D1764928

LIVERPOOL JMU LIBRARY

3 1111 01461 5783

MOSQUITO ECOLOGY

Mosquito Ecology

Field Sampling Methods
Third Edition

by

JOHN B. SILVER
New York

 Springer

A C.I.P. Catalogue record for this book is available from the Library of Congress.

ISBN 978-1-4020-6665-8 (HB)
ISBN 978-1-4020-6666-5 (e-book)

Published by Springer,
P.O. Box 17, 3300 AA Dordrecht, The Netherlands.

www.springer.com

Printed on acid-free paper

All Rights Reserved
© 2008 Springer Science+Business Media B.V.
No part of this work may be reproduced, stored in a retrieval system, or transmitted
in any form or by any means, electronic, mechanical, photocopying, microfilming, recording
or otherwise, without written permission from the Publisher, with the exception
of any material supplied specifically for the purpose of being entered
and executed on a computer system, for exclusive use by the purchaser of the work.

For Mel

Preface to Third Edition

When Mike Service first approached me and asked me if I would be interested in producing a third edition of Mosquito Ecology: Field Sampling Methods I was initially surprised and somewhat daunted by the prospect, given my relative inexperience in field-based entomology compared to Mike and many of the researchers whose work is quoted liberally in the pages that follow. This initial feeling of surprise soon gave way to a feeling of pride at having been asked, but this emotion in turn rapidly evaporated as I contemplated the immensity of the task ahead. Mike stated in the preface to the second edition of this book that a first edition is relatively easy to produce and a second edition somewhat more difficult. I propose that writing a third edition of a book written by another author is perhaps even more difficult, especially when the initial author is as illustrious and highly-regarded as Mike Service. In writing this edition, I was forced to admit that I had only a passing acquaintance with the previous editions that was gained during my time as a student of Mike's, and so I did not have the detailed knowledge of the content and structure that comes with having written two previous editions and so my first task was to become better acquainted with the second edition. Consequently, I suspect that I have now become only the third person (after Mike and his wife Wendy) to have read the whole of the second edition from cover to cover on more than one occasion.

My basic approach to this mammoth task was to maintain as much as possible of the original content, and to adopt a similar style to the first two editions, while undertaking significant restructuring and reorganising of the text into several smaller, and consequently I hope more easily digestible chapters, while adding new and deleting old material as appropriate. Perhaps the hardest task was to identify material that could be removed without devaluing the book as a comprehensive resource to field workers interested in both "state of the art" mosquito sampling, as well as historical developments in the field.

In describing apparatus and techniques I have retained the units of measurement given in the original descriptions. Although this results in some descriptions being in metric and others in non-metric units it avoids the 'odd' amounts which result from converting original measurements, frequently chosen for convenience, to the metric scale.

A note on mosquito nomenclature adopted in this edition

In 1999 and 2000 John Reinert began to publish a series of papers dealing with some of the issues relating to the phylogeny, classification and taxonomy of the very large and diverse tribe the Aedini. In the first paper, the subgenus *Verrallina* of genus *Aedes* was restored to generic rank (Reinert, 1999). This paper was rapidly followed by restoration of a second subgenus (*Ayurakitia*) of *Aedes* to generic rank (Reinert, 2000*a*). It is probably true to say that these two papers created relatively little emotion among all but the most dedicated of mosquito taxonomists, as both *Verrallina* and *Ayurakitia* are groups of relatively few species that are not known vectors of disease, and in the case of *Ayurakitia* their known distribution is restricted to Thailand. However, the publication of a third paper (Reinert, 2000*b*) in 2000 made wholesale changes to the classification of the tribe Aedini, including the elevation of subgenus *Ochlerotatus* to generic status, with important consequences for many researchers, control specialists and publishers, as *Ochlerotatus* is a large group containing many well-known species and several important human and animal disease vectors. Despite much discussion and some reluctance (see for example Savage & Strickman (2004) and the response by Black (2004)) this change in classification and nomenclature has been adopted in the literature almost universally. Subsequently, Reinert et al. (2004) proposed wholesale changes and a major re-classification of the whole tribe Aedini based on the examination of 172 morphological characters across all life-stages from egg to adult. The results of this study revealed that the tribe Aedini is a polyphyletic assemblage of *Aedes* and *Ochlerotatus*, in which either seven or eight additional genera (depending on analysis method) were embedded. In the light of failure of either analytical method to unequivocally resolve the relationships within *Aedes* and *Ochlerotatus*, the authors proposed a 'reasonable and conservative compromise classification' involving recognition as genera of those groups that were 'weighting independent', i.e. those that were common to the results of both analyses. This 'conservative' reclassification resulted in proposed generic status for 32 subgenera within the tribe, including the restoration of *Stegomyia* to generic rank. *Aedes* (*Stegomyia*) includes perhaps the most well-known of all mosquitoes, the yellow fever and dengue vector *Aedes aegypti* and the proposed changes would necessitate a change in name of this species to *Stegomyia aegypti*. Perhaps even more controversially, another important vector of human disease, *Aedes albopictus* would require a name change to *Stegomyia albopicta*, which includes a change in gender of the termination of the species name in order for it to agree with the gender of the generic name. The changes proposed in Reinert et al. (2004) have not been so readily accepted as the elevation of *Ochlerotatus* to generic rank. Indeed, the Editor-in-Chief and Subject Editors of the Journal of Medical Entomology have resisted these latest changes, taking the position that more research and interpretation is required in order to develop a consensus on the classification and nomenclature of the tribe Aedini (Anon, 2005). The editors of Journal of Medical Entomology encourage authors to persist with the traditional nomenclature, except when they can present valid reasons for using the classification and nomenclature of Reinert et al. (2004). The apparent arbitrary way in which these various proposed changes have been either accepted or resisted is somewhat surprising in my opinion, and much of the resistance appears to have arisen primarily because of the proposed name changes to well-known vectors of human disease. However, this reaction to the change in name of *Aedes aegypti* is perhaps somewhat surprising, as this species has undergone many name changes since its original description as *Culex aegypti* by Carl Linné. Indeed, it was also known as *Stegomyia fasciata* for many years. My own view from reading the various publications and having an amateur interest in mosquito identification and classification, is that the tribe Aedini is indeed somewhat chaotic, containing many undetermined phylogentic relationships and artificial groupings of probably unrelated species. Zavortink (1990) has suggested that the number of subgenera recognised is currently too low for a family as large as the Culicidae, when compared with other groups with better-resolved taxonomies, and estimated the total number of genera to be around 225. Harbach & Kitching (1998) also suggested that the recognition of more, smaller subgenera within the Culicidae would ultimately lead to a more stable classification and nomenclature, and ultimately ease the task of identifying specimens. It is likely that clarifying these relationships within the tribe Aedini will potentially be fertile ground for taxonomic studies for many years to come. In the meantime, I have adopted a somewhat conservative approach to the use of mosquito names in this volume, reflecting the changes proposed in Reinert (1999, 2000*a,b*), but not those in Reinert et al. (2004), such that *Ochlerotatus* replaces *Aedes* (*Ochlerotatus*), but *Aedes aegypti* and *Aedes albopictus* retain their familiar names. Although I support the changes proposed by Reinert et al. (2004), maintenance of the pre-Reinert et al. (2004) nomenclature will keep the names used in this publication consistent with recent editorial decisions of some of the major entomological journals (e.g. Anon, 2005), and hopefully will facilitate cross-referencing of species names with other publications, at least in the short-term until these issues are fully resolved to the satisfaction of all.

As regards the Anophelinae, I use the phylogeny and classification proposed by Sallum et al. (2000), which was based on a morphological analysis of 64 species, and formally synonymized the genus *Bironella* and the subgenera *Stethomyia*, and *Lophopodomyia* with *Anopheles*, subgenus *Anopheles*. Inclusion of *Bironella* within *Anopheles* remains slightly controversial, as it was not supported by the analyses of Harbach & Kitching (1998) or Foley et al. (1998), although it was supported by morphological character analysis and subsequent rDNA and combined rDNA and mtDNA analyses (Sallum et al., 2002), although Sallum et al. (2002) did conclude that the proposed subgeneric status of *Bironella* requires further testing.

Anon (2005) Journal Policy on Names of Aedine Mosquito Genera and Subgenera. J Med Entomol 42: 511

Black WC (2004) Learning to use *Ochlerotatus* is just the beginning. J Am Mosq Control Assoc 20: 215–216

Foley DH, Bryan JH, Yeats D, Saul A (1998) Evolution and systematics of *Anopheles*: insights from a molecular phylogeny of Australian mosquitoes. Molecular Phylogenetics and Evolution 9: 262–275

Harbach RE, Kitching IJ (1998) Phylogeny and classification of the Culicidae (Diptera). Systematic Entomology 23: 327–370

Polaszek A (2006) Two words colliding: resistance to changes in the scientific names of animals – *Aedes* vs *Stegomyia.* Trends Parasitol 22: 8–9

Reinert JF (1999). Restoration of *Verrallina* to generic rank in tribe Aedini (Diptera: Culicidae) and descriptions of the genus and three included subgenera. Contributions of the American Entomological Institute 31: 1–83

Reinert JF (2000) Restoration of *Ayurakitia* to generic rank in tribe Aedini and a revised description of the genus. J Am Mosq Control Assoc 16: 57–65

Reinert JF (2000) New classification for the composite genus *Aedes* (Diptera: Culicidae: Aedini), elevation of subgenus *Ochlerotatus* to generic rank, reclassification of the other subgenera, and notes on certain subgenera and species. J Am Mosq Control Assoc 16: 175–188

Sallum MAM, Schultz TR, Wilkerson RC (2000). Phylogeny of Anophelinae (Diptera: Culicidae) based on morphological characters. Ann Entomol Soc Am 93: 745–775

Sallum MAM, Schultz TR, Foster PG, Aronstein K, Wirtz RA, Wilkerson RC (2002) Phylogeny of Anophelinae (Diptera: Culicidae) based on ribosomal and mitochondrial DNA sequences. Systematic Entomology 27: 361–382

Savage HM (2005) Classification of mosquitoes in tribe Aedini (Diptera: Culicidae): paraphylyphobia, and classification versus cladistic analysis. J Med Entomol 42: 923–927

Savage HM, Strickman D (2004) The genus and subgenus categories within Culicidae and placement of *Ochlerotatus* as a subgenus of *Aedes*. J Am Mosq Control Assoc 20: 208–214

Zavortink TJ (1990) Classical taxonomy of mosquitoes – a memorial to John N. Belkin. J Am Mosq Control Assoc 6: 593–599

John B Silver
New York, USA
August 2007

Preface to Second Edition

It is relatively easy to produce a first edition, all that one needs is self-discipline, an inexhaustible supply of paper, a good library and an understanding spouse. A second edition requires something more. You have to read through the entire first edition and in so doing begin to realise that probably no one else has ever done so. You then begin to be overwhelmed by the amount of new information that you have managed to discover on your subject and ponder on just how selective you should be in integrating it into the new edition. At this point you may reflect that Dr Samuel Johnson said 'A man may turn over half a library to make one book'. After which you may try and console yourself with another of his quotations, namely, 'What is written without effort is in general read without pleasure'.

I have generally adopted the layout of the previous edition but have expanded sections that, although not directly concerned with sampling, are nevertheless associated with it. For example, I have given more details on oviposition attractants, larval growth retardant factors, and aggregation of populations, and now include summaries of methods to identify blood-meals, and accounts of vectorial capacity and the calculation of inoculation rates. I have added a small final chapter, which one of my students described as containing 'junk data'. But after having written all other chapters I was left with information on recent books on mosquito ecology and behaviour that contain information relevant to sampling, accounts of attempts to model mosquito populations, and publications containing important mathematical and statistical considerations, that did not easily fit into any one chapter, but were applicable to many. So Chapter 12 was created for this type of data.

In the 17 years since the First Edition was published there has been a proliferation of publications on mosquito ecology, especially on subjects such as estimating adult survivorship and on mosquitoes colonising tree-holes and other natural container-type habitats. During the intervening years many mosquito workers have broadened their outlook and adopted or modified ecological principles and techniques used by ecologists for their own studies on mosquitoes. At the same time mosquitoes have attracted the attention of 'pure ecologists' who have brought new ideas to the study of mosquito ecology, although admittedly they have usually concentrated on species breeding in discrete container habitats. Such tractable microhabitats make it easier to sample and study mosquitoes living in them. Several of these papers are highly mathematical and are frightening to those who have no love for mathematics. I have tried to understand such papers and present the basics so that entomologists can apply the procedures and analyse their data without needing to read the original papers; which for some may be difficult to obtain. It has proved a daunting task reading the enormous volume of literature for this Second Edition, which contains some 1800 additional references. Only a few of the older ones could be deleted.

Finally, it is a pleasure to thank my wife, Wendy, for typing and retyping the manuscript, checking the references, making the indices, proof-reading, and encouraging me to continue when I felt I was losing the battle with such a wealth of new information.

M. W. Service
Liverpool
December 1991

Preface to First Edition

This book has been written for the field worker, whether he is mainly concerned with mosquito control and wishes, for example, to learn more about the use of light-traps for monitoring pest mosquitoes or maybe improve his larval sampling procedures, or whether he is more interested in ecology and population dynamics of mosquitoes. Descriptions of traps and procedures, and methods of analysing results, have been described mostly in sufficient detail to enable those with limited access to the original publications to understand and use the methods. It has proved impractical to describe all the many different types of traps and methods that have been used to collect mosquitoes. I have tried therefore to present a selection of the more useful and interesting techniques while at the same time introducing others that, although little used in mosquito studies, appear to have potential.

In describing apparatus and techniques I have retained the units of measurement given in the original descriptions. Although this results in some descriptions being in metric and others in non-metric units it avoids the 'odd' amounts which result from converting original measurements, frequently chosen for convenience, to the metric scale.

M. W. Service
Liverpool

Contents

Preface to Third Edition……………………………………..vii

Preface to Second Edition…………………………………xiii

Preface to First Edition……………………………………xv

Acknowledgements………………………………………....xix

Useful Conversion Data………………………………......xxi

Chapter 1
DESIGNING A MOSQUITO SAMPLING PROGRAMME……………..1

Chapter 2
SAMPLING THE EGG POPULATION……………………………....25

Chapter 3
SAMPLING THE LARVAL POPULATION……………………..137

Chapter 4
SAMPLING THE EMERGING ADULT POPULATION…………..…339

Chapter 5
SAMPLING THE ADULT RESTING POPULATION………………...373

Chapter 6
SAMPLING ADULTS BY ANIMAL BAIT CATCHES AND BY
ANIMAL-BAITED TRAPS………………………………………493

Chapter 7
BLOOD-FEEDING AND ITS EPIDEMIOLOGICAL
SIGNIFICANCE………………………………………………...677

Chapter 8
SAMPLING ADULTS WITH NON-ATTRACTANT TRAPS………...771

Chapter 9
SAMPLING ADULTS WITH LIGHT-TRAPS……………………......845

Chapter 10
SAMPLING ADULTS WITH CARBON DIOXIDE
TRAPS...947

Chapter 11
SAMPLING ADULTS WITH VISUAL ATTRACTION TRAPS,
SOUND TRAPS AND OTHER MISCELLANEOUS ATTRACTION
TRAPS...1027

Chapter 12
ESTIMATION OF THE MORTALITIES OF THE IMMATURE
STAGES..1049

Chapter 13
METHODS OF AGE-GRADING ADULTS AND ESTIMATION OF
ADULT SURVIVAL RATES...1161

Chapter 14
ESTIMATING THE SIZE OF THE ADULT
POPULATION...1273

Chapter 15
MEASURING ADULT DISPERSAL.......................................1377

Chapter 16
EXPERIMENTAL HUT TECHNIQUES..................................1425

Chapter 17
INDICES OF ASSOCIATION AND SPECIES DIVERSITY
INDICES..1445

Index...1469

Species Index..1479

Acknowledgements

Grateful acknowledgements are given to the authors and publishers of figures, tables and formulae that I have used in original or modified form. In all instances references have been given to the authors in the text, tables or figures.

I would like to thank the editors and publishers of the following for permission to reproduce figures and published tables: American Journal of Hygiene, American Journal of Tropical Medicine and Hygiene, American Midland Naturalist, Annals of Applied Biology, Annals of the Entomological Society of America, Annals of Tropical Medicine and Parasitology, Australian Journal of Entomology, Bulletin of Entomological Research, Bulletin of the World Health Organization, Bulletin of the Society for Vector Ecology, California Vector Views, Canadian Entomologist, Canadian Journal of Zoology, Ecological Entomology, Ecology, Endemic Disease Bulletin of Nagasaki University, Entomologia Experimentalis et Applicata, Entomologist's Gazette, Environmental Entomology, Hydrobiologia, Japanese Journal of Sanitary Zoology, Journal of Animal Ecology, Journal of Applied Ecology, Journal of the Australian Entomological Society, Journal of Communicable Diseases, Journal of Economic Entomology, Journal of the American Mosquito control Association, Journal of the Entomological Society of Southern Africa, Journal of Medical Entomology, Journal of Tropical Medicine and Hygiene, Journal of Vector Ecology, Medical and Veterinary Entomology, Mosquito News, Nature, Pacific Entomologist, Parassitologia, Oecologia, Researches on Population Ecology, Southeast Asian Journal of Tropical Medicine and Public Health, Transactions of the Royal Entomological Society of London, Tropical Biomedicine, Tropical Medicine (Nagasaki), and to Springer, Methuen & Co, W. B. Saunders & Co., the Entomological Society of America, Indian Council of Medical Research, and the World Health Organization.

I am greatly indebted to the following, who kindly gave permission to use original photographs: J. E. Freier, B. H. Kay, M. W. Service, J. A. Shemanchuk.

This work would not have been possible without access to several libraries, including those at: Liverpool School of Tropical Medicine, Liverpool, UK; International Centre for Insect Physiology and Ecology, Nairobi, Kenya; Swiss Tropical Institute, Basel, Switzerland, Columbia

University, New York, USA; American Museum of Natural History, New York, USA; United Nations Environment Programme, Nairobi, Kenya; and the World Health Organization, Geneva, Switzerland.

There are instances where we have been unable to trace or contact the copyright holder. If notified the publisher will be pleased to rectify any errors or omissions at the earliest opportunity.

I wish to thank Mike Service for his encouragement and for helpful comments on an earlier draft of this book.

Useful Conversion Data

To convert	Multiply by
Inches to centimetres	2.540
Centimetres to inches	0.394
Yards to metres	0.914
Metres to yards	1.094
Miles to kilometres	1.609
Kilometres to miles	0.621
Square inches to square centimetres	6.452
Square centimetres to square inches	0.155
Square yards to square metres	0.836
Square metres to square yards	1.196
Acres to hectares	0.405
Hectares to acres	2.471
Ounces to grams	28.350
Grams to ounces	0.035
Pounds to kilograms	0.454
Kilograms to pounds	2.205
Fluid ounces to millilitres	28.413
Millilitres to fluid ounces	0.035
Pints to litres	0.568
Litres to pints	1.760
Gallons to litres	4.546
Litres to gallons	0.220
Cubic feet to cubic metres	0.028
Cubic metres to cubic feet	35.314
Cubic yards to cubic metres	0.765
Cubic metres to cubic yards	1.308
Imperial gallons to US gallons	1.201
US gallons to Imperial gallons	0.833

Chapter 1 Designing a Mosquito Sampling Programme

Interest in mosquito ecology and the use of appropriate sampling methods commenced at the beginning of the nineteenth century with the discovery that mosquitoes could act as vectors of diseases to humans and domestic animals. Since those early days, the science and technology of mosquito sampling have developed apace, as evidenced by the increase in size and scope of successive editions of this book.

As stated by Southwood and Henderson (2000), arguably the primary factor in selecting an appropriate sampling methodology is careful formulation of the specific ecological problem that the investigator wishes to address. This will determine the timing and duration of the sampling programme, whether it is an extensive study that attempts to describe large-scale distribution and abundance of a species or population, or an intensive study of a specific population, usually with a view to determining its absolute or relative size and the factors that regulate its size. Southwood and Henderson (2000) have also proposed a useful stepwise decision-making process that can be applied to the design of any sampling programme, including for mosquitoes. Firstly, initial studies should be undertaken to determine species richness and the required levels of taxonomic discrimination required. This is of particular relevance to the study of anopheline mosquitoes, many of which exist as groups of several morphologically indistinguishable sibling species that sometimes exhibit marked differences in behaviour and ecology. The relatively high costs of the techniques available to separate these sibling species may preclude their use on more than a sample of the total individuals collected, resulting in studies that treat sibling species groups as a single taxonomic and ecological entity. The second decision involves selection of the sampling universe: should the investigator focus on a single discrete unit of habitat (pond, rice field, house, animal stable, etc.) or should she/he examine representatives of a specific habitat type over a wider geographical area? This will depend on whether an extensive or intensive study is desired. The next decision concerns the required accuracy and precision of the population estimates obtained, where accuracy refers to the closeness of an estimate to the true population size, and precision refers to the reproducibility of that estimate. This decision will depend on the variability of the system being studied, which in the case of mosquitoes and other insects may be very high, even over a short sampling timeframe, such that the levels of accuracy and precision achieved may be lower. Githeko et al. (1996) emphasised the importance of taking the often substantial spatial variability (often up to a factor of 10) of entomological parameters into account when designing large-scale malaria entomological sampling programmes, usually through sampling stratification. In extensive studies, sampling of a specific life-stage from multiple locations may be appropriate, for example in the case of mosquitoes, the life-stage responsible for transmitting the infection or causing the nuisance, namely the adults. Alternatively, if the objective of the study is to develop predictors for outbreaks, then sampling of a different life-stage may be more appropriate. An individual habitat may require further sub-division based on prior knowledge of likely differences in environmental parameters, such that sampling of mosquito immatures from a pond, for example, would be stratified according to water depth or distance from the shore, etc. It is often advisable to take samples of different sizes from the same habitat in order to increase precision and to use more than one method of sampling to ensure consistency of estimates and identify potential bias in sampling.

Population sampling is employed as a means of estimating population size when a direct count of all the animals constituting a population is impossible or impractical. Any estimate obtained should be as accurate as possible given the resource constraints, be they time, money or available personnel.

Githeko et al. (1996) provide a useful description of the main steps involved in designing an entomological sampling programme for monitorring the effects of large-scale vector control interventions in Africa. The steps proposed are as follows: (1) describe the pattern of disease transmission in the study area (seasonality and intensity), stratified accord-ing to ecological zones; (2) determine the main vector species present, at least to sibling species level, but preferably to cytotype; (3) measure vector densities, parity rates, human blood indices, and vector infectivity rates using appropriate techniques in order to determine any 'mass effect'. For monitoring the density of malaria vectors, indoor and outdoor light-traps or pyrethrum spray catches alongside window exit-traps are recommended. Minimum levels of sampling recommended by Githeko et al. (1996) for this type of monitoring are at least ten villages (five intervention and five control) and up to 10% of all villages in the study area, and 10–30 rooms per village per month. If possible, control and intervention villages should be matched for transmission intensity at the start of the study. Pit traps were recommended as a suitable method for collecting outdoor-resting blood-fed female mosquitoes in order to determine the human blood index. It was noted that sample sizes used in the determination of vector infectivity rates are often inadequate, and a minimum of 200–500 mosquitoes is recommended in areas of intense transmission. Collection of the above data facilitates the calculation of the entomological inoculation rate; (4) in the case of insecticide-based control interventions, bioassay tests to determine the efficacy of the insecticide are required monthly or at least quarterly, and insecticide susceptibility testing should be carried out annually on a minimum sample of 400 mosquitoes of the same age; (5) an estimate of the coverage of personal protection methods of vector control in use in the study area is desirable; (6) human behaviour, including sleeping patterns, etc. should also be monitored.

Hurlbert (1984) presents a detailed critique of how ecologists design and analyse their field experiments, including a discussion of some of the problems encountered in randomising experiments (e.g. Latin Squares). He points out that if: (1) experiments are spatially or temporarily segregated; (2) replicate experiments are somehow interconnected; or (3) if replicates are only samples from a single experimental unit, then such replicates are not in fact independent. If data from such experimental designs are used then one is committing pseudo-simplification. This is a very important and readable paper. As noted by Reisen and Lothrop (1999) mosquito sampling intensity is more often dictated by practical constraints that include landscape, access to sites, availability of human and other resources, time required for sample processing, etc. As a result, sample sites are almost never randomly distributed and are frequently stratified, based on prior knowledge of distribution of the target species, the occurrence of human disease cases, or other parameters. This type of sampling, termed best-estimate sampling by Reisen and Lothrop (1999) tends to overestimate abundance across wide geographical areas that consist of a mosaic of favourable and unfavourable habitats.

Sample size

Many publications, mainly in statistical journals, describe methods for estimating sample size, and several computer software programmes are available to calculate sample sizes. A selection of these methods have been presented by Elliott (1977), Southwood (1978), and Southwood and Henderson (2000) and some of the more common method are described below. Unfortunately, these methods have not always been used consistenly by mosquito workers. Practical information on sampling and the problems of undertaking surveys and censuses is presented by Yates (1981), while Prepas (1984) gives a detailed and easy to read account of the statistics needed in the design of sampling experiments and the analysis of the results. A relatively simple account of bootstrapping, a non-parametric method used to estimate sampling distribution, is given by Solow (1989), while Buonaccorsi and Liebhold (1988) explain the methods for estimating ratios, such as densities, with particular reference to confidence intervals, and describe jackknifing and boot-strapping techniques. They also describe how to determine sample size. Altman (1982) presents a concise account of how to determine sample size, together with a nomogram for a two-sample comparison of a continuous variable, which relates the power of the test to detect a difference, total sample size, the expected difference in sample measurements, and the 0.01 and 0.05% significance levels. A useful book by Lemeshow et al. (1990), published on behalf of the World Health Organization, aims to guide epidemiologists on the sample sizes that need to be taken in surveys. More than half the book consists of tables, for example to determine sample size necessary to estimate the proportion of a population with a certain characteristic within 0.01–0.25 absolute percentage points with 99% confidence. This publication should be of interest to medical entomo-logists. Smith and Morrow (1996) also present several useful equations for determining sample size in health intervention field trials.

All sampling methods are selective in catching certain elements of a population to a greater extent than others, for example bait catches collect predominantly, or completely, hungry female mosquitoes orientated to blood-feeding. Catches from non-attractant suction traps may sample a more representative proportion of the population, but again these traps will be biased towards active individuals that fly near to the traps. These types of bias need to be understood and quantified in sample data if the results are to be used to get a better understanding of the overall population dynamics of a species.

Perry (1989) gives an informative statistical review of sampling, mainly as has been applied to agricultural entomology. Elliott (1977) gives a very readable account of the application of parametric and non-parametric sta-tistical tests to the analysis of benthic invertebrates and applicable to mos-quito sampling, written with the biologist not the mathematician in mind.

Insect populations, including vector populations, tend to exhibit spatial heterogeneity in their densities, and this spatial heterogeneity, together with temporal changes in abundance have important consequences for insect ecology, and in particular the design of sampling strategies and control interventions. Taylor (1984) and Kuno (1991) discuss the statistical concepts of spatial distribution in insects and how this can affect sampling. Two criteria are important for determining the appropriate sample size, namely the desired precision, which relates to the width of the confidence intervals (or more correctly fiducial limits [Southwood and Henderson 2000]) around the estimate; and secondly the power of the study to detect differences between populations or to detect the effect of a control intervention. Sampling errors make it impossible to guarantee obtaining a significant result, even where one exists, and so researchers have to rely on a probability of detecting a significant difference. The power of a study depends on: (1) the value of the true difference between the two samples; (2) the size of the samples, such that the larger the sample, the greater the power; and (3) the probability level selected as being indicative of a significant result (Smith and Morrow 1996). The decision to use precision or alternatively power to determine sample size ultimately depends on the study objective. For example, where we are certain that a control intervention will definitely have an effect, then the priority should be on determining the magnitude of that effect with the greatest precision. If we are not sure of the outcomes of an intervention, then choosing a sample size with sufficient power to detect a difference would be the most appropriate option, however, this method results in lower precision of the estimates obtained. Where multiple outcomes are under investigation, then the largest sample size calculated for each of the outcomes should be adopted across the whole study, where resources permit.

The minimum sample size (n) required for detecting the population mean abundance with a targeted precision level and at a significance level α can be determined based on prior knowledge of the spatial distribution of mosquito populations. The precision level is defined as $D = SE / \overline{x}$, where $\overline{x}$ is the sample mean abundance and SE is the standard error of the mean abundance. The allowable precision level in ecological research is typically 10–25% (Southwood and Henderson 2000). Where the population distribution is essentially random and can be described by the Poisson distribution, the minimum sample size (n) can be calculated using the formula $n = (t / D)^2 / \overline{x}$, where t is the critical value for the t distribution at the fixed type I error with the degree of freedom that sample mean ($\overline{x}$) was calculated. If the mosquito spatial distribution is clumped or aggregated and described by a negative binomial distribution, n can be calculated using $n = (t / D)^2 (\overline{x} + k) / \overline{x}k$, where k is the parameter describing the negative binomial distribution. Several empirical methods are available to allow for the formulation of sampling plans that do not require prior knowledge of the form of the spatial distribution of the population. One such method is to use Taylor's power law to determine sample size and to design a sampling plan based on the information obtained on sample mean and variance using the formula $n = a\bar{x}^{(b-2)}(t/D)^2$, where a and b are the parameters obtained from Taylor's power law (Taylor 1961). Several authors have used this method in determining sample size for mosquito population studies (Service 1971; Mackey and Hoy 1978; Sandoski et al. 1987; Ritchie and Johnson 1991; Pitcairn et al. 1994; Reisen and Lothrop 1999; Lindblade et al. 2000).

The pattern of spatial distribution has important consequences for the sample size required to achieve the desired level of precision in an abundance estimate, as well as influencing the effectiveness of control measures. Resisen and Lothrop (1999), for example, have shown that a stratified random design based on the known spatial distribution pattern was the most accurate method for estimating *Culex tarsalis* adult abundance in California, USA.

Relative variation (RV) is a useful statistic to compare the efficiency of various sampling methods, in fact RV = 100 $S\bar{x}/\bar{x}$, where $S\bar{x}$ is the standard error of the mean ($\bar{x}$). According to Southwood (1978) a RV < 25 is usually adequate for most extensive sampling surveys, although in certain intensive programmes an RV < 10 may be required. A highly aggregated population will likely produce a higher RV than an underdispersed or randomly distributed population. To try to overcome this the following formula takes into consideration k, which is a parameter of the negative binomial distribution reflecting the degree of aggregation

$$D = \sqrt{\frac{1}{n\bar{x}} + \frac{1}{nk}}, \quad \text{and} \quad n = \frac{1}{D^2}\left(\frac{1}{\bar{x}} + \frac{1}{k}\right) \tag{1.1}$$

where D = fixed proportion of the mean expressed as a decimal.

Another common statistic to compare efficiencies is the Coefficient of Variability (CV), that is CV = 100 $S/\bar{x}$, or the Coefficient of Variation i.e. CV = $S/\bar{x}$. There are two ways of expressing reliability, one is to define the standard error as equal to a fixed constant such as the half-length of the standard error interval, or as a fraction or percentage of the mean. The other approach is to express reliability as a probability that the estimated mean should be within a certain value of the true population mean. Again values can be given as a fixed numerical quantity or as a percentage of the mean. Determining reliability aids in deciding how many samples (n) should be taken to achieve this.

Karandinos (1976) clearly summarised methods for determining optimum sampling size (n) and presented simple formulae to obtain the numbers of samples needed for reliability as defined by the Coefficient of Variability (CV). Thus a general formula is

$$n = \left(\frac{S}{\bar{x}\,\mathrm{CV}}\right)^2 \qquad (1.2)$$

McArdle et al. (1990) pointed out that calculating the standard deviation of log $(n + 1)$, which is widely done, can be seriously biased. This is because if there are zero counts then the calculated variance will be an underestimate of the true value, and the more zeros the worse this bias will be. Also this bias is most marked with means of 0–20. These authors suggest greater use is made of the Coefficient of Variation (CV), because this does not require any transformation of data and so there are no associated problems with zeros. Now, plotting log CV against log mean density will result in a horizontal line if variability as measured by CV is independent of the mean, if variability increases or decreases then the plot will have the appropriate slope. This very readable paper reiterates difficulties of choosing the correctly sized sampling unit to measure spatial distribution and variability. Some of the problems, such as bias associated with Taylor's power law are discussed.

Another procedure for defining reliability is in terms of formal probabilistic statements, such as setting a confidence interval for the mean where D = a set proportion of the mean. Thus a general formula is

$$n = \left(\frac{Z_\alpha/2}{D}\right)^2 \frac{S^2}{\bar{x}^2} \qquad (1.3)$$

where $Z/2_\alpha$= is the upper $\alpha/2$ point of the normal distribution, that is 1.96.

Finally, another approach using probabilistic statements is to set reliability in terms of a fixed number (h), in order to have an acceptable error of say ± 10 mosquitoes (i.e. $h = 10$), for this the general formula is

$$n = \left(\frac{Z_\alpha/2}{h}\right)^2 S^2 \qquad (1.4)$$

These and formulae for other distributions are given in Table 1.1.

The number of samples (n) required for estimating the mean density of eggs of *Wyeomyia smithii* per pitcher (*Sarracenia purpurea*) with a set level of precision (D = standard error/mean) was calculated by Mogi and Mokry (1980) using the method of Iwao and Kuno (1968), namely

$$n = \frac{t^2}{D^2}\left(\frac{\alpha+1}{\overline{x}}\right) + \beta - 1 \tag{1.5}$$

where t is Student t-test α and β are the regression coefficients (intercept and slope) of the plot of mean crowding on mean density. Because it is unlikely that the distribution of larvae and pupae are more aggregated than eggs, Mogi and Mokry (1980) thought this sampling plan could be used for them as well.

Table 1.1. Formulae giving the optimum sample size n derived under three definitions of the estimate's reliability (after Karandinos 1976)

Biological distribution	Coefficient of Variability (CV)	Definition of reliability by Probalisitic statement Confidence interval equal to: % of Paraneter (D)		Fixed positive number (h)
General	$n = \left(\dfrac{S}{\overline{x}\,CV}\right)^2$	$n = \left(\dfrac{Z_\alpha/2}{D}\right)^2 \dfrac{S^2}{\overline{x}^2}$		$n = \left(\dfrac{Z_\alpha/2}{h}\right)^2 S^2$
Negative binomial	$n = \dfrac{\dfrac{1}{\overline{x}}+\dfrac{1}{k}}{CV^2}$	$n = \dfrac{(Z_a/2)^2\left(\dfrac{1}{\overline{x}}+\dfrac{1}{k}\right)}{D^2}$		$n = \left(\dfrac{Z_a/2}{h}\right)^2\left(\dfrac{k\overline{x}+\overline{x}^2}{k}\right)$
Poisson	$n = \dfrac{1}{\overline{x}\,CV^2}$	$n = \left(\dfrac{Z_\alpha/2}{D}\right)^2 \dfrac{1}{\overline{x}}$		$n = \left(\dfrac{Z_\alpha/2}{h}\right)^2 \overline{x}$
Binomial	$n = \dfrac{q}{p\,CV^2}$	$n = \left(\dfrac{Z_\alpha/2}{D}\right)^2 \dfrac{q}{p}$		$n = \left(\dfrac{Z_\alpha/2}{h}\right)^2 pq$

The relationship between the mean and variance, which is required to determine the minimum number of samples needed in a sampling programme, is given approximately (Iwao and Kuno 1968) by

$$S^2 = (a+1)\,\overline{x} + (b-1)\,\overline{x}^{\,2} \tag{1.6}$$

The smallest number of samples (n) required for a predetermined level of precision (C), such as 0.1, 0.2, 0.3, with confidence level ($1-\alpha$) is

$$n = \left(\frac{t_\alpha / 2S}{C\overline{x}} \right)^2 \quad \text{or} \quad = t_\alpha / 2^2 S^2 \overline{x}^{-2} C^{-2} \quad (1.7)$$

or using the power law relationship, $(S^2 = a\overline{x}^b)$, and substituting this for S^2 (Elliott and Drake 1981) we have $n = t^2 a\overline{x}^b \overline{x}^{-2} C^{-2}$ or $n = t^2 a\overline{x}^{(b-2)} C^{-2}$ where $t_\alpha/2$ is the standard deviate of Student's t-test corresponding to α. When sample size is >30, t can be replaced by the standard normal deviate z. Many investigators express precision in terms of the standard error of the mean, which in practice is approximately equivalent to using 65% confidence intervals, especially with large sample sizes where $t \approx z = 1$. Karandinos (1976) derived optimal sample size equations for several common distributions, but according to Régnière and Sanders (1983) failed to recognise this fact in discussing the use of the standard error of the mean.

Iwao and Kuno (1968) showed that for randomly dispersed insects sample size could be derived from the regression of Lloyd's (1967) mean crowding, m^*, on the mean, $\overline{x}$, thus

$$m^* = \alpha + \beta \overline{x} \quad (1.8)$$

the number of samples, n, needed for a magnitude of error relative to the mean, C (i.e. $d/\overline{x}$, where d is half-width of confidence interval) can be estimated as follows

$$n = \frac{t^2}{C^2} \left(\frac{\alpha + 1}{\overline{x}} + \beta - 1 \right) \quad (1.9)$$

where t is students t-test and is 2 for 95% confidence probability if $n > 30$. Since $\alpha = 0$ and $\beta = 1$ in a Poisson distribution, this equation can be simplified to

$$n = \frac{t^2}{C^2} \frac{1}{\overline{x}} \quad (1.10)$$

when larvae are more or less randomly distributed. Having obtained a regression of $m^* - m$ this formula was used by Ikemoto (1978) in sampling *Anopheles sinensis* larvae in rice fields. If, however, the distribution mimics a negative binomial with a common k, that is $\alpha = 0$ and $\beta = 1 + 1/k$, then the appropriate equation is

$$n = \frac{t^2}{C^2} \left(\frac{1}{\overline{x}} + \frac{1}{k} \right)$$ (1.11)

Reuben et al. (1978) applied this formula to determine the number of samples they needed to take to estimate the mean number of *Aedes aegypti* pupae per house. Setting C at 0.10 more than 3000 houses would have to be sampled, and this was not feasible with the resources available, but it was possible to search 1500–2000 houses, in which case C would be between 0.10 and 0.20. Kuno (1976) also used the regression of mean crowding on the mean in his development of multistage sampling, that is selecting primary sampling units and then subdividing these units into secondary sampling units, and sometimes even dividing these into further units. As with the method of Iwao and Kuno (1968) with set levels of precision (C), the numbers of samples needed can be estimated.

The optimal sample size for monitoring larval abundance in rice fields was investigated in two ways by Pitcairn et al. (1994). One method examined the number of dips needed to estimate population abundance with a fixed level of precision given specified amounts of aggregation in the sampled population, while the second method involved calculating the minimum number of dips necessary to ensure collection of at least one mosquito larva in 95 of 100 samples. Sampling was conducted by three individuals at 104 rice field sites using standard mosquito dippers (400 ml) at 10 points along linear transects. Results for each dip were recorded separately and one sample consisted of 90 dips. In the first part of the study, Pitcairn et al. (1994) used the modification of Wilson and Room (1983) to the formula of Karandinos (1976), which incorporates Taylor's power law, namely:

$$n = \left(\frac{Z_\alpha /2}{D} \right)^2 ax^{b-2}$$ (1.12)

where n is the number of samples, $Z_{\alpha/2}$ is the upper $\alpha/2$ part of the standard normal distribution, α is the confidence level, and D is a fixed proportion of the mean. For a 95% CI, $\alpha = 0.05$, for which $Z_{0.025} = 1.96$ when $n > 30$. Using estimates for a and b of Taylor's power law calculated from pooled data for all instars, Pitcairn et al. (1994) calculated that for fixed precision levels of 10% and 20% at least 1800 and 461 dips respectively are necessary to estimate the larval population of a field with a density of 0.05 larvae per dip, a density frequently encountered during the study (Fig. 1.1).

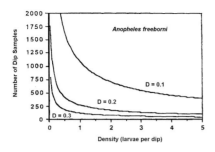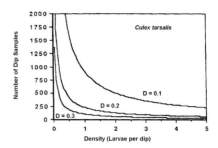

Fig. 1.1. Number of dips required to estimate the mean number of *Anopheles freeborni* and *Culex tarsalis* larvae at three fixed precision (D) levels after Pitcairn et al. (1994)

In the second part of the study, the authors considered that a useful objective for a sampling programme relating to control operations would be to determine if the population of larvae in a rice field exceeds some treatment threshold, and in this case an alternative sampling programme, based on the minimum number of dips required to collect at least one mosquito larva may be appropriate. When larval abundance is low, it can be difficult to collect any larvae, even when larvae are present. Given an arbitrary density and an assumed degree of aggregation in the larval population, the minimum number of dips necessary to collect at least one larva in 95 out of 100 samples can be estimated. For the Poisson distribution the variance is assumed to be equal to the mean so the probability of finding at least one larva depends only on density. For the negative binomial, the degree of aggregation of the sampled population, k was estimated using Taylor's *a* and *b* values for all instars. The number of dips necessary to collect at least one larva in 95 out of 100 samples was estimated as

$$1 - [P_{(0)}]^n = 0.95 \tag{1.13}$$

where $P_{(0)}$ is the probability of collecting no larvae in a dip and n is the sample size. Solving for n gives

$$n = \log_e(1 - 0.95) / \log_e(P_{(0)}) \tag{1.14}$$

where $\log_e$ is the natural logarithm. The authors estimated $P_{(0)}$ using the Poisson and the negative binomial probability distributions. For the Poisson

$$P_{(0)} = e^{-\bar{x}} \tag{1.15}$$

and for the negative binomial distribution

$$P_{(0)} = (1 + \bar{x}/k)^{-k} \tag{1.16}$$

where k is the dispersion pattern of the sampled population. Wilson and Room (1983) used Taylor's model to relate k to the sample mean as $k = \bar{x}^2/(s^2 - \bar{x})$. Substituting for k gives

$$P_{(0)} = (1 + (s^2 - \bar{x})/\bar{x})^{-\bar{x}^2/(s^2 - \bar{x})} \tag{1.17}$$

using Taylor's model, $s^2 = a\bar{x}^b$, and substituting for s^2 produces

$$P_{(0)} = (1 + (a\bar{x}^b - \bar{x})/\bar{x})^{-\bar{x}^2/(a\bar{x}^b - \bar{x})} \tag{1.18}$$

The minimum number of dips necessary to collect at least one larva, given a 5% error rate, was estimated using $P_{(0)}$ described by equation 1.15 for the Poisson distribution and equation 1.18 for the negative binomial distribution. For *Anopheles freeborni* and *Culex tarsalis*, 61% and 68% respectively, of field samples did not have a variance to mean ratio significantly greater than 1 and the greatest proportion of these occurred at low population densities, suggesting that a Poisson distribution may be an appropriate description of the spatial distribution, at least early in the season. The estimated sample sizes required when larval densities are between 0 and 0.3 larva per dip are shown in Figure 1.2. The minimum number of samples increased with decreasing densities, but sample sizes were substantially lower than those required for a fixed-precision estimate. For example, for a density of 0.05 *Anopheles freeborni* larvae per dip, 461 dips were required to obtain a sample value within 20% of the population mean; whereas only 60 samples were required to ensure collection of at least one larva in 95 out of 100 samples. Sampling to detect the presence of larvae was recommended by the authors as a realistic alternative for providing an efficient and cost-effective method for monitoring larval mosquito abundance for control purposes in northern California rice fields.

Frequently determination of the numbers of dips (samples) to be taken is based on more arbitrary considerations. For example, in Pakistan in studying the associations between larvae of many different species in habitats with different environmental qualities of the water, Reisen et al. (1981) took 10–20 dips from each breeding place. But more dips were taken from large heterogeneous habitats where the numbers collected varied considerably from dip to dip, while fewer dips were taken from smaller more homogeneous habitats, where larvae appeared to be more evenly dispersed.

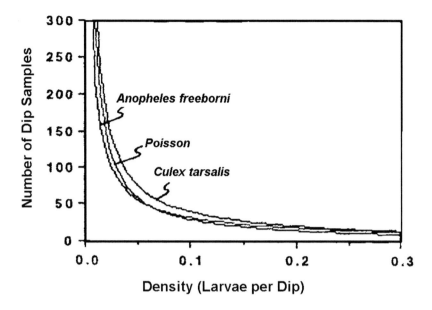

Fig. 1.2. Estimated minimum number of dips required to collect at least one mosquito larva in 95 of 100 samples for *Anopheles freeborni* (negative binomial) and *Culex tarsalis* (negative binomial). A Poisson distribution curve is illustrated for comparison (Pitcairn et al. 1994)

In Louisiana in studying the distribution of anopheline larvae in rice fields McLaughlin et al. (1987) selected fields for sampling from a table of random numbers. They then took 10 samples, each sample comprising two dips taken at 10-step intervals from each side of levees that subdivided the field. In Sri Lanka Amerasinghe and Ariyasena (1990) varied the number of dips taken from rice fields in accordance with the area of the breeding places (Chap. 3).

Turning to pyrethrum spray catches as a method for estimating adult indoor-resting resting abundance, a village hut is usually considered to be the sampling unit, and several huts (sampling units) must be sampled before the relative size of the indoor resting population in an area or village can be reliably indicated by the mean hut density. An obvious difficulty is that huts differ considerably in both size and attractiveness to mosquitoes, consequently the sampling unit is not standardised. This usually results in large variations between the numbers of mosquitoes caught from different huts, and leads to a very large variance of the mean hut density. This variability means that a large number of samples (huts) must be taken to obtain a reliable mean value. This limitation is very frequently ignored, and in many surveys the comparison of mean hut densities between differtent areas is unreliable because it is based on too small a sample. The problem of sample size determination in the study of the spatial distribution of resting adult anopheline mosquitoes among village huts has been considered in detail by Zhou et al. (2004) in the highlands of western Kenya. These authors conducted pyrethrum spray catches (see Chap. 5) in randomly selected samples of village huts in an initial area 3 km × 3 km, later expanded to 4 km × 4 km. The spatial distribution pattern of adult resting mosquitoes varied with season and with species. The mean abundance of *Anopheles gambiae* (PCR testing of 100 randomly selected specimens of *Anopheles gambiae* revealed 100% to be *Anopheles gambiae* s.s.) was 8.9 mosquitoes per house in the long rainy season of May 2002, but only 1.6–3.2 mosquitoes per house during the dry (August 2002) and short rainy seasons (November 2002 and February 2003). *Anopheles funestus* (of 50 specimens examined by PCR, all were *Anopheles funestus* s.s.) abundance was 0.6 mosquitoes per house in May 2002, and peak abundance of 2.1 mosquitoes per house was observed in August 2002, immediately following the long rainy season. For both species, the coefficient of variation was >1, indicating that the distribution of mosquito adults was aggregated. Houses harbouring the greatest number of mosquitoes were situated within 400 m of the Yala River. None of the four datasets could be fitted to the Poisson distribution. *Anopheles gambiae* was distributed according to the negative binomial in August 2002 only, while *Anopheles funestus* distributions fitted the negative binomial in both May and August 2002. Log-transformed sampling variances and means from all four sampling occasions and both species gave a good fit with Taylor's power law ($r^2 > 95\%$, $P < 0.001$), with values of $b > 1.0$.

Using the negative binomial model, the estimated minimum sample size required to achieve a precision level of 30% with a type I error of 0.05 for the average mosquito abundance was calculated as 251 houses for *Anopheles gambiae* in August 2002 and 362 and 325 houses for *Anopheles funestus* in May and August 2002, respectively. Note that acceptable precision levels for ecological studies are usually between 10 and 25%. If the required precision was adjusted to 20%, then a minimum sample of 564 houses was required to determine *Anopheles gambiae* mean abundance in August 2002, and 814 houses and 732 houses for *Anopheles funestus* in May and August 2002, respectively.

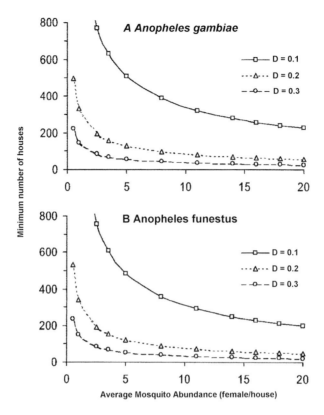

Fig. 1.3. Minimum sample size requirements for malaria vectors *Anopheles gambiae* (A) and *Anopheles funestus* (B) at three precision levels calculated from Taylor's power law based on parameters for all seasons combined (Zhou et al. 2004)

The authors used Taylor's power law to calculate the relationship between the minimum sample size and the mean mosquito abundance per house at several levels of precision and found that to achieve a precision level of 20%, required sample sizes were 69, 251, 274, and 184 houses for *Anopheles gambiae* and 391, 212, 672, and 603 houses for *Anopheles funestus* in May, August, and November 2002 and February 2003, respectively (Fig. 1.3). The extent of the required sampling programme based on these estimates is high and the resources required to conduct such surveys are probably beyond the means of many research programmes.

Lindblade et al. (2000) applied three approaches (direct, minimum sample size, and sequential sampling) to determine if indoor resting densities of *Anopheles gambiae* s.l. exceeded critical levels associated with epidemic transmission in Uganda. They observed that a density of 0.25 *Anopheles* mosquitoes per house was associated with epidemic transmission, whereas 0.05 mosquitoes per house was selected as the normal level expected during non-epidemic months. The direct approach to cal-culating mean *Anopheles* density with an allowable error of 20–50% of the mean would require the sampling of 102–116 houses, respectively. In contrast, with only 7 houses, the minimum sample size approach could be used to determine whether *Anopheles* density had exceeded the critical epidemic level. This method, however, would result in an overestimation of the risk of an epidemic at low *Anopheles* density. Finally, a sequential sampling plan could require as many as 50 houses to conclude that there was a risk of an epidemic, but the disadvantage of this relatively large sample size was offset by being able to preset the probabilities of con-cluding that risk of an epidemic exists at both the critical and normal *Anopheles* densities. The authors concluded that the study illustrated the feasibility and expediency of including monitoring of *Anopheles* resting density in highland malaria epidemic early warning systems.

Graham and Bradley (1961) concluded that if light-traps were used to measure changes in population size in an area, then more statistically reliable results would be obtained if the traps were sited randomly and not in a few selected sites. This, however, does not necessarily follow if information on relative changes in population size is required. It might be better to site traps in localities where they will give large catches than in unproductive areas where sample sizes will be small. It must be remem-bered that under these conditions it may not be appropriate to compare population sizes of different species, because specific sites may be heavily biased in favour of only some species. Deciding how many traps to use and where to locate them is frequently based on logistical capacity and resource availability rather than on the bionomics or spatial distribution of the target species. To address this issue, Ryan et al. (2004) conducted an intensive light trapping study to investigate the spatial pattern of four different mosquito species in Australia using an array of 81 carbon dioxide- and octenol-baited light traps. Spatial autocorrelation analyses were used to determine whether trap counts in neighbouring traps were correlated, and a spatial interpolation method (kriging) was used to produce weekly contour maps of mosquito densities. Kriging is a method of spatial prediction that incorporates a model of the covariance of the random function and uses a weighted moving average interpolation to produce the optimal spatial linear prediction (Cressie 1991). Eighty-one trap sites were distributed approximately in a grid pattern, with adjacent traps separated by a distance of approximately 1.5–3.0 km. Trap sites were divided into two groups of 41 and 40. Adjacent traps were allocated into different groups, and each group of traps was then set on alternate consecutive nights. Traps were set 2–4 h before sunset and retrieved the next morning between 800 and 1000 h. Moran's I analyses were used to assess the degree of spatial autocorrelation, or interdependence, in the numbers of each mosquito species collected in traps separated by different distances. The numbers of each species at unsampled locations throughout the mainland and Macleay Island areas of Redland Shire were estimated using universal kriging applied to the numbers of mosquitoes collected at nearby trap sites. Analyses were conducted using Arc-View GIS 3.2a (ESRI 1996*a*) and ArcView Spatial Analyst extension (ESRI 1996*b*). An estimated 502 571 mosquitoes were collected over 1194 trap nights. In terms of species composition, *Ochlerotatus vigilax* was the most commonly collected mosquito and represented 46.1% of the total collection, followed by *Culex annulirostris* (25.6%), *Coquillettidia linealis* (8.4%), *Culex sitiens* (5.8%), *Ochlerotatus notoscriptus* (3.8%), *Ochlerotatus procax* (2.7%), and *Verrallina funerea* (2.3%). Moran's I an-alyses of weekly mosquito counts were undertaken for four mosqueito species (*Coquillettidia. linealis, Culex annulirostris, Ochlerotatus notoscriptus*, and *Ochlerotatus vigilax*). All species, except *Ochlerotatus notoscriptus* exhibited positive spatial autocorrelation in trap counts. Auto-correlation was greatest in *Ochlerotatus vigilax* at a lag distance of 0–1.5 km, with Moran's I values ranging from 0.30 to 0.64. The existence of spatial autocorrelation suggested that the numbers of mosquitoes in unsampled areas could be estimated with greater accuracy using this method, as opposed to relying solely on the sample mean and variance, an important consideration where a species has a skewed population distribution.

A uniform sampling grid of 63 CDC light-traps baited with CO_2 was set up as part of operational research to develop a Geographical Information System for the Coachella Valley Mosquito and Vector Control District (Reisen and Lothrop 1999). Mosquito abundance data obtained from these 63 traps were analysed to determine the mean and variance, to determine the most appropriate transformation for least squares analysis, to verify abundance estimates obtained previously by best-estimate and transect sampling methods, and to evaluate the ability of different sampling schemes (a random sample of 8 traps, transects of 5 or 6 traps, use of the nine most productive traps, and stratified random sampling) to estimate abundance. Traps were positioned at 1-mile intervals across a 10×17 mile area. Conditional simulations were performed on transformed catch data using subsets of traps selected according to sampling designs that either presumed no knowledge of mosquito distribution (transect, uniform, or random) or required information from prior sampling (best-estimate or stratified). The arithmetic mean and variance of trap catches of *Culex tarsalis* were calculated for each trap site and a $\ln(y + 1)$ transformation was successful in controlling the variance, making it independent of the mean. This transformation also normalised the frequency distribution of catch size by trap nights. Variability among trap sites was significant, but contributed only 11.9% to the total sum of squares in the ANOVA. Variability by month contributed 70.1%. A random sample of 8 traps yielded a mean that did not differ significantly from the overall mean, but precision was markedly reduced and confidence intervals increased by 2.4 to 3.3 times compared to those obtained from the full 63-trap sample. One of two transects of 6 and 5 traps selected from the 63-trap array accurately estimated the mean, while the other transect, which crossed some unsuitable habitat, produced a mean that differed significantly from that of the total sample. The mean calculated from the nine most productive traps was significantly higher than the mean for the full 63-trap array and also had the widest confidence intervals of any sample comprising a similar number of traps. Stratified random samples of 11 traps from strata within 1 mile of the shore, an area known to yield high numbers of mosquitoes, produced means that were below the 95% confidence intervals of the overall mean in two trials, and in the third trial, the confidence intervals of the mean did not overlap the confidence intervals for the overall mean. Reduction of the number of traps within the uniform array led to decreased precision of the estimate of the mean, with the result that the mean for only one trial fell within the confidence intervals for the mean of the full 63-trap array. All sampling strategies appeared to depict the pattern of seasonal variation in abundance reasonably well, although the amplitude of the peak varied by a factor of more than 2, depending on sampling strategy. The authors concluded that a uniform distribution of traps is required to provide information on both seasonal and spatial distribution of abundance, with the number of traps required varying according to habitat heterogeneity and degree of dispersal of the target species. In this case, an array of traps separated by 2-mile intervals was ultimately selected as the optimum design for the Coachella Valley sampling programme.

Keating et al. (2003) described a sampling methodology based on a GIS platform that they considered appropriate for studying relationships between human activity and malaria vectors in urban Africa. First, study area maps were overlaid with a 270×270 m grid, corresponding to the resolution of 9×9 pixel LANDSAT Thematic Mapper remote sensing satellite images. The level of urban planning and drainage in each grid cell was assessed and used to stratify the data by assigning a value according to the following: 1) planned (incorporating information on house-spacing, engineered drainage, etc) and well drained; 2) planned, poorly drained; 3) unplanned, well drained; 4) unplanned, poorly drained; and 5) peri-urban. Individual grid cells within a stratum were selected by systematic sampling from a random start. The sampling interval was calculated for each stratum using the formula $I = F/S$, where I is the sampling interval for each stratum, F is the total number of grid cells in each stratum, and S is the desired sample size for each stratum, proportionate to the actual number of grid cells per stratum. A random number generator was used to select the first grid cell and subsequently every Ith grid cell in the respective strata was selected. Within each selected grid cell all aquatic habitats were sampled using a dipper. Pearson's correlation coefficient and multivariate regression were used to link entomological data with socio-economic data. Results indicated that the number of households per selected grid cell was a significant factor affecting the abundance of potential larval sites in both Kisumu and Malindi towns, Kenya. The number of larval habitats per cell increased with the number of households up to a threshold point, beyond which increasing density of households appeared to reduce the number of aquatic habitats.

A common problem facing ecologists is determining when a rare species is not present. Obviously this can be answered only by a complete and 100% search of a habitat, which in most cases is impractical. McArdle (1990), however, showed how to determine the probability (α) of rare species being recorded in a sampling programme.

$$\alpha = 1 - (1 - p)^N \tag{1.19}$$

where p = probability of the species appearing in a single sample, N = the number of random samples taken from the habitat, α = the probability or confidence that the species will be detected in a sampling programme of N samples, so $(1 - p)^N$ = the probability of the sampling programme not detecting it. If the most extreme level of rarity (p) worth detecting in a sampling programme is determined, and also the confidence level (α) that is desirable, then the number of samples (N) estimated to be needed to record this rare species is

$$N = \frac{\log(1 - \alpha)}{\log(1 - p)} \tag{1.20}$$

As an example, to detect a rare mosquito species where $p = 0.02$ with a probability of detecting it of 0.9, this would require as many as 114 samples.

In a most useful paper directed at entomologists Jones (1984) criticised the use of many multiple-comparison tests commonly used to compare three or more means, medians or proportions. He criticised the inadequate coverage given by most statistical text books on the appropriate statistical tests, e.g. Duncan's multiple range test and the Student–Newman–Keuls test. This is a paper that should be read by everybody. A cautionary tale is told by Cruess (1989) who in a review of statistical procedures found that in a survey of 201 papers in the *Journal of the American Society of Tropical Medicine and Hygiene* 73.5% had at least one detectable statistical error, the most common fault was the misuse of the concepts of standard deviation and standard error.

References

Altman DG (1982) Statistics and ethics in medical research. In: Gore SM, Altman DG (eds) Statistics in Practice. British Medical Association, London, UK, pp. 1–24

Amerasinge FP, Ariyasena TG (1990) Larval survey of surface water-breeding mosquitoes during irrigation development in the Mahaweli project, Sri Lanka. J Med Entomol 27: 789–802

Buonaccorsi JP, Liebhold AM (1988) Statistical methods for estimating ratios and products in ecological studies. Environ Entomol 17: 572–580

Cressie N (1991) Statistics for spatial data. Wiley, New York

Cruess DF (1989) Review of use of statistics in the American Journal of Tropical Medicine and Hygiene for January–December 1988. Am J Trop Med Hyg 41: 619–626

Elliott JM (1977) Some Methods for the Statistical Analysis of Samples of Benthic Invertebrates. Freshw biol Ass sci Publ No. 25

Elliott JM, Drake CM (1981) A comparative study of seven grabs used for sampling benthic macroinvertebrates in rivers. Freshwater Biology 11: 99–120

[ESRI] Environmental Systems Research Institute Inc. (1996a) ArcView GIS 3.2a for Windows. ESRI, Redlands, Australia

[ESRI] Environmental Systems Research Institute Inc. (1996b) Using the Arc-View Spatial Analyst. ESRI, Redlands, Australia

Githeko AK, Mbogo CNM, Curtis CF, Lines J, Lengeler C (1996) Entomological monitoring of large-scale vector-control interventions. Parasitol Today 12: 127–128

Graham JE, Bradley IE (1961) An evaluation of some techniques used to measure mosquito populations. Proc Utah Acad Sci Arts & Letters 39: 77–83

Hurlbert SH (1984) Pseudoreplication and the design of ecological field experiments. Ecol Monogr 54: 187–211

Ikemoto T (1978) Studies on the spatial distribution pattern of larvae of the mosquito, *Anopheles sinensis*, in rice fields. Researches in Population Ecology 19: 237–249

Iwao S, Kuno E (1968) Use of the regression of mean crowding and mean density for estimating sample size and the transformation of data for the analysis of variance. Researches in Population Ecology 10: 210–214

Jones D (1984) Use, misuse, and role of multiple-comparison procedures in ecological and agricultural entomology. Env Entomol 13: 635–649

Karandinos MG (1976). Optimum sample size and comments on some published formulae. Bull ent Soc Am 22: 417–421

Keating J, Macintyre K, Mbogo C, Githeko A, Regens JL, Swalm C, Ndenga B, Steinberg LJ, Kibe L, Githure JI, Beier JC (2003) A geographic sampling strategy for studying relationships between human activity and malaria vectors in urban Africa. Am J Trop Med Hyg 68: 357–365

Kuno E (1976) Multi-stage sampling for population estimation. Researches in Population Ecology 18: 39–56

Kuno E (1991) Sampling and analysis of insect populations. Annu Rev Entomol 36: 285–304

Lemeshow S, Hosmer DW, Klar J, Lwanga SK (1990) Adequacy of Sample Size in Health Studies. John Wiley & Sons, Chichester

Liebhold AM, Rossi RE, Kemp WP (1993) Geostatistics and geographic information systems in applied insect ecology. Annu Rev Entomol 38: 303–327

Lindblade KA, Walker ED, Wilson ML (2000). Early warning of malaria epidemics in African highlands using *Anopheles* (Diptera: Culicidae) indoor resting density. J Med Entomol 37: 664–674

Lloyd M (1967) Mean crowding. J Anim Ecol 36: 1–30

Mackey BE, Hoy JB (1978) *Culex tarsalis*: Sequential sampling as a means of estimating populations in Californian rice fields. J Econ Entomol 71: 329–334

McArdle BH (1990). When are rare species not there? Oikos 57: 276–277

McArdle BH, Gaston KJ, Lawton JH (1990) Variation in the size of animal populations: patterns, problems and artifacts. J Anim Ecol 59: 439–454

McLaughlin RE, Brown MA, Vidrine MF (1987) The sequential sampling of *Psorophora columbiae* larvae in rice fields. J Am Mosq Control Assoc 3: 423–428

Mogi M, Mokry J (1980) Distribution of *Wyeomyia smithii* (Diptera, Culicidae) eggs in pitcher plants in Newfoundland, Canada. Trop Med 22: 1–12

Perry JN (1989) Review: Population variation in entomology: 1935–1950. I. Sampling. Entomologist 108: 184–198

Pitcairn MJ, Wilson LT, Washino RK, Rejmankova E (1994) Spatial patterns of *Anopheles freeborni* and *Culex tarsalis* (Diptera: Culicidae) larvae in California rice fields. J Med Entomol 31: 545–553

Prepas EE (1984) Some statistical methods for the design of experiments and analysis of samples. In: Downing JA, Rigler FA (eds) A Manual on Methods for the Assessment of Secondary Productivity in Fish Waters, 2nd edn. IBP Handbook No. **17**. Blackwell Scientific Publications, Oxford, pp. 266–335

Régnière J, Sanders CJ (1983) Optimal sample size for the estimation of spruce budworm (Lepidoptera: Tortricidae) populations on balsam fir and white spruce. Can Entomol 115: 1621–1626

Reisen WK, Lothrop HD (1999). Effects of sampling design on the estimation of adult mosquito abundance. J Am Mosq Control Assoc 15: 105–114

Reisen WK, Siddiqui TF, Aslamkhan M, Malik GM (1981) Larval inter-specific associations and physico-chemical relationships of ground-water breeding mosquitoes of Lahore. Pak J sci Res 3: 1–23

Reuben R, Das PK, Samuel D, Brooks GD (1978) Estimation of daily emergence of *Aedes aegypti* (Diptera: Culicidae) in Sonepat, India. J Med Entomol 14: 705–714

Ritchie SA, Johnson ES (1991) Distribution and sampling of *Aedes taeniorhynchus* (Diptera: Culicidae) in a Florida mangrove forest. J Med Entomol 28: 270–274

Ryan PA, Lyons SA, Alsemgeest D, Thomas P, Kay BH (2004). Spatial statistical analysis of adult mosquito (Diptera: Culicidae) counts: an example using light trap data, in Redland Shire, southeastern Queensland, Australia. J Med Entomol 41: 1143–1156

Sandoski CA, Kring TJ, Yearian WC, Meisch MV (1987) Sampling and distribution of *Anopheles quadrimaculatus* immatures in rice fields. J Am Mosq Control Assoc 3: 611–615

Service MW (1971) Studies on sampling larval populations of the *Anopheles gambiae* complex. Bull World Health Organ 45: 169–180

Smith PG, Morrow RH (eds) (1996) Field Trials of Health Interventions in Developing Countries. A Toolbox. 2nd edn. Macmillan Education Ltd, London

Solow AR (1989) Bootstrapping sparsely sampled spatial point patterns. Ecology 70: 379–382

Southwood TRE (1978) Ecological Methods with Particular Reference to the Study of Insect Populations. Chapman & Hall, London

Southwood TRE, Henderson PA (2000) Ecological Methods. 3rd edn. Blackwell Science, Oxford

Taylor LR (1961) Aggregation, variance and the mean. Nature 189: 732–735

Taylor LR (1984) Assessing and interpreting the spatial distribution of insect populations. Annu Rev Entomol 29: 321–357

Wilson LT, Room PM (1983) Clumping patterns of fruit and arthropods in cotton, with implications for binomial sampling. Environ Entomol 12: 50–54

Yates F (1981) *Sampling Methods for Censuses and Surveys,* 4th edn. C. Griffin & Co., London

Zhou G, Minakawa N, Githeko A, Yan G (2004) Spatial distribution patterns of malaria vectors and sample size determination in spatially heterogeneous environments: a case study in the west Kenyan highland. J Med Entomol 41: 1001–1009

Chapter 2 Sampling the Egg Population

Oviposition strategies of mosquitoes can be classified into four broad categories. Species of *Anopheles, Toxorhynchites, Sabethes* and *Wyeomyia* hover above the water surface and deposit eggs singly onto the water surface, often without ever coming into physical contact with the water. Species of *Culex, Coquillettidia* and *Culiseta* lay egg-rafts onto the water surface. *Mansonia* species fix their eggs to the surface of vegetation, often below the water surface. *Aedes, Ochlerotatus* and *Psorophora* species lay eggs singly, not in water, but on substrates that are subject to intermittent flooding, such as leaf litter, soil at the edges of ponds, on the walls of man-made containers, or on the surfaces of plants, tree-holes and bamboo. Mosquito eggs are found in many different habitats, including small pools, large marshes, rock pools, tree-holes, plant axils, flower bracts, fallen leaves, fruit husks, empty snail shells, bromeliads and a variety of man-made containers.

Oviposition site location and selection are governed by the combined effects of a range of physical and chemical stimuli that act over different distances from a potential oviposition site. Long-range cues are thought to be primarily visual, and may include optical density of the water, reflectance, and colour. Short-range stimuli include the chemical and physical characteristics of the environment and a large number of oviposition attractants and stimulants have been identified, mostly in the laboratory. Readers are directed to the review by Bentley and Day (1989) for further information on the subject of chemical and behavioural aspects of mosquito oviposition. Recent studies have demonstrated that preferences for certain oviposition media can be conditioned by exposure to those conditions as immatures (e.g. Kaur et al. 2003; McCall and Eaton 2001) and McCall and Cameron (1995) present a recent review of the role of pheromones in oviposition by insect vectors.

The study of egg biology and ecology can reveal much useful information. For example, the detection of eggs in aquatic habitats gives more reliable information on the oviposition sites selected by females than the presence or absence of larvae, as not all habitats selected for oviposition will prove suitable for development to the larval stages. Eggs laid in unsuitable habitats nevertheless represent part of the reproductive input of the adults. Egg surveys are particularly useful for species that remain in the egg state for many months, because potential larval habitats can be identified and enumerated without waiting for the larvae to appear. The ability to sample eggs and derive population estimates in natural habitats is of paramount importance in ecological studies concerning population dynamics. If there is also information on the size of the emergent adult population, then the probability of a viable egg giving rise to an adult mosquito can be estimated. The importance of this parameter in predicting population size and the impact of genetic control measures has been stressed by Cuéllar (1969*a,b*). Lopp (1957) emphasised the usefulness of egg surveys in predicting the potential size of pest populations of mosquitoes.

In genetic control programmes based on the production of sterile eggs by field populations, the ability to sample the egg population enables the proportions of sterile eggs laid at varying distances from the centre of control operations to be assessed. Sampling the egg population is usually more difficult and time consuming than larval surveys, especially when eggs have to be extracted from samples of soil and debris.

Specific identification of the eggs obtained in surveys may sometimes be difficult because eggs have been described for comparatively few species (although the situation has improved in recent years with the publication of electron micrographs of the eggs of many species, primarily by John Linley and colleagues), and some species cannot be separated on egg morphology. This can usually be overcome by either identifying 1st instar larvae dissected out from the eggs or by soaking the eggs and identifying the resultant 4th instar larvae or adults.

Because of the diversity of oviposition sites utilised by mosquitoes many different sampling techniques would be required to adequately sample the eggs of all species. However, relatively few methods have been developed for sampling mosquito egg populations, besides ovitraps. Apart from sampling eggs already present in natural habitats, useful information can be obtained by collecting eggs from artificial oviposition sites. Such techniques have frequently been used in surveillance of *Aedes aegypti* and *Aedes albopictus* (Chadee and Corbet 1987; Evans and Bevier 1969; Fay and Eliason 1966; Freier and Francy 1991; Jakob and Bevier 1969*a,b*; Jakob et al. 1970; Marques et al. 1993; Pratt and Jakob 1967; Subra and Mouchet 1984; Thaggard and Eliason 1969) and with other aedine species ovipositing in small container habitats such as domestic utensils, tree-holes and snail shells (Buxton and Hopkins 1927; Corbet 1963, 1964*a*; Dunn 1927; Goettel et al. 1980; Kitron et al. 1989; Lambrecht and Zaghi 1960; Lewis and Tucker 1978; Philip 1933; Service 1965; Tikasingh and Laurent 1981; Yates 1979).

Other types of traps have been developed to sample *Culex* (Haeger and O'Meara 1983; O'Meara et al. 1989b; Reiter 1983; Reiter et al. 1986; Strickman 1988; Surgeoner and Helson 1978), *Haemagogus* (Chadee et al. 1984), *Toxorhynchites* (Schuler and Beier 1983), *Eretmapodites* (Lounibos 1980) and *Trichoprosopon* (Lounibos and Machado-Allison 1986).

Natural oviposition sites

Direct collection methods

Anopheles

Few methods have been developed to sample *Anopheles* eggs, but Barber (1935) seems to have been the first to have seriously proposed a collecting method. He successfully collected eggs by skimming the water surface of larval habitats with a collecting bowl and straining the contents through a white muslin bag or mitten placed over the hand. Sometimes several hundred *Anopheles* eggs were collected by this method. Both Bates (1940) and Lewis (1939) successfully used this technique to collect eggs of the *Anopheles maculipennis* complex from natural habitats in Albania. Bates (1941) proposed a modification, whereby the mitten is replaced by a piece of muslin stretched over a small wooden hoop. Several such sieves can be made and placed at the edge of larval habitats and a known number of dips strained through them. A plastic wash bottle is used to wash fine silt through the sieves and also to wash off any eggs stuck to pieces of wood or debris. Eggs can be collected from the sieve, or the contents floated off in water. An alternative method is to use a metal dipper with the bottom removed and replaced by a fine metal gauze. After several dips, or sweeps through the water, the dipper is turned upside down and the contents washed into a bowl. Individual eggs can be picked out with fine forceps or with a glass pipette and sorted into tubes for later counting and identification.

Earle (1956) described an automatic strainer for concentrating larval collections and as he mentioned that *Anopheles* eggs were also retained this might prove a useful piece of apparatus for removing eggs from collections made with a dipper. A description of the method is given in Chap. 3.

Swellengrebel and de Buck (1938), Aitken (1948) and Rozeboom and Hess (1944) found it unnecessary to strain their samples and simply counted *Anopheles* eggs on the surface of the water collected in a dipper. In Holland Swellengrebel and de Buck (1938) collected as many as 242 eggs of the *Anopheles maculipennis* complex from 10 dips. Muirhead-Thomson (1940 *a,b*) found that eggs of certain Indian and African malaria vectors, such as *Anopheles minimus* and *Anopheles funestus*, were so small that they were washed through the usual type of muslin mitten or sieve. He therefore collected eggs by skimming the surface with a white enamel tray and reported that *Anopheles* eggs were easily seen against the white background of the tray. The method proved successful in still waters but in streams, even where *Anopheles minimus* was breeding abundantly, no eggs were collected. To try to overcome this difficulty Muirhead-Thomson (1940 *a,b*) removed pieces of vegetation and scraped surface mud from the edges of streams, and washed the material in a bowl. Although very successful in still waters, this technique was still only partially successful with stream-breeding *Anopheles* (Muirhead-Thomson, 1940*a*). With pool breeders Muirhead-Thomson (1940*a*) found a good correlation between the abundance of eggs and larvae of different *Anopheles* species collected from the same habitats.

Similarly, Rozeboom and Hess (1944) found very good correlations between the numbers of eggs and larvae of *Anopheles quadrimaculatus* s.l. collected by skimming the water surface of reservoirs containing differing amounts and types of vegetation. Aitken (1948) compared the incidence of both eggs and larvae in various habitats in Albania. He found that of the 546 collection stations having immature stages of *Anopheles* both eggs and larvae were collected from 81%, while eggs but no larvae were collected from 13% of them. These results further demonstrate the value of egg surveys.

In Sierra Leone Muirhead-Thomson (1945) collected eggs of both *Anopheles gambiae* and *Anopheles melas* from pools, puddles and partially dried up streams by using either a white enamel scoop or bowl, or directly from the water with a wire loop. In India both Muirhead-Thomson (1940*a*) and Russell and Rao (1942) found that dipping was unnecessary, and instead collected *Anopheles* eggs by lifting them from the water surface with a small wire loop.

In Thailand it appears that *Anopheles dirus* sometimes lays her eggs on damp soil above the water line and Rosenberg (1982) obtained eggs by sluicing the banks of a larval habitat with 5–10 litres of clean water and then quickly ladling the water draining back into the centre of the pool and passing it through a 150-μm cloth sieve. Examination under a microscope revealed unhatched eggs of *Anopheles dirus*. A high proportion of eggs

(5/7 and 21/33) recovered up to 10–12 days after a breeding site was drained remained viable and hatched in the laboratory after flooding.

In Kenya Beier et al. (1990) collected dry soil from habitats and obtained *Anopheles gambiae* and *Anopheles arabiensis* larvae when the samples were flooded with water in the laboratory. The authors suggested that there may be greater tolerance to desiccation in populations of a species living in dry areas than exhibited by eggs of the same species collected from wetter areas.

It is unlikely, however, that the total egg population can be removed and counted from many natural habitats. In most instances *Anopheles* egg surveys will only detect the presence of eggs in a habitat, or give relative population indices such as the number of eggs per dip or the number collected with a wire loop within unit time. However, even with these methods there may be sampling problems. For example, if the water surface in one habitat is clean most eggs will be stranded along the edges, but if in another site floating vegetation and debris are present a number of eggs will cling to these objects and occur away from the edges. The distribution pattern of the eggs will consequently differ in the two habitats and this will most likely be reflected in the numbers caught, although the actual number of eggs present in both habitats could be the same. Furthermore, if the same number of eggs are present in two different sized pools, it is likely that more will be collected per dip, or unit time, from the smaller pool. Such factors have to be taken into consideration if the numbers caught from different habitats are to be compared.

Aedes and Ochlerotatus

Occasionally aedine eggs have been recovered from the field by locating the sites in which they are laid. Corbet (1964*b*, 1965, 1966), for example, found that in the Arctic *Ochlerotatus nigripes* lays her eggs on the mossy slopes of the northern banks of ponds, whereas eggs of *Ochlerotatus impiger* are deposited some 5–20 mm below the soil surface in cracks. Corbet (1964*b*) reported that oviposition sites of *Ochlerotatus nigripes* were often rendered conspicuous by the corpses of individuals that died after oviposition. Wesenberg-Lund (1921) located eggs of *Ochlerotatus communis* amongst leaf litter of a dried up pond. Service found that eggs of *Ochlerotatus cantans* and *Ochlerotatus rusticus* are mainly deposited on the undersides of the top layer of leaves resulting from the previous autumn, with very few eggs occurring in soil beneath the leaf litter. Smith (1904), quoted by Mattingly (1969), was able to detect eggs of various aedine species in cut out sods of earth by examining the cut edges with a hand lens.

James (1966) also observed the location of aedine eggs in soil samples by visual inspection.

Mansonia (Mansonioides)

Species of the subgenus *Mansonioides* and some species of *Mimomyia* (e.g. *Mimomyia hybrida*) lay their eggs in clusters glued onto the undersides of leaves of floating aquatic plants, such as *Pistia stratiotes*. Dyar and Knab (1916) collected egg masses of *Mansonioides* from *Pistia* plants, and Wanson (1944) reported that it was comparatively easy to collect eggs of *Mansonioides* from their natural habitats. Bonne-Wepster and Brug (1939) also had little difficulty in finding *Mansonioides* eggs on the underside of leaves.

In Sri Lanka Laurence and Samarawickrema (1970) recorded daily the presence or absence of *Mansonioides* egg masses on specific plant leaves and compared their distribution, resulting from overnight oviposition, with a Poisson model to determine whether the egg masses were randomly distributed. They were found to exhibit a distinctly aggregated distribution, and there was a marked preference for oviposition on leaves that had already been selected for egg laying. In addition to selection for individual leaves there was also preference for specific areas of the habitat. Laboratory experiments, however, failed to confirm these field observations.

In Thailand Gass et al. (1983) successfully collected egg masses of *Mansonia annulifera*, *Mansonia indiana* and *Mansonia uniformis* by removing plants including *Pistia*, *Eichhornia*, *Salvinia*, *Jussiaea*, *Nymphaea* and *Marsilia* from either randomly selected 15 × 15-cm or 1 × 1-m plots within selected areas of 100–200 m^2. The distribution of eggs was aggregated and fitted a negative binomial distribution.

Other culicine species

Egg rafts of *Culex, Culiseta, Uranotaenia* and *Coquillettidia* and also some other genera are usually readily seen on the water surface, and are often collected in larval surveys. Buxton and Breland (1952) successfully collected egg rafts of *Culiseta morsitans* from natural habitats, while Barr (1958) used his hand to submerge aquatic plants so that egg rafts floating on the water could be more easily seen. To collect egg rafts of *Coquillettidia perturbans* in Canada, Allan et al. (1981) constructed floating oviposition frames by bending plastic tubing into a circle to enclose about 0.05 m^2 of water. Care was taken to ensure that no egg rafts were present at the beginning of the exposure period, then at weekly intervals over a period of about 10 weeks all egg rafts were removed from a series of 15 frames. Only nine of the oviposition frames contained egg rafts, the mean number being 2.73 rafts/frame, 86% of which were in circles near or enclosing *Typha latifolia* and 13% in frames with and near *Carex* spp.

Extraction methods for soil samples

The detection of eggs in either natural habitats or in samples removed from oviposition sites gives some information regarding the actual site in which the eggs are laid, but the method is very time consuming and cannot be used quantitatively. One of the commoner techniques for both detecting and determining the relative abundance of aedine eggs in oviposition sites is to count the numbers of larvae that hatch when soil and leaf litter samples from oviposition sites are soaked in water (Bidlingmayer and Schoof 1956; Bradley and Travis 1942; Breeland and Pickard 1963; Buxton and Breland 1952; de Szalay and Resh 2000; Dunn 1926; Elmore and Fay 1958; Enfield and Pritchard 1977; James 1966; Micks and McNeil 1963; Ritchie and Addison 1991; Service 1965, 1970; Wallace et al. 1990; Wilkins and Breland 1949).

Fallis and Snow (1983*b*) reported that with *Ochlerotatus punctor* and *Ochlerotatus cantans* a slow reduction in oxygen content, achieved by immersing eggs in 0.1% Bacto Nutrient broth (Novak and Shroyer 1978) induced hatching, whereas transferring eggs directly to deoxygenated water failed to stimulate hatching.

Filsinger (1941) studied the vertical distribution of eggs of *Aedes vexans* in sod samples removed from oviposition sites. Vegetation that grew above 2 in from soil level was cut and discarded, then vegetation that reached from 1–2 in was cut and soaked in water, and vegetation that grew up to an inch from the sod samples was similarly cut and soaked. Then the sods were sliced and separated into the 1st inch below soil level, the 2nd inch and finally the remaining 4 in. All samples were soaked in water for 24 h and the larvae that hatched counted. Based on larval counts he concluded that only 4% of the eggs were located above the soil, 14% were contained within the 1st inch of soil, 20% in the 2nd inch and 47% below this. 'Trimmings', which represented material remaining after the sod samples had been cut up, contained about 12% of the eggs. Filsinger (1941) then placed sod samples on the top of a nest of 20-, 40-, 60-, 80- and 100-mesh sieves and washed them for 1 h with water from a sprinkler. None of the eggs was retained by the top sieve, most were collected by the second sieve. From this simple experiment he concluded that in the field, rain was important in washing eggs down to lower depths.

In studies on *Ochlerotatus taeniorhynchus* and *Ochlerotatus sollicitans* Bradley and Travis (1942) removed sod samples with an iron ring, 1 in deep and 3.3 in in diameter, mounted on a 3-ft hoe handle. The 8-in square samples obtained were soaked in water and the number of larvae that hatched after 24 h used to indicate the relative abundance of these two species. Elmore and Fay (1958) also studied the oviposition sites selected by these mosquitoes, but undertook a more critical evaluation of the method for determining their relative abundance. They conditioned soil samples at either 15, 21, or 26°C (60, 70 or 80°F) for 1–12 days before flooding them with water at the temperature at which they were conditioned. After soaking, the water was poured off and each sample stored for 10 days at 26°C (80°F) and 70% R.H. before being resoaked. It was found that temperature greatly influenced egg hatching. Fewer eggs hatched on the first soaking from samples that were stored at 15°C (60°F) than those that were held at higher temperatures. Furthermore, the proportions of the two species obtained varied according to temperature and storage time (1–12 days). Less than 3% of the larvae that hatched from samples stored and flooded at 15°C (60°F) were *Ochlerotatus taeniorhynchus*, and samples conditioned at 21°C (70°F) for 1–3 days before flooding produced no larvae of *Ochlerotatus taeniorhynchus*. At 26°C (80°F) very few larvae of *Ochlerotatus taeniorhynchus* hatched from samples for 1–2 days, but with increasing storage time the number of *Ochlerotatus taeniorhynchus* that hatched increased and rapidly exceeded those of *Ochlerotatus sollicitans*. They also found that the prevalence of *Ochlerotatus sollicitans* calculated from the numbers of 4th instar larvae obtained from soaking soil samples was greater than when identification was based on 1st instar larvae. Evidently their rearing procedures favoured *Ochlerotatus sollicitans* more than *Ochlerotatus taeniorhynchus*. Bidlingmayer and Schoof (1956), however, did not experience this difficulty. Soil samples from salt marshes were held for a week at 26°C (80°F) then flooded and the numbers of larvae, mainly *Ochlerotatus taeniorhynchus* and *Ochlerotatus sollicitans*, that hatched within 24 h were counted. The proportions of these species were about the same when based on identification of 1st instar larvae or reared adults.

Ritchie and Johnson (1991*a,b*) used a 10-cm diameter golf-core sampler to take 10-cm deep soil cores in mangrove forests to study the distribution of *Ochlerotatus taeniorhynchus* eggs. Soil samples were soaked in water for at least 3 days to allow newly oviposited eggs to mature, after which the core samples were flooded with dilute yeast solutions to promote hatching. In other surveys Ritchie and Addison (1991) collected soil samples from mangrove forests with modified 6- and 60-ml plastic syringes. The tips of the syringes were cut off and the front edge of the barrel bevelled to form sharp cutting edges. It was apparently necessary to cut off the tip of the plunger to enable it to be inserted into the barrel with minimum resistance. When the syringes were pushed 2.5 cm into the soil, cores of 3- and 15-ml volume were obtained by the smaller and larger syringes.

De Szalay and Resh (2000) collected soil samples from dry wetlands from among stands of saltgrass (*Distichlis spicata*), pickleweed (*Salicornia virginica*), alkali bulrush (*Scirpus robustus*) and cattail (*Typha* spp.). Plant stems more than 10 cm above-ground were cut off and discarded and soil samples were collected using a core sampler (11 cm diameter) that was driven about 15 cm into the soil. Samples were maintained in the laboratory at ambient temperature with a 14:10 light:dark cycle and were flooded with a dilute nutrient broth made from crushed corn (1% by volume) and filtered water collected from irrigation channels. The broth was decanted after 24 h, and aquatic macroinvertebrates were collected. To allow for the fact that some mosquito eggs do not hatch the first time they are flooded, the samples were maintained for an additional 2–6 weeks and then were reflooded. Only aedine mosquito eggs wrere collected from all substrate types, at densities ranging from 9–174 eggs m-2.

Buxton and Breland (1952) collected and flooded samples of mud, soil leaf litter and debris from a wide range of different types of habitat to detect oviposition sites. Although they recorded over 19 species, less than a third of their samples yielded larvae, and they concluded that this procedure was not a suitable routine method for detecting mosquito breeding sites. However, although samples from tree-holes and rock pools were repeatedly soaked, larger sod samples were soaked only once, and this would increase the likelihood of negative results. In Panama Stone and Reynolds (1939) did not collect samples, but flooded small damp depressions in natural sites, and by this procedure identified the oviposition sites of several *Culex, Anopheles* and *Psorophora* species.

Not all eggs hatch on the initial soaking of soil or other samples, and this is presumably an adaptation to oviposition sites that undergo frequent flooding and drying out. About 80% of the eggs of *Ochlerotatus detritus* hatched from mud collected from salt marshes on the 2nd–5th soakings, but a few remained unhatched until the 18th soaking (Service, 1968*a*). When gourds (used in Nigeria to study mosquitoes breeding in tree-holes) were repeatedly soaked a few aedine eggs failed to hatch until the 7th soaking. Similarly, in Panama a small number of eggs of *Haemagogus* species in bamboo pots remained unhatched until the 10th flooding (Galindo et al. 1955). Buxton and Breland (1952) soaked tree-hole litter 13 times and obtained an egg hatch of *Ochlerotatus triseriatus* on 12, and of *Ochlerotatus zoosophus* on 9 occasions. With the rock pool species *Aedes vittatus*, most of the eggs contained in mud samples hatched during the 2nd and 3rd soakings, but a few hatched on the 6th soaking (Service, 1970). Although the detection and estimation of egg populations by soaking soil samples might appear an attractive and simple procedure a number of difficulties exist. For reliable results each sample must be flooded a relatively large number of times, and this makes the procedure time consuming. Another disadvantage is that there is no guarantee that the proportions of two, or more, aedine species present in an oviposition site will be accurately measured by the species composition of the larvae that hatch. Eggs of some species may hatch more readily than others. Despite these limitations the method can still in many instances be usefully employed in mosquito surveys.

In studying the spatial distribution of the immature stages of *Culicoides variipennis* Vaughan and Turner (1987) used a simple plastic sampler that was thrust into the mud and which had sliding flexible partitions inserted to divide the sample from the top of the mud downwards into sections that were 0–1, 1–2, 2–3 and 3–5 cm deep. The sampler was then eased out of the mud and the various sections washed into separate containers. If certain modifications were made to this apparatus, such as giving the sliding partitions of metal a cutting edge, it might prove useful for studying the depth distribution of aedine eggs.

The earliest published description of an egg separating machine appears to be given by Gjullin (1938), who adapted a 24-in wide commercial grain cleaner to sift dried soil samples passed over a series of 14-, 30- and 40-mesh shaker sieves to remove particles. Eggs and fine debris which passed through these sieves were collected on an 80-mesh sieve from which they were shaken onto a 60-mesh roller sieve and finally collected underneath in a small pan. Gjullin (1938) successfully used this equipment to extract eggs of *Ochlerotatus sticticus, Aedes vexans* and *Ochlerotatus dorsalis*, and although employed by Stage et al. (1952) in their studies on the mosquitoes of the northwestern states of America the apparatus has been little used by others.

Horsfall (1956) described a method of wet sieving to remove eggs of floodwater mosquitoes (*Psorophora, Aedes* and *Ochlerotatus*) from soil and leaf litter samples. Samples are removed from oviposition sites by a 'cutting square'. This consists of a sharpened metal band bent into a 6-in square fixed to a wooden square with a handle on top. It is pushed into the ground to a depth of about 1 in and the sample cut from the soil below with a spade after which it is placed in a bag and taken to the laboratory. The sod of soil is then placed in the inner of three concentric cylindrical metal screens, having 4, 8 and 18 meshes per in respectively. The lower halves of these cylinders are immersed in a water bath. A central shaft runs through the middle of the inner cylinder and when its handle is turned the three cylindrical sieves rotate and pass through the water bath (Fig. 2.1).

The operator turns the handle at a rate of about 50 rev./min, first in one direction then in the opposite direction to complete about 125 revolutions.

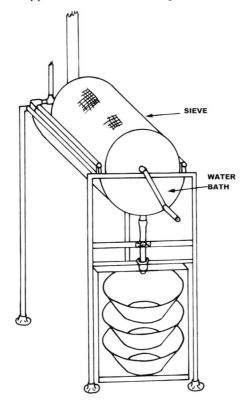

Fig. 2.1. Horsfall's soil washing machine (after Horsfall 1956)

This treatment breaks up the soil sample in the sieves and flushes the eggs, and other comparatively small particles, through into the water bath. During the final 25 turns, a bottom tap on the water bath is opened and the contents empty into the first of three metal sieves (40, 60 and 100 mesh/in) placed one above the other. A strong jet of water washes the eggs through the top two sieves onto the screen of the bottom sieve, from which they are washed on to a small cylindrical fine mesh 'transfer' screen. Further separation from soil particles is achieved by flotation. Eggs are washed with about 1.5 litres of saturated sodium chloride solution from the transfer screen into a 2-litre conical funnel. The solution is stirred for 1–2 min with a glass tube through which air is passed from a pump. This causes the eggs, together with other fine organic particles, to float to the top, while soil particles sink to the bottom and are removed by opening a drain tap.

The eggs are then filtered through a fine sieve, and washed with tap water into a small dish. Floating debris and most of the water is decanted as waste and the residue, which contains the eggs, reflooded with saturated salt solution. Eggs float to the top and are poured on to another fine mesh sieve, from which they are washed with water into a dish which is scanned with a microscope and the eggs removed and identified. Horsfall (1956, 1963) reported a recovery rate of eggs of 81–89% and 80% respectively using this method.

An alternative method for separating eggs from other organic debris after extraction using the sieving method of Horsfall (1956) involves pouring the water containing eggs and other organic debris into a porcelain dish, then adding about 100 ml warm water (60–70°C). After a few seconds the contents are poured through a 100-mesh sieve. The retained eggs and organic debris are washed with the minimum amount of water into a dish lined with paraffin wax. About 100 ml of a 0.3% (or stronger) hydrogen peroxide solution, prepared from commercial 3% stock solution, is poured into the wax-lined dish. The hydrogen peroxide results in bubbles forming within debris particles, causing them to float to the surface. After such separation the debris can be poured off, and the eggs which remain on the bottom removed with a pipette. Recovery of aedine eggs ranges from 94–100%. If viable eggs are required the pre-separation heat treatment with warm water should be omitted, but this may reduce the recovery rate to 89–98%. Although Service (1993) reported using this method he preferred the much simpler method of flotation in salt or magnesium sulphate solution.

Chambers et al. (1979) took 15 × 15 × 2.54-cm soil samples from Louisiana rice fields to collect eggs of *Psorophora columbiae, Psorophora ciliata, Psorophora discolor* and *Ochlerotatus sollicitans*. A modified version (no details of the modification were given) of the egg separator of Horsfall (1956) and Meek and Olson (1976) was used to extract eggs from the samples. This was then followed by the salt-water flotation method. Lopp (1957) tested the efficiency of the machine, which he mechanised to cope with large numbers of samples, by placing a single mosquito egg in each of five soil samples. An egg was recovered from four of the samples after they had passed through the machine.

Pausch and Provost (1965) used Horsfall's extraction method to calculate the average number of eggs of *Ochlerotatus taeniorhynchus* per sod sample so that approximate estimates could be made of the total eggs present in different areas. The machine was also used by McDaniel and Horsfall (1963) and Horsfall (1963) to study the local distribution of eggs of floodwater mosquitoes. McDaniel and Horsfall (1963) investigated the location of eggs at different levels in the soil. They selected areas known to contain concentrations of eggs and which were free from sticks and stones which could interfere with the removal of the samples. A 6-in square metal frame with sharpened cutting edges was placed on the ground and 25 metal tubes, 1 in in diameter and 3 in long with the bottom edges sharpened, were placed within the frame (Fig. 2.2).

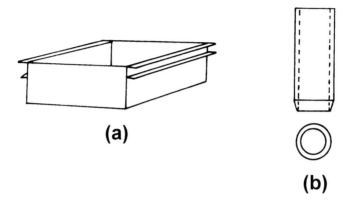

(a)

(b)

Fig. 2.2. Sampling square (*a*); sampling tubes (*b*) (after McDaniel and Horsfall 1963)

Both tubes and the frame were carefully hammered into the soil. The soil outside the frame was removed on three sides to allow a sheet of galvanised steel to be driven horizontally underneath the frame. This prevented the soil falling out of the tubes when the frame together with the tubes was carefully lifted and transported to the laboratory. A square of plywood was placed on top of the tubes to enable the frame to be removed without disturbing the tubes. The soil cores were expelled from the upper end of the tubes by slowly inserting a cylindrical cork plunger. Slices 5-mm thick were cut and isolated, and eggs extracted by sieving and flotation. Most eggs of *Ochlerotatus stimulans* and *Aedes vexans* were found to be within about the upper 25 mm, but a few eggs of *Aedes vexans* were recovered from a depth of 71–75 mm. Checks were made to ensure that the metal tubes had not forced the eggs down to unnatural depths.

In studying the effect of tillage on the distribution of *Aedes vexans* eggs in floodplains, an 8.3-cm diameter commercial grass plugger was used to take core samples to a maximum depth of 12.7 cm (Cooney et al. 1981). In the field the core samples were forced from the plugger into quart-sized cylindrical milk cartons of the same diameter as the corer. In the laboratory a hand-operated mechanical jack was calibrated so that each stroke of the piston forced the core 6 mm up into the carton and out at the top. An electric carving knife was used to slice off these 6-mm sections, which were sealed in plastic bags until processed by a modification of the Horsfall (1956) method. Eggs recovered from 0–6.0 mm represented 61.2% of the total retrieved, from 6.0–12.0 mm 22.4%, from 12.0–18.0 mm 11.2%, from 18.0–25.0 mm 4.4%, from 25.0–31.0 mm 1%, no eggs were recovered from a depth of 31.0–106.0 mm.

Scotton and Axtell (1979) used a 15 × 15-cm stainless steel tray with 3 sides upturned to a height of 5 cm, with the remaining side protruded as a 5-cm lip to take soil samples from dredge spoil. To sample surface soil the lip was pushed down 2 cm into the soil and then the tray thrust horizontally to obtain 15 × 15 × 2-cm (450 cm^3) samples. Soil samples were tipped into water and when necessary broken up using an electric blender with the blades covered with rubber tubing to reduce the risk of damaging the eggs for 15 seconds. Wet sieving and flotation methods modified from Horsfall (1956) and Service (1968b) were applied, and eggs of *Ochlerotatus taeniorhynchus* and *Ochlerotatus sollicitans* floated off in a 1.1 sp. gr. solution of magnesium sulphate. The recovery rate was 71 ± 8%, with only 8.5% of recovered eggs being damaged.

Metge and Hassaïne (1998) used a metallic cylinder (5 cm long × 25 cm^2 cross-section) open at both ends and with a cutting blade on the lower end to take soil samples along transects in temporarily flooded saline depressions in Algeria. The full length of the cylinder was thrust into the soil and the sample was divided into 1 cm long sections using a square metal plate. The authors do not describe the method used to extract the eggs from the 1 cm soil sections. Vertical egg distribution was measured in three vegetational zones: *Salicornia radicans, Arthrocnemum glaucum,* and *Suaeda fruticosa,* representing three distinct soil types. Across all vegetation types, the majority (76.6%) of *Ochlerotatus caspius and Ochlerotatus detritus* viable eggs was found in the first 1 cm and the depth to which the eggs were buried was related to soil texture and structure. For example, in the clay soil-type (*Salicornia radicans*), 90% of eggs were found in the first 1 cm of the soil sample, while in the sandy soils (*Suadea fruticosa*) only 11.1% of eggs were found in the top 1 cm.

A criticism of Horsfall's method is that it necessitates the construction of a special, and fairly elaborate, piece of apparatus. The extraction technique is also time consuming. Despite these limitations, the Horsfall (1956) method, or a modification of it, has been the most commonly used system for extracting aedine eggs from soil. To simplify and speed up the removal of aedine eggs from soil samples a Salt–Hollick soil washing machine (Salt and Hollick, 1944) was used by Service (1968b). The machine is available commercially as a standard piece of equipment used by soil zoologists to extract nematodes from soil samples, but if it has to be constructed it is more easily made than Horsfall's machine.

Soil samples are removed from oviposition sites and transported to the laboratory in plastic bags. They can be processed immediately or stored in a refrigerator until required. Freezing followed by thawing may be useful for helping break up lumps of clay in samples and Fisher (1981) believed that freezing soil samples for at least a day helped to break up soil aggregates. Alternatively, chemical dispersing agents such as sodium citrate (d'Aguilar et al. 1957), sodium hexametaphosphate and sodium carbonate (Raw, 1955) or sodium oxalate (Seinhorst, 1962) can be used. Fisher (1981) emptied his samples into a bucket of warm water containing the water softener Calgon.

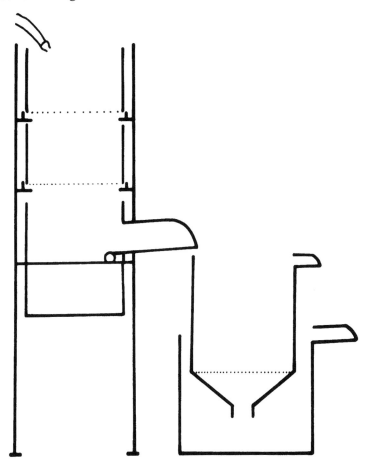

Fig. 2.3. Salt–Hollick type of soil washing machine (after Salt and Hollick 1944)

In extracting *Culicoides* larvae from soil samples Mullens and Rodriguez (1984) found that the addition of a commercial flocculating agent (2 drops, i.e. about 0.05 ml of 0.5% solution, of 'Separan NP10' from Dow Chemical Co.) speeded up settling of mud particles and made sorting and counting easier when there was flotation in NaCl or MgSO$_4$. Following any preliminary treatment the sample is placed in a white bucket and flushed with a strong jet of water. When about three-quarters full the contents are vigorously stirred to break up the sample and dislodge eggs from plant debris, after which it is washed with a strong jet of water from a hose through 7- and 2.5-mm sieves which are mounted above each other in the Salt–Hollick machine (Fig. 2.3).

Small particles, together with the eggs, are washed through the finer sieve into the settling can beneath, then the sample is tipped into the 'Ladell can' which has a 0.2-mm phosphor bronze mesh screen at the bottom. After all the water has drained through, a rubber bung is inserted into the bottom of the can which is then filled with a solution of magnesium sulphate, sodium chloride or almost any other solution, including cane sugar, having a specific gravity of 1.2. The contents are stirred, and after allowing about 5 min for soil particles to sink and organic matter and the eggs to float to the surface, the top half is decanted through a small 0.2 mm sieve (Fig. 2.4).

Fig. 2.4. Decanting eggs and surface debris from Ladell can into a fine sieve

Eggs are dislodged from the sides of the can and from the debris with a washbottle filled with magnesium sulphate. Finally the material collected on the sieve is washed into a convenient container. Samples can be stored in a refrigerator or a deep freeze, or processed immediately. Each sample is tipped into a white porcelain evaporating dish and examined under a stereoscopic microscope, and the eggs, which float to the top, removed and counted. Sometimes a lot of plant debris floats on top of the sample and further separation is required. The sample is poured into a 200-ml narrow-mouthed centrifuge bottle containing saturated sodium chloride and spun at $700 \times g$ for 10 min. Mosquito eggs settle out at the top and are decanted into an evaporating dish. Very occasionally when the samples contain excessive amounts of plant litter, such as small seeds and pieces of leaves, it may be necessary to tip the sample into a glass jar and place it under partial vacuum in a vacuum desiccator for 2–3 min, prior to centrifuging.

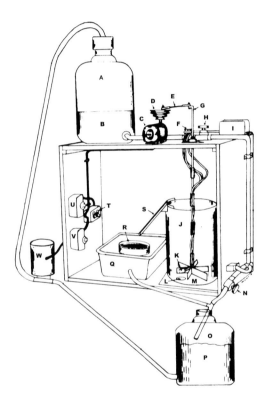

Fig. 2.5. Modified Salt–Hollick machine (Lawson and Merritt 1979) A—storage reservoir, B—magnesium sulphate solution, C—agitator motor, D—three step sheave, E—connecting rod, F—gang valve, G—agitator shaft, H—solution flow valve, I—air pump, J—flotation cylinder, K—air stone, L—drain hole, M—agitator blade, N—cylinder drain valve, O—aspirator bottle, P—filter floss, Q—collection basin, R—retention sieve, S—sluice gate, T—on-off switch, U—agitator speed control, V—filter pump speed control, W—filter pump.

As pressure is reduced plant debris sinks to the bottom. Care must be taken not to reduce the pressure too much otherwise the desiccator may implode;

for this reason it is recommended that the desiccator is placed within a strong metal wire cage.

Fallis and Snow (1983*a*) found that placing samples under a vacuum was ineffective in recovering eggs of *Ochlerotatus punctor* from leaf litter because many became entangled on sinking organic matter and were not recovered. They also reported that flotation and centrifugation did not separate eggs from organic matter; they washed their leaf litter samples with a water jet through a number of sieves (5.6 mm, 710 μm and 80 μm).

Ritchie and Johnson (1989) pointed out that even with wet sieving and flotation methods it can sometimes still be difficult to identify aedine eggs from background debris. This is especially so in mangrove soils rich in peat deposits, because low density peat fragments are not adequately separated from the eggs. To make separation easier, they placed the filtered material in 2.5% sodium hypochlorite (50% commercial bleach) for 3–5 min, stirring occasionally until the soil and peat particles turned brownish yellow. The solution was then poured through a 0.15-mm sieve and washed for 30 s with water to remove much of the peat, which had been partially dissolved by the bleach. The darker unbleached eggs should be removed or counted immediately because with time they will tend to bleach, making separation more difficult.

It has been shown that the Salt–Hollick soil washing machine removes about 83% of the eggs from soil samples (Service, 1968*b*). Eggs of many *Aedes* and *Ochlerotatus* species and also those of *Culiseta morsitans* are still viable after samples have been stored in a refrigerator, processed through the soil washing machine and centrifuged in sodium chloride. This method has been used to extract eggs of several *Aedes* and *Ochlerotatus* species from hundreds of soil and leaf litter samples collected from woodland habitats and fresh and salt water marshes.

Lawson and Merritt (1979) described modifications to the Salt–Hollick soil washing machine (Fig. 2.5). These modifications comprised an electric motor to drive a paddle to agitate the soil sample when flooded with magnesium sulphate, and an electric pump to bubble air through the sample to cause further agitation. Another modification allows the magnesium sulphate flotation solution to be recycled after filtering, so as to flood a subsequent soil sample.

Because eggs of some *Ochlerotatus* species, such as those of *Ochlerotatus nigromaculis* readily hatch on the first flooding (Husbands 1952), Miura (1972) considered that methods using water for separating eggs from soil samples might not be reliable, presumably because some eggs might hatch during processing. To overcome this he used a sonic sifter (Allen–Bradley, Model L3P), originally designed for particle size analysis, to separate *Ochlerotatus nigromaculis* eggs from air-dried soil samples.

Samples measuring 25 × 25 mm were cut to a depth of 10 mm from ovi-position sites and processed through a sonic sifter having 14-, 40-, 60-, 80- and 100-mesh sieves stacked on top of each other. The amplitude and pulse of the sifter were set at values of 5 and 4, respectively and the samples sifted for 5 min. Most eggs of *Ochlerotatus nigromaculis* (which were 0.664 ± 0.004 mm in length and 0.182 ± 0.001 mm in width) were retained by the 80-mesh sieve, but about 10% passed through to the 100-mesh sieve. The efficiency of extraction was tested and found to be on average 91.68 ± 1.34%, but this varied according to the operator. Miura (1972) pointed out that the number and mesh sizes of the sieves, the amplitude and pulse rate of the sifter, and the time required to sift a sample largely depends on the type of soil being processed. With fairly clear sandy loams about 1.75 hr was needed to examine each sample. There was no detect-able effect on the viability of about 8000 eggs of *Ochlerotatus nigromaculis* sifted by the machine for 1–10 min.

Ritchie and Addison (1991) processed soil samples collected using a modified 60 ml plastic syringe from a mangrove forest by nested sieving, through 0.185 and 0.170 mm sieves, of wet material or by coarse sieving and flotation. For the latter soil samples were passed through nested 0.30- and 0.15-mm sieves and retained soil rinsed on to 0.15-mm screening. The screen was placed on paper towelling to remove water and then completely dried in an oven for 24 h at 50°C. The dry soil was gently broken up in a mortar and pestle, rinsed through a 0.15-mm sieve and flushed into a 1-litre separating funnel containing about 100 ml water. After 1 min the stopcock was opened and the settled soil drained out. A wash bottle rinsed down debris clinging to the inside of the funnel, and the stopcock was reopened to let out more settled soil. The residue was then filtered through a 0.15-mm sieve and the debris flushed into containers for identification of eggs of *Ochlerotatus taeniorhynchus* under a microscope. Ritchie and Addison (1991) concluded that their flotation method recovered more eggshells (62.0%) than the sieving procedure (33.8%). They also used it to recover eggshells of *Ochlerotatus infirmatus* and *Aedes vexans*. However, they nevertheless believed that the sieving and bleaching method of Ritchie and Johnson (1989) and hatching method of Bidlingmayer and Schoof (1956) require less labour and are more efficient for recovering aedine eggs.

Ritchie (1994) used Ritchie and Addison's (1991) method to investigate the spatial stability of eggshells of *Ochlerotatus vigilax* in southeastern Queensland and concluded that relatively few eggshells were flushed from the preferred oviposition sites (pond banks) by tidal movements, making this method suitable for locating oviposition sites.

Ritchie et al. (1992) used 6 ml and 60 ml modified syringes and nested sieving to sample eggshells of *Ochlerotatus taeniorhynchus* in mangrove swamps in Florida in order to determine the spatial distribution of eggshells within the habitat, their degree of clumping, movement of eggshells as a result of water run-off, and the relationship of eggshell density to soil depth. Eggshells were found to be concentrated in the sloping banks of ponds, hummocks, and pot-holes, with a mean density of 1.45 ± 0.75 SE.

Ritchie and Addison (1992) used 3 cm^3 soil samples obtained with modified 6 ml plastic syringes and compared eggshell density with elevation (mean, standard deviation, and range and basin spillover elevation), soil characteristics (litter standing crop, soil bulk density, and percentage organic matter), and vegetation (total absolute dominance and vegetative association). Fifty-eight per cent of eggshell density variation was attributable to basin spillover elevation and elevation range, the only statistically significant variables as determined by piecewise regression. Eggshells were absent from sites exposed to tidal flushing with a frequency in excess of 15%, and this was attributed to oviposition avoidance in these areas, rather than flushing of eggshells by receding tidal waters.

Using the same methods, Addison et al. (1992) investigated the relationship of *Ochlerotatus taeniorhynchus* eggshell density to larval production in the same Florida mangrove swamps. Soil samples were taken from line transects within a 31×37 m grid, which encompassed both ponds and flood plain. Larval sampling was carried out in the same grids follo-wing flood events and consisted of 50 random samples per grid per ses-sion using a 300-ml dipper. Detritus and litter were sampled using a 13-cm diameter plastic corer and all detritus was removed from within the corer until firm soil was reached. In order to increase sample size, larval production was also estimated using three categories of larval production, namely light broods, 1 to 2 moderate broods upon initial flooding and many broods per year. Significant relationships between number of eggshells per cm^3 of soil and larval production were obtained, both for the quantitative data and the qualitative data, however the authors acknowledged that the relationship is complicated by soil characteristics.

Ritchie and Jennings (1994) later modified the above method by introducing the use of an electric blender (Phillips HR1375/A) to break up soil samples, rather than performing this step manually. The blender processing of soil samples was found to significantly reduce sample processing time, especially for clay soils, but resulted in reduced *Ochlerotatus vigilax* eggshell recovery rates from sandy soil. In addition, the technique favoured the recovery of new eggshells rather than old ones.

Turner and Streever (1997a) compared the use of a flotation method and direct examination of sediment as methods to determine the numbers of eggshells of *Ochlerotatus vigilax* in soil samples of approximately 15 cm^3 volume. Soil samples were collected from a saltmarsh border area in New South Wales, Australia, using a 2.9 cm diameter plastic corer pushed into the soil to a depth of about 2.2 cm. Sixteen samples were collected from 1m^2 areas at each of 14 sampling stations. The 16 core samples from each station were pooled and added to 250 ml of water, shaken vigorously for 1 minute and then washed through 3 sieves of 300-, 250-, and 150 μm mesh size. Any remaining lumps of soil were blended for 10 sec and passed through the sieves. Twenty subsamples were extracted from each of the 14 sieved samples as follows: material from the 150 μm sieve was poured into a 2-litre beaker and topped up to a volume of 200 ml with tap water. The solution was stirred to keep soil particiles in suspension and twenty 2-ml subsamples were removed with a glass pipette. For the direct examination method, subsamples were examined under a stereo micro-scope and the numbers of hatched eggshells, unhatched eggshells and unhatched eggs were recorded. Subsampling precision was determined according to Southwood (1978) as:

$$n = (st_\alpha / Em)^2 \qquad (2.1)$$

where *n* is the required subsample size, *m* is the mean of the 20 subsamples, *s* is the standard deviation of the 20 subsamples, *E* is the halfwidth of the confidence interval as a decimal of the mean and t$_a$ is from the Student's *t* distribution. *E* was set at 10% of the mean and values selected for a were 0.10, 0.05 and 0.01, giving an estimated sample size with 90%, 95% and 99% confidence that the true mean is within 10% of the estimated mean. The finite population correction of Krebs (1989) (n* = n/ [1 + (n − 1)/N], where *n** is the corrected subsample value, *n* derives from the equation above and *N* is the maximum number of subsamples (100), was applied as the subsampling procedure removed a significant proportion of the total sample. The flotation procedure, which was applied to the same 14 samples as the direct observation method was as follows: twenty 2-ml subsamples were extracted from each of the 200 ml samples as described above and air-dried on the 150 μm sieve for at least 24 h. The dried material was brushed from the sieve onto a petri dish and broken down manually. The material was then returned to the 150 μm sieve, wetted, and washed from the sieve into a 500-ml beaker. The beaker was placed into the sieve and a fine jet of water was applied for 60–120 sec to overfill the beaker so that light material overflowed into the sieve and dense material remained in the beaker. The material remaining in the sieve was then examined for eggs and eggshells. Recovery of eggshells using the two methods was compared by Wilcoxon's matched pairs test. Eggshell counts were cube-root transformed to stabilise variances prior to regression analysis. The mean number of hatched eggshells recovered using the flotation method was 407.3 (SE = 223.3), and for the direct examination method it was 1134.6 (SE = 692.2) and the difference was statistically significant (P < 0.002). Mean eggshell recovery using the flotation method was 44%. The flotation method took about 92 min per sample to process, compared with 278 min for the direct examination method. The high value for the coefficient of determination (r^2 = 0.97) for the regression of eggshell counts obtained by direct examination on counts obtained by flotation, suggests that little information is lost by using the more rapid flotation method and determining the true count from the regression equa-tion. At the eggshell densities obtained, 20 subsamples were deemed sufficient to give a good indication of true eggshell density, although a larger number of subsamples would be required at lower eggshell densities.

The relationship of *Ochlerotatus vigilax* eggshell density and environ-mental factors on Kooragang Island, Australia was also determined by Turner and Streever (1997*b*) using 15 cm³ core samples, as described above (Turner and Streever 1997*a*). Eggshells were extracted from soil samples using the water flotation method (Turner and Streever 1997*a*) and densities were correlated with elevation, and presence or absence of different vegetation types using multiple regression. Sampling was con-ducted at two spatial scales. For the macroscale study, 16 core samples were taken from each of 40 sampling stations. For the mesoscale study, 100 soil samples were taken from each of two 400 m² plots. The distribution of eggs and eggshells was highly aggregated in the macroscale study and density was significantly correla-ted with distance from a drainage culvert, *Sarcocornia quinqueflora* and *Sporobolus virginicus* per-centage cover, elevation, depression, pond, and proximity of a creek. These factors explained 53% of the variation in eggshell density. The distribution of eggshells was also highly aggregated in the mesoscale study. Eggshell density in Plot 1 was highest at a relative elevation of 8.5 cm and in the presence of vegetation. In Plot 2, maximum densities were associated with low elevations and the presence of *Sporolobus virginicus* or mixed vegetation.

Dale et al. (1999) used Ritchie and Addison's (1991) method to extract *Ochlerotatus vigilax* eggs from intertidal soil samples in Australia in order to compare quantitative and qualitative (presence/absence) methods for determining the distribution of eggshells among different soil and vegetation types. Eggshell counts per cm³ were determined for samples taken from 3 habitat types and 4 classes of vegetation cover. Relationships between eggshell numbers/cm³ and site type or vegetation cover type were analysed using 1-way ANOVAs. Data were converted to qualitative presence-absence data and were analysed using the likelihood ratio Chi-squared value.

Comparison of the two sets of results revealed similar patterns in relation to habitat and vegetation cover. The authors concluded that the qualitative method is useful from a vector control perspective as it allows the same conclusions to be drawn regarding oviposition site selection as the quantitative assessment, whilst being less laborious and time-consuming.

Dale et al. (2002) used the methods described above to investigate the effects of habitat modification on eggshell distribution of *Ochlerotatus vigilax* in salt marshes in Queensland, Australia. Habitat modification in the form of runnelling and open marsh water management tended to reduce eggshell numbers, but not in all cases, probably as a result of significant heterogeneity over small spatial scales.

A simple extraction technique can be employed with samples collected from marshes, muddy ground pools, rock pools and tree-holes etc., so long as they contain little leaf litter and vegetation. The sample is placed in a beaker of water vigorously stirred and after any lumps have been broken up it is poured through ¼-in, ⅛-in and a phosphor bronze sieve stacked on top of each other. The sample is washed through the sieves with a jet of water, and the eggs retained on the phosphor bronze sieve are floated off in a solution of sodium chloride (sp. gr. 1.2). Service has reported successfully extracting eggs of *Aedes vittatus* from many rock pool samples with this procedure as well as eggs from soil samples from salt marshes and woodland pools (Service 1993).

Montgomery et al. (1979) present a relatively simple washing–flotation method for extracting insect eggs and larvae from various types of soil. The soil sample is placed in a wash tank (26-cm diameter and 30-cm deep) and water is passed into the bottom of the tank through a short length of rubber hose fitted to a pipe fixed to an inlet fitting, which is a threaded end-cap with three 0.32-cm holes drilled in the sides. This results in directing jets of water in an upward and circular motion so as to break up and mix the soil sample. The floating material is then passed from the overflow of the wash tank into a stacked series of graded sieves. Material collected on the sieves is flushed into another container with a small hand-sprayer connected to the mains water supply. Flotation is in magnesium sulphate having a specific gravity of 1.15. The recovery of eggs of *Otiorhynchus sulcatus* (black vine weevil) from 1 litre of sandy silt loam was 95.0%, but was reduced to 87.3% when 2.4-litre volumes were processed. Recovery of the smaller eggs of *Diabrotica longicornis* (northern corn rootworm) from 0.5-litre samples was 96.4%. After extraction and flotation eggs were still viable.

Whatever extraction method is used not all eggs will be recovered from samples. The efficiency of any extraction method should therefore be assessed by determining the percentage recovery of a known number of eggs processed through the apparatus. The number of eggs extracted from samples can then be corrected for 100% efficiency. For example, in a population study of *Ochlerotatus cantans* in England conducted by Service over several years, one hundred 10 × 10-cm soil samples, representing about 5% of the total oviposition area of a habitat, were collected in September, when all the eggs of the year had been laid. The number of eggs extracted by processing through a Salt–Hollick machine was adjusted for total recovery and multiplied by 20 to get an estimate of the total egg population of the habitat. Another egg estimate was made in late December, just prior to egg hatching to determine egg loss during the intervening months. This second estimate was then corrected for the percentage of eggs that fail to hatch, due to sterility or other factors, to give an estimate of the number of viable eggs available for hatching.

Methods applicable to other habitats

Tree-holes are probably the most widespread class of natural mosquito habitats and species of several genera of mosquitoes breed in them. It is well known that *Aedes* and *Ochlerotatus* can be collected from tree-holes by removing dry debris from them (Buxton and Breland 1952; Dunn 1926; Lounibos et al. 1985; Trpis 1972; Wilkins and Breland 1951). In addition to collecting material from the bottom of tree-holes Dunn (1926) carefully scraped the inside walls and bottom with a metal spoon to recover the maximum number of eggs. Arnell and Nielsen (1967) also obtained eggs by scraping the walls of tree-holes. In Nigeria Lambrecht and Peterson (1977) used various sized spoons to scrape the debris from the inside of tree-holes, and after soaking obtained larvae of *Aedes aegypti, Aedes stokesi, Aedes bromeliae, Aedes luteocephalus, Ochlerotatus ingrami, Aedes apicoargenteus, Aedes africanus*, and *Aedes dendrophilus*. In both collections *Aedes aegypti* was by far the most common species.

Although eggs are often collected by such methods it is difficult to obtain quantitative results. Kitching (1971), however, standardised the collecting method by developing a small core sampler for collecting semi-fluid substrates from tree-holes. One part consists of a 35-cm long piece of brass tubing 1.85 cm in internal diameter and with the distal end sharpened. A loosely fitting pierced cap fits over the opposite end (Fig. 2.6). The other part fits into the brass tubing and consists of a commer-cially produced steel drill bit with the small screw part of the tip and squa-red upper part of the shaft cut off. The remaining section is fitted into the lower half of a solid brass collar, the upper half of which is fitted by a brass shaft to a wooden handle.

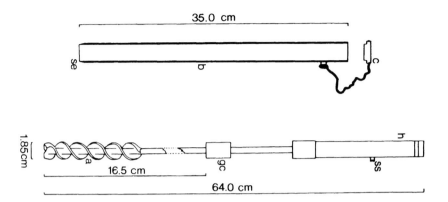

Fig. 2.6. Core sampler for tree-holes, a—auger, b—barrel of corer, c—cap, gc—guiding collar, h—wooden handle, se—sharpened edge; ss—stop screw (Kitching 1971)

Both the collar and drill bit fit closely into the length of brass tubing. A screw projects out of the side of the wooden handle so that when the auger is pushed through the tubing as far as the screw permits, the auger and shaft project from the bottom of the tubing with the bottom of the brass collar level with the sharpened lower edge of the tubing. In taking a sample the tube is placed in a tree-hole and first worked in by hand, then with the cap on the top it is hammered through the substrate until hard underlying wood is reached. The auger is then carefully screwed down the tubing. In practice it was found that when the auger had penetrated part way into the tree-hole the whole apparatus could be lifted out together with the sample. When the auger is pushed through the tubing the brass collar scrapes the side and ejects the sample into a plastic bag. Tree-hole debris still attached to the auger is washed into the bag. Knowing the diameter and length of the core, its volume and surface area sampled are readily calculated.

When attempting to compare or estimate the egg population in different sized tree-holes the area of the bottom of the tree-holes being sampled must be known. Furthermore, many mosquitoes lay at least some of their eggs on the inner walls of tree-holes, and these will be missed unless the walls in addition to the bottom debris is sampled (Jenkins and Carpenter 1946). It is more difficult to collect eggs from the walls of tree-holes, especially when they are deposited in cracks and crevices. In tree-holes with narrow openings it will even be difficult to collect eggs from the bottom.

Several tropical mosquitoes oviposit in water-filled sections of bamboo, and these are generally more easily sampled than tree-holes. It is not so difficult to remove bottom debris, and eggs can usually be collected more easily from the smooth walls than from those of tree-holes. Sides of growing bamboo are sometimes punctured by insects and birds, and in some regions, especially Latin America and Malaysia when these bamboo sections become filled with rainwater, certain mosquito species lay their eggs in them. It is difficult to remove debris from these habitats unless the bamboo is cut across, thereby destroying the habitat. In tree-holes and bamboo where it is difficult or impossible to remove bottom debris, eggs can sometimes be collected by filling habitats with water and then siphoning or pumping out the contents. This method is still not very effective in collecting eggs adhering to the inner walls. The presence of mosquito eggs in debris collected from tree-holes and bamboo is usually detected by soaking it in water and collecting the larvae that hatch out, but as discussed previously, this may not give a reliable indication of either the number of eggs present or species composition. A better approach is to extract the eggs from the debris by sieving and flotation.

Some mosquitoes oviposit in plant axils such as those formed in banana plants, pineapples, *Ravanela*, bromeliads, *Nepenthes*, grass, and in cavities of pitcher plants (Lounibos et al. 1985). There is little information on the recovery of eggs from these habitats, but eggs can sometimes be located in situ by pulling the plants apart. For example, in Canada eggs of *Wyeomyia smithii* were collected from pitcher plants (*Sarracenia purpurea*) by dissecting the plants under a stereoscopic microscope. Alternatively, accumulated debris in the axils can be flushed out, sieved and eggs recovered by flotation techniques, or their presence detected by soaking the debris and removing the larvae that hatch.

Large numbers of eggs of *Aedes vittatus* have been recovered from mud collected from rock pools by sieving and flotation (Service, unpublished data). Eggs have also been detected in rock pools by repeatedly soaking mud samples and identifying larvae that hatched. *Aedes vittatus* eggs have been collected by covering the walls of small water-filled rock pools with pink blotting paper, which acts as an oviposition substrate (Service 1970). In coastal areas of East Africa where *Aedes aegypti* breeds in coral rock holes, eggs have been detected by soaking soil and detritus from the rock holes and identifying the resultant larvae (Trpis et al. 1971).

Evans (1962) collected a few eggs of *Psorophora confinnis* by scraping the walls of burrows made by crawfish.

Mosquito larvae, but rarely eggs, are often collected from a range of natural containers such as split fruit husks, coconut shells, dead leaves lying on forest floors, flower sheaths, rotting fallen tree trunks, and snail shells, among others.

With species that lay eggs in rafts or masses, both the numbers of egg rafts and total number of eggs can be counted. The number of egg rafts may not, however, represent the number of gravid females that oviposited, because while females usually deposit all eggs in a single raft, a number of smaller rafts will be laid in the same or different containers if the female is interrupted during egg laying. Egg rafts are also easily broken and incomplete rafts may be recorded as complete ones. Because they break easily rafts collected in the field should be stranded on wet filter paper in individual tubes when they are transported to the laboratory. Simple population estimates can be made by multiplying the number of 'intact' rafts by the mean number of eggs per raft; intact rafts will represent the number of females that have oviposited in the habitat.

Aedine eggs can sometimes be directly observed in containers. For example, Service (1970) was able to count the eggs of *Aedes vittatus* laid on the walls of small rock pools and emergent plants.

Chadee and Small (1988) used a small scoop for collecting the hydrophobic eggs of *Toxorhynchites moctezuma* (= *Toxorhynchites theobaldi*), from small natural and artificial container habitats. The scoop was made from a white plastic teaspoon (122 mm long with a bowl 40 mm long, 30 mm wide and 8 mm deep) from which a 20-mm diameter hole was removed from the centre by pressing down with a piece of copper piping heated in a flame. The bowl of the spoon was placed in chloroform for a few seconds and then while still soft a small piece of nylon mesh (aperture 660 μm) was stuck to the underside. This spoon was also useful for collecting adult Trichoprosopon digitatum guarding her egg raft after a glass tube had been placed over her. Such a simple scoop may be useful for collecting other mosquito eggs that float on the water surface.

With small container habitats, e.g. snail shells, fruit husks, fallen leaves, etc., the presence of eggs can be detected by immersing the containers in water and counting the larvae that hatch.

Spatial distribution of aedine eggs in habitats

In Canada Enfield and Pritchard (1977) took frozen core samples (15 × 15 cm and 2 cm deep) at 2-m intervals along transects radiating from the centre of a pond when it was not flooded. When, however, the pond was flooded core samples were taken by cutting the earth from inside 15- and 10-cm diameter PVC pipes, which were used as templates. The numbers of eggs of *Aedes cinereus* and *Aedes vexans* per sample were estimated by repeated flooding

(3) of the core samples and identification of the larvae. The pond was divided into six strata, and the mean numbers of eggs per sample unit was divided by the sample area (0.0225 m^2 or 0.00785 m^2) to give the mean density of eggs within each stratum. From these values, and the areas of the strata, estimates of overall mean egg densities and their standard errors were calculated for each stratum, and also the total egg population in the entire pond was estimated. In their pond Enfield and Pritchard (1977) could find no evidence that the distribution of eggs was related to any physical features. Precision, as judged by the size of the standard errors, was reasonably good (7.8 and 10.8%) but there were larger standard errors when egg densities were low and many samples contained no eggs. A larger core sampler might have resulted in better estimates at these low densities. A disadvantage of this, and related methods, is that estimating egg numbers from larvae hatched on repeated soakings can be laborious, and of course will not work if the eggs are in diapause. Moreover, if the soil is frozen hard it can be very difficult to cut samples from the ground.

In Florida, USA citrus groves Curtis and Frank (1981) removed 100-cm^2 soil samples from three zones, namely the bottom of furrows, from a distance of 1–2 m from the bottom on sloping banks and from an area between 2 m to the crown of the furrow. The samples were processed according to Horsfall (1956). The mean numbers of eggs increased from 0.3, 2.0 to 43.3 in the three zones, increasing with distance from the bottom of the furrows.

Leftkovitch and Brust (1968) studied the distribution of *Aedes vexans* eggs in a pond to determine their distribution and the best procedures for egg sampling. The pond was divided into 10-ft squares and a 1-ft square soil sample was cut to a depth of 1 in and taken from each square. Eggs were extracted from the samples by Horsfall's method. The heights below the level at the edge of the pond from which the samples were taken were measured so that an inverted contour map of the pond could be drawn, taking the highest point as zero. The mean number of eggs per sample was 33.92, the variance was 2890.31. In a random distribution the mean is an estimate of the theoretical variance, i.e. the mean and variance are equal. Now, because in this instance the variance was much greater than the mean it shows that the eggs were highly aggregated, that is they occurred in clumps in the pond. Since the eggs were not evenly distributed it was decided to find out whether their distribution was related to topographical features of the pond. First, the numbers of eggs extracted from the samples were grouped into categories corresponding with the successive heights (in inches) in the pond from which they were collected. There was a significant relationship between the mean numbers of eggs in each group and its variance. For statistical reasons this relationship must be removed by a suitable transformation before the data can be analysed. The transformation $z = 103 (1 - y - 0.352)$, where y is number of eggs obtained by the power law of Healy and Taylor (1962) and then a scale factor, was found to remove this association. The mean values of the transformed data for each depth group were calculated, and the mean level (5.07 in) and standard error (0.876 in) calculated by the summation method described by Elderton (1953). The 95% confidence limits of the mean were calculated as 3.11–7.01 in. These results show that with the correct sampling procedure relatively few samples need to be collected to elucidate the vertical distribution of eggs in a habitat.

While it is not anticipated that such a mathematical approach to sampling will be generally adopted by mosquito workers, the data clearly show that the distribution of *Aedes* eggs in natural habitats is likely to be highly aggregated. This necessitates the use of suitable transformation before the results can be statistically analysed. In practice, however, it may not be worth the effort to obtain precise transformations; in many instances converting field counts to log $(x + 1)$ will suffice. Miura (1972), however, used a modified square root transformation, $\sqrt{(x + 0.5)}$, for comparing the mean number of eggs of *Ochlerotatus nigromaculis* in different parts of oviposition sites.

Fay and Perry (1965) were among the first to demonstrate that *Aedes aegypti* females do not lay all eggs in a single location, presumably to avoid sibling competition among immature stages. The initial laboratory results of Fay and Perry (1965) were later confirmed in the field in Trinidad, West Indies by Chadee and Corbet (1987). This phenomenon was termed 'skip oviposition' by Mogi and Mokry (1980), working on *Wyeomyia smithii* in Newfoundland, Canada and is presumed to occur in several species that use small containers for oviposition. Corbet and Chadee (1993) have developed a method for detecting substrate preferences among species that exhibit 'skip oviposition', which is based on introducing a single test substrate into an array of eight substrates, of which seven are controls. The test substrate is then rotated among all eight positions to remove any positional effects. Harrington and Edman (2001), however, presented an alternative view, which states that skip oviposition uses greater maternal energy reserves, may increase the risk of adult female mortality and does not reduce other forms of competition, and therefore may not be a universally beneficial strategy. These authors used an alternative, indirect approach to determine if skip oviposition occurs in nature. It was reasoned that if adult females do not deposit all their eggs in one site, then field-collected samples of females should include some females that had deposited some of their eggs but still retained some mature eggs. Therefore the mean number of mature eggs (Christophers' stage V) would be expected to be significantly lower than the mean number of developing oocytes in females with eggs in Christophers' stages IIIa–IVb. Collections were made of mosquitoes resting on walls, clothing, water jars, ant traps, and other surfaces inside houses, between 0800 and 1600 h on 17 dates in October and November 1998, using a modified vacuum aspirator (National MC3500 Airpower 170, 1 000 W, Bangkok, Thailand). A total of 384 female *Aedes aegypti* was dissected and evaluated. Of those females with oocytes in stages late IIIa V, most had eggs at stage V (37.5%); fewest were at late stage IIIa (7.2%). No significant differences were found among oocyte number by stage (F = 1.72, d.f. = 4, P = 0.143). No significant differences were found between the mean number of combined immature (Christophers' stages less than V) and mature (Christophers' stage V) oocytes per female (t = −1.31, d.f. = 382, P = 0.19). This was true even when the data for those immature oocytes that may degenerate before maturing (Christophers' stages IIIa - IIIb) were removed from the analysis (t = −1.34, d.f. = 360, P = 0.18). No significant differences were detected when the number of mature versus immature oocytes were compared within three female size classes. Harrington and Edman (2001) ascribed the results of earlier studies that showed small numbers of eggs laid in several small containers as being due to the small size of the surveillance container and its location in an area where large containers may be preferred oviposition sites.

Artificial oviposition sites

Direct sampling

Aedine eggs can be sampled from artificial containers directly, a method used by Lambrecht and Peterson (1977) in Nigeria. These authors used ladles to scrape mud and debris from earthen water-storage pots, and on soaking the materials hatched out larvae of *Aedes aegypti, Aedes fowleri, Aedes bromeliae, Aedes luteocephalus, Aedes apicoargenteus* and *Aedes unilineatus* in that order of abundance. They also used various sized spoons to scrape the debris from the inside of tree-holes, and after soaking obtained larvae of *Aedes aegypti, Aedes stokesi, Aedes bromeliae, Aedes luteocephalus, Ochlerotatus ingrami, Aedes apicoargenteus, Aedes africanus*, and *Aedes dendrophilus*. In both collections *Aedes aegypti* was by far the most common species.

In a Mexican cemetery Arredondo-Bernal and Reyes-Villanueva (1989) collected eggs of *Toxorhynchites theobaldi* from containers with a simple plastic scoop. By firstly removing all the eggs in this manner from 25 artificial oviposition containers, followed by the collection of eggs at 2-h intervals (0600–2100 h) from these containers the diel pattern of oviposition was determined. The lowest mean number of eggs per container (approx. 8) was recorded at 1500 h, while a peak mean of 80.9 eggs was recorded at 1900 h, just 1 h before twilight. The numbers of eggs laid in a container showed a positive correlation ($r = +0.70$) with surface area, and was expressed by the regression line $y = -14.12 + 0.126x$, where y is the mean number of eggs per sample and x is surface area in cm^2. The slope of the equation ($b = 0.1266$) means that for each 1-cm^2 of water the oviposition rate increased by 0.13 eggs, thus an increase of 7.9 cm^2 allows 1 more egg to be laid in the container.

Artificial pools and ponds

Small water-filled borrow pits have sometimes been used as artificial oviposition sites for *Anopheles* species. In India Russell and Rao (1942) studied the effect of mechanical obstructions and shade on egg laying by *Anopheles culicifacies* by digging a number of oviposition pits, 9–12 in deep and 2 or 3 ft square. They filled the pits with seepage water to about 2–3 in from the top and kept them free of macroscopic vegetation. Eggs of *Anopheles culicifacies, Anopheles subpictus* and *Anopheles vagus* were collected from the water surface of the pits by lifting them off with a wire loop. In studying the oviposition behaviour of *Anopheles melas* in relation to salinity in West Africa, Muirhead-Thomson (1945) also dug a number of artificial pits to attract ovipositing females.

Christie (1958) devised complicated artificial oviposition sites for *Anopheles gambiae*. A metal box containing suitable pool water was placed at the bottom of a shallow pool about 18–20 in square. Metal boxes in several pools were connected by piping via a cistern and ballcock to a reservoir of water thus ensuring that they remained flooded. A number of small soil trays were made by tacking wooden slats to 1-in wide iron banding and covering the bottoms with 16-mesh plastic mosquito gauze. These trays were filled with soil and placed side by side in the metal box with peripheral trays positioned at an angle of about 30°. Flanges connected the trays together and prevented the soil from being washed down between them (Fig. 2.7*a,b,c*).

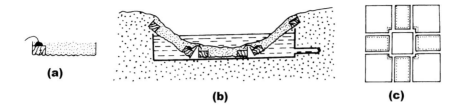

Fig. 2.7. Artificial oviposition trap for *Anopheles* eggs; (*a*) a soil tray; (*b*) soil trays in metal box sunk in small pool; (*c*) exploded plan of soil tray arrangement, dotted lines show flanges joining trays (Christie 1958)

After a night's exposure water was drained from beneath the trays, first by emptying the cistern, then by lifting one of the corner trays and pumping the remaining water out of the metal box. As the water drained through the trays *Anopheles* eggs were retained on the water-logged soil. Eggs were either removed from the trays in the field (Christie 1958) or the trays were placed in individual plastic bags and taken to the laboratory for egg extraction. Eggs were recovered by gently lowering the tray into a metal box containing water (A). More water was carefully added so that it lapped the lip (B) at one end of the box. A plastic boom (C) with rubber flanges (D) at the ends was used to sweep floating debris and eggs into a gently sloping gutter (E) from where they were flushed into a small can (F) having a 100-mesh bottom (Fig. 2.8).

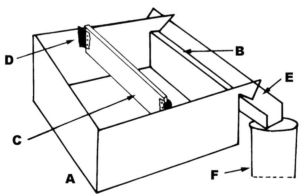

Fig. 2.8. Apparatus for skimming *Anopheles* eggs from water surface showing: A—box-like water container, B—lip, C—plastic boom, D—rubber flanges, E—sloping gutter, F—collecting can (after Christie 1958)

Material collected at the bottom of the can was washed through a 16-mesh sieve onto a 100-mesh one, and then finally into a conical vessel, which had an 8-cm opening and a 2.5-cm diameter base covered with 100-gauge mesh. This inverted conical vessel was lowered into a container of water so that debris and eggs floated to within about 1 cm from its rim. Floating debris containing eggs was transferred by a small paint brush to a small piece of paper waterproofed with cellulose paint and folded up like a concertina to give a series of gutters. Debris in the gutters was flooded with clean water, examined under a stereo-microscope and the eggs lifted out by a fine wire loop. After all eggs had been collected they were transferred to filter paper having a 3-mm grid to facilitate counting.

In three field trials 373, 392 and 579 eggs of *Anopheles gambiae* were recovered from the trays. A recovery rate of 60–69% was obtained when the efficiency of the extraction technique was tested by placing a known number of eggs of *Anopheles gambiae* in the artificial oviposition sites in the field (Christie, 1958).

Christie's procedure of producing artificial oviposition sites and extracting the eggs appears unnecessarily complicated and could be simplified by excavating small shallow depressions, lining them with plastic sheeting, placing small amounts of soil on the bottom and flooding with suitable pool water. After mosquito oviposition the water can be siphoned or baled out and passed through a fine mesh sieve to retain any eggs. The water-logged soil at the bottom of the pool together with the plastic lining can be removed and taken to the laboratory for washing through a series of graded sieves. Final separation of the eggs could be achieved by flotation in magnesium sulphate or sodium chloride (sp. gr. 1.2).

In Florida, USA Smith and Jones (1972) constructed artificial oviposition pools for *Culex nigripalpus* by stapling black plastic cloth to a wooden frame 30 in long, 18 in wide and 3 in deep. These artificial pools were embedded in the ground in shaded sites near large collections of water, with the tops of the frames level with the ground. About 2.5% of the egg rafts collected from them failed to hatch, but of those that did about 89% were *Culex nigripalpus* and the remainder other *Culex* species. There was no difference between the numbers of egg rafts laid in pools containing water or hay infusion, but about three times as many eggs were laid in pools containing crushed 40% hog supplement ('Purina') which was added at the rate of approximately 8 g/gal of water.

In studying the influence of soil fermentation on the selection of oviposition sites in California Gerhardt (1959) dug a series of shallow 2-ft square pits and lined them with polythene sheeting. A 6-in layer of soil was placed at the bottom of each pit and covered with 6 in of tap water. These pits failed to attract ovipositing mosquitoes, but eggs of *Culex stigmatosoma* and a few also of *Culex tarsalis* and *Anopheles freeborni* were laid in pits which were supplemented with 2 lb prepared dog meal.

Egg rafts of *Culex stigmatosoma* were also collected from pits to which 11 lb of either sucrose, casein or hydrogenated vegetable oil was added.

De Meillon et al. (1967) studied the oviposition cycle of *Culex quinque-fasciatus* in Yangon by creating an attractive artificial oviposition site consisting of a shallow galvanised tray (0.9 × 0.6 m) containing septic tank water together with scum and floating debris. To study the diel pattern of oviposition two collectors worked 3-h periods and removed egg rafts from the water with a piece of stiff white paper as soon as they were laid. A roof of plastic sheeting and palm leaves was erected about 1.2 m over the artificial pool to protect the collectors from rain. In calm weather peak oviposition was around sunset and sunrise, corresponding to the principal arrival times of gravid females, but wind and heavy rain delayed oviposition and caused an irregular cycle. The arrival time of gravid females was investigated by placing on a septic tank a very simple trap consisting of a wooden frame (53 × 51 cm) 71 cm high at the front having a sloping plywood roof. The top compartment was covered with mosquito gauze while the sides of the lower section were made of stiff plastic sheeting. The bottom was covered with mosquito gauze. A 1.3-cm wide louvre-type entrance between the two compartments allowed gravid mosquitoes to enter the trap.

From field trials testing the larvicidal properties of n-capric (decanoic) acid on mosquitoes (Maw and House 1971) it was discovered that the acid eventually turned pools into abnormally attractive oviposition sites (Maw, 1970). The acid acted as a 'fertiliser' to bacteria of the family Pseudomonodaceae, which then made the pools attractive to gravid mosquitoes. Maw and Bracken (1971) used this information in developing effective artificial oviposition sites for *Culex restuans*. Eight 1-m square pools made from 5 × 15-cm sections of wood and having a bottom of 4-mm thick hardboard were lined with polythene plastic sheeting. A series of 21 smaller pools measuring 30 cm^2 and 10 cm deep were also constructed. The pools were filled with water collected from nearby streams or temporary pools. To each large pool was added 40 ml capric acid dissolved in 15 ml 95% ethanol, 10 ml saturated ammonium nitrate, and 1 litre of water collected from pools that had been treated with capric acid the previous year and which were known to be attractive. This water, which was kept frozen until required, contained the necessary bacteria. The smaller pools had proportionally less attractants added. When the water in the pools lost its turbidity, reflecting a decrease in bacteria level, further amounts of capric acid were added to maintain maximum attractiveness. Egg rafts were collected from the pools daily from June to September.

A total of 7115 egg rafts, presumably all of *Culex restuans*, were collected from the eight large pools ($\bar{x} = 889.4$) and 1962 rafts from the smaller pools ($\bar{x}$ = 94.4). Initially no eggs were deposited in a series of 16 untreated pools, but from mid-August to the conclusion of the trials in September 11 egg rafts were collected from these pools. It was thought that eggs found in early June were from females that had overwintered. If so, then these pools were probably more effective in assessing the re-appearance of hibernating populations of *Culex restuans* than were the light-traps employed by Belton and Galloway (1966) as they failed to notice overwintering females in their catches. Maw and Bracken (1971) found that the seasonal incidence of the egg rafts was very similar in both the large and small pools.

In Canada Brust (1990) constructed artificial pools to collect egg rafts of *Culex tarsalis* and *Culex restuans*. Each pool was 1 m^2 and made from 4-cm thick wood to form a 1 × 1 × 0.2-m frame lined with black polyethylene sheeting. A 70 × 70 × 2-cm thick sod of lawn grass was placed in each pool as an oviposition attractant, and water added to a depth of 10 cm. Six holes (2-cm) were drilled 3 cm from the top of the pool and covered with fine netting to allow excess water to drain out. Best results were obtained when the sods were changed at 3-week intervals for *Culex tarsalis*, and every 1–3 weeks for *Culex restuans*. Buth et al. (1990) made similar pools to collect egg rafts of these two species and also *Culiseta inornata*. Madder et al. (1980) used inflated plastic paddling pools (84 cm diameter) lined with a layer of sods and filled with tap water as oviposition sites in Canada for *Culex pipiens* and *Culex restuans*. From a series of pools they collected 13 606 egg rafts over about 3.5 months. They found that the addition of decanoic acid, 95% ethanol, and ammonium nitrite to pool water in the proportions described by Maw and Bracken (1971) did not provide any additional attractant for these for two *Culex* species.

In Israel, Stav et al. (2000) created artificial oviposition pools to study the effects of a predator (nymphs of the dragonfly *Anax imperator*) and crustaceans on oviposition by *Culiseta longiareolata*. Plastic tubs with a trapezoidal cross-section and dimensions of 23 cm in height, 34 × 59 cm at the base, and 37 × 62 cm at a height of 20 cm were filled with 53 litres of aged tap water, to which was added a finely ground combination of cat food, mouse chow, yeast and fish flakes. Also in Israel, Eitam and Blaustein (2004) used similar artificial oviposition sites to study the interactions of *Notonecta* predators on oviposition by *Culex* and *Culiseta* mosquitoes Oviposition sites were constructed as follows: 20 plastic tubs (volume approximately 30 litres, dimensions 48 × 27 cm at bottom, 55 × 33 cm at top, height 19 cm) were arranged in a 10 × 2 array under a canopy of *Pinus halepensis*. Pools were filled with tap water and 250 cm^3 of *Quercus* leaf litter, plus 10 cm^3 of soil were added to each pool. All egg rafts obtained were from either *Culex laticinctus* or *Culiseta longiareolata*.

Reiskind and Wilson (2004) created artificial oviposition habitats for *Culex restuans* in Michigan, USA, using 63.6-litre blue plastic tubs filled with 10 litres of unchlorinated well water. A nylon netting sac containing 100 g of pesticide-free, organic hay (100 g) was bound with plastic cable ties, placed into the water and weighted down with silica pebbles. Egg clutches were collected and counted daily. Reiskind et al. (2004) used these same plastic tubs to study the responses of ovipositing *Culex restuans* to artificial pools that were filled with high-, low-, and no-nutrient media and placed outside in natural woodlots. Significantly more eggs were laid in containers holding high- or low-nutrient water, compared with the containers holding plain water ($P < 0.001$). Pupal weight was also greater in high nutrient containers, compared with low- and no-nutrient containers.

Ovitraps

Ovitraps are relatively easy to construct and are sensitive at detecting the presence of gravid females, even at low population densities, making them ideal for surveillance activities. They do, however suffer from a number of disadvantages: they cannot be used to determine absolute population density; the infusions commonly used as oviposition attractants are made from natural ingredients such as hay infusion, grass sod infusion, etc., and as such will have variable chemical properties which may affect their attractiveness for different species, making it difficult to standardise between traps and trapping occasions. The utility of a standardised synthetic attractant or range of attractants has been noted by Millar et al. (1992). Additionally, maintenance and servicing of traps is labour intensive.

In Trinidad Chadee (1986) compared the efficiency of human landing catches, larval surveys and ovitraps for detecting relatively low levels of *Aedes aegypti*. As was reported by both Fay and Eliason (1966) and Tanner (1969) ovitraps were the most sensitive sampling method for *Aedes aegypti*, but did not identify larval habitats. Only one ovitrap contained eggs of another species (*Haemagogus janthinomys*), whereas another seven species were caught in bait collections, and four in larval surveys. Giglioli (1979) and Slaff et al. (1983) found that bait catches were inadequate in monitoring *Aedes aegypti* populations. Furlow and Young (1970) found ovitrap surveys about equally as sensitive as larval surveys in detecting *Ochlerotatus triseriatus*. However, in Jakarta Nelson et al. (1976) found they were less sensitive than human bait catches or larval surveys for monitoring *Aedes aegypti*, and in Bangkok Pant et al. (quoted by Nelson et al., 1976) also found ovitraps the least sensitive method of detecting low populations of *Aedes aegypti* after control operations.

In a recent review, Focks (2003) critically examined sampling methods for dengue vectors and concluded that ovitraps are a useful and cost-effective method for demonstrating the presence/absence of dengue vectors, particularly when populations are low, but emphasised that ovitrap data cannot be reliably used to estimate differences in adult population abundance. In addition, ovitrap data cannot be used to estimate transmission risk unless a statistical relationship between the two variables has previously been demonstrated for the particular location under investigation. Numbers of positive ovitraps or numbers of eggs per ovitrap do not bear a simple relationship with adult population density or trans-mission due to the complex interaction of many variables, including stage-specific mortality, productivity of different types of container, spatial distribution of containers, seroprevalence, etc. Ovitrap data are useful in determining the efficacy or otherwise of anti-adult vector control operations, as both the number of positive ovitraps and numbers of eggs per ovitrap should decrease with reduced adult population size, but the data cannot be used to assess the outcome of oviposition container removal measures, as oviposition may simply be concentrated into fewer containers following this intervention.

Construction of ovitraps for Aedes aegypti and Aedes albopictus

Ovitraps were initially developed for detecting the presence of *Aedes aegypti* during the first 3 years of the US *Aedes aegypti* Eradication Program (Fay and Eliason 1966; Fay and Perry 1965; Jakob and Bevier 1969*a*; Pratt and Jakob 1967), which began in 1964 (Schliessmann 1964). Field evaluations showed that the ovitrap was potentially a sensitive and efficient technique for detecting populations of *Aedes aegypti* (Chadee and Corbet 1987, 1990; Evans and Bevier 1969; Fay and Eliason 1966; Frank and Lynn 1982; Furlow and Young 1970; Hoffman and Killingsworth 1967; Nayar 1981; Ritchie 1984*a*; Subra and Mouchet 1984; Tanner 1969), and is a particularly useful method when population densities are low, (Braga et al. 2000; Jakob and Bevier 1969*b*; Focks 2003), giving more reliable results than visual inspection of containers in and around premises (Rawlins et al. 1998). Early ovitraps were usually constructed from 1 pint capacity glass jars, about 5 in tall, with tapered sides and with a diameter of about 3 in at the top, and painted glossy black on the outside. Water to a depth of about 1 in was added to the jar and a 3/4-in wide, 5-in long hardboard paddle having a smooth and a rough surface was attached vertically with a paper clip to the inside of the jar (Fig. 2.9).

Fig. 2.9. *Aedes aegypti* black glass jar ovitrap (Service 1993)

Paddles should preferably be made of the hardboard used for interior decorating as this is more absorbent than the exterior-type hardboard, and thus presents a more suitable oviposition surface (Thaggard and Eliason 1969). Identification marks can be written on the smooth side of the paddle. Eggs of *Aedes aegypti* are usually deposited just above the water line on the rough side of the paddle that faces towards the centre of the jar. Originally a small glass vial of ethyl acetate was suspended within the oviposition jar, supposedly acting as an attractant for gravid females, but in 1967 this practice was discontinued when it was discovered that eggs of *Aedes aegypti* were obtained just as frequently in jars without ethyl acetate (Hoffman and Killingsworth 1967; Thaggard and Eliason 1969).

Chadee (2004) used this design of ovitrap to investigate the vertical distribution of oviposition by *Aedes aegypti* in an apartment block in Trinidad. At each site, two ovitraps were placed at ground level and at heights of 0–12 m (1st floor), 13–24 m (2nd floor), 25–36 m (3rd floor), 37–48 m (4th floor), and 49–60 m (5th floor). On each floor one ovitrap was placed indoors and one outdoors. Significantly more eggs ($G = 14.4$; d.f. $= 3$; $P < 0.01$) were collected at the 2nd floor height of 13–24 m, compared with the other heights, for both indoor and outdoor ovitraps. It was suggested that *Aedes aegypti* in Trinidad had become progressively more adapted to high-rise living as previous studies (Chadee, 1991) reported only 1.2% of eggs collected at heights above 4.5 m.

In the laboratory, Yap et al. (1995) studied the oviposition preferences of *Aedes albopictus* for different coloured glass ovitraps. Glass jars (15 cm in height and with a diameter of 7.5 cm at the open end) painted black, red or blue were preferred oviposition sites when compared with white, yellow, green, or clear glass. Compared with other types of container frequently used as oviposition sites by *Aedes albopictus*, including tyres, bamboo stumps, coconut shells, etc. black glass jar ovitraps containing a 12.5 × 2.0 cm strip of Whatman No. 1 chromatography paper were preferred (26.1 ± 3.4% of total eggs laid), along with coconut shells (17.5% ± 4.1) and leaf axils (16.2% ± 4.2).

Vezzani and Schweigmann (2002) examined the use of different types of flower vases as oviposition sites for *Aedes aegypti* in cemeteries in Buenos Aires, Argentina. The majority of vases are provided by local municipal authorities and comprise black plastic cubes or cones and metal cones. Other containers commonly used included ceramic vases, glass flasks and cans of different shapes and sizes and all containers have capacities of between 0.25 and 2 litres. In a preliminary study, sampling of containers was conducted in 40 squares of 50 × 50 m each, which were selected at random on a weekly basis, using a virtual grid that covered the whole cemetery. In the main study, containers were sampled from four distinct environments: (1) open field, exposed to permanent sunlight and containing microenvironments sheltered from the sun by vegetation; (2) wall 1, sunlight from midday to sunset and absence of vege-tation cover; (3) wall 2, sunlight from dawn to midday and absence of vegetation cover; (4) underground gallery, shade provided by roof and presence of vegetation. A random sample of 100 containers was taken weekly from each patch and the container index was calculated. The CIs calculated for each of the studied patches were as follows: open field = 4.8% (78/1,605), wall 1 = 3.1% (12/388), wall 2 = 2.7% (17/632), gallery = 1.3% (10/747). Of 117 containers positive for mosquito immatures, plastic containers were the most frequently positive (82.1%), followed by glass (8.5%), metal (6%) and ceramic (3.4%) containers. However, when considering the availability of each type of container per patch, the use of containers was observed to be proportional to availability in both open fields and gallery (correlation coefficient = 0.99 and $P < 0.05$ in both cases). By contrast, in both wall habitats, plastic containers were most often used, while metal containers were the least used in relation to their availability.

Vezzani et al. (2003) used 330 ml capacity black glass jars containing 100 ml of a 3 × 10 cm hardboard paddle resting against the upper rim to investigate the seasonal occurrence of *Aedes aegypti* in two Buenos Aires cemeteries.

In São Paulo state, Brazil, Cardoso Junior et al. (1997) used ovitraps constructed from black glass jars 12 cm high with a diameter of 5 cm and containing strips of eucatex 12 cm long and 2 cm wide as the oviposition substrate.

Glass jars are increasingly being replaced with plastic cups (e.g. Chadee 1992; Goettel et al. 1980; Lourenço-de-Oliveira et al. 2004; Mogi et al. 1996; O'Meara et al. 1989a 1995; Savage et al. 1993).

In Fiji Goettel et al. (1980) used ovitraps made from black plastic cups containing hardboard paddles which were removed at 3- or 4-day intervals. Each paddle was soaked in water for 2 weeks in the laboratory and when a few were soaked for 3 weeks an extra 6.4% *Aedes pseudoscutellaris* and 5.2% *Aedes aegypti* hatched.

In comparative trials in Louisiana Kloter et al. (1983) found that black glass jars and black plastic beakers were equally attractive to ovipositing *Aedes aegypti*. A potential problem with ovitraps is the risk of flooding with rainwater, but this can be more easily prevented with plastic ovitraps by drilling an overflow hole in them.

Corbet and Chadee (1990) substituted glass jars with black plastic jars having a top diameter of 8 cm. The oviposition liquid in the traps was a yeast mixture (15 mg dry yeast/350 ml water), and an overflow hole was drilled 7.6 cm from the top.

Chadee (1992) used 9 cm diameter black plastic cups containing 450 ml of tap water, and with an overflow hole 3 cm from the top, to survey *Aedes aegypti* oviposition along transects in urban Trinidad.

Bellini et al. (1996) compared 550 ml plastic, 590 ml glass and 440 ml metal ovitraps for monitoring *Aedes albopictus* in northern Italy. Ovitraps were painted gloss black, filled with 200 ml dechlorinated tap water, and a 2 × 12 cm strip of Masonite was used as the oviposition substrate. Plastic and glass traps were equally effective, and collected significantly more eggs than the metal traps. Also in Italy, Toma et al. (2003) used a network of 500 black plastic flower-pot ovitraps to monitor *Aedes albopictus* presence in Rome. A strip of Masonite (3 × 15 cm) was used as the oviposition substrate and pots were filled with 350ml of water.

Mogi et al. (1996) used 40 ovitraps constructed from 9 cm diameter and 8 cm depth black plastic containers lined with paper toweling and half-filled with either well water or stored rainwater to study the occurrence of Aedes mosquitoes in villages in Indonesia. Ovitraps were collected after 5 days of exposure. In Saur village, ovitraps placed outdoors collected 478 eggs and indoor traps collected 26 eggs. In Kao village, 69 eggs were collected outdoors and 26 indoors. Of the 187 adults reared from eggs laid in ovitraps, 185 were *Aedes scutellaris* and two were *Aedes aegypti*.

Ovitraps constructed from dark plastic containers of 15 cm diameter and 12 cm depth with a 28 cm × 8 cm piece of cloth covering the inside surface, were used in a survey of dengue vectors in central Lao PDR by Tsuda et al. (2002). These traps yielded an average of 74.0 eggs per week per trap when placed indoors and 51.4 outdoors.

In Salto City, Uruguay, black plastic containers of variable size are used in the *Aedes aegypti* monitoring programme. Each ovitrap contains an oviposition substrate consisting of a strip of wood that is positioned vertically, with the bottom third inserted into the water at the bottom of the trap. Holes for mosquito entry are made in the sides of the container at regular intervals just above the level of the water (da Rosa et al. 2003).

Lourenço-de-Oliveira et al. (2004) studied the invasion of a forest by urban dengue vectors in Brazil using ovitraps constructed from black plastic 400 ml jars. Each ovitrap contained 300 ml of sieved stream water plus washed and broken fallen leaves. A 12 × 2 cm wooden paddle acted as the oviposition substrate. Ovitraps were numbered and set in the shade on the ground, usually fixed to tree trunks. Ovitraps were set 1 m, 10 m, 100 m, 500 m, and 1 km into the forest from two houses located beside a paved road adjacent to the forest. Five ovitraps were set at each distance. Additionally, four plastic bottles, designed to simulate treeholes, were hung 15 m high in the forest, two at 500 m and two at 1 km from the houses. Ovitraps were exposed for 15 d each month and examined for mosquito immatures after seven and fifteen days, from July 1997 to June 1998. A total of 25 783 immature mosquitoes was collected. The majority (45% and 32% respectively) were collected 1 m and 10 m from the houses. *Aedes albopictus* was the most abundant species at all distances and accounted for 83.7% of the total mosquitoes collected. *Aedes aegypti, Limatus durhami*, and *Culex dolosus* accounted for 6.4, 4.3, and 4.0% respectively. *Aedes aegypti* was scarce at distances >10 m and was never recovered at 500 m and 1 km. *Aedes albopictus* and *Limatus durhami* were more commonly trapped close to houses, but were present at all distances. In contrast, *Culex dolosus* was more numerous at 1000 m than at the other distances.

Badano and Regidor (2002) compared the preference of *Aedes aegypti* for ovipositing in plastic ovitraps presenting different surface areas (177 cm^2 or 57 cm^2) and colours under conditions of direct sunlight and 90% shade. Filter paper was used as the oviposition substrate. Under conditions of direct sunlight, red and black ovitraps were preferred over green or white. No significant differences were observed between the two diameters of trap. Under shaded conditions, black and red traps were again preferred, and green traps were the least preferred. Ovitraps with the larger surface area were preferred over those with the smaller surface area (Mann-Whitney $U = 36.0$, $P < 0.01$). Taking colour and surface area together, it was observed that under direct sunlight red and black traps with a surface area of 177 cm^2 were the most successful, while under shade, black traps with a surface area of 177 cm^2 were preferred. Collins and Blackwell (2000) reported a preference for black containers over white, red, yellow, green, or blue containers as oviposition sites for gravid female *Toxorhynchites moctezuma* and *Toxorhynchites amboinensis* in the laboratory.

Vezzani et al. (2004) studied the preference of *Aedes aegypti* for different sized larval habitats in three cemeteries in Buenos Aires, Argentina. A total of 21 244 containers were examined, of which 10 254 held water and 615 were positive for *Aedes aegypti* larvae and/or pupae. Total container availability, water-holding container availability, and breeding sites were classified according to their capacity into four categories: 0–0.5, >0.5–1, >1–3, and >3 litres. Containers of all capacity categories were positive for *Aedes aegypti*. Highest mosquito abundance was recorded for category >0.5–1 litres (45%), followed by >1–3 litres (35.6%), 0–0.5 litres (11.2%), and >3 litres (8.2%). However, use of containers of different capacity was proportional to container availability, suggesting that *Aedes aegypti* does not exhibit a preference for any particular container capacity.

Kaw Bing Chua et al. (2004) conducted a field study to determine the characteristics of man-made containers that made them most attractive to ovipositing *Aedes aegypti* and *Aedes albopictus* in the city of Petaling Jaya, peninsular Malaysia. Ovitraps were constructed from 1-gallon plastic containers normally used to store medicinal syrups, with varying numbers, shapes and sizes of holes cut from the containers. Some containers had palm (*Ptychosterma macathurii*) leaflets inserted into the water and others had their exterior surfaces painted black. Each container was filled with 1.5 litres of rainwater and pairs of containers were placed at random in shady positions within the garden of a single house. Pairs of containers were moved to a different random location after 2 study cycles. A study cycle was considered completed when the first pupae had appeared. In a second study, three sets comprising two containers each with two 12 × 8 cm oval holes in the top surface, one with a palm leaf (A1 type) and one without (A2), plus a standard Municipal Council of Kuala Lumpur ovitrap were placed at 3 fixed locations in the garden. In the third study eight pairs of containers with two 12 × 8 cm holes were placed in fixed locations. One container of each pair was painted black on its outside surface, excluding the base. In the first study, over 17 oviposition cycles, containers with the two 12 × 8cm holes and a palm leaf (type A1) produced the most larvae (171). The addition of a palm leaf made each design of trap more efficient than the same design without the palm leaf. In the second study, the A1 type traps and the standard ovitrap were both more efficient than the type A2 trap (no palm leaf). Mean number of larvae was higher with the Type A1 traps than for the standard ovitrap, although both were positive on approximately the same number of occasions. Containers painted black were positive on more occasions (79:65) and contained more larvae than white containers (1769:693), despite both containers being of the same shape and size and both containing a palm leaf. The authors did not attempt to differentiate between *Aedes aegypti* and *Aedes albopictus* in this study.

Forattini et al. (1998) in Brazil described ovitraps constructed from large plastic boxes in which a constant volume of about 70 litres of water was maintained. A small, sloping corrugated iron roof structure was constructed over the traps to keep out rainfall. Although the traps were primarily designed for *Aedes albopictus*, the authors reported the unexpected recovery of several larvae and pupae of *Anopheles bellator*.

Tikasingh and Martinez (1983) developed a multipaddle trap to collect eggs of *Haemagogus equinus*, other *Haemagogus* species and *Aedes aegypti*. The modified trap consists of a 1.5-litre plastic ice cream carton about 16.5 cm in diameter and 10 cm deep. Twelve hardboard paddles (2.5 × 13.0-cm) are positioned vertically in wire hoops around the inside wall of the carton. In field trials in Trinidad with four traps exposed from May 1981 to February 1982, 148 (15.8%) of the 936 paddles examined had eggs. Total eggs were 1013, giving a mean of 6.8 per positive paddle. There appeared to be no difference between the attractiveness of red cartons and those painted black inside and out. A. B. Knudsen (Tikasingh and Martinez, 1983) found that in Anguilla multipaddle traps attracted *Aedes aegypti*. The authors believed the employment of many paddles increased the numbers of eggs obtained. An alternative to using multiple paddles would be to line the inside of the carton with strong brown paper towels, or embossed benchkote paper died grey-black, as used by Yates (1974).

Chan et al. (1971) found that the most common out-of-door habitat of *Aedes albopictus* in Singapore was discarded tin cans, and Chan (1971) made use of them as convenient ovitraps. Empty condensed milk tins were painted black and placed at ground level in shaded sites, such as under bushes and banana clumps. The oviposition surface consisted of a piece of hardboard (Bristol Board is quoted in the publications but this was due to confusion of terms) measuring 1 × 4 ½ × 1/18 in.

Iriarte et al. (1991) used a cylindrical metal ovitrap with a volume of 350 ml to investigate the distribution of mosquitoes in relation to human dwellings in Nagasaki, Japan. Species collected in 27 ovitraps included *Aedes japonicus, Aedes albopictus, Tripteroides bambusa, Armigeres subalbatus* and *Orthopodomyia anopheloides*.

Swanson et al. (2000) used 340 ml capacity aluminium cans painted black on the outside to monitor *Aedes albopictus* after its introduction to Illinois, USA. Overflow holes were drilled 2.5 cm from the top of the trap and pre-soaked balsa wood paddles were used as the oviposition substrate.

On Samui Island, Thailand, Thavara et al. (2001) used modified Pratt and Jakob (1967) style ovitraps to monitor the oviposition behaviour of

Aedes aegypti and *Aedes albopictus*. Traps were 450 ml capacity flower pots (9 cm high and 10.5 cm in diameter at the top) that had no drain holes. Eighty ovitraps filled with 300 ml of rainwater were set 15 m away from houses on the ground in shady areas protected from intense rain and wind. The oviposition substrate was a strip of white filter paper. On inspection, 60% of the 80 ovitraps contained *Aedes* eggs. The average number of eggs deposited per trap per three days was 26, with a range of 11 to 59. All larvae reared to adults were identified as *Aedes albopictus*.

Ovitraps consisting of 500 ml capacity tin cans (11.5 cm high with a 8.5 cm diameter) painted black on the outside and filled with approximately 400 ml of water were used to study the seasonal abundance of dengue vectors in Manila, Republic of the Philippines (Schultz, 1993). The oviposition substrate was a tongue depressor (15.5 cm × 2.0 cm) wrapped in brown paper toweling. Two ovitraps were placed indoors and two outdoors at each of 10 houses in the first year and 12 in the second year. Traps were left in position for 7 days before being re-set. Outdoor catches ranged from 13.3 to 140.2 eggs/trap/-week at the Sampaloc site and 21.8 to 123.7 eggs/trap/week at the Tondo site. More than 99% of all eggs deposited were of *Aedes aegypti*, the remainder were *Aedes albopictus*.

Abu Hassan et al. (1996) used 10 tin can ovitraps, 105 mm high and 772 mm in diameter, and painted black inside and out to sample *Aedes albopictus* in Penang, Malaysia. Traps were filled with 200 ml of water and a cardboard paddle with a rough surface was used as the oviposition substrate. A hole in the side of the can at a height of 90 mm prevented the traps from overflowing. Trap paddles were removed every hour over six 24-h periods. Peak oviposition was observed to occur at 1600 h.

In Samoa Buxton and Hopkins (1927) used artificial test containers to study the factors controlling egg laying in *Aedes pseudoscutellaris* and *Aedes aegypti*. Their artificial containers, or pots as they were called, consisted of glass vessels 15 cm in diameter and about 10 cm tall, and were half-filled with different types of water. Each oviposition pot was covered with a 4-gal petrol tin, 36 cm high and 24 in square which had the bottom removed and a 7.5-cm diameter hole cut in the top. The pots were inspected at weekly intervals, the *Aedes* eggs removed with a small paint brush, and the water level maintained by adding distilled water.

In Florida, USA after mass release of *Aedes aegypti* in an otherwise aegypti-free area ovitraps comprising twenty 3-litre half-filled buckets lined with filter paper and eleven 55-gal barrels lined with white cloth were set out. Over a 22-day period 96 652 eggs were collected from one release site. At another release site 61 954 eggs were collected from these ovitrap-buckets (Seawright et al. 1977).

Oviposition substrates. Other materials besides hardboard have been evaluated as oviposition substrates in ovitraps. After testing more than 50 different materials Jakob et al. (1970) concluded that brown or grey velour paper paddles were about as efficient as hardboard paddles. More than 98% of *Aedes aegypti* eggs were deposited on the face of the velour paddles, whereas only about 81% were deposited on the rough side of hardboard paddles, 19% being laid along the edges of the paddles.

Many, if not most, *Aedes* species apparently prefer to lay their eggs in cracks and crevices or at least on rough, in preference to smooth, surfaces (Beckel 1955; Dunn 1927; Fay and Perry 1965; O'Gower 1955, 1957, 1958, 1963; Penn 1947; Wallis 1954).

O'Meara et al. (1989*a*) used red velour paper paddles of Kloter et al. (1983) in paired glossy black polypropylene plastic jars to collect eggs of *Ochlerotatus bahamensis* in Florida. O'Meara et al. (1995) later used the same design of ovitrap alongside surveys of used tires and other containers to monitor the spread of *Aedes albopictus* and the decline of *Aedes aegypti* in Florida, USA.

Several investigators have used balsa paddles, for example Hanson et al. (1988) used balsa wood paddles in their ovitraps for monitoring *Ochlerotatus triseriatus* populations, while Kitron et al. (1989) attached their balsa strips (15 cm long, 2.5 cm wide) with a clip to black-painted can-type ovitraps to minimise animal damage.

Schuler and Beier (1983), Beehler and DeFoliart (1990) and Beier et al. (1982) used presoaked balsa paddles in black aluminium cans to collect eggs of *Ochlerotatus triseriatus, Ochlerotatus hendersoni, Toxorhynchites rutilus rutilus* and *Toxorhynchites brevipalpis.*

In Japan Toma et al. (1982) used paper towels as an oviposition substrate in their survey of *Aedes albopictus*, while in Tanzania Trpis (1972) lined his pots with paper towelling in addition to using hardboard paddles. Ballard et al. (1987) used ovitraps similar to those of Novak and Peloquin (1981) with tongue depressor blades (15 cm) which had been scratched with a saw blade as the oviposition substrate.

Ritchie (2001) compared ovitraps containing roughened wooden tongue depressors, Masonite™ hardboard paddles and seed germination paper and reported that the mean number of positive traps did not differ among the three oviposition substrates, but the lighter coloured tongue depressors and the seed germination paper facilitated visual inspection and identification of positive traps.

Thavara et al. (2004) compared 27 × 6 cm strips of Whatman No. 1 Filter paper, 2mm thickness yellow sponge sheets, and 12 × 2 cm hardboard paddles as oviposition substrates for *Aedes albopictus* in the laboratory in Thailand. The mean number of eggs laid by colony reared females and collected from filter paper was significantly lower than from hardboard paddles and sponge sheets, but there was no significant difference between the mean numbers of eggs collected from hardboard paddles and sponge. The mean (± S.E.) number of eggs collected was: 169.7 ± 21.6, 277.9 ± 27.5, and 364.1 ± 40.9 from filter paper, sponge sheet, and hardboard paddle, respectively. The average hatching rate of *Aedes albopictus* eggs obtained on the hardboard paddles was higher than the hatches on filter paper and sponge sheets, with mean (± S.E.) hatching of 54 ± 3.7%, 31.6 ± 2.8%, and 3.7 ± 0.8%, respectively, with significant differences observed among all three substrates. The apparent inhibitory effect of sponge sheeting on the hatching of *Aedes albopictus* eggs was felt by the authors to be a factor recommending its use as an oviposition substrate in ovitraps used for *Aedes* monitoring.

Madon et al. (2003) constructed ovitraps from clear-plastic water bottles (9–20 oz) with the tops cut off, 2 holes punched approximately 1 in below the top opening for attachment of an "S" hook and prevention of overflow. Traps were painted black on the outside. Strips of seed-germination paper (Anchor Paper Company, St. Paul, Minneapolis, USA) were used as the oviposition medium.

In the USA Berry (1986) used muslin cloth strips attached to the rims of ovitraps with paper clips as an oviposition surface.

Savage et al. (1992) fitted their 700 ml plastic cups with cloth liners as the oviposition substrate during a survey for *Aedes albopictus* in Nigeria. *Aedes albopictus* comprised 18.1% of the 271 *Aedes* specimens collected.

Rozeboom et al. (1973) omitted paddles from their traps, instead lining them with rough brown paper. They found that only about 17% of the pots contained more than 61 eggs of *Aedes albopictus*, whereas laboratory observations showed that the average egg batch size was 63 eggs. Similar observations were made on the oviposition behaviour of *Aedes polynesiensis*. They concluded that these species did not discharge all their eggs in a single oviposition site, however, it is also possible that the traps were not very attractive and thus females only deposited a few eggs in these, whereas they normally laid all their eggs in a single natural habitat. In India

Reuben et al. (1977) found that a brown cloth strip placed in black glass ovitraps was considerably more attractive to ovipositing *Aedes aegypti* than jars having velour paper or hardboard strips. They also found that more eggs were laid on a green cloth strip (1332) than red (627), yellow (672), brown (698) or blue cloth strips (752). In addition to *Aedes aegypti* eggs of *Aedes albopictus, Aedes vittatus, Aedes unilineatus* and *Aedes micropterus* were laid in jars having green strips of cloth. Ovitraps with brown cloth strips were also successfully used in later studies (Reuben et al. 1978).

Reiter et al. enhanced CDC ovitrap. Reiter et al. (1991) found that 10% hay infusion in ovitraps was a good oviposition attractant for *Aedes aegypti* in Puerto Rico. The use of paired ovitraps, one containing 100% hay infusion and one containing 10% infusion collected the highest numbers of eggs (92.2/collection), which was 8.1 times more than a single ovitrap with water. Most eggs were laid in the pots containing the 10% hay infusion, suggesting that a concentrated hay infusion acts as a powerful attractive olfactory stimulus, but on arrival gravid females prefer to oviposit in traps containing a less concentrated infusion. Other useful combinations were 100%/water in which most eggs were laid in the ovitrap containing just water and the paired concentration 100%/100%. Apostol et al. (1996) used the enhanced CDC ovitrap to carry out a population genetic structure analysis of *Aedes aegypti* collected from sixteen barrios in six cities in Puerto Rico. The 100%/10% hay infusion combination was used in the traps.

In Tucson, Arizona, USA, Hoeck et al. (2003) modified the Reiter et al. (1991) trap by attaching a fibreglass screen held in place by the lid-ring closure to deter lizards, which were attracted to ovitraps containing the 100% hay infusion.

Hay infusion-enhanced CDC ovitraps of Reiter et al. (1991) were compared with ovitraps containing 400 ml tap water plus 0.5 g or 1.0 g of lupin-based small animal chow (compressed seeds of the leguminous genus Lupinus) or ovitraps containing 400 ml tap water plus 0.25 g or 0.5 g of alfalfa pellets (Ritchie, 2001). Neither trap performed as well as the enhanced CDC traps when tested using a Latin square experimental design, and the ones containing lupin-based animal feed pellets developed a film on the water surface which killed mosquito larvae.

The enhanced attractiveness of hay infusion-enhanced ovitraps has also been demonstrated in Trinidad, where Rawlins et al. (1998) found that both *Aedes aegypti* and *Culex quinquefasciatus* preferred ovitraps containing hay infusion to either a suspension of brewer's yeast (1 g/20 ml) or aged tap water, however the authors considered that brewer's yeast was a preferable medium for survey work as it was easier to prepare and gave similar results to the hay infusion baited ovitraps.

Kay et al. (2002) used 650 ml capacity black plastic ovitraps containing 350 ml of a 10% hay infusion to evaluate the effects of a control intervention against *Aedes aegypti* breeding in wells in Queensland, Australia. The oviposition substrate was a wooden tongue depressor. To prevent predation of eggs by ants and cockroaches, each trap was placed in a clear rectangular plastic container filled with water.

In the laboratory, *Aedes aegypti* has demonstrated a preference for oviposition in lagoon water or waste water (collected from a University waste water reservoir) over swimming pool water, tap water (both chlorinated), distilled water, and sea water. The preference for the waste water and lagoon water appeared to be related to the presence of coliform bacteria (Navarro et al. 2003).

Lee and Kokas (2004) evaluated lawn sod and rabbit chow infusions as attractants for West Nile virus vectors in ovitraps and gravid traps in New York Sate, USA. Pairs of ovitraps, one containing lawn sod infusion and one containing rabbit chow infusion were placed under the eaves of buildings at two sites and egg rafts were counted and removed daily. Overall, higher daily mean numbers of *Culex* egg rafts were laid in traps containing lawn sod infusion, compared with rabbit chow infusion, however, this relationship did not persist throughout the trial. In the ninth week of the study, equal numbers of egg rafts were recovered from each infusion, and in the tenth week more egg rafts were recovered from rabbit chow infusion-baited traps. The authors did not consider this to be due to an increase in the relative size of the *Culex pipiens* population, which Lampman and Novak (1996*a*) demonstrated as having a slight preference for rabbit chow infusion, as identification of larvae reared from egg rafts laid during the weeks when oviposition was high in rabbit chow infusion traps revealed that almost 100% of egg rafts in both traps were laid by *Culex restuans*. The changes in attractiveness of the two infusions over time was ascribed to physical or chemical changes in the infusions caused by changes in external environmental condtions. Jackson et al. (2005) also observed that relative attractiveness of infusions changed during the course of an evaluation of gravid trap and ovitrap infusions in Virginia, USA. Manure infusion-baited traps yielded more *Culex restuans* and *Culex pipiens* egg rafts during three of the first 4 weeks of the study, while hay infusion baited-traps yielded more egg rafts during weeks 5–9. These authors reached a similar conclusion to Lee and Kokas (2004) and postulated that the higher temperatures experienced during outdoor incubation of the infusions later in the season were responsible for the increased attractiveness of hay infusions over time.

Madon et al. (2003) reported the use of water obtained after rinsing black tiger shrimps (*Penaeus monodon*) to enhance attractiveness of CDC ovitraps to *Aedes albopictus*. The attractant was prepared by rinsing 3–5 lbs of tiger shrimp with small amounts of tap water about 8–10 times until approximately 1–3 gal of rinse water was obtained. The rinse water was stored in a refrigerator and used as needed.

Occasionally ovitraps have been placed in cement half-blocks, painted black, to prevent them tipping over (Anon 1979; Ritchie 1984*a*). O'Meara et al. (1989*a*) attached black polypropylene ovitrap jars to pieces of white plywood to stabilise them, and also used a wire bar across the entrance to prevent animals drinking from them.

Sticky ovitraps. Ordóñez-Gonzalez et al. (2001) constructed sticky ovitraps from 3.8 litre black plastic containers. A 6 cm wide strip of thin black cardboard was attached with paperclips around the inside surface of the container, above the water line. The black cardboard strip was coated with a layer of adhesive used for rodent trapping (Trapper™, Bell Laboratories, Inc., Madison, Wisconsin, USA), which remains sticky under a range of humidities and temperatures. Effectiveness of the traps was determined by releasing marked adult *Aedes aegypti* females in an area of 300m diameter containing 100 sticky ovitraps. 7.7% of the 401 females marked and released were recaptured by the sticky ovitraps.

Ritchie et al. (2003) developed a similar sticky ovitrap by attaching an arc of overhead transparency plastic (5.5 cm × 45 cm top and 36 cm bottom) coated with a thin layer of polybutylene adhesive to the inside surface of 1.2 litre black plastic buckets. The sticky ovitraps were baited with 50% hay infusion (Reiter and Gubler 1997), and supplemented with two 0.2-g pellets of lucerne (alfalfa) to enhance the infusion (Ritchie 2001). Field trials were conducted in dengue endemic residential areas of Cairns in coastal northern Queensland, Australia. The authors reported that the sticky ovitrap was as sensitive as the standard ovitrap in detecting *Aedes aegypti*, but also collected other species, including *Culex quinquefasciatus* and *Ochlerotatus notoscriptus*. Although adult mosquitoes attached to the glue are often damaged and are found in positions that obstruct the view of important characters used for identification, most specimens could be identified to species using a hand lens. A major advantage of this method is that ovipositing mosquitoes can be identified immediately, without requiring hatching of the eggs and rearing of larvae. The sticky ovitraps were positive for *Aedes aegypti* more often than a combination of standard ovitraps and a "sentinel" car tyre used to monitor exotic and epidemiologically significant mosquitoes at Cairns International Airport. In addition, the ovipositing mosquitoes using the sticky ovitrap were retained and killed.

Ritchie et al. (2004) used the Ritchie et al. (2003) sticky ovitraps to undertake entomological investigations during a dengue virus serotype 2 outbreak, which occurred in a suburb of Cairns, Queensland, Australia during January-April 2003. The sticky ovitraps used were slightly modified from the earlier versions, being fitted with two 15 × 5.5 cm strips of overhead transparency plastic coated in adhesive and fixed to opposite sides of the 1.2 litre plastic bucket with 50 mm paper clips, rather than an arc of adhesive coated plastic. Each ovitrap contained 500 ml of 33% hay infusion plus 0.5 g of alfalfa pellets. Two s-methoprene pellets were added to prevent larval development. Two hundred and eighty-six sticky ovitraps were placed adjacent to houses in a 0.49 km^2 area of a suburb of Cairns and collected a total of 222 female

Aedes aegypti between 21 January and 08 May, with a mean of more than 3 females per trap per week before initiation of control operations, falling to below 1 per trap per week 2–3 weeks after control activities began. The authors reported that the sticky ovitrap is a valuable tool for surveillance in dengue control campaigns, including for measurement of the efficacy of control operations. In addition, sticky ovitraps collect gravid females that can be assessed for viral RNA using PCR-based techniques. The authors proposed a sticky ovitrap index (mean number of female *Aedes aegypti* per trap per week) and suggested that this may be easier to obtain than other *Aedes aegypti* indices, such as the Breteau Index (BI), House Index (HI), and Pupal Index.

Russell and Ritchie (2004) also used the Ritchie et al. (2003) sticky ovitrap (with black buckets replaced by dark red ones) to undertake surveillance and behavioural investigations of *Aedes aegypti* and *Aedes polynesiensis* on Moorea, French Polynesia. The effects of adding infusions to the traps, timing of trap exposure, degree of trap exposure, and trap orientation on oviposition by the two species were investigated in a series of trials. Results of the trials indicated that both species tended to avoid traps containing leaf infusions, compared with plain water or grass infusions. Significantly larger numbers of both species were collected from traps set and collected at midday, compared with traps set and collected at midnight. Both species were more frequently caught in traps in sheltered and shady, rather than exposed, positions. More female *Aedes aegypti* were collected from traps against west-facing walls compared to east-facing walls. There was no significant difference in numbers of *Aedes polynesiensis* by trap orientation.

A disadvantage of sticky ovitraps is that individual mosquitoes may be damaged during removal from the adhesive, hampering species identification or subsequent classification of ovarian development stage.

Automatic recording ovitraps. To study the time of oviposition of *Aedes albopictus* in Japan Tsuda et al. (1989) used an ovitrap incorporating an automatic recorder (Fig. 2.10). A 1.5-litre water tank allows water to drain down into a water cup A to maintain a constant depth of 2 cm, from here water drains into a cup B (50 ml) which when full overflows into one of the small cups of the water-mill wheel. This is connected to a strip of filter paper (4 × 120 cm) which has one end in the ovitrap, and advances it 11 cm, thus exposing a new area for oviposition. The part of the filter paper on which eggs have been deposited is then lightly sandwiched between two strips of plastic (5 × 130 cm) to prevent further oviposition.

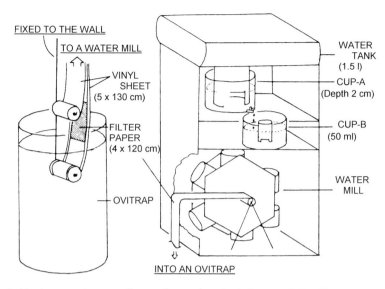

Fig. 2.10. Automatic recording ovitrap, the shaded part of the filter paper is the part that becomes exposed to ovipositing females for 1 h (Tsuda et al. 1989)

Autocidal traps. In Singapore Chan et al. (1977) and Lok et al. (1977) designed autocidal *Aedes aegypti* ovitraps that had one to two hardboard paddles inserted through a floating doughnut shaped ring with nylon mesh covering the centre hole. This arrangement caused larvae to suffocate, or prevented adults from escaping from the trap. The modifications to this trap made by Cheng et al. (1982) for use in the USA to control *Aedes aegypti* are described here. The ovitrap jar consists of a dark bottle (approx. 10 cm high and 8 cm in diameter) and two expanded polystyrene rings having a piece of nylon mesh glued between them, which is then placed on the water in the bottle. Two short pieces of hardboard (3.2 × 6.4 × 0.3 mm) acting as oviposition paddles are inserted into the polystyrene ring with their lower edges in contact with water in the ovitrap (Fig. 2.11).

As in conventional ovitraps paddles can be periodically replaced, but if left undisturbed so that eggs laid on them hatch 2nd instar larvae are unable to squeeze through the nylon mesh and eventually drown. These traps have proved to be a sensitive and reliable method of detecting and monitoring not only *Aedes aegypti* but also *Ochlerotatus triseriatus*.

A simplified lethal ovitrap was developed and laboratory-tested by Zeichner and Perich (1999). The trap consists of a 473 ml capacity black polyethylene cup, with an 11cm × 2.5 cm red velour strip treated with insecticide as the oviposition substrate attached to the rim of the cup with a paper clip. The Zeichner and Perich (1999) lethal ovitrap was evaluated by Perich et al. (2003) against *Aedes aegypti* in the field in Rio de Janeiro State, Brazil.

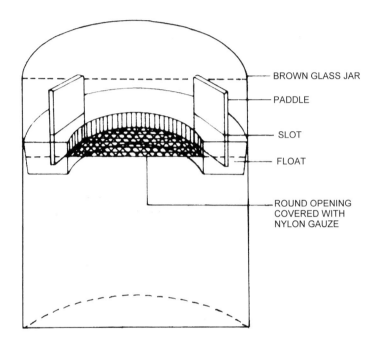

Fig. 2.11. Autocidal *Aedes aegypti* ovitrap (Cheng et al. 1982)

In each of two municipalities, 10 lethal ovitraps per house, five indoors and five outdoors, were used in a group of 30 houses. The 30 intervention houses were compared with 30 similar, non-intervention houses. The oviposition substrates were treated with 1.0 mg a.i./strip of deltamethrin insecticide. *Aedes aegypti* density was monitored using three methods: (i) percentage of containers positive for larvae and/or pupae; (ii) total pupae/house; (iii) total adult females/house collected by aspirator indoors. The ovitraps had significant impact on the *Aedes aegypti* population in the first locality (Areia Branca), as measured by all three sampling parameters, including adult females per house, 30 days after introduction of the traps. The traps also reduced all three population indices in the second location (Nilopolis), but only after 90 days of intervention and this was attributed to the availability of a higher number of alternative oviposition sites in the second locality.

To minimise the risk of larvae developing in ovitraps, particularly tyre ovitraps, insecticide can be added to the water in the traps, but this may affect the attractiveness of the trap to ovipositing females. Ritchie and Long (2003) examined the effect of adding s-methoprene to black plastic ovitraps and tyre ovitraps on oviposition by *Aedes aegypti* and *Ochlerotatus notoscriptus* in Queensland, Australia. The control ovitrap consisted of a 1.2-litre black plastic bucket filled with ~500 ml of tap water to which a 0.5-g alfalfa pellet was added to create an infusion to enhance oviposititon. A roughened wooden tongue depressor served as the oviposition substrate. Treatment ovitraps were identical, except for the addition of a single Altosid™ pellet (~0.13 g of 40 g/kg s-methoprene). The addition of a single Altosid™) pellet had no significant effect upon oviposition by *Aedes aegypti* or *Ochlerotatus notoscriptus* at either of 2 sites, as determined by a Mann-Whitney rank-sum test. The authors concluded that the addition of s-methoprene pellets to ovitraps, particularly sentinel tyre traps, should be considered in ovitrapping programmes to minimise the threat of mosquito development in the traps.

Ovitrap surveys

Ovitraps are usually serviced every 7 days, and Ritchie (1984*a*) concluded that in most surveys a weekly exposure period was suitable. To overcome the problem of predation Frank and Lynn (1982) suggested having the shortest possible time between ovipaddle collections, but this makes surveillance labour intensive. Shorter trap exposure periods, however, have been used, including 1-day periods (Frank and Lynn 1982; Nayar 1981) and 48 h periods in Nigeria (Savage et al. 1992). In comparative trials in Louisiana Kloter et al. (1983) found that fibreboard paddles were better than velour paper ones because snails and cockroaches sometimes destroyed the paper ones. Even so predators may still remove eggs from fibreboard paddles without destroying the paddles. In Trinidad ovitrap paddles were changed every 2 h to study the diel oviposition cycle of Aedes aegypti. One to 43 eggs were oviposited on a paddle during this interval (Chadee and Corbet, 1990). Ovitrap size can also be important. For instance Berry (1986) found that 12-oz can-type traps collected about seven times more *Aedes aegypti* eggs compared with 10-oz ovitraps.

Ovitraps have been used by many workers in North America as a routine surveillance tool. Fay and Eliason (1966) found that one mosquito inspector could cover a three to five times larger area if oviposition surveys were made instead of larval surveys, and the costs were halved, or even quartered. Jakob and Bevier (1969*b*) reported a 17-fold decrease in working days when ovitraps were substituted for larval surveys.

Dibo et al. (2005) reported the results of a study to determine the best location for siting ovitraps when conducting *Aedes* surveys in and around houses in Brazil. Ovitraps consisting of 1-litre black plastic pots filled with 500 ml of tap water, and with a wooden paddle measuring 12.5 cm by 2 cm placed inside the pot to act as the oviposition substrate, were placed in the bedroom, in the living room, in a sheltered outdoor site (veranda or service area), and in an exposed site (under leaves on tree branches or in bushes). From the 100 residences in which traps were installed, a total of 4400 paddles were obtained, of which 38.6% (95% CI: 37.1–40.0) were positive for eggs. Outdoor traps were more often positive for *Aedes aegypti* eggs than indoor traps. Exposed sites were more frequently positive than sheltered sites outdoors. No significant difference was observed between the bedroom and living room sites.

Collection of other species

Although primarily developed for *Aedes aegypti* surveillance, ovitraps have attracted ovipositing adults of other *Aedes* and *Ochlerotatus* species when used in the United States, including *Ochlerotatus triseriatus, Ochlerotatus atropalpus, Ochlerotatus mediovittatus, Ochlerotatus zoosophus* and *Aedes albopictus*, as well as *Orthopodomyia signifera* (Beehler and DeFoliart 1990; Pratt and Kidwell 1969).

In a study of the occurrence of *Ochlerotatus triseriatus* and other in a La Crosse virus endemic area of North Carolina, USA, Szumlas et al. (1996) used ovitraps constructed from tin cans (volume approximately 4 litres) painted glossy black inside and out. Drainage holes were drilled 2.5 cm from the top of the can and dry oak leaf litter was added to a depth of 3–5 cm and a piece of 4-gauge galvanised metal screen was placed on top of the leaf litter. Sufficient tap water was added to half-fill the cans and a red velour paper oviposition strip was clipped to the inside of each trap. Each ovitrap was covered with 2 cm gauge wire screen to prevent entry by large animals. Although five species of mosquito were detected in the traps, 98% of all eggs deposited were of *Ochlerotatus triseriatus*. Other species collected included *Culex restuans, Ochlerotatus hendersoni, Aedes albopictus*, and *Culex pipiens*.

In Wisconsin ovitraps with balsa wood paddles sometimes contained eggs of *Orthopodomyia signifera* (Beehler and DeFoliart 1990; Loor and DeFoliart 1970), while in Tahiti *Aedes aegypti* ovitraps (Fay and Eliason 1966; Fay and Perry 1965) were used to attract *Toxorhynchites amboinensis* (Rivière 1985).

Use of standard black plastic ovitraps containing red velour oviposition strips and a 1:1 mixture of freshwater and local seawater set at 37 locations within 50 m of the high-tide line and adjacent to natural or man-made rock formations along the Pacific coast of Washington State, USA, failed to detect the presence of *Ochlerotatus togoi*, although 15 specimens were recovered by active surveillance of rock holes (Sames et al. 2004).

In Trinidad ovitraps of the Fay and Eliason (1966) design were found to be suitable for collecting eggs of *Haemagogus equinus* (Tikasingh and Laurent 1981). Of 6678 oviposition paddles exposed 69% were positive and 24 445 eggs were collected. The number of eggs deposited per paddle per week ranged from 1 to 150, with an average of 35.

Also in Trinidad ovitraps consisting of 500-ml capacity polystyrene cups (height 135 mm, basal diameter 60 mm, mouth 90 mm) painted black and filled with 275–300 ml water were evaluated as ovitraps for *Toxorhynchites moctezuma*. To prevent eggs being displaced by rain some traps had an inverted plastic petridish supported 60 mm above the cups on three wire supports, lid and supports were also painted black (O'Malley et al. 1989). However, only 1.4% of the pots were colonised, compared to 6.0% of natural oviposition sites comprising fruits of the tree *Lecythis zapucajo*.

Eggs of *Culex (Melanoconion)* Group Pilosus were recovered from three oviposition traps set in the toilet of a Bus Station in Joinville city, State of Santa Catarina, Brazil, during *Aedes aegypti* surveillance (Gomes et al. 1998).

In the Western pacific region ovitraps have been used in Taiwan, Guam and Okinawa, and apart from attracting ovipositing females of *Aedes aegypti* and *Aedes albopictus*, eggs of *Ochlerotatus aureostriatus okinawanus, Aedes riversi, Aedes pandani* and *Aedes nocturnus* were collected on filter paper paddles used in the traps (Reisen and Basio 1972). In Tanzania Trpis (1972) used ovitraps to study oviposition of *Aedes bromeliae* in different ecological zones.

Chadee and Tikasingh (1989) used the modified ovitraps described by Chadee and Corbet (1987) to study diel oviposition by *Haemagogus janthinomys*. Paddles were removed at 2-h intervals for 24 h on 1 day/week for 53 weeks. They also studied diel oviposition of *Haemagogus equinus* (Chadee and Tikasingh 1990) using the same method. Only 175 eggs of *Haemagogus janthinomys* were collected over the entire period, compared with 820 eggs of *Haemagogus equinus*. In other trials in Tobago Chadee et al. (1984) recovered eggs of *Haemagogus equinus* and *Haemagogus celeste*, as well as those of *Ochlerotatus taeniorhynchus* and *Ochlerotatus berlini*, from their paddles.

Alencar et al. (2004) constructed ovitraps for sampling *Haemagogus janthinomys* in an Atlantic forest area of Espírito Santo State, Brazil. Traps comprised 1 litre black plastic buckets without lids, which were half-filled with rainwater and leaf litter. The oviposition substrate was provided in the form of four strips (2.5 cm × 14 cm) of hardboard attached vertically to the interior surface of the buckets by way of wire clips. Traps were hung from trees at heights of between 2.5 m and 6 m. 96 of 768 (12.5%) oviposition strips were positive, with a total of 734 eggs collected. Larvae of *Haemagogus janthinomys* and *Ochlerotatus terrens* were also collected. Traps without oviposition strips and placed on the ground were colonised by *Limatus durhami*.

In São Paulo, Brazil, Marques et al. (1993) compared larval traps and ovitraps consisting of 15 cm diameter and 15 cm deep plastic bottles for surveying the dengue and yellow fever vectors *Aedes aegypti* and *Aedes albopictus* in rural and peri-urban areas. The authors found that ovitraps were more efficient in determining the presence of *Aedes albopictus* compared with larval traps. *Aedes aegypti* was not sampled and *Ochlerotatus terrens* was the only other species collected.

Bamboo pot ovitraps. Water-filled sections of bamboo, usually termed bamboo pots or cups, have commonly been used as artificial oviposition sites to attract mosquitoes that breed in tree-holes and bamboo (Bang et al. 1979; Causey and dos Santos 1949; Corbet 1963, 1964*a*; Dunn 1927; Galindo et al. 1951, 1955; Harris 1942; Harrison et al. 1972; Kemp and Jupp 1993; Laarman 1958; Lambrecht and Zaghi 1960; Lounibos 1979, 1981; Petersen and Willis 1971; Philip 1933; Sempala 1983; Service 1965, 1970; Yates 1979). Bamboo ovitraps, however, may not always sample the same species as natural tree-holes. For example, in the Democratic Republic of Congo Laarman (1958) found that larvae of *Toxorhynchites brevipalpis* and *Culex albiventris* were rarely collected from bamboo pots although they were common in tree-holes.

If suitable bamboo does not grow in areas in which the pots are to be used, it may have to be 'imported', although this may present its own problems. In the dry savanna areas of Nigeria bamboo pots obtained from the rain forests of the south split in the severe dry season. It was discovered, however, that cylindrical gourds of *Lagenaria siceraria* did not crack, even when exposed to direct sunlight, and when used instead of bamboo pots they accurately reflected the mosquito species breeding in tree-holes (Service, 1965). In England, to prevent bamboo pots from splitting, their outsides were coated with embedding wax (Yates 1974). Furthermore, to obtain the maximum number of pots from a limited supply of bamboo, lengths which were open at both ends had a 5-mm thick piece of cork glued to one end, which was also coated with wax (Yates 1974).

The conventional method of sampling bamboo pots is to tip out the contents and identify the mosquito larvae, but McClelland (1956) in studying *Aedes aegypti* in Uganda pointed out that *Aedes* eggs might remain undetected on pot walls without hatching for relatively long periods, during which time larval inspections would be negative. He also considered that other factors such as unfavourable conditions in the water, competition with larvae of other species, and possibly selective predation by *Toxorhynchites* larvae might result in high, or complete larval mortality of *Aedes aegypti* before the pots were examined. By inserting a 4¾-in cylinder of filter paper attached to a cellulose acetate sheet into each pot he was able to collect *Aedes aegypti* eggs on the filter paper just above the water line. Corbet (1963) dispensed with the cellulose sheet and lined the insides of bamboo pots with filter paper which was dyed grey (30 ml black 'Pelikan' waterproof ink in 4 litres water) to make the surface more attractive as an oviposition site. In Tanzania, Trpis (1972) used bamboo pots lined with paper towelling, but in addition introduced a 20 × 120-mm hardboard 'paddle' in the pots, which were placed in different ecological areas and also at different heights. *Aedes* eggs were laid on both oviposition surfaces. Pink or green blotting paper can be substituted for dyed filter paper.

Aedine species appear to prefer to lay their eggs in cracks or crevices or on rough surfaces. For example, under laboratory conditions *Ochlerotatus triseriatus* laid more eggs on paper towelling that had been embossed with a pattern from a 16-mesh hardware cloth than on smooth towelling (Wilton 1968), ans it is usually considered appropriate to line bamboo pots with paper having an embossed surface. This is particularly important in studies on the distribution of eggs in relation to the water level in the pots, because if smooth paper is used nearly all *Aedes* eggs are deposited along the vertical edges where the ends of the paper strip overlap.

A useful technique for using bamboo pots as oviposition sites has been developed by Yates (1974) working in England. Sections of bamboo are cut across obliquely at an angle of about 50° because it was found that more eggs of *Ochlerotatus geniculatus* were laid in these pots than those having a horizontal opening (Fig. 2.12). Such oviposition preferences may not be shown by other tree-hole species. When blotting paper or filter paper is left in water-filled pots for any length of time it usually becomes difficult to remove without tearing or disintegrating. Yates overcame this by using 'laboratory bench paper' which is commercially available in large sheets (e.g. Benchkote™) and consists of absorbent white filter-type paper backed with a thin sheet of plastic paper. This is dyed grey in an alcoholic solution of black drawing ink (1 part 'Pelikan' ink: 50 parts water); then the wet paper is passed through a domestic mangle which has been modified by slipping a cylinder of hard wire mesh over one of the rollers to give an embossed pattern on the paper. Before placing the paper lining in the pot it is dried and alcohol removed by placing it in an incubator. The paper

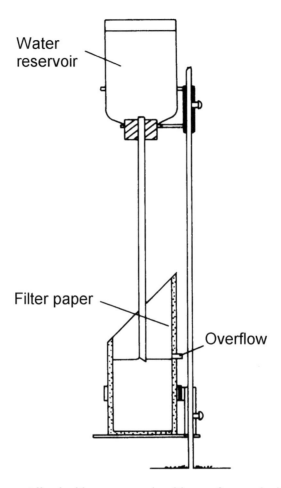

Fig. 2.12. Bamboo pot lined with paper as oviposition surface, and with water reservoir above (Yates 1974)

linings are numbered, placed in the bamboo pots in the field and can be replaced at regular intervals.

To study the vertical deposition of eggs in relation to the water line a constant water level must be maintained in the pots. Yates (1974) achieved this by making a water reservoir from an inverted polythene bottle. A length of 14-mm diameter glass tubing is placed through a rubber bung inserted into the neck of the bottle, and its lower end cut at 45°. A 13-mm hole is drilled through the bamboo pot and paper lining at the required water level and a short length of tube inserted. This serves as an overflow. With this arrangement a drop of only 2 mm in water level is compensated by water descending the glass tube. Yates (1979) used his bamboo pots in England to study oviposition behaviour of *Ochlerotatus geniculatus*. In summary pots were fixed to trees at heights, 0.5, 2.0, 4.0 and 8.0 m and to a tower at 9 heights, from ground level to 11.2 m at 1.4-m intervals.

Regression coefficients (b) of the linear relationships between height and $\log_e$ (eggs + 1) were calculated. Because regression coefficients varied between individual trees, two estimates of the overall slope were made. Firstly a weighted overall estimate was made by averaging the regression estimates from all trees with the reciprocals of their estimated variance being used as weights. The other estimate was obtained after the number of eggs laid in the pots was adjusted for differences in surface area available for oviposition due to slightly different-sized pots (Fig. 2.13a). Corrected egg numbers were then transformed and plotted against height. Pots of similar shape and size were used on the tower, so allowing the number of eggs in pots at the same height to be pooled and the resulting total egg numbers to be transformed to logs and plotted against height (Fig. 2.13b).

In order to get the seasonal incidence of oviposition the numbers of eggs laid on the linings of the bamboo pots were standardised to correct for differences in the length of exposure. This was done by calculating the number of eggs per day for each period and then expressing this number as a percentage of the yearly total. By replacing the paper lining in the pots at regular, though variable, intervals Yates (1979) found that 97.8% of the eggs (1663) were laid during the daytime, with peak oviposition being 2–3 h before sunset.

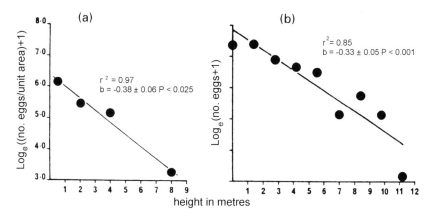

Fig. 2.13. The vertical distribution of *Ochlerotatus geniculatus* ovipositing in bamboo pots as shown by the relationship between (a) $\log_e$ ((no. of eggs/unit area of oviposition surface) + 1) and height in 1972; and (b) $\log_e$ (no. of eggs + 1) and height in 1973 (Yates 1979)

In Nigeria Bang et al. (1979) found that the numbers of *Aedes (Stegomyia)* species collected in bamboo cups nailed to trees at a height of 1.5 m in rural habitats were much greater than found in tin-can type ovitraps nailed alongside them. They also found that all eggs of all species hatched on three soakings in April (during the rainy season), but in November during the dry season three soakings produced only the following hatches, 8% *Aedes bromeliae*, 27% *Aedes dendrophilus*, 53% *Aedes luteocephalus*, 60% *Aedes africanus*, 61% *Aedes apicoargenteus* and 94% *Aedes aegypti*. At least nine repeated soakings were required to cause all eggs to hatch, and consequently to enable the true proportions of the species ovipositing in the ovitraps to be determined.

Before pots are used to sample the local population of mosquitoes they should be matured for 1–2 weeks in the field by filling them with filtered tree-hole water or rain water and adding a few dead leaves. If this conditioning is omitted they may be unattractive when they are first used (Harris 1942). In a yellow fever vector survey in southern Nigeria CDC-type ovitraps (Fay and Eliason, 1966) and bamboo cups, which had been weathered for at least 4 weeks and boiled, were set up in six vegetation types. The commonest vector species was *Aedes aegypti* and about the same percentage of bamboo pots and ovitraps were positive in the six different vegetation zones studied (Bown et al. 1980).

Paper linings in bamboo pots can be changed at regular intervals and the numbers of eggs counted and identified. When specific identification of the eggs is impossible the linings must be repeatedly soaked and the resultant larvae identified. Larval identification is usually necessary when pots are used without linings and eggs are deposited on the pot walls. Some of the difficulties associated with interpreting oviposition results by larval identification have been outlined previously. In Thailand Harrison et al. (1972) used unlined pots. After exposure in the field larvae were removed from the pots, which were then returned to the laboratory and soaked in water and the numbers of larvae hatching recorded daily. To prevent extraneous oviposition the pots were covered with paper towels. Before pots were returned to the field they were cleaned with a wire brush, sandpaper and boiling water to remove any residual unhatched eggs.

Certain species, such as some of those belonging to the genera *Haemagogus, Sabethes* and *Armigeres*, oviposit in closed sections of bamboo that have a small hole in the side, usually made by certain beetles or birds. To attract ovipositing females that lay eggs in these habitats bamboo pots covered by a lid and having a small hole bored in the side near the top can be used (Carpenter et al. 1952; Galindo et al. 1951, 1955). The pots are sampled by removing the top and tipping the contents into a white enamel bowl. These pots could also be lined with filter paper so long as the entrance hole was not obscured. More recently in Sri Lanka Amerasinghe and Alagoda (1984) compared mosquito oviposition in two types of bamboo pots (45–50 cm long, 10–12 cm in diameter), one having the top open, and the other having the cut end covered with a removable piece of hardboard, and a 2-cm hole bored in the side for entry of ovipositing mosquitoes. Coarse blotting paper dyed grey was placed inside the pots. In some pots the modification devised by Yates (1974) was used to maintain a constant water level, so that the height eggs were laid above the water surface could be obtained. Traps of both types were placed at ground level and at heights of 3.5 and 7.0 m. The most common species were *Aedes albopictus, Armigeres subalbatus, Aedes novalbopictus* and *Culex quinquefasciatus*. The two *Aedes* species exhibited a clear preference for ovipositing in traps with open tops, the preference was not quite so marked with *Armigeres subalbatus* and *Culex quinquefasciatus* showed little choice between the two types of traps.

In southern Africa over 6 years a total of 19 species were collected from bamboo pots (Jupp and McIntosh 1990). The most common species were *Aedes aegypti, Aedes ledgeri, Aedes metallicus, Ochlerotatus fulgens, Culex nebulosus* and *Culex horridus. Aedes furcifer/cordellieri* were known to be common in the area, but few were collected from bamboo pots during 1976–1977 and 1977–1978 (3–4% only of pots positive) or bottles in 1980 (1%). It was thought that this might have been because the openings were too large, and this supposition was substantiated when the openings of the bamboo pots were made smaller in 1980 and 1981, and the percentage of samples with *Aedes furcifer/cordellieri* increased to 14 and 48%, respectively. This agrees with findings in Senegal that *Aedes furcifer/cordellieri* prefers to oviposit in tree-holes with small openings (Raymond et al. 1976).

The horizontal and vertical distribution of different species can be studied by placing bamboo pots, or cylindrical gourds, lined with paper in different ecological zones and at different heights (Bang et al. 1979; Causey and dos Santos 1949; Corbet 1964*a*; Galindo et al. 1951; Harris 1942; Laarman 1958; Lounibos 1979, 1981; Service 1965; Surtees 1959; Yates 1979). Seasonal incidence and diel periodicities of egg laying can be investigated by regularly replacing the oviposition papers in pots. Alternatively the pots can be covered with lids, a few of which are removed each hour throughout the 24-h day (Corbet 1964*a*; Lambrecht and Zaghi 1960). It will be more difficult to use bamboo pots to measure or compare the population size of mosquitoes in different areas or habitats, because the incidence of egg laying in the pots not only depends on population size but also on the number of alternative natural oviposition sites that are available. This difficulty is not limited to the use of bamboo pots, but is inherent in most types of sampling programmes where artificial habitats are created.

Other artificial tree-hole ovitraps. Loor and DeFoliart (1969) used an oviposition trap made from a beer can to detect the presence of the tree-hole mosquito *Ochlerotatus triseriatus*. The top of a 12-oz beer can was removed and its outside covered in beige coloured masking tape. The can was then filled to 1 in of the top with distilled water. The relative attractiveness was evaluated of cans containing: (1) only water; (2) water plus organic debris consisting of 75% dry and 25% green oak leaves; (3) a black muslin sleeve lining the interior; and finally (4) organic debris and a black muslin sleeve. Cans were attached to trees at heights of 2.5 and 5 ft. They were examined weekly and organic debris and the sleeve removed and the eggs counted. Of the total of 2394 eggs of *Ochlerotatus triseriatus* that were recovered 69% were from cans with organic debris and the black muslin sleeve, and 26% from cans with only the black sleeve. A few eggs of *Orthopodomyia signifera*, a species that normally breeds in tree-holes, were also collected from the cans.

Because of difficulties in locating sufficient numbers of easily accessible natural tree-holes of the same type and size in their studies in California, USA Lewis and Tucker (1978) made artificial ones based on an earlier model (Lewis and Christenson 1975). Traps were made of ¾-in thick wooden boards about 8 × 10 or 10 × 12-in cut and double bevelled at 45° angles. To prevent water leakage silicone seal ('silicone glue and seal', or 'silicone chalk and seal') was used between the joints as a gasket. Two lengths of 4.76-mm diameter galvanised steel wire with each end formed into a loop were passed round the assembled boards and tension obtained by tightening the nuts on bolts passed through these loops. A piece of plywood was then screwed onto the bottom of the trap. Another similar piece of plywood with a 2-in diameter hole cut from the middle was hinged to the top with a length of webbed strapping, and the trap kept closed except when samples were being removed. Water and alfalfa pellets were added to the traps as necessary. Oviposition paddles (ovisites) consisted of 2 × 3-in pieces of wood abutted with cork strips over which thin grooved layers of balsa wood were stapled. Adjustments were made of the amount of cork to ensure the ovisites floated on the water surface and the balsa wood remained moist but not covered with water. Ovipositions were removed weekly. These traps proved attractive to both *Ochlerotatus sierrensis* and *Orthopodomyia signifera*.

During the 1980s there were several ecological studies on *Ochlerotatus triseriatus*, many using various ovitraps, some of which are described here.

Beier and Trpis (1981) used the ovitraps of Loor and DeFoliart (1969) to monitor *Ochlerotatus triseriatus* breeding at the Baltimore Zoo. They concluded that ovitraps competed with natural tree-holes as oviposition sites, because fewer eggs were collected from traps placed near beech trees with water-filled holes, than those placed near beeches lacking holes. Clark et al. (1986) also used the ovitraps of Loor and DeFoliart (1969) but lined them with black flannel to collect eggs of *Ochlerotatus triseriatus*.

Kitron et al. (1989) used 12-oz lidless aluminium tins painted black on the outside as oviposition traps for *Ochlerotatus triseriatus*. They were half-filled with oak leaf infusion, provided with an overflow hole, and attached to the bases of trees. Even weekly topping up failed to prevent the traps sometimes drying out between weekly inspections. Balsa strips, as advocated by Novak and Peloquin (1981), 2.5 cm wide and 15 cm long served as oviposition paddles. Paddles were attached to the can with a 'binder clip' to minimise damage from animals. Ovitraps of Novak and Peloquin (1981) were used by Walker et al. (1987) to collect *Ochlerotatus triseriatus* and *Ochlerotatus hendersoni* in Indiana ovipositing at three different heights.

In Texas, USA Aziz and Hayes (1987) placed 400-ml plastic beakers lined with paper towels and filled with 300 ml of a mixture of tree-hole and rainwater at heights of 0.6, 1.2, 1.8, 2.7 and 3.7 m in trees to collect eggs of *Ochlerotatus triseriatus*. Tongue depressors wrapped in paper towelling and towelling lining the beakers served as oviposition sites. Although eggs were obtained at all heights most were collected from the lower ovitraps (0.6–1.2 m).

In a test of the theory that more productive sites have more species, Srivastava and Lawton (1998) used artificial tree-holes constructed from transparent plastic boxes 11 cm × 16 cm × 5.5 cm deep set up in a mixed birch (*Betula pendula*) and beech (*Fagus sylvatica*) woodland in the United Kingdom. Each trap contained a measured amount of leaf litter and was filled to a volume of 750 ml with tap water. A piece of birch bark was added as the oviposition substrate. Garden netting with a 1 cm × 1 cm mesh size prevented the entry of additional leaf litter but allowed for oviposition by insect colonisers. Artificial tree-holes were colonised by *Ochlerotatus geniculatus, Culex torrentium, Anopheles plumbeus*, and *Culiseta annulata*.

In Nigeria, Dunn (1927) found that tin cans were much less attractive to *Aedes aegypti* and other tree-hole species than bamboo pots, but apart from the addition of a few leaves no attempt was made to make the cans more attractive to gravid females. Because in many areas tin cans are more readily obtained than bamboo pots, it would consequently be worthwhile assessing the effectiveness of suitably prepared tins cans as artificial breeding sites for tree-hole mosquitoes, much as has been done for *Ochlerotatus triseriatus* in the USA. Size can be an important factor as shown in Indiana, USA where Hanson et al. (1988) found that large metal can ovitraps—3100 ml capacity 18 cm tall, 16 cm in diameter and painted black—collected 3.19 times more eggs of *Ochlerotatus triseriatus* per positive trap than smaller traps—350 ml, 12 cm tall, 6.5 cm in diameter— and moreover 4.86 times as many of the larger traps were positive.

Weinbren and O'Gower (1966) constructed an ovitrap from a 4¼-in diameter, 6¾-in high tin can for studying tree-hole breeding mosquitoes in Puerto Rico. A circular metal pie dish, with sloping sides and having a basal diameter of 5½ in and an opening of 7¼ in, is held some 4 in above the tin by three stout equally spaced wires to serve as a cover (Fig. 2.14). At least one wire support is easily detachable from the cover for access to the contents in the tin. Both the insides and outsides of the tin and cover are painted matt black. Two holes about 2 in apart are punched in the tin 1¾ in from the bottom, and two more holes are punched diametrically opposite. Two pieces of stout wire (e.g. plastic covered copper wire) are passed through these holes to provide a support for a 4⅛-in diameter platform of very fine mesh. A small quantity of 2-week-old horse manure infusion is placed in the bottom of the can to attract ovipositing females to the traps. To prevent the infusion mixture rising too high in the can a -in drain hole is drilled in the can 1¼ in from the bottom. Non-absorbent cotton wool is dipped in water and placed in a 90-mm plastic petri-dish and then covered with a circle of coarse paper which has been dyed black ('Tintex' dye). This is the oviposition substrate. The petri-dish is lowered onto the wire mesh screen situated near the base of the tin. An eye-bolt is passed through the centre of the pie dish cover so that the trap can be suspended amongst vegetation.

In a study of a La Crosse encephalitis outbreak in eastern Tennessee, USA, Gottfried et al. (2002) attached four sets of ten 473-ml black (inside and outside) plastic cups to the north side of trees approximately 0.5 m from the base and spaced approximately 10 m apart, following the methods of Loor and DeFoliart (1969). Two holes were cut below the rim of the cup to allow for drainage and each cup was filled with 100 ml of tap water. Strips of 76-lb seed germination paper (5 × 25.5 cm served as the oviposition substrate. *Ochlerotatus triseriatus* contributed 77% (126 670) of the total eggs collected, the remainder being eggs of *Aedes albopictus* (38 230). The mean number of *Aedes albopictus* eggs was 18.0 and 26.0 per trap per week at the two sites studied. A similar trapping methodology was used by Barker et al. (2003) to describe the spatial and temporal distributions of *Ochlerotatus triseriatus* in a La Crosse virus transmission area in southwestern Virginia, USA. Ovitraps consisting of 450 ml black plastic cups, with drainage holes approximately halfway up were nailed to trees or posts. Strips of seed germination paper, approximately 5 cm wide, served as the oviposition substrate. *Aedes albopictus* eggs were also collected at all sites, but in lower numbers than those of *Ochlerotatus triseriatus*.

In Illinois, USA Lang (1990) compared oviposition by *Ochlerotatus triseriatus* in 2.8-litre can ovitraps, painted black and containing an oak leaf-litter infusion, having either horizontal (open top) and vertical (side hole) entrances. For the latter type of trap the can was closed with a lid and a 9 × 10-cm hole was cut in the side of the upper part of the can. Drain holes were punched in all traps 8 cm from the bottom. A 7.5-cm wide strip of muslin cloth attached by a paper clip to the rim of the traps and extending into the water served as the oviposition surface. Significantly more eggs were obtained from ovitraps having the standard horizontal top openings.

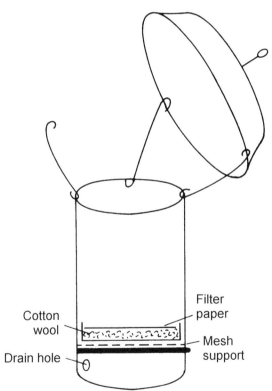

Fig. 2.14. *Aedes aegypti* oviposition trap of Weinbren and O'Gower (1966)

Beier et al. (1982) systematically placed 36 ovitraps made from 350-ml black aluminium cans fitted with a partial lid to keep out rain and debris in a wood to study the spatial distribution of *Ochlerotatus triseriatus*. Presoaked balsa paddles (Novak and Peloquin 1981) were used as the oviposition substrate. Each ovitrap was partially filled with 200 ml of a 1:3 dilution of oak tree stemflow and distilled water. Traps were fixed to trees at a height of 50 m. From 7 weekly collections 13 311 eggs were collected, and based on identification of larvae hatched from eggs it was estimated that 98.4% were *Ochlerotatus triseriatus* and 1.6% *Ochlerotatus hendersoni*. There was no correlation between the numbers of eggs in the different traps and the numbers of *Ochlerotatus triseriatus* collected from the surrounding area by aspirator collections.

In studying oviposition behaviour of *Ochlerotatus triseriatus* in 300 ovitraps with balsa wood paddles in Illinois, USA Kitron et al. (1989) used several measures to define the dispersion of eggs in ovitraps. They calculated: (i) prevalence, that is proportion of ovitraps with eggs; (ii) mean intensity, that is mean number of eggs per positive ovitrap; (iii) mean density (md), which is the product of prevalence and mean intensity (or total eggs in traps divided by number of traps with eggs); (iv) Lloyd's (1967) mean crowding (mc), which can be calculated as mean density (md) plus the variance (var.) divided by mean density (md) minus one, thus $mc = md + \mathrm{var}/md - 1$, and (v) patchiness, which is mean density (md) divided by mean crowding (mc). The regression of mean crowding on mean density (Iwao 1968, 1970) was plotted to separate the effect on aggregation of numbers of eggs per oviposition and dispersion of oviposition events among the ovitraps. The intercept measures the numbers of eggs per oviposition and is zero when a single egg comprises an oviposition. The slope measures the degree of aggregation of oviposition events and equals 1 when the distribution is random. They found that most eggs were deposited on balsa paddles without eggs and not on paddles with eggs that had been returned to the ovitraps. The dispersion pattern was highly aggregated, so some traps had many eggs whereas many had none. Frequently the numbers of ovipositions per trap could be fitted to a negative binomial distribution. Non-random, but selective, oviposition occurred not only spatially within weekly samples but also temporally among weekly samples. Ribeiro, Mather and DeFoliart (see Kitron et al. 1989) found that the dispersion pattern of eggs among ovitraps fitted a logarithmic distribution and oviposition events were distributed spatially in a multinominal fashion among the traps.

Whereas in laboratory experiments 80–130 eggs were laid by single *Ochlerotatus triseriatus* (Mather and DeFoliart 1983), eggs appeared to be laid in ovitraps in clumps of 29–47, suggesting that gravid females scatter their eggs in 2–4 ovitraps (Kitron et al. 1989).

In Florida, USA Mortenson et al. (1978) used conventional glass jar ovitraps, with hardboard paddles, fixed to trees to monitor the tree-hole species *Ochlerotatus sierrensis*. The tops of the jars were covered with ¼-in screening fitted to a Kilner (Mason) jar screw-cap ring to exclude rodents. Sometimes up to 81.3% of the ovitraps were positive after a week's exposure, the maximum number of eggs in a single trap was 495, recorded in late May.

Landry and DeFoliart (1987) wanted to age-grade female *Ochlerotatus triseriatus* by ovariole dilatations, after they had laid eggs, so they designed an ovitrap that retained females after oviposition (Fig. 2.15).

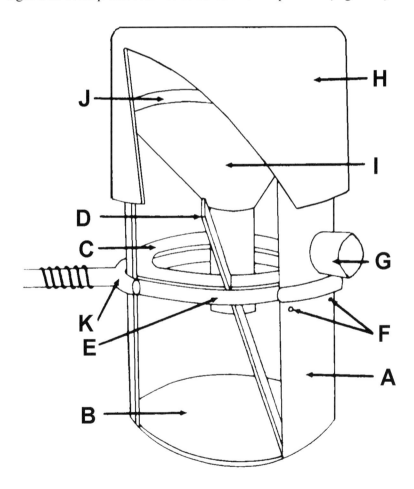

Fig. 2.15. Female-retaining *Aedes* ovitrap. See text for explanation of parts (Landry and DeFoliart 1987)

LIVERPOOL JOHN MOORES UNIVERSITY
LEARNING SERVICES

Their trap consists of a 20.4-cm length of 10.2-cm diameter PVC tubing (A) closed at the bottom with a circular piece of plexiglas (B) stuck on with ethyl dichloride. A plywood (1.3-cm thick) ring (C) having an outer diameter of 10 cm and an inner diameter of 5 cm, and with a 60° slit cut through to accommodate an oviposition paddle (D), was positioned on a ring (E) made of the original PVC tubing with a 2.5-cm piece removed so that it could be glued inside the trap body 6.35 cm from the bottom. This ring-shaped platform served as a resting site for mosquitoes. Eight over-flow holes (F) were drilled below this supporting ring. Three holes (G) (2.54-cm diameter) were spaced around the body to allow insertion of an aspirator to remove adults trapped after oviposition, these are normally plugged with rubber stoppers. The entire trap is painted black except for the underside of the plywood platform, which was painted white to make eggs more easily detectable. Gravid mosquitoes enter through the lid (H) made from a PVC pipe-coupler (10.2-cm diameter) and pass through a funnel (I) of metal mosquito screening secured to it by a plywood platform support (similar to part E) inside the screen funnel (J), and attached to the coupler with screws. The stem of the funnel (2.54 cm diameter) was posi-tioned 1.9 cm above the overflow holes. A metal ring (K) was welded to a lag-screw which can be screwed into a tree, and the trap slipped through the metal ring to rest on the three rubber stoppers. These traps were baited with either oak infusion water made by adding oak leaves to distilled water in the bottom of the trap, or from a laboratory stock of oak leaf water, or were filled with filtered tree-hole water.

A total of 1715 ovipositing *Ochlerotatus triseriatus* were collected over 4 years from 149 traps, which were checked three to five times a week. Live adults were removed with aspirators and kept for at least 28–30 h to allow ovariole sacs to contract and for the formation of dilatations. Many mosquitoes escaped from the traps after oviposition, and several traps con-tained eggs but no adults. Uniparous adults formed 79.9–92.7% (mean 84.1%) of the trapped mosquitoes, means of 13.2% were 2-parous, 2.1% 3-parous, 0.5% 4-parous and 0.1% 5-parous. These female-retaining ovitraps were used later by Landry et al. (1988) to collect ovipositing *Ochlerotatus triseriatus*.

Beehler and DeFoliart (1990) studied the spatial dispersion of *Ochlero-tatus triseriatus* eggs in ovitraps in a wood. Using Taylor's (1961) power law they calculated b as 1.4, which indicates a clumped distribution of eggs among traps. Lloyd's (1967) mean crowding index was calculated and regressed against the mean, and a slope of greater than one was ob-tained, again indicating a clumped distribution. Like Beier et al. (1982) they were unable to explain why some traps were more attractive or less attractive. The intercept of the regression line (mean crowding vs mean) of

30.3 suggested that individual eggs are clumped in groups of 31 ± 9.8, which is taken as an estimate of the egg batch size of *Ochlerotatus triseriatus*. Beehler and DeFoliart (1990) point out that their estimate is in between estimates of 29.3 ± 16.9 and 46.9 ± 25.3 derived by Kitron et al. (1989). They argue that their estimate is likely to be more precise as it was calculated from data obtained from daily, not weekly, sampling as undertaken by Kitron et al. (1989). They concluded that despite a contagious distribution of oviposition in their ovitraps, only a few traps were needed for detecting and monitoring populations of *Ochlerotatus triseriatus*.

Field trials in Puerto Rico showed that the Weinbren and O'Gower (1966) ovitrap was effective in collecting eggs of *Aedes aegypti*, which was particularly useful as the species was not caught at light or in bait catches. However, in later trials (Haber and Moore, 1973) in the same area *Aedes aegypti* eggs were not collected in these traps when they were baited with either horse manure or rabbit food infusion, but neither were they collected from tyres, bamboo pots nor the traps of Fay and Eliason (1966). The mosquito was apparently absent from the area, possibly due to changes in the environmental conditions since the previous survey of Weinbren and O'Gower (1966). Eggs of *Ochlerotatus mediovittatus, Culex antillummagnorum* and *Culex secutor* were retrieved from glass ovitrap jars, but none of these nor any other species was collected from the Weinbren and O'Gower trap.

In California, USA Woodward et al. (1998) used artificial tree-hole oviposition traps to monitor oviposition activity of two tree-hole breeding species, *Orthopodomyia signifera* and *Ochlerotatus sierrensis*. Traps were constructed using a translucent polyethylene cup (473 ml capacity) filled with 380 ml of water obtained from natural blue oak tree holes or an infusion prepared from blue oak leaves. The oviposition substrate consisted of a 10×27 cm strip of towel (Teri-wiper™). Individual cups were placed inside a plywood box (29.2 cm height, 15.2 cm width, and 17.8 cm depth), painted gloss black on all surfaces. An 11.4×11.4 cm vertical entrance hole, backed with 2.5 cm mesh hardware cloth, was cut into the centre of the front of the box, 7.7 cm below the top. The box lid was hinged to allow access. Traps were staked to the ground on the north side of tree trunks, with the opening facing north and positioned approximately 20 cm above the ground. A series of six traps was also attached to a living tree trunk at heights of 0.2, 1.0, 3.0, 5.0, 7.5 and 10 m above ground to investigate the vertical distribution of oviposition by the two species. Totals of 4241, 6083, 4135, and 750 eggs of *Orthopodomyia signifera* were collected from the ovitraps in 1991, 1993, 1995 and 1997 respectively, with peak oviposition occurring in August and September, corresponding to the period when the majority of natural tree holes in the area had dried out. In the 1994 study of vertical distribution of oviposition, a total of 1293 *Orthopodomyia signifera* eggs were recovered, all from traps located 3.0m or more above ground, with 74% of the total obtained from the traps at 5.0 m and 7.5 m. Peak oviposition by *Ochlerotatus sierrensis* occurred earlier in the season (May-June), with the majority of oviposition occurring before the natural tree holes had dried out. Total eggs collected per year amounted to 38 725 in 1991, 25 358 in 1993, 65 054 in 1995 and 27 761 in 1997. A total of 21 538 *Ochlerotatus sierrensis* eggs were collected from the six vertically distributed ovitraps in 1994, with no significant difference in the vertical distribution of egg numbers among the six ovitraps. These same authors later studied the seasonal population dynamics of *Ochlerotatus sierrensis* in northern California, USA, using ovitraps (Woodward et al. 1996) consisting of a black plywood box (29 crn high, l5 cm wide, 18 cm deep) with a hinged lid and a screened (2.5 cm mesh) vertical entrance (11 × 11 cm) near the top of the front panel that held a polyethylene cup (473 ml) lined with a Terri-wiper towel strip (10 × 27 cm), which acted as the oviposition substrate. Each cup held blue oak (*Quercus douglasii*) tree hole water (380 ml). When operated alongside Fay-Prince CO_2 traps, no direct relationship was observed between egg counts and adult female abundance in traps. The fact that the use of ovitraps by *Ochlerotatus sierrensis* was dependant on the availability of natural oviposition sites precluded, in the opinion of the authors, their use as a means of measuring or predicting female population size.

To study the dispersal of dengue vectors from urban areas into adjacent forest in Brazil, Lourenço-de-Oliveira et al. (2004) used plastic jar ovitraps placed on the ground at varying distances from houses as well as four artificial treeholes. These artificial treeholes were made from brown 1.5 litre bottles with a round opening of 10 cm diameter in the middle, which contained 400 ml of sieved treehole water with fallen leaves that had previously been washed and broken into pieces. The inner wall of these bottles was covered with brown paper soaked with the same treehole water. In total, the artificial treeholes yielded only 44 immature mosquitoes: 12 at a distance of 500 m and 32 at 1 km into the forest. A few eggs of *Aedes albopictus* (one at 500 m, eight at 1 km), but none of *Aedes aegypti*, were recovered. Other mosquitoes collected were *Culex (Microculex)* spp., *Toxorhynchites* spp., *Sabethes* spp., and *Limatus durhami*.

Car tyre ovitraps. In the USA Bradshaw and Holzapfel (1985) placed car tyres at the base of trees and put in two handfuls of sterilised tree detritus to establish ovitraps to monitor breeding mosquitoes (Fig 2.16). They found that the relative abundance of *Orthopodomyia signifera, Ochlerotatus triseriatus, Anopheles barberi* and *Toxorhynchites rutilus* in these sentinel traps closely approximated that in actual tree-holes. In New Orleans, USA Freier and Francy (1991) evaluated tyre traps for ovipositing *Aedes albopictus*. Firstly, an oviposition medium was made by incubating 1 g rabbit pellets for 3 days at 27°C in 3.8 litres of water.

Fig. 2.16. Car tyre ovitrap (photograph reproduced with permission of M. W. Service)

A tyre was then placed horizontally on the ground and six 7-cm diameter equally spaced holes made around the tread.

A plastic container containing 1 litre of oviposition water was placed on the ground within the tyre centre, and a plywood board covering the tyre opening was placed on top of the tyre. A 2-m length of 10-cm diameter plastic tubing projected from an opening cut from the middle of the board. A motor and fan placed at the end of this tubing sucked mosquitoes that had entered the tyre trap through the six peripheral holes, into a 0.5 litre screened carton inserted in the tubing near the tyre.

A vertical tyre trap was also made. This consisted of three tyres with their sidewalls attached together and placed with the tread on the ground. About 1 litre of oviposition medium was placed in the middle tyre. Neither arrangement was very successful because more *Aedes albopictus* (mean 4.2/6-h trap-day) were caught in the gravid trap of Reiter (1983), than in the horizontal (1.5) and vertical (0.8) tyre traps. Similar decreases in numbers were observed across these three trap types for *Ochlerotatus triseriatus* and *Culex salinarius*.

Micieli and Campos (2003) used car tyres cut in two across the diameter to monitor oviposition activity and seasonality of *Aedes aegypti* in sub-tropical Argentina. Half tyres were placed approximately 1m above ground level in the shade of private houses. Wooden paddles (8 × 6 cm) covered with thick brown paper were attached to the inner wall of the tyre, at the water level and were removed weekly, along with the liquid contents of the traps. The authors did not record if they used any oviposition attractants in the traps. *Culex* spp. and *Toxorhynchites* spp. were also collected from the tyre traps.

Pena et al. (2003, 2004) developed and used a trap constructed from a motorcycle tyre to monitor *Aedes albopictus* populations in the Dominican Republic. Traps were constructed as follows: a motorcycle tyre measuring 165 cm in circumference was cut into 50 cm sections. A 6 cm long brass screw with an eye was inserted into each end of the cut section. A 100 cm length of nylon cord was attached to the brass screws and used to suspend the trap from a nail hammered into a tree at a height of 1.2 m above ground level. Traps were filled with 300 ml of tap water and exposed for one week. Traps set at 100 m along a transect through the centre of a public park in Santo Domingo caught a total of 518 *Aedes albopictus* from 63/1 664 (3.8%) positive traps. Species collected included: *Aedes aegypti* (74.7%), *Aedes albopictus* (10.5%), *Culex corniger* (3.6%), *Ochlerotatus albonotatus* (8.7%), and *Toxorhynchites* sp. (2.5%).

Miscellaneous ovitraps.   In studying the seasonal variations in relative abundance of *Aedes albopictus* and *Aedes aegypti* in Thailand Mogi et al. (1990*b*) used greenish dark-grey ceramic ant traps that formed a circular trough as ovitraps. Diameters of the inner and outer rims were 7 and 15 cm, respectively. The trough held about 400 ml of water. The inner side of the outer rim was lined with brown paper towelling that had a rough surface and which remained intact for at least a week after soaking. However, a disadvantage of using such towelling was that, unlike hardboard paddles, eggs hatched without submersion, presumably because the paper gets much wetter. There was a 5-day exposure between removal of the paper towels. The distribution of eggs in these ovitraps was distinctly contagious but the data did not fit a negative binomial model with a common *k* (Mogi et al. 1990*a*).

Lounibos and Machado-Allison (1986) successfully used split cocoa pods as oviposition traps for *Trichoprosopon digitatum*.

Several species of *Eretmapodites* preferentially oviposit in the water-filled shells of *Achatina fulica*, and Lounibos (1980) used them as oviposition traps in Kenya. He half-filled clean shells with spring water and placed them on the ground in the shade. After a 6-day exposure the shells were collected, the larvae removed and reared to adulthood, and the water discarded. Two to 3 days later the dry shells were immersed for 24 h in water containing liver powder to stimulate hatching of unhatched eggs. Finally, the snail shells were placed in boiling water for 5 min to kill any remaining eggs and to sterilise them before they were returned to the field as oviposition traps. In the Shimba hills a total of 539 *Eretmapodites silvestris conchobius*, 569 *Eretmapodites quinquevittatus* and 58 *Eretmapodites subsimplicipes* were identified from snail ovitraps. Other species occasionally found in the traps were *Aedes calceatus*, *Aedes aegypti*, *Aedes bromeliae*, *Aedes soleatus*, *Aedes heischi* and *Culex nebulosus*. In Kombeni forest 164 *Eretmapodites quinquevittatus* were collected from similar snail traps.

A simple trap for collecting the eggs of *Ochlerotatus taeniorhynchus* in South Carolina, USA, was developed by Wallace (1996) and consists of two black plastic MacCourt tubs, measuring $20 \times 60 \times 90$ cm, available from home supply stores. A cotton towel, soaked in water, and contaminated with bacteria and other micro-organisms was placed in the bottom tub and covered with wheat straw or locally available plant material. The other tub was inverted and supported above the lower tub using steel rods (120 cm $\times 1.27$ cm) that were pushed about 80 cm into the ground. The clearance between the top and bottom tubs was about 20 cm. Contaminated towels were prepared by exposing 100% cotton towels to airborne micro-organisms over a period of 2–3 weeks in the laboratory. Following 24 h of exposure, plant material was removed from the trap, placed in another plastic tub and washed with water from a garden hose with a high-pressure nozzle. Washings were then sieved through 45 and 100 mesh standard USA testing sieves. Towels were draped over a glass-drying rack, which was then placed into a plastic tub. The towel was washed with water in the same manner as the vegetation. Use of contaminated towels and the tub that acted as the trap cover both significantly increased the number of eggs recovered compared with clean towels and no trap cover. Colour of the towel and the trap did not appear to influence the number of eggs laid. The trap was also successful at obtaining *Ochlerotatus sollicitans* eggs, with a maximum of approximately 5 000 eggs being laid in a single trap during a 24 h period.

Smith and Enns (1967) floated artificial oviposition blocks of 'Styrofoam' plastic on oxidation lagoons in Missouri. Four 6-in square pieces were cut from an 18-in square and 3-in thick block of 'Styrofoam', leaving a 2-in margin between the cut-out portions and between the outer edges. The total oviposition area in each block was 1 ft^2. A length of nylon rope was tied to two 4½-in eye-bolts inserted through two opposite ends of the block to secure it to a cement anchor. From an exposure period from April to August, 7715 egg rafts of the *Culex pipiens* complex, 79 of *Culiseta inornata* and 27 of *Culex tarsalis* were collected. Larval collections of the *Culex* species were comparable to the results of the egg survey, except that about 0.1% of the larvae collected were *Culex salinarius*, a species not represented in egg collections.

Scott and Crans (2003) proposed the use of small blocks of expanded polystyrene added to container habitats as a method for monitoring *Ochlerotatus japonicus* oviposition in the northeastern United States. Commercially available sheets (1.9 × 34.6 × 121.9 cm) of expanded polystyrene were broken into 12-cm-wide strips, and the strips were further broken into 5-cm-long pieces to produce a float measuring 12 × 5 × 2 cm. Sixty similarly-sized EPS floats could be obtained from a single panel. Breaking, rather than cutting the sheets was found to be quicker and provided a more suitable rough surface for oviposition. Floats were placed in existing containers known to be habitats of immature *Ochlerotatus japonicus*, including buckets, car tyres, and cement catch basins (Fig. 2.17).

Fig. 2.17. Typical habitats suitable for use of expanded polystyrene floats. (A) Plastic tray; (B) discarded tire; (C) trash can; and (D) gravid trap (Scott and Crans 2003)

In large water bodies, lengths of monofilament line were tied to the floats to prevent them being washed away and to facilitate retrieval. Retrieved floats were collected and placed into individual resealable plastic bags for transport to the laboratory. Floats were stored in the plastic bags for 7–14 days at room temperature and under a 16:8 h light:dark cycle. Individual floats wexdrtre then submerged in pans containing l litre of 24-h aged tap water plus 10 mg of powdered rat chow. Floats remained submerged for 7 days, and emerging larvae were identified and counted. Eggs of *Ochlerotatus japonicus*, *Ochlerotatus hendersoni*, *Ochlerotatus triseriatus*, and *Aedes albopictus* were collected using the floats. Field observations showed that mosquitoes tended to deposit their eggs in the crevices between the cells of expanded polystyrene on the rough surfaces created by breaking the larger sheets. Floats were found to be very durable, and withstood exposure to sunlight, a wide range of temperatures (–l2 to 37°C) and repeated drying without any apparent deterioration.

Oliver and Howard (2005) made use of these styrofoam oviposition blocks to examine the fecundity of field-collected *Ochlerotatus japonicus* in the laboratory in New York State, USA. A total of 289 females oviposited 32 970 eggs on the styrofoam blocks or on the water surface. Most eggs were laid on the blocks, with only 4802 eggs not laid on blocks. These authors reported that *Ochlerotatus japonicus* readily used styrofoam blocks as an oviposition substrate and tended to oviposit on the four vertical sides of the block at or just above the water line. Eggs were never deposited on the horizontal surface of the block.

Culex ovitraps

Clay pots, generally found outside, but sometimes also inside houses, or other receptacles such as jars, can be used as artificial oviposition sites both in (Southwood et al. 1972) and away from houses (Service 1970). Yasuno et al. (1973) poured about 2 litres of 1% yeast infusion into clay pots to improve their efficiency in attracting gravid *Culex quinquefasciatus*. Other types of man-made receptacles such as bottles, tin cans and tyres can be placed in different habitats to detect the presence of ovipositing females (Bond and Fay 1969; Dunn 1927).

In Japan rice-straw infusions in earthen jars provided attractive oviposition sites for *Culex pipiens* form *pallens*. By collecting egg rafts from a series of pots at hourly intervals Oda (1967) was able to study the oviposition cycle of a natural population.

On the island of Seahorse Key, off the Gulf Coast of Florida, USA, artificial oviposition sites for *Culex quinquefasciatus* consisted of 2.5-gal plastic wash tubs containing an infusion of equal parts of liver powder, brewer's yeast and hog food supplement. Egg rafts were removed daily, and every 3rd or 4th day the tubs were emptied and refilled to prevent the formation of surface scum or the establishment of predators (Lowe et al. 1973).

In India 4- or 5-litre clay pots holding 2 litres of water containing 1% baker's yeast infusion were placed as *Culex quinquefasciatus* ovit-raps in courtyards of houses (Sharma et al. 1976). Preliminary experiments had shown this yeast infusion to be better than hay infusions, dog biscuit infusion and water from drains, but in the cool season because yeast fermentation is reduced, larval waters from laboratory colonies were added to some traps. Ovitraps appeared to be efficient from June to September with the highest mean number of egg rafts per trap, 18.5, being recorded in August. Peak densities correlated well with rainfall, but not with adult densities.

Oviposition traps have often been used to monitor the seasonal abundance of *Culex* vectors of arboviruses (Leiser and Beier 1982; Lee and Rowley 2000; Madder et al. 1980; Reiter 1986), and they have also been employed to catch older (gravid or recently oviposited) *Culex* to increase the chance of obtaining virus-infected mosquitoes (Reiter 1987; Reiter et al. 1986).

In Texas, USA Strickman (1988) collected egg rafts of *Culex quinque-fasciatus* from oviposition traps consisting of 6-litre plastic rubbish cans containing foul-smelling water produced by putting 32 g alfalfa pellets/ 4 litres water and maintaining at about 27°C for 11 days. In field trials in Indiana, USA (Hoban and Craig 1981; Hoban et al. 1980) it was found that fresh cow manure diluted in water was a better attractant than horse ma-nure or commercial dehydrated cow manure for attracting ovipositing *Culex restuans*. This led to the development of a simple bucket ovitrap with the lid propped 2–3 cm open and containing cow manure in a cheese cloth sack. Altosid was added to the traps to prevent adult emergence. In some situations over 100 egg rafts, mainly *Culex restuans* and *Culex pipiens*, were recorded each day (Hoban and Craig, 1981). The numbers of egg rafts collected corresponded with the numbers of adults caught in light-traps.

Leiser and Beier (1982) compared oviposition traps with New Jersey light-traps in Indiana, USA for monitoring *Culex pipiens* and *Culex restuans*. The Hoban and Craig (1981) type ovitrap was used, consisting of a 5-litre plastic bucket three-quarters full of water and with the lid propped partially open (10 cm) with a clothes peg. A cloth bag of 300 g of fresh cow manure is added; rocks placed in the bag ensure it stays submerged. An Altosid tablet is added to the water. Sixteen ovitraps were placed in the shade at various locations 5 m from a New Jersey light-trap, which ope-rated from 2200 – 0600 h daily, from May to October. A total of 365 992 mosquitoes were collected from the light-traps, including 25 232 female and 35 051 male *Culex* spp. At the same time 4193 egg rafts of *Culex pipiens* or *Culex restuans* were retrieved from the ovitraps. Both sampling methods showed that *Culex* populations peaked in July, when in a single week the maximum catch was 18 856 adults in the light-traps, and a week later a peak of 960 egg rafts were collected. There was a positive correlation ($r = 0.63$) between the numbers of *Culex* taken in light-traps and egg rafts found in ovitraps for 14 of the 16 collecting stations. It was thought that in the two locations showing a negative correlation this was likely due to the existence of a multitude of alternative breeding sites. Leiser and Beier (1982) concluded that although both methods adequately monitor changes in population size, many more ovitraps can be operated than light-traps with the same man-hours. Moreover, ovitraps are less expensive.

In California, USA, Beehler et al. (1993) studied the spatial and circadian oviposition patterns of *Culex quinquefasciatus* using black plastic tubs ($50 \times 40 \times 18$ cm) containing 300 ml of Bermuda grass infusion (Millar et al. 1992), 750 ml of alfalfa hay infusion (Reiter 1986) and 8 litres of tap water, placed along a 200 m transect. On one night each month, egg rafts were removed and counted every two hours throughout the night. On a second night, egg rafts were only collected once the following morning. Eighty-eight percent of the egg rafts collected were of *Culex quinquefasciatus. Culex stigmatosoma* contributed 3%, *Culex tarsalis* 5%, and *Culiseta incidens* 4%. There appeared to be no significant effect of trap position on the number of egg rafts recovered.

In Florida, USA ovitraps have collected egg rafts of *Culex nigripalpus, Culex quinquefasciatus, Culex salinarius* and *Culex restuans* (Haeger and O'Meara, 1983). The relative abundance of these species appears to be influenced by seasonal and geographical variations (Lowe et al. 1974; Nayar 1982; O'Meara et al. 1989*b*) and by the type of trap (O'Meara et al. 1989*b*; Smith and Jones 1972). In Florida O'Meara et al. (1989*b*) compared two types of ovitraps for attracting ovipositing *Culex* mosquitoes, namely 1-quart (0.95-litre) Kilner (Mason) jars inserted into concrete blocks (19.5-cm cubes). The outside of the jar and its block were painted black, and two jars in their blocks were placed side by side. The other trap consisted of a rectangular (56×44 cm and 8 cm deep) plastic tub painted black. An oviposition infusion was prepared by adding about 2.5 kg of oak leaf litter to a 76-litre container filled with tap water and fitted with a lid. This infusion was left for at least 1 week before being placed in the two types of ovitrap. A few hours before sunset 0.5 litres of the infusion was poured into the jars and 3.8 litres into the plastic tubs; ovitraps were restocked with infusion on each of three consecutive nights a week over a year. Six tubs and six jar-type oviposition traps were placed alternately along transect lines.

From a total of 3720 trap-nights 4540 egg rafts were collected. Significantly greater numbers of egg rafts were recovered from tubs than from

jars for *Culex nigripalpus* (7.7 ×), and *Culex restuans* (2.9 ×), but more
(1.4 ×) egg rafts of *Culex quinquefasciatus* were recorded from jars than
tubs. Although the number of *Culex salinarius* egg rafts collected from jars
was also larger (1.2 ×) the difference was not significant. Not only did the
type of trap (jar or tub) affect oviposition by *Culex quinquefasciatus* and
Culex nigripalpus but, whereas the former species showed no preference
for ovitraps placed in shaded or unshaded situations, fewer *Culex quinquefasciatus* laid eggs in shaded traps. This emphasises the effects of environmental conditions and trap location on the size and species composition
of catches.

Lampman and Novak (1996*a*) compared several types of infusion in
ovitraps used to monitor *Culex pipiens* and *Culex restuans* mosquitoes in
Illinois, USA, in order to maximise trap efficacy and investigate differences in attractancy of the various infusions for the two species. Rabbit
chow infusion was the standard infusion used in *Culex* surveillance in Illinois and was used as the reference against which a range of infusions were
tested, namely mixed grass clippings, alfalfa, oak leaves, maple leaves,
lawn sod, and wheat straw. The ovitraps used in the study were those of
Steinly and Novak (1990), consisting of 19 litre plastic buckets with a series of six 7.5 cm holes evenly spaced around the circumference of the
bucket about 15 cm below the top. Ovitraps were placed in 2 woodlots. Infusions were prepared by placing 150 g of each material in a cheesecloth
bag and soaking the bag in 6 litres of water for 48 h. For sod-baited traps, a
25 × 25-cm section of sod was placed in the ovitrap, where it remained
throughout each test. Continuous lawn sod infusions showed the least variability in attractiveness and became active in a relatively short time in both
previously used ovitraps and unused ovitraps. Other infusions took longer
to become active, especially in previously unused ovitraps. Sod infusions
were also found to be attractive to female and male *Aedes albopictus* in
field studies and attraction to sod infusion was confirmed in the laboratory
using 2-choice oviposition preference tests and an olfactometer (Lampman
and Novak, 1996*b*). Approximately 5% of female *Aedes albopictus* collected in gravid traps near a waste tyre site were gravid and 30% of the
total mosquitoes caught were male.

As a contribution to St. Louis encephalitis virus surveillance in Iowa,
USA, Lee and Rowley (2000) used oviposition traps to monitor the temporal changes in ovipositing populations of *Culex salinarius, Culex restuans,*
and *Culex pipiens*. Ovitraps were constructed from 20-litre buckets containing 8 litres of tap water flooding a section of 30 by 45-cm Kentucky
bluegrass (*Poa pratensis*) lawn sod. Two ovitraps were placed under the
eaves of a wooden warehouse near a cornfield. Traps were inspected daily
from Monday to Friday throughout the summer over three years and

replaced weekly with fresh traps. A total of 4180 egg rafts were collected from the ovitraps. Five species oviposited in the ovitraps: *Culex restuans, Culex pipiens, Culex salinarius, Culex tarsalis*, and *Culiseta inornata*. The proportions of the total number of egg rafts contributed by each species varied among years. Ageing of the traps from Monday–Friday did not significantly affect the number of egg rafts collected.

Gravid traps

In Sri Lanka Samarawickrema (1967) caught gravid females of *Culex quinquefasciatus* from 1800–2000 h as they alighted on the walls of an open cesspit to lay eggs; but De Meillon et al. (1967) appear to be among the first to have constructed a trap to monitor the arrival of gravid *Culex quinquefasciatus*. Frequently mosquitoes caught in light-traps with or without carbon dioxide, are predominantly nulliparous (Magnarelli 1975; Morris and DeFoliart 1971), so the probability of collecting infected mosquitoes in arbovirus studies is relatively small. This can be overcome by employing traps that catch gravid females, and several such traps have been designed (De Meillon et al. 1967; Lewis et al. 1974; Surgeoner and Helson 1978). However, these traps are not very portable and for these reasons Reiter (1983) developed what has become known as the 'gravid trap'.

Recently there has been a surge of interest in the use of gravid traps to monitor populations of *Culex* species coinciding with their incrimination as the primary vectors of West Nile virus in the United States (e.g. Sardelis et al. 2001). The introduction of West Nile virus to North America from Africa resulted in an outbreak of the disease in humans in New York City in 1999 (Lanciotti et al. 1999) and the virus has assumed public health significance in much of the western hemisphere.

Surgeoner and Helson's gravid trap

In Canada, Surgeoner and Helson (1978) built a trap consisting of an 84-cm diameter plastic inflatable paddling pool, the middle of which was placed in a hole so that when filled with water the depth at the centre was about 23 cm. The pool was lined with sods of earth, and 40 ml n-capric (decanoic) acid diluted in 15 ml of ethanol was added to improve the attractiveness of the water as a *Culex* oviposition site. In addition about 0.6 litres of water from a nearby highly productive source of *Culex restuans* was added. About every 3 weeks a further amount of 40 ml capric acid was added to the water. A 30-cm metal container, 26 cm high and weighted with stones was placed in the centre of the pool, having the rim of the container about 3 cm above the water. A CDC light-trap with the collecting bag replaced by a pint-sized plastic container with two 3 × 2-cm windows and bottom covered with netting was placed on the stones. The lid of the CDC trap was about 6 cm above the water. An altosid briquette was added to the pool water to prevent development of mosquitoes.

Surgeoner and Helson (1978) compared the numbers of mosquitoes collected in five of these traps, five CDC traps and five cone-traps baited with dry ice. The total numbers and proportions of female *Culex pipiens* and *Culex restuans* caught from about 40 trap-nights were: 1199 (94.7%) in the oviposition traps, 7340 (37.4%) in the CDC traps, and just 387 (72.4%) in the CO_2 traps.

Although the CDC traps caught more *Culex* adults (2748) than the oviposition trap (1136) the authors believed that the former caught substantially higher proportions of nulliparous mosquitoes, and that together with the tedium of sorting the *Culex* from other mosquitoes meant that the more selective oviposition trap was better for catching *Culex* for virus isolation studies. The oviposition trap also caught a few *Culiseta inornata, Aedes vexans*, and *Ochlerotatus triseriatus*. This trap has, however, been largely superseded by the Reiter oviposition traps (Reiter 1983, 1987; Reiter et al. 1986).

Reiter's gravid traps

The original trap consists of a 3-in diameter PVC inlet tube housing a 6-V d.c. motor, as used in CDC traps, on which is mounted a four-bladed 3-in counter-clockwise fan. (Alternatively an upward flow of air can be produced by reversing the terminals of a CDC fan and motor, but this eliminates the aerodynamic efficiency of the fan and specimens may be damaged.) The inlet tube is clamped between two vertical wooden boards that fit over a black plastic box (18.5 × 14.0 × 6.5 in). A plastic 12-in long PVC chimney slots into the upper end of the inlet tube. The top half consists of three struts as shown in Fig. 2.18 that fit into and support a netting collecting bag. For this the middle of the collecting bag is reinforced with a circular patch of denim cloth. The oviposition attractant is made by adding 1 lb of hay and 1 oz each of dried brewer's yeast and lactalbumen powder to 30 gal of tap water. This infusion is allowed to mature for 5 days. Traps are placed in position 1 h before sunset, and trapped mosquitoes removed the next morning with an aspirator. New oviposition media are used for each trap-night.

In trials in California, USA Reisen and Pfuntner (1987) reported that surprisingly the gravid trap of Reiter (1983) performed poorly, in catching only a mean (% of total catch in parentheses) of 2.7–16.0/trap-night (0–13%) of *Culex quinquefasciatus* in an area where considerable numbers of adults were caught in carbon dioxide traps. In fact the mean number of gravid *Culex* was greater from collections from walk-in red boxes (3–7) than from

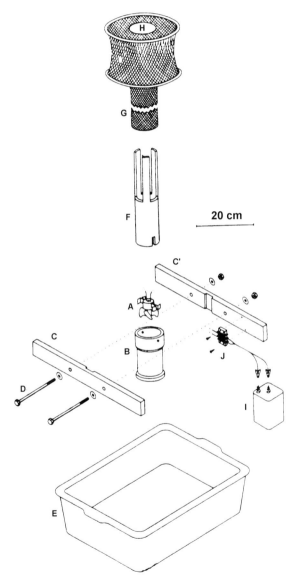

Fig. 2.18. Reiter's (1983) gravid trap. A—motor/fan assembly, B—inlet tube, C—& C'—cross bars, F—chimney, G—collecting bag, H—reinforced support for bag, I—6-V battery, J—connector block

the gravid traps (0.4). This contrasts with reports of Reiter et al. (1986) of 142.3 females/trap-night, of which 95% were gravid, and Ritchie (1984*b*) of 405.2 females/trap-night of which 57% were gravid.

The original gravid mosquito trap of Reiter (1983) suffers from certain limitations, namely up to 10% of the catch of adults is damaged by passing through the fan blades, and adults tend to die of desiccation. Moreover, ants, racoons and birds can damage the traps and the mosquitoes. Reiter (1987) therefore redesigned the trap to consist of a rectangular box 41 cm long, 27 cm high and 13 cm wide made of 1.1-cm plywood. The trap is composed of an upper, middle and lower compartment (Fig. 2.19) held together with suitcase latches.

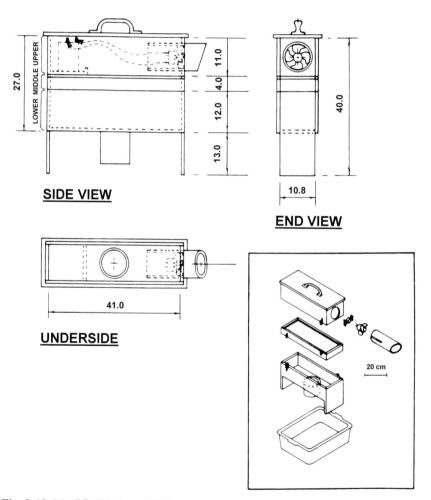

Fig. 2.19. Modified Reiter (1987) trap

The upper part is 11 cm high and has a carrying handle screwed on the top, and a small shelf inside on which batteries are placed. The 4-cm high middle compartment has an 8 mesh/cm screen fastened across the entire top. The lower compartment, 12 cm high, has solid ends that extend down 13 cm and support the trap when it is not resting on the oviposition pan. A 6-V d.c. CDC-type motor and a 4-bladed 7.6-cm diameter fan is mounted in a bracket that fits into a 9-cm slot cut in an 18-cm length of 7.6-cm diameter PVC tubing. Dry ice or an anaesthetic is placed on the screen top of the middle compartment to anaesthetise or kill the catch. The two compartments are then tipped upside down and mosquitoes that have fallen onto the screen are removed.

Reiter (1986) described a routine for making an oviposition attractant for *Culex* and several other raft-ovipositing genera. The equipment consists of two tapered 120-litre plastic dustbins (garbage cans) stacked one in the other, with the top bin retaining its lid. The inner top dustbin has numerous 0.6-cm diameter holes drilled in the bottom, while the outer dustbin has a tap towards the base and is mounted on a 4-wheel dolly. Grass-hay (0.5 kg) and 5 g each of dried brewer's yeast and lactalbumen and 114 litres of water is put in the inner dustbin and left to mature for 6 days. The inner dustbin is then hoisted out by an overhead pulley system while the bottom dustbin with the attractant oviposition water can be rolled up into the back of a pick-up truck and transported to field sites.

In Tennessee, USA 135 724 mosquitoes belonging to at least 25 species were collected in CDC gravid mosquito traps of Reiter (1983) in 954 trapnights (Reiter et al. 1986), of which 98.78% were *Culex pipiens* s.l. and *Culex restuans*, important vectors of St. Louis encephalitis and West Nile viruses. These traps also caught reasonable numbers of *Aedes aegypti* (236), *Ochlerotatus triseriatus/hendersoni* (251) and *Culex erraticus* (544). The average catch was 142.3 mosquitoes/trap-night. At least 95% of the females were gravid and usually 80–95% were alive when the traps were emptied. This preponderance of gravid mosquitoes should increase the likelihood of catching disease-infected mosquitoes. The gravid traps caught 88 times more *Culex* than were collected by mechanical aspiration of outdoor mosquitoes resting in culverts, and underground shelters, and 96 times more *Culex*/person-hour. One operator can service at least 20–30 gravid traps/day compared to just 8–10 resting sites.

Niebylski and Meek (1992) used Reiter (1983) gravid traps to collect blood-fed *Culex quinquefasciatus* mosquitoes in Louisiana, USA. A total of 1104 individuals were captured from 85 collections in three areas (adjacent to houses, adjacent to dog kennels, and in forest). Dogs were the most frequent source of blood meals in both urban sites, while passeriform birds were the most common source of bloodmeal in the forest.

Savage et al. (1993) used the gravid trap of Reiter (1983), alongside CDC light-traps supplemented with CO_2 and with or without light, and hand-held aspirator collections of resting mosquitoes in an entomological investigation of an epidemic of St. Louis encephalitis in Arkansas, USA. The number of female *Culex pipiens quinquefasciatus* caught in gravid traps was significantly greater than the number caught in CDC light-traps supplemented with CO_2 ($t = 2.76$; $P < 0.01$).

Similarly, Nasci et al. (2002) used the Reiter (1983) gravid trap along-side CDC light-traps supplemented with CO_2 to investigate the transmission of West Nile virus in the New York City metropolitan area, USA. Traps were set up in the vicinity of areas from which dead birds had been reported. A total of 17 220 mosquitoes belonging to the genus *Culex* and 50 individuals of other species, primarily *Aedes vexans*, was collected from three sites.

Reisen et al. (1999) compared Reiter's (1987) trap, as modified by Cummings (1992), with NJ light-traps and CO_2-baited CDC traps for *Culex* surveillance in California, USA. These authors found that gravid traps baited with bulrush infusion collected the greatest numbers of *Culex pipiens* complex females and males, especially in urban situations. The bulrush infusion used out-performed an alfalfa infusion in rural agricultural areas, collecting significantly more females than the alfalfa infusion in 3 habitats in Orange County and a greater diversity of species. However, the bulrush infusion failed to collect large numbers of gravid female *Culex tarsalis* compared with CO_2 traps.

In a comparison of methods for surveillance of West Nile virus vectors in Delaware, USA, Gingrich and Casillas (2004) observed that *Culex restuans* was most frequently collected from infusion-baited gravid traps, compared with CO_2-baited CDC light-traps, CO_2-baited omnidirectional Fay traps, and human landing catches. Other species collected in reasonable numbers in gravid traps included *Culex pipiens pipiens*, *Culex salinarius*, and *Ochlerotatus triseriatus*.

McCardle et al. (2004) also compared CDC gravid traps with Miniature CDC light-traps with and without CO_2, Fay-Prince traps baited with CO_2 and 1.8 m^3 partially open unbaited cages for sampling gravid mosquitoes in Maryland, USA. Of the total 12720 females collected, 15.3% were gravid. Fay-Prince traps caught the highest mean number of individuals per night (411.8) and gravid traps the lowest (22.3%). However, only 2.5% of the females caught in Fay-Prince traps were gravid, compared to 71.7% in gravid traps.

Allan and Kline (2004) evaluated three commercial gravid trap models for collecting *Culex* in rural habitats in Florida, USA: Gravid CDC traps A (Hausherr's Machine Works, Toms River NJ, USA) and B (model 1712,

John W. Hock Company, Gainesville, FL, USA), both based on the original Reiter (1983) design, and one based on the revised Reiter (1987) gravid box trap (BioQuip, Rancho Dominguez, CA, USA). Traps were baited with 2 litres (small pans) or 4 litres (large pans) of hay infusion (450 g dried hay, 5 g brewer's yeast, 20 g lactalbumen hydrolysate, 75 litres water) and operated with 6-V 10-ampere-h rechargeable gel cell batteries. Efficiency of each model of trap was determined by running individual traps overnight in screened field cages (1.83 m wide × 1.83 m long × 1.83 m tall), each cage containing 100 laboratory-reared gravid *Culex quinquefasciatus*. In an early summer comparison, a total of 12 029 Culex were collected over 36 trap-nights, with *Culex quinquefasciatus* females the most commonly collected (77.6%), followed by *Culex nigripalpus* females (7.4%). Significant differences were observed between trap models. The largest collections were obtained with CDC A and B traps as compared to the box-type gravid traps. There were no significant differences between collections in the CDC gravid A and CDC gravid B traps. The proportions of gravid *Culex* females caught by the three trap models did not differ significantly. In the field cage trials of trap efficiency, numbers of gravid females collected in CDC gravid A and B traps were significantly greater than in the box gravid traps ($F = 53.4$; df = 2, 35; $P < 0.0001$). Traps fitted with large (surface area 1496 cm^2) oviposition pans caught 2.05 to 2.73 times more gravid *Culex* than traps fitted with small pans (748 cm^2). Use of darker coloured pans (black and green) resulted in larger collections than use of lighter-coloured pans (tan or white).

The effects of pan size and colour on gravid trap catches were evaluated by Rapaport et al. (2005) in Illinois, USA. Pan sizes evaluated were 23 × 33 cm internal dimensions (759 cm^2 surface area and 3.0 litre infusion-holding capacity) and 35 × 47 cm internal dimensions (1645 cm^2 surface area and 5.3 litre infusion-holding capacity). Colurs tested were green, blue and black. No effect of pan colour on trap catch was observed, but large pans increased catches by approximately 3 times compared with the small pans ($P = 0.0033$).

Scott et al. (2001) investigated the potential for using a commercially available version of the Reiter (1983) gravid trap (Hausherr's Machine Works) to trap *Ochlerotatus japonicus*, a species that is only rarely caught by standard or CO_2-baited light-traps. Traps were powered by either a rechargeable 6-V gel cell or four 1.5-V D-cell batteries. Both grass and hay infusions were used as attractants. The infusions were made by placing approximately 0.9 kg (2 pounds) of fresh grass clippings or hay into a 121-litre (32 gal) plastic trash can, with 5 g of brewer's yeast and 114 litres (30 gal) of tap water. The lid was then securely fitted and the container was left in a sunny location and allowed to ferment for 5–7 days before use. Traps were operated overnight for periods of from 12 to 26 h. A total of 125 *Ochlerotatus japonicus* was collected over 20 trap nights. All but 2 of the specimens collected were females, of which approximately two thirds were gravid. The average number collected per trap per night was 6.25, with a range of 0 to 31. Other species collected were *Culex restuans, Culex pipiens* and *Ochlerotatus triseriatus*.

Oliver and Howard (2005) used a commercially-available version of the Reiter gravid trap (J. W. Hock Co., Gainesville, Florida, USA) baited with a mixture of well water and composted manure to sample *Ochlerotatus japonicus* in New York State, USA. Two traps were set on the ground against out-buildings in an area of minimal disturbance, partial shade and sunlight, and protected from wind and artificial lighting. Traps were typically set between 1800 and 2000 h and retrieved by 2130 h, and then reset and retrieved the next morning between 0800 and 0900 h. Most *Ochlerotatus japonicus* were collected between 1800 and 2130 h.

Reisen and Meyer (1990) laboratory- and field-tested eight different potential oviposition attractants for *Culex tarsalis*, namely tap water, a slightly modified Reiter (1983) medium, the modification proposed by Ritchie (1984*b*) of adding isopropanol to Reiter's (1983) medium, leaf infusion, alfalfa infusion, steer manure infusion, and water which had contained either larvae or pupae of *Culex tarsalis*. The gravid traps of Reiter (1983) were baited with these solutions, and in addition sod-baited traps of Maw and Bracken (1971) were evaluated. It was concluded that none of these attractants was of any use for trapping field populations of gravid *Culex tarsalis*, although in the laboratory there was some indication that steer manure was somewhat attractive. They also reported that the numbers of egg rafts of *Culex quinquefasciatus* collected per trap were very variable and seemed to be strongly influenced by the numbers of natural competitive oviposition sites, as well as by trap placement. This emphasises the importance of trap location in sampling.

In Sri Lanka Jayanetti et al. (1988) baited Reiter (1983) type oviposition traps with water that had 5 days previously had 250 g alfalfa pellets and 0.2 g yeast added to about 18 litres of water. From a total of 119 trap collections seven species of mosquitoes were caught, but *Culex quinquefasciatus* (83%) and *Armigeres subalbatus* (16%) comprised most of the catch. The mean numbers trapped per night was 32.17 *Culex quinquefasciatus* and 6.83 *Armigeres subalbatus*, of which 95 and 77% respectively were gravid females. The authors considered that in terms of collecting effort and cost, the gravid traps were much more effective than catching mosquitoes indoors with aspirators, especially as *Armigeres subalbatus* is partially, or mainly, exophilic.

Using basically the Reiter gravid trap Ritchie (1984*b*) in Florida, USA suspended a CDC light-trap over a 29 × 34 × 12-cm deep brown plastic pan containing 6 litres of three different oviposition attractants. The basic solution was produced by adding 0.9 kg hay, 10 g brewer's yeast and 114 litres of water to a bucket and leaving it covered to mature for a week. The other attractants consisted of a 2:1 mixture of this hay infusion and industrial grade isopropyl alcohol, and isopropyl alcohol without the hay infusion. With traps having just isopropyl alcohol the mean catch of *Culex* mosquitoes, predominantly *Culex nigripalpus*, was 168.8, the hay infusion trap caught a mean of 227.7 *Culex*, while the mixture of both attractants caught a mean of 405.2 *Culex* mosquitoes. These increases were accompanied by increases in the numbers of gravid females collected. In paired trials, while a carbon dioxide-baited CDC light-trap caught almost 5 times as many *Culex* as the hay-isopropyl infusion trap (of 1562.7 vs 335.1) the latter collected almost 50 times (3.0 vs 147.6) as many gravid females, and moreover the parity of unfed females was almost double that recorded from the carbon dioxide light-traps.

Burkett et al. (2004) evaluated five different gravid trap infusions in the commercially-available version of Reiter's (1983) gravid trap in Georgia, USA. The five infusions were: the standard infusion described by Reiter (1983); an experimental set-up consisting of paired gravid traps, one containing standard infusion and the other containing 5 ml of standard infusion in 3.5 litres of rain water; sod infusion obtained by soaking a 30 × 30 cm section of Bermuda grass sod in 11.5 litres of water for 7 days; oak infusion prepared by adding 95 g of oak leaves to 10.5 litres of water and ageing for 7 days; and dilute hay infusion prepared from 5 ml of standard hay infusion diluted in 3.5 litres of rain water. Infusions were compared using a 5 × 5 Latin square experimental design. Traps differed significantly ($P = 0.0001$) in the total number of mosquitoes captured by infusion type as follows: sod = standard hay infusion = hay adjacent to dilute hay = dilute hay adjacent to hay = oak > dilute hay. Oak infusion captured the greatest number of *Aedes albopictus*, while sod infusions captured significantly higher numbers of *Culex restuans*. *Culex quinquefasciatus* was captured in higher numbers in traps baited with standard hay infusion, hay adjacent to dilute hay, and sod infusion, compared with the other infusions.

Further evaluations of gravid trap attractants were conducted by Lee and Kokas (2004) in New York State, USA. Two attractants were compared in CDC gravid traps: a rabbit chow infusion (Lampman and Novak 1996) and a lawn sod infusion (Lee and Rowley 2000). Traps were baited with 3.5 litres of 4-day old infusions and operated in pairs from 2 h before sunset until the following morning. Over 13 nights, the two pairs of traps collected a total of 671 mosquitoes: 582 in lawn sod infusion-baited traps and 89 in rabbit chow infusion-baited traps. Lawn sod infusion-baited traps caught 447 *Culex restuans/Culex pipiens*, 104 *Ochlerotatus japonicus*, 23 *Ochlerotatus triseriatus*, and eight individuals of other species. Rabbit chow infusion-baited traps caught 54 *Ochlerotatus japonicus* and 25 *Culex restuans/Culex pipiens*, and 10 individuals of other species.

Allan et al. (2005) also tested several infusions as attractants to gravid *Culex* mosquitoes in the laboratory and field in Florida, USA. Infusions were prepared according to published protocols as follows: hay infusion (Reiter 1983); cow manure infusion (Leiser and Beier 1982); oak leaf infusion (O'Meara et al. 1989*b*); alfalfa pellet rabbit food infusion (Strickman 1988); and cattail (*Typha latifolia*) infusion (after Du and Millar, 1999). Water, from a commercial dairy effluent lagoon, was also used as a treatment. The field studies were carried out adjacent to an effluent water lagoon at a commercial dairy farm. Trap pans were conditioned by filling them with water and letting them sit for at least one week, after which they were scrubbed, rinsed and dried prior to use. Each day traps were baited with 2 litres of freshly-thawed infusion and operated for 22–24 h from 1000–1200 h. A total of 11 604 *Culex* was collected over 72 trap nights, with *Culex quinquefasciatus* the most commonly collected species (58.7%), followed by *Culex nigripalpus* (31.9%). Traps with cow manure and *Typha* leaf infusions caught significantly more gravid female *Culex quinquefasciatus* than alfalfa hay infusions, but not more than Bermuda grass hay, dairy effluent water or oak leaf infusion. Cow manure infusion also attracted more *Culex nigripalpus* (Table 2.1). Traps with alfalfa hay infusions consistently caught the fewest numbers of *Culex* mosquitoes.

Further evaluations of gravid trap and ovitrap infusions were carried out by Jackson et al. (2005) in Virginia, USA. Gravid traps consisted of large plastic pans (50.8 L × 38.1 W × 20 cm H) with a 37.9 litre capacity, to which 3.8 litres of infusion were added. The infusions tested were: cow manure, mixed grass clippings, wheat straw, and rabbit chow. Initially, catches of *Culex* mosquitoes in manure infusion-baited traps were highest (weeks 1 and 2). Later in the season, however, the relative attractiveness of hay infusion appeared to increase, and there were no significant differences between treatments during weeks 3–6.

Table 2.1. Average mean daily collection (± SE) of *Culex* mosquitoes in CDC gravid traps baited with different infusions in Florida, USA (Allan et al. 2005)

	Alfalfa hay	Bermuda hay	Cow manure	Dairy effluent	Oak leaves	Typha leaves
Total female	68.3	146.3	178.7	133.2	143.1	135.0
Culex	(8.6)b	(19.6)ab	(35.2)a	(29.8)ab	(30.7)ab	(30.3)ab
Culex	40.6	98.6	117.4	85.0	90.7	107.0
quinquefasciatus	(5.1)b	(16.9)ab	(30.3)a	(19.5)ab	(20.1)ab	(27.2)ab
Culex	27.2	47.6	70.8	48.2	52.5	30.0
nigripalpus	(4.3)b	(5.9)ab	(13.8)a	(11.8)ab	(11.4)ab	(3.5)b
Total gravid female						
Culex	54.8	104.8	123.8	103.4	109.2	114.8
	(8.3)b	(14.2)a	(27.5)a	(23.1)a	(22.9)a	(25.5)a
Culex	31.0	72.6	98.6	68.9	68.9	90.4
quinquefasciatus	(5.6)b	(13.0)ab	(25.5)a	(15.2)ab	(13.8)ab	(22.7)a
Culex	24.8	32.2	43.4	34.6	37.5	24.4
nigripalpus	(3.4)b	(3.0)ab	(9.7)a	(9.2)ab	(8.9)ab	(3.4)b
Bloodfed	1.0	1.8	0.8	0.6	0.8	1.8
females	(0.4)a	(0.6)a	(0.3)a	(0.4)a	(0.4)a	(0.6)a
Male *Culex*	7.6	12.4	20.5	11.4	16.7	9.4
	(1.5)a	(2.9)a	(5.8)a	(2.5)a	(6.9)a	(3.3)a

Row means followed by the same letter are not significantly different ($P < 0.05$, Tukey's standardised test).

Presence–absence technique/binomial sampling

Mogi et al. (1990*a*) applied for the first time presence–absence sampling, a technique previously used with agricultural pests (Wilson and Room, 1983), to *Aedes* ovitrap surveys in Thailand. They also combined it with sequential sampling procedures. The model used for presence-absence sampling was as follows

$$\log_e \bar{x} = \log_e a + b \log_e \{-\log_e (1-p)\} \qquad (2.2)$$

where $\bar{x}$ = the mean, p = proportion of positive samples, and a and b are constants which can be determined by plotting the linear regression of $\log \bar{x}$ against $\log_e \{-\log_e (1-p)\}$. Mogi et al. (1990*a*) determined the regression equation to be $\log \bar{x} = 3.38 + 0.99 \log_e \{-\log_e (1-p)\}$, where $a = 29.3$ and $b = 0.99$. Now the number of samples (n_0) needed for a set and predetermined level of precision, D is given by

$$n_0 \frac{p}{D^2(1-p)} \left(\frac{b}{-\log_e(1-p)} \right)^2 \tag{2.3}$$

In their situation they calculated that 100 ovitraps could keep $D < 0.3$ for $0.11 < p < 0.99$, and be sufficient to study *Aedes aegypti* populations in their area.

Mogi et al. (1990*a*) concluded that for the implementation of presence–absence sampling it may be necessary to proceed stepwise by making: (i) a preliminary survey to establish the m - p relationship; and (ii) a trial survey to compare estimates and actual counts, and then if the agreement is good routine surveys based on presence–absence sampling can be undertaken. They also combined presence–absence sampling with sequential sampling using computer simulations to decide when population levels of *Aedes aegypti*, as determined by ovitraps, had reached a size that needed to be controlled to prevent potential dengue outbreaks.

Bellini et al. (1996) evaluated the presence-absence or binomial sampling methodology of Mogi et al. (1990*a*) during *Aedes albopictus* surveillance in an urban area of northern Italy. The determination coefficient (r^2) was high for all three trap types used (plastic, $r^2 = 0.96$; glass, $r^2 = 0.88$ and metal, $r^2 = 0.80$), indicating that field data fitted the binomial model well. The authors also presented sample size curves for the three different trap types for precision levels of $D = 0.2$, $D = 0.3$, and $D = 0.4$ and concluded that fewer plastic traps are required to obtain the same level of precision, based on their increased efficacy.

References

Abu Hassan A, Adanan CR, Rahman WA (1996) Patterns in *Aedes albopictus* (Skuse) population density, host-seeking, and oviposition behavior in Penang, Malaysia. J Vector Ecol 21: 17–21

Addison DS, Ritchie SA, Webber LA, Van Essen F (1992) Eggshells as an index of aedine mosquito production 2: relationship of *Aedes taeniorhynchus* eggshell density to larval production. J Am Mosq Control Assoc 8: 38–43

Aitken THG (1948) Recovery of anopheline eggs from natural habitats, an aid to rapid survey work. Ann Ent Soc Am 41: 327–9

Alencar J, Gil-Santana H-R, Lopes CM, dos Santos JS, Guimarães AE (2004) Utilização de armadilha "ovitrampa" para monitoramento de *Haemagogus janthinomys* (Diptera: Culicidae) em area de Mata Atlântica. Entomología y Vectores 11: 369–374

Allan SA, Kline D (2004) Evaluation of various attributes of gravid female traps for collection of *Culex* in Florida. J Vector Ecol 29: 285–294

Allan SA, Surgeoner GA, Helson BV, Pengelly DH (1981) Seasonal activity of *Mansonia perturbans* adults (Diptera: Culicidae) in southwestern Ontario. Can Entomol 113: 133–9

Allan SA, Bernier UR, Kline DL (2005) Evaluations of oviposition substrates and organic infusions on collection of *Culex* in Florida. J Am Mosq Control Assoc 21: 268–273

Amerasinghe FP, Alagoda TSB (1984) Mosquito oviposition in bamboo traps, with special reference to *Aedes albopictus, Aedes novalbopictus* and *Armigeres subalbatus*. Insect Science and its Application 5: 493–500

Anon (1979) Vector Topics Number 4: Biology and Control of *Aedes aegypti*. U.S. Dept. Hlth Educ. Welfare, Publ. Hlth Serv.

Apostol BL, Black WC IV, Reiter P, Miller BR (1996) Population genetics with RAPD-PCR markers: the breeding structure of *Aedes aegypti* in Puerto Rico. Heredity 76: 325–334

Arnell JH, Nielsen LT (1967) Notes on the distribution and biology of tree hole mosquitoes in Utah. Proc Utah Mosq Abatement Assoc 20: 28–9

Arredondo-Bernal HC, Reyes-Villanueva F (1989) Diurnal pattern and behavior of oviposition of *Toxorhynchites theobaldi* in the field. J Am Mosq Control Assoc 5: 25–8

Aziz N, Hayes J (1987. Oviposition and biting patterns of *Aedes triseriatus* in the flood plains of Fort Bend county, Texas. J Am Mosq Control Assoc 3: 397–399

Badano EI, Regidor HA (2002) Selección de hábitat de oviposición en *Aedes aegypti* (Diptera: Culicidae) mediante estímulos físicos. Ecología Austral 12: 129–134

Ballard EM, Waller JH, Knapp FW (1987) Occurrence and ovitrap site preference of tree hole mosquitoes: *Aedes triseriatus* and *Aedes hendersoni* in eastern Kentucky. J Am Mosq Control Assoc 3: 42–44

Bang YH, Bown DN, Onwubiko AO, Lambrecht FL (1979) Prevalence of potential vectors of yellow fever in the vicinity of Enugu, Nigeria. Cah ORSTOM, sér Entomol méd Parasit 17: 139–147

Barber MA (1935) Malaria studies in Greece. A method of detecting the eggs of *Anopheles* in breeding places and some of its applications. Riv Malar Sez 1 14: 146–149

Barker CM, Paulson SL, Cantrell S, Davis BS (2003) Habitat preferences and phenology of *Ochlerotatus triseriatus* and *Aedes albopictus* (Diptera: Culicidae) in Southwestern Virginia. J Med Entomol 40: 403–410

Barr RA (1958) The mosquitoes of Minnesota (Diptera: Culicidae: Culicinae). Tech. Bull. Minn. agric. Exp. Stn, 228

Bates M (1940) Oviposition experiments with anopheline mosquitoes. Am J Trop Med Hyg 20: 569–583

Bates M (1941) Field studies of the anopheline mosquitoes of Albania. Proc Ent Soc Wash 43: 37–58.

Beckel WE (1955) Oviposition site preference of *Aedes* mosquitoes (Culicidae) in the laboratory. Mosquito News 15: 224–8

Beehler JW, DeFoliart GR (1990) Spatial distribution of *Aedes triseriatus* eggs in a site endemic for La Crosse encephalitis virus. J Am Mosq Control Assoc 6: 254–257

Beehler JW, Webb JP, Mulla MS (1993) Spatial and circadian oviposition patterns in an urban population of *Culex quinquefasciatus*. J Am Mosq Control Assoc 9: 383–388

Beier JC, Trpis M (1981) Local distribution of *Aedes triseriatus* (Diptera: Culicidae) at the Baltimore Zoo. Mosquito News 41: 447–454

Beier JC, Berry WJ, Craig GB (1982) Horizontal distribution of adult *Aedes triseriatus* (Diptera: Culicidae) in relation to habitat structure, oviposition, and other mosquito species. J Med Entomol 19: 239–247

Beier JC, Copeland R, Oyaro C, Masinya A, Odago WO, Oduour S, Koech DK, Roberts CR (1990) *Anopheles gambiae* complex egg-stage survival in dry soil from larval development sites in western Kenya. J Am Mosq Control Assoc 6: 105–109.

Bellini R, Carrieri M, Burgio G, Bacchi M (1996) Efficacy of different ovitraps and binomial sampling in *Aedes albopictus* surveillance activity. J Am Mosq Control Assoc 12: 632–636

Belton P, Galloway MM (1966) Light-trap collections of mosquitoes near Belleville, Ontario, in 1965. Proc Entomol Soc Ontario 96: 90–96

Bentley MD, Day JF (1989) Chemical ecology and behavioral aspects of mosquito oviposition. Annu Rev Entomol 34: 401–421

Berry RL (1986) Ovitrap sampling for *Aedes triseriatus*. Proc Annu Mtg New Jers Mosq Control Assoc 72: 98–104

Bidlingmayer WL, Schoof HF (1956) Studies on the viability of salt-marsh mosquito eggs. Mosquito News 16: 298–301

Bond HA, Fay RW (1969) Factors influencing *Aedes aegypti* occurrence in containers. Mosquito News 29: 113–116

Bonne-Wepster J, Brug SL (1939) Observations on the breeding habits of the subgenus *Mansonioides* (genus *Mansonia*, Culicidae). Tijdschr Entomol 82: 81–90

Bown DN, Bang YH, Knudsen AB, Arata AA, Fabiyi A (1980) Forest types and arbovirus vectors in the Mamu river forest reserve of eastern Nigeria. Mosquito News 40: 91–102

Bradley GH, Travis BV (1942) Soil sampling for studying distribution of mosquito eggs on salt marshes in Florida. Proc New Jers Mosq Exterm Assoc 29: 143–146

Bradshaw WE, Holzapfel CM (1985) The distribution and abundance of treehole mosquitoes in eastern North America; perspectives from north Florida. In: Lounibos LP, Rey JR, Frank JH (eds) Ecology of Mosquitoes: Proceedings of a Workshop. Florida Medical Entomology Laboratory, Vero Beach, Florida, pp. 3–23

Braga IA, Gomes AC, Nelson M, Mello RCG, Bergamaschi DP (2000) Comparação entre pesquisa larvária e armadilha de oviposição, para detecção de *Aedes aegypti*. Rev Soc Bras Med Trop 33: 347–353

Breeland SG, Pickard E (1963) Life history studies on artificially produced broods of floodwater mosquitoes in the Tennessee valley. Mosquito News 23: 75–85

Brust RA (1990) Oviposition behavior of natural populations of *Culex tarsalis* and *Culex restuans* (Diptera: Culicidae) in artificial pools. J Med Entomol 27: 248–255.

Burkett DA, Kelly R, Porter CH, Wirtz RA (2004) Commercial mosquito trap and gravid trap oviposition media evaluation, Atlanta, Georgia. J Am Mosq Control Assoc 20: 233–238

Buth JL, Brust RA, Ellis RA (1990) Development time, oviposition activity and onset of diapause in *Culex tarsalis, Culex restuans* and *Culiseta inornata* in southern Manitoba. J Am Mosq Control Assoc 6: 55–63

Buxton JA, Breland OP (1952) Some species of mosquitoes reared from dry materials. Mosquito News 12: 209–214

Buxton PA, Hopkins GHE (1927) Researches in Polynesia and Melanesia—Entomology. IV. Experiments Performed on *Aedes variegatus* and *Aedes argenteus*. Mem Ser Lond Sch Hyg Trop Med No. 1, 125–256

Cardoso Junior RP, Scandar SAS, de Mello NV, Ernandes S, Botti MV, Nascimento EMM (1997) Detecção de *Aedes aegypti* e *Aedes albopictus*, na zona urbana do município de Catanduva-SP, após controle de epidemia de dengue. Rev Soc Bras Med Trop 30: 37–40

Carpenter SJ, Galindo P, Trapido H (1952) Forest mosquito studies in an endemic yellow fever area in Panama. Mosquito News 12: 156–164

Causey OR, dos Santos GV (1949) Diurnal mosquitoes in an area of small residual forests in Brazil. Ann Entomol Soc Am 42: 471–482

Chadee DD (1986) A comparison of three *Aedes aegypti* sampling methods in Trinidad, W. I. Cah ORSTOM sér Entomol méd Parasitol 24: 199–205

Chadee DD (1991) Seasonal incidence and vertical distribution patterns of oviposition by *Aedes aegypti* in an urban environment in Trinidad, W. I. J Am Mosq Control Assoc 7: 383–386

Chadee DD (1992) Seasonal incidence and horizontal distribution patterns of oviposition by *Aedes aegypti* in an urban environment in Trinidad, West Indies. J Am Mosq Control Assoc 8: 281–284

Chadee DD (2004) Observations on the seasonal prevalence and vertical distribution patterns of oviposition by *Aedes aegypti* (L.) (Diptera: Culicidae) in urban high-rise apartments in Trinidad, West Indies. J Vector Ecol 29: 323–330

Chadee DD, Corbet PS (1987) Seasonal incidence and diel patterns of oviposition in the field of the mosquito, *Aedes aegypti* (L.) (Diptera: Culicidae) in Trinidad, West Indies: a preliminary study. Ann Trop Med Parasitol 81: 151–161

Chadee DD, Corbet PS (1990) A night-time role of the oviposition site of the mosquito *Aedes aegypti* (L.) (Diptera: Culicidae). Ann Trop Med Parasitol 84: 429–433

Chadee DD, Small GJ (1988) A simple spoon device for collecting eggs of *Toxorhynchites* from small containers in the laboratory and field. J Fla Anti-Mosq Assoc 59: 5–6

Chadee DD, Tikasingh ES (1989) Observations on the seasonal incidence and diel oviposition periodicity of *Haemagogus mosquitoes* (Diptera: Culicidae) in Trinidad, W. I.: Part I. *Haemagogus janthinomys* Dyar. Ann Trop Med Parasitol 83: 507–516

Chadee DD, Tikasingh ES (1990) Observations on the seasonal incidence and diel oviposition periodicity of *Haemagogus mosquitoes* (Diptera: Culicidae) in Trinidad, W. I. II. *Haemagogus equinus* Theobald. Ann Trop Med Parasitol 84: 267–275

Chadee DD, Connell NK, Le Maitre A, Ferreira SB (1984) Surveillance for *Aedes aegypti* in Tobago, West Indies (1980–82). Mosquito News 44: 490–492

Chambers DM, Steelman CD, Schilling PE (1979) Mosquito species and densities in Louisiana ricefields. Mosquito News 39: 658–668

Chan KL (1971) Life table studies of *Aedes albopictus* (Skuse). In: Sterility Principles for Insect Control or Eradication. Proc Symp IAEA & FAO (Athens, September, 1970), STI/PUB/265, pp. 131–44

Chan KL, Ho BC, Chan YC (1971) *Aedes aegypti* (L.) and *Aedes albopictus* (Skuse) in Singapore City. 2. Larval habitats. Bull World Health Organ 44: 629–633

Chan KL, Kiat NS, Koh TK (1977) An autocidal ovitrap for the control and possible eradication of *Aedes aegypti*. Southeast Asian J Trop Med Public Health 8: 56–62

Cheng M-L, Ho B-C, Bartnett RE, Goodwin N (1982) Role of a modified ovitrap in the control of *Aedes aegypti* in Houston, Texas, USA. Bull World Health Organ 60: 291–296

Christie M (1958) Improved collection of anopheline eggs and analysis of oviposition behaviour in *A. gambiae*. J Trop Med Hyg 61: 282–286

Clark GG, Crabbs CL, Elias BT (1986) Absence of La Crosse virus in the presence of *Aedes triseriatus* on the Delmarva peninsula. J Am Mosq Control Assoc 2: 33–37

Collins LE, Blackwell A (2000) Colour cues for oviposition behaviour in *Toxorhynchites moctezuma* and *Toxorhynchites amboinensis* mosquitoes. J Vector Ecol 25: 127–135

Cooney JC, Pickard E, Upton JW, McDuff BR (1981) Tillage—a non-chemical method for the control of floodwater mosquitoes. Mosquito News 41: 642–649

Corbet PS (1963) The oviposition-cycles of certain sylvan culicine mosquitoes (Diptera, Culicidae) in Uganda. Ann Trop Med Parasitol 57: 371–381

Corbet PS (1964a) Observations on mosquitoes ovipositing in small containers in Zika forest, Uganda. J Anim Ecol 33: 141–164

Corbet PS (1964b) Autogeny and oviposition in arctic mosquitoes. Nature 203: 668

Corbet PS (1965) Reproduction in Mosquitoes of the High Arctic. Proc. Int. Congr. Ent. XIIth (1964), pp. 817–818

Corbet PS (1966) Diel patterns of mosquito activity in a high arctic locality: Hazan camp, Ellesmere Island, NWT. Can Entomol 98: 1238–1252

Corbet PS, Chadee DD (1990) Incidence and diel pattern of oviposition outdoors of the mosquito, *Aedes aegypti* (L.) (Diptera: Culicidae) in Trinidad, W. I. In relation to solar aspect. Ann Trop Med Parasitol 84: 63–78

Corbet PS, Chadee DD (1993) An improved method for detecting substrate preferences shown by mosquitoes that exhibit 'skip oviposition'. Physiol Entomol 18: 114–118

Cuéllar CB (1969a) A theoretical model of the dynamics of an *Anopheles gambiae* population under challenge with eggs giving rise to sterile males. Bull World Health Organ 40: 205–212

Cuéllar CB (1969b) The critical level of interference in species eradication in mosquitoes. Bull World Health Organ 40: 213–219

Cummings RC (1992) Design and use of a modified Reiter gravid mosquito trap for mosquito borne encephalitis surveillance in Los Angeles County, California. Proc California Mosq Vector Control Assoc 60: 110–116

Curtis GA, Frank JH (1981) Establishment of *Aedes vexans* in citrus groves in southeastern Florida. Environ Entomol 10: 180–182

d'Aguilar J, Benard R, Bessard A (1957) Une méthode de lavage pour l'extraction des arthropodes terricoles. Annls Epiphyt C, 8: 91–99

da Rosa EG, Lairihoy R, Leivas JC, González W, Paulino D (2003) Monitoreo de *Aedes aegypti* mediante el use de ovitrampas. Entomología y Vectores, 10: 451–456

Dale PER, Ritchie SA, Chapman H, Brown MD (1999) Eggshell sampling: quantitative or qualitative data? J Am Mosq Control Assoc 15: 74–76

Dale PER, Chapman H, Brown MD, Ritchie SA, Knight J, Kay BH (2002) Does habitat modification affect oviposition by the saltmarsh mosquito, *Ochlerotatus vigilax* (Skuse) (Diptera: Culicidae)? Aust J Entomol 41: 49–54

De Meillon B, Sebastian A, Khan ZH (1967) Time of arrival of gravid *Culex pipiens fatigans* at an oviposition site, the oviposition cycle and the relationship between time of feeding and time of oviposition. Bull World Health Organ 36: 39–46

de Szalay FA, Resh VH (2000) Factors influencing macroinvertebrate colonization of seasonal wetlands: responses to emergent plant cover. Freshwater Biology 45: 295–308

Dibo MR, Chiaravalloti-Neto F, Battigaglia M, Mondini A, Favaro EA, Barbosa AAC, Glasser CM (2005) Identification of the best ovitrap installation sites for gravid *Aedes (Stegomyia) aegypti* in residences in Mirassol, State of São Paulo, Brazil. Mem Inst Oswaldo Cruz 100: 339–343

Du Y, Millar JG (1999) Electroantennogram and oviposition bioassay responses of *Culex quinquefasciatus* and *Culex tarsalis* (Diptera: Culicidae) to chemicals in odors from Bermuda grass infusions. J Med Entomol 36: 158–166

Dunn LH (1926) Mosquitos bred from dry material taken from holes in trees. Bull Entomol Res 17: 183–187

Dunn LH (1927) Mosquito breeding in 'test' water-containers. Bull Entomol Res 18: 17–22

Dyar HG, Knab F (1916) Eggs and oviposition in certain species of *Mansonia*. Insecutor Inscit Menstr 4: 61–68

Earle HH (1956) Automatic device for the collection of aquatic specimens. J Econ Entomol 49: 261–262

Eitam A, Blaustein L (2004) Oviposition habitat selection by mosquitoes in response to predator (*Notonecta maculata*) density. Physiol Entomol 29: 188–191

Elderton WP (1953) Frequency Curves and Correlation. 4th edn. Cambridge University Press. Cambridge

Elmore CM, Fay RW (1958) *Aedes sollicitans* and *A. taeniorhynchus* larval emergence from sod samples. Mosquito News 18: 230–233

Enfield MA, Pritchard G (1977) Estimation of population size and survival of immature stages of four species of *Aedes* (Diptera: Culicidae) in a temporary pond. Can Entomol 109: 1425–1434

Evans BR (1962) Survey for possible mosquito breeding in crawfish holes in New Orleans, Louisiana. Mosquito News 22: 255–257

Evans BR, Bevier GA (1969) Measurements of field populations of *Aedes aegypti* with the ovitrap in 1968. Mosquito News 29: 347–353

Fallis SP, Snow KR (1983a) Distribution of the eggs of *Aedes punctor* (Diptera: Culicidae). Ecol Entomol 8: 139–144

Fallis SP, Snow KR (1983b) The hatching stimulus for eggs of *Aedes punctor* (Diptera: Culicidae). Ecol Entomol 8: 23–28

Fay RW, Eliason DA (1966) A preferred oviposition site as a surveillance method for *Aedes aegypti*. Mosquito News 26: 531–535

Fay RW, Perry AS (1965) Laboratory studies of ovipositional preferences of *Aedes aegypti*. Mosquito News 25: 276–281

Filsinger C (1941) Distribution of *Aedes vexans* eggs. Proc New Jers Mosq Exterm Assoc 28: 12–19

Fisher JR (1981) System for extracting corn rootworm larvae from soil samples. J Econ Entomol 74: 103–105

Focks DA (2003) A review of entomological sampling methods and indicators for dengue vectors. World Health Organization. TDR/ IDE/Den/03.1, 40 pp.

Forattini OP, Kakitani I, Marques GRAM, de Brito M (1998) Formas imaturas de anofelíneos em recipientes artificiais. Rev Saúde Pública 32: 189–191

Frank JH, Lynn HC (1982) Standardizing oviposition traps for *Aedes aegypti* and *Culex quinquefasciatus*: Time and medium. J Fla Anti-Mosq Assoc 53: 22–27

Freier JE, Francy DB (1991) A duplex cone trap for the collection of adult *Aedes albopictus*. J Am Mosq Contr Assoc 7: 73–79

Furlow BM, Young WW (1970) Larval surveys compared to ovitrap surveys for determining *Aedes aegypti* and *Aedes triseriatus*. Mosquito News 30: 468–470

Galindo P, Carpenter SJ, Trapido H (1951) Ecological observations on forest mosquitoes of an endemic yellow fever area in Panama. Am J Trop Med 31: 98–137

Galindo P, Carpenter SJ, Trapido H (1955) A contribution to the ecology and biology of tree hole breeding mosquitoes in Panama. Ann Entomol Soc Am 48: 158–164

Gass RF, Deesin T, Surathin K, Vutikes S, Sucharit S, Harinasuta C (1983) Studies on oviposition characteristics of *Mansonia (Mansonioides)* mosquitoes in southern Thailand. Ann Trop Med Parasitol 77: 605–614

Gerhardt RW (1959) The influence of soil fermentation on oviposition site selection by mosquitoes. Mosquito News 19: 151–159

Giglioli MEC (1979) *Aedes aegypti* programs in the Caribbean and emergency measures against the dengue pandemic of 1977–1978: a critical review. In: Dengue in the Caribbean. Proceedings of a Workshop, Montego Bay, Jamaica (8–11 May 1978), pp. 133–152. Pan-Am. Hlth Org. sci. Publ., No. 375

Gingrich JB, Casillas L (2004) Selected mosquito vestors of West Nile virus: comparison of their ecological dynamics in four woodland and marsh habitats in Delaware. J Am Mosq Contr Ass 20: 138–145

Gjullin CM (1938) A Machine for Separating Mosquito Eggs from Soil. U.S. Dept. agric. Bur. Ent. & Pl. Quar., ET-135

Goettel MS, Toohey MK, Pillai JS (1980) The urban mosquitoes of Suva, Fiji: seasonal incidence and evaluation of environmental sanitation and ULV spraying for their control. J Trop Med Hyg 83: 165–171

Gomes A de C, da Conceição MBE, Sallum MAM, Portes M da GT, Machado JP, da Silva IJ (1998) Observação sobre característica natural de oviposição de *Culex (Melanoconion)* Grupo Pilosus (Diptera: Culicidae). Rev Saúde Pública 32: 370–371

Gottfried KL, Gerhardt RR, Nasci RS, Crabtree MB, Karabatsos N, Burkhalter KL, Davis BS, Panella NA, Paulson DJ (2002) Temporal abundance, parity, survival rates, and arbovirus isolation of field-collected container-inhabiting mosquitoes in eastern Tennessee. J Am Mosq Control Assoc 18: 164–172

Haber WA, Moore CG (1973) *Aedes aegypti* in the Puerto Rican rain forest: Results of a one-year survey. Mosquito News 33: 576–578

Haeger JS, O'Meara GFO (1983) Separation of first-instar larvae of four *Culex (Culex)*. Mosquito News 43: 76–77

Hanson SM, Song M, Craig GB (1988) Urban distribution of *Aedes triseriatus* in northern Indiana. J Am Mosq Control Assoc 4: 15–19

Harrington LC, Edman JD (2001) Indirect evidence against delayed "skip-oviposition" behavior by *Aedes aegypti* (Diptera: Culicidae) in Thailand. J Med Entomol 38: 641–645

Harris WV (1942) Notes on culicine mosquitos in Tanganyika territory. Bull Entomol Res 33: 181–193

Harrison BA, Boonyakanist P, Mongkolpanya K (1972) Biological observations on *Aedes seato* Huang in Thailand with notes on rural *Aedes aegypti* (L.) and other *Stegomyia* populations. J Med Entomol 9: 1–6

Healy MJR, Taylor LR (1962) Tables for power-law transformations. Biometrika 49: 557–559

Hoban B, Craig GB (1981) Effectiveness of the ovitrap for monitoring *Culex* mosquitoes. Proc Indiana Vector Control Assoc 5: 10–16

Hoban B, Fish D, Craig GB (1980) The influence of organic substrates upon oviposition site-selection in the mosquito *Culex restuans*. Proc Indiana Acad Sci 89: 208

Hoeck PAE, Ramberg FB, Merrill SA, Moll C, Hagedorn HH (2003) Population and parity levels of *Aedes aegypti* collected in Tucson. J Vector Ecol 28: 65–73

Hoffman BL, Killingsworth BF (1967) The egg-laying habits of *Aedes aegypti* (Linnaeus) in central Texas. Mosquito News 27: 466–469

Horsfall WR (1956) A method for making a survey of floodwater mosquitoes. Mosquito News 16: 66–71

Horsfall WR (1963) Eggs of floodwater mosquitoes (Diptera: Culicidae). IX. Local distribution. Ann Entomol Soc Am 56: 426–441

Husbands RC (1952) Some techniques used in the study of *Aedes* eggs in irrigated pastures in California. Mosquito News 12: 145–150

Iriarte WLZ, Tsuda Y, Wada Y, Takagi M (1991) Distribution of mosquitoes on a hill of Nagasaki city, with emphasis to the distance from human dwellings. Tropical Medicine (Nagasaki) 33: 55–60

Iwao S (1968) A new regression method for analyzing the aggregation pattern of animal populations. Researches in Population Ecology 10: 1–20

Iwao S (1970) Problems of spatial distribution in animal population ecology. In: Patil GP, Pielou, EC (eds) Random Counts in Biomedical and Social Sciences, Vol. 2. Pennsylvania State University Press, University Park, pp. 117–149

Jackson BT, Paulson SL, Youngman RR, Scheffel SL, Hawkins B (2005) Oviposition preferences of *Culex restuans* and *Culex pipiens* (Diptera: Culicidae) for selected infusions in oviposition traps and gravid traps. J Am Mosq Control Assoc 21: 360–365

Jakob WL, Bevier GA (1969a) Application of ovitraps in the U.S. *Aedes aegypti* eradication program. Mosquito News 29: 55–62

Jakob WL, Bevier GA (1969*b*) Evaluation of ovitraps in the U.S. *Aedes aegypti* eradication program. Mosquito News 29: 650–653

Jakob WL, Fay RW, von Windeguth DL, Schoof HF (1970) Evaluation of materials for ovitrap paddles in *Aedes aegypti* surveillance. J Econ Entomol 63: 1013–1014

James HG (1966) Location of univoltine *Aedes* eggs in woodland pool areas and experimental exposure to predators. Mosquito News 26: 59–63

Jayanetti SR, Perera HAS, Wijesundara M de S (1988) Evaluation of the CDC gravid mosquito trap for sampling periodomestic mosquito filarial vectors. Mosq-Borne Dis Bull 5: 18–21

Jenkins DW, Carpenter SJ (1946) Ecology of the tree hole breeding mosquitoes of nearctic North America. Ecol Monogr 16: 32–47

Jupp PG, McIntosh BM (1990) *Aedes furcifer* and other mosquitoes as vectors of chikungunya virus at Mica, northeastern Transvaal, South Africa. J Am Mosq Control Assoc 6: 415–420

Kaur JS, Lai YL, Giger AD (2003) Learning and memory in the mosquito *Aedes aegypti* shown by conditioning against oviposition deterrence. Med Vet Entomol 17: 457–460

Kaw Bing Chua, I-Ly Chua, I-Ee Chua, Kerk Hsiang Chua (2004) Differential preferences of oviposition by *Aedes* mosquitos in man-made containers under field conditions. Southeast Asian J Trop Med Public Health 35: 599–607

Kay BH, Ryan PA, Lyons SA, Foley PN, Pandeya N, Purdie D (2002). Winter intervention against *Aedes aegypti* (Diptera: Culicidae) larvae in subterranean habitats slows surface recolonization in summer. J Med Entomol 39: 356–361

Kemp A, Jupp PG (1993) Results of a preliminary ovitrap survey for *Aedes albopictus* (Skuse) (Diptera: Culicidae) at two tyre retreading companies in Durban, South Africa. African Entomol 1: 63–65

Kitching RL (1971) A core sampler for semi-fluid substrates. Hydrobiologia 37: 205–209

Kitron UD, Webb DW, Novak RJ (1989) Oviposition behavior of *Aedes triseriatus* (Diptera: Culicidae): Prevalence, intensity, and aggregation of eggs in oviposition traps. J Med Entomol 26: 462–467

Kloter KO, Bowman DD, Carroll MK (1983) Evaluation of some ovitrap materials used for *Aedes aegypti* surveillance. Mosquito News 43: 438–441

Krebs CJ (1989) Ecological Methodology. Harper and Row, New York

Laarman JJ (1958) Research on the ecology of culicine mosquitoes in a forest region of the Belgian Congo. Acta Leidensia 28: 94–98

Lambrecht FL, Peterson RD (1977) The hatching of mosquito larvae from material collected in dry tree holes and in dry water storage jars Anambra State, Nigeria, WHO/VBC/76.649: 6 pp. (mimeographed)

Lambrecht FL, Zaghi A (1960) Observations sur la ponte des culicides dans la forêt ombrophile d'Irangi, Congo Belge. Rev Zool Bot Afr 61: 87–97

Lampman RL, Novak RJ (1996*a*). Oviposition preferences of *Culex pipiens* and *Culex restuans* for infusion-baited traps. J Am Mosq Control Assoc 12: 23–32

Lampman RL, Novak RJ (1996*b*) Attraction of *Aedes albopictus* adults to sod infusion. J Am Mosq Control Assoc 12: 119–124

Lanciotti RS, Roehrig JT, Deubel V, Smith J, Parker M, Steele K, Crise B, Volpe KE, Crabtree MB, Scherret JH, Hall RA, Mackenzie JS, Cropp CB, Panigrahy B, Ostlund E, Schmitt B, Malkinson M, Banet C, Weissman J, Komar N, Savage HM, Stone W, McNamara T, Gubler DJ (1999) Origin of the West Nile virus responsible for an outbreak of encephalitis in the northeastern United States. Science 286: 2333–2337

Landry SV, DeFoliart GR (1987) Parity rates of *Aedes triseriatus* (Diptera: Culicidae) collected in a female-retaining ovitrap. J Med Entomol 24: 282–285

Landry SV, DeFoliart GR, Hogg DB (1988) Adult body size and survivorship in a field population of *Aedes triseriatus*. J Am Mosq Control Assoc 4: 121–128

Lang JT (1990) Ovipositional response of *Aedes triseriatus* females to horizontally and vertically open ovitraps in southern Illinois. J Am Mosq Control Assoc 6: 530–531

Laurence BR, Samarawickrema WA (1970) Aggregation by ovipositing *Mansonioides* mosquitoes. J. Med. Entomol., 7: 594–600

Lawson DL, Merritt RW (1979) A modified Ladell apparatus for the extraction of wetland macroinvertebrates. Can Entomol 111: 1389–1393

Lee Joon-Hak, Kokas JE (2004) Field evaluation of CDC gravid trap attractants to primary West Nile virus vectors, *Culex* mosquitoes in New York State. J Am Mosq Control Assoc 20: 248–253

Lee Joon-Hak, Rowley WA (2000) The abundance and seasonal distribution of *Culex* mosquitoes in Iowa during 1995–97. J Am Mosq Control Assoc 16: 275–278

Leftkovitch LP, Brust RA (1968) Locating the eggs of *Aedes vexans* (Mg.) (Diptera: Culicidae). Bull Entomol Res 58: 119–122

Leiser LB, Beier JC (1982) A comparison of oviposition traps and New Jersey light traps for *Culex* population surveillance. Mosquito News 42: 391–395

Lewis DJ (1939) The seasonal and geographical distribution of *Anopheles maculipennis* in Albania. Riv Malar Sez 1 18: 237–248

Lewis LF, Christenson DM (1975) Residual activity of temephos, chlorpyrifos, DDT, fenthion, and malathion against *Aedes sierrensis* (Ludlow) in fabricated tree-holes. Mosquito News 35: 381–384

Lewis LF, Tucker TW (1978) Fabrication of artificial treeholes and their performance in field tests with *Aedes sierrensis* and *Orthopodomyia signifera*. Mosquito News 38: 132–135

Lewis LF, Clark TB, O'Grady JJ, Christenson DM (1974) Collecting ovigerous *Culex pipiens quinquefasciatus* Say near favorable resting sites with louvered traps baited with infusions of alfalfa pellets. Mosquito News 34: 436–439

Lloyd M (1967) Mean crowding. J Anim Ecol 36: 1–30

Lok CK, Kiat NS, Koh TK (1977) An autocidal ovitrap for the control and possible eradication of *Aedes aegypti*. Southeast Asian J Trop Med Public Health 8: 56–62

Loor KA, DeFoliart GR (1969) An oviposition trap for detecting the presence of *Aedes triseriatus* (Say). Mosquito News 29: 487–488

Loor KA, DeFoliart GR (1970) Field observations on the biology of *Aedes triseriatus*. Mosquito News 30: 60–64

Lopp OV (1957) Egg sampling as an index of mosquito breeding. Proc New Jers Mosq Exterm Assoc 44: 60–65

Lounibos LP (1979) Temporal and spatial distribution, growth and predatory behaviour of *Toxorhynchites brevipalpis* (Diptera: Culicidae) on the Kenya coast. J Appl Ecol 48: 213–236

Lounibos LP (1980) The bionomics of three sympatric *Eretmapodites* (Diptera: Culicidae) at the Kenya coast. Bull Entomol Res 70: 309–320

Lounibos LP (1981) Habitat segregation among African treehole mosquitoes. Ecol Entomol 6: 129–154

Lounibos LP, Machado-Allison CE (1986) Mosquito maternity: Egg brooding in the life cycle of *Trichoprosopon digitatum*. In: Taylor F, Karban R (eds) The Evolution of Insect life Cycles. Springer-Verlag, New York pp. 173–84

Lounibos LP, Rey JR, Frank JH (eds) (1985) Ecology of Mosquitoes: Proceedings of a Workshop. Florida Medical Entomology Laboratory, Vero Beach, Florida

Lourenço-de-Oliveira R, Castro MG, Braks MAH, Lounibos LP (2004) The invasion of urban forest by dengue vectors in Rio de Janeiro. J Vector Ecol 29: 94–100

Lowe RE, Ford HR, Smittle BJ, Weidhaas DE (1973) Reproductive behavior of *Culex pipiens quinquefasciatus* released into a natural population. Mosquito News 33: 221–227

Lowe RE, Ford HR, Cameron AL (1974) Seasonal abundance of *Culex* species near Cedar Key, Florida. Mosquito News 34: 118–119

Madder DJ, MacDonald RS, Surgeoner GA, Helson BV (1980) The use of oviposition activity to monitor populations of *Culex pipiens* and *Culex restuans* (Diptera: Culicidae). Can Entomol 112: 1013–1017

Madon MB, Hazelrigg JE, Shaw MW, Kluh S, Mulla MS (2003) Has *Aedes albopictus* established in California? J Am Mosq Control Assoc 19: 297–300

Magnarelli LA (1975) Relative abundance and parity of mosquitoes collected in dry-ice baited and unbaited CDC miniature light traps. Mosquito News 35: 350–353

Marques CCA, Marques GRAM, de Brito M, dos Santos Neto LG, Ishibashi VC, Gomes FA (1993) Estudo comparativo de eficácia de larvitrampas e ovitrampas para vigilância de vetores de dengue e febre amarela. Rev Saúde Pública 27: 237–241

Mather TN, DeFoliart GR (1983) Effect of host blood source on the gonotrophic cycle of *Aedes triseriatus*. Am J Trop Med Hyg 32: 182–93

Mattingly PF (1969) Mosquito eggs. V. Mosq Syst Newsletter 1: 70–80

Maw MG (1970) Capric acid as a larvicide and an oviposition stimulant for mosquitoes. Nature 227: 1154–1155

Maw MG, Bracken GK (1971) The use of artificial pools in assessing populations of the mosquito *Culex restuans* Theobald. Proc Entomol Soc Ontario 102: 78–83

Maw MG, House HL (1971) On capric acid and potassium capricate as mosquito larvicides in laboratory and field. Can Entomol 103: 1435–1440

McCall PJ, Cameron MM (1995) Oviposition pheromones in insect vectors. Parasitol Today 11: 352–355

McCall PJ, Eaton G (2001) Olfactory memory in the mosquito *Culex quinquefasciatus*. Med Vet Entomol 15: 197–203

McCardle PW, Webb RE, Norden BB, Aldrich JR (2004) Evaluation of five trapping systems for the surveillance of gravid mosquitoes in Prince Georges County, Maryland. J Am Mosq Control Assoc 20: 254–260

McClelland GAH (1956) Field Studies on *Aedes aegypti*. In: Haddow AJ (ed) E. Afr. Virus Res. Inst. Rep. No. 6, 1955–56, Government Printer, Nairobi, pp. 38–42

McDaniel IN, Horsfall WR (1963) Bionomics of *Aedes stimulans* (Diptera: Culicidae). I. Effect of moisture on the distribution of eggs. Am Midl Nat 70: 479–489

Meek CL, Olson JK (1976) Oviposition sites used by *Psorophora columbiae* (Diptera: Culicidae) in Texas ricefields. Mosquito News 36: 311–315

Metge G, Hassaïne K (1998) Study of the environmental factors associated with oviposition by *Aedes caspius* and *Aedes detritus* along a transect in Algeria. J Am Mosq Control Assoc 14: 283–288

Micieli MV, Campos RE (2003) Oviposition activity and seasonal pattern of a population of *Aedes (Stegomyia) aegypti* (L.) (Diptera: Culicidae) in subtropical Argentina. Mem Inst Oswaldo Cruz 98: 659–663

Micks DW, McNeil JC (1963) Preliminary egg sampling studies of salt marsh mosquitoes. Proc New Jers Mosq Exterm Assoc 50: 339–342

Millar JG, Chaney JD, Mulla MS (1992) Identification of oviposition attractants for *Culex quinquefasciatus* from fermented Bermuda grass infusions. J Am Mosq Control Assoc 8: 11–17

Miura T (1972) Laboratory and field evaluation of a sonic sifter as a mosquito egg extractor. Mosquito News 32: 433–436

Mogi M, Mokry J (1980) Distribution of *Wyeomyia smithii* (Diptera, Culicidae) eggs in pitcher plants in Newfoundland, Canada. Trop Med 22: 1–12

Mogi M, Choochote W, Khambooruang C, Suwanpanit P (1990*a*). Applicability of presence–absence and sequential sampling for ovitrap surveillance of *Aedes* (Diptera: Culicidae) in Chiang Mai, northern Thailand. J Med Entomol 27: 509–514

Mogi M, Khamboonruang C, Choochote W, Suwanpanit P (1990*b*). Ovitrap surveys of dengue vector mosquitoes in Chiang Mai, northern Thailand; seasonal shifts in relative abundance of *Aedes albopictus* and *Ae. aegypti*. Med Vet Entomol 2: 319–324

Mogi M, Miyagi I, Toma T, Hasan M, Abadi K, Syafruddin (1996) Occurrence of *Aedes (Stegomyia)* spp. mosquitoes (Diptera: Culicidae in Halmahela villages, Indonesia. J Med Entomol 33: 169–172

Montgomery ME, Musick GJ, Polivka JB, Nielsen DG (1979) Modifiable washing-flotation method for separation of insect eggs and larvae from soil. J Econ Entomol 72: 67–69

Morris CD, DeFoliart GR (1971) Parous rates in Wisconsin mosquito populations. J Med Entomol 8: 209–212

Mortenson EW, Rotramel GL, Prine JE (1978) The use of ovitraps to evaluate *Aedes sierrensis* (Ludlow) populations. California Vector Views 25: 29–32

Muirhead-Thomson RC (1940*a*). Studies on the behaviour of *Anopheles minimus*. Part 1. The selection of the breeding place and the influence of light and shade. J Malar Inst India 3: 265–322

Muirhead-Thomson RC (1940*b*). Mosquito Behaviour in Relation to Malaria Transmission and Control in the Tropics. Edward Arnold, London

Muirhead-Thomson RC (1945). Studies on the breeding places and control of *Anopheles gambiae* and *A. gambiae* var. *melas* in coastal districts of Sierra Leone. Bull Entomol Res 36: 185–252

Mullens BA, Rodriguez JL (1984). Efficiency of salt flotation for extraction of immature *Culicoides variipennis* (Ceratopogonidae) from mud substrates. Mosquito News 44: 207–211

Nasci RS, Komar N, Marfin AA, Ludwig GV, Kramer LD, Daniels TJ, Falco RC, Campbell SR, Brookes K, Gottfried KL, Burkhalter KL, Aspen SE, Kerst AJ, Lancioti RS, Moore CG (2002) Detection of West Nile virus-infected mosquitoes and seropositive juvenile birds in the vicinity of virus-positive dead birds. Am J Trop Med Hyg 67: 492–496

Navarro DMAF, de Oliveira PES, Potting RPJ, Brito AC, Fital SJF, Sant'Ana AEG (2003) The potential attractant or repellent effects of different water types on oviposition in *Aedes aegypti* L. (Dipt., Culicidae). J Appl Entomol 127: 46–50

Nayar JK (1981) *Aedes aegypti* (L.) (Diptera: Culicidae): Observations on dispersal, survival, insemination, ovarian development and oviposition characteristics of a Florida population. J Fla Anti-Mosq Assoc 52: 24–40

Nayar JK (1982) Bionomics and physiology of *Culex nigripalpus* (Diptera: Culicidae) of Florida: an important vector of diseases. Fla Agric Exp Stn Techn Bull No. 827

Nelson MJ, Usman S, Pant CP, Self LS (1976) Seasonal abundance of adult and immature *Aedes aegypti* (L.) in Jakarta. Bull Penel Keseh Health Std Indonesia 4: 1–8

Niebylski ML, Meek CL (1992) Blood-feeding of *Culex* mosquitoes in an urban environment. J Am Mosq Control Assoc 8: 173–177

Novak RJ, Peloquin JJ (1981) A substrate modification for the oviposition trap used for detecting the presence of *Aedes triseriatus*. Mosquito News 41: 180–181

Novak RJ, Shroyer DA (1978) Eggs of *Aedes triseriatus* and *Ae. hendersoni*: a method to stimulate optimal hatch. Mosquito News 38: 515–521

Oda T (1967) Hourly and seasonal distribution of the number of egg rafts of *Culex pipiens pallens* deposited in earthen jars. Trop Med 9: 39–44

O'Gower AK (1955) The influence of the physical properties of a water container surface upon its selection by gravid females of *Aedes scutellaris* (Walker) for oviposition (Diptera, Culicidae). Proc Linn Soc N.S.W. 79: 211–218

O'Gower AK (1957) The influence of the surface on oviposition by *Aedes aegypti* (Linn.) (Diptera, Culicidae). Proc Linn Soc N.S.W. 82: 240–244

O'Gower AK (1958) The influence of the surface on oviposition by *Aedes albopictus* (Skuse) and *Aedes scutellaris katherinensis* Woodhill (Diptera, Culicidae). Proc Linn Soc N.S.W. 82: 285–288

O'Gower AK (1963) Environmental stimuli and the oviposition behaviour of *Aedes aegypti* var. *queenslandensis* Theobald (Diptera, Culicidae). Anim Behav 11: 189–197

Oliver J, Howard JJ (2005) Fecundity of naturally blood-fed *Ochlerotatus japonicus*. J Med Entomol 42: 254–259

O'Malley SLC, Hubbard SF, Chadee DD (1989) Oviposition habitat preferences of *Toxorhynchites moctezuma* mosquitoes in four types of tropical forest in Trinidad. Med Vet Entomol 3: 247–252

O'Meara GF, Larson VL, Mook DH, Latham MD (1989a). *Aedes bahamensis*: its invasion of south Florida and association with *Aedes aegypti*. J Am Mosq Control Assoc 5: 1–5

O'Meara GF, Vose FE, Carlson DB (1989b). Environmental factors influencing oviposition by *Culex (Culex)* (Diptera: Culicidae) in two types of traps. J Med Entomol 26: 528–534

O'Meara GF, Evans LF Jr, Gettman AD, Cuda JP (1995) Spread of *Aedes albopictus* and decline of *Ae. aegypti* (Diptera: Culicidae) in Florida. J Med Entomol 32: 554–562

Ordóñez-Gonzalez JG, Mercado-Hernandez R, Flores-Suarez AE, Fernández-Salas I (2001) The use of sticky ovitraps to estimate dispersal of *Aedes aegypti* in northeastern Mexico. J Am Mosq Control Assoc 17: 93–97

Pausch RD, Provost MW (1965) The dispersal of *Aedes taeniorhynchus* IV. Controlled field production. Mosquito News 25: 1–8

Pena CJ, Gonzalvez G, Chadee DD (2003) Seasonal prevalence and container preferences of *Aedes albopictus* in Santo Domingo City, Dominican Republic. J Vector Ecol 28: 208–212

Pena CJ, Gonzalvez G, Chadee DD (2004) A modified tire ovitrap for monitoring *Aedes albopictus* in the field. J Vector Ecol 29: 374–375

Penn GH (1947) The larval development and ecology of *Aedes (Stegomyia) scutellaris* (Walker 1859) in New Guinea. J Parasitol 33: 43–50

Perich MJ, Kardec A, Braga IA, Portal IF, Burge R, Zeichner BC, Brogdon WA, Wirtz RA (2003) Field evaluation of a lethal ovitrap against dengue vectors in Brazil. Med Vet Entomol 17: 205–210

Petersen JJ, Willis OR (1971) Effects of salinity on site selection by ovipositing tree hole mosquitoes in Louisiana. Mosquito News 31, 352–355

Philip CB (1933) Mosquito species breeding in 'test' water containers in West Africa. Bull Entomol Res 24: 483–491

Pratt HD, Jakob WL (1967) Oviposition Trap Reference Handbook. *Aedes aegypti* Handbook Series No. 6, National Communicable Disease Center

Pratt HD, Kidwell AS (1969) Eggs of mosquitoes found in *Aedes aegypti* oviposition traps. Mosquito News 29: 545–548

Rapaport AS, Lampman RL, Novak RJ (2005) Evaluation of selected modifications to CO_2 and infusion-baited mosquito traps in Urbana, Illinois. J Am Mosq Control Assoc 21: 395–399

Raw F (1955) A flotation extraction process for soil micro-arthropods. In: McE Kevan DK (ed) Soil Zoology. Butterworths Science Publishers, London pp. 341–346

Rawlins SC, Martinez R, Wiltshire S, Legall G (1998) A comparison of surveillance systems for the dengue vector *Aedes aegypti* in Port of Spain, Trinidad. J Am Mosq Control Assoc 14: 131–136

Raymond HL, Cornet M, Dieng PY (1976) Études sur les vecteurs sylvatiques du virus amaril. Inventaire provisoire des habitats larvaires d'une forêt-galerie dans le foyer endèmique du Senegal oriental. Cah ORSTOM sér Entomol Méd Parasitol 14: 301–306

Reisen WK, Basio RG (1972) Oviposition trap surveys conducted on four USAF installations in the Western Pacific. Mosquito News 32: 107–108

Reisen WK, Meyer RP (1990) Attractiveness of selected oviposition substrates for gravid *Culex tarsalis* and *Culex quinquefasciatus* in California. J Am Mosq Control Assoc 6: 244–250

Reisen WK, Pfuntner AR (1987) Effectiveness of five methods for sampling adult *Culex* mosquitoes in rural and urban habitats in San Bernardino County, California. J Am Mosq Control Assoc 3: 601–606

Reisen WK, Boyce K, Cummings RC, Delgado O, Gutierrez A, Meyer RP, Scott TW (1999) Comparative effectiveness of three adult mosquito sampling methods in habitats representative of four different biomes of California. J Am Mosq Control Assoc 15: 24–31

Reiskind MH, Wilson ML (2004) *Culex restuans* (Diptera: Culicidae) oviposition behavior determined by larval habitat quality and quantity in southeastern Michigan. J Med Entomol 41: 179–186

Reiskind MH, Walton ET, Wilson ML (2004) Nutrient-dependent reduced growth and survival of larval *Culex restuans* (Diptera: Culicidae): laboratory and field experiments in Michigan. J Med Entomol 41: 650–656

Reiter P (1983) A portable, battery-powered trap for collecting gravid *Culex* mosquitoes. Mosquito News 43: 496–498

Reiter P (1986) A standardized procedure for the quantitative surveillance of certain *Culex* mosquitoes by egg raft collection. J Am Mosq Control Assoc 2: 219–221

Reiter P (1987) A revised version of the CDC gravid mosquito trap. J Am Mosq Control Assoc 3: 325–327

Reiter P, Gubler DJ (1997) Surveillance and control of urban dengue vectors. CAB International, New York

Reiter P, Jakob WL, Francy DB, Mullenix JB (1986) Evaluation of the CDC gravid trap for the surveillance of St. Louis encephalitis vectors in Memphis, Tennessee. J Am Mosq Control Assoc 2: 209–211

Reiter P, Amador MA, Colon N (1991) Enhancement of the CDC ovitrap with hay infusions for daily monitoring of *Aedes aegypti* populations. J Am Mosq Control Assoc 7: 52–55

Reuben R, Panicker KN, Das PK, Kazmi SJ, Suguna SG (1977) A new paddle for the black jar ovitrap for surveillance of *Aedes aegypti*. Indian J Med Res 65 (Suppl.): 115–119

Reuben R, Das PK, Kazmi SJ, Brooks GD (1978) Seasonal changes in egg-laying activity of *Aedes* species in Sonepat by the use of the black jar ovitrap. Indian J Med Res 67: 763–766

Ritchie SA (1984*a*) The production of *Aedes aegypti* by weekly ovitrap survey. Mosquito News 44: 77–79

Ritchie SA (1984*b*). Hay infusion and isopropyl alcohol-baited CDC light trap; a simple, effective trap for gravid *Culex* mosquitoes. Mosquito News 44: 404–407

Ritchie SA (1994) Spatial stability of *Aedes vigilax* (Diptera: Culicidae) eggshells in southeastern Queensland. J Med Entomol 31: 920–922

Ritchie SA (2001). Effect of some animal feeds and oviposition substrates on *Aedes* oviposition in ovitraps in Cairns, Australia. J Am Mosq Control Assoc 17: 206–208

Ritchie SA, Addison DS (1991) Collection and separation of *Aedes taeniorhynchus* eggshells from mangrove soil. J Am Mosq Control Assoc 7: 113–115

Ritchie SA, Addison DS (1992). Oviposition preferences of *Aedes taeniorhynchus* (Diptera: Culicidae) in Florida mangrove forests. Environ Entomol 21: 737–744

Ritchie SA, Jennings CD (1994) Dispersion and sampling of *Aedes vigilax* eggshells in southeast Queensland, Australia. J Am Mosq Control Assoc 10: 181–185

Ritchie SA, Johnson ES (1989) Use of sodium hypochlorite to detect aedine mosquito eggs in mangrove soils and insect feces. J Am Mosq Control Assoc 5: 612–615

Ritchie SA, Johnson ES (1991*a*) Distribution and sampling of *Aedes taeniorhynchus* (Diptera: Culicidae) eggs in a Florida mangrove forest. J Med Entomol 28: 270–274

Ritchie SA, Johnson ES (1991*b*). *Aedes taeniorhynchus* (Diptera: Culicidae) oviposition patterns in a Florida mangrove forest. J Med Entomol 28: 496–500

Ritchie SA, Long S (2003) Does s-methoprene affect oviposition by *Aedes aegypti* in an ovitrap? J Am Mosq Control Assoc 19: 170–171

Ritchie SA, Addison DS, Van Essen F (1992) Eggshells as an index of aedine mosquito production 1: distribution, movement and sampling of *Aedes taeniorhynchus* eggshells. J Am Mosq Control Assoc 8: 32–37

Ritchie SA, Long S, Hart A, Webb CE, Russell RC (2003) An adulticidal sticky ovitrap for sampling container-breeding mosquitoes. J Am Mosq Control Assoc 19: 235–242

Ritchie SA, Long S, Smith G; Pyke A, Knox TB (2004) Entomological investigations in a focus of dengue transmission in Cairns, Queensland, Australia, by using the sticky ovitraps. J Med Entomol 41: 1–4

Rivière F (1985) Effects of two predators on community composition and biological control of *Aedes aegypti* and *Aedes polynesiensis* In: Lounibos LP, Rey JR, Frank (eds) Ecology of Mosquitoes: Proceedings of a Workshop. Florida Medical Entomology Laboratory, Vero Beach, Florida pp. 121–35

Rosenberg R (1982) Forest malaria in Bangladesh. III. Breeding habits of *Anopheles dirus*. Am J Trop Med Hyg 31: 192–201

Rozeboom LE, Hess AD (1944) The relation of the intersection line to the production of *Anopheles quadrimaculatus*. J Natn Malar Soc 3: 169–179

Rozeboom LE, Rosen L, Ikeda J (1973) Observations on oviposition by *Aedes (S.) albopictus* Skuse and *A. (S.) polynesiensis* Marks in nature. J Med Entomol 10: 397–399

Russell PF, Rao TR (1942). On relation of mechanical obstruction and shade to ovipositing of *Anopheles culicifacies*, J Exp Zool 91: 303–329

Russell RC, Ritchie SA (2004) Surveillance and behavioral investigations of *Aedes aegypti* and *Aedes polynesiensis* in Moorea, Polynesia, using a sticky ovitrap. J Am Mosq Control Assoc 20: 370–375

Salt G, Hollick FSJ (1944) Studies of wireworm populations. 1. A census of wireworms in pasture. Ann Appl Biol 31: 53–64

Samarawickrema WA (1967) A study of the age-composition of natural populations of *Culex pipiens* fatigans Wiedemann in relation to the transmission of filariasis due to *Wuchereria bancrofti* (Cobbold) in Ceylon. Bull World Health Organ 37: 117–137

Sames WJ, Herman WE, Florin DA, Maloney FA (2004) Distribution of *Ochlerotatus togoi* along the Pacific coast of Washington. J Am Mosq Control Assoc 20: 105–109

Sardelis MR, Turell MJ, Dohm DJ, O'Guinn ML (2001) Vector competence of selected North American *Culex* and *Coquillettidia* mosquitoes for West Nile virus. Emerg Infect Dis 7: 1018–1022

Savage HM, Ezike VI, Nwankwo ACN, Spiegel R, Miller BR (1992) First record of breeding populations of *Aedes albopictus* in continental Africa: implications for arboviral transmission. J Am Mosq Control Assoc 8: 101–103

Savage HM, Smith GC, Moore CG, Mitchell CJ, Townsend M, Marfin AA (1993) Entomologic investigations of an epidemic of St. Louis encephalitis in Pine Bluff, Arkansas, 1991. Am J Trop Med Hyg 49: 38–45

Schliessmann DJ (1964) The *Aedes aegypti* eradication program of the U.S. Mosquito News 24, 124–132

Schuler TC, Beier JC (1983) Oviposition dynamics of two released species of *Toxorhynchites* (Diptera: Culicidae) and potential prey species. J Med Entomol 20: 371–376

Schultz GW (1993) Seasonal abundance of dengue vectors in Manila, Republic of the Philippines. Southeast Asian J Trop Med Public Health 24: 369–375

Scott JJ, Crans SC, Crans WJ (2001) Use of an infusion-baited gravid trap to collect adult *Ochlerotatus japonicus*. J Am Mosq Control Assoc 17: 142–143

Scott JJ, Crans WJ (2003) Expanded polystyrene (EPS) floats for surveillance of *Ochlerotatus japonicus*. J Am Mosq Control Assoc 19: 376–381

Scotton GL, Axtell RC (1979). *Aedes taeniorhynchus* and *Ae. sollicitans* (Diptera: Culicidae) oviposition on coastal dredge spoil. Mosquito News 39: 97–110

Seawright JA, Dame DA, Weidhaas DE (1977) Field survival and ovipositional characteristics of *Aedes aegypti* and their relation to population dynamics and control. Mosquito News 37: 62–70

Seinhorst JW (1962). Extraction methods for nematodes inhabiting soil In: Murphy PW. Progress in Soil Zoology (ed) Butterworths, London, pp. 243–256

Sempala SDK (1983) Seasonal population dynamics of the immature stages of *Aedes africanus* (Theobald) (Diptera: Culicidae) in Zika forest, Uganda. Bull Entomol Res 73: 11–18

Service MW (1965) The ecology of the tree-hole breeding mosquitoes in the northern guinea savanna of Nigeria. J Appl Ecol 2: 1–16

Service MW (1968a) The ecology of the immature stages of *Aedes detritus* (Diptera: Culicidae). J Appl Ecol 5: 513–630

Service MW (1968b). A method for extracting mosquito eggs from soil samples taken from oviposition sites. Ann Trop Med Parasitol 62: 478–480

Service MW (1970). Studies on the biology and taxonomy of *Aedes (Stegomyia) vittatus* (Bigot) (Diptera: Culicidae) in Northern Nigeria. Trans R Entomol Soc Lond 122: 101–143

Service MW (1993) Mosquito Ecology. Field Sampling Methods. 2nd edn. Chapman and Hall, London

Sharma VP, Patterson RS, LaBrecque GC, Singh KRP (1976) Three field release trials with chemosterilized *Culex pipiens fatigans* Wied. in a Delhi village. J Commun Dis 8: 18–27

Slaff M, Crans WJ, McCuiston LJ (1983) A comparison of three mosquito sampling techniques in northwestern New Jersey. Mosquito News 43: 287–290

Smith WL, Enns WR (1967) Laboratory and field investigations of mosquito populations associated with oxidation lagoons in Missouri. Mosquito News 27: 462–466

Smith WW, Jones DW (1972) Use of artificial pools for determining presence, abundance, and oviposition preferences of *Culex nigripalpus* Theobald in the field. Mosquito News 32: 244–245

Southwood TRE (1978) Ecological Methods with Particular Reference to the Study of Insect Populations. Chapman and Hall, London

Southwood TRE, Murdie G, Yasuno M, Tonn RJ, Reader PM (1972) Studies on the life budget of *Aedes aegypti* in Wat Samphaya, Bangkok, Thailand. Bull World Health Organ 46: 211–226

Srivastava DS. Lawton JH (1998) Why more productive sites have more species: an experimental test of theory using tree-hole communities. Am Nat 152: 510–529

Stage HH, Gjullin CM, Yates WW (1952) Mosquitoes of the Northwestern States. U.S. Dept Agric Handbook, No. 46

Stav G, Blaustein L, Margalit Y (2000) Influence of nymphal *Anax imperator* (Odonata: Aeshnidae) on oviposition by the mosquito *Culiseta longiareolata* (Diptera: Culicidae) and community structure in temporary pools. J Vector Ecol 25: 190–202

Steinly BA, Novak JR (1990) *Culex restuans* and *Culex pipiens* oviposition frequency characteristics during a drought year (1988) in east-central Illinois. Proc Illinois Mosq Vector Control Assoc 1: 16–24

Stone WS, Reynolds FHK (1939) Hibernation of anopheline eggs in the tropics. Science 90: 371–372

Strickman D (1988) Rate of oviposition by *Culex quinquefasciatus* in San Antonio, Texas, during three years. J Am Mosq Control Assoc 4: 339–344

Subra R, Mouchet J (1984) The regulation of preimaginal populations of *Aedes aegypti* (L.) (Diptera: Culicidae) on the Kenya coast. II. Food as a main regulatory factor. Ann Trop Med Parasitol 78: 63–70

Surgeoner GA, Helson BV (1978) An oviposition trap for arbovirus surveillance in *Culex* sp. mosquitoes (Diptera: Culicidae). Can Entomol 110: 1049–1052

Surtees G (1959) On the distribution and seasonal incidence of culicine mosquitoes in southern Nigeria. Proc R Entomol Soc Lond (A) 34: 110–120

Swanson J, Lancaster M, Anderson J, Crandell M, Haramis L, Grimstad P, Kitron U (2000) Overwintering and establishment of *Aedes albopictus* (Diptera: Culicidae) in an Urban La Crosse virus enzootic site in Illinois. J Med Entomol 37: 454–460

Swellengrebel NH, de Buck A (1938) Malaria in the Netherlands. Scheltema and Holkema, Amsterdam

Szumlas DE, Apperson CS, Powell EE (1996) Seasonal occurrence and abundance of *Aedes triseriatus* and other mosquitoes in a La Crosse virus-endemic area in western North Carolina. J Am Mosq Control Assoc 12: 184–193

Tanner GD (1969). Oviposition traps and population sampling for the distribution of Aedes aegypti (L.). Mosquito News 29: 116–121

Taylor LR (1961) Aggregation, variance and the mean. Nature 189: 732–735

Thaggard CW, Eliason DA (1969) Field evaluation of components for an *Aedes aegypti* (L.) oviposition trap. Mosquito News 29: 608–612

Thavara U, Tawatsin A, Chansang C, Kong-ngamsuk W, Paosriwong S, Boon-Long J, Rongsriyam Y, Komalamisra N (2001) Larval occurrence, oviposition behavior and biting activity of potential mosquito vectors of dengue on Samui Island, Thailand. J Vector Ecol 26: 172–180

Thavara U, Tawatsin A, Chompoosri J (2004) Evaluation of attractants and egg-laying substrate preference for oviposition by *Aedes albopictus* (Diptera: Culicidae). J Vector Ecol 29: 66–72

Tikasingh ES, Laurent E (1981) Use of ovitraps in monitoring *Haemagogus equinus* populations. Mosquito News 41: 677–679

Tikasingh ES, Martinez R (1983) A multipaddle ovitrap for collecting *Haemagogus* and *Aedes aegypti* eggs. Mosquito News 43: 358–359

Toma L, Severini F, di Luca M, Bella A, Romi R (2003) Seasonal patterns of oviposition and egg hatching rate of *Aedes albopictus* in Rome. J Am Mosq Control Assoc 19: 19–22

Toma T, Sakamoto S, Miyagi I (1982) The seasonal appearance of *Aedes albopictus* in Okinawajima, the Ryuku archipelago, Japan. Mosquito News 42: 179–183

Trpis M (1972) Breeding of *Aedes aegypti* and *A. simpsoni* under the escarpment of the Tanzanian plateau. Bull World Health Organ 47: 77–82

Trpis M, Hartberg WK, Teesdale C, McClelland GAH (1971) *Aedes aegypti* and *Aedes simpsoni* breeding in coral rock holes on the coast of Tanzania. Bull World Health Organ 45: 529–531

Tsuda Y, Takagi M, Wada Y (1989) Field observations on oviposition time of *Aedes albopictus*. Trop Med 31: 161–165

Tsuda Y, Kobayashi J, Nambanya S, Miyagi I, Toma T, Phompida S, Manivang K (2002) An ecological survey of dengue vector mosquitos in central Lao PDR. Southeast Asian J Trop Med Public Health 33: 63–67

Turner PA, Streever WJ (1997*a*). Test of a mosquito eggshell isolation method and subsampling procedure. J Am Mosq Control Assoc 13: 43–46

Turner PA, Streever WJ (1997*b*). The relationship between the density of *Aedes vigilax* (Diptera: Culicidae) eggshells and environmental factors on Kooragang Island, New South Wales, Australia. J Am Mosq Control Assoc 13: 361–367

Vaughan JA, Turner EC (1987) Seasonal microdistribution of immature *Culicoides variipenniis* (Diptera: Ceratopogonidae) at Saltville, Virginia. J Med Entomol 24: 340–346

Vezzani D, Schweigmann N (2002) Suitability of containers from different sources as breeding sites of *Aedes aegypti* (L.) in a cemetery of Buenos Aires City, Argentina. Mem Inst Oswaldo Cruz 97: 789–792

Vezzani D, Velázquez SM, Schweigmann N (2003) Seasonal pattern of abundance of *Aedes aegypti* (Diptera: Culicidae) in Buenos Aires city, Argentina. Mem Inst Oswaldo Cruz 99: 789–792

Vezzani D, Velázquez SM, Schweigmann N (2004) Containers of different capacity as breeding sites of *Aedes aegypti* (Diptera: Culicidae) in the cemeteries of Buenos Aires, Argentina. Entomología y Vectores 11: 305–316

Walker ED, Copeland RS, Paulson SL, Munstermann LE (1987) Adult survivorship, population density, and body size in sympatric populations of *Aedes triseriatus* and *Aedes hendersoni* (Diptera: Culicidae). J Med Entomol 24: 485–493

Wallace FL (1996) Construction of a field trap for initiating an ovipositional response in *Aedes taeniorhynchus*. J Am Mosq Control Assoc 12: 491–493

Wallace FL, Tidwell MA, Williams DC, Jackson KA (1990) Effects of controlled burning of *Aedes taeniorhynchus* eggs in an abandoned rice impoundment in South Carolina. J Am Mosq Control Assoc 6: 528–529

Wallis RC (1954) Observations on oviposition of two *Aedes* mosquitoes (Diptera, Culicidae). Ann Entomol Soc Am 27: 393–396

Wanson M (1944) Elevage du *Taeniorhynchus (Coquillettidia) metallicus* Theobald. East Afr Med J 21: 269–272

Weinbren MP, O'Gower AK (1966) A simple artificial tree-hole for recovering mosquito eggs, with a note on the recovery of Aedes aegypti eggs from rain forests in Puerto Rico. Mosquito News 26: 522–526

Wesenberg-Lund C (1921) Contributions to the Biology of the Danish Culicidae. K. Danske Vidensk Selsk Skr 8th Ser, 7, No. 1

Wilkins OP, Breland OP (1949) Recovery of the mosquito. *Culiseta inornata* (Williston) from dry material. Proc Entomol Soc Wash 51: 27–28

Wilkins OP, Breland OP (1951) The larval stages and the biology of the mosquito, *Orthopodomyia alba* Baker (Diptera: Culicidae). J N.Y. Entomol Soc 59: 225–240

Wilson LT, Room PM (1983) Clumping patterns of fruit and arthropods in cotton, with implications for binomial sampling. Environ Entomol 12: 50–54

Wilton DP (1968) Oviposition site selection by the tree-hole mosquito, *Aedes triseriatus* (Say). J Med Entomol 5: 189–194

Woodward DL, Colwell AE, Anderson NL (1996) Temporal and spatial distribution of *Aedes sierrensis* oviposition. Proc Mosq Vector Control Assoc California 64: 51–62

Woodward DL, Colwell AE, Anderson NL (1998) Surveillance studies of *Orthopodomyia signifera* with comparison to *Aedes sierrensis*. J Vector Ecol 23: 136–148

Yap HH, Lee CY, Chong NL, Foo AES, Lim MP (1995) Oviposition site preference of *Aedes albopictus* in the laboratory. J Am Mosq Control Assoc 11: 128–132

Yasuno M, Kazmi SJ, LaBrecque GC, Rajagopalan PK (1973) Seasonal Change in Larval Habitats and Population Density of *Culex fatigans* in Delhi Villages. WHO/VBC/73.429, 12 pp. (mimeographed)

Yates M (1974) An artificial oviposition site for tree-hole breeding mosquitoes. Entomol Gaz 25: 151–154

Yates MG (1979) The biology of the tree-hole breeding mosquito *Aedes geniculatus* (Olivier) (Diptera: Culicidae) in southern England. Bull Entomol Res 69: 611–628

Zeichner BC, Perich MJ (1999) Laboratory testing of a lethal ovitrap for *Aedes aegypti*. Med Vet Entomol 13: 234–238

Chapter 3 Sampling the Larval Population

Mosquito larvae and pupae are found in a great variety of habitats, ranging from large expanses of water such as swamps, marshes and rice fields to small collections of water in tyres, domestic utensils, tree-holes, plant axils, snail shells and fallen leaves. A number of often arbitrary classifications of larval habitats have been used (Almirón and Brewer 1996; Bates 1949; Boyd 1930; Hopkins 1952; Mattingly 1969; Mogi 1981). Almirón and Brewer (1996) coded habitat characteristics using a binary presence/absence system and related these characteristics to the species of mosquito collected in the different habitats in Córdoba, Argentina. A basic matrix of data comprising 19 rows (species) and 86 columns (habitat characteristics) was prepared and species and habitat associations were determined. Other systems have been proposed by Service (1993*a*), and by Laird (1988) who gives a useful review of past classifications and presents much detail on the community ecology of mosquito larval habitats. Joy and Hildreth-Whitehair (2000) developed a classification of the larval habitats of *Ochlerotatus triseriatus*, the vector of LaCrosse encephalitis, in West Virginia, USA that was based on Laird's (1988) proposed standard system, but with the addition of sunlit or shaded as criteria. Joy and Clay (2002) presented a classification of mosquito larval habitats also in West Virginia, USA comprising seven major sunlit categories and eight shaded categories.

Some mosquitoes exhibit considerable plasticity in their selection of breeding places, such as many *Culex*, *Ochlerotatus* and *Aedes* species which occur in a variety of ground collections of water, whereas other species are very restricted in their choice of breeding places like those colonising tree-holes, bromeliads or pitcher plants.

Until the mid 1940s larval collections formed an essential part of malaria surveys and control programmes, but with the advent of spraying houses with residual insecticides, attention was focused more on the biology and behaviour of adults. In the 1970s, a few entomologists began to study the natural mortalities of *Anopheles* larvae, but despite this renewed interest in larval ecology and continued efforts to sample mosquito larvae, it remains difficult to accurately monitor changes in larval population size. One of the major problems is the fluctuating size of the larval habitat, due to mainly rainfall and desiccation, which makes it difficult to standardise sampling techniques. Other problems such as unequal dispersion of larvae in the habitat and changes in vegetative cover compound the difficulties. In discussing larval surveys of *Culex tarsalis* in rice fields over 30 years ago Loomis (1959) concluded that even the best sampling techniques lacked standardisation and the necessary reliability to make results meaningful on any large scale. Reisen and Reeves (1990) concurred with this, and identified the aggregated dispersion patterns of the immature stages as one of the complications. The intensity of larval aggregation, however, usually decreases as a function of age, and the probability of collecting each instar consequently often increases. They pointed out that most mosquito control agencies in California, USA relied on qualitative, not quantitative, surveys to record breeding, and used light-traps to monitor adult mosquito population size. In fact the main objective of larval surveys is usually the detection of larval habitats and assessment of any gross changes in larval density caused by control measures. Only rarely are attempts made to obtain quantitative estimates of either larval density or population size.

While spraying the interior surfaces of houses with residual insecticides can effectively control anopheline mosquitoes, this method is less effective at controlling culicine mosquitoes. Larval control, whether by insecticides, biological control agents, habitat modifications or elimination, has therefore remained a useful method for reducing culicine populations and consequently larval surveys have continued to be important in assessing population size and the impact of control measures for these species. A good knowledge of pre-imaginal ecology is paramount for understanding the dynamics of mosquito populations, but larval sampling has frequently been replaced by routine light-trap catches and biting collections of adults.

In the following accounts reference is made almost entirely to sampling mosquito larvae. No special methods have been devised specifically for collecting pupae, but most of the techniques described for the collection of larvae are also applicable to pupae.

Remote sensing of larval habitats

In the 14 years since the publication of the second edition of this book, there has been an explosion of interest in the use of remotely sensed data to predict vector distribution and disease transmission by a range of arthropod vectors, including *Anopheles* mosquitoes (Roberts et al. 1996; Sharma et al. 1996; Thomson et al. 1996, 1997; Wood et al. 1991*a, b*; 1992; Rejmankova et al. 1995, 1998), *Aedes* (Linthicum et al. 1990), *Ochlerotatus*

(Dale et al. 1993, 1998; Beck et al. 1994, 1997), *Culex* (Linthicum et al. 1990; Sharma et al. 1996), *Culiseta* (Moncayo et al. 2000), *Culicoides* biting midges (Baylis et al. 1998), *Ixodes* ticks (Dister et al. 1997; Kitron and Kazmierczak 1997; Kitron et al. 1992), *Glossina* tsetse flies (Kitron et al. 1996, Rogers and Randolph, 1991) and *Phlebotomus* sand flies (Cross et al. 1996; Thomson et al. 1999). The distribution and abundance of arthropod vectors of disease are determined by the complex interaction of a range of ecological, climatic and environmental variables. All mosquitoes undergo a facultative aquatic stage in their life-cycle and many authors have identified associations between various types of plant or plant communities and larval habitats of certain mosquito species (e.g. Cousserans et al. 1969; Jolivet et al. 1974; Pautou et al. 1973; Provost 1975; Rioux et al. 1968). Accurate identification of the preferred larval habitats of vectors of human and animal disease is desirable as it can assist in designing focused and cost-efficient control programmes within a resource-limited environment. Traditional land-based larval habitat surveillance systems involving sampling with dippers or nets are generally impractical where these habitats cover a wide geographic area, requiring significant human and other resources to effectively monitor them. The need for a rapid and simple tool for identifying larval mosquito habitats prompted several workers to investigate the utility of black-and-white, colour and colour-infrared aerial photography to delineate such habitats (e.g. NASA 1973 [cited in Washino and Wood 1994]; Wagner et al. 1979; Welch et al. 1989*a, b*). Oviposition sites and larval habitats such as ditches, low lying areas and tyre tracks can be detected, even on photographs at a scale of 1:42 000 (Welch et al. 1989*a*) *Ochlerotatus sollicitans* larval habitats are saltmarshes intermittently flooded with freshwater, and 90% of them are dominated by *Spartina patens* and *Juncus roemerianus* vegetation, which could be reliably identified from visual inspection of colour-infrared photographs in Louisiana, USA (Cibula 1976, cited in Washino and Wood 1994 and in Hay et al. 1996). Later use of a multi-spectral scanner improved the discrimination of the plant communities and allowed identification of larval habitats across the entire Mississippi delta (Barnes and Cibula 1979). Fleetwood et al. (1981) modified the combined aerial surveillance and mapping techniques developed by Gooley and Lesser (1975) and Kent and Sutherland (1977) for identifying breeding places of *Ochlerotatus sollicitans*, to detect breeding of *Psorophora columbiae* in Louisiana rice fields. Meek and Olson (1991) in reviewing the ecology of riceland mosquitoes pointed out the value of aerial colour infrared (CIR) remote survey techniques in monitoring changes in land use and thus identifying potential breeding places for mosquitoes such as *Psorophora columbiae*.

Enlargements at a scale of 1:10 000 obtained from 1:30 000 scale colour aerial photographs have been used to develop a rapid technique for the identification of urban habitats for the aquatic stages of *Culex annulirostris* in Brisbane, Australia (Dale and Morris 1996). Likely habitats were identified from the aerial photographs of two areas, one a coastal area, and the other further inland and more heavily urbanised, and these habitats were checked by ground inspection to confirm whether or not they harboured juvenile mosquitoes. Larval habitats identified during ground inspections were marked on the aerial photographs. Data on site variables, including shade, aquatic and terrestrial vegetation cover were assessed and tabulated. The 1:10 000 photographs were converted to transparencies, producing a 3-layer RGB image, which was transformed using the MicroBRIAN image analysis software package (Jupp et al. 1985), to produce an image with a pixel resolution of 10 × 10 m. Actual and potential larval habitats were identified on the photographs and used to create training themes in the image processor. Significant habitat relationships for *Culex annulirostris* were with site type and whether the site was bare or grassy. Of 78 sites initially surveyed, 68 (87%) were correctly identified by the image analysis. Of 71 sites identified by the image analysis package as grassy field, 53 (75%) were confirmed by field inspection as being actual or likely larval habitats. Many of the positive sites identified were smaller than the 100 m^2 pixel size of the digitised photographs, but the image analysis software was able to identify small depressions of the type inhabited by *Culex annulirostris*, even when these were not distinguishable on the aerial photographpahs. The authors concluded that digitised aerial photographs taken at a time when surface water conditions were optimum for mosquito breeding, and processed using an image analysis package could be used to identify *Culex annulirostris* larval habitats.

Moloney et al. (1998) in Queensland, Australia, tested the usefulness of aerial photography as a surveillance tool in identifying residential premises at high risk of *Aedes aegypti* infestation. In 1995, 360 premises were inspected for the presence of immature *Aedes aegypti* and Premise Condition Index (PCI2) scores (incorporating assessments of yard condition and shade) were assigned. PCI2 values were also estimated from visual inspection of 1:3000 colour and infrared aerial photographs. Shade level estimates from both the colour and infrared photography resulted in the highest level of agreement with ground estimates (75.5% and 77.4%, respectively). The overall PCI2 estimates from the colour images were less accurate than those from the infrared images (49% versus 56% matched pairs, respectively). Infrared and colour PCI2 estimates were lower than ground estimates in 24% and 35% of cases, respectively, and higher in 20% and 16% of cases. Yard conditions, however, could not be accurately identified from either type of photograph. The PCI2 values obtained from photographs did not significantly correlate with indices of mosquito infestation (proportion of positive premises or the mean number of positive containers per premise), and logistic regression further demonstrated that neither aerial photograph type allowed the accurate prediction of Aedes aegypti infestation risk and it was concluded that the ability of low-level aerial photography to enhance *Aedes aegypti* larval site surveillance is at present limited.

Black-and-white film records reflected light in the visible and near-infrared regions of the spectrum between 0.4 μm and 0.9 μm and colour film records in the range between 0.4 μm and 0.7 μm. The range of the spectrum that can be detected is enhanced by the use of multispectral scanners that use electronic detectors to record emitted or reflected energy across several spectral regions simultaneously (Wood and Washino 1994). The shift to digital data collection facilitates quantitative analysis and standardisation.

Since the 1970s multispectral scanners have been used in earth-orbiting satellites to transmit spectral data obtained from the Earth's atmosphere and surface. Satellites are either geostationary or polar-orbiting. Geostationary satellites orbit the Earth at the equator and travel at the same speed as the Earth's rotation, and therefore record data from a fixed area of the Earth's surface. Polar-orbiting satellites orbit the Earth repeatedly and record data from different sections of the Earth's surface as it rotates (Hay et al. 1996). The spectral, temporal and spatial resolutions of satellite data represent a compromise arising from the characteristics of the on-board detectors, on-board storage capacity, the speed of orbit in the case of polar-orbiting satellites, the height of orbit, and the transmission characteristics of the atmosphere at any given time. In general, a satellite detector will provide data at either high spatial or high temporal resolution, but not both. Satellite detectors with high spatial resolution (Landsat Multi-Spectral-Scanner [MSS], Landsat Thematic Mapper [TM], and Satellite Pour l'Observation de la Terre [SPOT]) provide images with a spatial resolution down to around 10 m, but these images are only updated every 16 or 26 days. Cloud cover also results in a relatively low number of clear images in any year. The geostationary Meteosat satellite produces images every 30 minutes, with a resolution of 2.5 to 5 km depending on the spectral waveband used. By contrast, the polar-orbiting satellites of the National Oceanographic and Atmospheric Administration (NOAA) complete 14.1 orbits per day and produce two global-coverage images every 24 h. NOAA satellites carry a five-channel Advanced Very High Resolution Radiometer (AVHRR) producing a nominal resolution of 1.1 km directly beneath the satellite track, but this is degraded with distance along and across the track. Spatial resolution is further reduced by the need to average data across blocks of 5 × 3 pixels in order to reduce data storage capacity requirements. This produces images with a resolution of 4 × 4 km, and is termed global area coverage or GAC data. Prior to their use for ecological and environmental analysis, spectral data obtained by sensors need to be converted into indices that describe relevant and meaningful characteristics. One of the first, and subsequently most widely used indices is the normalised-difference vegetation index or NDVI, which exploits the fact that chlorophyll and carotenoids in active, photosynthesising plant tissues absorb light in the visible red wavelength (AVHRR channel 1), while mesophyll tissue reflects light in the near-infra red wavelength (AVHRR channel 2), so that a healthy photosynthesising plant will appear darker than a dying or dead plant. The ratios of reflectances in channels 1 and 2, adjusted for reflectance from dark soil backgrounds is the NDVI and is calculated as:

$$NDVI = \frac{(Ch_2 - Ch_1)}{(Ch_2 + Ch_1)} \tag{3.1}$$

NDVI is a particularly useful measure in areas of sparse vegetation, as values tend to saturate where there is high vegetation coverage, for example forest canopy. NDVI exhibits a linear relationship with saturation deficit (Rogers and Randolph 1991) and so can be used as a proxy measure of this variable. Another commonly used satellite-derived environmental variable is cold-cloud duration (CCD), which is used as an estimator of rainfall. In the tropics, where rainfall is predominantly convective in nature, clouds with a cloud-top temperature of below $-40°C$ are more likely to be producing rainfall (Hay et al. 1996). Cloud-top temperatures are recorded by the Meteosat satellite, the Geostationary Operational Environmental Satellite (GEOS), the Geostationary Operational Meteorological Satellite (GOMS), and Geosynchronous Meteorological Satellite (GMS). The relationship between threshold temperatures and associated rainfall has to be determined for each location, but is linear. The African Real-Time Environmental Monitoring and Information System (ARTEMIS) of the Food and Agriculture Organisation of the United Nations uses this data to produce freely-available 10-day (decadal) and monthly images of cold-cloud duration for Africa. (metart.fao.org). Thomson et al. (1996) present a useful table that describes the type of remotely sensed data and the methods that can be applied to these data to provide information on ecological variables of relevance to malaria transmission. One of the first uses of multi-spectral-scanner satellite data in vector studies was conducted by NASA and the Mexican Government on screw-worm flies (Washino and Wood 1994).

Table 3.1. Ecological variables, remote sensing methodologies and relevance for malaria transmission (Thomson et al. (1996)

Ecological variable	Remotely sensed data and methodology	Relevance for malaria transmission
Surface water	AVHRR NDVI (Linthicum et al. 1990; Thomson et al. 1996) AVHRR day/night thermal difference (Malone et al. 1994)	Larval habitat
Rainfall quantity	Meteosat thermal infrared Cold-cloud duration (Milford and Dugdale 1990)	
Evapotranspiration	Meteosat Thermal infrared and NDVI (Lamblin and Strahler 1994; Kogan 1995)	
Water flow	Meteosat Thermal infrared and NDVI (Grimes et al. 1995)	
Water temperature	AVHRR sea-surface temperatures (SST) (Coll et al. 1994)	Egg and larval development rates
Ambient temperature	AVHRR land-surface temperatures (LST) (Coll et al. 1994)	Oviposition cycle length, extrinsic incubation cycle of parasite
Saturation deficit	NDVI (Rogers and Randolph 1991; Thomson et al. 1996)	Vector survival
Vegetation type and biomass	Vegetation indices (Flasse and Verstraete 1994)	Vector resting sites
Seasonality (mean phase amplitude)	Fourier time series analysis of GAC archived NOAA AVHRR and Meteosat datasets (Rogers and Williams 1994)	Transmission season length

Hayes et al. (1985) made use of Landsat 1 and 2 satellites to identify and map mosquito larval habitats in Nebraska and South Dakota, USA, based on associations with freshwater plant communities. Field data were used to calibrate the satellite data in a process known as supervised classification and this classification subsequently resulted in 95% accuracy in predicting larval habitats. In Africa, Linthicum et al. (1987) demonstrated how Advanced Very High Resolution Radiometer (AVHRR) instruments could be used to monitor ecological parameters associated with Rift Valley Fever viral activities. Subsequently Linthicum et al. (1990) described how such instruments on board the polar-orbiting NOAA satellites 7 and 9 were used to measure NDVI dynamics in Kenya from which ground moisture and rainfall patterns were derived, and used to monitor the transient shallow larval habitats of the *Aedes,*

Ochlerotatus and *Culex* vectors of Rift Valley Fever. Unfortunately flooded and dry habitats could not be distinguished using this method. Later, Pope et al. (1992) used Landsat data and airborne Synthetic Aperture Radar to discriminate between the flooded and dry larval habitats.

In southern Chiapas, Mexico, Beck et al. (1994) used remotely sensed data from the Landsat Thematic Mapper to identify and map landscape elements within the coastal plain area, focusing primarily on landscape elements previously shown to be associated with *Anopheles albimanus* abundance. The spatial resolution of the images was 28.5×28.5 m. Accuracy of the landscape classification using the digital data was checked by field observations and comparison with colour infrared photographs. Mosquito abundance data for 40 randomly selected villages were calculated as mean numbers captured between 1800 and 0600 h by CDC ultra-violet light-traps operated over a 16-week period. The ten villages with the highest mosquito abundance were assigned to a high-abundance category, while the remaining 30 villages were classified as low abundance. A stepwise discriminant analysis was used to identify the landscape elements that most contributed to the separation of high-abundance from low-abundance categories of village. Stepwise linear regression was also used to identify the best subset of landscape elements that predicted mosquito abundance. Both analyses revealed that the proportions of transitional swamp and unmanaged pasture were the most important landscape elements ($P < 0.001$ in both cases). Later, Beck et al. (1997) applied the above model to predicting malaria transmission risk in villages in Chiapas, using a blind test. Landscape elements in the study area obtained from Landsat Thematic Mapper images were classified according to the characteristics identified by Beck et al. (1994). The proportions of each landscape element within a 1 km buffer zone around each of 40 villages were determined and the values for transitional swamp and unmanaged pasture, the two most important variables identified by Beck et al. 1994, were input into the previously developed discriminant function equations in order to classify villages as having either high or low *Anopheles albimanus* abundance. The classifications were then compared with actual abundance data collected from CDC ultraviolet light-traps. The discriminant function equations identified 28 of the 40 villages as high abundance villages and the remaining 12 as low abundance. Of the 28 villages predicted to have high vector abundance, 22 were ranked in the top 28 villages for abundance by light-trapping and six of the 12 villages classified as low abundance had their statuse confirmed by the light-trap data, giving a predictive accuracy of 79% for high abundance villages and 50% for low abundance villages. Further analysis revealed that in this study area, transitional swamp was the only remotely-sensed landscape element that was significantly associated with mosquito abundance ($P = 0.005$). The authors recommended preliminary calibrations of the model based on small-scale field studies to identify differences in landscape elements and vector ecology in order to improve the predictive accuracy of the model.

Rejmankova et al. (1995) and Roberts et al. (1996) undertook similar studies of remotely-sensed data and their association with *Anopheles albimanus* abundance and distribution on the Pacific coast of northern Belize. In contrast to the situation described above for the Atlantic coastal plain of Mexico, *Anopheles albimanus* in Belize is primarily associated with marshes containing cyanobacterial mats and river margins and this example illustrates the importance of local knowledge of vector ecology prior to the application of predictive models based on remotely-sensed data. Rejmankova et al. (1995) processed a 60 km × 60 km SPOT image, with a resolution of 20 m × 20 m using an unsupervised classification process that resulted in the identification of 23 distinct classes of landscape element, and these classifications were confirmed by field observation. Predictions of mosquito abundance in houses were based on the known flight range of *Anopheles albimanus* and the distance of houses from suitable larval habitats. Actual mosquito abundance was determined for 16 sites using indoor and outdoor human landing catches between 1830 and 2000 h. Outdoor landing densities exceeded 0.5 per human per minute in eight of the nine villages predicted to have high mosquito densities, equivalent to a predictive ability of 89%. All sites predicted to have low mosquito densities had landing rates below 0.5 per human per minute. Also in Belize, Roberts et al. (1996) used the methods of Beck et al. (1994) and Rejmankova et al. (1995) to predict the distribution of *Anopheles pseudopunctipennis*. This species is associated primarily with increasing altitude and the presence of sun-exposed pools or waterways containing filamentous algae. Indoor and outdoor human landing catches yielded *Anopheles pseudopunctipennis* in 50% of the locations predicted to have a high probability of *Anopheles pseudopunctipennis* presence. *Anopheles pseudopunctipennis* was not collected from houses predicted to have low probability of presence and no significant difference was observed between predicted and actual distributions.

Also in Belize, Rejmankova et al. (1998) conducted extensive larval surveys to identify the larval habitat of *Anopheles vestitipennis* and *Anopheles punctimacula*. Canonical discriminant analysis of *Anopheles vestitipennis*-positive, *Anopheles punctimacula*-positive, and all negative sites, revealed that while larval habitats of *Anopheles punctimacula* were clustered in the area dominated by trees, larval habitats of *Anopheles vestitipennis* were found in both tree-dominated and tall dense macrophyte dominated environments. Data collected from the ground-based larval survey and data from existing vegetation and land-use maps were then compared with a dry season (February 1992–93) SPOT multispectral satellite image. Ten land cover types were identified, including swamp forest and tall dense marsh. The classifications based on the SPOT image did not indicate the presence of larval habitats where there were none. However, confirmed larval habitats in many locations were not identified and could have been more abundant than indicated. The omission errors were most severe for the swamp forest classification, for which the error rate was as high as 50%.

Based on these studies in Central America, a stepwise methodology has been proposed (Roberts et al. 1996) for conducting similar studies, and comprising the following steps: step 1 involves defining the environmental and ecological characteristics of the species under investigation; step 2 defines the spatial scale for analysing the identified environmental determinants using satellite-derived data; step 3 validates remotely-sensed data through field observation; step 4 develops predictions based on the associations identified in steps 1 to 3 and finally step 5 tests those predictions against field data.

In India, Sharma et al. (1996) used remotely-sensed data obtained from Indian satellites to test the feasibility of using such data to identify larval mosquito habitats in 6 study sites in and around Delhi. Multi-date IRS 1A and B, LISS-II satellite data (similar to Landsat TM) were collected and false colour composite (FCC) images generated, with a spatial resolution of 36.5 m. Land cover was assessed using a supervised classification based on training sets developed from field observation. Larval samples were obtained from water bodies by dipping and adult mosquito collections were made using resting catches from cattle sheds and human dwellings. Seasonal changes in mosquito population densities were compared with changes in water and vegetation coverage and human settlements. The results indicated that suitable habitats can be identified albeit with a limitation of resolution determined by the images obtained, in this case 36.5 m. Seasonal and spatial changes in mosquito density were positively correlated with changes in the environmental variables of water and vegetation cover, but not at all study sites. It was noted that peak mosquito densities occurred in the post-monsoon season, however, it is during this period that extensive cloud cover prevents clear satellite images being obtained, restricting land cover analysis during this period. Diuk-Wasser et al. (2004) encountered a similar constraint in Mali, where early stages of rice growth are known to coincide with peak larval production, but during this period clouds often obstruct Landsat sensor data.

Satellite imagery was combined with a GIS to investigate the possibility of identifying areas at risk of eastrern equine encephalomyelitis virus transmission in Massachusetts, USA, as determined by abundance of the enzootic vector *Culiseta melanura* and of six possible epidemic-epizootic vector mosquitoes (Moncayo et al. 2000). Data from the New England Gap Analysis Project, which has developed base vegetation maps from Landsat Thematic Mapper data combined with aerial videography data, were used to generate a map of landscape elements at epidemic-epizootic foci. GIS technology was used to determine the proportion of landscape elements surrounding 15 sites from where human and horse cases had been reported. Abundance data for *Culiseta melanura, Ochlerotatus canadensis, Aedes vexans, Culex salinarius, Coquillettidia perturbans, Anopheles quadrimaculatus,* and *Anopheles punctipennis* were obtained using ABC traps (American Biophysics Corporation). Traps were equipped with a photosensitive flickering light that responds to changes in light intensity and starts operating at dusk, and compressed CO_2, which was continuously emitted at 500 ml/min from a storage tank. Two ABC traps were operated at each case site during two consecutive nights each week. The relationships between vector abundance and landscape proportions were analysed using stepwise linear regression, which indicated that wetlands were the most important landscape element, accounting for up to 72.5% of the observed variation in the host-seeking populations of *Ochlerotatus canadensis*. Conifer dominant was the landscape element that accounted for 52.5% of the observed variation in *Aedes vexans* abundance. The conifer element was negatively correlated with mosquito abundance. When wetland was added to the model, the R^2 value increased from 0.525 to 0.700, and the r value increased from 0.725 to 0.836. Variation in abundance of *Culiseta melanura* populations was best accounted for by wetlands ($R^2 = 0.482$). Stepwise linear regression was unable to devise models to predict the abundance of *Coquillettidia perturbans, Culex salinarius, Anopheles quadrimaculatus,* or *Anopheles punctipennis*.

Barker et al. (2003) used a combination of geo-referenced ovitraps and remotely sensed satellite imagery to develop a Bayesian decision-rule model to predict oviposition activity and abundance of *Aedes albopictus* and *Ochlerotatus triseriatus* in Virginia, USA at a spatial resolution of 28.5 m. The authors concluded that the model was moderately successful. The fact that the model was both developed and tested using the same dataset, rather than developed from a previously existing independent data set was cited as a possible reason for the moderate success of the model. Overall, county-wide accuracy of prediction of oviposition activity ranged from 55 to 79% for *Ochlerotatus triseriatus*, with the highest probabilities for high oviposition activity in forested areas, the preferred oviposition site for this species; and 70–94% for *Aedes albopictus*, with highest accuracies in urban areas, the preferred habitat of this species. Accuracy of predictions tended to decrease as the season progressed and oviposition activity increased.

Diuk-Wasser et al. (2004) assessed the usefulness of remotely sensed data for the identification of malaria vector breeding habitats in an irrigated rice growing area in Mali. Landsat data are frequently obscured by cloud cover during the early part of the rice-growing season, when larval abundance is highest, so attempts were made to base a classification on two Landsat Enhanced Thematic Mapper (ETM)+ scenes acquired in the middle of the season and at harvesting times, extrapolating back to infer which rice growth stages were present earlier in the season. A maximum likelihood supervised classification was performed using on-the-ground classification of 100 m strips of land in the study area. Rice was distinguished from other land uses with 98% accuracy. Three age cohorts of rice crop could be discriminated with 84% accuracy and two age cohorts with 94% accuracy.

In Kenya, Jacob et al. (2005) assessed the usefulness of multispectral thermal imager (MTI) satellite data in the visible spectrum at 5 m resolution to enhance predictions of urban mosquito densities. Larval surveys were conducted in Malindi and Kisumu towns and georeferenced data were compared with data obtained from the MTI images. Poisson and logistic regression models were developed using *Anopheles* larval counts. For the Poisson regression, shade was the most important factor for the Malindi and combined Malindi and Kisumu datasets, while presence of domestic animals was the best predictor for Kisumu larval data. In the logistic regression models, habitat size was the most important factor for Malindi, domestic animals for Kisumu, and shade for the combined data set. All significant determinants of larval abundance were obtained from field-collected data, rather than from the satellite data, and the authors concluded that satellite-derived data at a resolution of 5 m should not replace field surveys as a method of determining larval habitat distribution and larval abundance.

Shaman et al. (2002) developed a dynamic hydrology model, to predict flood and swamp water mosquito abundances in watersheds in New Jersey, USA. The model resulted in a mosaic of surface wetness depicting the spatial variability of terrain (topography, vegetation, and soil type) and meteorologic conditions at a spatial resolution as high as 10 m × 10 m. A time series of local mean water depth could be calculated for any pixel within the watershed. The model was compared with New Jersey light trap catch records of adult mosquitoes from northern New Jersey, consisting of a 13-year time series of daily mosquito counts (May through September) from seven sites, and a 15-year time series of daily mosquito counts (June through September) from a single site in the Great Swamp National Wildlife

Refuge. Trap locations were identified with a global positioning sensor. Water table depths were used to create indices of local wetness, and time series measures of the surface wetness of the immediate area surrounding each of the mosquito collection sites. Time series regression analyses were used to determine the tendency and strength of the association between mosquito abundances and indices of local wetness at each site. For the full 15-year record of trap catches at the Great Swamp site, the index of local wetness was positively correlated with a 10- day lag ($P <$ 0.0001) for both *Aedes vexans* and *Anopheles walkeri*. While statistically significant, the variation in the indices of local wetness failed to explain more than 12% of the variance in abundance of any of the mosquito species analysed. Logistic regression was also performed to investigate if mass emergence ($\geq$ 128 individuals per trap per day) of *Aedes vexans* could be predicted. Model fit was significant ($P < 0.0001$) and yielded the following equation for the probability of a mass emergence of *Aedes vexans* 10 days later: p ($\geq$ 128 *Aedes vexans*) = (1 + exp (4.701 – 0.00804* index of local wetness)). A significant positive correlation was also obtained for *Anopheles walkeri* and a significant negative correlation for *Culex pipiens*. The authors proposed several advantages of this methodology over related satellite-data derived models, including: the dynamic model models the actual aquatic environment used by the mosquitoes, not a filtered proxy; it offers continuous real-time prediction of mosquito abundance; it identifies the location of potential breeding habitats at a very fine scale; and the model is readily coupled to global climate models, allowing additional medium- and long-range forecasts of mosquito abundances.

Much recent work has focused on extending the utility of satellite data beyond predicting larval habitats and mosquito abundance to predicting disease transmission risk. Thomson et al. (1996) used meteorological data, cold-cloud duration data from Meteosat, and NDVI data, to examine the environmental variation in relation to relatively small-scale spatial and temporal variations in malaria epidemiology in The Gambia. NDVI values of –0.015 to 0.2 were found to be associated with alluvial soils and tributaries of the River Gambia, breeding sites for the vectors *Anopheles gambiae* s.l. and *Anopheles melas*. NDVI values also clearly identified a 600 ha rice-growing area that was known to be an important larval habitat for *Anopheles arabiensis*. In Kenya, Hay et al. (1998) attempted to correlate NOAA-AVHHR and Meteosat HRR data with previously collected data on childhood malaria cases. Temporal Fourier processing revealed the NDVI lagged by one month to be the variable most consistently and significantly correlated with malaria admissions (mean adjusted $r^2 = 0.71$, range = 0.61 – 0.79).

Elnaiem et al. (2003) used a GIS to study variations in visceral leishmaniasis transmission in relation to environmental variables, including rainfall, vegetation status, soil type, altitude, distance from river, topography, wetness indices, and average rainfall estimates in Sudan. Logistic binary regression analysis produced a model that correctly predicted all observed positive sites as visceral leishmaniasis-endemic villages, however, the model failed to predict any of the 33 villages negative for visceral leishmaniasis.

Hay et al. (1997) reviewed the impact of remote sensing on the study and control of invertebrate intermediate host and vectors for disease. A review of the potential use of satellite imagery in describing and predicting malaria epidemiology, as well as some of the challenges, including the relative paucity of available high quality malaria epidemiological data for training the models based on remotely-sensed data has recently been produced by Rogers et al. (2002). This paper provides a useful summary of the types of satellite sensors and their resolutions, as well as the types of data processing and analysis used to interpret the images. Ceccato et al. (2005) have recently reviewed the usefulness of remote sensing for assessing and monitoring malaria risk, and they include a review of the factors involved in malaria transmission and how these can be tracked using remote sensing technologies.

Hugh-Jones (1989) discusses the application of remote sensing in the study of disease vectors, paying particular attention to the resolution obtained from the various satellites. Examples are given of the technique for detecting breeding places of *Ochlerotatus sollicitans, Psorophora columbiae*, and malaria vectors, including *Anopheles freeborni*. Dale et al. (1998) discuss the application of several remote sensing techniques, including colour infrared photography, thermal imaging and synthetic aperture radar to the study of mosquito vectors, focusing primarily on *Ochlerotatus vigilax*, the vector of Ross River virus in Australia. Advantages and disadvantages of the methods are usefully tabulated. Additional reviews of the application of remote-sensing techniques to the study of arthropod vectors have been published recently, including Hay et al. (1996, 1998), Thomson and Connor (2000), Thomson et al. (1996), and Washino and Wood (1994).

The recent development of several computer software packages for the storage and analysis of spatial data, alongside the availability of relatively cheap, handheld global positioning system (GPS) satellite receivers, has allowed for the combination of remote sensing data with other georeferenced data within a geographic information system. Geographic Information Systems include both spatial data (locations) and descriptive data (attributes). Use of a GIS has the potential to further strengthen the usefulness of remotely sensed data for studying disease epidemiology and guiding disease and vector management and control. Geographical information systems (GIS) incorporating climatological information have been used to model the distribution of *Aedes albopictus* in Japan (Kobayashi et al. 2002). More than 200 sites in 26 cities, towns, and villages in the Tohoku district of Japan were surveyed for mosquito larvae and pupae and locations of collection sites were input into the GIS. Local climate data for annual mean temperatures and January mean temperatures were obtained and compared with the larval distribution records. A strong correspondence was observed between areas with mean annual temperatures above 11°C and presence of *Aedes albopictus*.

Kutz et al. (2003) conducted a geospatial analysis of the location of used tyre dumps, presence of West Nile virus-infected mosquito pools obtained from light-trap data, location of infected birds, humans and horses, and land cover and land use, within a GIS in Maryland, USA. Data analysis confirmed the presence at the same locations of infected mosquitoes, birds, humans, horses, and other vertebrate hosts, presence of *Aedes albopictus* and *Ochlerotatus japonicus japonicus*, and susceptible vertebrate hosts. Land cover within 1.6 km of positive mosquito pools was dominated by urban and suburban development.

Liebhold et al. (1993), Clarke et al. (1996) and Kitron (1998) have all reviewed the application of GIS and geostatistical methods for describing spatial distributions to insect ecology and disease epidemiology. Many examples of the application of GIS and spatial statistics-based approaches to disease and vector monitoring have been published since the second edition of this book, including for malaria vector larval habitats (Sithiprasasna et al. 2003*a, b*), distribution of adult malaria vectors (Sithiprasasna et al. 2003*b*), arbovirus surveillance (Kitron et al. 1997; Lothrop and Reisen 1999), dengue (Ali et al. 2003; Sithiprasasna et al. 2004), among others.

Other methods of predicting species distributions

Levine et al. (2004*a*) describe a method for predicting the distribution of member species of the *Anopheles gambiae* complex using ecological niche modelling. Fourteen environmental data layers were initially assessed for inclusion in the model: annual mean temperature, annual mean maximum temperature, annual mean minimum temperature, daily temperature range, frost days, topographic aspect, flow accumulation, topographic index, annual mean precipitation, wet days, elevation, and vapour pressure, tree cover, and land-use/land-cover. Of these elevation and vapour pressure were excluded from the final model. An existing *Anopheles gambiae* s.l. dataset (Coetzee et al. 2000) was supplemented with additional data from published literature to give a total dataset of 581, 501, and 86 unique occurrence points for *Anopheles gambiae, Anopheles arabiensis,* and *Anopheles quadriannulatus* respectively, where an occurrence was defined as a georeferenced collection site for a particular species. Fourteen African countries were selected at random from the total set of 28 countries for which sufficient data were available. Data from these fourteen countries were then used to build models and to predict species' distributions in the remaining 14 countries. Ecologic niche models were created using a Genetic Algorithm for Rule-set Prediction (GARP), an algorithm specifically designed for the study. After creating 100 GARP models, a final predicted map was created by summing the five best-subset models. Test data, which were not used for model building, were overlaid on predicted distributions, and model predictions assessed using a chi-square analysis. The final maps produced from the model were in broad agreement with the distributions modelled by Lindsay et al. (1998) based on climate suitability and predicted general sympatry of *Anopheles arabiensis* and *Anopheles gambiae*, with *Anopheles arabiensis* being more widely distributed, particularly in south-central Africa, than is usually accepted. The distributions produced differ from those published by the WHO (White 1989) in that they predict the presence of *Anopheles arabiensis* across much of central Africa (e.g., Angola, Congo, Democratic Republic of Congo, and Gabon). The Levine et al. (2004*a, b*) distributions also predict the presence of *Anopheles gambiae* in northeastern Africa, at lower elevations in Ethiopia, northern Kenya, southern coastal Somalia, and southeastern Sudan. The authors noted that predictions of the distribution of *Anopheles gambiae* and *Anopheles arabiensis* produced from the GARP analysis also differed from those published by Rogers et al. (2002), which relied on satellite imagery techniques and ascribed those differences to the inability of regression-based inferences as used by Rogers et al. (2002) to predict distributions in broad areas for which limited samples are available. Levine et al. (2004*b*) also attempted to predict the likely distribution of members of the *Anopheles quadrimaculatus* complex in the United States using the same GARP method of Stockwell and Peters (1999) as applied to *Anopheles gambiae* above. Preliminary analysis demonstrated that environmental variables with the greatest effects on the distribution were daily temperature range, precipitation, mean minimum temperature, mean maximum temperature, mean temperature, vapour pressure, aspect, elevation, and slope. Preliminary distribution maps were prepared and additional mosquito samples were analysed from states considered by the authors to be boundary and undersampled regions. Using the maps, a previously undetected population of

Anopheles diluvialis was found in coastal North Carolina and Massachusetts, an unsampled region in the northernmost part of the prediced range of this species, demonstrating the model's accuracy.

Bayoh et al. (2001) attempted to map the distribution of chromosomal forms of *Anopheles gambiae* s.s. in West Africa, using climate data. Published distribution records for the chromosomal forms were linked to climate data obtained from a computerised environmental database (Corbett and O'Brien 1997). The percentage of sites correctly classified as absent or present for the different chromosomal forms in the pilot model (utilising half of the dataset of 144 sites) was 94.4% for Forest, 87.5% for Mopti, 95.8% for Savanna and 84.7% for Bissau (Bamako chromosomal form was excluded from the study due to relative lack of data). When tested against the 72 sites not used to construct the model, presence was accurately predicted for 87.5% of Forest, 90.3% of Mopti, 80.6% of Savannah and 80.6% of Bissau forms. When the data were combined the models correctly predicted 87.5% of Forest, 88.9% Mopti, 87.5% of Savanna and 86.8% of Bissau.

Lindsay and Bayoh (2004) later attempted to map the distribution of *Anopheles gambiae* and *Anopheles arabiensis* using climate data and the known differences in preference of these two species for more humid or more arid environments respectively. Existing distribution maps of the two species in Africa (Davidson and Lane 1981; White 1989) were digitised and incorporated into a Geographical Information System (GIS) containing climate surfaces for the continent. The climate surfaces were produced from meteorological data collected from 621 African weather stations and interpolating between the stations. Summaries of climate variables (e.g. temperature, precipitation (P) and potential evapotranspiration (PE)) in the wettest 5 months of the year were used in preference to annual rainfall. The relationship between P/PE and relative abundance of the two species was used to map habitat suitability and the model was tested against a dataset from Tanzania, achieving a correct prediction of relative abundance at all 14 sites. The approach was also extended to mapping the chromosomal forms of *Anopheles gambiae* s.s. Logistic regression models derived from a stratified random sample of half of a dataset of characterised sites in West Africa correctly classified over 80% of sites for the presence or absence of each chromosomal form. When these models were tested against the sites excluded from model development, they were correct at over 80% of sites. The combined data produced models that were correct at over 86% of sites. Limitations of the approach acknowledged by the authors included the fact that the model only predicted relative, not absolute abundance, and could only make predictions at relatively coarse spatial scales, whereas mosquito ecology and malaria transmission are known to be highly variable at small spatial scales.

Collecting techniques

Dippers

The dipper is undoubtedly the most commonly used tool for collecting mosquito larvae and pupae from a wide variety of habitats. Dippers vary in size and shape but are often white inside to facilitate detection of larvae. Local availability often dictates the size of dipper used, which unfortunately is not always clearly recorded with the results of surveys. Soup ladles, 9–10 cm in diameter and holding 100–150 ml water, have been used (Mori 1989; Service 1968) (Fig. 3.1a) as have dippers about 15 cm in diameter and 2–3 cm deep (Russell et al. 1963; Trapido 1951; Wada et al. 1971a, b), or those having capacities of 80 ml (Fischer and Schweigmann 2004), 240 ml (Klinkenberg et al. 2003), 350 ml (Bøgh et al. 2003; Sandoski et al. 1987), 375 ml (Bailey et al. 1980), 400 ml (Andis et al. 1983; Claborn et al. 2002; Walton and Workman 1998), or 450 ml (Jaal and Macdonald 1993; Campos et al. 2004). Many workers, especially those in North America, have used what is termed the 'standard pint dipper' (Amerasinghe and Ariyasena 1990; Amerasinghe and Indrajith 1994; Dixon and Brust 1972; Downing 1977; Hagstrum 1971a; Lemenager et al. 1986; Markos 1951; Roberts and Scanlon 1974; Scholefield et al. 1981), whose capacity is recorded as varying from 390 to 500 ml, although more usually it is given as 473 ml.

Forattini et al. (1993, 1994a, 1994b) used a 500 ml dipper to sample mosquitoes in rice fields in Brazil. Larger dippers such as those measuring 18 cm in diameter and 10 cm deep (Imai et al. 1988), or having capacities of 1 litre (Croset et al. 1976; Maire 1982; Papierok et al. 1975) or 1 quart (Shemanchuk 1959) have also been employed. Grillet (2000) used a long-handled dipper with a capacity of 1.5 litres to sample mosquito larvae in Venezuela. Sometimes the bottoms of dippers have plastic or metal screen bottoms (Rajagopalan et al. 1975) (Fig. 3.1b). To reach relatively inaccessible water long handles can be attached to dippers. The standard pint dipper is particularly well suited for this because it has a hollow handle, into which a wooden stake, piece of dowling or bamboo cane can be inserted. Short lengths of 16-mm diameter dowling can be joined together with plastic couplings as used by plumbers to extend the length of the dipper handle. Readers interested in the design of dippers used by workers in the early part of the twentieth century (e.g. Boyd 1930; Russell and Baisas 1935; Russell and Santiago 1932) should consult the second edition of this book.

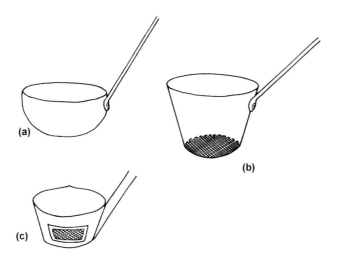

Fig. 3.1. Larval dippers: (a) Small soup ladle; (b) 'pint dipper' with screen bottom; (c) dipper with metal guard protecting mesh screen (Service 1993)

There is little standardisation in how the dipper is used to take samples. It can be skimmed at an angle quickly through the water and removed before it overflows, or alternatively be lowered gently at one point and water allowed to flow into it. In very shallow water the dipper must be pressed firmly into the bottom mud and debris and water allowed to flow into it.

To minimise damage to rice fields from trampling Markos (1951) used wooden planks to construct elevated paths that stretched across flooded fields from bank to bank. Samples with a pint dipper attached to a 5-ft handle could therefore be taken along the banks (levees) and also from the middle areas of rice fields. By taking samples from certain collecting stations in the morning and from others in the afternoon Markos (1951) avoided casting shadows on the water surface, which are liable to adversely affect the numbers of larvae caught. In studying the abundance and distribution of larvae of *Culex tarsalis* and *Anopheles freeborni* in Californian rice fields Lemenager et al. (1986) walked around a rice paddy field taking three samples (473-ml dipper) every 10 paces (6 m), except for along one edge where no samples were taken. Thus in effect two samples were taken along each of three sides and one at each of the two corners making a total of eight sampling points.

In a study of the effects of artificial wastewater processing marshes in southern California, USA Walton and Workman (1998) conducted an initial larval survey by taking five 400 ml dips at each of 32 points along the periphery of the marsh. After the initial survey, the number of sampling stations per marsh was reduced to 16 and samples were taken weekly using 3 dips from a 400 ml dipper along 2 m transects at each of the 16 collecting stations. Larval *Culex* populations in vegetated habitats of one-phase marshes were significantly larger than in the vegetated zones of 3-phase marshes, with numbers per dip differing by 5- to 10-fold during July, and 2- to 3-fold thereafter.

Claborn et al. (2002), working in the Republic of Korea, measured the circumference of larval habitats, divided this measurement by 30 and then took 30 dips from 30 equidistant points around the circumference using a 400 ml dipper. In northern Thailand, Overgaard et al. (2002) attempted to characterise *Anopheles minimus* larval habitats by correlating larval density and features of the habitat. The dipper used was 13 cm in diameter and 6.5 cm deep. The stream sampled was divided into three zones and each zone was divided into sections 10 m long. Larvae were sampled from each section on 2 consecutive days per month. On each sampling day, one dip per meter was taken from the edge on each side of the stream in each section. Thus, for each collection day a total of 20 dips was taken in each section. The density of *Anopheles minimus* and other *Anopheles* species were $\sqrt{x}$ transformed and tested for normality using the Shapiro-Wilks statistic, W. Because larval densities were not normally distributed, a new data set was created by combining three adjacent sections. Thus, the 200 sections were reduced to 67 sections that were each 30 m long. The mean density of the three sections was $\sqrt{x}$ transformed and habitat variables were averaged. Pearson's multiple correlation analysis of larval and habitat data were performed for dry and wet seasons separately. Habitat variables were selected for the regression analysis by removing one of two correlated variables. Finally, stepwise regression with forward selection was used to find the best prediction model. The mean density of *Anopheles minimus* per 30 m section was 23.0 larvae per 120 dips (19.1–27.0, 95% CI) in the dry season and 12.8 larvae per 120 dips (10.6–15.0, 95% CI) in the wet season. The best prediction model for *Anopheles minimus* density in the dry season was: y_{min} = 0.1980–0.1733* (water current velocity)–0.0317* (height of large-leaved emergent aquatic plants)–0.0249* (height of riparian small-leaved plants) + 0.0192* (height of riparian ferns)–0.0170* (height of left and right stream bank) + 0.0017* (mean water depth of left and right sides of stream)–0.0016* (% of water edge with wood) + 0.0011* (% of water edge with debris), ($R^2 = 0.51$, $P < 0.001$). The R^2 value of the fitted model for *Anopheles minimus* in the wet season was only 34.5%. The most important single variable associated with larval density was water current velocity (partial $R^2 = 17\%$).

Bøgh et al. (2003) took samples using a 350 ml dipper at 100 m intervals along four transects positioned in flooded alluvial soils and among mangrove forest along the banks of the river Gambia. A delay of 3 minutes following arrival at each collecting site and prior to dipping was used to allow mosquito larvae to return to the water surface. At each point along the transect, 10 dips were taken from among plants and roots within a 5 m radius. Klinkenberg et al. (2003) constructed two transects across each of ten rice fields in Mali using lengths of rope. Ten dips were taken at five points along each transect giving a total of 100 dips per rice field. The position of the transects was changed daily. Larval density data were log-transformed and analysed in relation to time since planting, water management regime, plant height and density.

Sampling for *Anopheles gambiae* mosquitoes in the Kenyan highlands was conducted along a 600 m linear transect crossing river, swamp, cultivated land and flooded ditches by Minakawa et al. (2004). Sampling for larvae was carried out at 1 m intervals using a 350 ml dipper, and a presence/absence classification was used for each site. Anopheline larvae were encountered in 191 discrete sites and *Anopheles gambiae* s.s. was present in 104 of these sites.

A variety of other utensils have also been used to dip for larvae, e.g. rectangular enamel pie dishes, photographic trays, frying pans, plastic or enamel bowls, and household metal strainers. In very shallow pools and hoof prints conventional dippers may be too large, and spoons may be used to collect larvae (Knowles and Senior White 1927; Russell and Baisas 1935; Russell et al. 1963).

Larval concentrators

Larvae collected from a number of dips are frequently pooled, and various simple devices have been designed to strain the larvae from the water of the samples to concentrate the catch. An obvious disadvantage of any method that combines catches from several samples is that the variances cannot be calculated.

Earle (1956) reported that even with experienced collectors as much as 23% of the younger instars and 4 and 8% of the later instars and pupae respectively of mixed populations of *Anopheles, Culex* and *Aedes* could be missed when the immature stages were pipetted from dippers. To overcome this and to reduce the time needed to remove the catch from dippers he constructed an automatic device for collecting aquatic specimens, termed ADCAS. A 6-in diameter metal funnel is soldered into the top of a 7-in high metal cylinder (Fig. 3.2*a*). The 1-in diameter opening at the opposite end of the funnel is soldered to a metal screw cap of a bottle of about 120-ml capacity. Two rectangular 1 × 2¼ in openings are cut from the sides of the funnel and covered with 96-mesh screen. A 7/16-in hole is drilled into the side of the funnel and a copper inverted U-tube acting as a siphon is inserted, with one end placed inside the collecting bottle. A detachable cylinder of 96-mesh screen 9/16 in in diameter is soldered onto a bronze collar, which is screwed on to the threaded end of the copper siphon. Four bronze wire vertical guards are soldered around the outside of the cylinder to help protect it from damage. A removable 6½-in diameter sieve made from about 8-mesh gauze is also constructed. Successive samples collected by a dipper are poured through the funnel and when the collecting bottle fills up the siphon automatically lowers the water in the bottle to the level of its intake. The coarse circular sieve can be placed over the top of the funnel to exclude stones and debris. When sufficient samples have been poured into the bottle it can be replaced with another. In addition to 1st instar larvae, mosquito eggs, including those of *Anopheles* species, are retained in the strainer.

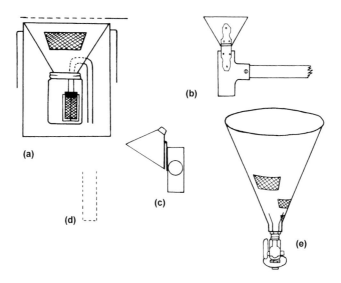

Fig. 3.2. Larval concentrators: (a) 'ADCAS' automatic siphoning type (after Earle, 1956); (b)–(d) Husbands' (1969) concentrator, (b) wooden handle supporting brass tee and funnel; (c) funnel folded back; (d) mesh screen vial; (e) cone fitted to gas valve tap (after Fanara 1973)

Warren and Eddleman (1965) poured their samples through a plastic strainer, which was sufficiently light to be held in the hand. The concentrator of Warren and Eddleman (1965) consists of a 1-gal white plastic bottle with one end removed and replaced with a ⅛-mesh sieve held in position by three 1¾-in bolts. A length of heavy insulating wire is looped through the sieve to form a convenient handle. A hole is cut in one side of the plastic bottle and covered with 35-in mesh screen, which can be fixed in position by melting the plastic along the edges of the hole with a hot soldering iron. The contents of several dips are poured through the coarse sieve, which retains large unwanted material while mosquito larvae and fine debris are retained in the bottle as the water drains out through the mesh window. Husbands (1969) developed a much smaller larval concentrator (Fig. 3.2b) consisting of a metal funnel attached to a hollow brass tee by a hinge that can be folded away from the upright part of the tee (Fig. 3.2c). A tubular vial of 60-mesh wire (Fig. 3.2d) is dropped into the tee to retain the larvae when the sample is poured through the funnel. The opening of the mesh vial can be closed with a cork and placed in a tube of alcohol. The contents from a number of dips are poured through into a single vial and the larvae counted and identified on return to the laboratory. A coarse sieve can be placed just below the entrance of the funnel to exclude debris (Miura et al. 1970; Reed 1970). When a comparison was made between removing larvae and pupae from each dip with a glass pipette and using the concentrator Miura et al. (1970) found that both methods gave the same species compositions and proportions of different instars, but the concentrator speeded up sampling almost threefold. Reed and Husbands (1970) found the concentrator particularly useful in larval surveillance surveys in California, USA. Nagamine et al. (1979) modified the larval concentrator of Husbands (1969) to fit on an upright pipe thrust into the mud of rice fields.

Hagstrum (1971b) used the very simple procedure of pouring the contents from 10 dips through a muslin cloth, which was then tied up and placed in 5% formaldehyde. Reisen et al. (1989) also used cloth to concentrate larvae. They found that only about 19% of 1st instar larvae of *Culex tarsalis* were lost when dipper samples were poured through organdy cloth.

Fanara (1973) considered that specimens could be lost from the concentrator of Husbands (1969) if algae occluded the screened containers. Because of this and because he found it was often difficult and time consuming to collect material from the muslin cloth concentrator of Hagstrum (1971b), a new concentrator comprising a funnel and gas valve assembly was developed (Fig. 3.2e). The upper part of the concentrator consists of a wide-mouthed (10½-in) polyethylene funnel which has three rectangular sections cut out and replaced with 80-mesh screening. The stem of the funnel is cut to leave about a 1-in length, which fits tightly over the 1⅛-in diameter end of a plastic coupling. Four brass pins are cut so as to pass through the funnel and into the plastic coupling, but not penetrate it. They are heated red hot before insertion. The gas valve is firmly screwed into position. A plastic lid of a screw-on container (e.g. 40 dram vial) has a hole cut from the middle and is held onto the base of the valve assembly by two nuts. In operation the valve is closed and water samples poured into the funnel, which is gently swirled to throw out water through the mesh screens. The valve is then opened and the residue collected in a container, which is screwed to the plastic lid. A small quantity of 95% alcohol can be used to flush out any larvae remaining in the funnel.

Stokes and Payne (1976) describe a simple and inexpensive larval and pupal concentrator for collecting large numbers of specimens for laboratory and other uses. Fig. 3.3a shows an 8-in diameter metal funnel whose opening is covered with wide spaced mesh (hardware cloth) and with the lower end, with the spout removed, soldered to the lid of a Kilner (Mason) jar. A 16-in length of ⅜-in diameter copper tubing is bent into a J-shape and passed through a hole in the metal funnel near the top of the Kilner jar lid. Other details are as shown in the figure. The concentrator can, if required, be fixed to a stake in the field.

Fleetwood et al. (1978) made a larval concentrator from a Mazola oil bottle. The bottom was removed and covered with coarse (0.64 cm) screening and became the top of the concentrator. Three 2.5 × 3.8-cm pieces were cut from the sides of the plastic bottle near the handle and covered with fine (32-mesh) netting. Dipper samples were poured into the top of the container; the position of the fine netting permitted only 50 ml of water to be retained. The bottle cap was then removed and the plastic slide of a WHO adult mosquito insecticide susceptibility kit was fixed in position. When this transfer slide was moved to the open position the water (50 ml in the bottle) passed into a 100-ml vial fixed below which contained 50 ml 95% ethanol. In sampling mosquito larvae in rice fields Chambers et al. (1979) poured the contents of five standard dippers through this concentrator of Fleetwood et al. (1978), but having the following modifications. The wide end of a cut down plastic funnel was glued to the underside of a WHO insecticide adult plastic exposure tube. A lid from a 35-ml vial with a large hole drilled through the middle was then glued to the narrower end of the plastic funnel. This allowed the use of smaller (35-ml) tubes than those used by Fleetwood et al. (1978). These tubes contained about 10 ml 95% ethanol.

Driggers et al. (1978) although appreciating the value of previous larval concentrations developed over the preceding years (Earle 1956 to Fanara 1973) considered there was room for improvements.

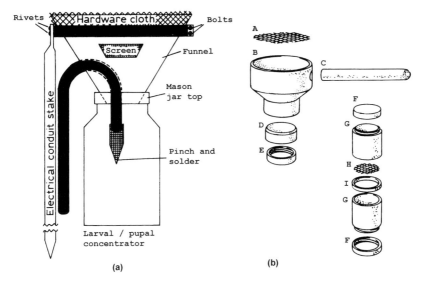

Fig. 3.3. Larval concentrators: (a) Stokes and Payne (1976) type; (b) Driggers et al. (1978) exploded diagram showing A–E concentrator base unit with B—funnel, C—handle, D—bushing, E—thread, F–I collection and storage assembly with F—screw cap, G—tube body, H—collection screen, I—connection ring

They wanted a compact, light-weight, portable device that could be used with the standard pint-dipper. Their system comprised a funnel concentrator (Fig. 3.3b A–E) and a collecting and storage assembly unit (Fig. 3.3b F–I). They made their concentrator from items commercially available in the USA, alternatives will have to be found for construction in other countries. They used a PVC 3-in to 1.5-in reduced coupling (B) to act as the receptacle funnel, to which a 6-in long PVC rod is fixed as the handle (C). Coarse metal screening (A) is press-fitted into the mouth of the funnel. A PVC 1.5-in diameter bush (D) is bevelled at the top (48°) so that it can be glued into the base of the funnel. A 0.453-in deep, 1.734-in diameter hole is drilled into the bottom of the bush to form a seat for the thread insert (E). This is made by cutting the top from the screw-cap of a dollar-sized coin storage tube, after which it is glued into the bottom of the bush (D).

The collection unit consists of two 2-in clear plastic tubes (G) with screw caps (F) (commercially available silver dollar storage tubes), a 0.278-in diameter fine mesh (40 × 40-mesh gauze made of 0.010-in diameter wire) collecting screen (H), and a 0.375-in connector ring (I) cut from a clear plastic tube. The various components are glued together. The top and bottom screw caps (F) are removed from the collecting tube and the tube screwed into the concentrator at E. After the required number of dips have been poured through and larvae become stranded on the screen (H), the top cap (F) of the collection tube is screwed onto the bottom of the collection tube and water poured through the funnel to float the larvae. The tube is then unscrewed from the concentrator and capped. Both laboratory and field tests showed that virtually all larvae (1st–4th instars) poured through were collected and retained in the collecting tubes.

In Canada, Scholefield et al. (1981) made use of a very simple technique of just pouring larvae from a pint dipper into plastic cups with a mesh screen near the top to concentrate their samples. In West Virginia, USA, Joy and Hildreth-Whitehair (2000) poured samples from a dipper through a handheld strainer with a mesh size of 1 mm. Trapped larvae were removed from the mesh using fine jeweller's forceps and transferred to bottles containing 70% ethanol.

Campos and Garcia (1993) describe a simple apparatus for separating live mosquito larvae and pupae from murky water or water containing a lot of debris. The centre of the screw-cap from a wide-mouthed glass jar is removed and an inverted plastic funnel is glued to the remaining portion of the screw-cap using waterproof glue. Small holes are made around the base of the funnel using a heated nail. The screw-cap with attached funnel can then be easily screwed onto and released from the glass jar, to which the water sample is added. The whole apparatus containing the water sample is placed in a large bucket or similar container, which is then filled slowly with clean water to a depth at least 5 cm greater than the height of the top of the inverted funnel. The clean water enters the apparatus through the holes in the base of the plastic funnel. Pupae and larvae exit the apparatus through the narrow end of the funnel and are collected from the surrounding clean water. In trials, 81% (± 2.71) of larvae and pupae were successfully extracted from five samples obtained from tyres after 15 minutes. This proportion increased to 99.0% after 60 minutes.

Containers for larvae

Larvae and pupae collected by dipping or by any other method can conveniently be pipetted into small glass or plastic screw-cap tubes (2 × 9 cm, 30-ml capacity) for transportation to the laboratory. Alternatively larvae can be placed in 'Whirl-Pak' plastic bags (7.5 × 18 cm, 150-ml capacity) (Janousek and Lowrie 1989), or in snap-sealing plastic bags of a similar size. When large numbers of larvae and pupae are collected they can be placed in a variety of large plastic or glass bottles or plastic cups with snap-on lids (Scholefield et al. 1981). Care must be taken that larvae are not placed in the sun or allowed to heat up by any other means. A commercial, or home-made, cool box is useful for keeping collections cool. Prolonged shaking of the water in the tubes or bottles during transportation may cause larvae or pupae to drown. To overcome this, immature stages can be placed on very wet filter paper overlying cotton wool in petridishes, or on wet mud. In Canada Scholefield et al. (1981) placed their larvae in a cool box, and made 5-min stops every 30 min to allow larvae to surface and breathe.

Recording and analysing results from dipping

Sometimes the number of larvae/dip is recorded separately, but frequently only the total number of larvae collected from a known number of dips is recorded, such as usually happens when a larval concentrator is used. More information is obtained by the former procedure, because in addition to being able to calculate the mean number of larvae per dip ($\overline{x}$), the degree of variability between the samples is measured. This variability is usually expressed as the sample variance (S^2) or standard error of the mean. For example if 100 larvae are collected in 100 samples (i.e. dips) and the unlikely situation occurs in which one larva is caught in each dip, the mean number of larvae per dip is 1 and there is no variability. In this case the sample mean ($\overline{x}$) would be a very accurate estimate of the true, but unknown, population mean (μ). However, if all 100 larvae are collected in a single dip the mean is still 1, but the standard error is 10, and the variance (S^2) is 100. A variance greater than the mean indicates a highly aggregated or clumped larval distribution. Because there is so much variability between sample counts more samples must be taken if any reliability is to be credited to the sample mean ($\overline{x}$) as an estimate of the population mean (μ). The number of larvae in each sample, whether it is from dips, quadrats, cylinders, or aquatic net collections, must be recorded separately if statistical comparisons are required between the mean numbers of larvae obtained from different collections. To be able to associate a run of low counts with adverse habitat conditions the numbers of larvae in each dip must be recorded separately and also in chronological order. A common error in dipping is the tendency to omit younger instars from larval counts.

Russell et al. (1945) made a careful evaluation of the various ways in which the number of *Anopheles* larvae caught by dippers could be expressed. Over a 33-month period they collected larvae of *Anopheles culicifacies*, and other Indian *Anopheles* species, from a variety of habitats, which included wells, irrigation canals, borrow pits and water tanks. Although this work is almost 50 years old their analysis of their data is presented here in some detail because it represents one of the relatively few attempts to critically examine results from dipping. They plotted logarithms of the total number of *Anopheles* larvae caught each month against; (1) time in minutes spent dipping, (2) areas sampled in square feet, (3) geometric means of minutes × square feet, and (4) number of collections (Fig. 3.4). With any of these two variables there was a significant positive correlation indicating that a geometric increase in the number of *Anopheles* larvae results from an increase in collection time, or area sampled. The highest correlation was between catch size and minutes ($r = +0.943$). The relationship is exponential and may be expressed as:

$$y = ab^x \text{ or } \log y = \log a + x(\log b) \tag{3.2}$$

where y = number of larvae caught, x = number of minutes spent dipping (or area, number of collections etc.), a = a constant and b = the regression coefficient.

A multiple equation in which the catch was computed in terms of both minutes and square feet was set up:

$$\log y = \log a + x(\log by_{x.z} + z(\log by_{z.x}) \tag{3.3}$$

where z = square feet sampled, $by_{x.z}$ = regression coefficient of the proportional increase in number of larvae caught per unit increase in minutes for a given number of square feet and $by_{z.x}$ = regression coefficient indicating proportional increase in number of larvae caught per unit increase in square feet for a given number of minutes. The three variables, number of larvae, minutes and square feet were plotted to give a three-dimensional figure and a regression plane (Fig. 3.5). Although the number of *Anopheles* larvae caught depended on both time and area, the best correlation was between larvae and times. The association between larvae and area arises because minutes and feet are themselves correlated, and in fact if minutes are held constant the association between the numbers of *Anopheles* caught and area sampled is lost.

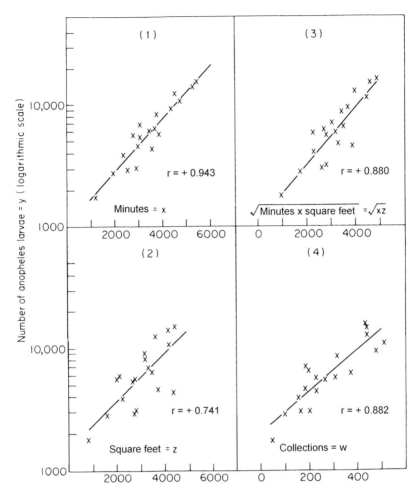

Fig. 3.4. Monthly captures of *Anopheles* larvae in terms of (1)—minutes spent dipping, (2)—square feet sampled, (3)—geometric means of minutes and square feet, and (4)—number of collections (after Russell et al. 1945)

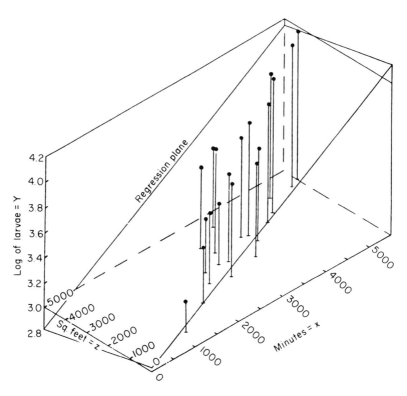

Fig. 3.5. The regression plane of monthly captures of *Anopheles* larvae on minutes and square feet (after Russell et al. 1945)

The data from these collections were then examined in greater detail. The degree of correlation between the numbers of *Anopheles culicifacies* larvae caught and time spent dipping, and also the area sampled was the same ($r = + 0.938$). A slightly greater correlation existed between the numbers caught and the geometric mean of time × area ($r = +0.945$). When the collections of *Anopheles culicifacies* larvae were analysed according to habitat it was discovered that there was a strong positive correlation between the numbers caught from wells and area, but in other habitats the best correlation was between numbers of larvae caught and time. It was finally concluded that minutes were the most universally applicable method for measuring numbers of *Anopheles culicifacies*, in other words when larval densities are high more larvae are caught per unit time. If logarithms of the numbers caught are divided by minutes the ratios become more or less constant due to the close exponential relationship between the two variables. For practical purposes the simple ratio of the number of larvae caught to minutes spent dipping is a good measure of larval density. Larval densities between two or more habitats can be compared by the numbers caught within unit time. In long term surveys there must be care that collecting efficiency does not improve with time, otherwise there may be an increase in numbers caught in unit time not associated with changes in mosquito density.

In Japan Wada et al. (1971a) investigated the best and quickest procedure to measure the relative densities of *Culex tritaeniorhynchus* larvae in different areas containing a number of paddy fields. Using a 15-cm diameter dipper they recorded the number of larvae in each of 10 dips taken from 22 paddy fields. Not surprisingly they found that the larvae were aggregated. Previous sampling had shown that the distribution of larvae mimicked that of the negative binomial model, and Wada et al. (1971b) were able to calculate a common k (see pp. 97–8 for explanation of this exponent). The number of larvae caught in each dip (x) was converted to log ($x + 1$), which normalised the data and permitted the variances of the mean catches within (S_s^2) and between (S_p^2) paddy fields to be calculated. The optimum number of samples (n) to take from a paddy field is derived from the ratio of the within (S_s^2) and between (S_p^2) variances and also the cost (i.e. time involved) of moving to another paddy field to take a dip (C_p) or the cost of taking another dip from the same paddy field (C_s) (Hansen et al. 1953).

$$n = \sqrt{\frac{S_s^2}{S_p^2} \times \frac{C_p}{C_s}} \qquad (3.4)$$

As it was considered that both costs were about the same the optimum number of dips depends only on the ratio of the two variances, S_s^2/S_p^2, which in this instance had the values of 0.0152 and 0.0116, giving a ratio of approximately unity. Consequently only one dip per paddy field was required. The number of paddy fields (N_p) to be sampled within an area to get a required level of accuracy (D), expressed as a decimal, usually 0.1 (i.e. 10%) is given by:

$$N_p = \frac{(S_s^2 / ns) + S_p^2}{\bar{x}^2 \times D^2} \qquad (3.5)$$

which in this instance simplifies to

$$N_p = \frac{0.0268}{\bar{x}^2 D^2} \qquad (3.6)$$

LIVERPOOL JOHN MOORES UNIVERSITY
LEARNING SERVICES

where ns = the number of samples required per paddy field, already calculated in this case as 1, $\bar{x}$ = the mean number of larvae per dip, calculated from transformed values, and S_s^2 and S_p^2 as above. To get estimates of the mean number of larvae from different areas within an accuracy of 10%, so that their relative population densities can be compared, only 10 paddy fields need to be sampled if the mean number of larvae/dip is 0.5. But if the mean number of larvae/dip is 0.05 then more than 1000 paddy fields must be sampled to get the same level of precision. In general Wada et al. (1971a) found that more accurate results were obtained if only one dip/paddy field was taken, but the number of paddy fields sampled in the area increased. If the same number of paddy fields are sampled in two areas then obviously the larger the mean numbers of larvae per dip the more accurate the estimates.

In later studies Mogi and Wada (1973) studied the distribution and density of larvae of *Culex tritaeniorhynchus* in 100 rice fields in 11 groups scattered over an area of about 15 ha. Three dips with a ladle were taken from each rice field, the numbers of larvae recorded and the density level (low, moderate or high) of each rice field classified from tables of sequential sampling. They found that high larval densities tended to occur in fields in certain areas, although the density levels sometimes differed markedly between rice fields in the same area. They concluded that the best sampling procedure for monitoring population trends was to take a single dip from each rice field, but to sample as many as feasible.

Mogi et al. (1995) examined the abundance of mosquitoes and potential predators in Indonesian rice fields by taking 30 dips from the edge of rice fields using a dipper with a diameter of 15 cm and a depth of 3 cm. Numbers of mosquitoes and other aquatic insects collected per 30 dips were transformed to $\log_{10}(y + 1)$. *Culex vishnui, Culex tritaeniorhynchus* and *Anopheles paeditaeniatus* were the three most common species in rain-fed and irrigated fields.

In experiments for evaluating the efficiency of a dipper for sampling the various instars of *Culex tritaeniorhynchus* Wada and Mogi (1974) used four concrete tanks holding water to a depth of 10 cm and having a surface area of 0.7225 m². One hundred pupae and larvae of each instar, that is 2000 pre-adults, were placed in each tank, to give a density of 692.0 immature stages/m². When 160 dips were taken by four people from each tank a mean of 3.73 immature stages/dip was recorded. The collecting efficiency was calculated by dividing the numbers of immature stages collected/dip by the numbers/m², that is 3.73/692, which is 0.0054. The reciprocal of the dipping efficiency is the expected number of larvae/m² when one larva is collected in a single dip, in this instance 186 larvae. Now, the mean number/dip can be used to estimate the likely number of larvae/m^2, e.g. 0.5 larvae/dip would be expected to indicate a population density of 93 larvae/m^2. Wada and Mogi (1974) pointed out that in practice a mean of 10 larvae of *Culex tritaeniorhynchus* per dip is frequently encountered in rice fields during the summer; this would indicate a density of 1860 larvae/m^2 or 18.6×10^5 larvae/ha.

From the results of Russell et al. (1945) it appears that one of the better methods of comparing anopheline, and possibly culicine, larval densities is to record catch size against time. Alternatively the numbers of larvae caught within unit time can be compared. These indices measure larval density, not absolute population size and if larval densities in different *sized* habitats are compared problems arise. For example, the same number of larvae may be present in a small and large habitat, but because there is a higher larval density in the smaller habitat the mean number per dip will be greater. One of the commoner methods of trying to overcome this difficulty is to adjust the catch for surface area. Such a procedure was used in sampling *Ochlerotatus detritus* larvae in a salt marsh where the area of the water surface was constantly changing. The marsh was usually about half full of water and the surface area during this state of flooding was taken as the standard. The numbers of larvae caught by dipping was multiplied by the quotient of the area existing at the time of sampling (Service 1968). This procedure is similar to the 'Mosquito Production Index', which is the mean number of larvae per dip × the breeding area in square feet, as used by the Mosquito Investigation Unit for mosquito surveys in irrigation areas in Montana, USA (Anon 1953, quoted by Knight 1964). Shemanchuk (1959) used a similar index but specified that 20 dips with a quart-sized dipper should be made in each habitat and that only the number of 4th instar larvae caught should be multiplied by habitat area in square yards. Siverly and DeFoliart (1968) obtained crude estimates of the population size of *Ochlerotatus communis* larvae by multiplying the mean number/dip by the volume of water in the habitat.

Belkin (1954) developed a 'Mosquito Breeding Index' (*BI*) for survey work. First, 20 dips are taken for every 100 ft^2 of effective breeding surface; areas of open deep standing water or open flowing water which experience shows do not contain mosquito larvae are omitted. If, when the appropriate number of dips (*ND*) are taken from a habitat none contains larvae then the results are omitted. No record of negative dips is made until the first positive one is obtained. Thereafter the number of dips required to sample the habitat according to its size are taken. As an illustration, if larvae are first recorded in the 8th dip taken from a habitat measuring 200 ft^2 then this is considered the first effective dip and another 39 dips are taken. Both the number of positive dips and total number of larvae collected are recorded. A breeding place (*BP*) is defined as habitat in which 1–20 positive dips are obtained. If there are more than 20 positive dips, then the number of breeding places recorded is one for each 20 dips or fraction thereof. The average number of larvae and pupae per dip (*APD*) is obtained by dividing the total numbers caught (*TLP*) by the number of dips (*ND*). The breeding index (*BI*) is the average number of pre-adults per dip × the number of breeding places (i.e. *BI* = *APD* × *BP*). Belkin (1954) stated that whereas there was practically no correlation between the average number per dip (*APD*) and the breeding index, there was a positive correlation between this index and the total number of larvae and pupae collected. This suggested that if the number of dips per unit is standardised, then the total catch of larvae and pupae can be used directly for comparing population sizes. He also proposed a very simple breeding index for rapid survey work. The approximate surface area of each habitat is measured in square feet, and a known number of dips are taken (remember that generally the more samples collected the more reliable the results), and both the number of positive dips and total catch of larvae and pupae recorded. The breeding index for each separate habitat is then given by:

$$BI = \frac{SA \times PD \times TLP}{(ND)^2} \tag{3.7}$$

where *SA* = surface area in square feet and *PD* = number of positive dips. Belkin (1954) introduced a factor of 10 or 100 in the denominator to produce a more easily handled index. The breeding index for an area is obtained by summation of the indices calculated for individual habitats. This index is essentially based on the mean number per dip (*TLP/ND*) multiplied by the surface area (*SA*), but with the addition of a correcting factor (*PD/ND*) to take into account the patchiness of the larvae within a habitat. If larvae are found in each dip then no correction is needed (*PD/ND* = 1), but if larvae are aggregated fewer dips will contain larvae, resulting in a smaller breeding index. This is an extremely crude and simplified approach, but at least it makes some attempt to take into consideration the patchy distribution of mosquito larvae.

In studying larval populations in irrigated pastures Husbands (Knight 1964) used the Belkin breeding index but omitted the correction factor for patchiness. He confirmed Belkin's observations that if the number of dips per unit area are standardised then the equation is simplified and the breeding index approximately equals the total number of larvae and pupae collected (*TLP*). He divided the effective breeding area in irrigated pastures into different ecological sub-areas, and took a number of dips from each sub-area that was proportional to their contribution to the total area.

Wooster and Rivera (1985) in the Philippines used Belkin's (1954) definition of a breeding point, except they modified it to 2-m stretches of stream bank and required at least 3 consecutive positive dips within 20 dips for the site to be considered as a major mosquito breeding place. This modification was done to eliminate any negative dips and minor breeding places.

In routine surveys in Utah, Graham and Bradley (1969) recorded the mean number of larvae/dip and then coded the results into different indices (1–7) so that the data, together with other relevant information such as water temperature, could be stored on punch cards for computer analysis. Although the limitations of the method were recognised it was concluded that, with *Culex tarsalis* at least, a very useful and simple index which reflected population size was the total number of pools with larvae recorded for each half-month period throughout the year (Graham and Bradley 1969). It was also found that the mean number/dip when larvae of any particular species occurred alone was greater than when it coexisted with another species (Graham and Bradley 1962, 1969; Graham and Collett 1969). That is the mean number of larvae/dip of a species from habitats in which another species occurred was always less than the sum of the mean numbers/dip when the two species occurred alone.

Population estimates from dipping

In studying pool breeding aedine mosquitoes in Canada Dixon and Brust (1972) gently stirred the water prior to sampling. Using a pint dipper three samples were taken from small pools but 10–20 samples from pools measuring >50 ft^2. By calculating the volume of water in the pools, the numbers caught with the dipper were related to the capacity of the pools and larval population estimates made. Mori (1989) placed 600 larvae of all instars and pupae of *Ochlerotatus togoi* in water jars holding 2 litres of water. He then sampled the jars by taking three dips after the water had been stirred, and another three dips when larvae were at the water surface and the water had not been stirred. Three dips with his 100-ml capacity dipper were equivalent to 0.15 of the 2 litres of water in the jars, and by simple proportion he should have obtained 90 larvae of each instar and 90 pupae. When he dipped after stirring the water his actual numbers ranged from 88 3rd stage larvae to 103 4th stage larvae. He concluded that three dips, or even one, gave a reliable sample of the larvae and pupae and could be used to estimate the total numbers present. When, however, samples were taken at the water surface from jars without prior stirring this greatly overestimated larval and pupal numbers, especially of 1st and 3rd stage larvae.

Several workers have tried and failed to relate the numbers of larvae in a dipper of known volume to the numbers in a breeding place of known size (Boyd 1930; Goodwin and Eyles 1942; Service 1971). Others (Belkin 1954; Farlow et al. 1978; Horsfall 1946; Knight 1964; Russell et al. 1945; Shemanchuk 1959) have used area samplers (static quadrats) to estimate absolute population size by relating their sample area to the total area of the habitat. However, this is based on the assumption that mosquito larvae are randomly distributed and this is rarely so. Nevertheless, both Hess (1941) and Boyd (1949) considered that quantitative results could be obtained with a dipper by multiplying the diameter of the dipper at the line of intersection with the water by the length of the sweep. For example, 10 separate sweeps of a 1-m length by a 10-cm diameter dipper sampled 1 m^2 of water surface. Boyd (1930) also calculated that by allowing water to gently flow into a dipper, 18.7 cm in diameter, an area of about 500 cm^2 would be sampled.

In France Papierok et al. (1975) systematically took a total of 118 dips with a 1-litre ladle at 1.5-m intervals along transects through a marsh. The number of larvae of *Ochlerotatus cataphylla* caught in 118 litres of water was converted by simple proportion into the expected total number of larvae in the marsh, the volume of water in it having been estimated. In a series of four dipping experiments they estimated that the larval population varied from 1 573 131 to 2 009 632 with a mean population size of 1 793 024. This mean estimate was only 8.1% smaller than population estimates of 1 938 972 ± 24 447 obtained by mark—recapture studies made on the same day—but reported two years earlier (Papierok et al. 1973). They concluded that if sufficient numbers of dips were taken dipping could give reliable estimates of total larval populations. Croset et al. (1976) similarly believed that dipping could be used to give absolute population estimates (*P*) of larvae by using simple proportions, such as capacity of the dipper (*c*), number of dips taken (*n*) and volume of water in the larval habitat (*v*).

$$P = vn/c \qquad (3.8)$$

While theoretically this is true, obtaining reliable results with this approach is very difficult and dipping is unlikely to be a good sampling method for obtaining absolute estimates of population size. In their work Croset et al. (1976) concluded that at least 95% of samples should contain larvae and for this they had to use a 1-litre ladle for *Ochlerotatus cataphylla* and a 1.5-litre ladle for *Ochlerotatus detritus*. Ladles were attached to 1.5-m handles and they undertook stratified sampling, taking dips at intervals of 3 m or less along the perimeter of the habitat, both near and away from the edges. They acknowledged that accuracy is exponentially related to the intensity of sampling (Papierok et al. 1975). Also that the numbers caught depended on the weather. For example, after snowfall there is less larval activity and population size is underestimated.

In Kenya an area of 1800 m^2 of a natural low-lying land was partitioned off by earthen dykes and artificially flooded with water pumped in from a river 200 m away (Linthicum et al. 1985). The area was maintained flooded for about 17 days, to allow large numbers of *Aedes lineatopennis* eggs in the soil to hatch, and for adult emergence to occur. Pupae were systematically sampled along predetermined transect lines with a 0.47-litre capacity dipper according to the protocol described by Linthicum et al. (1983, 1984). Based on pupal sampling the flooded area was estimated to contain a population of 1.3 × 10^6 female and 1.2 × 10^6 male pupae. Sabesan et al. (1986) sampled a variety of *Anopheles culicifacies* breeding sites on Rameswaram island (56 km^2) with a dipper. They calculated the surface area of all habitats, and by relating this to larval density in the sample concluded that coconut pits, having a surface area varying from 6.8–36.7 ha, produced an estimated 3.3–140 million *Anopheles culicifacies* a month, while casuarina pits (2–6 ha) produced some 2.7–33.5 million *Anopheles culicifacies*. However, because of the very variable numbers of pools and fluctuations in their size it proved impossible to calculate their contribution to the *Anopheles culicifacies* population.

In a long paper on the population dynamics of *Culex tritaeniorhynchus* in Japanese rice fields Mogi (1978) obtained total population estimates of larvae and pupae in his study area by simple proportional arithmetic. For example, previous studies (Wada and Mogi 1974) had shown that one larva or pupa obtained with a standard dipper actually represented 186 larvae or pupae/m^2 of water surface in rice fields. Mogi (1978) proposed that total number of larvae (or pupae) in rice fields surveyed = (σ (No. per dip × Water-logged rate)) × 186 × 300. Where water-logged rate represents the proportion of a rice field with water and varies from 0 to 1, while 300 is the mean area of a rice field in m^2. As he surveyed 200 of the 500 rice fields in his area he multiplied the product of the above equation by 2.5 to obtain an estimate of pre-adults in the whole of his study area. Also in Japan in developing mean-generation life-tables for *Culex tritaeniorhynchus* Chubachi (1979) multiplied the number of pre-adults/dip by a conversion factor, which was the reciprocal of the efficiency of the dipper (Wada and Mogi 1974), to estimate the density of larvae and pupae/m^2 of rice fields.

Sunish and Reuben (2001, 2002) studied the effects of abiotic and biotic factors on the density and abundance of the immature stages of Japanese encephalitis vectors in rice fields in India. Twenty dips at equal intervals along the margins of rice fields were taken using a 450 ml dipper attached to a long handle. The population density of mosquito immatures was calculated as the number of larvae and pupae per dip, adjusted by a correction factor equal to the proportion of the breeding habitat covered by water.

Stewart and Schaefer (1983) compared dipper counts of *Culex tarsalis* and absolute population densities in Californian rice fields. In one series of experiments 12 aluminium cylinders, 1.12 m in diameter, 30 cm tall and enclosing 1 m^2 of water surface were thrust into rice fields. 1st, 2nd, and 3rd instar larvae and pupae were placed within the cylinders at densities of 20, 50, 100, 200, and 400/cylinder; with 4th instars an additional density of just 10 larvae/cylinder was used. Eight samples were taken from each cylinder on each sampling date using 473-ml capacity enamel dippers attached to 1.5-m wooden handles. There were significant regressions of mean numbers of larvae and pupae per dip and densities present within the enclosed cylinders ($r^2 = 0.9652$–0.9978). Statistical analysis showed that in all cases the slopes were not different, nor were the levels of regression. In a second experiment the cylinders enclosed unknown numbers of a natural population of *Culex tarsalis*. The total collected from five dips from each cylinder represented one sampling unit, and six units were taken from each enclosure. Numbers of larvae were 'analysed by methods described by Southwood (1978)' to estimate population size by the removal method. Results were added to the data from the first experiment, and regression coefficients recalculated. Except for 2nd instar larvae ($r^2 = 0.3960$) regression coefficients were much the same, no pupae were collected from the field population. It was concluded that the mean number per dip could be used to estimate absolute population size (Table 3.2).

In a study of the influence of rainfall and temperature on the population dynamics of *Ochlerotatus albifasciatus* in Buenos Aires, Argentina, Fontanarrosa et al. (2000) sampled immature mosquitoes with 80 ml plastic dippers. The number of dips taken per habitat was related to the estimated size of the flooded areas, using simple proportions. Between 5 and 130 dips were taken from each habitat.

Chavasse et al. (1995) studied the population of immature *Culex quinquefasciatus* in Dar-es-Salaam, Tanzania, prior to implementing a control intervention using expanded polystyrene beads. Pupal densities in enclosed breeding sites were estimated by making two to five dips with a soup ladle attached to a pole and assigning the number of pupae sampled to one of five points on a scale (0, 1–5, 6–25, 26–125, 126–625). An average number per dip for each site was calculated by taking the geometric mean of the upper and lower limits of the relevant scale point for each site. A dip containing 10 pupae for example would be assigned to the 6–25 scale point and have a geometric mean of 12.25 ($\sqrt{6 \times 25}$). The mean number of pupae per site was calculated from the arithmetic mean of the above dipscores and the ratio of the area of the ladle opening to the surface area of the water body sampled.

Table 3.2. Estimated numbers of *Culex tarsalis* immatures based on regression statistics of 1980–1981 enclosure studies (Stewart and Schaefer 1983)

Stage	No./dip	No./hectare
	0.10	407 700
I	0.50	941 400
	1.00	1 608 400
	5.00	6 944 712
	0.10	29 100
II	0.50	1 984 600
	1.00	4 429 000
	5.00	23 984 418
	0.10[a]	250 104
III, IV, Pupa	0.50	1 013 951
	1.00	1 968 759
	5.00	9 607 228

[a] Cumulative total for all 3 stages.

In India Rajagopalan et al. (1975) obtained population estimates of *Culex quinquefasciatus* by several methods, and concluded that its population could be estimated by the number/dip (Fig. 3.6). Somewhat similarly in the USA Sandoski et al. (1987) showed that the numbers of *Anopheles quadrimaculatus* larvae/dip could be related to absolute population densities over a range of high densities, using linear regression, but at low densities, which were the most commonly encountered densities in the field, precision was poor.

In Japan Okazawa and Mogi (1984) studied the relationship between the numbers of larvae collected by dippers (density) and absolute population size by enclosing varying numbers (25–800) of all instar larvae of *Anopheles sinensis* in 1 × 1-m enclosures in rice fields. Irrespective of rice height they obtained good regression lines of mean numbers of larvae plotted against their total numbers, that is larval density was proportional to larval population size, at least within the range of 25–800 larvae of each instar. Rather different results were obtained by Sandoski et al. (1987). They released reared larvae of different instars and also pupae of *Anopheles quadrimaculatus* in 1-m diameter (0.72 m^2) enclosures made of sheet aluminium about 30 cm high in Arkansas, USA rice fields. Because Stewart and Schaefer (1983) observed the necessity of using non-reflective surfaces for such enclosures, aged enclosures were used.

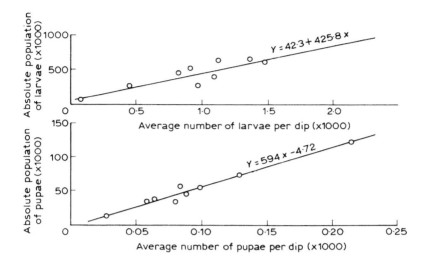

Fig. 3.6. Correlation between relative and absolute densities of larvae and pupae of *Culex quinquefasciatus* (Rajagopalan et al. 1975)

Each enclosure had two 1.5-cm diameter openings covered with 100-mesh aluminium below the water level to allow water movements between the enclosures and rice fields. Densities of the introduced larvae were 10, 50, 100, 200, 500, 1000 and 1500/0.72 m². Ten samples with a 350-ml dipper were taken from each enclosure and the mean number/dip calculated for all instars at all densities. Regression equations were developed using the mean values/10 dips to estimate total populations at the known densities. This regression method was unreliable in predicting populations when there were 200 or less larvae per 0.72 m², which corresponds to a mean of 9.55 *Anopheles quadrimaculatus* larvae per dip, or less. The authors also estimated that a prohibitive number of 6424 dipper samples would be needed in their rice fields to detect differences within 10% of the true mean with 5% and 10% probability of Types I and II errors, respectively, when the mean number/dip was 0.62. It should be noted that their larvae approximated a Poisson distribution; if they had mimicked an aggregated distribution it is likely that sample size needed would have been even greater.

In El Salvador Bailey et al. (1980) investigated whether larval sampling of *Anopheles albimanus* could reflect changes in its adult population size. For this 40 dips with a 375-ml capacity ladle were taken weekly from 18 different breeding sites. The numbers of 3rd and 4th instars in each dip were recorded, and as a comparative measure their number/100 dips was calculated and averaged for each type of habitat (rivers, marshes, irrigation ditches etc.). In addition, numbers were averaged for each sampling site at 4-weekly intervals throughout the study period (June 1976 to March 1978). Adult *Anopheles albimanus* were collected with aspirators for one hour from 14 cattle sheds. Numbers were expressed as 1 man-h collections, and as with larval collections numbers were averaged at 4-week intervals. In the rainy seasons the ratios of the average catches of adults and larvae (adults/man-hr and larvae/100 dips) varied from 2.9 to 72.0 and showed no similar trends in population size; in the dry seasons these ratios varied much less, from 1.3 to 5.3. It was concluded that larval sampling did not accurately reflect changes in adult population size, especially during the rainy season.

Efficiency of dipping

Hagstrum (1971*a*) investigated the reliability of dipping in sampling all larval instars of *Culex tarsalis*. A 20 × 25-cm aluminium frame, 25 cm high, and open at both ends was thrust rapidly down into the bottom of a breeding place. After an interval of 15 min a single sample with a pint dipper having a surface area of 95 cm^2 was taken from the water enclosed by the frame. The water was then pumped out of the quadrat and all the immature stages counted. There were nine replicate experiments. The surface area of the dipper was 19% of that enclosed by the quadrat, and a mean of 18, 21, 22 and 26% of the total numbers of 1st–4th larval instars that were enclosed by the quadrat were caught in a dip. Hagstrum (1971*a*) concluded that dipping sampled all larval instars as effectively as the metal quadrat. Very few pupae were collected (a mean of 0.9/quadrat and none by dipping); consequently no conclusions could be made regarding the efficiency of dipping in sampling pupal populations. In the same paper he concluded that the quadrat was more efficient in collecting 1st instar larvae than dipping.

Wada and Mogi (1974) conducted an instructive series of experiments to test the efficiency of a dipper in sampling the immature stages of *Culex tritaeniorhynchus*. One hundred larvae of all larval instars and pupae of *Culex tritaeniorhynchus* and 100 4th instar larvae of *Culex pipiens* form *pallens* were released into four 85 × 85 cm water tanks holding water to a depth of 10 cm. Each tank had 20 equally spaced rice plants. Using a 15-cm diameter dipper four people took 10 random samples from each tank. After recording the numbers of larvae in each dip the larvae were returned to the tank and the next dip taken only after a short interval to allow the larvae to redistribute themselves. A total of 160 dips were taken. There was little difference between the numbers of larvae in the samples taken from the different tanks. Combining all samples, the numbers of pre-adults collected in the 160 dips were 87 1st instars, 95 2nd instars, 133 3rd instars, 125 4th instars and 157 pupae of *Culex tritaeniorhynchus* and as few as 39 4th instar larvae of *Culex pipiens* form *pallens*. Clearly the efficiency of the dipper differed considerably for the different age classes of *Culex tritaeniorhynchus* and also for 4th instar larvae of the two species. The efficiency of a dipper can be obtained by dividing the mean number of larvae per dip by their absolute density. Wada and Mogi (1974) defined efficiency as the probability of one larva in 1 m^2 of water to be collected by one dip. The reciprocal of the efficiency is the expected number of larvae/m^2 when one larva is collected in one dip, and can be regarded as a conversion factor. Using these criteria Chubachi (1976) concluded that the efficiency of an 8.4-cm dipper was about the same (0.002 10–0.003 00) for all larval instars and pupae of *Culex tritaeniorhynchus* in rice fields, contrasting with the findings of Wada and Mogi (1974), who found early instars undersampled. With *Anopheles sinensis*, however, Chubachi (1976) found that the efficiency of the dipper was less for pupae (0.0079) and early instars (0.000 67–0.000 88) than later instar larvae (0.003 00–0.003 27).

In England once or twice a week, for 17 weeks, larvae and pupae of *Ochlerotatus rusticus* and *Ochlerotatus punctor* were collected from 12 open-ended metal cylinders each of which enclosed 459 cm^2 of water surface when thrust into woodland pools (Service, unpublished results). In addition 100 dips, with a dipper having an area of 63 cm^2, were taken from the same area of the habitat but not from within the cylinders. Compared with the proportions of the age classes enclosed by the cylinders both 4th instar larvae and pupae of *Ochlerotatus rusticus* and *Ochlerotatus punctor* were seriously underestimated by dipping. It is also possible that the cylinders failed to give representative samples of the different instars, although it is likely that they gave a more accurate representation of the instar proportions in the pools.

Some factors influencing dipping and sampling bias

Larval and pupal alarm reactions and time spent at water surface

Dipping will usually only catch larvae and pupae at the water surface and since, both in the presence and absence of alarm reactions, different species and also different instars of the same species may remain at the water surface for varying periods, it follows that dipping may frequently be biased for particular species or instars. For example, dipping will not usually sample

Anopheles and culicine species equally, because whereas the latter often submerge for long periods to feed at the bottom of habitats *Anopheles* larvae rarely leave the surface unless disturbed. Apart from these difficulties, some larvae will escape capture by dipping by swimming away, and their ability to escape may differ between species and also instars. After an eradication programme in Sardinia Trapido (1951) undertook extensive larval surveys to locate any remaining larvae of *Anopheles atroparvus*. It proved exceedingly difficult to find *Anopheles* larvae by dipping when the population density was low. Where the main objective is to detect residual breeding sites requiring further insecticidal treatment, rather than obtain quantitative estimates of population density or size, aquatic nets are more useful, as they sample a greater volume of water than conventional dipping. Trapido (1951) produced interesting information on the durations of submergence for larvae of four *Anopheles* species after they were disturbed at the water surface. About 50% of *Anopheles atroparvus* larvae returned to the water surface within about 3 s and all within 4 min, whereas it was 40 s before about 50% of the larvae of *Anopheles claviger* surfaced, and one larva remained submerged for as long as 14 min. Even more surprising, one larva of *Anopheles hispaniola* remained below the water surface for 35 min. It is evident that although *Anopheles* larvae normally occur at the water surface, they may remain submerged for comparatively long periods. Differences in alarm reactions and submergence times can lead to bias in sampling different species in the same habitat. For example, Wada and Mogi (1974) and Okazawa and Mogi (1984) found that dipping efficiency for *Anopheles sinensis* was about twice that for *Culex tritaeniorhynchus*, probably because the latter species is quicker in its escape responses.

Rather surprisingly Duhrkopf and Benny (1990) found that although larvae of different strains of both *Aedes aegypti* and *Aedes albopictus*, from various localities in the USA, spent different times submerged after being disturbed at the water surface, there were no differences in the alarm reactions between these two species.

Thomas (1950) found that regular repetition of changes in light intensity, such as shadows on the water surface, caused a decrease in the numbers of larvae of *Culex quinquefasciatus* that dived to the bottom. He also obtained evidence that their alarm reaction was density-dependent, that is the more larvae present the more likely they were to submerge when disturbed. Furthermore, older larvae did not react to alarm stimuli by diving so much as younger instars, and in the absence of alarm stimuli younger instar larvae remained submerged much longer than older instars. Thomas (1950) considered that this was because the ratio of the body surface: volume was greater in younger instar larvae thus enabling them to obtain more of their oxygen requirements by cuticular respiration of dissolved oxygen. Nielsen and Nielsen (1953) also observed that 1st instar, and to a lesser extent 2nd instar, larvae of *Ochlerotatus taeniorhynchus* came up to the water surface much less frequently than older larvae.

Le Sueur (1991) recorded the numbers of all instar and pupae of *Anopheles merus* in 11 consecutive dips with a ladle. Instead of a decline in numbers, the number per dip tended to increase with increasing dips, especially with 1st instar larvae. Le Sueur (1991) argued that this was because of larval escape reactions and that 1st instar larvae stayed submerged longer. With pupae, however, there was no such pattern and the numbers/dip for the 11 dips were more or less random. But he also realised that because of the very small size of his pools there was a concentrating effect as water was being removed, making it increasingly easier to catch the remaining larvae.

Workman and Walton (2003) observed that in the laboratory, larvae of *Culex erythrothorax, Culex tarsalis, Culex quinquefasciatus*, and *Culex stigmatosoma* all tended to dive less often under conditions of low food availability.

Tuno et al. (2004) demonstrated that larvae of *Anopheles gambiae* tended to dive deeper in muddy water columns in the laboratory as compared to clean water, and mortality of larvae resulting from an inability to swim back to the surface was higher in deeper water compared to shallow water.

Differences between submersion times of the different immature stages will probably result in sampling bias when the age-structure of the population is derived from dipping. In fact dippers have been reported as being less efficient for 1st instar larvae of various species than for older instars (Andis et al. 1983; Chubachi 1976; Hagstrum 1971a; Okazawa and Mogi 1984; Wada and Mogi 1974). In constructing vertical life-tables to estimate pre-adult mortality of *Culex tarsalis* from sampling with a dipper, Reisen et al. (1989) sometimes found that even after correction for instar duration, the numbers of 2nd instar larvae collected exceeded those of the 1st instar; the same was found by Mogi et al. (1986) for larvae of *Anopheles sinensis* in Thailand. Similarly in Canada Pritchard and Scholefield (1983) found that the numbers of 1st instars of several aedine species were sampled with lower efficiency than later instars. In laboratory experiments Shogaki and Makiya (1970) found that with both an ordinary dipper and one having a wire mesh bottom, fewer 1st–3rd instar larvae of *Culex pipiens* form *pallens* were caught than 4th instar larvae when equal numbers of all instars were present. They thought that this bias in sampling might be due to differences in vertical distribution of the different instars, and this belief seems to be substantiated by the following observations in Canada. Using a pint dipper with a 4-ft dowel handle Downing (1977) sampled all instars of *Ochlerotatus canadensis* at different depth (means 6.7–11 in) in small woodland pools. He found no differences between the numbers of 1st instar larvae collected in different water depths, but larger numbers of 2nd and 3rd instars were collected from the shallower water, and there were slightly more 4th instar larvae collected from the deeper waters. Since older instar larvae may spend more time at the water surface than earlier instars (Thomas 1950) the chance of collecting older instars with a dipper is greater. Downing (1977) believed that in deeper waters increased depth allowed early instars, which spend less time at the water surface, and moreover are more easily frightened into diving to escape from dipper collections. Thus there is a bias towards older larvae in deeper waters.

Vegetation coverage, stage of plant growth, and environment type

Early workers (Hess and Hall 1943; Rozeboom and Hess 1944) found a good correlation between densities of *Anopheles* larvae and the amount of floating and emergent vegetation. They developed the intersection line value—'the number of meters of intersection line per square meter of water surface', where the intersection line is defined as the 'line of intersection between three interfaces', water–air, water–plant and plant–air. Later workers (Balling and Resh 1984; Collins and Resh 1989; Walker et al. 1988) have corroborated that anopheline larval densities are higher when there is plant coverage. Orr and Resh (1989) concluded that emergent or submerged macrophytes favoured the survival of *Anopheles* larvae, mainly by providing a refuge from predators.

Sen (1948) caught 11 *Anopheles* species in rice fields in India, and recorded a succession in the dominance of individual species with time and this was dependent on the maturity and height of the rice crop and age and condition of the water. Takagi et al. (1995) in Japan studied the seasonal occurrence of *Culex tritaeniorhynchus* larvae in rice fields under three distinct types of cultivation. Larval abundance, defined as density (=log-transformed number of larvae per 10 dips) × percentage of rice field inundated with water, was determined by the mean number of larvae per ten dips using a dipper with a diameter of 13 cm and a depth of 7.5 cm. Samples were concentrated using small plastic containers with a fine (0.138 mm) stainless steel mesh on the sides. Differences in rice culturing practices appeared to affect the seasonal occurrence of *Culex tritaeniorhynchus* immatures. Takagi et al. (1996, 1997) used the same methods described by Takagi et al. (1995) to estimate the densities of *Culex tritaeniorhynchus* and *Anopheles sinensis* in rice fields in

Japan and *Culex* spp. in Thailand. Densities were highest in the unplanted sections of the fields and population densities were also found to be higher in sections of rice fields planted with short varieties of rice (mean height 45.0 cm) than in sections planted with tall varieties (mean height 98.5 cm) (Takagi et al. 1996). Species succession has been reported for several mosquitoes, particularly in rice fields. As the rice plants grow taller sun-loving species decrease in numbers while populations of shade-loving mosquitoes increase (Chambers et al. 1979; Chandler and Highton 1975, 1976; Mogi 1984*b*; Rao 1984; Robert et al. 1988; Snow 1983; Surtees 1970; Surtees et al. 1970; and see Lacey and Lacey 1990 for short review on rice plant height and mosquito density). In Louisiana, USA McLaughlin et al. (1987) studied the relative weekly and seasonal distribution and possible succession of larvae of *Anopheles quadrimaculatus* and *Anopheles crucians* by comparing the numbers of 3rd and 4th stage larvae collected with dippers from rice fields. They calculated the numbers of larvae of *Anopheles quadrimaculatus: Anopheles crucians* and then, omitting 0:1 and 1:0 ratios, plotted a linear regression of the ratio against days post-flooding of the rice fields. The regression clearly showed that the ratio of these species changed with time, although some fields were dominated by one species over most of the season.

In comparing mosquito breeding in ground waters at different phases of the Mahaweli irrigation development project in Sri Lanka, Amerasinghe and Ariyasena (1990) used a standard pint (0.473-litre) dipper and generally took six dips/m^2 of water surface, but the actual number taken was proportional to the estimated area of the breeding places. With large habitats dips were taken from 5-m^2 quadrats along their perimeters. Relative density of the species in the habitats was computed as the geometric mean of the numbers of immatures per dip (after a log transformation) for habitats positive for mosquitoes × 1000. Habitats in which larvae were not found were excluded. Comparisons between mosquito breeding during different phases of the irrigation project were based on prevalence (*p*), which was calculated as the relative density (*d*) of a species in all habitats × percentage occurrence (*o*), (i.e. *p* = *d* × *o*). Comparisons between phases of development were expressed as the Index of Change (IC) where

$$IC = (p_2 - p_1)/(p_2 + p_1) \tag{3.9}$$

where p_1, and p_2 are the prevalence rates for a species in phases 1 and 2, respectively. The index will vary from +1, representing maximum increase when a species has invaded the habitat, to −1 when a species has been eliminated. A value of ± 0.33, which is 100% difference between p_1 and p_2, indicates a considerable change, while values greater than ± 0.71 (i.e. a 500% or greater difference) indicate major changes. The authors point out that one limitation to this analysis is that the index of change based on p values is dependent on density level (o), which may be difficult to calculate with reliability. Other limitations, mainly due to sampling difficulties, are discussed. Amerasinghe and Indrajith (1996) used the same sampling methodology to investigate the occurrence and abundance of mosquito larvae in different habitats associated with irrigated rice fields. Forattini et al. (1995*a*) used the index of change proposed above to investigate the relative abundance of different mosquito species in environments ranging from vestigial forest (phase 1) to an anthropic environment (phase 2) in southeast Brazil. Overall means were computed from values obtained from Shannon trap catches in vestigial forest, on open land and in the domiciliary environment. Indices of change were computed as 0.8846 for *Ochlerotatus scapularis* and –0.9930 for *Ochlerotatus serratus*, indicating that the relative prevalence of *Ochlerotatus scapularis* had increased 16.33 times between phase 1 and phase 2 of development towards a domestic environment, while the prevalence of *Ochlerotatus serratus* decreased 284.0 times between phase 1 and phase 2. Indices of change were also calculated for *Culex nigripalpus* (0.9853) and for the *Culex* Coronator Group (0.3433) (Forattini et al. 1995*b*).

Larval dispersion and aggregation

The dispersion of individuals in a population can be conveniently classified as following one of three basic patterns: (1) regular; (2) random; or (3) contagious. This is a useful but greatly simplified classification, since the three types may overlap resulting in compound distributions. Usually individuals, including mosquito larvae, do not conform to either a regular ($\sigma^2 < \mu$) or random ($\sigma^2 = \mu$) dispersion but are contagious ($\sigma^2 > \mu$), that is they exhibit a patchy type of dispersion in which individuals are aggregated or clumped. As most environments are not completely uniform, it is not surprising that certain areas will be favoured and occupied at the expense of less attractive ones. In Canada, Hocking (1953) observed that larvae of *Ochlerotatus communis* aggregated round the edges of a pool. Up to 1003 larvae were collected in a single sample taken with a 400-ml dipper from the clumps of larvae, whereas elsewhere in the pool an average of only 20 larvae/dip were caught. Nielsen and Nielsen (1953) recorded the existence of 'balls' of *Ochlerotatus taeniorhynchus* some 10–15 cm in diameter, and also reported that other people had seen clusters of larvae some 20–25 cm in diameter and 0.5 m deep. These are instances of extreme aggregation, often controlled it appears by temporary crowding, temperature, availability of food and photonegative responses (Nayar and Sauerman 1968). Less marked but nevertheless distinctly aggregated populations of mosquito larvae are common, if not the usual phenomenon (Andis and Meek 1984; Chubachi 1979; Ikemoto 1978; Reisen and Siddiqui 1979; Stewart and Schaefer 1983; Stewart et al. 1983*b*).

Several mathematical models have been proposed to describe contagious distributions of natural populations, but the negative binomial model appears to be the most widely applicable, probably because it is so flexible. It has been fitted to the dispersion of larvae and/or pupae of several mosquito species, for example *Anopheles arabiensis* and *Anopheles gambiae* complex (Service 1971, 1985). *Culex tarsalis* (Mackey and Hoy 1978; Stewart and Schaefer 1983). *Ochlerotatus cantans* (Service 1968, 1985; Renshaw et al. 1995), *Ochlerotatus poicilius* and *Ochlerotatus flavipennis* (Lang and Ramos 1981), *Aedes flavopictus miyarai* and *Malaya genurostris* (Mogi 1984*a*), *Aedes aegypti* (Reuben et al. 1978), several Canadian *Ochlerotatus* species (Wada 1965), *Coquillettidia perturbans* (Morris et al. 1990), *Toxorhynchites brevipalpis* (Lounibos 1979), and *Culex tritaeniorhynchus* (Wada et al. 1971*a*). However, Walton et al. (1990) found that larvae of *Culex tarsalis* in a freshwater marsh exhibited more a log-normal than a negative binomial distribution ($k = 0.16$), while Sandoski et al. (1987) found that *Anopheles quadrimaculatus* larvae in Arkansas, USA rice fields had a dispersion approaching a Poisson distribution, and rather surprisingly Downing (1977) calculated that larvae of *Ochlerotatus canadensis* were uniformly distributed in pools.

The negative binomial distribution is described by two parameters, the arithmetic mean (μ) and exponent k, which is a measure of aggregation. The smaller the value of k the more aggregated the population. Taylor et al. (1979) showed that exponent k of the negative binomial is related to the variance (S^2) as follows $S^2 = a \overline{x}^{\,b}$, since $1/k = ax^{b-2} - \overline{x}^{-1}$, and that if $b > 2$ then $1/k$ approaches α as the mean ($\overline{x}$) approaches α. If $1 < b < 2$ then $1/k$ approaches 0 as $\overline{x}$ approaches α and there is a turning point. The minimum value of k depends on parameters a and b. As noted with mosquito pre-adults (Service 1985) and in phytophagous insects (Wellings 1987) k is not necessarily constant in all parts of a species density range. Service (1985) found with both *Ochlerotatus cantans* and *Anopheles gambiae* that the value of k varies according to the age class of the pre-adults sampled, even from the same habitats (Service 1985). In other words the degree of aggregation differs among the various larval instars and pupae, usually decreasing with older instars which are less numerous than younger ones. Changes in k are often taken as indicating changes in the degree of aggregation, which may affect the degree of intraspecific competition, as well as interspecific competition between apparently cohabiting competitors. However, Perry and Taylor (1986) pointed out that contrary to usual expectations aggregation did not increase with k, moreover, there was often no constant value (k_c) over all population densities. They concluded that the use of k as a measure of aggregation should be abandoned. This paper should be read for a discussion on the interactions of two populations (such as host and parasitoid or prey and predator), their stabilities, aggregation, behaviour and the negative binomial parameter k. Useful descriptions of the negative binomial and other mathematical distribution models are given by Elliott (1977).

Because mosquito larvae and pupae are aggregated there are often large differences between the numbers caught in different samples. On account of this variability more samples usually have to be taken to get a reliable estimate of the population mean than if the population was randomly dispersed. Thus, a large number of dips are usually necessary if the mean number of larvae per dip is to be a meaningful indicator of larval density. Kuno et al. (1963) and Rojas (1964) independently showed that if a contagious distribution mimics a negative binomial model then the required number of samples depends on the dispersion characteristics of the population, as shown by parameter k. However, this can be of limited practical value, especially when the minimum number of samples are collected and the shape of the distribution cannot be defined. Usually *at least* 50 samples will be needed before a mathematical distribution can be fitted, to determine whether it mimics a negative binomial, and k can be calculated. The value of k is also taken into consideration in sequential sampling if the distribution of individuals mimics a negative binomial model.

Recognising that mosquito larvae were highly aggregated within even apparently more or less homogeneous habitats, Fanara and Mulla (1974) tried to reduce the large variability in numbers of larvae per dip, resulting from inadvertently sampling areas with clumps of larvae and areas with few larvae, by restricting sampling to areas with high densities. This was achieved by placing a handful of clean straw tied to a wire in the water at each corner of artificial ponds, and only dipping from these sites which attracted large numbers of *Culex tarsalis* and *Culiseta inornata* larvae. This procedure did not alter species composition, but helped to standardise sampling by lowering the variability of the numbers of larvae per sample. In this instance population estimates were not required, just a comparison of the mean number of larvae per dip in ponds receiving different insecticidal treatments.

Indices of spatial dispersion

The spatial distribution of animals has been much studied both experimentally and theoretically and clearly most animals, including mosquitoes, have an aggregated distribution.

Two regression models have been proposed to describe density-dependent relationships in spatial distributions, namely Taylor's (1961) power law, and Iwao's (1968) model. The former is based on the relationship between the sample mean ($\bar{x}$) and variance (S^2) and is as follows

$$S^2 = a\,\bar{x}^{\,b}, \log_e S^2 = \log_e a + b \log_e x \qquad (3.10)$$

where a and b are constants, the latter being an index of aggregation and, when $b > 1$ this shows that the degree of aggregation is density-dependent (Taylor et al. 1978). Originally b was believed to be a species-specific parameter not affected by the environment, but Downing (1986) considered it could be variable within a species. His analysis was, however, criticised by Taylor et al. (1988). Sawyer (1989) and Yamamura (1990) have studied the variability of b in simulation models, while Kuno (1991) again argues that b is not species-specific and reviews the merits and drawbacks of Taylor's power law. Yamamura (1990) concluded that the shape of the regression of the power law, b, is inversely related to the movements of individuals being sampled, and that b is expected to be larger as the scale of sampling increases.

Iwao's (1968) model is as follows

$$m^* = \alpha + \beta\,\bar{x} \qquad (3.11)$$

where m^* is Lloyd's mean crowding statistic, that is $m^* \approx \bar{x} + (S^2/x - 1)$, and α and β are constants which Iwao claimed were biological realities, and this consequently led him to believe his equation gave more information on spatial distribution than Taylor's equation. However, several biologists have found that their data do not fit his model, although they frequently fit Taylor's model and produce meaningful results (see Drake 1983).

Taylor et al. (1978) critically reviewed Iwao's method and suggested that the equation should be rewritten as a polynomial,

$$S^2 = a + b\,\bar{x} + c\,\bar{x}^{\,2} \dots i\,\bar{x}^{\,-i-1} \qquad (3.12)$$

and the data plotted on transformed axes.

Weseloh (1989) used a mathematically sophisticated technique called two-dimensional spectral analysis to study insect spatial distribution, and found the results generally complementary to Lloyd's patchiness index.

Those interested in the theory, evolution and mathematics of spatial dynamics of insects should read the paper by Taylor and Taylor (1979). They have argued that constant *b* in Taylor's power law is due to opposing movements of individuals to or away from their population centres. The mathematical model designed to explain this is the Delta model (Taylor 1981*a, b*), which is based on the assumption that the distance an individual moves towards (immigration) or away (emigration) from the population centre is proportional to a fractional power of the density of the population. Emigration is treated as a positive force while immigration is negative, and these two interacting density-dependent behaviours are expressed as Delta. There have been several papers criticising this variance-mean relationship on theoretical (Anderson et al. 1982; Dye 1983), ecological (Hanski 1982), and statistical (Downing 1986) grounds, and also on its internal logic (Thórarinsson 1986). Hanski (1987) points out the importance of cross-correlations in the dynamics of local populations, and that standard population dynamic theories can explain the spatial distribution of insect populations and the properties of the regression slope *b*. Taylor (1984) presents a masterly review of spatial distribution of insect populations, discussing in depth his own power law, Lloyd's mean crowding, Morista's index, Iwao's regression between mean crowding and density, and the negative binomial distribution. Taylor (1984) argues that the validity of *k* of the negative binomial as a biologically meaningful parameter of aggregation is dubious, pointing out that *k* is not a predictable parameter except over small ranges of population density. He also points out the similarity, at least for practical sampling purposes, of $1/k + 1$, Morista's index and mean crowding of Lloyd. Those interested in these indices, or related ones, should read this paper. For other papers on aggregation and animal movement, which also have many references to the subject, the reader is referred to Taylor (1986), Turchin (1989) and Kuno (1991).

Application of aggregation indices to mosquito larvae. In sampling larvae of *Coquillettidia perturbans* from a marsh in Florida, USA Morris et al. (1990) found an aggregated distribution in all of their 12 samples. The parameter *k* of the negative binomial distribution was estimated by maximum likelihood methods (Elliott 1977) to range from 0.073–0.0323, with a common *k* calculated as 0.205. They also used Taylor's (1961) power law and estimated *b* (slope of the regression line) for each larval instar. Values ranged from 2.44–3.39 with a value of 2.19 for all instars combined. This is just above a value of 2 that would indicate a log-normal distribution. Using the calculated value for each instar the appropriate power transformations were calculated, which showed a downward trend from $x^{-0.7}$ for 1st instar larvae to $x^{-0.2}$ for 4th instar larvae. This indicates that older instars are less aggregated than younger ones. Mackey and Hoy (1978) found most of their data on larvae of *Culex tarsalis* fitted Taylor's (1961) power law, and that the best fit of a common k of the negative binomial distribution was 0.09602 (Fig. 3.7).

Renshaw et al. (1995) studied the spatial distribution of larvae and pupae of *Ochlerotatus cantans* in temporary woodland pools in northern England. Data obtained from collections made using a 120 ml dipper were compared with the Poisson distribution and the negative binomial using χ^2 tests. Lloyd's (1967) index of mean crowding ($m*$) was used to perform Iwao's patchiness regression of $m*$ on the mean (Iwao 1968) and the data were used to derive a Taylor's power law equation (Taylor 1961). The variance to mean ratios of the sampling data for all larval instars and pupae were greater than one, indicating a contagious distribution, although the ratio tended to decrease for later instars. Data for none of the larval instars fitted the Poisson distribution, although they did fit the negative binomial, with values of k varying between 0.05 and 0.96. Correlation coefficients for Taylor's power law ranged from 0.89 for pupae to 0.97 for second and third instar larvae and the values of a and b ranged from 0.70 (pupae) to 1.13 (first instars) and 1.34 (pupae) to 1.88 (first and second instars) respectively. Parameter b showed a consistent decrease in aggregation from first instar larvae through to pupae. Iwao's patchiness regression gave a strong correlation between mean crowding and the mean with values of β all in excess of 1, indicating that the population was aggregated in relation to habitat heterogeneity. The relationship of larval density to water depth and temperature was not consistent between the two ponds studied and between the two years of data collection.

Mogi et al. (1985) applied Taylor's (1961) and Iwao's (1968) spatial distribution models to study the distribution of aquatic fauna in axils of aroids (*Alocasia* spp.) in the Philippines. They found that *Malaya genurostris, Topomyia dubitans* and *Armigeres baisasi* all exhibited varying degrees of contagion, but not *Toxorhynchites splendens,* which was more or less uniformly distributed in plant axils. The distribution of *Ochlerotatus flavipennis* and *Ochlerotatus poicilius* in banana axils in the Philippines (Lang and Ramos 1981), and *Malaya genurostris* and *Aedes flavopictus miyarai* in aroid axils in the Ryukyus (Mogi 1984*a*) also showed a negative binomial distribution. In East Africa Lounibos (1979) found the distribution of *Toxorhynchites brevipalpis* in bamboo sections fitted a negative binomial distribution. Mogi and Suzuki (1983) found that the distribution of *Topomyia yanbarensis* in bamboo in Japan changed from an aggregated distribution in the eggs to a uniform one in the 4th instar larvae.

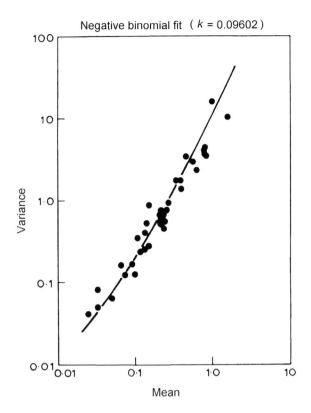

Fig. 3.7. The relationship between the means and variances of *Culex tarsalis* larvae plotted against a series of negative binomial distribution with a $k_c = 0.09602$ (Mackey and Hoy 1978)

In the USA Sandoski et al. (1987) used Taylor's (1965) index of aggregation *b*, obtained as the regression slope, in determining the distribution of *Anopheles quadrimaculatus* larvae in Arkansas rice fields. Their calculated value of 0.821 was significantly different from 0, but not from the value of 1 which would indicate a Poisson distribution. They also calculated the exact transformation (*z*) needed to normalise their data

$$Z = x^{1-0.5b} \tag{3.13}$$

and found that the value of *z* was 0.59, not very different from the 0.5 value (square root) recommended for a Poisson distribution. Thus, they concluded that although their data did not fit a Poisson distribution nevertheless they approximated it. Later Walker et al. (1988) used 0.5-cm diameter (0.1963 m²) plastic tubular hoops floated on the water surface to sample larvae of *Anopheles quadrimaculatus* in a marsh. Larvae enclosed by the hoops were removed with a pipette. They studied larval dispersal using Taylor's (1961) power law (Fig. 3.8a) and Lloyd's (1967) mean crowding index ($m^* = x^*$). They regressed m^* on the mean numbers of larvae ($\overline{x}$) per sample (Fig. 3.8b) to obtain 'an index of basic contagion' (α, the intercept), and the 'density-contagiousness coefficient' (β, the regression slope) according to the method of Iwao (1968, 1970). The ratio of $m^*:\overline{x}$ which is a measure of patchiness (Iwao, 1970) was also calculated. The regression in Taylor's power law (Fig. 3.8a) had a positive slope of <1.0, and similarly Iwao's density-contagiousness coefficient was positive and <1 (Fig. 3.8b), thus corroborating that larvae were aggregated in response to habitat heterogeneity. In fact there was evidence that larvae positioned themselves at the vegetation interface at the water surface, that is they clustered around floating and emergent vegetation. However, Iwao's index of basic contagion (α) was negative (Fig. 3.8b), which is theoretically impossible because it implies that samples with small means have negative variances (Taylor et al. 1978). Walker et al. (1988) interpreted this as indicating that larvae do not aggregate in relation to each other, that is they do not live in groups.

Andis and Meek (1984) used an area sampler (0.1 m^2) to study the spatial dispersion of mosquito larvae in Louisiana, USA rice fields. Numbers per sample were converted to a square root transformation. Estimates of mean crowding (m^*) were calculated using the equation $m^* = \overline{x} + (S^2/\overline{x} - 1)$ (Lloyd 1967), and mean densities were calculated for all instars of *Psorophora columbiae* and *Anopheles crucians*. Mean crowding was then plotted against mean density to generate a linear regression ($m^* = \alpha + \beta x$), where α = intercept on m^* axis and β = the slope. As calculated values of α were greater than zero, this suggested larvae occurred in groups, especially the younger instars. The fact that values of α were larger than unity indicated that these groups of larvae were themselves aggregated. Values of both parameters decreased with later classes.

Before rice harvesting, 93.4% of larvae of *Psorophora columbiae* are within 1 m of the leaves indicating that for more efficient larval surveillance, sampling could be restricted to along the contour levees. Within this 1-m zone, however, the distribution of larvae was very patchy. After harvest when ruts form over the rice fields larvae occur more uniformly throughout the rice fields, but again to save time sampling can be restricted to water adjacent to the levees. In contrast the distribution of larvae of *Anopheles crucians* was independent of harvest, and larvae were distributed at all distances from the levees. The regression of mean crowding on mean density gave values of 0.01 for α and 1.19 for β indicating that larvae did not occur in groups, although individual larvae were contagiously distributed within the rice field.

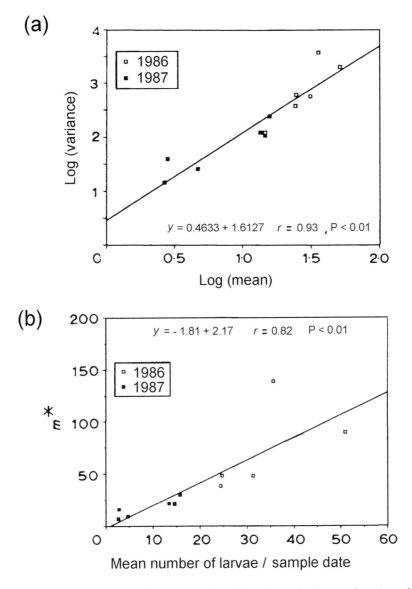

Fig. 3.8. (a) Taylor's power law regression of log (sample date variance) on log (sample date mean) for *Anopheles quadrimaculatus* larvae collected on six dates in 1986 and six dates in 1987; (b) lwao's regression of Lloyd's mean crowding (m*) on mean number of *Anopheles quadrimaculatus* larvae/sample date, for six dates in 1986 and six dates in 1987 (Walker et al. 1988)

The authors stress that such studies as these provide the statistical foundation for a better sampling strategy and a basis for interpreting the results. However, in the vast majority of instances there are no statisticcal considerations given as to the number of samples required for meaningful results.

Pitcairn et al. (1994) studied the spatial distribution of *Anopheles freeborni* and *Culex tarsalis* in California, USA rice fields. Sampling of larvae was carried out as follows: three individuals spaced approximately 10 m apart each took three dips using a 400 ml dipper at 10 sites, 3–5 m apart, along a linear transect. One 'sample' therefore consisted of 90 dips. Taylor's power law and Iwao's patchiness regression were used to quantify the relationship between the sample mean and variance. For *Anopheles freeborni*, Taylor's model showed that the level of aggregation was highest among the first instars and then decreased, as demonstrated by a decrease in both *a* and *b* values. Taylor's model applied to the spatial distribution of *Culex tarsalis* larvae also showed that aggregation was highest among the first instars and then declined. Iwao's model also showed that early instar *Anopheles freeborni* larvae were more aggregated than older instars, as indicated by the decline in β with age. The value of α was significantly different from 0 for first instars only. For *Culex tarsalis*, β was highest among first instars and lowest among third and fourth instars.

Mogi and Mokry (1980) used Iwao's (1968) method of plotting the regression of mean crowding (m^*) on mean density (m) of larvae of *Wyeomyia smithii* in small (11–20-ml) and large (21–30-ml) capacity pitchers of *Sarracenia purpurea*. They found that the index of basic contagion (α), that is the intercept, was positive, indicating an aggregated distribution, similarly the density-contagiousness coefficient (β), the slope of the regression line, was greater than unity again indicative of contagion.

Ikemoto (1978) also plotted mean crowding against mean density to study the spatial distribution of *Anopheles sinensis* larvae in Japanese rice fields. All larval instars and pupae were contagiously distributed, and the pattern did not vary significantly between age classes or between rice fields on 22 sampling days.

Bradshaw (1983) and Bradshaw and Holzapfel (1983) made good use of Lloyd's (1967) mean crowding index to study the association and coexistence of tree-hole mosquitoes in Florida, USA. As they pointed out Lloyd's indices are especially suited to discrete habitats such as tree-holes where mean crowding can be taken as the mean numbers of species *y* encountered per litre of water by the average individual of species *x*. Now if x_i represents the numbers of species *x*, y_i the numbers of individuals of species *y*, and the volume of water in the *i*th tree-hole is v_i, then the mean *inter*specific crowding of *x* by species *y* is given by

$$\sum \frac{x_i y_i}{v_i} / \sum x_i \qquad (3.14)$$

Mean *intra*specific crowding is obtained by substituting $(x_i - 1)$ for y_i above. Fig. 3.9 summarises interspecific and intraspecific crowding as found by Bradshaw and Holzapfel (1983).

In a later study of *Anopheles plumbeus, Ochlerotatus geniculatus* and *Culex torrentium* breeding in tree-holes in England Bradshaw and Holzapfel (1991) also calculated mean crowding of one species on another.

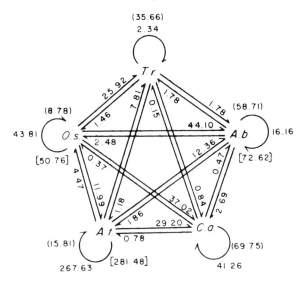

Fig. 3.9. Mean intra- and interspecific crowding/litre of actual tree-hole volume in 35 holes sampled 17 times, where T.r. = *Toxorhynchites rutilus*, A.b. = *Anopheles barberi*, C.a. = *Corethrella appendiculata*, A.t. = *Ochlerotatus triseriatus*, O.s. = *Orthopodomyia signifera*. Straight arrows show extent of interspecific crowding where y → x is the crowding of *y* on *x*. Circular arrows designate intraspecific crowding. Numbers in parentheses indicate the total numbers of individuals of all other species encountered/litre by an individual of a given species. Numbers in brackets indicate the total crowding/litre, including intra- and interspecific, of filter-feeders on the given species (Bradshaw and Holzapfel 1983)

Costanzo et al. (2005) used mean crowding as used by Bradshaw (1983) to investigate the average number of *Aedes albopictus* that one *Culex pipiens* individual encounters within discarded automoblie tyres in Illinois, USA. The equation for mean crowding used was:

$$\sum_{i=1}^{5}(a_i * c_i) / \sum_{i=1}^{5}c_i \qquad\qquad (3.15)$$

where a is total number of *Aedes albopictus* larvae found within a tyre and c is total number of *Culex pipiens* found within a tyre. These authors disregarded the volume of water in the tyres and so results were expressed per tyre rather than per unit volume. In May, June, and July, *Culex pipiens* larvae encountered a median of <6 *Aedes albopictus* larvae per tyre, but in August and September, they encountered >25 *Aedes albopictus* larvae per tyre.

Iwao's (1968) regression analysis has been used to describe dispersion of a number of mosquito larval populations (Andis and Meek 1984; Chubachi 1979; Ikemoto 1978; Service 1985); the regression coefficients were generally low, which may be due to the difficulty of deciding on an appropriate sample unit for a larval population. With container breeders Kitron et al. (1989) suggested that ovitraps provide convenient, albeit artificial sampling units. They presented results showing that the egg distribution of *Ochlerotatus triseriatus* in 300 ovitraps were highly aggregated.

Chubachi (1979) studied the relationship between the density of *Culex tritaeniorhynchus* larvae and the distances samples were taken from the edge of paddy fields, in order to estimate absolute density over his entire census area (279 paddy fields). Counts were transformed according to the procedures of Iwao and Kuno (1971) so as to stabilise the variances. The actual form of transformation needed is derived from the relationship between mean crowding (m^*) and mean density of larvae and pupae (m) calculated from

$$m^* = \alpha + \beta m \qquad\qquad (3.16)$$

He found that the appropriate transformations were $\log (x + 1)$ for 1st instar larvae, $\sqrt{x}$ for 4th instars, and $\sinh^{-1}\sqrt{((\beta-1)/(\alpha+1))x}$ for 2nd and 3rd instars and pupae. Rather surprisingly Chubachi (1979) found that the constant α was not significantly greater than zero except for 4th instar larvae, which suggests that all other age classes do not occur in groups.

Service (1985) and Walker et al. (1988) present several useful references to papers that have examined the mathematical distribution of mosquito larvae, and to those that contain calculations of various indices of aggregation (e.g. Taylor's, Iwao's, Lloyd's).

Aquatic Nets

Large or small nets mounted on a frame and attached to a handle can be drawn through the water to sample mosquito larvae in relatively large habitats, e.g. borrow pits, ponds and marshes. The bag should be made from relatively coarse material because if the weave is too fine it is likely to become clogged with silt and fine particles, preventing the flow of water through it. Because of the relatively large volume of water sampled, nets are very useful in collecting large numbers of larvae, and also in detecting the presence of breeding when populations are very small. Drawing a net through the water causes considerable disturbance and mosquito larvae within a comparatively large area submerge. Consequently, further samples should be taken some distance away otherwise the numbers caught per sweep will be considerably reduced. When floating debris and vegetation are caught in the bag, the net should be partially submerged in the water and the debris vigorously shaken to dislodge larvae before it is discarded. The contents are then tipped into a large white bowl containing clean water, and after allowing mud to settle, the larvae are removed with a pipette. It is useful to have more than one bowl so that while the sediment is settling in the first bowl further samples can be tipped into another bowl. Some nets have a glass, or preferably plastic, vial about 4 × 10–12 cm inserted through a hole in the bottom (Figs 3.10a, b) to concentrate the catch in a small volume of water. With these nets positive sweeps, that is those that have collected larvae, can often be identified without tipping the contents out into a dish for examination. However, larvae may become stranded on the net and not collect in the bottle unless flushed through with further water. The narrow opening of the bottle may become blocked if floating debris or vegetation is present.

Nets are most useful when the objective of the exercise is to identify the presence or absence of certain species as a relatively large volume of water can easily be sampled. Lardeaux et al. (2002) working in French Polynesia used small plankton nets with a 300µm mesh to sample mosquito immatures. Because of the difficulties encountered in sampling some water tanks and cisterns, net sampling could not be quantified and so a presence/absence system with four abundance categories was used: no larvae, low (<10 larvae), medium (<100 larvae), and high (>100 larvae). While obtaining quantitative estimates of population density with nets is problematic, when nets of known diameter are drawn through the water for measured distances, attempts can be made to relate the catch to the total population. For example, in Arkansas, USA Meisch et al. (1982) used both nets and dippers to quantitatively sample mosquito larvae (e.g. *Psorophora columbiae, Anopheles quadrimaculatus, Culex erraticus*) in rice fields. They used a 30.5-cm D-frame net and made 1-m sweeps through the water. Ten such sweeps, sampling 405 litres of water, were combined as a sampling unit. Similarly, 10 dipper samples, 4.2 litres of water, were combined for a sampling unit. However, in the analysis of the results the dipper and net samples were combined by calculating the number of larvae/dip from the volume of water sampled.

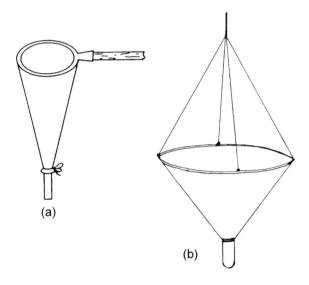

(a)

(b)

Fig. 3.10. (*a*) aquatic net with plastic/glass collecting bottle (Service 1993); (*b*) well net with plastic collecting bottle (World Health Organization 1992)

Marten et al. (1996) used a plankton net (120 μm mesh) attached to a 20 cm-sided square wooden frame attached to a pole to sample mosquito immatures and associated fauna and flora from potential *Anopheles albimanus* larval habitats in Colombia. These nets were usually used to sample between 10 and 50 m, along a linear transect where possible.

Fischer and Schweigmann (2004) sampled *Culex* immature stages from temporary pools in urban Buenos Aires, Argentina, by dragging a 350 μm mesh hand net (10 × 8 cm) once along the bottom of each pond along its longest axis. Mosquito abundance was related to pond surface area, depth, temperature, shade, among others. The surface area of each pool was estimated by measuring the area of a rectangle containing the pool, and multiplying this by the estimated proportion of the area covered by water. Fischer et al. (2002) used the same technique to study the seasonal dynamics of *Ochlerotatus albifasciatus* in temporary pools in Buenos Aires. In this study, the number of samples taken per water body was proportional to the surface of the water body and ranged from 5 to 160 dips. The relative abundance of larvae in each pool was calculated by multiplying the surface area times the mean number of larvae per dip.

Doran and Lewis (2003) describe a technique for sampling invertebrates including mosquitoes from snowmelt and rain-fed pools using a small aquarium net measuring 15 × 12.5 cm and a D-frame dip net measuring 25 × 31 cm, in suburban Montreal, Canada. The larger D-frame dip net was used when the water depth exceeded 15 cm and the smaller net was used when the maximum water depth was less than 15 cm. A maximum of 10 sweeps one metre in length was made in each pool, and these sweeps were performed both near the edge and in the middle of the pool. As pools became smaller in area, the number of sweeps decreased so as to sample one half of the surface area. Nets were dragged close to the bottom of each pool, ensuring the collection of organisms from the pond bottom and from the water column. The authors recognised the difficulties of quantitatively analysing the data obtained from different sized ponds using different nets, so to compare the relative abundance of mosquitoes as a measure of sampling effort, the total number of specimens collected per week was converted to number of insects per dip, by dividing the total number of larvae collected in all pools in a given week by the total number of sampling dips in that week. Dip values for sweeps with the smaller aquarium net were converted by multiplying these values by a factor of 2.07, as the small aquarium net sampled 2.07 times less than the D-frame dip net. Approximately 0.0375 m^3 of water was sampled by every 1 m sweep with the D-frame dip net.

Lopes and Lozovei (1995) used a D-shaped aquatic net (maximum diameter 53.5 cm) attached to a three-piece aluminium pole capable of extension to a total length of 3 m to sample mosquito larvae along a river bank in Paraná, Brazil. The net was made from cotton and had a maximum depth of 20 cm. At each sampling site, 20 net sweeps were performed. A total of 2764 mosquitoes of 18 species were collected using this method.

Hatfield et al. (1985) used what they termed a dredge net to quantify larval sampling of riceland mosquitoes. They made a 38-cm tall and 10-cm wide rectangular frame from 0.64-cm diameter steel rods, and then welded on the bottom two angled 20-cm long steel rod 'skids' 4 cm below the frame to prevent the net from dragging on muddy bottoms of ponds and rice fields (Fig. 3.11). A wooden handle was fixed into a locking ferrule fixed to the top of the frame. The muslin collecting bag had the apical part made of 100-mesh nylon netting. Two marks 1 m apart were painted on the handle so that when the handle was pulled through the operator's hand from the upper to the lower mark, a 1-m length of water was sampled. The mean number of larvae/area sampled (10 cm × 1 m = 0.1 m^2) or per volume

(10 cm × 1 m × water height in the entrance of the frame) can be easily obtained.

Montgomery and Ritchie (2002) used a small aquatic net to sample *Aedes aegypti* and *Ochlerotatus notoscriptus* from deeply flooded roof guttering in Queensland, Australia. A linear meter was randomly selected in any flooded gutter and small sandbags placed at either end to prevent dispersion of the pupae. Leaf litter was removed by hand. Four sweeps of a 10 × 12.5 cm net through one linear metre of gutter caught 71% of the *Aedes aegypti* pupae ($n = 10$, mean ± SE = 7.1 ± 0.53) in tests of the method. The number of pupae/gutter was then estimated by extrapolation incorporating a sampling efficiency of 71% and the proportion of flooded guttering that was sampled. A total of 212 pupae were collected from 13 roof gutters at 11 premises, giving an estimated total population of 1472 pupae. Pupae collected were either *Aedes aegypti* (57.3%) or *Ochlerotatus notoscriptus* (42.7%), with 49.8% and 48.2% of the standing crop (roof gutter plus yard survey) originating from roof gutters. No *Culex quinquefasciatus* were collected from roof gutters.

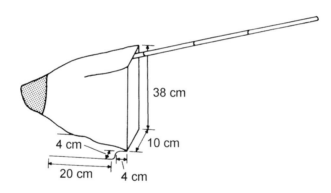

Fig. 3.11. Dredge with 20-cm long 'skids' 4 cm below net frame (Hatfield et al. 1985)

Hamad et al. (2002) report using a dip-net attached to a 15 m length of rope to sample larvae from wells in eastern Sudan. In Myanmar, Oo et al. (2002) reported sampling *Anopheles dirus* larvae and pupae using a well net that consisted of a conical shaped, white cloth net with a diameter of 35 cm and length of 46 cm. It was held at an angle by four strings and controlled by a long string or rope. The net was introduced into the well with the lower side of the net just under the surface of the water and its opening at an angle of about 45°. The net was moved slowly around the side of the well twice, then quickly withdrawn and inverted into a bowl of water. Ten dips per well, per week were taken and the mean number of larvae and pupae per dip were determined. The mean number of larvae and pupae per dip ranged from 2.5 in March 1999 to 58.4 in October 1999 and numbers appeared to be positively correlated with rainfall.

Tun Lin et al. (1994) describe a quantitative method for sampling *Aedes aegypti* immatures from large metal drums (200 litre capacity, 57 cm diameter) using a cotton (100 μm gauze) sweep net 20 cm long and attached to a 10 cm wide × 20 cm high metal frame and a 180 cm extendable handle. The net was immersed gently into the water against the side of the drum and swept rapidly (~10 s) once around the circumference. An estimated 29.5 litres of water were sampled. Experiments were carried out to compare the efficiency of the net and a 350 ml, 14 cm diameter dipper at sampling from drums containing known numbers of immature mosquitoes. Percentage recovery from drums containing 200 larvae in either 67, 133, or 200 litres of water was similar for all volumes when using the net (25.7% to 29.3%). With the dipper, the proportion of larvae recovered decreased significantly with decreasing water volume and in 30 trials of 10 dips each, positive dips were only recorded in 20% of trials when the volume of water was 67 litres, compared with 80% positive when the water volume was 133 litres. From 225 trials sampling from populations of 100–400 immatures in a full drum, at least one immature was collected by sweeping the surface or dipping on 93.3% and 72.9% of occasions respectively. The net was twice as efficient as the dipper when the full drum contained only 100 immatures. Pearson's correlations of numbers of fourth instar larvae collected with a single net sweep against actual numbers of larvae present, gave coefficients of 0.99 and 0.98 from field data in 1989 and 1990 respectively. The above method was later modified (Tun-Lin et al. 1995c) to improve the efficiency of removal of larvae. In the modified method, an initial sweep with the net was carried out as described above. This was followed approximately 1 to 2 minutes later by a second sweep made at the bottom-centre of the drum, directed at collecting the larvae accumulating in the tail of the whirlpool created by the first sweep. Third and fourth sweeps were carried out approximately five minutes after the first sweep, repeating the action of the first and second sweeps (Fig. 3.12). This method was tested in the laboratory in drums containing 67, 133, or 200 litres and 200 of each of the four larval instars and pupae. Mean percentage recovery irrespective of water volume and larval stage was 50.6 ± 12.1% for a drum with rust and 61.9 ± 12.6% for a drum without rust. This difference was not significant. The four-sweep method was most effective at sampling fourth instar larvae (66.2 ± 19.4% for the rusted drum and 73.0 ± 17.0 for the drum without rust). Field sampling from 28 drums containing

known numbers of immatures showed that the four-sweep method removed
85.4 ± 10.7% of the immatures. This method was more effective when the
drums were full (93.8 ± 6% removal).

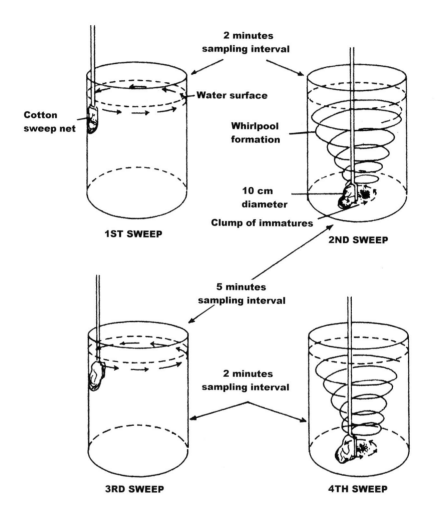

Fig. 3.12. Diagrammatic representation of the method of Tun-Lin et al. (1995*c*)
for sampling *Aedes aegypti* immatures from a 200 litre drum using a cotton
sweep net

Kay et al. (2000) used the quantitative sweep-netting method of Tun-Lin
et al. (1994) to sample mosquito immatures in subterranean water drainage
manholes in Australia. The net was swept around the perimeter of the
manhole just below the water surface, and at the middle and bottom of the

water column. In trials, the net sweeping method recovered at least one larva in 25–75% of replicates when larval populations were between 25 and 100 larvae per manhole. At populations in excess of 200 larvae per manhole, presence of larvae was always detected. Net sampling recovered 5.8% (95% CI 0–11.7%) of the larval population.

In Puerto Triunfo, Colombia, a method similar to that described by Tun-Lin et al. (1994) was used by Romero-Vivas et al. (2002) to determine the productivity of large tanks and drums for *Aedes aegypti*. Trials were conducted to relate the numbers obtained from a single sweep of the surface layer of water and the total number of 4th instar larvae and pupae in the water container. One hundred, 500, and 1000 4th instar larvae and 100, 200 and 500 pupae were added to a 220 litre metal drum, a 446 litre tank and a 1498 litre tank. Each tank type was tested at three water levels: one-third full, two-thirds full and full. Ten individual sweeps at 10 minute intervals were performed for each combination of tank, water depth and number of larvae and pupae. Regression analysis was used to determine calibration factors between numbers in a single sweep and numbers in the container. The estimated recovery of 4th instars of *Aedes aegypti* by a single sweep varied between 20.6% and 30.7% depending on the size of container and the water level. The recovery of pupae of *Aedes aegypti* was estimated at between 31.6% and 50.8%, depending on the container type and water level.

Kubota et al. (2003) also used a sweeping method to sample *Aedes* larvae from 80-litre drums in Brazil. Two hundred fourth instar larvae were added to the drum and 10 sweeps were made with a plastic sieve with a 16 cm diameter and a 60 cm plastic handle. Eight sweeps with the sieve proved sufficient to collect up to 72% of the total larvae in the drum.

Area samplers and quadrats

Cambournac (1939) and Bates (1941) appear to have been the first to have used quadrats in the form of metal cylinders or boxes to sample mosquito larvae. In Portugal mosquitoes breeding in rice fields were collected by imbedding rectangular metal tins, open at both ends, into the bottom mud so that an area of 0.1 m^2 was enclosed (Cambournac 1939). In Albania Bates (1941) thrust open-ended petrol tins down into the mud to sample larvae of the *Anopheles maculipennis* group. Each tin enclosed about 500 cm^2 of water surface. Because it is easier to find and collect larvae enclosed within quadrats when the water level is near the top, a series of tins of various heights were made for use in different depths of water. More recently, Victor and Reuben (1999, 2000a, 2000b) used metal quadrats with dimensions of 33 × 33 × 15 cm and open at both ends to sample mosquito larvae inhabiting rice fields in India. The quadrats were pushed into the rice field soil at random locations and the water contained within was removed using an enamel bowl.

Horsfall (1946) considered it difficult to quantify results obtained from larval dipping due to variations in efficiency of the collectors, and so developed what he termed 'area samplers' to collect mosquito larvae. Originally (Horsfall 1942) these consisted of metal cylindrical screen cages, 13 in diameter, 13 in tall and open at both ends, circumscribed by a band of metal at each end which were joined together by four vertical metal struts. Later, area samplers were made from 16-in square metal boxes (Horsfall 1946). They were dropped at random in larval habitats and the enclosed larvae counted and removed. To test the efficiency of this method larvae of *Psorophora confinnis* were introduced into two experimental plots at densities of 16 and 64/ft^2 (ratio of 1:4) (Horsfall 1946). When 100 dips with a conventional dipper were taken from these two plots 2 and 17 larvae respectively (ratio 1:8.5) were collected. Whereas when 10 quadrats (area samplers) were used, which gave the same total area sampled as 100 dips, 10 and 35 larvae were caught from the two plots (ratio 1:3.5). From this simple trial it was concluded that quadrats appeared to give a more accurate comparative index of larval populations than was obtained by dipping. In natural habitats, Horsfall (1946) compared the efficiency of 10 dips and a quadrat, both of which sampled about 1 ft^2 of water surface. In areas with low population densities larvae were detected in 27% of the different catching stations by dipping and in 50% by the quadrat. In areas with higher larval densities breeding was indicated in 73% of localities by dipping and in 95% by quadrats, again showing that quadrats were more efficient.

In Sarawak Heathcote (1970) successfully obtained quantitative samples by sinking 0.5 m × 0.5 m metal frames into shallow areas of rice fields and removing larvae with a fine mesh net. Curtis and Frank (1981) used this method to sample *Aedes vexans* in Florida. They used two 500-ml capacity dippers with screen mesh bottoms to remove larvae from the quadrats. One dipper had a 250 μm mesh bottom and was used for 1st instar larvae, but was prone to clogging with detritus so a 500 μm dipper was used for older larvae. In trials with known numbers of larvae 88 ± 5% of 1st and 94 ± 3% of older instar larvae were recovered.

Rajendran et al. (1995) used square (31.6 cm × 31.6 cm), 15 cm tall metal frames open at both ends to sample mosquitoes in rice fields in Madurai, southern India. The 0.1 m^2 quadrats were thrust into the mud and larvae and pupae counted, however, counting all immatures present in the quadrat was found to be too time-consuming, and so only pupae were counted in order to make comparisons between fields.

In America Darrow (1949) used metal quadrats to detect the presence of *Anopheles quadrimaculatus* larvae stranded on mud after the water had receded from breeding places. She thrust the metal tins into the mud, filled them with water and collected any larvae that swam to the surface.

Nielsen and Greve (1950) used what they termed a 'larvascope', but is essentially a quadrat, to record the numbers of mosquito larvae in temporary pools. It consists of 6 × 6 in plate painted white on its upper surface and fixed at right angles to a stick, so that when it is lowered in the water larvae and pupae can be counted against its white background. According to Nielsen and Nielsen (1953) it is possible to get good population estimates from a series of such counts when larvae are more or less evenly distributed.

In Turkey, Simsek (2004) studied the larval population dynamics of *Culex theileri* in a lake ecosystem. Thirteen quadrats, each enclosing an area of $1m^2$ were set up in seven habitat categories. Samples comprising of a single dip with a standard long-handled dipper (350 ml) were taken from every quadrat on a monthly basis. Samples were always collected by the same individual at the same time in the morning (0900–1200 h) or afternoon (1400–1700 h), and each sample was maintained separately. Seasonal changes in larval populations were determined by taking the averages of the numbers obtained monthly from the 13 sampling quadrats.

Working in Russia Nikolaeva and Ol'shvang (1978) concluded that only about 56–61% of the *Anopheles* larvae present in an area are actually caught with an aquatic net, and that the capture rate is always higher with 4th instar larvae, whereas when using a dipper the capture rate is about 73%. For culicines the capture rates are considerably smaller. Because of such inefficiencies they developed a 'biocoenometer'. This is a metal square quadrat of varying length (30–70 cm) with 25-cm long sides which is pushed into the bottom mud of larval breeding places. The quadrat has a base area of 0.0625 m^2. A small net (12 × 12 cm) is used to remove the mosquitoes entrapped within the quadrat. In water with relatively few larvae, e.g. 50 per sample, 2–5 min are needed to remove all larvae, whereas at higher densities (400 larvae/sample) it takes 10–12 min; apparently at least 90–92% of the larvae enclosed by the biocoenometer are removed. In small pools (10–15 m^2) 1–2 quadrats are sufficient for larval sampling, but in larger pools (200–400 m^2) 5–7 quadrats are needed. The authors believe that this approach gives a highly accurate estimate of the numbers of larvae/1 m^2.

Hagstrum (1971*a*) collected larvae and pupae enclosed within metal quadrats by pumping out the water. This was done by connecting the inlet nozzle of a hand tyre pump with a length of rubber tubing to an empty gallon bottle, and placing another piece of tubing from the bottle into the quadrat. On pumping, pressure is reduced in the bottle causing water to be sucked into it from the quadrat. Larvae, and also debris, are filtered out before the water enters the bottle. In a series of experiments Hagstrum (1971a) pumped out each quadrat three times and recorded the total catch of larvae and pupae. The proportions of all instars of *Culex tarsalis* obtained by emptying the quadrats and also by dipping from within them were similar, but when dips were taken outside the quadrats a smaller proportion of 1st instars was obtained.

In experiments in Japan known numbers (5, 10, 15, 20) of all larval instars of *Anopheles sinensis* and *Culex quinquefasciatus* were introduced into a 25 × 25 × 25 cm metal quadrat, and the water sucked into a bottle by means of a portable air cleaner-type suction pump. In evacuating the water five times Ikemoto (1976) found that virtually all larvae were removed by the third sample. Although the method was considered suitable for field work it is too tedious for extensive surveys.

In ponds in Louisiana, USA Holck et al. (1988) used a 106-cm length of PVC irrigation pipe (700 cm^2 opening) and a similarly long plexiglas box with a 1000 cm^2 opening, modified from the area sampler of Farlow et al. (1978), to collect mosquito larvae. Each sampler had two handles near the top and bevelled edges on the bottom to help them embed firmly into bottom mud. A manually operated portable bilge pump with a 100-mesh nylon bag fitted to the outlet end of the bilge hose pumped out the enclosed water and larvae. In addition, about 2–3 cm of bottom mud was removed with a 450-ml plastic cup attached to a 1-m wooden dowel and placed in the nylon bag, which was shaken in pond water to remove the excess fine soil and other filterable debris. The sample was then tipped from the bag into a 0.95-litre flask containing 95% ethanol and larvae and pupae removed in the laboratory.

In estimating the numbers of *Ochlerotatus taeniorhynchus* larvae and pupae in experimental plots Pausch and Provost (1965) stirred up the water to get a more even distribution of the immature stages before dropping narrow plastic cylinders into the water. The total catch removed from the cylinders was related to the volume of water sampled so that estimates of the total population in the habitat could be obtained.

Several difficulties may be encountered when area samplers—cylindrical or square-shaped—are pushed down into the bottom mud. In thrusting a sampler into position some larvae may escape from the volume of water that is to be enclosed before the sampler is firmly bedded down. Consequently the numbers per unit volume or area will be underestimated. Submerged debris sometimes prevents cylinders being pushed firmly down into the bottom of habitats, and may even buckle the lower edges of thin metal cylinders. Area samplers therefore need to be made of strong plastic or metal, with the lower edge sharpened, bevelled or reinforced to enable it to be pushed firmly into mud and cut through submerged leaf litter and twigs. Sometimes it may be necessary to bolt a band of saw-like teeth along the bottom edge so that when the sampler is rotated first in one, then in the opposite, direction it saws through any bottom debris (Wilding 1940). For this reason it is more convenient to use cylinders than square-shaped tins.

Because larvae may take a long time to resurface, it is better to make several repeated collections from a series of samplers, than to try to collect all larvae from one sampler before examining the others. Larvae from each sampler can be kept separate by placing them in a small bottle attached to the outside of the cylinder. It is sometimes helpful to use cylinders that are slightly tapered so that they can be stacked within each other, thus making it easier to carry several to the field.

Walker et al. (1988) analysed the distribution and abundance of *Anopheles quadrimaculatus* larvae in a marsh using a point-quadrat system (Southwood 1978). In 1986 30 wooden dowel rods were inserted 1 m apart in a north-south transect across the marsh, and perpendicular to this 15 similar dowels marked off an east–west transect. In 1987 the transects consisted of 10 parallel lines of 10 dowels, thus forming a square of 100 dowels 1 m apart. Sampling points (dowel posts) were chosen randomly according to the method of Schoen and Fruchter (1983) and a 0.5-cm diameter (0.1963 m^2) hoop of plastic tubing was placed over the dowel. The larva within this floating hoop nearest the dowel was identified, then the distance to its nearest neighbour measured with a ruler. This enabled the nearest-neighbour estimate of larval density to be calculated, using the simple formula of $1/4r^2$ (Southwood 1978), where r is the mean distance to the nearest neighbouring larva. Then 15–60 min were spent pipetting all larvae from within the hoop. From this the absolute number of larvae/m^2 was estimated. In 1986 the nearest-neighbour estimate of larvae/m^2 was 166.5, compared to 152.8 ± 77.1 based on complete removal of all larvae within the area samplers (hoops). In 1987, the nearest-neighbour estimate was 37.1 larvae/m^2 and the estimate based on removal of larvae was 44.5 ± 21.3 larvae/m^2. Nearest-neighbour estimates require populations to be randomly dispersed, so it is surprising that in the present case, where *Anopheles quadrimaculatus* larvae were aggregated, that estimates of population size agreed so well with estimates based on larval removal. Walker et al. (1988) discuss the possible reason for this, and stress that the validity of this method should be evaluated against other methods of population estimation.

Lesser (1977) combined sampling with quadrats with the removal method of estimating population size by plotting, either by eye or by least mean squares, the regression of numbers caught against cumulative catches. The procedure was to take about five dips with a ladle from each of a number of quadrats (0.018 m^2) and to establish total mosquito population within each quadrat. A mean population value for a quadrat was obtained and related to quadrat area, for example an estimated 64 *Ochlerotatus sollicitans* larvae/0.018 m^2, and by simple proportion the total population in the breeding sites could be estimated. Under laboratory conditions checks made by introducing known numbers of larvae of *Ochlerotatus sollicitans* and *Culex pipiens* into cylinders showed an average accuracy of 96 and 93% respectively for the two species.

Area sampler of Roberts and Scanlon

This consists of two plexiglas cylinders (Roberts and Scanlon 1974). The larger one is 5 in. in diameter and 11 13/16 in. tall, while the smaller is 4 7/16 in. in diameter, 12 in. tall and has an inverted plastic cone with a 67° slope inserted into the bottom end (Fig. 3.13).

Fig. 3.13. Plexiglas area sampler (Roberts and Scanlon 1974)

A weighted rubber stopper is secured to a length of nylon string attached to a plastic rod, which is placed across the opening at the opposite end of the cylinder. In use, after the larger cylinder has been firmly thrust into the bottom of a pool, the smaller cylinder is inserted and carefully pushed down to the bottom. After a predetermined time, say 15–20 min, to allow mosquito larvae enclosed by the cylinders to pass through the smaller opening of the cone, the stopper is pulled up to seal the inner cylinder. This is then withdrawn and trapped larvae emptied into a dish and sorted.

Roberts and Scanlon (1974) reported that this sampler was exceptionally useful, but when it was used in woodland pools leaves and debris floated to the top and obstructed the narrow opening of the cone. To overcome this, plant debris is removed from the enclosed area after the outer cylinder has been positioned but before the inner one is introduced.

This trap is cheap, easy to construct and has the advantage that it can be used in pools with less than 1-in depth of free water. For use in such shallow habitats after it has been thrust as deep as possible into the muddy substrate, water is added to the inner chamber until it rises above the small opening in the cone.

Area sampler of Downing

Downing (1977) developed a relatively simple area sampler based on previous models (Roberts and Scanlon 1974; Welch and James 1960). His sampler consists of a plastic outer open-ended cylinder with height graduations in inches marked on the outside. A slightly smaller cylinder with an inverted funnel attached to the base is pushed down and into the outer cylinder. After an appropriate exposure period—found with *Ochlerotatus canadensis* breeding in shallow pools to be 15 min for 50% of the trapped larvae to swim up through the cylinder into the inner cylinder—a rubber bung is inserted into the funnel. The cylinder is then pulled out by a wooden dowel handle fixed across its top, and the contents tipped into a white tray for sorting. The water surface sampled is 18 in^2.

When this sampler was placed in shallow (6.7-in.), intermediate (9.6-in.) and deeper (11-in.) parts of 7.5-ft diameter pools Downing (1977) found no differences between the numbers of larvae collected in these areas. Graduation marks on the outer cylinder, allowed the volume of water sampled to be obtained and the numbers of larvae/unit volume, taking 100 in^3 as a convenient standard unit, to be calculated. Again there were no differences between numbers of larvae estimated/100 in^3 in different areas of the pools. He concluded that larvae of *Ochlerotatus canadensis* were not aggregated but uniformly distributed throughout the pools.

Enfield and Pritchard sampler

A full description of this sampler and its use in ponds containing aedine larvae is given by Enfield and Pritchard (1977*a*), and its further use is described by Enfield and Pritchard (1977*b*). They thrust, as quickly as possible, a 10- or 20-cm diameter PVC cylinder into the mud at the bottom of flooded larval habitats. A galvanised so-called trapping-vessel was then lowered within the cylinder with a polythene funnel fitted at the bottom in the raised position (Fig. 3.14). The water level should be just above the inner funnel. Larvae swim through the gap between the inner funnel and an outer funnel to the water surface, but do not swim back again. After 20–60 min the inner polythene funnel is lowered to close the gap between it and the outer funnel, and then the trapping vessel is withdrawn from the PVC cylinder. Larvae can be filtered off and counted, or preserved. In the laboratory the sampler recovered 95–100% of any instar of *Aedes vexans*.

This sampler has been used in Canada to sample several aedine species (Enfield and Pritchard 1977*a*, *b*). It may tend to undersample 1st instar larvae because they may be able to stay submerged and obtain most of their oxygen through cuticular respiration. Enfield and Pritchard (1977*b*) also identified contagious larval distribution as a possible problem in getting reliable population estimates. Increasing the numbers of samples and/or using larger diameter cylinders may help overcome some of the sampling problems. They concluded that they needed a minimum of about 40 samples, and the size of the samples (cylinder diameter) must be appropriate to the density of larvae being sampled.

In one experiment Enfield and Pritchard (1977a) thrust a graduated post into the lowest point, which was about in the middle, of a pond. Eight transect lines radiated out from the post in such a way as to divide the pond into eight sectors of approximately equal areas. Each transect line was marked off at 0.5-m intervals allowing the area of the flooded part of the pond to be calculated. When the pond was extensively flooded intermediate points were marked off so that the flooded area could be more accurately measured. The larval populations of *Aedes cinereus* and *Aedes vexans* were sampled by thrusting the Enfield and Pritchard samplers into the pond at either random or regular sites along the transect lines. The number of samples taken depended on the extent of flooding of the pond. The samplers were left in position for 20–60 min and samples were taken at the same time each day (1000–1030 h) so that any changes in the spatial distribution of the larvae caused by increasing water temperatures were minimised. For each stratum the density of the larvae was calculated by dividing the mean number/sample by the area of the cylinder being used (0.0314 m^2 or 0.00785 m^2). Population estimates were obtained by multiplying these densities by the appropriate areas of the strata.

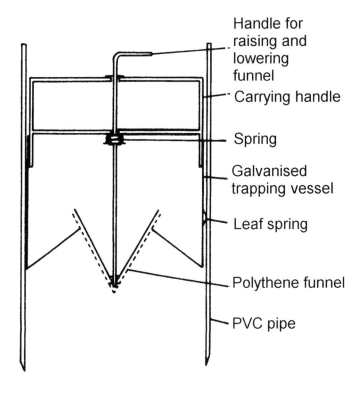

Fig. 3.14. Sampler of Enfield and Pritchard (1977*a*)

Area sampler of Taylor

Area samplers usually suffer from two major problems. Firstly time is wasted in waiting for larvae to surface and be collected after placement of the sampler, and secondly unless the substrate forms a good seal with the sampler it is inoperable in water less than about 5 cm deep. To some extent Roberts and Scanlon (1974) overcame the first problem by using 10 samplers simultaneously, while Lesser (1977) reduced the time spent on each sampler by applying the removal method for about 5 min and then estimating the number of larvae enclosed.

To overcome such deficiencies Taylor (1979) constructed a sampler comprising of a 33-cm long, 12.0-cm internal diameter plexiglas cylinder (Fig. 3.15) with a serrated steel band extending 3 cm on the lower end (H).

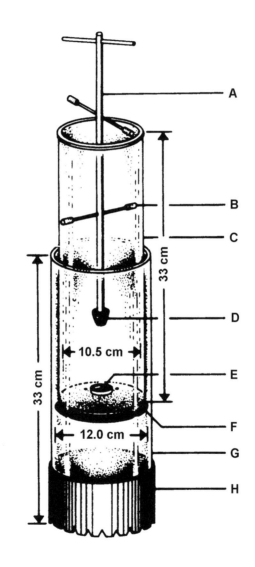

Fig. 3.15. Sampler of Taylor (1979) showing A—steel rod supporting stopper (D), B—cross-bar, F—rubber gasket, G—outer cylinder, H—serrated bottom steel band

There is a similar length of plexiglas 10.5 cm in diameter (C) with the bottom closed with two discs of plexiglas in between which is sandwiched a rubber ring that forms a seal with the outer cylinder (F). A 3.6-cm diameter hole (E) is made in this 'perspex-rubber sandwich' which is bevelled on the lower side to facilitate water flowing into the cylinder, and bevelled into the middle on the interior rim to facilitate seating of a rubber stopper (D) mounted at the end of a steel rod (A). This rod has two plexiglas-tipped cross bars (B) to assist in locating the stopper in the hole in muddy waters.

To operate, the outer cylinder is thrust into the bottom substrate and then the inner cylinder is pushed down so that the water enters through the bottom hole. The rubber stopper is then pushed into the hole and the sample poured through a 10-cm funnel mounted over a screen cylinder (6 cm $\times$ 2.5 cm) similar to that used by Husbands (1969). These cylinders are placed in vials of water or alcohol for later sorting of the larvae. The sampler can be used in pools as shallow as 2 cm and it only takes about 2 min to take each sample.

In simple trials with 2nd and 4th instar larvae of *Ochlerotatus infirmatus* the sampler appeared to collect both age classes about equally, and worked well with both high and low larval densities, but was less efficient in shallow waters. Taylor (1979) presented some calibration factors in terms of the proportion of the larvae in the outer cylinder that are collected from the inner cylinder. Depths with conversion factors in parentheses were 2 cm (0.50), 5 cm (0.64), 8 cm (0.78), 10 cm (0.84), 13 cm (0.89) and 15 cm (0.95). Hence the mean numbers of larvae collected in waters of these depths should be corrected by dividing through by these conversion factors (c), while the variances are multiplied by $1/c^2$. It would be advisable to check the efficiency of this method for different species in various depths of water.

Comparison of dippers and area samplers (quadrats)

Christensen and Washino (1978) in comparing an aquatic net, a 50-in^2 plastic tub that was thrust into the water, and a standard pint (390 ml) dipper, found that the greatest number of larvae of *Culex tarsalis* and *Anopheles freeborni*/sample was obtained by the tub followed by the net and then dipper. As a continuation of this work Nagamine et al. (1979) compared the efficiency and effectiveness of a dipper, an area sampler (69 cm tall, 20 cm in diameter) and a plastic tub (34.5 cm long, 26 cm wide and 11.5 cm high). The long edge of the tub was thrust into the water and the tub allowed to become approximately ¾ full of water (about 4430 ml). When the mean number of larvae/ml was calculated, the dipper consistently gave the highest density, followed by the tub and area sampler. But for any one sample the tub caught the largest number of larvae.

In a comparison of dippers, cylindrical minnow traps and an area sampler for collecting aquatic fauna in Californian rice fields, Takahashi et al. (1982) concluded that the area sampler collected both greater numbers and more taxa than dippers and more taxa than the minnow trap. Their area sampler consisted of an open-ended 50-cm tall 6.53-mm thick plastic box covering an area of 0.1 m^2. A similar, but slightly smaller yet taller, plastic box fitted tightly within the outer box. The inner box was closed at the bottom but incorporated a plastic funnel entrance. In operation the outer box was quickly thrust into the water and bottom mud, and emergent or floating vegetation removed before the inner box was pushed down inside it. A large rubber bung then closed its funnel entrance, the inner box was then withdrawn and the contents filtered through a sieve. The inner box was graduated so the volume of the sample was known. Although the area sampler was more efficient for most organisms, an average of only 1.3 larvae of *Anopheles* and *Culex* were caught per 200 litres of water compared to 37.5 per 200 litres obtained with a 400-ml capacity dipper.

In Texas, USA Roberts and Scanlon (1979) sampled the immature stages of *Ochlerotatus atlanticus, Ochlerotatus tormentor, Psorophora ferox, Psorophora longipalpus* and *Psorophora howardii* with a standard dipper attached to a 4-ft handle so as to cause minimum disturbance to the larvae, and with an area sampler (Roberts and Scanlon, 1974). Area samplers were placed along two transects in two pools and 30 min later a dipper was used to collect larvae from the sample area. The next collecting day the dipper was used first followed by the area sampler. About twice as many larvae of both genera were caught by area samplers than by dippers. A greater proportion of 1st instar aedine species were obtained with dippers, about equal proportions of 2nd and 3rd instar larvae (combined) were taken by the dipper and area samplers, but significantly smaller proportions of 4th instar larvae and pupae were collected with the dipper. With *Psorophora ferox* and *Psorophora longipalpus* there appeared to be no difference in the proportions collected by the two methods. As pointed out by the authors their results do not agree with the findings of Downing (1977), who found that dippers obtained most larvae of *Ochlerotatus canadensis* from the shallow regions of pools, whereas area samplers indicated larvae were equally dispersed throughout the pools.

Andis et al. (1983) compared the numbers of larvae of *Psorophora columbiae* and *Anopheles crucians* in Louisiana, USA rice fields collected in 10 dips with a ladle (capacity 400 ml) taken adjacent to an area sampler (0.1 m^2) on 18 sampling occasions.

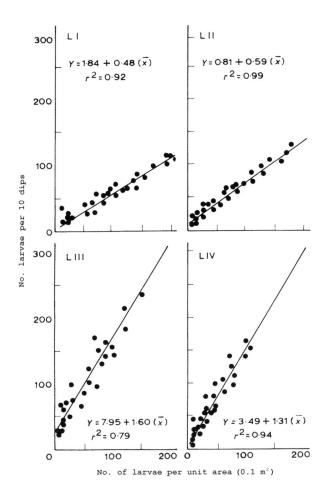

Fig. 3.16. Relationship between number of different instars (I–IV) of *Anopheles crucians*/10 dips and number/larvae/area sampler (0.1 m²) (Andis et al. 1983)

The regression coefficients for all four larval instars combined, obtained by plotting numbers of larvae/10 dips against numbers of larvae/area sampler, resulted in coefficients of 0.77 for *Psorophora columbiae* and 0.69 for *Anopheles crucians*, but to obtain more precise relationships regression lines were plotted separately for all four larval instars (Fig. 3.16).

A limitation of such regressions is that the mean number of larvae/10 dips cannot be directly converted into larval densities. So, to overcome this, mathematical models were derived from the regression analyses to estimate larval densities (larvae/0.1 m²) from the numbers collected in 10 dips. The predictive models are given in Table 3.3. The area sampler was not just more precise in detecting low density breeding, but yielded more species and a greater proportion of early instar larvae. Moreover, the coefficient of variability was usually lower from the area sampler than from dippers, and decreased rapidly as population size increases (Fig. 3.17).

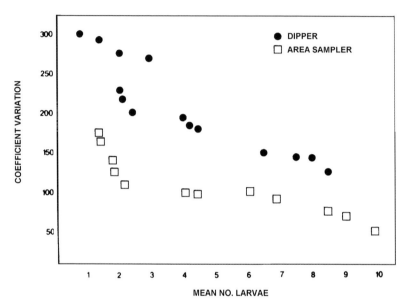

Fig. 3.17. Comparison of coefficient of variation using an area sampler and dipper methods for estimating relative abundance of *Anopheles crucians* (Andis et al. 1983)

Both Andis and Meek (1984) and Ikemoto (1978) believed that area samplers could allow considerable precision in estimating population density and the analysis of larval spatial dispersion patterns. However, the main disadvantage of the area sampler is that it is usually time-consuming to operate.

Awono-Ambéné and Robert (1999) compared tray dippers with static quadrats for sampling *Anopheles arabiensis* from market-gardener wells in Senegal. Tray dippers were 22 cm long, 11 cm wide and 1 cm deep with a capacity of 500 ml. A total of 50 dips were taken in each well, 25 in the centre and 25 at the periphery and all larvae were returned after each dip. The wooden quadrat had dimensions of 1 × 1 × 1 m and was sufficiently deep to reach the bottom of each well. All larvae enclosed by the quadrat were counted. Quadrat sampling was performed once at the centre and once at the periphery of each well. In larger wells, a third quadrat sample was taken from a point between the centre and the periphery. Larvae were returned to the well after each individual quadrat sample. No significant differences were observed in the total numbers nor in mean densities per square metre of each larval stage sampled using the two methods when analysed by Wilcoxon test for two paired groups. Quadrats were more effective at sampling pupae than dippers ($z = -1.99$, $P = 0.046$), collecting approximately twice as many per square metre sampled.

Table 3.3. Predictive models for the calculation of larval *Psorophora columbiae* and *Anopheles crucians* density from dipper sample values (Andis et al. 1983)

Species	Instar	Predictive model[a] $1 - r^{2b}$
Ps. columbiae	I	$X = -0.27 + 1.92$ (Y) 0.08
	II	$X = -0.93 + 1.69$ (Y) 0.01
	III	$X = 1.70 + 0.49$ (Y) 0.21
	IV	$X = -1.63 + 0.72$ (Y) 0.06
	I–IV[c]	$X = -16.29 + 1.48$ (Y) 0.23
An. crucians	I	$X = 1.11 + 2.56$ (Y) 0.08
	II	$X = 0.63 + 1.04$ (Y) 0.07
	III	$X = 1.33 + 0.55$ (Y) 0.28
	IV	$X = 0.14 + 0.66$ (Y) 0.14
	I–IV[c]	$X = 1.20 + 1.08$ (Y) 0.31

[a] X = number of larvae/unit area of 0.1 m^2.
Y = mean number of larvae/10 dips.

Breev et al. (1983) presented a mathematical paper comparing the sampling efficiencies of a photographic dish (100, 300 or 500 cm^2 areas), 'nets' constructed from the three sizes of photographic dish with the bottom removed and replaced with netting, and a cylinder enclosing an area of 300 cm^2. Fourth instar larvae and pupae of *Aedes vexans* were placed at varying densities in wooden water butts 34 cm tall having a top diameter of 70 cm (63 cm at the bottom) and containing water to a depth of 24 cm. It was concluded that the best sampling tool was the net with an area of 500 cm^2 because smaller nets gave unreliable results. They also suggested that natural breeding places should be divided into quadrats, and a sample taken from each, but this is not usually practical. In their experiments, the dish or net was firstly placed on the bottom of the water butt and then slowly lifted through the water, which will allow many mosquito immatures to escape, especially when the dish is used. Apparently, this is not standard practice for sampling mosquitoes in Russia (Service 1993*b*). Also, Breev et al. (1983) calculated surface areas of their dishes from bottom measurements, but as the dishes have sloping sides the top measurements are bigger, so for example, although a 500-cm^2 area dish is described this in fact would have been 560 cm^2.

In India Thenmozhi and Gajanana (1990) compared square-frame samplers having an enclosed area of 0.1 m^2 with larval sampling with a dipper. Because it took so long to remove all mosquito larvae from the quadrats only pupae were collected. After transforming counts to logarithms a good linear regression was found between the numbers of culicine pupae/m^2 with the number/dip, but, presumably because they were less aggregated, the better regression for anopheline pupae was between untransformed counts.

Clearly sampling with dippers is considerably quicker and easier than with area samplers (quadrats), and dipping is much more versatile because it can be used in a greater variety of breeding places. However, it would be expected that dipping would show more bias in sampling different age classes and species, because of different amounts of time spent at the water surface and differences in larval escape reactions. But from the comparative studies described above there is no consensus as to whether dippers or area samplers give the more reliable results. Problems with such comparative studies are that the efficiency of dippers and area samplers likely vary according to the types of breeding places and larval dispersion patterns within a habitat, as well as the precise method by which the samples are taken, this applies especially to dipping. There is considerable scope for more studies comparing the efficiencies of different sampling methods.

Floating traps

For sampling *Anopheles* larvae Goodwin and Eyles (1942) used quadrat frames made from 1 × 6 in wooden planks to enclose 0.5–1 m^2 of water surface. *Anopheles* larvae within the floating frames were removed by dipping and according to the authors the number of larvae escaping from them was very small, if not negligible. By using 10–20 quadrats systematically placed in a habitat they were able to estimate the total larval population (Goodwin and Eyles 1942). Similar floating quadrats were used in studying the numbers and distribution of larvae (Hess and Hall 1943) and eggs and larvae (Rozeboom and Hess 1944) of *Anopheles quadrimaculatus* in relation to the total perimeter of plants that intersected the water surface. Hess and Hall (1943) found that stirring up the mud facilitated the collection of larvae. They also concluded that *Anopheles* larvae did not escape from the quadrats, even when vegetation was removed prior to collecting. Despite these statements Service (1993*b*) thought it extremely probable that there is some interchange of larvae between areas within and outside the quadrats, and this is likely to be greater with culicine larvae because of their repeated ascents and descents in the water. Many years later Walker et al. (1988) sampled larvae of *Anopheles quadrimaculatus* in a marsh using 0.5-cm diameter (0.1963 m^2) plastic tubular hoops which floated on the water surface. Larvae enclosed by the hoops were collected with a pipette.

A small, lightweight and easily constructed floating larval trap was described by Undeen and Becnel (1994). The trap comprises a 100 ml transparent plastic cup with a convex bottom, in the centre of which a 6-mm diameter hole was cut. Four pieces of cork were cemented around the outside edge of the cup, to act as floats. The device floated at the water surface, with the hole in the convex bottom situated about 5 mm below the surface (Fig. 3.18). These floating traps were evaluated in field studies in which they were placed in tyres containing leaf litter and three litres of water. To determine if the traps concentrated larvae, a "trapping ratio" was calculated by dividing the number of larvae/cm^2 in the sampler by the number of larvae/cm^2 in the tyre. On average, the traps sampled about 22% of the total larvae in the tyres. Trapping ratios were significantly in excess of 1, indicating that the traps did indeed trap and concentrate larvae.

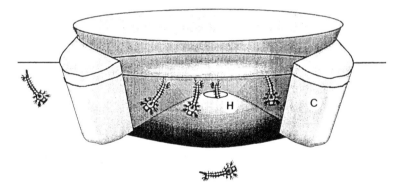

Fig. 3.18. Floating larval sampling device of Undeen and Becnel (1994). H, entry hole; C, cork float

Funnel traps

Harrison et al. (1982) devised a floating trap, which they termed the AFRIMS trap (Armed Forces Research Institute of Medical Sciences), to obtain more accurate estimates of *Aedes aegypti* larval densities. Their trap was based on the one used by Muul et al. (1975) for collecting larvae of *Culiseta melanura* colonising ground (stump) holes. The trap consists of a

13-cm diameter 6-cm deep plastic kitchenware container with a removable lid. A white or light-coloured 10-cm diameter funnel, about 12 cm long and ending in a 1-cm diameter stem, is inserted into a hole cut from the middle of the bottom of the container (Fig. 3.19a, b). A 44-cm length of plastic tubing is wrapped round the middle of the container and joined together by a plug of wood, this serves as a float. A hole is cut towards the edge of the snap-on lid of the container and a rubber stopper with a screw hook is pushed in. A length of nylon fishing line tied to the screw hook is threaded through the stem of the funnel and tied to a screw hook in a similar rubber stopper.

To operate, both rubber stoppers are pulled free and the trap is slowly pushed into the breeding place until the water rises to about the top of the plastic tubing float, then the stopper is pushed down into the lid to trap air. The trap now floats and larvae swim up through the funnel into the container. The trap can be removed after 12 or 24 h, and for this the top stopper is removed and pulled up so that the lower stopper is drawn into the neck of the funnel. The contents can then be tipped out through the hole in the lid or the snap-on lid removed to facilitate emptying. In laboratory trials traps caught means of 37–69% of *Aedes aegypti* and 44–79% of *Culex quinquefasciatus* larvae contained in earthen water-storage jars. In field trials 42 541 larvae and pupae were collected from traps placed in 1322 containers positive for *Aedes aegypti*, and over 98% of pre-adults trapped were *Aedes aegypti*. Only 21 traps failed to collect immature stages after a 24-h exposure period, indicating a trap sensitivity (Dixon and Massey 1969) of 98.4%.

Although the trap can be used in breeding places with a depth of 14 cm or more of water, they are too large for use in small or shallow habitats such as ant traps, vases and tyres, and so consequently cannot be used on their own to estimate the classical *Aedes aegypti* indices such as container or house indices. Another disadvantage is that two trips are required to sample a container, namely to set and then to remove the trap. The major advantage, however, is that human bias in sampling is more or less eliminated. Also the trap is cheap and easily constructed.

Kay et al. (1992) simplified the funnel traps described above constructing theirs from an inverted 500 ml polystyrene jar with a screw cap (Corning cat. no. 25628–500) that acted as the reservoir, a 22 cm diameter plastic funnel inserted through a hole cut in the lid of the plastic bottle, a 50 g weight to sink the trap and cord for lowering and retrieving the trap from wells. The trap as described cost US$ 4 and can be used in water depths over 40 cm. The polystyrene bottle is filled with 400 ml of water so that when inverted an air space of 3–4 cm is left, sufficient to float the device in the water. On immersion the weight of the sinker causes the trap to invert so that the reservoir bottle is uppermost. Pulling the attached cord to retrieve the trap causes it to invert again, trapping the mosquitoes in the reservoir bottle.

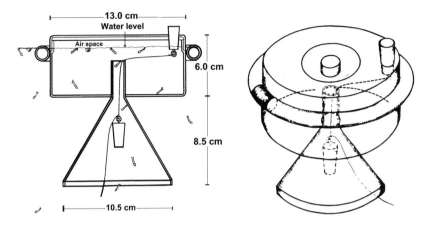

Fig. 3.19. AFRIMS *Aedes aegypti* trap. (a) lateral view with dimensions; (b) assembled trap (Harrison et al. 1982)

The effectiveness of the trap for collecting larval mosquitoes and copepods was tested in several water bodies in Ceará and 285 wells in Fortaleza, Brazil. In order to determine the sensitivity and efficacy of the funnel traps, the following numbers (and number of replicates) of mixed third and fourth instar *Culex quinquefasciatus* were introduced into negative wells: 10 (8), 25 (12), 50 (8), 100 (9), 200 (8), 400 (6), 600 (4), 1000 (3), 1500 (6) and 2000 (4). On the next day the following numbers (and number of replicates) of the copepod *Mesocyclops aspericornis* were released into the wells: 25 (4), 50 (4), 100 (4), 200 (4), 400 (4), 500 (2) and 800 (4). Traps were left in place from 1700 to 0800 h. Trap captures of *Culex quinquefasciatus* and *Mesocyclops aspericornis* averaged 6.1 and 3.6% respectively. It was observed that 12% to 29% of trapped *Culex quinqefasciatus* larvae were able to swim out of the trap through the funnel tube. Reduction in size of the funnel tube to 1 cm diameter did not reduce the proportion escaping. The sensitivity of the overnight funnel traps for detecting the presence of *Culex* larvae or *Mesocyclops* was dependent on the number of animals released. Detection success reached 100% for 100 or more larvae or 400 or more copepods. In the 285 wells, the mean numbers of larvae and pupae recovered from the funnel traps were 188.5 and 17.5 respectively, although one trap did collect >5000 larvae and >500 pupae.

Funnel traps of Kay et al. (1992) were used by Russell et al. (1996) in Charters Towers, Queensland, Australia, a town with a history of dengue fever, to sample mosquito larvae and potential predators from abandoned mine shafts and wells. Funnel traps were placed in water bodies for approx. 24 h per sample. A fishing rod was used to lower the traps into unstable or inaccessible mine shafts and shafts with a depth to water of up to 45 m were sampled using this method. Nine of ten wells sampled contained *Aedes aegypti* larvae or pupae, and were therefore identified as key breeding sites for this species. Disused mine shafts were not important sources of *Aedes aegypti*. Russell et al. (2002) later demonstrated the epidemiological significance of this type of larval habitat for dengue virus transmission in the 1993 Charters Towers epidemic by analysing the spatial distribution of dengue seropositive residents and subterranean larval habitats and demonstrating that residents within 160 m of a well or service manhole were at 2.47 times (95% CI 1.88–3.24) higher risk of being seropositive than residents living more than 160 m from these habitats.

Later, Russell and Kay (1999) described a calibrated funnel trap for use in subterranean larval habitats. The trap consists of a plastic funnel (185 mm diameter) attached to a plastic screw-top container. The spout of the funnel is inserted through a circular hole in the container cap and held in place with glue or self-tapping screws. A counter-balance manufactured from a 20-mm long section of galvanised water pipe is slipped onto the funnel spout before attachment to the screw-top container. The screw-cap container is filled two-thirds full with water prior to the whole apparatus being lowered into the larval habitat. The counter-balance causes the trap to invert and the air in the screw-top container causes it to float. The trap is retrieved by means of a cord attached to the rim of the funnel mouth. The trap detected the presence of larvae at all larval densities in the range 0.0022 to 0.0884 larvae per cm^2 habitat surface area. Development of a regression model ($y = 0.182x + 5.35$, $r^2 = 0.992$) relating trap catch to absolute numbers of larvae allowed the estimation of absolute population size with an accuracy of 70–95% accuracy in field trials.

Kay et al. (2000) used these calibrated funnel traps alongside a calibrated sweep-netting method to sample mosquito larvae from wells and manholes in Australia. In Yogyakarta, Indonesia, Gionar et al. (1999) designed a floating funnel trap to assist in the sampling of wells as part of a prospective seroepidemiological dengue study in an urban area. Their trap comprised a 20-cm-diameter white plastic funnel, a 1-litre polystyrene bottle (reservoir) with screw cap, a 420 g metal ring (sinker) fixed at the point of attachment of the reservoir to the neck of the funnel, and a plastic cord attached to the funnel apparatus to allow the trap to be lowered into the well and subsequently removed. The traps were operated as follows: the reservoir was filled to half capacity with clean fresh water and attached to the funnel. The trap, reservoir uppermost, was then lowered into the well using the cord. On contact with the water surface, the weight of the sinker naturally inverts the trap into the collecting position and the air pocket in the reservoir allows the funnel to float freely. Each well was sampled for approximately 24 h from between 1000 and 1200 h. Traps were retrieved by inverting the trap using the cord and then raising the trap from the well. Water contained in the reservoir was filtered through a 100-mesh screen and all mosquito larvae and copepods were pipetted into plastic bags. Thirty-one (33%) wells contained *Aedes aegypti* larvae, and 4 (4.3%) contained *Culex quinquefasciatus*. The average number of *Aedes aegypti* larvae (all instars) per positive well trap was 8.8 (range 1–63) per 24-h period.

Kay et al. (2002) used calibrated funnel traps of Russell and Kay (1999) to identify wells that were important larval habitats for mosquitoes in Charters Towers, Australia, prior to testing a winter control intervention comprising container removal or treatment with s-methoprene.

Nam et al. (2003) constructed funnel traps from a 185-mm-diameter plastic funnel a 500-ml polystyrene jar (reservoir) with a screw cap and a 20-mm section of galvanised water pipe that acted as a counterbalance. To set the trap, the reservoir was filled to two-thirds capacity with clean water and placed on the water surface. The weight of the metal counterbalance caused the trap to invert and float on the water surface. Traps were used to sample mosquitoes from the type of water storage jars commonly used in Vietnam. The funnel trap sampling rnethod was 100% sensitive for detecting the presence of immature *Aedes aegypti* in 0.28 to 0.18 m diameter jars with larval densities as low as 25 larvae per jar. In the 0.52 m-diameter jars, funnel traps were 100% sensitive at larval densities of 25 per jar: however, at densities of 50 larvae per jar the sensitivity was 70, 80, and 90% at 8, 16, and 24 h respectively after the traps had been set. The decreased sensitivity of the trap in sampling 50 larvae compared with 25 larvae was unexpected, but significant (Fisher exact test $P = 0.012$). Funnel traps were 100% sensitive at larval densities ≥ 100 per jar. Jar size had the largest effect on the percentage recapture of larvae (ANOVA, $F = 166.3$, $P < 0.001$). The duration of sampling had a smaller but still significant effect on the recapture rate (ANOVA, $F = 5.238$, $P = 0.006$).

Aquatic light traps

Submerged light-traps have occasionally been used to collect aquatic insects (Aiken 1979; Apperson and Yows 1976; Carlson 1971; Faber 1981), but as yet they have proved of limited value in catching mosquito larvae. Further investigations, however, may show that some species are strongly attracted to certain types of light and can be adequately sampled by them.

It might also be worthwhile investigating the use of flexible fibre optic light-guides, which could be used to beam light from a source on land to a submerged trap. Very little light is lost during its reflected passage through the fibre, and the light can be bent round corners and obstructions. In the first and second editions of this book Service (1976, 1993*b*) described several aquatic light-traps, but because they rarely, if ever, caught mosquitoes, descriptions of only a few of the more successful and more recently developed traps are included here. Readers interested in older models of aquatic light-traps or those that are more successful in trapping other species than mosquitoes should consult Service (1976, 1993*b*).

A very simple light-trap was used by Husbands (1967) to collect mosquito larvae and other aquatic insects in Californian rice fields. It consisted of a widemouthed screw-cap glass jar (Mason or Kilner) having a translucent plastic funnel retained in its mouth by the 'screw on' metal ring of the cap. Illumination was by a 6-V torch bulb suspended inside a test tube inserted through a 15-mm diameter hole drilled in the bottom of the jar (Fig. 3.20). The trap can be operated from a dry cell battery. Husbands (1967) incorporated a photoelectric cell in the electrical circuit to switch the trap on and off at around sunset and sunrise. The trap was two-thirds filled with water and either suspended in the water by a rope or supported by floats; a cork soaked in chloroform was placed in the bottle to kill the catch.

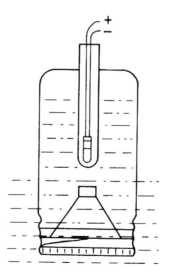

Fig. 3.20. Subaquatic light-trap with torch bulb (after Husbands 1967)

Washino and Hokama (1968) modified the trap by using a 9-cm diameter and 20-cm long glass cylinder. About 5 ml of chloroform in a glass tube was placed in the upper half of the trap.

In preliminary trials more larvae of *Culex tarsalis* and *Anopheles freeborni* were usually caught by dipping than by the light-trap but on one occasion two traps caught 72 aedine larvae, whereas none was caught by dipping. Washino and Hokama (1968) concluded that the light-trap was not as useful as dipping for collecting *Culex tarsalis* and *Anopheles freeborni* from rice fields.

Bertram et al. (1970) used a betalight in their aquatic traps. These lights are commercially available and consist of sealed small glass tubes, discs, spheres and other shapes coated on the inside with a phosphor and filled at less than atmospheric pressure with tritium, a radionuclide of hydrogen. On decay this emits exclusively beta particles, which are absorbed by the phosphor causing it to emit light in the visible spectrum. The betalight tube was mounted on rubber bungs in a strong test tube clipped onto the inside of the lid of a cylindrical tin (about 19 × 14 cm) (Fig. 3.21a, c). A wide slit entrance was cut in the opposite end of the tin (Fig. 3.21b).

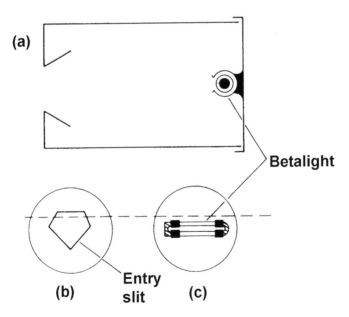

Fig. 3.21. Aquatic light-trap with betalight (after Bertram et al. 1970). (*a*) side view of trap; (*b*) view of entrance; (*c*) view of rear end

The trap, painted white inside and black outside, was placed horizontally and partially submerged in water, and was either supported by floats or secured by anchoring.

In preliminary trials in Sarawak all three *Culex* species present in the larval habitats were caught in the trap. The size of the catch varied greatly, but in ditches 15–95 *Culex* larvae, which included *Culex tritaeniorhynchus* were caught compared with 0.2 larvae in control traps without betalights. Also, when the trap was placed in forest pools about six times as many larvae and pupae of *Culex mimulus* were caught in betalight traps than in control traps (Bertram et al. 1970). Heathcote (1970) reported that in Sarawak the traps were successful in water 18–36 cm deep, but generally of little use in paddy fields. In laboratory trials the trap caught considerable numbers of *Culex quinquefasciatus* and *Ochlerotatus togoi* but few larvae of *Anopheles atroparvus* and *Aedes aegypti* (Bertram et al. 1970). In contrast Ree (1971) had better success in the laboratory in catching larvae of *Anopheles atroparvus*, he also trapped larvae of *Ochlerotatus togoi* and *Anopheles gambiae*. In field tests in southern England Ree (1971) caught larvae and pupae of both *Ochlerotatus cantans* and *Culiseta morsitans*. He found that the traps caught more larvae in relation to pupae than were obtained by dipping. He also confirmed findings of Bertram et al. (1970) that more larvae and pupae are collected from traps if short exposure periods are used, presumably due to some larvae and pupae escaping. A more efficient one-way entrance is therefore needed to prevent the immature stages that have entered during the night, from escaping before the traps are examined in the morning.

Betalights normally pose no radiation health hazards, but if the tritium is under pressure in the tubes (e.g. 2.5 times atmospheric) to increase brightness, 2 Ci of tritium may be exceeded and their use may be controlled by local government safety regulations. Betalights will continue to emit light for about 10–15 years, by which time the light will be reduced to about 50%. They are available in several different shapes and sizes, and in different colours, such as green, yellow, white, orange, deep orange, blue and red, in order of decreasing brightness.

Chemical light-traps

Chemical light-tubes, as sold in camping and other shops, were first used in light-traps by Service and Highton (1980) for sampling adult haematophagous insects, and later evaluated in aquatic light-traps (Service et al. 1983). The Cyalume lightsticks used in these traps consist of a 152-mm long translucent, pliable plastic tube containing 6 ml of a greenish fluorescing solution and a small glass ampoule containing 1.5 ml of a colourless energy-producing solution. When the lightstick is bent, it breaks the inner ampoule and releases its contents, which mix with the fluorescent liquid to produce a bright light of about 15-m lamberts. Brightness decreases to about 4.5-m lamberts after about 8 h, the rate of decrease depends upon temperature. Although the light appears greenish, spectrometer measurements show that there is a single emission peak at 510 nm, meaning that there is a considerable amount of light at the red end of the visible spectrum.

In trials in England and Ghana a lightstick was suspended in the middle of the same type of cylinder as used for betalights (Service et al. 1983; Service 1993b). In England, these traps were placed for 24 h on the bottom of a woodland pond and a ditch, while in Accra, Ghana they were put in a concrete drain, an ornamental pond and in a freshwater pond, but in more detailed trials traps were placed from 1800–2000 h in five saltwater ponds. In England all larval instars and pupae of *Ochlerotatus cantans* were caught. There was considerable variation in the numbers trapped on different nights, ranging from 41–675 immatures of *Ochlerotatus cantans*, 0–430 *Ochlerotatus rusticus* and 0–52 *Ochlerotatus punctor*. The largest single catch consisted of 1748 larvae of all three species. In Ghana, one trap caught 1631 *Aedes vittatus* larvae and pupae when placed overnight in a concrete drain. In the saltwater ponds because so many larvae and pupae of *Culex thalassius* were caught, their numbers were usually estimated. On two occasions, an estimated 18 500 and 3000 larvae and pupae of *Culex thalassius* were caught overnight in a single trap, in another pond 1776 and 2033 *Culex thalassius* were caught from 1800–2000 h in a trap on each of two evenings. *Culex thalassius* larvae were very abundant in the five saltwater ponds ($\bar{x}$ = 25.6 ± 34.9 per dip). In these ponds unlit control traps caught 33–126 larvae from 1800–2000 h, whereas traps with chemical lights caught 416–4560 pre-adults (Service et al. 1983).

A commercial aquatic light-trap for mosquitoes that uses lightsticks consists of a heavy duty black plastic (ABS) cylinder (21 cm tall, 17 cm diameter) painted white inside and having four equally spaced 'portholes' around the middle of the cylinder. These have transparent plastic inverted 10-cm diameter cones through which mosquito larvae enter the trap. A lead weight, or stone, inside the trap ensures it rests on the bottom of habitats, alternatively without a weight it can be floated partially submerged.

There seems to be little doubt that chemical lights provide the simplest and most convenient source of light in submerged aquatic light-traps. Cyalume lightsticks are available in a range of different colours (yellow, red, orange, blue, white) but the original greenish ones produce the most light. Reasonable light output continues for about 12 h with the 152-mm long green, yellow, orange and red lightsticks, but for only 8 h for the white and blue ones. Longer lightsticks (381 mm) are also available as are high intensity ones, but these only last for about 30 min. Because light is produced chemically it is temperature-dependent, thus above 28°C light intensity increases but its duration decreases, while below 21°C the light is less bright but lasts longer. If lightsticks are kept in their individual metal foil wrappers shelf-life is 2–4 years. After breaking the inner ampoule by bending the lightstick, the plastic tube can be cut and the light-producing liquid which is quite safe to handle, tipped into suitable glass or plastic containers, or the contents of more than one lightstick combined in a trap. In fact, the chemiluminescent liquid was removed from lightsticks for trials in England and Sierra Leone with sticky light-traps (see below).

Aquatic chemical lightsticks merit further evaluation in collecting mosquito larvae. They also collect a variety of other aquatic organisms (Service et al. 1983).

Sticky light-traps

In laboratory experiments Sulaiman (1982) found that certain adhesives remained tacky even when submerged in water and caught mosquito larvae including those of *Ochlerotatus cantans*. However, for some unexplained reason they caught very few *Ochlerotatus cantans* in field trials.

A chemical light was combined with an adhesive to form a sticky light-trap (Service, 1984). For this a pair of plastic petri-dishes (9 cm) were sealed together with waterproof sealant and a small (5-mm diameter) hole cut and covered with strong rubber sheeting. The mixed contents of a Cyalume lightstick (see under chemical light-traps) were syringed into the petri-dish; after trials the liquid can be syringed out. The water-resistant adhesive chosen was rat varnish—as sold to trap rodent pests in warehouses—because in laboratory tests this proved the best adhesive for trapping mosquito larvae (Sulaiman 1982). The varnish was smeared onto a $160 \times 160 \times 1.5$-mm sheet of transparent plastic stuck by 'Velcro' fastener onto the top of the larger of the pairs of petri-dishes.

In Sierra Leone these petri-dishes were placed overnight on the bottom of four village water-storage jars containing known numbers of *Aedes aegypti*. About 72–85% of the larvae (all instars) in the pots were caught, but only about 18% of the pupae were trapped on the adhesive. In one series of trials in England extending over 10 days only 0–29 mosquito larvae were caught per night in five unilluminated traps, whereas the mean numbers of *Ochlerotatus cantans* caught overnight on illuminated petri-dishes placed in ponds ranged from 73.4 ± 14.2–163.0 ± 19.9. In the same pond dipping with a ladle gave only 6.9 ± 5.5–11.8 ± 7.7 larvae/dip, and area samplers (cylinders) caught means of 83.8 ± 17.0–139.4 ± 37.9 larvae. The sum of the coefficients of variation was calculated by expressing the standard deviation as a percentage of the mean number of larvae caught over 10 different days. In two separate trials the values were 785 and 1223 for collections with a dipper, whereas for area samplers the coefficients of variation were 32.5% and 24.7% of these values respectively, and 22.5% and 29.5% for the sticky light-traps. This shows that least daily variations in numbers caught occur when the light-traps are used. In one series of trials the proportions of larval instars caught by these sticky light-traps and dippers did not differ significantly, although the proportions of 1st and 2nd instar larvae were slightly larger in an area sampler than in the light-traps, while the proportion of 3rd instars was smaller. However, in other trials the sticky traps caught disproportionately more 3rd and 4th instar larvae than either area samplers or dippers. These sticky light-traps have also been successful in trapping larvae of *Culiseta morsitans* (Service 1984). They are inefficient at collecting mosquito pupae.

It was concluded that, at least for *Ochlerotatus cantans*, these sticky light-traps were about as efficient as cylindrical traps incorporating a light-stick, and were simpler to make and had the advantage of being usable in shallow waters. Sticky light-traps may prove to be useful in sampling low population densities of larvae. It was also suggested that if chemiluminescent liquid was poured into small tubes (e.g. test tubes) coated with rat varnish adhesive, these could be dipped into larval habitats having restrictive opening such as tree-holes, crabholes and defective soak-away pits.

Evacuation of habitats

All larvae can be collected from various small container habitats such as pots, tin cans, bamboo pots, snail shells and sometimes even tree-holes, by either tipping or siphoning out the water. This procedure is not usually practical with larger habitats but Christie (1954) developed a technique for removing water from small ground pools so that larvae of *Anopheles gambiae* and other pool-breeding species could be counted. A semi-rotary pump, delivering about 150 gal/h, pumps out the water from pools into a 100-gal capacity galvanised drum. The distal end of the pump's 12-ft intake hose is covered with 100-mesh gauze to prevent mosquito larvae being sucked up. Pumping is stopped when the water level has been reduced to about 3–4 in. This residual water, in which the mosquito larvae are concentrated, is carefully baled out and poured through a series of 5-, 16-, 60- and 80-mesh sieves stacked on top of each other. These sieves are surrounded by a metal water jacket so that the water level can be maintained about 1 in above the mesh of the top sieve. This helps prevent the larvae being damaged as they are tipped through the sieves. A dish and dipper are used to remove the last few ounces of pool water, which because it is very muddy is not passed through the sieves but examined separately for larvae. When the pool has been emptied it is washed down with about 2–3 gal of water from the drum, and when the mud has resettled the contents are again baled out and passed through the sieves. The process is then repeated. Christie (1954) used this method in pools about 4 ft in diameter and 2–3 ft deep, formed in black alluvial soil. He considered that in pools with a firmer bottom where the mud is less easily disturbed, pumping would be unnecessary and that all water could be removed by baling.

When the efficiency of this extraction method was tested by introducing a known number of *Anopheles gambiae* larvae into pools and extracting all the water, Christie (1954) concluded that at least 95% of all stages, except 1st instar larvae, are recovered. Christie (1954) pointed out that the method was impracticable in pools in which water rapidly seeped. In later experiments Christie (1958) claimed to have overcome this difficulty by baling out the residual water (10–15 gal) as quickly as possible into a 40-gal oil drum which was cut in half longitudinally and painted white inside. As the mud settled larvae were removed from the water surface, and when most of the sediment had settled out the water passed through the series of sieves. Finally the small quantity of muddy water remaining in the oil drum was examined for larvae.

Although this method was first used almost 40 years ago it has rarely been employed, mainly because it is time consuming and consequently cannot be used in routine larval surveys or collections, and moreover is applicable only in relatively small pools. Furthermore, it is difficult to ensure that all larvae are extracted from the mud at the bottom of pools and from vegetation. Nevertheless, despite these limitations Le Sueur (1991) in estimating larval mortalities of *Anopheles merus* in South Africa, used a method similar to that of Christie (1954) and collected and sieved all water from small pools. By adding approximately 60 larvae of each instar and pupae to five pools, he estimated recovery rates were 96.9% for 1st instar larvae, 90.0% for 2nd, 89.1% for 3rd, and 94.0% for 4th instar larvae, and 90.0% for pupae.

Sampling procedures for special habitats

Wells

It is usually difficult to collect mosquito larvae from wells. In some the water is at a great depth, but even in shallower wells it is rarely sufficiently near the top to enable larvae to be seen at the water surface. Sampling is consequently performed at a distance with long-handled dippers or more usually with nets attached to rope. Floating funnel traps suitable for use in wells have been described earlier. Dippers mounted on jointed handles, which allow the cup to be adjusted to suitable angles have been found useful (Russell and Santiago 1932). Since larvae usually congregate around the walls of wells dippers should be drawn up against them.

Boyd (1930) preferred using a specially adapted bucket, to which a bail with one tip weighted is secured just above the centre of gravity, thus allowing the bucket to be gently lowered into the water at an angle. In India Menon and Rajagopalan (1979) collected *Anopheles* larvae from wells using an ordinary bucket. Four samples were taken along the sides and one from the middle. Sabesan et al. (1986) using buckets to sample *Anopheles culicifacies* took two samples from the sides of a well and one from the middle. They left an interval of 3–4 min between each bucket sample to allow larvae and pupae to resurface. Panicker et al. (1982) also collected *Aedes aegypti* larvae from wells in buckets. Aquatic nets, however, are more frequently used.

Senior White (Knowles and Senior White 1927) used a muslin net stretched over a 12-in diameter brass ring which had three chains attached at equal distances and joined together about 18 in above the net. A length of string was tied at this point and another to the edge of the metal ring so that the net was lowered into the well edge-on in a more or less closed position. When submerged it was gently opened and pulled up. In India Rajagopalan et al. (1976) and Yasuno et al. (1977) sampled *Culex quinquefasciatus* in wells with nets made of 1-cm thick aluminium frames supporting a flat square-shaped (33 × 33 cm, 10-cm deep) bag. The bottom of the bags were made of PVC netting having 15 mesh/cm, which allowed water to flow through quickly without overflowing when the nets were withdrawn. However, most 1st and 2nd instar larvae escaped, although egg rafts as well as older instar larvae were retained. Rajagopalan et al. (1976) lowered simultaneously into each well four nets along the sides and one in the middle, they remained submerged 30 cm below the water surface for 5 min before they were pulled up. The surface area of the wells was about 3.14 m^2 and therefore the five nets sampled 15.9% of the water surface.

Rajnikant et al. (1993) in Delhi, India used a 25 cm diameter well net to sample mosquitoes from wells. Eight species of anophelines were sampled, of which *Anopheles barbirostris* was the most common (45.4%), followed by *Anopheles stephensi* (36.1%), *Anopheles subpictus* (13.5%) and *Anopheles culicifacies* (3.5%). No anophelines were collected at depths in excess of 12 m from the ground surface.

Thu et al. (1985) used a conical net made of fine cotton to collect larvae of *Anopheles balabacensis* from wells. In Myanmar Tun-Lin et al. (1986) sampled *Anopheles dirus* larvae from wells by gently lowering into the water a white cotton cloth net measuring 36 cm in diameter and 46 cm long, and with a stone at the bottom. After 2 min the net was carefully pulled up against the side of the well and the contents tipped into a white dish.

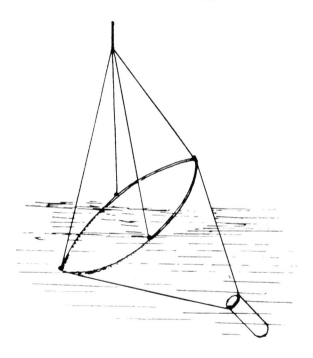

Fig. 3.22. Well net in use (World Health Organization 1992)

Later in collecting larvae from wells Tun-Lin et al. (1988) made a well net according to the description of WHO (1975), namely a nylon bag about 30 cm deep which is sewn onto a 20–25-cm diameter metal ring. The upper part of the bag around the ring is reinforced with stronger material, while a transparent plastic tube (4 × 10 cm) is fitted through a hole in the bottom of the net (Fig. 3.22). Nylon string is attached to four points on the ring in such a way that the ring is at an angle of 30°, and then a rope is attached. A small stone or metal weight (about 50 g) is placed in the net to ensure it stays submerged. However, because of a lot of surface debris and shrubs growing out from the sides of their wells, which were also >3 m deep, Tun-Lin et al. (1988) found that the recommended protocol of waiting 2–3 min after the net was submerged before dragging it two or three times slowly round the edges of the well, inappropriate. Instead they lowered the net into the centre of the well until it touched the water, and waited a minute before sinking it to a depth of 60 cm and drawing it to the edge of the well having the most shade. The net was then slowly raised until it was 15 cm from the water surface, when it was jerked out of the water and the contents tipped into a white dish. In comparing the WHO and modified procedure in 10 wells they reported that in all instances the mean number of *Anopheles dirus* larvae per dip (2.3–15.3) by their method was significantly greater than with the WHO procedure (0–9.3). They found that three similar dips per well were sufficient to estimate relative larval densities, but present no details explaining how this was judged.

Phytotelmata (tree-holes, bamboo, bromeliads, leaf axils, etc).

The term phytotelmata refers to small collections of water held in any parts of plants, such as tree-holes, bamboo, bromeliads, leaf axils, or pitcher plants. The term has come into popular usage by entomologists since the publication of the book 'Phytotelmata: terrestrial plants as hosts for aquatic insect communities,' (Frank and Lounibos 1983). An advantage of studying the population dynamics of mosquitoes breeding in phytotelmata is that it is often possible to census all individuals in their habitats, facilitating studies on competition, food webs, predation and density-dependent factors.

Many important papers on the ecology of mosquitoes breeding in tree-holes, pitcher plants and plant axils have been published in recent years by several authors, including Bradshaw and Holzapfel (1983, 1984, 1985, 1986*a, b*, 1988, 1989), Fish and Carpenter (1982), Fisher et al. (1990), Hard et al. (1989), Istock et al. (1976*a, b*), Lounibos (1979, 1981), Lounibos et al. (1987), Seifert (1980), Seifert and Barrera (1981) and by these and other authors in the books edited by Frank and Lounibos (1983) and Lounibos et al. (1985).

Methods employed to collect mosquitoes from these diverse habitats are described below.

Tree-holes. The collecting method will depend to some extent on the size and shape of the tree-hole. In very large tree-holes small dippers can be used to collect larvae (Sakakibara 1960; Siverly 1966), and in those in which the larvae can be seen they can be pipetted directly from the water surface. A useful tool is a pipette made from a 20–25 cm length of glass tubing with a 50-ml capacity rubber suction bulb fitted to one end. This sucks up relatively large volumes of water. Flexible plastic tubing is a useful substitute for the glass part of the pipette, as it allows the pipette to be used in tree-holes formed at awkward angles. Stirring up the water prior to sampling by repeatedly sucking up and discharging the contents with a pipette may increase the likelihood of collecting larvae. Lounibos (1981) used a large bore (5-mm) glass pipette with a large bulb to collect larvae from tree-holes. To check on the efficiency of this method, after all water had been collected, the tree-hole was refilled with an equal volume of water and larvae again collected. This procedure was repeated twice. A total of 3397 larvae and pupae were collected from 74 tree-holes, of these 68.3% were collected from the original water and 20.4 and 11.3% from the two subsequent floodings. Rosenberg (1982) collected *Anopheles dirus* larvae from containers such as tree-holes with rubber tubing attached to a 50-ml syringe. Water removed with the syringe was poured through 150-μm nylon netting and the restrained larvae washed into pans for counting.

In studying *Ochlerotatus triseriatus* Haramis (1984) removed water from tree-holes with a squeeze-bulb powered pump as used for cleaning aquaria. The sides of the tree-holes were then rinsed with 150 ml of water to flush any remaining immatures into the bottom, and the remaining liquid removed with a meat baster. Kruger and Pinger (1981) also used a suction meat baster to sample larvae from tree-holes. Arnell and Nielsen (1967) strained water collected in pipettes through a sieve to ensure that all larvae were removed from dark-coloured tree-hole water. Larvae can also be collected by siphoning out all, or part, of the water in tree-holes with a large bottle (pint capacity or larger) having a tight fitting top with two pieces of rubber or plastic tubing inserted through it (Fig. 3.23*a*). One length of tubing is placed in the tree-hole while the other is sucked to start water siphoning out. Water continues to flow out so long as the bottle is held below the opening of the tubing in the tree-hole. Another type of siphon recommended by WHO (1975) is illustrated in Fig. 3.23*b*.

In a study of the composition of tree-hole communities in Pennsylvania, USA, Barrera (1996) decided to completely remove the entire contents of the tree holes, in order to ensure that all insects were collected. Preliminary collections had shown that samples comprising only leaf litter contained aedine mosquitoes (47%), ceratopogonids (28%) and scirtids (25%) and similar proportions were obtained from sampling sediment. Aedine mosquitoes (86%) were also dominant in pipetted samples of free water, obtained using a 5 mm diameter pipette, although ceratopogonids (6%) and scirtids (8%) were also collected using this method. The final sampling method adopted was as follows: firstly all leaf litter and coarse materials were removed. Standing water was then siphoned off and sediment was removed to a depth of at least 3 cm. Finally, tree holes were flushed out at least 3 times with aerated water.

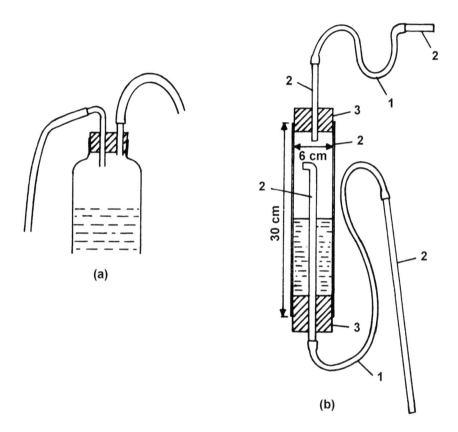

Fig. 3.23. Larval samplers: (*a*) bottle for siphoning larvae from tree-holes and other phytotelmata; (*b*) siphon for sampling tree-holes, 1—rubber tube, 2—plastic tubing, 3—cork stopper (World Health Organization 1975)

Waters and Slaff (1987) described the simple apparatus shown in Fig. 3.24, by which mouth suction is applied to the end of the uppermost plastic tubing to suck water and mosquito larvae from tree-holes and other rather inaccessible habitats. This sampler seems to have no advantages over using a bottle with rubber tubing to siphon water from tree-holes.

In Florida, USA larvae were pipetted or siphoned from tree-holes and the samples transported in quart-sized (0.95-litre) plastic bags to the laboratory where they were maintained in an incubator set at the tree-hole temperature. Counting the mosquitoes in the samples could take 3–4 h, and so each sample was tipped into a plastic pan surrounded by a dish of ice to keep the larvae at temperatures close to those in the field. The samples were eventually returned to the tree-holes (Bradshaw and Holzapfel 1983).

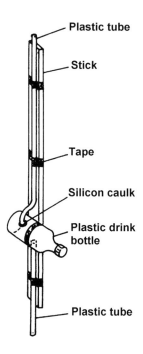

Fig. 3.24. Larval collecting siphon of Waters and Slaff (1987)

Larvae collected by pipetting should be recorded separately to enable the mean number of larvae per pipette to be calculated ($\bar{x} \pm SE$). But the mean number per pipette will depend not only on the number of larvae present but also on the volume of water in the tree-hole. In other words larval density and not population size is measured. If larval populations in different sized tree-holes are to be compared then the water in each tree-hole should be removed, measured and then replaced. The number of larvae per pipette sample can then be related to volume of water present in the tree-hole at the time of sampling. This refinement is not always necessary; it depends on the information required from sampling. In using pipettes care should be taken that larvae do not become stranded in the bulb and get flushed out with the following sample (Davis 1944). This can be avoided by placing a cotton wool plug in the glass tubing just below the bulb.

In Tahiti the breeding of *Aedes polynesiensis* in both tree-holes and co-conut husks was studied by Bonnet and Chapman (1956). In an effort to quantify breeding in tree-holes they counted the numbers of different tree species in an area of 100 m × 2000 m, and recorded the number having larvae of *Aedes polynesiensis*. The flame tree was found to have the highest incidence of breeding (45.4%) but only a few trees were found, whereas the breadfruit was the most important tree for harbouring mosquitoes (19.2% of 291 trees examined being positive).

Hirian and Tyagi (2004) observed that *Aedes albopictus* eggs and larvae could be recovered from cocoa (*Theobroma caco*) pods on the ground and still attached to trees in Kerala State, India. The pods had holes gnawed in them by rodents. Larvae and eggs were recovered from pods by filling them with water and catching the overflow.

Bamboo. Mosquitoes which breed in water-filled cut sections of bamboo are generally more easily sampled than those breeding in tree-holes, because larvae can be more readily collected with pipettes or by siphoning out the water. Sunahara and Mogi (1998) used a 5 ml pipette to detect the presence of *Tripteroides bambusa* in water-filled bamboo stumps in Saga, south-western Japan. In these types of habitats the collecting methods described for tree-hole sampling can be used. Larvae of some species (notably sabethines), however, occur in bamboo that is not open at the top, the only access being a small hole in the side of the plant. *Sabethes chloropterus* has been observed in the laboratory to hover near a hole in the bamboo stem and expel eggs through the air into the hole (Galindo 1957). These larval habitats are more difficult to sample. The bamboo can either be cut across to gain access but this destroys the habitat, or a length of rubber tubing can be inserted through the small hole in the side of the bamboo and the contents siphoned out. Water and larvae can then be replaced with a large capacity pipette or poured back through a funnel and rubber tubing. Marcondes and Mafra (2003) used a battery-operated drill to bore up to 300 holes with a diameter of 5 mm in bamboo stems in Santa Catarina, Brazil. Holes were drilled in each internode from the base to a height of 3.5–4.2 m and 20 ml of water was injected into each hole using a plastic syringe. Every 14 days, two bamboo stems were cut into sections at each internode just above the drilled holes and the water was emptied into plastic trays. The bamboo sections were rinsed with 150–200 ml of water to ensure all eggs and larvae were collected. Using this method more than 200 individual mosquitoes comprising *Sabethes* *aurescens, Wyeomyia limai, Toxorhynchites* sp. and one specimen of *Sabethes melanonymphe* were obtained, with a mean of 2.87 mosquitoes per internode.

The use of bamboo pots as oviposition sites for mosquitoes breeding in bamboo and tree-holes has been described in the preceding chapter.

Plant axils, bromeliads and pitcher plants. Pipettes with small or large bulbs are commonly used to collect mosquito larvae from plant axils. Those with the glass barrel drawn out into a narrow opening are useful in removing the small quantities of water that collect at the base of plant axils. Larvae living at the bottom of very small pockets of water can be floated to the surface for easier collection by adding water to the axils.

In Canada Paterson (1971) investigated the mortality of *Wyeomyia smithii* larvae during the winter, when they remain frozen for about 4 months in the small quantities of water collected at the base of the leaves of the pitcher plant, *Sarracenia purpurea*. On each sampling day a few leaves were removed from several plants and taken to the laboratory where they were thawed. The contents of the leaf axils were flushed out and the numbers of larvae counted. The axils were then filled with water to measure their potential capacity. When the larvae in the leaf axils were expressed as numbers/ 10 ml potential water there was significant correlation between larval density and leaf axil capacity. In Venezuela Seifert (1980) and Seifert and Barrera (1981) removed all immatures, including eggs, of *Culex bihaicola, Trichoprosopon digitatum* and *Wyeomyia felicia* from bracts of *Heliconia aurea* by repeatedly flushing them with water using a hydraulic pooter and pipette. Removal was considered complete when three successive flushings yielded no mosquitoes. Dissection of the bracts rarely showed the presence of any remaining eggs. To prevent oviposition the bracts of an inflorescence were sometimes covered with a plastic bag, which neither harmed the plants nor mosquitoes (Seifert and Barrera 1981). In Tanzania Trpis (1972*a*) collected larvae of *Aedes aegypti* from leaf axils by direct pipetting and also by taking the plants to the laboratory, removing the leaves and washing out the contents. Larval density was expressed as both the percentage of leaf axils with larvae and as the mean number of larvae per positive plant. In Nigeria *Aedes simpsoni* group larvae and pupae were collected with a long stemmed glass pipette from leaf axils of banana/plantain plants, from cocoyams, *Dracaena* and pineapple plants (Bown and Bang 1980). The largest numbers of immatures/axil were obtained from cocoyam plants followed by banana/plantain plants (no distinction between the two was made). Lutwama and Mukwaya (1994, 1995) used a plastic pipette attached to a rubber bulb to collect larvae of *Aedes simpsoni* complex from plant axils in Uganda. Sampling efficiency was estimated by twice rinsing the plant axils with a similar volume of water to that originally removed from the axil. Although the method appeared to show different collecting efficiencies for different instars the differences were not statistically significant.

In the Philippines Lang and Ramos (1981) developed the following device for collecting immatures of *Ochlerotatus poicilius* and *Ochlerotatus flavipennis* from banana leaf axils. An eyedropper-type pipette was joined to a length of 8-mm diameter glass tubing strapped by tape to a section of thin bamboo (Fig. 3.25).

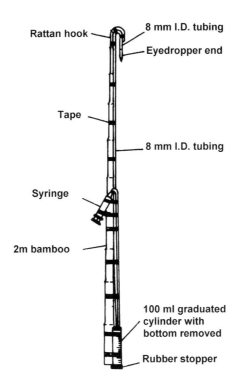

Fig. 3.25. Siphon for collecting larvae from leaf axils (Lang and Ramos 1981)

The other end of the glass tubing was inserted through a rubber bung into the top of a 100-ml plastic graduated measuring cylinder. Another section of glass tubing passing through this bung was connected to a plastic syringe, and when its plunger was slowly pulled out water was siphoned from the leaf axil into the cylinder. The bottom of this cylinder had been removed and closed by a rubber bung, which when taken out discharged the contents into a suitable container. Axils were not flushed out after the naturally occurring water was withdrawn. Placing known numbers of immature stages in leaf axils showed that the proportions removed were, 79%

for pupae, 58% for 4th instar larvae, 56% for 3rd instars and 31% for 2nd instars. Efficiency with regards 1st instar larvae was not measured, but based on the decreasing efficiency for other instars it was estimated as being only about 20%.

Boyd (1930) collected larvae from bromeliads by removing them from the trees, tipping out the water held in their axils and flushing out any larvae adhering to the leaves with a pipette. Bates (1949) advocated similar methods for collecting larvae from aroids and *Heliconia* flowers. In Trinidad detailed ecological investigations were made by Pittendrigh (Downs and Pittendrigh 1946; Pittendrigh 1948, 1950*a,b,c*) on the *Anopheles* species breeding in bromeliads on cocoa plantations. In routine surveys a climber cut the leaves from a plant just above the level of the water in the interfoliar 'tanks' while the plant was still attached to the tree. The plant was then carefully pried away from the tree and the water that poured out collected in a bucket. Finally the plant was flushed out with its own water, which was then examined for larvae against a white background. The humus, decaying leaves, and other debris washed out from the bromeliads may make the water very dark, making it difficult to ensure that all the larvae are removed, especially those that stay submerged for long periods. To speed up survey work Pittendrigh sometimes limited collecting to 15 min, after which any remaining uncaught larvae were discarded.

To compare the size of larval populations between different areas, surveys were made of the species composition and abundance of the various bromeliads in each area (Pittendrigh 1948, 1950*b*). The breeding incidence in the different species was measured in experimental plots having randomly placed equal numbers of different bromeliad species. *Gravisia aquilegia* was the commonest bromeliad in the plantations and nearly always contained *Anopheles* larvae; consequently it constituted the most important larval habitat. Pittendrigh (1950*b*) therefore expressed the Breeding Index as a decimal fraction of the number of larvae that a bromeliad species supported relative to the number supported by *Gravisia aquilegia*. The total breeding capacity of a tree with bromeliads was obtained by summation of the product of the mean numbers of each bromeliad species per tree and the appropriate Breeding Indices. Because the trees in the plantations were evenly spaced it was relatively easy to calculate the size of the breeding ground per unit area and estimate the size of the larval population (Pittendrigh 1950*b*). Quantitative evaluation of breeding was only possible because both the number of trees and bromeliads could be accurately counted. It is unlikely that similar quantitative population estimates can be made of species breeding in tree-holes or leaf axils because of the great difficulties of enumerating them in any area.

The bromeliad *Tillandsia utriculata* does not require attachments by its roots to grow, so in Florida, USA Frank et al. (1976) were able to suspend plants from trees and shrubs by nylon cords. Such bromeliads make ideal sampling units. To obtain its larvae a bromeliad is removed and inverted over an aluminium cone (Fig. 3.26*a*) (4) and dunked 20 times in a drum of water (1). The cone (4) together with the attached plexiglas tube (6) are removed and its water drained through the mesh screen (9) at the bottom of the tube (Fig. 3.26*b*). The plexiglas tube is then removed and the contents washed into a petri-dish and the larvae collected. In other experiments Frank et al. (1977) collected bromeliads, numbered them with plastic tags, suspended them 140 cm from the ground on a nylon cord, and regularly sampled them.

O'Meara et al. (1995, 2003) used a 1 m long pipette or meat baster to sample mosquitoes from native and exotic bromeliads in Florida, USA. Water was added to the plant axils to facilitate collection when axils were small. The exotic tank bromeliads *Aechmea fasciata, Aechmea distichantha, Billbergia pyramidalis, Neoregelia carolinae* and *Neoregelia spectabilis* yielded the following species: *Wyeomyia mitchellii, Wyeomyia vanduzeei, Ochlerotatus bahamensis, Aedes aegypti* and *Aedes albopictus* (O'Meara et al. 1995). Later studies (O'Meara et al. 2003) revealed *Culex biscaynensis* to be the dominant species in exotic bromeliads. Low numbers of *Aedes aegypti, Aedes albopictus* and *Culex quinquefasciatus* were also found in exotic bromeliads. *Wyeomyia mitchellii* was the most common mosquito encountered in native bromeliad species (*Tillandsia utriculata* and *Tillandsia fasciculata*).

In studying predation by *Toxorhynchites haemorrhoidalis* on the fauna of phytotelmata in Venezuela, Lounibos et al. (1987) removed bromeliads (*Aechmea nudicaulis* and *Aechmea aquilega*) from their natural habitat and suspended them on a rack 1.5 m from the ground. These 'artificial' oviposition sites were periodically and vigorously shaken in a bucket of water to dislodge mosquito larvae and other fauna.

Hutchings (1994) studied the use of fallen palm bracts as breeding sites for *Toxorhynchites haemorrhoidalis* in an upland forest of the Central Amazon, Brazil. Larvae were collected from the fallen bracts by emptying the entire contents of the bract into a container.

Ferreira et al. (2001) in Brazil, sampled mosquito larvae and other insects from water contained in the cup-like structures formed by the fungus *Aquascypha hydrophora* using pipettes. Species collected included *Limatus durhamii, Limatus pseudomethysticus, Limatus flavisetosus, Culex urichii, Culex bonnei*, members of the *Culex coronator* Group, and members of the subgenus *Melanoconion* of *Culex*.

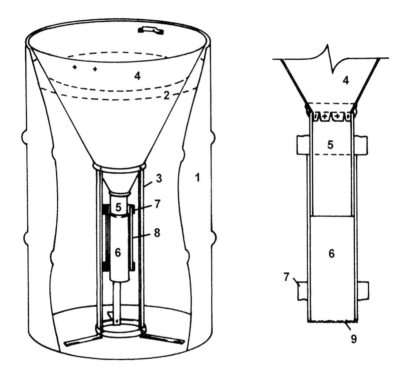

Fig. 3.26. Apparatus for sampling contents from bromeliads. (a) 1—water drum, 2—water level, 3—stand, 4—aluminium cone with 5—tubular plexiglas extension, 6—removable plexiglas tube, 7—plexiglas lug to hold 8—rubber band; (b) lower part of aluminium cone with items numbered as in (a) and showing 9—end of removable plexiglas tube with 100-mesh stainless steel hardware cloth (Frank et al. 1976)

Rock Pools

The collection of mosquito larvae from rock pools is usually conducted either with a dipper or pipette. Alternatively the entire contents of small pools can be siphoned out; the method used by Nakamura et al. (1988) to sample *Ochlerotatus togoi* in Okinawa, Japan, for example.

Wada et al. (1993) considered the method of completely emptying rock pools to be too time-consuming. Mori (1989) proposed thoroughly stirring the water prior to taking samples with a dipper and estimating absolute density of immature *Ochlerotatus togoi*. Wada et al. (1993) report the use of a modified removal method (Wada 1962*b*) to determine the absolute numbers of *Ochlerotatus togoi* larvae in a pool to within 10% of the true number. Results of this work are described under the section on removal sampling.

In a survey of *Aedes aegypti* breeding in rock holes on a coral peninsular in Tanzania Trpis (1972*b*) firstly estimated the number of rock holes that held water suitable for mosquito breeding by quadrat sampling. In April, 1970, during the rainy season the mean number of water holding rock holes was 34.8/ha, but only 9.0/ha in July during the dry season. At about weekly or monthly intervals all water was removed with a pipette from some of the holes, the larvae counted and identified to species and instar and the water returned. By determining the mean number of larvae per rock pool Trpis (1972*b*) estimated the number of *Aedes aegypti* larvae per hectare on a number of different sampling days during April to September. Occasionally, all holes were dry, but when at least some were flooded the larval population was estimated to vary from 2–5290/ha, which gave an estimated population on the coral peninsular of 510–715 575 larvae on the different sampling days.

Crab Holes

Crab holes can be dug out to gain access to the water and larvae at the bottom (e.g. Dunn 1928) but usually the water is siphoned or pumped out; and Dunn (1928) modified a car tyre pump to remove water from crab holes. In Japan Mogi et al. (1984) collected mosquitoes from crab holes with a 2-m length of 5-mm diameter silicone tubing fixed to a 0.5-litre plastic bottle which in turn was connected to a 12-cm long, 2.5-cm diameter hand pump —originally used for suction removal of venom from snake bites. After removal of all the water two further pumpings were made after seepage water had reached about the original level. If, however, seepage was slow then about the same volume of water was added as was originally extracted. After the third pumping the procedure was repeated until two successive samples contained no immatures.

In the South Pacific area, by reversing the cup washer (A, B, C) and the check valve (D, E) assembly Goettel et al. (1981) converted a garden pressure sprayer into a suction pump for removing water from crab holes (Fig. 3.27*a, b*). The hose and spray nozzle attached to the base of the 4.5-litre plastic container were removed. A 1.5-m length of 7-mm diameter plastic tubing was reinforced along its length by stiff but pliable electrical wire bound to the plastic tubing with binding wire (Fig. 3.27*c*). This allowed the tubing to be twisted and turned while pushing it down a crab hole. To prevent the end of the tubing becoming blocked with mud the electrical wire was twisted round and inserted into the end of the tubing, and 1-cm square hole cut from the side of the tubing. Blowing down the tubing alerted the operator when it had reached the water in the crab hole. The tubing was then connected to the nozzle at the base of the plastic container bottle and water pumped out.

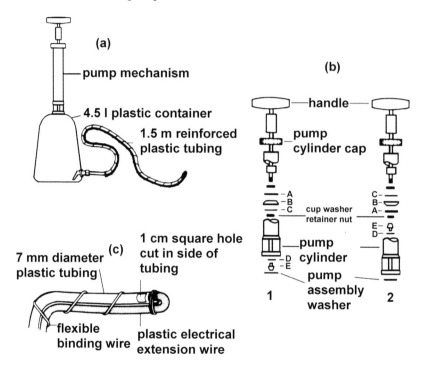

Fig. 3.27. (*a*) Modified garden pressure sprayer for pumping water from crab holes; (*b*) conversion of a commercial pump mechanism (1) to a suction pump (2) A—cup backing washer, B—cup washer (leather), C—cup spread washer, D—check valve O ring, E—check valve assembly: details for conversion will vary between different models of pump; (*c*) modifications and reinforcement of tubing used with the pump (Goettel et al. 1981)

The water can be sieved or tipped directly into suitable containers. As pointed out by the authors this simple extraction pump can be used to sample other container-habitats, such as tree-holes.

This method of Goettel et al. (1981) was also used by Takagi and Narayan (1988) to pump water from crab holes colonised by *Aedes polynesiensis*, but they experienced difficulties in getting their tubing down to the bottom of the meandering tunnels. In Samoa, Samarawickrema et al. (1993)

used Goettel's pump design to sample 244 crab holes in Vailu'utai village on Upolu island, but found none positive for *Aedes polynesiensis* or *Aedes aegypti*.

In Nigeria Bruce-Chwatt and Fitzjohn (1951) found that tunnels made by crabs may be many feet long and contain up to about 1 gal water, and in Louisiana, USA Evans (1962) reported that crawfish holes were sometimes so tortuous that unless the water was near the entrance it was impossible to siphon it out.

Natural underground habitats

Woodrow and Howard (1994) used a similar device to that described by Walker and Crans (1986) to sample larvae of *Culiseta melanura* from typical larval habitats under tree stumps, root masses, etc., in New York State, USA. Their sampling device was a modified 3.8 × 96.5 cm chemical transfer pump (Fig. 3.28). The pump was modified by wrapping a 15.9 × 13.9 cm piece of 0.6 cm mesh hardware cloth screen around the inlet end of the pump and securing it with a hose clip. The open end of the hardware cloth screen was closed off with a 4 cm diameter plastic cap. A 3.8 to 2.5 cm reducing coupler was added to the outlet end of the hose to allow samples to be pumped into 1-litre capacity narrow-mouthed polyethylene bottles. Field trials were conducted by taking samples along 70 m transects from the edge of a swamp towards its centre. The inlet hose of the pump was introduced into the habitat through existing holes, or where no holes were present in the surface substrate, a folding pruning saw was used to make one. Approximately 21% of closed habitats yielded larvae. Most positive samples contained 5 or fewer larvae, and the majority of larvae collected were 1st, 2nd, or 3rd instars, suggesting a bias of this sampling device against the capture of older instars.

Underground storm drains

Kroenenwetter-Koepel et al. (1995) used a hand-pumped siphon to transfer water directly from underground concrete stormwater catch basins into a collection container in a study of the role of *Culex* mosquitoes in the transmission of West Nile virus in Wisconsin, USA. Using this method a total of 1 244 larvae were collected from catch basins, representing 4 genera and 9 species, with members of the genus *Culex* accounting for 94% of all immatures collected.

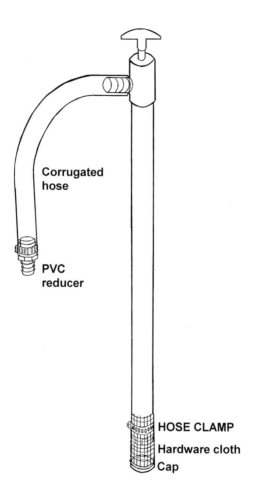

Fig. 3.28. Modified chemical transfer pump of Woodrow and Howard (1994)

Automobile tyres

In a car dump in Tanzania Trpis (1972*b*) counted all the different types of *Aedes aegypti* larval habitats, such as discarded tyres, tins, coconut shells and snail shells (*Achatina fulica*). Larvae were removed and counted from representative samples of each habitat. To facilitate the removal of larvae from tyres a 15-mm hole was made in the top of each tyre, which was turned through 180° when samples were required and the contents emptied into a bowl. After removing the larvae the tyre was returned to its original position and the water replaced. The population of larvae in the car dump was calculated from the mean number of larvae in each type of container and the percentage holding water. During April to September 1970 the estimated total population of *Aedes aegypti* varied from 0–22 477 on different sampling days. Monthly estimates of total mosquito productivity were obtained by summation of the estimated larval populations calculated each week (in fact every 6–8 days). Similar procedures were used with larvae collected from coral rock pools. These values, however, will not be reliable unless it is certain that none of the larvae was counted twice, such as a young instar being counted on the subsequent sampling day as an older instar. Also, no larvae should have hatched and then pupated between sampling occasions.

In a survey of 232 abandoned tyre dumps in West Virginia, USA, Joy et al. (2003) sampled mosquito immatures by repeatedly sweeping a small hand-held strainer with a mesh size of approximately 1.0 mm through the water that had collected in the tyres. Using this method 12 species were collected, of which *Ochlerotatus triseriatus* was the most common in terms of proportion of visits positive for this species, followed by *Culex restuans* and *Culex territans*. Other species collected were: *Ochlerotatus atropalpus*, *Toxorhynchites rutilus*, *Aedes albopictus*, *Anopheles punctipennis*, *Anopheles barberi*, *Culex pipiens*, *Orthopodomyia signifera*, *Anopheles quadrimaculatus*, and *Aedes vexans*. Larvae of *Culex restuans* ($\chi^2 = 13.12$, d.f. = 1, $P < 0.05$), *Anopheles punctipennis* ($\chi^2 = 5.87$, d.f. = 1, $P < 0.05$), *Culex territans* ($\chi^2 = 5.63$, d.f. = 1, $P < 0.05$) and *Aedes albopictus* ($\chi^2 = 4.38$, d.f. = 1, $P < 0.05$) were all more likely to be encountered in peridomestic sites, while *Ochlerotatus triseriatus* was more common in forested sites ($\chi^2 = 4.59$, d.f. = 1, $P < 0.05$). Joy (2004) used the same methods in 112 abandoned tyre sites in West Virginia, but on this occasion found no species significantly associated with peridomestic sites. *Ochlerotatus triseriatus* was significantly associated with habitats classified as non-peridomestic. Additional species collected were *Ochlerotatus japonicus* and *Ochlerotatus canadensis*.

Because the volume of water varied considerably in different sized tyres in a tyre yard in Indiana, USA Beier et al. (1983) tried to overcome any bias this might introduce by stirring the contents with a ladle and removing 300 ml water from each tyre. In a study of the distribution of *Ochlerotatus triseriatus* in Louisiana, USA Nasci (1988) used wooden stakes to prop tyres at 45° angles in different types of habitat. Immature stages were searched for with the aid of a torch, pupae removed and reared to adults for identification. Focks et al. (1980) tied tyres vertically to the sides of trees, and forced a wooden stick between the rims of each tyre to spread it out, thus facilitating observations on larvae in the collected water.

Gettman and Hall (1989) cut the rim (bead) of a tyre diametrically opposite each other with bolt cutters, then the top wall of the tyre was sliced round so that it could be lifted up like a flap. Two small sections of the wall (2.5 cm) were left uncut to serve as hinges. The tyre is positioned on its side at an angle of 23–30°. When larvae were to be collected the hinged flap (side wall) of the tyre was lifted. A drain hole can be bored into the opposite (underneath wall) directly below one of the upper hinges of the flap to standardise the maximum amount of water held by the tyre. With practice an old tyre can be modified as described in about 10 min. These modifications cannot, however, be used with steel-belted tyres.

Suwonkerd et al. (1996) sampled used tyres placed on the ground to protect young tree saplings to investigate the seasonal occurrence of *Aedes aegypti* and *Aedes albopictus* in Chiangmai, Thailand. Larvae were collected weekly from 47 tyres using a siphon. The number of tyres yielding *Aedes albopictus* was consistently higher than the number yielding *Aedes aegypti*. The average number of larvae sampled per tyre over the collecting period (July 1992 to August 1994) was 0.12 *Aedes aegypti* and 0.49 *Aedes albopictus* ($t = 12.06$, $P < 0.001$). However, when samples containing no larvae were excluded from the analysis, the mean densities were not significantly different (2.65 for *Aedes aegypti* and 2.85 for *Aedes albopictus*).

Zhen and Kay (1993) compared the sampling efficacy of sweeping with a small net (63 cm diameter) and conventional dipping (350 ml dipper) for collecting *Aedes aegypti* larvae from tyres. They found that when sampling known numbers of larvae, between 2.0 and 6.8 times more larvae were recovered from 10 sweeps with the net compared with a similar number of dips. Sweeping also gave more consistent results between sampling occasions. When low numbrs of larvae were present (20), sweeping with the net detected the presence of larvae in 100% of 158 samples. Dipping on the other hand returned 20 negative samples from the same number of trials. The presence of sediment in the tyres had a greater effect on recovery rates obtained by sweeping than on recovery rates by dipping.

Methods for sampling *Mansonia* and *Coquillettidia* larvae

Larvae and pupae of *Mansonia* and *Coquillettidia*, and a few other mosquitoes such as *Mimomyia hybrida* and *Mimomyia pallida*, are difficult to sample by ordinary methods because they do not normally rise to the water surface, but remain submerged and attached to plants from which they obtain their oxygen. Because of this, various methods have been devised to collect them, several of which have been briefly reviewed by Lounibos and Escher (1983).

Uprooting and washing

Several workers have pulled up host plants and vigorously shaken their roots and other submerged parts in a bowl or bucket of water. After mud and debris have settled larvae and pupae are collected either from the water surface or are pipetted from the bottom mud (van den Assem and Metselaar, 1958). McNeel (1931) uprooted plants and washed their roots in a bucket of water having a screen bottom and collected larvae of *Coquillettidia perturbans* from amongst the debris collected on the screen. This extraction procedure can be performed either at the edge of larval habitats or plants placed in plastic bags and taken to the laboratory for processing. In East Africa Gillett (1946) uprooted handfuls of aquatic vegetation and thrust a large enamel basin under the roots to collect some of the larvae which were disturbed and swimming sluggishly in search of new roots. After allowing about 30 s for mud in the bowl to settle, the water was poured off until only about 1 cm remained from which the larvae were easily removed. This is basically similar to one of the collecting methods used by Bonne-Wepster and Brug (1939) in Java. Although this technique is often successful in Africa in shallow waters, Wharton (1962) found it did not work in deep water in Malaysia.

Laurence (1960) pointed out that *Mansonia* larvae are often not easily detached from plants even by vigorous shaking, Moreover, both van den Assem and Metselaar (1958) and Ingram (1912) noted that detached larvae may remain hidden and submerged for long periods; according to van den Assem and Metselaar (1958) up to almost 1 h with larvae of *Mansonia uniformis*. Bidlingmayer (1968) reported that in the absence of plants *Coquillettidia perturbans* larvae could remain submerged for about 5 h during the day and 3.6 h at night, and that *Mansonia dyari* could remain submerged for 2 h during the day and for 1 h at night. In Florida, USA Bailey (1984) collected *Pistia stratiotes* plants from a *Mansonia* breeding place 2 h after sunrise and 1 h after sunset and recorded the numbers of larvae attached. He found that substantial numbers of larvae of *Mansonia dyari* and *Mansonia titillans* appeared to detach at night; the ratios of daytime collections of larvae to night-time collections ranged from 5.0:1 to 1.3:1. This type of behaviour makes it more difficult to collect larvae by uprooting plants and shaking them to detach the larvae.

Lounibos and Escher (1983) uprooted plants (cattails) and vigorously shook their roots in buckets of tap water to detach larvae of *Coquillettidia perturbans*. Then an insert, originally designed as a part of the Bidlingmayer trap, was fitted into the bucket and tap water added until the level rose to just a few cm above the cone-shaped aperture of the insert. In the absence of roots larvae surfaced through the aperture of the apparatus and were siphoned off with the water after a 24 h holding period. To check the efficiency of this method the inserts were placed over seven samples for a further 24 h. The numbers of larvae recovered varied between 8.3 and 29.9% ($\bar{x}$ = 16.0%) of the original sample. To determine whether many larvae were lost when plants were uprooted from field sites, larvae were allowed to attach to roots of wheat that had been grown hydroponically (Guille 1973) in a bucket of water. Twenty-four hours later the wheat plants were lifted out and transferred to clean water and the numbers of larvae remaining attached to the roots counted. Only 52.3% of *Coquillettidia perturbans* larvae had successfully transferred onto the wheat roots, whereas 95.7% of larvae of *Mansonia dyari* remained attached. Some 30 years earlier Hagmann (1953) had reported 50% detachment of *Coquillettidia perturbans* when plants were uprooted. These results illustrate how differences in behaviour of species can introduce bias in sampling programmes. Lounibos and Escher (1983) also found that 1st instar larvae readily detach from plants and were infrequent in their samples, whereas pupae were very firmly attached and even vigorous shaking did not always dislodge them.

Because of the failure of all larvae to swim through inserts placed in cylinders and differences between detachment rates of both instars and species from plants, quantitative sampling of larval populations of *Mansonia* and *Coquillettidia* mosquitoes by uprooting remains difficult.

McDonald (1970) found that when plants of *Eichhornia* with attached larvae of *Mansonia uniformis* were shaken vigorously in water only four to five larvae were dislodged, and not all these swam to the surface. When, however, plants were immersed in a 5% solution of sodium hydroxide apparently all larvae and pupae detached themselves and rose to the surface. Similar but less dramatic results were achieved if plants were immersed in solutions of sodium chloride stronger than 15%. It is likely that other species will react the same way. Wharton (1962) found that Malaysian species of *Mansonia* detached themselves from plants when they were soaked overnight in pyrethrum emulsion. Chemical treatments, however, are now rarely undertaken to collect *Mansonia* or *Coquillettidia* preadults.

Cylinders

Bonne-Wepster and Brug (1939) had limited success in collecting *Mansonia* larvae by using open-ended 16-in diameter and 30-in high cylinders. These were pressed down into the mud and after 15–20 min all *Pistia* plants enclosed by the cylinders carefully removed and searched for larvae. The water within the cylinders was then stirred up and scooped out as rapidly as possible and examined for larvae, and finally pieces of roots in the mud were pulled up and inspected. Larvae were collected from all three subdivisions of the sample, i.e. attached to the free floating *Pistia*, to its roots and swimming in the water. In Malaysia Wharton (1962) constructed galvanised metal cylinders, 14 in in diameter and 18 or 30 in high, having a pair of handles near the top. The lower edge was reinforced with a sharpened band of metal to facilitate cutting through plant roots. Roots and plants were pulled up and washed vigorously within the cylinders, then discarded, while the water was baled out and strained first through a coarse hexagonal chicken wire sieve and then through a 16-mesh sieve placed underneath. The contents retained by the second sieve were washed into white dishes containing clean water or saturated magnesium sulphate solution. Larvae were collected as they surfaced. While the sample was being flushed through the sieves the cylinder was refilled with water and this was again baled out and sieved, and then the process repeated a third time. Wharton (1962) concluded that when the material collected on the sieves was flooded with saturated magnesium sulphate more larvae were extracted than when water was used.

In Russia Morozov (1965) designed a special piece of apparatus for sampling larvae of *Coquillettidia richiardii*. This consisted of a 22-cm length of a 12-cm diameter cylinder into the bottom of which is fitted an inverted funnel having three small openings (windows). A 1-m rod passes down the cylinder and connects with three sharp shutters that fit over the three windows. In use the cylinder is thrust into the water with the windows open so that grass and vegetation projects through the windows, then by rotating the rod the shutters slice off these pieces of vegetation and close the windows. The cylinder is withdrawn and the shutters reopened to allow the water and plant pieces to empty into a bucket, from which any captured larvae can be removed.

Bidlingmayer's larval trap

This trap works on the principle that when *Mansonia* and *Coquillettidia* larvae are shaken free from plants they eventually rise to the water surface, although this may take a long time (Bidlingmayer 1968; Bonne-Wepster and Brug 1939; Ingram 1912; van den Assem and Metselaar 1958). The trap consists of two basic parts. One part is a 30-in tall, 13⅞-in diameter galvanised cylinder fitted with a pair of handles, which samples an area of 1 ft². The other part of the trap is inserted in the cylinder, and one of two types is used (Fig. 3.29*a–c*). The simplest is a 12-in high metal cone with a 1¾-in apical opening and a basal diameter slightly less than 13⅞ in, so that it is a sliding fit within the cylinder. A small metal cylinder 12½ in high and 9 in in diameter is placed over the cone and riveted in position. The alternative insert for the trap consists of a 10-in high cylinder also with a diameter slightly less than 13⅞ in. The floor is about 4 in from the base and is composed of a series of small pyramids (14 complete or partially complete pyramids) each 2¾ in high, with a base of 3½ in square. Each small pyramid has a ½-in apical opening.

To use the trap the outer cylinder is thrust into the mud and enclosed plants pulled up and vigorously shaken in the water within the cylinder. It is essential that all pieces of plant, either rooted or free floating, are removed. After some 5–10 min, to allow coarser material to settle, either type of insert is pushed down into the cylinder until its opening is about 1–2 in below the water surface. Because larvae are denied access to plants they surface and pass through the single cone, or pyramids, and are trapped in the space above the opening surrounding the insert. To facilitate removal of the larvae Bidlingmayer (1954) recommended that some of the muddy water above the cone, or pyramids, is baled out and replaced with clean water. The trap can be left in position for 24 h, or removed sooner so long as sufficient time has been allowed for the larvae to surface. Larvae are recovered from the trap by baling or siphoning out the water surrounding the cone, or pyramids. The water can be emptied into a white dish and the larvae collected. A more rapid larval recovery might be obtained if the contents were sieved and floated in saturated magnesium sulphate.

Bidlingmayer (1954) found that with the cone type of insert an average of 83% of *Coquillettidia perturbans* larvae introduced into the trap were recovered when the trap had been kept in position for 24 h. When the pyramid type of insert was used only 71% of the larvae were recovered. The pyramid insert is less efficient and considerably harder to construct, but Bidlingmayer (1954) considered it the better choice because it permits the trap to be used in only 4 in of water whereas the cone insert requires a depth of at least 14 in, and it also results in more sediment contaminating the collection.

One disadvantage of the method is that not all larvae are dislodged from the weed before it is discarded, and pupae are rarely caught (Bidlingmayer 1954), as they appear to attach themselves more tenaciously than the larvae (Bonne-Wepster and Brug 1939; Dorer et al. 1950; Lounibos and Escher 1983). It may also be difficult to ensure that all fragments of plants are uprooted and removed; if any are left detached larvae may reattach themselves. Sometimes so much plant material and debris has to be removed before the cone can be inserted that the water level within the cylinder drops drastically, and more water must be added so the level rises above the pyramid or cones.

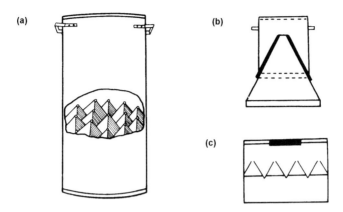

Fig. 3.29. Bidlingmayer's (1954) larval trap; (*a*) cylinder; (*b*) cone-type inset; (*c*) pyramid-type inset

The trap is also apparently less efficient in cold waters because larvae are sluggish and relatively few pass through the cones. However, Bidling-mayer (1954, 1968) considered that all four larval instars, but not the pupae, were sampled equally by the trap.

Scraping plants

In Malaysia Wharton (1962) managed to collect *Mansonia* larvae from some habitats by dragging a net through the water against the roots of host plants. The contents were tipped out into white dishes and larvae removed. A similar approach was used in Canada by Batzer and Sjogren (1986). They needed to survey 86 marshes for larvae of *Coquillettidia perturbans* and considered that using either cylinders or uprooting plants too time-consuming. Instead they pulled a 14-cm diameter dipper, mounted on a 90-cm handle, from the base of emergent plants (e.g. cattails—*Typha* spp.) upwards scraping larvae from the roots. Other times they thrust the dipper under floating mats of plants and scraped along 50 cm of roots. The bottom of the dipper was screened with mesh (7.9 wires/cm), and larvae were removed from it with an ordinary larval pipette. A second scraping of roots did not collect significant numbers of larvae.

Batzer (1993) describes a method for collecting *Coquillettidia perturbans* larvae from plant roots in Minnesota, USA, using a dipper constructed from a 12.7 cm diameter, 4-cm high 1-mm mesh screen. For sampling among emergent plants, this dipper was oriented with the distal edge of the dipper against the bottom of the plant and was then drawn up vertically through the root mass and along the plant stem to the water surface. To sample floating plants, the dipper was oriented horizontally beneath the plant at a distance of 50 cm below the plant. The edge of the dipper was then pushed firmly into the root mass and moved outwards along the underside of the floating plant. Batzer (1993) considered this method more suitable for large-scale monitoring programmes than other methods described here.

Suction pump of Morris et al.

Morris et al. (1985) considered that the methods developed by various workers (Allan et al. 1981; Barton 1964; Bidlingmayer 1954; Gozhenko 1978; Guille 1975; McNeel 1931; Morozov 1965; Rademacher 1979) for collecting *Mansonia* and *Coquillettidia* larvae were either laborious, time-consuming, or inappropriate for quantitative studies, so they developed their own system for collecting *Coquillettidia perturbans* larvae in Florida, USA. Their apparatus comprises a probe, a pump, a sieve and connecting hose (Fig. 3.30*a*). The probe (B) is a 30-in length of 0.5-in diameter electrical conduit with four ⅜-in holes drilled perpendicular to each other near the end. This end section of the conduit is enclosed in a 2-in diameter 10-in long sleeve of 18-gauge stainless steel perforated with numerous 5/16-in holes (C) acting as a coarse filter. The end of the conduit is closed with a bolt-on conical cap (D). A length of garden hose connects the upper part of the conduit to a 12-V d.c. self-priming pump operated from a 12-V car battery. About 4.24 gal of water are sucked up per minute. Alternatively a 12-V-6-A gel cell battery can be used to make the system more portable, but this will only power about 30 min of pumping. A second length of conduit of variable length is bolted on the outside of the first piece to form a handle (A). The probe is placed at the bottom of a habitat or amongst mats of vegetation and water pumped up into an inverted 8-in diameter metal funnel (Fig. 3.30*b*) having a 350-μm mesh nylon sieve (B) attached to the bottom. The narrow end of the funnel (A) is placed in a 1-quart Kilner (Mason) jar, and material deposited on the sieve flushed through the jar. A series of Kilner jars containing material are then taken to the laboratory for further processing with the separation chamber (Fig. 3.30*c*). This consists of a 10-in length of 3-in diameter PVC cylinder glued to a 4.5-square plexiglas base. Three ⅜-in holes drilled through the cylinder 6.5 in from the base allow 3/16-in self-tapping bolts (seating pins) (C) to be inserted. An inverted plastic circular funnel (A), trimmed to fit into the cylinder and with its stem removed to create a ⅝-in diameter hole, is pushed down to rest on the three bolts, and held there by a 9.25-in length of 3/16-in diameter coiled plastic tubing (B). This is in reality a miniaturisation of the apparatus devised by Lounibos and Escher (1983). The water sample from the Kilner jar is tipped into the cylinder and water added until it rises to the level of the bolts. The funnel and tubing are then inserted and extra water added until it is about ½ in above the funnel opening. The samples are held overnight, after which the water above the funnel (about 200 ml) is poured into a white tray and *Coquillettidia* larvae pipetted out.

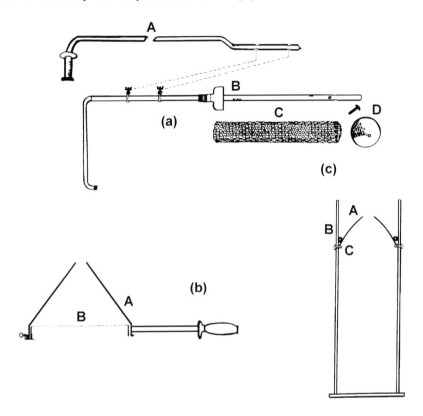

Fig. 3.30. (*a*) suction trap, A—handle, B—probe conduit, C—filter, D—end cap; (*b*) sieve unit with A—galvanised funnel, B—350-μm mesh screen; (*c*) separation chamber. A—inverted plastic funnel, B—coiled plastic tubing. C—self-tapping bolts (Morris et al. 1985)

Apparently two people can take 75–100 samples/day. Because of the probe's small diameter and the ability to vary the length of the handle the probe can be used in almost any type of *Coquillettidia* habitat, including areas with deep water or thick roots. The team can work from the bank or from flat-bottomed boats. The electric pump and the separation system standardise the sampling procedure, and this eliminates human errors and variabilities. This apparatus was subsequently used by Callahan and Morris (1987) and Morris et al. (1990) to collect *Coquillettidia perturbans* larvae.

Larvae of *Culex, Anopheles* and *Uranotaenia* in addition to *Mansonia* and *Coquillettidia* have been collected, suggesting that the technique can be used to collect surface-breathing mosquitoes. The authors suggest that with some modifications the system can be used to sample tree-holes, crab holes, rock pools and artificial containers.

Bilge pump of Walker and Crans

Walker and Crans (1986) developed a hand-operated boat bilge pump for collecting larvae and pupae. The bilge pump is firstly modified by removing the intake valve at the bottom of the pump shaft, which converts the pump into a syringe. If this valve is not removed debris will clog the pump and prevent larvae from being sucked up. The outlet hose located at the top of the pump can be discarded because it now serves no useful purpose. The end of the modified bilge pump is thrust into a breeding place and located at the base of vegetation, such as cattails, but trying to avoid pushing it into thick mud. Water is sucked up the shaft by pulling up the handle, after which the pump must be withdrawn quickly from the habitat and the collected column of water allowed to drain into a bowl or through a fine sieve. Because the valve has been removed the water will drain out of the pump by gravity. The sample should be covered with water to allow larvae and pupae to swim to the surface; if large amounts of detritus have been sucked up larvae may not be apparent for 30–60 s. For more quantitative results the contents of the sieve should be flushed into a bowl and washed through a series of stacked graded sieves (mesh sizes 5 × 5, 2.5 × 2.5 and 1 × 1 mm).

Olds et al. (1989) used this bilge pump technique to collect larvae of *Coquillettidia perturbans* from cattail marshes. They then followed the procedures of Morris et al. (1985) of using inverted plastic funnels to separate and sort the larvae in the laboratory. They recovered 94% of the larvae within 20 h. A limitation of this method, but not that of Morris et al. (1985), is that it cannot be used in water deeper than about 1 m.

Use of artificial container habitats for sampling larvae

The use of small container habitats, such as clay pots, glass jars and bamboo pots, for studying and detecting mosquito breeding has been outlined in Chap. 2 in connection with artificial oviposition sites. In larval surveys eggs are not looked for, but the contents of the containers are usually emptied into a white dish and the larvae of the different species counted, and then discarded or returned. The addition of more water may be necessary to compensate for evaporation. The possible differences between species prevalence based on eggs and larvae have been outlined in the previous chapter.

There are several obvious advantages in using small artificial containers. For example, both the total population of larvae and predators, as well as other associated fauna, can frequently be counted with the minimum of effort, returned and recounted on successive sampling occasions. Artificial containers can be placed in different ecological niches, at different heights and in different types of vegetation etc. to study the abundance and prevalence of different mosquitoes in relation to habitat.

Apart from small container habitats larger artificial ones, such as shallow pools or pits, have sometimes been dug to attract pool-breeding species (Bates and de Zulueta 1949; Brust 1990; Christie 1954, 1958; Darrow 1949; Madder et al. 1980; Russell and Rao 1942a, b). In Sardinia Trapido and Aitken (1953) dug two pools, 2 m in diameter and 20 cm deep, near the edge of a stream so that they readily filled with seepage water. Algae and a few aquatic plants collected from the stream were introduced into the pools from which regular larval collections were made to study the seasonal succession and density of *Anopheles atroparvus* and *Anopheles hispaniola*.

Artificial habitats are sometimes used to monitor changes in the seasonal abundance of mosquitoes breeding in natural sites in an area, but there may be severe limitations on their ability to reflect true population changes. When water is maintained in artificial habitats when natural ones are drying out they may, because of their availability, attract abnormally large numbers of ovipositing females, and consequently not reflect the decrease in population size that is occurring within the area. It is also difficult to compare population size in different areas by using artificial habitats. For example, the absolute mosquito populations may be the same in two areas, but if in one area there are twice as many breeding places as in the other, then artificial habitats will indicate that the population is about half that of the other. A correct interpretation would only be possible if the total number of available habitats in each area was known. If, as usual, the size and attractiveness of the natural larval habitat varies, the problem becomes increasingly difficult.

Aedes aegypti indices

Classical methods

Because of the importance of *Aedes aegypti* as a vector of yellow fever in Africa and the Americas and of dengue and haemorrhagic dengue in the wider tropics and subtropics, considerable attention has been paid to larval surveys. So-called *Stegomyia* indices were developed in the early 1920s (Connor and Monroe 1923) in conjunction with the Rockefeller Foundation's efforts to eradicate yellow fever from coastal cities in Brazil. The first two indices proposed were:

1. House (=Premise) Index = percentage of houses (including examination of surrounding compound) examined that have larvae of *Aedes aegypti* in at least some containers.
2. Container (=Receptacle) Index = percentage of water-holding containers examined that contain larvae of *Aedes aegypti*.
3. These two indices were supplemented with a third, the Breteau Index, some thirty years later. Breteau Index = total number of containers with larvae of *Aedes aegypti* per 100 houses.

These three indices remain the most commonly used measurements in *Aedes aegypti* surveys to this day.

The house index has been used for many years and it is probably the most widely employed single index, although increasing use is being made of the Breteau index. For example, the Breteau index has recently been used to study the spread of *Aedes albopictus* in São Paulo state, Brazil (Neto 1996, 2002), in investigations of a dengue outbreak in Taiwan (Lien et al. 1993), and in a study of large water drums as *Aedes aegypti* larval habitats in Trinidad (Chadee and Rahaman 2000).

In the original type of survey the presence of *Aedes aegypti* larvae in water-holding containers found in or outside houses was recorded. Sometimes, especially in small hamlets and villages, all houses are inspected for *Aedes aegypti* breeding, but usually only a percentage of the houses are searched. The selection of houses that are sampled can either be systematic, e.g. say every 2nd or 3rd house (Anon 1956) or every other house in every 4th block of houses (Tinker 1967) or at random (Reid 1954; Tinker and Hayes 1959). Whatever method is used samples should be taken from houses in different suburbs or areas of towns, from different types of houses and from houses whose occupiers come from a range of socio-economic backgrounds in order to get a representative incidence of breeding in the whole town or village. Vezzani et al. (2001) used a similar survey method to calculate infestation levels (container index) for *Aedes aegypti* in cemeteries in Argentina and attempted to relate these to environmental variables. Chambers et al. (1986) gave a useful account of how to undertake a survey of potential and actual breeding places of mosquitoes in people's backyards in relation to their income levels. They showed the usefulness of using census tracts (i.e. small geographical more or less homogenous political subdivisions of a town) and their accompanying economic and social statistics, to objectively select areas for artificial container mosquito production studies. More recently Focks et al. (1993) proposed a methodology for conducting household surveys that incorporates a brief demographic survey alongside collections of adult and immature mosquitoes from inside and outside the household.

The method of Focks et al. (1993) was used to undertake an analysis of the spatial pattern of *Aedes aegypti* adults and immatures in two areas of Iquitos, Peru (Getis et al. 2003). Households were sampled by assigning a unique code to each property, which was painted on the front of 550 houses in Maynas zone and 510 houses in Tupac Amaru zone. Households were surveyed in sequence daily along the block from a designated start house between 0700 and 1300 h. Unoccupied or closed houses and houses where residents did not provide permission for the survey, businesses, offices, and schools were not sampled. 95% of houses were surveyed in each zone. Collecting teams were rotated among blocks each day in order to reduce bias. Access to closed houses was attempted a minimum of three times. In each household, one member of the collecting team administered the demographic survey questionnaire and the other member carried out the entomological survey. Adult mosquitoes were collected from indoors and outdoors using a backpack aspirator (John W. Hock Company, Gainesville, Florida, USA). Collections were made from walls, under furniture, inside closets, and other likely adult mosquito resting sites in all rooms of the house (when permitted). Outdoor collections were made from exterior walls, under eaves, vegetation, and in and around outdoor stored materials. Larval and pupal collections were also made from inside and outside houses. The spatial patterns of four variables were examined, namely adult *Aedes aegypti*, pupae, all water-holding containers, and water-holding containers positive for larvae and/or pupae. Global K-functions, point and weighted, were used to identify clustering in each of the two zones and the local statistic, Gi^*, was used to determine the magnitude of each variable in each household for each neighborhood. These statistics are components of the suite of spatial statistical programs available as part of the Point Pattern Analysis (PPA) program, an online version of which is available. To investigate the presence or absence of a variable in a household for each neighbourhood, chi-square tests were used. The K-function describes the number of pairs of observations between a point, which is located at the centre of a disk and other points that are distance d away. For a stationary, isotropic process, $\lambda(d)$ is the expected number of points within distance d of an arbitrary point. The estimator of λ is N/A where N is the number of points in the study area A. The estimator of K(d) is:

$$\hat{K}(d) = A/N^2 \sum_i \sum_j u_{ij}^{-1} I_d(d_{ij} \le d), i \ne j \qquad (3.17)$$

where d_{ij} is the distance between the ith and jth observed points and $I_d(d_{ij} \le d)$ is an indicator function that is 1 if d_{ij} is less than or equal to d and 0 otherwise. For a circle centered on i passing through point j, u_{ij} is the proportion of the circumference of the circle that lies within A. When d_{ij} is less than the distance from i to one or more borders of the study area, u_{ij} is 1. The "border correction" makes $\hat{K}(d)$ an approximately unbiased estimator of K(d) provided that d is less than the circumference of A. A square-root scale makes the function linear and stabilizes the variance.

$$\hat{L}(d) \equiv \sqrt{[\hat{K}(d)/\pi]} \qquad (3.18)$$

which is the estimator of $L(d) \equiv \sqrt{[K(d)/\pi]}$. The mean of $L(d)$ is d and the approximate variance is $\frac{1}{2}(\pi N^2)$. The expectation of $L(d)$ given the hypothesis of complete spatial randomness (CSR) is d. CSR is a homogenous planar Poisson process where all points are independent of all other points and all locations are equally likely to contain a point. For CSR, a plot of $\hat{L}(d)$ against d on similarly scaled axes yields a 45-degree line beginning at the natural origin. A clustered pattern occurs when $\hat{L}(d)$ is greater than d and a dispersed pattern can be identified when $\hat{L}(d)$ is less than d. Statistical significance at the $P < 0.05$ level is assumed to exist when the observed $\hat{L}(d)$ function falls outside of an envelope containing 19 permutations of the location of the N objects where each permutation is based on CSR. $\hat{L}(d)$ is usually calculated for a series of distances d. Instead of considering each point as a nominal scale variable, points can be weighted according to some measure of size or intensity,

$$\hat{L}_w(d) = [\{A \sum_i \sum_j u_{ij}^{-1} I_d(d_{ij} \le d) x_i x_j\} / \{\pi [(\sum_i x_i)^2 - \sum_i x_i^2]\}]^{1/2}, i \ne j \qquad (3.19)$$

where X is a random variable having values x for adult mosquitoes in houses at sites i. Equation (3.19) is the estimator for $L_w(d)$, which is equal to $E[\hat{L}_w(d)]$. In the cases discussed in their paper, the weights are in turn numbers of adult mosquitoes, pupae, water-holding containers, and positive containers. For each x_i, there are $(N - 1)$ values x_j. In this case, the numerator of $\hat{L}_w(d)$ represents the product of the pairs of values $x_i\, x_j$ within distance d of each x. The denominator is scaled such that if all x are of equal value, then $\hat{L}(d)$ will be approximately equal to $\hat{L}_w(d)$. Thus, equation (3.19) represents a measure of clustering or dispersion identified in equation (3.18). If the number of adult mosquitoes, for example, is independently distributed within the plots of houses, $\hat{L}(d)$ will be approximately equal to $\hat{L}_w(d)$. Upper and lower significance boundaries for $\hat{L}_w(d)$ can be determined by a permutation procedure in which the various observed values for number of adult mosquitoes, x_i, are permuted among the house locations a specified number of times. The authors also explored the increments to $\hat{L}(d)$ and $\hat{L}_w(d)$ observed for each equal increase of distance. In a CSR pattern of adult mosquitoes, these successive values will be the same for each equal increase of d. The focus is on the noncumulative properties of these pattern indicators. When the change in $\hat{L}(d)$ is greater or less than the change in $\hat{L}_w(d)$ for a given distance band, the adult mosquitoes are less concentrated or more concentrated, respectively, than that expected in the observed pattern, no matter how clustered the pattern of houses. That is, the number of adult mosquitoes is not randomly distributed among the houses. In essence, $\Delta \hat{L}(d)$ with $\Delta \hat{L}_w(d)$ are compared for a given small change in d.

In addition to $L(d)$, the local statistic, $Gi*$ was used to identify individual members of clusters. For $Gi*$ each house is taken as a centre, one at a time, and the nearby area is searched for occurrences of more or fewer adult mosquitoes than expected. In this way, specific houses are identified as members or non-members of clusters. The $Gi*$ statistic is defined as:

$$G_i*(d) = [\sum_j w_{ij}(d)x_j W_i * \bar{x}]/[s\{[NS_{1i}* - W_i*^2]/(N-1)\}^{1/2}]\ \text{all } j \qquad (3.20)$$

where $w_{ij}(d)$ is the i, jth element of a one/zero spatial weights matrix with ones if the jth house is within d of a given ith house; all other elements are zero; $Wi* = \sum_{ij}(d)$, where w_{ii} is included, and $S_{1i}* = \sum w^2_{ij}$ (all j).

The mean of the adult mosquitoes in houses is x and s is the standard deviation. The value of $Gi*(d)$ is given in normal standard deviates. Note that this statistic has as its expectation, $Wi\overline{x}$, which controls for the number of houses within d of each house. Also note that $Gi*(d) = 0$ where adult mosquitoes are randomly distributed within d of house i. For this study, values greater than an arbitrary value of 2.575 (the 0.01 level of confidence) were defined as representing houses that were members of clusters of adult mosquitoes. Results of the survey indicated that adults clustered strongly within houses and weakly to a distance of 30 m beyond the household. Clustering was not detected beyond 10 m for positive containers or pupae. One hundred and sixty-four (31.1%) of the houses had one or more adult mosquitoes present; however, only 35 of them (21.3%) were members of statistically significant house clusters. This indicated that clusters were made up mainly of household concentrations, and that 79.7% of the households with mosquitoes were located randomly among all households. Pupae were observed to cluster strongly within houses, but households infested with pupae were dispersed evenly throughout the neighbourhood.

Morrison et al. (2004a) reported the results of a long-term large-scale *Aedes aegypti* survey in Iquitos, Peru, in which daily mosquito surveys were carried out over a period of 3 years and seven months between January 1999 and August 2002. The entomological sampling was one component of a prospective longitudinal cohort study of dengue virus infections in approximately 2400 students and their families. Entomological survey circuits were established that ensured that each serological study participant's household as well as neighbouring households were sampled. Each circuit contained 17 short (1–2-week) surveys. The duration of a complete circuit was approximately 4 months and around 6000 households (range 5721–6466) were sampled per circuit. The methodology of the entomological surveys was as described by Getis et al. (2003) and Morrison et al. (2004b) All containers were described and categorised by measuring their diameter, length, width, and height, and assigning them to 64 container types comprising 14 broad categories. Containers were also scored for solar exposure (proportion of day with direct sunlight) method by which container was filled with water (manual, rain-filled, rain-filled with aid of a rain gutter or roof), and whether the container had a lid. For *Aedes aegypti*-positive containers, the number of larvae was estimated (1–10, 11–100, and > 100), and the total number of pupae was counted. All entomological data were linked to specific houses and entered into a GIS. Differences in mean abundance of water-holding, positive, and pupae positive containers and absolute numbers of pupae were analsed using analysis of variance

(ANOVA) with the general linear models procedure (PROC GLM) of SAS. A total of 289 941 wet containers were observed and characterized and the mean number of wet containers per household surveyed was 4.85 (308 containers per hectare). Of these, 7.3% contained *Aedes aegypti* and 3.6% were positive for pupae. Most wet containers were found inside houses (57.4%), did not have lids (68.0%), and were filled manually by occupants of the household with water from a pipe or well (85.0%). However, only 17.9% of positive containers were located inside houses, 97.5% did not have a lid, and only 19.5% were manually filled. Wet container, positive container, pupae-positive container, pupal abundance, and house density all varied significantly among the eight geographic zones studied. Using a definition of a key container as one that contained >500 pupae, approximately 2% of pupae-positive large tanks and non-traditional containers were classified as key containers. Of these, three of seven contained > 1000 pupae on the day of the survey. Plastic containers accounted for 37% of all positive and pupae-positive containers and 25% of all pupae. However, the plastic container category was not considered an appropriate category for directing control efforts at the infestation rate (3.6%) was low and there were very many plastic containers present in the area.

Also in Iquitos, Peru, Morrison et al. (2004b) used a survey methodology based on that proposed by Focks et al. (1993) and enabling the traditional larval indices (PAHO 1994), and pupal indices (Focks and Chadee 1997) to be obtained. In addition, adult mosquito collections were made using a backpack aspirator (John W. Hock Company, Gainesville, Florida, USA). Six hundred houses on 20 city blocks in Maynas and on 14 city blocks in Tupac Amaru were surveyed twice, the first from mid-November to mid-December 1998 and the second from mid-December 1998 to mid-January 1999. One member of each two-person entomology team conducted a brief demographic survey to determine key features of the property, including the number of occupants, property size, construction materials, method of cooking, water use patterns, type of sewage disposal, and insecticide use. The other member of the team began collecting adult mosquitoes using the backpack aspirator from all rooms to which access was granted. Aspiration collections were also made outside the house from external walls, under eaves, and amongst vegetation and materials stored outside. After completion of the demographic questionnaire all potential *Aedes aegypti* containers both inside and outside the house were examined for water, larvae, and pupae. All pupae and a sample of larvae collected were placed in a twist-top plastic bag and labelled with a house and container code. The following entomological indices were calculated in each of the four surveys: house index (HI); container index (CI); Breteau index (BI); pupae per hectare; pupae per person; pupae per house; adult index

(AI = number of houses positive for adult *Aedes aegypti* divided by the number of houses inspected multiplied by 100); adults per hectare; adults per person; and adults per house. Results of the demographic and entomological census surveys were as follows: in the first survey, the house index was 28.8% (142/562) in Tupac Amaru and 44.7% (260/582) in Maynas ($\chi^2 =$ 30.8, d.f. = 1, $P < 0.0001$). During the second survey period, the house indices were 22.7% (124/545) and 38.1% (216/567) in Tupac Amaru and Maynas respectively ($\chi^2 = 30.8$, d.f. = 1, $P < 0.0001$). All other indices were higher in Maynas than in Tupac Amaru. Larval indices (HI, CI, and BI) were 1.5 to 1.9 times higher in Maynas than in Tupac Amaru. Differences in pupal indices were even greater, ranging from 2.2 to 2.8 times higher in Maynas than Tupac Amaru. Adult indices reflected an intermediate magnitude of difference between the two study areas. Simulations of a standard rapid assessment survey used by the Ministry of Health in Peru were conducted on the demographic and entomological data maintained within a GIS. The sampling methodology used by the Ministry of Health consists of a systematic sampling strategy carried out from a randomly selected starting block and continuing along a transect running from northwest to southeast. The sample size used for each simulation was 10% of the total houses, as recommended by the Pan American Health Organization (1994), and in this case equivalent to 60 houses. The means of the 10 simulated surveys were similar to the indices calculated for each index using the full census data sets for both periods and in both neighbourhoods, but the range of estimates among the individual simulations was wide. Indices calculated from census data and from simulations are presented as table 3.4. Morrison et al. (2004*a*) concluded that individual households are the appropriate spatial unit for entomological surveys and the rapid assessment transect sampling technique is an efficient and appropriate study design. They further recommended that, logistical considerations aside, pupal or adult counts per hectare should be used as the best estimate of immediate risk for dengue virus transmission. A potential drawback of the rapid assessment transect approach is that random selection of starting blocks can result in overlapping routes among collecting groups.

Use of a transect sampling methodology relies on a random distribution *of Aedes aegypti* infested households within the sampled neighbourhoods. As demonstrated by the work of Getis et al. (2003) described above, this assumption appears to hold true in Iquitos at scales in excess of 30 m.

In an eight-city larval survey in the USA Moore et al. (1990) used both random and non-random sampling methods. Although the study focused on *Aedes albopictus* the methodology applies also to *Aedes aegypti*, consequently a brief description of their strategy is merited. Non-random sites included specially selected high risk premises such as tyre dumps and salvage yards, while a random cluster sampling technique was used to sample residential premises. Census tracts were chosen randomly from information obtained from the local census bureau on the basis of probability proportional to size, that is estimated number of premises. With this procedure census tracts with large numbers of residences had a higher chance of being sampled.

Table 3.4. *Aedes aegypti* indices calculated from entomological surveys and sampling simulations carried out during Nov.– Dec. 1998 and Dec. 1998–Jan. 1999 in the Maynas neighborhood of Iquitos, Peru (after Morrison et al. 2004*b*)

Date	Index	Census survey	Simulation surveys				
			Mean	SD	Range	CV	CV/mean
Nov–Dec	HI	44.67	47.84	5.50	38.3–56.1	11.49	0.24
1998	CI	12.78	12.98	1.92	10.7–16.7	14.81	1.14
	BI	83.16	88.33	10.80	71.9–110.3	12.23	0.14
	Pu/ha	432.26	479.83	90.11	249.9–778.8	39.62	0.08
	Pu/per	1.23	1.29	0.5	0.7–2.2	38.51	29.85
	Pu/hse	9.09	8.70	3.37	5.0–14.2	38.78	4.46
	AI	30.07	33.11	5.54	26.2–45.8	16.73	0.51
	Ad/ha	37.54	41.90	7.04	32.2–52.0	16.81	0.40
	Ad/per	0.11	0.11	0.02	0.08–0.15	18.13	164.82
	Ad/hse	0.70	0.76	0.11	0.61–0.93	15.64	20.58
Dec 1998	HI	38.10	39.89	2.02	36.4–42.4	5.06	0.13
– Jan 1999	CI	9.35	9.33	0.66	7.9–10.0	7.04	0.75
	BI	60.49	63.10	4.68	54.1–69.0	7.42	0.12
	Pu/ha	354.36	315.58	141.87	131.1–650.8	44.96	0.14
	Pu/per	0.98	0.80	0.39	0.3–1.8	48.52	60.65
	Pu/hse	6.57	5.52	2.60	2.2–11.8	47.11	8.53
	AI	27.16	26.52	4.41	19.7–32.8	16.62	0.63
	Ad/ha	36.99	33.38	7.96	23.3–48.0	23.85	0.71
	Ad/per	0.10	0.08	0.02	0.06–0.13	39.03	487.88
	Ad/hse	0.69	0.58	0.16	0.4–0.9	26.77	46.16

Census tracts with less than 640 premises were combined with an adjacent tract. A minimum of 160 premises were inspected in each selected tract, and the numbers of tracts chosen for a city depended on the size of the city. Within the selected census tracts 40 random numbers were used to identify the block locations of the target premises. The randomly selected premises on that block formed the cluster of premises to be inspected. A total of 5728 premises were inspected in the eight cities, and 24.4% harboured larvae of *Aedes albopictus* and/or other mosquito species. Problems of interpreting the results of larval surveys from high-risk non-randomly selected premises are discussed, and it is concluded that random larval surveys give the most reliable estimates of infestation rates of cities. This approach will likely have to be modified for use in other countries.

Premise Condition Index

In reviewing *Aedes aegypti* surveillance programmes in the Caribbean, Nathan (1993) made observations on some of the difficulties in sustaining such programmes over time, namely lack of resources, and weak motivation of staff due to the boring and repetitive nature of house-to-house surveys. Following initial studies that identified key premises for *Aedes aegypti* production as those with ≥ three positive containers (Tun-Lin et al. (1995*a*) a Premise Condition Index has recently been proposed by Tun-Lin et al. (1995*b*) as a rapid assessment tool that can be used to increase the efficiency of detecting premises positive for *Aedes aegypti* in surveys in Australia. House-to-house surveys were conducted in four localities in northern Queensland where dengue and *Aedes aegypti* were known to have occurred. Surveys for infested containers were restricted to the outside of premises and each premise was classified according to house condition, yard condition, and degree of shade, based on a series of standard photographs. Questionnaires were also sent to 168 positive and 168 negative households to determine the number of persons in each premises, their age, gender, occupation, employment status, whether the property was owned or rented, etc. The authors then used Poisson regression to produce two types of generalised linear models (univariate and multivariate). The univariate models tested the associations between number of positive containers and individual variables. The adjusted, or multiple generalised linear models examined whether the associations observed from the univariate model could be explained by confounding with other variables. Significant associations were found for the three variables house condition, yard condition, and amount of shade and PCI_3 scores were calculated by summing the scores for each variable. A PCI_2 score was also calculated for the variables yard and shade. The correlation coefficient for the relationship between PCI_3 score and the proportion of positive houses was $r = 0.94$, d.f. $= 7$, $P < 0.002$ and for PCI_3 score and the number of positive containers per house, $r = 0.91$, d.f. $= 7$, $P < 0.02$. Correlations between PCI_2 scores and the two variables were also significant ($r = 0.92$, d.f. $= 4$, $P < 0.02$ for the proportion of positive houses, and $r = 0.89$, d.f. $= 4$, $P < 0.04$ for the number of positive containers per house). Characteristics of the household occupants were not significantly correlated with either variable. By only inspecting premises with a PCI_3 score of 8 or 9 the probability of the premise being positive was increased by 2.7 and the probability of finding a positive container increased by 3.7. Premises with a PCI_3 score of 8 or 9 represented only 9.5% of the total premises but accounted for 34.7% of positive containers. The authors concluded that the method is a useful tool for rapidly selecting premises for further survey as it does not require entry into the premises or collection of containers or larvae. It was acknowledged that the grading of premises is essentially subjective and would require training of staff to ensure consistency. Kay et al. (2002) used the Premise Condition Index to categorise wells in Queensland Australia, prior to testing an intervention against *Aedes aegypti*. The Premise Condition Index of Tun-Lin et al. (1995*b*) was later used by Nogueira et al. (2005) to conduct an *Aedes* survey in Botacatu city, São Paulo, Brazil. The city was divided into 105 quadrants and one house per quadrant was randomly selected for the survey. One black plastic ovitrap was set per house, usually in a shaded location in the garden, or alternatively on the patio. House and yard condition and amount of shade were assigned PCI scores (3–9) and these scores were compared with the presence or absence of *Aedes* eggs. *Aedes albopictus* eggs were recovered from 41 of 105 ovitraps that were distributed across the city. Two ovitraps contained only *Aedes aegypti* eggs, and 11 ovitraps were positive for eggs from both species. Sixty-five per cent of houses with PCI scores of 8 and 9 (representing poor condition) yielded positive ovitraps, while only 19% of the properties with PCI scores of 3 and 4 (representing well-maintained houses) yielded positive ovitraps. There was a positive correlation ($r = 0.9684$, $P < 0.01$) between house condition and percentage of houses with positive ovitraps.

One-larva-per-container method

The one-larva-per-container method was developed by the World Health Organization *Aedes* Research Unit in Thailand as an alternative to the time-consuming classical indices for conducting *Aedes aegypti* surveys throughout the country. In this approach only a single larva was collected from each container (Sheppard et al. 1969). As the larval habitats of *Aedes aegypti* in Thailand could be conveniently divided into three main indoor and three main outdoor categories, namely, water-jars, ant guards and miscellaneous containers, only six appropriately labelled collecting bottles were needed by a collector. One larva is collected from each positive container associated with each house and placed in the appropriate bottle. A record is kept of the number of containers with and without larvae, and it follows that the numbers of each type of container with larvae should correspond to the number of larvae in the respective bottle, providing a useful check on the reliability of the collectors.

The single larval method measures the prevalence of *Aedes aegypti* relative to other species and gives the minimum number of positive breeding habitats per house. When *Aedes aegypti* is the only container-breeding species present then the one-larva-per-container survey gives the normal house, container, and Breteau indices, but when there are other species these indices will be underestimated. For this reason, the one-larva-percontainer method is not used in Australia, where *Aedes aegypti* and *Ochlerotatus notoscriptus* occupy the same habitats (Tun-Lin et al. 1995). The discreepancy in calculated indices arising from use of the different methods, may not, however, be very great. In a survey of Nigerian villages for example, although several mosquito species occurred in village pots, there was very little difference between the various *Aedes aegypti* indices calculated from conventional methods, and from a single-larva survey (Service 1974). In surveys of *Aedes aegypti* breeding in tyre dumps in India Mahadev and Geevarghese (1978) compared the single larva method against removing about 20 larvae/tyre, and found that both produced similar container indices. However, they believed that if the water was turbid and inhabited by both *Aedes aegypti* and *Culex quinquefasciatus* then larvae of the latter being slower moving were more likely to be collected. In Malaysia Hii (1979) reported that the one-larva-per-container method was very useful in evaluating control measures directed against *Aedes aegypti*. In Nigeria Bang et al. (1981) found that in Ogui, a suburb of Enugu town, there was no difference between the classical indices and the single-larva/container method, but in two villages (Egede, Abor) the latter method gave container indices that were 1.8% and 2.7% smaller. They concluded, however, that the single-larva method was justified in monitoring *Aedes aegypti* larval populations.

The single-larva survey, however, cannot give information on the degree of association or coexistence between larvae of *Aedes aegypti* and other species. In the single-larva survey the larva taken from each container should be collected at random, but this may be difficult if larvae of other species that are noticeably different from *Aedes aegypti* are present. The collector is faced with the choice of collecting a larva that is possibly *Aedes aegypti* and one that is definitely not. In India Reuben and Panicker (1975) conducted one-larva/container surveys but instead of collecting a single larva/pot irrespective of genus, they chose to collect a single *Aedes* larva. Apart from *Aedes aegypti*, larvae of *Aedes vittatus* and *Aedes albopictus* were collected, so the surveys gave a relative index of pots infested with *Aedes aegypti* larvae in relation to the other two species. Similarly Geevarghese et al. (1975) used the single larva survey method in India, but gave preference to selecting an *Aedes* larva.

The main advantage of the one-larva/container survey is that much larger areas can be surveyed than by conventional methods. The choice between the two methods depends on the information required and the resources available, but the information provided by the two methods may not be dissimilar (Mahadev and Geevarghese 1978; Service 1974; Sheppard et al. 1969). One larva/container surveys have principally been used in Asia (Geevarghese et al.

1975; Gould et al. 1971; Hii 1979; Macdonald and Rajapaksa 1972; Mahadev and Geevarghese 1978; Rao et al. 1973; Reuben and Panicker 1975; Sheppard et al. 1969; Tonn and Bang 1971).

Other larval indices

Other indices which are sometimes used include the larval density index, which is the mean number of *Aedes aegypti* larvae per house and is obtained by counting all larvae in the containers. Chan et al. (1971*b*) found a good positive correlation between the larval density index and both the house index and the infested receptacle index (average number infested receptacles/housing unit), but not between the larval density and the container indices. They concluded that for practical purposes the infested receptacle index was the most useful and convenient measure of *Aedes aegypti* populations. In Thailand, Strickman and Kittayapong (2002) used a larval index based on that of Chan et al. (1971b). Abundance of *Aedes* larvae in a container was recorded in 1 of 4 categories: absent, low density (1–9 larvae), moderate density (10–50 larvae), or high density (> 50 larvae). The larval index was calculated as the sum of larvae in all containers, with individual containers considered as having no larvae, 5 larvae (low density), 25 larvae (moderate density), or 51 larvae (high density), and was an estimate of the number of *Aedes* larvae per house. The number of containers per house was not correlated to larval index ($r^2 = 0.216$, d.f. $= 1/7$, $F = 1.93$, $P = 0.207$). However, the number of positive containers was strongly correlated both to the larval index ($r^2 = 0.915$, d.f. $= 1/7$, $F = 75.39$, $P = 0.0001$) and the number of high density sites per house ($r^2 = 0.782$, d.f. $= 1/7$, $F = 25.07$, $P = 0.0016$).

In Nigeria Bang et al. (1981) introduced the *Stegomyia* index, which is the number of positive containers/1000 people. Although epidemiologically this is probably a better index than the Breteau index, it is difficult in practice to obtain because of the need to have an accurate human population census. Working in Singapore Chan (1985) proposed two other indices, namely the

$$\textit{Stegomyia larval density index} = \frac{\text{Number of larvae in an area}}{\text{Number of people in the area}} \times 1000$$

and the

$$\text{Larvitrap density index} = \frac{\text{Number of larvae}}{\text{Number of positive ovitraps}}$$

The former relates the numbers of larvae directly to people, while the latter makes use of ovitraps. Tun Lin et al. (1996) calculated a modified *Stegomyia* Larval Density Index (MSLDI) defined as mean (rather than absolute) number of larvae per 1000 persons. These authors also calculated a modified MSLDI using the mean number of fourth instar larvae only per 1000 persons, termed SFDI.

Another index is the block index (Tinker 1967). This is the percentage of blocks of houses that have houses where breeding is occurring, and can be especially useful in control campaigns where areas may be divided into blocks for the convenience of spraying.

Usefulness of Aedes aegypti indices

Interpretation of the various indices in relation to epidemic risk, however, can be difficult. For example, there may be only a few containers with larvae but they may be producing very large numbers of adults, in such a situation the container index will be low although the adult biting rate may be high. The World Health Organization (Anon. 1973) tabulated a series of density figures (1–9) which were derived from the principal larval indices by averaging the data from a number of localities where two or three types of indices were obtained simultaneously. This allowed the values of the three separate indices to be equated and a conversion table derived. With this information tables of the prevalence of *Aedes aegypti* throughout the world were made (Anon. 1973). The following ranges of larval indices have been associated with epidemic risk—house index of 4–>35, container index of 3–>20, and a Breteau index of 5–>50. A density figure above 5 is taken to indicate that the population size of *Aedes aegypti* has reached a level which represents a threat of urban transmission of yellow fever. In a survey of *Aedes aegypti* in northern Nigeria the density figures calculated separately from the house, container and Breteau indices were usually the same, or at least very similar (Service 1974), thus supporting the claim that any one index can be used to obtain a reliable density figure but Focks and Chadee (1997) found weak correlations among the classical indices and inconsistencies between the indices when comparing them with the WHO density table. Lien et al. (1993) used the density figure of >5 to indicate villages at risk of dengue transmission in Taiwan and observed a reduction in the proportion of villages with a Breteau index >35 (equivalent to density >5) and the number of reported and confirmed cases of dengue following application of indoor ultra-low volume (ULV) spraying with 3.6% permethrin or 1.5% alphacypermethrin.

Sulaiman et al. (1996) examined the relationship between the BI and HI and cases of dengue/dengue haemorrhagic fever in Kuala Lumpur,

Malaysia. Human cases of dengue ranged from 0 to 21 cases per month across the six municipal zones studied. The BI ranged from 1–11 and the HI from 0.1–7.5. The correlation coefficients for the BI and the monthly number of human dengue cases were not significant in five of the six zones of the city. A significant correlation was only obtained in the City Centre zone ($r = 0.60$, $P < 0.05$). Similar results were obtained with the HI, although a correlation significant at the 2.5% level was observed in the City Centre zone.

Sharma et al. (2005) reported that in India, all cases of dengue/dengue haemorrhagic fever outbreaks were associated with container indices exceeding 20. Chadee (2003) recorded Breteau indices of >5 in all four districts of Tobago during a survey of the whole island. The majority of dengue cases (66, or 63.5%) occurred in the Central district, where the Breteau indices ranged from 7 to 44. Thavara et al. (2001) calculated house, container and Breteau indices for *Aedes aegypti* and *Aedes albopictus* on Samui Island in the Gulf of Thailand. In each of nine surveys a minimum of 50 houses were sampled indoor and outdoor larval containers. A total of 18 937 containers from 3233 houses were inspected and of these containers, a total of 7514 containers from 2425 houses were infested with *Aedes* larvae. The Breteau index exceeded 100 in eight of nine surveys. At its peak, the BI was approximately six times higher than than the national control BI target set by the Ministry of Public Health for the Dengue Haemorrhagic Fever control program. The house and container indices ranged from 43 to 89 and 16 to 50, respectively. The HI and CI remained relatively more stable over the collecting perion than the BI. In many areas of West Africa, density figures much greater than 5 have been recorded without disease transmission (Hamon et al. 1971; Pichon et al. 1969; Service 1974; Shidrawi et al. 1973). Although there may be large populations of *Aedes aegypti* closely associated with man it does not necessarily follow that this constitutes a risk of yellow fever transmission, as much depends on the existence of virus reservoirs such as monkeys (which may be absent or scarce) or infected humans entering the area.

In a recent major review commissioned by the Special Programme for Research and Training in Tropical Diseases (TDR) of the World Health Organization, Focks (2003) discussed the relative merits and shortcomings of the indicators and indices currently available for dengue vector surveillance and monitoring. The classical *Stegomyia* indices continue to be the chief surveillance tool used by control programmes, however, they are generally considered to be inadequate for the measurement of risk of transmission, or the effectiveness of control operations. Neither do they provide programmes with nformation that could assist in targeting of control interventions. The CI is considered to be perhaps the weakest of the three commonly used indices as it describes only the proportion of containers in an area that are positive for *Aedes aegypti*. It gives no information on the number of positive containers per house, per person, or per area. Soper (1967) considered that the house index did not adequately measure the intensity of *Aedes aegypti* breeding, because this was governed not only by the number and distribution of infested houses but also by the number and size of the larval habitat. The HI offers a marginal improvement over the CI but it does not provide information on the number of positive containers per positive house. Teixeira et al. (2002) for example, found that the HI for *Aedes aegypti* in the city of Salvador in north-east Brazil ranged from 0.27% to 25.6% between surveyed areas (mean = 7.4% and median = 5.2%). Only a weak, non-significant positive correlation between HI and incidence of dengue infection was observed ($r^2 = +0.17$; $P = 0.36$). When the sampling areas were re-grouped according to HI level, in the areas with the lowest HI, the observed seroincidence (adjusted for age and mean seroprevalence) was the lowest but was nevertheless very high (54.6%). In the areas with higher HI levels, the seroincidences were similar. A HI of 3 was considered as the threshold level, below which the transmission rate starts to decrease. The BI provides more information as it combines information on both containers and houses, but as with the other two classical indices it does not take into account the differential productivity of the large variety of container types utilized by *Aedes aegypti* and other vector species. The omission of a single, highly productive house can lead to a low BI and result in under-estimation of the transmission risk. In Queensland, Tun-Lin et al. (1995) estimated that if a single premise with three or more positive containers was missed during a standard sample of 50 premises, then the BI could change from 0 to 6, changing the assessment of risk from zero to possible. For example a high BI could be calculated from a sample of 50 containers each containing 10 larvae, but a "safe" BI could be obtained from a sample of five wells each containing 5,000 larvae (Tun-Lin et al. 1996). In recognition of the fact that the classical Breteau Index does not take into account differential productivity, Fauran (1996) in Martinique proposed disaggregating the index and expressing its components separately by container type. A software programme was also developed to automatically calculate revised Breteau indices by incorporating a coefficient to represent differential productivity of the main container types. The coefficient used varied from 1 for flower vases to 5 for drums. The three indices do tend to be correlated with one another and often yield similar density figures when reading from the WHO table, as for example in northern Nigeria (Service 1974), however this is not always the case, even in the same country (Bang et al. 1981). From a practical perspective, it should be noted that in some areas larvae of other

Aedes species and short-siphoned *Culex* mosquitoes which superficially resemble *Aedes aegypti* may also occur in containers, making it difficult to record the numbers of containers with *Aedes aegypti* by visual inspection in the field. Usually only a few larvae are collected from each container for laboratory examination and identification, but in some surveys all larvae are removed from at least the smaller types of habitats and identified (Chan et al. 1971*a,b*). Other problems include the difficulty in deciding up to what distance from a house a container can be said to belong to that house, whether a pot is located inside or outside a house, or even what constitutes a house or housing unit (Chan et al. 1971*b*; Macdonald 1956; Tonn et al. 1969). Perhaps the major shortcoming of all three classical indices is that they do not provide information on key variables that are known to be linked to transmission risk, such as vector density per person or per area.

Kittayapong and Strickman (1993) conducted a container index survey of 186 homes at Ban Laem Hin village approximately 85 km east of Bangkok. 44% of all containers examined were positive, with a mode of two containers infested with *Aedes* larvae per house. Indoor containers were more frequently infested than outdoor containers. The house index was 92% and the Breteau index was 411 positive containers per 100 houses. Somewhat surprisingly, water storage jars kept outdoors and covered with commercially available aluminium lids were infested significantly more often than uncovered jars. This observation was confirmed in laboratory tests which demonstrated that gravid *Aedes aegypti* females were able to enter and oviposit in 200 l capacity ceramic water storage jars fitted with aluminium covers, although the number of eggs deposited was reduced by 77% (Strickman and Kittayapong (1993). A 2-cm wide strip of foam rubber glued to the inside flange of the lid effectively prevented oviposition.

In India, Dutta et al. (1998) reported container indices of 30.0 to 88.0 for used tyres in major townships situated along the national highways and trunk roads in the north east of the country. Both *Aedes aegypti* and *Aedes albopictus* were collected, with the former species dominating the collections.

Comparisons of larval indices

Tinker (1967) found that the population levels of *Aedes aegypti* as shown by the container, house, infested receptacle, and block indices were usually similar. In fact, there was a good positive correlation between all four indices, the best being between the house and block indices. Using a logarithmic transformation two regression formulae were devised: (1) House Index = antilog (0.0861 + 0.01586 × Block Index) and

(2) Block Index = 1.50 + log 52.40 × log House Index. Thus if one index is obtained a very good approximation of the other can be calculated. Later Tinker (1978) showed in surveys in Jamaica and El Salvador that the house and Breteau indices were highly correlated in a quarctic relationship. For example, at low infestation rates (i.e. indices less than 5%) the house and Breteau indices were essentially the same, but with increasing infestation rates the two diverged due to the greater number of houses with multiple breeding places.

In India Mahadev et al. (1978) obtained a good correlation between the container and Breteau indices in their *Aedes aegypti* larval surveys. Generally the Breteau index was 2.5 times greater than the container index in the dry season, but about 3 times more during the rains. Later Mahadev (1983) tried to see whether the patchy distribution of water receptacles containing water was associated with the container index. For this he calculated an index of patchiness for the water receptacles having water. Using Lloyd's mean crowding parameter he found that there was a poor correlation between this index and the number of pots with *Aedes aegypti* larvae. During a survey of dengue vectors in Lao PDR, Tsuda et al. (2002) found the *Aedes aegypti* container index to be 51.8% in Nongbok and 40.2% in Thakhek. The house index was 94.9% in Nongbok and 75.2% in Thakhek. Chareonviriyaphap et al. (2003) conducted larval surveys inside and outside 50 houses in each of five geographical regions of Thailand. Of the 272 houses investigated, 206 were positive for *Aedes* larvae. Dry season (January to April 2002) values for the House, Container and Breteau indices differed considerably from one another, and no consistent pattern in the indices was observed across the five regions. For example, the lowest HI (52) was obtained from the north, but the north had the highest BI (190). Similarly, the lowest BI (99) was obtained in the central region, which had the highest HI (96). These observations concur with those of Tinker (1978) who observed that at high infestation levels, the two indices tend to diverge. In all regions, the BI exceeded the national target set by the Vector Control Program of the Ministry of Public Health of less than 50. Nagao et al. (2003) conducted household larval surveys in 18 provinces in the northern region of Thailand and calculated House container, and Breteau indices and attempted to correlate these indices with a variety of climatic and social risk factors. Socio-economic characteristics were obtained from rural databases which contain data collected through routine biannual questionnaire surveys. Data for 38 variables were correlated with the values for larval indices. Larval indices were transformed to approximately normal distributions, by using arcsine (square root) transformation for HI, and CI, and taking the natural logarithm of BI. Data analysis comprised an initial univariate analysis, from which socio-economic variables that showed a significant association ($P < 0.05$) with all the entomological indices were selected and subjected to multivariate analysis. All larval indices were observed to be highly correlated with each other ($r > 0.83$). Public water wells, transport services, and tin houses were all significantly positively correlated with each of HI, CI and BI. Private water wells, health insurance, ethics education, health education, religious rites, rice bank, kindergarten, primary school, community centre, library, use of firewood, thatched houses and house density were all significantly negatively correlated. The authors advised caution in assigning a direct causal relationship to the socio-economic variables and the larval indices. Sharma et al. (2005) surveyed 103 778 houses in Delhi, India and calculated House, Container, and Breteau indices for *Aedes aegypti*. All three indices showed the same seasonality pattern. Water coolers and tires were found to be the preferred larval habitats.

In a larval survey of *Aedes aegypti* in Puerto Rico Moore et al. (1978) found that fluctuations in container and house indices were compatible with those of the Breteau index. Generally, however, Breteau indices in Puerto Rico were higher (for a given house or container index) than the international averages presented by Brown (1971), indicating that the average positive house in Puerto Rico had more containers with *Aedes aegypti* than those in many other countries. Larval surveys conducted in a public park and one hundred surrounding houses in Santo Domingo in the Dominican Republic identified 1982 containers as potential breeding sites with Styrofoam lunch containers (1372), plastic glasses (201), plastic buckets (139), rock holes (50), and tree holes (67) the most common. The House index was $2/100 = 2.0\%$ positive with 11 positive containers (Container index $11/1982 = 0.6\%$) and the Breteau index was $11/100 = 11.0$ (Pena et al. 2003). In a container survey carried out in Trinidad, Chadee and Rahaman (2000) examined all natural and artificial containers at more than 100 houses and compounds in each of nine sites. The mean number of large water storage drums containing *Aedes aegypti* immatures ranged from 1.6 to 4.9 while the Breteau index varied from 6.2 in St. Andrew/St. David to 52.5 in Victoria East, with a median of 29.5. At the end of the 1978–1979 yellow fever epidemic in The Gambia Germain et al. (1980) carried out *Aedes aegypti* surveys incorporating the Breteau and container indices, and by integrating these into a scale of WHO density figures obtained density values of 0–9. A potential for yellow fever transmission exists at a value of 2, while a density value of 6 or higher indicates a high risk of an epidemic. Because the survey was carried out in the dry season when many of the water containers were dry, Germain et al. (1980) derived a 'potential peridomestic potential breeding index', defined as the number of potential peridomestic habitats/100 houses. In one area values ranged from 65–1270 but could not be correlated with the WHO density values; for example both the minimum and the maximum index were found in localities having a WHO density index of 0.

In northern Nigeria Service (1974) found that the average number of containers per house varied from 2–9, and in this situation found that container, house, and Breteau indices gave similar WHO density figures. In contrast, in southern Nigeria Bang et al. (1981) reported that the average number of containers/house ranged from 19–30, and in this situation the regression of the container index on the house index (Fig. 3.31) and the Breteau index did not correspond with the relationship demonstrated by the WHO density figures. However, the numerical relationship between the house index and Breteau index fits that on which the WHO density figure was based (Fig. 3.31). Bang et al. (1981) point out that if there are large numbers of containers associated with houses, then reliable density figures cannot be derived from the container index.

Tidwell et al. (1990) determined the *Aedes aegypti* premise, container and Breteau indices in Santa Domingo, Dominican Republic. They also obtained a 'female adult density index'. This was the average number of females caught in 5 min by two men from randomly selected houses in various areas of the city. Collections were made with 12-in sweep-nets with particular attention to searching under tables, chairs, beds and in cupboards. Occasionally one of the collectors used a 6-V hand-held battery-powered aspirator, but they preferred both collectors to use nets since this did not require a constant supply of batteries. In addition an 'adult positive house index' was used to indicate the percentage of houses having *Aedes aegypti* adults. The female adult density ranged from 1.22 to 15.04, but in some areas it was not uncommon to collect more than 20 *Aedes aegypti* from a house, and in one house as many as 134 females were collected. There was, however, no significant relationship between these adult densities and any of the three larval indices, or with the numbers of eggs collected by ovitraps. Tidwell et al. (1990) concluded that their results demonstrated the difficulty of estimating adult populations from larval indices. They believed that measuring adult densities was more appropriate in evaluating the effectiveness of control measures than larval surveys. They point out that Fox and Specht (1988) suggested that 5-min bait catches might be practical for evaluating the presence of *Aedes aegypti* in areas of high density. But they stress that there will most probably have to be several catching stations in an area to obtain reliable data on changes in relative population size. If it is not possible to catch adults, then a weekly pupal density index would be a relatively good indicator of fluctuations in adult populations.

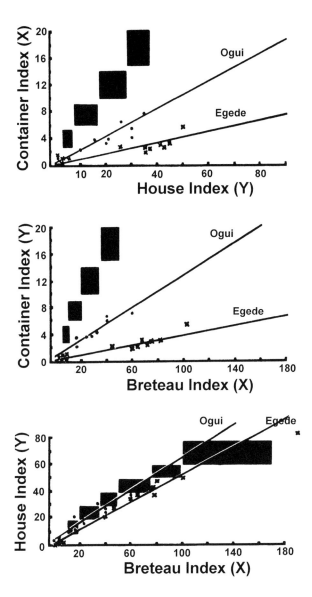

Fig. 3.31. Relationships of the three standard *Aedes aegypti* larval indices and the WHO density figures (shaded areas) for the Nigerian villages of Ogui and Egede (Bang et al. 1981)

Tun-Lin et al. (1996) undertook a critical examination of the following *Aedes* indices in Queensland, Australia: Breteau, House, Container, Modified *Stegomyia* Larval Density Index (MSLDI), *Stegomyia* Fourth instar

Density Index (SFDI), Fourth Instar Density Index (FDI), and an Adult Productivity Index (API) using Spearman's correlation analysis. Immature stage indices were compared with adult abundance determined by collecting outdoor resting individuals for a 10 min period using battery-powered aspirators and expressing the catch as number of female *Aedes aegypti* collected outdoors per 100 houses, and the average number of immature *Aedes aegypti* collected outdoors per premise. Spearman's correlations of indices against immature and adult density are shown in table 3.5. Indices that incorporated measurements of abundance of immatures (MSLDI, API, SFDI, FDI) gave higher positive correlations than the Breteau, the House, or the Container index. Significant positive correlations were obtained with adult abundance for Breteau, House, MSLDI, and FDI, but not for the Container index. Correlations with API were generally lower but still significant. These authors also conducted a computer simulation using pooled data from all of the localities sampled in 1990. Thirty samples were selected from 50 randomly chosen houses with replacement. Values for the indices and their coefficients of variation were calculated and those with low coefficients of variation compared with the mean were considered to be more reliable. Values for the different indices calculated from the computer simulation varied widely. The coefficient of variation indicated that the Breteau, adult productivity index based on the arithmetic mean (API_{am}), House, adult density and Container indices were the most robust. It was concluded that the modified *Stegomyia* larval indices showed high correlations with adult density but their usefulness is constrained by the difficulties in obtaining an accurate census of the human population and in counting individual mosquito immatures.

Romero-Vivas and Falconar (2005) compared *Aedes aegypti* classical indices, as well as egg, larval, pupal and adult density indices in a dengue endemic urban area of Colombia. Two indices were calculated from ovitrap survey data: an ovitrap premise index (OPI), equal to the number of premises positive for *Aedes aegypti* eggs × 100/number of premises with ovitraps; and an ovitrap density index (ODI), equal to number of *Aedes aegypti* eggs recovered during survey/no.of premises. Larval and pupal density indices were calculated as follows: larval premise index (LPI) = no. of premises positive for *Aedes aegypti* L4 larvae × 100/no. of premises inspected; larval density index (LDI) = no. of L4 larvae of *Aedes aegypti*/no. of premises; pupa premise index (PPI) = no. of premises positive for pupae × 100/no. of premises inspected; and pupa density index (PDI) = no. of *Aedes aegypti* pupae/no. of premises. Similarly, adult density indices were also calculated, where the adult premise index (API) = no. of premises positive for adult female *Aedes aegypti* × 100/no. of premises inspected; and adult density index (ADI) = number of adult female

Aedes aegypti collected (by 2 collectors in 15 minutes)/no. of premises. A comparison of the geometric mean monthly density indices of eggs, larvae, pupae and adults revealed a significant positive correlation over time between LDI and PDI ($r = 0.90$. d.f. $= 5$, $P < 0.005$). A negative correlation was observed between ODI and LDI. No other significant correlations were obtained among the density indices. In addition, no significant correlation was observed between density indices and the number of dengue cases reported. As regards the various premise indices, the ODI was found to be the most sensitive for assessing the presence of *Aedes aegypti*. The other three premise indices displayed marked variation over time. No significant correlations were found among premise indices and dengue cases. The BI and HI were found to be significantly correlated with each other, but not with any of the other premise or density indices.

Productivity estimates

In Kenya Subra (1983) estimated the productivity of different domestic containers by removing, counting and then returning the pupae of *Aedes aegypti*. Surtees (1959) in Nigeria and Pichon and Gayral (1970) in Burkina Faso counted all the *Aedes aegypti* larvae in a selected number of village pots so that they could calculate changes in larval population size. Although capable of providing useful information on population dynamics, the method is too time-consuming to be used in routine surveys.

In *Aedes aegypti* surveys in New Orleans, USA Focks et al. (1981) estimated productivity in the various city blocks by taking account of the numbers of different types of larval habitats and the mean numbers of early and late instar larvae and pupae they contained. They also calculated the mean daily emergence of *Aedes aegypti* per block as follows:

$$\text{Mean emergence of } \female \text{ /block/day} = \frac{(\text{No. pupae}) \, (A) \, (0.5)}{(\text{No. blocks}) \, (2 \text{ days})}$$

where A is the proportion of pupae collected that are *Aedes aegypti*, 0.5 is the proportion of females and 2 days represents the pupal period. This means that half of the pupae observed would be expected to emerge as adults the following day.

If the daily survival of adult females (p) is constant and independent of age (t) then the absolute population of *Aedes aegypti* can be estimated by

$$\text{No. } \female/\text{block} = \text{Mean no. emerging } \female/\text{block/day} \int_{o}^{a} p^{t} dt \qquad (3.21)$$

Table 3.5. Spearman's correlations and significance levels of different *Aedes aegypti* indices with immature and adult densities in Townsville (1989, 1990) and Charters Towers, Mingela, and Ravenswood (1990), Australia (Tun-Lin et al. 1996)

Index	Definition	Immature density† 1989	Immature density† 1990	Adult density† 1990
Breteau	Positive containers/100 houses	0.75	0.81	0.78
House	Positive houses (%)	0.79	0.68	0.70
Container	Positive containers (%)	0.89	0.38‡	0.35‡
Stegomyia	Positive containers/1000 population	0.78	0.80	0.90
MSLDI	Mean immature stages/1000 population	1.00	0.98	0.83
SFDI	Mean 4th instar larvae/1000 population	0.96	0.79	0.65
API$_{gm}$	Sum (container type x mean larvae)	0.89	0.83	0.52§
API$_{am}$	Sum (container type x mean larvae)	0.86	0.90	0.64
FDI	Mean 4th instar larvae/house	0.96	0.89	0.71

MSLDI = modified *Stegomyia* larval density index; SFDI = *Stegomyia* 4th-stage density index; APIgm=adult productivity index based on geometric mean; APIam = adult productivity index based on arithmetic mean; FDI = 4th stage density index.
† Unless otherwise stated, P < 0.005 – 0.001
‡ Not significant
§ P < 0.05

Forattini et al. (1997) used a slightly modified version of the above formula to determine the productivity of *Aedes albopictus* found in a single abandoned concrete water tank structure with a capacity of about 70 litres in São Paulo state, Brazil. Productivity of *Aedes albopictus* was calculated as:

$$\text{mean pupae per collection (A) (0.5)}/2$$

$$= (31.13 \times 0.269 \times 0.5)/2 = 2.1$$

This figure was multiplied by 7 to obtain the total productivity, as the volume sampled on each occasion was around 10 litres, or one-seventh of the total volume of water in the tank.

In an analysis of two sequential surveys of three locations in Queensland, Australia, conducted in 1989 and 1990, and involving more than 1300 premises, Tun-Lin et al. (1995) found that a small proportion of premises was responsible for the majority of production, and these foci were categorised as key premises. As an example, <2% of inspected premises in Townsville accounted for 47% of all positive containers in 1989; in 1990, 3% of all premises accounted for 53%. The authors also determined that some container types were more productive than others and categorised these as key containers. Examples of especially productive containers included wells and rainwater tanks. Spatially and temporally stable estimates of the productivity of each class of container could be calculated and the product of this average and the abundance of the class could be used to reliably estimate relative importance. In Trinidad, Chadee (2004) also identified both key premises and key containers that were responsible for a disproportionate share of *Aedes aegypti* productivity. For example in Curepe only 51 of 390 (13%) of premises were positive, while only 11 premises accounted for 43% of positive containers. Drums with a mean surface area of 0.28 m^2 accounted for less than 13% of total positive containers but produced more than 44% of the *Aedes aegypti* pupae. Similarly, buckets with a mean surface area of 0.06 m^2 accounted for 20% of the positive containers but produced 33% of the pupae. Small, miscellaneous containers including pots, pans, discarded cans, etc. while numerous contributed less than 1% of pupae. It was proposed that control efforts concentrate on the identification of key premises and the removal of key containers, supplemented with education of householders.

In Australia, Tun-Lin et al. (1994) attempted to take the productivity of large containers into account by sampling larvae and pupae with sweep nets and correlating these standing crop data with adult densities. Correlations of between 0.92 and 0.96 were obtained, allowing for productivity of large containers to be estimated relatively reliably without the need to count the larvae and pupae present. Tun-Lin et al. (1996) calculated an Adult Productivity Index (API) using both geometric (gm) and arithmetic mean (am) numbers of immature stages per container type. Scores were derived by summation of the frequency of each container type $\sum C_{i-\infty}$ weighted by the product of the mean number of larvae in each container type $L_{i-\infty(am)or(gm)}$. Correlations of API with adult density were found to be lower than for the Breteau index (Tun-Lin et al. 1996).

Local customs vary, but pots containing drinking water are often emptied every 2–3 days, and although these may contain larvae they will not contribute to the adult population. By attracting ovipositing females they are in fact helping to decrease the adult population. The presence of pupae or pupal exuviae may be more appropriate criteria in indicating breeding places that produce adult *Aedes aegypti*. Broken and abandoned pots are often important man-made larval habitats, but during the dry season they are more likely to dry out than pots containing drinking water. In surveys attention should also be paid to special breeding foci such as collections of discarded tins, containers in warehouses, tyre dumps and other places where breeding may be localised but intense.

In many areas *Aedes aegypti* breeds almost exclusively in peridomestic man-made containers but in other areas larvae are frequently found in natural habitats, such as tree-holes and plant axils. It is very difficult to quantify breeding in these natural habitats. The best procedure is probably to record these habitats separately from artificial ones, but not attempt to define the intensity of breeding in them (Mouchet 1972*a, b*).

Pupal survey

In light of the difficulties in demonstrating clear, empirical relationships between the classical *Aedes aegypti* indices and absolute mosquito density or risk of transmission, Focks (2003) recently proposed the pupal survey as an alternative method for determining risk, partly based on observations from Trinidad (Focks and Chadee 1997). During May and June 1994, all natural and artificial containers, both indoors and outdoors, for 100 houses in each of 16 locations across all eight counties of Trinidad were inspected. All pupae found in wet containers were counted and container types classified. A total of eleven container types were identified, ranging from flower vases and tin cans, to 207 litre metal drums. The relative importance of each container type in terms of the standing crop of *Aedes aegypti* pupae varied considerably (Fig. 3.32). Island-wide mean density of containers positive for *Aedes aegypti* pupae was 287 ± 33 (SE) per hectare. The mean standing crop of pupae was 9.1 ± 0.9 (SE) per wet container. The standing crop in the most productive containers (flower pots) was 12 times greater than for the least productive containers (indoor flower vases). However, the flower-pot type of container was only found in one district and when this container type was excluded from the analysis, the difference between highest and lowest standing crop was reduced to 6-fold. Outdoor drums, laundry tubs, buckets and small containers were responsible for >90% of all pupae, with buckets and small containers being responsible for more than 70%.

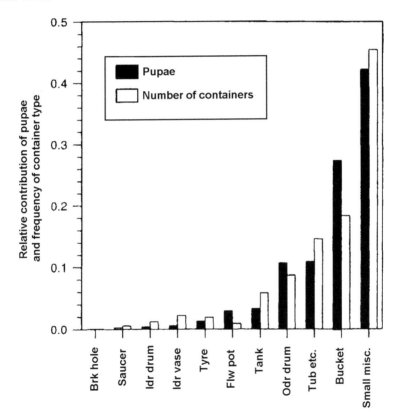

Fig. 3.32. Relative importance of container types in terms of standing crop of *Aedes aegypti* pupae and numbers of containers. Data represent island-wide averages for Trinidad (Focks and Chadee 1997)

Both the numbers and types of containers varied significantly among the 16 sites studied, and correspondingly, standing crop of pupae also varied considerably, with a 39-fold difference between the site with the lowest standing crop and the highest (both sites were urban). Correlations between the traditional *Aedes aegypti* indices and the standing crop of *Aedes aegypti* either per hectare or per person showed no significant relationships, except for CI and standing crop, although in this case the relationship was negative (table 3.6). Efforts to reduce the ubiquitous small, discarded receptacles, such as tin cans, drinks bottles, and car tyres would reduce *Aedes aegypti* production by an estimated 43%, however, it was acknowledged that water storage containers could not be eliminated until the supply of piped water to many communities is improved. The authors concluded that their study served to illustrate the weak relationship between the traditional indices and actual *Aedes aegypti* production, and ultimately therefore transmission risk, and proposed absolute vector density measurement, including pupal survey as a preferred alternative.

Later, Focks et al. (2000) attempted to determine transmission thresholds for dengue in terms of the number of *Aedes aegypti* pupae per person, taking into account the roles of ambient temperature and herd immunity. Their estimates of transmission thresholds were based on a pair of interrelated simulation models (the Container-Inhabiting Mosquito Simulation Model [CIMSiM] and the Dengue Simulation Model [DENSiM]) (Focks et al. 1993*a*, 1993*b*, 1995).

Table 3.6. Pearson product moment correlations between traditional *Aedes aegypti* indices and pupal survey estimates for Trinidad (Focks and Chadee 1997)

Measure	Pupae per hectare	Pupae per person	Breteau index	Container index	House index
Pupae per hectare	1.000	0.963†	−0.119	−0.522†	0.121
	1.000	0.000	0.660	0.038	0.660
Pupae per person	0.963†	1.000	−0.181	−0.535†	0.117
	0.000	1.000	0.502	0.033	0.665
Breteau index	−0.119	−0.181	1.000	0.590†	0.683†
	0.660	0.502	1.000	0.016	0.004
Container index	−0.522†	−0.535†	0.590†	1.000	0.552†
	0.038	0.033	0.016	1.000	0.027
House index	0.121	0.117	0.683†	0.552†	1.000
	0.660	0.665	0.004	0.027	1.000

Top figure in each row is the correlation coefficient, lower figure is the P value
† Significant correlation

The effects of pre-existing antibody levels in human populations, ambient air temperatures, and size and frequency of viral introductions on human and vector components of transmission were modelled. Two scenarios of viral introduction were considered: 1, 2, 4, or 8 viraemic person(s) introduced into an area on day 90 of the year or a single viraemic person introduced once-a-month throughout the year, termed Single Introduction 1, 2, 4, and 8 and Monthly Introductions.

Resulting threshold levels were estimated to range between about 0.5 and 1.5 *Aedes aegypti* pupae per person for ambient air temperatures of 28°C and initial seroprevalences ranging between 0% to 67% (Tables 3.7 and 3.8).

Table 3.7. Estimated number of *Aedes aegypti* pupae per person required to result in a 10% or greater increase in seroprevalence of antibody to dengue during the course of a year under conditions of a single viral introduction of 1 or 2 viremic individual(s) on day 90 of the year; the estimates for 2 individuals are in parentheses* (Focks et al. 2000)

Temp °C	Transmission thresholds by initial seroprevalence of antibody		
	0%	33%	67%
22	9.57 (9.16)	14.10 (12.83)	30.55 (29.15)
24	2.92 (2.68)	4.47 (4.21)	9.22 (8.68)
26	1.42 (1.23)	2.03 (1.98)	4.26 (4.01)
28	0.53 (0.48)	0.75 (0.72)	1.69 (1.38)
30	0.13 (0.12)	0.19 (0.18)	0.38 (0.35)
32	0.07 (0.07)	0.10 (0.10)	0.26 (0.18)

* In a series of simulations in the dengue simulation model, these values resulted in a 10% or greater increase in prevalence approximately 50% of the time.

Table 3.8. Estimated number of *Aedes aegypti* pupae per person required to result in a 10% or greater increase in seroprevalence of antibody to dengue during the course of a year under conditions of a single viral introduction of 4 or 8 viremic individuals on day 90 of the year; the estimates for 2 individuals are in parentheses* (after Focks et al. 2000)

Temp °C	Transmission thresholds by initial seroprevalence of antibody		
	0%	33%	67%
22	8.02 (7.13)	11.66 (10.69)	24.66 (22.11)
24	2.52 (2.20)	3.69 (3.27)	7.76 (7.02)
26	1.09 (1.08)	1.80 (1.57)	3.79 (3.24)
28	0.47 (0.41)	0.63 (0.62)	1.33 (1.27)
30	0.11 (0.09)	0.18 (0.15)	0.33 (0.31)
32	0.06 (0.06)	0.09 (0.09)	0.18 (0.16)

* In a series of simulations in the dengue simulation model, these values resulted in a 10% or greater increase in prevalence approximately 50% of the time

Focks et al. (2000) concluded that in light of the fact that the ratio of *Aedes aegypti* pupae to humans has been observed to range between 0.34

and >60 in 25 sites in dengue-endemic or dengue-susceptible areas in the Caribbean, Central America, and Southeast Asia, then the results of their analysis were not particularly encouraging for those charged with dengue control through vector habitat source reduction. Examples were given, whereby if an initial seroprevalence of 33% is assumed for Puerto Rico, Honduras, and Bangkok, Thailand, then the degree of suppression required to essentially eliminate the possibility of summertime transmission was estimated to range between 10% and 83%; in Mexico and Trinidad, reductions of >90% would be required. This was translated into the proportion of containers that would need to be eliminated to further illustrate the potential difficulties. In Puerto Rico, where the ratio of observed standing crop of *Aedes aegypti* and and the calculated threshold was 1.7, this would mean that approximately 7 of every 17 breeding containers would have to be eliminated. For Reynosa, Mexico, with a ratio of approximately 10, 9 of every 10 containers would have to be eliminated, and for sites in Trinidad with ratios averaging approximately 25, the elimination of 24 of every 25 containers would be required. The authors argued that the strength of the thresholds presented lies in ability to provide targets, in terms of the upper bound on the number of pupae per person to ensure that viral introductions would result in very little or no transmission.in a specific location. The use of thresholds could also be used to estimate the relative contribution of different types of container to that threshold, and hence allowing for control efforts to be directed at the most important containers, and also to determine the relative effectiveness of different rates of elimination of containers.

Strickman et al. (2003) applied a modified version of the Focks et al. (2000) sampling method in Thailand and compared it with results obtained from traditional surveillance methods (Kittayapong and Strickman 1993; Strickman and Kittayapong 2002). Two sampling protocols were followed. Firstly, small fish nets (approximately 20 × 15 cm with a long handle) were swirled through all containers in 3 villages in the hot season (February–April 1990), wet season (May–October 1990), and cool season (November 1990–January 1991). Secondly, complete filtering of all water was conducted by emptying every container in a house through a strainer and capturing all immature mosquitoes. The complete filtering approach was applied in 10 houses per month for 1 year in one village, comprising 120 houses. Due to the time taken to conduct this kind of survey, only 1 or 2 houses could be sampled in a day and so the 10 houses were sampled over the whole month. The same house was never resampled during the year.

A monthly estimate of pupae per house was calculated as follows:

$$(P/C)_m = (TP_m)/(TC_m) \qquad (3.22)$$

where: P/C_m = pupae per container during 1 month, TP_m = total pupae collected in 10 houses during 1 month, and TC_m = total containers sampled in 10 houses during 1 month

$$P/C = (\Sigma TP_m)/(\Sigma TC_m) \tag{3.23}$$

where: P/C = pupae per container for entire year

$$Dm = (P/Cm)/(P/C) \tag{3.24}$$

Where: D_m = deviation in pupal count for a given month

$$P/H = (\Sigma TP_m)/(TH) \tag{3.25}$$

where: P/H = pupae per house for the year, ΣTP_m = total number of pupae collected during the year, and TH = total number of houses in the village (=120)

$$P/H_m = D_m([P/H]/12) \tag{3.26}$$

where: P/H_m = estimate of pupae per house in a given month.

Pupal counts for container types that yielded more than 100 replicates during a year's sampling were compared statistically using the Mann-Whitney U-test. Smaller containers were found to be relatively more productive in terms of pupae per litre of water than larger ones (0.007 for standard jars, 0.05 for small jars, and 4.8 for ant traps). Comparison of the estimate of mosquito abundance to the critical threshold for dengue transmission, assuming a household size of 5 persons and a seroprevalence of 67%, showed that an outbreak could have occurred at almost any time during the year. December and February were the only months when the estimated number of female *Aedes aegypti* was lower than the Focks et al. (2000) threshold for a 10% increase in transmission. In this case, the authors found that the Breteau index was sensitive enough to detect the higher *Aedes aegypti* abundance in May and the lower abundance in December through February.

Montgomery and Ritchie (2002) conducted pupal surveys using either turkey basters or small hand nets (see section 3.9 above for a description of the method used) of roof gutters and house yards in Queensland, Australia during the dry and wet seasons. In the wet season, 527 gutters totalling a length of 3398 meters were inspected. Of these, 77 (14.6%) were flooded, with a total submerged length of 464.5 meters. Mosquito pupae were found in 11 (13.7%) flooded gutters. Although relatively uncommon, flooded gutters were the primary source of *Aedes aegypti* and *Ochlerotatus notoscriptus* pupae (52.6% and 66.8%, respectively) in the wet season. In the dry season, only two roof gutters contained mosquito pupae but these accounted for 39.5% of the *Aedes aegypti* pupae collected. In the wet season house yard survey, 634 *Aedes aegypti* pupae were collected from 70 containers within seven categories. Garden objects (plant pot bases, birdbaths) were the most abundant (*n* = 28) positive containers, followed by rubbish (12), discarded household items (11), and domestic commercial usage containers (9). Garden objects, discarded household items, and rubbish contained the largest proportion of *Aedes aegypti* pupae (36.4%, 28.1%, and 20.6%, respectively). However, the highest mean number of *Aedes aegypti* pupae per container occurred in discarded household items, followed by rubbish, garden accoutrements, and domestic-commercial usage containers (16.2, 10.9, 8.2, and 7.0, respectively). These four categories generated 95% of the *Aedes aegypti* pupae in yards. A total of 312 *Ochlerotatus notoscriptus* pupae was collected in 25 containers, with garden objects, natural breeding sites, and domestic-commercial usage containers accounting for most positive containers. These three categories generated 98.2% of the pupae, with garden accoutrements the most productive (mean = 20.5 pupae). *Aedes aegypti* pupal indices were highest in the wet season, ranging from 2.36 to 0.59 for the wet and dry season surveys, respectively.

Pupal surveys were also used by Montgomery et al. (2004) in studies to determine the relative productivity of drain sumps as key containers for *Aedes aegypti* in Cairns, Australia. Pupal sampling was conducted by what the authors termed a "vortex technique", and appears similar to that described by Tun-Lin et al. (1995*c*). Any debris was removed from the surface of the water and the water was then stirred ten times with a 30 cm ruler. A small aquarium net was then used to take a sample from the central column of the vortex. A total of 285 pupae were collected from 18 water-containing drain sumps. Drain sumps constituted 13 of 50 large containers sampled during the survey, and contributed 33.4% to the total pupae sampled. As regards *Aedes aegypti* standing crop, approximately 14.7% was contributed by four *Aedes aegypti*-positive drain sumps.

Pupal surveys are a useful tool in epidemiological risk assessment for dengue, as they can be used as a proxy measure to determine in absolute terms the ratio of *Aedes aegypti* females to humans. This ratio is required in order to apply the mass action principal, which states that the course of an epidemic is dependent on the rate of contact between susceptible hosts and infectious vectors, and threshold theory, whereby the introduction of a few infectious individuals into a community of susceptible individuals will not give rise to an outbreak unless the density of vectors exceeds a certain critical level (Anderson and May 1991) to dengue risk assessment and control. Pupae are used as a proxy measure for female mosquito density for the following reasons (Focks 2003): 1) unlike any of the other life stages, it is possible to actually count the absolute number of *Aedes aegypti* pupae in most domestic environments; 2) container-inhabiting *Stegomyia* pupae are easily and inexpensively separated from other genera and identified to species either as pupae or as emerged adults; 3) because pupal mortality is low and well-described, the number of pupae is highly correlated with the number of adults. Importantly, counting pupae also permits evaluation of the relative and absolute contributions of the various classes of containers. Larval and egg counts are not useful proxies are they are logistically difficult to obtain and due to delayed hatching and density-dependent interactions, they are almost impossible to relate consistently with adult population size. An additional feature of pupal indices that makes them preferable to egg or larval indices is the fact that the pupa is the life stage that directly precedes the virus-transmitting adult, making abundance of pupae a theoretically more direct measure of transmission risk than larval abundance. Focks et al. (2000) have developed a table of dengue transmission thresholds in terms of *Aedes aegypti* pupae per person (Table 3.9), based on dynamic life tables and a dengue simulation model (Focks et al. 1993*a*; 1993*b*; 1995).

Table 3.9. Estimated number of *Aedes aegypti* pupae per person required to result in a 10% or greater rise in seroprevalence of antibody to dengue during the course of a year resulting from 12 monthly viral introductions of a single viremic individual, the Monthly Introduction threshold* (Focks et al. 2000)

Temperature °C	Transmission threshold by initial seroprevalence of antibody		
	0%	33%	67%
22	7.13	10.70	23.32
24	2.20	3.47	7.11
26	1.05	1.55	3.41
28	0.42	0.61	1.27
30	0.10	0.35	0.30
32	0.06	0.09	0.16

Population estimates

Mosquito larval collections are usually made to determine the presence or absence of various species in different habitats, or to monitor relative changes in numbers associated with seasonal abundance or control measures, but occasionally there are attempts to estimate the size of the absolute population in a habitat from the number of larvae per dip, or the number enclosed by quadrats. Some of the difficulties of interpreting the results of these catches in terms of real population size have already been discussed in this chapter. For reliable population estimates, as may be required in ecological studies, better methods must be used. Several are in fact available for obtaining population estimates of insects and other animals, and these are well described by Begon (1979), Blower et al. (1981) and Southwood and Henderson (2000), while Otis et al. (1978) and Seber (1973) describe the statistical and theoretical aspects of getting population estimates. Methods that have been applied to obtain population estimates of mosquito larvae, together with a few others that might be applicable, are described in this section.

Removal methods

When samples are taken from a population the numbers of individuals caught are related to the size of the population present. Now, if a number of successive samples are removed and the individuals not returned to the parent population, then theoretically there should be a gradual decline in the numbers caught in each sample due to the population being slowly depleted. The number of individuals caught in samples should decrease in geometrical regression (Le Pelley 1935), and the rate of decline in numbers be directly related to the population present. Moran (1951) outlined the conditions that must be satisfied if this method is to produce reliable results, but not all his criteria are important in sampling mosquito populations. For example, an important and common sense criterion is that the population remains stable over the sampling period, which in mammal trapping may extend over many days or even weeks, but with mosquito populations sampling is normally over a very short period. It is essential that the chances of larvae being caught do not increase or decrease during the sampling programme. Obviously if the probability of capture increases with time, due possibly to a diminishing sensitivity to alarm reaction, then more larvae will be caught in later samples and the population over-estimated. Populations will be underestimated, however, if during sampling larvae become increasingly difficult to catch. Care must also be exercised that changes in weather conditions, such as cloudiness, or changes in diel cyclical activities of the larvae do not affect their behaviour and chances of capture.

   If the number of larvae caught in a dip is proportional to the number present in the habitat then the following equation, which was proposed by Zippin (1956) for estimating small mammal populations, applies. The notation is that of Wada (1962a) who used this method with *Culex pipiens* form *pallens* in Japan.

$$A_n = a(S - Y_{n-1})$$
(3.27)

where A_n = the number of larvae caught by the nth dip, Y_{n-1} = the accumulated number of larvae collected in all dips till the $(n-1)$th dip, in other words the total catch from all previous dips, S = the total population originally present and which is to be estimated and a = a proportional constant, and is the ratio of the catch to those yet to be caught, i.e. the slope of the regression line of A_n on Y_{n-1}. Now, the ratio of larvae collected in the nth dip, i.e. A_n, to the number caught in the $(n-1)$th dip, i.e. A_{n-1}, is:

$$\frac{A_n}{A_{n-1}} = \frac{a(S - Y_{n-1})}{a(S - Y_{n-2})} \tag{3.28}$$

which simplifies to:

$$\frac{A_n}{A_{n-1}} = 1 - a \tag{3.29}$$

Wada (1962a) used the removal method to estimate the larval population of *Culex pipiens* form *pallens* in a 2.2-m diameter fertiliser pit. He took 50 and 100 dips with a 15-cm diameter dipper on two consecutive days, but combined the catch from a number of dips. On the first day he combined the catches from five consecutive dips, while on the second day he added together the numbers of larvae caught in 10 consecutive dips. These combined samples were termed 'super-unit catches'. Hence one dip with a dipper represented a unit catch, but a number of successive dips (t) (which in this case was 5 or 10) were termed the super-unit catch. Thus $A(t)_n$ represented the number of larvae collected in the nth superunit catch, and from this Wada (1962a) derived the following equation:

$$A(t)_n = \{1 - (1 - a)^t\}(S - Y(t)_{n-1}) \tag{3.30}$$

where $Y(t)_{n-1}$ = the total number of larvae caught up to the $(n-1)$th super-unit catch. This is basically of the same form as eqn (1). The ratio of larvae caught in the nth super-unit catch to those caught in the previous catch, $A(t)_n / A(t)_{n-1}$, is $(1-a)^t$, which is similar to the ratio of $(1-a)$ obtained for larvae in single dips. From eqns (3.27) or (3.29) a regression can be obtained between the numbers caught in the nth sample and those caught in all previous samples, i.e. till the $(n-1)$th sample. Figure 3.33 shows the plots of the numbers of larvae of *Culex pipiens* form *pallens* obtained in the nth super-unit catch and the accumulated number till the $(n-1)$th super-unit catch (Wada 1962a). The regression line can be fitted visually or its plot calculated by the method of weighted or unweighted least squares (Zippin 1956; Wada 1962a). The estimated total population of larvae is given by the intercept of the regression line on the abscissa.

When Wada (1962a) used the values obtained on the first day of dipping the estimated larval population of *Culex pipiens* form *pallens* in the fertiliser pit was calculated as 113 058 but only 56 318 on the second day (Fig. 3.33). However, on the first day a total of 75 730 larvae were removed by dipping, and if these are added to the second estimate the estimated population is 132 048. Finally, by combining the results of dippings obtained on the 2 days the estimated population was 122 048 larvae. The similarity of the three estimates indicates that it is reasonable to conclude that the removal method satisfactorily estimated the size of the true population in the pit. A very large number of samples, however, had to be collected. In fact a common disadvantage of the removal method is that often many samples have to be taken before sufficient numbers are removed from the population and the decline in the catch rate becomes apparent (see also removal trapping applied to adults resting in houses in Chap. 5).

The removal method can also be applied to the decline in catch rate with time. Wada (1962b), for example, recorded the numbers of larvae of *Ochlerotatus togoi* collected each minute from rock pools with a glass pipette. This constituted his unit catch and the combined number of larvae from five consecutive 1-min catches formed the super-unit catch. When the numbers of larvae caught in the nth unit (A_n) or super-unit ($A(t)_n$) catch were plotted against the total catch (Y_{n-1}, or $Y(t)_{n-1}$) a series of good regression lines were obtained thus enabling the population size in a number of rock pools to be estimated. There was a very close agreement between population size estimated by 1-min unit catches (A_n/Y_{n-1}) and 5-min super-unit catches ($A(t)_n/Y(t)_{n-1}$). Furthermore, the estimates of population size were very similar to the actual numbers of larvae known to be present in the pools, but all regression lines, and hence population estimates, were based on larval collections in which about 99% of the total larvae were collected from the pools. Obviously to be of any practical use it is essential that estimates can be calculated from a much smaller proportion of the total population. Wada (1962b) therefore checked the efficiency of the method against a varying number of unit (1-min) and super-unit (5-min) catches. He found that if there were 50–300 larvae in the pools and population estimates within an error of 10% were required then about 3–8 min should be spent pipetting. Zippin (1956, 1958) also concluded that comparatively large proportions of a population must be collected if reliable estimates were to be obtained. He concluded that to obtain a coefficient of variation (population estimate/SE × 100) of 10% about 70–75% of the individuals must be removed from a population of about 500 or less, 60% from a population of 1000, 35% from a population of 10 000 and 20% from a population of about 100 000 (Zippin 1958).

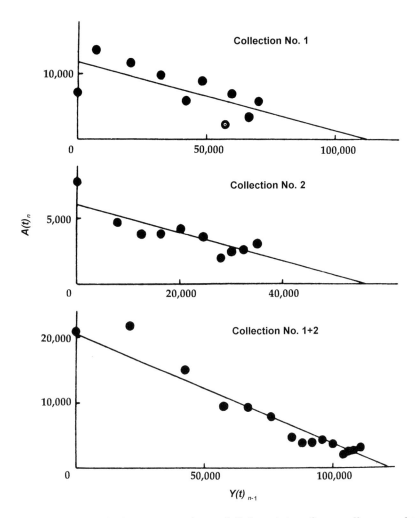

Fig. 3.33. relationship between numbers of *Culex pipiens* form *pallens* caught in nth super-unit catch ($A(t)_n$) and the accumulated number till the $(n-1)$th super-unit catch ($Y(t)_{n-1}$) (after Wada 1962*a*)

The slopes of the regression lines in Fig. 3.33 are measures of the constant *a*, i.e. the rate at which larvae are captured in a catch. It can be regarded as the collecting efficiency. Working with *Ochlerotatus togoi* larvae Wada (1962*b*) investigated the relationship between collecting efficiency and (1) surface area (cm²) of water in the rock pools, (2) surface area × depth of water, (3) number of larvae collected in first 5 min, and, finally, (4) the product of three parameters, surface area, depth and number of larvae collected in first 5 min. When values of constant *a* were plotted against each of these variables a rectangular hyperbolic curve resulted, and this was most pronounced when *a* was plotted against the product of the three parameters (Fig. 3.34*a*). It was found that the larger and deeper the pools and the more larvae present, then in general collecting efficiency was reduced.

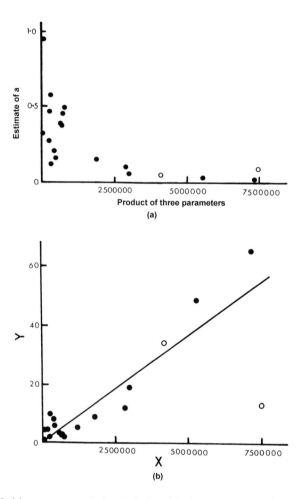

Fig. 3.34. *Ochlerotatus togoi.* (*a*) Relationship between collecting efficiency represented by the estimate of a and the product of the parameters–surface area, water depth and number of larvae caught in first 5 min; (*b*) relationship between the reciprocal of the estimate for the constant (a) (i.e. *Y*) and the product of the parameters–surface area, water depth and number of larvae caught in first 5 min (i.e. *X*). In both graphs solid and open circles represent pipette and dipper collections, respectively (after Wada 1962*b*)

When the reciprocal of a was plotted against the product of the three variables a linear regression line passing through the origin was obtained (Fig. 3.34b). A regression equation in the form of $Y = kX$ was produced by the method of least squares where Y = the reciprocal of a, X = the product of the three parameters and k = a constant. This relationship was represented by

$$Y = 0.007956\,X \qquad (3.31)$$

Now if a 5-min catch with a pipette is made ($t = 5$) and the value of X is determined for the specific habitat, then Y can be readily calculated from eqn (4), and its reciprocal is another estimate of a. In eqn (3) $t = 5$ and by letting $n = 1$ and $A(t)_n$ = the larvae caught in the first 5 min of pipetting the following is derived:

$$S = \frac{A(5)_1}{1 - (1 - a)^5} \qquad (3.32)$$

where $A(5)_1$ = the number of larvae collected within the first 5 min. Hence another estimate of larval population size (S) is obtained. Wada (1962b) considered that the population estimates that this formula produced compared favourably with those obtained by plotting the regression of about 15 1-min unit or three 5-min super-unit catches, against total catch. However, if more catches are plotted to give a regression line then more accurate population estimates are obtained, but accuracy will have been achieved at the expense of a time-consuming collecting method. Wada et al. (1993) applied the methods of Wada (1962b) to the estimation of seasonal changes in absolute abundance of *Ochlerotatus togoi* inhabiting rock pools on Fukue Island, Nagasaki, Japan in relation to precipitation, pool surface area and depth. Numbers of larvae tended to vary in line with pool surface area, and were also affected by water quality and the presence of predators.

Despite Wada's careful work removal trapping has rarely been used with mosquitoes, mainly because of the large proportion of the population that has to be removed for an accurate population estimate. The few attempts that have been made are reported here.

Croset et al. (1976) in France used the removal method of Zippin (1956, 1958) to estimate the population of *Ochlerotatus communis* in a very small pool (about 1.5 × 1 m and 0.15 m deep). Five 0.5-h spaced samples were taken with an aquatic net. The regression method (Fig. 3.35) gave a total population of 1006 larvae, while a maximum likelihood method resulted in an estimate of 1066 ± 56 larvae. A mark–recapture estimate using radioactive ^{32}P estimated the population as 1170, and dipping gave a population estimate of 1387.

The authors concluded that because of the necessity of removing large proportions of larvae the removal method could be used only in very small habitats.

Miura (1980) used the removal method to estimate the population of *Culex quinquefasciatus*. When sampling was from a 1-m^2 pond five samples with a dipper represented the 'catch unit' that was plotted against previous total catch, but a 15-dip unit was used when collections were from a catch basin. In addition estimates were made of the numbers of larvae in 4-litre capacity fabricated tree-holes and 1-litre cemetery vases, and on these occasions the numbers collected every 3 min with a pipette were regressed against total catch. In all instances there was very good agreement between population estimates obtained by the regression method and graphically (Zippin 1956).

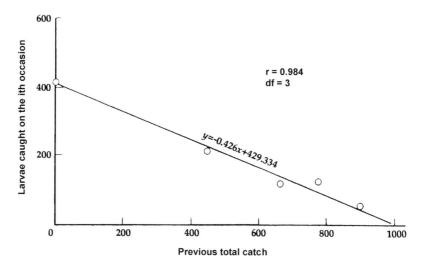

Fig. 3.35. Removal method. Estimated larval population of *Ochlerotatus communis* is 1006. Zippin's method gave an estimation of 1066 ± 56, dipping an estimation of 1387, and a capture–recapture method an estimation of 1170) (Croset et al. 1976)

In Japan Mogi et al. (1984) compared four removal methods for estimating the total populations of immatures of *Ochlerotatus baisasi*, *Culex tuberis* and *Uranotaenia ohamai* in crab holes, namely by linear regression, and the methods of Kono (1953), Zippin (1956) and Seber and Le Cren (1967) (Chap. 5). They also repeatedly pumped out the crab holes to obtain a real population estimate. With the Seber and Le Cren method only the first two samples were used to estimate population size, whereas with the other methods the first three samples were needed. Both the Zippin and Seber and Le Cren methods gave 95% confidence limits of the estimate. They concluded that all methods often gave realistic population estimates, but Zippin's method may be preferred, except that estimation is impossible when the numbers in the second sample are just twice those in the third sample. Seber and Le Cren's method does not require a third sample, but has several disadvantages. For example, the range of the 95% confidence limits can be very large, estimation is impossible when first and second samples contain the same numbers of individuals, and estimates are negative when the first sample is smaller than the second sample.

Mogi et al. (1984) discovered that the numbers of larvae and pupae caught in the first sample were directly proportional to the real population, consequently just a single sample may be satisfactory to estimate relative population size, provided of course that all water in the crab hole is pumped out during the first sample.

Kono's method of time unit collecting

In most of the foregoing examples the population estimates were obtained by regression, the line being fitted either by eye or its plot calculated, such as by least squares. A simpler and quicker method of analysing catch data to get an approximate population estimate is to consider the numbers caught (n_1, n_2, n_3) at only three points of time (t_1, t_2, t_3), where $t_3 = \frac{1}{2}(t_1 + t_2)$ (Kono 1953). As an example let the numbers of larvae collected during three intervals of time, by either pipetting or dipping, be n_1, n_2, and n_3, these being the accumulated catches. The population estimate (P) is given by:

$$P = \frac{n_1 n_2 - (n_3)^2}{n_1 + n_2 - 2n_3} \qquad (3.33)$$

The method has been used by Mogi et al. (1984) to calculate the population size of several mosquito species.

Zippin's method of estimating population size

A more accurate method of estimating population size by removal methods is to use maximum likelihood methods in analysing the data (Moran 1951; Zippin 1956, 1958). An added advantage is that it also provides an estimate of the standard error. In this method the numbers caught in all samples (e.g. numbers of larvae caught in all dips, pipettes or small units of time) are added to give the total catch (T) which can be expressed as follows:

$$T = \sum_{i=1}^{k} y_i = y_1 + y_2 + y_3 + ... y_k \tag{3.34}$$

where y_1, y_2, y_3 and y_i = the numbers of larvae collected during the 1st, 2nd, 3rd and ith (i.e. last) catch and k = the number of catches. The following expression must also be calculated:

$$\sum_{i=1}^{k}(i-1)y_i = (1-1)y_1 + (2-1)y_2 + (3-1)y_3 + ...(k-1)y_k \tag{3.35}$$

Next the ratio (R) of this calculated value to the total catch (T) is determined:

$$R = \frac{\sum_{i=1}^{k}(i-1)y_i}{T} \tag{3.36}$$

Now the estimated proportion of the population caught throughout the sampling programme is:

$$(1-(1-\hat{p})^k) = (1-\hat{q}^k) \tag{3.37}$$

where $\hat{P} = (y_1 - y_2/y_1)$, that is the estimated probability of capture of larvae during a single sampling (y_1 and y_2 are numbers of larvae caught in 1st and 2nd samplings), and $\hat{q} = 1 - \hat{p}$, the probability of a binomial distribution. Values of $\hat{p}$ and $\hat{q}$ can be calculated from the following equation:

$$R = \frac{\hat{q}}{\hat{p}} - \frac{k\hat{q}^k}{(1-\hat{q}^k)} \tag{3.38}$$

and the estimate of the population ($\hat{P}$) is given by:

$$\hat{P} = \frac{T}{(1-\hat{q}^k)} \tag{3.39}$$

which is the total catch (T) divided by the estimated proportions of the larval population that is caught. The standard error of this multinomial population estimate is given by:

$$SE(\hat{P}) = \sqrt{\dfrac{\hat{P}(\hat{P}-T)T}{T^2 - \hat{P}(\hat{P}-T)\dfrac{(k\hat{p})^2}{(1-\hat{p})}}} \tag{3.40}$$

The 95% confidence limits are approximately twice this calculated value of the standard error when the population is 200 or more, and when 90% or less of it is caught. When populations are less than 200 but more than 50 then $2 \times SE$ gives 90% not 95% confidence limits. Santos and Oliva (1991) also presented equations for numerical solutions of Zippin's removal method.

To simplify the mathematics of solving P and its standard error Zippin (1958) presented a number of graphs in which the values of R were plotted against $(1 - \hat{q}^k)$ and also against $\hat{p}$, for $k = 3, 4, 5$ and 7. It is unlikely, however, that these graphs will be of value in estimating mosquito larval populations, unless the numbers of larvae caught in a large number of dips are pooled, to represent catches on the 1st, 2nd and 3rd occasions, or else much larger sampling units are used.

Two-catch removal method

The method was devised by Seber and Le Cren (1967) for situations where a large proportion of the population can be removed. They used it to study fish populations but considered that it had other applications. It has not been used to estimate mosquito larval populations, but might prove useful in certain situations, such as where marking is difficult and where a high proportion, but not all, larvae can be readily removed, e.g. larvae from some tree-holes. Two successive catches, C_1 and C_2 are made with equal effort (i.e. equal time, or number of samples). An estimate of population size (P) is then obtained from the following very simple equation:

$$P = (C_1)^2 /(C_1 - C_2), \quad \text{var. } P = [(C_1)^2(C_2)^2(C_1 + C_2)]/(C_1 - C_2)^4 \tag{3.41}$$

Robson and Regier (1968) suggested that a less biased population estimate would be obtained if the enumerator was $(C_1^2 - C_2)$ not $(C_1)^2$. Seber and Le Cren (1967) pointed out that the method might be useful when a sample requires lengthy and tedious sorting to remove all wanted individuals, such as the removal of mosquito eggs from soil samples or adults from light-trap collections. Catches C_1 and C_2 would represent the numbers of individuals removed in two time intervals and the total catch could be estimated, thus saving considerable time and effort. It is clear that the two-catch technique is a removal method. Seber and Le Cren (1967) also give methods by which approximate population estimates can be made from a single catch, and describe how the two-catch method can be combined with mark-recapture methods. Mogi et al. (1984) used this last method with mosquito larvae.

Kelker's selective removal method

Following Service (1993) this method is referred to as Kelker's method. The method uses any natural marks or distinguishing features that occur on a percentage of the individuals in a population (Kelker 1940). The proportions of the two different forms in the population are recorded and a known number of one form removed, so that their relative proportions are changed. The new ratio of the two forms is then recorded, and population size estimated from the magnitude of the difference between these two ratios. This and related methods are known as CIR (Change in Ratio) methods.

The two contrasting components in the population whose ratio is determined before and after removal can be different morphs of a polymorphic species, different sexes, or different age classes. For example, the collection of different larval instars and then the selective removal of one instar could be used to calculate population size. The mathematical theory of the method together with formulae for calculating the variance of the population estimate are presented by Chapman (1955), and good reviews of CIR methods are given by Paulik and Robson (1969), Smirnov (1967) and Seber (1973). In comparing the method with the mark–recapture method Chapman (1955) concludes that the latter method will usually give more information for the same effort. A procedure is also given by which the two methods can be combined, the individual estimates acting as a check on each other.

The method was used in Kenya to estimate the population size of 4th instar larvae of the *Anopheles gambiae* complex in pools ranging from 64–921 m^2 in surface area (Service 1971). Three distinct larval mutants were recognised in addition to normal larvae, but for purposes of estimating their population size larvae were classified as either normal (n) or mutant (m), and after the first collection to determine the proportions of normal and mutant larvae a number of normal larvae were removed and the proportions reassessed. The estimate of the population (P) is obtained as follows:

$$P = Kn \div \left[Dn_1 - \frac{(Dm_1 \ Dn_2)}{Dm_2} \right] \qquad (3.42)$$

where Kn = the number of normal larvae that are removed after the first collections, Dn_1 and Dn_2 = the proportions of normal larvae as a decimal of the total numbers of larvae in the two collections, and Dm_1 and Dm_2 = the proportions of mutant larvae in the two collections. To check on the efficiency of the technique the populations of *Anopheles gambiae* in the pools were also estimated by a simple mark–recapture method using Rhodamine dye to stain the larvae. The populations in the different pools were ranked in the same order of magnitude as shown by Kelker's selective removal method, but there was no good agreement between the actual estimates obtained by the two methods. In some instances Kelker's method gave a larger population estimate; in others the reverse was true. More studies are needed to assess the usefulness and practicalities of this method. One advantage of this method is that there is no danger of affecting larval survival or behaviour by marking them.

Visual classification methods

To avoid the inherent problems of manually counting mosquito larvae to determine population size, Carron et al. (2003) have proposed a visual estimation method, based on the use of a visual "abacus" representing 5 or 10 abundance classes of mosquito larvae or pupae, which when used with a standardised collecting tray can be used to determine the abundance and density of mosquito immatures. Graphical representations of 5 or 10 classes of abundance are printed on sheets of A4 paper. Samples of mosquito immatures are placed in a standardised tray, and the abundance pattern observed is allocated to one of the 5 or 10 classes, and is then converted to an abundance range. This method was tested in the laboratory and in a field situation. Laboratory results revealed no significant differences between abundance estimates calculated using the two abacuses and manual counting of larvae. Differences in abundance estimates among different users of the technique when sampling from high density populations were observed, and this was attributed to their differing experience in mosquito larval sampling. The abacus method was also able to detect changes in abundance following treatment of a natural habitat with *Bti*. The authors note several potential restrictions to the widespread use of this method, including its unsuitability for turbid water or water containing a lot of leaf litter, or other vegetable matter. The abacus is also unsuitable where abundance of different species or life-stages is required, e.g. for life-table studies.

Mark–recapture methods

Mark–recapture methods for the determination of absolute population size have been used to estimate the size of larval mosquito populations and some examples of their use are described in this chapter. Mark-release recapture techniques are generally more applicable and have been more frequently utilised to study populations of adult mosquitoes, and readers are therefore referred to Chapter 12 for a full description of the range of mark-recapture techniques available, many of which can be successfully applied to the study of larval populations.

Marking methods

Radionuclides. The use of radionuclides to mark mosquito larvae was first proposed by Bugher and Taylor (1949), but the method was only infrequently used until the 1970s. No recent examples of the radioactive marking of mosquito larvae are available and so this technique is not discussed further here. Readers interested in this method from a historical perspective are advised to refer to the first and second editions of this book.

Vital dyes and stains. The most useful dyes for marking mosquitoes are probably methylene blue, Giemsa and rhodamine B, but Nile blue A and neutral red may also be useful. All these dyes are usually characterised by low insect toxicity (Gast and Landin 1966; Heron 1968; Vail et al. 1966). Useful accounts of the effect of vital dyes on living organisms are given in the publications of Barbosa and Peters (1971) and David (1963).

Mosquito larvae can be stained by immersing them in solutions of water-soluble dyes and stains, but in early experiments this sometimes caused high larval mortalities (Bailey et al. 1962; Chang 1946; Reeves et al. 1948; Weathersbee and Hasell 1938). It has also been reported that vital dyes such as Nile blue A, methylene blue and neutral red can, at least in *Aedes aegypti*, cause retardation of larval growth, decrease in pupation rate and increase in larval mortality (Barbosa and Peters 1970; Chevone and Peters 1969; Peters and Chevone 1968). It is therefore essential that the toxicities of dyes and stains are investigated before use and only those causing negligible mortality should be used in mark–recapture studies. Another adverse property of some dyes is the prolongation of larval development but this will be of little practical importance when mark–recapture experiments are made during a short period, say 24 h. Over much longer periods any extension of the larval period may result in underestimating population size, but this may to some extent be counter-balanced if there is increased mortality of marked larvae between the interval of marking and recapture. An additional hazard of using dyes is that some are photodynamic, that is their detrimental effects on larvae may be increased by exposure to bright light. The chemical properties of dyes may also alter with changes in pH. Because of the adverse effects dyes may have on larvae the sooner the second sample is taken the more reliable the population estimate, but only if sufficient time has elapsed to allow the marked larvae to mix with the parent population.

Despite the various adverse affects reported stains have successfully been used to estimate larval population size. For example larvae of *Ochlerotatus detritus* and *Anopheles gambiae* were marked by immersing them for 12–24 hr in solutions of 0.1 mg/litre rhodamine B with little or no mortality, and population estimates obtained (Service 1968, 1971). More recently in England, larvae of *Ochlerotatus cantans* and *Ochlerotatus punctor* have been marked by placing them in 1–5 ppm of rhodamine B for 12 h with very little mortality (Service 1993*b*). In India Rajagopalan et al. (1975) stained larvae of *Culex quinquefasciatus* in 0.1% methylene blue for 8 h, with insignificant mortality, and Kaur and Reuben (1981) successfully marked larvae of *Anopheles stephensi* by keeping them overnight in 0.05% methylene blue. In the USA Fish and Joslyn (1984) marked *Ochlerotatus communis* larvae by placing them in 100 ppm of Giemsa stain, and there was no detectable mortality. In Myanmar 3rd and 4th instar larvae of *Anopheles dirus* were placed in 2% Giemsa stain for 1.5 h. Shorter periods or weaker stains marked fewer larvae, whereas the present concentration ensured that all larvae were stained and there were no mortalities after 6 h, and only 5% after 24 h.

Population estimates of mosquito larvae

In India Rajagopalan et al. (1975) stained larvae of *Culex quinquefasciatus* with 0.1% methylene blue for 8 h, and then released 10 000–20 000 larvae in 30-m sections of town drains. The water was then gently stirred to mix the stained and unstained larvae and the sections sampled 2 h later with dippers. The total population of larvae and pupae was estimated using Bailey's modification of the Lincoln Index (see Chap. 12). The relative densities of both larvae and pupae, as measured by the mean numbers of larvae per dip, were correlated with the absolute population estimates by regression analysis (Fig. 3.6). It was concluded that it was possible to estimate total population density from the numbers of pre-adults per dip. Also in India Kaur and Reuben (1981) released either 200 or 400 *Anopheles stephensi* larvae into wells, and using Bailey's correction to the Lincoln Index calculated the population of the larvae from the ratio of marked and unmarked larvae, while the number of pupae was estimated from their proportion in the sample.

From the data a straight regression line was obtained of the number of larvae in five bucket samples (y) on the absolute number of larvae in a well (x), $Y=0.1242x - 0.392$. This equation was then used to obtain estimates of population size in 20 wells in the area using buckets.

The population of the pre-adults of *Anopheles dirus* was estimated in a well in Myanmar by staining 3rd and 4th instar larvae in Giemsa. After marking, larvae were released into the well and after 6 h the larvae were sampled by lowering a net into the well and the combined population of 3rd and 4th stage larvae estimated by the Lincoln Index (Tun-Lin et al. 1986). Previous samplings had given the proportions of 1st and 2nd instar larvae, 3rd and 4th instar larvae, and pupae in the well, and using these proportions the total numbers of 1st and 2nd instars were calculated as 385, 3rd and 4th instars as 435, and pupae as 83. The total immatures in this one well was therefore estimated as 903.

Fish and Joslyn (1984) marked batches of 1000 *Ochlerotatus communis* larvae by placing them for 1–3 h in a 5-litre bucket containing 100 ppm of Giemsa stain, and after rinsing in clear pond water they were returned to their five separate pools of differing size. To ensure thorough mixing they walked through the pools as larvae were released. After about 18 h some 500 larvae were collected and examined by eye against a white background. Population estimates, using Bailey's (1951) modified Lincoln Index, ranged from 8416 for the smallest pool to 194 000 for the largest pool. There was a positive correlation between these estimates and both the surface area of the pools and the volume of water they contained, but surface area was the better predictor of size ($y = 22047 + 478.7x$; $r = 0.96$) (In their paper there are errors in Table 1. In columns 5 and 6 the 1st and 2nd entries should be reversed as should the 4th and 5th entries.)

Other examples of mark–recapture methods being used to obtain population estimates of mosquito larvae include *Ochlerotatus cataphylla* and *Ochlerotatus communis* (Croset et al. 1976), *Ochlerotatus communis* (Fish and Joslyn 1984), *Ochlerotatus sollicitans* (Joslyn and Conrad 1984), *Anopheles dirus* (Tun-Lin et al. 1986), *Ochlerotatus detritus* (Service 1968), and *Anopheles gambiae* (Service 1971).

Sequential sampling

This type of sampling was developed in 1943 for quality control in factories and was first classified as restricted, but was cleared for more general use in 1945. In the biological field it is generally used to classify populations into one of usually three predefined population levels, low, medium or high. For example, we might consider that a mean number of larvae per dip of <1 reflects a low population, a mean of >5 larvae a high population, and intermediate values a medium sized population. Most sequential sampling plans are used to determine whether pest levels have reached a density where control measures are required.

The number of samples taken is variable and depends on whether the cumulative numbers collected in a series of samples provide the necessary information to allow the population to be classified into one of the three population levels. Once this has been achieved sampling stops.

Although the potential usefulness of sequential sampling in mosquito surveys was pointed out by Knight (1964) many years ago the method has not been widely adopted, having been used by only a few entomologists working with mosquitoes (Ikemoto 1978; Mackey and Hoy 1978; McLaughlin et al. 1987; Stewart et al. 1983a, b; Wada et al. 1971b). A difficulty with employing a sequential sampling programme is that the biological distribution (dispersion) of the animal in the habitat must be known because this determines the manner in which the data are analysed (Iwao 1975; Onsager 1976; Wald 1947; Waters 1955). Determination of the dispersion pattern requires a rather large number of preliminary samples, and moreover the dispersion pattern must not change appreciably during the sampling timeframe, which may extend over weeks or months. It may prove difficult to fit the dispersion of mosquito larvae to any general mathematical model (Service 1971), although Wada (1965) found the larval dispersion of several Canadian species fitted the negative binomial model, and in Japan the dispersion of *Culex tritaeniorhynchus* larvae also mimicked this model (Wada et al. 1971b).

Onsager (1976) gives a good clear account of sequential sampling and formulae for use when the data fits: (i) a binomial distribution, such as classifying counts into two categories, e.g. males and females, or infested and uninfested; (ii) a Poisson distribution which can be used as a model for a random distribution; and (iii) a negative binomial distribution which is applicable to aggregated sample data. Further details on the use of sequential sampling in entomology are given in chapters 4 and 5 in Kogan and Herzog (1980). Readers interested in the use of sequential sampling methods are advised to consult Wada et al. (1971b) or Service (1993) for detailed descriptions of the methods.

References

Aiken RB (1979) A size selective underwater light trap. Hydrobiologia 65: 65–68

Ali M, Wagatsuma Y, Emch M, Breiman RF (2003) Use of a geographic information system for defining spatial risk for dengue transmission in Bangladesh: role for *Aedes albopictus* in an urban outbreak. Am J Trop Med Hyg 69: 634–640

Allan SA, Surgeoner GA, Helson BV, Pengelly DH (1981) Seasonal activity of *Mansonia perturbans* adults (Diptera: Culicidae) in southwestern Ontario. Can Entomol 113: 133–139

Almirón WR, Brewer ME (1996) Classification of immature stage habitats of Culicidae (Diptera) collected in Cordoba, Argentina. Mem Inst Oswaldo Cruz 91: 1–9.

Amerasinghe FP, Ariyasena TG (1990) Larval survey of surface water-breeding mosquitoes during irrigation development in the Mahaweli project, Sri Lanka. J Med Entomol 27: 789–802

Amerasinghe FP, Indrajith NG (1994) Postirrigation breeding patterns of surface water mosquitoes in the Mahaweli Project, Sri Lanka, and comparisons with preceding developmental phases. J Med Entomol 31: 516–523.

Anderson RM, Gordon DM, Crawley MS, Hassell MS (1982) Variability in the abundance of animal and plant species. Nature 296: 245–248

Andis MD, Meek CL (1984) Bionomics of Louisiana ricefield mosquito larvae, II. Spatial dispersion patterns. Mosquito News 44: 371–376

Andis MD, Meek CL, Wright VL (1983) Bionomics of Louisiana riceland mosquito larvae, I. A comparison of sampling techniques Mosquito News 43: 195–203

Anon. (1956) Manual de Normas Tecnicas y Administratives de la Campana de Erradicacion de *Aedes aegypti*. Pan-Am Sanit Bur 244 pp. (mimeographed)

Anon. (1973) WHO Computer Survey of *Stegomyia* Mosquitoes, 1972. VBC/73.11, 2 pp., 50 maps (mimeographed)

Apperson CS, Yows DG (1976) A light trap for collecting aquatic organisms. Mosquito News 36: 205–206

Arnell JH, Nielsen LT (1967). Notes on the distribution and biology of tree hole mosquitoes in Utah. Proc Utah Mosq Abatement Assoc 20: 28–29

Awono-Ambéné HP, Robert V (1999) Survival and emergence of immature *Anopheles arabiensis* mosquitoes in market-gardener wells in Dakar, Senegal. Parasite 6: 179–184

Bailey DL (1984) Comparison of diurnal and nocturnal *Mansonia* larval populations of water lettuce plants. Mosquito News 44: 548–552

Bailey DL, Lowe RE, Kaiser PE, Dame DA, Fowler JE (1980) Validity of larval surveys to estimate trends of adult populations of *Anopheles albimanus*, Mosquito News 40: 245–251

Bailey NTJ (1951) On estimating the size of mobile populations from recapture data. Biometrika 38: 293–306

Bailey NTJ (1952) Improvements in the interpretation of recapture data. J Anim Ecol 21: 120–127

Bailey SF, Eliason DA, Iltis WG (1962) Some marking and recovery techniques in *Culex tarsalis* Coq. flight studies. Mosquito News 22: 1–10

Balling SS, Resh VH (1984) Seasonal patterns of pondweed standing crop and *Anopheles occidentalis* densities in Coyote Hills Marsh. Proc California Mosq Vector Control Assoc 52: 122–125

Bang YH, Bown DN, Onwubiko AO (1981) Prevalence of larvae of potential yellow fever vectors in domestic water containers in south-east Nigeria. Bull World Health Organ 59: 107–114

Barbosa P, Peters TM (1970) Dye-induced changes in the developmental physiology of *Aedes aegypti* larvae. Entomol Exp Appl 13: 293–299

Barbosa P, Peters TM (1971) The effects of vital dyes on living organisms with special reference of methylene blue and natural red. Histochemistry J 3: 71–93

Barker CM, Brewster CC, Paulson SL (2003). Spatiotemporal oviposition and habitat preferences of *Ochlerotatus triseriatus* and *Aedes albopictus* in an emerging focus of La Crosse virus. J Am Mosq Control Assoc 19: 382–391

Barnes CM. Cibula WG (1979) Some implications of remote sensing technology in insect control programs including mosquitoes. Mosquito News 39: 271–282

Barrera R (1996) Species concurrence and the structure of a community of aquatic insects in tree holes. J Vector Ecol 21: 66–80

Barton WI (1964) A survey technique for *Mansonia perturbans*. Mosquito News 24: 224–225

Bates M (1941). Field studies of the anopheline mosquitoes of Albania. Proc Entomol Soc Wash 43: 37–58

Bates M (1949). The Natural History of Mosquitoes. The MacMillan Co., New York

Bates M, de Zulueta J (1949) The seasonal cycle of anopheline mosquitoes in a pond in eastern Colombia. Am J Trop Med 29: 129–150

Batzer DP (1993) Technique for surveying larval populations of *Coquillettidia perturbans*. J Am Mosq Control Assoc 9: 349–351

Batzer DP, Sjogren RD (1986) Larval habitat characteristics of *Coquillettidia perturbans* (Diptera: Culicidae) in Minnesota. Can Entomol 118: 1193–1198

Baylis M, Bouayoune H, Touti J, El Hasnaoui H (1998) Use of climatic data and satellite imagery to model the abundance of *Culicoides imicola*, the vector of African horse sickness virus, in Morocco. Med Vet Entomol 12: 255–266

Bayoh MN, Thomas CJ, Lindsay SW (2001) Mapping distributions of chromosomal forms of *Anopheles gambiae* in West Africa using climate data. Med Vet Entomol 15: 267–274

Beck LR, Rodriguez MH, Dister SW, Rodriguez AD, Rejmankova E, Ulloa A, Meza RA, Roberts DR, Paris JF, Spanner MA, Washino RK, Hacker C, Legters LJ (1994) Remote sensing as a landscape epidemiologic tool to identify villages at high risk for malaria transmission. Am J Trop Med Hyg 51: 271–280

Beck LR, Rodriguez MH, Dister SW, Rodriguez AD, Washino RK, Roberts DR, Spanner MA (1997) Assessment of a remote-sensing based model for predicting malaria transmission risk in villages of Chiapas, Mexico. Am J Trop Med Hyg 56: 99–106

Begon M (1979) Investigating Animal Abundance: Capture-recapture for Biologists. Edward Arnold, London

Beier JC, Travis M, Patricoski C, Kranzfelder J (1983) Habitat segregation among larval mosquitoes (Diptera: Culicidae) in tire yards in Indiana, USA. J Med Entomol 20: 76–80

Belkin JN (1954) Simple larval and adult mosquito indexes for routine mosquito control operations. Mosquito News 14: 127–131

Bertram DS, Varma MGR, Page RC, Heathcote OHU (1970) A betalight trap for mosquito larvae. J Med Entomol 7: 267–270

Bidlingmayer WL (1954) Description of a trap for *Mansonia* larvae. Mosquito News 14: 55–58

Bidlingmayer WL (1968) Larval development of *Mansonia* mosquitoes in central Florida. Mosquito News 28: 51–57

Blower JG, Cook LM, Bishop JA (1981) Estimating the Size of Animal Populations. Allen & Unwin, London

Bøgh C, Clarke SE, Jawara M, Thomas CJ, Lindsay SW (2003) Localized breeding of the *Anopheles gambiae* complex (Diptera: Culicidae) along the River Gambia, West Africa. Bull Entomol Res 93: 279–287

Bonnet D, Chapman H (1956) The importance of mosquito breeding in tree holes with special reference to the problem in Tahiti. Mosquito News 16: 301–305

Bonne-Wepster J, Brug SL (1939) Observations on the breeding habits of the subgenus *Mansonioides* (genus *Mansonia*, Culicidae). Tijdschr Entomol 82: 81–90

Bown DN, Bang YH (1980) Ecological studies on *Aedes simpsoni* (Diptera: Culicidae) in southeastern Nigeria. J Med Entomol 17: 367–374

Boyd MF (1930) An Introduction to Malariology. Harvard University Press, Cambridge, Massachussets

Boyd MF (1949) Malariology. A Comprehensive Survey of All Aspects of this Group of Diseases from a Global Standpoint. Volume 1. W. B. Saunders Co. London

Bradshaw WE (1983) Interaction between the mosquito *Wyeomia smithii*, the midge *Metriocnemus knabi*, and their carnivorous host *Sarracenia purpurea*, In: Frank JH, Lounibos LP (eds) Phytotelemata: Terrestrial Plants as Hosts for Aquatic Insect Communities. Plexus Publishing Inc., Medford, New Jersey, pp 161–189

Bradshaw WE, Holzapfel CM (1983) Predator-mediated, non-equilibrium coexistence of tree-hole mosquitoes in southeastern North America. Oecologia (Berl.) 57: 239–256

Bradshaw WE, Holzapfel CM (1984) Seasonal development of tree-hole mosquitoes (Diptera: Culicidae) and Chaoboridae in relation to weather and predation. J Med Entomol 21: 366–378

Bradshaw WE, Holzapfel CM (1985) The distribution and abundance of tree-hole mosquitoes in eastern North America: perspectives from north Florida. In: Lounobos LP, Rey JR, Frank JH (eds) Ecology of Mosquitoes: Proceedings of a Workshop. Florida Medical Entomology Laboratory, Vero Beach, Florida, pp 3–23

Bradshaw WE, Holzapfel CM (1986a) Habitat segregation among European tree-hole mosquitoes. Natn Geogr Res 2: 167–178

Bradshaw WE, Holzapfel CM (1986b). Geography of density-dependent selection in pitcher-plant mosquitoes. In: Taylor F, Karban R (eds) The Evolution of Insect Life Cycles. Springer-Verlag, New York, pp 48–65

Bradshaw WE, Holzapfel CM (1988) Drought and the organization of tree-hole mosquito communities. Oecologia (Berl.) 74: 507–514

Bradshaw WE, Hopzapfel CM (1989) Life-historical consequences of density-dependent selection in the pitcher-plant mosquito, *Wyeomyia smithii*. Am Nat 133: 869–887

Bradshaw WE, Hopzapfel CM (1991) Fitness and habitat segregation of British tree-hole mosquitoes. Ecol Entomol 16: 133–144

Breev KA, Kornikov VV, Bart AG, Boyko VI, Kalinin OM (1983) A method of calibrating catching facilities using as an example the quantitative estimation of mosquitos in a region of the Baikal-Amur railway (beta-scheme). Parazitol. Sbornik., 31: 48–61 (In Russian, English summary)

Brown AWA (1971) World wide surveillance of *Aedes aegypti*. Proc California Mosq Control Assoc 42: 20–25

Bruce-Chwatt LJ, Fitzjohn RA (1951) Mosquitoes in crab-burrows on the coast of West Africa and their control. J Trop Med Hyg 54: 116–121

Brust RA (1990) Oviposition behavior of natural populations of *Culex tarsalis* and *Culex restuans* (Diptera: Culicidae) in artificial pools. J Med Entomol 27: 248–255

Bugher JC, Taylor M (1949) Radiophosphorous and radiostrontium in mosquitoes. Preliminary report. Science 110: 146–147

Callahan, JL, Morris CD (1987) Habitat characteristics of *Coquillettidia perturbans* in central Florida. J Am Mosq Control Assoc 3: 176–180

Cambournac FJC (1939) A method for determining the larval *Anopheles* population and its distribution in rice fields and other breeding places. Ric Malar 18: 17–22

Campos RE, Garcia JJ (1993) A simple apparatus to separate mosquito larvae from field collected samples. J Am Mosq Control Assoc 9: 100–101

Campos, RE, Fernández LA, Sy VE (2004) Study of the insects associated with the floodwater mosquito *Ochlerotatus albifasciatus* (Diptera: Culicidae) and their possible predators in Buenos Aires Province, Argentina. Hydrobiologia 524: 91–102

Carlson D (1971) A method for sampling larval and emerging insects using an aquatic black light trap. Can Entomol 103: 1365–1369

Carron A, Duchet C, Gaven B, Lagneau C (2003) An easy field method for determining the abundance of culicid larval instars. J Am Mosq Control Assoc 19: 353–360

Ceccato P, Connor SJ, Jeanne I, Thomson MC (2005) Application of Geographical Information Systems and Remote Sensing technologies for assessing and monitoring malaria risk. Parassitol 47: 81–96

Chadee DD (2003) Surveillance for the dengue vector *Aedes aegypti* in Tobago, West Indies. J Am Mosq Control Assoc 19: 199–205

Chadee DD (2004) Key premises, a guide to *Aedes aegypti* (Diptera: Culicidae) surveillance and control. Bull Entomol Res 94: 201–207

Chadee DD, Rahaman A (2000). Use of water drums by humans and *Aedes aegypti* in Trinidad. J Vector Ecol 25: 28–35

Chambers DM, Steelman CD, Schilling PE (1979) Mosquito species and densities in Louisiana ricefields. Mosquito News 39: 658–668

Chambers DM, Young LF, Hill HS (1986) Backyard mosquito larval habitat availability and use as influenced by census tract determined resident income levels. J Am Mosq Control Assoc 2: 539–544

Chan KL (1985) Methods and indices used in the surveillance of dengue vectors. Mosq-Borne Dis Bull 1: 79–88

Chan KL, Ho BC, Chan YC (1971a) *Aedes aegypti* (L.) and *Aedes albopictus* (Skuse) in Singapore City. 2. Larval habitats. Bull World Health Organ 44: 629–363

Chan YC, Chan KL, Ho BC (1971b) *Aedes aegypti* (L.) and *Aedes albopictus* (Skuse) in Singapore City. 1. Distribution and density. Bull World Health Organ 44: 617–627

Chandler JA, Highton RB (1975) The succession of mosquitoes species in ricefields in the Kisumu area of Kenya, and their possible control. Bull Entomol Res 65: 295–302

Chandler JA, Highton RB (1976) The breeding of *Anopheles gambiae* Giles (Diptera; Culicidae) in rice fields in the Kisumu area of Kenya. J Med Entomol 113: 211–215

Chang HT (1946) Studies on the use of fluorescent dyes for marking *Anopheles quadrimaculatus* Say. Mosquito News 6: 122–125

Chapman DG (1955) Population estimation based on change of composition caused by a selective removal. Biometrika 42: 279–290

Chareonsook O, Sethaputra S, Singklang K, Yawwa T, Purahong S, Suwankiri P (1985) Prevalence of *Aedes* mosquito in big cement jars and rain water tanks. Commun Dis J 11: 247–263 (In Thai, English Summary)

Chareonviriyaphap T, Akratanakul P, Nettanomsak S, Huntamai S (2003) Larval habitats and distribution patterns of *Aedes aegypti* (Linnaeus) and *Aedes albopictus* (Skuse) in Thailand. Southeast Asian J Trop Med Public Health 34: 529–535

Chavasse DC, Lines JD, Ichimori K, Marijani J (1995) Mosquito control in Dar es Salaam. I. Assessment of *Culex quinquefasciatus* breeding sites prior to intervention. Med Vet Entomol 9: 141–146

Chevone BI, Peters TM (1969) Retardation of larval development of *Aedes aegypti* (L.) by the vital dye, nile blue sulphate (A). Mosquito News 29: 243–251

Christensen JB, Washino RK (1978) Sampling larval mosquitoes in a rice field: a comparison of three techniques. Proc California Mosq Vector Control Assoc 46: 46

Christie M (1954) A method for the numerical study of larval populations of *Anopheles gambiae* and other pool-breeding mosquitoes. Ann Trop Med Parasitol 48: 271–276

Christie M (1958). Predation on larvae of *Anopheles gambiae* Giles. J Trop Med Hyg 61: 168–176

Chubachi R (1976) The efficiency of the dipper in sampling of mosquito larvae and pupae under differing conditions. Sci. Rep. Tohoku Univ., Ser. IV (Biol.), 37: 145–149

Chubachi R (1979) An analysis of the generation-mean life table of the mosquito, *Culex tritaeniorhynchus summorosus*, with particular reference to population regulation. J Anim Ecol 48: 681–702

Claborn DM, Hshieh PB, Roberts DR, Klein TA, Zeichner BC, Andre RG (2002) Environmental factors associated with larval habitats of malaria vectors in northern Kyunggi province, Republic of Korea. J Am Mosq Control Assoc 18: 178–185

Clarke KC, McLafferty SL, Tempalski BJ (1996) On epidemiology and Geographic Information Systems: a review and discussion of future directions. Emerg Infect Dis 2: 85–92

Coetzee M, Craig M, le Sueur D (2000) Distribution of African malaria mosquitoes belonging to the *Anopheles gambiae* complex. Parasitol Today 16: 74–77

Coll C, Casselles V, Sobrino JA, Valor E (1994) On the atmospheric dependence of the split-window equation for land surface temperature. Int J Remote Sensing 15: 105–122

Collins JN, Resh VH (1989) Guidelines for ecological control of mosquitoes in non-tidal wetlands of the San Francisco Bay area. California Mosquito and Vector Control Association, Sacramento, California

Corbett JD, O'Brien RF (1997) The Spatial Characterization Tool - Africa. Texas A&M University, Texas

Costanzo KS, Mormann K, Juliano SA (2005) Asymmetrical competition and patterns of abundance of *Aedes albopictus* and *Culex pipiens* (Diptera: Culicidae). J Med Entomol 42: 559–570

Cousserans J, Gabinaud A, Simonneau P, Sinègre G (1969) Les bases écologiques de la démoustification: méthodes de réalisation et d'utilisation de la carte phytoécologique. Vie et Mielieu C 20: 1–20

Croset H, Papierok B, Rioux JA, Gabinaud A, Cousserans J, Arnaud D (1976) Absolute estimates of larval populations of culicid mosquitoes: comparison of 'capture–recapture', 'removal' and 'dipping' methods. Ecol Entomol 1: 251–256

Cross ER, Newcomb WW, Tucker CJ (1996) Use of weather data and remote sensing to predict the geographic and seasonal distribution of *Phlebotomus papatasi* in southwest Asia. Am J Trop Med Hyg 54: 530–536

Curtis GA, Frank JH (1981) Establishment of *Aedes vexans* in citrus groves in southeastern Florida. Environ Entomol 10: 180–182

Dale PER, Morris CD (1996) *Culex annulirostris* breeding sites in urban areas: using remote sensing and digital image analysis to develop a rapid predictor of potential breeding areas. J Am Mosq Control Assoc 12: 316–320

Dale, PER, Ritchie SA, Territo BM, Morris CD, Muhar A, Kay BH (1998) An overview of remote sensing and GIS for surveillance of mosquito vector habitats and risk assessment. J Vector Ecol 23: 54–61

Darrow EM (1949) Factors in the elimination of the immature stages of *Anopheles quadrimaculatus* Say in a water level fluctuation cycle. Am J Hyg 50: 207–235

David J (1963) Les effets physiologiques de l'intoxication par le bleu de methylene sur la Drosophile. Bull Biol Fr Belg 97: 515–530

Davidson G, Lane J (1981) Distribution maps for the *Anopheles gambiae* complex. Mosquito Studies at the London School of Hygiene and Tropical Medicine. Progress Report 40: 2–4

Davis DE (1944) Larval habitats of some Brazilian mosquitoes. Rev Entomol Rio de Janeiro 15: 221–235

Dister SW, Fish D, Bros SM, Frank DH, Wood BL (1997) Landscape characterization of peridomestic risk for lyme disease using satellite imagery. Am J Trop Med Hyg 57: 687–692

Diuk-Wasser MA, Bagayoko M, Sogoba N, Dolo G, Touré MB, Traoré SF, Taylor CE (2004) Mapping rice field anopheline breeding habitats in Mali, West Africa, using Landsat ETM + sensor data. Int J Remote Sensing 25: 359–376

Dixon RO, Brust RA (1972) Mosquitoes of Manitoba. III. Ecology of larvae in the Winnipeg area. Can Entomol 104: 961–968

Dixon WJ, Massey FJ (1969) Introduction to Statistical Analysis, McGraw-Hill Book Co., New York

Doran BR, Lewis DJ (2003) The species composition and seasonal distribution of mosquitoes in vernal pools in suburban Montreal, Quebec. J Am Mosq Control Assoc 19: 339–346

Dorer RE, Carter RG, Bickley WE (1950) Observations on the pupae of *Mansonia perturbans* (Walk.) in Virginia. Proc New Jers Mosq Exterm Assoc 37: 110–113

Downing JA (1986) Spatial heterogeneity: evolved behaviour or mathematical artefact? Nature 323: 255–257

Downing JD (1977) A comparison of the distribution of *Aedes canadensis* larvae within woodland pools using the cylindrical sampler and the standard pint dipper. Mosquito News 37: 362–366

Downs WG, Pittendrigh CS (1946) Bromeliad malaria in Trinidad, British West Indies. Am J Trop Med 26: 47–66

Drake CM (1983) Spatial distribution of chironomid larvae (Diptera) on leaves of the bullrush in a chalk stream. J Anim Ecol 52: 421–437

Driggers DP, Cranford HB, Parsons RE, Desrosiers RE, Kardatzke JT (1978) Development and evaluation of the army improved portable immature mosquito concentrator system. Mosquito News 38: 480–485

Duhrkopf RE, Benny H (1990) Differences in the larval alarm reactions in populations of *Aedes aegypti* and *Aedes albopictus*. J Am Mosq Control Assoc 6: 411–414

Dunn LH (1928) Further observations on mosquito breeding in tree-holes and crabholes. Bull Entomol Res 18: 247–250

Dutta P, Khan SA, Sharma CK, Doloi P, Hazarika NC, Mahanta J (1998) Distribution of potential dengue vectors in major townships along the national highways and trunk roads of northeast India. Southeast Asian J Trop Med Public Health 29: 173–176

Dye C (1983) Insect movement and fluctuations in insect population size. Antenna 7: 174–178

Earle HH (1956). Automatic device for the collection of aquatic specimens. J Econ Entomol 49: 261–262

Elliott JM (1977) Some Methods for the Statistical Analysis of Samples of Benthic Invertebrates. Freshwater Biological Association Scientific Publication No. 25

Elnaiem DA, Schorscher J, Bendall A, Obsomer V, Osman ME, Mekkawi AM, Connor SJ, Ashford RW, Thomson MC (2003) Risk mapping of visceral leishmaniasis: the role of local variation in rainfall and altitude on the presence and incidence of kala-azar in eastern Sudan. Am J Trop Med Hyg 68: 10–17

Enfield MA, Pritchard G (1977a) Methods for sampling immature stages of Aedes spp. (Diptera: Culicidae) in temporary ponds. Can Entomol 109: 1435–1444

Enfield MA, Pritchard G (1977b). Estimation of population size and survival of immature stages of four species of Aedes (Diptera: Culicidae) in a temporary pond. Can Entomol 109: 1425–1434

Evans BR (1962) Survey for possible mosquito breeding in crawfish holes in New Orleans, Louisiana. Mosquito News 22: 255–257

Faber DJ (1981) A light-trap to sample littoral and limnetic regions of lakes. Verh Int Ver Limnol 21: 776–781

Fanara DM (1973) An efficient aquatic sample concentrator. Mosquito News 33: 603–604

Fanara DM, Mulla MS (1974) Population dynamics of larvae of Culex tarsalis Coquillett and Culiseta inornata (Williston) as related to flooding and temperature of ponds. Mosquito News 34: 98–104

Farlow JE, Breaud TP, Steelman CD, Schilling PE (1978). Effects of the insect growth regulator diflubenzuron on non-target aquatic populations in a Louisiana intermediate marsh. Environ Entomol 7: 199–204

Fauran P (1996) Lutte contre Aedes aegypti en Martinique. Rapport des études entomologiques. Bull Soc Pathol Exot 89: 161–162

Ferreira RLM, Oliveira AF, Pereira ES, Hamada N (2001) Occurence of larval Culicidae (Diptera) in water retained in Aquascypha hydrophora (Fungus: Stereaceae) in Central Amazônia, Brazil. Mem Inst Oswaldo Cruz 96: 1165–1167

Fischer S, Schweigmann N (2004) Culex mosquitoes in temporary urban rain pools: Seasonal dynamics and relation to environmental variables. J Vector Ecol 29: 365–373

Fischer S, Marinone MC, Schweigmann N (2002) *Ochlerotatus albifasciatus* in rain pools of Buenos Aires: seasonal dynamics and relation to environmental variables. Mem Inst Oswaldo Cruz 97: 767–773

Fish D, Carpenter SR (1982) Leaf litter and larval mosquito dynamics in tree-hole ecosystems. Ecology 63: 283–288

Fish D, Joslyn DJ (1984) Larval population estimates of *Aedes communis* using Giemsa marking. Mosquito News 44: 565–567

Fisher IJ, Bradshaw WE, Kammeyer C (1990) Fitness and its correlates assessed by intra- and interspecific interactions among tree-hole mosquitoes. J Anim Ecol 59: 819–829

Flasse SP, Verstraete MM (1994). Monitoring the environment with vegetation indices: comparison of NDVI and GEMI using AVHRR data over Africa. In: Veroustraete F, Ceulemans R (eds) Vegetation, Modelling and Climate Change Effects. SPB Academic, The Hague, Netherlands, pp 107–135

Fleetwood SC, Steelman CD, Schilling PE (1978) The effects of waterfowl management practices on mosquito abundance and distribution in Louisiana coastal marshes. Mosquito News 38: 105–111

Fleetwood SC, Chambers MD, Terracina L (1981) An effective and economical mapping system for monitoring *Psorophora columbiae* in rice and fallow fields in southwestern Louisiana. Mosquito News 41: 174–177

Focks DA (2003) A review of entomological sampling methods and indicators for dengue vectors. World Health Organization TDR, Geneva

Focks DA, Chadee DD (1997) Pupal survey: an epidemiologically significant surveillance method for *Aedes aegypti*: an example using data from Trindad. Am J Trop Med Hyg 56: 159–167

Focks DA, Dame DA, Cameron AL, Boston MD (1980) Predator-prey interaction between insular populations of *Toxorhynchites rutilus rutilus* and *Aedes aegypti*. Environ Entomol 9: 37–42

Focks DA, Sackett SR, Bailey DL, Dame DA (1981) Observations on container-breeding mosquitoes in New Orleans, Louisiana, with an estimate of the population density of *Aedes aegypti* (L.). Am J Trop Med Hyg 30: 1329–1335

Focks DA, Haile DG, Daniels E, Mount GA (1993a). Dynamic life table model for *Aedes aegypti* (Diptera:Culicidae): Analysis of the literature and model development. J Med Entomol 30: 1003–1017

Focks DA, Haile DG, Daniels E, Mount GA (1993b). Dynamic life table model for *Aedes aegypti* (Diptera:Culicidae): Simulation results and validation. J Medl Entomol 30: 1018–1028

Focks DA, Daniels E, Haile, DG, Keesling, JE (1995). A simulation model of the epidemiology of urban dengue fever: Literature analysis, model development, preliminary validation, and samples of simulation results. Am J Trop Med Hyg 53: 489–506

Focks DA, Brenner RJ, Hayes J, Daniels E (2000) Transmission thresholds for dengue in terms of *Aedes aegypti* pupae per person with discussion of their utility in source reduction efforts. Am J Trop Med Hyg 62: 11–18

Fontanarrosa MS, Marinone MC, Fischer S, Orellano PW, Schweigmann N (2000) Effects of flooding and temperature on *Aedes albifasciatus* development time and larval density in two rain pools at Buenos Aires University City. Mem Inst Oswaldo Cruz 95: 787–793

Forattini OP, Kakitani I, Massad E, Marucci D (1993) Studies on mosquitoes (Diptera: Culicidae) and anthropic environment. 2-Immature stages research at a rice irrigation system location in South-Eastern Brazil. Rev Saúde Pública 27: 227–236

Forattini OP, Kakitani I, Massad E, Marucci D (1994*a*) Studies on mosquitoes (Diptera: Culicidae) and anthropic environment. 5-Breeding of *Anopheles albitarsis* in flooded rice fields in South-Eastern Brazil. Rev Saúde Pública 28: 329–331

Forattini OP, Kakitani I, Massad E, Marucci D (1994*b*). Studies on mosquitoes (Diptera: Culicidae) and anthropic environment. 6-Breeding in empty conditions of rice fields in South-Eastern Brazil. Rev Saúde Pública 28: 395–399

Forattini OP, Kakitani I, Massad E, Marucci D. (1995*a*). Studies on mosquitoes (Diptera: Culicidae) and anthropic environment. 9-Synanthropy and epidemiological vector role of *Aedes scapularis* in south-eastern Brazil. Rev Saúde Pública 29: 199–207

Forattini OP, Kakitani I, Massad E, Marucci D (1995*b*). Studies on mosquitoes (Diptera: Culicidae) and anthropic environment. 10-Survey of adult behaviour of *Culex nigripalpus* and other species of *Culex (Culex)* in south-eastern Brazil. Rev Saúde Pública 29: 271–278

Forattini OP, Kakitani I, Sallum MAM, de Rezende L (1997) Produtividade de criadouro de *Aedes albopictus* em ambiente urbano. Rev Saúde Pública 31: 545–555

Fox I, Specht P (1988) Evaluating ultra-low volume ground applications of malathion against *Aedes aegypti* using landing counts in Puerto Rico, 1980–84. J Am Mosq Control Assoc 4: 163–167

Frank JH, Curtis GA, Evans HT (1976) On the bionomics of bromeliad-inhabiting mosquitoes. I. Some factors influencing oviposition by *Wyeomyia vanduzeei*. Mosquito News 36: 25–30

Frank JH, Curtis GA, Evans HT (1977) On the bionomics of bromeliad-inhabiting mosquitoes. II. The relationship of bromeliad size to the numbers of immature *Wyeomyia vanduzeei* and *Wy. medioalbipes*. Mosquito News 37: 180–192

Frank JH, Lounibos LP (eds) (1983). Phytotelmata: Terrestrial Plants as Hosts for Aquatic Insect Communities. Plexus Publishing Inc., New Jersey

Gast RT, Landin M (1966) Adult boll weevils and eggs marked with dye fed in larval diet. J Econ Entomol 59: 474–475

Geevarghese G, Kaul HN, Dhanda V (1975) Observations on the re-establishment of *Aedes aegypti* population in Poona city and suburb, Maharashtra State, India, Indian J Med Res 63: 1155–1163

Germain M, Francy DB, Monath TP, Ferrara L, Bryan J, Salaun J-J, Heme G, Renaudet J, Adam C, Digoutte J-P (1980) Yellow fever in The Gambia, 1978–1979; Entomological aspects and epidemiological correlations. Am J Trop Med Hyg 29: 929–940

Getis A, Morrison AC, Gray K, Scott TW (2003) Characteristics of the spatial pattern of the dengue vector, *Aedes aegypti*, in Iquitos, Peru. Am J Trop Med Hyg 69: 494–505

Gettman AD, Hall DW (1989) A modification of scrap automobile tires for field studies of artificial container-breeding mosquitoes. J Am Mosq Control Assoc 5: 439

Gillett JD (1946) Notes on the subgenus *Coquillettidia* Dyar (Diptera: Culicidae). Bull Entomol Res 36: 425–438

Gionar YR, Rusmiarto S, Susapto D, Bangs MJ (1999) Use of a funnel trap for collecting immature *Aedes aegypti* and copepods from deep wells in Yogya-karta, Indonesia. J Am Mosq Control Assoc 15: 576–580

Goettel MS, Toohey MK, Engber BR, Pillai JS (1981) A modified garden sprayer for sampling crab hole water. Mosquito News 41: 789–790

Goodwin MH, Eyles DE (1942) Measurements of larval populations of *Anopheles quadrimaculatus* Say. Ecology 23: 376

Gooley BR, Lesser FH (1975) Formulation and interpretation of aerial mapping of breeding areas. Proc New Jers Mosq Control Assoc 63: 106–110

Gould DJ, Mount GA, Scanlon JE Sullivan MF, Winter PE (1971) Dengue control on an island in the gulf of Thailand. I. Results of an *Aedes aegypti* control program. Am J Trop Med Hyg 20: 705–714

Gozhenko VA (1978) Biotopes and times of development of *Mansonia richiardii* (Ficalbi) 1889 in the conditions of the Ukrainian steppes. Medskaya Parazi-tol., 47: 36–40 (In Russian)

Graham JE, Bradley IE (1962) The effects of species on density of mosquito larval populations in Salt Lake County, Utah. Mosquito News 22: 239–247

Graham JE, Bradley IE (1969) Mode of action of factors responsible for increases in *Culex tarsalis* Coq. populations in Utah. Mosquito News 29: 678–687

Graham JE, Collett GC (1969) Ten years of surveillance for western equine en-cephalitis in Utah. Mosquito News 29: 451–456

Greenstone MH (1979) A sampling device for aquatic arthropods active at the wa-ter surface. Ecology 60: 642–644

Grillet EM (2000) Factors associated with distribution of *Anopheles aquasalis* and *Anopheles oswaldoi* (Diptera: Culicidae) in a malarious area, northeastern Venezuela. J Med Entomol 37: 231–238

Grimes D, Bonifacio R, Dugdale G, Diop M (1995) Flow forecasting of a semi-arid catchment with METEOSAT and NOAA AVHRR data. Proceedings of the Meteorological Satellite Data Users' Conference, Winchester, 4–8 September 1995

Guille G (1973) Mode d'alimentation de la larve de *Coquillettidia (Coquillettidia) richiardii* (Diptera: Culicidae). Bull Biol Fr Belge 107: 265–269

Guille G (1975). Recherches éco-éthologiques sur *Coquillettidia (Coquillet-tidia) richiardii* (Ficalbi), 1889 (Diptera-Culicidae) du littoral méditerranéen Français, I.—Techniques d'étude et morphologie. Ann Sci Nat Zool Paris 17: 229–272

Hagmann LE (1953) Biology of '*Mansonia perturbans*' (Walker). Proc New Jers Mosq Exterm Assoc 40: 20–22

Hagstrum DW (1971*a*). Evaluation of the standard pint dipper as a quantitative sampling device for mosquito larvae. Ann Entomol Soc Am 64: 537–540

Hagstrum DW (1971*b*). A model of the relationship between densities of larvae and adults in an isolated population of *Culex tarsalis* (Diptera: Culicidae). Ann Entomol Soc Am 64: 1074–1077

Hamad AA, Nugud AEHD, Arnot DE, Giha HA, Abdel-Muhsin AMA, Satti GMH, Theander TG, Creasey AM, Babiker HA, Elnaiem DEA (2002) A marked seasonality of malaria transmission in two rural sites in eastern Sudan. Acta Trop 83: 71–82

Hamon J, Pichon G, Cornet M (1971) La transmission du virus amaril en Afrique occidentale. Ecologie, répartition, fréquence et contrôle des vecteurs, et observations concernant l'épidémiologie de la fièvre jaune. Cah ORSTOM sér. Entomol Méd Parasitol 9: 3–60

Hansen MH, Hurwitz WN, Madow WG (1953). Sample Survey Methods and Theory, Vol. 1. John Wiley & Sons, London

Hanski I (1982) On patterns of temporal and spatial variation in animal populations. Ann Zool Fenn 19: 21–37

Hanski I (1987) Cross-correlation in population dynamics and the slope of spatial variance-mean regressions. Oikos 50: 148–151

Haramis LD (1984) *Aedes triseriatus*: A comparison of density in tree holes vs. discarded tires. Mosquito News 44: 485–489

Hard JJ, Bradshaw WE, Malarkey DJ (1989) Resource- and density-dependent development in tree-hole mosquitoes. Oikos 54: 137–144

Harrison BA, Callahan MC, Watts DM, Panthusiri L (1982) An efficient floating larval trap for sampling *Aedes aegypti* populations. J Med Entomol 19: 722–727

Hatfield LD, Riner JL, Norment BR (1985) A dredge sampler for mosquito larvae. J Am Mosq Control Assoc 1: 372–373

Hay SI, Tucker CJ, Rogers DJ, Packer MJ (1996) Remotely sensed surrogates of meteorological data for the study of the distribution and abundance of arthropod vectors of disease. Ann Trop Med Parasitol 90: 1–19

Hay SI, Packer MJ, Rogers DJ (1997) The impact of remote sensing on the study and control of invertebrate intermediate host and vectors for disease. Int J Remote Sensing 18: 2899–2930

Hay SI, Snow RW, Rogers DJ (1998). From predicting mosquito habitat to malaria seasons using remotely sensed data: practice, problems and perspectives. Parasitol Today 14: 306–313

Hayes RO, Maxwell EL, Mitchell CJ, Woodzick TL (1985) Detection, identification, and classification of mosquito larval habitats using remote sensing scanners in earth-orbiting satellites. Bull World Health Organ 63: 361–374

Heathcote OHU (1970) Japanese encephalitis in Sarawak: studies on juvenile mosquito populations. Trans R Soc Trop Med Hyg 64: 483–488

Heron RJ (1968) Vital dyes as markers for behavioral and population studies of the larch sawfly, *Pristophora erichsonii* (Hymenoptera: Tenthredinidae). Can Entomol 100: 470–475

Hess AD (1941) New Limnological Sampling Equipment. Special Publications of the Limnological Society of America 6

Hess AD, Hall TF (1943) The intersection line as a factor in anopheline ecology. J Natn Malar Soc 2: 93–98

Hii JLK (1979) Evaluation of an *Aedes* control trial using the one-larva-per-container method in Labuan island, Sabah, Malaysia. Jap J Sanit Zool 30: 127–134

Hiriyan J, Tyagi BK (2004) Cocoa pod (*Theobroma caco*) – a potential breeding habit (sic) of *Aedes albopictus* in dengue-sensitive Kerala State, India. J Am Mosq Control Assoc 20: 323–325

Hocking B (1953) Notes on the activities of *Aedes* larvae. Mosquito News 13: 77–81

Holck AR, Puissegur WJ, Meek CL (1988) Mosquito productivity of crawfish ponds and irrigation canals in Louisiana ricelands. J Am Mosq Control Assoc 4: 82–84

Hopkins GHE (1952) Mosquitoes of the Ethiopian Region. 1. Larval Bionomics of Mosquitoes and the Taxonomy of Culicine Larvae. British Museum (Natural History), London

Horsfall WR (1942) Biology and Control of Mosquitoes in the Rice Area. Bull Agric Exp Stn Arkansas 427

Horsfall WR (1946) Area sampling of populations of larval mosquitoes in rice fields. Entomol News 57: 242–244

Hugh-Jone, M (1989) Applications of remote sensing to the identification of parasites and disease vectors. Parasitol Today 5: 244–251

Husbands RC (1967) A subsurface light trap for sampling aquatic insect populations. California Vector Views 14: 81–82

Husbands RC (1969) An improved technique of collecting mosquito larvae for control operations. California Vector Views 16: 67–69

Hutchings RSG (1994) Palm bract breeding sites and their exploitation by *Toxorhynchites (Lynchiella) haemorrhoidalis haemorrhoidalis* (Diptera: Culicidae) in an upland forest of the central Amazon. J Med Entomol 31: 186–191

Ikemoto T (1976) A method, using a standard quadrat device and a small suction pump for sampling of the immature stages of mosquitoes in rice fields. Jap J Sanit Zool 27: 153–156

Ikemoto T (1978) Studies on the spatial distribution pattern of larvae of the mosquito, *Anopheles sinensis*, in rice fields. Res Popul Ecol 19: 237–249

Imai C, Ikemoto T, Takagi M, Yamugi H, Pohan W, Hasibuan H, Sirait H, Panjaitan W (1988) Ecological studies of *Anopheles sundaicus* larvae in a coastal village of North Sumatra, Indonesia, I. Topography, land use, and larval breeding. Jap J Sanit Zool 39: 293–300

Ingram A (1912) Notes on the mosquitos observed at Bole, Northern Territories, Gold Coast. Bull Entomol Res 3: 73–78

Istock CA, Zisfeln J, Vavra KJ (1976a) Ecology and evolution of the pitcherplant mosquito: 2. The substructure of fitness. Evolution 30: 535–547

Istock CA, Vavra KJ, Zimmer H. (1976*b*). Ecology and evolution of the pitcherplant mosquito: 3. Resources tracking by a natural population. Evolution 30: 548–557

Iwao S (1968) A new regression method for analyzing the aggregation pattern of animal populations. Res Popul Ecol 10: 1–20

Iwao S (1970) Problems of spatial distribution in animal population ecology. In: Patil GP, Pielou EC, Water WE (eds) Random Counts in Biomedical and Social Sciences, Vol. 2. Pennsylvania State University Press, University Park, pp. 117–149

Iwao S (1975) A new method of sequential sampling to classify populations relative to a critical density. Res Popul Ecol 16: 281–288

Iwao S, Kuno E (1971) An approach to the analysis of aggregation pattern in biological populations. In: Patil GP, Pielou EC, Water WE (eds) Random Counts in Biomedical and Social Sciences, Vol. 2. Pennsylvania State University Press, University Park, pp. 461–513.

Jaal Z, Macdonald WW (1993) The ecology of anopheline mosquitos in northwest coastal Malaysia: larval habitats and adult seasonal abundance. Southeast Asian J Trop Med Public Health 24: 522–529

Jacob BG, Arheart KL, Griffith DA, Mbogo CM, Githeko AK, Regens JL, Githure JI, Novak R, Beier JC (2005) Evaluation of environmental data for identification of *Anopheles* larval habitats in Kisumu and Malindi, Kenya. J Med Entomol 42: 751–755

Janousek TE, Lowrie RC (1989) Vector competency of *Culex quinquefasciatus* (Haitian strain) following infection with Wuchereria bancrofti. Trans R Soc Trop Med Hyg 83: 679–680

Jolivet P, Yi BG, Ree HI, Lee KW (1974) Application of phytoecological cartography to detect the mosquito breeding places on an island in the Yellow Sea, Korea. WHO/VBC/74.485, 14 pp. (mimeographed)

Joslyn DJ, Conrad LB (1984) Larval density estimates of *Aedes sollicitans* (Walker) using the Giemsa self-marker. Proc New Jers Mosq Control Assoc 71: 72–73

Joy JE (2004) Larval mosquitoes in abandoned tire pile sites from West Virginia. J Am Mosq Control Assoc 20: 12–17

Joy JE, Clay JT (2002) Habitat use by larval mosquitoes in West Virginia. Am Midl Nat 148: 363–375

Joy JE, Hildreth-Whitehair A (2000) Larval habitat characterization for *Aedes triseriatus* (Say), the mosquito vector of LaCrosse encephalitis in West Virginia. Wilderness Environ Med 11: 79–83

Joy JE, Hanna AA, Kennedy BA (2003) Spatial and temporal variation in the mosquitoes (Diptera: Culicidae) inhabiting waste tires in Nicholas County, West Virginia. J Med Entomol 40: 73–77

Jupp DLB, Heggan SJ, Mayo KK, Kenwal SW (1985) The BRIAN handbook - an introduction to Landsat and the BRIAN system for users. CSIRO Division of Land and Water Resources, Natural Resources Series, 3

Kaur R, Reuben R (1981) Studies of density and natural survival of immatures of *Anopheles stephensi* Liston in wells in Salem (Tamil Nadu). Ind J Med Res 73 (Suppl.): 129–135

Kay BH, Cabral CP, Araujo DB, Ribeiro ZM, Braga PH, Sleigh AC (1992) Evaluation of a funnel trap for collecting copepods and immature mosquitoes from wells. J Am Mosq Control Assoc 8: 372–375

Kay BH, Ryan PA, Russell BM, Holt JS, Lyons SA, Foley PN (2000) The importance of subterranean mosquito habitat to arbovirus vector control strategies in North Queensland, Australia. J Med Entomol 37: 846–853

Kay BH, Ryan PA, Lyons SA, Foley PN, Pandeya N, Purdie D (2002) Winter intervention against *Aedes aegypti* (Diptera: Culicidae) larvae in subterranean habitats slows surface recolonization in summer. J Med Entomol 39: 356–361

Kelker GH (1940) Estimating deer populations by a differential hunting loss in the sexes. Proc Utah Acad Sci 17: 65–69

Kent RB, Sutherland DJ (1977) A compact map-panel system for large acreage. Mosquito News 37: 765–767

Kitron U (1998) Landscape ecology and epidemiology of vector-borne diseases: tools for spatial analysis. J Med Entomol 35: 435–445

Kitron UD, Webb DW, Novak RJ (1989) Oviposition behavior of *Aedes triseriatus* (Diptera: Culicidae): Prevalence, intensity, and aggregation of eggs in oviposition traps. J Med Entomol 26: 462–467

Kitron U, Kazmierczak JJ (1997) Spatial analysis of the distribution of Lyme disease in Wisconsin. Am J Epidemiol 145: 558–566

Kitron U, Otieno LH, Hungerford LL, Odulaja A, Brigham WU, Okello OO, Joselyn M, Mohamed-Ahmed MM, Cook E (1996) Spatial analysis of the distribution of tsetse flies in the Lambwe valley, Kenya, using Landsat TM satellite imagery and GIS. J Anim Ecol 65: 371–380.

Kitron U, Michael J, Swanson J, Haramis L (1997) Spatial analysis of the distribution of LaCrosse encephalitis in Illinois, using a geographic information system and local and global spatial statistics. Am J Trop Med Hyg 57: 469–475

Kittayapong P, Strickman D (1993) Distribution of container-inhabiting *Aedes* larvae (Diptera: Culicidae) at a dengue focus in Thailand. J Med Entomol 30: 601–606

Klinkenberg E, Takken W, Huibers F, Touré YT (2003) The phenology of malaria mosquitoes in irrigated rice fields in Mali. Acta Trop 85: 71–82

Knight KL (1964) Quantitative methods for mosquito larval surveys. J Med Entomol 1: 109–115

Knowles R, Senior White R (1927) Malaria. Its Investigation and Control with Special Reference to Indian Conditions. Thacker, Spink & Co., Calcutta

Kobayashi M, Nihei N, Kurihara T (2002) Analysis of northern distribution of *Aedes albopictus* (Diptera: Culicidae) in Japan by geographical information system. J Med Entomol 39: 4–11

Kogan F (1995) AVHRR data for detection and analysis of vegetation stress. Proceedings of the Meteorological Satellite Data Users' Conference, Winchester, 4–8 September 1995

Kogan M, Herzog DC (eds) (1980) Sampling Methods in Soybean Entomology. Springer-Verlag, New York

Kono T (1953) On the estimation of insect populations by time unit collecting. Res Popul Ecol 2: 85–94 (In Japanese, English summary)

Kronenwetter-Koepel TA, Meece JK, Miler CA, Reed KD (2005) Surveillance of above- and below-ground mosquito breeding habitats in a rural midwestern community: baseline data for larvicidal control measures against West Nile virus vectors. Clin Med Res 3: 3–12

Kruger RM, Pinger RR (1981) A larval survey of the mosquitoes of Delaware county, Indiana. Mosquito News 41: 484–489

Kubota RL, de Brito M, Voltolini JC (2003) Método de varredura para exame de criadouros de vetores de dengue e febre amarela urbana. Rev Saúde Pública 37: 263–265

Kuno E (1991) Sampling and analysis of insect populations. Annu Rev Entomol 36: 285–304

Kuno E, Yamamoto S, Satomi H, Outi Y, Okada T (1963) On the assessment of insect populations in a large area of paddy field based on the negative binomial distribution. Proc Assoc Pl Prot Kyushu 9: 33–36 (In Japanese, English summary)

Kutz FW, Wade TG, Pagac BB (2003) A geospatial study of the potential of two exotic species of mosquitoes to impact the epidemiology of West Nile virus in Maryland. J Am Mosq Control Assoc 19: 190–198

Lacey LA, Lacey CM (1990) The medical importance of riceland mosquitoes and their control using alternatives to chemical insecticides. J Am Mosq Control Assoc 6 (Suppl. No. 2): 1–93

Laird M (1988) The Natural History of Larval Mosquito Habitats. Academic Press, London

Lamblin EF, Strahler AH (1994) Indicators in land cover change for change vector analysis in multi-temporal space at coarse spatial scales. Int J Remote Sensing, 15: 2099–2119

Lang JT, Ramos AC (1981) Ecological studies of mosquitoes in banana leaf axils on Central Luzon, Philippines. Mosquito News 41: 665–673

Lardeux F, Sechan Y, Loncke S, Deparis X, Cheffort J, Faaruia M (2002) Integrated control of peridomestic larval habitats of *Aedes* and *Culex* mosquitoes (Diptera: Culicidae) in atoll villages of French Polynesia. J Med Entomol 39: 493–498

Laurence BR (1960) The biology of two species of mosquito, *Mansonia africana* (Theobald) and *Mansonia uniformis* (Theobald), belonging to the subgenus *Mansonioides* (Diptera: Culicidae). Bull Entomol Res 51: 491–517

Lemenager DC, Bauer SD, Kauffman EE (1986) Abundance and distribution of immature *Culex tarsalis* and *Anopheles freeborni* in rice fields of the Sulter-Yuba M. A. D.: 1. Initial sampling to detect major mosquito producing rice fields, augmented by adult light trapping. Proc California Mosq Vector Control Assoc 53: 101–104

Le Pelley RH (1935) Observations on the control of insects by hand-collection. Bull Entomol Res 26: 533–541

Lesser CR (1977) A method to estimate populations of mosquito larvae in shallow water. Mosquito News 37: 517–519

Le Sueur D (1991) The ecology, overwintering and population dynamics of the preimaginal stages of the *Anopheles gambiae* Giles complex (Diptera: Culicidae) in northern Transvaal, South Africa. Ph.D. thesis, University of Natal

Levine RS, Peterson AT, Benedict MQ (2004*a*). Geographic and ecologic distributions of the *Anopheles gambiae* complex predicted using a genetic algorithm. Am J Trop Med Hyg 70: 105–109

Levine RS, Peterson AT, Benedict MQ (2004*b*). Distribution of members of *Anopheles quadrimaculatus* Say s.l. (Diptera: Culicidae) and implications for their roles in malaria transmission in the United States. J Med Entomol 41: 607–613

Liebhold AM, Rossi RE, Kemp WP (1993) Geostatistics and geographic information systems in applied insect ecology. Annu Rev Entomol 38: 303–327

Lien JC, Lin TH, Huang HM (1993) Dengue vector surveillance and control in Taiwan. Trop Med (Nagasaki) 35: 269–276

Lindsay SW, Bayoh MN (2004) Mapping members of the *Anopheles gambiae* complex using climate data. Physiol Entomol 29: 204–209

Lindsay SW, Parson L, Thomas CJ (1998) Mapping the range and relative abundance of the two principle African malaria vectors, *Anopheles gambiae* sensu stricto and *An. arabiensis*, using climate data. Proc R Soc London Series B 265: 847–854

Linthicum KJ, Davies FG, Bailey CL, Kairo A (1983) Mosquito species succession in a dambo in an East African forest. Mosquito News 43: 464–470

Linthicum KJ, Davies FG, Bailey CL, Kairo A (1984) Mosquito species encountered in a flooded grassland dambo in Kenya. Mosquito News 44: 228–232

Linthicum KJ, Bailey CL, Davies FG, Kairo A (1985) Observations on the dispersal and survival of a population of *Aedes lineatopennis* (Ludlow) (Diptera: Culicidae) in Kenya. Bull Entomol Res 75: 661–670

Linthicum KJ, Bailey CL, Davies FG, Tucker CJ (1987) Detection of Rift Valley fever viral activity in Kenya by satellite remote sensing imagery. Science 235: 1656–1659

Linthicum KJ, Bailey CL, Tucker CJ, Mitchell KD, Logan TM, Davies FG, Kamau CW, Thande PC, Wagateh JN (1990) Application of polar-orbiting, metereological satellite data to detect flooding of Rift Valley Fever virus vector mosquito habitats in Kenya. Med Vet Entomol 4: 433–438

Lloyd M (1967) Mean crowding. J Anim Ecol 36: 1–30

Loomis EC (1959) The function of larval surveys in the California encephalitis surveillance program. Proc California Mosq Control Assoc 21: 64–67

Lopes J, Lozovei AL (1995) Ecologia de mosquitos (Diptera: Culicidae) em criadouros naturais e artificiais de área rural do norte do Estado do Paraná, Brasil. I - Coletas ao longo do leito de ribeirão. Rev Saúde Pública 29: 183–191

Lothrop HD, Resisen WK (1999) A geographical information system to manage mosquito and arbovirus surveillance and control data in the Coachella valley of California. J Am Mosq Control Assoc 15: 299–307

Lounibos LP (1979) Mosquitoes occurring in the axils of *Pandanus rabaiensis* Rendle on the Kenya coast. Cah ORSTOM sér Entomol Méd Parasitol 17: 25–29

Lounibos LP (1981) Habitat segregation among African tree-hole mosquitoes. Ecol Entomol 6: 129–154

Lounibos LP, Escher RL (1983) Seasonality and sampling of *Coquillettidia perturbans* (Diptera: Culicidae) in south Florida. Environ Entomol 12: 1087–1093

Lounibos LP, Rey JR, Frank JH (eds) (1985) Ecology of Mosquitoes: Proceedings of a Workshop. Florida Medical Entomology Laboratory, Vero Beach, Florida

Lounibos LP, Frank JH, Machado-Allison CE, Ocanto P, Navarro JC (1987) Survival, development and predatory effects of mosquito larvae in Venezuelan phytotelmata. J Trop Ecol 3: 221–242

Lutwama JJ, Mukwaya LG (1994) Studies on some of the physical and biological factors affecting the abundance of the *Aedes simpsoni* (Diptera: Culicidae) complex-larvae and pupae in plant axils. Bull Entomol Res 84: 255–263

Lutwama JJ, Mukwaya LG (1995) Estimates of mortalities of larvae and pupae of the *Aedes simpsoni* (Theobald) (Diptera: Culicidae) complex in Uganda. Bull Entomol Res 85: 93–99

Macdonald WW (1956) *Aedes aegypti* in Malaya. 1. Distribution and dispersal. Ann Trop Med Parasitol 50: 385–398

Macdonald WW, Rajapaksa N (1972) A survey of the distribution and relative prevalence of *Aedes aegypti* in Sabah, Brunei and Sarawak. Bull World Health Organ 46: 203–209

Mackey BE, Hoy JB (1978) *Culex tarsalis*: Sequential sampling as a means of estimating populations in Californian rice fields. J Econ Entomol 71: 329–334

Madder DJ, MacDonald RS, Surgeoner GA, Helson BV (1980) The use of oviposition activity to monitor populations of *Culex pipiens* and *Culex restuans* (Diptera: Culicidae). Can Entomol 112: 1013–1017

Mahadev PVM (1983) A case study of *Aedes aegypti* prevalence by settlement types in Dehu town group of Maharashtra state. Indian J Med Res 78: 537–546

Mahadev PVM, Geevarghese G (1978) Comparison of single larva and conventional pool methods for the study of *Aedes aegypti* in tyre dumps. Indian J Med Res 68: 934–939

Mahadev PVM, Dhanda V, Shetty PS (1978) *Aedes aegypti* (L.) in Maharashtra state—distribution and larval habitats. Indian. J Med Res 67: 562–580

Maire A (1982) Selectivity by six snow-melt mosquito species for larval habitats in Quebec subarctic string bogs. Mosquito News 42: 236–243

Malone JB, Huh OK, Fehler DP, Wilson PA, Wilensky DE, Holmes RA, Elmagdoub AI (1994). Temperature data from satellite imagery and the distribution of schistosomiasis in Egypt. Am J Trop Med Hyg 50: 714–722

Marcondes CB, Mafra H (2003) Nova técnica para o estudo da fauna de mosquitos (Diptera: Culicidae) em internódios de bambus, com resultados preliminares. Rev Soc Bras Med Trop 36: 763–764

Markos BG (1951) Distribution and control of mosquitoes in rice fields in Stanislaus County, California. J Natn Malar Soc 10: 233–247

Marten GG, Suarez MF, Astaeza R (1996) An ecological survey of *Anopheles albimanus* larval habitats in Colombia. J Vector Ecol 21: 122–131

Mattingly PF (1969) The Biology of Mosquito-Borne Disease. George Allen & Unwin, London

McDonald JL (1970) Preliminary results on experimental detection of *Mansonia uniformis* (Theob.) mosquito immatures. Mosquito News 30: 614–619

McLaughlin RE, Brown MA, Vidrine MF (1987) The sequential sampling of *Psorophora columbiae* larvae in rice fields. J Am Mosq Control Assoc 3: 423–428

McNeel TE (1931) A method for locating the larvae of the mosquito *Mansonia*. Science 74: 155

Meek CL, Olson JK (1991) Determination of riceland mosquito population dynamics. In: Heinricks EA, Miller TA (eds) Rice Insects: Management Strategies. Springer-Verlag, New York, pp. 107–139

Meisch MV, Anderson AL, Watson RL, Olson L (1982) Mosquito species inhabiting ricefields in five growing regions of Arkansas. Mosquito News 42: 341–346

Menon PKB, Rajagopalan PK (1979) Seasonal changes in the density and natural mortality of immature stages in the urban malaria vector, *Anopheles stephensi* (Liston) in wells in Pondicherry. Indian J Med Res 70 (Suppl.): 123–127

Minakawa N, Sonye G, Mogi M, Yan G (2004) Habitat characteristics of *Anopheles gambiae* s.s. larvae in a Kenyan highland. Med Vet Entomol 18: 301–305

Miura T (1980) Estimation of mosquito population size in confined natural breeding sites. Proc California Mosq Vector Control Assoc 48: 108–112

Miura T, Husbands RC, Reed DE (1970) Field evaluation of the concentrator-dipper technique for sampling mosquito larvae. Mosquito News 30: 448–452

Mogi M (1978) Population studies on mosquitoes in the rice field area of Nagasaki, Japan, especially on *Culex tritaeniorhynchus*. Trop Med 20: 173–263

Mogi M (1981) Population dynamics and methodology for biocontrol of mosquitoes. In: Laird M (ed) Biocontrol of Medical and Veterinary Pests. Praeger Science, New York, pp. 140–172

Mogi M (1984*a*) Distribution and overcrowding effects in mosquito larvae (Diptera: Culicidae) inhabiting taro axils in the Ryukyus, Japan. J Med Entomol 21: 63–68

Mogi M (1984*b*) Mosquito problems and their solution in relation to paddy rice production. Protect Ecol 7: 219–240

Mogi M, Mokry J (1980) Distribution of *Wyeomyia smithii* (Diptera: Culicidae) eggs in pitcher plants in Newfoundland, Canada. Trop Med 22: 1–12

Mogi M, Suzuki H (1983) The biotic community in the water-filled internode of bamboos in Nagasaki, Japan, with special reference to mosquito ecology. Jap J Ecol 33: 271–279

Mogi M, Wada Y (1973) Spatial distribution of larvae of the mosquito *Culex tritaeniorhynchus summorosus* in a rice field area. Trop Med 15: 69–83

Mogi M, Miyagi I, Okazawa T (1984) Population estimation by removal methods and observations on population-regulation factors of crab-hole mosquito immatures (Diptera: Culicidae) in the Ryukus, Japan. J Med Entomol 21: 720–726

Mogi M, Horio M, Miyagi I, Cabrera BD (1985) Succession, distribution, overcrowding and predation in the aquatic community in aroid axils, with special reference to mosquitoes. In: Lounibos LP, Rey JR, Frank JH (eds) Ecology of Mosquitoes: Proceedings of a Workshop. Florida Medical Entomology Laboratory, Vero Beach, Florida, pp. 95–119

Mogi M, Okazawa T, Miyagi I, Sucharit S, Tumrasvin W, Deesin T, Khamboonruang C (1986) Development and survival of anopheline immatures (Diptera: Culicidae) in rice fields in northern Thailand. J Med Entomol 23: 244–250

Mogi M, Memah V, Miyagi I, Toma T, Sembel DT (1995) Mosquito (Diptera: Culicidae) and predator abundance in irrigated and rain-fed rice fields in north Sulawesi, Indonesia. J Med Entomol 32: 361–367

Moloney JM, Skelly C, Weinstein P, Maguire M, Ritchie S (1998) Domestic *Aedes aegypti* breeding site surveillance: limitations of remote sensing as a predictive surveillance tool. Am J Trop Med Hyg 59: 261–264

Moncayo AC, Edman JD, Finn JT (2000) Application of geographic information technology in determining risk of eastern equine encephalomyelitis virus transmission. J Am Mosq Control Assoc 16: 28–35

Montgomery B, Ritchie SA (2002) Roof gutters: a key container for *Aedes aegypti* and *Ochlerotatus notoscriptus* (Diptera: Culicidae) in Australia. Am J Trop Med Hyg 67: 244–246

Montgomery BL, Ritchie SA, Hart AJ, Long SA, Walsh ID (2004) Subsoil drain sumps are a key container for *Aedes aegypti* in Cairns, Australia. J Am Mosq Control Assoc 20: 365–369

Moore CG, Cline BL, Ruiz-Tibén E, Lee D, Romney-Joseph H, Rivera-Correra E (1978) *Aedes aegypti* in Puerto Rico; Environmental determinants of larval abundance and relation to dengue virus transmission. Am J Trop Med Hyg 27: 1225–1231

Moore CG, Francy DB, Eliason DA, Bailey RE, Campos EG (1990) *Aedes albopictus* and other container-inhabiting mosquitoes in the United States; results of an eight-city survey. J Am Mosq Control Assoc 6: 173–178

Moran PAP (1951) A mathematical theory of animal trapping. Biometrika 38: 307–311

Mori A (1989) A simple method for sampling the immature stages of *Aedes togoi*. Trop Med 31: 171–174

Morozov VA (1965) Spread of *Mansonia richiardii* Fic. in the Krasnodar region and methods of collection of their larvae. Medskaya Parazitol. 34: 514–517 (In Russian, English summary)

Morris CD, Callahan JL, Lewis RH (1985) Devices for sampling and sorting immature *Coquillettidia perturbans*. J Am Mosq Control Assoc 1: 247–250

Morris CD, Callahan JL, Lewis RH (1990) Distribution and abundance of larval *Coquillettidia perturbans* in a Florida freshwater marsh. J Am Mosq Control Assoc 6: 452–460

Morrison AC, Gray K, Getis A, Astete H, Sihuincha M, Focks D, Watts D, Stancil JD, Olson JG, Blair P, Scott TW (2004a). Temporal and geographic patterns of *Aedes aegypti* (Diptera: Culicidae) production in Iquitos, Peru. J Med Entomol 41: 1123–1142

Morrison AC, Astete H, Chapilliquen F, Ramirez-Prada G, Diaz G, Getis A, Gray K, Scott TW (2004*b*). Evaluation of a sampling methodology for rapid assessment of *Aedes aegypti* infestation levels in Iquitos, Peru. J Med Entomol 41: 502–510

Mouchet J (1972*a*) Etude préliminaire sur les vecteurs potentiels de fièvre jaune au Ghana. Cah ORSTOM sér Entomol Méd Parasitol 10: 177–188

Mouchet J (1972*b*). Prospection sur les vecteurs potentiels de fièvre jaune en Tanzanie. Bull World Health Organ 46: 675–684

Muul I, Johnson BK, Harrison BA (1975) Ecological studies of *Culiseta melanura* (Diptera: Culicidae) in relation to eastern and western equine encephalomyelitis viruses on the eastern shore of Maryland. J Med Entomol 11: 739–748

Nagamine LR, Brown JK, Washino RK (1979) A comparison of the effectiveness and efficiency of three larval sampling techniques. Proc California Mosq Vector Control Assoc 47: 79–82

Nagao Y, Thavara U, Chitnumsup P, Tawatsin A, Chansang C, Campbell-Lendrum D (2003) Climatic and social risk factors for *Aedes* infestation in rural Thailand. Trop Med Int Health 8: 650–659

Nam VS, Ryan PA, Yen NT, Phong TV, Marchand RP, Kay BH (2003) Quantitative evaluation of funnel traps for sampling immature *Aedes aegypti* from water storage jars. J Am Mosq Control Assoc 19: 220–227

Nasci RS (1988) Biology of *Aedes triseriatus* (Diptera: Culicidae) developing in tires in Louisiana. J Med Entomol 25: 402–405

Nathan MB (1993) Critical review of *Aedes aegypti* control programs in the Caribbean and selected neighbouring countries. J Am Mosq Control Assoc 9: 1–7

Nayar JK, Sauerman DM (1968) Larval aggregation formation and population density interrelations in *Aedes taeniorhynchus*, their effects on pupal ecdysis and adult characteristics at emergence. Entomol Exp Appl 11: 423–442

Nelson MJ, Usman S, Pant CP, Self LS (1976) Seasonal abundance of adult and immature *Aedes aegypti* (L.) in Jakarta. Bull Penel Keseh Health Std Indonesia 4: 1–8

Neto FC, Costa AIP da Soares MRD, Scandar SAS, Cardoso Junior RP (1996) Descrição da colonização de *Aedes albopictus* (Diptera: Culicidae) na região de São José do Rio Preto, SP, 1991–1994. Rev Soc Bras Med Trop 29: 543–548

Neto FC, Dibo MR, Anália A, Barbosa C, Battigaglia M (2002) *Aedes albopictus* (S) na região de São José do Rio Preto, SP: estudo da sua infestação em área já ocupada pelo Aedes aegypti e discussão de seu papel como possível vetor de dengue e febre amarela. Rev Soc Bras Med Trop 35: 351–357

Nielsen ET, Greve H (1950) Studies on the swarming habits of mosquitoes and other Nematocera. Bull Entomol Res 41: 227–258

Nielsen ET, Nielsen AT (1953) Field observations on the habits of *Aedes taeniorhynchus*. Ecology 34: 141–156

Nikolaeva NV, Ol'shvang VN (1978) Simple biocenometer for counting aquatic organisms in small water bodies. Ekologiya, 5: 93–95 (In Russian)

Nogueira LA, Gushi LT, Miranda JE, Madeira NG, Ribolla PEM (2005) Short report: application of an alternative *Aedes* species (Diptera: Culicidae) surveillance method in Botucatu City, São Paulo, Brazil. Am J Trop Med Hyg 73: 309–311

Okazawa T, Mogi M (1984) Efficiency of the dipper in collecting larvae of *Anopheles sinensis* (Diptera: Culicidae) in rice fields. Jap J Sanit Zool 35: 367–371

Olds EJ, Merritt RW, Walker ED (1989) Sampling, seasonal abundance, and mermithid parasitism of larval *Coquillettidia perturbans* in south-central Michigan. J Am Mosq Control Assoc 5: 586–592

O'Meara GF, Evans LF Jr, Gettman AD, Patteson AW (1995) Exotic tank bromeliads harboring immature *Aedes albopictus* and *Aedes bahamensis* (Diptera: Culicidae) in Florida. J Vector Ecol 20: 216–224

O'Meara GF, Cutwa MM, Evans LF Jr (2003) Bromeliad-inhabiting mosquitoes in south Florida: native and exotic plants differ in species composition. J Vector Ecol 28: 37–46

Onsager JA (1976). The rationale of sequential sampling, with emphasis on its use in pest management. Tech Bull Agr Res Serv USDA, No. 1526

Oo TT, Storch V, Becker N (2002) Studies on the bionomics of *Anopheles dirus* (Culicidae: Diptera) in Mudon, Mon State, Myanmar. J Vector Ecol 27: 44–54

Orr BK, Resh VH (1989) Experimental test on the influence of aquatic macrophyte cover on the survival of *Anopheles* larvae. J Am Mosq Control Assoc 5: 579–585

Otis DL, Burnham KP, White GC, Anderson DR (1978). Statistical inferences from capture data on closed animal populations. Wld Monogr 62: 1–135

Overgaard HJ, Tsuda Y, Suwonkerd W, Takagi M (2002) Characteristics of *Anopheles minimus* (Diptera: Culicidae) larval habitats in Northern Thailand. Environ Entomol 31: 134–141

Pan American Health Organization (1994) Dengue and dengue hemorrhagic fever in the Americas. Guidelines for prevention and control. Scientific Publication, 548

Panicker KN, Bai MG, Kalyanasundaram M (1982) Well breeding behaviour of *Aedes aegypti*. Indian J Med Res 76: 689–691

Papierok B, Croset H, Rioux JA (1973) Estimation de l'effectif des populations larvaires d'*Aedes (O.) cataphylla* Dyar, 1916 (Diptera-Culicidae). 1.— Méthode de 'capture–marquage–recapture'. Cah ORSTOM sér Entomol Méd Parasitol 11: 243–249

Papierok B, Croset H, Rioux JA (1975) Estimation de l'effectif des populations larvaires d'*Aedes (O.) cataphylla* Dyar, 1916 (Diptera, Culicidae), II, Méthode utilisant le 'coup de louche' ou 'dipping'. Cah ORSTOM sér Entomol Méd Parasitol 13: 47–51

Paterson CG (1971) Overwintering ecology of the aquatic fauna associated with the pitcher plant *Sarracenia purpurea* L. Can J Zool 49: 1455–1459

Paulik GJ, Robson DS (1969) Statistical calculations for change-in ratio estimators of population parameters. J Wildl Mgmt 33: 1–27

Pausch RD, Provost MW (1965) The dispersal of *Aedes taeniorhynchus*. IV. Controlled field production. Mosquito News 25: 1–8

Pautou P, Aïn G, Gilot B, Cousserans J, Gabinaud A, Simonneau P (1973) Cartographie écologique appliquée à la démoustication. Doc Cartogr Écol Univ Scient Méd. Grenoble, France, 11: 1–16

Pena CJ, Gonzalvez G, Chadee DD (2003) Seasonal prevalence and container preferences of *Aedes albopictus* in Santo Domingo City, Dominican Republic. J Vector Ecol 28: 208–212

Perry JN, Taylor LR (1986) Stability of real interacting populations in space and time: implications, alternatives and the negative binomial kc. J Anim Ecol 55: 1053–1068

Peters TM, Chevone BI (1968) Marking *Culex pipiens* Linn. larvae with vital dyes for larval ecological studies. Mosquito News 28: 24–28

Pichon G, Gayral P (1970) Dynamique des populations d'*Aedes aegypti* dans trois villages de savane d'Afrique de l'Ouest. Cah. ORSTOM, sér. Entomol. méd. Parasit., 8: 49–68

Pichon G, Hamon J, Mouchet J (1969) Groupes ethniques et foyers potentiels de fièvre jaune dans les états francophone d'Afrique occidentale; considerations sur les methodes de lutte contre *Aedes aegypti*. Cah. ORSTOM, sér. Entomol. méd. Parasit., 7: 39–50

Pipitkool V, Srisawangwonk T, Sithithaworn P, Tesna S (1984) The prevalence of *Aedes* mosquito in Khon Khaen University. Bull. Knon Khaen Univ. Hlth Sci., 7: 6–11 (In Thai, English summary)

Pitcairn MJ, Wilson LT, Washino RK, Rejmankova E (1994) Spatial patterns of *Anopheles freeborni* and *Culex tarsalis* (Diptera: Culicidae) larvae in California rice fields. J Med Entomol 31: 545–553

Pittendrigh CS (1948) The bromeliad–*Anopheles*–malaria complex in Trinidad. 1. The bromeliad flora. Evolution, 2: 58–89

Pittendrigh CS (1950*a*) The quantitative evaluation of *Kerteszia* breeding grounds. Am. J. trop. Med., 30: 457–468

Pittendrigh CS (1950*b*) The ecoclimatic divergence of *Anopheles bellator* and *A. homunculus*. Evolution, 4: 43–63

Pittendrigh CS (1950*c*) The ecotopic specialization of *Anopheles homunculus*; and its relation to competition with *A. bellator*. Evolution, 4: 64–78

Pope KO, Sheffner EJ, Linthicum KJ, Bailey CL, Logan TM, Kasischke ES, Birney K, Njogu AR, Roberts CR (1992) Identification of central Kenyan rift valley fever virus vector habitats with Landsat TM and evaluation of their flooding status with airborne imaging radar. Remote Sensing and the Environment, 40: 185–196

Pritchard G, Scholefield PJ (1983) Survival of *Aedes* larvae in constant area ponds in southern Alberta (Diptera: Culicidae). Can. Ent., 115: 183–188

Provost MW (1975) Needle rush as an indicator of breeding of salt-marsh mosquitoes. Ann. Proc. Fla Anti-Mosq. Contr. Ass., 46: 23–28

Rademacher RE (1979) Studies of overwintering larvae of *Coquillettidia perturbans* mosquitoes in Minnesota. Mosquito News, 39: 135–136

Rajagopalan PK, Menon PKB, Mani TR, Brooks GD (1975) Measurement of density of immatures of *Culex pipiens fatigans* in effluent drains. J. Commun. Dis., 7: 327–337

Rajagopalan PK, Yasuno M, Menon PKB (1976) Density effect on survival of immature stages of *Culex pipiens fatigans* in breeding sites in Delhi villages. Indian J. Med. Res., 64: 688–708

Rajendran R, Reuben R, Purushothaman S, Veerapatran R (1995) Prospects and problems of intermittent irrigation for control of vector breeding in rice fields in southern India. Annals of Tropical Medicine and Parasitology, 89: 541–549

Rajnikant, Bhatt RM, Gupta DK, Sharma RC, Srivastava HC, Gautam AS (1993) Observations on mosquito breeding in wells and its control. Indian Journal of Malariology, 30: 215–220

Rao TR (1984) The Anophelines of India, (Revised edition), Malaria Research Centre, Indian Council of Medical Research, Delhi

Rao TR, Trpis M, Gillett JD, Teesdale C, Tonn RJ (1973). Breeding places and seasonal incidence of *Aedes aegypti*, as assessed by the single-larva survey method. Bull. Wld Hlth Org., 48: 615–622

Ree HI (1971) The use of betalights in trapping mosquito larvae. M.Sc. Project Rept. University of London, London School of Hygiene and Tropical Medicine

Reed DE (1970) Operational use of an improved mosquito larvae concentrator. Mosquito News, 30: 274

Reed DE, Husbands RC (1970) Integration of larval surveillance techniques in the operational program of the Fresno Westside mosquito abatement district. Proc. Calif. Mosq. Contr. Ass., 37: 98–101

Reeves WC, Brookman B, Hammon WM (1948) Studies on the flight range of certain *Culex* mosquitoes, using a fluorescent-dye marker, with notes on *Culiseta* and *Anopheles*. Mosquito News, 8: 61–69

Reid JA (1954) A preliminary *Aedes aegypti* survey. Med. J. Malaya, 9: 161–168

Reisen WK, Reeves WC (1990) Bionomics and ecology of *Culex tarsalis* and other potential mosquito vector species. In: Reeves WC (ed) Epidemiology and Control of Mosquito-borne Arboviruses in California, 1943–1987. California Mosquito and Vector Control Association, Sacramento, pp. 254–329

Reisen WK, Siddiqui TF (1979) Horizontal and vertical estimates of immature survivorship for *Culex tritaeniorhynchus* (Diptera: Culicidae) in Pakistan. J Med Entomol 16: 207–218

Reisen WK, Meyer RP, Shields J, Arbolante C (1989) Population ecology of preimaginal *Culex tarsalis* (Diptera: Culicidae) in Kern county, California. J Med Entomol 26: 10–22

Rejmankova E, Roberts DR, Pawley A, Manguin S, Polanco J (1995) Predictions of adult *Anopheles albimanus* densities in villages based on distances to remotely sensed larval habitats. Am J Trop Med Hyg, 53: 482–488

Rejmankova E, Pope KO, Roberts DR, Lege MG, Andre R, Greico J, Alonzo Y (1998) Characterization and detection of *Anopheles vestitipennis* and

Anopheles punctimacula (Diptera: Culicidae) larval habitats in Belize with field survey and SPOT satellite imagery. Journal of Vector Ecology, 23: 74–88

Renshaw M, Silver JB, Service MW, Birley MH (1995) Spatial dispersion patterns of larval *Aedes cantans* (Diptera: Culicidae) in temporary woodland pools. Bulletin of Entomological Research, 85: 125–133

Reuben R, Panicker KN (1975) *Aedes* survey in five districts of Rajasthan, India. J. Commun. Dis., 7: 1–9

Reuben R, Das PK, Samuel D, Brooks GD (1978) Estimation of daily emergence of *Aedes aegypti* (Diptera: Culicidae) in Sonepat, India. J Med Entomol 14: 705–714

Rioux J-A, Croset H, Corre J-J, Simonneau P, Gras G (1968) Phyto-ecological basis of mosquito control: cartography of larval biotopes. Mosquito News, 28: 572–582

Robert V, Ouari B, Ouedraogo V, Carnevale P (1988) Étude écologique des Culicidae adultes et larvaires dans une rizière en Vallée du Kou, Burkina Faso. Acta trop., 45: 351–359

Roberts DR, Scanlon JE (1974) An area sampler for collecting mosquito larvae in temporary woodland and field pools. Mosquito News, 34: 467–468

Roberts DR, Scanlon JE (1979) Field studies on the population biology of immature stages of six woodland mosquito species in the Houston, Texas area, Mosquito News, 39: 26–34

Roberts DR, Paris JF, Manguin S, Harbach RE, Woodruff R, Rejmankova E, Polanco J, Wullschleger B, Legters LJ (1996) Predictions of malaria vector distribution in Belize based on multispectral satellite data. Am J Trop Med Hyg, 54: 304–308

Robson DS, Regier HA (1968) Estimations of population number and mortality rates. In: Ricker WE (ed) Methods for Assessment of Fish Production in Fresh Waters. IBP Handbook No. 3, Blackwell, Oxford, pp. 124–158

Rogers DJ, Randolph SE (1991) Mortality rates and population density of tsetse flies correlated with satellite imagery. Nature, 351: 739–741

Rogers DJ, Williams BG (1994) Tsetse distribution in Africa: seeing the wood and the trees. In: Edwards, PJ, May RM, Webb NR (eds) Large-scale Ecology and Conservation Biology. Blackwell Science, Oxford, pp 247–272

Rogers DJ, Randolph SE, Snow RW, Hay SI (2002) Satellite imagery in the study and forecast of malaria. Nature, 415: 710–715

Rojas BA (1964) La binomial negativa y la estación de intensidad de plagas en el suelo. Fitotecnia Latinamer, 1: 27–36

Romero-Vivas CME, Wheeler JG, Falconar AKI (2002) An inexpensive intervention for the control of larval *Aedes aegypti* assessed by an improved method of surveillance and analysis. J Am Mosq Control Assoc 18: 40–46

Romero-Vivas CME, Falconar AKI (2005) Investigation of relationships between *Aedes aegypti* egg, larvae, pupae, and asult density indices where the main breeding sites were located indoors. J Am Mosq Control Assoc 21: 15–21

Rosenberg R (1982). Forest malaria in Bangladesh. III. Breeding habits of *Anopheles dirus*. Am. J. trop. Med. Hyg., 31: 192–201

Rozeboom LE, Hess AD (1944) The relation of the intersection line to the production of *Anopheles quadrimaculatus*. J. natn. Malar. Soc., 3: 169–179

Russell BM, Kay BH (1999) Calibrated funnel trap for quantifying mosquito (Diptera: Culicidae) abundance in wells. J Med Entomol 36: 851–855

Russell BM, Muir LE, Weinstein P, Kay BH (1996) Surveillance of the mosquito *Aedes aegypti* and its biocontrol with the copepod *Mesocyclops aspericornis* in Australian wells and gold mines. Medical and Veterinary Entomology, 10: 155–160

Russell BM, McBride WJH, Mullner H, Kay BH (2002) Epidemiological significance of subterranean *Aedes aegypti* (Diptera: Culicidae) breeding sites to dengue virus infection in Charters Towers, 1993. J Med Entomol 39: 143–145

Russell PF, Baisas FE (1935) The technic of handling mosquitoes. Philipp. J. Sci., 56: 257–294

Russell PF, Rao TR (1942*a*) On the ecology of larvae of *Anopheles culicifacies* Giles in borrow-pits. Bull Entomol Res 32: 341–361

Russell PF, Rao TR (1942*b*). On relation of mechanical obstruction and shade to ovipositing *Anopheles culicifacies*. J. exp. Biol., 91: 303–329

Russell PF, Santiago D (1932) *Anopheles minimus* larvae from wells in Laguna province, Philippine Islands. Philipp. J. Sci., 49: 219–223

Russell PF, Rao TR, Putnam P (1945) An evaluation of various measures of *Anopheles* larva density. Am. J. Hyg., 42: 274–298

Russell PF, West LS, Manwell RD, Macdonald G (1963) Practical Malariology. Oxford University Press, Oxford

Sabesan S, Krishnamoorthy K, Jambulingam P, Rajendran G, Kumar NP, Rajagopalan PK (1986) Breeding habitats of *Anopheles culicifacies* on Rameswaram island. Indian J. med. Res., 84: 44–52

Sakakibara M (1960) On the seasonal distributions of the larvae of *Anopheles (A) omorii* and nine other mosquito species found in a tree-hole. Endem. Dis. Bull. Nagasaki. Univ., 2: 236–242

Samarawickrema WA, Sone F, Kimura E, Self LS, Cummings RF, Paulson GS (1993) The relative importance and distribution of *Aedes polynesiensis* and *Aedes aegypti* larval habitats in Samoa. Medical and Veterinary Entomology, 7: 27–36

Sandoski CA, Kring TJ, Yearian WC, Meisch MV (1987) Sampling and distribution of *Anopheles quadrimaculatus* immatures in rice fields. J Am Mosq Control Assoc 3: 611–615

Santos JLF, Oliva WM (1991) Sobre o precedimento de Zippin para estimar populaçöesde animais. Rev. Saúde públ., 25: 53–55

Sawyer AJ (1989) Inconstancy of Taylor's b: simulated sampling with different quadrat sizes and spatial distributions. Res. Popul. Ecol., 31: 11–24

Schoen DJ, Fruchter D (1983) A calculator-assisted method of random sampling. Ecology, 64: 205–206

Scholefield PJ, Pritchard G, Enfield MA (1981) The distribution of mosquito (Diptera, Culicidae) larvae in southern Alberta, 1976–1978. Quaest. ent. 17: 147–168

Seber GAF (1973) The Estimation of Animal Abundance and Related Parameters. Griffen, London

Seber GAF, Le Cren ED (1967) Estimating population parameters from catches large relative to the population. J. Anim. Ecol., 36: 631–643

Seifert RP (1980) Mosquito fauna of *Heliconia aurea*. J. Anim. Ecol., 49: 687–697

Seifert RP, Barrera R (1981) Cohort studies on mosquito (Diptera: Culicidae) larvae living in the water-filled floral bracts of *Heliconia aurea* (Zingiberales: Musaceae). Ecol. Ent., 6: 191–197

Sen P (1948) *Anopheles* breeding in the rice fields of lower Bengal: Its relation with the cultural practices and with the growth of rice plants. Indian J. Malar., 2: 221–237

Service MW (1968) The ecology of the immature stages of *Aedes detritus* (Diptera: Culicidae). J. appl. Ecol., 5: 613–630

Service MW (1971) Studies on sampling larval populations of the *Anopheles gambiae* complex. Bull. Wld Hlth Org., 45: 169–180

Service MW (1974) Survey of the relative prevalence of potential yellow fever vectors in north-west Nigeria. Bull. Wld Hlth Org., 50: 487–494

Service MW (1984). Evaluation of sticky light traps for sampling mosquito larvae. Entomologia exp. appl., 35: 27–32

Service MW (1985) Population dynamics and mortalities of mosquito preadults. In: Lounibos LP, Rey, JR, Frank, JH (eds) Ecology of Mosquitoes: Proceedings of a Workshop. Florida Medical Entomology Laboratory, Vero Beach, Florida, pp. 185–201

Service MW (1993*a*) Culicidae. In: Lane RP, Crosskey RW (eds) Medical Insects and Arachnids. Chapman and Hall, London

Service, M. W. (1993*b*). Mosquito Ecology. Field Sampling Methods. 2nd edn. Chapman and Hall, London

Service MW, Highton RB (1980) A chemical light trap for mosquitoes and other biting insects. J Med Entomol 17: 183–185

Service MW, Sulaiman S, Esena R (1983) A chemical aquatic light trap for mosquito larvae (Diptera: Culicidae). J Med Entomol 20: 659–663

Shaman J, Stieglitz M, Stark C, le Blancq S, Cane M (2002) Using a dynamic hydrology model to predict mosquito abundances in flood and swamp water. Emerging Infectious Diseases, 8: 6–13

Sharma RS, Kaul SM, Sokhay J (2005) Seasonal fluctuations of dengue fever vector, *Aedes aegypti* (Diptera: Culicidae) in Delhi, India. Southeast Asian Journal of Tropical Medicine and Public Health, 36: 186–190

Sharma VP, Dhiman RC, Ansari MA, Nagpal BN, Srivastava A, Manavalan P, Adiga S, Radhakrishnan K, Chandrasekhar MG (1996) Study on the feasibility of delineating mosquitogenic conditions in and around Delhi using Indian Remote Sensing Satellite data. Indian Journal of Malariology, 33: 107–125

Shemanchuk JA (1959) Mosquitoes (Diptera: Culicidae) in irrigated areas of southern Alberta and their seasonal changes in abundance and distribution. Can. J. Zool., 37: 899–912

Sheppard PM, Macdonald WW, Tonn RJ (1969) A new method of measuring the relative prevalence of *Aedes aegypti*. Bull. Wld Hlth Org., 40: 467–468

Shidrawi GR, Clarke JL, Boulzaguet JR, Ashkar TS (1973). Culicine Mosquitoes with Particular Reference to *Aedes aegypti* and Their Prevalence in a Rural Area of the African Sudan Savannah, Garki District, Kano State, Nigeria.' WHO/VBC/73.420, 10 pp. (mimeographed)

Shogaki Y, Makiya K (1970) An improved device for quantitative sampling of mosquito larvae. Jap. J. sanit. Zool., 21: 172–178 (In Japanese, English summary)

Simsek FM (2004). Seasonal Larval and Adult Population Dynamics and Breeding Habitat Diversity of *Culex theileri* Theobald, 1903 (Diptera: Culicidae) in the Golbasi District, Ankara, Turkey. Turkish Journal of Zoology, 28: 337–344

Sithiprasasna R, Linthicum KJ, Liu G-J, Jones JW, Singhasivanon P (2003*a*). Use of GIS-based spatial modeling approach to characterize the spatial patterns of malaria mosquito vector breeding habitats in northwestern Thailand. Southeast Asian Journal of Tropical Medicine and Public Health, 34: 517–528

Sithiprasasna R, Linthicum KJ, Liu G-J, Jones JW, Singhasivanon P (2003*b*) Some entomological observations on temporal and spatial distribution of malaria vectors in three vilages in northwestern Thailand using a geographic information system. Southeast Asian Journal of Tropical Medicine and Public Health, 34, 505–516.

Sithiprasasna R, Patpoparn S, Attatippaholkun W, Suvannadabba S, Srisuphanunt, M (2004). The geographic information system as an epidemiological tool in the surveillance of dengue virus-infected Aedes mosquitos. Southeast Asian Journal of Tropical Medicine and Public Health, 35: 919–926

Siverly RE (1966) Occurrence of *Culiseta melanura* (Coquillett) in Illinois. Mosquito News, 26: 95–96

Siverly RE, DeFoliart GR (1968) Mosquito studies in northern Wisconsin. 1. Larval studies. Mosquito News, 28: 149–154

Smirnov VS (1967) The estimation of animal numbers based on the analysis of population structure, Vol. 1. In: Petrusewicz K (ed) Secondary Productivity of Terrestrial Ecosystems. Inst. Ecology, Polish Acad. Sci., Warsaw., pp. 199–223.

Snow WF (1983) Mosquito production and species succession from an area of irrigated rice fields in The Gambia, West Africa. J. trop. Med. Hyg., 86: 237–245

Soper FL (1967) Dynamics of *Aedes aegypti* distribution and density. Bull. Wld Hlth Org., 36: 536–538

Southwood TRE (1978) Ecological Methods with Particular Reference to the Study of Insect Populations. Chapman and Hall, London

Stewart RJ, Schaefer CH (1983) The relationship between dipper counts and the absolute density of *Culex tarsalis* larvae and pupae in rice fields. Mosquito News, 43: 129–135

Stewart RJ, Miura T, Parman RB (1983*a*) Comparison of sample patterns for *Culex tarsalis* in rice fields. Proc. Calif. Mosq. & Vect. Contr. Ass., 51: 54–58

Stewart RJ, Schaefer CH, Miura T (1983*b*) Sampling *Culex tarsalis* (Diptera: Culicidae) immatures in rice fields treated with combinations of mosquitofish and *Bacillus thuringiensis* H-14 toxin, J. econ. Ent., 76: 91–95

Stockwell DRB, Peters D (1999) The GARP modelling system: problems and solutions to automated spatial prediction. International Journal of Geographic Information Science, 13: 143–158

Stokes GM, Payne D (1976) Larval/pupal concentrator. Mosquito News, 36: 200–202

Strickman D, Kittayapong P (1993) Laboratory demonstration of oviposition by *Aedes aegypti* (Diptera: Culicidae) in covered water jars. J Med Entomol 30: 947–949

Strickman D, Kittayapong P (2002) Dengue and its vectors in Thailand: introduction to the study and seasonal distribution of *Aedes* larvae. Am J Trop Med Hyg, 67: 247–259

Strickman D, Kittayapong P (2003) Dengue and its vectors in Thailand: calculated transmission risk from total pupal counts of *Aedes aegypti* and association of wing-length measurements with aspects of the larval habitat. Am J Trop Med Hyg, 68: 209–217

Subra R (1983) The regulations of preimaginal populations of *Aedes aegypti* L. (Diptera: Culicidae) on the Kenya coast. 1. Preimaginal population dynamics and the role of human behaviour. Ann. trop. Med. Parasit., 77: 195–201

Sulaiman S (1982) The ecology of *Aedes cantans* (Meigen) and biology of *Culex pipiens* in hibernation sites in northern England Ph.D. thesis, University of Liverpool

Sulaiman S, Pawanchee ZA, Arifin Z, Wahab A (1996) Relationship between Breteau and House indices and cases of dengue/dengue hemorrhagic fever in Kuala Lumpur, Malaysia. J Am Mosq Control Assoc 12: 494–496

Sunahara T, Mogi M (1998) Distribution and turnover of a mosquito (*Tripteroides bambusa*) metapopulation among bamboo groves. Ecological Research, 13: 291–299

Sunish IP, Reuben R (2001) Factors influencing the abundance of Japanese encephalitis vectors in ricefields in India – I. Abiotic. Medical and Veterinary Entomology, 15: 381–392

Sunish IP, Reuben R (2002). Factors influencing the abundance of Japanese encephalitis vectors in ricefields in India – II. Biotic. Medical and Veterinary Entomology, 16: 1–9

Surtees G (1959) Influence of larval population density on fluctuation in mosquito numbers. Nature, Lond., 183: 269–270

Surtees G (1970) Large-scale irrigation and arbovirus epidemiology, Kano Plain, Kenya, I. Description of the area and preliminary studies on mosquitoes. J Med Entomol 7: 509–517

Surtees G, Simpson DIH, Bowen ETW, Grainger WE (1970) Rice field development and arbovirus epidemiology, Kano Plain, Kenya. Trans. R. Soc. trop. Med. Hyg., 64: 511–518

Suwonkerd W, Tsuda Y, Takagi M, Wada Y (1996) Seasonal occurrence of *Aedes aegypti* and *Ae. albopictus* in used tires in 1992–1994, Chiangmai, Thailand. Tropical Medicine (Nagasaki), 38: 101–105

Takagi M, Narayan D (1988) Relative importance of crab holes, tree holes, and coconut husks as breeding sources of *Aedes polynesiensis* in a riverside delta in Fiji. Jap. J. sanit. Zool., 39: 151–153

Takagi M, Sugiyama A, Maruyama K (1995) Effect of rice culturing practices on seasonal occurrence of *Culex tritaeniorhynchus* (Diptera: Culicidae) immatures in three different types of rice-growing areas in central Japan. J Med Entomol 32: 112–118

Takagi M, Sugiyama A, Maruyama K (1996) Effect of rice plant covering on the density of mosquito larvae and other insects in rice fields. Applied Entomology and Zoology, 31: 75–80

Takagi M, Suwonkerd W, Tsuda Y, Sugiyama A, Wada Y (1997) Effects of rice culture practices on the abundance of *Culex* mosquitoes (Diptera: Culicidae) in northern Thailand. J Med Entomol 34: 272–276

Takahashi RM, Miura T, Wilder WH (1982) A comparison between the area sampler and the two other sampling devices for aquatic fauna in rice fields. Mosquito News, 42: 211–216

Taylor LR (1961) Aggregation, variance and the mean. Nature, Lond., 189: 732–735

Taylor LR (1965) A Natural Law for the Spatial Disposition of Insects. Proc. int. Congr. Ent. 12th, pp. 396–397

Taylor LR (1984) Assessing and interpreting the spatial distributions of insect populations. A. Rev. Ent., 29: 321–357

Taylor LR (1986) Synoptic dynamics, migration and the Rothamsted insect survey. J. Anim. Ecol., 55: 1–38

Taylor LR, Woiwod IP, Perry JN (1978) The density-dependence of spatial behaviour and the rarity of randomness. J. Anim. Ecol., 47: 383–406

Taylor LR, Woiwod IP, Perry JN (1979) The negative binomial, as an ecological model and the density dependence of k. J. Anim. Ecol., 48: 289–304

Taylor LR, Perry JN, Woiwod IP, Taylor AR (1988) Specificity of the spatial power-law exponent in ecology and agriculture. Nature, Lond., 332: 721–722

Taylor NJ (1979) A rapid, efficient area sampler for estimating absolute abundance of floodwater mosquito larvae. Env. Ent., 8: 1004–1006

Taylor RAJ (1981*a*) The behavioural basis of redistribution. I. The Δ-model concept. J. Anim. Ecol., 50: 573–586

Taylor RAJ (1981*b*). The behavioural basis of redistribution. II. Simulations of the Δ-model. J. Anim. Ecol., 50: 587–604

Taylor RAJ, Taylor LR (1979) A behavioural model for the evolution of spatial dynamics. In: Anderson RM, Turner BD, Taylor LR (eds) Population Dynamics. The 20th Symposium of the British Ecological Society. Blackwell Science Publishers, London, pp. 1–27

Teixeira M da G, Barreto ML, Costa M da CN, Ferreira LDA, Vasconcelos PFC, Cairncross S (2002) Dynamics of dengue virus circulation: a silent epidemic in a complex urban area. Tropical Medicine and International Health, 7: 757–762

Thavara U, Tawatsin A, Chansang C, Kong-ngamsuk W, Paosriwong S, Boon-Long J, Rongsriyam Y, Komalamisra N (2001) Larval occurrence, oviposi-tion behavior and biting activity of potential mosquito vectors of dengue on Samui Island, Thailand. Journal of Vector Ecology, 26: 172–180

Thenmozhi V, Gajanana A (1990) Comparison of quadrat and dipper sampling for density measurements in rice fields. In: Annual Report (1989–90). Centre for Research in Medical Entomology, Indian Council for Medical Research, Madurai, pp. 24–28

Thomas IM (1950) The reactions of mosquito larvae to regular repetitions of shadows as stimuli. Aust. J. scient. Res. (B), 3: 113–123

Thomson MC, Connor SJ, Milligan PJM, Flasse SP (1996) The ecology of malaria – as seen from earth-observation satellites. Annals of Tropical Medicine and Parasitology, 90: 243–264

Thomson MC, Connor SJ, Milligan PJM, Flasse SP (1997). Mapping malaria risk in Africa – what can satellite data contribute? Parasitology Today, 8: 313–318

Thomson MC, Connor SJ (2000) Environmental information systems for the control of arthropod vectors of disease. Medical and Veterinary Entomology, 14: 227–244

Thorarinsson K (1986) Population density and movement: a critique of models. Oikos, 46: 70–81

Thu MM, Lin WT, Sebastian A, Kanda T, May K (1985) Preliminary biological study on two taxa of the *Anopheles balabacensis* complex in Burma. Jap. J. Genet., 60: 373–380

Tidwell MA, Williams DC, Tidwell TC, Peña CJ, Gwinn TA, Focks DA, Zaglul A, Mercedes M (1990) Baseline data on *Aedes aegypti* populations in Santo Domingo, Dominican Republic. J Am Mosq Control Assoc 6: 514–522

Tinker ME (1967) Measurements of *Aedes aegypti* populations. J. econ. Ent., 60: 634–637

Tinker ME (1978) Relationship of the house index and the Breteau index for *Aedes aegypti*. PAHO/WHO Newsletter on Dengue, Yellow Fever and *Aedes aegypti* in the Americas 1978. 7: 11–13

Tinker ME, Hayes GR (1959) The 1958 *Aedes aegypti* distribution in the United States. Mosquito News, 19: 73–78

Tonn RJ, Bang YH (1971) One-larva-per-container mosquito surveys conducted in Bangkok-Thonburi, Thailand, in 1969. Bull. Wld Hlth Org., 45: 270–274

Tonn RJ, Sheppard PM, Macdonald WW, Bang YH (1969) Replicate surveys of larval habitats of *Aedes aegypti* in relation to dengue haemorrhagic fever in Bangkok, Thailand. Bull. Wld Hlth Org., 40: 819–829

Trapido H (1951) Factors influencing the search for anopheline larvae in Sardinia. J. natn. Malar. Soc., 10: 318–326

Trapido H, Aitken T (1953) Study of a residual population of *Anopheles labran-chiae* Falleroni in the Geremeas valley, Sardinia. Am. J. trop. Med. Hyg., 2: 658–676

Trpis M (1972a) Breeding of *Aedes aegypti* and *A. simpsoni* under the escarpment of the Tanzanian plateau. Bull. Wld Hlth Org., 47: 77–82

Trpis M (1972*b*). Seasonal changes in the larval populations of *Aedes aegypti* in two biotopes in Dar es Salaam, Tanzania. Bull. Wld Hlth Org., 47: 245–255

Tsuda Y, Kobayashi J, Nambanya S, Miyagi I, Toma T, Phompida S, Manivang K (2002) An ecological survey of dengue vector mosquitos in central Lao PDR. Southeast Asian Journal of Tropical Medicine and Public Health, 33: 63–67

Tun-Lin W, Moe M, Aung H, Paing M. (1986) Larval density estimation of well-breeding anophelines by mark–release–recapture techniques in Burma. Mosq.-Borne Dis. Bull., 3: 1–4

Tun-Lin W, Myo-Paing, Zaw-Myint (1988) A modification of the W. H. O. dipping procedure for well-breeding anophelines in Burma. Trop. Biomed., 5: 51–55

Tun-Lin W, Kay BH, Burkot TR (1994) Quantitative sampling of immature *Aedes aegypti* in metal drums using sweep net and dipping methods. J Am Mosq Control Assoc 10: 390–396

Tun-Lin W, Kay BH, Barnes A (1995*a*). Understanding productivity, a key to *Aedes aegypti* surveillance. Am J Trop Med Hyg, 53: 595–601

Tun-Lin W, Kay BH, Barnes A (1995*b*) The premise condition index: a tool for streamlining surveys of *Aedes aegypti*. Am J Trop Med Hyg, 53: 591–594

Tun-Lin W, Maung Maung Mya, Sein Maung Than, Tin Maung Maung (1995*c*). Rapid and efficient removal of immature *Aedes aegypti* in metal drums by sweep net and modified sweeping method. Southeast Asian Journal of Tropical Medicine and Public Health, 26: 754–759

Tun-Lin W, Kay BH, Barnes A, Forsyth S (1996) Critical examination of *Aedes aegypti* indices: correlations with abundance. Am J Trop Med Hyg, 54: 543–547

Tuno N, Miki K, Minakawa N, Githeko A, Yan G, Takagi M (2004) Diving ability of *Anopheles gambiae* (Diptera: Culicidae) larvae. J Med Entomol 41: 810–812

Turchin P (1989) Population consequences of aggregative movement. J. Anim. Ecol., 58: 75–100

Undeen AH, Becnel JJ (1994) A device for monitoring populations of larval mosquitoes in container habitats. J Am Mosq Control Assoc 10: 101–103

Vail PV, Howland AF, Henneberry TJ (1966) Fluorescent dyes for mating and recovery studies with cabbage looper moths. J. econ. Ent., 59: 1093–1097

Van den Assem J, Metselaar D (1958) Host-plants and breeding places of *Mansonia (Mansonioides) uniformis* in Netherlands New Guinea. Trop. geogr. Med., 10: 51–55

Vezzani D, Velázquez SM, Soto S, Schweigmann NJ (2001) Environmental characteristics of the cemeteries of Buenos Aires city (Argentina) and infestation levels of *Aedes aegypti* (Diptera: Culicidae). Memórias do Instituto Oswaldo Cruz, 96: 467–471

Victor TJ, Reuben R (1999) Population dynamics of mosquito immatures and the succession in abundance of aquatic insects in rice fields in Madurai, South India. Indian Journal of Malariology, 36: 19–32

Victor TJ, Reuben R (2000*a*) Effect of plant spacing on the population of mosquito immatures in rice fields in Madurai, South India. Indian Journal of Malariology, 37: 18–26

Victor TJ, Reuben R (2000*b*) Effects of organic and inorganic fertilisers on mosquito populations in rice fields of southern India. Medical and Veterinary Entomology, 14: 361–368

Wada Y (1962*a*). Studies on the population estimation of insects of medical importance. 1. A method of estimating the population size of mosquito larvae in a fertilizer pit. Endem. Dis. Bull. Nagasaki Univ., 4: 22–30

Wada Y (1962*b*) Studies on the population estimation for insects of medical importance. II. A method for estimating the population size of larvae of *Aedes togoi* in the tide-water rock-pool. Endem. Dis. Bull. Nagasaki Univ., 4: 141–156

Wada Y (1965) Population studies on Edmonton mosquitoes. Quaest. ent., 1: 187–222

Wada Y, Mogi M (1974) Efficiency of the dipper in collecting immature stages of *Culex tritaeniorhynchus summorosus*. Trop. Med., 16: 35–40

Wada Y, Mogi M, Nishigaki J (1971*a*) Studies on the population estimation for insects of medical importance. IV. A method for the estimation of the relative density of *Culex tritaeniorhynchus summorosus* larvae in whole paddy-fields of an area. Trop. Med., 13: 86–93

Wada Y, Mogi M, Nishigaki J (1971*b*) Studies on the population estimation for insects of medical importance. III. Sequential sampling technique for *Culex tritaeniorhynchus summorosus* larvae in the paddy-field. Trop. Med., 13: 16–25

Wada Y, Ito S, Oda T (1993) Seasonal abundance of immature stages of *Aedes togoi* at Fukue island, Nagasaki (Diptera: Culicidae) southwestern Japan. Tropical Medicine (Nagasaki), 35: 1–10

Wald A (1947) Sequential Analysis. John Wiley & Sons, New York

Walker ED, Crans WJ (1986) A simple method for sampling *Coquillettidia perturbans* larvae. J Am Mosq Control Assoc 2: 239–240

Walker ED, Merritt RW, Wotton RS (1988) Analysis of the distribution and abundance of *Anopheles quadrimaculatus* (Diptera: Culicidae) larvae in a marsh. Env. Ent., 17: 992–999

Walton WE, Schreiber ET, Mulla MS (1990) Distribution of *Culex tarsalis* larvae in a freshwater marsh in Orange County, California. J Am Mosq Control Assoc 6: 539–543

Walton WE, Workman PD (1998) Effect of marsh design on the abundance of mosquitoes in experimental constructed wetlands in southern California. J Am Mosq Control Assoc 14: 95–107

Warren ME, Eddleman CD (1965) A self-straining larval concentrator. Mosquito News, 25: 486–487

Washino RK, Hokama Y (1968) Quantitative sampling of aquatic insects in a shallow-water habitat. Ann. ent. Soc. Am., 61: 785–786

Washino RK, Wood BL (1994) Application of remote sensing to arthropod vector surveillance and control. Am J Trop Med Hyg, 50: 134–144

Waters BT, Slaff M (1987) A small habitat, larval mosquito sampler. J Am Mosq Control Assoc 3: 514

Waters WE (1955) Sequential sampling in forest insect surveys. Forest Sci., 1: 68–79

Weathersbee AA, Hasell PG (1938) Mosquito studies. On the recovery of stain in adults developing from anopheline larvae stained in vitro. Am. J. trop. Med., 18: 531–43

Welch HE, James HG (1960) The Belleville trap for quantitative samples of mosquito larvae. Mosquito News, 20: 23–26

Welch JB, Olson JK, Hart WG, Ingle SG, Davis MR (1989a). Use of aerial color infrared photography as a survey technique for *Psorophora columbiae* oviposition habitats in Texas ricefields. J Am Mosq Control Assoc 5: 147–160

Welch JB, Olson JK, Yates MM, Benton AR, Baker RD (1989b). Conceptual model for the use of aerial color infrared photography by mosquito control districts as a survey technique for *Psorophora columbiae* oviposition habitats in Texas ricefields. J Am Mosq Control Assoc 5: 369–373

Wellings PW (1987) Spatial distribution and interspecific competition. Ecol. Ent., 12: 359–362

Weseloh RM (1989) Evaluation of insect spatial distributions by spectral analysis, with particular reference to the Gypsy moth (Lepidoptera: Lymantriidae) and *Calosoma sycophanta* (Coleoptera: Carabidae). Env. Ent., 18: 201–207

Wharton RH (1962) The Biology of *Mansonia* Mosquitoes in Relation to the Transmission of Filariasis in Malaya. Bull. Inst. med. Res. Fed. Malaya (N.S.) No. 11

White GB (1989) Malaria. Geographical Distribution of Arthropod-Borne Diseases and their Principal Vectors. World Health Organization, Geneva

Wilding JL (1940) 'A New Square-Foot Aquatic Sampler.' Spec. Publs limnol. Soc. Am., 4

Wood BL, Beck LR, Washino RK, Hibbard KA, Salute JS (1992) Estimating high mosquito-producing rice fields using spectral and spatial data. International Journal of Remote Sensing, 13: 2813–2826

Woodrow RJ, Howard JJ (1994) Use of a modified chemical transfer pump for sampling *Culiseta melanura* larvae. J Am Mosq Control Assoc 10: 427–429

Wooster MT, Rivera D (1985) Breeding point and larval association of anopheline mosquitoes of northwest Mindoro, Philippines. Southeast Asian J. trop. Med. publ. Hlth., 16: 59–65

Workman PD, Walton WE (2003) Larval behavior of four *Culex* (Diptera: Culicidae) associated with treatment wetlands in the southwestern United States. Journal of Vector Ecology, 28: 213–228

World Health Organization (1975) Manual on practical entomology in malaria. Part II Methods and techniques. WHO Offset Publication No. 13, World Health Organization, Geneva

Yamamura K (1990). Sampling scale dependence of Taylor's power law. Oikos, 59: 121–125

Yasuno M, Rajagopalan PK, Kazmi SJ, LaBrecque GC (1977). Seasonal changes in larval habitats and population density of *Culex fatigans* in Delhi villages. Indian J. Med. Res., 65 (Suppl.): 52–64

Zhen Tian-Min, Kay BH (1993) Comparison of sampling efficacy of sweeping and dipping for *Aedes aegypti* larvae in tires. J Am Mosq Control Assoc 9: 316–320

Zippin C (1956) An evaluation of the removal method of estimating animal populations. Biometrics, 12: 163–189

Zippin C (1958) The removal method of population estimation. J. Wildl. Mgmt, 22: 82–90

Chapter 4 Sampling the Emerging Adult Population

Emergence traps have been widely used for sampling aquatic insects, especially chironomids, in both deep and shallow waters, and these traps can be conveniently divided into two broad categories. First, those such as funnel traps, which are completely submerged in the water, and secondly those which either float on, or are positioned over, the water, such as floating conical box traps, cages erected over the water and sticky traps (Jónasson 1954; Kimerle and Anderson 1967; Morgan 1971; Morgan et al. 1963; Mundie 1956, 1971; and see Southwood 1978, and Merritt et al. 1984). Only those emergence traps that are positioned over the water surface are used to sample mosquitoes. Davies (1984) discusses several topics of relevance to emergence trapping of aquatic insects, including the effects of transparent or opaque traps, trap size, temperature, tests for sampling efficiency and methods for predicting emergence.

Emergence traps can be used to detect mosquito breeding in inaccessible habitats such as crab holes, pit latrines and deep wells, to study diel and seasonal patterns of emergence and to obtain estimates of adult productivity. Ross (1910) erected a mosquito net over marshes in Mauritius to estimate daily emergence rates of *Anopheles gambiae* as long ago as 1908, and they have been used ever since, though perhaps not as widely as other sampling techniques.

The use of emergence traps to calculate the proportions of the population emerging daily or to estimate total productivity is problematic, as habitat size may change rapidly due to desiccation or flooding and numbers caught in the trap must be related to the total surface area of the breeding place producing emerging adults. This is not necessarily equivalent to the area of free water, because some pupae may survive amongst water-logged leaf litter and give rise to adults (Roberts and Scanlon 1979). In laboratory-type experiments Castleberry et al. (1989) found that the ratio of the numbers of mosquitoes entering a jar at the apex of a pyramidal emergence trap, to the number of adults that emerged from different numbers of pupae placed under the trap did not vary according to pupal densities.

This indicates that traps were sampling emerging adults with equal efficiency at different pupal densities, but it would be unwise to extrapolate these results to field situations without further experiments. Corbet (1965) considered that emergence traps could not be used to derive population estimates because of bias in sampling, but in spite of the difficulties in obtaining representative samples, emergence trapping can, at least in some situations, provide estimates of the total emergent population. Lakhani and Service (1974), for example, found that the probabilities of viable eggs of *Ochlerotatus cantans* giving rise to adults based on field estimates of the egg and emergent adult populations were similar to the probabilities calculated from sampling the different age classes of the immature stages. Moreover, similar survival rates of eggs to adults were obtained when an emergence cage erected over a small pond caught all emergent adults (Service 1977). Furthermore, Smith and McIver (1984) felt justified in extrapolating the numbers of pupae caught in their traps to estimate total emerging population size of several mosquito species.

Other potential difficulties with the use of emergence traps include the attractive or repulsive effects on pupae of shading and reduced illumination of the water surface caused by the traps and the influence of wind on pupal distribution in the habitat. Kimerle and Anderson (1967) showed in both laboratory and field experiments that emergence traps covered with clear transparent plastic caught 4–5 times more midges (chironomids and chaoborids) than traps covered with black plastic. Scott and Opdyke (1941) reported reduced catches of insects in traps covered with opaque materials. In contrast, Pritchard and Scholefield (1980) found shade did not affect the numbers of pupae under their traps, while in Canada Smith and McIver (1984) concluded that aedine pupae neither accumulated under their floating traps nor avoided them. In comparative studies on patterns of mosquito emergence Corbet (1964, 1965) considered that during high winds pupae tended to seek shelter under emergence traps and as a result, samples would be unrepresentative of the emerging population. It seems reasonable to assume that when emergence occurs predominantly at night or during twilight periods the effect of shadows cast by the trap will be less important. Furthermore, traps in shaded woodland habitats will provide less shelter from wind and cast less shadow than those located in exposed situations.

In studying the seasonal emergence of an odonatan, Taketo (1960) reported the time elapsed from the start of emergence to the point when 50% (EM_{50}) of the population had emerged. Corbet and Danks (1973) pointed out that although this is a useful statistic, it gives no information on the spread of emergence over time. In their studies on *Ochlerotatus impiger* and *Ochlerotatus nigripes* in which emergence was mainly completed within 7–10 days, Corbet and Danks (1973) plotted cumulative percentage emergence against time (Fig. 4.1) when, 0, 10, 50, 90 and 100% of the population had emerged (EM_0, EM_{10}, EM_{50}, EM_{90}, EM_{100}); the first and last values being the day before emergence began and the last day of emergence, respectively. This method was also used to describe the emergence pattern of *Ochlerotatus cantans* (Renshaw 1991) and *Ochlerotatus rusticus* (Andreasen et al. 1997) in northern England. The EM_{10} and EM_{90} usually approximated to the points of inflection on the sigmoid emergence curve. The more or less straight line between the two points shows that during this interval emergence is most rapid. A log–probit plot would probably convert the sigmoid curve into a straight line.

Slaff (1986) used a degree-week (DW) model to estimate the adult emergence of *Coquillettidia perturbans* from overwintering larvae, as follows: DW = [Max T °C + Min T °C]/2–threshold temp. of development × weeks. Accumulated degree-weeks are calculated, and the weeks when emergence begins and the weeks and resultant degree-weeks when 50% of adults have emerged are recorded.

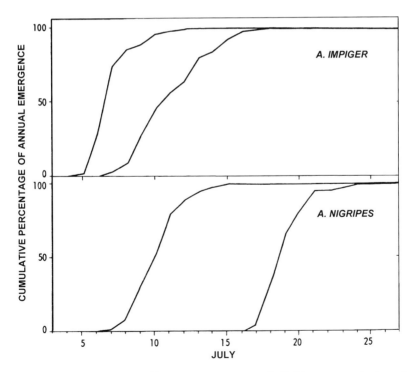

Fig. 4.1. Emergence of *Ochlerotatus impiger* and *Ochlerotatus nigripes* from ponds on Ellesmere Island, Canada, which had the earliest and latest emergences in 1963 (after Corbet and Danks 1973)

Floating traps

Bradley's trap

From 1921–24 Bradley (1926) studied the emergence of North American *Anopheles*, mainly *Anopheles quadrimaculatus* s.l., by using 3-ft square pyramidal floating traps that were anchored to overhanging vegetation. In shallow waters 4-ft high screen cages, either 3- or 4-ft square, with a door, and which rested on the bottom of the pools were used. Adults were removed daily from both types of traps and all predators, such as spiders, were killed.

Mundie's trap (modified)

Mundie (1956, 1971) presents useful reviews of emergence traps used to sample aquatic insects in both deep and shallow waters. One of the traps he illustrates for use in shallow waters has been modified and successfully used for several years to sample mosquitoes, mainly *Ochlerotatus cantans*, breeding in small pools and ditches (Service 1977). The modified version consists of a conical framework made of duraluminium, or some other light alloy, having a 40-cm diameter base and covered with white mosquito netting. An inverted glass collecting jar with a small plastic funnel taped into its entrance is pushed into the trap's 8-cm diameter apical opening (Fig. 4.2). The trap's three main struts extend horizontally and support two round floats. Copper floats should be avoided as an electrochemical reaction occurs in the water between the copper and duraluminium resulting in the supports slowly disintegrating. Plastic ballcocks such as those used in lavatory cisterns make convenient floats. Mundie (1956) points out that condensation may occur in the collecting jar and spoil the catch. He suggests that small holes are drilled into the jar to create ventilation, and that similar holes are made around the base of the funnel to drain off any condensation that collects. Alternatively the bottom part of the collecting jar can be removed and replaced with pale transparent plastic mesh. The traps are designed so that the base of the cone is just submerged. This is important because if the trap is raised a few centimetres above the water surface low flying predacious flies get trapped, and these may eat the catch of mosquitoes.

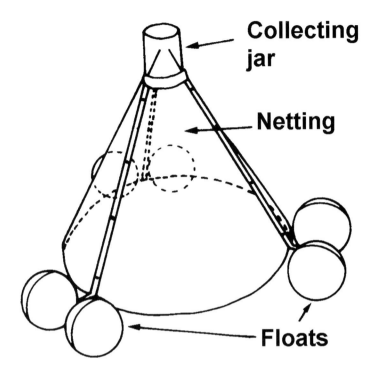

Fig. 4.2. Conical floating emergence trap (Service 1977)

If the trap fails to 'seat' on the water it can be tied down. Mosquitoes that have not entered the collecting jar are collected from the trap by inserting an aspirator through a small sleeve opening.

Pyramidal traps

Aubin et al. (1973) developed a pyramidal trap to function effectively in very shallow waters and to kill mosquitoes and preserve them in good condition for taxonomic studies. It consists of an 18-cm high, truncated pyramid of 3-mm plexiglas with a 25-cm square base supported on a styrofoam framework (Fig. 4.3*a*). An 8.5-cm square plexiglas plate with a large hole (5.8 cm) removed from its centre is glued on top of the pyramid. The cap of a wide-mouth 8-oz glass jar with a 4-cm diameter hole removed from the centre is fixed within the hole in the plexiglas plate. A plastic funnel is glued to the inside of this cap to both guide mosquitoes into the

collecting jar and to retain the 25 ml of formalin that are poured around
the edges of the cap. The collecting jar is then screwed into position.
Aubin et al. (1973) reported that it was very successful in catching
mosquitoes in the Quebec area of Canada, and that it remained stable
on the water surface even in storm conditions.

Smith and McIver (1984) used pyramidal floating emergence traps
modified from the design of Aubin et al. (1973). Their traps had a wooden
frame with a base measuring 1 × 1 m and covered with polyethylene sheeting. Each trap was tied to a pole driven into the substrate. Mosquitoes that
failed to fly up into the collecting bottle were removed with a Black and
Decker 'Mod 4' car vacuum cleaner inserted through a netting sleeve sewn
into one of the sides of the trap. From 20 trap-nights in 1979 a total of
4373 mosquitoes were caught, including 2018 *Ochlerotatus communis* and
1364 *Ochlerotatus excrucians*, while in 1980 a total of 3959 mosquitoes
were caught in 25 traps, including 1979 *Ochlerotatus communis*, 966
Ochlerotatus abserratus and 566 *Ochlerotatus excrucians*, along with several other *Ochlerotatus* species. Based on visual inspection the authors
concluded that there was no tendency for pupae to either accumulate under
the floating traps, or to avoid them, and consequently, they felt justified in
extrapolating the numbers caught in all their traps to obtain estimates of
the total emerging populations of the different species.

Appleton and Sharp (1985) developed a trap based on the design of
Aubin et al. (1973) consisting of a 520-mm square resin-coated polystyrene base with the middle section (400 mm square) removed (Fig. 4.3*b*).

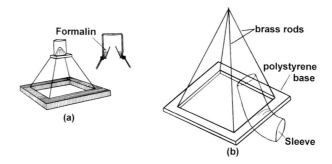

Fig. 4.3. (*a*) Styrofoam trap (after Aubin et al. 1973); (*b*) floating emergence trap
of Appleton and Sharp (1985) for *Mansonia* spp.

The square-framed base supports four brass rods forming a 550-mm
high pyramid covered with netting, to which a sleeve is attached for the
removal of mosquitoes. A total of 72 traps, each sampling 160 cm^2 of water, were used to collect *Mansonia uniformis* emerging from different areas

of a marsh. The mean catch/trap-night ranged from 1–20, with a median of one *Mansonia uniformis*/trap. These traps were later used by Sharp et al. (1987) to study the breeding places of *Mansonia uniformis, Mansonia africana, Culex theileri, Aedes durbanensis*, and various other *Culex, Aedes* and *Anopheles* species occurring in different vegetation zones of a swamp.

In brackish, forested wetland in Australia, Ryan et al. (2000) used modified pyramidal emergence traps based on those of Aubin et al. (1973). Each trap consisted of a 12 mm by 1200 mm wooden stake that was driven into the ground, on top of which a metal sleeve, a mesh cloth tent 1 m^2 at the base, and a plastic trap lid were fixed (Fig. 4.4). The four corners of the skirt of the tent were then fixed in position with metal spikes and covered with *Casuarinas* branches. A white plastic bottle (100 mm by 150 mm) containing a sticky cardboard insert (Austech International Pty Ltd, Mount Hawthorne, WA, Australia) was then screwed onto the lid atop the emergence trap to immobilise collected mosquitoes and to protect them from ants and other scavengers. Ten traps were set in each of two study areas and mosquito collections were retrieved weekly. A total of 825 adult mosquitoes were collected from both study areas over a 12-week period. Overall, *Ochlerotatus procax* was the most abundant species collected (55%), followed by *Verrallina funerea* (16%), *Culex halifaxii* (9%), *Uranotaenia pygmaea* (5%) and *Culex quinquefasciatus* (5%). The five other species collected made up only 9% of the total trap catch.

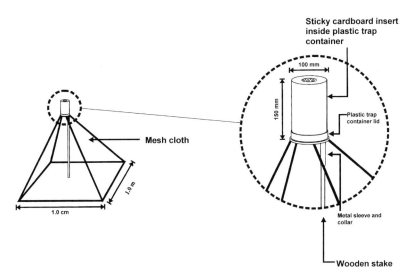

Fig. 4.4. Emergence trap of Ryan et al. (2000)

Walton et al. (1998) describe an emergence trap constructed from a frame of 1.3-cm-diameter PVC tubing fitted with fibreglass mesh window screening and a wide mouth (473-ml) Mason jar fitted with a plastic funnel placed at the apex of the trap to collect the emerging adults. The surface area of water enclosed by the traps was 0.25 m^2. Eighty-four traps were set in experimental constructed wetland in southern California, USA. Bulrush (*Schoenoplectus* spp.) plants were cut back to just above the water surface prior to placement of the traps. On average, the number of mosquitoes emerging declined from approximately 1500 adults/m^2/week in early July to 300 adults/m^2/week in early August. Application of the microbial insecticide *Bacillus thuringiensis* var. *israelensis* pellets at a rate of 19 kg/ha did not appreciably affect the mosquito emergence pattern. In a subsequent study (Workman and Walton 2000), traps were positioned over emergent vegetation at the shore (maximum depth 5 cm), in shallow water (maximum 30 cm depth) and deeper water (maximum depth 60 cm). Ninety-four per cent of the emerging mosquitoes collected were *Culex erythrothorax* at an average collection rate of 59 emergent adults per day per square metre. The other two most common species collected were *Culex quinquefasciatus* and *Culex tarsalis*. None of the three species demonstrated any preference for a specific water depth. Total numbers of emerged *Culex erythrothorax* adults had a strong positive correlation with emergent vegetation density. Placement of the traps amongst emergent vegetation was considered by the authors to have mitigated any potential effects of decreased light intensity under the traps on pupal avoidance or concentration.

LeSage and Harrison (1979) described two types of pyramidal traps for use on standing or lotic water. One type consists of a square base (0.37 m^2 opening) made of wood and fixed to four strips of polystyrene to provide extra buoyancy; but to provide a stronger base for use in running water and because eventually polystyrene loses it buoyancy, other traps had four 51-cm lengths of 10-cm diameter plastic drainage tubing joined to form a square. It was necessary to ensure that all joints were waterproof. One design of trap had to be emptied daily while another collected emerging insects in alcohol and was emptied about weekly. Predators, such as spiders, were reportedly a problem when traps were left unattended for several days. The mesh used to cover these pyramidal traps was very pale blue synthetic muslin. It was considered that traps made of polyethylene (Sublette and Dendy 1959) were inappropriate because condensation occurred within the traps, damaging the collection. The muslin apparently only reduced the light within the trap slightly, and therefore probably did not reduce the catch.

Floating trap of Morgan et al.

This is a floating box trap designed to minimise shadows cast on the water and to reduce wave damage to the catch (Morgan et al. 1963). It consists of a white painted narrow wooden framework supporting a plastic mesh collecting cage having a clear plexiglas roof (Fig. 4.5). Plexiglas panels (A) at the sides and a submerged apron of plexiglas (B) reduce wave action; the apron also retains floating larval and pupal exuviae. Service (1993) used this type of trap (modified by having a cloth sleeve in the roof (C) for the insertion of an aspirator) on a large pond to trap emerging adults of *Ochlerotatus annulipes*, but found it to be less than satisfactory as it was bulky, easily damaged and unless staked into position drifted to the sides of the pond in the slightest wind. Similar traps used by Edwards et al. (1964) in England to sample chironomids emerging from large ponds also caught adults of *Culex pipiens, Anopheles maculipennis* complex and *Culiseta annulata*, species that were not recovered in larval surveys.

In Finland 41 153 mosquitoes belonging to 18 species were caught over 3 years in about 20 floating box-like emergence traps each covering 0.25 m^2 in area (Brummer-Korvenkontio et al. 1971). These traps were basically of the type used by Hirvenoja (1960) and originally described by Mundie (1956) as a tent-like trap.

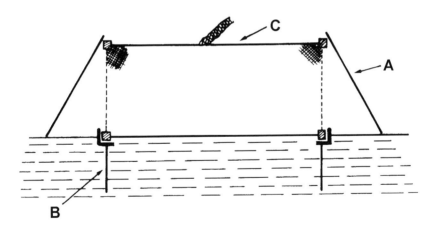

Fig. 4.5. Box trap showing A—plexiglas side panels, B—apron, C—roof with sleeve (after Morgan et al. 1963)

Non-floating traps and cages

These consist of various sized and shaped traps that are fixed in position with the lower edges below the water surface. Some traps rest firmly on the bottom of habitats and isolate an enclosed volume of water, others have the bottom edges raised thus permitting free movement of immature stages to and from under the trap.

Pyramid traps

A pyramidal trap used in Canada by James and Smith (1958) to sample *Chaoborus* species also caught *Ochlerotatus* mosquitoes. It consists of a 3-ft square wooden framework mounted on four pointed stakes driven into the mud to partially submerge the cage. The short upright sides of the cage were made of translucent plastic while the inwardly sloping upper sections were of mesh screening. Adults were collected in an inverted glass bottle placed over the apical opening. Larger box-like mesh screen cages (James 1957) were also used.

Pyramidal wooden frames covered with nylon screening were used as emergence traps for collecting *Ochlerotatus taeniorhynchus* and *Ochlerotatus sollicitans* in the USA (Vorgetts et al. 1980). The apex of the trap terminated in a plywood platform in the middle of which was inserted a 1-quart polyethylene ice-cream carton, the top of which was covered with netting. A hole was cut from the bottom to receive an inverted plastic funnel.

Lounibos and Escher (1983) made some useful emergence traps for sampling adult *Coquillettidia perturbans* in a phosphate pit that was covered with a mat of vegetation. Each trap consisted of a wooden or aluminium pyramidal frame about 1 m high, covered with plastic netting and with a base enclosing 4 m^2. The concentrating device of LeSage and Harrison (1979) and termed WEEK was incorporated in the trap thus allowing flying mosquitoes in the trap to fall into liquid preservative in a bottle that was replaced weekly. Maximum total catches from three traps were over 100 adults/week during mid-June and early September. Although *Coquillettidia perturbans* formed 73.9% of the catch, a few other species were also trapped, such as *Anopheles crucians* (17.5%) *Culex nigripalpus* (6.2%), *Uranotaenia sapphirina* (1.5%), *Culex erraticus* (0.4%) and *Mansonia titillans* (0.2%).

In another area of Florida, USA four similar traps were used to sample mosquitoes breeding amongst *Pistia stratiotes*. A total of 45 932 mosquitoes was caught over 3 years, mostly *Mansonia dyari* (89.7%), *Mansonia titillans* (6.2%) and *Culex erraticus* (1.5%), but a further 11 species were also trapped.

Prism traps

Ettinger (1979) describes a collapsible triangular prism-shaped trap (Fig. 4.6) that rests on the bottom of shallow waters.

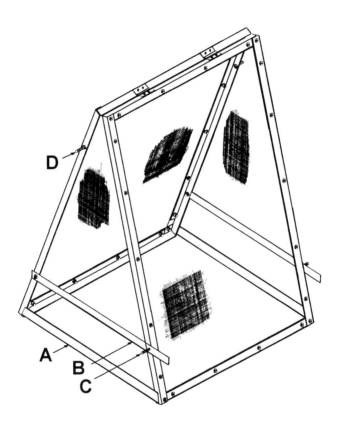

Fig. 4.6. Prism-type emergence trap showing A—rubber tape reinforcing folding side of net, B—aluminium strip used to keep trap open, C—bolt used to secure aluminium strip with trap open, D—bolt used to immobilise aluminium strip with trap closed (Ettinger 1979)

The two rectangular sides (31.5 × 47.3 cm) are made of angle-aluminium and flat aluminium strips are used to fix netting to cover these frames. The triangular sides (31.5 × 47.3 × 47.3 cm) are formed of netting reinforced along the the bottom with a strip of rubber tape (A). The trap, which is held open by bolting two strips of aluminium (B) to the frame, encloses 0.1 m^3 of water. After sampling emergent insects, the aluminium strips (B) are disengaged and immobilised with bolts (D), and the triangular netting pieces are folded into the traps, which is then closed and the insects taken to the laboratory enclosed within the trap. Alternatively, a netting sleeve could easily be fitted into one of the sides to allow the catch to be aspirated out. This trap was made to collect a variety of aquatic insects (e.g. Diptera, Ephemeroptera and Odonata), and because of its portability it might prove useful for trapping emerging mosquitoes.

Cylindrical traps

Large copper mesh screen cylinders with open ends were fastened to two iron stakes and thrust into marshes in Canada (Hocking et al. 1950). Each cylinder was provided with a detachable screen cover. To study the periodicities and succession of emergence of different *Ochlerotatus* species, larvae collected from elsewhere in the marshes were added to the cages to ensure large numbers of emerging adults. Emerging adults of *Ochlerotatus* and *Psorophora* species breeding in small woodland ponds were caught by thrusting into the bottom substrate a 5-gal metal can from which the top and bottom were removed and the top covered with netting. Emerging mosquitoes were collected by carefully lifting a small section of the netting and aspirating out the mosquitoes (Roberts and Scanlon 1979).

Box traps

Kimerle and Anderson (1967) reported that a box-like trap with three sides and roof of clear plastic and the remaining side of 'Nitex' screening, staked with the bottom edge about 3 in below the water surface was more efficient than a floating pyramid trap for sampling chironomids and chaoborids. A strong plastic base plate was slotted into grooves near the bottom of the trap to enclose adults before it was removed from the water. The catch was anaesthetised and emptied out of the box.

In studying *Coquillettidia perturbans* in Florida, USA Bidlingmayer (1968) placed 4-ft square bottomless cages over vegetation so that the lower edges were submerged. Adult mosquitoes were removed through a small cloth screen.

In Sweden a large plastic rectangular translucent box-like structure placed over a pool caught 1899 newly emerged adults of *Ochlerotatus communis, Ochlerotatus hexodontus* and *Ochlerotatus punctor*. All three species exhibited diurnal emergence (0600–1800 h), with less than 10% emerging between 2400–0600 h (Dahl 1973).

In Java Martono (1987) successfully used small (50 × 50 cm square, 100 cm tall) and large (1 × 1 m square, 1.5 m tall) emergence cages made of aluminium frames and covered with nylon netting, which were easily erected on site, to trap emerging adult *Anopheles aconitus*.

Bucket traps

In Canada Allan et al. (1981) made emergence cages for collecting *Coquillettidia perturbans* from domestic 2-gal capacity buckets, with a part of their sides removed and covered with plastic netting. They were inverted over water near, or enclosing, *Typha latifolia* or *Carex* spp., and were held in position by wire thrust into the bottom mud. Each trap covered 0.05 m^2 of water. Catches ranged from 0–14 from 41 traps, and the mean catch of *Coquillettidia perturbans* was 0.87 ± 0.34.

Corbet's emergence trap

Corbet (1965) considered the floating conical trap of Mundie (1956) unsatisfactory for several reasons, including the fact that in the high arctic mosquitoes, and other trapped insects, do not readily fly into collecting jars placed on the tops of such traps. Also the arrangement of the floats prevented the trap from being lowered to the bottom of ponds when the water level was very low. To compensate for decreasing water level a trap should be adjustable allowing it to be raised or lowered so that it stands over a standard depth of water. Corbet (1965) designed a trap specifically for sampling mosquitoes emerging from shallow water (Fig. 4.7). His trap consisted of a 19¼-in high aluminium cone, having a basal diameter of 14 in and an apical 5-in opening, covered with coloured saran mesh supported on four 18-in long struts. A petridish was tightly secured by a sprung cord over the top opening of the trap, which was provided with a rubber or plastic foam collar to ensure a tight fit. Insects in the trap are viewed from above through the glass petri-dish and are collected by inserting the long, curved, pliable copper collecting piece of a specifically designed aspirator through a slit in a double-rubber seal fixed into the netting of the trap. The trap is maintained with its bottom rim about 3 cm below the water surface by seating the trap on adjustable sliding brackets fixed to three rod-like legs. As the water level fluctuates trap height is adjusted, but when the water depth falls to about 5 cm the trap can be moved to a deeper part of the habitat.

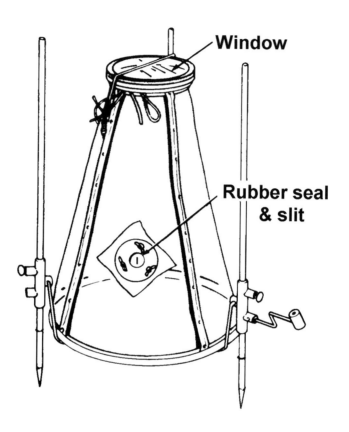

Fig. 4.7. Adjustable conical emergence trap (after Corbet 1965)

Using these traps to study the sex ratios of the high arctic mosquitoes, *Ochlerotatus impiger* and *Ochlerotatus nigripes*, Danks and Corbet (1973) found a marked predominance of females of both species emerging from temporary pools, and a slightly bigger overall proportion of females from permanent pools. The most likely cause was thought to be different mortality of the two sexes in the immature stages. Using the same type of traps Corbet and Danks (1973) compared the seasonal pattern of emergence of these two species from shallow ponds on Ellesmere Island during 1962–1966. Emergence began about 3 days earlier in *Ochlerotatus impiger* than in *Ochlerotatus nigripes*. In both species emergence was completed within 7–10 days and males emerged 1–2 days before the females.

Pritchard and Scholefield trap

Pritchard and Scholefield (1980) found that in many types of emergence traps mosquitoes remained in the cones and failed to enter the collecting bottles, so they developed a cone trap in which the cone itself acts as the collecting vessel. A 1-m pointed metal stake is passed through the centre of a 25-cm diameter aluminium alloy rim, the position of which on the stake can be adjusted by a clamp screw (Fig. 4.8).

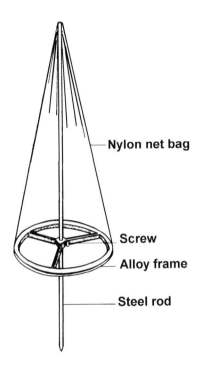

Nylon net bag

Screw

Alloy frame

Steel rod

Fig. 4.8. Cone-type emergence trap (Pritchard and Scholefield 1980)

A conical nylon bag having an elastic base is fitted tightly over the rim. The trap is pushed into the larval habitat until the metal rim is just below the water surface. To collect mosquitoes the bag is removed from the rim underwater and slipped over the metal stake. In the field the catch can be killed by placing the collecting bag in a plastic bag with chloroform, then the bag can be replaced on the trap. In Canada Pritchard and Scholefield (1980) found the trap gave good quantitative collections of emerging adult *Culiseta inornata*. Shade did not appear to affect the numbers of pupae under the traps and hence the numbers of mosquitoes emerging into it.

Large emergence cages

In England two different types of large cages have been used to collect large numbers of *Ochlerotatus cantans* emerging from heavily shaded woodland ditches. One type consists of natural coloured plastic mesh screening glued onto a pyramidal frame made of light alloy, 5 ft high and having a 4-ft square base. This cage rests on the bottom of the ditch but is raised about 2–3 ft each morning by pulling on a nylon cord attached to its apex and passed over a horizontal cross-bar. The collector stoops down and enters the cage to collect the mosquitoes with an aspirator. Predators, such as spiders, are also removed. The other type of cage is a 6-ft high, 4-ft wide and 8-ft long tubular steel framework joined at the corners with 'Klee-klamp' connections. A natural coloured plastic mesh cover, with canvas reinforced corners, is fitted over the framework. A heavy duty zip fastener, sewn along the entire length of one vertical edge, is unzipped to allow the collector to enter the cage.

Both types of cages, but especially the latter, proved exceptionally useful in ecological studies on *Ochlerotatus cantans*. They have provided absolute population estimates of emergent adults (Lakhani and Service 1974), adults of known age for mark–recapture experiments, and phenological data on emergence (Service 1977).

Packer and Corbet (1989) used similar large rectangular cages (2 × 1 × 1.5 m) to trap emerging *Ochlerotatus punctor* in England.

Mosquito nets

Since 1975 mosquito ecologists have frequently used single or double rectangular-type mosquito nets to sample emerging adults. The nets can be suspended by string from tree branches or from vertical poles thrust into ponds, or from a wooden frame. Nets having the bottom parts made of fabric, not netting, are preferable, as the fabric tends to snag less on vegetation in the ponds and also sinks in the water more easily than netting. In windy conditions the lower edges may have to be secured with string to short wooden stakes pushed into the bottom substrate to prevent the nets lifting and blowing in the wind.

Emergence traps for cesspits and sewerage systems

Several simple traps have been designed to catch mosquitoes emerging from cesspits, septic tanks, and underground sewerage systems. For example, Saliternik (1960) described a very simple, but apparently effective, trap to catch adult *Culex pipiens* form *molestus* emerging from cesspits. The trap is a 45-cm galvanised framework cube covered with mosquito netting and provided with a cord handle (Fig. 4.9). The base of the trap consists of a galvanised sheet extending 6 cm on two opposite sides. A 30-cm tall inverted mesh cone is placed over a 25-cm diameter hole cut from the floor of the trap. The apical opening of the cone is 2.5 cm in diameter. A sleeve in one side of the cage enables mosquitoes to be removed with an aspirator. When used in Israel the traps were usually placed over cracks and openings in cesspits about an hour before sunset.

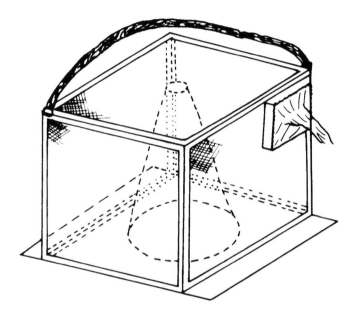

Fig. 4.9. Cesspit trap (after Saliternik 1960)

In Yangon, Myanmar, De Meillon et al. (1967) placed a wooden frame covered with mosquito gauze over a manhole cover of a septic tank that contained large numbers of larvae of *Culex quinquefasciatus*. Adults emerging from the tank were caught in the trap and removed hourly over a 24-h period. The peak times of emergence from the pupae were between 1800–1900 h and between 2000–2100 h, but there was only one peak of adults flying into the emergence trap, between 1800–1900 h. This suggested that adults emerging after about 1900 h remained within the septic tank until the following night. Batra et al. (1995) placed 60 × 60 × 60 cm nylon-net cages over partially covered or broken sewer manhole lids in Delhi, India, in order to study the potential of underground sewerage systems to act as a source of adult mosquitoes. The majority of emergence occurred between 1900 and 2000 h. The mean number of mosquitoes emerging per trap ranged from 5 to 1461 according to locality, with an overall average of 328 individuals per trap across all 21 localities. Only *Culex* mosquitoes were collected.

In Fiji Goettel et al. (1980) placed wire-framed cages (15-cm cube) over terminal air vents of septic tanks, and caught *Culex quinquefasciatus* in 7 of their 22 cages. In Papua New Guinea Charlwood and Galgal (1985) fitted a 50-cm cube wooden box (Fig. 4.10) over the vent pipe of a septic tank to trap newly emerging *Armigeres milnensis* escaping from the tank, as well as parous females that were leaving after ovipositing and also gravid females attempting to enter to lay eggs.

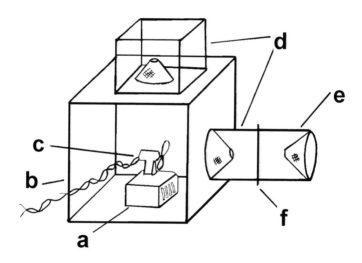

Fig. 4.10. Entrance/exit trap, a—septic tank vent, b—wooden box, c—fan and motor assembly, d—exit traps with wire mesh cones, e—entrance trap, f—mesh screen (Charlwood and Galgal 1985)

A metal exit trap (d) with a fine mesh cone is placed over a 14-cm diameter hole cut from the top of the box, and a cylindrical exit trap also with a mesh conical entrance (d) is placed over a similar hole cut from one side of the wooden box. The end of this second exit trap is covered with fine mesh (f), and a similar metal cylindrical trap with a cone mesh entrance (e) is fitted to the mesh end of this exit trap. A small fan run from a 4-V power supply (a) is placed inside the wooden box on top of the septic tank vent. This arrangement gently blows 'polluted air' through the side exit trap into the attached entry trap. The trap was operated for 30 min every hour, at other times the trap was removed to allow free entry and egress of mosquitoes. The maximum mean catch of female *Armigeres milnensis* entering the trap per 30 min was about 17, while the mean maximum catches of newly emerged females and those that had recently oviposited caught in the exit trap were about 19 and 8, respectively.

In Nigeria Irving-Bell et al. (1987) used much simpler so-called bucket traps to sample mosquitoes emerging from septic tanks. The top of the plastic bucket is covered with plastic mosquito screening and a 10-cm diameter hole is cut in the side to which is attached a netting sleeve. The sleeve is placed over a cardboard funnel fixed on top of the air vent of a pit latrine. In addition to *Culex quinquefasciatus* five other species were collected; the numbers of mosquitoes per trap-night ranged from 0 to over 900.

In Tanzania Curtis (1980) used exit traps fitted to pit latrines to catch and kill *Culex quinquefasciatus* and *Chrysomya putoria*. A box (about 1 m³) made from a wooden frame and covered with plastic or metal gauze was fixed to a plywood baseplate fitted with handles. A 15-cm hole was cut in the baseplate and bottom of the trap and covered with mosquito gauze. A 2-cm entry hole is made in the middle and is surrounded by a 3-cm high tube of gauze (Fig. 4.11). To use these traps as a sampling device, rather than to kill emerging insects, one or more cloth sleeves need to be incorporated in the box to allow the insertion of a battery-powered aspirator to remove the catch. There is the danger that ants, cockroaches and other pests may enter and eat the catch.

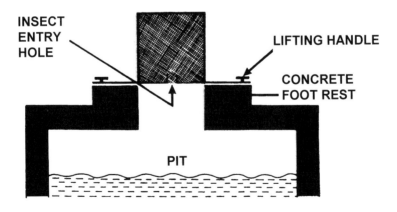

Fig. 4.11. Box-type pit latrine emergence trap (Curtis 1980)

In later studies in Tanzania Curtis and Hawkins (1982) converted 4-litre metal, or preferably plastic, paint buckets into emergence traps, by cutting a large hole from the bottom and covering it with plastic netting using strong contact adhesive. A similar hole was cut from the paint container lid and an inverted mesh funnel inserted (Fig. 4.12). The traps were fitted over vent pipes of pit latrines, over squatting plates or on pedestal latrine seats.

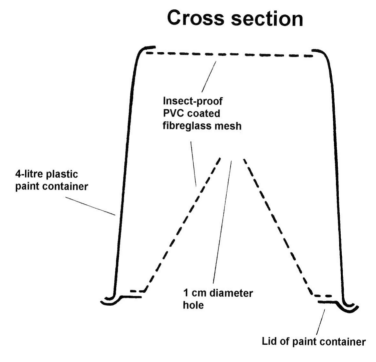

Fig. 4.12. Paint bucket-type pit latrine emergence trap (Curtis and Hawkins 1982)

Trapped insects were killed by spraying the trap with pyrethrum or a pyrethroid insecticide, after which the trap was washed before being reused. As many as 1248 mosquitoes, mainly *Culex quinquefasciatus*, were caught during a single day, and up to 6111 mosquitoes from 3 trap-days. When some pits had the trap reversed to act as an entry trap 141 gravid *Culex quinquefasciatus* were caught from 8 trap-days. These traps were later used by Chavasse et al. (1995) to assess the effectiveness of applying expanded polystyrene beads to the surface water of pit latrines and other enclosed water bodies as an intervention to control the breeding of *Culex quinquefasciatus* in Dar-es-Salaam, Tanzania.

In Kenya Subra and Dransfield (1984) used very simple traps (Fig. 4.13) to catch *Culex quinquefasciatus* and *Culex cinereus* emerging or entering pit latrines hourly over 24-h periods. The entry trap, consisting of a cylindrical wire cage and entrance funnel covered with mosquito netting, is placed in one of two holes in the top of the pit latrine (*a*), the other hole is left unobstructed for emerging mosquitoes. On removal of the trap adults are collected by inserting an aspirator into the sleeve at the end of the trap. Conversely with the exit trap one hole is left open for gravid mosquitoes to enter, while the other is covered with a 40-cm cube wire cage covered with mosquito netting (*b*). A piece of plywood can be slid under the trap to close it, and mosquitoes can be aspirated out through the netting sleeve.

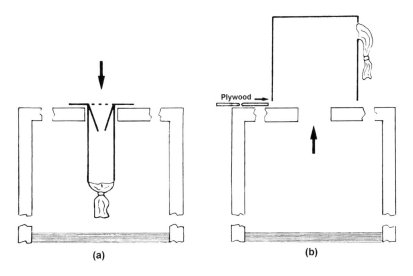

Fig. 4.13. Entry (*a*) and exit (*b*) pit latrine emergence traps of Subra and Dransfield (1984)

Girikumar and Rao (1984) constructed a cheap and simple emergence trap for use in pit latrines and ditches, consisting of two conical plastic electric lamp shades about 17-cm in diameter at their base, fixed one above the other by two 25-cm long wooden supports (Fig. 4.14). The top shade has a small hinged flap-like door, through which an aspirator, or tubing delivering carbon dioxide to anaesthetise the catch, can be inserted. A plastic mesh cone with a 1-cm apical opening is placed on the top of the bottom lamp shade. The area between the two lamp shades is covered with cotton or nylon mosquito netting. The base of the trap, which weighs about 150 g, is fixed to expanded polystyrene to enable it to float on the water. The trap is about 23 cm tall and 54 cm in circumference at its base.

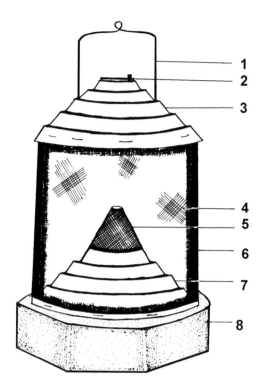

Fig. 4.14. Lantern-type pit latrine emergence trap, 1—handle, 2—hinged door, 3—upper lamp shade, 4—white mosquito netting, 5—cone, 6—wooden support, 7—lower lamp shade, 8—styrofoam float (Girikumar and Rao 1984)

Emergence trap for crab hole mosquitoes

Evans (1962) designed a trap to catch mosquitoes emerging from crawfish burrows that were so tortuous that breeding could only be detected by siphoning out water when this was only a few feet from the entrances. A large Kilner (Mason) jar with an inverted waxed paper cone inserted into its mouth is fixed over an 8-in copper mesh cylinder, which has the other end placed in the entrance of a crab or crawfish hole. The main purpose of the cylinder is to raise the collecting bottle from the ground to prevent submersion when the area is flooded. Evans (1962) only caught one *Culex quinquefasciatus* and seven *Psorophora confinnis* from 18 traps. He pointed out that some mosquitoes caught in the traps may not represent emerging adults but could be adults that have sought shelter in the burrows.

This could be confirmed by dissections to determine whether they are inseminated and the state of ovarian development.

In French Guyana Rivière et al. (1979) fitted small exit traps over crab holes to collect *Aedes polynesiensis*. Their trap consisted of a plastic funnel, fixed in position over the hole with an iron rod, with the opposite end leading to a small removable netting mosquito cage. After 9 days only 104 females, most of which were parous, and 94 male *Aedes polynesiensis* were caught. A major problem encountered was that crabs upset and damaged the traps as they emerged at night.

Emergence trap for tree-hole mosquitoes

Emergence traps were developed by Yates (1974) to obtain regular samples of emerging mosquitoes breeding in tree-holes. A 9-cm high, 7-cm diameter, glass collecting bottle with an inverted plastic cone taped into the mouth is slipped into a spring collar. The small end of a terylene mesh conical sleeve is glued onto the inside of the collar while the other end is tacked round the tree-hole with either drawing pins or staples, using a large stapler normally used for attaching labels to wooden crates (Fig. 4.15). To prevent the mesh cone from tearing and to get a better seal round the hole, the staples are passed through a shoe lace that encircles the tree-hole.

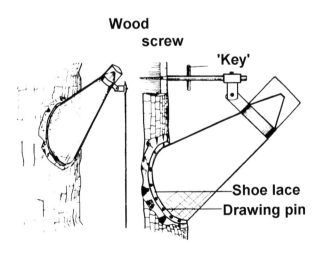

Fig. 4.15. Tree-hole emergence trap (after Yates 1974)

The trap is fixed by a sliding bush to a 7/16-in diameter steel rod that is driven into the ground when the trap is used on tree-holes near the base of the tree. For use on tree-holes high above the ground a ¼-in wood screw is brazed onto the end of the rod which, by passing a 'key' through it, can be screwed horizontally into the tree. A simpler arrangement for horizontal tree-holes near the ground is to screw the metal collar supporting the collecting bottle onto a wooden stake that is driven into the ground. To prevent corrosion all metal parts are painted with a proprietary cold galvanising paint. These traps have been successfully employed in England to study the seasonal emergence of *Ochlerotatus geniculatus* and other mosquitoes (Yates 1979).

Washburn et al. (1989) made very simple traps to collect and study temporal patterns of adult *Ochlerotatus sierrensis* emerging from tree-holes. Two layers of nylon netting are wrapped round a supporting wire framework, which has a foam sponge base to ensure a tight fit when fitted over a tree opening. The apex of the trap tapers into a 8.5-cm diameter short length of corrugated plastic hose sealed at the end with nylon netting. From 17 traps used over 149 days 8 444 adult *Ochlerotatus sierrensis* were caught.

Morris (1984) used simple mesh cone emergence traps with a large cardboard carton over the top (Pierson and Morris 1982) fitted to holes of tree-root hummocks to sample *Culiseta melanura* and *Culiseta morsitans*.

Light-traps for sampling emergent insects

Chandler and Highton (1975, 1976) used light-traps to catch mosquitoes emerging from rice fields in Kenya. The 11.2-cm diameter open end of an elbowed metal cylinder is positioned about 10 cm above the water surface (Fig. 4.16). The light source consists of a 3.8-V torch bulb operated through a 33-Ω resistor from a 12-V car battery placed on the bank of the rice field. A three-bladed fan with a displacement of 11.06 m^3 air/min through the trap sucks up mosquitoes and delivers them into a large white terylene netting bag. Traps were run from 1900–0630 h.

Using this trap in Kenya Chandler and Highton (1975) caught 10 782 mosquitoes belonging to 35 species in 655 trap-nights. The six most common being *Culex poicilipes, Culex antennatus, Anopheles ziemanni, Mansonia uniformis, Culex aurantapex* and *Anopheles arabiensis*. Since only 0.9% of the females caught were blood-fed and only 5.1% were gravid, and as more than 85% of the males in the traps had incompletely rotated hypopygia, it was concluded that the traps sampled adults shortly after emergence and attracted very few older individuals. Careful checks are required that this type of trap is sampling newly emerged adults and not attracting large numbers of older individuals.

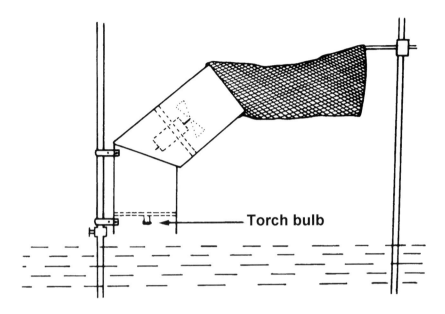

Torch bulb

Fig. 4.16. Emergence light-trap as used by Chandler and Highton (1975)

Emergence wheel of Corbet

Although not strictly an emergence trap it is appropriate to describe this piece of apparatus, which was used by Corbet (1966) to study natural diel periodicities of emergence of *Ochlerotatus impiger* and *Ochlerotatus nigripes* in the field. These species exhibit closely synchronised emergence, about 75% of which is completed within 4–6 days. In addition, wind can complicate matters by disturbing or postponing emergence. To eliminate these difficulties, the 'emergence wheel' was invented. Individual pupae are placed in pond water in glass vials (18 × 65 mm), which are immersed in ponds so that the water levels in and outside the vials are the same. The vials are placed in holes in a wooden or metal disc (wheel), about 80 cm in diameter, mounted on a short central axle to facilitate controlling their position and allowing for inspection. Two hundred and eighty-seven glass vials containing individual pupae are inserted in the wheel along 24 radii in alternating groups of six and nine, the remaining 107 vials being positioned

364 Chapter 4 Sampling the Emerging Adult Population

around the perimeter. At 3-h intervals the wheel is raised, the number of emerged adults noted, and the wheel lowered again but rotated 45° clockwise (15° for each hour). This prevents shadows of the central axle falling on each line of vials more than once a day, thus eliminating any effect of a time cue. Corbet (1966) found that both sexes of *Ochlerotatus impiger* exhibited a marked peak in emergence near or soon after solar noon, and a depression near solar midnight, times when temperatures were highest and lowest, respectively. Female *Ochlerotatus nigripes* showed somewhat similar but less pronounced diel periodicity of emergence, but this was not evident in the males. Corbet concluded that on Ellesmere Island emergence of both species was governed mainly by temperature.

Sticky emergence traps

In Massachusetts, Armstrong (1941) constructed sticky traps to sample emerging adults of *Coquillettidia perturbans*. Each trap consisted of a four-sided wire mesh enclosure which was thrust into the water with 'about a foot or more' of the cage projecting above the water. A horizontal cloth roof coated with 'Tanglefoot' was placed on top of the cage. As many as 1783 adults were caught in one trap during 'the season'.

Slaff et al. (1984) used a sticky emergence trap to sample *Mansonia titillans* and *Mansonia dyari* because they considered that not all mosquitoes passed through the inverted cones and baffles incorporated in most emergence traps. They designed a pyramidal trap having a square wooden framework base enclosing 36 in^2 of water. Aluminium angle side supports rise from each corner and slope inwards to support a 10 in^2 plastic top. The sides of the trap are covered with plastic screening. Clear self-adhesive sheets of plastic film are applied to the inside of the four sloping sides, and then painted with 'Tack-trap'. Emerging mosquitoes are trapped on this adhesive, and the plastic sheeting can easily be removed. Laboratory and field tests showed a sticky emergence trap caught 2.5–3.3 times as many mosquitoes as a baffle-type trap, and moreover was about half the cost of the baffle trap. In Florida, USA Slaff and Haefner (1985) used these pyramidal emergence traps to sample *Mansonia dyari, Mansonia titillans* and *Coquillettidia perturbans*. Areas of water containing the water lettuce (*Pistia stratiotes*) produced more emergent adults of both *Mansonia* species than water covered with water hyacinths (*Eichhornia crassipes*) or cattails (*Typha* spp.).

Kay et al. (2000) designed a simple sticky pipe trap for sampling mosquitoes entering and exiting drainage manholes in Australia (see Chap. 6 for a detailed description).

Sampling the Total emergent population

The total emergent population of mosquitoes could be fairly easily obtained from small habitats, such as tree-holes, crab holes and rock holes, although in fact this has very rarely been attempted. It is occasionally feasible to cover larger habitats with cages to catch all the emerging adults. These cages will prevent oviposition in the habitats, although this may not always be a serious disadvantage. In southern England the egg population of *Ochlerotatus cantans* was estimated from soil and leaf litter samples collected from a 5-m diameter pond in December, a time when the whole population was in the egg state. The complete habitat was then covered with a cage of clear plastic netting tacked to a wooden framework and all adults emerging the following April–June from the overwintering population were caught. These two procedures enabled the mortality from egg to emergent adult to be estimated. The cage remained permanently over the habitat to prevent further oviposition, and during subsequent years was used to collect adults arising from eggs that had failed to hatch in previous years (Service 1977).

Indirect methods of estimating emergence

Estimates of adult emergent populations can sometimes be made by the daily counting and removal of pupae exuviae. For example, this method has been employed with dragonflies (Corbet 1957, Pajunen 1962) but their exuviae are usually attached to vegetation and are larger and more conspicuous than those of mosquitoes. With mosquitoes the method is more feasible in small container habitats such as water butts and tanks, concrete drains, and rock pools, than in ground collections where exuviae tend to become stranded at the edge of habitats and overlooked. Counts of pupal skins give the number of mosquitoes that have emerged, but not necessarily those that have successfully emerged, because some adults may die during or very shortly after emergence. Observations, for example, have indicated that predation by various adult Diptera on emerging mosquitoes may be more important than generally realised (Service 1973*a,b*).

An even more indirect, and rarely satisfactory, method is the deduction of emergence rates from the capture of newly-emerged adults, principally from resting places near larval habitats. Newly emerged individuals can often be detected by incomplete rotation of male terminalia, presence of meconium, stage of ovarian development and possibly absence of insemination, but in some species copulation may occur within hours of emergence. Apart from lacking any precision these methods can rarely identify adults as emerging from specific habitats.

In India Reuben et al. (1978) searched 1500–2000 houses and collected all pupae of *Aedes aegypti* breeding in domestic containers, from this they estimated the total population of pupae (*T*) by simple proportion as:

$$T = \frac{\text{Total number pupae} \times \text{total number houses}}{\text{No. houses sampled}} \tag{4.1}$$

The estimated number of pupae divided by mean pupal duration gave an estimate of the daily numbers of emerging adults, which ranged from 316/day in February to 23 022/day in June.

In California, USA Nelson et al. (1978) estimated daily emergence (E_t) of *Culex tarsalis* by the differences between daily population estimates (N_t) obtained by mark–recapture methods, and the estimated survivors from the previous day (sN_{t-1}), where survival rates (s) were estimated from mark–recapture results, or from parous rates (Davidson 1954), thus

$$E_t = N_t - sN_{t-1} \tag{4.2}$$

References

Allan SA, Surgeoner GA, Helson BV, Pengelly DH (1981) Seasonal activity of *Mansonia perturbans* adults (Diptera: Culicidae) in southwestern Ontario. Can Entomol 113: 113–9

Andreasen MH, Silver JB, Renshaw M (1997) Emergence pattern, size variation and reproductive success of *Aedes rusticus* (Diptera: Culicidae) in northern England. The Entomologist 116: 153–162

Appleton CC, Sharp BL (1985) A preliminary study on the emergence of *Mansonia uniformis* (Diptera: Culicidae) from swamps at Richards Bay, Natal, South Africa. J Entomol Soc Southern Afr 48: 179–184

Armstrong RL (1941) *Mansonia perturbans* (Walk.) on Cape Cod. Proc New Jers Mosq Exterm Assoc 28: 184–188

Aubin A, Bourassa JP, Pellissier M (1973) An effective emergence trap for the capture of mosquitoes. Mosquito News 33: 251–252

Batra CP, Mittal PK, Adak T (1995) A study on the mosquito emergence from the underground sewerage system in some areas of Delhi. Indian J Malariol 32: 85–88

Bidlingmayer WL (1968) Larval development of *Mansonia* mosquitoes in central Florida. Mosquito News 28: 51–57

Bradley GH (1926) Observations on the emergence of *Anopheles* mosquitoes. Am J Trop Med 6: 283–297

Brummer-Korvenkontio M, Korhonen P, Hameen-Antilla R (1971) Ecology and phenology of mosquitoes (Dipt., Culicidae) inhabiting small pools in Finland. Acta Entomol Fenn 28: 51–73

Castleberry DT, Cech JJ, Kristensen AB (1989) Evaluation of the effect of varying mosquito emergence on the efficiency of emergence traps over enclosed environments. J Am Mosq Control Assoc 5: 104–105

Chandler JA, Highton RB (1975) The succession of mosquito species (Diptera: Culicidae) in rice fields in the Kisumu area of Kenya, and their possible control. Bull Entomol Res 65: 295–302

Chandler JA, Highton RB (1976) The breeding of *Anopheles gambiae* Giles (Diptera: Culicidae) in rice fields in the Kisumu area of Kenya. J Med Entomol 13: 211–215

Charlwood JD, Galgal K (1985) Observations on the biology and behaviour of *Armigeres milnensis* Lee (Diptera: Culicidae) in Papua New Guinea. J Aust Entomol Soc 24: 313–19

Chavasse DC, Lines JD, Ichimori K, Majala AR, Minjas JN, Marijani J (1995) Mosquito control in Dar es Salaam. II. Impact of expanded polystyrene beads and pyriproxyfen treatment of breeding sites on *Culex quinquefasciatus* densities. Med Vet Entomol 9: 147–154

Corbet PS (1957) The life-history of the Emperor Dragonfly *Anax imperator* Leach (Odonata: Aeshnidae). J Anim Ecol 26: 1–69

Corbet PS (1964) Temporal patterns of emergence in aquatic insects. Can Entomol 96: 264–279

Corbet PS (1965) An insect emergence trap for quantitative studies in shallow ponds. Can Entomol 97: 845–858

Corbet PS (1966) Diel patterns of mosquito activity in a high-arctic locality: Hazan Camp, Ellesmere Island, N.W.T. Can Entomol 98: 1238–1252

Corbet PS, Danks HV (1973) Seasonal emergence and activity of mosquitoes (Diptera: Culicidae) in a high-arctic locality. Can Entomol 105: 837–872

Curtis C (1980) Insect traps for pit latrines. Mosquito News 40: 626–628

Curtis CF, Hawkins P (1982) Entomological studies of on-site sanitation systems in Botswana and Tanzania. Trans R Soc Trop Med Hyg 76: 99–108

Dahl C (1973) Emergence and its diel periodicity in *Aedes (O.) communis* (DeG.), *punctor* (Kirby) and *hexodontus* Dyar in Swedish Lapland. Aquilo: Seria Zool 14: 34–45

Danks HV, Corbet PS (1973) Sex ratios at emergence of two species of high-arctic *Aedes* (Diptera: Culicidae). Can Entomol 105: 647–651

Davidson G (1954) Estimation of the survival-rate of anopheline mosquitoes in nature. Nature 174, 792–793

Davies IJ (1984) Sampling aquatic insect emergence. In: Downing JA, Rigler FH (eds) Manual on Methods for the Assessment of Secondary Productivity in Fresh Waters (2nd edn). IBP Handbook, No. 17. Blackwell Scientific Publications, Oxford, pp 161–227

De Meillon B, Sebastian A, Khan ZH (1967) Exodus from a breeding place and time of emergence from the pupa of *Culex pipiens fatigans*. Bull World Health Organ 36: 163–167

Edwards RW, Egan H, Learner MA, Maris PJ (1964) The control of chironomid larvae in ponds using TDE (DDD). J Appl Ecol 1: 97–117

Ettinger WS (1979) A collapsible insect emergence trap for use in shallow standing water. Entomol News 90: 114–117

Evans BR (1962) Survey for possible mosquito breeding in crawfish holes in New Orleans, Louisiana. Mosquito News 22: 255–257

Girikumar A, Rao PV (1984) A low cost floating cage to trap emerging mosquitoes under urban or rural conditions. Mosquito News 44: 416–417

Goettel MS, Toohey MK, Pillai JS (1980) The urban mosquitoes of Suva, Fiji; seasonal incidence and evaluation of environmental sanitation and ULV spraying for their control. J Trop Med Hyg 83: 165–171

Hirvenoja M (1960) Ökologische Studien über die Wasserinsekten in Riihimäki (Südfinnland). I. Chaoborinae (Dipt., Culicidae). Ann Entomol Fenn 26: 31–44

Hocking B, Richards WR, Twinn CR (1950) Observations on the bionomics of some northern mosquito species (Culicidae: Diptera). Can J Res (D), 28: 58–80

Irving-Bell RJ, Okoli EI, Diyelong DYEO, Onyia OC (1987) Septic tank mosquitoes: competition between species in central Nigeria. Med Vet Entomol 1: 243–250

James HG (1957) *Mochlonyx velutinus* (Ruthe) (Diptera: Culicidae), an occasional predator of mosquito larvae. Can Entomol 89: 470–480

James HG, Smith BC (1958) Observations on three species of *Chaoborus* Licht. (Diptera: Culicidae) at Churchill, Manitoba. Mosquito News 18, 242–248

Jónasson PM (1954) An improved funnel trap for capturing emerging aquatic insects with some preliminary results. Oikos 5: 179–188

Kay BH, Sutton KA, Russell BM (2000). A sticky entry-exit trap for sampling mosquitoes in subterranean habitats. J Am Mosq Control Assoc 16: 262–265

Kimerle RA, Anderson NH (1967) Evaluation of aquatic insect emergence traps. J Econ Entomol 60: 1255–1259

Lakhani KH, Service MW (1974) Estimated mortalities of the immature stages of *Aedes cantans* (Meigen) (Dipt., Culicidae) in a natural habitat. Bull Entomol Res 64: 265–276

LeSage L, Harrison AD (1979) Improved traps and techniques for the study of emerging aquatic insects. Entomol News 90: 65–78

Lounibos LP, Escher RL (1983) Seasonality and sampling of *Coquillettidia perturbans* (Diptera: Culicidae) in south Florida. Environ Entomol 12: 1087–1093

Martono (1987). An experiment on mosquito capturing technique using a de-mountable cage. Bull Penel Keseh Health Std Indonesia 15: 29–31

Merritt RW, Cummins KW, Resh VH (1984) Collecting sampling and rearing methods for aquatic insects. In: Merritt RW, Cummins KW (eds) An Introduction to the Aquatic Insects of North America. 2nd edn. Kendall/Hunt Publishing Co., Dubuque, Iowa, pp 11–26

Morgan NC (1971) Factors in the design and selection of insect emergence traps. In: Edmondson WT, Winberg GG (eds) A Manual on Methods for the Assessment of Secondary Productivity in Fresh Waters. IBP Handbook, No. 17. Blackwell Scientific Publications, Oxford, pp 93–108

Morgan NC, Waddell AB, Hall WB (1963) A comparison of the catches of emerging aquatic insects in floating box and submerged funnel traps. J Anim Ecol 32: 203–219

Morris CD (1984) Phenology of trophic and gonobiological states in *Culiseta morsitans* and *Culiseta melanura* (Diptera: Culicidae). J Med Entomol 21: 38–51

Mundie JH (1956) Emergence traps for aquatic insects. Mitt Int Verein Theor Angew Limnol 7: 1–13

Mundie JH (1971) Techniques for sampling emerging aquatic insects. In: Edmondson WT, Winberg GG (eds) A Manual on Methods for the Assessment of Secondary Productivity in Fresh Waters. IBP Handbook, No. 17. Blackwell Scientific Publications, Oxford, pp 80–93

Nelson RL, Milby MM, Reeves WC, Fine PEM (1978) Estimates of survival, population size, and emergence of *Culex tarsalis* at an isolated site. Ann Entomol Soc Am 71: 801–808

Packer MJ, Corbet PS (1989) Seasonal emergence, host-seeking activity, age composition and reproductive biology of the mosquito *Aedes punctor*. Ecol Entomol 14: 433–442

Pajunen VI (1962) Studies on the population ecology of *Leucorrhinia dubia* v.d. Lind. (Odon., Libellulidae). Ann Zool Soc Zool-Bot Vanamo 24: 1–79

Pierson JW, Morris CD (1982) Epizootiology of eastern equine encephalomyelitis virus in upstate New York, USA. IV. Distribution of *Culiseta* (Diptera: Culicidae) larvae in a freshwater swamp. J Med Entomol 19: 423–428

Pritchard G, Scholefield PJ (1980) An adult emergence trap for use in small shallow ponds. Mosquito News 40: 294–296

Renshaw M (1991) Population dynamics and ecology of *Aedes cantans* (Diptera: Culicidae) in England. PhD thesis, University of Liverpool

Reuben R, Das PK, Samuel D, Brooks GD (1978) Estimation of daily emergence of *Aedes aegypti* (Diptera: Culicidae) in Sonepat, India. J Med Entomol 14: 705–714

Rivière F, Pichon G, Chebret M (1979) Écologie d'*Aedes (Stegomyia) polynesiensis* Marks, 1951 (Diptera, Culicidae) en Polynésie Française. I Lieux de repos des adultes. Application dans la lutte antimoustique à Bora-Bora. Cah ORSTOM sér Entomol Méd Parasitol 17: 235–241

Roberts DR, Scanlon JE (1979) Field studies on the population biology of immature stages of six woodland mosquito species in the Houston, Texas area. Mosquito News 39: 26–34

Ross R (1910) The Prevention of Malaria. John Murray, London

Ryan PA, Kay BH (2000) Emergence trapping of mosquitoes (Diptera: Culicidae) in brackish forest habitats in Maroochy Shire, south-east Queensland, Australia, and a management option for *Verrallina funerea* (Theobald) and *Aedes procax* (Skuse). Aust J Entomol 39: 212–218

Saliternik Z (1960) A mosquito light trap for use on cesspits. Mosquito News 20: 295–296

Scott W, Opdyke DF (1941) The emergence of insects from Winona Lake. Invest Ind Lakes 2: 3–14

Service MW (1973a) Mortalities of the larvae of the *Anopheles gambiae* complex and detection of predators by the precipitin test. Bull Entomol Res 62: 359–369

Service MW (1973b) Study of the natural predators of *Aedes cantans* (Meigen) using the precipitin test. J Med Entomol 10: 503–510

Service MW (1977) Ecological and biological studies on *Aedes cantans* (Meig.) (Diptera: Culicidae) in southern England. J Appl Ecol 14: 159–196

Sharp BL, Appleton CC, Thompson DL, Meenehan G (1987) Anthropophilic mosquitoes at Richards Bay, Natal, and arbovirus antibodies in human residents. Trans R Soc Trop Med Hyg 81: 197–201

Slaff M (1986) Predicting the spring emergence of *Coquillettidia perturbans*. J Am Mosq Control Assoc 2: 227–228

Slaff M, Haefner JD (1985) Seasonal and spatial distribution of *Mansonia dyari*, *Mansonia titillans*, and *Coquillettidia perturbans* (Diptera: Culicidae) in the central Florida, USA, phosphate region. J Med Entomol 22: 624–629

Slaff M, Haefner JD, Parsons RE, Wilson F (1984) A modified pyramidal emergence trap for collecting mosquitoes. Mosquito News 44: 197–199

Smith BP, McIver SB (1984) The patterns of mosquito emergence (Diptera: Culicidae; *Aedes* spp.): their influence on host selection by parasitic mites (Acari: Arrenuridae; *Arrenurus* spp.). Can J Zool 62: 1106–1113

Southwood TRE (1978) Ecological Methods with Particular Reference to the Study of Insect Populations. Chapman and Hall, London

Sublette JE, Dendy JS (1959) Plastic materials for simplified tent and funnel traps. S West Nat 3: 220–223

Subra R, Dransfield RD (1984) Field observations on competitive displacement, at the preimaginal stage, of *Culex quinquefasciatus* Say by *Culex cinereus* Theobald (Diptera: Culicidae) at the Kenya coast. Bull Entomol Res 74: 559–568

Taketo A (1960). Studies on the life-history of *Tanpteryx pryeri* Selys (Odonata, Petaluridae). 1. Observations on adult dragonflies. Kontyû 28: 97–109 (In Japanese, English summary)

Vorgetts J, Ezell WB, Campbell JD (1980) Species composition of mosquitoes produced in dredged material, wildlife management, and natural saltmarsh habitats of the south Carolina coast. Mosquito News 40: 501–506

Walton WE, Workman PD, Randall LA, Jiannino JA, Offill YA (1998) Effectiveness of control measures against mosquitoes at a constructed wetland in southern California. J Vector Ecol 23: 149–160

Washburn JO, Anderson JR, Mercer DR (1989) Emergence characteristics of *Aedes sierrensis* (Diptera: Culicidae) from California treeholes with particular reference to parasite loads. J Med Entomol 26: 173–182

Workman PD, Walton WE (2000) Emergence patterns of *Culex* mosquitoes at an experimental constructed treatment wetland in southern California. J Am Mosq Control Assoc 16: 124–130

Yates M (1974) An emergence trap for sampling adult tree-hole mosquitoes. Entomologist's Monthly Magazine 109: 99–101

Yates MG (1979) The biology of the tree-hole breeding mosquito *Aedes geniculatus* (Olivier) (Diptera: Culicidae) in southern England. Bull Entomol Res 69: 611–628

Chapter 5 Sampling the Adult Resting Population

Animal bait catches (see Chap. 6) usually catch only unfed females in search of a blood-meal and the choice of bait species will in many cases determine the species of mosquito sampled. Attractant traps (Chap. 11) also predominantly sample host seeking females, and may often attract only certain species. Non-attractant traps (Chap. 8) give less biased collections of mosquitoes and, at least theoretically, should sample all species more or less equally. However, they only sample the proportion of the population that is active and airborne, which again mainly comprises unfed females, although these may not all be actively host seeking. Non-attractant traps also sample the active proportion of the male population better than attractant traps. Adult mosquitoes probably spend a majority of their time resting in natural or man-made shelters, and are thus unavailable to be sampled using bait catches, or attractant and non-attractant traps. Resting collections sample unfed females, including those not actively host seeking, both blood-fed and gravid females and also males. Resting collections therefore provide much additional information and will yield samples of the population that more accurately represent the sex ratio, age structure and physiological condition of the population as a whole. The collection of resting blood-engorged adult females can be used to study natural host preferences, for example.

Only very few mosquito species commonly rest in man-made shelters, most rest in natural shelters, such as amongst vegetation, in hollow trees, animal burrows, and crevices in the ground, etc. Searches for outdoor resting mosquitoes have frequently proved time-consuming and unrewarding (e.g. Bahang et al. 1984; Bown and Bang 1980; Downe 1960; Magnarelli 1977b; Muirhead-Thomson 1956; Muirhead-Thomson and Mercier 1952; Ree et al. 1976; Rehn et al. 1950). However, worthwhile numbers of mosquitoes have sometimes been caught from vegetation with oral and motorised aspirators or by sweep-netting and successes have been achieved through the use of several designs of artificial resting shelter (Breeland 1972a,b; Chandler et al. 1976; Cordellier et al. 1983; Day et al. 1990; Komar et al. 1995; McHugh 1989; Mutero et al. 1984; Nasci and Edman 1984; Nayar 1982; Senior White 1951; Wharton 1950).

The indoor resting population

Although comparatively few mosquito species regularly rest in human and animal habitations many of those that do are important vectors of malaria, filariasis and more rarely arboviruses, and hence have been the subject of considerable study.

Mosquitoes are usually caught from houses and animal quarters using manual or mechanical aspirators or by knock-down pyrethrum spray collections.

Hand-held or head-mounted torches are frequently used to both locate indoor resting mosquitoes and also to aid the collection of adults that have been knocked down by space spraying. The light source should not be so powerful that it causes resting mosquitoes to take flight when they are illuminated. When collecting mosquitoes from African huts Haddow (1942) placed greaseproof paper over the torch to diffuse the light. Grimstad and DeFoliart (1974) collected mosquitoes visiting flowers at night with a torch strapped to the head that used a red filter over the bulb. The use of red filters or bulbs has also often been used in bait collections to reduce the likelihood of disturbing mosquitoes settling on the bait.

Collecting tubes

The simplest, but not necessarily the most efficient, method of collecting resting adults is to carefully place a plastic or glass tube, such as a test tube, over them, causing them to fly upwards into the tube, which can then be sealed with a piece of cotton wool (Russell and Baisas 1935; Russell et al. 1963, World Health Organization 1992) (Fig. 5.1). Several adults can be collected in each individual tube by pushing the cotton wool down to confine the mosquito to the end of the tube. Subsequently captured individuals are isolated between additional cotton wool plugs. Collecting tubes can be placed in a domestic freezer to kill the mosquitoes collected. Alternatively, a collecting tube can be made into a killing tube by placing cut-up pieces of rubber bands or other small pieces of rubber in the bottom of a suitable sized test tube (e.g. 150 × 18 or 24 mm) and flooding them with chloroform. After allowing the liquid to be soaked up the excess is poured away and a layer of absorbent cotton wool inserted, which is then covered with 1–2 layers of white blotting or filter paper. Care should be taken to keep the paper layers dry. If the paper becomes wet due to too much chloroform remaining in the tube, then the tube should be left uncorked until the excess has evaporated.

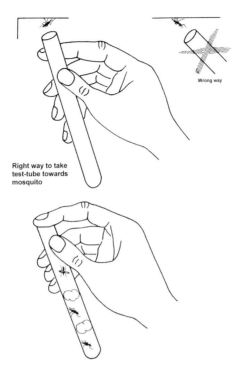

Fig. 5.1. Collecting resting mosquitoes with a test-tube (World Health Organization 1992)

This type of killing tube remains effective for several weeks, if not months. To recharge it the cotton wool and filter paper discs are removed and the rubber bands resoaked with chloroform. Alternatively, crushed potassium or sodium cyanide is placed in the bottom 6 mm of the tube and covered with 1.3-cm layer of dry plaster of Paris or sawdust and 2–3 filter paper discs. Finally a wet mixture of plaster of Paris is added to the tube. A roll of paper, which can afterwards be withdrawn, can be inserted into the tube while adding the wet plaster of Paris to keep the sides of the tube clean. A small plastic conical funnel fixed into the mouth of the tube helps to reduce the evaporation of the killing agent and prevent mosquitoes from flying or falling out (King et al. 1939). A tube with cyanide lasts much longer than one with chloroform but is not rechargeable and is dangerous if broken. Furthermore, there may be difficulties in safely disposing of it when no longer required. Burton (1954) made a killing tube from 6 × 1¼ in. 'Lusteroid plastic' test tubes. Killing agents such as carbon tetrachloride or chloroform were soaked up by either cut-up rubber bands placed at the bottom and held in place by a wedge of cellulose sponge or directly by the sponge. Chloroform, however, may gradually damage plastic test tubes and ether must be avoided as it dissolves both the sponge and test tube. Practice may be needed in positioning a tube over a mosquito because toxic vapours escaping from it may cause the mosquito to fly away before it can be caught. Mosquitoes can be collected directly into killing tubes or blown into them from aspirators.

Oral aspirators

Mosquitoes found resting on various surfaces are frequently collected with an aspirator, sometimes referred to as a 'pooter'. Such collections are commonly referred to as 'hand-catches'. The simplest and most widely used type of aspirator is made from a 30–45-cm length of 8–12-mm internal diameter plastic or glass tubing. A piece of mosquito netting is taped either directly over one end, or over a short piece of smaller diameter tubing, which is inserted into the end of the larger tubing. A 50-cm length of rubber or pliable plastic tubing is slipped over the end of the glass tubing, and a small section of plastic or glass tubing is inserted into the opposite end to form a mouthpiece (Fig. 5.2).

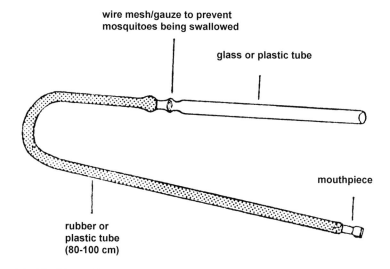

Fig. 5.2. World Health Organization oral aspirator (WHO 1992)

Several mosquitoes can be sucked up into the aspirator before they are gently blown into a suitable storage container. Blood-engorged adults must be sucked up gently otherwise their abdomens may be damaged by hitting the netting barrier at the end of the tube. Captured mosquitoes are prevented from flying out by stopping the open end of the tube with a thumb, finger or piece of cotton wool. Aspirators with reservoirs (Fig 5.3 *a–d*) can be used to reduce the likelihood of mosquitoes escaping and to enable larger numbers to be collected before the aspirators need emptying. In these aspirators the collecting tube has to be removed before the adults can be extracted. Ryan (1989) suggested a small modification to reservoir-type pooters to allow insects to be stunned or killed. He placed together citric acid crystals and sodium hydrogen bicarbonate at the bottom of the pooter and covered them with a thin plug of cellulose wadding. Insects were sucked up in the normal manner and then a few drops of water were carefully introduced down the glass entry tube of the pooter, resulting in the production of carbon dioxide. When stunned, the insects were either tipped out and sorted, or the reservoir removed from the pooter and capped with a cork, and replaced with another tube containing citric acid and bicarbonate.

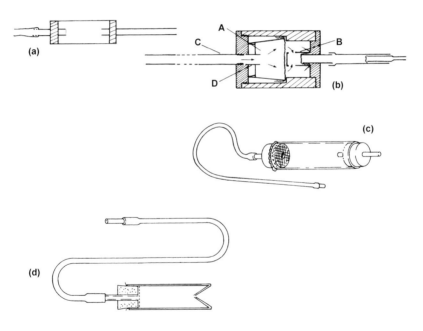

Fig. 5.3. (*a*) Aspirator with reservoir; (*b*) aspirator with A—removable cardboard cup showing B—plastic tubing with holes near its end, C—plastic collecting tube, D—tubing glued to bottom of the cup (after Coluzzi and Petrarca 1973); (*c*) and (*d*) reservoir-type aspirators (after World Health Organization 1975)

Various sized plastic or paper cups covered with mosquito netting are widely used as inexpensive and expendable small holding containers for mosquitoes. Coluzzi and Petrarca (1973) developed an aspirator in which mosquitoes are collected directly into such cups (Fig. 5.3b). A plexiglas cylindrical chamber in which the cup (A) is housed forms the body of the aspirator. A short length of plastic tubing with a ring of five to six small holes near its end (B) is glued over a small central hole in the permanently fixed plastic base of the cylinder. A smaller diameter piece of tubing is inserted through this hole and is attached to a length of rubber tubing, which forms the mouth-piece. A piece of fine barrier netting is inserted between the tubing with holes and the smaller diameter section. A plastic collecting tube (C) is threaded into the top of the cylinder, which when screwed into position retains the paper cup within the cylinder. A short section of plastic tubing (D) glued to the inside of the top projects into a hole made in the bottom of the paper cup. By having the collecting tube threaded into the aspirator different shapes and lengths of tubing can be used, while the function of the five to six small holes in the tubing in contact with the netting of the cup is to ensure a more even aspiration, thus minimising damage to the mosquitoes. After mosquitoes have been sucked into the paper cup, the top of the plexiglas cylinder is unscrewed and the paper cup removed after the hole in its bottom has been plugged with cotton wool.

Cartridge-type aspirators were also described by Barnes and Southwick (1967). Different sized and shaped cartridges can be used. Sholdt and Neri (1974) described how an aquarium dip tube of the type used to remove organic debris can be easily modified to form a mouth aspirator incorporating a holding cage.

The practice of collecting mosquitoes by sucking them up into aspirators can become tiring when practised for long periods, and moreover the prolonged inhalation of mosquito scales, dust and other fine debris may cause, or aggravate, bronchial allergies (Bellas 1990; Douglas 1984). A cotton wool plug can be inserted into mouth aspirators to minimise these hazards, but this can reduce suction power. As long ago as 1939 Woodbury and Barnhart described an aspirator using the venturi principle to avoid the risk of inhaling foreign particles when collecting insects by sucking. Spielman (1964) described and illustrated an aspirator operating from a source of compressed air, which employed a venturi made from glass tubing. Insertion of a wad of very fine plastic mesh as used in oxygen and anaesthetic lines in hospitals to deliver clean air to patients does not reduce suction to any appreciable extent and The Liverpool School of Tropical Medicine inserts this type of 'filter' (Thermo Vent 1200, Portex Co.) directly into the rubber tubing of its simple aspirator. To collect phlebotomine sandflies Warburg (1989) modified a standard disposable bottle-top filler, as used to sterilise laboratory cultures, by removing the Millipore filter and inserting a double-layered paper tissue, so when the filter was placed into an oral aspirator it prevented dust being sucked up into the mouth.

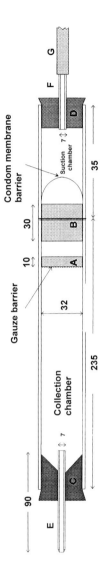

Fig. 5.4. Barrier aspirator of Tang (1996). A—cardboard ring with gauze barrier stretched across collection chamber; B—cardboard ring holding piece of condom as a barrier across the suction chamber; C,D—rubber bungs; E—glass inlet tube (7 mm internal diameter) with one end tapered; F—glass tube (7 mm internal diameter) connected to; G—length of flexible tube leading to the mouthpiece

Tang (1996) proposed the use of a barrier formed from a section of latex condom to reduce the risk of inhaling dirt and other fine debris (Fig. 5.4). Aspirators incorporating a 0.3 µm HEPA filter are available commercially.

Several more ingenious methods have been developed to overcome the problem of inhaling dust and dirt. In the aspirator illustrated in Figure 5.5*a* below both the aspirator (A) and blowing (B) bulbs have valves.

Another type of aspirator (Fig. 5.5*b*) that avoids the inhalation of dust particles was developed for the collection of agricultural insects (Wiens and Burgess 1972). A thin plastic bag (A) is positioned in a 3-pint capacity cardboard carton with its open end folded over the rim, the lid is then re-placed and the joint between the lid and carton sealed with silicone rubber or adhesive tape. A one-way flap valve (B) is made in the bottom of the carton by cutting out a 1-cm hole and loosely taping a small piece of poly-thene plastic over it. A suitable mouthpiece (C) is inserted through the lid and a length of rubber tubing used to connect the bottom of the carton to a reservoir-type pooter (D). In operation air is first blown into the plastic bag to inflate it and this results in expelling air from the carton through the flap valve. Air is then sucked out of the bag, which on deflation draws air, and insects, into the pooter.

Sometimes the end of the glass tubing of an aspirator is bent at an angle to facilitate collecting from awkward corners and surfaces.

Indoor resting collections using an oral aspirator and a torch for illumi-nation are widely used in entomological and epidemiological studies, espe-cially in the Indian sub-continent, where catches are frequently expressed as 'man hour density'. In Pakistan Reisen and Milby (1986) caught more than 14 species of mosquitoes resting inside houses and cattle sheds, the most common of which were *Anopheles stephensi, Culex quinquefasciatus, Anopheles culicifacies, Anopheles annularis* and *Anopheles subpictus*. In another study in Pakistan Reisen et al. (1982) caught 13 mosquito species from houses, and 17 species from cattle sheds. In both sites the most com-mon two species were *Culex quinquefasciatus* and *Culex tritaeniorhynchus*.

In India Yasuno and Rajagopalan (1977) employed 15-min aspirator collections in store rooms of houses and in cattle sheds to monitor sea-sonal changes in *Culex quinquefasciatus* adult densities. In Nepal, Reisen et al. (1993) collected indoor-resting *Anopheles* mosquitoes from the same eight houses, four mixed human-animal dwellings and four cattle sheds. Additionally, collections were made from four randomly selected human dwellings and four cattle sheds. Collections were made by two collectors for 10 minutes per structure using oral aspirators and commencing just after dawn.

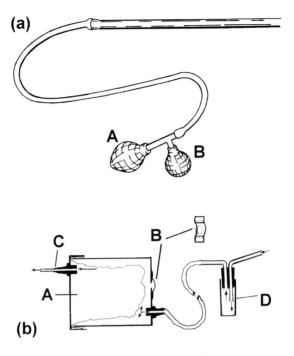

Fig. 5.5. (a) Aspirator with aspirating bulb A—with valve, B—blowing bulb; (b) aspirator after Wiens and Burgess (1972) showing A—plastic bag, B—cardboard carton having a one-way flap valve, C—mouthpiece, D—reservoir-type pooter

Resting mosquitoes were also collected from vegetation and two to four pit shelters. Collections were expressed as females per person hour per month. A total of 20 species of *Anopheles* were collected using these methods, the commonest species being *Anopheles maculatus, Anopheles vagus, Anopheles splendidus, Anopheles tessellatus* and *Anopheles annularis*. Indoor residual spraying with DDT reduced the indoor resting collections of all species except *Anopheles fluviatilis*. Over a two-year period, Chand et al. (1993) collected 15 species of *Anopheles*, 9 species of *Culex,* and one species each of *Aedes, Armigeres,* and *Mansonia* in 15-min indoor-resting collections using a manual aspirator and torch in a broken-forest ecosystem in north-west Orissa, India. *Anopheles culicifacies* was present throughout the year, at highest densities per man hour in cattlesheds. *Anopheles subpictus* and *Anopheles vagus* both occurred with two annual peaks and were more commonly captured in cattle sheds. Further studies of indoor resting mosquitoes in Orissa, India were conducted by Nutan Nanda et al. (2000). Indoor resting mosquito collections were made on a fortnightly basis between 0600 and 0800 h. In each village, mosquitoes were collected for 15 minutes from each of 4 cattle sheds and 4 human dwellings using an aspirator and torch. In forested villages, high man-hour densities (mean = 2.9 to 13.1) of *Anopheles fluviatilis* were observed during the cooler months (October–January), but very low densities (mean = 0.0 to 0.1) were observed in summer (May–June). By contrast, densities of *Anopheles culicifacies* in forested villages were relatively lower during winter (November–January). Densities of *Anopheles fluviatilis* in human dwellings were significantly higher than those in cattle sheds, but the reverse was observed in the case of *Anopheles culicifacies*. In the riverine villages in the deforested area high densities of *Anopheles culicifacies*, reaching up to 290 per person-hour were recorded. The density of *Anopheles culicifacies* was higher in cattle sheds.

De and Chandra (1994) in West Bengal, India, performed 12-minute resting collections from 10 human dwellings on a weekly basis, giving a total of 8 hours of collection effort per month. Manual aspirator and torch collections carried out between 0600 h and 0800 h on a fortnightly basis from 3 human dwellings and 3 cattle sheds at each of six sites in India over the period from January 1987 to December 1995 yielded 17 786 anophelines of six species, of which *Anopheles subpictus* (67.7%) was the most frequently collected. Neeru Singh et al. (1996) collected indoor-resting anophelines from houses and cattle sheds in central India during the early morning on four consecutive days twice per month, using oral aspirators. Results were expressed as per man-hour resting counts. In forest villages, the mean hourly catch of *Anopheles culicifacies* ranged from 0.0 in May to 70.5 in August. For *Anopheles fluviatilis*, mean catch per hour ranged from 0.0 in April–June to 3.5 in October. In villages located away from forest, a maximum of 34.4 *Anopheles culicifacies* and 2.0 *Anopheles fluviatilis* were collected per hour. Almost 29 000 *Culex quinquefasciatus* were collected (Dua et al. 1997).

Tiwari et al. (1997) collected 25 019 anopheline mosquitoes of 14 species resting indoors through the use of oral aspirator collections. Collections were carried out from two fixed and two randomly selected human dwellings and cattle sheds in each of four villages in Uttar Pradesh, India. Mosquito collections were performed from 0600 to 0800 h on a fortnightly basis. The four dominant species collected were *Anopheles subpictus* (40.8%), *Anopheles culicifacies* (36.7%), *Anopheles annularis* (17.0%) and *Anopheles pallidus* (4.4%).

Sampath et al. (1998) used indoor resting collections to evaluate the effectiveness against *Anopheles culicifacies* populations of lambda-cyhalothrin-treated mosquito nets in Orissa, India. Two single-room huts and two cattle sheds acted as fixed collecting stations, to which an additional four randomly selected huts and cattle sheds were added on each sampling occasion. Collections were carried out for 15 min at each sampling station using an oral aspirator and torch, and results were expressed as numbers per hour of collection effort. In Bihar, India, Das et al. (2000), in contrast to many workers, conducted their indoor resting catches in the evening between 1800 and 2100 h. Catches were performed using an oral aspirator in human dwellings, cattle sheds and goat cabins. Indoor resting mosquitoes were collected fortnightly between 0600 and 0900 h using mechanical aspirators (Hausherr's Machine Works, New Jersey, USA) over a 12 month period in Andhra Pradesh, India by Murty et al. (2002*a*). Results of collections were expressed as mean per man hour densities. Peak man hour densities of *Culex quinquefasciatus* in rural areas were highest around December and lowest around June. Murty et al. (2002*b*) performed aspirator collections in and around cattle sheds and indoor dwellings in 11 villages in Andhra Pradesh, India. Collections began at 0800 h and continued for exactly 1 h. These authors consider this to be a simple method that provides a stable index, and is repeatable. A total of 15 512 mosquitoes of 2 genera and 5 species was collected. Members of the *Culex vishnui* subgroup dominated the collections, comprising 42.6% of the total. Other species collected were *Anopheles subpictus* (40.4%), *Anopheles hyrcanus* (12.8%), *Culex gelidus* (3.5%), and *Anopheles barbirostris* (0.8%).

In Ethiopia, five species of anophelines were collected from inside village huts using oral aspirators, with *Anopheles pharoensis* and *Anopheles gambiae* s.l. the two commonest species (Nigatu et al. 1994).

In El Salvador Lowe and Bailey (1979) found that one-hour collections between 0800–1000 h or 1845–2020 h in stables provided good numbers of female *Anopheles albimanus*. In general more were collected in the mornings than in the evening collections, and morning catches provided a higher proportion of blood-fed individuals. In Mexico in an evaluation of insecticidal residual house-spraying, aspirator collections of *Anopheles albimanus* for 15 min/house were made weekly between 0900–1100 h in 32 selected village houses (Bown et al. 1984). Similar collections were also made from animal corrals and from vegetation, but from 2000–2200 h.

Flores-Mendoza and Lourenço-de-Oliveira (1996) made five-minute aspirator catches of mosquitoes resting on the walls of a house in Rio de Janeiro state, Brazil, twice per month at the following times: 1855 to 1900 h, 1955 to 2000 h and 2100 to 2105 h. A 10-minute collection was made from a horse stable at 2105 to 2115 h. Collections were expressed as number of mosquitoes per man hour.

Small battery-powered aspirators

Small battery operated hand vacuum cleaners can be converted into aspirators (Dell'Uomo 1967; Dyce et al. 1972; Husbands 1958; Husbands and Holten 1967; Jackson and Grothaus 1971; McCreadie et al. 1984; Meek et al. 1985; Nelson and Chamberlain 1955; Saliternik 1963; Spencer 1962; Sudia and Chamberlain 1967; Trpis 1968). Suction is usually produced by high speed rotation of a plastic or metal fan, usually having two or more blades, which is driven by a 3- or 6-V d.c. motor powered by dry-cell torch batteries, often of the rechargeable type. Several models include additional nozzle attachments or tools, which can be adapted to improve the utility of the aspirator. As model specifications and names frequently change, I have tried to keep the description as generic and therefore applicable to other models, as possible. A version developed at the London School of Hygiene and Tropical Medicine is shown in Fig. 5.6a. After removal of the bristles forming a brush around the air intake a small plastic funnel is cut to the correct size and glued in position. The short spout of the funnel is covered with a piece of mosquito netting and projects into the glass or plastic collecting tube. The small dust bag originally present is replaced with a length of rubber tubing so that with the batteries switched off the catch of mosquitoes can be blown into a suitable container. An alternative method consists of gluing or screwing a 12-cm long, 4-cm diameter clear plastic cylinder (e.g. a plastic holding tube supplied with the WHO mosquito susceptibility test kit) over the air intake to form a collecting reservoir. An aspirator of this design is shown in Fig. 5.6d. Because of its large size, mosquitoes may get damaged by being repeatedly blown around inside the reservoir. Insertion of three approximately equally spaced horizontal stiff cardboard baffles into the collecting reservoir reduces air turbulence and the risk of damage to the collected mosquitoes. Carver (1967) made an aspirator from a battery operated vacuum cleaner by clipping the bristles of the brush and fixing mosquito netting over the inlet with plastic bands. A plastic funnel about 7 cm in diameter was taped over the inlet with adhesive tape. Mosquitoes sucked into the funnel through its spout were removed by blowing through the outlet hole.

Another modification was described by Husbands and Holten (1967). They removed the brush assembly of a battery vacuum cleaner and fixed a plastic plate over the air intake. A short stub of plastic tubing (28 mm internal diameter) with fine netting over the free end was cemented in the middle of the plate (Fig. 5.6b). A 20-cm extension tube of glass or rigid plastic is inserted over the short plastic stub. A number of small collecting tubes, that also serve as holding tubes, are made by cutting the end off a conical centrifuge tube and covering it with mosquito netting. One of these tubes is inserted into the end of the extension tube. After one or more mosquitoes have been collected the end of the centrifuge tube is closed with cotton wool and the tube removed and another inserted. For collecting in inaccessible places the extension tube is made of rubber or flexible plastic, but two hands are then required to hold the aspirator. A larger type of portable car vacuum cleaner that operates from two 6-V dry-cell batteries was modified by Harden et al. (1970) for collecting mosquitoes at bait catches. A length of 1-in diameter plastic tubing with mosquito netting at one end is screwed into the suction hose of the cleaner; the end is closed periodically with a cork and the tube removed and replaced with another.

Trpis (1968) described a more complicated aspirator (Fig. 5.6c) constructed as follows: after removing the brush a circular plastic base plate with five 10-mm diameter holes is permanently fixed over the air intake. Two 3-mm thick and 3-mm^2 spacers (A) are glued on to the face of the base plate. A rectangular acrylic plastic outer casing, $180 \times 40 \times 37$ mm (B), is cemented on to the base plate. The opposite end is fitted with a rubber bung through which a length of 9-mm diameter glass tubing is inserted. Rectangular holding cages, $155 \times 30 \times 25$ mm, are made with the two ends and upper and lower sections of 3-mm acrylic plastic but with the two sides of plastic mosquito screening (C). A 9-mm diameter hole is cut from one end of each holding tube. The holding tube is inserted into the larger permanently fixed plastic casing to rest on the two spacers on the base plate. When sufficient numbers of mosquitoes have been sucked up into the holding tube it is removed, corked and replaced with another.

McCreadie et al. (1984) modified a commercial 3.6 V hand vacuum cleaner, by removing the filter from the distal end of its detachable nozzle and internal flap near the apex and inserting a rectangular cloth collecting bag (8×15 cm) into the apex of the nozzle, and holding it in place by way of elastic bands.

Meek et al. (1985) described two very useful battery operated aspirators also made from adapting commercially-available handheld vacuum cleaners, one of which is described here. The supplied crevice tool (B) is cut and two triangular-shaped pieces of plexiglas (C) are stuck on either side with hot glue (Fig. 5.7). A rectangular hole the same size as the cross-section of the modified crevice tool is cut into the screw-top lid (D) of a 20-dram transparent plastic prescription vial (F), which is then permanently attached to the end of the modified crevice tool. Collecting tubes are constructed from two prescription vials joined together with epoxy cement. A 4.45-cm circular piece of plastic screening (E) is attached over one of the threaded ends of the collecting tube with cement. This screened end of the tube is screwed into the vial cap (D) attached to the modified crevice tool. When sufficient mosquitoes have been caught the open end of the tube is closed, and the tube can be replaced with another.

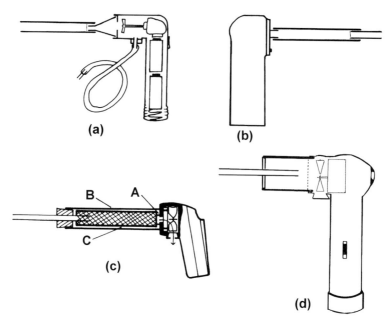

Fig. 5.6. Battery operated aspirators using modified small hand vacuum cleaners: (*a*) Type developed by London School of Hygiene and Tropical Medicine; (*b*) after Husbands and Holten (1967); (*c*) after Trpis (1968) showing A—spacers, B—acrylic outer casing, C—holding tube; (d) battery-powered aspirator of the World Health Organization (1975)

An alternative type of collecting tube can be made using the transparent cylinders and screw caps comprising the exposure tubes of the WHO adult mosquito insecticide susceptibility kits.

Governatori et al. (1993) describe a portable mechanical aspirator constructed from a computer cooling fan and transparent plastic tubing (Fig. 5.8). The suction fan used measures 80 × 80 × 25 mm and is driven by a brushless, direct current motor that consumes only 2.6 W, and can be powered by 6–12 V batteries. The holding cage portion of the device is constructed from a 50 mm diameter × 200 mm long transparent plastic tube, which is fitted into the motor using a PVC bell reducer glued to a plastic retaining ring fitted to the base of the bell reducer. The base of the holding cage is covered with 0.7 mm nylon mesh to prevent the mosquitoes being sucked into the fan. The holding cage section of the apparatus can be removed and replaced with a new, empty cartridge as required.

The suction tube is constructed from a transparent plastic cylinder with a diameter of 38 mm and a variable length between 400 and 800 mm. The various tubes are connected one to the other using plastic rings. The almost silent operation of the motor and fan unit was found to be a distinct advantage in field trials.

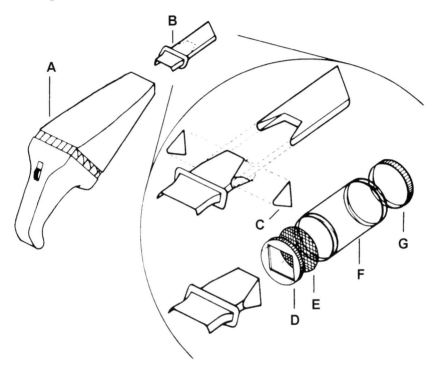

Fig. 5.7. Black and Decker 'Dustbuster Plus' cordless vacuum cleaner modified for aspirating mosquitoes: A—vacuum cleaner, B—crevice tool, C—triangular piece of plexiglas, D—screw cap, E—50-mesh plastic screen, F—transparent plastic vial, and G—screw cap (Meek et al. 1985).

If vehicles can be parked near the collecting site battery powered aspirators that operate from 12 V can be plugged directly into the standard cigarette lighter socket located on the instrument panel of some cars, or into the sockets for inspection lamps (Harden et al. 1970).

McGavin and Furlong (1981) designed a cheap electronic counter that could be attached to an aspirator to count insects. In the field it operates from rechargeable batteries providing about 10.8 V, in the laboratory mains output of 8–12 V d.c. at approximately 300 mA can be used. In tests it proved reliable for counting objects sucked up, ranging in size from wheat grains to *Calliphora* puparia.

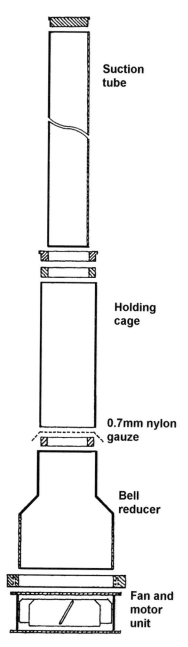

Suction
tube

Holding
cage

0.7mm nylon
gauze

Bell
reducer

Fan and
motor
unit

Fig. 5.8. Mechanical aspirator of Governatori et al. (1993)

A non-portable, solid-state, electronic insect counter for use in the laboratory was described by Pearson et al. (1975), and Bennett (1980) used it to count *Ochlerotatus sierrensis* adults prior to release in mark–release experiments.

In a comprehensive and integrated analysis of the *Anopheles quadrimaculatus* species complex in Florida, USA, Reinert et al. (1997) aspirated adult mosquitoes from daytime resting sites using the battery-powered mechanical aspirator of Meek et al. (1985). The aspirator was further modified by using 2 extension tubes (approximately 46 and 86 cm long) fabricated from 3.3-cm (outside diameter) white polyvinyl chloride (PVC) pipe. By using 1 or a combination of the 2 PVC tubes, most adult resting sites (e.g., in large rot cavities of trees, livestock barns, wooden resting boxes, and outdoor shelters, and under bridges, culverts, and eaves of buildings) could be easily reached. Although the authors acknowledged that the method may be biased, they found that on at least some occasions 4 species (*Anopheles inundatus, Anopheles maverlius, Anopheles quadrimaculatus,* and *Anopheles smaragdinus*) of the complex were collected from a single resting site.

Pålsson et al. (2004) collected indoor-resting mosquitoes three times per year from 30 houses in a suburban area in Guinea Bissau using battery-powered aspirators, and related resting abundance to environmental characteristics. They observed that significantly greater numbers of mosquitoes were collected from rooms that had open eaves and from houses with a well located in the compound. Presence of pigs was also associated with increased resting density of mosquitoes.

Resting counts

This method has been used in studies on the effect of residual insecticides on house resting mosquitoes (Smith 1964), in studies of hibernating mosquitoes, and in other investigations of mosquito resting habits. The mosquitoes resting on the walls and roof of a hut are carefully counted but not caught, so that both the behaviour of the mosquitoes and the toxicity of the insecticide can be assessed. However, in East Africa Smith (1964) found that *in situ* counting was actually only able to record about 47 and 61% of the total *Anopheles gambiae* population resting in grass and mud-roofed huts, respectively. When directly counting resting mosquitoes it is important not to disturb them, causing them to resettle elsewhere and potentially resulting in double-counting. Hibernating females are more easily counted than non-hibernating mosquitoes as they are not so readily disturbed. Weekly counts have been made of hibernating populations of *Culex pipiens* and/or *Culiseta annulata* in England in brick-built shelters (Service 1969; Sulaiman and Service 1983) and caves

LIVERPOOL JOHN MOORES UNIVERSITY
LEARNING SERVICES

to show their seasonal build-up and decline in numbers, and also to study the rate of spread of fungal infections in *Culex pipiens* populations. In Japan populations of *Culex pipiens* form *pallens*, and a few other species, were studied by counting the resting adults in a cave every 7–10 days for 16 months (Shimogama and Takatsuki 1967). The position of resting adults on a wall or roof can be marked with a pencil, so that the approximate proportion of the population that moves in-between sampling days can be determined (Service 1969). In Japan Makiya and Taguchi (1982) studied the movement of overwintering *Culex pipiens* form *pallens* in underground air-raid shelters by marking off 37 squares (50 × 50 cm) and photographing adults resting in each square. Their numbers and position were checked weekly with their photographs. It was observed that about 70% of the adults changed their position from week to week. Also in Japan Natuhara et al. (1991) measured seasonal fluctuations of *Culex tritaeniorhynchus* by counting the numbers resting on grey sheets (100 × 145 cm) marked with grid lines and pinned to the walls of cow sheds. Temporal patterns of abundance agreed well with collections in light-traps.

Chadee et al. (2002) describe a preliminary protocol for collecting blood-engorged *Culex quinquefasciatus* mosquitoes for examination for the presence of microfilariae, as used in the filariasis prevalence monitoring programme in Trinidad and Tobago. The so-called xenomonitoring protocol comprises five stages, namely: notification of householders, mapping, preparation of equipment, house collections, and finally laboratory processing. The fourth stage, house collections, is described in more detail here. The target sample size was set at 5% (1000) of households in each of Trinidad and Tobago's ten counties. Two teams, of three individuals each, arrive at the first selected house between 0500 and 0600 h. The teams collect resting *Culex quinquefasciatus* from all the bedrooms, using a CDC backpack aspirator (John W. Hock Co., Gainesville, Florida, USA) and mouth aspirators. Brooms made from coconut palm leaves are used to flush out resting, bloodfed mosquitoes from difficult-to-reach areas. Flying mosquitoes are collected using hand nets. All mosquitoes collected are transferred into pre-prepared jars or used plastic ice-cream containers lined with plaster of Paris. All collections are conducted between 0600 and 1000 h. In the laboratory, all fully engorged *Culex quinquefasciatus* are dried for 2 h in an oven at 90°C. In preliminary testing of the protocol, more than 3000 fully engorged *Culex quinquefasciatus* were collected from 1400 houses across two counties. The authors noted that freshly blood-fed specimens were fragile and their abdomens often ruptured when being collected using the aspirator or hand nets and they therefore recommended that collections be undertaken 2–4 h after peak feeding activity.

Resting sites of mosquitoes in houses

The walls, ceilings, roof, clothing, furniture, including the underside of beds, and other objects in huts should be examined with a torch to locate resting mosquitoes. In some areas where houses are raised on stilts mosquitoes may also rest underneath them (Spencer 1965). In Tanzania Smith (1955) found that in both round and rectangular village huts about 62–66% of the females and 60–70% of the males of *Anopheles gambiae* rested below the thatched roof, mainly on the mud walls, but also on various objects and under the beds. Similarly, in later studies Smith (1962*a,b*) found that about 56–75% of *Anopheles gambiae* rested on hut walls and household objects.

In Java most *Anopheles aconitus* found indoors were collected resting on walls just above the floor, especially in crevices and in dark sheltered corners, very few rested higher up the walls or on the ceilings (Joshi et al. 1977). Adults were also collected from hay and straw used as fodder and stored in animal sheds. In contrast in India Batra et al. (1979) had to use ladders in houses and other buildings to collect *Anopheles stephensi*, because they tended to rest high up on the walls and ceilings beyond the reach of the collectors. Also in India, Chatterjee et al. (1993) studied the resting habits of *Anopheles stephensi* in temporary shelters (jhupries), cattle-sheds and brick-built rooms in Calcutta. Mosquitoes were collected for 10 minutes per shelter between 0600 and 0800 h using aspirators and test-tubes and numbers converted to density per man hour. Higher densities were observed in temporary structures (0.42 per man hour) and cattlesheds (0.68), compared with brick-built rooms (0.01). In temporary shelters, 42.8% of mosquitoes were collected resting on hanging objects and 22.4% on furniture. In cattle sheds, 48.3% of mosquitoes collected were resting on the ceiling and 14.9% on hanging objects. Gunasekaran et al. (1995) reported on the vertical distribution of *Anopheles fluviatilis* resting on the internal walls of dwellings in Orissa, India. Resting collections were carried out using oral aspirators indoors in the morning (0600 to 0900 h) and from verandas and external walls in the early evening (1800 to 2100 h). Six day-time collections from three houses yielded 689 female *Anopheles fluviatilis*. The preferred resting sites were the walls of the room, followed by the roof and a sizeable proportion were collected resting on surfaces unsuitable for indoor residual spraying, namely umbrellas, clothes, baskets, bags, etc. The greatest proportion (50.2%) of the females resting on walls was collected from a height above four feet. Evening collections also revealed a preference for upper walls and veranda roofs as resting sites for this species. In Pakistan Reisen et al. (1979) using a mechanical aspirator (Davis and Gould 1973) collected *Anopheles subpictus* resting on the floor and lower walls of animal sheds and houses. Also in Pakistan, Hewitt et al. (1995) studied the distribution of resting mosquitoes inside tents as part of an evaluation of the usefulness of treating the tents with residual insecticide as a malaria control measure. Anopheline mosquitoes (primarily *Anopheles stephensi* and *Anopheles culicifacies*) preferred to rest on the apex of the tent roof, while culicines preferred to rest briefly on the ground before exiting the tent, while those that did rest on the tent (4%) preferred the walls.

The local distribution of adults resting in huts may be affected by the type of hut construction, the building materials used, the availability of alternative resting places, such as clothing and furniture, and the presence of fires. There may also be differences between the resting pattern during the day and night, and also seasonal variations. Smith et al. (1966) found that in huts with grass roofs about 80% of *Anopheles gambiae* rested on the roofs during the day and night, but in huts with corrugated iron roofs, although 80% rested on them at night less than 10% rested on the roofs during the day, when temperatures were higher.

In the D'Entrecastleaux Islands, Spencer (1965) found that about 60% of *Anopheles farauti* rested on the hut walls below the level of 3 ft.

Ameneshewa and Service (1996) also studied the movements of indoor-resting anophelines in Ethiopia using a mark-release-recapture method. They marked and released 199 female *Anopheles arabiensis* with lunar yellow A27 Dayglow fluorescent powder and released them early in the morning in a hut sprayed with DDT. Released mosquitoes were recaptured during the same afternoon using aspirators and portable, battery-powered ultra-violet lights. The majority (53.1%) of the 32 females recaptured were found resting on objects within the room, including pots, clothes, under beds, etc., all of which would not have been sprayed with DDT. 155 females were marked with blue A60 fluorescent powder and released in an unsprayed hut. Fifty-three females were recaptured, most of which (35.8%) were found resting on the walls less than 0.5m above ground.

In Kenya, Mutinga et al. (1995) studied the preference of *Anopheles gambiae* for resting on bare walls or on different coloured cloth attached to the internal walls of rural houses at different heights. *Anopheles gambiae* preferred to rest on cloth-covered walls rather than on bare walls and demonstrated a preference for white coloured cloth, followed by red, yellow, black, blue and green in order of preference. There also appeared to be a preference for the lower section of wall as a resting site, for both cloth-covered and bare walls and for the walls which received the least ambient light.

In a coastal area of southern Mexico, Arredondo-Jiménez et al. (1995) used knowledge of the preferred resting sites of the malaria vector *Anopheles albimanus* to design a control intervention relying on the selective application of insecticide to two 1m wide horizontal bands just above the base of the walls and at the base of internal roofs. The preferred resting sites of blood-fed females were determined by marking females in situ, just as they had finished feeding on a collector seated in the open doorway of a house. Marked females were then followed for 1 hour by a second technician using an ultraviolet lamp. Mean resting height of this species was found to be 1.25 m on walls and 0.5 m from the base of internal roofs. The selective application of insecticide to the preferred resting sites was found to be as effective as spraying the total interior surfaces of walls and roofs, but was 67% less expensive.

In The Gambia although Bryan (1979) caught some adults of the *Anopheles gambiae* complex from walls and ceilings of village houses, she caught most mosquitoes inside mosquito nets which are widely used in that country. Also in The Gambia, Lindsay et al. (1995) studied the spatial and temporal distribution of the numbers of mosquitoes collected from mosquito nets in village houses in order to estimate exposure of children to *Anopheles gambiae*. Mosquitoes resting inside mosquito nets were collected from a stratified random sample of 140 nets occupied by children under the age of seven years. Nets were searched between 0700 and 1000 h each day. The mean and variance of numbers of mosquitoes per mosquito net were calculated for each daily collection and compared with Poisson and negative binomial model distributions and an index of dispersion I (Southwood and Henderson, 2000) was calculated for each study night.

$$I = (\text{variance}) (n - 1)/\text{mean} \tag{5.1}$$

Overall, 74% (203) of samples gave a good fit to the negative binomial model, with 38% (105) fitting the Poisson model. In 20% of samples the variance was less than the mean. The Poisson model was found to be most appropriate during the low season, when mean geometric counts of mosquitoes per net were <0.3, while the negative binomial model gave a better fit when mosquito numbers were high (daily geometric mean up to 7.0 per net). Values for the index of dispersion ranged from 26.0 in the cold dry season to 109.0 in the hot dry season. The authors concluded that since the number of mosquitoes per net was overdispersed during the malaria season, most people received only a few, or no, bites per night, whilst a few individuals received more bites and hence were at greater risk of contracting malaria.

Indoor attractant resting box of Edman

During aspirator collections of *Aedes aegypti* from inside houses, Edman et al. (1997) noticed that resting adults were more frequently found in dark locations, and on dark, non-reflective surfaces, such as clothing in closets. This led the authors to test three designs of indoor attractant resting stations in eastern Thailand. The first design was a medium-weight brown cardboard box 90 cm tall × 30 cm wide × 30 cm deep, from which a 50 cm × 30 cm section was cut from the centre of one of the long sides. Following initial trials, all surfaces of the box and the panel traps were covered with black muslin cloth attached using a staple gun. The second design consisted of two 90 cm × 30 cm flat cardboard panels arranged in a cross design (Fig 5.9).

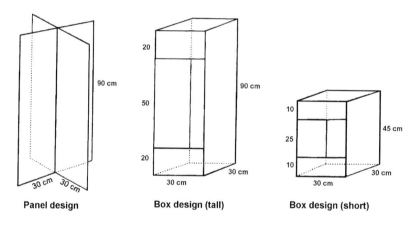

Fig. 5.9. Resting attractant panel and tall and short attractant resting boxes for *Aedes aegypti* (Edman et al. 1997)

The third design was also a box, measuring 45 cm × 30 cm × 30 cm. Collections from the traps were made twice each day, between 0930 and 1100 h and again between 1330 and 1530 h, using a mechanical aspirator. Resting boxes were sampled for 30 seconds per box, and this was followed by 8–10 minutes searching the rest of the house. Collections were made for eight consecutive days using 2 box and two panel traps per house for 12 houses, with trap types and locations rotated daily. Resting boxes and panels accounted for 40% of all males and 46% of all females collected.

More females were collected from the box design (374 v 231; $\chi^2 = 29.38$, $P < 0.00001$), but more males were collected from the panel design (297 v 252). Given the much shorter time required to empty the traps, compared with searching the whole house, traps were found to be nearly 4 times as efficient as house searches.

Ponlawat and Harrington (2005) reported that these resting boxes were relatively unsuccessful in capturing *Aedes albopictus* in Thailand.

Other indoor resting shelters

In Kenya indoor resting densities of *Anopheles gambiae* complex and *Anopheles funestus* were based partly on weekly collections of adults resting on a 5 × 6 ft reed mat placed in houses to form a ceiling (Sexton et al. 1990).

In India Yasuno et al. (1973b, 1977) placed plywood boxes (50 × 30 × 28 cm) having one-half of the lid hinged and left partially open overnight—for entry of mosquitoes—in dark corners of bedrooms in 20 houses. The boxes were lined with black cloth and contained a sponge saturated with water to increase humidity. On one occasion as many as 325 *Culex quinquefasciatus* were collected overnight from a single trap, while the mean catch reached a peak of 40.7 in April. These traps were effective only when mosquito densities were large and humidity low.

In Trinidad over about 16 months Nathan (1981) collected 1720 female *Culex quinquefasciatus* and substantial numbers of *Aedes aegypti* from resting shelters similar to those used by Yasuno et al. (1973a) placed inside houses. These shelters comprised open-ended 30-cm square plywood boxes painted white outside and black inside and provided with a screened jar of water to improve their attractiveness.

Collecting efficiency

The numbers of mosquitoes caught from huts are usually expressed as the mean number per hut, or hut density. In India the number of *Anopheles culicifacies* caught resting in houses is frequently expressed as the mean hour density or man hour density (MHD)

$$MHD = \frac{n \times 60}{t \times p} \tag{5.2}$$

where n = numbers caught, t = collection time in minutes, and p = number of collections (Subbarao et al. 1988) and this method is commonly used on the sub-continent (e.g. Konradsen et al. 1998; Malakar et al. 1995; Murty et al. 2002*b*; Neeru Singh and Mishra 2000; Yadav et al. 1989), and has also been used by Forattini et al. (1997) in Brazil.

Mosquitoes caught from under over-hanging eaves of huts, from underneath huts or from their outside walls are sometimes included in the calculation of hut densities although they are not resting in the huts. Hut densities will nearly always represent an underestimate of the numbers of mosquitoes in a hut because of the inability to catch the entire resting population. Aspirator collections will recover a smaller proportion of the mosquitoes resting in huts than pyrethrum spray catches, but are essential if live individuals are required.

In a series of 30-min catches from huts in two unsprayed villages in India, Viswanathan et al. (1950) found that the mean female density of *Anopheles culicifacies, Anopheles fluviatilis* and *Anopheles stephensi* calculated from morning catches (0700–1100 h) was about 30% greater than afternoon (1100–1500 h) catches. The relationship between the two densities was shown to be: Mean afternoon density = 0.6455 × mean morning density + 1.3. In two villages that were sprayed with DDT the relationship was: Mean afternoon density = 0.488 × mean morning density + 1.27. Because of their greater activity it was more difficult to collect adults at night (2000–2100 h), but despite this about twice as many were caught in both unsprayed and DDT sprayed huts than in morning collections (0700–1100 h). In HCH sprayed huts, however, only about 40% more were caught during the night than in the morning.

The efficiency of hand-catching was assessed from 620 hut collections in six Indian villages (Viswanathan et al. 1952). Aspirator collections were made in each hut with the doors and windows shut and all openings and cracks closed with paper and cloth. After completion of these collections white sheets were placed on the floor and the hut space sprayed with 0.2% pyrethrum in kerosene. Since all openings had been carefully closed it was considered that the pyrethrum collections recovered all the mosquitoes resting in the huts. Although this is unlikely, it would probably not be a serious source of error. A total of 7785 anophelines and 4940 culicines were caught in these collections. Hand collections terminated after 30 min irrespective of whether or not more mosquitoes remained. About 32% of the *Anopheles culicifacies* and 26% of *Anopheles subpictus* actually present in the huts were caught in the hand-catches. Of the 10 other *Anopheles* species and culicines caught from the huts 28% and 31%, respectively, of the total numbers resting in the huts were collected in the hand-catches. If collecting for a standard time (e.g. 30 min) catches a known proportion of the mosquitoes in huts, then so long as the collector's efficiency does not change over a sampling programme the actual numbers present can be estimated. The proportion of the total population caught by hand-catches from different types of huts may vary, and this should be accounted for in calculating mean hut densities. Viswanathan et al. (1952) concluded that when aspirator collections were made for 30 min and the results expressed as numbers caught per man-hour this gave a useful index for comparing mosquito densities between different catching stations.

Batra et al. (1979) found that aspirator catches collected 31.3% of the males and 28.9% of the females of *Anopheles stephensi* resting in houses, whereas with *Anopheles subpictus* only 8.8% of the males and 7.1% of the females were caught; the remainder being collected by pyrethrum spray catches.

In West Africa Ribbands (1946a) found that even when attempts were made to catch with aspirators all the *Anopheles* resting in a hut, pyrethrum catches still yielded about a further 28%. In India Senior White and Rao (1946) also found that despite attempts to catch with aspirators all anophelines resting in huts about 28% remained uncaught, and in Morocco pyrethrum collections in houses following aspirator collections showed that only about 47% of the *Anopheles labranchiae* were caught with aspirators (Bailly-Choumara 1973). In Jamaica as little as about 10% of the indoor resting population of *Anopheles albimanus* were caught in hand collections (Muirhead-Thomson and Mercier 1952). Ribbands (1946a) reported that the proportion of the total catch obtained in aspirator collections varied greatly (50–90%) on different days. Because male and unfed females are generally more difficult to locate, he concluded that they were more likely to be underestimated in aspirator collections than blood-fed and gravid individuals. Unfortunately he caught only 494 mosquitoes in his experiments, and consequently his results are not as meaningful as those of Viswanathan et al. (1952) who found no evidence that males were underestimated more than females.

Undoubtedly a greater proportion of the absolute numbers of mosquitoes are caught from huts if aspirator collections are continued until no further adults can be found (i.e. catching to completion), but in practice the method is too time-consuming for routine catches. It may take 2–3 h for all the mosquitoes to be collected from a single hut (Rao, in Viswanathan et al. 1952). Catching to completion is also probably subject to greater individual bias than collecting for a specific time, because of the wide variations in the abilities and conscientiousness of the collectors. The ease with which mosquitoes are found also varies between huts depending on the availability of different types of resting sites.

In Ethiopia Krafsur (1977) treated the successive monthly proportions of unfed, partly fed, blood-fed and all other gonotrophic categories of

Anopheles caught resting in houses each in turn as a set of independent random variables having a binomial distribution. Then, the probabilities that monthly proportions in the different gonotrophic stages were homogeneous were estimated from the chi-squared statistic

$$\chi^2 = \sum \frac{(x_i - n_i\theta)^2}{n_i\theta(1-\theta)} \tag{5.3}$$

where $n_i\theta$ = expected proportion of any gonotrophic condition, n_i = sample size, θ = population mean, and x_i = observed sample mean. He found considerable heterogeneity, which was caused in part by the proportional increase in unfed and partly fed females when the mosquito population was growing, and a decrease when the population was declining (See Krafsur (1977) for further details of his methods of calculation.)

Nagasawa (1976) using three years' data on the collection of *Culex quinquefasciatus* resting in houses (2000–2100 h) and from human landing catches (2100–2400 h) in the same seven catching stations in Myanmar firstly transformed the data into logs and calculated the mean numbers caught/man-hour. Then by harmonic analysis, following the method of Bliss (1970), two-term Fourier curves were plotted to show seasonal patterns of abundance obtained by the two methods, and estimates were derived of maximum and minimum numbers of mosquitoes caught by the two sampling methods.

It is not only often difficult to find outdoor resting populations of *Anopheles*, but generally even more problematic to estimate the proportion of the population that rests out of doors. In Nigeria Molineaux and Gramiccia (1980) estimated the proportion of *Anopheles gambiae* complex that fed on man indoors and then rested indoors as follows:

Let *IRD* = the true indoor resting density

$I\hat{R}D$ = the indoor resting density estimated by pyrethrum spray-sheet collections

$I\hat{R}D$ = $b1 \times IRD$

b_1 = $I\hat{R}D/IRD$ = the bias of the estimated indoor resting density

MBR = the true man-biting rate

$M\hat{B}R$ = the man-biting rate estimated by night-biting collections

$M\hat{B}R$ = $b_2 \times MBR$

b_2 = $M\hat{B}R/MBR$ = the bias of the estimated man-biting rate

HBI = the proportion of blood-meals in the pyrethrum spray-sheet collection positive for man (human blood index)

N = the number of persons per hut (the population of the village, divided by the number of huts)

x = the proportion of blood-meals followed by resting indoors (at least until the next morning when pyrethrum spray-sheet collections are made).

T = period in days of resting indoors after feeding.

Then:

$$IRD \times HBI = MBR \times N \times x \times T \qquad (5.4)$$

Or

$$\frac{\hat{IRD}}{b_1} \times HBI = \frac{\hat{MBR}}{b_2} \times N \times x \times T \qquad (5.5)$$

$\hat{IRD}$, $\hat{MBR}$, HBI, N are measured directly. T is estimated by

$$T = 1 + \frac{G}{F}, \qquad (5.6)$$

where G and F are the proportions of gravid and blood-fed individuals, respectively, caught in the pyrethrum spray-sheet collection (if the maturation time is 2 days, as is suggested by the clear-cut bimodal distribution of the pyrethrum spray-sheet collection by abdominal appearance); x and b_2/b_1 (the relative bias of the two sampling methods) are unknown. If we know the one, we can compute the other from eqn (5.4):

$$x = \frac{\hat{IRD}}{\hat{MBR}} \times \frac{b_2}{b_1} \times \frac{HBI}{N \times T} \qquad (5.7)$$

$$\frac{b_2}{b_1} = \frac{\hat{MBR}}{\hat{IRD}} \times x \times \frac{N \times T}{HBI} \qquad (5.8)$$

The values, $\hat{MBR}$, $\hat{IRD}$, N, T and HBI were readily obtained by Molineaux and Gramaccia (1980) and inserted in eqns (5.7) and (5.8). With *Anopheles funestus* past experience shows that nearly all blood-fed individuals rest indoors, thus x can be given the value of 1, thus enabling the value of b_2/b_1 to be computed from formula 3. It is found to be 1.16, that is in comparison with the indoor-resting density, the man-biting rate is overestimated by 16%. Now, assuming that the relative bias of the two sampling methods is the same for both *Anopheles funestus* and *Anopheles gambiae* s.l., then the value of *x* for *Anopheles gambiae* s.l. can be derived from formula 2, and is found to be 0.47. That means that only about half the blood-meals taken on man by *Anopheles gambiae* are followed by resting indoors.

Lines et al. (1986) studied the mixing of indoor and outdoor-resting adults of the *Anopheles gambiae* complex and *Anopheles funestus* in Tanzania by marking them with different coloured fluorescent powders. Out of 568 female *Anopheles gambiae* caught outside in pit shelters and marked and released just 31 were recaptured, all from inside houses. An estimate of the proportion of outdoor-resting females that rest exclusively out of doors is 0/31. The 95% upper confidence limit (f), can be calculated from the binomial distribution by

$$(1-f)^{31} = 0 \cdot 05 \quad \text{i.e.} \ f = 0 \cdot 092 \tag{5.9}$$

Now, if the outdoor resting sample had contained 9.2% or more exclusively outdoor-resting mosquitoes, at least one should have appeared in the sample of 31 in 95% of the observations. But this estimate is based on unbiased sampling with respect to the endophilic and exophilic populations and fails to take into account various biological variables. A full analysis of the recapture data taking into account the probabilities that the outdoor resting collections consist of endophilic females that had 'strayed outside' together with a few exclusively exophilic females, and other mixes of the indoor and outdoor collections, is presented by Lines et al. (1986).

The numbers of *Anopheles gambiae* entering village houses was shown in The Gambia to be about 43% fewer in those with closed eaves, that is houses with no or very little space between the tops of the wall and the roof (Lindsay and Snow 1988). Clearly these types of houses should be avoided if large collections of mosquitoes are required.

Hand-net catches

Occasionally small hand-nets, about 15 cm in diameter, made of fine mosquito netting have been used to catch adult mosquitoes resting in human and animal habitations. In Taiwan for example Rosen et al. (1989) used a net once a week from July 1980 to December 1983 for 1 hour, or slightly longer, after sunset and collected 142 434 female *Culex tritaeniorhynchus* resting in a single shed used to rear pigs. Obviously very large numbers of mosquitoes must have been resting in this shed. In the Dominican Republic Tidwell et al. (1990) found that fluctuations in adult population levels of both *Aedes aegypti* and *Culex quinquefasciatus* could be monitored by two men using 12-in hand-nets to catch mosquitoes in houses for 5-min periods. Particular attention was paid to searching underneath beds, chairs and tables as well as in cupboards. They found the mean number of female *Aedes aegypti* per house ranged from 1.22–15.04, with several houses having more than 20 females, and one as many as 134. The average numbers of female and male *Culex quinquefasciatus* collected were 3.2 ± 4.8 and 3.0 ± 4.3, respectively; the highest catches were 68 females and 39 males.

In determining the seasonal abundance of *Aedes aegypti* in Indonesian huts, adults were caught from some huts by 10-min aspirator collections and from others by sweep-netting for 5 min, but no details of the latter technique were given (van Peenen et al. 1972). More *Aedes aegypti* were caught with aspirators than by sweep-netting, but no males were collected, whereas substantial numbers of males were caught with the sweep-nets.

In India Rajagopalan et al. (1977) were unable to use the 'hand-catch index' to sample indoor resting *Culex quinquefasciatus* in urban houses because their densities were too high, mosquitoes were therefore caught in 10-min collections from houses using an 8-in sweep-net (Yasuno et al. 1973b). The mean numbers caught per house by this method varied from 56.4 (June) to 354.3 (October) during 1972–1973, and from 11.5 (January) to 117.9 (October) in 1973–1974. They also used a removal method to estimate the total population of *Culex quinquefasciatus* (see section 1.9 below).

Sometimes net-catches are preceded by gently spraying houses with non-toxic oils (e.g. Risella or citronella oil) paying particular attention to cracks and crevices. This causes adults that have escaped capture to take flight, and these are caught in the net.

Pyrethrum spray collections

Pyrethrum spraying of habitations was originally used as a malaria control measure (Covell et al. 1938; De Meillon 1936; Eddey 1944; Russell and Knipe 1939, 1940, 1941), but the practice was discontinued in the 1940s, when it was replaced by spraying huts with DDT and other residual insecticides. Knock-down space spraying with pyrethrum is a standard, quick and easy method of catching mosquitoes resting in huts and animal shelters. It is usually the most efficient of the available methods for collecting mosquitoes, but its efficiency depends on the type of hut construction.

The routine procedure is as follows: all occupants, animals, easily removable objects such as small tables and chairs, exposed food and drinking water are firstly removed from a hut that is to be sprayed. However, as reported by Dossou-yovo et al. (1995), working in Côte d'Ivoire, this is not always easy. These authors were unable to carry out regular PSC catches in their study village of Alloukoukro, due to the bedrooms being cluttered with very many items. When water is stored in pots that are embedded in the floor and cannot be removed, the pot opening should be covered with a lid. Furniture should be returned to a house after a spray catch has been completed, otherwise householders may refuse permission for their houses to be sprayed on future occasions.

After removing furniture and other objects, two people carefully lay white sheets made of calico or some other strong fabric over the entire floor, over the bed, and over furniture and miscellaneous objects that have not been removed. Where fires are made in hut floors a small area of floor space has to remain uncovered. It is convenient to have a good supply of both small (2×1 m) and large (2×2 m) spray sheets available. Sheets must be regularly washed, because mosquitoes are easily over-looked when collecting from dirty sheets. In Egypt sheets are frequently not placed over the floor in pyrethrum spray catches, instead a 1-m^2 white sheet supported on two sticks is moved around a room and under furniture while someone sprays a kerosene solution of 0.2% pyrethrum (El Said et al. 1986; Gad et al. 1995; Kenawy et al. 1986, 1990). This so-called index-sheet method is also used in animal shelters. Clearly consid-erably fewer mosquitoes than obtained by conventional spray sheet methods will be collected, and there seems little to recommend this pro-cedure, unless for some reason householders forbid sheets being spread over the floor and furniture. All doors and windows are closed and the hut space-sprayed with 0.1–0.2% pyrethrum in kerosene (paraffin) usu-ally synergised with piperonyl butoxide. In Egypt Kenawy et al. (1990) used 0.2% neopybuthrin to spray houses. In Tanzania Bushrod (1979) added 0.3% chloroform to her 0.3% pyrethrum spray as she considered this prevented mosquitoes that were knocked down from recovering. Cer-tain proprietary brands of pyrethrum concentrate, such as Pycon 819E contain, in addition to piperonyl butoxide, emulsifiers enabling the con-centrate to be mixed with water instead of kerosene. For this the diluent (water) should be reduced by 20% from that needed with kerosene. Strong spray-guns of medium capacity and preferably with a long plunger stroke are best. A pin or other suitable implement is used to clear the nozzle, when the pump becomes blocked, which happens frequently. The small, cheap pumps that can often be bought locally should generally be avoided, because they usually leak, are easily broken and often lack sufficient power to direct the spray to the top of high-roofed rooms or houses. Some operators in preference to buying cheap locally made 'flit-guns' have resorted to using larger sprayers, usually with a lance, that are sold for agricultural spraying. In India Batra et al. (1979) space-sprayed houses, huts, cattle sheds, firewood sheds and other shelters with a swingfog machine. Because of the open nature of the structures they had to use 1% pyrethrum, not 0.2% as generally recommended, to get a good knock-down.

Whatever type of spray gun is used the spray is first directed at all potential escape routes, such as closed doors, windows and eaves; it is then aimed towards the roof or ceiling. To reduce the numbers of mosquitoes escaping from the large gaps that exist in some huts between the tops of the walls and roof, one or two technicians walk round the outside of the hut spraying these gaps either before or while the inside is sprayed. In small huts, however, it may be possible to fill up these gaps with surplus spray sheets. If a hut has either no door or a badly-fitting one, then a spray sheet can be held over the doorway both during and after spraying. In Sri Lanka Büttiker (1958) pinned or pegged cotton sheets over the palm matting walls of improvised huts and shelters to prevent mosquitoes escaping. Use can be made of the mosquitoes' attraction to light when irritated by insecticides. The door of a hut is left open and mosquito netting held over the doorway. On spraying adults fly towards the doorway and eventually fall on to the floor sheets placed at the base of the doorway. Whatever method is employed sprayers leave the hut after spraying. Usually about 90% of the mosquitoes fall on to the spray sheets within the first 7 min after spraying, and after about 10 min the sprayers re-enter the hut to retrieve the mosquitoes.

Mosquitoes can be collected from the spray sheets while they remain in position on the hut floor, but it is usually better to carefully pick up each sheet by the corners and shake the contents on to a single sheet. Weather permitting this sheet is taken outside the hut and examined by two collectors. If it is windy or raining, mosquitoes are collected from the sheet inside the hut, using a torch for illumination (Fig. 5.10).

Mosquitoes which have been carefully picked up from the sheets with forceps or fingers are placed in small convenient containers lined with damp filter paper or lint. Plastic petri-dishes, tobacco or cigarette tins make useful receptacles. In hot weather the containers holding the mosquitoes can be placed in wide-mouthed vacuum flasks or in commercial cold boxes to keep them cool. On return to the laboratory mosquitoes will still be in a sufficiently good condition to allow dissections for malarial and filarial parasites, ovarian age-grading, and the preparation of gut smears for immunological (e.g. ELISA) tests etc. If they cannot be dissected the same day mosquitoes can be stored overnight in a domestic refrigerator, or in a deep freeze for months for age-grading dissections (see Chap. 13). It is usually convenient to count, sex and record the gonotrophic stages and species of mosquitoes by individual hut, prior to pooling collections from several huts in a compound or village.

Fig. 5.10. Collecting mosquitoes from a spray sheet, using a torch and oral aspirator (photograph reproduced with permission of J. B. Silver)

Commercial aerosol containers containing pyrethrum or pyrethroids can be used to space spray huts, but the method is usually more expensive than spraying with pyrethrum for routine use. Krafsur (1971, 1977) used aerosols containing 0.6% pyrethrins synergised with 1.4% piperonyl butoxide to spray houses in Ethiopia, and commercially-available aerosols are frequently used to conduct PSCs in many African countries. For example, Faye et al. (1998) used a commercially available pyrethroid spray (Yotox®) to collect resting mosquitoes from houses in Wassadou village in southeast Senegal in order to assess the impact of ITNs on several entomological indices of malaria transmission. Many aerosol preparations contain residual insecticides and these should be avoided.

Factors influencing pyrethrum spray collections

The numbers of mosquitoes resting in different huts in the same compound may vary considerably. Unfed mosquitoes tend to be attracted to huts in numbers related to the number of human occupants (Haddow 1942) but no general arithmetic relationship has been established between catch size and number of occupants. Small village huts occupied by several people generally harbour large numbers of resting adults. These types of huts can be selected and sprayed when large numbers of mosquitoes are required, such as for determining infection rates, or seasonal changes in numbers. In addition to the number of people in a hut many other variables such as the presence of fires, the roofing materials used, whether or not the inhabitants use insecticide-treated nets, large gaps between the eaves, whether or not the doorway is left open, proximity to larval habitats, etc. also affect the numbers of mosquitoes resting in houses. Selecting houses containing disproportionally large numbers of mosquitoes introduces a pronounced sampling bias. To obtain a reliable estimate of the mean hut density of mosquitoes in a village, or area, representative samples of the different types of huts must be sprayed, and not just huts known or suspected to contain large numbers of mosquitoes.

It is more difficult to efficiently spray very large huts, especially those that are divided into a number of rooms, and also those that have high roofs. A smaller proportion of the mosquito population is caught in huts with an incomplete ceiling below the roof, because some of the adults that are knocked down fall on to the ceiling partition and are not collected. In East Africa Gillies (1955) considered that pyrethrum collections underestimated the mosquitoes resting in huts by about 10–20%; but because of their more sluggish flight blood-fed individuals were less likely to escape capture than other categories. The ability to escape capture may vary between different species as well as between different gonotrophic stages. If so, then pyrethrum catches may not give representative samples of either the relative hut densities of different species, or of the gonotrophic stages within a single species.

Pyrethrum spray can have a repellent effect on indoor-resting mosquitoes. For example, Swellengrebel and de Buck (1938) observed that heavy spraying with pyrethrum reduced the numbers of the *Anopheles maculipennis* complex resting in houses, and in India Senior White et al. (1945) noted that pyrethrum spraying exerted a repellent effect on *Anopheles minimus* resting in houses. In Assam Ribbands (1946*b*) observed a reduction of about 90% of *Anopheles minimus* resting in huts the night after they had been sprayed with 0.1% pyrethrum at the rate of 25 ml/1000 ft^3. Some degree of repellency persisted up to the fourth day. In West Africa Muirhead-Thomson (1948) found that when village huts were sprayed about every other day it reduced the house resting *Anopheles* to a third of their original numbers. In unsprayed huts, for example, only about 20% of the *Anopheles* left as blood-fed individuals whereas in regularly sprayed huts there was an exodus of about 76%. It was concluded that the reduced catch of mosquitoes in sprayed huts was mainly due to the exodus of blood-fed females. In East Africa Haddow (1942) also reported that pyrethrum spraying had a repellent effect on the numbers of mosquitoes resting in huts. There is also evidence that the kerosene solvent is itself repellent. Whatever the cause of the repellency, it is generally agreed that because the effect is short lived huts can be sprayed weekly, and possibly even twice a week, but not at shorter intervals, without there being a reduction in the number of mosquitoes caught.

It is often recommended that pyrethrum or aspirator collections are started before sunrise so as to include in the catch mosquitoes that leave huts with the onset of daylight. If the species or population under investigation tends to exhibit a dawn exodus then catching should be completed, not just started, before dawn. However, unless huts are occupied by vector control or research staff members it is usually difficult or impossible, to gain access to huts before dawn. If it is not possible to sample all structures before dawn then it is better that all collections are made after dawn, although in Nigeria no significant differences were observed between the numbers of *Anopheles gambiae* or *Anopheles funestus* caught in huts sprayed at different times from 0430–0730 h (Service 1964). Similarly, in Kenya there was no significant difference between the mean hut densities (70.9 and 68.4) of *Anopheles gambiae* calculated from pyrethrum catches made at 0730 h or 1400 h (Joshi et al. 1973). In contrast far fewer adults were caught at 1400 h in huts that had been sprayed with fenitrothion. The difference in results appears to have been due to adults that had fed outside on cattle entering the huts around dawn, just before the morning collections were made. In the fenitrothion sprayed huts most of these were killed by the residual insecticide before the afternoon catches were made and consequently the mean catch was smaller.

Lutwama et al. (1999) used PSC and indoor and outdoor light-traps to undertake entomological investigations following an outbreak of o'nyong-nyong fever in south central Uganda. PSC catches caught more mosquitoes than light traps and more than 94% of all the anophelines caught (5094) were obtained by PSC. Light traps caught high numbers of culicines. All 34 houses that were sprayed were positive for anophelines and collections of between 100 and 300 anophelines per house were obtained from 65% of houses. The mean and SE of anophelines per house was 204 ± 22. *Anopheles funestus* was the most abundant mosquito species collected, comprising about 45% of the total.

Malaria epidemiological studies were undertaken by Ravoahangimalala et al. (2003) in Madagascar using indoor and outdoor human landing catches and pyrethrum spray catches. PSC were conducted in a minimum of 10 houses per study site and per collection. PSC's caught more *Anopheles funestus* (1200) than either indoor (198) or outdoor (160) human landing catches, but more *Anopheles gambiae* s.l. were caught using human landing catches (43 indoors, 64 outdoors) than PSC (4).

Interpretation of house catches

House catches are usually expressed in terms of mean hut densities, but aspirator collections can also be expressed as the numbers caught/man-hour. Hut densities will nearly always be smaller than the real densities of resting adults, although under ideal conditions knock-down spray catches may recover almost all the mosquitoes resting in huts.

Hut densities are commonly used to measure changes in the seasonal and annual abundance of mosquitoes, to compare house resting densities in different villages or areas, and to assess the impact of control measures on endophilic species. Only rarely have attempts been made to estimate the total population of an endophilic mosquito in a village from hut densities. In an East African village Gillies (1955) calculated the total indoor populations of *Anopheles gambiae* (15 577) and *Anopheles funestus* (15 002) not by taking a sample of huts but by spraying all the huts (119) on five separate days over a period of 11 days. Because the percentage of both species that rested out of doors was known (about 5%), Gillies was able to estimate their total populations in the village, an area comprising 380 acres. Regular human landing collections were made in a hut throughout a year to measure seasonal changes in relative population size, and by using these values Gillies (1955) was able to adjust his single population estimates of *Anopheles gambiae* and *Anopheles funestus* to give estimated figures for total populations throughout the year.

The ratios of the number of mosquitoes caught in indoor night human landing catches to the number of human blood-fed mosquitoes caught in pyrethrum spray sheet collections can be used as an index of exophily; the higher the index the greater the degree of exophily. In northern Nigeria Molineaux et al. (1976) found that this index increased in the rainy season. This could be due to: (i) progressive increases in anthropophagy (i.e. decrease in non-human feeds found in pyrethrum catches); (ii) decrease in time mosquitoes rested indoors; (iii) changes in the relative bias of the two sampling methods; or (iv) a decrease of the proportion of mosquitoes resting indoors after feeding (i.e. endophily). They concluded that the latter was the most likely explanation and was associated with an increase in vegetation during the rains which provided more outdoor resting places.

Resting mosquito sampling data have been fitted to mathematical distribution models by several researchers. Nedelman (1983) tested data obtained from the Garki project (Molineaux and Gramiccia, 1980) against the negative binomial, Neyman Type A and Polya-Aeppli distributions. The numbers of adult *Anopheles gambiae* s.l. caught from houses in pyrethrum spray sheet collections were best described by a negative binomial distribution with temporally varying means but a constant k.

Taking k to indicate heterogeneity among the houses, then this is largest for pyrethrum collections, then for exit traps fitted to the huts and finally for indoor human bait collection. Molineaux and Gramiccia (1980) reported that the relative efficiencies of these three sampling methods, and also out of door collections, varied over time. Nedelman (1983), however, believed that except in one of eight villages these trapping methods maintained constant relative efficiencies throughout the rainy season, and in all but two of the villages the relative efficiencies of the four sampling methods were the same.

Ribeiro et al. (1996) used hand-held oral aspirators and pyrethrum spray catches to determine the spatial and temporal distribution of mosquitoes in a village in Arba Minch region, Ethiopia. A team of six collectors each sampled an average of 54 houses per month. Mosquitoes were collected using an oral aspirator and in every sixth house a pyrethrum spray collection was carried out. Data were analysed using Taylor's power regression (Taylor 1961), Iwao's mean crowding (Iwao 1968) and kriging, which calculates the spatial moving average of neighbouring points in a plane. Taylor's power plots and Iwao's plots both yielded slopes significantly in excess of unity, indicating a clumped or overdispersed distribution pattern of mean *Anopheles gambiae* s.l. (presumed to be *Anophlees arabiensis*) hut densities. Low k values and a poor fit to the negative binomial model were obtained. In most months, less than 20% of the houses contained more than 50% of the mosquitoes collected.

Shililu et al. (2003) used PSC to undertake a study on the spatial distribution of anopheline mosquitoes resting inside huts in Eritrea and also found that the spatial distribution of *Anopheles arabiensis* was distinctly aggregated. More than 80% of individuals collected originated from 1.8% of houses in northern Red Sea zone, from 2.0% of houses in Debub, from 8.5% in Anseba, and 10.2% in Gash Barka.

Keating et al. (2005) recently examined the spatial distribution of *Anopheles gambiae* s.l. and *Anopheles funestus* among village houses at the Kenya coast. Mosquitoes were sampled from 10 houses within a 2 km radius of the local primary school in each of 30 villages, using pyrethrum spray catches, every two months. Mean household resting mosquito density and the corresponding variance were calculated for each village for each sampling period. In total, 5476 *Anopheles gambiae* s.l. and 3461 *Anopheles funestus* mosquitoes were collected. The total number of mosquitoes collected from any one house ranged from 0 to 121 for *Anopheles gambiae* s.l. and from 0 to 152 for *Anopheles funestus*. Mosquito densities were overdispersed, with 70% of *Anopheles gambiae* s.l. and 80% of *Anopheles funestus* collected from approximately 20% of the total number of houses sampled. In most sampling periods, zero mosquitoes were collected from around 50% of the houses sampled. Clustering of mosquitoes did not appear to be restricted to any specific houses across the study period. The relationship between the variance/mean ratio and the mean mosquito density per village for each sampling period revealed strong positive correlations in all sampling periods, and this indicates that variability in mosquito distribution increases with increasing mean mosquito density per village.

In Haiti Hobbs et al. (1986) compared the numbers of *Anopheles albimanus* recorded inside houses in human bait collections, with the numbers of blood-fed mosquitoes caught during 15-min search periods performed hourly throughout the night. Because considerably fewer blood-feds were caught resting in houses than were caught as unfeds in all night bait collections, it was concluded that engorged *Anopheles albimanus* rested for only short periods in houses after feeding. However, a problem with such comparisons is that often only a small percentage of mosquitoes actually resting in houses is caught. Nevertheless, in Jamaica Muirhead-Thomson and Mercier (1952) concluded that only about 10% of *Anopheles albimanus* feeding in houses remained in them during the day.

Estimation of the indoor population by the removal method

The application of the removal method to estimate population size is discussed in detail with reference to larval populations in Chap. 3. Yasuno et al. (1973b, 1977) applied the method to the collection of adult *Culex quinquefasciatus* resting in urban houses in Delhi. The principle is that the numbers of individuals caught in a sample are related to the size of the population, and that if a number of successive samples are taken without replacement, then the population decreases in size and consequently the numbers caught in the samples diminish. In their study Yasuno et al. (1973b) collected *Culex quinquefasciatus* from 10 houses using either four people per house collecting with aspirators or two people catching adults in hand-nets. The total numbers caught by the two methods were separated into seven 5-min units of time, but in one house collecting was extended for 18 time units (90 min) and 605 mosquitoes were caught. Figure 5.10 shows that in this catch there was a reasonably good regression of the numbers caught against time. The estimate of total population present ($\hat{N}$) is given by:

$$\hat{N} = \sum_{i-1}^{k}(i-1)y_i \div (1-q)^k \qquad (5.10)$$

where y_i = numbers caught in ith time unit collection, k = number of time unit collections, p = probability of capture during a single time unit and hence $q = 1 - p$. A more complete explanation of the procedure is given in Chap. 3.

The average probability of capture per unit time (p) based on collections from 10 houses was 0.124 for aspirator collections and 0.297 for hand-net collections. This difference might have been the cause of the greater standard error attached to the first type of collection (6.1–23.1% of The average probability of capture per unit time (p) based on collections from 10 houses was 0.124 for aspirator collections and 0.297 for hand-net collections. This difference might have been the cause of the greater standard error attached to the first type of collection (6.1–23.1% of The average probability of capture per unit time (p) based on collections from 10 houses was 0.124 for aspirator collections and 0.297 for hand-net collections. This difference might have been the cause of the greater standard error attached to the first type of collection (6.1–23.1% of $\hat{N}$) than to the latter (1.4–10.4% of $\hat{N}$). It was concluded that even four people collecting in a single house with aspirators was inefficient and did not give an accurate population estimate. However, as the standard error of the probability of capture (p) was quite small (0.17–0.035) it seemed that if more time unit collections were performed a more reliable population estimate could be obtained.

For example, from Fig 5.11 seven unit collections give the estimate of the population of *Culex quinquefasciatus* in the house as 752 ± 145, whereas from 18 unit collections the population estimate has a much smaller error, 711.8 ± 21.6. When two men caught mosquitoes with hand-nets the confidence limits were generally within ±10% of the estimates, and under the conditions of the experiment about seven time unit collections were sufficient. It was concluded that the sweep-net method was the better of the two in collecting a large proportion of the mosquitoes resting in houses.

In Papua New Guinea Charlwood and Bryan (1987) collected indoor resting *Anopheles punctulatus* from a house with aspirators and segregated the catch into twelve 5-min intervals; they then estimated the absolute population by the regression of numbers caught in the ith catch against previous total catch, and also by maximum likelihood methods (Zippin 1956, 1958). The regression method estimated the population as 192 and the Zippin method as 282 ± 53.1, whereas the actual total catch was 280. These very similar results indicated that aspirator collections were efficient in collecting most if not all *Anopheles punctulatus* resting in houses—or at least in this house.

In Tanzania Charlwood et al. (1995) estimated the total number of mosquitoes resting in a house by fitting an exponential decay curve to the numbers of mosquitoes collected in each five-minute period over a total collecting period of 35 minutes and then integrating the fitted curves to estimate the total number that would have been caught had collecting continued indefinitely. The number of mosquitoes removed from the population in the *i*th time interval was defined as n_i, which was assumed to be proportional to the total number of mosquitoes remaining uncollected. Each value of n_i was expressed as a proportion of n, the total number collected in a 35-minute period, such that:

$$n_i = n\beta_0 \exp(\beta_1 i) \tag{5.11}$$

where β_0 is proportional to the initial rate of capture of individual mosquitoes and β_1 is the decline over time in the rate of capture. The relationship between these parameters was investigated by fitting Poisson regression models by maximum likelihood methods.

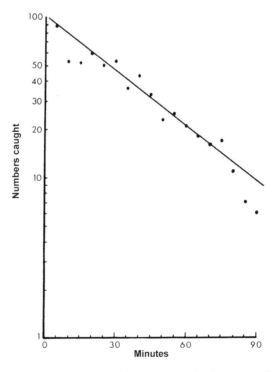

Fig. 5.11. Number of *Culex quinquefasciatus* caught by successive 5-min collections with aspirators from a house in India (after Yasuno et al. 1973*b*)

The proportion of the total number of mosquitoes resting in the house that were removed in the first j intervals was then estimated as $1-\exp(\beta_1 j)$ and the total number resting in the house was estimated from:

$$N = n/(1 - \exp(7\beta_1)) \qquad (5.12)$$

The analysis indicated a highly significant decrease with time in the numbers of mosquitoes removed ($F = 417.9$, d.f. $= 1.8277$, $P < 0.001$) with a stronger relationship at lower total density, such that the proportion of the total mosquitoes collected in 35 minutes depends on the total resting population in the house.

The ability of removal trapping to provide an accurate estimate of the total population of mosquitoes resting in a collection of houses, or in a whole village or town, obviously depends on the number of houses sampled. Although removal trapping can be used to obtain the total numbers of mosquitoes resting in houses, in most instances equally, if not more, reliable results could be obtained more conveniently by carefully performed pyrethrum space spraying.

The outdoor resting population

Most mosquito species rest exclusively out of doors in natural resting places, and only a relatively few species rest in man-made shelters. However, it is usually more difficult to find those mosquitoes that rest out of doors than those that rest in buildings such as houses, shelters, barns and animal quarters. This is because outdoor populations are usually widely distributed over large areas and not concentrated in discrete units. Even with a species such as *Anopheles gambiae*, which is regarded as highly endophilic, a certain proportion of the population may be found resting out of doors (Gillies 1954) even when suitable indoor resting sites are available. The size and importance of the exophilic population of species that commonly rest inside houses is probably often overlooked.

Gillies (1956) considered that three basic types of exophily could be exhibited by malaria vectors: (1) *obligatory*, where because of the absence of man-made shelters *Anopheles* were compelled to rest in natural outdoor shelters; (2) *facultative*, where human and animal habitations were available but adults rested in either natural outdoor shelters or in houses; and (3) *deliberate*, where mosquitoes specifically avoided resting in habitations although they were available. Mosquitoes in the third category were further divided into: (a) those that in fact fed in houses (endophagic) and rested outside; and (b) those that both fed (exophagic) and rested out of doors. Gillies (1956) was principally concerned with the epidemiological importance of exophily exhibited by malaria vectors, and was little interested in outdoor resting populations of unfed females, or males. From an ecological point of view, however, it may be important to adequately sample all gonotrophic stages and age classes of mosquito populations.

It has usually proved very difficult to get reliable comparisons between either the relative or absolute sizes of the indoor and outdoor resting populations of mosquitoes. Molineaux and Gramiccia (1980), however, tried to estimate the proportion of the *Anopheles gambiae* complex and *Anopheles funestus* resting out of doors in northern Nigeria by an indirect method which is summarised in section above.

Exophilic mosquitoes shelter in many types of habitats, such as animal burrows, hollow trees, holes at the base of termite mounds, cracks and crevices in the ground, on tree trunks, under bridges, on fencing and brick walls, in abandoned mines, and amongst a variety of vegetation. These different habitats provide a wide range of microclimatic conditions. Generally, adults resting amongst vegetation will be afforded less protection from wind, sunlight and desiccation than those sheltering in rodent holes and tree-holes, etc. Different species may exhibit marked preferences for resting in particular habitats. Harwood (1962) found *Anopheles freeborni* overwintered in animal burrows, whereas *Culex tarsalis* occurred mainly amongst vegetation and rock-holes and fissures. Breeland (1972*a*) gave a detailed list and description of the favoured natural diurnal resting sites in El Salvador of *Anopheles quadrimaculatus* s.l. and *Anopheles pseudopunctipennis*, both species favouring rock crevices and fissures. Also in El Salvador *Anopheles albimanus* can be found resting under bridges, in culverts and rock crevices and between the buttress roots of trees. In England *Coquillettidia richiardii* and several *Aedes* and *Ochlerotatus* species rest primarily amongst vegetation, whereas various *Culiseta* spp., *Anopheles plumbeus* and *Anopheles claviger* mainly occur on tree trunks (Service 1969, and unpublished results). In Guatemala Cupp et al. (1986) collected few *Mansonia titillans* until they discovered they rested in tall grasses growing in bunches around bases of trees in open fields. During surveys in Tennessee over a million *Anopheles quadrimaculatus* s.l. and numerous other *Anopheles* species were caught from barns, but although *Anopheles walkeri* comprised about 30% of the *Anopheles* caught at bait, at first none, and later only a few, were found in the barns (Snow and Smith 1956). It was later discovered that *Anopheles walkeri* rested during the day on the stems of various swamp plants.

Not all vegetation is equally attractive to mosquitoes. During collections of mosquitoes from various resting sites in England none was caught resting in bracken (*Pteris aquilinum*) although the plant would appear to offer attractive shelter (Service 1971*a*). In Malaysia, however, bracken (*Glichenia* sp.) proved to be a favourite resting site of several *Anopheles* species (Wharton 1950). In Tanzania the outdoor population of *Anopheles gambiae* was found sheltering amongst the dense growths of salt bushes and also in crevices in the ground, whereas *Anopheles pharoensis* occurred in less dense growths of salt bush (Smith 1961). In some areas of Kenya *Anopheles arabiensis*, and to a lesser extent *Anopheles gambiae*, are commonly found in small granaries made of maize stalks (Clarke et al. 1980; Githeko 1992), while in other areas lacking such granaries adults of *Anopheles arabiensis* can be found in cracks and crevices of brick pits and in cracks in the ground (Service 1985). In these highly favoured resting sites adults of *Anopheles pretoriensis, Anopheles pharoensis* and *Anopheles rufipes* can also be found, but surprisingly not *Culex quinquefasciatus* although this species is very common in nearby houses. In the Awash valley, Ethiopia, relatively large numbers of *Anopheles arabiensis* were collected from the earth banks around hot springs and from ditches shaded by low bushes (Ameneshewa and Service 1996). In Uttar Pradesh State, India, *Culex epidesmus* was most commonly found resting in bushy undergrowth of mango and teakwood plantations during the day, but no individuals were recovered resting amongst the abundant vegetation of surrounding sugarcane fields (Kanojia 2003).

The distribution of mosquitoes resting amongst vegetation may change during the day so that they avoid direct sunlight (Senior White et al. 1945; Service 1971*a*). There may also be differences between day-time and night-time resting sites; for example in El Salvador *Anopheles albimanus* rests during the day in rock crevices, culverts and other natural sites, but at night adults accumulate in large numbers on walls and fences enclosing cattle corrals (Breeland 1972*a,b*, 1974; Breeland et al. 1974). Apart from movements within a habitat, changes in weather conditions may cause some species to seek shelter in more protected habitats. In an area in England where *Ochlerotatus cantans* normally rested amongst vegetation, numerous adults were found in rodent burrows during an exceptionally dry period. They were previously absent from these burrows and vacated them again after heavy rainfall (Service 1973). In Tennessee *Anopheles walkeri* rests almost exclusively amongst vegetation, but during the hot summer months a few adults are found in barns (Snow and Smith 1956). Similarly in Pakistan *Anopheles culicifacies* is found in exposed dryish resting sites during the cooler months of the year, whereas they tend to switch to resting in damp cattle sheds during hotter periods. In Turkey, *Anopheles sacharovi* rests principally in shelters during the winter months, but in more exposed places in the hotter months (Service 1989).

Indoor spraying with residual insecticides can also affect the behaviour of mosquitoes, such as *Anopheles minimus,* which was formerly an endophilic malaria vector over much of the Oriental region. In Myanmar after some 35 years of DDT house-spraying this species is now predominantly exophilic. Similarly, in India *Anopheles minimus* was formerly predominantly endophilic and endophagic, but it is now frequently found resting out of doors, and in Thailand this vector is now mainly exophilic and exophagic (Ismail et al. 1974, 1975, 1978). In contrast, in Nepal, house-spraying has not selected out exophily, but resulted in the virtual elimination of the species (Parajuli et al. 1981). Other cases of increased exophily have been reported for the *Anopheles gambiae* complex (Muirhead-Thomson 1960), *Anopheles sundaicus* (Kalra 1980; Kalra in Bang 1985), *Anopheles philippinensis* (Bang 1985), and *Anopheles farauti* (Taylor 1975).

Natural resting sites

Vegetation

Oral aspirators. Oral aspirators can be used to collect mosquitoes resting among vegetation. For example, Mani et al. (1991), in studies on Japanese encephalitis in India, collected adult female *Culex tritaeniorhynchus, Culex vishnui* and *Culex pseudovishnui* from in and around cattle sheds in the hour after sunset and multiplied the average number caught by the proportion parous to obtain a 'dusk index'. When this index was compared with human landing catches (1800–0600 h) both exhibited a sharp increase after transplantation of rice seedlings, but the dusk index remained high after biting counts had decreased to a low level. This was due to the paucity of *Culex tritaeniorhynchus* at human bait, and its abundance in dusk collections around cattle sheds. The dusk index was therefore routinely used to monitor vector densities. Also in southern India, Arunachalam et al. (2004) collected 150 454 female mosquitoes representing 6 anopheline and 12 culicine species using oral aspirators from vegetation around cattle sheds and pigsties during monthly 1 h collections conducted immediately after dusk. However, due to the relative difficulty in locating natural outdoor resting sites and the dispersed distribution of resting mosquitoes, alternative methods to oral aspiration of individual mosquitoes are more frequently employed.

Sweep-netting. Many mosquito species rest amongst grassy and shrubby vegetation and on the foliage of bushes and shrubs. Mosquitoes have sometimes been collected by slowly walking through vegetation and capturing them in small hand-nets as they are disturbed and fly out (McClelland 1957; McClelland and Weitz 1963; Teesdale 1959). McClelland (1957), however, realised that this procedure was likely to be biased in favour of collecting unfed females of species that bite man, because having been disturbed they will tend to be attracted to the collector. In one series of catches in East Africa the catchers used an insect repellent (dimethyl-phthalate) to try and prevent unfed females of *Aedes aegypti* being attracted to them after they had been disturbed from their day-time resting sites (McClelland 1957).

In the USA Copeland (1986) caught adults of *Ochlerotatus thibaulti* from a wood by disturbing vegetation with a stick and collecting them with a sweep-net. He also used two green resting boxes, but most adults were collected with mechanical aspirators (Nasci 1981). In South Africa mosquitoes that were flushed out by walking through grassy vegetation were caught in test tubes as they resettled on nearby vegetation (De Meillon et al. 1957).

Some species may not be readily flushed out by walking through vegetation but can be caught if the vegetation is vigorously sweep-netted. In Kenya, van Someren et al. (1958) caught 65 species of mosquitoes by this method. The most suitable sweep-net consists of a strong white calico bag fastened by pop studs over a D-shaped metal frame to which a 2–3-ft wooden handle is attached. A number of swift forward and backward strokes are made without interruption through the vegetation, then the bag, which is quickly folded over to prevent the catch from escaping, is lightly sprayed with chloroform and placed in a large plastic bag. After 1–2 min the net is removed from the bag and the contents tipped into a white photographic dish. Leaves and debris are discarded and the mosquitoes collected. If mosquitoes caught in sweepnet collections are required alive the sweep-net bag can be placed for about 2–4 min in a cold box containing dry ice. The net is then removed and the contents tipped into a white dish, and the mosquitoes aspirated into paper or plastic cups.

Sweep-net bags made of calico or similar material are extremely strong and can be used to sweep holly bushes (*Ilex aquifolium*) and even bramble (*Rubus* spp.) without tearing. When, however, vegetation is wet the bags become soaked with water and the collection becomes a mass of sodden leaf litter and damaged specimens. Bags made of mosquito netting are preferable for sweeping wet vegetation because they do not become so sodden, but of course they do tear more easily. Sweep-net bags become dirty and sticky due to accumulation of sugary plant secretions and should be removed periodically from their frames and washed.

Some entomologists use a sweep-net having a detachable small bag fixed with velcro 'touch and close' fastener to the bottom (Fig. 5.12). After a predetermined number of sweeps the bag is transferred to a carton and a new one substituted. Mosquitoes caught in sweep-nets tend to become denuded of scales and consequently not too many sweeps should be made before the catch is removed from the bag. In Texas, USA standardised sampling, consisting of three series of 20 sweeps through rice fields with a 30.5-cm diameter sweep-net, were made every 30 min from 1900–2400 h to study insemination and temporal abundance of newly emerged *Psorophora columbiae* (Robert and Olson 1986). Magnarelli (1977*a*) collected *Ochlerotatus canadensis* by taking two 30-min sweep-net samples from the edge of its larval habitats each week during the adult season. Sweep-netting has been used in England to study the distribution of mosquitoes resting in various types of vegetation (Service 1971*a*) and in many parts of the world to collect blood-engorged females for blood-meal identification (Aitken et al. 1968; Ardö 1958; Cordellier et al. 1983; McClelland and Weitz 1963; Pajot 1977; Service 1971*b*; Takahashi et al. 1971; Wharton 1950; Williams et al. 1958).

Fig. 5.12. Sweep-net with detachable bag at bottom (photograph reproduced with permission of M. W. Service)

Sweep nets can also be used to sample mosquitoes in flight. Service (1994) used a sweep net to sample from swarms of *Culex torrentium* in West Kirby, England in order to determine the sex and species composition of the swarming mosquitoes. Rajavel (1995) also used a sweep net to sample swarming *Armigeres subalbatus* in Pondicherry, south India. Three sweeps were made every 15 minutes at either dusk or dawn and continued until no more mosquitoes were collected. Dusk swarming commenced around 1600 h with peak densities obtained 45–90 minutes later. Peak density at dusk coincided with a mean light intensity of 116 lux (SD 50.01). Dawn swarming commenced around 0530 h and peaked 30–45 minutes later, when light intensity was 19 lux (SD 9.88).

In Louisiana, USA Holck and Meek (1991) obtained absolute density measurements of *Psorophora columbiae* and *Culex quinquefasciatus* by placing 32-cm diameter 1-m long plastic area samplers with netting tops on the ground to enclose an area of 0.1 m^2. Adults were removed with aspirators from a side-hole fitted with a netting sleeve. At the same time D-vac and sweep-net samples were taken over 5-m transects. Using regression analysis it was concluded that both D-vac and sweep-net sampling could be used to estimate absolute population density (as obtained by the area sampler), but because the coefficient of variation was less for net samples and as less time was involved in taking these samples, sweep-netting was the method of choice for estimating population size.

Sometimes relatively few mosquitoes are caught by sweep-netting. For example, Magnarelli (1977*b*) encountered difficulties in finding outdoor resting blood-fed mosquitoes in the USA. Only 489 engorged females of *Psorophora ferox, Coquillettidia perturbans* and *Ochlerotatus* were caught in sweep-nets, while searches underneath fallen trees and collections from resting boxes of Goodwin (1942) yielded just 81 engorged *Culex* and *Culiseta* species. In the Ivory Coast although Cordellier et al. (1983) caught large numbers of *Aedes* and *Culex* mosquitoes (9555 females and about 10 670 males) by sweep-netting vegetation in forested areas very few females (82) of known or potential yellow fever or dengue vectors, species in which they were interested, were collected.

In mosquito surveys in Bali Lee et al. (1983) collected 17 species during the evenings in sweep-nets and by aspirator collections of out of door resting adults, whereas 20 mosquito species (*Ochlerotatus, Anopheles, Armigeres, Culex* and *Mansonia*) were caught in CDC light-traps placed in or near animal shelters. However, 2.6 times as many mosquitoes were caught by sweep-nets and aspirator collections than in the light-traps.

Relative densities of different mosquito species resting in different types of vegetation, and changes in densities associated with different times of the day and year, can be obtained if the collecting technique is standardised and a known number of sweeps are taken; then the average number caught per sweep, or ten sweeps, can be calculated. In Nigeria, for example sweep-net collections were made in four different biotopes around compounds, mainly to collect *Aedes simpsoni* (Bown and Bang 1980). But from 25 collections of 100 sweeps just 32 female and 22 male *Aedes simpsoni* were caught, mostly from bushes and from under trees, with fewer adults being caught in cocoyam and banana plantations where the species bred. In addition 47 females and 21 males belonging to other *Aedes* species, and 352 female and 881 male *Culex* species, were collected, most of which were caught from cocoyam and banana plantations.

Tonkyn (1980) quantified the process of sweep-netting as follows: the collector moves forward at a constant rate while swinging the net through a circular arc. The volume of air thus sampled is

$$\beta(\pi r + \frac{2\Delta y}{\pi}) \tag{5.13}$$

where β = area of net opening, Δy = distance the collector moves during the sweep, and r = radius of net. So, with a sweep of 180° of radius 1.25 m with a net area of 0.073 m^2 (12-in diameter), made by a collector who takes two steps totalling 1.7 m with each sweep, the volume of air sampled (V) will be

$$V = 0.073(1.25\pi + \frac{2(1.7)}{\pi}) \tag{5.14}$$

$$= 0.366 \text{ m}^3$$

Consequently, 25 sweeps will, for example, sample 9.15 m^2 of air. For sweep angles different from 180° the reader should consult this paper for the appropriate formula. Southwood (1978) gives other references on the efficiency of sweeping for non-medical insects. However, there are many variables that can make sweep-netting an imprecise sampling method. For instance the efficiency of sweeping is likely to vary considerably between different workers; some species may be dislodged and collected more easily than others; there may be differences between vertical distribution of different species in vegetation; differences between the availability of species due to diel cyclical activities; and variations in collecting efficiency due to weather conditions. Nevertheless, despite these limitations Service (1993) believed that sampling mosquitoes by sweep-netting has been underused, and that in several, if not many, situations sweeping vegetation would prove rewarding.

Drop-net cages. De Zulueta (1950) in Colombia and Rehn et al. (1950) in Puerto Rico were amongst the first to collect mosquitoes resting in vegetation by drop-nets. Rehn et al. (1950) caught very few mosquitoes but de Zulueta (1950) had more success. His procedure consisted of driving four vertical poles into the ground, taking care not to disturb the vegetation between them, and suspending a 2-m high and 2-m square muslin net (or tent) from the poles, the sides are then quickly pulled down. Two men enter the tent and spray the vegetation with citronella oil to disturb the mosquitoes, which are collected with aspirators from the sides of the net. De Zulueta (1950) found it took only about 10 min to catch all the mosquitoes flushed up from the 4 m^2 of ground enclosed by the net. From 104 drop-net catches performed from 0700–1700 h 1106 culicines (mainly *Culex chrysonotum, Culex (Melanoconion) spp. and Psorophora confinnis)* and 128 anophelines (*Anopheles braziliensis, Anopheles peryassui* and *Anopheles parvus*) were collected. This gave an average of 11.9 mosquitoes/4 m^2, which is equivalent to about 3 million mosquitoes/km^2. About 23% of the anophelines and 52% of the culicines caught in the cages were males; blood-fed females formed about 11% of the catch.

To shorten the time taken to make a catch de Zulueta (1952) constructed a portable cage made from a wooden framework covered with mosquito screening. The cage, which was 1.73 m square and 2 m high, enclosed an area of about 3 m^2, and was fitted with a door and handles to facilitate carrying. Using this already erected cage, collections could be made about every 10 min. These collections were made in savannah grassland in an inhabited area, as distinct from the previous catches, which were made in an uninhabited area (de Zulueta 1950). Of those collected 40.0–68.8% of the female *Anopheles* and 24.8–31.9% of the culicines were blood-fed. There was a higher proportion of engorged specimens in catches made near houses. From 732 collections 2396 culicines and 220 *Anopheles* were caught, of which about 59 and 48% respectively were males. The smaller density of mosquitoes, 1.2/m^2, obtained in these catches was thought to reflect a real diminution in their population size.

In Jamaica Muirhead-Thomson and Mercier (1952) placed a mosquito net over vegetation but waited until sunset to collect adults that emerged of their own accord from the vegetation. They also used a large canvas tent having a single opening facing the sun and covered with a mosquito netting exit trap to collect mosquitoes resting in vegetation. In Trinidad Senior White (1952) found that drop-nets were useful for collecting mosquitoes resting amongst grassy vegetation, but not for *Anopheles aquasalis*, as adults did not respond by flying out of the vegetation in the presence of sunlight to be caught on the walls of the net.

Modified cages consisting of light wooden frames, 6-ft square and 5-ft high, constructed in four sections to facilitate carrying, were used by McClelland in East Africa (McClelland 1957; McClelland and Weitz 1963). A net was suspended within the framework in the manner of a Barraud cage (Fig. 5.13).

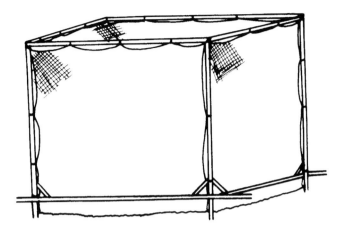

Fig. 5.13. Large drop-net cage (Service 1993)

The top of the net and lower part of each side panel was made of calico while the upper sections were of white mosquito netting. Mosquitoes were flushed from the vegetation with either a bee smoker in which a mixture of charcoal and coarse sawdust was burnt, or with a fine spray of citronella and sesame oil. A single catch took 5–20 min depending on the numbers of mosquitoes present. Smoke was superior to spraying in driving mosquitoes out from the vegetation. Larger cages enclosing about 10 m^2 of ground were made by suspending nets within a hexagonal frame, 2 m high and 2 m wide, constructed from 18 sections that were assembled in the field by five people. Three assistants entered the cage by a sleeve on one side to collect the mosquitoes. Smoke was produced outside the cage and blown into it by a fan through a 15-ft length of reinforced rubber hose. Unfortunately the fan rapidly became heavily coated with resinous deposits and required frequent washing. Over 24 mosquito species belonging to eight genera were collected from vegetation by these cages (McClelland 1957). About 23% of the females were blood-fed whereas only about 8% of the adults collected by sweeping vegetation were engorged. McClelland (1957) suggested that the lower proportion of engorged adults obtained by sweep-netting was due to the dilution effect of unfed mosquitoes attracted to the catchers and included in the collections.

A useful lightweight drop-net cage can be made from a framework of light alloy ½-in tubing, with a nylon net suspended within it. It is convenient to have a zip fastener sewn down one of the vertical edges of the cage so that collectors can enter it without having to crawl in underneath. The erected cage can be easily carried short distances and re-sited elsewhere by two to four people, or dismantled in about 15 min and carried to other areas.

At the Kenya coast drop-net cages 2 m square and 2 m high and having cloth loops at the corners to enable them to be easily manoeuvred by two people were placed over coconut husks, fallen coconut fronds and other debris which formed attractive resting places for *Anopheles merus*. The enclosed debris was disturbed with a stick or by kicking and adults caught for 15-min periods with powered aspirators (Mutero et al. 1984). Up to 100 *Anopheles gambiae* complex, including *Anopheles merus*, were caught per man-hour. Of 724 females trapped 64% were unfed, 27.4% blood-fed, 4.8% half-gravid and 3.7% were gravid. In contrast no mosquitoes were caught in box shelters of Gillies (1954).

Anil Prakash et al. (1998) used drop-nets between 0630 and 0730 h to sample mosquitoes resting outdoors among vegetation in Tripura state, India. From five attempts using this method they collected 123 mosquitoes representing five species and four genera. *Armigeres subalbatus* comprised 94.3% of the total catch.

Drop-net cages were used to study the effects of insecticide-treated cattle ear-tags on blood-feeding by *Ochlerotatus dorsalis* and *Ochlerotatus melanimon* in Wyoming, USA, by Lloyd et al. (2002). Drop-nets (2.4 × 2.4 × 2.7 m) were lowered over individual restrained cattle during the evening period of peak host-seeking activity. Mosquitoes were collected from inside the trap approximately 10 min later.

Drop-net cages may not always be suitable for collecting mosquitoes. For example in Guyana, Symes and Hadaway (1947) found that when adults of *Anopheles darlingi* were introduced into a 6 × 12 × 7-ft cage positioned over low scrub vegetation and grass, very few could be collected on the sides of the cage after the vegetation had been beaten. They discovered that adults were resting not on the vegetation but on the ground and on fallen leaves. In Sardinia, although a number of culicines were caught from vegetation enclosed within a mosquito net, *Anopheles* were not collected (Trapido and Aitken 1953), and in Indonesia Bahang et al. (1984) tried collecting out of door resting mosquitoes by dropping a net over bushes, but very few were caught.

Drop-nets cannot be easily used in wooded areas with emergent scrubby vegetation, although there may be concentrations of day-time resting mosquitoes in such habitats. Sweep-netting is more economical in manpower than using drop-nets and can be carried out in a greater variety of habitats, but is less easy to quantify.

Plastic tent. To collect adults of *Culex modestus*, which in the Carmague overwinter in reed (*Phragmites*) piles and amongst the dense vegetation in reed swamps, plastic tents have been used (Mouchet et al. 1969). These are made from tough white transparent plastic sheeting pulled over a 1.5-m high metal or wooden frame erected over swamp vegetation. An area of 3 × 3 m is enclosed by the plastic tent. The interior becomes very humid and hot, and this combined with the action of disturbing the enclosed vegetation results in mosquitoes flying out from their resting places and settling on the tent walls or roof. In addition to *Culex modestus*, adults of *Uranotaenia unguiculata, Culex pipiens, Culex impudicus, Culiseta annulata, Culiseta subochrea* and *Anopheles hyrcanus* and a variety of other Diptera including *Culicoides* spp. have been collected in France in these plastic tents.

In attempts to collect overwintering *Culex tritaeniorhynchus* in Korea plastic tents were placed over rock piles of stones that were dismantled. A temperature of about 10–12°C was needed to induce mosquitoes to fly and settle on the roof and walls of the tent, and when the sun failed to raise the temperature to this level a gas stove was used to heat the interior. Plastic tents were also placed over piles of straw, reeds and rock walls. Ree et al. (1976) considered the technique was not very effective in disturbing mosquitoes resting near the ground because the temperature there remained low. The only mosquitoes collected by this method were *Anopheles sinensis* (15), *Culex pipiens* (31), *Culex orientalis* (4) and *Anopheles pullus* (1).

Plastic tents have not been widely used, but this method might prove useful in collecting non-hibernating mosquitoes resting in vegetation, especially species such as *Anopheles darlingi*, which normally rest on or near the ground and are not retrieved with drop-net cages.

Crab holes

In Panama large numbers of several *Deinocerites* species were collected from crab holes by placing fine mesh cages over their entrances and then either dislodging the mosquitoes by blowing in smoke, or by forcing them up into the cages by flooding burrows with water (Tempelis and Galindo 1970). In Jamaica mosquitoes were also collected from crab holes by lightly spraying with pyrethrum and catching the escaping adults in cages placed over their entrances (Muirhead-Thomson and Mercier 1952).

In Malaysia Rudnick (1986) reported the capture of 257 male and 666 female adult mosquitoes from nine aspirator collections from crab holes. All the males, except one, were *Uranotaenia lateralis*, and 55% of the females were also of this species.

Rajavel et al. (2000) reported gently blowing air into crab holes in a mangrove forest in south India and rapidly collecting the disturbed mosquitoes using an aspirator. Adult density of *Aedes portonovensis* was expressed as number per 10 crab holes.

Caves

Mosquitoes resting in caves can usually be caught directly with aspirators, but in Israel Saliternik (1965) found it more convenient to catch *Anopheles sergentii*, which rested during the day in caves, crevices and fissures in limestone rocks, in exit cages. The cages, which consisted of a 20 × 20 × 18 cm wooden framework covered with mosquito netting, were placed over or close to the opening of crevices or caves during the day, and the catch retrieved the following morning. Large numbers of *Anopheles* were also collected from caves in Palestine by Shapiro et al. (1944).

Tree trunks, fencing, culverts, banks etc.

Mosquitoes resting on relatively exposed surfaces can be collected with aspirators. Their distribution may be very patchy, but with practice favoured resting places, sometimes the basal 0.5–1 m of tree trunks, may be identified. Gently tapping or prodding tree trunks, bromeliads, and earthen banks etc. with a stick usually disturbs resting mosquitoes, which can then be caught in a small hand-net.

In Zika forest, Uganda from 20 collections performed throughout the 24-h day from tree trunks McCrae et al. (1976) succeeded in catching 1328 male and 1777 female *Anopheles implexus*. Anil Prakash et al. (1997) collected 22 adult *Anopheles dirus* (8 males, 14 females) from outdoor resting sites, of which moist, dark areas of large tree trunks yielded the majority of specimens (20). No *Anopheles dirus* were collected from searches of the interior of human dwellings or cattle sheds.

Mosquitoes resting on branches and foliage of trees have sometimes been collected by spraying the trees with insecticide. For example, in Australia Kay (1983) used a swingfog machine to spray 0.1% pyrethrins synergised with 0.6% piperonyl butoxide into trees. The operator had to sometimes stand on a ladder to ensure the insecticidal fog reached the tops of the trees. Mosquitoes and other insects were collected beneath the trees on plastic sheets. This method demonstrated that considerable numbers of *Culex squamosus* were resting in the trees, a species poorly represented in aspirator collections from vegetation and resting box catches. In the USA Simmons et al. (1989) used a backpack sprayer fitted with a ULV nozzle to spray the lower canopy of trees with 3.3% resmethrin insecticide. Resting simuliid flies were knocked down and collected from sheets spread on the ground under the trees. Shaking the trees helped dislodge insects that had fallen onto leaves. From 32 person-hours of collecting, 84 simuliid blackflies ($\bar{x} = 2.63$) were obtained, but truck trapping for 450 person-hours yielded 8730 blackflies ($\bar{x} = 19.4$).

Yamashita and Ishii (1977) gave a description in English of the use of commercial insecticidal smoke 'bombs' (canisters) to collect arthropods from trees, originally published in Japanese in the period 1968–1971. Basically the smoke canisters were ignited and either hand-held or placed in metal containers supported on a variety of extension tubes to reach high up in the tree canopy. Smoking lasted 1–3 min and insects knocked down were collected on white trays or sheets placed on the ground underneath the trees and bushes. Diptera formed 7.0–21.8% of the arthropods caught by this method.

Frank and Curtis (1977) give an interesting description of their frustrations in trying to catch large numbers of both sexes of *Wyeomyia vanduzeei* for mark–release experiments. The creation of artificial resting sites and 'honey pot' feeding stations proved unsuccessful. The only way of collecting adults was to aspirate them from their natural resting sites on the rough bark of buttonwood trees. With experience they were able to catch about 80 adults, with approximately equal numbers of males and females, in 2 h. Catching for 2 h daily, 5 days/week for 42 weeks yielded 5920 male and 3836 female *Wyeomyia vanduzeei*.

Tree-holes, rodent burrows, termite mounds, crevices etc.

Aspirators. Adults resting in these types of recesses can be located with a torch and collected with aspirators. Büttiker (1958) collected 32 female and 22 male *Anopheles culicifacies* including blood-fed individuals from a tree-hole in Sri Lanka. In a survey of tree-hole habitats in Orissa, India Yadav et al. (1997) collected resting adult mosquitoes from tree-holes and hollow tree trunks using an oral aspirator and flashlight. In villages, 74 of 387 hollow tree trunks harboured resting adult mosquitoes. In an uninhabited forest area, resting adults were collected from 92 of 421 trees. Adults of 20 species belonging to six genera, including five species of *Anopheles*, eight *Culex*, three *Aedes*, one *Ochlerotatus*, one *Armigeres*, 1 *Orthopodomyia*, and one *Toxorhynchites* were collected resting in hollow tree-trunks. The most commonly collected species were *Culex brevipalpis* (220/452), *Aedes albopictus* (58/452), and *Culex tritaeniorhynchus* (38/452). In El Salvador Breeland (1972*b*) collected with aspirators an average of 42.4 adults of *Anopheles albimanus* and 12.9 *Anopheles pseudopunctipennis* per hour from natural diurnal resting sites such as rock crevices, tree-holes and ground holes. In other surveys, 7670 *Anopheles albimanus* and 2344 *Anopheles pseudopunctipennis* were caught in 181 collections from natural shelters. These day-time collections of resting adults were much less time consuming than night captures, and were probably less influenced by climatic conditions. Furthermore, they were of particular value in catching adults of *Anopheles pseudopunctipennis*, which are infrequently caught in light-traps, at human bait, in stable captures or in collections from houses. Breeland (1972*a,b*) concluded that the collection of adults from day-time resting sites provided reliable information on changes in population levels of *Anopheles albimanus* and *Anopheles pseudopunctipennis*. In southern California and Mexico hibernating adults of *Anopheles freeborni* and several other mosquitoes were collected from nests of wood rats (*Neotoma fuscipes*) (Ryckman and Arakawa 1951, 1952).

In India, Rajendran et al. (1995) collected mosquitoes resting amongst vegetation in rice fields using hand-held oral aspirators. In conventionally irrigated rice fields, as many as 188.8 individuals of *Culex tritaeniorhynchus* and 101.6 *Culex vishnui* were caught per person-hour of collection effort. Numbers collected were reduced to 35.8 and 26.2 individuals per person-hour respectively in fields under intermittent irrigation. Rajendran et al. (2003) used the same method during the first hour after dusk to collect vectors of Japanese encephalitis resting in vegetation in and around cattle sheds, also in India. Biweekly collections conducted between June 1998 and May 2000 yielded 72 958 females of 17 species. Mean catches per person-hour ranged from 16.5 to 654.4. *Culex tritaeniorhynchus* was the most commonly collected species (80.46%) followed by *Culex gelidus* (12.38%), *Culex vishnui* (2.39%), *Culex fuscocephala* (1.23%), and *Culex infula* (0.47%). Bansal and Singh (1993) performed resting collections in 12 villages in Rajasthan, India. Collections were performed at dusk and dawn from inside human dwellings and cattle sheds, from cracks and crevices in the banks of ponds, from wells and clay water containers, using oral aspirators. 1559 anophelines of six species were collected. *Anopheles pulcherrimus* and *Anopheles barbirostris* were only collected from cracks and crevices in pond banks.

In Canada Hudson (1978) collected overwintering mosquitoes from man-made rock piles that were covered with snow. Firstly he removed the snow and then turned over the stones by hand and aspirated out adults. Twelve searches in seven rock piles (116.5 man-hour) yielded 108 *Anopheles earlei*, 102 *Culex territans*, 1 *Culiseta alaskaensis* and 1 *Culiseta minnesotae*.

I have collected several species of culicines from the entrances of animal burrows in Kenya, using a torch and oral aspirator.

Fumigation and spraying. In many situations relatively few mosquitoes can be collected by direct searches, but more may be caught if they are disturbed from their resting sites and caught as they fly out. Tobacco smoke, smoke from a beehive fumigator or produced by burning corrugated paper soaked in potassium chlorate can be blown in, or Risella or citronella oil or a weak solution of pyrethrum (0.5%) can be sprayed into, the resting sites. Alternatively commercially available insecticidal aerosols can be used (Service 1963). Büttiker (1958) collected *Anopheles culicifacies* from large cavities in tree-holes in Sri Lanka by placing spray sheets over the bottom and also over the opening and then spraying with pyrethrum. Zukel (1949*a*) collected adults from hollow trees by fumigating them with sulphur dioxide produced by burning cheese cloth impregnated with a paste of sulphur and fuel oil wrapped round a wire frame. This was lighted and placed in a hollow tree, which had all openings covered with canvas, and the mosquitoes collected on a white sheet placed at the base of the hollow. Zukel (1949*a*) reported that whereas only a single female of *Anopheles quadrimaculatus* s.l. was collected from hollow trees prior to smoking, substantial numbers of *Anopheles quadrimaculatus* s.l., *Uranotaenia sapphirina, Culex erraticus*, and a few *Anopheles punctipennis, Culex quinquefasciatus* and *Culex peccator* were caught following fumigation. He also found that smoking was more efficient in driving them out of their hiding places than fumigating with hydrogen cyanide or pyrethrum. Love and Goodwin (1961) also found fumigation useful in collecting adults from tree-holes. Mosquitoes have also been caught from hollow trees, rodent burrows and other natural resting sites by spraying them with chloroform or acetone and catching the escaping adults in nets or small cages placed over their openings (Loomis and Green 1959; Mortenson 1953; Trapido and Aitken 1953; Zukel 1949*b*). In California, USA Reisen et al. (1989) lightly sprayed 605 rodent burrows with atomised chloroform, but only four female and two male escaping *Culiseta inornata* were collected in sweep-nets.

In India Batra et al. (1979) tried several methods to collect exophilic mosquitoes, including lowering a canvas cloth fixed to an umbrella frame down wells and then space spraying above with 1% pyrethrum. Nine or more mosquito species, including 50 *Anopheles stephensi* and 924 *Culex quinquefasciatus*, were caught from 102 urban wells, and 214 *Culex quinquefasciatus* from rural wells. Most of these mosquitoes, however, were newly emerged individuals. Mosquitoes resting in culverts and in some other miscellaneous places were smoked out by lighting mosquito coils and catching escaping adults in hand-nets. In other shelters, such as tree-holes, bamboo, holes in walls and also other culverts, adults were caught in hand-nets after they had been flushed out by spraying with 1% citronella oil. From 273 culverts 45 *Culex quinquefasciatus* were caught when mosquito coils were lit, but none was caught from 60 culverts sprayed with citronella oil, and only seven mosquitoes were caught from all the other resting sites sprayed with citronella oil.

Traps. Harwood and Halfhill (1960) developed a simple trap for collecting mosquitoes, mainly *Culex tarsalis* and *Anopheles freeborni*, from amongst vegetation, crevices in the ground and from animal burrows. Their trap consists of a 2-lb coffee can with a fine wire mesh inverted cone fixed into the base, and with a large circular hole cut from the top and covered with fine mesh. The can rests on a metal flange soldered to a circle of ½-in wire mesh (Fig. 5.14). The trap is placed over an animal burrow, and earth is placed around the base of the trap so that light only enters through the top of the can. Mosquitoes fly out of the burrow through the inverted cone of the trap. A similar trap was used in Australia for collecting biting Diptera emerging from vertebrate burrows (Dyce et al. 1972).

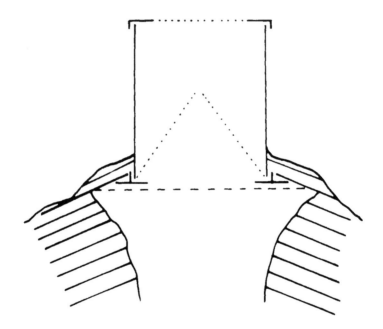

Fig. 5.14. Trap used by Harwood and Halfhill (1960) to collect mosquitoes resting in crevices in the ground

Wire mesh over the entrance of the can prevents rodents entering the trap. The trap can be modified for collecting mosquitoes from vegetation and rocky outcrops and crevices in the ground (Harwood 1962; Harwood and Halfhill 1960). Canvas, black heavy duty polythene or tarpaulin is fixed to the trap's metal flange which is supported on a tripod having 1–1½-ft long legs. The bottom edges of the tarpaulin are held down with rocks so that a tent-like structure is erected over the area to be sampled.

The tripod legs are coated with motor oil to prevent spiders ascending and spinning webs inside the trap. When 118 traps were placed in position over rodent holes during the mid-afternoon and recovered the following day an average of 2.0 female and 3.0 male *Culex tarsalis* and 12.7 female and 8.2 male *Anopheles freeborni* were collected per night. Some burrows gave consistently high catches. One that was probably inhabited, or used by wood rats, produced an average of 9.0 female and 13.1 male *Culex tarsalis* and 8.7 female and 11.0 male *Anopheles freeborni*, while a marmot burrow gave an average of 4.5 female and 7.0 male *Culex tarsalis* and 210 female and 46.5 male *Anopheles freeborni* per night. This type of trap was also used to study hibernating populations of *Culex tarsalis* and *Anopheles freeborni* in the USA (Harwood 1962), and overwintering adults of *Culex tarsalis*, *Culiseta inornata* and *Anopheles earlei* in Canada (Shemanchuk 1965).

Hudson (1978) also working in Canada used modified traps of Harwood and Halfhill (1960) to collect mosquitoes emerging from burrows of the badger (*Taxidea taxus*). Traps consisted of 15-cm long, 10-cm diameter cans, with the bottom removed and replaced with a black 1-mm mesh cone funnel, ending in a 1.5-cm diameter hole at the apex. A 2.5-cm hole was cut from the metal push-on lid and covered with 2-mm metal mesh screening. The base of the can was inserted through a hole cut from a 20-cm square piece of plywood, having a skirt of black plastic attached to its edges to help seal the trap over the badger's burrow. To prevent rodent damage traps were enclosed in protective 6-mm wire screening. Every 7–10 days the push-on lid was removed and the catch aspirated out, but 10–20% escaped; clearly a more efficient removal method is required. A total of 127 *Anopheles earlei* and 1 *Culex territans* were caught from 53 burrows, but the mean numbers of *Anopheles earlei* caught per burrow varied from 0.5 to 7.0, and in fact no mosquitoes were recovered from 49% of the burrows.

In Sri Lanka a few mosquitoes were collected by placing muslin gauze in and over holes in termite mounds and stimulating mosquitoes to fly out by pouring about 5 ml of mosquito repellent containing a benzoate derivative into some of the openings (Büttiker 1958). Mosquitoes resting in nests of kingfishers, sand martins and bee-eaters, which consist of narrow burrows in earthen banks, were examined for mosquitoes by the insertion of a narrow strip of filter paper and spraying with 2% pyrethrum. The idea was that mosquitoes knocked down by the spray were caught on the paper strip, which was then withdrawn. However, only four female culicines were recovered from 12 birds' nests by this method (Büttiker 1958).

Granaries

Clarke et al. (1980) were the first to show that in the Kisumu area of Kenya substantial numbers of the *Anopheles gambiae* complex (subsequently shown to be both *Anopheles arabiensis* and *Anopheles gambiae*) rested in grain stores, whose walls and roofs were made of maize or millet stalks plastered with mud. Mosquitoes resting in these granaries can be caught by pyrethrum spray sheet collections. For this a sheet is placed over the stored grain inside the granary while another is held over the entrance of the granary. Alternatively a person with a torch enters the granary and collects mosquitoes with an aspirator. In the study by Clarke et al. (1980) Muirhead–Thomson-type pit shelters were dug in the same compounds as those with grain stores. The mean monthly number of the *Anopheles gambiae* complex they collected per house was 17.0 times greater than those from pit shelters, but just 2.5 times more than those collected from granaries. Grain stores gave collections 6.8 times greater than from pit shelters. *Anopheles funestus* is a more endophilic species and few adults were collected from either grain stores or pits, the numbers in houses being 53.3 and 34.3 times greater than those found in pit shelters and granaries. The numbers in granaries were 1.6 times greater than those in pit shelters. These and later studies (Githeko 1992) have shown that *Anopheles arabiensis* more commonly rests in granaries and other out of door structures than does *Anopheles gambiae* s.s. Githeko (1992) constructed simple walk-in box-like shelters made of papyrus fronds in areas lacking granaries, and found they were very attractive out of door resting sites for *Anopheles arabiensis*.

Storm drains

Underground storm drain systems in metropolitan areas in Orange County, California, USA and elsewhere have been found to constitute important adult resting habitats for domestic and peridomestic mosquitoes and other diptera (Chanda and Shisler 1980; Giorgio et al. 1994; Mulligan and Schaefer 1981; Pfuntner 1978; Smith and Schisler 1981). Tianyun Su et al. (2003) used non–attractant CDC traps to sample adult mosquitoes resting in these drains. *Culex quinquefasciatus* dominated at all sampling sites. Adult female mosquitoes caught in traps were either unfed or gravid. Only 3 blood-fed females were found in all collections over the 2-year sampling period, however, this could also have been due to blood-fed females not being efficiently sampled by the traps.

Mechanical suction sweepers

Dietrick's machine ('D-vac')

Many of the earlier models of suction machines developed to collect arthropods from vegetation required either mains supply of electricity or large cumbersome generators. Many of these suction machines have narrow diameter (2.5–7.5 cm) suction hoses, convenient for extracting insects from short grassland but which are totally unsuitable for collecting insects, such as mosquitoes, resting in tall grass and herbage. The models developed by Dietrick et al. (1959) and Dietrick (1961) have much larger diameter suction hoses. The latter, which is available commercially under the name of the 'D-vac', is the more useful model as it is portable. It operates from a single-cylinder, air-cooled, two-cycle petrol engine. By altering the engine throttle air displacement can vary from about 500–1500 ft^2/min. Total weight is about 27 lb. A 5-ft length of 8-in diameter flexible wire ribbed canvas air ducting is connected to the fan intake. The other end is joined by an 18-in length of tapering plastic-coated nylon ducting to a fibre glass collecting head 11 in in diameter and about 11 in long (Fig. 5.15). A fine 18-in long tapered organdie bag, into which the insects are sucked, is positioned within the fibre glass collecting head.

Service successfully used this machine in England to collect mosquitoes resting in grassy and shrubby vegetation and it was employed by Mitchell and Chen (1973) in Taiwan and by Tempelis et al. (1970) in Hawaii to collect blood-fed mosquitoes for precipitin testing. Kuntz et al. (1982) successfully used D-vac aspirators to collect reasonable numbers of blood-fed *Anopheles crucians* and *Psorophora columbiae*, as well as a few engorged adults of six other species. In North Carolina, USA Irby and Apperson (1992) used a D-Vac to investigate the spatial and temporal distribution of resting mosquitoes. Operators walked through the habitat at a speed of approximately 1 m/s, sweeping the nozzle of the the D-Vac through a 1 m arc with each step, estimated to sample about 43 m^2/min.

Counts were transformed to $\sqrt{y} + 0.5$ to stabilise variance and resting densities were expressed per 214 m^2, equivalent to 5 min collecting, for statistical analyses, with backtransformed means used to describe densities per 1000 m^2. Over a period of two years, 24 830 mosquitoes were collected, representing 28 species and seven genera.

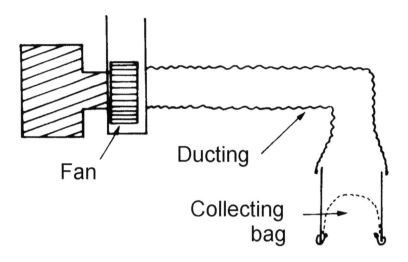

Fig. 5.15. Mechanical suction sweeper of Dietrick (1961)

Due to the relatively high cost of the D-Vac and its unavailability in some regions Sheldahl (1974) converted a petrol-driven, backpack insecticide mistblower into an efficient and powerful aspirator. The petrol tank and sprayer hose are removed and the air intake connected to a flexible 4-in diameter, 8-ft length of hose ending in an aluminium collection chamber which houses the collection carton. The conversion of a standard mistblower to power the aspirator apparently takes only some 15–20 min. The machine should be operated on about quarter throttle. Unwanted insects are excluded by placing a mesh screen over the collecting tube, and also by varying the force of the air intake. Thornhill (1978) described how a machine much cheaper than the D-vac, but very similar, could be made by using a lawn mower 4 h.p. air-cooled two-cycle 98 cc engine.

Summers et al. (1984) found the D-vac heavy, cumbersome and expensive and developed an improved sampler called the UC-VAC, and described it in detail. The basic unit consists of an Echo-PB-400 Power Blower with a special venturi attachment—Echo-PBAV-400—that uses energy of the exhaust to create suction in a separate section of piping. The metal collection canister has an internal diameter of 30.48 cm and is 21.10 cm long, with 'lips' 6.03 cm in diameter to which suction pipes are attached. These measurements are important in ensuring that there is maximum suction. As in the D-vac there is a removable cloth bag in the collection canister to retain the insect catch.

The total weight of the D-vac is 19.5 kg, the UC-VAC weighs 15.0 kg, at full throttle the rev/min are 3000 and 8000 respectively, in addition the UC-VAC is less noisy, vibrates less, and is less than half the price of the D-vac.

Moreby (1991) described a very simple modification that can be made to motorised vacuum samplers of the Dietrick (1961) and Thornhill (1978) type to facilitate removing the catch from the collecting bag (Fig. 5.16).

This consists of taping the net collecting bag on to the apical removable nozzle ring A, thus dispensing with the hooks or clips that normally secure the bag to nozzle B. Ring A now fits tightly into ring B, without the need of hooks or clips. The insect collection can be easily emptied into a plastic bag by turning the bag inside out. Only one collecting bag is therefore needed.

In England attempts to use a smaller type of motorised machine developed by Arnold et al. (1973) and known as the 'Univac' to collect mosquitoes from woodland scrub vegetation have not proved successful.

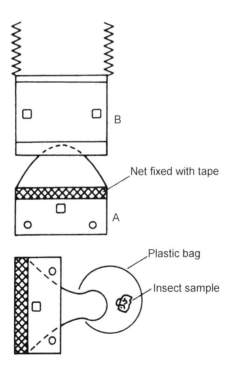

Fig. 5.16. Modification to Dietrick mechanical suction aspirator proposed by Moreby (1991)

When placed near the ground the strong air-intake through the 6.3–15-cm diameter nozzle leading to a 6.3-cm suction hose caused much ground debris and soil to be collected, which ruined the collected mosquitoes. Even when it was not placed near the ground, the few mosquitoes that were caught were badly damaged (Renshaw 1991). Southwood (1978) described how two bags can be used in the machine, one of nylon netting to retain insects and an outer larger cloth bag to prevent dust passing through the fan.

Suction sweeper of de Freitas

This machine (de Freitas et al. 1966) is composed of three main parts: (1) a 19.5-cm length of 9-cm diameter plastic tubing carrying on its lower side a 12-cm long side port made of the same material (A); (2) a 9-cm long nylon netting sleeve (B); and (3) a 26-cm long, 15-cm diameter collecting chamber (C) made from 32 × 32 mesh wire screening (Fig. 5.17). A small 3–4½-V d.c. motor, such as used in CDC light-traps, having an eight-bladed propeller mounted at the rear is positioned within the upper section of the plastic tubing. The tube's opening is covered with 14 × 14 mesh wire to prevent mosquitoes entering. These are collected through the lower side port, which is covered with 2 × 2 screening to prevent leaves and other debris being sucked in. The complete machine is suspended underneath a 1-m long wooden handle.

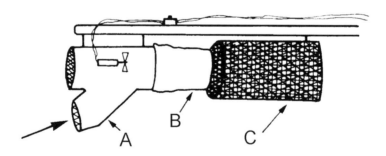

Fig. 5.17. Mechanical suction sweeper model of de Freitas et al. (1966). A—collecting tube, B—netting sleeve, C—collecting chamber

The motor is wired through an on/off switch to a 3-V power supply consisting of torch batteries or a rechargeable wet cell battery attached to the belt of the operator. This suction sweeper has been successfully used in Brazilian forests, to collect mosquitoes resting amongst vegetation, from rodent burrows and from the bases of trees. Mosquitoes are first gently disturbed by prodding vegetation or rodent burrows with a stick and then caught as they take flight. When sufficient mosquitoes have been caught, the nylon sleeve is closed with a clip or piece of string and the collection chamber and sleeve removed. Mosquitoes collected by this machine are apparently in a good enough condition for taxonomic study (de Freitas et al. 1966). A greater number of species (at least 63) were caught by this apparatus than obtained by light-traps (27) and a variety of other collecting methods (42–59). The machine was used to collect mosquitoes for virus isolation studies, to obtain engorged females for blood-meal determination, and to study the day-time distribution of mosquitoes resting in different types of vegetation. The inventors stressed that much of the success of collecting adults depended on the ingenuity of the operator in locating natural resting places.

In South Africa Jupp and McIntosh (1987) constructed an aspirator, basically the same as that of de Freitas et al. (1966), to collect *Aedes circumluteolus* from vegetation. It consists of an 18-cm length of 15-cm diameter perspex tubing housing a 12-V, 8-W d.c. motor with a three-bladed rubber propeller. The end of the perspex tube is covered with 13 × 13-mm wire mesh to prevent debris being sucked up. Their procedure was to disturb vegetation with a stick held in one hand and then with the aspirator held in the other about 10 cm from the ground to aspirate insects in flight. About 10% of the mosquitoes that were sucked up through the fan blades into a 20-cm^3 collecting cage were damaged. Over a 16-month period the numbers of *Aedes circumluteolus* caught per trap-hour in different months varied from 0–416 males and <1–561 females. Other relatively common mosquitoes collected were *Culex neavei, Culex zombaensis, Culex insignis* and *Culex antennatus*.

CDC sweeper

Like the de Freitas machine, suction is provided by a 3–4½-V d.c. motor, but this operates from a 6-V not a 3-V battery (Hayes et al. 1967). A 3-in diameter two-blade aluminium propeller is fixed to a motor housed in a 3½-in outside diameter 4-in length of clear cast acrylic tubing (A) (Fig. 5.18). The basal ⅜ in of a 1-in wide, 3½-in inner diameter ring of acrylic plastic is cemented over the opposite end of the fan housing. A pint-sized cardboard carton (C), 3⅜ in in outer diameter, with the bottom end removed and covered with no. 24 nylon mesh, is seated within the ⅝-in projecting rim of this plastic ring. This comprises the removable collection chamber. A collecting tube (B) made from acrylic tubing of the same shape and size as the fan housing is fitted to the top of the cardboard carton by a similar 1-in wide 3½-in inside diameter ring of plastic. A 15-in length of 1-in aluminium angle is fitted along the outside of the complete collecting unit, being permanently fastened to the fan housing by a circular metal ring and to the collecting tube by elastic bands. The 3-in length of angle that extends from the back of the fan housing is riveted to a 24-in long handle of 1-in diameter aluminium tubing.

After mosquitoes have been collected the motor is kept running while the front tube is removed and a lid is placed on the cardboard carton, which is then removed and replaced with another. A quart-size carton or two 1-pint ones taped together can be used for larger catches, but increasing the capacity of the collection chamber decreases the suction power at the intake. The original aim was to collect mosquitoes resting amongst vegetation by slowly moving the apparatus backwards and forwards in a sweeping motion, hence its name. The motor, however, provides insufficient suction for this and the machine is more usually used as a powered aspirator for collecting mosquitoes resting in shelters and traps. In Hawaii Tempelis et al. (1970) found this machine very useful in collecting mosquitoes, including blood-fed individuals.

Herrel et al. (2004) used the commercially available version of the CDC mechanical aspirator (model 1412 John W. Hock Co., Gainesville, Florida, USA) to sample mosquitoes resting among vegetation between 0700 and 1300 h in South Punjab, Pakistan. Collections were carried out for 15 minutes per crop or vegetation type. *Anopheles cullicifacies* was only rarely collected from vegetation, whereas *Anopheles peditaeniatus* was only collected from vegetation, albeit in small numbers, and not from inside houses or animal shelters.

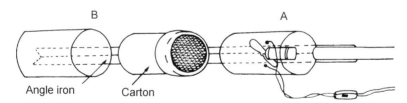

Fig. 5.18. CDC-type mechanical suction sweeper, A and B—cylindrical tubing, C—holding carton in middle (Hayes et al. 1967)

Yamashita and Ishii (1977) described an aspirator for collecting insects, which is based on the CDC sweeper of Hayes et al. (1967) but also somewhat resembles the aspirator made by Hall et al. (1968) to collect ants.

Suction sweeper of Garcia

A long torch provides both a convenient handle and housing for five 1.5-V dry-cell batteries which are needed to power a small fan placed at the rear of a metal cylinder mounted on brackets beneath the torch. A small conical section forms the intake and is a sliding fit over the cylindrical chamber (Fig. 5.19). A detachable fine muslin collecting bag is tied on to an inner metal tube projecting from the conical section. A metal cap covered with fine mesh is placed over the end of the intake to prevent mosquitoes escaping when the collecting bag is withdrawn from the cylinder. Mosquitoes are sucked into the collecting bag, which is tied across with string and removed.

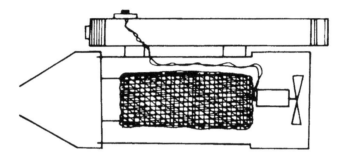

Fig. 5.19. Mechanical suction sweeper model of Garcia et al. (1988)

Davis and Gould's aspirator

This small portable, battery operated aspirator was developed in Thailand to collect mosquitoes resting amongst vegetation (Davis and Gould 1973). It consists of four basic parts, a collecting tube (A), a collecting chamber (B), a housing for the collecting chamber (C) and a fan unit (D) (Fig. 5.20). The collecting tube consists of a 36-in length of 3½-in diameter plastic water pipe. Half-inch wire mesh can be taped over the entrance to prevent leaves and large insects being sucked up. This tube fits into a 3½-in diameter metal sleeve fitted to the hinged lid of the aluminium collecting chamber housing, which is 8 in in diameter and length. A 1-gal paper ice cream carton with the bottom replaced with nylon netting and with an 18-in length of 6-in wide surgical stockinette taped around the other end forms the collecting cage and is inserted into the aluminium housing. The free end of the stockinette is slipped over the section of the collecting tube that extends about 4 in through the metal sleeve in the lid of the metal housing. Suction is provided by a 'squirrel cage' fan blower from a Volkswagen car heater unit. This is bolted through a gasket, made from a car inner tube to the bottom of the metal housing, which has a 3½-in diameter hole cut from the middle. The blower fan operates from a 6-V motorcycle battery, which can be attached to a belt worn by the operator. This aspirator was successfully used to collect mosquitoes from a variety of habitats during a study of Japanese encephalitis in Thailand (Gould et al. 1974).

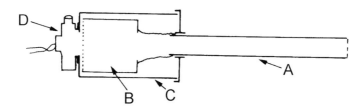

Fig. 5.20. Mechanical suction sweeper model of Davis and Gould (1973), A—mailing tube, B—collecting chamber, C—housing for collecting chamber, D—fan unit

This aspirator or a modified version was used in several studies in Pakistan by Reisen and colleagues. Using it, Reisen et al. (1982) caught 12 mosquito species resting in fields and 14 species resting in the forest, including *Culex tritaeniorhynchus, Culex quinquefasciatus, Aedes lineatopennis, Culex pseudovishnui, Culex fuscocephala, Aedes culicinus, Verrallina indica* and *Verrallina yusafi*. Adults of all species were more common in forest collections, and in fact *Verrallina yusafi* was collected only in the forest. In other studies the Davis and Gould (1973) aspirator caught *Culex tritaeniorhynchus* and a few adults of 11 other species from fields in Pakistan (Reisen et al. 1978), while Reisen and Milby (1986) used the aspirator to collect more than 14 species of exophilic mosquitoes, the most common again being *Culex quinquefasciatus*, while *Culex pseudovishnui* was also common.

Suction sweeper of Kay

Kay (1983) made an aspirator from a 1-m length of 10-cm diameter PVC drain pipe (Fig. 5.21). A 12-V motor drawing 37 W and a fan are powered by a 9-Amphr motorcycle battery carried in a bag on the hip. The velocity of air through the intake is 5 m/s and mosquitoes are collected in terylene bags inserted into the intake of the PVC tubing. These bags can be secured with rubber bands and are removed every 15 min. This aspirator is similar to that of Davis and Gould (1973) but is simpler in that the enlarged holding cage is omitted. In Australia mosquitoes were collected from inside houses and from natural out of door resting places. In one site 21 591 mosquitoes, mainly *Culex quinquefasciatus*, were caught, the mean catch being 35.2 mosquitoes. In another site the total catch was 55 961 and the mean catch was 48.6 mosquitoes. Using this aspirator Kay (1983) collected some 7552 blood-engorged mosquitoes of 13 species, but mainly *Culex annulirostris* from 2119 collections from vegetation in sylvan and urban areas of Queensland.

Fig. 5.21. Mechanical aspirator of Kay (1983) (photograph reproduced by permission of B.H. Kay)

Aspirator of Nasci

To address some of the disadvantages of the D-vac machine in terms of weight, noise, need for refuelling with petrol, and cost, Nasci (1981) constructed an aspirator from locally available materials that was some seven times cheaper, and weighed less than 5 kg. It consists of sheet metal (aluminium) rolled round to form a 61-cm long and 34.3-cm diameter tube (Fig. 5.22). A 12-V d.c. motor with a 25-cm fan blade is mounted on a 14-cm square piece of plywood screwed onto a 34.3-cm length of 4 × 10-cm strip of wood rounded at the ends to fit into the base of the metal tube. Transparent sheet plastic is rolled and bolted to form a 61-cm and 34.3-cm diameter tube, which can be inserted 10.1 cm into the metal tube. A D-vac collecting bag is folded over the other end of the plastic tube and held in position by S-shaped clips attached to large rubber bands. Two metal gate handles are fixed on the metal cylinder. Wire from the motor passes to a push-button switch on the metal tube and then to two 6-V gel batteries connected in series, which will power the aspirator for at least 6 h.

The Nasci (1981) aspirator has proved successful in collecting a variety of mosquitoes from amongst vegetation, and appears to be one of the more widely used aspirators in the Americas.

Using this type of aspirator Beier et al. (1982) collected 2220 female and 1127 male mosquitoes belonging to more than nine species from 15 daily collections of 5-min duration from 15 different areas in a wood. *Aedes vexans* formed 60.4%, and *Culex* spp. formed 23.9% of the catch. In Florida, USA Nayar (1982) succeeded in using this type of aspirator to collect large numbers of male and female *Wyeomyia mitchellii* resting on trunks of oak trees. Savage et al. (1993) used this aspirator alongside hand-held aspirators, CDC light-traps supplemented with CO_2 and human landing catches to collect blood-fed *Aedes albopictus* in Missouri, USA to investigate blood-feeding patterns. Also in the USA weekly sampling over 3 months from woodlots collected 841 *Ochlerotatus triseriatus* (Nasci 1982). Forattini et al. (1993) used the Nasci (1981) battery-powered aspirator in the Ribeira valley region of south-east Brazil to collect mosquitoes resting in vegetation within a wooded area and in the ecotone between the woodland and open land, which contained grasses and shrubs up to 1.0 metre in height. A series of 5 or 10 min collections were made during a period of one hour during the day on a fortnightly basis for one year (January 1992–January 1993). A total of 22 140 individuals were collected (12 790 females and 9350 males), comprising 11 species. The most common species at the residual woodland sites and at the secondary bush site was *Ochlerotatus serratus*, whch comprised 39.0% and 24.8% of the total catch respectively at each site. At the primitive forest site, *Anopheles cruzii* was marginally the most commonly collected mosquito, comprising 16% of the total catch at this site. Eighty-two species were recorded from the first woodland site, 83 from the secondary bush site, and 65 from the primitive forest.

Later, Forattini et al. (1997) used the Nasci (1981) aspirator to collect mosquitoes resting among vegetation in the vicinity of a flooded concrete structure that acted as a productive source of *Aedes albopictus* pupae. Collections were performed for 30 minutes from 1800 to 1830 h and mean hour densities calculated. A total of 984 individuals of 38 species (plus a further nine partially identified species or groups) in nine genera were collected during 15 collections. Forattini et al. (2000) used the same aspirator to collect *Aedes albopictus* and *Ochlerotatus scapularis* in Brazil. A total of 441 female *Aedes albopictus* were collected in the outdoor collections, with a peak in numbers observed between January and May, corresponding with the peak in the biting population.

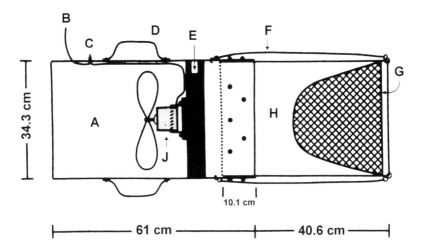

Fig. 5.22. Nasci-type aspirator, A—aluminium tube, B—wire to battery, C—push-button switch, D—heavy duty gate handle, E—wooden, motor support, F—heavy rubber band with 'S' clips at either end to secure collecting bag. G—is a D-vac bag folded over rim of plastic tube, H—plastic tube partially inserted and bolted inside aluminium tube, J—motor mount assembly (Nasci 1981)

A total of 99 *Ochlerotatus scapularis* were obtained from outdoor collections, with peak numbers also occurring between January and May, however, this did not coincide with the peak in biting activity, which occurred between July and October. Gomes et al. (2003) also used the Nasci aspirator to sample resting mosquitoes from areas shaded by large trees, within households, and near materials of domestic origin deposited immediately surrounding the household, in Brazil. A total of 10 751 blood-fed female mosquitoes was captured using the aspirator and hand-held nets, comprising *Ochlerotatus scapularis, Culex nigripalpus, Culex quinquefasciatus*, and *Aedes albopictus*. Of these 1092 specimens were fully engorged and later tested by the precipitin method.

The distribution of resting female *Aedes vexans* in metropolitan Minneapolis-St. Paul, USA, was studied by Boxmeyer and Palchick (1999) using the Nasci mechanical aspirator. Aerial photographs were used to randomly select 4 sampling sites of each of four habitats: wooded, garden, crop, and residential yards. Each sample comprised a 2.5 min collection and sampling was carried out between 0800 and 1330 h at least weekly for five consecutive weeks. The swath of the aspirator extended from ground level to approximately 1 m above ground, and was estimated to sweep out a volume of 22 m^3. Female *Aedes vexans* comprised 89% of the total collection of 2899 female mosquitoes. The wooded habitats yielded a higher mean density (24.33 mosquitoes per 57.6 m^3) than the garden (2.95 mosquitoes per 57.6 m^3), crop (6.00 mosquitoes per 57.6 m^3), or yard (2.95 mosquitoes per 57.6 m^3) habitats.

The Nasci (1981) aspirator was used by Ashford et al. (2003) to undertake entomological sampling as part of investigations into an outbreak of dengue fever in Palau, western Pacific. The aspirator was used for periods of 10 to 15 minutes to sample mosquitoes both indoors and resting amongst vegetation outdoors. Five species were collected, namely *Aedes aegypti* ($n = 22$), *Aedes albopictus* ($n = 13$), *Aedes hensilli* ($n = 16$), *Aedes* sp. ($n = 2$), and *Culex quinquefasciatus*.

Arbovirus Field Station (AFS) Suction sweeper of Meyer

Meyer et al. (1983) found the aspirators of Davis and Gould (1973) and Nasci (1981) unsuitable, mainly because of their size and/or lack of power, so they designed a lightweight more powerful machine termed the AFS (Arbovirus Field Station) sweeper. Basically it consists of a lightweight L-shaped metal backpack adapted to hold a plywood platform (8 × 13 in) supporting two 6-V d.c. rechargeable gel cell batteries (Fig. 5.23). These are connected in series to deliver 12-V at a minimum of 16 A, which is needed to operate the blower for about 1 h. The blower is a 12-V Dayton model (2C646) and is mounted underneath the plywood platform on the horizontal section of the backpack. Air displacement of the blower is about 2.12 m^3/min, which represents improved suction over most previous battery operated aspirators.

The suction tube is composed of various sections of 4-in diameter black PVC piping (as used by plumbers and called ABS tubing), and about 4-ft of 4-in diameter automobile defroster hose, making altogether a tube about 8 ft long. Mosquitoes are sucked into a 1-pint sized paper carton having a nylon-screened bottom which is inserted where the handle section and nozzle join (Fig. 5.23). This collection carton can be rapidly removed, its lid placed on and a new carton inserted. A screen of ¼ or ½-in netting is inserted over the collecting tube to prevent leaves and twigs being sucked into the collection carton. This aspirator was used in California, USA to catch mosquitoes resting in various outdoor shelters and amongst vegetation (Reisen et al. 1988).

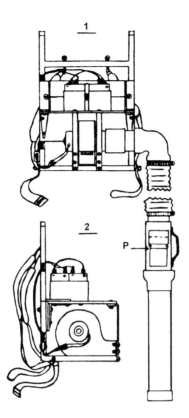

Fig. 5.23. AFS-sweeper, 1—rear view of sweeper showing collection carton (P) inserted into handle part of tubing, 2—side (left) view (Meyer et al. 1983)

Generally substantially more *Culex quinquefasciatus, Culex tarsalis, Culex stigmatosoma* and *Culiseta incidens* were collected from shrubbery and flower beds than from house eaves, porches, miscellaneous buildings and debris littering yards. Resting sites were characterised by high humidity and shade provided by trees or buildings. The numbers of *Culex quinquefasciatus, Culex tarsalis, Culex stigmatosoma* and *Culiseta incidens* caught by aspirators, and also in gravid mosquito traps, in different house compounds was highly aggregated and fitted a negative binomial distribution. In later studies in California vegetation was sampled with the AFS sweeper and with the more powerful D-vac machine, but these collections generally proved unrewarding, except when vegetation near larval habitats was sampled when large numbers of males and fewer newly emerged females of *Culiseta inornata* were caught (Reisen et al. 1989). From 31 collections with the sweeper, mainly near larval habitats, 217 females ($\bar{x} = 7.0$) and 12 150 males ($\bar{x} = 391.9$) were collected, while from five collections with the D-vac 189 female ($\bar{x} = 37.8$) and 1527 male ($\bar{x} = 305.4$) *Culiseta inornata* were caught. The AFS was also used to collect outdoor-resting *Culex tarsalis* for diapause studies (Reisen et al. 1995).

In a study of the daytime resting sites of adult mosquitoes at an urban residence in southern California, USA, Schreiber et al. (1993) used the AFS sweeper to sample 30 outdoor resting sites fortnightly from June to October 1989. Biotic factors, including presence of con-specifics, other mosquito species, chironomds, other non-dipteran flying insects, and predators, were assessed as were three abiotic factors: light intensity ambient air temperature and relative humidity. *Culex quinquefasciatus* was the most commonly collected mosquito throughout the study period. Overall, males predominated in all samples, although the proportion of males did decline over the study period. Total numbers of mosquitoes collected varied over the five months, the distribution of mosquitoes among the five resting habitats sampled remained consistent, with the favoured sites being on shaded, vertical surfaces situated above decaying vegetation (a compost pile, or other areas of the garden treated with mulch). Significantly more mosquitoes were collected at heights of 0–1 m above ground than at 4–6 m. Abundance of adult resting mosquitoes was higher in midday collections than either morning or late afternoon collections, in all months except September. Abiotic factors did not appear to adequately describe the observed distributions, except for late afternoon samples collected in August, in which both temperature and relative humidity were significantly correlated with abundance.

The AFS sweeper was used by Szumlas et al. (1996) to sample mosquitoes resting among vegetation and leaf litter at a La Crosse virus endemic area in North Carolina, USA. Four 15-min samples were taken between 1000 and 1400 h by walking slowly within a 100 m radius of the residence from which a case of La Crosse virus infection had been reported. Aspirator samples were conducted weekly from mid-May to late October or early November in 1989 and 1990, and yielded a total of 6046 adult mosquitoes comprising 11 species. *Ochlerotatus triseriatus* was the most commonly collected species in sweeper samples, representing 88.4% of the total catch. The mean number of adult mosquitoes collected per 15-min sample was 8.4 in 1989 and 3.6 in 1990.

CDC Backpack Aspirator

Clark et al. (1994) modified the AFS sweeper of Meyer et al. (1983) as follows: the shelf for the 6-V batteries was removed; a 3/8 in. thick piece of plywood, with a hole for the exhaust, was attached to the base of the frame; the blower position was changed to the operator's right side; the net-covered collecting carton was placed at the end of the collecting tube, rather than the middle; the blower was rotated 90° so that the exhaust was directed downwards; the 6-V batteries were replaced with a single 12-V motorcycle battery; a wire mesh battery compartment was added (Fig. 5.24).

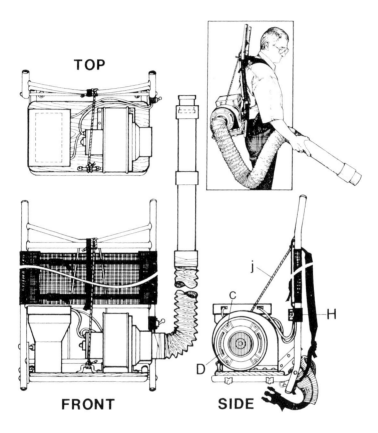

Fig. 5.24. The CDC backpack aspirator (Clark et al. 1994)

The apparatus was tested in indoor collections of *Aedes aegypti* from rooms and cupboards in Puerto Rico. Full details of the construction of this aspirator are provided as an appendix to the original paper, for those interested in constructing one.

Perdew and Meek aspirator

Perdew and Meek (1990) argued that their aspirator, made by modifying a commercial cordless 'broom' blower, provides greater suction and requires less modification than many previous battery powered aspirators. The basic equipment is a blower-type broom that consists of a plastic body incorporating a fan and a motor, and in its handle rechargeable cadmium batteries, and an extension tube tapered and curved distally. To convert the broom into an aspirator the wires to the motor are reversed and the tapered end of the extension tube cut off.

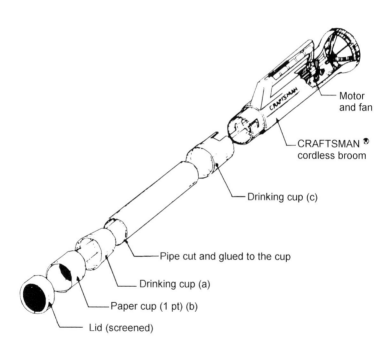

Motor and fan

CRAFTSMAN ® cordless broom

Drinking cup (c)

Pipe cut and glued to the cup

Drinking cup (a)

Paper cup (1 pt) (b)

Lid (screened)

Fig. 5.25. Cordless broom aspirator (after Perdew and Meek 1990)

Two rigid plastic tapered beakers (474 ml) having a minimum 9-cm inner diameter at the top are required. To make an elongated version of the aspirator a selected length of the bottom of the plastic cup (*a*) is cut off to leave a 6.9-cm opening that can be glued to the inside of the extension tube (Fig. 5.25). A 473-ml paper cup (*b*) has a 6.4-cm diameter hole cut from both its lid and its bottom, and both holes covered with screening of suitably sized mesh. This forms the collection carton, and without its lid, is pushed tightly into the plastic cup stuck in the end of the extension tube. This extended version of the aspirator is 118-cm long. A shorter (65-cm) version can be made by cutting the bottom from a similar plastic cup (*c*) to leave a 7.4-cm diameter opening. After two opposing rectangles are cut out the cup is fitted over the end of the main body of the cordless broom and secured with four screws. The paper cup (collection carton) is fitted into the end of the plastic cup, the extension tube is not used. After using either the extended or short aspirator the screened lid is pushed on the paper cup, which is then removed and replaced with another.

The aspirator can operate efficiently for about 45 min, with a minimum air flow of 21 m^3/min.

Battery-powered aspirators

In Florida, USA Day and Curtis (1989) monitored the abundance of *Culex nigripalpus* by undertaking two 10-min collections with battery powered aspirators from vegetation three times a week for 3 years. In some months more than 5000 females were collected. A time series analysis of the numbers caught and rainfall revealed a significant cross correlation between the numbers of blood-fed females and rainfall. In 1985 there was best correlation between engorgement rates 2–13 days after rain, but in 1986 trends were not so clear, and in 1987 the only positive association was on the day of rainfall. In Brazil Natal and Marucci (1984) constructed a simple 3-in diameter aspirator from PVC drainpipe operating from a 6-V battery attached to a belt from around the waist to collect mosquitoes resting either indoors or amongst vegetation. Flores-Mendoza and Lourenço-de-Oliveira (1996) used a slightly modified version of the Natal and Marucci aspirator to collect mosquitoes resting among vegetation in Rio de Janeiro State, Brazil.

Rubio-Palis and Curtis (1992) used a large battery-operated aspirator consisting of a 14 cm diameter × 125 cm long PVC collecting tube, a small fan and two 6V batteries to collect anophelines resting among vegetation in western Venezuela. A total of 2470 anophelines of eight species were collected. About 13% of the sample was unidentifiable. Catches were significantly correlated with human landing catches for the four commonest species collected. Relative sampling efficiency compared with human landing catch was determined by calculating the ratio of mean resting catch to mean landing catch. The aspirator was relatively efficient at collecting *Anopheles triannulatus*, but less efficient for *Anopheles oswaldoi* and the dominant species in the landing catches, *Anopheles nuneztovari*. In one site the aspirator caught more *Anopheles albitarsis* than human landing catches.

Vehicle-mounted aspirators

Service (1993) described two vehicle-mounted aspirators developed in the 1960s for collecting mosquitoes, and Southwood (1978) refers to a few others that have been used to sample non-medical arthropods. As these types of machine appear to have been only rarely used, they are not described here and interested readers are advised to consult Service (1993) for details.

Artificial resting places

Because of the difficulties inherent in locating outdoor mosquito populations, which may be distributed over wide areas of vegetation, specially constructed artificial shelters have been used to try to attract mosquitoes to specific sites from which they can conveniently be collected. Nuttall and Shipley (1902) appear to have been the first to have suggested that traps could be used to attract and catch resting adults. As a result various types of box-like shelters were used both in and outdoors by later workers in efforts to control mosquitoes (see Service 1993 for references). It was not until Russell and Santiago (1934) constructed their earth-lined box trap that it was fully appreciated that artificial resting places could be used to study and sample exophilic mosquitoes.

Since then several different types of artificial resting shelters have been made and evaluated. Some, such as the box shelters of Edman et al. (1968), have been used mainly to collect specific species or genera, such as *Culiseta melanura*, whereas others, such as pit shelters, are used to collect a broader range of species and genera.

Keg shelters

Field observations in the Tennessee Valley area of the USA showed that in general more *Anopheles quadrimaculatus* s.l. entered unbaited traps with the door left open than baited traps with a closed door (Smith 1942). This together with the discovery that adults readily entered empty, small box-like structures led to the development of the keg shelter as an artificial diurnal resting site for *Anopheles quadrimaculatus*.s.l. The shelter consists of an ordinary, untreated wooden barrel-shaped nail keg with the lid removed, placed on its side on the ground in a deeply shaded area, especially near the edge of larval habitats. Its entrance should be protected from wind and the sun by siting the keg at the base of a tree, alongside or under a log, under bushes or amongst scrub vegetation.

It is now very difficult to obtain wooden nail kegs. Because of this and the effectiveness of other simple and cheaper shelters (see below) kegs are now little used as artificial resting sites.

Small red box shelters

Goodwin (1942) reported that nail kegs were not successful in Georgia, USA, for collecting *Anopheles quadrimaculatus* s.l. so he experimented with various wooden boxes as alternative resting sites. Boxes open at one end and measuring 1 ft^3, or 1 ft^2 in cross section, and 2 ft deep caught more *Anopheles quadrimaculatus* s.l. than smaller boxes, and more adults were caught in a 1-ft^3 box placed below 6 ft than in those placed higher. Boxes facing east always caught fewer mosquitoes than those whose opening faced other directions. Goodwin (1942) found that the mean catch of *Anopheles quadrimaculatus* s.l. was greater in boxes painted red inside (24.13) than those painted white (0.04), yellow (0.28), blue (0.43), black (12.79) or green (0.47). Moreover, mosquitoes were more easily seen and collected from red boxes than from black ones where torches were sometimes needed. It was concluded that the best shelter was a 1-ft^3 wooden box painted dull black on the outside, red inside and positioned on the ground in a sheltered position, preferably not facing east. In comparative trials these red boxes were on average about 2.8 times more attractive than nail kegs (Goodwin 1942).

Red boxes were used by Zukel (1949*a*) to collect *Anopheles quadrimaculatus* s.l. in Georgia, USA and by Breeland (1972*a*) to collect *Anopheles albimanus* and *Anopheles pseudopunctipennis* in El Salvador. Both Burbutis and Jobbins (1958) and Moussa et al. (1966) found they were useful in collecting *Culiseta melanura*, and in fact could replace light-traps for monitoring seasonal changes in population size of this mosquito (Burbutis and Jobbins 1958). Boxes on the ground facing approximately due west caught more mosquitoes than those facing other directions, or placed at a height of about 4 ft. Few mosquitoes were found in boxes inspected between 0600–0700 h. Burbutis and Jobbins (1958) considered that most entered between 0800–0900 h. In addition to *Culiseta melanura*, adults of *Anopheles crucians, Anopheles quadrimaculatus* s.l., *Anopheles punctipennis, Culex salinarius, Culex restuans, Culex pipiens, Ochlerotatus canadensis, Ochlerotatus sollicitans, Coquillettidia perturbans*, and *Uranotaenia sapphirina* were caught in the boxes. Loomis and Sherman (1959) found that the boxes were very useful for collecting and measuring population changes of

Anopheles freeborni and *Culex tarsalis* in California, USA. In fact both they and Bradley (1943) caught more *Anopheles freeborni* in red boxes than in light-traps. Yuval and Fritz (1994) used resting boxes constructed from red-painted plywood and measuring 1 × 1 × 1 m to collect *Anopheles freeborni* in California, USA. Hayes et al. (1958) also caught large numbers of *Culex tarsalis* in box shelters and Reisen et al. (1995) used them to collect *Culex tarsalis* in California, USA.

Although Carpenter et al. (1946) reported better catches of mosquitoes in light-traps than in the resting boxes of Goodwin, Gusciora (1961) in 33 comparable tests in New Jersey, USA caught 13 240 mosquitoes in box shelters but only 6260 in light-traps. A ¼-in mesh screen was placed over the openings of the box shelters to reduce the large numbers of tipulids that sought shelter in them. Collections were made by placing a cloth net bag with elastic around the opening over the entrance of the box and spraying with chloroform. Gusciora (1961) suggested that increased catches might be obtained if ½-in thick plastic foam pressed into 3-in deep folds was wetted and tacked to the floor of the box, so as to increase humidity and the resting area available for mosquitoes. In India Yasuno et al. (1973*a*) did in fact place a sponge soaked in water in boxes placed inside houses to attract endophilic *Culex quinquefasciatus*.

Pletsch (1970) reported that the red boxes of Goodwin were useful in El Salvador for collecting *Anopheles*, but as they were very bulky and cumbersome he developed a collapsible box shelter made of ¼-in plywood. The four sides (12-in square) are joined together with strips of canvas, leather or plastic, stapled or nailed to the sides. The back of the box (12 × 13½ in) is joined to the bottom by a canvas hinge, and a transverse strip of wood (12 × 1½ × ¾ in) fastened to the inner face ½ in from the top gives rigidity to the box when the back is folded into position. This back panel is held in position with string or cord. Sixteen or more of these traps can be transported or stored in the space required to stack two conventional red boxes of Goodwin.

Walk-in red boxes

In California, USA Nelson and Spadoni (1972) developed much larger (6 ft tall, 4 ft wide, 6 ft deep) boxes painted red inside and out and called walk-in red boxes. These were later modified by Meyer (1985) to have a curtain that can be pulled across the entrance when a person enters to aspirate the catch (Fig. 5.26). These generally trap more mosquitoes than the smaller red boxes having 1 ft^2 entrances.

Fig. 5.26. Walk-in red box (photograph reproduced with permission of M.W. Service)

In California, USA Reisen et al. (1983) compared the numbers of male *Culex tarsalis* caught in standard (0.3 × 0.3 × 0.0.03 m) red boxes (Goodwin 1942) and larger (2 × 1 × 2 m) walk-in red boxes with those caught in CDC light-traps, truck traps and those caught from swarms. The red boxes, especially the larger ones, caught the most males, and were considered the best method for monitoring their seasonal abundance and Reisen et al. (1995) later used them to collect *Culex tarsalis* for diapause studies in California, USA. Resisen and Lothrop (1995) used the Meyer (1985) modification of the walk-in red box to sample *Culex tarsalis* as part of a mark-release-recapture study using 49 CO_2-baited CDC light-traps in the

Coachella valley, California, USA. Only 116 females (7% of a total 1780 recaptures) and 9 males were recaptured in two walk-in red boxes. Lothrop and Reisen (2001) successfully used walk-in red boxes of Meyer (1985) to collect *Culex tarsalis* for bloodmeal identification studies in the Coachella Valley, California. These boxes have been used to sample other mosquitoes such as *Culex quinquefasciatus* and *Culex stigmatosoma* (Reisen et al. 1990), and *Anopheles freeborni* (McHugh 1989). In California 30 827 male and 22 813 female *Anopheles freeborni* were collected from 33 daily collections from 15 walk-in red boxes (1.8 × 1.2 × 1.3 m). Blood-fed females formed 19.6% of the catch, but most females (75.3%) were unfed (McHugh 1989). Niebylski and Meek (1992) used walk-in red boxes (Meyer 1985) and hay infusion-baited ovitraps (Reiter 1983 and see Chap. 2) to collect blood-fed *Culex* in an urban environment in Louisiana, USA. Weekly sampling between 24 May and 15 October 1987 yielded 159 *Culex* specimens. A higher proportion of the mosquitoes collected in the walk-in red boxes located in forest had fed on humans and dogs than mosquitoes collected by ovitraps in the same forest. This was explained by the red boxes acting as a visual attractant for engorged females dispersing from houses and dog pens in the vicinity. Wekesa et al. (1996, 1997) used walk-in red boxes of Meyer (1985) to collect blood-fed *Anopheles freeborni* and *Culex tarsalis* in California, USA. Wekesa et al. (1997) reported collecting a total of 63 373 female *Anopheles freeborni* and 25 687 female *Culex tarsalis* from 12 red boxes operated during the period June to September in 1991 and 1992 and emptied weekly. Overall, 28% of the females collected were blood-fed. Nelson et al. (1978) found that Goodwin (1942) type red boxes and larger walk-in ones sited near larval habitats caught almost exclusively newly emerged *Culex tarsalis*, whereas older adults were collected by carbon dioxide-baited light-traps. Possibly older mosquitoes used other resting sites, or those further away from the emergence sites.

Reisen et al. (1989) made concerted efforts to collect outdoor resting *Culiseta inornata* in California. From 314 collections from 1-ft^3 red boxes 190 female ($\bar{x} = 0.61$) and 30 male ($\bar{x} = 0.10$) *Culiseta inornata* were obtained, compared to 867 females ($\bar{x} = 0.49$) and 846 males ($\bar{x} = 0.47$) from 1784 collections from walk-in red boxes. In this instance the larger boxes were no better in collecting *Culiseta inornata* than the smaller ones.

Robertson et al. (1993) constructed resting boxes 2.1 m tall, 1 m wide and 1 m deep from plywood with the exterior surfaces painted dark green. Dark-coloured cloth was fixed across the top 50 cm of the opening. Traps were positioned with the opening facing west when located amongst trees and with the opening facing the water when located close to a swamp. Mosquitoes were collected weekly using hand-held aspirators between 1000 and 1400 h. In total, 7009 females and 9344 males of *Anopheles quadrimaculatus* s.l. were collected from 14 shelters between May and the beginning of December 1985. In 1986, 9228 females and 11 657 males were collected from 16 shelters over the same period. The mean number of female *Anopheles quadrimaculatus* s.l. per shelter ranged from 2.4 to 109.3 in 1985 and 1.3 to 67.5 in 1986. The other common species collected using this method was *Culex erraticus*, of which 2678 females and 772 males were collected in 1985 and 2517 females and 600 males in 1986. Mean numbers per shelter ranged from 1.1 to 18.4 in 1985 and 0.11 to 13.4 in 1986.

McNelly and Crans (1982) rightly emphasised that catches from different walk-in red boxes should be recorded separately for statistical tests, and not all box catches in an area combined.

Red cloth resting shelter

Breeland and Glasgow (1967) devised a cheaper, lighter and more portable version of the red box shelter of Goodwin (1942). A 52-in length of red broadcloth which is 36 in wide is folded down the middle and stitched across at each end, so that when the material is cut between the middle seams two sacks are formed. Each sack is stapled to a 1-ft^2 unpainted wooden frame, which is nailed to a tree (Fig. 5.27*a*). The sack is held out horizontally by tying the ends with string to a convenient tree or stake. In two separate series of trials in Alabama, USA these red cloth shelters caught significantly more *Anopheles quadrimaculatus* s.l. than the conventional red boxes.

McCardle et al. (2004) used un-baited tan-coloured Saran cages with an internal volume of 1.8 m^3 to collect presumably unbiased samples of mosquitoes in Maryland, USA, as part of studies to determine the best method for collecting gravid female mosquitoes. The entrance slit was kept open to allow mosquito entry and was only closed during emptying of the nets with a hand-held mechanical aspirator, which was undertaken between 0900 and 1100 h twice per week. Total catch per night in these cages was 77.7 mosquitoes, of which 41.3% were gravid or blood-fed. *Culex restuans* was the most commonly collected species, followed by *Ochlerotatus canadensis, Culex pipiens, Culex territans, Aedes vexans, Ochlerotatus triseriatus, Culex erraticus, Anopheles crucians, Coquillettidia perturbans, Anopheles quadrimaculatus* s.l., *Psorophora ferox, Anopheles punctipennis, Aedes hendersoni*, and *Uranotaenia sapphirine*.

Resting box of Edman

This resting box was developed by Edman et al. (1968) specifically for collecting blood-fed adults of *Culiseta melanura* in Florida, USA. The final version of their trap consisted of a plywood box 30 in long, 18 in wide and 12 in high, painted matt grey on the outside and matt black inside. A red cotton collecting bag slightly smaller in width and height than the box but 8 in longer was put in the box and held in position by placing elastic loops sewn at the bottom four corners over hooks screwed into the bottom of the box. The open end of the cloth bag which projected from the box was folded back over its opening, and secured by elastic tape threaded through its end seam. A plywood concave frame, painted matt grey outside and matt black inside, with a 28 × 52-in opening tapering to a 10 × 16 base was placed just inside the opening of the box (Fig. 5.27*b*). Mosquitoes were collected from the box shelters by first carefully withdrawing the concave frame and then pulling the turned over end section of the collecting bag from the box and closing it.

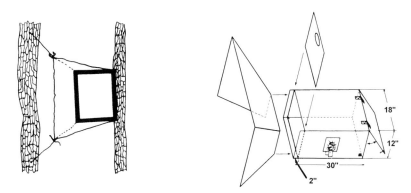

Fig. 5.27. Artificial outdoor resting shelters: (*a*) Red cloth shelter of Breeland and Glasgow (1967); (*b*) resting box of Edman et al. (1968)

Boxes without a cloth bag caught similar numbers of mosquitoes as those with the bag, but it was more time consuming to remove them with an aspirator than by removing the entire catch within the bag. Larger numbers of *Culiseta melanura*, especially males, were caught in boxes facing west than east. Boxes with a concave frame caught about 44 mosquitoes per day, whereas those without a frame caught only about nine per day but the proportion of blood-fed females increased from about 11 to 30%. However, because of the larger catch the absolute numbers of engorged females was higher in the boxes having a concave frame. It appeared that when the frame was positioned in the entrance of a box mosquitoes were attracted from a greater distance. Most *Culiseta melanura* entered the boxes before 0830 h, but there was some flight activity during most of the day. Optimum collection time was influenced by weather conditions; in general, the drier and windier the weather the earlier adults entered the boxes. In addition to *Culiseta melanura*, more than 14 other mosquito species were collected from the boxes.

Nasci and Edman (1981*b*) used these resting boxes to collect *Culiseta melanura*; 177 adults were caught, about 15% of which were blood-fed, from just four nights. In more prolonged collections the boxes were placed with their open-bottoms on damp soil at the bases of trees, with their openings facing west to avoid the morning sun. Several thousand blood-engorged *Culiseta melanura* were caught over 2 years from 10 boxes (Nasci and Edman 1981*a*). Nasci et al. (1993) used two resting boxes (presumably of the same design) to sample *Culiseta melanura* as part of a surveillance programme for eastern equine encephalitis virus in Ohio, USA. Over four nights in September 1991 the two boxes caught 47 *Anopheles quadrimaculatus* s.l. and two *Anophles punctipennis*. No *Culiseta melanura* was caught. In Massachusetts, USA Nasci and Edman (1984) caught 12 094 female *Culiseta melanura* from 10 resting boxes. The mean number of females per trap night varied from less than 10 to about 75; blood-engorged adults ranged from less than 10% to about 30%.

Resting box of Morris

A great variety of very simple (Goodwin 1942) and more complicated (Edman et al. 1968) resting shelters, including walk-in red boxes (Meyer 1985; Nelson and Spadoni 1972) have been used. It is difficult, however, to compare their efficiencies because they have been operated in different areas to collect different mosquito species, and have varied in colour and in location. Morris (1981) undertook experiments in New York, USA to study the effect of size and shape, location, and colour on mosquitoes caught.

In his experiments the standard shelter was a 30-cm wooden box painted matt black. Experimental designs included 10-, 20- and 40-cm cubes, the 30-cm box with the front opening partially closed, a box partitioned with one or two vertical or horizontal partitions, large boxes (20 cm high, 40 cm wide, 20 cm deep) on 10-cm high legs, and a less deep box (10 × 40 × 20 cm) on 20-cm high legs. Boxes were made of plywood or masonite, painted black, red, blue, brown or green, and had their openings facing in different directions. Morris (1981) concluded that a shelter 30 cm high, 40 cm wide and 20 cm deep caught about the same numbers of mosquitoes

(*Culiseta melanura, Culiseta morsitans, Anopheles punctipennis, Anopheles quadrimaculatus* s.l. and *Culex territans*) as the standard 30-cm cube box, but smaller shelters caught fewer mosquitoes. Making boxes from masonite reduces cost and weight, and prevents damage by plywood-eating porcupines. There were no differences between the numbers caught in shelters painted black, red, brown, blue or green, but shelters painted white, grey or partially white and black attracted fewer mosquitoes. Presence of vertical or horizontal partitions had no effect on the catch. West-facing shelters consistently had more mosquitoes than east-facing ones when collections were made in the morning, but in the evenings there was no such difference. Obviously catches are increased by employing more shelters, but (Morris 1981) considered they should be placed at least 1.5 m apart.

Morris et al. (1980) placed shelters 3 m apart with one of the longer sides on the ground and the entrance facing west. Trapped mosquitoes were killed by placing a chloroform-soaked pad in a shelter and closing it with a lid. From 640 shelter-days in New York, USA 1403 female *Culiseta melanura* were collected. Depending on the collecting area, means of 32, 36 and 42% were parous, while blood-fed and gravid females formed 38.5, 52.0 and 84.9% of the catches. Later Howard et al. (1989) found these resting shelters useful in collecting unfed, blood-fed, gravid and males of *Culiseta melanura* and *Culiseta morsitans* in their mark–recapture studies on flight range and dispersal.

Oliver et al. (1996) used these resting shelters to collect blood-fed *Culiseta melanura* in Oswego County, New York, USA for studies on fecundity.

Resting boxes of Morris (1981) are routinely used alongside CDC miniature light-traps supplemented with dry ice to collect *Culiseta melanura* and *Culiseta morsitans* as part of the eastern equine encephalitis surveillance programme in upstate New York, USA (Howard et al. 1996).

Resting box of Weathersbee

Although box-type resting stations of Edman et al. (1968) have proved useful in collecting *Anopheles quadrimaculatus* s.l. (Weathersbee et al. 1986), they were considered cumbersome and rather expensive to make, consequently a more portable and cheaper resting unit was designed (Weathersbee and Meisch 1988). It consists of a 113.5-litre plastic refuse container (72.3 × 41.9 × 80.6 cm) placed on its side on the ground, to the front entrance of which is attached a tent-like 3-ft extension made from a sheet of black plastic (3 × 5 ft) (Fig. 5.28). The narrower end was fixed with waterproof insulating tape to the refuse box, while cord threaded through holes along the ridge of the plastic tent and its two sloping ends (all reinforced with insulating tape) were attached to a centre 3-ft dowel. Two large nails passed through loops in the cord from the sloping ends and secured the canopy to the ground.

Fig. 5.28. Resting box of Weathersbee and Meisch (1988) (photograph reproduced with permission of A.A. Weathersbee)

The mean numbers of *Anopheles quadrimaculatus* s.l. per day per box ranged from 118.3 ± 22.8 to 347.3 ± 93.4 males, and from 107.8 ± 27.1 to 227.5 ± 46.1 females. Numbers compared favourably with mean catches per day of 316–423 of the same species recorded by Snow (1949) from privy-type shelters. These boxes of Weathersbee and Meisch (1988) were used in later studies, again to collect *Anopheles quadrimaculatus* s.l. (Weathersbee and Meisch 1990).

Resting box of Kay

Kay (1983) used brown cardboard cartons ($33 \times 27 \times 25$ cm) which had the top flaps folded and tapered to form an ingress aperture as outdoor resting boxes in Australia. The boxes were stacked in groups of 10–25 on top of each other. In the dry season wet cheesecloth was sometimes introduced into the boxes, while during the wet season they were sheltered against rain. In one site the mean catch was 26.3 mosquitoes and in another just 4.1. The most common mosquito species collected was *Culex quinquefasciatus*.

Resting box of Charlwood

In Papua New Guinea Charlwood et al. (1985) reported that at dusk they caught *Anopheles farauti*, in all gonotrophic conditions, in a portable resting trap. This consisted of a 2-m long, 1-m wide, 1.5-m high wooden frame covered with black plastic with an exit trap fitted on one side and another on top. No mosquitoes were caught in the trap during the day.

Fibre pot resting box of Komar et al.

Commercially available, nestable, fibre plant-pots were used by Komar et al. (1995) to sample potential arbovirus vectors as part of an arbovirus surveillance programme in Massachusetts, USA. The pots used were constructed from recycled wood pulp, formed into hollow, truncated pyramids, 28 cm high, with a 28×28 cm square open end, tapering to a 15×15 cm closed end. Fibre pots were compared with conventional plywood resting boxes, as described by Edman et al. (1968). Pots and resting boxes were placed on the ground with their openings randomly oriented. More than 2000 specimens, representing 10 different species were collected from 295 fibre pots over a period of about 10 weeks. *Culiseta melanura* was the most commonly collected species, comprising 75% of the total. More than half of the females collected contained blood. Fibre pots were about as efficient at collecting mosquitoes as plywood boxes, but the addition of baffles to the openings of the plywood boxes made them much more efficient than the fibre pots.

Garvin et al. (2004) used 40 of the Komar et al. (1995) fibre pots to sample *Culiseta melanura* adults during an interepizootic period of eastern equine encephalomyelitis transmission in Ohio, USA. Boxes were arranged in groups of four in woodland with openings facing north, south, east and west and adult mosquitoes were collected weekly from mid-June to mid-August using hand-held aspirators. Only 5 *Culiseta melanura* adults were collected using this method, all from a single set of resting boxes during the first week of July (none was collected by CDC light-trap).

Rolled up mattresses

Khin Maung (1964) reported that in Myanmar during residual house-spraying programmes it had proved difficult to collect outdoor-resting *Anopheles culicifacies* until artificial resting shelters were made by rolling bamboo mattresses (6×4 ft) into hollow cylinders. These were stood vertically on the ground and the open tops covered with thatch. From 21 man-hours 535 *Anopheles culicifacies* were collected from these artificial shelters.

Plastic dustbin (trash) bags

In very dry regions of Dubai with scant vegetation Service (1986) made resting shelters by placing black dustbin (trash bin) bags over 1-ft cube wire frames, leaving the bag partially open to form an entrance. Cotton wool soaked in water was placed in the bag to increase humidity. These shelters were placed in position at 1700 h and inspected the following day at 0800 h. From 22 trap-nights 84 male and 264 female *Ochlerotatus caspius* were collected, 12.9% of the females were blood-fed, and from 44 trap-nights 614 males and 1745 female *Culex quinquefasciatus* were obtained, of which 12.6% were blood-fed. In northern England in the late autumn I have observed large numbers of *Ochlerotatus detritus* resting inside open heavy-duty plastic bags containing potting compost and stored in a wooden garden shed. Plastic dustbin bag traps, however, caught no mosquitoes in wooded areas in England (Service 1993) and Ponlawat and Harrington (2005) also found these traps to be unsuccessful in capturing *Aedes albopictus* in Thailand, however, the simplicity and cheapness of these traps means that they probably deserve further evaluation in different habitats.

Earth-lined box shelters

Earth-lined boxes were first used by Russell and Santiago (1934) in the Philippines to collect *Anopheles flavirostris* and other *Anopheles* species. Their trap consists of a 3-ft long and 2-ft square, or larger, wooden framework with a 1-in thick layer of soil held in place against the inside walls and roof by 16-mesh/in screening. There is no bottom to the trap, which is placed directly on the ground. A black cloth hangs down over the entrance to within about ½ ft of the ground. Drip cans full of water keep the earth lining the box moist (Fig. 5.29). When similar boxes without an earth lining were used the average catch of mosquitoes per night per box was 4.4–12.4, whereas in earth-lined boxes an average of 18.8–69.5 mosquitoes was caught. Further, none or very few *Anopheles flavirostris* were caught from the unlined boxes. Placing a light in the earth-lined box produced a maximum mean catch of 137.3 mosquitoes per night but no adults of *Anopheles flavirostris* entered the boxes.

In India Rao (1984) collected 'reasonable numbers' of *Anopheles culicifacies* from earth-lined wooden boxes placed out of doors. Also in India,

Sadanandane et al. (2004) describe the use of earth-lined boxes 60 × 60 × 90 cm in size, constructed by inserting wooden planks into natural hillsides.

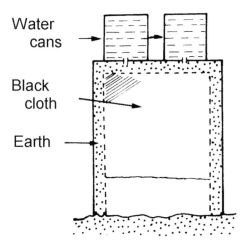

Fig. 5.29. Earth-lined box shelter (Rao 1984)

In studies on the outdoor resting populations of *Anopheles gambiae* and *Anopheles funestus* in a humid sub-coastal belt of Tanzania Gillies (1954) used a modified version of the earth-lined shelter of Russell and Santiago (1934). His box consists of a 3-ft long, 3-ft high and 2-ft wide wooden frame covered with plastic mosquito gauze, with a black cloth hanging down from the front entrance to leave a 6–8 in entry gap underneath. These boxes are buried in a suitably shaded earth bank, and mosquitoes collected from them early in the morning. None of these boxes caught large numbers of *Anopheles*, the mean catch varied from 0.3–3 *Anopheles* per shelter per day, although occasionally up to 20 females were collected from a shelter. A comparison of the mosquitoes caught in these artificial shelters with those from natural outdoor resting sites showed a similar composition of unfed (32.7%), blood-fed (7.5%) and gravid (59.8%) individuals of *Anopheles gambiae*. In contrast a higher proportion of unfed females of *Anopheles funestus* were caught from natural (80.0%) than artificial shelters (40.3%). Exophily in *Anopheles gambiae* was also studied in an arid area of Tanzania that was interspersed with large swamps and irrigation areas. Cattle were more abundant in this area than the coastal region and larval habitats were larger but more concentrated. During peak populations 'well over 3000 *Anopheles gambiae*' were caught from 23 box shelters over a 10-day period

(Gillies 1956). Occasionally over 100 *Anopheles gambiae* were collected from a single shelter. A considerably high proportion of the females caught in these shelters were blood-fed than in those in the coastal areas. This apparent higher degree in exophily was probably due to the combination of a larger mosquito population, a scarcity of suitable natural resting sites and the greater availability of cattle in the area (Gillies 1956).

Gillies (1954) thought that the box shelters sited near houses might compete with them in offering suitable alternative resting places, and in fact contain some adults which might otherwise have entered houses. A comparison of the ratio of half-gravid:gravid females caught in boxes placed near houses and in boxes some distance from houses with the ratio of half-gravid:gravid adults caught in houses did in fact indicate that boxes near houses in part reflected the endophilic population. Moreover, the construction of a hut near a box shelter caused a very marked reduction in the numbers of mosquitoes resting in it. It was concluded that whereas catches of mosquitoes resting in shelters away from houses gave a valid sample of the exophilic population, those collected from boxes placed near houses might not do so.

As with other artificial resting shelters the numbers of mosquitoes caught in these box shelters is greatly influenced by their location, with the result that identical boxes often catch greatly different numbers of mosquitoes. But, whereas portable resting shelters can be easily moved around so that locations giving the highest catches can be selected, the box shelters of Gillies are not easily resited. To overcome this difficulty portable mud-lined boxes similar to those of Russell and Santiago (1934) were used in Nigeria (Service unpublished data). They were 3 ft long and 2 ft^2 in cross-section and covered on the inside with plastic mosquito gauze but unlike the original earth-lined boxes they had a bottom. Earth was packed on all four sides and bottom in between the 2-in thick wooden framework and held in place against the mosquito gauze by plywood panels nailed to the box. A black cloth hung down over the entrance to leave a gap of about 6 in. Water was sprinkled through the gauze sides of the box to dampen the earth and the boxes sited in shaded places in village compounds. During the wet season in Nigeria a mean of 4.7 female and 1.9 male *Anopheles gambiae* and 2.2 female and 2.6 *Anopheles funestus* were collected per night from eight such boxes. In the dry season the mean catch was reduced to 1.4 female and 0.4 male *Anopheles gambiae*, and 1.1 female and 0.8 male *Anopheles funestus*. The maximum overnight catch was 37 female *Anopheles gambiae* from eight boxes, recorded in the wet season.

Some of the difficulties that may be experienced in using these artificial resting sites include theft of the black cloth hanging over the openings of the boxes, and the destruction of the boxes by termites, goats and vandalism, and the building of wasp nests in them.

Pipe traps

Nelson (1980) developed an artificial rodent burrow termed the pipe trap for collecting *Culex tarsalis*. The device consists of three parts, a pipe, a plunger and a collecting carton (Fig. 5.30). The pipe consists of a 91-cm length of 18.5-cm outside diameter asbestos-cement sewer pipe, both ends of which taper to 16.2 cm in diameter. The pipe absorbs moisture. The plunger is a 91-cm length of 4.76-mm steel rod having a 14-cm wooden disc fixed at the end, to which is glued a piece of 16.5-cm diameter foam rubber. The collecting component consists of a 16.5-cm long, 3.8-litre cardboard carton. The bottom is removed and a 30.5-cm length of tubular stockinette is taped over the outside of the carton. The other end (top) is removed or modified so that the sleeve section of a CDC collecting bag is fitted over the end. Alternatively a sleeve of stockinette or mosquito netting is fixed over the top end of the carton. A hole is dug in the ground (by a post-hole digger if available) and the pipe inserted at an angle of 20–25° from the horizontal so that about 7.5 cm projects from the hole. The piston-like plunger is pushed down to the bottom of the pipe. The following day any mosquitoes that have rested in the pipe are removed, by firstly fitting the stockinette sleeve on the bottom of the carton over the end of the pipe. The collector then reaches through the top sleeve and slowly pulls up the plunger, while squeezing the top sleeve to prevent mosquitoes escaping.

When the foam rubber disc is flush with the end of the pipe the stockinette tube is eased off the pipe and pinched off across the face of the disc. The carton is then slipped off from the rod (Fig. 5.30) and both ends tied. The trap is then reset by pushing the plunger down to the bottom of the pipe. Rodents sometimes enter pipe traps, and possibly scorpions and snakes may enter them, so caution may be needed in withdrawing the plunger.

Fig. 5.30. Pipe trap and its operation. 1—components, 2–8—process of removing catch (Nelson 1980)

In California, USA Nelson (1980) collected a mean of 13.8 male and 11.2 female (unfeds, blood-feds and gravids) *Culex tarsalis* per pipe compared to means of 28.8 and 24.6 from 1-ft red box shelters (Goodwin 1942). Other species resting in the traps were *Culex quinquefasciatus, Culiseta inornata* and *Anopheles franciscanus*. However, Reisen et al. (1989) found these traps ineffective in their studies, as only one male and one female were collected from 160 collections, but then only a single female was collected from 100 collections from cone traps fitted to rodent burrows, whereas previously such traps had caught considerably more *Culex tarsalis* (Reisen et al. 1985).

It seems that pipe traps could prove useful in other areas of the world, especially in hot dry regions, for a variety of mosquito species that normally rest in rodent burrows and other holes, or in cracks and crevices in the ground.

Metal drums, pots, etc

Empty petrol drums (Laarman 1959), village clay pots and other containers may also serve as useful resting places if they are buried in an earth bank or partially covered with a pile of earth. Vale (1971) described a useful resting place for the tsetse, *Glossina morsitans*, that he used in Zimbabwe, and which might prove attractive to mosquitoes. It consists of a metal drum covered in earth, and a box with a thatch roof positioned over the entrance. Dichlorvos (DDVP) strips were placed in a small flask to kill the catch. Girod et al. (1999) used old metal drums with a capcity of 200 litres and laid on their side to collect outdoor-resting *Anopheles arabiensis* on the island of La Réunion. A total of 1823 individual *Anopheles gambiae* s.l. mosquitoes were collected from these external shelters from 3348 person-hours of collection effort. Blood-fed female *Anopheles gambiae* s.l. accounted for 234 of the total.

Pit shelters

Blin (1908) first described how *Anopheles* could be caught, and he considered controlled, by digging small holes in the ground. Many years later Muirhead-Thomson (1951) used a similar method to sample outdoor resting mosquitoes in Tanzania. After initially failing to collect any *Anopheles gambiae* or *Anopheles funestus* resting on natural earth banks, he discovered that if horizontal pits or channels were dug into them, unfed, blood-fed and gravid females of both species sought refuge in these dark niches. He also created attractive resting places for both species by undercutting the wall near the bottom of an abandoned excavation pit. These observations led to the development of artificial pit shelters (Muirhead-Thomson 1958).

These pits are 5–6 ft deep and if possible are dug under trees or large bushes so that their openings (4–5 × 3–4 ft) are shaded from above; failing this a suitable cover should be placed partially over the pit entrance. About 1½–2 ft from the bottom of the pit one or two small cavities, about 30 cm deep, are dug out horizontally from each of the four sides (Fig. 5.31). Mosquitoes are collected from both these small cavities or from the wall of the pit itself. It is advisable to encircle the pits with a thorn or fence enclosure to prevent cattle or young children falling into them, or them being used as toilets. In Tanzania four pit shelters sited 30–150 m from a village yielded, after about a month, 674 male and 626 female *Anopheles gambiae* complex and 130 male and 150 female *Anopheles funestus*. This represented just 20.8 and 4.6%, and 20.0 and 7.6%, respectively, of the total catches of females and males of these two species collected from inside houses and in pit shelters combined. No anophelines were caught by searching vegetation and earth banks, nor in box-type traps made from tea chests (Lines et al. 1986). Muirhead-Thomson (1958) found pit shelters very useful in Zimbabwe for collecting *Anopheles gambiae*, the *Anopheles funestus* group and also several other *Anopheles*. In Zimbabwe Mpofu (1985) found that pit traps caught both *Anopheles quadriannulatus* and *Anopheles funestus*, 87.9% of the latter found in these traps were blood-fed. Nevertheless, only 589 *Anopheles funestus* and 404 *Anopheles quadriannulatus* were collected from 12–13 pits inspected on 14 days/month for a year.

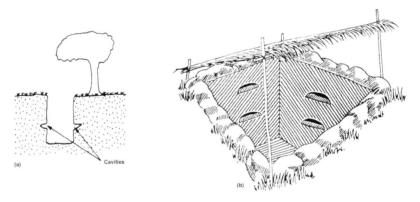

Fig. 5.31. (*a*) Muirhead-Thomson pit shelter; (*b*) pit shelter with roof (World Health Organization 1992)

Pit shelters have proved an effective method for sampling *Anopheles arabiensis* in the Awash valley of Ethiopia, where Ameneshewa and Service (1996) collected 451 individuals, representing 66.7% of the total outdoor-resting catch from four pits operated over a period of 22 days. Pit shelters were 2m deep, with a 1.5 m × 1.5 m opening. In Eritrea, Shililu et al. (2004) caught a total of 1359 anophelines resting in outdoor pit shelters, of which 87.3% were *Anopheles arabiensis*. Other species collected included *Anopheles cinereus* (10.4%), *Anopheles pretoriensis* (0.3%), *Anopheles d'thali* (0.9%), *Anopheles squamosus* (0.2%), *Anopheles demeilloni* (0.8%), *Anopheles garnhami*, and *Anopheles rupicola* (0.1%). The outdoor resting collection of *Anopheles arabiensis* females comprised 22.6% unfed, 42% blood-fed, 12.8% half-gravid, and 22.6% gravid.

Mnzava et al. (1994) used three outdoor pit-traps of the Muirhead-Thomson (1951) design located approximately 6 m from houses to collect *Anopheles arabiensis* for bloodmeal analysis and determination of chromosomal inversion frequencies in Kenya. The frequencies of the 2Rb and 3Ra inversions (Coluzzi et al. 1979) differed between the indoor mechanical-aspirator resting collections and the outdoor pit-shelter collections in one of the two study sites, as determined by a χ^2 test for heterogeneity, however, the authors concluded that the differences may have arisen as a result of combining data from two years' collections. Significant differences in the bloodmeal sources of indoor and outdoor collections was observed ($\chi^2 = 246$, $P < 0.001$), but it was felt that this could have reflected differences in host availability rather than genetic differences between the two populations, given that the chromosomal inversion frequencies were similar in both collections.

La Grange and Coetzee (1997) found pit shelters very effective at catching both males and females of the *Anopheles funestus* complex in Northern Province, South Africa. From 12 pit trap collections per month between June 1987 and August 1989 9234 *Anopheles funestus* females and 6807 males were collected. The majority of individuals of both male and female *Anopheles funestus* group entered the pits between 0600 and 0900 h. Two samples revealed approximately equal numbers of unfed, fed, and gravid (Christophers' stages IV–V) *Anopheles funestus* group females in the pit trap collections (115:146:146 respectively). Other species collected included *Anopheles gambiae* complex (45 females, 20 males), *Anopheles coustani* (3 females, 3 males), and *Anopheles rufipes* (41 females, 22 males). Small numbers of *Anopheles pretoriensis, Anopheles marshallii* group, *Anopheles longipalpis*, and *Anopheles demeilloni* were also captured using this method.

In Tanzania, Mnzava et al. (1995) used pit shelters 3 m deep with eight 15 cm diameter horizontal tunnels dug into the face of each of the four walls of the pit to collect *Anopheles gambiae* and *Anopheles arabiensis*. No *Anopheles gambiae s.s.* was collected in the traps, but they did catch large numbers of *Anopheles arabiensis*.

Ijumba et al. (2002), also in Tanzania, studied malaria transmission in three different agroecosystems (rice cultivation, sugarcane and savannah) using indoor and outdoor human-landing collections, light traps, pyrethrum spray catches and pit shelters. Pit shelters were examined fortnightly and mosquitoes were removed from the pits using manual aspirators over a period of 15 min between 0630 and 0730 h. A total of 2612 *Anopheles arabiensis* was collected from pit shelters and this species was the dominant species collected by this method in all three agroecosystems. *Anopheles funestus* was also collected in pit shelters, but at much lower numbers.

In Côte d'Ivoire, three pit shelter collections in August 1992 produced 217 *Anopheles gambiae* s.l. and 29 *Anopheles funestus*, while five days of hand collections inside 38 bedrooms during the same month produced only 47 *Anopheles gambiae* s.l. and 7 *Anopheles funestus* (Dossou-yovo et al. 1995).

In India Shalaby (1971, 1972) dug pit shelters which were 160 cm deep, 130 × 120 cm wide and had 30-cm deep pockets dug in all four walls about 50 cm from the bottom of the pits. Whenever possible these were located under a tree, but when not, shade was provided by erecting thatch roofs over the pit entrances. These shelters proved very successful, a total of 12 *Anopheles* species were collected from them. Also in India Mani et al. (1984) dug pit traps (1 × 1 × 1 m) in riverine villages and over 2 years caught 12 *Anopheles* species, the most common being *Anopheles varuna, Anopheles subpictus* and *Anopheles culicifacies*, the maximum number of any species caught was 30.61 *Anopheles varuna*/10 trap-nights during July, 1981. The pits were made more attractive to mosquitoes by watering them the evening before collections were made. In Gujarata state, India Bhatt et al. (1989) caught 4998 anophelines belonging to 10 different species and 1855 culicines from 20 pit shelters (1.2 × 1 m, and 1.5 m deep) that had 15-cm wide and 30-cm deep hollows cut in each side some 30–40 cm from the pit bottom. These shelters were examined fortnightly for a year, so in total 480 pit shelter collections were made. The most common anophelines collected from these pits were *Anopheles tessellatus* (32.32%), *Anopheles subpictus* (22.37%) and *Anopheles culicifacies* (11.44%).

A series of more than 100 pit shelters dug at 100 m intervals radiating out from a central village were used by Sadanandane et al. (1993) to study dispersal of malaria vectors in Orissa State, India. Pits were 45–60 cm deep and 30–45 cm across and were dug into the sides of well-shaded mounds of earth. 1,668 anophelines were collected from these shelters, including *Anopheles fluviatilis* (45% of total catch), *Anopheles jeyporiensis* (24%), *Anopheles tessellatus* (16%) and occasional individuals of *Anopheles culicifacies* and *Anopheles annularis*. 14% of the *Anopheles fluviatilis* collected from the pit shelters had fed on human blood, the rest on bovine blood. Sadanandane et al. (2004) used similar pit shelters to study the relative efficiency of different collecting methods within the National Malaria Surveillance Programme in India. Together with earth-lined box shelters, these outdoor resting traps caught about 2% of the total female mosquito catch.

In contrast, Amerasinghe et al. (1999) found pit shelters (1.5 m × 1.5 m × 1.5 m) covered with woven coconut thatch to be very inefficient at collecting anophelines in a dry zone in Sri Lanka. Only 6 individuals of five species were collected during the period August 1994 to February 1997.

In Iran shelters (120 × 90 and 150 cm deep with 30 cm deep cavities dug horizontally into the walls of the pit about 45 cm above the bottom have been used with some success to catch *Anopheles stephensi, Anopheles fluviatilis, Anopheles dthali*, and *Anopheles culicifacies* (Zaim et al. 1986; 1995), although only 4.2% of the total 1658 *Anopheles culicifacies* females collected were obtained from pit shelters (Zaim et al. 1995).

In Central Java Barodji and Supratman (1983) found pit shelters very useful in collecting substantial numbers of *Anopheles aconitus*, the lowest mean catch from four to eight pits was 2.9 in November, and the highest mean catch was 34.6 *Anopheles aconitus* in June. More adults/person-hour were collected from the traps in the dry season than in collections made from a variety of natural shelters, but the reverse was true in the rainy season. They concluded therefore that pit shelters did not accurately reflect population changes of exophilic adults.

Fernandez-Salas et al. (1993) used ten uncovered pit shelters (1.5 × 1.2 × 1.0 m) to increase the numbers of outdoor-resting *Anopheles pseudopunctipennis* in southern Mexico in host selection studies. Fernandez-Salas et al. (1994) later used similar pit shelters alongside human landing catches, a horse-baited trap and collections from inside houses and natural resting sites in a study to determine the bionomics and vector status of *Anopheles pseudopunctipennis*. Ten pit shelters were established in each of four villages and inspected each morning for one week per month for six months. A total of 1298 female *Anopheles pseudopunctipennis* were caught in the pit shelters during the 1991 dry season, compared with 341 and 220 collected from natural resting sites.

Muirhead-Thomson was little troubled with problems of the pits becoming flooded, but Service found that in both Kenya and Nigeria most pits become inoperative each rainy season because of flooding. This can be difficult to prevent. The erection of suitable roofing over the pits may help to keep out rainwater, but it does not alleviate flooding by seepage water. Pits are not easily dug or maintained in areas where the soil is either rocky or sandy.

Privy-type shelters

In California, USA Schoof (1944) constructed 6-ft high wooden structures with 3-ft^2 bases and overhanging wooden roofs, which he placed in cool dark places near larval habitats and termed 'anopheline houses.' He considered that they were better than red boxes or nail kegs in collecting *Anopheles crucians, Anopheles punctipennis* and *Anopheles quadrimaculatus*.s.l.

A small privy-type shelter made of wood and about 7 ft high, 4 ft square with a sloping roof and partially open front was mounted on wooden legs and used by Carpenter et al. (1946) to catch *Anopheles quadrimaculatus*. S.l. A modified privy shelter was designed and used by Snow (1949) in the Tennessee Valley, USA. It consists of a wooden framework 6.5 ft high with a 4-ft^2 base and with the roof and four sides covered with weatherproof cellulose board ('Celotex'). The inner surfaces are lined with a moisture proof backing of black asphalt-like material. Each side is composed of four panels, any of which can be removed to either leave an open space or provide an opening for a Bates (1944) type ingress baffle (Fig. 6.23*a*).

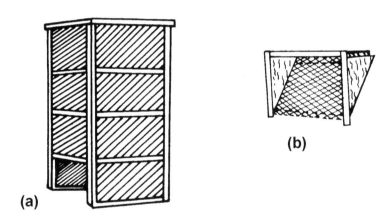

(a)

(b)

Fig. 5.32. (*a*) Privy-type shelter trap; (*b*) ingress baffle fitted to privy-type shelter of Snow (1949)

The trap is easily dismantled for transportation to the field. The biggest catch of *Anopheles quadrimaculatus* s.l. was obtained when the lowest panel on the west face of the trap was removed and the third panel on the opposite side replaced by an ingress baffle. The mean catch per day then varied from 316–423 *Anopheles quadrimaculatus* s.l., some 15–24 times greater than that from nail keg barrels. When the lowest panel was not removed only about 4% of the *Anopheles quadrimaculatus* s.l. attracted to the trap actually entered it (i.e. through the baffle), 96% rested on the outside, but with the lower panel removed 71% of the catch was collected from inside the trap. The behaviour of *Anopheles crucians* differed from that of *Anopheles quadrimaculatus* s.l. More adults were collected from the trap if the two panels were removed from all sides. In addition to *Anopheles quadrimaculatus* s.l., *Anopheles crucians* and *Anopheles pictipennis* about 15 culicine species were caught in the shelters.

In Texas, USA Hayes et al. (1973) used privy-type shelters and other shelters to collect outdoor resting mosquitoes, including blood-fed specimens.

General considerations

Different mosquito species and different gonotrophic stages of the same species may require different types of resting sites, consequently artificial shelters are unlikely to give unbiased samples of the exophilic population of all species in an area. The efficiency of an artificial shelter in catching a particular species may vary in different areas, or in the same area at different times of the year. An artificial resting shelter must compete with natural outdoor resting sites. In areas, where natural resting-sites these are minimal, artificial shelters may attract larger numbers of mosquitoes than similar shelters located in areas with abundant natural resting sites. It follows that these larger catches do not necessarily reflect a larger exophilic population. In fact during periods of dry weather mosquito populations may be greatly reduced, but because of the reduction of suitable alternative outdoor resting sites greater numbers of mosquitoes may seek refuge in artificial shelters.

Artificial shelters provide resting places not only for mosquitoes but also for numerous other animals such as lizards, spiders and scorpions, and as these may be predators of mosquitoes they should be removed every time collections are made from the shelters.

References

Aitken THG, Worth CB, Tikasingh ES (1968) Arbovirus studies in bush bush forest, Trinidad, W.I., September 1959–December 1964, III. Entomological studies. Am J Trop Med Hyg 17: 253–268

Ameneshewa B, Service MW (1996) Resting habits of *Anopheles arabiensis* in the Awash river valley of Ethiopia. Ann Trop Med Parasitol 90: 515–521

Amerasinghe PH, Amerasinghe FP, Konradsen F, Fonseka KT, Wirtz RA (1999) Malaria vectors in a traditional dry zone village in Sri Lanka. Am J Trop Med Hyg 60: 421–429

Anil Prakash, Bhattacharyya DR, Mohapatra PK, Mahanta J (1997) Breeding and day resting habitats of Anopheles dirus in Assam, India. Southeast Asian J Trop Med Public Health 28: 610–614

Anil Prakash, Bhattacharyya DR, Mohapatra PK, Mahanta J (1998) Investigation on malaria vectors and mosquito fauna in South Tripura district, Tripura state. Indian J Malariol 35: 151–159

Ardö P (1958) On the feeding habits of Scandinavian mosquitoes. Opusc Entomol 23: 171–191

Arredondo-Jiménez JI, Bown DN, Rodriguez MH, Loyola EG (1995) Control of *Anopheles albimanus* mosquitos in southern Mexico by spraying their preferred indoor resting sites. Bull World Health Organ 73: 329–337

Arnold AJ, Needham PH, Stevenson JH (1973) A self-powered portable insect suction sampler and its use to assess the effects of azinphos methyl and endosulfan on blossom beetle populations on oil seed rape. Ann Appl Biol 75: 229–233

Arunachalam N, Samuel PP, Hiriyan J, Thenmozhi V, Gajanana A (2004) Japanese encephalitis in Kerala, south India: can *Mansonia* (Diptera: Culicidae) play a supplemental role in transmission? J Med Entomol 41: 456–461

Ashford DA, Savage HM, Hajjeh RA, McReady J, Bartholomew DM, Spiegel RA, Vorndam V, Clark GG, Gubler DG (2003) Outbreak of dengue fever in Palau, Western Pacific: risk factors for infection. Am J Trop Med Hyg 69: 135–140

Bahang Z, Saafi L, Bende N, Kirnowardoyo S, Lim Boo Liat (1984) Malaysian filariasis studies in Kendari regency, southeast Sulawesi, Indonesia II: surveillance of mosquitoes with reference to two *Anopheles* vector species. Bull Penel Keseh Health Std Indonesia 12: 8–20

Bailly-Choumara H (1973) Étude comparative de différentes techniques de récolté de moustiques adultes (Diptera, Culicidae) faite au Maroc, en zone rurale. Bull Soc Sci Nat Phys Maroc 53: 135–187

Bang YH (1985) Implication in the control of malaria vectors with insecticides in tropical countries of south-east Asian region. Part II—Consequences of insecticide use. J Commun Dis 17: 300–310

Bansal SK, Singh KV (1993) Prevalence and seasonal distribution of anopheline fauna in district Bikaner (Rajasthan). Indian J Malariol 30: 119–125

Barnes WW, Southwick JW (1967) AEHA cartridge-type aspirators. Mosquito News 27: 521–522

Barodji, Supratman S (1983) Evaluation of pit shelters as a monitoring device for outdoor populations of malaria vector *Anopheles aconitus* Donitz. Bull Penel Keseh Health Std Indonesia 11: 20–24

Bates M (1944) Notes on the construction and use of stable traps for mosquito studies. J Natn Malar Soc 3: 135–145

Batra CP, Reuben R, Das PK (1979) Studies of day-time resting places of *Anopheles stephensi* Liston in Salem (Tamil Nadu). Indian J Med Res 69: 583–588

Beier JC, Berry WJ, Craig GB (1982) Horizontal distribution of adult *Aedes triseriatus* (Diptera: Culicidae) in relation to habitat structure, oviposition, and other mosquito species. J Med Entomol 19: 239–247

Bellas TE (1990) Occupational inhalant allergy to arthropods. Clin Rev Allergy 8: 15–29

Bhatt RM, Sharma RC, Yadav RS, Sharma VP (1989) Resting of mosquitoes in outdoor pit shelters in Kheda district, Gujarat. Indian J. Malariol., 26: 75–81

Blin G (1908) Déstruction des moustiques par le procédé des trous-pièges. Bull Soc Pathol Exot 1: 100–103

Bliss CI (1970) Statistics in Biology, McGraw-Hill Book Co., New York

Bown DN, Bang YH (1980) Ecological studies on *Aedes simpsoni* (Diptera: Culicidae) in southeastern Nigeria. J Med Entomol 17: 367–374

Bown DN, Rios JR, del Angel Cabañas G, Guerrero JC, Méndez JF (1984) Evaluation of chlorphoxim used against *Anopheles albimanus* on the south coast of Mexico: 1; Results of indoor chlorphoxim applications and assessment of the methodology employed. Bull Pan-Am Health Organ 18: 379–388

Boxmeyer CE, Palchick SM (1999) Distribution of resting female *Aedes vexans* (Meigen) in wooded and nonwooded areas of metropolitan Minneapolis-St. Paul, Minnesota. J Am Mosq Control Assoc 15: 128–132

Bradley GH (1943) Determination of densities of populations of *Anopheles quadrimaculatus* on the wing. Proc New Jers Mosq Exterm Assoc 30: 22–27

Breeland SG (1972a) Studies on the diurnal resting habits of *Anopheles albimanus* and *A. pseudopunctipennis* in El Salvador. Mosquito News 32: 99–106

Breeland SG (1972b) Methods for measuring anopheline densities in El Salvador. Mosquito News 32: 62–72

Breeland SG (1974) Population patterns of *Anopheles albimanus* and their significance to malaria abatement. Bull World Health Organ 50: 307–315

Breeland SG, Glasgow JW (1967) An improved portable resting station for *Anopheles quadrimaculatus* Say. Mosquito News 27: 5–9

Breeland SG, Jeffery GM, Lofgren CS, Weidhaas DE (1974) Release of chemosterilized males for the control of *Anopheles albimanus* in El Salvador. I. Characteristics of the test site and the natural population. Am J Trop Med Hyg 23: 274–281

Bryan JH (1979) Observations on the member species of the *Anopheles gambiae* complex in The Gambia, West Africa. Trans R Soc Trop Med Hyg 73: 463–466

Burbutis PP, Jobbins DM (1958) Studies on the use of a diurnal resting box for the collection of *Culiseta melanura* (Coquillett). Bull Brooklyn Entomol Soc 53: 53–58

Burton GJ (1954) Suggested improvements for an unbreakable aspirator and killing tube. Mosquito News 14: 27–30

Bushrod FM (1979) Studies on filariasis transmission in Kwale, a Tanzanian coastal village, and the results of mosquito control measures. Ann Trop Med Parasitol 73: 277–285

Büttiker W (1958) Notes on exophily in anophelines in south-east Asia. Bull World Health Organ 19: 1118–1123

Carpenter SJ, Middlekauff WW, Chamberlain RW (1946) The Mosquitoes of Southern United States East of Oklahoma and Texas. Am Midl Nat Monogr 3. The University Press of Notre Dame, Indiana

Carver HD (1967) A portable aspirator for collecting mosquitoes. Mosquito News 27: 428–429

Chadee DD, Williams SA, Ottesen EA (2002) Xenomonitoring of *Culex quinquefasciatus* mosquitoes as a guide for detecting the presence or absence of lymphatic filariasis: a preliminary protocol for mosquito sampling. Ann Trop Med Parasitol 96 (Supplement 2): S47–S53

Chand SK, Yadav RS, Sharma VP (1993) Seasonality of indoor resting mosquitoes in a broken-forest ecosystem of North-Western Orissa. Indian J Malariol 30: 145–154

Chanda DA, Shisler JK (1980) Mosquito control problems associated with storm water control facilities. Proc New Jers Mosq Control Assoc 67: 193–200

Chandler JA, Highton RB, Boreham PFL (1976) Studies on some ornithophilic mosquitoes (Diptera: Culicidae) of the Kano Palin, Kenya. Bull Entomol Res 66: 133–143

Charlwood JD, Bryan JH (1987) A mark–recapture experiment with the filariasis vector *Anopheles punctulatus* in Papua New Guinea. Ann Trop Med Parasitol 81: 429–436

Charlwood JD, Dagoro H, Paru R (1985) Blood-feeding and resting behaviour in the *Anopheles punctulatus* Donitz complex (Diptera: Culicidae) from coastal Papua New Guinea. Bull Entomol Res 75: 463–475

Charlwood JD, Smith T, Kihonda J, Heiz B, Billingsley PF, Takken W (1995) Density independent feeding success of malaria vectors (Diptera: Culicidae) in Tanzania. Bull Entomol Res 85: 29–35

Chatterjee KK, Biswas D, Choudhuri DK, Mukherjee H, Hati AK (1993) Resting sites of *Anopheles stephensi* Liston in Calcutta. Indian J Malariol 30: 109–112

Clark GG, Seda H, Gubler DJ (1994) Use of the "CDC backpack aspirator" for surveillance of *Aedes aegypti* in San Juan, Puerto Rico. J Am Mosq Control Assoc 10: 119–124

Clarke JL, Pradhan GD, Joshi GP, Fontaine RE (1980) Assessment of the grain store as an unbaited outdoor shelter for mosquitoes of the *Anopheles gambiae* complex and *Anopheles funestus* (Diptera: Culicidae) at Kisumu, Kenya. J Med Entomol 17: 100–102

Coluzzi M, Petrarca V (1973) Aspirator with paper cup for collecting mosquitoes and other insects. Mosquito News 33: 249–250

Copeland RS (1986) The biology of *Aedes thibaulti* in northern Indiana. J Am Mosq Control Assoc 2: 1–6

Cordellier R, Bouchité B, Roche J-C, Monteny N, Diaco B, Akoliba P (1983) Circulation sylvatique du virus dengue 2 en 1980, dans les savanes sub-soudaniennes de Côte d'Ivoire. Données entomologiques et considérations épidémiologiques. Cah ORSTOM sér. Entomol Méd Parasitol 21: 165–179

Covell G, Mulligan HW, Afridi MK (1938) An attempt to control malaria by the destruction of adult mosquitoes with insecticidal sprays. J Malar Inst India 1: 105–113

Cupp EW, Scherer WF, Lok JB, Brenner RJ, Dziem GM, Ordonezi JV (1986) Entomological studies at an enzootic Venezuelan equine encephalitis virus focus in Guatemala, 1977–1980. Am J Trop Med Hyg 35: 851–859

Das NG, Bhuyan M, Das SC (2000) Entomological and epidemiological studies on malaria in Rajmahal range, Bihar. Indian J Malariol 37: 88–96

Davies JB (1973) A simple battery operated suction trap for insects attracted to animal, light or chemical bait. Mosquito News 33: 102–104

Davis EW, Gould DJ (1973) A portable suction apparatus for collecting mosquitoes. Mosquito News 33: 246–247

Day JF, Curtis GA (1989) Influence of rainfall on *Culex nigripalpus* (Diptera: Culicidae) blood-feeding behavior in Indian River county, Florida. Ann Entomol Soc Am 82: 32–37

Day JF, Curtis GA, Edman JD (1990) Rainfall-directed oviposition behavior of *Culex nigripalpus* (Diptera: Culicidae) and its influence on St. Louis encephalitis virus transmission in Indian River county, Florida. J Med Entomol 27: 43–50

De SK, Chandra G (1994) Studies on the filariasis vector - *Culex quinquefasciatus* at Kanchrapara, West Bengal (India). Indian J Med Res 99: 255–258

de Freitas EN, Shope RE, Causey OR (1966) A portable suction apparatus for capturing insects. Mosquito News 26: 368–372

Dell'Uomo G (1967) Un aspiratore portatile per cattura di zanzare, azionato a batteria. Riv Parassitol 28: 221–223

De Meillon B (1936) II. The control of malaria in South Africa by measures directed against the adult mosquitoes in habitations. Q Bull Health Organ L of N 5: 134–137

De Meillon B, Paterson HE, Muspratt J (1957) Studies on arthropod-borne viruses of Tongaland, II. Notes on the more common mosquitoes. S Afr J Med Sci 22: 47–53

de Zulueta J (1950) A study of the habits of the adult mosquitoes dwelling in the savanna of eastern Colombia. Am J Trop Med 30: 325–339

de Zulueta J (1952) Observations on mosquito density in an endemic malarious area in eastern Colombia. Am J Trop Med Hyg 1: 314–329

Dietrick EJ (1961) An improved backpack motor fan for suction sampling of insect populations. J Econ Entomol 54: 394–395

Dietrick EJ, Schlinger EI, van den Bosch R (1959) A new method for sampling arthropods using a suction collecting machine and modified Berlese funnel separator. J Econ Entomol 52: 1085–1091

Dossou-yovo J, Doannio JMC, Riviere F, Chauvancy G (1995) Malaria in Cote d'Ivoire wet savannah region: the entomological input. Trop Med Parasitol 46, 263–269

Douglas RB (1984) The hazard of pooting insects. Antenna 8, 193–194

Downe AER (1960) Blood-meal sources and notes on host preferences of some *Aedes* mosquitoes (Diptera: Culicidae). Can J Zool 38, 689–699

Dua VK, Sharma SK, Srivastava A, Sharma VP (1997) Bioenvironmental control of industrial malaria at Bharat Heavy Electricals Ltd., Hardwar, India - results of a nine-year study (1987–95). J Am Mosq Control Assoc 13: 278–285

Dyce AL, Standfast HA, Kay BH (1972) Collection and preparation of biting midges (Fam. Ceratopogonidae) and other small Diptera for virus isolation. J Aust Entomol Soc 11: 91–96

Eddey LG (1944) Spray-killing of mosquitoes in houses—A contribution to malaria control on the Gold Coast. Trans R Soc Trop Med Hyg 38: 167–197

Edman JD, Evans FDS, Williams JA (1968) Development of a diurnal resting box to collect *Culiseta melanura* (Coq.). Am J Trop Med Hyg 17: 451–456

Edman J, Kittayapong P, Linthicum K, Scott T (1997) Attractant resting boxes for rapid collection and surveillance of *Aedes aegypti* (L.) inside houses. J Am Mosq Control Assoc 13: 24–27

El Said S, Beier JC, Kenawy MA, Morsy ZS, Merdan AI (1986) *Anopheles* population dynamics in two malaria endemic villages in Faiyum governorate, Egypt. J Am Mosq Control Assoc 2: 158–163

Faye O, Konate L, Gaye O, Fontenille D, Sy N, Diop A, Diagne M, Molez JF (1998) Impact de l'utilisation des moustiquaires pré-imprégnées de perméthrine sur la transmission du paludisme dans un village hyperendémique du Sénégal. Méd Trop (Mars) 58: 355–360

Fernandez-Salas I, Roberts DR, Rodríguez MH, Rodríguez MC, Marina-Fernandez CF (1993) Host selection patterns of *Anopheles pseudopunctipennis* under insecticide spraying situations in southern Mexico. J Am Mosq Control Assoc 9: 375–384

Fernandez-Salas I, Rodriguez MH, Roberts DR, Rodriguez MC, Wirtz RA (1994) Bionomics of adult *Anopheles pseudopunctipennis* (Diptera: Culicidae) in the Tapachula foothills area of southern Mexico. J Med Entomol 31: 663–670

Flores-Mendoza C, Lourenco-de-Oliveira R (1996) Bionomics of *Anopheles aquasalis* Curry 1932, in Guarai, State of Rio de Janeiro, southeastern Brazil - I. Seasonal distribution and parity rates. Mem Inst Oswaldo Cruz 91: 265–270

Forattini OP, Kakitani I, Massad E, Marucci D (1993) Studies on mosquitoes (Diptera: Culicidae) and anthropic environment. 4 - Survey of resting adults and synanthropic behaviour in South-Eastern Brazil. Rev Saúde Pública 27: 398–411

Forattini OP, Kakitani I, Sallum MAM, de Rezende L (1997) Produtividade de criadouro de *Aedes albopictus* em ambiente urbano. Rev Saúde Pública 31: 545–555

Forattini OP, Kakitani I, dos Santos RLC, Kobayashi KM, Ueno HM, Fernandez Z (2000) Comportamento de *Aedes albopictus* e de *Ae. scapularis* adultos (Diptera: Culicidae) no Sudeste do Brasil. Rev Saúde Pública 34: 461–467

Frank JH, Curtis GA (1977) On the bionomics of bromeliad-inhabiting mosquitoes. V. A mark–release–recapture technique for estimation of population size of *Wyeomyia vanduzeei*. Mosquito News 37: 444–452

Gad AM, Riad IB, Farid HA (1995) Host-feeding patterns of *Culex pipiens* and *Cx. antennatus* (Diptera: Culicidae) from a village in Sharqiya Governorate, Egypt. J Med Entomol 32: 573–577

Garcia R, des Rochers BS, Voigt WG (1988) A bait/carbon dioxide trap for the collection of the western tree hole mosquito *Aedes sierrensis*. J Am Mosq Control Assoc 4: 85–88

Garvin MC, Ohajuruka OA, Bell KE, Ives SL (2004) Seroprevalence of eastern equine encephalomyelitis virus in birds and larval survey of *Culiseta melanura* Coquillett during an interepizootic period in central Ohio. J Vector Ecol 29: 73–78

Gillies MT (1954) Studies of house leaving and outside resting of *Anopheles gambiae* Giles and *Anopheles funestus* Giles in East Africa. I. The outside resting population. Bull Entomol Res 45: 361–373

Gillies MT (1955) The density of adult *Anopheles* in the neighbourhood of an East African village. Am J Trop Med Hyg 4: 1103–1113

Gillies MT (1956) The problem of exophily in *Anopheles gambiae*. Bull World Health Organ 15: 437–449

Girod R, Salvan M, Simard F, Andrianaivolambo L, Fontenille D, Laventure S (1999) Evaluation de la capacité vectorielle d'*Anopheles arabiensis* (Diptera: Culicidae) à l'île de La Réunion: une approche du risque sanitaire lié au paludisme d'importation en zone d'éradication. Bull Soc Pathol Exot 92: 203–209

Githeko AK (1992) The behaviour and ecology of malaria vectors and malaria transmission in Kisumu district of Western Kenya. Ph.D. thesis, University of Liverpool

Gomes AC, Silva NN, Marques GRAM, Brito M (2003) Host-feeding patterns of potential human disease vectors in the Paraíba Valley Region, State of São Paulo, Brazil. J Vector Ecol 28: 74–78

Goodwin MH (1942) Studies on artificial resting places of *Anopheles quadrimaculatus* Say. J Natn Malar Soc 1: 93–99

Gould DJ, Edelman R, Grossman RA, Nisalak A, Sullivan MF (1974). Study of Japanese encephalitis virus in Chiangmai Valley, Thailand. IV. Vector studies. Am J Epidemiol 100: 49–56

Governatori M, Bulgarini C, Rivasi F, Pampiglione S (1993) A new portable aspirator for Culicidae and other winged insects. J Am Mosq Control Assoc 9: 460–462

Grimstad PR, DeFoliart GR (1974) Nectar sources of Wisconsin mosquitoes. J Med Entomol 11: 331–334

Gunasekaran K, Jambulingam P, Das PK (1995) Distribution of indoor-resting *Anopheles fluviatilis* in human dwellings and its implication on indoor residual spray. Indian J Malariol 32: 42–46

Gusciora WR (1961) The resting box technique for the sampling of *Culiseta melanura* (Coquillett). Proc New Jers Mosq Exterm Assoc 48: 122–125

Haddow AJ (1942) The mosquito fauna and climate of native huts at Kisumu, Kenya. Bull Entomol Res 33: 91–142

Hall H, Drew WA, Eisenbraun EJ (1968) A portable battery-operated aspirator (ant collector). Ann Entomol Soc Am 61: 1348–1349

Harden FW, Poolson BJ, Bennett LW, Gaskin RC (1970) Analysis of CO_2 supplemented mosquito adult landing rate counts. Mosquito News 30: 369–374

Harwood RF (1962) Trapping overwintering adults of the mosquitoes *Culex tarsalis* and *Anopheles freeborni*. Mosquito News 22: 26–31

Harwood RF, Halfhill JE (1960) Mammalian burrows and vegetation as summer resting sites of the mosquitoes *Culex tarsalis* and *Anopheles freeborni*. Mosquito News 20: 174–178

Hayes RO, Bellamy RE, Reeves WC, Willis MJ (1958) Comparison of four sampling methods for measurement of *Culex tarsalis* adult populations. Mosquito News 18: 218–227

Hayes RO, Kitaguchi GE, Mann RM (1967) The 'CDC sweeper', a six-volt mechanical aspirator for collecting adult mosquitoes. Mosquito News 27: 359–363

Hayes RO, Tempelis CH, Hess AD, Reeves WC (1973) Mosquito host preference studies in Hale County, Texas. Am J Trop Med Hyg 22: 270–277

Herrel N, Amerasinghe FP, Ensink J, Mukhtar M, Van der Hoek W, Konradsen F (2004) Adult anopheline ecology and malaria transmission in irrigated areas of South Punjab, Pakistan. Med Vet Entomol 18: 141–152

Hewitt S, Rowland M, Muhammad N, Kamal M, Kemp E (1995) Pyrethroid-sprayed tents for malaria control: an entomological evaluation in Pakistan. Med Vet Entomol 9: 344–352

Hobbs JH, Sexton JD, St. Jean Y, Jacques JR (1986) The biting and resting behavior of *Anopheles albimanus* in northern Haiti. J Am Mosq Control Assoc 2: 150–153

Holck AR, Meek CL (1991) Comparison of sampling techniques for adult mosquitoes and other Nematocera in open vegetation. J Entomol Sci 26: 231–236

Howard JJ, White DJ, Muller SL (1989) Mark–recapture studies on the *Culiseta* (Diptera: Culicidae) vectors of eastern equine encephalitis virus. J Med Entomol 26: 190–199

Howard JJ, Grayson MA, White DJ, Oliver J (1996) Evidence for multiple foci of eastern equine encephalitis virus (Togaviridae: Alphavirus) in central New York State. J Med Entomol 33: 421–432

Hudson JE (1978) Overwintering sites and ovarian development of some mosquitoes in central Alberta, Canada, Mosquito News 38: 570–579

Husbands RC (1958) An improved mechanical aspirator. California Vector Views 5: 72–73

Husbands RC, Holten JR (1967) An improved mechanical method of aspirating insects. California Vector Views 14: 78–80

Ijumba JN, Mosha FW, Lindsay SW (2002) Malaria transmission risk variations derived from different agricultural practices in an irrigated area of northern Tanzania. Med Vet Entomol 16: 28–38

Irby WS, Apperson CS (1992) Spatial and temporal distribution of resting female mosquitoes (Diptera: Culicidae) in the costal plain of North Carolina. J Med Entomol 29: 150–159

Ismail IAH, Notananda V, Schepens J (1974) Studies on malaria and responses of *Anopheles balabacensis balabacensis* and *Anopheles minimus* to DDT residual spraying in Thailand. Part I—Pre-spraying observations. Acta Trop 31: 129–164

Ismail IAH, Notananda V, Schepens J (1975) Studies on malaria and responses of *Anopheles balabacensis balabacensis* and *Anopheles minimus* to DDT residual spraying in Thailand. Part II Post-spraying observations. Acta Trop 32: 206–231

Ismail IAH, Pinichpongse S, Boonrasri P (1978) Responses of *Anopheles minimus* to DDT residual spraying in a cleared forest hill area in central Thailand. Acta Trop 35: 69–82

Iwao S (1968) A new regression method for analyzing the aggregation pattern of animal populations. Res Popul Ecol 10: 1–20

Jackson SC, Grothaus RH (1971) A combination aspirator and killing tube for collecting mosquitoes and other insects. Mosquito News 31: 112–113

Joshi GP, Fontaine RE, Thymakis K, Pradhan GD (1973) The cause of occasional high counts of *An. gambiae* in morning pyrethrum spray collections in huts sprayed with fenitrothion, Kisumu, Kenya. Mosquito News 33: 29–38

Joshi GP, Self LS, Usman S, Pant CP, Nelson MJ, Supalin (1977) Ecological studies on *Anopheles aconitus* in the Semarang area of central Java, Indonesia. WHO/VBC/77.677; 155 pp. (mimeographed)

Jupp PG, McIntosh BM (1987) A bionomic study of adult *Aedes (Neomelaniconion) circumluteolus* in northern Kwazulu, South Africa. J Am Mosq Control Assoc 3: 131–136

Kalra NL (1980) Emergence of malaria zoonosis of simian origin as natural phenomenon in Greater Nicobars, Andaman & Nicobar islands—A preliminary note. J Commun Dis 12: 49–54

Kanojia PC (2003) Bionomics of *Culex epidesmus* associated with Japanese encephalitis virus in India. J Am Mosq Control Assoc 19: 151–154

Kay BH (1983) Collection of resting adult mosquitoes at Kowanyama, northern Queensland and Charleville, south–western Queensland. J Aust Entomol Soc 22: 19–24

Keating J, Mbogo C, Mwangangi J, Nzovu JG, Gu W, Regens JL, Yan G, Githure JI, Beier JC (2005) *Anopheles gambiae* s.l. and *Anopheles funestus* mosquito distributions at 30 villages along the Kenyan coast. J Med Entomol 42: 241–246

Kenawy M, Zimmerman JH, Beier JC, El Said S, Abbassy MM (1986) Host-feeding patterns of *Anopheles sergentii* and *An. multicolor* (Diptera: Culicidae) in Siwa and el Gara oases, Egypt. J Med Entomol 23: 576–577

Kenawy M, Beier JC, Asiago CM, El Said S (1990) Factors affecting the human-feeding behavior of anopheline mosquitoes in Egyptian oases. J Am Mosq Control Assoc 6: 446–451

Khin Maung, Kyi U (1964) Rapid and efficient methods for sampling anopheline populations in insecticide treated areas. Burmese Med J 12: 130–134

King WV, Bradley GH, McNeel TE (1939) The Mosquitoes of the South–eastern States. Miscellaneous Publications of the U.S. Department of Agriculture No. 336

Komar N, Pollack RJ, Spielman A (1995) A nestable fiber pot for sampling resting mosquitoes. J Am Mosq Control Assoc 11: 463–467

Konradsen F, Stobberup KA, Sharma SK, Gulati OT, van der Hoek W (1998) Irrigation water releases and *Anopheles culicifacies* abundance in Gujarat, India. Acta Trop 71: 195–197

Krafsur ES (1971) Malaria transmission in Gambela, Illubabor province. Ethiop Med. J 9: 75–94

Krafsur ES (1977) The bionomics and relative prevalence of *Anopheles* species with respect to the transmission of Plasmodium to man in western Ethiopia. J Med Entomol 14: 180–194

Kuntz KJ, Olson JK, Rade BJ (1982) Role of domestic animals as hosts for blood-seeking females of *Psorophora columbiae* and other mosquito species in Texas ricefields. Mosquito News 42: 202–210

la Grange JJP, Coetzee M (1997) A mosquito survey of Thomo village, Northern Province, South Africa, with special reference to the bionomics of exophilic members of the *Anopheles funestus* group (Diptera: Culicidae). African Entomol 5: 295–299.

Laarman JJ (1959) A new species of *Anopheles* from a rain-forest in eastern Belgian Congo. Acta Leidensia 29: 200–211

Lee VH, Atmosoedjono S, Rusmiarto S, Aep S, Semendra W (1983) Mosquitoes of Bali island, Indonesia: common species in the village environment. Southeast Asian J Trop Med Public Health 14: 298–307

Lindsay SW, Snow RW (1988) The trouble with eaves; house entry by vectors of malaria. Trans R Soc Trop Med Hyg 82: 645–646

Lindsay SW, Armstrong Schellenberg JRM, Zeiler HA, Daly RJ, Salum FM, Wilkins HA (1995) Exposure of Gambian children to *Anopheles gambiae* malaria vectors in an irrigated rice production area. Med Vet Entomol 9: 50–58

Lines JD, Lyimo EO, Curtis CF (1986) Mixing of indoor- and outdoor-resting adults of *Anopheles gambiae* Giles s.l. and *A. funestus* Giles (Diptera: Culicidae) in coastal Tanzania. Bull Entomol Res 76: 171–178

Lloyd JE, Schmidtmann ET, Kumar R, Bobian RJ, Waggoner JW, Legg DE, Hill DC (2002) Suppression of bloodfeeding by *Ochlerotatus dorsalis* and *Ochlerotatus melanimon* on cattle treated with Python® ear tags. J Am Mosq Control Assoc 18: 207–209

Loomis EC, Green DH (1959) Ecological observations on *Culex tarsalis* Coquillett and other mosquitoes in the delta region of the Central Valley of California, 1953–1956 (Diptera: Culicidae). Ann Entomol Soc Am 52: 524–533

Loomis EC, Sherman EJ (1959) Comparison of artificial shelters and light traps for measurement of *Culex tarsalis* and *Anopheles freeborni* populations. Mosquito News 19: 232–237

Lothrop HD, Reisen WK (2001) Landscape affects the host-seeking patterns of *Culex tarsalis* (Diptera: Culicidae) in the Coachella Valley of California. J Med Entomol 38: 325–332

Love GJ, Goodwin MH (1961) Notes on the bionomics and seasonal occurrence of mosquitoes in southwestern Georgia. Mosquito News 21: 195–215

Lowe RE, Bailey DL (1979) Comparison of morning and evening captures of adult female *Anopheles albimanus* from stables in El Salvador. Mosquito News 39: 532–535

Lutwama JJ, Kayondo J, Savage HM, Burkot TR, Miller BR (1999) Epidemic o'nyong-nyong fever in southcentral Uganda, 1996–1997: entomologic studies in Bbaale village, Rakai district. Am J Trop Med Hyg 61: 158–162

Magnarelli LA (1977*a*) Seasonal occurrence and parity of *Aedes canadensis* (Diptera: Culicidae) in New York state, USA. J Med Entomol 13: 741–745

Magnarelli LA (1977*b*) Host feeding patterns of Connecticut mosquitoes (Diptera: Culicidae). Am J Trop Med Hyg 26: 547–552

Makiya K, Taguchi I (1982) Ecological studies on overwintering populations of *Culex pipiens* patterns. 3. Movement of the mosquitoes in a cave during overwintering. Jap J Sanit Zool 33: 335–343

Malakar P, Das S, Saha GK, Dasgupta B, Hati AK (1995) Indoor resting anophelines of North Bengal. Indian J Malariol 32: 24–31

Mani TR, Tewari SC, Reuben R, Devaputra M (1984) Resting behaviour of anophelines & sporozoite rates in vectors of malaria along the river Thenpennai (Tamil Nadu). Indian J Med Res 80: 11–17

Mani TR, Rao CVRM, Rajendran R, Devaputra M, Prassana Y, Hanumaiah Gajanana A, Reuben R (1991) Surveillance for Japanese encephalitis in villages near Madurai, Tamil Nadu, India. Trans R Soc Trop Med Hyg 85: 287–291

McCardle PW, Webb RE, Norden BB, Aldrich JR (2004) Evaluation of five trapping systems for the surveillance of gravid mosquitoes in Prince Georges County, Maryland. J Am Mosq Control Assoc 20: 254–260

McClelland GAH (1957) Methods of Collection of Blood-fed Females in the Field. E. Afr. Virus Res. Inst. Rep., 1956–1957. Government Printer, Nairobi, pp. 47–55

McClelland GAH, Weitz B (1963) Serological identification of the natural hosts of *Aedes aegypti* (L.) and some other mosquitoes (Diptera, Culicidae) caught resting in vegetation in Kenya and Uganda. Ann Trop Med Parasitol 57: 214–224

McCrae AWR, Boreham PFL, Ssenkubuge Y (1976) The behavioural ecology of host selection in *Anopheles implexus* (Theobald) (Diptera: Culicidae). Bull Entomol Res 66: 587–631

McCreadie JW, Colbo MH, Bennett GF (1984) A trap design for the collections of hematophagous Diptera from cattle. Mosquito News 44: 212–216

McGavin GC, Furlong J (1981) An electronic counter for use in quantitative biology. J Appl Ecol 18: 481–485

McHugh CP (1989) Ecology of a semi-isolated population of adult *Anopheles freeborni*: abundance, trophic status, parity, survivorship, gonotrophic cycle length, and host selection. Am J Trop Med Hyg 41: 169–176

McNelly J, Crans WJ (1982) Limitations in the use of resting boxes to assess populations of the mosquito, *Culiseta melanura*. Proc New Jers Mosq Control Assoc 69: 32

Meek CL, Meisch MV, Walker TW (1985) Portable battery-powered aspirators for collecting adult mosquitoes. J Am Mosq Control Assoc 1: 102–105

Meyer RP (1985) The "walk-in" type red box for sampling adult mosquitoes. Proc New Jers Mosq Control Assoc 72: 104–105

Meyer RP, Reisen WK, Hill BR, Martinez VM (1983) The "AFS sweeper", a battery powered back pack mechanical aspirator for collecting adult mosquitoes. Mosquito News 43: 346–350

Mitchell CJ, Chen PS (1973) Ecological studies on the mosquito vectors of Japanese encephalitis. Bull World Health Organ 49: 287–292

Mnzava AEP, Mutinga MJ, Staak C (1994) Host blood meals and chromosomal inversion polymorphism in *Anopheles arabiensis* in the Baringo District of Kenya. J Am Mosq Control Assoc 10: 507–510

Mnzava AEP, Rwegoshora RT, Wilkes TJ, Tanner M, Curtis CF (1995) *Anopheles arabiensis* and *An. gambiae* chromosomal inversion polymorphism, feeding and resting behaviour in relation to insecticide house-spraying in Tanzania. Med Vet Entomol 9: 316–324

Molineaux L, Gramiccia G (1980) The Garki Project. Research on the Epidemiology and Control of Malaria in the Sudan Savanna of West Africa. World Health Organization, Geneva

Molineaux L, Shidrawi GR, Clarke JL, Boulzaguet R, Ashkar T, Dietz F (1976) The impact of propoxur on *Anopheles gambiae* s.l. and some other anopheline populations, and its relationship with some pre-spraying variables. Bull World Health Organ 54: 379–389

Moreby S (1991) A simple time-saving improvement to the motorized insect suction sampler. The Entomologist 110: 2–4

Morris CD (1981) A structural and operational analysis of diurnal resting shelters for mosquitoes (Diptera: Culicidae). J Med Entomol 18: 419–424

Morris CD, Zimmerman RH, Edman JD (1980) Epizootiology of eastern equine encephalomyelitis virus in upstate New York, USA. II. Population dynamics and vector potential of *Culiseta melanura* (Diptera: Culicidae) in relation to distance from breeding site. J Med Entomol 17: 453–465

Mortenson EW (1953). Observations on the overwintering habits of *Culex tarsalis* Coquillett in nature. Proc California Mosq Control Assoc 21: 59–60

Mouchet J, Rageau J, Chippaux A (1969) Hibernation de *Culex molestus* (Ficalbi) (Diptera: Culicidae) en Camargue. Cah ORSTOM sér. Entomol Méd Parasitol 7: 35–37

Moussa MA, Gould DJ, Nolan MP, Hayes DE (1966) Observations on *Culiseta melanura* (Coquillett) in relation to encephalitis in southern Maryland. Mosquito News 26: 385–393

Mpofu SM (1985) Seasonal vector density and disease incidence patterns of malaria in an area of Zimbabwe. Trans R Soc Trop Med Hyg 79: 169–175

Muirhead-Thomson RC (1948) The effects of house spraying with pyrethrum and with DDT on *Anopheles gambiae* and *A. melas* in West Africa. Bull Entomol Res 38: 449–464

Muirhead-Thomson RC (1951) Studies on salt-water and fresh-water *Anopheles gambiae* on the East African coast. Bull Entomol Res 41: 487–502

Muirhead-Thomson RC (1956) The part played by woodland mosquitoes of the genus *Aedes* in the transmission of myxomatosis in England. J Hyg Camb 54: 461–471

Muirhead-Thomson RC (1958) A pit shelter for sampling outdoor mosquito populations. Bull World Health Organ 19: 1116–1118

Muirhead-Thomson RC (1960) The significance of irritability, behaviouristic avoidance and allied phenomena in malaria eradication. Bull World Health Organ 22: 721–734

Muirhead-Thomson RC, Mercier EC (1952) Factors in malaria transmission by *Anopheles albimanus* in Jamaica. Part 1. Ann Trop Med Parasitol 46: 103–116

Mulligan FS, Schaefer CH (1981) The breeding of *Culex quinquefasciatus* within the Fresno urban storm drain system. Proc California Mosq Vector Control Assoc 49: 101–103

Murty US, Sai KSK, Kumar DVRS, Sriram K, Rao KM, Krishna D, Murty BSN (2002*a*) Relative abundance of *Culex quinquefasciatus* (Diptera: Culicidae) with reference to infection and infectivity rate from the rural and urban areas of east and west Godavari districts of Andhra Pradesh, India. Southeast Asian J Trop Med Public Health 33: 702–710

Murty US, Satyakumar DVR, Sriram K, Rao KM, Singh TG, Arunachalam N, Samuel PP (2002*b*). Seasonal prevalence of *Culex vishnui* sub-group, the major vectors of Japanese encephalitis virus in an endemic district of Andhra Pradesh, India. J Am Mosq Control Assoc 18: 290–293

Mutero CM, Mosha FW, Subra R (1984) Biting activity and resting behaviour of *Anopheles merus* Donitz (Diptera: Culicidae) on the Kenya coast. Ann Trop Med Parasitol 78: 43–47

Mutinga MJ, Odhiambo TR, Kamau CC, Odulaja A, Amimo FA, Wachira DW (1995) Choice of resting sites by *Anopheles gambiae* (Diptera: Culicidae) in Mwea Rice Irrigation Scheme, Kirinyaga District, Kenya. East Afr Med J 72: 170–175

Nagasawa S (1976) An analysis of seasonal pattern in a population of *Culex pipiens fatigans* Wiedemann (Diptera: Culicidae). Kontyû Tokyo 44: 102–107

Nasci RS (1981) A lightweight battery-powered aspirator for collecting resting mosquitoes in the field. Mosquito News 41: 808–811

Nasci RS (1982) Differences in host choice between the sibling species of treehole mosquitoes *Aedes triseriatus* and *Aedes hendersoni*. Am J Trop Med Hyg 31: 411–415

Nasci RS, Edman JD (1981*a*) Blood feeding patterns of *Culiseta melanura* (Diptera: Culicidae) and associated sylvan mosquitoes in southeastern Massachusetts eastern equine encephalitis enzootic foci. J Med Entomol 18: 493–500

Nasci RS, Edman JD (1981*b*). Vertical and temporal flight activity of the mosquito *Culiseta melanura* (Diptera: Culicidae) in southe–eastern Massachusetts. J Med Entomol 18: 501–504

Nasci RS, Edman JD (1984) *Culiseta melanura* (Diptera: Culicidae) population structure and nectar feeding in a freshwater swamp and surrounding areas in southeastern Massachusetts, USA. J Med Entomol 21: 567–572

Nasci RS Berry RL, Restifo RA, Parsons MA, Smith GC, Martin DA (1993) Eastern equine encephalitis virus in Ohio during 1991. J Med Entomol 30: 217–222

Natal D, Marucci D (1984) Apareho de sucção tipo aspirador par captura de mosquitos. Rev Saúde Pública 18: 418–420

Nathan MB (1981) Bancroftian filariasis in coastal north Trinidad, West Indies: Intensity of transmission by *Culex quinquefasciatus*. Trans R Soc Trop Med Hyg 75: 721–730

Natuhara Y, Takagi M, Maruyama K, Sugiyama A (1991) Monitoring *Culex tritaeniorhynchus* (Diptera: Culicidae) abundance in cow sheds by in situ counting. J Med Entomol 28: 551–552

Nayar JK (1982) *Wyeomyia mitchellii*: Observations on dispersal, survival, nutrition, insemination and ovarian development in a Florida population. Mosquito News 42: 416–427

Nedelman J (1983) A negative binomial model for sampling mosquitoes in a malaria survey. Biometrics 39: 1009–1020

Neeru Singh, Mishra AK (2000) Anopheline ecology and malaria transmission at a new irrigation project area (Bargi Dam) in Jabalpur (Central India). J Am Mosq Control Assoc 16: 279–287

Neeru Singh, Singh OP, Sharma VP (1996) Dynamics of malaria transmission in forested and deforested regions of Mandla District, central India (Madhya Pradesh). J Am Mosq Control Assoc 12: 225–234

Nelson DB, Chamberlain RW (1955) A light trap and mechanical aspirator operating on dry cell batteries. Mosquito News 15: 28–32

Nelson RL (1980) The pipe trap, an efficient method for sampling resting adult *Culex tarsalis* (Diptera: Culicidae). J Med Entomol 17: 348–351

Nelson RL, Spadoni RD (1972) Nightly pattern of biting activity and parous rates of some California mosquito species. Proc California Mosq Control Assoc 40: 72–76

Nelson RL, Milby MM, Reeves WC, Fine PE (1978) Estimates of survival, population size, and emergence of *Culex tarsalis* at an isolated site. Ann Entomol Soc Am 71: 801–808

Niebylski ML, Meek CL (1992) Blood-feeding of *Culex* mosquitoes in an urban environment. J Am Mosq Control Assoc 8: 173–177

Nigatu W, Petros B, Lulu M, Adugna N, Wirtz R (1994) Species composition, feeding and resting behaviour of the common anthropophilic anopheline mosquitoes in relation to malaria transmission in Gambella, south west Ethiopia. Insect Science and its Application 15: 371–377

Nutan Nanda, Yadav RS, Subbarao SK, Hema Joshi, Sharma VP (2000) Studies on *Anopheles fluviatilis* and *Anopheles culicifacies* sibling species in relation to malaria in forested hilly and deforested riverine ecosystems in northern Orissa, India. J Am Mosq Control Assoc 16: 199–205

Nuttall GHF, Shipley AE (1902) Studies in relation to malaria. II. The structure and biology of Anopheles. J Hyg Camb 1: 58–84

Oliver J, Howard JJ, Morris CD (1996) Fecundity of naturally bloodfed *Culiseta melanura*. J Am Mosq Control Assoc 12: 664–668

Pajot F-X (1977) Préférence trophiques, cycle d'activité et lieux de repos d'*Aedes (Stegomyia) simpsoni* (Theobald, 1905) (Diptera: Culicidae). Cah ORSTOM sér Entomol Méd Parasitol 15: 73–91

Pålsson K, Jaenson TGT, Dias F, Laugen AT, Björkman A (2004) Endophilic *Anopheles* mosquitoes in Guinea Bissau, West Africa, in relation to human housing conditions. J Med Entomol 41: 746–752

Parajuli MB, Shrestha SL, Vaidya RG, White GB (1981) Nationwide disappearance of *Anopheles minimus* Theobald, 1901, previously the principal malaria vector in Nepal. Trans R Soc Trop Med Hyg 75: 603

Perdew PE, Meek CL (1990) An improved model of a battery-powered aspirator. J Am Mosq Control Assoc 6: 716–719

Pfuntner AR (1978) The development and control of *Culex quinquefasciatus* Say and *Cx. peus* Speiser in urban catch basins. Proc California Mosq Control Assoc 46: 126–129

Pletsch DJ (1970) A collapsible model of the 'red box' for measuring mosquito population density. Mosquito News 30: 646–648

Ponlawat A, Harrington LC (2005) Blood feeding patterns of *Aedes aegypti* and *Aedes albopictus* in Thailand. J Med Entomol 42: 844–849

Rajagopalan PK, Menon PKB, Brooks GD (1977) A study on some aspects of *Culex pipiens fatigans* population in an urban area, Faridabad, northern India, Indian J Med Res 65 (Suppl.): 65–76

Rajavel AR (1995). Field observations on the swarming behavior of *Armigeres subalbatus* (Coq) (Diptera: Culicidae). Southeast Asian J Trop Med Public Health 26: 168–171

Rajavel AR, Natarajan R, Vaidyanathan K, Munirathinam A (2000) Seasonal incidence of *Aedes (Rhinoskusea) portonovoensis* in a mangrove forest of South India. J Am Mosq Control Assoc 16: 340–341

Rajendran R, Reuben R, Purushothaman S, Veerapatran R (1995) Prospects and problems of intermittent irrigation for control of vector breeding in rice fields in southern India. Ann Trop Med Parasitol 89: 541–549

Rajendran R, Thenmozhi V, Tewari SC, Balasubramanian A, Ayanar K, Manavalan R, Gajanana A, Kabilan L, Thakare JP, Satyanarayana K (2003) Longitudinal studies in South Indian villages on Japanese encephalitis virus infection in mosquitoes and seroconversion in goats. Trop Med Int Health 8: 174–181

Rao TR (1984). The Anophelines of India (Revised edition), Malaria Research Centre, Indian Council of Medical Research, Delhi

Ravoahangimalala RO, Rakotoarivony HL, le Goff G, Fontenille D (2003) Écoéthologie des vecteurs et transmission du paludisme dans la région rizicole de basse altitude de Mandritsara, Madagascar. Bull Soc Pathol Exot 96: 323–328

Ree HI, Wada Y, Jolivet PHA, Hong HK, Self LS, Lee KW (1976) Studies on over-wintering *Culex tritaeniorhynchus* Giles in the Republic of Korea. Cah ORSTOM sér Entomol Méd Parasitol 14: 105–109

Rehn JWH, Maldonado Capriles J, Henderson JM (1950) Field studies on the bionomics of *Anopheles albimanus*. Parts II and III: diurnal resting places—Progress report. J Natn Malar Soc 9: 268–279

Reinert JF, Kaiser PE, Seawright JA (1997) Analysis of the *Anopheles (Anopheles) quadrimaculatus* complex of sibling species (Diptera: Culicidae) using morphological, cytological, molecular, genetic, biochemical, and ecological techniques in an integrated approach. J Am Mosq Control Assoc 13 (Supplement): 1–102

Reisen WK, Lothrop HD (1995) Population ecology and dispersal of *Culex tarsalis* (Diptera: Culicidae) in the Coachella Valley of California. J Med Entomol 32: 490–502

Reisen WK, Milby MM (1986) Population dynamics of some Pakistan mosquitoes: Changes in adult relative abundance over time and space. Ann Trop Med Parasitol 80: 53–68

Reisen WK, Aslam Y, Siddiqui TF, Khan AQ (1978) A mark–release–recapture experiment with *Culex tritaeniorhynchus* Giles. Trans R Soc Trop Med Hyg 72: 167–177

Reisen WK, Mahmood F, Parveen T (1979) *Anopheles subpictus* Grassi: observations on survivorship and population size using mark–release–recapture and dissection methods. Res Popul Ecol 21: 12–29

Reisen WK, Hayes CG, Azra K, Niaz S, Mahmood F, Parveen T, Boreham PFL (1982) West Nile virus in Pakistan. II. Entomological studies at Changa Manga national forest, Punjab province. Trans R Soc Trop Med Hyg 76: 437–448

Reisen WK, Milby MM, Meyer RP, Reeves WC (1983) Population ecology of *Culex tarsalis* (Diptera: Culicidae) in a foothill environment in Kern county, California: Temporal changes in male relative abundance and swarming behavior. Ann Entomol Soc Am 76: 809–815

Reisen WK, Milby MM, Reeves WC, Eberle MW, Meyer RP, Schaefer CH, Parman RB, Clement HL (1985) Aerial adulticiding for the suppression of *Culex tarsalis* in Kern county, California, using low volume propoxur. 2. Impact on natural populations in foothill and valley habitats. J Am Mosq Control Assoc 1: 154–163

Reisen WK, Meyer RP, Martinez VM, Gonzalez O, Spoehel JJ, Hazelrigg JE (1988) Mosquito abundance in suburban communities in Orange and Los Angeles counties, California, 1987. Proc California Mosq Vector Control Assoc 56: 75–85

Reisen WK, Meyer RP, Milby MM (1989) Studies on the seasonality of *Culiseta inornata* in Kern county, California. J Am Mosq Control Assoc 5: 183–195

Reisen WK, Pfuntner AR, Milby MM, Tempelis CH, Presser SB (1990) Mosquito bionomics and the lack of arbovirus activity in the Chino area of San Bernardino county, California. J Med Entomol 27: 811–818

Reisen WK, Pradhan P, Shrestha JP, Shrestha SL, Vaidya RG, Shrestha JD (1993) Anopheline mosquito (Diptera: Culicidae) ecology in relation to malaria transmission in the inner and outer Terai of Nepal, 1987–1989. J Med Entomol 30: 664–682

Reisen WK, Smith PT, Lothrop HD (1995) Short-term reproductive diapause by *Culex tarsalis* (Diptera: Culicidae) in the Coachella Valley of California. J Med Entomol 32: 654–662

Reiter P (1983) A portable, battery-powered trap for collecting gravid *Culex* mosquitoes. Mosquito News 43: 496–498

Renshaw M (1991) Population dynamics and ecology of *Aedes cantans* (Diptera: Culicidae) in England. Ph.D. thesis, University of Liverpool

Ribbands CR (1946*a*) Moonlight and house-haunting habits of female anophelines in West Africa. Bull Entomol Res 36: 395–417

Ribbands CR (1946*b*) Repellency of pyrethrum and Lethane sprays to mosquitos. Bull Entomol Res 37: 163–172

Ribeiro JMC, Seulu F, Abose T, Kidane G, Teklehaimanot A (1996) Temporal and spatial distribution of anopheline mosquitos in an Ethiopian village: implications for malaria control strategies. Bull World Health Organ 74: 299–305

Robert LL, Olson JK (1986) Temporal abundance and percent insemination of newly emerged adult female *Psorophora columbiae* near the larval habitat. J Am Mosq Control Assoc 2: 485–489

Robertson LC, Prior S, Apperson CS, Irby WS (1993) Bionomics of *Anopheles quadrimaculatus* and *Culex erraticus* (Diptera: Culicidae) in the Falls Lake basin, North Carolina: seasonal changes in abundance and gonotrophic status, and host-feeding patterns. J Med Entomol 30: 689–698

Rosen L, Lien J-C, Lu L-C (1989) A longitudinal study of the prevalence of Japanese encephalitis virus in adult and larval *Culex tritaeniorhynchus* mosquitoes in northern Taiwan. Am J Trop Med Hyg 40: 557–560

Rubio-Palis Y, Curtis CF (1992) Evaluation of different methods of catching anopheline mosquitoes in western Venezuela. J Am Mosq Control Assoc 8: 261–267

Rudnick A (1986) Dengue virus ecology Malaysia In: Rudnick A, Lim TW (eds) Dengue Fever Studies in Malaysia. Bull Inst Med Res Malaysia 23, pp. 51–53

Russell PF, Baisas FE (1935) The technic of handling mosquitoes. Philipp J Sci 56: 257–294

Russell PF, Knipe FW (1939) Malaria control by spray-killing adult mosquitoes. First season's results. J Malar Inst India 2: 229–237

Russell PF, Knipe FW (1940) Malaria control by spray-killing adult mosquitoes. Second season's results. J Malar Inst India 3: 531–541

Russell PF, Knipe FW (1941) Malaria control by spray-killing adult mosquitoes. Third season's results. J Malar Inst India 4: 181–197

Russell PF, Santiago D (1934) An earth-lined trap for anopheline mosquitoes. Proc Entomol Soc Wash 36: 1–21

Russell PF, West LS, Manwell RD, Macdonald G (1963) Practical Malariology. Oxford University Press, London

Ryan R (1989) A practical method for the use of carbon dioxide as an entomological killing agent in the field. Antenna 13: 16–17

Ryckman RE, Arakawa KY (1951) *Anopheles freeborni* hibernating in wood rats' nests (Diptera: Culicidae). Pan-Pacific Entomol 27: 172

Ryckman RE, Arakawa KY (1952) Additional collections of mosquitoes from wood rats' nests. Pan-Pacific Entomol 28: 105–106

Sadanandane C, Gunasekaran K, Jambulingam P, Das PK (1993) Studies on dispersal of malaria vectors in a hilly tract of Koraput District, Orissa State, India. Southeast Asian J Trop Med Public Health 24: 508–512

Sadanandane C, Jambulingam P, Subramanian S (2004) Role of modified CDC miniature light-traps as an alternative method for sampling adult anophelines (Diptera: Culicidae) in the National Mosquito Surveillance Programme in India. Bull Entomol Res 94: 55–63

Saliternik Z (1963) Catching of adult mosquitoes by the aid of a flashlight. Mosquito News 23: 351

Saliternik Z (1965) A simple, practical method of collecting samples of *Anopheles sergentii* mosquitoes in a cave with the aid of a standard mosquito cage. Mosquito News 28: 218

Sampath TRR, Yadav RS, Sharma VP, Adak T (1998) Evaluation of lambdacy-halothrin-impregnated bednets in a malaria endemic area of India. Part 2. Impact on malaria vectors. J Am Mosq Control Assoc 14: 437–443

Savage HM, Niebylski ML, Smith CG, Mitchell CJ, Craig GB (1993) Host-feeding patterns of *Aedes albopictus* (Diptera: Culicidae) at a temperate North American site. J Med Entomol 30: 27–34

Schoof HF (1944) Adult observation stations to determine effectiveness of the control of *Anopheles quadrimaculatus*. J Econ Entomol 37: 770–779

Schreiber ET, Walton WE, Mulla MS (1993) Mosquito utilization of resting sites at an urban residence in southern California. Bull Soc Vector Ecol 18: 152–159

Senior White RA (1951) Studies on the bionomics of *Anopheles aquasalis* Curry, 1932 (contd.) Part II. Indian J Malariol 5: 465–512

Senior White RA (1952) Studies on the bionomics of *Anopheles aquasalis* Curry, 1932 (concld.). Indian J Malariol 6: 29–72

Senior White RA, Rao VV (1946) On the relative efficiency of hand and spray catching of mosquitoes. J Malar Inst India 6: 411–416

Senior White RA, Ghosh AR, Rao JVV (1945) On the adult bionomics of some Indian anophelines: with special reference to malaria control by pyrethrum spraying. J Malar Inst India 6: 129–245

Service MW (1963) The ecology of the mosquitos of the northern guinea savannah of Nigeria. Bull Entomol Res 54: 601–632

Service MW (1964) An analysis of the numbers of *Anopheles gambiae* Giles and *A. funestus* Giles (Diptera: Culicidae) in huts in Northern Nigeria. Bull Entomol Res 55: 29–34

Service MW (1969) Observations on the ecology of some British mosquitoes. Bull Entomol Res 59: 161–194

Service MW (1971a) The daytime distribution of mosquitoes resting in vegetation. J Med Entomol 8: 271–278

Service MW (1971*b*) Feeding behaviour and host preferences of British mosquitoes. Bull Entomol Res 60: 653–61

Service MW (1973) Flight activities of mosquitoes with emphasis on host seeking behaviour. In: Hudson A (ed) Biting Fly Control and Environmental Quality Proc Symp Univ Alberta, Canada, May, 1972, pp. 125–32

Service MW (1985) *Anopheles gambiae*: Africa's principal malaria vector, 1902–1984. Bull Entomol Soc Am Fall issue, 8–12

Service MW (1986) The biologies of *Aedes caspius* (Pallas) and *Culex quinquefasciatus* Say (Diptera: Culicidae) in Dubai. Insect Science and its Application 7: 11–18

Service MW (1989) The importance of ecological studies on malaria vectors. Bull Soc Vector Ecol 14: 26–38

Service MW (1993) Mosquito Ecology. Field Sampling Methods. 2nd ed. Chapman & Hall , London

Service MW (1994) Male swarming of the mosquito *Culex (Culex) torrentium* in England. Med Vet Entomol 8: 95–98

Sexton JD, Ruebush TK, Brandling-Bennett AD, Breman JG, Roberts JM, Odera JS, Were JBO (1990) Permethrin-impregnated curtains and bed-nets prevent malaria in western Kenya. Am J Trop Med Hyg 43: 11–18

Shalaby AM (1971) Sampling of outdoor resting populations of *Anopheles culicifacies* and *Anopheles fluviatilis* in Gujarat State, India. Mosquito News 31: 68–73

Shalaby AM (1972) A study of the outdoor population of anopheline mosquitoes in Gujarat State of India. Bull Soc Entomol Egypte 56: 369–388

Shapiro JM, Saliternik Z, Belferman S (1944) Malaria survey of the Dead Sea area during 1942, including the description of a mosquito flight test and its results. Trans R Soc Trop Med Hyg 38: 95–116

Sheldahl JA (1974) A simple conversion of a back mist-blower into an efficient power aspirator. Mosquito News 34: 166–169

Shemanchuk JA (1965) On the hibernation of *Culex tarsalis* Coquillett, *Culiseta inornata* Williston and *Anopheles earlei* Vargas (Diptera: Culicidae). Mosquito News 25: 456–462

Shililu J, Ghebremeskel T, Mengistu S, Fekadu H, Zerom M, Mbogo C, Githure J, Gu WD, Novak R, Beier JC (2003) Distribution of anopheline mosquitoes in Eritrea. Am J Trop Med Hyg 69: 295–302

Shililu J, Ghebremeskel T, Seulu F, Mengistu S, Fekadu H, Zerom M, Asmelash GE, Sintasath D, Mbogo C, Githure J, Brantly E, Beier JC, Novak RJ (2004) Seasonal abundance, vector behavior, and malaria parasite transmission in Eritrea. J Am Mosq Control Assoc 20: 155–164

Shimogama M, Takatsuki Y (1967) Seasonal changes in the distribution and abundance of mosquitoes, especially *Culex pipiens pallens* in a cave in Nagasaki City. Endem Dis Bull Nagasaki Univ 8: 159–165

Sholdt LL, Neri P (1974) Mouth aspirator with holding cage for collecting mosquitoes and other insects. Mosquito News 34: 236

Simmons KR, Edman JD, Bennett SR (1989) Collection of blood-engorged black flies (Diptera: Simuliidae) and identification of their source of blood. J Am Mosq Control Assoc 5: 541–546

Smith A (1955) The distribution of resting *A. gambiae* Giles and *A. funestus* Giles in circular and rectangular mud walled huts in Ukara Island, Tanganyika. East Afr Med J 32: 325–329

Smith A (1961) Resting habits of *Anopheles gambiae* and *Anopheles pharoensis* in salt bush and in crevices in the ground. Nature 190: 1220–1221

Smith A (1962*a*) The preferential indoor resting habits of *Anopheles gambiae* in the Umbugwe area of Tanganyika. East Afr Med J 39: 631–635

Smith A (1962*b*) Studies on domestic habits of *A. gambiae* that affect its vulnerability to insecticides. East Afr Med J 39: 15–24

Smith A (1964) A review of the origin and development of experimental hut techniques used in the study of insecticides in East Africa. East Afr Med J 41: 361–374

Smith A, Obudho WO, Esozed S (1966) Resting patterns of *Anopheles gambiae* in experimental huts treated with malathion. Trans R Soc Trop Med Hyg 60: 401–408

Smith CM, Shisler JK (1981) An assessment of storm water drainage facilities as sources of mosquito breeding. Mosquito News 41: 226–230

Smith GE (1942) The keg shelter as a diurnal resting place of *Anopheles quadrimaculatus*. Am J Trop Med 22: 257–269

Snow WE (1949) Studies on portable resting stations for *Anopheles quadrimaculatus* in the Tennessee valley. J Natn Malar Soc 8: 336–343

Snow WE, Smith GE (1956) Observations on *Anopheles walkeri* Theobald in the Tennessee valley. Mosquito News 16: 294–298

Southwood TRE (1978) Ecological Methods with Particular Reference to the Study of Insect Populations. Chapman & Hall, London

Southwood TRE, Henderson PA (2000) Ecological Methods. 3rd ed. Blackwell Science, Oxford

Spencer M (1965) Malaria in the d'Entrecasteaux islands, Papua, with particular reference to *Anopheles farauti* Laveran. Proc Linn Soc N.S.W 90: 115–127

Spencer TET (1962) Notes on a suction device for catching mosquitoes. Papua New Guinea Med J 6: 32

Spielman A (1964) Two mechanical aspirators for the manipulation of mosquitoes. J Parasitol 50: 585

Stern VM, Dietrick EJ, Mueller A (1965) Improvements on self-propelled equipment for collecting, separating, and tagging mass numbers of insects in the field. J Econ Entomol 58: 949–953

Subbarao SK, Vasantha K, Raghavendra K, Sharma VP, Sharma GK (1988) *Anopheles culicifacies*: siblings species composition and its relationship to malaria incidence. J Am Mosq Control Assoc 4: 29–33

Sudia WD, Chamberlain RW (1967) Collection and Processing of Medically Important Anthropods for Virus Isolation, U.S. Dept Hlth Educ National Disease Center, Atlanta, Georgia

Sulaiman S, Service MW (1983) Studies on hibernating populations of the mosquito *Culex pipiens* in southern and northern England. J Nat Hist 17: 849–857

Summers CG, Garrett RE, Zalom FG (1984) New suction device for sampling arthropod populations. J Econ Entomol 77: 817–823

Swellengrebel N, de Buck A (1938) Malaria in the Netherlands. Scheltema & Holkema, Amsterdam

Symes CB, Hadaway AB (1947) Initial experiments in the use of DDT against mosquitoes in British Guiana. Bull Entomol Res 37: 399–430

Szumlas DE, Apperson CS, Powell EE (1996) Seasonal occurrence and abundance of *Aedes triseriatus* and other mosquitoes in a La Crosse virus-endemic area in western North Carolina. J Am Mosq Control Assoc 12: 184–193

Takahashi M, Yabe S, Shimizu Y (1971) Observations on the feeding habits of some mosquitoes in Gunma prefecture, Japan. Jap J Med Sci Biol 24: 163–169

Tang Y (1996) Condom barrier in a mouth-operated aspirator prevents inhalation of debris when handling small insects. Med Vet Entomol 10: 288–290

Taylor B (1975) Observations on malaria vectors of the *Anopheles punctulatus* complex in the British Solomon Islands Protectorate. J Med Entomol 11: 677–687

Taylor LR (1961) Aggregation, variance and the mean. Nature 189: 732–735

Teesdale C (1959) Observations on the mosquito fauna of Mombasa. Bull Entomol Res 50: 191–208

Tempelis CH, Galindo P (1970) Feeding habits of five species of *Deinocerites* mosquitoes collected in Panama. J Med Entomol 7: 175–179

Tempelis CH, Hayes RO, Hess AD, Reeves WC (1970) Blood-feeding habits of four species of mosquitoes found in Hawaii. Am J Trop Med Hyg 19: 335–341

Thornhill EW (1978) A motorised insect sampler. PANS 24: 205–207

Tianyun Su, Webb JP, Meyer RP, Mulla MS (2003) Spatial and temporal distribution of mosquitoes in underground storm drain systems in Orange County, California. J Vector Ecol 28: 79–89

Tidwell MA, Williams DC, Tidwell TC, Peña CJ, Gwinn TA, Focks DA, Zaglul A, Mercedes M (1990) Baseline data on *Aedes aegypti* populations in Santo Domingo, Dominican Republic. J Am Mosq Control Assoc 6: 514–522

Tiwari SN, Anil Prakash, Ghosh SK (1997) Seasonality of indoor resting anophelines in stone quarry area of District Allahabad, U.P. Indian J Malariol 34: 132–139

Tonkyn DW (1980) The formula for the volume sampled by a sweep net. Ann Entomol Soc Am 73: 452–453

Trapido H, Aitken THG (1953) Study of a residual population of *Anopheles l. labranchiae* Falleroni in the Geremeas valley, Sardinia. Am J Trop Med Hyg 2: 658–676

Trpis M (1968) A suction apparatus for collecting mosquitoes and other insects. Mosquito News 28: 647–648

Vale GA (1971) Artificial refuges for tsetse flies (Glossina spp.). Bull Entomol Res 61: 331–350

van Peenen PFD, Atmosoedjono S, Lien JC, Saroso S (1972) Seasonal abundance of *Aedes aegypti* in Djakarta, Indonesia. Mosquito News 32: 176–179

van Someren ECC, Heisch RB, Furlong M (1958) Observations on the behaviour of some mosquitos of the Kenya coast. Bull Entomol Res 49: 643–660

Viswanathan DK, Rao TR, Bhatia SC (1952) The validity of estimation of *Anopheles* densities on the basis of hand collection on a timed basis from fixed catching stations. Indian J Malariol 6: 199–213

Viswanathan DK, Rao TR, Halgeri AV, Karandikar VS (1950) Observations on *Anopheles* densities in indoor shelters during the forenoon, afternoon and night. Indian J Malariol 4: 533–547

Warburg A (1989) An improved air filter for sandfly aspirators. Med Vet Entomol 3: 325–326

Weathersbee AA, Meisch MV (1988) An economical lightweight portable resting unit for sampling adult *Anopheles quadrimaculatus* populations. J Am Mosq Control Assoc 4: 89–90

Weathersbee AA, Meisch MV (1990) Dispersal of *Anopheles quadrimaculatus* (Diptera: Culicidae) in Arkansas ricefields. Environ Entomol 19: 961–965

Weathersbee AA, Meisch MV, Sandoski CA, Finch MF, Dame DA, Olson JK, Inman A (1986) Combination ground and aerial adulticide applications against mosquitoes in an Arkansas riceland community. J Am Mosq Control Assoc 2: 456–460

Wekesa JW, Yuval B, Washino RK (1996) Spatial distribution of adult mosquitoes (Diptera: Culicidae) in habitats associated with the rice agroecosystem of northern California. J Med Entomol 33: 344–350

Wekesa JW, Yuval B, Washino RK, de Vasquez AM (1997) Blood feeding patterns of *Anopheles freeborni* and *Culex tarsalis* (Diptera: Culicidae): effects of habitat and host abundance. Bull Entomol Res 87: 633–641

Wharton RH (1950) Daytime resting places of *Anopheles maculatus* and other anophelines in Malaya, with results of precipitin tests. Med J Malaya 4: 260–271

Wiens JE, Burgess L (1972) An aspirator for collecting insects from dusty habitats. Can Entomol 104: 1557–1558

Williams MC, Weitz B, McClelland GAH (1958) Natural hosts of some species of *Taeniorhynchus* Lynch Arribalzaga (Diptera: Culicidae) collected in Uganda, as determined by the precipitin test. Ann Trop Med Parasitol 52: 186–190

Woodbury EN, Barnhart CS (1939) Tests on crawling insects. Soap Sanit Chem 15: 93–113

World Health Organization (1975) Manual on practical entomology in malaria. Part II. Methods and techniques. WHO Offset Publication, Geneva, No. 13

World Health Organization (1992) Entomological Field Techniques for Malaria Control. Part I. Learner's Guide. World Health Organization, Geneva, Switzerland

Yadav RS, Sharma RC, Bhatt RM, Sharma VP (1989) Studies on the anopheline fauna of Kheda district and species specific breeding habitats. Indian J Malariol 26: 65–74

Yadav RS, Sharma VP, Chand SK (1997) Mosquito breeding and resting in treeholes in a forest ecosystem in Orissa. Indian J Malariol 34: 8–16

Yamashita Z, Ishii T (1977) Smoking method as a survey method of the arboreal arthropod fauna. Ecological studies on the arboreal arthropod fauna. 2. Rept. Environ Sci Mie Univ 2: 69–94

Yasuno M, Rajagopalan PK (1977) Population estimation of *Culex fatigans* in Delhi villages. J Commun Dis 9: 172–183

Yasuno M, Kazmi SJ, LaBrecque GC, Rajagopalan PK (1973*a*) Seasonal Change in Larval Habitats and Population Density of *Culex fatigans* in Delhi Villages.' WHO/VBC/73.429, 12 pp. (mimeographed)

Yasuno M, Russel S, Rajagopalan PK (1973*b*) An Application of the Removal Method to the Population Estimation of *Culex fatigans* Resting Indoors. WHO/VBC/73.458, 9 pp. (mimeographed)

Yasuno M, Rajagopalan PK, Russel S (1977) An application of the removal method to the population estimate of *Culex fatigans* resting indoors. Indian J Med Res 65 (Suppl.): 34–42

Yuval B, Fritz GN (1994) Multiple mating in female mosquitoes - evidence from a field population of *Anopheles freeborni* (Diptera: Culicidae). Bull Entomol Res 84: 137–139

Zaim M, Ershadi MRY, Manouchehri AV, Hamdi MR (1986) The use of CDC light traps and other procedures for sampling malaria vectors in southern Iran. J Am Mosq Control Assoc 2: 511–515

Zaim M, Manouchehri AV, Motabar M, Emadi AM, Nazari M, Pakdad K, Kayedi MH, Mowlaii G (1995) *Anopheles culicifacies* in Baluchistan, Iran. Med Vet Entomol 9: 181–186

Zippin C (1956) An evaluation of the removal method of estimating animal populations. Biometrics 12: 163–189

Zippin C (1958) The removal method of population estimation. J Wildl Mgmt 22: 82–90

Zukel JW (1949*a*) A winter study of *Anopheles* mosquitoes in southwestern Georgia, with notes on some culicine species. J Natn Malar Soc 8: 224–233

Zukel JW (1949*b*) Observations on ovarian development and fat accumulation in *Anopheles quadrimaculatus* and *Anopheles punctipennis*. J Natn Malar Soc 8: 234–237

Chapter 6 Sampling Adults by Animal Bait Catches and by Animal-Baited Traps

Female mosquitoes are sampled primarily through the use of a suitable bait to attract hungry host-seeking individuals. Human bait catches or landing catches have been used for many years and remain the most useful single method to collect anthropophagic species. Variations on the simple direct bait catch have included enclosing humans or other bait animals in nets, cages or traps which, in theory at least, permit the entrance of mosquitoes but prevent their escape. Although bait catches are not completely free from sampling bias they are usually more so than most other collecting methods that employ an attractant. Direct human landing catches are easily performed and require no complicated or expensive equipment. In some areas, especially in North America, light-traps, with or without carbon dioxide as a supplement, have more or less replaced human and animal baits as a routine sampling method for several species, and the use of light-traps in conjunction with a human volunteer sleeping inside a mosquito net as an alternative to the human landing catch has been investigated in several countries and is discussed further in Chapter 9. However, despite intensive studies on host-seeking behaviour no trapping method has been identified that is sufficiently effective to completely replace the use of human landing catches or animal-baited traps.

Mosquito attraction to hosts is mediated by intrinsic, genetic factors as well as extrinsic factors, including heat, water vapour, carbon dioxide and various odours emanating from hosts. Some authors have proposed that host attraction may also be mediated by associative learning and experience (Charlwood et al. 1988; Mwandawiro et al. 2000; McCall and Eaton 2001; McCall et al. 2001; McCall and Kelly 2002).

Sampling the human biting population

The relative size of the human biting population of mosquitoes is an important variable in determining the entomological inoculation rate (EIR) or the vectorial capacity, important parameters for studying the transmission of human pathogens, particularly malaria (see below for a description of these two parameters and techniques for their estimation).

Stationary direct landing (bait) catches

Human landing catches are the most direct method for estimating the human biting rate (ma), or the number of bites/person/day, which is the product of adult mosquito density in relation to the human population and the proportion of mosquitoes feeding on humans. The human biting rate is an important component of the entomological inoculation rate. In direct bait catches man acts both as bait and collector. Kerr (1933) working in West Africa was largely responsible for developing the collection of mosquitoes from man as a routine sampling method, the technique being later modified and standardised by Kumm and Novis (1938) in South America. A common procedure for conducting a human landing catch is for a person to sit on the ground, or a stool, and allow hungry unfed mosquitoes to alight on her/his clothing or exposed skin. Shirt sleeves or trouser legs may be rolled up or shirts removed, but frequently there is no need to expose bare skin to attract mosquitoes. In fact, it is sometimes advantageous to wear 'protective' clothing. For example, collectors studying arbovirus transmission by mosquitoes in a forest in Iquitos, Peru, wore hooded, screened jackets to prevent mosquitoes and other biting insects from feeding on the upper parts of the body. Only the lower portions of the collectors' legs were exposed (Jones et al. 2004). Whenever possible landing mosquitoes should be caught before they have had a chance to insert their mouthparts, thereby reducing the risk of acquiring mosquito-borne infections and avoiding irritation and discomfort due to the act of biting. In practice, however, it may prove impossible to collect all adults before they have bitten and it may therefore be advisable that collectors take prophylactic antimalarial medicines (e.g. Charlwood et al. 1995a; Voorham 2002) and are immunised against yellow fever (e.g. Dossou-Yovo et al. 1998, 1999; Diallo et al. 2003). Mendis et al. (2000) by contrast did not provide their paid collectors with malaria prohylaxis as they presumed that mosquitoes would be collected before they bit the collector and hence the collector was not considered to be at increased risk. Simard et al. (2005) working in Cameroon, opted to exclude human landing catches as a technique for sampling *Aedes aegypti* and *Aedes albopictus*, preferring larval sampling, due to the risks of contracting dengue and the lack of a vaccine or effective treatment. Mosquitoes that have settled can be caught with small battery operated aspirators, or in oral aspirators and then blown into suitable containers. Bates (1944a) reported that the use of oral aspirators in bait catches was abandoned because of the possibility of getting the tongue infected with larvae of *Dermatobia* through sucking up mosquitoes harbouring their eggs. Suitable containers for captured mosquitoes consist of small Barraud cages (Fig. 6.1*a*), cardboard cartons with one or both ends removed and replaced with netting, or glass cylinders or lantern globes which have a piece of rubber sheeting with a 1-in slit stretched over one end (Fig. 6.1*c*). This slit allows the end of the aspirator to be inserted to discharge the catch, but conveniently closes on withdrawal. Another useful container consists of a quart-sized Kilner or Mason jar having a screw cap lid. One or two 1-in wide vertical strips of adhesive tape are fixed to the inside walls, and then the bottom and walls coated with plaster of Paris. When this is dry the strips are removed to reveal windows through which the catch can be observed (Fig. 6.1*b*).

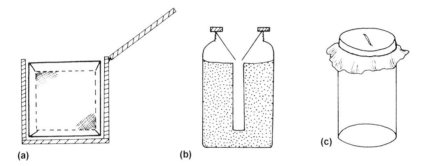

Fig. 6.1. (*a*) small Barraud cage placed in wooden box for transportation; (*b*) Kilner (Mason) jar lined with plaster of Paris; (*c*) container with self-closing slit in rubber top

Mosquitoes are blown from an aspirator into the jar through a plastic funnel which is inserted into its mouth (Aitken et al. 1968*a*). If the plaster is dampened before use mosquitoes can be held in these jars with very little mortality for a considerable time. They are also useful for keeping blood-fed mosquitoes alive in the laboratory. Alternatively, mosquitoes that have landed can be captured by carefully placing a test tube, or some other suitable transparent vial, over them. Some four to five mosquitoes separated by cotton wool plugs can be conveniently caught in a 13-cm long test tube. Occasionally mosquitoes are caught in chloroform killing tubes (Beadle 1959; Galindo et al. 1950) or in small vials charged with ethyl acetate placed over mosquitoes landing on an exposed human forearm (Butts 2001). Landing catches may be carried out by a single person or a group of people and this will depend on the resources available, the type of landing catch and the information required. Much useful information can be obtained from landing catches performed under standardised conditions by a single person (Service 1969a; 1971a,b). When more than one person participates it is essential to minimise individual bias in both the skill and collecting efficiency of the catchers and in their attractiveness to mosquitoes, which is likely to vary from person to person (Freyvogel 1961; Khan et al. 1965, 1971; Shidrawi et al. 1974; von Rahm 1956, 1958; Woke 1962). Individual catchers or groups of catchers should be 'rotated' so that they do not perform catches during the same specific time interval during a continuous landing-catch or from a single locality when catches from more than one locality are being compared. Haddow (1954) gives an explicit account of how he reduced this type of sampling bias. The sources of variation in catch size that may be encountered in human landing collections have been well described by Kettle and Linley (1967, 1969a,b) for Ceratopogonidae.

When catches are performed at night, or inside dark houses or shelters, subdued light from torches or hurricane lamps is used to locate mosquitoes settling on the body. In Kenya head-mounted torches have been employed during all night collections (Chandler et al. 1975). Murphey and Darsie (1962) used a red lens over their torches which only transmitted light of 6720–6869 Å, which they considered did not disturb feeding mosquitoes. Red cellophane was placed over a torch by Aitken (1967) for catching mosquitoes at night in Trinidadian rain forests, and used by Grimstad and DeFoliart (1974) in Wisconsin, USA so that the light had a wavelength of about 6800 Å and was thus invisible to the mosquitoes. In Surinam during landing catches Hudson (1984) covered torches with red plastic to make the light less visible to *Anopheles darlingi*.

Lindsay et al. (1995) collected mosquitoes landing on eight randomly selected village children in The Gambia. Mosquitoes landing on the children were collected by adult field-workers using aspirators and torches. Mosquitoes were collected from children outdoors from 1900 h until the child retired to bed at 2200 h. On retiring, the child slept in a double-layered mosquito net, the outer net having 5 × 5 cm holes cut in it to represent a torn net. Mosquitoes were collected from these nets at 0700 h the next morning. The movement of children was not restricted during the collection period and the authors stated that the study depended on the children's exposure to biting being no different from usual, but how this was assessed is not discussed.

Anil Prakash et al. (2000) and Dev et al. (2001) report collecting mosquitoes from human volunteers as they relaxed or slept.

Location of landing catches

Human landing collections can be performed in a variety of environments, such as in houses, animal shelters, caves, on verandahs and in buildings and natural shelters, or outdoors in village compounds (Fig. 6.2), cleared bush, banana plantations, and farms and forests.

Typically human landing catches are performed concurrently both indoors and outdoors, often by two-person teams in each location in order to effectively sample both the endophagic and exophagic proportions of the population. On Samoa, for example, Samarawickrema et al. (1987) used two two-men teams to collect *Aedes polynesiensis* in a series of 10-min human landing catches at different sites from 0830–1030 h and from 1600–1800 h, and for *Aedes samoanus* from 1930–2130 h. One team, consisting of bait and collector, caught indoors while the other team worked out of doors. More rarely catches were performed indoors and outdoors from 0600–1800 h, and just indoors from 1800–0600 h. In Jakarta *Aedes aegypti* were caught by a team of three people catching adults inside houses for half-hour periods every hour, from 0600–1800 h. At the end of every hour one of the collectors was replaced by another, and by employing a team of six people each collector worked 3 h then had 3 h rest (Nelson et al. 1978). From biting collections in 72 houses during both the wet and dry seasons the largest catches were obtained in houses that had containers positive for larvae. It was tentatively suggested that adults may be attracted to human bait in the same houses from where they had emerged, and/or that after feeding adults may tend to oviposit in the same houses.

Sithiprasasna et al. (2003) conducted only outdoor human landing catches to study the temporal and spatial distribution of malaria vectors in three villages in northwestern Thailand. Collecting teams were located adjacent to houses in the centre of the village and worked in two 6 h shifts from 1800 to 0600 h. More than 18 000 anophelines were collected over 68 trap nights and *Anopheles minimus* was the most frequently captured species, comprising 15 662 individuals.

In a study to characterise the behavioural heterogeneities displayed by anopheline species in southeast Asia, Trung et al. (2005) reported the use of human landing collections carried out by one collector inside and two collectors outside each house from 1800 to 0600 h working in two shifts at two houses per village. In addition, one cattle-shed was selected for cattle-bait collections performed by two collectors, from 2100 to 2400 h each night, and indoor morning resting collections were conducted in another 10 houses from 0600 to 1000 h by two collectors.

Fig. 6.2. Typical night-time human landing catch outside a house (photograph reproduced with permission of M. W. Service)

Each night-time mosquito collection was followed by an indoor morning resting collection. Indoor and outdoor human landing rates were calculated as number of mosquitoes per man per night. Cattle biting rates were estimated as the numbers of mosquitoes caught on cattle by a collector. The degree of anthropophily of the *Anopheles* species was evaluated by the ratio of outdoor human landing rates to cattle landing rates, and the endophagic trend was estimated by the ratio of indoor human landing rates to outdoor human landing rates. The ratio of the number of mosquitoes collected by indoor morning resting collections to total number of mosquitoes collected by indoor, outdoor human landing and cattle landing collections was used to evaluate the endophilic trend. Intraspecific behavioural differences were observed among populations of most *Anopheles* species. However, *Anopheles dirus* A showed a high degree of anthropophily at all study sites where it was found. The preference of *Anopheles minimus* A to feed on humans by contrast was only found in the study sites of central Vietnam and in Laos, whereas *Anopheles minimus* A from northern Vietnam and Cambodia were more attracted to cattle, and this could have been related to differences in cattle abundance among sites.

Gass et al. (1982) pointed out that with mosquitoes such as *Mansonia* a comparison of only indoor and outdoor biting did not always identify the true degree of vector-man contact. For example, in Thailand they found that the ratios of the numbers caught biting in forests, on verandahs and indoors was 7.2:2.2:1 for *Mansonia annulata*, and 9.5:3.9:1 for *Mansonia uniformis*, demonstrating that although both species were reluctant to enter houses a substantial number bit people on verandahs. Moreover, peak biting occurred just after sunset, a time when people relax on their verandahs. Das et al. (1983) studied daytime biting by *Armigeres subalbatus* in India by having collectors catch mosquitoes from a man lying on a bed inside a house, and from another person on a bed outside a house.

In Australia Kay (1985) compared the numbers of mosquitoes biting human baits indoors in darkness, indoors with a 75-W bulb, and out of doors in natural darkness. Catches were made from two humans simultaneously in two paired situations from 1900–1945 h on 12 nights on each of four visits, in a randomised block design so that each paired situation (e.g. bait 1 indoors in darkness and bait 2 indoors in light) was replicated. To minimise collector bias each situation evaluated contained an equal number of collections by each collector. For example, for the 12 nights, each situation (indoors/dark; indoors/light; outdoors/dark) was analysed on the basis of eight collections each, i.e. four each by collectors 1 and 2. Catches were transformed to log (n + 1) and results analysed by 2-way ANOVA to detect any differences between situations (2 degrees of freedom (d.f.)), changes in abundance during the four trips (3 d.f.) and their interactions (6 d.f.). This paper provides useful guidance on the statistical design of experiments to compare vector-man contact and the risk of being infected with pathogens in different situations.

In Mexico in an evaluation of the effect of insecticidal residual house-spraying Bown et al. (1984) organised simultaneous indoor and outdoor human landing collections from 1800–0600 h. The indoor and outdoor collectors changed places every 3 h and each collected for 6 h before being relieved. In addition *Anopheles albimanus* engorging on a person seated near the door of a house were dusted with fluorescent powders and their movements into the house followed with the aid of an ultraviolet lamp. Their activities, such as number of landings, types of resting surfaces in the house and duration of resting times, were recorded for 1 h. At the end of this period the mosquitoes were collected and kept for 24 h to determine their survival. In Brazil a series of human landing collections were made 10, 20 and 40 m from a house. Only *Anopheles darlingi* was caught in collections made nearest the house, but six *Anopheles* species, *Ochlerotatus fulvus* and *Psorophora cingulata* were caught 20 m from the house, and six *Anopheles* species, *Ochlerotatus fulvus, Psorophora cingulata* and *Culex spissipes* were caught biting 40 m distance (Roberts et al. 1987). In Brazil, Rosa-Freitas et al. (1992) performed 12-h outdoor landing catches to collect specimens of *Anopheles darlingi* for isoenzyme electrophoresis and cuticular hydrocarbon analysis. Indoor human landing catches were not performed as no unsprayed or unscreened houses were available in the localities surveyed.

In Ethiopia Krafsur (1977) employed two people to collect mosquitoes attempting to bite a sleeping person inside a house from 1800–0700 h. He compared the numbers of *Anopheles* caught biting per man-hour inside and outside houses, and also estimated the biting rate deduced from pyrethrum spray collections in houses. For example, the mean number of bites per man per night estimated by the total catch over a year of *Anopheles wellcomei* biting indoors divided by the number (in this instance 276) of man-nights was 1.95. But the mean number biting a man per night obtained by dividing the mean number caught in a house by the proportion of those that were blood-fed, and then dividing this by the numbers of people in the house was 0.005. The ratio of 1.95:0.005 shows that the estimated average man-biting rate inside houses was in fact 390 greater than suggested by collections from space-spray collections made in the early morning. This clearly shows that while *Anopheles wellcomei* may be partially endophagic, it is strongly exophilic.

Bockarie et al. (1995) conducted both indoor and outdoor human landing catches in southern Sierra Leone to determine vectorial capacity and entomological inoculation rates. Outdoor catches were carried out from 1800 to 0600 h. Interestingly, the first part of the 'indoor' catches was actually conducted outside between 1800 and 2200 h, the period when inhabitants were still active outside. At 2200 h the collection shifted indoors to a bedroom and continued until 0600 h.

Ameneshewa and Service (1997) studied the biting habits of *Anopheles arabiensis* in central Ethiopia using indoor and outdoor human landing catches. Collections were carried out from 1800 h to 0600 h two nights per month over a period of 23 months by two-person teams working in four-hour shifts. A total of 2326 females were collected in human landing catches. In the first year the indoor collection (795) was slightly higher than the outdoor collection (723), but in the second year the outdoor collection (497) was greater than the indoor collection. No consistent pattern of endophagy:exophagy was observed even when separating the catches between wet and dry seasons. The overall degree of exophagy was estimated to be 57.7%. Biting activity commenced between 1845 and 1930 h and peaked between 2000 and 2100 h both indoors and outdoors. Outdoor biting tended to exhibit a second lower peak around midnight, while indoor biting showed a small peak just before dawn at around 0400 h.

In south Cameroon, Manga et al. (1993) performed indoor human landing catches in four houses located along a transect between the larval habitat and a small hill several hundred metres distant. Human landing catches were found to be higher in houses located closer to larval habitats. For example, in Pongo, 86.5% of mosquitoes caught were collected in the two houses situated closest to the larval habitat. None was collected in house 4 on the hill, 300 m distant. In studying malaria transmission in Bangladesh Rosenberg and Maheswary (1982) undertook human landing catches inside a house and simultaneously had two men seated outside 15 ft away under a 4-m^2 tarpaulin sheet. Each night's catch consisted of hourly 45-min collections from 1800–0600 h. The 15 'free' minutes each hour were used to record rainfall, wind velocity, cloud cover, and wet and dry bulb temperatures; and also to collect mosquitoes from exit traps and light-traps. One team of two collectors worked from 1800–2400 h, and the other from 2400–0600 h, and indoor and outdoor teams exchanged positions hourly. From 198 collections over 21 months 6098 *Anopheles* of 15 species or species groups were caught. By far the commonest species was *Anopheles dirus*, of which in 1975 81.4% were caught biting outdoors, whereas the following year only 57.5% were caught outdoors. Other anophelines included *Anopheles philippinensis, Anopheles maculatus, Anopheles karwari, Anopheles annularis, Anopheles vagus* and the *Anopheles hyrcanus* group.

Abu Hassan et al. (2001) used 2 groups of 2 persons stationed either indoors or outdoors to determine species composition and biting cycles of *Anopheles* mosquitoes near the Malaysia-Thailand border. One-hour biting catches were used between sunset and sunrise, each catch comprising 50 minutes collecting and 10 minutes rest. Nine species of *Anopheles* were collected, the majority in outdoor collections. The major malaria vector in the area, *Anopheles maculatus,* was captured in equal numbers indoors and outdoors and peak indoor biting activity was observed around 2130 h. *Anopheles aconitus* peak biting was just after sunset, with a smaller peak between 2230 and 0130 h. Neeru Singh et al. (1999) also used this method of conducting landing catches for 50 minutes each hour, with a ten-minute rest period, during which meteorological measurements were taken.

In China using a so-called special human bait hut to sample anthropophagic mosquitoes was considered much more realistic than landing collections on mosquito workers. In 1982 and 1983 the mean number of bites per night by *Anopheles sinensis* in the human bait hut was 1.7 and 1.0, whereas in more artificial human bait collections the mean numbers were 14.6 and 20.9 (Guan et al. 1986).

Catches can be performed at ground level or at various heights up to the tree canopy by fixing wooden ladders and platforms to forest trees (Aitken et al. 1968*b*; Bates 1944*a*; Bugher et al. 1944; Deane et al. 1948; Galindo et al. 1950; Garnham et al. 1946; Haddow et al. 1947; Happold 1965; Lourenço-de-Oliveira and Luz 1996; Mattingly 1949*a,b*; Novak et al. 1981; Trapido and Galindo 1955, 1957). Alternatively, steel or wooden towers can be erected which allow catches to be performed not only at various heights up to the forest canopy but also beyond it (Haddow 1964; Haddow et al. 1961; Haddow and Ssenkubuge 1965; Rickenbach et al. 1971). To enable catches to be made at different sites, but at the same height a small bridge, or walkway, can be suspended between trees. Some species, e.g. *Ochlerotatus ingrami, Aedes africanus, Coquillettidia aurites, Coquillettidia pseudoconopas*, have been shown to exhibit pronounced daily vertical migrations within the forest (Germain et al. 1972, 1973; Haddow 1954, 1961*a,b*; Haddow and Ssenkubuge 1965; Mattingly 1949*a*). The percentage biting at different levels can vary according to time. The species composition and biting cycles of the same species may differ according to locality, habitat and height (Galindo et al. 1950; Germain et al. 1972; Haddow 1945*b*, 1961*a,b*; Haddow and Ssenkubuge 1965; Happold 1965; Lumsden 1958*a*; Rickenbach et al. 1971; van Someren and Furlong 1964). In Zika forest, Uganda McCrae et al. (1976) showed by 24-h human landing catches that biting times of *Anopheles implexus* varied greatly according to different ecological zones. For example, within the forest about 66% of the mosquitoes were caught during the day whereas at the exposed forest edge only 3% bit during the daytime.

Lourenço-de-Oliveira and Luz (1996) investigated the biting habits of anopheline mosquitoes inside and outside of two forest habitats in the Brazilian Amazon, using human landing catches at ground level and from platforms erected at heights of 14–17 m in the tree canopy. Collections were carried out from 1800 to 2100 h at one site and from 1800 to 2100 h and 0500 to 0700 h at a second site. Ten species of anopheline mosquitoes were collected, with *Anopheles nuneztovari* and *Anopheles oswaldoi* being the most common species overall, although *Anopheles mediopunctatus* was the dominant species at the Rondônia site. Nine species were collected within the forest and only two species outside. Of the two species collected outside the forest, *Anopheles darlingi* was not collected from inside the forest. At Balbina in Amazonas State, most species were collected in approximately equal numbers at ground level and in the canopy. Almost 72% of *Anopheles mediopunctatus* were collected from the canopy, while *Anopheles mattogrossensis* was only captured at ground level. At the Samuel site in Ronodônia State, almost 80% of all anophelines were captured in the forest canopy. *Anopheles mediopunctatus* exhibited a strong preference for canopy feeding, with 85.5% of this species collected in the canopy.

In common with many other sampling techniques, landing collections in certain localised sites may result in larger catches than similar catches made only a short distance away (Service 1969*a*, 1971*b*). For example, in Papua New Guinea Charlwood et al. (1984) found that the numbers of *Anopheles farauti* caught in human landing catches in nearby areas in a village differed considerably, emphasising that the number of bites people receive depends on their whereabouts in the village. Again this stresses the difficulties of realistically estimating biting rates. Furthermore, species composition may also differ over relatively short distances (Service 1971*b*). As with other sampling procedures it is therefore important that collecting procedures are strictly standardised.

Duration of catches

Haddow (1945*a*) appears to have been the first to have introduced and emphasised the importance of a continuous 24-h human landing catch for collecting representative samples of all anthropophagic mosquitoes in an area. A detailed appraisal of the 24-h catch technique, which has been widely adopted though sometimes with modification, is given by Haddow (1954). Cheong et al. (1988) provide a good example of 24-h landing catches undertaken in Malaysia, in which at least 13 culicine species were collected. Smits et al. (1996) in Burundi used two sets of collectors, one working from 1800 to 2400 h and a second from 2400 to 0600 h both indoors and outdoors. Collections were carried out at fixed stations for five consecutive nights. Almeida et al. (2005) used several different lengths of human landing catches to sample *Aedes albopictus* on the Chinese island of Macao, ranging from 5–24 h in the wet season. In the dry season, three teams of two people simultaneously collected during the period from 0600–1900 h. Because an unusually high biting rate may be experienced at the start of any catch (Haddow 1954; Service 1969*a*, 1971*b*) 25-h catches are sometimes performed and the numbers caught during the first hour excluded from the results (de Kruijf 1972). In a Trinidadian forest for example, bait catches were undertaken for 25 h because it was noticed that at the start of each catch abnormally large counts were encountered (Aitken et al. 1968*b*). In addition, the arrival, every 2 h, of the relief team of catchers also resulted in an increased catch. This was probably caused by mosquitoes being attracted to the movements of the catchers and following them as they walked to the catch site. To overcome this, catching commenced at even and odd hours on alternative working days. Chadee et al. (1993) used the methods of Haddow (1954) and Aitken et al. (1968*b*) to investigate the biting activity of *Haemagogus equinus* over a 24 h period in Trinidad, using six collectors at each site working in three 8 h shifts. The monthly mean of the daily number of female *Haemagogus equinus* landing on human bait was 73.7 per pair of collectors per day in the wet season and 5.5 in the dry season. The same methods were also used to study the diurnal landing activity of *Haemagogus janthinomys* and *Haemagogus leucocelaenus* (Chadee et al. 1992, 1995).

Continuous catches of longer duration than 25 h, even lasting several days, are sometimes performed, such as the 96-h landing catches performed by Linthicum et al. (1984) in Kenya.

In many situations, a 12-h landing catch is used, especially in studies related to malaria epidemiology, where the *Anopheles* vectors are predominantly active at night. The 12-h catch has been used by Dossou-yovo et al. (1995, 1998, 1999), for example, in Côte d'Ivoire, where teams of four persons worked in pairs indoors and outdoors from 1800 to 0600 h, the first pair being replaced by the second pair at midnight. Over 576 nights of capture, 13 654 female mosquitoes of eleven species were collected (Dossou-Yovo et al. 1995). *Anopheles gambiae* s.l. was the most frequently collected mosquito. In Cameroon, Quakyi et al. (2000) conducted indoor human landing catches on two consecutive nights from 1800 to 0600 h the following morning in representative village houses to estimate the biting rate per person per night and the nightly inoculation rate. Githeko et al. (1996) used indoor and outdoor human landing catches from 1900 to 0600 h to investigate the biting behaviour of *Anopheles gambiae s.s., Anopheles arabiensis* and *Anopheles funestus* in western Kenya. Mendis et al. (2000) reported the use of pairs of paid collectors working indoors and outdoors in 6 h shifts from 1800 to 0600 h in Mozambique. Adults of *Anopheles funestus* and *Anopheles gambiae* (85% *Anopheles arabiensis* and 15% *Anopheles merus*) were obtained both indoors and outdoors in every month of the year. Bangs et al. (1996) conducted indoor human landing catches from 1800 to 0600 h and outdoor collections from 1800 to 2100 h once per week for 88 weeks in and adjacent to two sentinel houses in Irian Jaya, Indonesia. Of the 2577 anophelines collected, 97.7% were *Anopheles punctulatus*. In Papua New guinea Attenborough et al. (1997) mainly conducted human landing catches from 1800 to 0600 h, except in Beitafip, where negligible numbers of mosquitoes were collected prior to 2200 h. More than 93% of the anophelines captured below 240 m elevation were *Anopheles koliensis*, while at 650 m altitude all anophelines collected were *Anopheles punctulatus*. Bockarie et al. (2002) conducted all-night (1800 to 0600 h) human landing-catches to collect *Anopheles vectors* of *Wuchereria bancrofti* filariae in Papua New Guinea. In Sri Lanka Ramasamy et al. (1994) conducted human-landing collections from 1800 to 0600 h using pairs of collectors working in 6 h shifts inside and outside houses. Chadee and Martinez (2000) used human landing captures according to the methods of Haddow (1954) to determine the landing periodicity of *Aedes aegypti* in Trinidad. Collections were made on one day each week between 0400 and 2400 h in two houses, with one collector sitting outside on the porch and one sitting indoors in the living room. Each collector worked 10 h shifts. Mosquitoes were caught with hand nets or aspirators from the catcher's lower legs and ankles and transferred into jars lined with moistened plaster of Paris. Diel landing periodicity was trimodal, with peaks at 0700, 1100 and 1700 h. Landing periodicity at the urban site was predominantly diurnal (90%), but with a nocturnal component (10%). At the rural site, no nocturnal component was recorded. In Haiti, Molez et al. (1998) studied the biting habits of *Anopheles albimanus* by conducting indoor and outdoor human landing catches from 1700 to 0600 h on two nights per month over a 12 month period. At one of the two sites studied outdoor biting intensity of *Anopheles albimanus* was estimated at between 3.0 and 3.7 bites per person per hour, and indoor biting was estimated to be between 0.4 and 1.3. At the other site, indoor and outdoor biting densities were similar. Xavier and Rebêlo (1999) conducted human landing catches indoors and outdoors from 1800 to 0600 h once per month for one year in Maranhão, Brazil. They collected a total of 1407 anophelines of six species (*Anopheles aquasalis* [82%], *Anopheles galvaoi* [10.2%], *Anopheles albitarsis* [6.4%], *Anopheles evansae*, *Anopheles nuneztovari*, and *Anopheles triannulatus davisi* [1.4% combined]). In Amapá State, Brazil, Voorham (2002) employed adults from the local community to undertake human landing collections on three consecutive nights per month during eight months The observation period commenced 30 minutes before sunset and ended 30 minutes after sunrise, which corresponded to a collecting period of 13 h as the study site was located almost at 0° latitude. Each collection was split into two shifts of six and seven hours. Among the 27 458 anophelines collected, *Anopheles. darlingi* (74.7%) and *Anopheles marajoara* (22.7%) were the most abundant. In a multi-country (Guatemala, El Salvador, Honduras, Nicaragua, Costa Rica, Belize, and Panama) study of the ecology and biology of *Anopheles albimanus* in Central America human landing catches were carried out indoors and outdoors for 12 hours between 1800 and 0600 h (except in El Salvador, where collections were performed for only 6 h between 1800 and 2400 h (Pan American Health Organization 1996). Results were expressed as mosquitoes/human/hour. Vythilingham et al. (2003) in Lao PDR conducted all-night, human landing collections inside and outside two houses per village. Collections were carried out from 1800 to 0600 h using two collectors working in shifts from 1800 to 2400 h and from 2400 to 0600 h in a study of the seasonality of *Anopheles*. If biting starts around dusk and continues during the night, it is usually more informative to continue collecting throughout the night rather than just for a part of it. However, in El Salvador, where *Anopheles albimanus* could be caught biting throughout the night, a 2-h catch in the early evening was found to be sufficient for monitoring changes in biting density (Rachou et al. 1965).

Although landing catches may be made for 12 or 24 h, mosquitoes may not be collected continually throughout this period. For instance, in California, USA Cope et al. (1986) collected an average of 828 mosquitoes/night, mainly *Culex erythrothorax*, by two people catching mosquitoes at human bait for 15 min in each hour, from 1900–0600 h. In Texas, USA in 14-h human bait collections one person exposed his bare legs and arms to mosquitoes while the rest of his body was protected by mosquito netting, heavy clothing and boots. Collections were made for only 15-min periods each hour (Roberts and Scanlon 1975). The biting behaviour of *Anopheles darlingi* in Brazil was investigated by having two human baits collecting together for 30 min every hour throughout the night in a house, while at the same time a similar collection was performed outside 20 m away. Sometimes, however, because of high biting densities just a single person caught mosquitoes, inside or outside, for 10–15 min of each hour (Roberts et al. 1987). In the Dominican Republic Mekuria et al. (1990) performed human bait catches for 45-min periods each hour from 1800–0600 h.

In evaluating the effects on *Anopheles pseudopunctipennis* of spraying houses with DDT or bendiocarb in southern Mexico, Casas et al. (1998) conducted a combination of human landing catches from 1800 to 0600 h for six consecutive nights prior to spraying and six consecutive nights post-spraying. One member of the team of two collectors performed a human-landing catch outside the house and the other conducted a catch indoors. Both collectors collected mosquitoes for 45 min after which time the outdoors collector entered the house and both members of the team collected indoor-resting mosquitoes for the remaining 15 minutes in each hour.

In a study to determine the geographic distribution of *Anopheles darlingi* in Peru, Schoeler et al. (2003) described the use of trained volunteers who undertook landing collections from 1800 to 0600 h, during which period collections were made for 50 min, with a 10-min rest/break period each hour. A total of 60 585 anopheline mosquitoes of 12 species and two subgenera (*Nyssorhynchus* and *Anopheles*) were obtained using single night human-landing collections at 93 sites. Also in Peru, Jones et al. (2004) report the use of 24 h human landing catches conducted in two 12 h shifts by volunteers who collected mosquitoes for 40 min each hour and rested for 20 mins. Similarly, Rattanarithikul et al. (1996) used a 50 minute collection and 10 minutes rest cycle from 1800 to 0600 h for collecting anophelines in southern Thailand.

When peak biting activity periods are well known, landing catches may be restricted to 1–2 h around the peak biting time. In Tanzania, for example, Corbet and Smith (1974) concluded that because of the consistency of diel landing rates of *Aedes aegypti* it was unnecessary to catch this species over its entire biting cycle (about 0500–1900 h) but for only a part of this time, 0600–0900 and 1500–1900 h, to obtain reliable measurements of density. However, the dangers of making assumptions about peak biting times were demonstrated by Diarrassouba and Dossou-Yovo (1997), who conducted human landing catches from 1600 to 0600 h in Côte d'Ivoire and observed a somewhat atypical biting pattern for *Aedes aegypti*, normally a day-biting species. Females were collected continuously throughout the night, with peak biting occurring around midnight. Lardeux et al. (1995) carried out human landing catches for 15 min at each of 247 sites in Tahiti, French Polynesia to sample *Aedes polynesiensis*, the local vector of *Wuchereria bancrofti*. Catches were performed in the morning between 0630 and 1030 h and in the afternoon from 1500 to 1800 h as this species exhibits peaks in biting activity at dawn and dusk. Russell et al. (2005) conducted preliminary trials to determine the most appropriate times to undertake human landing catches to sample *Aedes aegypti* and *Aedes polynesiensis* in French Polynesia, and selected a 2 h period prior to sunset for subsequent studies. In Bangladesh, Renshaw et al. (1996) reported using indoor and outdoor human landing catches for a duration of 3 h directly after sunset. Hii et al. (1995) used nightly outdoor human landing catches from 1815 to 2215 h over a period of 30 consecutive nights to determine the effect of mosquito nets treated with permethrin and DDT spraying on survival rate and oviposition interval of *Anopheles farauti* No.1 in the Solomon Islands. In Britain Packer and Corbet (1989) studied seasonal abundance of host-seeking *Ochlerotatus punctor* with human baits always facing downwind of any breeze and collecting was restricted to 1.5 h before to 0.5 h after sunset. In São Tomé and Príncipe Pinto et al. (2000) conducted outdoor human landing catches from 2100 to 2300 h coinciding with peak biting activity. In São Tomé, 2259 female *Anopheles gambiae* s.l. were collected in 108 h of collecting and in Príncipe, 112 female *Anopheles gambiae* s.l. were collected in 22 h of collecting. Fontenille and Toto (2001) reported collecting *Aedes albopictus* for the first time in Cameroon, using human landing catches conducted by volunteers from 1700 to

1830 h. Zeze (1996) working in Abidjan, Côte d'Ivoire used human landing catches from 1600 to 0100 h to sample *Culex quinquefasciatus* during the period when human inhabitants were presumed to be most exposed to the risk of biting. Faye et al. (1994) used one team of two to conduct indoor human landing catches from 1900 to 0100 h and a second team to collect from 0100 to 0700 h in a malaria epidemiology survey in Senegal. Collections were carried out once per month from June to November and once every two months from December to May and for two consecutive nights per village. Also in Senegal, Diallo et al. (2003) report the use of human landing catches in forest canopy and at ground level from 1730 to 2030 h in order to study the sylvatic cycle of dengue virus type 2. Govere et al. (2000) conducted human landing catches in Mpumulanga, South Africa using two four-person collecting teams. Catches were performed from 1800–2200 h for four consecutive nights per month. *Anopheles coustani* was the most abundant species, comprising 55.8% of the total catch. A total of 435 female *Anopheles gambiae* s.l. were collected and 425 were identified using PCR, of which 56.0% were *Anopheles merus*, 30.4% *Anopheles quadriannulatus* and 13.6% *Anopheles arabiensis*. During entomological investigations of the first recorded outbreak of yellow fever in Kenya in 1992–1993, Reiter et al. (1998) report that human landing collections were made by pairs of collectors at 22 sites representing the full range of forest biotopes in the transmission area. Mosquito collections were made on 19 consecutive days for 1–3 h after dawn and in the last 3–4 h of daylight. Human-landing collections were also made from platforms constructed in the forest canopy (6 m above the ground) at two sites at the edge of a mountain stream. Marrama et al. (1999), working in Madagascar, conducted human landing catches indoors from 1700 to 0500 h and outdoors from 1700 to 2400 h. *Anopheles mascarensis* was observed to be the most frequently captured anopheline both indoors and outdoors, but was rare in PSC catches.

Relatively short duration human landing catches are a common technique used in South America. Rebêlo et al. (1997) conducted human landing catches from 1730 to 2000 h on three consecutive nights per month in the Maranhão state of Brazil and collected 269 individuals of seven species of *Anopheles*. Conn et al. (2002) collected biting adults from 1900 to 2100 h in Amapá state, Brazil. Barbosa et al. (2003) performed human landing catches between the hours of 1700 and 2000 one night per month in a botanical garden in the municipality of Curitiba, Paraná, Brazil. Luz and Lourenço-de-Oliveira (1996) conducted human landing catches from 1800 to 2100 h three times per week in the vicinity of a hydroelectric plant in Rondônia, Brazil. A total of 3769 mosquitoes, comprising nine genera and

21 species were collected over the period August 1990 to July 1991. Charlwood et al. (1995*b*) employed local people to carry out human landing catches using oral aspirators and torches inside and outside their own houses between 1800 h and 2200 h for seven consecutive nights per month for one year in Rondonia, Brazil. Eleven species of *Anopheles* plus *Chagasia bonneae* were collected in outdoor catches and seven species of *Anopheles* were collected in indoor catches. *Anopheles darlingi* was the dominant species both indoors and outdoors. Póvoa et al. (2003) used human landing catches between 1800 and 2100 h to sample anophelines in the city of Belém in Brazil. Forattini et al. (1993*a*) used 2 h crepuscular human biting catches indoors and outdoors to sample *Anopheles (Kerteszia)* in south-eastern Brazil. From a total of 25 samples 11 373 *Anopheles* females were obtained, comprised of *Anopheles cruzii* (92.3%) and *Anopheles bellator* (7.7%). In further studies of the biting activity of anophelines in Brazil, Forattini et al. (1996*a*) conducted human landing catches for 2 h starting at sunset. On this occasion, 7561 *Anopheles (Kerteszia)* were collected *Anopheles bellator* was at least four times more common than *Anopheles cruzii* in indoor collections. *Anopheles bellator* was also successfully sampled using this method (Forattini et al. 1999). Forattini and colleagues used similar methods to sample *Culex* mosquitoes in Brazil (Forattini et al. 1995*b*). Human landing catches were carried out from 1730 to 1930 h or from 1800 to 2000 h. A total of 6631 adult mosquitoes (6344 females and 287 males) of the *Culex* Spissipes Section was collected, of which 764 (11.5%) were collected from human landing catches and 5602 (84.5%) using Shannon traps. Souza-Santos (2002) described the seasonality of malaria vectors in Rondônia State, Brazil, as determined by indoor and outdoor human landing catches conducted by the Fundação Nacional de Saúde (National Health Foundation (FUNASA). In this case, human landing catches were conducted for a 3 h period from 1800 h. Póvoa et al. (2001) carried out human landing catches for a 4 h period from 1800 to 2200 h as part of malaria vector studies in Amapá State, Brazil. Guimarães et al. (1997) also conducted human landing catches over a two-hour period (1800 to 2000 or 1900 to 2100 h) to investigate the anopheline fauna associated with a hydroelectric plant in Paraná State, Brazil. Da Silva-Vasconcelos et al. (2002) conducted 2 h human landing catches from 1800–2000 h for 1 to 5 or 7 to 10 consecutive nights per month between May 1996 and May 1998 in Brazil. In addition, 12 h catches were performed on one night per month to determine the biting cycle throughout the night. Guimarães et al. (2000*a, b, c*) conducted bi-weekly human landing collections in three sites in the Serra do Mar State Park, São Paulo State, Brazil. Catches were performed during three periods each day: 1000–1200 h, 1400–1600 h, and 1800–2100 h. From 622

collections, a total of 24 943 adult mosquitoes comprising 57 species were collected, of which 34 species were from the Sabethini (Guimarães et al. 2000*a*). Guimarães et al. (2000*d*) performed human landing catches at the same three times in a study of the mosquito fauna of the Serra do Bocaina National Park, also in Brazil. On this occasion, a total of 11 808 specimens representing 28 species were collected from 576 hours of human landing catches. *Runchomyia reversa* and *Anopheles cruzii* were the predominant species, accounting for 52.5% and 17.9% of total captures. Dégallier et al. (1998) performed daily human landing catches in the forest canopy at heights of 25–30 m in Para State, Brazil between 1000 and 1400 h. Captured *Haemagogus* spp. and *Sabethes* spp. mosquitoes were marked with Dayglo fluorescent dust and released for subsequent recapture.

In Sri Lanka Amerasinge and Munasingha (1988) compared the numbers of species collected in diurnal (1400–1700 h) and nocturnal (sunset plus 6 h) human landing collections, in CDC light-traps placed outside houses, and from searches inside thatched temporary huts. A total of 1 172 male and 14 071 female mosquitoes belonging to 71 species were identified, and although most (50%) were caught in the light-traps they trapped large numbers of only a few species (*Mimomyia hybrida, Culex pseudovishnui*). In human landing catches 38 species were collected biting during the afternoons, the most common being *Aedes albopictus, Aedes w-albus, Aedes novalbopictus* and *Aedes jamesi*, whereas 48 species were caught in nighttime collections, the most common being *Mansonia annulifera, Culex gelidus, Culex fuscocephala, Mansonia uniformis* and *Mansonia indiana*. Thirty-two species were collected resting indoors, the most common being *Mansonia annulifera* and *Culex quinquefasciatus*.

Landing catches are frequently performed for about 1–3 h, but sometimes for much shorter periods lasting only a few minutes. Many diurnal species can be adequately sampled by short daytime bait catches. For example, in studying the dispersal of domestic and peridomestic *Aedes aegypti* by mark–recapture methods Trpis and Haüsermann (1986) caught adults during 15-min biting collections in 16 houses in a Kenyan village. In Florida, USA Freier and Francy (1991) had a collector attract *Aedes albopictus* for 2 min, but did not allow any to land on clothing or exposed lower legs, then for the next 5 min those mosquitoes that landed on the body were collected. Thavara et al. (2001) collected day-biting *Aedes* mosquitoes from inside and outside houses on Samui Island, Thailand using teams of three volunteers, with each team collecting for 20 minutes per house, between 0800 and 1200 h. An alternative methodology used by these authors in houses known to have a high population of mosquitoes was to station one volunteer in each of three houses. Mosquitoes were collected from the volunteers' bared legs for a 20-minute period, followed by a 10-minute break, followed by a second 20-minute collection. All collections were made between 0800 and 1700 h. Indoor catches comprised 75.4% *Aedes aegypti* and 24.6% *Aedes albopictus*, while the outdoor catches (15 m from houses) comprised only 1% *Aedes aegypti* and 99% *Aedes albopictus*. The indoor biting rate of *Aedes aegypti* ranged from 1.5 to 8.1 mosquitoes/person-hour, whereas the outdoor biting rate of *Aedes albopictus* varied between 5 and 78 mosquitoes/person-hour.

Butts (2001) performed weekly human landing catches for 20 min per site in the vicinity of beaver (*Castor canadensis*) ponds in upstate New York, USA and collected 15 species of mosquito.

Analysis of results from landing catches

It is usually better to transform the numbers of mosquitoes caught in landing catches to log $(n + 1)$ to allow more realistic means to be calculated, and the application of parametric statistical tests. Results are often plotted as total catches, means or percentages against unit time, e.g. hour, week or month. Downing (1976) emphasised the need to transform data to logarithms to calculate geometric (William's) means, and also pointed out how daily fluctuations in numbers caught could be smoothed out by taking moving (running) means, such as over 5-day periods. Both procedures are explained and illustrated step-wise in his paper. Hawley (1985) used 11-day running means on daily catches of *Ochlerotatus sierrensis* at human bait.

In analysing the relationship between the numbers of individuals and the numbers of mosquito species caught in a series of catches in Tanzania, Nagasawa (1973) found that the results fitted both the logarithmic series of Fisher (1943) and the truncated lognormal distribution introduced by Preston (1948). There was, however, a slightly better fit to the log-normal model. An advantage of having data fit this distribution is that it enables the number of uncaptured species to be estimated. In this instance Nagasawa (1973) predicted that about 11 species (29.7%) were missed in the catches, but should eventually have been caught if catching had been continued. The accuracy of such predictions is attested for by the good agreement between the number of tabanid species estimated to be available for capture by this method and the actual results obtained from field collections (Nagasawa 1967).

Ability to predict relative abundance of insect pests would be an enormous benefit in control programmes, and in fact various attempts have been made to forecast insect abundance. For example, time-series analyses (Hacker et al. 1973) and stochastic probability models (Moon 1976) have been tried with mosquitoes, while thermal summation models (Ring and Harris 1983; Toscano et al. 1979) have been used in attempts to predict outbreaks of agricultural pests. Milby (1985) used logarithmic transformed mean monthly numbers of female *Culex tarsalis* caught in a previous month in New Jersey light-traps, and meteorological variables such as: (i) monthly rainfall; (ii) cumulative rainfall since January 1; (iii) mean temperature in °F; (iv) number of day-degrees above 65°F; and (v) cumulative day-degrees since January 1, to develop regression equations for predicting *Culex tarsalis* numbers in future months. Only limited success was achieved. She found (predictably!) that the more recent the data of the variables were to the month that she wished to predict, the better the prediction. Often the only variables needed were (i), (iii) and (vi), and for some months the only variable that mattered in predicting the size of *Culex tarsalis* populations was its population size the previous month.

Strickman and Hagan (1986) analysed the results of human bait collections of *Chrysops variegatus* for seasonal distribution, and effect of meteorological conditions on the numbers caught. A modified form of time-series analyses as used by Hayes and Downs (1980) on catches of *Culex quinquefasciatus* was employed to determine any seasonal periodicity of biting. In this approach the numbers of flies caught (Y) is given as

$$Y = A + \{B \cdot \cos[(C \cdot X) - D]\} \tag{6.1}$$

where A = mean number of flies throughout the study period; B = degree of amplitude of the periodic waves; C = length of periods, where the period in weeks is equal to $(2\pi)/C$; X = week number; and D = shift in phase. The Kolmogorov–Smirnov test of difference was used to test the significance of the results from 'white noise', together with the asymptotic 95% confidence limits of the calculated period.

The effect of meteorological variables on biting activity was analysed using multiple linear regression, as follows:

$$Y = A + (B \cdot X_1) + (C \cdot X_2) + (D \cdot X_3) \tag{6.2}$$

where Y and A as already defined, and X_1 = dry bulb temperature in °C; X_2 = wind speed on a scale of 0–4; and X_3 = % relative humidity.

These statistical procedures are relevant to mosquito ecology, and the paper by Strickman and Hagan (1986) should be consulted by those interested in applying this approach.

The relative abundances of mosquitoes, such as in landing catches, are often shown by just ranking the species by their absolute numbers, but problems arise when this approach is used with data from several distinct collection sites, because no weight is given to the within site spatial distribution of populations. The alternative approach of scoring the presence or absence of species in different sites ignores among site variability in numbers of the species present. Because of these limitations Roberts and Hsi (1979) proposed a new Index of Species Abundance (ISA) that is calculated for individual species, and which takes into account numerical abundance and spatial distribution. The first step is to tabulate in a column all species caught and then to fill in the numbers of each species caught in the different collection sites in rows. Then keeping the species in the column, another table is made ranking each species in order of abundance at each collecting site, (the most common species being ranked 1), calculating mean ranks for tied scores, then the statistic C is calculated, where C = single largest assigned rank + 1, in other words the highest number in the table. This value of C is then multiplied by the number of zero cells in all columns for each row (i.e. species) to derive the statistic a. In other words the number of collecting sites without a particular species × C. Then R_j is calculated for each row (species) by adding all the rank numbers in that row. Then for each species the ISA is

$$ISA = \frac{a + R_j}{K} \tag{6.3}$$

where K = number of collection sites.

Clearly the range of ISA will be determined by the largest rank number (i.e. number assigned to rarest species), and will differ for different sets of data. To overcome this and to be able to compare the index on a scale of 0–1 the standardised ISA can be calculated as follows

$$ISA = \frac{C - ISA}{C - 1} \tag{6.4}$$

where as already defined C = largest assigned rank number + 1. The most common abundant species at all sites will have the smallest ISA values, whereas they will have the largest standardised ISA values. The variance of the ISA is derived as follows

$$VAR_{ISA} = 1/12K \; [N^2 p^3 (4-3p) + 6Np^2 (2 + q - 2C) \tag{6.5}$$

$$+ \; 3pq(1 + q - 2C)^2 - p^2(6-5p)]$$

where p = the overall proportion of cells in the table of rankings which have non-zero counts at all sites, where $q = 1 - p$ and N = total numbers of species.

Roberts et al. (1984), used the ISA in a study of mosquitoes caught biting human baits and resting inside houses in Bolivia. Later, in Belize,

Roberts et al. (2002) used indoor and outdoor human landing catches to investigate the spatial distribution of the malaria vectors *Anopheles darlingi* and *Anopheles albimanus*. Adult mosquitoes were collected from 1830 to 2000 h by one collector sitting outdoors within 10 m of the house and one collector sitting indoors. Oral aspirators were used to capture mosquitoes as they landed on the exposed feet and legs of collectors. Regression analysis was used to estimate outdoor to indoor ratios of the numbers collected. The index of species abundance (ISA) of Roberts and Hsi (1979) described above was used to calculate an epidemiological parameter comprising presence or absence and numerical abundance of different species at multiple houses in order to define the relative dominance of one species versus other species. Overall, a total of 63 collections were conducted in 42 separate houses. All collections were used in calculating outdoor:indoor ratios for the two species. The outdoor:indoor ratio for *Anopheles albimanus* was 1: 0.211 ($r^2 = 0.68$), and for *Anopheles darlingi* it was 1:0.60 ($r^2 = 0.95$). *Anopheles albimanus* was most often present (at 18 of 22 sites) and most abundant in outdoor collections, with an ISA value of 2.02 (± 0.29). *Anopheles darlingi* was the next most widely distributed (present at 16 of 22 sites) and abundant species, with an ISA value of 2.36 (± 0.29). *Anopheles darlingi* was most often present (12 of 22 sites) and most abundant in indoor collections, with an ISA value of 1.98 (± 0.15). *Anopheles albimanus* was the next most widely distributed and abundant species with an ISA value of 2.27.

Campos et al. (2004) used the ISA of Roberts and Hsi (1979) to study the seasonal abundance of immature stages of *Ochlerotatus albifasciatus* in floodwater pools in Buenos Aires, Argentina. Standardised ISA values for *Ochlerotatus albifasciatus* were highest in autumn (early: 0.99; late: 1.00), followed by summer (ISA = 0.98) and spring (early: 0.96; late: 0.86).

In studying population fluctuations of mosquitoes caught in landing catches in Panama (Wolda and Galindo 1981) seasonal fluctuations were distinguished from yearly fluctuations by calculating the Annual Variability parameter (AV) of Wolda (1978). For example, the numbers of mosquitoes of species i caught in year 1 is termed N_1; and in the second year N_2; so change in abundance from year 1 to year 2 is the ratio of $N_{2i}/N_{1i} = R_i$. It is better to use logarithms and obtain log R_i. In a catch of n species a total of n values of log R_i are obtained and can be plotted as a frequency distribution which will approximate normality. The mean of log R_i gives information on whether all species together tend to increase (positive mean) or decrease (negative mean) in abundance from one year to another. The variance of log R_i, termed Annual Variability (AV), reflects differences between species. If all species change more or less the same way, then AV will be small, but if some species increase greatly in abundance while others decrease, AV will be large. Consequently AV can be used as a measure of stability of the species being collected. Only those species that are represented by at least five individuals each year should be used in determining R_i.

Wolda and Galindo (1981) analysed data for seven mosquito species and found that AV varied from 0.105–0.525, denoting large variations in abundance.

In studying changes in abundance of different species at bait in Sri Lanka Amerasinghe and Ariyasena (1991) applied the Index of Change (IC), previously devised to study changes in abundance of larval populations. Amerasinghe and Ariyasena (1990) also applied the procedure for collections of mosquitoes in light-traps and those resting indoors.

Alternatives to human landing catches

Given the inherent risks of conducting human landing catches in areas where mosquitoes transmit harmful or deadly infections to humans, several workers have attempted to correlate catches obtained using other methods, most notably light-traps alongside mosquito nets containing a sleeping human volunteer, with human landing catches. The results of these trials appear to vary according to the mosquito species studied and no universally satisfactory alternative to human landing catches has been developed to date. Results of investigations to find alternatives to direct human landing catches are presented in Chapter 9.

Hand-net collections

Small, hand-held nets can be successfully used to catch mosquitoes that have not landed on a collector but are flying in the immediate vicinity. When mosquitoes are overwhelmingly numerous, as often occurs in sub-arctic environments, the best procedure may be to make several figure-of-eight sweeps with a small hand-net around the head of the collector, or colleague, and thus standardise sampling in this manner (Gjullin et al. 1961). In fact in the sub-arctic mosquitoes may be so numerous that it is impossible to perform conventional stationary bait catches. In these situations it may only be possible to sample mosquitoes, which are attracted in clouds, by sweeping the air around the head and body and then retreating to the safety of a car, tent or building to sort out the catch. Mosquito repellents may be useful in reducing the numbers attracted. Alternatively, the collector can wear protective clothing, including a mosquito net over, but away from, the face, and restrict the collection of adults to only a few minutes from a specific site on the body, such as below the knee. However, biting catches under these conditions are frequently unsatisfactory. One reason is that collections over very short periods may give unrepresentative samples of the very large mosquito populations that are present. In Sweden Andersson (1990) used hand-nets for 5–10 mins at 30-min intervals to collect 16 species of mosquitoes arriving at bait, either throughout the 24-h day or from 1600–0300 h; the shorter time being used when biting densities were high. Biting activity was recorded as the numbers of mosquitoes, mainly *Ochlerotatus communis* (63%), caught per minute.

A combination of sweeping the air around the host and aspirating settled adults was the method employed by Ho et al. (1971) to collect *Aedes albopictus*. In Canada Hocking et al. (1950) caught mosquitoes 'on the wing' by making 20 sweeps through the air with a net immediately upon arrival at a catch site, in addition to those landing on them to feed. Also in Canada Lewis and Bennett (1979, 1980) collected mosquitoes attracted to humans by performing once during the day and again at dusk, or later during darkness, 40 standardised figure-of-eight sweeps with a 30-cm diameter net. Similarly Taylor et al. (1979) collected host-seeking Canadian aedines and *Mansonia* by having a collector make 40 figure-of-eight sweeps about himself, this took about 1 min and comprised a single sample. To collect *Culex salinarius* in New Jersey, USA Slaff and Crans (1981) walked to a specific catching station, waited for 1 min, then swept the air around the body with a sweep-net for 5 min. This routine was repeated every 30 min for 3 h after sunset (i.e. time of maximum host-seeking activity), and then hourly until 1 h after sunrise. Most adults were caught during the first 30–60 min after sunset.

In a study of the biting habits of Canadian mosquitoes Lewis and Webber (1985) collected adults by three methods. Namely, catching mosquitoes that landed on a 0.09-m^2 blue cloth placed on the person's lap during 2-min intervals; those biting or attempting to bite the left forearm during 2-min intervals; and thirdly by 40 figure-of-eight sweeps made around the collector with a 30-cm diameter net. Consecutive counts and collections were made by a single person. Most mosquitoes, of which *Ochlerotatus punctor* and *Ochlerotatus communis* were the commonest species, were obtained by sweep-netting, followed by landing catches, and then landing counts on the blue cloth. All three collecting methods showed the same seasonal trends in abundance.

In Bolivia because some *Sabethes* were difficult to catch even while feeding, Roberts et al. (1984) found it necessary to use small hand-nets in addition to aspirators to collect adults arriving at human bait. But they pointed out that this may have meant that some other mosquitoes were caught in the net, such as a few *Uranotaenia*, that were not actually attempting to feed on people. A common procedure in Russia is that after a predetermined number of sweeps (10–100) over a period of 15 min or longer—depending on the density of mosquitoes—the sack of a butterfly net containing the catch is removed and replaced with a new sack. Rasnitsyn and Kosovskikh (1979) described and figured an improved butterfly net which had the upper 20 cm made of cloth and the lower 40 cm of mosquito netting, to the end of which is attached a removable small (12-cm long, 6-cm diameter) bag, into which the mosquitoes are collected. Masalkina and Kachalova (1989) found that they need make only 50, not 100, sweeps to get reliable data. In Britain Packer and Corbet (1989) caught mosquitoes arriving at bait with a 40-cm diameter butterfly net, and Renshaw (1991) successfully used a 15cm diameter hand net constructed from a loop of wire covered with mosquito netting to catch *Ochlerotatus cantans* arriving at human bait.

Hilburn et al. (2003) caught *Anopheles quadrimaculatus* s.s. females in Arkansas, USA, for allozyme analysis by sweep-netting immediately after dark any mosquitoes attracted to the collector, who was standing in the beam of the headlights of an automobile.

In Taiwan, Ming-Hui Weng et al. (2005) collected mosquitoes for Japanese encephalitis virus surveillance by using a sweep net around a human bait for a period of 1.5 h commencing 30 min before sunset.

Drop-net catches

Although mosquito densities are only rarely so high as to preclude the use of landing catches, there may sometimes be too many mosquitoes for the collector to catch in test tubes, aspirators or hand-nets. In other situations 'nervous' or 'shy' ('dilettante') species may be encountered. These are species which although attracted to the collector hover around for a considerable time before settling, and even then may be very easily disturbed. In ordinary landing catches such species are more difficult to collect than those which readily settle on bait and consequently their numbers may be underestimated. One method of partly overcoming these difficulties is to use a drop-net. This commonly takes the form of a cylindrical or bell-shaped tube of cloth weighted along the bottom edge with a metal hoop and rolled up and suspended just above the head of the collector (Fig. 6.3*a*). A cord causes the net to descend rapidly to enclose the bait and mosquitoes (Fig. 6.3*b*) in the immediate vicinity (Blagoveshenskii et al. 1943; Dyce and Lee 1962; Minar 1959; Mohrig 1969; Monchadskiy and Radzivilovskaya 1947; Rasnitsyn and Kosovskikh 1979, 1983).

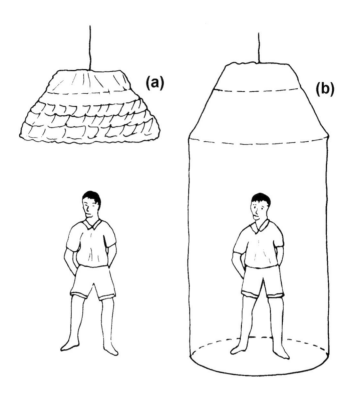

Fig. 6.3. (*a*) Drop-net rolled up; (*b*) drop-net descended to enclose bait (Service 1993)

The dark bell of Monchadskiy or Berezantev, described and figured by Monchadskiy and Radzivilovskaya (1947), has been commonly used in human landing collections in Russia since about 1937. There are several variations, but basically the trap consists of a bell-shaped net made of dark cloth that is suspended about 2 m above the bait. After an exposure time, ranging from 2–15 min or longer, a cord is pulled and the collector becomes enclosed within the drop-net. In some designs the trapped mosquitoes accumulate in a small transparent dome-shaped chamber at the top of the dark bell. Mosquitoes are killed with chloroform or aspirated alive from this apical chamber. In comparative trials Rasnitsyn and Kosovskikh (1979) reported that there was no real difference between the mosquitoes caught with the dark bell and those flying around a human bait which were collected in 25- or 30-cm diameter butterfly nets. Their dark bell caught 1153 mosquitoes belonging to 20 species, while 1214 mosquitoes belonging to 15 species were collected with the butterfly net having removable sacks. With both methods *Aedes vexans nipponii* (38.5 and 49.5%) and *Ochlerotatus punctor* (14.6 and 11.9%) were the two most common species.

In four different localities in Russia Masalkina (1979) found there were no differences between the coefficients of similarity (Jaccard 1912; Shorygin 1939) between the numbers of mosquitoes, and also species, caught by the dark bell net, butterfly nets with removable sacks, and catches from the forearm. Nine species were identified, the most common being *Aedes vexans*. Collections from the dark bell and the forearm were considered generally to give the most complete picture of the species present. These two methods, together with catches with a butterfly net, gave similar results as regards the percentage species composition. In contrast light-traps collected almost only *Aedes vexans*, but they were tested in only two of the four areas. In later evaluations Masalkina (1981) found that there was a linear relationship between the numbers of mosquitoes, mainly *Aedes vexans*, caught with the dark bell and on a forearm, and with a butterfly net with removable sacks and those biting the forearm, but only up to densities where it was no longer possible to collect all those biting the forearm. At densities up to 300 per sample with the dark bell, and up to 150 per sample with the net, the ratio was constant and a linear regression could be obtained between the numbers caught by the two methods.

$$\log y = \log 47.76 + 0.34 \log x \qquad (6.6)$$

$$\log x = 2.94 \log y - 4.94 \qquad (6.7)$$

where x and y = the numbers caught by the net (10 sweeps) and bell, respectively. From this it was concluded that the numbers caught by the net (x) could be converted to the numbers that would have been collected by the dark bell ($y = 2.64x$), and *vice versa* ($x = 0.38y$), a relationship confirmed by Masalkina and Kachalova (1989) in later studies ($y = 2.6x \pm 43$, and $x = 0.4y \pm 17$). They recorded a total of 13 species, the most common being *Ochlerotatus communis, Ochlerotatus punctor, Culiseta alaskaensis, Ochlerotatus pullatus* and *Ochlerotatus cantans. Culiseta bergrothi* was caught by sweep-netting and not in the dark bell net, while *Ochlerotatus hexodontus* was caught only in the dark bell net. To save time they suggested that the exposure period in their net could be reduced from 5 to 2 min, and the number of sweeps from 100 to 50. Their sweep-net had interchangeable bags. The numbers of mosquitoes caught by either method could be converted to the other by simple regression, as was done by Masalkina (1981).

In Kenya a drop-net having at the top an 82-cm diameter ring of wire covered in canvas was supported on four 2.5-m right-angled poles which held the net in the 'up' position (Chandler et al. 1976*b*). At 30 min past each hour throughout the night a man stood under the raised net for 3 min before allowing it to descend and enclose him and the mosquitoes. The lower half of the net was held in position by a 120-cm diameter steel ring which rested on the ground. From 35 12-h night collections 3571 mosquitoes belonging to 12 species were trapped, the commonest being *Mansonia uniformis, Mansonia africana, Anopheles arabiensis, Anopheles ziemanni* and *Aedes circumluteolus*.

Klock and Bidlingmayer (1953) described an umbrella-type drop-net consisting of a 4-ft diameter wooden or metal circular frame covered with mosquito netting and supported on a 7½-ft centre pole. The netting extends down over the outside of the frame to reach the ground, but is initially held in a rolled-up position by four equally spaced cloth straps. The free ends of these straps are loosely held in position over four steel pins, behind each of which is a metal washer tied to a cord, which in turn is connected to a handle. When mosquitoes have been attracted to a man standing on a canvas or plastic sheet underneath the trap, the handle is pulled and the net rapidly drops to enclose the bait. Mosquitoes are collected from within the drop-net by aspirators, or are knocked down onto the floor sheet with a pyrethrum space-spray.

Rooker et al. (1994) further developed the principle of the design of Klock and Bidlingmayer (1953) into what they termed a whole-person bag sampler (WPBS). The WPBS is a free-standing cylindrical drop-net with a sampling volume of approximately $4m^3$ (Fig. 6.4). A 1.5 m diameter spoked top frame is supported by a detachable 2.1 m centre pole with a tripod base. A reel and pulley system allows the rapid raising and lowering of the trap. The WPBS was compared with a stationary sweep-net catch and a catch made with a Nasci (1981) mechanical aspirator. The numbers of mosquitoes collected in the WPBS did not vary with length of exposure of the collector (2, 4 and 8 minutes), in contrast with the sweep-netting and aspirator collections. Catches did not appear to be affected by the use of different individuals as collectors.

Tent traps

The orange tent method of Trpis consists of one or more people sitting in an orange coloured canvas gable-type tent (about 2 m × 2 m × 2 m) which has one end wide open. All mosquitoes entering the tent are captured.

Tent traps have been used in both what was then Czechoslovakia (Trpis 1962a,b, 1971) and Tanzania (Tonn et al. 1973; Trpis et al. 1973).

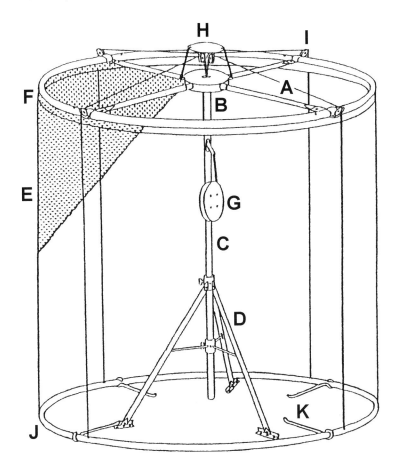

Fig. 6.4. Whole-person bag-sampler (WPBS) of Rooker et al. (1994). A, 70.2 cm long sections of 1.3 cm diameter metal conduit attached to B, a central hub. A 15 cm length of metal conduit extends from the bottom of the central hub and fits into C, a 213 cm tall piece of 1.9 cm diameter metal conduit, supported by D, a 91 cm tall tripod.E, "no-see-um" netting covers the framework, and is supported by 152 cm diameter hoops (F). A 17 cm wooden reel (G) attached to the central pole and dacron tow-line (68 kg test) runs from the reel through an angled nylon hose adapter tapped into a hole in the centre pole, from where the tow-line runs inside the centre pole up through the central hub to swivel eye pulleys (H), which are riveted to a base plate suspended above the hub. The line then passes through fixed eye pulleys (I) at the distal end of each spoke and down to J, the weighted bottom hoop. K, heavy wire covered with plastic tubing to act as net gatherers during raising of the net

In the then Czechoslovakia the procedure was to open the tent entrance for 15 min every 2 h over the period of a 24-h catch. At the end of each 15-min catching period the tent was closed and the mosquitoes caught in a small hand-net. During darkness a small 3-W bulb, operated from a 6-V battery, was suspended inside the tent, not to attract mosquitoes, but to facilitate catching.

Mosquito net traps

One of the earlier references to the use of a trap to catch host-seeking mosquitoes is by Ross (1902) who considered that the number of mosquitoes found in a room during the day was not necessarily a reliable guide to the number that had fed on the occupants during the night. To overcome this disparity, and to measure the numbers seeking a blood-meal, Drs H. E. Annett and J. E. Dutton suggested that a servant slept under a mosquito-net with holes in it, as this would retain many of the mosquitoes that fed on him during the night. Much more recently Bryan (1983) considered that as many people in The Gambia used mosquito nets these could provide a useful method for collecting female *Anopheles gambiae* and *Anopheles melas*. It is usually difficult to compare the relative abundance of these two species in houses from house-resting collections, because *Anopheles melas* is more exophilic than *Anopheles gambiae*, but it is likely that biting a person under a net prevents many from leaving after feeding. Thomson et al. (1995) used the bed-net catch (BNC) method alongside pyrethrum spray catches, exit trap catches and outdoor human landing catches to evaluate the entomological component of the Gambia's National Impregnated Bed-net Programme. A minimum of 300 mosquito nets were searched early in the morning in each study village and counts were log $(n + 1)$ transformed. Fonseca et al. (1996) also collected resting, blood-fed mosquitoes from inside mosquito nets habitually used by the population inhabiting Bissau City in Guinea Bissau. Mosquitoes were collected using an oral aspirator at around 0730 h each morning. *Anopheles gambiae* sensu stricto and *Anopheles melas* were collected using this method in the ratio of approximately 65:35.

In Panama Le Prince and Orenstein (1916) caught large numbers of *Anopheles* in mosquito nets which were placed over a man, dog or chickens and which had the lower edges pulled up and pinned back. However, it was Gater (1935) who was responsible for developing and popularising the bed-net technique. The following account is mainly concerned with human baits under mosquito nets, their use with animal baits is described later, as are the general limitations of mosquito net traps, whether enclosing people or animals.

Starting in 1928 Manalang (1931) devised and experimented with several types of human-baited nets in the Philippines, but it was Gater (1935) who introduced the method to Malaysia where it has now been used for many years (Colless 1959; Hodgkin 1956; Jaal 1990; Reid 1961; Wharton 1953). The original trap consisted of a large net 10 ft long and both 7 ft wide and high with a 3-ft wide flap on each longer side which was rolled up to leave two entrances. A person entered the net around sunset and slept enclosed within a smaller protective net. At about sunrise he was able by pulling on cords to unroll the two flaps of the outer net without leaving his own net. With both entrances closed the entrapped unfed hungry mosquitoes were collected. The original design has undergone various modifications to suit specific purposes. For example, smaller nets are frequently used, especially when they are baited with birds or small mammals, and the inner protective net is often omitted to allow trapped mosquitoes to engorge on the bait, which if an animal is usually tied to a stake or placed in a cage. Removal of the inner net usually results in a larger catch of mosquitoes. Mosquitoes also tend to escape too easily if the original type of net is used; consequently the entrances by which mosquitoes enter, have frequently been modified. Very often a single door-like opening is used (Colless 1959) (Fig. 6.5a). Sometimes one or more sides of the net are partly rolled up and pinned or tucked into place (Sasa and Sabin 1950; Service 1963), or a horizontal slit made (Davidson 1949). Alternatively, the entire net can be raised a few inches from the ground (Akiyama 1973; Hamon 1964; Laarman 1958) to give access to hungry mosquitoes (Fig. 6.5b).

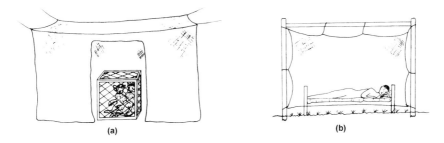

(a)   (b)

Fig. 6.5. (a) Bait-net with cage for small mammals; (b) mosquito-net raised from ground (Service 1993)

Baited mosquito-nets have been used by several Japanese entomologists (e.g. Omori 1942; Takeda et al. 1962; Wada et al. 1967, 1970).

Those used by Wada and his colleagues consist of a double net (400 × 240 × 240 cm) having a single opening (240 cm long and 120 cm high) usually with one, but occasionally two, men enclosed within an inner net (200 × 110 × 150 cm).

In Nigeria Bown and Bang (1980) compared the numbers of *Aedes* species caught biting a man underneath a mosquito net to a man outside the net. In the net collections 162 *Aedes africanus*, 38 *Aedes aegypti* and 17 other *Aedes* mosquitoes were trapped, whereas outside the net 234 *Aedes africanus*, 10 *Aedes aegypti* and 23 other *Aedes* were caught. This emphasises that there may be differences in the relative proportions of species caught by direct landing catches and by mosquito-net collections.

In Sarawak, Malaysia, Seng et al. (1999) used human-baited net catches, human landing catches and CDC light-trap catches to investigate aspects of malaria transmission. Human-baited nets were more effective than either human landing catches or CDC light-traps. For both *Anopheles leucosphyrus* and *Anopheles donaldi*, a significant positive correlation was observed between $\ln(x + 1)$ transformed human landing catches and human-baited net catches ($r = 0.85$, $P < 0.001$ and $r = 0.89$, $P < 0.001$) respectively. Charlwood et al. (1986*a*) reported that in Papua New Guinea mosquito nets raised about 8 cm from the ground allowed a considerable number of unfed *Anopheles punctulatus* to escape, irrespective of whether the human bait was enclosed within an inner protective net. To overcome this problem they placed a person within an inner protective net and surrounded him with a larger net raised approximately 8 cm from the ground, but which had an inverted CDC light-trap suspended from the top of it (Fig. 6.6). Bait catches caught significantly more female *Anopheles farauti* (593 ± 228/night) than mosquito nets having CDC traps with (348 ± 138) or without a light (39 ± 22). Differences between numbers collected from the two net traps were considered to be due to the light attracting mosquitoes within the net into the light-trap, whereas in the unlit trap many mosquitoes escaped. A rather similar method was employed in Kenya by Mutero and Birley (1987). They placed an updraft trap without any light about 40 cm above the head of a person sleeping under a net. Their trap was made from an ordinary plastic bucket, 30-cm tall with diameters of 30 and 24 cm at the top and bottom, respectively. A 10-cm length of plastic, or metal, tubing was thrust halfway through a hole cut from the centre of the bottom of the bucket. A 12-V, 0.17-A d.c. electric motor, powered by a 12-V, 5.7-A d.c. sealed lead-acetate battery, and carrying a small 3-bladed plastic propeller was placed inside the tubing. Mosquitoes hovering around a host in bed were sucked up into the bucket which had the top covered with netting. Moist cotton wool placed inside the bucket prevented mosquitoes from desiccation. The trap was operated from 1800–0600 h and caught mainly hungry unfed female *Anopheles gambiae, Anopheles arabiensis, Anopheles merus* and *Anopheles funestus*. The battery was recharged daily from a solar panel (12-V, 1-A), which was often placed on the roof of the house.

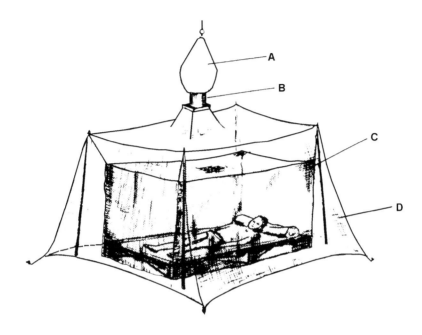

Fig. 6.6. Light-trap type of bed-net, A—cage for collecting mosquitoes, B—updraft CDC suction trap, C—inner bed-net protecting sleeping bait, D—outer bed-net with edges raised (Charlwood et al. 1986*a*)

Parsons (1977) described a bait trap for collecting anthropophagic mosquitoes in which a person is protected from bites by being enclosed in an inner compartment; non-human hosts could also be placed in the trap. Basically his trap consisted of a plastic mesh rectangular cage (76 × 76, 52-in high) supported at the four corners by vertical poles connected at the top of the trap by horizontal ones. A zippered flap in the middle of the trap provides the entrance for a person (bait) to enter. The trap is divided into two 76-in long, 52-in high but just 12-in wide outer chambers, each having a 2-in wide entrance slit. Mosquitoes attracted to the bait in the inner compartment pass through the two horizontal slits into the outer compartments and are collected by the bait with an aspirator inserted through six circular openings fitted with sleeves arranged in two rows on the two walls separating the inner and outer compartments. The trap can apparently be dismantled by two men in 10 min, and is easily transported. It is somewhat reminiscent of a Shannon net trap, except that hungry mosquitoes enter through slits and not through a gap separating the two outer compartments from the ground. No details are presented of the numbers caught, except that in Panama it attracted *Anopheles albimanus*.

Le Goff et al. (1997) compared the traditional human landing catch with single-net and double-net traps (Gater 1935) in Cameroon. For the human landing catches, the catcher sat on a chair indoors with his bare legs exposed and caught landing mosquitoes into a haemolysis tube. One catcher worked from 2000 to 0100 h and a second catcher worked from 0100 to 0600 h. In the single net catches, the catcher slept on a bed under a single mosquito net (1.6 m long, 0.7 m wide and 1.4 m high. One side of the net was lifted 10 cm above the ground. Mosquitoes were removed from the net at 0100 h and at dawn using an oral aspirator. The double-net collections were conducted by completely enclosing a single net similar to the one described above with a larger double net (2 m long, 1.5 m wide and 1.8 m high). One side of the outer net was raised 10 cm above the ground and mosquitoes were collected from the space between the two nets at 0100 and 0500 h. The three collection methods were compared on three nights per month for six months in four houses per method. Human landing catches caught 2227 mosquitoes during 70 person-nights, of which 260 were *Anopheles gambiae*, 1900 *Anopheles nili*, and 13 *Anopheles funestus*. The remainder were culicines or *Mansonia* spp. The single net method yielded 1316 mosquitoes from 68 person-nights, of which 152 were *Anopheles gambiae*, 1143 were *Anopheles nili*, and 14 were *Anopheles funestus*. Human-landing catches captured 1.66 times as many *Anopheles gambiae* and 1.61 times as many *Anopheles nili* as single net catches. Parous rates for *Anopheles gambiae* were higher for the landing catch samples than for the single-net samples (69.7% v 57.9%, $\chi^2 = 4.99$, $P = 0.025$). In contrast, parous rates of *Anopheles nili* were lower in the landing catches than in the single-net catches (66.2% v 74.6%, $\chi^2 = 22.6$, $P < 0.00001$). No significant differences in sporozoite rates were observed between human landing catches and single-net catches for either species. In the comparison between single and double-net catches, it was observed that the single-nets caught 1007 mosquitoes over 38 person-nights, while the double-nets caught 238 mosquitoes over 36 person-nights, equivalent to a ratio of 4.7:1.0 for anophelines. The proportions of *Anopheles gambiae* and *Anopheles nili* in the two catches was similar at approximately 4% and 96% respectively. No significant differences in parous rates or sporozoite rates were observed between the two methods.

The Mbita trap. In response to evidence that light traps hung next to occupied mosquito nets may demonstrate bias in terms of the specific cohorts of the mosquito population sampled, particularly as regards parous rates, as well as the relatively high cost of light-traps, Mathenge et al. (2002) developed several mosquito bed-net traps, the most successful of which they named the Mbita trap. The final design was a pyramidal net 200 cm long by 111 cm wide and approximately 200 cm high at the apex made from white cotton cloth. At the apex of the net, a funnel 30 cm tall, with a 60 cm diameter upper opening narrowing to a 5 cm diameter lower opening and manufactured from 156 mesh polyester netting was affixed. 70 cm below the upper end of the funnel a horizontal panel of polyester netting is fitted to the inside of the net, creating a barrier between the human sleeping in the net and the mosquitoes entering through the funnel. The horizontal netting panel has a hole in the centre, to which a collecting bag is attached (Fig. 6.7).

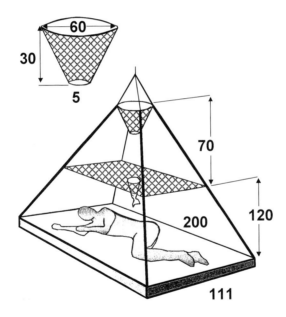

Fig. 6.7. Mbita trap, with detail of funnel (redrawn from Mathenge et al. 2002)

The Mbita net trap was evaluated against model 512 CDC light-traps and human landing collections using laboratory-reared, uninfected *Anopheles gambiae* s.s. mosquitoes. The three trapping methods were tested in experimental huts constructed inside a modified metal-framed glasshouse, in which the glass walls were replaced by mosquito netting. This experimental set-up was designed to mimic as much as possible natural conditions without exposing the experimenters to the potential of becoming infected by wild anopheline mosquitoes. Known numbers of laboratory-reared mosquitoes were released inside the glasshouse at a point equidistant from the two experimental huts. In the first experiment, one hut contained an Mbita trap with a human occupant, while the other hut contained an un-treated mosquito net with a CDC light-trap hung nearby at a height of approximately 1 m. In a second experiment the Mbita trap was compared against an individual who performed a human landing-catch while seated on a chair in the second experimental hut. Human volunteers and traps were rotated between the experimental huts. Experiments were conducted from 2200 h until 0600 the following morning. In addition to the experiments in which different traps were directly compared one against the other simultaneously, a series of experiments comparing the efficiency of one type of trap against the other on different nights was conducted. Performance of the Mbita trap in relation to the other traps was assessed using logistic regression. The Mbita trap caught $45.3 \pm 3\%$ of the mosquitoes released, compared with the net-light-trap combination which trapped $11.8 \pm 1.5\%$ of the released mosquitoes when the two traps were operated together. No difference in size of the mosquitoes caught by each trap was observed. When the Mbita trap was evaluated directly against the human-landing catch, it was observed that the Mbita trap yielded $22.4 \pm 1.3\%$ of the mosquitoes released compared with $52 \pm 2.9\%$ for the human landing catch. Again, no significant size differences in the mosquitoes captured by the two methods were observed. In the non-competitive experiments, the Mbita traps caught $71.1 \pm 3.9\%$ of the mosquitoes released, compared to $79.6 \pm$ for the human landing catch. In experiments designed to determine the rate of capture of the three trapping methods, it was reported that the human landing catch caught 50% of the released mosquitoes in 1.8 ± 0.5 h, compared with 4.2 ± 0.1 h for the Mbita trap, although the overall proportion captured during the whole experimental period was similar for both methods.

Mathenge et al. (2004) later compared the efficiency of the Mbita trap in relation to CDC light-traps hung adjacent to an un-treated, occupied mosquito net and a human landing-catch under field conditions. Each of the three sampling methods was allocated to one of three houses on a given night in a 3 × 3 randomized Latin square experimental design replicated 18 times. Sampling was carried out from 2000 to 0600 h the following morning. Over the 54 nights of the trial, the Mbita trap, human landing collection, and the CDC light trap-mosquito net system caught 592, 1215, and

2162 *Anopheles gambiae* s.l,. and 291, 390, and 742 *Anopheles funestus*, respectively. The Mbita trap caught approximately 50% and 70% of the number of *Anopheles gambiae* s.l. and *Anopheles funestus* respectively caught by human landing collection. In contrast to other studies on CDC light-trap-mosquito net combinations (see Chapter 9), the CDC light trap-mosquito net system in this study caught almost twice the number of *Anopheles gambiae* s.l. and *Anopheles funestus* caught by human landing collection. The authors found that the numbers of mosquitoes sampled by the three methods were more or less consistently proportional to each other enabling an estimate of the human landing catch on any given night from the catch in either a Mbita trap or in a CDC light trap-mosquito net system. The species composition of *Anopheles gambiae* s.l. did not vary according to the sampling method used. Sporozoite prevalence did not vary according to species. Similarly, parity of *Anopheles gambiae* s.l. was independent of the sampling method used and was estimated to be $78.0 \pm 1.2\%$ (170 of 218).

In Madagascar, where many of the *Anopheles* vectors are zoophilic and exophagic, Laganier et al. (2003) observed that the Mbita trap collected on average only 0.55 *Anopheles* mosquitoes per trap-night, compared with 10.30 per night for human landing catch. Additionally, Mbita traps sampled only three species of anopheline and five culicines, compared with 10 and 16 species respectively captured in human landing catches.

In northern Tanzania, Braimah et al. (2005) compared Mbita nets with CDC light-traps and observed that in almost all cases the CDC traps caught more mosquitoes than the Mbita traps. In comparison with human landing catches, the authors observed that Mbita traps caught only 20% and 30% respectively of the numbers of *Anopheles gambiae* s.l. and *Anopheles funestus* caught in human landing catches. Braimah et al. (2005) also found no consistent relationship in catch sizes between the Mbita and CDC traps and evidence of systematic differences among villages and also for species of mosquito, concluding that the Mbita traps were probably not sufficiently efficient to replace CDC light-traps or the human landing catch.

Moving landing catches

Most human landing catches are performed with a stationary bait, but occasionally collections are made by the person slowly walking through vegetation and periodically stopping to catch mosquitoes that have alighted on her/himself or her/his companions. Catches of this sort, termed roving catches, have been made in forests of Trinidad (Aitken et al. 1968*a,b*).

The same type of catch but called a walking–landing method was used in Tanzania (Tonn et al. 1973). The method consists of slowly walking through the bush for 1 min then sitting on the ground and collecting mosquitoes as they land for 5 min, or until no more land. The collector then moves on and repeats the process. To sample *Ochlerotatus cantator* in Maryland, USA a collector upon arrival at a site stood still for 1 min and counted the mosquitoes landing below the waist, then he moved around the area for 5 min aspirating host-seeking mosquitoes. Another 1-min count was then taken before proceeding to the next catching station (Weaver and Fashing, 1981). In Uganda, Henderson et al. (1972) employed mosquito catchers to walk at least 25 yd then stop and catch for 10 min, after which they proceeded for another 25 yd or more. In Kenya more adults of *Aedes aegypti* and other *Aedes* species were caught by catchers slowly walking through the bush than in stationary catches (Teesdale 1955, 1959). Jupp and Kemp (1998) found that the most rewarding human landing catch methods for sampling potential vectors of Wesselsbron virus in South Africa were as follows: an individual operator walked a few paces through the sedge to disturb the resting mosquitoes, halted, and then collected the mosquitoes that alighted on his body; alternatively 2 collectors stood close to one another and collected mosquitoes from themselves and one another. During the period from late afternoon until sunset as many as 182 mosquitoes could be collected per man hour using this method. Unfortunately, the risk to the collectors of becoming infected with Wesselsbron virus is high using this method, and in fact one of the authors did contract the virus during the study. In São Paulo State, Brazil, Marques et al. (1997) used both stationary landing catches and a moving catch method to sample *Aedes albopictus*. For the moving catch, collectors visited 48 different points for five minutes each over a total period of four hours in the morning and the afternoon, once per month. Of the 637 females of *Aedes albopictus* collected, 54 (8.4%) were obtained from stationary landing catches and 583 (91.6%) from moving landing catches.

Renshaw (1991) used a moving landing catch method in northern England to sample host-seeking *Ochlerotatus cantans* feeding in fields containing livestock. The success of this type of collecting method depends largely on the mosquito fauna and the manner in which they are attracted to baits. In England, for example, Service found that mosquitoes (including *Ochlerotatus cantans*) are not usually caught by slowly walking through vegetation during the day, although it may be harbouring very large mosquito populations. The explanation is that a moving bait does not stimulate the resting population in time for them to fly out, locate and settle on the collector. On the other hand, Renshaw (1991) believed that *Ochlerotatus cantans* followed her when she walked to a catch site, and this was one of the reasons for a high initial catch.

Trap huts

During the 1950s and 1960s various so-called trap huts were developed to catch mosquitoes that enter houses to feed. These traps are now rarely used, and will not be described here. Readers wishing to obtain more information are referred to previous editions of this book (Service 1976; 1993).

Exit and entry traps fitted to huts

Experimental huts fitted with various exit traps have been widely used since the 1950s to assess the impact of insecticides on endophilic mosquitoes. A description of the use of these huts is given in Chapter 16. The present account is concerned only with the use of exit and entry traps fitted to huts to catch mosquitoes attracted to human or animal occupants. The idea is that natural cracks and crevices, open doorways, windows and eave gaps allow mosquitoes to both enter and leave houses, but when one or more traps are inserted into the walls, windows or door of a house a sample of the mosquitoes leaving, or entering, are caught.

Muirhead-Thomson exit trap

The most widely used exit trap is probably that developed by Muirhead-Thomson (1947, 1948), or one of its modifications. The original trap consisted of a cage made from a 1-ft cube framework of wire covered with white mosquito netting. One side was inverted to form an entrance funnel narrowing to about a ¼-in diameter opening. The funnel was supported within the cage by string tied from its narrow end to the four corners of the trap (Figs 6.8*a*). One or two small cloth sleeves incorporated in the sides of the cage enabled aspirators to be inserted to remove the catch. The trap is usually placed in the middle of a piece of black cloth which is secured over a hut window. A large proportion of the mosquitoes which seek to leave the hut at dawn are attracted by light entering the window, and are consequently caught in window traps as they try to escape. It is sometimes possible, or even necessary, to partially block the eaves and various cracks and crevices (Mpofu et al. 1988) to allow mosquitoes to enter, but to discourage them from leaving by the eaves (Fig. 6.8*b,c*).

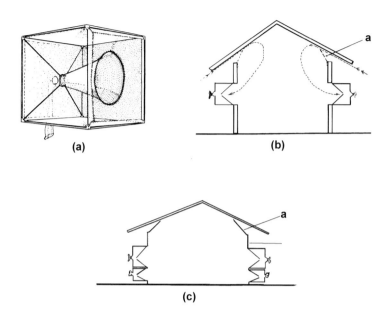

Fig. 6.8. Exit traps: (*a*) Muirhead-Thomson type; (*b*) and (*c*) exit traps fitted to houses showing a—partially blocked eaves (World Health Organization 1975*b*)

Various modifications have been made to this original design to take into account local building materials and variations in hut construction as well as the behaviour of the mosquitoes entering and leaving houses. Githeko (1992) found that in Kenya if the small circular opening was too close to the facing posterior wall, many *Anopheles* failed to fly through and enter the trap, but instead turned round and flew out. He found the best distance was 10.2 cm. Instead of using cones, traps having horizontal entry slits can be used. For example, WHO (1975*b*) recommended a rectangular prism-shaped trap about 1–2 m long, 35 cm deep and 40–50 cm high, with a long horizontal slit (Fig. 6.9). One or more such traps can be inserted into house walls, and Service (1993) considered this design to be an improvement over those with circular openings. In experimental hut studies in Belize, Grieco et al. (2000) removed the entry funnel from their exit and entry traps after initial trials, as they felt it impeded mosquito entry into the trap.

In Nigeria a 1-ft cube cloth-covered window trap was used as both an entry and exit trap when fitted to ordinary village huts (Fig. 6.8*a*). Because normal window openings in the Kaduna area were too small to accommodate the traps larger windows, or well fitting doors, were constructed.

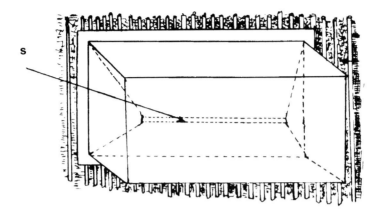

Fig. 6.9. Wall-type trap showing entry S—entry slit (after World Health Organization 1975*b*)

They were provided with a 1-ft square flap-like door hinged along the top edge, which when not bolted in place to close the opening was pushed upward to rest on the top of the exit trap. This afforded some protection from rain, but the main purpose of the hinged door was to provide the hut owner with a means of closing the exit trap space when a trap was not fitted, thus giving security to his hut (Service 1963). Any spaces between the cage and opening in the doors or window were filled with cotton wool, foam rubber or leaves. When the cages were collected a plug of cotton wool was inserted into the narrow opening of the entrance funnel. In some areas most, if not all, huts have no suitable door over their entrances in which exit traps can be inserted. To overcome this, mosquito netting exit traps are sewn into a large piece of dark coloured cloth which is nailed in position over the outside of the doorway (Pant et al. 1969). A disadvantage, however, is that this prevents the occupants having free access to their huts; they have to retire early and stay inside. Although nobody should be entering or leaving any hut with an exit trap after it has been placed in position, there is invariably some degree of movement. This can, and must often, be tolerated if close fitting doors are fitted and these are shut every time a person enters or leaves.

In Zimbabwe lobster-type (Muirhead-Thomson) traps were used to monitor the exodus of mosquitoes from houses (Mpofu et al. 1988). But before they could be fitted, doors had to be removed and replaced by an adjustable retractable door frame, having black calico sheeting pinned to it to block the doorway. A plywood insert with a hole was fixed into each frame to allow a window-type exit trap to be fitted. All eaves and crevices were effectively plugged with cotton waste to maximise the catch of *Anopheles arabiensis* in the exit traps.

Up to four window traps, one on each wall, have occasionally been used (Hadjinicolau in Muirhead-Thomson 1968). In Malaysia Wharton (1951*a*) constructed wooden huts with thatched roofs which were raised a foot or two from the ground. They resembled typical village huts. The ceilings and inside walls were lined with smooth brown opaque paper to prevent light entering the huts and also to make the collection of mosquitoes easier. In Malaysia *Anopheles maculatus* is not so markedly orientated to feeding on man and entering houses as is *Anopheles gambiae* in Africa, and the natural cracks and openings in these experimental huts were insufficient for the entry of adults into the huts. Consequently, special louvre openings were constructed in the two opposite longer walls of the huts for the access of *Anopheles maculatus*. The louvres consisted of a series of black ½-in thick wooden slats 2.5 ft long, fitted one above the other at an angle of 30° with the vertical so as to leave eighteen 1¾-in wide longitudinal entrance gaps. Mosquitoes leaving these experimental huts were caught in a Muirhead-Thomson-type mosquito netting exit trap fitted to a window (Reid and Wharton 1956; Wharton 1951*a,b,c*). To check on the proportion of mosquitoes that escaped via the louvres and not through the window trap, Wharton (1951*a,b*) placed an exit trap over about a quarter of each louvre surface. In unsprayed huts he estimated that about 44% of the *Anopheles maculatus* that entered the hut left via the louvres.

In comparing the relative attractiveness of different hosts to *Anopheles* (Wharton 1951*b*) and culicines (Wharton 1951*c*) the routine procedure was to bait one hut with a man and another some 30 ft away with a calf, goats or dogs. In the early evening, prior to introducing the bait into the huts and inserting the window traps, blinds which covered the louvres during the day were rolled up. Before sunrise these blinds were pulled down to prevent mosquitoes escaping, and the bait taken out, then about 1–1.5 h after sunrise the exit cages, which contained a representative sample of the mosquitoes leaving the huts were removed. Wharton (1951*b*) also attempted to determine the times that mosquitoes left the huts by counting the numbers in the exit traps every hour at around dawn. This, however, was not very successful because it was difficult to count mosquitoes in the traps due to their movements.

Specially designed huts built in Tanzania and baited with two men and provided with Muirhead-Thomson-type window traps were used to study the exodus of *Anopheles gambiae* and *Anopheles funestus* (Gillies 1954).

Mosquitoes entered the huts through a line of 2½-in high slit shutters placed in all hut walls just below the eaves. They were closed before dawn so that the only available exits for mosquitoes were the window traps. However, when the eave shutters were fitted with mosquito netting cages it was found that of the small numbers (4–8%) that left the hut as blood-feds, a high proportion escaped through the eaves before the shutters were closed just before dawn.

On Bioko Island, Equatorial Guinea, Cano et al. (2004) reported collecting relatively high numbers of blood-engorged females of *Anopheles funestus* and *Anopheles gambiae* s.s. in window exit-traps, indicating a high degree of exophily, although the authors acknowledged that further investigations would be required to confirm this observation.

In certain areas much larger numbers of *Anopheles* are sometimes caught when huts are baited with large mammals instead of man. In Java, for example, insufficient adults were attracted to experimental huts baited with men for an assessment of the effect of insecticides on the population of *Anopheles aconitus*. More attractive hosts, such as bovids, were needed. Existing thatched-roofed cattle sheds were therefore completely surrounded with bamboo canes and walls of finely woven bamboo. These incorporated horizontal 2-cm wide slits to allow mosquitoes to enter. Mosquitoes leaving the cattle sheds were caught in two window traps placed in the walls (Soerono et al. 1965). When the sheds contained Zebu cattle a mean of 35 *Anopheles aconitus* was collected from the traps after 5 days, but when water buffalo were kept in the sheds 586 mosquitoes were caught in the traps. In Indonesia Barodji et al. (1986) compared catches from exit traps fitted to cattle sheds with collections of indoor resting adults. Although more *Anopheles aconitus* were caught by the latter method, exit traps proved useful in catching blood-fed females and thus confirmed that the vector was basically exophilic. In Iran, Zaim et al. (1995) used exit and entry traps to sample *Anopheles culicifacies* from houses and animal shelters. Exit traps caught four times as many *Anopheles culicifacies* than entry traps. Thirty-three percent and 60% of the total exit trap catch were caught between 1800 and 2100 h and 0300 and 0600 h respectively, while approximately 80% of the total caught by entry traps was caught between 0300 and 0600 h.

In Nigeria cloth exit traps were fitted to horse and cow stables and also to chicken huts. Although few mosquitoes (28) were caught in traps placed in five chicken huts, larger catches of mosquitoes belonging to about 33 species entered traps fitted to both a cow (952) and two horse (3943) stables (Service 1964). Precipitin tests on 53 blood-fed mosquitoes caught in the traps showed that only five females had fed on animals other than those in the stables.

Rachou exit traps

Working in El Salvador, Rachou et al. (1965) considered that any kind of restrictive entrance to an exit trap, whether a slit-like baffle or lobster-type funnel, probably hindered the entry of mosquitoes. Consequently, an exit cage without any kind of one-way entrance was used. At short intervals a partition was slid across the window cage to enclose the catch and enable the cage to be removed and another inserted. It was thought that if the cage was frequently changed very few mosquitoes would escape by flying back into the hut. Although such an exit trap providing unimpeded entry *may* catch more mosquitoes than one with a restrictive entrance, the necessity of regularly replacing the cages and removing the catch involves considerable manpower.

Collapsible window trap

Because of their bulkiness window traps are diifcult to transport in large numbers to the field. To try to overcome these difficulties Shidrawi (1965) described a collapsible window trap (Fig. 6.10).

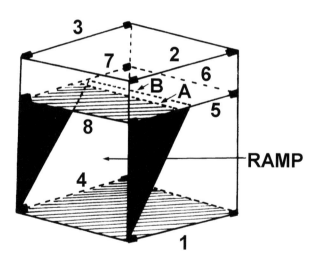

Fig. 6.10. Collapsible type of exit trap, 1–8—rods, A and B—rods forming a 2-cm gap (Shidrawi 1965)

This consists of two lengths of 3-mm thick wire rods bent and soldered to form two rectangular frames (55 × 40 cm). These are joined together at the corners by four lengths (Nos. 1–4) of 40-cm wire which are fixed into bushes welded or screwed on to the rectangular frames. Four similar rods (Nos. 5–8) are fixed between the framework 15 cm from the top to divide the trap into a small upper (40 × 15 cm) and a large lower (40 × 40 cm) section. The 'floor' of the upper section is divided about 24 cm from its front by two horizontal rods (A, B) having about a 2-cm gap between them. A mosquito netting and cloth cage made to fit the shape of the trap is suspended within the frame-work by tapes. A ramp of mosquito netting is made to stretch from the bottom edge of the cage to the posterior wire rod (B). A rectangular piece of dark cloth covers the floor of the upper section from the front to the first rod (A). Mosquitoes entering the lower half of the trap are guided up the ramp through the narrow slit between rods A and B into the upper section and lower section behind the ramp. Triangular pieces of dark cloth are used for the sides of the entrance. Shidrawi (1965) claimed that these traps could be dismantled and reconstructed within 10 min, and that 10 such traps when collapsed occupy less space than a conventional cloth window trap.

Veranda traps of Smith

Although window traps are very useful in catching mosquitoes which are attracted by light entering windows at dawn, they trap only a relatively small proportion of such mosquitoes, and moreover not all species are attracted by early morning sunlight. Also, when huts are sprayed with certain insecticides a large proportion of the mosquitoes that would normally be caught in window traps may be stimulated to leave before sunrise, and many of these escape via the eaves. To sample such mosquitoes Smith (1965) developed a veranda trap. This is made by extending the thatch roof beyond the four walls of a square or rectangular hut to form a roof over a veranda, which is enclosed on two sides by mesh screening. Mosquitoes enter the hut through the eaves of the two walls having unscreened verandas. These huts are discussed and illustrated in Chap. 16.

In preliminary trials in Tanzania 51% of the *Anopheles gambiae* population left a man-occupied hut each night and of these 85% left via the windows and 15% via the eaves. In marked contrast 90% of *Mansonia uniformis* left each night, of which 69% left via the eaves and 31% through the windows. In huts with iron roofs the exodus of *Anopheles gambiae* increased to 63% and most escaped through the eaves, but there was no significant increase in the percentage of *Mansonia uniformis* that left (Smith et al. 1967). Although Smith fitted the veranda cages to specially constructed huts, normal village huts can sometimes be adapted for fitting these, or simpler, veranda cages.

In Burkina Faso Coz et al. (1965) used both veranda-type exit traps and window traps with horizontal entrance slits to catch mosquitoes leaving huts sprayed with insecticides. They also placed boxes having Bates-type entrance slits in the mud walls of huts, so that mosquitoes could enter but not leave. These boxes could be closed by lowering a hinged lid over their openings. In an evaluation of permethrin and fenitrothion residual spraying in houses in Kenya Taylor et al. (1981) used veranda-type exit traps (WHO 1975a), but caught relatively few mosquitoes in them.

Coz (1971) in comparing the efficiency of window and veranda traps concluded that although the former were more easily managed and could be changed several times a day they tended to delay mosquitoes leaving huts when compared with catches in veranda traps. With sprayed huts this would mean increased insecticidal contact and higher mortalities.

In Brazil Roberts et al. (1987) fitted window-type traps to window spaces and to the gables of houses to sample *Anopheles darlingi* both leaving and entering houses. In addition two modified veranda traps were constructed. One trap enclosed a relatively large area of wall (1.8 × 2.95 m) which included an exit trap in one window, whereas the other veranda trap enclosed just the corner of a house (1.8 × 0.2 × 0.2 × 1.8 m) at the meeting of two walls. These traps were quite successful in catching *Anopheles darlingi*, which were collected at 2-h intervals so as to measure times of entry and exodus. The numbers of *Anopheles darlingi* caught in entry traps fitted to windows were greater than in similar traps fitted to the gables.

Curtain traps

Because many houses in Latin America have walls made of loose-fitting bamboo or matting, the collection of mosquitoes in exit traps fitted to windows or doors, or even by veranda traps, is inefficient. To overcome these problems Elliott (1972), devised a technique by which a house was completely encircled from the lower edge of the roof to the ground with a nylon mosquito netting curtain (Fig. 6.11). The curtain was raised for the first 30 min of each hour to allow ingress of mosquitoes and then mosquitoes resting on the inside and outside surfaces of the netting collected. Interpretation of the data was complicated because the numbers caught leaving were 2–3 times greater than those caught entering. Elliott (1972) concluded that the net curtain trap was better at trapping exiting rather than entering mosquitoes. Another problem was that when the curtain was raised it was impossible to evaluate mosquito movements because they were able to enter and leave freely.

Fig. 6.11. Colombian curtain in place and enclosing a house in Kenya (photograph reproduced with permission of M. W. Service)

In Mexico Bown et al. (1985, 1986) used Elliott's technique, and a modification of it, to study the behaviour and mortality of *Anopheles albimanus*. Later only the modified approach was used, as follows. The curtain was made by sewing several pieces of mosquito netting to form a rectangle, 3 × 40 m, which was attached to the eaves of the roof and dropped down to reach the ground. The bottom edges were bordered with calico to afford protection against wear. The ends of the wrap-round curtain overlapped considerably to allow house occupiers to enter without having to raise the curtain. Firstly all mosquitoes were removed from the house in the late afternoon, then again after dusk (1800 h), and then once every hour any mosquitoes resting on the outside of the curtain were collected, and unfed ones released into the house. This process of collection and release continued hourly until 0600 h. In addition mosquitoes resting on the inside surface of the curtains were collected hourly at half past each hour and sorted into unfed and blood-fed females. Alternatively mosquitoes collected at human or animal baits were sometimes released into the house at the start of the night and since they were released together, the time they stayed in the house before being collected on the inside of the curtain was determined. Such collections gave useful information on mosquitoes entering and leaving houses. Dead and moribund mosquitoes were also collected from inside the house and in the space between the wall and the curtain. To facilitate this a 1-m wide strip of white sheeting was placed on the floor inside the house around the walls, and in the ground space between the curtain and outside of the walls.

Bown et al. (1987) using this modified method collected *Anopheles albimanus* from the inside and outside of curtains surrounding unsprayed and sprayed houses (bendiocarb and deltamethrin) on alternate hours from 1800–0600 h. For one hour all mosquitoes caught on the outside of the curtain were collected, but only the unfed females were released into the house. During the subsequent hour mosquitoes resting inside were collected and classified according to their physiological condition. Later Bown et al. (1991) used basically the same method; that is, with the curtain lowered to the ground, the inside of a house was searched for 45 min at 1715 h to remove all live and dead mosquitoes. From 1830–2100 h four people collected a minimum of 150 unfed female *Anopheles albimanus* from other houses in the village, and released them in the curtained house which had five to seven people in it. At hourly intervals from 2200–0600 h mosquitoes resting between the interior of the curtain and outside house wall were recaptured and their gonotrophic conditions recorded. Collections of dead and moribund mosquitoes were removed from the floor sheets placed between the walls, and at 0600 h a final collection of live indoor resting mosquitoes as well as dead ones was made.

Villarreal et al. (1995) used the net of Bown et al. (1985, 1986) to assess the movements of mosquitoes into and out of a house in Mexico. White sheets were placed on the floor alongside the net both inside and outside of the house to facilitate location of dead or moribund mosquitoes after spraying the house with lambda-cyhalothrin.

Curtain traps were also used by Casas et al. (1998) to study the effects of house spraying with DDT or bendiocarb on the behaviour of *Anopheles pseudopunctipennis* in southern Mexico. Two technicians collected mosquitoes resting on the outside of the net between 1800 and 2045 h and marked them with fluorescent powder before releasing them inside the house enclosed by the net, which contained two volunteers. Marked mosquitoes resting on the interior surface of the net were assumed to be exiting the house and the numbers of mosquitoes exiting the house from 2100 to 2200 h and from 2200 to 2400 h were significantly greater in huts treated with insecticide, compared with controls.

Muirhead-Thomson type entry traps

Window traps have also occasionally been used as entry traps to sample mosquitoes entering a hut to feed. The much smaller numbers usually caught in entry, as compared to exit, traps clearly show that they are not very efficient in sampling mosquitoes entering huts.

In Kenya large window traps, more than 1-ft cube and having an inverted funnel entrance of mosquito netting, caught over 14 000 mosquitoes belonging to 30 species when they functioned as exit traps, but when used as entry traps only 289 mosquitoes belonging to six species were caught (van Someren et al. 1958). Similarly Teesdale (1955) caught considerably fewer mosquitoes in window traps used as entry rather than exit traps, but in Nigeria entry traps fitted to village huts caught substantial numbers of *Anopheles gambiae*, *Anopheles funestus* and even *Anopheles nili* (Service 1963). They showed that about 24, 16 and 52% respectively of these three species entered huts as blood-fed individuals. They also demonstrated that there was some movement of half-gravid and gravid females of *Anopheles gambiae* and *Anopheles funestus* into huts.

In Korea entry window traps fitted to pig-baited portable sheds have caught large numbers of *Culex tritaeniorhynchus* (Ree et al. 1969). A common procedure in Japan is to place a bed-net, either with one side partially rolled up or with a section cut out, over the entrance of stables and cattle sheds to catch mosquitoes entering these animal quarters. By removing the catch at hourly intervals the biting times can be studied (Katô and Toriumi 1950).

Man-biting rates

In epidemiological studies man-biting rates, based on 12- or 24-h catches during different seasons, are often used to obtain a theoretical number of bites a person could receive in a year, and to calculate the inoculation risk with malarial sporozoites, microfilariae or arboviruses. A common procedure involves the collection of mosquitoes throughout the night that are attracted to a human bait sitting outdoors in a village compound. The numbers caught outside may have little bearing on the average number of bites a person normally sleeping inside might receive. Even if catches are made indoors there may still be differences between the numbers caught and those normally biting a hut occupant. In India, in trying to get realistic estimates of the biting density of *Culex quinquefasciatus*, Gubler and Bhattacharya (1974) employed two people working for 2-h shifts both inside and outside houses throughout most of the night to catch mosquitoes from a person who slept, acted and dressed normally. They appreciated that estimating biting densities was difficult, but calculated that in a Kolkata suburb a person would receive over 115 000 bites a year from *Culex quinquefasciatus*, and moreover that this was probably an underestimate!

From their estimated biting densities and from filarial infection rates they calculated the number of infective bites a person receives over a year, and also the numbers of larvae of *Wuchereria bancrofti* deposited on such a person.

In Liberia Kuhlow and Zielke (1978) had two people sitting during the night in a partitioned part of a village hut that was otherwise used normally by the occupants. From the numbers of vectors caught per man-night in different months they estimated by simple proportions: (i) the numbers of bites a person would receive in a year; (ii) the number of bites with mosquitoes infective with *Wuchereria bancrofti*; and (iii) the number of infective larvae deposited on a person during a year. They were careful to use these values as comparative indices and not as absolute estimates of biting and transmission potential.

Nathan (1981) studied the intensity of bancroftian filariasis transmission in Trinidad by weekly collections over a year of indoor biting (1900–0600 h) *Culex quinquefasciatus*. Because very few people were active outdoors during darkness, catches were made only indoors. Houses were selected randomly without replacement from a list of 200 random numbers. If a house was unoccupied, or could not be used for some other reason, then the nearest house was used as a catching station. Residents of houses were asked to carry out their normal domestic duties while the collector sat on the floor and caught mosquitoes from his exposed legs and feet. Torches were used for only about 2-min durations to minimise any affect torchlight might have on the normal behaviour of the mosquitoes. To prevent ovarian development before the mosquitoes could be dissected they were placed in glass jars lined with damp plaster of Paris and kept in an ice chest. Nathan (1981) estimated that a person was exposed to 17 948 bites a year from *Culex quinquefasciatus*, including 14 infective bites. This transmission potential is very low when compared with estimates of 1850 infective bites for Kolkata (Gubler and Bhattacharya 1974) and 1106 for Pondicherry (Rajagopalan et al. 1977), although in Tanzania (White 1971) and in Kenya (Wijers and Kiilu 1977) estimates of numbers of infective bites a person received per year were only 23 and 46, respectively.

Effect of environmental conditions

Wind and rain, though not necessarily light drizzle, usually drastically reduce the numbers of mosquitoes caught biting, and catches have sometimes to be abandoned because of bad weather. It is important to know whether adverse weather conditions have prevailed during any part of a bait catch, but have not been reported, or perhaps not even noticed, by the collectors, so that the entire catch, or perhaps only part of it, can be excluded from the results. Snow (1980) recorded that above a wind speed of 1.2 m/s biting by *Anopheles melas* and *Culex thalassius* in The Gambia virtually ceased.

In South Africa Sharp (1983) investigated the effect of environmental factors, such as temperature, wind speed and rain on the biting cycle of *Anopheles merus*. Not surprisingly both an increase in wind speed or rain decreased, or sometimes stopped, biting activity. Subra (1972) gives a useful account of how weather conditions affect outdoor biting by *Culex quinque-fasciatus*, while Service (1980) briefly discusses the effect of wind on suppressing biting behaviour, pointing out that whereas winds of about 8 km/h or less usually prevent host-seeking activities, arctic species seem to continue biting in such winds. In Canada for example, Haufe (1966), reported that only speeds of about 29 km/h or more deterred mosquito flight. Although wind and low temperature can inhibit biting, it must be remembered that some temperate and subarctic species may continue to fly in winds of 2–8 m/s and temperatures as low as around 4°C (Jaenson 1988). Temperature changes may also cause shifts in peak biting times. For example, in Pakistan Reisen and Aslamkhan (1978) found that *Anopheles* bit mainly during the evening in the cool season, but later at night in the warm season. This observation was confirmed by Bhatt and Kohli (1996) for *Anopheles culiciaficies, Anopheles varuna, Anopheles aconitus*, and *Anopheles tessellatus* in Gujarat, India. In East Africa *Anopheles merus* bites mainly after midnight, whereas in South Africa Sharp (1983) found that females may bite earlier in the night when minimum temperatures drop to 16°C.

An analysis of mosquitoes, and other haematophagous insects, biting man in Panamanian forests demonstrated that biting activity is largely dependent on temperature and vapour pressure, and that most activity is concentrated in quite a narrow range of these two environmental factors (Read et al. 1978). For example, the greatest numbers of *Haemagogus lucifer* were caught biting both at ground level and in the canopy when temperatures were 26.5–28.1°C, and the vapour pressure (millibars) was 31.4–32.3 (forest floor) and 29.4–30.3 (canopy). In fact, during 1973 and 1974 86–96% of both *Haemagogus lucifer* and *Haemagogus equinus* were caught biting when temperatures were in excess of 24.7°C. Both species are day-biters, and during the day temperatures are usually higher and relative humidities lower than at night, but it must be remembered that both are influenced by rainfall and wind. For instance, heavy rain will tend to lower ground temperature for hours or even days, which in turn may reduce evaporation and result in air near the ground or even in the canopy, being near saturation. Later Read and Adames (1980) investigated the relationship of air temperature, dew point temperature and evaporation on numbers

(>435 000) of *Mansonia dyari* biting human baits and derived the following empirical regression equation:

$$Y = -522 + 1035\,X \qquad (6.8)$$

where Y = the number of *Mansonia dyari* caught biting man/24 h and X = evaporation (mm)/48 h. The minimum value limiting the applicability of this equation is 0.5 mm/48 h of evaporation. It was calculated that 66% of all variations in the numbers biting is accounted for by the evaporation rate.

Read and Adames (1980) believed that the above equation could be used to predict times when biting densities should be high, but because of the complexity of factors affecting biting activity they cautioned it was best to regard predictions in terms of probabilities.

Charlwood and Galgal (1985) calculated polynomial regressions of the percentage of the total catch of *Armigeres milnensis* caught every 5 min in human landing catches against both light intensity and time. The multiple r^2 for a polynomial regression of degree 2 was 0.383 for percentage biting against light (log lux), and 0.510 for percentage biting against time, thus showing that time was a better predictor of biting activity than light intensity.

Light intensity is often the most important environmental factor influencing mosquito activity. *Anopheles neivai* biting activity, for example, occurs between 0.1 and 10 lux, peaking at 3.0 lux (Murillo et al. 1988).

Moonlight

Moonlight has a biological effect on the behaviour of many animals, including mosquitoes. Bowden (1973*a*) has shown that at Kampala, Uganda, light from the full moon at zenith (0.2 lux) is about the same as that experienced 30 min after sunset (about civil twilight) on a clear moonless evening.

At full moon the decrease in illumination in the hour following sunset is much less than on a moonless night. On nights immediately following a full moon illumination may decrease to starlight (0.009 lux) before moonrise, but thereafter increase sharply. Muirhead-Thomson (1940) gives some interesting light readings of moonlight measured in India.

Many nocturnal mosquitoes are more numerous in bait catches on nights of full moon (Bidlingmayer 1964; Charlwood et al. 1986*b*; Pandian and Chandrashekaran 1980), it being suggested that this is due to moonlight enhancing the mosquito's ability to locate hosts, and also oviposition sites (Allan et al. 1987; Charlwood et al. 1988). In Senegal Hervy et al. (1986)

found that *Aedes taylori* increased in numbers at human bait during moonlit nights, and in China Wang and Chang (1957) found that biting and flight activities of *Anopheles sinensis* were greater during moonlit nights. In the Republic of Korea, Strickman et al. (2000) reported that hourly landing collections of *Anopheles sinensis* in 1997 were negatively correlated to the phase of the moon present during the hour of collection. Controlling for effects of wind (zero order partial correlation = –0.1133, d.f. = 202, $P = 0.107$ and temperature (zero order partial correlation = 0.1161, d.f. = 202, $P = 0.001$), between phase of the moon (or absence if it had not yet risen or if it had already set) and hourly landing collections was -0.2245 (d.f. = 200, $P = 0.001$). This result supported the apparent negative correlation between moon phase and total nightly landing collections of *Anopheles sinensis*.

Guimarães et al. (2000c) observed no influence of the lunar cycle on overall populations of nocturnal species collected by human landing catches in Paraná State, Brazil. However, individual species did appear to show a preference for biting on nights with a new moon. *Anopheles fluminensis* showed the strongest preference for biting on dark nights, while *Anopheles oswaldoi* and *Anopheles mediopunctatus* mean captures did not change throughout the lunar cycle. *Anopheles cruzii* was more frequently collected during the new moon but was also captured in significant numbers during the first and last quarter moons. *Ochlerotatus scapularis* was the only species of *Ochlerotatus* that was collected more frequently during the full moon.

It might be expected that the illumination from the moon on nights just before and after full moon might change the shape of the crepuscular biting profile of mosquitoes on these nights. However, Haddow (1964) failed to find that a full moon affected the timing of the crepuscular biting peaks of mosquitoes inhabiting the forest canopy near the equator. Furthermore, although Corbet (1964) found that the numbers of mosquitoes caught in light-traps above the forest canopy may be *less* on nights with a full moon, there was no evidence that their times of appearance significantly differed from those recorded on nights with little moonlight. In human landing collections in India the biting cycle was exactly the same whether catches were performed in houses or out of doors (Pandian and Chandrashekaran 1980). Similarly Krafsur (1977) reported that endophagic and exophagic *Anopheles* in Ethiopia had similar biting cycles. McClelland (1960) found that in coastal Kenya the indoor biting cycle of *Aedes aegypti* exhibited no pronounced peaks, although in Uganda an outdoor biting population showed pronounced peaks in biting (McClelland 1959).

In contrast to many of the foregoing examples van Someren and Furlong (1964) showed that on Pate Island, just off the Kenyan coast, moonlight had a pronounced effect on the biting times of *Aedes pembaensis* and *Aedes mombasaensis*. With *Aedes pembaensis* biting appeared to be enhanced by moonlight and inhibited by darkness and so biting was most intense in the early evening at new moon and during the first quarter, times when the moon rose before sunset. When the moon rose after sunset, such as at full moon and in the last quarter, the early evening wave of biting was depressed, but the early morning peak was bigger. With *Ochlerotatus fryeri* moonlight appeared to modify times of biting at spring tides, whereas at neap tides the phases of the moon did not appreciably alter the biting pattern. Also more females of both *Aedes pembaensis* and *Aedes mombasaensis* were collected at bait during spring tides. Gillies and Furlong (1964) found that there was a tendency for a higher proportion of *Anopheles parensis* to bite later in the night when the moon rose late, such as during the last quarter. At new moon there appeared to be a slight increase in biting just after 1800 h, and just before 0600 h. In other words there was a slight, but significant, tendency for increased biting activity during periods of moonlight, compared with hours of greater darkness. In Papua New Guinea *Anopheles farauti* bites mainly in the middle and latter parts of the night during full moon periods, but on moonless nights most activity is in the early evenings (Charlwood et al., 1986*b*). In Colombia there were earlier peaks of biting by *Anopheles punctimacula* and *Anopheles nuneztovari* during light moon phases (crescent and full), whereas by contrast *Anopheles darlingi* showed the same degree of biting during the new and crescent phases (Elliott 1972). Davies (1975) found that at new moon in Trinidad there were biting peaks of *Culex portesi* and *Culex taeniopus* on mice during twilight of the evenings and mornings, although the dawn peak was not very pronounced with *Culex taeniopus*. At full moon, however, the evening and dawn peaks were replaced with increased activity at moon rise and during the middle of the night.

In Bangladesh the most concentrated biting by *Anopheles dirus* with respect to time (i.e. sharpest peak) occurred during the first quarter of the moon, when about half of the moon was above the horizon as early as sunset (approx. 1842 h) (Rosenberg and Maheswary 1982). As the quarters advanced moonrise was progressively later, until during the last quarter moonrise was near midnight (Fig. 6.12*b*) and biting peaked at midnight (Fig. 6.12*a*). During new moon when there was no moonlight, biting peaked at 2200 h and remained high until 0145 h, and this activity was believed to represent the intrinsic biting pattern.

In Borneo, Malaysia, Colless (1957) found that in his catches (2000–2300 h) more *Anopheles balabacensis* were caught out of doors between half and full moons than during other moon phases.

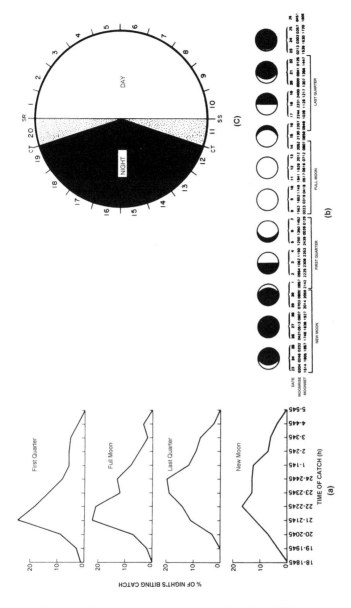

Fig. 6.12. (a) Influence of moon phase on *Anopheles dirus* biting man outdoors (Rosenberg and Maheswary 1982); (b) phases of moon and times of moonrise and moonset at 20°N latitude, 23 June to 26 July 1976. Time of sunset was 1943 hr on 23 June and 1840 hr on 26 July (Rosenberg and Maheswary 1982); (c) division of diel cycle into 20 periods, based upon times of sunrise (SR), sunset (SS), and civil twilight (CT). (Lillie et al. 1987)

During the first half of the night in Uganda biting catches of *Anopheles implexus* were similar at all phases of the moon, but in the latter half of the night, greater numbers were caught at full moon and activity was prolonged (McCrae et al. 1976). Provost (1958) found that swarming in *Psorophora confinnis* was extended for at least an extra hour at full moon because illumination had not dropped to the critical level (0.02 lux). In studying the diel and seasonal flying activities of *Culicoides* species in Florida, USA Lillie et al. (1987) divided the 24-h day into 20 periods based on times of sunset, sunrise, and twilight so that catches at different times of the year could be compared when there were changes in the duration of photophase and scotophase. Photophase was represented by 10 equal periods (which ranged from 62 to 84 min depending on the time of year). Period 1 started at sunrise, while period 10 ended at sunset (Fig. 6.12*c*). Evening twilight comprising the time from sunset to the end of civil twilight was period 11 (52–60 min). Scotophase comprised 8 equal periods (60–90 min) starting with period 12 at the end of twilight and ending with period 19 at the beginning of morning twilight. The duration of these periods will of course not just vary seasonally but in different parts of the world.

Further information concerning moonlight and its possible effects on insect behaviour is to be found in the publications of Brown and Taylor (1971), Bowden (1973*a,b*), Bowden and Church (1973), Beck (1968) and Bidlingmayer (1967, 1985), see also the effect of moonlight on light-trap catches in Chapter 9.

Twilight and crep units

Frequently the numbers of mosquitoes caught biting each hour, or more usually the transformed counts obtained from a series of similar catches, are expressed as percentages of the total 24-h catch to give diel biting profiles of the species (Fig. 6.13). Sometimes the numbers caught in smaller time intervals, such as every 5, 10 or 15 min, or even each minute (Haddow 1956, 1964) are recorded (Fig. 6.14).

The beginning or cessation of biting in many species appears to be initiated by changes in illumination, and biting profiles are often correlated with times of sunset and sunrise. Such correlations may be difficult because these times alter according to both locality and season. The importance of adjusting catch times in relation to exact times of sunset and sunrise was stressed by Lumsden (1952, 1957) and Haddow (1954).

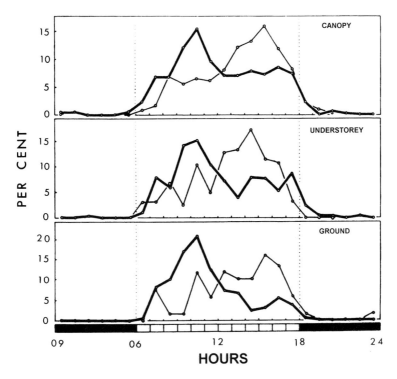

Fig. 6.13. Biting cycle of *Aedes apicoargenteus* by hour and level in Zika forest (thick line) and Bwamba (thin line), Uganda (after Haddow 1961*a*)

Even at the equator a variation of 31 min can occur between times of sunset at different times of the year. Much greater variations in sunset and sunrise times are encountered further from the equator. During a series of landing catches in May to September in England the difference between times of sunset and sunrise was about 2.5 h. It was shown that in August the peak biting times of *Ochlerotatus detritus* and *Coquillettidia richiardii* were an hour earlier and later than during July (Service 1969*a*), corresponding to the new times respectively of sunset and sunrise.

Another difficulty is that away from the equator day length varies and consequently it is difficult to adjust the catch clock to times of both sunset and sunrise. One method, at least near the equator, is to use sun times, that is to start each continuous 12-, 18- or 24-h catch at the time of sunset, which in fact will entail catches beginning at different clock times (Lumsden 1952).

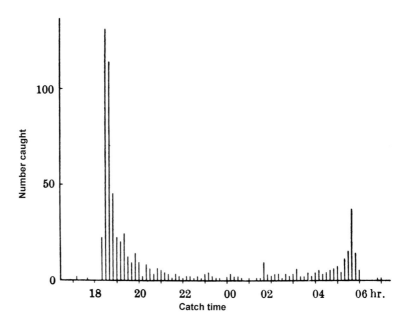

Fig. 6.14. Females of *Coquillettidia fuscopennata* caught in Uganda in a series of 24-h human bait catches expressed as numbers caught each 10-min period. Sunset at 1800 h (from Haddow 1956)

Nielsen (1963) pointed out that illumination differs in two important respects from environmental measurements, such as temperature, humidity and wind speed, in that the daily variations are enormous. Secondly, under standard conditions, that is excluding the effects of cloud cover, haze, etc., the level of illumination at any locality at any time can be precisely calculated. This is because illumination depends only on the sun's altitude which is completely predictable. The sun's altitude is measured as the angle (As) between a line from the centre of the sun to the observer and from the observer to the horizon. The angle of elevation depends on latitude (f), the date and hour, and is calculated as follows:

$$\sin(As) = (\cos t \times \cos f \times \cos d) + (\sin f \times \sin d) \qquad (6.9)$$

where t = the hour angle and d = the declination of the sun. Values for each day of the year and for each hour of local time are found in navigational tables (e.g. The American Ephemeris, The Nautical Almanac, The Air Almanac). Local time is derived from standard time (e.g. Middle European time, Eastern Standard Time) by adding 4 min for each degree of longitude west on which standard time is calculated. Similarly 4 min are subtracted for each degree east of the longitude on which standard time is based.

Nielsen (1963) defined civil twilight as the period when the sun passes from 0°.50′ to 6°00′ below the horizon, and on this basis nautical and astronomical twilight would end when the sun was 12° and 18° respectively below the horizon. The Smithsonian Physical Tables also define sunset (astronomical) as lasting until the sun is about 18° below the horizon, but some authorities (Nautical Almanac) give twilight not as a duration but as the moment when the sun is 6°, 12° or 18° below the horizon. Nielsen's definition is retained here as it is convenient to talk about a twilight period. Using it in this sense, the duration of twilight varies considerably according to locality and season. At the equator the sun sets at 1800 h and twilight lasts only 20–23 min, but its duration increases progressively further from the equator. At latitudes greater than 50° twilight lasts until midnight, above a latitude of 61° it lasts all night at midsummer and at latitudes above 67° there is no twilight as the sun never sets. It is important to realise that the same changes in light intensity are experienced during twilight irrespective of locality. At the moment of sunset light intensity is 395 lux but by the end of civil twilight it has been reduced to 3.55 lux. The only difference is that nearer the equator the changes in illumination are faster. It is therefore clear that it is not meaningful to compare biological phenomena, such as biting cycles, with clock times in localities having widely different latitudes. To facilitate comparisons at different latitudes and also in different seasons, Nielsen (1961) introduced the crep unit, which is defined as the interval between sunset and the end of civil twilight, i.e. the period when light is decreasing to the level of 3.55 lux. Crep values are calculated as follows:

$$\text{crep value} = \frac{\text{time of day} - \text{time of sunset}}{\text{duration of civil twilight}} \qquad (6.10)$$

Or

$$\frac{\text{time of sunrise} - \text{time of day}}{\text{duration of civil twilight}} \qquad (6.11)$$

Positive crep values refer to periods when the sun is below the horizon and negative values when it is above the horizon. A value of 0 corresponds to times of sunset and sunrise, and + 1 indicates the end of twilight in the evening or the beginning at dawn. Figure 6.15 shows the crepuscular biting cycles of *Coquillettidia richiardii* and *Ochlerotatus detritus* in England plotted against minutes before and after sunset and sunrise, and also against corresponding crep values (Service 1969*a*). Relatively few workers have used crep units, apart from Forattini and Gomes (1988) in the study of biting cycles of *Culex ribeirensis* (Fig. 6.16) and *Ochlerotatus scapularis* in Brazil. Forattini et al. (1996*b*) performed human landing catches every two weeks at a rice field margin. The sunset crepuscular period was divided into five crep units according to the methods of Nielsen (1961, 1963). A unimodal peak of landing activity of *Anopheles albitarsis* s.l. occurred between the 2nd and 3rd crep unit.

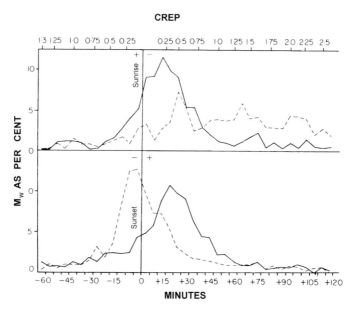

Fig. 6.15. Crepuscular biting cycle of *Coquillettidia richiardii* (solid line) and *Ochlerotatus detritus* (broken line) in relation to times before (−) and after (+) sunset and sunrise, and also crep units (from Service 1969*a*)

Useful tables of illumination in log lux + 10 (this does away with the use of a negative index) corresponding to both positive and negative crep values, of the relationship between crep and the sun's altitude and also the correction in log lux that must be added in different months for latitudes greater than 40°, are presented by Nielsen (1963). Haddow (1964) gives some minute to minute zenith light readings taken near the equator in Uganda and discusses these in relation to mosquito biting activity and the readings of illumination calculated by Nielsen (1963). Haddow et al. (1968) have shown that even at, or near, the equator when twilight is of such a constant duration the conversion of catch times to crep values is still well worthwhile, as it shows better than clock times the forms of waves of biting activity. One of the more interesting facts to emerge from this study was that peak biting activities were governed more by the sun's altitude than clock times. This important paper should be consulted for an account of a detailed analysis of catch data with reference to crep units, and also for data on light values at sunset, including the rate of decay of illumination. In the USA Wright and Knight (1966) studied the effect of decreasing light intensity (and also changes in temperature and relative humidity) around sunset on the biting density of *Aedes vexans* and *Ochlerotatus trivittatus*.

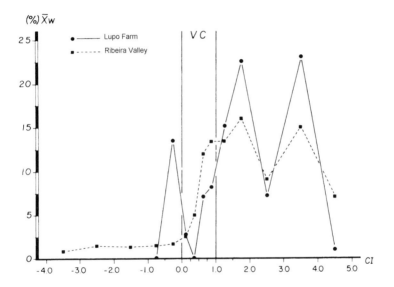

Fig. 6.16. Vespertine crepuscular and pericrepuscular biting by *Culex ribeirensis* at human bait at two localities (Lupo farm, Ribeira valley) in Brazil, CI—crep intervals corresponding to 1700–2000 h; VC—vespertine crepuscular period (Forattini and Gomes 1988)

The effect of polarised light on mosquitoes arriving at bait was studied in North America by Wellington (1974). He found that nearby resting adults of various *Aedes* and *Culex* species arrived at human bait irrespective of whether polarised light was present, but mosquitoes arrived at bait from further afield only when there was natural polarised light present. Polarisation is most intense near sunset and sunrise. Passing clouds reduced polarisation and this was associated with a reduction in the numbers of mosquitoes arriving at bait.

Influence of physiological status on host-seeking

Unfed females invariably predominate in human landing catches and in most cases they are the only category caught. The capture of unfed females that have either alighted on, or are hovering around, the bait is commonly taken as showing that they have been attracted for the purpose of taking a blood-meal, but this may not always be true. *Uranotaenia, Ficalbia* and *Hodgesia* species are sometimes caught at bait (Haddow et al. 1951; Macdonald 1957; Mattingly 1949*b*), but there is often no clear evidence that they would bite if given the opportunity. In Canada, Hocking et al. (1950) and Haufe (1952) recorded separately the landing and biting rates of *Aedes* species attracted to man, and Haddow and Ssenkubuge (1963) emphasised the importance of distinguishing between these two phenomena, i.e. arrival at bait and the intention of feeding. Even with a highly anthropophagic species such as *Aedes aegypti* collection in bait catches does not prove that all individuals are orientated to blood feeding. As much as 18% of the female *Aedes aegypti* collected in a series of catches in Tanzania refused to take a blood-meal (McClelland and Conway 1971). In addition to unfed females, blood-fed, partially and fully gravid females (Gould et al. 1970) and also unfed females with fat reserves are occasionally caught at bait (Service 1969*a*).

Body size is a key determinant of host-seeking and blood-feeding success. Klowden et al. (1988) showed that host-seeking avidity was greater in large *Aedes aegypti* than in smaller adults, while Nasci (1991) showed that large females were significantly more persistent biters than smaller ones. In England Renshaw (1991) and Renshaw et al. (1994) found an increase in the size of *Ochlerotatus cantans* at human bait later in the season, and individuals that were still nulliparous many weeks after emergence were small. These results support the idea that larger adults are more successful in getting a blood-meal. Host seeking activity in *Aedes aegypti* continues until a threshold volume of blood is ingested and Xue et al. (1995) demonstrated that in larger and older female *Aedes aegypti* a larger blood meal is generally required before subsequent feeding activity is inhibited (Table 6.1).

This increase in threshold blood meal size may partially explain why a greater proportion of older females take multiple blood meals.

Takken et al. (1998) observed that responsiveness of *Anopheles gambiae* s.s. females to emanations from a human hand in an olfactometer varied according to body size. One-day old females did not respond to host odours. Host responsiveness reached a maximum when females were 6 days old. Mean maximum response among small females was 61.9% compared with 77.2% for large females.

Table 6.1. Effect of age and body size of female *Aedes aegypti* with different sized blood meals on percentage responding to host (Xue et al. 1995)

Age[a]	Body size	Blood meal size (mg/female)											
		0.1	0.2	0.3	0.4	0.5	0.6	0.7	0.8	0.9	1.0	1.1	≥1.2
5[b]	Small[c]	100	100	85	79	57	25	25	0	0	0	0	0
10[b]	Small[d]	100	100	100	100	87	75	50	33	33	0	0	0
5[e]	Large[e]	96	100	89	80	70	50	50	25	33	0	0	0
10[e]	Large[d]	100	96	100	100	88	67	65	67	50	20	20	0

Large females (mean wing length = 2.98 ± 0.034 mm, n = 30); small females (mean wing length = 2.64 ± 0.134 mm, n = 30)
[a]Five-day-old nulliparous, 10-day-old parous, n = 300 females for each age per body size group
[b]Five v 10-day small, χ^2 = 108.5, df = 9, P < 0.001
[c]Small v large 5 day, χ^2 = 62.2, df = 8, P < 0.001
[d]Small v large 10 day, χ^2 = 30.1, df = 10, P < 0.001
[e]Five v 10 day large, χ^2 = 49.3, df = 10, P < 0.001

Chronological age post-emergence has been shown to affect blood-feeding activity from an artificial membrane feeding system in *Aedes aegypti* and *Aedes albopictus* in the laboratory. However, age was not significantly associated with increased blood-feeding for either species when presented with a restrained chicken (Alto et al. 2003).

Renshaw et al. (1995) demonstrated that lipid reserves may be an important factor in the initiation of host-seeking activity in at least some species of mosquito. Newly-emerged females of *Ochlerotatus cantans* in England given free access to 10% w/v sucrose solution did not attempt to feed from a human volunteer until 192 h post-emergence when the mean lipid levels were found to be similar to those found in field-collected host-seeking mosquitoes. By contrast, *Ochlerotatus punctor* initiated blood-feeding only 48 h post-emergence when lipid levels were again similar to those determined for field-collected host-seeking females. *Ochlerotatus cantans* appears to require approximately double the lipid reserves of *Ochlerotatus punctor* before blood-feeding is initiated.

Serotonin concentrations in mosquitoes may also affect blood-feeding success, as demonstrated by laboratory investigations undertaken by Novak and Rowley (1994). Two amine-depleting drugs (α-methyl-tyrosine [AMT] and α-methyl-tryptophan [AMTP]) were administered to *Ochlerotatus triseriatus*. Host-seeking behaviour, as determined by response to a rabbit in an olfactometer was not altered by treatment with either agent, however, fewer mosquitoes treated with the serotonin depleting agent AMTP fed on the ears of a restrained rabbit, fewer fed to repletion and the time taken to feed to repletion increased, compared to controls.

Jong-Jin Lee and Klowden (1999) identified a peptide synthesised in the accessory glands of male *Aedes aegypti* that in conjunction with a peptide synthesised in the neurosecretory cells of the female inhibits host-seeking during oogenesis.

In Sweden Andersson (1990) found that many mosquitoes caught at bait, especially *Ochlerotatus communis*, had fed on nectar (fructose). It was concluded that nulliparous females commenced nectar-feeding earlier than parous ones, but parous individuals contained most fructose. Fructose, an indicator of nectar-feeding, was found in females in all gonotrophic conditions, and moreover, Andersson (1990) observed blood-engorged mosquitoes feeding on flowers. Van Handel and Day (1990) caught *Ochlerotatus taeniorhynchus* attracted to humans and by quantitatively testing them for fructose concluded that it appeared that nectar-feeding occurred mostly after the onset of darkness, with very little or any feeding during the daytime.

Based on semi-field studies in large screened field cages, Impoinvil et al. (2004) demonstrated that *Anopheles gambiae* males and females both feed on plant sugars when these are available and laboratory tests showed an increase in survival of mosquitoes feeding on sugars obtained from the castorbean plant, compared with those feeding on water or glucose solution. Foster (1995) comprehensively reviews sugar feeding by mosquitoes.

Sugar-feeding in facultatively autogenous mosquitoes, defined as those that can mature eggs with or without a blood meal tends to diminish host-seeking activity of nulliparous females, with the effect that they are often not responsive to a host until after their first gonotrophic cycle. Under laboratory conditions, O'Meara et al. (1993) demonstrated that the majority of field-collected and colony-reared *Ochlerotatus bahamensis* females developed eggs autogenously when they had access to a sugar source. A reduced proportion of small females, or those starved of sugar exhibited autogeny and clutch size was reduced. The predominance of parous females of *Ochlerotatus bahamensis* among host-seeking individuals captured in the field led the authors to conclude that blood-feeding did not occur until after the first egg batch had been laid. Sugar-feeding was considered to prevent blood-feeding and promote autogeny. Further laboratory experiments on sugar-feeding and host-seeking by this species by Bowen and Romo (1995*a*) led to the conclusion that autogenous gravid, autogenous parous and anautogenous females of *Ochlerotatus bahamensis* are all responsive to host stimuli and furthermore, autogenous gravid females readily took a blood meal when given access to a host. Reduced sugar-feeding by autogenous gravid females may increase host-seeking activity in *Ochlerotatus bahamensis* (Bowen and Romo 1995*a*). By contrast, sugar-feeding appears to have no effect on host-seeking behaviour in parous females of *Ochlerotatus bahamensis* (Bowen and Romo 1995*b*). Hancok and Foster (1997) investigated the relationship between larval and adult nutrition and sugar-feeding and blood-feeding by adult *Culex nigripalpus*. Unfed mosquitoes were found to be less responsive to a bird host in the laboratory compared with sugar-fed mosquitoes of the same age. In an olfactometer assay, unfed females preferred the scent of a sugar source to host odours, but this preference switched to the bird host after sugar feeding and this appeared to be related to an accumulation of lipid and follicular growth. If previously sugar-fed mosquitoes were subsequently starved their preference switched back to a sugar source. Crowding of larvae appeared to increase the preference for sugar over a blood meal and together with the above results this suggested that energy reserves were the key determinant of responsiveness to host stimuli. In contrast to aedine mosquitoes, sugar feeding in *Anopheles* does not appear to inhibit blood feeding (Klowden and Briegel 1994). Straif and Beier (1996) demonstrated that the chronologically oldest females (≥ 20 days) of *Anopheles gambiae* took more blood meals and fed more frequently when deprived of access to sugar compared with females of a similar age that had access to sugar. Among younger cohorts, sugar availability did not appear to affect blood feeding.

Host-seeking is often inhibited during egg development, presumably by physiological processes linked to blood meal digestion and follicle development, including abdominal distension (Klowden and Lea 1979). Inhibition of host-seeking tends to diminish with chronological and physiological age of the female, with older females more likely to be attracted to hosts when gravid. Male accessory gland substances (MAG) transferred during mating appear to be the inhibitory factor (Fernandez and Klowden 1995). In the laboratory, mating has been shown to reduce responsiveness to a host. When females of the same chronological age were mated on either day 3 post-emergence or on day 22 post-emergence and subsequently exposed to a human host on day 25, responsiveness of the late-mated females was lower than in the group which mated earlier (Klowden and Fernandez 1996). Further experiments showed that females that have taken multiple blood meals are more responsive to hosts than females of the same chronological age that have not previously fed. Females injected with a MAF homogenate also showed inhibition of host-seeking (Klowden and Fernandez 1996).

In Sri Lanka Samarawickrema (1967, 1968) found small differences between the age composition of *Culex quinquefasciatus* and *Mansonia uniformis* biting at different times, while in Trinidad, Senior White (1953) found slight differences between the biting times and age of *Anopheles aquasalis*. In England there was a small but significant decrease between the proportion of parous *Ochlerotatus detritus* biting between 0300–0500 h than at other hours (Service 1969*a*). Yajima et al. (1971) found a higher parous rate in *Culex tritaeniorhynchus* caught in the latter half of the night from pig-baited traps. Furthermore, older uniparous females with contracted dilatations were commoner in the early part of the night, whereas adults with uncontracted or partly contracted sacs were commoner later in the night. In Brazil Charlwood and Wilkes (1979) found that based on 755 female *Anopheles darlingi* caught at human bait there was a preponderance of nulliparous individuals biting at dusk (64.4%) and again at dawn (71.0%). Differences in the composition of field-derived bait catches have been noted in West Africa, where Coz (1964) reported a small difference between the proportions of parous and nulliparous *Anopheles gambiae* biting at different times of the night, and Hamon (1963*a*) found a slight tendency for older *Anopheles* to bite more in the middle of the night and in the early morning than in the early evening. Bockarie et al. (1996) reported that in Sierra Leone nulliparous females of the forest form of *Anopheles gambiae* s.s. bit earlier in the evening than parous individuals, and similarly nulliparous females of *Anopheles punctulatus* predominated in the early hours of all-night human landing catches in Papua New Guinea. Further investigations revealed that only one of 76 *Anopheles punctulatus* females positive for *Plasmodium falciparum* sporozoite antigen was captured in the early evening between 1800 and 2100 h. The *Plasmodium falciparum* sporozoite rate of mosquitoes captured between 0200 and 0600 h was significantly higher than that for the period 1800 to 2200 h as determined by Yates's corrected χ^2 ($P = 0.037$). Apart from these examples, few differences have been found between the biting times of parous and nulliparous mosquitoes, as have been found in some species of *Simulium* (Davies 1963; Le Berre 1966; McCrae et al. 1969) and *Chrysops* (Duke 1960). By contrast, many field investigations in Africa (Charlwood et al. 2003; Corbet 1961, 1962; Corbet and Smith 1974; Germain et al. 1973; Gillett 1957; Gillies 1957; Gillies and Wilkes 1963, 1965; Hamon et al. 1959, 1961; McCrae 1972), Trinidad (Nathan 1981), Malaysia (Chiang et al. 1984*a*) and in Myanmar (De Meillon and Sebastian 1967) have failed to show any difference between the age composition of mosquitoes biting at different times.

Working in the Congo Carnevale and Molinier (1980) generated a general formula for determining the average number of times a parous anopheline bites in one day (L) based on its gonotrophic cycle and its behaviour before and after oviposition, the formula for *Anopheles gambiae* for example is

$$L = \frac{1}{4 - A - \alpha} \qquad (6.12)$$

where A = the proportion of females which bite on the night that eggs are laid, and α = the proportion of females which oviposit 2 days after their blood-meal. The value of 4 is derived from these patterns of behaviour, and for *Anopheles nili* for example which has an extended gonotrophic cycle the figure is 5. When calculated values of L are multiplied by the anthropophagic index the parameter a of Macdonald (1957) is obtained. Carnevale and Molinier's paper gives graphical illustrations of the biting and oviposition rhythms of both these malaria vectors. Canyon et al. (1999) observed that for *Aedes aegypti* in the laboratory, host biting is largely opportunistic, such that biting frequency increases as more opportunities to bite a host are presented. In their experiments, females presented with a host at intervals of 6 h took five or six blood meals over a 3 day period.

In addition to females, males of a few species are not infrequently encountered at bait (Cordellier and Geoffroy 1974; Hamon 1963*b*). Substantial numbers of male *Aedes aegypti* have been collected in human landing catches (Boorman 1960; Corbet and Smith 1974; Hartberg 1971; Lumsden 1957; McClelland 1960; Pillai and Rakai 1976; Soman 1978), and both De Meillon and Sebastian (1967) and Lumsden (1957) caught males of *Culex quinquefasciatus* on man. Bates (1944*a*) reported that male *Aedes aegypti* settle on a bait and await the opportunity to pounce on females coming to feed. It seems that the host can be a focal point for mating in *Aedes aegypti* (Hartberg 1971) as has been shown for *Ochlerotatus varipalpus* and *Ochlerotatus sierrensis* (Lee 1971; Peyton 1956). Other examples of male mosquitoes being attracted to hosts for sexual encounters are *Mansonia uniformis* (McIver et al. 1980), *Aedes vittatus* (Cordellier and Geoffroy 1974), *Ochlerotatus triseriatus* and *Aedes albopictus* (Reeves 1951), *Aedes furcifer/taylori* (Jupp 1978; McIntosh et al. 1977), *Eretmapodites chrysogaster* (Gillett 1971) and *Armigeres subalbatus* (Das et al. 1983).

In most catches the periodicity of males arriving at bait is similar to that of the females. Trpis et al. (1973) thought that the arrival at bait of males might better indicate the underlying endogenous activity rhythm of a species than the arrival of females. They argued that an abundance of suitable hosts in the morning might result in a smaller percentage of unfed *Aedes aegypti* remaining in the local population to bite in the afternoon, thus causing a depression in the late afternoon biting peak.

Sampling the animal biting population

As with humans, increase in host size, or numbers, seems to increase the numbers of mosquitoes attracted (Edman and Webber 1975) to a bait animal. If the conditions conducive to interrupted blood-feeding are also conducive to multiple feeding, then the dynamics of multiple feeding may be influenced by the interaction of many factors such as host density, host species, host behaviour, mosquito density, and infection of host and/or mosquitoes with disease organisms (Day and Edman 1983, 1984; Edman and Scott 1987; Klowden and Lea 1979; Walker and Edman 1985*a,b*, 1986).

Tethered animals

Collection of tsetse flies from a tethered host, usually an ox, is a common sampling method for *Glossina* (see Glasgow and Phelps in Mulligan (1970) for references), but catching mosquitoes from bait animals not enclosed in any kind of trap has been less widely employed. When, however, the method is used a common procedure is for one or more collectors to visit the tethered animal at intervals to collect with aspirators or test tubes mosquitoes that have settled on it and may in fact be feeding. In addition, mosquitoes that have not settled on the bait but are hovering around are sometimes caught in small hand-nets. An objection to direct catches from animals is that mosquitoes that may be attracted to the collectors while they are catching mosquitoes from the bait are likely to get included within the catch. The use of repellents may usefully reduce the numbers of mosquitoes attracted to the collectors, but at the same time may deter them from being attracted to the bait animal. Mosquitoes will normally be collected more quickly from bait animals during the day than during the night, consequently the likelihood of collecting those attracted to the collectors will be less.

Direct bait catches from tethered docile water buffalo or oxen have been used to sample mosquitoes in Taiwan (Hu and Grayston 1962). Collections for a 3-h period starting at sunset provided information on the relative abundance and seasonal prevalence of *Culex fuscocephala*,

Culex tritaeniorhynchus, Culex vishnui, Anopheles sinensis and *Anopheles tessellatus*. In addition to these mosquitoes, 16 other species were caught. In Pakistan Aslam et al. (1977) and Reisen and Aslamkhan (1978) caught *Culex tritaeniorhynchus* and several other mosquito species for 15-min periods each hour of the night as they arrived to feed on buffaloes or cattle tethered to feed troughs. In later collections mosquitoes were caught from a tethered buffalo with the aid of torches and mouth aspirators, for a 30-min period starting 20 min after sunset (Reisen and Milby 1986). In Assam, India, teams of two collectors aspirated mosquitoes from a cow kept outside from dusk to dawn in a study of the biting habits of potential vectors of Japanese encephalitis (Bhattacharyya et al. 1995). Using this method 22 species from 5 genera were collected, including *Culex fuscocephala, Culex pseudovishnui, Culex vishnui, Culex tritaeniorhynchus*, and *Anopheles peditaeniatus*. Also in India, Neeru Singh and Mishra (2000) performed animal landing catches as follows: one collector caught anophelines attracted to buffalo baits (1 inside the house and 1 outside the house) for 15 min each hour. Catches of *Culex tritaeniorhynchus* from cows were made from both inside and outside cow sheds in Korea (Ree et al. 1969). In Malaysia, Wharton (1951*b,c*) collected mosquitoes attracted to tethered cattle, while in Trinidad Senior White (1952) made observations on *Anopheles aquasalis* feeding on a tethered ox, goat and a horse. In Jamaica Muirhead-Thomson and Mercier (1952) made routine collections of *Anopheles albimanus* from a tethered donkey, while in the USA Jones et al. (1977) collected mosquitoes from tethered horses and donkeys. In South Africa several mosquito species were successfully caught from tethered oxen (De Meillon et al. 1957). In England in a study to identify the potential mosquito vectors of myxomatosis, mosquitoes were collected at 10-min intervals from a rabbit tied by a 1-m lead to a tree (Service 1971*c*). In Australia Myers (1956) also collected mosquitoes attracted to rabbits, but immobilised them by pinioning them firmly to a board. Mosquitoes attracted to the rabbit were caught by carefully lowering a cone-shaped trap over the bait at intervals. In Canada Hudson (1983) used aspirators and torches to collect mosquitoes biting unrestrained calves, but when mosquito densities were high the cattle became restless and sometimes stampeded, and so they had to be tethered for mosquito collections.

In California, USA Barnard and Mulla (1977) studied the diel feeding patterns, from 1400 or 1600 to 0745 or 0700 h, of *Culiseta inornata* on a calf tethered to a stake in an open area and supplied with food and water. Throughout the night the bait was approached at 15-min intervals and the numbers of mosquitoes feeding counted; a procedure taking just 60–90 s. A 25-W lamp provided illumination. During daylight hours records of feeding were made hourly.

In Mali and Burkina Faso Touré and Coluzzi (1986) encircled cattle corrals with a fence of white mosquito netting. This was done by pushing vertically into the ground at 3-m intervals a series of 1.8-m tubular metal posts and attaching 1-m sections of metal tubing at an angle of 120° onto the tops of these posts. Mosquito netting was then hung from a series of rings placed at 30-cm intervals on both the vertical and slanting metal tubes. This netting barrier was erected about 1 m from the edge of a fence enclosing the cattle. Mosquitoes flew over this 2.3–2.5 m high barrier, and also through a 10–20 cm gap formed between the ground and the bottom edge of the netting, to feed on the encircled cattle. Large numbers of blood-fed females of the *Anopheles gambiae* complex, and other mosquitoes, were collected resting on the inside surfaces of the upright and inwardly leaning netting barrier.

In El Salvador periodic collections of *Anopheles albimanus* comprising mostly blood-fed individuals, were made from vegetation and from the beams and poles of corrals, having a roof but no walls, in which cattle were tethered. Regular 2-h collections of mosquitoes attracted to these animals provided information both on the seasonal incidence of *Anopheles albimanus* and the variations in densities in different localities (Rachou et al. 1965). Crans and Rockel (1968) caught mosquitoes attracted to turtles by tying nylon cord to a ring passed through a hole drilled in the edge of the carapace.

Other less manageable animals my need to be anaesthetised. In the Ivory Coast Cordellier et al. (1983) placed an anaesthetised monkey in a widely spaced mesh cage and had a person collect, with an aspirator, mosquitoes landing on it.

Comtois and Berteaux (2005) report the use of 5-min landing counts on a blue cushion (30 cm × 33 cm), every day, at sunset, from 23 May to 26 July 2002 to assess the abundance of biting mosquitoes and blackflies in a study of the effects of insect biting on porcupine (*Erethizon dorsatum*) behaviour and habitat use. The cushion-landing counts were undertaken during observations of individual porcupines with the observer sitting within 30 m of the porcupine being observed. In addition, these authors also used 5-min landing counts on a hemisphere (30 cm long × 21 cm wide × 30 cm high) designed to simulate the silhouette of a porcupine. The silhouette was covered with a blue cloth and a plastic tube (200 cm long, diameter 0.79 cm) permitted the observer to introduce exhaled breath into the silhouette, serving as a source of CO_2. Landing counts were conducted using the silhouettes placed at the entrance of and approximately 1 m inside rock dens used by porcupines. In addition, the silhouettes were suspended from branches at a height of approximately 5 m in five species of trees and biting abundance was estimated. Landing counts were also conducted at the foot of each tree. While an interesting approach, the methods used clearly did not directly measure attraction of biting flies to porcupines, rather they measured abundance of species attracted to the collector and potentially to porcupines.

Animal-Baited Net Traps

Mosquito nets have also been widely used to collect mosquitoes attracted to a variety of animals. The simplest arrangement consists of positioning an ordinary single or double-sized mosquito net over a host and raising it a few cm from the ground to allow entry of host-seeking mosquitoes.

In Okayama City, Japan, when Sasa and Sabin (1950) baited standard US army mosquito nets with a man and a variety of animals the most common mosquito caught was *Culex tritaeniorhynchus* followed by *Culex pipiens* form *pallens, Anopheles hyrcanus* and *Anopheles sinensis*. Surprisingly the mean catch of mosquitoes in a net baited with both a chicken and a rabbit (588) was consistently greater than the mean catch obtained when it was baited with only a chicken (84) or rabbit (239), and also greater than the combined mean catches (323). Sasa et al. (1950) considered baited nets the most convenient and accurate collecting method for studying host preferences and temporal fluctuations in population size of the local mosquitoes. By not enclosing the bait animals within an inner net the mosquitoes were able to feed on the bait, and the rates of engorgement by different species on various baits were used to provide additional information on host acceptability.

In studies on vectors of simian malaria in Malaysia, nets baited with monkeys (*Macaca irus*) were used to sample mosquitoes (Wharton et al. 1963, 1964). The monkeys were kept in a small cage covered with chicken wire, which did not prevent the mosquitoes from feeding on them. Small nets measuring 4 × 3 × 3 ft, having a 14-in wide 3-ft high opening on either side held two monkeys and larger nets measuring 6 × 5 × 4 ft with two 22-in wide and 4-ft high openings were used to hold four monkeys. The traps were positioned either at ground level or at various heights in trees, and mosquitoes collected from them every 1–2 h throughout the night. The larger nets with four monkeys caught about twice as many mosquitoes as did the smaller nets, and 5.6 times as many adults of *Anopheles hackeri*, the principal vector of *Plasmodium knowlesi*. More adults of *Anopheles hackeri* were caught in traps placed in a mangrove swamp where wild monkeys slept, than in traps sited immediately above the vector's larval habitats (Wharton et al. 1963). In later trials as many as 45 culicine and 18

Anopheles species were caught in monkey-baited net traps (Wharton et al. 1964).

In South Africa Jupp and McIntosh (1967) placed an animal bait (goat, sheep) in a 2-in wire mesh cage in the centre of a mosquito net 7 ft long, 5.5 ft wide and 6 ft tall supported on poles and raised 8 in from the ground. A large zippered panel allowed a collector to enter. Ten mosquito species were caught, the commonest being *Culex pipiens* (53%), *Culex theileri* (28%) and *Culex univittatus* (12%), this being the same order of abundance as obtained with CO_2-baited lard-can traps. Jupp and Kemp (1998) later used net traps of Jupp and McIntosh (1967) baited with CO_2 to investigate the vectors of Wesselsbron virus in South Africa. A total of 8117 mosquitoes were collected, with *Culex theileri* accounting for 86.9%. Other species collected in descending order of frequency included *Ochlerotatus juppi/caballus, Aedes mcintoshi/luridus, Anopheles squamosus, Culex univittatus, Ochlerotatus juppi, Culex pipiens, Ochlerotatus caballus, Aedes mcintoshi, Aedes dentatus*, and a single *Aedes mixtus/microstictus*. These authors however, considered the human landing catch to be the most productive in relation to effort. The large numbers of *Culex theileri* collected in the net trap and CO_2-baited light traps resulted in damage to the specimens of floodwater *Aedes* and *Ochlerotatus* collected, making identification difficult. Jupp et al. (2002) also used this type of net trap baited with CO_2 to sample potential Rift Valley fever vectors following an outbreak in Saudi Arabia in 2000. The traps collected as many as 18 600 mosquitoes (predominantly *Culex tritaeniorhynchus, Aedes vexans arabiensis*, and *Ochlerotatus caballus*) per hour when set near a sedge bed at the periphery of a large dam. In Nigeria Bown and Bang (1980) suspended a mosquito net about 30 cm above the ground over cages baited with one goat, two chickens, two monkeys and three rats. From eighteen 24-h collections only 70 females belonging to six *Aedes* species were collected.

Chevalier et al. (2004) used a mosquito net trap adapted from that used by Bown and Bang (1980) to investigate mosquito feeding on sheep in a Rift Valley fever virus transmission zone in Senegal. The trap consisted of a metal frame covered with a mosquito net, which was raised slightly above the ground to allow mosquitoes to enter the trap at its base. The trap was baited with a sheep and mosquitoes flying upwards after biting were trapped in a removable box positioned on top of the trap. Traps were operated from 1800 to 06.00 h. A total of 8122 female mosquitoes were collected, with the three presumed RVF vectors (*Aedes vexans, Culex poicilipes* and *Aedes ochraceus*) accounting for 94.2% of the total. A Poisson model (MacCullagh and Nelder 1989) incorporating three variables (study period, pond, and capture site and a quasi-likelihood algorithm method were used to analyse the data for *Aedes vexans*.

In a study of the effects of use of treated mosquito nets and DDT on survival rate and oviposition interval of *Anopheles farauti* No.1 in the Solomon Islands, Hii et al. (1995) used a pig-baited net trap. The trap consisted of a bamboo fence enclosure covered by mosquito netting and with a 10 cm gap left between the bottom of the net and the ground. The trap was sheltered by a palm thatch roof erected above the net. These traps caught 21.3% of the total catch in one site, but only 4% in another, when operated alongside human landing catches and indoor CDC light-trap collections.

In Malaysia Reid (1961) compared the mosquitoes attracted by two men, a calf and two goats enclosed in 10-ft long, 7-ft high and 7-ft wide nets which had a 3-ft wide entrance in each long side which could be closed by a flap. The nets were positioned under a shelter without sides, and mosquitoes collected hourly from 1900–2300 h. Results were expressed in terms of attraction ratios, calculated as the ratio of the numbers of a particular species caught on man compared with the numbers caught on a calf or two goats. Reid (1961) considered that although experimental conditions (e.g. sampling error, size of baits, different types of trap) might change the values of the ratios, their order of magnitude for different species would nevertheless remain more or less the same, and could therefore be used to compare results from experiments made at different times, or in different localities. For example, the man:calf ratio calculated for *Culex quinquefasciatus* and *Mansonia uniformis* caught in bait nets was 1.7:1 and 1:2.8 respectively. The man:calf ratios for the same species calculated from window trap huts were very different in magnitude (4.0:1 and 1:36), but the two species were ranked in the same order of attractiveness by both sampling methods.

In India Reuben (1971) caught over 21 mosquito species attracted to a man, a buffalo and a bullock and confirmed Reid's (1961) observations on attraction ratios. *Culex quinquefasciatus* preferred man, *Culex bitaeniorhynchus* was about equally attracted to all three baits, while the other species preferred animal baits to man.

In Sabah, Chiang et al. (1984*b*) placed a mosquito net (300 × 240 × 210 cm) having a 90-cm wide gap in each of the longer sides over a calf. Mosquitoes were collected for 15-min periods every 2 h from 1800–0600 h, or for 15 min every hour from 1800–2100 h. Seventeen species were caught, including large numbers of *Culex gelidus, Aedes vexans* and *Culex tritaeniorhynchus*. The use of cow-baited net traps has been widely practiced in Malaysia in recent years. For example, Jaal and Macdonald (1992, 1993) used a cow-baited mosquito net based on Gater's 1935 design to sample *Anopheles* mosquitoes in Penang, Malaysia. A 2.4 × 2.4 × 1.6 m net having a 1.2-m wide door panel closed by zippers on one side and a tarpaulin placed over the net to protect against rain (Fig. 6.17) was used.

Fig. 6.17. An animal-baited bed-net trap (photograph reproduced with the permission of Zairi bin Jaal)

Every hour, three collectors entered the net for 15 min to remove the anophelines captured, which included *Anopheles peditaeniatus, Anopheles sinensis, Anopheles subpictus, Anopheles lesteri paraliae* and *Anopheles vagus*. The cow-baited traps were much more successful than human-baited traps operated concurrently. The human-baited net consisted of a double mosquito net, the dimensions of the outer net being 2.1 m long, 2 m wide and 1.8 m high. The inner net contained a camp bed or chair on which the subject rested. The outer net was raised 0.3 m above ground. Similar cow-baited net traps have been used to undertake surveys of adult anophelines in peninsular Malaysia by Abu Hassan et al. (1997) and Rahman et al. (1995, 2002) and along the Malaysia-Thailand border (Rahman et al. 1993). Rahman et al. (1995) collected twelve species of anophelines from traps operated from 1900 to 0700 h on one night per month for 12 months. The predominant species was *Anopheles maculatus*, which comprised 55% of the total catch. In their 2002 paper, Rahman et al. reported collecting 14 species of *Anopheles*, the dominant species on this occasion being *Anopheles aconitus* and *Anopheles barbirostris*. *Anopheles maculatus*, the most important malaria vector in the area comprised 9% of the catch. The cow-baited trap was considered to be a relatively inefficient method for sampling *Anopheles aconitus* on the Malaysia-Thailand border as only 25.5% of the total catch (1043) were of this species, despite it being both exophagic and zoophilic (Rahman et al. 1993). A $5 \times 5 \times 2$ m double net containing a tethered buffalo was used by Somboon et al. (1998) to study entomological aspects of malaria transmission in Karen villages in forest areas of north west Thailand. The nets were suspended from a bamboo frame and the outside net was raised 25 cm above the ground, with a 60 cm gap between the inner and outer nets. Rwegoshora et al. (2002) used a cow-baited trap (4 m $\times$ 4 m $\times$ 3 m) alongside indoor and outdoor human landing catches to collect *Anopheles minimus* in Thailand. The cow-baited trap caught a total of 479 *Anopheles minimus* species A and 1859 species C when conducted on two consecutive nights per month over a 12-month period (excluding June, July and August). The proportion of species A in the total catch was slightly higher for the cow-baited traps compared with the human landing catches.

Thin Thin Oo et al. (2003) used nets baited with either a cow or water buffalo, alongside indoor and outdoor human landing catches and indoor resting catches to investigate the transmission of malaria in Myanmar. Nets used had dimensions of $330 \times 330 \times 180$ cm and contained a tethered bait animal. Mosquitoes were collected from the net using an aspirator at intervals of three hours (1800, 2100, 2400, 0300, and at sunrise). Human landing catches caught more *Anopheles dirus* than cow-baited net traps.

Kobayashi et al. (1997) used a large $4 \times 4 \times 2$ m net enclosing a cow or buffalo to collect anopheline mosquitoes in Lao PDR. Also in Lao PDR, Toma et al. (2002) used $4 \times 4 \times 2$ m double nets, with a cow or buffalo tethered within the inner net. The bottom edge of the outer net was raised to facilitate the entry of mosquitoes. Using this method, the authors captured 19 species of *Anopheles* in the wet season (12 in subgenus *Cellia*, and 7 in subgenus *Anopheles*), the dominant species being *Anopheles nivipes*. In the dry season, 16 species of subgenus *Cellia* were captured and 6 species of subgenus *Anopheles*. *Anopheles philippinensis* was the dominant species at one dry season site, *Anopheles minimus* at the two remaining sites.

In India Kulkarni (1987) caught 3007 female mosquitoes comprising 19 species of anophelines in mosquito nets raised 30 cm from the ground and enclosing a cow tied to a stake. Mosquitoes were removed hourly and the most numerous were *Anopheles karwari, Anopheles jamesii, Anopheles maculatus* and *Anopheles splendidus*. Bhatt and Kohli (1996) used the same methods to sample biting anophelines in Gujarat, India. Fully-fed females were collected from the inside surface of the net during the last 15 minutes of every hour from 1800–0600 h. Using white-coloured adult bullocks as bait animals, 41 552 anophelines were captured from 70 catches. *Anopheles subpictus* comprised 61.7% of the total catch, followed by *Anopheles aconitus* (8.9%), *Anopheles nigerrimus* (6.4%), *Anopheles culicifacies* (5.5%), and *Anopheles annularis* (5.5%).

In the Philippines, Walker et al. (1998) used carabao (water buffalo) baited mosquito net traps to collect *Ochlerotatus poicilius* as part of an investigation into the components of vectorial capacity for *Wuchereria bancrofti* transmission. The carabao was tethered to a stake and covered with a mosquito net, which had one side left open. The trap was operated from 1800 to 0600 h. Mosquitoes resting on the inside of the net were collected with an aspirator. A total of 7856 female mosquitoes of 27 species were collected in carabao-baited traps during 18 nights of sampling. The catch was dominated by *Culex tritaeniorhynchus* (21.4%) and only 105 *Ochlerotatus poicilius* were captured. A comparison of the total mean number of *Ochlerotatus poicilius* collected in the carabao-baited trap with human landing collections showed no significant difference when subjected to a Mann-Whitney U test. However, as the carabao-baited traps were operated throughout the night, while the human landing catches were conducted only from 1800 to 2400 h the numbers collected from the animal-baited trap were divided by 2 and after this adjustment, it was found that the human landing catches caught significantly more *Ochlerotatus poicilius*.

In a study on western equine encephalomyelitis in Argentina Mitchell et al. (1985) used large (360 × 360, 210 cm tall) net-type traps made of nylon tulle and baited with horses. The trap was raised by ropes 30–46 cm from the ground to allow hungry mosquitoes to enter, and was lowered when mosquitoes were being collected. The bait animal, which was usually exposed from dusk to 1 or 2 h after sunrise, was led in through a 1-m wide door-flap made in the middle of one side of the net. From just five trap-nights in one locality and three in another, 2752 and 6929 mosquitoes belonging to at least 20 species were caught; *Culex (Culex)* predominated at the two sites (45.8 and 95.7%). The commonest species were the *Culex pipiens* complex, *Ochlerotatus albifasciatus, Psorophora ciliata* and *Psorophora pallescens*. The authors reported that the trap performed better than a portable Magoon trap. In later studies carried out over three years using the same traps, Mitchell et al. (1987) collected 20 697 mosquitoes belonging to at least 28 species, the majority being unidentified *Culex (Culex)*.

Arredondo-Jiménez et al. (1996) used large nylon nets baited with a horse to collect *Anopheles vestitipennis* in southern Mexico. The trap was raised 20–30 cm from the ground to facilitate mosquito entry. After observing that

Anopheles vestitipennis tended to rest on the net before and after feeding, the technique was modified by hanging a net from a tree adjacent to the horse. Five-minute searches for mosquitoes resting on the net were conducted every 15 min throughout the night and catches were expressed as mean number collected per hour.

Similar horse-baited traps were used by Balenghien et al. (2005) to study the vectors of West Nile virus in the Camargue region of southern France. A horse was maintained inside a large (4 × 4 × 3 m) net, with a gap of approximately 10 cm left between the bottom of the net and the ground to allow mosquitoes to enter the trap. Traps were set up at noon and operated for 24 h and mosquitoes were collected every 4 h using a backpack aspirator (Modified CDC Backpack Aspirator model 1412, John W. Hock, Gainesville, Florida, USA). At the Tour du Valat site in the delta, *Ochlerotatus caspius* (39,347), *Aedes vexans* (34,897), *Anopheles maculipennis* sensu lato (38 000), and *Anopheles hyrcanus* sensu lato (5881) represented 98.8% of engorged females collected on the horse. At the Lunel-Viel site (a riding stables in a neighbouring dry zone) *Culex pipiens* (251), *Ochlerotatus caspius* (159), and *Culiseta annulata* (72) represented 95.4% of engorged females collected on the horse.

In order to study the protective effects of treating canvas tents with pyrethroid insecticide to protect the inhabitants of Afghan refugee camps in Pakistan from malaria, Hewitt *et al.* (1995) erected wooden cow pens (2 × 1 m) in the centre of purpose-built raised platforms (6 × 5 m) situated within mud-walled compounds. Ridge-pole tents (3.5m width × 4.0 m length × 2.1 m height) made from a double sheet of cotton with a water-proofed canvas outer fly-sheet were erected over each cow pen and a 100 denier white polyester mosquito net was suspended over the tent. One hour before sunset a cow was tethered within the pen inside the tent and the walls of the mosquito net were lowered, leaving a gap of 30 cm between the bottom of the net and the ground to allow mosquitoes to enter. The nets were closed one hour before sunrise and all live and dead mosquitoes were collected. The preferred resting sites of mosquitoes within the tents were evaluated by making resting collections at 3 h intervals over three consecutive nights using two collectors. Untreated nets caught a total of 1619 culicines and 790 anophelines, of which 664 were *Anopheles stephensi* and 80 *Anopheles culicifacies*. 98% of *Anopheles stephensi* and 96% of *Anopheles culicifacies* were blood-fed. More *Anopheles stephensi* were caught in the net trap (65%) than in the tent (45%), with the reverse being true for *Anopheles culicifacies*, of which 26% were caught in the net, and 74% in the tent. Hewitt et al. (1995) proposed a method to calculate the collecting efficiency of their trap which compares the numbers of mosquitoes caught actually biting the bait and the number of blood-fed mosquitoes caught in the open net traps. The method used is as follows:

$$a_o = d_o + e_o \qquad (6.13)$$

where a_o = number of mosquitoes attracted to the bait, d_o = number collected from the net trap at dawn, and e_o = the number that escaped from the net. If there is a constant ratio between captured and escaped mosquitoes, E, then

$$e_o = E(d_o) \qquad (6.14)$$

And

$$a_o = d_o + E(d_o) \qquad (6.15)$$

Similarly, for the biting collections, in which mosquitoes were collected directly from the cow baits by two collectors working in shifts through the night,

$$a_b = b + d_b + E(d_b) \qquad (6.16)$$

where a_b is the number of mosquitoes attracted to the bait, b is the number caught feeding on the bait, and d_b is the number collected from the net trap at dawn.

If a_o and a_b are equal, as would seem likely where the experiments were undertaken on consecutive nights, then

$$d_o + E(d_o) = b + d_b + E(d_b) \qquad (6.17)$$

and

$$E = \frac{b + d_b - d_o}{d_o - d_b} \qquad (6.18)$$

and the percentage capture efficiency (%C) is derived from

$$\%C = 1/(1 + E(100)) \qquad (6.19)$$

and

$$\%C = ((d_o - d_b)(100))/b \qquad (6.20)$$

Using this formula, the authors calculated the efficiency of the traps as 11% for *Anopheles stephensi*, 26% for *Anopheles subpictus*, and 85% for culicines.

Wright and DeFoliart's net trap

In Wisconsin, USA Wright and DeFoliart (1970) used a modified Gater-type bed-net made of saran cloth which is supported at each corner by wooden stakes and measures 6 ft in length, width and height. A heavy duty nylon zipper is sewn into one side and after the bait is introduced this is opened to about three-quarters of the distance to the top, and the flaps tied back to leave an opening about one-third of the total area of one side. Twenty species of bait animals were exposed in a variety of cages made of 1 or 2-in mesh and ranging in size from $5 \times 5 \times 16$ in to $11 \times 12 \times 20$ in; frogs, snakes and small rodents were enclosed in cylindrical cages 9 in in diameter and length made of 5/16-in hardware cloth. None of the cages prevented mosquitoes feeding on the bait animals. When deer were the bait, a $3 \times 2 \times 2$-ft hutch was placed in a corner of a chicken wire pen enclosing the deer, and the mosquito net suspended from the roof of the pen over the hutch. Bait animals were usually placed in the traps 2–3 h before sunset and mosquitoes removed 2–3 h after sunrise by a collector wearing protective clothing, including gloves and a helmet fitted with a bee net. In seven trials with unbaited traps 674 mosquitoes were collected, but only about 1.9% were engorged, from which Wright and DeFoliart (1970) concluded that very few blood-fed mosquitoes entered the traps.

Shannon's net-trap

This net was developed by Shannon (1939) in South America apparently independent of the more simple net of Gater (1935) and was usually baited with a donkey to catch day-time biting forest mosquitoes. It consisted of a large central compartment, which in the original design was 130 cm wide, 200 cm high and 300 cm long, and two identical smaller lateral compartments, each 60 cm wide, 300 cm long but only 135 cm deep, thus leaving a 65 cm gap along the bottom (Fig. 6.18).

The tops and ends of all compartments were made of strong white muslin cloth, while the lateral panels of the central and two outer compartments were of mosquito netting. In addition, a small window of mosquito netting, 130 cm wide and 50 cm high, was inserted into one of the end panels about 25 cm from the top. A bait animal, usually confined to a pen or cage, is placed in the middle compartment, the lower edges of which are secured to the ground to prevent mosquitoes from entering it.

Mosquitoes are collected from the two outer compartments. Alternatively, bait animals may be replaced by a light source.

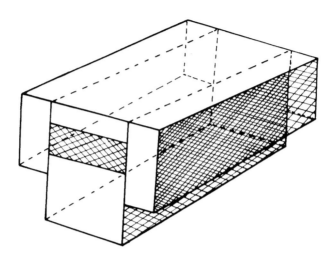

Fig. 6.18. Shannon-type net with central compartment for bait (Shannon 1939)

Shannon-type nets made completely of cotton sheeting and with shorter side compartments are often used. Figure 6.19 shows such a net in operation in Brazil, erected under a roof to protect against rain.

Shannon (1939) also described a one-compartment trap consisting of only the middle section of the typical Shannon net, with the roof made of muslin, but all four sides of mosquito netting. Mosquitoes are collected either resting on the outside of the trap or inside it when the lower edges are raised. This later modification gives a trap very similar to the net of Gater (1935). Vexanet et al. (1986) developed another modification of the Shannon trap, in which the central compartment was replaced by a vertically hanging piece of cloth, with the lateral compartments retained, although these were reduced in size. Shannon net traps can also be used with a light source as an alternative to a bait animal.

Use of the Shannon net has been mainly restricted to South America (e.g. Barbosa et al. 1993; Cardoso et al. 2004; Christensen et al. 1996; Forattini et al. 1981, 1993*b*, 1995*b,c,d*, 2000; Need et al. 1993; Teodoro et al. 1995). Forattini et al. (1993*b*) used Shannon traps with two-person collecting teams to sample mosquitoes in residual woodland, primitive forest and open land adjacent to rice fields in south-eastern Brazil. CDC light-traps baited with dry ice were also used.

Fig. 6.19. Collecting mosquitoes from Shannon net trap with battery-powered aspirator (photograph reproduced with permission of O. P. Forattini)

Traps were operated from 1700 to 2200 h. Of the total specimens collected, 34 843 females were obtained from the Shannon traps and 17 546 from the CDC traps. *Ochlerotatus scapularis* was the commonest species collected, comprising 15.3% of the total, followed by *Anopheles albitarsis* (12.1%), *Culex nigripalpus* (10.8%), *Mansonia indubitans* (8.5%), *Coquillettidia venezuelensis* (8.3%), and *Culex ribeirensis* (7.3%). Anopheline mosquitoes were virtually absent from the Shannon net collections in residual woodland. In the primitive forest, *Anopheles cruzii* and *Anopheles bellator* were dominant (91.7% and 3.4% respectively) in the Shannon trap catches. These authors used the Shannon trap captures and results of human landing catches to calculate indices of synanthropy (Nuorteva 1963) as follows:

$$s = \frac{2a + b - 2c}{2}$$ (6.21)

where:

a = the percentage collected in the anthropic settlement; b = percentage collected in rural surroundings; and c = percentage collected in the wild. The index ranges from +100 to –100, the former corresponding to the highest degree of synanthropy.

Shannon traps located adjacent to the rice paddy were considered to represent the anthropic environment (a). Data for rural surroundings (b) corresponded to Shannon trap catches from residual woodland, and c was determined from Shannon trap catches from primitive forest. *Anopheles albitarsis* and *Ochlerotatus scapularis* showed the highest degrees of synanthropy using this method (99.9 and 71.0 respectively). Forattini et al. (1993c) also calculated indices of synanthropy for resting mosquitoes collected from residual woodland and secondary bush (a), peridomestic environment (b) and primitive forest (c) using a Nasci (1981) mechanical aspirator. High synanthropy was exhibited by *Aedes nubilus* (98.89), *Ochlerotatus scapularis* (85.82) *Ochlerotatus serratus* (88.84), *Anopheles cruzii* (94.59), *Culex bidens* (96.23), among others. Later, Forattini et al. (1995a) used Shannon traps operated by two persons in forest remnants approximately 1 km from rice fields and close to the Pariquiera-Açu river in southeast Brazil. On this occasion traps were operated from 1700 to 2000 h on a fortnightly basis. Shannon traps yielded 1123 female and 3 male *Anopheles albitarsis* species A and 20 189 females and 147 males of species B during the period from February to December 1993. Forattini and colleagues used the same methods to sample *Culex* mosquitoes in Brazil (Forattini et al., 1995b). A total of 6631 adult mosquitoes (6344 females and 287 males) of the *Culex* Spissipes Section was collected, of which 5602 (84.5%) originated from Shannon traps and 764 (11.5%) from human landing catches. *Culex sachettae* (49.3%) and *Culex ribeirensis* (46.6%) were the most commonly collected species. Shannon traps also proved successful in sampling *Culex (Culex)* mosquitoes (Forattini et al., 1995d). A total of 1812 adult mosquitoes (1787 females and 93 males) belonging to the subgenus *Culex* were obtained from Shannon traps, compared with only 68 from human landing catches. The commonest species was *Culex nigripalpus* (77.9%), followed by *Culex coronator/Culex usquatus* (10.1%). Also in Brazil, Forattini et al. (1999) used Shannon traps and human landing catches to sample *Anopheles bellator* Shannon traps were operated fortnightly from 1700 to 2000 h. A total of 2382 *Anopheles bellator* and 4374 *Anopheles cruzii* were caught from traps operated at three sites between May 1996 and September 1997. Forattini et al. (2000) used Shannon traps alongside other methods to compare the behaviours of sympatric populations of *Aedes albopictus* and *Ochlerotatus scapularis* in southeastern Brazil. Two nets were operated from 1500–1800 h and one trap from 1700–2000 h. Traps captured a total of 1253 specimens (1141 *Ochlerotatus scapularis* and 103 *Aedes albopictus*), corresponding to mean hourly collections of 2.7 and 0.3 respectively. Guimarães et al. (1997) used Shannon traps illuminated with propane gas lamps between 1800 and 2000 h to collect anopheline mosquitoes in Paraná State, Brazil. *Anopheles albitarsis* s.l. (61.6%) and *Anopheles galvaoi* (35.2%) were the most frequently collected species. Guimarães et al. (2003) later used the same type of Shannon traps illuminated with propane gas lamps alongside human landing catches to study the mosquito fauna of the Iguaçu National Park in Brazil. Together, these two methods resulted in the capture of 20 273 adult mosquitoes belonging to 44 species. *Ochlerotatus serratus* (10.3%), *Haemagogus leucocelaenus* (9.7%), *Mansonia titillans* (9.6%), and *Chagasia fajardoi* (8.8%) were the most frequently captured species. Data were not presented separately for the two methods used.

Teodoro et al. (1995) used Shannon gas-lamp illuminated light-traps in conjunction with human landing catches to sample mosquitoes in the Itaipu Lake region of southern Brazil. Two traps were operated once per month from January to December 1991. The Shannon traps captured 21 280 mosquitoes representing 41 species and eight genera, compared to 1010 mosquitoes captured in human landing catches. The most commonly captured species was *Coquillettidia shannoni*, followed by *Mansonia humeralis*, and *Anopheles triannulatus*.

Need et al. (1993) found that a human-baited Shannon net trap yielded the most species-diverse mosquito collections in the Iquitos area of Peru, with 39 of the 40 taxa captured using eight different methods caught in this type of trap. Trapido and Aitken (1953) introduced the Shannon trap to Sardinia in a mainly unsuccessful attempt to catch *Anopheles* with an unbaited net containing a light. In Malaysia, Shannon nets were less effective than Gater-type nets baited with monkeys (Wharton et al. 1963). In South Africa Sharp et al. (1988) used Shannon nets baited with a man, goat or bovid, and at about 2100, 2400 and 0300 h aspirated mosquitoes from the inside walls of the net. Only 133 mosquitoes, however, were caught from three night's collection. Burkett et al. (2001) used 6-foot × 4-foot × 6-foot Shannon traps in a comparison of several baited traps for mosquito surveillance in the Republic of Korea. The Shannon traps were baited with two human collectors and a propane lantern. Mosquitoes landing on the inside or outside of the trap were collected with an aspirator. A total of 8653 mosquitoes was collected over a period of 11 days in June 2000. *Aedes vexans* was the most abundant species, followed approximately equally by *Anopheles sinensis* and *Anopheles yatsushiroensis*. *Anopheles lesteri* was more frequently captured with the Shannon trap, than with the other traps.

Galati et al. (2001) observed that Shannon nets of the Vexanet et al. (1986) type, when constructed from black cloth, were more successful at catching phlebotomine sand flies in Brazil than white traps. A fluorescent lamp (15 W, 12 V) hanging between the two traps, and a manual lantern (0.15 A, 6 V) were used as light sources. Traps were set side by side near the entrance of a cave, and operated in the early evening over a total period of 44 h, comprising 12 observations. A total of 889 sand flies, representing 13 species were collected: 801 on the black traps and 88 on the white traps. Overall, hourly Williams means for black and white traps were 8.67 and 1.24 respectively.

Mpofu and Masendu net trap

In Zimbabwe Mpofu and Masendu (1986) developed an ox-baited trap that is essentially a modified mosquito net. It consists of a framework of light metal tubing to form a tent-like structure (Fig. 6.20) that is covered with either mosquito netting or a khaki lightweight canvas.

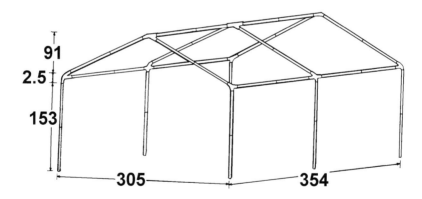

Fig. 6.20. Tubular framework of animal-baited trap, measurements in cm (Mpofu and Masendu 1986)

The latter is preferable in rainy weather as it helps protect the bait animal and mosquito catch. For extra stability guy ropes can be fitted to the four corners of the trap.

One of the smaller ends has a vertical slit in the netting, or canvas, to allow the bait and collectors to enter. A gap (measurements not given, but probably about 25 cm) is left between the bottom of the net or canvas and the ground for entry of host-seeking mosquitoes. Over a 3-month period in an area with apparently low mosquito densities 147 *Anopheles gambiae* complex were trapped from one such trap.

Drop-net cages

Drop-net cages consisting of a framework of PVC or aluminium tubing covered with plastic screening to form a box-like trap (2–3 m diameter, 2–2.5 m high) have been used for the capture of *Culicoides* attracted to animal baits (Schmidtmann et al. 1980; Zimmerman and Turner 1983; and Hayes et al. 1984). The cages are suspended from a wooden crossbeam 5 m above the ground over an enclosed bait animal (Fig. 6.21). A pulley system and winch attached to the upright poles supporting the crossbeam is used to lower and raise the trap.

Fig. 6.21. Drop-net trap (Schmidtmann et al. 1980)

Limitations of mosquito net-traps for human and animal baits

While mosquito net traps have been used extensively over many years in several countries they do suffer from several limitations. Because the nets are often raised above the ground to facilitate mosquito entry a proportion of the catch can escap, and this is exacerbated if bait animals are kept in a separate net, as is often the case, thus preventing mosquitoes from feeding. During use of Gater-type nets in Malaysia it was observed that the mosquitoes caught in the outer nets remained active because they were continuously being stimulated by a bait upon which they could not feed, and this caused a number to escape (Colless 1959). To try to reduce this exodus, nets having only a single entrance were used, and the bait was not enclosed within an inner protective net. When a man was employed as the bait he collected throughout the night mosquitoes from the inside walls of the net on which they rested before attempting to bite. Catches of mosquitoes from human baited nets from which mosquitoes were continuously collected were compared with those from unbaited nets from which they were collected for about 20 min every hour. In one series of trials the mosquitoes which entered the net were given the opportunity to feed on the sleeping occupant. Surprisingly about 39–49% of the total catch of mosquitoes were collected from unbaited nets.

When a Shannon net was baited with monkeys and baboons in Uganda, Haddow (1945*a*) found that although large numbers of *Coquillettidia fuscopennata* entered the nets, most escaped, especially around sunrise. About three times as many mosquitoes entered the net as were retrieved from the nets in the morning. Moreover, more than twice as many mosquitoes were caught biting a boy outside a net as were observed to enter it when it was baited with a boy. Hamon (1964) also reported that in Mauritania most of the mosquitoes that entered baited net traps in the early part of the night left before sunrise.

Working in Malaysia Moorhouse and Wharton (1965) also found that trapped mosquitoes escaped from their nets. They also made the observation that net traps did not give an accurate reflection of biting rates. The ratios of the numbers of mosquitoes in hourly collections from net traps to those caught in direct bait catches was 9:1 for *Anopheles letifer*, 1.8:1 for *Anopheles donaldi*, 2.3:1 for *Anopheles campestris* and 6:1 for *Anopheles maculatus*. Net traps caught more individuals of all of these species than did direct landing catches on man. These authors considered that the Gater-type net acted in part as an outdoor resting place for mosquitoes, especially for host seeking ones. For example, instead of mosquitoes resting on nearby vegetation prior to landing on the bait and feeding they rested in or on the nets.

Net traps have been shown to collect not only host-seeking mosquitoes, but also half-gravid, gravid and blood-fed females, as well as some males (Colless 1959; Hamon 1964 and Hodgkin 1956) and hence catches from these net traps do not necessarily represent the composition of the true host-seeking population. Blood-fed individuals caught within a baited net may not necessarily have fed from that species of host. In addition, net traps may not sample the same range of species or in the same proportions as direct landing catches. Colless (1959) found that when mosquitoes could feed on the bait there was often a disparity between the proportions of the different species caught in net traps and the percentage that fed on the bait. For example, large numbers of *Culex tritaeniorhynchus* were caught in nets baited with man but few took a blood-meal, whereas only a few *Culex quinquefasciatus* were caught but a high percentage of them were blood-fed. In Nigeria more adults of *Anopheles gambiae* and *Anopheles nili* were caught in direct bait catches on man than in mosquito-net collections irrespective of whether the catches were performed in village huts or outside in the compounds (Service 1963). There was no significant difference between the numbers of *Anopheles funestus* caught by the two methods. In a series of outdoor net trap collections species other than the above three *Anopheles* formed only 1.4% of the total catch, whereas in direct bait catches 'other species' formed as much as 49.1% of the catch. Furthermore, *Anopheles brohieri*, which was not collected from net traps, constituted 13.8% of the mosquitoes in direct catches. Clearly different species may react differently to a bait enclosed within a net. In comparative trials in Burkina Faso more individuals of most *Anopheles* species, particularly *Anopheles broheiri* and *Anopheles flavicosta*, were caught on human bait outside a net than from a man-baited net. With culicines about the same number of *Aedes* were attracted to man in and outside a net, but for other genera more mosquitoes were caught in net traps than in direct landing catches (Hamon 1964). Fewer adults of most species were retrieved when collections were made every 2 h instead of hourly. In Morocco, except for *Aedes* mosquitoes, Bailly-Choumara (1973) collected the same species from human-baited nets as in direct landing catches, but in nearly all instances the numbers caught in nets were fewer. In Pakistan Akiyama (1973) found that in five trap-nights human-baited net traps raised a few inches from the ground caught 1570 female and 139 male *Anopheles*, including 22 female and 15 male *Anopheles culicifacies*. Half-gravid, gravid and blood-fed females were retrieved from the baited nets, but the engorged individuals all contained bovid blood. On the same nights, however, no *Anopheles culicifacies* and only 16 female *Anopheles hyrcanus* group and 10 female *Anopheles pulcherrimus* were caught in direct landing catches. Unbaited nets caught 172 mosquitoes. Clearly entry into the nets was either mainly accidental or to find shelter, and not due to host attraction. The presence of large numbers of mosquitoes in unbaited nets is probably often due to them acting as efficient Malaise traps and consequently in these situations the catch comprises individuals that have flown in indiscriminately or in search of a resting site, as well as those attracted by the bait. When unbaited control nets catch only few mosquitoes, then almost all those collected from bait-nets can be considered to have been attracted by the bait.

The biting cycles of mosquitoes may differ slightly between net traps and direct bait catches (Hamon 1964). Net traps may also in some cases deter or repel mosquitoes from biting a normally acceptable host, and this was first observed by Haddow et al. (1948) in East Africa. *Aedes africanus* avidly attacked monkeys in the open forest, but did not readily bite monkeys enclosed in cages. As a consequence sentinel monkeys had to be tethered to wooden posts fixed to tree platforms in the open, rather than kept in cages.

In spite of the limitations described above, net traps can in many situations provide a cheap and easy method of catching mosquitoes attracted to man and other animals.

Stable traps

Stable traps were introduced in the study of mosquito biology later than net traps, and have mainly been used in the Americas. They were originally used to catch mosquitoes attracted to equines and bovids, but have since been baited with man, a wide variety of mammals, birds and even cold-blooded vertebrates. They are heavier and more permanent structures than net traps but lightweight models can be made, which are readily dismantled and easily transported.

Magoon trap

The first use of stable traps is usually credited to Payne, who used them in Haiti in 1923, but earlier than this Metz (1920) caught *Anopheles crucians* and *Anopheles quadrimaculatus* in Florida, USA using small wooden shelters baited with pigs and having a longitudinal slit entrance on two sides tapering from an 8-in to a 1-in opening. Stable traps were introduced into Puerto Rico in 1926 where they were modified by Earle (Earle and Howard 1936), after which their popularity spread to other West Indian Islands, Colombia and Panama, and they became known as Caribbean traps. Only after Magoon had visited Puerto Rico was a detailed description of a modified and more portable trap published (Magoon 1935). The trap in its original or modified form has now been used in many countries. Its size, construction and the materials from which it is made can be adapted to suit local conditions and requirements.

A useful trap is one with a base measuring about 3½ × 6½ ft, with one of the two longer upright sides about 7 ft high joined by a sloping and waterproof roof to the opposite vertical side which is about 5½ ft high (Fig. 6.22).

When animals such as donkeys and calves are used there is no need for a floor, but unless animals such as pigs and rodents are enclosed within a cage, a floor may be necessary to prevent them from burrowing and escaping. A door is fitted at one end of the trap, and another may be fitted at the opposite end so that larger animals can be removed without turning them round. Apart from the roof as much as possible of the trap should be made of plastic or wire mesh mosquito gauze to allow bait odours to escape, but the lower half usually has to be made of stout plywood or sheet metal on a wooden frame, to prevent the animals from kicking it to pieces.

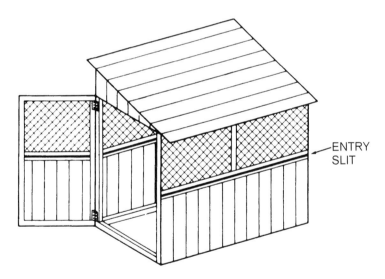

Fig. 6.22. Magoon-type bait stable trap (World Health Organization 1975*b*)

Mosquitoes enter the trap through a horizontal entrance slit (Fig. 6.23) placed about halfway up the trap. This entrance slit may extend completely round the trap, or be confined to the two longer sides. The construction of the slit is the most critical part in building the trap. It is usually made from

4-in wooden planks placed to form a V-shaped trough with a 6–8 in wide opening to the outside converging to leave a ¾–1-in slit-like opening in the trap. The walls of the trap above the entrance baffle are made of mosquito gauze and those below of wood.

It is often convenient to make the trap demountable with the various sections fixed into position with bolts and wing nuts, enabling it to be readily transported to different sites (Fletcher et al. 1988). Strips of plastic foam stuck along the edges of the different sections eliminate gaps through which the mosquitoes might escape. Wood chemically treated with insecticides or fumigant preservatives should be avoided as this may deter mosquitoes from entering the trap, or kill those that have already entered. A clear non-toxic varnish or paint can be used to protect the outside of the trap.

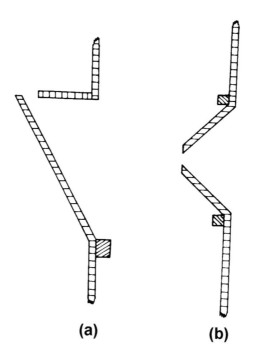

(a)   **(b)**

Fig. 6.23. (*a*) Bates (Egyptian)-type (i.e. vertical) entry baffle for stable traps; (*b*) Magoon-type (i.e. horizontal) baffle for stable traps

Various predators such as lizards and spiders may enter the trap and will destroy the catch unless removed. Both Bates (1944*b*) and Bradley et al. (1949) recommended that permanent trap sites should have a concrete base surrounded by a water trough to prevent ants entering the trap and destroying the catch. The normal routine is to expose a bait animal in the trap for 12 or 24 h, possibly with food or water, after which a collector enters the trap and carefully removes all mosquitoes resting on the walls and roof.

In Texas, USA because the numbers of mosquitoes biting cattle were so great Kuntz et al. (1982) modified the Magoon trap to prevent trapped mosquitoes feeding on the bait animals. This was done by fixing a wooden collection box (91 × 182 × 31 cm) covered with plastic mesh screening over the lower part of each side of the mosquito-proofed stanchion housing a host animal (horse, calf, pig, dog, sheep). Two horizontal louvre openings ending in a 2.5-cm wide slit were positioned at heights of 7.3 and 48 cm on each box. Louvres were fitted with door panels which were closed at the end of the trapping period. Mosquitoes were removed through sleeve-type armholes fitted in the top and two ends of each box. Two traps were operated simultaneously 15 m apart for 2-h periods after sunset, one in a flooded plain site and the other at a rice-farm site. Five mosquito genera were caught, and the most common species was *Psorophora columbiae* which formed 67.0 and 85.6% of the total catches from these two locations. At the floodplain site the next most common mosquitoes were *Culex salinarius* (12.8%), *Anopheles quadrimaculatus* (5.9%), and *Ochlerotatus taeniorhynchus* (3.8%), while on the rice-farm the next most common mosquitoes were *Anopheles quadrimaculatus* (6.4%) and *Anopheles crucians* (4.5%). In addition to the animals listed above, the stable traps were sometimes baited with 3 litres/min of carbon dioxide, or 4 CDC light-traps. Horse- and calf-baited traps attracted most mosquitoes, and except for CO_2 traps at the farm site, carbon dioxide and light-traps were not very attractive. No mosquitoes were caught in unbaited traps.

Kay and Bulfin (1977) described a cheap and easily made sectional trap constructed mainly from angle-iron (Dexion) that can be readily transported and fitted to cattle crushes in remote areas to convert them to small Magoon-type traps. The design prevents the entrapped mosquitoes from feeding on the cattle or other livestock bait.

In Tennessee, USA Hribar and Gerhardt (1986) compared the mosquitoes attracted to dogs in a modified Magoon trap and a Shemanchuk (1978) trap (see section 3.4 and Fig. 6.29 below). The Magoon trap measured 1.2 × 2.44 m and was 1.83 m high, with entry louvres on the two longer sides. Areas above the louvres were made of mesh screening while the part below the louvres was made of wood. The trap was mounted on a trailer to enable it to be easily transported to different areas. The Shemanchuk trap consisted of a 1.8 × 1.8 and 1.2 m wooden frame with a 1-m high sloping gable on top. The entire trap was covered with plastic mosquito screening. The four sides could be raised to allow hungry mosquitoes to enter. At about 1600 h two dogs were placed in the Magoon trap and at about 2230 h all trapped mosquitoes were removed, after which one of the dogs was tethered under the Shemanchuk trap, while the other remained in the Magoon trap. The sides of the Shemanchuk trap were lifted for 10 min to allow mosquitoes to enter, then they were lowered and 5 min allowed for entrapped mosquitoes to feed on the host. The subsequent collection of mosquitoes took 5 min so the sides of the trap were repeatedly raised every 20 min from 1830–2230 h.

Although the same five species were collected from both these traps there were some important differences. For example, *Ochlerotatus trivittatus* was collected about equally from both traps, but *Psorophora ferox* and *Aedes vexans* were more common in the Shemanchuk trap, while *Ochlerotatus triseriatus* and *Culex salinarius* were more commonly caught in the Magoon trap. The two dog-baited traps therefore ranked the species differently in abundance, and because of this the authors stressed the need for more than one trapping technique in epidemiological studies.

Ernst and Slocombe (1984) used a Magoon trap modified to take a dog (Ernst 1982). The roof was made of plywood and most of the upper part of the sides consisted of mosquito screening. Entrance baffles were small V-shaped longitudinal slits placed directly beneath the screening on three sides of the trap. The dog was restrained in a wire cage. From 1980 to 1981 when the trap was operated on an unspecified number of occasions from evening to midday, 3310 mosquitoes belonging to 10 species were caught, including 1114 *Culex pipiens/restuans*, 739 *Aedes vexans*, 510 *Ochlerotatus trivittatus* and 569 *Coquillettidia perturbans*. In India Mahadev et al. (1978) using pig-baited Magoon traps caught a total of 33 mosquito species, but only 0.9 mosquitoes/trap-night. Their traps had two 0.75-cm horizontal entrance slits which unfortunately let some mosquitoes escape. They also used the portable bait trap of Rao (1957), which when containing a chicken caught a mean of 1.4 mosquitoes/trap-night.

In El Salvador Magoon traps were successfully employed to study the seasonal variations in the numbers of *Anopheles albimanus* and *Anopheles pseudopunctipennis* (Kumm and Zúniga 1944), but Lofgren et al. (1974) were unsuccessful in their attempts to catch *Anopheles albimanus* in El Salvador in calf-baited traps. Lowe and Bailey (1981), however, developed a simple Magoon-type trap that proved very successful in catching this vector. Basically the trap consists of a wooden frame (1.2 × 2.4 m and 1.8 m high) with the bottom half on three sides covered with plywood and the upper halves covered with plastic mesh netting. Between the top and bottom sections there is a horizontal 15-cm wide slit-like entrance tapering to 2.5 cm, to allow entry of hungry mosquitoes. The roof is covered with plastic sheeting to protect against rain, and a door is sited at one end. A calf is tethered to a wooden support inside the trap. Two people could lift the trap into the back of a pick-up truck. The average catch of female *Anopheles albimanus* per night from four traps varied from 12–157, and was about 1.7 greater than the numbers collected per man-hour from stables. In Venezuela Gabaldon et al. (1940) found stable traps useful for sampling *Anopheles albimanus*, but not *Anopheles darlingi*. Sasse and Hackett (1950) used stable traps in Peru to study the host preference of *Anopheles pseudopunctipennis*.

Magoon traps were used by Ulloa et al. (2002) in Mexico to capture *Anopheles vestitipennis* for mark-release-recapture studies. Two traps were placed 5 m apart and were operated over 12-h periods from (1800 to 0600 h) for 20 consecutive nights. In the first experimental series, traps were baited with 2 adult male humans in one trap and one horse in the other, while in a second series, traps contained either two men or two pigs, and in a third series, two men or one cow were used as baits. Mosquitoes were collected by oral aspirator for 45 minutes per hour in human-baited traps, and in animal-baited traps, mosquitoes were collected for the final 15 minutes of each hour. In later studies, Ulloa et al. (2005) also used Magoon traps containing either two human volunteers or one horse to collect *Anopheles vestitipennis* for gonotrophic cycle duration and survival rate estimations. In human-baited traps, mosquitoes were collected for the first 45 minutes of each hour. In animal-baited traps, mosquitoes resting on the interior surfaces of the trap were collected during the final 15 minutes of each hour.

In studying host preferences in Nigeria one Magoon trap was baited with a goat while another 45 ft away was baited alternatively with a sheep, a pig and two monkeys (*Erythrocebas patas patas*). About 33 mosquito species were caught, including 15 *Anopheles* species which formed about 85–93% of the catches of mosquitoes from the trap when it was baited with sheep, goat, and a pig. Only 16 mosquitoes were collected from the monkey-baited trap and 10 of these were *Culex quinquefasciatus* (Service 1964).

In dengue studies in Malaysia Rudnick (1986) used relatively large Magoon traps which were baited with 4–6 monkeys. Traps on the ground contained macaque monkeys (*Macaca* spp.) while those raised by winch and pulley to 75 ft in the tree canopy were baited with leaf-monkeys (*Presbytis* spp.). Numerous mosquito species (35) and genera (17) were collected in the high canopy traps, e.g. species of *Anopheles, Culex, Mansonia, Coquillettidia, Zeugnomyia, Orthopodomyia, Armigeres, Heizmannia, Tripteroides, Aedes* and seven other genera. The most common mosquito was *Culex cinctellus*, but of the 3194 females caught in one locality only three had fed on the monkeys, whereas 16 of the 45 female *Ochlerotatus niveus* group were engorged. In some localities at least 53 species belonging to 22 genera were caught in *Macaca*-baited traps at ground level. In dengue studies in Malaysia Garcia et al. (1988) modified this small Magoon trap by constructing the roof, door, upper and lower sides separately from mitred aluminium screen moulding. The base of the trap measured 60 × 60 × 56 cm, while the panels were formed of plastic mesh screening held together by four vertical angle-iron corner supports. Four triangular pieces of mesh screening stitched together and fitted within an aluminium frame formed the roof (60 × 60 × 50-cm high). A hook and four lines supported the configuration of the roof (Fig. 6.24).

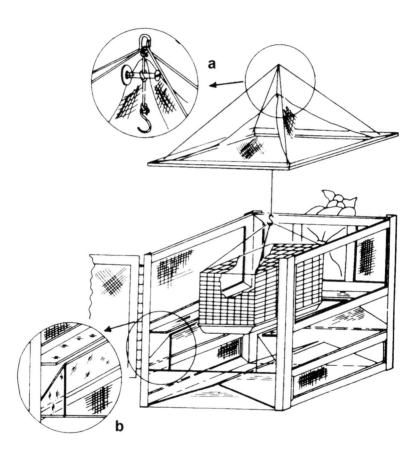

Fig. 6.24. Miniature Magoon trap for rabbits with door open, and roof raised to give better view of the inside of the trap, a—details of roof and system for suspension of rabbit cage, b—details of entry slits (Garcia et al. 1988)

An animal bait cage was suspended in the centre of the trap by a nylon line from the roof hook, and 3 kg dry-ice placed in a 15 × 20 × 30-cm styrofoam box insulated with newspapers was placed on the roof and held in place by slipping the corners of the box under the four support lines. A 2-cm hole in the lower side of the box allowed carbon dioxide to flow down into the trap. Mosquitoes entered the trap through 1-cm wide baffles made of plexiglas and fitted to all sides of the trap, except on the end door. Another point of entry was the slit formed by two sloping (20°) panels of plexiglas that formed the bottom of the trap (Fig. 6.24). Mosquitoes were removed with aspirators inserted through a cloth sleeve fitted to the rear end of the trap. In California, USA these traps were baited with a rabbit and carbon dioxide and operated from 1400–1000 h the following day, and mean catches of female *Ochlerotatus sierrensis* were 39.1, 79.9 and 355.4 depending on ecological location (Garcia et al. 1989). The maximum overnight catch was 901 mosquitoes. Males hovered around the trap but relatively few were collected from inside it, the male:female ratio varied from 1:21 to 1:95. A Fay-Prince trap (see Chap. 11) augmented with carbon dioxide caught similar numbers of female *Ochlerotatus sierrensis* as the rabbit-baited trap, but was more convenient to use because it was not encumbered with a live bait. Moreover, it was 15–20 times better at catching males. This is one of the few traps that employs both a bait animal and carbon dioxide. Landry and DeFoliart (1986), however, used mice and dry ice in a CDC-type trap but found that the addition of mice did not improve their catch of *Ochlerotatus triseriatus*.

Bates' (Egyptian) stable trap

Although Gabaldon et al. (1940) reported that in Venezuela less than 3% of the mosquitoes caught in Magoon traps escaped before they were collected shortly after sunrise, Bates (1944*b*) working in Egypt found that the trap was inefficient. A large proportion of mosquitoes caught escaped back through the entrance slits before collections were made in the early morning. He therefore made a new type of trap which he called the Egyptian trap, but which has subsequently also become known as Bates' stable trap. The principal difference between the Bates and Magoon traps is in the design of the entrance slit (Fig. 6.23*a*). This was modelled on the entrance baffle used by van Thiel et al. (1939) in their experimental work on host attractiveness. A further difference is that unlike the original Magoon trap entrance slits were incorporated only in the two longer sides and not in the ends. Various sized traps were made, but finally standardised to measure 2 m long, 1 m wide and 1.75 m high; they were placed on a concrete base. A transparent roof was placed on the trap but this has not generally been used by later workers. The upper part of the entrance baffle consists of a 20-cm wide length of wood set at right angles to the side of the trap. The lower section of the baffle is 44 cm wide and is positioned 38 cm below and slopes upwards to leave a vertical 2-cm opening between the upper and lower lengths of wood.

In Egypt Bates (1944*b*) caught 4000–5000 mosquitoes/trap-night, including up to 1000 *Anopheles*. In comparative trials in Colombia Bates (1944*b*) clearly showed that many more mosquitoes were caught in Bates than Magoon traps, e.g. the catch of *Anopheles* (mostly *Anopheles rangeli*) was about 10 times greater. De Zulueta (1950) reported that very few mosquitoes escaped from these traps.

In Australia Kay et al. (1979*a*) baited Bates' traps with a man, a feral pig, a dog, two domestic fowls, a grey kangaroo and a calf. They used the Feeding Index of Kay et al. (1979*b*) to study host preferences. At one site 44 626 mosquitoes belonging to at least 35 taxa were trapped from 360 collections, from another site 26 215 mosquitoes belonging to 15 taxa were caught in 90 collections. The most common species were *Culex annulirostris, Culex quinquefasciatus, Anopheles bancroftii, Ochlerotatus normanensis* and *Anopheles annulipes*. Although the traps were placed in very similar topographical situations there were considerable differences between catches in differently located traps. This paper provides a good interpretation of feeding preferences from baited traps and from the collection of blood-engorged mosquitoes.

Nelson et al. (1976) baited Bates' traps simultaneously with a jackrabbit and a chicken or pheasant, and after some 90–110 trap-nights more than 21 000 mosquitoes were collected, of which nearly 90% were *Culex tarsalis*.

In Canada Hudson (1983) caught 14 mosquito species from 12 nights operation of a Bates' trap, the most numerous being *Aedes vexans* (1325), *Ochlerotatus communis* group (358), *Culiseta inornata* (110), and *Culiseta alaskaensis* (94).

More recently, a slightly modified horse-baited Bates' stable trap was used by Samui et al. (2003) in Louisiana, USA. The trap was 3.1 m long × 1.5 m wide × 2.4 m high, and had mosquito wire screen covering the longer sides, the roof, three-quarters of the rear door, and one quarter of the front door to maximise diffusion of bait odours. The stable trap had no floor and the horse was tethered to ring bolts. Horses were kept in the traps from 1800 or 1900 to 2300 or 0000 h and mosquitoes were collected by hand aspirator. 28 581 individual mosquitoes, representing 26 species, were caught in the traps, the dominant species being *Psorophora columbiae, Culex (Melanoconion)* sp., and *Culex salinarius*. Important horse-feeding species were classified as those that constituted more than 1% of the total catch at a site and had an engorgement rate in excess of 50% (source of blood meal was not determined in the study). Species considered important horse feeders at all three sites included *Culex salinarius, Culex (Melano-conion)* spp., *Aedes vexans* and *Anopheles crucians*, species from which Eastern Equine Encephalitis virus has been isolated. Other potential EEE vectors that were collected include *Anopheles quadrimaculatus* s.l., *Co-quillettidia perturbans*, and *Psorophora columbiae*. A comparison between CO_2 baited CDC traps and horse traps showed that a stable catch of 500 mosquitoes could be converted to a CDC trap catch of 120 000.

Roberts' stable trap

Roberts (1965) designed a modified Magoon-type trap which had better air circulation thus enabling bait odours to disseminate more efficiently from the trap. Another feature of the trap was that mosquitoes could enter it at several heights. The trap consists of a wooden framework, 7 ft long, 5 ft wide, 6 ft high, with a flat roof and a door at one end (Fig. 6.25).

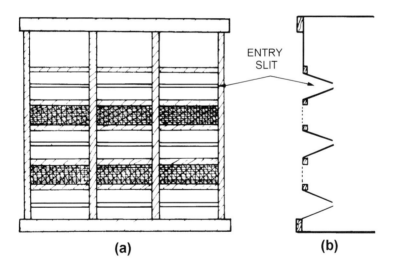

ENTRY SLIT

(a)   **(b)**

Fig. 6.25. (*a*)–(*b*) stable trap with entrance at three levels (after Roberts 1965)

The bait animal is confined in a stanchion to keep it in the centre and prevent it from damaging the trap. About the top 1 ft of all sides and door are covered with polythene sheeting, while the rest of the door and sections in between the baffles are covered with fine copper wire mesh screen. The Magoon-type horizontal baffles are built into the two long sides of the trap at ground level and at heights of 2 and 4 ½ ft. There are no baffle entrances in the end sections. In a comparison of the mosquitoes caught in light-traps and in steer-baited traps in Mississippi, USA 7381 mosquitoes belonging to 16 species were caught at light and 105 387 representing 23 species were caught in the stable traps. Only *Culex territans* and *Uranotaenia sapphirina* were caught at light and not in the stable traps.

In Wyoming, USA Pennington and Lloyd (1975) baited Roberts' trap with a heifer, and collections were made at four irregular intervals throughout the diel. The numbers of mosquitoes in each collecting period were determined by either counting, or by weighing and estimating the numbers from the weight of a 100-mosquito subsample. From sixteen 24-h collections 71 440 *Ochlerotatus melanimon* representing 44.34% of the total catch, 56 388 *Ochlerotatus dorsalis* (35.06%) and another six species were caught.

In the USA Jones and Lloyd (1985) found that about equal numbers of nine mosquito species were attracted to a 5–6-month-old ewe held in a Roberts' trap and to a CDC light-trap placed 300 m away and baited with 4.5 kg dry ice placed in a 1.37-kg coffee can with four 3-mm holes drilled in the bottom. This arrangement resulted in the dry ice releasing about the same amount of carbon dioxide as an adult bovine-sized animal (Morris and DeFoliart 1969). Hayakawa et al. (1990) used a modified version of the Roberts' trap in which the entrance slits (10 cm) were much wider than the original 1.9-cm slits. Their cattle-baited trap caught more mosquitoes than did carbon dioxide-baited bed-nets.

Loftin et al. (1996) used four stable traps (2.2 × 1.2 × 1.6 m) based on Roberts' (1965) design, constructed from 1.9 cm diameter aluminium tubing and covered with aluminium screen material to test the attractiveness of cattle to mosquitoes following treatment of the cattle with different insecticides. The trap was modified from the original design by having horizontal openings 1.9 cm wide running the length of the traps at heights of 22.8, 77.6 and 121.9 cm above ground level. Nylon netting was placed behind each door to reduce mosquito escapes during experimental manipulations. The traps collected a total of 20 738 mosquitoes comprising 6 species. Mean catches of 348 *Aedes vexans* and 291 *Psorophora confinnis* were obtained per animal per night, and these two species represented 98.6% of the total catch. Similar traps, constructed from 1.9 cm square steel tubing were used by Loftin et al. (1997) to investigate the host preferences of mosquitoes in New Mexico, USA. Traps were baited with a cow, a horse, a dog, and six chickens and the side slots were opened at 1900 h and closed the following morning at 0700 h. Mosquitoes were removed from the traps using a battery-powered aspirator (Meek et al. 1985) after allowing 30 min for any unfed mosquitoes in the trap to feed. Trapping was conducted for four consecutive nights each month from June to September at two sites. Ten species of five genera were collected, with *Aedes vexans, Ochlerotatus dorsalis, Culex quinquefasciatus, Culex tarsalis*, and *Culiseta inornata* the most abundant. *Aedes vexans* demonstrated a host preference for cattle and horses, as shown by numbers collected, proportion engorged, and proportions of engorged individuals collected from each host. *Ochlerotatus dorsalis* also preferentially fed on cattle and horse. *Culex quinquefasciatus* was most frequently collected in traps containing either dogs or chickens. The dog-baited trap caught more individuals than the chicken-baited trap, but the proportion of mosquitoes that was engorged was higher in the chicken-baited trap (88%). *Culex tarsalis* also appeared to have a preference for feeding on small animals, but this was less marked than for *Culex quinquefasciatus*. Engorgement proportions on all hosts were similar. *Culiseta inornata* was predominantly collected in the traps baited with large animals.

Wright and DeFoliart's stable trap

In Wisconsin, USA Wright and DeFoliart (1970) constructed two sizes of small Magoon-type traps which were baited with squirrels, rodents, reptiles and amphibians. The smaller size trap consisted of an $18 \times 20 \times 20$-in wooden framework with the top and bottom made of 3/8-in plywood, and all four sides, but excluding the door ($11 \times 11\frac{1}{2}$ in), covered with 52-in mesh natural colour saran screening. A 4-in wide V-shaped plywood baffle leaving a 3/4-in entrance slit was placed 4 in from the top of the trap and extended the entire length of all four sides. The larger type of trap was similar in design except that it was 27 in long, and the door measured 12×13 in. In 17 checks with these small Magoon traps 190 mosquitoes were trapped but none was blood-fed.

Shannon's stable trap (= dawn trap)

Although sometimes referred to as a stable trap it is really a hybrid between a stable trap and an experimental hut containing a window-type exit trap. It consists of two parts, a wooden stable and a dawn trap (Fig. 6.26). The stable has specially constructed lightproof overhanging eaves which leave a 6-in gap on three sides of the stable, through which mosquitoes can enter to feed on the bait animal. A close fitting door is provided to the stable which is dark inside, and may be painted black. The wooden dawn trap which measures about $3\frac{1}{2} \times 3\frac{1}{2}$ ft and 5 ft high is painted white inside except for the rear wall which is made of mosquito screening. It is fitted to the wall of the stable that faces the rising sun. Mosquitoes that have entered the stable at night to feed on the bait are attracted to the light coming from the dawn trap. They consequently fly though the 30-in wide opening of a screen mesh baffle fitted into the rear wall of the stable and pass through a 1½-in wide longitudinal slit into the dawn trap. The original description of the trap by Shannon (1943) is not very accessible, but a good account together with a diagram and photograph is presented by Earle (1949). These traps have mainly been used in Central and South America. Shannon (Earle 1949) caught as many as 7145 *Anopheles aquasalis* in one night from a dawn trap, and in Trinidad Senior White (1952) collected mosquitoes in traps baited with man, ox, goats, horse and a pig.

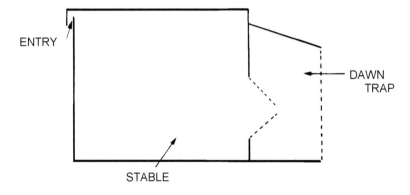

Fig. 6.26. Shannon stable trap (dawn trap) (after Earle 1949)

Russell's calf-baited hut

In studying the flight range of *Anopheles culicifacies* in India Russell et al. (1944) built 80 calf-baited trap huts which they placed at various distances from the release point of marked mosquitoes. Each trap consisted of a small hut, 7 ft high, 6 ft long and 5 ft wide, built from a framework of *Casuarina* scantlings covered with palm matting. A thatched sloping roof reached to the ground on both long sides, and both ends were also covered with thatch. A doorway (21/2 × 11/2 ft) was left at one end for the entry of mosquitoes and a bamboo gate was fixed across this entrance to prevent jackals entering. Two strips of dark cloth were suspended from the roof to provide additional resting places for mosquitoes, which were collected by aspirators after a canvas curtain had been pulled across the doorway.

Despite the success of the trap in catching 65 893 *Anopheles culicifacies* and 141 928 other species, including *Culex*, they have been little used as bait traps. Similar cattle-baited traps were used by Amerasinghe and Amerasinghe (1999) and Amerasinghe et al. (1999) in a study of malaria vectors in Sri Lanka. Hut dimensions were 3 m × 3 m × 3 m and were constructed from coconut thatch. A gap of 6cm was left at the bottom of the walls to allow mosquitoes to enter. A host animal was tethered within each hut at 1800 h and removed at 0500 h the next morning, at which time all mosquitoes within the trap hut were collected using battery powered mechanical aspirators (Hausherr's Machine Works, U.S.A.). The most commonly captured mosquitoes in these huts were *Anopheles varuna* and *Anopheles culicifacies*. The trap huts were much more effective in catching *Anopheles culicifacies* than human bait catches, cattle bait catches, indoor-resting collections and collections from pit traps.

General considerations of stable traps

As with bait-nets and most types of animal-baited traps certain mosquito species may be reluctant to enter them. A check on the efficiency of a stable trap in catching mosquitoes that feed on a certain bait can be made by comparing the mosquitoes caught biting the bait outside and inside the trap. The efficiency and usefulness of stable traps may vary in different localities and for different mosquito species. Although de Zulueta (1950) found that with donkey-baited traps the smell from a previous night's bait did not attract any mosquitoes, this is a difficulty that may be encountered with stable traps, especially those with wooden floors. They tend to become contaminated with animal excreta and urine, so that the 'smell' of one animal persists after it has been removed and another introduced. Scherer et al. (1959) found that large numbers of mosquitoes were still caught in stable traps the day after a pig was removed; the numbers decreased on the following 2 days. The likelihood of mosquitoes being attracted to a lingering smell of a bait animal can be reduced by covering the floor with disposable plastic sheeting, or confining the bait to a cage that is afterwards removed. Alternatively the trap can be scrubbed out with clean water and left open for a few days before a different animal is introduced, or if a trap is used without a floor it can be moved to a new site after the conclusion of trials with one type of bait animal. As previously noted by Gabaldon et al. (1940) in Venezuela, Bates (1944*b*) found that in 35 of 54 weekly catches the numbers caught on the first night were greater than those on the third consecutive night. When using Magoon traps Gabaldon et al. (1940) found that in any one location more *Anopheles* were caught on the first night than on either the second or third nights, but when the trap was moved only about 3 m a 'first night's catch' was obtained. The most likely explanation is that when a stable trap is first introduced into an area it catches, on the first night, both the local resting population that has built up over several days in addition to mosquitoes flying in from further afield. On subsequent nights the local population of hungry females having been depleted, only those flying into the area are caught. A change of trap location is likely to result in another high catch on the first night. In Colombia stable traps placed amongst grassy vegetation caught several times as many mosquitoes as traps no more than 10 m away, but sited in an area of cleared grass (de Zulueta 1952). Presumably in the latter area there were fewer resting places for mosquitoes. Surprisingly, however, more *Anopheles darlingi* were caught in traps in the cleared area than in traps amongst grass. De Zulueta (1952) suggested that this was related to differences between the host seeking behaviour of *Anopheles darlingi* and the other species.

Although an average 1470 mosquitoes were caught per night in a donkey-baited trap in Colombia placed in savannah areas the species composition differed markedly from that obtained by drop-net collections (de Zulueta 1952). For example, drop-net collections showed that *Anopheles* comprised 19% of the total mosquito population resting amongst the grass, whereas in stable traps *Anopheles* constituted as much as about 73% of the catch. Clearly culicines were inadequately sampled. Baiting the traps with a calf or fowls had no effect on species composition. During four trials with unbaited traps, in which the door was sometimes left open, only 12 mosquitoes were caught, clearly demonstrating that mosquitoes were not just seeking shelter in the traps but were attracted to the bait animals. There was no evidence that the smell of a bait animal, such as a donkey, which might have lingered on from a previous night's trial attracted any mosquitoes into the trap; the bait had to be within the trap.

Murphey et al. (1967) caught 6803 mosquitoes belonging to 14 genera in 2-ft cube Magoon traps baited with 19 different vertebrate hosts enclosed in wire mesh cages. It was considered that although the numbers caught might reflect the attractiveness of the different hosts, they were dependent on the size of the population and flight activities of the mosquitoes, whereas the percentage that actually fed on the bait was less dependent on these factors.

Stable traps usually contain mammals, but in the USA Blackmore and Dow (1958) used a small trap (4 × 4 × 4 ft) with Egyptian-type baffles baited with adult and nestling birds. A much higher proportion of *Culex tarsalis* caught in the traps had fed on nestling birds than on adults, but the difference varied according to the species of bird used. Mosquitoes probably found it easier to get blood-meals from nestling birds because they have fewer feathers and are more quiescent. As a result of these experiments Blackmore and Dow (1958) thought that birds nesting during the encephalitis season might be important reservoirs of infection.

When different animals are placed close together hungry mosquitoes are given a choice of hosts and host preferences can be studied. If, however, they are widely separated there is no such choice, and a measure of only the relative attractiveness to mosquitoes is obtained. If the animals are placed too far apart, they may be in different ecological environments containing different mosquito populations. It has been shown that mosquitoes caught in human landing catches performed at different sites in a more or less homogenous habitat may differ (Service 1971*b*). Under these conditions it may not be valid to make close comparisons between trap catches. In Japan small stable traps and other small traps baited with birds were used to collect *Culex tritaeniorhynchus* and *Culex pipiens* (Scherer et al. 1959). Pronounced differences sometimes occurred between the numbers of mosquitoes caught in identical traps due to both trap location and microhabitat around the traps. In addition some traps consistently caught many more mosquitoes than others, even despite changing both baits and trap positions. Even more surprising was the fact that although similar numbers of *Culex tritaeniorhynchus* were caught in six bird-baited traps, two traps never caught mosquitoes infected with Japanese encephalitis virus (Scherer et al. 1959). These same workers made some interesting observations on the attractiveness of traps without baits. They found that after a pig had been removed from a stable trap about the same numbers of *Culex tritaeniorhynchus* were caught on the following day. However, when a pig was removed for three consecutive days, the catch progressively decreased from over 2000 mosquitoes/night to less than 50. Similar but less well documented reductions occurred when man and birds were removed from stable traps for 3 days.

Other traps for large animals

McCreadie et al. trap

Because mosquitoes sometimes do not enter animal-baited traps employing baffles and cones, and because the suspension of drop-nets over a bait may deter some mosquitoes from approaching and biting Jones (1961) devised a tent-like net that was operated by springs and which flipped over an enclosed small animal, such as a sheep. For animals such as cattle

McCreadie et al. (1984) developed a larger tent-like trap, in which the animal is placed in a wooden pen surrounded on the ground by a rectangular frame (2.4 × 3.0 m) of 4.3-cm diameter galvanised steel piping. Three hoops, each consisting of two straight and one curved section, of 2.1-cm diameter piping are fixed to this framework to form a 2.1-m high arc over the enclosed bait. A fine plastic netting tent is made to have a 30-cm canvas lower border and three 8-cm wide canvas strips sewn into the netting to lie directly underneath each metal hoop. Brass eyes and 5-cm diameter metal rings are sewn into these strips so that the net-tent is fixed to a series of rings which surround the three hoops arising from the basal metal frame (Fig. 6.27). In operation the tent is collapsed on the ground along one side of the bait animal, then after a 10-min exposure period two collectors walk to the trap and rapidly pull the tent up and over the three metal hoops to enclose the animal and mosquitoes attacking it. This operation takes about 3 s and only a few mosquitoes hovering around the bait may escape. If required a further 10 min can be allotted for entrapped mosquitoes to feed on the host. A zipper sewn into one end of the tent allows a collector to enter and remove the catch with aspirators. McCreadie et al. (1984) used modified commercial Black and Decker Dustbuster hand vacuum cleaners as aspirators. In Canada when baited with a calf the trap caught 26 species of biting flies, including six mosquito species, the most common of which were *Culiseta impatiens, Ochlerotatus abserratus* and *Ochlerotatus punctor*.

Fig. 6.27. Bait-net of McCreadie et al. (1984) (photograph reproduced with permission of J. W. McCreadie)

The authors suggested that smaller versions of the trap can be made for smaller animals, and that perhaps less conspicuous pens could be made so as to allow a more visible silhouette of the baits.

Later McCreadie et al. (1985) found that about double the numbers of simuliids (10 747 vs 5720) were caught in their trap baited with a calf than in the carbon dioxide suction trap of Trueman and McIver (1981) (see Chapter 10). Only a few mosquitoes were caught, the most common species being *Ochlerotatus abserratus, Culiseta impatiens* and *Ochlerotatus punctor*, of which the calf-baited trap trapped 363 whereas the carbon dioxide trap caught just 117 adults.

Because of reports of bias in the collection of haematophagous insects from tethered baits (Bennett 1960; Zimmerman and Turner 1983), Fletcher et al. (1988) preferred to collect biting insects from a horse-baited trap, which used the same principles as the one designed by McCreadie et al. (1984). Namely, the bait (horse) is restricted in a wooden stanchion surrounded by a wooden frame (4 × 10 ft and 7.5 ft high) (Fig. 6.28a). The upper and lower parts of the frame are covered with 40-mesh plastic screening, except that the upper 30 cm is covered with clear polythene sheeting and contains on each long side three collecting holes (6 × 8 in) with lengths of stockinette (Fig. 6.28b). The edges of the two long sides of the trap (left and right) are sealed with plastic foam weather stripping and when erected over the horse are held in position by elastic straps. Each side of the trap is hinged to a base frame of wooden boards (2 × 8 in and 6 ft long) thus allowing the two sides to be opened to expose the horse and when required raised and closed. The roof slopes down towards the middle to form a V-shape, this causes the entrapped mosquitoes to gather along the upper edges of the trap and facilitates their removal, when aspirators are inserted through the entrance ports. For ease of construction the trap is made in sections. The upper part is divided into two sections while the lower part comprises four 'L'-shaped sections. The horse is exposed to biting flies for about 15 min and is then enclosed within the trap, and after allowing 30 min for entrapped mosquitoes to feed they are collected. Although most biting flies (*Simulium*, tabanids, *Culicoides*) fly to the top of the trap and are readily collected, blood-engorged mosquitoes tend to rest on the sides so additional collection holes should be made along the middle of the trap and at the two ends.

Another somewhat similar trap is that of Shemanchuk (1978) designed especially for trapping *Simulium arcticum*, but which in Canada also caught *Aedes vexans, Ochlerotatus flavescens, Ochlerotatus fitchii, Ochlerotatus excrucians, Ochlerotatus punctor* and *Culiseta inornata*. The bait animal (100–450 kg steer or heifer) was closely confined in a metal stanchion bolted (Fig. 6.29) to a plywood white-painted floor and surrounded by a wooden framework (170 × 245 cm and 180 cm tall). Nylon screening was permanently fixed to the roof but fixed to the sides with Velcro, so that it could be folded back onto the roof to fully expose the bait but quickly pulled down to enclose the animal and flies biting it.

Fig. 6.28. *a–b.* Bait trap of Fletcher et al. (1988) (photograph reproduced with permission of M. G. Fletcher)

Fig. 6.29. Steer-baited trap of Shemanchuk (1978) (photograph reproduced with permission of J. A. Shemanchuk)

The distance between the stanchion and two longer sides of the wooden framework was 45 cm, while 52.5 cm separated the frame and stanchion at the two ends. After a suitable exposure period all sides were closed as quickly as possible, and after allowing 10 min for the flies to engorge on the bait, a person entered and collected the insects.

Portable traps of Mitchell et al. and Wilton et al.

The Mitchell et al. (1985) trap is quite large (360 × 360 cm, 210 cm tall) and comprises stitched panels of nylon tulle suspended by ropes tied to the four corners and to trees or some other suitable supports. The bait animal is introduced through a 1-m wide door flap made in the middle of one side, and then secured to a stake. The ropes are pulled to raise the trap 30–46 cm from the ground. The following morning the trap is lowered to touch the ground and the bait removed, after which mosquitoes are collected.

When used in Argentina 2752 and 6929 mosquitoes belonging to at least 18 species in six genera were caught after five and three trap-nights, respectively. *Culex (Culex)* species (45.8%) and *Ochlerotatus albifasciatus* (21.7%) predominated. The trap proved better than a Magoon-type trap. In later trials with a horse as bait, Mitchell et al. (1987) caught over a year 2752 mosquitoes belonging to at least 23 species, but mostly unidentified *Culex (Culex)* species.

Wilton et al. (1985) considered that the usefulness of this portable trap of Mitchell et al. (1985) was limited because it was not self-supporting, but had to be suspended by ropes from nearby trees. To overcome this Wilton et al. (1985) modified two commercially available portable summer 'screen rooms' normally used to provide insect-proof living facilities out of doors. The smaller screen room measured 3.6 × 3.6 × 2.2 m and weighed about 14 kg, the larger room was 3.0 × 4.2 × 2.3 m and weighed some 15.8 kg. All four sides were made of mesh screen and were supported by tubular metal frames, the rooms had a nylon zipper entrance. To convert them to mosquito traps two 41-cm vertical slits were cut at the bottom of all four corners to enable the sides to be raised 20–30 cm above the ground and secured by spring clips for entry of mosquitoes. In Colorado, USA when the larger trap was baited with a horse for six nights 2776 mosquitoes belonging to 11 species were caught, the most common of which were *Aedes vexans* (68.8%), *Culex tarsalis* (9.9%), *Ochlerotatus dorsalis* (9.5%) and *Ochlerotatus melanimon* (5.8%). The mean catch was 462.5 mosquitoes/trap-night compared to 367/trap-night with CDC light-traps supplemented with dry ice. A person exposed under the protection of a mosquito net in the smaller net on five nights from 1 h before, to 1 h after sunset caught 464 mosquitoes belonging to five species, again the most common was *Aedes vexans* (64.6%).

Dog-baited traps

In a study of the potential vectors of *Dirofilaria immitis* in dogs in Minnesota, USA Bemrick and Sandholm (1966) constructed modified Magoon traps, in which the plywood sides (6 × 5 and 5 × 5 ft) were made in two sections to make the trap more portable. A V-shaped baffle with a ¾-in slit opening extended completely round the trap at a height of 28 in from the ground. The large numbers of mosquitoes, especially *Coquillettidia perturbans* and *Aedes vexans*, which were caught in the traps were removed by the collector entering the trap through a door at one end.

A more portable dog-baited trap was made by Villavaso and Steelman (1970). This measured 36 in wide, 46 in long and 30 in high, but when dismantled and the sections stacked on top of each other was only 24 × 30 × 36 in. Four dismantled traps could be transported in a ½-ton pick-up truck and each trap only took about 15 min to assemble. The basic design was that the dog was confined in a centrally placed compartment, about 38 in long and 30 in in width and height. Two removable mosquito collecting boxes (30 × 36 in and 8 in deep) formed the two long sides of the trap. The outside panel of each collecting box was made of 16-in mesh screen wire incorporating two ¼–½-in wide longitudinal entry baffles fitted at 10 and 20 in from the base of the trap. The inner wall of each box was covered with 16-mesh screening protected by a layer of ½-in wire mesh to prevent the dog tearing the screening. Mosquitoes which entered the two collecting boxes through the horizontal openings were prevented by the mesh screening from feeding on the dogs and were collected by inserting aspirators through 1-in square openings cut in the outside mesh screening. Villavaso and Steelman (1970) reported that in the USA 16 mosquito species were caught in dog-baited traps, and that as relatively large numbers of *Culex quinquefasciatus* (90% of the total catch) were obtained, the traps might prove useful for monitoring relative population densities of this mosquito in urban areas.

In Louisiana, USA Cupp and Stokes (1973) also caught 16 mosquito species in this type of dog-baited trap, the most common two species being *Culex salinarius* and *Culex quinquefasciatus*. Their procedure for removing mosquitoes from the collecting boxes was to place them in a plywood box and introduce car exhaust fumes for 5–6 min. About 90–95% of the mosquitoes which were knocked down and collected by aspirators recovered. Brock and Crans (1977) caught very few mosquitoes when using collapsible dog-baited traps of Villavaso and Steelman (1970) in New Jersey, USA. By comparing the relative abundance of mosquitoes in light-traps with those of dog traps they concluded that the trap was biased in favour of *Culex* species, such as *Culex salinarius*, but was inefficient at collecting aedines such as *Aedes vexans* and *Ochlerotatus sollicitans*. They postulated that these species may find it more difficult than *Culex* species to enter the trap via their louvre-type openings.

In studying potential vectors of *Dirofilaria immitis* Lewandowski et al. (1980) constructed a special lightweight wooden dog-baited trap, having louvres of the Bates-type mounted on two opposite sides (Fig. 6.30). Few details other than the figure were given. During two field seasons 2166 mosquitoes belonging to 14 species were caught; the most common of which were *Coquillettidia perturbans* (44%), *Culex pipiens* (17%), *Aedes vexans* (12%) and *Anopheles walkeri* (11%). In other studies on *Dirofilaria immitis* Walters and Lavoipierre (1982) built dog kennels in California, USA measuring 90 × 120 cm and 120 high to the apex of the sloping roof.

There were two Bates-type baffles on three sides at heights of 45 and 72 cm. Only relatively few *Aedes vexans* and *Ochlerotatus sierrensis* entered the traps.

In Indiana, USA Pinger (1985) used modified kennel traps of Klowden and Lea (1979) to study mosquitoes attracted to dogs. An inner restraining cage (69 cm long, 41 cm wide and 53 cm high) made from 2.54-mesh wire had a removable top which when in place was fixed by ties of insulated copper wire.

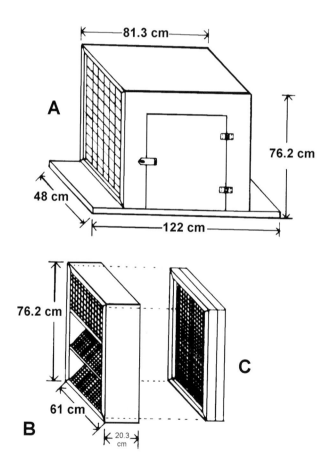

Fig. 6.30. Dog-baited trap, A—dog compartment, B—mosquito collection chambers which are mounted on either side of dog compartment, C—removable screen frame which is removed to collect mosquitoes (Lewandowski et al. 1980)

The holding cage was placed inside an outer cage 91-cm long, 61-cm wide and 61-cm high having a plywood floor and 0.14-cm mesh screen sides. Two sheets of 97–61-cm plexiglas taped together and supported on wooden stakes formed a sloping roof over the cage. The ends of the outer cage had sleeves made from plastic dustbin bags with their ends removed, for replacement of the inner cage. Mosquitoes entered either through a 2-cm eave gap between the plexiglas roof and outer cage or through a 2-cm slit on each side of the outer cage positioned 24.5-cm from the ground. The dog was exposed in a trap from 1800–2100 h. Over 2 years 14 species of mosquitoes belonging to five genera were caught, the most common were *Ochlerotatus trivittatus* (66%), *Culex pipiens/restuans* (5%), *Culex erraticus* (72%) and *Aedes vexans* (26%); the numbers in parentheses are the percentages engorging on the dog bait. Results confirmed other studies that *Aedes vexans* is not very attracted to dogs, whereas dogs are a good source of blood-meals for *Ochlerotatus trivitattus*.

Pinger (1985) also noticed that mosquitoes often hovered around the dog-baited kennel traps as if experiencing difficulties in entering them through the horizontal 2-cm wide slits; other times adults were seen exiting the traps.

Labarthe et al. (1998a,b) used cat- or dog-baited traps to collect potential *Dirofilaria immitis* vectors in Rio de Janeiro State, Brazil. The traps consisted of a 75 × 100 × 120 cm wooden-framed outer cage covered with mosquito netting and an inner restraining wire cage (50 × 50 × 50 cm), also covered by mosquito netting, and suspended 45 cm from the outer cage floor. The outer cage was designed with two "Egyptian type baffles" on each side to allow host odours to escape, plus horizontal slots to allow mosquitoes to enter the trap. Traps were baited with either a 7 kg female dog or a 3.5 kg female cat and were operated for 3 h from daybreak and again for 3 h in the evening, commencing 30 min before sunset for four days per month between March 1995 and February 1996. Human landing catches were also carried out concurrently. The trapping programme yielded 3888 individual mosquitoes representing 21 species. The commonest mosquito collected was *Ochlerotatus taeniorhynchus*, comprising 30% of the total. Dogs were found to be significantly more attractive than cats to *Ochlerotatus taeniorhynchus*, *Culex quinquefasciatus*, *Ochlerotatus scapularis*, *Culex declarator* and *Culex nigripalpus*, while more *Aedes albopictus* and *Wyeomyia bourrouli* were collected on cats, although insufficient numbers of the latter two species were collected to allow for statistical analysis. Dissections of the specimens collected revealed *Ochlerotatus scapularis*, *Ochlerotatus taeniorhynchus* and *Culex quinquefasciatus* to be the only species harbouring infective dirofilaria immitis larvae (Labarthe et al. 1998b). The Labarthe et al. (1998a) dog-baited trap has also been successfully used by Ahid and Lourenço-de-Oliveira (1999) to collect potential dirofilariasis vectors in northeastern Brazil.

Small animal-baited traps

Numerous traps designed to catch mosquitoes attracted to small mammals and birds, and more rarely to amphibia and reptiles were developed in the 1960s and 1970s, but relatively few of them have been used by investigators other than those that first designed them. For this reason I have included here only accounts of traps that have been widely used, or were designed since the publication of the second issue of this book. Readers interested in descriptions of these traps are advised to consult earlier editions of this book (Service 1976, 1993).

Buescher et al. bird-baited trap

These traps were baited with various birds and used in Japan in a study of Japanese encephalitis (Buescher et al. 1959). They are made of wood and resemble small Magoon stable traps, and measure 36 × 44 in and 40 in high, with a centrally pitched wooden roof. A 5-cm deep entrance extends the length of the two longer sides parallel to and 10 in from the base of the trap. The lower part of the entrance consists of an upwardly slanting strip of wood which leaves a 1-in vertical opening, as found in the Bates-type stable trap, about 18 in from the trap base. Because birds ate mosquitoes entering these traps a ½-in wire mesh frame is inserted inside the traps just below the slit-like entrance to confine the birds to the lower half of the trap. A small door in the upper part of the trap at each end allows aspirators to be inserted for the removal of the catch. A third small door is made near the base of the trap for introducing the bait animals and for facilitating feeding and cleaning out the trap.

Control traps without birds attracted no mosquitoes, but large numbers of *Culex tritaeniorhynchus* (4170) were caught from traps containing a Black-crowned Night Heron, a bird which also attracted large numbers of *Culex pipiens* (8712). However, the highest numbers of this mosquito were caught from chicken-baited traps (11 679) (Scherer et al. 1959). Both trap location and the microhabitat around the traps considerably affected the numbers of mosquitoes caught. Also, identical traps sometimes caught greatly different numbers of mosquitoes, a difference that persisted even when both baits and trap positions were changed. In Japan, Flemings (1959) attached ropes to the top of small stable traps (36 × 36 × 24 in) baited with birds and pulled them up to various heights. Adults of *Culex tritaeniorhynchus, Culex quinquefasciatus* and *Culex bitaeniorhynchus* were caught in traps at ground level and all four heights, but *Armigeres subalbatus* was only caught in traps on the ground. Scherer et al. (1959) also used small stable-type traps baited with birds and suspended at various heights in a study of Japanese encephalitis.

Balenghien et al. (2005) used similar wooden-framed traps covered with netting and baited with a duck to investigate the potential vectors of West Nile virus in the Camargue, southern France. Dimensions of the trap were (0.5 × 0.5 × 0.7 m) with a gutter-shaped opening and a mosquito-collecting chamber at the top of the cage. The bait animal was kept in the lower portion of the cage and mosquitoes are removed from the top portion, which acts as a holding cage. At each of two sites, two bird-baited traps containing one duck per trap were suspended in a tree approximately 1 m and 8 m above the ground. Traps were installed at noon and operated for 24 h and mosquitoes were removed every 4 h by way of hand aspirators. *Culex pipiens* was the dominant species collected at both heights.

Blower trap of de Freitas et al.

The collection in the morning of mosquitoes from under the hood of a sentinel trap is not very satisfactory as mosquitoes that are attracted to the bait but do not remain under the trap until the morning are missed. To get more representative collections a trap was devised in which mosquitoes attracted to sentinel mice were automatically trapped at intervals (de Freitas et al. 1966).

Mice are exposed in a 10 × 10 × 12-cm cage of wire mesh suspended from an aluminium hood, 62 cm square and 32 cm high (Fig. 6.31*a,b*). A 26-cm tall, 21-cm square, chimney having the lower half made of nylon netting and the upper half of aluminium is placed underneath the hood and surrounds the bait cage. A 19-cm fan blade is fixed to the spindle of a 6-V d.c. motor enclosed in a plastic bag and placed under the centre of the bait cage. It operates from a 6-V car or gel cell battery. The updraft of air produced opens a pair of 11.5-cm square plastic foam trap doors fitted to the top of the aluminium hood, and which fall shut when the fan is not operating. Originally the minute hand of a clock, which was housed in a small box underneath an aluminium roof, was wired to the positive lead from the car battery to the fan motor, so that electrical contact was made for a period of 90 s every 15 min, but the timer described by Kimsey and Brittnacher (1985) could be used. During the short exposure period mosquitoes attracted to the sentinel mice are blown up through the open doors into a 23-cm square, 30-cm high, plastic mesh collecting cage placed on top of the hood. The complete trap is suspended by wires under a large flat protective aluminium roof, which if attached to pulleys allows the traps to be used at various heights. The trap can also be mounted on metal rods. Wires or rods supporting the trap need to be greased to keep out ants. The collecting technique differs from that employed in the Lumsden (1958*b*) trap in that mosquitoes are blown and not sucked into the trap.

In Brazil these traps have proved exceptionally useful in collecting large numbers of mosquitoes. For example, de Freitas et al. (1966) caught 41 825 mosquitoes from a blower trap containing sentinel mice (mother and young) operated for 84 daily periods. The average catch was 498 mosquitoes per night. The identification of representative samples showed that about 96% of the catch comprised *Culex* species, of which about two-thirds were of the subgenus *Melanoconion*. The numbers caught were only slightly less when the fan functioned every 30 instead of 15, min, but the numbers of blood-fed mosquitoes increased from about 7 to 20%.

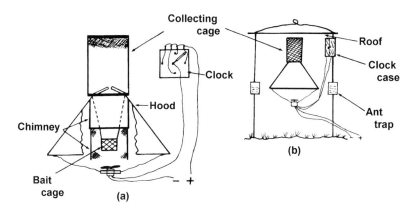

Fig. 6.31. (*a*) and (*b*) blower trap (after de Freitas et al. 1966)

With the conventional type of hooded sentinel bait cage, without a fan, a number of mosquitoes undoubtedly return to the neighbouring forest after feeding on the mice. De Freitas et al. (1966) pointed out that if sentinel mice become infected and develop viraemia they serve as artificial reservoirs and mosquitoes feeding on them will become potential vectors. The blower trap reduces the likelihood of any such transmission occurring in the forest. In studies in Bush Bush forest in Trinidad this risk was reduced by never exposing sentinel mice for more than 14 h in No. 10 Trinidad traps, thus preventing them becoming viraemic (Aitken et al. 1968*a*).

This trap with little or no modification could be baited with a variety of other small animals or dry ice, and if a bigger fan was fitted and the trap modified it could be used to collect mosquitoes attracted to larger animals such as rabbits or even monkeys. Such a trap would probably be simpler to construct than that of Lumsden (1958*b*).

Baited suction trap (Davies)

One of the important features of this trap is that mosquitoes do not have to pass through any baffles or restrictive entrances to get at the bait since it is relatively exposed. The trap is made from readily available materials, and is relatively cheap. The motor cycle battery needed to operate could be attractive to thieves. The trap was briefly referred to by Davies (1971), but a complete description together with diagrams did not follow until some 2 years later (Davies 1973).

The trap consists of four basic components, a cylindrical net collecting cage, a small metal tubular fan housing, a wire mesh bait cage and a time switch which can be made from a cheap clock (Fig. 6.32). The collecting cage is made from nylon or terylene netting and is 20 in long and 12 in in diameter except that at both ends a curtain wire threaded through a ½-in hem reduces the diameter to about 10 in. A 12-in circular piece of ¼-in plywood with a wire loop handle is inserted in the top of the cage as a convenient lid. A short length of plastic tubing, such as a vial with the bottom removed, is cemented into the middle of the cage so that an aspirator can be inserted for removal of the catch. An 8-in long, 6-in diameter metal tube (e.g. a 5-lb dried milk tin with both ends removed) is mounted in a 12-in disc of plywood placed in the bottom of the trap. A 6-V d.c. motor is held by aluminium brackets or straps midway inside the tube. Ferrous mountings should be avoided as they may interfere with the magnets of the motor and hence its operation.

A 2- or 3-bladed 6-in diameter plastic propeller with a 4-in pitch, such as used in model aircraft, is fixed to the spindle of the motor. The top of the metal tube is closed by a circular flap of ¼-in polystyrene foam, cemented only at one point, so that when the motor operates, the updraft of air forces the flap open. It falls back into place when the motor stops. A bait cage about 3 in in diameter and length made from ½-in galvanised wire mesh is supported at the bottom of the fan housing by two long meat skewers running through it at right angles. Wiring the motor to a 6-V motor cycle or gel cell battery through a clock allows the fan to automatically operate for 30–45 s every 7–10 min. Alternatively a more simple and cheap electronic timer could be used (Kimsey and Brittnacher 1985). A metal or plastic roof can be placed over the complete trap.

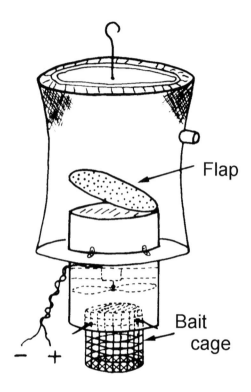

Fig. 6.32. Baited suction trap (after Davies 1973)

Tikasingh and Davies (1972) compared the efficiency of this trap, the Trinidad No. 10 trap (Worth and Jonkers 1962), the CDC light-trap and the No. 17 trap (Davies 1971) in the rain forests of Trinidad. The baited suction trap caught more than twice as many mosquitoes (*Culex* species, mainly *Culex portesi*) as the No. 17 trap and about four and eight times as many as the CDC and No. 10 traps. Although the suction trap caught more mosquitoes, because of its comparative bulkiness, and the fact that under field conditions it sometimes broke down, the No. 17 trap with no working parts was considered the best practical trap.

Davies (1973) pointed out that trap dimensions are not critical, in fact bigger traps can be made for baiting with larger animals and in fact he used a large version in studying the attraction of *Culex portesi* and *Culex taeniopus* to various hosts in Trinidad (Davies 1978). For this the bait cage

(15.2 cm diameter, 8.9 cm high) was made of wire mesh having 1.3-cm squares. Half of this cylindrical bait cage protruded below the overhead fan housing into which mosquitoes were sucked up by a 22.9-cm diameter propeller powered by a 6-V d.c. motor. A time switch activated these motors for approximately 45 s every 7.5 min. Power to each trap was standardised by connecting them in parallel to a single 12-V battery by 15.24 m of identical cable; the resistance of the cable reduced the 12-V to 6-V at the outlet. Six traps were suspended 90 cm from the ground 13.5 m apart along the circumference of a 27-m diameter circle. Bait cages were boiled in 10% bleach solution after use to eliminate any residual odours.

In all trials one trap was baited with two white mice to serve as a standard, and an adjustment factor, based on the ratio between the mean catch on the two mice and 50 mosquitoes/night, was calculated for each experiment to compensate for seasonal variations in mosquito population size. The mean catch, as well as its standard error, was then multiplied by this correction factor. Neither *Culex portesi* nor *Culex taeniopus* were attracted to crabs, toads or lizards exposed in the traps, but all rodents, bats and birds exposed attracted varying numbers of mosquitoes. The best bait for *Culex taeniopus* was the opossum (*Dildelphis marsupialis*), while *Culex portesi* was attracted to a wide range of rodents and marsupials. High attraction to a bait did not necessarily result in high feeding success, the most extreme case was the high degree of attraction of *Culex portesi* to the grass mouse (*Akodon urichii*), yet only 6.08% engorged on the mouse. Need et al. (1993) found hamster-baited suction traps of Davies to be a relatively inefficient method of collecting mosquitoes in the Iquitos area of Peru, and their use was discontinued after two years, although they did sample 16 of the 40 taxa collected using eight different trapping methods.

IMR bait-trap

This trap (IMR = Institute for Medical Research, Kuala Lumpur) consists of a rectangular cage 45 × 30 cm and 38 cm high having a sheet metal base. The sides are covered with copper mesh except that one of the smaller ends (30 × 38 cm) has a nylon netting sleeve attached for introducing the baits and removing the catch (Fig. 6.33). The two longer sides are bent inwards at 7 cm from the metal base plate making an angle of 145°, and along the apices of these angles are horizontal 1-cm wide slits for mosquitoes to enter. When these traps were baited with either a chicken or pigeon near Kuala Lumpur, Malaysia a mean of 9.1 and 5.3 *Coquillettidia crassipes* were collected, respectively. There were no statistical differences between the numbers caught in these traps, in No. 10 Trinidad traps or in lard-can bait traps (Chiang et al. 1986). A number of other species were collected including *Culex quinquefasciatus* and *Aedomyia catasticta*. Klinkaewnarong et al. (1985) trapped 172 mosquitoes during seven nights when one trap was baited with a chicken and another with a gerbil in their studies on vectors of *Cardiofilaria nilesi*.

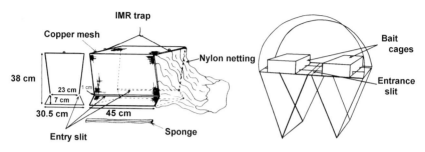

Fig. 6.33. (*a*) IMR bait trap (Chiang et al. 1986); (*b*) Trinidad No. 10 trap (Service 1969*b*)

No. 10 Trinidad trap

This has proved to be one of the most versatile animal- or CO_2-baited traps, although it appears to have been used somewhat less widely than merited. The frame of this double-baited trap is made from $\frac{1}{8}$-in galvanised or stainless steel wire. The various sections can be soldered or tied together, or with some ingenuity the entire framework of the trap can be made by bending a single length of wire into the appropriate shape (Worth nd Jonkers, 1962). Each end section of the trap is formed by bending the wire to form a 'W' and then the outer limbs of the 'W' are connected overhead by two semicircular pieces or wire (Fig. 6.33*b*). The two end sections are linked by two 16-in horizontal lengths of wire running from the outer top ends of the 'W'-shaped wire frame, and by two similar strips separated by a 1-in gap connecting the central apex of the two 'W'-shaped frames. This gap forms the entrance for the mosquitoes. The framework of the lower part of the trap is covered with rustless wire mesh while the upper part supported by the two semi-circles of wire is covered with white nylon mosquito netting which is sewn on to the framework except at the front end. A 'touch and close' fastener such as 'Velcro' is sewn on to the wire mesh at the front end of the trap and also along the bottom edge of the mosquito netting cover. Alternatively the top cover can be made as a bag that is dropped over the cage and held in position by a large strip of rubber such as cut from an old motor car inner tube. However the cover is fitted, it should not be made of cotton mosquito netting as this tends to become mouldy and fluffy, causing the holes to become occluded, and poor ventilation results in poor catches.

The trap is normally baited with small vertebrates contained in a small mesh cage supported on wire supports in each section of the trap, but the trap has also been successful when baited with dry ice contained in plastic bags or polystyrene boxes (Service 1969b). The trap is normally suspended from a tree and protected from rain by a horizontal sheet (18 × 24 in) of metal. Mosquitoes enter from below through the long vertical centrally placed slit. They are collected by inserting an aspirator underneath the bottom edge of the mosquito netting cage covering the trap. Worth and Jonkers (1962) found that over a test period of several days only about half a dozen mosquitoes entered a trap containing bait cages and food, but no animals. When the trap was completely covered with green plastic mesh no mosquitoes were caught even when baited with mice.

These traps have mainly been used in Trinidad where they have proved to be exceptionally useful, although to some extent they have been replaced with the simpler No. 17 Trinidad trap (below). In preliminary trials in Trinidad they caught up to 20 different species, the maximum catch being 1929 mosquitoes (Worth and Jonkers 1962). Traps baited with mice caught more mosquitoes than Shannon-type traps baited with chickens (Aitken et al. 1963). In later studies traps containing mice and White Leghorn chicks were used at both ground level and at 55 ft up in the tree canopy. At ground level chick-baited traps caught 21 species and mice-baited ones 25 species, but chick-baited traps caught more mosquitoes; 33 species were collected at human bait. Fewer mosquitoes were caught in traps in the canopy than at ground level (Aitken 1967). The traps have proved very useful in virus isolation studies in Trinidadian rain forests where Aitken et al. (1968b) caught 25–50% of the mosquitoes used in virus isolation experiments in these traps. When baited with various rodents, lizards and chicks, 42 mosquito species, including those of *Limatus, Aedes, Mansonia, Psorophora* and *Wyeomyia* were caught, but *Culex* species, especially *Culex nigripalpus* formed the bulk of the catches (Aitken et al. 1968b). In contrast very few mosquitoes were caught in traps baited with mice in England (Service 1969b), although when about 0.5 lb of dry ice was placed in a polystyrene box in each section of the trap nine species, including *Anopheles*, were caught. The maximum overnight catch from a single trap was 105 females.

In Kenya Chandler et al. (1976a,b) baited Trinidad traps with six white mice, and from 34 trap-nights caught 787 mosquitoes belonging to 19 species, of which *Mansonia uniformis, Mansonia africana, Culex poicilipes, Culex univittatus* and *Culex antennatus* formed the bulk of the catch. From 56 trap-nights at another site 7696 mosquitoes were caught, with again the first three species comprising most (91%) of the catch. In other trials traps contained two young chickens in each V-shaped section and were suspended from trees in a heronry near a rice irrigation scheme at heights of 1.9, 7.4 and 11.7 m. From 54 nocturnal trap catches 364 mosquitoes belonging to 12 species were caught, of which the *Culex univittatus* group formed 77.7% of the total (Chandler et al. 1976a).

Ferro et al. (2003) used a slightly modified version of the No. 10 Trinidad traps baited with golden Syrian hamsters to sample vectors of Venezuelan equine encephalitis virus in Colombia. The modifications were as follows: 1) the metal can comprising the trap opening was replaced by a polyvinyl chloride pipe, 10 cm in diameter; 2) the cylindrical animal cage was enlarged to 11 cm in diameter and 12 cm in height; 3) the roof was constructed from plexiglass; and 4) the opening for mosquito aspiration was a simple buttonhole sewn into the polyester collection net. Baited traps were suspended approximately 1.5 m above the ground and placed along transects at 10-m intervals. Traps were checked each morning between 0600 and 0800 h, and occasionally also between 1700 and 1900 h. Mosquitoes were removed from the traps with an aspirator. Thirteen species of mosquito were collected, including the two main vectors: *Culex pedroi* and *Culex vomerifer*, and *Culex adamesi* was identified as a new vector. These modified traps were also used by Yanoviak et al. (2005) to collect mosquitoes in an enzootic focus of Venezuelan equine encephalitis in Peru. Six traps were baited with Syrian golden hamsters and hung from low branches approximately 1.2 m above ground level. A total of 352 mosquitoes comprising 20 species were collected over a three-day trapping period. Mean ± 1SD mosquito numbers per trap were per day were 19.6 ± 16.3. One hamster seroconverted during the study and virus-positive pools of *Culex gnomatos* and *Culex amazonensis* were obtained.

In Peru, Need et al. (1993) used chicken-baited Trinidad traps as one of eight trapping methods in a survey in the Iquitos area, but found them to be inefficient and their use was discontinued after two years of sampling.

Worth and Jonkers (1962) considered that only few of the mosquitoes caught escaped by 'blundering out', but Service (1969b) found that despite mosquitoes entering No. 10 traps much more readily than they did cylindrical traps with conical entrances a greater proportion escaped, although a few of these subsequently re-entered the trap. If mosquitoes are allowed to feed on the baits then they become less active and fewer escape, but it is not always desirable to let them engorge.

No. 17 Trinidad trap

This trap was developed as a small, simple and cheap trap to collect mosquitoes attracted to small rodents (Davies 1971). It consists of four distinct parts—lid, net cage, bait cage and spreader ring (Fig. 6.34). The lid is a ¼-in thick 12-in diameter piece of plywood with a small central hole through which string or wire is fixed for suspending the trap. The upper surface can be painted, usually black, to protect the wood, while the lower surface is painted white. The net cage is made from a piece of 38 × 18 in terylene netting with a ½-in hem along both longer edges, by sewing the two shorter sides together to form a tube. A 28-in length of flexible curtain wire is threaded through the top hem and a 9-in length of ¼-in wide elastic through the bottom hem. A short section of a 1-in diameter plastic tube having a removable top or cork is cemented into a slit made in the side of the bag. This provides a simple but efficient opening for inserting an aspirator. Alternatively a slit-like opening furnished with a touch and close fastening such as 'Velcro' could be used.

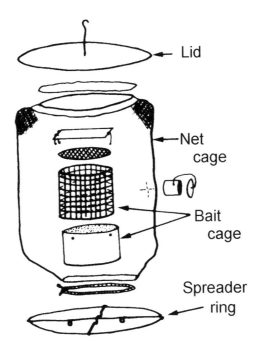

Fig. 6.34. Trinidad No. 17 trap (after Davies 1971)

The bait cage is made from a 1 US quart motor oil tin cut in half and with both ends removed to give a 4-in diameter, 2 ¾-in long cylinder. Half-inch galvanised wire mesh screening rolled round to form a 4-in diameter, 13-in long cylinder is slipped down over the bait tin for about ½ in and held in position by adhesive tape, small bolts or self-tapping screws. A disc of ¼-in mesh screen is wired inside the wire cylinder about 1½ in above the top of the tin to form a floor for the bait cage, and a piece of plywood or galvanised metal is hooked on top of the wire cylinder to form a roof. A piece of 16-gauge galvanised wire is placed underneath the middle of the tin that forms the bottom of the bait cage, and is bent to form two U-shaped slots to accommodate the rim of the tin. Two such pieces of wire are then placed at right angles to each other across the spreader ring, this being a 12-in diameter circle of stiff wire, which supports the cage within the trap. Mice, or other small baits, are confined within the 2-in high space between the roof and the wire mesh floor of the bait cage.

Davies (1971) found that there was a greatly reduced catch if the trap bag was made of nylon mosquito netting with round holes. The most efficient traps had a bag of terylene netting with square holes and 22 meshes to the inch. It was also important that the floor of the wire mesh cylinder was made of finer mesh than its sides. If two or more adult mice are used the bait chamber should be divided to prevent fighting. One man can carry 10 net bags and trap lids in one box and 10 bait cages packed in another box to the field.

Davies (1971) reported that in Trinidadian forests a trap baited with two adult mice catches up to 200–300 mosquitoes per night belonging to about 30 species; this represents about half the catch obtained by a No. 10 trap of Worth and Jonkers (1962) containing four mice. The portability, simplicity and cheapness of the No. 17 trap together with its efficiency have been responsible for its popularity in Trinidad and the almost complete phasing out of the double baited No. 10 trap. In comparative trials of four different traps the No. 17 trap has been identified as the best (Tikasingh and Davies 1972). This trap was regularly employed in the extensive trapping programmes undertaken at Belém, Brazil, and has also been used for short periods in British Honduras, but since then has rarely been used.

Lard-can traps

Birds or small mammals instead of dry ice can be placed in the metal cylindrical traps of Bellamy and Reeves (1952) (Dow et al. 1957; Downing and Crans 1977; Lounibos and Escher 1985; Lounibos and Linley 1987). A description of the construction of these traps is included in Chapter 10.

In trials in England a rabbit or young chicken not enclosed in any restraining cage was placed directly into a metal cylinder, 35 cm long and 25 cm in diameter, with inverted wire mesh funnels at both ends (Fig. 6.35). With rabbits nine mosquito species were caught, the most common being *Coquillettidia richiardii* but relatively large catches of *Ochlerotatus detritus* were also obtained. At night relatively large numbers of unfed females of both *Culiseta morsitans* and *Culex pipiens* were trapped, species known to be almost entirely ornithophagic in the area, but few (0.6–2.3%) fed on the rabbits (Service 1969*b*). In contrast when the traps were baited with pullets very few mosquitoes entered them, and moreover none engorged on the birds. The reasons why mosquitoes which normally feed on birds entered a rabbit-baited trap are not understood, but the results emphasise the caution needed in interpreting trap catches in terms of natural host preferences.

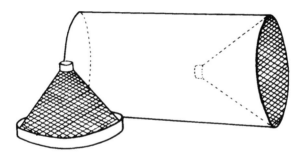

Fig. 6.35. Cylindrical lard-can trap (after Service 1969*b*)

A better technique of exposing bait animals in cylindrical traps is to restrict them to a small section of the trap. For example, bait animals can be placed in a small screen cage soldered to the floor of a trap and provided with a small door underneath to allow their easy insertion and removal (Dow and Morris 1972). Ehrenberg (1966) introduced a pigeon through a hinged flap door in the side of the cylinder, and the two mesh cone entrances at the ends opened into a screened cage, thus preventing biting on the host. Downing and Crans (1977) found these traps very useful in catching *Culex* mosquitoes in New Jersey, USA recording a mean catch of 55.9 per night. In addition to *Culex pipiens*, which formed 73.0% of the *Culex* caught, small numbers of *Culex salinarius* and *Culex restuans* were obtained, as well as a very few *Aedes, Culiseta* and *Coquillettidia*. Slaff and Crans (1981) using a pigeon-baited Ehrenberg trap found that most host-seeking *Culex salinarius* were trapped during the first 2 h after sunset.

Anderson et al. (2004) compared lard-can traps of Bellamy and Reeves (1952), CDC light-traps and mosquito magnet experimental traps in a study of the prevalence of West Nile virus in tree canopy-inhabiting mosquitoes in Connecticut, USA. Slightly-modified lard-can traps constructed from a can measuring 63.5 cm long with a diameter of 34.3 cm with a screen cone leading inward into the can from each end were also used. The quail or hamster bait was placed in a side door and was protected with a screen from feeding mosquitoes. Lard-can traps baited with quail and hamster caught similar numbers of *Culex pipiens*. Significantly more *Culex pipiens* were captured in the quail/hamster baited lard-can traps placed in the canopy, while those placed close to the ground captured more *Culex salinarius*.

Mitchell and Millian (1981) made lard-can traps in three sections. The two longer end sections had conical entrance funnels which were separated by a shorter removable circular bait cage of equal diameter having both ends made of mesh and held in position by 'snap-on' clips. A curved hinged flap in the side of the circular cage allowed the bait (a chicken) to be introduced and removed.

Nayar et al. (1980) also modified lard-can traps. They cut a 15 × 25-cm hole in the side and screened it from the interior by a fine mesh cage (16 × 26 cm and 12 cm deep). The bait animals, 1–3-week-old chicks, were confined to a wire cage attached to the lid, which was fitted within the mesh cage, and then the lid was inserted into the can and fastened. Thus, it was possible to introduce and remove the bait without opening the trap. The outside of the trap was sprayed with matt black paint. A 6-cm cube (approx. 500 g) of dry ice was placed in a 12 × 12-cm styrofoam box hung outside the baited trap. When these traps were suspended 1–2 m above the ground many hundreds of *Culex nigripalpus* were caught.

In California, USA Dow et al. (1957) confined birds and other bait animals to cages placed in a fine mesh screen recess inserted in the top of a cylindrical trap. The outside opening of the recess was closed with a hinged cover. In later experiments mosquitoes were given the opportunity of feeding on the birds, which were placed in a ½ × 1-in wire mesh cage introduced into the trap through a hinged trap door provided with a 3-ft cloth sleeve. To minimise the effect of trap position, four identical cylindrical traps were suspended at right angles to each other about 5 ft from the ground on 3-ft booms, which slowly rotated horizontally. The testing procedure was set out as a Latin Square. Each individual bait was exposed in a different trap each successive night, until each bait had been placed in all the traps. The numbers caught in the different traps were transformed to log $(x + 1)$ for an analysis of variance. The percentages which had engorged on the baits were subjected to an inverse sine transformation (Bartlett 1947). It was found that the numbers of *Culex tarsalis* attracted to different birds were directly correlated with the size of the birds and not species, whereas engorgement rates were independent of attraction rates and size of the birds, but were related to the species and also to different birds of the same species. A density-dependent phenomenon was observed, namely an increase in catch size of *Culex tarsalis* resulted in a decrease in the proportion feeding.

Lord and Day (2000) used pairs of lard-can traps (Dow et al. 1957) baited with either live domestic leghorn chickens (*Gallus sallus*) or bobwhite quail (*Colinus virginianus*) to study the host preferences of *Culex nigripalpus* for chickens, which are routinely used as sentinel hosts for St. Louis encephalitis surveillance. Traps were placed at 2 sites (>2–5 m apart) in a southern live oak (*Quercus virginiana*) and cabbage palm (*Sabat palmetto*) hammock near Vero Beach, Florida, USA. Paired traps were hung approximately 1 m above ground and 5 m apart, with entry holes oriented in the same direction. Traps were operated from 1600–1700 h until 0800–0900 h the following day 4 nights per week from September 22 to October 10, 1997. More *Culex nigripalpus* were caught in the chicken-baited traps (signed rank test, $S = 135.5$, $n = 24$, $P < 0.0005$) indicating a significant preference for chickens over quail. No significant effect was observed for the other variables measured.

In studying the host preferences of mosquitoes in Massachusetts, USA Hayes (1961) baited cylindrical 120-lb capacity lard-tins with 25 different vertebrate species, including birds, bats, rabbits, squirrels, snakes, turtles, frogs, toads and salamanders. Small animals were placed in a small hammock, 10½ in wide, 6½ in deep, made of nylon mesh (28 strings/in) and edged with a 2-in muslin collar with four button holes. The hammock was inserted through a 6-in diameter hole cut from the top of the cylinder and held in position by spreading the collar over the outside of the cylinder and covering it with a 1½-in wide metal gasket bolted to the cylinder. A metal cover was bolted over the pocket on the outside of the trap to prevent mosquitoes reaching the bait without entering the trap. Except for the turtles, which were placed unrestrained within the hammock, baits were immobilised. Birds with their feet bound together were placed in the toe of a nylon stocking and then placed on their backs in the hammocks; amphibians were also held in a stocking. A wad of cotton wool on the floor of the trap underneath the hammock absorbed urine which was generally produced by the animals. Moderately sized mammals and snakes were restrained in galvanised ½-in wire mesh cages before being placed in the hammock, but animals such as rabbits, rats and squirrels which were too large for the hammock were placed in cages on the floor of the trap. Following an exposure period the metal cover over the hammock was removed and the baits taken out, after which mosquitoes in the trap were anaesthetised and collected. After use the entire trap, including the nylon hammock, was washed to remove bait odours.

In response to the continuing need for small animal-baited mosquito traps and the discontinuation of metal cans of the type used in earlier traps, LePore et al. (2004) developed a lard-can trap constructed from materials readily available from suppliers to the ventilation industry. Traps consist of a 24 in long section of 16 in diameter galvanised ducting, fitted with collecting chambers at each end of the central tube. The body of these chambers is constructed from a 6 in length of 10 in diameter galvanised tubing designed to fit over the crimped end of a 10 in end cap. The collecting chamber bodies are fitted to the inner faces of 16 in diameter end plates using a crimped collar. One collecting chamber slides over the crimped end of the main trap body, the other chamber is crimped to fit inside the main trap body. Mosquitoes enter the collecting chambers through inward-facing mesh funnels with a 1 in diameter opening, and are prevented from feeding on the bait animal by a 16-mesh aluminium screen attached to the inside end of the collecting chambers (Fig. 6.36). The entire trap weighs about 15 lbs and cost less than US$80 (2002). Removal of mosquitoes from the collecting chambers is achieved as follows: the chambers are removed from the trap body and placed with the larger diameter end facing down onto an inflatable ring, as used by children to aid buoyancy when swimming, taped to a board. A plastic shower-cap with a 1 cm hole is placed over the narrow end of the collecting chamber. Carbon dioxide gas is introduced into the chamber from a gas cylinder via a tube threaded through a hole in the board to which the inflatable ring is attached. Anaesthetised mosquitoes can then be removed from the collecting chamber by removing the end cap and aspirating them. Traps baited with rock doves (*Columba livia*) and European starlings (*Sturnus vulgaris*) enclosed in wire cages were evaluated during field trials in Massachussets, USA, where they yielded 6623 mosquitoes, primarily *Culex pipiens* and *Culex restuans* females. Nightly trap catches ranged from 0 to 100 per trap. More mosquitoes were caught in traps positioned in the canopy of tall trees at heights of around 15 m.

Deegan et al. (2005) used modified (LePore et al. 2004) lard-can traps baited with pigeons for West Nile virus surveillance in New York, USA. Each trap was constructed from a 61 cm long cylinder of steel duct pipe with a diameter of 41 cm. A removable chicken-wire cage with dimensions of 30.5 cm × 30.5 cm × 35.6 cm was fitted inside the traps and the ends of the pipe were covered by screened cones that allowed entry of mosquitoes.

Detachable inner screens could be fitted to prevent mosquitoes feeding on the enclosed pigeons. Traps were set at nine sites in each of three locations. Three of the sites at each location were in open habitat, three in forest-edge habitat and three in forested habitat. At each site one trap was installed at a height of 1.5 m and another at a height of 7.6–9.1 m. During 654 collections, a total of 4985 adult female mosquitoes were collected, comprising predominantly *Culex* species (99.8%). Three-factor ANOVA revealed a significantly higher number of mosquitoes captured at the higher elevation ($P < 0.0001$) and revealed significant differences in the numbers captured among the different habitat types. Significantly more mosquitoes were collected from forested habitat than open habitat ($P < 0.01$).

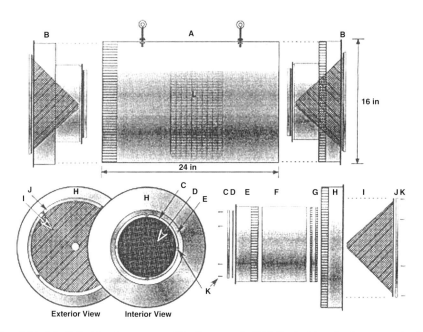

Fig. 6.36. Modified lard-can trap of LePore et al. (2004). A. Trap body, comprised of a length of galvanized duct (24-in long × 16-in diameter) and with 2 eyebolts inserted for mounting. B. Mosquito-collecting chambers, assembled from 9 parts (C-J) and fastened by screws (K). C. Ring, galvanized sheet metal (9.5-in outer diameter × 8.5-in inner diameter). D. Panel, 16-mesh aluminum screen (11 in). E. End cap, galvanized duct (10-in diameter with a 9-in diameter opening cut in its centre). F. Duct, galvanised (6-in length × 10-in diameter). G. Collar, galvanized (10-in inner diameter). H. End cap, galvanised duct (l6-in diameter with a 10-in hole cut in its centre). I. Funnel, 16-mesh aluminum screen (12-in diameter with 1-in diameter hole at its apex). J. Ring, galvanized sheet metal (9.5-in outer diameter ×8.5-in inner diameter). K. Sheet metal screws (0.25 in). L. Animal cage

In addition, the seroconversion rate of pigeons in traps at the higher elevation was significantly greater than at the lower elevation, and the authors proposed setting traps at heights of several metres above ground to improve surveillance. Because Blackmore and Dow (1958) had found that if birds were relatively active in cylindrical traps they inhibited mosquitoes from getting a blood-meal, Reeves et al. (1961) restrained bait chickens in nylon stocking sleeves before placing them in a bait cage on the floor of the trap. When the chickens were not enclosed within a stocking restrainer an average of only 30.3–47.3% fed on them, whereas when a restrainer was used engorgement rates increased to over 90%.

In Texas, USA Easton et al. (1968) compared the mosquitoes collected in cylindrical traps, made from 110-lb capacity lard-cans, with those caught in Malaise traps of Townes (1962) (See Chap. 8). Larger bait animals, such as Californian jackrabbits and Audubon's cottontail rabbits, were enclosed in 6-in diameter, 15-in long cylindrical wire restraining cages, while smaller animals were placed in a 15-in long rectangular box made of 1/4-in wire mesh. Each trap had the screw top of a Mason (Kilner) jar soldered over a 2-in diameter hole cut from one side of the trap. A plastic cone was cemented inside the screw lid so that when the glass jar was screwed in it projected into it. The baited traps were placed amongst shade, which usually resulted in mosquitoes caught in the traps flying into the collecting jars. This procedure enabled the catch to be easily removed. Because artificial light failed to attract mosquitoes into the jar at night, catches were removed only during the day. With the exception of *Culex erraticus*, and possibly *Culex quinquefasciatus* and *Ochlerotatus trivittatus*, the Malaise traps were much more efficient in collecting mosquitoes and other haematophagous flies than the cylindrical traps. When the cylindrical traps were baited with jackrabbits all trapped mosquitoes fed on them, but none fed on smaller mammals enclosed in the 1/4-in mesh cages, possibly because of the reduced aperture of these cages, or the greater movement afforded to the smaller animals. To avoid bias resulting from trapping in different sites, the procedure that had been used by Dow et al. (1957) was largely followed, i.e. the traps were suspended from horizontal arms and slowly rotated. The test was also based on the experimental layout of a Latin Square. By exposing 25 vertebrates in 113 trials, 15 182 mosquitoes belonging to 13 species were caught. The catch consisted mainly of *Culiseta melanura, Culex salinarius, Culex pipiens* and *Ochlerotatus canadensis*, all of which were attracted to amphibians although only *Culex salinarius* and *Ochlerotatus canadensis* fed on them. *Culiseta melanura* and *Culex pipiens*, but not *Culex salinarius* and *Ochlerotatus canadensis*, were much more common in bird-baited traps than in those containing mammals.

The most critical evaluation of wind as a factor in operating cylindrical bait traps was made in Florida, USA by Dow and Morris (1972). Their traps which were baited with two Leghorn pullets, and divided into two equal parts by a vertical wire mesh partition, were suspended horizontally about 1 m from the ground. By an arrangement of pulleys, the traps could be orientated parallel to the wind, at right angles to it, at a random fixed angle to it, or made to continuously rotate at a speed of 0.2 rev./min about a vertical axis. Results showed that revolving traps caught the most *Culex nigripalpus* and *Psorophora confinnis*, while those facing into the wind caught the fewest *Culex nigripalpus* and traps at right angles to it the least *Psorophora confinnis*. In discussing practical considerations Dow and Morris (1972) concluded that if large numbers of mosquitoes are required two ordinary traps set in any random direction should catch more mosquitoes than a trap maintained in any of the orientations they tested. However, if suitable power was available a revolving trap would probably catch most mosquitoes.

In Florida, USA lard-can-type traps were baited with chickens on eight nights near a drainage canal where the predominant species in emergence traps were *Mansonia dyari* (89.7%) and *Mansonia titillans* (6.2%). A total of 3993 female *Mansonia*, of which 78.3% were *Mansonia dyari* and 21.7% were *Mansonia titillans*, were caught in the bait traps (Lounibos and Escher 1985). The differences in the proportion caught as emerging adults and as blood-seeking females may indicate that *Mansonia titillans* is more strongly attracted than *Mansonia dyari* to chickens, that the former species entered the traps more readily than *Mansonia dyari*, or that the bait traps were sampling over a larger area than the drainage canal.

Almirón and Brewer (1995) used lard-can traps baited with either chickens, frogs, rabbits or tortoises to assess mosquito host feeding preferences in central Argentina, but the authors provide no details of the construction of the traps or whether the bait animals were restrained or immobilised. Thirteen species of mosquito were collected, the majority of individuals belonging to ten species within the genus *Culex*. 68.7% of the total catch were from chicken-baited traps, 29.9% from rabbit-baited traps, 0.8% from tortoise traps and 0.5% from frog traps.

Lard-can traps baited with a chicken have also been used alongside CDC light-traps supplemented with CO_2 in St. Louis Encephalitis surveillance in the suburbs of Córdoba city in Argentina (Díaz et al. 2003). No details of the design or construction of the traps was provided by the authors. The traps used caught a total of 443 mosquitoes during the period November 2001 to March 2002, of which 433 were *Culex quinquefasciatus*.

In contrast *Ochlerotatus albifasciatus* was the commonest species (4302, 78%) captured in the CDC light-traps, but only one specimen was caught in the can traps.

In South Africa lard-can traps baited with a rodent caught 232 *Culex rubinotus* during 17 trap-nights, compared to 17 and 14 when baited with pigeons and bats (Jupp et al. 1976). Also in South Africa, in addition to using lard-can traps containing birds enclosed within a nylon mesh restrainer McIntosh et al. (1972) constructed larger cylindrical traps, 76 cm long and 43 cm in diameter, baited with monkeys restrained in wire cages (20 × 20 × 25 cm). Traps were exposed at both ground level and at a height of 12 m in the gallery forest. Twenty culicine species were caught in the monkey-baited traps, and 22 culicines and one anopheline in the fowl-baited traps; in human bait catches 25 culicine and seven anopheline species were collected. The traps were not very efficient in catching *Aedes* or *Anopheles* mosquitoes, and some species caught at human bait which failed to enter the traps readily fed on monkeys in the laboratory. Similar large cylinders (44-gal drums) baited with monkeys contained in expanded metal cages were hauled up 20ft in Malaysia, but caught relatively few mosquitoes (Wharton et al. 1963). In Zaire, Laarman (1959) caught 91 *Anopheles theileri* in large empty petrol drums baited with porcupines, but only six and one adult when they contained a monkey or rabbit.

Emord and Morris (1982) reported that they often observed female *Culiseta melanura* resting on the funnel entrances of lard-can traps, but not entering them, while Mitchell and Millian (1981) consistently reported the entry of blood-fed mosquitoes into their chicken-baited traps.

Emord and Morris trap

Animal-baited lard-can traps have not always attracted many mosquitoes (Emord and Morris 1982; Main et al. 1966; Stamm et al. 1962), and because of this Emord and Morris (1982) designed an alternative small animal-baited trap, based on the CDC light-trap. They took a standard CDC trap and removed the light source and made the following modifications. A semi-cylindrical cage (6 × 4.5 × 4 in) made from 0.5-in mesh wire screening wrapped round a plywood D-shaped base was attached by a metal hook to a 8 × 0.25-in threaded metal rod fixed to one side of the CDC trap. The hinged top of the bait cage was also made from wire screening. A 1.5-in thick layer of household sponge was placed on the floor of the bait cage to protect the bait (sparrow) from the cold, and also to facilitate cleaning the cage. (Clearly other bait animals could be put in the cage). Traps operated from four 1.5-V torch batteries fixed to the CDC trap (Fig. 6.37).

The collecting container consisted of a plastic 1-quart frozen food storage container with a section from the bottom and three sections from the side removed and replaced by plastic mosquito screening. These screenings, and a cloth sleeve fixing the container to the trap, were embedded into the plastic with a soldering iron.

The trap was evaluated against a sparrow-baited lard-can trap and a CDC light-trap with 3–4 lb dry ice suspended nearby in a cloth bag. Significantly more *Culiseta melanura, Culiseta morsitans* and *Culex pipiens/restuans* group were collected in the bird-baited CDC trap, whereas more mammalian-feeding *Aedes vexans* and *Anopheles* spp. and *Coquillettidia perturbans* were caught in the CO_2-CDC trap; the lard-can trap performed the worst. Emord and Morris (1982) concluded that the greater exposure of bait in their trap over the lard-can trap, and the difficulty some species have in entering the latter, made their trap more effective in collecting several species.

Several others have found the Emord and Morris trap effective. For example, Howard et al. (1983) in studies on the vectors of eastern equine encephalomyelitis baited these traps with house sparrows, and placed them in 12 different types of habitats, at ground level, at heights of 5 and 10 m and in the tree canopy. From a total of 641 trap-nights spread over 32 nights, 15 077 female mosquitoes belonging to at least 15 species were caught, the most common of which were *Culiseta melanura* (32.0%), *Coquillettidia perturbans* (18.9%), *Culex pipiens/restuans* group (12.8%), *Ochlerotatus canadensis* (11.1%) and *Culiseta morsitans* (10.8%).

In studying the feeding patterns of Swedish mosquitoes Jaenson (1985) and Jaenson and Niklasson (1986) used Emord and Morris (1982) traps placed 1 m above ground, and the net traps of Jupp and McIntosh (1967) which were made of white netting measuring 1.5 m³, and having the sides rolled up to provide a 20-cm opening all round. Both sets of traps were baited with a rabbit, guinea pig, hen or a dove, and in addition there were unbaited traps. The traps were operated for three 24-h periods each week for 5 months. At least 17 species belonging to *Anopheles, Culex, Culiseta* and *Coquillettidia* were caught; the most common species were *Ochlerotatus communis, Ochlerotatus excrucians* s.1., *Ochlerotatus diantaeus, Ochlerotatus intrudens, Aedes cinereus, Ochlerotatus cantans* and *Culex pipiens*. The Emord and Morris traps proved to be more efficient than the net traps.

The percentage of engorged mosquitoes in the net traps (13.5%) was on average nine times greater than engorged females collected from the suction baited traps (1.5%). Surprisingly there was considerable movement of mosquitoes between traps. For example, of the 17 blood-meals identified serologically as guinea pig blood, only five were collected from guinea pig traps, furthermore only 8% of blood-fed *Aedes* tested from the bird-baited net trap contained avian blood possibly indicating that birds are difficult to feed upon, and from 23 bloodfed *Aedes* mosquitoes collected from rabbit-baited Emord and Morris traps only five had fed on rabbits. Some mosquitoes caught in the traps had fed on cervids and/or cattle. Another problem in interpreting the results was that relatively large numbers of mosquitoes were caught in unbaited traps.

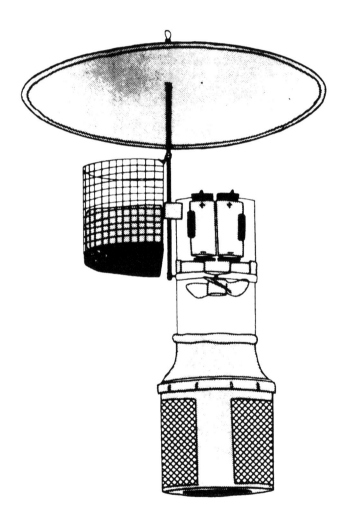

Fig. 6.37. Host-baited CDC trap (Emord and Morris 1982)

For example, large numbers of male *Ochlerotatus diantaeus* and small numbers of male *Ochlerotatus communis* and *Ochlerotatus intrudens* were caught in the Emord and Morris trap; males were also collected in the baited net trap of Jupp and McIntosh but generally in smaller numbers.

The Emord and Morris traps caught considerable numbers of *Culex pipiens* and *Culiseta morsitans* confirming the view of Emord and Morris (1982) that these traps are efficient in catching these species. In contrast the net traps caught many fewer *Culex* and *Culiseta* mosquitoes.

Using rather similar traps, that is modified CDC light-traps normally baited with carbon dioxide (Pfuntner 1979), but baited with a white mouse, dry ice or with both attractants, Landry and DeFoliart (1986) working in Iowa, USA found that location of the traps was very important in determining the numbers of *Ochlerotatus triseriatus* caught.

Howard et al. (1989) successfully used a chicken-baited Emord and Morris trap to catch *Culiseta melanura* and *Culiseta morsitans*.

Meyer and Bennett trap

This is a duck-baited trap, but can be adapted to hold other animals. The bait cage consists of a 30 × 30 and 35-cm high chicken wire cage nailed to a wooden board (35 × 45 cm), there is no wire bottom. A hole is cut from the top to allow the bait to be introduced, after which it is covered with a piece of chicken wire netting. Food and water dishes are wired to one of the sides of the cage. The mosquito collecting cage is a 60-cm cube wooden frame covered with fine plastic mosquito netting on five sides, the bottom being left open (Fig. 6.38). Four polythene 25-cm diameter funnels with their stems cut off to leave about 5-cm diameter holes are glued to fit into four 25-cm holes cut from two panels (30 × 60 cm) of plywood fixed at the top of the cage and on two opposite sides. Finally, a mosquito netting sleeve is sewn into the top of the cage for removal, with an aspirator, of the mosquitoes.

In Canada during 15 weeks when traps were exposed from about 1800–0900 or 1100 h on 37 trap-nights they caught 7235 mosquitoes belonging to at least 13 species. The mean catch per night was 188 mosquitoes (range 1–1445). The most common species were *Coquillettidia perturbans* (6333), *Anopheles walkeri* (330), *Culiseta morsitans* (257) and *Ochlerotatus cantator* (132). Species not found in larval surveys nor in human bait catches were also caught. About 30% of the mosquitoes had engorged on the duck, the only species not feeding on the bird was *Culex territans* (Meyer and Bennett 1976).

In their studies in the USA on the vector of *Plasmodium elongatum* Beier and Trpis (1981) used a slightly enlarged version of this trap to accommodate a penguin. From 63 trap-nights in 1978 they caught 739 female mosquitoes, while in 1979 they caught 455 mosquitoes from 69 trap-nights, all but two were *Culex pipiens/restuans*.

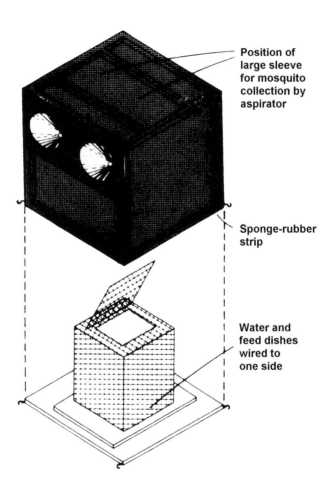

Position of large sleeve for mosquito collection by aspirator

Sponge-rubber strip

Water and feed dishes wired to one side

Fig. 6.38. Duck-baited trap (Meyer and Bennett 1976)

Anderson and Brust box trap

In a study of multiple host contacts during blood-feeding by three species of *Culex*, Anderson and Brust (1995) designed a box trap to hold a pair of Japanese quail (*Coturnix japonica*) contained in cylindrical wire cages. Traps were 30 by 30 by 30 cm and had baffled, slotted entrances (narrowing from 30 by 8 cm to 30 by 2 cm) on the underside. Baffles were constructed from fine mesh to permit downward movement of host odours. On one side of the trap was a hole with a stocking sleeve to permit introduction and removal of the quails and to facilitate emptying of the trap (Fig. 6.39). Perspex and plywood traps were constructed. Traps were suspended 1 m above the ground. In Manitoba, Canada, in 1991 box traps containing pairs of quail and operated over 165 trap nights collected 13 857 female mosquitoes, of which 5218 were *Culex tarsalis*. In Florida, USA, in 1992 perspex box traps containing pairs of northern bob-white (*Colinus virginianus*) captured 2110 female mosquitoes, of which 2041 were *Culex nigripalpus*.

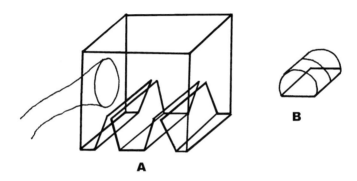

Fig. 6.39. Box trap used to collect blood-fed mosquitoes attracted to marked quail. (A) Trap design, (B)Wire cage for quail. Quail cage is placed inside main cage, between entry baffles (Anderson and Brust 1995)

Sentinel animals

Smithburn et al. (1949) introduced the method of exposing immigrant animals, such as monkeys, into an area and regularly bleeding them to detect the presence of circulating arboviruses. Rudnick (1986), for example used sentinel monkeys in modified Magoon traps during dengue studies in Malaysia. This procedure is now widely adopted in arbovirus studies, and in addition to monkeys, various rodents, marsupials, deer, baby chicks, adult chickens, bobwhite quail, and mice are used as sentinels (Andre et al. 1985; Artsob et al. 1983; Crans 1986; LeDuc 1978; Reisen et al. 1990;

Vigliano and Carlson 1986), and in India Mani et al. (1991) used donkeys as sentinels for Japanese encephalitis and West Nile viruses.

Apart from bleeding the sentinels, the mosquitoes attracted to them are frequently caught and tested for arboviruses.

A common procedure involves placing sentinel mice, usually consisting of a mother and a number (six) of new-born infants, in a $10 \times 10 \times 12$-cm wire test tube basket which has the top covered with wire mesh and the bottom and lower quarter of the sides with aluminium sheeting (Causey et al. 1961). Wood shavings are provided for bedding, and dry pelleted food and a water bottle are added. The basket containing the mice is suspended by a wire hook underneath a 56-cm square aluminium hood which tapers to a 10-cm square top, and is 38 cm deep. To prevent ants entering the trap the wire that suspends the hood from a tree branch should be coated with grease, oil or a permanently sticky adhesive. Some of the mosquitoes attracted to the sentinel mice rest on the basket or underside of the aluminium roof and can be periodically removed with aspirators. Alternatively a single collection can be made in the mornings by carefully fitting a screen over the bottom edge of the hood, and removing the catch by inserting an aspirator through a cloth sleeve in the middle of the screen. Obviously only a fraction of the mosquitoes attracted during the exposure period are caught in the morning.

Sentinel chicken shed

Domestic chickens are susceptible to various viral encephalides, such as St. Louis encephalitis, western equine encephalomyelitis and eastern equine encephalomyelitis, and show good antibody responses to all three. They are suitable as sentinels because of their widespread distribution, and because haemagglutination-inhibition, fluorescent antibody complement fixation and ELISA techniques can be used for determining their antibodies (see Monath 1988). Viral activity can be studied by making antibody surveys of farmyard chickens, but it is usually better if variables such as flock size and types of poultry shelters are controlled. Furthermore, transmission indices may be required in areas where there are no farm flocks. For these reasons the sentinel chicken shed was developed in Colorado, USA by Rainey et al. (1962). The complete trap consists of three basic parts, a shed, two removable mosquito traps and a chicken wire pen (Fig. 6.40). A 20-in wide, 33-in high doorway is made in front of the shed and a 10-in high door, that is normally raised to allow free entry of chickens into the shed from the pen, is hinged along its upper edge and extends across the entire back of the shed.

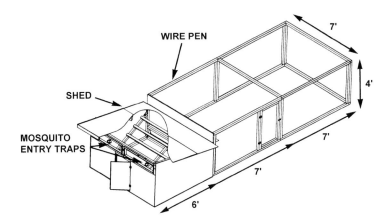

Fig. 6.40. Sentinel chicken shed and pen (Rainey et al. 1962)

A screened opening, 47-in wide and 24½-in high, is fixed above the hinged door to increase ventilation through the shed, which has been shown to increase the catch of *Culex tarsalis* fivefold. It also results in large catches of other *Culex* species, *Mansonia* and *Culiseta* mosquitoes, but there is no such marked increase in *Aedes* species.

Two removable mosquito entry traps are fitted to the top of the front wall of the shed and incorporate Bates'-type baffles (Fig. 6.23*a*). The pen, which is pushed up against the back of the shed, is 14 ft long, 7 ft wide and 4 ft high and is made from chicken wire fixed to a wooden framework. To prevent predators entering the pen the floor can also be covered with chicken wire. A 23 × 43-in doorway is placed in one side of the pen to provide access. If the shed and pen are made in prefabricated panels, two men can apparently assemble, or dismantle, a trap in about an hour (Rainey et al. 1962).

Hayes et al. (1967) found these traps useful in studying the ecology of arboviruses in Texas, USA while Shemanchuk (1969) in Canada used similar traps, but modified to include an entry trap to catch mosquitoes before they could feed on the birds, and an exit trap to collect those leaving the trap. Reisen et al. (1995*a,b*) used sentinel flocks of 10 or 20 white leghorn laying hens to monitor arbovirus activity in southern California, USA. Sentinel chickens were bled by jugular venipuncture every two or four weeks, however, bleeding every four weeks was found to be unsatisfactory as it may delay detection of transmission of WEE virus by up to eight weeks. The location of the sentinel flocks was found to be more critical than flock size in detecting arbovirus activity, with flocks located close to *Culex tarsalis* breeding sites and vegetation suitable for host-seeking mosquitoes had higher seroconversion rates. The Florida, USA arbovirus surveillance programme uses flocks of six hens and bleeds them on a weekly basis (O'Bryan and Jefferson 1991). In Malaysia a sentinel chicken shed trap collected a mean of 12.02 *Culex vishnui* and smaller numbers of *Culex quinquefasciatus* (6.15) and other *Culex* species per day (Wallace et al. 1977).

Odour-baited traps

Costantini et al. (1993) have developed an odour-baited entry trap (OBET) that attempts to address some of the problems associated with landing catches, baited net traps and drop-nets. The key feature of the trap is that the host odours are separated from other cues associated with the host, such as heat, visual cues, etc. The OBET consists of a steel wire frame 40 cm high × 40 cm wide × 60 cm long, which is covered with mosquito netting. A 50 cm long, 10 cm diameter plastic tube is inserted through the centre of the rear panel of the cage and protrudes about 10 cm into the cage. The tube is supported by two 10 cm diameter plastic rings and the end of the tube which is inside the cage is covered with mosquito netting to prevent mosquitoes entering it.

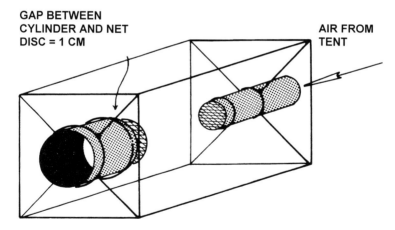

Fig. 6.41. Stylised drawing of the OBET showing the plastic tube to which the lay-flat tubing is connected (narrow cylinder) and the foam rubber tube which is the entrance for mosquitoes (wider cylinder). Cross-hached areas represent discs of mosquito netting. Steel wire rings, netting around the cage and cotton sleeve around the plastic tube are not shown (Costantini et al. 1993)

The rear end of the tube is attached to a length of 'lay-flat' polythene tubing, through which the host odours are delivered into the cage. Inserted into the face of the cage opposite the plastic tube is a foam rubber hollow cylinder with an external diameter of 20 cm and walls 1 cm thick (Fig. 6.41). This foam rubber cylinder protrudes about 14 cm into the cage and is supported by 2 plastic rings. Approximately 1 cm further into the cage another plastic ring (16 cm diameter) supporting a disc of mosquito netting is inserted. This leaves a gap of approximately 1 cm between the internal end of the entry tube and the mosquito netting disc, through which the mosquitoes enter the trap (Fig. 6.41). In the study described, Costantini et al. (1993) introduced host odours to the trap from a tent containing bait animals. The tent was constructed from heavy-duty polythene sheeting suspended from poles. The dimensions of the tent can be adjusted to suit the size of the bait animals being used. A 10 cm diameter hole was cut into one face of the tent and a 12-V DC fan powered by a rechargeable battery was inserted and connected to the 'lay-flat' polythene tube, which was in turn connected to the plastic cylinder inserted into the rear of the trap. The authors found that the air stream produced by the fan was sufficient to fully inflate the polythene tube over a length of at least 12 m. A mean air speed of 5 ms^{-1} was used. In trials of the trap in the Sudanese savanna of Burkina Faso, Costantini et al. (1993) caught about 1500 mosquitoes during the period October–December 1992. *Anopheles gambiae s.l., Anopheles funestus* and *Mansonia* sp. were the most commonly collected species, along with *Anopheles pharoensis, Anopheles rufipes, Anopheles ziemanni, Anopheles wellcomei, Anopheles squamosus, Culex poicilipes, Culex quinquefasciatus* and other *Culex* species. More than 85% of mosquitoes caught were un-fed or partially fed, 5% were fully fed and the rest were gravid. The authors successfully used the trap outdoors, mounted on top of an empty oil drum, and as entry trap fitted to the windows of village houses. Over a period of three nights, a human, a calf, or 3 goats were placed in each of three separate tents, with positions of the baits being rotated among the three tents each night. Total catches were relatively low for both the human-baited and the animal-baited traps but the species composition of the catches was highly significantly different between traps emitting human or animal odours. When no bait animals were in the tent, the trap caught only one unfed *Anopheles funestus* in 26 replicates, indicating that the mosquitoes trapped are responding to the host odour rather than to any visual or other cues produced by the trap itself. The influence of air speed produced by the fan and concentration of host odours on the size and composition of the catch were not investigated. In a later experiment, Costantini et al. (1996) compared the attractiveness to mosquitoes of whole host odour from a human and CO_2 alone using OBETs. In the first experiment two OBETs were set up side-by-side near the window of an experimental hut. One OBET was baited with the whole host odour of a single male human sleeping inside a tent within the experimental hut, with host odour being delivered through approximately 9 m of lay-flat polythene tubing. Carbon dioxide was delivered to the second OBET from gas cylinders secreted behind the hut on the opposite side to the OBET. A similar concentration of CO_2 to that produced by the human bait was supplied. A CO_2/human catch index was calculated as the log-transformed ratio of the CO_2-baited trap catch +1/human-baited trap catch +1. The antilogarithm of the mean catch index, weighted by the nightly sample size gave an estimate of the mean proportion of mosquitoes caught with CO_2 alone relative to whole host odour. Departures of the catch index from 1 were tested using the Mann-whitney U test. Two further experiments were conducted to examine the effect of separating the two trap types by 20 m and also the dose response to CO_2. For the dose response experiment, five concentrations of CO_2 were administered: 0.04%, 0.06%, 0.15%, 0.30% and 0.60%. In the first experiment *Anopheles gambiae* s.l. was the only mosquito species that showed a significant preference for one trap over the other as determined by comparison with a binomial distribution with an expectation of random 'choice' (probability 50%). 33% (geometric mean 13.5, 95% CI 8.9–20.2) were caught in the CO_2-baited trap and 67% (geometric mean 27.3, 95% CI 19.8–37.5) in the human-baited trap. When traps were separated by 20 m the mean catch index indicated that numbers of *Anopheles gambiae* s.l. and *Anopheles funestus* attracted by the CO_2 traps were less than half of those caught by the human-baited trap (mean catch index 0.48 and 0.39 respectively). In contrast, more than twice as many *Anopheles pharoensis* were attracted to the CO_2 trap compared with the human-baited trap (mean catch index 2.09). *Anopheles gambiae* s.s. and *Anopheles arabiensis* were more attracted to whole host odour than to CO_2, even at high concentrations. *Mansonia uniformis* appeared to be strongly attracted to CO_2, with increasing concentrations resulting in higher catches of this species. Later experiments by Costantini et al. (1998*a*) compared the relative attractiveness of human and cattle host odours presented by means of two adjacent OBETs operated under field conditions in Burkina Faso. Human host odour was obtained from either of two volunteers sleeping in a polyethylene tent approximately 7 m upwind of the trap. Cattle host odour was obtained from a tethered calf of similar mass to the human volunteers also kept inside a polyethylene tent. In order to reduce the influence of differential CO_2 production among hosts on mosquito catch, on half of the nights where differences in concentration were observed, supplemental CO_2 was provided from a compressed gas cylinder in order to equalize the release rates from the two traps. Experiments commenced at 2100 h and were ended at 0500 h. Catch results were subjected to maximum likelihood generalized linear modeling. Differences between species in their degree of anthropohily were statistically significant ($G = 88.2$, df = 3, $P < 0.001$). Species with an anthropophily index in excess of 0.5, representing no preference, included *Anopheles gambiae* s.l. (index = 0.96, $t_{18} =$ 6.59, $P < 0.0001$), and *Anopheles pharoensis* (index = 0.68, $t_{18} = 2.74$, $P <$ 0.02). *Culex antennatus* had an index of 0.25, indicating a significant preference for the bovine host odour ($t_{11} = 2.95$, $P < 0.02$). Subsequent molecular identification of members of the *Anopheles gambiae* species complex revealed that *Anopheles arabiensis* comprised 92% of the specimens identified from the cattle-baited trap, and 52% of those caught in the human-baited trap.

Comparisons between CO_2 and human host odours in their 'attractiveness' to mosquitoes were also conducted in the Sudan savanna belt of Burkina Faso by Costantini et al. (1998b). Carbon dioxide at concentrations of 0.04–0.6% (cf. 0.03% ambient concentration) was supplied via OBETs. Carbon dioxide emitted at a rate equivalent to that released by a human bait was also compared with the 'attractiveness' of whole human odour in direct choice tests with two OBETs placed side-by-side. In this case, the number of *Anopheles gambiae* s.l. entering the trap with human odour was double the number trapped with CO_2 alone when emitted at the human equivalent rate. No significant differences between OBET catches were observed for *Anopheles funestus, Anopheles pharoensis, Culex quinquefasciatus* and *Mansonia uniformis*. The dose-response for all mosquito species was essentially similar: a linear increase in catch with increasing dose on a log-log scale. The slopes of the dose-response curves were not significantly different between species, although there were significant differences in the relative numbers caught. In relation to human landing catches, however, it was observed that even the highest dose of CO_2 did not catch more *Anopheles gambiae* s.l. than one human landing catch. By contrast, the 3 highest doses of CO_2 caught significantly more *Mansonia uniformis* than a single human landing catch.

In Senegal, Costantini and Diallo (2001) used OBETs (Costantini et al. 1993) baited with either two monkeys (*Cercopithecus aethiops*) or a human child to investigate the behavioural response of culicine and anopheline mosquitoes to the odours of alternative hosts. By pairing the OBETs, mosquitoes were exposed to odours from both hosts before selecting which trap to enter. A total of 192 mosquitoes belonging to 12 species and 4 genera were collected over 8 nights of trapping. All species were more abundant in the human-baited trap (except for a single individual of *Anopheles fowleri* caught in the monkey-baited trap. In excess of 90% of

634   Chapter 6 Sampling Adults by Animal Bait Catches and Traps

the human malaria vector species *Anopheles gambiae* s.l., *Anopheles funestus* and *Anopheles nili* were caught in the human-baited trap. In Madagascar, Duchemin et al. (2001) used OBETs to study the host choice of wild female anopheline mosquitoes. Two odour-baited entry traps (Costantini et al. 1993) were set up side by side with their entrances facing down-wind, and a choice of odours from two alternative hosts was presented to approaching mosquitoes. One adult man 25–40-years-old and a calf of similar mass were concealed in two separate tents. The mean speed of the air current carrying the host odour exiting from the OBETs was adjusted to approximately 0.6 m/s. OBETs were operated from 1900 to 0500 h. Trap positions were swapped nightly to compensate for any position effects. In each of four villages four replicate tests were conducted, totalling 16 trap nights (i.e. one pair of nights × 2 traps × 4 replicates) per village.

In 48 trap nights the OBETs caught 1930 female mosquitoes belonging to 22 species and four genera (*Anopheles, Mansonia, Aedes* and *Culex*). *Mansonia uniformis* was the most frequently caught species ($n = 767$), followed by *Anopheles gambiae* s.l. ($n = 362$) and *Anopheles funestus* ($n = 266$). Other species included *Culex antennatus* ($n = 169$), *Anopheles coustani* ($n = 143$) and *Culex tritaeniorhynchus* ($n = 94$). The local malaria vector *Anopheles mascarensis* was occasionally caught from all villages but only in the calf-baited OBETs. *Anopheles funestus* females exhibited a 'preference' for the OBET with human odour in all three villages. Conversely, *Anopheles gambiae* s.l. 'preferred' the OBET baited with calf odour. The majority of variability in the anthropophilic index (defined here as the mean proportion p of the total number collected nightly from both traps) was due to temporal and position effects.

Also in Senegal, Dia et al. (2005) compared the effectiveness of OBETs with human landing catches for sampling malaria vectors in arid and forested zones. Each trapping method was used both indoors and outdoors. Four houses were selected in each village, and the following sampling method was used: on the first night, an indoor human landing catch was carried out in the first house, an indoor OBET in the second house, an outdoor human landing catch in the third house, and an outdoor OBET in the fourth house. On the subsequent three nights, the four sampling procedures were rotated among houses, such that each house was sampled once by each of the four collection methods. Four consecutive nights represented one replicate. Overall, 2695 females *Anopheles* mosquitoes comprising 10 species were collected. Human landing catches caught more species than the OBETs in both villages, the additional species caught by landing catch being *Anopheles pharoensis* and *Anopheles rufipes* in the dry zone and *Anopheles hancocki, Anopheles squamosus,* and *Anopheles ziemanni* in the forested zone. In both villages, the numbers of mosquitoes collected in OBETs indoors were always lower than those obtained by landing catch. The authors concluded that OBETs were a useful collection method, removing the need to expose humans to the risk of infection, but were less useful for studying the anopheline fauna in an area as they tended to catch fewer species than human landing catch.

Mboera et al. (1997) used a similar method to compare the efficiency of whole host odour and CO_2 as attractants for mosquitoes in Tanzania. Whole host odour or CO_2 was introduced into a tent trap containing a mosquito net and a CDC miniature light-trap and fitted with two Muirhead-Thomson exit traps. Whole host odour was obtained from a human subject seated in an underground pit dug below the tent and odour was blown into the mosquito net inside the tent through a PVC tube containing a fan from a CDC light-trap. Carbon dioxide was pumped into the mosquito net from a cylinder at rates of 300 ml min^{-1} or 1500 ml min^{-1} (Fig. 6.42).

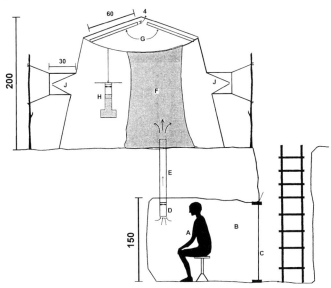

Fig. 6.42. Tent trap as used by Mboera et al. (1997). Test subject (A) seated in underground pit (B), sealed by a polythene sheet (C). Host odour was pumped by a fan (D) through a PVC tube (E) into a mosquito net (F). Mosquitoes entered through entry slits (G) and were caught by a CDC light-trap (H) or window exit traps (J)

Tent traps into which human host odour was pumped caught significantly more *Anopheles gambiae* s.l. and *Anopheles funestus* than traps into which CO_2 was pumped. Raising the flow rate of CO_2 did not affect the *Anopheles gambiae* s.l. catch, but did increase the *Anopheles funestus* catch to around 70% of that obtained using human odour.

References

Abu Hassan A, Rahman WA, Che Salmah MR, Rashid MZR, Jaal Z, Adanan CR, Shahrem MR (1997) The distribution of common *Anopheles* mosquitoes in northwestern peninsular Malaysia. J Vector Ecol 22: 109–114

Abu Hassan A, Rahman WA, Rashid MZA, Shahrem MR, Adanan CR (2001) Composition and biting activity of *Anopheles* (Diptera: Culicidae) attracted to human bait in a malaria endemic village in peninsular Malaysia near the Thailand border. J Vector Ecol 26: 70–75

Ahid SMM, Lourenco-de-Oliveira R (1999) Mosquitos vetores potenciais de diro-filariose canina na regiao nordeste do Brazil. Rev Saúde Pública 33: 560–565

Aitken THG (1967) The canopy-frequenting mosquitoes of Bush Bush forest, Trinidad, West Indies. Atas do Simpósio sôbre a Biota Amazónica (Patologia) 6: 65–73

Aitken THG, Jonkers AH, Worth CB (1963) Bush Bush Bonanza. A Study of Virus-Vector Relationships in a Trinidadian Forest. Proc Int Congr Trop Med Malaria 7th in Anais Microbiol 11: 67–77

Aitken THG, Worth CB, Jonkers AH, Tikasingh ES, Downs WG (1968*a*) Arbovirus studies in Bush Bush forest, Trinidad, W. I., September 1959–December 1964. II. Field program and techniques. Am J Trop Med Hyg 17: 237–252

Aitken THG, Worth CB, Tikasingh ES (1968*b*). Arbovirus studies in Bush Bush forest, Trinidad, W. I., September 1959–December 1964. III. Entomological studies. Am J Trop Med Hyg 17: 253–268

Akiyama J (1973) Interpretation of the results of baited trap net collections. J Trop Med 76: 283–284

Allan SA, Day JF, Edman JD (1987) Visual ecology of biting flies. Annu Rev Entomol 32: 297–316

Almeida APG, Baptista SSSG, Sousa CAGCC, Novo MTLM, Ramos HC, Panella NA, Godsey M, Simões MJO, Anselmo ML, Komar N, Mitchell CJ Ribeiro H (2005) Bioecology and vectorial capacity of *Aedes albopictus* (Diptera: Culicidae) in Macao, China, in relation to dengue virus transmission. J Med Entomol 42: 419–428

Almirón WR, Brewer MM (1995) Preferencia de hospedadores de Culicidae (Diptera) recolectados en el centro de la Argentina. Rev Saúde Pública 29: 108–114

Alto BW, Lounibos LP, Juliano SA (2003) Age-dependent bloodfeeding of *Aedes aegypti* and *Aedes albopictus* on artificial and living hosts. J Am Mosq Control Assoc 19: 347–352

Ameneshewa B, Service MW (1997) Blood-feeding behaviour of *Anopheles arabiensis* Patton (Diptera: Culicidae) in central Ethiopia. J Afr Zool 111: 235–245

Amerasinghe PH, Amerasinghe FP (1999) Multiple host feeding in field populations of *Anopheles culicifacies* and *An. subpictus* in Sri Lanka. Med Vet Entomol 13: 124–131

Amerasinghe FP, Ariyasena TG (1990) Larval survey of surface water-breeding mosquitoes during irrigation development in the Mahaweli project, Sri Lanka. J Med Entomol 27: 789–802

Amerasinghe FP, Ariyasena TG (1991) Survey of adult mosquitoes (Diptera: Culicidae) during irrigation development in the Mahaweli Project, Sri lanka. J Med Entomol 28: 387–393

Amerasinghe FP, Munasingha NB (1988) A predevelopment mosquito survey in the Mahaweli development project area, Sri Lanka: Adults. J Med Entomol 25: 276–285

Amerasinghe PH, Amerasinghe FP, Konradsen F, Fonseka KT, Wirtz RA (1999) Malaria vectors in a traditional dry zone village in Sri Lanka. Am J Trop Med Hyg 60: 421–429

Anderson JF, Andreadis TG, Main AJ, Kline DL (2004) Prevalence of West Nile virus in tree canopy-inhabiting Culex pipiens and associated mosquitoes. Am J Trop Med Hyg 71: 112–119

Anderson RA, Brust RA (1995) Field evidence for multiple host contacts during blood feeding by *Culex tarsalis, Cx. restuans*, and *Cx. nigripalpus* (Diptera: Culicidae). J Med Entomol 32: 705–710

Andersson HI (1990) Nectar feeding activity of *Aedes* mosquitoes, with special reference to *Aedes communis* females. J Am Mosq Control Assoc 6: 482–489

Andre RG, Rowley WA, Wong YW, Dorsey DC (1985) Surveillance of arbovirus activity in Iowa, USA, 1978–1980. J Med Entomol 22: 58–63

Anil Prakash, Bhattacharyya DR, Mohapatra PK, Mahanta J (2000) Preliminary observations on man-mosquito contact in Soraipung village of district Dibrugarh, Assam (India). Indian J Malariol 37: 97–102

Arredondo-Jimenez JI, Gimnig J, Rodriguez MH, Washino RK (1996) Genetic differences among *Anopheles vestitipennis* subpopulations collected using different methods in Chiapas state, southern Mexico. J Am Mosq Control Assoc 12: 396–401

Artsob H, Spence L, Surgeoner G, Th'ng C, Lampotang V, Grant L, McCreadie J (1983) Studies on a focus of California group virus activity in southern Ontario. Mosquito News 43: 449–455

Aslam Y, Reisen WK, Aslamkhan M (1977) The influence of physiological age on the biting rhythm of *Culex tritaeniorhynchus* Giles (Diptera: Culicidae). Southeast Asian J Trop Med Public Health 8: 364–367

Attenborough RD, Burkot TR, Gardner DS (1997) Altitude and the risk of bites from mosquitoes infected with malaria and filariasis among the Mianmin people of Papua New Guinea. Trans R Soc Trop Med Hyg 91: 8–10

Bailly-Choumara H (1973) Etude comparative de differentes techniques de récolté de moustiques adultes (Diptera Culicidae) faite au Maroc, en zone rurale. Bull Soc Sci Nat Phys Maroc 53: 135–187

Balenghien T, Fouque F, Sabatier P, Bicout DJ (2005) Horse-, bird-, and human-seeking behavior and seasonal abundance of mosquitoes in a West Nile virus focus of southern France. J Med Entomol 42: 936–946

Bangs MJ, Rusmiarto S, Anthony RL, Wirtz RA, Subianto DB (1996) Malaria transmission by *Anopheles punctulatus* in the highlands of Irian Jaya, Indonesia. Ann Trop Med Parasitol 90: 29–38

Barbosa OC, Teodoro U, Lozovei AL, Filho VLS, Spinosa RP, de Lima EM, Ferreira MEMC (1993) Nota sobre culicídeos adultos coletados na região sul do Brasil. Rev Saúde Pública 27: 214–216

Barbosa AA, Navarro-Silva MA, Calado D (2003) Atividade de Culicidae em remanescente florestal na região urbana de Curitiba (Paraná, Brasil). Rev Bras Zool 20: 59–63

Barnard DR, Mulla MS (1977) Diel periodicity of blood feeding in the mosquito *Culiseta inornata* in the Coachella valley of southern California. Mosquito News 37: 669–673

Barodji, Sularto, Bamlang Haryanto, Supratman S, Supalin (1986) Manfaat penggunaan 'exit-trap' dalam penilaian padat populasi vektor malaria *Anopheles aconitus* di kandang pada malam hari. Bull Penel Keseh Health Std Indonesia 14: 18–24 (In Indonesian, English summary)

Bartlett MS (1947) The use of transformations. Biometrics 3: 39–52

Bates M (1944a) Observations on the distribution of diurnal mosquitoes in a tropical forest. Ecology 25: 159–70

Bates M (1944b) Notes on the construction and use of stable traps for mosquito studies. J Natn Malar Soc 3: 135–145

Beadle LD (1959) Field observations on the biting habits of *Culex tarsalis* at Mitchell, Nebraska, and Logan, Utah. Am J Trop Med Hyg 8: 134–140

Beck SD (1968) Insect Photoperiodism. Academic Press, New York

Beier JC, Trpis M (1981) Incrimination of natural culicine vectors which transmit *Plasmodium elongatum* to penguins at the Baltimore zoo. Can J Zool 59: 470–475

Bellamy RE, Reeves WC (1952) A portable mosquito bait-trap. Mosquito News 12: 256–258.

Bemrick WJ, Sandholm HA (1966) *Aedes vexans* and other potential mosquito vectors of *Dirofilaria immitis* in Minnesota. J Parasitol 52: 762–767

Bennett GF (1960) On some ornithophilic blood-sucking Diptera in Algonquin Park Ontario, Canada. Can J Zool 38: 377–389

Bhatt RM, Kohli VK (1996) Biting rhythms of some anophelines in central Gujarat. Indian J Malariol 33: 180–190

Bhattacharyya DR, Dutta P, Khan SA, Doloi P, Goswami BK (1995) Biting cycles of some potential vector mosquitos of Japanese encephalitis of Assam, India. Southeast Asian J Trop Med Public Health 26: 177–179

Bidlingmayer WL (1964) The effect of moonlight on the flight activity of mosquitoes. Ecology 45: 87–94

Bidlingmayer WL (1967) A comparison of trapping methods for adult mosquitoes: species response and environmental influence. J Med Entomol 4: 200–220

Bidlingmayer WL (1985) The measurement of adult mosquito population changes—some considerations. J Am Mosq Control Assoc 1: 328–348

Blackmore JS, Dow RP (1958) Differential feeding of *Culex tarsalis* on nestling and adult birds. Mosquito News 18: 15–17

Blagoveshenskii DI, Sregetova NG, Monchadskiy AS (1943) Activity in mosquito attacks under natural conditions and its diurnal periodicity. Zool Zh Ukr 22: 138–153 (In Russian)

Bockarie MJ, Service MW, Barnish G, Touré YT (1995) Vectorial capacity and entomological inoculation rates of *Anopheles gambiae* in a high rainfall forested area of southern Sierra Leone. Trop Med Parasitol 46: 164–171

Bockarie MJ, Alexander N, Bockarie F, Ibam E, Barnish G, Alpers M (1996) The late biting habit of parous *Anopheles* mosquitoes and pre-bedtime exposure of humans to infective female mosquitoes. Trans R Soc Trop Med Hyg 90: 23–25

Bockarie MJ, Tavul L, Kastens W, Michael E, Kazura JW (2002) Impact of untreated bednets on prevalence of *Wuchereria bancrofti* transmitted by *Anopheles farauti* in Papua New Guinea. Med Vet Entomol 16: 116–119

Boorman J (1960) Studies on the biting habits of the mosquito *Aedes (Stegomyia) aegypti*, Linn., in a West African village. West Afr Med J (N. S.) 9: 111–122

Bowden J (1973a) The significance of moonlight in photoperiodic responses of insects. Bull Entomol Res 62: 605–612

Bowden J (1973b) The influence of moonlight on catches of insects in light-traps in Africa. Part I. The moon and moonlight. Bull Entomol Res 63: 113–128

Bowden J, Church BM (1973) The influence of moonlight on catches of insects in light-traps in Africa. Part II. The effect of moon phase on light-trap catches. Bull Entomol Res 63: 129–142

Bowen MF, Haggart D, Romo J (1995a) Long-distance orientation, nutritional preference, and electrophysiological responsiveness in the mosquito *Aedes bahamensis*. J Vector Ecol 20: 203–210

Bowen MF, Romo J (1995b) Host-seeking and sugar-feeding in the autogenous mosquito *Aedes bahamensis* (Diptera: Culicidae). J Vector Ecol 20: 195–202

Bown DN, Bang YH (1980) Ecological studies on *Aedes simpsoni* (Diptera: Culicidae) in southeastern Nigeria. J Med Entomol 17: 367–374

Bown DN, Rios JR, del Angel Cabañas G, Guerrero JC, Méndez JF (1984) Evaluation of chlorphoxim used against *Anopheles albimanus* on the south coast of Mexico: 1; Results of indoor chlorphoxim applications and assessment of the methodology employed. Bull Pan-Am Health Organ 18: 379–388

Bown DN, Rios JR, del Angel Cabañas G, Guerrero JC, Méndez JF (1985) Evaluation of chlorphoxim used against *Anopheles albimanus* on the south coast of Mexico: 2. Use of two curtain-trap techniques in a village-scale evaluation trial. Bull Pan-Am Health Organ 19: 61–68

Bown DN, Rios JR, Frederickson C, del Angel Cabañas G, Méndez JF (1986) Use of an exterior curtain-net to evaluate insecticide/mosquito behavior in houses. J Am Mosq Control Assoc 2: 99–101

Bown DN, Frederickson EC, del Angel Cabañas G, Méndez JF (1987) An evaluation of bendiocarb and deltamethrin applications in the same Mexican village and their impact on populations of *Anopheles albimanus*. Bull Pan-Am Health Organ 21: 121–135

Bown DN, Rodriguez MH, Arrendo-Jiménez JI, Loyola EG, del Carman Rodriguez M (1991) Age structure and abundance levels in the entomological evaluation of an insecticide used in the control of *Anopheles albimanus* in southern Mexico. J Am Mosq Control Assoc 7: 180–187

Bradley GH, Goodwin MH, Stone A (1949) Entomologic technics as applied to anophelines. In: Boyd MF (ed) Malariology, vol. I. W. B. Saunders Co., Philadelphia & London, pp 331–378

Braimah N, Drakeley C, Kweka E, Mosha F, Helinski M, Pates H, Maxwell C, Massawe T, Kenward MG, Curtis C (2005) Tests of bednet traps (Mbita traps) for monitoring mosquito populations and time of biting in Tanzania and possbile impact of prolonged insecticide treated net use. Int J Trop Insect Sci 25: 208–213

Brock B, Crans WJ (1977) A field study of the collapsible dog-baited mosquito trap. Proc New Jers Mosq Control Assoc 64: 72–77

Brown ES, Taylor LR (1971) Lunar cycles in the distribution and abundance of airborne insects in the equatorial highlands of East Africa. J Anim Ecol 40: 767–779

Bryan JH (1983) *Anopheles gambiae* and *A. melas* at Brefet, The Gambia and their role in malaria transmission. Ann Trop Med Parasitol 77: 1–12

Buescher EL, Scherer WF, Rosenberg MZ, Gresser I, Hardy JL, Bullock HR (1959) Ecologic studies of Japanese encephalitis virus in Japan. II. Mosquito infection. Am J Trop Med Hyg 8: 651–664

Bugher JC, Boshell-Manrique J, Roca-Garcia M, Osorno-Mesa E (1944) Epidemiology of jungle yellow fever in eastern Colombia. Am J Hyg 39: 16–51

Burkett DA, Lee WJ, Lee KW, Kim HC, Lee HI, Lee JS, Shin EH, Wirtz RA, Cho HW, Claborn DM, Coleman RE, Klein TA (2001) Light, carbon dioxide and octenol-baited mosquito trap and host-seeking activity evaluations for mosquitoes in a malarious area of the Republic of Korea. J Am Mosq Control Assoc 17: 196–205

Butts WL (2001) Beaver ponds in upstate New York as a source of anthropophilic mosquitoes. J Am Mosq Control Assoc 17: 85–86

Campos RE, Fernández LA, Sy VE (2004) Study of the insects associated with the floodwater mosquito *Ochlerotatus albifasciatus* (Diptera: Culicidae) and their possible predators in Buenos Aires Province, Argentina. Hydrobiologia 524: 91–102

Cano J, Berzosa PJ, Roche J, Rubio JM, Moyano E, Guerra-Neira A, Brochero H, Mico M, Edú M, Benito A (2004) Malaria vectors in the Bioko Island (Equatorial Guinea); estimation of vector dynamics and transmission intensities. J Med Entomol 41: 158–161

Canyon D, Hii JLK, Muller R (1999) The frequency of host biting and its effect on oviposition and survival in *Aedes aegypti* (Diptera : Culicidae). Bull Entomol Res 89: 35–39

Cardoso J da C, Corseuil C, Barata JMS (2004) Anophelinae (Diptera, Culicidae) occorentes no estado do Rio Grande do Sul, Brasil. Entomología y Vectores 11: 159–177

Carnevale P, Molinier M (1980) Le cycle gonotrophyque et le rythme quotidien des piqures d'*Anopheles gambiae* (Giles), 1902 et *Anopheles nili* (Theobald), 1904. Parassitologia 22: 173–185

Casas M, Torres JL, Bown DN, Rodríguez MH, Arredondo-Jiménez JI (1998) Selective and conventional house-spraying of DDT and bendiocarb against *Anopheles pseudopunctipennis* in southern Mexico. J Am Mosq Control Assoc 14: 410–420

Causey OR, Causey CE, Maroja OM, Macedo DG (1961) The isolation of arthropod-borne viruses, including members of two hitherto undescribed serological groups in the Amazon region of Brazil. Am J Trop Med Hyg 10: 227–249

Chadee DD, Martinez R (2000) Landing periodicity of *Aedes aegypti* with implications for dengue transmission in Trinidad, West Indies. J Vector Ecol 25: 158–163

Chadee DD, Tikasingh ES, Ganesh R (1992) Seasonality, biting cycle and parity of the yellow fever vector mosquito *Haemagogus janthinomys* in Trinidad. Med Vet Entomol 6: 143–148

Chadee DD, Ganesh R, Hingwan JO, Tikasingh ES (1995) Seasonal abundance, biting cycle and parity of the mosquito *Haemagogus leucocelaenus* in Trinidad, West Indies. Med Vet Entomol 9: 372–376

Chadee DD, Hingwan JO, Persad RC, Tikasingh ES (1993) Seasonal abundance, biting cycle, parity and vector potential of the mosquito *Haemagogus equinus* in Trinidad. Med Vet Entomol 7: 141–146

Chandler JA, Highton RB, Hill MN (1975) Mosquitoes of the Kano Plain, Kenya. I. Results of indoor collections in irrigated and nonirrigated areas using human bait and light traps. J Med Entomol 12: 504–510

Chandler JA, Highton RB, Boreham PFL (1976*a*). Studies on some ornithophilic mosquitoes (Diptera: Culicidae) of the Kano Plain, Kenya. Bull Entomol Res 66: 133–143

Chandler JA, Highton RB, Hill MN (1976*b*) Mosquitoes of the Kano Plain, Kenya. II. Results of outdoor collections in irrigated and nonirrigated areas using human and animal bait and light traps. J Med Entomol 13: 202–207

Charlwood JD, Galgal K (1985) Observations on the biology and behaviour of *Armigeres milnensis* Lee (Diptera: Culicidae) in Papua New Guinea. J Aust Entomol Soc 24: 313–319

Charlwood JD, Wilkes TJ (1979) Studies on the age-composition of samples of *Anopheles darlingi* Root (Diptera: Culicidae) in Brazil. Bull Entomol Res 69: 337–342

Charlwood JD, Paru R, Dagoro H, Lagog M, Kelepak L, Yabu S, Krimbo P, Pirou P (1984) Factors affecting the assessment of man biting rates of malaria vectors. In: Bryan JH, Moodie PM (eds) Malaria. Proceedings of a Conference to Honour Robert H. Black, Sydney, February 1983. Australian Government Publishing Services, Canberra, pp 143–151

Charlwood JD, Paru R, Dagara (sic) H (1986*a*) A new light-bed net trap to sample anopheline vectors of malaria in Papua, New Guinea. Bull Soc Vector Ecol 11: 281–282

Charlwood JD, Paru R, Dagoro H, Lagog M (1986*b*) The influence of moonlight and gonotrophic age in the biting activity of *Anopheles farauti* (Diptera: Culicidae) from Papua New Guinea. J Med Entomol 23: 132–135

Charlwood JD, Graves PM, Marshall TFde C (1988) Evidence for a 'memorized' home range in *Anopheles farauti* females from Papua New Guinea. Med Vet Entomol 2: 101–108

Charlwood JD, Smith T, Kihonda J, Heiz B, Billingsley PF, Takken W (1995*a*) Density independent feeding success of malaria vectors (Diptera: Culicidae) in Tanzania. Bull Entomol Res 85: 29–35

Charlwood JD, Alecrim WD, Fe N, Mangabeira J, Martins VJ (1995*b*) A field trial with lambda-cyhalothrin (Icon) for the intradomiciliary control of malaria transmitted by *Anopheles darlingi* Root in Rondonia, Brazil. Acta Trop 60: 3–13

Charlwood JD, Pinto J, Sousa CA, Ferreira C, Gil V, do Rosário VE (2003) Mating does not affect the biting behaviour of *Anopheles gambiae* from the islands of São Tomé and Príncipe, West Africa. Ann Trop Med Parasitol 97: 751–756

Cheong WH, Chiang GL, Loong KP, Mahadevan S, Samarawickrema WA (1988) Biting activity cycles of some mosquitoes in the Bengkoka peninsula, Sabah state with notes on their importance. Trop Biomed 5: 27–31

Chevalier V, Mondet B, Diaite A, Lancelot R, Fall AG, Ponçon N (2004) Exposure of sheep to mosquito bites: possible consequences for the transmission risk of Rift Valley Fever in Senegal. Med Vet Entomol 18: 247–255

Chiang GL, Samarawickrema, WA, Cheong WH, Sulaiman I, Yap HH (1984*a*) Biting activity, age composition and survivorship of *Mansonia* in two ecotypes in Peninsular Malaysia. Trop Biomed 1: 151–158

Chiang GL, Cheong WH, Samarawickrema WA, Mak JW, Kan SKP (1984b) Filariasis in Bengkoka peninsula, Sabah, Malaysia: vector studies in relation to the transmission of filariasis. Southeast Asian J Trop Med Publ Health 15: 179–189

Chiang GL, Samarwickrema WA, Eng KL, Cheong WH, Sulaiman I, Yap HH (1986) Field studies on the surveillence of *Coquillettidia crassipes* (Van der Wulp) and the isolation of Cardiofilaria in peninsular Malaysia. Ann Trop Med Parasitol 80: 235–244

Christensen HA, Vasquez AM, Boreham, MM (1996) Host-feeding patterns of mosquitoes (Diptera: Culicidae) from central Panama. Am J Trop Med Hyg 55: 202–208

Colless DH (1957) Components of the catch curve of *Culex annulus* in Singapore. Nature, Lond 180: 1496–1497

Colless DH (1959) Notes on the culicine mosquitoes of Singapore. VI. Observations on catches made with baited and unbaited trap-nets. Ann Trop Med Parasitol 53 251–258

Comtois A, Berteaux D (2005) Impacts of mosquitoes and black flies on defensive behaviour and microhabitat use of the North American porcupine (*Erethizon dorsatum*) in southern Quebec. Can J Zool 83: 754–764

Conn JE, Wilkerson RC, Segura MNO, de Souza RTL, Schlichting CD, Wirtz RA, Póvoa MM (2002) Emergence of a new Neotropical malaria vector facilitated by human migration and changes in land use. Am J Trop Med Hyg 66: 18–22

Cope SE, Barr AR, Bangs MJ, Morrison AC, Guptavanij P (1986) Human bait collections of mosquitoes in a southern California freshwater marsh. Proc California Mosq Vector Control Assoc 54: 110–112

Corbet PS (1961) Entomological studies from a high tower in Mpanga forest, Uganda. VIII. The age-composition of biting mosquito populations according to time and level. Trans R Entomol Soc Lond 113: 336–345

Corbet PS (1962) The age-composition of biting mosquito populations according to time and level: a further study. Bull Entomol Res 53: 406–416

Corbet PS (1964) Nocturnal flight activity of sylvan Culicidae and Tabanidae (Diptera) as indicated by light-traps: a further study. Proc R Entomol Soc Lond (A) 39: 53–57

Corbet PS, Smith SM (1974) Diel periodicities of landing of nulliparous and parous *Aedes aegypti* (L.) at Dar es Salaam, Tanzania (Diptera, Culicidae). Bull Entomol Res 64: 111–121

Cordellier R, Geoffroy B (1974) Contribution á l'étude des culicides de la République Centrafricaine, rythmes d'activités en secteur préforestier. Cah ORSTOM sér Entomol Méd Parasitol 12: 19–48

Cordellier R, Bouchité B, Roche J-C, Monteny N, Diaco B, Akoliba P (1983) Circulation selvatique du virus dengue 2 en 1980, dans les savanes sub-soudaniennes de Côte d'Ivoire. Données entomologiques et considérations épidémologiques. Cah ORSTOM sér Entomol Méd Parasitol 21: 165–179

Costantini C, Diallo M (2001) Preliminary lack of evidence for simian odour preferences of savanna populations of *Anopheles gambiae* and other malaria vectors. Parassitologia 43: 179–182

Costantini C, Gibson G, Brady J, Merzagora L, Coluzzi M (1993) A new odour-baited trap to collect host-seeking mosquitoes. Parassitologia 35: 5–9

Costantini C, Gibson G, Fale Sagnon N, della Torre A, Brady J, Coluzzi M (1996) Mosquito responses to carbon dioxide in a West African Sudan savanna village. Med Vet Entomol 10: 220–227

Costantini C, Sagnon NF, della Torre A, Diallo M, Brady J, Gibson G, Coluzzi M (1998a) Odor-mediated host preferences of West African mosquitoes, with particular reference to malaria vectors. Am J Trop Med Hyg 58: 56–63

Costantini C, Sagnon NF, Sanogo E, Merzagora L, Coluzzi M (1998b) Mosquito responses to carbon dioxide in a West African Sudan savanna village. Bull Entomol Res 88: 503–511

Coz J (1964) Étude des variations de l'âge physiologique d'*A. gambiae*, Giles et *A. mascarensis*, De Meillon, au cours de captures de nuit. Bull Soc Pathol exot 57: 619–626

Coz J (1971) Étude comparative des fenêtres et des verandas-pièges, comme moyen des sortie pour les moustiques, Koumbia (Haute-Volta). Cah ORSTOM sér Entomol Méd Parasitol 9: 239–246

Coz J, Eyraud M, Venard P, Attiou B, Somda D, Ouedraogo V (1965) Expérience en Haute-Volta sur l'utilisation de cases pièges pour la mesure d l'activité du DDT contre les moustiques. Bull World Health Organ 33: 435–52

Crans WJ (1986) Failure of chickens to act as sentinels during an epizootic of Eastern Equine Encephalitis in southern New Jersey, USA. J Med Entomol 23: 626–629

Crans WJ, Rockel EG (1968) The mosquitoes attracted to turtles. Mosquito News 28: 332–337

Cupp EW, Stokes GM (1973) Identification of bloodmeals from mosquitoes collected in light traps and dog-baited traps. Mosquito News 33: 39–41

da Silva-Vasconcelos A, Kató MYN, Mourão EN, de Souza RTL, Lacerda RNda L, Sibajev A, Tsouris P, Póvoa MM, Momen H, Rosa-Freitas MG (2002) Biting indices, host-seeking activity and natural infection rates of anopheline species in Boa Vista, Roraima, Brazil from 1996 to 1998. Mem Inst Oswaldo Cruz 97: 151–161

Das P, Bhattacharya S, Chakraborty S, Palit A, Das S, Ghosh KK, Hati AK (1983) Diurnal man-biting of *Armigeres subalbatus* (Coquillett, 1898) in a village in West Bengal. Indian J Med Res 78: 794–798

Davidson G (1949) A field study on 'Gammexane' on malaria control in the Belgian Congo. 1. The anophelines of Yaligimba and their bionomics. Ann Trop Med Parasitol 43: 361–372

Davies JB (1971) A small mosquito trap for use with animal or carbon dioxide baits. Mosquito News 31: 441–443

Davies JB (1973) A simple battery operated suction trap for insects attracted to animal, light or chemical bait. Mosquito News 33: 102–104

Davies JB (1975) Moonlight and the biting activity of *Culex (Melanoconion) portesi* Senevet & Abonnenc and *C. (M) taeniopus* D. & K. (Diptera: Culicidae) in Trinidad forests. Bull Entomol Res 65: 81–96

Davies JB (1978) Attraction of *Culex portesi* Senevet & Abonnenc and *Culex taeniopus* Dyar & Knab (Diptera: Culicidae) to 20 animal species exposed in a Trinidad forest. I. Baits ranked by numbers of mosquitoes caught and engorged. Bull Entomol Res 68: 707–719

Davies L (1963) Seasonal and diurnal changes in the age-composition of adult *Simulium venustum* Say (Diptera) populations near Ottawa. Can Entomol 95: 654–667

Day JF, Edman JD (1983) Malaria renders mice susceptible to mosquito feeding when gametocytes are most infective. J Parasitol 69: 163–170

Day JF, Edman JD (1984) Mosquito engorgement on normally defensive hosts depends on host activity patterns. J Med Entomol 21: 732–740

Deane LM, Causey OR, Deane MP (1948) Notas sôbre a distribuição e a biologia dos anofelinos das regiões Nordestina e Amazônica Mo Brasil. Rev Serv Esp Saúde Pública Rio de Janeiro 1: 827–965

Deegan CS, Burns JE, Huguenin M, Steinhaus EY, Panella NA, Beckett S, Komar N (2005) Sentinel pigeon surveillance for West Nile virus by using lard-can traps at differing elevations and canopy cover classes. J Med Entomol 42: 1039–1044 de Freitas EN, Shope RE, Toda A (1966) A blower trap for capturing mosquitoes. Mosquito News 26: 373–377

Dégallier N, Sá Filho GC, Monteiro HAO, Castro FC, da Silva OV, Brandão RCF, Moyses M, da Rosa APAT (1998) Release-recapture experiments with canopy mosquitoes in the genera *Haemagogus* and *Sabethes* (Diptera: Culicidae) in Brazilian Amazonia. J Med Entomol 35: 931–936

de Kruijf HAM (1972) Aspects of the ecology of mosquitoes in Surinam. Studies on the fauna of Suriname and other Guyanas. 13, No. 51, Martinus Nijhoff, The Hague

De Meillon B, Sebastian A (1967) Qualitative and quantitative characteristics of adult *Culex pipiens fatigans* populations according to time, site and place of capture. Bull World Health Organ 36: 75–80

De Meillon B, Paterson HE, Muspratt J (1957) Studies on arthropod-borne viruses of Tongaland, II. Notes on the more common mosquitoes. S Afr J Med Sci 22: 47–53

Dev V, Ansari MA, Hira CR, Barman K (2001) An outbreak of *Plasmodium falciparum* malaria due to Anopheles minimus in central Assam, India. Indian J Malariol 38: 32–38

de Zulueta J (1950) A study of the habits of the adult mosquitoes dwelling in the savannas of eastern Colombia. Am J Trop Med 30: 325–339

de Zulueta J (1952) Observations on mosquito density in an endemic malarious area in eastern Colombia. Am J Trop Med Hyg 1: 314–329

Dia I, Diallo D, Duchemin J-B, Ba Y, Konate L, Costantini C, Diallo M (2005) Comparisons of human-landing catches and odor-baited entry traps for sampling malaria vectors in Senegal. J Med Entomol 42: 104–109

Diallo M, Ba Y, Sall AA, Diop OM, Ndione JA, Mondo M, Girault L, Mathiot C (2003) Amplification of the sylvatic cycle of dengue virus type 2, Senegal, 1999–2000: entomologic findings and epidemiologic considerations. Emerg Infect Dis 9: 362–367

Diarrassouba S, Dossou-Yovo J (1997) Rythme d'activité atypique chez *Aedes aegypti* en zone de savane sub-soudanienne de Côte d'Ivoire. Bull Soc Pathol Exot 90: 361–363

Díaz LA, Almirón WR, Ludueña Almeida F, Spinsanti LI, Contigiani MS (2003) Vigilancia del virus Encefalitis de San Luis y mosquitos (Diptera: Culicidae) en la Provincia de Córdoba, Argentina. Entomología y Vectores 10: 551–566

Dossou-yovo J, Doannio JMC, Riviere F, Chauvancy G (1995) Malaria in Cote d'Ivoire wet savannah region: the entomological input. Trop Med Parasitol 46: 263–269

Dossou-Yovo J, Doannio JMC, Diarrassouba S, Chauvancy G (1998) Impact d'aménagements de rizières sur la transmission du paludisme dans la ville de Bouaké, Côte d'Ivoire. Bull Soc Pathol Exot 91: 327–333

Dossou-Yovo J, Diarrassouba S, Doannio J, Darriet F, Carnevale P (1999) Le cycle d'agressivité d'*Anopheles gambiae* s.s. à l'intérieur des maisons et la transmission du paludisme dans la région de Bouaké (Côte d'Ivoire). Intérêt de l'utilisation de la moustiquaire imprégnée. Bull Soc Pathol Exot 92: 198–200

Dow RP, Morris CD (1972) Wind factors in the operation of a cylindrical bait trap for mosquitoes. J Med Entomol 1: 60–66

Dow RP, Reeves WC, Bellamy RE (1957) Field tests of avian host preferences of *Culex tarsalis* Coq. Am J Trop Med Hyg 6: 294–303

Downing JD (1976) Statistical analysis and interpretation of mosquito light trap data. Proc New Jers Mosq Control Assoc 63: 127–133

Downing JD, Crans WJ (1977) The Ehrenberg pigeon trap as a sampler of *Culex* mosquitoes for St Louis encephalitis surveillance. Mosquito News 37: 48–53

Duchemin J-B, Leong Pock Tsy J-M, Rabarison P, Roux J, Coluzzi M, Costantini C (2001) Zoophily of *Anopheles arabiensis* and *An. gambiae* in Madagascar demonstrated by odour-baited entry traps. Med Vet Entomol 15: 50–57

Duke BOL (1960) Studies on the biting habits of *Chrysops*. VII. The biting cycle of nulliparous and parous *C. silacea* and *C. dimidiata* (Bombe form). Ann Trop Med Parasitol 54: 147–155

Dyce AL, Lee DJ (1962) Blood-sucking flies (Diptera) and myxomatosis transmission in a mountain environment in New South Wales. II. Comparison of the use of man and rabbits as bait animals in evaluating vectors of myxomatosis. Aust J Zool 10: 84–94

Earle WC (1949) Trapping and deflection of anopheline mosquitoes. In: Boyd MF (ed) Malariology, vol. II. W. B. Saunders Co., Philadelphia & London, pp 1221–1231

Earle WC, Howard HH (1936) The determination of *Anopheles* mosquito prevalence. Boln Asoc Méd P Rico 28: 233–240

Easton ER, Price MA, Graham OH (1968) The collection of biting flies in west Texas with Malaise and animal baited traps. Mosquito News 28: 465–469

Edman JD, Scott TW (1987) Host defensive behaviour and feeding success of mosquitoes. Insect Science and its Application 8: 617–622

Edman JD, Webber LA (1975) Effect of vertebrate size and density on host-selection by caged *Culex nigripalpus*. Mosquito News 35: 508–512

Ehrenberg HA (1966) Some comparisons of a pigeon-baited trap with a New Jersey light trap. Proc New Jers Mosq Exterm Assoc 53: 175–182

Elliott R (1972) The influence of vector behavior on malaria transmission. Am J Trop Med Hyg 21: 755–763

Emord DE, Morris CD (1982) A host-baited CDC trap. Mosquito News 42: 220–224

Ernst J (1982) *Dirofilaria immitis*: its vectors in southwestern Ontario and tolerance to low temperatures in *Aedes triseriatus*. MSc Thesis, University of Guelph

Ernst J, Slocombe JOD (1984) Mosquito vectors of *Dirofilaria immitis* in southwestern Ontario. Can J Zool 62: 212–216

Faye O, Gaye O, Faye O, Diallo S (1994) La transmission du paludisme dans des villages éloignés ou situés en bordure de la mangrove au Sénégal. Bull Soc Pathol Exot 87: 157–163

Fernandez NM, Klowden MJ (1995) Male accessory gland substances modify the host-seeking behavior of gravid *Aedes aegypti* mosquitoes. J Insect Physiol 41: 965–970

Ferro C, Boshell J, Moncayo AC, Gonzalez M, Ahumada ML, Kang WL, Weaver SC (2003) Natural enzootic vectors of Venezuelan equine encephalitis virus, Magdalena Valley, Colombia. Emerg Infect Dis 9: 49–54

Fisher RA (1943) The relation between the number of species and the number of individuals in a random sample of an animal population. Part 3. A theoretical distribution for the apparent abundance of different species. J Anim Ecol 12: 54–58

Flemings MB (1959) An altitude biting study of *Culex tritaeniorhynchus* (Giles) and other associated mosquitoes in Japan. J Econ Entomol 52: 490–492

Fletcher MG, Turner EC, Hansen JW, Perry BD (1988) Horse-baited insect trap and mobile insect sorting table used in a disease vector identification study. J Am Mosq Control Assoc 4: 431–435

Fonseca LF, di Deco MA, Carrara GC, Dabo I, do Rosario V, Petrarca V (1996) *Anopheles gambiae* complex (Diptera: Culicidae) near Bissau City, Guinea Bissau, West Africa. J Med Entomol 33: 939–945

Fontenille D, Toto JC (2001) *Aedes (Stegomyia) albopictus* (Skuse), a potential new dengue vector in Southern Cameroon. Emerg Infect Dis 7: 1066–1067

Forattini OP, Gomes A de C (1988) Biting activity patterns of *Culex (Melanoconion) ribeirensis* in southern Brazil. J Am Mosq Control Assoc 4: 175–178

Forattini OP, Gomes A de C, Santos JLF, Galati EAB, Rabello EX, Natal D (1981) Observações sôbre actividade de mosquitos Culicidae, em mata residual no Vale do Ribeira, S. Paulo, Brasil. Rev Saúde Pública 15: 557–586

Forattini OP, Kakitani I, Massad E, Gomes A de C (1993*a*) Studies on mosquitoes (Diptera: Culicidae) and anthropic environment. 1 - Parity of blood seeking *Anopheles (Kerteszia)* in South-Eastern Brazil. Rev Saúde Pública 27: 1–8

Forattini OP, Kakitani I, Massad E, Marucci D (1993*b*) Studies on mosquitoes (Diptera: Culicidae) and anthropic environment. 3 - Survey of adult stages at the rice irrigation system and the emergence of *Anopheles albitarsis* in South-Eastern, Brazil. Rev Saúde Pública 27: 313–325

Forattini OP, Kakitani I, Massad E, Marucci D (1993*c*) Studies on mosquitoes (Diptera: Culicidae) and anthropic environment. 4 - Survey of resting adults and synanthropic behaviour in South-Eastern Brazil. Rev Saúde Pública 27: 398–411

Forattini OP, Kakitani I, Massad E, Marucci D (1995*a*) Studies on mosquitoes (Diptera: Culicidae) and anthropic environment. 7 - Behaviour of adult *Nyssorhynchus* anophelines with special reference to *Anopheles albitarsis* s.l. in south-eastern Brazil. Rev Saúde Pública 29: 20–26

Forattini OP, Sallum MAM, Kakitani I, Massad E, Marucci D (1995*b*) Studies on mosquitoes (Diptera: Culicidae) and anthropic environment. 8 - Survey of adult behaviour of Spissipes section species of *Culex (Melanoconion)* in south-eastern Brazil. Rev Saúde Pública 29: 100–107

Forattini OP, Kakitani I, Massad E, Marucci D (1995*c*) Studies on mosquitoes (Diptera: Culicidae) and anthropic environment. 9 - Synanthropy and epidemiological vector role of *Aedes scapularis* in south-eastern Brazil. Rev Saúde Pública 29: 199–207

Forattini OP, Kakitani I, Massad E, Marucci D (1995*d*) Studies on mosquitoes (Diptera: Culicidae) and anthropic environment. 10 - Survey of adult behaviour of *Culex nigripalpus* and other species of *Culex (Culex)* in south-eastern Brazil. Rev Saúde Pública 29: 271–278

Forattini OP, Kakitani I, Massad E, Marucci D (1996*a*) Studies on mosquitoes (Diptera: Culicidae) and anthropic environment. 11 - Biting activity and blood-seeking parity of *Anopheles (Kerteszia)* in south-eastern Brazil. Rev Saúde Pública 30: 107–114

Forattini OP, Kakitani I, Massad E, Marucci D (1996*b*) Studies on mosquitoes (Diptera: Culicidae) and anthropic environment. 12. Host-seeking behaviour of *Anopheles albitarsis* s.l. in south-eastern Brazil. Rev Saúde Pública 30: 299–303

Forattini OP, Kakitani I, dos Santos R La C, Ueno HM, Kobayashi KM (1999) Role of *Anopheles (Kerteszia) bellator* as malaria vector in Southeastern Brazil (Diptera: Culicidae). Mem Inst Oswaldo Cruz 94: 715–718

Forattini OP, Kakitani I, dos Santos R La C, Kobayashi KM, Ueno HM, Fernandez Z (2000) Comportamento de *Aedes albopictus* e de *Ae. scapularis* adultos (Diptera: Culicidae) no Sudeste do Brasil. Rev Saúde Pública 34: 461–467

Foster WA (1995) Mosquito sugar feeding and reproductive energetics. Annu Rev Entomol 40: 443–474

Freier JE, Francy DB (1991) A duplex cone trap for the collection of adult *Aedes albopictus*. J Am Mosq Control Assoc 7: 73–79

Freyvogel TA (1961) Ein Beitrag zu den Problemen um die Blutmahlzeit von Stechmücken. Acta Trop 18: 201–251

Gabaldon A, Ochoa-Palacios M, Perez-Vivas YMA (1940) Estudios sôbre anofelinos. Serie 1. Observaciones sôbre lecturas de trampas-stablo con cebo animal. Publication no. 5. Caracas, Venezuela, Div. de Malariol

Galati EAB, Nunes VLB, Dorval MEC, Cristaldo G, Rocha HC, Gonçalves-Andrade RM, Naufel G (2001) Attractiveness of Black Shannon Trap for Phlebotomines. Mem Inst Oswaldo Cruz 96: 641–647

Galindo P, Trapido H, Carpenter SJ (1950) Observations on diurnal forest mosquitoes in relation to sylvan yellow fever in Panama. Am J Trop Med 30: 533–573

Garcia R, des Rochers BS, Voigt WG (1988) A bait/carbon dioxide trap for the collection of the western tree hole mosquito *Aedes sierrensis*. J Am Mosq Control Assoc 4: 85–88

Garcia R, Colwell AE, Voigt WG, Woodward DL (1989) Fay-Prince trap baited with CO_2 for monitoring adult abundance of *Aedes sierrensis* (Diptera: Culicidae). J Med Entomol 26: 327–331

Garnham PCC, Harper JO, Highton RB (1946) The mosquitoes of the Kaimosi forest, Kenya Colony, with special reference to yellow fever. Bull Entomol Res 36: 473–496

Gass RF, Deesin T, Surathin K, Vutikes S, Sucharit S (1982) Observations on the feeding habits of four species of *Mansonia (Mansonioides)* mosquitoes in southern Thailand. Southeast Asian J Trop Med Public Health 13: 211–215

Gater BAR (1935) Aids to the Identification of Anopheline Imagines in Malaya. Govt. Straits Settlement & Malar. Adv. Bd., F. M. S., Singapore

Germain M, Eouzan JP, Ferrara L, Button JP (1972) Observations sur l'écologie et le comportement particuliers d'*Aedes africanus* (Theobald) dans le nord du Cameroun occidental. Cah ORSTOM sér. Entomol Méd Parasitol 10: 119–126

Germain M, Eouzan JP, Ferrara L, Button JP (1973) Données complémentaires sur le comportement et l'écologie d'*Aedes africanus* (Theobald) dans le nord du Cameroun occidental. Cah ORSTOM sér Entomol Méd Parasitol 11: 127–146

Gillett JD (1957) Age analysis of the biting-cycle of the mosquito *Taeniorhynchus (Mansonioides) africanus* Theobald, based on the presence of parasitic mites. Ann Trop Med Parasitol 51: 151–158

Gillett JD (1971) Mosquitos. Weidenfeld & Nicholson, London

Gillies MT (1954) Studies in the house leaving and outside resting of *Anopheles gambiae* Giles and *Anopheles funestus* Giles in East Africa. Bull Entomol Res 45: 375–387

Gillies MT (1957) Age-groups and the biting cycle in *Anopheles gambiae*. A preliminary investigation. Bull Entomol Res 48: 553–559

Gillies MT, Furlong M (1964) An investigation into the behaviour of *Anopheles parensis* Gillies at Malindi on the Kenya coast. Bull Entomol Res 55: 1–16

Gillies MT, Wilkes TJ (1963) Observations of nulliparous and parous rates in a population of *Anopheles funestus* in East Africa. Ann Trop Med Parasitol 58: 204–213

Gillies MT, Wilkes TJ (1965) A study of the age-composition of populations of *Anopheles gambiae* Giles and *A. funestus* Giles in north-eastern Tanzania. Bull Entomol Res 56: 237–262

Githeko AK (1992) The behaviour and ecology of malaria vectors and malaria transmission in Kisumu district of Western Kenya. Ph.D. thesis, University of Liverpool

Githeko AK, Adungo NI, Karanja DM, Hawley WA, Vulule JM, Seroney IK, Ofulla AVO, Atieli FK, Ondijo SO, Genga IO, Odada PK, Situbi PA, Oloo JA (1996) Some observations on the biting behaviour of *Anopheles gambiae* s.s., *Anopheles arabiensis*, and *Anopheles funestus* and their implications for malaria control. Exp Parasitol 82: 306–315

Gjullin CM, Sailor RI, Stone A, Travis BV (1961) The mosquitoes of Alaska. U. S. Department of Agriculture Handbook

Gould DJ, Mount GA, Scanlon JE, Ford HR, Sullivan MF (1970) Ecology and control of dengue vectors on an island in the gulf of Thailand. J Med Entomol 7: 499–508

Govere J, Durrheim DN, Coetzee M, Hunt RH, la Grange JJ (2000) Captures of mosquitoes of the *Anopheles gambiae* complex (Diptera: Culicidae) in the lowveld region of Mpumalanga Province, South Africa. Afr Entomol 8: 91–99

Grieco JP, Achee NL, Andre RG, Roberts DR (2000) A comparison study of house entering and biting behavior of *Anopheles vestitipennis* (Diptera: Culicidae)

using experimental huts sprayed with DDT or deltamethrin in the southern district of Toledo, Belize, C. A. J Vector Ecol 25: 62–73

Grimstad PR, DeFoliart GR (1974) Nectar sources of Wisconsin mosquitoes. J Med Entomol 11: 331–334

Guan D, Shi W, Deng D (1986) On the surveying method for man-biting rate of *Anopheles sinensis*. J Parasitol Parasitol Dis 4: 28–31 (In Chinese, English summary)

Gubler DJ, Bhattacharya NC (1974) A quantitative approach to the study of bancroftian filariasis. Am J Trop Med Hyg 23: 1027–1036

Guimarães AÉ, de Mello RP, Lopes CM, Alencar J, Gentile C (1997) Prevalência de anofelinos (Diptera: Culicidae) no crepúsculo verspertino em áreas da Usina Hidrelétrica de Itaipu, no município de Guaíra, Estado do Paraná, Brasil. Mem Inst Oswaldo Cruz 92: 745–754

Guimarães AÉ, de Mello RP, Gentile C, Lopes CM (2000*a*) Ecology of mosquitoes (Diptera: Culicidae) in areas of Serra do Mar State Park, State of São Paulo, Brazil. I - Monthly frequency and climatic factors. Mem Inst Oswaldo Cruz 95: 1–16

Guimarães AÉ, Gentile C, Lopes CM, de Mello RP (2000*b*) Ecology of mosquitoes (Diptera: Culicidae) in areas of Serra do Mar State Park, State of São Paulo, Brazil. II - habitat distribution. Mem Inst Oswaldo Cruz 95: 17–28

Guimarães AÉ, Gentile C, Lopes CM, de Mello RP (2000*c*) Ecology of mosquitoes (Diptera: Culicidae) in areas of Serra do Mar State Park, State of São Paulo, Brazil. III - daily biting rhythms and lunar cycle influence. Mem Inst Oswaldo Cruz 95: 753–760

Guimarães AE, Gentile C, Lopes CM, Sant'Anna A, Jovita AM (2000*d*) Ecologia de mosquitos (Diptera: Culicidae) em áreas do Parque Nacional da Serra da Bocaina, Brasil. I - Distribuição por habitat. Rev Saúde Pública 34: 243–250

Guimarães AÉ, Lopes CM, de Mello RP, Alencar J (2003) Ecologia de mosquitos (Diptera, Culicidae) em áreas do Parque Nacional do Iguaçu, Brasil. 1 - Distribuição por hábitat. Cad Saúde Pública 19: 1107–1116

Hacker CS, Scott DW, Thompson JR (1973) A forecasting model for mosquito population densities. J Med Entomol 10: 544–551

Haddow AJ (1945*a*) The mosquitoes of Bwamba county, Uganda. II. Biting activity with special reference to the influence of microclimate. Bull Entomol Res 36: 33–73

Haddow AJ (1945*b*) The mosquitoes of Bwamba county, Uganda. III. The vertical distribution of mosquitoes in a banana plantation and the biting-cycle of *Aedes (Stegomyia) simpsoni* Theo. Bull Entomol Res 36: 297–304

Haddow AJ (1954) Studies of the biting-habits of African mosquitoes. An appraisal of methods employed with special reference to the twenty-four-hour catch. Bull Entomol Res 45: 199–242

Haddow AJ (1956) Rhythmic biting activity of certain East African mosquitoes. Nature 177: 531–532

Haddow AJ (1961*a*) Entomological studies from a high tower in Mpanga forest, Uganda. VII. The biting behaviour of mosquitoes and tabanids. Trans R Entomol Soc Lond 113: 315–335

Haddow AJ (1961*b*) Studies on the biting habits and medical importance of East African mosquitos of the genus *Aedes*. II. Subgenera *Mucidus, Diceromyia, Finlaya* and *Stegomyia*. Bull Entomol Res 52: 317–351

Haddow AJ (1964) Observations on the biting-habits of mosquitoes in the forest canopy at Zika, Uganda, with special reference to the crepuscular periods. Bull Entomol Res 55: 589–608

Haddow AJ, Ssenkubuge Y (1963) Studies on the biting habits of East African mosquitoes in the genera *Uranotaenia, Ficalbia* and *Hodgesia*. Bull Entomol Res 53: 639–652

Haddow AJ, Ssenkubuge Y (1965) Entomological studies from a steel tower in Zika forest, Uganda. Part I. The biting activity of mosquitoes and tabanids as shown by twenty-four-hour catches. Trans R Entomol Soc Lond 117: 215–243

Haddow AJ, Gillett JD, Highton RB (1947) The mosquitoes of Bwamba county, Uganda, V. The vertical distribution and biting cycle of mosquitoes in rainforest, with further observations on microclimate. Bull Entomol Res 37: 301–330

Haddow AJ, Smithburn KC, Dick GWA, Kitchen SF, Lumsden WHR (1948) Implication of the mosquito *Aedes (Stegomyia) africanus* Theobald in the forest cycle of yellow fever in Uganda. Ann Trop Med Parasitol 42: 218–223

Haddow AJ, van Someren ECC, Lumsden WHR, Harper JO, Gillett JD (1951) The mosquitoes of Bwamba county, Uganda. VIII. Records of occurrence, behaviour and habitat. Bull Entomol Res 42: 207–238

Haddow AJ, Corbet PS, Gillett JD (1961) Entomological studies from a high tower in Mpanga forest, Uganda. I. Introduction. Trans R Entomol Soc Lond 113: 249–256

Haddow AJ, Casley DJL, O'Sullivan JP, Ardoin PML, Ssenkubuge Y, Kitama A (1968) Entomological studies from a high steel tower in Zika forest, Uganda. Part II. The biting activity of mosquitoes above the forest canopy in the hour after sunset. Trans R Entomol Soc Lond 120: 219–236

Hamon J (1963*a*) Étude de l'âge physiologique des femelles d'*Anophèles* dans les zones traitées au DDT, et non traitées de la région de Bobo-Dioulasso, Haute-Volta. Bull World Health Organ 28: 83–109

Hamon J (1963*b*) Les moustiques anthropophiles de la région de Bobo-Dioulasso (République de Haute-Volta). Cycles d'aggressivité et variations saisonnieres. Ann Soc Entomol Fr 132: 85–144

Hamon J (1964) Observations sur l'emploi des moustiquaires-pièges pour la capture semi-automatique des moustiques. Bull Soc Pathol Exot 57: 576–588

Hamon J, Chauvet G, Thélin L (1961) Observations sur les méthodes d'évaluation de l'âge des femelles d'*Anophèles*. Bull World Health Organ 24: 437–443

Hamon J, Choumara R, Adam JP, Bailly H (1959) Le paludisme dans la zone pilote de Bobo Dioulasso, Haute-Volta. Cah ORSTOM 1: 37–98

Hancock RG, Foster WA (1997) Larval and adult nutrition effects on blood/nectar choice of *Culex nigripalpus* mosquitoes. Med Vet Entomol 11: 112–122

Happold DCD (1965) Mosquito ecology in Central Alberta. II. Adult populations and activities. Can J Zool 43: 821–846

Hartberg WK (1971) Observations on the mating behaviour of *Aedes aegypti* in nature. Bull World Health Organ 45: 847–850

Haufe WO (1952) Observations on the biology of mosquitoes (Diptera: Culicidae) at Goose Bay, Labrador. Can Entomol 84: 254–263

Haufe WO (1966) Synoptic correlation of weather with mosquito activity. In: Biometeorology, Vol. 2, Part 2, Proc. 3rd Int. Cong. Biomet. 1963. Pergamon Press, Oxford, pp 523–540

Hawley WA (1985) Population dynamics of *Aedes sierrensis*. In: Lounibos LP, Rey JR, Frank, JH (eds) Ecology of Mosquitoes: Proceedings of a Workshop Florida Medical Entomology Laboratory, Vero Beach, Florida, pp 167–184

Hayakawa H, Yamashita N, Hasegawa T, Tsubaki Y, Iwane K, Shinjo G (1990) Notes on the construction and use of animal-baited traps for sampling haematophagous Diptera. Jap J Sanit Zool, 41: 121–125 (In Japanese, English summary)

Hayes J, Downs TD (1980) Seasonal changes in an isolated population of *Culex pipiens quinquefasciatus* (Diptera: Culicidae): A time series analysis. J Med Entomol 17: 63–69

Hayes ME, Mullen GR, Nusbaum KE (1984) Comparison of *Culicoides* spp. (Diptera: Ceratopogonidae) attracted to cattle in an open pasture and bordering woodland. Mosquito News 44: 368–370

Hayes RO (1961) Host preferences of *Culiseta melanura* and allied mosquitoes. Mosquito News 21: 179–187

Hayes RO, LaMotte LC, Holden P (1967) Ecology of arboviruses in Hale county, Texas, during 1965. Am J Trop Med Hyg 16: 675–687

Henderson BE, McCrae AWR, Kirya BG, Ssenkubuge Y, Sempala SDK (1972) Arbovirus epizootics involving man, mosquitoes and vertebrates at Lunyo, Uganda. Ann Trop Med Parasitol 66: 343–355

Hervy J-P, Legros F, Ferrara L (1986) Influence de la clarté lunaire sur l'activité trophique d'*Aedes taylori* (Diptera, Culicidae). Cah ORSTOM sér Entomol Méd Parasitol 24: 59–65

Hewitt S, Rowland M, Muhammad N, Kamal M, Kemp E (1995) Pyrethroid-sprayed tents for malaria control: an entomological evaluation in Pakistan. Med Vet Entomol 9: 344–352

Hii JLK, Birley MH, Kanai L, Foligeli A, Wagner J (1995) Comparative effects of permethrin-impregnated bednets and DDT house spraying on survival rates and oviposition interval of *Anopheles farauti* No. 1 (Diptera: Culicidae) in Solomon Islands. Ann Trop Med Parasitol 89: 521–529

Hilburn LR, Parrack JW, Cooksey LM (2003) Allozyme diversity in *Anopheles quadrimaculatus* (sensu stricto) populations in northeastern Arkansas. J Am Mosq Control Assoc 19: 6–12

Ho BC, Chan KL, Chan YC (1971) *Aedes aegypti* (L.) and *Aedes albopictus* (Skuse) in Singapore City. Bull World Health Organ 44: 635–641

Hocking B, Richards WR, Twinn CR (1950) Observations on the bionomics of some northern mosquito species (Culicidae: Diptera). Can J Res (D) 28: 58–80

Hodgkin EP (1956) The Transmission of Malaria in Malaya. Stud Inst Med Res Malaya, No. 27

Howard JJ, Emord DE, Morris CD (1983) Epizootiology of eastern equine encephalomyelitis virus in upstate New York, USA. V. Habitat preference of host-seeking mosquitoes (Diptera: Culicidae). J Med Entomol 20: 62–69

Howard JJ, White DJ, Muller SL (1989) Mark–recapture studies on the *Culiseta* (Diptera: Culicidae) vectors of eastern equine encephalitis virus. J Med Entomol 26: 190–199

Hribar LJ, Gerhardt RR (1986) Mosquitoes attacking dogs in Knox county, Tennessee. J Am Mosq Control Assoc 2: 552–553

Hu SMK, Grayston JT (1962) Encephalitis on Taiwan. II. Mosquito collection and bionomic studies. Am J Trop Med Hyg 11: 131–140

Hudson JE (1983) Seasonal succession and relative abundance of mosquitoes attacking cattle in central Alberta. Mosquito News 43: 143–146

Hudson JE (1984) *Anopheles darlingi* Root (Diptera: Culicidae) in the Suriname rain forest. Bull Entomol Res 74: 129–142

Impoinvil DE, Kongere JO, Foster WA, Njiru BN, Killeen GF, Githure JI, Beier JC, Hassanali A, Knols BGJ (2004) Feeding and survival of the malaria vector *Anopheles gambiae* on plants growing in Kenya. Med Vet Entomol 18: 108–115

Jaal ZB (1990) Studies on the Ecology of the Coastal Anopheline Mosquitoes of Northwestern Peninsular Malaysia. Ph.D. thesis, University of Liverpool

Jaal Z, Macdonald WW (1992) Anopheline mosquitoes of northwest coastal Malaysia. Southeast Asian J Trop Med Public Health 23: 479–485

Jaal Z, Macdonald WW (1993) The ecology of anopheline mosquitos in northwest coastal Malaysia: host preferences and biting-cycles. Southeast Asian J Trop Med Public Health 24: 530–535

Jaccard P (1912) The distribution of the flora in the alpine zone. New Phytol 11 37–50

Jaenson TG (1988) Diel activity of blood-seeking anthropophilic mosquitoes in central Sweden. Med Vet Entomol 2: 177–187

Jaenson TGT (1985) Attraction to mammals of male mosquitoes with special reference to *Aedes diantaeus* in Sweden. J Am Mosq Control Assoc 1: 195–198

Jaenson TGT, Niklasson B (1986) Feeding patterns of mosquitoes (Diptera: Culicidae) in relation to the transmission of Ockelbo disease in Sweden. Bull Entomol Res 76: 375–383

Jones CJ, Lloyd JE (1985) Mosquitoes feeding on sheep in southeastern Wyoming. J Am Mosq Control Assoc 1: 530–532

Jones JW, Turell MJ, Sardelis MR, Watts DM, Coleman RE, Fernandez R, Carbajal F, Pecor JE, Calampa C, Klein TA (2004) Seasonal distribution, biology, and human attraction patterns of culicine mosquitoes (Diptera: Culicidae) in a forest near Puerto Almendras, Iquitos, Peru. J Med Entomol 41: 349–360

Jones RH (1961) Some observations on biting flies attacking sheep. Mosquito News 21: 113–115

Jones RH, Hayes RO, Potter HW, Francy DB (1977) A survey of biting flies attacking equines in three states of the southwestern United States. J Med Entomol 14: 441–447

Jong-Jin Lee, Klowden MJ (1999) A male accessory gland protein that modulates female mosquito (Diptera: Culicidae) host-seeking behavior. J Am Mosq Control Assoc 15: 4–7

Jupp PG (1978) A trap to collect mosquitoes attracted to monkeys and baboons. Mosquito News 38: 288–289

Jupp PG, Kemp A (1998) Studies on an outbreak of Wesselsbron virus in the Free State province, South Africa. J Am Mosq Control Assoc 14: 40–45

Jupp PG, McIntosh BM (1967) Ecological studies on Sindbis and West Nile viruses in South Africa. II. — Mosquito bionomics. S Afr J Med Sci 32: 15–33

Jupp PG, McIntosh BM, Anderson D (1976) *Culex (Eumelanomyia) rubinotus* Theobald as vector of Banzi, Germiston and Witwatersrand viruses. IV. Observations on the biology of *C. rubinotus*. J Med Entomol 12: 647–651

Jupp PG, Kemp A, Grobbelaar A, Leman P, Burt FJ, Alahmed AM, Al-Mujalli D, Al-Khamees M, Swanepoel R (2002) The 2000 epidemic of Rift Valley fever in Saudi Arabia: mosquito vector studies. Med Vet Entomol 16: 245–252

Katô M, Toriumi M (1950) Studies in the associative ecology of insects. I. Nocturnal succession of a mosquito association in the biting activity. Sci Rep Tôhoku Univ Ser 4 (Biology) 18: 467–472

Kay BH (1985) Man-mosquito contact at Kowanyama northern Queensland, Australia. J Am Mosq Control Assoc 1: 191–194

Kay BH, Bulfin ET (1977) Modifications of a livestock crush into a stable trap for mosquito collection. J Med Entomol 13: 515–516

Kay BH, Boreham PFL, Williams GM (1979*a*) Host preferences and feeding patterns of mosquitoes (Diptera: Culicidae) at Kowanyama, Cape York Peninsula, northern Queensland, Bull Entomol Res 69: 441–457

Kay BH, Boreham PFL, Edman JD (1979*b*) Application of the 'feeding index' concept to studies of mosquito host-feeding patterns. Mosquito News 39: 68–72

Kerr JA (1933) Studies on the abundance, distribution and feeding habits of some West African mosquitos. Bull Entomol Res 24: 493–510

Kettle DS, Linley JR (1967) The biting habits of *Leptoconops bequaerti*. I. Methods; standardization of techniques; preferences for individuals, limbs and positions. J Appl Ecol 4: 379–395

Kettle DS, Linley JR (1969*a*) The biting habits of some Jamaican Culicoides. I. *C. barbosai* Wirth & Blanton. Bull Entomol Res 58: 729–753

Kettle DS, Linley JR (1969*b*) The biting habits of some Jamaican Culicoides. II. *C. furens* (Poey). Bull Entomol Res 59: 1–20

Khan AA, Maibach HI, Strauss WG, Fenley WR (1965) Screening humans for degrees of attractiveness to mosquitoes. J Econ Entomol 58: 694–697

Khan AA, Maibach HI, Strauss WG (1971) A quantitative study of variation in mosquito response and host attractiveness. J Med Entomol 8: 41–43

Kimsey RB, Brittnacher JG (1985) A simple electronic timer for animal-baited intermittent suction insect traps. J Am Mosq Control Assoc 1: 14–16

Klinkaewnarong W, Chiang GL, Eng KL (1985) Studies of *Coquillettidia (Co-quillettidia) crassipes* (Vanderwulp, 1881) in relation to transmission of *Cardiofilaria nilesi*. Southeast Asian J Trop Med Public Health 16: 10–14

Klock JW, Bidlingmayer WL (1953) An adult mosquito sampler. Mosquito News 13: 157–159

Klowden MJ, Briegel H (1994) Mosquito gonotrophic cycle and multiple feeding potential: contrasts between *Anopheles* and *Aedes* (Diptera: Culicidae). J Med Entomol 31: 618–622

Klowden MJ, Fernandez NM (1996) Effects of age and mating on the host-seeking behavior of *Aedes aegypti* mosquitoes. J Vector Ecol 21: 156–158

Klowden MJ, Lea AO (1979) Effect of defensive host behavior on the blood meal size and feeding success of natural populations of mosquitoes. J Med Entomol 15: 514–517

Klowden MJ, Blackmer JL, Chambers GM (1988) Effects of larval nutrition on the host-seeking behavior of adult *Aedes aegypti* mosquitoes. J Am Mosq Control Assoc 4: 73–75

Kobayashi J, Mambanya S, Miyagi I, Vanachone B, Manivong K, Koubouchan T, Amano H, Nozaki H, Inthakone S, Sato Y (1997) Collection of anopheline mosquitos in three villages endemic for malaria in Khammouane, Lao PDR. Southeast Asian J Trop Med Public Health 28: 615–620

Krafsur ES (1977) The bionomics and relative prevalence of *Anopheles* species with respect to the transmission of *Plasmodium* to man in western Ethiopia. J Med Entomol 14: 180–94

Kuhlow F, Zielke E (1978) Dynamics and intensity of *Wuchereria bancrofti* transmission in the savannah and forest regions of Liberia. Tropenmed Parasitol 29: 371–381

Kulkarni SM (1987) Feeding behaviour of anopheline mosquitoes in an area endemic for malaria in Bastar district, Madhya Pradesh. Indian J Malariol 24: 163–171

Kumm HW, Novis O (1938) Mosquito studies on the Ilha de Marajó, Pará, Brazil. Am J Hyg 27: 498–515

Kumm HW, Zúniga H (1944) Seasonal variations in the numbers of *Anopheles albimanus* and *A. pseudopunctipennis* caught in stable traps in Central America. Am J Hyg 39: 8–15

Kuntz KJ, Olson JK, Rade BJ (1982) Role of domestic animals as hosts for blood-seeking females of *Psorophora columbiae* and other mosquito species in Texas ricefields. Mosquito News 42: 202–210

Laarman JJ (1958) Research on the ecology of culicine mosquitoes in a forest region of the Belgian Congo. Acta Leidensia 28: 94–98

Laarman JJ (1959) A new species of *Anopheles* from a rain forest in eastern Belgian Congo. Acta Leidensia 29: 200–211

Labarthe N, Serrão ML, Melo YF, de Oliveira SJ, Lourenço-de-Oliveira, R (1998) Mosquito frequency and feeding habits in an enzootic canine dirofilariasis area in Niterói, State of Rio de Janeiro, Brazil. Mem Inst Oswaldo Cruz 93: 145–154

Labarthe N, Serrão ML, Melo YF, de Oliveira SJ, Lourenço-de-Oliveira R (1998) Potential vectors of *Dirofilaria immitis* (Leidy, 1856) in Itacoatiara, oceanic region of Niterói Municipality, State of Rio de Janeiro, Brazil. Mem Inst Oswaldo Cruz 93: 425–432

Laganier R, Randimby FM, Rajaonarivelo V, Robert V (2003) Is the Mbita trap a reliable tool for evaluating the density of anopheline vectors in the highlands of Madagascar? Malar J 2: 42

Landry SV, DeFoliart GR (1986) Attraction of *Aedes triseriatus* to carbon dioxide. J Am Mosq Control Assoc 2: 355–357

Lardeux F, Nguyen NL, Cartel JL (1995) *Wuchereria bancrofti* (Filariidea: Dipetalonematidae) and its vector *Aedes polynesiensis* (Diptera: Culicidae) in a French Polynesian village. J Med Entomol 32: 346–352

Le Berre R (1966) Contribution à l'étude biologique et écologique de *Simulium damnosum* (Theobald, 1903) (Diptera, Simuliidae). Memoires ORSTOM No. 17

LeDuc JW (1978) Natural transmission of keystone virus to sentinel rabbits on the Del Mar Va peninsula. Am J Trop Med Hyg 27: 1041–1044

Lee D (1971) The role of the mosquito, *Aedes sierrensis*, in the epizootology of the deer body worm, *Setaria yehi*. Thesis, University of California, Berkeley

Le Goff G, Carnevale P, Fondjo E, Robert V (1997) Comparison of three sampling methods of man-biting anophelines in order to estimate the malaria transmission in a village of south Cameroon. Parasite 4: 75–80

LePore TJ, Pollack RJ, Spielman A, Reiter P (2004) A readily constructed lard-can trap for sampling host-seeking mosquitoes. J Am Mosq Control Assoc 20: 321–322

Le Prince JA, Orenstein AJ (1916) Mosquito Control in Panama. G. P. Putnam's London

Lewandowski HB, Hooper GR, Newson HD (1980) Determination of some important natural potential vectors of dog heartworm in central Michigan. Mosquito News 40: 73–79

Lewis DJ, Bennett GF (1979) Biting flies of the eastern maritime provinces of Canada. II. Culicidae. Mosquito News 39: 633–639

Lewis DJ, Bennett GF (1980) Observations on the biology of *Mansonia perturbans* (Walker) (Diptera: Culicidae) in the Nova Scotia-New Brunswick border region. Can J Zool 58: 2084–2088

Lewis DJ, Webber RA (1985) Species composition and relative abundance of anthropophilic mosquitoes in subarctic Quebec. J Am Mosq Control Assoc 1: 521–523

Lillie TH, Kline DL, Hall DW (1987) Diel and seasonal activity of *Culicoides* spp. (Diptera: Ceratopogonidae) near Yankeetown, Florida, monitored with a vehicle-mounted insect trap. J Med Entomol 24: 505–511

Lindsay SW, Armstrong Schellenberg JRM, Zeiler HA, Daly RJ, Salum FM, Wilkins HA (1995) Exposure of Gambian children to *Anopheles gambiae* malaria vectors in an irrigated rice production area. Med Vet Entomol 9: 50–58

Linthicum KJ, Davies FG, Kairo A (1984) Observations of the biting activity of mosquitoes at a flooded dambo in Kenya. Mosquito News 44: 595–598

Lofgren CS, Dame DA, Breeland SG, Weidhaas DE, Jeffery G, Kaiser R, Ford HR, Boston MD, Baldwin KF (1974) Release of chemosterilized males for the control of *Anopheles albimanus* in El Salvador III. Field methods and population control. Am J Trop Med Hyg 23: 288–297

Loftin KM, Byford RL, Craig ME, Steiner RL (1996) Evaluation of cattle insecticide treatments on attraction, mortality, and fecundity of mosquitoes. J Am Mosq Control Assoc 12: 17–22

Loftin KM, Byford RL, Loftin MJ, Craig ME, Steiner RL (1997) Host preference of mosquitoes in Bernalillo County, New Mexico. J Am Mosq Control Assoc 13: 71–75

Lord CC, Day JF (2000) Attractiveness of chickens and bobwhite quail for *Culex nigripalpus*. J Am Mosq Control Assoc 16: 271–273

Lounibos LP, Escher RL (1985) Mosquitoes associated with water lettuce (*Pistia stratiotes*) in southeastern Florida. Fla Entomol 68: 169–178

Lounibos LP, Linley JR (1987) A quantitative analysis of underwater oviposition by the mosquito *Mansonia titillans*. Physiol Entomol 12: 435–443

Lourenço-de-Oliveira R, Luz SLB (1996) Simian malaria at two sites in the Brazilian Amazon - II. Vertical distribution and frequency of anopheline species inside and outside the forest. Mem Inst Oswaldo Cruz 91: 687–694

Lowe RE, Bailey DL (1981) Calf-baited traps as a method for selective sampling of adult populations of *Anopheles albimanus* Weidemann. Mosquito News 41: 547–551

Lumsden WHR (1952) The crepuscular biting activity of insects in the forest canopy in Bwamba, Uganda. A study in relation to the sylvan epidemiology of yellow fever. Bull Entomol Res 42: 721–760

Lumsden WHR (1957) The activity cycle of domestic *Aedes (Stegomyia) aegypti* (L.) (Dipt., Culicid.) in Southern Province, Tanganyika. Bull Entomol Res 68: 769–782

Lumsden WHR (1958*a*) Periodicity of Biting behaviour of some African Mosquitoes.' Proc Int Congr Entomol Xth (1956) vol. 3, pp 785–790

Lumsden WHR (1958*b*) A trap for insects biting small vertebrates. Nature 181: 819–820

Luz SLB, Lourenco-de-Oliveira R (1996) Forest Culicinae mosquitoes in the environs of Samuel Hydroelectric Plant, State of Rondônia, Brazil. Mem Inst Oswaldo Cruz 91: 427–432

MacCullagh P, Nelder JA (1989) Generalized Linear Model. Chapman & Hall, London

Macdonald WW (1957) Malaysian Parasites. XVI. An Interim Review of the Non-anopheline Mosquitoes of Malaya. Stud Inst Med Res Malaya No. 28, pp 1–34

Magoon EH (1935) A portable stable trap for capturing mosquitos. Bull Entomol Res 26: 363–369

Mahadev PVM, Dhanda V, Geevarghese G, Mishra AC, Deshukh PK, Kaul HN, Modi GB, Shetty PS, George PJ, Guttikar SN, Dhanapal J (1978) Studies on the mosquitoes of Bankura district, West Bengal: adult populations. Indian J Med Res 68: 248–263

Main AJ, Tonn RJ, Randall EJ, Anderson KS (1966) Mosquito densities at heights of five and twenty-five feet in southeastern Massachusetts. Mosquito News 26: 243–248

Manalang C (1931) Malaria transmission in the Philippines. I. The natural vector. Philipp J Sci 45: 241–248

Manga L, Fondjo E, Carnevale P, Robert V (1993) Importance of low dispersion of *Anopheles gambiae* (Diptera: Culicidae) on malaria transmission in hilly towns in south Cameroon. J Med Entomol 30: 936–938

Mani TR, Rao CVRM, Rajendran R, Devaputra M, Prasanna Y, Hanumaiah Gajanana A, Reuben R (1991) Surveillance for Japanese encephalitis in villages near Madurai, Tamil Nadu, India. Trans R Soc Trop Med Hyg 85: 287–291

Marques GRAM, Gomes A de C (1997) Comportamento antropofílico de *Aedes albopictus* (Skuse) (Diptera: Culicidae) na região do Vale do Paraíba, Sudeste do Brasil. Rev Saúde Pública 31: 125–130

Marrama L, Laventure S, Rabarison P, Roux J (1999) *Anopheles mascarensis* (De Meillon, 1947), major vector of malaria in Fort-Dauphin (south–east Madagascar). Bull Soc Pathol Exot 92: 136–138

Masalkina TM (1979) A comparative evaluation of methods of capturing blood-sucking mosquitoes. Communication I. Species composition and ratio of mosquito species caught by different methods. Medskaya Parazit 48: 47–52 (In Russian, English summary)

Masalkina TM (1981) A comparative evaluation of methods of capturing blood-sucking mosquitoes. Communication III. A statistical analysis of data on mosquito abundance obtained by different methods of capture. Medskaya Parazit 50 (in error given as 60 on reports), 58–61 (In Russian, English summary)

Masalkina TM, Kachalova NA (1989) On the dark bell and net with removable sacks methods for counting blood-sucking mosquitoes. Medskaya Parazit 58: 27–30 (In Russian, English summary)

Mathenge EM, Killeen GF, Oulo DO, Irungu LW, Ndegwa PN, Knols BGJ (2002) Development of an exposure-free bednet trap for sampling Afrotropical malaria vectors. Med Vet Entomol 16: 67–74

Mathenge EM, Omweri GO, Irungu LW, Ndegwa PN, Walczak E, Smith TA, Killeen GF, Knols BGJ (2004) Comparative field evaluation of the Mbita trap, the Centers for Disease Control light trap, and the human landing catch for sampling of malaria vectors in western Kenya. Am J Trop Med Hyg 70: 33–37

Mattingly PF (1949a) Studies on West African forest mosquitos. Part I. The seasonal distribution, biting cycle and vertical distribution of four of the principal species. Bull Entomol Res 40: 149–168

Mattingly PF (1949b) Studies on West African forest mosquitoes. Part II. The less commonly occurring species. Bull Entomol Res 40: 387–402

Mboera LEG, Knols BGJ, Takken W, della Torre A (1997) The response of *Anopheles gambiae* s.l. and *A. funestus* (Diptera: Culicidae) to tents baited with human odour or carbon dioxide in Tanzania. Bull Entomol Res 87: 173–178

McCall PJ, Eaton G (2001) Olfactory memory in the mosquito *Culex quinquefasciatus*. Med Vet Entomol 15: 197–203

McCall PJ, Kelly DW (2002) Learning and memory in disease vectors. Trends Parasitol 18: 429–433

McCall PJ, Mosha FW, Njunwa KJ, Sherlock K (2001) Evidence for memorized site-fidelity in *Anopheles arabiensis*. Trans R Soc Trop Med Hyg 95: 587–590

McClelland GAH (1959) Observations on the mosquito, *Aedes (Stegomyia) aegypti* (L.) in East Africa. I. The biting cycle in an outdoor population at Entebbe, Uganda. Bull Entomol Res 50: 227–235

McClelland GAH (1960) Observations on the mosquito, *Aedes (Stegomyia) aegypti* (L.) in East Africa. II. The biting cycle in a domestic population on the Kenya coast. Bull Entomol Res 50: 687–696

McClelland GAH, Conway GR (1971) Frequency of blood feeding in the mosquito Aedes aegypti. Nature 232: 485–486

McCrae AWR (1972) Age-composition of man-biting *Aedes (Stegomyia) simpsoni* (Theobald) (Diptera: Culicidae) in Bwamba county, Uganda. J Med Entomol 9: 545–550

McCrae AWR, Manuma P, Mawejje C (1969) The *Simulium damnosum* Species Complex. (c). Studies on the Relationship Between Microclimate and the Biting, Rhythms of Nulliparous and Parous S. damnosum (Sebwe A or Sanje form). East Afr Virus Res Inst Rep 1968, No. 18, pp 106–108

McCrae AWR, Boreham PFL, Ssenkubuge Y (1976) The behavioural ecology of host selection in *Anopheles implexus* (Theobald) (Diptera: Culicidae). Bull Entomol Res 66: 587–631

McCreadie JW, Colbo MH, Bennett GF (1984) A trap design for the collections of hematophagous Diptera from cattle. Mosquito News 44: 212–216

McCreadie JW, Colbo MH, Bennett GF (1985) The seasonal activity of hematophagous Diptera attacking cattle in insular Newfoundland. Can Entomol 117: 995–1006

McIntosh BM, Jupp PG, de Sousa J (1972) Mosquitoes feeding at two horizontal levels in gallery forest in Natal, South Africa, with reference to possible vectors of chikungunya virus. J Entomol Soc S Afr 35: 81–90

McIntosh BM, Jupp PG, dos Santos I (1977) Rural epidemic of Chikungunya in South Africa with involvement of *Aedes (Diceromyia) furcifer* (Edwards) and baboons. S Afr J Sci 73: 267–269

McIver SB, Wilkes TJ, Gillies MT (1980) Attraction to mammals of male *Mansonia (Mansonioides)* (Diptera: Culicidae). Bull Entomol Res 70: 11–16

Meek CL, Meisch MV, Walker TW (1985) Portable battery-powered aspirators for collecting adult mosquitoes. J Am Mosq Control Assoc 1: 102–105

Mekuria Y, Tidwell MA, Williams DC, Mandeville JD (1990) Bionomic studies of the *Anopheles* mosquitoes of Dajabon, Dominican Republic. J Am Mosq Control Assoc 6: 651–657

Mendis C, Jacobsen JL, Gamage-Mendis A, Bule E, Dgedge M, Thompson R, Cuamba N, Barreto J, Begtrup K, Sinden RE, Hogh B (2000) *Anopheles arabiensis* and *An. funestus* are equally important vectors of malaria in Matola coastal suburb of Maputo, southern Mozambique. Med Vet Entomol 14: 171–180

Metz CW (1920) On the possibilities of using mosquito traps in antimalarial work. Public Health Rep Wash 35: 1974–1977

Meyer CL, Bennett GF (1976) Observations on the sporogony of *Plasmodium circumflexum* Kikuth and *Plasmodium polare* Manwell in New Brunswick. Can J Zool 54: 133–141

Milby MM (1985) Predicting *Culex tarsalis* abundance in Kern county. Proc California Mosq Vector Control Assoc 52: 153–155

Minar J (1959) The influence of meteorological factors to the activity of mosquitoes in south-western Bohemia. Čslká Parasitol 6: 57–74

Ming-Hui Weng, Jih-Ching Lien, Dar-Der Ji (2005) Monitoring of Japanese encephalitis virus infection in mosquitoes (Diptera: Culicidae) at Guandu Nature Park, Taipei, 2002–2004. J Med Entomol 42: 1085–1088

Mitchell CJ, Millian KY (1981) Continued host seeking by partially engorged *Culex tarsalis* (Diptera: Culicidae) collected in nature. J Med Entomol 18: 249–250

Mitchell CJ, Darsie RF, Monath TP, Sabattini MS, Daffner J (1985) The use of an animal-baited net trap for collecting mosquitoes during western equine encephalitis investigations in Argentina. J Am Mosq Control Assoc 1: 43–47

Mitchell CJ, Monath TP, Sabattini MS, Christensen HA, Darsie RF, Jakob WL, Daffner JF (1987) Host-feeding patterns of Argentine mosquitoes (Diptera: Culicidae) collected during and after an epizootic of western equine encephalitis. J Med Entomol 24: 260–267

Mohrig W (1969) Die Culiciden Deutschlands, Untersuchungen sur Taxanomie, Biologie, und Ökologie der Eineimischen Stechmücken. Parasit. SchrReihe, 18. Fisher-Verlag, Jena

Molez JF, Desenfant P, Jacques JR (1998). Bio-écologie en Haïti d'*Anopheles albimanus* Wiedemann, 1820 (Diptera: Culicidae). Bull Soc Pathol Exot 91: 334–339

Monath TP (ed) (1988) The Arboviruses: Epidemiology and Control, Vols I–V. CRC Press, Boca Raton, Florida

Monchadskiy AS, Radzivilovskaya ZA (1947) A new quantitative method of estimating activity of attacking bloodsuckers. Mag. Parasitol 9: 147–65 (In Russian)

Moon TE (1976) A statistical model of the dynamics of a mosquito vector (*Culex tarsalis*) population. Biometrics 32: 355–368

Moorhouse DE, Wharton RH (1965) Studies on Malayan vectors of malaria; methods of trapping, and observations on biting cycles. J Med Entomol 1: 359–370

Morris CD, DeFoliart GR (1969) A comparison of mosquito catches with miniature light traps & CO_2-baited traps. Mosquito News 29: 424–426

Mpofu SM, Masendu HT (1986) Description of a baited trap for sampling mosquitoes. J Am Mosq Control Assoc 2: 363–365

Mpofu SM, Taylor P, Govere J (1988) An evaluation of the residual lifespan of DDT in malaria control. J Am Mosq Control Assoc 4: 529–535

Muirhead-Thomson RC (1940) Studies on the behaviour of *Anopheles minimus*. Part 1. The selection of the breeding places and the influence of light and shade. J Malar Inst India 3: 265–294

Muirhead-Thomson RC (1947) The effects of house spraying with pyrethrum and with DDT on *Anopheles gambiae* and *A. melas* in West Africa. Bull Entomol Res 38: 449–464

Muirhead-Thomson RC (1948) Studies on *Anopheles gambiae* and *A. melas* in and around Lagos. Bull Entomol Res 38: 527–558

Muirhead-Thomson RC (1968) Ecology of Insect Vector Populations. Academic Press, London

Muirhead-Thomson RC, Mercier EC (1952) Factors in malaria transmission by *Anopheles albimanus* in Jamaica. Part I. Ann Trop Med Parasitol 46: 103–116

Mulligan HW (1970) (ed) The African Trypanosomiases. George Allen & Unwin, London

Murillo C, Astaiza R, Fajardo P (1988) Biología de *Anopheles (K.) neivai* Howard, Dyar & Knab, 1913 (Diptera: Culicidae) en la Costa Pacífica de Colombia. III. Medidas de luminosidad y el comportamiento del picadura. Rev Saúde Pública 22: 109–112

Murphey FJ, Darsie RF (1962) Studies on the bionomics of *Culex salinarius* Coquillett. I. Observations on the crepuscular and nocturnal activities of adult females. Mosquito News 22: 162–171

Murphey FJ, Burbutis PP, Bray DF (1967) Bionomics of *Culex salinarius* Coquillett. II. Host acceptance and feeding by adult females of *C. salinarius* and other mosquito species. Mosquito News 27: 366–374

Mutero CM, Birley MH (1987) Estimation of the survival rate and oviposition cycle of field populations of malaria vectors in Kenya. J Appl Ecol 24: 853–863

Mwandawiro C, Boots M, Tuno N, Suwonkerd W, Tsuda Y, Takagi M (2000) Heterogeneity in the host preference of Japanese encephalitis vectors in Chiang Mai, northern Thailand. Trans R Soc Trop Med Hyg 94: 238–242

Myers K (1956) Methods of sampling winged insects feeding on the rabbit *Oryctolagus cuniculus* (L.) C.S.I.R.O. Wildl Res 1: 45–58

Nagasawa S (1967) Seasonal prevalence of horse-flies attacking cattle. Jap J Sanit Zool 18: 259–269

Nagasawa S (1973) Fitting the Logarithmic and Truncated Lognormal Distributions to a Mosquito Trap Record. WHO/VBC/73.459, 6 pp (mimeographed)

Nasci RS (1981) A lightweight battery-powered aspirator for collecting resting mosquitoes in the field. Mosquito News 41: 808–811

Nasci R (1991) Influence of larval and adult nutrition on biting persistence in *Aedes aegypti* (Diptera: Culicidae). J Med Entomol 28: 522–526

Nathan MB (1981) Bancroftian filariasis in coastal north Trinidad, West Indies: Intensity of transmission by *Culex quinquefasciatus*. Trans R Soc Trop Med Hyg 75: 721–730

Nayar JK, Provost MW, Hansen CW (1980) Quantitative bionomics of *Culex nigripalpus* (Diptera: Culicidae) populations in Florida. 2. Distribution, dispersal and survival patterns. J Med Entomol 17: 40–50

Need JT, Rogers EJ, Phillips IA, Falcon R, Fernandez R, Carbajal F, Quintana J (1993) Mosquitoes (Diptera: Culicidae) captured in the Iquitos area of Peru. J Med Entomol 30: 634–638

Neeru Singh, Mishra AK (2000) Anopheline ecology and malaria transmission at a new irrigation project area (Bargi Dam) in Jabalpur (Central India). J Am Mosq Control Assoc 16: 279–287

Neeru Singh, Mishra AK, Chand SK, Sharma VP (1999) Population dynamics of *Anopheles culicifacies* and malaria in the tribal area of central India. J Am Mosq Control Assoc 15: 283–290

Nelson MJ, Self LS, Pant CP, Usman S (1978) Diurnal periodicity of attraction to human bait of *Aedes aegypti* (Diptera: Culicidae) in Jakarta, Indonesia. J Med Entomol 14: 504–510

Nelson RL, Tempelis CH, Reeves WC, Milby MM (1976) Relation of mosquito density to bird:mammal feeding ratios of *Culex tarsalis* in stable traps. Am J Trop Med Hyg 25: 644–654

Nielsen ET (1961) Twilight and the 'crep' unit. Nature 190: 878–879

Nielsen ET (1963) Illumination at twilight. Oikos 14: 9–21

Novak MG, Rowley WA (1994) Serotonin depletion affects blood-feeding but not host-seeking ability in *Aedes triseriatus* (Diptera: Culicidae). J Med Entomol 31: 600–606

Novak RJ, Peloquin J, Rohrer W (1981) Vertical distribution of adult mosquitoes (Diptera: Culicidae) in a northern deciduous forest in Indiana. J Med Entomol 18: 116–122

Nuorteva P (1963). Synanthropy of blowflies (Dipt., Calliphoridae) in Finland. Ann Entomol Fenn 29: 1–49

O'Bryan PD, Jefferson HJ (1991) The year of the chickens: the good and bad of a sentinel chicken flock during the 1990 SLE epidemic. J Florida Mosq Control Assoc 62: 59–63

O'Meara GF, Larson VL, Mook DH (1993) Blood feeding and autogeny in the peridomestic mosquito *Aedes bahamensis* (Diptera: Culicidae). J Med Entomol 30: 378–383

Omori N (1942) Observations on the nocturnal activities of the anopheline mosquitoes in Formosa. I. Preliminary Report. Acta Nippon Med Trop 4: 59–67

Packer MJ, Corbet PS (1989) Seasonal emergence, host-seeking activity, age composition and reproductive biology of the mosquito *Aedes punctor*. Ecol Entomol 14: 433–442

Pan American Health Organization (1996) Biology and ecology of *Anopheles albimanus* Wiedemann in Central America. Technical Paper, 43: 1–44

Pandian RS, Chandrashekaran MK (1980) Rhythms in the biting behaviour of a mosquito *Armigeres subalbatus*. Oecologia (Berl.) 47: 89–95

Pant CP, Joshi GP, Rosen P, Pearson JA, Renaud P, Ramasamy M, Vandekar M (1969) A village-scale trial of OMS-214 (Dicapthon) for the control of *Anopheles gambiae* and *Anopheles funestus* in Northern Nigeria. Bull World Health Organ 41: 311–315

Parsons RE (1977) An improved bait trap for mosquito collecting. Mosquito News 37: 527–528

Pennington RG, Lloyd JE (1975) Mosquitoes captured in a bovine-baited trap in a Wyoming pasture subject to river and irrigation flooding. Mosquito News 35: 402–428

Peyton EL (1956) Biology of the pacific coast tree hole mosquito *Aedes varipalpus* (Coq.). Mosquito News 16: 220–228

Pfuntner AR (1979) A modified CO_2-baited miniature surveillance trap. Bull Soc Vector Ecol 4: 31–35

Pillai JS, Rakai IM (1976) Mosquito-borne infections in Fiji: VI. Diel periodicity in the landing of *Aedes aegypti* on man. Mosquito News 36: 186–189

Pinger RR (1985) Species composition and feeding success of mosquitoes attracted to caged dogs in Indiana. J Am Mosq Control Assoc 1: 181–185

Pinto J, Sousa CA, Gil V, Ferreira C, Gonçalves L, Lopes D, Petrarca V, Charlwood JD, do Rosário VE (2000) Malaria in São Tomé and Príncipe: parasite prevalences and vector densities. Acta Trop 76: 185–193

Póvoa MM, Wirtz RA, Lacerda RNL, Miles MA, Warhurst D (2001) Malaria vectors in the Municipality of Serra do Navio, State of Amapá, Amazon Region, Brazil. Mem Inst Oswaldo Cruz 96: 179–184

Póvoa MM, Conn JE, Schlichting CD, Amaral JCOF, Segura MNO, da Silva ANM, dos Santos CCB, Lacerda RNL, de Souza RTL, Galiza D, Santa Rosa EP, Wirtz RA (2003) Malaria vectors, epidemiology, and the re-emergence of *Anopheles darlingi* in Belém, Pará, Brazil. J Med Entomol 40: 379–386

Preston FW (1948) The commonness, and rarity, of species. Ecology 29: 254–283

Provost MW (1958) Mating and Male Swarming in *Psorophora* Mosquitoes. Proc Int Congr Entomol Xth (1956), vol. 2, pp 553–561

Quakyi IA, Leke RGF, Befidi-Mengue R, Tsafack M, Bomba-Nkolo D, Manga L, Tchinda V, Njeungue E, Kouontchou S, Fogako J, Nyonglema P, Harun LT, Djokam R, Sama G, Eno A, Megnekou R, Metenou S, Ndoutse L, Same-Ekobo A, Alake G, Meli J, Ngu J, Tietche F, Lohoue J, Mvondo JL, Wansi E, Leke R, Folefack A, Bigoga J, Bomba-Nkola C, Titanji V, Walker-Abbey A, Hickey MA, Johnson AH, Wallace Taylor D (2000) The epidemiology of *Plasmodium falciparum* malaria in two Cameroonian villages: Simbok and Etoa. Am J Trop Med Hyg 63: 222–230

Rachou RG, Lyons G, Moura-Lima M, Kerr JA (1965) Synoptic epidemiological studies of malaria in El Salvador. Am J Trop Med Hyg 14: 1–62

Rahman WA, Abu Hassan A, Adanan CR (1993) Seasonality of *Anopheles aconitus* mosquitoes, a secondary vector of malariain a village near the Malaysia Thailand border. Acta Trop 55: 263–265

Rahman WA, Hassan AA, Adanan CR, Razha RM (1995) A report of *Anopheles* (Diptera: Culicidae) attracted to cow bait in a malaria endemic village in peninsular Malaysia near the Thailand border. Southeast Asian J Trop Med Public Health 26: 359–363

Rahman WA, Adanan CR, Abu Hassan A (2002) Species composition of adult *Anopheles* populations and their breeding habitats in Hulu Perak district, Peninsular Malaysia. Southeast Asian J Trop Med Public Health 33: 547–550

Rainey MB, Warren GV, Hess AD, Blackmore JS (1962) A sentinel chicken shed and mosquito trap for use in encephalitis field studies. Mosquito News 22: 337–342

Rajagopalan PK, Karmi SJ, Mani TR (1977) Some aspects of transmission of *Wuchereria bancrofti* and ecology of the vector *Culex pipiens fatigans* in Pondicherry. Indian J Med Res 66: 200–215

Ramasamy MS, Kulasakera R, Srikrishnaraj KA, Ramasamy R (1994) Population dynamics of anthropophilic mosquitoes during the northeast monsoon season in the malaria epidemic zone of Sri Lanka. Med Vet Entomol 8: 265–274

Rao TR (1957) Description of a portable baited mosquito trap. Bull Nat Soc Ind Malar 5: 86–88

Rasnitsyn SP, Kosovskikh VL (1979) Improved method of sampling of mosquito abundance around man with a butterfly net and comparison of this with results observed by a dark bell. Medskaya Parazit 48, 18–24 (In Russian, English summary)

Rasnitsyn SP, Kosovskikh VL (1983) Influence of the exposure duration on the number of blood-sucking Diptera captured with a dark bell-trap. Medskaya Parazit 52, 69–72 (In Russian, English summary)

Rattanarithikul R, Konishi E, Linthicum KJ (1996) Observations on nocturnal biting activity and host preference of anophelines collected in southern Thailand. J Am Mosq Control Assoc 12: 52–57

Read RG, Adames AJ (1980) Atmospheric stimulation of man-biting activity in tropical insects. Environ Entomol 9: 677–680

Read RG, Adames AJ, Galindo P (1978) A model of microenvironmental and man-biting tropical insects. Environ Entomol 7: 547–552

Rebêlo JMM, da Silva AR, Ferreira LA, Vieira JA (1997) *Anopheles* (Culicidae, Anophelinae) e a malária em Buriticupu-Santa Luzia, Pré-Amazõnia maranhense. Rev Soc Bras Med Trop 30: 107–111

Ree HI, Chen YK, Chow CY (1969) Methods of sampling population of the Japanese encephalitis vector mosquitoes—preliminary report. Med J Malaya 23: 293–295

Reeves WC (1951) Field studies on carbon dioxide as a possible host stimulant to mosquitoes. Proc Soc Exp Biol Med 77: 64–66

Reeves WC, Bellamy RE, Scrivani RP (1961) Differentiation of encephalitis virus infection rates from transmission rates in mosquito vector populations. Am J Hyg 73: 303–315

Reid JA (1961) The attraction of mosquitoes by human or animal bait in relation to the transmission of disease. Bull Entomol Res 52: 43–62

Reid JA, Wharton RH (1956) Trials of residual insecticides in window-trap huts against Malayan mosquitos. Bull Entomol Res 47: 433–468

Reisen WK, Aslamkhan M (1978) Biting rhythms of some Pakistan mosquitoes (Diptera: Culicidae). Bull Entomol Res 68: 313–330

Reisen WK, Milby MM (1986) Population dynamics of some Pakistan mosquitoes: Changes in adult relative abundance over time and space. Ann Trop Med Parasitol 80: 53–68

Reisen WK, Hardy JL, Reeves WC, Presser SB, Milby MM, Meyer RP (1990) Persistence of mosquito-borne viruses in Kern county, California, 1983–1988. Am J Trop Med Hyg 43: 419–437

Reisen WK, Lothrop HD, Presser SB, Milby MM, Hardy JL, Wargo MJ, Emmons RW (1995*a*). Landscape ecology of arboviruses in southern California: temporal and spatial patterns of vector and virus activity in Coachella Valley, 1990–1992. J Med Entomol 32: 255–266

Reisen WK, Hardy JL, Lothrop HD (1995*b*) Landscape ecology of arboviruses in southern California: patterns in the epizootic dissemination of western equine encephalomyelitis and St. Louis encephalitis viruses in Coachella Valley, 1991–1992. J Med Entomol 32: 267–275

Reiter P, Cordellier R, Ouma JO, Cropp CB, Savage HM, Sanders EJ, Marfin AA, Tukei PM, Agata NN, Gitau LG, Rapuoda BA, Gubler DJ (1998) First recorded outbreak of yellow fever in Kenya, 1992–1993. II. Entomologic investigations. Am J Trop Med Hyg 59: 650–656

Renshaw M (1991) Population dynamics and ecology of *Aedes cantans* (Diptera: Culicidae) in England. Ph.D. thesis, University of Liverpool

Renshaw M, Service MW, Birley MH (1994) Size variation and reproductive success in the mosquito *Aedes cantans*. Med Vet Entomol 8: 179–186

Renshaw M, Silver JB, Service MW (1995) Differential lipid reserves influence host-seeking behaviour in the mosquitoes *Aedes cantans* and *Aedes punctor*. Med Vet Entomol 9: 381–387

Renshaw M, Elias M, Maheswary NP, Hassan MM, Silver JB, Birley MH (1996) A survey of larval and adult mosquitoes on the flood plains of Bangladesh, in relation to flood-control activities. Ann Trop Med Parasitol 90: 621–624

Reuben R (1971) Studies on the mosquitoes of North Arcot district, Madras State, India. Part 2. Biting cycles and behaviour on human and bovine baits at two villages. J Med Entomol 8: 127–134

Rickenbach A, Ferrara L, Germain M, Eouzan JP, Button JP (1971) Quelques données sur la biologie de trois vecteurs potentiels de fièvre jaune *Aedes (Stegomyia) africanus* (Theo.), *A. (S.) simpsoni* (Theo.), et *A. (S.) aegypti* (L.) dans la région de Yaundé (Cameroun). Cah ORSTOM sér Entomol Méd Parasitol 9: 285–299

Ring DR, Harris MK (1983) Predicting pecan nut casebearer (Lepidoptera: Pyralidae) activity at College Station, Texas. Environ Entomol 12: 482–486

Roberts DR, Hsi BP (1979) An index of species abundance for use with mosquito surveillance data. Environ Entomol 8: 1007–1013

Roberts DR, Scanlon JE (1975) The ecology and behavior of *Aedes atlanticus* D. & K. and other species with reference to Keystone virus in the Houston area, Texas. J Med Entomol 12: 537–546

Roberts DR, Peyton EL, Pinheiro FP, Balderrama F, Vargas R (1984) Association of arbovirus vectors with gallery forests and domestic environments in south-eastern Bolivia. Bull Pan-Am Health Organ 18: 337–350

Roberts DR, Alecrim WD, Tavares AM, Radke MG (1987) The house-frequenting, host seeking and resting behavior of *Anopheles darlingi* in south-eastern Amazonas, Brazil. J Am Mosq Control Assoc 3: 433–441

Roberts DR, Manguin S, Rejmankova E, Andre R, Harbach RE, Vanzie E, Hakre S, Polanco J (2002) Spatial distribution of adult *Anopheles darlingi* and *Anopheles albimanus* in relation to riparian habitats in Belize, Central America. J Vector Ecol 27: 21–30

Roberts RH (1965) A steer-baited trap for sampling insects affecting cattle. Mosquito News 25: 281–285

Rooker JR, Read NR, Smith ME, Leach SC (1994) A whole-person sampler for assessing numbers of host-seeking adult mosquitoes. J Am Mosq Control Assoc 10: 127–130

Rosa-Freitas MG, Broomfield G, Priestman A, Milligan PJM, Momen H, Molyneux DH (1992) Cuticular hydrocarbons, isoenzymes and behavior of three populations of *Anopheles darlingi* from Brazil. J Am Mosq Control Assoc 8: 357–366

Rosenberg R, Maheswary NP (1982) Forest malaria II. Transmission by *Anopheles dirus*. Am J Trop Med Hyg 31: 183–191

Ross R (1902) Mosquito Brigades and How to Organise Them. George Philip & Son, London

Rudnick A (1986) Dengue virus ecology in Malaysia. In: Rudnick A, Lim TW (eds) Dengue Fever Studies in Malaysia. Bull Inst Med Res Malaysia 23, pp 51–53

Russell RC, Webb CE, Davies N (2005) *Aedes aegypti* (L.) and *Aedes polynesiensis* Marks (Diptera: Culicidae) in Moorea, French Polynesia: a study of adult population structures and pathogen (*Wuchereria bancrofti* and *Dirofilaria immitis*) infection rates to indicate regional and seasonal epidemics. J Med Entomol 42: 1045–1056

Russell PF, Knipe FW, Rao TR, Putnam P (1944) Some experiments on flight range of *Anopheles culicifacies*. J Exp Zool 97: 135–163

Rwegoshora RT, Sharpe RG, Baisley KJ, Kittayapong P (2002) Biting behavior and seasonal variation in the abundance of *Anopheles minimus* species A and C in Thailand. Southeast Asian J Trop Med Public Health 33: 694–701

Samarawickrema WA (1967) A study of the age composition of natural populations of *Culex pipiens fatigans* Wiedemann in relation to the transmission of *Wuchereria bancrofti* (Cobbold) in Ceylon. Bull World Health Organ 37: 117–137

Samarawickrema WA (1968) Biting cycles and parity of the mosquito *Mansonia (Mansonioides) uniformis* (Theo.) in Ceylon. Bull Entomol Res 58: 299–314

Samarawickrema WA, Sone F, Cummings RF (1987) Seasonal abundance, diel biting activity and parity of *Aedes polynesiensis* Marks and *A. samoanus* (Grünberg) (Diptera: Culicidae) in Samoa. Bull Entomol Res 77: 191–200

Samui KL, Gleiser RM, Hugh-Jones ME, Palmisano CT (2003) Mosquitoes captured in a horse-baited stable trap in Southeast Louisiana. J Am Mosq Control Assoc 19: 139–147

Sasa M, Sabin AB (1950) Ecological studies on the mosquitoes of Okayama in relation to the epidemiology of Japanese B encephalitis. Am J Hyg 51: 21–35

Sasa M, Kano R, Hayashi K, Kimura M, Miura A, Oyama K, Sato K (1950) Two years' observation on the seasonal activities and zoophilism of mosquitoes in Tokyo, by animal trap method. Jap J Exp Med 20: 509–517

Sasse BE, Hackett LW (1950) Note on the host preference of *Anopheles pseudopunctipennis*. J Natn Malar Soc 9: 181–182

Scherer WF, Buescher EL, Flemings MB, Noguchi A, Scanlon J (1959) Ecologic studies of Japanese encephalitis virus in Japan. III. Mosquito factors. Zootropism and vertical flight of *Culex tritaeniorhynchus* with observations on variations in collections from animal-baited traps in different habitats. Am J Trop Med Hyg 8: 665–677

Schmidtmann ET, Abend JF, Valla ME (1980) Nocturnal blood-feeding from pastured calves by the ceratopogonid midge, *Culicoides venustus*, in New York State. Mosquito News 40: 571–577

Schoeler GB, Flores-Mendoza C, Fernández R, Reyes Davila J, Zyzak M (2003) Geographical distribution of *Anopheles darlingi* in the Amazon Basin Region of Peru. J Am Mosq Control Assoc 19: 286–296

Seng CH, Matusop A, Sen FK (1999) Differences in *Anopheles* composition and malaria transmission in the village settlements and cultivated farming zone in Sarawak, Malaysia. Southeast Asian J Trop Med Public Health 30: 454–459

Senior White RA (1952) Studies on the bionomics of *A. aquasalis* Curry, 1932 (concld.) Part III. Indian J Malariol 6: 29–72

Senior White RA (1953) On the evening biting activity of three neotropical *Anopheles* in Trinidad, British West Indies. Bull Entomol Res 44: 451–460

Service MW (1963) The ecology of the mosquitos of the northern Guinea savannah of Nigeria. Bull Entomol Res 54: 601–632

Service MW (1964) The attraction of mosquitoes by animal baits in the northern Guinea savannah of Nigeria. J Entomol Soc Sthn Afr 27: 29–36

Service MW (1969a) Observations on the ecology of some British mosquitoes. Bull Entomol Res 59: 161–194

Service MW (1969b) The use of traps in sampling mosquito populations. Entomol Exp Appl 12: 403–412

Service MW (1971a) Feeding behaviour and host preferences of British mosquitoes. Bull Entomol Res 60: 653–661

Service MW (1971b) The daytime distribution of mosquitoes resting amongst vegetation. J Med Entomol 8: 271–278

Service MW (1971c) A reappraisal of the role of mosquitoes in the transmission of myxomatosis in Britain. J Hyg Camb 69: 105–111

Service MW (1976) Mosquito Ecology. Field Sampling Methods. Applied Science Publishers

Service MW (1980) Effects of wind on the behaviour and distribution of mosquitos and blackflies. Int J Biomet 24: 347–353

Service MW (1993) Mosquito Ecology. Field Sampling Methods. 2nd ed. Chapman & Hall , London

Shannon RC (1939) Methods for collecting and feeding mosquitoes in jungle yellow fever studies. Am J Trop Med 19: 131–140

Shannon RC (1943) Trinidad Government-Rockefeller Foundation: Malaria, Annual Report of the Cooperative Work in Trinidad and Tobago, B.W.I. Government Printer, Port-of-Spain, Trinidad

Sharp BL (1983) *Anopheles merus* (Dönitz) its biting cycle in relation to environmental parameters. J Entomol Soc Sthn Afr 46: 367–374

Sharp BL, Le Sueur D, Ridl FC (1988) Host preference studies on *Aedes (Aedimorphus) durbanensis* (Theobald). J Entomol Soc Sthn Afr 51: 137–150

Shemanchuk JA (1969) Epidemiology of western encephalitis in Alberta: response of natural populations of mosquitoes to avian hosts. J Med Entomol 6: 269–275

Shemanchuk JA (1978) A bait trap for sampling the feeding population of bloodsucking Diptera on cattle. Quaest Entomol 14: 433–439

Shidrawi GR (1965) A Modified Window-Trap for Mosquito Collection. WHO/Mal/509.65, 6 pp. (mimeographed)

Shidrawi GR, Boulzaguet JR, Ashkar TS, Bröger S (1974) Night-Bait Collection: The Variation Between Persons Used as Collector-baits. MPD/TN/74.1 Technical Note 17: 9–18 (WHO mimeographed document)

Shorygin AA (1939) Food and food preference of some Gobidae of the Caspian Sea, Zool Zh 18: 27–53 (In Russian, English summary)

Simard F, Nchoutpouen E, Toto JC, Fontenille D (2005) Geographic distribution and breeding site preference of *Aedes albopictus* and *Aedes aegypti* (Diptera: Culicidae) in Cameroon, Central Africa. J Med Entomol 42: 726–731

Sithiprasasna R, Linthicum KJ, Liu G-J, Jones JW, Singhasivanon P (2003) Some entomological observations on temporal and spatial distribution of malaria vectors in three vilages in northwestern Thailand using a geographic information system. Southeast Asian J Trop Med Public Health 34: 505–516

Slaff M, Crans WJ (1981) The host seeking activity of *Culex salinarius.* Mosquito News 41: 443–447

Smith A (1965) A verandah-trap hut for studying the house-frequenting habits of mosquitoes and for assessing insecticides. I. A description of the verandah-trap hut and of studies on the egress of *Anopheles gambiae* Giles and *Mansonia uniformis* Theo. from an untreated hut. Bull Entomol Res 56: 161–167

Smith A, Obudho WO, Esozed S (1967) The egress of *Anopheles gambiae* and *Mansonia uniformis* from experimental huts with corrugated iron roofs. East Afr Med J 44: 169–172

Smithburn KC, Haddow AJ, Lumsden WHR (1949) An outbreak of sylvan yellow fever in Uganda with *Aedes (Stegomyia) africanus* Theobald as principal vector and insect host of the virus. Ann Trop Med Parasitol 43: 74–89

Smits A, Roelants P, van Bortel W, Coosemans M (1996) Enzyme polymorphisms in the *Anopheles gambiae* (Diptera: Culicidae) complex related to feeding and resting behavior in the Imbo Valley, Burundi. J Med Entomol 33: 545–553

Snow WF (1980) Field estimates of the flight speed of some West African mosquitoes. Ann Trop Med Parasitol 74: 239–242

Soerono M, Badawi AS, Muir DA, Soedono A, Siran M (1965) Observations on doubly resistant *Anopheles aconitus* Donitz in Java, Indonesia, and on its amenability to treatment with malathion. Bull World Health Organ 33: 453–459

Soman RS (1978) Studies on diel periodicity in the landing of *Aedes aegypti* on man in Bangalore city. Indian J Med Res 67: 937–941

Somboon P, Aramrattana A, Lines J, Webber R (1998) Entomological and epidemiological investigations of malaria transmission in relation to population movements in forest areas of north-west Thailand. Southeast Asian J Trop Med Public Health 29: 3–9

Souza-Santos R (2002) Seasonal distribution of malaria vectors in Machadinho d'Oeste, Rondônia State, Amazon Region, Brazil. Cad Saúde Pública 18: 1813–1818

Stamm DD, Chamberlain RW, Sudia WD (1962) Arbovirus studies in south Alabama. Am J Hyg 76: 61–81

Straif SC, Beier JC (1996) Effects of sugar availability on the blood-feeding behavior of *Anopheles gambiae* (Diptera: Culicidae). J Med Entomol 33: 608–612

Strickman D, Hagan DV (1986) Seasonal and meteorological effects on activity of *Chrysops variegatus* (Diptera: Tabanidae) in Paraguay. J Am Mosq Control Assoc 2: 212–216

Strickman D, Miller ME, Heung-Chul Kim, Kwan-Woo Lee (2000) Mosquito surveillance in the demilitarized zone, Republic of Korea, during an outbreak of *Plasmodium vivax* malaria in 1996 and 1997. J Am Mosq Control Assoc 16: 100–113

Subra R (1972) Études écologiques sur *Culex pipiens fatigans* Weidemann, 1828, (Diptera, Culicidae) dans une zone urbaine de savane soudanienne ouest-africaine. Tendances end-exophages et cycle d'agressivité. Cah ORSTOM sér Entomol Méd Parasitol 10: 335–345

Takeda U, Kurihara T, Suzuki T, Sasa M, Miura A, Matsumoto K, Tanaka H (1962) Studies on a collection method of mosquitoes by dry ice and a mosquito net. Jap J Sanit Zool 13: 31–35 (In Japanese, English summary)

Takken W, Klowden MJ, Chambers GM (1998) Effect of body size on host seeking and blood meal utilization in *Anopheles gambiae* sensu stricto (Diptera: Culicidae): the disadvantage of being small. J Med Entomol 35: 639–645

Taylor DM, Bennett GF, Lewis DJ (1979) Observations on the host-seeking activity of some Culicidae in the Tantramar marshes, New Brunswick. J Med Entomol 15: 134–137

Taylor RN, Hill MN, Stewart DC, Slatter R, Gichanga M (1981) A field evaluation of permethrin (OMS 1821) and NRDC 161 (OMS 1998) for residual control of mosquitoes. Mosquito News 41: 423–434

Teesdale C (1955) Studies on the bionomics of *Aedes aegypti* (L.) in its natural habitats in a coastal region of Kenya. Bull Entomol Res 46: 711–742

Teesdale C (1959) Observations on the mosquito fauna of Mombasa. Bull Entomol Res 50: 191–208

Teodoro U, Guilherme ALF, Lozovei AL, Filho VLS, Fukushigue Y, Spinosa RP, Ferreira MEMC, Barbosa OC, de Lima EM (1995) Culicídeos do lago de Itaipu, no rio Paraná, Sul do Brasil. Rev Saúde Pública 29: 6–14

Thavara U, Tawatsin A, Chansang C, Kong-ngamsuk W, Paosriwong S, Boon-Long J, Rongsriyam Y, Komalamisra N (2001) Larval occurrence, oviposition behavior and biting activity of potential mosquito vectors of dengue on Samui Island, Thailand. J Vector Ecol 26: 172–180

Thin Thin Oo, Storch V, Becker N (2003) *Anopheles dirus* and its role in malaria transmission in Myanmar. J Vector Ecol 28: 175–183

Thomson MC, Adiamah JH, Connor SJ, Jawara M, Bennett S, D'Alessandro U, Quinones M, Langerock P, Greenwood BM (1995) Entomological evaluation of The Gambia's National Impregnated Bednet Programme. Ann Trop Med Parasitol 89: 229–241

Tikasingh ES, Davies JB (1972) Comparative tests of four mosquito traps. Mosquito News 32: 623–627

Toma T, Miyagi I, Okazawa T, Kobayashi J, Saita S, Tuzuki A, Keomanila H, Nambanya S, Phompida S, Uza M, Takakura M (2002) Entomological surveys of malaria in Khammouane Province, Lao PDR, in 1999 and 2000. Southeast Asian J Trop Med Public Health 33: 532–546

Tonn RJ, Bang YH, Pwele A (1973) Studies on *Aedes simpsoni* and *Aedes aegypti* in Three Rural Coastal Areas of Tanzania. WHO/VBC/73.442, 16 pp (mimeographed)

Toscano NC, van Steenwyk RA, Sevacherian V, Reynolds HT (1979) Predicting population cycles of the pink bollworm by thermal summation. J Econ Entomol 72: 144–147

Touré YT, Coluzzi M (1986) Barrière de tulle moustiquaire pour l'énchatillonnage des culicides zoophiles. Parassitologia 28: 79–82

Townes H (1962) Design for a Malaise trap. Proc Entomol Soc Wash 64: 253–262

Trapido H, Aitken THG (1953) Study of a residual population of *Anopheles (1.) labranchiae* Falleroni in the Geremeas valley, Sardinia. Am J Trop Med Hyg 2: 658–676

Trapido H, Galindo P (1955) The investigation of a sylvan yellow fever epizootic on the north coast of Honduras, 1954. Am J Trop Med Hyg 4: 655–674

Trapido H, Galindo P (1957) Mosquitoes associated with sylvan yellow fever near Almirante, Panama. Am J Trop Med Hyg 6: 114–144

Trpis M (1962a) Neue Erkenntnisse über eine Forschungsmethodik der Bestimmung der Aktivität von Stechmücken. Biologia Bratisl 17: 123–129

Trpis M. (1962*b*) Aktivita a sezónna dynamika komárov v meistach ich úkrytov vo vegetách dunajských luzných lesov. Biologia Bratisl 17: 263–282 (In Slovak, German summary)

Trpis M (1971) Seasonal Variation in Adult Populations of *Aedes aegypti* in the Dar es Salaam Area, Tanzania. WHO/VBC.71.291, 29 pp (mimeographed)

Trpis M, Haüsermann W (1986) Dispersal and other population parameters of *Aedes aegypti* in an African village and their possible significance in epidemiology of vector-borne diseases. Am J Trop Med Hyg 35: 1263–1279

Trpis M, McClelland GAH, Gillett JD, Teesdale C, Rao TR (1973) Diel periodicity in the landing of *Aedes aegypti* on man. Bull World Health Organ 48: 623–629

Trueman DW, McIver SB (1981) Detecting fine-scale temporal distributions of biting flies: A new trap design. Mosquito News 41: 439–443

Trung HD, Van Bortel W, Sochantha T, Keokenchanh K, Briet OJT, Coosemans M (2005) Behavioural heterogeneity of *Anopheles* species in ecologically different localities in Southeast Asia: a challenge for vector control. Trop Med Int Health 10: 251–262

Ulloa A, Arredondo-Jiménez JI, Rodriguez MH, Fernández-Salas I (2002) Mark-recapture studies of host selection by *Anopheles (Anopheles) vestitipennis*. J Am Mosq Control Assoc 18: 32–35

Ulloa A, Rodriguez MH, Arredondo-Jiménez JI, Fernández-Salas I (2005) Biological variation in two *Anopheles vestitipennis* populations with different feeding preferences in southern Mexico. J Am Mosq Control Assoc 21: 350–354

van Handel E, Day JF (1990) Nectar-feeding habits of *Aedes taeniorhynchus*. J Am Mosq Control Assoc 6: 270–273

van Someren ECC, Furlong M (1964) The biting habits of *Aedes (Skusea) pembaensis* Theo. and some other mosquitoes of Faza, Pate island, East Africa. Bull Entomol Res 55: 97–124

van Someren ECC, Heisch RB, Furlong M (1958) Observations on the behaviour of some mosquitoes of the Kenya coast. Bull Entomol Res 49: 643–660

van Thiel PH, Reuter J, Sautet J, Bevere L (1939) On zoophilism and anthropophilism of *Anopheles* biotypes and species. Riv Malar 18: 95–125

Vexanet JA, Barreto AC, Cuba CC, Marsden PD (1986) Características epidemiológicas da leishmaniose tegumentar americana em uma região endêmica do estado da Bahia. III. Fauna flebotomínica. Mem Inst Oswaldo Cruz 81: 293–301

Vigliano RR, Carlson DB (1986) An adjustable restrainer for sentinel chickens used in encephalitis surveillance. J Am Mosq Control Assoc 2: 357–359

Villarreal C, Rodriguez MH, Bown DN, Arredondo-Jiménez JI (1995) Low-volume application by mist-blower compared with conventional compression sprayer treatment of houses with residual pyrethroid to control the malaria vector *Anopheles albimanus* in Mexico. Med Vet Entomol 9: 187–194

Villavaso EJ, Steelman CD (1970) A collapsible dog-baited mosquito trap. Mosquito News 30: 39–42 von Rahm U (1956) Zum problem der Attracktion von Stechmücken durch den Menschen. Acta Trop 13: 319–344

von Rahm U (1958) Die Attraktive Wirkung der von Menschen agergebenen Dufstaffe auf *Aedes aegypti* L. Z. Tropenmed Parasitol 9: 146–156

Voorham J (2002) Intra-population plasticity of *Anopheles darlingi*'s (Diptera, Culicidae) biting activity patterns in the state of Amapá, Brazil. Rev Saúde Pública 36: 75–80

Vythilingam I, Phetsouvanh R, Keokenchanh K, Yengmala V, Vanisaveth V, Phompida S, Hakim SL (2003) The prevalence of *Anopheles* (Diptera: Culicidae) mosquitoes in Sekong Province, Lao PDR in relation to malaria transmission. Trop Med Int Health 8: 525–535

Wada Y, Kawai S, Itô S, Oda T, Nishigaki J, Omori N, Hayashi K, Mifune K, Shichijo A (1967) Ecology of vector mosquitoes of Japanese encephalitis, especially of *Culex tritaeniorhynchus*. 1. Results obtained in 1965. Trop Med 9: 45–57

Wada Y, Kawai S, Itô S, Oda T, Nishigaki J, Suenaga O, Omori N (1970) Ecology of vector mosquitoes of Japanese encephalitis, especially of *Culex tritaeniorhynchus*. 2. Nocturnal activity and host preference based on all-night-catches by different methods in 1965 and 1966 near Nagasaki City. Trop Med 12: 79–89

Walker ED, Edman JD (1985*a*) Feeding-site selection and blood-feeding behavior of *Aedes triseriatus* (Diptera: Culicidae) on rodent (Sciuridae) hosts. J Med Entomol 22: 287–294

Walker ED, Edman JD (1985*b*) The influence of host defensive behavior on mosquito (Diptera: Culicidae) biting persistence. J Med Entomol 22: 370–372

Walker ED, Edman JD (1986) Influence of defensive behavior of eastern chipmunks and gray squirrels (Rodentia: Sciuridae) on feeding success of *Aedes triseriatus* (Diptera: Culicidae). J Med Entomol 23: 1–10

Walker ED, Torres EP, Villanueva RT (1998) Components of the vectorial capacity of *Aedes poicilius* for *Wuchereria bancrofti* in Sorsogon province, Philippines. Ann Trop Med Parasitol 92: 603–614

Wallace HG, Rudnick A, Rajagopal V (1977) Activity of Tembusu and Umbre viruses in a Malaysian community; Mosquito studies. Mosquito News 37: 35–42

Walsh JF, Davies JB, Le Berre R (1978) Standardization of criteria for assessing the effect of *Simulium* control on onchocerciasis control programmes. Trans R Soc Trop Med Hyg 72: 675–676

Walters LL, Lavoipierre MMJ (1982) *Aedes vexans* and *Aedes sierrensis* (Diptera: Culicidae): Potential vectors of *Dirofilaria immitis* in Tehama county, northern California, USA. J Med Entomol 19: 15–23

Wang H, Chang C (1957) The outdoor nocturnal biting activity of *Anopheles hyrcanus* var. *sinensis* in relation to the factors of temperature, humidity, light and rain. Acta Entomol Sin 7: 473–480

Weaver SC, Fashing NJ (1981) Dispersal behavior and vector potential of *Aedes cantator* (Diptera: Culicidae) in southern Maryland. J Med Entomol 18: 317–323

Wellington WG (1974) Changes in mosquito flights associated with natural changes in polarized light. Can Entomol 106: 941–948

Wharton RH (1951*a*) The behaviour and mortality of *Anopheles maculatus* and *Culex fatigans* in experimental huts treated with DDT and BHC. Bull Entomol Res 42: 1–20

Wharton RH (1951*b*) The habits of adult mosquitoes in Malaya. I. Observations on anophelines in window-trap huts and at cattle-sheds. Ann Trop Med Parasitol 45: 141–154

Wharton RH (1951*c*) The habits of adult mosquitoes in Malaya. II. Observations on culicines in window-trap huts and at cattle-sheds. Ann Trop Med Parasitol 45: 155–160

Wharton RH (1953) The habits of adult mosquitoes in Malaya. III. Feeding preferences of anophelines. Ann Trop Med Parasitol 47: 272–284

Wharton RH, Eyles DE, Warren McW (1963) The development of methods for trapping the vectors of monkey malaria. Ann Trop Med Parasitol 57: 32–46

Wharton RH, Eyles DE, Warren McW, Cheong WH (1964) Studies to determine the vectors of monkey malaria in Malaya. Ann Trop Med Parasitol 58: 56–77

White GB (1971) Studies on transmission of bancroftian filariasis in north-eastern Tanzania, Trans R Soc Trop Med Hyg 65: 819–829

Wijers DJR, Kiilu G (1977) Bancroftian filariasis in Kenya. III. Entomological investigations in Mambrui, a small coastal town, and Jaribuni, a rural area more inland (Coast province). Ann Trop Med Parasitol 71: 347–359

Wilton DP, Darsie RF, Story R (1985) Trials with portable screen rooms modified for use as animal-bait traps for mosquito collection. J Am Mosq Control Assoc 1: 223–226

Woke PA (1962) Observations on differential attraction among humans for *Mansonia nigricans* and *Anopheles albimanus* (Diptera: Culicidae). Proc New Jers Mosq Exterm Assoc 49: 173–181

Wolda H (1978) Fluctuations in abundance of tropical insects. Am Nat 112: 1017–1045

Wolda H, Galindo P (1981) Population fluctuations of mosquitoes in the non-seasonal tropics. Ecol Entomol 6: 99–106

World Health Organization (1975*a*) Manual of practical entomology in malaria. Part I. Vector bionomics and organisation of anti-malaria activities. WHO Offset Publication, No. 13. World Health Organization, Geneva

World Health Organization (1975*b*) Manual on practical entomology in malaria. Part II. Methods and techniques. WHO Offset Publication, No. 13. World Health Organization, Geneva

Worth CB, Jonkers AH (1962) Two traps for mosquitoes attracted to small vertebrates. Mosquito News 22: 15–21

Wright RE, DeFoliart GR (1970) Associations of Wisconsin mosquitoes and woodland vertebrate hosts. Ann Entomol Soc Am 63: 777–786

Wright RE, Knight KL (1966) Effects of environmental factors on biting activity of *Aedes vexans* (Meigen) and *Aedes trivittatus* (Coquillett). Mosquito News 26: 565–578

Xavier MM dos SP, Rebêlo JMM (1999) Espécies de *Anopheles* (Culicidae, Anophelinae) em área endêmica de malária, Maranhão, Brasil. Rev Saúde Pública 33: 535–541

Xue R-D, Edman JD, Scott TW (1995) Age and body size effects on blood meal size and multiple blood feeding by *Aedes aegypti* (Diptera: Culicidae). J Med Entomol 32: 471–474

Yajima T, Yoshida S, Watanabe T (1971) Ecological studies on the population of adult mosquito, *Culex tritaeniorhynchus summorosus* Dyar; the diurnal activity in relation to physiological age. Jap J Ecol 21: 204–214

Yanoviak SP, Aguilar PV, Lounibos LP, Weaver SC (2005) Transmission of a Venezuelan equine encephalitis complex alphavirus by *Culex (Melanoconion) gnomatos* (Diptera: Culicidae) in northeastern Peru. J Med Entomol 42: 404–408

Zaim M, Manouchehri AV, Motabar M, Emadi AM, Nazari M, Pakdad K, Kayedi MH, Mowlaii G (1995) *Anopheles culicifacies* in Baluchistan, Iran. Med Vet Entomol 9: 181–186

Zeze GD (1996) Variations saisonnieres de l'agressivite des femelles et dynamique des populations de *Culex quinquefasciatus* Say 1823 a Abidjan (Republique de Cote d'Ivoire). Bull Soc Pathol Exot 88: 187–193

Zimmerman RH, Turner EC (1983) Host-feeding patterns of *Culicoides* (Diptera: Ceratopogonidae) collected from livestock in Virginia, USA. J Med Entomol 20: 514–519

Chapter 7 Blood-feeding and its Epidemiological Significance

Identification of blood-meals

Formerly mosquito blood-meals were identified mostly by the interfacial precipitin (ring) test (described in Chapter 12), and it is sometimes still used (Alencar et al. 2005; Anderson and Gallaway 1988; Andrade and Lorosa 1998; Irby and Apperson 1988; la Grange and Coetzee 1997; Pates et al. 2001; Savage et al. 1993), as is the microcapillary precipitin test variant of Tempelis and Lofy (1963), which uses antisera produced in chickens (e.g. Christensen et al. 1996; Gad et al. 1995; Gomes et al. 2003). A variety of other techniques have been employed, including complement fixation, latex agglutination, and especially in China (Huang and Luo 1986; Shihai and Jun 1989; Wang 1986) and also India (Kumari et al. 1993) cellulose acetate or agar gel counter immunoelectrophoresis. More recently, antibody methods and DNA methods have become the preferred techniques for blood-meal identification.

ELISA methods

One of the most commonly used methods from the 1980s to the present has been the enzyme-linked-immunosorbent assay (ELISA), either direct (Beier et al. 1988, Savage et al. 1993), indirect (Burkot et al. 1981, 1988), combined direct and indirect (Linthicum et al. 1985), and sandwich (Service et al. 1986). The direct method developed by Beier et al. (1988) involves a two-step protocol in which antigen in the test sample is adsorbed onto a microtitre plate well. A primary antibody conjugated with an enzyme is added to the plate well and results are visualised by use of an enzymatic substrate. Beier et al. (1998) used antihuman peroxidase conjugate and antibovine phosphate conjugate to test a single mosquito for two hosts using the same microtitre plate well. ELISA protocols have also been developed to allow a single mosquito to be tested for both the presence of malarial sporozoites and for identification of its blood-meal. Beier et al. (1988) argued that because host-specific antisera are not required the direct method is easier than indirect or sandwich methods such as that of Service et al. (1986). Indirect ELISA is usually used when no primary labelled antibody is available. The indirect ELISA differs from the direct version in that antibody rather than antigen is adsorbed onto the microtitre plate and this captures antigen from the test sample. Antigen-specific enzyme conjugated antibodies are used to detect the bound antibody-antigen complex. The sandwich ELISA method is most suitable for the detection of small amounts of host antigen (Chow et al. 1993). Service (1993) considered the indirect method to be much cleaner, and moreover should not require any blocking agents such as milk powder, gelatin, BSA or casein, although the Beier et al. (1988) method appears to have been more frequently used in recent years.

Some authors (e.g. Rubio-Palis et al. 1992) recommend not using the wells at the edge of the microtitre plates as these may be subject to an 'edge effect' as a result of thermal gradients across the plate (Oliver et al. 1981).

Mwangangi et al. (2003) recently used the Beier et al. (1988) ELISA method to identify blood meal sources of members of the *Anopheles gambiae* complex at 30 sites along the Kenya coast. Blood meals were identified by direct single-step ELISA using antihost (IgG) conjugate against human and cow proteins. Non-reacting samples were then tested against chicken and goat IgG. A total of 1480 (77.1%) *Anopheles gambiae* s.l. and 439 (22.9%) *Anopheles funestus* were collected and tested. The ELISA identified bloodmeal sources for 92% and 81% of *Anopheles gambiae* s.l. and *Anopheles funestus* respectively. Human IgG was detected in 98.97% of *Anopheles gambiae* s.l. and 99.48% of *Anopheles funestus*. Subsequent identification of individual members of the *Anopheles gambiae* complex revealed no significant differences in the proportion of blood meals taken from human hosts among *Anopheles gambiae* s.s., *Anopheles arabiensis*, and *Anopheles merus*.

Savage et al. (1993) used precipitin tests (Tempelis and Lofy 1963) and a novel direct ELISA technique to identify blood-meals of field-collected *Aedes albopictus* in Missouri, USA. Precipitin tests were used to initially screen samples against mammalian or avian antiserum. Samples identified as avian blood were further tested against antisera of different bird groups using precipitin tests. Samples testing positive for mammalian blood were tested using the authors' own plate ELISA test or further precipitin tests. Of 172 blood-fed specimens collected, 139 were positive for either mammalian (110, 64.0%) or avian (29, 16.9%) blood respectively. The remaining blood-meals were unidentified despite screening against reptilian blood and the authors ascribed these negative results to small blood-meal volumes. Results of experiments on the effect of blood-meal digestion on ELISA activity indicated that host-specific IgG can be detected up to 30 h after full engorgement for human blood-meals but activity was reduced to about half at 36 h post-engorgement. These results were similar to those reported by Service et al. (1986) and Beier et al. (1988). Activity may also be reduced under field conditions, or where a proportion of the mosquito population takes incomplete or small blood-meals.

In Korea, Han-Il Ree et al. (2001) used the ELISA method of Beier et al. (1988), as modified by Loyola et al. (1990) to identify the sources of blood meals in field-collected *Anopheles sinensis*. At all locations, *Anopheles sinensis* was found to have fed almost exclusively on bovines (89.8%). Other hosts included humans (0.7%), swine (3.3%), dogs (0.7%), and chickens (1.6%), with 3.6% of hosts unidentified.

Gomes et al. (2001) recently compared the specificity, sensitivity and accuracy of precipitin and ELISA tests to identify blood meal sources in *Ochlerotatus fluviatilis* and *Aedes aegypti* in the laboratory. Precipitin tests demonstrated higher specificity and accuracy than ELISA, whereas ELISA was more sensitive. Both tests produced better results if mosquitoes were killed and frozen within 48 h of taking a blood meal.

Chow et al. (1993) compared sandwich-A ELISA, sandwich-B ELISA and direct ELISA protocols for the identification of *Aedes aegypti* blood meals and found the sandwich-B ELISA method to be the most specific. This protocol does not require cross-adsorption with heterologous sera, in contrast with the sandwich-A protocol which requires overnight cross-adsorption at 4°C. Reaction with ≥ 1 host was achieved with the sandwich-B ELISA in 70 of 80 field-collected *Aedes aegypti*. By contrast, the direct ELISA identified the source of blood meals for only 33 of the 80 samples. The sandwich-B protocol of Chow et al. (1993) has become quite widely used since. For example, Scott et al. (1993*a*) used this method to identify blood meals of *Aedes aegypti* in a rural village in Thailand. Host determination was possible for 73% of the 1230 field-collected specimens. Of the meals identified, 88% had come from humans only, and 2% from dogs only. Overall 7% of all blood meals contained blood from two hosts, including human, and six meals contained blood from three hosts (cryptic multiple meals from the same host species are not identified using this method).

Walker et al. (1998) used the method of Chow et al. (1993) to identify blood meals in 296 specimens of *Ochlerotatus poicilius* in the Philippines and reported that 67.7% reacted positively with human antisera, 2.4% reacted with dog antisera, 1.0% with goat, and 0.7% with pig. Human and dog mixed blood meals were observed in six specimens, human and pig in five specimens, human and chicken in three specimens, human and goat in three specimens, and chicken and pig in one specimen.

The Chow et al. (1993) method was also used by Ponlawat and Harrington (2005) to identify the blood meal sources from *Aedes aegypti* and *Aedes albopictus* collected from inside and underneath houses in Thailand using CDC backpack aspirators. Host identity was successfully determined for 85% (869/1021) of the blood meal samples.

Gingrich and Williams (2005) also used the Chow et al. (1993) method to undertake a comprehensive survey of host-feeding patterns among mosquitoes collected using CO_2-baited CDC light-traps, CO_2-baited omni-directional Fay-Prince traps, human landing catches, and gravid traps, in Delaware, USA.

The Service et al. (1986) ELISA can be used without a plate reader. Blood-engorged mosquitoes can be smeared on to filter paper and stored for months or even years over a desiccant (e.g. phosphorous pentoxide, silica gel) or kept in a refrigerator or deep-freeze. Alternatively blood-fed mosquitoes can be stored in gelatine capsules (Tempelis and Lofy 1963), although this is not necessary in this instance because contrary to the findings of Eligh (1952) and Roy and Sharma (1987) no proteins appeared to be lost when blood-engorged mosquito abdomens were squashed on to filter paper (Service et al. 1986). The sandwich technique is sufficiently sensitive to identify blood in half-gravid mosquitoes, and also in most three-quarter gravids. Microdot and dipstick methods have been evaluated (Abdulaziz and Pal 1989; Lombardi and Esposito 1986; Roy and Sharma 1987), but they have not proved as efficient and reliable as the ELISA microtitre plate method. However, Hunter and Bayly (1991) working with simuliids described an interesting modified ELISA test using a biotinylated second antibody and a streptoavidin–biotinylated peroxidase complex. Using this approach sensitivity was considerably increased. The method deserves evaluation in situations where the blood content in mosquitoes is small. Blackwell et al. (1994, 1995) have improved the sensitivity of the indirect ELISA procedure, including through reading absorbance at 490 nm and incorporating continuous plate shaking into the protocol. This method allowed the authors to screen *Culicoides* midges against up to nine potential host IgGs per assay. Blood meals were identified for 437 of the 494 females tested (Blackwell et al. 1995).

Collins et al. (1986) described a modified gel diffusion method for identifying mosquito blood meals in which 16 blood meals on a microscope slide (7.6 cm) can be tested simultaneously against two hosts (e.g. cow and man), and using a 10-cm long slide up to 25 blood-meals can be tested. Van den Hurk et al. (2003) used a slight modification of the gel diffusion method of Collins et al. (1986) to identify the sources of blood meals in engorged mosquitoes captured in CDC light-traps supplemented with CO_2 and octenol in northern Australia and Papua New Guinea. Abdomens of engorged mosquitoes were separated from the thorax and homogenised individually in 100–140 μl phosphate-buffered saline, centrifuged at 9510 g for 5 min and stored at 20°C. Each blood meal extract was tested against a panel of nine commercially available polyclonal antisera for the identification of human, pig, cow, horse, dog, cat, kangaroo, chicken and non-specific avian hosts. Eight microlitres of each blood meal extract was loaded into one of six wells cut from a 0.9% agarose gel layer placed on microscope slides. The six wells containing blood-meal extracts surrounded a central well that contained 8 μl of the undiluted antiserum to be tested against. Slides were incubated at 23°C and read after 18 h and 36 h. Formation of a white precipitate between the antigen and antiserum indicated a positive reaction. From a total of 570 trap collections 3658 blood-fed mosquitoes of 15 species were obtained, with *Culex annulirostris* comprising 87.4%. Positive reactions were obtained with 2368 of 3197 (74.1%) *Culex annulirostris* blood meals and the most common source was identified as mammalian. Only 7.2% of blood meals originated from avian hosts. Overall, pigs accounted for only 9.1% of *Culex annulirostris* bloodmeals identified. In the Gulf Plains region, Cape York Peninsula and Daru area of Papua New Guinea, marsupials accounted for 77.8%, 59.7% and 63.8% respectively of blood meals identified. The relatively high proportion of unidentified blood-meals was ascribed to either feeds from hosts for which antisera were not available or alternatively the small blood volumes present in the females attracted to the type of trap used. This method has also been used in India (Chandra et al. 1996; Reuben et al. 1992; Samuel et al. 2004).

Antibody methods of identifying blood-meals suffer from several disadvantages, including the need to prepare antibodies against each potential host, which can be a time-consuming process. Antibody-based assays are often capable of identifying blood meals only to the order level and cannot distinguish between individual species of host.

Boorman (1986) adapted the pyridine haemochromogen test methods used by Heller and Adler (1980) to detect the presence of dried blood on the Turin Shroud, to detect blood in Ceratopogonidae preserved in either alcohol or formalin. Insects, or just their guts, are ground up with a single drop of 99.5% hydrazine in the shallow wells of a porcelain plate, after about 5 min a single drop of formic acid is added. This results in a puff of white 'smoke'. When the reaction has subsided two more drops of acid are added and the plate examined under ultraviolet light. If blood is present there is a bright pinkish–red fluorescence, the intensity of which increases after 10–30 s irradiation. Harrington (1990) adapted this test for detecting the presence of blood in old museum Hemiptera.

Rubidium chloride has been used as a marker in studies on several herbivorous insects (Fleischer et al. 1986; Pearson et al. 1989; Stimmann 1974; van Steenwyck et al. 1978, 1979) and caesium has also occasionally been used (Moss and van Steenwyck 1982). Kimsey and Kimsey (1984) were the first to mark mosquitoes with rubidium to detect arthropod bloodmeals. They injected mice, chickens and lizards intraperitoneally with rubidium chloride, and found that a dose of Rb^+ of 500 mg/kg had no adverse effects on the hosts, nor the mosquitoes feeding on them. Bloodengorged mosquitoes were prepared by an acid digestion method involving the wet ashing technique of Smith (1953), except that there was no need to add a vanadium catalyst, and also that if mosquito samples were left for 3 days they could be digested without heat. Rubidium was detected by flame spectrophotometry using an atomic emission mode at 779.6 nm. All *Culex tarsalis* fed on rubidium-marked quail remained identifiable for up to 6–7 days after they had fed. If atomic absorption machines are available (e.g. in medical and research laboratories) then the method is relatively cheap, and analytical procedures are simple and safe. No special preservation methods are necessary to keep blood-fed mosquitoes prior to testing and their shelf-life is indefinite. The authors point out that the technique is not a replacement for serological detection of hosts, but can be used as an adjunct, and where for some reason other tests are not practical. For example, Anderson et al. (1990) injected chickens with rubidium and caesium in order to study multiple feeding by natural populations of *Culiseta melanura*. These two metals are easily distinguished by their emission wavelengths, and can be detected in mosquitoes for up to 3 days, but for this the dosage of caesium has to be higher than that of rubidium. Anderson and Brust (1996) injected individuals of pairs of Japanese quail with either rubidium or caesium in a laboratory study of the effects of host defensive behaviour on mosquito feeding success. This method allowed the authors to identify which of the two hosts had been fed on by individual mosquitoes and revealed the individual differences in host defensive reaction and therefore blood-feeding success. The authors stress the value of the method in quantifying multiple feeding on hosts that are serologically indistinguishable.

Xue et al. (1995) used the caesium and rubidium marking method of Anderson et al. (1990) to determine the frequency of multiple bloodfeeding in *Aedes aegypti*. These authors demonstrated that chronologically older females (age 15 days) demonstrated a higher frequency of multiple feeding than did 5 day old nullipars or 10 day old parous females. In addition, larger females in each age cohort were more likely to engage in multiple blood feeding.

Similarly, Hodgson et al. (2001) marked the blood of robins (*Turdus migratorius*) and European starlings (*Sturnus vulgaris*) with either rubidium or caesium salts to investigate if *Culiseta melanura* takes multiple blood meals. Birds were injected with a single dose of either 205.4 mg/kg body weight of RbCl or 320.5 mg/kg body weight of CsCl, or alternatively with 500 mg/kg RbCl administered in two doses 1.5 h apart, or 780 mg/kg CsCl also in two doses 1.5 h apart. Birds were individually caged with five birds from each species arranged adjacent to each other and alternating according to species. Birds marked with caesium in month 1 were marked with rubidium in month 2 and vice versa. Mosquitoes were collected in the field from black-painted resting boxes. Of 49 mosquitoes containing marker, seven contained both markers, indicating a mixed meal. Six of the seven individual mosquitoes that had taken a blood meal from both bird species contained a relatively larger amount of robin blood than starling blood. The other mosquito contained small amounts of blood from both species. No mosquitoes contained a large amount of starling blood and a small amount of robin blood.

Investigations of the potential use of radiosodium (^{24}Na and ^{22}Na) in the identification of blood meal sources in insects have been conducted by Knaus et al. (1993), who concluded that the method could be useful in estimating blood meal volumes when these are in the nanolitre range. The short half-life of radiosodium also means that there is no need for disposal of radioactive material and all apparatus and equipment can be re-used after only one week.

Simple agglutination tests can be performed on blood-fed mosquitoes up to 20 h after feeding to identify whether they had fed on humans having A, B or O blood-groups (Bryan and Smalley 1978). Boreham et al. (1978) used gradient-gel electrophoresis to identify different haptoglobin types in studies on host preferences between humans sleeping in the same houses. Identification of actual hosts can be useful in behavioural studies. Although feeds on individuals can be identified by these methods the number of genotypes for these loci are few, and this makes the method of limited practical use.

Handling blood is a potentially dangerous activity, and Boreham (1976) proposed a variety of methods to deal with the potential of infection arising from pathogens contained in insect blood-meals squashed and dried on filter paper. He proposed immersing the papers in diethyl ether for 1 h, exposing them to temperatures of 60°C for 1 h, or exposing them to UV light. However, these treatments may be ineffective against pathogens in whole insects sent for blood-meal identification. Infection with HIV from blood smears, or blood-engorged mosquitoes that have been dried for a day or more is highly unlikely, as the virus does not remain viable in a dry state.

DNA 'fingerprinting' methods

Coulson et al. (1990) first demonstrated the possibility of using DNA finger-printing techniques to identify blood from individual humans in mosquito blood meals. Test mosquitoes were stored in one of three ways: frozen at $-70°C$, or maintained on silica gel, or in 100% isopropanol, both at room temperature. Blood meal analysis was performed using DNA extraction, *Alu*-polymerase chain reaction (PCR) amplification (Nelson et al. 1989) of extracted DNA and southern blotting electrophoresis on agarose gels, as described by Sambrook et al. (1989). Human DNA in the blood meals was identified using a ^{32}P-labelled total human DNA probe (pλg3). Human DNA was detectable in frozen mosquitoes killed up to 20 h post blood meal. Specimens stored in isopropanol also gave good results, although desiccated specimens only gave strong results at 0 h. The authors concluded that the techniques were suitable for extraction, amplification and hybridization of human DNA extracted from mosquito blood meals up to ten hours post-feeding as long as the samples are frozen or stored in isopropanol. The *Alu*-PCR technique did not produce a sufficiently specific pattern to enable identification of which host the meal had originated from. Gokool et al. (1993) used the methods of Coulson et al. (1990) to identify the human sources of blood meals obtained from individual field-collected female *Anopheles gambiae* in Tanzania. One hundred and twenty-five female *Anopheles gambiae* collected from veranda-trap huts were analysed using variable numbers of dinucleotide repeats (VNDR) microsatellite markers and from these 39 readable DNA profiles were obtained (15 contained no human DNA, 58 did not produce profiles, and 13 of the profiles were unreadable). Obtaining a readable result did not appear to be related to the quantity of DNA in the samples. Analysis of the 39 readable profiles revealed that 35 of the blood meals had originated from the occupants of the veranda-trap hut, even though the occupants slept under insecticide-treated nets.

A non-radioactive DNA dot blot hybridisation method was designed by Sato et al. (1992) to identify human blood in mosquitoes. The method is essentially as follows: DNA is extracted from blood samples and applied to nitrocellulose membranes, which were then hybridised with a biotin-labelled probe. A minimum of 10 ng of DNA was required to produce a positive result. Positive results were obtained with blood samples obtained from *Aedes aegypti* females that had fed on a human volunteer up to 100 h earlier. The long time period over which a blood meal can be identified compares favourably with ELISA tests, which typically can only detect blood meals up to a maximum of 40 h post-feeding (e.g. Service et al. 1986). Jensen et al. (1996) used the quick-blot DNA probe methodology of Sato et al. (1992) to identify human blood in different members of the *Anopheles quadrimaculatus* species complex in Florida, USA. Nineteen of 177 blood-engorged specimens of species A hybridized with the human DNA probe, compared with 0 of 62 specimens of species B (*Anopheles smaragdinus*, Reinert et al. 1997) and 4 of 327 specimens of species C (*Anopheles diluvialis*, Reinert et al. 1997). The human blood-feeding rate for species A (*Anopheles quadrimaculatus* s.s.) was significantly higher than for either species B (*Anopheles smaragdinus*) or C (*Anopheles diluvialis*).

Ansell et al. (2000) developed a modified DNA extraction and PCR methodology which is suitable for field use and allows for mosquito blood meals to be identified to individual human hosts, including members of the same family. Field-collected mosquitoes are squashed onto filter paper, which is then dried at room temperature and placed in a plastic bag with desiccant and stored in air-tight containers in a refrigerator at 4°C. Dried blood meals and finger-prick blood samples are cut out of the filter paper and left to soak in 0.5 ml of phosphate-buffered saline (PBS) overnight to elute the samples. To increase the chance of finding differential genotypes between a woman and child(ren) sharing the same mosquito net and make it possible to distinguish one blood meal from another five different highly polymorphic microsatellite markers were used.

Federal Bureau of Investigation protocols for the PCR-based genetic typing of humans were slightly modified by Chow-Shaffer et al. (2000) to identify individual humans as sources of *Aedes aegypti* blood meals in Thailand. In the field study, mosquitoes were collected daily from houses using modified vacuum cleaner aspirators (Scott et al. 1993*a*, 2000) between 0900 and 1600 h. Allele patterns derived from blood samples obtained from mosquitoes were compared with patterns obtained from oral swabs of human inhabitants of the study houses. Complete profiles at all four loci were obtained from 43 of 50 individuals, and when all four loci were used unique profiles were obtained for each individual. Of 20 blood samples from mosquitoes that were completely fingerprinted, 9 were multiple blood meals, and 13 contained blood from individuals who did not reside in the house from which the mosquito was collected. Seven of 10 meals in which all hosts were identified contained only blood from residents of the house from which the mosquitoes were collected. DNA in 33 of 40 blood samples obtained from mosquitoes was sufficient for amplification and profiling up to 24 h after feeding at ambient temperature of

29°C. After 24 h identification was possible in only one of 58 samples. While it was possible to identify mixed meals originating from members of the same family, the error rate (8%) is higher in this situation than when considering a larger population of related and non-related individuals (1%).

In Pondicherry, south India, Michael et al. (2001) used a multiplex PCR method to identify the individual human sources of blood meals taken by *Culex quinquefasciatus*. Female *Culex quinquefasciatus* were collected for 10 min per house using battery-powered aspirators between 0730 and 0930 h. Mosquitoes from each house were preserved individually within an hour in 100% isopropanol at room temperature. The authors developed and optimized a 9-locus multiplex short-tandem repeat (STR) system to undertake the PCR analysis. A total of 361 mosquitoes were collected from 19 randomly selected households, with the number of blood-fed *Culex quinquefasciatus* ranging from 2 to 77 per house per night. More than 75% of the blood meal samples from the study households provided unambiguous PCR results. Blood meal sources could not be definitively identified when the mosquitoes tested contained small blood meals (classified by the authors as quarter-full). Similarly to the results of Chow-Shaffer et al. (2000) cited above, 27.5% of the mosquitoes caught within a household did not contain alleles of the inhabitants of that household. Overall, 13.0% of the mosquitoes collected resting indoors had taken multiple meals. The numbers of bites on people were highly overdispersed (variance to mean ratio = 5.5) with >55% of the total bites received by <20% of the people. There was no significant difference in the overall number of bites received by males and females (quasi-likelihood Poisson generalized linear model, $F = 0.97$, d.f. $= 1,74$, $P = 0.327$) In the female population, there was a clear nonlinear pattern for biting density to increase in girls up to the age of approximately 15 years before decreasing significantly among women. Biting densities were not age-related in males.

De Benedictis et al. (2003) used DNA profiling of four polymorphic human loci to examine three aspects of the interaction between *Aedes aegypti* and its human host in Florida, Puerto Rico: namely, whether multiple blood-feeding involved different individual human hosts; to test the hypothesis that mosquito bites are not uniformly distributed across the human host population; and to determine to what extent household visitors are bitten by resident mosquitoes. Mosquitoes were collected from 22 of 25 selected houses using backpack aspirators (John W. Hock Co.). Mosquito abdomens from blood-fed mosquitoes were removed and squashed onto filter paper and dried. After drying, pieces of filter paper with the dried bloodstains were excised, individually placed into 1.5 ml screw-capped plastic tubes, and held at ambient temperature until tested. Oral swabs were taken from 82 of the 84 residents of the 22 households included in the study and complete profiles were obtained for 70 individuals. In addition, swabs were obtained from 19 members of the field teams, and five visitors. Of the 217 completely or partially profiled meals from wild mosquitoes, at least one source was categorized as resident (R) (n = 174) in 164 meals (76%), neighbour (N) (n = 34) in 32 meals (15%), visitor (V) (n = 2) in 2 meals (1%), and Unknown (U) (n = 41) in 41 meals (19%). Totals exceed 100% due to some meals containing blood from two sources. Specific individuals were identified as the source of 200 (80%) of the 250 completely and partially profiled blood meals from wild mosquitoes. Observed frequencies indicate that feeding was not random (χ^2 = 185.4, d.f. = 6, P = 2.42 × 10^{-17}). Meals were not taken in proportion to the age groups of the 75 Yanes III residents (χ^2 = 74.9, d.f. = 3, P = 3.81 × 10^{-16}), and there was a tendency to feed more frequently on males than females (P = 0.013). This non-random pattern was due to three people accounting for 56% (110 of 198) of all blood meals. The time limit for allele detection under natural field conditions was estimated at two days, based on analysis of meals obtained from recaptured laboratory-reared mosquitoes that had been fed human blood just prior to release.

The probability of obtaining a successful PCR amplification of human DNA from blood meals has been shown to decrease with time since ingestion in laboratory colonies of *Anopheles gambiae* s.s. (Mukabana et al. 2002). Between 8 and 32 h post digestion there was a significant reduction in the probability of obtaining useful PCR products as determined by logistic regression. Approximately 50% of samples were successful at 15 h post ingestion. There was no reduction in success during the first 8 h following ingestion of a blood meal. Meal size did not significantly affect the success of the PCR amplification process. The smallest blood meal from which DNA was successfully amplified at both loci under investigation was 0.08 mg, which was estimated to contain approximately 2 ng of human DNA. Only 50% of blood meals obtained from ten *Anopheles gambiae* complex mosquitoes maintained for 24 h after feeding on cattle in a cattle-baited trap in Zimbabwe could be identified successfully using a PCR technique, compared with 17/18 specimens processed immediately (Prior and Torr 2002).

When multiple feeding in a single gonotrophic cycle is common, as it is for *Aedes aegypti*, identification of the individual sources of blood meals becomes more complicated. When there are as many as three sources in a single blood meal, there are always instances when all hosts cannot be specifically identified using DNA profiling (e.g., the presence of a child's blood cannot be detected in a meal containing the blood of both biologic parents).

Heteroduplex analysis of cytochrome b DNA

Cytochrome b is a useful marker for identifying arthropod blood meals as it is present in high copy number as a mitochondrial gene and demonstrates sufficient genetic variation at the primary sequence level to permit reliable identification of vertebrate taxa. Boakye et al. (1999) developed a polymerase chain reaction (PCR)-heteroduplex assay (HDA) to distinguish blood meals derived from several species of vertebrates. Blood meal-derived cytochrome b sequences were amplified and this was followed by heteroduplex analysis of the resulting products. This method allowed vertebrate blood meals in black flies and tsetse flies to be identified to the genus level.

Lee et al. (2002) modified the methods of Boakye et al. (1999) to enable them to identify the individual bird species that served as hosts to *Culex tarsalis* in Mississippi and Alabama, USA. Genomic DNA from avian blood samples and blood-fed mosquitoes was extracted. Sensitivity of the avian cytochrome PCR assay in detecting blood meals was determined by testing DNA extracted from individual *Culex tarsalis* that had fed on a quail and maintained for different periods post-ingestion. It was possible to detect quail cytochrome b sequence within individual mosquitoes up to seven days post-feeding. The post-feeding period over which blood-meal analysis was still successful was relatively long compared with results obtained by authors working on human blood (Boakye et al. 1999; Kent and Norris 2005; Mukabana et al. 2002), and this was ascribed as being at least partially due to the fact that avian erythrocytes contain nuclei and hence a larger amount of DNA than mammal-derived blood-meals (Lee et al. 2002). Tests on field-collected *Culex tarsalis* produced detectable PCR products in more than 90% (19/21) of fully engorged mosquitoes. The sensitivity of the assay was reduced in partially engorged mosquitoes (83%, $n = 24$) and in gravid specimens (71%, $n = 7$).

A combined indirect ELISA and PCR heteroduplex analysis method was used by Apperson et al. (2002) to investigate host-feeding habits of mosquitoes in New York, USA. Blood-fed mosquitoes were initially subjected to the indirect ELISA method of Apperson and Irby (1988), in which blood meal extracts were screened against antisera with broad anti-bird or anti-mammal activity. Samples positive for mammal blood were then screened against a range of specific antisera in order to identify the host. Blood meal extracts positive for bird sera were subjected to a PCR-heteroduplex analysis. The method used was based on that of Boakye et al. (1999), but with the addition of a nested amplification step. The PCR analysis revealed that *Culex* spp. and *Coquillettidia perturbans* had fed on

12 species of birds. All three species of *Culex* tested had fed on Northern cardinal, American robin, and Brown-headed cowbird.

Cupp et al. (2004) used the combined ELISA and PCR-heteroduplex method of Apperson et al. (2002) to identify the sources of blood meals from mosquitoes collected as part of an arbovirus activity study in northern Mississippi, USA. Genomic DNA prepared from blood-fed mosquitoes was used as a template in a nested PCR amplification reaction by using primers designed to specifically amplify vertebrate cytochrome b sequences as per the method of Lee et al. (2002) described above. Blood meal data and avian abundance data derived from point counts of all birds seen or heard during three-minute counts at 20 stops along a trail were used to derive the feeding index of Kay et al. (1979). As well as using unadjusted point count data, feeding indices were also calculated in terms of the overall biomass of each species at the site obtained by multiplying the number of individuals of each species of bird observed by the average body mass of each species. A total of 101 blood engorged mosquitoes representing three likely vectors of eastern equine encephalomyelitis were collected and analysed, namely: *Aedes vexans* (*n* = 21), *Coquillettidia perturbans* (*n* = 20), and *Culex erraticus* (*n* = 60). Of the 101 mosquitoes tested, 54 (54%) produced a positive result in the PCR assay. Of these, 29 were avian in origin with eight species represented. The great blue heron, *Ardea herodias*, was the major source (55%) of the avian blood meals identified. Feeding indices calculations from unadjusted abundance data revealed the great blue heron, northern mockingbird, greenbacked heron, wild turkey, brown-headed cowbird, and blue jay to have a feeding index in excess of 10, indicating strong preference for taking blood meals from these birds. When the data were adjusted for biomass, only the blue jay, brown-headed cowbird, and northern Mockingbird were highly overrepresented in the blood meals Table 7.1).

In a further development of this approach, Meece et al. (2005) used a Terminal Restriction Fragment Length Polymorphism (T-RFLP) profile analysis of the cytochrome b gene to identify mosquito blood meal sources. This method is considered to be both accurate and capable of generating reproducible results. A second advantage is the availability of a searchable database of RFLP products that can aid in the specific identification of the source of a blood meal. T-RFLP also has the potential to identify mixed blood meals originating from two or more species of host. Fluorescent primers were incorporated to amplify 358 bases of a region of the cytochrome b gene and a database of sequence-predicted and experimentally determined terminal restriction fragment profiles from 55 avian, 13 mammalian, and one amphibian species. Blood meal sources of 59 field-collected mosquitoes representing five genera and seven species were tested. Thirty-two of the 59 blood meals were identified as having originated from mammalian hosts and 18 from avian hosts. Blood meal sources for nine of the mosquito samples could not be determined as no matches were found in the database.

Table 7.1. Feeding indices (FI) for avian blood meal hosts in mosquitoes collected from the Iuka study site in 2002 (Cupp et al. 2004)

Species	FI based on point count	FI based on body mass
Wild turkey	13.4	0.1
Carolina chickadee	0.9	7.1
Brown-headed cowbird	27.8	47.9
Northern cardinal	0.6	1.0
Blue jay	14.4	12.5
Great blue heron	160.0	3.7
Northern mockingbird	13.4	20.9
Green-backed heron	13.4	4.7

Kent and Norris (2005) recently developed a vertebrate specific multiplexed primer set based on mitochondrial cytochrome b to identify the mammalian blood hosts of field collected mosquitoes. The primer set removes the need for the added time and expense of a restriction enzyme digestion or heteroduplex analysis step following the PCR and allows mammalian blood hosts to be identified directly from size specific fragments separated by agarose gel electrophoresis. 15 ng (0.5 μl) of extracted DNA from whole blood was found to be sufficient template for use in positive control reactions. In the laboratory 100% of human, cow, dog, pig, and goat blood meals were correctly identified. Testing of the method against blood-fed mosquitoes collected by PSC in Zambia and stored dry on silica desiccant for 2–7 months prior to DNA extraction resulted in positive blood meal identifications from 100 (81%) of 123 mosquito samples: 61 of 74 *Anopheles gambiae* s.s., 19 of 20 *Anopheles arabiensis*, 8 of 15 *Anopheles funestus* s.s., 8 of 8 *Anopheles leesoni*, 1 of 1 *Mansonia uniformis*, and 3 of 5 *Culex quinquefasciatus*. The considerably smaller body size and therefore smaller blood meal size of *Anopheles funestus* resulted in a lower success rate with this species, especially when the blood meals were either incomplete or partially digested. Blood meals taken from all five animal hosts could be identified from laboratory mosquitoes killed 24–30 h post-feeding, but at 48 h post-ingestion, only goat blood was identifiable.

Histologic techniques for identifying multiple blood-feeding

Romoser et al. (1989) appear to have been the first to use histologic parameters to identify mixed blood meals in *Culex nigripalpus*. Blood-fed specimens were killed in alcoholic Bouin's fixative for 10 min and left to fix at room temperature for 24 h, after which they were stored in 70% ethyl alcohol. Specimens were prepared for microscopy by infiltrating them with Paraplast, mounting them in Tissue-Tek blocks, and cutting into 7 µm sagittal sections, stained using the modified Azan trichrome method of Hubschman (1963). The most useful histologic parameters for identifying multiple blood meals were: the peritrophic membrane, a plug of material forming between anterior and posterior mid-gut, a zone of digested blood surrounding undigested blood, the presence of the remnants of a previous fully digested blood meal, and distinct physical separation of blood meals. In specimens processed less than 84 h post-feeding, two blood meals could be observed in 92.9% of cases. The Romoser et al. (1989) technique was later applied to *Aedes aegypti* by Scott et al. (1993*b*). When laboratory-reared and fed mosquitoes were fixed immediately following their second blood meal, multiple feeding could be identified using histologic parameters in 88% (30/34) of cases. Ability of the method to detect multiple feeding in *Aedes aegypti* decreased markedly if the interval between successive meals exceeded 24 h and specimens were fixed more than 12 h after feeding, contrasting with the results of Romoser et al. (1989) with *Culex nigripalpus*, in which multiple meals could be detected up to 84 h post-feeding. If meals were separated by 48 h then only 25% of specimens could be identified as containing multiple blood meals. When the technique was applied to field-collected specimens of *Aedes aegypti*, 50% (48/96) were identified as containing multiple blood meals. In California, USA Wekesa et al. (1995) determined the proportion of field-collected *Anopheles free-borni* that had taken more than one blood meal to be 9.7% ($n = 134$), based on an assessment of several histologic parameters, including presence of haem, peritrophic membrane/peritrophic plug, partially digested blood meal, physical separation of blood meals and ovarian development. Histologic methods are useful as they can detect cryptic double feeds, when two meals are taken from the same host species. Even this method may underestimate the proportion of multiple feeds if the time elapsed between feeds is short enough for the blood to mix, or alternatively that the period between blood meals (within the same gonotrophic cycle) is sufficiently long for the first meal to have been digested beyond detection or digested and excreted. In later studies, Wekesa et al. (1997*b*), using the modified precipitin test of Tempelis and Lofy (1963), determined the proportion of mixed blood-meals in *Anopheles freeborni* in California, USA to be 1.9%.

Further histologic studies of *Anopheles freeborni* and *Culex tarsalis* in California, revealed multiple blood meals in 44 of 333 *Anopheles freeborni* and in 4 of 41 *Culex tarsalis*. 1.2% of the *Anopheles freeborni* examined showed evidence of at least 3 blood meals during a single gonotrophic cycle (Wekesa et al., 1997*a*). Lorenz and Scott (1996) also used the histologic technique of Romoser et al. (1989) and Scott et al. (1993*b*) to identify multiple blood feeding in *Culiseta melanura*. In the laboratory, females that had been induced to take two blood meals could not be identified by the histologic technique as having taken multiple meals when the meals were separated by less than 24 h or by more than 30 h. This was in contrast with the results of Scott et al. (1993*b*) in which multiple blood meals separated by only 1 h could be identified. Dissection of 654 specimens collected in the field using resting boxes (Edman et al. 1968; see Chapter 5) failed to reveal any histologic evidence of multiple blood-feeding, emphasising the need to refine the techniques and conduct preliminary studies for each mosquito species under investigation.

Mahmood and Crans (1997) examined 532 bloodfed female *Culiseta melanura* and classified them as having fed on more than one host when they were observed to contain fresh blood as well as partially digested blood surrounded by a separate peritrophic membrane and eggs in stage III or IV of ovarian development. Less than 1% (5 of 532) of those examined showed physical evidence of having taken multiple blood meals.

In Sri Lanka Amerasinghe and Amerasinghe. (1999) found, using the histological techniques of Scott et al. (1993*b*) and Wekesa et al. (1997*a*), that among 3306 *Anopheles culicifacies* 34.4% had taken multiple blood meals and of 871 *Anopheles subpictus* 30.4% had taken more than one blood meal. Of the multiple feeds, double meals accounted for 92.7% and 89.5% and triple feeds for 7.3% and 10.5% of feeds for *Anopheles culicifacies* and *Anopheles subpictus* respectively.

Attraction to hosts

Compounds used by mosquitoes to locate their hosts are known as kairomones, that is substances from the emitters (hosts) are favourable to the receiver (mosquitoes) but not to themselves. Wright (1975) considered warmth and humidity to be the main attractants of mosquitoes to humans, and doubted whether there was any skin odour involved in host attraction, but Khan (1977) believed that in addition to skin temperature and skin colour, body odour and other factors were involved. Price et al. (1979) concluded that female *Anopheles quadrimaculatus* were mainly attracted to humans by chemicals emanating from the skin, while studies by Schreck et al. (1981, 1990) showed that there were unidentified attractants to mosquitoes in the sweat from human subjects. There are two types of sweat, eccrine sweat which comes from most body surfaces but especially from the palms of the hands and soles of the feet, and apocrine sweat from the axillary, perigenital and perianal regions. [S]-Lactic acid (formerly called L-lactic acid) is in fact produced by glycolysis in the eccrine sweat glands, and excess remains in the final secretion—sweat. Most mammals have apocrine-type sweat glands; birds lack sweat glands. Schreck et al. (1990) found that sweat from the face and hands generally elicited the greatest response from mosquitoes, and there were significant differences between the attractiveness of sweat from the hands of different people. However, of the 12 mosquito species tested four showed no response, while another four species were only weakly attracted to human sweat: the species most attracted was *Aedes aegypti*, followed by *Aedes albopictus* and *Anopheles albimanus*. Cork and Park (1996) examined the composition of human sweat for electrophysiologically important substances. Sweat samples were chemically fractionated into acid and non-acid components and the most abundant volatiles were identified by linked gas chromatography mass spectrometry as aliphatic carboxylic acids of C_2–C_{18} carbon atoms. Lactic acid was not identified in the samples, probably due to the inefficiency of the linked gas chromatography mass spectrometry technique used and in particular degradation of lactic acid during gas chromatography. Activities of compounds were tested by electronantennography against intact *Anopheles gambiae* antennae and the largest response for a given dose was achieved with methanoic acid. Two non-acidic compounds, 1-octen-3-ol and 4-methylphenol elicited significant dose-dependent responses.

Bernier et al. (2002) compared the chromatograms of skin emanations desorbed off of glass beads handled by two human volunteers with markedly different attractiveness to *Aedes aegypti*. The chromatograms of the two subjects contained the same peaks but significant differences in the abundances of compounds were observed. Forty-five compounds were identified as being present in higher relative concentrations in the "attractive" volunteer, compared with the less "attractive" volunteer.

In wind tunnel studies, filter papers impregnated with human sweat were observed to elicit a landing response in *Anopheles gambiae* (Healy and Copland 2000). Equivalent doses of lactic acid as found in the human sweat samples failed to elicit a response, again suggesting that lactic acid is not the compound in sweat that stimulates landing. 2-oxypentanoic acid also elicited a landing response and provides further evidence of the importance of carboxylic acids in mosquito attraction to and feeding from hosts. Shirai et al. (2001) demonstrated in the laboratory that L-lactic acid acted as a repellent to *Aedes albopictus* when applied to the skin of mice or humans.

Using a dual-choice olfactometer in the laboratory, Dekker et al. (2002) observed that L-lactic acid alone was not attractive to *Anopheles gambiae* s.s., but it significantly augmented the attractiveness of CO_2, skin odour, and skin rubbing extracts from humans and other vertebrate animals. These authors also reported that human skin has a relatively high concentration of L-lactic acid, as compared with cow skin, for example. Blends of two or three kairomones, comprising L-lactic acid plus either acetone, dichloromethane, or dimethyl disulphide were shown to elicit a synergistic attraction response in laboratory-reared female *Aedes aegypti* (Bernier et al. 2003). Two terms were used to describe the mode of action of these synergistic compounds: base attractant and activator. A base attractant was defined as a compound that causes minimal excitation in the mosquitoes but results in attraction of those mosquitoes that do become activated to flight. In the study described, L-lactic acid acted as a base attractant for *Aedes aegypti*, while CO_2 functioned as an activator, causing high excitation but resulting in only low attraction. The authors noted that the results did not mean that the combination could, in direct competition, attract *Aedes aegypti* better than emanations from a live human.

Takken and Kline (1989), Takken (1991) and Lehane (1991) have briefly summarised what is known about substances that attract mosquitoes to baits and odour-baited traps, while the role of carbon dioxide in host attraction has been reviewed by Gillies (1980). Gillett (1979) discussed possible mechanisms by which mosquitoes orientate upwind to hosts in the absence of visual cues; he also presents some pertinent physical characteristics of wind speed near the ground. Gibson and Torr (1999) published a broader review of the visual and olfactory responses of haematophagous diptera to host stimuli.

McIver (1982) listed five types of stimuli that have been shown to elicit host responses in mosquitoes, namely vision, heat, water vapour, carbon dioxide and host odours. A major activator in host location is the concentration of carbon dioxide emitted by hosts, which mosquitoes detect by capitate pegs on their palps. An increase of only 0.01% in carbon dioxide concentration may be detectable, and the response is almost logarithmic to a saturation level of 0.05–0.5%. As the biological range of carbon dioxide concentration emanating from animals is between 3–5%, it is not surprising that artificial concentrations as great as 10% from dry ice or gas cylinders elicit little additional response. Carbon dioxide from animals is therefore 100 or more times greater than the background concentration of 0.02–0.04%, but yet much less than the concentration of about 100% at the release point emitted from gas cylinders or dry ice. Although discharge rates can be altered, the mixing of the gas in the air at various distances from the trap—that is its concentration—will depend on local environmental conditions, which can be very variable in both time and space, and usually remain largely unknown in trapping experiments.

In addition to carbon dioxide expired breath contains several organic compounds (Teranishi et al. 1972) including acetone (Crofford 1976) and 300–400 compounds have been identified from human body emanations (Cork and Park 1996), some of which may be attractive to mosquitoes. Sutcliffe (1986) gives a good review of how blackflies locate their hosts and in a later article (Sutcliffe 1987) gives a more general review of the orientation of biting flies to their hosts, much of which is of interest to those concerned with host orientation by mosquitoes. Takken and Kline (1989) reported for the first time from field experiments that octenol had potential as a mosquito attractant. Later Kline et al. (1990) conducted field trials in Florida, USA with unlit CDC-type light-traps baited with various combinations of attractants including: (1) CO_2; (2) octenol; (3) octenol + CO_2; (4) octenol + butanone + CO_2; (5) lactic acid + CO_2; (6) lactic acid + octenol + CO_2; (7) honey; (8) phenols; and (9) phenols + octenol. Not surprisingly different mosquito species sometimes responded differently to these chemicals. Very few species are attracted in any numbers to octenol alone, but when octenol and carbon dioxide are used together there appears to be a synergistic effect and Kline et al. (1990) obtained a twofold or greater catch with most species of *Aedes, Psorophora, Anopheles, Coquillettidia* and *Mansonia* encountered in the area. With *Culex* species, however, there was little attraction to either chemical alone or in combination. But in contrast to these generalisations *Ochlerotatus taeniorhynchus* and *Coquillettidia perturbans* seemed to respond to octenol alone. Honey (500 ml diluted with 300 ml 29% sodium chloride, then extracted overnight with 250 ml hexane in a liquid/liquid extractor, followed by concentration over a steam bath) was very attractive to *Ochlerotatus taeniorhynchus* (not *Coquillettidia perturbans* as stated in the paper's abstract). The presence of butanone seemed to decrease collections of all species.

Laboratory experiments with *Aedes aegypti* and other species have suggested that mosquitoes might selectively feed on hosts having a rise in temperature due to viral or other parasitic infections (Gillett and Connor 1976; Mahon and Gibbs 1982; Turell et al. 1984). In laboratory experiments Day and Edman (1983) reported that mice were more susceptible to feeding mosquitoes when they were infected with malaria, but in later experiments hypothermia had no significant impact on numbers of mosquitoes feeding on mammals (Day and Edman 1984*a*). If infected hosts are more susceptible to biting mosquitoes, and this applies to human malaria, then there could be epidemiological consequences. This topic and other aspects of blood-feeding and host location are reviewed by Edman and Spielman (1988), while Bowen (1991) provides a good review of host finding cues.

Alekseev et al. (1977) were the first to describe a possible 'invitation effect' whereby additional females of *Ochlerotatus communis* were attracted to feed on a human hand in proportion to the number of females already feeding on it and the phenomenon was later demonstrated in *Ochlerotatus sierrensis* in a series of laboratory choice-chamber experiments by Ahmadi and McClelland 1985). Charlwood et al. (1995*b*) further investigated the phenomenon in the European mosquito *Ochlerotatus cantans* and several species of anophelines and culicines in Tanzania. In 12 replicates with *Ochlerotatus cantans*, legs on which females were already feeding attracted 160 females compared with 68 females attracted to legs on which no insects were feeding (*t*-test, $P < 0.0001$). In contrast in Tanzania almost equal numbers of mosquitoes were attracted to legs on which other mosquitoes were feeding or legs on which no mosquitoes were feeding. The existence of an invitational effect or aggregation pheromone has implications in trap design and interpretation of results. For instance, it might be expected that an animal bait-trap which prevented mosquitoes feeding on the host would attract fewer mosquitoes than one in which blood-feeding was allowed. In fact Emord and Morris (1982) reported that with bird-baited traps double-screening to prevent mosquitoes feeding on the birds was accompanied by a considerable reduction in numbers of mosquitoes caught. They considered that the extra screening diminished host odours escaping from the trap, but it is possible that the reduction in mosquitoes caught was due, at least in part, to the prevention of host feeding and elimination of any invitational effect.

Diurnal species of mosquito also respond to visual characteristics of potential hosts, including colour, brightness, pattern, shape, and movement (Allan et al. 1987). Even nocturnal species respond to conspicuous objects and may be attracted to them from distances up to 15–20 m in the absence of other cues (Bidlingmayer 1994).

Arrival at bait

Species which normally feed at twilight or during the night will often bite during the day if a suitable host is present. In England species which were essentially crepuscular and nocturnal were caught in large numbers during the day whenever bait catches were performed in sheltered sites, where unfed females were resting among the vegetation (Service 1969, 1971*b*). In contrast few or no mosquitoes were caught during day-time catches in exposed areas such as in fields or on pathways. At night, however, baits in both sheltered and exposed areas were bitten (Service 1971*b*). It was concluded that although during the day mosquitoes resting among vegetation were not actively orientated to host feeding, they would nevertheless readily feed if a host was in the immediate area. At night under the influence of an endogenous biting rhythm adults actively flew in search of blood-meals and were consequently encountered in both sheltered and exposed areas. The same phenomenon has been described for *Aedes africanus* in West Africa (Germain et al. 1973). The role of endogenous daily rhythms in vector-host-parasite interactions has recently been reviewed by Barrozo et al. (2004), including several examples pertaining to mosquitoes.

Several workers have reported a burst of biting activity during the first 15–20 min in daytime bait catches followed by a decline in numbers (Germain et al. 1973; Mogi and Yamamura 1981; Nishimura 1982; Roberts and Scanlon 1975; Service 1969, 1971*b*; Wellington 1974), but McCrae et al. (1976) working on *Anopheles implexus* in Uganda undertook the best analyses of this type of behaviour. They pointed out that during passive (opportunistic) biting by mosquitoes formerly resting amongst vegetation, there would be two principal categories; namely a static component (*s*) representing mosquitoes already present within the host's area of attraction, and a component of arrivals (*a*) flying into the host's attractant area after the bait had arrived. But because different species exhibit very different catch curves when collected from the same hosts at the same time, then clearly there must also be different intrinsic behaviours in addition to environmental stimuli affecting their sequence of arrival at bait.

Roberts and Scanlon (1975) observed a depletion effect in their series of 15-min catches, that is high initial catches of *Ochlerotatus atlanticus, Ochlerotatus tormentor* and *Psorophora ferox* during the first 5 min followed by a decline during the next 10 min. The initial high biting rate was said to be caused by host movement, supporting the contention of Gillett (1972) that movement attracts diurnally active mosquitoes to their hosts. Roberts and Scanlon (1975), however, failed to observe any obvious depletion effect with nocturnal species such as *Culex salinarius* and *Ochlerotatus fulvus*. An initial high catch was also reported in *Aedes aegypti* in catches performed both outdoors and in huts in Kenya (Teesdale 1955).

Colless (1956, 1957) considered that mosquitoes arrived at bait by a process of random wandering, or at least by a process not directly orientated to the bait. In Singapore he found that the numbers of *Culex annulus* caught each hour (1900 – 0600 h) declined progressively with time, and fitted the following linear relationship:

$$\log (K - C) = a + bt \tag{7.1}$$

where K = the initial population, C = the cumulative catch, t = time, and a and b are constants. This is in fact an example of removal trapping, i.e. the numbers caught depends on the population still available for capture. Colless (1957) stated that in Singapore the biting cycles of most *Culex* and *Mansonia* species were characterised by a depletion of catch with time. This implies that there is no marked temporal cycle of attraction to the bait, a theory that does not really explain the arrival patterns of most mosquitoes to a host.

In Japan Nishimura (1982) obtained high initial catches of *Aedes albopictus*, and *Ochlerotatus japonicus* during the first 10 min of human landing catches, but not with *Tripteroides bambusa*. It seemed that this was because *Tripteroides bambusa* caught at bait consisted of only actively host-seeking females, and not those resting amongst vegetation. Often a shift of only a few metres can result in another high initial catch (Gillett 1967; Service 1969, 1971*b*). By modifying the method of Service (1971*b*) Nishimura (1982) concluded that the range of attraction for female *Aedes albopictus* was 4.0 m while for *Ochlerotatus japonicus* it was 9.0 m. However, Mogi and Yamamura (1981) saw this paper by Nishimura (1982) before it was published and criticised some of the methodology. They also adopted Service's method, and after performing a 30-min human landing catch undertook a second catch at distances of 2, 4, 6, 8, 10 or 12 m. They analysed the results by applying a new type of removal method similar to that of Kono (1953), and concluded that the range of attraction for *Aedes albopictus* was 4–5 m.

Because of such opportunistic feeding it may be necessary to perform a preliminary bait catch for about 30 min, or even 1 h, to clear an area of hungry unfed mosquitoes before diel biting cycles can be studied (Service 1969; Teesdale 1955). In England, however, Renshaw (1991) believed the high initial catch of *Ochlerotatus cantans* was due to hungry females following her through a wood to the catch site. Another characteristic of some bait catches is that adults may arrive in waves (Haddow 1954; Service 1969), that is the sudden appearance of several individuals followed by short periods when few or no individuals arrive. This is possibly explained by slight changes in the drift of host odours causing the simultaneous stimulation of small groups of resting mosquitoes, which then arrive at the bait more or less together.

There may be a marked interval between the appearance of adults on nearby vegetation and their alighting on the bait. Such pre-biting resting behaviour has been reported in *Mansonia* species (Haddow 1961; Service 1969; Wharton 1962) and *Anopheles* (Colless 1956; Hudson 1984; Lee et al. 1980; Moorhouse and Wharton 1965; Ribbands 1946; Senior White 1953; Smith 1958).

A delay of several weeks between adult emergence to blood-feeding has been recorded independently several times in *Ochlerotatus cantans* (Renshaw 1991; Renshaw et al. 1995; Service 1977; Sulaiman 1982), in *Culiseta morsitans* (Service 1969), in *Ochlerotatus sierrensis* (Bennett 1978; Garcia et al. 1975; Lee 1971), and in *Ochlerotatus thibaulti* (Shields and Lackey 1938). Renshaw et al. (1995) proposed a possible explanation for the delay observed in *Ochlerotatus cantans* as a requirement for a build-up of lipid reserves through sugar-feeding. The possible reasons for such a delay in other species remain unexplained and uninvestigated.

One method of determining the range of insect attraction to animals is to arrange non-attractive interception traps at various distances and direction from a host-baited trap (e.g. Gillies and Wilkes 1970, 1972). In Florida, USA Edman (1979) used ramp traps similar to those of Gillies (1969) (see Chap. 8) to study host orientation to animal baits and carbon dioxide. Contrary to the observation of Gillies and Wilkes (1974) he found no evidence that host-seeking mosquitoes fly downwind, in fact *Culex nigripalpus* exhibited strong upwind flight. It appeared that hosts, or carbon dioxide, stimulated host-seeking at a distance of 15 m. Another approach is to place two identical traps at varying distances from each other, and then determine at what distance the size of the catch per trap is no longer decreased by the other, this would then be equal to twice the trap's range of attraction. Alternatively the numbers caught in traps set at increasing distances from a concentrated source of adults (e.g. isolated breeding sites) can be recorded. However, the decrease in numbers caught in the traps must be distinguished from the natural decline in numbers dispersing associated with increasing distance. A description and detailed account of how these latter two approaches were used to measure the distance of attraction of *Glossina pallidipes* is given by Dransfield (1984).

Tuno et al. (2003) studied the spatial distribution of pre- and postprandial resting mosquitoes in relation to a tethered cow bait in Thailand as follows: a 25 m^2 grassy plot was selected several hundred metres distant from any human or domestic animal habitation. A single cow was tethered in the centre of the plot from 1800 h to the following morning. Forty bamboo stakes (0.7 m high × 4 cm wide) were arranged at the four main compass points in arrays of 10 stakes per point, with stakes spaced 1 m apart in a cross pattern. Mosquitoes resting on the stakes were collected by 2 persons using aspirators for 20 min. at 1900, 2000, 2100, 2200, and 2300 h. Control collections followed the same procedure without the presence of a tethered cow. Canonical correspondence analysis (CCA) was used to determine the spatial relationship of resting mosquitoes to the host. A total of 1566 mosquitoes of 25 species and five genera was collected. *Anopheles aconitus* (638) was the commonest species collected, followed by *Anopheles peditaeniatus* (173), *Culex pseudovishnui* (152) and *Culex vishnui* (133). Not surprisingly, the presence of a cow strongly influenced the distribution of resting mosquitoes, as did distance from the cow and orientation in the east–west direction (but not north–south). Orientation in the east–west direction was primarily influenced by *Mansonia uniformis*, which attacked preferentially from the east (possibly the direction in which its larval habitat was located). Both fed and unfed mosquitoes were concentrated within 5 m from the host. Analysis of the ratio of fed to unfed females and distance from the host by species was conducted using seven clustering methods (weighted and unweighted, average, complete, Ward minimum variance, single, centroid and median). These analayses revealed two distinct groups of species, which the authors termed the 'slow' and 'rapid' groups. The proportion of females that had fed in the rapid group was higher at every distance from the cow compared with the slow group, suggesting that rapid species tended to have a shorter pre-attack resting phase and completed feeding more rapidly. The slow group comprised *Anopheles minimus, Mansonia uniformis, Culex tritaeniorhynchus, Anopheles nivipes, Anopheles aconitus, Culex pseudovishnui*, and *Culex vishnui*. The rapid group comprised *Aedes vexans, Anopheles peditaeniatus, Culex gelidus, Aedes lineatopennis*, and *Armigeres* spp. The authors acknowledged that it was not possible to completely remove the potential influence of the collectors on the distribution.

After a mosquito has landed it may 'freeze' for a few seconds and during this period it is readily disturbed by the host's movements. Following this short initial period of apparent inactivity there is usually a short exploratory period before the mosquito actually probes the skin (Gillett, 1967). Service (1971*a*) studied the feeding behaviour of several British species, while Yates (1979) investigated the feeding behaviour of *Ochlerotatus geniculatus*. They recorded the total time (116–240 s) spent on the host but also divided this into three behavioural pauses, the exploratory period (7–31 s), penetration period (25–92 s), and the feeding period (82–150 s). Mosquitoes generally spent at least 1 min hovering in the vicinity (1–2 m away) of the host prior to alighting.

Preferred biting site

Many haematophagous Diptera have well-defined preferred biting regions. As long ago as 1921 Wesenberg-Lund reported that in Europe *Aedes cinereus* normally bit the legs, but also the hands if they were placed down amongst the grassy vegetation harbouring the insects. Haddow (1954, 1956) found that 97–98% of the adults of *Eretmapodites chrysogaster* bit a standing man below the knees, almost entirely from the ankles upwards, i.e. in a well-defined band of 6–18 in from the forest floor. A man lying horizontally on the ground was rarely bitten, but if he was raised some 6 in all parts of the body were attacked. *Aedes simpsoni* feeds mainly on the head (Haddow 1945), and Aitken has observed that *Sabethes belisarioi* bites almost exclusively on the nose (Gillett 1971). In India Das et al. (1983) collected 58.3% of *Armigeres subalbatus* from people's lower extremities while only 10.07% and 7.48% bit the abdomen and face respectively. Shirai et al. (2002*a*) investigated the preferred biting sites of *Aedes albopictus* on semi-naked volunteers exposed to 120 adult females with their proboscis removed. The preferred biting site on upright and supine individuals was the foot. The proportion of mosquitoes landing on the foot when the volunteer was lying down was lower than for a standing volunteer. The second most preferred landing site was the hand, followed by the face. No correlation was found between preferred landing sites and body temperature. In field studies in Papua New Guinea, Cooper and Frances (2000) demonstrated that *Anopheles koliensis* bit humans more frequently on the feet and ankles rather than the leg. Convection currents around the body of a human volunteer were thought to be the primary factors in determining preferred biting site for three members of the *Anopheles gambiae* complex in laboratory studies (Dekker et al. 1998). When the volunteer was seated, the legs and feet were preferentially bitten by *Anopheles gambiae* s.s., whereas when the volunteer lay on the ground with legs and feet raised, the head, trunk and arms (i.e. the body parts closest to the ground were most frequently bitten. *Anopheles quadriannulatus* also bit the head of the test subject more frequently than either *Anopheles gambiae* s.s. or *Anopheles arabiensis*. De Jong and Knols (1996) reviewed mosquito preferences for specific biting sites and concluded that most species studied do appear to show definite preferences but these tend to differ among species.

Host preference

Tempelis (1975) postulated that there are nine host feeding patterns exhibited by mosquitoes, namely: species that feed almost exclusively on mammals; almost exclusively on birds; species that feed readily on mammals and birds; species that feed almost exclusively on amphibians; those that feed predominantly on reptiles and occasionally on homeothermic animals; species that feed exclusively on fish; species that feed readily on poikilothermic and homeothermic animals; species that feed preferentially on birds in spring, but later in the year shift their feeding habits onto mammals; and species that feed exclusively on birds in one geographical location, and mammals in another. Observations that adults of many species of mosquitoes feed on more than one species of host have posed the question as to whether host preference is genetically determined and heritable or dependent on ecological factors relating to numbers and availability of specific hosts. Rawlings and Curtis (1982) studied the feeding behaviour of *Anopheles culicifacies* species B in Sri Lanka by a series of bait catches conducted from 1830–2100 h on a cow in a hut, and on five men sitting in an adjacent house. Mosquitoes caught biting the cow were marked with magenta fluorescent dusts while those caught in human bait catches were dusted yellow: all mosquitoes were released outside the huts at the end of the catch period. On six subsequent evenings mosquitoes caught on either bait were checked for markings, and unmarked mosquitoes marked with the appropriate colour, after which all were released. On the final seventh evening the mosquitoes were collected and killed. A total of 1150 and 188 mosquitoes were caught biting the cow and men, respectively. Recaptures of marked adults were small. Only two originally biting a cow were caught at human bait, and just four firstly caught on men were later caught biting the cow. It was tentatively concluded that there were no distinct anthropophagic and zoophagic populations of *Anopheles culicifacies* species B, at least in their area.

Using the same methods in Malaysia Loong et al. (1990) found there were no separate populations of *Anopheles maculatus* feeding on cattle and people.

Arredondo-Jimenez et al. (1992) studied the basis of population differences in host preference shown by *Anopheles albimanus* by a series of experiments in which progeny of females were allowed to feed on either a cow or human host and then released into a choice chamber containing humans and a cow. Approximately 65% were attracted to the cow bait and 35% to the human bait and this proportion did not vary according to geographic source of the mosquito parents, nor to the original bloodmeal source, suggesting that learned host associations are unlikely in this species.

In Pakistan, Hewitt et al. (1994) reported that sleeping close to cattle or goats significantly increased the human-biting rate of zoophilic anophelines. Similarly, Seyoum et al. (2002) in Ethiopia, noted that the human biting rate (HBR) of *Anopheles arabiensis* tends to increase when cattle are kept in the same dwellings as humans, though the differences in biting rates were not statistically significant due to relatively large confidence intervals of the estimated HBRs.

Hadis et al. (1997) looked at the host preferences of indoor-resting *Anopheles arabiensis* using the direct ELISA method of Beier et al. (1988). Mosquitoes were collected from human dwellings, cattle sheds and mixed dwellings between 0500 and 0700 h, using oral aspirators. Of 611 *Anopheles arabiensis* collected, 325 (53.2%) gave a positive reaction to either human or bovine blood, or both. The highest proportion of human blood was found in mosquitoes collected from human dwellings.

Also in Ethiopia, Habtewold et al. (2001) investigated differences in feeding behaviour of *Anopheles* mosquitoes in relation to sleeping and livestock keeping arrangements in three different areas of the country. Resting mosquitoes were collected from inside houses, cattle enclosures and granaries between 0530 and 0700 h using oral aspirators. The proportion of mosquitoes feeding on livestock (96.9%) as compared to humans was much higher in Konso district, where inhabitants sleep on wooden platforms in acacia trees approximately 3–4 m above ground and livestock are kept in enclosures beneath the platforms. In the other two sites, where inhabitants either slept in houses and livestock were maintained in separate enclosures, or where people shared their sleeping places with their livestock, there was no significant difference in the proportions of livestock or human feeds. *Anopheles arabiensis* was the only member of the *Anopheles gambiae* complex collected inside houses in this part of Ethiopia.

Killeen et al. (2000) developed an availability-based model of host choice where the availability of a given host to the vector population was defined as the mean rate at which a typical single hostseeking vector encounters and feeds upon that host in a single feeding cycle. Thus the availability (a) of a given host (j) of a particular species (s) can be envisaged as the product of the rate at which individual vectors encounter that host while host seeking ($\varepsilon_{s,j}$) and the likelihood that, once encountered, the vector will acquire a bloodmeal from that host ($\phi_{s,j}$), and

$$a_{s,j} = \varepsilon_{s,j}\phi_{s,j} \tag{7.2}$$

Host species availability (As) was defined as the sum of the availabilities of all hosts ($j = 1, 2,. N,$) of that species (s), such that:

$$A_s = \sum_{j}^{N_s} a_{s,j} \tag{7.3}$$

where N_s is the total number of hosts of species s within the transmission sion focus. Furthermore, the total host availability (A) is defined as the sum of the availabilities of all host species present which represent potential bloodmeal sources ($s = 1, 2,. S$):

$$A = \sum_{s}^{S} A_s = \sum_{s}^{S} \sum_{j}^{N_s} N_{s,j} \qquad (7.4)$$

The mean length of time vectors spend in locating a host (η) is equivalent to the inverse of total host availability (A). The proportion of blood-meals taken from a host species (Q_s) also depends on availability of that host species, such that $Q_s = A_s/A$. Using the following notation, where h signifies human, c cattle and o other, then the human blood index (Q_h) is given by:

$$Q_h = A_h / A_h + A_c + A_o \qquad (7.5)$$

If human and cattle population sizes are known (N_h and N_c respectively), then Q_h can be expressed in terms of mean availability of these two hosts

$$Q_h = N_h \bar{a}_h / N_h \bar{a}_h + N_c \bar{a}_c + A_o \qquad (7.6)$$

which in terms of relative availability of hosts can be rewritten as:

$$1/Q_h - 1 = \rho + \lambda N_c / N_h \qquad (7.7)$$

$$Q_h = 1/(1 + \rho + \lambda(N_c / N_h)) \qquad (7.8)$$

where $\rho = A_o/Ah$ and $\lambda = \bar{a}_c / \bar{a}_h$. The authors then fitted data on host and cattle populations available from two published studies in sub-Saharan Africa using least squares linear regression and least squares non-linear regression

Minakawa et al. (2002) studied the influence of livestock and human host availability on the distribution and abundance of malaria vectors in the Lake Victoria region of western Kenya. Larval mosquitoes were collected by dipping and adult mosquitoes by pyrethrum spray collection. Distance to the nearest larval habitat was measured and the locations of all cowsheds were mapped, and the number of cows in each cowshed recorded. Cow density around each larval habitat was estimated by averaging the number of cows in the five nearest cowsheds. Similarly, human density around a larval habitat was estimated by averaging the number of residents in the five nearest houses. Multiple regression analysis of the arc-sine transformed relative abundance of *Anopheles gambiae* s.s. as a proportion of the total number of *Anopheles gambiae* s.l. collected against the ratio of average resident number to average cow number around each larval habitat, the ratio of average house distance to average cowshed distance from the larval habitats, and the distance among larval habitats. Geographic distances among larval habitats were represented by a distance matrix. A similar method was used to determine the relationship between the absolute densities of *Anopheles gambiae* and *Anopheles arabiensis* adults and human/livestock distributions. In the early rainy season, the ratio of human density to cow density and the ratio of distance to a house from a larval habitat to distance to a cowshed from a larval habitat were significantly associated with the relative abundance of *Anopheles gambiae* larvae, such that if a larval habitat was farther away from a house but closer to a cowshed, fewer *Anopheles gambiae* larvae were found. In the late rainy season, only one variable (the ratio of distance to a house from a larval habitat to distance to a cowshed from a larval habitat) showed a significant and negative association with the relative abundance of *Anopheles gambiae* larvae. For adult mosquitoes collected in both March and May, distance from a house to the nearest larval habitat was the only variable significantly associated with *Anopheles gambiae* density. More than 90% of *Anopheles gambiae* adults were found in the houses within 300 metres from the nearest larval habitat in both sampling periods. Interestingly, *Anopheles gambiae* density in a house was not correlated with either human and cow densities in the homestead or with the distances to the cowshed from the house.

Crans et al. (1996) investigated the feeding habits of *Ochlerotatus sollicitans* in New Jersey, USA, in order to determine if the relatively low proportion of bird feeds (1–2%) identified by precipitin tests was primarily a result of host availability or sampling bias, rather than an innate preference. When offered a restrained bird (Japanese quail) or mammal (guinea pig) in separate cages, no significant differences in host preference were observed, with 54.9% feeding on the quail and 52.9% feeding on the guinea pig. When presented with the two hosts simultaneously, 66.4% of the mosquitoes that took a blood meal took it from the mammal 33.6% from the bird.

Learning and host attraction

McCall and Kelly (2002) present an interesting paper on learning and memory in disease vectors, reviewing some of the evidence in relation to host finding and oviposition. Short-term memory and the use of spatial maps would confer clear benefits on any animal living in environments in which resources (oviposition sites, resting sites, food sources) demonstrate heterogeneity over space and time. There is increasing evidence that mosquitoes may select hosts based on prior experience. In Sabah Hii (1985) and Hii and Vun (1987) used mark-recapture methods to study the feeding preferences and behaviour of *Anopheles balabacensis* and *Anopheles donaldi* on buffaloes and people. Mosquitoes feeding on four men were caught, dusted with blue powder and released, while those feeding on a buffalo some 33 m away were marked green and then released. On subsequent nights when returning to feed mosquitoes were recaptured and treated as previously. It was found that adults of both species tended to return to the same types of host. Similarly in Thailand Nutsathapana et al. (1986) found that there was a statistically significant tendency for adult *Anopheles minimus* to return to the hosts on which they were first caught, thus showing host-preference heterogeneity in the population. Mwandawiro et al. (2000) recently demonstrated that *Culex* mosquitoes were more likely to feed on a host on which they had previously fed successfully. In their study in Chiang Mai, Thailand, Mwandawiro et al. (2000) placed either a single cow or a single pig in large mosquito nets erected 7 m apart. Mos-quitoes attempting to feed on the enclosed animals were collected and maintained on water overnight before marking them with fluorescent dye: blue for the mosquitoes attracted to the cow and red for those attracted to the pig. Blue marked mosquitoes were then released into the net containing the cow and allowed to feed before being collected again. Red-marked mosquitoes were released into the net containing the pig and allowed to feed. Blood meals were identified using direct ELISA. In a second experiment mosquitoes attracted to the cow were released into the net containing the pig and those attracted to the pig were released into the net containing the cow. Mosquitoes attracted to cows fed more successfully when released into the net containing a cow than when released into a net containing a pig. This was true for *Culex tritaeniorhynchus*, *Culex gelidus* and *Culex vishnui*, but not for *Aedes vexans*. Of those mosquitoes originally attracted to the pig, only *Culex vishnui* fed more successfully on the pig than the cow. When marked mosquitoes were released into a net containing both a cow and a pig, it was observed that they fed preferentially on the host to which they were first attracted. Blood fed mosquitoes were also collected from a cow shed or a pigsty and allowed to lay eggs. Offspring from these eggs were raised to adults, marked according to the host preference of their parent and released into a net containing both a pig and a cow. No significant association between parent host preference and that of the offspring was observed, and in fact offspring of parents collected from either the cow shed or the pigsty exhibited a significant preference for feeding on the cow, suggesting that mosquitoes may develop host preferences according to prior experience. Learning to select less defensive hosts would be expected to confer distinct survival advantages on individuals.

In northern Tanzania, McCall et al. (2001) used a mark-release-recapture methodology to investigate whether *Anopheles arabiensis* exhibits site and/or host fidelity. Four groups of mosquitoes were collected: human-fed at house 1 (marked orange), human-fed at house 2 (green), cattle-fed at house 1 (blue), and cattle-fed at house 2 (magenta). Determination of whether or not mosquitoes were human-fed or cattle-fed was based on the site from which they were collected, such that mosquitoes collected from inside mosquito bed nets were considered to have fed on humans, while those collected resting on the walls of cattle shelters were considered to be 'cattle-fed'. This assumption was subsequently refuted as only half of bloodmeals tested from mosquitoes caught resting inside mosquito nets in which humans slept did in fact contain only human blood. All marked mosquitoes were released from an uninhabited house over a period of five days. Mosquitoes were recaptured in resting collections, using oral aspirators. Pyrethrum spray catches were also conducted on two days. Mosquitoes recaptured on the walls of bedrooms were classified as returning to humans, whilst those caught in the animal shelters were classified as returning to cattle. A total of 4382 anopheline mosquitoes was caught, marked and released. Over the subsequent 8 days. 17 099 were caught and examined, of which 44 were marked. No marked mosquitoes were recovered from neighbouring houses. Fifty-nine percent of *Anopheles arabiensis* initially caught at house 1 were recaptured at house 1, while 82% of those first caught at house 2 were recaptured at house 2. This result was significant ($\chi^2 = 5.81$, $P = 0.007$) and suggested that *Anopheles arabiensis* females tended to return to the house from which they were originally captured prior to the next bloodmeal, behaviour termed 'site-fidelity'. Thirty-eight of fourty-four (86%) *Anopheles arabiensis* returned to cattle sheds, regardless of where they were first caught.

Ulloa et al. (2002) conducted similar studies with *Anopheles vestitipennis* originally captured in Magoon traps baited with humans, a horse, pigs, or a cow. Host fidelity of mosquitoes was estimated by recapturing marked mosquitoes returning for a second blood meal. In the first experiment, 66% (23 of 35) of mosquitoes collected on humans returned to humans, whereas 85% (22 of 26) of mosquitoes collected on the horse returned to the horse. In a second experiment, 60% (15 of 25) of mosquitoes showed fidelity to human hosts, whereas 33% (1 of 3) returned to feed from a pig. Numbers of mosquitoes recaptured were small in several cases. Pooled data revealed that overall mosquitoes returned to feed from the same host to which they had initially been attracted,and this host fidelity appeared to be stronger for non-human baits than for human baits. In later studies, Ulloa et al. (2004) undertook further investigations on the nature of host preference in wild-caught *Anopheles vestitipennis* and their F_1 progeny in experimental hut trials in southern Mexico. Wild mosquitoes were collected using human landing catches or from horses, cows, or pigs and were marked with fluorescent powders before being released into huts with three compartments. Marked mosquitoes were released into the central compartment, while the two end compartments contained either 2 human volunteers, 1 horse, 1 cow, or 2 pigs so as to control for differences in host surface area. Wild-caught mosquitoes showed no apparent preference for subsequently feeding on the hosts from which they were originally collected, in contrast with the results of their earlier studies (Ulloa et al. 2002), described above. For example, 25% of mosquitoes that were originally collected from humans fed on humans and 31% of mosquitoes originally collected from a horse also selected human hosts for their next blood meal. Similar patterns were observed for females originally collected from cows and pigs. Interestingly, 67% (174/260) of F_1 progeny raised from females that selected a human host in the first hut experiment selected the human-containing compartment when offered the choice of humans or horse. By contrast, only 14% (25/179) of F_1 females that had selected horse in the first experiment entered the human compartment in a human:horse choice test. Similar patterns were again observed for the other host combinations. Overall, 59% of F_1 females raised from females that had originally been collected from a human host also selected a human host. Only 20% of F_1 females raised from females that originally selected a non-human host subsequently selected a human host. The authors postulated that the differences in results of the two studies could have arisen as a result of the original host preferences of the wild-caught females being more influenced by host availability, or opportunistic feeding rather than any innate preference, such that the original sample comprised a collection of mosquitoes with a mixture of innate preferences.

Alonso et al. (2003) studied the ability of *Aedes aegypti* to associate unconditional stimuli with particular odours or visual stimuli, using an olfactometer and a "visual arena". These authors found no evidence of a learning ability, but acknowledged that the experimental apparatus and procedures used may have been insufficient to detect such a learning response.

Seasonal, physiological and geographical heterogeneities in feeding behaviour

In studying seasonal changes in population size attraction to specific hosts should not change over the sampling period, however, such changes in host preference do occur and can be due to genetic and behaviourial (Boreham and Garrett-Jones 1973; Gillies 1964), or environmental (Edman 1974), factors or to changes in vector or host abundance (Chandler et al. 1977; Reeves 1971). In Florida, USA for example *Culex nigripalpus* feeds more on birds than mammals during the cooler months of the year, but the reverse occurs in the warmer months (Edman and Taylor 1968). *Culex tarsalis* exhibits a similar preference for feeding on mammals as opposed to birds during the summer months (Tempelis et al. 1967). Several other examples of shifts in host preference from birds to mammals and *vice versa* have been reported (Bertsch and Norment 1983; Hayes et al. 1973; Reeves 1971; Suyemoto et al. 1973; Tempelis 1975). In Iowa, USA Ritchie and Rowley (1981), reported an apparent midsummer increase in the proportions of *Culex pipiens, Culex restuans* and *Culex salinarius* feeding on mammals in preference to birds. In Massachusetts, USA Nasci and Edman (1981) found a seasonal feeding shift in *Culiseta melanura*, from almost exclusively feeding on passerines at the beginning of the year (June) to feeding on non-passerines, and to a much lesser extent mammals, reptiles and amphibia, later in the summer (August–September). In Kenya Chandler et al. (1977) recorded seasonal changes in feeding in the *Culex univittatus* group, which was believed to be due to changes in the availability of hosts. This shift involved predominantly feeding on ciconiform birds early in the year to feeding exclusively on passerines at the end of the year. In addition to changes in feeding on different types of birds, *Culex univittatus* also tended to shift towards feeding more on mammals, mainly cattle, at the end of the year. Nasci (1984) gives several references of seasonal changes in the feeding patterns of *Culex nigripalpus, Culex tarsalis, Culex univittatus* and *Culiseta melanura*. He found that in Indiana, USA *Aedes vexans* and *Ochlerotatus trivittatus* exhibited considerable daily variability in the types of host fed upon both within and between different types of habitats.

In India *Culex quinquefasciatus* feeds more on man in the hotter months than on cattle (Kaul and Wattal 1968). The relative proportions of indoor and outdoor biting may also vary according to season. In Jamaica Muirhead-Thomson and Mercier (1952) noted a marked increase in the proportion of *Anopheles albimanus* biting indoors in the rainy season. A similar shift to indoor biting during the monsoon, and increased proportions biting out of doors during the cooler months, has been observed in *Culex quinquefasciatus* in India (Gubler and Bhattacharya 1974). In Colombia, Elliott (1968) found that during periods when population densities were at a maximum, outdoor biting by *Anopheles darlingi* could be less important than indoor biting, whereas at other times outdoor biting was often more important. Seasonal shifts in biting times have also been recorded. For example, in Pakistan Reisen and Aslamkhan (1978) recorded that in warm weather *Anopheles culicifacies, Anopheles stephensi* and *Anopheles subpictus* fed on cattle mainly late at night, but with the onset of cooler months they became crepuscular feeders. However, no such seasonal shifts in biting activity were recorded in *Anopheles nigerrimus, Culex tritaeniorhynchus* and *Culex pseudovishnui.*

Bowen et al. (1995*a,b,c*) report that both autogenous and anautogenous individuals of *Ochlerotatus bahamensis*, a facultatively autogenous mosquito, are preferentially attracted to plant volatiles rather than vertebrate host volatiles as newly-emerged females. The development of sensitivity to vertebrate host volatiles develops as the number of lactic-acid-sensitive receptors increases with age. In anautogenous females the number of sensitive receptors is never high and correspondingly responsiveness to vertebrate hosts is diminished. This could be due to the fact that anautogenous females, which are always smaller, have relatively fewer receptors. Alternatively, recent sugar feeding may affect haemolymph composition and suppress receptor sensitivity. In a wind-tunnel olfactometer study of preference for sugar volatiles or human volatiles Foster and Takken (2004) demonstrated that newly-emerged male and female *Anopheles gambiae* responded more strongly to sugars during the period 24–36 h post emergence. Females that fed on sugar during the night of emergence responded more strongly to human volatiles, and after five days of access to a sugar source females responded almost exclusively to human volatiles.

Dia et al. (2001) observed significant variations in the percentage of blood meals taken from different hosts by *Anopheles funestus* collected in PSCs conducted in villages along a sampling transect in Senegal. The proportion of feeds identified as human by direct ELISA varied from 32.3% in Wassadou, a village in the sudano-guinean biogeographical zone, to 100% in Ndiop, a village in the northern Sudanese zone. 52% of feeds were from horses in Wassadou, despite the number of horses in this village being similar. Chromosomal inversion polymorphisms in *Anopheles funestus* are known to differ from west to east Senegal (Dia et al. 2000; Lochouarn et al. 1998) and this could explain the differences in host preference. Konaté et al. (1999) observed a similar pattern in degree of anthropophily amongst indoor-resting *Anopheles funestus* from east to west in Senegal, with 69.9% of 69 blood meals identified as non-human in the village of Wassadou.

Feeding preferences for individuals and sub-populations of hosts

Another practical difficulty in epidemiological studies is preferential biting, in which biting is biased in favour of certain individuals, age-classes, host size, sex, health, or other factors (Ansell et al. 2002; Day and Edman 1983; Elliott 1968; Lindsay et al. 1993; Port et al. 1980; Smith 1961; Spencer 1967). The pattern of contact between vectors and individual hosts is in reality non-random and heterogenous (Woolhouse et al. 1997), in stark contrast to the assumptions made by traditional models of transmission, including vectorial capacity (see Chap. 7), which assume that the vector population is evenly distributed among hosts. Experiments with *Aedes aepypti* have also shown that a person's attractiveness can vary over short periods (Khan et al. 1971). In Nigeria Shidrawi et al. (1974) found a four-fold difference between the numbers of *Anopheles gambiae* and *Anopheles funestus* caught by different men aged 16–25. Carnevale et al. (1978) in the Congo compared biting rates in teams of different ages (0–2, 2–10, 10–20, >20 years) and sex, and discovered that the number of bites received from *Anopheles gambiae* s.s. increased proportionally as 1:2:2.5:3 for the four age-groups. Males and females were bitten indiscriminately. In Sierra Leone Thomas (1951) reported that 59.3 and 79.2% of the variation in numbers of *Anopheles gambiae* s.l. biting people in two families was apparently due to age, fewer bites being on younger people. In Jamaica Muirhead-Thomson (1951) concluded that the large variations (65.4, 81.7 and 91.3%) in biting rates of *Anopheles albimanus* in three families were also due to age, most biting being on adults. In Kenya Boreham et al. (1978) showed that *Anopheles gambiae* s.l. and *Culex quinquefasciatus* fed more frequently on mothers than babies. In The Gambia, by typing human blood in engorged mosquitoes into A, B and O blood groups, and by identifying different haptoglobins, Port et al. (1980) attributed the larger number of bites on adults than young children to their greater size, i.e. both weight and estimated surface area of skin. In fact with both infants (less than 18 months) and adults they obtained significant regressions of the numbers of *Anopheles gambiae*, and other mosquitoes, biting with increasing weight of the baits. Lindsay et al. (1993) observed that certain individual human subjects consistently attracted more mosquitoes into the experimental hut in which they slept, over four trials, spanning two-and-a-half years in The Gambia. The differential attractiveness persisted even when the individuals were rotated among the experimental huts. No relationship was observed between the number of mosquitoes entering a hut and the number blood feeding, suggesting that individuals that attract large numbers of mosquitoes do not necessarily suffer more bites. Also in The Gambia, Ansell et al. (2002) demonstrated that pregnant women were more attractive to *Anopheles gambiae* mosquitoes than non-pregnant women sleeping under an untreated mosquito net. The number of mosquitoes entering the nets each night was 1.7–4.5 times higher in the pregnant group ($P = 0.02$) and pregnant women also received a higher proportion of bites than did non-pregnant women (70% vs 52%, $P = 0.001$). Studies in eastern Sudan carried out by Himeidan et al. (2004) also demonstrate an apparent preference for pregnant women. Nine pairs of women, each pair consisting of one pregnant woman and one non-pregnant woman of roughly similar age, weight and parity, were identified and gave their informed consent to participate. Each woman was asked to stay outdoors between 1900 and 0600 h on each of four consecutive nights/week, in the yard of her house, on a bed inside an untreated mosquito net in which three triangular holes (each side measuring 10 cm) had been cut at low level. Mosquitoes were collected from inside the nets each morning at 0600 h. Compared with the non-pregnant, the pregnant women attracted similar numbers of culicines ($P = 0.902$) but almost twice as many *Anopheles arabiensis*. The mean *Anopheles arabiensis* biting rates (and 95% confidence intervals) were 0.94 (0.33–1.54) bites/woman-night for the pregnant women and 0.49 (0.12–0.98) bites/woman-night for the non-pregnant ($P = 0.005$). The differential attractiveness of pregnant women is thought to be due to a combination of factors, including an increase in the amount of exhaled breath, increases in size and surface area, and increased production of body heat. In Tanzania Knols et al. (1995) used tent traps having two Muirhead-Thomson type exit traps fitted to opposite sides and containing a single individual enclosed within a mosquito net to investigate differences in attractiveness among three individual male Tanzanians aged 21, 22 and 24 years. A randomised block design was used to allocate the individuals to tents over a nine-day test period. Catches differed significantly among the three individuals for *Anopheles gambiae* s.l., *Anopheles funestus*, and *Culex quinquefasciatus*, but not for *Mansonia*. sp. Gass et al. (1982) showed that adolescents and adults were more attractive to *Mansonia annulata* than children, although this bias was not found in the other three *Mansonia* species they collected.

Prior and Torr (2002) applied a similar approach to investigate differences in attractiveness of individual cattle to *Anopheles gambiae* complex mosquitoes in Zimbabwe. A single adult cow, a single calf, an adult and a calf, two adults, two calves, or no animals were introduced into a trap hut

($3 \times 4 \times 3$ m) fitted with an exit trap. The exit trap comprised two funnels constructed from mosquito netting, with an entrance diameter of 50 cm and an exit diameter of 10 cm. The two funnels were joined together at the narrow ends and a gauze-covered box was attached to the exit side of the trap to retain the mosquitoes attempting to exit the hut. Blood-meal samples extracted using the methods of Torr et al. (2001) were amplified using the OLADRB primer set (Paterson et al. 1998). Blood-meal analysis of mosquitoes collected from traps containing an adult and a calf revealed that adults were consistently fed upon more frequently than calves (ratio of adult:calf meals 2.1:1) in all four test replicates. If likelihood of being bitten was dependent on host body mass then the expected ratio would have been 5.2:1 as the mean weight of adult cattlewas 385 kg, compared to 74 kg for calves. The ratio of approximate surface areas was 2.8:1. No significant differences were observed in the proportions of meals taken from the different hosts by *Anopheles arabiensis* and *Anopheles quadriannulatus*, the two species captured. These two species did however differ in the proportions of blood-meals that contained blood from both animals in the hut, with 20% of blood meals from *Anopheles arabiensis* containing blood from both individual hosts, compared with 9% for *Anopheles quadriannulatus*.

Quiñones et al. (2000) described a study to determine whether permethrin-treated mosquito nets achieve their effects on child morbidity and mortality through a diversion of biting mosquitoes away from children to adults. An interesting method of differentiating blood meals taken from adults from those taken from children was employed. All children in the study were vaccinated against rabies and detection of antibodies to the rabies vaccine was used as a marker to identify blood meals originating from children. These authors observed that the proportion of blood-fed mosquitoes collected indoors that had fed on children fell from 30.8% to 9.2% and from 28.0% to 6.9% in two villages following the introduction of insecticide-treated nets.

In an attempt to answer the question whether parasite-infected hosts are more attractive to their insect vectors than non-infected hosts, Lacroix et al. (2005) undertook a study of host preference of *Anopheles gambiae* in Kenya. The study was conducted under semi-natural conditions on 12 groups of three schoolchildren. Each group of three children comprised one uninfected child, one asymptomatic child with asexual stage parasitaemia, and one child with gametocytes. The relative attractiveness of the children in each group was measured with a three-way olfactometer consisting of a central chamber attached via PVC tubes to three tents where the three children of the group were resting or sleeping. Mosquitoes were released in the central chamber and given the opportunity to enter any of the three chambers. After the experiment, all children with detectable parasitaemia were treated with sulphadoxine-pyrimethamine and used in a repeat assay two weeks after the first assay in order to identify any differences in innate attractiveness to mosquitoes. While it is known that sulphadoxine-pyrimethamine treatment may enhance gametocytogenesis, each child was confirmed gametocyte-negative by microscopy, prior to the second assay. Before treatment, gametocyte carriers attracted an average of 10.2 mosquitoes, uninfected children attracted 5.3 mosquitoes, and children with asexual parasitaemia attracted 5.4. Following antimalarial treatment, the children who had previously harboured gametocytes attracted a similar number of mosquitoes to those who had previously been uninfected or infected with asxual stages.

Wood et al. (1972) reported an apparent prefence exhibited by *Anopheles gambiae* to feed on humans of blood group O, as demonstrated by characterising bloodmeals post-feeding. Recently, Shirai et al. (2004) demonstrated that the mean relative percent of *Aedes albopictus* that landed on blood group O secretors (83.3%) was significantly higher than landed on group A secretors (46.5%), however, the authors considered their results to be inconclusive.

There is some evidence that ingestion of alcohol may make human hosts more attractive to *Aedes albopictus* in the laboratory (Shirai et al. 2002*b*). The proportion of mosquitoes landing on human volunteers who had ingested 350 ml of beer was significantly increased compared with attraction prior to alcohol ingestion. Ethanol content of sweat was not correlated with alcohol ingestion or mosquito landing.

Host defensive behaviour and feeding success

Edman pioneered the study of host-defensive behaviour and feeding success of mosquitoes over 22 years ago (Edman and Kale 1971), and since then there have been several interesting papers on host responses, including defensive reactions to being fed upon by mosquitoes (e.g. Charlwood et al. 1995*b*; Cully et al. 1991; Day and Edman 1984*b*; Downes et al. 1986; Edman et al. 1972, 1985; Edman and Scott 1987; Kale et al. 1972; Kelly (2001); Klowden 1983; Klowden and Lea 1979; Molyneux and Jefferies 1986; Scott et al. 1988, 1990; Walker and Edman 1985*a,b*, 1986). Host-defensive reactions and other aspects of host location are reviewed by Edman and Spielman (1988). Downes et al. (1986) review the effects of insects, including mosquitoes, on population movement of caribou. Those interested in the protective measures adopted by animals against biting flies should also read the papers by Waage (1979, 1981).

Anderson and Brust (1996) conducted studies on the effects of host defensive behaviour exhibited by Japanese quail on blood feeding success of *Aedes aegypti* and *Culex nigripalpus* in the laboratory. Five behaviours exhibited by quail not exposed to biting mosquitoes, but considered to be potential defensive reactions all increased in frequency when hungry mosquitoes were introduced to the bird cages. In addition, the intensity of three key defensive behaviours was observed to differ significantly among individual birds.

Following further analysis of the data obtained using the box traps described in section 3.5.13 above (Anderson and Brust, 1995), Anderson and Brust (1997) reported on the feeding success of *Culex tarsalis, Culex nigripalpus* and *Culex restuans* from individual quails. In 18 of 25 trap-night collections of *Culex tarsalis*, the distribution of blood meals between the host marked with caesium and the host marked with rubidium differed significantly from 0.5. For *Culex nigripalpus* the proportions of feeds from individual birds only differed from 0.5 in 7 of 20 collections. For combined *Culex restuans* and *Culex tarsalis* data, proportions differed from 0.5 in 23 of 25 collections. Incomplete feeding, as evidenced by decreased blood meal volume was between 2 and 10 times more likely from one bird of a pair, compared with the other. The authors ascribed the differences in the probability of individual birds being fed upon and differences in the incidence of incomplete feeding between individual birds to differences in individual host defensive reactions.

Charlwood et al. (1995*a*) investigated the relationship between feeding success of *Anopheles gambiae* and *Anopheles funestus* and indoor resting population density in Tanzania. Feeding success (*Ps*) was calculated as:

$$Ps = (c + d)/(a + b + c + d) \tag{7.9}$$

where a = proportion unfed, b = proportion part-fed, c = proportion fed, and d = proportion semi-gravid. Resting collections were made on 109 different dates from 48 houses, resulting in 423 collections that contained *Anopheles funestus* and 405 that contained *Anopheles gambiae*. The proportion of gravid mosquitoes and feeding success both decreased with resting population density, however, most of the variation in feeding success was accounted for by highly significant differences among houses, as demonstrated by logistic regression. Inter-house differences in apparent feeding success could have arisen through non-uniform dispersal of mosquitoes among houses and may not have been the result of host avoidance at high biting density. The relationship between feeding success and density for individual households was only significant for *Anopheles gambiae* (rank correlation, $r = -0.41$, $P = 0.006$). Similarly, a significant relationship between feeding success and density on different nights in the same house was only observed for *Anopheles gambiae* (Likelihood ratio test, $\chi^2 = 75.9$, d.f. $= 1$, $P < 0.0001$). The authors compared the ratio of P_s for a given collection in a specific house to P_s for all collections from that same house in order to control for inter-house variation. No consistent pattern in the ratio was observed for *Anopheles funestus*, but for *Anopheles gambiae*, the ratio tended to exceed 1 (indicating greater than expected feeding success) more frequently at high densities. The authors concluded that these two species of highly synanthropic mosquitoes had evolved a late-biting habit coinciding with their host's sleeping patterns in order to minimize density-dependent host avoidance behaviours.

Kelly (2001) discusses the influence of host defensive behaviour as a selection pressure on vectors to feed on hosts that are more amenable to being bitten. The author argues that blood-feeding insects should evolve strategies that minimize contact with more defensive hosts (host X) in favour of less defensive hosts (host Y). Evidence suggests that in most cases host defensiveness is dependent on the density of biting and as a result, the distribution of biting insects among host X and host Y will eventually stabilise when host defensiveness is equal between the two hosts. This point of equilibrium is termed the ideal free distribution, and at this point a vector that changes host will achieve no benefit, as the increased vector density will raise the level of host defensiveness, reducing feeding success. The ideal free distribution (IFD) predicts that small differences in host defensiveness will drive large differences in host choice, leading to a highly skewed distribution of vectors among hosts. The IFD for host preference with a two host system is given by:

$$h_x = \frac{N_{x,i}.Hx}{N_{x,i}.Hx + \left[\left(\dfrac{Q_y}{Q_x} \right)^{1/m_y} . N_{x,i}^{m_x/m_y} . H_y \right]} \tag{7.10}$$

where h_x is the proportion of vectors feeding on host x, $N_{x,i}$ is the number of vectors on an individual, I, of host x; H_x and H_y are the numbers of hosts x and y; Q_x and Q_y are the innate defensiveness of the hosts (the probability that a single biting insect will obtain a bloodmeal when feeding in isolation); and m_x and m_y determine the rate at which defensiveness of hosts x and y increases with biting density. Kelly (2001) used data from Fujito et al. (1971) to illustrate how the biting preference of *Culex tritaeniorhynchus* would rapidly shift from a cow, which has a low innate defensiveness but rapid increase in defensiveness with biting rate and a pig, which has a higher innate defensiveness but a lower rate of increase with increasing biting intensity.

Comtois and Berteaux (2005) reported the results of studies on the effects of biting flies (blackflies and mosquitoes) on the behaviour and habitat use by American porcupines (*Erethizon dorsatum*). The frequency of five defensive movements was counted by direct observation of individual porcupines: head shaking, body shaking, leg shaking, scratching, and other body movements apparently showing discomfort. These movements were compared with measures of mosquito and blackfly abundance obtained from 5-min counts of insects landing on a blue cushion (30 cm × 33 cm) placed on the observer's legs, who sat within 30 m of the porcupine under observation. The numbers of blackfly bites received by individual porcupines wwere directly determined by immobilizing the animal and counting the number of bites contained within circles drawn with a nontoxic pen on three sites on the abdomen: around the left teat (50.5 cm^2), at the centre of the abdomen (50.5 cm^2), and inside the right hind leg (23.0 cm^2). The number of bites observed on the teats, abdomen, and hind legs of porcupines increased with abundance of blackflies as measured in the week prior to the counts of bites on porcupines. Porcupines generally performed more repelling movements when insects were present than when they were absent, but the effect of insect presence was only significant for a small portion of repelling-movement categories, that is scratching (mosquitoes: $F_{[9,123]} = 1.450$, $P = 0.016$; blackflies: $F_{[9,123]} = 2.359$, $P < 0.001$) and head shaking (blackflies: $F_{[9,123]} = 1.683$, $P = 0.044$). The authors concluded that porcupines are relatively tolerant of biting insects and this could be largely dependent on their relatively large size. Although differences were observed among microhabitats in terms of biting pressure, porcupines did not appear to change their use of available microhabitats when insects were most abundant.

The observed positive association between blood-meal volume and the size of egg clutches suggests that mosquitoes should benefit from behaviours that maximise bloodmeal volume, but these behaviours should be limited by the risks associated with host defensive reactions. Anderson and Roitberg (1999) used stochastic simulation and dynamic programming models to examine the trade-offs between maximising bloodmeal volume while avoiding being killed by the host. Two models were developed, comparing a population consisting of persistent feeders with a population of non-persistent individuals that did not make a second attempt to feed if disturbed. Fitness was determined as the average total number of eggs laid per female during her lifetime. Simulations revealed that persistent mosquitoes tended to exhibit higher fitness under conditions of low probability of mortality due to host defensive reaction. When the probability of this type of mortality was high, the nonpersistent strategy resulted in higher fitness. A dynamic programming model, which included allowances for variation in individual behaviour, depending on circumstances, mirrored the results of the simulations and revealed that the advantage gained in taking a second meal increases as a function of the probability of surviving that meal, up to about 25%. Size of the initial meal was also shown to influence the adaptive value of making a subsequent feeding attempt, such that it was only worthwhile when probability of survival during feeding is high and host availability is low.

Gonotrophic dissociation and gonotrophic discordance

Swellengrebel (1929) and De Buck and Swellengrebel (1934) were the first to observe that female *Anopheles* sometimes took blood meals without ovipositing and this was termed gonotrophic dissociation in the case of females about to enter hibernation and gonotrophic discordance where more than one blood meal was required to complete a single gonotrophic cycle. Field data for several other species of *Anopheles* (Amerasinghe and Amerasinghe 1999; Gillies and De Meillon 1968; Resisen and Aslamkhan 1976) and also some culicines (Bang and Reeves 1942; Edman and Downe 1964; Scott et al. 1993*b*; Trpis and Häusermann 1986) have confirmed these observations. Briegel and Hörler (1993) investigated multiple blood-feeding in *Anopheles albimanus, Anopheles gambiae* and *Anopheles stephensi* in the laboratory and found that all three species readily took successive blood meals at 6, 12 or 24 h during a single oviposition cycle, even after feeding to repletion on each previous occasion. The initial blood meal is used primarily for the synthesis of maternal reserves, with subsequent meals being used to increase fecundity through synthesis of additional yolk protein and lipid. Mixed meals do not, at least in *Anopheles*, necessarily indicate interrupted feeding as a result of host defensive behaviour, rather they are a well-adapted strategy to improve female survival through increasing reserves and to increase fecundity. Briegel (2003) in a review of the physiological basis of mosquito ecology, goes as far as to suggest that the concept of rigidly defined gonotrophic cycles does not apply to the Anophelini, only the Culicini.

Beier (1996) observed that gonotrophic discordance was common amongst both *Anopheles gambiae* s.l. and *Anopheles funestus* in western Kenya.

Overall, among blood-fed mosquitoes, 14.6% of 1064 *Anopheles gambiae* s.l. and 16.0% of 941 *Anopheles funestus* in Kisian and 12.7% of 2166 *Anopheles gambiae* s.l. and 18.2% of 1021 *Anopheles funestus* in Saradidi had oocytes in Stage V. Only 60.0% of 1287 *Anopheles gambiae* s.l. and 60.0% of *Anopheles funestus* oviposited after a single blood meal when held in individual oviposition chambers and provided with access to sugar.

Almirón et al. (2000) reported that 14% and 4% of female *Ochlerotatus albifasciatus* collected in human landing collections in Argentina in 1996 and 1997 respectively had ovarian follicles in Christophers' stages III–IV, indicating gonotrophic discordance.

Interrupted feeding

Klowden and Lea (1978) cited several studies that indicated mosquitoes may take multiple blood-meals during a single gonotrophic cycle, while Magnarelli (1977) showed that 4.9% of mosquitoes caught at human bait in Connecticut, USA already contained small amounts of blood. Cupp and Stokes (1976) found that 12.5% of blood-meals identified from *Culex salinarius* were from a mixture of hosts. In Colorado, USA Mitchell and Millian (1981) found that 1.5% of *Culex tarsalis* caught in animal-baited lard-can traps already had some blood. Canyon et al. (1998) studied the role of host activity, host defensive behaviour and biting persistence on multiple host-feeding on human volunteers by *Aedes aegypti* in the laboratory. They found that the median number of hosts bitten or landed upon per mosquito was constant and independent of host activity, host defensive behaviour and biting persistence.

The proportion of blood-meals that are taken from more than one host depends on the probability of two or more hosts being selected by a hungry mosquito and the probability of the blood-meal being interrupted. Burkot et al. (1988) studied mixed feeding by species in the *Anopheles punctulatus* complex in Papua New Guinea. They elaborated on the model proposed by Boreham and Garrett-Jones (1973) for estimating the proportion of cryptic mixed blood-meals. The proportion of cryptic mixed blood-meals can be derived from the proportion of unmixed and patent mixed meals and the probability of feeding on these hosts (Boreham and Lenahan 1976; Boreham et al. 1978, 1979; Bryan and Smalley 1978; Port et al. 1980). It can also be measured more directly by ABO blood groups or by serum protein haptoglobins, because from this the probability of a meal on a host species being interrupted can be estimated (Boreham and Lenahan 1976; Boreham et al. 1978, 1979; Bryan and Smalley 1978). For simplicity

Burkot et al. (1988) considered just a two-host situation, a human and a non-human host. If it is assumed that blood-feeding is interrupted just once, then the proportion of blood-meals that are patent mixed will be

$$Q(1-Q)(l_H + l_N) \tag{7.11}$$

in which Q = probability of humans being the host; I_H = probability of a human feed being interrupted; I_N = probability of a non-human feed being interrupted. The proportion of patent mixed feeds increases as the probability of interruption increases. However, the greatest proportion of patent

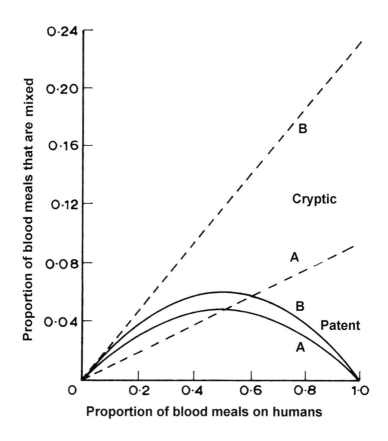

Fig. 7.1. Influence of interrupted feeding on the proportion of patent- and cryptic-mixed blood meals. Solid lines show proportion of all blood meals that are patent-mixed meals; dashed lines show the proportion of human blood meals that are cryptic-mixed meals (Burkot et al. 1988)

mixed feeds will be when there is an equal likelihood of the mosquito selecting a human or non-human host (i.e. when $Q = 0.5$) (Fig. 7.1).

The proportion of all blood-meals that are cryptic mixed on humans will be Q^2I_H, and dividing by Q gives the proportion of human feeds that are cryptic mixed (QI_H). That is the proportion of cryptic mixed feeds on humans increases linearly with Q for a given I_H (Fig. 7.1). Now, although the proportion of both cryptic and patent mixed meals increases as the probability of interruption increases, the overall proportion of meals taken on humans remains unchanged, although the proportion of mixed meals increases. What actually changes is the absolute number of blood-meals taken. Consequently if none of the feeds is interrupted then the HBI is a direct estimate of Q.

If, however, interrupted feeds occur on both human and non-human hosts, then the HBI (as usually measured as the total numbers of meals positive for human blood) will overestimate Q. This estimate, however, can be corrected as follows:

Q = proportion of feeds with only human blood + $[I_H / (I_H + I_N)] \times$ total proportion of patent mixed human feeds.

So, if $I_H = I_N$ then Q is obtained by adding half of the proportion of patent mixed human feeds to the proportion of only human feeds.

Now, if the proportion of mixed feeds is measured, then the probability of a feed being interrupted can be estimated. But we must know either I_H or I_N to estimate the other, or else assume that the two are equal (as Burkot et al. 1988 did in their study) and so obtain a common estimate that is applicable to both. If we assume the latter; then from eqn (1) we have

$$I_H = I_N = \text{proportion of patent mixed blood-meals}/2Q(1 - Q) \qquad (7.12)$$

Scott et al. (1993a) used the above methods to calculate the probability of *Aedes aegypti* taking two or more meals within a single gonotrophic cycle in a rural village in Thailand. The probability fluctuated between 0.0 in September, October and December 1990 and 1.0 in July 1990 and February 1991, although the authors noted that sample sizes were low in September–December. Overall, the probability of a multiple meal was 0.450.

Forage ratios and feeding index

In most studies on host preferences, as determined by identification of mosquito blood-meals, there is little or no information on the numbers of available hosts present in an area. Murphey et al. (1967) assessed host preferences of different mosquito species for bait animals in small Magoon traps by using a host attractive index, obtained by multiplying the percentage that fed on the baits by the numbers caught in the traps and dividing by 100. An index of 8 or more was considered to reflect an attractive host, 3–7 a moderately attractive one and below this a poorly attractive host. This is a very simplified approach of comparing relative host attractiveness. It overlooks the fact that some species may find it difficult to enter the traps although they may be strongly attracted to the bait. Furthermore, the percentage feeding on a host confined within a trap may have little bearing on the proportion successfully obtaining a blood-meal from the same animal under natural conditions. In order to better quantify feeding patterns, Hess et al. (1968) introduced the 'forage ratio' into mosquito studies. This ratio is obtained by dividing the percentage of blood-fed mosquitoes (caught from natural resting places) that have been shown by serological methods to have fed on a particular animal species or group of species by the percentage it comprises of the total available population of hosts in the area, i.e. availability of hosts is taken into consideration. A forage ratio of approximately 1 indicates neither preference nor avoidance of the indicated host, whereas ratios significantly greater than 1 indicate selective host feeding. The main difficulty of trying to apply this technique is that it is rarely possible to obtain reliable estimates of the relative proportions of the different hosts in an area. Furthermore, there may be seasonal changes in the proportions of available hosts. This approach was used by Hayes et al. (1973) to investigate possible seasonal shifts in feeding patterns in Texas, USA. In studying host preferences of *Psorophora columbiae* and *Anopheles crucians* Kuntz et al. (1982) carried out a census of the proportions of available hosts in a rice field area in Texas, and applied the forage ratio technique to results of precipitin tests performed on wild-caught adults.

In southern Mexico, Fernandez-Salas et al. (1993) used an indirect ELISA technique (Beier et al. 1988) to determine the sources of blood meals in indoor and outdoor resting collections of *Anopheles pseudopunctipennis*. The proportion of females that had fed on human blood was higher among mosquitoes captured resting indoors, compared with those captured from artificial pit-shelters and other outdoor resting sites. Weighted mean Human Blood Index values of 34.0% and 29.5% were obtained in 1990 and 1991 respectively. Of all available hosts, dogs were the second most frequent source of blood meals for *Anopheles pseudopunctipennis* collected resting outdoors. Humans were three times more abundant than domestic animals (excluding chickens), but the forage ratio was less than 1.0 in both years. Forage ratios in excess of 1.0 were calculated for dogs, pigs, and horses.

In Central America, in a multi-country study (Pan American Health Organization 1996) on the biology and ecology of *Anopheles albimanus*, censuses of humans and domestic animals were conducted in each study site and biomass of each type of host was estimated using weight-for-age tables for humans and by on-the-spot estimates for domestic animals. Blood meal sources, as determined by ELISA were then related to the numbers, proportions, and biomass of each host present in the study area. Human biomass was dominant in Guatemala, Costa Rica and Nicaragua, while in El Salvador and Panama, bovines predominated in terms of biomass. Sufficient numbers of blood meals to permit statistical analysis of host preference were only obtained in Guatemala. The proportion of blood meals from humans was 54.5% and humans comprised 67% of the biomass. By contrast, 13.1% of blood meals were from bovines, although these animals represented only 7.5% of the biomass, indicating a preference for feeding on bovines. In Egypt Beier et al. (1987), Kenawy et al. (1990) and Zimmerman et al. (1988) used the forage ratio to study feeding preferences of various mosquitoes, including sometimes those caught as blood-feds in light-traps. Kumari et al. (1993) calculated forage ratios for *Anopheles sundaicus* on Car Nicobar island in the Bay of Bengal as follows: human/pig = 2.42, human/goat = 0.37, human/bovine 5.77, human/domestic animal = 0.09, indicating that preferential feeding on pigs and bovines was occurring. Loyola et al. (1993) determined the forage ratio of Hess et al. (1968) for *Anopheles albimanus* on the coastal plains of Chiapas, southern Mexico (Table 7.2). Chickens accounted for more than 50% of all hosts, but only 0.01% of blood meals were positive for chicken and so this host was removed from the analysis. Forage ratios in excess of 1, indicating host preference, were obtained for bovine, equine, porcine and canine hosts.

Ponlawat and Harrington (2005) used the forage ratio approach to study the preferred hosts of *Aedes aegypti* and *Aedes albopictus* in Thailand. Forage ratios for *Aedes aegypti* of 3.4, 0.6, and 0.2 were obtained for humans, dogs, and pigs from one village (Nakhon Sawan), but in a second village (Nakhon Ratchasima) a value of 5 was obtained for pigs and 2.9 for humans, however, the authors were unable to identify any specific cause for these differences.

It is often difficult, or more usually nearly impossible, to enumerate the numbers of possible hosts in an area, and the failure to take into consideration their ecology and availability to mosquitoes. Edman (1971) pointed out these as well as other deficiencies in trying to use the forage ratio. Attempting to overcome these difficulties Kay et al. (1979) proposed the 'Feeding Index' (*FI*). This is defined as the proportion of feeds on one

Table 7.2. Forage ratios and human blood indices for female *Anopheles albimanus* resting indoors and outdoors in three ecological zones on the coastal plains of Chiapas, Mexico. (Loyola et al. 1993)

Host	Proportion blood meals	Proportion total hosts	Forage ratio	HBI	HBI$_{adj}$
Banana plantation					
Human	0.25	0.54	0.46	0.21	0.19
Bovine	0.43	0.11	3.95	—	—
Equine	0.08	0.02	3.61	—	—
Porcine	0.17	0.18	0.99	—	—
Canine	0.06	0.15	0.41	—	—
Total	242	1153	—	—	—
Mixed agriculture					
Human	0.14	0.45	0.31	0.11	0.13
Bovine	0.35	0.19	1.78	—	—
Equine	0.14	0.06	2.43	—	—
Porcine	0.22	0.16	1.37	—	—
Canine	0.16	0.14	1.10	—	—
Total	725	2772			
Cattle-raising					
Human	0.16	0.49	0.32	0.12	0.12
Bovine	0.45	0.21	2.10	—	—
Equine	0.10	0.04	2.26	—	—
Porcine	0.15	0.12	1.19	—	—
Canine	0.15	0.13	1.15	—	—
Total	826	1174			

host with respect to another divided by the expected proportion of feeds on these two hosts based on factors affecting feeding.

These factors include host abundance and size, their temporal and spatial concurrence with the mosquito species, and the mosquitoes' feeding success. Thus

$$FI = \frac{Ne / Ne^1}{Ef / Ef^1} \tag{7.13}$$

where Ne = numbers of feeds identified from host 1; Ne^1 = number of identified feeds from host 2; Ef = expected proportion of feeds on host 1; and Ef^1 = expected proportion of feeds on host 2. An index of 1 indicates equal feeding on both host species being compared, while smaller or larger values indicate a decrease or increase of feeds on host 1 compared to host 2.

As an example, Kay et al. (1979) presented data on *Culex annulirostris* feeding on dogs and fowl in an Australian village. The estimated dog population was 100 and the fowl population 80, thus the expected ratio of feeds based just on their abundance would be 1:0.8, or 1.25. The actual analysis of blood-meals showed 35 dog feeds and 8 fowl feeds, that is observed feeding (Ne/Ne^1) is 35/8 = 4.38. Now, ignoring all other environmental factors the feeding index is calculated as 4.38/1.25 = 3.50, i.e. greater feeding on dogs than fowl. However, ecological and behavioural studies were able to measure the concurrence of these two hosts and *Culex annulirostris*. During the feeding times of the mosquito the proportions of dogs outdoors and indoors were 0.9 and 0.1, the proportions for fowls were 1.0 and 0; and finally for *Culex annulirostris* 0.92 of the population were feeding outdoors and 0.08 feeding indoors. Thus the concurrence for dogs to fowls is

$$\frac{(0.9 \times 0.92) + (0.1 \times 0.08)}{(1.0 \times 0.92)} = 0.91 \tag{7.14}$$

showing that *Culex annulirostris* is more likely to encounter fowl than dogs. In animal-baited stable-trap experiments 96% of *Culex annulirostris* fed on dogs and 83% on fowl. So there is the following adjustment to be made to account for feeding success, 96/83 = 1.16 for dogs relative to fowl. Finally, the fact that a dog is about five times the weight of a fowl is taken into consideration. So, the expected comparative feeding rates for dogs with respect to fowl would be

$$1.25 \times 0.91 \times 1.16 \times 5.0 = 6.59 \tag{7.15}$$

So the true feeding index (FI) = 4.38/6.59 = 0.66. A very different value from the crude feeding index of 3.5 obtained when environmental factors were ignored. Such calculations require much information on host availability, host size, and concurrence as well as the feeding success of mosquitoes on different hosts, and moreover the estimation of these parameters will likely be inaccurate. These difficulties were recognised by Kay et al. (1979), who admitted that factors determining host selection were so complex that perhaps any such above analysis was of limited value. However, they believed that they had provided a framework on which a better understanding of mosquito host-feeding patterns might be built, and Kay et al. (1985) used the feeding index in later studies on *Culex annulirostris* and other species. Burkot et al. (1988) used it to study feeding preferences of the *Anopheles punctulatus* complex in Papua New Guinea, and Renshaw et al. (1994) calculated the feeding index for *Ochlerotatus cantans* feeding on cattle, horses and sheep in northern England. An equal preference for feeding on cattle and sheep was observed in both 1989 and 1990. In 1989 mosquitoes showed a strong preference for feeding on horses, compared with cattle and sheep, however this pattern was reversed in 1990. It was postulated that the strong preference for feeding from a horse in 1989 was due to the fact that one of the two horses present in the field at that time was sick and displayed almost no defensive behaviour against mosquito bites. The sick horse died later that year, leaving only a healthy animal that displayed strong host avoidance behaviours.

Bockarie et al. (1995) estimated the feeding index for *Anopheles gambiae* s.s. in Sierra Leone using the method of Kay et al. (1979). Blood meal ELISAs were only positive for humans (163/165) and dogs (5/165) and the populations of humans and dogs in the study area were 48 and 5 respectively. The calculated feeding index was therefore 32.6/9.6 = 3.2, indicating positive selection of humans as hosts.

Rubio-Palis et al. (1994) used an ELISA method incorporating a TMB peroxidase substrate to investigate host choice of anopheline mosquitoes in western Venezuela and calculated both the Human Blood Index and the Feeding Index. A census of domestic animals within a 2 km radius of experimental huts in three villages was conducted using a household questionnaire. A total of 2497 mosquitoes were collected resting among vegetation in the vicinity of the experimental huts and subjected to ELISA testing. The majority of blood meals were from bovine sources. Up to 12.5% of blood meals were from more than one host. HBI values varied among species of mosquito and among the three study villages. The highest HBI observed was 57.1% in *Anopheles oswaldoi*, the lowest 15.2% in *Anopheles albitarsis*. Human-bovine FIs varied among mosquito species and study villages. The FI for *Anopheles nuneztovari* ranged from 0.6 to 2.6, for *Anopheles albitarsis* from 0.21 to 5.5, for *Anopheles triannulatus* from 0.8 to 5.3, and for *Anopheles oswaldoi* from 1.2 to 3.3. In the village with the highest cattle populations, the FI for all four species exceeded 1.0, indicating that there was preferential feeding on humans relative to bovines. However, in the village of Jabilos there were relatively fewer cattle, but the FIs for *Anopheles nuneztovari, Anopheles albitarsis* and *Anopheles triannulatus* were all below 1.0 indicating a preference for feeding on cattle. The authors suggested that proximity of cattle enclosures to human habitations could have explained these contrasting results, but this study emphasises some of the difficulties in interpreting blood meal source data.

Gomes et al. (2003) identified by precipitin test the blood meal sources of 1092 mosquitoes collected in Brazil and calculated the feeding index of Kay et al. (1979). A census of 249 properties revealed that 649 people had resided in the households the previous night. Additionally, 318 dogs, 85 horses, 120 cows, and 922 chickens were recorded. Feeding indices were as follows: *Aedes albopictus* 41.90 (human/chicken), 1.45 (human/cattle), 9.64 (human/dog), 3.86 (human/horse); *Ochlerotatus scapularis* 5.01 (human/chicken), 0.2 (human/chicken), 0.24 (human/cattle), 0.11 (human/dog), 2.37 (human/horse) *Culex nigripalpus* 1.35 (human/chicken), 0.80 (human/cattle), 1.00 (human/dog), 0.11 (human/horse); and *Culex quinquefasciatus* 3.57 (human/chicken), 13.70 (human/cattle), 1.31 (human/dog), 4.60 (human/horse).

In Belize, Grieco et al. (2002) calculated the feeding index for several species of *Anopheles* captured by oral aspirator, a 12V DC backpack aspirator and a truck trap. Sources of bloodmeals were determined using the ELISA method of Chow et al. (1993). The standard feeding index equation was modified to take account of differences in body size of hosts by multiplying the denominator by the proportional difference in body size between pairs of hosts. Cattle were taken to be 5 times the size of a human, a pig was taken to be the same size as a human, and a dog was taken to be 1/3 the size of a human. A total of 1298 blood-engorged *Anopheles* were collected and tested. Feeding indices were calculated separately for each collection method and for each species. For *Anopheles vestitipennis*, for example, the human:cow feeding indices were 24.8 unadjusted and 124 adjusted for body size from outdoor oral aspiration collections, 3.4 unadjusted and 17.2 adjusted for the backpack aspirator collections and 0.9 unadjusted and 4.6 adjusted from vehicle-mounted trap collections. These results indicate a clear preference for feeding on humans over cows in this species.

The feeding index for *Aedes aegypti* in a rural village in Thailand was calculated by Scott et al. (1993*a*) and demonstrated a clear preference for feeding on humans (FI range = 10.31 to 32.27).

Wekesa et al. (1997*b*) calculated the feeding index for *Anopheles freeborni* and *Culex tarsalis* in four habitats in California, USA. Both species preferentially fed on leporids relative to humans (FI ≤ 0.1) in all four habitats (riparian, rice, pasture, mixed). *Anopheles freeborni* females fed preferentially on bovids relative to humans (FI ≤ 0.7) in all habitats except riparian (FI = 1.3). By contrast, *Culex tarsalis* fed preferentially on humans relative to bovids (FI ≥ 1.2).

With the development of vertebrate-specific polymerase chain reaction-heteroduplex analysis (Lee et al. 2002) it has become possible to identify the source of avian blood meals obtained from mosquitoes to the species level. Hassan et al. (2003) used this technique to study the host preferences of vectors of eastern equine encephalomyelitis (EEE) virus in Macon County, Alabama, USA. Mosquito collections were made from resting boxes and from natural resting sites twice per week between 0830 and

1000 h from May to September. Feeding indices were calculated using the methods of Kay et al. (1979) with populations of each bird species being determined from a series of 12 point count censuses. Bird abundance data were corrected for biomass and surface area using published figures on the mass of bird species and defining surface area as body mass raised to the two-thirds power. 447 blood-fed mosquitoes were collected, of which *Culex erraticus* was the most common, comprising 63% of the total. 198 of 264 blood samples produced detectable PCR products. For *Aedes vexans*, 25% of blood

Table 7.3. Bird species over-represented or under-represented in blood meals when compared to the observed abundance of each species at the Tuskegee National Forest (Hassan et al. 2003)

Species	Count feeding index	Mass feeding index	Area feeding index
Carolina chickadee	4.70 (P = 0.00001)	51.56 (P = 0.00001)	17.48 (P = 0.00001)
Northern mocking-bird	>32.81 (P = 0.00001)	>71.44 (P = 0.00001)	>40.99 (P = 0.00001)
Yellow-crowned night heron	255.94 (P = 0.00001)	16.97 (P = 0.00001)	22.23 (P = 0.00001)
Wild turkey	>19.65 (P = 0.00001)	>0.29 (P = 0.00001)	>0.93 (NS)
Orchard oriole	3.72 (NS)	16.4 (P = 0.00001)	7.46 (P = 0.009)
Great blue heron	13.98 (P = 0.00001)	0.47 (P = 0.00001)	1.14 (NS)
Acadian flycatcher	<0.19 (P = 0.026)	<1.66 (NS)	<0.61 (NS)
Northern cardinal	0.34 (P = 0.031)	0.87 (NS)	0.47 (NS)
White-eyed vireo	<0.19 (P = 0.036)	<1.88 (NS)	<0.66 (NS)
Red-bellied wood-pecker	<0.20 (P = 0.039)	<0.36 (NS)	<0.22 (NS)
American crow	<0.22 (NS)	<0.04 (P = 0.00001)	<0.05 (P = 0.00001)
Great egret	<1.83 (NS)	<0.21 (P = 0.02)	<0.32 (NS)
Pileated wood-pecker	<0.60 (NS)	<0.22 (P = 0.024)	<0.22 (NS)

Feeding indices and statistical significance were calculated from the raw point count data and the point count data adjusted for biomass and surface area. Only species found to be significantly over- or under-represented in one or more of the analyses are shown. NS = not significant.

meals were positive for avian hosts, whilst this figure was 48% in *Culex erraticus* and 33% in *Coquillettidia perturbans*. 83 blood samples were derived from a single avian host and there were 18 different PCR product patterns, indicating 18 species of avian host and 14 of these could be matched to published DNA sequences. The most common hosts were the Carolina chickadee (*Poecile carolinensis*) comprising 19% of avian feeds from *Culex erraticus* and the yellow-crowned night heron (*Nyctanassa violacea*) from which 27% of *Culex erraticus* avian feeds derived. The yellow-crowned night heron was significantly over-represented in the feeding indices adjusted for bird population density, biomass, and surface area. This species represented less than 1% of the adult birds recorded at the site, but the calculated feeding indices were 255.94 (population abundance); 16.97 (biomass) and 22.23 (surface area) indicating a very strong preference for blood meals from this host.

Vectorial capacity

An important concept in the epidemiology of disease is the basic reproduction rate (R_0), which is the average number of secondary cases of a disease (e.g. malaria) arising from each primary infection in a defined population of susceptible individual hosts. In other words R_0 represents the maximum reproductive rate per generation, leaving aside complications such as host immunity and super-infection. If $R_0 > 1$ the disease is maintained, with the level of transmission depending on the size of R_0 but if $R_0 < 1$ the disease decreases and will eventually disappear from the population. Vectorial capacity is the entomological component of the basic reproduction rate of malaria. It is the average number of inoculations from a single case of malaria in unit time, usually a day, that the vector population transmits to man, where all vectors biting an infected person become infective. Reducing vectorial capacity reduces R_0. The usual formula for vectorial capacity (C), in terms of a daily rate, as derived by Garrett-Jones (1964) is

$$C = \frac{ma^2 P^n}{-\log_e P} \qquad (7.16)$$

but a parameter V (sometimes written as c or b) can be inserted to describe the inability of all mosquitoes that become infected to become infective, in other words V is the proportion of mosquitoes with sporozoites actually infective to man, thus

$$C = \frac{ma^2 V P^n}{-\log_e P} \qquad (7.17)$$

where C = new infections disseminated per person per day by each mosquito, ma = the number of bites/man/day, a = the proportion of females feeding on man divided by the duration of the gonotrophic cycle in days. a is multiplied by ma because re-feeding is necessary for transmission. P = probability of daily survival, estimated vertically (sometimes called cross-sectional as it is based on acquisition of infection with age) from the population age structure if the duration of the gonotrophic cycle is known, or horizontally (longitudinal, based on time) from the daily loss rate of identified cohorts over time, and n = time from infection to infectivity in days and is usually estimated from the ambient temperature using a degree-day relationship. Thus, P^n = probability of a mosquito surviving to become infective, and the expected duration of life in days = $1/-\log_e P$.

Molineaux et al. (1978) rewrote the definition of vectorial capacity as

$$C = ma(P/F)e^{-n/E}E \qquad (7.18)$$

where ma = number of mosquito bites/man/day, i.e. the biting rate, P = the proportion of blood-meals taken on man, F = the interval between feeding and refeeding in days, n = duration of extrinsic incubation period (e.g. of malaria) in the mosquito, and E = life expectancy of the mosquito calculated from $1/-\log_e P$. The term ma (P/F), like ma^2 defines mosquito–man contact, while $e^{-n/E}E$ or its equivalent $P^n/(-\log_e P)$ is the expectation of infective life, which is compounded in terms of probability of survival to a later age $(x + n)$ and the life expectancy of survivors at that age.

In Pakistan Reisen and Boreham (1982) estimated malaria vectorial capacity by this modified formula of Molineaux et al. (1978) where ma was calculated as $(P_t a)/(gc/H)$; where P_t = daily population size of female vectors—estimated in this instance by the Lincoln Index with Bailey's (1952) correction, a = proportion of blood-meals positive for human blood amongst those tested from representative collection sites (note, confusion can arise here because a is being used to mean something different to a in the Garrett-Jones' formula), gc = duration of gonotrophic cycle in days; H = human population—determined in this instance by a household census, n = duration of sporogony of *Plasmodium vivax* and *Plasmodium falciparum*—calculated by temperature summation as described by Detinova (1962), and E = life expectancy of the female vector, that is where P = daily female vector survivorship—estimated in this instance by the regression of numbers of ovarian dilatations against age in days.

Because of a very low biting rate on humans during the monsoon season and relatively low life expectancy, the pooled vectorial capacity for *Anopheles culicifacies* and *Anopheles stephensi* was less than 1.34×10^{-2} for *Plasmodium vivax* and 8.58×10^{-3} for *Plasmodium falciparum*, very low figures when compared to estimates (0.006–22.25) for the *Anopheles gambiae* complex and *Anopheles funestus* in Africa (Reisen and Boreham 1982).

In deforested areas of western Kenya, Afrane et al. (2005) observed that ambient temperatures tended to be 0.5°C higher than in neighbouring forested areas and this was associated with a decrease in the length of the first and second gonotrophic cycles of *Anopheles gambiae* of 1.7 d (59%) and 0.9 d (43%) in the dry season. During the rainy season, the average indoor temperature of houses located in the deforested area was 1.2°C higher than in houses in the forested area and consequently the duration of the first and second gonotrophic cycles was shortened by 1.5 d (17%) and 1.4 d (27%), respectively, at the deforested highland site. These reductions in gonotrophic cycle length imply increased daily biting frequency and thus increased vectorial capacity.

In Sichian Province, China entomological surveys showed that the vectorial capacity of *Anopheles lesteri anthropophagus* was 0.654, and for *Anopheles sinensis* 0.019, similarly the entomological inoculation rate (see below) calculated as the product of the man-biting rate and the sporozoite rate was 0.003367 for *Anopheles lesteri anthropophagus* and 0.000185 for *Anopheles sinensis*. Using Krafsur and Armstrong's (1978) formula for estimating risk of infection (R) defined as the probability of receiving one or more sporozoite inoculations per unit time, then $R = 1 - e^{-sn/t}$, where s = sporozoite rate, n = number of bites in t days, so sn/t is the entomological inoculation rate, the probability (risk = R) of inoculation with sporozoites was 0.1829 and 0.0110, respectively for both vectors, that is 94.3% of local malaria transmission is by *Anopheles lesteri anthropophagus* and 5.7% by *Anopheles sinensis* (Liu et al. 1986).

Almeida et al. (2005) calculated vectorial capacity of *Aedes albopictus* in relation to dengue viruses on the island of Macao, China, using the formula of Garrett-Jones (1964). Both field-collected data, and data on feeding frequency and gonotrophic cycle duration obtained from a laboratory colony were used. The value of C was calculated using daily survival rate based on the parous rate (Davidson, 1954), and on survival observed in the laboratory. The corresponding values of vectorial capacity calculated using these two methods were 482 and 880 new infections per respectively in the wet season, based on a biting rate of 314 per person per day. In the dry season, vectorial capacity was calculated as 144 and 263, using the two estimates of daily survival and a human biting rate of 94 per person per day.

Calculations of vectorial capacity are usually based on random biting and the man-biting rate (ε^2 V/B) is usually based on the average biting rate per person among a team of B bait collectors. The resulting estimate of vectorial capacity ($\varepsilon^2 V/\delta H$) is at best proportional to the true vectorial capacity. When, however, there is non-random biting by mosquitoes on hosts, then, as shown by Dye and Hasibeder (1986), the vectorial capacity is likely greater than when calculated on the assumption of random biting. Taking into consideration the heterogeneity factor (summation part of equation below), the definition of vectorial capacity in their model can be written as

$$C = \frac{\varepsilon^2 V}{\delta H} \sum_i \frac{\gamma_i^2}{h_i} \qquad (7.19)$$

where V and H = the numbers of vectors and hosts (man) in the area, of which a proportion of hosts (h_i) reside in area i and are bitten by a proportion (γ_i) of all mosquitoes, ε = number of bites taken on man by one mosquito per day, and $1/\delta$ = expectation of infective mosquito life. Woolhouse et al. (1997) further considered the effects of non-random host-vector contact and observed that typically, 20% of the host population contributes at least 80% of the net transmission potential, as measured by the basic reproduction number, R_0 Heterogeneities in vector-host contact rates lead to increases in the value of R_0, by 2 to 4 times for vector-borne infections. In reality, however, it is usually impossible to estimate h_i and γ_i, and so the cruder estimate of vectorial capacity (Garrett-Jones 1964) has to be used, which, however, is likely to change proportionally with the true but unknown vectorial capacity. Dye and Hasibeder (1986) showed that when this crude estimate of vectorial capacity is reduced (e.g. by vector control or chemotherapy) then calculations based on random mixing of biting on people will at first produce a conservative estimate of the success of any control programme, but when transmission is much reduced, predictions on reduced transmission rates will be over optimistic. For further explanations see Dye and Hasibeder (1986) and Hasibeder and Dye (1988).

In the laboratory, Gary and Foster (2001) demonstrated that *Anopheles gambiae* s.s. females that fed on blood and sugar lived longer and had higher daily fecundities than those fed on blood alone. However, total fecundity and intrinsic rate of increase were not decreased in the absence of sugar. Moreover, biting frequency tended to increase among females that fed only on blood, and theoretically at least could result in increased vectorial capacity. *Anopheles gambiae* is thought to use sugar only facultatively, if at all, in nature.

Despite the relative sophistication of models for estimating vectorial capacity, it appears that this measure is only sometimes marginally better correlated with parasitological data on malaria transmission than the very much simpler measure of challenge based on the man-biting rate (*ma*) (Dye 1986). It therefore seems questionable whether the extra work involved in calculating vectorial capacity is justified, and as Dye (1986) has pointed out 'methods based on untested assumptions are used to estimate parameters with unknown errors'. An interesting paper on measuring the vectorial capacity of simuliid blackflies as vectors of onchocerciasis concluded, that if there was little variation in the proportion of infective flies, then estimating their survival rate for computing the vectorial capacity was redundant, and that the easiest parameter to obtain, namely the biting rate, could account for variations in inoculation rate recorded in different areas and at different times (Dye and Baker 1986).

Entomological Inoculation Rate (EIR)

Vectorial capacity is an indirect method of estimating transmission rate by a vector, a more direct way is to use the entomological inoculation rate (EIR) or infective biting rate (IBR), or as it is often called the inoculation rate (*h*), which is simply the product of the (human) biting rate and the infection rate, or sporozoite rate.

Figure 7.2 illustrates seasonal changes in human biting rate and the EIR for three malaria vectors in Dielmo Senegal (Fontenille et al. 1997).

Malaria transmission intensity in Africa is highly variable with annual EIRs ranging from <1 to >1,000 infective bites per person per year. Beier et al. (1999) examined the relationships between EIRs and Plasmodium falciparum malaria prevalence in sub-Saharan Africa from published data from 31 sites (Fig. 7.3).

The EIR data ranged from 0 to 702 infective bites per person per year, while the malaria prevalence data ranged from 7.0% to 94.5%. The majority of sites fitted a linear $(r^2 = 0.712)$ relationship between malaria prevalence and the logarithm of annual EIR, such that: malaria prevalence (%) = 24.2 (95% confidence interval [CI] = 18.1 – 30.2) $\log_{10}$ (annual EIR) + 24.68

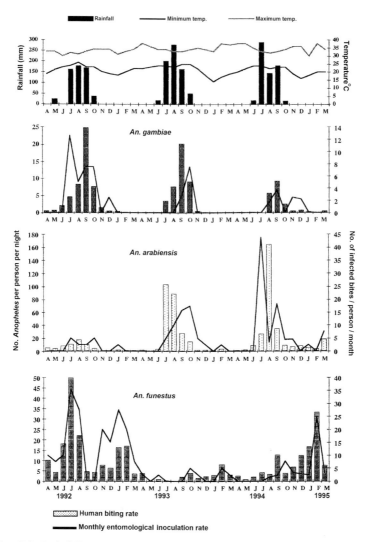

Fig. 7.2. Rainfall, temperature, monthly human biting rate, and monthly entomo-logical inoculation rate calculated using enzyme-linked immunosorbent assay for all three *Plasmodium* species for three main vector species in Dielmo, Senegal (Fontenille et al. 1997)

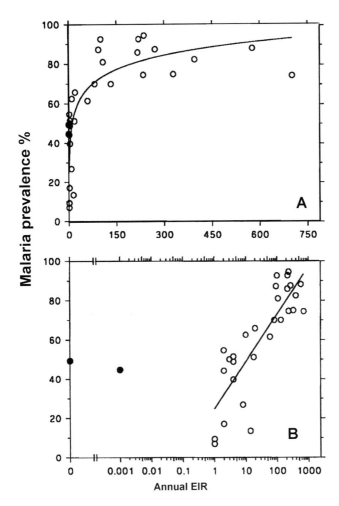

Fig. 7.3. Relationship between annual entomologic inoculation rates (EIRs) and the prevalence of *Plasmodium falciparum* malaria based on data obtained from 31 sites throughout Africa plotted on A, linear and B, logarithmic axes. Two sites with undetectable or scarcely detectable annual EIRs but malaria prevalence > 40% (•) at Mikingirini and Ufouni on the Kenyan coast are clear outliers (Beier et al. 1999)

(95% CI = 14.1–35.3). In addition to large differences in the EIR over large spatial scales across the continent of Africa, EIR and transmission intensity can also vary over smaller spatial scales and also seasonally. Shililu et al. (2003) observed high geographical and seasonal variation in the EIR in Eritrea, a semi-arid country in the horn of Africa. EIRs were calculated from human landing catch data collected one night per month for 24 months and monthly pyrethrum spray catches. Biting density of *Anopheles arabiensis* expressed as bites per person per night varied significantly among study villages and across months. In the highlands and western lowlands peak biting occurred between July and November. Biting rates were lower in coastal villages, where peak biting was observed between December and February. The EIRs differed significantly among the study villages ($F = 2.884$, d.f. = 7, 95, $P = 0.009$). The highest levels of transmission were recorded in Hiletsidi in the Gash Barka zone and in Maiaini in the Debub zone, with mean annual EIR figures of 70.6 and 32.1 infective bites per person, respectively, over the two-year period of the study (Tables 7.3 and 7.4).

Mbogo et al. (2003) observed similarly high variations in the EIR and in malaria transmission among different sites along the Kenyan coast. Houses were sampled by PSC in the afternoons (noon to 1500 h) and at least 15 sites (five from each district) were sampled each month from June 1997 to May 1998. Analysis of 1961 specimens of *Anopheles gambiae* complex mosquitoes by PCR revealed the relative proportions of each species to be 81.9% *Anopheles gambiae* s.s., 12.8% *Anopheles arabiensis*, and 5.3% *Anopheles merus*. The relationships between mosquito density, EIR, and area-specific rainfall were analysed using cross-correlation with various time lags. Analysis of variance was used to determine differences in mosquito abundance and EIR among the sites. Daily human biting rates ranged from 0.25 to 2.45 bites per person per night in Malindi, 0.00 to 1.07 in Kilifi, and 0.14 to 2.7 in Kwale. The mean annual EIR at the 30 sites ranged from 0 to 120 infective bites per person. In an earlier paper, Mbogo et al. (1999) were unable to demonstrate a direct link between vector abundance and the risk of developing severe malaria in a matched case-control study at the Kenyan Coast. Under conditions of low vector densities, as encountered in the study, it appeared that the risk of severe malaria was multifactorial and not strictly associated with transmission intensity.

Lindblade et al. (2000) calculated the EIR for *Anopheles gambiae* s.l. mosquitoes in eight villages situated along the borders of natural and man-made papyrus swamps in the Ugandan highlands. EIR was calculated for each house sampled, an arithmetic mean was computed for each village, and a summary value was calculated for each swamp type as the arithmetic average of all house-captures. The EIR was calculated to be 0 infected bites/person/night (*ib/p/n*) for many villages, due to the scarcity of sporozoite positive mosquitoes. The highest EIR observed (0.0071) was obtained from a village adjacent to a cultivated swamp and the summary EIR was almost an order of magnitude higher in villages located along cultivated swamps than in villages located along natural swamps but the difference was not statistically significant (Wilcoxon rank-sum test, *P* = 0.1307). With an assumed transmission period of only 14 weeks, and assuming that transmission rates remained relatively constant, the calculated EIR corresponded to each individual living adjacent to a cultivated swamp receiving a total of 0.29 infective bites during that period, while those living along natural swamps would have received 0.04 infective bites per person.

Drakeley et al. (2003) estimated the EIR at Ifakara in the Kilombero valley of Tanzania using both the standard method whereby the product of the sporozoite rate and the man-biting rate is determined, as well as a second approach in which the number of positive sporozoite ELISAs is divided by the number of catches and multiplied by 365. This second approach has the advantage that it allows for the calculation of confidence intervals around the resulting EIR estimates, an assumed negative binomial distribution of sporozoite positive mosquitoes. As light-trap plus mosquito nets were used to collect the mosquitoes, rather than human landing-catches, a conversion factor of 1.605 was used to convert the light-trap catches to a human landing-catch equivalent as per the method of Lines et al. (1991). A total of 26 134 *Anopheles gambiae* complex, and 615 *Anopheles funestus* were caught. Polymerase chain reaction (PCR) analysis revealed that 91.5% of the *Anopheles gambiae* were *Anopheles arabiensis*. Estimates of the EIR obtained were 31 using the standard method and 29 infective bites per person per year (95% CI: 19, 44) for the alternative method. Seasonal variation in the EIR was observed, with a maximal EIR of 108 infective bites per person per year (95% CI: 69, 170) in the wet season and a minimum of four (95% CI: 1, 17) in the cool season.

Dossou-Yovo et al. (1998) have also demonstrated the seasonal and relatively small-scale spatial variation of the EIR associated with rice fields and market gardens in Côte d'Ivoire. While the numbers of biting *Anopheles gambiae* were higher in the rice growing areas, transmission was similar, due to the lower proportion of older females and hence infective bites received, compared with the market garden areas.

Doannio et al. (2002) recorded variations in EIR of *Anopheles gambiae* s.s. in Côte d'Ivoire that corresponded broadly with the stage in the rice-growing season. Mean EIR was around one infective bite per person per night, and ranged from 0.2 in February to a high of 6.2 in July, corresponding to the stage just prior to harvesting.

Table 7.4. Mean annual entomologic inoculation rates by *Anopheles arabiensis* over 24 months in eight villages in Eritrea* (Shililu et al. 2003)

Zone	Village	Latitude	Longitude	Elevation (m)	Rainfall (mm)	Year 1 Biting rate (b/p/n) ± SE	Year 1 SR‡ (%) ± SE	Year 1 1/EIR§	Year 1 Annual EIR	Year 2 Biting rate (b/p/n) ± SE	Year 2 SR‡ (%) ± SE	Year 2 1/EIR§	Year 2 Annual EIR
Gash Barka	Hiletsidi	15.07.050	36.39.091	570	286.9	6.40 ± 2.60	0.6 ± 0.003	27.9	13.1	17.2 ± 6.10	2.0 ± 0.003	2.8	128.1
	Dasse	14.55.346	37.29.050	850	196.9	1.00 ± 0.70	3.0 ± 0.021	32.8	11.1	1.0 ± 0.29	1.3 ± 0.009	76.3	4.8
Anseba	Adibosqual	15.41.725	38.38.956	1560	381.2	6.40 ± 4.48	0.2 ± 0.002	64.2	5.7	7.0 ± 2.87	0.2 ± 0.002	71	5.1
	Hagaz	15.41.902	38.16.784	860	241.1	0.04 ± 0.04	0	-	0	0.3 ± 0.21	1.8 ± 0.009	203.1	1.8
Debub	Maiaini	14.48.510	39.05.765	1540	378.1	6.90 ± 3.50	1.3 ± 0.005	11.3	32.2	6.7 ± 3.89	1.3 ± 0.005	11.5	31.9
	Shekae-yamo	14.42.410	38.50.884	1870	657.7	0.50 ± 0.45	0	-	0	0.7 ± 0.61	0	-	0
NRS	Gahtelay	15.31.350	39.09.193	295	51.6	0.02 ± 0.02	0	0	0	0.7 ± 0.31	0	-	0
	Ghinda	15.26.327	39.06.108	850	358.7	0.08 ± 0.06	0	-	0	0.1 ± 0.06	0	-	0

* b/p/n = bites/person/night; SR = sporozoite rate; EIR = entomologic inoculation rate.

† Mean rainfall over the 24 months of the study

‡ Sporozoite rate based on enzyme-linked immunosorbent assay determination of *Plasmodium falciparum* circumsporozoite antigen in the head and thorax of female *Anopheles arabiensis*.

The standard error (SE) for sporozoite rates was calculated according to a binomial distribution.

§ Number of days it would take to receive at least a single infective bite from *Anopheles arabiensis*.

Table 7.5. Monthly variation of *Anopheles arabiensis* human biting rates, sporozoite rates, and entomological inoculation rates in eight villages in Eritrea* (Shililu et al. 2003)

	Hiletsidi			Dasse			Ad-ibosqual			Hagaz			Maiaini			Shekaeyamo			Gahtelay			Ghinda		
	HBR	SR	EIR	HBR	SR	EIR	HBR	SR	EIR	HBR	SR	EIR	HBR	SR	EIR	HBR	SR	EIR	HBR	SR	EIR	HBR	SR	EIR
Jan	1.8	0	0	0.1	0	0	0	0	0	0	0	2.3	0	0	0	0	0	0	1	0	0	0.5	0	0
Feb	1.4	0	0	0.5	0	0	0	0	0	0	0	0	0.1	0	0	0	0	0	1.1	0	0	0.3	0	0
Mar	21.3	0.2	1.4	0	0	0	0	0	0	0	0	0	0.1	0	0	0	0	0	0.4	0	0	0	0	0
Apr	5.1	0	0	0	0	0	0	0	0	0	0	0	0	0	0	0	0	0	0	0	0	0.3	0	0
May	0.3	12.5	0.9	0	0	0	0	0	0	0	0	0	0	0	0	0	0	0	0	0	0	0	0	0
Jun	0	0	0	0	0	0	0	0	0	0	0	0	0	0	0	0	0	0	0	0	0	0	0	0
Jul	13	1.5	5.7	1.8	0	6.4	1.2	0	0	0	0	0	5.6	5.2	8.7	0	0	0	0	0	0	0	0	0
Aug	17.6	4.1	21.9	2.8	3.4	5.5	0	0	0	1.3	0	0	21.1	1.4	8.6	0.1	0	0	0	0	0	0	0	0
Sep	46.3	3.1	43.6	2.6	2.4	21.8	0	0	0	1.9	0.5	0.3	40.6	1.3	16.2	6.5	0	0	0.3	0	0	0	0	0
Oct	28.6	0.6	5	4.9	0	39.8	0.2	0	0	0.1	0	2.7	12.9	0	0	0.3	0	0	0	0	0	0	0	0
Nov	12.9	0	0	1.6	0	5.6	0	0	0	0	0	0	0.8	0	0	0.4	0	0	1	0	0	0.5	0	0
Dec	0.3	0	0	0.4	6.3	0.7	0	0	0	0	0	0	0.5	0	0	0	0	0	3.3	0	0	0.3	0	0

* HBR = human biting rate; SR = sporozoite rate; EIR = entomologic inoculation rate. The sporozoite rate was derived from indoor resting collection and human-landing catches. EIR = infective bites/person/month.

Elissa et al. (1999) observed that the EIR showed considerable variation between rural and suburban areas around Franceville, Gabon. The observed EIRs were one infective bite per person every 6 and 17 days for *Anopheles funestus* and *Anopheles gambiae* s.l. respectively, at the suburban site (Akou), while at the rural site (Benguia), the EIRs were one infective bite per person every 2, 3, 6, and 19 days for *Anopheles funestus, Anopheles gambiae* s.l., *Anopheles nili,* and *Anopheles moucheti* respectively. The EIR was also observed to vary seasonally, with a mean value of 1.001 infective bites per person per night, rising to 2.691 in October 1994 and reaching a low of 0.177 in April 1994.

Patz et al. (1998) demonstrated that *Anopheles* biting rates and EIR could be relatively well correlated with soil moisture in western Kenya, thereby allowing for predictions of seasonal and inter-annual variations in the EIR to be made. A soil moisture model of surface-water availability was used to combine multiple weather parameters with land cover and soil features to improve disease prediction. Modelling soil moisture substantially improved prediction of biting rates compared to rainfall; soil moisture lagged by 2 weeks explained up to 45% of variation in *Anopheles gambiae* biting rate, compared with 8% for raw rainfall data. For *Anopheles funestus*, soil moisture explained 32% variability, peaking after a 4-week lag. Modelled soil moisture accounted for up to 56% variability of *Anopheles gambiae* EIR, peaking at a lag of 6 weeks.

In malariology the man biting rate (*ma*) is multiplied by the sporozoite rate *s* to give

$$h = mabs \qquad (7.20)$$

where *b* is the proportion of mosquitoes containing sporozoites that are actually infective. This is not an easy parameter to measure, but *b* is little if at all affected by changes in the indirect factors. However, when transmission is at a low level the sporozoite rate is usually low, and the confidence intervals at the 95% probability level vary considerably with sample size of mosquitoes dissected. For example, if 2000 mosquitoes are dissected and the sporozoite rate is 0.10% the sporozoite rate could be 0.01–0.36%, and even if 8000 mosquitoes were dissected the true rate could be as low as 0.04% or double the calculated value (0.20%) (Onori and Grab 1980). But as the sporozoite rate is a function of the mosquito survival rate, the sporogonic cycle and the gametocyte rate, then the inoculation rate can be estimated without resource to the sporozoite rate as follows

$$h = \frac{ma^2 bgxP^n}{agx - \log_e P} \qquad (7.21)$$

where gx = the gametocyte rate, n = the duration of the sporogonic cycle, m = vector density in relation to man, a = the man-biting rate, b = proportion of vectors with sporozoites actually infective, and P = the daily survival rate. The inoculation rate is very sensitive to changes in P and n.

The paper by Krafsur (1977) on the calculation of sporozoite inoculation rate and the probabilities of a person receiving one or more inoculations per year is worth reading.

Killeen et al. (2000) describe the adaptation of a relatively simple cyclical model (Saul et al. 1990) (see also Chap. 13) to allow the calculation of transmission intensity as a function of its three fundamental components: the infectiousness of the human reservoir; the capacity of individual mosquitoes to transmit malaria; and the mosquito emergence rate relative to human population density. The model is based on the life histories of individual mosquitoes, rather than on individual humans and the vector biting densities they experience. When combined with mosquito emergence rates relative to human population size and infectiousness, these components define the transmission intensity experienced by any given human population. The model assumes that the length of the feeding cycle (f), the probability of surviving per feeding cycle (P_f), the number of days required for parasite development (n), and the susceptibility of the vector to infection as well as ability to become infectious (κ) do not change with age. Mathematically, this means that f, P_f, n, and κ are independent of the number of feeding cycles a given mosquito has completed and whether the mosquito is infected or not, at any cycle. The model also assumes that each vector feeds from a single host during each feeding cycle and that the vector and human populations mix homogeneously. Killeen et al. (2000) define b_h as the mean number of human blood meals a vector will acquire during its lifetime. During its lifetime, a typical mosquito will bite humans in proportion to its proba-bility of reaching each possible feeding cycle and Q, its preference for human hosts:

$$b_h = Q\sum_f (P_f)^i \tag{7.22}$$

where P_f is the survivorship per feeding cycle and i = 1,2,3, etc. It is therefore possible to estimate b_h by summing the probabilities of surviving to each feeding cycle, up to feeding cycle 20 or less and multiplying this sum by Q, the preference for feeding on humans.

Assuming that multiple infections are negligible, the probability of a mosquito being infectious at a given feeding cycle is related directly to the number of infectious blood meals taken from humans that have had sufficient time to allow infectious sporozoites to appear in the salivary glands

(n). This interval is expressed as F, the mean number of feeding cycles required for sporogonic development of the parasite:

$$F = n/f \qquad (7.23)$$

The probability of becoming infected is directly related to the proportion of human hosts that are infectious (x), the susceptibility of the vectors to infection (k), and the proportion of vectors that progress from being infected to infectious if they live long enough (v). The model of Saul et al. (1990) treated x and k together as K and v was assumed to be 1 and ignored. In the Killeen et al. (2000) model therefore, the product of x, k, and v (κ) is used instead, which reflects the overall capacity of a human reservoir to produce infectious vectors. If super-infections of mosquitoes are assumed to be negligible, S_i, the probability of a mosquito being infectious after surviving to a given feeding cycle i, can be expressed as a function of the number of blood meals it has taken which have the potential to result in infectious status:

$$S_i = \kappa Q(i - \delta) \qquad (7.24)$$

for $i - \delta = 1,2,3$, etc. Thus, for any given age group, i, the probability of a mosquito surviving from emergence and being infective (I_i) is the product of these two possible occurrences:

$$I_i = S_i P_i \qquad (7.25)$$

The cyclical nature of this model allows the probable mean number of infectious bites transmitted by a typical emerging mosquito over the course of its lifetime (β) to be calculated as the product of the human biting preference and the sum of the probabilities being alive and infectious at each cycle:

$$\beta = Q \sum I_i \qquad (7.26)$$

for $i - \delta = 1,2,3$, etc. By substituting the above equations, β can be expressed as:

$$\beta = Q \sum Q\kappa(i - \delta)(P_f)^i \qquad (7.27)$$

which by rearrangement yields

$$\beta = \kappa Q^2 \sum (i - \delta)(P_f)^i \qquad (7.28)$$

For the purpose of resolving malaria transmission by individual mosquitoes into two distinct contributing factors, this equation can be broken down into κ and L, a life–history function for the local vector population:

$$\beta = \kappa L \tag{7.29}$$

where

$$L = Q^2 \sum (i - \delta)(P_f)^i \tag{7.30}$$

Thus, the transmission capacity of individual mosquitoes becomes the product of two separate variables with real meaning in the field. κ reflects the effective infectiousness of the human reservoir and the physiologic compatibility of the local vector and parasite populations. L reflects the ability of an individual vector to transmit malaria from infectious human hosts over its lifetime, based on its longevity (P_f) and blood-feeding habits (Q). The model offers an alternative form for the calculation of EIR, which is derived from the fact that the EIR experienced by an individual person in a discrete transmission focus is the product of the mean number of infections transmitted by individual mosquitoes over their lifetimes (β) and the mean rate at which vector mosquitoes emerge (E) divided by the number of humans available for them to feed upon (N_h):

$$EIR = \beta E / Nh \tag{7.31}$$

Which, with substitution gives:

$$EIR = \kappa L E / Nh \tag{7.32}$$

S, the mean proportion of adult vectors which are infectious, reflects the proportion of bites that are infectious over the lifetime of an individual mosquito and is equal to the number of infectious bites divided by the number of total bites for an individual mosquito over its lifetime:

$$S = \beta / bh \tag{7.33}$$

Emergence rates of vectors can be estimated from the overall human biting rate (HBt) and average number of bites per lifetime (bh). Combining the equations for EIR above

$$H_{Bt}S = \beta E / N_h \tag{7.34}$$

And by substituting $S = \beta / b_h$ and rearranging

$$E = HBtNh/bh \tag{7.35}$$

and

$$E/N_h = H_{Bt}/b_h \qquad (7.36)$$

Allowing for the estimation of the emergence rate of mosquitoes per human host. The model was applied to four *Plasmodium falciparum*-endemic sites, Kankiya and Kaduna in Nigeria, Namawala in Tanzania, and Butelgut in Papua New Guinea for which relevant data were available. The probable number of infectious bites transmitted by a mosquito was defined by the overlap of the two functions, mortality and infectiousness, over the course of the mosquito's lifetime. Predicted EIR ± SD values were 1.13 – 0.37 times those measured in the field (range = 0.84–1.59) and did not differ significantly (df = 3, $t = -0.81$, $P = 0.48$, by paired t-test). The authors expressed concern that they were only available to identify four sites for which all of the necessary parameters (κ, P_f, F, Q, and H_{Bt}) have been reported and two of these (Kaduna and Kankiya) required values for the length of the feeding cycle and the sporogonic incubation period to be assumed or calculated because direct estimates were unavailable, despite standard methods for measuring P_f and Q being well established.

Birley & Boorman (1982) showed that the expected infective life of a mosquito (V) may be estimated as

$$V = P^{d/u} /(1 - P) \qquad (7.37)$$

where P = survival rate per oviposition cycle, d = duration of the extrinsic incubation period of the parasite in the mosquito and u = estimated length of the interval between blood-feeding and oviposition, i.e. the oviposition cycle. Clearly V is extremely sensitive to small changes in P, the survival rate. In this approach survival rate is calculated over discrete time-intervals, the oviposition cycles, and does not represent the daily survival rate which is a more continuous measurement.

Smith (1987) presented a modification of the malaria reproduction rate formula of Macdonald (1952) to estimate the reproductive rate of an arbovirus (R), such as western equine encephalomyelitis, where R is defined as the average number of vertebrate maintenance hosts infected by mosquitoes infected from a single vertebrate maintenance host, thus

$$R = \frac{mbhs_m Vs_v P^i}{-\log_e P} \qquad (7.38)$$

where m = bites/bird/night, b = number of feeds by a mosquito each day (if gonotrophic cycle is 4 days then the value is 0.25), h = proportion of blood-meals taken from birds (say 0.85), s_m = vector competence for western equine encephalomyelitis (WEE) (say 0.67), V = duration of infective viraemia in birds (3 days), s_v = proportion of birds susceptible to infection (say 0.67), P = mosquito daily survival (say 0.8) and i = intrinsic incubation period of WEE (say 6 days). Using these values $R = m$ (0.25 × 0.85) 0.67 × 3 × 0.67 (0.8^6) (1/−log$_e$ 0.8) which equals m × 0.336 (Reisen 1989). So if $R = 1$, which is necessary for WEE maintenance then, $m = 2.98$ bites/bird/night. Now when m is greater than this then R is >1, and represents the numbers of new infections/infected bird. However, this approach, based on so many untested assumptions, has been criticised by Dye (1992).

There is a series of six interesting papers on the estimation of vectorial capacity, mainly orientated to arbovirus infections, published in the *Bulletin of the Society of Vector Ecology* (1989) 14, 39–70, and an excellent and readable account of vectorial capacity is presented by Dye (1992). Boudin et al. (1998) provide an interesting description in French of the parameters involved in the measurement of malaria transmission.

The fifth report of the WHO Expert Committee on filariasis (WHO 1992) recently recommended the use of annual biting rate (ABR), annual infective biting rate (AIBR) and annual transmission potential (ATP) as the key entomological parameters for the measurement of transmission intensity for lymphatic filariasis. The use of resting vector collections in human dwellings and determination of mosquito parasite infection rates has been proposed as an alternative monitoring tool to the dangerous and unethical human-landing catch.

Transmission intensity index (TII) = number of resting females collected per man-hour (PMH) × proportion of females containing L3 larvae × average number of L3 per female.

ATP is estimated as MBR × total L3/mosquitoes dissected and is usually calculated from the sum of monthly or more frequent collections

Sunish et al. (2003) calculated these indices for *Culex quinquefasciatus* in Tamil Nadu, South India to monitor filariasis control programmes. These authors observed a significant positive linear relationship between the transmission intensity index calculated from resting collections and the annual transmission potential calculated from human landing catches, such that ATP = 173.9 + 658.3 TII ($r = 0.809$; $P < 0.001$). Sunish et al. (2003) noted that the validity of TII as an evaluation tool depends upon the endophilic and endophagic nature of the vector and studies to validate the relationship between TII and ATP need to be conducted in each geographical area prior to adopting resting collections alone to monitor effectiveness of large-scale filariasis control programmes.

Problems of estimating man-biting rates

Non-random biting on people violates the assumption made in nearly all mathematical models on disease transmission, that is that everybody is at equal risk from mosquito bites (e.g. Bailey 1975, 1982). Dye and Hasibeder (1986) found that when mosquitoes selectively feed on certain people this results in the vectorial capacity and the basic reproductive rate of malaria being larger than, or equal to, their estimated values under homogeneous mixing; a result anticipated by Dietz (1980). In fact, the results of Muirhead-Thomson (1951) indicate that non-random host-biting by *Anopheles albimanus* can result in a basic reproduction rate more than 2.5 times that which would occur with uniform exposure. Dye and Hasibeder (1986) emphasise the limitations of estimating vectorial capacity from field collected entomological data. Burkot (1988) and Dye (in Burkot 1988) present a mini-review of non-random host selection and its epidemiological implications in malaria transmission.

The influence of human behaviour on estimates of biting rate are illustrated by an example from Papua New Guinea. The human blood index of *Anopheles farauti* in Maraga village, Papua New Guinea, was estimated to be about 5% (Charlwood et al. 1986*a*; and P. M. Graves and T. Burkot quoted by Saul 1987), the majority of feeds being from pigs. Mark–recapture studies estimated the biting population per night to be about 46 000, which would equate to approximately 2300 feeds per night on humans. As the human population in the area was 125, the mean number biting a person per night would be expected to be around 20. However, in human bait collections some 500 *Anopheles farauti* were caught per person per night. Using this capture rate at bait to calculate vectorial capacity, or malaria inoculation rates, gives estimates 25 times greater than those obtained from the human blood index and population estimates of both mosquitoes and man. This example emphasises the danger of uncritical use of man-biting rates. In bait catches the collector is trying to collect as many mosquitoes as possible and is therefore likely to catch more than would have bitten him under natural conditions. In areas where mosquito nets are used the real biting rate will be further reduced. On the other hand, interrupted multiple feeding by individual mosquitoes during a single night will increase the biting rate, though this would not be reflected in the results of routine bait collections.

Clearly the assessment of man-biting rates is based on an artificial system. With some species an alternative approach is to base man-biting rates on the numbers of freshly blood-fed mosquitoes resting in, and leaving, a house (Garrett-Jones 1968, 1970; Garrett-Jones and Shidrawi 1969). However, it is unlikely that all the mosquitoes that have fed on hut occupants will be collected the following morning. Gubler and Bhattacharya (1974) for example believed that basing biting rates on the numbers of blood-fed female *Culex quinquefasciatus* found resting indoors during early morning collections (about 0700–0830 h) seriously under-estimated biting rates, because they had observed, by using exit traps, that a substantial number of adults left houses at 0400–0600 h. Furthermore, the problem of interrupted feeding remains. Theoretically a better approach to determine the degree of man–mosquito contact would be to collect all the outdoor and indoor resting mosquitoes in a small area (Ungureanu 1947), but this is very rarely possible. However, Brady (1974) attempted to estimate the biting rate of *Anopheles* on man from the numbers of blood-fed and gravid mosquitoes found in houses in early morning pyrethrum spray-sheet collections and the proportions of blood-fed and gravid mosquitoes found resting outside. But the formula he derived is applicable only if the gonotrophic cycle lasts 48 h (in which case there should be no half-gravids in early morning pyrethrum catches), also relatively high outdoor resting densities must be discovered, and confirmation that blood-engorged mosquitoes have in fact fed on man. Because of these, and other limitations, his approach has rarely been used.

Thomson et al. (1995) discuss some of the problems in estimating the EIR in insecticide-treated net (ITN) programmes. Where the human population protects itself from bites using mosquito nets, then the EIR calculated from light-trap catches or human landing catches will overestimate the EIR. Additionally, the repellent effects of permethrin treatment of mosquito nets will affect the estimates of human-biting rate obtained from numbers of indoor-resting mosquitoes. A review of some of the difficulties of correctly assessing man-biting rates and the epidemiological problems involved is given by Garrett-Jones (1970). Nájera (1974) also discusses some of these problems and emphasises the importance of obtaining reliable estimates of biting rate for use in models of malaria transmission.

Despite all the above limitations human landing catches remain of paramount importance in both epidemiological and ecological studies, and in assessing nuisance biting.

References

Abdulaziz HHQ, Pal RA (1989) Dipstick ELISA for the identification of mosquito bloodmeal, Mosq-Borne Dis Bull 6: 5–6

Afrane YA, Lawson BW, Githeko AK, Yan G (2005) Effects of microclimatic changes caused by land use and land cover on duration of gonotrophic cycles of *Anopheles gambiae* (Diptera: Culicidae) in western Kenya highlands. J Med Entomol 42: 974–980

Ahmadi A, McClelland GAH (1985) Mosquito-mediated attraction of female mosquitoes to a host. Physiol Entomol 10: 251–255

Alekseev AN, Rasnitsyn SP, Vitilin LM (1977) Group attack by females of blood-sucking mosquitoes (Diptera: Culicidae). Part I. Discovery of the 'invitation effect'. Medskaya Parazit 46: 23–24. (In Russian, English summary)

Alencar J, Lorosa ES, Dégallier N, Serra-Freire NM, Pacheco JB, Guimarães AE (2005) Feeding patterns of *Haemagogus janthinomys* (Diptera: Culicidae) in different regions of Brazil. J Med Entomol 42: 981–985

Allan SA, Day JF, Edman JD (1987) Visual ecology of biting flies. Annu Rev Entomol 32: 297–316

Almeida APG, Baptista SSSG, Sousa CAGCC, Novo MTLM, Ramos HC, Panella NA, Godsey M, Simões MJO, Anselmo ML, Komar N, Mitchell CJ, Ribeiro H (2005) Bioecology and vectorial capacity of *Aedes albopictus* (Diptera: Culicidae) in Macao, China, in relation to dengue virus transmission. J Med Entomol 42: 419–428

Almirón WR, Almeida FFL, Brewer M (2000) Relative abundance and gonotrophic status of *Aedes albifasciatus* (Diptera: Culicidae) during the autumn-winter period in Córdoba Province, Argentina. J Med Entomol 37: 16–20

Alonso WJ, Wyatt TD, Kelly DW (2003) Are vectors able to learn about their hosts? A case study with *Aedes aegypti* mosquitoes. Mem Inst Oswaldo Cruz 98: 665–672

Amerasinghe PH, Amerasinghe FP (1999) Multiple host feeding in field populations of *Anopheles culicifacies* and *An. subpictus* in Sri Lanka. Med Vet Entomol 13: 124–131

Anderson RA, Brust RA (1995) Field evidence for multiple host contacts during blood feeding by *Culex tarsalis, Cx. restuans*, and *Cx. nigripalpus* (Diptera: Culicidae). J Med Entomol 32: 705–710

Anderson RA, Brust RA (1996) Blood feeding success of *Aedes aegypti* and *Culex nigripalpus* (Diptera: Culicidae) in relation to defensive behavior of Japanese quail (*Coturnix japonica*) in the laboratory. J Vect Ecol 21: 94–104

Anderson RA, Brust RA (1997) Interrupted blood feeding by *Culex* (Diptera: Culicidae) in relation to individual host tolerance to mosquito attack. J Med Entomol 34: 95–101

Anderson RA, Gallaway WJ (1988) Hosts of *Anopheles earlei* Vargas (Diptera: Culicidae) in southwestern Manitoba. J Med Entomol 25: 149–150

Anderson RA, Roitberg BD (1999) Modelling trade-offs between mortality and fitness associated with persistent blood feeding by mosquitoes. Ecology Letters 2: 98–105

Anderson RA, Edman JD, Scott TW (1990) Rubidium and cesium as host blood-markers to study multiple blood feeding by mosquitoes (Diptera: Culicidae). J Med Entomol 27: 999–1001

Andrade RE, Lorosa ES (1998) Identification of food sources of mosquitoes in the Municipality of Nova Iguaçú, RJ, Brazil, using the precipitin reaction technique. Entomología y Vectores 5: 85–92

Ansell J, Hu J, Gilbert SC, Hamilton KA, Hill AVS, Lindsay SW (2000) Improved method for distinguishing the human source of mosquito bloodmeals between close family members. Trans R Soc Trop Med Hyg 94: 572–574

Ansell J, Hamilton KA, Pinder M, Walraven GEL, Lindsay SW (2002) Short-range attractiveness of pregnant women to *Anopheles gambiae* mosquitoes. Trans R Soc Trop Med Hyg 96: 113–116

Apperson CS, Harrison BA, Unnasch TR, Hassan HK, Irby WS, Savage HM, Aspen SE, Watson DW, Rueda LM, Engber BR, Nasci RS (2002) Host-feeding habits of *Culex* and other mosquitoes (Diptera: Culicidae) in the Borough of Queens in New York City, with characters and techniques for identification of *Culex* mosquitoes. J Med Entomol 39: 777–785

Arredondo-Jiménez JI, Bown DN, Rodríguez MH, Villarreal C, Loyola EG, Frederickson CE (1992) Tests for the existence of genetic determination or conditioning in host selection by *Anopheles albimanus* (Diptera, Culicidae). J Med Entomol 29: 894–897

Bailey NTJ (1952) Improvements in the interpretation of recapture data. J Anim Ecol 21: 120–127

Bailey NTJ (1975) The Mathematical Theory of Infectious Diseases and its Applications. (2nd edn) Charles Griffin & Co. Ltd, London

Bailey NTJ (1982) The Biomathematics of Malaria. Charles Griffin & Co. Ltd. London

Bang FB, Reeves WC (1942) Mosquitoes and encephalitis in the Yakima Valley, Washingtom. III. Feeding habits of *Culex tarsalis* Coq., a mosquito host of the viruses of western equine and St. Louis encephalitis. J Infect Dis 70: 273–274

Barrozo RB, Schilman PE, Minoli SA, Lazzari CR (2004) Daily Rhythms in Disease-Vector Insects. Biological Rhythm Research 35: 79–92

Beier JC (1996) Frequent blood-feeding and restrictive sugar-feeding behavior enhance the malaria vector potential of *Anopheles gambiae* s.l. and *An. funestus* (Diptera: Culicidae) in western Kenya. J Med Entomol 33: 613–618

Beier JC, Zimmerman JH, Kenawy M, El Said S, Abbassy MM (1987) Host-feeding patterns of the mosquito community (Diptera: Culicidae) in two Faiyum governorate villages, Egypt. J Med Entomol 24: 28–34

Beier JC, Perkins PV, Wirtz RA, Koros J, Diggs D, Gargan TP, Koech DK (1988) Bloodmeal identification for direct enzyme-linked immunosorbent assay (ELISA), tested on *Anopheles* (Diptera: Culicidae) in Kenya. J Med Entomol 25: 9–16

Beier JC, Killeen GF, Githure JI (1999) Short report: entomologic inoculation rates and *Plasmodium falciparum* malaria prevalence in Africa. Am J Trop Med Hyg 61: 109–113

Bennett SR (1978) Use of thermal summations to predict host-feeding rates of *Aedes sierrensis* females in northern California. Proc California Mosq Vector Control Assoc 46: 46

Bernier UR, Kline DL, Schreck CE, Yost RA, Barnard DR (2002) Chemical analysis of human skin emanations: comparison of volatiles from humans that differ in attraction of *Aedes aegypti* (Diptera: Culicidae). J Am Mosq Control Assoc 18: 186–195

Bernier UR, Kline DL, Posey KH, Booth MM, Yost RA, Barnard DR (2003) Synergistic attraction of *Aedes aegypti* (L.) to binary blends of L-lactic acid and acetone, dichloromethane, or dimethyl disulfide. J Med Entomol 40: 653–656

Bertsch ML, Norment BR (1983) The host-feeding patterns of *Culex quinquefasciatus* in Mississippi. Mosquito News 43: 203–206

Bidlingmayer WL (1994) How mosquitoes see traps: role of visual responses. J Am Mosq Control Assoc 10: 272–279

Birley MH, Boorman JPT (1982) Estimating the survival and biting rates of haematophagous insects with particular reference to the *Culicoides obsoletus* group (Diptera, Ceratopogonidae) in southern England. J Anim Ecol 51: 135–148

Blackwell A, Mordue (Luntz) AJ, Mordue W (1994) Identification of bloodmeals of the Scottish biting midge, *Culicoides impunctatus*, by indirect enzyme-linked immunosorbent assay (ELISA). Med Vet Entomol 8: 20–24

Blackwell A, Brown M, Mordue W (1995) The use of an enhanced ELISA method for the identification of *Culicoides* bloodmeals in host-preference studies. Med Vet Entomol 9: 214–218

Boakye DA, Tang J, Truc P, Merriweather A, Unnasch TR (1999) Identification of bloodmeals in haematophagous Diptera by cytochrome B heteroduplex analysis. Med Vet Entomol 13: 282–287

Bockarie MJ, Service MW, Barnish G, Touré YT (1995) Vectorial capacity and entomological inoculation rates of *Anopheles gambiae* in a high rainfall forested area of southern Sierra Leone. Trop Med Parasitol 46: 164–171

Boorman J (1986) A test for blood in haematophagous insects and its application to apparent blood feeding in insectivorous Ceratopogonidae (Diptera). Ann Trop Med Parasitol 80: 649–651

Boreham PFL (1976) Sterilization of arthropod bloodmeals prior to bloodmeal identification. Mosquito News 36: 454–457

Boreham PFL, Garrett-Jones C (1973) Prevalence of mixed blood meals and double feeding in a malaria vector (*Anopheles sacharovi* Favre). Bull World Health Organ 48: 605–614

Boreham PFL, Lenahan JK (1976) Methods for detecting multiple blood meals in mosquitoes (Diptera: Culicidae). Bull Entomol Res 66: 671–679

Boreham PFL, Chandler JA, Jolly J (1978) The incidence of mosquitoes feeding on mothers and babies at Kisumu, Kenya. J Trop Med Hyg 81: 63–67

Boreham PFL, Lenahan JK, Boulzaguet R, Storey J, Ashkar TS, Nambiar R, Matsushima T (1979) Studies on multiple feeding by *Anopheles gambiae* s.l. in a sudan savanna area of north Nigeria. Trans R Soc Trop Med Hyg 73: 418–423

Boudin C, Bonnet S, Tchuinkam T, Gouagna LC, Gounoue R, Manga L (1998) L'évaluation des niveaux de transmission palustre: méthodologies et paramètres. Méd Trop 58: 69–75

Bowen MF (1991) The sensory physiology of host-seeking behaviour in mosquitoes. Annu Rev Entomol 36: 139–158

Bowen MF, Haggart D, Romo J (1995*a*) Long-distance orientation, nutritional preference, and electrophysiological responsiveness in the mosquito *Aedes bahamensis*. J Vector Ecol 20: 203–210

Bowen MF, Romo J (1995*b*) Host-seeking and sugar-feeding in the autogenous mosquito *Aedes bahamensis* (Diptera: Culicidae). J Vector Ecol 20: 195–202

Bowen MF, Romo J (1995*c*) Sugar deprivation in adult *Aedes bahamensis* females and its effects on host seeking and longevity. J Vector Ecol 20: 211–215

Brady J (1974) Calculation of anopheline man-biting densities from concurrent indoor and outdoor resting samples. Ann Trop Med Parasitol 68: 359–361

Briegel H (2003) Physiological bases of mosquito ecology. J Vect Ecol 28: 1–11

Briegel H, Hörler E (1993) Multiple blood meals as a reproductive strategy in *Anopheles* (Diptera: Culicidae). J Med Entomol 30: 975–985

Bryan JH, Smalley ME (1978) The use of ABO blood groups as markers for mosquito biting studies. Trans R Soc Trop Med Hyg 72: 357–360

Burkot TR (1988) Non-random host selection by anopheline mosquitoes. Parasitol Today 4: 156–162

Burkot TR, Goodman WG, DeFoliart GR (1981) Identification of mosquito blood meals by enzyme-linked immunosorbent assay. Am J Trop Med Hyg 30: 1336–1341

Burkot TR, Graves PM, Paru R, Lagog M (1988) Mixed blood feeding by the malaria vectors in the *Anopheles punctulatus* complex (Diptera: Culicidae). J Med Entomol 25: 205–213

Canyon DV, Hii JLK, Muller R (1998) Multiple host-feeding and biting persistence of *Aedes aegypti*. Ann Trop Med Parasitol 92: 311–316

Carnevale P, Frézil JL, Bosseno MF, Le Pont F, Lancien J (1978) Étude de l'agressivité d'*Anopheles gambiae* A en fonction de l'âge et du sexe des sujets humains. Bull World Health Organ 56: 147–154

Chandler JA, Parsons J, Boreham PFL, Gill GS (1977) Seasonal variations in the proportions of mosquitoes feeding on mammals and birds at a heronry in western Kenya. J Med Entomol 14: 233–240

Chandra G, Chatterjee KK, Hati AK (1996) Feeding behaviour of *Anopheles stephensi* in Calcutta. Indian J Malariol 33: 103–105

Charlwood JD, Graves PM, Birley MH (1986) Capture–recapture studies with mosquitoes of the group of *Anopheles punctulatus* Dönitz (Diptera: Culicidae) from Papua New Guinea. Bull Entomol Res 76: 211–227

Charlwood JD, Smith T, Kihonda J, Heiz B, Billingsley PF, Takken W (1995*a*) Density independent feeding success of malaria vectors (Diptera: Culicidae) in Tanzania. Bull Entomol Res 85: 29–35

Charlwood JD, Billingsley PF, Hoc TQ (1995*b*) Mosquito-mediated attraction of female European but not African mosquitoes to hosts. Ann Trop Med Parasitol 89: 327–329

Chow E, Wirtz RE, Scott TW (1993) Identification of blood meals in *Aedes aegypti* by antibody sandwich enzyme-linked immunosorbent assay. J Am Mosq Control Assoc 9: 196–205

Chow-Shaffer E, Sina B, Hawley WA, De Benedictus J, Scott TW (2000) Laboratory and field evaluation of polymerase chain reaction-based forensic DNA profiling for use in identification of human bloodmeal sources of *Aedes aegypti* (Diptera: Culicidae). J Med Entomol 37: 492–502

Christensen HA, de Vasquez AM, Boreham MM (1996) Host-feeding patterns of mosquitoes (Diptera: Culicidae) from central Panama. Am J Trop Med Hyg 55: 202–208

Colless DH (1956) The *Anopheles leucosphyrus* group. Trans R Entomol Soc Lond 108: 37–116

Colless DH (1957) Components of the catch curve of *Culex annulus* in Singapore. Nature, Lond 180: 1496–1497

Collins RT, Dash BK, Agrawala RS, Dhal KB (1986) An adaption of the gel diffusion technique for identifying the source of mosquito blood meals. Indian J Malariol 23: 81–89

Comtois A, Berteaux D (2005) Impacts of mosquitoes and black flies on defensive behaviour and microhabitat use of the North American porcupine (*Erethizon dorsatum*) in southern Quebec. Can J Zool 83: 754–764

Cooper RD, Frances SP (2000) Biting sites of *Anopheles koliensis* on human collectors in Papua New Guinea. J Am Mosq Control Assoc 16: 266–267

Cork A, Park KC (1996) Identification of electrophysiologically-active compounds for the malaria mosquito, *Anopheles gambiae*, in human sweat extracts. Physiol Entomol 10: 269–276

Coulson RMR, Curtis CF, Ready PD, Hill N, Smith DF (1990) Amplification and analysis of human DNA present in mosquito bloodmeals. Med Vet Entomol 4: 357–366

Crans WJ, Sprenger DA, Mahmood F (1996) The blood-feeding habits of *Aedes sollicitans* (Walker) in relation to eastern equine encephalitis virus in coastal areas of New Jersey. II. Results of experiments with caged mosquitoes and the effects of temperature and physiological age on host selection. J Vector Ecol 21: 1–5

Crofford OB (1976) Acetone in breath. Trans Am Clin Clim Assoc 88: 128–139

Cully JF, Grieco JP, Kissel D (1991) Defensive behavior of eastern chipmunks against *Aedes triseriatus* (Diptera: Culicidae) J Med Entomol 28: 410–416

Cupp EW, Stokes GM (1976) Feeding patterns of *Culex salinarius* Coquillett in Jefferson parish. Mosquito News 36: 293–302

Cupp EW, Tennessen KJ, Oldland WK, Hassan HK, Hill GE, Katholi CR, Unnasch TR (2004) Mosquito and arbovirus activity during 1997–2002 in a wetland in northeastern Mississippi. J Med Entomol 41: 495–501

Das P, Bhattacharya S, Chakraborty S, Palit A, Das S, Ghosh KK, Hati AK (1983) Diurnal man-biting of *Armigeres subalbatus* (Coquillett, 1898) in a village in West Bengal. Indian J Med Res 78: 794–798

Davidson G (1954) Estimation of the survival-rate of anopheline mosquitoes in nature. Nature 174: 792–793

Day JF, Edman JD (1983) Malaria renders mice susceptible to mosquito feeding when gametocytes are most infective. J Parasitol 69: 163–170

Day JF, Edman JD (1984*a*) The importance of disease-induced changes in mammalian body temperature to mosquito blood-feeding. Comp Physiol Biochem 77: 447–452

Day JF, Edman JD (1984*b*) Mosquito engorgement on normally defensive hosts depends on host activity patterns. J Med Entomol 21: 732–740

De Benedictis J, Chow-Shaffer E, Costero A, Clark GG, Edman JD, Scott TW (2003) Identification of the people from whom engorged *Aedes aegypti* took blood meals in Florida, Puerto Rico, using polymerase chain reaction-based DNA profiling. Am J Trop Med Hyg 68: 437–446

De Buck A, Swellengrebel NH (1934) Behaviour of Dutch *Anopheles atroparvus* and *messeae* in winter under artificial conditions. Riv Malariol 13: 404–416

de Jong R, Knols BGJ (1996) Selection of biting sites by mosquitoes. In: Cardew G (ed) Olfaction in mosquito-host interactions. John Wiley and Sons, Chichester, UK, pp 89–103

Dekker T, Takken W, Knols BGJ, Bouman E, van de Laak S, de Bever A, Huisman PWT (1998) Selection of biting sites on a human host by *Anopheles gambiae* s.s., *An. arabiensis* and *An. quadriannulatus*. Entomol Exp Appl 87: 295–300

Dekker T, Steib B, Cardé RT, Geier M (2002) L-lactic acid: a human-signifying host cue for the anthropophilic mosquito *Anopheles gambiae*. Med Vet Entomol 16: 91–98

Detinova TS (1962) Age-grading methods in Diptera of medical importance. World Health Organ Monogr Ser 47

Dia I, Lochouarn L, Boccalini D, Costantini C, Fontenille D (2000) Spatial and temporal variations of the chromosomal inversion polymorphism of *Anopheles funestus* in Sénégal. Parasite 7: 179–184

Dia I, Lochouarn L, Diatta M, Sokhna CS, Fontenille D (2001) Préférences trophiques des femelles endophiles d'*Anopheles funestus* au Sénégal. Bull Soc Pathol Exot 94: 210–213

Dietz K (1980) Models for vector-borne parasitic diseases. Lect Notes Biomath 39: 264–277

Doannio JMC, Dossou-Yovo J, Diarrassouba S, Rakotondraibé ME, Chauvancy G, Chandre F, Rivière F, Carnevale P (2002) La dynamique de la transmission du paludisme à Kafiné, un village rizicole en zone de savane humide de Côte d'Ivoire. Bull Soc Pathol Exot 95: 11–16

Dossou-Yovo J, Doannio JMC, Diarrassouba S, Chauvancy G (1998) Impact d'aménagements de rizières sur la transmission du paludisme dans la ville de Bouaké, Côte d'Ivoire. Bull Soc Pathol Exot 91: 327–333

Downes CM, Theberge JB, Smith SM (1986) The influence of insects on the distribution, microhabitat choice, and behavior of the Burwash caribou herd. Can J Zool 64: 622–629

Drakeley C, Schellenberg D, Kihonda J, Sousa CA, Arez AP, Lopes D, Lines J, Mshinda H, Lengeler C, Schellenberg JA, Tanner M, Alonso P (2003) An estimation of the entomological inoculation rate for Ifakara: a semi-urban area in a region of intense malaria transmission in Tanzania. Trop Med Int Health 8: 767–774

Dransfield RD (1984) The range of attraction of the biconical trap for *Glossina pallidipes* and *Glossina brevipalpis*. Insect Science and its Application 5: 363–368

Dye C (1986) Vectorial capacity: Must we measure all its components? Parasitol Today 2: 203–209

Dye C (1992) The analysis of parasite transmission by bloodsucking insects. Annu Rev Entomol 37: 1–19

Dye C, Baker RHA (1986) Measuring the capacity of blackflies as vectors of onchocerciasis: *Simulium damnosum* s.l. in southwest Sudan. J Appl Ecol 23: 883–893

Dye C, Hasibeder G (1986) Population dynamics of mosquito-borne disease: effects of flies which bite some people more frequently than others. Trans R Soc Trop Med Hyg 80: 69–77

Edman JD (1971) Host-feeding patterns of Florida mosquitoes. 1. *Aedes, Anopheles, Coquillettidia, Mansonia* and *Psorophora*. J Med Entomol 8: 687–695

Edman JD (1974) Florida mosquitoes. III. *Culex (Culex)* and *Culex (Neoculex)*. J Med Entomol 11: 95–104

Edman JD (1979) Orientation of some Florida mosquitoes (Diptera: Culicidae) towards small vertebrates and carbon dioxide in the field. J Med Entomol 15: 292–296

Edman JD, Downe AER (1964) Host-blood sources and multiple-feeding habits of mosquitoes in Kansas. Mosquito News 24: 154–160

Edman JD, Kale HW (1971) Host behavior: its influence on the feeding success of mosquitoes. Ann Entomol Soc Am 64: 513–516

Edman JD, Scott TW (1987) Host defensive behaviour and feeding success of mosquitoes. Insect Science and its Application 8: 617–622

Edman JD, Spielman A (1988) Blood-feeding by vectors: physiology, ecology, behavior, and vertebrate defense. In: Monath TP (ed) Volume I. The Arboviruses: Epidemiology and Ecology. CRC Press, Boca Raton, Florida, pp 153–189

Edman JD, Taylor DJ (1968) *Culex nigripalpus*: seasonal shift in the bird–mammal feeding ratio in a mosquito vector of human encephalitis. Science 161: 67–68

Edman JD, Evans FDS, Williams JA (1968) Development of a diurnal resting box to collect *Culiseta melanura* (Coq.). Am J Trop Med Hyg 17: 451–456

Edman JD, Webber LA, Kale HW (1972) Effect of mosquito density on the interrelationship of host behavior and mosquito feeding success. Am J Trop Med Hyg 21: 487–491

Edman J, Day J, Walker E (1985) Vector-host interplay: factors affecting disease transmission. In: Lounibos LP, Rey JR, Frank JH (eds) Ecology of Mosquitoes: Proceedings of a Workshop. Florida Medical Entomology Laboratory, Vero Beach, Florida, pp 273–285

Eligh GS (1952) Factors influencing the performance of the precipitin test in the determination of blood meals of insects. Can J Zool 30: 213–218

Elissa N, Karch S, Bureau P, Ollomo B, Lawoko M, Yangari P, Ebang B, Georges AJ (1999) Malaria transmission in a region of savanna-forest mosaic, Haut-Ogooué, Gabon. J Am Mosq Control Assoc 15: 15–23

Elliott R (1968) Studies on man-vector contact in some malarious areas in Colombia. Bull World Health Organ 38: 239–253

Emord DE, Morris CD (1982) A host-baited CDC trap. Mosquito News 42: 220–224

Fernandez-Salas I, Roberts DR, Rodríguez MH, Rodríguez MC, Marina-Fernandez CF (1993) Host selection patterns of *Anopheles pseudopunctipennis* under insecticide spraying situations in southern Mexico. J Am Mosq Control Assoc 9: 375–384

Fleischer SJ, Gaylor MJ, Hue NV, Graham LC (1986) Uptake and elimination of rubidium, a physiological marker, in adult *Lygus lineolaris* (Hemiptera: Miridae). Ann Entomol Soc Am 79: 19–25

Fontenille D, Lochouarn L, Diagne N, Sokhna C, Lemasson JJ, Diatta M, Konate L, Faye F, Rogier C, Trape JF (1997) High annual and seasonal variations in malaria transmission by anophelines and vector species composition in Dielmo, a holoendemic area in Senegal. Am J Trop Med Hyg 56: 247–253

Foster WA, Takken W (2004) Nectar-related vs. human-related volatiles: behavioural response and choice by female and male *Anopheles gambiae* (Diptera: Culicidae) between emergence and first feeding. Bull Entomol Res 94: 145–157

Fujito S. et al. (1971) Effect of the population density of *Culex tritaeniorhynchus* Giles on bloodsucking rates in cow sheds and pigpens in relation to its role in the epidemic of Japanese encephalitis. Jap J Sanit Zool 22: 38–44

Gad AM, Riad IB, Farid HA (1995) Host-feeding patterns of *Culex pipiens* and *Cx. antennatus* (Diptera: Culicidae) from a village in Sharqiya Governorate, Egypt. J Med Entomol 32: 573–577

Garcia R, des Rochers BS, Telford AD (1975) Surveillance and detection of *Aedes sierrensis* (Ludlow). Proc California Mosq Control Assoc 43: 156

Garrett-Jones C (1964) The human blood index of malaria vectors in relation to epidemiological assessment. Bull World Health Organ 30: 241–261

Garrett-Jones C (1968) Epidemiological Entomology and its Application to Malaria. WHO/MAL/68.672, 17 pp. (mimeographed)

Garrett-Jones C (1970) Problems of epidemiological entomology as applied to malariology. Miscellaneous Publications of the Entomological Society of America 7: 168–180

Garrett-Jones C, Shidrawi GR (1969) Malaria vectorial capacity of a population of *Anopheles gambiae*. An exercise in epidemiological entomology. Bull World Health Organ 40: 531–545

Gary RE Jr, Foster WA (2001) Effects of available sugar on the reproductive fitness and vectorial capacity of the malaria vector *Anopheles gambiae* (Diptera: Culicidae). J Med Entomol 38: 22–28

Gass RF, Deesin T, Surathin K, Vutikes S, Sucharit S (1982) Observations on the feeding habits of four species of *Mansonia (Mansonioides)* mosquitoes in southern Thailand. Southeast Asian J Trop Med Public Health 13: 211–215

Germain M, Eouzan JP, Ferrara L, Button JP (1973) Données complémentaires sur le comportement et l'écologie d'*Aedes africanus* (Theobald) dans le nord du Cameroun occidental. Cah ORSTOM sér Entomol Méd Parasitol 11: 127–146

Gibson G, Torr SJ (1999) Visual and olfactory responses of haematophagous Diptera to host stimuli. Med Vet Entomol 13: 2–23

Gillett JD (1967) Natural selection and feeding speed in a blood-sucking insect. Proc R Soc (B) 167: 316–329

Gillett JD (1971) Mosquitos. Weidenfeld & Nicholson, London

Gillett JD (1972) The Mosquito: Its life Activities and Impact on Human Affairs. Doubleday & Co., Garden City, New York

Gillett JD (1979) Out for blood: flight orientation up-wind in the absence of visual clues (sic) (although printed as clues, the word cues is intended). Mosquito News 39: 221–229

Gillett JD, Connor J (1976) Host temperature and transmission of arboviruses by mosquitoes. Mosquito News 36: 472–477

Gillies MT (1964) Selection for host preferences in *Anopheles gambiae*. Nature 203: 852–854

Gillies MT (1969) The ramp-trap, an unbaited device for flight studies of mosquitoes. Mosquito News 29: 189–193

Gillies MT (1980) The role of carbon dioxide in host-finding by mosquitoes (Diptera: Culicidae): a review. Bull Entomol Res 70: 525–532

Gillies MT, De Meillon B (1968) The Anophelinae of Africa South of the Sahara. Publication of the South African Institute for Medical Research, 54

Gillies MT, Wilkes TJ (1970) The range of attraction of single baits for some West African mosquitoes. Bull Entomol Res 60: 225–235

Gillies MT, Wilkes TJ (1972) The range of attraction of animal baits and carbon dioxide for mosquitoes. Studies in a freshwater area of West Africa. Bull Entomol Res 61: 389–404

Gillies MT, Wilkes TJ (1974) The range of attraction of birds as baits for some West African mosquitoes (Diptera: Culicidae). Bull Entomol Res 63: 573–581

Gingrich JB, Williams GM (2005) Host-feeding patterns of suspected West Nile virus mosquito vectors in Delaware, 2001–2002. J Am Mosq Control Assoc 21: 194–200

Gokool S, Curtis CF, Smith DF (1993) Analysis of mosquito bloodmeals by DNA profiling. Med Vet Entomol 7: 208–215

Gomes AC, Silva NN, Marques GRAM, Brito M (2003) Host-feeding patterns of potential human disease vectors in the Paraíba Valley Region, State of São Paulo, Brazil. J Vect Ecol 28: 74–78

Gomes LAM, Duarte R, Lima DC, Diniz BS, Serrão ML, Labarthe N (2001) Comparison between precipitin and ELISA tests in the bloodmeal detection of *Aedes aegypti* (Linnaeus) and *Aedes fluviatilis* (Lutz) mosquitoes experimentally fed on feline, canine, and human hosts. Mem Inst Oswaldo Cruz 96: 693–695

Grieco JP, Achee NL, Andre RG, Roberts DR (2002) Host feeding preferences of *Anopheles* species collected by manual aspiration, mechanical aspiration, and from a vehicle-mounted trap in the Toledo district, Belize, Central America. J Am Mosq Control Assoc 18: 307–315

Gubler DJ, Bhattacharya NC (1974) A quantitative approach to the study of bancroftian filariasis. Am J Trop Med Hyg 23: 1027–1036

Habtewold T, Walker AR, Curtis CF, Osir EO, Thapa N (2001) The feeding behaviour and *Plasmodium* infection of *Anopheles* mosquitoes in southern Ethiopia in relation to use of insecticide-treated livestock for malaria control. Trans R Soc Trop Med Hyg 95: 584–586

Haddow AJ (1945) The mosquitoes of Bwamba county, Uganda. II. Biting activity with special reference to the influence of microclimate. Bull Entomol Res 36: 33–73

Haddow AJ (1954) Studies of the biting-habits of African mosquitoes. An appraisal of methods employed with special reference to the twenty-four-hour catch. Bull Entomol Res 45: 199–242

Haddow AJ (1956) Observations on the biting-habits of African mosquitos in the genus *Eretmapodites* Theobald. Bull Entomol Res 46: 761–72

Haddow AJ (1961) Entomological studies from a high tower in Mpanga forest, Uganda. VII. The biting behaviour of mosquitoes and tabanids. Trans R Entomol Soc Lond 113: 315–335

Hadis M, Lulu M, Makonnen Y, Asfaw T (1997) Host choice by indoor-resting *Anopheles arabiensis* in Ethiopia. Trans R Soc Trop Med Hyg 91: 376–378

Han-Il Ree, Ui-Wook Hwang, In-Yong Lee, Tae-Eun Kim (2001) Daily survival and human blood index of *Anopheles sinensis*, the vector species of malaria in Korea. J Am Mosq Control Assoc 17: 67–72

Harrington J (1990) Detecting evidence of hematophagy in dry museum specimens of *Clerada apicornis* (Hemiptera: Lygaeidae: Rhyparochrominae). Ann Entomol Soc Am 83: 545–548

Hasibeder G, Dye C (1988) Mosquito-borne disease dynamics: persistence in a completely heterogeneous environment. Theor Popul Biol 33: 31–53

Hassan KH, Cupp EW, Hill GE, Katholi CR, Klinger K, Unnasch TR (2003) Avian host preference by vectors of eastern equine encephalomyelitis virus. Am J Trop Med Hyg 69: 641–647

Hayes RO, Tempelis CH, Hess AD, Reeves WC (1973) Mosquito host preference studies in Hales county, Texas. Am J Trop Med Hyg 22: 270–277

Healy TP, Copland MJ (2000) Human sweat and 2-oxopentanoic acid elicit a landing response from *Anopheles gambiae*. Med Vet Entomol 14: 195–200

Heller JH, Adler AD (1980) Blood on the Shroud of Turin. Appl Opt 19: 2742–2744

Hess AD, Hayes RO, Tempelis CH (1968) The use of the forage ratio technique in mosquito host preference studies. Mosquito News 28: 386–389

Hewitt S, Kamal M, Muhammad N, Rowland M (1994) An entomological investigation of the likely impact of cattle ownership on malaria in an Afghan refugee camp in the North West Frontier Province of Pakistan. Med Vet Entomol 8: 160–164

Hii JLK (1985) Evidence of the existence of genetic variability in the tendency of *Anopheles balabacensis* to rest in houses and to bite man. Southeast Asian J Trop Med Public Health 16: 173–182

Hii JLK, Vun YS (1987) The influence of a heterogeneous environment on host feeding behaviour of *Anopheles balabacensis* (Diptera: Culicidae). Trop Biomed 4: 67–70

Himeidan YE, Elbashir MI, Adam I (2004) Attractiveness of pregnant women to the malaria vector, *Anopheles arabiensis*, in Sudan. Ann Trop Med Parasitol 98: 631–633

Hodgson JC, Spielman A, Komar N, Krahforst CF, Wallace GT, Pollack RJ (2001) Interrupted blood-feeding by *Culiseta melanura* (Diptera: Culicidae) on European starlings. J Med Entomol 38: 59–66

Huang W, Luo M (1986) A cellulose acetate membrane counter immunoelectrophoresis test for identification of host source of mosquito blood meals. J Parasitol Parasitol Dis 4: 186–188. (Chinese, English summary)

Hubschman J H (1962) A simplified Azan process well suited for crustacean tissue. Stain Technol 37: 379–380

Hudson JE (1984) *Anopheles darlingi* Root (Diptera: Culicidae) in the Suriname rain forest. Bull Entomol Res 74: 129–142

Hunter FF, Bayly R (1991) ELISA for identification of blood-meal source in black flies (Diptera: Simuliidae). J Med Entomol 28: 527–532

Irby WS, Apperson CS (1988) Hosts of mosquitoes in the coastal plain of North Carolina. J Med Entomol 25: 85–93

Jensen T, Cockburn AF, Kaiser PE, Barnard DR (1996) Human blood-feeding rates among sympatric sibling species of *Anopheles quadrimaculatus* mosquitoes in northern Florida. Am J Trop Med Hyg 54: 523–525

Kale HW, Edman JD, Webber RH (1972) Effect of behavior and age of individual ciconiform birds on mosquito feeding success. Mosquito News 32: 343–350

Kaul HN, Wattal BL (1968) Studies on culicine mosquitoes. IV. Influence of climate on the feeding behaviour of *Culex fatigans* females in village Arthala, near Delhi. Bull Ind Soc Mal Commun Dis 5: 45–52

Kay BH, Boreham PFL, Edman JD (1979) Application of the 'feeding index' concept to studies of mosquito host-feeding patterns. Mosquito News 39: 68–72

Kay BH, Boreham PFL, Fanning ID (1985) Host-feeding patterns of *Culex annulirostris* and other mosquitoes (Diptera: Culicidae) at Charleville, southwestern Queensland, Australia. J Med Entomol 22: 529–535

Kelly DW (2001) Why are some people bitten more than others? Trends Parasitol 17: 578–581

Kenawy M, Beier JC, Asiago CM, El Said S (1990) Factors affecting the human-feeding behavior of anopheline mosquitoes in Egyptian oases. J Am Mosq Control Assoc 6: 446–451

Kent RJ, Norris DE (2005) Identification of mammalian blood meals in mosquitoes by a multiplexed polymerase chain reaction targeting cytochrome b. Am J Trop Med Hyg 73: 336–342

Khan AA (1977) Mosquito attractants and repellents. In: Storey HH, McKelvey JJ (eds) Chemical Control of Insect Behavior Wiley, New York, pp 305–325

Khan AA, Maibach HI, Strauss WG (1971) A quantitative study of variation in mosquito response and host attractiveness. J Med Entomol 8: 41–43

Killeen GF, McKenzie FE, Foy BD, Schieffelin C, Billingsley PF, Beier JC (2000) A simplified model for predicting malaria entomologic inoculation rates based on entomologic and parasitologic parameters relevant to control. Am J Trop Med Hyg 62: 535–544

Kimsey RB, Kimsey PB (1984) Identification of arthropod blood meals using rubidium as a marker; a preliminary report. J Med Entomol 21: 714–719

Kline DL, Takken W, Wood JR, Carlson DA (1990) Field studies on the potential of butanone, carbon dioxide, honey extract, 1-octen-3-ol, L-lactic acid, and phenols as attractants for mosquitoes. Med Vet Entomol 4: 383–391

Klowden MJ (1983) The physiological control of mosquito host-seeking behavior. In: Harris KF (ed) Current Topics in Vector Research. Vol. I. Praeger Scientific, New York, pp 93–116

Klowden MJ, Lea AO (1978) Blood meal size as a factor affecting continual host-seeking by *Aedes aegypti* (L.). Am J Trop Med Hyg 27: 827–831

Klowden MJ, Lea AO (1979) Effect of defensive host behavior on the blood meal size and feeding success of natural populations of mosquitoes. J Med Entomol 15: 514–517

Knaus RM, Foil LD, Issel CJ, Leprince DJ (1993) Insect blood meal studies using radiosodium 24Na and 22Na. J Am Mosq Control Assoc 9: 264–268

Knols BGJ, de Jong R, Takken W (1995) Differential attractiveness of isolated humans to mosquitoes in Tanzania. Trans R Soc Trop Med Hyg 89: 604–606

Konaté L, Faye O, Gaye O, Sy N, Diop A, Diouf M, Trape JF, Molez JF (1999) Zoophagie et hôtes alternatifs des vecteurs du paludisme au Sénégal. Parasite 6: 259–267

Kono T, (1953) On the estimation of insect populations by time unit collecting. Researches in Population Ecology 2: 85–94 (In Japanese, English summary)

Krafsur ES (1977) The bionomics and relative prevalence of *Anopheles* species with respect to the transmission of Plasmodium to man in western Ethiopia. J Med Entomol 14: 180–194

Krafsur ES, Armstrong JC (1978) An integrated view of entomological and parasitological observations on falciparum malaria in Gambela, western Ethiopian lowlands. Trans R Soc Trop Med Hyg 72: 348–356

Kumari R, Joshi H, Giri A, Sharma VP (1993) Feeding preferences of *Anopheles sundaicus* in Car Nicobar Island. Indian J Malariol 30: 201–206

Kuntz KJ, Olson JK, Rade BJ (1982) Role of domestic animals as hosts for blood-seeking females of *Psorophora columbiae* and other mosquito species in Texas ricefields. Mosquito News 42: 202–210

Lacroix R, Mukabana WR, Gouagna LC, Koella JC (2005) Malaria infection increases attractiveness of humans to mosquitoes. PLOS Biol 3: e298

la Grange JJP, Coetzee M (1997) A mosquito survey of Thomo village, Northern Province, South Africa, with special reference to the bionomics of exophilic members of the *Anopheles funestus* group (Diptera: Culicidae). Afr Entomol 5: 295–299

Lee D (1971) The role of the mosquito, *Aedes sierrensis*, in the epizootology of the deer body worm, *Setaria yehi*. Thesis, University of California, Berkeley

Lee JH, Hassan H, Hill G, Cupp EW, Higazi TB, Mitchell CJ, Godsey MS, Unnasch TR (2002) Identification of mosquito avian-derived blood meals by polymerase chain reaction-heteroduplex analysis. Am J Trop Med Hyg 66: 599–604

Lee VH, Atmosoedjono S, Aep S, Swaine CD (1980) Vector studies and epidemiology of malaria in Irian Jaya, Indonesia. Southeast Asian J Trop Med Public Health 11: 341–347

Lehane MJ (1991) Biology of Blood-Sucking Insects. Harper-Collins Academic, London

Lindblade KA, Walker ED, Onapa AW, Katungu J, Wilson ML (2000) Land use change alters malaria transmission parameters by modifying temperature in a highland area of Uganda. Trop Med Int Health 5: 263–274

Lindsay SW, Adiamah JH, Miller JE, Pleass RJ, Armstrong JRM (1993) Variation in attractiveness of human subjects to malaria mosquitoes (Diptera: Culicidae) in The Gambia. J Med Entomol 30: 368–373

Lines JD, Curtis CF, Wilkes TJ, Njunwa KJ (1991) Monitoring human-bait mosquitoes (Diptera: Culicidae) in Tanzania with light-traps hung beside mosquito nets. Bull Entomol Res 81: 77–84

Linthicum KJ, Kaburia HFA, Davies FG, Lindqvist KJ (1985) A blood meal analysis of engorged mosquitoes found in Rift Valley Fever epizootics areas in Kenya. J Am Mosq Control Assoc 1: 93–95

Liu C, Qian H, Gu Z, Pan J, Zheng X, Peng Z (1986) Quantitative study on the role of Anopheles lesteri anthropophagus in malaria transmission. J Parasitol Parasitol Dis 4: 161–164. (Chinese, English summary)

Lochouarn L, Dia I, Boccolini D, Coluzzi M, Fontenille D (1998) Bionomical and cytogenetic heterogeneities of *Anopheles funestus* in Senegal. Trans R Soc Trop Med Hyg 92: 607–612

Lombardi S, Esposito F (1986) A new method for identification of the animal origin of mosquito bloodmeals by the immunobinding of peroxidase–antiperoxidase complexes on nitrocellulose. J Immunol Methods 86: 1–5

Loong KP, Chiang GL, Eng KL, Chan ST, Yap HH (1990). Survival and feeding behaviour of Malaysian strains of *Anopheles maculatus* Theobald (Diptera: Culicidae) and their role in malaria transmission. Trop Biomed 7: 71–76

Lorenz LH, Scott TW (1996) Detection of multiple blood feeding in *Culiseta melanura* using a histologic technique. J Am Mosq Control Assoc 12: 135–136

Loyola EG, Rodriguez MH, Gonzalez L, Arredondo JI, Bown DN, Vaca MA (1990) Effect of indoor residual spraying of DDT and bendiocarb on the feeding patterns of *Anopheles pseudopunctipennis* in Mexico. J Am Mosq Control Assoc 6: 635–640

Loyola EG, González-Cerón L, Rodríguez MH, Arredondo-Jiménez JI, Bennett S, Bown DN (1993) *Anopheles albimanus* (Diptera: Culicidae) host selection patterns in three ecological areas of the coastal plains of Chiapas, southern Mexico. J Med Entomol 30: 518–523

Macdonald G (1952) The analysis of equilibrium in malaria. Trop Dis Bull 49: 813–829

Magnarelli LA (1977) Physiological age of mosquitoes (Diptera: Culicidae) and observations on partial blood-feeding. J Med Entomol 13: 445–450

Mahmood F, Crans WJ (1997) Observations on multiple blood-feeding in field-collected *Culiseta melanura*. J Am Mosq Control Assoc 13: 156–157

Mahon R, Gibbs A (1982) Arbovirus infected hens attract more mosquitoes. In: MacKenzie, JS (ed) Viral Diseases in South East Asia and the Western Pacific. Academic Press, New York, pp 502–505

Mbogo CNM, Kabiru EW, Glass GE, Forster D, Snow RW, Khamala CPM, Ouma JH, Githure JI, Marsh K, Beier JC (1999) Vector-related case-control study of severe malaria in Kilifi District, Kenya. Am J Trop Med Hyg 60: 781–785

Mbogo CM, Mwangangi JM, Nzovu J, Gu W, Yan G, Gunter JT, Swalm C, Keating J, Regens JL, Shililu JI, Githure JI, Beier JC (2003) Spatial and temporal heterogeneity of *Anopheles* mosquitoes and *Plasmodium falciparum* transmission along the Kenyan coast. Am J Trop Med Hyg 68: 734–742

McCall PJ, Kelly DW (2002) Learning and memory in disease vectors. Trends Parasitol 18: 429–433

McCall PJ, Mosha FW, Njunwa KJ, Sherlock K (2001) Evidence for memorized site-fidelity in *Anopheles arabiensis*. Trans R Soc Trop Med Hyg 95: 587–590

McCrae AWR, Boreham PFL, Ssenkubuge Y (1976) The behavioural ecology of host selection in *Anopheles implexus* (Theobald) (Diptera: Culicidae). Bull Entomol Res 66: 587–631

McIver SB (1982) Review article: Sensillae of mosquitoes, J Med Entomol 19: 489–535

Meece JK, Reynolds CE, Stockwell PJ, Jenson TA, Christensen JE, Reed KD (2005) Identification of mosquito bloodmeal source by terminal restriction fragment length polymorphism profile analysis of the cytochrome b gene. J Med Entomol 42: 657–667

Michael E, Ramaiah KD, Hoti SL, Barker G, Paul MR, Yuvaraj J, Das PK, Grenfell BT, Bundy DAP (2001) Quantifying mosquito biting patterns on humans by DNA fingerprinting of bloodmeals. Am J Trop Med Hyg 65: 722–728

Minakawa N, Seda P, Yan G (2002) Influence of host and larval habitat distribution on the abundance of African malaria vectors in western Kenya. Am J Trop Med Hyg 67: 32–38

Mitchell CJ, Millian KY (1981) Continued host seeking by partially engorged *Culex tarsalis* (Diptera: Culicidae) collected in nature. J Med Entomol 18: 249–250

Mogi M, Yamamura N (1981) Estimation of the attraction range of a human bait for *Aedes albopictus* (Diptera: Culicidae) adults and its absolute density by a new removal method applicable to populations with immigrants. Researches in Population Ecology 23: 328–343

Molineaux L, Dietz K, Thomas A (1978) Further epidemiological evaluation of a malaria model. Bull World Health Organ 56: 565–571

Molyneux DH, Jefferies D (1986) Feeding behaviour of pathogen-infected vectors. Parasitology 92: 721–736

Moorhouse DE, Wharton RH (1965) Studies on Malayan vectors of malaria; methods of trapping, and observations on biting cycles. J Med Entomol 1: 359–370

Moss JI, van Steenwyk RA (1982) Marking pink bollworm (Lepidoptera: Gelechiidae) with cesium. Environ Entomol 11: 1264–1268

Muirhead-Thomson RC (1951) The distribution of anopheline mosquito bites among different age groups, a new factor in malaria epidemiology. BMJ 1: 1114–1117

Muirhead-Thomson RC, Mercier EC (1952) Factors in malaria transmission by *Anopheles albimanus* in Jamaica. Part I. Ann Trop Med Parasitol 46: 103–116

Mukabana WR, Takken W, Seda P, Killeen GF, Hawley WA, Knols BGJ (2002) Extent of digestion affects the success of amplifying human DNA from blood meals of *Anopheles gambiae* (Diptera: Culicidae). Bull Entomol Res 92: 233–239

Mwandawiro C, Boots M, Tuno N, Suwonkerd W, Tsuda Y, Takagi M (2000) Heterogeneity in the host preference of Japanese encephalitis vectors in Chiang Mai, northern Thailand. Trans R Soc Trop Med Hyg 94: 238–242

Mwangangi JM, Mbogo CM, Nzovu JG, Githure JI, Yan GY, Beier JC (2003) Blood-meal analysis for anopheline mosquitoes sampled along the Kenyan coast. J Am Mosq Control Assoc 19: 371–375

Nájera JA (1974) A critical review of the field application of a mathematical model of malaria eradication. Bull World Health Organ 50: 449–457

Nasci RS (1984) Variations in the blood-feeding patterns of *Aedes vexans* and *Aedes trivittatus* (Diptera: Culicidae). J Med Entomol 21: 95–99

Nasci RS, Edman JD (1981) Blood feeding patterns of *Culiseta melanura* (Diptera: Culicidae) and associated sylvan mosquitoes in southeastern Massachusetts eastern equine encephalitis enzootic foci. J Med Entomol 18: 493–500

Nelson DL, Ledbetter SA, Corbo L, Victoria MF, Ramirez-Solis R, Webster TD, Ledbetter DH, Caskey CT (1989) Alu PCR: a method for rapid isolation of human-specific sequences from complex DNA sources. Proc Natl Acad Sci USA 86: 6686–6690

Nishimura M (1982) How mosquitoes fly to man. Researches in Population Ecology 24: 58–69

Nutsathapana S, Sawasdiwongphorn P, Chitprarop U, Cullen JR, Gass RF, Green CA (1986) A mark–release–recapture demonstration of host preference heterogeneity in *Anopheles minimus* Theobald (Diptera: Culicidae) in a Thai village. Bull Entomol Res 76: 313–320

Oliver DG, Sanders AH, Douglas Hogg R, Hellman JW (1981) Thermal gradients in microtitration plates effects on enzyme-linked immunoassay. J Immunol Methods 42: 195–201

Onori E, Grab B (1980). Indicators for the forecasting of malaria epidemics. Bull World Health Organ 58: 91–98

Pan American Health Organization (1996) Biology and ecology of *Anopheles albimanus* Wiedemann in Central America. Technical Paper 43

Paterson S, Wilson K, Pemberton JM (1998) Major histocompatability complex variation associated with juvenile survival and parasite resistance in a large unmanaged ungulate population (*Ovis aries* L.). Proc Natl Acad Sci USA 95: 3714–3719

Pates HV, Takken W, Curtis CF, Huisman PW, Akinpelu A, Gill GS (2001) Unexpected anthropophagic behaviour in *Anopheles quadriannulatus*. Med Vet Entomol 15: 293–298

Patz JA, Strzepek K, Lele S, Hedden M, Greene S, Noden B, Hay SI, Kalkstein L, Beier JC (1998) Predicting key malaria transmission factors, biting and entomological inoculation rates, using modelled soil moisture in Kenya. Trop Med Int Health 3: 818–827

Pearson AC, Ballmer GR, Sevacherian V, Vail PV (1989) Interpretation of rubidium marking levels in beet armyworm eggs (Lepidoptera: Noctuidae). Environ Entomol 18: 844–848

Ponlawat A, Harrington LC (2005) Blood feeding patterns of *Aedes aegypti* and *Aedes albopictus* in Thailand. J Med Entomol 42: 844–849

Port GR, Boreham PFL, Bryan JH (1980) The relationship of host size to feeding by mosquitoes of the *Anopheles gambiae* Giles complex (Diptera: Culicidae). Bull Entomol Res 70: 133–144

Price GD, Smith N, Carlson DA (1979) The attraction of female mosquitoes (*Anopheles quadrimaculatus* Say) to stored human emanations in conjunction with adjusted levels of relative humidity, temperature, and carbon dioxide. J Chem Ecol 5: 383–395

Prior A, Torr SJ (2002) Host selection by *Anopheles arabiensis* and *An. quadriannulatus* feeding on cattle in Zimbabwe. Med Vet Entomol 16: 207–213

Quiñones ML, Drakeley CJ, Muller O, Lines JD, Haywood M, Greenwood BM (2000) Diversion of *Anopheles gambiae* from children to other hosts following exposure to permethrin-treated bednets. Med Vet Entomol 14: 369–375

Rawlings P, Curtis CF (1982) Tests for the existence of genetic variability in the tendency of *Anopheles culicifaces* species B to rest in houses and to bite man. Bull World Health Organ 60: 427–432

Reeves WC (1971) Mosquito vector and vertebrate host interaction: the key to maintenance of certain arboviruses. In Fallis AM (ed) The Ecology and Physiology of Parasites. University of Toronto Press, Toronto, pp 223–230

Reisen WK (1989) Estimation of vectorial capacity: relationship to disease transmission by malaria and arbovirus vectors. Bull Soc Vector Ecol 14: 67–70

Reisen WK, Aslamkhan M (1976) Observations on the swarming and mating behaviour of *Anopheles culicifacies* Giles in nature. Bull World Health Organ 54: 155–158

Reisen WK, Aslamkhan M (1978) Biting rhythms of some Pakistan mosquitoes (Diptera: Culicidae). Bull Entomol Res 68: 313–330

Reisen WK, Boreham PF (1982) Estimates of malaria vectorial capacity for *Anopheles culicifacies* and *Anopheles stephensi* in rural Punjab province, Pakistan. J Med Entomol 19: 98–103

Renshaw M (1991) Population dynamics and ecology of *Aedes cantans* (Diptera: Culicidae) in England. Ph.D. thesis, University of Liverpool

Renshaw M, Service MW, Birley MH (1994). Host finding, feeding patterns and evidence for a memorized home range of the mosquito *Aedes cantans*. Med Vet Entomol 8: 187–193

Renshaw M, Silver JB, Service MW (1995) Differential lipid reserves influence host-seeking behaviour in the mosquitoes *Aedes cantans* and *Aedes punctor*. Med Vet Entomol 9: 381–387

Reuben R, Thenmozhi V, Samuel PP, Gajanana A, Mani TR (1992) Mosquito blood feeding patterns as a factor in the epidemiology of Japanese encephalitis in southern India. Am J Trop Med Hyg 46: 654–663

Ribbands CR (1946) Moonlight and house-haunting habits of female anophelines in West Africa. Bull Entomol Res 36: 395–415

Ritchie SA, Rowley WA (1981) Blood-feeding patterns of Iowa mosquitoes. Mosquito News 41: 271–275

Roberts DR, Scanlon JE (1975) The ecology and behavior of *Aedes atlanticus* D. & K. and other species with reference to Keystone virus in the Houston area, Texas. J Med Entomol 12: 537–546

Romoser WS, Edman JD, Lorenz LH, Scott TW (1989) Histological parameters useful in the identification of multiple blood meals in mosquitoes. Am J Trop Med Hyg 41: 737–742

Roy A, Sharma VP (1987) Microdot ELISA: Development of a sensitive and rapid test to identify the source of mosquito blood meals. Indian J Malariol 24: 51–58

Rubio-Palis Y, Wirtz RA, Curtis CF (1992) Malaria entomological inoculation rates in western Venezuela. Acta Trop 52: 167–174

Rubio-Palis Y, Curtis CF, Gonzales C, Wirtz RA (1994) Host choice of anopheline mosquitoes in a malaria endemic area of western Venezuela. Med Vet Entomol 8: 275–280

Sambrook J, Fritsch EF, Maniatis T (1989) Molecular Cloning: A Laboratory Manual, 2nd ed

Samuel PP, Arunachalam N, Hiriyan J, Thenmozhi V, Gajanana A, Satyanarayana K (2004) Host-feeding pattern of *Culex quinquefasciatus* Say and *Mansonia annulifera* (Theobald) (Diptera: Culicidae), the major vectors of filariasis in a rural area of south India. J Med Entomol 41: 442–446

Sato C, Furuya Y, Harada M, Suguri S (1992) Identification of human blood in mosquitoes (Diptera: Culicidae) using non-radioactive DNA dot blot hybridization. J Med Entomol 29: 1045–1048

Saul A (1987) Estimation of survival rates and population size from mark-recapture experiments of bait-caught haematophagous insects. Bull Entomol Res 77: 589–602

Saul A, Graves PM, Kay BH (1990) A cyclical model of disease transmission and its application to determining vectorial capacity from vector infection rates. J Appl Ecol 27: 123–133

Savage HM, Niebylski ML, Smith CG, Mitchell CJ, Craig GB (1993) Host-feeding patterns of *Aedes albopictus* (Diptera: Culicidae) at a temperate North American site. J Med Entomol 30: 27–34

Schreck CE, Smith N, Carlson DA, Price GD, Haile D, Godwin DR (1981) A material isolated from human hands that attracts female mosquitoes. J Chem Ecol 8: 429–438

Schreck CE, Kline DL, Carlson DA (1990) Mosquito attraction to substances from the skin of different humans. J Am Mosq Control Assoc 6: 406–410

Scott TW, Edman JD, Lorenz LH, Hubbard JL (1988) Effects of disease on vertebrates' ability behaviorally to repel host-seeking mosquitoes. In: Scott TW, Grumstrupp-Scott J (eds). Proceedings of a Symposium: The Role of Vector-Host Interactions in Disease Transmission. Miscellaneous Publications of the Entomological Society of America 68: pp 7–17

Scott TW, Lorenz LH, Edman JD (1990) Effects of house sparrow age and arbovirus infection on attraction of mosquitoes. J Med Entomol 27: 856–863

Scott TW, Chow E, Strickman D, Kittayapong P, Wirtz RA, Lorenz LH, Edman JD (1993*a*) Blood-feeding patterns of *Aedes aegypti* (Diptera: Culicidae) collected in a rural Thai village. J Med Entomol 30: 922–927

Scott TW, Clark GG, Lorenz LH, Amerasinghe PH, Reiter P, Edman JD (1993*b*) Detection of multiple blood feeding in *Aedes aegypti* (Diptera: Culicidae) during a single gonotrophic cycle using a histologic technique. J Med Entomol 30: 94–99

Scott TW, Morrison AC, Lorenz LH, Clark GG, Strickman D, Kittayapong P, Zhou H, Edman JD (2000) Longitudinal studies of *Aedes aegypti* (Diptera: Culicidae) in Thailand and Puerto Rico: population dynamics. J Med Entomol 37: 77–88

Senior White RA (1953) On the evening biting activity of three neotropical *Anopheles* in Trinidad, British West Indies. Bull Entomol Res 44: 451–460

Service MW (1969) Observations on the ecology of some British mosquitoes. Bull Entomol Res 59: 161–194

Service MW (1971*a*) Feeding behaviour and host preferences of British mosquitoes. Bull Entomol Res 60: 653–661

Service MW (1971*b*) The daytime distribution of mosquitoes resting amongst vegetation. J Med Entomol 8: 271–278

Service MW (1977) Ecological and biological studies on *Aedes cantans* (Meig.) (Diptera: Culicidae) in southern England. J Appl Ecol 14: 159–196

Service MW (1993) Mosquito Ecology. Field Sampling Methods. 2nd edn. Chapman & Hall, London

Service MW, Voller A, Bidwell DE (1986) The enzyme-linked immunosorbent assay (ELISA) test for the identification of blood-meals of haematophagous insects. Bull Entomol Res 76: 321–330

Seyoum A, Balcha F, Balkew M, Ali A, Gebre-Michael T (2002) Impact of cattle keeping on human biting rate of anopheline mosquitoes and malaria transmission around Ziway, Ethiopia. East Afr Med J 79: 485–490

Shidrawi GR, Boulzaguet JR, Ashkar TS, Bröger S (1974) Night-Bait Collection: The Variation Between Persons Used as Collector-baits. MPD/TN/74.1 Technical Note 17, 9–18 (WHO mimeographed document)

Shields SE, Lackey JB (1938) Conditions affecting mosquito breeding with special reference to *Aedes thibaulti* Dyar and Knab. J Econ Entomol 31: 95–102

Shihai W, Jun ZH (1989) Blood meals of mosquito vectors of filariasis in Guizhou province, China. Southeast Asian J Trop Med Public Health 20: 175–177

Shililu J, Ghebremeskel T, Mengistu S, Fekadu H, Zerom M, Mbogo C, Githure J, Novak R, Brantly E, Beier JC (2003) High seasonal variation in entomologic inoculation rates in Eritrea, a semi-arid region of unstable malaria in Africa. Am J Trop Med Hyg 69: 607–613

Shirai Y, Kamimura K, Seki T, Morohashi M (2001) L-lactic acid as a mosquito (Diptera: Culicidae) repellent on human and mouse skin. J Med Entomol 38: 51–54

Shirai Y, Funada H, Kamimura K, Seki T, Morohashi M (2002*a*) Landing sites on the human body preferred by *Aedes albopictus*. J Am Mosq Control Assoc 18: 97–99

Shirai Y, Tsuda T, Kitagawa S, Naitoh K, Seki T, Kamimura K, Morohashi M (2002*b*) Alcohol ingestion stimulates mosquito attraction. J Am Mosq Control Assoc 18: 91–96

Shirai Y, Funada H, Takizawa H, Seki T, Morohashi M, Kamimura K (2004) Landing preference of *Aedes albopictus* (Diptera: Culicidae) on human skin among ABO blood groups, secretors or nonsecretors, and ABH antigens. J Med Entomol 41: 769–799

Smith A (1958) Outdoor cattle feeding and resting of *A. gambiae* Giles and *A. funestus* Theo. in the Pare-Taveta area of East Africa. East Afr Med J 35: 559–567

Smith A (1961) Observations on the man-biting habits of some mosquitoes in the South Pare area of Tanganyika. East Afr Med J 38: 246–255

Smith CEG (1987) Factors influencing the transmission of western equine encephalomyelitis virus between its vertebrate maintenance hosts and from them to humans. Am J Trop Med Hyg 37 (suppl): 33s–39s

Smith GF (1953) The wet ashing of organic matter employing concentrated perchloric acid. The liquid fire reaction. Anal Chim Acta 5: 397–421

Spencer M (1967) Anopheline attack on mother and infant pairs. Fergusson Island. Papua N Guinea Med J 10: 75

Stimmann WW (1974) Marking insects with rubidium: cabbageworm marked in the field. Environ Entomol 3: 327–328

Sulaiman S (1982) The Ecology of *Aedes cantans* (Meigen) and Biology of *Culex pipiens* in Hibernation Sites in Northern England. Ph.D. thesis, University of Liverpool

Sunish IP, Rajendran R, Mani TR, Munirathinam A, Tewari SC, Hiriyan J, Gajanana A, Reuben R, Satyanarayana K (2003) Transmission intensity index to monitor filariasis infection pressure in vectors for the evaluation of filariasis elimination programmes. Trop Med Int Health 8: 812–819

Sutcliffe JF (1986) Black fly host location; a review. Can J Zool 64: 1041–1053

Sutcliffe JF (1987) Orientation of biting flies to their hosts. Insect Science and Its Application 8: 611–616

Suyemoto W, Schiefer BA, Eldridge BF (1973) Precipitin tests of blood-fed mosquitoes collected during the VEE surveillance survey in the southern United States in 1971. Mosquito News 33: 392–395

Swellengrebel NH (1929) La dissociation des fonctions sexuelles et nutritives (dissociation gonotrophique) d'*Anopheles maculipennis* comme cause du paludisme dans les pays-bas et ses rapports avec "l'infection domiciliaire". Ann Inst Pasteur (Paris) 43: 1370–1389

Takken W (1991) The role of olfaction in host-seeking of mosquitoes: a review. Insect Science and its Application 12: 287–295

Takken W, Kline DL (1989) Carbon dioxide and 1-octen-3-ol as mosquito attractants. J Am Mosq Control Assoc 5: 311–316

Teesdale C (1955) Studies on the bionomics of *Aedes aegypti* (L.) in its natural habitats in a coastal region of Kenya. Bull Entomol Res 46: 711–742

Tempelis CH (1975) Host-feeding patterns of mosquitoes, with a review of advances in analysis of blood meals by serology. J Med Entomol 11: 635–653

Tempelis CH, Lofy MF (1963) A modified precipitin method for identification of mosquito blood-meals. Am J Trop Med Hyg 12: 825–831

Tempelis CH, Francy DB, Hayes RO, Lofy MF (1967) Variations in feeding patterns of seven culicine mosquitoes on vertebrate hosts in Weld and Larimer counties. Colorado. Am J Trop Med Hyg 16: 111–119

Teranishi R, Mon TR, Robinson AB, Cary P, Linus P (1972) Gas chromatography of volatiles from breath and urine. Analyt Chem 44: 18–20

Thomas TCE (1951) Biting activity of *Anopheles gambiae*. BMJ 2: 1402

Thomson MC, Adiamah JH, Connor SJ, Jawara M, Bennett S, D'Alessandro U, Quiñones M, Langerock P, Greenwood BM (1995) Entomological evaluation of The Gambia's National Impregnated Bednet Programme. Ann Trop Med Parasitol 89: 229–241

Torr SJ, Wilson PJ, Schofield S, Mangwiro TNC, Akber S, White BN (2001) Application of DNA markers to identify the individual-specific hosts of tsetse feeding on cattle. Med Vet Entomol 15: 78–86

Trpis M, Haüsermann W (1986) Dispersal and other population parameters of *Aedes aegypti* in an African village and their possible significance in epidemiology of vector-borne diseases. Am J Trop Med Hyg 35: 1263–1279

Tuno N, Tsuda Y, Takagi M, Suwonkerd W (2003) Pre- and postprandial mosquito resting behavior around cattle hosts. J Am Mosq Control Assoc 19: 211–219

Turell MP, Rossignol PA, Spielman A (1984) Enhanced arboviral transmission by mosquitoes that concurrently ingest microfilaria, Science 225: 1039–1041

Ulloa A, Arredondo-Jiménez JI, Rodriguez MH, Fernández-Salas I, González-Cerón L (2004) Innate host selection in *Anopheles vestitipennis* from southern Mexico. J Am Mosq Control Assoc 20: 337–341

Ungureanu EM (1947) Contribution to the study of anophelini in relation to the transmission of malaria in northern Romania. Ann Acad Romậne 22, Mem 7 (In Romanian)

van den Hurk AF, Johansen CA, Zborowski P, Paru R, Foley PN, Beebe NW, Mackenzie JS, Ritchie SA (2003) Mosquito host-feeding patterns and implications for Japanese encephalitis virus transmission in northern Australia and Papua New Guinea. Med Vet Entomol 17: 403–411

van Steenwyk RA, Ballmer GR, Page AL, Reynolds HT (1978) Marking pink bollworm with rubidium. Ann Entomol Soc Am 71: 81–84

van Steenwyk RA, Henneberry TJ, Ballmer GR, Wolf WW, Sevacherian V (1979) Mating competitiveness of laboratory-cultured and sterilized pink bollworm for use in sterile moth release program. J Econ Entomol 72: 502–505

Waage JK (1979) The evolution of insect/vertebrate associations. Biol J Linn Soc 12: 187–224

Waage JK (1981) How the zebra got its stripes—biting flies as selective agents in the evolution of zebra coloration. J Entomol Soc Sthn Afr 44: 351–358

Walker ED, Edman JD (1985a) Feeding-site selection and blood-feeding behavior of *Aedes triseriatus* (Diptera: Culicidae) on rodent (Sciuridae) hosts. J Med Entomol 22: 287–294

Walker ED, Edman JD (1985b) The influence of host defensive behavior on mosquito (Diptera: Culicidae) biting persistence. J Med Entomol 22: 370–372

Walker ED, Edman JD (1986) Influence of defensive behavior of eastern chipmunks and gray squirrels (Rodentia: Sciuridae) on feeding success of *Aedes triseriatus* (Diptera: Culicidae). J Med Entomol 23: 1–10

Walker ED, Torres EP, Villanueva RT (1998) Components of the vectorial capacity of *Aedes poicilius* for *Wuchereria bancrofti* in Sorsogon province, Philippines. Ann Trop Med Parasitol 92: 603–614

Wang S (1986) Study on counterimmunoelectrophoresis and cellulose acetate diffusion techniques to identify mosquito blood meals. Chin J Epidemiol 7: 215

Wekesa JW, Yuval B, Washino RK (1995) Multiple blood feeding in *Anopheles freeborni* (Diptera: Culicidae). Am J Trop Med Hyg 52: 508–511

Wekesa JW, Yuval B, Washino RK (1997a) Multiple blood feeding by *Anopheles freeborni* and *Culex tarsalis* (Diptera: Culicidae): spatial and temporal variation. J Med Entomol 34: 219–225

Wekesa JW, Yuval B, Washino RK, de Vasquez AM (1997*b*) Blood feeding patterns of *Anopheles freeborni* and *Culex tarsalis* (Diptera: Culicidae): effects of habitat and host abundance. Bull Entomol Res 87: 633–641

Wellington WG (1974) Changes in mosquito flights associated with natural changes in polarized light. Can Entomol 106: 941–948

Wesenberg-Lund C (1921) Contributions to the Biology of the Danish Culicidae. D. Kgl. Danske Vidensk. Selsk. Skrifter, Copenhagen, series 8, 7

Wharton RH (1962) The Biology of *Mansonia* Mosquitoes in Relation to the Transmission of Filariasis in Malaya. Bull Inst Med Res Fed Malaya (N.S.) No. 11

Wood CS, Harrison GA, Dore C, Weiner JS (1972) Selective feeding of *Anopheles gambiae* according to ABO blood group status. Nature 239: 165

Woolhouse MEJ, Dye C, Etard JF, Smith T, Charlwood JD, Garnett GP, Hagan P, Hii JLK, Ndhlovu PD, Quinnell RJ, Watts CH, Chandiwana SK, Anderson RM (1997) Heterogeneities in the transmission of infectious agents: implications for the design of control programs. Proc Natl Acad Sci USA 94: 338–342

World Health Organization (1992) Lymphatic Filariasis: The Disease and its Control. Fifth report of the WHO Expert Committee on Filariasis. Technical Report Series No. 82. World Health Organization, Geneva

Wright RH (1975) Why mosquito repellents repel. Sci Am 233: 104–111

Xue R-D, Edman JD, Scott TW (1995) Age and body size effects on blood meal size and multiple blood feeding by *Aedes aegypti* (Diptera: Culicidae). J Med Entomol 32: 471–474

Yates MG (1979) The biology of the tree-hole breeding mosquito *Aedes geniculatus* (Olivier) (Diptera: Culicidae) in southern England. Bull Entomol Res 69: 611–628

Zimmerman JH, Abbassy MM, Hanafi HA, Beier JC, Dees WH (1988) Host-feeding patterns of mosquitoes (Diptera: Culicidae) in a rural village near Cairo, Egypt. J Med Entomol 25: 410–412

Chapter 8 Sampling Adults with Non-attractant Traps

Adults of haematophagous Diptera, especially mosquitoes, are usually caught by using human or animal baits or in light or carbon dioxide traps. Most of these attractant traps collect predominantly unfed females orientated to host-feeding. The use of different baits or attractants usually results in different groups of species being collected, e.g. anthropophilic or ornithophilic species. Often attractant traps are used specifically to collect certain species for isolation of viruses or other pathogens, and in this case the absolute numbers collected rather than the efficiency of collection of different species takes priority. Sometimes, however, especially in ecological investigations, more representative samples of mosquito populations are needed. Because of the virtual impossibility of finding an attractant trap that will sample equally all species, non-attractant traps are preferred. A disadvantage of non-attractant traps, however, is that because they catch mosquitoes only in their immediate area the numbers obtained are small unless mosquito populations are large. All traps discussed in this chapter sample the aerial population, hence the numbers caught depend not only on population density but also on the flight activities of the individuals. Unfed females in fact usually comprise the greatest element of the catch in non-attractant traps, mainly because they are normally the most active. A non-attractant trap is unlikely to be completely free from sampling bias although every effort is made to minimise this. For example, the physical presence of a trap may promote visual responses, causing mosquitoes to be either attracted or repelled by it; similarly the hum of the motor in suction traps may influence the numbers caught. Despite these limitations aerial populations of insects can usually be more efficiently sampled, and with less bias, than most aquatic or terrestrial invertebrate populations. Suction traps probably give the most reliable estimates of both relative and absolute population size. To better understand the ecology of the total mosquito population in an area, the non-active resting populations should also be sampled (see Chap. 5).

771

Malaise traps

This trap, which was originally invented by Dr René Malaise (1937) has been modified and simplified many times and used to catch a variety of insects. Breeland and Pickard (1965), however, appear to have been the first to specifically report on its usefulness in catching mosquitoes. They used a modified Malaise trap designed by Townes (1962). The original paper should be consulted for a detailed step by step construction of the trap, but the four basic parts are as follows: (1) a wooden frame about 76 in square and 50 in tall held in position with guy ropes and within which the trap is supported (this is not an essential component, and is not needed if the trap can be suspended between conveniently placed trees etc.); (2) a lower part of the trap, which consists of four baffles set at right angles to each other and made from two pieces of 102 × 42 in, *black* mosquito netting; (3) a pyramid of *white* netting divided by four white baffles fixed over the framework and lower baffles; (4) a collecting jar consisting of a curved metal cone leading to a transparent plastic funnel, underneath which is a killing bottle, which is fitted over the opening in the apex of the pyramid (Fig. 8.1a,b). Insects flying into the trap are prevented from escaping by the lower baffles and fly upwards into the lightest part of the trap, and eventually pass into the killing jar. Pinger et al. (1975) used Townes-type (1962) Malaise traps in the USA and over about 4.5 months, a period when mosquitoes were active, trapped 8604 mosquitoes, of which 36% were *Aedes vexans*, 13% were *Ochlerotatus trivittatus* and another 13% were *Culiseta inornata*, while *Culex pipiens, Culex restuans* and *Culex salinarius* formed 31% of the catch, and just 3% were *Culex tarsalis*. In other studies Pinger and Rowley (1975) caught 385 blood-fed mosquitoes in Malaise traps, as against only 215 in CDC light-traps supplemented with dry ice.

Marston (1965) described a useful trap made by suspending the net part of a Malaise trap within a framework of tubular aluminium having telescopic legs, such as used in tents. Apart from being very light it is claimed that the trap can be erected in about 10 min. More insects appeared to be caught when the cage was made of 'Visqueen' polystyrene than when 'Saran' cloth was used, but unfortunately this type of polystyrene deteriorates rapidly in sunlight.

Breeland and Pickard (1965) found that, of the 29 mosquito species they collected in Tennessee, USA 27 were collected in Townes-type Malaise traps compared with 19 in light-traps, and about 3½ times as many females as males were caught. They considered that these traps gave more representative samples of mosquito populations than light-traps. In Texas, USA

Easton et al. (1968) used a trap similar to that designed by Townes, except that the framework was constructed of aluminium and not wood; and they also concluded that a Malaise trap could be a useful survey tool for mosquitoes.

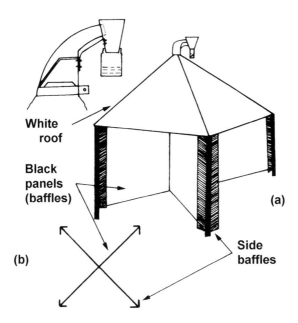

White roof

Black panels (baffles)

(a)

(b)

Side baffles

Fig. 8.1. (*a*) Townes-type Malaise trap, and details of collecting bottle; (*b*) plan of Townes-type trap (Townes 1962)

Both Gressitt and Gressitt (1962) and Butler (1965) used much simpler traps. Of the three rather similar nets described by Gressitt and Gressitt (1962) for use as Malaise traps the one that should prove most suitable for catching mosquitoes is as follows. The net is made from black nylon organdie and consists of a central median panel 7 m long and 3.6 m high, with two end-panels 1.8 m wide and set at right angles (Fig. 8.2). The roof is made to slope downwards on either side of the central panel. Panels 20 cm wide are sewn to the edges of the roof and the side panels and slant inwards to help retain the catch. At the two ends of the central panel the nylon netting of the roof and the end panels is extended to form a cone of netting leading to a plastic cylindrical collecting tube. A straight length of rope is run along under the centre of the roof adjacent to the central panel, to emerge through the conical extensions. The two plastic collecting tubes are fixed to this rope, which is slanted upwards and tied to a tree. Each collecting tube is 22 × 10 cm with an inverted funnel at the entrance.

The opposite end of the tube is removable and contains a small container with a perforated removable lid. A killing agent, such as cyanide wrapped in cotton wool or absorbent paper, is placed in the small tube (Fig. 8.2). Two thinner pieces of rope support the two outer edges of the roof panel, and are tied to a tree or staked to the ground some distance from the trap. Rope is used to tie down the four bottom corners of the side panels and the two corners of the median panel. Finally, a 30-cm high double strip of heavy black cotton poplin is sewn on to the lower edge of the median panel to help weight it down.

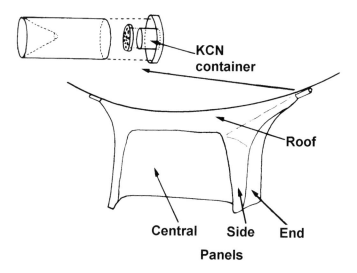

Fig. 8.2. Gressitt and Gressitt-type malaise trap, with detail of catching chamber containing killing agent (after Gressitt and Gressitt 1962)

Butler's trap (1965) is even simpler. It is made from a mosquito bed-net by cutting out one of the longer side panels, but leaving a 1-ft deep strip along the top edge. For greater strength it is advisable, though not essential, to sew a 10-in square piece of cloth into the middle of the roof panel before a hole is cut out from it and a metal cylinder (coffee tin with both ends removed), is inserted. A transparent plastic bag is tied by an elastic band to the top end of the cylinder. A killing agent can be placed at the bottom of the plastic bag. Two light pieces of wood (dowels, bamboo canes, etc.) are inserted across the inside of the two short sides of the roof panel to hold the net open. A long loop of nylon cord is attached to each of the projecting ends of the wooden supports. A piece of wire or string is passed under each nylon loop and attached in the middle to the top of the cylinder and its end secured to a tree. This arrangement results in pulling the centre of the roof panel upwards about 18 in so that a funnel-shaped roof is formed that leads to the collecting bag.

Roos (1957) used a trap divided vertically into two equal parts, each with its own collecting bottle. Thus insects from two opposite directions were caught and retained separately. He positioned his traps over streams to study the upstream migration of aquatic insects. Pruess and Pruess (1966) also used a directional trap but had a separate collecting bottle for each of the four sides, which faced the cardinal points of the compass. A total of 104 mosquitoes were caught.

Malaise traps are normally used at ground level but they can also be suspended at various heights.

It is often assumed that provided shadows are not cast over their entrance Malaise traps give unbiased collections of insects (Breeland and Pickard 1965; Graham 1969; Gunstream and Chew 1967), but it seems likely that at least some insects, including mosquitoes, will either be attracted to, or repelled by, Malaise traps. Townes (1962) for example recognised that different trap designs might affect the relative proportions of different species of insects caught in the trap. Roberts (1970) concluded that tabanids did not just blunder at random into Malaise traps but responded to both trap colour and its light reflectance. Vision plays a very important part in host location by tabanids, so it is not surprising that they can respond to the presence of Malaise traps. Roberts (1976) provides photographs of several types of Malaise traps, as well as the plexiglas trap (Schreck et al. 1970), the canopy trap (Catts 1970) and a Manning trap (Hansens et al. 1971), as used in an evaluation of their attractiveness to tabanids. Roberts (1978) also discusses the effects that modifications, such as introduction of internal horizontal baffles and addition of carbon dioxide, has on the tabanids collected in Malaise traps. Vision is not generally so important in mosquitoes for host seeking, consequently Malaise traps may give more representative population samples, but this needs to be critically evaluated.

On Ellesmere Island in Canada, Corbet and Danks (1973) found that site position can markedly effect the catch of mosquitoes in Malaise traps. They concluded that although the traps were unlikely to have given reliable data on the relative numbers of *Ochlerotatus nigripes* and *Ochlerotatus impiger*, they were nevertheless useful in providing phenological information on the emergence, periods of flight activity and reproduction in these two species.

In California, USA Gunstream and Chew (1967) compared the mosquitoes collected in Townes-type Malaise traps and CDC light-traps using a 0.15-A, radio light. The same seven species of mosquitoes were collected by both traps. The relative proportions of *Psorophora confinnis, Culex tarsalis, Ochlerotatus dorsalis* and *Aedes vexans* varied greatly according to the trapping method. For example, the ratio of the total female catch of these four species in Malaise traps was about 24.7:67.3:2.5:1, whereas in the light-trap the ratios were 1322:86.0:58.0:1.0. In all species a higher proportion of blood-fed females were caught in the Malaise traps, and it was concluded that this trap probably gave more realistic relative measures of population size of the different species than light-traps.

Malaise traps were used alongside CDC light traps, New Jersey Light traps and Shannon traps to collect mosquitoes in Panama, for determination of host preference (Christensen et al. 1996).

Although they can be cheap, easy to make and operate, and require the minimum of attention, Malaise traps have been relatively little used for collecting mosquitoes.

Window trap

In Iceland Jónsson et al. (1986) used, very successfully, a new type of window trap for catching large numbers of chironomids and *Simulium vittatum*. The trap, which might be useful for catching mosquitoes, consists of a 16 × 20-cm perspex box, 16 cm high and divided in the middle by a 20 × 36-cm sheet of perspex (window). The two compartments of the box are filled to a depth of about 12 cm with 4–6% formalin containing a few drops of detergent; in winter ethylene glycol can be added to prevent freezing. One or two holes drilled in one side of the box at 12 cm and covered with fine netting prevent the trap overflowing after heavy rain. Traps are mounted on aluminium poles. Flying insects on hitting the transparent vertical plastic window fall into the formalin.

Ramp traps

Gillies (1969) devised a directional trap, variously called the ramp, flight or intervention trap. These traps allowed the entry of mosquitoes from only one direction, but unlike the traps of Nielsen (1960) or Provost (1960) they were not used to investigate the exodus of mosquitoes from larval habitats but to study the flight direction of hungry unfed females. The traps were used at, or just above, ground level. Each trap is composed of two separate parts, a ramp unit and a detachable collecting cage. The ramp consists of a wooden frame 6 ft long and 3 ft wide, covered with plastic or glass fibre netting and fixed by upright supports at an angle of 135°. Two triangular sections make up the ramp frame. Initially each side section has the two equal sides 4 ft 6 in long, but the upper inner corners are cut off to leave an upper side about 4 ft 2 in long. These two sections are mounted on either side of the ramp platform and a 4 ft 2-in long, 3-ft wide roof rests horizontally across them (Fig. 8.3a). When these side frames are fitted together, a 4-in gap, through which mosquitoes pass into the collecting cage, is left between the top edge of the ramp and the inner edge of the roof. The framework of the entire ramp unit is covered with netting. The collecting cage, which is 3 ft 5 in wide, 1 ft 5 in deep and 2 ft high, fits tightly against the vertical supports of the ramp. A horizontal strut fixed some 7 in from the top of the cage enables it to rest across the roof of the ramp. The section of the front wall that fits over the ramp is covered with wood, while the rest of the cage is covered with netting. Gillies (1969) found that the section of wooden frame of the ramp unit that projected into the cage had to be tapered so that when the two were fitted together a clearance was left between the sides of the cage and the netting on either side of the entry slit. Without this modification some mosquitoes rested in inaccessible parts of the cage and were difficult to collect. Any gaps can be filled in with cotton wool or foam rubber. When the trap is not in use a removable mesh screen is placed over the ramp entrance to prevent various insects entering the trap. There is nothing critical about the dimensions of the trap; all that is required is a suitably inclined ramp that will guide mosquitoes into a collecting cage. Gillies (1969) pointed out that on moonlit nights traps would be more readily seen than on moonless nights, and since some mosquitoes might be either attracted or repelled by them, catches on these nights might differ. He also observed that a ramp trap could reduce wind speed on the leeward side by about 50%.

These traps were successfully used in West Africa to study the orientation of several mosquito species, including *Anopheles melas, Culex thalassius, Culex tritaeniorhynchus* and the *Culex decens* group, to carbon dioxide and animal baits (calf and man) (Gillies and Wilkes 1969, 1970).

In a further study on the range of attraction of mosquitoes (mainly *Anopheles ziemanni, Mansonia africana, Mansonia uniformis, Culex thalassius, Culex decens* group and *Aedes* spp.) to both carbon dioxide and bait animals a modified trap was devised (Gillies and Wilkes 1972). The most important change in design was the separation of the ramp from the collecting cage. This cage is 2 ft high, 3 ft wide, 1 ft 6 in deep and covered with plastic netting and is mounted on a 4-ft high stand. A 3-in entry slit with a hinged wooden lid is made in the cage as shown in Fig. 8.3b,c). The ramp unit consists of a sloping rectangular frame covered with wide-mesh nylon netting which is hooked on to the top of the cage stand. Its bottom edge rests on the ground 6 ft in front of the stand supporting the collecting cage. The tops of two 6-ft high vertical stakes inserted into the ground on either side of the bottom edge of the ramp are connected to the cage by horizontal bamboo canes. The funnel-shaped framework that results from this construction is covered with wide-mesh, double weave cotton netting. This netting was chosen in preference to ordinary mosquito netting as it presents less wind resistance. The funnel entrance to the trap is about 6 ft high and a little less than 6 ft wide. In general these traps caught larger numbers of mosquitoes than the older type, presumably due, at least in part, to the larger cross-sectional area of the ramp entrance. There was, however, little increase in the catch of certain *Anopheles* species.

When the traps were not in use insects were prevented from entering them by simply closing the slit entrance with the hinged lid. The traps were prone to damage by strong winds. To reduce this, the horizontal bamboo canes were loosely attached to the vertical supports, and the netting loosely tied to the vertical bamboo canes inserted lightly into the ground. Consequently, in the face of strong winds the funnel entrance part of the trap collapsed but the rest of the trap was left intact.

To study the vertical distribution of mosquitoes in a coastal area of The Gambia, Snow (1975) used the ramp traps of Gillies (1969) and Gillies and Wilkes (1972), and also 22.9-cm diameter 'Vent-Axia' suction traps. In five trials the ramp traps were set at heights of ground level–1.37 m, 0.69–2.06 m, 1.45–2.82 m, 2.13–3.51 m and 2.90–4.27 m, while in one series of experiments suction traps were placed at heights of 0.68, 2.13, 3.51 m, and in another at 0.91, 3.05, 6.10 and 9.15 m (Fig. 8.4). Most mosquitoes, including *Anopheles melas, Anopheles squamosus, Aedes pseudothoracis, Culex decens* group, *Culex tritaeniorhynchus, Mansonia uniformis* and *Mansonia africana* flew near the ground and were collected in the lower traps. Only the ornithophagic *Culex neavei* and *Culex weschei* were commonest in the higher traps. On moonlit nights ramp trap catches of the *Culex decens* group and *Culex thalassius* were larger, probably because the traps were more visible on these nights. In contrast, catches in suction traps were not more numerous on moonlit nights, indicating that mosquitoes did not respond visually to these smaller traps. More *Anopheles melas* were caught in the suction traps than in the ramp traps, which they seemed to avoid. However, in later experiments, suction traps on moonlit nights yielded reduced catches of *Anopheles melas*, suggesting trap avoidance (Snow 1982). Although ramp traps have been used in The Gambia to study mosquito flight levels and direction (Snow 1975 1976, 1977; Snow and Wilkes 1977), it seems that at least some species respond visually to the traps. For example, the *Culex decens* group and *Culex thalassius* were commonly caught in ramp traps, whereas *Anopheles melas* tended to avoid them. Ramp traps therefore may not give unbiased data. Suction traps were later preferred in studies on flight behaviour (Gillies and Wilkes 1976, 1981; Snow 1977).

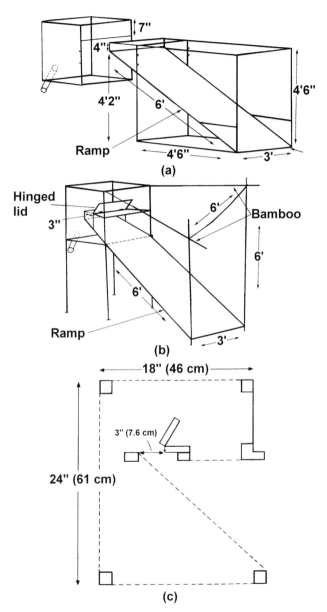

Fig. 8.3. Ramp traps: (*a*) original model (Gillies 1969); (*b*) modified model (Gillies and Wilkes 1972); (*c*) entrance of modified model (after Gillies and Wilkes 1972)

Fig. 8.4. Scaffold tower in The Gambia with Vent-Axia-type suction traps at different heights (photograph reproduced with permission of W. F. Snow)

Stationary nets

Fixed stationary nets

Mosquitoes have occasionally been sampled by horizontal nets in which the opening, which is perpendicular to the ground, is permanently fixed in one direction. Nielsen (1960) used this type of net to catch windborne and migrating mosquitoes. The trap consists of dark blue nylon netting made into the shape of a pyramid with a 2 × 2-m entrance narrowing at the opposite end to an 18-cm diameter opening. The four corners of the funnel-like net are fastened to curtain rings that can be slid up and down two vertical metal supporting rods. A 2½-m long, 18-cm diameter, cylindrical white nylon sleeve is fixed to the apex of the funnel and held out horizontally by tying the distal end to an upright support (Fig. 8.5). Nielsen (1960) used a unit consisting of two nets mounted one above the other to study the dispersal of mosquitoes from their breeding places. Individuals not having a strong directional flight will not readily enter this type of net, or if they do, not many are retained. Provost (1960) studying the dispersal of *Ochlerotatus taeniorhynchus* positioned a pair of such nets at approximately the four cardinal points of the compass facing inwards to the larval habitat. The lower net was centred about 6 ft above ground level and the upper one at about 15 ft. A marked downwind dispersal of mosquitoes was found.

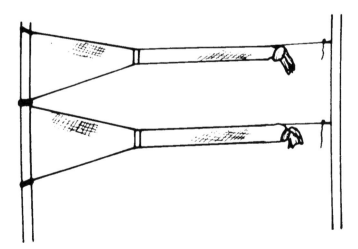

Fig. 8.5. Stationary net (Nielsen 1960)

Wind orientated stationary nets

Although mosquitoes have been sometimes successfully caught in fixed stationary nets as described under the previous heading, nets used to sample aerial populations of insects are usually pivoted so that their openings always face into the wind (Broadbent 1948; Davies 1965; Farrow and Dowse 1984; Freeman 1945; Gorham 1946; Hardy and Milne 1938; Johnson 1950*b*). Such nets are commonly referred to as 'tow nets', which can be misleading as they are not pulled or towed through the air. They do not sample mosquitoes flying in one particular direction, but those flying with the wind. The volume of wind passing through any kind of stationary net is less than would pass through the same areas without a net. Holzapfel and Harrell (1968) measured the speed of the wind passing through net traps used on board ship by placing an anemometer in their entrances. At low speeds both fixed and wind-orientated nets may not be very efficient because mosquitoes may avoid entering them, but when wind speeds are in excess of the mosquito's flight speed, mosquitoes can be regarded as inanimate objects that are blown into the trap.

The numbers of mosquitoes caught in tow nets depends on the numbers of mosquitoes passing through unit area in unit time, such as m^2/s (i.e. flux or migration). Thus catches of insects at low density but travelling fast may be similar to catches of insects at higher density but travelling slower. With insects as small as mosquitoes their travelling speed can be equalised to wind speed. Consequently to compare aerial samples taken with different sized tow nets, in different wind speeds and for different durations, their densities can be computed as follows: Density = catch per unit area/ volume of air sampled (m^3).

Kites

Numerous kite-type trap designs have been described (Pelham 1976) and many readymade kites can be purchased (Jenkins 1981). Because aerial densities of insects can be low the cross-sectional area of the entrance of the tow nets was made as large (1 m^3) as the lifting power of the kite in light winds would allow, while the terylene 0.5-mm mesh net was made as light as possible (1050 g).

With winds of 10–20 km/h or greater, kites were launched on 100–200 m of braided nylon line, using a winch or short tow by a vehicle driven as far as practical (2–5 km) at 5–30 km/h. These kites were pulled to an operating height of 100–500 m. In light to moderate winds (15–50 km/h) the parafoil kite with a drogue gave the best lift, whereas with winds in excess of 50 km/h the smaller of the conyne kite with a drogue was best. A commercial radio-controlled system for model aircraft, using a tow-line trigger, enabled the tow nets to be opened when the kites were at the desired height and then closed. The aim was to get an airflow of 20–30 km/h through the net to retain the catch and prevent strongly flying insects from avoiding the tow nets. Tow nets were kept aloft for 1–3 h, sometimes for 5 h at night. In calmer weather, however, the kites remained airborne for just short periods (10–30 min). Nets could be raised and lowered independently of the kite, permitting the kite to remain aloft. A better arrangement, however, is to use helium-filled balloons to support nets which can remain in the air for much longer, and moreover can ascend to greater heights.

Kay and Farrow (2000) provided circumstantial evidence for the movement of mosquitoes, including disease vectors, from Papua New Guinea to Australia across the Torres Strait. A net with a 1 m^2 cross-section opening and a detachable terylene bag 50 cm long was attached to a kite and raised to various heights above ground using the methods of Farrow and Dowse (1984). Forty of 368 collections contained mosquitoes, yielding a total of 294 specimens, of which 221 were identified to species. Species collected and their relative proportions were as follows: *Culex australicus* (58.8%), *Culex annulirostris* (21.3%), *Anopheles annulipes* s.l. (10.4%), *Ochlerotatus theobaldi* (7.2%), *Ochlerotatus rubrithorax* (1.4%), and *Ochlerotatus sagax* (0.9%). These preliminary findings were further researched by Johansen et al. (2003) who also used a Farrow and Dowse (1984) type net with a 1 m^2 opening suspended at altitudes of 70–150 m from a helium-filled kytoon (Advanced Inflatable Products, Ltd, Essex, UK) to sample wind-borne insects crossing the Torres Strait from New Guinea in order to determine if this was a potential route of invasion of Australia by Japanese encephalitis vectors. In contrast to the Kay and Farrow (2000) study, only three male mosquitoes were captured – two *Verrallina funerea* and one *Ochlerotatus vigilax* and these were thought to have originated locally. Reynolds et al. (1996) used a net and kytoon apparatus described in Riley et al. (1991) to sample airborne insects at heights of 150 m in West Bengal, India. Mosquito species collected during three sampling occasions in November 1992, and March and October 1994 were *Anopheles vagus, Anopheles hyrcanus, Culex fuscocephala, Culex gelidus, Culex quinquefasciatus, Culex tritaeniorhynchus, Culex vishnui, Culex pseudovishnui,* unidentified *Culex* spp., *Mansonia annulifera, Mansonia indiana,* and 17 unidentified specimens. A total of 130 individuals were collected, the most common species being *Culex gelidus* (25), *Culex quinquefasciatus* (22) and *Culex tritaeniorhynchus* (19). 129 of the mosquitoes were captured between dusk and dawn when there were no vertical air movements that could have carried them to this height, suggesting that they had been engaged in active flight to reach and maintain the observed height. Mean wind speeds at 150 m above ground were between 3.4 and 5.1 ms^{-1} which would result in considerable horizontal displacements if the insects remained at this height for several hours. The observation that at least during October and March peak densities were observed from dusk (1630–1830) until midnight or beyond suggests that long distance dispersal could occur. Physiological age of the specimens was not determined.

For catching Ceratopogonidae in Jamaica, Davies (1965) used nylon nets with a 2-ft square opening suspended on a wire frame, and by the use of a wind vane ensured that they always pointed into the wind (Fig. 8.6). The catching container was lined with sticky paper to retain the midges.

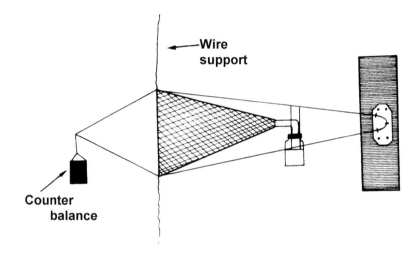

Fig. 8.6. Wind orientated trap (Davies 1965)

Taylor (1962a) described a useful isokinetic insect net; that is one with the inlet so designed that the airflow neither diverges nor converges at the edges of the inlet, the air-flow lines being straight. He concluded that such nets should sample small insects with an error of very much less than 10% in wind speeds above 8 mph, and that 100% efficiency is approached in winds of about 14 mph. Used in conjunction with an anemometer, isoki-netic nets can be used to measure aerial densities, and would probably be useful in sampling aerial populations of mosquitoes in exposed situations, being especially valuable in places where there is no electricity to operate suction traps. Estimates can be made of absolute population size, and with this trap the appropriate formula is:

$$\text{No. mosquitoes/}10^6 \text{ ft}^3 \text{ of air} = \frac{\text{No. caught} \times 10^6}{1 \cdot 53 \times \text{outside wind run in ft}} \qquad (8.1)$$

An estimate of aerial density (D) can be obtained from stationary nets from the following general formula devised by Johnson (1950b):

$$D = (T/(Ax)3600)K_x/100 \qquad (8.2)$$

where T = number of mosquitoes caught per h, A = area of net opening in in ft^2, x = wind speed in ft/s, K_x = percentage of air passing through the net at a wind speed of x ft/s. To use this formula the wind speed measured both outside and through the net must be obtained. The higher the wind speed then the greater the proportion that passes through the net.

Isard et al. (1990) described and illustrated an isokinetic net which was mounted on a helicopter to sample aphids at various heights, later the system was modified to allow discs to be electronically released by the pilot to segregate catches into 12 samples, in terms of elevation, location, time and volume of air sampled (Hollinger et al. 1991).

Radar

Radar provides instantaneous information on the numbers of insects in a volume of air about 10^7 m^3, whereas tow nets sample much less air, for example in moderate winds a net with a 1 m^3 opening samples about 5×10^4 m^3. However, although radar has been successfully employed to demonstrate insect migration it neither identifies the insects (Riley 1979; Schaefer 1976) nor can it usually detect insects smaller than about half the wavelength of the radar (i.e. about 15 mm for conventional 3-cm radar). Because of these and other problems associated with radar Farrow and Dowse (1984) used tow nets carried on kites in Australia for sampling insects from the upper air. However, 8.8-mm wavelength pulsed incoherent radar has recently been evaluated to study flights of insects weighing just 2 mg. Such radar can detect single planthoppers (body length 1.6 mm) at ranges of about 1.3 km (Riley 1992).

Ming et al. (1993) used this type of radar to guide the positioning of a net with a 0.64 m^2 aperture suspended from a kytoon to sample *Culex tritaeniorhynchus* at different heights in China. Fifteen individuals were captured at heights up to 300 m and 29 from heights of 300 m or above.

Chapman et al. (2003) have reviewed the use of vertical-looking radar as a tool for studying high-altitude insect migration. These authors describe the advantages of radar for entomological studies as: its unique capacity to detect insects simultaneously at a range of altitudes that can reach more than 1 km above ground level; and the large sampling volume that it provides relative to traditional sampling methodologies. Analysis of the data received can be used to estimate the body shape of insects passing through the beam, as well as body mass, and even wing beat frequencies. The system currently in use at the Rothamstead Research centres in England can reliably detect movements of relatively large insects, i.e. those with a body mass in excess of 1 mg. To detect relatively small insects like aphids (or mosquitoes) shorter wavelengths than 3.2 cm are required, however, at present this would probably require certain components to be individually manufactured, thereby raising the cost considerably.

Rotary traps

In North America during the 1940s through to about 1970, various designs of rotary traps, petrol driven or powered by electricity, were used by a few mosquito workers such as Chamberlin and Lawson (1945), Graham (1969), Horsfall (1942), Love and Smith (1957), Provost (1957) and Snow and Pickard (1957), but rotary traps have very rarely been used to catch mosquitoes since and are not discussed further here.

Movable traps

These traps are of various shapes and sizes and can be pulled or towed through the air by aeroplanes (Glick 1939; Glick and Noble 1961; Gressitt et al. 1961; Odinstov 1960; Yoshimoto et al. 1962a), by ships (Holzapfel and Harrell 1968; Yoshimoto and Gressitt 1959, 1963; Yoshimoto et al. 1962a,b), or on land by various vehicles (see truck traps below). By adjusting the speed of the vehicle it is possible to control the volume of air sampled, but sampling is not from a point source but from a transect. Ships or aeroplanes have only rarely been used to sample aerial populations of mosquitoes, but Bidlingmayer and Schoof (1957), Nielsen and Greve (1950) and Provost (1960) have all used nets pulled by aeroplanes in abortive attempts to catch mosquitoes. Holzapfel and Harrell (1968) list 43 Culicidae as being caught by traps on ships at sea, and illustrate the various types of traps in which insects were caught, such as fixed and wind orientated windsocks, sticky traps, ventilation traps and suction traps.

Ascending nets

Reference must be made to the nets used by Nielsen and Greve (1950) in Denmark to sample swarms of mosquitoes and other Nematocera. They are not drawn horizontally through the air but sample mosquitoes by rapidly ascending. The net is made of tulle and weighs only 125 g; it is 250 cm long, 60 cm across at the opening but tapers to a narrow cylindrical section only 20 cm in diameter. An 8-m length of line from a hydrogen-filled balloon is attached to three equally spaced strings fixed to the edge of the opening. Another fine length of line is attached to the rim of the net and threaded through a loop made 50 cm from the net in the balloon line (Fig. 8.7a–c). With this arrangement the opening of the net is placed vertically and does not sample the air as the line is slackened to let the net and balloon rise to the desired height.

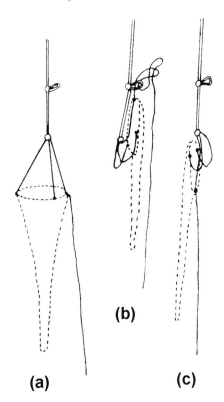

(b)

(a)

(c)

Fig. 8.7. Ascending nets raised by balloons, (*a*) net open for catching on ascent; (*b*) net closed for ascending to catch height; (*c*) net closed for descending (after Nielsen and Greve 1950)

Then by tugging on the line the net is released and its opening now points directly upwards. By releasing the brake on a drum, on to which the line is wound, the net is allowed to ascend rapidly, about 200 m/min, to sample the air. Finally, by pulling on the line the net is closed and slowly hauled down.

Vehicle-mounted traps

In mosquito studies, moving traps have usually been fixed to motor vehicles (Barnard 1979; Bidlingmayer 1966, 1967, 1974; Holbrook and Wuerthele 1984; Johansen et al. 2003; Loy et al. 1968; Provost 1952, 1957; Sommerman and Simmet 1965; Stage et al. 1952; Steelman et al. 1968; de Zulueta 1950) and are called truck or vehicle traps. Hill (1971), however, used a bicycle mounted trap, while Provost (1960) and Steelman et al. (1968) used traps mounted on power boats.

Some of the earliest traps attached to motor vehicles consisted of large cones mounted on either the near or offside front wing (Chamberlin and Lawson 1945; Stage and Chamberlin 1945; Stage et al. 1952). De Zulueta (1950), however, used two muslin conical nets with 60-cm openings mounted on poles that extended 2 m beyond the side of the vehicle and were 1½ m above the ground. When the car was driven at 30 km/h over open savannah country about 20 min after sunset both culicines and anophelines were collected. None of these trapping methods was referred to as a truck trap collection; this term was only introduced later.

Several different designs of truck traps have been developed. Bidlingmayer (1961) used a modification of the trap designed by Chamberlin and Lawson (1945), to catch *Culicoides furens*. He later added a few minor alterations (Bidlingmayer 1966). The trap consists of a pyramidal frame-work of 2 × 2-in wood covered on the inside with glass fibre mesh screening. The opening is 2 ft high and 7 ft wide, and the trap tapers back, about 10 ft to end in a 4 × 4-in opening. The leading edges are made of tempered hardboard. A projection about 15 in long is attached to the top of the trap to support the end of the cloth collecting bag which is tied to the end of the trap. The trap is mounted a few inches above the roof of a vehicle with the front edge projecting over the windscreen (Fig. 8.8). The vehicle is driven at 20–25 mph and at the end of a run, usually of several miles, the bag is quickly removed and replaced by an empty one. It would obviously be an advantage to fit a cone into the sleeve to prevent the mosquitoes escaping when the vehicle stopped.

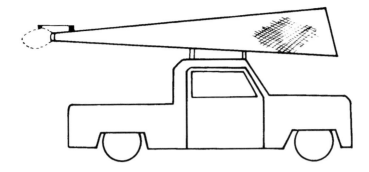

Fig. 8.8. Truck trap (Bidlingmayer 1974)

In Japan a truck trap similar to that used by Bidlingmayer (1966) was mounted on a vehicle driven at 25–30 km/h to sample mosquito populations in villages and rice fields. From regular collections made about 3 h after sunset the seasonal incidence of the four commoner species in the area was obtained. In addition to unfed individuals a number of blood-fed and gravid females were collected, and in the case of *Culex pipiens* these constituted a major element of the catch (Shimizu et al. 1969).

Yoshimoto and Gressitt (1959) soaked the nets in 5% endrin and then sprayed the insides with DDT and pyrethrum to kill the captured insects and prevent them from escaping. Provost (1952) used conical nets mounted both on the roof and front bumper of a vehicle to sample dispersing *Ochlerotatus taeniorhynchus*; from a total of 735 truck trap collections 33 259 female and 8432 males were caught. In later collections Provost (1957) used only one trap which was placed on the roof and was 18 in wide and made in the form of an inverted scoop to catch mosquitoes that were swept upwards from the front of the vehicle. This trap caught 344 148 females and 82 287 males of *Ochlerotatus taeniorhynchus* in 1176 collections.

Loy et al. (1968) designed a lightweight trap weighing less than 25 lb to enable it to be taken as personal baggage on passenger airlines. The trap was constructed of ¾-in aluminium tubing, and the opening of their final model was 5 ft wide and 2 ft high. To reduce the bulkiness of the trap, each of the longer sides was provided with a 2-ft removable section, held in position with sliding pins (split pins). When dismantled and folded the maximum dimensions were 1½ × 2 ft. Nylon netting was used to make the pyramidal collecting funnel, but the anterior 12 in into which the frame was sewn, was made from smooth fabric. A 4-in diameter fabric sleeve was sewn into the apical opening of the trap and glued into a 4-in diameter collecting tube. A standard CDC trap collecting bag was fitted over the other end of the collecting tube. The trap was mounted on top of a frame made of telescopic aluminium tubing. Suction cups and straps were used to secure the trap to the top of almost any vehicle. Guy ropes were attached from the top edge of the trap opening to a convenient structure on the vehicle to help hold the trap in position. In assessing the relative size of mosquito populations the trap was driven over carefully mapped out routes, usually of 5 miles, during crepuscular periods when mosquito flight activities were greatest. Care was taken to sample the population continuously from the beginning to the end of the crepuscular flight period.

In Australia, Dyce et al. (1972) used a modified design of this trap mounted on top of a vehicle driven at 15 mph over preselected routes. The small sleeve-like terylene voile collecting bags were removed at half-hour intervals.

The Dyce et al. (1972) modified truck trap was used by Kettle et al. (1998) to sample *Culicoides* in coastal Queensland, Australia. Collections were made over a period of 27 months between November 1973 and February 1976. The rectangular trap, with an opening 0.6 m high and 1.5 m wide was mounted on a vehicle with the base of the trap 1.35 m above ground level. The truck was driven at 25 km/h over a 3 km course. Collections were made hourly covering the period two hours before sunset and three hours after sunset for 95 days. The trap caught 29 378 *Culicoides* of nine species, of which *Culicoides brevitarsis* was the most common, comprising 35.2% of the total catch. Catches were influenced by time of day, with peak collections occurring at sunset. Some species exhibited a lunar cycle of abundance in trap catches, while intertidal breeding species exhibited a tidal cycle.

Janousek and Olson (1994) used a funnel-shaped 16-mesh nylon net secured to the roof of a motor vehicle to undertake studies of flight activity during a lunar eclipse in Texas, USA. The opening of the trap measured 122 × 76 cm, tapering to a 10 cm circular opening at the posterior end, where a collecting bag was attached. The mobile trap was operated at speeds of 19–24 km/h along a 2.8 km road in a northeast to southwest direction. Three 15-min collections were made each hour from one hour before sunset to one hour after sunrise. The vehicle-mounted trap collected *Ochlerotatus sollicitans, Ochlerotatus taeniorhynchus, Anopheles crucians, Anopheles quadrimaculatus, Coquillettidia perturbans, Culex erraticus, Culex salinarius, Psorophora columbiae* and *Uranotaenia* sp. Collections of each of the four species collected in sufficient numbers for statistical analysis decreased as the eclipse reached totality.

In Belize, Grieco et al. (2002) used a truck-mounted trap similar to that of Bidlingmayer (1966, 1967) to sample *Anopheles* mosquitoes. The frame of the net was constructed from ½ inch PVC tubing covered with green polyester netting, with a mesh size of 530 μm. The trap opening measured 4 ft by 3 ft and was positioned with the base of the opening 6 in above the roof of the vehicle. A plastic funnel encased in a PVC pipe-coupler was fitted to the tapered end of the net. Collecting cups consisted of ice-cream containers with the bottoms removed and replaced with two alternating pieces of dental dam with single 1 in slits cut in their centres. The constant forward motion of the vehicle created sufficient airflow to keep mosquitoes in the collection cup. The vehicle was driven at a speed of 10 mph along a one-mile section of road. Collection runs were conducted every 15 min throughout the night beginning 1 h before sunset and continuing 1 h after sunrise. Only the flashing hazard lights of the vehicle were used during the collection in order to minimise the possibility of mosquitoes being attracted to the headlights. Approximately 1.85 blood-fed mosquitoes were collected in the trap per person-hour of collection.

Steelman et al. (1968) constructed a lightweight trap weighing about 35 lb made from a framework of angular aluminium and covered with glass fibre netting. An 18-in long sleeve at the end of their trap folded on itself when the vehicle stopped, thus preventing the catch escaping. The trap was used on a variety of motor vehicles, and also mounted on the front of a power boat.

This trap proved unsuccessful in catching blood-engorged mosquitoes in Texan rice fields, more were collected by D-vac aspirators (Kuntz et al. 1982). Similarly, although Williams and Meisch (1983) found it collected more mosquitoes and more species (14) than collections with aspirators from vegetation (5), and from artificial resting shelters (6), the proportions of blood-engorged mosquitoes was very low (0.0–2.4%). For example, only 2.2% of *Psorophora columbiae*, the most common species in all collections, were blood-fed compared with 37.1% from D-vac collections and 40.4% from resting stations. In contrast relatively large numbers of blood-fed mosquitoes, especially *Culex annulirostris* and *Anopheles annulipes* were caught in truck traps operating in Australia (Muller et al. 1981).

Sommerman and Simmet (1965) developed a new type of vehicle mounted trap in Alaska, USA in which the mosquitoes were directed into a collecting cage mounted inside the vehicle. The driver was thus able to follow the pattern of collecting as the trap was in use, and could remove and replace the collecting cage without having to get out. The original paper should be consulted for a detailed account of its construction, but the general features are as follows.

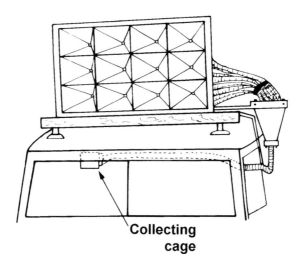

Fig. 8.9. Truck trap of Sommerman and Simmet (1965)

There is a rectangular funnel opening 36 in wide and 21 in high, but because the catch is diverted down into the front of the vehicle the trap only
extends backwards 21 in. Because of this a single funnel is not suitable for
catching and retaining the mosquitoes. Instead the opening is composed of
12 small rectangular funnels (7 × 9 in, about 15 in deep) having steep sides
(76–78°) and arranged in three rows (Fig. 8.9). Mosquitoes collected in
these separate funnels are conducted by 12 flexible plastic tubes into a
funnel mounted vertically outside and just above the driver's window.
A flexible tube from this 'concentrating funnel' guides the mosquitoes
into the collecting cage attached to the sun visor inside the vehicle. The
trap was designed to be sufficiently light for one person to handle, yet
strong enough to withstand rough roads.

Truck traps have rarely been used outside the USA, but in India
Rajagopalan et al. (1977) used a trap on top, and another mounted on the
side, of a jeep driven at 30–35 kph on seven nights along a 11-km stretch of
road. A total of 7309 male and 5825 female *Culex quinquefasciatus* were
collected, most females were unfed, but there were a few blood-fed, half-
gravid and gravid individuals.

Roberts and Kumar (1994) and Roberts (1996) used a 1 m high × 1.5 m
wide × 2 m long net tapering to a 5cm cylinder mounted on the roof of a
vehicle to collect sandflies in northern Oman. A removable collecting bag
was attached to the cylinder at the narrow end of the trap and the vehicle
was driven at 40 km/h for 15 km along a road. Trapping was carried out

from 1700 to 0700 h. It was estimated that the trap filtered approximately 22 500 m^3 of air in a thirty minute period. Seven species of sandfly were collected using the traps, with the four predominant species being *Sergentomyia clydei, Phlebotomus alexandri, Sergentomyia tiberiadis*, and *Sergentomyia fallax*. Truck traps caught proportionately more *Phlebotomus alexandri* (29% of the catch) compared to sticky traps (4%) and roof-mounted nets caught more sand flies than a net mounted on the side of the vehicle, 0.3 m above ground (Roberts and Kumar 1994).

In Louisiana, USA a truck trap with a 1.48-m^2 opening positioned 2 m above the ground was driven at 21 km/h over a 7.1-km route for 30 min each hour from 1830–0600 h. The trap sampled 10 480 m^3 of air during each run, and from three such runs caught 527 male and 3597 female mosquitoes belonging to 15 species. The commonest were *Culex salinarius* (58.6%), *Ochlerotatus taeniorhynchus* (16.3%), *Ochlerotatus sollicitans* (7.8%), *Anopheles crucians* (2.7%) and *Uranotaenia lowii* (2.3%). Most species exhibited peak flight activities just after sunset and before sunrise (Carroll and Bourg 1977).

Barnard (1979) reviewed the use of truck traps and considered most were complicated and expensive to make, and that the Bidlingmayer (1966) model could not withstand hard use. As a consequence Barnard (1979) described a simple, inexpensive, portable, durable trap that one person could assemble on the roof of a car in about 15 min. The disassembled trap fits into the back of a station wagon-type car or pick-up. The frame of the trap is made of a 2.5-cm diameter thin wall steel conduit with the ends hammered flat and bent at 45° angles and bolted together to form a 190 × 72-cm entrance leading to a rectangular cone of 12.5-mesh/cm of nylon marquisette (Fig. 8.10). The collection cage consists of a 10-cm diameter and 20-cm long galvanised tube connected with a hose clamp to a 50-cm long, 15-cm diameter nylon finely woven cloth bag. The entire trap is about 310 cm long. Driving the vehicle at 40 km/h over a 4-km course takes 7 min and about 5500 m^3 of air are sampled. In Colorado, USA in addition to catching 13 species of *Culicoides*, the main target insects, *Ochlerotatus dorsalis, Ochlerotatus nigromaculis, Ochlerotatus trivittatus, Aedes vexans, Culex tarsalis* and *Culiseta inornata* were collected.

Holbrook and Wuerthele (1984) also constructed a lightweight (19 kg) portable vehicle-mounted trap. It was fixed in front of a vehicle to avoid turbulence caused by air passing over the front end of the car (Bidlingmayer 1966). The frame and supporting upright mounts fixing the trap to the front bumper are made of lightweight metal tubing about 1.9-cm in diameter. The forward and rearward sections of the trap are of light canvas. The 256-cm long middle section is made of fine mosquito netting with an entrance of 195 × 85 cm tapering to a 10-cm diameter opening at the rear which is connected to a removable 30-cm long netting bag. Apparently one person can mount the trap in position within 15 min and dismantle it in 10 min. The longest section of the trap is 112-cm, and all parts are easily packed into a fabric carrying case. Using this trap in Colorado, USA Tsai et al. (1989) found that per unit effort, in this instance 1 h operation per night, truck traps were more effective in catching *Culex pipiens* (mean of 21.6) than either CDC light-traps baited with CO_2 (6.4) or Reiter-type (1983) gravid traps (12.9). Disappointingly only three blood-fed *Culex pipiens* were collected by truck traps. However, truck traps were less effective in catching *Culex tarsalis* (3.4) than light-traps (5.0), but superior to gravid traps (1.1). The ratio of the *Culex tarsalis*: *Culex pipiens* collection with truck traps was 1:6.40, with gravid traps 1:12.10 and 1:1.27 with light-traps.

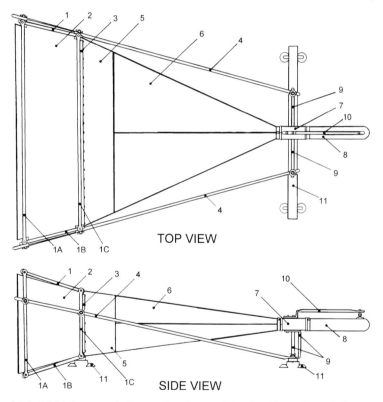

Fig. 8.10. Vehicle-mounted trap: 1—frame, 1A—leading edge of frame, 1B — cross-member, 1C—trailing edge of frame, 2—collecting funnel, 3—retaining bar, 4—frame support, 5—net attachment band, 6—net, 7—concentrator, 8—receiving bag, 9—concentrator support, 10—receiving bag support, 11—luggage rack (Barnard 1979)

It was considered that light-traps seriously underestimated the population of *Culex pipiens*, although species diversity was greatest in light-traps (15 spp.) compared to eight species caught by both gravid and truck traps. These studies were actually comparing collections made at static points (light-traps and gravid traps) and collections made by the truck traps along a transect (3.2 km). To study the flight activities and population size of various British simuliids Davies and Roberts (1973) designed a useful trap fitted to a roof rack mounted on a van. The trap has a 91.5-cm wide, 61-cm high entrance which tapers to 10.2 cm diameter at the opposite end. To ensure a smooth air-flow through the trap it is covered with polyester netting having 13.3 meshes/cm and a 50% open area. The lower edge of the trap is positioned 23 cm in front of the leading edge of the van's roof so as to minimise the effect of the slipstream of air over the windscreen. The volume of air sampled depends not only on the van's speed and the cross-sectional area of the trap, but also on wind speed and direction. To measure the actual volume of air sampled an anemometer is fixed in the entrance of the trap. Insects collected by the trap are forced through tubing and delivered into perspex collecting tubes contained in the back of the van, which is driven at 48 km/h. At lower speeds large insects such as muscids are not forced down into the tubes, while at higher speeds a back pressure is set up which prevents a smooth flow of air through the trap.

The collecting tubes into which the insects are delivered are mounted in a turntable, placed in the back of the van, which is rotated by an electric motor connected to the van's 12-V battery and also to an auxiliary one. At the end of each kilometre run the driver presses a switch on the van's instrument panel which causes the turntable to advance and position a new collecting tube beneath the end of the delivery tube.

Johansen et al. (2003) during a study of wind-borne dispersal of insects across the Torres Strait between New Guinea and Australia used a truck trap consisting of a net with a 1 m^2 opening mounted on a 2 m pole fixed to the back of a pick-up truck. No further details of the net design were given. The truck was driven at 15 km/h over a distance of 5 km along an airstrip between the times of 19:10 and 19:30. Over 14 evenings, this trap collected 9802 arthropods representing 10 orders, with Diptera being the most commonly collected group. A total of 1984 mosquitoes was collected, including *Uranotaenia* spp. (47.4%), *Culex sitiens* sub-group (21.5%), *Coquillettidia crassipes* (9.0%), *Verrallina funerea* (9.6%), *Ochlerotatus kochi* (2.3%), *Ficalbia* spp. (2.2%), *Mimomyia* spp. (1.7%), *Culex quinquefasciatus* (0.9%), *Mansonia uniformis* (0.7%), *Aedeomyia catasticta* (0.3%), *Anopheles farauti* (0.2%), *Culex gelidus* (0.2%), *Anopheles annulipes* (0.1%), plus other unidentified culicid species.

The paper by Davies and Roberts (1980) provides a good example of the analysis that should be applied to truck trap data.

Limitations of truck traps

Non-attractant traps only sample mosquitoes that occur in the specific location occupied by the trap, and this also applies to vehicle-mounted traps, which sample the aerial populations of mosquitoes only in the terrain and volume of air covered by the trap. Thus large populations of certain species may occur in the general area but be largely missed unless the truck trap traverses their flight paths. For example, species which prefer woodland habitats will not be sampled by truck traps. Attractant traps, in contrast, sample mosquitoes not just from the immediate vicinity of the trap but also from a considerably greater area.

The numbers of mosquitoes caught in a truck trap will obviously depend on the time of catching and the mosquito flight times. Bidlingmayer (1966, 1967) proposed that the night might be divided into eight approximately equal periods, the first being from sunset to the end of astronomical twilight. Each of the two periods from the end of twilight to midnight, and from midnight to the beginning of morning twilight are divided into three equal periods, and the final period is from the beginning of astronomical twilight to sunrise. Truck trap collections may have to be made during all eight periods, and also possibly during daylight hours if mosquito populations are to be adequately sampled. However, in many instances collections can be restricted to around sunset, when many mosquito species are most active. During twilight periods truck trap collections can also be divided into short intervals, e.g. 10 min, to study the build up and subsequent decline of flight activities (Bidlingmayer 1966).

Meteorological conditions, especially wind, will affect the efficiency of catching. Bidlingmayer (1967) found that with winds of 0.1–0.9 mph the catches of female *Ochlerotatus taeniorhynchus* and *Culex nigripalpus* were reduced by about 58% of the numbers caught on still nights (wind <0.1 mph), and by about 80% in winds of more than 1 mph. It is recognised that there is increased flight activity of many species on moonlit nights, consequently, truck trap collections are likely to be greater on these nights, although light-trap catches will usually be less (Chapter 9). However, if collections are restricted to crepuscular periods there may be little, if any, significant difference between catch size according to the phase of the moon (Bidlingmayer 1967). Bidlingmayer (1974) in an informative paper on the influence of environmental factors, studied the effects of light level, wind speed, humidity, and temperature, as well as characteristics of the trap site, on the numbers of mosquitoes caught in truck traps, and also on the numbers caught by vehicle-mounted aspirators, in suction, bait and light-traps. At least 10 species were caught; all were crepuscular and were two to four times more active on moonlit than moonless nights. Wind velocity of 0.45–0.89 mph reduced the catches to about a third of that obtained in winds of <0.45 mph; temperatures of 16–18°C about halved the catch that was obtained when temperatures were 19–21°C. Only *Culex nigripalpus* and *Aedes vexans* showed a response to higher relative humidities. Apart from environmental conditions, the speed of the vehicle and the shape of its contours will also influence the size of the catch.

In a later paper Bidlingmayer (1985) again discusses the effect of meteorological conditions on truck trap catches. He stresses the great difficulty of measuring day-to-day population fluctuations because of variables such as weather conditions. He also points out that most sampling methods collect mosquitoes at only one or at the most a few specific sites, not along a transect as do truck traps, and because of environmental variations, it is difficult to interpret catches in terms of the overall population in the area.

Bicycle-mounted trap

Hill (1971) found that a trap mounted on a bicycle was useful for collecting mosquitoes during both the day and night from terrain unsuitable for motor vehicles in Sarawak.

The trap consists of a 60-cm long wooden frame covered with fine wire mesh mosquito netting having an opening 50 cm square tapering to the

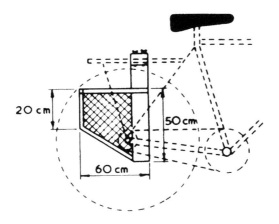

Fig. 8.11. Bicycle trap (Hill 1971)

rear end, which is 20 cm square. The trap is suspended from the luggage carrier of a bicycle and positioned adjacent to the rear wheel about 15–20 cm from the ground (Fig. 8.11). By cycling measured distances along footpaths through villages, rice fields and areas of scrub vegetation at a speed of about 5.8 mph Hill collected about 250 females of *Culex tritaeniorhynchus* in about 27 min, during which time a distance of 4320 m was covered. He also expressed his catch as the number caught/1000 m per 10 min. Most mosquitoes were unfed females, but some blood-fed and gravid females were caught, as were a few males. Hill (1971) found that mosquitoes collected in the trap could not be forced into a collecting bag at its rear, consequently the catch was removed at the end of each run with a battery operated aspirator. Two modifications were suggested, namely that if the trap was mounted on the front wheel then a larger one could be used and there would be less turbulence around the mouth of the trap; secondly, if the trap was mounted on a small motorcycle it would be easier to handle and maintain at a constant speed.

Remote-controlled planes

Large model planes can be used to spray insecticides, and to trap insects (Benzon et al. 1986; Kaniuka 1985). Gottwald and Tedders (1986) described a 2.04-m long model biplane having a 2.44-m wingspan which was powered by a 4-hp chainsaw motor, controlled in flight by a 7-channel FM radio. A conical 19-cm diameter, 35.6-cm long net, which could be opened and closed by remote ground control, was placed on either side of the fuselage between the upper and lower wings. Flying on three occasions in Georgia, USA at heights of 2–54 m over pecan and peach trees the plane caught a large variety of insects including two mosquitoes (Tedders and Gottwald 1986).

Reling and Taylor (1984) described a collapsible 68.6-cm diameter net for trapping aerial populations that sampled some 647 m^3 air/min. The net is inexpensive and can be fitted, with virtually no modification, to a light monoplane aircraft. It proved useful in the USA for catching leafhoppers.

In attempts to catch phlebotomine sandflies in France Killick-Kendrick (1986) flew a radio-controlled model plane having a 4-stroke engine and wingspan of 2.2 m at heights of 10–39 m. A 155-cm diameter 12-cm long pod supporting 30- or 75-cm long gauze nets was fixed under each wing. Although sandflies were not caught, six Diptera, including a *Culicoides* female, were collected in 13 short flights. In some situations there is probably some potential for sampling mosquitoes at heights of about 10–50 m by radio-controlled model planes, as well as it being considerable fun! But as Killick-Kendrick (1986) points out an experienced pilot is needed for take-off and landing if the plane is not to crash.

Non-attractant sticky traps

Sticky traps can be divided into two basic types, attractant and non-attractant. Examples of the former are those employing carbon dioxide (Gillies and Snow 1967; Wilson et al. 1966), a bait animal (Disney 1966) and traps of a particular colour or shape that attract insects (Allan and Stoffolano 1986; Broadbent 1948; Snoddy 1970). Some of these traps are described elsewhere. This section is devoted to non-attractant traps, which rely on insects being caught either as they alight or are blown on to sticky surfaces.

A variety of sticky compounds have been used including commercial tree banding resins ('Bentley's Tree Grease', 'Tanglefoot', 'Stop Moth', 'Stickem', 'Deadline', 'Tack Trap', etc), various greases, castor oil and other oils, mixtures of oils and greases and commercial sticky adhesives. Greases and oils are not such strong adhesives as resins and usually only relatively small insects will be trapped, but if the correct formulation is used mosquitoes can be caught (e.g. Provost 1960). Bird repellents such as 'Beacon' (Walsh 1980), and 'Roost-No-More' (West et al. 1971) have also been used on sticky insect traps. Ryan and Molyneux (1981) compared the efficiency of 23 different adhesives in the laboratory and found that poly-butene adhesives (e.g. 'Oecotak', 'Hyvis') were the best. They present tables giving the physical, chemical and handling properties of the adhesives, their solvents, and numerous references to the 23 adhesives they tested, as well as to three untested ones ('Tangletrap', 'Tack Trap' and 'Stickem'). In England Service (1984) found that 'Hyvis 2000' and 'Rat Varnish' are very suitable for retaining mosquitoes—and also heavier insects, and moreover can be used underwater to trap mosquito larvae.

Although tree banding resins are the most efficient for catching a wide variety of different sized insects the catch is difficult to remove, and with many resins insects can be satisfactorily removed only by heating the adhesive and then scraping them off. With greases and oils insects can usually be picked off and cleaned in benzene, petrol, methanol, isopropanol or a variety of other solvents. One of the most common sticky materials used in commercial adhesives is polyisobutylene, which being non-polar is not easily dissolved in polar solvents such as acetone. Murphy (1985) lists the following solvents as best for removing insects from sticky traps using polyisobutylene adhesives: toluene, heptane, hexane, xylene, ethyl acetate, and various concoctions of these. Other suitable substances are fingernail polish remover and methylchloroform (1,1,1-trichloroethane), which is the modern substitute for carbon tetrachloride as a domestic grease remover. Less effective solvents are petrol and kerosene, which linger on insects for a long time. After dissolving the adhesive with one of these solvents the excess solvent should be removed with filter paper and the insects washed in ethylene glycol ethyl ether (cellosolve) for 1 h or longer (even overnight) to remove the solvent. The cellosolve is then removed by putting the insects in xylene for 30–60 min, after which the specimens can be placed on blotting paper, allowed to dry and then pinned.

It is often difficult to get an adhesive of the correct viscosity and tackiness. If the adhesive is too thin it tends to run down the coated surface and get washed off by rain. High temperatures may also cause greases and oils to become fluid. When the mixture is too thick many insects alighting on the sticky surface are not held and fly off again, only those blown forcibly on to the surface are trapped. In field experiments Browne and Bennett (1981) found that coating trap surfaces with 'Tanglefoot' was ineffective in catching mosquitoes (*Coquillettidia perturbans*, *Ochlerotatus cantator* and *Ochlerotatus punctor*) because they tended to hover around the target and to 'test' it with an extended leg, which on encountering a sticky surface caused them to reverse flight and escape. In certain situations there may be a problem with traps becoming covered with extraneous material such as dust, sand, seeds and even unwanted insects.

Strong (1987) described a very cheap and simple system for dipping cardboard, wooden or metal panels in adhesives, such as 'Stickem Special' and 'Sticky Stuff' to coat them. Adhesives could be applied to as many as 140 panels per hour.

Sticky adhesives can be applied to flat, horizontal or vertical surfaces, e.g. glass plates, boards and screens. Such sticky traps are usually directional, but if mounted at right angles to a wind vane they will trap windborne insects. Furthermore if four sticky flat surfaces are mounted at right angles to each other then mosquitoes from all directions will be caught. As long ago as 1916 Le Prince and Orenstein figured and described such a trap, which was conceived by Mr Quimby, to study flight direction of mosquitoes.

Sticky traps can be placed in various places, such as under the eaves of village huts, near or over rodent holes and amongst vegetation to sample mosquitoes resting in these sheltered situations. More usually, however, they are used to sample aerial populations and most mosquitoes are caught as a result of wind impaction. In Texas, USA Gordon and Gerberg (1945) used 20 × 20-in screens of 18-mesh copper netting mounted in wooden frames to study mosquito dispersal. To determine flight direction four similar screens were slid into grooved arms set at right angles to each other on a post. The centre of each screen was about 5½ ft above the ground. Only a light coating of 'Tanglefoot' was applied to avoid closing the holes in the screens. The five mosquito species caught in order of abundance were *Ochlerotatus sollicitans, Anopheles quadrimaculatus, Ochlerotatus taeniorhynchus, Psorophora confinnis* and *Culex quinquefasciatus*; 88% came from the southeast which was the direction of the prevailing winds and the largest breeding area in the vicinity.

Kay et al. (2000) developed a sticky entry-exit trap for sampling mosquitoes entering or exiting subterranean drainage systems in Australia. The trap was constructed from a 310-mm long section of 55-mm-outside-diameter polyvinyl chloride (PVC) pipe with a spring steel clip punch-riveted to the top. Red-coloured sticky cardboard (Austech International Pty. Ltd., Mt. Hawthorne, Western Australia 6016, Australia), with a 310 × 15.5-mm area of proprietary adhesive on 1 side, was inserted into the tube, with the adhesive facing inwards. One end of the trap was covered with gauze and two tubes were attached to the underside of a manhole cover. One trap was positioned with the open end facing downwards and with the gauze-covered end against the underside of the manhole cover, covering the keyhole. This acted as the exit trap. The second pipe trap was positioned with the open end against the underside of the manhole cover and the gauze-covered end facing downwards. This trap collected mosquitoes entering the sewerage system through the keyholes in the manhole cover. The efficiency of the entry traps was tested by releasing a known number of *Aedes aegypti* into a cage constructed on top of a manhole cover over a mosquito-free drainpipe. Exit trap efficiency was tested by releasing a known number of *Aedes aegypti* into the drain below the manhole cover. A larger proportion of the released mosquitoes were captured in the entry traps than in the exit traps (87% vs 63%). In field trials the following species were collected: *Aedes aegypti, Aedes tremulus, Aedes notoscriptus, Culex quinquefasciatus*, and *Ochlerotatus vigilax*. The majority of mosquitoes entering the manholes were gravid females. Few nulliparous and parous empty females and no blood-fed females were collected suggesting that females are entering the manholes primarily to oviposit. This trap could prove useful for collecting *Culex* mosquitoes from storm and sanitary sewerage systems in cities prone to West Nile virus outbreaks.

Similar traps were used by Montgomery et al. (2004) to determine the relative contribution of subsoil drain sumps to the *Aedes aegypti* standing crop in Cairns, Australia. The trap, termed a Sticky Emergent Adult Trap, or SEAT, consisted of a black plastic sheet with a hole in the centre. A clear plastic cup (250 ml) with the bottom removed and replaced with a 1.0 × 0.5 mm mesh screen was inserted through the hole, forming an airtight seal. The mesh screen allowed for proper functioning of the drain. The black plastic sheet was held in place by the sump cover. A sheet of clear plastic cut to size was coated in isobutylene adhesive and wrapped around the outside surface of the plastic cup and secured in place using a single staple. Traps were left in place for 48 h before they were recovered and the sticky plastic sheet removed. A total of 866 adult mosquitoes (21% *Aedes aegypti* and 79% *Culex quinquefasciatus*) were collected from 162 drain sumps. A disadvantage of this design of sticky trap noted by the authors is the fact that due to the limited surface area available for trapping mosquitoes, in highly productive sites, not all mosquitoes will be sampled.

In Florida, USA Haeger (Provost 1960) used 22-in cylindrical nylon nets coated with an adhesive (1½ lb amber gear grease, 12 pints No. 20 motor car oil and 1 pint mineral spirits) and caught 355 adults of *Ochlerotatus taeniorhynchus* during the short period of evening exodus from larval habitats. The following evening rectangular sections of mosquito netting were mounted in 1 × 2 ft aluminium frames, coated on both surfaces with adhesive, and suspended in pairs at right angles to each other at heights of 10, 20 and 30 ft. From 34 nets a total of 176 *Ochlerotatus taeniorhynchus* were caught during mass dispersal of adults from their larval habitats (Provost 1960). Males represented 26% of the catch, although on the previous night using cylindrical sticky traps they formed 79% of the catch. On both occasions the proportions of males on the sticky nets were less than obtained by collecting resting mosquitoes, thus indicating that not all emerging males dispersed.

A disadvantage of solid flat surfaces for trapping is that eddies usually develop around their edges (Fröhlich 1956) and consequently not all insects blown towards them are caught. Cylinders or mesh screens are more efficient in catching windborne insects. Cylindrical sticky traps, as originally developed by Broadbent (1946) to sample aphids, are more efficient than most flat surfaces in sampling windborne insects because the air flows more smoothly past them than with solid flat surfaces; in general they become more efficient as their diameters decrease. They do not appear to have been used to sample mosquito populations but might prove useful in certain situations. The best procedure is not to coat the trap with an adhesive but to apply this to a sheet of paper or plastic that is wrapped around the trap, which can consist of a test tube, tin can, glass jar or a length of plastic or drain pipe. Such sticky surfaces can be readily removed and replaced with new ones.

Rohitha and Stevenson (1987) made an automatic sticky trap for sampling aphids that changed a sticky cylinder daily for 7 days. Basically a vertically mounted long section of plastic tubing houses small (188-mm long) sticky cylinders, one of which is dropped down on a central rod into an exposed situation where it catches aphids. After 24 h a simple cog and notch arrangement operated by a 7-day clock (ex thermohydrograph) allows this sticky cylinder to drop into a lower section of plastic tubing, while at the same time another sticky cylinder descends to replace it.

Gregory (1951) showed that the efficiency of a vertical sticky cylinder for catching fungus spores was related to a non-dimensional function k:

$$k = \frac{V_s V_0}{rg} \tag{8.3}$$

where V_s = wind speed, V_0 = terminal velocity of the spores, r = radius of of the cylindrical trap and g = gravitational acceleration, i.e. 9.81 m/s^2. Taylor (1962b) showed that small insects, up to about ¼ in long, behave as inert particles in wind speeds of more than 2 mph. Apparently only in winds of less than this do small insects exert any control over whether they are caught on cylindrical sticky traps. Aerodynamic efficiency increases steeply with increasing wind speed and approaches a constant value in winds over 6 mph. Taylor presented a table of conversion factors (in logs) which when added to the log catch per hour of insects caught on a 5-in diameter, 12-in long non-attractive sticky cylinder gave log density per 10^6 ft^3 of air. These conversion factors apply only to traps of this size and only to insects up to about ¼-in in length, but this would include a number of mosquito species.

Water traps

These consist of metal, glass or plastic receptacles of various shapes, sizes and colours containing water to which a wetting agent has been added so that when insects rest on, or fall into, the water they are wetted and trapped. Small amounts of preservatives, such as formalin, can be added to the water to help preserve the catch. It seems unlikely that as such these traps will be of much value in catching mosquitoes. But Grigarick (1959) used floating water traps to help catch insects settling on water. Possibly, such traps, or even the addition of a wetting agent direct to the water of small larval habitats (e.g. containers, small ground pools), might wet mosquitoes that rest on the water surface to oviposit.

Suction traps

Insect suction traps measure the aerial density of insects at the site of the trap. Because they do not employ any attractant the differences between numbers of various insects caught should reflect their natural densities. They are thus usually considered to give unbiased samples (Service 1969; Taylor 1962*a*), but with some insects this may not be completely true. For example, Banks (1959) considered that small suction traps might be selective in catching syrphids as some of the larger species are strong fliers and might consequently escape capture. Way and Banks (1968) in fact found that the number of syrphids caught in 'Johnson–Taylor'-type traps (Johnson 1950*a*; Taylor 1951) could not be correlated with the numbers caught on sticky traps. Also, the proportions of *Chrysops caecutiens* and *Haematopota pluvialis* caught in suction traps differed markedly from their proportions in human landing catches (Service 1973*a*). It was thought that the actual traps might differ in their visual attractiveness to the two species. In the USA Bidlingmayer and his colleagues have conducted several field experiments showing that their rather large L-shaped suction traps (Fig 8.12), are visible to mosquitoes, as are surrounding bushes and trees. They concluded that trap catches are influenced by these visible cues (Bidlingmayer and Hem 1979, 1980; Bidlingmayer et al. 1985). Similarly Snow (1975) believed that some, but not all, species of mosquitoes tended to avoid the much smaller 'Vent-Axia' suction traps he used in The Gambia. Bidlingmayer also considered that even moderate winds reduced the numbers of mosquitoes caught in his suction traps (Bidlingmayer 1974; Bidlingmayer and Hem 1980; Bidlingmayer et al. 1985). Nevertheless despite these possible limitations suction traps probably give less biased samples of mosquitoes (Bidlingmayer 1967; Gillies and Wilkes 1976; Service 1969, 1971*a*; Snow 1975) than other sampling methods. However, truck traps will also give 'unbiased' catches, the only important variables affecting their catch size, apart from mosquito numbers *per se*, will be weather conditions that affect flight behaviour.

In addition to measuring absolute aerial densities suction traps can be used to study vertical distribution, diel flight activities, seasonal incidence, and flight direction. Because they are non-attractant, visual cues excepted, few mosquitoes will be caught in the traps when populations are low, consequently prolonged sampling may be necessary if statistically reliable numbers are to be obtained.

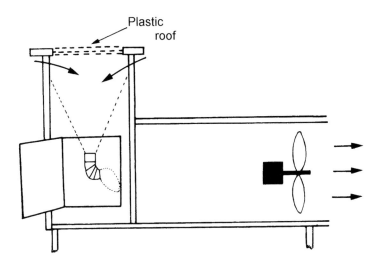

Fig. 8.12. Suction trap of Bidlingmayer (Bidlingmayer 1974)

It must be emphasised that suction traps do not measure or sample adult populations of mosquitoes as such, but only that fraction which is actively flying; it is the aerial population that is sampled. Consequently, the numbers of mosquitoes caught in the trap depends both on the size of their population and their flight activity. It is therefore not surprising that unfed females constitute the greatest proportion of mosquitoes that are usually collected by suction traps.

The volume of air sampled by a trap depends on both the diameter and speed of the fan and also the angle of pitch of its blades. The same air speed can be obtained with a steep pitch blade angle and slow motor as with a small pitch angle and a fast motor, but as a fast motor uses less power and is consequently lighter, the best combination is a small pitch angle and a fast motor. When traps are operated from a long length of cable sufficiently thick wire should be used to minimise voltage drop, which may otherwise slow down the fan and result in a smaller air sample. The total impedance is greater in a 3-phase than 1-phase motor of the same power, thus the proportion of voltage lost is less with a 3-phase motor. They also have the advantage of being considerably lighter than 1-phase motors. All traps should be efficiently earthed. Lightweight materials are essential when large fan blades are used.

Although the volume of air sampled by specific fans may be given in publications it is advisable to check the air displacement of a fan to be used in a trap, either with the manufacturer or in the laboratory (e.g. Loomis

1959; Macauley et al. 1988; Mulhern 1953; Taylor 1955, 1962*a*; Wainhouse 1980), because of the variability between similar traps, and because a new model may have been produced which gives an increased air flow. In certain studies, however, it is unnecessary to know the volume of air sampled so long as all traps are sucking in the same, but unknown, quantity of air.

Two very different designs of suction traps have been mainly used to sample mosquitoes. Namely, vertical models employing a 22.9-cm 'Vent Axia' fan based on the traps of Johnson (1950*a*), Taylor (1951) and Service (1971*a*), which have been predominantly used in Britain and The Gambia, and the much larger L-shaped traps of Bidlingmayer (1964, 1967) used in the USA in which the 24-in fan is mounted on its side (Fig. 4.8*e*). A few other traps that have occasionally been used to sample mosquitoes, or have the potential to do so, are described towards the end of this chapter.

'Johnson-Taylor' exposed cone traps

These traps were designed at Rothamsted Experimental Station, England, to sample aphids mainly near the ground or in areas with little wind. The original trap was devised by Johnson (1950*a*) but its design and construction was improved by Taylor (1951) (Fig. 8.13*a,b,c*), and a commercially-manufactured version is available in the UK (Burkard Manufacturing Co., Rickmansworth). The papers of Johnson (1950*a*) and Taylor (1951) should be consulted if a trap with dropping discs to segregate the catch into time intervals is to be constructed. The essential components are as follows: a 22.9-cm diameter 'Vent-Axia' fan, with the grill over the opening removed is mounted vertically. It is important that an impeller-type fan is used, i.e. a fan with reversed blades, so that the air is sucked down into the trap and not blown out. The unit is mounted with the fan blades uppermost and the motor underneath. Attached to the bottom of the fan is a copper, or monel metal, 21-in long cone made of 26-mesh gauze (Fig. 8.13*b*). Three equally spaced metal struts run down the outside of the cone and extend beyond it to hold in position the metal cylindrical casing that houses the collecting magazine. A central 24-in long rustproof metal guide rod is bolted to the base of the motor casing and projects downwards through the mesh cone to extend about an inch beyond the bottom of the collecting magazine. A solenoid and disc release mechanism is mounted just beneath the motor. Each segregating disc is made of brass or stainless steel, is 1¼ in in diameter and has a central collar ¼ in high and $^5/_{16}$ in in diameter to keep the discs apart (Fig. 8.13*c*). When a disc is released and falls down the central guide rod this fringe brushes against the walls of the collecting magazine ensuring that small insects are not missed but trapped between successive discs. The discs should be soaked in 1–3% pyrethrum solution or other killing agent. Insects are drawn into the trap by the fan and are gently blown down the collecting cone and deposited on to an oil film smeared on the upper exposed surface of the last fallen disc. They are killed by the killing agent and are sealed off by the next disc that descends the rod.

The traps are operated from mains electricity and a control box allows the fans to work at slow, normal or boost speeds. Unless mosquito densities are high, traps are best operated on the fastest speed as this samples a greater volume of air. An electromagnetically operated time switch allows the discs to be released at pre-set time intervals. Burgess and Muir (1970) described how a simple piece of additional mechanism can be fitted to the 'Sangamo–Weston' time switch, which is usually used with the complete trap unit, to enable the catch to be divided into 12-h intervals.

Several simple precautions are needed for the efficient operation of the traps. Great care must be taken not to bend the central rod in the trap otherwise the discs will not fall freely. The interior of the fan casing must be kept clean otherwise from time to time large numbers of insects may become squashed on the inner surface of the casing surrounding the fan. This may build up to such an extent that the fan is stopped, which results in the motor overheating and may cause it to burn out. In wooded areas leaves may lodge in the trap and either prevent the fan from turning or stop the discs descending. It is essential that the monel mesh cone of the trap is kept clean. If dirt is allowed to accumulate on the mesh this may affect the smooth flow of air through the trap, which is essential if the catch is to be related to the volume of air sampled. These types of traps having an exposed gauze collecting cone should not be used in situations with cross winds of more than about 10 mph, otherwise there is a significant reduction in the intake of air. This can be minimised, however, by enclosing the cones in a cylindrical tube. Southwood and Henderson (2000) consider these traps to be most suited to sheltered conditions, for example between plants in a crop field.

These traps have been used in England to obtain the seasonal incidence and diel flight periodicities of several British mosquitoes (Service 1971a). They have also been valuable in detecting the presence of adults of several anthropophagic species before they are orientated to blood feeding and are caught in bait catches (Service 1973b).

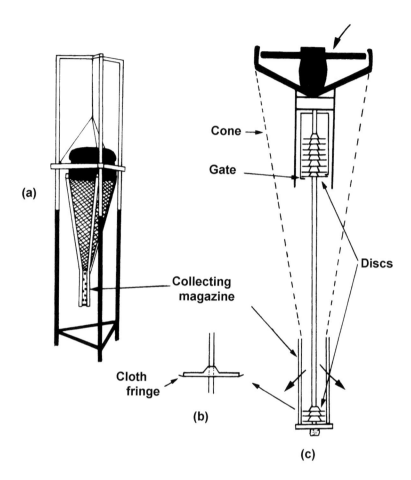

Fig. 8.13. (*a*) Johnson–Taylor suction trap (Johnson 1950*a*; Taylor 1951); (*b*) Johnson–Taylor trap showing arrangement of fan, cone and discs; (*c*) single disc with cloth fringe

Larger traps with enclosed cones (aerofoil and airscrew traps)

Larger and more powerful traps have been designed for sampling insects in exposed situations such as high up in the air where wind speeds are more likely to affect the sampling efficiency of the trap. More powerful traps, sampling a greater volume of air, are also needed in situations where insect density is low. Four large traps have been described by Johnson and Taylor

(1955*a–c*), the most useful being the 'Enclosed' 12 and 30-in traps. The 12-in trap has an aerofoil fan mounted in a 14-in diameter, 4-ft long steel cylinder and delivers about twice the volume of air as the 12-in 'Vent-Axia' fan. The larger trap employs a 30-in light-weight airscrew fan mounted in a 2½-ft diameter, 7½-ft long cylinder and delivers about five times the air as the 12-in aerofoil trap. Because insects get damaged if they pass through the fan blades of these high speed fans, the fan is positioned below the collecting cone and segregating device. An essential feature of the trap is that the collecting cone is enclosed within a cylindrical duct made either from sheet metal or rubberised fabric stretched over a light metal frame. By shielding the cones the effect of cross winds on sampling efficiency is greatly reduced.

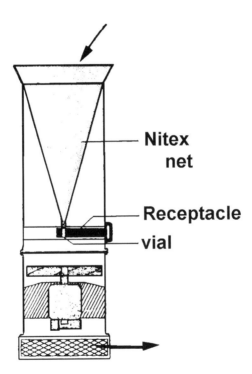

Fig. 8.14. Large diameter suction trap (from Holzapfel and Harrell 1968)

Holzapfel and Harrell (1968) described a 90-cm diameter suction trap with a conical nitex funnel-shaped net surrounded by an aluminium cylinder (Fig. 8.14). A 1-h.p. 1-phase electric motor operating a tube axial fan was placed below the collecting cone and displaced some 6800 m^3 air/h.

Insects did not pass through the fan but collected in a vial held in place by a plastic receptacle which was removed by sliding it horizontally from the trap. When used at sea the fan tended to reverse its direction when the wind was more than about 22 mph. To increase the volume of air sampled, a metal scoop provided with a wind vane was fitted on top of the trap.

In Texas, USA an enclosed suction trap of the Johnson–Taylor (1955a) design and having an 18-in diameter fan, was modified to incorporate a turntable containing pint-capacity glass jars half-filled with ethylene glycol (car antifreeze) as a preservative. This allowed the catch to be separated into 2-h intervals (Goodenough et al. 1983). The trap used a more powerful tube axial fan (1.1 kW) (i.e. 1.5 h.p.) than the Johnson–Taylor trap and sampled about 4500 m^3 air/2-h sampling period, as against about 3000 m^3 for the original Johnson–Taylor trap.

Allison and Pike (1988) describe a home-made enclosed type of suction trap having an 85-cm long, conical net leading to a collecting jar, both of which are enclosed within a 1.5-m length of 38-cm diameter PVC tubing. On top of this a 30-cm diameter 4-m length of PVC tubing is fixed to act as a chimney and sample insects about 8 m from the ground. A three-bladed fan attached to a motor draws the catch down into a collecting jar.

Rothamsted insect survey trap

This trap was designed to enable insects to be sampled at considerable heights without having to construct any complex supporting mechanisms, such as a steel tower (Taylor and Palmer 1972). It consists of a 30-ft tall, 10-in diameter plastic 'chimney' mounted on a 10-ft high, 30-in square wooden box, which contains a centrifugal fan at its base. This type of fan is chosen because it ensures an almost constant air intake in most wind speeds, and also when the gauze collecting cone at the bottom of the trap becomes partially blocked by dust and dirt. Insects sucked down the tall chimney are collected in a bottle attached to a collecting cone. To reduce the speed of insect impaction, which can be as much as 35 mph at the base of the trap, the cone has a large surface area. The trap samples 101 040 ft^3 air/h.

Macauley et al. (1988) modified this trap. Their trap consists of a 9.2-m length of 244-mm diameter plastic pipe with a flared inlet positioned 12.2 m from the ground, while the other end is placed on top of a 3-m high box containing a centrifugal fan, and the necessary filtering and storage devices. The inlet air speed greatly exceeds insect flight speed, i.e. >50 km/h, to give a sample volume of 45 m^3/min. Air flow measurements can be made with a Lambrecht direction velocity metre which has a Prandtl-type pitot static tube. When readings are used in conjunction with tables supplied by the fan's manufacturer air flow through the trap in m^3/min is obtained. But this process likely overestimates air flow by more than 10%. According to Winternitz and Fischl (1957) a more accurate measurement is derived from

$$V_m = 1/5[V_{0.5} + 2(V_{0.081} + V_{0.919})]$$ (8.4)

where V_m = mean velocity and $V_{0.081}$, etc. are velocities at distances of 0.081, 0.5 and 0.919 diameters from one wall of the trap. Errors are approximately within ±1%.

Schaefer et al. (1985) used radar methods and the remote sensing IRADIT infra-red system to measure aerial densities of insects near the Rothamsted Insect Survey (10-in diameter) traps positioned with their inlet at 12.2 m, and aerofoil traps (12-in diameter). The aim was to study the effectiveness of these suction traps in relation to increasing wind speeds, such as when traps are used at elevations or in exposed areas.

Simple 'Vent-axia' traps

When it is not necessary to divide the catch of mosquitoes into time intervals a much simpler trap than the 'Johnson–Taylor' one, without any segregating mechanisms, can be used; the catch being removed daily or at longer intervals (Gillies and Wilkes 1976; Service 1971a,b, 1974, Snow 1975, 1977). A simple but efficient suction trap can be constructed from a 6, 7½, 9 or 12-in 'Vent-Axia' fan made for window mounting; at boost speeds these fans sample 12 000, 18 000, 30 000 and 62 000 ft^3 of air/h. First, the outer grill is removed, then any recesses or ledges near the fan blades are covered over and the fan unit inserted into a circular hole cut out from a piece of plywood (Fig. 8.15). A monel gauze collecting cone (B) is attached to the board underneath the fan and a small fine mesh bag (C) is tied to the bottom of the cone to collect the mosquitoes. Two 1-in holes are drilled at opposite ends of the board and lengths of ⅝-in galvanised tubing inserted and pushed into the ground. By securing the board to the tubing by small adjustable clamps (e.g. 'Klee-Klamps') its height can be easily altered, thus allowing the trap to be used at various heights. Wire from the trap enters a control box, which allows the use of three fan speeds. Control boxes from a number of fans can be mounted within a waterproof metal box. A length of flexible tubing (A), having a diameter several inches larger than the base of the collecting cone (B), can be attached underneath the plywood board to enclose the cone and thus reduce the effect of cross-winds on sampling efficiency. However, if the traps are used in sheltered positions, such as near the ground or amongst the shelter of vegetation this may not be necessary.

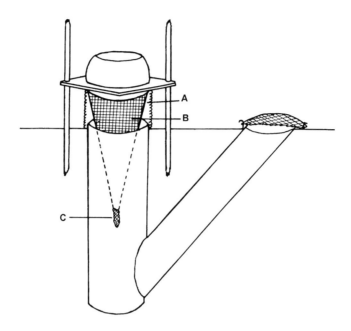

Fig. 8.15. A 9-in diameter suction trap with cone projecting into plastic tubing imbedded in the ground: A—flexible tubular ducting, B—mesh collecting cone, C—collecting bag (Service 1971a)

Trap inlets must be positioned near the ground to sample mosquitoes flying near the ground. In very dry areas this can be achieved by just digging a hole and lowering the trap into it, in which case care must be taken that the filtered air exhausted from the trap is conducted away from the fan inlet and not resampled, thereby reducing the sample volume. For example, Snow (1982) and Gillies and Wilkes (1976) were able to place their 22.9-cm 'Vent-Axia' traps in a 1-m^3 pit during the dry season in The Gambia. Two shallow trenches from the pit allowed exhaust gases to escape. But in many areas the ground is waterlogged and holes rapidly fill with seepage water. To overcome this, the collecting cone can be placed in a length of plastic or metal tubing. To ensure a smooth flow of air through the trap (Service 1971a) another section of smaller diameter tubing is welded on to it near the base before it is placed in a hole dug in the ground (Fig. 8.15). It may be necessary to hold the tubing in position with ropes while the earth is tightly packed around it. The opening of the exhaust vent should be covered with plastic mesh to prevent small animals slipping down the tubing. A small amount of rain water may accumulate at the bottom of the tubing but the air blowing through the trap helps to evaporate this; however if too much forms it can be pumped out. It is essential that any water that does collect at the bottom does not become contaminated with organic debris, otherwise large numbers of gravid flies may be attracted and become squashed on the inside of the trap.

If the shape of the collecting cone is altered or enclosed within a cylinder it is advisable to check that there is still a smooth flow of air through the trap. This is very simply done by producing smoke at the entrance of the trap and watching its passage through the trap. Titanic chloride streamers or Bee Keeper's smoke generators are useful. Alternatively smoke can be produced by burning corrugated paper soaked in potassium chlorate. The manufacturers' specifications of the volume of air displaced by a trap may be reduced when a collecting cone is added, but Taylor (1955) found that with the 9-in trap this is usually negligible.

Simple 'Vent-Axia' suction traps without any segregating device were used for several years to study the behaviour of British mosquitoes (Service 1971*a*, 1974) and *Culicoides* (Service 1971*b*, 1974), while in The Gambia they have been used to catch mosquitoes flying in different directions and at different heights (Gillies and Wilkes 1976; Gillies et al. 1978; Snow 1975, 1977, 1982). They have also been used to study the behaviour of mosquitoes attracted to human baits (Gillies and Wilkes 1978). Their 'Vent-Axia' fans were sometimes placed on metal scaffolding towers at different heights (Fig. 8.4) (Gillies and Wilkes 1976; Snow 1975, 1982), or very near the ground (10–20 cm) (Gillies et al. 1978; Snow 1982).

Suction trap of Bidlingmayer

The Bidlingmayer suction trap differs from the traps developed by Johnson and Taylor in having the fan mounted on its side, not vertically. The body of the trap is made of plywood over a wooden frame and is shaped like the letter 'L' lying on its side (Fig. 8.12). At the distal end of the section lying horizontal to the ground is a 24-in diameter ¼-h.p. fan with a displacement of about 216 000 ft^3 air/h. The intake of the trap, which measures 31 × 31 in is situated 5 ft above the ground at the top of the short vertical section. Air drawn through the intake passes through a large mesh cone and is discharged through a metal tube attached at right angles to the cone. A cloth bag is fitted to the end of this metal tube to retain the mosquitoes, which in this trap do not pass through the fan blades. To prevent rain and debris entering the trap a clear plastic flat roof is positioned about 8 in above the intake (Fig. 8.12).

These traps have been extensively used in Florida, USA by Bidlingmayer to study the flight behaviour of mosquitoes in relation to topography and mete-orological conditions (Bidlingmayer 1964, 1967, 1971, 1974; Bidlingmayer and Evans 1987; Bidlingmayer and Hem 1979, 1980; Bidlingmayer et al. 1985, 1995). In a brief re-description of the trap Bidlingmayer and Hem (1979) gave the same measurements as detailed originally by Bidlingmayer (1964), but in metric, and stated that it sampled 102 m^2 air/min (velocity = 2.74 m/s). Measurements made at 5 cm above and horizontal to the edge of the trap's entrance showed that in still air ve-locities were 347, 238, 145, 60 and 25 cm/s at distances of 0, 10, 20, 30 and 40 cm, respectively.

Employing traps which were somewhat similar to Bidlingmayer's trap, Dow and Gerrish (1970) found that day-to-day differences in the catch size of *Culex nigripalpus*, but not *Ochlerotatus taeniorhynchus*, were signifi-cantly positively correlated with day-to-day differences in relative humid-ities measured 1 h after sunset.

Bidlingmayer and Evans (1985) describe a time interval sampler that can divide the catch from Bidlingmayer-type suction traps into 27 samples. It consists of two concentric acrylic plastic cylinders (6, 10), about 15 in high (Fig. 8.16). The inner cylinder (6) is mounted around a central stainless steel helix screw (20) and contains a stacked series of 14 paired semicircular collection chambers (cups) (17). A motor (115-V, 5-W, 12 rev/min) and drive assembly turns the helix screw and screws the inner cylinder down into the outer one (10), and in so doing positions a cup op-posite an entrance port (16) through which the air and insects of the suc-tion trap are blown. The overall height of the cylinder is 32 in. Although this rather complicated device is built to receive insects from a horizontal air-stream, the authors point out it could be mounted horizontally to re-ceive a vertical airstream. Moreover, if three cups were positioned at 120° apart on the inner cylinder instead of two at 180° the height could be reduced to just 22 in.

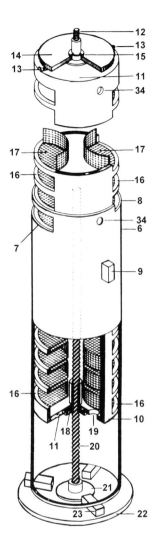

Fig. 8.16. Cylinder assembly of telescopic collection chamber for suction traps and other traps: 6—outer cylinder, 7—inlet port, 8—notch, 9—key, 10—inner cylinder, 11—end plates, 12—upper spindle, 13—eared tabs, 14—gear, 15—washer, 16—cup openings, 17—collecting cups, 18—helix nut, 19—drain holes, 20—helix screw, 21—flange, helix screw, 22—basal disc, 23—spacer, 34— locking holes. For full details of other parts see Bidlingmayer and Evans (1985)

Directional suction trap of Horsfall

Because stationary nets with vertical openings mainly collect mosquitoes flying with the wind Horsfall (1961) used directionally orientated suction traps to study flight direction. He constructed a group of four suction traps into a single unit. Each trap consisted of an 8-in fan mounted at the top of a 10-in diameter, 18-in long cylinder, which had a copper mesh (14 × 18 meshes/in) collecting cone at the base leading to a small collecting cage. A jointed and elbowed 10-in diameter metal cylinder, with a flared opening 13½ in in diameter at right angles to the fan, was mounted on top of the trap. The openings of the four traps comprising a single unit faced different directions. Horsfall (1961) considered that when an 8-in diameter fan was used insects were drawn into the traps from a distance of up to 6 in from the openings. However, this obviously will depend, among other factors, on wind speed and insect size. Because of the limited distance from which mosquitoes were sucked into the traps Horsfall (1961) considered that those collected were mainly individuals flying towards the traps and consequently directional flight was measured.

Trap construction could be simplified by using flexible tubing (e.g. 'Flexitube') instead of a jointed metal cylinder; alternatively the traps could be positioned on their sides with their openings perpendicular to the ground.

In Indiana, USA Novak et al. (1981) modified the Horsfall trap to study the vertical distribution of mosquitoes in a wood. Their trap consisted of a 25.4-cm diameter metal cylinder elbowed at the top and flared to a 58.4-cm diameter intake (Fig. 8.17), which was painted black with contrasting white stripes. A mesh screen funnel leading to a mesh collecting cylinder (14 × 18-mesh) having a removable screen end was positioned within the top of the straight section of tubing. A Dayton duct-type 28-cm diameter fan was mounted near the bottom of the tubing.

Mosquitoes up to 25.4 cm from the intake were sucked into the trap and were not damaged by passing through the fan blades. To position traps at different heights in the wood a lead pellet (70 g) fitted to a spool of 36-kg nylon line was catapulted over a branch capable of supporting the 9-kg trap. The nylon line was then attached to a rope on which a pulley was mounted that allowed traps to be raised and lowered. From 40 days sampling with two such traps 924 mosquitoes of both sexes and belonging to

20 species were caught, the most common being unidentified *Culex* species, *Ochlerotatus hendersoni*, and *Anopheles barberi*. These traps were particularly useful in catching *Culex* species and *Anopheles barberi*, because they were rarely collected at human bait. In other collections from five different microhabitats considerable numbers of *Ochlerotatus triseriatus* and *Ochlerotatus hendersoni* were collected.

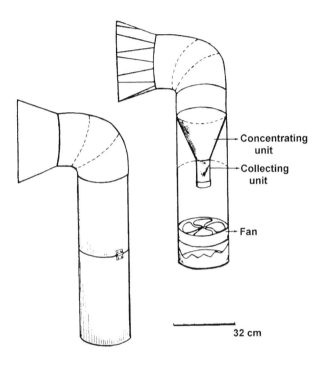

Fig. 8.17. Directional suction trap (Novak et al. 1981)

Snow (1977) modified his earlier suction trap (Snow, 1975) into a directional suction trap (Fig. 8.18). Each trap had a 1.2 × 1.2-m entrance with an inclined floor of green plastic netting (Netlon) and sides and roof made of 1-cm mesh plastic netting, dyed dark green to minimise visual contrast. Mosquitoes were thus guided up this funnel-type entrance to a horizontally-mounted 22.9-cm 'Vent-Axia' suction fan. These traps were used to determine the height and direction of flight of Gambian mosquitoes.

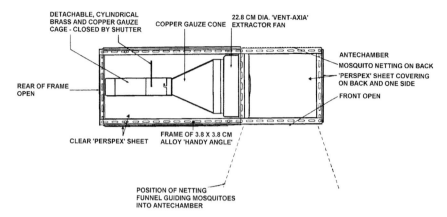

Fig. 8.18. Directional suction trap using Vent-Axia fan (Snow 1977)

Koch et al. trap

Koch et al. (1977) described and illustrated an inexpensive suction trap having a turntable device for separating catches of biting flies into hourly samples. Basically the trap consists of a 120-V 9-in diameter 6-bladed duct fan attached to a 1/70 h.p. kitchen fan motor, secured in a short length of 12-in diameter metal tubing. A bronze mesh cone underneath delivers the insects to collecting jars fixed under a notched turntable. A 2-lb lead weight tied to a rope advances the turntable when a solenoid causes a timer to close for about 15 s every hour, and this releases a sliding bar (slider-bar) resting against one of the notches in the turntable. This trap is rather similar to the Johnson–Taylor trap, but with a different sorting mechanism.

Wainhouse suction trap

Wainhouse (1980) described a battery-operated suction trap for catching small insects that would also be suitable for mosquitoes. The fan consists of an 'Air-max' type PR-Y4393 in an aluminium casing, 15.2 cm in internal diameter and 12 cm deep. The fan blades are inverted and the wires to the motor brushes switched to reverse the direction of rotation, needed for the arrangement shown in Fig. 8.19. A flared inlet made from a 5-cm strip cut from the top of a 60° plastic funnel is fitted to the top of the fan housing. A 6-cm deep flange bolted to the bottom of the fan casing holds in position a nylon mesh cone tapering to a mesh collecting bag. The trap weighs about 1.6 kg. The fan has a 31-W motor and operates at 2900 rev/min from two 25-Amphr 'Alcad' alkaline batteries connected in parallel, the combined weight of which is about 37 kg. The batteries run the trap for about 16 h. These batteries can be rapidly recharged, and are less sensitive to overcharging than lead-acid batteries and can be left uncharged for short periods without harm. The volume of air sampled by two traps measured with a Metrovic velometer was 329 and 355 m^3/h, which is about 60% of the volume sampled by 9-in 'Vent Axia' mains-operated traps (Johnson 1950*a*).

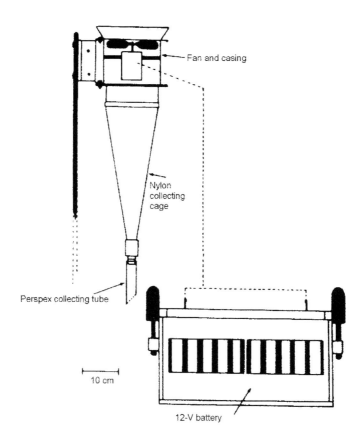

Fig. 8.19. Battery-operated suction trap (Wainhouse 1980)

Barnard and Mulla suction trap

Barnard and Mulla (1977) constructed an inexpensive and simple suction trap. It consists of a 36-in long and 21-in diameter galvanised metal cylinder supported on three ½-in angle-iron legs, adjusted to position the top of the cylinder 54 in from the ground. A 20-in, 5-bladed fan attached to a 115-V, 2.6-A motor (McGraw-Edison model 7327) is mounted on a metal plate supported by three 1-in wide metal arms in the lower part of the cylinder (Fig. 8.20). The motor sucks air through the trap at the rate of 2600 ft^3/min. A nylon netting cone is pulled over a metal band that is placed over the top of the trap, and a plastic vial with a mesh screen bottom is fixed with an elastic band on to the bottom of the cone. This is removed at the end of each sampling period.

In California, USA Barnard and Mulla (1978) used these suction traps, New Jersey light-traps, a D-vac aspirator and diurnal resting boxes for

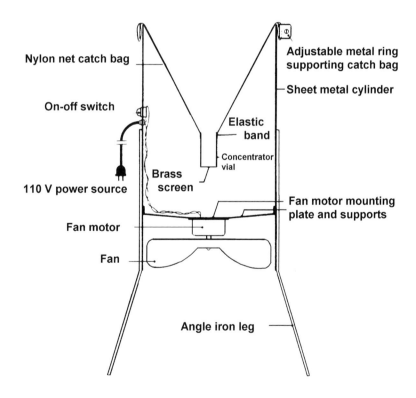

Fig. 8.20. Suction trap of Barnard and Mulla (1977)

sampling *Culiseta inornata*. The light-trap caught most females, followed by the D-vac, resting boxes and the suction trap.

Lumsden–Goma suction trap

Goma (1965) modified the suction trap developed by Lumsden (1957; 1958; see Chapter 7) for catching mosquitoes attracted to bait animals, by removing the transparent louvres at the sides to allow easier entry of mosquitoes. These modified traps without any bait were placed at ground level and on six platforms at 20-ft intervals on a steel tower in Zika forest, Uganda. From a series of forty 24-h continuous catches 4151 mosquitoes belonging to 34 species or species groups were caught, of which males only formed 3.5% of the catch. Several *Mansonia* and *Coquillettidia* species and *Aedes apicoargenteus* were sufficiently common for their vertical distributions and hourly flight periodicities to be analysed and plotted.

New Jersey-type suction traps

In Puerto Rico New Jersey light-traps were converted into suction traps for trapping *Aedes aegypti* by painting them black and removing the light bulb and cover. When placed in buildings they caught large numbers of adults of both sexes (Anon 1979).

Trap efficiency and absolute densities

Taylor (1962*a*) made a critical and very valuable evaluation of the absolute efficiency of insect suction traps. Among the factors considered were wind speed, insect size and size of the traps. By taking these factors into consideration Taylor (1962*a*) was able to derive the following formula for calculating the efficiency of any trap with regard to the insects being caught:

$$E = (w+3)(0.0082\ C_e - 0.123) + (0.104 - 0.159 \log i) \qquad (8.5)$$

where E = log efficiency, w = wind speed in mph, i = insect size in mm^2 (obtained by multiplying body length by wing span), and C_e = coefficient of efficiency of the trap, calculated by dividing the cube root of the volume of air sampled per hr by the square root of the fan diameter (in inches) thus:

$$C_e = (\text{vol. air sampled in ft}^3 / hr)^{\frac{1}{3}} \div (\text{inlet diam. in inches})^{\frac{1}{2}} \qquad (8.6)$$

Now the catch of mosquitoes can be converted to aerial density by dividing the numbers caught by the volume of air sampled, but remembering that an adjustment to the standard air-flow through the trap may be necessary if traps are used in exposed conditions where wind speeds exceed about 5 mph (Taylor 1955). This value is then corrected for the inefficiency of extraction of mosquitoes by the above formula, which gives a negative value in logs by which the catch of mosquitoes is less than the real density. This calculated value, known as the conversion factor (log f), is added to the log catch to give estimated absolute aerial density thus:

$$\text{Log density per } 10^6 \text{ ft}^3 \text{ air} = \text{log catch} + \text{conversion factor (in logs)} \quad (8.7)$$

The use of anitlogs gives actual densities.

Taylor (1962a) gave tables of conversion factor (log f) for different sized insects and wind speeds (0–10 mph) for 9, 12, 18 and 30-in traps. Unfortunately, because the manufacturers have increased the air-flow through the traps, the tables are no longer applicable. Correct values, kindly supplied by Dr L. R. Taylor, are given here for the 9- and 12-in suction traps (Tables 8.1 & 8.2). These values are added to the log catch to give aerial densities per 10^6 ft^3 air, thus eliminating the need to work out the values by the above formulae. It must be remembered that these values refer to fans working at normal speed. If a fan operates at a higher speed then the difference between the logs of the volume of air sampled per hour at normal and boost speeds must be *subtracted* from the published conversion factor. Fans used at lower speeds have the differences between the logs of the volumes of air sampled *added* to the conversion factor.

Table 8.1. Conversion factor (log f) for 12-in 'Vent-Axia' fan trap at normal speed

Insect size (mm^2)	Wind speed (mph)				
	0–2	2–4	4–6	6–8	8–10
1–3	1.58	1.69	1.81	1.92	2.03
3–10	1.66	1.77	1.89	2.00	2.11
10–30	1.74	1.85	1.97	2.08	2.19
30–100	1.82	1.93	2.05	2.16	2.27
100–300	1.90	2.01	2.13	2.24	2.35
300–1000	1.98	2.09	2.21	2.32	2.43

Table 8.2. Conversion factor (log f) for 9-in 'Vent-Axia' fan trap at normal speed

Insect size (mm^2)	Wind speed (mph)				
	0–2	2–4	4–6	6–8	8–10
1–3	1.81	1.92	2.04	2.15	2.26
3–10	1.89	2.00	2.12	2.23	2.34
10–30	1.97	2.08	2.20	2.31	2.42
30–100	2.05	2.16	2.28	2.39	2.50
100–300	2.13	2.24	2.36	2.47	2.58
300–1000	2.21	2.32	2.44	2.55	2.66

As is to be expected individual traps of the same size may differ slightly in the volume of air they collect (Taylor 1955) and this can produce small, but often significant, differences between the numbers of mosquitoes they collect. Taylor (1962a) considered that with 9- and 12-in 'Vent-Axia' fans a difference in catch size of about 6 and 2%, respectively, could not be attributed to real differences in population size. If it is not possible to measure the air-flow through individual traps, then when a number are used their positions should be alternated to minimise individual differences in the amount of air they sample.

It may not always be necessary to convert the catches in the traps to absolute densities. For example, flight periodicities can be studied simply by using the numbers caught each hour; similarly, relative change in population size can be recorded without calculating absolute population size. Since doubling the size of an insect only changes the efficiency of a trap by about 5% (Taylor 1962a) direct comparisons can be made between the numbers of different mosquitoes caught if they do not differ greatly in size. Wind speed will be the most important single factor influencing the size of the catch.

Vertical distribution of mosquitoes

Ecological studies in wooded areas in England clearly showed that the densities of unfed *Ochlerotatus cantans* and female *Culex pipiens* with fat reserves (Service 1971a) and *Culicoides* species (Service 1971b) decreased rapidly with small increases in height (23–550 cm) (Fig. 8.21). It seemed likely that this was correlated with low level flights for host seeking and oviposition. This assumption was supported by the discovery that during the gonoactive season *Culex pipiens*, an ornithophagic species, was commonest in the highest suction trap (550 cm), but when blood-feeding stopped and females sought hibernation sites adults were commonest in the lowest traps.

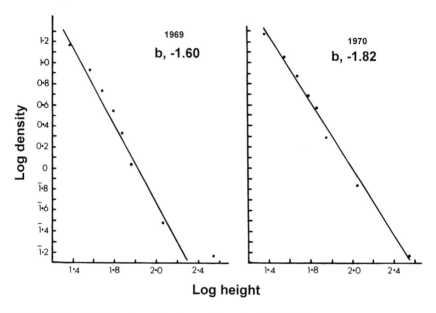

Fig. 8.21. Log height × log density profile of unfed *Ochlerotatus cantans* caught in suction traps at different heights (Service 1971*a*)

Snow (1975) initiated studies in The Gambia on the vertical distribution of mosquitoes. In a coastal region he used 22.9-cm diameter 'Vent-Axia' suction traps positioned at heights of 0.68–9.15 m on short and taller scaffolding towers; ramp traps were also used. The commonest species trapped were *Anopheles melas* and *Culex thalassius*. Later Gillies and Wilkes (1976) studied the vertical distribution of mosquitoes in a savannah area of The Gambia, employing suction traps at heights of 0.5–6.0 m (Fig. 8.22). Three main patterns of flight behaviour were recognised, namely: (1) species such as *Mansonia uniformis* and *Mansonia africana* (not separated), *Aedes* spp. and *Anopheles pharoensis* and *Anopheles ziemanni* which mostly fly below 1 m; (2) species such as *Anopheles funestus, Anopheles gambiae* and *Culex neavei* with flight levels more or less evenly distributed near the ground, but decreasing above 2–4 m; and (3) the high-fliers comprising *Culex antennatus, Culex thalassius* and *Culex poicilipes* with densities at 6 m greater or much greater than at 1 m, (Fig. 8.22). In a final series of observations on vertical distribution of Gambian mosquitoes Snow (1982) placed suction traps at heights of 0.1, 0.25, 0.5, 1.0, 2.1, 3.9

and 7.9 m in savannah areas near a swamp and near a village. Their vertical distribution was categorised into four main groups: (1) species whose densities progressively decreased with increasing height, e.g. *Anopheles, Aedes* and *Mansonia (Mansonioides)*; (2) species such as *Culex thalassius* whose densities increased to heights of 0.5 or 1 m, then decreased; (3) species such as *Culex poicilipes* and *Culex weschei* which were most common in the highest trap (7.9 m); and (4) some uncommon species such as *Culex neavei* that appeared to be common at all heights. This, however, is a simplified summary because within a species flight levels sometimes differed according to sex and the gonotrophic state of the females. Moreover there was an increase in the proportion of mosquitoes taken in the lowest trap (inlet at ground level) during the latter part of the night (2300–0500 h) (Gillies and Wilkes 1976; Snow 1982). Snow (1982) thought that such changes in flight levels might have been associated with falling ambient temperature.

In most of the above trials, the positions of the traps were changed after each night's catch to avoid bias caused by variations in their efficiencies. The distribution of mosquitoes in the air will be determined in part by selection of their own flight level. When, however, wind speed is greater than their flight speed they will have less control over their flight, and some will be swept into the upper air. Species breeding in exposed areas are more likely to be subjected to wind dispersal than those in sheltered sites, such as woods. Glick (1939) recorded mosquitoes up to 1530 m, and *Culex tarsalis* has been caught at 610 m (Glick and Noble 1961). In the tropics some nocturnal sylvan species both swarm and feed high up in the forest canopy, but because air turbulence and convection is usually at a minimum during the night, night flying insects are less likely to be affected by air current than day fliers.

In studying the distribution of insects at low levels Taylor (1960) concluded that the density of most small insects decreased markedly with increasing height but larger insects tended to select their own flight levels. This seems to be the case with *Chrysops caecutiens*, which showed no general decrease in density with increasing height, but a definite flight level was selected, below and above which the population was smaller (Service 1973a). Taylor (1958) introduced the term 'boundary layer' to describe a hypothetical layer of air near the ground within which insects could control their flight because this was greater than the wind speed. Above the boundary layer the type of flight will depend largely on the degree of protection provided by vegetation. Further evidence of the existence of a boundary layer was given by Taylor (1974).

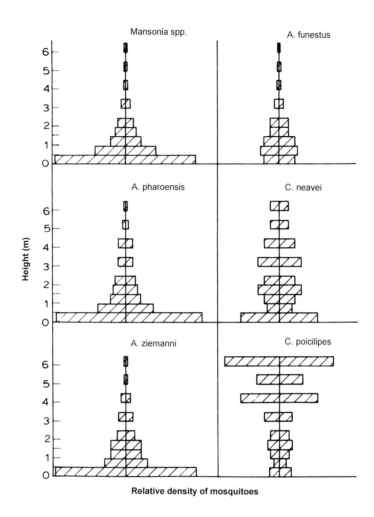

Fig. 8.22. Vertical distribution of six species, or groups, of mosquitoes over open farmland showing low-flying species on the left, and intermediate and high-level species on the right (Gillies and Wilkes 1976)

Johnson (1957) found that the diminishing density of insects with height could be fitted to the equation:

$$f(z) = C(z + z_e)^{-\lambda} \tag{8.8}$$

where $f(z)$ = density at height z, λ = an index of the gradient of density on height, C = a scale factor or constant related to population size and z_e = a constant, which is a measure of the departure from linearity of the curve obtained by plotting log values of aerial density against log height.

Total aerial population

In estimating the total aerial population of mosquitoes between any two heights the first step is to plot log density (log $f(z)$) against log heights (log z) and obtain a curvilinear graph (Fig. 8.23). Then by trial and error a constant (z_e) is found which when added to each of the values of the heights (z) converts the curve to a straight line when log ($z + z_e$) is plotted against log $f(z)$.

The densities of the mosquitoes (fz_1, fz_2) are then read off the graph at two heights (z_1, z_2) near the ends of the plotted line, and the value of λ (the regression coefficient) can be calculated from the following formula:

$$\lambda = \log \frac{(fz_1)}{(fz_2)} \div \log \frac{(z_2 + z_e)}{(z_1 + z_e)} \tag{8.9}$$

Alternatively the value of λ can be calculated by normal regression methods.

The value of C can be obtained from the following equation:

$$\log C = \log f(z) + \lambda \log(z + z_e) \tag{8.10}$$

Having calculated the values of Z_e, λ and C, then the number of mosquitoes (P) estimated to be dispersed in a column of air between any two heights (z_1, z_2), is obtained by integrating density on height (i.e. $f(z) = C(z + z_e)^{-\lambda}$):

$$P = \frac{C}{1 - \lambda}[(z_2 + z_e)^{1-\lambda} - (z_1 - z_2)^{1-\lambda}] \tag{8.11}$$

Less accurate, but nevertheless good, approximations of the total population of mosquitoes between two heights can be obtained from simple graphical methods (Johnson et al. 1962). The first step is to plot log densities of mosquitoes against log heights, and to read off from the visually fitted curve density estimates at various selected heights. These values are then plotted against heights on arithmetic graph paper and the area under the curve represents the estimated total population of mosquitoes.

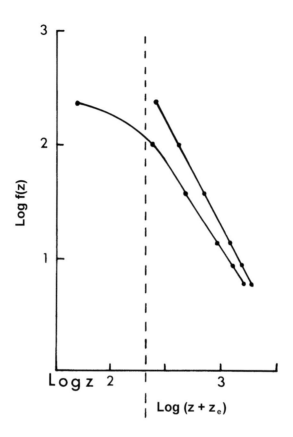

Fig. 8.23. Log density (*f(z)*) of insect catch in suction trap plotted against log height (*z*) to give a curvilinear plot. A constant value (z_e) is then added to each height to produce a straight line (after Johnson 1957)

Flight behavioural studies of Bidlingmayer and others

In Florida, USA, using his L-shaped suction traps, Bidlingmayer (1975) studied mosquito flight paths of several mosquitoes, in particular *Anopheles crucians, Anopheles quadrimaculatus, Aedes vexans, Coquillettidia pertur-bans, Culex nigripalpus, Culiseta melanura* and *Psorophora confinnis*. Their response to the visual effects of vertical and horizontal barriers made of mesh netting and placed adjacent to or over the traps, but with sufficient space for mosquitoes to fly through the mesh holes, was investigated. When traps having vertical netting were placed in an exposed position in a field the catches were larger (1.4–3.9 ×) than catches in similar traps without netting (except in one series where the mean catch of *Aedes vexans* remained the same). In contrast, in a wooded swamp the only species whose catch was significantly increased was *Culiseta melanura* (1.4–1.9 ×). In fact the number of *Anopheles crucians* was significantly reduced in traps with netting (0.6–0.8 ×). When horizontal netting was placed over the traps catches were reduced for all species, except that in some trials the numbers of *Culiseta melanura* (1.2 ×) and *Culex salinarius* (1.1 ×) were slightly greater. At full moon the response to both vertical and horizontal netting was substantially reduced, suggesting that the netting may have been perceived at a greater distance and avoidance action taken.

In a wood in England Service (1974) had netting with 3.8-cm diameter openings radiating out for 2 m in three directions from his 'Vent-Axia' suction traps. Catches of all mosquito species (*Ochlerotatus cantans, Culex pipiens, Culex torrentium. Anopheles plumbeus, Culiseta morsitans, Ochlerotatus geniculatus, Anopheles claviger* and *Coquillettidia richiardii*) were greater in traps with than without netting, but only in the first five species was the increase (1.78–3.34 ×) significant. These results suggest that, at least under certain circumstances, mosquitoes are guided by vertical structures and fly alongside 'barriers'. Differences in behaviour of woodland mosquitoes to vertical barriers reported by Bidlingmayer and Service may be due to differences in the species sampled, and/or differences in vegetative cover in the two woods.

Also in Florida, USA Bidlingmayer and Hem (1979) found that mosquitoes must approach within 30-cm or less of the suction trap intake to be captured. Among the 14 species caught, the more common were *Culex nigripalpus, Uranotaenia lowii, Uranotaenia sapphirina, Anopheles crucians, Anopheles atropos, Deinocerites cancer* and *Culiseta melanura*. More adults of all species, except *Ochlerotatus sollicitans* (an uncommon species in the study area), were collected in traps covered with black panels, than in traps without these panels, but both these traps caught more mosquitoes, except *Uranotaenia lowii*, than an acrylic transparent trap. It appeared that increasing trap visibility could either increase or decrease the numbers caught, depending on the behaviour of different species. For example, several species were attracted from some distance to conspicuous objects, but in closing in they avoided them by flying over or round them.

Bidlingmayer (1971) showed that some mosquitoes, which he referred to as field species, rest in exposed areas of grassland during the day and feed in these areas at night, in contrast to commuter species which shelter in woods during the day but fly out to seek hosts at night. Finally there are the so-called woodland species that rest and feed primarily in woods. Bidlingmayer and Hem (1979) believed that woodland mosquitoes come nearer (<30 cm) to objects before avoiding them than do field species. They argued that a larger percentage of woodland species will be captured in suction traps than will field species, and concluded that large suction traps in open habitats cannot be regarded as non-attractant sampling devices. It might be pointed out, however, that the suction traps used by Bidlingmayer are much bigger than the Johnson–Taylor traps or the 'Vent-Axia' traps used in England and The Gambia. Later Bidlingmayer and Hem (1981) showed that field species such as *Psorophora columbiae* and *Psorophora ciliata*, were caught at night in about equal numbers in suction traps sited 11 or 87 m from the edge of a woodland. In contrast woodland species, such as *Aedes vexans, Anopheles crucians, Culiseta melanura* and *Culex nigripalpus*, were caught in reduced numbers in traps sited 87 m from the woods. That is, the population of woodland species declined with increasing distance from the woods. It was suggested that these species may maintain visual contact with the silhouette of the woodland edge.

The effect of nearby traps on catches was investigated by placing 4–20 identical L-shaped traps 15 or 30 m apart in variously spaced configurations in a large open field (Bidlingmayer and Hem 1980). It was concluded that most species responded visually to traps at approximately 15.5–19 m. *Aedes vexans* and *Psorophora columbiae* responded to traps from the greatest distance, followed by *Culex nigripalpus, Culiseta, melanura. Anopheles crucians, Psorophora ciliata* and *Uranotaenia lowii*. Only *Uranotaenia sapphirina* and *Culex quinquefasciatus* appeared to respond to traps from just 7.5 m or less. The numbers of female mosquitoes caught in a trap decreased on average by 33% with each additional competing trap, which acted as a visual target. The ratio of the numbers caught from suction traps surrounded by a group of traps (inside), and traps placed in an outer row (edge), at a corner and beyond a corner are shown in Fig. 8.24. The estimated catch ratio for a trap placed beyond the visual competitiveness of other traps is also shown, and by extrapolating the curve in Fig. 8.24 such a trap would likely catch about five times as many mosquitoes as a trap surrounded by competing traps. To avoid trap interference (i.e. competition between traps) they should be placed at least 40 m apart. However, the actual distance that different mosquito species perceive objects (traps) varies.

Bidlingmayer and Hem (1980) considered that their L-shaped traps caught more mosquitoes than traps discharging air downwards and outwards (e.g. Johnson–Taylor traps, 'Vent-Axia' traps). This was because the discharge from their traps formed an angle of only 23° with the centre of the trap intake, and they believed that only over this zone would mosquitoes have difficulty in flying to the trap. Furthermore, because the exhaust air is emitted 1.6 m from the trap intake, and mosquitoes are only caught within a 30-cm range of the air intake, they argued that their traps would cause less disturbance to flying mosquitoes than traps discharging air below them horizontally in all directions. Bidlingmayer and Hem (1980) believed that the catch in a suction trap was dependent on two major factors, namely: (1) the physical features of the surrounding terrain such as natural objects (trees, bushes etc) and artificial objects (other traps) which may visually compete with it; and (2) even slight variations in wind speed. According to them, catches will be most variable when traps are surrounded by many various sized objects in an irregular distribution.

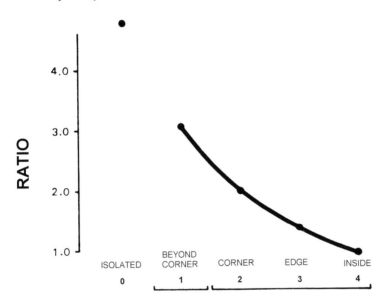

Fig. 8.24. Mean trap catch ratios for female mosquitoes between beyond corner, corner, edge, and the inside traps (inside traps = 1.0) with one, two, three and four adjacent traps, respectively, serving as competing visual attractants. $Y = -1.65$ (log x) + $3.28 = 2.38$, 2.14, 1.47 and 0.99, respectively, when $x = 1, 2, 3$ and 4. The estimated catch ratio for a trap without visual competition is shown also (Bidlingmayer and Hem 1980)

Bidlingmayer (1974) found that the numbers of *Culex nigripalpus* caught in his suction traps decreased by 50% in winds of 0.45–0.89 m/s and by 73% in winds of >0.89 m/s. In Florida when some suction traps were placed on land and others on a raft moored in a water-filled borrow pit 107.5 m from the shoreline, Bidlingmayer et al. (1985) found that wind speeds up to 0.24 m/s had no discernable effect on the numbers of *Aedes, Anopheles, Coquillettidia, Culex, Culiseta, Psorophora,* or *Uranotaenia* caught. However, wind speeds of 0.25–0.49 m/s reduced catches of *Culex nigripalpus* and *Culex erraticus* by about 75%, and when wind speeds were about 0.50 m/s or more catches were reduced by 90%. There was no evidence of downwind flight at any velocities. It seemed that high winds greatly reduced the numbers of mosquitoes captured, which is contrary to findings obtained with suction traps in grassland. In later studies, Bidling-mayer et al. (1995) observed that suction trap catches in relatively exposed sites were reduced by about 50% by winds of 0.5 m/sec and 75% by winds of 1.0 m/sec. No wind velocity threshold was observed below which mos-quito flight activity as measured by suction trapping was not affected. In The Gambia, Snow (1980), using suction traps, found that host-seeking flights of *Anopheles melas* and *Culex thalassius* did not appreciably de-crease until the wind increased to 120 cm/s.

Because of the often quite large day-to-day variations in the numbers of mosquitoes caught by most sampling methods, mainly caused by meteoro-logical conditions, Bidlingmayer (1985) proposed that the percentage in-crease or decrease in catch size caused by moonlight, temperature, relative humidity and wind speed should be taken into account. He illustrated the approach using collections of *Culex nigripalpus* in suction traps. From re-gression equations he calculated the increases and decreases of *Culex ni-gripalpus* in suction traps caused by moonlight and temperature, and more directly the percentage change caused by wind and relative humidities. Us-ing these in simple formulae he derived correction factors for each meteoro-logical factor. For example, for moonlight correction (k), $k = 1 - (np/1 - np)$, where n = days and p = mean percentage change. Then he multiplied trap catches by the product of all the correction factors to give an adjusted trap catch. The validity of this approach has still to be tested in the field for dif-ferent species and in different weather conditions, before its usefulness can be determined.

In summary, when wind speed is below the mosquito's flight speed mosquitoes generally fly upwind (Gillies et al. 1978; Service 1980; Snow 1975), but conspicuous objects can cause mosquitoes to deviate from a strictly upwind flight (Bidlingmayer and Hem 1980; Snow 1975, 1976). The silhouette of an object can be discerned at a distance, but as a mos-quito approaches the outline and shape of the object is lost (Mazokhin-Porshnyakov and Vishnevskaya in Browne and Bennett, 1981). However, as the closeness of approach to objects differs among species (Bidlingmayer and Hem 1979), the mosquito's appreciation of the object will differ and they will respond differently when at close quarters (Bidlingmayer and Evans 1987). In this last paper the authors point out that differences in be-haviour at close range (e.g. a metre or less) to visual targets can affect their orientation to hosts, and also the measurement of mosquito populations with traps when visual stimuli play a role. It appears that some species will tend to fly up and over an object while others will fly around it. They consider that trees, shrubs and other barriers may affect feeding patterns, and that physical features in the vicinity of a trap may affect the composition of the catch.

Electric grids and flight direction

Gillies et al. (1978) used high-voltage electric grids to screen suction trap (22.9-cm diameter) sunk into the ground to study low level mosquito flight in response to wind direction. These electrocuting grids, which are very similar to those designed to study tsetse flies (Vale 1974), could be used with other sampling methods and so are described here.

Steel wires (0.15 mm) 5 mm apart were tied at one end with nylon thread and tensioned at the opposite end with a steel spring to 44-cm wide and 46-cm high aluminium frames. A step-up transformer, as used in commercial electrocuting fly traps, boosted the voltage from a 240-V a.c. generator to about 5000 V a.c. A series of resistors inserted into the primary circuit reduced the input voltage until there was no spontaneous sparking in the grid; the working output voltage was then about 2000 V. A switch was inserted in the circuit to bypass these resistors so that when the trap was switched on sparking showed it to be working, after which the switch was operated. The authors believed that the current was about 1 mA, which is harmless to accidental human contact, but either knocks down or burns up mosquitoes hitting the wires. Apparently the electrocuting grid killed 75–80% of the insects attempting to fly through it. In later experiments (Gillies and Wilkes 1981) to help prevent mosquitoes passing through the screen a second grid of wires 3 mm apart, but not electrified, was placed behind the live grid. A problem that may arise is that some of the electrocuted insects may be so damaged that they are difficult to identify to species (Gillies and Wilkes 1981).

The following equations (Gillies et al. 1978) show the effect of grid efficiency (75–80%) on the relationship between the true proportion of mosquitoes flying upwind (a) and the estimated proportion flying upwind (x).

$$x = \frac{a(1 - Z_d) + Z_d}{1 + Z_d + a(Z_u - Z_d)} \quad \text{and} \quad a = \frac{x(1 + Z_d) - Z_d}{1 - Z_d - x(Z_u - Z_d)} \tag{8.12}$$

where Z_d = proportion of downwind flying mosquitoes passing through the grid, and Z_u = proportion of upwind flying mosquitoes passing through the grid. It is reasonable to assume that $Z_d = Z_u$, so using just Z the equations simplify to

$$x = \frac{a(1-Z)+Z}{1+Z} \quad \text{and} \quad a = \frac{x(1+Z)-Z}{1-Z} \tag{8.13}$$

From the above equations it can be calculated that with a grid efficiency of 75–80% (Z = 0.2–0.25), a total catch of say 64% (x) in traps facing downwind would indicate that 72–73% of the mosquitoes were in reality flying upwind (a).

References

Allan SA, Stoffolano JG (1986) Effects of background contrast on visual attraction and orientation of *Tabanus nigrovittatus* (Diptera: Tabanidae). Environ Entomol 15: 689–694

Allison D, Pike KS (1988) An expensive suction trap and its use in an aphid monitoring network. J Agric Entomol 5: 103–107

Anon (1979) Vector Topics Number 4: Biology and Control of *Aedes aegypti*. U.S. Dept. Health Education and Welfare, Public Health Service

Banks CJ (1959) Experiments with suction traps to assess the abundance of Syrphidae (Diptera), with special reference to aphidophagous species. Entomol Exp Appl 2: 110–124

Barnard DR (1979) A vehicle-mounted insect trap. Can Entomol 111: 851–854

Barnard DR, Mulla MS (1977) A non-attractive sampling device for the collection of adult mosquitoes. Mosquito News 37: 142–144

Barnard DR, Mulla MS (1978) The ecology of *Culiseta inornata* in the Colorado desert of California: Seasonal abundance, gonotrophic status, and oviparity of adult mosquitoes. Ann Entomol Soc Am 71: 397–400

Benzon GL, Lake RW, Murphey FJ (1986) A remotely piloted vehicle (RPV) for ULV experimentation. J Am Mosq Control Assoc 2: 86–87

Bidlingmayer WL (1961) Field activity studies of adult *Culicoides furens*. Ann Entomol Soc Am 54: 149–156

Bidlingmayer WL (1964) The effect of moonlight on the flight activity of mosquitoes. Ecology 45: 87–94

Bidlingmayer WL (1966) Use of the truck trap for evaluating adult mosquito populations. Mosquito News 26: 139–143

Bidlingmayer WL (1967) A comparison of trapping methods for adult mosquitoes: Species response and environmental influence. J Med Entomol 4: 200–220

Bidlingmayer WL (1971) Mosquito flight paths in relation to the environment. I. Illumination levels, orientation, and resting places. Ann Entomol Soc Am 64: 1121–1131

Bidlingmayer WL (1974) The influence of environmental factors and physiological stage on flight patterns of mosquitoes taken in the vehicle aspirator and truck, suction, bait and New Jersey light traps. J Med Entomol 11: 119–146

Bidlingmayer WL (1975) Mosquito flight paths in relation to the environment. Effect of vertical and horizontal visual barriers. Ann Entomol Soc Am 68: 51–57

Bidlingmayer WL (1985) The measurement of adult mosquito population changes—some considerations. J Am Mosq Control Assoc 1: 328–348

Bidlingmayer WL, Evans DG (1985) A telescopic collecting cup changer for insect traps. J Am Mosq Control Assoc 1: 33–37

Bidlingmayer WL, Evans DG (1987) The distribution of female mosquitoes about a flight barrier. J Am Mosq Control Assoc 3: 369–377

Bidlingmayer WL, Hem DG (1979) Mosquito (Diptera: Culicidae) flight behaviour near conspicuous objects. Bull Entomol Res 69: 691–700

Bidlingmayer WL, Hem DG (1980) The range of visual attraction and the effect of competitive visual attractants upon mosquito (Diptera: Culicidae) flight. Bull Entomol Res 70: 321–342

Bidlingmayer WL, Hem DG (1981) Mosquito flight paths in relation to the environment. Effect of the forest edge upon trap catches in the field. Mosquito News 41: 55–59

Bidlingmayer WL, Schoof HF (1957) The dispersal characteristics of the salt-marsh mosquito, *Aedes taeniorhynchus* (Wiedemann) near Savannah, Georgia. Mosquito News 17: 202–212

Bidlingmayer WL, Evans DG, Hansen CH (1985) Preliminary study of the effects of wind velocities and wind shadows upon suction trap catches of mosquitoes (Diptera: Culicidae). J Med Entomol 22: 295–302

Bidlingmayer WL, Day JF, Evans DG (1995) Effect of wind velocity on suction trap catches of some Florida mosquitoes. J Am Mosq Control Assoc 11: 295–301

Breeland SG, Pickard E (1965) The Malaise trap—an efficient and unbiased mosquito collecting device. Mosquito News 25: 19–21

Broadbent L (1946) Alate aphids trapped in north-west Derbyshire, 1945. Proc R Entomol Soc Lond (A) 21: 41–46

Broadbent L (1948) *Aphis* migration and the efficiency of the trapping method. Ann Appl Biol 35: 379–394

Browne SM, Bennett GF (1981) Response of mosquitoes (Diptera: Culicidae) to visual stimuli. J Med Entomol 18: 502–521

Burgess RJ, Muir RC (1970) 'A Modification of the Johnson–Taylor Suction Trap to provide a Twelve-Hour Segregation of the Catch.' Rep East Malling Res Stn (1969), pp 169–170

Butler GD (1965) A modified Malaise insect trap. Pan-Pacific Entomol 41: 51–53

Carroll MK, Bourg JA (1977) The night-time flight activity and relative abundance of fifteen species of Louisiana mosquitoes. Mosquito News 37: 661–664

Catts EP (1970) A canopy trap for collecting Tabanidae. Mosquito News 30: 472–474

Chamberlin JC, Lawson FR (1945) A mechanical trap for the sampling of aerial insect populations. Mosquito News 5: 4–7

Chapman JW, Reynolds DR, Smith AD (2003) Vertical-Looking Radar: a new tool for monitoring high-altitude insect migration. Bioscience 53: 503–511

Christensen HA, de Vasquez AM, Boreham MM (1996) Host-feeding patterns of mosquitoes (Diptera: Culicidae) from central Panama. Am J Trop Med Hyg 55: 202–208

Corbet PS, Danks HV (1973) Seasonal emergence and activity of mosquitoes (Diptera: Culicidae) in a high-arctic locality. Can Entomol 105: 837–872

Davies JB (1965) Studies on the dispersal of *Leptoconops bequaerti* (Kieffer) (Diptera: Ceratopogonidae) by Means of Wind Traps. Proc Int Congr Entomol XIIth (1964), pp 754–755

Davies L, Roberts DM (1973) A net and a catch segregating apparatus mounted in a motor vehicle for field studies on flight activity of Simuliidae and other insects. Bull Entomol Res 63: 103–112

Davies L, Roberts DM (1980) Flight activity of female black-flies (Diptera: Simuliidae) studies with a vehicle-mounted net in northern England. J Nat Hist 14: 1–16

de Zulueta J (1950) A study of the habits of the adult mosquitoes dwelling in the savannas of Eastern Colombia. Am J Trop Med 30: 325–339

Disney RHL (1966) A trap for Phlebotominae sandflies attracted to rats. Bull Entomol Res 56: 445–451

Dow RP, Gerrish GM (1970). Day-to-day change in relative humidity and the activity of *Culex nigripalpus* (Diptera: Culicidae). Ann Entomol Soc Am 63: 995–999

Dyce AL, Standfast HA, Kay BH (1972) Collection and preparation of biting midges (Fam. Ceratopogonidae) and other small Diptera for virus isolation. J Aust Entomol Soc 11: 91–96

Easton EM, Price MA, Graham OH (1968) The collection of biting flies in West Texas with Malaise and animal-baited traps. Mosquito News 28: 465–469

Farrow RA, Dowse JE (1984) Method of using kites to carry tow nets in the upper air for sampling migrating insects and its application to radar entomology. Bull Entomol Res 74: 87–95

Freeman JA (1945) Studies in the distribution of insects by aerial currents. The insect population of the air from ground level to 300 feet. J Anim Ecol 14: 128–154

Fröhlich G (1956) Methoden zur Bestimmung der Befalls-bzw. Bekämpfungstermine verschiedener Rapssch ädlinge, insbesondere des Rapsstengelrüsslers (*Ceuthorrhynchus napi* Gyll). NachrBl dt Pflschutzdienst Berl 10: 48–53

Gillies MT (1969) The ramp-trap, an unbaited device for flight studies of mosquitoes. Mosquito News 29: 189–193

Gillies MT, Snow WF (1967) A CO_2-baited sticky trap for mosquitoes. Trans R Soc Trop Med Hyg 61: 20

Gillies MT, Wilkes TJ (1969) A comparison of the range of attraction of animal baits and of carbon dioxide for some West African mosquitoes. Bull Entomol Res 59: 441–456

Gillies MT, Wilkes TJ (1970) The range of attraction of single baits for some West African mosquitoes. Bull Entomol Res 60: 224–235

Gillies MT, Wilkes TJ (1972) The range of attraction of animal baits and carbon dioxide for mosquitoes. Studies in a freshwater area of West Africa. Bull Entomol Res 61: 389–404

Gillies MT, Wilkes TJ (1976) The vertical distribution of some West African mosquitoes (Diptera: Culicidae) over open farmland in a freshwater area of the Gambia. Bull Entomol Res 66: 5–15

Gillies MT, Wilkes TJ (1978) The effect of high fences on the dispersal of some West African mosquitoes (Diptera: Culicidae). Bull Entomol Res 68: 401–408

Gillies MT, Wilkes TJ (1981) Field experiments with a wind tunnel on the flight speed of some West African mosquitoes (Diptera: Culicidae). Bull Entomol Res 71: 65–70

Gillies MT, Jones MDR, Wilkes TJ (1978) Evaluation of a new technique for recording the direction of flight of mosquitoes (Diptera: Culicidae) in the field. Bull Entomol Res 68: 145–152

Glick PA (1939) The distribution of Insects, Spiders, and Mites in the Air. Tech Bull U.S. Dept Agric No. 673

Glick PA, Noble LW (1961) Airborne Movement of the Pink Bollworm and Other Arthropods. Tech Bull U.S. Dept Agric Agric Res Serv No. 1255

Goma LKH (1965) The flight activity of some East African mosquitoes (Diptera: Culicidae). 1. Studies on a high steel tower in Zika forest, Uganda. Bull Entomol Res 56: 17–35

Goodenough JL, Jank PC, Carroll LE, Sterling WL, Redman EJ, Witz JA (1983) Collecting and preserving airborne arthropods in liquid at timed intervals with a Johnson–Taylor-type suction trap. J Econ Entomol 76: 960–963

Gordon WM, Gerberg EJ (1945) A directional mosquito barrier trap. J Natn Malar Soc 4: 123–125

Gorham RP (1946) The use of flight traps in the study of aphid movement. Acadian Naturalist 2: 106–111

Gottwald TR, Tedders WL (1986) MADDSAP-1, a versatile remotely piloted vehicle for agricultural research. J Econ Entomol 79: 857–863

Graham P (1969) A comparison of sampling methods for adult mosquito populations in central Alberta, Canada. Quaest Entomol 5: 217–261

Gregory PH (1951) Deposition of air-borne *Lycopodium* spores on cylinders. Ann Appl Biol 38: 357–376

Gressitt JL, Gressitt MK (1962) An improved malaise trap. Pacific Insects 4: 87–90

Gressitt JL, Sedlacek J, Wise KAJ, Yoshimoto CM (1961) A high speed airplane trap for air-borne organisms. Pacific Insects 3: 549–555

Grieco JP, Achee NL, Andre RG, Roberts DR (2002) Host feeding preferences of *Anopheles* species collected by manual aspiration, mechanical aspiration, and from a vehicle-mounted trap in the Toledo district, Belize, Central America. J Am Mosq Control Assoc 18: 307–315

Grigarick AA (1959) A floating pan trap for insects associated with the water surface. J Econ Entomol 52: 348–349

Gunstream SE, Chew RM (1967) A comparison of mosquito collection by Malaise and miniature light traps. J Med Entomol 4: 495–496

Hansens EJ, Bosler EM, Robin JW (1971) Use of traps for study and control of saltmarsh greenhead flies. J Econ Entomol 64: 1481–1486

Hardy AC, Milne PS (1938) Studies in the distribution of insects by aerial currents. Experiments in aerial two-netting from kites. J Anim Ecol 7: 199–229

Hill MN (1971) A bicycle-mounted trap for collecting adult mosquitoes. J Med Entomol 8: 108–109

Hocking B (1970) Insect flight and entomologists' inheritance Entomol News 81: 269–278

Holbrook FR, Wuerthele W (1984) A lightweight, hand-portable vehicle-mounted insect trap. Mosquito News 44: 239–242

Hollinger SE, Sivier KR, Irwin ME, Isard SA (1991) A helicopter-mounted isokinetic aerial insect sampler. J Econ Entomol 84: 476–483

Holzapfel EP, Harrell JC (1968) Transoceanic dispersal studies of insects. Pacific Insects 10: 115–153

Horsfall WR (1942) Biology and Control of Mosquitoes in the Rice Area. Univ. Arkansas Agric Exp Stn Bull 427

Horsfall WR (1961) Traps for determining direction of flight of insects. Mosquito News 21: 296–299

Hudson JE (1981) Studies on flight activity and control of mosquitoes in the Edmonton area, 1971–73. Quaest Entomol 17: 179–188

Isard SA, Irwin, ME, Hollinger SE (1990) Vertical distribution of aphids (Homoptera: Aphididae) in the planetary boundary layer. Environ Entomol 19: 1473–1484

Janousek TE, Olson JK (1994) Effect of a lunar eclipse on the flight activity of mosquitoes in the upper Gulf coast of Texas. J Am Mosq Control Assoc 10: 222–224

Jenkins GJ (1981) Kites and meteorology. Weather 36: 294–300

Johansen CA, Farrow RA, Morrisen A, Foley P, Bellis G, Van Den Hurk AF, Montgomery B, Mackenzie JS, Ritchie SA (2003) Collection of wind-borne haematophagous insects in the Torres Strait, Australia. Med Vet Entomol 17: 102–109

Johnson CG (1950a) A suction trap for small airborne insects which automatically segregates the catch into successive hourly samples. Ann Appl Biol 37: 80–91

Johnson CG (1950b) The comparison of suction trap, sticky trap and tow-net for the quantitative sampling of small airborne insects. Ann Appl Biol 37: 268–285

Johnson CG (1957) The distribution of insects in the air and the empirical relation of density to height. J Anim Ecol 26: 479–494

Johnson CG, Taylor LR (1955a) The development of large suction traps for airborne insects. Ann Appl Biol 43: 51–62

Johnson CG, Taylor LR (1955b) The measurement of insect density in the air. Part 1. Laboratory Practices 4: 187–192

Johnson CG, Taylor LR (1955c) The measurement of insect density in the air. Part II. Laboratory Practices 4: 235–239

Johnson CG, Taylor LR, Southwood TRE (1962) High altitude migration of *Oscinella frit* L. (Diptera: Chloropidae). J Anim Ecol 31: 373–383

Jónsson E, Gardarsson A, Gislason G (1986) A window trap used in the assessment of flight periods of Chrironomidae and Simuliidae (Diptera). Freshwater Biol 16: 711–719

Kaniuka R (1985) Biocontrol takes off in a pilotless miniplane. Agr Res 33: 8–9

Kay BH, Farrow RA (2000) Mosquito (Diptera: Culicidae) dispersal: implications for the epidemiology of Japanese and Murray Valley encephalitis viruses in Australia. J Med Entomol 37: 797–801

Kay BH, Sutton KA, Russell BM (2000) A sticky entry-exit trap for sampling mosquitoes in subterranean habitats. J Am Mosq Control Assoc 16: 262–265

Kettle DS, Edwards PB, Barnes A (1998) Factors affecting numbers of *Culicoides* in truck traps in coastal Queensland. Med Vet Entomol 12: 367–377

Killick-Kendrick R (1986) Sampling aerial populations of insects with a radio-controlled model aircraft. Antenna 10: 8–11

Koch HG, Axtell RC, Baughman GR (1977) A suction trap for hourly sampling of coastal biting flies. Mosquito News 37: 674–680

Kuntz KJ, Olson JK, Rade BJ (1982) Role of domestic animals as hosts for blood-seeking females of *Psorophora columbiae* and other mosquito species in Texas ricefields. Mosquito News 42: 202–210

LePrince JA, Orenstein AJ (1916) Mosquito Control in Panama. G.P. Putnam's Sons, London

Loomis EC (1959) A method for more accurate determination of air volume displacement of light traps. J Econ Entomol 52: 343–345

Love GJ, Smith WW (1957) Preliminary observations on the relation of light trap collections to mechanical sweep net collections in sampling mosquito populations. Mosquito News 17: 9–14

Loy VA, Barnhart CS, Therrien AA (1968) A collapsible, portable vehicle-mounted insect trap. Mosquito News 28: 84–87

Lumsden WHR (1957) Further Development of Trap to Estimate Biting Insect Attack on Small Vertebrates. East Afr Virus Res Inst Rep July 1956–June 1957. Government Printer, Nairobi, pp 33–35

Lumsden WHR (1958) A trap for insects biting small vertebrates. Nature 181: 819–820

Macaulay EDM, Tatchell GM, Taylor LR (1988) The Rothamsted Insect Survey '12-metre' suction trap. Bull Entomol Res 78: 121–129

Malaise R (1937) A new insect-trap. EntomolTijdskr 58: 148–160

Marston N (1965) Some recent modifications in the design of Malaise insect traps with a summary of the insects represented in collections. J Kans Entomol Soc 38: 154–162

Ming J-G, Jin H, Riley JR, Reynolds DR, Smith AD, Wang RL, Cheng J-Y, Cheng X-N (1993) Autumn southward 'return' migration of the mosquito *Culex tritaeniorhynchus* in China. Med Vet Entomol 7: 323–327

Montgomery BL, Ritchie SA, Hart AJ, Long SA, Walsh ID (2004) Subsoil drain sumps are a key container for *Aedes aegypti* in Cairns, Australia. J Am Mosq Control Assoc 20: 365–369

Mulhern TD (1953) The use of mechanical traps in measuring mosquito populations. Proc California Mosq Control Asso 21: 64–66

Muller MJ, Murray MD, Edwards JA (1981) Blood-sucking midges and mosquitoes feeding on mammals at Beatrice Hill, N. T. Aust J Zool 29: 573–588

Murphy WL (1985) Procedure for the removal of insect specimens from sticky-trap material. Ann Entomol Soc Am 78: 881

Nielsen ET (1960) A note on stationary nets. Ecology 41: 375–376

Nielsen ET, Greve H (1950) Studies on the swarming habits of mosquitos and other Nematocera. Bull Entomol Res 41: 227–258

Novak RJ, Peloquin J, Rohrer W (1981) Vertical distribution of adult mosquitoes (Diptera: Culicidae) in a northern deciduous forest in Indiana. J Med Entomol 18: 116–122

Odinstov VS (1960) Air-catch of insects as a method of study upon entomofauna of vast territories. Entomol Obozr 39: 227–230 (In Russian)

Pelham D (1976) The Penguin Book of Kites. Penguin, London

Pinger RR, Rowley WA (1975) Host preferences of *Aedes trivittatus* (Diptera: Culicidae) in central Iowa. Am J Trop Med Hyg 24: 889–893

Pinger RR, Rowley WA, Wong YW, Dorsey DC (1975) Trivittatus virus infections in wild mammals and sentinel rabbits in central Iowa. Am J Trop Med Hyg 24: 1006–1009

Provost MW (1952) The dispersal of *Aedes taeniorhynchus*. I. Preliminary studies. Mosquito News 12: 174–190

Provost MW (1957) The dispersal of *Aedes taeniorhynchus*. II. The second experiment. Mosquito News 17: 233–247

Provost MW (1960) The dispersal of *Aedes taeniorhynchus*. III. Study methods for migrating exodus. Mosquito News 20: 148–161

Pruess KP, Pruess NC (1966) Note on a Malaise trap for determining flight direction of insects. J Kans Entomol Soc 39: 98–102

Rajagopalan PK, Brooks GD, Menon PKB, Mani TR (1977) Observations on the biting activity and flight periodicity of *Culex pipiens fatigans* in an urban area. J Commun Dis 9: 22–31

Reiter P (1983) A portable, battery-powered trap for collecting gravid *Culex* mosquitoes. Mosquito News 43: 496–498

Reling D, Taylor RA (1984) A collapsible tow net used for sampling arthropods by airplane. J Econ Entomol 77: 1615–1617

Reynolds DR, Smith AD, Mukhopadhyay S, Chowdhury AK, De BK, Nath PS, Mondal SK, Das BK, Mukhopadhyay S (1996) Atmospheric transport of mosquitoes in northeast India. Med Vet Entomol 10: 185–186

Riley JR (1979) Radar as an aid to the study of insect flight. In: Amlaner CJ, Macdonald DW (eds) A Handbook on Biotelemetry and Radar Tracking. Pergamon, New York, pp 131–139

Riley JR (1992) A millimetric radar to study the flight of small insects. Electr Commun Eng J 4: 43–48

Riley JR, Cheng X-N, Zhang X-X, Reynolds DR, Xu G-M, Smith AD, Cheng J-Y, Bao A-D, Zhai B-P (1991) The long-distance migration of *Nilaparvata lugens* (Stal) (Delphacidae) in China: radar observations of mass return flight in the autumn. Ecol Entomol 16: 471–489

Roberts DM (1996) Circadian flight activity of Arabian sandflies (Diptera: Psychdidae) using a vehicle-mounted net. Bull Entomol Res 86: 61–66

Roberts D, Kumar S (1994) Using vehicle-mounted nets for studying activity of Arabian sand flies (Diptera: Psychodidae). J Med Entomol 31: 388–393

Roberts RH (1970) Color of Malaise trap and the collection of Tabanidae. Mosquito News 30: 567–571

Roberts RH (1976) The comparative efficiency of six trap types for the collection of Tabanidae (Diptera). Mosquito News 36: 530–535

Roberts RH (1978) Effect of Malaise trap modifications on collections of Tabanidae. Mosquito News 38: 382–385

Rohitha BH, Stevenson BE (1987) An automatic sticky trap for aphids (Hemiptera: Aphididae) that segregates the catch daily. Bull Entomol Res 77: 67–71

Roos T (1957) Studies on upstream migration in adult stream-dwelling insects. I. Rep Inst Freshwater Res Drottningholm 38: 167–193

Ryan L, Molyneux DH (1981) Non-setting adhesives for insect traps. Insect Science and its Application 1: 349–355

Schaefer GW (1976) Radar observations of insect flight. In: Rainey C (ed) Insect Flight. Symp R Entomol Soc Lond 7. Blackwell Scientific Publications, Oxford, pp 157–196

Schaefer GW, Bent GA, Allsopp K (1985) Radar and opto-electronic measurements of the effectiveness of Rothamsted insect survey suction traps. Bull Entomol Res 75: 701–715

Schreck CE, Gouck HK, Posey KH (1970) An experimental plexiglas mosquito trap utilizing carbon dioxide. Mosquito News 30: 641–645

Service MW (1969) The use of insect suction traps for sampling mosquitoes. Trans R Soc Trop Med Hyg 63: 656–663

Service MW (1971a) Flight periodicities and vertical distribution of *Aedes cantans* (Mg.), *Ae. geniculatus* (Ol.), *Anopheles plumbeus* Steph. and *Culex pipiens* L. (Dipt., Culicidae) in southern England. Bull Entomol Res 60: 639–651

Service MW (1971b) Adult flight activities of some British *Culicoides* species. J Med Entomol 8: 605–609

Service MW (1973a) Observations on the flight activities of *Chrysops caecutiens* L. Ann Trop Med Parasitol 67: 445–454

Service MW (1973b) Flight Activities of Mosquitoes with Emphasis on Host Seeking Behaviour. Proceedings of a Symposium on Biting Fly Control and Environmental Quality, May 1972, Edmonton, Canada. Defence Research Board, Ottawa, No. DR. 217

Service MW (1974) Further results of catches of *Culicoides* and mosquitoes from suction traps. J Med Entomol 11: 471–479

Service MW (1980) Effects of wind on the behaviour and distribution of mosquitoes and blackflies. Int J Biometeorol 24: 347–353

Service MW (1984) Evaluation of sticky light traps for sampling mosquito larvae. Entomol Exp Appl 35: 27–32

Shimizu Y, Takahashi M, Yabe S (1969) Use of the truck trap for the survey of mosquito population and the physiological age composition. Jap J Sanit Zool 20: 76–80 (In Japanese, English summary)

Snoddy EL (1970) Trapping deer flies with colored weather balloons (Diptera: Tabanidae). J Ga Entomol Soc 5: 207–209

Snow WE, Pickard E (1957) Correlation of vertical and horizontal flight activity of *Mansonia perturbans* with reference to marked changes in light intensity (Diptera, Culicidae). Ann Entomol Soc Am 50: 306–311

Snow WF (1975) The vertical distribution of flying mosquitoes (Diptera: Culicidae) in West African savanna. Bull Entomol Res 65: 269–277

Snow WF (1976) The direction of flight of mosquitoes (Diptera: Culicidae) near the ground in West African savanna in relation to wind direction, in the presence and absence of bait. Bull Entomol Res 65: 555–562

Snow WF (1977) The height and direction of flight of mosquitoes in West African savanna, in relation to wind speed and direction. Bull Entomol Res 67: 271–279

Snow WF (1980) Field estimates of the flight speed of some West African mosquitoes. Ann Trop Med Parasitol 74: 239–242

Snow WF (1982) Further observations on the vertical distribution of flying mosquitoes (Diptera: Culicidae) in West African savanna. Bull Entomol Res 72: 695–708

Snow WF, Wilkes TJ (1977) Age composition and vertical distribution of mosquito populations in The Gambia, West Africa. J Med Entomol 13: 507–513

Sommerman KM, Simmet RP (1965) Car-top insect trap with terminal cage in auto. Mosquito News 25: 172–182

Southwood TRE, Henderson PA (2000) Ecological Methods. 3rd ed. Blackwell Science, Oxford

Stage HH, Chamberlin JC (1945) Abundance and flight habitats of certain Alaskan mosquitoes, as determined by means of a rotary-type trap. Mosquito News 5: 8–16

Stage HH, Gjullin CM, Yates WW (1952) Mosquitoes of the Northwestern States. U.S. Dept Agric Handb No. 46

Steelman CD, Richardson CG, Schaefer RE, Wilson BH (1968) A collapsible truck-boat trap for collecting blood-fed mosquitoes and tabanids. Mosquito News 28: 64–67

Strong WB (1987) A new method of adhesive application for sticky insect traps. J Econ Entomol 80: 525–526

Taylor LR (1951) An improved suction trap for insects. Ann Appl Biol 38: 582–591

Taylor LR (1955) The standardization of air-flow in insect suction traps. (Coleman, W.S.—Appendix. Comments on the measurement of air-flow in the smaller traps, pp 406–408). Ann Appl Biol 43: 390–408

Taylor LR (1958) Aphid dispersal and diurnal periodicity. Proc Linn Soc Lond 169: 67–73

Taylor LR (1960) The distribution of insects at low levels in the air. J Anim Ecol 29: 45–63

Taylor LR (1962a) The absolute efficiency of insect suction traps. Ann Appl Biol 50: 405–421

Taylor LR (1962b) The efficiency of cylindrical sticky insect traps and suspended nets. Ann Appl Biol 50: 681–685

Taylor LR (1974) Insect migration, flight periodicity and the boundary layer. J Anim Ecol 43: 225–238

Taylor LR, Palmer JMP (1972) Aerial sampling. In van Emden HF (ed) Aphid Technology With Special Reference to the Study of Aphids in the Field. Academic Press, London, pp 189–234

Tedders WL, Gottwald TR (1986) Evaluation of an insect collecting system and an ultra-low-volume spray system on a remotely piloted vehicle. J Econ Entomol 79: 709–713

Townes H (1962) Design for a Malaise trap. Proc Entomol Soc Wash 64: 253–262

Tsai TF, Smith GC, Happ CM, Kirk LJ, Jakob WL, Bolin RA, Francy DB, Lampert KJ (1989) Surveillance of St. Louis encephalitis virus vectors in Grand Junction, Colorado, in 1987. J Am Mosq Control Assoc 5: 161–165

Vale GA (1974) The response of tsetse flies (Diptera: Glossinidae) to mobile and stationary baits. Bull Entomol Res 64: 545–588

Wainhouse D (1980) A portable suction trap for sampling small insects. Bull Entomol Res 70: 491–494

Walsh JF (1980) Sticky trap studies on *Simulium damnosum* s.l. in northern Ghana. Tropenmed Parasitol 31: 479–486

Way MJ, Banks CJ (1968) Population studies on the active stages of the black bean aphid, *Aphis fabae* Scop., on its winter host *Euonymus europaeus* L. Ann Appl Biol 62: 177–197

West AS, Baldwin WF, Gomery J (1971) A Radioisotopic-Sticky Trap-Autoradiographic Technique for Studying the Dispersal of Black-flies. WHO/ONCHO/71.84, 18 pp. (mimeographed)

Williams DC, Meisch MV (1983) Collection methods for a blood host study of riceland mosquitoes. Mosquito News 43: 355–356

Wilson BH, Tugwell NP, Burns EC (1966) Attraction of tabanids to traps baited with dry ice under field conditions in Louisiana. J Med Entomol 3: 148–149

Winternitz FAL, Fischl CF (1957) A simplified integration technique for pipeflow measurement. Water Power 9: 225–234

Yoshimoto CM, Gressitt JL (1959) Trapping of air-borne insects on ships in the Pacific (Part II). Proc Hawaii Entomol Soc 17: 150–155

Yoshimoto CM, Gressitt JL (1963) Trapping of air-borne insects in the Pacific–Antarctic area, 2. Pacific Insects 5: 873–883

Yoshimoto CM, Gressitt JL, Mitchell CJ (1962*a*) Trapping of air-borne insects in the Pacific–Antarctic area, 1. Pacific Insects 4: 847–858

Yoshimoto CM, Gressitt JL, Wolff T (1962*b*) Air-borne insects from the Galathea expedition. Pacific Insects 4: 269–291

Chapter 9 Sampling Adults with Light-Traps

Light-traps, the earliest using paraffin or acetylene lamps, have been used for many years to catch insects and many papers have been published dealing with various trap designs, different light sources and factors influencing catch size.

Some types of insects, including certain mosquito species appear to be strongly 'attracted' to light, whereas others seem to be disorientated rather than attracted by light. Some insects, including mosquitoes, may initially exhibit positive phototaxis and then at a certain distance, varying according to the light intensity of the trap, negative phototaxis. As long ago as 1937 Headlee realised that mosquitoes could be repelled by a bright light and in fact, mosquitoes are rarely caught unless they are sucked into a trap with a fan. The distance from a light source at which mosquitoes are 'influenced' will vary according to many factors, including the design of the trap, the type of light and of course the species concerned.

Bowden and Church (1973) showed that the illumination produced by a lamp of known intensity at a specified distance is determined by the inverse square law. It is calculated from the following simple equation:

$$I = \frac{L}{D^2} \tag{9.1}$$

where I = illumination, D = distance and L the intensity of the lamp in the trap. The boundary region of the trap's influence, that is where illumination from the light source equals that from the background (e.g. moonlight), is found by rewriting the equation above, so

$$D^2 = \frac{L}{I} \tag{9.2}$$

Huffaker and Back (1943) considered that the distance from which mosquitoes were attracted or repelled by light would most likely vary not only according to species but also their physiological condition. From laboratory experiments and field observations Nielsen and Nielsen (1953) considered females of *Ochlerotatus taeniorhynchus* to be positively attracted to light on about the seventh day after emergence; thereafter they

845

appeared to be attracted cyclically on about every fifth day. Field studies on *Ochlerotatus taeniorhynchus* by Provost (1952) substantiated this, except that they found that females initially responded to light 5 days after emergence, rather than 7 days. In contrast males appeared to be attracted to light only during the first 3 days of their life. Bidlingmayer and Schoof (1957) also obtained evidence of a cyclic attraction of females to light, but they also found that females were first attracted 5 days post-emergence.

Not all mosquito species are caught at light, and, those that are, are often caught in disproportionate numbers to the size of their populations. For example, Fox (1958) concluded that light-traps at the Puerto Rico International Airport failed to reflect the true abundance of *Ochlerotatus sollicitans* and probably also *Culex quinquefasciatus*. Bradley (1943) drew attention to the inefficiency of New Jersey traps for collecting *Anopheles quadrimaculatus* in southeastern USA, and Pippin (1965) reported that a trap operated for a year in the Philippines caught only about 30% of the *Anopheles* species, subspecies and varieties known to be present in the area. Moreover, Miller et al. (1969) found that CDC light-traps were not attractive to four *Anopheles* species in Thailand. Huffaker and Back (1943) concluded that light-traps did not give unbiased samples of mosquito populations.

In Florida, USA Bidlingmayer (1967) caught mosquitoes in non-attractant suction traps, and in suction traps and New Jersey light-traps both incorporating a 60-W light-bulb. Because suction traps do not employ any attractant they were considered to give the least biased catches (see Chapter 8). More than half the species were considerably more numerous (2.3–14.6 times) in a suction trap with than without a light, while a few species were only slightly more numerous (1.4–2.0 times). Males of *Anopheles quadrimaculatus*, and females of *Culiseta melanura* and *Coquillettidia perturbans* were less common (0.5–0.7 times) in suction traps with, rather than without, a light. In catches with the New Jersey trap some species were more, and others less, numerous than in the suction traps without a light. It was concluded that different species responded differently to light, some being actually repelled by it, and the response was not necessarily the same for both sexes of a single species. Species inhabiting wooded areas in preference to exposed habitats appeared to be the least attracted to light. In England light-traps do not usually catch mosquitoes except in September–October, when *Culex pipiens* and *Culex torrentium* with well-developed fat reserves are seeking hibernation sites.

Perhaps not surprisingly light-traps are usually of little use for sampling day flying mosquitoes. Haufe and Burgess (1960) failed to catch any in New Jersey light-traps in sub-arctic regions of Canada. They pointed out that this could not be related to species composition, because *Ochlerotatus communis,* which is common in the sub-arctic, was absent from light-trap catches although it was frequently caught in light-traps further south. They considered that the non-attractiveness of their light-traps was because there was no period of total darkness.

Employing light as an attractant in a trap can be considered to be a more artificial situation than the use of most other attractants in traps, because the light disrupts the insect's normal behaviour.

Use of light-traps

Light-traps are frequently used to catch large numbers of mosquitoes for laboratory studies, such as for virus isolations (Artsob et al. 1983; Chamberlain et al. 1964; Malainual et al. 1987; Olson et al. 1979; Reeves 1968; Srihongse et al. 1980).

In Connecticut, USA 157 646 mosquitoes belonging to over 24 species were caught over 10 years in CDC light-trap surveillance programmes (Main et al. 1979); in New York State, USA as many as 123 464 mosquitoes, mainly *Ochlerotatus* species, were caught in CDC surveillance traps (Srihongse et al. 1979), and over 5 years 918 047 mosquitoes were caught (Srihongse et al. 1980). In a 4-year study in Iowa 136 807 mosquitoes belonging to 28 species were caught in CDC-type traps (Wong et al. 1978), and in Japan Natuhara et al. (1991) caught 432 287 mosquitoes in just 27 light-trap-nights, of which 96.4% were *Culex tritaeniorhynchus*.

The type of trap selected for use in mosquito studies is often based on the need to catch as many mosquitoes, of the desired species, as possible. Light-traps are also commonly employed to record changes in abundance of mosquitoes before and after control campaigns, and to compare annual or seasonal fluctuations in population size (Chandler et al. 1975*b*; Miller et al. 1977). Light-traps have frequently been used to study dispersal (Brust 1980; Gillies and Wilkes 1978; LaSalle and Dakin 1982; Provost 1952, 1957). Although light-traps can give relative indices of mosquito abundance (Chandler et al. 1976*b*; Graham and Bradley 1961; Loomis and Hanks 1959; Milby and Reeves 1986; Siverly and DeFoliart 1968), Collett et al. (1964) concluded that for more reliable results larval collections should be made. Breeland (1974), however, reported good agreement between light-trap catches and larval collections when monitoring changes in population size of *Anopheles albimanus* in El Salvador. Moreover, in Okinawa, Japan McDonald and Savage (1973) found that there were good correlations between larval dipping and the catches in New Jersey light-traps over 7 years for many, but not all, mosquito species.

Light-traps are sometimes used to study the vertical distribution of mosquitoes (Chandler et al. 1976a; Mitchell 1982; Mitchell and Rockett 1979; Nasci and Edman 1981) but caution is required in interpreting results because the presence of a light-trap may cause mosquitoes to fly at heights outside of their normal flight range, as a result of attractancy of the trap. Similar care is required in interpreting flight times.

Although light traps predominantly sample un-fed females, they may also attract relatively large numbers of blood-fed individuals (Bailly-Choumara 1973a; Carnevale and Boreham 1978; Christensen et al. 1996; Hii et al. 1986; Service and Highton 1980; Sun 1964; Vervent and Coz 1969).

In routine surveys traps are frequently operated nightly, but sufficient information on seasonal changes in population size etc. may be obtained from light-traps that are used on only a few nights a week (Bradley and Travis 1943; Loomis and Hanks 1959; Miller et al. 1977; Silvain and Pajot 1981). In Japan Ishii (1971c) compared the numbers of *Culex tritaeniorhynchus, Aedes vexans nipponii* and *Culex pipiens* form *pallens* caught in a light-trap per night (trap index) computed on one, two, three, four and seven nights of operation. Not unexpectedly trap indices based on one and two nights fluctuated much more widely than indices calculated from four and seven nights of operation.

To summarise, light-traps may be very useful in catching large numbers of certain mosquito species and in measuring relative changes in abundance of these species in time and space, but as an ecological tool they are of strictly limited value because they sample different species unequally.

Some factors influencing light-trap catches

Motors and fans

A fan is usually necessary to draw mosquitoes into a light-trap that would otherwise tend to fly around the light source without entering the trap (e.g. Breyev 1958). However, traps without fans have sometimes been successfully employed to catch mosquitoes (Corbet 1961, 1964; Corbet and Haddow 1961; LaSalle and Dakin 1982; Loomis 1959a; McDonald 1970). Occasionally New Jersey traps have used fans larger than the normal 8-in ones (Mulhern 1953a; Pincus 1938), but Barr et al. (1963) found that there was very little difference between catch size when 7½, 8, 8½ or 9-in fans were used. However, in laboratory experiments Wilton and

Fay (1972*a*) found that with the much smaller CDC traps a reduced air flow was accompanied by a significant reduction in the numbers of *Anopheles albimanus* caught, although the percentage that remained alive in the trap increased.

The design and performance of the miniature 6-V and 12-V motors used in CDC-type and Monks Wood light-traps has been continually improving over many years. A great variety of both expensive and inexpensive motors are available from toy and model shops and electrical stores. Generally the more costly the motor, the greater is its life expectancy. Some precision motors last in excess of 20 000 h, thus allowing them to function for about 1600 nights. The efficiency of a motor may be around 40% for cheap models but 80% or more for precision built motors. The less efficient motors have a higher current rating and require more power, consequently the drain on a battery is greater. The number of revolutions per minute will also vary greatly between different types of motors. Choosing a suitable motor is something of a compromise, between cost and efficiency; in general those with the lowest current rating are best. The best motor has silver-graphite replaceable brushes and stainless steel bearings, and has an input of 150–200 mA at about 6-V. The fan fixed to the motor can have two, three or four blades, or be multi-bladed. Frequently a model aeroplane plastic propeller can be used, but if the blades need cutting to shorten them, or if blades are made from plastic or aluminium, care must be taken to ensure the propeller is balanced around the centre, otherwise it will damage the motor. The pitch of the blade is important in getting maximum airflow through a trap (Rohe and Fall 1979). It may be possible to measure this (Frommer et al. 1976; Loomis 1959*b*; Mulhern 1953*a*), but if not, smoke can be used to ensure that there is a good flow of air through the trap. Mosquitoes are rarely damaged by passing through the fan of a trap, but damage is more likely if there is virtually no space between the tips of the fan blades and walls of the trap; Sandoski et al. (1983) had to modify their CDC-type traps because they considered mosquitoes were often damaged.

Electricity supply

Traps such as the New Jersey or American light-trap can be operated from 115 or 220V a.c. mains voltage, or alternatively, modified versions can be operated from 12V car batteries (Hurlbut and Weitz 1956; Standfast 1965). Reliance on mains electricity greatly limits the usefulness of light-traps, and so considerable attention has focused on developing battery-operated traps.

There have been considerable improvements in the performance and maintenance of lead acetate 12-V car batteries, such as the development of sealed units that do not need topping-up with distilled water. The most important characteristic of car batteries is their amp-hour rating—the greater this is (e.g. heavy-duty batteries), the longer a battery will run a trap without recharging. There have also been vast improvements in rechargeable 6- and 12-V batteries. For example, in addition to four 1.25-V nickel-cadmium rechargeable batteries wired in series, a 6-V sealed gel-cell lead-dioxide battery can operate CDC-type light-traps. With recharging 6-V or 12-V gel-cell batteries should last for 2–5 years. There are also efficient rechargeable 6-V lantern-type batteries, and a variety of inexpensive a.c. battery chargers. Note that the voltage of the nickel-cadmium Dcell batteries is 1.25, not the normal 1.5-V of the non-rechargeable equivalents. These 'nicad' batteries can be recharged at 500 mA.

Light source characteristics

Various linear units are used to describe wavelength, and their suitability varies according to different regions of the spectrum. Some of the more common units and their equivalents are as follows, 1 micron (μ) = 1000 nanometres (nm) = 10 000 Ångstrom units (Å. U.) = 10^{-6} metres.

The wavelength of ultraviolet emission extends up to 400 nm, visible light is in the region of 400–700 nm and infrared beyond 700 nm; wavelengths beyond about 5000 nm are absorbed by the glass of the lamp. The output of a light source is measured in energy units such as watts or milliwatts, or in luminous energy units or lumens. The (perceived) brightness of visible light from a light source is termed its luminance and this is measured in candelas (cd), and 1 candela = 1 lumen/steradian = 1 metre-candle/steradian, or equals 3.1426 lumens/cm^2. With circular bulbs output in lumens is divided by 12.57 (i.e. 4π) to obtain brightness in candelas, with pear-shaped bulbs the divisor is 10 and with fluorescent tubes 9.25. Another unit of brightness is the foot-lambert which = $1/\pi$ candles/ft^2, or 1 lumen/ft^2.

Illumination is the amount of light falling on a surface and is measured in lux, 1 lux = 1 lumen/m^2 = 0.0929 foot-candles. Full sunlight with zenith sun produces an illumination of about 92 900 lux (10 000 foot-candles), full moonlight about 0.19 lux (0.02 foot-candles) and adequate illumination for reading is about 93.0 lux (10 foot-candles). These measurements, however, are based on spectral sensitivity of the human eye and not on what insects perceive, many of which have good sensitivity in the blue and ultraviolet end of the spectrum. So, what appears bright to an insect may appear very dim to the human eye and register very few lux (Burkhardt 1977; Green and Cosens 1983). Young et al. (1987) considered it was often more relevant, as well as easier, to measure illumination, i.e. the amount of light falling onto a surface, as opposed to luminance, which is the perceived brightness. They used two units of illumination, W/m^2, which is a measure of incident light, and einsteins/s/m^2 which measures photons impinging on a surface. This is a large quantity (6×10^{23}) so microeinsteins (μE) are used as a convenient unit. The energy of photons is inversely proportional to wavelength, consequently an increasing number are necessary to carry the same amount of light as wavelength increases. There is no universal conversion factor because the relative abundance of different wavelengths in a light source determines the numbers of photons that correspond to 1 W of total light power, but Table 9.1 shows equivalents of these two measurements.

Table 9.1. Conversion Factors for Light Units (after Young et al. 1987)

Light source	Unit		
	W/m^2	$\mu E/m^2$	lux (lumens/m^2)
Daylight (sun + sky)	1	4.55	355
Tungsten (240 V, 100 W)	1	5.13	142
Quartz–halogen (24 V, 150 W)	1	4.95	180
Fluorescent (warm white)	1	4.74	362
Fluorescent (daylight)	1	4.55	355
400–450 nm (violet)	1	3.57	5
450–500 nm (blue)	1	3.95	73
500–550 nm (green)	1	4.40	456
550–600 nm (yellow)	1	4.80	622
600–650 nm (orange)	1	5.24	266
650–700 nm (red)	1	5.64	31

Young et al. (1987) describe a portable light-meter and a laboratory mains-operated one that will measure illumination ranging from starlight to bright sunlight. Measurements are in W/m^2 and the photocell used responds equally to the same amounts of energy of wavelengths between 420–850 nm (violet–infrared). This range misses near ultraviolet light, but by having an appropriate filter in front of the detector and using the photocell's sensitivity curve this type of radiation can be detected. Moreover, ultraviolet sensitivity can be extended if needed by adding a second photo-detector, namely a UV enhanced photocell.

Sometimes the radiation intensity of a particular part of the wave band, such as mW/cm^2 of ultraviolet light, is more important than that for the entire spectral emission. Incandescent lamps produce light by heating a tungsten filament. Their 'continuous' spectral emission includes a small amount of ultraviolet and considerable quantities of visible light especially rich in yellow and red, but peak radiation is in the infrared region, which comprises about three-quarters of the total lamp output. Discharge lamps are filled with various gases under pressure, each gas giving its own specific 'non-continuous' spectrum of a number of peak emissions at various wavelengths. The most commonly used discharge lamp is the mercury vapour lamp, which produces mainly invisible ultraviolet radiation, some blue and green radiation but very little red, or infrared.

Fluorescent lamps

Fluorescent lamps are basically low pressure discharge lamps filled with mercury vapour, but with the inner surface of the lamp coated with various phosphors. These absorb shortwave ultraviolet radiation and reradiate the energy at longer wavelengths. Certain phosphors fluoresce in the ultraviolet region while others do so in the visible spectrum. These lamps give a 'continuous' spectrum from the fluorescent coating plus the bright peaks that characterise the mercury vapour discharge lamp. Some fluorescent lamps are coated with a phosphor that converts the energy of the basic mercury discharge lamp to longer ultraviolet wavelengths (320–380 nm). Such lamps are often referred to as 'UV actinic blue 05' or 'blacklights'. In some of these lamps the envelope is made of dark cobalt glass or a Wood's glass filter, which lets through ultraviolet energy but filters off most of the visible light. These are often referred to as 'blacklight' lamps, whereas similar lamps without a dark glass envelope are termed 'blacklight-blue' lamps, because they appear blue when lit. For many years the common phosphor coating was magnesium germanate, but in the early 1970s the coating was changed to yttrium vanadate, which is more efficient at reclaiming wasted UV radiation. It seems that long-wavelength ultraviolet (300–400 nm) output from high pressure mercury lamps is decreased from about 3% to 1%, and that phosphor coatings have virtually eliminated the ultraviolet output of blended lights. Not surprisingly differences in coatings of the lamps can affect their attractiveness to insects (Hill 1977; Upton 1973a,b).

Fluorescent lamps may produce a flickering effect. Belton and Pucat (1967) reported that the blacklight fluorescent tube they used in Canada had a flicker rate of 120/s. Flies have a temporal resolution, with flicker-fusion frequencies of about 200–300 Hz, which enables them to track small fast moving objects in close proximity. Flicker is due to the frequency of the a.c. voltage applied to the tubes, and fluorescent lamps, and to a lesser extent incandescent ones, flicker at twice the mains a.c. frequency, i.e. 100 or 120 Hz. Syms and Goodman (1987) showed that houseflies were more attracted to a UV fluorescent tube that flickered than one operated from a flicker-free d.c. source.

It should be remembered that the longer the wavelength emitted by a phosphor coating of a lamp the slower it takes for the luminescence to fade. For example, actinic (BL)-type fluorescent lamps decay almost to zero before the next rise in a.c. voltage, thus flickering twice for each cycle of the main power supply, whereas a phosphor emitting at longer wavelengths, such as used for various white fluorescent tubes, may only decay to about 60% before the next cycle (Syms and Goodman 1987). Mercury vapour lamps, as used sometimes in mosquito traps, also flicker at twice the mains frequency, their intensity decays to around 12% peak emission before the following rise in supply voltage.

Care is consequently needed in interpreting behavioural responses in terms of just spectral emission, because mosquitoes could also be responding to flicker and decay characteristics of the light source. In recent years 'apparent flicker' has been much reduced in fluorescent lamps by using improved phosphor coatings and improved transistor gear; tubes having a frequency of 20 000 cycles produce no visible flicker. Incandescent lamps usually show very little flicker.

Low temperature may tend to increase the voltage required for the ignition of fluorescent tubes by an inverter. To increase the chance of strike Wagner et al. (1969) taped a 5-mm wide metal foil strip along the length of the fluorescent tube in their light-trap to act as a coupling capacitor.

It is not always appreciated that although an ultraviolet fluorescent tube may light up for many months, its UV output may be greatly reduced after 2000–2500 h operation. Consequently after this period the spectral emission will be altered. 'Over-running' a filament lamp increases its intensity and wattage but greatly shortens its life. The higher the wattage the higher the light intensity. Increasing the intensity of a light source by using higher wattage bulbs may change the spectral emission, by altering the colour temperature in an upwards direction, i.e. there is a colder appearance. Changes in emission characteristics can be avoided if the luminous intensity is reduced by covering up part of the light source, such as with metal foil (Belton and Kempster 1963; Hollingsworth et al. 1968; Taylor and Brown 1972).

The complexities of quantifying the characteristics of artificial light were stressed by Philogène (1982), who among other things pointed out that the amount of solar UV light reaching the earth is less than 1% whereas the UV content of ordinary artificial white light is often about 3–6%.

Polarised light

The ability to perceive polarised light is common to insects, and is generally associated with short wavelength receptors, but there is no evidence for wavelength discrimination. Some insects such as bees and ants use polarised light for orientation and navigation (Dethier 1963; von Frisch 1950; Waterman 1951), but there is little evidence as to whether mosquitoes can perceive polarised light. Kalmus (1958) claimed that insects cannot detect polarised light directly, but only when reflected from a dark background which produces an increase in brightness, but Clements (1963) considered more experimental data were needed to resolve the issue. Wellington (1974) pointed out that although mosquitoes which inhabit woods and forests may not be exposed to polarised light, those living in more exposed and lighter habitats can hardly avoid it.

In Russia, Kovrov and Monchadskiy (1963) used modified New Jersey light-traps with 1000-W mercury quartz bulbs to compare catches of mosquitoes, and other insects, attracted to both polarised and non-polarised lights. Metal shields having 45–50-mm wide apertures were built around the traps to restrict the light to a fairly narrow beam. Two traps were arranged so that their beams of light crossed at about 15 m. Polarisation of light from one trap was achieved by reflecting it from a highly polished 200×80 mm duraluminium mirror placed at an angle of $45°$ to the beam of light. Although the catch of *Culicoides* was about 3.7 times greater in the traps emitting polarised light, there was no difference between the numbers of mosquitoes caught with polarised and non-polarised light. Goldsmith (1964) gives a concise review of the physiological evidence for and against the capabilities of various insects to detect polarised light.

Light source and mosquito response

Differences between the numbers and species of mosquitoes caught with different types of light may be due not only to differences between spectral emission but also changes in light intensities. Even when colours of the same brightness are used in tests to evaluate their attractiveness to mosquitoes, the physiological intensities may be different due to different wavelength absorption by the mosquito's eye (Roeder 1953). Thus it is extremely difficult to determine whether mosquitoes can in fact differentiate between coloured lights. The effect of light intensity on mosquito catches has been investigated by several people. Barr et al. (1960) reported that trap efficiency for collecting *Culex tarsalis* increased with light intensity of the bulb, while Pfuntner (Reisen and Pfuntner 1987) found the same for *Culex stigmatosoma*. By comparing catches in New Jersey traps having 20-, 50-,

75- and 100-W bulbs, Barr et al. (1963) found that during periods of new moon there were significant differences between catch size of both sexes of *Culex tarsalis* and *Anopheles freeborni* and females of *Ochlerotatus melanimon* when different wattage light-bulbs were used. At full moon only catches of females of these three species increased significantly with increasing light intensities. In Puerto Rico, Pritchard and Pratt (1944) caught more mosquitoes with a 40 than a 25-W bulb, but there was no appreciable difference between catches when the wattage was increased to 50- or 60-W. Pratt (1944; 1948) showed that when 25-, 50- and 100-W bulbs were used the numbers of *Anopheles albimanus* and *Anopheles vestitipennis* were approximately proportional to the wattage of the bulbs. However, although about twice as many *Anopheles grabhamii* were collected with a 50-W than with a 25-W bulb, there was no significant increase in numbers caught when a 100-W bulb was used. Reisen et al. (1999) in California, USA, found no significant differences in NJ light-trap catches using 25-watt incandescent or 5-watt fluorescent lamps that yielded comparable brightness of ~250 lumens. In Malaysia Wharton et al. (1963) reported that the species composition was about the same in a New Jersey trap with 25- and 100-W bulbs, but the latter light attracted about twice as many mosquitoes. Among the species caught were *Aedes amesii, Verrallina butleri, Culex gelidus* and large numbers of *Uranotaenia*. In preliminary trials in Japan Ikeuchi (1967) recorded that greater numbers of *Culex pipiens* form *pallens* were caught in traps using a 6-W blacklamp fluorescent tube (BL-FL) that emitted ultraviolet and only a little visible light than in traps employing a 6-W blacklamp blue tube (BLB-FL) that emitted ultraviolet light but also considerable amounts of visible light, a 60-W daylight fluorescent tube (DL-FL) or a 20-W incandescent lamp. In later trials, although more species of mosquitoes were caught in traps having an ultraviolet light-tube than in those with an incandescent bulb, the latter was only a little less effective in catching *Culex tritaeniorhynchus* and had the advantage that it did not attract so many unwanted insects. In field trials in Georgia more *Aedes vexans* and *Culex salinarius* were caught in New Jersey traps having blue lamps peaking at 4470 Å than in traps with red lamps peaking at 6700 Å, with yellow lamps peaking at 5700 Å and with traps having white lamps with a slight peak at 6490 Å, when all were operated at equal light intensity. In contrast more *Culex quinquefasciatus* were attracted to red, yellow and white lamps than to blue lamps. There was no significant difference between the attractiveness of these lamps to *Culex nigripalpus* (Bargren and Nibley 1956).

Laboratory studies into the attractiveness of lights of different colours and intensities fitted to New Jersey traps, to both sexes of *Culex tarsalis, Culex quinquefasciatus* and males of *Ochlerotatus sierrensis*, used ceramic-dipped red, green, blue, orange and white incandescent bulbs and also a 4-W ultraviolet light (Gjullin et al. 1973). More males of all three species were attracted to a 7.5-W red lamp than to a green, blue, or orange bulb, or to white or ultraviolet lights, and moreover this lamp was more attractive than a 40-W red one. Although more females than males were caught in these experiments there were no convincing differences between their responses to these different coloured lights. In laboratory experiments Wilton and Fay (1972b) convincingly demonstrated that when adults of *Anopheles stephensi* were exposed to light from a 7.5-W clear incandescent lamp and monochromatic light of different wavelengths (290–740 nm), both sexes were strongly attracted to wavelengths of 290 and 365 nm in the ultraviolet region, and surprisingly also to 690 nm in the red end of the spectrum. Light in the blue, green and yellow regions (490, 540 and 590 nm) appeared repellent in comparison with light from the incandescent lamp. In contrast Lewis and Teller (1967) found that green fluorescent tubes in modified New Jersey traps attracted females of *Culex nigripalpus*, *Psorophora confinnis* and *Uranotaenia sapphirina*.

In response to the relatively recent availability of several colours of highly efficient, low-cost, ultra-bright light-emitting diodes (LEDs), Burkett et al. (1998) tested their usefulness against standard incandescent light sources in CDC light-traps. LEDs use significantly less power (approximately 1 mA/8 h) than incandescent bulbs and also have a longer life-span. The LED was secured into a piece of 2×2-cm plexiglass and fastened to the screen atop the lid assembly 3 cm below the aluminium trap lid. Four different colours of LED (Toshiba Toshbright®) were compared with no light and the standard incandescent bulb in traps with (trial 1) or without (trial 2) the addition of CO_2 from a compressed gas cylinder at a constant rate of 200 ml min^{-1}. The diodes tested were red (613 ± 50 nm, 1600 milicandela [mcd], 22°), orange (605 ± 50 nm, 2 000 mcd, 22°), yellow (587 ± 50 nm, 2 300 mcd, 22°) and green (567 ± 50 nm, 2 400 mcd, 8°). In a later trial, infrared (940 ± 50 nm, 22°, Martech Optoelectronics) and blue (450 ± 50 nm, 800 mcd, 22°, Panasonic) LEDs were also tested. Each LED was powered by 2 alkaline D-cell batteries at 2.8 ± 0.2V and 18 ± 2 mA. Traps were hung at heights of 1.65 m above ground level and were separated by 30 m, along the banks of a seasonal forest stream in Florida, USA and operated from 1800 to 0600 h for six days. Traps were tested using a 6×6 or 8×8 Latin square experimental design. No significant differences were observed in the total number of mosquitoes captured in traps with different coloured light sources in either the CO_2-baited or unbaited trials $P = 0.08$, $P = 0.24$ respectively). However, differences were observed in the response of individual species. In the trial with 6 colours of

LED, plus incandescent and no light, overall catch was greatest with the standard white broad-spectrum incandescent light, followed by blue, green, orange, yellow, red, no light control, and infrared, respectively. As regards individual species, white light captured the most *Anopheles crucians* s.l.; the greatest numbers of *Culiseta melanura* were captured in traps with white, green, and orange LEDs, and the traps emitting blue or white light caught the most *Psorophora columbiae*. Significantly more *Uranotaenia sapphirina* were captured in traps with standard white light or a blue LED.

A better method of studying the responses of insects to coloured lights is to undertake electrophysiological (ERG) investigations on their vision. This allows precise control over a wide range of light bandwidths and intensities, without the problems of behavioural anomalies introduced by normal test conditions (Pickens 1991; Pickens et al. 1987).

Ludueña Almeida and Gorla (1995) observed that both temperature and illumination were significantly correlated with catches of *Ochlerotatus albifasciatus* in CDC light-traps baited with CO_2 in Argentina. Trap catches were highly correlated with illumination when ambient temperatures exceeded a threshold of $6^{\circ}C$.

The presence of a bait animal may significantly alter the response of a mosquito to light, as suggested by the results obtained by Costantini et al. (1998) from West Africa, for both *Anopheles gambiae* s.l. and *Anopheles funestus.* When a human bait was enclosed within a mosquito net adjacent to a CDC light-trap, whether the light was on or off did not significantly affect the catch either indoors or outdoors, suggesting the dominant importance of host cues over light as an attractant for these species.

Killing agents

In some traps, such as CDC-types and Monks Wood light-traps, mosquitoes are not purposely killed, but every effort made to reduce mortality to a minimum by introducing wet cotton wool covered with filter or blotting paper into the bottom of the collecting bag, and by protecting the bag from direct sunlight. Some designs of CDC traps have side pockets to hold moistened cotton wool balls. In some traps, however, the aim is to kill the catch. A few insect traps employ electric grids but mosquitoes are very rarely killed by electrocution. In some traps insects, including mosquitoes, are collected directly into alcohol (McDonald 1970, 1980), aqueous 0.01% Teepol and 0.02% Hibitane (Taylor et al. 1982), 0.8% saline with 0.1% household detergent, into 50% benzoic acid (Nielsen et al. 1980) or into phosphate buffered saline (Lindsay et al. 1991), but these procedures hamper the subsequent identification of many species. Walker and Boreham

(1976) found that blood in engorged *Culicoides* collected in saline (7.5 g NaCl/litre with 1:2000 antiseptic added as a wetting and bactericidal agent) could still be distinguished by the precipitin test, often after storage for 48 h, but thereafter there was a dramatic reduction in those that could be identified.

A wide variety of chemicals that emit toxic vapours have been employed in light-traps to kill mosquitoes. Both potassium and sodium cyanides are used, the latter being better in humid climates as it does not deliquesce so quickly (Frost 1964). A pint or quart-sized killing jar, preferably plastic, having 0.4–0.75 oz of granulated cyanide covered with 0.25–0.5 in of sawdust and then with an equally thick layer of plaster of Paris makes a convenient killing jar, and should remain effective for 3–6 months, depending on how often it is used. Calcium cyanide is also a very effective killing agent but it has to be replenished almost nightly. Chloroform, ethyl acetate, diethyl ether, and carbon tetrachloride can be used but they evaporate very quickly and it is more practical to use less volatile agents such as trichloroethylene, tetrachloroethylene and trichloroethane. Because their heavy vapours evaporate more slowly a single charge of killing fluid, which can conveniently be poured onto a block of plaster of Paris placed in the traps, can last through one or two nights. Contact with the skin and prolonged inhalation of these vapours should be avoided as they can be carcinogenic, but even chloroform and carbon tetrachloride should be handled with caution because of (low risk) carcinogenic properties. Other killing agents include resin strips impregnated with dichlorvos (DDVP) insecticide (Bartnett and Stephenson 1968; Easton et al. 1986; Herbert et al. 1972; McDonald et al. 1964; Mitchell 1982; Pennington 1967; Preiss et al. 1970).

Automatic time switches and segregating devices

A variety of automatic timing devices to switch traps on at dusk and off at dawn can be incorporated into the circuits of those operating from mains electricity or from batteries (e.g. Ross and Service 1979; Turner and Earp 1968). For battery-operated traps a light-activated switch employing a photosensitive cell can be used (Buckley and Stewart 1970; Elston and Apperson 1977; Service 1970). Clockwork mechanical timers (Sholdt et al. 1974) remain useful.

Various segregating devices have been introduced into light-traps to divide the catch into hourly samples, or when the trap is left unattended for several days, into daily samples. Holub (1983) described a simple mechanism that segregates catches in a New Jersey light-trap into three 24-h samples, which was found useful over the weekends in routine surveillance programmes. The mechanisms involved can be divided basically between, those which employ a series of containers fixed to a revolving table which are advanced under the base of the trap at predetermined time-intervals (Frost 1952; Hutchins 1940; King et al. 1965; Lafferty and Murphy 1948; Nagel and Granovsky 1947; Siddorn and Brown 1971; Standfast 1965; Strickland 1967; Taylor et al. 1982; Williams 1935), and those which use a series of dropping discs to partition the catch (Harcourt and Cass 1958; Haufe and Burgess 1960; Horsfall 1962; Johnson 1950; Taylor 1951). Bast (1960), however, applied a different principle, using a metal cylinder, 15½ in tall and 3 in internal diameter, having seven sloping metal louvres positioned into the base of a New Jersey trap. A series of seven solenoid-actuated plates are automatically drawn into and across the vertical cylinder to seal off the catch at regular intervals. Nielsen et al. (1980) described a mains operated (220-V) light-trap for catching small insects which had a turntable carrying 24 bottles which could be placed under the trap's funnel to receive and segregate the catch into 10, 20, 30 or 60-min intervals. Mitchell (1982) slightly modified Sudia and Chamberlain's (1962) CDC trap so that it could be used with a turntable-type hourly segregating mechanism. Collecting vials (1.25 × 5 in) were mounted beneath a 10-in diameter acrylic turntable and advanced by a 1-lb weighted pulley after activation by the timer. This consisted of a magnet attached to the bronze hour-arm of a wind-up clock, which activated a mercury switch, which in turn activated a solenoid and released the turntable. A 6-V 30-A motorcycle battery powered the fan, light and solenoid.

Meyer et al. (1984) described a time-segregating device (TSS = time-segregated sampler) used in conjunction with a CDC light-trap and the release of carbon dioxide from a cylinder. The authors claim that an advantage of their trap over those of Standfast (1965) and Mitchell (1982) is that large numbers of mosquitoes can be retained alive in their modified 1 quart-sized collection cartons. Basically a counterweight moves the light-trap and CO_2 delivery system, which are mounted on a trolley, along a horizontal aluminium platform until the 'T' arm of a solenoid contacts the front edge of a window lock and stops the trolley over a collection carton. After a time interval of an hour the solenoid releases the trolley and the counterweight system pulls the trap along to be stopped over the next carton by another window lock. (Non-Americans may find the author's description of the trap difficult to follow because many of the parts, e.g. aluminium angle equal leg, kick plate, clothes dryer reducer, will be unfamiliar to them). In California, USA the trap enabled the flight periodicities of host-seeking female *Culex tarsalis, Culex quinquefasciatus* and *Culiseta inornata* to be determined. Reisen et al. (1997) used the time-segregated device of Meyer et al. (1984) to study the host-seeking behaviour of *Culex tarsalis* in California, USA. Carbon dioxide was released at a rate of 1 litre/min in 1983, 1984 and 1994 and 0.5 litres/min in 1995. Host-seeking in this species commenced at the hottest and driest time of night, close to sunset.

Linley and Evans (1968) described the construction of a turntable device for laboratory time-lapse studies, in which insects are deposited at hourly intervals into a cold box. This type of apparatus can be used with a New Jersey trap not only to segregate catches into hourly samples but also to keep them in good condition for later virus isolations or dissection.

In the USA Hendricks (1985) mounted an infrared light emitting diode having peak emission in the 880-nm wavelength and two phototransistor infrared detectors at the entrance of a light-trap to record the periodicity and numbers of moths entering it. The infrared detection system operated from a 6-V d.c. battery. This type of approach could probably be adapted to monitor mosquito flight times. In fact Moore et al. (1986) built equipment that triggered the recording of 512 light intensity measurements within about 0.05 s, when an insect flew into a light beam, due to reflected light from the insect's wings. Spectral analysis of wingbeat frequencies allowed all female *Ochlerotatus triseriatus* and all male *Aedes aegypti* that flew into the light beam to be identified, 93% of the male *Ochlerotatus triseriatus* were also recognised, but only 43% of female *Aedes aegypti* were correctly identified. The authors believed that for field operation the employment of low power red light or infrared laser light would allow for a considerable distance between the light source and its detectors. This type of sophisticated electronic equipment could possibly be incorporated into a light-trap, but there might be problems with variations in wingbeat frequency due to the mosquito's size, age, and its physiological condition as well as ecological variables such as temperature, which can affect wingbeat frequency.

Screens

Hemmings (1959) followed the recommendation of Mulhern (1953*b*) and excluded larger unwanted insects from New Jersey traps by substituting a vertical screen between the conical roof and trap body for the traditional horizontal one. Although these screens reduced the numbers of unwanted insects, they also reduced the mosquito catch by about 50%. Barr et al. (1963) also found that a ¼-in mesh vertical screen reduced the catch of mosquitoes more than a horizontal one. However, Gjullin et al. (1973) found in laboratory experiments that a ¼-in mesh vertical screen did not reduce the catches of male or female *Culex tarsalis, Culex quinquefasciatus* or male *Ochlerotatus sierrensis*. Bartnett and Stephenson (1968) substituted a $\frac{1}{4}$-in mesh screen for the standard $\frac{5}{16}$-in one normally used in a New Jersey trap. This reduced the catch of unwanted insects, such as beetles and moths, which were causing damage to the mosquitoes collected, and of the five mosquito species caught only the catch of *Psorophora ciliata*, a large mosquito, was reduced by the decrease in mesh size. If a live catch is required from New Jersey traps Floor and Grothaus (1971) advised that the horizontal mesh screening be removed, because debris left by larger insects hitting the screen tended to damage or kill mosquitoes being drawn through it. Mosquitoes can be collected in a 12-diameter, 6-in long nylon netting bag, similar to that used on a CDC trap.

Screens can be made of galvanised wire, plastic or from fishing nets. With CDC type traps Service (1993) reported preferring to omit horizontal screens unless there are problems with large insects entering the trap.

Physiological condition of mosquitoes caught in light-traps

Most mosquitoes caught in light-traps comprise unfed females (Bryan 1986; Chamberlain et al. 1964; Chandler et al. 1976*a*; Fox 1958; Husbands and Reed 1970; Ikeuchi 1967, 1970; Loomis and Hanks 1959; Odetoyinbo 1969; Pritchard and Pratt 1944, etc.), but sometimes substantial numbers of males are caught (Belton and Galloway 1966; Corbet 1964; Corbet and Haddow 1961; Itô 1964; Provost 1952; Sun 1962), and occasionally relatively high proportions of blood-fed females are caught (Bailly-Choumara 1973*a*; Carnevale and Boreham 1978; Hii et al. 1986; Service and Highton 1980; Sun 1964; Vervent and Coz 1969). Since light-trap collections are often made in situations where there are known, or suspected, concentrations of biting adults it is not surprising that unfed females predominate.

Moonlight

There is considerable difference in light intensity on nights with and without a moon, for example Kuiper (1938) says that a night sky without a moon gives about 140 times less radiation than one with a full moon. Moonlight increases both the intensity and duration of illumination, and some insects fly for longer periods on moonlit nights. For those working between 10°N and 10°S the table of average illumination during four 3-hr periods after sunset and for 34 different moon phases presented by Bowden (1973) is useful. Greatest illumination, namely 2009×10^4 lux is at full moon and at 4–6 h after sunset, compared to no moon when light from the stars is only 9×10^4 lux. Half moon gives just about 10% of the light of a full moon. A characteristic of moon phases is the rapid rise in illumination at full moon and its rapid decline afterwards. Illumination also varies considerably both seasonally and latitudinally and this can have a great influence on trap effectiveness.

Numerous papers by Bowden and colleagues have done much to help understand how light-traps sample insect populations, although none of them refers to mosquitoes. Bowden and Morris (1975) using estimates of nocturnal illumination presented by Bowden (1973) and an inverse relationship drew curves of theoretical trap effectiveness, which generally agreed very well with observed catches of insects. It was suggested how light-trap catches could be corrected for variations in trap effectiveness at times of different moonlight (Bowden and Church 1973; Bowden and Morris 1975), and this led to the relationship

$$\text{catch} = \text{constant} \times \sqrt{W / I} \qquad (9.3)$$

where W = trap illumination and I = background light intensity. But if an insect is active for only part of a night then variations in background illumination, and thus of trap effectiveness, outside its flight times are irrelevant. Bowden and Morris (1975) consequently constructed curves of theoretical trap effectiveness based on changes of radius (i.e. distance at which trap illumination equals background light levels), caused by changes in illumination during different parts of the night.

Special difficulties in studying the effects of lunation on light-trap catches are encountered in high latitude (i.e. temperate) regions. First, the marked seasonality of insects makes their study over a long period of time in relation to lunar periods difficult. Secondly, the apparent movement of the moon causes large variations in the distribution and quantity of moonlight at these high latitudes; see Bowden (1981) for a brief explanation on retardation of moonrise and moonset and the effects of this on changes of declination of the moon.

Strictly speaking the efficiency of a light-trap is measured by the percentage of insects within the trap's attraction zone that are actually caught. But just because a trap catches more insects than another it is not necessarily more efficient, because the trap may be sampling a larger volume of air. For these reasons McGeachie (1988) expressed the efficiency of a 125-W mercury vapour light-trap as the percentage of those insects detected with a video camera within a volume of 9.9 m^3 around the trap that were caught.

McGeachie (1989) discussed the difficulties of determining whether nocturnal insects respond to changes in moonlight based on light-trap catches made at different moon phases. One of the many problems is the difficulty of accurately estimating incident moonlight. McGeachie (1989) overcame some of these difficulties by developing an algorithm to estimate moon illumination plus residual starlight. He was able to compare catches of moths with moonlight, temperature and wind speed. This paper should be referred to for other details. The 28-day lunar cycle not only increases illumination from new to full moon, but also increases the duration of moonlight, from a brief period at new moon to light almost throughout the night at full moon. Moreover, there is a sharp increase in the angle of illumination to a max of 90° when the full moon is at its zenith. In temperate regions, away from the equator, added complications are the large differences between the lengths of day and night during summer and winter.

Other useful papers on the physics of trap illumination and factors affecting catches of insects are those of Bowden (1982, 1984). See also the effects moon-light has on human landing catches (Chap. 6).

Turning now to mosquitoes, both Ribbands (1945) and Provost (1958) considered that moonlight enabled the increase in flight activity associated with twilight in many mosquito species to be extended into the night at a higher level than in the absence of moonlight. Bidlingmayer (1964) found that the flight activity of *Ochlerotatus taeniorhynchus* was increased by 95% at quarter moon and 546% at full moon, over that occurring during moonless periods. Similar, but less pronounced increases in flight activity with increase in moonlight were recorded for several other mosquito species. The numbers of mosquitoes caught in light-traps have often been found to be less at full than new moon (Bargren and Nibley 1956; Barr et al. 1963; Bidlingmayer 1967; Bradley and McNeel 1935; Costantini et al. 1998; Horsfall 1943; Miller et al. 1970; Neeru Singh et al. 1996; Onishi 1959; Pajot et al. 1977; Pratt 1948; Provost 1959; Reisen et al. 1983). Huffaker and Back (1943) noted that more mosquitoes were caught in New Jersey traps on dark than light nights. The presence of moonlight reduces the contrast between illumination from the light-trap and the background resulting in an apparent reduction in the brightness of the bulb. It is generally accepted that this reduction is responsible for the decrease in catch size on moonlit nights. Bidlingmayer (1967) found that catches of species found in open areas at night, where light intensities are relatively high, were more influenced by moonlight than those inhabiting wooded areas. He suggested that for woodland species a smaller type of bulb, such as used in the CDC trap (Sudia and Chamberlain 1962), might be more effective in catching mosquitoes than the more powerful bulbs employed by New Jersey traps. Provost (1959) pointed out that as there are many factors influencing the size of the catch in light-traps the effect of moonlight on catch size is often masked. Similar conclusions were reached by

Bidlingmayer (1964). Barr et al. (1963) found significant nightly variations in catch size, and unless comparative trials were made on the same night, real differences could be overshadowed.

In *Aedes vexans* peak flight activity, as shown by light-traps, occurred at different times on different nights (Platt et al. 1958). Evidence suggested that more were caught as relative humidity rose to about 70% and thereafter any increase in humidity resulted in a decrease in catch size. There was no correlation between numbers caught and temperature. In Delaware, USA peak numbers of *Culex salinarius* were caught in New Jersey traps with 25-W bulbs in the sixth 15-min interval after sunset (Murphey and Darsie 1962). Flight activity appeared to be governed mainly by decrease in natural illumination, but on nights with a minimum fall in temperature flight activity, as measured by light-traps and human bait catches, was prolonged. It was suggested that measurements of population size of *Culex salinarius* should be restricted to sampling during peak flight activity periods, thus reducing the variability in numbers caught due to fluctuations in temperature during the night. In Thailand Miller et al. (1970) collected more *Anopheles* and *Culex*, including *Anopheles subpictus, Anopheles vagus, Culex gelidus* and *Culex tritaeniorhynchus*, in light-traps during periods of new moon than on nights with a full moon, but the moon appeared to have no effect on catches of *Mansonia uniformis, Mansonia indiana, Mansonia annulifera* and *Aedes lineatopennis*. In Australia catches of *Anopheles annulipes* in light-traps from dusk to 2130 h were greater on moonlit nights (Foley and Bryan 1991).

In Madhya Pradesh, central India, Neeru Singh et al. (1996) studied the association between size of catch from indoor and outdoor CDC light traps and the phase of the moon. Catches of *Anopheles culicifacies* were significantly higher on moonless nights, but the proportion of *Anopheles culiciafcies* in the total catch remained approximately the same. The total number of culicines collected was approximately three times greater on moonless nights compared with moonlit nights.

In Venezuela, Rubio-Palis (1992) investigated the effect of moon phase on indoor CDC light-trap catches of anopheline mosquitoes. Light-traps were set in three experimental huts and operated for 12 h per night, two nights per week, three weeks per month for 15 months. In each hut a human volunteer slept within a mosquito net. A total of 7636 anophelines, comprising nine species was collected, but only *Anopheles nuneztovari* was collected in sufficient numbers to permit further analysis. Analysis of variance of log-transformed data revealed significant effects for month of capture ($P = 0.0001$) and moon phase ($P = 0.002$). The mean number of *Anopheles nuneztovari* collected on nights with no moon was 1.86 times greater than on nights with a full moon. Flores-Mendoza and Lourenço-de-Oliveira (1996) reported no discernible differences in average hourly catches (human landing catch and light-traps) of *Anopheles aquasalis* during new or full moons in Brazil, although this finding contradicted that of Lourenço-de-Oliveira (1984) (cited in Flores-Mendoza and Lourenço-de-Oliveira (1996)) who observed an average capture of 2.9 *Anopheles aquasalis* per 10 man-hours during nights with a new moon, compared with 1.6 during a crescent moon, 1.4 during a full moon and 1.0 during a waning moon.

Santos de Marco et al. (2002) examined the effects of moonlight on catches of *Lutzomyia* sandflies in Shannon and Falcão light-traps in Brazil. Of the ten species collected, significant differences in abundance among lunar phases were observed only for *Lutzomyia whitmani*, which exhibited reduced abundance at the time of the crescent moon. The activity of female *Lutzomyia intermedia* was lower during full and crescent moon phases.

In Texas, USA Janousek and Olson (1994) studied the effect of a lunar eclipse on flight activity of mosquitoes, assessed using a vehicle-mounted trap and a stationary CDC light-trap supplemented with CO_2. The light-trap was supplemented with approximately 1.5 kg of dry ice in a padded envelope suspended near the trap entrance. Three 15-min collections per hour were made with each trap from one hour before sunset to one hour after sunrise. The lunar eclipse commenced at 2021 h, was total between 2120 and 2256, and ended at 2356 h. Light intensity during the total phase of the eclipse was 0.60 lux, compared with 1.18 lux during the remainder of the night. Collections of *Ochlerotatus sollicitans, Anopheles crucians, Culex salinarius* and *Psorophora columbiae* females decreased in the vehicle-mounted trap as the eclipse reached totality, and increased in the light-trap. The decreased flight activity as measured by the vehicle-mounted trap was considered by the authors to be consistent with the reduced numbers observed during new moon phases (e.g. Bidlingmayer 1964).

Other environmental factors affect trap numbers, for instance some, but not all, species show marked increases in catch size after rainfall. In Indonesia Olson et al. (1983) obtained good positive correlations between the log mean numbers of female *Culex tritaeniorhynchus* and *Culex gelidus* caught per night in different months in CDC light-traps with log monthly mm of rainfall. Porter and Gojmerac (1970) found that in North America there was some evidence to suggest an increase in temperature during the night was associated with an increase in the numbers of *Aedes vexans* caught in light-traps. In South Africa the numbers of *Culex rubinotus* mosquitoes caught in CO_2-baited light-traps decreased on one night from a mean catch of 500 to only 12 the following night, which was very windy (Jupp et al. 1976).

Trap colour

To test the effect of trap colour Barr et al. (1963) covered the cylindrical body of a New Jersey trap with metal strips painted red, yellow, green or white. In all trials the outer surface of the roof was painted dark green and the inner surface white. They reported that the numbers of male and female *Culex tarsalis* and female *Anopheles freeborni* caught were influenced by trap colour, but the catch of *Ochlerotatus melanimon* was unaffected. To discover whether heat from the light source attracted mosquitoes, catches were compared from the traps in which the 25-W bulb was painted black with catches from traps with inoperative bulbs. Too few *Ochlerotatus melanimon* were caught for any conclusions, but there was no significant difference between the catches of *Culex tarsalis* and *Anopheles freeborni* from the two types of trap.

CDC light-trap

The CDC light-trap was first conceived by Sudia and Chamberlain (1962). The original model was made from a 1-ft length of 3¼-in internal diameter plexiglas tubing, slotted on each side to receive a metal bracket (Fig. 9.1*a*). A 4.5-V d.c. motor was mounted on the bracket and a 2⅝-in two-bladed fan made from plastic or aluminium was fixed on its spindle. A 4-V torch bulb was mounted directly above the motor at the top of the trap body. The motor and light-bulb drew 0.15 A each, and by connecting a 75-Ω resistor in series in the circuit the trap operated from a 6-V battery. A 30-Amphr motorcycle battery allowed the trap to be used for about five nights before the battery needed recharging. A detachable wire mesh screen was placed over the entrance of the trap body to exclude large insects. A fine cloth mesh bag with a long narrow neck was fitted to the end of the trap body to receive the catch. Finally, a large plastic or metal flat cover was placed above the trap to protect it from rain when used outdoors.

There have been many changes to the original design of the CDC trap, some of the most important involve substituting 1.5-V D-cell torch batteries, or 1.25-V nickel–cadmium rechargeable batteries, or more recently a 6-V gel-cell battery, for the original motorcycle battery. Other changes include the employment of much more efficient miniature motors and 4-bladed or multi-bladed fans to ensure a better flow of air through the trap, incorporation of automatic timers to switch traps on and off, segregating mechanisms for sorting the catch into time-intervals, and frequently the use of less powerful lights, such as 'grain of wheat bulbs', although this is usually accompanied by supplementation with dry ice.

Some years ago Johnston et al. (1973) removed the 7.5-Ω resistor from their CDC trap and substituted a 2-V light-bulb which used only 65–70 mA/h instead of 130–150 mA/h of the 4-V bulb originally used. This modified trap was either run from two 1.5-V heavy duty alkaline batteries wired in series, or from four standard 1.5-V carbon-zinc batteries, three of which were wired in series but with the fourth in parallel. With two heavy duty batteries the trap could operate for four nights, after which the batteries needed to be recharged. The motor ('AristoRev' or 'Barber–Coleman') was 'underrun' resulting in reduced air flow through the trap, but this did not apparently appreciably reduce the numbers of mosquitoes caught. Using four batteries the motor operates from 4.5 V but the bulb is wired into the circuit to work on 3 V, this causes the 2-V bulb to burn brighter than normal and slightly shortens its life, but in practice this is not a serious problem. If two heavy duty alkaline batteries are substituted the trap can operate for about four nights before recharging is required, but it is usually more convenient to use ordinary batteries and discard them after one night's use. Batteries are held in place on two opposite sides of the plexiglas trap body by metal brackets that contain the electrical contact points. The electrical circuit is completed by inserting a three-pronged plug from the batteries into a socket from the bulb and motor. Singh et al. (1993) used CDC light-traps fitted with a black light indoors to sample mosquitoes in Madhya Pradesh, India.

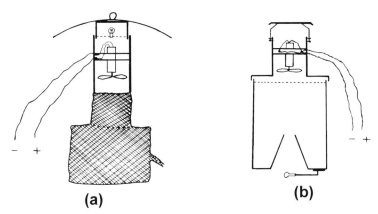

(a)   **(b)**

Fig. 9.1. (*a*) CDC light-trap; (*b*) modified CDC light-trap (after Chaniotis and Anderson 1968)

Chaniotis and Anderson (1968) used a modified type of CDC trap to collect phlebotomines, but one that might also be useful for mosquitoes. The maind difference between this and the normal CDC trap is that insects are sucked up through the bottom and are retained in a 15-cm diameter plastic cylinder placed below the fan. A nylon mesh screen underneath the fan prevents the catch from passing up through the fan blades, while an inverted 5-cm diameter plastic cone in the bottom of the cylinder prevents the catch escaping (Fig. 9.1b). The trap's 3-V motor and torch bulb are run from two 1.5-V dry cell batteries connected in series.

Several models of the CDC trap hold the D-cell batteries in place in metal brackets fixed to the trap body, but Service (1993) found that this arrangement is often not very robust and the electrical connections may be temperamental. Service (1993) prefers to fix the four batteries in plastic battery holders screwed into the bottom of a plastic sandwich box (approx. 16 × 11 cm, 6 cm deep), having a snap-on lid, which makes the box rainproof. In operation the box can be placed on the ground, or on a convenient piece of furniture when traps are used in houses. Two wires connected to banana plugs in the side of the box lead to two similar banana plugs in the trap body. If 'nicad' batteries are used these can be recharged without removing them from the box. Although a transparent plastic cylinder (e.g. plexiglas) is usually used for the trap body of a CDC-type trap, it is considerably cheaper if opaque plastic tubing, as sold by builder's merchants for water pipes, is used. Such tubing is more readily available in many developing countries than perspex or plexiglas.

Sometimes the opening from the cylinder part of the trap to the collecting bag has a plastic funnel or an air-actuated gate system, such as a weight biased butterfly check valve, inserted to prevent mosquitoes escaping if the power source fails. However, this is often unnecessary because the light normally fails before the fan, which although often only continuing to run at a low speed nevertheless prevents mosquitoes flying out of the bag.

Githeko et al. (1994) constructed a modified CDC miniature light-trap in western Kenya from locally available materials as follows: the body of the trap consisted of a 15.2 cm length of 10.2 cm diameter plastic drainage pipe. A 6-V motor from an audio cassette player was mounted inside the plastic pipe about halfway along, using a 2.5 cm wide aluminium bracket. A 0.5 mm gauge steel sheet was shaped into a 4.5 cm radius fan blade and soldered onto the motor axis. A 6.2-V standard light-bulb was fitted at the top of the trap using an aluminium bracket. A four-cell stainless steel battery holder was attached to the outside of the plastic tube.

In India, Gunasekaran et al. (1994) constructed modified CDC light-traps from a 14 cm long piece of PVC pipe with an internal diameter of 9 cm. A 6-V d.c. motor with a 7.5 cm diameter three-bladed fan was attached to the top of the tube with a metal bracket. The trap used a 2.5-V bulb mounted just above the motor. A 25 cm length of PVC pipe was attached to the bottom of the trap to prevent mosquitoes escaping in the event of failure of the fan. A 0.03 m^3 cage constructed from a rigid frame covered with cloth was attached to the bottom of the trap to act as the collecting vessel.

Sadanandane et al. (2004) also constructed CDC-type miniature light-traps from locally available materials for use in the National Mosquito Surveillance Programme in India. These authors omitted the wire mesh screen from the entrance to the standard trap and attached a 25 cm length of PVC tubing as per Gunasekaran et al. (1994). These modified traps caught 77.6% of the total mosquitoes collected using five methods (light-traps, indoor and outdoor resting catches, and indoor and outdoor landing catches). A total of 16 species of anophelines were collected using the five methods, and all 16 species were represented in the light-trap samples.

Parsons et al. (1981) describe how to fit two 14.5-cm vertical welding rods to the frame of the CDC collecting bag to prevent it collapsing and ruining the catch during transportation. In Brazil, to prevent catches being spoiled by rain Gomes et al. (1985) and Forattini et al. (1987) replaced the typical, and rather large, collection bag of the CDC trap with a cylindrical mesh cage with a removable bottom to facilitate removal of the catch. This was then enclosed within a 30-cm length of 15-cm diameter plastic tubing. Ginsberg (1986, 1988) often caught too many mosquitoes in CDC light-traps to be counted. He therefore estimated the numbers present by counting the numbers (twice) resting within a 12-cm band drawn on the net collection bag 14 cm from the top. He found there was a very good regression ($R^2 = 0.887$) when the square root of the log numbers counted within the band were plotted against log numbers of the total catch in the bag.

A parabolic aluminium reflector can be fixed to the inside of the roof of a CDC trap to increase light output.

Evans et al. (2005) describe an easily-constructed T-bar apparatus that can be used to support a CDC light-trap on one arm and a CO_2 source on the other (Fig. 9.2). The apparatus is constructed from ¾-in PVC tubing with an inner tube of ½-in aluminium pipe for strengthening. The internal aluminium pipe is also pushed into the ground to anchor the support. A T-shaped coupling is attached to the top of the central PVC pipe and two side arms (6 and 8-in long) are connected to the coupling. An S hook is attached to each arm by way of a ¼-in × 2 ½-in eyebolt. The CDC light-trap is hung from the 8-in arm and the CO_2 container is hung from the shorter arm. The authors described two lengths of support – one of 60-in and one of 20-in. The 20-in support was designed so that the collection net rests just above the ground, making it suitable for collecting low-flying species, such as *Aedes albopictus*. Holes can be drilled in the internal aluminium pole of the central support to allow the height of the T-bar to be raised or lowered.

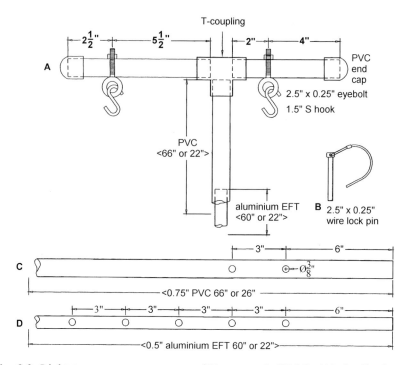

Fig. 9.2. Light-trap support apparatus of Evans et al. (2005). (A) detail of top of pole showing attachments for hanging traps. (B) wire lock pin for securing internal aluminium tube to external PVC pipe and adjusting height. (C) PVC pipe detail. (D) Aluminium pipe detail

The outer PVC pipe is secured at the desired height by inserting a $\frac{1}{4} \times 2\,\frac{1}{2}$-in wire lock pin into the appropriate hole drilled into the aluminium tube.

Sandoski et al. (1983) considered that mosquitoes were often damaged in passing through the fan of CDC-type light-traps, so they altered the trap as follows. The fan is mounted upside down above the light-bulb and the wiring reversed so that air is blown downwards (Fig. 9.3).

The trap lid is placed directly on the top of the trap, and screening closes off the top of the trap. Mosquitoes enter through four 5-cm diameter holes cut near the base of the trap. A 2-cm cone-shaped baffle, made from an ordinary laboratory funnel, is fitted in the trap to ensure the down draft of air blows the mosquitoes into the collection cage and not out through the four entry holes. Wet cloth is placed over the catch bags during transportation. In addition to these modifications the CDC trap has spawned several other miniature battery-operated light-traps, such as the ACIS trap, US Army solid state surveillance traps, and the EVS and Pfuntner traps.

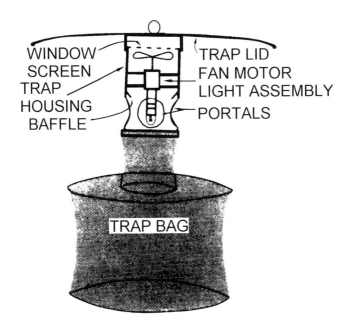

Fig. 9.3. Modified CDC light-trap (Sandoski et al. 1983)

Results obtained with CDC light-traps

Odetoyinbo (1969) was the first to show that the CDC light-trap could be successfully employed to catch *Anopheles gambiae*. In The Gambia the mean catch of *Anopheles gambiae* was as high as 1217.5/trap-night when the trap was hung inside huts on a level with the eaves. Traps situated inside village compounds caught a mean of 76.3 females/night, but when placed near the larval habitats the catch dropped to 5.8/trap-night. The maximum catch of *Anopheles gambiae* exceeded 3000/trap-night. Substantial numbers of *Anopheles melas, Anopheles ziemanni, Anopheles funestus, Anopheles pharoenis* and *Anopheles squamosus* were also caught. A mean catch of only 10.0 culicines was obtained when a trap was used near their breeding places, but when it was placed under the outside eaves of huts, the catch increased to 267.3, and inside rooms on level with the eaves to 409.5. However, the largest mean catch 610.2, was obtained when the trap was located in an open field. The maximum catch of culicines exceeded 2000/trap-night, the most abundant species being *Culex tritaeniorhynchus,* but *Aedes irritans, Aedes furcifer, Aedes aegypti* and *Mansonia* species were also commonly caught.

Odetoyinbo (1969) concluded that maximum numbers of endophilic mosquitoes were caught when the trap was placed as near as possible to the host, numbers caught decreasing with increasing distance from the host. He considered that the effective range of the CDC miniature light-trap was less than 5 m, and that the presence of a trap in a hut was unlikely to attract more mosquitoes into it.

Since these early studies CDC-type traps have been evaluated in several African countries, and are now probably the most widely used light traps for mosquito sampling. For example, when Chandler et al. (1975*b*) placed light-traps in village houses on a rice irrigation area they caught, from 40 trap-nights over a period of 11 months, 22 522 mosquitoes belonging to 17 species, although about 74% of the catch comprised *Anopheles arabiensis* and 16% *Anopheles funestus*. In other studies in Kenya whereas only 10 species were collected in day time searches in village huts, 15 species were caught in CDC light-traps placed in huts, including endophagic but exo-philic species such as *Mansonia uniformis, Mansonia africana, Culex an-tennatus* and *Culex univittatus* (Chandler et al., 1975*a*). When CDC traps using a 3.8-V torch bulb and modified to operate from 12-V car batteries were placed for 107 trap-nights in a heronry near a rice irrigation scheme at heights of 1.9, 7.4 and 11.7 m they caught 5434 mosquitoes belonging to 20 species (Chandler et al. 1976*a*). *Mansonia uniformis* formed 30.1% of the total catch of females, *Culex antennatus* 20.3%, *Mansonia africana* 14.2% and the *Culex univittatus* group 9.5%. Unfed females predominated (91.8%), while small numbers of gravid (5.4%) and blood-fed (2.8%) indi-viduals were also caught; males formed 9.6% of the catch.

Also in Kenya, Johnson et al. (1981) caught 21 809 mosquitoes belong-ing to eight species, when CDC traps were placed for a week in houses near a rice field. The *Anopheles gambiae* complex comprised 53.4% of the catch, *Mansonia uniformis* and *Mansonia africana* 21.5 and 1.8%, respec-tively, while *Anopheles funestus*, from which they isolated o'nyong-nyong virus, formed 21.1% of the catch.

Highton (1981) in a careful study of mosquitoes on an irrigation scheme in Kenya concluded that CDC-type light-traps placed inside houses gave reliable and unbiased collections of the *Anopheles gambiae* complex and could be used to monitor population changes. Both Vervent and Coz (1969) and Coz et al. (1971) successfully used CDC traps to catch *Anopheles gambiae* and *Anopheles funestus* in huts in Burkina Faso, but whereas Odetoyinbo (1969) found that only about 1.2–3.6% of *Anopheles gambiae* were blood-fed, Vervent and Coz (1969) found that over 50% of those caught indoors were engorged. Similarly Coz et al. (1971) found that 57%

of *Anopheles gambiae* were blood-fed, and that 36% of *Anopheles funestus* and 78% of *Anopheles nili* caught in indoor light-traps were also blood-fed. In traps placed outside the huts the proportion of blood-fed individuals decreased to 18, 8 and 35% respectively for these three species. Coz et al. (1971) considered that samples of the malaria vectors provided by the CDC light-traps could be used to study their biting rates, physiological age and sporozoite rates. Contrary to these and other observations in Burkina Faso, Hamon et al. (1969) and Carnevale and Le Pont (1973) in the Congo, reported that *Anopheles gambiae* s.s. was only moderately attracted to CDC light-traps. The trap seemed more efficient in collecting *Anopheles funestus, Anopheles nili, Anopheles paludis* and *Anopheles flavicosta*, but Hamon et al. (1969) considered that these species were in fact attracted into huts by the light. In the Congo Carnevale (1974) and Carnevale and Boreham (1978) also found that CDC light-traps placed inside houses were useful for catching *Anopheles nili*, including blood-fed individuals.

CDC light-traps have been used alongside human landing catches in investigations of outbreaks of yellow fever and chikungunya arboviruses in Senegal (Thonnon et al. 1998, 1999).

In Madagascar when CDC traps were placed in cow-sheds they caught over a 3-month period, some 19 432 *Anopheles coustani* and *Anopheles squamosus/cydippis*, as well as 2496 *Culex antennatus* and lesser numbers of 14 other species including small numbers of *Anopheles gambiae* s.l. and *Anopheles funestus* (Fontenille and Rakotoarivony 1988).

Charlwood (1997) operated a standard CDC miniature light-trap for 45 consecutive nights in a single house in southeastern Tanzania. The trap was powered by a cadmium dry-cell battery and hung near the foot of the two occupants' bed at a height of 1.5 m. Light emitted from the trap was not visible from outside of the house. The householders slept under a mosquito net. A total of 48 049 anophelines belonging to 12 species or species groups were collected. *Anopheles gambiae* s.l. was the commonest species collected (17 595), followed by *Anopheles funestus* (12 397).

Twenty species were caught by CDC light-traps placed inside houses in Gabon, including *Anopheles obscurus, Anopheles hancocki, Culex perfuscus* and most surprisingly the *Eretmapodites chrysogaster* group, which are species that were not caught in human bait catches nor in Monks Wood light-traps placed out of doors (Service 1976). In Ethiopia Sholdt et al. (1974) reported that some 41 400 mosquitoes were collected over 24 trap-nights in CDC traps, and one trap caught over 12 000 mosquitoes during a single night. In Morocco Bailly-Choumara (1973*a,b*) found CDC traps caught more anophelines and culicines than were obtained in human bait collections; they were also useful for sampling outdoor populations.

From 85 trap-nights in one locality in Sabah, Malaysia 56 *Anopheles balabacensis*, 208 other anophelines and 3134 culicines (91.4% of total catch) were caught (Hii et al. 1986). In another locality, 313 *Anopheles balabacensis*, 283 other anophelines and 3133 culicines (84.0%) were obtained from 92 trap-nights. Their other anophelines were *Anopheles donaldi* (16.3 and 19.6% of the total anophelines from the two catch sites respectively), *Anopheles hyrcanus* (4.7 and 12.9%), *Anopheles tessellatus* (3.9 and 26.1%), *Anopheles vagus* (5.7 and 18.2%) and *Anopheles philippinensis* (3.8 and 13.6%). The proportions of blood-engorged adults varied considerably both between and within species depending on location and date, for example from 0–53.6% for *Anopheles donaldi* and from 44.4–79.3% for *Anopheles balabacensis*. Hii et al. (1986) reported that in other collections elsewhere in Malaysia *Anopheles flavirostris* was caught in CDC light-traps, and also in one locality four *Anopheles sundaicus*. They concluded that indoor light-traps were very valuable in catching malaria vectors. When CDC traps and modified versions were used in piggeries in Sarawak, very large numbers of *Culex tritaeniorhynchus* were collected (Hill 1970).

Singh et al. (1993) collected 1711 anophelines of 14 species from 44 trap nights using CDC light-traps fitted with a black light and hung indoors at a height of 5 ft 6 in. *Anopheles culicifacies* was the dominant species comprising 64.3% of the anophelines caught, followed by *Anopheles subpictus* (21%), *Anopheles annularis* (5.3%) and *Anopheles fluviatilis* (3.1%). Males comprised 15% of the 1101 *Anopheles culicifacies* collected.

In Iran, although CDC traps caught *Anopheles stephensi* and *Anopheles fluvialitis* from a variety of situations, most were collected when traps were placed in animal shelters. Also rather surprisingly more were collected from traps placed in the compound of a house and in the middle of a village, than when placed inside a house (Zaim et al. 1986). These light-traps caught all known anophelines in the area, namely in addition to *Anopheles stephensi* and *Anopheles fluviatilis*, adults of *Anopheles dthali*, *Anopheles multicolor, Anopheles pulcherrimus, Anopheles sergentii, Anopheles superpictus* and *Anopheles turkhudi* were also caught. Large numbers of culicines were also collected, again mainly from animal shelters.

Pajot et al. (1977) found that when CDC light-traps were placed inside houses, on porches, in compounds, or even in stables in French Guyana they collected *Anopheles aquasalis*, *Anopheles braziliensis* and *Anopheles darlingi*, the species composition varied greatly between the collecting sites. The numbers of *Anopheles aquasalis* and *Anopheles darlingi* caught on porches were greater on moonless nights, but moonlight had little effect on the catch of *Anopheles braziliensis*. Later, in French Guyana, Silvain and Pajot (1981) operated three CDC traps in different out of door situations for two nights a week for about a year, and caught 41 019 female *Anopheles aquasalis*. As previously noted by Pajot et al. (1977) the numbers caught not only varied greatly between the three traps but within each trap on different nights. For instance, one trap caught six *Anopheles aquasalis* during one night, then six nights later the same trap in the same position caught 2486 females.

Bryan (1986) did not find CDC traps very useful in Papua New Guinea. From nine collections with traps placed indoors few mosquitoes were caught (49 *Aedes nocturnus*, 19 *Anopheles punctulatus* and 14 *Culex annulirostris*), and most of them comprised unfed females.

Mitchell (1982) used CDC traps with a turntable-segregating mechanism in Ohio, USA to study nightly activity cycles and vertical stratification of *Ochlerotatus canadensis*, *Ochlerotatus stimulans*, *Ochlerotatus sticticus*, *Aedes vexans* and *Ochlerotatus trivittatus*. The first three species exhibited maximum activity from about 1800–2200 h, *Ochlerotatus trivittatus* was also most active during these hours, but substantially more were collected throughout the remainder of the night than the other species. *Aedes vexans* showed peak activity from 2100–2300 h and from 0400–0600 h. Most *Culex pipiens* were caught from 2200–0700 h, and about 80% were caught in a trap positioned at 30 ft and 20% in a trap at 5 ft, substantiating the preference for high elevations found by Mitchell and Rockett (1979), whereas about 84% of *Aedes vexans* were caught in a trap placed at 5 ft.

CDC-type light-traps manufactured by American Biophysics Corporation (Rhode Island, USA) supplemented with CO_2 and octenol were used by Bosak et al. (2001) to study the preferred habitats of *Coquillettidia perturbans*, a potential vector of eastern equine encephalomyelitis, in New Jersey, USA. Three traps were set up at a height of 1.0m above ground at each of three sites representing different habitats: forest, open field, and marsh and operated from late afternoon to late morning the following day. Traps were also set at 1.0, 2.5 and 6.0 m in a forest habitat. Host-seeking mosquitoes were more common in the forest than in the marsh or open fields. *Coquillettidia perturbans* were also more frequently collected at a height of 1.0m above ground than at 2.5 or 6.0 m.

Light-traps used in conjunction with mosquito nets

Garrett-Jones and Magayuka (1975) working in Tanzania were the first to exploit the use of light-traps in houses in which the human hosts were enclosed within a protective mosquito net. Since then several others have used this method to catch endophagic mosquitoes.

Lines et al. (1991) undertook a detailed comparison in Tanzania of the numbers of *Anopheles gambiae* s.l., *Anopheles funestus* and *Culex quinquefasciatus* caught in a CDC light-trap hung in a bedroom 1.5 m from the floor and about 50 cm from a person enclosed within a mosquito net, with those caught during other nights by two people catching biting adults from 1830–0630 h. When the numbers of females caught by three CDC light-traps were plotted against numbers biting two people indoors (Fig. 9.4*a*) there was a clear correlation between the two methods ($r = 0.85$, $P < 0.0001$). To determine whether the relative sampling efficiency (which was shown not to vary according to species or villages where catches were made) depended on mosquito densities, the ratios were plotted between light-trap catches and human-bait catches against geometric means of the two catches. The vertical scatter of the observations in Fig. 9.4*b* shows little or no relationship with mosquito densities. The mean log ratio was 0.0315, the antilog of which was 1.07 and is shown by the horizontal solid line, which means that the numbers caught from three light-traps were about the same as the numbers caught by paired bait collections. The two broken lines (Fig. 9.4*b*) indicate the 95% range. The numbers of ovariolar dilatations in *Anopheles gambiae* caught by the two methods were the same, and there were no significant differences between parous or sporozoite rates in the two *Anopheles* species in the two collections. *Culex quinquefasciatus* females were not dissected. It was concluded that light-traps sampled the same elements of the population as bait collections, but only so long as the hosts were fully protected by mosquito nets. Without a net it appears that indoor light-traps can sample different endophilic species with different bias, and there can be bias for age of females caught as well as for species (Carnevale and Le Pont 1973; Coz et al. 1971; Shidrawi et al. 1973).

Maxwell et al. (1990), also in Tanzania provided mosquito nets to the inhabitants of houses in which CDC traps were used. The average number of *Culex quinquefasciatus* caught per night in a trap placed in a room that usually contained two sleeping occupants, was multiplied by the number of nights in each month, and after adding the monthly estimates together, the total was multiplied by 1.5, the conversion factor later published by Lines et al. (1991). The end product was the estimated number of bites received by a person a year.

Faye et al. (1992) compared the use of CDC light-traps with human landing catches in Senegal. CDC light-traps were set up indoors and outdoors and were run from 1900 to 0700 h. Traps were hung at a height of 60 cm, slightly above the level of the beds used by the village inhabitants.

While catches using both methods were somewhat similar indoors, when operated outdoors CDC light-traps caught fewer *Anopheles gambiae* s.l. and *Anopheles pharoensis* than human landing catches. In terms of the mean number of *Anopheles gambiae* s.l. females captured per night seasonal figures were similar for both capture methods indoors. Outdoors, light-traps caught slightly more *Anopheles gambiae* s.l. females than human landing catches in the period September to October 1984 (58.5 v 43.3 per night respectively), but caught only one quarter of the human landing catch from June to October the following year (25.3 v 118.3). The high initial catch outdoors was ascribed to the location of the trap close to animal stables. More *Anopheles ziemanni* were captured in CDC light-traps both indoors and outdoors. *Anopheles rufipes* was only sampled by CDC light-traps. No differences were observed in the physiological age or sporozoite rate of female *Anopheles gambiae* s.l. caught using the two methods.

The potential use of light-traps to estimate human exposure to malaria vectors in Kilifi, Kenya, an area with lower vector abundances than Muheza, Tanzania, was investigated by Mbogo et al. (1993). In each house, light traps, operated alongside an occupant sleeping in an untreated mosquito net, were used for two consecutive nights, followed on the third night by an all-night human landing catch. Log $(x + 1)$-transformed catches were compared using the graphical and parametric analyses of Altman and Bland (1983), which regresses $z = \log(x + 1) - \log(y + 1)$ on $[\log(x + 1) + \log(y + 1)]/2$. A significant correlation was obtained for the numbers of *Anopheles gambiae* s.l. captured using the two methods ($r = 0.64$, d.f. = 260, $P < 0.0001$). The ratio of human-landing catches to light-trap catches showed a tendency to increase with increasing mosquito abundance.

Light-traps were less efficient at detecting the presence of *Anopheles funestus* compared with human landing catches. Light-traps on the other hand caught about twice the proportion of infected mosquitoes compared to human landing catches. Smith (1995) re-examined the apparently conflicting results of Lines et al. (1991) and Mbogo et al. (1993), in which the latter study purported to show that the ratios in the trapping efficiencies of light-traps and human landing catches were not independent of mosquito abundance.

Smith (1995) noted that if the ratio of the numbers of mosquitoes caught in light traps (x) to the number caught in human landing catches (y) is constant, then so too is the relationship $\log(x) - \log(y)$, but this does not apply when the $\log x + 1$ transformation $(\log (x +1) - \log (y + 1))$ is used, as this is highly dependent on mosquito abundance at low values of x and y. This was the transformation used by Mbogo et al. (1993) to correct for the high proportion of zero catches in their data. Smith (1995) therefore concluded that the high proportion of zero catches in the data from Kilifi, Kenya, were disproportionately influential, giving the false impression that at high densities, z is independent of abundance.

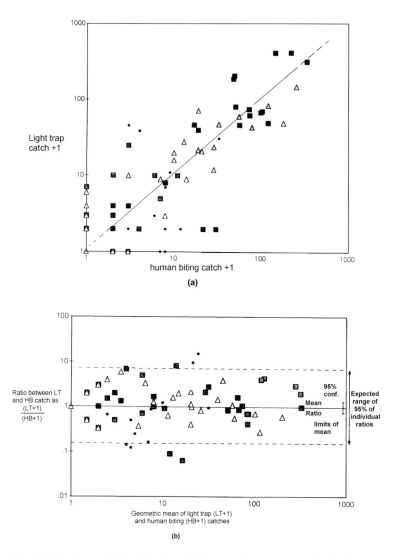

(a)

(b)

Fig. 9.4. (*a*) Numbers of female *Anopheles gambiae* s.l., *Anopheles funestus* and *Culex quinquefasciatus* caught in three CDC light-traps inside a house plotted against numbers of mosquitoes caught by two people in indoor human bait collections. The line shows predicted relationship between the two sampling methods; (*b*) same data as in (*a*) plotted as the ratio between light-trap catches (LT) and human bait catches (HB) against the geometric mean of the two catches. The solid line shows the mean ratio of 1.07, the two dotted lines show the range in which 95% of the ratios are expected to be (Lines et al. 1991)

Githeko et al. (1994) compared catches obtained using locally-manufactured CDC light-traps and human landing catches in a rice irrigation area and a sugarcane belt in western Kenya. CDC light-traps were set up indoors in houses in which mosquito nets were in use. The traps were suspended from the roof, about 30 cm below the eaves and close to the mosquito net. Some traps were also operated outdoors, where they were suspended from the eaves of houses or from the fence of a cattle kraal. Indoor human landing catches were conducted in houses where the occupants did not use mosquito nets. The relationship between the numbers of female *Anopheles arabiensis* caught in CDC light-traps and in human landing collections was log $y = 0.425 + 0.82(\log x)$, and for *Anopheles funestus* the relationship was described as log $y = 1.1 + 0.53(\log x)$. The relationship was not statistically significant for *Anopheles arabiensis* ($r = 0.611$, d.f. = 8, $P > 0.05$). For *Anopheles funestus* and *Anopheles gambiae* s.l., the relationships were significant ($r = 0.88$, d.f. = 9, $P < 0.05$; $r = 0.9$, $P = 0.05$, respectively). These authors concluded that CDC light-traps could potentially substitute for human landing collections for *Anopheles funestus*, but not for *Anopheles arabiensis*. They ascribed the strong correlation obtained by Lines et al. (1991) in Tanzania as being due to the fact that in that part of Tanzania, *Anopheles gambiae* s.s. comprises 90% of the *Anopheles gambiae* population. *Anopheles arabiensis* is known to feed more frequently outdoors and on animals than *Anopheles gambiae* s.s.

Davis et al. (1995) also compared CDC miniature light-traps with human landing collections for sampling anopheline mosquitoes indoors in Tanzania. Human biting catches were made for 30 min each hour from 1800 to 0600 h by two collectors working indoors in the vicinity of a mosquito net occupied by a sleeping individual. CDC light-trap catches were performed in the same houses, either the night before or the night after the human landing catch. A CDC trap was hung 1.5 m above the ground, adjacent to the occupied mosquito net. A total of 2042 anophelines were captured (9.3% *Anopheles funestus*, 90.7% *Anopheles gambiae* s.l.). Numbers of each species collected by the two methods did not differ between study villages or households. Light-traps collected geometric means 1.18 times larger than human bait catches for *Anopheles gambiae* and 1.32 times larger for *Anopheles funestus*. Light-traps caught more gravid females than did the human bait catches (*Anopheles gambiae* $\chi^2 = 91.12$, d.f. = 1, P < 0.001; *Anopheles funestus* $\chi^2 = 27.75$, d.f. = 1, P < 0.001). Human landing catches caught a higher proportion of *Anopheles gambiae* that tested positive for human blood than light-traps, but the proportions of human blood-feds among *Anopheles funestus* were the same for the two collection methods. Also in Bagamoyo, Tanzania Shiff et al. (1995) used the combined light-trap and mosquito net method of Lines et al.

(1991) and the correction factor determined by Davis et al. (1995) to convert CDC light-trap catches to human biting rate and determine the entomological inoculation rates of *Anopheles gambiae s.s., Anopheles arabiensis* and *Anopheles funestus*. Ellman et al. (1998) also used the 1.5 conversion factor of Lines et al. (1991) to calculate entomological inoculation rates for malaria in the Muheza district of Tanzania.

Aranda et al. (2005) used the combined light-trap and human-baited mosquito net method in ten houses to collect mosquitoes as part of a study to characterise malaria transmission in Manhiça, southern Mozambique. *Anopheles tenebrosus* was caught using this method, but was not sampled using the PSC method. These authors did not appear to use a correction factor prior to estimating the overall EIR for Manhiça, at around 15 infective bites per person per year.

In Burkina Faso, West Africa, Costantini et al. (1998) examined the role of CDC light-traps in combination with human bait protected by mosquito nets as a sampling technique. CDC light-traps fitted with 150 mA incandescent bulbs or alternatively 4 W fluorescent UV 'black-light' tubes were positioned 30 cm to one side of an untreated mosquito net suspended over a sleeping adult male volunteer. These authors also used a double net system whereby the volunteer slept inside a mosquito net, which was itself covered by a second larger net, and which had one side raised approximately 30 cm from the ground to allow mosquitoes to enter. In this case, the light trap was suspended in between the two nets and near to the ceiling of the outer net. Traps were also set up outdoors in a rectangular shelter constructed from 2-m long poles covered with a thatched or corrugated metal roof. Indoor and outdoor traps were operated from 2100 to 0500 h. Light-trap collections were compared with human landing collections performed by individual catchers using an electrically-operated aspirator (Coluzzi and Petrarca 1973). The following eight treatments, plus the landing catch, were compared in a 9 × 9 randomized Latin square design: A = no light + no net; B = incandescent bulb + no net; C = no light + single net; D = incandescent bulb + single net; E = no light + double net; F = incandescent bulb + double net; G = UV tube + no net; H = UV tube + single net. In addition, treatments B, C, and D performed outdoors plus an outdoor landing catch were compared in a 4 × 4 Latin square design. Analysis of Williams' means was carried out using a Generalised Linear Modelling (GLIM) package (Payne 1987; Crawley 1993). The UV fluorescent tube caught the most *Anopheles gambiae* s.l., while the incandescent bulb caught the most *Anopheles funestus*. The catch increased three-fold for both species, when the mosquito net containing a human bait was used. There was no significant difference between incandescent and UV fluorescent light sources. In outdoor catches, addition of a human bait sleeping within a mosquito net significantly increased the Williams' mean catch with or without a light. On average, the catch from a single indoor light-trap plus mosquito net was 1.08 times that of a human landing collection (Fig. 9.5).

Cano et al. (2004) compared the effectiveness of human-landing catches against CDC light-traps operated alongside human volunteers protected by mosquito nets, indoor resting collections, tent traps, and window exit traps as rapid assessment techniques for estimating malaria vector dynamics and transmission intensity on Bioko Island, Equatorial Guinea. Human landing catches were the most effective at trapping *Anopheles funestus*, yielding 55% of the total catch, followed by light-traps (18%), indoor resting collections (14%), tent traps (7%), and window exit traps (6%). For *Anopheles gambiae* s.s., the proportions captured using the different methods were as follows: human landing catch (54%), tent traps (20%), light traps (10%), indoor resting collections (8%) and exit traps (8%).

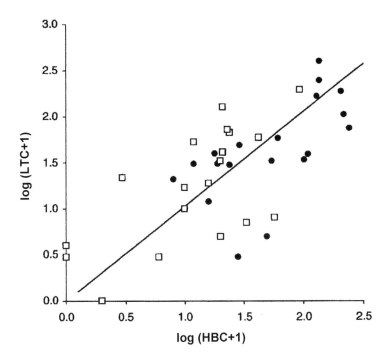

Fig. 9.5. Scatter plot for the relationship between indoor CDC light-trap catches (LTC) and human biting collections (HBC) of *Anopheles gambiae* s.l. (closed circles) and *Anopheles funestus* (open squares). The Pearson correlation coefficient for the pooled data was $r = 0.69$ ($P < 0.0001$). (Costantini et al. 1998)

Hii et al. (1995) in the Solomon Islands, by contrast, found no correlation between the numbers of *Anopheles faruati* No. 1 captured in indoor CDC light-traps hung close to sleeping occupants and human-landing collections. The sampling efficiency of CDC light-traps and mosquito nets containing a human bait was later compared with indoor and outdoor landing collections in Papua New Guinea by Hii et al. (2000). Human landing catches were carried out by pairs of collectors working in 6 h shifts between 1800 and 0600 h over five consecutive nights per month. CDC light-traps were set up in four houses located as close as possible to the landing catch sites while not being visible from them. CDC traps were equipped with a 100 mA incandescent bulb and were positioned 20–100 cm from an untreated net, with the trap inlet positioned approximately 1.8 m above the ground. Traps were operated from 2000 to 0600 h.

Following the observations of Smith (1995), Hii et al. (2000) used a method for comparing the catches obtained that assumes that the overdispersed nature of the variability in catch data can be parameterized as a gamma mixture of Poisson distributions (Venables and Ripley 1994). Observed landing catches were assumed to arise from an underlying gamma or log normal distribution of densities with a distinct value, $E(x_i)$ for each matched set of one landing catch plus four light traps, such that:

$$E(x_i) \sim Gamma(\alpha_0, \alpha_1) \, or \, \log(E(x_i)) \sim Normal(\mu, \sigma^2) \qquad (9.4)$$

where α_0, α_1, μ, σ^2 are unknown parameters of the distribution of $E(x_i)$. The sampling variation about $E(x_i)$ was assumed to follow a Poisson distribution, i.e.:

$$x_i \sim Poisson(E(x_i)) \qquad (9.5)$$

where the distribution of $E(x_i)$ was assumed to be gamma, counts were modelled as being distributed according to the negative binomial. Where the distribution of $E(x_i)$ was assumed to be log normal, very similar values for $E(x_i)$ were obtained. Zero counts are handled by this model with a probability of $x_i = 0$, given by $\exp(-E(x_i))$.

Two distinct models for the numbers of mosquitoes caught in the light traps ($y_{i,j}$, where $j = 1,2,3,4$ for the four light traps) and corresponding with the landing catch were fitted:

$$E(y_i) = \beta_0 E(x_i) \qquad (9.6)$$

which assumes proportionality, and

$$E(y_i) = \beta_1 E(x_i)^{\beta_2} \qquad (9.7)$$

which allows for a curvilinear relationship between the numbers of mosquitoes caught using the different methods. $y_{i,j}$ was assumed to be

$$y_{i,j} \sim \text{Poisson } (E(v_i)) \qquad (9.8)$$

The independence of the terms x_i and $y_{i,j}$ allows for the posterior distributions of the two parameters to be sampled using a (Bayesian) Gibbs Sampling approach (Thomas et al. 1992). Agreement between the sampling methods was assessed by:

1. calculation of deviance statistics and determination of goodness of fit using:

$$d_x = 2\sum_i \left[x_i \ln((x_i + 0.5)/E(x_i + 0.5)) - (x_i - E(x_i)) \right] \qquad (9.9)$$

and

$$d_y = 2\sum_{i,j} \left[y_{i,j} \ln((y_{i,j} + 0.5)/E(y_i + 0.5)) - (y_{i,j} - E(y_i)) \right] \qquad (9.10)$$

The sum ($dx + dy$) should approximate to the degrees of freedom if the fit of the lines is good and the sampling error is approximately distributed as Poisson.

2. by comparing the two different regression lines to see if the proportionality assumption holds. If the two methods are sampling the same proportion of the mosquito population then the fitted line for the curvilinear model should be close to that for the proportionality model.

Anopheles koliensis was the most frequently captured mosquito, followed by *Anopheles punctulatus*, *Anopheles karwari*, *Anopheles farauti* s.l., *Anopheles bancroftii* and Anopheles *longirostris*. All species, except *Anopheles koliensis*, were more frequently captured by landing catches than by light traps. For all species, except *Anopheles longirostris* (indoor landing catch) and *Anopheles bancroftii* (outdoor landing catch), the numbers caught by the different methods were not proportional, although the deviation from proportionality was small in most cases. *Anopheles farauti* s.l. showed a strong curvilinear relationship between sampling methods. Catches of *Anopheles bancroftii* in light traps decreased as the indoor landing catch increased. Only the results for *Anopheles longirostris* indicated that the two sampling methods were proportional over the full range of mosquito densities observed (Fig 9.6 and Table 9.2).

The *Plasmodium falciparum* circumsporozoite rate was significantly higher in light trap collections compared with indoor or outdoor landing catches for both *Anopheles punctulatus* and *Anopheles farauti* s.l.

The authors concluded that the high levels of extra-Poisson variation in sampling error between the different sampling methods would make predictions of biting rates from light-trap catches very imprecise in the case of the anopheline fauna of Papua New Guinea. The lack of proportionality also suggests that the different methods were sampling different proportions of the mosquito population.

To investigate the effects on the catch of position of the light-trap relative to the human host within the mosquito net, Mboera et al. (1998) hung a standard miniature model 512 CDC light trap with an incandescent light bulb beside the net and the trap entrance in one of four positions: the head end or foot end of the bed, at either 70 cm or 150 cm from the floor (approximately 25 cm or 105 cm above the bed, respectively). Traps were operated between 2100 h and 0600 for *Culex quinquefasciatus* and between 2000 h 0600 h for *Anopheles gambiae*. Female catches were log $(x + 1)$ transformed and subjected to Latin square ANOVA. A total of 13 402 *Anopheles gambiae* and 4051 *Culex quinquefasciatus* were caught during the eight- and 12-day experiments, respectively. Analysis of trap position effects showed that when the CDC trap was positioned at the foot end of the bed and at a height of 150 cm, this resulted in 1.3–2.1 times higher catches ($P < 0.05$) for *Anopheles gambiae* and 2.3–3.1 times higher catches ($P < 0.001$) for *Culex quinquefasciatus*. Pooled results showed that for both species traps near the top of the net caught significantly more parous mosquitoes than traps near the bottom ($\chi^2 = 4.36$, d.f. $= 1$, $P = 0.017$ for *Anopheles gambiae*; $\chi^2 = 15.50$, d.f. $= 1$, $P = 0.0001$ for *Culex quinquefasciatus*). To date, there is no clear consensus on the best position to place the CDC light trap relative to the person occupying the net.

Table 9.2. Numbers of anopheline mosquitoes caught in matched landing and light-trap catches in Papua New Guinea (Hii et al. 2000)

	Indoor landing		Outdoor landing		Light traps		Ratio of light-trap : landing catch	
	Total	Rate (person^{-1} n^{-1})	Total	Rate (person^{-1} n^{-1})	Total	Rate (trap^{-1} n^{-1})	Indoors	Outdoors
Anopheles punctulatus	5453	11.5	6007	12.6	2031	1.11	0.1	0.09
Anopheles koliensis	9347	19.6	9777	20.5	10346	5.68	0.29	0.28
Anopheles karwari	3007	6.32	2981	6.27	1595	0.88	0.14	0.14
Anopheles farauti	603	1.27	673	1.41	174	0.1	0.08	0.07
Anopheles bancroftii	111	0.23	158	0.33	87	0.05	0.2	0.14
Anopheles longirostris	197	0.41	341	0.72	78	0.04	0.1	0.06
No. of collections	476		476		1822			

Includes only CDC traps in houses where all sleepers used mosquito nets, and matched landing catches

n = night

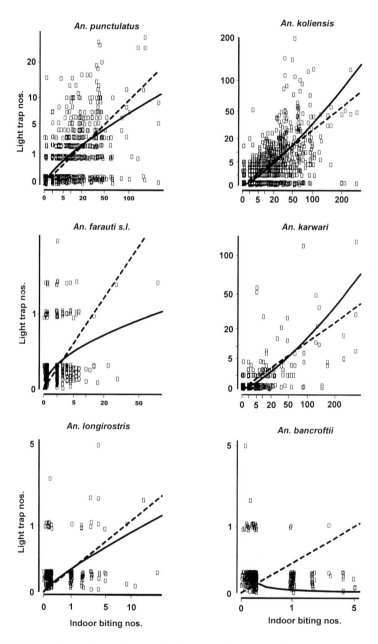

Fig. 9.6. Plots of numbers of anopheline mosquitoes caught in 1250 CDC light-trap collections, against those caught in 476 matched indoor landing catches. Dashed line is fitted from model (i) and solid line is fitted from model (ii) (Hii et al. 2000)

Comparisons of light-trap catches with other methods

In India, Gunasekaran et al. (1994) compared the efficiency of modified CDC miniature light-traps with daytime resting, human landing and night-time resting collections for sampling *Anopheles fluviatilis*. Light traps were operated in two randomly-selected houses and two cattle sheds between 1800 and 0600 h. Traps were suspended at heights of 60 cm above ground level in the human dwellings and at a height of 120 cm above ground in the cattle shed. Daytime indoor resting collections were performed from 0600 to 0800 h for 10 minutes in each of nine human dwellings and nine cattle sheds. Outdoor resting collections were made from natural resting sites, including tree holes, bushes, roots, etc and also artificial pits and pot shelters. Night landing collections were made for 15 minutes per hour between 1800 and 0600 h in each of three houses. The mean light-trap catch in cattle sheds (7.70 ± 1.85 per trap) was significantly higher than in human dwellings (0.85 ± 0.24). The mean number collected in indoor hand catches did not differ between cattle sheds and human dwellings. A significant correlation was obtained between the number per trap in light trap catches and the number collected per man hour in indoor ($r = 0.625$, d.f. = 20, $P = 0.0019$) and outdoor resting catches ($r = 0.603$, d.f. = 20, $P = 0.0029$). No significant correlation was obtained between light-trap catches and human landing catches or nocturnal resting catches and the authors ascribed this to the possible presence of sub-populations of *Anopheles fluviatilis* with different host preferences.

In order to identify a satisfactory method of sampling anopheline mosquitoes that would reduce reliance on the use of human landing catches, Rubio-Palis and Curtis (1992), working in Venezuela, compared human landing catches with CDC light-traps, a human-baited double net trap, a calf-baited trap and outdoor resting collections. Human landing catches were performed between 1900 and 0700 h indoors and outdoors for two nights per week per location. Teams of six catchers worked in pairs in shifts of 4 h and were rotated among shifts and indoor or outdoor collections. CDC light-traps were also operated for 12 h in occupied huts in which the occupants slept under mosquito nets. The double net catches were also performed over the same 12 h period. Searches for trapped mosquitoes were made hourly using a torch and specimens were collected with a manual aspirator. The calf-baited trap comprised a wooden pen (180×120 cm) covered with a netting roof and canvas walls. A gap of 20 cm was left between the bottom of the walls and the ground. Resting mosquitoes were collected using a large battery-operated aspirator. Collections were performed between 0610 and 0800 h. Indoor human landing catches yielded 21 748 mosquitoes representing 11 species of *Anopheles*. The commonest species collected was *Anophles nuneztovari*, which comprised more than 70% of the total collection. Light traps caught fewer individuals, but the four most abundant species were the same as collected by human landing catch. Significant correlations were obtained between the two methods for each of the four commonest species. The ratios of mean light trap catch to mean human landing catch were calculated for the five commonest species in order to estimate relative efficiency of the two methods. Light traps caught only 10% of the *Anopheles nuneztovari* caught by human landing catch, but caught relatively higher proportions of the other four species. The parous rate of females collected by light-trap was significantly lower than that for the human landing catch. The double net was found to be ineffective and its use was discontinued. The calf-baited trap caught only 69 anophelines during 13 nights of operation. In one site, more *Anopheles albitarsis* were collected resting on vegetation than in human landing catches. The authors concluded that the only reliable and available method for monitoring anopheline populations in western Venezuela was the human landing catch.

Standard New Jersey light-trap

As early as 1922 Headlee suggested that because human collectors varied in their degree of attractiveness to mosquitoes and also in their aptitude in catching them, a mechanical device should be developed to sample mosquitoes. This eventually led to the development of the New Jersey light-trap (Mulhern 1934, 1942).

The New Jersey trap consists of a 12-in vertical metal cylinder, 9 in in diameter which is covered by a conical metal roof 16 in in diameter, supported 2 in from the top of the cylinder by three equally spaced metal struts ($^3/_{16}$ in × 1¼ in). These struts are attached to the outside of the cylinder and extend 8 in beyond it to form legs (Fig. 9.7*a*). In later models the body was lengthened and the legs eliminated (Mulhern 1953*a*). A 25-W light-bulb is inserted into a socket fixed to the underside of the apex of the conical roof. A block of rubber or sponge foam can be placed between the socket and roof to reduce vibrations that tend to shorten the life of the bulb. The entrance to the cylinder is usually covered with a $^5/_{16}$ - or ¼-in mesh horizontal screen to exclude large insects. Mosquitoes attracted to the light are drawn into the cylinder by an 8-in diameter fan mounted near the top of the cylinder. A removable funnel of 16-wire mesh is placed within the cylinder just below the fan to guide the catch down into a collecting jar, which is usually screwed into position. This jar is frequently charged with a killing agent such as potassium or calcium cyanide, tri- or tetrachloroethylene or tetrachloroethane absorbed on small blocks of plaster of Paris. These fluids are much more suitable than highly volatile killing agents such as chloroform, ether or ethyl acetate, which evaporate too quickly. Alternatively strips of plastic impregnated with DDVP (dichlorvos) can be used. To prevent mosquitoes becoming contaminated with the killing agent a waxed paper cup with small holes punctured in its sides, or a fine wire mesh container, can be fitted into the entrance of the killing jar to receive the catch.

A useful killing jar consisting of two separated containers was described by Hanson (1959). A metal lid with a large hole in the centre (e.g. Kilner jar ring) is soldered around the base of the screen funnel of the light-trap, and the top of a pint plastic container is riveted to it, so that the pint-sized container can be screwed into it (Fig. 9.7b). The lid of a ½-pint plastic container is now riveted to the bottom of the pint container, and the smaller container which contains the killing agent is screwed into its lid. Two or three small holes are made through the lid of the ½-pint container and bottom of the pint container to allow vapour of the killing agent to enter the pint container. Foam rubber or sponge is fixed to the underside of the lid to prevent mosquitoes coming into contact with the killing agent. White (1964) also devised a method by which captured insects were prevented from coming into contact with the killing agent (potassium cyanide). Occasionally insects are collected directly into methanol (Miller et al. 1977). Gjullin and Brandl (1978) describe a New Jersey type-light-trap that automatically sterilises adults with tepa and then releases them. The outside of the trap is usually painted green or black and the inner surface of the roof white to reflect the light. Bulbs of different light intensity can be used, but in many surveys frosted 25- or 40-W bulbs are preferred. Electrical time switches can be used to automatically switch the trap on and off.

In Australia McDonald (1980) used a 240-V New Jersey light-trap (Mulhern 1942) modified to have a neon pilot globe (0.1 lux) as the light source, and a time clock to switch the trap on and off at dusk and dawn. The insects were killed and retained in 70% methanol and the trap emptied weekly or twice weekly. From 1975–1979, five *Aedes* species, five *Culex* species, of which *Culex annulirostris* and *Culex australicus* were the most common mosquitoes, and *Coquillettidia linealis* and *Anopheles annulipes* were caught. Because of the high failure rate of bulbs in New Jersey light-traps West and Cashman (1985) converted the electricity to the bulbs from a.c. to d.c. by detaching one of the leads to the lamp socket and attaching a 1-A axial lead diode rated at 1000 PIV (peak inverse voltage) to the lamp socket. The other lead of the diode was soldered to the wire that had been detached.

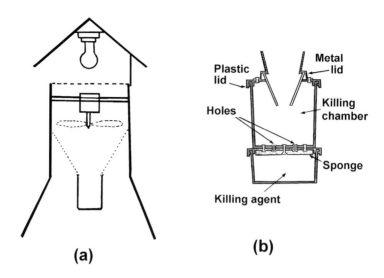

Fig. 9.7. (*a*) New Jersey light-trap; (*b*) killing bottle for New Jersey light-trap (after Hanson 1959)

The 120-V a.c. current is reduced to 54-V d.c., the current from about 0.108 A to 0.056 A, and the wattage from 12.9 to 3 W. This results in greatly diminishing the light output, so if a 15-W bulb was formerly used then to achieve approximately the same brightness a 60-W bulb should be substituted. The exposed electrical leads should be taped to prevent any electrical shorts to the trap body.

In studying the effect of dry ice on collecting *Culicoides* species Wieser-Schimpf et al. (1991) described how by incorporating an inverter ballast and a two-way toggle switch in the circuit, a standard New Jersey light-trap can be converted to operate either the normal light-bulb or a 15-W black fluorescent tube.

A battery operated New Jersey trap with a turntable segregating device was successfully used in Australia (Standfast 1965), but nowadays most battery operated traps are of the CDC-type.

New Jersey traps can be operated from ground level or at different heights (Bast and Rehn 1963; Bast et al. 1964; Blakeslee et al. 1959; Love et al. 1963; MacCreary 1941; Main et al. 1966; Meyers 1959, etc.). As a generalisation insect density decreases with height, consequently there is usually a marked reduction in mosquitoes caught with increasing height, but the density of species feeding on birds may in fact increase with height

(Main et al. 1966; Service 1971). In routine surveillance, traps are often positioned so that the lower edge of the roof is 175 cm above the ground.

In Thailand Miller et al. (1977) found that operating New Jersey light-traps every day was burdensome because of the large numbers of mosquitoes that had to be sorted and identified. They therefore compared catches from one, three and five nights' operation per week, and the numbers collected at six 2-h intervals throughout the night. Eleven species were caught, the most common being *Culex gelidus, Culex tritaeniorhynchus, Mansonia uniformis, Culex annulus, Culex fuscocephala* and *Mansonia annulifera*. It was concluded that just one night's collection per week would monitor mosquito seasonal abundance, but clearly this depends on the numbers caught and nightly variations in catch size. With the more common species different numbers were caught during the six different 2-h periods, and therefore it might be unwise to compare species abundance based on 2-h collection periods.

For many years New Jersey traps have proved extremely useful in North America, especially in surveillance programmes, and also elsewhere in the world. Fox (1958) used six light-traps for about 1 year in Puerto Rico and caught 176 053 mosquitoes representing 26 species, including those of *Aedes, Psorophora, Culex, Uranotaenia, Mansonia* and *Anopheles. Deinocerites cancer* was the third most common mosquito caught in his traps. Pritchard and Pratt (1944) and Fox and Capriles (1952) also successfully used light-traps in Puerto Rico, the latter workers catching as many as 73 010 mosquitoes in 276 trap-nights. Pippin (1965) caught large numbers of mosquitoes belonging to seven genera in traps operated in the Philippines, 51% of the catch consisted of *Culex* species and 46% *Anopheles* species. The three most common *Anopheles* caught were *Anopheles indefinitus, Anopheles limosus* and *Anopheles peditaeniatus*; the principal malaria vector *Anopheles flavirostris* was not collected, confirming the findings of Thurman and Thurman (1955) that in Thailand this species is not caught at light. In Taiwan Hu and Grayston (1962) caught as many as 27 313 mosquitoes in a single trap. In Malaysia 395 female and 95 male mosquitoes belonging to 32 species were caught in four trap-nights when traps using 25- or 100-W bulbs were used (Wharton et al. 1963). Kim et al. (2003a,b) successfully used New Jersey light-traps to survey mosquitoes on military installations in the Republic of Korea. Traps incorporating a 25-watt, white frosted light bulb set approximately 1.5 metres above ground level were used. Traps were operated between the hours of dusk and dawn (1800 to 0800 h) 1–4 times weekly. In 1999–2000, a total of 68 051 and 62 526 adults were collected in 1999 and 2000 respectively, comprising 19 species from 7 genera. The most commonly collected species were *Anopheles sinensis* (34.2%), *Culex tritaeniorhynchus* (29.4%), *Aedes vexans nipponii* (18.2%) and *Culex pipiens* (16.8%). Control operations were launched when the "trap index", corresponding to ten females per trap per night of the malaria vector *Anopheles sinensis*, or the Japanese encephalitis vector *Culex tritaeniorhynchus*, was exceeded. A total of 56 656 adults were collected during 2001. In contrast to these successes very few mosquitoes were caught in a New Jersey trap using a 25-W lamp over 5 years in Nigeria (Service and Boorman 1965). In southern England a New Jersey trap with a 150-W light caught considerable numbers of *Ochlerotatus detritus*, but despite the presence of large populations of *Culiseta morsitans, Culiseta annulata* and *Coquillettidia richiardii* very few were caught (Service 1969). However, when a Monks Wood battery operated light-trap was used in the same locality a substantial number of *Culiseta morsitans* and its sibling species *Culiseta litorea* were trapped (Service 1970).

Milby and Reeves (1989) reported that catch size with New Jersey light-traps was reduced markedly in the presence of competing light sources and in California, USA this has led most mosquito control agencies to incorporate supplemental sampling methods into their mosquito surveillance programs (Reisen et al. 2002), including CDC light-traps (Sudia and Chamberlain 1962), encephalitis virus surveillance (EVS) traps (Pfuntner 1979) augmented with dry ice, and CDC (Reiter 1987) or Cummings (Cummings 1992) gravid female traps baited with alfalfa infusion (Reiter 1983).

American mosquito light-trap

Following reports on the use of badly worn, and sometimes incorrect motors in New Jersey traps Mulhern (1953a) made a series of modifications to the standard New Jersey trap, the more important of which are as follows. The new model has vertical, not horizontal, ¼-in screening to exclude large insects, the cylindrical trap body is enlarged from 9 to 10 in in diameter to take an 8¾-in motor which replaces the standard 8-in motor and finally, the trap body is lengthened to eliminate the three legs. The new model became known as the American mosquito light-trap. However, not all these modifications have proved popular and most workers still use what they term the New Jersey trap, although this may incorporate some of the modifications such as removal of the legs, lengthening of the body or replacement of the metal bracket by which the trap is suspended by a chain. Jewell (1981) modified the American model light-trap to prevent vandalism and the potential danger of people tampering with the cyanide killing jar, by extending the trap and constructing a hinged lockable bottom screen.

To prevent wind-damage to traps in Arizona McDonald et al. (1974) constructed a stabiliser from two 4-ft long iron pipes. One is ⅜ in in outside diameter and the other slightly greater in internal diameter. A simple platform is made of reinforcing rod to hold the base of the trap and is welded to the larger diameter pipe. The smaller pipe is pushed into the ground beneath the trap. The larger pipe is then slipped over it and held in position by screws so that the correct heights can be obtained.

ACIS trap

Boobar et al. (1987) described the construction of a collapsible light-trap that they believed would replace the standard New Jersey trap, and which they christened the Army Collapsible Insect Surveillance (ACIS) trap. The trap consists of two main sections made of thick polyvinyl chloride tubing. The upper part has an outside diameter of 20 cm that collapses into the lower part having a diameter of 20.2 cm. A 35.6-cm diameter aluminium lid serves as a rain shield and is supported on three adjustable rods and has three height settings. Three telescopic aluminium legs attached to the lower part of the trap serve as a tripod base. A 16.8-cm rigid thermal plastic (RTP) fan is mounted on a totally enclosed 1/60 h.p. motor and produces an airflow of 12 m^3/min. To extend bulb life an axial-lead diode, rated at 1000 peak inverse voltage (PIV) at 1 A, was installed according to the procedure described by West and Cashman (1985). Insects pass into a plastic collecting jar, which is inserted into a larger killing jar containing a suitable killing agent. A 20.3-cm deep plastic mesh collecting funnel with the killing jar attached is fixed to the main body of the trap by a 1.9-cm wide velcro fastener. A photocell allows the trap to be automatically switched on and off. The trap operates from 110-V a.c. power. It weighs 4.4 kg compared to the 6.6 kg of a New Jersey trap, and when collapsed it is only 24.1 cm high. Its robustness was shown by its failure to break when the authors repeatedly dropped it from 4 ft onto a hard surface. The developers' desire to see the ACIS trap replace the New Jersey light-trap do not appear to have materialised and it is not currently used.

U.S. Army Miniature Solid State light trap (AMSS)

The AMSS (or SSAM) is essentially an improved CDC light-trap (Driggers et al. 1980). The 6-V d.c. motor is a Mabuchi RF510T, chosen because it gives optimum rev./min over a wide range of operating voltages. It also has the highest rating in terms of rev./min per mA of current used, and high life expectancy. The light-bulb is either a Chicago Miniature Lamp No.503 or 1490, chosen because they have a long life and relatively high light output. The battery of choice is a gel-cell 6-V, 7.5-Amphr rechargeable sealed battery, which provides a relatively constant operating voltage for a long time. A 14-in diameter aluminium pizza pan is used as a lid. A light-cell automatically switches the lamp and fan on at dusk, while at dawn increasing light switches the lamp off, but the fan continues to run, thus preventing mosquitoes from escaping. The circuitry is designed so that the trap will operate only with correct polarity, and when 6 V are applied the motor operates at 5.25 V and the lamp at 4.25 V. The trap can run at peak efficiency for more than 24 h before there is battery failure. The trap can also operate on five or even four 1.5-V torch batteries, but the catch of mosquitoes with the gel-cell battery is 2.67 and 6.60 times greater, and moreover more species are caught. When a CDC trap using four 1.5-V batteries was compared to the AMSS trap operating from five similar batteries, the AMSS trap caught 3.48 times as many mosquitoes. Finally, by using an a.c. converter the trap can operate from a 12-V a.c. source.

From 13 AMSS traps positioned 1 m from the ground near a large breeding place, baited with 2 kg dry ice, and operated from 1600–0830 h on 45 nights in Kenya, some 30 669 female *Aedes lineatopennis* were collected (Linthicum et al. 1985). Also in Kenya, Murphy et al. (2001) used AMSS traps baited with dry ice, hexanoic acid, L-lactic acid and synthetic cheese volatiles. Attractants were delivered from a cotton wick 3.8 cm long, to which the attractants, dissolved in paraffin, were added by way of a pipette. Wicks were attached to the traps approximately 3 cm from the trap inlet. Traps baited with CO_2 + lactic acid caught significantly more *Anopheles gambiae* and *Anopheles funestus* than traps baited with CO_2 alone or with CO_2 + an alternative attractant. No other attractant combination yielded catches in excess of those obtained with CO_2 alone.

In Egypt Zimmerman et al. (1988) caught large numbers of blood-fed *Culex pipiens* in AMSS traps baited with dry ice placed in 3.8-litre cans having perforated bottoms which were hung over the traps. In Senegal when baited with dry ice 2.2% of the 62 055 mosquitoes caught in these traps contained blood (Gordon et al. 1991).

Pagac et al. (1992) used AMSS traps baited with CO_2 and sentinel quails (*Colinus virginianus*) to investigate eastern equine encephalomyelitis activity at the Patuxent Wildlife Research Center, Laurel, Maryland, USA. Traps were positioned approximately 700 m apart and within a 3.2 km radius of pens containing a captive breeding population of whooping cranes (*Grus americana*), 54% of which showed evidence of EEE virus infection. The mean number of *Culiseta melanura* captured per trap per night during August and September for the years 1985–1990 ranged from a

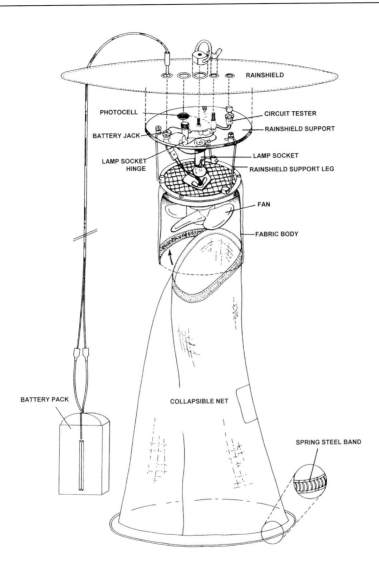

Fig. 9.8. Fabric-body light-trap of Collier et al. (1992)

minimum of 1.6 in 1987 to 11.7 in 1989. In four of the six years, mean catch per trap per night was less than 10.

In the Dominican Republic, Perich et al. (1993) successfully used two AMSS and a Fay-Prince light trap to evaluate the effectiveness of a barrier spraying method to control *Anopheles albimanus*.

Strickman et al. (2000) reported the use of AMSS traps for mosquito surveillance following an outbreak of *Plasmodium vivax* in the demilitarized zone in the Republic of Korea in 1996 and 1997. Traps were hung from a tree or fence about 1.5 m above the ground. Each trap was operated with the light on and carbon dioxide was provided by way of dry ice contained in an insulated, 1-litre jug with an open pour spout. Rankings of the three most abundant species collected was similar for light trap and landing collections and absolute numbers of *Anopheles sinensis* collected were also similar in light trap and landing collections. A significant result was obtained when the log of landing collections and the log of the mean catch from all 5 light traps were compared ($y = 1.14(x) - 0.297$, $r^2 = 0.386$, $F = 10.05$, d.f. $= 1.16$, $P = 0.0059$).

Collier et al. (1992) modified the standard AMSS trap, converting it to a collapsible fabric body light trap to reduce weight and facilitate transport and storage. Plain weave 3.8 oz nylon cloth was used for the 15.24 cm long and 8.89 cm diameter body of the trap and Herculite™ 80 fabric for the folding 35.56 cm diameter rain-shield. The trap net is attached to the body of the trap using a 1.27 cm wide Velcro strip. A 4.7mm thick aluminium disc was attached to the top of the cloth body of the trap to provide support, while a spring steel band stitched around the opening of the net keeps the mouth of the net open and allows it to be collapsed when not in use. The dimensions of the collapsed trap are 12.7 × 12.7 × 4 cm and the weight is 0.5 kg compared with 1 kg for the standard AMSS trap (Fig. 9.8). In comparisons at three field sites, the modified trap showed no significant differences from the standard trap in terms of numbers or species of mosquitoes collected.

EVS light-trap

The top part of the encephalitis virus surveillance trap (EVS) consists of a 1-gal paint tin painted black on the outside and lined inside with polyurethane foam (Rohe and Fall 1979). About 1.45 kg of dry ice is placed in the tin after which a plastic snap-on polyethylene lid closes it. Four to six holes punctured in the bottom of the tin allow carbon dioxide to flow downwards over the lower part of the trap, which consists of a vertical plastic plate having holders for two 1.5-V dry cell batteries (Fig. 9.9) that power a Mabuchi RE-260 motor (about 5000 rev./min) and a sub-miniature 1.5-V, 70 mA 'grain-of-wheat' light-bulb (C. G. Electronics E2-374). The fan was originally a two-bladed plastic propeller that was boiled for a few minutes to soften the plastic, and while still hot the two blades were twisted to increase their pitch to about 45°, after which it was held under cold water so that it set to this configuration. Later models, however, had a simple fan blade cut from sheet metal. The blade is mounted upside down on the spindle of the motor, which is fixed inside a 7-cm long piece of 7.6-cm diameter plastic tubing. A switch near the batteries on the vertical section of plastic allows the trap to be turned on and off.

Mosquitoes attracted by the carbon dioxide disseminating down from the dry ice container and by the light-bulb (but apparently mainly the former) are sucked through a 5-mm plastic or metal mesh screen down into a simple 30-cm long nylon netting catch bag. A plastic circular lid is dropped into the bottom of the bag to make it open out as a cylindrical sleeve. The trap weighs only about 0.95 kg, and when not in use the entire trap can be placed within the carbon dioxide bait-tin. The trap can be wired so that both motor and light-bulb can be replaced in the field without tools. The batteries must be changed nightly.

Barker-Hudson et al. (1993) modified the EVS trap of Rohe and Fall 1979) by adding a photosensitive switching device to turn the traps on and off at dusk and dawn. A baffle was also installed between the fan unit and the collection bag to prevent the escape of captured mosquitoes once the light had been switched off.

Pfuntner (1979) reported that the EVS trap caught a mean of 176 female mosquitoes/night (88% *Culex tarsalis*), whereas a CDC-type trap caught just 53 mosquitoes. In California, USA Walters and Smith (1980) used the EVS trap baited with about 1 kg dry ice to measure seasonal fluctuations of potential virus vectors. To minimise injury to the collected mosquitoes a 1-litre carton was attached to the net collecting bag to receive the mosquitoes, the most common being *Culex erythrothorax* and *Culex tarsalis*. The catch of *Culex quinquefasciatus* remained small, but unresponsiveness of this species in California to light-traps has already been recorded by Magy et al. (1976) and by Webb et al. (1977). However, in July when the trap was supplemented with 1 kg dry ice and placed in deep-water sections of storm drains it caught a mean of 3722 ± 1615 *Culex quinquefasciatus*/trap-night. But when placed in household gardens the trap caught just 37 ± 9 adults/night (Mulligan and Schaefer 1982). Also, in a sampling programme in California Cope and Hazelrigg (1989) found that five to eight EVS light-traps baited with carbon dioxide caught 85% of the 65 824 adults of *Culex erythrothorax* (88% of total catch), *Culex tarsalis*, *Culex quinquefasciatus*, *Culex stigmatosoma*, *Culiseta incidens* and *Culiseta inornata* that were trapped. In contrast, one New Jersey light-trap caught 14% and human bait collections just 1% of these species, demonstrating the effectiveness of the EVS trap for *Culex erythrothorax*.

Fig. 9.9. EVS trap (photograph reproduced with permission of M. W. Service)

Because of the weak light output few insects other than mosquitoes are caught, but up to 10 150 *Aedes vexans* have been collected during a single trap-night (Rohe and Fall 1979). In California, USA two EVS traps caught 433 *Culex tarsalis* in eight nights compared with 246 adults in two CDC-type traps, but only 1048 *Culex erythrothorax* compared to 2232 in the CDC traps.

EVS traps have been used in St. Louis encephalitis virus surveillance in Florida, USA (Zyzak et al. 2002). EVS traps supplemented with 1.5 kg of dry ice were placed at a height of 1.5 m above ground in five forest sites on a weekly basis and operated from 1500 to 0730 h the following day. *Culex nigripalpus* and *Culex salinarius* comprised 81% of the 1991 catch, 84% of the 1992, and 66% of the 1993 catch.

Although the EVS trap has proved popular in the USA it has not been widely used outside America, except in Australia, where Foley and Bryan (1991) found it useful for catching *Anopheles annulipes*. Also in Australia Broom et al. (1989) used a modified form to suit their weather conditions to trap *Ochlerotatus normanensis* and *Culex annulirostris*, and other mosquitoes. In their trap a detachable 50-cm aluminium veranda was attached to the underneath of the trap-can to protect the catch against excessive rain. The trap bag was replaced with a 10-cm tube of nylon netting glued to an inverted 2-litre plastic container. A circular 12-cm diameter section of the lid was replaced with fly-screen netting so that airflow would not be affected. This modified bag was attached with elastic bands and adhesive tape to the bottom of the fan housing. These modifications increased survival during periods of low humidity.

Also in Australia, Barton et al. (2004) used CO_2-baited EVS traps to define the spatial distribution of *Ochlerotatus camptorhynchus*, the commonest mosquito species inhabiting brackish waters in coastal Victoria, and the probable major vector of enzootic Ross River and Barmah Forest viruses. A total of 303 901 *Ochlerotatus camptorhynchus* was recorded from 918 trap nights in the Shire of Wellington (1991–2001), and 53 771 over 270 trap nights in East Gippsland (1990–94). *Ochlerotatus camptorhynchus* was the dominant mosquito species (80–98% of total catch) recorded at all trap localities, except Marlay Point (43%) and Matheson Swamp (43%) where *Culex australicus* and *Culex globocoxitus* were dominant.

Wishart (1999) used EVS traps baited with dry ice over a 13-week period to sample potential virus vectors in and around Melbourne, Australia. Some of the traps malfunctioned prior to the end of the survey. Over the 13 week survey, 6396 mosquitoes, representing 14 species in five genera were trapped. *Culex molestus* was the most abundant species caught, and 41% of the entire catch of this species was recorded in just one week. *Aedes camptorhynchus* was the second most abundant species collected. In a later study, Wishart (2002) used the same traps to undertake mosquito surveys in Victoria, Australia. On this occasion, a total of 43 039 female mosquitoes, representing 23 species from five genera were collected during April to November, over a three-year period. *Culex annulirostris* was the most abundant species in the trap catches, followed by *Anopheles annulipes*.

In a comparison of EVS traps and human landing collections in Australia a total of 21 mosquito species were collected, 20 in the EVS trap and 19 biting (Jones et al. 1991). Regressions of biting collections against EVS catches showed that the relative abundance of the five most common species, namely *Culex annulirostris* ($r = 0.61$), *Anopheles annulipes* ($r = 0.69$), *Ochlerotatus vigilax* ($r = 0.84$), *Mansonia uniformis* ($r = 0.92$) and *Mansonia septempunctata* ($r = 0.86$), were ranked similarly by both methods. However, a more detailed analysis of the data suggested that the EVS trap was better suited for sampling *Ochlerotatus vigilax*, while human (or animal) baits were better for *Culex annulirostris*.

Van Essen et al. (1994) studied the effects of adding 1-octen-3-ol (octenol) to EVS traps on their efficiency at collecting mosquitoes, particularly *Ochlerotatus vigilax*, a major vector of Ross River virus in Australia. The following combinations were tested: 2200 g of CO_2 alone, CO_2 plus octenol released at low, medium and high rates, and CO_2 supplemented with light from a 6V incandescent bulb. Traps were arranged in a 5 × 5 Latin square design and were placed 200m apart, perpendicular to the prevailing wind direction. Traps were operated for an 18 h period (1500–0900 h) with a different daily bait combination at each trap location. Octenol was released from microreaction vials fitted with plastic lids and neoprene septa and the rate of release was varied according to the following method: a low release rate was obtained by inserting a bent 11 cm pipe cleaner into the vial so that it pressed against the underside of the septum, which was pierced with two pin-holes. Medium release rates were obtained by having approximately 0.5 cm of the pipe cleaner extending through the lid of the vial. Similarly, a high release rate was obtained by having 2 wicks extending 2 cm through the vial lid. During a five-day trial at Brisbane airport, a total of 7192 mosquitoes were collected, of which 92.6% were *Ochlerotatus vigilax*. Other species collected included *Ochlerotatus alternans, Aedes funereus, Anopheles annulipes, Coquillettidia xanthogaster, Culex annulirostris* and *Culex sitiens*. At a second site, 47 467 mosquitoes were collected with *Ochlerotatus vigilax* again the dominant species, comprising 94.9% of the catch. Additional species to those collected at the airport site were *Ochlerotatus kochi, Ochlerotatus notoscriptus, Ochlerotatus procax, Anopheles atratipes, Culex australicus* and *Mansonia uniformis*. At both sites, addition of octenol to the standard CO_2 baited traps significantly increased the numbers of *Ochlerotatus vigilax* caught, compared with CO_2 alone. Medium and high octenol with CO_2 gave higher catches of *Ochlerotatus vigilax* than the CO_2 + light combination. No significant differences were observed in the numbers of *Ochlerotatus vigilax* caught among the three different octenol release rates. Addition of octenol did not significantly increase the numbers of *Culex* trapped.

Ryan et al. (1997) used EVS traps supplemented with CO_2 and octenol to trap mosquitoes in an investigation of the role of gray-headed flying foxes (*Pteropus poliocephalus*) in the ecology of Ross River virus in Brisbane, Australia. Traps were set at four sites at heights of 1.2 m and 8 m. and were operated on 14 occasions between January and August. A total of 272 014 mosquitoes representing nine species were collected from the four sites. *Aedes funereus* comprised 84.2% of the total catch, followed by *Culex annulirostris* (9.1%) and *Ochlerotatus vigilax* (6.6%). Each of these three species was significantly less common in traps set at a height of 8 m, although 35% of *Aedes funereus* were trapped at a height of 8 m, and this species was more common at sites in the immediate vicinity of flying fox camps.

Kimsey and Chaniotis trap

Kimsey and Chaniotis (1984) found that in Panamanian forests rain spoilt the catches of insects when any of the standard designs of light-traps were used, and moreover metal parts corroded and caused frequent electrical failure. To overcome these difficulties they designed a completely new trap made almost entirely of transparent plexiglas (Fig. 9.10) and consisting of three principal parts. The trap head (A), in which there is a standard CDC light-trap motor and four-bladed plastic fan, is mounted under the rain cover (20.32-cm diameter). A vertical 5-cm baffle (c) positioned below the fan and across the trap prevents vortexing and so reduces damage to the catch. The middle holding cage (B) is made of a 20.32-cm length of 15.24-cm diameter plexiglas, and has a fine mesh screen (i) fitted at the top. The light-bulb (h) is maintained below the disc (bb) of the bottom part (C) of the trap. This disc and central entrance chimney (I) are painted black to prevent light diffusing up through the trap. Insects attracted to the light source are sucked up by the fan through the small chimney into the trap body, where they are retained.

Permanent bonding of plexiglas is achieved with ethylene dichloride, and all soldered connections are made with non-acid core flux and then coated with a thin layer of silicone sealant, which is also used to make parts of the trap water-proof, such as the screw fixing the circular rain cover to the trap body. Routine maintenance after each trap-night consists of spraying electrical connectors (alligator clips, banana plugs and the bulb socket) with WD-40 silicone aerosol lubricant.

Insects are undamaged and mainly alive when traps are emptied. The entrance chimney is plugged with cotton wool and the catch killed by putting the holding cage in a freezer.

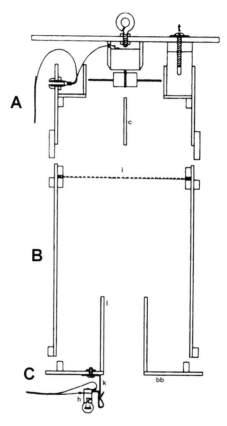

Fig. 9.10. Light-trap of Kimsey and Chaniotis (1984) showing A — trap head, B — holding cage, C — bottom part with bb — black-painted disc and I — chimney-type entrance

Pfuntner light-trap (CO$_2$-4 trap)

Pfuntner (1979) modified the EVS trap of Rohe and Fall (1979) to obtain larger samples. The main modifications were that the short plastic cylinder housing the motor and fan was increased in diameter from 7.6 to 10.8 cm. The batteries were fixed to a cylindrical metal disc above the cylinder (Fig. 9.11), and the mesh screen over the cylinder was omitted. A 10.2-cm

Thimboe-Drome propeller was mounted onto a Mabuchi RE-26 motor, but apparently any 1.5–3.0-V d.c. motor can be used. As in some models of the EVS trap, the blades were twisted to a 45° pitch. The light source was a sub-miniature 25-mA bulb.

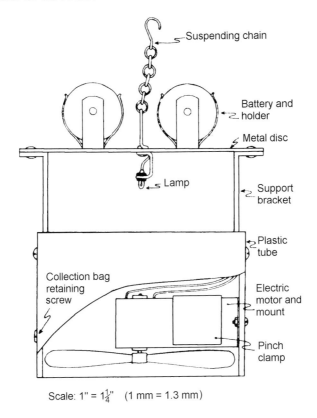

Scale: 1" = $1\frac{1}{4}$"   (1 mm = 1.3 mm)

Fig. 9.11. Carbon dioxide baited light-trap (CO_2-4) of Pfuntner (1979)

A cloth collecting bag was fixed to the end of the plastic cylinder. This part of the trap is, as in the EVS trap, suspended under a 1-gal capacity tin containing dry ice, which flows out through four 6.4-mm diameter holes drilled in its bottom. An optional extra is a light sensitive phototransistor wired into an electronic time-circuit to automatically switch the trap on at dusk and off at dawn. The trap can operate from two carbon–zinc 1.5-V D-cell batteries wired in parallel, or from one alkaline D-cell battery. In comparative trials this trap caught a mean of 269 female mosquitoes (88% *Culex tarsalis*) a night compared with 42 in CDC-type traps, and in other trials the trap caught 153 females/night, while the EVS trap of Rohe and Fall (1979) caught 68 mosquitoes/night.

Updraft light-traps

A disadvantage of conventional light-traps having a fan below the light is that despite the presence of a screen, large unwanted insects tend to drop down on the catch and cause damage. The use of smaller screens to eliminate these unwanted insects may significantly reduce the catch of mosquitoes. Rupp and Jobbins (1969) were apparently the first to publish an account of a trap in which the fan is mounted above the light source to provide an updraft of air to draw mosquitoes into the trap. Unfortunately, although they presented a photograph of the trap together with some constructional details, they gave neither a complete description, nor any worthwhile results concerning its efficiency. It was left to Wilton and Fay (1972a) to develop and evaluate the updraft light-trap. They had found that despite the use of a 0.6-cm screen numerous scarab beetles entered the ultraviolet light-traps they used in El Salvador and destroyed much of the catch. To try to overcome this problem laboratory experiments were conducted with a trap having a horizontal 4-W black fluorescent light tube and a 6-V d.c. motor fitted with aluminium fan blades mounted above the light source. Air was sucked upwards and expelled at right angles, through the collecting cage (Fig. 9.12a). This modified trap caught significantly more (42–78%) *Anopheles albimanus* and *Anopheles stephensi* than a conventional trap having a downward displacement of air (up to 28%). By connecting a 0–100 Ω variable resistor in series with the motor, fan speed, and consequently volume of air sampled, could be varied. When fan speed was reduced there was a substantial increase in the survival of mosquitoes in the two traps, but whereas in the conventional trap this was accompanied by a marked reduction in numbers caught, there was no such decrease with the updraft trap.

They concluded from observations on the movements of mosquitoes marked with fluorescent powders that when a mosquito encounters an air stream produced by a light-trap it attempts to evade it by vigorous flight activity. As well as a forward thrust an upward flight movement is involved and this tends to help mosquitoes escape capture in conventional traps, but increases their likelihood of capture in updraft traps. Ultraviolet updraft light-traps of Wilton and Fay (1972a) have proved useful in collecting *Anopheles albimanus* in Mexico (Rodriguez et al. (1996), Haiti (Taylor et al. 1975) and in El Salvador (Wilton 1975a). In other trials in El Salvador about 1 lb of dry ice was wrapped in newspaper and placed near updraft ultraviolet light-traps and CDC traps, other traps operated without dry ice. Nine or more species were caught in both CDC and UV traps, with or without dry ice. With the UV traps the catch of females was increased

1.2–2.4 times in those with dry ice, the biggest difference being with *Mansonia titillans*. With the CDC traps dry ice increased catches 1.2–23.0 times, the biggest difference being in the attraction of *Ochlerotatus taeniorhynchus* (23.0 ×), *Mansonia titillans* (14.1 ×) and *Ochlerotatus scapularis* (12.0 ×). Although Wilton (1975*a*) concluded that dry ice was a worthwhile adjunct to CDC traps, the trouble and expense of getting carbon dioxide for use with UV traps was probably not justified. He could not explain the 3.1 times increased attraction of male *Mansonia titillans* to UV traps with dry ice, and the 114.9 times increase in CDC traps having dry ice.

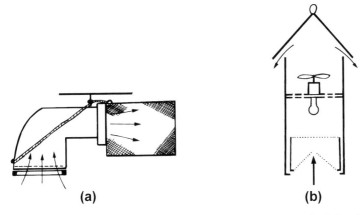

Fig. 9.12. (*a*) Updraft trap (Wilton and Fay 1972a); (*b*) bottom draft (updraft) trap (Grothaus and Jackson 1972)

The ultraviolet light-trap of Wilton (1975*b*) consists of a 4-W blacklight fluorescent tube (peak radiation about 3650 Å) operated from a 12-V car battery through an inverter and a 6-V d.c. motor with a 2-bladed fan connected through a 75-Ω resistor to the same battery. A 1.2-cm square mesh screen is placed behind the lamp. The collection cage consists of 'Tubeguaz' stretched over two 10-in wooden embroidery hoops held apart by three 7-in long metal struts. The trap is constructed so that it can operate as a downdraft trap with the light above, or by inverting it as an updraft trap with the light below. Field trials in El Salvador found that when operated in the updraft position the trap caught 2.4 times the numbers of female *Anopheles albimanus* as when used in the conventional down-draft position.

The updraft ultraviolet trap caught about 4.4 times as many mosquitoes when placed inside houses than when used out of doors, confirming findings in El Salvador (Wilton 1975*a*) that catches were larger with indoor traps. With human landing collections the ratio of numbers caught outside and inside was 3.1:1, whereas with CDC light-traps the ratio of the numbers of *Anopheles albimanus* caught outside to indoors was 0.3:1. The updraft ultraviolet trap caught most *Anopheles albimanus* (7682), followed by biting collections (2207) and CDC light-traps (1343). It was concluded that the ultraviolet light-trap was very efficient in collecting female *Anopheles albimanus*, and moreover there was very good survival of mosquitoes caught in the trap, allowing dissections for malaria infectivity and age-grading. Rodriguez et al. (1996) employed a commercially available UV trap (John W. Hock, Gainesville, Florida, USA) to sample adult *Anopheles albimanus* in 14 villages in a malarious area of southern Mexico. Mosquito abundance data were correlated with several landscape characteristics using a generalized linear modelling approach.

In Australia, van den Hurk et al. (1997) found that the updraft CDC trap did not significantly increase trap collections of *Anopheles farauti* complex when compared with the standard CDC trap.

Grothaus and Jackson (1972) also designed an updraft trap (Fig. 9.12*b*) and found the updraft principle seemed to enhance mosquito collections while reducing the catch of large unwanted insects. The updraft principle can be applied to CDC and similar traps.

Monks wood light-trap

This trap combines the best features of the CDC and the Pennsylvania light traps (Service 1970). Light is provided by a 23-cm 6-W fluorescent tube mounted vertically between three white plastic or metal baffles (Fig. 9.13). Light tubes giving either ultraviolet or white light with emissions of different spectral power distributions can be used. These tubes consume only 0.5 A and have a life of about 5000 h. They operate through a transistor ballast (inverter) from a 12-V d.c. battery. The baffles housing the tube are slotted into a 19-cm length of clear acrylic 10-cm diameter tubing. A removable wire mesh screen is held in position at the entrance of the trap body by a metal ring secured by three small clips. A detachable metal cover rests on top of the baffles to protect it against rain. Suction to draw mosquitoes into the trap is provided by either a 6-V motor connected to a resistor or by a 12-V motor. A netting bag similar in shape to that used with the CDC trap, but differing in that the top and upper sections of its walls are made of cotton cloth not netting, is fitted to the trap body. A slit-like opening in the middle extends along a third of its circumference to allow damp cotton wool, or plastic foam sponge covered with filter paper, to be placed over the floor of the bag to help keep the catch alive overnight. The edges of the slit opening are covered with 'Velcro'. A small transistorised photosensitised light cell, which permits the trap to be automatically switched on and off around sunset and sunrise, can be incorporated into the top of the metal box housing the ballast. To prevent the catch escaping when the trap is automatically switched off, a smooth 11-cm diameter plastic funnel can be inserted into the base of the neck of the bag. A small switch on the ballast box permits the operator to choose between manual or automatic operation.

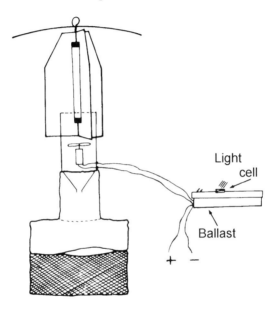

Fig. 9.13. Monks Wood light-trap (Service 1993)

The trap can run for about four nights from a heavy duty 12-V battery, but in practice it is advisable to recharge the battery every two to three nights. For protection against rain the ballast box is enclosed within a plastic bag. In England the trap caught both sexes of *Culiseta morsitans* and *Culiseta litorea*, species not caught in a New Jersey-type trap using a 150-W bulb (Service 1969). Other species caught included *Culiseta annulata, Culex pipiens, Ochlerotatus detritus, Coquillettidia richiardii* and *Anopheles claviger* (Service 1970). In preliminary trials in Nigeria and Kenya 20 mosquito species, including a few *Aedes aegypti* were caught. When the Monks Wood light-trap with either white light or ultraviolet light was used inside Nigerian village huts about three to four times more

Anopheles gambiae, Anopheles funestus and *Culex quinquefasciatus* were caught than with CDC traps. In 14 catches from village huts in Kenya the mean catch of females per night was 69.6 *Anopheles gambiae*, 14.4 *Anopheles funestus* and 16.3 *Culex quinquefasciatus;* there was no difference between the numbers caught when white or ultraviolet light tubes were used. Both *Anopheles arabiensis* and *Anopheles gambiae* were caught. Several other species were trapped including *Mansonia uniformis, Mansonia africana, Coquillettidia fuscopennata, Anopheles pharoensis* and *Anopheles ziemanni*, none of which was common in early morning pyrethrum catches in the huts. The ultraviolet light tube attracted considerably more males of *Anopheles gambiae* than the white light tube when the traps were used in huts in Nigeria, but there was no evidence of this in Kenya. A blacklight tube operated on five nights in Nigeria caught no mosquitoes.

In The Gambia in studying the dispersal of mosquitoes over high fences Gillies and Wilkes (1978) placed a calf inside a mosquito-proof stall positioned along the wall of a village hut, and a Monks Wood trap was suspended just below the eaves on the opposite wall of the hut. Few mosquitoes were caught in the light-trap until aluminium foil was used to mask all but the lower 2.5 cm of the tube just above the fan. In two series of experiments the mean catch per trap-night was 38.3 and 40.9 for *Mansonia africana/uniformis*, and 43.7 and 57.4 for *Anopheles gambiae* and *Anopheles funestus* combined.

In Gabon Monks Wood light-traps with daylight fluorescent tubes placed in village compounds caught 35 mosquito species, the most common mosquito being *Malaya taeniarostris. Hodgesia cuptopous* was also quite common, and several species of *Anopheles, Coquillettidia* and *Urano-taenia* were also collected. These catches constituted 81.4% of all the mosquitoes collected as adults in the area by different techniques and included 17 species not caught biting man nor in indoor CDC light-traps (Service 1976).

In Thailand Malainual et al. (1987) found the Monks Wood light-trap useful for collecting *Culex tritaeniorhynchus, Culex gelidus* and *Culex fusco-cephala* in Japanese encephalitis studies.

A modified Monks Wood light-trap was designed (Ross and Service 1979) that incorporated circuitry to allow the fluorescent tube to flash on and off for adjustable time periods, or for the tube to operate continuously. The trap also included a photocell that allowed it to cut in and off at predetermined light levels. A possible advantage of a flashing light is that it reduces power consumption and can extend intervals before the 12-V car battery needs to be recharged. Although in Ghana this trap caught large numbers of *Simulium squamosum* (6520 in a single night) (Service 1979), and caught more mosquitoes of certain species in Colombia when flashing than when the tube was not flashing, catches of other species were smaller (Ross and Service 1979).

Braverman and Linley (1993) investigated the influence of the height at which Monks Wood suction light-traps fitted with blacklight tubes (Philips TL 6 W/O5) were mounted on the catch of *Culicoides* in Israel. Traps were suspended at heights of 1.4 m or 26 m from a Eucalyptus tree approximately 100 m from a cattle shed. The higher trap caught significantly more *Culicoides circumscriptus* than the lower trap. Numbers of *Culicoides imicola* and *Culicoides schultzei* species group did not differ significantly between the two trap heights.

Japanese commercial light-traps

Mains-operated light-traps have been used in Japan both inside animal shelters and outside for collecting vectors of Japanese encephalitis. The original 'Nozawa' type of trap is no longer manufactured and has been replaced by a similar one ('AS' trap) that has a 15-W circular fluorescent tube (Fig. 9.14). Mosquitoes attracted to the light in either type of trap are drawn down into a wire mesh collecting cage housed within the trap body above the fan. When the trap is automatically switched off at dawn a pair of flap-like doors close above the collecting cage to prevent mosquitoes from escaping. To remove the catch the lower part of the trap body holding the fan is swung aside and the collecting cage withdrawn. Other ultraviolet light-traps, including the MC-4100, which uses a 6-W UV tube, and the MC-5100, which uses a UV 30-W circular tube have also been used, but less often than the Nozawa or AS traps because mosquitoes pass through the fan and tend to get damaged. These types of traps have proved successful in catching large numbers of *Culex tritaeniorhynchus* (Self et al. 1973) and other mosquitoes such as *Anopheles sinensis, Aedes vexans nipponii* and *Culex pipiens* form *pallens* (Ishii 1970, 1971a,b), and also other mosquitoes (Ikeuchi 1967).

Similar traps were used in light-trap surveys in Korea (Han-Il Ree et al. 1973); the most common species caught over 3 years were *Culex pipiens, Culex tritaeniorhynchus, Aedes vexans nipponii* and *Anopheles sinensis*. Han-Il Ree et al. (2001) also used 'Nozawa-type' light-traps fitted with a black light and operated from 1900 to 0600 h in a half-open cowshed in Korea in order to sample *Anopheles sinensis*, however, no details on the efficiency or usefulness of the method were provided. Self et al. (1976) using Nozawa blacklight (60-W) light-traps in a semi-urban area of Jakarta and in four rural rice-growing sites in Java caught 22 mosquito species.

The most common species were *Culex tritaeniorhynchus* and *Culex gelidus*, of which some 2000 were caught per night; the next most common species were *Anopheles barbirostris* and *Culex quinquefasciatus*.

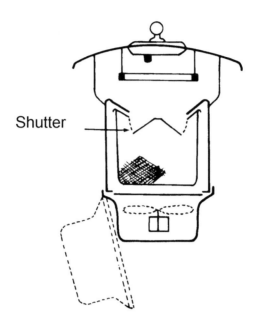

Shutter

Fig. 9.14. 'Nozawa'-type light-trap

Takagi et al. (1995) report the use of a light trap manufactured by the Fujihira Kogyo Co. Ltd., of Tokyo to catch potential vectors of bancroftian filariasis in Nagate prefecture. The species most frequently collected in the light trap was *Culex tritaeniorhynchus*. Larger numbers of the filariasis vector *Culex pipiens* form *pallens* were collected by light traps than by human landing catch, although overall numbers of this species were small, presumably as a result of removal of the principal larval habitats, which had previously been open drains.

Miscellaneous ultraviolet light-traps

In addition to the above Japanese traps and the occasional use of ultraviolet fluorescent tubes in the Monks Wood trap and in the updraft traps of Wilton, other types of trap have occasionally employed UV lights.

For example in Russia light-traps employing 375-W mercury vapour lamps having spectral emission of 2483–6907 Å, but with peak emission in the 3022–5770 Å range, have caught very large numbers of mosquitoes, mainly *Anopheles hyrcanus, Anopheles messeae, Aedes vexans, Culex pipiens, Culex modestus* and *Coquillettidia richiardii*. The same proportions of different species were collected by light-traps and bait catches, but whereas average catches as high as 6207 mosquitoes/5 min were recorded in light-traps, an average of only 655 mosquitoes were caught from human bait in this time (Breyev 1958). Catch size was not reduced when a black filter letting only ultraviolet light through was used, but traps without a fan caught an average of only 45 mosquitoes in 5 min, clearly demonstrating the need for a fan in light-traps.

Comparing the effect of different light sources on mosquito catches Breyev (1963) showed that within the range of 300–620 nm the numbers of mosquitoes caught varied in proportion to radiation output. It was concluded that for the collection of *Anopheles hyrcanus, Anopheles messeae, Aedes vexans, Culex pipiens* and *Culex modestus*, a high intensity light was best, and also that because different species appeared to be attracted to different wavelengths, lights such as mercury lamps giving a broad spectrum were best. Incandescent lamps were more species selective.

In Uganda traps incorporating several features of New Jersey and Robinson light-traps (Robinson and Robinson 1950) were successfully used at ground level and at various heights up to 120 ft to catch forest mosquitoes (Corbet 1961, 1964; Corbet and Haddow 1961). The trap's 125-W mercury vapour bulb was operated via a choke from a generator; there was insufficient power for a fan. The outside of the trap was painted black and the inner surfaces with aluminium paint (Fig. 9.15). Mosquitoes and other small insects, which passed through the ⅜-in mesh vertical screen surrounding the trap entrance, were collected in a bottle containing filter paper impregnated with trichloroethylene. When the traps were operated in two forest localities over 67 mosquito species were caught, including large numbers of *Coquillettidia* spp., *Culex insignis*, and *Culex subrima*, but few *Anopheles* or *Aedes* spp. Both sexes were well represented in the collections.

In discussing the response of synanthropic flies to light-traps using black light Tarry et al. (1971) mention that in field trials in England *Ochlerotatus* mosquitoes were commonly attracted to 20-W lamps emitting light ranging from 3100–4400 Å and peaking at 3510 Å. No further details of the mosquitoes caught are given. In Japan, Kitaoka and Itô (1964) reported that generally more mosquitoes, especially *Culex tritaeniorhynchus* and *Anopheles hyrcanus*, were attracted to black light than white light.

In Thailand Somboon et al. (1989) caught more than 30 mosquito species in UV light-traps of unspecified design, operated out of doors from about

1900–2100 h. In contrast only about a third of these species were obtained in human bait catches. In both collections the predominant species was *Culex tritaeniorhynchus*, others caught included *Culex gelidus* and *Culex fuscocephala*.

Jaenson et al. (1986) suspended miniature ultravioletlight-traps (unspecified design) 1.8 m from the ground in a Swedish forest. Only 142 mosquitoes belonging to 12 species were caught from two such traps operated for usually three nights a week for 5 months. The most common species were *Ochlerotatus communis* (28.6%), *Ochlerotatus excrucians* (19.7%), *Ochlerotatus intrudens* (8.5%), mosquitoes which were also common in animal-baited traps, and *Anopheles maculipennis* s.1. (12.0%), *Culiseta morsitans* (8.5%) and *Culiseta alaskaensis* (6.3%), all of which were rare or uncommon in bird and mammal-baited traps. *Aedes cinereus* was not caught, although it formed 6.5–7.0% of mosquitoes caught in animal-baited traps.

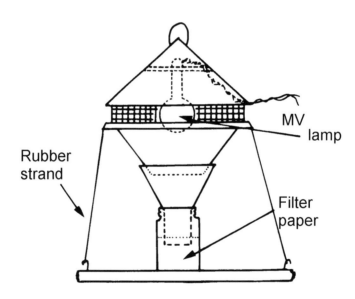

Fig. 9.15. Mercury vapour light-trap (Corbet, 1961)

Flashing light-traps

Vavra et al. (1974*a,b*) and Ross and Service (1979) have employed flashing lights to trap mosquitoes, while Service (1979) has used them to collect simuliids. Vavra et al. (1974*a,b*) used a neon glow lamp connected through a 0.1 mf capacitor and a 9.5-MΩ resistor to a 90-V battery to produce 2 flashes per second. In Panama the mean catch of mosquitoes per night was just eight, but in Maryland, USA the flashing light caught a mean of 318 per night (Vavra et al. 1974*b*). In Texas, USA Lang (1984) found a 75-W incandescent black light-bulb in a New Jersey light-trap with flashing on/off cycles of 1/1, 5/5, 20/10, 30/30, 45/15 and 39/1 s, did not increase the numbers of mosquitoes, or species, attracted compared with a trap operating a continuous light-bulb. Unlike Ross and Service (1979) who caught several Colombian mosquitoes in a Monks Wood light-trap flashing on and off 20–30 times a minute, Lang (1984) caught no mosquitoes with the 1/1 on/off cycle. The only advantage was that selecting an appropriate on/off cycle, such as in the range of 10/20 or 20/10, resulted in reducing the catch size without sacrificing species diversity. This might be advantageous in reducing the numbers of mosquitoes to be sorted when their densities are high.

Non-electrical light-traps

Vavra et al. (1974*a,b*) appear to have been the first to have evaluated betalights and chemical lights for catching adult mosquitoes, both of which are described in Chapter 3 in connection with larval sampling. Their betalights had a brightness of 390–1300 microlamberts, and in Panama when used in CDC traps caught just 3–30 mosquitoes per trap-night (Vavra et al. 1974*a*), but in Maryland, USA the catch was 426–965 per trap-night (Vavra et al. 1974*b*). Baldwin and Chant (1975) in comparing light-traps having betalights and 3-V torch bulbs found that in Canada traps with betalights caught 20 mosquito species, whereas those with bulbs caught 22 species. But the traps with blue betalights attracted more mosquitoes than those with torch bulbs. In 1972, 1740 mosquitoes were collected from about 14 trap-nights with a blue betalight, and 927 mosquitoes in traps having a white betalight, whereas only 915–1068 adults were caught in traps fitted with torch bulbs. The most commonly collected species were *Ochlerotatus punctor, Ochlerotatus communis, Ochlerotatus excrucians* and *Coquillettidia perturbans*, but adults of *Anopheles, Culex* and *Culiseta* species were also collected (Baldwin and Chant 1975).

Fig. 9.16. Updraft chemical light-trap employing a light-stick (photograph reproduced with permission of M. W. Service)

Using chemical lights that initially had a brightness of 10-ft lamberts, but which decreased to 2-ft lamberts after 10 h, 114 mosquitoes/trap-night were caught in Panama (Vavra et al. 1974*b*). Service and Highton (1980) used chemical light-sticks having an initial output of 50-ft lamberts, but which slowly decreased to 15-ft lamberts after 8 h. There was a single emission peak at 510 nm, with light intensity declining to 10% of the maximum on either side of the peak at wavelengths of 490 and 570 nm.

There is therefore considerable illumination at the red end of the visible spectrum, although this is not apparent by looking at the lights. A light-stick was suspended below a CDC-type light-trap without a bulb, to make an updraft light-trap (Fig. 9.16). In Ghana 170 mosquitoes were collected in 23 trap-nights when traps were placed out of doors. When the traps were placed in village houses 648 female and 27 male mosquitoes, belonging to 12 species were caught, including 116 *Anopheles gambiae* complex, 395 *Anopheles nili* and smaller numbers of *Anopheles funestus, Mansonia uniformis, Mansonia africana, Aedes aegypti* and *Aedes simpsoni*. Although these traps do not appear to catch as many mosquitoes as CDC light-traps they do seem particularly effective in catching *Anopheles nili*, of which 97.5% were blood-fed.

In Kenya from six trap-nights inside houses these chemical traps caught 251 *Culex quinquefasciatus* compared to 1384 *Culex quinquefasciatus* caught by CDC traps (Service and Highton 1980).

Light-sticks are also available in different colours such as yellow, orange, red, blue and white in addition to the original greenish ones. In addition to attracting mosquitoes they have been used to collect *Culicoides* and phlebotomine sandflies. They deserve further evaluation.

Comparisons of different light-traps

Many comparative trials of different models of light-traps and trials of light-traps against other traps and sampling methods have been undertaken and the results of some of these studies are described below. The selection of a particular type of light-trap will depend on the species of mosquito under investigation, the type of investigation being undertaken (faunal survey, monitoring of a particular species, etc.), and availability of resources, among other factors, and no one model is likely to be preferred under all circumstances. The variability of results obtained with different species, and in different ecological situations makes forming general conclusions regarding the use of different trap types particularly problematic.

In the USA Matsumoto and Maxfield (1985) reported that although New Jersey light-traps caught the most *Culiseta melanura*, CDC traps nevertheless effectively measured relative changes in population density. As they are smaller and easier to transport, they suggested CDC traps could replace the larger New Jersey traps in surveillance programmes.

In California, USA Meyer et al. (1984) compared the numbers of *Ochlerotatus melanimon, Anopheles freeborni* and *Culex tarsalis* collected in a standard red resting box, a New Jersey light-trap (25-W bulb) and a CDC trap supplemented with about 2.5 kg of dry ice, both operating from 1900–0800 h. At each trap site the three traps were placed 30 m apart. The largest mean catch of female *Culex tarsalis* (249.9) was from the CDC trap, and the largest mean catch of female *Anopheles freeborni* (74.1) was from the New Jersey trap. Both light-traps caught about the same numbers of *Ochlerotatus melanimon* (14.4 CDC, 9.8 NJ). The red boxes performed poorly, except in collecting male and female *Anopheles freeborni* (23.6 ♂, 28.8 ♀), but New Jersey light-traps were about as equally effective for this species (20.7 ♂, 74.1 ♀). In Dakota, USA New Jersey light-traps with a frosted 25-W bulb caught fewer mosquitoes (3329) than either CDC (6433) or EVS traps (7719). The most common species captured were *Aedes vexans* and *Culex tarsalis* which were caught in about equal proportions by both the New Jersey (1:0.96) and CDC (1.10:1) traps, whereas the EVS trap caught 13.49 times as many *Culex tarsalis* as *Aedes vexans* (Easton et al. 1986).

In Iowa, USA Acuff (1976) compared the mosquitoes caught in a New Jersey light-trap having a 40-W frosted bulb, a CDC light-trap baited with 2 kg dry ice, a Malaise trap (Townes 1962) and those caught at human bait. A total of 18 species were collected, of which both light-traps caught 15 species whereas only seven species were collected by the Malaise trap and at bait. The carbon dioxide-baited CDC trap caught 1.96 times more mosquitoes than the New Jersey trap. The most common species in all traps was *Aedes vexans*. Olson et al. (1968) reported that CDC traps (without CO_2) caught 16 species of mosquitoes in Utah, USA as compared to just seven mosquito species in steer-baited Magoon traps, and also 1.5 times more mosquitoes than the bait traps. Similarly Lloyd and Pennington (1976) caught over 5.5 times as many mosquitoes in a CO_2-baited CDC trap than in bovine-baited traps of Roberts (1965). As many as 49 935 *Ochlerotatus melanimon* and 20 159 *Ochlerotatus dorsalis* were caught in just two nights in their CDC traps supplemented with carbon dioxide.

Vaidyanathan and Edman (1997*a*) compared different collection methods for estimating human exposure to potential vectors of eastern equine encephalomyelitis (EEE) virus. They compared human landing catches with collections from CDC, American Biophysics Corporation (ABC) and New Jersey (NJ) light-traps and resting boxes in Massachusetts, USA. Human biting collections caught significantly more host-seeking females (mean catch per night = 125) than resting boxes (mean = 3) or unbaited light traps (CDC = 24, ABC with flickering light source = 7, ABC trap with steady light source = 4, NJ light-trap = 4). The numbers of mosquitoes captured by human landing catch and ABC flicker trap + CO_2 were not significantly different. Regression analysis of human landing collections and ABC traps supplemented with CO_2 indicated that 61.2% of the human biting risk from *Aedes vexans, Ochlerotatus canadensis, Ochlerotatus stimulans s.l., Ochlerotatus abserratus, Ochlerotatus triseriatus, Ochlerotatus trivittatus, Ochlerotatus aurifer* and *Coquillettidia perturbans* can be predicted from ABC + CO_2 trap catches. ABC traps supplemented with CO_2 and octenol were capable of predicting 65% of biting risk due to *Anopheles punctipennis, Anopheles quadrimaculatus*, and *Anopheles walkeri*. No single sampling method was accurate for predicting human biting risk by *Culex salinarius*, and the authors concluded that no single method could predict human biting risk by all potential vectors surveyed.

In further studies to determine which of 11 trapping methods most successfully sampled populations of 6 potential vectors of eastern equine encephalitis (EEE) virus in Massachusetts, USA, Vaidyanathan and Edman (1997*b*) compared several commercially available light-traps, with and without CO_2 and other trap additives. Traps compared were: New Jersey (NJ) light traps, CDC light traps, CDC traps with octenol, CDC traps with CO_2, CDC traps with CO_2 plus octenol, the American Biophysics Corporation light trap with a flickering light source (ABC flicker), ABC trap with a steady light source, ABC flicker trap with octenol, ABC flicker trap with CO_2, ABC flicker trap with CO_2 plus octenol, and 3 resting boxes. Octenol was emitted from traps at a rate of 2.0 mg/h. Trap collections for the eight light traps in 1994/95 and 1996, and results of significance tests for pairwise comparisons between traps, are presented as Tables 9.3 and 9.4 below. No significant differences were observed between ABC traps with a steady light source and those with a flickering light source, and between ABC flicker and NJ light-traps. CDC light-traps collected a significantly greater mean number of females per night than either ABC flicker traps or ABC traps with a steady light source. The addition of octenol slightly increased trap catch with both CDC and AB light traps, but not significantly. The addition of CO_2 to ABC and CDC traps increased the catch significantly. Addition of CO_2 + octenol to ABC flicker traps significantly increased the catch compared with octenol alone, but not compared with CO_2 alone. The use of octenol + CO_2 increased collections of *Anopheles* species. Resting boxes were found to be the best method for sampling bloodfed or parous *Anopheles* spp. The NJ, ABC flicker and ABC steady light traps were ineffective at sampling potential vectors of EEE virus. Both ABC light-traps and CDC light-traps performed better than NJ light-traps. Given that ABC flicker traps use less electricity than the ABC traps with a steady light source, ABC flickers were preferred by the authors, who concluded that ABC flicker traps or CDC traps were better than NJ light-traps for prolonged surveillance of potential EEE vectors.

Table 9.3. Numbers of mosquitoes collected by various sampling methods in Massachussets, USA (after Vaidyanathan and Edman 1997b)

Species	CDC	ABC steady	ABC flicker	New Jersey	ABC flicker + CO_2	ABC flicker + octenol	ABC flicker + CO_2 + octenol	Resting boxes
1994/95								
Aedes vexans	3	0	0	0	6	0	4	0
Ochlerotatus canadensis	23	3	2	5	154	22	147	0
Ochlerotatus spp.[1]	344	5	113	7	1230	340	794	1
Anopheles spp.	29	1	8	0	98	17	127	63
Culiseta melanura	170	5	27	1	51	52	50	16
Coquillettidia perturbans	212	17	42	7	659	41	210	0
Culex spp.	90	15	16	18	85	24	145	1
Total	871	46	208	38	2283	496	1477	81
Total mean catch	24	4	7	4	74	15	55	3

Species	CDC	CDC + octenol	CDC + CO_2	CDC + CO_2 + octenol	ABC flicker	ABC flicker + octenol	ABC flicker + CO_2	ABC flicker + CO_2 + octenol
1996								
Aedes vexans	0	2	1	10	0	0	4	4
Ochlerotatus canadensis	9	9	290	335	8	13	426	529
Ochlerotatus spp.	19	35	1374	1180	29	54	1207	1120
Anopheles spp.	6	10	164	211	3	20	113	257
Culiseta melanura	0	0	5	0	1	0	0	7
Coquillettidia perturbans	145	302	2855	1523	93	190	1253	2166
Culex spp.[3]	9	14	102	27	5	10	84	85
Total	188	372	4791	3286	139	287	3087	4168
Total mean catch	13	25	319	365	9	19	206	321

[1] Ochlerotatus stimulans, Ochlerotatus fitchii, Ochlerotatus excrucians, Ochlerotatus triseriatus, Ochlerotatus absserratus, Aedes cinereus, Ochlerotatus aurifer, Ochlerotatus trivittatus
[2] Anopheles quadrimaculatus, Anopheles punctipennis, Anopheles walkeri
[3] Culex salinarius, Culex restuans, Culex territans

Table 9.4. Significance values for pairwise comparisons of sampling methods using Wilcoxon signed rank test for mean trap catch for all species, Massachussets, USA, 1994/95 and 1996. *P* values two-tailed tests are for a normal approximation with continuity correction (Vaidyanathan and Edman 1997*b*)

Pairwise comparison	P value
1994/95	
Significant difference (trap 1 > trap 2)	
CDC v ABC flicker	0.00
ABC flicker + CO_2 v ABC flicker	0.00
ABC flicker + CO_2 v ABC flicker + octenol	0.00
ABC flicker + CO_2 + octenol v ABC flicker + octenol	0.01
CDC v ABC steady	0.00
ABC flicker + CO_2 v CDC	0.00
ABC flicker + CO_2 + octenol v CDC	0.00
No significant difference	
ABC flicker v ABC steady	0.18
ABC flicker v NJ	0.10
ABC flicker v ABC flicker + octenol	0.07
ABC flicker + CO_2 v ABC flicker + CO_2 + octenol	0.39
1996	
Significant difference (trap 1 > trap 2)	
CDC + CO_2 + octenol v ABC flicker + CO_2 + octenol	0.03
CDC + CO_2 v CDC	0.00
CDC + CO_2 + octenol v CDC	0.01
No significant difference	
CDC v ABC flicker	0.53
CDC + octenol v ABC flicker + octenol	0.85
CDC + CO_2 v ABC flicker + CO_2	0.84
CDC v CDC + octenol	0.24

In California, USA Milby et al. (1978) undertook a comparison over 60–61 trap-nights between the mosquitoes caught in American-type New Jersey light-traps employing a 25-W bulb and CDC traps augmented with 4-lb blocks of dry ice wrapped in paper. Eleven species were caught in the New Jersey traps and 12 in the CDC traps but the only common species were *Culex tarsalis, Ochlerotatus melanimon* and *Anopheles freeborni*; the numbers of females of the former two species were 2.8 and 2.6 times greater in the CDC traps, whereas *Anopheles freeborni* was 2.6 times more common in the New Jersey traps. The authors believed they could use these conversion factors to convert catches from one type of trap to that of another.

Milby and Reeves (1989) showed that in urban areas of California, USA CDC traps with lights and baited with carbon dioxide caught considerably more *Culex tarsalis* and *Culex quinquefasciatus* than New Jersey traps (although the reverse was true for *Anopheles freeborni*). This prompted them to question the suitability of New Jersey light-traps in surveillance programmes and they suggested that CDC traps baited with CO_2 could replace New Jersey traps for monitoring virus vectors. However, to get the same number of mosquitoes about twice as many traps are needed, and males are rarely caught in traps without light.

To investigate the potential for replacing NJ light-traps with CDC traps in the California mosquito monitoring program in light of the reducing catches obtained by NJ light-traps (Wegbreit & Reisen, 2000), Reisen et al. (2002) compared NJ light-traps with CO_2-baited CDC traps without a light. CDC traps without lights or rain covers and supplemented with 1–2 kg of dry ice were compared directly against standard NJ light-traps over a period of 16–21 weeks. Traps were set up in pairs approximately 25–50 m apart and were operated concurrently. The CDC traps always collected more females of all species compared with NJ light-traps, except in the case of *Anopheles freeborni* which was collected equally well by both trap types. Males of *Culex tarsalis* and *Anopheles freeborni* were more frequently collected in NJ light-traps. Correlations between abundance of catches in paired traps were significant over time and space for 12 of 14 groups of females, but for only 2 of 7 groups of males, however the difference in catch size among sites was also highly significant. A summary of the features of CDC and NJ traps in relation to mosquito abundance monitoring in California is shown as Table 9.5 below.

Table 9.5. Comparative features of Centers for Disease Control and New Jersey light-traps (Reisen et al. 2002)

Feature	CDC Light-Trap	New Jersey Light-Trap
Sensitivity, females	High	Low
Sensitivity, males	None	Moderate
Impact of weather	High	Low
Processing time	Fast	Slow
Virus surveillance	Yes	No
Cost of operation	High	Low
Duration of available historical data	Short	Long

A more detailed multivariate ANOVA investigation of *Culex tarsalis* catches revealed that the magnitude of the difference in catches between the two traps increased as a function of population size and it was not possible to calculate a consistent conversion factor to translate the catch sizes between the two trap types. Regression analyses failed to provide a universal functional relationship between trap types and although both trap types generally measured similar trends in abundance of females, no consistent pattern in the shape of the regression functions was observed: the relationship was linear in 2 districts and curvilinear in 2 other districts.

Reisen et al. (1999) compared New Jersey light-traps with CO_2 enhanced CDC traps and gravid traps of Reiter (1987) as modified by Cummings (1992) as *Culex* surveillance tools in California, USA. In the Coachella valley, gravid traps collected significantly more *Culex quinquefasciatus* females and males than either NJ light traps or CO_2 traps. The catch in CO_2 traps was greatest in agricultural habitat, whereas gravid trap catches were highest in agricultural and golf course habitats. Gravid traps also caught significantly more *Culex tarsalis* than NJ light-traps. In the San Joaquin valley, CO_2 traps caught the highest numbers of *Culex tarsalis*. Catches of *Anopheles freeborni* were highest in CO_2 and NJ light traps operated in agricultural habitat in the Sacramento valley. As less than 20% of the variation in NJ light-trap catches could be described by CO_2 trap catches, a regression approach to estimating the catch in NJ light traps from the catch in CO_2 traps was not considered to be a useful method. The authors concluded that the NJ light-traps were efficient at sampling *Anopheles freeborni, Culex tarsalis*, and *Psorophora columbiae* in agricultural and riparian habitats where competing illumination was minimal. Carbon dioxide traps had the advantage that they were unaffected by light pollution and they collected large numbers of *Culex pipiens* complex females. It was recommended that gravid traps should be used in urban and suburban environments to monitor *Culex pipiens*, while NJ light-traps should be replaced by CO_2 traps in urban and suburban habitats and in those areas where light pollution precludes the effective operation of NJ light traps.

In Pakistan Suleman et al. (1977) compared catches from CDC and New Jersey traps with catches from buffaloes. They concluded that both light-traps were suitable for sampling local mosquitoes, although they were much less productive than buffalo biting collections.

Caglar et al. (2003) compared NJ light-traps and CO_2-baited CDC traps for collecting mosquitoes in the Belek area of Turkey. New Jersey light-traps were fitted with 40-W bulbs and CO_2 was supplied from a 0.7 kg block of dry ice. Traps were set 1.5 m above ground and were operated from 1800 to 0600 h, one night per week for 30 weeks between May and December. A total of 4529 mosquitoes of six species was collected, of which *Ochlerotatus caspius* (23.74%) was the most frequently collected, followed by *Culex pipiens* (26.89%), *Culex tritaeniorhynchus* (23.74%), *Aedes cretinus* (10.69%), *Ochlerotatus dorsalis* (8.74%) and *Culiseta annulata* (6.20%). The CO_2 traps were significantly more attractive to *Aedes cretinus* ($P < 0.01$; $F = 12.19$), compared with NJ light-traps. The NJ light-traps were significantly more efficient in trapping *Ochlerotatus caspius* and *Ochlerotatus dorsalis*. There was no statistically significant difference in numbers of *Culex pipiens, Culex tritaeniorhynchus*, or *Culiseta annulata* caught in the two trap types. Also in Turkey, Simsek (2004) operated seven New Jersey light traps once a month from 1900 to 0700 h to investigate the seasonal population dynamics of *Culex theileri* and other species associated with a lake ecosystem.

In Haiti Sexton et al. (1986) compared a CDC trap, a modified updraft ultraviolet light-trap and human landing collections for sampling *Anopheles albimanus*. Their updraft trap consisted of the cylindrical plastic body of a CDC trap with its motor and fan mounted upside down to create an updraft. A 15.2-cm long 4-W blacklight fluorescent strip (peak emission near 3650 Å) was positioned horizontally across the bottom of the cylindrical body. The circular metal lid of the CDC trap had two 'foldback' (binder) paper clips attached with bolts and nuts opposite each other near the rim of the lid, in such a position that the bottom of the CDC collecting bag could be firmly held in position in the clips. The neck of the bag was slipped over the top of the CDC plastic body and tied in position with string. The fluorescent tube operated through an inverter ballast from a 12-V motorcycle battery with a power of 7 or 6 A, while a 75-Ω resistor allowed the 6-V motor to run from the same 12-V battery.

A commercially available updraft UV light-trap (John W. Hock & Co., Gainesville, Florida, USA) was compared with human landing catches for sampling *Anopheles albimanus* in southern Mexico by Ulloa et al. (1997). A single human landing catch caught more *Anopheles albimanus* than a single updraft light-trap in villages (34.3 ± 6.3 versus 15.9 ± 3.3) and close to larval habitats (14.6 ± 2.9 versus 2.4 ± 0.3), but catches using the two methods were significantly correlated in both habitats ($r^2 = 0.78$, $P < 0.0001$ in villages and $r^2 = 0.51$, $P < 0.0001$ close to larval habitats). At sampling sites close to larval habitats, human landing catches yielded higher proportions of nulliparous and parous females, but fewer gravid females, whereas at village collection sites, light traps and human landing catches yielded a similar proportion of parous individuals (23% versus 26%). The authors concluded that updraft UV light-traps could be successfully used as an alternative to human landing catches for population monitoring of *Anopheles albimanus*, as both methods were able to detect differences in relative abundance between sampling sites, and at least in village habitats, both methods yielded similar proportions of parous females.

When used in animal or human habitations in Morocco Monks Wood light-traps using ultraviolet light caught more mosquitoes, including *Culex pipiens* and *Anopheles labranchiae* than CDC traps, although out of doors this species was more abundant in CDC traps (Bailly-Choumara 1973a,b).

There were large mortalities of mosquitoes caught in the traps. This could, however, have probably been reduced if damp cotton wool had been placed in the bottom of the trap cages.

In Kenya both more species (21) and more blood-engorged mosquitoes were caught in Monks Wood light-traps having a white fluorescent tube and hung under the outside eaves of village huts than were collected by indoor CDC light traps (15 species) and by indoor resting catches (10). Although 21 species of exophilic adults were collected by battery operated aspirators there were many fewer blood-fed specimens (Chandler et al. 1975a). The most common mosquitoes in the Monks Wood light traps were *Anopheles pharoensis, Culex antennatus, Mansonia uniformis, Aedes circumluteolus, Anopheles ziemanni, Aedes ochraceus* and *Culex univittatus*.

In Thailand, Ismail et al. (1982) compared the efficiency of CDC and Monks Wood light-traps. When placed in houses the CDC traps caught 3.8 times as many *Anopheles balabacensis*, 2.9 times as many *Anopheles maculatus*, 2.2 times as many *Anopheles minimus* and 2.4 times as many *Culex* species as did the Monks Wood trap. However, in out of door collections the Monks Wood trap caught 3.0 times as many *Anopheles philippiensis*. The use of white or ultraviolet fluorescent tubes in the Monks Wood trap did not appear to significantly alter the catches.

Analyses of results from light-traps

Hacker et al. (1973a) used time-series analysis to analyse objectively and quantitatively the densities of mosquitoes recorded by daily or weekly catches from light-traps operated over long-periods, such as 3–5 years. Use was made of the statistical techniques of cross variance, autovariance functions and power spectrum analysis. They also used time-series analyses on light-trap data to predict mosquito densities up to 6 days in the future (Hacker et al. 1973b). Later Hacker et al. (1975) presented two other heuristic models belonging to a class of time-series models termed transfer function models. From knowledge about past mosquito densities, and observations on meteorological variables, mosquito densities can be forecast. One such model uses rainfall, maximum daily temperature and minimum daily temperature at preceding time intervals; while the second model is simpler and associates mosquito densities with only rainfall and minimum daily temperature. A detailed account of these two models using light-trap catches of female *Culex tarsalis* is presented by Hacker et al. (1975).

In South Africa light-traps were established at four stations. Mean weekly numbers of different species of mosquitoes caught in the traps and mean climatic variables, such as temperature and relative humidity were pooled from the four catching stations to get overall means (Hewitt et al. 1982). To try and establish a quantitative predictive model multiple regression analyses were performed on transformed weekly counts obtained over the entire catching season against weekly climatological data. However, neither this approach nor the autoregressive model used by Hacker et al. (1973a,b, 1975) proved satisfactory, consequently a circular approach, as developed by de Waal and Hewitt (1979), was used. This involves presenting mean weekly counts on the circumference of a unit circle and fitting a calculated distribution without taking into account any independent variables. The appendix in the original paper should be consulted for details of the method.

Roberts and Conner (1979) proposed that the following demographic equation could be used on light-trap collections of *Culex pipiens*

$$N_t = N_0 \, e^k \, t \qquad\qquad (9.11)$$

where t = time in months, N_t = number of *Culex pipiens* caught after time t (as adults/trap night), N_0 = number of adults caught at the beginning, k = constant for growth rate (birth rate – death rate), and e = base of natural logarithms. After solving k for each month in 1 year when adults are caught in light-traps, these values can then be inserted into the equation for a month the following year to predict population size in subsequent months. But this simplistic approach will likely only work if growth processes are the same during consecutive years.

McLaughlin and Focks (1990) studied the effect of cattle density on the numbers of *Psorophora columbiae, Anopheles crucians, Culex salinarius/Culex erraticus* and *Anopheles quadrimaculatus* caught in New Jersey light-traps in two agricultural rice-growing areas having seven traps (363 trap-nights) and 11 traps (530 trap-nights). They discovered that night-to-night catches often varied by an order of 4 logs (base 10) due to the interaction of a number of environmental variables that were not necessarily synchronised across the entire sampling areas. They concluded that the only way of comparing light-trap catches against such a variable background was to use the mean numbers caught per trap-night at each location, calculated from the total year's collection. They estimated the number of cattle per hectare in five concentric annuli of 0.8, 1.6, 2.4, 3.4 and 4.8 km, and plotted the mean numbers of mosquitoes caught in light-traps in these annuli against cattle density within them, so as to obtain estimates of the mean annual capture of a species in a light-trap. The most abundant mosquito was *Psorophora columbiae*, averaging 1063/night. To illustrate this approach the regression equation for this species for one of their two areas was: Expected captures of *Psorophora columbiae* = $312.3 + 4240.1 \times HD_{0.8-km}$ $(R^2 = 97\%)$, where 312.3 = the intercept, $HD_{0.8}$ = the model coefficient for the 0.8-km annulus and R^2 (the regression coefficient) = 0.97, which means that 97% of variations in the capture of this species is attributable to cattle density. There was no significant relationship for *Culex salinarius/Culex erraticus*, whereas 74% of the variation in catches of *Anopheles quadrimaculatus* and 68% of the variations in *Anopheles crucians* were due to cattle densities.

By analysing the numbers of mosquitoes in New Jersey light-traps from several sites in California, USA from 1953 to 1973, Olson et al. (1979) were able to show a positive correlation between the numbers of *Culex tarsalis* caught and the incidence of St Louis encephalitis (SLE) and western equine encephalomyelitis (WEE) in humans. This provided a reliable means for forecasting years of highest virus incidence. Dividing the total numbers of female *Culex tarsalis* caught in all traps in a district by the product of the total numbers of traps and nights each trap was operated, gave what they termed a Light-Trap Index (LTI). The ratio of rural to urban indices was 4.2:1.0 for female *Culex tarsalis*. It was believed justifiable to use this ratio as a conversion factor, so that collections obtained in earlier studies in rural areas could be approximately converted to compare with levels obtained in later studies with urban traps. The authors concluded that the critical level of *Culex tarsalis* in urban sites, below which human cases of SLE and WEE were not detected, was a Light-Trap Index of 0.1. Other LTIs relating to transmission thresholds in rural areas, and in areas with large and small human populations, were also estimated. For example, peaks of weekly incidence of SLE and WEE were associated with LTIs of 21 and 81, respectively from urban light-traps.

In California, USA Nelson et al. (1978) found a good correlation between routine light-trap catches of *Culex tarsalis* and mark–recapture population estimates.

In an analysis of parity rates of phlebotomine sandflies in light-traps Gibb et al. (1988) warned against making incorrect conclusions that light-traps may be biased in favour of various age-classes. They pointed out that differences between parity of flies caught in light-traps and in other collections may reflect more trap location and the non-random distribution of parous and nulliparous insects than a real difference in parity.

New Jersey Light-trap surveillance data from West Central Florida, USA, were analyzed by Zhong et al. (2003) using ecological parameters consistent with a non-interactive community, under the assumption that the population densities and numbers of individual species were too low to cause competition for resources. The number of species in a given community, or species richness (Price 1984), was summarised for each trap site using 2 years of data. Adult mosquito populations also were analysed for dominant species, seasonal abundance, community structure (indices of aggregation, diversity, and evenness), and trap site clustering. The dominant species index (d) was estimated using the Berger-Parker equation (Southwood and Henderson 2000):

$$d = \frac{N_{max}}{N_T} \tag{9.12}$$

which expresses the proportion of the total catch N_T comprised by the most common species N_{max}.

Diversity was calculated using the Shannon-Wiener diversity index H

$$H = -\sum_{i=1}^{S_{obs}} p_i \log_e p_i \tag{9.13}$$

and an evenness index (J) was estimated from the formula.

$$J = \frac{H}{\log S} \tag{9.14}$$

where H is the Shannon-Wiener index and $\log S$ is the total number of species collected at each trap site. Cluster analysis was conducted by nearest-neighbour clustering. The analysis revealed that seven species accounted for 84% of the 157 600 individuals collected and the dominance indices for these seven species ranged from 0.054 to 0.227. The most dominant species was *Anopheles crucians*.

References

Acuff VR (1976) Trap biases influencing mosquito collecting. Mosquito News 36, 173–176

Altman DG, Bland JM (1983) Measurement in medicine - the analysis of method comparison studies. Statisitician 32: 307–317

Aranda C, Aponte JJ, Saute F, Casimiro S, Pinto J, Sousa C, do Rosario V, Petrarca V, Dgedge M, Alonso P (2005) Entomological characteristics of malaria transmission in Manhiça, a rural area in southern Mozambique. J Med Entomol 42: 180–186

Artsob H, Spence L, Surgeoner G, Th'ng C, Lampotang V, Grant L, McCreadie J (1983) Studies on a focus of California group virus activity in southern Ontario. Mosquito News 43: 449–455

Bailly-Choumara H (1973a) Étude préliminaire d'une récolte d'*Anopheles labranchiae* par piège CDC réalisée dans la région de Larache, Maroc. Bull World Health Organ 49: 49–55

Bailly-Choumara H (1973b) Étude comparative de différentes techniques de récolté de moustiques adultes (Diptera, Culicidae) faîte au Maroc, en zone rurale. Bull Soc Sci Nat Phys Maroc 53: 135–187

Baldwin WF, Chant GD (1975) Seasonal succession of the mosquitoes (Diptera: Culicidae) of the Chalk River area. Can Entomol 107: 947–952

Bargren WC, Nibley C (1956) Comparative Attractiveness of Colored Lights of Equal Intensity to Specific Species of Mosquitoes'. Research Report of Third Army Area Medical Laboratory, SU 3004, Fort McPherson (mimeographed)

Barker-Hudson P, Kay BH, Jones RE, Fanning ID, Smythe LD (1993) Surveillance of mosquitoes and arbovirus infection at the Ross River Dam (Stage 1), Australia. J Am Mosq Control Assoc 9: 389–399

Barr AR, Smith TA, Boreham MM (1960) Light intensity and the attraction of mosquitoes to light traps. J Econ Entomol 53: 876–880

Barr AR, Smith TA, Boreham MM, White KE (1963) Evaluation of some factors affecting the efficiency of light traps in collecting mosquitoes. J Econ Entomol 56: 123–127

Bartnett RE, Stephenson RG (1968) Effect of mechanical barrier mesh size on light trap collections in Harris county, Texas. Mosquito News 28: 108

Barton PS, Aberton JG, Kay BH (2004) Spatial and temporal definition of *Ochlerotatus camptorhynchus* (Thomson) (Diptera: Culicidae) in the Gippsland Lakes system of eastern Victoria. Aust J Entomol 43: 1–2

Bast TF (1960) An automatic interval collector for the New Jersey light trap. Proc New Jers Mosq Exterm Assoc 47: 95–104

Bast TF, Rehn JWH (1963) Vertical distribution of mosquitoes as indicated by light trap collections in two environments. Proc New Jers Mosq Exterm Assoc 50: 219–229

Bast TF, Rehn JWH, Stockwell WE (1964) Density of mosquitoes at two elevations. Proc New Jers Mosq Exterm Assoc 51: 146–152

Belton P, Galloway MM (1966) Light-trap collections of mosquitoes near Belleville, Ontario in 1965. Proc Entomol Soc Ontario 96: 90–96

Belton P, Kempster RH (1963) Some factors affecting the catches of Lepidoptera in light traps. Can Entomol 95: 832–837

Belton P, Pucat A (1967) A comparison of different lights in traps for *Culicoides* (Diptera: Ceratopogonidae). Can Entomol 99: 267–272

Bidlingmayer WL (1964) The effect of moonlight on the flight activity of mosquitoes. Ecology 45: 87–94

Bidlingmayer WL (1967) A comparison of trapping methods for adult mosquitoes: species response and environmental influence. J Med Entomol 4: 200–220

Bidlingmayer WL, Schoof HF (1957) The dispersal characteristics of the salt-marsh mosquito, *Aedes taeniorhynchus* (Wiedemann) near Savannah, Georgia. Mosquito News: 17: 202–212

Blackeslee TE, Axtell R, Johnston L (1959) *Aedes vexans* and *Culex salinarius* light trap collections at five elevations. Mosquito News 19: 283

Boobar LR, Sardelis MR, Nelson JH, Brown WM (1987) A new type of collapsible insect-surveillance light trap for sampling Diptera. Med Vet Entomol 1: 215–218

Bosak PJ, Reed LM, Crans WJ (2001) Habitat preference of host-seeking *Coquittettidia perturbans* (Walker) in relation to birds and eastern equine encephalomyelitis virus in New Jersey. J Vector Ecol 26: 10–09

Bowden J (1973) The influence of moonlight on catches of insects in light-traps in Africa. Part I. The moon and moonlight. Bull Entomol Res 63: 113–128

Bowden J (1981) The relationship between light- and suction-trap catches of *Chrysoperla carnea* (Stephens) (Neuroptera: Chrysopidae), and the adjustment of light-trap catches to allow for variation in moonlight. Bull Entomol Res 71: 621–629

Bowden J (1982) An analysis of factors affecting catches of insects in light-traps. Bull Entomol Res 72: 535–556

Bowden J (1984) Latitudinal and seasonal changes of nocturnal illumination with a hypothesis about their effect on catches of insects in light-traps. Bull Entomol Res 74: 279–288

Bowden J, Church BM (1973) The influence of moonlight on catches of insects in light-traps in Africa. Part II. The effect of moon phase on light-trap catches. Bull Entomol Res 63: 129–142

Bowden J, Morris MG (1975) The influence of moonlight on catches of insects in light-traps in Africa. III. The effective radius of a mercury-vapour light-trap and the analysis of catches using effective radius. Bull Entomol Res 65: 303–348

Bradley GH (1943) Determination of densities of *Anopheles quadrimaculatus* on the wing. Proc New Jers Mosq Exterm Assoc 30: 22–27

Bradley GH, McNeel TE (1935) Mosquito collections in Florida with the New Jersey light trap. J Econ Entomol 28: 780–786

Bradley GH, Travis BV (1943) Time-saving methods for handling mosquito light-trap collections. J Econ Entomol 36: 51–53

Braverman Y, Linley JR (1993) Effect of light trap height on catch of *Culicoides* (Diptera: Ceratopogonidae) in Israel. J Med Entomol 30: 1060–1063

Breeland SG (1974) Population patterns of *Anopheles albimanus* and their significance to malaria abatement. Bull World Health Organ 50: 307–315

Breyev KA (1958) On the use of ultra-violet light-traps for determining the specific composition and numbers of mosquito populations. Parazitol Sb 18: 219–238 (In Russian, English summary)

Breyev KA (1963) The effect of various light sources on the numbers and species of blood-sucking mosquitoes (Diptera, Culicidae) collected in light traps. Entomol Obozr 42: 280–303 (In Russian, English summary)

Broom AK, Wright AE, Mackenzie JS, Lindsay MD, Robinson D (1989) Isolation of Murray Valley encephalitis and Ross River viruses from *Aedes normanensis* (Diptera: Culicidae) in western Australia. J Med Entomol 26: 100–103

Brust RA (1980) Dispersal behaviour of adult *Aedes sticticus* and *Aedes vexans* (Diptera: Culicidae) in Manitoba. Can Entomol 112: 31–42

Bryan JH (1986) Vectors of *Wuchereria bancrofti* in the Sepik provinces of Papua New Guinea. Trans R Soc Trop Med Hyg 80: 123–131

Buckley DJ, Stewart WWA (1970) A light-activated switch for controlling battery-operated light traps. Can Entomol 102: 911–912

Burkett DA, Butler JF, Kline DL (1998) Field evaluation of colored light-emitting diodes as attractants for woodland mosquitoes and other Diptera in north central Florida. J Am Mosq Control Assoc 14: 186–195

Burkhardt D (1977) On the vision of insects. J Comp Physiol 120: 33–50

Caglar SS, Alten B, Bellini R, Simsek FM, Kaynas S (2003) Comparison of nocturnal activities of mosquitoes (Diptera: Culicidae) sampled by New Jersey light traps and CO_2 traps in Belek, Turkey. J Vector Ecol 28: 12–22

Cano J, Berzosa PJ, Roche J, Rubio JM, Moyano E, Guerra-Neira A, Brochero H, Mico M, Edu M, Benito A (2004) Malaria vectors in the Bioko Island (Equatorial Guinea); estimation of vector dynamics and transmission intensities. J Med Entomol 41: 158–161

Carnevale P (1974) Variations saisonnières d'une population d'*Anopheles nili* (Theo.), 1904 en République Populaire du Congo. Cah ORSTOM sér Entomol Méd Parasitol 12: 165–174

Carnevale P, Boreham PFL (1978) Études des preferences trophique d'*Anopheles nili* (Theo.), 1904. Cah ORSTOM sér Entomol Méd Parasitol 16: 17–22

Carnevale P, Le Pont F (1973) Epidemiologie du paludism humain en République Populaire du Congo. II. Utilisation des pièges lumineux 'C.D.C'. comme moyen d'échantillonnage des populations anophèliennes. Cah ORSTOM sér Entomol Méd Parasitol 11: 263–273

Chamberlain RW, Sudia WD, Coleman PH, Beadle LD (1964) Vector studies in the St. Louis encephalitis epidemic, Tampa Bay area, Florida, 1962. Am J Trop Med Hyg 13: 456–461

Chandler JA, Boreham PFL, Highton RB, Hill MN (1975*a*) A study of the host selection patterns of the mosquitoes of the Kisumu area of Kenya. Trans R Soc Trop Med Hyg 69: 415–425

Chandler JA, Highton RB, Hill MN (1975*b*) Mosquitoes of the Kano Plain, Kenya. I. Results of indoor collections in irrigated and nonirrigated areas using human bait and light traps. J Med Entomol 12: 504–510

Chandler JA, Highton RB, Boreham PFL (1976*a*) Studies on some ornithophilic mosquitoes (Diptera: Culicidae) of the Kano Plain, Kenya. Bull Entomol Res 66: 133–143

Chandler JA, Highton RB, Hill MN (1976*b*) Mosquitoes of the Kano Plain, Kenya. II. Results of outdoor collections in irrigated and nonirrigated areas using human and animal bait and light traps. J Med Entomol 13: 202–207

Chaniotis BN, Anderson JR (1968) Age structure, population dynamics and vector potential of *Phlebotomus* in northern California. Part II. Field population dynamics and natural flagellate infections in parous females. J Med Entomol 5: 273–292

Charlwood JD (1997) Vectorial capacity, species diversity and population cycles of anopheline mosquitoes (Diptera: Culicidae) from indoor light-trap collections in a house in southeastern Tanzania. Afr Entomol 5: 93–101

Christensen HA, de Vasquez AM, Boreham MM (1996) Host-feeding patterns of mosquitoes (Diptera: Culicidae) from central Panama. Am J Trop Med Hyg 55: 202–208

Clements AN (1963) The Physiology of Mosquitoes. Pergamon Press, London

Collett GC, Graham JE, Bradley IE (1964) Relationship of mosquito light trap collection data to larval survey data in Salt Lake county. Mosquito News 24: 160–162

Collier BW, Solberg VB, Brown MW, Boobar LR (1992) A fabric body light trap for sampling mosquitoes. J Am Mosq Control Assoc 8: 413–415

Coluzzi M, Petrarca V (1973) Aspirator with paper cup for collecting mosquitoes and other insects. Mosquito News 33: 249–250

Cope SE, Hazelrigg JE (1989) Evaluations of trapping methods, biting behavior, and parity rates for mosquitoes—Harbor lake, Wilmington, California, U.S.A. Bull Soc Vector Ecol 14: 277–281

Corbet PS (1961) Entomological studies from a high tower in Mpanga forest, Uganda. VI. Nocturnal flight activity of Culicidae and Tabanidae as indicated by light-traps. Trans R Entomol Soc Lond 113: 301–314

Corbet PS (1964) Nocturnal flight activity of sylvan Culicidae and Tabanidae (Diptera) as indicated by light traps: a further study. Proc R Entomol Soc Lond (A) 39: 53–67

Corbet PS, Haddow AJ (1961) Observations on nocturnal flight activity in some African Culicidae (Diptera). Proc R Entomol Soc Lond (A) 36: 113–118

Costantini C, Sagnon NF, Sanogo E, Merzagora L, Coluzzi M (1998) Relationship to human biting collections and influence of light and bednet in CDC light-trap catches of West African malaria vectors. Bull Entomol Res 88: 503–511

Coz J, Hamon J, Vervent G, Sales S (1971) Contribution à l'étude du piège lumineux 'C.D.C. miniature light trap' comme moyen d'échantillonnage des populations anophèliennes dans le Sud-Ouest de la Haute-Volta. Cah ORSTOM sér Entomol Méd Parasitol 9: 417–430

Crawley MJ (1993) GLIM for Ecologists. Blackwell Scientific Publications, Oxford

Cummings RC (1992) Design and use of a modified Reiter gravid mosquito trap for mosquito borne encephalitis surveillance in Los Angeles County, California. Proc California Mosq Vector Control Assoc 60: 110–116

Davis JR, Hall T, Chee EM, Majala A, Minjas J, Shiff CJ (1995) Comparison of sampling anopheline mosquitoes by light-trap and human-bait collections indoors at Bagamoyo, Tanzania. Med Vet Entomol 9: 249–255

Dethier VG (1963) The Physiology of Insect Senses. Methuen & Co., London

De Waal DJ, Hewitt PH (1979) A circular approach to cyclical data. Dept Math Stat Univ Orange Free State Technical Rept No. 45

Driggers DP, O'Connor RJ, Kardatzke JT, Stup JL, Schiefer BS (1980) The U.S. army miniature solid state mosquito light trap. Mosquito News 40: 172–178

Easton ER, Coker RS, Ballinger R (1986) Occurrence and seasonal incidence of mosquitoes on Indian reservations in Iowa, Nebraska and South Dakota during 1983. J Am Mosq Control Assoc 2: 190–195

Ellman R, Maxwell C, Finch R, Shayo D (1998) Malaria and anaemia at different altitudes in the Muheza district of Tanzania: childhood morbidity in relation to level of exposure to infection. Ann Trop Med Parasitol 92: 741–753

Elston R, Apperson C (1977) A light-activated on-off switch for the C.D.C. light trap. J Med Entomol 14: 254–255

Evans CL, Wozniak A, McKenna B, Vaughan DR, Dowda MC (2005) Design of a mosquito trap support pole for use with CDC miniature light traps. J Am Mosq Control Assoc 21: 114–116

Faye O, Diallo S, Gaye O, Ndir O, Faye O (1992) Efficacité comparée de l'utilisation des pièges lumineux du type CDC et des sujets humains pour l'échantillonnage des populations anophéliennes. Bull Soc Pathol Exot 85: 185–189

Floor TG, Grothaus RH (1971) Conversion of the New Jersey light trap for collecting live mosquitoes in Da Nang, Vietnam. Mosquito News 31: 221–222

Flores-Mendoza C, Lourenco-de-Oliveira R (1996) Bionomics of *Anopheles aquasalis* Curry 1932, in Guarai, State of Rio de Janeiro, southeastern Brazil - I. Seasonal distribution and parity rates. Mem Inst Oswaldo Cruz 91: 265–270

Foley DH, Bryan JH (1991) *Anopheles annulipes* Walker (Diptera: Culicidae) at Griffith, New South Wales. 2. Biology and behaviour of two sibling species. J Aust Entomol Soc 30: 113–118

Fontenille D, Rakotoarivony I (1988) Reappearance of *Anopheles funestus* as a malaria vector in the Antananarivo region, Madagascar. Trans R Soc Trop Med Hyg 82: 644–645

Forattini OP, Gomes A de C, Natal D, Kakitani I, Marucci D (1987) Preferências alimentares de mosquitoes Culicidae no Vale do Ribeira, São Paulo, Brazil, Rev Saúde Pública 21: 171–187

Fox I (1958) The mosquitoes of the international airport, isla verde, Puerto Rico, as shown by light traps. Mosquito News 18: 117–124

Fox I, Capriles JM (1952) Light trap studies on mosquitoes and *Culicoides* in western Puerto Rico. Mosquito News 13: 165–166

Frommer RL, Schiefer BA, Vavra RW (1976) Comparative effects of CO_2 flow rates using modified CDC light traps on trapping adult black flies (Simuliidae; Diptera). Mosquito News 36: 355–358

Frost SW (1952) Light traps for insect collection, survey and control. Bull Agric Exp Stn Pennsylvania 550

Frost SW (1964) Killing agents and containers for use with insect light traps. Entomol News 75: 163–166

Garrett-Jones C, Magayuka S (1975) Studies on the natural incidence of *Plasmodium* and *Wuchereria* infections in *Anopheles* in rural East Africa: I—Assessment of densities by trapping hungry female *Anopheles gambiae* Giles; species A. WHO/MAL 75.851 & WHO/VBC 75.541 17 pp. (mimeographed)

Gibb PA, Anderson TJC, Dye C (1988) Are nulliparous sandflies light-shy? Trans R Soc Trop Med Hyg 82: 342–343

Gillies MT, Wilkes TJ (1978) The effect of high fences on the dispersal of some West African mosquitoes (Diptera: Culicidae). Bull Entomol Res 68: 401–408

Ginsberg HS (1986) Dispersal patterns of *Aedes sollicitans* (Diptera: Culicidae) at the east end of the fire island national seashore, New York, USA. J Med Entomol 23: 146–155

Ginsberg HS (1988) Survivorship of mosquitoes (Diptera: Culicidae) captured in CDC miniature light traps. Proc Ann Mtg New Jers Mosq Control Assoc 75: 86–92

Githeko AK, Service MW, Mbogo CM, Atieli FA, Juma FO (1994) Sampling *Anopheles arabiensis, A. gambiae* sensu lato and *A. funestus* (Diptera: Culicidae) with CDC light-traps near a rice irrigation area and a sugarcane belt in western Kenya. Bull Entomol Res 84: 319–324

Gjullin CM, Brandl DG (1978) An automatic chemosterilizing insect light trap. J Med Entomol 14: 585–588

Gjullin CM, Brandl DG, O'Grady JJ (1973) The effect of colored lights and other factors on the numbers of *Culex pipiens quinquefasciatus, C. tarsalis* and *Aedes sierrensis* entering light traps. Mosquito News 33: 67–71

Goldsmith TH (1964) The visual system in insects. In Rockstein M (ed) The Physiology of Insecta, Vol. 1. Academic Press, London, pp. 397–462

Gomes A de C, Rabello EX, Natal D (1985) Uma nova câmara coletora para armadilha CDC-miniatura. Rev Saúde Pública 19: 190–191

Gordon SW, Tammeriello RF, Linthicum KJ, Wirtz RA, Digoutte JP (1991) Feeding patterns of mosquitoes collected in the Senegal river basin. J Am Mosq Control Assoc 7: 424–432

Graham JE, Bradley IE (1961) An evaluation of some techniques used to measure mosquito populations. Proc Utah Acad Sci Arts Letters 39: 77–83

Green CH, Cosens D (1983) Spectral response of the tsetse fly *Glossina morsitans morsitans*. J Insect Physiol 29: 795–800

Grothaus RH, Jackson SC (1972) A new bottom-draft light trap for mosquito studies. Mosquito News 32: 634–635

Gunasekaran K, Jambulingam P, Sadanandane C, Sahu SS, Das PK (1994) Reliability of light trap sampling for *Anopheles fluviatilis*, a vector of malaria. Acta Trop 58: 1–11

Hacker CS, Scott DW, Thompson JR (1973*a*) Time series analysis of mosquito populations. J Med Entomol 10: 533–543

Hacker CS, Scott DW, Thompson JR (1973*b*) A forecasting model for mosquito population densities. J Med Entomol 10: 544–551

Hacker CS, Scott DW, Thompson JR (1975) A transfer function forecasting model for mosquito populations. Can Entomol 107: 243–249

Hamon J, Sales S, Gayral P (1969) Evaluation de L'efficacité des Pièges Lumineux C.D.C. pour L'échantillonnage des Populations des Moustiques dans le Sud-ouest de la Haute-Volta, Afrique Occidentale. I. Evaluation des Pièges á L'interieur des Habitations.' OCCGE Centre Muraz, Lab Entomol No. 18/Ent./69 (mimeographed)

Han-Il Ree, Ui-Wook Hwang, In-Yong Lee, Tae-Eun Kim (2001) Daily survival and human blood index of *Anopheles sinensis*, the vector species of malaria in Korea. J Am Mosq Control Assoc 17: 67–72

Hanson JC (1959) A new insect killing container for use with the American mosquito light trap. California Vector Views 6: 85–87

Harcourt DG, Cass LM (1958) A controlled-interval light trap for microlepidoptera. Can Entomol 90: 617–622

Haufe WO, Burgess L (1960) Design and efficiency of mosquito traps based on visual response to patterns. Can Entomol 92: 124–140

Headlee TJ (1922) The problem of evaluating mosquito density and the advantages to be realised from its solution. Proc New Jers Mosq Exterm Assoc 9: 48–56

Headlee TJ (1937) Some facts underlying the attraction of mosquitoes to sources of radiant energy. J Econ Entomol 30: 309–312

Hemmings RJ (1959) Observations on the operation of mosquito light traps with a cylindrical vertical screen. Mosquito News 19: 101

Hendricks DE (1985) Portable electronic detector system used with inverted-cone sex pheromone traps to determine periodicity and moth captures. Environ Entomol 14: 199–204

Herbert EW, Meyer RP, Tubes PG (1972) A comparison of mosquito catches with CDC light traps and CO_2-baited traps in the republic of Vietnam. Mosquito News 32: 212–214

Hewitt PH, van der Linde TG de K, van Pletzen R, Kok DJ, Fourie S, Mostert DJ, Nel A (1982) Temporal fluctuations in the numbers of female mosquitoes trapped at a state in the western Orange Free State. J Entomol Soc Sthn Afr 45: 69–92

Hii J, Chin KF, MacDonald M, Vun YS (1986) The use of CDC light traps for malariometric entomology surveys in Sabah, Malaysia. Trop Biomed 3: 39–48

Hii JLK, Birley MH, Kanai L, Foligeli A, Wagner J (1995) Comparative effects of permethrin-impregnated bednets and DDT house spraying on survival rates and oviposition interval of *Anopheles farauti* No. 1 (Diptera: Culicidae) in Solomon Islands. Ann Trop Med Parasitol 89: 521–529

Hii JLK, Smith T, Mai A, Ibam E, Alpers MP (2000) Comparison between anopheline mosquitoes (Diptera: Culicidae) caught using different methods in a malria endemic area of Papua New Guinea. Bull Entomol Res 90: 211–219

Highton RB (1981) The evaluation of CDC light traps and human bait collections for sampling *Anopheles arabiensis* Patton. M.Sc. thesis, University of Liverpool

Hill MN (1970) Japanese encephalitis in Sarawak: studies on adult mosquito populations. Trans R Soc Trop Med Hyg 64: 489–496

Hill RL (1977) Lamps for light-trapping. N.Z. Entomol 6: 314–315

Hollingsworth JP, Hartstack AW, Lindquist DA (1968) Influence of near-ultra-violet output of attractant lamps on catches of insects by light traps. J Econ Entomol 61: 515–521

Holub RE (1983) Modification of New Jersey light trap for multiple sample collection. Mosquito News 43: 241–242

Horsfall WR (1943) Some responses of the malaria mosquito to light. Ann Entomol Soc Am 36: 41–45

Horsfall WR (1962) Trap for separating collections of insects by interval. J Econ Entomol 55: 808–811

Hu SMK, Grayston JT (1962) Encephalitis on Taiwan. II. Mosquito collection and bionomic studies. Am J Trop Med 11: 131–140

Huffaker CB, Back RC (1943) A study of methods of sampling mosquito populations. J Econ Entomol 36: 561–569

Hurlbut HS, Weitz B (1956) Some observations on the bionomics of the common mosquitoes of the Nile Delta. Am J Trop Med Hyg 5: 901–908

Husbands RC, Reed DE (1970) A comparison of larval mosquito species occurrence and light trap data. Proc California Mosq Control Assoc 37: 101–108

Hutchins RE (1940) Insect activity at a light trap during various periods of the night. J Econ Entomol 33: 654–657

Ikeuchi M (1967) Ecological studies on mosquitoes collected by light traps. Trop Med 9: 186–200

Ikeuchi M (1970) Observations on the physiological age of *Culex tritaeniorhynchus* collected by light traps. Jap J Sanit Zool 21: 209–212 (In Japanese, English summary)

Ishii T (1970) Seasonal abundance of mosquitoes in Kyoto prefecture in 1969. Ann Rep Kyoto Pref Inst Public Health 15: 1–18 (In Japanese)

Ishii T (1971*a*) Seasonal prevalence of several species of mosquitoes caught in one light trap. Ann Rep Kyoto Pref Inst Public Health 16: 51–54 (In Japanese, English summary)

Ishii T (1971*b*) Mosquito abundance surveyed with light trap: a comparison of operation for various numbers of nights a week (preliminary report). Ann Rep Kyoto Pref Inst Public Health 16: 55–62 (In Japanese, English summary)

Ishii T (1971*c*) Mosquito abundance surveyed with light trap: A comparison of operation for various numbers of nights a week (preliminary report). Ann Rep Kyoto Pref Inst Public Health 16: 55–61

Ismail IAH, Pinichpongse S, Chitprarop V, Prasittisuk C, Schepens J (1982) Trials with CDC and Monks Wood light-traps for sampling malaria vectors in Thailand. WHO/VBC/82,864, 7 pp. (mimeographed)

Itô S (1964) Collection of mosquitoes by light traps at four stations of Nagasaki city. Endem Dis Bull Nagasaki Univ 6: 231–241

Jaenson TGT, Niklasson B, Henriksson B (1986) Seasonal activity of mosquitoes in an Ockelbo disease endemic area in central Sweden. J Am Mosq Control Assoc 2: 18–28

Janousek TE, Olson JK (1994) Effect of a lunar eclipse on the flight activity of mosquitoes in the upper Gulf coast of Texas. J Am Mosq Control Assoc 10: 222–224

Jewell D (1981) A security modification for the 'American model' mosquito light trap. Mosquito News 41: 183–184

Johnson BK, Gichogo A, Gitau G, Patel N, Ademba G, Kirui R, Highton RB, Smith DA (1981) Recovery of O'nyong-nyong virus from *Anopheles funestus* in western Kenya. Trans R Soc Trop Med Hyg 75: 239–241

Johnson CG (1950) A suction trap for small airborne insects which automatically segregates the catch into successive hourly samples. Ann Appl Biol 37: 80–91

Johnston JG, Weaver JW, Sudia WD (1973) Flashlight batteries as a power source for CDC miniature light traps. Mosquito News 33: 190–194

Jones RE, Barker-Hudson P, Kay BH (1991) Comparison of dry ice baited light traps with human bait collections for surveillance of mosquitoes in northern Queens-land, Australia. J Am Mosq Control Assoc 7: 387–394

Jupp PG, McIntosh BM, Anderson D (1976) *Culex (Eumelanomyia) rubinotus* Theobald as vector of Banzi. Germiston and Witwatersrand viruses. IV. Observations on the biology of *C. rubinotus*. J Med Entomol 12: 647–651

Jupp PG, McIntosh BM, Nevill EM (1980) A survey of the mosquito and *Culicoides* faunas at two localities in the Karoo region of South Africa with some observations on bionomics. Onderstepoort J Vet Res 47: 1–6

Kalmus H (1958) Responses of insects to polarized light in the presence of dark reflecting surfaces. Nature 157: 512

Kim HC, Lee KW, Richards RS, Schleich SS, Herman WE, Klein TA (2003*a*) Seasonal prevalence of mosquitoes collected from light traps in Korea (1999–2000). Korean J Entomol 33: 9–16

Kim HC, Friendly OS, Pike JG, Schuster AL, O'Guinn ML, Klein TA (2003) Seasonal prevalence of mosquitoes collected from light traps in the Republic of Korea, 2001. Korean J Entomol 33: 189–199

Kimsey RB, Chaniotis BN (1984) A light trap for biting Nematocera in moist environments. Mosquito News 44: 408–412, and correction slip, J Am Mosq Control Assoc (1985) 1: 114

King EW, Pless CD, Reed JK (1965) An automatic sample-changing device for light-trap collecting. J Econ Entomol 58: 170–172

Kitaoka S, Itô K (1964) Attractiveness of black light to biting midges and mosquitoes. Jap J Sanit Zool 15: 208–209 (In Japanese, English summary)

Kovrov BG, Monchadskiy AS (1963) The possibility of using polarized light to attract insects. Entomol Rev Wash 42: 25–28

Kuiper GP (1938) The magnitude of the sun, the stellar temperature scale, and bolometric corrections. Astrophys J 88: 429–471

Lafferty AL, Murphy R (1948) An automatic device for changing daily the collection receptacle on mechanical mosquito traps. Proc New Jers Mosq Exterm Assoc 35: 88–89

Lang JT (1984) Intermittent light as a mosquito attractant in New Jersey light traps. Mosquito News 44: 217–220

LaSalle MW, Dakin ME (1982) Dispersal of *Culex salinarius* in southwestern Louisiana. Mosquito News 42: 543–550

Lewis LA, Teller LW (1967) The use of fluorescent tubes in a modified New Jersey light trap. Proc New Jers Mosq Exterm Assoc 54: 163–170

Lindsay SW, Wilkins HA, Zieler HA, Daly RJ, Petrarca V, Byass P (1991) Ability of *Anopheles gambiae* mosquitoes to transmit malaria during the dry and wet seasons in an area of irrigated rice cultivation in The Gambia. J Trop Med Hyg 94: 313–324

Lines JD, Curtis CF, Wilkes TJ, Njunwa KJ (1991) Monitoring human-bait mosquitoes (Diptera: Culicidae) in Tanzania with light-traps hung beside mosquito nets. Bull Entomol Res 81: 77–84

Linley JR, Evans FDS (1968) A simple device to aid time-lapse studies with insects. Ann Entomol Soc Am 61: 775–777

Linthicum KJ, Bailey CL, Davies FG, Kairo A (1985) Observations on the dispersal and survival of a population of *Aedes lineatopennis* (Ludlow) (Diptera: Culicidae) in Kenya. Bull Entomol Res 75: 661–670

Lloyd JE, Pennington RG (1976) Mosquitoes collected in a CO_2-baited CDC miniature light trap and a bovine-baited trap in Wyoming. Mosquito News 36: 457–459

Loomis EC (1959*a*) Selective response of *Aedes nigromaculis* (Ludlow) to the Minnesota light trap. Mosquito News 19: 260–263

Loomis EC (1959*b*) A method for more accurate determination of air volume displacement of light traps. J Econ Entomol 52: 343–345

Loomis EC, Hanks SG (1959) Light trap indices of mosquito abundance: a comparison of operation for four and seven nights a week. Mosquito News 19: 168–171

Love GJ, Platt RB, Goodwin MH (1963). Observations on the spatial distribution of mosquitoes in south-western Georgia. Mosquito News 23: 13–22

Ludueña Almeida FF, Gorla DE (1995) Daily pattern of flight activity of *Aedes albifasciatus* in central Argentina. Mem Inst Oswaldo Cruz 90: 639–644

MacCreary D (1941) Comparative density of mosquitoes at ground level and at an elevation of approximately one hundred feet. J Econ Entomol 34: 174–179

Magy HI, Work TH, Thomas CV (1976) A reassessment of *Culex pipiens* as a potential St Louis encephalitis vector in Imperial county. Proc California Mosq Control Assoc 44: 41–45

Main AJ, Tonn RJ, Randall EJ, Anderson KS (1966) Mosquito densities at heights of five and twenty-five feet in southeastern Massachusetts. Mosquito News 26: 243–248

Main AJ, Brown SE, Wallis RC (1979) Arbovirus surveillance in Connecticut, II. California serogroup. Mosquito News 39: 552–559

Malainual A, Chansang C, Thavara U, Phan-Urai P (1987) Time intervals and location of biting and flying activities of JE vectors. Bull Dept Med Sci 29: 103–111 (In Thai, English summary)

Matsumoto BM, Maxfield HK (1985) A comparison of female *Culiseta melanura* captured in New Jersey and CDC light traps in southern Massachusetts. J Am Mosq Control Assoc 1: 90–92

Maxwell CA, Curtis CF, Haji H, Kisumku S, Thalib AI, Yahya SA (1990) Control of bancroftian filariasis by integrating therapy with vector control using poly-styrene beads in wet pit latrines. Trans R Soc Trop Med Hyg 84: 709–714

Mboera LEG, Kihonda J, Braks MAH, Knols BGJ (1998) Influence of Centers for Disease Control light trap position, relative to a human-baited bed net, on catches of *Anopheles gambiae* and *Culex quinquefasciatus* in Tanzania. Am J Trop Med Hyg 59: 595–596

Mbogo CNM, Glass GE, Forster D, Kabiru EW, Githure JI, Ouma JH, Beier JC (1993) Evaluation of light traps for sampling anopheline mosquitoes in Kilfi, Kenya. J Am Mosq Control Assoc 9: 260–263

McDonald G (1980) Population studies of *Culex annulirostris* Skuse and other mosquitoes (Diptera: Culicidae) at Mildura in the Murray Valley of southern Australia. J Aust Ent Soc 19: 37–40

McDonald JL (1970) A simple, inexpensive alcohol light trap for collecting *Culicoides*. Mosquito News 30: 652–654

McDonald JL, Savage LB (1973) Seasonal abundance of mosquito larval counts and adult light trap catches in Okinawa. Mosquito News 33: 105–107

McDonald JL, Granger J, Olton GS (1974) A mosquito light-trap stabilizer. Mosquito News 34: 234

McDonald WA, Naseir A, Fulmer AC, Ryan JM (1964) DDVP as a killing agent in New Jersey light trap. Mosquito News 24: 225

McGeachie WJ (1988) A remote sensing method for the estimation of light-trap efficiency. Bull Entomol Res 78: 379–385

McGeachie WJ (1989) The effects of moonlight illuminance, temperature and wind-speed on light-trap catches of moths. Bull Entomol Res 79: 185–192

McLaughlin RE, Focks DA (1990) Effects of cattle density on New Jersey light trap mosquito captures in the rice/cattle agroecosystem of southwestern Louisiana. J Am Mosq Control Assoc 6: 283–286

Meyer RP, Washino RK, McKenzie TL, Fukushima CK (1984) Comparison of three methods for collecting adult mosquitoes associated with rice field irri-gated pasture habitats in northern California. Mosquito News 44: 315–320

Meyer RP, Reisen WK, Eberle MW, Milby MM, Martinez VM, Hill BR (1984) A time segregated sampling device for determining nightly host-seeking pat-terns. Proc California Mosq Vector Control Assoc 52: 162–166

Meyers EG (1959) Mosquito collections by light traps at various heights above ground. Proc California Mosq Control Assoc 27: 61–63

Milby MM, Reeves WC (1986) Changes in the relative abundance of *Aedes nigromaculis, Aedes melanimon* and *Culex tarsalis* in the central valley of California. Proc Conf California Mosq Vector Control Assoc 54: 96–100

Milby MM, Reeves WC (1989) Comparison of New Jersey light-traps and CO_2-baited traps in urban and rural areas. Proc California Mosq Vector Control Assoc 57: 73–79

Milby MM, Kauffman EE, Harvey JF (1978) Conversion of CDC light trap indices to New Jersey light trap indices for several species of Californian mosquitoes. Proc California Mosq Vector Control Assoc 46: 58–60

Miller TA, Stryker RG, Wilkinson RN, Esah S (1969) Notes on the use of CO_2-baited CDC miniature light traps for mosquito surveillance in Thailand. Mosquito News 29: 688–689

Miller TA, Stryker RG, Wilkinson RN, Esah S (1970) The influence of moonlight and other environmental factors on the abundance of certain mosquito species in light-trap collections in Thailand. J Med Entomol 7: 555–561

Miller TA, Stryker RG, Wilkinson RN, Esah S (1977) The influence of time and frequency of collection on the abundance of certain mosquito species in light-trap collections in Thailand. J Med Entomol 14: 60–63

Mitchell L (1982) Time-segregated mosquito collections with a CDC miniature light trap. Mosquito News 42: 12–18

Mitchell L, Rockett CL (1979) Vertical stratification preferences of adult female mosquitoes in a sylvan habitat (Diptera: Culicidae). Great Lakes Entomol 12: 219–223

Moore A, Miller JR, Tabashnik BE, Gage SH (1986) Automated identification of flying insects by analysis of wingbeat frequencies. J Econ Entomol 79: 1703–1706

Mulhern TD (1934) A new development in mosquito traps. Proc New Jers Mosq Exterm Assoc 21: 137–140

Mulhern TD (1942) New Jersey Mechanical Trap for Mosquito Surveys. New Jers Agric Exp Stn Circ 421

Mulhern TD (1953a) Better results with mosquito light traps through standardizing mechanical performance. Mosquito News 13: 130–133

Mulhern TD (1953b) The use of mechanical traps in measuring mosquito populations. Proc California Mosq Control Assoc 21: 64–66

Mulligan FS, Schaefer CH (1982) A physical barrier for controlling mosquitoes which breed in urban storm drains. Mosquito News 42: 360–365

Murphey FJ, Darsie RF (1962) A mechanical aspirator for collecting adult mosquitoes. Proc New Jers Mosq Exterm Assoc 49: 115–117

Murphy MW, Dunton RF, Perich MJ, Rowley WA (2001) Attraction of *Anopheles* (Diptera: Culicidae) to volatile chemicals in western Kenya. J Med Entomol 38: 242–244

Nagel RH, Granovsky PA (1947) A turntable light trap for taking insects over regulated periods. J Econ Entomol 40: 583–586

Nasci RS, Edman JD (1981) Vertical and temporal flight activity of the mosquito *Culiseta melanura* (Diptera: Culicidae) in southwestern Massachusetts. J Med Entomol 18: 501–504

Natuhara Y, Takagi M, Maruyama K, Sugiyama A (1991) Monitoring *Culex tritaeniorhynchus* (Diptera: Culicidae) abundance in cow sheds by in situ counting. J Med Entomol 28: 551–552

Neeru Singh, Mishra AK, Curtis CF, Sharma VP (1996) Influence of moonlight on light-trap catches of the malaria vector *Anopheles culicifacies* (Diptera: Culicidae) in central India. Bull Entomol Res 86: 475–479

Nelson RL, Milby MM, Reeves WC, Fine PE (1978) Estimates of survival, population size, and emergence of *Culex tarsalis* at an isolated site. Ann Entomol Soc Am 71: 801–808

Nielsen ET, Nielsen AT (1953) Field observations on the habits of *Aedes taeniorhynchus*. Ecology 34: 141–156

Nielsen SA, Siewertz-Poulsen KE, Nielsen BO (1980) A time-sorting insect light-trap. Entomol Meddr 48: 29–32

Odetoyinbo JA (1969) Preliminary investigation on the use of a light-trap for sampling malaria vectors in The Gambia. Bull World Health Organ 40: 547–560

Olson JG, Reeves WC, Emmons RW, Milby MM (1979) Correlation of *Culex tarsalis* population indices with the incidence of St. Louis encephalitis and western equine encephalomyelitis in California. Am J Trop Med Hyg 28: 335–343

Olson JG, Atmosoedjono S, Lee VH, Ksiazek TG (1983) Correlation between population indices of *Culex tritaeniorhynchus* and *Cx. gelidus* (Diptera: Culicidae) and rainfall in Kapuk, Indonesia. J Med Entomol 20: 108–109

Olson JK, Elbel RE, Smart KI (1968) Mosquito collections by CDC miniature light traps and livestock-baited stable traps at Callao, Utah. Mosquito News 28: 512–516

Onishi A (1959) Influence of moonlight on mosquito collection with the light trap. Shikoku Acta Med 15: 1993–1998 (In Japanese, English summary)

Pagac BB, Turell MJ, Olsen GH (1992) Eastern equine encephalomyelitis virus and *Culiseta melanura* activity at the Patuxent wildlife research center, 1985–90. J Am Mosq Control Assoc 8: 328–330

Pajot F-X, Le Pont F, Molez J-F (1977) Utilisation des pièges lumineux "C.D.C. miniature light trap" comme moyen d'énchantillonnage des populations anophèlienes dans une village du littoral de la Guyane françiase. Cah ORSTOM sér Entomol Méd Parasitol 15: 233–240

Parsons RE, Champion E, Wilson F (1981) An inexpensive support rod for the CDC light trap collecting bag. Mosquito News 41: 796–797

Payne CD (1987) The GLIM System Release 3.77 Manual. 2nd ed. Numerical Algorithms Group Ltd, Oxford

Pennnington NE (1967) Comparison of DDVP and cyanide as killing agents in mosquito light traps. J Med Entomol 4: 518

Perich MJ, Tidwell MA, Dobson SE, Sardelis MR, Zaglul A, Williams DC (1993) Barrier spraying to control the malaria vector *Anopheles albimanus*: laboratory and field evaluation in the Dominican Republic. Med Vet Entomol 7: 363–368

Pfuntner AR (1979) A modified CO_2-baited miniature surveillance trap. Bull Soc Vector Ecol 4: 31–35

Philogène BJ (1982) Experiments with artificial light: necessity for properly identifying the source. Can Entomol 114: 377–379

Pickens LG (1991) Colorimetric versus behavioral studies of face fly (Diptera: Muscidae) vision. Environ Entomol 19: 1242–1252

Pickens LG, Carroll JF, Azad AF (1987) Electrophysiological studies of the spectral sensitivities of cat fleas, *Ctenocephalides felis*, and oriental rat fleas, *Xenopsylla cheopis* to monochromatic light. Entomol Exp Appl 45: 193–204

Pincus S (1938) Mosquito control in greater New York. Proc New Jers Mosq Exterm Assoc 25: 115–121

Pippin WF (1965) Notes on the operation of a light trap in central Luzon, Philippine islands. Mosquito News 25: 183–187

Platt RB, Love GJ, Williams EL (1958) A positive correlation between relative humidity and the distribution and abundance of *Aedes vexans*. Ecology 39: 167–169

Porter CH, Gojmerac WL (1970) Temperature; its influence on light trap catches of *Aedes vexans* (Meigen). Mosquito News 30: 54–56

Pratt HD (1944) Studies on the comparative attractiveness of 25-, 50- and 100-watt bulbs for Puerto Rican *Anopheles*. Mosquito News 4: 17–18

Pratt HD (1948) Influence of the moon on light trap collections of *Anopheles albimanus* in Puerto Rico. J Natn Malar Soc 7: 212–220

Preiss FJ, Barefoot HL, Stryker RG, Young WW (1970) Effectiveness of DDVP as a killing agent in mosquito killing jars. Mosquito News 30: 417–419

Pritchard AE, Pratt HD (1944) I. A comparison of light trap and animal trap anopheline mosquito collections in Puerto Rico. II. A list of the mosquitoes of Puerto Rico. Public Health Rep Wash 59: 221–233

Provost MW (1952) The dispersal of *Aedes taeniorhynchus*. I. Preliminary studies. Mosquito News 12: 174–190

Provost MW (1957) The dispersal of *Aedes taeniorhynchus*. II. The second experiment. Mosquito News 17: 233–247

Provost MW (1958) Mating and Swarming in *Psorophora* Mosquitoes. Proc Int Congr Entomol Xth (1956), Vol. 2, 553–561

Provost MW (1959) The influence of moonlight on light-trap catches of mosquitoes. Ann Entomol Soc Am 52: 261–271

Ree HI, Self LS, Hong HK, Lee KW (1973) Mosquito light trap surveys in Korea 1969–1971. Southeast Asian J Trop Med Public Health 4: 382–386

Reeves WC (1968) A review of the development associated with the control of western equine and St. Louis encephalitis in California during 1967. Proc California Mosq Control Assoc 36: 65–70

Reisen WK, Pfuntner AR (1987) Effectiveness of five methods for sampling adult *Culex* mosquitoes in rural and urban habitats in San Bernardino county, California. J Am Mosq Control Assoc 3: 601–606

Reisen WK, Milby MM, Meyer RP, Reeves WC (1983) Population ecology of *Culex tarsalis* (Diptera: Culicidae) in a foothill environment in Kern county, California: Temporal changes in male relative abundance and swarming behavior. Ann Entomol Soc Am 76: 809–815

Reisen WK, Lothrop HD, Meyer RP (1997) Time of host-seeking by *Culex tarsalis* (Diptera: Culicidae) in California. J Med Entomol 34: 430–437

Reisen WK, Boyce K, Cummings RC, Delgado O, Gutierrez A, Meyer RP, Scott TW (1999) Comparative effectiveness of three adult mosquito sampling methods in habitats representative of four different biomes of California. J Am Mosq Control Assoc 15: 24–31

Reisen WK, Eldridge BF, Scott TW, Gutierrez A, Takahashi R, Lorenzen K, DeBenedictis J, Boyce K, Swartzell R (2002) Comparison of dry ice-baited Centers for Disease Control and New Jersey light traps for measuring mosquito abundance in California. J Am Mosq Control Assoc 18: 158–163

Reiter P (1983) A portable, battery-powered trap for collecting gravid *Culex* mosquitoes. Mosquito News 43: 496–498

Reiter P (1987) A revised version of the CDC gravid mosquito trap. J Am Mosq Control Assoc 3: 325–327

Ribbands CR (1945) Moonlight and house-haunting habits of female *Anopheles* in West Africa. Bull Entomol Res 36: 395–415

Roberts FC, Conner GE (1979) Use of a mathematical model to predict levels of adult *Culex pipiens* in Almeda county. Proc California Mosq Vector Control Assoc 47: 114–115

Roberts RH (1965) A steer-baited trap for sampling insects affecting cattle. Mosquito News 25: 281–285

Robinson HS, Robinson PJM (1950) Some notes on the observed behaviour of Lepidoptera in flight in the vicinity of light-sources together with a description of a light-trap designed to take entomological samples. Entomol Gaz 1: 3–20

Rodriguez AD, Rodriguez MH, Hernandez JE, Dister SW, Beck LR, Rejmankova E, Roberts DR (1996) Landscape surrounding human settlements and *Anopheles albimanus* (Diptera: Culicidae) abundance in southern Chiapas, Mexico. J Med Entomol 33: 39–48

Roeder KD (1953) Insect Physiology. John Wiley & Sons, New York

Rohe DL, Fall RP (1979) A miniature battery powered CO_2 baited light trap for mosquito borne encephalitis surveillance. Bull Soc Vector Ecol 4: 24–27

Ross D, Service MW (1979) A modified Monks Wood light trap incorporating a flashing light. Mosquito News 39: 610–616

Rubio-Palis Y (1992) Influence of moonlight on light trap catches of the malaria vector *Anopheles nuneztovari* in Venezuela. J Am Mosq Control Assoc 8: 178–180

Rubio-Palis Y, Curtis CF (1992) Evaluation of different methods of catching anopheline mosquitoes in western Venezuela. J Am Mosq Control Assoc 8: 261–267

Rudolfs W (1922) Chemotropism of mosquitoes. Bull New Jers Agric Exp Stn No. 367, 4

Rupp HR, Jobbins DM (1969) Equipment for mosquito surveys: two recent developments. Proc New Jers Mosq Exterm Assoc 56: 183–188

Ryan PA, Martin L, Mackenzie JS, Kay BH (1997) Investigation of gray-headed flying foxes (*Pteropus poliocephalus*) (Megachiroptera: Pteropodidae) and mosquitoes in the ecology of Ross River virus in Australia. Am J Trop Med Hyg 57: 476–482

Sadanandane C, Jambulingam P, Subramanian S (2004) Role of modified CDC miniature light-traps as an alternative method for sampling adult anophelines

(Diptera: Culicidae) in the National Mosquito Surveillance Programme in India. Bull Entomol Res 94: 55–63

Sandoski CA, Meisch MV, Case DA, Olson JK (1983) Effects of collection and handling techniques on riceland mosquitoes used in laboratory and field insecticide susceptibility tests. Mosquito News 43: 445–448

Santos De Marco T, Gaia MC de M, Brazil RP (2002) Influence of the lunar cycle on the activity of phlebotomine sandflies (Diptera: Psychodidae). J Am Mosq Control Assoc 18: 114–118

Self LS, Shin HK, Kim KH, Lee KW, Chow CY, Hong HK (1973) Ecological studies on *Culex tritaeniorhynchus* as a vector of Japanese encephalitis. Bull World Health Organ 49: 41–47

Self LS, Usman S, Nelson MJ, Saroso JS, Pant CP, Fanara DM (1976) Ecological studies on vectors of malaria, Japanese encephalitis and filariasis in rural areas of West Java. Bull Penel Keseh Health Std Indonesia 4: 41–55

Service MW (1969) The use of traps in sampling mosquito populations. Entomol Exp Appl 12: 403–412

Service MW (1970) A battery-operated light-trap for sampling mosquito populations, Bull World Health Organ 43: 635–641

Service MW (1971) Flight periodicities and vertical distribution of *Aedes cantans* (Mg.), *Ae. geniculatus* (Ol.), *Anopheles plumbeus* Steph. and *Culex pipiens* L. (Diptera: Culicidae) in southern England. Bull Entomol Res 61: 639–651

Service MW (1976) Contribution to the knowledge of mosquitoes (Diptera: Culicidae) of Gabon. Cah ORSTOM sér Entomol Méd Parasitol 14: 259–263

Service MW (1979) Light trap collections of ovipositing *Simulium squamosum* in Ghana. Ann Trop Med Parasitol 73: 487–490

Service MW (1993) Mosquito Ecology. Field Sampling Methods. 2nd ed. Chapman & Hall, London

Service MW, Boorman JPT (1965) An appraisal of adult mosquito trapping techniques used in Nigeria, West Africa. Cah ORSTOM sér Entomol Méd 3 & 4: 27–33

Service MW, Highton RB (1980) A chemical light trap for mosquitoes and other biting insects. J Med Entomol 17: 183–185

Sexton JD, Hobbs JH, St. Jean Y, Jacques JR (1986) Comparison of an experimental updraft ultraviolet light trap with the CDC miniature light trap and biting collections in sampling for *Anopheles albimanus* in Haiti. J Am Mosq Control Assoc 2: 168–173

Shidrawi GR, Clarke JL, Boulzaguet JR (1973) Assessment of the CDC miniature light trap for sampling malaria vectors in Garki district, northern Nigeria. World Health Organization. Technical Note, No. 11, MPD/TN/73.1, 44–9 (mimeographed)

Shiff CJ, Minjas JN, Hall T, Hunt RH, Lyimo S, Davis JR (1995) Malaria infection potential of anopheline mosquitoes sampled by light trapping indoors in coastal Tanzanian villages. Med Vet Entomol 9: 256–262

Sholdt LL, Neri P, Seibert DJ (1974) Flashlight body and timer unit for powering CDC light traps. Mosquito News 34: 237–238

Siddorn JW, Brown ES (1971) A Robinson light trap modified for segregating samples at predetermined time intervals, with notes on the effect of moonlight on the periodicity of catching insects. J Appl Ecol 8: 69–75

Silvain J-F, Pajot F-X (1981) Écologie d'*Anopheles (Nyssorhynchus) aquasalis* Curry, 1932 en Guyane Française. 1. Dynamique des populations imaginales. Caractérisation des gîtes larvaires. Cah ORSTOM sér Entomol Méd Parasitol 19: 11–21

Simsek FM (2004) Seasonal larval and adult population dynamics and breeding habitat diversity of *Culex theileri* Theobald, 1903 (Diptera: Culicidae) in the Golbasi District, Ankara, Turkey. Turkish J Zool 28: 337–344

Singh N, Mishra AK, Singh OP (1993) Preliminary observations on mosquito collections by light traps in tribal villages of Madhya Pradesh. Indian J Malariol 30: 103–107

Siverly RE, DeFoliart GR (1968) Mosquito studies in Northern Wisconsin II. Light trapping studies. Mosquito News 28: 162–167

Smith T (1995) Proportionality between light trap catches and biting densities of malaria vectors. J Am Mosq Control Assoc 11: 377–378

Somboon P, Choochote W, Khamboonruang C, Keha P, Suwanphanit P, Sukontasan K, Chaivong P (1989) Studies on the Japanese encephalitis vectors in Amphoe Muang, Chiang Mai, northern Thailand. Southeast Asian J Trop Med Public Health 20: 9–17

Southwood TRE, Henderson PA (2000) Ecological Methods. 3rd ed. Blackwell Science, Oxford

Srihongse S, Grayson MA, Bosler EM (1979) California encephalitis complex virus isolations from mosquitoes collected in northeastern New York, 1976–1977. Mosquito News 39: 73–76

Srihongse S, Woodhall JP, Grayson MA, Weibel R (1980) Arboviruses in New York state: surveillance in arthropods and nonhuman vertebrates, 1972–1977. Mosquito News 40: 269–276

Standfast HA (1965) A miniature light trap which automatically segregates the catch into hourly samples. Mosquito News 25: 48–53

Strickland RE (1967) Insect suction trap for collecting segregated samples in a liquid. J Agric Engineerig Res 12: 319–321

Strickman D, Miller ME, Heung-Chul Kim, Kwan-Woo Lee (2000) Mosquito surveillance in the demilitarized zone, Republic of Korea, during an outbreak of *Plasmodium vivax* malaria in 1996 and 1997. J Am Mosq Control Assoc 16: 100–113

Sudia WD, Chamberlain RW (1962) Battery-operated light trap, an improved model. Mosquito News 22: 126–129

Suleman M, Reisen WK, Aslamkhan M (1977) Observations on the time of attraction of some Pakistan mosquitoes to light traps. Mosquito News 37: 531–533

Sun WKC (1962) A study of the seasonal succession of some medically important insects in the Taipei area. Biol Bull Tunghai Univ Taichung Taiwan 10: 1–19

Sun WKC (1964) The seasonal succession of mosquitoes in Taiwan. J Med Entomol 1: 277–284

Syms PR, Goodman LJ (1987) The effect of flickering U-V light output on the attractiveness of an insect electrocutor trap to the house-fly. *Musca domestica.* Entomol Exp Appl 43: 81–85

Takagi M, Tsuda Y, Wada Y, Takafuji A (1995) Decrease of vector mosquitoes of bancroftian filariasis in a village on Fukue island, Nagasaki, southwestern apan. Trop Med (Nagasaki) 37: 159–163

Tarry DW, Kirkwood AC, Herbert VCN (1971) The response to 'black-light' radiation of some common flies of economic importance. Entomol Exp Appl 14: 23–29

Taylor J, Padgham DE, Perfect TJ (1982) A light-trap with upwardly directed illumination and temporal segregation of the catch. Bull Entomol Res 72: 669–673

Taylor LR (1951) An improved suction trap for insects. Ann Appl Biol 38: 582–591

Taylor LR, Brown ES (1972) Effects of light-trap design and illumination on samples of moths in the Kenya highlands. Bull Entomol Res 62: 91–112

Taylor RT, Solis M, Weathers DB, Taylor JW (1975) A prospective study of the effects of ultralow (ULV) aerial applications of malathion on epidemic *Plasmodium falciparum* malaria. II. Entomologic and operational aspects. Am J Trop Med Hyg 24: 188–193

Thomas A, Spiegelhalter DJ, Gilks WR (1992) BUGS: a program to perform Bayesian inference using Gibbs sampling. In: Bernardo JM, Berger JO, David AP, Smith AFM (eds) Bayesian Statistics Volume 4. Clarendon Press, Oxford, pp. 837–842

Thonnon J, Spiegel A, Diallo M, Sylla R, Fall A, Mondo M, Fontenille D (1998) Yellow fever outbreak in Kaffrine, Senegal 1996: epidemiological and entomological findings. Trop Med Int Health 3: 872–877

Thonnon J, Spiegel A, Diallo M, Diallo A, Fontenille D (1999) Epidémies à virus Chikungunya en 1996 et 1997 au Sénégal. Bull Soc Pathol Exot 92: 79–82

Thurman DC, Thurman EB (1955) Report of the initial operation of a light trap in northern Thailand. Mosquito News 15: 218–224

Townes H (1962) Design for a Malaise trap. Proc Entomol Soc Wash 64: 253–262

Turner EC, Earp UF (1968) A timing device for direct current New Jersey light traps. Mosquito News 28: 75–76

Ulloa A, Rodriguez MH, Rodriguez AD, Roberts DR (1997) A comparison of two collection methods for estimating abundance and parity of *Anopheles albimanus* in breeding sites and villages of southern Mexico. J Am Mosq Control Assoc 13: 238–244

Upton MS (1973*a*) Collecting lamps—a warning. Aust Entomol Soc News Bull 9: 17

Upton MS (1973*b*) Collecting lamps—a further note. Aust Entomol Soc News Bull 10: 57–58

Vaidyanathan R, Edman JD (1997*a*) Sampling with light traps and human bait in epidemic foci for eastern equine encephalomyelitis virus in southeastern Massachusetts. J Am Mosq Control Assoc 13: 348–355

Vaidyanathan R, Edman JD (1997*b*) Sampling methods for potential epidemic vectors of eastern equine encephalomyelitis virus in Massachusetts. J Am Mosq Control Assoc 13: 342–347

Van den Hurk AF, Beebe NW, Ritchie SA (1997) Responses of mosquitoes of the *Anopheles farauti* complex to 1-octen-3-ol and light in combination with carbon dioxide in northern Queensland, australia. Med Vet Entomol 11: 177–180

Van Essen PHA, Kemme JA, Ritchie SA, Kay BH (1994) Differential responses of *Aedes* and *Culex* mosquitoes to octenol or light in combination with carbon dioxide in Queensland, Australia. Med Vet Entomol 8: 63–67

Vavra RW, Carestia RR, Frommer RL, Gerberg EJ (1974*a*) Field evaluation of alternative light sources as mosquito attractants in the Panama Canal zone. Mosquito News 34: 382–384

Vavra RW, Frommer RL, Carestia RR, Harding FL, Linehan DD (1974*b*) Field evaluation of chemical, radioactive and blinking light sources as mosquito attractants. Proc California Mosq Control Assoc 42: 93–95

Venables WN, Ripley BD (1994) Modern Applied Statistics with S-plus. Springer-Verlag, New York

Vervent G, Coz J (1969) Contribution à L'étude des Pièges Lumineux Comme Moyen de Capture des Anopheles. OCCGE Centre Muraz, Lab Entomol No. 482/69; 9 pp. (mimeographed)

von Frisch K (1950) Perception of polarized light by insects. Experimentia 6: 210–221

Wagner RE, Barnes MM, Ford GM (1969) A battery-operated timer and power supply for insect light traps. J Econ Entomol 62: 575–578

Walker AR, Boreham PFL (1976) Saline as a collecting medium for *Culicoides* (Diptera, Ceratopogonidae) in blood feeding and other studies. Mosquito News 36: 18–20

Walters LL, Smith TA (1980) Bio-ecological studies of *Culex* mosquitoes in a focus of western equine and St. Louis encephalitis virus transmission (New River Basin, Imperial Valley, California), I. Larval ecology and trends of adult dispersal. Mosquito News 40: 227–235

Waterman TH (1951) Polarized light and orientation by insects. Trans NY Acad Sci 14: 11–14

Webb JP, Work TH, McAndrews TP, Jacobson D (1977) A preliminary comparative study of *Culex tarsalis* and *Culex pipiens quinquefasciatus* from the New River, Imperial Valley, California. Proc California Mosq Vector Control Assoc 45: 16–17

Wegbreit J, Reisen WK (2000) Relationships among weather, mosquito abundance, and encephalitis virus activity in California: Kern County 1990–98. J Am Mosq Control Assoc 16: 22–27

Wellington WG (1974) Change in mosquito flight associated with natural changes in polarized light. Can Entomol 106: 941–948

West HW, Cashman DL (1985) New Jersey light trap modification to extend bulb life. J Am Mosq Control Assoc 1: 378–379

Wharton RH, Eyles DE, Warren McW (1963) The development of methods for trapping the vectors of monkey malaria. Ann Trop Med Parasitol 57: 32–46

White EG (1964) A design for the effective killing of insects caught in light traps. N.Z. Entomol 3: 25–27

Wieser-Schimpf L, Foil LD, Holbrook FR (1991) Effect of carbon dioxide on the collection of adult *Culicoides* spp. (Diptera: Ceratopogonidae) by a new modification of black light New Jersey light traps. J Am Mosq Control Assoc 7: 462–466

Williams CB (1935) The times of activity of certain nocturnal insects, chiefly Lepidoptera, as indicated by a light trap. Trans R Ent Soc Lond (A) 83: 523–555

Wilton DP (1975a) Mosquito collections in El Salvador with ultra-violet and CDC miniature light traps with and without dry ice. Mosquito News 35: 522–525

Wilton DP (1975b) Field evaluation of three types of light traps for collection of *Anopheles albimanus* Wiedeman (Diptera: Culicidae). J Med Entomol 12: 382–386

Wilton DP, Fay RW (1972a) Air flow direction and velocity in light trap design. Entomol Exp Appl 15: 377–386

Wilton DP, Fay RW (1972b) Responses of adult *Anopheles stephensi* to light of various wavelengths. J Med Entomol 9: 301–304

Wishart E (1999) Adult mosquito (Diptera: Culicidae) and virus survey in metropolitan Melbourne and surrounding areas. Aust J Entomol 38: 310–313

Wishart E (2002) Species composition and population studies of mosquitoes (Diptera: Culicidae) in the Mildura district in the Murray Valley of southern Australia. Aust J Entomol 41: 45–48

Wong YW, Rowley WA, Dorsey DC, Hausler WJ (1978) Surveillance of arbovirus activity in Iowa during 1972–1975. Mosquito News 38: 245–251

Young S, David CT, Gibson G (1987) Light measurement for entomology in the field and laboratory. Physiol Entomol 12: 373–379

Zaim M, Ershadi MRY, Manouchehri AV, Hamdi MR (1986) The use of CDC light traps and other procedures for sampling malaria vectors in southern Iran. J Am Mosq Control Assoc 2: 511–515

Zhong H, Yan ZC, Jones F, Brock C (2003) Ecological analysis of mosquito light trap collections from West Central Florida. Environ Entomol 32: 807–815

Zimmerman JH, Abbassy MM, Hanafi HA, Beier JC, Dees WH (1988) Host-feeding patterns of mosquitoes (Diptera: Culicidae) in a rural village near Cairo, Egypt. J Med Entomol 25: 410–412

Zyzak M, Loyless T, Cope S, Wooster M, Day JF (2002) Seasonal abundance of *Culex nigripalpus* Theobald and *Culex salinarius* Coquillett in north Florida, USA. J Vector Ecol 27: 155–162

Chapter 10 Sampling Adults with Carbon Dioxide Traps

As early as 1922 Rudolfs reported that carbon dioxide was an attractant for mosquitoes, and that carbon dioxide produced by breathing was an important factor in attracting mosquitoes to their hosts. It is now generally accepted that carbon dioxide is, in concert with other olfactory cues, an attractant to virtually all haematophagous flies. In most of the West African species studied by Gillies and Wilkes (1969) carbon dioxide was a middle range factor in host orientation, activating and attracting mosquitoes from about 15–30 m, but the actual distance over which it acted depended on the species. Ornithophagic species appeared to be little attracted by carbon dioxide (Gillies and Wilkes 1972). In later trials Gillies and Wilkes (1974), using ramp traps, found that whereas birds attracted *Anopheles melas* and *Culex thalassius* from at least 7 m, carbon dioxide (50 ml/min) attracted these ornithophagic species from up to only 4 m. In contrast, in Uganda Henderson et al. (1972) caught large numbers of ornithophagic mosquitoes in CDC light-traps when they were supplemented with dry ice, seemingly indicating that carbon dioxide was an important attractant.

Gillies (1980) gave a useful, albeit brief, review of the role of carbon dioxide in host-seeking by mosquitoes, while Sutcliffe (1986, 1987) discusses the distances over which biting flies are attracted to their hosts. Kline et al. (1990*a*) carried out field trials to study the role of carbon dioxide, octenol and other possible host attractants in attracting mosquitoes. Further information on carbon dioxide and host orientation is also given in Chapter 7.

Different types of animals release different amounts of carbon dioxide. Roberts (1972) cited some interesting unpublished data supplied by P. W. Moe and H. F. Tyrrell on the rate of carbon dioxide produced by cattle. Fasted dairy cows produced an average of 1617 litres/24 h, while lactating cows produced 5005 litres, 'beef heifers' growing slowly 1727 litres and those on 'full feed' 2639 litres carbon dioxide/24 h. In contrast five small chickens released about 72 litres/24 h (Gillies and Wilkes 1974).

Human exhaled breath contains about 4.5% carbon dioxide, and a host odour plume would remain above background atmospheric levels

(0.03–0.04%) until carried downwind and diluted by a factor of about 100. Field trials with human subjects breathing normally or wearing a breathing apparatus that removed 95.5% of the expired carbon dioxide showed that significantly fewer mosquitoes were attracted to the subjects wearing the apparatus (Snow 1970). However, once adults had arrived at the host there was no significant difference between the numbers attempting to feed on people with and without the apparatus; thus appearing to confirm that carbon dioxide is usually a medium- or long-range attractant.

McIver (1982) included carbon dioxide as one of the five main stimuli concerned with host orientation, while in Canada McIver and McElligott (1989) used ramp traps of Gillies (1969) to determine the range of attraction (3, 7, 11, 15 and 19 m) of various release rates (250, 500, 1000 and 4000 ml/min) of carbon dioxide from a cylinder. In general increased levels of gas attracted more *Aedes vexans*, and increasing the discharge from 1000–4000 ml/min extended the range of attraction from between 3–7 m to 7–11 m for *Aedes vexans*, other aedine species and *Anopheles walkeri*. However, increased rates of release had no significant effect on catches of *Culiseta inornata, Culiseta morsitans, Culex restuans* and *Culex pipiens*.

Traps supplemented with carbon dioxide

The role of carbon dioxide as a mosquito attractant has led to its use in a variety of traps. For example, as long ago as 1934 Headlee reported that delivering carbon dioxide gas over a New Jersey light-trap for only 2 h each evening increased the mosquito catch by 400–500%. Headlee (1941) later reported that up to 19 times more mosquitoes were collected in New Jersey light-traps supplemented with dry ice in addition to light. Huffaker and Back (1943) caught about eight times as many mosquitoes in New Jersey light-traps baited with about 3 lb of dry ice than in traps without it, but they noted that the relative order of abundance of the different species was altered by the addition of dry ice. Furthermore, not all species were attracted equally by the addition of dry ice. Iha (1971) found that in Okinawa, Japan, dry ice traps collected a greater number of species than other trapping methods. *Culex tritaeniorhynchus* and *Aedes vexans nipponii* were the two most abundant species, and *Anopheles tessellates*, which was almost entirely absent from other collections, was caught in dry ice traps.

Reeves and Hammon (1942) were among the first to seriously propose the addition of carbon dioxide to light-traps to increase mosquito catches, but their use was limited by the relative bulkiness of early light-traps and the difficulty in obtaining CO_2. The introduction of battery operated CDC miniature light-traps (Sudia and Chamberlain 1962) and the increased availability of dry ice stimulated renewed interest in combining light with carbon dioxide in mosquito traps.

Since the mid 1970s there has been increasing usage of carbon dioxide, usually in the form of dry ice, as a supplement in other traps, occasionally in animal-bait traps but more usually in light-traps. In many areas of the USA CDC-type light-traps incorporating CO_2 are routinely operated in surveillance programmes, sometimes with the light source removed. These and other traps are described below. Descriptions of the construction of light-traps that are sometimes supplemented with carbon dioxide, such as CDC traps and those originally designed for use with carbon dioxide, for example the EVS (Rohe and Fall 1979) and the CO_2-4 trap (Pfuntner 1979), are presented in Chapter 9.

Source of carbon dioxide

Carbon dioxide can be added to traps in the form of gas, dry ice, or from CO_2-producing granular sachets. An obvious advantage of using cylinders for releasing carbon dioxide into traps is that its discharge can be regulated, an important consideration if it is suspected that different species are attracted to different emission rates. Uniform release of gas, however, necessitates a sensitive regulatory valve system and meters to control and measure flow rates, and these together with the cylinders, are more costly and bulky than dry ice. These factors can prove a serious disadvantage if several traps are used.

Carestia and Savage (1967) discharged carbon dioxide from a cylinder into four CDC light-traps at mean flow rates of 240, 548, 945 and 2000 ml/min. In another trap without light, gas was emitted at the rate of 470 ml/min, while a sixth trap had a light source but no carbon dioxide. About four times as many mosquitoes were caught in the trap discharging 240 ml of gas/min as in the trap without carbon dioxide, and the size of the catch increased with larger discharge rates. The trap without a light, but releasing 470 ml of carbon dioxide/min, caught almost 10 times as many mosquitoes as the unbaited trap. An advantage of not using light in the traps was that few other insects were caught, thus simplifying sorting and identification.

Parker et al. (1986) connected a two-way solenoid switch, powered by a 6-V battery, to a standard single-stage 20-lb gas cylinder to facilitate the delivery of carbon dioxide to light-traps and (other traps) only at times when the gas was needed. A standard CDC-type photocell was wired to the solenoid. Using this procedure the authors were able to obtain five nights of 9–10 h trapping with a 20-lb cylinder delivering about 2500 ml/min.

A useful alternative to using individual gas cylinders for each trap is to take into the field a 25-lb gas cylinder and several truck or tractor tyre inner tubes, which can be inflated with gas at the trapping sites. The inner tubes can be fitted with a valve mechanism to regulate the flow of gas (20–2000 ml/min). This arrangement has been used successfully in Scotland for trapping simuliids (Coupland 1991).

Saitoh et al. (2004) describe the novel use of an age-old method for generation of gaseous carbon dioxide for use in mosquito traps, namely the fermentation of sugar by yeast. The apparatus comprises two 2-litre plastic bottles containing different concentrations of dry yeast and sugar. One bottle contains 150 g sugar + 12 g dry yeast, made up to 1500 ml with water. The second bottle contains 100 g sugar and 6 g yeast, made up to 1750 ml. The two-bottle system combines the relatively high output of CO_2 obtained in the high concentration set-up with the more prolonged output of the lower concentration set-up. The two bottles are connected one to the other with polypropylene tubing and to a third small plastic bottle (500 ml), from which CO_2 is released via a 5 mm hole in one side. The small bottle is hung close to the trap entrance (Fig. 10.1). Stabilisation of CO_2 output was determined to take approximately 1.5 h from initial set-up of the apparatus. Maximum CO_2 output was 40.6 ± 2.1 ml/min and was achieved after approximately 10 h. Output then decreased gradually to a rate of 28.0 ± 0.6 ml/min at the end of the experiment (28.5 h). Mean output was 32.4 ml/min. The efficacy of the yeast CO_2 generation system was tested in field experiments against traps without CO_2 and traps baited with 1kg of dry ice. Traps baited with yeast-generated CO_2 consistently trapped higher numbers of mosquitoes than unbaited traps. Traps baited with 1 kg of dry ice produced CO_2 at a rate of 387 ml/min and probably as a result of this higher output, trapped more mosquitoes than traps baited with yeast-generated CO_2. The most commonly trapped species in traps baited with yeast-generated CO_2 were *Culex pipiens pallens* and *Aedes albopictus*, the same as trapped by dry ice-baited traps. Other species collected by the traps baited with yeast-generated CO_2 included *Armigeres subalbatus, Culex halifaxii, Ochlerotatus japonicus*, and *Tripteroides bambusa*.

Dry ice is relatively cheap and light, although in certain areas it may be more difficult to obtain than cylinders. Simple and small pieces of commercially available laboratory apparatus can be fitted to a gas cylinder to produce small blocks (about 100 mm across and 75 mm thick) of dry ice weighing about 500 g.

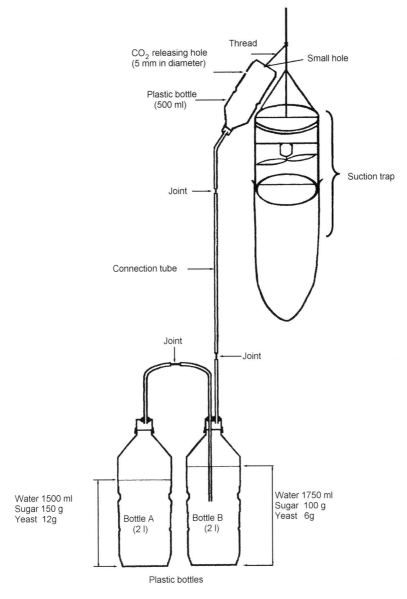

Fig. 10.1. Apparatus for supplying carbon dioxide from yeast fermentation of sugar to a suction trap (Saitoh et al. 2004)

The amount of dry ice produced is approximately half that of the weight of gas in the cylinder. This apparatus enables the rapid production of a number of uniform-sized blocks, which can conveniently be made in the field for immediate use. It overcomes the need to transport large blocks of dry ice to the field for breaking up into irregular lumps. Other equipment can make 800-g blocks (70 × 87 × 190 mm) of dry ice, a 50-lb cylinder can produce eight or nine such blocks. This useful piece of equipment would appear to have considerable potential in areas where supplies of dry ice are limited.

Dry ice can be wrapped in paper or metal foil (Bailey et al., 1965; Bellamy and Reeves 1952; Fraissignes et al. 1968; Ginsberg 1988; Nelson et al. 1978; Newhouse et al. 1966; Siverly and DeFoliart 1968), placed in perforated plastic bags or in bags with the openings closed with elastic bands to allow some control over the release of the gas (Herbert et al. 1972; Janousek and Kramer 1999; Service 1969), or in polystyrene or other insulated containers (Garcia et al. 1989; Jupp and McIntosh 1990; Landry and DeFoliart 1986; Pfuntner 1979; Pinkovsky and Sutton 1977; Service 1969).

Janousek and Kramer (1999) supplied dry ice to miniature CDC light-traps for a survey of mosquitoes in Nebraska, USA using the following method. A 1.2- to 1.4-kg piece of dry ice was placed in 3.8-litre cans lined with a layer of open cell foam 1.3 cm thick. The can and foam liner had 3-mm holes in the bottom. A non-insulated metal lid was placed on top of the can, and the light trap was suspended 10 cm below the can bottom. Another useful arrangement is to place about 4–5 kg of dry ice in a styrofoam ice chest with the lid sealed with tape. A length of 0.32-cm diameter polyethylene tubing from the chest allows carbon dioxide to be released into a suitable trap. Dry ice in such a chest can last for several days, and up to a week in cool temperate climates. Nevertheless, despite these arrangements there is relatively little control over the release rate of gas.

The warmer the weather, the faster the sublimation and the higher the local concentration of gas, which if excessive, may deter some mosquitoes from entering traps. Furthermore, host-seeking mosquitoes are generally attracted to warmth, whereas dry ice will usually lower the temperature in the vicinity of the trap. In fact Raymond (1977) using Manitoba traps for higher Diptera in France found that CO_2 produced by dry ice was much less effective than that from cylinders, probably because of the lower temperatures of the gas sublimating from the dry ice.

Because of the expense and difficulty sometimes encountered in getting dry ice or gas cylinders Hoy (1970) used a 4-cylinder engine adapted to operate on liquid propane gas. The carbon dioxide and carbon monoxide fumes generated by the engine were blown out by a squirrel-type fan fitted to the crankcase and carried to a Malaise trap. Trials with pure carbon monoxide showed that this did not cause a decrease in catch size of mosquitoes, so it was concluded that the production of this gas in addition to carbon dioxide did not have a deterrent effect. Shipp (1985) describes a portable and compact (1.5 kg) commercial generator that produces carbon dioxide, heat and moisture. It uses a platinum catalyst that emits gas at approximately 264–297 ml/min and operates from a 0.5-kg tank of propane; a 6 V-dry cell battery is used to ignite the catalyst. In trials in Canada this generator did not attract as many simuliids as did a 3.0–3.5-kg block of dry ice, but was nevertheless considered useful in remote areas where dry ice was unavailable. This principle has been adopted in the recently-developed commercial Counterflow Geometry Traps and the Mosquito Magnet (see below for a description of these traps).

The recent development of CO_2-producing granular sachets has the potential to make adding CO_2 to mosquito traps considerably easier. The sachets contain two different types of granules in separate compartments. The two compounds are mixed by breaking the separator and shaking the contents together. Carbon dioxide and water vapour are released for up to 24 h. Rapaport et al. (2005) evaluated the use of commercially available CO_2-producing granular sachets as a source of CO_2 in CDC light-traps in Illinois, USA. Traps baited with granular sachets caught significantly fewer individual mosquitoes (mean = 10.2 per trap, SE = 4.0) than traps baited with 2.3 kg of dry ice in padded envelopes (mean = 201.7, SE = 52.5). Traps baited with granular sachets also caught fewer species than dry-ice-baited traps. However, traps baited with granular sachets did collect significantly more individuals than un-baited CDC traps with the light removed. The CO_2 release rate of the sachets was determined using a bubble-flow meter to be about 43 ml/min initially, but slowed to around 10 ml/min at 40 min and 4.5 ml/min at 180 min. Combining sachets with octenol released from a wick did not significantly increase catch size.

CDC light-traps

Newhouse et al. (1966) suspended dry ice, usually 1–2 lb wrapped in newspaper, over CDC light-traps during the late afternoon, and emptied the traps the following morning. When smaller pieces of dry ice were used, and also on rainy nights, the dry ice was wrapped in aluminium foil to increase insulation. In trials in Florida, Georgia and North Carolina, USA catches from 72 light-trap-nights with dry ice were compared with those from 116 trap-nights without dry ice. In the three areas the catches of mosquitoes from carbon dioxide-baited light-traps were 4, 6 and 6 times more than from unbaited light-traps. In some species the increase in numbers was very marked, for example about 20 times more *Culex salinarius, Wyeomyia mitchellii, Ochlerotatus atlanticus* and *Ochlerotatus tormentor*, about 33 times more *Psorophora ferox* and 53 times more *Ochlerotatus canadensis* were caught. In addition to increasing the catch size dry ice also resulted in a 19–25% increase in the number of species in the traps. Another benefit of incorporating dry ice in light-traps was that their location was not so critical in ensuring a good catch. Moreover, by baiting traps several hours before dusk certain diurnal species, rarely caught in light-traps, were collected. Newhouse et al. (1966) considered that carbon dioxide helped to attract mosquitoes into the operating range of the trap, and that the fan provided a more efficient method of collecting mosquitoes than in more conventional carbon dioxide traps, such as lard-cans (Chapter 6), where mosquitoes have to make their own way into the traps.

In France, Fraissignes et al. (1968) suspended about 1 kg of dry ice wrapped in newspaper and aluminium foil above a CDC light-trap, which had the top cover removed and the plastic body made opaque so that light only came from the top of the trap. On a warm summer night it took about 13–15 h for the ice to completely sublime. It was thought that low temperatures produced locally by the dry ice and an excessive discharge of carbon dioxide might be repellent to mosquitoes. In field trials, however, large numbers of *Culex pipiens* and *Culex modestus* were caught, but very few *Anopheles hyrcanus* and relatively few *Ochlerotatus caspius*, although human bait catches showed them to be common in the area.

In mosquito surveillance programmes in Thailand CDC light-traps were unsatisfactory because they caught so few mosquitoes, but the catch was increased by about 100-fold when dry ice was incorporated in the traps. A direct comparison between the attractiveness of light and dry ice showed that carbon dioxide traps caught about 30 times as many mosquitoes as light-traps (Miller et al. 1969). In Malaysia Parsons et al. (1974) found that the addition of carbon dioxide to light-traps greatly increased the catch of mosquitoes.

In the Philippines CDC light-traps, operated for 17 nights, were compared with CDC traps without a light but with about 2.2 kg dry ice placed in a styrofoam container suspended near the trap and run for 14 nights (Pinkovsky and Sutton 1977). A 15-mm diameter plastic tube extending from the container delivered carbon dioxide to the trap. A total of 10 099 adults belonging to 51 species were collected. Of these 97.3%, comprising 47 species, were caught by the dry ice trap of which 22 species were caught only in these traps. CDC traps caught 29 species of which four species were caught only in these traps. By far the most common species in both traps were *Culex tritaeniorhynchus* (22.7%), *Culex bitaeniorhynchus* (20.9%) and the *Culex vishnui* group (20.1%). It was concluded that the best trap would probably use a combination of light and dry ice, as previously shown by Magnarelli (1975) in New York State, USA, where dry ice-baited CDC traps caught more mosquitoes than light-traps, but dry ice added to CDC traps with a light generally yielded the largest catches.

In Vietnam the addition of about 4–5 lb of dry ice in a perforated plastic bag increased the catch of mosquitoes in a CDC light-trap about 26 times (Herbert et al. 1972). *Culex* species, including *Culex tritaeniorhynchus* and *Culex quinquefasciatus* constituted the bulk of the catch. A few *Anopheles*, including *Anopheles peditaeniatus*, and *Aedes albopictus* and *Mansonia uniformis* were also caught. Surprisingly the catch of mosquitoes was almost doubled when the light was removed and the traps baited with only dry ice. In this instance it appeared that light repelled some mosquitoes.

The efficiency of CDC light-traps with and without carbon dioxide was reassessed in trials by Carestia and Horner (1968). Flow rates of 25, 50, 125, 250 and 500 ml/min from a gas cylinder were selected although in practice there were considerable variations between discharge rates due to difficulties of maintaining a constant flow. Again, all light-traps baited with carbon dioxide caught more mosquitoes than those without it, and generally traps using both light and gas caught both more individual mosquitoes and species than those just having carbon dioxide, in contrast to the findings of Reisen et al. (1983) and Magnarelli (1975). The numbers caught by Carestia and Horner (1968) usually increased with greater carbon dioxide levels, but it was concluded that in practice a minimum discharge of about 125 ml/min was necessary to produce a significant increase in catch size.

Frommer et al. (1976) found that catches of simuliids (*Cnephia mutata* and *Prosimulium hirtipes*) increased in CDC traps (minus light) when carbon dioxide flow rate increased from 50 to 500 ml/min, but increases above this were wasteful as they did not necessarily increase the catch. There was, however, considerable variation and overlap between catches at different rates.

In Senegal Cornet and Chateau (1971) released carbon dioxide at the rate of 2 litres/min into a CDC light-trap. Many *Aedes* species, especially those of the subgenus *Aedimorphus*, were caught, but others such as *Aedes taylori, Aedes furcifer, Aedes vittatus* and *Aedes luteocephalus* were less common. Most known or suspected yellow fever vectors were caught only in small numbers, but a reduction in the flow rate of the gas might have increased the catch of these species.

Large numbers of mosquitoes can be caught in CDC-type dry ice traps. For example, in 1971 LeDuc et al. (1975) processed for virus isolation 77 000 female mosquitoes caught in CDC light-traps baited with dry ice, and over 106 000 in 1972. A total of 13 species were collected by the traps, the more common being *Ochlerotatus canadensis, Ochlerotatus cantator, Ochlerotatus atlanticus, Coquillettidia perturbans, Psorophora ferox* and

Ochlerotatus sollicitans. Hayes et al. (1976) supplemented CDC traps with 2 kg of dry ice in six states in the USA, and over a year caught 173 074 mosquitoes belonging to 41 species. Light-traps baited with dry ice caught relatively large numbers of *Culiseta melanura, Culiseta morsitans, Ochlerotatus canadensis* and *Coquillettidia perturbans* in upstate New York, USA (Howard et al. 1988); the mean number of females/night in two traps varied from 219.5–805.5. Also in the USA as many as 14 803 male and 104 500 female *Culex tarsalis* were caught over 48 trap-days with CDC light-traps supplemented with a 1.8-kg block of dry ice wrapped in white paper and suspended 30 cm from the trap. In northern Wisconsin, USA Siverly and DeFoliart (1968) suspended about 5 lb of dry ice wrapped in thick paper a few feet from, but on a level with, the hood of a light-trap. Some 21 790 mosquitoes belonging to 26 species, the most abundant being *Ochlerotatus communis, Ochlerotatus punctor, Ochlerotatus canadensis, Aedes vexans* and *Coquillettidia perturbans*, were recovered from the traps. In Ecuador Calisher et al. (1981) caught more than 770 000 mosquitoes in carbon dioxide-baited CDC traps and 44 646 from similar traps in Argentina.

Merdić and Boca (2004) used CO_2-baited CDC light-traps to sample *Anopheles maculipennis* complex in Croatia, and trapped a total of 3508 specimens during an eight-year study.

In New Jersey, USA Ginsberg (1988) found that overnight survival of mosquitoes in a CDC light-trap having dry ice wrapped in newspaper hung alongside, was greatest in wooded areas than when traps were placed in residential areas. However, survival rates varied greatly from day to day and appeared to be positively correlated with nightly minimum temperatures, but not with wind speed.

Crans and Sprenger (1996) used CDC light-traps baited with 3 kg of dry ice and with the light source removed to determine the habitat preference, vertical distribution and diel periodicity of host-seeking *Ochlerotatus sollicitans* in New Jersey, USA as part of studies to investigate the reasons for the relatively low prevalence of blood meals originating from birds in this species. Vertical distribution was determined by suspending traps from tree branches at heights of 0.5, 2.0 and 6.0 m along the boundary between an open field and a deciduous woodland. Mosquitoes were collected from the traps every hour over a 25-hour period. The first hour's catch was excluded from the analysis to ensure that the effects of disturbance and initial attraction of resting mosquitoes was discounted. A total of 8875 individual *Ochlerotatus sollicitans* was collected from the traps set at different heights. The traps set at a height of 0.5 m caught 81.9% of the total catch ($P < 0.001$).

Larval dipping and light-trapping with CDC miniature light-traps supplemented with CO_2 were used to study the elevational distribution of mosquitoes in Wyoming, USA, by Denke et al. (1996). Light-traps were supplied with CO_2 from a 2 kg block of dry ice and set up between 1930 and 2100 h. A sub-sampling methodology was used on the large catches obtained in traps operated in river valleys at low altitudes, as follows: the catch was spread onto a painted grid and all unusual specimens removed for identification. A further 35 individuals were removed from each sample for identification. In general, species diversity of catches decreased with increasing elevation, although the site at the lowest elevation did not yield the greatest number of species. The majority of specimens was collected from elevations between 2134 and 2591 m, and were predominantly species associated with flood pool habitats (*Ochlerotatus campestris, Ochlerotatus dorsalis, Ochlerotatus flavescens*, and *Ochlerotatus melanimon*). Two species were found predominantly at higher elevations, above 3200 m: *Ochlerotatus impiger* and *Ochlerotatus punctor*.

Reisen et al. (1990*b*) caught considerably more *Culex tarsalis* and *Culex stigmatosoma* in CDC carbon dioxide traps hung at 5 m in trees than in traps at ground level (1.5–2.0 m), whereas *Culiseta incidens* was most common in the ground level traps. It is worth noting that both *Culex* species fed mainly on birds, whereas *Culiseta incidens* fed primarily on dogs.

Carbon-dioxide-baited CDC traps with the light removed were also used to determine the flying height of *Ochlerotatus caspius* and *Culex pipiens* in Italy (Bellini et al. 1997). Traps were set at heights of 1.5, 3, 4, 5, 6, and 7 m in a wooded area and at 1.5, 3, 4, and 5 m in an open area. Traps were baited with 500 g of dry ice releasing approximately 200–300 ml/min of CO_2. Traps were operated one night per week from 1 h before sunset until the following morning. *Ochlerotatus caspius* was collected predominantly at a height of 1.5 m at both wooded (577/631) and open sites (2954/3486). Catches of *Culex pipiens* were more evenly distributed across trap heights, with 46.9% collected at 1.5 m and 10.5% at 7 m in the wooded area.

During investigations to identify the vectors of *Wuchereria kalimantani* in east Kalimantan, Indonesia, Atmosoedjono et al. (1993) used CO_2-baited CDC light-traps positioned at heights of 2, 5 and 10 m in forest. Thirteen species from five genera were caught, with *Anopheles balabacensis* and *Culex quinquefasciatus* the most commonly collected species, each representing about 21% of the total catch. *Anopheles balabacensis* were captured at each of the three trap heights.

In Quebec, Canada CDC traps baited with 2 kg dry ice wrapped in newspaper, which was torn to facilitate the sublimation of gas, were operated twice weekly in six locations from May to October and trapped 61 712 female and 2098 male mosquitoes belonging to 26 species in seven genera. The most common were females of *Coquillettidia perturbans* (59.1%), *Ochlerotatus aurifer* (32.0%), and *Anopheles walkeri* (4.2%). Allan et al. (1981) also caught *Coquillettidia perturbans* in similar traps in Canada.

In southern Africa Jupp & McIntosh (1990) employed CDC-type light-traps (Jupp et al., 1980) incorporating a tin insulated with corrugated cardboard containing about 2 kg dry ice. A 3-mm bore plastic tube delivered the gas to the entrance of the trap. From 63 trap-nights 3590 mosquitoes belonging to at least 17 species were collected. The most common mosquitoes were 2498 various *Anopheles* species, 334 *Culex poicilipes*, 265 male *Aedes furcifer*, 174 female *Aedes furcifer/cordellieri*, and 149 assorted *Culex* species.

Maciá (1997) used CO_2-baited CDC light-traps (with the light retained) to study the physiological age structure of female mosquito populations in the Buenos Aires area of Argentina. A total of 4466 individuals comprising seven species were dissected and classified as nulliparous (71%), parous (27%) or engorged, half-gravid, and gravid (2%). All 86 individuals of *Runchomyia paranensis* dissected were parous, indicating that this species may be autogenous. Carbon dioxide-baited CDC light-traps have also been used to successfully sample mosquitoes in the Iquitos area of Peru (Need et al. 1993).

Ludueña Almeida and Gorla (1995) compared human landing catches and CO_2-baited CDC light-traps to investigate the daily flight activity of *Ochlerotatus albifasciatus* in Argentina. Density estimates derived from the two sampling methods were highly correlated ($R^2 = 74.9\%$, $P = 0.001$, $n = 22$), but the proportion of the total catch comprised of *Ochlerotatus albifasciatus* was greater in human landing catches (0.75) than in CDC light-traps (0.32).

Also in Argentina, Gleiser et al. (2000, 2002) used CDC light-traps baited with CO_2 to characterise the spatial pattern of abundance of *Ochlerotatus albifasciatus*. In the earlier study, seven traps were located along an 80 km transect and were operated from 1900 to 0700 h. Carbon dioxide was supplied from a 500 g block of dry ice wrapped in cardboard paper. Spatial patterns of abundance were studied using three methods. In the first method, mosquito abundance data from one site (i) was correlated with that from the next site ($i + d$, where d = distance between the paired sites). The second method used a semivariogram to plot semivariance of abundance against distance between sites (Jongman et al. 1995). The third method used was principal component analysis. The semivariogram indicated that changes in mosquito abundance at inter-site distances of 10 or more km were not spatially related. Principal component analysis revealed three groups of trap sites associated with distance to larval habitats and rainfall. In the later study (Gleiser et al., 2002) abundance was described in relation to surface water dynamics and vegetation coverage, obtained from a Landsat image. Densities of mosquitoes were also linked to cattle densities, elevation, and rainfall. A significant model that explained 49% of the variation in spatial abundance of this species was obtained using the variables of *Salicornia* and *Sesuvium* prairie coverage, distance to woodland and slope of land.

Jensen et al. (1993) used three inverted updraft CDC traps baited with 2 kg of dry ice to sample mosquito populations over 18 consecutive 24 h periods in a central Florida, USA swamp. Updraft CDC traps were considered to be more efficient at sampling *Anopheles* than the standard CDC light-traps. A total of 3818 female and 48 male *Anopheles quadrimaculatus* s.l. were collected (identified by DNA probe or electrophoresis as either *Anopheles quadrimaculatus* or *Anopheles diluvialis*) as well as 1900 female and 41 male *Anopheles crucians*, and 12 female *Anopheles perplexens*.

In their studies in California, USA on *Culex tarsalis, Culex quinquefasciatus* and *Ochlerotatus nigromaculis* Meyer et al. (1991) stressed that the placement of CDC-type carbon dioxide traps in the right type of vegetation where hosts, small mammals and birds in this instance, lived was important in maximising catches. Meyer (1991) also reported that these traps were inefficient in collecting *Culex quinquefasciatus* in urban areas (1.4–3.1 females/trap per night) compared to rural areas (31.8–111.2 females/trap per night). In fact catch size was inversely correlated with housing density.

Use of CO_2-baited CDC light-traps in arbovirus surveillance

Light-traps supplemented with carbon dioxide have been most widely used in the USA, especially in arbovirus surveillance programmes. CDC light-traps supplemented with CO_2 and with the light removed are routinely used to monitor *Culex tarsalis* populations in relation to arbovirus activity in California, USA (e.g. Reisen et al. 1995a,b; 1996). Carbon dioxide-baited CDC traps and gravid traps are used for mosquito surveillance in relation to transmission of West Nile (WNV), St. Louis (SLV) and western equine encephalomyelitis (WEE) viruses in California, USA and a description of their use in tracking WNV is given in Reisen et al. (2004). In rural southeastern California, WNV was reportedly tracked best by testing pools of *Culex tarsalis* collected by CO_2 traps and by monitoring sentinel chicken sera. Few mosquitoes were collected by either CO_2 or gravid traps in urban or peri-urban areas of Los Angeles. In urban neighbourhoods generally, gravid traps tend to perform better than CO_2 traps and other methods. Takeda et al. (2003) describe the results obtained from the use of CDC light-traps baited with CO_2 in eastern equine encephalomyelitis surveillance in Rhode Island, USA, between 1995 and 2000. Over the six-year period, a total of 185 537 mosquitoes were collected, yielding 193 arbovirus positive pools (3%), including 71 positive for EEE. *Coquillettidia perturbans* was the most frequently collected species and comprised 21% of the total. Wozniak et al. (2001) describe the use of CO_2-baited CDC light-traps for surveillance of eastern equine encephalitis virus, St. Louis encephalitis virus, Flanders virus, Tensaw virus, and a variant of Jamestown Canyon virus in South Carolina, USA. More than 85 000 mosquitoes representing 34 taxonomic units were collected over 1431 trap-nights. Thirty-seven of 300 mosquito pools tested positive for arboviruses.

Over 236 trap-nights in 1987 13 099 mosquitoes were collected in a western equine encephalitis surveillance programme in northern Colorado, USA (Smith et al. 1993), using CDC light-traps. Species collected included *Culex tarsalis, Culex pipiens, Aedes vexans, Aedes cinereus, Ochlerotatus dorsalis, Ochlerotatus melanimon, Ochlerotatus nigromaculis, Ochlerotatus sticticus, Ochlerotatus trivittatus, Ochlerotatus hendersoni, Anopheles earlei,* and *Culiseta inornata.* In 1991, 13 493 mosquitoes were collected over 90 trap-nights (Smith et al. 1993).

In eastern equine encephalomyelitis virus surveillance studies in Ohio, USA in 1991 Nasci et al. (1993) collected 22 095 mosquitoes of 17 species including *Coquillettidia perturbans* (43%), *Aedes vexans* (22%) and *Culex salinarius* (18%).

In arbovirus studies in northern Thailand Leake et al. (1986) caught 353 042 mosquitoes belonging to 59 species from 115 CDC light-trap collections made near houses. The most common species were *Culex tritaeniorhynchus, Culex vishnui, Culex gelidus* and *Culex fuscocephala*, which together formed 97.6% of the total collection. Bloodfed mosquitoes formed only about 1–8% of the catches. There were 623 collections from carbon dioxide-baited traps and 52 from traps without carbon dioxide, the mean catches were about double (1.9–2.2 ×) when carbon dioxide was incorporated. Interestingly 53 virus isolations were made from adults caught in traps emitting carbon dioxide, but only 10 isolations from those without gas.

As part of investigations on Ockelbo virus, a north European sub-type of Sindbis virus, Lundström et al. (1996) used dry-ice baited miniature CDC light-traps to study the mosquito fauna in southern and central Sweden. CDC light-traps were suspended at various heights in the tree canopy of a deciduous forest and were supplemented with CO_2 released from a 1–2 kg block of dry ice kept in an insulated envelope and suspended alongside the light-trap. Traps were placed in trees using a modified slingshot method

(Novak et al. 1981) in which a 90 g lead weight attached to a length of nylon monofilament line was fired over a suitable branch. A rope was attached to the other end of the filament line and the trap and CO_2 were raised into the canopy (Fig. 10.2).

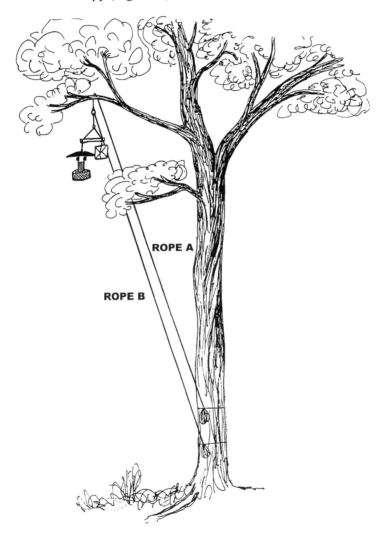

Fig. 10.2. the method used by Lundström et al. (1996) to raise and lower a CDC miniature light-trap and an envelope containing dry ice into the tree canopy

A total of 15 186 mosquitoes were collected over 43 trap nights at two study sites using this method, including 93 *Culiseta morsitans* and 182 *Culex pipiens/torrentium* (the two species were not distinguished). *Culiseta morsitans* and *Culex pipiens/torrentium* were most commonly collected in the canopy at heights of 12 to 15.5 m at the Norra Åssum site and at 14 to 18 m at the Tärnsjö site. *Aedes cinereus* was predominantly collected at ground level (1.5 m) or at heights of 6 to 9 m.

In Senegal, Diallo et al. (2000) used CDC light-traps baited with CO_2 adjacent to potential larval habitats and CDC light-traps without CO_2 in sheep-folds or cow sheds. In total, 31 944 mosquitoes belonging to 6 genera and 20 species were captured. *Culex poicilipes* and *Mansonia uniformis* were the most commonly collected species and the presence of RVF virus was demonstrated in *Culex poicilpes* for the first time.

Moncayo and Edman (1999) and Moncayo et al. (2000) used ABC traps (CDC type light-traps) equipped with a photosensitive flickering light and compressed CO_2 delivered from a storage tank at a continuous flow rate of 500 ml/min to sample putative vectors of eastern equine encephalomyelitis virus in Massachusetts, USA. The CO_2 flow rate selected was intended to mimic the average CO_2 discharge from an adult human. Two ABC traps were placed at each of the 15 sites from where human or animal cases had been reported and were operated on two consecutive nights each week from mid-July to mid-September 1996. Moncayo and Edman (1999) used a relatively complex, two-stage ranking method to determine which species would be the most likely vector(s) on the basis of abundance. Species were ranked from most to least abundant at each site; a score of 1 was given to the most abundant and a score of up to 6 for the least abundant. If less than 6 species were present at a site, rankings would range from 1 to the number of species present. Absent species received a score of 7. Once the relative order of abundance was determined for the 6 species, another ranking scheme based on quartiles was used, in which the range from the minimum to the maximum score from the first scheme was divided into four quartiles, with each quartile assigned a score of 0, 1, 2, or 3 respectively. Abundance of between 15 and 37.5 was assigned a score of 0 and abundance of between 82.5 and 105 was assigned a score of 3. The relative abundances of the 6 potential vectors varied with trap site. In general, the most abundant of the 6 species was *Ochlerotatus canadensis*, followed by *Coquillettidia perturbans, Culex salinarius, Aedes vexans, Anopheles.punctipennis*, and *Anopheles quadrimaculatus*. The abundance of *Ochlerotatus canadensis* was skewed by the enormous population present at one site.

Hribar et al. (2003) also used ABC traps supplemented with CO_2 and gravid traps to monitor populations of suspected West Nile virus in Florida, USA. Over 30 000 mosquitoes comprising 22 taxa were collected during a three month period, including virus positive *Anopheles atropos, Deinocerites cancer*, and *Ochlerotatus taeniorhynchus*. All positive mosquito pools were captured in CO_2-baited traps. More mosquitoes were collected in ABC traps than in gravid traps for all species analysed ($t = 4.21$, d.f. = 162, $P = 0.00004$).

Comparisons of CO_2-baited CDC light-traps and other traps

In trials in New Jersey, USA comparing a New Jersey light-trap, a CDC trap supplemented with dry ice, and human bait catches, Slaff et al. (1983) found that most species, such as *Ochlerotatus abserratus, Ochlerotatus canadensis, Ochlerotatus excrucians, Ochlerotatus trivittatus, Aedes vexans, Anopheles quadrimaculatus, Anopheles punctipennis* and *Coquillettidia perturbans*, were caught in the CO_2-CDC trap, and generally there was a good agreement with numbers and species of mosquitoes caught at human bait. For example, Horn's (1966) similarity index calculated for different months varied from 0.58–0.92 for the comparison of the CO_2-CDC trap and biting counts. In contrast indices for the New Jersey light-trap and biting counts were 0 (complete dissimilarity) to 0.65, showing that the New Jersey trap was a poor indicator of nuisance mosquitoes.

Reisen et al. (1983) caught significantly more female *Culex tarsalis* in CDC traps baited with 1–2 kg dry ice when the bulb was removed than in traps retaining the bulb, and more than in either large ($2 \times 1 \times 2$ m) walk-in red boxes or standard ($0.3 \times 0.3 \times 0.3$ m) red boxes. In Vietnam Herbert et al. (1972) found that catches were almost doubled when light was removed from their traps.

A single CDC light-trap baited with CO_2 was used alongside human landing catches to determine the infection rate of mosquitoes with the filarial worm *Dirofilaria immitis* in North Carolina, USA (Parker, 1993). Over 28 trapping occasions between June 1985 and August 1987 a total of 70 786 adult females were collected comprising four genera and eleven species. Of these, 97.1% were collected using the CO_2-baited light-trap. *Anopheles bradleyi* was the commonest species collected (66.6%), followed by *Culex salinarius* (15.5%), *Ochlerotatus taeniorhynchus* (8.1%) and *Ochlerotatus sollicitans* (7.4%).

Reisen et al. (2000) compared CDC-type light-traps and the EVS trap (Pfuntner 1979; Rohe and Fall 1979) for their suitability for *Culex* surveillance in California, USA. Four trap designs and three CO_2 presentation methods were evaluated. The four traps used were an Arbovirus Field Station (AFS) trap, an Orange County Vector Control District trap, an American Biophysics Corporation (ABC) trap and a model 1012 miniature CDC light-trap (John W. Hock Company). In the first two experiments CO_2 was supplied to each of the four traps in the form of a 1.5 kg block of dry ice, or alternatively 1.5 kg of pelletised dry ice. In the third experiment, CO_2

was supplied to AFS traps either from a 1.5 kg block of dry ice, as a constant flow of CO_2 gas from a cylinder, or alternatively pulsed flow from a CO_2 cylinder. In the constant flow experiment CO_2 was released at rates of 0.5, 1.0 and 1.5 l/min and in the pulsed release experiment a flow of 0.5 l/min was released 15 times per minute to simulate respiration of a host animal. In both experiments 1 and 2 the AFS trap without a light source collected significantly more *Culex tarsalis* than the other three trap types. The CDC and AFS traps were the most efficient at sampling *Culex erythrothorax*. In the third experiment, 95% of the catch comprised *Culex tarsalis* and the highest numbers were caught in the devices releasing the most CO_2. Constant flow CO_2 traps caught more mosquitoes than dry ice-baited traps, and both caught more than the traps operated with pulsed release of CO_2. No significant difference was observed in the catches when dry ice was kept in a Styrofoam container or a grey plastic bucket with a lid, even though the release rate for the grey plastic bucket holder was higher at 0.5 l/min than the Styrofoam container at 0.4 l/min. It was concluded that CDC type traps without a light source were effective at collecting *Culex* mosquitoes and were preferable to traps with a light as they collected far fewer chironomids. Pulsed release of CO_2 was not found to be an effective trapping method

Comparisons of different CO_2-baited traps and other trapping methods were also undertaken by Gingrich and Casillas (2004) for West Nile virus surveillance in Delaware, USA. The collection methods compared were: human-landing catch, CO_2-baited commercial CDC light-trap, CO_2-baited omnidirectional Fay trap, and infusion-baited gravid traps. Abundance was calculated as the number of mosquitoes collected per person or per trap per collection period. A total of 29 species of mosquito were collected. Two suspected West Nile virus vectors (*Aedes vexans* and *Culex pipiens pipiens*) were most abundant in light trap collections. The third suspected vector (*Culex salinarius*) was captured using both light-traps and human landing catches. Omnidirectional Fay traps caught fewer mosquitoes than human landing collections or CDC light-traps, however, these traps did collect slightly higher numbers of *Aedes albopictus, Anopheles crucians, Culiseta melanura, Ochlerotatus mitchellae, Culex restuans,* and *Psorophora columbiae*. Mean abundance estimates obtained from light-traps tended to exhibit higher standard deviations than human-landing catches, indicating much higher variability between catching sites and occasions. The overall proportion parous was highest in human landing catches on 19 of 31 trapping occasions and highest in CDC traps on 9 of 31 occasions.

Jones et al. (2004) also used both human landing catches and CO_2-baited CDC light-traps to sample mosquitoes in a forest in Iquitos, Peru. A total of seventy species from 14 genera were collected using the two methods.

During daylight hours, 62% of the total mosquitoes collected were from human landing catches, whereas during the night 68% of the total was collected in light-traps. Several species (*Johnbelkinia longipes, Haemagogus janthinomys, Haemagogus baresi, Sabethes chloropterus, Ochlerotatus argyrothorax, Mansonia amazonensis,* and *Mansonia humeralis*) were collected almost exclusively in human landing catches.

Jupp et al. (2002) used CO_2-baited ABC suction traps (CDC type light-traps manufactured by American Biophysics Corporation) with or without a light source, alongside human landing catches and a CO_2-baited tent trap to carry out entomological investigations following an outbreak of Rift Valley fever in Saudi Arabia in September 2000. The ABC traps were operated from 1800 to 2030 h inside and outside of a hut used to house sheep during the night. The light traps caught as many as 800 mosquitoes per hour at a flooded agricultural site and a total of 74 500 mosquitoes were collected using all methods. The three main species collected were *Culex tritaeniorhynchus, Aedes vexans arabiensis* and *Ochlerotatus caballus.*

In the Belem area of Antalya province, Turkey, Alten et al. (2000) used CDC light-traps baited with 0.7 kg dry ice alongside New Jersey light traps with 40 W bulbs to study mosquito population dynamics and seasonality following the introduction of integrated vector control interventions. One trap of each type was placed 200 m apart at each of four sampling sites representing different ecosystems. *Culex pipiens* was the most frequently collected species and was caught in approximately equal numbers in CO_2-baited traps and NJ light-traps. *Culex tritaeniorhynchus* and *Ochlerotatus caspius* were more common in the New Jersey light-traps.

For collecting *Culex* mosquitoes Reisen and Pfuntner (1987) compared the efficiencies of a New Jersey light-trap with a 25-W bulb, a CDC-type trap without a light but baited with about 1 kg dry ice (Pfuntner 1979), a battery-powered trap fitted with a high intensity 12-V car tail light bulb (Pfuntner 1979), walk-in red resting boxes (2 × 2 × 1.3 m) fitted with a screened door (Meyer 1985; Nelson 1966), and a gravid trap of Reiter (1983). *Culex quinquefasciatus* formed 94.9% of the total catch of 23 159 adults, *Culex stigmatosoma* 4.2% and *Culex tarsalis* 0.9%. Female mosquitoes were collected in largest numbers by the carbon dioxide trap (84.9%), followed by red boxes (9.7%), the New Jersey light-trap (2.0%), the high intensity light-trap (1.8%) and the gravid trap (1.2%). In rural situations the red boxes were the most effective method of catching male mosquitoes. In later trials in California Reisen et al. (1990*a*) collected *Culex tarsalis, Culex quinquefasciatus, Culex stigmatosoma, Culiseta incidens* and *Culiseta inornata* in that order of abundance, in the carbon dioxide traps of Pfuntner (1979).

In Uganda Henderson et al. (1972) caught large numbers of ornithophagic *Coquillettidia* mosquitoes with CDC traps incorporating dry ice, whereas very few of them were caught in human landing catches. This trap also proved useful for collecting various species of *Culex, Ficalbia* and *Uranotaenia*, but was ineffective in collecting *Aedes* and *Hodgesia* species.

Pfuntner traps baited with CO_2

Because of the poor response of *Ochlerotatus triseriatus* to most animal-bait traps and light-traps, as well as other sampling problems, Landry and DeFoliart (1986) tested a trap (Fig. 10.3) based on the CO_2-baited light-trap of Pfuntner (1979) (Fig. 9.11) but with the trap body painted matt black. When traps were operated during the day dry ice attracted more *Ochlerotatus triseriatus* (0.09–11.43/trap-day) than when traps were baited with a white mouse (0.04–1.45/trap-day).

In conclusion they proposed that dry ice suspended in a styrofoam box directly over the trap intake would provide a suitable system for monitoring adult biting activity. In later studies they removed the plastic box containing dry ice, lowered the trap lid, and delivered carbon dioxide gas at approximately 1 litre/min. Gas cylinders were also used by Pfuntner et al. (1988) in comparing catches of *Culex tarsalis, Culex quinquefasciatus* and *Culex stigmatosoma* in traps (Pfuntner 1979) placed at different heights (2, 5, 10 m) and receiving varying amounts of carbon dioxide (250, 500, 1000 ml/min). Only with *Culex tarsalis* did the numbers increase with increasing discharge rates. In rural areas catches of the three species did not differ significantly at different heights, but in urban locations there was some evidence to suggest that catches were larger at 5 m than at 2 m.

Encephalitis Virus Surveillance (EVS) traps baited with CO_2

Encephalitis Virus Surveillance traps of Rohe and Fall (1979) baited with dry ice are routinely used for arbovirus surveillance in Australia. In New

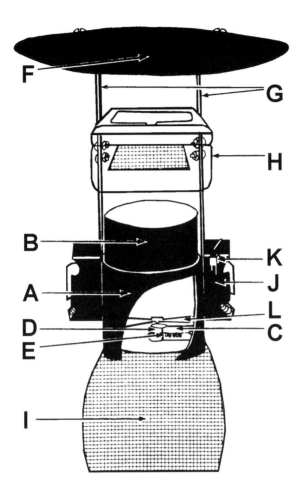

Fig. 10.3. Modified Pfuntner trap, A—black trap body, B—sewage pipe insert, C—motor-mount, D—motor, E—spring to hold motor, F—trap lid, G—lid supports, H—plastic crisper box, I—collection bag, J—battery holder, K—toggle switch, L—propeller (Landry and DeFoliart 1986)

South Wales at least two traps are operated at each of approximately 30 locations weekly from November to April each year (Russell 1998).

In Australia Russell (1985) found that EVS (Fig. 9.9) traps baited with 500 g of dry ice, caught substantially more mosquitoes than EVS traps containing a chicken, rabbit or a guinea pig, than in EVS traps without either dry ice or animals, and in CDC light-traps. The CDC trap occasionally caught male *Culex annulirostris*, but the other traps never trapped males. Moreover, a single EVS trap caught 4–8 times as many *Culex annulirostris* as did five 1-ft cube resting boxes (Goodwin 1942). In other experiments Russell (1986a) set dry ice-baited EVS traps 1–2 m above the ground and caught 12 species, including *Culex annulirostris, Culex australicus, Ochlerotatus sagax, Ochlerotatus bancroftianus, Ochlerotatus theobaldi, Ochlerotatus alboannulatus, Anopheles annulipes*, and more rarely *Coquillettidia linealis*. In other localities Russell and Whelan (1986) caught 24 different species in these traps, the most common being *Culex annulirostris, Ochlerotatus vigilax, Anopheles bancroftii, Anopheles farauti* s.l., and *Anopheles hilli*. In different trials Russell (1986c) used EVS light-traps baited with dry ice and CDC light-traps without any dry ice. Only *Culex australicus* was more common in the CDC traps (20) than the dry ice traps (4), and whereas the dry ice traps caught 628 mosquitoes the light-traps caught only 141 mosquitoes. Russell (1986b) concluded from this, and other field trials, that dry ice traps were much superior to unbaited light-traps.

Kay et al. (1996) used six EVS traps supplemented with CO_2 to sample mosquitoes at the Ross River reservoir in Queensland, Australia. Traps had a photo-resistor fitted so that they automatically switched on and off at dusk and dawn respectively. The mean number of mosquitoes captured per light trap was 72.4, 72.6, 116.5, and 191 for the four study sites.

Cooper et al. (1996) used CO_2-baited EVS light-traps to investigate the distribution of anopheline mosquitoes in northern Australia.

Fay-Prince carbon-dioxide trap

In California, USA Garcia et al. (1989) compared the bait-carbon dioxide trap of Garcia et al. (1988) containing a rabbit, with the visual attraction trap (Fig. 11.2) of Fay and Prince (1970) having 1 litre/min of carbon dioxide discharging above the intake fan, for catching *Ochlerotatus sierrensis*. In another series of experiments catches in a Fay–Prince trap having 2.5 kg of dry ice placed in a styrofoam box fixed on top of the trap and with a 2.5-cm hole made for escape of the gas, were compared with catches in similar traps without dry ice. There was no statistical difference between the mean numbers of female *Ochlerotatus sierrensis* caught in Fay–Prince traps with carbon dioxide gas (11.2–48.5) and rabbit-baited carbon dioxide traps (23.5–63.8), but 15–20 times more males were attracted to the Fay–Prince traps. Fay–Prince traps with dry ice caught a mean of 33.8 ± 39.6 male and 21.3 ± 14.3 female *Ochlerotatus sierrensis*, compared to just 0.7 ± 1.2 males and 0.2 ± 0.4 females in traps without dry ice. It was concluded that a Fay–Prince trap supplemented with dry ice or carbon dioxide was an efficient method for sampling both sexes of *Ochlerotatus sierrensis*, but without carbon dioxide the trap was inefficient for this species. Also in California, Woodward et al. (1998) found that while CO_2-baited Fay-Prince traps were effective at sampling *Ochlerotatus sierrensis*, they did not sample adults of *Orthopodomyia signifera.*

Washburn et al. (1992) compared catches of *Ochlerotatus sierrensis* from CO_2-baited Fay-Prince traps, duplex cone traps (Freier and Francy, 1991), and human landing catches using an AFS suction sweeper (Meyer et al. 1983). At one site, two Fay-Prince and two duplex cone traps, plus one human landing catch were operated. At two other forested sites, three Fay-Prince and two duplex cone traps plus a human landing catch were set up. Traps were baited with 1 kg of dry ice and operated between 1700 and 2100 h for a total of 19 evenings. Human landing catches were made by two people wearing white clothing and seated at the catching station between 1930 and 2030 h. Catch data from the traps ($n + 1$) were plotted against landing catch ($n + 1$) on a log-log scale for each of the nineteen nights. Significant correlations ($P < 0.001$) were obtained between numbers of females collected by landing catch and in the two traps, with values of R^2 equal to 88% for the Fay-Prince traps and 57% for the duplex cone traps. The *y* intercept for correlations between females caught in Fay-Prince traps and landing catches did not differ significantly from zero, indicating their potential usefulness as replacements for human landing catches. For males, a significant correlation was obtained for the landing catch and duplex cone traps, but R^2 in this case was only 27%. No significant correlation was observed with the landing catches for males collected using the Fay-Prince traps. Overall, the Fay-Prince traps caught significantly more of both sexes than the duplex cone traps ($t = 2.738$, d.f. = 18, $P < 0.001$ for females and $t = 3.887$, d.f. = 18, $P < 0.001$ for males). The Fay-Prince trap was also more efficient when operated against the duplex cone trap in paired tests ($t = 4.348$, d.f. = 11, $P < 0.001$ for females).

Jones et al. (2003) also evaluated the efficiency of the omnidirectional Fay-Prince trap (Fig. 11.2), alongside the Wilton and Kloter (1985) trap and sticky lures in comparison with human landing catches as methods for capturing adult *Aedes aegypti* in Thailand. The omnidirectional Fay-Prince trap consists of two standard Fay-Prince traps mounted back to back.

Carbon dioxide was supplied to both the Fay-Prince and Wilton and Kloter traps from a 1 kg block of dry ice in a plastic container. The sticky lure consisted of an adhesive material applied to a red coloured cardboard background, with the addition of a proprietary mosquito attractant. Each of the collection methods were used daily in four houses over 16 days, using a Latin square experimental design. Traps were operated from 0800 to 1700 h. Human landing collections were made for 50 minutes per hour. A total of 1272 adult *Aedes aegypti* was collected over two 16-day trials. No other mosquito species was collected. All three traps collected significantly fewer *Aedes aegypti* than the human landing catch, but there were no significant differences in mean daily catches among the three traps tested. Human landing catches were estimated to be approximately 10 times more effective at collecting *Aedes aegypti* than either the Fay-Prince or Wilton and Kloter traps, with a mean of 33.5, 3.4. and 2.9 mosquitoes per trap per day, respectively. The sticky lure failed to capture any *Aedes aegypti*.

Woodward et al. (2003) studied the seasonal variability of *Ochlerotatus sierrensis* populations in northern California, USA using Fay-Prince CO_2 traps and ovitraps. Fay-Prince traps baited with 3.2 kg of dry ice were hung from trees and operated 1.25 m above ground level at the centre of each of a series of 100 m transects. Eight ovitraps (Chapter 2) were placed on the ground on the north side of trees and spaced approximately 25 m apart. Ovitraps were operated continuously at the same locations (4 on each transect) as the other traps and over a similar period. Yearly mean capture dates of females in the Fay-Prince traps occurred 21-28 days after males. Eggs were recovered from ovitraps 2 weeks after the first collections of host-seeking females in both 1991 and 1996.

The effects of adding octenol to Fay-Prince traps on catches of *Aedes albopictus* was studied by Shone et al. (2003) in 12 locations in Maryland, USA. Each trap was suspended at a height of approximately 0.9 m above ground level and supplemented with octenol, CO_2, octenol + CO_2, or left unbaited. Traps were set each day between 1100 and 1400 h and operated for 24 h. Carbon dioxide was released from a 1.9-litre plastic beverage cooler suspended directly above the trap and containing approximately 1.4 kg of dry ice. The approximate release rate of CO_2 was 53 g/h. Three millilitres of octenol was supplied from a 5 ml microreaction vial attached 5 cm below the trap entrance and fitted with a plastic cap and neoprene septum with a 1.5 mm hole bored through it. A pipe-cleaner wick was folded over and inserted into the vial so that approximately 1.5 cm protruded. The estimated release rate of octenol was 44 mg/h. Over 144 trap nights a total of 1121 female *Aedes albopictus* was collected. The overall ranking of response to each bait was as follows: CO_2 + octenol > CO_2 > no bait > octenol. However, the response to CO_2 + octenol did not differ significantly from the response to CO_2 alone.

Counter-flow geometry (CFG) traps

Kline (1999) evaluated a counter-flow geometry (CFG) trap developed by the commercial company American Biophysics Corporation (East Greenwich, Rhode Island, USA) and compared it with a modified CDC-type light-trap developed by the same company (ABC Standard PRO trap) in both outdoor cages and in field studies. The counter-flow geometry trap operates on the principle of producing two air streams flowing in opposite directions. A downward flow produces a plume of carbon dioxide or other attractants from the bottom of the trap. Mosquitoes are attracted to the CO_2/odour plume and are caught in an updraft generated by a second fan. The original trap was constructed from a clear PVC pretzel container (ca. 11.4 litre), litre), modified by removing the bottom and adding a mounting flange and a 10.16-cm (4-in.) diameter × 17.78-cm (7-in.) length of PVC thin-walled pipe. A 5.08-cm (2-in.) diameter × 30.48-cm (12-in.) length of PVC pipe was fitted inside the larger diameter pipe, extending 7.62 cm (3 in.) beyond each end of the 10.16-cm pipe. An 80-mm fan was attached to the top of the PVC container. A 40-mm fan was mounted so as to seal the 5.08-cm tube from the outer container and produce a downwards airflow which exits from the bottom of the 5.08-cm pipe. A tube with an inner diameter of 0.32-cm (0.125-in.) passes through the outer container and into the side of the 5.08-cm pipe, above the 10.16-cm pipe, and provides an entrance port for CO_2 to enrich the exit plume produced by the 40-mm fan. A small lid was attached above the 80-mm fan to prevent rain damage to the fan and to the mosquito collection (Fig. 10.4). The CFG trap was hung from a pole so that the bottom of the attractant plume was approximately 50 cm (19.69 in.) above ground level. Carbon dioxide was supplied from 9-kg (20-lb) compressed gas cylinders. A flow rate of 500 ml/min was used for all trap-bait combinations. Control of CO_2 flow rate was achieved with ABC's FLOWSET1 pressure regulator with an output fixed at 15 psig, a 10-µm line filter, a 500-ml/min flow control orifice, and quick-connect luer fittings. American Biophysics Corporation's OCT1 Slow Release 1-octen-3-ol packets were used to supply the octenol. Each packet contained a crushable glass vial containing 1 ml of octenol. The odour of octenol permeated the inner pouch and was detectable about 1 h after crushing the vial at room temperature (21°C). After that time, the packet released octenol at a rate of approximately 0.5 mg/h until the supply was exhausted (around 2 months).

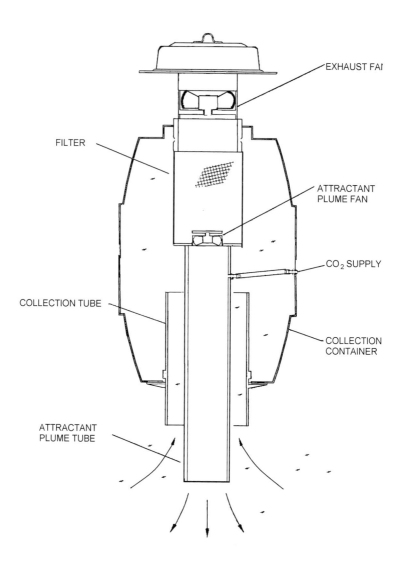

EXHAUST FAN

FILTER

ATTRACTANT
PLUME FAN

CO_2 SUPPLY

COLLECTION TUBE

COLLECTION
CONTAINER

ATTRACTANT
PLUME TUBE

Fig. 10.4. Original design of the Counter-flow Geometry (CFG) trap (Kline 1999)

In experimental cage studies, 1000 *Ochlerotatus taeniorhynchus* were released 90 minutes before sunset in batches of 250 in each of the four corners of an outdoor cage. A trap was located in the centre of each cage and was operated with either no bait, 500 ml/min CO_2 only, or 500 ml/min CO_2 + 1 package of ABC octenol for a minimum of 3 nights. Traps were retrieved 90 minutes after sunrise the following day. Field studies were conducted in a wooded area adjacent to bay and cypress swamps. Two traps, positioned 50-m apart, were used for the field studies. Each trap was baited with 500 ml/min CO_2.and run for 12 nights between April 9 and May 7, 1997. Traps were alternated each night between the 2 trapping stations. In the cage studies, CO_2-baited CFG traps caught significantly greater numbers of *Ochlerotatus taeniorhynchus* compared to the CO_2-baited standard PRO traps. In the field studies 5220 mosquitoes from eighteen species were collected over the 12 nights trapping period. The CFG trap collected 4627 mosquitoes, compared with 493 collected by the PRO trap. The species collected, in order of abundance, were *Anopheles crucians, Coquillettidia perturbans, Ochlerotatus canadensis, Culex salinarius, Culex erraticus, Ochlerotatus infirmatus, Anopheles quadrimaculatus, Aedes vexans, Ochlerotatus atlanticus, Anopheles punctipennis, Psorophora columbiae, Ochlerotatus triseriatus, Ochlerotatus dupreei, Orthopodomyia signifera, Ochlerotatus mitchellae, Psorophora ciliata, Aedes aegypti*, and *Culiseta melanura. Aedes aegypti, Ochlerotatus dupreei*, and *Culiseta melanura* were not collected by the PRO trap; *Ochlerotatus mitchellae, Orthopodomyia signifera*, and *Psorophora ciliata* were absent from the CFG trap collections. The total catch by the CFG trap was significantly greater than the PRO trap ($P = 0.0006$) and the CFG also collected significantly more individuals of six of the eight most abundant species collected. The main advantage of the CFG trap over other updraft traps, is that the CFG trap produces a downwards flow of air that can be enhanced with CO_2 or other attractants to increase the catch.

Mboera et al. (2000a) compared the effectiveness of the CFG trap of Kline (1999) with CDC light-traps and electric nets (Knols et al. 1998) for collecting mosquitoes outdoors in Tanzania. Traps were supplemented with CO_2 as follows: for the CDC trap and the CFG trap CO_2 was provided from a pressurized cylinder at a rate of 300 ml/min through silicon tubing situated at the entrance to the CDC trap (approximately 1 m above ground level) and via the inlet tube of the CFG trap. The CFG trap expelled CO_2 from the bottom of its outlet pipe, approximately 20 cm above ground level. For the electric nets, CO_2 was also supplied via the cylinder and silicon tubing, which was attached to a 25 cm long perforated plastic tube itself attached across the centre of the electric net. Comparisons between the CFG trap and a CDC trap with a light revealed that the CFG trap caught significantly more *Anopheles gambiae* than the CDC trap (mean ± SD = 40.2 ± 1.2; 5.0 ± 0.4 respectively, $P < 0.05$). A similar result was achieved when comparing the CFG trap with the CDC trap with the light off (36.5 ± 1.0; 3.8 ± 3.3 respectively, $P < 0.05$). Electric nets baited with CO_2 caught more *Anopheles gambiae* than a CDC trap with the light off (geometric mean = 23.0 and 8.6 respectively, $P < 0.05$). Catch size for the CFG and the CO_2-baited net traps were not significantly different. The CFG traps also caught more *Culex quinquefasciatus* than CDC traps with either the light on or off.

Also in Tanzania, Mboera et al. (2000c) compared the effectiveness of CFG traps and CDC light-traps baited with grass infusion or the synthetic oviposition pheromone (5R,6S)-6-acetoxy-5-hexadecanolide for collecting gravid *Culex quinquefasciatus*. Pheromone was provided according to the methods of Dawson et al. (1990) in which 0.1 ml of the mosquito oviposition pheromone preparation in hexane (200 mg/ml) was placed on a blank effervescent tablet. This method delivers a pheromone dose of 5 mg. Grass infusion was prepared by fermenting 2 kg of *Digitaria* sp. grass cuttings in 10 litre of tap water for 5 days at room temperature. The resulting solution was filtered through fine netting and frozen until used. To compare the two trap types, a pheromone-treated tablet was dissolved in 800 ml of tap water and an open glass vial containing 4 ml of the pheromone solution was attached near the entrance of the CDC trap or fixed in a small plastic tube attached at the odour entrance port of the CFG trap. To compare grass infusion and oviposition pheromone attractants in CFG traps, white plastic jars were filled with 800 ml of tap water treated with either a tablet impregnated with pheromone or hexane (control), grass infusion, or plain tap water. The jars were covered with black netting material to prevent mosquitoes ovipositing on the solution. Counter-flow geometry traps baited with pheromone and set outdoors were more efficient in collecting gravid *Culex quinquefasciatus* than CDC traps baited with pheromone. Significantly more gravid *Culex quinquefasciatus* were caught in traps baited with either pheromone or grass infusion than in traps baited with tap water. The number of gravid *Culex quinquefasciatus* collected in the CFG trap baited with pheromone did not significantly differ from that collected in the trap baited with grass infusion.

The original CFG trap of Kline (1999) was also compared with CDC light-traps and American Biophysics Corporation PRO traps for collecting adult biting midges in Florida, USA by Cilek and Kline (2002). Treatments were CO_2 delivered from 9-kg compressed gas cylinders (release rate 500 ml/min), a 4:1:8 mixture of octenol: 3-*n*-propylphenol: 4-methylphenol, and a combination of CO_2 + the octenol-phenol mixture. The octenol-phenol mixture was released from 16 ml screw-capped glass vials via a single wick. Traps were located 33 m apart and arranged in a 12×12 Latin square experimental design. Four species of ceratopogonids were captured in all traps and treatments: *Culicoides mississippiensis, Culicoides barbosai, Culicoides melleus*, and *Culicoides furens*. Regardless of trap type, significantly more *Culicoides mississippiensis, Culicoides barbosai*, and *Culicoides melleus* were collected when traps were supplemented with either CO_2 or CO_2 plus the octenol-phenol mixture compared with the mixture alone or no treatment. *Culicoides mississippiensis* were collected in significantly greater numbers in ABC PRO traps baited with CO_2 plus the octenol-phenol mixture compared with similarly baited CDC and CFG traps, when all were operated with a 6-V battery. Significantly more *Culicoides barbosai* were collected from the CFG trap when baited with CO_2 and the octenol-phenol mixture compared with similarly baited CDC and ABC PRO traps. Midge collections from traps powered with 12-V batteries were not significantly greater than those from traps powered with 6-V batteries ($F = 1.38$, d.f. = 43,131, $P = 0.24$; 12-V mean = 517.1 ± 59.1, 6-V mean = 447.9 ± 48.4), contrasting with the recommendation of Kline (1999) to use 12-V batteries with the CFG trap, at least for mosquito collections.

In response to the limitations of the original CFG traps in terms of their requirement for an external power source and a separate source of the attractants CO_2 or heat, several commercially available propane-powered models have been developed and have been tested by Kline (2002). The combustion of propane produces CO_2, heat and water vapour and by addition of a thermoelectric generator, also produces power to drive the trap's fans. Kline (2002) compared the original CFG trap to a propane-1 trap, a portable-CO_2 trap, a portable propane trap, Mosquito Magnet Beta-1 and Beta-2 traps, the commercially available Mosquito Magnet trap and the Nicosia trap (Chapter 9). The propane-1 trap is an original CFG trap, with CO_2, heat and power for operating the trap's fans supplied by the combustion of propane. The fans are powered by 12-V d.c. gel batteries and CO_2 is provided from a compressed gas cylinder. In the portable propane trap heat, CO_2 and power to run the unit's fans are provided by propane combustion. The Mosquito Magnet Beta-1 trap is the same as the Counterflow 2000™ trap (Kline 1999). The Mosquito Magnet Beta-2 trap incorporates a chamber within the CO_2 outflow tube to which other attractants, such as octenol can be added. Propane traps were tested in large outdoor field cage studies as described in Kline (1999), with the addition of releases of *Aedes aegypti* when available. Released mosquitoes were marked with a different colour of fluorescent powder for each release day and trap collections were retrieved approximately 90 min after sunrise. Field collections were also carried out in Gainesville, Florida, USA, using a 4 × 4 Latin square experimental design in which the ABC PRO trap, CFG trap, portable-CO_2 and portable propane traps were compared. Carbon dioxide was supplied to the first three traps from a compressed gas cylinder at a rate of 500 ml/min. Octenol was supplied at a rate of 0.5 mg/h from a commercially available slow release packet taped to each trap.

In the large cage studies, the traps caught between 19.8% and 62.6% of the released *Aedes aegypti* and 26.1% to 71.2% of the released *Ochlerotatus taeniorhynchus*. The commercially available Mosquito Magnet was the most efficient at trapping both species in cages. In the field studies, the CFG trap caught the highest numbers of the five most frequently collected species (*Ochlerotatus canadensis, Anopheles crucians, Coquillettidia per-turbans, Culex erraticus*, and *Culex salinarius*) in each replicate and the PRO trap the lowest. The portable-CO_2 and portable propane traps were intermediate in their efficiency and one type did not consistently out-perform the other across all replicates. A total of 21 different species were collected, 19 by the CFG trap and 17 by the propane traps. Some species were only collected in three or fewer trap types (Tables X to X). The propane traps appeared to perform less well than the CFG in collecting *Anopheles crucians*, but slightly better for *Culex erraticus* and *Culex salinarius*.

Table 10.1. Mean (±SD) responses of laboratory reared *Ochlerotatus taeniorhynchus* and *Aedes aegypti* to various traps in large outdoor cage studies (Kline 2002)

Trap type	n	Ochlerotatus taeniorhynchus Landing count	Net	n	Aedes aegypti Landing count	Net
Propane-1	2	90.0 (48.1)	261.0 (166.9)	2	28.0 (39.6)	197.5 (82.7)
PC	4	123.0 (30.9)	314.2 (149.6)	1	0.0	229
PP	30	70.0 (63.6)	413.5 (192.9)	19	12.9 (14.7)	469.2 (226.3)
MMB-1	25	100.2 (71.1)	371.0 (234.4)	6	55.2 (53.4)	378.5 (232.1)
MM-1	11	38.3 (36.8)	520.1 (200.9)	11	11.0 (15.3)	378.3 (168.7)
MM-2	8	13.0 (8.4)	712.3 (141.4)	9	2.6 (3.5)	626.1 (198.0)

The original CFG trap was developed into a commercially available and patented CFG trap, named the American Biophysics Counterflow 2000™. This trap has been tested in field studies in Sweden by Blackmore and Dahl (2002), who, contrary to Kline (1999) found no difference between

Table 10.2. Mean ($\pm$SD)[1] response of total mosquitoes collected and 5 most commonly collected species to various trap types located in a woodland wetland in Gainesville, Florida, USA, in Spring 1997 (Kline 2002)

Species	Trap type			
	CFG	PP	PC	PRO
Total	422.6 (200.9) a	153.1 (174.5) bc	211.8 (174.5) b	56.4 (49.3) c
Ochlerotatus canadensis	40.9 (44.6) a	9.5 (10.5) b	13.3 (12.7) b	9.3 (11.0) b
Anopheles crucians	328.3 (156.9) a	91.4 (61.3) bc	173.3 (169.1) b	27.2 (27.0) c
Coquillettidia perturbans	21.3 (19.0) a	17.2 (16.2) a	12.8 (12.4) ab	5.1 (4.7) b
Culex erraticus	5.3 (6.2) a	5.7 (7.6) a	4.1 (5.9) a	1.1 (1.9) a
Culex salinarius	19.7 (12.7) ab	23.5 (16.3) a	5.2 (3.4) c	9.7 (13.7) bc

[1]n = 12 nights. Means in the same row followed by the same letter are not significantly (P > 0.05) different; Ryan-Einot-Gabriel-Welsh multiple range test.

Table 10.3. Mean ($\pm$SD)[1] response of total mosquitoes collected and 5 most commonly collected species to various trap types located in a woodland wetland in Gainesville, Florida, USA, in Spring 1998 (Kline 2002)

Species	Trap type			
	CFG	PP	PC	PRO
Total	653.3 (253.4) a	197.5 (86.9) b	148.0 (73.9) b	41.5 (13.2) b
Ochlerotatus canadensis	10.5 (5.1) a	6.5 (5.3) a	6.3 (3.5) a	4.3 (3.9) a
Anopheles crucians	496.8 (175.9) a	78.3 (63.9) b	50.3 (32.8) b	8.5 (5.5) b
Coquillettidia perturbans	83.0 (49.4) a	57.5 (10.4) ab	38.0 (25.4) ab	20.5 (9.0) b
Anopheles quadrimaculatus	4.3 (3.8) a	1.3 (1.3) a	0.8 (0.9) a	3.5 (3.1) b
Culex salinarius				

[1]n = four nights. Means in the same row followed by the same letter are not significantly (*P* > 0.05) different; Ryan-Einot-Gabriel-Welsh multiple range test.

the Counterflow 2000™ and CO_2-baited miniature light-traps with the light switched off in terms of total catch, although the Counterflow 2000™ trapped a greater diversity of species than the CDC traps. A total of 12 726 mosquitoes was collected in 64 trap days. Seventeen species belonging to 6 genera were collected, with *Coquillettidia richiardii* being the most abundant species (39% of total), followed by *Ochlerotatus sticticus* (25%), *Aedes rossicus* (19%) and *Aedes cinereus* (11%). Catches were significantly associated with trap location and this was thought to be primarily due to variations in wind exposure, although this variable was not measured during the study. The Counterflow 2000™ trap appeared to be more effective than the CDC traps in wind-exposed sites or on windy days, presumably because the CO_2 and heat plume is more closely associated with the trap entrance in windy conditions than is the case with the CDC traps. Simuliidae, Chironomidae, Psychodidae, and Mycetophilidae were also collected by both traps, although the CDC traps collected 7 times more Simuliidae than the Counterflow 2000™ traps.

The CFG trap was evaluated against ABC light traps and Shannon traps for mosquito surveillance in a malarious area of the Republic of Korea by Burkett et al. (2001). Light, CO_2, octenol, and combinations of these attractants were used in the traps. Traps employing counterflow technology (Mosquito Magnet and CFG traps) captured significantly greater numbers of mosquitoes compared to other baited ABC or CDC light traps. In contrast to other reports (e.g. Becker et al. 1995; Mboera et al. 2000b), Burkett et al. (2001) found that all octenol-baited traps were repellent to certain species, resulting in significantly lower catches of *Culex pipiens* compared with un-baited traps. No *Culex orientalis* were caught in octenol-baited traps.

Similarly, in Thailand Sithiprasasna et al. (2004) compared human landing catches with CDC light-traps baited with dry ice, the ABC standard light-trap with dry ice and octenol, the ABC CFG trap with dry ice, the ABC Mosquito Magnet with octenol, and the Nicosia and Reinhardt Company Mosquito Attractor Device as alternatives to human-landing catches for malaria vector surveillance. These authors observed that human landing catches were the most productive, yielding 5.2, 7.0, 7.3, 31.1, and 168.8 times as many mosquitoes as the CDC, ABC, CFG, Mosquito Magnet, and N&R traps respectively. This result contrasts with those obtained by Burkett et al. (2001), Kline (1999), and Mboera et al. (2000) who all observed the CFG trap to be effective at sampling anopheline mosquitoes, and this was ascribed by Sithiprasasna et al. (2004) to be possibly due to the use of dry ice instead of CO_2 delivered from a compressed gas cylinder and also the use of 6 V, as opposed to 12 V d.c. batteries, in the Thailand study. In the Thailand study, *Anopheles minimus* was the most commonly collected mosquito species, comprising 82.5% of the total (27 104) collection. *Anopheles kochi* (not a malaria vector in Thailand) was only rarely collected in human-landing catches, comprising only 0.7% of the total catch using this method, but accounted for 11.5% of the total catch from traps and human-landing catches.

Experimental Mosquito Magnet counterflow traps were evaluated against cylindrical lard-can traps (Bellamy and Reeves 1952) and CDC light-traps in a study of West Nile virus prevalence in tree canopy-inhabiting mosquitoes in Connecticut, USA by Anderson et al. (2004). The counter-flow trap used was constructed from an 11.4-liter clear polyvinyl chloride container with a fan blowing CO_2 out of the bottom of the trap and another fan providing airflow into the bottom of the trap. Carbon dioxide was supplied from a 20-lb compressed gas cylinder at a flow rate of 500 ml/min. The counterflow traps captured significantly more *Culex pipiens* than either the CDC or lard-can traps placed in the canopy. Counterflow traps placed in the tree canopy caught more mosquitoes than those situated close to the ground.

Cilek and Hallmon (2005) reported mixed results when assessing the effectiveness of the Mosquito Magnet trap at reducing biting midge populations in residential gardens in Florida, USA. Midge populations, as assessed by CDC light-traps were lower in premises using traps compared with control premises in only 2 of 45 weeks. Owners of premises using Mosquito Magnet traps also reported that the levels of midge reduction experienced by them were variable.

Ritchie et al. (2003) recently proposed that the Mosquito Magnet Pro trap could be used in place of the current system of using sentinel pigs for Japanese encephalitis surveillance in remote areas of Australia. Pools of mosquitoes containing an individual mosquito known to be infected with JE virus were identifiable after 7 or 14 d of being kept in MM Pro traps. Potential disadvantages of the proposed trap-based surveillance system identified by the authors included the expense of the MM Pro traps, their susceptibility to vandalism, and the fact that the propane tanks required to power the traps cannot be transported in passenger planes, limiting the potential for rapid deployment of traps in remote and inaccessible areas.

Also in Australia, Johansen et al. (2003) compared the commercially available Mosquito Magnet and CFG traps with EVS and CDC traps. The CFG, CDC and EVS traps were baited with a 1-kg block of dry ice and all 4 traps were supplemented with octenol at a release rate of approximately 4.5 mg/h. Trap efficacy was tested using a 4×4 Latin square experimental design. The MM trap performed significantly better than the EVS trap in terms of total mosquitoes collected as well as the number of *Culex sitiens* sub-group mosquitoes (the likely vectors of JE in northern Australia) collected. The MM trap performed significantly worse than the CFG trap in collecting *Ochlerotatus* spp. mosquitoes during one of three trials. These authors concluded that the MM trap is equal to or superior than the CDC or EVS traps routinely used for virus vector surveillance in Australia.

Commercially available Mosquito Magnet traps were compared with American Biophysics Corporation standard downdraft and u-v updraft models of CDC-type light-traps, and commercially available omnidirectional Fay-Prince traps for mosquito surveillance in Georgia, USA, by Burkett et al. (2004). In terms of total catch size, traps performed as follows: u-v updraft (no CO_2) $\geq$ Mosquito Magnet $\geq$ omnidirectional Fay-Prince trap with CO_2 $\geq$ CDC style downdraft trap with CO_2. Significantly greater numbers of *Aedes vexans* were caught in the u-v updraft traps.

Dennett et al. (2004) compared three commercially available traps with three standard mosquito surveillance traps, and a new design of swinging trap for collection of *Aedes aegypti* and *Aedes albopictus* in Texas, USA. The three commercial traps evaluated were the Mosquito Magnet Liberty trap; the Mosquito Deleto trap (manufactured by the Coleman Company of Kansas, USA) that uses open-flame combustion of propane to produce CO_2, moisture, and heat to attract mosquitoes to land on adhesive panels; and the Dragonfly Biting Insect Trap (Biosensory Incorporated, Connecticut, USA), which produces CO_2 from a 9 kg gas cylinder and electrocutes any insects attracted to it by way of an electrostatic grid. The standard traps used were CDC light-traps with an incandescent bulb, CDC traps with no light source, and an omnidirectional Fay-Prince trap, all operated on 3V d.c. The moving-target trap is essentially a motorised fan suction trap mounted on a motorised child's swing, so that it moves backwards and forwards. The moving-target trap is fully described in Chapter 11. Carbon dioxide was supplied to the three standard traps and the moving-target trap by way of 2 kg of dry ice placed in 1.9 litre beverage coolers with five 0.6 cm holes drilled in the bottom. Traps were tested in a 7×7 Latin square experimental design over 21 days. Traps were run continuously for 23 h per day, allowing 1 h to service traps and rotate their position within the Latin square. Significantly higher mean numbers of *Aedes aegypti* and *Aedes albopictus* (55 and 269 respectively) were captured in the Mosquito Magnet trap, compared with all other traps. The omnidirectional Fay-Prince trap caught significantly more *Aedes albopictus* than the Dragonfly™ trap, while the Dragonfly™ trap caught significantly more *Aedes aegypti* than the omnidirectional trap. The moving-target trap caught a mean of 40 *Aedes albopictus* and 6 *Aedes aegypti*, and catches did not differ significantly from those obtained with omnidirectional and Dragonfly™ traps. The Mosquito Deleto™ trap caught relatively few of either species. CDC light-traps with and without light did not differ significantly from one another in terms of numbers of *Aedes aegypti* and *Aedes albopictus* captured.

Mosquito Magnet™, omnidirectional and moving target traps all performed better than CDC traps for *Aedes albopictus*. For *Aedes aegypti*, Mosquito Magnet™ and Dragonfly™ traps performed better than CDC traps.

Comtois and Berteaux (2005) used a commercial Mosquito Magnet™ garden edition trap (American Biophysics Corp., East Greenwich, Rhode Island, USA), without octenol, to determine the abundance of mosquitoes and blackflies during a study on the effect of biting insects on behaviour and habitat use by American porcupines (*Erethizon dorsatum*).

Sticky CO$_2$-baited traps

Gillies and Snow (1967) produced a sticky dry ice trap by coating four 18-in square plastic (perspex) panels with 'Bostik Permanently Tacky Adhesives' (No. 1170 or S.96.102) and slotting them into a metal frame to form a square. Emission rates of 1200–1800 ml gas/min produced from either cylinders or dry ice in a tray situated at the top of the trap were aimed for. In trials in England large numbers of the mosquitoes attracted by the gas alighted and took off again from a variety of surfaces near the trap, with the result that many were eventually trapped on the sticky panels. Gillies and Snow (1967) considered that the traps might be useful for assessing mosquito densities in areas where populations were too great to be sampled by bait catches.

Wilson et al. (1966) used cylinders coated on the outside with an adhesive compound and baited with about 5.5 kg of dry ice to catch tabanids. A somewhat similar sticky cylindrical trap was used by Whitsel and Schoeppner (1965) to trap Ceratopogonidae. Their procedure was to wrap sheets of 'kraft' paper coated with an adhesive compound ('Tanglefoot') around cylinders containing 2–3 lb of dry ice. Because at relative humidities above 60% condensation formed on the sticky paper and reduced its tackiness, the inside of the metal cylinder had to be insulated with an ice-cream bag.

Large numbers of *Simulium* were caught on a trap consisting of a dark blue cylindrical lard-can baited with 1 kg dry ice wrapped in several layers of paper towelling and placed on top of the trap. The outside of the cylinder was covered with strips of plastic coated with 'Roost-no-more' or 'Tanglefoot' adhesive (Baldwin and Gross, 1972). Although none of these traps was designed for catching mosquitoes they might be useful in certain circumstances.

Mosquito nets baited with CO_2

Mosquito bed-nets baited with dry ice and with one or more sides raised have been used by a number of Japanese workers to collect mosquitoes, especially *Culex* species. Over many years Japanese entomologists have caught substantial numbers of *Culex tritaeniorhynchus, Culex pipiens* form *pallens, Anopheles sinensis* and *Aedes vexans nipponii*, as well as several other species, in a very simple trap consisting of a mosquito net suspended about 30 cm over a bowl of dry ice (Buei et al. 1986). The dry ice lasts about 3–4 h. Up to 5000 female mosquitoes, consisting mainly of *Culex tritaeniorhynchus*, but including *Culex pipiens* and *Anopheles* species, have been collected in one net during a single night (Takeda et al. 1962). These traps in fact caught more mosquitoes than light-traps or traps baited with goats or chickens. Regular collections throughout the night enabled the flight times of different mosquito species to be studied.

Omori et al. (1965) produced a stream of carbon dioxide by heating a small can of water containing dry ice that had been placed under a mosquito net. Adults of *Culex tritaeniorhynchus* attracted by the gas were collected both inside and outside the net using small hand-nets. Using a similar technique Mogi et al. (1970) produced a continuous discharge of carbon dioxide from about 2 h before to 2 h after sunset. When the carbon dioxide-baited nets were used near paddy fields considerable numbers of *Culex tritaeniorhynchus* and other species including *Culex pseudovishnui, Aedes vexans nipponii* and *Anopheles sinensis*, were collected. In Korea mosquito nets baited with dry ice have also been successful in catching *Culex tritaeniorhynchus* (Ree et al. 1969).

In Florida, USA Nayar (1982) attracted host-seeking *Wyeomyia mitchellii* by bubbling carbon dioxide through a beaker of water placed under a net.

Stable traps

Reeves (1951, 1953) used carbon dioxide as an attractant in Bates (1944) type stable traps. In preliminary trials traps were baited with 7–9 lb of dry ice wrapped in thick paper and placed in a large tin suspended on a level with the trap's baffles. More mosquitoes, mostly comprising *Culex tarsalis*, were caught in traps with dry ice than in those baited with three to five chickens or a single calf (Reeves 1951). In later trials by selecting release rates of about 25, 250 and 2500 ml carbon dioxide/min from gas cylinders the average amounts of carbon dioxide expired by a chicken, a man and a horse or cow, respectively were simulated (Reeves 1953).

Adults of *Ochlerotatus nigromaculis*, a zoophagic species, were attracted in increasing numbers to the highest concentration of gas, whereas more adults of *Culex quinquefasciatus*, an ornithophagic mosquito, were caught with the smallest discharge rate. *Culex tarsalis*, which commonly feeds on birds, was attracted in large numbers by all three concentrations, although there was a slight increase in numbers associated with increased discharge rates. It is possible that *Culex quinquefasciatus* was not so much attracted by low emission rates as repelled by the higher rates.

In Malaysia the addition of dry ice to a chicken-baited Magoon trap increased the numbers of *Culex* mosquitoes from 8.34 to 84.47/collection; numbers of *Culex vishnui*, the most common species caught, increased from 1.19 to 51.26/collection (Wallace et al. 1977).

Cylindrical lard-can traps

A cheap, portable and easily made trap has been developed in which carbon dioxide was produced from dry ice. This trap was originally made from a cylindrical 50-lb lard-can, 14½-in long and 12 in in diameter (Bellamy & Reeves, 1952). An 8¾-in hole was cut out from both ends of the can and a 5½-in long funnel of fine mesh gauze was soldered to the rims of the cut ends. The ¼-in diameter apical opening of each funnel was directed towards the middle of the can, which contained about 3 lb of dry ice wrapped in thick paper. These traps resemble those described in Chapter 6, which were baited with small animals. Traps baited at about mid-day with dry ice were suspended vertically or horizontally at different heights (ground level to 25 ft) amongst vegetation, and carbon dioxide was slowly released during most of the night. Trap location was often critical, and differences of only a few metres sometimes increased the catch many times. Horizontally placed traps were only slightly more effective than vertical ones. Ants sometimes destroyed the catch when traps were placed on the ground. In California, USA *Culex tarsalis* comprised about 99% of the catch, most of which were unfed females; males were extremely rare. On one occasion as many as 947 females were caught in a single trap baited overnight with about 5 lb of dry ice.

In a comparison of different collecting procedures for sampling *Culex tarsalis* in California, USA collections were made from natural shelters, from red box artificial shelters, from light-traps and from cylindrical traps baited with about 2.5 lb dry ice. It was concluded that all methods were satisfactory for measuring the abundance of females, and all except the carbon dioxide traps adequately sampled males (Hayes et al. 1958). This was a not unexpected result since carbon dioxide traps essentially catch females orientated to host feeding.

When Brockway et al. (1962) used dry ice as bait they found that if four cylindrical traps were mounted at right angles to each other the trap facing downwind caught 61.3% of the total catch of mosquitoes. This was about four times as many as caught in the upwind trap. Further experiments confirmed that traps facing into the wind caught most mosquitoes (Bailey et al. 1965). To investigate this phenomenon further, and to improve trapping efficiency for capturing marked individuals in dispersal studies, a wind directional trap was constructed. A cylinder baited with 4 lb of dry ice was fixed to each of four horizontal arms that were attached, at right angles to each other, to a vertical spindle. A wind vane and counter weight ensured that one trap always pointed into the wind, irrespective of direction, another trap pointed away from the wind and two were at right angles to it. Results clearly showed that downwind (leeward) trap caught about four times as many mosquitoes as the other three traps.

In their studies in California, USA on the dispersal of *Culex tarsalis* Bailey et al. (1965) used either 4-gal capacity metal cylinders with copper mesh entrance funnels or cardboard 3-gal ice cream canisters with plastic mesh funnels. The cardboard traps proved just as effective as the metal ones, but did not last so long. Initially traps were baited with 3–4 lb of dry ice wrapped in newspaper, but in later experiments the dry ice was placed in a plastic bag with the opening loosely constricted with an elastic band. Sublimation of the dry ice caused the bag to inflate and the trapped gas helped to insulate the remaining dry ice from the warm evening air. When traps were baited with dry ice just before sunset it lasted about 6 h. In the morning, before predators had a chance to destroy the catch, one of the metal cones was replaced with a cloth screen and the mosquitoes blown into an 18 × 18 × 26 in wire holding cage. Each side of the cage was divided into quarters and the total catch estimated by counting the numbers of mosquitoes resting on one quarter section of each side. Although very large numbers of female *Culex tarsalis* were caught, relatively few other species were collected. The next most common was *Anopheles freeborni*. In addition a few *Aedes, Culex, Culiseta* and other *Anopheles* species were caught.

The numbers of mosquitoes caught in an ordinary dry ice trap were compared with those caught in a trap in which gas from a dry ice canister was passed through a 5-gal can of hot water, so that warm, moist carbon dioxide was delivered into the trap. In two experiments only 7 and 157 *Culex tarsalis* were caught in the normal trap, compared with 48 and 72 in the trap with warm moist gas. Although it seemed that *Culex tarsalis* preferred traps emitting warm carbon dioxide, the combined catches in the 'cold' and 'warm' traps were 'little if any higher' than the normal catch obtained in a 'cold' trap at a similar site (Bailey et al. 1965). These results are inconclusive because differences between the numbers caught could have been due, at least in part, to trap location. These experiments need repeating. In dry ice traps the attractiveness of carbon dioxide may to some extent be offset by a deterrent effect of low temperatures produced around the trap. Using carbon dioxide from a cylinder overcomes this possible adverse effect.

In Taiwan Mitchell and Chen (1973) wrapped about 1 kg of dry ice in aluminium foil and when this was placed in cylindrical traps 9193 mosquitoes were caught in 39 trap-nights; 95.1% were *Culex annulus* and 3.3% *Culex tritaeniorhynchus*.

When Dow et al. (1965) baited 35 cylindrical traps with about 3 lb dry ice wrapped in brown paper an average of more than 8000 mosquitoes was caught per night. About 95% of these were *Culex tarsalis*, others comprised mostly *Culex erythrothorax* and *Anopheles franciscanus*. Dow et al. (1965) developed a technique for removing mosquitoes from traps that avoided the use of anaesthetics or a strong flow of air, procedures that can injure mosquitoes. A trap containing mosquitoes (Fig. 10.5a) is slowly lowered into a large drum of water and the lower inwardly projecting collecting cone replaced under water with an outwardly projecting 'transfer cone', which has a threaded metal ring from the top of a Kilner (Mason) jar soldered around its wide apical opening (Fig. 10.5b). The trap is then carefully removed from the water, and a hand placed over the apical opening to prevent mosquitoes escaping while it is inserted through the opening in the top of a cylindrical holding carton. The trap is then inverted so that the carton is at the top. The other end of the trap is now lowered into the water to force the mosquitoes into the 'transfer cone' and finally into the carton (Fig. 10.5c). This is carefully separated from the trap while still underwater and its opening covered with a threaded screw cap fitted with a mesh screen (Fig. 10.5d).

The number of mosquitoes in the carton is estimated by counting those resting between two vertically ruled zones; as each zone represents 1/20 of total wall area the counts are added and multiplied by 10. Because adults also rest on the ends of the cartons a better estimate would take these into consideration. The accuracy of the method can be checked by getting different workers to make estimates after which the mosquitoes are killed and the total catch counted. Dow et al. (1965) calculated a regression equation for each collector so that personal bias in estimating the size of catches could be corrected.

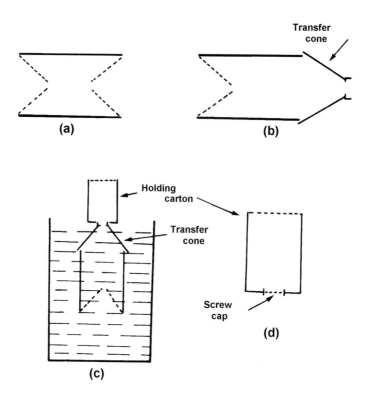

Fig. 10.5. Method of removing mosquito catch from cylinders which have been baited with dry ice (Dow et al. 1965), see text for explanation

In Florida, USA the attractiveness of young Leghorn chicks, 3.5 lb of dry ice and a combination of the two was compared by placing the baits in cylindrical lard-can-type traps (Vickery et al. 1966). When dry ice was used the insides of the metal cans were insulated with corrugated cardboard and the outsides by paper towelling. From 12 nights of trapping the mean catch of *Culex nigripalpus* in chick- and dry ice-baited traps was 141 and 318 respectively, whereas a combination of the two baits increased the mean catch to 930, which is more than twice the combined means from single-baited traps. Two-way analysis of variance showed that trap location and individual trap-nights contributed little variation, whereas individual chicks were a highly significant source of variation. Also in Florida Edman and Lea (1972) caught between 8000–10 000 blood-fed *Culex* spp. in 3 days when cylindrical traps were baited with both chicks and dry ice.

In southern California, USA Mian et al. (1990) placed lard-can carbon dioxide traps from 1900–0900 h at ground level (0.6 m) and at heights of 3 and 6 m. From 170 trap-nights 53 211 female mosquitoes belonging to at least eight species were caught, the most common being *Culex quinquefasciatus* (30.3%), *Culex tarsalis* (29.2%) and *Culex erythrothorax* (27.7%), with mean numbers/trap-night of 94.9, 91.4 and 86.7, respectively. Most mosquitoes were caught in the highest trap and fewest in the trap at 3 m. These dry ice traps proved convenient for monitoring seasonal abundance and spatial distribution of mosquitoes in different areas.

When Jupp and McIntosh (1967) used lard-can traps baited with 5 lb broken dry ice, they caught 12 mosquito species, and frequently more than 30 000 mosquitoes/trap-night. *Culex pipiens* (41%), *Culex theileri* (33%) and *Culex univittatus* (22%) were the most commonly collected species. Catches were sometimes so great as to hinder sorting and processing, and it was difficult to remove the catch from the can, especially when mosquitoes were wanted alive. To overcome this they developed their so-called 'number 2' trap. This is a 12-in long, 13-in diameter cylinder of galvanised metal positioned vertically, with an inverted mesh cone in the lower end terminating in a 1-in hole. Mosquitoes enter this cone, being attracted by dry ice (or a bait animal) placed in the cylinder just above the cone, and fly up and through an outwardly projecting plastic cone fitted to the top of the cylinders to enter a collecting cage. This trap proved very useful in catching *Culex pipiens* and *Culex univittatus*, but few other species entered the trap in significant numbers.

No. 10 Trinidad trap

These traps, which are described in Chap. 6, were designed for baiting with small animals, such as mice (Worth and Jonkers 1962) but can also be used with dry ice. In England when about 0.5 kg of dry ice in a polystyrene box was placed in each V-section of the trap, nine mosquito species, including an anopheline was caught, the maximum catch in a single trap being 105 mosquitoes. In contrast only six culicine species were caught in cylindrical traps baited with about 1 kg of dry ice, and the maximum catch was 30 mosquitoes (Service 1969). In further comparisons the Trinidad trap caught 226 mosquitoes compared with 24 in the cylindrical traps. Most species, including aedines, *Culiseta annulata, Culiseta morsitans* and *Culex pipiens*, were more numerous in traps put out overnight, but *Coquillettidia richiardii* was much more common in traps used during the day-time. In the cylindrical lard-can type of trap carbon dioxide diffuses out from the two ends whereas in the Trinidad trap there is a more general exodus over a larger area, although as CO_2 is heavier than air most will tend to diffuse downwards. Wind, however, will tend to more or less evenly disperse the gas (Barynin and Wilson 1972). Many mosquitoes were observed resting on the outside of the mesh funnels of cylindrical traps but not to enter them, whereas mosquitoes appeared to enter the Trinidad traps more readily, although because they remained in an activated state within the traps a number flew out again.

Trueman and McIver ramp trap

In their study on the temporal distribution of biting flies in Canada, Trueman and McIver (1981) designed a new type of trap employing multidirectional ramps, radiating vertical baffles and carbon dioxide. Their trap consists of a 4-ft high wooden-framed pyramid with an 8-ft square base covered with heavy-duty black plastic. Baffles (4 ft high, 8 ft long) made of burlap fabric supported by vertical wooden stakes radiate from the base of the trap. A 12-in 'Vent-Axia' fan mounted at the apex of the trap (Fig. 10.6) sucks insects through a fine nylon mesh cone and delivers them into collecting bags positioned on a turntable. Dichlorvos (DDVP) in the turntable box kills the insects in the bags, and needs replacing monthly. Cylinders deliver carbon dioxide at 700 ml/min to the top of a 4-ft square wooden roof positioned 8 in above the trap entrance.

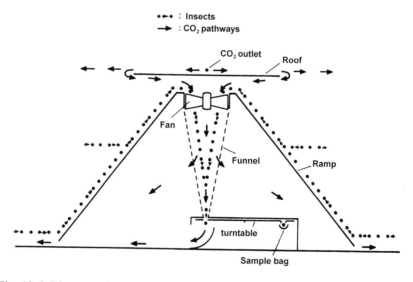

Fig. 10.6. Diagram of pyramid-type ramp trap (Trueman and McIver 1981)

The roof prevents rain and detritus being sucked into the trap. Apart from the fact that 20 mosquito species were caught, and about half in 'good numbers' no further details were given, however, the same traps were later used by Trueman and McIver (1984) to study the seasonal incidence and daily activity patterns of mosquitoes. From two traps operating over 2 years more than 16 000 mosquitoes belonging to 22 species were collected, the most common being *Ochlerotatus canadensis, Ochlerotatus communis, Ochlerotatus diantaeus, Ochlerotatus intrudens, Ochlerotatus punctor, Ochlerotatus abserratus* and *Coquillettidia perturbans*.

Plexiglas trap

This trap is constructed from 0.63-cm thick panels of plexiglas (polymethyl methacrylate) 45.7 cm high and 15.2 cm wide fitted together to form an octagon 39.8 cm in diameter (Fig. 10.7a,b).

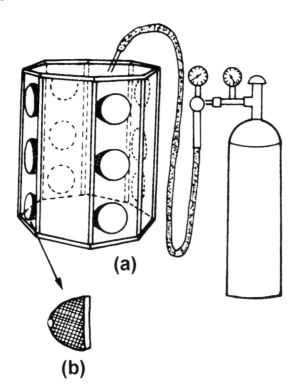

(a)

(b)

Fig. 10.7. (*a*) Plexiglas carbon dioxide trap and (*b*) single cone (after Schreck et al. 1970)

Three 10.2-cm diameter openings are cut from alternate panels to accommodate a total of 12 wire mesh funnels (Fig. 10.7*b*) having a small opening at their apices. The plastic top of the trap is fitted with a handle and is removable to allow access to trapped mosquitoes. Carbon dioxide released from a cylinder, via a regulator, at a rate of 1 litre/min is passed through a length of plastic tubing inserted into the top of the trap (Schreck et al. 1970).Several of these traps were evaluated in Florida, USA against field populations of *Culex salinarius, Coquillettidia perturbans, Ochlerotatus taeniorhynchus, Ochlerotatus sollicitans, Anopheles quadrimaculatus* and *Psorophora confinnis*. They were used from dusk to early morning at heights of 25 and 50 cm. Very large catches were obtained, with more than 102 000 *Psorophora confinnis* caught in a single trap operated for five nights. In one series of trials the number of *Culex salinarius* per trap ranged from 332–3366 while in other tests a single trap caught 1212 *Ochlerotatus sollicitans*, 4718 *Ochlerotatus taeniorhynchus*, 1427 *Psorophora confinnis*, 231 *Culex* spp. and also a few adults of other species. Apparently *Anopheles quadrimaculatus* was not caught, and only negligible numbers of male mosquitoes were trapped.

A plexiglas trap lined with brown paper and other experimental traps made to the same design but of plywood or glass were also tested. Surprisingly, the plexiglas trap without brown paper caught about twice as many mosquitoes as did the glass trap, and 6–14 times as many as traps made of plywood or lined with paper (Schreck et al. 1970). When a 15-W incandescent light was placed in two of the plywood traps the size of the catch was increased but was still much smaller than that in the plexiglas trap. In later trials ramp traps of Gillies (1969) were positioned along the perimeters of five concentric circles, at 15, 30, 60, 90 and 120 ft from the centre where a plexiglas trap containing dry ice was positioned (Schreck et al. 1972). The predominant species were *Culex salinarius, Psorophora confinnis* and *Coquillettidia perturbans* and about 92% were caught in traps within 60 ft of the centre, 5% in traps at 90 ft and 2% in traps along the 120 ft perimeter. Clearly the limit of the range of attraction of the dry ice traps was around 120 ft. When an identical wooden dry ice trap replaced the plexiglas one, the ramp traps still caught considerable numbers of mosquitoes, but very few were actually trapped by the wooden trap. This confirms earlier observations that wooden traps are inefficient in catching mosquitoes attracted to them (Schreck et al. 1970).

Katô's dry ice trap

This is a modification of the 'animal baited trap of Katô' (1955) which is essentially an exit trap fitted to a wooden board placed in the window of an animal shed (Fig. 10.8). It consists of a cylindrical framework of galvanised wire, 20 cm in diameter and 49 cm long, covered with plastic or wire mosquito netting (A). A funnel of netting, having the smaller opening 6 cm in diameter, is fitted inside one end of the cylinder to provide a one-way entrance for mosquitoes (B). The other end is covered by a short cone (C) through which the catch is removed. The cylinder together with the inserted funnel are fitted to a wire netting cone (D), 22 cm deep and having the larger 50-cm diameter opening attached to a wooden base board (E). For use as a dry ice trap the cone end is fitted over the opening of a larger plastic wire netting cone, 64 cm deep and with a 150-cm diameter opening. The whole trap is supported on four vertical wooden legs fixed to the large cone and thrust into the ground. About 2 kg of dry ice, either wrapped in newspaper or put in a metal box with 70 1–5 mm diameter holes drilled into it, is placed in the trap on a wooden platform some 12 cm above the bottom edge of the lower cone. The lower edge of the cone is positioned 30–40 cm from the ground (Katô et al. 1966).

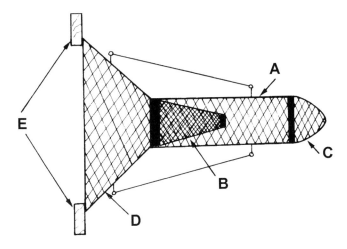

Fig. 10.8. Dry ice trap showing A—wire mesh collecting cylinder, B—entrance funnel, C—removable cone cover, D—basal wire netting cone, E—wooden base-board (after Katô et al. 1966)

In Japan more mosquitoes, especially *Culex tritaeniorhynchus*, were generally caught in these dry ice traps than in pig and cow-baited traps, although considerably fewer *Culex pipiens* form *pallens* were trapped by the dry ice trap. As many as 1415 mosquitoes, again mostly *Culex tritaeniorhynchus*, were caught overnight in one trap. Greater numbers of *Aedes vexans nipponii*, *Anopheles hyrcanus* and *Anopheles sinensis* were collected from dry ice than from animal-baited traps, but the differences were not very pronounced.

Conical trap of DeFoliart and Morris

This trap was designed to collect and retain a wide range of haematophagous flies in a frozen condition so that the trap need only be emptied every 2–3 days despite the fact that the specimens might be required for arbovirus isolations. The original design (DeFoliart and Morris 1967) was very complicated, but after two years' field use it was simplified (DeFoliart 1972). In addition to being easier to construct the later model has the advantage of retaining the catch alive, thus eliminating possible adverse effects that carbon dioxide may have on arbovirus infected flies. It is only necessary to empty the trap twice a week.

The modified trap consists of an insulating container 19.7 cm long, 14.7 cm wide and 17.8 cm high, in which about 4½ lb of dry ice is placed (Fig. 10.9). A 5-mm hole is drilled in each side of the box about 4 cm below the top to allow carbon dioxide to escape. The container rests on a small wooden rack placed on top of a 25-cm diameter plastic container called the 'crisper'. The only part of the 'crisper' not covered by the dry ice container is a small section at one end through which a hole is drilled to take a 2.5-cm diameter piece of tubing. A 75-cm high plastic cone, made by cutting out a semi-circle from a number of cellulose acetate sheets cemented together, is placed over the dry ice container and 'crisper'. Two 60-cm high triangular openings are cut out opposite each other from the base of the cone. An inverted cellulose acetate funnel (A), with the narrower opening 5 cm in diameter, is positioned over the apical opening of the large plastic cone. The short stem of this funnel is made of plastic tubing and leads upwards into another cellulose funnel (B), with its 25-cm diameter opening covered with a plastic cover. A length of plastic tubing (C) is inserted into the 2.5-cm diameter opening at the base of this second funnel and passes directly downwards through the small hole in the 'crisper', which is situated beneath it.

The trap works as follows. mosquitoes, and other haematophagous insects, attracted to the dry ice enter the plastic cone through the triangular openings, fly upwards through the inverted funnel into the top funnel and then finally slide down the vertical plastic tubing into the 'crisper', where they are retained. Pieces of wire mesh screening can be inserted in the 'crisper' to give mosquitoes additional resting places and to prevent larger insects, such as tabanids, from flying about and causing damage.

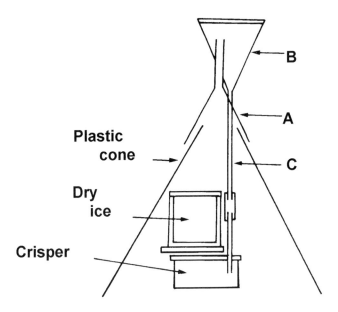

Fig. 10.9. Dry ice trap showing A—inverted funnel, B—upper plastic funnel, C—connecting plastic tubing to the collecting box (crisper) (after DeFoliart 1972)

DeFoliart (1972) reported that although many of the captured insects died during the 3–4-day collecting interval, considerable numbers remained alive. There was less mortality in mosquitoes than in simuliids or tabanids.

No details are given of the flies caught in the modified trap, but in the earlier model a wide variety of biting flies were collected (DeFoliart and Morris 1967; DeFoliart et al. 1967). However, the traps did not adequately sample *Anopheles, Culex* and *Culiseta* species, and although more *Aedes* were caught (e.g. maximum catches of some species were about 175/24 h) the catch was not as large as expected from the size of the populations prevailing at the time. In later studies Morris and DeFoliart (1969) found that in general more aedine mosquitoes were caught in dry ice traps than in light-traps, but the catches of *Anopheles, Culiseta* and *Coquillettidia* were considerably smaller. About equal numbers of *Culex salinarius* were caught in the two types of trap, but few of the other two *Culex* species caught during the trials were collected in the dry ice traps. Blood-fed specimens were 2.8 times more common in the dry ice traps than in the light-traps. There was some evidence to suggest that the parity of various aedine species obtained by the two sampling methods differed.

Malaise traps

A modified Malaise trap was baited with carbon dioxide by Blume et al. (1972) primarily for collecting tabanids, but mosquitoes were also caught. Three 2-m sections of aluminium tubing arranged to form a tripod frame are bent inwards at a height of 1.2 m at an angle of about 45° to form a pyramid. A 10-cm diameter metal ring is welded to three equally spaced steel rods bent at an angle so that they can easily be inserted into the open apical ends of the aluminium tubing of the pyramid. A collecting jar consisting of a clear 14-cm diameter plastic bottle with the bottom replaced by a screen cover is fitted on top of the trap. An inverted wire funnel is placed in the opening of this bottle. A 9-cm hole is cut out from the screw-on lid of the bottle, and the upper 3 cm of a 9-cm diameter plastic bottle, also having a screw lid, is glued on to the opening. Finally, a 7-cm hole is cut from the lid of the smaller (9-cm) bottle, which is fixed to the apex of the trap. The upper pyramidal part of the trap is covered with clear nylon reinforced plastic, while three pieces of black nylon reinforced plastic are joined together to form the baffles between the aluminium legs. The trap is steadied by placing its legs in tin cans three-quarters filled with concrete; oil is poured on top of the concrete to prevent ants climbing up the legs. Carbon dioxide from a cylinder released at a rate of 3 litres/min and introduced into the centre of the pyramid canopy about 20 cm below its apex, increased the catch of tabanids about four times that obtained in Malaise traps without carbon dioxide.

In Canada Graham (1969) released carbon dioxide from cylinders at an average rate of 5 litres/min into Malaise traps, but difficulties were experienced in controlling flow rates even though a flowmeter was used. Gas was also released from 25 and 50-lb cylinders without using a flowmeter. The smaller cylinder discharged carbon dioxide for about 16 h, and proved more useful because of its portability. These traps caught more mosquitoes than light-traps, visual attraction traps, rotary traps, or animal-baited traps, and about 51–125 times as many mosquitoes as unbaited Malaise traps. In all catches aedine species predominated.

In Mississippi, Roberts (1972) compared the relative attractiveness of carbon dioxide released from cylinders at the rate of 3500 ml/min (5005 litres/day) and a 1200-lb Hereford steer to mosquitoes and other haematophagous insects. The steer or gas cylinders were enclosed within an open 8-ft square wire strand pen surrounded by an 8-ft high fence of ¼-in mesh screening. Insects attracted to the steer or carbon dioxide were caught in four Townes (1962) type Malaise traps placed one on each side of the pen. Of the six mosquito species trapped only *Psorophora confinnis* was collected in large numbers, and almost twice as many were caught when the steer was used as bait. About equal numbers of *Aedes vexans* and *Anopheles quadrimaculatus* were caught when either the steer or gas was used as bait.

In Malaysia eight mosquito species were caught in Malaise traps containing blocks of dry ice, the mean catch of *Culex vishnui* was 10.76/trapnight (Wallace et al. 1977).

Human landing catches

In certain situations a quick, almost instant surveillance method may be required to assess the mosquitoes in an area. In surveys in Mississippi, USA Harden and Poolson (1969) used several different adult sampling techniques together with larval surveys, but considered that with the possible exception of CO_2-baited CDC traps none gave consistent and reliable results, and moreover use of CO_2-baited CDC traps required overnight operation, whereas an on the spot surveillance technique was really needed (Harden et al. 1970). After preliminary experiments they developed a method in which 5-min human landing catches were made at any time during the day but were supplemented by carbon dioxide produced from about 3 lb of dry ice placed in a perforated bucket within 2 ft of the collector. The addition of carbon dioxide increased the number of species caught from 18 to 25 and the total catch by a factor of 2.8, although different species responded differently. For example, the catch of *Ochlerotatus mitchellae* was only 1.1 times greater when dry ice was used whereas that of *Aedes vexans* was increased 22.1-fold. Despite the presence of large *Culex quinquefasciatus* populations none was caught, thus supporting the finding of Reeves (1953) that this species is little attracted by carbon dioxide.

Lactic acid, octenol and other trap additives

The report of Acree et al. (1968) that certain esters of lactic acid could increase mosquito activity stimulated Stryker and Young (1970) to undertake field trials in South Carolina, USA to discover whether the addition of these esters could increase catches of mosquitoes in light-traps. Three New Jersey light-traps employing 25-W bulbs and four CDC light-traps with 6-V bulbs were used in conjunction with three levels of carbon dioxide and five levels of L(+)-lactic acid (= [S]-lactic acid) (Table 10.4). Large numbers of individuals and species of mosquito were caught. The catch was usually larger when the discharge rate of carbon dioxide was increased from 400 ml/min to 1250 ml/min, but increasing the output to 2470 ml/min did not give an appreciably bigger catch. It was concluded that little was gained by the addition of light to traps using carbon dioxide, and that except for *Aedes vexans* lactic acid did not enhance the catch. The physiological aspects of host detection, especially responses to lactic acid, are discussed by Davis (1984), while a good review of the effect of lactic acid and carbon dioxide on host seeking behaviour is given by Bowen (1991).

Table 10.4. Indices[a] of catch size of mosquitoes caught in light-traps supplemented with carbon dioxide and lactic acid (after Stryker and Young 1970)

	Carbon dioxide	L(+) lactic acid level (ml)				
	ml/min	0.00	0.001	0.005	0.01	0.1
Lights on	2 470	43.25	28.34	b	13.26	
	1 250	24.50	17.49	2.82	16.64	4.90
	450	16.32	16.75	3.23	6.24	6.17
Lights off	2 470	16.89	7.49	b	13.25	4.14
	1 250	12.50	7.16	31.39	b	13.91
	450	10.82	3.92	11.50	1.05	14.33

[a]Regression indices obtained by comparing the catches in the light-traps against catches in a control trap by fitting a straight line forced through origin of the plot of each treatment versus its corresponding control.
[b]Insufficient data.

Tsetse flies are attracted to acetone, carbon dioxide and 1-octen-3-ol (octenol), all of which are contained in host odours emitted by cattle; in fact the latter is the most potent stimulant identified from cattle. In field trials all three chemicals attract tsetse flies, and the latter greatly enhances the attractive properties of carbon dioxide (Hall et al. 1984). Similarly French and Kline (1989) reported that the addition of octenol increased about threefold catches of tabanids in canopy traps, with and without dry ice. Kline and Wood (1988) also found that when octenol was placed in CDC traps they caught greater numbers of *Culicoides furens* and *Culicoides mississippiensis* than in traps baited with carbon dioxide gas. When octenol was combined with carbon dioxide the catch was more than doubled. In later experiments (Kline et al. 1990a) the combination of octenol and carbon dioxide at 200 ml/min increased catches of *Culicoides furens* about 100 times more than just carbon dioxide.

The successful results obtained using octenol as an attractant for *Culicoides* prompted investigators to determine if octenol was also an attractant to mosquitoes. So Kline and Wood (1989) used a large outdoor cage (30 × 60 ft, and 16 ft high at the centre) to compare the efficiency of 10 different trapping methods, many incorporating CO_2, for sampling adult *Ochlerotatus taeniorhynchus*. Grass and weeds, augmented with large potted shrubs, were allowed to grow within the cage to provide a selection of resting and nectar-feeding sites. Some 1000 3–4-day-old adults were released into the cage for each test run. A weather station was established outside the cage. It became clear in the experiments that environmental conditions, which could fluctuate from night to night, were very important in influencing the numbers caught in the different traps. The highest catches of *Ochlerotatus taeniorhynchus* were obtained with CDC light-traps baited with 200 ml/min CO_2 and octenol, with or without the light source (mean percentage recovered — 9.23–47.90). Light alone in either CDC traps (0.14–0.42%) or New Jersey light-traps with 25-W (0.88–1.45%) or 40-W (0.70–1.78%) bulbs caught very few mosquitoes, although the so-called CBS trap caught means of 15.03–28.71% of the released *Ochlerotatus taeniorhynchus*. (This is a carbon dioxide suction trap developed by C. D. Morris & R. H. Lewis (unpublished) having a 15-lb carbon dioxide gas cylinder placed in a pit dug in the ground with a gas line conducting the gas up into a cylindrical chamber into which mosquitoes are sucked up by a fan).

In field trials in Florida, USA Kline and Wood (1989) found that the addition of octenol to CDC light-traps baited with either 200 or 500 ml/min carbon dioxide increased the catch of *Coquillettidia perturbans* (approx. 2 ×), *Culex salinarius* (3–4 ×) and *Psorophora columbiae* (2 ×). Addition of octenol did not change the catches of *Anopheles crucians* and *Anopheles quadrimaculatus*, but appeared to slightly decrease the numbers of *Culex erraticus*.

Further experiments by Takken and Kline (1989) and Kline et al. (1990a,b) showed that octenol had a synergistic role in attracting mosquitoes when combined with the release of 200 ml/min CO_2, and this was confirmed in later studies (Kline et al. 1991a). In these studies in Florida, USA unlit CDC traps were arranged in a 12 × 12 Latin Square to test 12 treatments, comprising carbon dioxide released at rates of 0, 20, 200 and 2000 ml/min and octenol released at 0, low (approx. 3.0 ± 0.3 mg/h), and high (approx. 41.1 ± 3.0 mg/h) levels. For this octenol was placed in 5-ml microreaction vials fitted with plastic tops and neoprene septa. The tops had a 12-mm hole through the middle. For a low level of release a wick comprising a doubled-over length of pipe cleaner, a few millimetres longer than the distance from the bottom of the vial to the septum positioned in the top, was inserted. This allowed the wick to remain in contact with the permeable septum and allowed continual octenol release. For the higher level of release the wick protruded through a 1.5-mm hole and rose to about 2 cm above the septum. Release levels were monitored daily by weighing the vials before and after the exposure periods. Vials were fixed near the trap entrance, and when used in combination with CO_2 were placed adjacent to the gas release site.

From 12 days of trapping from just 1600–1900 h 300 234 mosquitoes belonging to 12 species were caught; *Ochlerotatus taeniorhynchus* formed 81.0% of the catch, while *Wyeomyia mitchelli* and *Wyeomyia vanduzeei* formed 14.5% and *Culex nigripalpus* 3.5% of the catch. Similar trials, also in Florida, showed that all octenol and carbon dioxide combinations increased collections of *Coquillettidia perturbans, Culex salinarius* and *Psorophora columbiae*, relative to the equivalent carbon dioxide release rates given above (Kline et al. 1991*b*). It was difficult to interpret the results with *Anopheles crucians* and *Anopheles quadrimaculatus*, because sometimes the addition of octenol to carbon dioxide appeared to enhance collections, while at other times it appeared to repel these species, confusing previous results (Kline et al. 1990*a,b*).

The effects of octenol were not uniform (Table 10.5). For example, catches of *Ochlerotatus taeniorhynchus* were increased by the addition of octenol, but with *Culex nigripalpus* the addition of a high level of octenol to 20 and 200 ml/min CO_2 discharge rates decreased catches, while a low octenol level increased catches when 200 ml/min of carbon dioxide were released. Various other combinations produced increased or reduced catches. With both *Wyeomyia* species octenol reduced the numbers caught. Kline et al. (1991*a*) showed that the numbers of *Ochlerotatus taeniorhynchus* in standard catches employing a 200-ml/min CO_2 discharge rate can be increased by addition of low levels of octenol (2.4 ×), by just increasing the discharge rate to 2000 ml/min (2.0 ×), or by adding either the low (5.0 ×) or high (6.4 ×) levels of octenol to a 2000-ml/min discharge rate.

Kline and Lemire (1998) investigated the use of type 512 CDC light-traps supplemented with CO_2 and octenol and cloth targets impregnated with an insecticide (lambda-cyhalothrin) and baited with the same combination and release rates of attractants as the traps, as a potential mosquito control method on Key Island, Florida, USA. Use of baited insecticidal targets has proved highly successful in tsetse control in parts of Africa (e.g. Torr 1994; Vale 1993). Carbon dioxide was supplied from a 9 kg compressed gas cylinder at a flow rate of 200ml/min using a double stage pressure regulator and was released approximately 5 cm from the trap entrance. Traps were suspended from metal poles so that the top of the trap was approximately 1.8 m above ground level. Octenol was released from a 15 cm pipe-cleaner wick folded into a v-shape and inserted into 5 ml micro-reaction vials. The octenol release rate was approximately

4 mg/h over a 24 h period. Octenol vials were attached near the trap entrance adjacent to the CO_2 release point. The baited targets consisted of collapsible cylinders (60 cm length × 53 cm diameter) of 60% polypropylene black shade cloth treated with an EC formulation (120gm/l) of lambda-cyhalothrin 0.2 g A.I./m^2 and left open at the bottom to allow mosquitoes to ener and rest on the internal surface. Targets were suspended from poles so that the base of the cylinder was approximately 15 cm above ground. The octenol vial and CO_2 delivery tube were taped to a short stake positioned centrally under each target, approximately 67 cm above ground. In 1993 using 42 light-traps, more than 6.5 million mosquitoes were collected between 21 June and 16 July comprising 16 species or species groups. The three commonest species collected were *Ochlerotatus taeniorhynchus* (84.7%), *Culex nigripalpus* (13.8%), and *Anopheles atropos* (1.4%). Based on 1993 data, a protective barrier of 52 CDC traps was operated from 14 June to 11 July 1994. These traps yielded approximately 3 million mosquitoes with the same three species dominant and in similar proportions. The trap barriers did not appear to have any significant effect on the mosquito populations within or outside the barrier, as determined by sentinel trapping either side of the barrier, however, the authors still felt that the method showed promise as a control tool with modifications to the trap density and placement.

Table 10.5. Geometric mean numbers of mosquitoes collected at traps baited with CO_2 (n = 36) and octenol (n = 48) in Everglades National Park during August 1998 (after Kline et al. 1991*a*)

Species	Octenol level (mg/hr)			CO_2 level (ml/min)			
	No	Low	High	0	20	200	2000
Aedes taenio-rhynchus	236.2	753.5	618.5	49.4	205.4	1379.2	3427.9
Wyeomyia spp.	105.9	128.4	49.6	7.4	55.3	248.6	506.8
Culex nigri-palpus	14.8	21.8	11.8	1.9	6.6	35.2	88.1
Culex (Melanoconion) spp.	4.3	2.5	1.8	0.6	1.6	4.5	7.0

Estimated standard deviations of the logarithms are *Aedes taeniorhynchus*, 0.665; *Culex nigripalpus*, 0.742; *Wyeomyia*, 0.778; *Culex (Melanoconion)* spp., 0.743.

Kline and Mann (1998) investigated the effects of adding CO_2, butanone, and octenol as attractants to model 512 CDC light-traps in cypress and bay swamps in Florida, USA. Carbon dioxide and octenol were supplied using the methods described above (Kline and Lemire 1998). Butanone was released from a wick extended 2 cm through the lid of a 125-ml amber bottle at an approximate rate of (781 mg/h). Different attractant combinations were tested using a series of Latin square experimental designs. In the CO_2 and light trials, 6321 mosquitoes of 17 species were collected. The largest collections were from light-plus-CO_2 traps, followed by CO_2 alone, then light. Four replicates comparing different CO_2 flow rates conducted over 100 trap nights yielded a total of l3 00l individuals of 22 species. Both the number of mosquitoes collected and species diversity increased with each increase in CO_2 release rate (Table 10.6), although some species responded differently to the changing concentrations. *Ochlerotatus atlanticus, Ochlerotatus canadensis* and *Anopheles crucians* were most attracted to CO_2 + octenol, while addition of butanone decreased the responses of *Culiseta melanura, Ochlerotatus atlanticus, Ochlerotatus canadensis* and *Anopheles crucians* (Table 10.7).

The effects of supplementing standard CDC miniature light-traps with CO_2 and octenol have also been tested in Queensland, Australia by van den Hurk et al. (1997). The light source used was a 6 V incandescent bulb. Carbon dioxide was provided from a 1 kg block of dry ice placed in a 4 litre Styrofoam box that had four 7 mm holes in its base, and which was suspended above the CDC trap. Octenol vapour was released from glass vials fitted with a protruding wick after the method of Kemme et al. (1993). Significant differences in trap catches by treatment were observed for *Anopheles farauti* No. 1, *Anopheles farauti* No. 2, and *Anopheles faruati* s.l. at Smithfield and for *Anopheles farauti* No. 2 and *Anopheles farauti* s.l. at Eubenangee swamp, but no differences were observed at Wyvari swamp. Addition of octenol significantly increased the catch of *Anopheles farauti* 1, 2 and s.l. at Smithfield when compared with CO_2 alone, but the effect was not significant at Wyvari swamp. The use of octenol alone resulted in capture of only a few *Anopheles farauti* No. 2 and no *Anopheles farauti* No. 3.

Ryan et al. (2000) successfully used CDC light-traps baited with CO_2 and octenol to collect potential Ross River virus vectors in Queensland, Australia. A total of 27 529 mosquitoes of fifteen species were collected from February to June and Ross River virus isolates were recovered from 6 of the 7 most abundant species. Van den Hurk et al. (2000) also in Queensland, Australia, used CDC light-traps with CO_2 and octenol (release rate 4.5 mg/h) to study the seasonal abundance of *Anopheles farauti* sibling species. A total of 45 041 individuals of the *Anopheles farauti* complex were collected, and based on PCR identification of a sample, the relative proportions of the three sibling species were estimated at 3.6% *Anopheles farauti* s.s., 65.1% *Anopheles faruati* No. 2, and 31.3% *Anopheles farauti* No. 3.

Table 10.6. Mean (SE) response of the most commonly collected mosquitoes to Model 512 Communicable Disease Center (CDC)-type traps, without light, baited with carbon dioxide (CO_2) at 5 different release rates.[1] (Kline and Mann 1998)

Species	CO_2 release rate (ml/min)					
	2	20	100	200	2,000	$P \leq$
Aedes atlanticus	14.05 (8.93)c	45.40 (41.07)c	59.00 (31.69)b	96.75 (67.99)b	195.15 (123.13)a	0.0001
Aedes canadensis canadensis	0.05 (0.05)b	0.05 (0.05)b	0.15 (0.15)b	0.04 (0.31)b	33.00 (30.65)a	0.005
Aedes dupreii	6.80 (6.75)a	5.85 (5.33)a	14.95 (14.53)a	10.75 (9.52)a	0.33 (0.46)a	0.25
Aedes infirmatus	0.15 (0.11)d	0.35 (0.13)c,d	3.45 (1.47)b,c	4.90 (1.78)b	5.60 (2.07)a	0.0001
Anopheles crucians	0.20 (0.12)c	0.35 (0.17)c	4.45 (2.00)b	5.05 (2.55)b	13.50 (4.22)a	0.0001
Coquillettidia perturbans	0.00 (0.00)b	0.00 (0.00)b	0.35 (0.25)b	0.65 (0.36)a,b	2.20 (1.16)a	0.0001
Culex erraticus	0.05 (0.05)b	0.10 (0.07)a,b	0.55 (0.36)a,b	0.75 (0.34)a,b	0.40 (0.15)a	0.05
Culex nigripalpus	1.10 (0.57)c	3.25 (0.93)b	5.05 (1.51)b	7.75 (2.85)b	11.40 (3.40)a	0.0001
Culex salinarius	0.15 (0.11)c	0.85 (0.39)b,c	1.00 (0.43)b	1.75 (0.73)a,b	3.00 (1.13)a	0.0001
Culiseta melamura	4.45 (1.77)b	18.75 (10.32)a,b	17.95 (6.15)a,b	23.70 (11.52)a,b	11.40 (4.23)a	0.001
Psorophora columbiae	0.00 (0.00)b	0.20 (0.16)b	0.45 (0.20)b	1.30 (0.92)b	5.15 (3.01)a	0.0001
Psorophora ferox	0.00 (0.00)b	0.00 (0.00)b	0.35 (0.26)b	0.90 (0.48)b	2.15 (1.10)a	0.0001

[1] n = 20 trap-nights. Means in the same row followed by the same letter are not significantly different; Ryan-Einot-Gabriel-Welsh multiple range test applied to ranked data.

Table 10.7. Mean (SE) response of freshwater swamp mosquitoes to Model 512 Communicable Disease Center (CDC)-type traps (no light) baited with combinations of butanone, 1-octen-3-01 (octenol), and carbon dioxide (CO_2) (n = 10) trap nights (Kline and Mann 1998)

Species	Butanone	Octenol	CO_2	Butanone + CO_2	Octenol + CO_2
Ochlerotatus atlanticus	0.80 (0.33)	5.50 (2.20)	6.20 (2.71)	5.20 (1.84)	37.10 (11.97)
Ochlerotatus canadensis	0.60 (0.31)	4.50 (1.42)	55.10 (12.95)	48.60 (l0.85)	224.30 (41.42)
Ochlerotatus infirmatus	0.00 (0.00)	0.10 (0.10)	0.80 (0.49)	2.20 (0.77)	4.10 (1.67)
Ochlerotatus mitchellae	0.00 (0.00)	0.00 (0.00)	0.10 (0.10)	0.00 (0.00)	0.00 (0.00)
Aedes vexans	0.00 (0.00)	0.00 (0.00)	0.10 (0.10)	0.10 (0.10)	0.20 (0.13)
Anopheles crucians	0.60 (0.27)	4.40 (1.86)	106.30 (47.90)	90.70 (31.00)	446.40 (122.32)
Anopheles punctipennis	0.00 (0.00)	0.00 (0.00)	0.00 (0.00)	0.00 (0.00)	0.10 (0.10)
Anopheles quadrimaculatus	0.00 (0.00)	0.00 (0.00)	0.00 (0.00)	0.00 (0.00)	0.80 (0.49)
Coquillettidia perturbans	0.10 (0.10)	0.30 (0.30)	4.50 (1.67)	8.50 (2.62)	20.50 (11.05)
Culex nigripalpus	0.00 (0.00)	0.00 (0.00)	0.10 (0.10)	0.00 (0.00)	0.00 (0.00)
Culex restuans	0.00 (0.00)	0.00 (0.00)	0.10 (0.10)	0.00 (0.00)	0.00 (0.00)
Culex salinarius	0.00 (0.00)	0.60 (0.40)	16.60 (4.93)	23.10 (13.78)	42.90 (10.05)
Culiseta melanura	1.10 (0.35)	1.00 (0.33)	13.50 (5.28)	8.30 (1.54)	1.20 (0.33)
Psorophora ferox	0.00 (0.00)	0.00 (0.00)	0.10 (0.10)	0.40 (0.31)	0.40 (0.27)

CDC light-traps baited with CO_2 and octenol were also used to undertake entomological investigations following an outbreak of Japanese encephalitis virus on Badu Island in the Torres Strait, which separates Australia and the island of New Guinea (Johansen et al. 2001). A total of twenty-five traps were operated over two nights, yielding 31 898 mosquitoes comprising 16 species. Virus isolations were obtained from *Culex sitiens* group mosquitoes.

Light-traps baited with CO_2 + octenol are commonly used to collect mosquitoes as part of arbovirus studies in the Australasian region. For example, Ritchie et al. (1997) used CDC light-traps supplemented with CO_2 and octenol to collect adult mosquitoes following an outbreak of poly-arthritis in suburban Brisbane, Australia, in 1994. From 181 trap nights, a total of 29 931 female mosquitoes were collected and 63 isolations of Ross River virus were obtained from 7 species of mosquito. Van den Hurk et al. (2003*a*) used CO_2-baited CDC light-traps to collect mosquitoes for Japanese encephalitis vector competence studies in Australia. Carbon dioxide was provided in the form of 1 kg of dry ice and 1-octen-3-ol was provided at a release rate of 4.5 mg/h. Traps were operated for 12 h, from 1800 to 0600 h. Seven species were collected, namely: *Aedes aegypti, Ochlerotatus notoscriptus, Ochlerotatus normanensis, Ochlerotatus purpureus, Culex sitiens, Mansonia septempunctata, Mansonia uniformis,* and *Verrallina carmenti.*

Van den Hurk et al. (2003*b*) used similar methods to those described above (van den Hurk et al. 2003*a*) to collect mosquitoes in a study of Japanese encephalitis virus transmission in northern Australia and Papua New Guinea. From a total of 570 trap collections between 1995 and 2001, a total of 3658 blood-fed mosquitoes of 15 species were collected, with the majority (87.4%) being *Culex annulirostris.*

Johansen et al. (2004) successfully used CDC light-traps baited with CO_2 and 1-octen-3-ol to collect mosquitoes for Flavivirus studies in the Torres Straits in Australia. Seventy-five traps operated over 15 nights yielded 84 210 adult mosquitoes, of which the most abundant were *Culex sitiens* subgroup (53.1%), *Coquillettidia crassipes* (18.7%), *Verrallina funerea* (18.4%), *Anopheles farauti* (4.1%), and *Ochlerotatus kochi* (3.2%).

Jeffery et al. (2002) used a 2×10 grid pattern consisting of 20 CDC light-traps supplemented with CO_2 and octenol after the method of Kemme et al. (1993) to investigate the spatial distribution of Ross River virus vectors in Australia. Traps were positioned approximately 1 km apart, and set 2 h before sunset and retrieved the next morning between 0800 and 0930 h. Universal kriging, a geostatistical method that interpolates values between observations when there is large variation in local means from different geographical areas, was applied to species counts obtained from weekly light trap collections. Mosquito counts were interpolated at 100 m × 100 m grid intervals. Analyses were performed using ArcView GIS 3.2a ([ESRI] Environmental Systems Research Institute Inc 1996a) and Arc View Spatial Analyst extension (ESRI 1996b). During 140 trap-nights on Russell Island, 1 365 564 female mosquitoes (mean 9754 per trap) comprising 25 species were collected. *Ochlerotatus vigilax* was the most commonly trapped mosquito, followed by *Coquillettidia lienalis*. Spatial analysis revealed significant clustering of higher than expected mosquito catches on Russell Island for *Coquillettidia linealis, Culex annulirostris, Ochlerotatus vigilax* and *Verrallina funerea*, and this was apparently associated with the presence of saltmarsh and mangrove areas.

In Germany, Becker et al. (1995) supplemented CDC light-traps with CO_2, octenol and host odours emanating from a caged hamster and examined the effects on mosquito catches. Three trap sites were established 200 m apart with four traps at each site, comprising a standard CDC light-trap, a CDC trap supplemented with CO_2, a trap with CO_2 and octenol, and a trap with octenol alone. The traps were positioned at the corners of a 50 m square and positions were rotated each night. Traps were operated over two series of four nights from 2130 – 0530 h during August 1991. Carbon dioxide-baited traps were supplemented with 1800 g of dry ice releasing 225g CO_2/h. Octenol was provided from 8 ml vials lined with filter paper soaked with 600 µl of octenol, which gave an average release rate of 107 mg/8 h. In the traps using a caged hamster as bait, the traps were fitted with an automatic timer that operated the fan for 5 min each hour during the trapping period. Of the 2551 mosquitoes collected, the four most abundant species were *Aedes vexans* (40%), *Aedes rossicus* (24%), *Aedes cinereus* (4.6%), and *Culex pipiens* (30%). Traps baited with CO_2 or CO_2 + octenol caught significantly more mosquitoes than unbaited traps or those baited with octenol alone. There was no synergistic effect between CO_2 and octenol in contrast to the results reported by Takken and Kline (1989) and Kline et al. (1990a, 1990b, 1991a, b). The attractiveness of the hamster as bait increased over the four days of the experiment, presumably as the smell from its cage increased. On days three and four the hamster was more attractive than the CO_2-baited traps.

Rueda et al. (2001) and Rueda and Gardner (2003) evaluated octenol, CO_2, and light as attractants in CDC light-traps in field trials in North Carolina, USA. The CO_2 was provided by way of a 2-litre plastic thermos containing about 1 kg of dry ice. The average CO_2 release rate was calculated to be 57 g/h. Octenol was provided by way of a patented wax-like medium (Biosensory Inc., Willimantic, Connecticut, USA) that released 1.5 mg/h at 27°C and was packaged in a crush-resistant plastic housing containing 3-mg octenol. The CO_2 container and octenol package were suspended near the trap intake. Traps were suspended approximately 1.3 metres above ground level. Traps were set between 1630 and 1700 h, and emptied the following morning between 0800 and 0830 h. The responses of mosquitoes to CDC traps baited with octenol + CO_2 + light, octenol + CO_2, octenol + light, and octenol alone were compared with traps baited with CO_2 + light using a 5×5 Latin square design. From 25 trap-nights in the salt marsh area, 56 855 mosquitoes comprising 12 species in 4 genera were collected. The general response pattern was for total collection size for octenol + CO_2 + light > octenol + CO_2 = CO_2 + light > octenol + light > octenol alone. *Ochlerotatus sollicitans*, *Ochlerotatus taeniorhynchus*, *Anopheles bradleyi*, and *Culex salinarius* were most attracted by octenol + CO_2 + light. The responses of *Ochlerotatus sollicitans*, *Ochlerotatus taeniorhynchus*, and *Culex salinarius* to octenol + CO_2 did not differ significantly from their responses to CO_2 + light. *Ochlerotatus sollicitans* and *Ochlerotatus taeniorhynchus* were least attracted by octenol + light and octenol alone. Octenol alone attracted fewer mosquitoes than any other treatment in most cases. *Anopheles punctipennis* and *Culex pipiens* were collected only from traps baited with octenol + CO_2 + light. At the flood plain site 19 498 mosquitoes (24 species in 7 genera) were collected over 50 nights. The response pattern was that total collection size for octenol + CO_2 + light = CO_2 + light > octenol + CO_2 > octenol + light > octenol alone. In the salt marsh area, the total mosquito catch was significantly greater ($P < 0.001$) in CO_2-baited light traps with octenol than those without octenol. In the creek flood plain, however, there was no significant difference in mosquito catches between CO_2-baited light traps with or without octenol.

The effects of supplementing CDC updraft UV light-traps with CO_2 and octenol on catches of anopheline mosquitoes in Venezuela was investigated by Rubio-Palis (1996). UV light-traps were operated without bait, with CO_2 alone, octenol alone, and CO_2 + octenol. Carbon dioxide was supplied from a 300 g block of dry ice wrapped in newspaper and hung approximately 50 cm from the traps. Octenol was supplied from 1 ml of octenol in a 2-dram vial attached close to the trap entrance. The estimated rate of evaporation of octenol was 92 ± 15 μl/h. Five species of *Anopheles* were collected: *Anopheles aquasalis*, *Anopheles albimanus*, *Anopheles apicimacula*, *Anopheles punctimacula*, and *Anopheles punctipennis*. Human landing catches caught more mosquitoes than any of the traps. To estimate the relative efficiency of the different traps at catching *Anopheles albimanus* and *Anopheles aquasalis*, a logistic regression model was used to estimate adjusted odds ratios from $[p_2/(1-p_2)]/[p_1/(1-p_1)]$, where p_2 is the proportion of either species in the total female catch of both species in a particular trap, and p_1 is the corresponding data for human landing catch.

Odds ratios for *Anopheles aquasalis* were significantly less than one, indicating low efficiency relative to the human landing catch. In contrast, the odds ratios for *Anopheles albimanus* were greater than 1 for all trapping methods. Addition of octenol, with or without CO_2 increased the catch of *Anopheles albimanus*, but had no effect on the numbers of *Anopheles aquasalis* caught.

The effects of carbon dioxide and 1-octen-3-ol on EVS trap catches was tested in Queensland, Australia, by Kemme et al. (1993). Carbon dioxide was supplied from a 6 kg gas cylinder at 200 ml/min using a pressure regulator and a fine flow regulator. Octenol was supplied from microreaction vials fitted with plastic lids and neoprene septa. Each lid had a 9-mm hole in it, and each septum had a 2-mm hole, through which a doubled-over cotton pipe cleaner was inserted, with 8–9 mm extending outwards. The mean octenol release rate obtained using this method was 5.12 ± 0.48 mg/h when operated without CO_2, and 6.60 ± 2.26 mg/h when CO_2 was also used. *Ochlerotatus vigilax* was the most commonly collected species, comprising 63.6% of the total catch. The number of mosquitoes collected in unbaited or octenol-baited traps was much lower than in CO_2-baited traps. Addition of octenol to CO_2 increased trap catches, but not significantly.

Ritchie et al. (1994) also used EVS traps (Rohe and Fall 1979) to study the effects of CO_2, light and octenol on trap catches of biting midges in Queensland, Australia. Traps were placed 200 m apart and perpendicular to the prevailing easterly wind. Traps were baited as follows: (1) no bait, (2) octenol, (3) CO_2, and (4) octenol + CO_2. Octenol was released at a mean ($\pm$ SEM) rate of 6.05 ± 0.31 mg/h from 4-ml microreaction vials with an exposed cotton pipe cleaner wick (Kemme et al. 1993). A release rate of 200 ml/min of CO_2 gas was obtained using a pressure regulator fitted with a fine flow regulator. Traps were operated from 1500 to 0900 h and the position of traps was rotated so that each treatment occupied each position for one night at each site. In total, 16 387 individual midges were trapped. The three most dominant species were *Culicoides molestus* (44.94%), *Culicoides histrio* (25.76%), and *Culicoides subimmaculatus* group (19.48%). Octenol by itself had no significant effect on trap catches. group (19.48%). Octenol by itself had no significant effect on trap catches. ($P > 0.05$). A mixture of CO_2 and octenol generally increased catches. Carbon dioxide significantly increased the mean number of *Culicoides histrio* and *Culicoides subimmaculatus* trapped. ($P < 0.05$). Of the five species or species groups collected in sufficient numbers for analysis, the addition of octenol to CO_2 significantly increased catches only of *Culicoides molestus*.

Miller et al. (2005) compared the use of CO_2- and octenol-baited EVS traps for the surveillance of Ross River virus, Barmah forest virus and dengue virus vectors at a military training camp in Queensland, Australia.

At one of three study sites, traps baited with both CO_2 and octenol caught a significantly higher mean number of species (3.67), compared with either CO_2-baited or octenol-baited traps. Traps baited with either CO_2 or octenol did not differ significantly in the mean number of species collected. At the Freshwater Beach site, traps baited with CO_2 + octenol captured more individuals of *Ochlerotatus vigilax, Mansonia uniformis*, and *Coquillettidia xanthogaster* than the traps baited with either CO_2 or octenol. At the other two study sites, addition of octenol to CO_2-baited traps did not significantly influence the number of species collected nor the particular species collected, however, it did appear to increase the total numbers of individuals trapped, albeit not significantly.

EVS traps were used by Cooper et al. (2004) to study the effectiveness of light, octenol, and carbon dioxide as attractants for anopheline mosquitoes in Papua New Guinea. Octenol was supplied to the traps using a wick that protruded 10 mm from a reservoir vial that was placed directly above the trap entrance. The mean release rate of octenol was 9.5 ± 0.73 mg/h. The light source was a 1.5 V bulb. A total of 1034 individual anopheline mosquitoes were collected, including *Anopheles koliensis* (439), *Anopheles farauti* 2 (232), *Anopheles farauti* 4 (32), *Anopheles longirostris* (223), and *Anopheles bancroftii* (114). Light traps caught very few anophelines. Carbon dioxide and octenol additives varied in their attractiveness according to species and trap location. Addition of octenol made the traps more effective than light alone at collecting *Anopheles koliensis, Anopheles bancroftii*, and *Anopheles farauti* s.l. at the Umun study site and *Anopheles farauti* s.l. only at the Pumpres study site. Carbon dioxide significantly increased trap collections of *Anopheles koliensis* at both sites and *Anopheles longirostris* at one site. The octenol + CO_2 combination had a significant synergistic effect on catch numbers for *Anopheles longirostris* only, and this was ascribed by the authors as being due to this species' zoophilic nature. The combination reduced the catches of *Anopheles bancroftii* at the Umun site and *Anopheles koliensis* at the Pumpres site.

Russell (2004) compared the efficiency of CDC miniature light-traps and EVS traps when supplemented with CO_2 and octenol on catches of *Aedes aegypti, Aedes polynesiensis* and *Culex quinquefasciatus* in Moorea, French Polynesia. CDC traps were set in groups of four in each of three localities, with one trap unbaited (i.e. light alone) and the other traps supplemented with either octenol in a glass vial with cotton pipe-cleaner as a wick, as per Van Essen et al. (1994), carbon dioxide (dry ice), or carbon dioxide plus octenol. The traps were placed in positions sheltered from wind, operated continuously for 24 h a day for four days with octenol continuously available and dry ice provided twice a day. Traps were rotated between the four positions on a daily basis to minimise any position effects. Comparisons between the two trap types were conducted as follows: two of each type of trap were operated in each of three localities. One trap of each type was baited with CO_2 alone and the other was baited with CO_2 plus octenol. Human landing collections were also undertaken in the vicinity of the traps for 15 min in the two hours prior to sunset. The relative abundance of the different species varied considerably between sites. The addition of octenol to light did not significantly increase the collections of any species. Carbon dioxide plus light significantly increased collections of each species at all three sites, but the supplementation of CO_2 with octenol did not increase collections further. For *Aedes aegypti* for example the mean number collected in CDC traps was as follows: light alone, 0.33 ± 0.58; light + octenol, 1.00 ± 1.00; light + CO_2, 15.00 ± 12.49; and light + octenol + CO_2, 11.00 ± 8.66. In the comparative trial, no significant differences in mean numbers of *Aedes aegypti* and *Aedes polynesiensis* collected by either CDC or EVS traps were observed, no matter which combination of attractants was used. The combination of octenol and carbon dioxide significantly reduced collections of *Culex quinquefasciatus* in CDC traps, but not in EVS traps.

In later studies, Russell and Hunter (2005) baited CDC light-traps (bulb removed) with bird uropygial gland odours in addition to CO_2, in an attempt to collect more *Culex pipiens* and *Culex restuans*, the ornithophilic potential West Nile virus vectors in Ontario, Canada. Baiting of the traps was conducted as follows: two frozen uropygial glands from the crow (*Corvus brachyrhynchus*) were thawed and the contents squeezed onto a petri dish. Ten cotton swabs were then rolled in the secretions and the swabs were individually placed into ten 14-ml Falcon tubes and transported to the field study site. Cotton swabs baited with the secretion were allocated randomly to half of the traps and were taped to the trap near the entrance to the fan. Traps were set at heights of 1.5 and 5 m above ground. From 280 catches, a total of 2482 mosquitoes was captured, of which *Culex pipiens/restuans* was the most abundant group (56.6%), followed by *Aedes vexans* (23.3%). Other species captured included *Culex territans, Ochlerotatus trivittatus, Ochlerotatus triseriatus, Ochlerotatus dorsalis, Anopheles quadrimaculatus, Anopheles punctipennis, Anopheles barberi,* and *Uranotaenia sapphirina.* No significant difference was observed in the overall median abundance of *Culex pipiens/restuans* in traps baited with bird odour compared with control traps. However, when taking into account trap height, a significantly greater number of *Culex pipiens/restuans* were captured in bird odour-baited traps set at 5 m versus control traps set at the same height ($\chi^2 = 9.33$, $P = 0.002$).

In laboratory studies Canyon and Hii (1997) observed that the addition of lactic acid and octenol to Fay-Prince traps decreased the catch of *Aedes aegypti* by about 50%. In field trials in Townsville, Australia, traps baited with lactic acid or octenol failed to trap any *Aedes aegypti*, although they were caught in human landing catches and in CO_2-baited traps. The authors ascribed the lack of attractancy to octenol to the fact that it is a by-product of rumination, and *Aedes aegypti* does not feed on ruminants.

Nilssen (1998) studied the effects of octenol on flight trap catches (Anderson and Nilssen 1996) of mosquitoes and tabanids in subarctic Norway. Traps were set up in four pairs, with one member of each pair acting as the test, the other as the control and were operated for 24 h. Paired traps were separated by approximately 50 m. Octenol was provided from an open 14 ml vial with a protruding cotton wick. The estimated rate of evaporation was 40 mg/h. In contrast to several other studies quoted here, addition of octenol increased mosquito (*Ochlerotatus communis, Ochlerotatus pionips*, and *Ochlerotatus punctor*) catches as well as the catch of *Hybomitra lundbecki*. Mean catches of mosquitoes were 770.5 per trap in octenol-baited traps, compared with 178.6 for control traps ($P = 0.0001$).

Mboera et al. (2000*b*) compared the effects of supplementing CDC light-traps and Counter-Flow Geometry (CFG) traps with CO_2, octenol, acetone, butyric acid and human foot odour on mosquito catches in Tanzania. Human foot odour was obtained from nylon stocking that had been worn by a human male volunteer for between four and seven days. The stocking was placed in a polythene bag and attached to the CO_2 supply point of the CFG trap and to the CDC trap using a 20 cm length of 5mm internal diameter silicon tubing. Carbon dioxide was delivered from a pressurised cylinder at a rate of 300 ml/min. Acetone was delivered through a silicon tube attached to a glass bottle containing 4 ml of acetone. Butyric acid was supplied from a 4-ml glass tube, also attached with silicon tubing. Traps were operated outdoors approximately 30 m from the nearest house.

A total of 1508 mosquitoes were collected by the different combinations of traps and attractants, including *Culex quinquefasciatus* (87.7%), *Anopheles gambiae* (2.9%), *Culex cinereus* (2.1%), *Anopheles coustani* (1.8%), *Anopheles funestus* (1.0%) and *Mansonia africana* (0.9%). Only *Culex quinquefasciatus* was collected in large enough numbers to permit statistical analysis. Counterflow Geometry traps baited with human foot odour caught significantly more mosquitoes than unbaited traps, but CFG traps baited with CO_2 caught more mosquitoes than traps baited with foot odour. Carbon dioxide-baited traps also yielded larger catches than those baited with acetone or butyric acid. CDC traps baited with CO_2 yielded 12 times more *Culex quinquefasciatus* than an unbaited trap and nine times more than a trap baited with octenol alone. The combination of CO_2 and octenol did not significantly affect the size of the catch.

Silva et al. (2005) recently tested a synthetic blend of chemicals comprising volatiles released by humans (acetone, L-lactic acid, and dimethyl disulphide) in CDC traps, a CFG trap, and a Fay-Prince trap in the laboratory. All traps when baited with the synthetic human odour caught significantly more *Aedes aegypti* females than un-baited traps. The baited CFG trap caught significantly more *Aedes aegypti* females than the other types of trap.

The highly variable results obtained from the use of different attractants in different traps in different geographical locations and under different experimental and microclimatic conditions described above, only serve to illustrate that different species appear to respond differently to single attractants and combinations of attractants, making it very difficult to generalise regarding the effect of attractants on mosquito trap catches. The only advice that can be given is to undertake investigations on the effects of attractants on trap catches using the chosen traps and the specific locations in which trapping will take place.

Parity

Reisen and Pfuntner (1987) pointed out that in their trials in California, USA, and at least in some other studies, most potential *Culex* virus vectors caught in carbon dioxide traps are unfed nullipars, thus greatly reducing the likelihood of detecting arboviruses. They pointed out that when 195 705 mosquitoes caught in a surveillance programme in 1985 were tested just 28 isolates of western equine encephalomyelitis and 30 isolates of St Louis encephalitis were recovered (Emmons et al. 1986). Reisen et al. (1995*b*), in recognition of the fact that in nature mature autogenous female *Culex tarsalis* are not attracted to CO_2 traps, corrected the parity rate for the autogeny rate as follows:

$$CPAR = [PAR \times (1 - AUT)], \text{ with } PAR = CPAR/[CPAR + NP] \quad (10.1)$$

CPAR = corrected parity rate, *PAR* = parity rate, *AUT* = autogeny rate (proportion of females collected as pupae that developed their first egg batch without a blood meal), *NP* = number of host-seeking nullipars.

Reisen and Pfuntner (1987) concluded that most trapping programmes are not very well suited for monitoring arbovirus infection rates. Also in

California, USA they found that in one locality more *Culex quinquefasciatus* were caught in dry ice traps than in gravid traps (Reiter 1983), but only 42.2% of them were parous. This meant that only 23.3 and 27.4 females/trap-night at ground level and at a height of 5 m could be potentially infected with viruses, and this was comparable to the numbers caught in their gravid trap (Reisen et al. 1988). In contrast in another locality in California, where mosquito populations were smaller, more *Culex quinquefasciatus* females were collected in gravid traps than in dry ice traps (Reisen et al. 1988). Morris et al. (1976) found that 56% of *Culiseta melanura* caught in carbon dioxide-baited CDC light-traps were parous, whereas only 18% of those found resting in artificial shelters (red boxes) were parous. The CDC trap was very inefficient in collecting *Culiseta morsitans* except during September, when almost all those caught were parous.

References

Acree F, Turner RB, Gouck HK, Beroza M, Smith N (1968) L-lactic acid: A mosquito attractant isolated from humans. Science 161: 1346–1347

Allan SA, Surgeoner GA, Helson BC, Pengelly DH (1981) Seasonal activity of *Mansonia perturbans* adults (Diptera: Culicidae) in southwestern Ontario. Can Entomol 113: 133–139

Alten B, Bellini R, Caglar SS, Simsek FM, Kaynas S (2000) Species composition and seasonal dynamics of mosquitoes in the Belek Region of Turkey. J Vector Ecol 25: 146–154

Anderson JR, Nilssen AC (1996) Trapping oestrid parasites of reindeer: the response of *Cephenemyia trompe* and *Hypoderma tarandi* to baited traps. Med Vet Entomol 10: 337–346

Anderson JF, Andreadis TG, Main AJ, Kline DL (2004) Prevalence of West Nile virus in tree canopy-inhabiting *Culex pipiens* and associated mosquitoes. Am J Trop Med Hyg 71: 112–119

Atmosoedjono S, Djoharti Purnomo, Bangs MJ (1993) *Anopheles balabacensis* (Diptera: Culicidae), a vector of *Wuchereria kalimantani* (Nematoda: Onchocercidae) in East Kalimantan (Borneo), Indonesia. Med Vet Entomol 7: 390–392

Bailey SF, Eliason DA, Hoffman BL (1965) Flight and dispersal of the mosquito *Culex tarsalis* Coquillett in the Sacramento valley of California. Hilgardia 37: 73–113

Baldwin WF, Gross HP (1972) Fluctuations in numbers of adult black flies (Diptera: Simuliidae) in Deep River, Ontario. Can Entomol 104: 1465–1470

Barynin JAM, Wilson MJG (1972) Outdoor experiments on smell. Atmos Environ 6: 197–207

Bates M (1944) Notes on the construction and use of stable traps for mosquito studies. J Natn Malar Soc 3: 135–145

Becker N, Zgomba M, Petric D, Ludwig M (1995) Comparison of carbon dioxide, octenol and a host-odour as mosquito attractants in the Upper Rhine Valley, Germany. Med Vet Entomol 9: 377–380

Bellamy RE, Reeves WC (1952) A portable mosquito bait-trap. Mosquito News 12: 256–258

Bellini R, Veronesi R, Draghetti S, Carrieri M (1997) Study on the flying height of *Aedes caspius* and *Culex pipiens* females in the Po Delta area, Italy. J Am Mosq Control Assoc 13: 356–360

Blackmore MS, Dahl C (2002) Field evaluation of a new surveillance trap in Sweden. J Am Mosq Control Assoc 18: 152–157

Blume RR, Miller JA, Eschle JL, Matter JJ, Pickens MO (1972) Trapping tabanids with modified Malaise traps baited with CO_2. Mosquito News 32: 90–95

Bowen MF (1991) The sensory physiology of host-seeking behavior in mosquitoes. Annu Rev Entomol 36: 139–158

Brockway PB, Eliason DA, Bailey SF (1962) A wind directional trap for mosquitoes. Mosquito News 22: 404–405

Buei K, Nakajima S, Itô S, Nakamura H, Yoshida M, Fujito S, Kunita N (1986) Ecological studies on the overwintering of mosquitoes, especially of *Culex tritaeniorhynchus* Giles, in Osaka prefecture. 1. Notes on the dry ice- and emergence-trapping in spring at terraced rice field areas, 1967–1975. Jap J Sanit Zool 37, 333–340 (In Japanese, English summary)

Burkett DA, Lee WJ, Lee KW, Kim HC, Lee HI, Lee JS, Shin EH, Wirtz RA, Cho HW, Claborn DM, Coleman RE, Klein TA (2001) Light, carbon dioxide and octenol-baited mosquito trap and host-seeking activity evaluations for mosquitoes in a malarious area of the Republic of Korea. J Am Mosq Control Assoc 17: 196–205

Burkett DA, Kelly R, Porter CH, Wirtz RA (2004) Commercial mosquito trap and gravid trap oviposition media evaluation, Atlanta, Georgia. J Am Mosq Control Assoc 20: 233–238

Calisher CH, Lazuick JS, Justines G, Francy DB, Monath TP, Gutierrez VE, Sabattini MS, Bowen GS, Jakob WL (1981) Viruses isolated from *Aedeomyia squamipennis* mosquitoes collected in Panama, Ecuador and Argentina: establishment of the Gamboa serogroup. Am J Trop Med Hyg 30: 219–223

Canyon DV, Hii JLK (1997) Efficacy of carbon dioxide, 1-octen-3-ol, and lactic acid in modified Fay-Prince traps as compared to man-landing catch of *Aedes aegypti*. J Am Mosq Control Assoc 13: 66–70

Carestia RR, Horner KO (1968) Analysis of comparative effects of selected CO_2 flow rates on mosquitoes using CDC light traps. Mosquito News 28: 408–411

Carestia RR, Savage LB (1967) Effectiveness of carbon dioxide as a mosquito attractant in the CDC miniature light trap. Mosquito News 27: 90–92

Cilek JE, Kline DL (2002) Adult biting midge response to trap type, carbon dioxide, and an octenol-phenol mixture in northwestern Florida. J Am Mosq Control Assoc 18: 228–231

Cilek JE, Hallmon CF (2005) The effectiveness of the Mosquito Magnet trap for reducing biting midge (Diptera: Ceratopogonidae) populations in coastal residential backyards. J Am Mosq Control Assoc 21: 218–221

Comtois A, Berteaux D (2005) Impacts of mosquitoes and black flies on defensive behaviour and microhabitat use of the North American porcupine (*Erethizon dorsatum*) in southern Quebec. Can J Zool 83: 754–764

Cooper RD, Frances SP, Waterson DGE, Piper RG, Sweeney AW (1996) Distribution of anopheline mosquitoes in northern Australia. J Am Mosq Control Assoc 12: 656–663

Cooper RD, Frances SP, Popat S, Waterson DGE (2004) The effectiveness of light, 1-octen-3-ol, and carbon dioxide as attractants for anopheline mosquitoes in Madang Province, Papua New Guinea. J Am Mosq Control Assoc 20: 239–242

Cornet M, Chateau R (1971) Intérêt du gaz carbonique dans les enquêtes sur les vecteurs sylvatiques du virus amaril. Cah ORSTOM sér Entomol Méd Parasitol 9: 301–305

Coupland JB (1991) The ecology of black flies (Diptera: Simuliidae) in the Scottish highlands in relation to control. Ph.D thesis, University of Aberdeen

Crans WJ, Sprenger DA (1996) The blood-feeding habits of *Aedes sollicitans* (Walker) in relation to eastern equine encephalitis virus in coastal areas of New Jersey. III. Habitat preference, vertical distribution, and diel periodicity of host-seeking adults. J Vector Ecol 21: 6–13

Davis EE (1984) Regulation of sensitivity in the peripheral chemoreceptor system for host-seeking behaviour by a haemolymph-borne factor in *Aedes aegypti*. J Insect Physiol 30: 179–183

Dawson GW, Mudd AL, Pickett JA, Pile MM, Wadhams LJ (1990) Convenient synthesis of mosquito oviposition pheromone and a highly fluorinated analog retaining biological activity. J Chem Ecol 16: 1779–1789

DeFoliart GR (1972) A modified dry ice-baited trap for collecting hematophagous Diptera. J Med Entomol 9: 107–108

DeFoliart GR, Morris CD (1967) A dry ice-baited-trap for the collection and field storage of hematophagous Diptera. J Med Entomol 4: 360–362

DeFoliart GR, Rao MR, Morris CD (1967) Seasonal succession of blood-sucking Diptera in Wisconsin during 1965. J Med Entomol 4: 363–373

Denke PM, Lloyd JE, Littlefield JL (1996) Elevational distribution of mosquitoes in a mountainous area of southeastern Wyoming. J Am Mosq Control Assoc 12: 8–16

Dennett JA, Vessey NY, Parsons RE (2004). A comparison of seven traps used for collection of *Aedes albopictus* and *Aedes aegypti* originating from a large tire repository in Harris County (Houston), Texas. J Am Mosq Control Assoc 20: 342–349

Diallo M, Lochouarn L, Ba K, Sall AA, Mondo M, Girault L, Mathiot C (2000) First isolation of the Rift Valley fever virus from *Culex poicilipes* (Diptera: Culicidae) in nature. Am J Trop Med Hyg 62: 702–704

Dow RP, Reeves WC, Bellamy RE (1965) Dispersal of female *Culex tarsalis* into a larvicided area. Am J Trop Med Hyg 14: 656–670

Edman JD, Lea AO (1972). Sexual behaviour of mosquitoes. 2. Large-scale rearing and mating of *Culex* for field experiments. Ann Entomol Soc Am 65: 267–269

Emmons RW, Milby MM, Walsh JD, Reeves WC, Bayer EV, Hui LT, Woodie JD, Murray RA (1986) Surveillance for arthropod-borne viral activity and disease in California during 1985. Proc California Mosq Vector Control Assoc 54: 1–8

[ESRI] Environmental Systems Research Institute Inc. (1996*a*) ArcView GIS 3.2a for Windows. ESRI, Redlands, USA

[ESRI] Environmental Systems Research Institute Inc. (1996*b*) Using the Arc-View Spatial Analyst. ESRI, Redlands, USA

Fay RW, Prince WH (1970) A modified visual trap for *Aedes aegypti*. Mosquito News 30: 20–23

Fraissignes B, Chippaux A, Mouchet J (1968) Captures de moustiques par des pièges lumineux associés à une source de gaz carbonique. Méd Trop (Mars) 28: 215–221

Freier JE, Francy DB (1991) A duplex cone trap for the collection of adult *Aedes albopictus*. J Am Mosq Control Assoc 7: 73–79

French FE, Kline DL (1989) 1-octen-3-ol, an effective attractant for Tabanidae (Diptera). J Med Entomol 26: 459–461

Frommer RL, Schiefer BA, Vavra RW (1976) Comparative effects of CO_2 flow rates using modified CDC light traps on trapping adult black flies (Simuliidae; Diptera). Mosquito News 36: 355–358

Garcia R, des Rochers BS, Voigt WG (1988) A bait/carbon dioxide trap for the collection of the western tree hole mosquito *Aedes sierrensis*. J Am Mosq Control Assoc 4: 85–88

Garcia R, Colwell AE, Voigt WG, Woodward DL (1989) Fay–Prince trap baited with CO_2 for monitoring adult abundance of *Aedes sierrensis* (Diptera: Culicidae) J Med Entomol 26: 327–331

Gillies MT (1969) The ramp-trap, an unbaited device for flight studies of mosquitoes. Mosquito News 29: 189–193

Gillies MT (1980) The role of carbon dioxide in host-finding by mosquitoes (Diptera: Culicidae): a review. Bull Entomol Res 70: 525–532

Gillies MT, Snow WF (1967) A CO_2-baited sticky trap for mosquitoes. Trans R Soc Trop Med Hyg 61: 20

Gillies MT, Wilkes TJ (1969) A comparison of the range of attraction of animal baits and of carbon dioxide for some West African mosquitoes. Bull Entomol Res 59: 441–456

Gillies MT, Wilkes TJ (1972) The range of attraction of animal baits and carbon dioxide for mosquitoes. Studies in a freshwater area of West Africa. Bull Entomol Res 61: 389–404

Gillies MT, Wilkes TJ (1974) The range of attraction of birds as baits for some West African mosquitoes (Diptera, Culicidae). Bull Entomol Res 63: 573–581

Gingrich JB, Casillas L (2004) Selected mosquito vectors of West Nile virus: comparison of their ecological dynamics in four woodland and marsh habitats in Delaware. J Am Mosq Control Assoc 20: 138–145

Ginsberg HS (1988) Survivorship of mosquitoes (Diptera: Culicidae) captured in CDC miniature light traps. Proc Ann Mtg New Jers Mosq Control Assoc 75: 86–92

Gleiser MR, Gorla DE, Schelotto G (2000) Population dynamics of *Aedes albifasciatus* (Diptera: Culicidae) south of Mar Chiquita Lake, Central Argentina. J Med Entomol 37: 21–26

Gleiser RM, Schelotto G, Gorla DE (2002) Spatial pattern of abundance of the mosquito, *Ochlerotatus albifasciatus*, in relation to habitat characteristics. Med Vet Entomol 16: 364–371

Goodwin MH (1942) Studies on artificial resting places of *Anopheles quadrimaculatus* Say. J Natn Malar Soc 1: 93–99

Graham P (1969) A comparison of sampling methods for adult mosquito populations in central Alberta, Canada. Quaest Entomol 5: 217–261

Hall DR, Beevor PS, Cork A, Nesbitt BF, Vale GA (1984) 1-octen-3-ol. A potent olfactory stimulant and attractant for tsetse isolated from cattle odour. Insect Science and its Application 5: 335–339

Harden FW, Poolson BJ (1969) Seasonal distribution of mosquitoes of Hancock county, Mississippi, 1964–1968. Mosquito News 29: 407–414

Harden FW, Poolson BJ, Bennett LW, Gaskin RC (1970) Analysis of CO_2 supplemented mosquito landing rate counts. Mosquito News 30: 369–374

Hayes RO, Bellamy RE, Reeves WC, Willis MJ (1958) Comparison of four sampling methods for measurement of *Culex tarsalis* adult populations. Mosquito News 18: 218–227

Hayes RO, Francy DB, Lazuick JS, Smith GC, Jones RH (1976) Arbovirus surveillance in six states during 1972. Am J Trop Med Hyg 25: 463–476

Headlee TJ (1934) Mosquito work in New Jersey for the year 1933. Proc New Jers Mosq Exterm Assoc 21: 8–37

Headlee TJ (1941) New Jersey mosquito problems. Proc New Jers Mosq Exterm Assoc 28: 7–12

Henderson BE, McCrae AWR, Kirya BG, Ssenkubuge Y, Sempala SDK (1972) Arborivus epizootics involving man, mosquitoes and vertebrates at Lunyo, Uganda. Ann Trop Med Parasitol 66: 343–355

Herbert EW, Meyer RP, Tubes PG (1972) A comparison of mosquito catches with CDC light traps and CO_2-baited traps in the republic of Vietnam. Mosquito News 32: 212–214

Horn H (1966) Measurement of 'overlap' in comparative ecological studies. Am Nat 100: 419–424

Howard JJ, Morris CD, Emord DE, Grayson MA (1988) Epizootiology of Eastern Equine Encephalitis virus in upstate New York, USA. VII. Virus surveillance 1978–85, description of 1983 outbreak, and series conclusions. J Med Entomol 25: 501–514

Hoy JB (1970) Trapping the stable fly by using CO_2 or CO as attractants. J Econ Entomol 63: 792–795

Hribar LL, Vlach JJ, Demay DJ, Stark LM, Stoner RL, Godsey MS, Burkhalter KL, Spoto MC, James SS, Smith JM, Fussell EM (2003) Mosquitoes infected with West Nile virus in the Florida Keys, Monroe County, Florida, USA. J Med Entomol 40: 361–363

Huffaker CB, Back RC (1943) A study of methods of sampling mosquito populations. J Econ Entomol 36: 561–569

Iha S (1971) Feeding preference and seasonal distribution of mosquitoes in relation to the epidemiology of Japanese encephalitis in Okinawa main island. Trop Med 12: 143–168 (In Japanese, English summary)

Janousek TE, Kramer WL (1999) Seasonal incidence and geographical variation of Nebraska mosquitoes, 1994–95. J Am Mosq Control Assoc 15: 253–262

Jeffery JAL, Ryan PA, Lyons SA, Thomas PT, Kay BH (2002) Spatial distribution of vectors of Ross River virus and Barmah Forest virus on Russell Island, Moreton Bay, Queensland. Aust J Entomol 41: 329–338

Jensen T, Kaiser PE, Barnard DR (1993) Short-term changes in the abundance and parity rate of *Anopheles quadrimaculatus* species C (Diptera: Culicidae) in a central Florida swamp. J Med Entomol 30: 1038–1042

Johansen CA, van den Hurk AF, Pyke AT, Zborowski P, Phillips DA, Mackenzie JS, Ritchie SA (2001) Entomological investigations of an outbreak of Japanese encephalitis virus in the Torres Strait, Australia, in 1998. J Med Entomol 38: 581–588

Johansen CA, Montgomery BL, Mackenzie JS, Ritchie SA (2003) Efficacies of the Mosquito Magnet TM and counterflow geometry traps in north Queensland, Australia. J Am Mosq Control Assoc 19: 265–270

Johansen CA, Nisbet DJ, Foley PN, Van Den Hurk AF, Hall RA, Mackenzie JS, Ritchie SA (2004) Flavivirus isolations from mosquitoes collected from Saibai Island in the Torres Strait, Australia, during an incursion of Japanese encephalitis virus. Med Vet Entomol 18: 281–287

Jones JW, Sithiprasasna R, Schleich S, Coleman RE (2003) Evaluation of selected traps as tools for conducting surveillance for adult *Aedes aegypti* in Thailand. J Am Mosq Control Assoc 19: 148–150

Jones JW, Turell MJ, Sardelis MR, Watts DM, Coleman RE, Fernandez R, Carbajal F, Pecor JE, Calampa C, Klein TA (2004) Seasonal distribution, biology, and human attraction patterns of culicine mosquitoes (Diptera: Culicidae) in a forest near Puerto Almendras, Iquitos, Peru. J Med Entomol 41: 349–360

Jongman RHG, Ter Braak CJF, Van Tongeren OFR (1995) Data Analysis in Community and Landscape Ecology Cambridge University Press, Cambridge

Jupp PG, McIntosh BM (1967) Ecological studies on Sindbis and West Nile viruses in South Africa. II.—Mosquito bionomics. S Afr J Med Sci 32: 15–33

Jupp PG, McIntosh BM (1990) *Aedes furcifer* and other mosquitoes as vectors of chikungunya virus at Mica, northeastern Transvaal, South Africa. J Am Mosq Control Assoc 6: 415–420

Jupp PG, McIntosh BM, Nevill EM (1980) A survey of the mosquito and *Culicoides* faunas at two localities in the Karoo region of South Africa with some observations on bionomics. Onderstepoort J Vet Res 47: 1–6

Jupp PG, Kemp A, Grobbelaar A, Leman P, Burt FJ, Alahmed AM, Al-Mujalli D, Al-Khamees M, Swanepoel R (2002) The 2000 epidemic of Rift Valley fever in Saudi Arabia: mosquito vector studies. Med Vet Entomol 16: 245–252

Katô M (1955) Ecology of Mosquitoes. DDT Kyokai, Tokyo (In Japanese)

Katô M, Ishii T, Watanabe T, Yoshida S (1966) A new dry ice trap for collecting mosquitoes. Jap J Sanit Zool 17: 83–88

Kay BH, Hearnden MN, Oliveira NMM, Sellner LN, Hall RA (1996) Alphavirus infection in mosquitoes at the Ross River reservoir, north Queensland, 1990–1993. J Am Mosq Control Assoc 12: 421–428

Kemme JA, Van Essen PHA, Ritchie SA, Kay BH (1993) Response of mosquitoes to carbon dioxide and 1-octen-3-ol in southeast Queensland, Australia. J Am Mosq Control Assoc 9: 431–435

Kline DL (1999) Comparison of two American Biophysics mosquito traps: the professional and a new counterflow geometry trap. J Am Mosq Control Assoc 15: 276–282

Kline DL (2002) Evaluation of various models of propane-powered mosquito traps. J Vector Ecol 27: 1–7

Kline DL, Lemire GF (1998) Evaluation of attractant-baited traps/targets for mosquito management on Key Island, Florida, USA. J Vector Ecol 23: 171–185

Kline DL, Mann MO (1998) Evaluation of butanone, carbon-dioxide, and 1-octen-3-ol as attractants for mosquitoes associated with north central Florida bay and cypress swamps. J Am Mosq Control Assoc 14: 289–297

Kline DL, Wood JR (1988) Natural and synthetic attractants for ceratopogonid biting midges. In: Olejníček J (ed) Medical and Veterinary Dipterology. Proc Int Conf November–December 1987 Cĕské Budĕjovice, Czechoslovakia, Dum Techniky Čsvts České Budĕjovice, pp 175–176

Kline DL, Wood JR (1989) Practical assessment and enhancement of some currently used adult surveillance techniques for Florida mosquitoes: Phase II. Final Rept Florida Department Health Rehabilitation Services, unpublished

Kline DL, Takken W, Wood JR, Carlson DA (1990a) Field studies on the potential of butanone, carbon dioxide, honey extract, 1-octen-3-ol, lactic acid, and phenols as attractants for mosquitoes. Med Vet Entomol 4: 383–391

Kline DL, Wood JR, Morris CD (1990b) Evaluation of 1-octen-3-ol as an attractant for Coquillettidia perturbans, Mansonia spp. and Culex spp. associated with phosphate mining operations. J Am Mosq Control Assoc 6: 605–611

Kline DL, Wood JR, Cornell JA (1991a) Interactive effects of 1-octen-3-ol and carbon dioxide on mosquito (Diptera: Culicidae) surveillance and control. J Med Entomol 28: 254–258

Kline DL, Dame DA, Meisch MV (1991b) Evaluation of 1-octen-3-ol and carbon dioxide as attractants for mosquitoes associated with irrigated ricefields in Arkansas. J Am Mosq Control Assoc 7: 165–169

Knols BGJ, Mboera LEG, Takken W (1998) Electric nets for studying odourmediated host-seeking behaviour of mosquitoes (Diptera: Culicidae). Med Vet Entomol 12: 116–120

Landry SV, DeFoliart GR (1986) Attraction of Aedes triseriatus to carbon dioxide. J Am Mosq Control Assoc 2: 355–357

Leake CJ, Ussery MA, Nisalak A, Hoke CH, Andre RG, Burke DS (1986) Virus isolations from mosquitoes collected during the 1982 Japanese encephalitis epidemic in northern Thailand. Trans R Soc Trop Med Hyg 80: 831–837

Le Duc JW, Suyemoto W, Keefe TJ, Burger JF, Eldridge BF, Russell PK (1975) Ecology of California encephalitis viruses on the Del Mar Va peninsular. I. Virus isolations from mosquitoes. Am J Trop Med Hyg 24: 118–123

Ludueña Almeida FF, Gorla DE (1995) The biology of Aedes (Ochlerotatus) albifasciatus Macquart, 1838 (Diptera: Culicidae) in central Argentina. Mem Inst Oswaldo Cruz 90: 463–468

Lundström JO, Chirico J, Folke A, Dahl C (1996) Vertical distribution of adult mosquitoes (Diptera: Culicidae) in southern and central Sweden. J Vector Ecol 21: 159–166

Maciá A (1997) Age structure of adult mosquito (Diptera: Culicidae) populations from Buenos Aires Province, Argentina. Mem Inst Oswaldo Cruz 92: 143–149

Magnarelli LA (1975) Relative abundance and parity of mosquitoes collected in dry-ice baited and unbaited CDC miniature light traps. Mosquito News 35: 350–353

Mboera LEG, Knols BGJ, Braks MAH, Takken W (2000a) Comparison of carbon dioxide-baited trapping systems for sampling outdoor mosquito populations in Tanzania. Med Vet Entomol 14: 257–263

Mboera LEG, Takken W, Sambu EZ (2000*b*) The response of *Culex quinquefasciatus* (Diptera: Culicidae) to traps baited with carbon dioxide, 1-octen-3-ol, acetone, butyric acid, and human foot odour in Tanzania. Bull Entomol Res 90: 155–159

Mboera LEG, Takken W, Mdira KY, Pickett JA (2000*c*) Sampling gravid *Culex quinquefasciatus* (Diptera: Culicidae) in Tanzania with traps baited with synthetic oviposition pheromone and grass infusions. J Med Entomol 37: 172–176

McIver SB (1982) Review article: Sensillae of mosquitoes. J Med Entomol 19: 489–535

McIver SB, McElligott PE (1989) Effects of release rates on the range of attraction of carbon dioxide to some southwestern Ontario mosquito species. J Am Mosq Control Assoc 5: 6–9

Merdic E, Boca I (2004) Seasonal dynamics of the *Anopheles maculipennis* complex in Osijek, Croatia. J Vector Ecol 29: 257–263

Meyer RP (1985) The 'walk-in' type red box for sampling adult mosquitoes. Proc New Jers Mosq Control Assoc 72: 104–105

Meyer RP (1991) Urbanization and the efficiency of carbon dioxide and gravid traps for sampling *Culex quinquefasciatus*. J Am Mosq Control Assoc 7: 467–470

Meyer RP, Reisen WK, Hill BR, Martinez VM (1983) The "AFS sweeper", a battery powered back pack mechanical aspirator for collecting adult mosquitoes. Mosquito News 43: 346–350

Meyer RP, Reisen WK, Milby MM (1991) Influence of vegetation on carbon dioxide trap effectiveness for sampling mosquitoes in the Sierra Nevada foothills of Kern county, California. J Am Mosq Control Assoc 7: 471–475

Mian LS, Mulla MS, Axelrod H, Chaney JD, Dhillon MS (1990) Studies on the bioecological aspects of adult mosquitoes in the Prado Basin of southern California. J Am Mosq Control Assoc 6: 64–71

Miller RJ, Wing J, Cope S, Davey RB, Kline DL (2005) Comparison of carbon dioxide- and octenol-baited encephalitis virus surveillance mosquito traps at Shoal Water Bay training area, Queensland, Australia. J Am Mosq Control Assoc 21: 497–500

Miller TA, Stryker RG, Wilkinson RN, Esah S (1969) Notes on the use of CO_2-baited CDC miniature light traps for mosquito surveillance in Thailand. Mosquito News 29: 688–689

Mitchell CJ, Chen PS (1973) Ecological studies on the mosquito vectors of Japanese encephalitis. Bull World Health Organ 49: 287–292

Mogi M, Kawai S, Oda T, Nishigaki J, Suenaga O, Itô S, Miyaga I, Wada Y, Omori N (1970) Ecology of vector mosquitoes of Japanese encephalitis, especially of *Culex tritaeniorhynchus*. 3. Seasonal changes in the time of being attracted to dry ice in the females of *Culex tritaeniorhynchus*. Trop Med 12: 122–127

Moncayo AC, Edman JD (1999) Toward the incrimination of epidemic vectors of eastern equine encephalomyelitis virus in Massachusetts: abundance of mosquito populations at epidemic foci. J Am Mosq Control Assoc 15: 479–492

Moncayo AC, Edman JD, Finn JT (2000) Application of geographic information technology in determining risk of eastern equine encephalomyelitis virus transmission. J Am Mosq Control Assoc 16: 28–35

Morris CD, DeFoliart GR (1969) A comparison of mosquito catches with miniature light traps and CO_2-baited traps. Mosquito News 29: 424–426

Morris CD, Zimmerman RH, Magnarelli LA (1976) The bionomics of *Culiseta melanura* and *Culex morsitans dyari* in Central New York State (Diptera: Culicidae). Ann Entomol Soc Am 69: 101–105

Morris CD, Zimmerman RH, Edman JD (1980) Epizootiology of eastern equine encephalomyelitis virus in upstate New York, USA. II. Population dynamics and vector potential of *Culiseta melanura* (Diptera: Culicidae) in relation to distance from breeding site. J Med Entomol 17: 453–465

Nasci RS, Berry RL, Restifo RA, Parsons MA, Smith GC, Martin DA (1993) Eastern equine encephalitis virus in Ohio during 1991. J Med Entomol 30 217–222

Nayar JK (1982) *Wyeomyia mitchelli*: Observations on dispersal, survival, nutrition, insemination and ovarian development in a Florida population. Mosquito News 42: 416–427

Need JT, Rogers EJ, Phillips IA, Falcon R, Fernandez R, Carbajal F, Quintana J (1993) Mosquitoes (Diptera: Culicidae) captured in the Iquitos area of Peru. J Med Entomol 30: 634–638

Nelson RL (1966) Newer collecting methods for vectors of arboviruses. Proc California Mosq Control Assoc 33: 65–66

Nelson RL, Milby MM, Reeves WC, Fine PE (1978) Estimates of survival, population size, and emergence of *Culex tarsalis* at an isolated site. Ann Entomol Soc Am 71: 801–808

Newhouse VF, Chamberlain RW, Johnston JG, Sudia WD (1966) Use of dry ice to increase mosquito catches of the CDC miniature light trap. Mosquito News 26: 30–35

Nilssen AC (1998) Effect of 1-octen-3-ol in field trapping *Aedes* spp. (Dipt., Culicidae) and *Hybomitra* spp. (Dipt., Tabanidae) in subarctic Norway. J Appl Entomol 122: 465–468

Novak RJ, Pelequin J, Rohrer W (1981) Vertical distribution of adult mosquitoes (Diptera: Culicidae) in a northern deciduous forest in Indiana. J Med Entomol 18: 116–122

Omori N, Wada Y, Kawai S, Itô S, Oda T, Suenaga O, Nishigaki J, Hayashi K, Mifune K (1965) Preliminary notes on the collection of hibernated females of *Culex tritaeniorhynchus* in Nagasaki. Endem Dis Bull Nagasaki Univ 7: 147–153

Parker BM (1993) Variation in mosquito (Diptera: Culicidae) relative abundance and *Dirofilaria immitis* (Nematoda: Filarioidea) vector potential in coastal North Carolina. J Med Entomol 30: 436–442

Parker M, Anderson AL, Slaff M (1986) An automatic carbon dioxide delivery system for mosquito light trap surveys. J Am Mosq Control Assoc 2: 236–238

Parsons RE, Dondero TJ, Hooi CW (1974) Comparison of CDC miniature light traps and human biting collections for mosquito catches during malaria vector surveys in peninsular Malaysia. Mosquito News 34: 211–213

Pfuntner AR (1979) A modified CO_2-baited miniature surveillance trap. Bull Soc Vect Ecol 4: 31–35

Pfuntner AR, Reisen WK, Dhillon MS (1988) Vertical distribution and response of *Culex* mosquitoes to differing concentrations of carbon dioxide. Proc California Mosq Vector Control Assoc 56: 69–74

Pinkovsky DD, Sutton DR (1977) A comparison of carbon dioxide and light as attractants for CDC mosquito traps at Clarke Air Base, Philippine islands. Mosquito News 37: 508–511

Rapaport AS, Lampman RL, Novak RJ (2005) Evaluation of selected modifications to CO_2 and infusion-baited mosquito traps in Urbana, Illinois. J Am Mosq Control Assoc 21: 395–399

Raymond HL (1977) Action d'anhydride carbonique et de facteurs visuels sur les performance de pièges "Manitobia" modifiés en milleu montagnard, Entomol Exp Appl 21: 121–129

Ree HI, Chen YK, Chow CY (1969) Methods of sampling populations of the Japanese encephalitis vector mosquitoes—a preliminary report. Med J Malaya 23: 293–295

Reeves WC (1951) Field studies on carbon dioxide as a possible host stimulant to mosquitoes. Proc Soc Exp Biol Med 77: 64–66

Reeves WC (1953) Quantitative field studies on a carbon dioxide chemotropism of mosquitoes. Am J Trop Med Hyg 2: 325–331

Reeves WC, Hammon WM (1942) Mosquitoes and encephalitis in Yakima Valley, Washington. IV. A trap for collecting live mosquitoes. J Infect Dis 70: 275–277

Reisen WK, Pfuntner AR (1987) Effectiveness of five methods for sampling adult *Culex* mosquitoes in rural and urban habitats in San Bernardino county, California. J Am Mosq Control Assoc 3: 601–606

Reisen WK, Milby MM, Reeves WC, Meyer RP, Bock ME (1983) Population ecology of *Culex tarsalis* (Diptera: Culicidae) in a foothill environment in Kern county, California: Temporal changes in female relative abundance reproductive status, and survivorship. Ann Entomol Soc Am 76: 800–808

Reisen WK, Meyer RP, Martinez VM, Gonzalez O, Spoehel JJ, Hazelrigg JE (1988) Mosquito abundance in suburban communities in Orange and Los Angeles counties, California, 1987. Proc California Mosq Vector Control Assoc 56: 75–85

Reisen WK, Hardy JL, Reeves WC, Presser SB, Milby MM, Meyer RP (1990a) Persistence of mosquito-borne viruses in Kern county, California, 1983–1988. Am J Trop Med Hyg 43: 419–437

Reisen WK, Meyer RP, Tempelis CH, Spoehel JJ (1990b) Mosquito abundance and bionomics in residential communities in Orange and Los Angeles counties, California. J Med Entomol 27: 356–367

Reisen WK, Lothrop HD, Presser SB, Milby MM, Hardy JL, Wargo MJ, Emmons RW (1995a) Landscape ecology of arboviruses in southern California: temporal and spatial patterns of vector and virus activity in Coachella Valley, 1990–1992. J Med Entomol 32: 255–266

Reisen WK, Lothrop HD, Hardy JL (1995*b*). Bionomics of *Culex tarsalis* (Diptera: Culicidae) in relation to arbovirus transmission in southeastern California. J Med Entomol 32: 316–327

Reisen WK, Hardy JL, Chiles RE, Kramer LD, Martinez VM, Presser SB (1996) Ecology of mosquitoes and lack of arbovirus activity at Morro Bay, San Luis Obispo County, California. J Am Mosq Control Assoc 12: 679–687

Reisen WK, Meyer RP, Cummings RF, Delgado O (2000) Effects of trap design and CO_2 presentation on the measurement of adult mosquito abundance using Centers for Disease Control-style miniature light traps. J Am Mosq Control Assoc 16: 13–18

Reisen WK, Lothrop H, Chiles R, Madon M, Cossen C, Woods L, Husted S, Kramer V, Edman J (2004) West Nile virus in California. Emerg Infect Dis 10: 1369–1378

Reiter P (1983) A portable, battery-powered trap for collecting gravid *Culex* mosquitoes. Mosquito News 43: 496–498

Ritchie SA, Van Essen PHA, Kemme JA, Kay BH, Allaway D (1994) Response of biting midges (Diptera: Ceratopogonidae) to carbon dioxide, octenol, and light in southeastern Queensland, Australia. J Med Entomol 31: 645–648

Ritchie SA, Fanning ID, Phillips DA, Standfast HA, McGinn D, Kay BH (1997) Ross River virus in mosquitoes (Diptera: Culicidae) during the 1994 epidemic around Brisbane, Australia. J Med Entomol 34: 156–159

Ritchie SA, Pyke AT, Smith GA, Northill JA, Hall RA, van den Hurk AF, Johansen CA, Montgomery BL, Mackenzie JS (2003) Field evaluation of a sentinel mosquito (Diptera: Culicidae) trap system to detect Japanese encephalitis in remote Australia. J Med Entomol 40: 249–252

Roberts RH (1972) Relative attractiveness of CO_2 and a steer to Tabanidae, Culicidae and *Stomoxys calcitrans* (L.). Mosquito News 32: 208–211

Rohe DL, Fall RP (1979) A miniature battery powered CO_2 baited light trap for mosquito borne encephalitis surveillance. Bull Soc Vector Ecol 4: 24–27

Rubio-Palis Y (1996) Evaluation of light traps combined with carbon dioxide and 1-octen-3-ol to collect anophelines in Venezuela. J Am Mosq Control Assoc 12: 91–96

Rudolfs W (1922) Chemotropism of mosquitoes. Bull New Jers Agric Exp Stn No. 367, 4

Rueda LM, Harrison BA, Brown JS, Whitt PB, Harrison RL, Gardner RC (2001) Evaluation of 1-octen-3-ol, carbon dioxide and light as attractants for mosquitoes associated with two distinct habitats in North Carolina. J Am Mosq Control Assoc 17: 61–66

Rueda LM, Gardner RC (2003) Composition and adult activity of salt-marsh mosquitoes attracted to 1-octen-3-ol, carbon dioxide, and light in Topsail Island, North Carolina. J Am Mosq Control Assoc 19: 166–169

Russell CB, Hunter FF (2005) Attraction of *Culex pipiens/restuans* (Diptera: Culicidae) mosquitoes to bird uropygial gland odors at two elevations in the Niagara region of Ontario. J Med Entomol 42: 301–305

Russell RC (1985) The efficiency of various collection techniques for sampling *Culex annulirostris* in southern Australia. J Am Mosq Control Assoc 1: 502–505

Russell RC (1986*a*) Seasonal abundance of mosquitoes in a native forest of the Murray valley of Victoria, 1979–1985. J Aust Entomol Soc 25: 235–240

Russell RC (1986*b*) Larval competition between the introduced vector of dengue fever in Australia, *Aedes aegypti* (L.), and a native container-breeding mosquito, *Aedes notoscriptus* (Skuse) (Diptera: Culicidae). Aust J Zool 34: 527–534

Russell RC (1986*c*) The mosquito fauna of Conjola state forest on the south coast of New South Wales. Part 1. Species composition and monthly prevalence. Gen Appl Entomol 18: 53–64

Russell RC (1998) Mosquito-borne arboviruses in Australia: the current scene and implications of climate change for human health. International J Parasitol 28: 955–969

Russell RC (2004) The relative attractiveness of carbon dioxide and octenol in CDC- and EVS-type light traps for sampling the mosquitoes *Aedes aegypti* (L.), *Aedes polynesiensis* Marks, and *Culex quinquefasciatus* Say in Moorea, French Polynesia. J Vector Ecol 29: 309–314

Russell RC, Whelan PI (1986) Seasonal prevalence of adult mosquitoes at Casuarina and Leanyer, Darwin. Aust J Ecol 11: 9–105

Ryan PA, Do K-A, Kay BH (2000) Definition of Ross River virus vectors at Maroochy Shire, Australia. J Med Entomol 37: 146–152

Saitoh Y, Hattori J, Chinone S, Nihei N, Tsuda Y, Kurahashi H, Kobayashi M (2004) Yeast-generated CO_2 as a convenient source of carbon dioxide for adult mosquito sampling. J Am Mosq Control Assoc 20: 261–264

Schreck CE, Gouck HK, Posey KH (1970) An experimental plexiglass mosquito trap utilizing carbon dioxide. Mosquito News 30: 641–645

Schreck CE, Gouck HK, Posey KH (1972) The range of effectiveness and trapping efficiency of a plexiglass mosquito trap baited with carbon dioxide. Mosquito News 32: 496–501

Service MW (1969) The use of traps in sampling mosquito populations. Entomol Exp Appl 12: 403–412

Shipp JL (1985) Evaluation of a portable CO_2 generator for sampling black flies. J Am Mosq Control Assoc 1: 515–517

Shone SM, Ferrao PN, Lesser CR, Glass GE, Norris DE (2003) Evaluation of carbon dioxide- and 1-octen-3-ol-baited Centers for Disease Control Fay-Prince traps to collect *Aedes albopictus*. J Am Mosq Control Assoc 19: 445–447

Silva IM, Eiras AE, Kline DL, Bernier UR (2005) Laboratory evaluation of mosquito traps baited with a synthetic human odor blend to capture *Aedes aegypti*. J Am Mosq Control Assoc 21: 229–233

Sithiprasasna R, Jaichapor B, Chanaimongkol S, Khongtak P, Lealsirivattanakul T, Tiang-Trong S, Burkett DA, Perich MJ, Wirtz RA, Coleman RE (2004) Evaluation of candidate traps as tools for conducting surveillance for *Anopheles* mosquitoes in a malaria-endemic area in western Thailand. J Med Entomol 41: 151–157

Siverly RE, DeFoliart GR (1968) Mosquito studies in Northern Wisconsin II. Light trapping studies. Mosquito News 28: 162–167

Slaff M, Crans WJ, McCuiston LJ (1983) A comparison of three mosquito sampling techniques in northwestern New Jersey. Mosquito News 43: 287–290

Smith GC, Moore CG, Davis T, Savage HM, Thapa AB, Shrestha SL, Karabatsos N (1993) Arbovirus surveillance in northern Colorado, 1987 and 1991. J Med Entomol 30: 257–261

Snow WF (1970) The effect of a reduction in expired carbon dioxide on the attractiveness of human subjects to mosquitoes. Bull Entomol Res 60: 43–48

Stryker RG, Young WW (1970) Effectiveness of carbon dioxide and L (+) lactic acid in mosquito light traps with and without light. Mosquito News 30: 388–393

Sudia WD, Chamberlain RW (1962) Battery-operated light trap, an improved model. Mosquito News 22: 126–129

Sutcliffe JF (1986) Black fly host location: a review. Can J Zool 64: 1041–1053

Sutcliffe JF (1987) Distance orientation of biting flies to their hosts. Insect Science and its Application 8: 611–616

Takeda U, Kurihara R, Suzuki T, Sasa M, Miura A, Matsumoto K, Tanaka H (1962) Studies on a collection method of mosquitoes by dry ice and a mosquito net. Jap J Sanit Zool 13: 31–35 (In Japanese, English summary)

Takeda T, Whitehouse CA, Brewer M, Gettman AD, Mather TN (2003) Arbovirus surveillance in Rhode Island: assessing potential ecologic and climatic correlates. J Am Mosq Control Assoc 19: 179–189

Takken W, Kline DL (1989) Carbon dioxide and l-octen-3-ol as mosquito attractants. J Am Mosq Control Assoc 5: 311–316

Torr SJ (1994) The tsetse (Diptera: Glossinidae) story: implications for mosquitoes. J Am Mosq Control Assoc 10: 258–265

Townes H (1962) Design for a Malaise trap. Proc Entomol Soc Wash 64: 253–262

Trueman DW, McIver SB (1981) Detecting time-scale temporal distributions of biting flies: A new trap design. Mosquito News 41: 439–443

Trueman DW, McIver SB (1984) Temporal patterns of host-seeking activity of mosquitoes in Algonquin park, Ontario. Can J Zool 64: 731–737

Vale GA (1993) Development of baits for tsetse flies (Diptera:Glossinidae) in Zimbabwe. J Med Entomol 30: 831–842

Van Essen PHA, Kemme JA, Ritchie SA, Kay BH (1994) Differential responses of *Aedes* and *Culex* mosquitoes to octenol or light in combination with carbon dioxide in Queensland, Australia. Med Vet Entomol 8: 63–67

van den Hurk AF, Beebe NW, Ritchie SA (1997) Responses of mosquitoes of the *Anopheles farauti* complex to 1-octen-3-ol and light in combination with carbon dioxide in northern Queensland, Australia. Med Vet Entomol 11: 177–180

van den Hurk AF, Cooper RD, Beebe NW, Williams GM, Bryan JH, Ritchie SA (2000) Seasonal abundance of Anopheles farauti (Diptera: Culicidae) sibling species in far North Queensland, Australia. J Med Entomol 37: 153–161 van den Hurk AF, Nisbet DJ, Hall RA, Kay BH, Mackenzie JS, Ritchie SA (2003*a*) Vector competence of Australian mosquitoes (Diptera: Culicidae) for Japanese encephalitis virus. J Med Entomol 40: 82–90

van den Hurk AF, Johansen CA, Zborowski P, Paru R, Foley PN, Beebe NW, Mackenzie JS, Ritchie SA (2003*b*) Mosquito host-feeding patterns and implications for Japanese encephalitis virus transmission in northern Australia and Papua New Guinea. Med Vet Entomol 17: 403–411

Vickery CA, Meadows KE, Baughman IE (1966) Synergism of carbon dioxide and chick as bait for *Culex nigripalpus*. Mosquito News 26: 507–508

Wallace HG, Rudnick A, Rajagopal V (1977) Activity of Tembusu and Umbre viruses in a Malaysian community: Mosquito studies. Mosquito News 37: 35–42

Washburn JO, Woodward DL, Colwell AE, Anderson JR (1992) Correlation of *Aedes sierrensis* captures at human sentinels with CO_2-baited Fay-Prince and duplex cone traps. J Am Mosq Control Assoc 8: 389–393

Whitsel RH, Schoeppner RF (1965) The attractiveness of carbon dioxide to female *Leptoconops torrens* Tns. and *L. kerteszi* Kieff. Mosquito News 25: 403–410

Wilson BH, Tugwell NP, Burns EC (1966) Attraction of tabanids to traps baited with dry ice under field conditions in Louisiana. J Med Entomol 3: 148–149

Wilton DP, Kloter KO (1985) Preliminary evaluation of a black cylinder suction trap for *Aedes aegypti* and *Culex quinquefasciatus* (Diptera: Culicidae). J Med Entomol 22: 113–114

Woodward DL, Colwell AE, Anderson NL (1998) Surveillance studies of *Orthopodomyia signifera* with comparison to *Aedes sierrensis*. J Vector Ecol 23: 136–148

Woodward DL, Colwell AE, Anderson NL (2003) Natural variability in the seasonal occurrence and densities of adult populations of *Ochlerotatus sierrensis*. J Am Mosq Control Assoc 19: 23–32

Worth CB, Jonkers AH (1962) Two traps for mosquitoes attracted to small vertebrates. Mosquito News 22: 15–21

Wozniak A, Dowda HE, Tolson MW, Karabatsos N, Vaughan DR, Turner PE, Ortiz DI, Wills W (2001) Arbovirus surveillance in South Carolina, 1996–98. J Am Mosq Control Assoc 17: 73–78

Chapter 11 Sampling Adults with Visual Attraction Traps, Sound Traps and Other Miscellaneous Attraction Traps

Visual attraction traps

Visual attraction traps are so efficient in catching tsetse flies that they are sometimes used as a control method. Tsetse entomologists have found that trap colour is a vital component of the attractiveness of these traps. Green and Flint (1986) have analysed the effects of colour on trap performance and their paper is of value to mosquito workers because it describes how reflectivity and wavelengths are measured for different coloured materials, and describes the differences between various spectral reflective curves. Agee and Patterson (1983) present a useful paper on the measurement of spectral sensitivity of muscoid flies and measurements of spectral reflectance, including UV light from different potential trap surfaces. Brach and Trimble (1985) give some useful graphs on the spectral reflectance of traps painted with different enamel paints and five fluorescent Day-Glo paints. They found that Tangle-Trap adhesive either decreased or increased reflectance from fluorescent coloured traps, and it appears that this adhesive acts as a barrier reducing the amount of incident UV radiation absorbed by the fluorescent pigment.

Browne and Bennett (1981) conducted interesting field experiments in Canada on the responses of mosquitoes, mainly *Coquillettidia perturbans* and *Ochlerotatus punctor*, to different colours and shapes. They designed 25.4-cm cube frames having five sides covered with coloured artboard (white, blue, yellow, red, black) and incorporating an entrance funnel; carbon dioxide was used as an attractant. The Munsell colour system was used in determining reflectance values. To test reaction to shape, cubic and pyramidal traps were made. They found that the percentage of luminous reflectance of light was inversely proportional to the numbers of mosquitoes attracted. Aedine species and *Coquillettidia perturbans* were attracted mostly to black, red and blue colours in preference to white and yellow.

Coquillettidia perturbans and *Ochlerotatus cantator* were attracted more to cubes than pyramids, whereas *Ochlerotatus punctor* was caught more often in pyramidal traps. Laboratory trials with Kodak colour filters (Kodak, 1990), selected to represent the electromagnetic spectrum from 300 nm (ultraviolet) to 720 nm (infrared and longer wavelengths), showed that *Coquillettidia perturbans* landed mostly on a filter transmitting 400–600 nm, while few landed on filters transmitting wavelengths longer than 600 nm.

Visual attraction traps have not been widely used for catching mosquitoes, the best known is probably the Fay–Prince trap, but this is commonly baited with carbon dioxide. Freier and Francy (1991) also designed a visual trap in trying to develop an efficient trapping method for the surveillance of *Aedes albopictus* in the USA, but it too is baited with carbon dioxide.

Haufe and Burgess trap

Haufe and Burgess (1960) considered that the failure to collect mosquitoes with light-traps in subarctic regions of Canada was due to the natural high light intensity of the nights and the absence of periods of total darkness in their working area. For these reasons, and because light-traps do not catch day-flying mosquitoes, and also because it is impossible to define the volume of air from which mosquitoes are caught, they designed traps that relied on a strong visual attraction. The final design was as follows. A fan is mounted about midway in a vertical metal cylinder, which is painted with 12 spiral $1\frac{5}{8}$-in wide black and white stripes. A horizontal disc considerably larger than the diameter of the cylinder is mounted above it to leave a small slit-like entrance gap (Fig. 11.1). A circular opening slightly smaller than the diameter of the cylinder is cut from the middle of the disc to allow a free flow of air through the middle of the disc down into the cylinder and out at the base. The disc is painted on both surfaces with 12 black and white stripes, which are expanded distally and taper towards the middle. Both cylinder and disc are rotated by a fan belt from a small motor mounted near the base of the trap. Mosquitoes attracted to the trap fly towards the black and white stripes on the cylinder, and their inwardly spiralling movement guides them to the top where they are prevented from 'overshooting' by the disc. They eventually fly through the gap between the disc and cylinder, and get sucked down into the trap and collected in a suitable net or killing bottle. Alternatively, the catch can be collected in a removable magazine into which a number of small discs drop at pre-determined time intervals thus enabling the catch to be divided into hourly samples (see Chap. 4 for a more detailed account of such segregating discs).

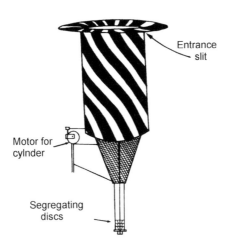

Fig. 11.1. Visual attractant trap of Haufe and Burgess (1960)

Because the attraction of mosquitoes to the trap depends on a minimum ommatidial (compound eye lens) angle being subtended by an individual stripe, the volume of air sampled depends on the width of the stripes. Changing the speed of rotation alters the flicker effect of the stripes. A stripe displacement rate of 160–180/min caused mosquitoes at a maximum distance of about 32 in from the trap to abruptly orientate towards it (Haufe and Burgess 1960). Mosquitoes approaching to within a few inches of the cylinder were often observed to circle it rapidly towards the top, but when rotation was increased to 200/min this circling flight increased and became erratic. At higher speeds many mosquitoes were disorientated and not collected by the trap. To increase the attractiveness of the rotating striped pattern at night when natural illumination is low, the trap can be illuminated by 60-W white light-bulbs placed 25 ft from the trap. These are shielded to ensure that only the trap cylinder is illuminated and also to reduce dazzle, which might repel mosquitoes near the trap.

Aedes and *Ochlerotatus* species have been caught in these traps in sub-arctic regions and more southern latitudes but the numbers have not been large, only about 10 times greater than catches obtained with non-attractant suction traps, the actual difference, however, varying according to species. Burgess and Haufe (1960) used three visual attraction traps at 5, 25 and 50 ft to study the vertical distribution of both prairie and forest mosquitoes. They did not illuminate them at night as this might have attracted mosquitoes from one level to another. In the prairie there was a great reduction in both sexes with increasing height, except that most males of *Aedes vexans* were caught at 50 ft. In the forest, females of three of the four species caught in any numbers were most common at 25 ft.

Further investigations are needed to determine the response of different mosquito species to this type of trap, and also whether the degree of attraction varies between day and night, although preliminary investigations suggest there is in fact little difference between day and night catches (Haufe and Burgess 1960). It would be informative to compare species composition and abundance obtained by this trap with that obtained by other traps, such as non-attractant suction traps and also highly selective traps, like New Jersey light-traps. Because this trap has an unusually short volume of influence it might prove suitable for comparing mosquito populations in closely situated areas, and studying vertical distributions. This trap has been little used, although Harwood (1961) operated a somewhat modified trap mounted on a mobile trailer, but no details of species caught are given. Corbet (1966) used visual attraction traps on Ellesmere Island to study diel flight periodicities of *Ochlerotatus nigripes* and *Ochlerotatus impiger* and to provide data on the seasonal build-up and decline of these two species. A mean maximum daily catch of around 160 females of *Ochlerotatus nigripes* was recorded. Corbet and Danks (1973) considered that although the traps provided representative samples of aerial populations, trap location could markedly affect the species prevalence in the catches. This is not surprising as this is a characteristic of most trapping techniques, and the shorter the range of attraction the greater the differences are likely to be between catches in different traps. Most traps only sample mosquitoes within a relatively localised area.

Fay-Prince trap

It is well known that female *Aedes aegypti* are attracted more to dark than light objects, and that shiny surfaces are more attractive than dull ones (Brett 1938; Brown 1951, 1954; Gjullin 1947; Peterson and Brown 1951; Sippell and Brown 1953), and also that males prefer to rest on glossy dark surfaces (Fay 1968). This has led to the evaluation of a number of simple traps for catching male as well as female mosquitoes, mainly *Aedes aegypti*. The final model has become known as the Fay–Prince trap (Fay and Prince 1970). It has an upper part, which is wind orientated, and a lower fixed part. The upper part consists of an aluminium trapezoid, 17½ in high (Fig. 11.2). A recessed shelf, 2¾ in deep and 2¼ in thick, is pushed to the rear of the trapezoid. A 4-in diameter semi-circle is cut from the centre of the front edge of this shelf. The rear end of the aluminium trapezoid above the shelf is covered with a transparent plastic window. This upper part of the trap is suspended by a wire attached to its middle. It has adjustable sliding panels to its front edge to counterbalance it and allow it to swing freely.

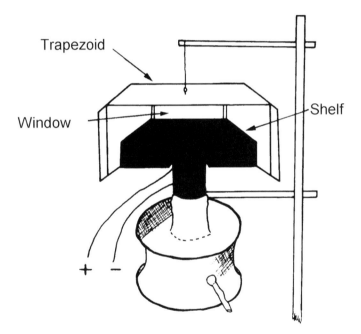

Fig. 11.2. Trapezoid visual attraction trap of Fay and Prince (1970)

The fixed part of the trap consists of a 6-in high, 3½-in diameter, metal cylinder fixed by a hose clamp to a horizontal support. A 4.5-V d.c. motor with 3⅛-in fan blades is mounted within the cylinder to draw mosquitoes down into a CDC collecting bag tied underneath. The recessed shelf, outside of the metal cylinder and upper half of its inside are painted gloss black. The rest of the trap, including the 6-V battery, is painted white. In operation the upper part of the trap is positioned so that the top of the recessed shelf is about ½ in above the top of the cylinder.

During 7 weeks' field trials in Mississippi, USA 12 of these traps caught 144 female and 277 male *Aedes aegypti*, and also 610 of approximately 2750 genetically marked and released males. They were also successful in other field studies on the dispersal of marked *Aedes aegypti* (Bond et al. 1970). Although Fay and Prince (1970) thought the traps were unlikely to be of any use for catching other mosquito species, the late Dr R. Fay later noticed that when the traps were used early in the morning *Culex quinquefasciatus* were caught exclusively until about 0800 h, but thereafter *Aedes aegypti* was caught. As a result of these observations Fay conducted a few trials with *Culex quinquefasciatus* in large outdoor cages. He found that substantially more adults were caught in traps working from 0600–0800 h than in ultraviolet light-traps operated from 2100–2300 h. Both traps caught more males than females.

Giglioli (1979) pointed out that sampling techniques for adult *Aedes aegypti* were inadequate, and partly for these reasons Kloter et al. (1983) decided to evaluate the following six traps for catching this important vector. A CDC light-trap (bulb EM47, General Electric), an ultraviolet light-trap (bulb F4T5, General Electric), a near infrared light-trap (10 diodes TIL32, Texas Instruments), a black cylinder trap (no light), a Fay–Prince (1970) trap (no light), and a so-called blank trap (CDC trap minus light). In all instances the 'collection cylinders' had the same inside diameter (8.5 cm), identical fans powered by 6-V d.c. motors (Pitman, P8512C375) operating at 1250 rev/min. Catch bags were made from light grey nylon netting. Evaluation was undertaken in a greenhouse, and during each experiment about 1500 nulliparous female and 500 male *Aedes aegypti* were released in the centre of a ring of traps.

A Fay–Prince trap caught significantly more males than any of the other traps, whether or not it was baited with 300 g dry ice, or operated during the daytime or at night. During the day the Fay–Prince trap, without any dry ice, was superior to other traps in catching females. At night the Fay–Prince and ultraviolet trap both without dry ice were equally efficient. In conclusion the Fay–Prince trap outperformed all other traps, except that at night the ultraviolet trap was about equally attractive. The poor performance of the CDC light-trap substantiated the findings of Brody (1977), that the Fay–Prince trap was superior to the CDC trap for trapping *Aedes aegypti*. Although the Fay–Prince has not often been tested for its attractiveness to other species, Garcia et al. (1989) caught *Ochlerotatus sierrensis* in it, and Wilton and Kloter (1985) collected *Culex quinquefasciatus*, although not as many as in their cylinder trap (see below).

Wilton and Kloter cylinder trap

Wilton and Kloter (1985) devised a trap for sampling *Aedes aegypti* which is lighter, more compact and more easily made than the Fay–Prince trap. It consists of an 18-cm length of plexiglas having an outside diameter of 9.5 cm and painted shiny black on the inside. It is fitted with machine bolts to the bottom of the plastic body of a CDC light-trap which has had the bottom part cut off to leave a 1-cm extension below the reinforced ring. The light-bulb is removed, and the trap inverted so that the CDC part is positioned below (Fig. 11.3). A 1-pint paper carton painted black on the inside and outside and with the bottom replaced with black netting is dropped into the top of the trap. A screen cone is fitted into the top of the carton to prevent mosquitoes escaping if there is a power failure. The trap operates from a 6-V battery.

In a series of twenty-one 24-h comparative trials with the Fay–Prince (1970) trap in New Orleans, USA the black cylinder trap caught a mean of 6.8 male and 2.4 female *Aedes aegypti*, while the Fay–Prince trap caught means of 3.4 male and 7.1 female *Aedes aegypti*. Surprisingly the black cylinder traps attracted the nocturnally active *Culex quinquefasciatus*; mean catches were 6.9 males and 6.1 females, compared to 4.6 male and 3.3 female *Culex quinquefasciatus* in the Fay–Prince trap.

Comparisons of the commercially available version of the Wilton and Kloter cylinder trap (John W. Hock & Co., Florida, USA) with omnidirectional Fay-Prince traps and American Biophysics Corporation ABC-PRO traps (essentially a CDC light-trap with a 2 litre container for dry ice attached to the top of the trap) for *Aedes aegypti* surveillance were recently undertaken in Peru by Schoeler et al. (2004). Efficacy of collections made with these traps was also compared with human landing catches and backpack aspirator collections. Trapping was undertaken in 6 premises from 0800 to 1000 h and again from 1600 to 1800 h three times per week every 2 weeks for 27 weeks. Backpack aspirator collections were the most efficient method, yielding 73% of the total 2411 mosquitoes obtained using all methods. Human landing catches caught 23% of the total. ABC-PRO traps were the most efficient of the four traps tested, yielding 1.7% of the total mosquitoes caught, followed by the Fay-Prince trap supplemented with CO_2 (0.9%), the Wilton and Kloter trap (0.8%) and the Fay-Prince trap without CO_2 (0.5%). *Aedes aegypti* represented 40.9% of the total collection, but

Fig. 11.3. Cylinder visual attraction trap of Wilton and Kloter (1985)

80% of the human landing catch, compared with only 23.8% *Aedes aegypti* in the backpack aspirator collections.

Duplex cone trap

The invasion of the USA by *Aedes albopictus* challenged American ento-mologists to devise a trap that could be used for surveillance of adults of this species. Carbon dioxide and light-traps have proved ineffective in southeast Asia for collecting *Aedes albopictus*, but it was noted in Louisiana, USA that adults often settled on shiny black plastic bags, and this eventually led Freier and Francy (1991) to develop a duplex cone trap.

The components of the trap are a 48-cm diameter, 41-cm tall black cone that has thirty 2.5-cm holes cut from around the base. This black cone is surmounted with a 50-cm diameter, 34-cm tall shiny galvanised cone supported 20 cm from the ground by four metal legs (Fig. 11.4).

Fig. 11.4. Duplex cone trap of Freier and Francy (1991) (photograph reproduced with permission of J. E. Freier)

A 10-cm diameter collar connects the top of the outer cone to a 46.5-cm tall, 8.3-cm diameter plastic cylinder which houses a CDC-type motor and fan powered by a 6-V battery. The trap is positioned over about 1 kg dry ice placed on the ground. Gas escapes through the holes around the base of the black cone and mosquitoes attracted by odour and the black cone are sucked up by the fan between the two cones into a CDC-type collecting bag placed on top of the trap. The duplex trap was evaluated in a small wood near New Orleans against the gravid trap of Reiter (1983), other oviposition traps, tyres, a CDC light-trap, an EVS trap, a Trinidad trap and a Malaise trap, all baited with dry ice, and a CDC-type trap having two hamsters as bait, and finally against human landing catches. The duplex trap proved the most efficient method for collecting female *Aedes albopictus* being almost three times better than the hamster trap and four times better than the Trinidad or gravid traps. A total of 12 mosquito species were collected, the most common in decreasing order of abundance were *Aedes albopictus, Culex salinarius, Coquillettidia perturbans, Ochlerotatus triseriatus, Ochlerotatus atlanticus, Anopheles crucians* and *Culex quinquefasciatus*. The mean catch per hour for these seven species ranged from 45.6 for *Aedes albopictus* to 4.5 for *Culex quinquefasciatus*. With the exception of *Culex quinquefasciatus* which was ranked 10th in human landing collections, the same species comprised the seven most common species at bait, but the exact order of the ranking differed, although *Aedes albopictus* was the most common in both the duplex cone catches and at human bait, where the mean catch was 45.0/min.

It was concluded that the duplex cone trap was an efficient trap for sampling female *Aedes albopictus*. It also caught a few males, whereas none of the other traps did. This trap deserves to be evaluated against other mosquito species.

Moving-target trap of Dennett et al.

Dennett et al. (2004) describe the construction of a visual attraction trap for *Aedes aegypti* and *Aedes albopictus* that incorporates a swinging motion. The trap itself is a suction trap constructed from a 10-cm length of thin-walled PVC pipe, a 6V d.c. motor and fan mounted onto a piece of sheet metal, and a PVC pipe collection container. This suction trap is mounted on a modified battery-powered baby swing, which provides the back-and-forth motion. The swing was modified by removing the seat and replacing it with four panels of 4-mm thick black polyethylene plastic sheeting, to form a box open at the top and bottom. The rear panel of this box measured 64 cm long × 23 cm wide and was attached to the top part of the swinging frame.

The front panel was 36 cm long × 25 cm wide and the two side panels were 36 cm long × 30 cm wide. The estimated cost of the apparatus, including purchase of a used baby swing was US$ 75.00. In trials against other traps, the moving-target trap baited with 2 kg of dry ice performed less well at catching *Aedes albopictus* than CDC light-traps (also baited with 2 kg of dry ice) and operated with or without a light source. No significant differences were detected in mean trap catches of *Aedes aegypti* from the moving-target traps, CDC traps, an omnidirectional Fay-Prince trap and a Dragonfly trap (see Chap. 8 for further details of the trial and results obtained).

Sound traps

In New York, USA in 1878, Sir Hiram Maxim noticed that large numbers of male mosquitoes were attracted to the whine of the commutator brushes of a new dynamo that had recently been installed (Maxim 1901). As he admitted in his letter to the newspaper, *The Times* (London), he was neither a naturalist nor an entomologist, but was able to recognise male mosquitoes by their feathery antennae and their much smaller size than females. On the basis of this statement Service (1993) thought it possible that chironomids and not mosquitoes may have been involved. Be that as it may, this and other reports of mosquitoes responding to sound led to field trials in Cuba to see whether mosquitoes could be attracted to sound. Kahn et al. (1945) suggested that sound might be incorporated into a mosquito trap, and later Kahn and Offenhauser (1949) reported that when the wing beat sound of a single *Anopheles albimanus*, of only about 7 s duration was repeatedly played at intervals of about 5 s near a swamp it attracted 'quite large numbers' of male *Anopheles albimanus*. A few females of *Anopheles albimanus* and other unspecified mosquito species were also caught in their sound traps, but the effective range of the trap was small, because if sound intensity was increased it actually repelled mosquitoes. Furthermore, sound did not attract inactive males such as those resting amongst vegetation.

Novak (1966) failed to stimulate male *Culex pipiens* by tape recordings of either female wing beat or pure sound ranging from 150–750 Hz. Other laboratory studies, however, have shown that in addition to responding to the sound produced by flying females, male *Aedes aegypti* also react to the same note from a tuning fork. Near the source of the sound the male mosquitoes exhibited seizing and clasping precursory mating behaviour (Roth 1948). Wishart and Riordan (1959) corroborated these findings, and discovered that the fundamental frequency of the flight tone was the important element of the female sound. Also, a simple sine wave was just as effective in attracting male *Aedes aegypti* as the more complex sound of a wing beat. When a male was offered a choice of sound of the correct frequency it selected the loudest, but having moved towards the source it appeared that high inten sity signals repelled it. Wishart and Riordan (1959) recommended the use of traps in which sounds can be radiated with decreasing intensity. Belton (1967) very briefly reported that sound frequencies of 180–350 Hz attracted male *Ochlerotatus stimulans* from a swarm when the sound was placed beneath it. Optimum intensity appeared in the region of 50–70 dB. No other relevant details are given in this paper, but he mentioned using a 'suction sound trap', so this must be one of the early examples of sound trapping. Later Belton and Costello (1979) analysed the fundamental wingbeat frequency of females of 13 Canadian mosquito species.

In Florida, USA sound of a pitch of 320 Hz produced by an audio-oscillator was amplified and emitted through a speaker placed just underneath the roof of a suction trap (Bidlingmayer 1967). It had a rating of 80 dB at 25 mm from the edge of the trap intake and was emitted repeatedly for 4 s at 6-s intervals. Results were not very encouraging because although 15 mosquito species were caught in the suction trap, in only three species was there any evidence that sound might have attracted them to the trap.

In the USA McKeever (1977) reported that *Corethrella* spp., which feed on tree frogs, were attracted to frog noises emanating from a cassette player. Later McKeever and Hartberg (1980) placed a cassette player broadcasting frog noises alongside a CDC light-trap. Nightly catches of *Corethrella* varied from 1–566, the largest catch was on a night when the bulb was removed; nine mosquitoes were also caught but were not identified. *Uranotaenia lowii* also feeds on amphibians and Bidlingmayer (1967) speculated that females might possibly use the calls of frogs and toads to locate their hosts, and so it would be interesting to see whether such mosquitoes can also be caught in a CDC-type sound trap.

There was little interest in sound trapping until the work of Ikeshoji and Kanda in the 1980s. Ikeshoji (1981, 1982, 1985) found that in the laboratory sound attracted males of *Aedes aegypti, Aedes albopictus, Culex pipiens* form *molestus* and *Anopheles stephensi*, and in cages their 'acoustic removal' system decreased insemination rates of female *Aedes aegypti* to 0–30% and of *Anopheles stephensi* to 0–20% (Ikeshoji 1981). However, under field conditions a sound trap *per se* is of little use, because male mosquitoes respond to sound only over very short distances, regardless of its intensity or frequency. Consequently some other attractant, such as swarm markers, needs to bring males within the effective distance of a sound trap.

To clarify what sounds males best respond to and the relationship of age and physiological condition to their attraction, Ikeshoji (1985) undertook laboratory experiments with *Culex pipiens* form *molestus, Culex pipiens* form *pallens, Aedes aegypti, Aedes albopictus* and *Anopheles stephensi* of known age, and field trials with *Culex tritaeniorhynchus* in Japan and *Culex tarsalis* and *Ochlerotatus melanimon* in California, USA. The age of field collected males was determined by counting daily cuticular growth layers of the mesothoracic furca (Schlein and Gratz 1972). In general, response to sound changed with age and was greatest in 2-day-old *Culex pipiens* form *molestus* and 3- and 4-day-old males of the other six species.

Peloquin and Olson (1986) used a solid state function (waveform) generator and loudspeaker to present emissions of various sounds in Texan ricefields to swarming male *Psorophora columbiae*. A 2 × 2 cloth swarm marker was positioned under a 2-m tall pole holding the loudspeaker. Video recorders monitored flight movements and showed that tones with frequencies of 200–425 Hz were attractive to males over at least 1 m. Males responded better to sine waveforms than to square or triangular waveforms.

Ikeshoji cylinder sound traps

In California, USA visual swarm markers consisting of a 1 × 2.5-m black cloth wrapped around the base of a 0.7-m tall tripod having 3-m long legs were used to attract male *Culex tarsalis* to cylindrical sound traps (Ikeshoji 1985; Ikeshoji et al. 1985). The sound trap was made from a rolled 30 × 32-cm transparent colourless polyethylene sheet into which was suspended a 9.3-cm diameter speaker (1 W, 8 Ω) (Fig. 11.5*a*). The inside of the sound cylinder was sprayed with Tangle Foot adhesive, and it was positioned horizontally on top of the tripod. Sound was provided by a hand-made sound generator consisting of a small audio-oscillator powered by four s1.5-V batteries and at dusk emitted sinusoidal sound of 370 or 400 Hz at 100 dB. When a trap was operated nightly for 20 min the batteries lasted several months. Similar traps were evaluated in Japan by the same workers.

The colour of the polyethylene cylinder did not appear to affect the catch, but more mosquitoes were caught in traps placed at heights of 0.7 m (52/trap-night) than in traps at 1.5 m (22/trap-night) (Ikeshoji et al. 1985). Best results were obtained when two sound traps reciprocally sounded for 5-s durations every 15 s. Intermittent sound was considered to be important in attracting males. In later field experiments with various waveforms

Ikeshoji et al. (1987) concluded that a synthesised repeat of one actual waveform of a female of 5-s duration but ending on a 1-s fading out vibration attracted most male *Culex quinquefasciatus*.

In studies on the spatial distribution of male *Culex tritaeniorhynchus* in Japanese rice fields a 9.3-cm speaker was enclosed in a 30-cm long transparent plastic cylinder fixed on a tripod 1 m above a 1 × 2-m black cloth placed on the ground that acted as a swarm marker (Ikeshoji 1986). A 370-Hz sound of either sinusoidal, rectangular or saw-tooth waves was emitted at 100 or 110 dB, with periods of 6 s on and 5 s off. Sound emission started at 1942 h and lasted 20 min.

Fig. 11.5. Sound traps: (*a*) cylindrical sound trap of Ikeshoji (1985) near a rice field; (*b*) cup trap of Ikeshoji and Ogawa (1988) sprayed with adhesive (both photographs reproduced with permission of T. Ikeshoji)

The numbers of male *Culex tritaeniorhynchus* varied from a mean of 36.7–248.9/trap-night. The numbers of females caught was 0.2–30.1/trap night.

Ikeshoji and Ogawa cup traps

In Japan Ikeshoji and Ogawa (1988) field-tested four different types of sound traps. The standard model was the cylindrical trap of Ikeshoji (1986). Their board trap consisted of a 5-cm diameter thin piezoplastic sheet speaker sandwiched between two 53 × 73 × 0.5-cm polystyrene foam boards painted black and sprayed with adhesive. Board traps were mounted vertically at 1 m and emitted 400 Hz, or horizontally at 30 cm and emitted 400 or 900 Hz. Their third and fourth traps consisted of 30- and 40-cm diameter halved polyethylene hollow balls (Fig. 11.5*b*) to form so-called cup traps. The outer convex surfaces were painted black and sprayed with adhesive. A 2-cm diameter piezoelectric metal disc, commonly used as a buzzer, was fixed with adhesive tape to the inside of the trap, and sound was emitted at 400 Hz or occasionally at 900 Hz. These traps were placed convex side uppermost on a tripod at a height of 30 cm. A hand-made generator (Ikeshoji 1986) was used to provide sound at 400 Hz, but 900 Hz sinusoidal sound was generated by a tape recorder; these are the mean wingbeat frequencies of female and male *Aedes albopictus* (Ikeshoji 1981). In all four types of traps sound intensity was kept to 90 dB at 1 cm from the speakers.

The larger (40-cm diameter) cup trap emitting 400 Hz caught the largest numbers of *Culex tritaeniorhynchus* (249.6 males/trap-night), whereas the standard 400 Hz cylindrical traps caught a mean of 89.8 males/trap-night. Most females were caught on board traps, but it was believed that these acted more as visual attraction traps than sound traps.

With *Aedes albopictus* most males and females were caught on 400 Hz vertical board traps (5.5 and 11.9/trap-night, respectively), but there were often large nightly variations in the numbers caught.

Kanda et al. cylinder and lantern traps

In Malaysia two types of sound trap were evaluated for attracting male *Mansonia uniformis*, other *Mansonia* species and *Aedes albopictus* (Kanda et al. 1987). The first trap consisted of two polythene cylinders held crosswise with a speaker in the middle of each (Fig. 11.6*a*). The inner surfaces of the cylinders were sprayed with an adhesive. The second trap consisted of a speaker placed in a paper lantern enclosed within a plastic bag sprayed with adhesive (Fig. 11.6*b*).

The largest number of mosquitoes was caught when emitted sound from a tape recorder had a frequency of 330 and 350 Hz. Sound by itself attracted a maximum of 14 *Mansonia* males to the cylinder trap, and two males to the lantern trap during exposure periods (1920–2030 h). When the cylinder trap was supplemented with dry ice the mean catch increased to 123.9 males, and the lantern trap catch to 28.8 males. But the largest numbers was obtained when both dry ice and a guinea pig supplemented sound in the cylinder traps, yielding a mean of 228.4 males and a maximum catch of 392 males.

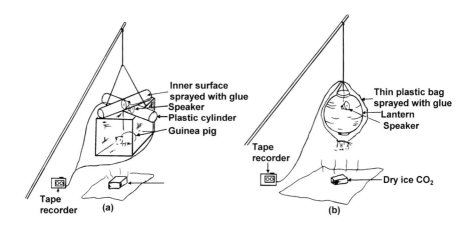

Fig. 11.6. (*a*) two cylindrical sound traps combined with odours from an animal bait and dry ice (Kanda et al. 1987); (*b*) paper lantern-type sound trap with dry ice (Kanda et al. 1987)

When sound at 480 Hz was combined with a guinea pig and dry ice in a cylinder trap, a mean catch of 17.8 male *Aedes albopictus* was obtained, the maximum catch being 23 males.

Kanda et al. (1988) again used this system in Malaysia to catch unspecified *Mansonia* species in sound traps. In these trials the lantern had holes to allow the emission of odours from a hamster and dry ice contained within it. The trapping period was for 1 h commencing 15 min after sunset. Different wingbeat frequencies were emitted from tapes in which the sine wave was recorded by the method of Ikeshoji (1985). Few *Mansonia* were collected in lantern traps (9–28 females, 1–11 males) when just one of the attractants was used. The addition of dry ice to sound traps (350 and 530 Hz) greatly increased the catches (601 and 1736 females, 197 and 575 males), but the largest catches were obtained when sound was added to lantern traps containing both hamsters and dry ice (1728 and 1766 females, 667 and 943 males). In other trials using six different wingbeat frequencies (300–1000 Hz) the most productive combination for trapping females was lantern traps having a hamster, dry ice and sound at a frequency of 500 Hz. The lantern-type trap caught many more female *Mansonia* than the cylindrical traps, but males were attracted more to the cylindrical trap, and the best frequency was 350 Hz.

In Thailand Leemingsawat (1989) modified the two above sound traps (lantern and cylinder) and baited them with both hamsters and dry ice, in addition to sound. The lantern trap proved superior, and when emitting sound at 800–1000 Hz attracted mean catches of up of 57.0 female *Anopheles minimus*/trap-night (1800–0600 h). A smaller peak in numbers occurred at frequencies of 500–600 Hz, but reduced numbers at 700 Hz. The reasons for attractancy at two frequency ranges is unclear, but might have been due to differences in mating behaviour of two divergent genetic groups. Much smaller numbers of *Anopheles maculatus* were caught, and they responded to all frequencies, but usually most were attracted to 800–1000 Hz. In the same area 16 *Anopheles* species were caught in buffalo-baited traps and seven species in human landing collections. Leemingsawat (1989) reported that buffalo traps caught four times as many *Anopheles minimus* and three times as many *Anopheles maculatus* as did human landing catches, and that sound traps caught one-third to one-quarter the numbers caught with buffaloes and three to four times more than those caught at human bait. He concluded that sound traps baited with hamsters and dry ice were efficient in sampling these two malaria vectors.

By contrast, female acoustic response has been less studied. However, in Thailand Leemingsawat et al. (1988) placed sound-emitting speakers in or on top of cages and compared the attraction of female *Culex tritaeniorhynchus* to hamsters and/or dry ice and tape-recorded sounds. At frequencies of 550–700 Hz relatively few females were caught, but the numbers greatly increased when traps also contained dry ice or a hamster, but the largest catches were made when all three attractants were combined. The most attractive frequency appeared to be 600 Hz, and the mean catch/trap-night rose to 71.7 when traps were baited with hamsters, dry ice and sound at this frequency.

Nicosia trap

The Nicosia trap produces an acoustical signal by way of pulsed liquid technology, which is designed to be similar to the signal produced by the circulatory system of host animals. This trap was compared against the ABC PRO trap, and various counter-flow geometry traps by Kline (2002). In field studies in Gainesville, Florida, USA, the Nicosia trap caught 13 species of mosquito but performed no better than an ABC PRO CDC-type light-trap supplemented with CO_2 in terms of total catch. *Ochlerotatus atlanticus, Orthopodomyia signifera* and *Psorophora columbiae* were only caught by the Nicosia trap, not by a counter-flow geometry trap, a Mosquito Magnet Beta-1 trap, nor an ABC PRO trap.

Conclusion

Leemingsawat (1989), Kerdpibule et al. (1989) and Ikeshoji et al. (1990) provide several useful references to the use of sound traps in the field to catch various species of mosquitoes. Nevertheless, it is most unlikely that sound traps will be widely used, because they mainly attract males, have a very short attraction range and usually have to be supplemented with host odours (animals or carbon dioxide).

References

Agee HR, Patterson RS (1983) Spectral sensitivity of stable, face, and horn flies and behavioral responses of stable flies to visual traps (Diptera: Muscidae). Environ Entomol 12: 1823–1828

Belton P (1967) Trapping mosquitoes with sound. Proc California Mosq Vector Control Assoc 35: 98

Belton P, Costello RA (1979) Flight sounds of the females of some mosquitoes of western Canada. Entomol Exp Appl 26: 105–14

Bidlingmayer WL (1967) A comparison of trapping methods for adult mosquitoes: species response and environmental influence. J Med Entomol 4: 200–220

Bond HA, Craig GB, Fay RW (1970) Field mating and movement of *Aedes aegypti*. Mosquito News 30: 394–402

Brach EJ, Trimble RM (1985) Effect of adhesive on the spectral reflectance of insect traps. Can Entomol 117: 1565–1568

Brett GA (1938) On the relative attractiveness to *Aedes aegypti* of certain coloured cloths. Trans R Soc Trop Med Hyg 32: 113–124

Brody MS (1977) Annual Report: *Aedes aegypti*. New Orleans Mosquito Control Board, pp 13–15

Brown AWA (1951) Studies on the responses of the female *Aedes* mosquito. Part IV. Field experiments on Canadian species. Bull Entomol Res 42: 575–582

Brown AWA (1954) Studies on the responses of the female *Aedes* mosquito. Part VI. The attractiveness of coloured cloths to Canadian species. Bull Entomol Res 45: 67–78

Browne SM, Bennett GF (1981) Response of mosquitoes (Diptera: Culicidae) to visual stimuli. J Med Entomol 18: 502–521

Burgess L, Haufe WO (1960) Stratification of some prairie and forest mosquitoes in the lower air. Mosquito News 20: 341–346

Corbet PS (1966) Diel patterns of mosquito activity in a high arctic locality; Hazan Camp, Ellesmere Island, N.W.T. Can Entomol 98: 1238–1252

Corbet PS, Danks HV (1973) Seasonal emergence and activity of mosquitoes (Diptera: Culicidae) in a high arctic locality. Can Entomol 105: 837–872

Dennett JA, Vessey NY, Parsons RE (2004) A comparison of seven traps used for collection of *Aedes albopictus* and *Aedes aegypti* originating from a large tire repository in Harris County (Houston), Texas. J Am Mosq Control Assoc 20: 342–349

Fay RW (1968) A trap based on visual responses of adult mosquitoes. Mosquito News 28: 1–7

Fay RW, Prince WH (1970) A modified visual trap for *Aedes aegypti*. Mosquito News 30 20–23

Freier JE, Francy DB (1991) A duplex cone trap for the collection of adult *Aedes albopictus*. J Am Mosq Control Assoc 7: 73–79

Garcia R, Colwell AE, Voigt WG, Woodward DL (1989) Fay–Prince trap baited with CO_2 for monitoring adult abundance of *Aedes sierrensis* (Diptera: Culicidae) J Med Entomol 26: 327–331

Giglioli MEC (1979) *Aedes aegypti* programs in the Caribbean and emergency measures against the dengue pandemic of 1977–78; a critical review. In: Dengue in the Caribbean. Proceedings of a Workshop, Montego Bay Jamaica (8–11 May 1978). Pan-Am Health Organ Sci Publ No. 375, pp 133–152

Gjullin CM (1947) Effect of clothing color on the rate of attack of *Aedes* mosquitoes. J Econ Entomol 40: 326–327

Green CH, Flint S (1986) An analysis of colour effects in the performance of the F2 trap against *Glossina pallidipes* Austen and *G. morsitans morsitans* Westwood (Diptera: Glossinidae). Bull Entomol Res 76: 409–418

Harwood RF (1961) A mobile trap for studying the behaviour of flying bloodsucking insects. Mosquito News 21: 35–39

Haufe WO, Burgess L (1960) Design and efficiency of mosquito traps based on visual response to patterns. Can Entomol 92: 124–140

Ikeshoji T (1981) Acoustic attraction of male mosquitoes in a cage. Jap J Sanit Zool 32: 7–15

Ikeshoji T (1982) Attractive sounds for autochemosterilization of male mosquitoes. Jap J Sanit Zool 33: 41–49

Ikeshoji T (1985) Age structure and mating status of male mosquitoes responding to sound. Jap J Sanit Zool 36: 95–101

Ikeshoji T (1986) Distribution of the mosquitoes, *Culex tritaeniorhynchus*, in relation to disposition of sound traps in a paddy field. Jap J Sanit Zool 37: 153–159

Ikeshoji T, Ogawa K (1988) Field catching of mosquitoes with various types of sound traps. Jap J Sanit Zool 39: 119–124

Ikeshoji T, Yap HH (1987) Monitoring and chemosterilization of a mosquito population, *Culex quinquefasciatus* (Diptera: Culicidae) by sound traps. Appl Entomol Zool 22: 474–481

Ikeshoji T, Yap HH (1990) Impact of the insecticide-treated sound traps on an *Aedes albopictus* population. Jap J Sanit Zool 41: 213–217

Ikeshoji T, Sakakibara M, Reisen WK (1985) Removal sampling of male mosquitoes from field populations by sound-trapping. Jap J Sanit Zool 36: 197–203

Ikeshoji T, Yamasaki Y, Yap HH (1987) Attractancy of various waveform sounds in modulated intensities to male mosquitoes, *Culex quinquefasciatus*, in the field, Jap J Sanit Zool 38: 249–252

Ikeshoji T, Langley P, Gomulski L (1990) Genetic control by trapping. In: Curtis CF (ed) Appropriate Technology in Vector Control. CRC Press, Boca Raton, Florida, pp 159–172

Kahn MC, Offenhauser W (1949) The first field tests of recorded mosquito sounds used for mosquito destruction. Am J Trop Med 29: 811–825

Kahn MC, Celestin W, Offenhauser W (1945) Recording of sounds produced by certain disease-carrying mosquitoes. Science 101: 335–336

Kanda T, Cheong WH, Loong KP, Lim TW, Ogawa K, Chiang GL, Sucharit S (1987) Collection of male mosquitoes from field populations by sound trapping. Trop Biomed 4: 161–166

Kanda T, Loong KP, Chiang GL, Cheong WH, Lim TW (1988) Field study on sound trapping and the development of trapping method for both sexes of *Mansonia* in Malaysia. Trop Biomed 5: 37–42

Kerdpibule V, Thongrungkiat S, Leemingsawat S (1989) Feasibility of wing beat sound trap for the control of mosquito vectors. Southeast Asian J Trop Med Public Health 20: 639–641

Kline DL (2002) Evaluation of various models of propane-powered mosquito traps. J Vector Ecol 27: 1–7

Kloter KO, Kaltenbach JR, Carmichael GT, Bowman DD (1983) An experimental evaluation of six different suction traps for attracting and capturing *Aedes aegypti*. Mosquito News 43: 297–301

Kodak Publications (1990) Handbook of Kodak Photographic Filters. Eastman Kodak Co., New York

Lang JT (1984) Intermittent light as a mosquito attractant in New Jersey light traps. Mosquito News 44: 217–220

Leemingsawat S (1989) Field trials of different traps for malaria vectors and epidemiological investigations at a foot-hill basin in Kanchanaburi, Thailand. Jap J Sanit Zool 40: 171–179

Leemingsawat S, Kerdpibule V, Limswan S, Sucharit S, Ogawa K-I, Kanda T (1988) Response of the female mosquitoes of Culex tritaeniorhynchus to sound traps of various wingbeat frequencies with hamsters and dry ice. Jap J Sanit Zool 39: 67–70

Maxim HS (1901) Mosquitoes and Musical Notes. The Times (London), Letter to Editor, 28th October

McKeever S (1977) Observations of *Corethrella* feeding on tree frogs (*Hyla*). Mosquito News 37: 522–523

McKeever S, Hartberg WK (1980) An effective method for trapping adult female *Corethrella* (Diptera: Chaoboridae). Mosquito News 40: 111–112

Novak D (1966) Note to the laboratory tests with the sounds of mosquitoes. Arch Roum Pathol Exp Microbiol 25: 849–852

Peloquin JJ, Olson JK (1986) Effects of sound on swarming male *Psorophora columbiae*. J Am Mosq Control Assoc 2: 507–510

Peterson DS, Brown AWA (1951) Studies on the responses of the female *Aedes* mosquito. Part III. The response of *Aedes aegypti* (L) to a warm body and its radiation. Bull Entomol Res 42: 535–541

Reiter P (1983) A portable, battery-powered trap for collecting gravid *Culex* mosquitoes. Mosquito News 43: 496–498

Roth LM (1948) A study of mosquito behavior. An experimental laboratory study of *Aedes aegypti* (Linnaeus). Am Midl Nat 40: 265–352

Schlein Y, Gratz NG (1972) Age determination of some flies and mosquitoes by daily growth layers of skeletal apodemes. Bull World Health Organ 47: 71–76

Schoeler GB, Schleich SS, Manweiler SA, Sifuentes VL (2004) Evaluation of surveillance devices for monitoring *Aedes aegypti* in an urban area of northeastern Peru. J Am Mosq Control Assoc 20: 6–11

Service MW (1993) Mosquito Ecology. Field Sampling Methods. 2nd edn. Chapman & Hall, London

Sippell WL, Brown AWA (1953) Studies on the responses of the female *Aedes* mosquito. Part V. The role of visual factors. Bull Entomol Res 43: 567–574

Wilton DP, Kloter KO (1985) Preliminary evaluation of a black cylinder suction trap for *Aedes aegypti* and *Culex quinquefasciatus* (Diptera: Culicidae). J Med Entomol 22: 113–114

Wishart G, Riordan DF (1959) Flight responses to various sounds by adult males of *Aedes aegypti* (L) (Diptera: Culicidae). Can Entomol 91: 181–191

Chapter 12 Estimation of the Mortalities of the Immature Stages

Preceding chapters have described various methods for sampling the different developmental stages and age groups of mosquito populations. This and subsequent chapters are concerned with analysing the numerical changes that occur in population size during the life-cycle of mosquitoes, identifying the causes of mortalities, and determining the age structure and survival rates of pre-imaginal and adult populations. A comprehension of the growth and regulation of mosquito populations is essential for understanding their population dynamics. Measurement of the mortality that necessarily occurs during the lifecycle from egg to ovipositing female has interested ecologists and statisticians alike, and many of the techniques used are founded on mathematical probabilities.

Density-dependent mortalities

MacArthur and Wilson (1967) in their book *The Theory of Island Biogeography* originated the terms r- and K-selection, defining them as the measure of fitness at very low and very high densities respectively. Mueller in a 1997 review contends that as no logical relationship exists between the logistic equation of the MacArthur and Wilson model and age-structured populations, then the theory of density-dependent natural selection cannot make predictions about the evolution of any trait other than population growth rates. Early models only considered populations without age structure, and therefore any life-stage could be considered as the population. As a result, it was observed that selection could increase, decrease, or have no effect on equilibrium numbers of the different life-stages, and in some cases, all three effects were observed according to which life-stage was considered (Prout 1980). Conventionally density-dependent mortality per generation is tested for by plotting percentage mortality, or the k-values of Varley and Gradwell (1960), against the population density on which mortality is acting. A positive correlation, that is a regression greater than 0, if statistically valid (Royama 1981*a,b*), is regarded as indicating the presence of density-dependent mortality (Varley and Gradwell 1968). Hassell et al. (1976) reviewed the role of density-dependent mortality in maintaining stability in several insect species, including *Aedes aegypti*, having little or no overlap between generations. They concluded that in most natural populations there was a return to a stable equilibrium after a disturbance, that is density-dependent regulation of population size was occurring. However, Dempster (1983) in analysing life-table data from Lepidoptera populations failed to find much convincing evidence that natural enemies are important in regulating insect populations in a density-dependent fashion. Hassell (1985) challenged this view by pointing out the difficulty in actually identifying density dependence when only average population densities of different life-stages are used and no or little account is given to spatial distribution of host populations, and other types of within-generation heterogeneity. May (1986) also argued that much density-dependent regulation arises from within-generation spatial heterogeneity, and that regulation depends only weakly on yearly variations in average population density. Hassell (1985) and May (1986) concluded that spatial heterogeneity and environmental stochasticity often make the detection of density dependence by conventional analysis of yearly fluctuations in average population density undetectable. This view is vigorously contested by Dempster and Pollard (1986) who proclaim that although spatial heterogeneity may modify the operation of density-dependent factors, in itself spatial heterogeneity cannot regulate a population. They stress that without temporal dependence there cannot be any population regulation, and that detection of density dependence still relies mainly on the analysis of year to year variations in average population size data. Hassell (1987) returned to the controversy and argued that not only should life-table studies take into account spatial distribution, but also other forms of within-generation variability in survival. Moreover, the problems of detecting within-generation density dependence may be compounded by sampling contagiously distributed populations. Mountford (1988) adopted a neutral stance and examined the facts from a purely mathematical viewpoint, and used simulation models. He found that spatial heterogeneity enhances the detection of density dependence, whereas population stochasticity is neutral in its effect, and that detailed analysis of within-generation mortalities may not always detect density dependence.

Pollard et al. (1987) urged caution in the use and interpretation of any test to detect density dependence from census data, because the tests depend on assumptions regarding population processes that may be incorrect, and also because errors of measurements may lead to spurious detection of density dependence (Itô 1972; Kuno 1973; Royama 1977, 1981a). Mueller (1997) notes if populations inhabiting different environments that differ only in terms of the density experienced the populations under study, then in principle any differences in the populations would be the result of density-dependent selection. However, this situation almost never occurs in field studies, where there are always multiple differences between populations, in addition to density, that confound simple interpretations. Consequently, any result, whether positive or negative, may have been due to one of the uncontrolled factors rather than differences in density per se, and hence carefully-controlled laboratory studies may prove more informative. Although most animal populations tend to fluctuate around a fixed level, statistical tests for density dependence often fail to show that such fluctuations in population size are not just random. Stiling (1988) reviewed 63 life-table studies involving 58 insect species and concluded that in 46.9% of them density dependence was not operating to regulate their populations. Hassell et al. (1989) re-examined the data presented by Stiling (1988) and found evidence for density dependence was more likely to be detected when data from large numbers of generations were considered. Solow and Steele (1990) were able to explain why this was so from a statistical viewpoint, and concluded that for even moderate density dependence it may be necessary to collect data on up to 30 generations before density independence can be firmly rejected. In fact, many ecologists believe there is no single and simple test to identify density-dependent or density-independent processes from census data. Pollard et al. (1987) proposed a distribution-free approach to detect density dependence from annual population censuses.

Kuno (1991) presents a brief review of the methods and pitfalls of detecting the presence, or otherwise, of density-dependent population regulation. He favours the classical approach of examining the slope b in the regression of total generation mortality (K) on initial density (N_t), but with some statistical modifications, to help overcome spurious density-dependent associations due to statistical quirks. He also explains how the fiducial limits of b can be obtained. Other interesting papers on density-dependent mortalities are those of May and Oster (1976), Bellows (1981), and Stewart-Oaten and Murdoch (1990).

More recent models of density-dependence, as reviewed by Mueller (1997) tend to be explanatory models that explicitly take into account specific components of the life cycle of some organism or group of organisms and try to model the response of these life-history components to density-dependence, rather than attempting to describe these complex processes with a single, simple equation. Theoretical predictions of these models are that density-dependent selection can lead to increased competitive ability, but will have no lasting impact on the adult equilibrium population size. These theories also predict that density-dependent selection can either increase or decrease adult body size, yielding no single or simple prediction about the evolution of body size due to density-dependent selection (Mueller 1997).

Berlow et al. (1999), following on from early work by Paine (1992) reviewed and summarised several methods of quantifying interaction strength between species, defined as the mean proportional change in the density of a non-manipulated population divided by the change in density of the manipulated (added or removed) species. This approach yields numerical estimates of the strength of interspecific interactions, or so-called 'interaction coefficients'. Abrams (2001) recently reviewed these approaches to describing and quantifying interspecifc interactions and concluded that such single numerical measures of the strength of interactions may be poor characterisations of dynamic, non-linear interactions, the form of which is poorly understood, in spite of many hundreds of field experiments involving population manipulations.

Density-dependence in mosquitoes

Field-based ecological studies have tended to focus predominantly on container-inhabiting species, which are easier to study and there have been many ecological investigations on mosquitoes inhabiting tree-holes (Bradshaw and Holzapfel 1983, 1984, 1985, 1986*a*, 1988, 1991; Fish and Carpenter 1982; Fisher et al. 1990; Hawley 1985*a,b*), pitcher plants and flower bracts (Bradshaw 1980, 1983; Bradshaw and Holzapfel 1986*b*; 1989; Istock et al. 1975, 1976*a,b*; Lounibos 1981, 1985; Lounibos et al. 2003; Machado-Allison et al. 1983), leaf axils and other phytotelmata (Lounibos 1979; Seifert 1980; Seifert and Barrera 1981; Sunahara and Mogi 2002*b*), as well as laboratory studies (Schneider et al. 2000; Smith et al. 1995). Studies on larval competition in larger habitats are relatively rare but Renshaw et al. 1993 undertook such studies on *Ochlerotatus cantans* in woodland pools. Many of these studies also attempted to determine the causes of immature mortalities, and moreover whether or not they are density-dependent. For example, evidence that *Wyeomyia smithii* inhabiting pitcher plants exhibited density-dependent mortalities was obtained byBeaver (1983), Bradshaw (1983), Bradshaw and Holzapfel (1986*b*) and Istock et al. (1976*b*), while Frank (1983) believed that bromeliad breedingmosquitoes were mainly regulated by density-dependent mortalities. In laboratory experiments with *Ochlerotatus triseriatus* larvae Fish and Carpenter (1982) found that increased larval densities resulted in decreased larval survivorship and pupation rates. Hard et al. (1989) also showed in laboratory experiments the existence of density-dependent fitness and mortality in *Ochlerotatus triseriatus* and *Ochlerotatus geniculatus*.

Gilpin and Langford (1978) obtained laboratory evidence that competition for food caused density-dependent growth rates and affected pupation success in *Ochlerotatus sierrensis*. Fisher et al. (1990) compared the fitness of *Ochlerotatus geniculatus* and *Ochlerotatus sierrensis* in the laboratory and found that at high larval densities *Ochlerotatus geniculatus* outcompeted *Ochlerotatus sierrensis*, again showing a density-dependent effect.

Smith et al. (1995) studied intraspecific and interspecific competition among larvae of *Culex tarsalis* and *Culex quinquefasciatus* reared at different densities in the laboratory. The distribution of these two species overlaps in California, USA and interspecific competition may have led to partial replacement of *Culex tarsalis* by *Culex quinquefasciatus* in some areas. Increasing larval density resulted in an increase in the median time from eclosion to emergence, decreased survivorship and reduced adult wing length. *Culex quinquefasciatus* displaced *Culex tarsalis* in a cage experiment within a single generation.

Costanzo et al. (2005) recently studied the competitive interactions between *Aedes albopictus* and *Culex pipiens*, two species commonly found co-occurring in used tyres in the United States, and observed that *Aedes albopictus* survivorship and developmental times were significantly affected by conspecific densities but not by the densities of *Culex pipiens*. In contrast, *Culex pipiens* survivorship and developmental times were significantly affected by densities of both conspecifics and *Aedes albopictus*.

In a study of intraspecific competition in *Aedes albopictus* under semi-natural conditions in Florida, USA, Lord (1998) reported that larval density ssignificantly affected the proportion surviving, wing length, and median time to emergence, while food availability significantly affected wing length only. The interaction between density and food availability only had a significant effect on median time to emergence and an index of population growth (r'), where $r' =$

$$r' = \frac{\ln \dfrac{1}{N_0} \sum_x A_x f(w_x)}{D + \left[\dfrac{\sum_x x A_x f(w_x)}{\sum_x A_x f(w_x)} \right]} \qquad (12.1)$$

and N_0 = the original number of females, D is the lag between emergence and oviposition in days, A_x is the number of females ovipositing on day x, w_x is the average wing length and $f(w_x)$ is fecundity associated with wing length, estimated from the relationship between wing length and the number of female offspring produced per female during the first gonotrophic cycle.

Gimnig et al. (2002) in western Kenya studied the growth and development of *Anopheles gambiae* larvae in artificial habitats constructed from plastic wash basins (35 cm diameter, 13 cm deep) into which five litres of local "black cotton" soil, 2–3 litres of water and either 20 or 60 larvae were added. The authors observed that larval development time was extended and emergent adults were smaller as rearing densities increased. Survival was not significantly affected. Addition of nutrients in the form of cow dung had no impact on larval growth and development. There was evidence of decreasing nitrogen levels associated with increasing larval densities indicating that nitrogen may be a limiting resource in the larval environment.

Schneider et al. (2000) conducted similar investigations of larval competition between the sympatric sibling species *Anopheles gambiae* and *Anopheles arabiensis* in the laboratory. *Anopheles gambiae* survival to the pupal stage was significantly higher than observed for *Anopheles arabiensis* when reared in single-species or in mixed-species experiments. At high larval densities (400 larvae per 200 cm^2 rearing tray) there was no significant difference in survival between the two species. Survival of *Anopheles arabiensis* decreased in mixed-species experiments, but survival of *Anopheles gambiae* was not affected by rearing in the presence of *Anopheles arabiensis*. There was no evidence of cannibalism or inter-specific predation.

Laboratory studies on the effects of priority in colonisation of a habitat on the outcome of intraspecific and interspecific competition revealed that survivorship and pupation success of cohorts of *Aedes albopictus* added to artificial container habitats on day 14 were affected negatively by cohorts of either *Aedes albopictus* or *Tripteroides bambusa* that had been added on day 0 (Sunahara and Mogi 2002*a*). Input of leaf-litter and also flushing of the habitats moderated the competitive effects of early cohorts on late cohorts suggesting that a toxic substance as well as competition for resources were factors involved in causing the negative effects of the early cohorts on survivorship of later cohorts.

A model which can qualitatively mimic the dynamics of a laboratory population of *Aedes aegypti* having discrete generations was built by Gilpin et al. (1976) and Gilpin and McClelland (1979). However, as pointed out by Dye (1984*a*) their model bears little resemblance to reality where there are overlapping generations of *Aedes aegypti* and eggs can be hatching daily. In contrast Dye (1984*a*) built analytical models for a real field population (Southwood et al. 1972) of adult *Aedes aegypti,* which were then used to develop to a multi-age class simulation model. It was shown that the population was monotonically stable, and that the adult population was more sensitive to changes in death rate than to changes in birth rate or numbers of larval habitats. Furthermore, the adult population was more sensitive to changes in adult survivorship than to changes in fecundity. It was concluded that density-dependent mortality acted on the eggs not the adults.

In Thailand Southwood et al. (1972) concluded that mortality between eggs and 2nd instar larvae of *Aedes aegypti* inhabiting water-storage pots was density-dependent, and was the most important mortality in regulating population size, but that mortality of 4th instar larvae appeared to be density-independent. Rogers (1983) reanalysed their data and concluded that in fact the greatest population loss was from the pupal stage to oviposition, but that mortality from egg hatching to 2nd instar larvae made the next most important contribution to total generation mortality.

Sunahara and Mogi (2002*b*) studied intra- and interspecific competition among larval mosquitoes and the effects on population performance of *Aedes albopictus* in bamboo groves. Their study also looked at the competitive interactions occurring at different spatial scales, both within a single bamboo grove and between separate groves. Experimental removal of competitors was the method chosen to examine the effects of competition at the scale of individual bamboo stumps, while comparison of population performance between bamboo groves with and without competitors was used to study the significance of competition over a larger spatial scale. Artificial oviposition sites were constructed from plastic cups (bottom and top diameter, 7 cm and 9 cm respectively; height, 10 cm) containing a cork sheet oviposition substrate measuring 3 × 10 cm in area and 2 mm thick. A smaller number of pupae and adults with lower body weight were observed at the plains site, where only *Aedes albopictus* occurred than at the hillside site, where *Aedes albopictus* occurred alongside other species, including *Tripteroides bambusa*. This suggests that the effect of intraspecific competition on *Aedes albopictus* was more severe than the combined effects of intra- and interspecific competition. A significant effect of leaf-litter on pupation of *Aedes albopictus* was also detected, confirming the importance of litter as a limiting factor for mosquito production. Drying of larval habitats removed the dominant effects of *Tripteroides bambusa*, presumably because *Aedes albopictus* has evolved high desiccation resistance in the egg stage (Sota and Mogi 1992). The authors concluded that drought-prone oviposition sites could benefit *Aedes albopictus* by acting as refuges from less drought-resistant competitive species.

Mpho et al. (2000) studied fluctuating asymmetry in overall wing length and other measures of wing size at late second instar larval densities of 25, 50, 100 and 200 *Culex quinquefasciatus* larvae per 140 ml of water. Fluctuating asymmetry is the random non-directional deviation from perfect bilateral symmetry with a mean distribution around zero (Van Valen 1962) and can be a useful indicator of environmental stress. All wing characters measured decreased in size for both sexes as larval density increased, as did body weight. Significant fluctuating asymmetry was observed, but it did not increase consistently with increasing larval density.

Lounibos et al. (2003) reported that fourth instar larvae of native *Wyeomyia* species negatively affected growth and survivorship of *Aedes albopictus* in exotic tank bromeliads in Florida, USA.

Hawley (1985*a,b*) and Lounibos (1985) have shown that density-dependent mortality arises in the tree-hole species, *Ochlerotatus sierrensis* and *Ochlerotatus triseriatus*. In South America *Trichoprosopon digitatum* lays her eggs in a variety of phytotelmata. In *Heliconia* bracts larval survivorship is not density-dependent (Seifert 1980), whereas in split cocoa pods, a more ephemeral habitat, there is density-dependent mortality (Lounibos and Machado-Allison 1986). *Aedes albopictus* has been shown to outcompete *Aedes aegypti* in resource-limited automobile tyres in a natural Florida, USA woodland (Juliano 1998).

Mori and Wada (1978) concluded that *Aedes albopictus* breeding in artificial containers was mainly regulated by density-dependent mortality. By calculating the *k*-values of Varley and Gradwell (1960) Mogi and Yamamura (1988) detected density-dependent larval mortality in *Armigeres theobaldi* in water-filled flower bracts, caused by contest competition, probably due to larvae being cannibalistic. Variations in pupal and adult size were considerably less than have been recorded for other species, and surprisingly were density-independent.

Most ecological studies on the population dynamics of mosquito larvae have concentrated on container breeders, because it is so much easier to census a population in small discrete habitats than in larger ones such as ground pools. Nevertheless, attempts have been made to determine the type of mortality experienced by species inhabiting more permanent collections of water. For example, Chubachi (1979) believed that *Culex tritaeniorhynchus* larvae in rice fields were mainly experiencing density-dependent mortalities. It appeared that the most important regulatory factor might have been intraspecific competition, probably for food. In India Rajagopalan et al. (1976*a*, 1977*b*) used Varley and Gradwell's (1963) method of plotting the regression of *k*-values (logarithm of the reciprocal of the survival rate) on egg density (N_e) to determine whether there was density-dependent mortality from

$$K = b_0 + b_1 N_e \qquad (12.2)$$

egg to pupa in *Culex quinquefasciatus* breeding in wells. In the above equation b_0 = the estimated intercept of *k* at zero density, and the reciprocal of the antilog gives estimated maximum survival, b_1 = slope of regression line and indicates change in survival probability. These regressions were tested for significance by an analysis of variance, and if the regression was significant then the maximum probability of survival (P_0) at zero density was calculated as

$$P_0 = e^{-b_0} \qquad (12.3)$$

where e = base of natural logarithms. Because k-values themselves are not independent of egg density Varley and Gradwell's (1968) test for density dependence was performed. This involved calculating the regression of log egg density on log pupal density and the regression of log pupal density on log egg density. Data collection from some of the wells showed that the slopes of both regressions were on the same side as 1.0, and differed significantly from 1.0, and so it was assumed that there was real density-dependent mortality. In other wells where larval densities were insufficiently high for such regulation, density-independent processes caused changes in population size. Nevertheless, they considered that as a whole the population of *Culex quinquefasciatus* breeding in wells in the area was under density-dependent regulation.

Renshaw et al. (1993) studied larval mortality, larval size and size of emergent adults of *Ochlerotatus cantans* in natural temporary woodland pools in northern England. In 1989, larval siphon length for all instars was significantly lower in a pond with a high larval population density compared with a pond with a lower population density. Adults emerging from the pond with a high larval population density had a mean winglength of 4.10 mm (SD $\pm$ 0.27), whereas those emerging from the pond with the lower population density had a mean winglength of 5.5 mm (SD $\pm$ 0.45) (Mann-Whitney, $P < 0.001$). In contrast in 1990, when larval densities in the two ponds were similar, there were no significant differences in larval siphon length nor adult wing length for the populations in the two ponds. Predator-exclusion cage experiments were also conducted in which 10, 50, 100, or 200 first instar larvae were maintained in inverted plastic plant pots covered with fine nylon mesh and floated in the same natural temporary woodland pools using expanded polystyrene sheets. The volume of water enclosed by each cage was approximately 100 ml. In the first experiment, larvae from each cage were killed after they had entered the second instar stage and larval siphon length was measured. Mean siphon length of larvae maintained at a density of 100 was significantly greater than for larvae maintained at a density of 50, which in turn was greater than siphon length of larvae maintained at a density of 10 larvae per 100 ml (ANOVA, $P < 0.001$). In a second experiment, survivorship curves and life-tables (according to the methods of Lakhani and Service 1974) were constructed for larvae maintained at densities of 10, 50, 100 or 200 per 100 ml. At the lowest larval density, the probability of an egg surviving to an adult was 0.448, compared with 0.350 at a density of 50, 0.149 at a density of 100, and 0.081 at a density of 200. Potential sources of larval mortality were identified as cannibalism (observed at high larval density in the laboratory) and contact inhibition of feeding among neighbouring individuals at high densities, as previously described by Broadie and Bradshaw (1991).

In an investigation as to whether or not kinship and rearing density influence cannibalism in *Ochlerotatus triseriatus*, Dennehy et al. (2001) placed F_1 larvae obtained from egg batches laid by adult females reared from wild-collected larvae into one of four treatment groups of the following rearing densities: 50 larvae/100 ml, 20 larvae/100 ml, 10 larvae/100 ml, and 5 larvae/100 ml, exposing them to different levels of stress. Five sibling or unrelated fourth instar larvae were added to vials containing first instar larvae. Kinship was assigned a value of 0 for non-kin prey and 1 for sibling prey. The rate at which fourth instar larvae consumed conspecifics apparently depended on the number of first instar prey available. Fourth instar larvae appeared able to discriminate and avoid consuming their siblings, but only under certain conditions. Cannibalistic larvae reared at high densities and exposed to high numbers of prey consumed fewer non-kin than when reared at low densities and presented with fewer prey. Fourth instar larvae exhibited a Type III functional response (Holling 1959) to non-kin first instar larvae, such that reduced rates of cannibalism are observed at low prey densities.

Edgerly et al. (1999) investigated intraguild predation among the native American tree-hole breeding *Ochlerotatus triseriatus* and the introduced container-inhabiting species *Aedes aegypti* and *Aedes albopictus* in the laboratory. Mortality of 1st instar larvae was highest in the presence of 4th instar *Ochlerotatus triseriatus*. Larvae of *Ochlerotatus triseriatus* were the least susceptible to predation of the three species studied, presumably due to their larger size.

Barr (1985) listed 34 published examples of 18 species of *Aedes, Culex, Culiseta* and *Anopheles* that show high pre-adult mortalities. In his study on *Culiseta incidens* he believed that the high mortality he observed was density-dependent, there being competition for food.

In contrast to the above evidence supporting the role of density-dependent mortality, Mogi et al. (1980a) believed that mortality of *Culex tritaeniorhynchus* larvae in fallow rice fields in Japan, caused by predators, was density-independent. This was partly because there were large numbers of many different predators, and mosquito larvae formed only a small component of their diets. However, in other studies Mogi (1978) concluded that mortality of *Culex tritaeniorhynchus* was density-dependent and was caused by larval overcrowding. Bradshaw and Holzapfel (1983) found no evidence of density-dependent regulation in tree-hole mosquitoes, *Orthopodomyia signifera, Anopheles barberi, Ochlerotatus triseriatus* and *Toxorhynchites rutilus*. Similarly Hawley (1985*a*) found no evidence of density-dependent mortality in *Ochlerotatus sierrensis*, but pupal weight was inversely correlated with larval density and thus so was adult weight. Seifert (1980) found no evidence of density-dependent mortality among larvae of *Wyeomyia felicia, Culex bihaicola* and *Trichoprosopon digitatum* colonising bracts of *Heliconia aurea*. But as argued by several ecologists it may be difficult to prove the existence, or otherwise, of density-dependent population regulation.

   In an interesting series of recent experiments Agnew and colleagues have studied the effects of density-dependent competition on selected life-history traits in *Culex quinquefasciatus* and *Aedes aegypti* using a novel experimental approach (Agnew et al. 2000, 2002). In the *Aedes aegypti* studies, larvae were reared at densities of one, two or three individuals only per 5 ml of rearing water under a range of food availabilities. Decreased food availability or increased larval density led to increased larval mortality, delayed pupation, the emergence of smaller adults and decreased time to starvation of those adults when deprived of a food source. The use of this simplified experimental approach allows for large numbers of independent replicates to be conducted.

Species Replacement in Mosquitoes

Both DeBach (1966) and Turnbull (1967) considered that competitive displacement was more common in the field than generally supposed and several examples occur in mosquitoes. There is evidence to suggest that in southeast Asia the indigenous *Aedes albopictus* has been replaced in many areas by invasions of *Aedes aegypti* (Gilotra et al. 1967; Hawley 1988; Macdonald 1956; Mattingly 1967; Rudnick 1965; Rudnick and Hammon 1960; Sucharit et al. 1978). Laboratory experiments with southeast Asian populations of *Aedes aegypti* and *Aedes albopictus* have shown that the former outcompetes *Aedes albopictus* (Chan et al. 1971; Macdonald 1956; Moore and Fisher 1969; Sucharit et al. 1978; Sucharit and Tumrasvin 1981) supporting the contention that *Aedes aegypti* might replace *Aedes albopictus* in the field. Hawley (1988), however, hypothesised that the apparent spread of *Aedes aegypti* in southeast Asia is caused by increased urbanisation, which favours breeding by this species which is also often more prevalent in indoor larval habitats, whereas *Aedes albopictus* breeds more in suburban and rural areas and tends to colonise outdoor habitats.

Aedes albopictus was first discovered in the USA in 1985, in Texas, and has since spread to 20 states. In certain areas of Texas and Louisiana there has been an observed increase in *Aedes albopictus* and decrease in *Aedes aegypti*, the reverse of what has been observed in southeast Asia. This led Black et al. (1989) to undertake laboratory studies on competition between these two species, but they could find no evidence to suggest that *Aedes albopictus* would displace *Aedes aegypti* in the USA. They thought that since tyres were the principal habitat of both species in the USA, and these are usually out of doors, this might help explain the spread of *Aedes albopictus*. But there are other possible explanations for the increase and spread of *Aedes albopictus* in the USA, some of which are discussed briefly by Black et al. (1989). In laboratory experiments and a few field trials in the USA and Singapore Ho et al. (1989) concluded that in field populations there was no evidence for the displacement of *Aedes aegypti* by *Aedes albopictus*; although it appeared that *Aedes albopictus* had some competitive advantages over *Ochlerotatus triseriatus* suggesting that it might replace *Ochlerotatus triseriatus* in the field.

Duhrkopf and Young (1979) believed that differences in the larval biology of *Aedes aegypti* and *Aedes albopictus*, such as time spent submerged, could contribute to the displacement of one species by the other. They argued that time spent on the bottom during an alarm reaction could be correlated with time spent feeding, so that larvae spending more time on the bottom eat more food, and also are more likely to avoid predation. If more food is obtained during diving episodes, then fewer dives are needed, and more resources could be diverted into growth, but on the other hand longer periods of submersion might indicate slower metabolic processes which might in turn reflect differences in fitness. However, later experiments (Duhrkopf and Benny 1990) showed no differences between the submergence times of *Aedes aegypti* and *Aedes albopictus*, in fact there were greater differences between strains of the species than between them. Their data therefore does not support the idea that this could account for the gradual displacement in many USA localities of *Aedes aegypti* by *Aedes albopictus*. Rai (1991) gives a review of *Aedes albopictus* in the Americas, and summarises competition experiments.

Following a series of laboratory investigations (Ali and Rozeboom 1971*a,b*, 1973); Gubler 1970*a,b*, 1971; Lowrie 1973*a,b*, Rozeboom 1971) indicating the ability of *Aedes albopictus* to replace *Aedes polynesiensis*, vector of non-periodic *Wuchereria bancrofti*, a field trial was conducted on Taiaro. This is a remote atoll about 5 km in diameter and about 40 km from the nearest land, another atoll (Rosen et al. 1976). Three different strains of *Aedes albopictus* were released on the atoll in the hope of replacing native populations of *Aedes polynesiensis*, but within 1 year one strain had disappeared and the remaining two disappeared within 4 years. The experiment was, however, rather unsatisfactory because it was not determined whether failure was due to: (i) the unsuitability of the selected strains, two of which had been colonised for 1–2 years; (ii) insufficient numbers being released; (iii) the environment being inhospitable (e.g. lack of suitable hosts for blood-meals, and/or hot, dry windy conditions); or (iv) whether the population of *Aedes polynesiensis* was so large that *Aedes albopictus* rarely managed to mate with its own species. These results emphasise how field results can contradict laboratory expectations, and also how difficult it can be to devise good ecological field tests.

Blackmore et al. (1995) have suggested that apparent competition, resulting from *Aedes aegypti* and *Aedes albopictus* having a shared enemy the protozoan parasite *Ascogregarina taiwanensis*, is a possible mechanism for the observed replacement of *Aedes aegypti*. To determine whether apparent competition or interspecific competition could explain species replacement in these two mosquito species, Juliano (1998) conducted a series of field experiments employing experimental cages. Each cage (height 221 mm × inside diameter 35 mm) consisted of a 150-µm nylon mesh cylinder attached to a 40 mm diameter jar lid in which a circular opening of 30 mm was cut. A plastic jar could then be screwed into the jar lid and acted as a receptacle for leaf litter and retained water when the cages were removed from car tyres, in which the experiments were conducted. Competition experiments were carried out with different densities of the two species under two different resource availabilities (amount of leaf litter added). Aedes albopictus was observed to be competitively superior to *Aedes aegypti* under simulated field conditions of high larval density, as evidenced by estimated finite rate of increase ($\lambda = \exp(r')$, where r' is as defined by Livdahl and Sugihara 1984 as:

$$r' = \frac{\ln\left[(1/N_0)\sum_x A_x f(w_x) \right]}{D + \left[\dfrac{\sum_x xAf(w_x)}{\sum_x A_x f(w_x)} \right]}$$ (12.4)

Where N_0 is the initial number of females in the cohort (assumed to be 50% of the cohort), A_x is the number of females eclosing on day x, w_x is a measure of the mean size of females eclosing on day x, $f(w_x)$ is a function relating production of female eggs to female size, and D is the time between adult eclosion and reproduction.

Resource constraints and increased density appeared to have a greater effect on *Aedes aegypti* compared with *Aedes albopictus*, and this effect was most pronounced in experimental treatments in which *Aedes albopictus* initially outnumbered *Aedes aegypti*, and where resources (leaf-litter) were limited. *Aedes albopictus* was relatively more successful at high densities, even in the absence of leaf-litter. Experiments to determine the effect of *Ascogregarina* parasitism appeared to indicate that this protozoan played very little role in determining the competitive disadvantage of *Aedes aegypti* as parasitism of *Aedes aegypti* was very low (1 of 117 larvae), compared with prevalence in *Aedes albopictus* (31/142). Juliano (1997) concluded that interspecific competition for resources and not apparent competition effectively explained the displacement of *Aedes aegypti*.

Insecticides have sometimes apparently been responsible for initiating species replacement in mosquitoes. For example, the widespread use of DDT in the malaria eradication campaign in Sardinia resulted in the virtual eradication of *Anopheles labranchiae*, but this was accompanied by a marked increase in the population of *Anopheles hispaniola* (Aitken and Trapido 1961). Surveys in Sardinia during 1980–1985, some 35 years after the end of the malaria eradication campaign in 1946, failed to detect *Anopheles hispaniola*, but *Anopheles labranchiae* was again common in several sites and was often the dominant anopheline. On the other hand it seems that the relative abundance of *Anopheles algeriensis* and *Anopheles petragnani* has switched since 1946, the former now being the more common species (Marchi and Munstermann 1987). In East Africa house-spraying with dieldrin seems to have caused the replacement of the endophilic *Anopheles funestus* with an exophilic species, *Anopheles rivulorum* (Gillies and Smith 1960). Similarly Service (1966) thought that repeated larviciding with DDT was probably the main reason for the apparent replacement of *Culex nebulosus* by *Culex quinquefasciatus* in many towns in West Africa. In South Africa, Anopheles arabiensis was effectively controlled by pyrethroid insecticides, but pyrethroid resistant Anopheles funestus, originating from Mozambique replaced it, and resulted in a major outbreak of malaria in 2000 (Coetzee 2004). Insecticides can also cause a change in breeding places. For example, the observed shift from container to tree-hole breeding by *Aedes aegypti* in Trinidad was thought possibly to have resulted from the repellent action of HCH applied to larval breeding places (Kellett and Omardeen 1957). If there has been selection of a population with a marked preference for tree-hole breeding, then if spraying ceases and containers again become suitable larval habitats, they may in the absence of *Aedes aegypti* larvae, become colonised by another species.

Irving-Bell et al. (1987) reported that in septic tanks in the Jos area of Nigeria *Culex decens* and *Culex cinereus* appeared to be competing with and replacing *Culex quinquefasciatus* in the wet season, but not during the long dry season. However, the actual timing of the displacement of *Culex quinquefasciatus*, which was sometimes total, varied yearly. In some interesting studies on the Kenya coast it was concluded that whereas *Culex quinquefasciatus* was a pioneer species successfully colonising flooded pit latrines in the early rains, after 3–4 months *Culex cinereus* invaded these larval habitats. Apparently as a result of competition-mediated succession, *Culex cinereus* annually displaced *Culex quinquefasciatus*, which reappeared only when the pits became re-flooded after drying out (Subra and Dransfield 1984). Whether larval competition for resources or interspecific interference was the cause was not determined, although the latter explanation was favoured. Elsewhere in Kenya the widespread pollution of urban breeding places by domestic detergents was believed to have caused the elimination of *Culex cinereus* from these sites. *Culex quinquefasciatus*, however, was not replaced, possibly because of its tolerance to detergent contamination. Alternatively it may be that habitats with detergents are unattractive oviposition sites for *Culex cinereus* (Subra et al. 1984).

Another possible example of species replacement is provided by the increase since about the 1950s in Britain of *Culex torrentium*, a species formerly rare in Britain that shares the same larval habitats as *Culex pipiens* and to which it is ecologically very similar (Service 1968; Gillies and Gubbins 1982).

Species replacement due to larval competition has possibly occurred on Guam. Surveys in 1948–1949 and in 1970 showed that during this interval the larval incidence of the indigenous mosquito, *Aedes guamensis*, has decreased in artificial container habitats by as much as 95% and by about 30% in more natural breeding sites. This coincides with the introduction around 1944 of *Aedes albopictus* on the island, a species which occurs in the same larval habitats and which has become increasingly abundant (Rozeboom and Bridges 1972).

Russell (1986) presented some experimental laboratory results suggesting that the disappearance of the introduced *Aedes aegypti* from parts of Australia was due to larval competition with the indigenous *Ochlerotatus notoscriptus* which has become increasingly more peridomestic.

Spielman and Feinsod (1979) reported that on Grand Bahamas Island *Ochlerotatus bahamensis* appeared to restrain colonisation by *Aedes aegypti*, while in Florida local populations of *Ochlerotatus bahamensis* seem to have displaced *Aedes aegypti* in some localities, although in others the two species coexist (O'Meara et al. 1989).

Steffan (1970) considered that *Toxorhynchites amboinensis* has probably replaced *Toxorhynchites brevipalpis* on the Hawaiian island of Oahu due to inter-specific larval competition.

Causes of larval mortalities

Limiting resources and growth retardants

Among the many causes of mortality in mosquitoes are competition for food or space, adverse climatic conditions, pathogens, parasites and predators. Barr (1985) concluded that the high mortality (97%) of *Culiseta incidens* larvae in a rain barrel was due to shortage of food. Bradshaw and Holzapfel (1986a) found good evidence of intraspecific competition in the tree-hole species, *Ochlerotatus geniculatus* and *Anopheles plumbeus*, in France and Britain; resources such as food seemed to be a limiting factor. Intraspecific competition was also shown to occur in *Ochlerotatus sierrensis* larvae (Hawley 1985a,b). In the Central African Republic Pajot (1975) believed that food was a limiting factor for *Aedes simpsoni* in banana axils, while in Nigeria Bown and Bang (1980) considered that shortage of food might have caused larval mortality in this species. Southwood et al. (1972) considered that in the absence of predators and pathogens food was likely the limiting factor for *Aedes aegypti* larvae in water-storage pots in Thailand.

Roberts (1998) showed that under conditions of high larval density of *Culex sitiens* in the laboratory, larval mortality increased to 99.6%. Changing larval rearing water daily increased the proportion pupating to 65% compared to 4% when the water was not changed, suggesting that chemical retardants played a role in population regulation. However, reducing mechanical interference between larvae by increasing surface area of the larval habitat increased pupation to 98%, suggesting that overcrowding leading to interference was the dominant density-dependent factor.

Laboratory studies have also shown that food availability is correlated with development time, resistance to starvation and size of emerging adults in *Aedes aegypti* (Arrivillaga and Barrera 2004). Adult body mass and resistance to starvation of *Aedes aegypti* collected from metal drums in the field in Anzoategui, Venezuela corresponded to laboratory rearing results at the lowest levels of feeding with liver powder and this was taken as evidence that immature *Aedes aegypti* in metal drums under natural conditions were subject to food limitation, possibly as a result of intra-specific competition.

Koenraadt et al. (2004) demonstrated in laboratory studies that limited food availability reduced *Anopheles gambiae* and *Anopheles arabiensis* larval development rates, but did not cause mortality *per se*. Space limitations, which presumably resulted in more encounters between individuals, significantly increased mortality of larvae.

In Kenya Subra (1983) found that food, accidentally introduced by humans, was the limiting factor determining mortality of *Aedes aegypti* in domestic water-storage pots. In Ethiopia, Ye-Ebiyo et al. (2003) investigated the development of larval *Anopheles arabiensis* in turbid water in relation to the proximity of flowering maize and larval density. In larval habitats remote from flowering maize, more pupae developed, and the resulting adults were larger. Larval crowding was observed to inhibit development in relatively clear water compared with turbid water. In close proximity to flowering maize, however, larval development was little affected by water turbidity or larval crowding, indicating that availability of a food source can counteract the effects of larval crowding.

The quality and quantity of food available may also affect larval development, as demonstrated in the laboratory for *Aedes albopictus* by Dieng et al. (2002). When larvae were reared in the presence of a rapidly decaying leaf (*Acer buergerianum*) (Angiospermae: Aceraceae) they developed faster than in the presence of a more slowly decaying leaf (*Cinnamomum japonicum*) (Angiospermae: Lauraceae). Larval mortality increased with population density in the *Cinnamomum* treatment, whereas mortality remained low even as densities increased in the *Acer* treatment.

Several workers have produced laboratory evidence suggesting that *Culex quinquefasciatus, Culex pipiens* and *Aedes aegypti* larvae produce so-called overcrowding factors (i.e. growth retardants), which assist in regulating their larval populations when they are overcrowded (Service 1985*b*). It appears that methyl branched hydrocarbons retard the growth and development of 1st and 2nd instar larvae, while bacterial decomposition of these chemicals results in 2- or 3-alkyl branched fatty acids which kill larvae at the point of ecdysis. Ikeshoji (1978) believes it likely that hydrocarbons are produced by aquatic organisms, not as originally believed by 3rd and 4th instar larvae (Hwang and Mulla 1975; Hwang et al. 1976; Ikeshoji 1977, Ikeshoji et al. 1977). Overcrowding can also result in decreased size of emerging adults, increase in the duration of the pre-feeding period, decrease in the blood volume ingested by females, increased dispersal, increase in the pre-gravid rate, reduced fecundity, reduced survivorship and reduced autogeny (see Mogi 1981; Mori 1979; and Service 1985*b* for references). Dye (1984*b*) reported some colony strains of *Aedes aegypti* produced growth retardants whereas others did not, and confirmed previous findings (Dye 1982) that growth retardants were of little importance in regulating larval populations.

Density dependence and intraspecific competition are bound closely together because whenever there is intraspecific competition its effects are density-dependent. Intraspecific competition in mosquito pre-adults can lead to reduced survivorship (Istock et al. 1976*b*, Reisen 1975; Reisen and Emory 1977), reduced fecundity (Reisen 1975), and increased development times (Frank and Curtis 1977*b*; Istock et al. 1975, 1976*a*; Moore and Whitacre 1972; Reisen 1975; Reisen and Emory 1977). Mori (1979) reported no difference in the survivorship of small and large adult *Aedes albopictus*, similarly Walker et al. (1987) found no difference in survival rates of small and large laboratory reared *Ochlerotatus hendersoni* and *Ochlerotatus triseriatus* released into the wild. The relationship between body size and survival is controversial, mainly because different studies have used dissimilar techniques leading to different conclusions (Landry et al. 1988). Peters and Barbosa (1977) give an overall review of the effect that population density has on the size, fecundity and develop-ment rates of insects, with some references to mosquitoes.

An interesting series of studies on the interactions between mosquito larvae and tadpoles in freshwater and brackish water habitats in Australia has been undertaken (Mokany and Shine 2002*a,b*, 2003*a,b*). Examination of gut contents of the tadpoles (*Limnodynastes peronii* and *Crinia signifera*) and the mosquito larvae that occur with them (*Culex quinquefasciatus* and *Ochlerotatus australis*) revealed that there is dietary overlap, with both groups consuming predominantly algae and bacteria. Competition experiments conducted in artificial ponds in the laboratory (Mokany and Shine 2003*a*) revealed the following: tadpole growth rates overall were depressed by the presence of mosquito larvae. Effects on mosquitoes were more complex, and differed between the two study systems. In freshwater ponds (*Limnodynastes peronii* and *Culex quinquefasciatus*), presence of tadpoles did not affect day to pupation or female wing size for *Culex quinquefasciatus* mosquitoes, but did affect male wing size, percentage pupation and percentage survival, all of which were lower in the presence of tadpoles. In brackish-water ponds (*Crinia signifera* and *Ochlerotatus australis*) presence of tadpoles did not affect day to pupation or percentage pupation of the mosquitoes, but wing sizes and percentage survival were reduced when the mosquitoes were raised with tadpoles. The effects of tadpoles on mosquito larvae were, perhaps unsurprisingly, most pronounced when food was limiting. However, substantial competitive effects were observed even at high food levels, suggesting that other competitive mechanisms were involved. The intensity of competitive suppression appeared to be influenced by attributes such as pond size (and hence, larval density), the location of food (on the water surface vs the substrate), and the extent of opportunities for direct physical interactions (Mokany and Shine 2002*a*). *Limnodynastes peronii* tadpoles suppressed *Culex quinquefasciatus* even when the two types of organisms were separated by a physical partition, suggesting that chemical or microbiological cues may be responsible. Later experiments (Mokany and Shine 2003*b*) revealed that water that had previously contained tadpoles suppressed the rates of survival and pupation of both species of mosquito. The addition of fungicides negated the suppression effect, suggesting that the inhibitors were fungal in origin, potentially occurring in tadpole faeces. Pond attributes also affected the impact of *Crinia signifera* tadpoles on *Ochlerotatus* larvae, but these effects disappeared when densities were lowered or when the tadpoles and mosquito larvae were physically separated. In other experiments, it was observed that increasing mosquito larval densities suppressed the growth of the mosquitoes as well as the tadpoles (Mokany and Shine 2002*b*).

Willems et al. (2005) concluded that tadpoles of four common species of Australian frog were not effective predators of *Culex annulirostris* under experimental conditions in the laboratory, as although they did consume some larvae (especially when starved for 24 h) and kill others, mosquito mortality was generally low.

Weather

Desiccation of habitats causes larvae and pupae to die and is a typical example of density-independent mortality. Although it cannot regulate population size in the sense of Varley and Gradwell (1963) drying out of breeding places may nevertheless cause large mortalities. Very heavy rain can result in habitat overflow, flushing out immature stages and also leading to population loss. In both Kenya (Teesdale 1957) and the Central African Republic (Pajot 1976) *Aedes simpsoni* s.l. suffers increased mortality during the rainy season because rain washes eggs out of banana axils. Similarly in the Americas eggs of *Wyeomyia vanduzeei* are easily washed from the axils of bromeliads (Frank and Curtis 1977*a*), while Focks and Sackett (1985) recorded that eggs of *Toxorhynchites amboinensis* were flushed out of man-made container habitats in urban areas of Louisiana, USA. The role of habitat overflow has also been de-monstrated for the tree-hole breeding mosquito *Ochlerotatus sierrensis* in both the field and laboratory by Washburn and Anderson (1993). In the laboratory, seven times more first instar larvae than fourth instars were found in overflow collection containers. In field studies, all larval stages and pupae were found in collections of overflow water following periods of rainfall, although it was acknowledged that a proportion of the larvae may have died prior to being flushed out and the pupae found in collecting vessels could have pupated after flushing out. In Venezuela, Lounibos and Machado-Allison (1986) have recorded rainfall as a mortality factor of eggs of *Toxorhynchites haemorrhoidalis* in *Heliconia* bracts. They also believe that the reason why female *Trichoprosopon digitatum* exhibit a maternal instinct and 'brood' their egg rafts by keeping them between the middle legs, is to prevent them from being washed out of cocoa husks and other small container habitats. Robert et al. (1999) used a nozzle system to direct drops of water vertically into the air, allowing them to reach terminal velocity and fall downwards, imitating a heavy tropical rainstorm. Mean drop diameter was 2.4 mm and mean kinetic energy of each drop was 23.5 $J/m^2/mm$ of rain. Wild *Anopheles arabiensis* larvae were placed in fine mesh sieves to prevent escape and water depth was maintained at a constant level of either 5 or 10 mm. Artificial rainfall delivered at 70 mm per hour for 40 minutes did not appear to result in larval mortality in the two experiments, although in the shallow water experiment four larvae were ejected from the sieve by the force of the rain.

Alto and Juliano (2001) investigated the effects of temperature and simulated precipitation regimes on caged *Aedes albopictus* populations in environmental chambers in the laboratory. In cages in which the larval habitat was allowed to completely dry out prior to refilling, higher temperatures were associated with reduced adult production. Conversely, where larval habitats were allowed to fluctuate between 25% and 90% of their original volume prior to refilling, higher temperatures led to higher production of adults. At temperatures of 22°C the effects of simulated precipitation on habitat fluctuation and drying were absent, suggesting that in cooler climates variability in precipitation is less of an influence on mosquito populations than in hotter climates, where adult production is increased as long as habitats do not completely dry out.

Temperature

Temperature is important for the aquatic stages of mosquitoes, not least because the immature stages of many species cannot survive temperatures in excess of 40°C for more than brief periods. Low temperatures may also have significant effects, with some species being susceptible, while the larvae of species from northern latitudes can survive even when the habitat is frozen. As well as the lethal effects of very high and very low temperatures on larvae, intermediate temperatures also affect development rates (e.g. Briegel and

Timmermann 2001; Fava et al. 2001; Tun-Lin et al. 2000), egg diapause, embryonic development and eclosion (de Carvalho et al. 2002; Lang 2003; Pritchard et al. 1996). Fluctuating temperatures during development may also affect development duration, adult longevity, fecundity, and blood-feeding behaviour (Joshi 1996; Tianyun Su and Mulla 2001). Zani et al. (2005) recently demonstrated that heat stress tolerance to a simulated warm-weather front decreased from embryo to larva to pupa to adult in *Wyeomyia smithii* in the laboratory. Juliano et al. (2002) have proposed that differential thermal tolerance of the eggs of *Aedes aegypti* and *Aedes albopictus* is a factor determining their competitive ability to utilize oviposition sites, with *Aedes albopictus* eggs being more susceptible to high temperatures and desiccation than those of *Aedes aegypti*. Lounibos et al. (2002) recently studied the role of temperature in modifying the effects of competition between *Aedes aegypti* and *Aedes albopictus* in urban areas in the United States. These authors compared growth and survivorship of the two species at controlled temperatures of 24° and 30°C in water-containing tires under conditions of intra- and interspecific competition and with or without leaf litter. When other variables were controlled statistically, the estimated finite rate of increase (λ') was significantly higher for both species at the higher temperature, and the proportional increases in λ' did not differ between species. The conclusion was that temperatures between 24° and 30°C would not alter the outcome of larval competition. The effects of temperature on mosquito development have been studied over many years for many species, including *Anopheles quadrimaculatus* s.l. (Huffaker 1944), *Anopheles gambiae* s.s. (Bayoh and Lindsay 2004), *Toxorhynchites brevipalpis* (Trpis 1972*b*), *Wyeomyia smithii* (Zani et al. 2005), *Aedes aegypti* (Bar-Zeev 1957; Tun-Lin et al. 2000), *Culex pipiens* molestus and *Culex quinquefasciatus* (Oda et al. 1999), *Culex tarsalis* (Reisen 1995), *Ochlerotatus albifasciatus* (Fava et al. 2001), *Culiseta melanura* (Mahmood and Crans 1998), *Aedes albopictus* and *Ochlerotatus triseriatus* (Teng and Apperson 2000), *Aedes vexans* (Read and Moon 1996), among others.

Parasites and pathogens

Many pathogens and parasites have been recorded from mosquitoes, but it is not within the scope of this book to discuss their detection, and estimation in field-collected larvae, as Goettel (1987) and Weiser (1991) provide comprehensive information on this subject. In addition, several publications have appeared over the last 30 years on parasites and predators of mosquitoes and their biological control, such as the books by Laird (1981), Laird and

Miles (1983, 1985), Chapman (1985), Dubitskij (1978) and Weiser (1991). Couch and Bland (1985) produced a seminal book on *Coelomomyces* species infecting mosquitoes, while de Barjac and Sutherland (1990) summarised most information on bacterial control of mosquitoes and blackflies, and Poinar (1979) discussed the use of nematodes for the biological control of insects. From 1979–1981 the World Health Organization produced 18 data sheets on various parasites and pathogens of mosquitoes (see Weiser 1991 for references). Jenkins (1964) provides a bibliography of pathogens, parasites and predators of mosquitoes and other medically important insects, while later publications have updated bibliographies on pathogens and parasites (Roberts and Castillo 1980; Roberts and Strand 1977; Roberts et al. (1983). Service (1983, 1985*a*) has written on the ecological considerations and limitations of biological control of mosquitoes. Parasitic infections may kill mosquito larvae or adults, or alternatively they may be non-lethal, causing larvae to be less tolerant to stress conditions such as adverse temperatures or insecticides. For example, although pupae and adults of *Ochlerotatus albifasciatus* were observed by Campos and Sy (2003) to be parasitised with the nematode worm *Strelkovimermis spiculatus*, no pupal mortality attributable to parasitism was recorded. Infections may also appreciably slow down escape reactions and hence make individuals more susceptible to predation, although Jackson (1953) found that predation by *Culex tigripes* was greatest on the more active mosquito larvae. If infected individuals exhibit slower movements they are probably more easily caught or trapped, which might introduce sampling bias into the estimation of the proportion of infected individuals in a population. When infected individuals are more easily detected, there may be a tendency to include disproportionate numbers of infected material in collections, so that sampling is not random.

Any increased rate of predation on individuals with a lethal infection will not add to population loss. If, however, the infection itself does not directly cause mortality but increases the likelihood of predation then this will contribute to population reduction. Infections may arise in larvae in a young instar but mortality may be delayed until later instars or even the adult stage. In England Service (1977*b*) and in Canada Goettel (1987) found that pathogens and parasites appeared to contribute little to the overall mortality of the mosquitoes they studied.

Factors which influence rates of parasitism, such as the efficiency of parasites in locating their prey, have been studied in the laboratory by several workers (Anderson 1979, 1982; Anderson and May 1978; Anderson et al. 1979; Fuxa 1987; Hassell 1971; Hassell and Rogers 1972; Hassell and May 1973, 1974; Hassell and Varley 1969; Sweeney and Becnel 1991; Varley and Edwards 1957; Varley and Gradwell 1970; Watt 1959). Royama

(1971) has critically reviewed the various models that have been proposed for predation and parasitism. Other interesting papers on mainly theoretical considerations of biological control and population dynamics, and with most examples drawn from agricultural science, include those by Hokkanen and Pimentel (1989), Murray (1979), Beddington et al. (1978) and Mackauer et al. (1990), and chapters in Huffaker and Rabb (1984).

Predators

Among the many causes of mortality in mosquitoes predation often seems to be the most important single factor determining population size. Predation may be on any stage of the life-cycle, but usually acts most heavily on the immature stages.

Andis and Meek (1984) and Marten et al. (2000) considered that predators, especially cyclopoid crustaceans, were a major source of larval mortality of *Psorophora columbiae* and *Anopheles quadrimaculatus* s.l. respectively in Louisiana rice fields. Knight et al. (2004) studied the interaction of *Anopheles quadrimaculatus* s.l. larvae with predators and other non-predatory species in semi-natural mesocosms in Missouri, USA. The mesocosms comprised 80 litre "keg buckets" 0.5 m in diameter × 1 m tall. Each bucket was filled with 25 g of dried tree leaves, to which approximately 60 litres of well water was added. Each mesocosm was covered with 1 mm Fiberglas® mesh "window-screening" in order to prevent colonization or emigration. Mesocosms were assigned to one of four treatments, each replicated eight times: (1) Control (no competitors or predators added), (2) Competitor (competitors present), (3) Predator (predators present), and (4) Both (competitors and predators present). Three days after establishment, Competitor and Both treatments received 2.5 litres of water containing zooplankton from nearby wetlands and six snails (*Physella gyrina* 3–6 mm in length). Three adult *Notonecta undulata* were also added to the Predator and Both treatments. Seventy-five 1st instar larvae of *Anopheles quadrimaculatus* s.l. were released into each mesocosm five days after setting up the kegs. Larvae were counted at the water surface by direct observation and classified according to instar. Two-way ANOVA results demonstrated that the presence of competitors ($F_{1,28} = 45.469$, $P < 0.001$) and predators ($F_{1,28} = 95.101$, $P < 0.001$) reduced the number of emerged adults, and that there was an interaction between these two factors ($F_{1,28} = 35.891$, $P < 0.001$). Larvae in the predator and competitor treatments also took longer to develop into adults than larvae in the control treatment.

Walton et al. (1990) concluded that predation, especially by aquatic beetles, was one of the main causes of larval mortality in *Culex tarsalis* in small ponds in southern California, USA. Morrison and Andreadis (1992) reported that the predatory chaoborid *Mochlonyx cinctipes* was a primary mortality factor among nearctic aedine mosquitoes in Connecticut, USA. This conclusion was based on population changes and spatial distribution of the predator in relation to the larval mosquito populations, rather than by testing gut contents, or other experimental methods.

In Sweden, Lundkvist et al. (2003) found that small dytiscids of the genus *Hydroporus* had no effect on the population size of mosquito larvae (mostly *Culex*) in artificial ponds in 2000. In 2001, *Hydroporus* beetles were less common, and were replaced by larger beetles of the genera *Ilybius, Rhantus* and *Agabus*. Beetle numbers significantly reduced mosquito population size, and ponds to which additional beetles were added had the lowest maximum mosquito population size. There was also a significant negative correlation between the mean number of dytiscid beetles in a pond and larval size.

Campos et al. (2004) studied the seasonal abundance of *Ochlerotatus albifasciatus* and associated insect fauna in temporary floodwater pools in Buenos Aires Province, Argentina. One hundred dips using a 450 ml dipper were taken each day from groundwater pools until all adult *Ochlerotatus albifasciatus* had emerged, or until the pools had become dry. The main predators were identified on the basis of relative abundance (ISA index), ecologically dominant groups, and the species association "I" index of Southwood (1978). Further details of the calculation of the indices are provided in Chap. 16. *Liodessus* sp. and *Rhantus signatus signatus* (Coleoptera: Dytiscidae) were the most abundant predators in the pools.

In the Philippines Mogi et al. (1985) found that *Toxorhynchites splendens* in aroids (*Alocasia* spp.) reduced other mosquitoes in the axils by about 70%. In the USA Lounibos (1983) thought that in some tree-holes *Corethrella appendiculata* was an important predator of *Ochlerotatus triseriatus*, noting it had a marked preference for young instar larvae. Chambers (1985) found that most mortality of *Ochlerotatus triseriatus* was caused by *Toxorhynchites rutilus*, which concentrated on the early instars.

Bradshaw and Holzapfel (1983) in studying the effect of the predator *Toxorhynchites rutilus* on *Orthopodomyia signifera, Ochlerotatus triseriatus* and *Anopheles barberi* cohabiting the same tree-holes, constructed 2 × 2 contingency tables of the presence and absence of predators and prey in them. They then calculated c^2, corrected for continuity as proposed by Pielou (1977), namely

$$\chi^2 = \frac{(ad - bc - (a+b+c+d)/2)^2(a+b+c+d)}{(a+b)(c+d)(a+c)(b+d)} \qquad (12.5)$$

where a = number of tree-holes where *Toxorhynchites rutilus* (predators) was present in a previous survey and mosquito larvae (prey) are present in the current-survey; b = number of tree-holes where predators were present in a previous survey and prey are absent in the current survey; c = number of tree-holes where predators were absent in a previous survey and prey are present in the current survey; d = tree-holes lacking predators in previous survey, and without prey in present survey. The type of effects, negative or positive, was determined by the sign of $(ad - bc)$. It was discovered that the presence of *Toxorhynchites rutilus* had no significant effect on the persistence of *Orthopodomyia signifera*, but *Ochlerotatus triseriatus* was more likely to be eliminated, while *Anopheles barberi* still persisted in the presence of predators. Schreiber et al. (1988) also used 2 × 2 contingency tables to determine the effect of predation by *Toxorhynchites rutilus septentrionalis* on tree-hole mosquitoes.

Nannini and Juliano (1997) investigated the potential role of developmental asynchrony between a prey species (*Ochlerotatus triseriatus*) and a predator (*Toxorhynchites rutilus*) on avoidance of predation in the laboratory. At relatively high densities, a developmental head start for *Ochlerotatus triseriatus* over its predator had a significant but weak effect on prey survivorship. The authors' interpretation was that variable developmental rates within a prey cohort meant that sufficient small prey individuals were always available to the larger predator, enabling the predator species to develop sufficiently rapidly to predate on all developmental stages of the prey and ultimately overcome most effects of developmental asynchrony. In a study on Moorea, French Polynesia, of predation by *Toxorhynchites amboinensis* on *Aedes polynesiensis* inhabiting coconut husks that had previously been chewed by rats, Mercer et al. (2005) observed that the presence of the predator significantly reduced survival of *Aedes polynesiensis*, with only 11 of 1550 surviving to the end of the experiment, compared with 347 survivors of 2450 in coconuts without predators. Neither winglength of emergent adults, nor time to emergence were significantly affected by the presence of the predator.

In the laboratory, young larvae of *Toxorhynchites splendens* have been shown to compulsively kill larvae of other species without eating them (Amalraj and Das 1994).

In Israel, Blaustein et al. (1995) studied the effects of the predatory backswimmer *Notonecta maculata* on *Culiseta longiareolata* in natural and artificial pools and observed that densities of all developmental stages of the mosquito (egg rafts, first and second instar larvae, third and fourth instar larvae, and pupae) were signicantly reduced by the presence of *Notonecta*. In pools containing *Notonecta*, no *Culiseta* reached the pupal stage. In later studies, Blaustein (1998) examined the impact of *Notonecta maculata* on invertebrate community structure in artificial and natural pools in Israel and observed that in artificial pools, *Notonecta* strongly reduced densities of *Culiseta longiareolata* and also ephydrid diptera.

Manrique-Saide et al. (1998) studied the effects of the copepod *Mesocyclops longisetus* on populations of *Aedes aegypti* in car tyres under semi-natural conditions in Mexico. Groups of fifty newly-emerged mosquito larvae were introduced into 20 of 40 car tyres along with 20 female copepods. Upon moulting to the next larval instar, surviving larvae were transferred to a new tyre that contained 20 copepods. All larval instars and pupae were counted daily until adult emergence had been completed and life tables were constructed. In the absence of copepods, *Aedes aegypti* larval mortality was estimated at 91%. Mortality in third and fourth instars was 54% and 44% respectively. In the presence of *Mesocyclops*, overall mortality was 99.8%, with mortality highest in the first and second instars.

In Uganda Sempala (1982) estimated that predation caused 83.2–91.0% of the mortality recorded in *Aedes africanus* breeding in tree-holes and bamboo. Mogi et al. (1980*a,b*) considered predators caused most larval mortality of *Culex tritaeniorhynchus* larvae in Japanese rice fields, while in Pakistan Reisen and Siddiqui (1979) found predation was an important mortality factor of *Culex tritaeniorhynchus* in a pond during the post-monsoon period, but not during the actual monsoon. Mogi et al. (1986) believed that predation was responsible for 19–54% of the estimated mortality of *Anopheles sinensis* in rice fields in Thailand, and Mogi et al. (1984) estimated that predators killed 48.7–87.0% of *Anopheles peditaeniatus* and *Culex vishnui* in Philippine rice fields. Miura and Takahashi (1988) considered that most of the 85.7–98.5% mortalities recorded in *Culex tarsalis* in Californian rice fields was due to predators, while Reisen et al. (1989) considered that predation accounted for 61.7–84.5% of mortality estimated in *Culex tarsalis*. In Australia McDonald and Buchanan (1981) estimated pre-adult mortality of *Culex annulirostris* as being 88.9%, and that predators caused most of it.

In contrast, Renshaw (1991) found no evidence from predator exclusion experiments that the high mortality (>90%) of *Ochlerotatus cantans* in woodland pool larval habitats studied over 3 years was caused by predators. It seemed that desiccation and interference competition resulting in *de facto* food shortages were important mortality factors. There are, however, many other examples of the proven, or suspected, importance of predators in reducing larval populations (Lacey and Lacey 1990; Service 1985*b* for references). Given the important role played by predation in population regulation of mosquitoes it is perhaps not unsurprising that some species appear to have evolved mechanisms that assist in the avoidance of predation. For example, Stav et al. (1999) studied the oviposition choices of *Culiseta longiareolata* in artificial ponds with or without predators. Significantly fewer egg rafts (52% fewer) were observed on the surface of pools containing *Anax imperator* dragonfly nymphs. Laboratory studies indicated that the reduction in the number of egg rafts was not due to predation on ovipositing females or directly on egg rafts. The precise mechanism by which female mosquitoes may be able to detect the presence of predators in oviposition sites is not known. Similar reductions in the number of egg rafts laid on water containing predators has been observed with *Bufo viridis* toad tadpoles (Blaustein and Kotler 1993) and *Notonecta maculata* (Blaustein et al. 1995; Blaustein 1998). Kiflawi et al. (2003*a*) attempted to discern which of the following possible alternative strategies to maximising the trade-off between the risk of predation and detrimental density dependence is adopted by *Culiseta longiareolata*. Alternative 1 describes a polymorphic scenario, in which a fixed proportion of females constantly avoid 'predator pools', while the remainder oviposits at random; the second alternative describes a monomorphic scenario, in which all females 'prefer' to oviposit in predator-free pools, but oviposit in predator pools with a certain probability. The experimental set-up utilised consisted of an array of 18 artificial pools (0.6 × 0.4 × 0.15 m) with a predator prevalence of either 0.25 or 0.75 of pools. Three *Notonecta* nymphs were introduced into each of nine pools. Six predator or non-predator pools were then sealed with plastic covers to give a predator prevalence across the pools of either 0.25 or 0.75. Oviposition was monitored for 22 nights, with the position of the covers shifting by one pool per night in a clockwise direction. Over the 13 nights in which predator prevalence was 0.75, 35 of the 42 egg rafts collected were from predator-free pools, Over the 9 days in which predator prevalence was 0.25, all of the 57 egg rafts collected were from predator-free pools. Comparing the likelihoods of the two scenarios showed the monomorphic scenario to be approximately twice as likely as the polymorphic scenario, suggesting that the lack of total predator avoidance by this mosquito is the result of a pure strategy of predator avoidance, but with a fixed probability of failing to detect the presence of predators. To further characterise the trade-off between predator avoidance and avoidance of high densities of conspecific immatures, Kiflawi et al. (2003*b*) examined oviposition patterns among three artificial habitats: predator pools, and non-predator pools with either low or high densities of *Culiseta* larvae. The majority (~88%) of females oviposited in low-density pools rather than in the predator- or high-density pools. Furthermore, a substantially higher proportion of females oviposited in predator pools when faced with the high-density alternative, however this was due largely to fewer females ovipositing in high- vs low-density pools. It was concluded that *Culiseta longiareolata* oviposition choice reflects an adaptive balance between risk of predation and density-dependent larval growth/survival; however, individual females are not adjusting their decisions to suit local hetero-geneity in habitat quality by making relative comparisons. Rather, indi-viduals appear to respond, in a probabilistic manner, to the absolute value of potential oviposition sites.

The effects of container surface area and depth on the population dy-namics of *Culex pervigilans* and associated predators was studied by Lester and Pike (2003) in New Zealand. Five container sizes ranging from 1.3 to 300 litres, and two water depths (high and low) were used. All the containers were circular and made of plastic, with the interior surface painted matt black to control for container colour. Mosquito densities were estima-ted by stirring the water and then removing a sample of 1 litre from the 10 to 300 litre containers, and a 100 ml sample from the 1.3 litre containers. The highest larval densities were observed in the smaller containers. Water depth did not significantly predict mosquito abundance. Predators were rare in all containers, but completely absent from the smallest containers.

Many accounts of predation refer either to more or less casual observa-tions or laboratory trials, and even when such trials are carefully designed and conducted, they have rarely provided much understanding of the im-portance of predation in nature. The number of mosquitoes eaten by a predator depends on a variety of factors, such as the size and number of both predator and prey, presence or absence of alternative prey, tempera-ture, interaction between different predator species, and presence or ab-sence of places of refuge for the prey and predators, and laboratory trials do not usually provide all these interactive possibilities. However, labora-tory trials may be of limited value in indicating whether certain suspected predators actually eat mosquito larvae and pupae (Dixon and Brust 1971; Lee 1967; Roberts et al. 1967; Watanabe and Wada 1968). But even nega-tive results may be misleading. For example, certain species of *Hydro-metra* (Hemiptera) are known predators of mosquito larvae, but frequently they cannot be induced to prey on larvae in the laboratory.

Data from field and laboratory studies have resulted in a theoretical frame-work for the study of predator-prey interactions and this is reviewed by Southwood and Henderson (2000). Predator-prey interactions are usually examined through the study of the death rate of the prey and the rate of in-crease of the predator, both of which are clearly interrelated. Death rate of the prey can most easily be measured in the laboratory under different prey densities. Two key components of the functional response of a predator to its prey are the attack rate (a) and the handling time (T_h), which measures saturation. Many functional responses are of the form:

$$N_a = \frac{aNTP}{1 + aT_h N} \tag{12.6}$$

where N_a = total number of prey attacked, N = total number of prey, P = predator density, and T = total time. These functional responses are termed type II of Holling (1959). This method will give inaccurate estimates of a and T_h where prey depletion is significant, and a revised formula is preferred in this case (Rogers 1972):

$$N_a = N[1 - \exp\{-1(PT - N_a T_h)\}] \tag{12.7}$$

Amalraj and Das (1998) used the above methods to study predation on *Aedes aegypti* by larval *Toxorhynchites splendens* in the laboratory. These authors found that the number of prey killed by the predator increased with prey density to a threshold of 40, after which it levelled out. Predation rates decreased with increasing prey size and *Toxorhynchites splendens* appeared to "prefer" second instar larvae.

Yasuda and Hashimoto (1995) observed that cannibalism in *Toxorhynchites towadensis* occurred despite the presence of prey larvae, however, no cannibalism was observed at high prey densities (20 prey per predator). Other studies on predation by *Toxorhynchites* include those by Russo (1983) and Sherratt and Tikasingh (1989).

Sunahara et al. (2002), based on the observation that small containers often lack predators, surmised that the significance of predation in mosquito populations depends on the habitat size. These authors surveyed 171 and 124 containers in summer and autumn, respectively and classified all containers into four size classes (<0.01 m^2, 0.01–0.1 m^2, 0.1–1 m^2, >1 m^2) by the area of the water surface. Mosquito immatures and other aquatic insects were collected using a 10-ml pipette, an aquarium net (opening area: 15 × 12 cm; mesh size: 0.3 mm), and a 1 litre capacity dipper of 17 cm diameter. Predators occurred predominantly in relatively large containers in open sites and infrequently in small containers and in bamboo groves. *Aedes (Stegomyia)* spp. and *Tripteroides bambusa*, species that inhabit small containers, showed negative correlations with all the predator groups, as predators rarely occur in containers with a surface area <0.01 m^2. In these small containers, intra- and interspecific competition for food, rather than predation, were thought to be responsible for population regulation.

Methods for detecting predation

Direct methods

Occasionally it is possible to identify predators directly in the field, for example when adult mosquitoes are found in spiders' webs inside village huts or amongst vegetation. A simple method is the detection of the hard parts of mosquitoes, such as the larval head capsule and siphon, or the head and wings of adults, in the predator's gut (Bay and Self 1972). This method is clearly unsuitable for identification of those predators that have sucking or piercing mouthparts, or species that have specialised mandibles and predigest their prey. Despite these restrictions James (1961, 1966) identified adult dytiscid beetles as predators of mosquito larvae by examining their crop contents. Another approach is to identify remains of mosquitoes in the excreta of predators, and regurgitated pellets of birds. Koenraadt and Takken (2003) were able to identify remains of first instar larvae (head capsules, abdominal hairs, and mouthparts) of *Anopheles arabiensis* in 11 of 43 samples of faeces of fourth instar larvae of the same species when reared together in a study to investigate cannibalism and predation among members of the *Anopheles gambiae* complex in the laboratory.Howerver, this method is unlikely to prove practical under field conditions.

Exclusion techniques

This involves comparing the population size of mosquitoes before and after predators have been removed or excluded from a habitat, and is more applicable to larvae than adults; it is not easy, however, to ensure that all predators and their eggs are removed, even from small isolated habitats. Even in very small ground pools predators may escape detection by burrowing into mud and leaf litter. Furthermore, maintaining habitats free of predators is usually difficult and frequently results in changing the environment, which may affect the survival of the mosquitoes.

Using predator-free cages in Louisiana, USA rice fields Andis and Meek (1983, 1984) concluded that most non-predator mortality of *Psorophora columbiae* occurred in the younger age classes. In later experiments Andis and Meek (1985) compared mortalities recorded from these cages with those experienced in the presence of natural predators, and found that predators increased larval mortality from 46.7–97.4%. This confirmed the importance of larval predators in regulating population size of rice field mosquitoes (Mogi et al. 1980*a,b*; Miura et al. 1978, 1984; Service 1977*a*). In California, USA Reisen et al. (1989) enclosed 1st instar larvae of *Culex tarsalis* in predator-exclusion cages placed in natural habitats and obtained horizontal survivorship rates from 1st instar larvae to pupa formation. By comparing mortality rates by the horizontal (predator free) with the vertical approach of Lakhani and Service (1974), Reisen et al. (1989) estimated that predation accounted for about 61.7–84.5% of pre-adult mortality. Also in California Miura and Takahashi (1988) estimated mortalities of 85.7–98.5% in natural populations of *Culex tarsalis*, but when predator-exclusion cages were used mortality dropped to just 25.3%.

In the Philippines Mogi et al. (1984) estimated the mortality from 1st instar larvae through to pupae of *Culex vishnui* and *Anopheles peditaeniatus* by placing 10 1st instar larvae in a series of small cages ($10 \times 10 \times 10$ cm) and 20 1st instar larvae in large ($30 \times 30 \times 10$ cm) cages, all of which were suspended in rice fields by styrofoam floats. Survivorship, as measured by numbers of pupal exuviae, varied in both species from 50.0–88.8%. In the same rice fields the method of Service (1971) was used to estimate survivorship of natural populations which varied from 0.0–1.8% (mean 0.9%) for *Culex vishnui* and from 1.1–4.7% (mean 2.4%) for *Anopheles peditaeniatus*. Crude estimates of mortality due to predation was inferred from the differences between mortalities obtained from predator-free cages and from natural populations.

James (1964) assessed the impact of predators of rock pool breeding mosquitoes by removing all mosquito larvae, predators and water, and then replacing the original water, and introducing a known number of mosquito larvae and predators. The ability of various predators to reduce larval populations of *Aedes vittatus* was studied in small rock pools in Nigeria (Service 1970). In more extensive experiments Christie (1958) evacuated all water and fauna from a series of paired shallow pits. He returned aquatic fauna including predators to one of each pair of pits but excluded all fauna from the other. He then introduced known numbers of 1st instar larvae of *Anopheles gambiae* to the pits which were covered with transparent plastic mesh to exclude natural invasions by both predators and *Anopheles gambiae*. After an interval of a few days to allow the young larvae to develop into 4th instars, the water was removed from the pits and the numbers of mosquito larvae surviving counted. The difficulties encountered in these trials are likely to apply to many similar situations, and are worth enumerating. One of the major problems was the impossibility of removing all the predators and their eggs. Larvae were up to 20 h old when introduced into the pits, consequently they were protected from predation during their early life, and moreover the experiments finished prior to pupation, therefore there was no information on predation on the last stages of their aquatic life. Because only one batch of larvae was introduced, predation caused a decrease in availability of food for predators, whereas under natural conditions there would be continual recruitment by further eggs hatching. Christie (1958) tried to overcome some of these difficulties in further trials. After evacuating the pits and returning the predators he introduced 400 1st instar larvae daily for 13 days, thus allowing time for the first batch of larvae to pupate and give rise to adults. Despite these improvements he was still unable to remove all predators from the pits. Nevertheless it was clear that predators were important in reducing the immature population of *Anopheles gambiae*.

In Tanzania, Trpis (1972*a*) studied the predation of *Toxorhynchites brevipalpis* on *Aedes aegypti* breeding in vehicle tyres by introducing known numbers of larvae of both species and recording the number of *Aedes aegypti* larvae consumed over 24 h. There is no mention of any alternative prey for the larvae of *Toxorhynchites brevipalpis* or whether oviposition in the tyres by *Aedes aegypti* was prevented. If this was not done then the predacious larvae could have fed on newly emerged larvae in addition to the older ones introduced.

In studying the development and mortality of the pre-adults of *Culex quinquefasciatus* populations in India metal frame containers (30 cm long, 8 cm wide and 9 cm tall, or 20 cm long, 20 cm wide and 28 cm tall) with the bottoms and sides covered with nylon cloth were floated partially submerged in wells and cesspools. One to eight egg rafts were placed in these containers and daily observations made on development times and numbers in the various instars. The number of adults emerging was determined from pupal exuviae. In the cool season the pre-adult mortality was 97.7% in the wells – probably mainly due to food shortages, whereas in the cesspools it was estimated to be 72.2%. In the hot season, mortality in the wells ranged from 60–96%; no observations were made in cesspools (Rajagopalan et al. 1975). It should be noted that these high mortalities were recorded in the absence of predators.

Mogi et al. (1980*a*) pushed 1-m^2 metal bottomless frames into the mud of fallow rice fields and placed known numbers of *Culex tritaeniorhynchus* eggs in them. Then starting 2 days later the immature stages were sampled by taking 10 or 20 dips every other day until none remained—usually 12–14 days later. The numbers of larvae per dip were converted into estimates of absolute numbers enclosed within the frames, using coefficients for the various age classes that had been calculated in earlier studies comparing the efficiency of dippers. For example, the number of 1st instar larvae per dip was multiplied by 256 to obtain an expected population/m^2, while 4th instars were multiplied by 77 (Wada and Mogi 1974). Mogi et al. (1980*a*) assumed that daily survival (*p*) was constant throughout the period of immature development, consequently the numbers of survivors (N_t) present on day t after the eggs had hatched was $N_0 p^t$, where $N_0 =$ the initial number of hatched larvae. Thus

$$\log N_t = \log N_0 + t \log p. \qquad (12.8)$$

This enabled log numbers of survivors to be plotted against time in days after release of the eggs, and the survival of *Culex tritaeniorhynchus* to be estimated. These survival rates, in the presence of natural predators entrapped within the 1 m² metal frames, were compared with the survival of *Culex tritaeniorhynchus* pre-adults released into cloth nets placed in the metal frames in the rice fields, and thus free from predation. Predators caused considerable mortality, but rates differed according to the types of predators present, for example fish caused much higher mortalities than insect predators. Also in Japan Mogi and Okazawa (1990) compared larval mortalities of *Culex pipiens* form *pallens* in predator-free cages and in natural field populations. They found that predation varied greatly in different areas, but generally accounted for more than 30% of total mortality.

Assessing the importance of predation by exclusion techniques is likely to be more reliable in small habitats, such as rock pools, village pots, wells and tree-holes than larger ones, because it is easier to ensure that all natural predators are removed from small habitats prior to the trials. However, the exclusion technique has limitations. The absence of predators may lead to greater competition for food and space by the mosquito larvae and might also increase the incidence of disease, and hence result in increased larval mortality. Sometimes all predators are removed, except those under test, but in such instances predation may be less intense than under natural conditions when a variety of predators are present. Moreover, if non-predator fauna, except for the mosquitoes, are removed this may result in increased mosquito predation due to absence of alternative food. On the other hand, removal of other fauna may reduce competition for food and space, and cause an increase in the survival rate of the mosquito larvae. There consequently needs to be care in interpreting results from exclusion experiments, as in any other experiments where field conditions are manipulated.

In Poland, Dabrowska-Prot and Łuczak studied spider predation on adult mosquitoes by outdoor cage experiments (Dabrowska-Prot 1966; Dabrowska-Prot et al. 1966; Łuczak and Dabrowska-Prot 1966). Although not strictly an exclusion technique, and applied to adult, rather than larval mosquitoes, the technique is most conveniently described here.

They constructed several small and large cages of plastic mesh stretched over a framework of metal rods, the larger ones measured 3 × 1.5 × 1 m and were divided into two equal size compartments (1.5 × 1.5 × 1 m). They were placed over natural plant cover and low lying scrub vegetation. About 50–180 adult mosquitoes of several species together with about 25–65 spiders belonging to four species were placed in the cages. In the control cages mosquitoes but not spiders were introduced. During the morning, and at mid-day and during the evening over a period of up to 2–3 weeks the numbers of mosquitoes seen resting on the vegetation and walls of cages were recorded. The difference between the numbers of mosquitoes counted in the control cages and those with spiders was taken to be the numbers killed by the spiders. Apparently no attempts were made to remove any natural mosquito predators that might have rested amongst the vegetation enclosed under the cages. If present, these natural predators would reduce the numbers of mosquitoes in cages with and without spiders, but not necessarily by the same amount because there might be interactions between the natural predators and the introduced spiders. Cages also prevented mosquitoes escaping predation by flying away.

Similar methods were used by Takagi et al. (1996) when studying the survival of newly-emerged *Culex tritaeniorhynhus* adults in field cages. Field cages were 2.0 m × 2.0 m × 1.5 m and were constructed from nylon gauze attached to a framework of iron pipes. Predators were removed from one of each set of paired cages by spraying with a 3% permethrin water dispersible powder. Two hundred field-collected pupae were introduced into each cage. The number of adults in the cages was counted every morning and adults were returned to the cage. The 45 minute collection period had previously been determined sufficient to collect 88.5% of 200 adults. Cumulative (S_n) and daily (S'_n) survival rates were calculated as follows:

$$S_n = 100N_n / (C_0 - \sum D_{n-1}) \qquad (12.9)$$

$$S'_n = 100N_n / (N_{n-1} - D_{n-1}) \qquad (12.10)$$

where C_0 = number of females that emerged and D_n = number of damaged females removed from the experiment. The number of surviving adults on day 4 averaged 14 in cages with predators and 83 in cages from which predators had been excluded. Average mortality rates over 4 days were 39.9 ± 28.2% and 76.7 ± 12.9% in cages without and with predators respectively. Lycosid spiders were considered to be the most important predators, but no specific investigations were conducted to confirm this.

Roitberg et al. (2003) conducted some interesting laboratory experiments on the potential increased risk of predation experienced by recently blood-fed *Anopheles gambiae* mosquitoes. These authors observed that engorged femals were three times less likely to escape predation by a single Salticid jumping spider (*Salticus scenicus*) compared with unfed females when placed in a small Perspex box with the spider. Whether or not this effect of blood-feeding is also seen in nature is unknown. By contrast, Canyon and Hii (1997) observed that unfed mosquitoes were more likely to be preyed upon by geckos than engorged ones, and this was ascribed to increased activity and hence visibility of un-fed mosquitoes.

Sunish and Reuben (2002) used a multiple regression method to associate the abundance of presumed predators to the density of mosquito immatures in rice fields in India. They observed notonectid density to be significantly negatively correlated with both larval and pupal population density. Dytiscids, anisopteran and zygopteran nymphs were also negatively correlated with density of mosquito immatures at different seasons.

Radionuclide techniques

Nowadays, there is relatively little interest in detecting or measuring predation by the use of radionuclides, with most studies on predation currently using serological techniques. Readers interested in the use of radionuclide techniques for detecting predation are referred to the first and second editions of this book (Service 1976; 1993).

Serological methods: interfacial precipitin test

The interfacial capillary ring precipitin test has been commonly used to identify the source of blood-meals in mosquitoes (Chap. 7), and also to study insect predator–prey relationships, especially by early workers (Dempster 1958, 1960, 1963; Dempster et al. 1959; Fox and MacLellan 1956; Frank 1967; Healey and Cross 1975; Loughton et al. 1963; Loughton and West 1962; Vickerman and Sunderland 1975). As long ago as 1946 Brooke and Proske using mosquito larvae and pupae as the prey, first showed that the precipitin test could be used to detect insect predators, but the method was overlooked in medical entomology until it was used to detect predators of *Ochlerotatus cantans* (Service 1973a), the *Anopheles gambiae* complex (Service 1973b, 1973c, 1977a) and *Simulium damnosum* (Service and Lyle 1975). More recently Sulaiman et al. (1990, 1996) used the interfacial precipitin test and identified several species of spider as predators of adult *Aedes albopictus* in Malaysia.

The test is based on the identification of prey, such as mosquito larvae, in a predator's gut by its reaction with the blood serum from a mammal, usually a rabbit, that has been sensitised against the prey by injections of prey antigens. Most antigens are proteins of animals or plants but polysaccharides and some artificial compounds can also function as antigens. When these are injected into an animal they stimulate the production of antibodies in the blood serum, which is called the antiserum. When this antiserum mixes with the appropriate antigens a flocculent white precipitate is produced due to the chemical combination of the two.

Numerous techniques have been developed for the production of antisera and readers are directed to the first and second editions of this book for details of the methods applied to mosquitoes and blackflies, as precipitin tests have largely been replaced by other techniques in recent years. Details of the method for testing predators are retained.

Predator testing

Guts of larger suspected predators are removed and squashed on to filter paper, while with small individuals the whole animal is squashed. Labelled smears are dried as quickly as possible and stored in a desiccator over silica gel or phosphorous pentoxide or at −40°C; they can be kept for many months, even as long as 2 years, without deterioration. For testing, the gut smear is cut out and soaked overnight in 0.1 ml normal saline, or in larger volumes of saline (e.g. 0.2–0.3 ml) if bigger smears are involved. It is advisable to centrifuge the extracts for about 15 min at $500 \times g$ before testing against the antiserum, otherwise tests may be spoilt by cloudiness or the formation of double rings. About 0.02 ml antigen is drawn up underneath an equal volume of gut extract in a small glass tube. Readings are taken after 2 h. The presence of a distinct precipitin ring at the interface shows that the predator's meal contained the mosquito under study.

For the interfacial ring test it is essential that both antigen and antiserum are completely clear. An alternative technique of double gel diffusion can be used if perfect clarity is not obtained by the first method, but the test is more time consuming. Wadsworth (1957) devised a very useful microtechnique that uses very small quantities of reagent and is conducted on a very thin agar film stuck on to a microscope slide. Modifications and improvements to the method were made by Crowle (1958). Basically it consists of placing the antiserum in the middle of a small well, drilled out of the centre of a plastic template, which rests on a very thin layer of agar gel on a microscope slide. Extracts from different gut smears are placed in eight equally spaced peripheral wells drilled in the template. After incubation for 1–2 days the template is removed and the slide washed and stained with a protein stain, e.g. thiazine red. Positive reactions are identified by one or more bands of precipitation between the central well, which contained the antiserum, and the peripheral wells which contained gut extracts from different predators. A good description of the method is given by Pickavance (1970).

Apart from detecting predators of mosquitoes, serological techniques can be used for analysing the food of mosquito larvae. Probable constituents of their diet, both animal and vegetable materials, are used to produce a number of specific antisera, which are then used to identify the gut contents of the larvae.

Period of detection and frequency of feeding

When digestion has proceeded to a certain stage the prey antigens become altered and no longer react with their homologous antibodies to form precipitates. Failure to detect prey can occur even when partially digested food is still present (West and Eligh 1952). It is important to know the period of time after feeding during which prey can be detected in the predator's gut. For example, if two distinct predators feed on the same number of prey, but the meal in one species is detectable for twice as long as the other, the precipitin test will indicate that it is the more important predator. The period of detection varies among different predator species because it depends on the size of the meal, temperature and also the rate of digestion and breakdown of protein antigens. To discover this period, individual mosquito larvae, or adults, are fed to various predators and a series of gut smears are made about every 6 h. Since temperature affects the rate of digestion, tests should be made at temperatures encountered in the field, and this may necessitate tests at relatively low and high temperatures to simulate winter and summer conditions. There is relatively little information on the detection period of meals in the guts of predators, but Dempster (1960) found that the broom beetle, *Phytodecta olivacea*, could be detected for up to about 24 h in the gut of its main predator, *Orthotylus adenocarpi* (Miridae), and Davies (1969) found that triclads could be detected in the guts of most predators for 24 h. Fourth instar larvae of *Ochlerotatus cantans* can be detected in dytiscid larvae for 24, but not 27 h and pupae for 20–22 h (Service 1973a).

Although the percentages of positive smears from two predators may be similar, one of them may be much commoner than the other and consequently it will be more important in determining the size of a mosquito population. By regular sampling the seasonal incidence of both predators and prey populations can be studied and, where appropriate, population estimates made of the predators so that their impact on the prey population can be assessed (Dempster 1960; Service 1973a). If there is reason to believe that a positive reaction represents one prey eaten, and if there is reliable data on population size of the predators (P), proportion of positive precipitin reactions (m), detection time (t) and population size of prey (T) then minimum number of prey consumed can be estimated (E). Thus:

$$E = \frac{pmT}{t}.\qquad(12.11)$$

Although the precipitin test can measure the proportion of a predator population feeding on mosquitoes within a known number of hours there is unfortunately no information on whether more than one individual has been eaten. Mosquitoes and in most cases their predators are very active and the immature stages of mosquitoes are usually much more numerous than their predators. Consequently a high proportion of most predators will probably have fed several times within the detection period, but also have fed on alternative prey. Laboratory experiments can be useful in indicating the likely number of feeds within the detection period. The precipitin test will also not distinguish between predators which cause a population loss and scavengers which have no direct effect.

Serological methods: latex agglutination test

Ohiagu and Boreham (1978) applied the latex agglutination test to identify insect predators. Their method is as follows.

IgG is separated from antiserum by adding 20 ml of acetate buffer (pH 4.0) to 10 ml of antiserum at 20°C, then adding 0.74 ml caprylic acid drop by drop with constant stirring for 30 min. The mixture was centrifuged at 18 000 rpm for 20 min and the supernatant removed and dialysed at 4°C against 0.1 m glycerine saline buffer (pH 8.2) for 24 h. Finally, the protein concentration of the IgG solution was determined by the folin phenol method of Lowry et al. (1951).

A 5% solution of latex (from BDH chemicals) was prepared in 0.1 m glycerine saline buffer. The latex and IgG solutions were then mixed together in the ratio of 11:4 and heated at 56°C for 30 min. Two drops of 2% bovine serum albumin was added to each ml of the mixture to stabilise the latex beads. The optimum concentration of IgG to add varies for the particular antiserum and has to be determined in a preliminary experiment in which a series of dilutions (neat to 1/32) are prepared, and that giving the best agglutination pattern determined. The sensitised latex should be stored at 4°C, and may be kept for up to about 3 months.

A single drop of the sensitised latex is added on a microscope slide and gently rocked with one drop of elute from the predator gut (antigen).

A positive reaction can be seen within 2–3 min by agglutination of the latex particles, and its strength recorded on a scale of 1–3.

Sensitivity is about the same as the interfacial precipitin test, but the actual testing time is rather short being 2–3 min as against 1–2 h.

Serological methods: passive haemagglutination inhibition test

Greenstone (1977) used a passive haemagglutination inhibition assay to measure the predation of spiders on *Ochlerotatus dorsalis* and other aquatic insects. Although more sensitive than the precipitin test it does not have the sensitivity or specificity of the ELISA technique which has become the usual method of choice.

Serological methods: ELISA test

The ELISA method is probably the widest used method to detect insect predators over the last 20 years or so (Crook and Sunderland 1984; Fitcher and Stephen 1984; Kapuge et al. 1987; Sunderland et al. 1987), including, predators of mosquito larvae (Palchick et al. 1986; Sulaiman 1982; Tempelis 1983). The method involves the immunisation of rabbits to produce antisera, followed by the separation of the IgG fraction by processes of precipitation and dialysis. For the sandwich method polyvinyl chloride plates are coated with IgG prepared against the mosquito species being studied, these can be stored for months in a refrigerator. Wells containing extracts of predator guts having mosquito antigen (that is positive feeds) turn a blue colour; this chemical reaction (hydrolysis), is stopped by the addition of 0.1 m HCl and the colour turns yellowish. The colour intensity depends on the number of mosquito larvae consumed and the time after feeding. A control well containing just PBS and another control well containing extract from a squashed mosquito larva are added to each plate as negative and positive checks. Ideally if a good balanced system is obtained positive wells can be detected by eye because the PBS control well should be completely colourless and any coloration in other wells—faint or strong, will indicate predation—as with the ELISA method of Service et al. (1986) used to identify mosquito blood-meals. However, it is likely that there will be some background coloration, and so it may be advisable to use a plate-reader to measure the intensity of the colour in the wells. An intensity of about 1.5 × the negative control well is usually taken as positive.

A good account of the methodology of the ELISA is given by Kapuge et al. (1987) who developed a system for detecting predators of *Pieris rapae*.

The ELISA method is generally more sensitive and often more specific than most other serological methods. Also, far less antisera are needed in the tests, for example just 50 or 100 μl/well compared to about 0.02–0.05 ml in the precipitin ring test.

Serological methods: radioimmune assay test

The method used is similar to the ELISA except that a radiolabelled reagent is used instead of an enzyme marker to identify positive reactions. Bailey et al. (1986) developed a radioimmunoassay, using ^{125}I-linked Protein A, as a tracer, to detect the predators of *Mansonia dyari*. The method consists of incubating gut extracts for 1–2 h at 37°C in the wells of polyvinyl chloride microtitre plates, followed by washing, and incubation for a further 1–2 h at 37°C with the anti-*Mansonia* serum (prepared in rabbits). Then after more washing, the radio-labelled [125]I Protein A tracer is added to each well and incubated for 2 h at 4°C. After these washings individual wells are cut out and placed in glass tubes and residual (i.e. bound) radioactivity measured using a gamma counter. A positive count was obtained up to 48 h after a *Toxorhynchites* larva had eaten a single 4th instar larva of *Mansonia dyari*. This is about twice the detection time reported for most precipitin test studies.

Other serological methods

McIver (1981) gives a useful comparison of the relative merits of the radial immunodiffusion test and the Ouchterlony double diffusion test for detecting arthropod predation. She also has useful notes on antigen production and general limitations of precipitin tests in predator–prey studies. Doane et al. (1985), working with predators of agricultural insects, present good descriptions of methods used to obtain antigens and methods for serological testing using the immuno-osmophoresis (IO) approach. They also give useful references to other relevant serological techniques. Boreham and Ohiagu (1978) present a review of the use of serology in predator–prey relationships.

Electrophoresis test

The detection of predators by serological methods necessitates the production of antisera in vertebrates. To obviate this Murray and Solomon (1978) introduced a new method using polyacrylamide gradient gel electrophoresis that detects prey enzymes within predator guts. For this, guts are removed and macerated, and then applied to the gels. Prey proteins (e.g. mosquitoes) are detected by staining with a mixture of 1-naphthyl acetate in acetone and fast blue RR in phosphate buffer. Predation is identified by comparing esterase bands of the prey, the predator being tested, and a starved predator. This technique has been applied in studies on red spider mites (Murray and Solomon 1978), terrestrial mites (Lister 1984) and notonectids (Giller 1982, 1984, 1986).

The method suffers the same limitations as other methods, such as inability to quantify numbers or size of prey eaten. However, in a brave attempt at quantification Lister et al. (1988) calculated field attack rates (A) of terrestrial mites on their prey using an exponential decay model of digestion incorporating the effect of meal size, where

$$A = -k\overline{Q}/(b\overline{M})$$ (12.12)

where $\overline{Q}$ = the mean quantity of esterase in sample, $\overline{M}$ = the mean weight of prey consumed per attacked prey, b = a constant of proportionality, and k = a negative constant that characterises the digestion curve. Laboratory experiments are needed to establish $\overline{Q}$ and $\overline{M}$ and then k and b are found by regression.

A simpler approach is that of Dempster (1967) namely predation rate is

$$\frac{\text{predator density} \times \text{proportion positive}}{\text{detection period}}$$ (12.13)

This assumes that a positive reaction represents a single prey eaten during the detection time. Rothschild (1966), however, takes a positive reaction to indicate that the predator has eaten a maximum number of prey (obtained from laboratory studies), and so calculates the predation rate as being the predator density × proportion positive × mean daily feeding rate as recorded in the laboratory. Both methods have serious limitations but the Rothschild approach is probably more appropriate when prey (mosquito larvae) are aggregated. In studying woodlice predators Sunderland and Sutton (1980) estimated predation rates by both methods and considered the true level of predation to be in between the two.

Urabe et al. (1982) used electrosyneresis to study predators of *Anopheles sinensis* in Japanese rice fields, and found the method better than immunodiffusion or micro-Ouchterlony methods.

DNA tests

Koenraadt and Takken (2003) used the rDNA-PCR method originally developed by Scott et al. (1993) for the identification of individual specimens of members of the *Anopheles gambiae* complex to detect predation by fourth instar larvae of *Anopheles gambiae* s.s. on first instar larvae of *Anopheles quadriannulatus* and vice versa. DNA of *Anopheles gambiae* s.s. was detected in 6 of 14 *Anopheles quadriannulatus* larvae that were known to have consumed at least one larva. *Anopheles quadriannulatus* DNA, on the other hand, was not detected in any of the 18 *Anopheles gambiae* s.s. larvae known to have consumed at least one first instar larva. Quantity of food available did not appear to have a significant effect on predation. Inability to identify DNA products in all larvae known to have consumed other larvae could be due to rapid digestion of prey DNA in the predator's stomach. The authors could not explain why the test failed to demonstrate the presence of *Anopheles quadriannulatus* DNA in fourth instar larvae of predatory *Anopheles gambiae* s.s.

Morales et al. (2003) used a PCR method to determine if the presence of *Anopheles gambiae* larval DNA could be detected in the gut of Libellulid dragonfly nymphs in a laboratory study in USA, following field observations on the frequency with which dragonfly nymphs were encountered in mosquito larval surveys in Kenya. These authors also used the ribosomal DNA method of Scott et al. (1993). Presence of mosquito larval DNA in dissected gut samples from a local species of dragonfly nymph was not detectable after periods of more than 1 hour had elapsed from prey ingestion, no matter how many prey individuals had been eaten. These results were in contrast to those obtained by Zaidi et al. (1999), who demonstrated that it was possible to identify prey DNA in predators more than 24 h post-ingestion.

Despite the mixed results obtained to date, it is likely that PCR-type methods will become increasingly useful for the demonstration of ingestion of prey species by predators.

Life-tables

The construction of life-tables, or budgets, was first applied to mosquitoes in the 1970s (Lakhani and Service 1974; Service 1971, 1973b; Southwood et al. 1972). The techniques have been more extensively used in agricultural and forest entomology (Begon et al. 1986; Morris 1963; Southwood and Henderson 2000; Varley and Gradwell 1970) and in fish studies (Beverton and Holt 1957; Ricker 1944, 1948; Wohlschlag 1954).

The purpose of a life-table is to summarise the survival and mortality rates of a population. Varley and Gradwell (1970) emphasised that the most instructive life-tables will usually be based on a continuous and intensive study of a population in a single habitat, not by sampling different populations in a number of similar habitats in different years. There are two basic types of life-tables. *Time-specific* or *vertical life-tables*, such as used by actuaries for assessing insurance premiums payable by people of different ages, measure the rate of an imaginary cohort by determining the age structure of the population at one given time. Both the age distribution of the population and deaths in the different age classes are recorded. This type of life-table is most useful with species having either overlapping generations or continual recruitment and was largely developed by Deevey (1947) and Ricker (1944, 1948). The second sort of life-table is the *age-specific* or *horizontal life-table* and is best suited for studying species with discrete non-overlapping generations. It measures the fate of a real cohort, such as the numbers of individuals of a single population. Not one, but a series of samples must be taken to estimate the absolute, or relative, numbers present at different time intervals during the progression of the population from the egg to adult. If there is a state of natural equilibrium in a population with the numbers of births just balancing out deaths and this does not change from year to year, then theoretically the two types of life-tables approach identity. But this rarely, if ever, happens. The actual construction and analysis of both types of life-table may be similar.

Laboratory studies

There have been several laboratory studies on the construction of life-tables of mosquitoes, the estimation of survival rates, intrinsic rate of increase, reproduction rates, etc. Such studies are of limited interest and generally outside the scope of this book, but a few examples are presented.

Lansdowne and Hacker (1975) constructed life-tables for five laboratory strains of *Aedes aegypti* in which fluctuating temperatures and humidities were incorporated. Liu et al. (1985) used a laboratory colony of *Aedes albopictus* to obtain life-table parameters, such as innate capacity for population increase, finite rate of increase, instantaneous birth and death rates, net reproductive rate and generation time.

Gómez et al. (1977) set up laboratory experiments with *Culex quinquefasciatus* to obtain age-specific mortality and fecundity data to determine various population parameters, such as intrinsic rate of increase, finite rate of increase, and net replacement rate. Suleman and Reisen (1979) extended this type of analysis to construct age-specific life-tables for three cohorts emerging from wild-caught pupae, while Reisen et al. (1979) studied life-table characteristics of different geographical populations of *Culex tritaeniorhynchus* from Asia. Mahmood (1997) used the methods of Reisen et al. (1979) to determine the life-table characteristics of *Anopheles albimanus* under controlled laboratory conditions. The daily survivorship of females at emergence was 0.97, and their net reproductive rate, age at mean cohort reproduction, instantaneous rate of increase, and mean generation time was 309.2, l0.8 days, 0.319, and 18 days, respectively.

Horizontal life-tables have been constructed for laboratory populations of *Anopheles culicifacies* and *Anopheles stephensi* (Reisen and Mahmood 1980), while Menon and Sharma (1981) compared life-table parameters of four colonies of *Anopheles stephensi* collected from different localities in India. Maharaj (2003) constructed horizontal life-tables for *Anopheles arabiensis* reared in programmable growth cabinets under conditions of fluctuating temperature designed to simulate field conditions during the four seasons experienced in KwaZulu Natal, South Africa. From the age-specific survivorship curve and the gonotrophic cycles, females were found to lay large numbers of eggs especially during spring and summer. Under simulated winter conditions mosquitoes did not oviposit in the laboratory. Trpis (1981) presented data on survivorship curves, fecundity and net reproductive rate of a laboratory colony of *Toxorhynchites brevipalpis*.

Laboratory conditions imperfectly mimic field situations, and so estimates of mortalities, and other life-history parameters, based on laboratory studies should probably not be extrapolated to natural populations.

Instar durations

For the construction of most life-tables the instar durations of the immature stages are required. There are several simple but not always very reliable, and some more sophisticated methods for determining these durations. For instance, larvae at the beginning of each instar can be put in small containers having water collected from natural breeding sites, which are then placed alongside or in, the breeding place. The time taken for 50% of the larvae to change into the next instar, that is the median time, can be taken as representing the instar duration in as near as possible field conditions. In California, USA Reisen et al. (1989) introduced 1st instar larvae of *Culex tarsalis* in predator-exclusion cages placed in a natural habitat to estimate the median durations of the different age classes. This approach was also adopted by Gomes et al. (1995) in a study of larval and pupal development and survival in natural and artificial containers in Brazil. A better, but more complicated, approach is to plot the regression of cumulative proportions of larvae changing into later instars, transformed to probits, against time, transformed to $\log_e$. This gives estimates of median instar development times. In Canada Enfield and Pritchard (1977) estimated the durations of the various larval instars of *Aedes cinereus* and *Aedes vexans* from the number of larvae collected from a pond over a 10-day period. Essentially they plotted the percentages of larvae in the different instars against time on $\log \times$ probability paper, and so estimated the time for 50% of the larvae (t_{50}) to have reached each successive instar. Campos and Sy (2003) used the same method for determining instar duration for *Ochlerotatus albifasciatus* sampled from floodwater pools in Argentina.

Birley (1979) pointed out the limitations of many simple methods for estimating the mean, or median, development time of the instars of mosquito larvae, such as their inability to estimate the range or variance. He applied a truncated log normal distribution model and also a skewed distribution model to the numbers of different instars of *Aedes aegypti* collected over 35 days from water pots in Bangkok (Southwood et al. 1972). A transfer function technique was developed to estimate the frequency distribution of developmental periods, and from this the mean, median or mode of the developmental periods could be calculated. With this method there is the ability to compare the effect of different temperatures or densities on these parameters.

In the laboratory Slater and Pritchard (1979) studied the relationship between the development times of all larval instars and pupae of *Aedes vexans* reared at five different constant temperatures (10–30°C), and expressed mortality at every 3 h of life. Their data were then used in developing a stepwise computer programme for the estimation of development times and survival rates at different temperatures that could be used with field populations. Their procedure, however, suffers from several limitations. For example, it does not take into account development at variable temperatures as experienced in the field, nor differences in laboratory and natural diets, and so seems to have limited use.

In the USA Hayes and Hsi (1975) studied larval mortality of *Culex quinquefasciatus* by placing individual egg rafts in 3.8-litre cans which had mesh netting on the bottoms and tops, and which were placed in a natural habitat. Every other day the entire contents were removed and all immature stages counted and their developmental stages recorded, after which they were returned to the can and replaced in the water. The mean duration of each immature stage (e) was estimated by standard life-table techniques based on cohort data (Chiang 1968). Thus

$$e = \frac{1}{1_0} \sum_{i=0}^{T} (X_i - X_0 + n_i/2) S_i \qquad (12.14)$$

where 1_0 = total number of surviving larvae entering the Nth instar or pupal stage, that is the sum of all the reductions in the surviving $N - 1$st immature stage over all time intervals, S_i = numbers of the Nth immature stages entering the Nth + 1st stage within the ith time interval (from time X_i to time X_{i+1}). That is (number of live Nth stage at time X_i) + (reduction in the $N - 1$st stage at time X_{i+1}) – (number of Nth stage whether dead or alive at time X_{i+1}). T = total number of observed intervals in the cohort study, and n_i = length of the ith interval in number of days (usually 2 in this instance). X_0 = time at which an $N - 1$st immature stage first entered into the Nth stage. The authors point out that although there may be some ambiguity in determining which of the larvae 'born' into the Nth stage at several X_0 becomes an $N + 1$st stage, the eventual estimate of e will remain unaffected. The mean developmental time of *Culex quinquefasciatus* based on four replicate cohorts each month ranged from 46.8 days in June to 10.8 days in September, while mortalities varied from 84% in January to 22% in July.

In their studies on the life-budget of *Aedes aegypti* Southwood et al. (1972) described how to construct life-tables on an age-specific basis, which involved calculating the expected numbers of each instar on each of their sampling days from the observed numbers of eggs laid in pots between an appropriate number of previous days. For this the instar durations were ex-trapolated backwards in time until all of the given instars must have been laid as eggs. Then a comparison of observed and expected numbers on each sampling day gave the total mortality up to each instar. A problem with this approach, however, is that instar duration depends considerably on tempera-ture, but this could be circumvented by using a physiological time scale (Hughes 1962, 1963).

Physiological time is the transformation of physical time so that tem-perature-dependent processes under study proceed at a constant rate. It eliminates the effect of varying temperatures, and simplifies analyses by expressing time in units directly related to the process being studied. Pajunen (1983, 1986) describes the usefulness of using physiological time in the analyses of stage-frequency field data of rock pool corixids. Pajunen (1986) explains how to use linear transformation of the physiological time scale and points out how the process is related to thermal summation methods using 'degree-hours' or 'degree-days' above developmental zero. The method is particularly suited to situations when use is made of areas under frequency curves (i.e. numbers in an age class multiplied by the time spent in that age class against time) to determine mortalities. The necessary transformation coefficients are found by the regression of larval size and developmental durations. This method could be applied to mosquitoes.

Atkinson (1977) pointed out that if instar durations were expressed in hour-degrees (hr°), then the expected number in each given stage could be estimated on a sampling day as follows. Let us say that the duration of 2nd instar larvae is 2000 hr°, and is achieved 5000 hr° after eggs were laid, then assuming that the youngest 2nd instar larvae on the sampling day entered that stage on that day, the date when the oldest entered that stage would be given by counting 2000 hr° back in time, such as from a thermograph trace or by a digital recorder. Counting 3000 hr° backward from both these days would then give the days when both the oldest and youngest 2nd instar larvae were laid as eggs. The numbers of eggs laid between these days would give the expected number of 2nd instar present on the given sampling day. A disadvantage of this approach is that it is difficult to count eggs laid, so instead Atkinson (1977) working with scale insects measured independently the daily production per female. This approach needs laboratory experiments to determine the threshold temperatures for development, and the hour- or day-degrees needed for each larval instar and pupa. For this the development time in hr° can be plotted against various temperatures, and the regression slope nearest to zero will indicate the threshold.

Both linear and non-linear functions have been proposed to estimate development rates; the former is in reality the day-degree (or hour) summation method. However, a linear relationship is not realistic because developmental rates cannot increase indefinitely with temperature. In fact insect development follows a logistic curve, which is relatively straight in the middle but curved at low and high temperatures. However, the thermal summation method can be quite accurate in predicting instar duration if field temperatures are in the middle of the developmental rate curve, that is on the straight line part of the relationship. But the logistic curve implies that the relationship is better described by a polynominal of the third degree. Harcourt and Yee (1982) therefore proposed the following cubic polynomial to predict development times of the alfalfa weevil.

$$r(x) = a_0 + a_1 x + a_2 x^2 + a_3 x^3 \qquad (12.15)$$

where x = temperature, and r = developmental rate at temperature x, the coefficients a_i were determined by solving the matrix equation

$$Aa = b \qquad (12.16)$$

where a = the vector of the coefficients, and A and b are from normal equations used to fit a polynomial to data in the sense of least squares. Harcourt and Yee (1982) determined the developmental time (t_d) for each stage as

$$\int_0^{t_d} \text{rate}_{(\text{stage})} \, [\text{temperature}(t)] \, dt = 1. \tag{12.17}$$

By using the trapezoid rule to approximate the integral at time intervals of Δt, the summation starts at $t = 0$ and stops when the approximate sum exceeds unity. The value of t at that point is the estimate for t_d.

Enfield, Pritchard and Scholefield method for estimating larval mortalities

In Canada Enfield and Pritchard (1977) obtained crude estimates of larval mortality of *Ochlerotatus euedes* and *Ochlerotatus mercurator* by plotting numbers of larvae caught on different days (irrespective of instars) against time on an arithmetic and also on a log scale, and calculating least squares regression lines. With *Ochlerotatus euedes* it appeared that mortality occurred in a linear fashion because the arithmetic plot gave the best fit, but in fact it is unlikely that mortality was constant, because if the linear rate continued the population would become extinct within 36 days, and this did not happen. No real mortality factors could be identified.

Using the sampling methods of Enfield and Pritchard (1977) and Pritchard and Scholefield (1980) the mortalities of *Aedes vexans, Ochlerotatus cataphylla* and *Ochlerotatus spencerii* were studied in two artificially flooded ponds in which the water levels were maintained constant (Pritchard and Scholefield 1983). Because: (i) 1st instar larvae were sampled with considerably lower efficiency than later instars; (ii) the two species could not be separated as 1st instars; and (iii) development rates were so rapid that insufficient samples were taken of the various instars, stage-specific mortality rates could not be estimated. To overcome the inability to identify 1st instar larvae, their numbers were estimated from the numbers and proportions of 2nd instar larvae present in the samples. This approach assumes that the proportion in each species was the same as when all species were 2nd instar larvae, and that the mortality of 1st instar larvae was the same for both species.

Linear ($y = a + bx$), exponential ($y = ae^b x$), logarithmic ($y = a + b \ln x$) and power ($y = ax^b$) regressions were calculated for each species. Generally the amount of variation in the estimates of population size obtained by the various regression models increased in the order Power < Linear

< Exponential < Logarithmic. Pritchard and Scholefield (1983) therefore concluded that mortality rates are higher in the early instars (the logarithmic model), rather than constant throughout (the exponential model), or decreasing (the linear model). However, while not necessarily giving the best fit to the data, the exponential model provides a single figure, in the form of exponent b, that can be used to compare mortality rates among or between species with time.

Service's method of time-specific life-tables

It has long been recognised that if the duration of the different age classes is taken into consideration then there is a relationship between the numbers of individuals collected in the different age classes and their survivorship. The example cited here is based on the method used by Service (1971) but refers to some of the data obtained from sampling larval populations of the *Anopheles gambiae* complex in Kenya and presented in a later paper (Service 1973b). Two hundred dips with a ladle were made daily of the pre-adults from a small pond for 10 days. It was considered reasonable to assume that during that relatively short sampling period the population of *Anopheles gambiae* was approximately stable, i.e. the number of eggs laid just balanced out death in all stages. Consequently the age distribution could be assumed to mimic the shape of the survivorship curve. The first step in constructing a life-table was to divide the total numbers of each instar collected by the appropriate instar durations and to plot these values against age in days of the immature stages (Fig. 12.1). The resultant histogram represents the *stage-specific* age distribution. Smooth curves are drawn through the mid-point of each histogram block, representing the mid-point in the life of each instar, to give the *age-specific* age distribution curve, which, so long as the steady state assumptions apply, will simulate the *time-specific* survivorship curve.

It follows that the numbers of pre-adults surviving to each age in days can be read off from this curve to give the numbers (n_x) surviving to age x (Table 12.1). From this basic data a life-table can be constructed. For comparative purposes it is often helpful to start a life-table with a population of 1000 individuals, therefore the figures in the n_x column are scaled up in the next column (l_x) which gives the numbers surviving at the beginning of age class x.

Successive subtractions of the (l_x) values give the numbers dying (d_x) between ages x and $x + 1$. The mortality rate (q_x), expressed here as the probability of an individual of age x dying before reaching age $x + 1$, is obtained by dividing the d_x values by the appropriate l_x values, or simply by subtracting the p_x values from 1, where p_x is the probability of survival from one age to the next.

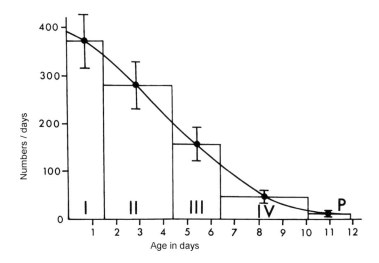

Fig. 12.1. Age distribution and survivorship curve for the immature stages of *Anopheles gambiae*. Vertical bars show 95% confidence limits of the mean total frequencies (Service 1973*b*)

The calculation of the expectation of life remaining (e_x) for individuals of age x is more complicated. Theoretically this is given by:

$$e_x = \frac{\int_x^w l_x d_x}{l_x},$$ (12.18)

but since the age intervals in the life-table are small the changes between ages x and $x + 1$ can be regarded as a linear function of x. It can therefore be assumed that the numbers of individuals (L_x) alive between the ages x and $x + 1$ which are exactly given by:

$$L_x = \int_x^{x+1} l_x d_x,$$ (12.19)

can in practice be given by:

$$L_x = \frac{l_x + l_{x+1}}{2}.$$ (12.20)

Now, successive values of L_x obtained in this way are summed from the bottom of the column up to each age of x to give values of T_x. The expectation of life in age units, in this example given in days, is now simply found by:

$$e_x = \frac{T_x}{l_x}, \tag{12.21}$$

values of L_x and T_x are frequently omitted from life-tables as they are only calculated to obtain values of e_x.

It must be understood that when, as in this instance, samples are taken from a population of unknown size the mortalities measured are relative, i.e. there is no information on the absolute numbers of larvae and pupae dying each day.

Table 12.1. Life-table for *Anopheles gambiae* pre-adults (after Service 1973*b*)

x	n_x	l_x	d_x	p_x	q_x	e_x
0	388	1000	59	0.941	0.059	4.814
1	365	941	109	0.884	0.116	4.085
2	323	832	136	0.837	0.163	3.554
3	270	696	134	0.807	0.193	3.151
4	218	562	124	0.779	0.221	2.783
5	170	438	116	0.735	0.265	2.429
6	125	322	98	0.696	0.304	2.124
7	87	224	93	0.585	0.415	1.835
8	51	131	49	0.626	0.374	1.782
9	32	82	30	0.634	0.366	1.538
10	20	52	18	0.654	0.346	1.154
11	13	34				0.500

Key: x = age in days, n_x = no. larvae surviving to age x, l_x = no. per 1000 larvae surviving to age x, d_x = mortality between ages x and $x + 1$, p_x = probability that a larva of age x survives to age $x + 1$, q_x = probability of a larva of age x dying before reaching age $x + 1$, e_x = expectation of life for individuals of age x.

A more simplified approach would be to estimate daily mortalities of the various instars assuming that mortality during any given instar is constant but not necessarily between instars, not an unreasonable assumption when instar durations are very short (e.g. 1.42–2.88 days in *Anopheles gambiae*). The numbers entering each instar are obtained from the survivorship curve and the calculations required to estimate instar mortalities are shown in Table 12.2.

Table 12.2. Instar mortalities of *Anopheles gambiae* (after Service 1973*b*)

Instars (i)	Age in days at beginning of instar	No. entering instar	Deaths in instar	Relative proportion dying in instar	Proportion dying daily in instars
	(t_{i-1})	(St_{i-1})	(D_i)	$\left(\dfrac{D_i}{St_{i-1}}\right)$	$1-\left(\dfrac{St_i}{St_{i-1}}\right)^{1/d*}$
I	0	388	48	0.1237	0.0795
II	1.42	348	143	0.4108	0.1678
III	4.30	205	89	0.4341	0.2568
IV	6.22	116	95	0.8190	0.3664
Pupa	9.97	21	10	0.4762	0.3034
Adult	11.76	11			

$d*$ = instar duration in days.

This approach has been used by several entomologists to estimate mortality of the immature stages of mosquitoes. Among the first to adopt the method were Rajagopalan et al. (1976*b*), who used it in India to construct survivorship curves and life-tables for the immature stages of *Culex quinquefasciatus* breeding in drains. In Thailand Apiwathnasorn et al. (1990) applied the method to estimate the mortalities (93.9–95.7%) of the pre-adults of *Culex tritaeniorhynchus* in paddy fields, of which more than half occurred in the pupal stage. In Uganda Sempala (1982) used time-specific survivorship curves to construct separate life-budgets for each month of the year for *Aedes africanus* breeding in tree-holes and bamboo pots. He also estimated and plotted *k*-values (Varley and Gradwell 1960) and found that overall mortality of the immature stages in tree-holes was 91.0%, whereas in bamboo pots a mortality of 83.2% was recorded; predation by *Toxorhynchites* species was thought to account for much mortality.

In Nigeria Bown and Bang (1980) sampled *Aedes simpsoni* larvae and pupae from cocoyam axils and using the method of Service (1971) constructed a survivorship curve, which showed that greatest mortality occurred in the 3rd and 4th instar larvae, with total immature mortality estimated as 97.9%. Daily output of adults/100 cocoyam plants was calculated by multiplying the mean number of pupae/100 plants by the pupal survival rate (66.7%) and then dividing by pupal duration (1.5 days). Another estimate of output was obtained from the density of the larvae and pupae and the estimated overall mortality rate of 97.9%. There was generally reasonable agreement between these two estimates. For example, the highest output was recorded in August, and based just on pupal numbers and their mortality was 10.9 *Aedes simpsoni*/100 plants per day, compared to an estimate of 8.83 adults based on all immature stages.

Lutwama and Mukwaya (1995) used the methods of Service (1973b, 1976) to estimate larval and pupal mortality for *Aedes simpsoni* s.l. developing in axils of *Colocasia esculenta, Xanthosoma sagittifolium*, and *Musa* spp. in Uganda. Instar durations were determined in the laboratory to be 2.48, 1.26, 2.12, 4.56, and 2.38 days for instars I-IV and pupae respectively. Life tables were constructed and mortality was observed to vary between dry and wet seasons and among the four sites sampled. Overall mortalities were highest in populations inhabiting banana plants (96.89%) and lower in populations inhabiting axils of *Colocasia esculenta* (47.11–88.37%). These mortalities are similar to those recorded for *Aedes simpsoni* by Bown and Bang (1980) in Nigeria.

The vertical method described above was also used by Edillo et al. (2004) to estimate *Anopheles gambiae* s.l. larval survivorship in rock pools, puddles and swamp in Mali. A total of 4174 larval *Anopheles gambiae* s.l. were collected from 2520 dips with a 350 ml dipper. Numbers of larvae from each habitat were as follows: rock pools, 2014 (48%), puddles, 1587 (38%), and swamp, 573 (14%). The numbers of larvae in each instar were divided by mean instar durations (derived from laboratory studies) to give the relative age structure. In some cases, they found that the relative numbers of earlier instars were less than later instars, a situation that could have arisen from sampling error related to the clumped spatial distribution of immatures. In these cases, the proportions were adjusted by setting the proportion in the earlier instar equal to the proportion in the later instar. The survivorship curve was constructed by regressing the corrected proportions with the midpoint of each instar mean duration and an exponential decay curve was superimposed on these data. A rough estimate of daily survival was obtained by regressing ln l_x on days, using a linear model with the intercept set at 0. Daily survival estimates obtained were 0.807 for rock pools, 0.899 for swamp, and 0.818 for puddles.

Survival of immature *Culex annulirostris* in three different breeding sites in Queensland, Australia was investigated by Mottram and Kettle (1997) using the Service (1971, 1973b) method. The three breeding sites chosen were a temporary pool, a semi-permanent pool, and flooded grassland. 50 first-instar larvae were introduced into $15 \times 15 \times 15$ cm cages constructed from a wire frame covered with 50-mesh nylon cloth to exclude predators. Three cages were floated in each breeding site using Styrofoam floats. Numbers and developmental stages were recorded at 24 h intervals for 15 days. Mortality in the breeding sites was determined by daily sampling consisting of 20 sub-samples with a 300 ml dipper from the two pools and fifty dips from the flooded grassland. As population stability is a requirement for use of the Service (1971, 1973b) method, Mottram and Kettle (1997) tested for this by dividing the data into early (days 1–6) and late (days 7–12) and calculating stage/group interactions on ln $(x+1)$ transformed data. The late group data (days 7–9 and 10–12) were found to be stable and were used for the construction of the life-table. Mean life expectancy on day 1 was found to be highest in the temporary pool (7.65 days), lower in the flooded grassland (4.95 days) and lowest in the semi-permanent pool (3.44 days). Daily percentage mortality was 7.9–8.5% in the temporary pool, 24.1–26.2% in the semi-permanent pool and 14.7–25.4% in the flooded grassland. Overall survival from first instar to adult in the semi-permanent pond was found to be similar at 2% to that calculated by Rae (1990), also for *Culex annulirostris*. Survival in the other two water bodies was higher at 4.2% and 34.8% from the flooded grassland and temporary pool respectively.

In Senegal, Awono-Ambéné and Robert (1999) used the method of Service (1971) to construct a survival curve for *Anopheles arabiensis* immatures in market-gardener wells from quadrat and dipper data. The equation for the curve that best fitted the data was: $y = 375 - 117.5 \, \mathrm{Log}_n x$.

In Argentina, Campos and Sy (2003) constructed life-tables for two cohorts of *Ochlerotatus albifasciatus* immatures, subject to parasitism by the nematode worm *Strelkovimermis spiculatus*. Percent survivorship of *Ochlerotatus albifasciatus* from first instars to adulthood was estimated as 1.4% in spring, 2% in summer, and 0.2–4.4% in fall. In cohort 1 (spring), no survival was recorded. The highest relative proportion dying (P_i) and daily proportion dying (P_{di}) occurred in the pupal stage in all cohorts. Life table analysis revealed low mortality in young instars, and higher in old instars, corresponding to an intermediate stage between Type III and I mortality curves of Slobodkin (1962).

Campos and Lounibos (2000) used the same method to study the population of *Toxorhynchites rutilus* inhabiting tree-holes and automobile tyres in Florida, USA. All immatures were removed from the containers prior to conducting the experiment. The entire liquid contents of tyres and tree-holes were removed every third day and numbers of each instar counted before samples were returned to their respective containers. Percent survivorship of *Toxorhynchites rutilus* from egg to adult was estimated at 1.8–3.4% in tyres and 4.5–5.6% in treeholes. Third instar mortality was higher in tyres compared with tree-holes. Survivorship in tyres in the autumn (fall) was significantly lower than in tyres or tree-holes in the summer. The shape of the survivorship curve was considered by the authors to most closely resemble Type IV of Slobodkin (1962), in which mortality risk is mainly concentrated in younger cohorts.

In Pakistan Reisen and Siddiqui (1979) constructed both horizontal and vertical life-tables for natural populations of *Culex tritaeniorhynchus*. For the former 63, and 625 and 2500 1st instar larvae were introduced into floating mesh cages (25×25 cm) placed in a village pond to give densities of 0.1, 1.0 and 4.0 larvae/cm^2. Emerged adults were counted daily, and the proportion of immatures surviving to adulthood calculated, as was the mean developmental time from 1st instar larvae to adult emergence.

For the estimation of vertical time-specific survivorship 15–25 dips were taken with a ladle twice weekly from the village pond, and larvae classified according to instars. Survivorship rates and life-table statistics were calculated according to the method of Service (1973*b*, 1977*a*), except that because their samples were taken within a 2-h interval the numbers collected in each instar were corrected by their probability of capture, rather than their instar durations. Probabilities of capture were obtained from other experiments which determined the hours of the day when larval moulting and pupation occurred, and then associating these with hours when samples were collected. For example, they reasoned that 1st, 2nd and 3rd instars would be present only during one sampling period, whereas 4th instar larvae and pupae would be present during two sampling periods, consequently the first three age-groups were divided by one and the numbers of 4th instar larvae and pupae by two to give stage-specific frequency distributions.

Using these corrected numbers of instars collected and instar durations, frequency histograms were constructed, and as suggested by Itô (1961) a segmented survivorship curve, not a smooth curve, was drawn through the midpoints (Fig. 12.2). By subtracting the overall mortality estimated from the horizontal approach with floating cages from the mortality estimated vertically, the crude mortality due to predation could be estimated, namely 13.4% during the pre-monsoon period and 97.1% post-monsoon.

Miura and Takahashi (1988) used the method of Service (1976) and Reisen and Siddiqui (1979) and constructed time-specific survivorship curves for *Culex tarsalis* in Californian rice fields, and found immature mortality ranged from different sample sites from 85.7–98.5%. In predator-free cages mortalities decreased to 25.3%, hence most mortality was probably caused by predators.

Reisen et al. (1982) applied the method of Service (1977*a*) as modified by Reisen and Siddiqui (1979) to obtain vertical life-tables for *Anopheles culicifacies* breeding in irrigation fields in Pakistan. Laboratory studies at different temperatures gave data on the duration of each age class and age at moulting, as well as the duration to total immature life from egg hatching to adult emergence. In addition to using some of these results in field studies on vertical life-tables the data were used to estimate the horizontal survivorship of the immature stages. Difficulties were encountered in sampling the age classes with equal efficiency. For example, there appeared to be an excess of 2nd instar larvae, as had been encountered by Service (1977a) with *Anopheles gambiae* in Kenya.

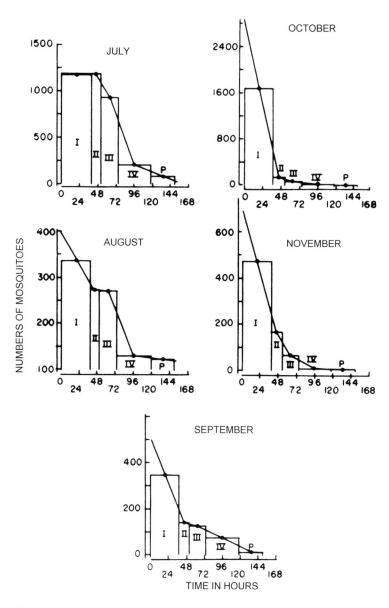

Fig. 12.2. Number of *Culex tritaeniorhynchus* collected in each stage and corrected by the probability of capture, and plotted as a function of instar duration in hours (Reisen and Siddiqui 1979)

In Thailand Southwood et al. (1972) working with *Aedes aegypti* and Mogi et al. (1986) working with *Anopheles* species overcame similar problems by simply pooling 1st and 2nd instar larvae in their calculations. Reisen et al. (1982) also reported low numbers of pupae of *Anopheles culicifacies*. Lakhani and Service (1974) also sometimes found difficulty in collecting all age classes without bias, for example they occasionally found more pupae than 4th instar larvae. These examples emphasise the difficulty of sampling all life stages equally, which probably arises from their different spatial dispersions and differences in their catchability.

Kaur and Reuben (1981) using the method of Service (1971) constructed a survivorship curve and time-specific life-table for *Anopheles stephensi* immatures breeding in wells in India. Instar durations at different times of the year were obtained by keeping larvae in wells in floating cages that excluded predators. Mortality occurring prior to pupation was very high in June (95.4%) and July–September (82.5–87.1%), the lowest mortality was recorded in March (42.0%). Mortality rates, termed k-values, were obtained by subtracting the logarithms of the relative numbers estimated to be pupating each day from the logarithm of the numbers of 1st instar larvae estimated to have entered the population each day 2 weeks previously. The time of 2 weeks was chosen because previous studies had shown that this was the development time from 1st instar to pupae. Because of the low larval densities in the wells the effect of density on mortality could be studied in only one well, and this was done by plotting k-values against logarithms of larval density. A straight line was obtained suggesting, according to Varley and Gradwell (1963), that mortality was density-dependent. Moreover, it was found that the slope of the regression of the logarithm of initial density on the logarithm of density of survivors, and the slope of the regression of the logarithm of the density of survivors on the logarithm of the initial density, were both on the same side of unity and differed significantly from unity, again indicating density-dependent mortality (Varley and Gradwell 1968).

In the Philippines Mogi et al. (1984) estimated the mortality of *Culex vishnui* and *Anopheles peditaeniatus* in rice fields by the method of Service (1971). However, they adopted the method used by Reisen and Siddiqui (1979) and Reisen et al. (1982) of drawing straight lines between the mid-points of the age of the various age classes to obtain a survivorship curve, in preference to the freehand curve preferred by Lakhani and Service (1974) and Service (1977*a*). Mortalities of natural populations in the rice fields varied from 98.2–100% (mean 99.1%) for *Culex vishnui*, and from 95.3–98.9% (mean 97.6%) for *Anopheles peditaeniatus*. Crude estimates of mortality due to predation were inferred from differences between mortalities obtained from predator-free cages and from sampling natural populations. Later Mogi et al. (1986) adopted the same procedures to study mortalities of *Anopheles peditaeniatus* and *Anopheles sinensis* in Thai rice fields. Although they divided the numbers of each age class caught by the median durations of each stage, the numbers of 2nd instar larvae were still sometimes greater than 1st instars, so they combined these two young instars in their histogram plots and straight line survivorship curves. Overall mortality was high, about 98% from egg hatching to adult emergence, and it was estimated that predators probably caused 19–54% of this.

Other examples of the use of Service's method are for *Culex annulirostris* in Australia (McDonald and Buchanan 1981; Rae 1990), for *Culex tarsalis* in the USA (Miura and Takahashi 1988; Stewart et al. 1985), for *Anopheles stephensi* in India (Menon and Rajagopalan 1979), and for *Culex tritaeniorhynchus* in Japan (Mogi 1978). Further data relating to the mortalities of the immature stages of the *Anopheles gambiae* complex are given by Service (1971, 1973b). A good explanation of the construction of time-specific life-tables, as applied to human populations, is given by Hill (1971). Other useful accounts describing how life-tables are constructed and used are presented by Begon et al. (1986), Deevey (1947), Southwood (1978), Southwood and Henderson (2000), and Varley et al. (1973).

Age-specific life-tables

For the construction of age-specific life-tables a series of samples of the population are taken at different times, so that the numbers entering the various age classes can be determined. The difference between successive estimates represents loss due either to mortality or dispersal. This approach is most suited to species having discrete non-overlapping generations. The method of analysis is similar to that for time-specific life-tables.

Southwood (1978) pointed out that heading the l_x column with 1000 can result in loss of important data, because there is no information on fluctuations in population size between different generations. It may also obscure density-dependent effects. When the actual estimates of the numbers in the various age classes are given in a table this can be referred to as a budget (Richards 1961; Southwood 1978).

In the ideal situation all members in a specific age class simultaneously change into the succeeding one; in practice, however, there is always some degree of overlap. This necessitates the integration of a number of estimates to obtain a reliable estimate of total population. Several methods have been proposed to estimate the population of each developmental stage on successive sampling occasions. One such method is to plot successive estimates of an age class from the beginning to the end of its life, and then divide the area beneath the graph by the mean instar duration to give the total numbers reaching the median age of that instar. This procedure can be repeated for each instar (Southwood and Jepson 1962). A big disadvantage of this method, and one that is common to many other methods, but not to those of Dempster (1961), Kiritani and Nakasuji (1967) and Kobayashi (1968), is its dependence on accurate determinations of instar durations. It may in practice be extremely difficult to obtain reliable measurements from field populations, especially under fluctuating conditions of temperature and food supply. Laboratory determinations may have little relationship to actual field durations.

Another relatively simple graphical approach applicable to instars having well-defined peaks of abundance is to plot logarithms of the accumulated total sample counts against time, and to fit a regression to the peak count and the following smaller counts. Extrapolation of this line back to the time when the stage was first found gives a population estimate of the numbers entering it (Richards and Waloff 1954). This method can only be used when there is a well-defined peak in numbers and a more or less steady mortality rate. The impact of mortality in successive instars will be compounded during the period of overlap of these stages. Now, if there are large differences between death rates then this could make the slope of the line meaningless and could give misleading estimates when the line is extrapolated back.

When recruitment and mortality greatly overlap so that there is no well-defined population peak this method is clearly inappropriate. However, Richards et al. (1960) showed that the total numbers of any stage collected in all samples (N) is given by:

$$N = P_0 \int_0^a S^t \mathrm{d}t = \frac{P_0(S^a - 1)}{\log_e S} \qquad (12.22)$$

where P_0 = the total number entering the stage, for example an accurate estimate of the egg population, S = fraction of this population that survives unit time and a = duration of the stage. Knowing the values of P_0, a and N from sampling, S can be calculated. The percentage mortality of the egg population is given by $100\,(1 - S^a)$, and therefore the numbers surviving and becoming 1st in-star larvae (that is a new P_0 value) can be calculated. So long as the duration of the 1st instar is known then an estimate of the numbers entering the 2nd instar can be calculated. By repeating this process the numbers entering each developmental stage can be obtained. The method, however, is not robust. The calculation of the numbers entering each successive instar is dependent on the initial calculation of P_0 for the egg population, hence there is the possibility of accumulation of errors. Furthermore, small errors in calculating the age duration (a) can result in severe discrepancies. Despite these difficulties this and the preceding method might prove useful for estimating the population size for various instars of mosquitoes.

Method of Southwood et al.

The life budget of *Aedes aegypti* was studied in Bangkok by Southwood et al. (1972). To estimate total numbers of larvae and pupae in their study area all pre-adults were removed three or four times a week from a few selected container habitats. After classifying into 1st and 2nd instars combined, 3rd and 4th instars combined and pupae, that is three age classes, all specimens were returned to the containers. To estimate egg production sample containers were exposed to wild ovipositing females for 48 h, after which the numbers of eggs laid were counted. The containers were then flooded and the number of eggs hatching determined by removing all 1st instar larvae over the next 10 days or so. The mean number of eggs per container multiplied by the number of containers that became flooded gave an estimate of the total number of eggs in the area available for hatching. The development pattern of each instar was studied daily by counting and removing larvae from a few sample containers; predators were excluded. Mean development times were calculated by recording the days taken for 50% of the numbers of each instar to change into the succeeding instar. Two different procedures were used to construct a life budget for *Aedes aegypti*.

One approach was the time-specific method, based on the daily recording of oviposition, hatching, emergence and the numbers in the different instars, which were estimated by dividing sample counts by instar durations. Survivorship curves of the log numbers in each instar ($\log x$), corrected for instar durations, were plotted against age classes for different months, and life budgets calculated for the different instars. Mortalities were expressed as k-values, these being logarithmic measures of the killing powers of a mortality factor, and are obtained by subtracting successive value of log survivors (l_x) (Varley and Gradwell 1960, 1970).

The other method of analysing the field data was an age-specific approach that consisted of determining the numbers of individuals that would be expected in the absence of mortality during the life cycle, i.e. the population of the different instars based on the daily egg input. It was assumed that all individuals had a constant and equal rate of development. For example, if the 3rd instar was reached on the 10th day after oviposition and lasted 3 days, then the total population of this stage expected on any given day (n) is the sum of the eggs laid on days $n - 9$, $n - 10$ and $n - 11$. However, because in practice instar durations are not constant, the population of any instar on a given day will be the sum of recruitment from oviposition over a number of days. In the absence of mortality the expected numbers of any instar on a sampling day can be calculated. The mortalities between the egg and different stages are found as the differences between those expected values and the actual numbers collected.

By using the stable age proportions of each instar, based on geometric mean development times, Southwood et al. (1972) obtained a second series of time-specific estimates of mortality to compare with those derived from the first method. A comparison of the monthly mortalities occurring between 2nd instar and pupae calculated by the two different methods showed they were ranked in the same order of magnitude. However, both methods were based on the same number of young instar larvae and consequently the two methods are not completely independent. It was decided that the first method, that is the time-specific approach, probably gave the most reliable results.

Southwood et al. (1972) found that mortality was most intense among the 1st, 2nd and 4th instar larvae of *Aedes aegypti*, and that in the 1st instar it was probably density-dependent. In the absence of predators and disease it was considered that mortality might have been caused by competition, such as for a limiting food supply. However, in their studies there was no direct relationship between pupae counted in one month and the eggs counted in the following month, because samples were not always collected in consecutive months, and moreover the egg and pupal populations were sampled differently. Dye (1984*a*) reanalysed both published and unpublished data of Southwood et al. (1972) and Sheppard et al. (1969) on populations of *Aedes aegypti* in Bangkok. He found that in addition to mortality of 1st–2nd instar larvae (k_1) being density-dependent, mortality during the 2nd–3rd instars (k_2) also appeared to be density-dependent. As a consequence this will increase the estimate of the finite rate of increase (λ) (net fecundity after life time density-independent mortalities), but decrease β, which is the maximum slope of the relationship between mortality—expressed here as a k-value—and log population size. Adult population size was more sensitive to changes in adult survivorship than to changes in fecundity.

Other methods used with Aedes aegypti and Aedes albopictus

Using both laboratory data and simple field observations Wijeyaratne et al. (1974) measured the proportion of *Aedes aegypti* eggs hatching and the proportion of 1st instar larvae surviving to the adult stage. They found that development time in natural habitats was much greater than in the insectary, presumably because of limited food supply. The survival rate of egg to adult was calculated as 0.42. Estimates of adult survival under field conditions were not attempted.

In Singapore Chan (1971) used milk tins with paddles as ovitraps for *Aedes albopictus*. They were covered over to exclude further oviposition as soon as eggs were found in them. Daily inspections established mean development times and mortalities of the various instars. A very simple life-table was constructed for the immature stages using these data. However, because the larval habitats were covered, preventing both further oviposition and excluding predators and leaf litter and other detritus from the containers, the mortalities recorded under these artificial conditions are likely to be an inaccurate representation of natural mortality.

Lakhani and Service method

A different approach was used to study the mortalities of the immature stages of *Ochlerotatus cantans* (Lakhani and Service 1974), which in southern England exists entirely as eggs from about September until the beginning of January when hatching starts. Larval development is slow and the first adults do not appear until late April or early May. The removal of numerous soil and leaf litter samples from a woodland ditch, and the extraction of eggs by wet sieving and flotation gave an estimate of the absolute egg population. This was corrected to take into account the likely number of eggs that would fail to hatch due to sterility or death of the embryos. The final estimate was of viable eggs. The adult population emerging in the field was sampled daily with emergence traps to give an estimate of its absolute population. From the estimates of egg input and adult output the probability of a viable egg giving rise to an adult was calculated for 3 years (Table 12.3). This gave an estimate of the total mortality over this period, but provided no information as to whether it acted most heavily on older or younger instars, or whether a constant number of larvae died per unit time. Having obtained a series of estimates of the number of each development stage on successive sampling days, and using independently obtained estimates of in-star durations, and also the estimates of the probability of a viable egg giving rise to an adult, Lakhani and Service (1974) fitted a generalised exponential survivorship model to the data (Fig. 12.3).

This was based on the assumption that the relative mortality rate during the duration of each instar of *Ochlerotatus cantans* was constant, although this rate was allowed to vary from instar to instar. This approach, and also an alternative graphical one given by Lakhani and Service (1974) agreed with the standard techniques of drawing a smooth curve through the stage-specific age-distributions.

All three methods indicated similar curves of type II of Slobodkin (1962), suggesting that the mortality was most intense in the younger instars, and that the relatively few individuals that survived to older age classes had a relatively high expectation of survival. Table 12.4 gives the final estimates of the instar mortalities of *Ochlerotatus cantans* obtained by these methods.

Table 12.3. Estimates of the total numbers of 'viable' eggs and emerging adults of *Ochlerotatus cantans* (after Lakhani and Service 1974)

	Estimated no. 'viable' eggs (no. entering 1st instar)	Estimated no. of emerging adults	P (probability of 'viable' egg resulting in an adult)
1969	197 058	13 044	0.066
1970	204 268	15 812	0.077
1971	454 187	47 427	0.104
Total	855 513	76 283	0.089

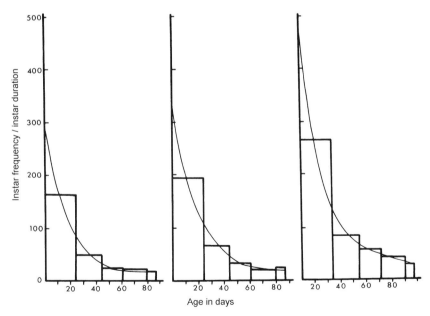

Fig. 12.3. Age distribution and survivorship curves of the immature stages of *Ochlerotatus cantans* (Lakhani and Service 1974)

Table 12.4. Estimated numbers of *Ochlerotatus cantans* entering each instar from freehand survivorship curves (after Lakhani and Service 1974)

I	Instar	Age in days at beginning of instar (t_{i-1})	Numbers entering instars $S(t_{i-1})$			
			1969	1970	1971	pooled data (all 3 years)
1	I	0	296	327	478	1 125
2	II	24	87	105	128	310
3	III	44.5	31	44	66	138
4	IV	61	22	25	46	94
5	pupa	80	19	20	34	76
6	adult	87	18	19	29	70

(Note: probability of a viable egg results in an adult = $S(87)/S(0) = 0.061$ for 1969, 0.058 for 1970, 0.061 for 1971, and 0.062 for pooled data).

Denoting the survivorship curve by $S(t)$, and noting that some 87 days elapse before newly hatched eggs give rise to adults the estimate of the probability of a viable egg resulting in an adult is $S(87)/S(0)$. The importance of this parameter in predicting population size and on the effects the release of sterile mosquitoes have on natural populations has been stressed by Cuéllar (1969a,b). Unlike egg sampling, larval sampling does not give an estimate of absolute population size, but measures the decreasing proportions in the various instars. Estimates of relative mortality rates (λ_i) are obtained from the following simple formula:

$$\lambda_i = \frac{1}{d_t} \log_e[S(t_{i-1})/S(t_i)]. \tag{12.23}$$

Thus, for example, for 1970 data the relative mortality rate of 2nd instar larvae is:

$$\lambda_2 = \frac{1}{20.5} \log_e[105/44] = 0.0424. \tag{12.24}$$

Another method of estimating relative instar mortalities would be to convert the numbers entering successive instars to $\log_{10}$ and to calculate k values (Varley and Gradwell 1970).

The accuracy of the approach used by Lakhani and Service (1974) relies on the ability to draw a smooth curve through the stage-specific age distribution, and also the accuracy of determining instar durations. Two alternative methods were proposed to overcome this subjective approach, and the paper by Lakhani and Service (1974) should be consulted for details. All three methods gave similar positively skewed survivorship curves of type

II of Slobodkin (1962), indicating that mortality was most intense in the younger instars, and that for the relatively few individuals that survived to older age classes there was a relatively high expectation of survival.

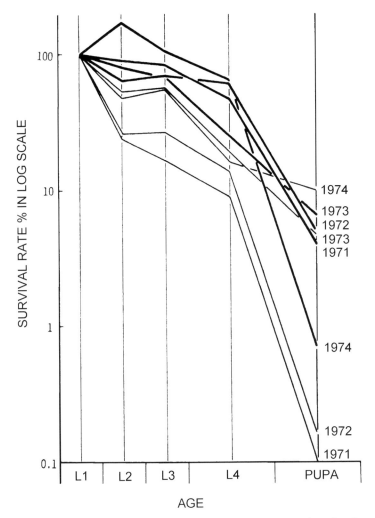

Fig. 12.4. Age-specific survivorship curves of *Culex tritaeniorhynchus* larvae and pupae; thick line—cultivated field, thin line—uncultivated field (Mogi 1978)

Fewer people have attempted to construct age-specific life-tables by this method, than by Service's time-specific method, probably because it is more difficult to determine instar durations, which can vary from the beginning to the end of the egg-adult season. However, Mogi (1978) con-structed survivorship curves for *Culex tritaeniorhynchus* in Japanese rice fields (Fig. 12.4) and Reisen et al. (1989) undertook the Lakhani and Service (1974) approach in California rice fields with *Culex tarsalis*. In both instances in preference to fitting curved survivorship lines straight lines between the midpoints of their histograms were fitted to give a stepped survivorship curve (Fig. 12.4) Reisen et al. (1989) estimated mortality from 1st instar larvae to adults as ranging from 66.7–98.9%. They also put 1st instar larvae of *Culex tarsalis* in predator-exclusion cages which were floated in natural habitats to estimate the median duration of the age classes, and obtain horizontal survivorship rates from 1st instar to pupal formation. Some of the cages were given ground rodent pellets as extra food. By comparing mortality rates by the horizontal (predator free) and vertical approaches, Reisen et al. (1989) estimated that predation accounted for about 61.7–84.5% of pre-adult mortality. They also considered that sampling with a dipper at fixed points along a transect, especially in a heterogeneous habitat, might have underestimated the proportions of 1st and perhaps 2nd instar larvae, because these are likely more aggregated than later instars (Mackey and Hoy 1978; Stewart et al. 1983). More recently in Australia Rae (1990) used this approach of Reisen et al. (1989) to construct life-tables of *Culex annulirostris*.

Mogi and Yamamura method

Mogi and Yamamura (1988) calculated larval mortalities (N_L) of *Amigeres theobaldi* inhabiting inflorescences having different numbers of water-filled bracts as follows.

$$N_L = 0.816(N_S + N_E)/B \tag{12.25}$$

where 0.816 = a hatch rate of 81.6% obtained from sampling the egg population, N_s = number of egg shells and dead eggs, N_E = number of eggs, and B = number of bracts per inflorescence. The k-value for each inflorescence was calculated as

$$k = \log_{10}((0.816\,N_S - N_I)/N_p) \tag{12.26}$$

where N_I = number of larvae alive and N_p = number of pupae and pupal exuviae at each sampling occasion. Now when k is density-independent it may relate to N_L as follows

$$k = c + \log_{10}(1 + (aN_L)^b) \tag{12.27}$$

where *a, b* and *c* are constants, indicating density-independent mortality (*a*), intensity of competition (*b*) and mode of competition (*c*) (Bellows, 1981). Mogi and Yamamura (1988) found that the correlation between

$\log_{10} N_L$ and k was significant ($r = 0.60$), and so they fitted the data to the above equation. They calculated that c was negative and so density-independent mortality was regarded as negligible, while b was 0.87, which being <1 indicated under-compensating contest competition and predicts a stable equilibrium.

Generation mean life-tables

Chubachi (1979) developed a new type of life-table specifically for insects with incompletely overlapping generations, which he termed the generation-mean life-table. Estimation of parameters required by this type of life-table is less affected by variations in age-distribution than that of time-specific life-tables. Hence the generation-mean life-table is more suited to populations having an unstable age distribution. He applied this approach to *Culex tritaeniorhynchus* breeding in Japanese rice fields. The pre-adult population from 279 paddy fields was sampled with a dipper once a week in 1966 from late May to early October and twice a week in 1967. Paddy fields varied from 100 to 1100 m² in size and the numbers of samples taken was roughly proportional to the area of water in the fields, eight or nine dips being taken per 1000 m².

The interval of mean generation time (G) is convenient to measure changes in population size with insects with overlapping generations, and is approximately given by

$$G = \frac{\sum_{x=1}^{\infty} G_x l_x m_x}{\sum_{x=1}^{\infty} l_x m_x} \tag{12.28}$$

where G_x = duration from beginning of egg state to the xth oviposition, l_x = probability of survival to the xth oviposition and m_m = mean number eggs laid by a female at the xth oviposition.

The population growth rate (R) during the mean generation time is

$$R = \frac{N_{t+G}}{N_t} \tag{12.29}$$

where Nt = population size at time t and $Nt + G$ = population size at time $t + G$. Now when lx and mx remain constant and the age distribution is stable the ratio R is approximated by

$$R = MR_0 M \sum_{x=1}^{\infty} l_x m_x \qquad (12.30)$$

where R_0 = net reproductive rate and M = ratio of population change as a result of migration. This equation is also applicable to insects with over-lapping generations, in fact the greater the overlap the more accurately this equation expresses population change. When parameters l_x and m_x are not constant then their mean values for all generations present during mean generation time can be used to give a good approximate value of R. This is the basis of the mean-generation life-table, and was proposed by Chubachi (1979) for populations with incompletely overlapping generations. For mosquitoes the following formula was used

$$R = M\overline{S}_E \overline{S}_{LP} \overline{F}_E \overline{P} \qquad (12.31)$$

where $\overline{S}_E$ = survival rate of eggs, $\overline{S}_{LP}$ = survival rates of larvae and pupae, $\overline{F}_E$ = expected total numbers of eggs a female will lay during her life time, i.e. fecundity, and $\overline{P}$ = proportion of females in the population. These symbols are mean values for all generations present during the interval t to $t + G$.

Estimation of the duration of the various age-classes is essential to the construction of most life-tables, and Chubachi (1979) used the summation temperature method to calculate these durations based on the following simple formula

$$1/Y = 1/K(X - C) \qquad (12.32)$$

where $1/Y$ = velocity of development, X = temperature, C = threshold temperature below which development ceases, and K = thermal constant. Having calculated instar durations and other intervals such as adult emer-gence to mating, mating to blood-feeding, blood-feeding to oviposition, and oviposition to blood-feeding by this method, G_x at any given tempera-ture was obtained by summation of these durations.

Now, mean generation time (G) depends not just on developmental du-rations but on survival of adults (l_x) and mean number of eggs at oviposi-tion (m_x). Although l_x and m_x probably change seasonally, Chubachi (1979) assumed that the ratios of l_x to l_1, and m_x to m_1, remained constant over the breeding season. The ratio of l_x to l_1, was estimated on the assumption of a stable age distribution from the age structure of adults collected during the breeding season. The ratio of m_x to m_1 was determined from previous labo-ratory experiments. Now although the assumption of constant ratios may not be quite true, and the estimated values may not be very accurate, these deficiencies may not produce any large error in the estimation of G, be-cause the $l_m \, m_x$ curve is monomodal. It should be mentioned that G, the mean generation time, is the most crucial parameter in this approach to constructing a life-table. Next G can be estimated by the method of tem-perature summation (Fig. 12.5). Because population size of each age group does not necessarily change at the same rate as that of all age groups, R should be estimated from sampling as many age groups as possible. How-ever, in his study Chubachi (1979) estimated R from just larval and pupal samples of *Culex tritaeniorhynchus*. As the duration of these pre-adults is 46–49% of the mean generation time, Chubachi (1979) considered that unless the age distribution of the population changed drastically then the estimated R values were probably quite accurate.

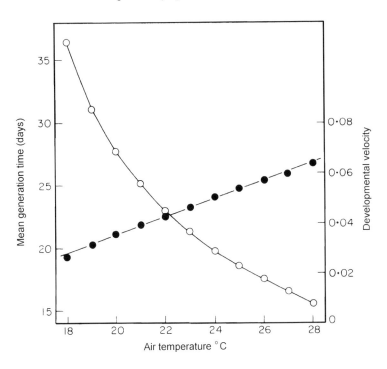

Fig. 12.5. Mean generation times (open circles) and developmental velocities (filled circles) of *Culex tritaeniorhynchus* at different air temperatures. The regression equation is $1/Y = 1/277.8(X - 10.1)$; $r^2 = 0.996$; $P < 0.001$ (Chubachi 1979)

If the age distribution in each stage is uniform and every individual has the same rate of development, then the survival rate of larvae and pupae

(S_{LP}) during the interval (I_n) between the nth and $n + 1$th sampling occasion (census) is estimated as

$$S_{LP,I_n} = \frac{\sum_{i=y+1}^{5} N_{n+1,i} + N_{n+1,y}\left(\sum_{i=1}^{y} T_i - I_n\right)/T_y}{\sum_{i=1}^{x-1} N_{n,i} + N_{n,x}\left(\sum_{i=1}^{y} T_i - I_n\right)/T_x} \qquad (12.33)$$

where $N_{n,i}$ = number of pre-adults in the ith stage (i = 1st, 2nd, 3rd, 4th instar larvae, and pupae) collected during the nth census. T_i = developmental duration of the ith stage, x = stage number of youngest individual present during the nth census and also expected to emerge as an adult before the $n + 1$th census, and y = stage number of oldest individual which is estimated to have hatched after the nth census and also alive at the $n + 1$th census.

The daily mean survival rate during the interval I_n is

$$\overline{S_{LP,D_n}} = S_{LP,I_n}^{1/I_n} \qquad (12.34)$$

The daily mean survival rate during the interval between any given day t and the day $(t + G)$ of one generation after t is estimated as

$$\overline{S_{LP,D_G}} = \left(\prod_{j=1}^{t+G-1} S_{LP,D_j}\right)^{1/G} \qquad (12.35)$$

where $S_{LP, Dj}$ = survival rate from any given day j to the next day, and which can be estimated from eqn (12.33). The mean survival rate from egg hatching to adult emergence during the interval from t to $t + G$ is approximately given by

$$\overline{S_{LP}} = \overline{S_{LP,D_G}}^{\overline{T_{LP}}} \qquad (12.36)$$

where $\overline{T_{LP}}$ = mean duration of larvae and pupae during the interval from t to $t + G$.

In his life-table studies on *Culex tritaeniorhynchus* Chubachi (1979) estimated the survival rate of the eggs (0.951) from previous laboratory experiments and the sex ratio by the proportion of female pupae (0.527) present in censuses. The life-table data were analysed by k-factor analysis (Varley and Gradwell 1960) both graphically and by regression of each k on K (Podoler and Rogers 1975). Chubachi (1979) found contrary to previous beliefs that the greatest mortality was due to pre-adult mortality other then drainage of the rice fields, although reduction in natality was also important, death due to drainage contributed least to total mortality (K).

Miscellaneous methods

A very simple method of estimating pre-adult mortality is to record numbers of 1st instar larvae and pupae in a habitat, and after correcting for instar durations, to take difference in numbers as representing mortality. This approach can be used to compare success rates in different habitats or in the same one at different times of the year.

Survivorship of overwintering larvae of *Ochlerotatus geniculatus* breeding in tree-holes in England was estimated by Bradshaw and Holzapfel (1991) by dividing the cumulative number of pupae, plus 3rd and 4th instar larvae, remaining at the end of the winter season by the numbers of larvae in the original cohort at the beginning of the observation period.

To determine the effects of different biotic and abiotic factors on *Anopheles gambiae* s.s. larval survivorship in forest in western Kenya, Tuno et al. (2005) set up four water basins (diameter 40 cm, depth 15 cm) in each of three habitat types: inside forest, 3–20 m from the forest edge (forest); under the forest canopy boundary but outside of the forest (forest edge); and outside the forest canopy, 5–30 m from the forest edge (open). Twenty second instars of *Anopheles gambiae* s.s. larvae from a laboratory colony (Kisumu strain) were introduced into each basin. Predators and other colonising animals were not prevented from entering the basins, but any new first instar larvae found in the basins were removed. Mortality was estimated from the number of larvae or pupae still alive on day 10 and redundancy analysis (Gauch 1982) was used to determine the key factors affecting survival. When data from all of the basins were analysed, redundancy analysis revealed that the factors significantly related to *Anopheles gambiae* survival were openness of the habitat, and the abundance of predatory animals, including odonate nymphs and *Culex tigripes*.

The Kiritani and Nakasuji (1967) method estimates stage-specific survival rates by sampling at regular intervals commencing when the insect enters the first stage and concludes when all in the last stage have died, but sampling the final life-stage is often difficult. Basically the method involves measuring areas under curves obtained from plotting stage-frequencies against time. Manly (1974a) demonstrated by simulation that this relatively simple method was about as good as some more mathematically complex models, but it has several restrictions limiting its use such as the need for populations to be regularly sampled. However, Manly (1976, 1977a) showed how the method could be modified to take into account sampling at infrequent intervals, how the durations of the stages can be derived from sampling the various age classes, and how jackknife techniques can be used to estimate the variances of various parameters (Manly 1977a). This last paper explains in a concise way the basic ideas of jackknife techniques and their limitations.

Life-table methods of Manly (1974b) and Kiritani and Nakasuji (1967) assume constant mortality rates for successive stages, in contrast the method proposed by Bellows and Birley (1981) allows mortality rates to vary between stages, although constant within the age class, and moreover recruitment to the initial stage need only be recorded. This is because recruitment to subsequent stages is predicted from the estimated developmental periods and mortalities of the various stages. Bellows et al. (1982) developed a model for the analysis of insect stage-frequency data which can be applied to a population with age-dependent mortality, for instance when the mortality rate of a population changes markedly at some fixed point in time.

Manly (1977b) proposed a new model for key factor analysis that can take into account 'circular populations', that is those in which adults in one generation (that is the survivors) produce all the individuals that are alive at the start of the next generation. It seems that the method of Kiritani and Nakasuji (1967) as modified by Manly (1976, 1977a) could be applied for estimating stage-specific survival of mosquito pre-adults.

Derr and Ord (1979) present a new method of estimating relative mortality and dispersal of insects with overlapping life-stages based on regular (e.g. weekly) sampling that might be applicable to immature stages of mosquitoes. Information is needed on instar duration. But only the mathematically brave need read their paper.

Walton and Workman (1998) used the relatively simple method of Aksnes and Ohman (1996) to estimate larval mortality in constructed wetlands in southern California, USA. The ratio of the numbers of individuals in younger (1st and 2nd instars) and older (3rd and 4th instars) subpopulations was used to estimate mortality (m), where

$$(r_i) = [\exp(ma_i) - 1] / [1 - \exp(-ma_{i-1})] \tag{12.37}$$

a, which refers to instar duration was determined from larval rearing in the field.

Life-table analysis

There are several ways of analysing life-table data to try to detect whether mortality at any given stage is directly, delayed or inversely density-dependent, or density-independent (Southwood 1978). Morris (1959) introduced the term 'key-factor' and used regression and correlation methods to identify the main factors responsible for important changes in total population density over a period of time. By extending and slightly modifying Haldane's (1949) logarithmic method for comparing successive mortalities Varley and Gradwell (1960) introduced a simple graphical method of key-factor analysis to determine whether mortality at any particular stage is in fact the key-factor. It should be noted that their method was specifically introduced for analysing age-specific life-tables of animals with non-overlapping generations. It uses estimates of mortalities of the various age classes in different years, thus providing information on mortality trends. Casanova and do Prado (2002) used the graphical method of Varley and Gradwell (1960) and the regression method of Podoler and Rogers (1975) to analyse life-tables for discrete generations of *Ochlerotatus scapularis* larvae in temporary pools in São Paolo State, Brazil. The key-factor most closely correlated with total mortality was pool dessication ($k_a = 1.081$) and the highest average mortality occurred in pupae ($k_5 = 0.316$) and fourth instar larvae ($k_4 = 0.201$). Larval population densities were calculated by sampling from 0.16 m^2 quadrats using an 11-cm diameter 500 ml capacity dipper and relating the mean number of individuals collected per sample to the total pool area, a method which rarely produces reliable results due to the over-dispersed distribution of mosquito immatures (see Chapter 3). Several authors (see Southwood and Henderson 2000) have pointed out the conceptual and statistical problems inherent in Varley and Gradwell's (1960) method and these have been addressed by Vickery (1991), Brown et al. (1993) and Sibley and Smith (1998). Sibley and Smith's (1998) method is termed λ contribution analysis and rather than determining the effects of various factors on mortality it identifies the elements in the life-table that most strongly affect population growth rates λ or *r*. Methods of analysing life-table data are described by Begon et al. (1986), Clark et al. (1967), Hassell and Huffaker (1969), Huffaker and Kennett (1969), Itô (1972), Kuno (1971), Luck (1971), Maelzer (1970), Morris (1959, 1963), Morris and Royama (1969), Pielou (1977), Smith (1973), Solomon (1968), Southwood (1967, 1978), Southwood and Henderson (2000), St Amani (1970) and Varley et al. (1973). The publications edited by Chapman and Gallucci (1981), Hazen (1970), Connell et al. (1970), May (1981) and McDonald et al. (1989) contain useful collections of papers on population dynamics.

In a three-year study on the population dynamics of *Ochlerotatus sierrensis* Hawley (1985*b*) obtained good regressions from plotting: (1) pupal weight against larval density; (2) number of eggs/egg batch against pupal weight; (3) pupal weight against wing length; and (4) percentage parous against pupal weight. Only the last plot gave a curvilinear line, showing that for some reason there was reduced survivorship for the largest adults. Hawley (1985*b*) used these regression equations to obtain *k*-values of the log reduction in fecundity at each log density. He also used the following equation of Hassell (1975), which describes a single species population with discrete generations

$$k = b \log(1 + aN_t) \tag{12.38}$$

where k = the log reduction in fecundity, N_t = population size and a and b = two parameters describing the shape of the relationship; b in fact is a measure of the density-dependent effect, and its value is taken as the maximum slope of the curvilinear plots of k (fecundity) against larval densities. Based on graphical analysis Hawley (1985b) found k to be between 0.26–0.80, showing that the reduction in fecundity due to size is under-compensating and the population necessarily exhibits exponential dampening. Hawley (1985b) points out that this type of analysis of life-table data has only been undertaken by Chubachi (1979) on *Culex tritaeniorhynchus*, who found undercompensating density-dependent mortality in the larvae, and by Dye (1984a) on *Aedes aegypti*, who reanalysed data collected in Bangkok by Southwood et al. (1972) and Sheppard et al. (1969). Dye (1984a) predicted that the population would have exponential dampening to equilibrium, pre-adult mortality being under-compensating ($b < 1$ in the above equation). This paper of Hawley (1985b) should be referred to for more details of procedures and analysis, and how it was concluded that the density-dependent size at metamorphosis was the main basis of population regulation in *Ochlerotatus triseriatus*.

Various models, some simplistic others highly mathematical and even philosophical have been developed in attempts to understand, or predict, population dynamics of mosquitoes. For example, Moore et al. (1990) constructed a deterministic age-specific life-table to track groups of individuals (30 cohorts) of *Culex pipiens* in different age-classes and physiological states through a period of 30 days following ULV application of insecticide. Immature stages were ignored. The limitations of the model, including sometimes insufficient field data, and the problem of pseudoreplication are discussed.

Moon (1976) developed a statistical model of the dynamics of *Culex tarsalis* that incorporated all the life-stages through which the mosquito passes. The model starts with the number of eggs deposited by a mosquito and takes into account the probabilities of survival during the subsequent larval and pupal stages. The effects of changes in survival rates on population dynamics are explored. A formula was derived to estimate the expected number of adult *Culex tarsalis* alive at any time during the spring or summer in Kern county, California, USA.

Ahumada et al. (2004) developed a discrete-time model for *Culex quinquefasciatus* populations in relation to temperature, rainfall and elevation in Hawaii. Developmental rate was included in the transition matrix, by using physiological ages of immatures instead of chronological age or stages, and the model also incorporated a consideration of intraspecific competition within larval habitats based on that described by Teng and Apperson (2000). Population projections from the model were compared with mosquito trap data collected from CO_2-baited CDC light-traps and oviposition traps at three representative sites. Model results for 2002–2003 showed a reasonable agreement with available mosquito data at three different elevations. At the high elevation station (1800 m) no mosquitoes were trapped during most trapping sessions, in line with the prediction of this altitude being above that suitable for *Culex quinquefasciatus*. At the intermediate elevation the model predictions replicated observed seasonal dynamics and explained approximately 75% of the variance observed in the data. At the low elevation site, the model predictions were out of phase with the seasonal dynamics in 2002, but consistent with 2003 observations, when they explained approximately 41% of the variance in the data.

Sawyer and Haynes (1984) point out that many of the methods for estimating stage-specific survival rates in populations with overlapping generations are computationally difficult (e.g. Birley 1977; Derr and Ord 1979; Ruesink 1975), others make various assumptions such as a constant daily survival rate for all stages (e.g. Kiritani and Nakasuji 1967; Manly 1974b), while other approaches need additional data such as estimates of recruitment to the first stage (e.g. Dempster 1961) or even worse, to each stage (e.g. Birley 1977), or require that sampling is made at equal intervals (e.g. Kiritani and Nakasuji 1967). All these restrictions impose severe limitations on the use of these methods on field populations. In contrast Southwood's (1978) graphical method is relatively simple, but is nevertheless applicable to real life situations where survival rates vary from stage to stage, or when data are unavailable for an entire generation. Furthermore, it does not require sampling to be made at equal intervals, and makes no assumption about the timing or distribution of recruitment to the population.

In Southwood's method data obtained at frequent sampling intervals are used to plot stage-frequency curves, and the area under each curve (total incidence) calculated graphically or numerically. When all mortality occurs at the end of a stage, an estimate of the number of individuals entering each stage (N) is obtained by dividing the area under each curve (A) by the duration of that stage (T). The survival rate for that stage is then estimated as

$$S_i = N_{i+1} / N_i \qquad\qquad (12.39)$$

where $N_i = A_i/T_i$ is the estimate of the number of individuals entering the ith stage.

Even this approach is subject to various errors: (1) systematic bias of the sampling procedures; (2) unequal sampling efficiency for successive life-stages; (3) random sampling error; (4) 'integration' error caused by sampling at finite levels; and (5) variations in stage durations. As Sawyer and Haynes (1984) point out the first four errors are due to sampling procedures and can, at least in theory, be overcome by increasing sampling size, sampling frequency or changing the sampling methods. The fifth error may have biological causes and involves inaccurate estimates of life-stage duration (T) and is more difficult to resolve, and in fact formed the basis of the paper by Sawyer and Haynes (1984). They concluded that although Southwood's method was fairly robust, often having acceptable errors in estimating survival rates of the first stage, precise estimates of stage-specific survival will nevertheless not usually be possible. Therefore direct measurements of mortality rates and survivorship patterns (Southwood 1978, p. 309) are strongly recommended, at least in preliminary sampling. For details on the need for or otherwise for correction procedures see Sawyer and Haynes (1984).

Reproductive potential

The life-tables just described are concerned mainly with estimating mortalities of the immature stages (optimistically from oviposition to adult emergence) but death continues. After emergence mosquitoes must live sufficiently long for mating, blood-feeding, maturation of the eggs and oviposition before there is any new population input from the emerging generation. During all these stages there will be further mortalities. The number of eggs laid and number of ovipositions by a female depend on a multitude of factors, including the availability of hosts and success in getting blood-meals, intrinsic longevity of the female and survival rates, climatic conditions and intervals between re-feeding. If sufficient field data are available on fecundity, number and duration of gonotrophic cycles and survival rates, then it may be possible to construct life and fertility tables for adults. The basic principles are summarised by Southwood (1978).

Net reproductive rate

The total number of female eggs a mosquito lays during her lifetime is sometimes termed the net reproductive rate (R_0), or in other words the number of times a population can multiply per generation. This is a function of fecundity, which is the number of eggs per female aged x, and life expectancy. That is R_0 is found as the summation of the mean fecundity at age x_1 (i.e. m_1, which is total natality divided by two to give number of female births) times the probability of survival to age x_1 (i.e. l_{x1}). Thus

$$R_0 = \Sigma l_x m_x. \tag{12.40}$$

Weidhaas (1974) considered that with insects such as mosquitoes which have immature stages developing in a completely different type of environment to the adults then it is easier to equate R_0 with the probability of adult survival (l_x) and the probability of survival of the immature stages (S_i) independently, thus:

$$R_0 = [\Sigma l_x m_x] S_i. \tag{12.41}$$

Another way of expressing the net reproductive rate is the ratio of individuals in the population at the start of one generation ($N_t + \tau$) to the numbers in a previous generation (N_t) thus

$$R_0 \frac{N_t + \tau}{N_t} \tag{12.42}$$

where τ = generation time. If $R_0 = 1$, then clearly the population is neither increasing nor decreasing.

Although since the 1970s many entomologists have tried to estimate pre-adult mortalities in several mosquito species, many fewer have attempted to calculate R_0 or its associated parameters, probably because there is generally a paucity of data to enable reliable estimates of reproductive rate to be made. However, Weidhaas (1974) used the last formula above to determine R_0 but substituted the ratio F_1/P, the ratio of female progeny to female parents and where $F_1 = [P\Sigma l_x\ m_x]S_i$ for $N_{t+\tau}/N_t$. He proposed that in control programmes which altered the reproductive potential of the field population, such as the production of sterile eggs, that a hatchability factor (h) be introduced, $h = 1 - S$ where S is the degree of sterility expressed as a decimal, thus:

$$F_1 = P[\Sigma l_x m_x](1 - S)(S_i). \tag{12.43}$$

Use was made of this approach in studying the effect of releasing chemosterilised male *Anopheles albimanus* into field populations in El Salvador (Weidhaas et al. 1974). The usefulness of estimating reproductive potential of populations before and during genetic control programmes is discussed by Weidhaas (1974) and by Weidhaas et al. (1972).

In his studies on the capacity for increase in *Wyeomyia smithii* Bradshaw (1980) calculated Lloyd's (1967) index of mean crowding (m^*) for larvae collected from pitchers of *Sarracenia purpurea*, while data from laboratory experiments enabled the capacity for increase (R_0), and mean generation time, which Bradshaw (1980) split into two components, namely pre-adult development (days from oviposition to adult emergence) and reproductive lag (days from emergence to oviposition), to be calculated. He found that capacity for increase was inversely correlated with mean crowding of the overwintering population.

Armbruster and Hutchinson (2002) demonstrated that pupal weight and wing length provided highly significant and equally accurate indicators of fecundity in *Aedes albopictus* reared in the laboratory, *Aedes geniculatus* reared in the laboratory, and *Aedes geniculatus* reared in its natural tree-hole larval habitat. Relative accuracy of pupal mass and wing length in predicting egg batch size was determined by comparing the coefficient of determination (r^2) for pupal mass versus egg batch size and wing length versus egg batch size regressions. For each regression, estimates of the standard error of r^2 were determined by performing 500 bootstrap itera- tions (i.e., resampling data points with replacement). For *Aedes albopictus*, wing length explained 79% (SE ± 4%) of the variation in fecundity, and pupal mass explained 83% (SE ± 3%) of the variation in fecundity. For *Aedes geniculatus* reared as larvae in the laboratory, wing length explained 60% (SE ± 9%) of the variation in fecundity, and pupal mass explained 61% (SE ± 10%) of the variation in fecundity. For *Aedes geniculatus* reared as larvae in their natural habitat, wing length explained 68% (SE ± 4%) of the variation in fecundity, and pupal mass explained 69% (SE ± 4%) of the variation in fecundity.

Intrinsic rate of increase

Another important parameter in population ecology is the intrinsic rate of natural increase or as it is sometimes called the innate capacity for in- crease. This is the natural growth rate and allows for some mortality in an unlimited environment. This was originally denoted by the symbol r (Birch 1948) but later as r_m (Andrewartha and Birch 1954). Now

$$N_t = N_0 e^r m^t \tag{12.44}$$

or in logs

$$\log_e N_t = \log_e N_0 + r_m t \tag{12.45}$$

where N_0 = the number of individuals at time t_0 and N_t = the number at time t. Exponent r_m is in fact the difference between the birth and death rate in a population. In a stable age population each age class grows at the same exponential rate, and each population has an intrinsic growth rate (r_m) for a given set of life-table parameters. The value of r_m can be calculated as follows

$$1 = \int_0^a e^{-r} m^x l_x m_x dx. \tag{12.46}$$

The solution for r_m is found by iteration. Useful methods for resolving this are presented by Birch (1948), Southwood (1978), Watson (1964) and Ricklefs (1973).

An approximate method of determining r_m was given by Andrewartha & Birch (1954), but Laughlin (1965) proposed that this approximate estimate should be distinguished from the more accurate estimate and be denoted as r_c and termed the capacity for increase. It is calculated as follows:

$$r_c = \frac{\log_e R_0}{T_c} \tag{12.47}$$

where T_c = the mean age of reproductive females in a cohort at oviposition, usually referred to as 'mean cohort generation time'. An approximate estimate of T_c independent of r_c can be obtained as follows:

$$T_c = \frac{\Sigma x l_x m_x}{\Sigma l_x m_x} \tag{12.48}$$

For a more detailed understanding of r, r_m, r_c, r_n and r_s refer to Southwood (1978).

As Southwood (1988) points out there has been some confusion as to the precise meaning of r_m, the rate of increase in a population, but generally field workers have started from the definition that

$$1 = \sum_t e^{-rt} l_t m_t \tag{12.49}$$

when the environment is unlimited and there is a stable age distribution this can be approximately expressed as the capacity for increase that is, r_c. So

$$r_c = \ln\left(\sum_t l_t m_t\right)/T_c \tag{12.50}$$

where l_t = number of female descendants alive at time t; m_t = number of female offspring produced at age t; and T_c = generation time.

Examples on mosquitoes

Crovello and Hacker (1972) conducted some interesting laboratory experiments on the life-table characteristics of 13 non-urban (ssp. *formosus*) and urban strains (ssp. *aegypti*) of *Aedes aegypti*. From cage colonies it was possible to determine the net reproductive rate (R_0) for each strain,

$$R_0 = a\sum_{x=0}^{w} l_x m_x \tag{12.51}$$

where w = the last interval (days or multiples of days) to which any females survived, a = proportion of females surviving to adult emergence, l_x = the proportion of adult females surviving to age x, and m_x = the mean number of female progeny produced for females of age x, i.e. the product of the production of mean number of eggs/female and the proportion that are female.

The intrinsic rate of increase (r_m) was calculated as follows:

$$1 = a\sum_{x=0}^{w} l_x m_x e^{-r_m(x+d)} \tag{12.52}$$

where x = age interval, e = base of natural logarithms, and d = duration for larval development from egg to age of egg production (a value of 12.5 days was in fact used). Although there was considerable variability with each strain of *Aedes aegypti* used in these experiments, there were nevertheless statistically significant differences for the mean expectation of life, net reproduction rate and intrinsic rate of increase between the strains. Non-urban (ssp. *formosus*) had smaller values of R_0 and r_m than did urban strains (ssp. *aegypti*). Interestingly two hybrid populations had very high net reproduction rates and intrinsic rates of increase, suggesting their usefulness in possible biological control programmes by the introduction of a sterility factor.

Yasuno (1974) reported on attempts in Delhi to measure the net reproductive rate of *Culex quinquefasciatus* at different times of the year, based on survival rates and duration of the immature stages, survival rates of adult females and duration of the gonotrophic cycle and frequency of oviposition and fecundity. In March R_0 was very high, 9.8, due to high survival rates of both larvae and adults, and clearly indicated an expanding population. Thereafter, despite fluctuations, R_0 tended to stabilise around unity, implying a more or less stationary population (Table 12.5). Presumably, later R_0 must have been <1, because over long periods $R_0 \approx 1$. The values obtained by Yasuno (1974), however, should be regarded with caution as the population studied in the different months changed as regards both locality and type of larval habitat. Moreover there was no information on the presence or absence of density-dependent factors, or whether there were marked seasonal changes in the intensity of predation.

Also in India Rajagopalan et al. (1977*a*) calculated R_0, generation time (T), number of ovipositions per female, and numbers of eggs laid per emerging female *Culex quinquefasciatus* breeding in a town south of Delhi. They also calculated the innate capacity of increase (r_m) and the finite capacity of increase (λ).

In El Salvador Weidhaas et al. (1974) attempted to measure the rates of increase (R_0) of natural populations of *Anopheles albimanus* when sterile males (S) had been released into the population.

$$R_0 = \frac{F}{P(1-S)} \tag{12.53}$$

where P = relative or absolute population at time t_1, F = relative or absolute population at time t_2, and S = proportion of introduced sterile males expressed as a decimal. They also equated the rate of increase of the population to the product of the number of egg batches per female (b), the number of eggs per batch giving rise to females (n) and the survival rate from egg to adult (S_i):

$$RI = bnS_i. \tag{12.54}$$

Because there was no direct information on S_i attempts were made to estimate this by comparing the numbers of 1st and 2nd instar larvae and the numbers of pupae caught in populations of immatures. At best this can only be a crude estimate. Furthermore, they assumed that 95% of all eggs hatched, that 60% of the 1st and 2nd instar larvae caught were in fact 1st instar, and that there was a 5% pupal mortality.

Table 12.5. Net Reproduction rate of *Culex quinquefasciatus* in a Delhi village in different months (after Yasuno 1974)

Months	Immature stages Survival rate	Adult stage					Cohort generation time (days)
		Days	Survival rate	Gonotrophic cycle[a]	Fecundity × 0.5	R_0	
Jan.-Feb.	0.054	21	0.650	13 & 6	113	0.022	36.2
March	0.043	11	0.932	5 & 3	63	9.80	27.6
May	0.016	11	0.854	5 & 3	63	1.21	20.8
	0.031	15	0.884	5 & 3	71	3.82	26.5
August	0.022	15	0.871	5 & 3	76	2.47	25.7
	0.014					1.57	
September	0.026	15	0.818	5 & 3	76	1.59	23.8
	0.016					0.98	

[a]Days from emergence to 1st oviposition and days between subsequent ovipositions.

No reasons are given for these assumptions and no consideration of instar duration is taken into account in trying to estimate the probability of survival from egg to adult.

In Florida, USA Wijeyaratne et al. (1974) placed known numbers of *Aedes aegypti* eggs in water tubs and estimated the probability of survival of the immature stages (S_i) simply by counting the numbers of different larval instars, pupae and emerging adults. They were unable to estimate directly the probability of adult survival, but obtained values for this by substituting various values of R_0 (0.5, 1, 2, 5 and 10) in the formula $R_0 = [\Sigma l_x\, m_x]$ (S_i). Values of m_x, expressed as the number of eggs laid by a female per day, were found by averaging the number of eggs laid daily over 39 days by 100 laboratory reared adults. Also in Florida, on Seahorse Key island, Weidhaas et al. (1971) estimated the reproductive potential ($R_0 = 1.4$–3.5) of *Culex quinquefasciatus* in successive generations by daily counting eggs from septic tanks which acted as ovitraps. Sterile males were then introduced into the population and all egg rafts collected from the septic tanks, which were considered to represent about half of those laid on the island, were destroyed. In the first generation following these control measures R_0 was 1.4 but as control continued R_0 increased to 5 and by the third or fourth generation to 10. It was thought that the decrease in population size caused by these control measures caused a reduction in larval competition, with the result that survival rates increased.

Suleman (1990) studied intraspecific variations in the reproductive potential and capacity for increase of *Anopheles stephensi* females under constant laboratory conditions. The standard equations for R_0, r_c, and T, as given in the preceding sections, were used with some modifications. For example, in the calculation of the reproductive rate

$$R_0 = a\sum_{x=1}^{n} l_x m_x \qquad (12.55)$$

a = the mean proportion of 1st instar larvae (hatched eggs) reaching the adult stage, x = the physiological age in terms of the gonotrophic cycle or oviposition cycle, l_x = proportion of females alive at each gonotrophic cycle, and m_x = the mean number of female larvae (assuming an equal sex ratio) produced per female during the gonotrophic cycle x. Then the capacity for increase (r_c) was calculated by the method of Laughlin (1965), and cohort generation time (i.e. mean age of females at birth of females offspring) was transformed from physiological age (T_{gc}) to chronological age (T) as follows

$$T\ \text{days} = [(T_{gc}) \times d_1] + d_2 + d_3 \qquad (12.56)$$

where d_1 = average gonotrophic cycle duration in days, d_2 = average adult life up to the beginning of the first gonotrophic cycle, and d_3 = mean development period for females from egg to adult emergence. It was difficult to estimate d_2 because it was not known how many females took a blood-meal when offered to adults 2 and 3 days old, so it was assured to be 3 days. R_0 was estimated as 25.7, r_c as 0.17, and T_{gc} was 1.96 gonotrophic cycles, or after transforming to chronological age 18.6 days.

References

Abrams PA (2001) Describing and quantifying interspecific interactions: a commentary on recent approaches. Oikos 94: 209–218

Agnew P, Haussy C, Michalakis Y (2000) Effects of density and larval competition on selected life history traits of *Culex pipiens quinquefasciatus* (Diptera: Culicidae). J Med Entomol 37: 732–735

Agnew P, Hide M, Sidobre C, Michalakis Y (2002) A minimalist approach to the effects of density-dependent competition on insect life-history traits. Ecol Entomol 27: 396–402

Ahumada JA, Lapointe D, Samuel MD (2004) Modeling the population dynamics of *Culex quinquefasciatus* (Diptera: Culicidae), along an elevational gradient in Hawaii. J Med Entomol 41: 1157–1170

Aitken THG, Trapido H (1961) Replacement Phenomenon Observed amongst Sardinian Anopheline Mosquitoes following Eradication Measures. Tech Mtg Un Conserv Nat 8th, 1960, pp. 106–114

Aksnes DL, Ohman MD (1996) A vertical life table approach to zooplankton mortality estimation. Limnology and Oceanography 41: 1461–1469

Ali SR, Rozeboom LE (1971a) Cross-insemination frequencies between strains of *Aedes albopictus* and members of the *Aedes scutellaris* group. J Med Entomol 8: 263–265

Ali SR, Rozeboom LE (1971b) Cross-mating between *Aedes (S.) polynesiensis* Marks and *Aedes (S.) albopictus* Skuse in a large cage. Mosquito News 31: 80–84

Ali SR, Rozeboom LE (1973) Comparative laboratory observations on selective mating of *Aedes (Stegomyia) albopictus* Skuse and *A. (S.) polynesiensis* Marks. Mosquito News 33: 23–28

Alto BW, Juliano SA (2001) Precipitation and temperature effects on populations of *Aedes albopictus* (Diptera: Culicidae): implications for range expansion. J Med Entomol 38: 646–656

Amalraj DD, Das PK (1994) Time to death from starvation and compulsive killing by the larvae of *Toxorhynchites splendens* (Diptera: Culicidae). Acta Trop 58: 151–158

Amalraj DD, Das PK (1998) Estimation of predation by the larvae of *Toxorhynchites splendens* on the aquatic stages of *Aedes aegypti*. Southeast Asian J Trop Med Public Health 29: 177–183

Anderson RM (1979) Parasite pathogenicity and depression of host population equilibria. Nature 279: 150–152

Anderson RM (1982) Epidemiology. In: Cox FEG (ed) Modern Parasitology. Blackwell Scientific Publications, Oxford, pp. 204–251

Anderson RM, May RM (1978) Regulation and stability of host-parasite population interactions. I. Regulatory processes. J Anim Ecol 47: 219–247

Anderson RM, Turner BD, Taylor LR (1979) Population Dynamics. Blackwell Scientific Publications, Oxford

Andis MD, Meek CL (1983) Estimated mortalities of the immature stages of rice-land mosquitoes in Louisiana. Proc Annu Mtg Texas Mosq Control Assoc 27: 13

Andis MD, Meek CL (1984) Survival of *Psorophora columbiae* larvae in Louisiana rice fields. Proc Annu Mtg Texas Mosq Control Assoc 28: 12–13

Andis MD, Meek CL (1985) Mortality and survival patterns of the immature stages of *Psorophora columbiae*. J Am Mosq Control Assoc 1: 357–362

Andrewartha HG, Birch LC (1954) The Distribution and Abundance of Animals. Chicago University Press, Chicago

Apiwathnasorn C, Sucharit S, Rongsriyam Y, Thongrungkiat S, Deesin T, Punavuthi N (1990) Survival of immature *Culex tritaeniorhynchus* in paddy fields. Mosq-Borne Dis Bull 7: 11–16

Armbruster P, Hutchinson RA (2002) Pupal mass and wing length as indicators of fecundity in *Aedes albopictus* and *Aedes geniculatus* (Diptera: Culicidae). J Med Entomol 39: 699–704

Arrivillaga J, Barrera R (2004) Food as a limiting factor for *Aedes aegypti* in water-storage containers. J Vector Ecol 29: 11–20

Atkinson PR (1977) Preliminary analysis of a field population of citrus red scale, *Aonidiella auranti* (Maske 11), and the measurement and expression of stage duration and reproduction for life tables. Bull Entomol Res 67: 65–87

Awono-Ambéné HP, Robert V (1999) Survival and emergence of immature *Anopheles arabiensis* mosquitoes in market-gardener wells in Dakar, Senegal. Parasite 6: 179–184

Bailey DL, Choate AL, Lawman JP (1986) A rapid radioimmunoassay for the detection of *Mansonia* antigen (Diptera: Culicidae): its potential use as a sensitive method for studying predator-prey relationships. Bull Entomol Res 76: 141–150

Barr AR (1985) Population regulation of immature *Culiseta incidens*. In: Lounibos LP, Rey JR, Frank JH (eds) Ecology of Mosquitoes: Proceedings of a Workshop. Florida Medical Entomology Laboratory, Vero Beach, Florida, pp. 147–154

Bar-Zeev M (1957) The effect of extreme temperatures on different stages of *Aedes aegypti* (L). Bull Entomol Res 48: 593–599

Bay EC, Self LS (1972) Observations on the guppy, *Poecilia reticulata* Peters, in *Culex pipiens fatigans* breeding sites in Bangkok, Rangoon, and Taipei. Bull World Health Organ 46: 407–416

Bayoh MN, Lindsay SW (2004) Temperature-related duration of aquatic stages of the Afrotropical malaria vector mosquito *Anopheles gambiae* in the laboratory. Med Vet Entomol 18: 174–179

Beaver RA (1983) The communities living in *Nepenthes* pitcher plants: fauna and food webs. In: Frank JH, Lounibos LP (eds) Phytotelmata: Terrestrial Plants as Hosts for Aquatic Insect Communities. Plexus Publishing Inc., Medford, New Jersey, pp. 129–159

Beddington JR, Free CA, Lawton JH (1978) Characteristics of successful natural enemies in models of biological control of insect pests. Nature 273: 513–519

Begon M, Harper JL, Townsend CR (1986) Ecology, Individuals, Populations and Communities. Blackwell Scientific Publications, Oxford

Bellows TS (1981) The descriptive properties of some models for density dependence. J Anim Ecol 50: 139–156

Bellows TS, Birley MH (1981) Estimating developmental and mortality rates and stage recruitment from insect stage-frequency data. Researches in Population Ecology 23: 232–244

Bellows TS, Ortiz M, Owens JC, Huddleston EW (1982) A model for analyzing insect stage-frequency data where mortality varies with time. Researches in Population Ecology 24: 142–156

Berlow EL, Navarette SA, Briggs CL, Power ME, Menge BA (1999) Quantifying variation in the strengths of species interactions. Ecology 80: 2206–2224

Beverton RJH, Holt SJ (1957) On the Dynamics of Exploited Fish Populations. Fishery Investigations, ser. 2, 19, Ministry of Agriculture Fisheries and Food, London, HMSO

Birch LC (1948) The intrinsic rate of natural increase of an insect population. J Anim Ecol 17: 15–26

Birley MH (1977) The estimation of insect density and instar survivorship functions from census data. J Anim Ecol 46: 497–510

Birley MH (1979) The estimation and simulation of variable developmental period, with application to the mosquito *Aedes aegypti* (L.). Researches in Population Ecology 21: 68–80

Black WC, Rai KS, Turco BJ, Arroyo DC (1989) Laboratory study of competition between United States strains of *Aedes albopictus* and *Aedes aegypti* (Diptera: Culicidae). J Med Entomol 26: 260–271

Blackmore, MS, Scoles, GA, Craig, GB (1995) Parasitism of *Aedes aegypti* and *Ae. albopictus* (Diptera: Culicidae) by *Ascogregarina* spp. (Apicomplexa: Lecudinidae) in Florida. J Med Entomol 32: 847–852

Blaustein L (1998) Influence of the predatory backswimmer, *Notonecta maculata*, on invertebrate community structure. Ecol Entomol 23: 246–252

Blaustein L, Kotler BP (1993) Oviposition and habitat selection by the mosquito, *Culiseta longiareolata*: effects of conspecifics, food and green toad tadpoles. Ecol Entomol 18: 104–108

Blaustein L, Kotler BP, Ward D (1995) Direct and indirect effects of a predatory backswimmer (*Notonecta maculata*) on community structure of desert temporary pools. Ecol Entomol 20: 311–318

Boreham PFL, Ohiagu CE (1978) The use of serology in evaluating invertebrate prey-predator relationships; a review. Bull Entomol Res 68: 171–194

Bown DN, Bang YH (1980). Ecological studies on *Aedes simpsoni* (Diptera: Culicidae) in southeastern Nigeria. J Med Entomol 17: 367–374

Bradshaw WE (1980) Blood-feeding and capacity for increase in the pitcher-plant mosquito, *Wyeomyia smithii*. Environ Entomol 9: 86–89

Bradshaw WE (1983) Interaction between the mosquito *Wyeomyia smithii*, the midge *Metriocnemus knabi*, and their carnivorous host *Sarracenia purpurea*. In: Frank JH, Lounibos LP (eds) Phytotelmata: Terrestrial Plants as Hosts for

Aquatic Insect Communities. Plexus Publishing Inc., Medford, New Jersey, pp. 161–189

Bradshaw WE, Holzapfel CM (1983) Predator-mediated, non-equilibrium coexistence of tree-hole mosquitoes in southeastern North America. Oecologia (Berl.) 57: 239–256

Bradshaw WE, Holzapfel CM (1984) Seasonal development of tree-hole mosquitoes (Diptera: Culicidae) and Chaoboridae in relation to weather and predation. J Med Entomol 21: 366–378

Bradshaw WE, Holzapfel CM (1985) The distribution and abundance of treehole mosquitoes in eastern North America: perspectives from north Florida. In: Lounibos LP, Rey JR, Frank JH (eds) Ecology of Mosquitoes: Proceedings of a Workshop. Florida Medical Entomology Laboratory, Vero Beach, Florida, pp. 3–23

Bradshaw WE, Holzapfel CM (1986a) Habitat segregation among European tree-hole mosquitoes. Natl Geogr Res 2: 167–178

Bradshaw WE, Holzapfel CM (1986b) Geography of density-dependent selection in pitcher-plant mosquitoes. In: Taylor F, Karban R (eds) The Evolution of Insect Life Cycles. Springer-Verlag, New York, pp. 48–65

Bradshaw WE, Holzapfel CM (1988) Drought and the organization of tree-hole mosquito communities. Oecologia (Berl.) 74: 507–514

Bradshaw WE, Holzapfel CM (1989) Life-historical consequences of density-dependent selection in the pitcher-plant mosquito, *Wyeomyia smithii*. Am Nat 133: 869–887

Bradshaw WE, Holzapfel CM (1991) Fitness and habitat segregation of British tree-hole mosquitoes. Ecol Entomol 16: 133–144

Briegel H, Timmermann SE (2001) *Aedes albopictus* (Diptera: Culicidae): physiological aspects of development and reproduction. J Med Entomol 38: 566–571

Broadie KS, Bradshaw WE (1991) Mechanisms of interference competition in the western tree-hole mosquito *Aedes sierrensis*. Ecol Entomol 16: 145–154

Brooke MM, Proske HO (1946) Precipitin test for determining natural predators of immature mosquitoes. J Natn Malar Soc 5: 45–56

Brown D, Alexander NDE, Marrs RW, Albon S (1993). Structured accounting of variance of demographic change. J Anim Ecol 62: 490–502

Campos RE, Lounibos LP (2000) Life tables of *Toxorhynchites rutilus* (Diptera: Culicidae) in nature in southern Florida. J Med Entomol 37: 385–392

Campos RE, Sy VE (2003) Mortality in immatures of the floodwater mosquito *Ochlerotatus albifasciatus* (Diptera: Culicidae) and effects of parasitism by *Strelkovimermis spiculatus* (Nematoda: Mermithidae) in Buenos Aires Province, Argentina. Mem Inst Oswaldo Cruz 98: 199–208

Campos RE, Fernández LA, Sy VE (2004) Study of the insects associated with the floodwater mosquito *Ochlerotatus albifasciatus* (Diptera: Culicidae) and their possible predators in Buenos Aires Province, Argentina. Hydrobiologia 524: 91–102

Canyon DV, Hii JLK (1997) The gecko: an environmentally friendly biological agent for mosquito control. Med Vet Entomol 11: 319–323

Casanova C, do Prado AP (2002) Key-factor analysis of immature stages of *Aedes scapularis* (Diptera: Culicidae) populations in southeastern Brazil. Bull Entomol Res 92: 271–277

Chambers RC (1985) Competition and predation among larvae of three species of treehole-breeding mosquitoes. In: Lounibos LP, Rey JR, Frank JH (eds) Ecology of Mosquitoes: Proceedings of a Workshop. Florida Medical Entomology Laboratory, Vero Beach, Florida, pp. 25–53

Chan KL (1971) Life table studies of *Aedes albopictus* (Skuse), IAEA-SM-138/19; pp. 131–44, In Sterility Principles for Insect Control or Eradication, IAEA, Vienna, STI/PUB/265

Chan KL, Chan YC, Ho BC (1971) *Aedes aegypti* (L.) and *Aedes albopictus* (Skuse) in Singapore city. 4. Competition between species. Bull World Health Organ 44: 643–649

Chapman DG, Gallucci VF (eds) (1981). Quantitative population dynamics. Stat Ecol Ser 13

Chapman HC (ed) (1985) Biological control of mosquitoes. Am Mosq Control Assoc Bull No. 6. Fresno, California

Chiang CL (1968) Introduction to Stochastic Processes in Biostatistics. John Wiley, New York

Christie M (1958). Predation on larvae of *Anopheles gambiae* Giles. J Trop Med Hyg 61: 168–176

Chubachi R (1979) An analysis of the generation-mean life table of the mosquito, *Culex tritaeniorhynchus summorosus* with particular reference to population regulation. J Anim Ecol 48: 681–702

Clark LR, Geier PW, Hughes RD, Morris RF (1967) The Ecology of Populations in Theory and Practice. Methuen, London

Coetzee M (2004). Distribution of the African malaria vectors of the *Anopheles gambiae* complex. Am J Trop Med Hyg 70: 103–104

Connell JH, Mertz DB, Murdoch WW (1970) Readings in Ecology and Ecological Genetics. Harper & Row, New York

Costanzo KS, Mormann K, Juliano SA (2005) Asymmetrical competition and patterns of abundance of *Aedes albopictus* and *Culex pipiens* (Diptera: Culicidae). J Med Entomol 42: 559–570

Couch JN, Bland CE (eds) (1985) The Genus *Ceolomomyces*. Academic Press, Orlando

Crook NE, Sunderland KD (1984) Detection of aphid remains in predatory insects and spiders by ELISA. Ann Appl Biol 105: 413–422

Crovello TJ, Hacker CS (1972) Evolutionary strategies in life table characteristics among feral and urban strains of *Aedes aegypti* (L.) Evolution 26: 185–196

Crowle AJ (1958) A simplified micro double-diffusion agar precipitin technique J Lab Clin Med 52: 784–787

Cuéllar CB (1969*a*) A theoretical model of the dynamics of an *Anopheles gambiae* population under challenge with eggs giving rise to sterile males. Bull World Health Organ 40: 205–212

Cuéllar CB (1969*b*) The critical level of interference in species eradication of mosquitoes. Bull World Health Organ 40: 213–219

Dabrowska-Prot E (1966) Experimental studies on the reduction of the abundance of mosquitoes by spiders. II. Activity of mosquitoes in cages. Bull Acad Pol Sci Cl II Sér Sci Biol 14: 771–775

Dabrowska-Prot E, Łuczak J, Tarwid K (1966) Experimental studies on the reduction of the abundance of mosquitoes by spiders. III. Indices of prey reduction and some controlling factors. Bull Acad Pol Sci Cl II Sér Sci Biol 14: 777–782

Davies RW (1969) The production of antisera for detecting specific triclad antigens in the gut contents of predators. Oikos 20: 248–260

DeBach P (1966) The competitive displacement and coexistence principles. Annu Rev Entomol 11: 183–212

De Barjac A, Sutherland DJ (eds) (1990) Bacterial Control of Mosquitoes and Black Flies. Biochemistry, Genetics and Applications of *Bacillus thuringienesis israelensis* and *Bacillus sphaericus*. Unwin Hyman, London

de Carvalho SCG, Martins Júnior A de J, Lima JBP, Valle D (2002) Temperature influence on embryonic development of *Anopheles albitarsis* and *Anopheles aquasalis*. Mem Inst Oswaldo Cruz 97: 1117–1120

Deevey ES (1947) Life tables for natural populations of animals. Quart Rev Biol 22: 283–314. Also reprinted in Hazen (1970)

Dempster JP (1958) A study of the predators of the broom beetle (*Phytodecta olivacea* Forster) using the precipitin test. Proc R Entomol Soc Lond (C) 23: 34

Dempster JP (1960) A quantitative study of the predators on the eggs and larvae of the broom beetle, *Phytodecta olivacea* Forster, using the precipitin test. J Anim Ecol 29: 149–167

Dempster JP (1961) The analysis of data obtained by regular sampling of an insect population. J Anim Ecol 30: 429–432

Dempster JP (1963) The natural prey of three species of *Anthocoris* (Heteroptera: Anthocoridae) living on broom (*Sarothamnus scoparius* L.) Entomol Exp Appl 6: 149–155

Dempster JP (1967) The control of *Pieris rapae* with D.D.T. 1. The natural mortality of the young stages of *Pieris*. J Appl Ecol 4: 485–500

Dempster JP (1983) The natural control of populations of butterflies and moths. Biol Rev 58: 461–481

Dempster JP, Pollard E (1986) Spatial heterogeneity, stochasticity and the detection of density dependence in animal populations. Oikos 46: 413–416

Dempster JP, Richards OW, Waloff N (1959) Carabidae as predators on the pupal stage of the chrysomelid beetle, *Phytodecta olivacea* (Forster). Oikos 10: 65–70

Dennehy JJ, Robakiewicz P, Livdahl T (2001) Larval rearing conditions affect kin-mediated cannibalism in a treehole mosquito. Oikos 95: 335–339

Derr JA, Ord K (1979) Field estimates of insect colonization. J Anim Ecol 48: 521–534

Dieng H, Mwandawiro C, Boots M, Morales R, Satho T, Tuno N, Tsuda Y, Takagi M (2002) Leaf litter decay process and the growth performance of *Aedes albopictus* larvae (Diptera: Culicidae). J Vector Ecol 27: 31–38

Dixon RD, Brust RA (1971) Predation of mosquito larvae by the Fathead Minnow. Manitoba Entomol 5: 68–70

Doane JF, Scotti PD, Sutherland ORW, Pottinger RP (1985) Serological identification of wireworm and staphylinid predators of the Australian soldier fly (*Inopus rubriceps*) and wireworm feeding on plant and animal foods. Entomol Exp Appl 38: 65–72

Dubitskij AM (1978) Biological Methods of Control of Blood Sucking Insects in the USSR. Alma-Ata (In Russian)

Duhrkopf RE, Benny H (1990) Differences in the larval alarm reactions in populations of *Aedes aegypti* and *Aedes albopictus*. J Am Mosq Control Assoc 6: 411–414

Duhrkopf RE, Young SS (1979) Some consequences of selection for fast and slow recovery from the larval alarm reaction of *Aedes aegypti*. Theor Appl Genet 55: 263–268

Dye C (1982) Intraspecific competition amongst larval *Aedes aegypti*: food exploitation or chemical interference? Ecol Entomol 7: 39–46

Dye C (1984a) Models for the population dynamics of the yellow fever mosquito, *Aedes aegypti*. J Anim Ecol 53: 247–268

Dye C (1984b) Competition amongst larval *Aedes aegypti*: the role of interference. Ecol Entomol 9: 355–357

Edillo FE, Touré YT, Lanzaro GC, Dolo G, Taylor CE (2004) Survivorship and distribution of immature *Anopheles gambiae* s.l. (Diptera: Culicidae) in Banambani village, Mali. J Med Entomol 41: 333–339

Edgerly JS, Willey MS, Livdahl T (1999) Intraguild predation among larval tree-hole mosquitoes, *Aedes albopictus, Ae. aegypti*, and *Ae. triseriatus* (Diptera: Culicidae), in laboratory microcosms. J Med Entomol 36: 394–399

Enfield MA, Pritchard G (1977) Methods for sampling immature stages of *Aedes* spp. (Diptera: Culicidae) in temporary ponds. Can Entomol 109: 1435–1444

Fava FD, Ludueña Almeida FF, Almirón WR, Brewer M (2001) Winter biology of *Aedes albifasciatus* (Diptera: Culicidae) from Córdoba, Argentina. J Med Entomol 38: 253–259

Fish D, Carpenter SR (1982) Leaf litter and larval mosquito dynamics in tree-hole ecosystems. Ecology 63: 283–288

Fisher IJ, Bradshaw WE, Kammeyer C (1990) Fitness and its correlates assessed by intra- and interspecific interactions among tree-hole mosquitoes. J Anim Ecol 59: 819–829

Fitcher BL, Stephen WP (1984) Time-related decay of prey antigens by arboreal spiders as detected by ELISA. Environ Entomol 13: 1583–1587

Focks DA, Sackett SR (1985) Some factors affecting interaction of *Toxorhynchites amboiensis* with *Aedes* and *Culex* in an urban environment. In: Lounibos LP, Rey JR, Frank JH (eds) Ecology of Mosquitoes: Proceedings of a Workshop. Florida Medical Laboratory, Vero Beach, pp. 55–64

Fox CJS, MacLellan CR (1956) Some Carabidae and Staphylinidae shown to feed on a wireworm, *Agnotes sputator* (L.) by the precipitin test. Can Entomol 88: 228–231

Frank JH (1967) A serological method used in the investigation of the predators of the pupal stage of the winter moth, *Operophtera brumata* (L.) (Hydriomenidae). Quaest Entomol 3: 95–105

Frank JH (1983) Bromeliad phytotelmata and their biota, especially mosquitoes. In: Frank JH, Lounibos LP (eds) Phytotelmata: Terrestrial Plants as Hosts for Aquatic Insect Communities. Plexus Publishing Inc., Medford, New Jersey, pp. 101–128

Frank JH, Curtis GA (1977a) On the bionomics of bromeliad-inhabiting mosquitoes. IV. Egg mortality of *Wyeomyia vanduzeei* caused by rainfall. Mosquito News 37: 239–245

Frank JH, Curtis GA (1977b) On the bionomics of bromeliad-inhabiting mosquitoes. III. The probable strategy of larval feeding in *Wyeomyia vanduzeei* and *Wyeomyia medioalbipes*. Mosquito News 37: 200–206

Fuxa JR (1987) Ecological considerations for the use of entomopathogens in IPM. Annu Rev Entomol 32: 225–251

Gauch HG (1982) Multivariate Analysis in Community Ecology. Cambridge University Press, Cambridge.

Giller PS (1982) The natural diets of waterbugs (Hemiptera-Heteroptera): electrophoresis as a potential method of analysis. Ecol Entomol 7: 233–237

Giller PS (1984) Predator gut state and prey detectability using electrophoretic analysis of gut contents. Ecol Entomol 9: 157–162

Giller PS (1986) The natural diet of the Notonectidae: field trials using electrophoresis. Ecol Entomol 11: 163–172

Gillies MT, Gubbins SJ (1982) *Culex (Culex) torrentium* Martini and *Cx. (Cx.) pipiens* L. in a southern English county, 1974–1975. Mosq Syst 14: 127–130

Gillies MT, Smith A (1960) The effect of a residual house-spraying campaign in East Africa on species balance in the *Anopheles funestus* group. The replacement of *A. funestus* Giles by *A. rivulorum* Leeson. Bull Entomol Res 51: 243–252

Gilotra SK, Rozeboom LE, Bhattacharya NC (1967) Observations on possible competitive displacement between populations of *Aedes aegypti* Linnaeus and *Aedes albopictus* Skuse in Calcutta. Bull World Health Organ 37: 437–446

Gilpin ME, Langford RP (1978) Evidence for density dependent growth regulation among larval *Aedes sierrensis* mediated by food competition. Proc California Mosq Vector Control Assoc 46: 42–45

Gilpin ME, McClelland GAH (1979) System analysis of the yellow fever mosquito *Aedes aegypti*. Fortschr Zool 25: 355–388

Gilpin ME, McClelland GAH, Pearson JW (1976) Space, time and stability of laboratory mosquito populations. Am Nat 110: 1107–1111

Gimnig JE, Ombok M, Otieno S, Kaufman MG, Vulule JM, Walker ED (2002) Density-dependent development of *Anopheles gambiae* (Diptera: Culicidae) larvae in artificial habitats. J Med Entomol 39: 162–172

Goettel MS (1987) Field incidence of mosquito pathogens and parasites in central Alberta. J Am Mosq Control Assoc 3: 231–238

Gomes A de C, Gotlieb SLD, Marques CC de A, de Paula MB, Marques GRAM (1995) Duration of larval and pupal development stages of *Aedes albopictus* in natural and artificial containers. Rev Saúde Pública 29: 15–19

Gomez C, Rabinovich JE, Machado-Allison CE (1977) Population analysis of *Culex pipiens fatigans* Wied. (Diptera: Culicidae) under laboratory conditions. J Med Entomol 13: 453–463

Greenstone MH (1977) A passive haemagglutination inhibition assay for the identification of stomach contents of invertebrate predators. J Appl Ecol 14: 457–464

Gubler DJ (1970*a*) Comparison of reproductive potentials of *Aedes (Stegomyia) albopictus* Skuse and *Aedes (Stegomyia) polynesiensis* Marks. Mosquito News 30: 201–209

Gubler DJ (1970*b*) Competitive displacement of *Aedes (Stegomyia) polynesiensis* Marks by *Aedes (Stegomyia) albopictus* Skuse in laboratory populations. J Med Entomol 7: 229–235

Gubler DJ (1971) Studies on the comparative oviposition behaviour of *Aedes (Stegomyia) albopictus* and *Aedes (Stegomyia) polynesiensis* Marks. J Med Entomol 8: 675–682

Haldane JBS (1949) Disease and evolution. In: Symposium sui fattori ecologici e genetici della speciazone negli animali. Ric Sci 19 (suppl.) 3–11

Harcourt DG, Yee JM (1982) Polynomial algorithm for predicting the duration of insect life stages. Environ Entomol 11: 581–584

Hard JJ, Bradshaw WE, Malarkey DJ (1989) Resource- and density-dependent development in tree-hole mosquitoes. Oikos 54: 137–144

Hassell MP (1971) Mutual interference between searching insect parasites. J Anim Ecol 40: 473–486

Hassell MP (1975) Density-dependence in single-species populations. J Anim Ecol 44: 283–295

Hassell MP (1985) Insect natural enemies as regulating factors. J Anim Ecol 54: 323–334

Hassell MP (1987) Detecting regulation in patchily distributed animal populations. J Anim Ecol 56: 705–713

Hassell MP, Huffaker CB (1969) The appraisal of delayed and direct density-dependence. Can Entomol 101: 353–361

Hassell MP, May RM (1973) Stability in insect host-parasitic models. J Anim Ecol 42: 693–726

Hassell MP, May RM (1974) Aggregation of predators and insect parasites and its effect on stability. J Anim Ecol 43: 567–587

Hassell MP, Rogers DJ (1972) Insect parasite responses in the development of population models. J Anim Ecol 41: 661–676

Hassell MP, Varley GC (1969) New inductive population model for insect parasites and its bearing on biological control. Nature 223: 1133–1137

Hassell MP, Lawton JH, May RM (1976) Patterns of dynamical behaviour in single-species populations. J Anim Ecol 45: 471–486

Hassell MP, Latto J, May RM (1989) Seeing the wood from the trees: detecting density dependence from existing life-table studies. J Anim Ecol 58: 883–892

Hawley WA (1985*a*) The effect of larval density on adult longevity of a mosquito, *Aedes sierrensis*: epidemiological consequences. J Anim Ecol 54: 955–964

Hawley WA (1985*b*) Population dynamics of *Aedes sierrensis*. In: Lounibos LP, Rey JR, Frank JH (eds) Ecology of Mosquitoes: Proceedings of a Workshop. Florida Medical Entomology Laboratory, Vero Beach, Florida, pp. 167–184

Hawley WA (1988) The biology of *Aedes albopictus*. J Am Mosq Control Assoc 4(suppl.): 1–39

Hayes J, Hsi BP (1975) Interrelationships between selected meteorologic phenomena and immature stages of *Culex pipiens quinquefasciatus* Say: study of an isolated population. J Med Entomol 12: 299–308

Hazen W (ed) (1970) Readings in Population and Community Ecology, 6th edn. WB Saunders, Philadelphia

Healey JA, Cross TF (1975) Immunoelectroosmorphoresis for serological identification of predators of the sheep-tick *Ixodes ricinus*. Oikos 26: 97–101

Hill AB (1971) Principles of Medical Statistics. Lancet Ltd, London

Ho BC, Ewert A, Chew L-M (1989) Interspecific competition among *Aedes aegypti, Ae. albopictus*, and *Ae. triseriatus* (Diptera: Culicidae): larval development in mixed cultures. J Med Entomol 26: 615–623

Hokkanen HMT, Pimentel D (1989) New associations in biological control: theory and practice. Can Entomol 121: 829–840

Holling CS (1959) The components of predation as revealed by a study of small mammal predation of the European pine sawfly. Can Entomol 91: 385–398

Huffaker CB (1944) The temperature relations of the immature stages of the malarial mosquito *An. quadrimaculatus* Say, with a comparison of the developmental power of constant and variable temperatures in insect metabolism. Ann Entomol Soc Am 37: 1–27

Huffaker CB, Kennett CE (1969) Some aspects of assessing efficiency of natural enemies. Can Entomol 101: 425–447

Huffaker CB, Rabb RL (eds) (1984) Ecological Entomology. John Wiley, New York

Hughes RD (1962) A method for estimating the effects of mortality on aphid populations. J Anim Ecol 31: 389–396

Hughes RD (1963) Population dynamics of the cabbage aphid, *Brevicoryne brassicae* (L.). J Anim Ecol 32: 393–424

Hurlbert SH (1984) Pseudoreplication and the design of ecological field experiments. Ecol Mongr 54: 187–211

Hwang Y-S, Mulla MS (1975) Overcrowding factors of mosquito larvae. Their potential for mosquito control. Proc California Mosq Control Assoc 43: 73–74

Hwang Y-S, Mulla MS, Majori G (1976) Overcrowding factors of mosquito larvae. VIII. Structure-activity relationship of methyl 2-alkylalkanoates against mosquito larvae. Agricultural Food Chemistry 24: 649–651

Ikeshoji T (1977) Self-limiting economes in the populations of insects and some aquatic animals. J Pest Sci 2: 77–89

Ikeshoji T (1978) Lipids self-limiting the populations of mosquito larvae. In: Symposium on the Pharmacological Effects of Lipids, AOCS Monograph No 5, pp. 113–121

Ikeshoji T, Ichimoto I, Ono T, Naoshima Y, Ueda H (1977) Overcrowding factors of mosquito larvae. X. Structure-bioactivity relationship and bacterial activation of the alkyl-branched hydrocarbons. Appl Entomol Zool 12: 265–273

Irving-Bell RJ, Okoli EI, Diyelong DY, Lyimo EO, Onyia OC (1987). Septic tank mosquitoes: competition between species in central Nigeria. Med Vet Entomol 1: 243–250

Istock CA, Wasserman SS, Zimmer H (1975) Ecology and evolution of the pitcher-plant mosquito: 1. Population dynamics and laboratory responses to food and population density. Evolution 29: 296–312

Istock CA, Zisfeln J, Vavra KJ (1976a) Ecology and evolution of the pitcher-plant mosquito. 2. The substructure of fitness. Evolution 30: 535–547

Istock CA, Vavra KJ, Zimmer H (1976b) Ecology and evolution of the pitcher-plant mosquito: 3. Resources tracking by a natural population. Evolution 30: 548–557

Itô Y (1961) Factors that effect the fluctuations of animal numbers, with special reference to insect outbreaks. Bull Natl Inst Agr Sci C13: 57–89

Itô Y (1972) On the methods for determining density-dependence by means of regression. Oecologia (Berl.) 10: 347–372

Jackson N (1953) Observations on the feeding habits of a predaceous mosquito larva, *Culex (Lutzia) tigripes* Grandpré and Charmoy (Diptera). Proc R Entomol Soc Lond (A) 28: 153–159

James HG (1961) Some predators of *Aedes stimulans* (Walk.) and *Aedes trichurus* (Dyar) (Dipt.: Culicidae) in woodland pools. Can J Zool 39: 533–540

James HG (1964) Insect and other fauna associated with the rockpool mosquito *Aedes altropalpus* (Coq.). Mosquito News 24: 325–329

James HG (1966) Insect predators of univoltine mosquitoes in woodland pools of the pre-Cambrian Shield in Ontario. Can Entomol 98: 550–555

Jenkins DW (1964) Pathogens, parasites and predators of medically important arthropods. Annotated list and bibliography. Bull World Health Organ 30 (Suppl.): 5–150

Joshi DS (1996) Effect of fluctuating and constant temperatures on development, adult longevity and fecundity in the mosquito *Aedes krombeini*. J Thermal Biol, 21: 151–154

Juliano SA (1998) Species introduction and replacement among mosquitoes: interspecific resource competition or apparent competition? Ecology 79: 255–268

Juliano SA, O'Meara GF, Morrill JR, Cutwa MM (2002) Desiccation and thermal tolerance of eggs and the coexistence of competing mosquitoes. Oecologia (Berl.) 130: 458–469

Kapuge SH, Danthanarayana W, Hoogenraad N (1987) Immunological investigation of prey-predator relationships for *Pieris rapae* (L.) (Lepidoptera: Pieridae). Bull Entomol Res 77: 247–254

Kaur R, Reuben R (1981) Studies of density and natural survival of immatures of *Anopheles stephensi* Liston in wells in Salem (Tamil Nadu). Indian J Med Res 73 (Suppl.): 129–135

Kellett FRS, Omardeen TA (1957) Tree hole breeding of *Aedes aegypti* (Linn) in Arima, Trinidad, B.W.I. W. Indian Med J 6: 179–188

Kiflawi M, Blaustein L, Mangel M (2003*a*) Predation-dependent oviposition habitat selection by the mosquito *Culiseta longiareolata*: a test of competing hypotheses. Ecol Letters 6: 35–40

Kiflawi M, Blaustein L, Mangel M (2003*b*) Oviposition habitat selection by the mosquito *Culiseta longiareolata* in response to risk of predation and conspecific larval density. Ecol Entomol 28: 168–173

Kiritani K, Nakasuji F (1967) Estimation of the stage-specific survival rate in the insect population with overlapping stages. Researches in Population Ecology 9: 143–152

Knight TM, Chase JM, Goss CW, Knight JJ (2004) Effects of interspecific competition, predation, and their interaction on survival and development time of immature *Anopheles quadrimaculatus*. J Vector Ecol 29: 277–284

Kobayashi S (1968) Estimation of the individual number entering each developmental stage in an insect population. Researches in Population Ecology 10: 40–44

Koenraadt CJM, Takken W (2003) Cannibalism and predation among larvae of the *Anopheles gambiae* complex. Med Vet Entomol 17: 61–66

Koenraadt CJM, Majambere S, Hemerik L, Takken W (2004) The effects of food and space on the occurrence of cannibalism and predation among larvae of *Anopheles gambiae* s.l.. Entomol Exp Appl 112: 125–134

Kuno E (1971) Sampling error as a misleading artifact in 'key factor analysis'. Researches in Population Ecology 13: 28–45

Kuno E (1973) Statistical characteristics of the density-dependent population fluctuations and the evaluation of density-dependence and regulation in animal populations. Researches in Population Ecology 15: 99–120

Kuno E (1991) Sampling and analysis of insect populations. Annu Rev Entomol 36: 285–304

Lacey LA, Lacey CM (1990) The medical importance of riceland mosquitoes and their control using alternatives to chemical insecticides. J Am Mosq Control Assoc 6 (Suppl. 2) 1–93

Laird M (ed) (1981) Biocontrol of Medical and Veterinary Pests. Praeger Publishers, New York

Laird M, Miles JW (eds) (1983) Integrated Mosquito Control Methodologies. Volume 1. Experience and Components from Conventional Chemical Control. Academic Press, London

Laird M, Miles JW (eds) (1985) Integrated Mosquito Control Methodologies. Volume 2. Biocontrol and Other Innovative Components, and Future Directions. Academic Press, London

Lakhani KH, Service MW (1974) Estimated mortalities of the immature stages of *Aedes cantans* (Meigen) (Dipt., Culicidae) in a natural habitat. Bull Entomol Res 64: 265–276

Landry SV, DeFoliart GR, Hogg DB (1988) Adult body size and survivorship in a field population of *Aedes triseriatus*. J Am Mosq Control Assoc 4: 121–128

Lang JD (2003) Factors affecting immatures of *Ochlerotatus taeniorhynchus* (Diptera: Culicidae) in San Diego County, California. J Med Entomol 40: 387–394

Lansdowne C, Hacker CS (1975) The effect of fluctuating temperature and humidity on the adult life table characteristics of five strains of *Aedes aegypti*. J Med Entomol 11: 723–733

Laughlin R (1965) Capacity for increase: a useful population statistic. J Anim Ecol 34: 77–91

Lee FC (1967) Laboratory observations on certain mosquito larval predators. Mosquito News 27: 332–338

Lester PJ, Pike AJ (2003) Container surface area and water depth influence the population dynamics of the mosquito *Culex pervigilans* (Diptera: Culicidae) and its associated predators in New Zealand. J Vector Ecol 28: 267–274

Lister A (1984) Predation in an Antarctic micro-arthropod community. Acarol 6: 886–892

Lister A, Block W, Usher MB (1988) Arthropod predation in an Antarctic terrestrial community. J Anim Ecol 57: 957–971

Liu Z, Zhang Y, Yang Y (1985) Population dynamics of *Aedes (Stegomyia) albopictus* (Skuse) under laboratory conditions. Acta Entomol Sin 28: 270–280 (In Chinese, English summary)

Livdahl TP, Sugihara G (1984) Non-linear interactions of populations and the importance of estimating per capita rates of change. J Anim Ecol 53: 573–580

Lloyd M (1967) Mean crowding. J Anim Ecol 36: 1–30

Lord CC (1998) Density dependence in larval *Aedes albopictus* (Diptera: Culicidae). J Med Entomol 35: 825–829

Loughton BG, West AS (1962) Serological assessment of spider predation on the spruce budworm, *Choristoneura fumiferana* (Clem.) (Lepidoptera: Tortricidae). Proc Entomol Soc Ontario 92: 176–180

Loughton BG, Derry C, West AS (1963) Spiders and the spruce budworm. Mem Entomol Soc Canada 31: 249–268

Lounibos LP (1979) Mosquitoes occurring in the axils of *Pandanus rabaiensis* Rendle on the Kenya coast. Cah ORSTOM sér Entomol Méd Parasitol 17: 25–29

Lounibos LP (1981) Habitat segregation among African treehole mosquitoes. Ecol Entomol 6: 129–154

Lounibos LP (1983) The mosquito community of treeholes in subtropical Florida. In: Frank JH, Lounibos LP (eds) Phytotelmata: Terrestrial Plants as Hosts for Aquatic Insect Communities. Plexus Publishing, Medford, pp. 223–246

Lounibos LP (1985) Interactions influencing production of tree-hole mosquitoes in south Florida. In: Lounibos LP, Rey JR, Frank JH (eds) Ecology of Mosquitoes: Proceedings of a Workshop. Medical Entomology Laboratory. Vero Beach, Florida, pp. 65–77

Lounibos LP, Machado-Allison CE (1986) Mosquito maternity: Egg brooding in the life cycle of *Trichoprosopon digitatum*. In: Taylor F, Karban R (eds) The Evolution of Insect Life Cycles. Springer-Verlag, New York, pp. 173–184

Lounibos LP, Suárez S, Menéndez Z, Nishimura N, Escher RL, O'Connell SM, Rey JR (2002) Does temperature affect the outcome of larval competition between *Aedes aegypti* and *Aedes albopictus*? J Vector Ecol 27: 86–95

Lounibos LP, O'Meara GF, Nishimura N, Escher RL (2003) Interactions with native mosquito larvae regulate the production of *Aedes albopictus* from bromeliads in Florida. Ecol Entomol 28: 551–558

Lowrie RC (1973*a*) The effect of competition between larvae of *Aedes (S) albopictus* Skuse and *A. (S) polynesiensis* Marks. J Med Entomol 10: 23–30

Lowrie RC (1973*b*) Displacement of *Aedes (S) polynesiensis* Marks by *A (S) albopictus* Skuse through competition in the larval stages under laboratory conditions. J Med Entomol 10: 131–136

Lowry OL, Rosebrough NJ, Farr AL, Randall RJ (1951) Protein measurement with the folin phenol reagent. J Biol Chem 193: 265–275

Luck RF (1971) An appraisal of two methods of analysing insect life tables. Can Entomol 103: 1261–1271

Łuczak J, Dabrowska-Prot E (1966) Experimental studies on the reduction of the abundance of mosquitoes by spiders. I. Intensity of spider predation on mosquitoes. Bull Acad Pol Sci Cl II Sér Sci Biol 14: 315–320

Lundkvist E, Landin J, Jackson M, Svensson C (2003) Diving beetles (Dytiscidae) as predators of mosquito larvae (Culicidae) in field experiments and in laboratory tests of prey preference. Bull Entomol Res 93: 219–226

Lutwama JJ, Mukwaya LG (1995) Estimates of mortalities of larvae and pupae of the *Aedes simpsoni* (Theobald) (Diptera: Culicidae) complex in Uganda. Bull Entomol Res 85: 93–99

MacArthur RH, Wilson EO (1967) The Theory of Island Biogeography. Princeton University Press. Princeton, New Jersey

Macdonald WW (1956) *Aedes aegypti* in Malaya. II. Larval and adult biology. Ann Trop Med Parasitol 50: 399–414

Machado-Allison CE, Rodriquez DJ, Barrera R, Cova CG (1983) The insect community associated with inflorescences of Heliconia caribea Lamarck in Venezuela. In: Frank JH, Lounibos LP (eds) Phytotelmata: Terrestrial Plants as Hosts for Aquatic Insect Communities. Plexus Publishing, Medford, pp. 247–270

Mackauer M, Ehler LE, Roland J (1990) Critical Issues in Biological Control. Intercept, Andover

Mackey BE, Hoy JB (1978) *Culex tarsalis*: Sequential sampling as a means of estimating populations in Californian rice field. J Econ Entomol 71: 329–334

Maelzer DA (1970) The regression of log Nn+1 on log Nn as a test of density dependence: an exercise with computer-constructed density-independent populations. Ecology 51: 810–822

Maharaj R (2003) Life table characteristics of *Anopheles arabiensis* (Diptera: Culicidae) under simulated seasonal conditions. J Med Entomol 40: 737–742

Mahmood F (1997) Life-table attributes of *Anopheles albimanus* (Wiedemann) under controlled laboratory conditions. J Vector Ecol 22: 103–108

Mahmood F, Crans WJ (1998) Effect of temperature on the development of *Culiseta melanura* (Diptera: Culicidae) and its impact on the amplification of eastern equine encephalomyelitis virus in birds. J Med Entomol 35: 1007–1012

Manly, BF J (1974a) A comparison of methods for the analysis of insect stage-frequency data. Oecologia (Berl.), 17: 335–48

Manly, BFJ (1974b) Estimation of stage-specific survival rates and other parameters for insect populations developing through several stages. Oecologia (Berl.), 15: 277–85

Manly, BF J (1976). Extensions to Kiritani and Nakasuji's method for analysing insect stage-frequency data. Res. Popul. Ecol., 17: 191–9

Manly, BFJ (1977a) A further note on Kiritani and Nakasuji's model for stage-frequency data including comments on the use of Tukey's jackknife techniques for estimating variances. Res. Popul. Ecol., 18: 177–86

Manly, BFJ (1977b) The determination of key factors from life table data. Oecologia (Berl.), 31: 111–7

Manrique-Saide, P., Ibáñez-Bernal, S., Delfín-González, H & Tabla, VP (1998) Mesocyclops longisetus effects on survivorship of Aedes aegypti immature stages in car tyres. Med Vet Entomol 12: 386–390

Marchi A, Munstermann LE (1987) The mosquitoes of Sardinia: species records 35 years after the malaria eradication campaign. Med. vet. Ent., 1: 89–96

Marten GG, Mieu Nguyen, Giai Ngo (2000) Copepod predation on Anopheles quadrimaculatus larvae in rice fields. J Vector Ecol 25: 1–6

Mattingly PF (1967) Aedes aegypti and other mosquitos in relation to the dengue syndrome. Bull World Health Organ 36: 533–55

May, RM (edit.). (1981) Theoretical Ecology: Principles and Applications (2nd edit.). Blackwell Scientific Publications, Oxford, ix + 489 pp.

May, RM (1986) The search for patterns in the balance of nature; advances and retreats. Ecology, 67: 1115–26

May, RM & Oster, GF (1976) Bifurcations and dynamic complexity in simple ecological models. Am. Nat., 110: 573–600

McDonald, G & Buchanan, GA (1981). The mosquito and predatory insect fauna inhibiting fresh-water ponds, with particular reference to Culex annulirostris Skuse (Diptera: Culicidae). Aust. J. Ecol., 6: 21–7.

McDonald, L., Manly, B., Lockwood, J & Logan, J (edit). (1989) Estimation and Analysis of Insect Populations. In Proc. Conf. Laramie, Wyoming, 25–29 January 1988. Springer-Verlag, Berlin, xiv + 492 pp.

McIver, JD (1981) An examination of the utility of the precipitin test for evaluation of arthropod predator-prey relationships. Can. Ent., 113: 213–22

Menon, PKB & Rajagopalan, PK (1979) Seasonal changes in the density and natural mortality of immature stages in the urban malaria vector, Anopheles stephensi (Liston) in wells in Pondicherry. Indian J. med. Res., 70 (Suppl.), 123–7

Menon, PKB & Sharma, VP (1981) Geographic variation in life table attributes of four populations of Anopheles stephensi Liston from India. Indian J. Malar., 18: 91–7

Mercer, DR., Wettach, GR & Smith, JL (2005) Effects of larval density and predation by Toxorhynchites amboinensis on Aedes polynesiensis (Diptera: Culicidae) developing in coconuts. Journal of the American Mosquito Control Association, 21: 425–431

Miura, T & Takahashi, RM (1988) Development and survival rates of immature stages of Culex tarsalis (Diptera: Culicidae) in central California rice fields. Proc. Calif. Mosq. & Vect. Contr. Ass., 56: 168–79

Miura, T., Takahashi, RM & Mulligan, FS (1978) Field evaluation of the effectiveness of predacious insects as a mosquito control agent. Proc. Calif. Mosq. & Vect. Contr. Ass., 46: 80–1

Miura, T., Takahashi, RM & Wilder, WH (1984) Impact of the mosquitofish (Gambusia affinis) on a rice field ecosystem when used as a mosquito control agent. Mosquito News, 44: 510–17

Mogi, M (1978) Population studies on mosquitoes in the rice field area of Nagasaki, Japan, especially on Culex tritaeniorhynchus. Trop. Med., 20: 173–263

Mogi, M (1981) Population dynamics and methodology for biocontrol of mosquitoes, pp. 140–172. In Biocontrol of Medical and Veterinary Pests, (edit. M. Laird). Praeger Scientific, New York, xx + 235 pp.

Mogi, M & Okazawa, T (1990) Factors influencing development and survival of Culex pipiens pallens larvae (Diptera: Culicidae) in polluted urban creeks. Res. Popul. Ecol., 32: 135–49

Mogi, M & Yamamura, N (1988) Population regulation of a mosquito Armigeres theobaldi with description of the animal fauna in zingiberaceous inflorescences. Res. Popul. Ecol., 30: 251–65

Mogi, M., Mori, A & Wada, Y (1980a) Survival rates of Culex tritaeniorhynchus (Diptera: Culicidae) larvae in fallow rice fields before summer cultivation. Trop. Med., 22: 47–59

Mogi, M., Mori, A & Wada, Y (1980b) Survival rates of immature stages of Culex tritaeniorhynchus (Diptera: Culicidae) in rice fields under summer cultivation. Trop. Med., 22: 111–26

Mogi, M., Miyagi, I & Cabrera, B D (1984) Development and survival of immature mosquitoes (Diptera: Culicidae) in Philippine rice fields. J Med Entomol 21: 283–91

Mogi, M., Horio, M., Miyagi, I & Cabrera, B D (1985) Succession, distribution, overcrowding and predation in the aquatic community in aroid axils, with special reference to mosquitoes, pp. 95–119. In Ecology of Mosquitoes: Proceedings of a Workshop (edit. LP Lounibos, JR Rey & JH Frank). Florida Medical Entomology Laboratory, Vero Beach, Florida, xix + 579 pp.

Mogi, M., Okazawa, T., Miyagi, I., Sucharit, S., Tumrasvin, W., Deesin, T & Khamboonruang, C. (1986). Development and survival of anopheline immatures (Diptera: Culicidae) in rice fields in northern Thailand. J Med Entomol 23: 244–50

Mokany, A & Shine, R (2002a) Pond attributes influence competitive interactions between tadpoles and mosquito larvae. Austral Ecology, 27: 396–404

Mokany, A & Shine, R (2002b) Competition between tadpoles and mosquitoes: the effects of larval density and tadpole size. Australian Journal of Zoology, 50: 549–563

Mokany, A & Shine, R (2003a) Competition between tadpoles and mosquito larvae. Oecologia, 135: 615–620

Mokany, A & Shine, R (2003b) Biological warfare in the garden pond: tadpoles suppress the growth of mosquito larvae. Ecological Entomology, 28: 102–108

Moon, TE (1976) A statistical model of the dynamics of a mosquito vector (Culex tarsalis) population. Biometrics, 32: 355–68

Moore CG, Fisher BR (1969) Competition in mosquitoes. Density and species ratio effects on growth, mortality, fecundity, and production of growth retardant. Ann. ent. Soc. Am., 62: 1325–31

Moore CG, Whitacre DM (1972) Competition in mosquitoes. 2. Production of Aedes aegypti larval growth retardant at various densities and nutrition levels. Ann. ent. Soc. Am., 65: 915–8

Moore CG, Reiter P, Eliason DA, Bailey RE, Campos EG (1990) Apparent influence of the stage of blood meal digestion on the efficacy of ground applied ULV aerosols for the control of urban Culex mosquitoes. III. Results of a computer simulation. J Am Mosq Control Assoc 6: 375–83

Morales, ME., Wesson, DM., Sutherland, IW., Impoinvil, DE., Mbogo, CM., Githure, JI & Beier, JC (2003) Determination of Anopheles gambiae larval DNA in the gut of insectivorous dragonfly (Libellulidae) nymphs by polymerase chain reaction. Journal of the American Mosquito Control Association, 19: 163–165

Mori, A (1979) Effects of larval density and nutrition on some attributes of immature and adult Aedes albopictus. Trop. Med., 21: 85–103

Mori, A & Wada, Y (1978) Seasonal abundance of Aedes albopictus in Nagasaki. Trop. Med., 20: 29–37

Morris, RF (1959) Single-factor analysis in population dynamics. Ecology, 40: 580–8.

Morris, RF (1963) Predictive population equations based on key factors. Mem. ent. Soc. Canada, 32: 16–21

Morris, RF & Royama, T (1969) Logarithmic regression as an index of responses to population density. A comment on a paper by MP Hassell and CB Huffaker. Can. Ent., 101: 361–4

Morrison, A & Andreadis, TG (1992) Larval population dynamics in a community of nearctic Aedes inhabiting a temporary vernal pool. Journal of the American Mosquito Control Association, 8: 52–57

Mottram, P & Kettle, DS (1997) Development and survival of immature Culex annulirostris mosquitoes in southeast Queensland. Med Vet Entomol 11: 181–186

Mountford, M D (1988) Population regulation, density dependence, and heterogeneity. J. Anim. Ecol., 57: 845–58

Mpho, M., Holloway, GJ & Callaghan, A (2000) Fluctuating wing asymmetry and larval density stress in Culex quinquefasciatus (Diptera: Culicidae). Bull Entomol Res 90: 279–283

Mueller, LD (1997) Theoretical and experimental examination of density-dependent selection. Ann Rev Ecol Syst, 28: 269–288

Murray, BG (1979) Population Dynamics. Academic Press, New York, ix + 212 pp.

Murray, RA & Solomon, MG (1978) A rapid technique for analysing diets of invertebrate predators by electrophoresis. Ann. appl. Biol., 90: 7–10

Nannini, MA & Juliano, SA (1997) Effects of developmental asynchrony between Aedes triseriatus (Diptera: Culicidae) and its predator Toxorhynchites rutilus (Diptera: Culicidae). J Med Entomol 34: 457–460

Oda, T., Uchida, K., Mori, A., Mine, M., Eshita, Y., Kurokawa, K., Kato, K & Tahara, H (1999) Effects of high temperature on the emergence and survival of adult Culex pipiens molestus and Culex quinquefasciatus in Japan. Journal of the American Mosquito Control Association, 15: 153–156

Ohiagu, CE & Boreham, PFL (1978) A simple field test for evaluating insect prey-predator relationships. Entomologia exp. appl., 23: 40–7

O'Meara GF, Larson VL, Mook DH, Latham MD (1989) Aedes bahamensis: its invasion of South Florida and association with Aedes aegypti. J Am Mosq Control Assoc 5: 1–5

Paine, RT (1992) Food web analysis through field measurement of per capita interaction strength. Nature, 355: 73–75

Pajot, F.-X (1975) Contribution a l'étude écologique d'Aedes (Stegomyia) simpsoni (Theobald, 1905) (Diptera: Culicidae). Études des gîtes larvaires en République Centrafricaine. Cah. ORSTOM, sér. Entomol. méd. Parasit., 13: 135–64

Pajot, F. -X (1976) Contribution a l'étude écologique d'Aedes (Stegomyia) simpsoni (Theobald, 1905) (Diptera: Culicidae). Observations concernant les stages préimaginaux. Cah. ORSTOM, sér. Entomol. méd. Parasit., 16: 129–50

Pajunen, VI (1983) The use of physiological time in the analysis of insect stage-frequency data. Oikos, 40: 161–5.

Pajunen, VI (1986) How to construct and use realistic physiological time scales: an analysis of larval mortality in rock pool corixids (Hemiptera). Oikos, 47: 239–50

Palchick, S., Schoof, DD., Tempelis, CH & Washino, RK (1986) Who is eating whom? An evaluation of an enzyme immunoassay for predator prey analyses. Proc. Calif. Mosq. & Vect. Contr. Ass., 53, 120

Peters, TM & Barbosa, P (1977) Influence of population density on size, fecundity, and development rate of insects in culture. A. Rev. Ent., 22: 431–50

Pickavance, JR (1970) A new approach to the immunological analysis of invertebrate diets. J. Anim. Ecol., 39: 715–24

Pielou, EC (1977) Mathematical Ecology. John Wiley & Sons, New York. x + 385 pp.

Podoler, H. & Rogers, D (1975) A new method for the identification of key factors from life table data. J. Anim. Ecol., 44: 85–114

Poinar, GO (1979) Nematodes for Biological Control of Insects. CRC Press, Boca Raton, 217 pp.

Pollard, E., Lakhani, KH & Rothery, P (1987) The detection of density-dependence from a series of annual censuses. Ecology, 68: 2046–55

Pritchard, G & Scholefield, PJ (1980) Efficiency of the Enfield sampler for estimates of larval and pupal mosquito populations. Mosquito News, 40: 383–7

Pritchard, G & Scholefield, PJ (1983) Survival of Aedes larvae in constant area ponds in southern Alberta (Diptera: Culicidae). Can. Ent., 115: 183–8

Pritchard, G., Harder, LD & Mutch, RA (1996) Development of aquatic insect eggs in relation to temperature and strategies for dealing with different thermal environments. Biological Journal of the Linnean Society, 58: 221–244

Prout, T (1980) Some relationships between density-independent selection and density-dependent population growth. Evolutionary biology, 13: 1–68

Rae, DJ (1990) Survival and development of the immature stages of Culex annulirostris (Diptera: Culicidae) at the Ross River dam in tropical eastern Australia. J Med Entomol 27: 756–62

Rai KS (1991) Aedes albopictus in the Americas. A. Rev. Ent., 36: 459–84

Rajagopalan, PK., Yasuno, M & Russel, S (1975) Studies on the development and survival of immature stages of Culex fatigans in nature. J. Commun. Dis., 7: 10–14

Rajagopalan, PK., Yasuno, M & Menon, PKB (1976a) Density effect on survival of immature stages of Culex pipiens fatigans in breeding sites in Delhi villages. Indian J. med. Res., 64: 688–708

Rajagopalan, PK., Brooks, GD & Menon, PKB (1976b) Estimation of natural survival rates of immatures of Culex pipiens fatigans in open effluent drains in Faridabad, northern India. J. Commun. Dis., 8: 11–17

Rajagopalan, PK., Menon, PKB & Brooks, GD (1977a) A study on some aspects of Culex pipiens fatigans population in an urban area, Faridabad, northern India. Indian J. med. Res., 65 (Suppl.), 65–76

Rajagopalan, PK., Curtis, CF., Brooks, GD & Menon, PKB (1977b) The density dependence of larval mortality of Culex pipiens fatigans in an urban situation and prediction of its effects on genetic control operations. Indian J. med. Res., 65 (Suppl.), 77–85

Read, NR & Moon, RD (1996) Simulation of development and survival of Aedes vexans (Diptera: Culicidae) larvae and pupae. Environmental Entomology, 25: 1113–1121

Reisen, WK (1975) Intraspecific competition in Anopheles stephensi Liston. Mosquito News, 35: 473–82

Reisen, WK (1995) Effect of temperature on Culex tarsalis (Diptera: Culicidae) from the Coachella and San Joaquin Valleys of California. J Med Entomol 32: 636–645

Reisen, WK & Emory, RW (1977) Intraspecific competition in Anopheles stephensi (Diptera: Culicidae). II. The effects of more crowded densities and the addition of antibiotics. Can. Ent., 109: 1475–80

Reisen, WK & Mahmood, F (1980) Horizontal life table characteristics of the malaria vectors Anopheles culicifacies and Anopheles stephensi (Diptera: Culicidae). J Med Entomol 17: 211–17

Reisen, WK & Siddiqui, TF (1979) Horizontal and vertical estimates of immature survivorship for Culex tritaeniorhynchus (Diptera: Culicidae) in Pakistan. J Med Entomol 16: 207–18

Reisen, WK., Siddiqui, TF., Aslam, Y & Malik, GM (1979) Geographic variation among the life table characteristics of Culex tritaeniorhynchus from Asia. Ann. ent. Soc. Am., 72: 700–779

Reisen, WK., Azra, K & Mahmood, F (1982) Anopheles culicifacies (Diptera: Culicidae): horizontal and vertical estimates of immature development and survivorship in rural Punjab province, Pakistan. J Med Entomol 19: 413–22

Reisen, WK., Meyer, RP., Shields, J & Arbolante, C (1989) Population ecology of preimaginal Culex tarsalis (Diptera: Culicidae) in Kern county, California. J Med Entomol 26: 10–22

Renshaw, M (1991) 'Population dynamics and ecology of Aedes cantans (Diptera: Culicidae) in England' Unpublished Ph.D. thesis, University of Liverpool, 186 pp.

Renshaw, M., Service, MW & Birley, MH (1993) Density-dependent regulation of Aedes cantans (Diptera: Culicidae) in natural and artificial populations. Ecological Entomology, 18: 223–233

Richards, OW. (1961). The theoretical and practical study of natural insect populations. A. Rev. Ent., 6: 147–62

Richards, OW & Waloff, N (1954) Studies on the biology and population dynamics of British grasshoppers. Anti-Locust Bull., 17, 182 pp.

Richards, OW., Waloff, N & Spradbery, JP (1960) The measurement of mortality in an insect population in which recruitment and mortality widely overlap. Oikos, 11: 306–10

Ricker, WE (1944) Further notes on fishing mortality and effort. Copeia, 1944: 23–44

Ricker, WE (1948) 'Methods of Estimating Vital Statistics of Fish Populations'. Indiana Univ. Publ. Sci. Ser., 15: 101 pp.

Ricklefs, RE (1973) Ecology. Nelson, London. x + 861 pp.

Robert, V., Planchon, O., Lapetite, JM & Esteves, M (1999) Rainfall is not a direct mortality factor for anopheline larvae. Parasite, 6: 195–196

Roberts, D (1998) Overcrowding of Culex sitiens (Diptera: Culicidae) larvae: population regulation by chemical factors or mechanical interference. J Med Entomol 35: 665–669

Roberts, DR., Smith, LW & Enns, WR (1967) Laboratory observations on predation activities of Laccophilus beetles in the immature stages of some dipterous pests found in Missouri oxidation lagoons. Ann. ent. Soc. Am., 60: 908–10

Roberts, DW & Castillo, JM (1980) Bibliography on pathogens of medically important arthropods: 1980. Bull World Health Organ 58 (Suppl.), 197 pp.

Roberts, DW & Strand, MA (1977) Pathogens of medically important arthropods. Bull World Health Organ 55 (Suppl.), 419 pp.

Roberts, DW., Daoust, RA & Wraight, SP (1983) Bibliography of pathogens of medically important arthropods: 1981. VBC/83.1, 324 pp.

Rogers, DJ (1972) Random search and insect population models. Journal of Animal Ecology, 41: 369–383

Rogers, DJ (1983) Interpretation of sample data, pp. 139–160. In Pest and Vector Management in the Tropics with Particular Reference to Insects, Mites and

Snails (edit. A Youdeowei & MW Service). Longman, London, xv + 399 pp.

Roitberg BD, Mondor EB, Tyerman JGA (2003) Pouncing spider, flying mosquito: blood acquisition increases predation risk in mosquitoes. Behavioral Ecology, 14: 736–740

Rosen L, Rozeboom LE, Reeves WC, Saugrain J, Gubler DJ (1976) A field trial of competitive displacement of Aedes polynesiensis by Aedes albopictus on a Pacific atoll. Am. J. trop. Med. Hyg., 25: 906–13

Rothschild, GHL (1966) A study of a natural population of Conomelus anceps Germar (Homoptera:Delphacidae) including observations on predation using the precipitin test. J. Anim. Ecol., 35: 423–33

Royama, T (1971) A comparative study of models for predation and parasitism. Res. Pop. Ecol., Suppl., 1, 91 pp.

Royama, T (1977) Population persistence and density dependence. Ecol. Monogr., 47: 1–35

Royama, T (1981a) Fundamental concepts and methodology for the analysis of animal populations dynamics, with particular reference to univoltine species. Ecol. Monogr., 51: 473–93

Royama, R (1981b) Evaluation of mortality factors in insect life table analysis. Ecol. Monogr., 51: 495–505

Rozeboom LE (1971) Relative densities of freely breeding populations of Aedes (S.) polynesiensis Marks and A. (S.) albopictus Skuse. A large cage experiment. Am. J. trop. Med. Hyg., 20: 356–62

Rozeboom LE, Bridges JR (1972) Relative population densities of Aedes albopictus and A. guamensis on Guam. Bull World Health Organ 46: 477–83

Rudnick A (1965) Studies on the ecology of dengue in Malaysia: A preliminary report. J. med. Entomol. 2: 203–8

Rudnick A, Hammon WM (1960) Newly recognized Aedes aegypti problems in Manila and Bangkok. Mosquito News, 20: 247–9

Ruesink, WG (1975) Estimating time-varying survival of arthropod life stages from population diversity. Ecology, 56, 244–7

Russell RC (1986) Larval competition between the introduced vector of dengue fever in Australia, Aedes aegypti (L.), and a native container-breeding mosquito, Aedes notoscriptus (Skuse) (Diptera: Culicidae). Aust. J. Zool., 34: 527–34

Russo RJ (1983) The functional response of Toxorhynchites rutilus rutilus (Diptera: Culicidae), a predator on container breeding mosquitoes. J Med Entomol 20: 585–590.

Sawyer, AJ & Haynes, DL (1984) On the nature of errors involved in estimating stage-specific survival rates by Southwood's method for a population with overlapping stages. Res. Popul. Ecol., 26, 331–51

Schneider, P., Takken, W. & McCall, P. J. (2000). Interspecific competition between sibling species larvae of Anopheles arabiensis and An. gambiae. Med Vet Entomol 14, 165–170

Schreiber, ET., Meek, CL & Yates, MM (1988) Vertical distribution and species coexistence of tree hole mosquitoes in Louisiana. J Am Mosq Control Assoc 4, 9–14

Scott, JA., Brogdon, WG & Collins, FH (1993) Identification of single specimens of the Anopheles gambiae complex by the polymerase chain reaction. American Journal of Tropical Medicine and Hygiene, 49, 520–529

Seifert, RP (1980) Mosquito fauna of Heliconia aurea. J. Anim. Ecol., 49, 687–97

Seifert, RP & Barrera, R(1981) Cohort studies on mosquito (Diptera: Culicidae) larvae living in the water-filled floral bracts of Heliconia aurea (Zingiberales: Musaceae). Ecol. Ent., 6: 191–7

Sempala, SDK (1982) Estimation of the mortality of the immature stages of Aedes (Stegomyia) africanus Theobald in a tropical forest in Uganda. Insect Sci. Applic., 2: 233–44

Service MW (1966) The replacement of Culex nebulosus Theo. by Culex pipiens fatigans Wied. (Diptera, Culicidae) in towns in Nigeria. Bull Entomol Res 56: 407–15

Service MW (1968) The taxonomy and biology of two sympatric sibling species of Culex, C. pipiens and C. torrentium (Diptera: Culicidae). J. Zool., Lond., 156: 313–23

Service MW (1970) Studies on the biology and taxonomy of Aedes (Stegomyia) vittatus (Bigot) (Diptera: Culicidae) in northern Nigeria. Trans. R. ent. Soc. Lond., 122: 101–43

Service MW (1971) Studies on sampling larval populations of the Anopheles gambiae complex. Bull World Health Organ 45: 169–80

Service MW (1973a) Study of the natural predators of Aedes cantans (Meigen) using the precipitin test. J Med Entomol 10: 503–10

Service MW (1973b) Mortalities of the larvae of the Anopheles gambiae complex and detection of predators by the precipitin test. Bull Entomol Res 62: 359–69

Service MW (1973c) Identification of predators of Anopheles gambiae resting in huts, by the precipitin test. Trans. R. Soc. trop. Med. Hyg., 67: 33–4

Service MW (1976) Mosquito Ecology. Field Sampling Methods. Applied Science Publishers, xii + 583 pp.

Service MW (1977a) Mortalities of the immature stages of species B of the Anopheles gambiae complex in Kenya: comparison between rice fields and temporary pools, identification of predators, and effects of insecticidal spraying. J Med Entomol 13: 535–45

Service MW (1977b) Ecological and biological studies on Aedes cantans (Meig.) (Diptera: Culicidae) in southern England. J. appl. Ecol., 14: 159–96

Service MW (1983) Biological control of mosquitoes—has it a future? Mosquito News, 43: 113–20

Service MW (1985a) Some ecological considerations basic to the biocontrol of Culicidae and other medically important arthropods, pp. 9–30 and 429–431 In Integrated Mosquito Control Methodologies, Volume 2. Biocontrol and Other Innovative Components and Future Directions (edit. M. Laird & J. W. Miles). Academic Press, London, xviii + 444 pp.

Service MW (1985b) Population dynamics and mortalities of mosquito preadults, pp. 185–201. In Ecology of Mosquitoes: Proceedings of a Workshop (edit. L. P. Lounibos, JR Rey & JH Frank). Florida Medical Entomology Laboratory, Vero Beach, Florida, 579 pp.

Service MW (1993) Mosquito Ecology. Field Sampling Methods. 2nd edn. Chapman & Hall, London

Service MW, Lyle P (1975) Detection of the predators of *Simulium damnosum* by the precipitin test. Ann Trop Med Parasitol 69: 105–108

Service MW, Voller A, Bidwell DE (1986) The enzyme-linked immunosorbent assay (ELISA) test of the identification of blood-meals of haematophagous insects. Bull Entomol Res 76: 321–330

Sheppard PM, Macdonald WW, Tonn RJ, Grab B (1969) The dynamics of an adult population of *Aedes aegypti* in relation to dengue haemorrhagic fever in Bangkok. J Anim Ecol 38: 661–702

Sherratt TN, Tikasingh ES (1989) A laboratory investigation of mosquito larval predation by *Toxorhynchites moctezuma* on *Aedes aegypti*. Med Vet Entomol 3: 239–246

Sibley RM, Smith RH (1998) Identifying key factors using lambda contribution analysis. J Anim Ecol 67: 17–24

Slater JD, Pritchard G (1979) A stepwise computer program for estimating development time and survival of *Aedes vexans* (Diptera: Culicidae) larvae and pupae in field populations in southern Alberta. Can Entomol 111: 1241–1253

Slobodkin LB (1962) Growth and Regulation of Animal Populations. Holt, Rinehart & Winston, New York

Smith PT, Reisen WK, Cowles DA (1995) Interspecific competition between *Culex tarsalis* and *Culex quinquefasciatus*. J Vector Ecol 20: 139–146

Smith RH (1973) The analysis of intra-generation change in animal populations. J Anim Ecol 42: 611–622

Solomon ME (1968) Logarithmic regression as a measure of population density response: comment on a report by G. W. Salt. Ecology 49: 357–358

Solow AR, Steele JH (1990) On sample size, statistical power, and the detection of density dependence. J Anim Ecol 59: 1073–1076

Sota T, Mogi M (1992) Interspecific variation in desiccation survival time of *Aedes (Stegomyia)* mosquito eggs is correlated with habitat and egg size. Oecologia 90: 353–358

Southwood TRE (1967) The interpretation of population change. J Anim Ecol 36: 519–529

Southwood TRE (1978) Ecological Methods with Particular Reference to the Study of Insect Populations. Chapman & Hall, London

Southwood TRE (1988) Tactics, strategies and templets. Oikos 52: 3–18

Southwood TRE, Henderson PA (2000) Ecological Methods. 3rd edn. Blackwell Science, Oxford

Southwood TRE, Jepson WF (1962) Studies on the populations of *Oscinella frit* L. (Dipt.: Chloropidae) in the oat crop. J Anim Ecol 31: 481–495

Southwood TRE, Murdie G, Yasuno M, Tonn RJ, Reader PM (1972) Studies on the life budget of *Aedes aegypti* in Wat Samphaya, Bangkok, Thailand. Bull World Health Organ 46: 211–226

Spielman A, Feinsod FM (1979) Differential distribution of peridomestic *Aedes* mosquitoes on Grand Bahama Island. Trans R Soc Trop Med Hyg 73: 381–384

St. Amani JLS (1970) The detection of regulation in animal populations. Ecology 51: 823–828

Stav G, Blaustein L, Margalith J (1999) Experimental evidence for predation risk sensitive oviposition by a mosquito, *Culiseta longiareolata*. Ecol Entomol 24: 202–207

Steffan WA (1970) Evidence of competitive displacement of *Toxorhynchites brevipalpis* Theobald by *T. amboinensis* Doleschall in Hawaii. Mosq Syst Newsletter 2: 68

Stewart RJ, Schaefer CH, Miura T (1983) Sampling *Culex tarsalis* (Diptera: Culicidae) immatures in rice fields treated with combinations of mosquitofish and *Bacillus thuringiensis* H-14 toxins. J Econ Entomol 76: 91–95

Stewart RJ, Schaefer CH, Miura T (1985) Age structure and survivorship of *Culex tarsalis* in central California ricefields. Proc California Mosq Vector Control Assoc 52: 148–152

Stewart-Oaten A, Murdoch WW (1990) Temporal consequences of spatial density dependence. J Anim Ecol 59: 1027–1045

Stiling P (1988) Density-dependent processes and key factors in insect populations. J Anim Ecol 57: 581–594

Subra R (1983) The regulation of preimaginal populations of *Aedes aegypti* L. (Diptera: Culicidae) on the Kenya Coast. 1. Preimaginal population dynamics and the role of human behaviour. Ann Trop Med Parasitol 77: 195–201

Subra R, Dransfield RD (1984) Field observations on competitive displacement, at the preimaginal stage, of *Culex quinquefasciatus* Say by *Culex cinereus* Theobald (Diptera: Culicidae) at the Kenya coast. Bull Entomol Res 74: 559–568

Subra R, Service MW, Mosha FW (1984) The effect of domestic detergents on the population dynamics of the immature stages of two competitor mosquitoes, *Culex cinereus* Theobald and *Culex quinquefasciatus* Say (Diptera: Culicidae) in Kenya. Acta Trop 41: 69–75

Sucharit S, Tumrasvin W (1981) Ovipositional attractancy of waters containing larvae of *Aedes aegypti* and *Aedes albopictus*. Jap J Sanit Zool 32: 261–264

Sucharit S, Tumrasvin W, Vutikes S, Viraboonchai S (1978) Interactions between larvae of *Aedes aegypti* and *Aedes albopictus* in mixed experimental populations. Southeast Asian J Trop Med Public Health 9: 93–97

Sulaiman S (1982) The ecology of *Aedes cantans* (Meigen) and biology of *Culex pipiens* in hibernation sites in northern England. Ph.D. thesis, University of Liverpool

Sulaiman S, Omar B, Omar S, Ghauth I, Jeffery J (1990) Detection of the predators of *Aedes albopictus* (Skuse) (Diptera: Culicidae) by the precipitin test. Mosq-Borne Dis Bull 7: 1–4

Sulaiman S, Pawanchee Z A, Karim MA, Jeffery J, Busparani V, Wahab A (1996) Serological identification of the predators of adult *Aedes albopictus* (Skuse) (Diptera: Culicidae) in rubber plantations and a cemetery in Malaysia. J Vector Ecol 21: 22–25

Suleman M (1990) Intraspecific variation in the reproductive capacity of *Anopheles stephensi* (Diptera: Culicidae). J Med Entomol 27: 819–828

Suleman M, Reisen WK (1979) *Culex quinquefasciatus*: Life table characteristics of adults reared from wild-caught pupae from north west frontier province, Pakistan. Mosquito News 39: 756–762

Sunahara T, Mogi M (2002*a*) Priority effects of bamboo-stump mosquito larvae: influences of water exchange and leaf litter input. Ecol Entomol 27: 346–354

Sunahara T, Mogi M (2002*b*) Variability of intra- and interspecific competitions of bamboo stump mosquito larvae over small and large spatial scales. Oikos 97: 87–96

Sunahara T, Ishizaka K, Mogi M (2002) Habitat size: a factor determining the opportunity for encounters between mosquito larvae and aquatic predators. J Vector Ecol 27: 8–20

Sunderland KD, Sutton SL (1980) A serological study of arthropod predation on woodlice in a dune grassland ecosystem. J Anim Ecol 49: 987–1004

Sunderland KD, Crook NE, Stacey DL, Fuller BJ (1987) A study of feeding by phytophagous predators on cereal aphids using ELISA and gut dissection. J Appl Ecol 24: 907–933

Sunish IP, Reuben R (2002) Factors influencing the abundance of Japanese encephalitis vectors in ricefields in India - II. Biotic. Med Vet Entomol 15: 1–9

Sweeney AW, Becnel JJ (1991) Potential of microsporidia for the biological control of mosquitoes. Parasitol Today 7: 217–220

Takagi M, Sugiyama A, Maruyama K (1996) Survival of newly emerged *Culex tritaeniorhynchus* (Diptera: Culicidae) adults in field cages with or without predators. J Med Entomol 33: 698–701

Teesdale C (1957) The genus *Musa* Linn. and its role in the breeding of *Aedes (Stegomyia) simpsoni* (Theo.) on the Kenya coast. Bull Entomol Res 48: 251–260

Tempelis CH (1983) Adaptation of the enzyme-linked immunosorbent assay for the study of predator-prey relationships. In: Fotaine RE (ed) Mosquito Control Research, Annual Report 1983. University of California, Davis, pp. 34–51

Teng HJ, Apperson CS (2000) Development and survival of immature *Aedes albopictus* and *Aedes triseriatus* (Diptera: Culicidae) in the laboratory: effects of density, food, and competition on response to temperature. J Med Entomol 37: 40–52

Tianyun Su, Mulla MS (2001) Effects of temperature on development, mortality, mating and blood feeding behavior of *Culiseta incidens* (Diptera: Culicidae). J Vector Ecol 26: 83–92

Trpis M (1972*a*) Predator-Prey Oscillations in Populations of Larvae of *Toxorhynchites brevipalpis* and *Aedes aegypti* in a Suburban Habitat in East Africa. WHO/VBC/72.399, 12 pp. (mimeographed)

Trpis M (1972*b*) Development and predatory behaviour of *Toxorhynchites brevipalpis* (Diptera: Culicidae) in relation to temperature. Environ Entomol 1: 537–546

Trpis M (1981) Survivorship and age-specific fertility of *Toxorhynchites brevipalpis* females (Diptera: Culicidae). J Med Entomol 18: 481–486

Tun-Lin W, Burkot TR, Kay BH (2000) Effect of temperature and larval diet on development rates and survival of the dengue vector *Aedes aegypti* in north Queensland, Australia. Med Vet Entomol 14: 31–37

Tuno N, Okeka W, Minakawa N, Takagi M, Yan G (2005) Survivorship of *Anopheles gambiae* sensu stricto (Diptera: Culicidae) larvae in western Kenya highland forest. J Med Entomol 42: 270–277

Turnbull AL (1967) Population dynamics of exotic insects. Bull Entomol Soc Am 13: 333–337

Urabe K, Sekijima Y, Ikemoto T, Aida C (1982) Studies on *Sympetrum frequens* (Odonata: Libellulidae) nymphs as natural enemies of the mosquito larvae, *Anopheles sinensis*, in the rice field. 1. Evaluation on an utilization of the electrosyneresis for the quantitative study of the prey–predator relationships. Jap J Sanit Zool 33: 55–60 (In Japanese, English summary)

Van Valen L (1962) A study of fluctuating asymmetry. Evolution 16: 125–142

Varley GC, Edwards RL (1957) The bearing of parasite behaviour on the dynamics of insect host and parasite populations. J Anim Ecol 26: 471–477

Varley GC, Gradwell GR (1960) Key factors in population studies. J Anim Ecol 29: 399–401

Varley GC, Gradwell GR (1963) The interpretation of insect population changes. Proc Ceylon AssocAdv Sci 18: 142–156

Varley GC, Gradwell GR (1968) Population models for the winter moth. In: Southwood TRE (ed) Insect Abundance. Symposium of the Royal Entomological Society of London. No 4. Blackwell Scientific Publications, Oxford, pp 132–142

Varley GC, Gradwell GR (1970) Recent advances in insect population dynamics. Annu Rev Entomol 15: 1–24

Varley GC, Gradwell GR, Hassell MP (1973) Insect Population Ecology, an Analytical Approach. Blackwell Scientific Publications, Oxford

Vickerman GP, Sunderland KD (1975) Arthropods in cereal crops: Nocturnal activity, vertical distribution and aphid predation. J Appl Ecol 12: 755–766

Vickery WL (1991) An evaluation of bias in k-factor analysis. Oecologia 85: 413–419

Wada Y, Mogi M (1974) Efficiency of the dipper in collecting immature stages of *Culex tritaeniorhynchus summorosus*. Trop Med 16: 35–40

Wadsworth C (1957) A slide microtechnique for the analysis of immune precipitates in gel. Int Archs Allergy 10: 355–360

Walker ED, Copeland RS, Paulson SL, Munstermann LE (1987) Adult survivorship, population density, and body size in sympatric populations of *Aedes triseriatus* and *Aedes hendersoni* (Diptera: Culicidae). J Med Entomol 24: 485–493

Walton WE, Workman PD (1998) Effect of marsh design on the abundance of mosquitoes in experimental constructed wetlands in southern California. J Am Mosq Control Assoc 14: 95–107

Walton WE, Tietz NS, Mulla MS (1990) Ecology of *Culex tarsalis* (Diptera: Culicidae): factors influencing larval abundance in mecocosms in southern California. J Med Entomol 27: 57–67

Washburn JO, Anderson JR (1993) Habitat overflow, a source of larval mortality for *Aedes sierrensis* (Diptera: Culicidae). J Med Entomol 30: 802–804

Watanabe M, Wada Y (1968) Studies on predators of larvae of *Culex tritaeniorhynchus summorosus* Dyar. Jap J Sanit Zool 19: 35–38 (In Japanese, English summary)

Watson TF (1964) Influence of host plant condition on population increase of *Tetranychus telarius* (Linnaeus) (Acarina: Tetranychidae). Hilgardia 35: 273–322

Watt KEF (1959) A mathematical model for the effect of densities of attacked and attaching species on the number attacked. Can Entomol 91: 129–144

Weidhaas DE (1974) Simplified models of population dynamics related to control technology. J Econ Entomol 67: 620–624

Weidhaas DE, Patterson RS, Lofgren CS, Ford HR (1971) Bionomics of a population of *Culex pipiens quinquefasciatus* Say. Mosquito News 31: 177–182

Weidhaas DE, LaBrecque GC, Lofgren CS, Schmidt CH (1972) Insect sterility in population dynamics research. Bull World Health Organ 47: 309–315

Weidhaas DE, Breeland SG, Lofgren CS, Dame DA, Kaiser R (1974) Release of chemosterilized males for the control of *Anopheles albimanus* in El Salvador. IV. Dynamics of the test population. Am J Trop Med Hyg 23: 298–308

Weiser J (1991) Biological Control of Vectors. Manual for Collecting, Field Determination and Handling of Biofactors for Control of Vectors. John Wiley, Chichester

West AS, Eligh GS (1952) The rate of blood digestion in mosquitoes. Precipitin test studies. Can J Zool 30: 267–272

Wijeyaratne P, Seawright JA, Weidhaas DE (1974) Development and survival of a natural population of *Aedes aegypti*. Mosquito News 34: 36–42

Willems KJ, Webb CE, Russell RC (2005) Tadpoles of four common Australian frogs are not effective predators of the common pest and vector mosquito *Culex annulirostris*. J Am Mosq Control Assoc 21: 492–494

Wohlschlag DE (1954) Mortality rates of whitefish in an arctic lake. Ecology 35: 388–396

Yasuda H, Hashimoto T (1995) Prey density effect on cannibalism by *Toxorhynchites towadensis* (Diptera: Culicidae). J Med Entomol 32: 650–653

Yasuno M (1974) Ecology of *Culex pipiens fatigans* in rural Delhi, India. J Commun Dis 6: 106–116

Ye-Ebiyo Y, Pollack RJ, Kiszewski A, Spielman A (2003) Enhancement of development of larval *Anopheles arabiensis* by proximity to flowering maize (*Zea mays*) in turbid water and when crowded. Am J Trop Med Hyg 68: 748–752

Zaidi RH, Jaal Z, Hawkes NJ, Hemingway J, Symondson WOC (1999) Can multiple-copy sequences of prey DNA be detected amongst the gut contents of invertebrate predators? Mol Ecol 8: 2081–2087

Zani PA, Cohnstaedt LW, Corbin D, Bradshaw WE, Holzapfel CM (2005) Reproductive value in a complex life cycle: heat tolerance of the pitcher-plant mosquito, *Wyeomyia smithii*. J Evol Biol 18: 101–105

Chapter 13 Methods of Age-grading Adults and Estimation of Adult Survival Rates

Methods for age-grading adults

Ability to determine the age structure and survival rate of female mosquitoes is of paramount ecological importance because longevity affects net reproduction rates and dispersal distances, and is often needed in analysing data from mark–recapture experiments, etc. A critical analysis of the age composition of a population is also crucial in epidemiological studies (Dye 1984*a*, 1992; Garrett-Jones 1970; Macdonald 1952, 1957; Molineaux and Gramiccia 1980; Molineaux et al. 1978; Nájera 1974), and knowledge of survival rates can help in assessing the impact of control measures (Molineaux et al. 1976, 1979).

All age determination methods estimate physiological age, except that cuticular growth banding and pteridine techniques initially relate to calendar age. Physiological age, however, can sometimes be converted to real age. Tyndale-Biscoe (1984) divided methods for age-grading insects into three main categories: (i) those based on external marks due to general wear and tear with age, such as wing fray, denudation of scales from the body, and although not included by Tyndale-Biscoe the presence of hygrobatid-type water mites; (ii) somatic changes occurring with age such as growth rings in the cuticle, presence or absence of meconium; and (iii) changes in the reproductive system, such as tracheation of the ovaries. Charlwood et al. (1980) and WHO (1975*a,b*) have reviewed methods for age grading mosquitoes, while more general reviews covering also non-vector insects are presented by Lehane (1985), Neville (1983) and Tyndale-Biscoe (1984).

External wear

Prior to the 1930s the degree of wear or rubbing on the wings was widely used to distinguish between old and young mosquitoes (Perry 1912). Corbet (1960, 1962a) reported that external wear of the abdominal sternites was a reliable method of detecting nulliparous *Coquillettidia fraseri* and *Coquillettidia metallica*; only 0.9 and 0.1%, respectively of these two species identified as nulliparous were shown by dissection to be parous. The reliability of this approach should be checked with each species. For example, with aedine mosquitoes rubbing of the scales from the wing margins provided a better indication of nulliparity than examination of the abdomen. Suzuki et al. (1993) observed that the degree of wing damage exhibited by *Aedes albopictus* females collected in Nagasaki, Japan differed significantly according to parity, with wings of parous individuals showing more damage than nulliparous mosquitoes. Corbet (1960, 1962a) also found that the coloration of the integument was of some use in certain species, such as *Coquillettidia pseudoconopas*, in recognising nulliparous individuals.

Aquatic mites

Larvae of water mites (Hydrachnidia = Hydrachnellae = Hydracarina) sometimes parasitise mosquitoes, and Gillett (1957) suggested that since they left female mosquitoes at the time of mosquito oviposition their presence indicated nulliparity. Although these mites are predominantly found on nulliparous females they have occasionally been reported on parous mosquitoes (Detinova 1962; Reisen and Mullen 1978; Wharton 1959). Corbet (1963) pointed out that nulliparity is only indicated if the 'parasitic water mites' in question are of the hygrobatid type and are alive. Hygrobtid mites are active beneath the water surface and attach to adult mosquitoes only as they emerge from pupae. Jalil and Mitchell (1972) reaffirmed this and stressed that in their experience most mites found on aedine species are of the genus *Thyas*, which attach during oviposition as well as during emergence. In England thyasid mites found on *Ochlerotatus cantans* were not indicators of nulliparity, but rather of parous females (Renshaw 1991). It is likely that the mites reported by Morris and DeFoliart (1970) and Graham (1969) on parous mosquitoes were probably thyasids and therefore not indicators of nulliparity (Corbet 1970). Good descriptions and figures illustrating the differences between hygrobatid and thyasid-type mites are given by Jalil and Mitchell (1972). Studies in India showed that although live *Arrenurus* mites were found on only nulliparous *Anopheles annularis* and *Anopheles stephensi*, 36.8% of *Anopheles culicifacies* infested with mites were parous (Biswas et al. 1980). Malhotra et al. (1983) found *Arrenurus* mites on 13 mosquito species (but not *Anopheles culicifacies*) collected resting in houses and other structures in India. All infested mosquitoes (547) were nulliparous except for one uniparous *Mansonia uniformis*. In Pakistan Reisen and Mullen (1978) found that 91.2% of 125 female *Culex tritaeniorhynchus* infested with live *Arrenurus madaraszi* larvae were nulliparous and 8.8% were parous; and when an additional 127 infested mosquitoes belonging to another nine species were included the proportion nulliparous was 91.7%. However, 19.9% of *Anopheles pulcherrimus* with mites were parous, as were smaller percentages of females of other species. The authors consequently did not believe that these mites were a reliable indicator of nulliparity.

McCrae (1976) describes a careful study of the association between the larvae of a limnesiid mite, probably *Limnesia* sp., and Arrenuridae, probably *Arrenurus* sp., on *Anopheles implexus* in Uganda. These two families belong to the so-called hygrobatid mites of Mitchell (1957). Limnesiid mites were found on 62.3% of all nulliparous *Anopheles implexus* with as many as 34 larvae/female, and on 6.8% of pars but with up to only two/mosquito. An index of mite association with nullipars (A_n) can be expressed as, the percentage by which infested nullipars as a proportion of infested nullipars plus infested pars, diverges from total unreliability, i.e. 50%, that is

$$A_n = \frac{100(N_i + P_i)}{2N_i} \tag{13.1}$$

where N_i = nullipars infested and P_i = pars infested. McCrae (1976) found that limnesiid mites indicated nulliparity with a reliability of 80.2%; by excluding mosquitoes with dead, as well as living, mites reliability increased to 85.6%. McCrae (1976) also divided his larval mites into four growth classes (*a–d*), and when only partially grown larval mites were used (*a–c*) there was total reliability between infestation with living mites and nulliparity. The much lower infestation rates with arrenurid mites prevented such a detailed analysis, but it appeared that these mites grew quicker than the limnesiids and therefore all larvae were ready to detach at a mosquito's first oviposition, and so the presence of live mites invariably indicated nulliparity.

Lanciani (1979*a*) found that *Arrenurus* mites sometimes detach from nulliparous *Anopheles crucians* before they have oviposited. This phenomenon has previously been reported, or suspected, such as by McCrae

(1976) with *Anopheles implexus*, and in earlier studies on *Anopheles crucians* (Lanciani and Boyt 1977). Laboratory and field studies appear to show that in at least some species (*Anopheles crucians, Anopheles quadrimaculatus* s.l.) adults with heavy infestations of aquatic mites die sooner than those with few mites (Lanciani 1979*b,c*, 1986, 1987; Lanciani and Boyett 1980; Lanciani and Boyt 1977; Reisen and Mullen 1978). Smith and McIver (1984) found that parasitism by *Arrenurus danbyesis* reduced egg production in *Coquillettidia perturbans* during the first gonotrophic cycle by 5% and in *Anopheles crucians* by 35%, but longevity was unaffected. On the other hand, Lanciani (1979*b*, 1986, 1987) found that as the numbers of *Arrenurus pseudotenuicollis* increased on *Anopheles quadrimaculatus* s.l. and *Anopheles crucians* their longevity decreased.

Lanciani (1979*a*) pointed out that aquatic mites can be found on male mosquitoes. For example, Reisen and Mullen (1978) found *Arrenurus madaraszi* on males as well as females of eight out of nine mosquito species harbouring mites, but the ratio of infested males:females was 1:9.76 in anophelines and 1:9.10 in *Culex* species. They tentatively suggested that this lower incidence might be because the survivorship of infested males was much lower than females, but believed the matter needed further investigation. Washburn et al. (1989) also reported that water mites (near *Euthyas*) were more common on female *Ochlerotatus sierrensis* emerging from tree-holes than males, which they thought suggested that mites could discriminate between the sexes. Unlike most situations males are not necessarily dead-end hosts, because they often seek shelter within tree-holes and here their mites can detach and find water.

The usefulness of mites as indicators of nulliparity relies very much on the percentage of infested mosquitoes in a population, which depends on the type of breeding places. Mosquitoes breeding in tree-holes, water-storage pots and polluted waters, habitats not normally colonised by aquatic mites, are unlikely to harbour mites, whereas those breeding in marshes and weedy ponds are more likely to be infested. *Mansonia* and *Coquillettidia* species frequently have relatively high infestation rates, for example as high as 34.91% in *Coquillettidia perturbans* (Lanciani and McLaughlin 1989). In certain situations *Anopheles* may also have high infestation rates. In Kenya Githeko (1992) found as many as 39 hygrobatid mites on a single *Anopheles arabiensis* originating from rice fields. If the larval mites are of the correct families and alive, then female mosquitoes can frequently be classified as nulliparous—thus saving dissection time.

Münchberg (1954) reviews the association of mites with mosquitoes, while Mitchell (1957) discusses their life-histories in terms of evolutionary trends. Prasad and Cook (1972) give a comprehensive account of the taxonomy of water mites (Hydracarina = Hydrachnellae) and Mullen (1974) gives an illustrated key for the identification to families and genera of mite larvae found on mosquitoes. Mullen (1975*b*) describes methods for the collection, preservation and rearing of these mites. Mullen (1975*a*) lists all known records of mosquitoes parasitised by mites, while Mullen (1976) presents morphological descriptions of the larvae of 13 species of *Arrenurus* (subgenus *Truncaturus*) mites parasitising North American mosquitoes. Cook et al. (1989) have written on the biology and strategies of 27 species of water mites of *Arrenurus*. Good reviews of the parasitic association of water mites with insects, including mosquitoes, are given by Smith and Oliver (1986) and Smith (1988).

Meconium and green mosquitoes

Freshly emerged adults of both sexes have a yellowish–brown to greenish coloured waste product in the stomach, termed meconium (billiverdin and/or insectoverdin). If mosquitoes have a pale integument this can be seen through the abdominal wall occupying segments IV–V (Detinova 1962; Mer 1936). However, as it is rapidly excreted, usually before or shortly after initial activity, it is of limited use in detecting nullipars or young males. Nevertheless Corbet (1963) was able to identify 34% of the resting population of *Coquillettidia fraseri* as nullipars by this method. In female *Ochlerotatus nigromaculis* maintained at 17°C, the meconium is egested within about 47 h and within 13 h if adults are kept at 28°C. With *Culex quinquefasciatus* the times for egestion at these temperatures are 49 and 25 h, respectively (Rosay 1961).

Although both Colless (1958) and Rosay (1961) recorded green coloration in adults of the *Culex pipiens* complex, it was De Meillon et al. (1967*d*) who recognised that it was a good indicator of newly emerged nullipars. Self and Sebastian (1971) found that in Yangon, Myanmar 99.7% of 5845 female *Culex quinquefasciatus* collected soon after emergence from either clean or heavily polluted water exhibited a distinct green coloration on the ventral and lateral aspects of the thorax and abdomen. The colour disappeared from the thorax after about 48 h and from the abdomen after 72 h, and could therefore be a reliable guide for identifying newly emerged females. Graham and Bradley (1972) confirmed that newly emerged females of *Culex quinquefasciatus* were green coloured for the first 3 days, and used this to calculate daily survival rates.

Kay (1979) found that green coloration in the abdomen of *Culex annulirostris* indicated they were nulliparous, and Day et al. (1990) used the presence of green adults of *Culex nigripalpus* to identify adults up to 24 h presence of green adults of *Culex nigripalpus* to identify adults up to 24 h after emergence, Morris (1984) found that all *Culiseta morsitans* with meconium were nulliparous, while 98% of *Culiseta melanura* with meconium were nulliparous.

Rotation of male genitalia

Newly emerged male mosquitoes may be recognised by the incomplete rotation of their genitalia (Haeger 1960; Nielsen 1958; Provost 1960). The time taken for various degrees of rotation can be experimentally determined in the field or laboratory. Rosay (1961) found that complete rotation in *Ochlerotatus nigromaculis* occurred within 58 h at a temperature of 17°C, after 12 h at 28°C, and within 53 and 17 h in *Culex quinquefasciatus* at the same two temperatures. De Meillon et al. (1967c) reported that in the field in Yangon, Myanmar the rotation in *Culex quinquefasciatus* was half completed in 7–8 h, and full rotation had occurred by 19 h.

Miscellaneous internal characters

In *Culicoides* a burgundy-red pigment, which is best seen in unfed females, develops under the abdominal wall during ovarian development and apparently reliably identifies parous *Culicoides*. Kay (1979) reported that this method could also be applied to *Culex annulirostris* in Australia. The pigment granules were situated in the parietal fat lining the abdomen and were best seen in sternites IV–V; occasionally pigmentation extended along the entire abdomen and was also found in fat bodies associated with the ovaries. He found that 'almost without exception' adults with a red pigment were parous; however, only 22.6% of the mosquitoes had the pigment.

The presence of muscle remnants of the immature stages in adults have also been investigated as a possible indicator of nulliparity in both sexes of mosquitoes (Hitchcock 1968; Rosay 1961). These degenerating muscles are visible as translucent tissue located between the gut and abdominal wall, but they disappear within 12–59 h, the rate of autolysis being greater at higher temperatures.

Gillies (1956) found that in recently mated female *Anopheles gambiae* there was a small translucent plug in the common oviduct, which was inserted by the male after copulation. This mating plug is normally absorbed within about 36 h, hence its presence only identifies newly emerged nullipars. Smiraglia et al. (1971) found that the mating plug disappeared after about 48 h in mosquitoes of the *Anopheles maculipennis* complex.

Pteridines

The increase with age of the quantity of several pigments in various structures of insects can sometimes be used to age grade them. The most widely known pigments associated with age are fluorescent substances called pteridines, which are degradation products of purine metabolism that function as light filters in eyes, as sexual colouration pigments or as enzyme cofactors (Penilla et al. 2002). Pteridines accumulate for at least 150 days in male tsetse flies and have been found to be a good indicator of chronological age in *Chrysomya bezziana* (Wall et al. 1990), *Cochliomyia hominivorax* (Thomas and Chen 1989) and *Musca autumnalis* (Moon and Krafsur 1995). Pteridines have been shown to accumulate with age in the eyes of *Stomoxys calcitrans*, *Simulium damnosum* complex, tsetse flies and other muscoid flies (Cheke et al. 1987; Cheke et al. 1990; Lehane and Hargrove 1988; Lehane and Mail 1985; Mail and Lehane 1988; Mail et al. 1983; Millest et al. 1992). Mullens and Lehane (1995) reported observing age- and physiologically-related changes in fluorescence in the biting midge *Culicoides variipennis sonorensis*, but the high variability in fluorescence among insects of a given age and sex reduced the usefulness of the technique in determining the age of individual specimens. For age grading, heads are homogenised in a suitable buffer (e.g. 0.06 m tris–HCl at pH 8), followed by centrifuging at $3000 \times g$ for 3 min, then the fluorescence of the supernatant is read using a fluorescence spectrometer. Flies can be kept dry at room temperature in the dark for at least 8 days without appreciable deterioration or loss of pteridines and Moon and Krafsur (1995) reported that pterin quantities in *Musca autumnalis* heads stored at room temperature in a dark, desiccating environment remained stable for two months. Stability of fluorescent pigment concentrations allows for examination of flies caught in remote areas to be postponed until there is access to a spectrofluorimeter. The studies by Cheke et al. (1987, 1990) and Millest et al. (1992) have demonstrated significant relationships among pteridine content in the head of *Simulium* spp. and fly age and size and these relationships are independent of temperature. Moon and Krafsur (1995) found that initial quantities of fluorescent pigments in teneral *Musca autumnalis* were proportional to head capsule width. Daily rates of accumulation of pigments were proportional to temperature for both females and males above a threshold temperature of 9.8°C, with rates of increase more rapid in males compared with females. These authors developed parsimonious degree-day models to estimate chronological age from pterin quantities in field-collected specimens. Adjustment for body size was recommended in order to improve accuracy of estimation of age. Chronological age did not correspond well with parity in field-collected specimens.

Lardeaux et al. (2000) considered spectrofluorometers to be inadequate for ageing *Aedes polynesiensis* and *Culex quinquefasciatus* in French Polynesia. Neither whole body nor head-only preparations gave results that exceeded the white noise level of the spectrofluorometer, defined as two standard deviations above the mean calculated from all negative controls.

Wu and Lehane (1999) achieved more success with the application of a reversed phase high performance liquid chromatography (HPLC) technique to measure total fluorescence and identify individual pteridines in laboratory-reared *Anopheles gambiae* and *Anopheles stephensi*. Pteridine extractions were carried out as follows: whole bodies, thoraces or head capsules of individual female mosquitoes or groups of ten or thirty individuals were homogenised in a Jencons microglass homogeniser with a buffer containing 0.1 M NaOH and 0.15 M glycine with a pH range of 6–11. Whole body fluorescence was found to be inversely proportional to age up to 30 days post-emergence for *Anopheles gambiae* and *Anopheles stephensi*. *Anopheles gambiae* fluorescence was only 50–70% that of *Anopheles stephensi*.

Penilla et al. (2002) identified several fluorescent compounds in the head of *Anopheles albimanus* using HPLC. Total pteridine concentrations were estimated using the method of Lehane and Mail (1985) in heads, body parts and whole bodies of insectary-reared and field-collected females in Mexico and investigated for their usefulness in age determination. Pteridine concentrations were observed to diminish with age in both laboratory and field-collected groups. Total pteridine concentrations (TPC) correlated with chronological age in insectary-reared sugar-fed females (heads: $r^2 = 0.35$; abdomens + legs + wings: $r^2 = 0.34$; $P < 0.001$). A lower correlation was observed with blood-fed females (heads: $r^2 = 0.22$; abdomens + legs + wings: $r^2 = 0.27$). TPC differed among females of the same age fed with blood at different times, indicating that blood-meals modify the rate of decrease of pteridine concentrations with age. A significant polynomial correlation was reported for TPC and the number of ovipositions (heads: $r^2 = 0.24$; abdomens + legs + wings: $r^2 = 0.27$; whole body: $r^2 = 0.52$; $P < 0.001$) in insectary-reared mosquitoes. The correlation was lower in field-collected mosquitoes (heads: $r^2 = 0.14$; abdomens + legs + wings: $r^2 = 0.10$; $P < 0.05$). Field-collected females showed a large increase in pteridines in one-parous females, followed by a reduction. The large initial increase, however, resulted in both physiologically young and old mosquitoes having the same pteridine concentrations. These authors concluded that TPC determination by spectrofluorometry was not a reliable method to estimate the age of wild-caught *Anopheles albimanus* females.

Cuticular growth rings

Neville (1963) and Dingle (1965) demonstrated the existence of daily growth rings in certain areas of cuticle in some insects. Neville (1983) reviewed the phenomenon of cuticular growth layers and pointed out that endopterygote insects can be best age graded by examination of the apodemes, but with exopterygotes by counting growth layers in the cuticle. Schlein and Gratz (1972, 1973) were the first to report the existence of growth layers in mosquitoes, and proposed that the daily deposition of cuticle on the skeletal apodemes could be used for age determination.

The rate of cuticle deposition depends on temperature and, at least in some species, on ultraviolet light. Growth ceases if temperature falls below the threshold value for deposition of cuticle, and increases with high temperatures. But if temperatures are too high full cuticle thickness is obtained quickly and the number of bands available for age determination is reduced (Tyndale-Biscoe and Kitching 1974). Growth bands are often clearest in insects subjected to fluctuating temperatures, with the minimum temperature just below that for cuticle growth. Schlein (1979) believed that insectary reared mosquitoes often failed to show cuticle growth lines because they are not normally exposed to fluctuating temperatures or UV light.

A limitation of the technique is that growth layers are deposited only for a short period after adults have emerged, for instance the mean maximum number of layers counted in *Anopheles* by Schlein (1979) was 10–14, and 10–11 in *Culex* (Schlein and Gratz 1972). Hence the method can be quite accurate during periods when growth layers are being deposited, but when the maximum number of layers have been deposited the insect may be that age or older. Schlein (1979) found that whereas in laboratory mosquitoes up to 14 daily growth layers can be counted, he only found two to four layers in many wild-caught adults. He attributed this to their dying before the cessation of cuticular growth. Up to eight daily bands were found in wild caught *Anopheles gambiae*, whereas 10–14 were counted in laboratory material (Schlein and Gratz 1973).

Schlein (1975, 1979) describes in some detail how mosquitoes should be processed for age grading. Basically they must be kept dry, or fixed in 3:1 ethanol/acetic acid and then stored in 70% ethanol, after which the tissues are macerated in 7–10% potassium or sodium hydroxide. The thoracic phragmata are dissected out and fixed in picric acid, and afterwards stained in 0.2% haematoxylin, aniline blue-orange G or some other stain. Growth layers can be seen under an ordinary light microscope, although some workers prefer phase contrast or interference microscopy (Neville 1983). Ikeshoji (1985) used the following modification of the Schlein and Gratz (1972) method to determine the age of male mosquitoes. Adults were boiled in 30% lactic acid for 15 min, and then the mesothoracic furcum was dissected out and stained on a slide in dilute potassium permanganate. The half-dried stained specimen was placed in immersion oil, covered with a coverslip and examined under $1000 \times$ magnification of an ordinary microscope. The number of daily growth layers were counted along the inside edge of the posterior part of the U-shaped furcum. Checks showed that about 70% of *Culex tarsalis, Culex tritaeniorhynchus* and *Ochlerotatus melanimon* were correctly aged in days.

Moore et al. (1986) used a slightly modified method of Schlein (1979) on the *Culex pipiens* complex, in which the apodemes were stained in mature Heidenhain's haematoxylin (0.2% in 70% ethanol) for up to 1 min. Examination under polarized light was found to be the best procedure, and growth lines furca were the most distinct. Moore et al. (1986) considered that the method was unreliable for ecological studies, at least on the *Culex pipiens* complex.

Despite several papers describing the application of this method to medical vectors, it has not proved very popular, probably because it can be used to calendar-age only young adults, and is no good under extremes of temperature—cold or very hot—because deposition of cuticle is inhibited. Moreover, the method is also rather time consuming so prohibiting the processing of large samples which are usually needed to determine the age-structure of populations.

Cuticular hydrocarbons

Desena et al. (1999*a,b*) developed and tested a gas-liquid chromatographic method for age determination in *Aedes aegypti*, based on the relative abundance of two cuticular hydrocarbons, pentacosane ($C_{25}H_{52}$) and nonacosane ($C_{29}H_{60}$). Hydrocarbons were extracted initially from whole insects (Desena et al. 1999*a*) and subsequently from just the legs of female *Aedes aegypti* (Desena et al. 1999*b*). In the initial study, five females were sampled from a laboratory colony of known age every 33 degree-days from 0 to 231 degree days. A 100 µg aliquot of n-hexane was added to a glass vial containing an individual mosquito, and was left to rest for five minutes for hydrocarbon extraction. After concentration to a volume of approximately 2 μl under a stream of nitrogen, the entire concentrated extract was injected into the gas chromatograph column. Analysis of hydrocarbon extracts was conducted using a Hewlett-Packard 5890 gas chromatograph and a flame ionisation detector. The chromatographic column used helium as the carrier gas at a flow rate of 3.8 ml/min. The temperature regime used was from 100 to 280°C at a rate of 7.5°C per minute, followed by 22 minutes at 280°C. Peaks 1 (pentacosane) and 5 (nonacosane) were used in the subsequent age-grading analysis. Relative abundance of pentacosane decreased over time, while that of nonacosane increased with time. Regression equations for peak 1 abundance were developed for un-transformed and logit transformed values, such that Degree-days = (P1 − 0.5985)/(−0.0037) ± 0.2432, and Degree-days = (logit P1 − 0.4619)/(−0.0177) ± 1.1959, with 95% confidence intervals in parentheses. Equation 1 predicted the actual age of laboratory-reared *Aedes aegypti* within 2 days (±0.1) and equation 2 within 2.1 days (±0.1). When applied to *Aedes aegypti* females of known age in a mark-recapture study in the field, the untransformed equation gave the same slope but a different intercept compared with the known actual ages of the recaptured individuals. The transformed equation yielded a different slope and intercept. Contamination of field samples with fatty acids, presumably resulting from damage to the insect cuticle in the wild, led to consistent underestimation of age. Desena et al. (1999*b*) subsequently tested just the legs of individual mosquitoes using the same extraction procedures as used for whole mosquitoes. In this study a third peak (Peak 4, identified as octacosane) was considered suitable for use in age-grading. Overall variability in cuticular hydrocarbon abundance from legs was reduced in comparison with whole body extractions, primarily as a result of decreased contamination by fatty acids, and the authors recommended the analysis of legs alone over analysis of whole body samples. Single-variable models described up to 75% of the variability in data series of up to 165 degree-days duration and 81% up to 132 degree-days. Degree-days = (P1 − 0.47)/(−1.93 × 10⁻³) (165 degree-days) and Degree-days = (logit P1 + 8.03 × 10⁻²)/(−9.24 × 10⁻³) (165 degree-days). Addition of peak 4 abundance data to a two-variable model with peak 5 increased the overall correlation by 9.5% for the 165 degree-day data set. This method would appear to show some promise where access to a gas chromatograh is available

Ampulla size

Mer (1932) found that the ampulla at the anterior end of the common oviduct increased in size during oviposition and did not return to its original size afterwards. Thus comparative measurements of the diameter of the ampulla could be used to differentiate between nulliparous and parous females. Polovodova (1941) confirmed this and also described qualitative changes in appearance of the ampulla prior to and after ovulation. Davidson (1955a) found that in nulliparous *Anopheles gambiae* the diameter of the ampulla varied according to the number of blood-meals taken by females and also the stage of ovarian development. Despite these variations he was able to use ampulla size to identify parous individuals. Hamon et al. (1961), however, considered this an unreliable and unsatisfactory method.

Tracheation of the ovaries

Undoubtedly of all the methods for age grading mosquitoes, two are outstanding. The simplest is based on the observation of Detinova (1945) that the terminations of the fine tracheoles covering the ovaries are tightly coiled into 'skeins' in females with ovaries that have not developed, but become stretched and uncoiled as the ovaries develop. This uncoiling is an irreversible change, so that the presence of skeins indicates nulliparity. This is a very simple method that demands no great dissecting skill and has become a routine and successful procedure that is widely used to determine the proportion of parous mosquitoes, and estimate survival. The method consists of dissecting out the ovaries in tap or distilled water (not saline) placing a line of 10–15 paired ovaries on a microscope slide and allowing them to dry out. It often helps if slides of dried ovaries are quickly rinsed under the tap and then shaken dry before examination under a compound microscope. So long as slides are protected against ants, cockroaches and other scavengers they can be kept more or less indefinitely before examination. Occasionally, intermediate conditions have been reported, in which a few but not all skeins are uncoiled (Blackmore and Dow 1962; Burdick and Kardos 1963; Kardos and Bellamy 1961). Kardos and Bellamy (1961) thought that this was probably due to the maturation of only a few eggs; this could be due to autogeny or an incomplete blood-meal. However, this intermediate condition is not always produced by autogeny, as shown by the absence of any skeins in adults of *Coquillettidia richiardii* that had laid the first batch of eggs autogenously (Service 1969).

Allan et al. (1981) were able to apply the tracheation method, as well as classifying the ovaries of *Coquillettidia perturbans* into the Christophers' developmental stages, after keeping females for a 'short time' at −70°C. Similarly *Ochlerotatus cantator* could be classified as nulliparous or parous after adults had been 'frozen' (Weaver and Fashing 1981). Packer and Corbet (1989) successfully identified *Ochlerotatus punctor* as parous or nulliparous after females had been stored in sealed vials for 2–12 weeks at at −35°C. They could also dissect these stored adults and count the ovariolar dilatations. Similarly Eldridge and Reeves (1990) were able to distinguish both parous and nulliparous *Ochlerotatus communis* by the tracheation method and count ovariolar dilatations after adults had been stored at −20°C. However, they noted that some specimens had tissue degradation making dissection of individual ovarioles difficult. Females of *Ochlerotatus cantans* can also be age-graded by the tracheation and dilatation methods after storing adults for months in a deep freeze (−70°C) (Renshaw 1991), as can *Psorophora ferox* (Magnarelli 1980). Haramis and Foster (1990) were also able to identify parous and nulliparous *Ochlerotatus triseriatus* after they had been stored at −40°C. They rehydrated any desiccated individuals by immersing them in 1% detergent solution for 4–12 h (Ungureanu 1972).

Ovariolar dilatations

The second important method of age determination is more complicated, requiring considerable dissection and interpretation skills, but is capable of distinguishing between nulliparous and parous females, and also allows determination of the numbers of separate egg batches laid. This is referred to as Polovodova's method (Polovodova 1949; Detinova 1949; Anufrieva and Artem'ev 1981).

Early interpretations considered an ovariolar dilatation to be caused by contraction of the sac-like distension of the ovariolar sheath formed by the passage of a mature egg down the elastic pedicel to the oviduct (Lehane 1985; Lehane and Laurence 1978). In this interpretation, during subsequent gonotrophic cycles the next egg follicle develops anterior to the small discrete dilatation, formed as a result of the first oviposition, and when it is eventually laid another dilatation is formed. Thus a two-parous mosquito has two dilatations in most ovarioles. Degenerating debris comprising epithelial cells and nurse cells is enclosed within each dilatation, and termed the follicular relic. These so-called follicular relics (corpus luteum) usually persist throughout life. Note that Detinova (1949) and many subsequent mosquito workers called the dilatations themselves follicular relics, whereas in fact these dilatations contain the follicular relics.

A different interpretation has been proposed by Khok (=Hoc) (1974), Lange and Khok (1981) and Sokolova (1981), and summarised in a more accessible publication by Hoc and Charlwood (1990) and see also Hoc and Schaub (1995) and Hoc and Wilkes (1995). These authors believe that ovariole sacs formed after oviposition do not contract to form dilatations, but remain as basal distentions of the pedicel and obscure any signs of previous gonotrophic activity within the ovariole. The number of such sacs shows the number of eggs laid during the last oviposition. They argue that dilatations are formed only when follicles degenerate at an early stage in the gonotrophic cycle. Diagnostic ovarioles are scarce in multipars, because the pedicel or pre-existing dilatations, or both, are destroyed during ovulation. Determination of parity therefore requires the examination of all ovarioles in order to identify the diagnostic ovarioles. Unfortunately, examination of all ovarioles increases the probability of detecting rogue ovarioles that have at least one more dilatation than diagnostic ovarioles (Fox and Brust 1994). In nulliparous *Culex tarsalis*, Fox and Brust (1994) observed that rogue ovarioles with two and three dilatations occurred after 12 and 18 days post-emergence respectively. An ovariole with a sac (representing successful egg development) followed by two dilatations (degenerating follicles) means that the mosquito is at least three-parous. If oviposition now occurs these dilatations are obliterated and the ovariole sheath remains stretched. But because oocytes degenerate in some ovarioles while eggs are developed and laid from others, the maximum number of dilatations still corresponds to the number of completed ovipositions. Fox and Brust (1996) by examining histological sections of whole ovaries of *Culiseta inornata* before and after ovulation in the first gonotrophic cycle confirmed that the pedicel is destroyed during ovulation. However, as the mosquito ages the numbers of ovarioles never having matured eggs decreases and so the chance of finding diagnostic ovarioles decreases, and the mosquito's age will tend to be underestimated.

Briegel and Graf (1994), based on observations primarily on *Anopheles* mosquitoes, consider the whole exercise of counting dilatations in order to elucidate reproductive history and subsequently epidemiological significance to be highly suspect. Briegel and Graf (1994) note that in *Anopheles* and also *Aedes aegypti*, frequent blood meals may be taken during a single gonotrophic cycle, with consequences for morphological analysis. These authors concluded that anatomical changes in mosquito ovaries cannot be relied upon to elucidate vectorial history.

Serial dilatations have been reported only in Diptera and are known to occur in many mosquito genera including *Anopheles, Aedes, Ochlerotatus, Culex, Culiseta, Mansonia* and *Psorophora* (Bertram and Samarawickrema 1958; Charlwood et al. 1985; Chiang et al. 1984*a,b*; Cristesco 1966; Detinova 1962, 1968; Fox and Brust 1996; Gillies and Wilkes 1965; Hitchcock 1968; Kay 1979; Loong et al. 1990; Magnarelli 1980; Reisen et al. 1979, 1980*a*, 1981, 1986; Rosay 1969; Samarawickrema 1962, 1968; Snow and Wilkes 1977; Spencer and Christian 1969).

Dissection of mosquitoes, especially small tropical species, is not easy. One of the problems is that the ovariole sheath is very fragile and liable to break when the ovariole stalk (pedicel) is being stretched to search for dilatations. Consequently a relatively large number of ovarioles have to be examined before the maximum number of dilatations counted can be taken as the number of gonotrophic cycles that the female has undergone. In Australia Russell (1986), working with *Culex annulirostris*, found that at least 10 ovarioles/ovary had to be examined to get a reliable count of the maximum number of dilatations. This makes dissections very time-consuming and so relatively few adults can be dissected from any population.

Another problem is that sometimes egg maturation is aborted with the result that the developing follicle is reabsorbed leaving behind a follicular relic, which is often indistinguishable from the follicular relic and dilatation formed after oviposition (Bellamy and Corbet 1974). Furthermore, if an egg then develops in the ovariole and oviposition occurs, the dilatation formed by a degenerating follicle may disappear (Lange and Khok 1981). The problem of degenerating follicles has been reported in *Aedes aegypti, Ochlerotatus nigromaculis, Culex annulirostris, Culex erraticus, Culex tarsalis* and many other *Culex* species, and also in *Anopheles atroparvus* and *Culiseta inornata*.

Another constraint is that in some species the sac formed after the first oviposition does not contract to form a discrete dilatation and further sacs or dilatations are not formed. Thus, even in multiparous females only a single sac-like dilatation is formed irrespective of the number of ovipositions. This condition has been reported in several species, including *Aedes aegypti, Anopheles atroparvus* and *Culex pipiens* and other *Culex* species (Detinova 1968; Kay 1979; Nayar and Knight 1981; Tyndale-Biscoe 1984 for references). For example, multiple dilatations are not always formed in *Culex tarsalis* (Nelson 1964), while abortive oogenesis may falsely indicate parity in *Culex tritaeniorhynchus* (Yajima 1970). Rosay (1969) reported that dilatations sometimes coalesced in *Culex quinquefasciatus* and *Culex tarsalis*, and aberrant dilatations were encountered in anautogenous adults of these two species and *Culiseta inornata* that had never taken blood. These dilatations are caused by resorption while new follicles are being produced by the germarium. These rogue ovarioles have been reported in field collected females of *Culiseta inornata* (Reisen et al. 1989). Rather similarly, Bellamy and Corbet (1973, 1974) found that in autogenous *Culex tarsalis* aberrant dilatations formed if the primary follicle failed to mature and degenerated, or was resorbed. Nayar and Knight

(1981) reported that 16–50% of laboratory reared *Culex nigripalpus* fed only 10% sucrose solutions had one aberrant or rogue dilatation 1 week after emergence, and during the next 2 weeks more than 90% had one to three aberrant dilatations. They concluded that the ovarian tracheation method may sometimes be more reliable for identifying parity in *Culex* mosquitoes. Mahmood and Crans (1998) reported that sugar feeding by nulliparous females of *Culiseta melanura* post-emergence leads to the formation of three types of ovarioles: 1) large primary follicles that subsequently develop into functional ovarioles; 2) small primary follicles that developed small amounts of yolk following a blood meal, but then degenerated; and 3) small primary follicles in which no yolk developed after blood-feeding, and in which uncontrolled cellular division occurred, giving rise to multiple false dilatations and becoming rogue ovarioles. Rogue ovarioles could be differentiated from true dilatations, as the former are attached to undeveloped tertiary follicles rather than fully developed secondary follicles. These authors concluded that the use of false dilatations could, however, be a useful tool for age-grading *Culiseta melanura* females, a species in which true dilatations are difficult to find. Nulliparous females with primary follicles at stage V have single false dilatations on degenerated ovarioles. One-parous gravid females have two false dilatations on each degenerating ovariole.

Fox and Brust (1996) reported that there was a decline in the number of 0-dilated ovarioles in nulliparous *Culiseta inornata* over a 34 day period. Rogue ovarioles with a single dilatation were observed after six days and rogue ovarioles with two dilatations were observed after 18 days. In addition, 10% of 2-pars lacked a diagnostic ovariole. The mean diagnostic index, that is the proportion of ovarioles that were diagnostic, was 46.0 ± 2.1% (12.3–100%) for 1-pars ($n = 47$) and 6.8 ± 1.0% (0.0–24.0%) for 2-pars ($n = 44$). These authors proposed the following parity-diagnostic criteria for anautogenous females: (1a) >20 0-dilated ovarioles = nullipar, (1b) ≥15 1-dilated ovarioles = 2-par, (2a) 0–4 2-dilated ovarioles = 1-parous or higher, (2b) ≥5 2-dilated ovarioles = 2-parous or higher.

Another problem is that some workers have recorded just a single dilatation in multiparous ovarioles of certain species, while others have counted serial dilatations in the same species. For example, in The Gambia Giglioli (1965) could find only one dilatation in *Anopheles melas*, whereas Snow and Wilkes (1977) working in the same country reported several distinct dilatations. Somewhat similarly Rosay (1969) found that in *Culex quinquefasciatus* there was coalescence of dilatations after repeated ovulations, but Samarawickrema (1967) had no difficulty in counting serial dilatations in this species. Similar difficulties in recognising dilatations have been found in *Culex tarsalis* (Nelson 1964), in several *Anopheles* species (Brady 1963) and *Ochlerotatus cantans* (Service 1993), but compare with Renshaw (1991) and Renshaw et al. (1994) who were able to discern a maximum of five dilatations in *Ochlerotatus cantans*. Multiple dilatations were reported in 18 aedine species examined by Carpenter and Nielsen (1965) and the method has been applied to several other aedine species (Hájková 1966; Shlenova and Bey-Bienko 1962; Volozina 1958; and for a review see Detinova 1968). Jensen et al. (1998) dissected *Anopheles punctipennis* caught in landing collections in California, USA and classified them as nulliparous or parous based on the condition of ovarian tracheoles. In 328 individuals, the tracheoles of one ovary were examined and the other ovary was stained, dissected, and number of dilatations on the pedicel of the ovarioles was counted. The overall parity rate observed was 0.82. Multiple dilatations were observed in the ovaries of 46% of the dissected mosquitoes. Individual females had as many as eight dilatations per ovariole indicating completion of eight gonotrophic cycles prior to collection, although laboratory studies were not carried out to ensure that multiple oviposition events can be reliably recognised by serial dilatations in this species.

In California, USA Reisen et al. (1995) were able to identify a maximum of three dilatations in 0.2–0.3% of *Culex tarsalis* caught in CDC light-traps supplemented with CO_2 and with the bulb removed. Reisen et al. (1995) also carried out blind tests to evaluate the accuracy of the dilatation method in *Culex tarsalis* and found that 96% of females that had oviposited once were correctly identified, while only 11% of those that had oviposited twice were correctly scored. Tests revealed that females were more frequently allocated to a lower parity category than the true parity (79%), rather than to a higher category. Dissection of 37 autogenous and anautogenous females of known oviposition history revealed that 15 of 15 1-pars exhibited a single dilatation, 12 of 16 2-pars had two dilatations and 2 of 6 3-pars had three dilatations.

The dissection of mosquito ovaries to count sequential dilatations formed during gonotrophic cycles remains a difficult technique requiring a certain 'knack'. In dissecting *Culex annulirostris* Kay (1979) found that the addition of Giemsa to the dissecting medium of physiological saline aided the detection of ovariole dilatations. He examined at least 6–10 ovarioles/female, and took the largest number of dilatations recorded as the number of ovipositions. Kay (1979) considered 60 dissections/day could be achieved, but Service (1993) reported taking 15–20 min to examine a single female.

Hoc and Charlwood (1990) describe an oil injection technique used to make the recognition of dilatations easier, that was originally developed in Moscow in 1973 and first published by Lange et al. (1981) in Russian.

Ovaries are dissected in Hayes saline containing neutral red stain (about 0.005%). A Pasteur pipette is drawn out to form a short length of very thin glass tubing which is filled with light paraffin or cooking oil. A small length of rubber tubing and a matchstick are added to the end (Fig. 13.1). The very fine glass pipette is introduced into the common oviduct and the rubber tubing squeezed to inject the oil. Lange et al. (1981) describe how the micropipette is obliquely introduced through the oviduct wall, and that in parous females both ovaries can be filled with oil, but with nulliparous ones they have to be filled separately because the paired oviducts are closed by an elastic membrane that requires some effort to pierce. Hoc (1996) suggested that inability to fill both ovaries via the common oviduct is indicative of nulliparity, avoiding any necessity for further dissection. If too much oil is injected the ovary splits apart. They suggest it is best to cut the ovary longitudinally with fine scissors. Whichever technique is used it inflates the ovaries and the ovarian sheath is rapidly cut with fine dissecting needles, which results in the ovarioles spreading out on the slide thus facilitating examination of their dilatations. Sokolova (1995) used a hanging drop preparation to examine the ovarioles. The oil injection technique is difficult and time-consuming and many unsuccessful attempts may ensue. A major problem is encountered in getting the oil to flow from the pipette into the oviducts. Hoc and Schaub (1996) later refined the oil-injection technique as follows: the micropipette apparatus is essentially the same as that illustrated in figure 13.1. A 1 ml syringe is used to fill a micropipette with oil (paraffin, vaseline, cooking oil) by inserting the syringe needle into the micropipette and withdrawing it slowly, avoiding introduction of air bubbles. The recommended diameter of the micropipette tip for examination of mosquito ovarioles is about 0.06 mm. Ovaries are dissected in physiological saline containing freshly prepared neutral red solution (1:5000 to 1:8000). Ovaries are filled with oil via the common oviduct to inflate it and separate out the ovarioles. The ovary is then cut lengthwise along the ventral surface using dissecting needles. A semi-permanent preparation can be made by replacing physiological saline with a sodium chloride-glycerol-formaldehyde (SGF) solution (25 parts Hayes saline, 10 parts glycerol and 1 part 3% formaldehyde) and sealing the edges of the coverslip with nail varnish. In the same paper, Hoc and Schaub (1996) also describe a new method, termed the ovariolar separation technique. In this technique ovaries are dissected from a mosquito and placed in a drop of 5% Carnoy's solution in distilled water and allowed to rest for 1 min, during which time the ovarian sheath loosens and the ovarioles separate. The Carnoy's solution is then removed with a piece of filter paper and replaced with SGF solution. After a coverslip is placed over the preparation, it can be examined, preferably using phase-contrast microscopy.

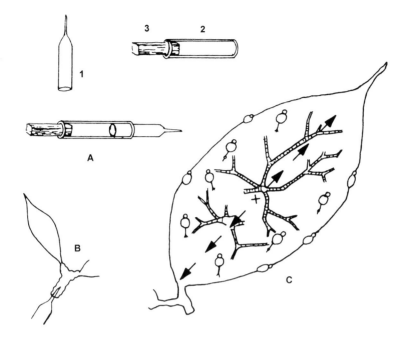

Fig. 13.1 Components of a micropipette used for injecting paraffin oil into mosquito ovaries and schematic representation of an expanded ovary. 1—glass pipette, 2—rubber tube, 3—matchstick, A—complete pipette, B— insertion of the pipette into the common oviduct, C—appearance of expanded ovary. The ovary is cut (in the directions shown by the arrows) with sharp needles from the point marked × (Hoc and Charlwood 1990)

This second technique was thought by the authors to be slightly easier to perform, but required accurate timing of the replacement of Carnoy's with SGF in order to avoid damage to the ovarioles.

Hoc (1996*a*) applied both the modified oil injection technique and the ovariolar separation technique to the examination of ovaries of several representatives of Culicidae, Simuliidae, and Tabanidae. The author observed that in the tabanid *Chrysops bicolor* and the blackfly *Simulium woodi*, the ovaries were robust and elastic and inflation with oil was often difficult. In addition, after cutting the ovarian sheath, the ovarioles in these two species often tended to clump together rather than spread out. In parous females, three different types of functional ovarioles were observed: 1) ovarioles with dilatations and a basal body (diagnostic ovarioles); 2) ovarioles possessing only a granular basal body; and 3) ovarioles with a granular basal body and a variable number of dilatations. The ovariolar separation technique produced similar results (as long as the timing of replacement of Carnoy's with SGF was accurate), although using this method the preparations are not stained. Hoc (1996a) proposed that parity could be directly determined by counting the number of dilatations in diagnostic ovarioles. Approximations of parity level can also be obtained by counting the number of dilatations plus 1 (the granular basal body) in those ovarioles that possess a granular basal body and a variable number of dilatations.

The presence of uncontracted sac-like dilatations was used by Detinova (1953) to determine whether re-feeding occurred within about 24 h after oviposition. Since then several workers have related the degree of sacciform contraction with different intervals following oviposition, for example in *Mansonia* species (Corbet 1964; Samarawickrema 1962, 1967, 1968), and *Anopheles gambiae* (Gillies and Wilkes 1965).

In summary it would appear that the anatomical changes associated with oogenesis and oviposition, while well described by several authors, remain open to interpretation. In addition, some authors question the relevance of these anatomical changes in determining physiological age or vectorial status of mosquitoes.

Granular basal bodies

Spencer (1979) described how she could age grade *Anopheles farauti* into nulliparous, 1-, 2-, 3-, 4- and 5- or more parous categories, not by counting serial dilatations but by examining intact ovaries and recording the changes in the appearance of granular material in the ovariole sacs and contracted dilatations, and the presence or absence of globules in the ovarian sheath. For example, ovarioles with only diffuse granulation are 1-parous, those with discrete masses of compact granulations occurring singly or in pairs are 2-parous.

Hoc and Schaub (1995) and Hoc (1996b) describe a simple method for distinguishing nulliparous from parous mosquitoes by staining dissected ovaries with vital red. Ovaries of parous females contain granular basal bodies, which stain red. These granular bodies are formed from the remnants of the egg sac, which result in a permanent granulation developing in the basal body cells. Fresh ovaries are dissected in a drop of neutral red solution in Hayes saline (0.6% NaCl solution) at a concentration of 1:3000–1:5000. For ovaries at Christophers' stages III and above, the ovarian sheath should be removed with a dissecting needle. The ovary preparation is allowed to stand for one to three minutes before examination under a stereo microscope (magnification 40X) against a white background. Preparations can also be examined under a compound microscope using a green filter, or alternatively a fluorescence microscope. In ovaries at Christophers' stages III, IV and V the granular basal bodies may be obscured by developing follicles and are best seen at the posterior end of the ovary or after manipulating the terminal follicles with a dissecting needle. Hoc and Schaub (1995) postulate that the increase in optical density of the granulation in the basal body with each successive oviposition could serve as a means to estimate parity.

Gryaznov (1995) observed both granular basal bodies and follicular relics (=ovariolar dilatations) in Russian and Canadian blackflies, when ovaries were prepared in neutral red solution. Absence of zones of granulation and follicular relics indicated nulliparity. In wild-collected females some contained either two zones of granulation or one follicular relic and one zone of granulation, and these individuals were presumed to be 2-parous. No apparently 3-parous females were encountered.

Female age

The product of the numbers of serial dilatations and intervals between successive gonotrophic cycles can provide an estimate of the age of females (Samarawickrema 1962; Gillies and Wilkes 1965). In north–eastern Tanzania these last two workers found that the first gonotrophic cycle in *Anopheles gambiae* lasted 3–4 days and later cycles 3 days. Newly emerged laboratory-reared females were marked and released so that recaptured individuals could be aged in days and also physiologically by the number of dilatations. A good correlation was found between the theoretical number of ovipositions that females of different ages could have made and the actual number as determined by dissections. It would, however, be rash to assume that in all species there is the minimum or even regular interval between feeding and oviposition and between oviposition and re-feeding. In fact this clearly did not occur in a population of *Ochlerotatus cantans* studied for several years in England (Service 1973*b*), in which some adults were nulliparous 6–7 weeks after emergence from natural habitats. Hamon et al. (1961) found that in natural populations of *Anopheles gambiae* the intervals between successive blood-meals were greater than the duration of the gonotrophic cycle. As shown in the section on estimation of adult survivorship delays between repetitive gonotrophic cycles can invalidate the estimation of survival rates based on parity.

Parasitic infections

Draper and Davidson (1953) introduced a method of estimating the survival rate of female mosquitoes based on immediate and delayed sporozoite rates. Later the age of filariae in mosquitoes, as determined by the morphology of their various developmental stages, was used by several workers to estimate survival rates of populations, e.g. *Culex quinquefasciatus* and *Anopheles peditaeniatus* in India (Laurence 1963), *Anopheles farauti* and *Ochlerotatus kochi* in Irian Jaya of Indonesia (van Dijk 1966), *Culex quinquefasciatus* in Sri Lanka (Samarawickrema 1967), *Culex quinquefasciatus* in Trinidad (Nathan 1981) and *Anopheles funestus* in Tanzania (Krafsur and Garrett-Jones 1977). In contrast Bryan (1986) could not use this approach with *Anopheles punctulatus* in Papua New Guinea because the filariae were causing mosquito mortality.

Estimation of adult survival rates

Mortality and age

Many studies have suggested that adult survivorship of mosquitoes is constant with age, that is resembling a type III survivorship curve of Slobodkin (1962). For example, both Laurence (1963) and De Meillon et al. (1967a) concluded that daily mortality was about constant with age, thus supporting Macdonald's (1952, 1957) theory that mosquitoes die too rapidly from a variety of factors than to die of old age. Similar conclusions were reached by Gillies (1961) in East Africa, Hitchcock (1968) in North America and Zalutskaya (1959) in Vietnam. There is little information on the causes of adult mortality, but Gillies (1961) suggested that in tropical species predators might be the most important regulators of population size, and this is also likely to be true in temperate regions. Graham and Bradley (1972), however, were not really convinced that in the population of *Culex quinquefasciatus* they were studying in Yangon, Myanmar, survival rate was constant. Clements and Paterson (1981) from a reanalysis of published survival data suggested that in many species mortality rates increased with age (Fig. 13.2) and thus fitted the Gompertz mortality function (i.e. a type I curve of Slobodkin (1962)).

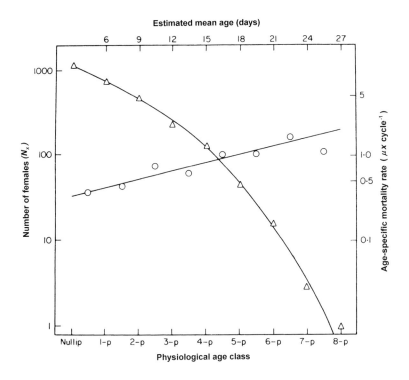

Fig. 13.2. Survival of female *Anopheles arabiensis* at Gonja, Tanzania. The size of the nulliparous sample has been adjusted. The fitted curves correspond to the Gompertz mortality and survival functions, α = 0.339, β = 0.225 (per cycle). From data of Gillies and Wilkes (1965) (Clements and Paterson 1981)

If true then the generally used exponential survivorship curve $S_t = S_0 e^{-\alpha t}$ is inappropriate, and the survivorship curve is in some, if not many, cases best obtained from the Gompertz model, thus

$$S_t = S_0 \exp\left(-\frac{\alpha}{\beta}(e^{-\beta t} - 1)\right) \tag{13.2}$$

where a and β = constants, e = base of natural logarithm. In practice the numbers of mosquitoes of a particular age N_x (e.g. 1-, 2-, 3-parous etc) are plotted as ln (N_x) against age to produce a survivorship curve. If an approximately straight line is obtained then the simple exponential model is probably appropriate, that is mortality is independent of age, and a straight line can be fitted by eye or by linear regression of ln (N_t) against age. The slope provides a good estimate of mortality rate $\mu(t)$, which is taken as constant (α). This can be calculated from

$$\alpha = \frac{\ln(y_1) - \ln(y_2)}{x_2 - x_1} \tag{13.3}$$

where x_1, y_1 and x_2, y_2 are two points on the regression line. The fitted curve is then calculated as follows.

$$\hat{N}_x = N_0 e^{-\alpha x} \tag{13.4}$$

If the points of the survivorship curve do not fall on a straight line but follow the Gompertz function on a curve that is concave below (Fig. 13.2) it indicates that mortality probably increases with age and the Gompertz curve describes mortality. A smooth curve is drawn by eye, and three or more tangents are drawn to this curve at well spaced intervals. Their slopes will represent instantaneous mortality rates (μ). These values of μ are plotted on a logarithmic scale against age on a linear scale, and if, as expected, they fall on a rising curve this confirms that mortality increases with the mosquito's age.

In studies in Pakistan Reisen et al. (1986) tested the presumptive Gompertz model by regressing the mortality of adults of *Anopheles culicifacies, Culex tritaeniorhynchus* and *Anopheles stephensi* as an exponential function of their age. Calculated regressions were not significantly different from 0, and the goodness of fit to the data for the model was generally poor. They considered that mortality was independent of age, and this substantiated results of previous studies on these three species, which in mark–release experiments showed recaptures decreased exponentially with time, strongly suggesting a constant loss rate (Reisen et al. 1978, 1980*a*; Reisen and Aslamkhan 1979). Amalraj and Das (1996) studied the life-table characteristics of *Toxorhynchites splendens* in the laboratory under different food regimens and concluded that female survivorship curves approximated Slobodkin Type-II curves, whereas males exhibited Slobodkin Type-II or Type-III curves.

Charlwood (1986) found that data on the survival rate of *Anopheles farauti* fitted the Gompertz model better than an exponential linear model, however, in Tanzania, population size of *Anopheles arabiensis*, as determined by CDC light-trap collections, declined exponentially and a log-linear model indicated a decline in catch of 0.191 per day, equivalent to a daily survival rate of 0.827 (Charlwood et al. 1995). Charlwood et al. (2000) fitted mark-release-recapture data to the Gompertz model and an exponential model in a study of the survival of anophelines during the dry season in Kilombero, Tanzania. An estimated 353 *Anopheles funestus* females and 155 males were collected resting in houses, marked with fluorescent powder and released in the house located closest to the presumed larval habitat. Survival curves were constructed and daily survival for this population, derived from the exponential model was 83.7%. In addition, the gonotrophic age of 46 *Anopheles arabiensis* and 80 *Anopheles gambiae* wase determined and survival curves constructed. The Gompertz survival model best described the *Anopheles arabiensis* and *Anopheles gambiae* data, indicating that survival decreased with age, but both Gompertz and exponential models fitted the *Anopheles funestus* data equally well. *Anopheles gambiae* was the most long-lived species, an estimated 6.3% surviving for four or more gonotrophic cycles, compared to 2.0% of the *Anopheles arabiensis* and 4.0% of the *Anopheles funestus*.

It is still debatable whether mortality of mosquitoes is exponential or follows the Gompertz model and in fact it may differ in different situations. If the latter model is suspected then the paper of Clements and Paterson (1981) should be consulted for further analysis and calculation of other parameters such as probability of a mosquito surviving n days, and the expectation of life at birth. They showed that there could be considerable discrepancies in the estimation of longevity factors based on exponential and Gompertz models, and clearly this has direct bearing on the epidemiology of disease transmission.

Percentage of green individuals in *Culex quinquefasciatus*

Graham and Bradley (1972) investigated the changes in age composition of adults of a population of *Culex quinquefasciatus* in Yangon, Myanmar after an intensive larviciding programme. They were not really convinced that the survival rate of adults was constant, but nevertheless made this assumption to compare the relative ages of adults in an area that had been sprayed, with adults in an untreated control area. If survival rate is in fact independent of age, and also relatively constant, then the numbers of survivors after t days, P_t, from an initial population at emergence, P_0, is estimated as follows:

$$\frac{dP}{dt} = -rP_0, \text{ or more conveniently } p_t = e^{-rt}P_0, \tag{13.5}$$

where e^{-rt} = the proportion of survivors after t days. Newly emerged females are green in colour and do not lose this coloration until 3 days old (see above). So if the proportion of non-green adults (Q) can be established in the population, and the number of adults emerging is more or less constant, then:

$$Q = \frac{P_3}{P_0} = e^{-3r}, \qquad (13.6)$$

Now, if reliable estimates are made of Q and P_0, then P_3 can be estimated and the daily death rate (r) calculated. The daily survival rate (s) is found as $s = 1 - r$. Once r is known then the proportion of the population surviving, e^{-rt}, for $t = 1, 2, 3, \ldots n$ days, can be solved. For example, in an untreated area Graham and Bradley (1972) estimated P_0 as 50 000 and Q as 0.267. Thus the numbers of survivors at 3 days was calculated as 13 350, whereas in a sprayed area P_0 was estimated as 1000 and Q as between 0.657–0.713, which gave the numbers of survivors after 3 days as between 657–713 adults. Clearly the adult population in the sprayed area had a higher survival rate so the proportion of old mosquitoes was larger than in the untreated area.

Davidson's parous rate method

Davidson (1954) and Davidson and Draper (1953) showed that daily survival rates can be calculated from the proportion of parous females in a population. If the gonotrophic cycle lasts 2 days then the daily survival rate is estimated as the square root of the parous rate, if it lasts 3 days then the cube root is used and so on. For example, if 74% of the population is parous and the gonotrophic cycle is 4 days then $p^4 = 0.74$ and daily survival rate is 0.927. The estimated daily mortality is 100 $(1 - p)$, in this example 7.96%. The probability of a mosquito surviving through two gonotrophic cycles both of 4 days is p^8, similarly the probability of a mosquito surviving through 3, 4 or n gonotrophic cycles etc. is p^{12}, p^{16} and p^{4n}. Davidson (1954) was careful to point out that this simple derivation only worked so long as the interval between successive gonotrophic cycles was accurately known in the field. This, however, may not be easy to obtain, and cannot always be extrapolated from laboratory observations on the time taken for a mosquito to become gravid after blood-feeding.

If females require two blood-meals for the first oviposition or have irregular feeding rhythms then the calculation of daily survival from the proportion parous is not quite so simple. Garrett-Jones and Grab (1964) computed daily survival rates for a variety of such feeding patterns (Table 13.1).

Table 13.1. Proportions of mosquitoes parous and corresponding daily survival rates for specified feeding rhythms and ovipositions[a] (After Garrett-Jones and Grab 1964)

Observed proportions parous, related to rhythm no.						Computable value of p
(1)	(2)	(3)	(4)	(5)	(6)	
0.898	0.833	0.815	0.807	0.740	0.730	0.95
0.793	0.708	0.656	0.632	0.546	0.520	0.90
0.689	0.575	0.522	0.482	0.400	0.362	0.85
0.587	0.457	0.410	0.357	0.291	0.247	0.80
0.491	0.353	0.316	0.257	0.208	0.165	0.75
0.402	0.268	0.240	0.181	0.146	0.108	0.70
0.323	0.198	0.179	0.124	0.101	0.068 1	0.65
0.252	0.142	0.130	0.082 1	0.068 0	0.042 0	0.60
0.192	0.098 8	0.091 5	0.052 5	0.044 2	0.024 9	0.55
0.143	0.066 6	0.062 5	0.032 3	0.027 8	0.014 1	0.50

[a]Rhythm no.	Supposed days of feeding and time of first oviposition (//)	Expression of proportion parous in terms of daily survival rate
1	2, // 5, 7, 9, 11, …	$\dfrac{p^3}{1-p^2+p^3}$
2	2, //6, 9, 12, 15, …	$\dfrac{p^4}{1-p^3+p^4}$
3	2, 4, // 6, 8, 10, …[b]	p^4
4	2, 4, // 7, 9, 11, …	$\dfrac{p^5}{1-p^4+p^5}$
5	2, 4, // 7, 10, 13, …	$\dfrac{p^5}{1+p^2-p^3}$
6	2, 4, // 7, 11, 14, …	$\dfrac{p^6}{1-p^2-p^3-p^5+p^6}$

[b]it should be noted that this rhythm would give the same proportion parous in the biting sample (i.e. a proportion equivalent to p^4) as would a regular rhythm with an oviposition interval of 4 days, such as:

$$2, // 6, 10, 14, 18, …$$

or

$$3, // 7, 11, 15, 19, …$$

They produced useful graphs for estimating daily survival rates from parous rates obtained from a biting population, with ovaries in stage I to late stage II, related to different feeding cycles (Fig. 13.3).

Due to its simplicity, both in determining parity and calculation of the survival rate, this is the most frequently used method of determining adult survival rates in field investigations, and some examples of its use are provided below.

Chandra et al. (1996) calculated daily survival rates of *Culex quinquefasciatus* in Calcutta, India using Davidson's (1954) method to be 0.87 overall and 0.86 in the rainy season, 0.91 in winter, and 0.81 in summer. In the Philippines Walker et al. (1998) calculated the daily survival rate of *Ochlerotatus poicilius*, a vector of *Wuchereria bancrofti*, to be 0.85 based on a parous rate of 0.62 and a gonotrophic cycle duration of 3 days. The daily survival rate of *Anopheles dirus* in Lao PDR was calculated by Sidavong et al. (2004), using the Davidson (1954) method, to range from 0.54 in Pier Geo village in August 2002 to 1.0 in the same village in October. In the majority of villages and throughout May to October, survival rates were between 0.84 and 0.92. In Sekong Province, Lao PDR, Vythilingham et al. (2003) calculated adult survival rates for *Anopheles minimus* and *Anopheles maculatus* using a gonotrophic cycle length of two days in the wet season and three in the dry season, and for *Anopheles jeyporensis* and *Anopheles dirus* a gonotrophic cycle of three days was used. For *Anopheles minimus* the daily probability of survival in the dry and wet seasons was 0.77 (parous rate 60.0% and 46.2% respectively). For *Anopheles maculatus* daily survival rates were 0.75 in the wet season and 0.86 in the dry season. Daily survival rates for *Anopheles jeyporiensis* and *Anopheles dirus* were 0.86 (wet) and 0.89 (dry) and 0.85 (wet) and 0.91 (dry) respectively. Han Il Ree et al. (2001) used the Davidson method to calculate daily survival of adult *Anopheles sinensis* females collected from cow-sheds using light-traps in Korea. These authors used a gonotrophic cycle duration of 3 days based on laboratory observations of developing follicles of field-collected fully blood-fed females. Parous rates ranged from 47.6% to 79.1%, equivalent to daily survival rates of between 0.781 and 0.925.

In Bangladesh Rosenberg and Maheswary (1982) used the proportion of parous *Anopheles dirus* caught in human bait collections and the duration of the gonotrophic cycle to estimate daily survival rates (p), and the percentage of the population expected to live sufficiently long to become infective with malarial sporozoites (p'), and finally the subsequent life-expectancy of this population ($p'/-\log_e p$).

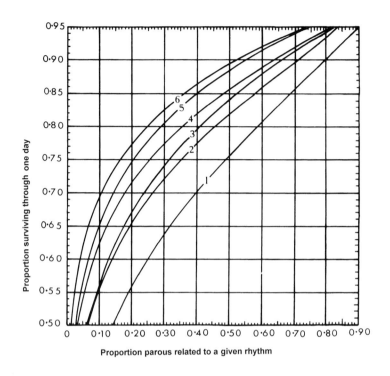

Fig. 13.3. Curves for deriving the proportion of mosquitoes surviving through 1 day (p) from the observed proportion parous, related to the various irregular feeding and oviposition rhythms given in Table 10.6. From data given by Garrett-Jones and Grab (1964)

It is not always easy to know accurately the duration of the gonotrophic cycle, because so much depends on the mosquito's ability to find hosts and oviposition sites. In their studies on survival rates Rosenberg and Maheswary (1982) took an arbitrary 4 days as the feeding cycle, but they point out this is a sensitive parameter, because if the parous rate is 70% and the cycle is 3 days then $p = 0.888$, but if the cycle is extended a day $p = 0.915$ and for a 5-day cycle $p = 0.931$. Using the formula proposed by Macdonald (1957) they also calculated the effective gametocyte rate (x) from the sporozoite rate (s), daily survival rate (p), and the average number of people bitten by a single *Anopheles dirus*/day (a), thus

$$x = \frac{s(-\log_e p)}{a(p^l - s)} \tag{13.7}$$

When the expected numbers of gametocyte carriers estimated from the monthly gametocyte rates (x) were compared with the actual numbers of gametocyte carriers identified by blood slides there was good agreement for 7 months, but statistical differences during the other months. Generally gametocyte numbers based on entomological data were fewer than those actually found, except in June 1978 when the expected number was 38.2 but just seven were found on blood slide examination.

Reisen et al. (1991) estimated the daily survivorship of *Culex quinque-fasciatus, Culex stigmatosoma* and *Culex tarsalis* vertically by Davidson's (1954) method. The survival rate of *Culex quinquefasciatus* was estimated as 0.84, which compared well with the horizontal method of regressing re-captured marked adults against cohort age in days which calculated the survival rate as 0.82. In California, USA, Reisen et al. (1995) used the par-ity method of Davidson (1954) to estimate survivorship of *Culex tarsalis*, using an adjusted parity rate to take into account the proportion of the population that was autogenous and not attracted to CO_2 traps. Survivor-ship varied on a monthly basis and ranged from 0.720 in March 1992 to 0.944 in February 1993 at one site and from 0.528 in September 1992 to 0.960 in January 1993 at a second site. In investigations of an outbreak of LaCrosse virus in eastern Tennessee, USA Gottfried et al. (2002) used the Davidson method to calculate adult survival rates of *Aedes albopictus*, us-ing published estimates of gonotrophic cycle lengths for this species of 4.6 days at 24°C and 10 days at 20°C (Hawley 1988). Calculated daily survival rates during the 4-month period of study ranged from 96% in September (mean temperature 24.9°C), using the estimated 4.6 day gonotrophic cycle length, to 99% during October (mean temperature 15.7°C), using the 10-day estimated gonotrophic cycle length.

The daily survival rate of *Anopheles gambiae* (a sample of which was identified by cytotaxonomy as *Anopheles gambiae* sensu stricto, forest form) females in rural Kinshasa was calculated by Coene (1993) using the parous rate method and an estimated period between two blood meals of 2.5 days. Daily survival rates varied between 0.63 and 0.94 in Kimbangu 3 (urban) and between 0.85 and 0.96 in Kwamuthu (rural). Touré et al. (1996) calculated the daily survival rates of *Anopheles gambiae* s.l. (in-cluding *Anopheles gambiae* s.s. [90% Mopti chromosomal form], and *Anopheles arabiensis*) in Mali from parous rates and an average gonotrophic cycle length of 2 days to be 0.89 in October (parous rate 80/99, 80.8%) and 0.61 in April (parous rate 57/154, 37.0%).

In Gabon, Elissa et al. (1999) used a gonotrophic cycle length of 2.5 days to calculate survival of several anopheline species from parity data using the formula of Davidson (1954). In a suburban zone of Franceville, the average parity rate of *Anopheles funestus* was about 93% ($n = 353$), equivalent to a daily survival rate of $p = 0.97$; for *Anopheles gambiae* s.l. parity was 90% ($n = 238$) and $p = 0.96$; and for *Anopheles nili* the parity rate was 95% ($n = 28$) and $p = 0.98$. In the rural zone, the average parity rate of *Anopheles funestus* was 87% ($n = 2998$), equivalent to a daily survival rate of $p = 0.95$; for *Anopheles gambiae* s.l. it was 84% ($n = 1643$), $p = 0.93$; for *Anopheles nili* the parity rate was 77% ($n = 833$) with $p = 0.90$; and for *Anopheles moucheti*, it was 72% ($n = 375$) and $p = 0.88$.

Daily survivorship rates of *Anopheles gambiae* s.l., *Anopheles funestus*, and *Anopheles pharoensis* in Ghana were calculated from parous rates, using a gonotrophic cycle length of three days (Appawu et al. 2001). At a coastal forest site, daily survival rates of *Anopheles gambiae* s.l. were estimated at 0.94 in 1992 and 0.76 in 1993. Survival rates of the same species in a savannah area were similar at 0.94 in 1992 and 0.74 in 1993. In south–eastern Tanzania, Charlwood (1997) estimated daily survival rates from the parous rate of anophelines collected by CDC light-trap. He was unable to use the preferred time series method of Birley and Boorman (1982) due to an absence of significant cross-correlations. Daily survival rate estimates were 0.788 for *Anopheles funestus*, 0.777 for *Anopheles gambiae* s.l., 0.794 for *Anopheles squamosus*, 0.737 for *Anopheles rivulorum*, and 0.815 for *Anopheles coustani*.

Daily survivorship of *Aedes vexans* and *Ochlerotatus melanimon* was estimated by Jensen and Washino (1991, 1994) by regressing the $\ln(y + 1)$ transformed number of marked mosquitoes collected each day as a function of days after release. Daily survivorship for *Aedes vexans* was estimated at 0.704 per day, and for *Ochlerotatus melanimon* it was estimated at 0.844 per day. The authors acknowledged that in this instance measurement of daily survivorship includes both mortality and emigration from the sampling area, however, the 0.844 survivorship estimate obtained for *Ochlerotatus melanimon* was only slightly lower than the 0.88 estimate obtained from a time series analysis of the number of parous and nulliparous females of unmarked *Ochlerotatus melanimon* conducted by Jensen and Washino (1991). Using Davidson's (1954) parous rate method and a gonotrophic cycle duration of 3 days, as determined by mark-release-recapture studies, Fernandez-Salas et al. (1994) estimated daily survival of *Anopheles pseudopunctipennis* in southern Mexico at 0.875 in the early dry season and 0.884 in the later dry season.

Vercruysse method

Obtaining an accurate estimate of daily survival (p) depends on unbiased sampling. For example, in East Africa Gillies and Wilkes (1963) showed that there was a deficit of nulliparous *Anopheles funestus* resting indoors, so indoor collections will underestimate this age-group. Consequently it seems better to use mosquitoes caught in landing catches for estimating their daily survival rates. So, in Senegal to estimate p for *Anopheles arabiensis* Vercruysse (1985) used only adults caught at human bait. He classified females with ovaries in Christopher's stage I or early II as pre-gravid nullipars, and those with ovaries at mid or late stage II as non-pre-gravids, while an examination of the ovariole tracheoles allowed this group to be classified as nullipars or pars. He firstly calculated the survival rate of each of these age-groups (pre-gravids, nullipars and pars) and then estimated the overall survival rate of the biting population. For this he used the following formulae

$$E(NP_1) = xp \tag{13.8}$$

i.e. Number of pre-gravids (at their first meal, ovaries stage I to early stage II).

$$E(NP_2) = xp^2 \tag{13.9}$$

i.e. Number of nullipars (at their second meal, 1 day after the first, ovaries mid or late stage II).

$$E(P) = xp^4 \frac{1}{1-p^2} \tag{13.10}$$

i.e. Numbers of parous females (representing 1st oviposition 4 days after emergence, subsequent ovipositions every 2 days)
   Consequently it follows that

$$E(NP_2) = pE(NP_1) \tag{13.11}$$

And

$$E(P) = p^2 E(NP_2) + E(P) \tag{13.12}$$

Now if n is the proportion of each age-group in the population being caught at bait, then

$$n_1 = NP_1/(NP_1 + NP_2 + P) \tag{13.13}$$

$$n_2 = NP_2/(NP_1 + NP_2 + P) \tag{13.14}$$

$$n_3 = P/(NP_1 + NP_2 + P) \tag{13.15}$$

and clearly $n_1 + n_2 + n_3 = 1$.

During his study Vercruysse (1985) caught 2370 female *Anopheles arabiensis* at human bait, of which 1123 were parous (P), 642 nullipars were pre-gravids (NP_1) and 605 were old nullipars (NP_2)

Thus

$$n_1 = 0.271, \; n_2 = 0.255 \text{ and } n_3 = 0.474 \tag{13.16}$$

$$E(NP_2)/E(NP_1) = 605/642 = 0.941, \text{ i.e. } p = 0.941 \tag{13.17}$$

and

$$E(P)/E(NP_2) + E(P) = 1123/1727 = 0.65, \text{ i.e. } p^2 = 0.65 \text{ and } p = 0.806 \tag{13.18}$$

Now

$$E(n_1) = \frac{1 - p^2}{1 - p^2 + p} = 0.83 \tag{13.19}$$

$$E(n_2) = \frac{p(1 - p^2)}{1 - p^2 + p} = 0.81 \tag{13.20}$$

$$E(n_3) = \frac{p^3}{1 - p^2 + p} = 0.82 \tag{13.21}$$

The best estimate of p is the one that minimises the sum of squares, so

$$\sum_{i=1}^{3} [n_i - E(n_i)]^2 \tag{13.22}$$

and p can be found by iteration and in this example is calculated as 0.818. The approximate standard error is

$$S.E.\,p = \sqrt{\dfrac{\displaystyle\sum_{i=1}^{3}[n_i - E(n_i)]^2}{\text{total mosquitoes caught}}} \qquad (13.23)$$

and in this case is 0.02. In this example the high survival rate of p in eqn (13.14) is because the pre-gravid rate was 100%, but if the pre-gravid is say 89% then

$$p = \dfrac{E(NP_2)G}{E(NP_1)} = 0.838 \qquad (13.24)$$

Also if 89% of females have a 2-day gonotrophic cycle and 11% have a 3-day cycle, then for the latter cycle the cube root, not square root, of p is calculated in eqn (13.12), and we get $p = 0.887$, and the overall population $p = (0.89 \times 0.82) + (0.11 \times 0.87) = 0.826$. This approach should only be applied to a population that is continuously reproducing, that is there are births and deaths with the result that the age-structure is more or less stable. However, with mosquitoes, such as *Anopheles arabiensis* breeding in a variety of ephemeral habitats it is likely that the population is rarely in a steady state. For example, observed monthly variations in age composition will result from unknown variations in emergence rates plus possibly unknown variations in survival rates, the two cannot be distinguished by the present, and related, methods of estimating p. This emphasises the need to usually estimate survival rates over an extended period.

The Vercruysse method was used by Kakitani and Forattini (2000) to calculate daily survival rates of *Anopheles albitarsis* A and B caught in Shannon traps in Brazil. The proportions of females with follicles in Stages I and II were high, at 97.5% and 98.2% for populations A and B respectively. The best estimate of p was calculated as 0.5339 ± 0.047 for population A and 0.5566 ± 0.015 for population B. Using these values for daily survival, the estimated lengths of the gonotrophic cycles calculated using the method of Davidson (1954) were 1.990 days for population A and 2.046 days for population B.

Problems associated with parous rate methods

Recruitment

To obtain a valid estimate of the survival rate from the proportion of parous females in a population there are a number of conditions which are commonly overlooked but which must be fulfilled. For example recruitment to the population must be constant throughout the sampling period. For example, the percentage of parous mosquitoes will be greatly reduced by large increases in emergence, and this will result in low estimates of survival. Conversely in an ageing population, i.e. where there are fewer births than deaths, the survival rate will be too high. Calculations of the proportion surviving through one day should be based only on populations having a relatively stable age distribution, a condition that may be more common in tropical mosquitoes, where, after an initial seasonal build-up, recruitment and loss may become more or less equal. However, even short-term fluctuations in recruitment can cause unstable parous rates. Garrett-Jones (1968) tried to overcome these difficulties by sampling only large and relatively stable populations. But a more satisfactory method to correct the distortions due to fluctuating numbers was proposed by Garrett-Jones (1973). Successive density indices of adults are obtained by a sampling method of constant efficiency, and the expression:

$$p^5 = \frac{V_d + 5}{M_d} \qquad (13.25)$$

is used, where M_d = overall density on a given day or its mean value over several days and V_d = the parous, or 'veteran', density index 5 days later. Thus the proportion sampled initially (d) is re-sampled after 5 days ($d + 5$) when all the survivors have become parous, i.e. 'veterans'. The mean ratio between these pairs of densities should represent the overall proportion of females, of all ages, surviving through 5 days.

   With univoltine species the population has to be sampled from zero recruitment at the beginning of the adult season through to zero recruitment at the end of the season in order for the parous rate to be equivalent to survivorship per gonotrophic cycle (Birley et al. 1983). Survivorship should be independent of age, which although for many years was believed to be true, may not in fact be so (see above).

Sampling bias and parity rates

The behaviour of nullipars and pars must be similar in order to avoid bias in sampling these two categories in a population, but nevertheless bias does sometimes occur. Gillies (1974) has discussed the difficulties of collecting representative samples of mosquitoes for determining their physiological age. He pointed out that the age-structure of a sample could vary according to the method and time of capture, and also location, as well as the physiological condition of the females. Several workers have reported a deficit of nullipars, and occasionally also 1-parous females in various collections, that is young females are not adequately represented. In Vietnam, for example samples of *Anopheles vagus* and *Anopheles minimus* made over 15 months showed a marked deficit of both these age classes (Zalutskaya 1959). Frequently *Anopheles maculipennis* populations sampled in Russia over an entire season have shown a deficit of nullipars, and sometimes of 1-parous and even 2-parous females (Detinova 1962). Such differences have often been ascribed to the different distributions of various age classes. For example, De Meillon et al. (1967*b*) found a lower parous rate in *Culex quinquefasciatus* collected from a variety of outdoor shelters than in females collected inside houses. They suggested that after emergence adults rested outside for some time before flying into houses to seek a blood-meal. In Madagascar Gruchet (1962) found marked differences in the parity of *Anopheles funestus* caught in night-biting collections performed in and out of doors, and when females were collected during the day resting in houses. Similarly in Tanzania a lower percentage of parous female *Anopheles funestus* were caught resting in houses during the day than were caught indoors biting at night (Gillies and Wilkes 1963). They, however, found that a greater percentage of nulliparous than parous females flew out of an experimental house after feeding. In later studies they also considered that there was a deficit of nullipars in their samples of *Anopheles gambiae*, *Anopheles arabiensis* and *Anopheles funestus* (Gillies and Wilkes 1965). They suggested that this might arise because a higher proportion rest out of doors than inside houses, but as they pointed out the evidence for this was conflicting. Now in these populations the duration of the first gonotrophic cycle was 4 days while subsequent durations were 3 days. So, in their collections from houses nullipars (comprising both pre-gravid and gravid females) represented a 4-day recruitment to the population, whereas all later age classes represented a 3-day recruitment. Clements and Paterson (1981) pointed out that in comparing mortality rates in different gonotrophic cycles an error may be introduced if the first cycle is not treated as being longer than subsequent cycles. They accordingly reduced the proportions of nullipars by 0.75 and reanalysed the data of Gillies and Wilkes (1965). They found that there was now no apparent deficit of nullipars in *Anopheles arabiensis* and *Anopheles funestus* when the exponential or Gompertz model was used, but there remained a deficit of nulliparous *Anopheles arabiensis* when using the exponential model, which they considered to be an inappropriate model for this species. Their reanalysis using the modified proportions of nullipars does not prove that in these three species all age classes were reliably sampled, but indicates that they may have been.

Charlwood et al. (1985) working in Papua New Guinea also found a deficit of nulliparous *Anopheles farauti* collected at bait and by other methods. They could not explain this under-sampling but considered it might be due to different dispersal and aggregation of different age classes.

Some workers have reported that CO_2 traps catch significantly greater numbers of nulliparous females than do light-traps (Feldlaufer and Crans 1979), while Barr et al. (1986) reported more nulliparous females in dry ice traps than in biting collections, and in Australia *Culex annulirostris* had a slightly lower parity rate when caught in carbon dioxide-baited light-traps than in animal bait traps (Russell 1985). However, Milby et al. (1983) and Nelson et al. (1978) found significantly more parous *Culex tarsalis* in carbon dioxide traps than in artificial resting shelters and high parity rates were reported for *Culex salinarius* caught in carbon dioxide traps (Slaff and Crans 1981). In California, USA Pfuntner et al. (1988) found that the parity of *Culex quinquefasciatus* was significantly higher in carbon dioxide traps (Pfuntner 1979) placed at a height of 10 m rather than at 5 or 2 m, but there were no such differences in the parity of *Culex tarsalis* and *Culex stigmatosoma*. In The Gambia Snow and Wilkes (1977) found that mean age of *Anopheles melas* and *Culex thalassius*, as determined by ovarian dilatations, caught in ramp traps and suction traps increased with trap height. This is similar to Corbet's (1961) findings with *Coquillettidia fuscopennata* (Corbet 1961) and *Coquillettidia aurites* (Corbet 1962*b*) in Uganda forests.

Some studies in California, USA have shown that CDC-type light-traps appear to collect a predominance of nulliparous mosquitoes (Schreiber et al. 1988, 1989). Morris and DeFoliart (1969) found considerable variability in parity among different species caught by light-traps and carbon dioxide baited traps. Furthermore, in comparative trials in California with red box shelters, carbon dioxide CDC light-traps, and New Jersey light-traps Meyer et al. (1984) reported that parous rates were highest in *Culex tarsalis* caught in the CDC traps (56%), highest in *Anopheles freeborni* caught in CDC traps (40%) or New Jersey light-traps (43%), and highest in *Ochlerotatus melanimon* sampled by New Jersey traps (57%). In Morocco, Bailly-Choumara (1973) found that the parous rate of *Anopheles labranchiae* collected by indoor light-traps was similar to the parous rates obtained by bait catches and collections from huts, but the parous rate of those caught in outdoor light-traps was much higher than those caught at human bait. In the Republic of the Congo the parous rate of *Anopheles gambiae* caught in huts with CDC light-traps was 46.6%, compared with 77.4% in adults caught biting (Carnevale and Le Pont 1973), whereas the parous rate of *Anopheles nili* was higher in adults caught in light-traps inside huts than those caught at bait.

In New York state, USA Emord and Morris (1982) found no differences in the parity of *Culiseta melanura* caught in their bird-baited CDC trap, a bird-baited lard-can trap, and a carbon dioxide-baited CDC trap, but in Japan Yajima et al. (1971) found that the proportion of parous *Culex tritaeniorhynchus* was greater in adults collected in pig-baited traps than in those from dry ice traps.

Gibb et al. (1988) cautioned against making incorrect assumptions that different traps are biased in favour of parous or nulliparous adults (sand-flies), because observed differences may be caused more by trap location and non-random distribution of parous and nulliparous adults than by real sampling differences.

Mostly there seems to be no difference in the biting times of nulliparous and parous mosquitoes, so sampling different periods of the 24 h diel should not lead to bias in parity rates, but there are some exceptions. For example, Charlwood et al. (1986*b*) in Papua New Guinea found that the parous proportion of *Anopheles farauti* increased in the latter part of the night, somewhat similarly Charlwood and Wilkes (1979) found very pronounced differences between the biting times of parous and nulliparous *Anopheles darlingi* in Brazil. In Malaysia Chiang et al. (1984*b*) found that in the forest a higher proportion of parous than nulliparous *Mansonia bonneae* bit from 1800–1900 h, but nulliparous females were more common after 2100 h. In Sri Lanka Samarawickrema (1968) recorded a higher proportion of parous *Mansonia uniformis* biting in the early part of the night, but in Japan Yajima et al. (1971) found a higher proportion of parous *Culex tritaeniorhynchus* biting pigs during the latter part of the night. It may therefore be necessary to sample biting populations over their entire biting cycle to avoid any bias in parity associated with biting times.

Statistical considerations concerning parous rates

The calculation of survival rates from the proportion that is parous clearly demands representative samples of the population, but in addition there may be statistical problems. The precision of a parous rate depends both on sample size and on the proportion parous. For example, if 400 females are dissected and the proportion that is parous is 0.40, the 95% confidence intervals put the range at 0.35–0.45. If the proportion that is parous represents p^3 then the survival rate lies somewhere between 0.705 and 0.765, which is rather a large interval on which to base expectation of life. De Meillon et al. (1967a) present useful tables of 95% confidence intervals for parity rates from different sample sizes. Some of these difficulties are discussed by Garrett-Jones (1970) and Garrett-Jones and Shidrawi (1969).

Gonotrophic cycle duration

The gonotrophic cycle is variously described as the interval between blood-feeding and time of the next blood-meal (Gillies and Wilkes 1963), or the time from just prior to one blood-meal to just prior to the next blood-meal (Hitchcock 1968). Other workers refer to the oviposition cycle, that is the interval between two consecutive acts of egg-laying, which should be of the same duration as the duration from blood-feeding to re-feeding. Frequently, however, the gonotrophic cycle is described as the interval between blood-feeding and oviposition, ignoring any time from oviposition and re-feeding.

These intervals of the gonotrophic cycle are often derived from laboratory observations, and not infrequently the time for blood-engorged females to become fully gravid is equated as the duration of the gonotrophic cycle. This, however, ignores the behaviour of gravid females in nature, which may experience some difficulty in rapidly locating a suitable oviposition site. The time involved from becoming gravid to actual deposition of eggs is part of the gonotrophic cycle. There may also be delays between oviposition and re-feeding. Moreover, the duration of the gonotrophic cycle may vary in the same species due to variations in ecological conditions (Birley and Charlwood 1989). Differences in parous rates of different sized mosquitoes have often been used as a measure of survival (Haramis and Foster 1983; Hawley 1985b; Landry et al. 1988; Nasci 1986a,b, 1988), but Pumpuni and Walker (1989) considered that the correlation between parous rate and longevity may not be reliable when the gonotrophic cycle is both variable and long. For instance Walker et al. (1987) found that female *Ochlerotatus triseriatus* seemed to have a mean life expectancy of just 11.6 days, but the interval from blood-feeding to oviposition varied from 8 to 17 days. So at least half the female *Ochlerotatus triseriatus* population would have died before their first oviposition and moreover nulliparous and parous samples would contain a wide range of young to old females. Slooff and Herath (1980) stressed the importance of correctly determining the gonotrophic cycle (feeding interval) in wild populations when survival rates are based on parity.

The more reliable methods for determining the duration of the gonotrophic cycle are based on field data. One way of determining the duration of the gonotrophic cycle is to use multiple linear regression on parity data obtained from sampling mosquitoes for 20 or more days. This method was developed by Birley and Boorman (1982) for *Culicoides* and used by Charlwood et al. (1985) in determining the gonotrophic cycle of *Anopheles farauti* in Papua New Guinea. Reisen et al. (1995) calculated gonotrophic cycle length for *Culex tarsalis* as the inverse of the oviposition rate, which was itself determined from a degree-day regression function developed by Reisen et al. (1992): OVR = –0.064 + 0.013 MOSQT, where OVR = oviposition rate and MOSQT = the microclimate temperature of mosquito resting sites, determined using the methods of Meyer et al. (1990).

Mahmood and Crans (1997) developed a thermal heat summation model based on that of Mahmood and Reisen (1981) to predict the duration of the gonotrophic cycle of *Culiseta melanura* under different temperature conditions. The model was calculated using regression analysis of temperature (t) and rate of development (V), such that $V = 1/g_c$ and g_c = mean time in days from females having eggs in Christophers' Stage I to oviposition. The regression of V on t must first be obtained as $V = a + b_t$, where a is the rate of development at time = 0, and b is the slope of the regression line. A thermal constant, k, equal to the number of degree days above t_0, which is the minimum below which ovarian development is arrested, was used. The value k is calculated as $1/b$ and t_0 is calculated as $-(a/b)$. For *Culiseta melanura* the empirical thermal minimum for ovarian development was determined as 6.4°C. The number of degree days above t_0 for completion of the gonotrophic cycle was 95.87.

An alternative and preferred approach is to use mark–release–recapture data (Charlwood et al. 1986*a*; Gillies and Wilkes 1965; McClelland and Conway 1971; Rawlings and Curtis 1982; Reisen et al. 1978; Slooff and Herath 1980; Suzuki 1977, 1978). Some examples of this approach are given below.

Gonotrophic cycle: mark–release-recapture methods

In Tanzania Gillies and Wilkes (1965) established from marking and releasing *Anopheles gambiae* females that the gonotrophic cycle in parous females was 3 days. Similarly in Bangkok by marking and releasing *Aedes aegypti* Sheppard et al. (1969) postulated that the gonotrophic cycle was probably 3 days. De Meillon et al. (1967*e*) found that when adults of *Culex quinquefasciatus* fed before 2400 h some two-thirds oviposited 2 days later, that is on the third night, but practically all females that fed after midnight oviposited on the fourth night. Not only did late feeding prolong the duration of the gonotrophic cycle, but De Meillon et al. (1967*d*) discovered that sugar feeding also delayed oviposition.

McClelland and Conway (1971) conducted a series of mark–recapture experiments in Tanzania to determine the interval between successive feedings of *Aedes aegypti*. On 23 consecutive days all females attempting to bite during a 5-h catch were caught, and during the first 11 days the mosquitoes were allowed to feed on the bait; these engorged individuals together with any bearing a previous mark were painted with a date specific mark and released. Thereafter *Aedes aegypti* were caught at bait and their markings recorded, but they were not released. Of 468 blood-fed females released, 58 (12.4%) were recaptured. The frequency distribution of the numbers recaptured at bait after various intervals after release is shown in Fig. 13.4.

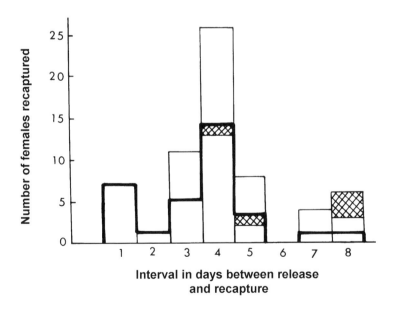

Fig. 13.4. Results of mark–recapture experiments with *Aedes aegypti* expressed as a frequency histogram of the intervals between release and recapture. Areas below thick lines represent females that were released after feeding and which engorged within 6 h after recapture. Shaded areas refer to intervals between first release and second recapture for females recaptured twice (after McClelland and Conway 1971)

The areas beneath the thick lines represent those released after feeding and which engorged within 6-h after recapture, while the shaded areas show the intervals between first release and second recapture for females recaptured twice. The marked 4-day peak can be taken to indicate a mean of 4-days between primary feeds, and that a proportion take a second feed on the second or third day of each gonotrophic cycle. This frequency of double feeding, however, decreases with each gonotrophic cycle (McClelland and Conway 1971; Conway et al. 1974). An alternative hypothesis is that the histogram shows that *Aedes aegypti* invariably takes a primary blood-meal every 4 days, but in each cycle a constant proportion return for a secondary feed on day 2 and a constant proportion re-feed on day 3. Conway et al. (1974) pointed out that the first hypothesis was more plausible, but as there were insufficient data from the experiment for a detailed analysis of the results they had to restrict their analyses to the second and simpler model.

They showed that the proportion of secondary feeders in the blood-feeding population was given by:

$$\frac{\phi(\alpha + \gamma\phi)}{1 + \alpha\phi + \gamma\phi^2} \tag{13.26}$$

where ϕ = the constant daily survival rate, α = the proportion of the population taking a supplementary feed on day 2 and γ = the proportion taking this second feed on day 3. Now, the proportion of the feeding population in the total female population is given by:

$$\frac{\phi(1 + \alpha\phi + \gamma\phi^2)(1 - \phi)}{1 - \phi^4} \tag{13.27}$$

The population available for capture is n_i/p_i, where n_i = number of mosquitoes caught on day i and p_i = the catch rate. It therefore follows that the total female population on day i, as derived from the second formula, is:

$$N = \frac{n_i(1 - \phi^4)}{p_i\phi(1 + \alpha\phi + \gamma\phi^2)(1 - \phi)} \tag{13.28}$$

In Thailand, Pant and Yasuno (1973) marked and released both newly emerged laboratory reared unfed virgin females and blood-fed and mated females of *Aedes aegypti* during the wet, cool–dry and hot seasons. Recaptures showed that the first blood–meal was taken 24–36 h after emergence, except in the cool–dry season when there was a longer interval. The minimum interval between emergence and oviposition was 81 h in the wet season and 93 h in the cool–dry and hot seasons, and second oviposition was recorded on the 8th day after emergence. From this data the interval between two consecutive ovipositions was calculated as 3 days. Similar mark–recapture studies have been made in three distinct seasons in India to determine the duration of the gonotrophic cycle in *Culex quinquefasciatus* (Singh and Yasuno 1972). In the Central African Republic, Germain et al. (1974) marked blood-fed adults of *Aedes africanus* with different coloured powders on different days and from a series of recaptures in human landing catches on consecutive days estimated that the gonotrophic cycle was on average as long as 7–8 days. They concluded that this long interval was mainly due to the time spent by females in trying to locate hosts.

In Florida, USA Lowe et al. (1973) released large numbers of *Culex quinquefasciatus* labelled with ^{32}P into the population. The first radioactive egg rafts were collected 4 or 5 days after release, and thereafter there was a peak in the number of radioactive egg rafts every 3 or 4 days, thus indicating an oviposition cycle of 3–4 days. Longevity was assessed by the length of time during which radioactive eggs were laid. Fernandez-Salas et al. (1994) marked and released both wild caught females and laboratory-reared progeny of wild caught females of *Anopheles pseudopunctipennis* in southern Mexico in order to determine gonotrophic cycle duration and survival rate. The first marked females with uncontracted ovarian sacs, indicating recent oviposition, were recaptured on day 3 post release, and thus a 3-day gonotrophic cycle duration was used in subsequent estimates of survival.

In Japan 24 608 laboratory blood-fed female *Culex tritaeniorhynchus* were marked with fluorescent powders and released near a cow shed. Over the following 15 days, two light-traps placed in the cow shed caught 194 females (0.79%), of which 59.8% were collected 48–60 h after release. From recording whether caught marked females were unfed or blood-fed, and the state of blood digestion, Buéi et al. (1980) estimated that the interval between the 1st and 2nd, and also the 2nd and 3rd, blood-meals was 2.9 days. In another experiment they released 62 838 unfed newly emerged marked *Culex tritaeniorhynchus*. Over 15 days 305 (0.49%) marked mosquitoes were caught in light-traps placed in a pigsty, horse stable, duck house and cow shed, 77.7% of which were caught 24–60 h after release. From an analysis of the physiological conditions of the marked females it was calculated that the interval between emergence and the 1st blood-meal was 2.6 days, and between the 1st and 2nd blood-meals 2.8 days.

In Sri Lanka Samarawickrema (1967) marked laboratory reared adults of *Culex quinquefasciatus* with different coloured paints and released them on different days. Recaptures were made of adults resting in houses and the days since release correlated with numbers of ovarian dilatations in order to estimate the duration of the gonotrophic cycle. Also in Sri Lanka Rawlings and Curtis (1982) marked with coloured dusts freshly fed *Anopheles culicifacies* (2443) caught in a cow-baited hut, and then released them back into the hut. On the next 4 days they recorded the gonotrophic condition of all marked mosquitoes caught in the hut. On the day after release there was a small but significant peak of half-gravids (50), while the biggest peak of blood-fed individuals (192) was found 2 days after release, with a smaller peak (78) on the third day. This suggested that the gonotrophic cycle was usually 2 days, but was sometimes 3 days, the mean duration was calculated as 2.3 days. The numbers of blood-engorged *Anopheles culicifacies* (76) caught on day 4 were assumed to have completed two cycles since their release (Rawlings and Curtis 1982). Using basically the same approaches Mori and Wada (1977) determined that the time from emergence of *Aedes albopictus* to the first blood-meal was 2 days and that the first gonotrophic cycle lasted 5 days; and in French Guyana the gonotrophic cycle of *Culex portesi* was determined as 8 days (Dégallier 1979). In Uganda Sempala (1981) estimated the duration of the gonotrophic cycle of *Aedes africanus* released as blood-feds and marked with powders and paints as 5–6 days. Reisen et al. (1983) using mark–recapture experiments determined the duration of the gonotrophic cycle of *Culex tarsalis* in California, USA was 4–5 days. In Papua New Guinea Birley and Charlwood (1989) showed by mark–recapture methods that the oviposition cycle of *Anopheles farauti* was 2.04 days at full moon, but was extended to 2.40 days during moonless periods. Also, when marked blood-fed females were released further away, the oviposition cycle was 3.0 days.

Measuring the oviposition cycle in the field by the interval from release of marked blood-fed mosquitoes to their recapture as unfeds at bait may give a smaller duration than release and subsequent recapture of marked unfed mosquitoes, because any delays in obtaining the initial blood-meal are eliminated. This limitation was appreciated by Haramis and Foster (1990) who in the USA marked and released 1.5–2.5-day-old laboratory reared *Ochlerotatus triseriatus* and determined the duration of the first gonotrophic cycle from the frequency distribution of marked females caught at human bait on subsequent days. The length of the first cycle would include times required for mating and sugar-feeding as well as times for blood-feeding and oviposition. The minimum and median times to complete this cycle were estimated from the first recapture of parous marked females and the time needed for 50% of the numbers of marked females to become parous, respectively. This median time was estimated by plotting a linear regression of the proportion parous recaptured at bait for consecutive 5-day periods against days after release.

In Fiji, based on numbers of blood-fed *Aedes polynesiensis* released after marking (449) that were recaptured at human bait (40), Suzuki (1977) found that the interval between successive blood-meals was 3–4 days; in later experiments in Japan Suzuki (1978) estimated this interval as 3–5 days. Suzuki (1977) also extended the mark–recapture method of estimating the duration of the gonotrophic cycle by taking into account parasite development in the mosquitoes. For example, on Tuvalu *Aedes polynesiensis* were fed on people exhibiting microfilaraemia of *Wuchereria bancrofti*, and then a few of the infected mosquitoes were dissected from days 1 to 13 post-feeding to measure the width and length of the larvae. A graph was drawn showing the relationship between both width and length over 13 days. In human landing catches 45 *Aedes polynesiensis* were caught with developing filarial larvae. The duration of infection in each mosquito was estimated from the graph, and this should be equal to *n* times ($n = 1, 2, 3$ ...) the interval of two successive blood-meals. Applying this approach the feeding interval was calculated as 3–4 days, which was in agreement with the more direct mark–recapture method.

Gonotrophic cycle: Charlwood et al. method

In Papua New Guinea Charlwood et al. (1986*a*) marked up to 500 *Anopheles* mosquitoes, caught on human and bovid baits and allowed to feed, with fluorescent powders and then released them the same day at sunrise. On subsequent nights adults were caught at bait and the numbers of marked ones recorded. They analysed their results by two different methods both of which are considerably more mathematically complex than those used by previous workers. Firstly a deterministic model was used. In this the numbers of recaptured marked mosquitoes (R_t) on night *t* after release is proportional to total numbers marked and released on day 0 (i.e. R_0) less the total already captured (B), times the probability of surviving and feeding on night *t* (i.e. S_t). (Unfortunately mortality and dispersal from the collection area cannot be separated.) The numbers recaptured on night *t* is thus

$$R_t = A(R_0 - B)S_t \qquad (13.29)$$

where A = the recapture rate. As the authors pointed out parameters A and R_0 are determined by experimental conditions, whereas S_t depends on natural factors, namely the survival rate and feeding frequency. Now, marked mosquitoes are available for recapture only at bait at set intervals—namely when they require a blood-meal, that is after oviposition. The duration of the oviposition cycle (u) is taken as being independent of the survival rate between blood-meals, because it seems logical to think that most mortality is likely to occur at feeding or at oviposition, two risky events in a mosquito's life, and not between them. It is also assumed that survival is independent of age. So let P_u be the proportion of mosquitoes that survive and have an oviposition cycle of u days. With *Anopheles punctulatus* the oviposition cycle was considered to last between 2–5 days and so the survival rate (S_t) can be estimated as follows

$$S_t = P_2 \times S_{t-2} + P_3 \times S_{t-3} + P_4 \times S_{t-4} + P_5 \times S_{t-5} \text{ and } S_0 = 1 \quad (13.30)$$

This is a renewal equation which is a special case of a convolution sum or transfer function (Birley 1977, 1979; Birley and Boorman 1982). The average survival per oviposition cycle (P), and average length of the oviposition cycle (u) can be defined as follows.

$$P = P_2 + P_3 + P_4 + P_5 \quad (13.31)$$

$$U = (2 \times P_2 + 3 \times P_3 + 4 \times P_4 + 5 \times P_5) / P \quad (13.32)$$

The average daily survival rate (p) is found as

$$p = P^{1/u} \quad (13.33)$$

Many different combinations of P_2, P_3, P_4 and P_5 will give the same value of p. Let f_u represent a unimodal frequency distribution of the oviposition cycle length, so that

$$P_u = P \times f_u \quad (13.34)$$

Then the survival rate (S_t) is

$$S_t = P(f_2 \times S_{t-2} + f_3 \times S_{t-3} + f_4 \times S_{t-4} + f_5 \times S_{t-5}) \quad (13.35)$$

The calculation of R_t and S_t by the above deterministic model suffers from assuming that the numbers recaptured represent a real or continuous variable that measures population density, whereas the data consist of small integers. Consequently the deterministic model does not adequately simulate reality.

Charlwood et al. (1986a) describe how a stochastic approach can be developed and more accurately analyse the data. A stochastic model can be constructed using the function BIN, which will simulate sampling from a binomial distribution. If there are R individual mosquitoes each having a probability P of surviving, then the numbers which actually do survive (X) may be obtained from computer simulation as follows

$$X = BIN(R, P). \tag{13.36}$$

The stochastic model is described by

$$T_t = BIN(T_{t-2}, P_2) + BIN(T_{t-3}, P_3) + BIN(T_{t-4}, P_4) + BIN(T_{t-5}, P_5) \tag{13.37}$$

$$R_t = BIN(T_t, A)$$

$$T_t = T_t - R_t (\text{BASIC notation})$$

where T_t = number of mosquitoes in the marked population which have not been caught t days after release, P_u = proportion of mosquitoes surviving to feed after u days, R_t = number of marked mosquitoes recaptured on day t, and A = recapture rate. Finally, since mosquitoes are not released again after capture at bait, T_t must be reduced by the numbers caught prior to T_t.

Charlwood et al. (1986a) questioned the validity of the widely used method of estimating the daily survival rate (p) from the regression of the log numbers of marked mosquitoes recaptured against time in which total numbers released on day 0 (R_0) and R_1, are ignored. They pointed out that drawing a straight line regression slope normally assumes there are no peaks or troughs in the data, except of course by chance, whereas with mosquitoes the intervals between re-feeding, and hence capture at bait, is characterised by a series of peaks. Furthermore, there may be some nights where $R_t = 0$ and of course log R_t cannot be found. They pointed out the problem could be overcome by either plotting log $(R_t + 1)$ or by ignoring days when $R_t = 0$. They tested these two methods with their stochastic simulation model, and found that the regression of log R_t against time does indeed provide a reasonable estimate of daily survival rate. Both of the above methods gave similar results, but if log $(R_t + 1)$ was used it was better to truncate recapture data so as not to include more than one or two trailing zeroes. In practice they suggested both plots should be tried and the one giving the most significant correlation coefficient selected. For some unexplained reason the intercept of the regression line does not estimate the recapture rate (A).

The approximate mean duration of the oviposition cycle (U') may be estimated as follows.

Firstly let $R' = R_2 + R_3 + R_4 + R_5$, and assume that the frequency with which a blood-meal is taken after u days from release (f_u) is

$$f_u = R_u / R'$$
(13.38)

If the oviposition cycle is distributed between 2 and 5 days, then U' and its standard deviation can be estimated from

$$U' = \sum_{u=2}^{5} u \times R_u / R'$$
(13.39)

$$SD = \sum_{u=2}(u - U')^2 \times R_u / (R' - 1)$$
(13.40)

where R_u = the numbers of unfed marked mosquitoes recaptured on day u after release. Note that SD is not a measure of the accuracy of the estimate U', but a measure of the variability of oviposition duration.

Charlwood et al. (1986a) found this estimate to be quite accurate so long as the minimum number of terms are included. For instance, if the peak oviposition cycle peak is at R_2 then the terms R_5 and possibly R_4 should be excluded. Using this method the duration of the gonotrophic cycle of *Anopheles punctulatus* in Papua New Guinea was estimated as 2.90 days (Charlwood and Bryan 1987).

Birley and Rajagopalan (1981) and Birley and Boorman (1982) in their studies of *Culex quinquefasciatus* in India and *Culicoides* in Britain presented a new method of calculating the duration of the gonotrophic cycle on data derived from captures over many consecutive days of both unfed and blood-fed females, which were classified as nulliparous and parous. However, as their approach was mainly to estimate survival rates it is more appropriate to consider it (Birley's method) when discussing this topic (see below).

Tsuda et al. method

In recognition of some of the difficulties inherent in determining parous rates, Tsuda et al. (1991) developed a model that attempts to describe the parous rate in terms of the following population parameters: daily adult survival rate, recruitment, adult longevity and gonotrophic cycle duration. In this model a narrower definition of the parous rate was used, namely the proportion of the feeding population that comprises parous females, rather than the whole population, which also includes resting and ovipositing females. The model assumed that newly emerged females only feed b days after emergence and subsequently re-feed n-1 times per gonotrophic cycle (g).

$$N_{t,x} = N_{t,b} + N_{t,b+g} + N_{t,b+2g} + N_{t,b+(n-1)g} \qquad (13.41)$$

where $N_{t,x}$ is the number of females aged x at time t. Assuming that daily survival rates, p, are constant, then

$$N_{t,x} = N_{t-x,0} P^{-x} \qquad (13.42)$$

and the total number of feeding females becomes

$$N_{t-b,0} P^b + N_{t-(b+g),0} P^{(b+g)} + N_{t-(b+2g),0} P^{(b+2g)} +$$
$$N_{t-(b+(n-1)g),0} P^{((b+(n-1)g)} = \sum_{x=0}^{n-1} N_{t-(b+gx),0} P^{(b+gx)} \qquad (13.43)$$

Assuming a constant finite rate of population increase r, then

$$N_{t,0} = r^x N_{t-x,0} \qquad (13.44)$$

and

$$N_{t-x,0} = r^{-x} N_{t,0} \qquad (13.45)$$

and the total number of feeding females becomes

$$\sum_{x=0}^{n-1} N_{t,0} r^{-(b+gx)} p^{(b+gx)} \qquad (13.46)$$

In this example, all females older than b days are parous and all females coming to feed on b days post-emergence are nulliparous and the total number of nulliparous females $=$

$$N_{t,b} = r^{-b} N_{t,0} p^b \qquad (13.47)$$

and the parous rate is defined as:

$$1 - \frac{N_{t,b}}{\sum_{x=0}^{n-1} N_{t,0} r^{-(b+gx)} p^{(b+gx)}} \qquad (13.48)$$

or

$$1 - N_{t,0}(p/r)^b / N_{t,0}(p/r)^b \sum_{x=0}^{n-1}(p/r)^{gx} \tag{13.49}$$

which is equivalent to:

$$\text{Parous rate} = (p/r)^g \{(p/r)^{g(n-1)} - 1\}/\{(p/r)^{gn} - 1\} \tag{13.50}$$

Using the survival rate per gonotrophic cycle S ($=pg$) and the finite rate of increase in the population per gonotrophic cycle, R ($=rg$) then the above equation for parous rate becomes:

$$\text{Parous rate} = (S/R\{(S/R)^{n-1} - 1\}\{(S/R)^n - 1\}) \tag{13.51}$$

When n is indefinite and $S/R < 1$, then

$$\text{Parous rate} = S/R = p^g/R \tag{13.52}$$

And if the population is stable, such that $R = 1$, then Parous rate $= S = p^g$, which is Davidson's formula for survival based on the parous rate. Tsuda et al. (1991) then conducted simulation studies to investigate the relationship of the parous rate with the length of the gonotrophic cycle, daily survival rate, the number of gonotrophic cycles per lifetime, and the rate of population increase per gonotrophic cycle. The effect of the number of lifetime gonotrophic cycles on estimated parous rates is greater when the rate of population increase is smaller and daily survival is higher. When the number of lifetime gonotrophic cycles equals 10, then the equation for parous rate (13.50) above can be used to give a relatively accurate estimate. The parous rate was found to be more sensitive to the length of the gonotrophic cycle and daily survival in decreasing populations, compared with stable or increasing populations. In most natural populations, where the lifetime number of gonotrophic cycles may approach 10, the number of gonotrophic cycles has little effect on the estimated parous rate.

Birley's method for survival rates

So long as recruitment is reasonably constant, the parous rate provides an estimate of average survival rate per average oviposition cycle, but is unreliable when the recruitment rate fluctuates rapidly and the population is observed for only short periods. An alternative formula for short runs of data, or rapidly fluctuating recruitment, was originally briefly mentioned by Garrett-Jones (1973), but the approach has been enlarged upon by Birley (1984), and was applied earlier by Birley and Rajagopalan (1981) to *Culex quinquefasciatus* and *Culicoides* (Birley and Boorman 1982). In essence a time-series analysis is performed.

Resting populations of *Culex quinquefasciatus* were sampled in a house daily at 0830 h, and mosquitoes categorised as unfed, freshly blood-fed, semi-gravid and gravid, and as nulliparous and parous by ovarian tracheation. The proposed formula for estimating survival, and also biting rates, is based on the following assumptions: (1) newly engorged females are sampled consistently at daily intervals and classified as nulliparous or parous; (2) nulliparous and parous mosquitoes are sampled with equal efficiency; (3) most females require just one blood-meal per gonotrophic, or oviposition cycle (note that the inverse of this period is the biting rate); (4) all females complete each cycle in the same number of days; and (5) all females repeat the cycle with the same probability of loss, through death or emigration. The underlying assumptions of the model are, however, difficult to test under field conditions, as the true survival rate is unknown, recruitment patterns may vary and cannot be controlled, and sampling biases may not be apparent. Variation in sampling efficiency from day to day may occur as the result of changes in wind speed or direction, rainfall and cloud cover, with consequences for the calculated survival rate (Lord and Baylis 1999). The concept underlying the model is that the number of parous flies biting at time t is equal to the total biting one gonotrophic cycle earlier multiplied by the proportion surviving per cycle.

Now

$$M_t = P_u \times T_{t-u}$$
$$T_t = N_t + M_t$$

(13.53)

where the variable given as a subscript (t) = time measured in integer days; N = number of nullipars in a sample, M = number of pars in a sample; u = duration of the oviposition cycle (taken here as time between successive blood-meals); and P = the survival rate per oviposition cycle. From eqn (1) it follows that an estimate of P_u is

$$P_u = \sum M_t / \sum T_{t-u}$$

(13.54)

Equation (13.54) is identical to the parous rate formula when $u = 0$, that is

$$P_0 = \sum M_t / \sum T_t \qquad (13.55)$$

Because the duration of the oviposition cycle was unknown the calculation was repeated for $u = 0, 1, 2, 3, 4, 5$ and 6 days. For each value of u the residual sum of squares is calculated and presented as a percentage goodness-of-fit or correlation index. The following formula can be used to calculate an approximate value for this index

$$R_u = (2 \times E - D \times P) \times P / B \qquad (13.56)$$

where E = sum of cross-products of the $M_t \times T_t - u$ pairs; D = sum of squares $T_t - u \times T_t - u$; B = sum of squares $M_t \times M_t$; and P = the estimate P_u.

An examination of the calculated values of R_u for all values of u will show whether the population is consistent from day to day (coherent) or is dominated by fluctuations (incoherent), and also whether recruitment is relatively variable. The value of R_u is between 0 and 1. Values near 1 indicate a high degree of correlation and consistency from day to day (coherency). R_u has a maximum at $u = 0$, and then declines as u is increased. It may have a second peak at about $u = 3, 4$ or 5, and this is the best estimate of the mean oviposition cycle. If all values of R_u are high and the peak poorly pronounced then this indicates recruitment is relatively constant. However, if all values of R_u are low (except $u = 0$), then this indicates that the population is incoherent, or the sampling protocol inadequate, and no survival estimate can be obtained by this approach.

The approximate standard error may be obtained for this estimated survival rate by

$$SE\ P_u = \sqrt{P_u \times [1 - P_u] / \text{total sample}}. \qquad (13.57)$$

To return to the catching of *Culex quinquefasciatus*, each day a new pair of 'data points' is obtained, that is the numbers of nullipars and pars in the sample, and these are added to the previous data. Calculations of P_u, SE P_u and R_u are made for $u = 0, 1, 2, 3, 4, 5, 6$ and 7. If the estimates stabilise and the standard error is acceptable, then the experiment is terminated, otherwise sampling is repeated on subsequent days until these criteria are met. Birley and Rajagopalan (1981) found that for *Culex quinquefasciatus* the correlation index of the survival rate was reasonably high, having peaks at $u = 0, 4$ and 6 days, which indicated a mean oviposition cycle of 4 days. It was believed that some individuals probably oviposited every

3 days. Secondary peaks arise at all time delays that are a multiple of the basic periodicity, hence the peak at 6 days. It was concluded that the best estimate of survival rate was 0.28 ± 0.02. However, McHugh (1989) argued that if teneral females were included in the samples used for determining parous rates, then this would have lowered the survival rate estimate relative to that derived from blood-fed mosquitoes. Moreover, the inclusion of teneral females would lead to overestimation of the length of the gonotrophic cycle. So McHugh (1989) considered that by combining data on unfed and blood-fed *Culex quinquefasciatus* Birley and Rajagopalan (1981) may have superimposed two estimates.

Birley and Boorman (1982) suggested that the estimation of survival rate and oviposition period by time series analysis be based on adults being sampled at bait for a minimum of 20–25 consecutive days. When sampling is performed for longer periods, such as over a season, then the various parameters can be estimated for different parts of the season. If for instance sampling is carried out for 160 consecutive days, the estimation procedures described above can be used on data collected on days 1–40, days 2–42, days 3–42 … and finally days 121–160. Thus there will be 120 separate estimates for each parameter forming a time series of running means. Each estimate is associated with a median date, for example for the above it would be for days, 20, 21 and so on until day 140. Running means may be calculated on shorter or longer runs, but Birley and Boorman (1982) suggested 40 days might be appropriate. More recently, Lord and Baylis (1999) investigated the minimum length of time series required and the effect of shorter series on estimates of the survival rate.

Holmes (1986) used Birley's method in Dubai to estimate the survival rate of *Culex quinquefasciatus* per oviposition cycle (0.301) sampled by light-traps. In California, USA McHugh and Washino (1986) using the method found that in unfed *Culex tarsalis* the gonotrophic cycle was 7 days and the survival rate/cycle was 0.86, whereas for blood-fed females the cycle was 5 days and the survival rate was 0.84. Similarly, in *Anopheles freeborni* the cycle and survival rates of unfed females were 6 days and 0.72, while for blood-fed individuals it was 4 days and 0.75. McHugh and Washino (1986) pointed out that Birley and Rajagopalan (1981) did not analyse their data on unfed and blood-fed females separately, although they had observed that in *Culex quinquefasciatus* the gonotrophic cycle peaked at 4 and 6 days. In retrospect this might have represented the durations for blood-fed and unfed females. Later McHugh (1989) again used this method on *Anopheles freeborni* collected over 23 days from red box shelters, and calculated the duration of the gonotrophic cycle based on unfed females as about 6 days, and a survival rate of 0.72, for blood-fed females the gonotrophic cycle appeared to be 4 days and the survival rate was computed as 0.74. The 6-day period was attributed to the inclusion of teneral females in the sample of resting adults. These estimates compared favourably with daily survival rates of 0.75 estimated 2 years later, from mortality of adults kept in an outdoor cage. McHugh (1990) applied the same approach to *Culex tarsalis* caught in walk-in red boxes in California, USA to estimate the daily survival rates and the duration of the gonotrophic cycle.

Holmes and Birley (1987) re-examined the Birley and Rajagopalan (1981) method and found that it was best to incorporate a generalised filter into cross-correlations between the time series so as to delete spurious cross-correlation peaks. They also improved the model by incorporating weighted linear regressions that stabilised the variance. For example, there are several ways of estimating the slope of a linear regression (b) through its origin when plotting paired observations ($x_i\ y_i$). If the variance increases proportionally to x then the weighted value (w_i) = k/x_i with k = var. ($e_{i}/_{xi}{}^{0.5}$), where the e_i values are the unweighted residuals. This gives the ratio of means estimator

$$b = \Sigma y_i / \Sigma x_i = \overline{y} / \overline{x}$$ (13.58)

with the SE calculated as either:

$$SE = \sqrt{(k / \sum_i x_i)}$$ (13.59)

or

$$SE = \sqrt{(b \times (1-b) / \sum_i x_i}$$ (13.60)

Jensen et al. (1993*a*) applied the Holmes and Birley (1987) method of cross-correlating time series of the numbers of nulliparous, parous and total females of *Anopheles quadrimaculatus* species C (*Anopheles diluvialis*, Reinert et al. 1997) collected by CDC light-trap supplemented with dry ice in Florida, USA. No statistically significant cross-correlation coefficients were obtained and these authors were unable to determine gonotrophic cycle length or survivorship using this method. Jensen et al. (1993*b*) applied the same methodology to study the short-term population dynamics of the salt marsh mosquito *Ochlerotatus dorsalis*. Analysis of filtered time series data yielded significant correlations for total females and parous females collected per day and for filtered parous and nulliparous time series with a lag of 5 days, indicating a gonotrophic cycle length of 5 days. Survivorship per gonotrophic cycle was estimated to be 14% (S.E. ± 0.7). In a later study, Jensen et al. (1998) used time-series analysis of *Anopheles puncti-pennis* parity in order to estimate gonotrophic cycle length in California, USA. Holmes and Birley's (1987) method of employing an autoregressive iterative moving average model was used to eliminate spurious yet seem-ingly significant cross-correlations from the time series. The three filtered time series were then cross correlated at time lags of zero to eight days. Cross-correlations were considered to be statistically significant if the correlation coefficient exceeded $\pm 2/\sqrt{n}$. Statistically significant cross correlation coefficients between time series with the same time lag were con-sidered to indicate the duration of the gonotrophic cycle. Statistically significant correlation coefficients were observed between the following time series: number of parous females collected per day and the number of nulliparous females collected per day; total females collected per day and the number of nulliparous females collected per day, with time lags of three days. The highest cross-correlation coefficient between time series parous and time series total was obtained with a time lag of three days, but this coefficient was not statistically significant. These results suggested that the duration of the gonotrophic cycle was approximately three days. These authors also used the Holmes and Birley (1987) method to estimate survival per gonotrophic cycle. Based on the assumption that the duration of the gonotrophic cycle was three days, survivorship per gonotrophic cycle was estimated to be 0.78. Daily survivorship was estimated using David-son's method (Davidson 1954) to be 0.92.

Hii et al. (1995) used the Holmes and Birley (1987) method applied to data from human landing catches, a pig-baited net trap and indoor CDC light-trap collections all operated over a 30 day period, to estimate the ef-fects of permethrin-treated mosquito nets and DDT house spraying on sur-vival of *Anopheles farauti* No. 1. Survival rates in villages with treated nets were significantly lower than in control villages (0.39, S.E. 0.01, and 0.53, S.E. 0.006 respectively).

Using the Birley and Rajagopalan (1981) method and the method of Holmes and Birley (1987), Ulloa et al. (2005) estimated the gonotrophic cycle lengths of two populations of *Anopheles vestitipennis* from southern Mexico that displayed different host preferences. No significant cross-correlations in parity over 15 consecutive nights were observed for females collected from either horse-baited or other animal-baited traps, but for the zoophilic population higher correlation indices were obtained on days 3 and 6, suggesting a 3-day gonotrophic cycle. Gonotrophic cycle duration for the anthropophilic population was estimated at 4 days.

Mutero and Birley method

The method of Birley and Rajagopalan (1981) and Birley and Boorman (1982) assumes that female mosquitoes require just one blood-meal for completion of the oviposition cycle, but some species require an additional blood-meal when they are nulliparous, that is, there is a pre-gravid stage, and so Mutero and Birley (1987) modified the equation to take this into account.

$$M_t = P \times x_{t-u} \qquad (13.61)$$

where $x_t = (1 - g) \times M_t + g \times T_t$ and $g = 1/(1 + c)$, and where c = approximate estimate of the numbers of pre-gravid nullipars. This is a more generalised form of equation $M_t = P \times T_{t-u}$ (Birley and Rajagopalan 1981).

In both equations there is a cross-correlation between the two time series M_t and T_t (and also N_t) which is equal to the duration of the oviposition cycle with a peak time-lag of u days. The cross-correlation function is a standard statistical formula, similar to a correlation function, but having a different value for each time lag. To overcome any spurious large peaks of cross-correlation both the parous P_t and total T_t series can be filtered using first-order autoregressive functions, having a time lag of 1 day. This procedure is explained by Holmes and Birley (1987), and can be summarised as follows

$$z_t = x_t - \alpha \times x_{t-1}. \qquad (13.62)$$

where x_t = time series to be filtered; z_t = filtered time series, and α = autoregression parameter estimated approximately as

$$\alpha = \frac{\Sigma\left((x_t - \overline{x}) \times (x_{t-1} - \overline{x})\right)}{\Sigma(x_{t-1} - \overline{x})^2}. \qquad (13.63)$$

The filtered cross-correlations R_u are considered to be significant if

$$R_u = \frac{2}{\sqrt{d}} \qquad (13.64)$$

where d = number of data points in the time series.

The survival rate per oviposition cycle is estimated by the normal Birley method (Birley and Boorman 1982; Birley and Rajagopalan 1981) as

$$P_u = \Sigma M_t / \Sigma T_{t-u} \qquad (13.65)$$

It should be noted that Birley and collaborators in a series of papers from 1979 to 1989 have often used different notations for the same parameter in different papers. The consistent notation adopted by Service (1993) is used here.

Zaim et al. (1993) compared the survival rates of *Anopheles culicifacies* s.l. and *Anopheles pulcherrimus* in sprayed and unsprayed villages in Baluchistan, Iran, using either the Mutero and Birley (1987) or the Davidson (1954) method. The daily survival rates calculated for *Anopheles pulcherrimus* from time series analysis were 0.83 in unsprayed villages and 0.78 in sprayed villages.

Bockarie et al. (1995) used the methods of Birley and co-workers described above to estimate the gonotrophic cycle duration of *Anopheles gambiae* s.s. in Sierra Leone. The time-series analysis showed peaks at intervals of three days, indicating a three-day cycle length, but the cross-correlation coefficients were not statistically significant, even after applying the weight linear regression and filtered cross-correlation recommended by Holmes and Birley (1987).

Lord and Baylis (1999) used a simulation model of Birley's method incorporating artificial population data to examine the accuracy and precision of survival rate estimates and the effects of violation of the assumptions listed above, by varying (i) the length of the data series; (ii) sampling efficiency; (iii) trends in recruitment of nulliparous individuals; and (iv) biases in sampling efficiency between nulliparous and parous flies. A series of 109 days of data was generated to allow for time series sample lengths up to 100 days. Four values of the input survival rate were used (0.2, 0.4, 0.6, 0.8 per cycle) to generate M_t. The gonotrophic cycle length was set at $u = 3$. Three relationships between sampling efficiencies for nulliparous and parous females were considered in the model: (i) equal efficiency, with no variation or bias between capture of parous and nulliparous individuals; (ii) independent efficiencies, where nulliparous and parous females are sampled independently from the same distribution, producing variation between types but no consistent bias; and (iii) biased efficiencies, where nulliparous and parous females are sampled from different distributions, such that there is variation between types and a consistent bias between nulliparous and parous individuals. Accuracy of the estimated survival rates was very high under conditions of equal sampling efficiency and independent sampling from the same distribution for all lengths of time series, except for the very shortest, while precision of the estimate increased substantially as the length of the time series increased. Lord and Baylis (1999) therefore concluded that when sampling is unbiased, time series of around 30 days give good accuracy and precision for estimates of survival rate. Under conditions of biased sampling efficiency, however, both the accuracy and precision of the survival rate estimate are severely reduced at the shorter time series lengths. When parous flies are sampled more efficiently than nulliparous flies, the survival rate is overestimated; when nulliparous flies are sampled more efficiently than parous flies the survival rate is underestimated.

Birley (1984) points out that most people attempt to estimate the *daily* survival rate of a mosquito population, and to do this they combine the survival rate per gonotrophic cycle with the estimate of the duration of the gonotrophic cycle, but the concept of a daily survival rate is biologically incorrect. During each phase of the gonotrophic cycle mosquitoes experience different mortality risks. Furthermore, vectorial capacity does not depend on daily survival, but on surviving the number of gonotrophic cycles needed to complete the intrinsic incubation period of the parasite (e.g. malaria). Birley (1990) and Hii et al. (1990) stressed that the survival rate/gonotrophic cycle, which is usually 0.4–0.6, varies relatively little between species, locations and seasons, whereas the duration of the gonotrophic cycle varies considerably with factors such as temperature, moonlight and distance to the breeding site.

Briët method

Briët (2002) developed a simplified method of estimating the survival rate of adult mosquito populations that are subject to seasonal variations in recruitment, which does not require the time-consuming daily sampling necessary for the time-series analysis methods described above. The method uses the proportion parous and the length of the first oviposition cycle, while correcting for population growth, and requiring sampling intervals of one or two weeks, rather than daily sampling. The model assumes an exponential population growth rate, such that when there is no fluctuation in the birth and mortality rates:

$$T_{(t)} = T_{(0)} e^{(\beta - \mu)t} \tag{13.66}$$

where $T(t)$ = population size of females at time t, β = birth rate, and μ = mortality rate. The model assumes that mortality is independent of age and that all age-groups contribute proportionally to the birth rate. If the total population is considered as consisting of nulliparous (N) and parous (P) females, and assumptions are made such that all females have the same oviposition cycle length and that the mortality rate is relatively constant during the first oviposition cycle (g), such that $\mu_{t-g} \approx \mu_t$, then the differential equation for the parous proportion of the population becomes:

$$P'_{(t)} = \beta_{(t-g)} T_{(t-g)} e^{-\mu_{(t)} g} - \mu_{(t)} P_{(t)} \qquad (13.67)$$

If the birth rate also changes little over the first oviposition period, then

$$P'_{(t)} = \beta_{(t)} T_{(t)} e^{-\beta_{(t)} g} - \mu_{(t)} P_{(t)} \qquad (13.68)$$

If the proportion parous $R_{(t)}$ is given by $R_{(t)} = P_{(t)}/T_{(t)}$, then the differential of R is

$$R'_{(t)} \frac{P'_{(t)} T_{(t)} - P_{(t)} T'}{T_{(t)}^2} \qquad (13.69)$$

and

$$R'_{(t)} = \beta_{(t)} (e^{-\beta_{(t)} g} - R_{(t)}) \qquad (13.70)$$

and the proportion of parous females is independent of mortality. If over short time periods, $R'(t) \sim 0$, then

$$R'_{(t)} = e^{-\beta_{(t)} g} \qquad (13.71)$$

which allows for estimation of β as follows:

$$\beta_{(t)} = -\frac{1}{g} \ln R_{(t)} \qquad (13.72)$$

Following the model of Tsuda $et\ al.$ (1991), which describes an age-structured population in which β and μ do not vary over time, then the number of females of age x at time t under conditions of a constant daily survival rate, $p = e^{-\mu}$, is given by:

$$T_{(t,x)} = T_{(t-x,0)} e^{-\mu x} \qquad (13.73)$$

and assuming a constant finite rate of population increase per day,

$$T_{(t,0)} = T_{(t-x,0)} e^{(\beta-\mu)x} \qquad (13.74)$$

Then

$$T_{(t-x,0)} = T_{(t,0)}e^{-(\beta-\mu)x} \tag{13.75}$$

and

$$T_{(t,x)} = T_{(t,0)}e^{-(\beta-\mu)x}e^{-\mu x} = T_{(t,0)}e^{-\beta x} \tag{13.76}$$

and the total number of females, is:

$$\sum_{x-0}^{\infty} T_{(t,0)}e^{-\beta x} = \frac{T_{(t,0)}}{1-e^{-\beta}} \tag{13.77}$$

In a population where all female mosquitoes require one blood meal to become parous and feed every φ days, such that $\varphi = g$, the proportion parous =

$$R = \frac{\displaystyle\sum_{x=1}^{\infty} T_{(t,0)}e^{-\varphi\beta x}}{\displaystyle\sum_{x=0}^{\infty} T_{(t,0)}e^{-\varphi\beta x}} = \frac{\dfrac{1}{1-e^{-\varphi\beta}}-1}{\dfrac{1}{1-e^{-\varphi\beta}}} = e^{-\varphi\beta} \tag{13.78}$$

The model can be adjusted for situations where a proportion of the population is pre-gravid, requiring more than one blood meal per oviposition cycle. If all members of the population are pre-gravid, requiring two bloodmeals per oviposition cycle, such that $2\varphi = g$, then the proportion parous is given by:

$$R = \frac{\displaystyle\sum_{x=2}^{\infty} T_{(t,0)}e^{-\varphi\beta x}}{\displaystyle\sum_{x=0}^{\infty} T_{(t,0)}e^{-\varphi\beta x}} = \frac{\dfrac{1}{1-e^{-\varphi\beta}}-1-e^{-\varphi\beta}}{\dfrac{1}{1-e^{-\varphi\beta}}} = e^{-2\varphi\beta} \tag{13.79}$$

If only a fraction (f) of the population needs two blood meals, then

$$R = fe^{-2\varphi\beta} + (1-f)e^{-\varphi\beta} \tag{13.80}$$

where $fe^{-\varphi\beta}$ is the proportion of pre-gravid mosquitoes in the population.

Briët (2002) applied the model to fortnightly mosquito capture data (Dossou-Yovo et al. 1995) with an assumed gonotrophic cycle length of 2 (minimum) or 4 (maximum) days. Birth rates were estimated from the proportion parous, using

$$\beta_{(t)} = -\frac{1}{g} \ln R_{(t)}$$

(13.81)

and birth and mortality rates were observed to correspond better with rainfall than with biting population dynamics, and this suggested that reproductive stress or humidity could be the factors influencing mortality.

Charlwood et al. comparisons

Charlwood et al. (1985) sampled *Anopheles farauti* in Papua New Guinea by man biting catches, pig-baited traps, CDC light-traps placed in houses, and from morning collections of out of door resting adults during both wet and dry seasons. Survival rates per oviposition cycle were estimated by four different methods: (1) the mean parous rate, where the proportion parous in a population with constant age-structure is equivalent to the survival rates per oviposition cycle; (2) log-linear regression of numbers in different parous age-classes (1-, 2-, 3-parous etc) against age classes; (3) a time series analysis of parous data as described by Birley and Boorman (1982) and Birley and Rajagopalan (1981) and where the numbers of parous mosquitoes (M_t) on night t is equal to the total (T_t) of nulliparous plus parous females sampled one oviposition earlier, multiplied by the survival rate P; and (4) finally by an extension of the time series methods that takes into account numbers in different parous age groups, thus equation $M_t = P \times T_{t-u}$ becomes

$$M_{i+1,t} = P_i \times M_{i,\,t-u}$$

(13.82)

where $M_{i,t}$ = number of females in parous group i collected on night t, nulliparous females will have $i = 0$, P_i = survival rate from age class i to age class $i + 1$. This fourth method was evaluated on *Anopheles farauti* for the first time.

Their conclusions were that survival rates based on multiparous dissections (methods 2 and 4) are the most reliable, but require considerable skill and are time-consuming, and so are not very practical for widespread use, and that the simplest method based just on the proportion parous (method 1) is really only applicable when there is constant recruitment. They found that overall method number 3, based on parous rate and time series analysis, was the best for estimating survival rates, but there must be care not to undersample the nulliparous population.

In Tanzania, Charlwood et al. (1997) calculated daily survival rates, survival rates per gonotrophic cycle, and survival rates per extrinsic incubation period for *Anopheles gambiae* s.l. using several methods. Mosquitoes were collected using CDC light-traps, 35-minute indoor resting catches, and 'net-with-holes' collections (mosquito nets with holes to allow entry of mosquitoes). The parous rate of mosquitoes was determined using examination of ovariolar stalks or alternatively looking for evidence of tracheolar coiling. The average duration of the gonotrophic cycle (u) was estimated using the methods of Charlwood et al. (1985) and is given as:

$$u = 3 - A \tag{13.83}$$

where A = the proportion of mosquitoes from light-trap collections with dilated ovariolar stalks, or so-called 'a-sacs'. Females with a-sacs are presumed to have returned to feed on the day of oviposition, whilst those without sacs are presumed to have delayed re-feeding for one day. Daily survival probabilities were first calculated from the rate of decline in a population of *Anopheles arabiensis* in the absence of recruitment using the method described in Charlwood et al. (1995), and denoted as P_d. A second estimate was determined from the decrease in daily recaptures of marked mosquitoes (P_c). Daily survival (P_M) was also estimated from the survival rate per feeding cycle as calculated using the method of Clements and Paterson (1981), combined with the estimate of u described above. Estimates of survival per gonotrophic cycle were calculated from the delayed oocyst rates obtained from resting catches and net-with-holes catches (P_r and P_n respectively), using the methods of Saul et al. (1990) and Burkot et al. (1990). Survival rates per extrinsic incubation period were also calculated from the estimated proportion infectious, S, and the delayed oocyst rate. A further estimate of survival rate per extrinsic incubation cycle was obtained from the parous rates and immediate oocyst rates ($P_{E.M}$). Estimates of the survival rates obtained using the different methods are provided as Table 13.2 below.

The three estimates of daily survival rate obtained were similar, with P_d the most precise. Survival rates per gonotrophic cycle varied considerably, depending on the method of calculation used. Survival rates per extrinsic incubation period (estimated to be 11.6 days) were similar for each of the methods used, except for $P_{E,n}$ and $P_{E,R}$, which gave higher estimates.

Table 13.2. Estimates of survival rates of *Anopheles gambiae* s.l. and *Anopheles funestus* in Tanzania, calculated using various methods (Charlwood et al. 1997)

	Namawala *Anopheles gambiae* s.l.	*Anopheles funestus*	Michenga *Anopheles gambiae* s.l.
Estimates of survival per day (p)			
p_d	0.827 (0.004)	–	–
p_c	0.813	0.645–0.730	–
$p_M = M^{(1/u)}$	0.839 (0.025)	0.813 (0.013)	0.77 (0.018)
Estimates of survival per gonotrophic cycle (P_u)			
M	0.623 (0.004)	0.611 (0.005)	0.49 (0.01)
$P_r = (R/D_r)^{0.5}$	0.427 (0.030)	0.665 (0.064)	–
$P_n = (R/D_n)^{0.5}$	0.604 (0.060)	0.849 (0.114)	–
Estimates of survival per extrinsic incubation period (P_E)			
$P_{E,d} = P_d^E$	0.110	–	–
$P_{E,c} = P_c^E$	0.091	0.006–0.026	–
$P_{E,M} = M^{E/u}$	0.130	0.091	0.046
$P_{E,r} = S/D_r$	0.152 (0.021)	0.163 (0.030)	–
$P_{E,n} = S/D_n$	0.304 (0.061)	0.367 (0.069)	0.324 (0.041)
$P_{E,R} = SM^2/R$	0.327 (0.041)	0.132 (0.018)	–

R = immediate oocyst rate.
S = sporozoite prevalence.
D_r = delayed oocyst rate (resting catch).
D_n = delayed oocyst rate (net-with-holes catch).
M = parous rate.

Ovariolar dilatations

Just as mosquito larvae can be partitioned into age classes (instars) and plotted to give survivorship curves so can adults, by using the numbers of serial ovariolar dilatations to represent the age classes. Among the first to use this approach were Reisen et al. (1979). Later Reisen et al. (1980a, 1982a) elaborated the method and estimated daily survival rates (S_d) by plotting the numbers of female *Anopheles culicifacies* with different numbers of dilatations (*i*) as a function of their chronological age at the midpoint of each gonotrophic cycle, established by dissecting marked females of known age. It seemed reasonable to assume a type II survivorship

(Deevey 1947), so survivorship rates were then estimated by fitting the following regression

$$\ln(y_i + 1) = \ln a - t_i \ln s_d \qquad (13.84)$$

y_i = numbers of females in each dilatation class, t_i = the chronological age in days at the midpoint of each gonotrophic cycle, and $a = M_0 r$ as defined by Nelson et al. (1978) (i.e. numbers marked mosquitoes released (M_0) and percentage of population being sampled (r)). Stage-specific (vertical) life-tables for *Anopheles culicifacies* were then calculated by solving the regression for y_t when t = female age at the beginning at each gonotrophic cycle.

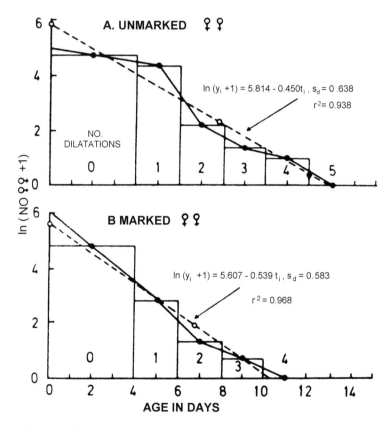

Fig. 13.5. Numbers of unmarked and marked female *Anopheles culicifacies* in each dilatation class (i), transformed to log (y + 1) and plotted as a function of the duration of the gonotrophic cycle in days (t). Included are least squares fits (----), with estimates of survivorship (s_d) and the coefficients of determination (r^2), and a graphic plot with a straight edge (———) through female age at the midpoints of each dilatation class (Reisen et al. 1980*a*)

For comparison a graphical method was used, consisting of drawing a straight line to the t_i values and estimating values of y_i (Fig. 13.5). Finally, stage-specific survivorship was estimated for each gonotrophic cycle, that is $y_i + 1/y_i$, both by regression and graphical methods (Reisen et al. 1980a, 1982a). Reisen et al. (1981, 1982c) calculated daily survival rates of *Anopheles culicifacies, Anopheles stephensi* and *Anopheles subpictus* using the same method, as did Chiang et al. (1984a,b) for *Mansonia uniformis, Mansonia annulata, Mansonia indiana, Mansonia annulifera, Mansonia bonneae* and *Mansonia dives* in Malaysia. Also in Malaysia Loong et al. (1990) obtained reasonable agreement between survival rates (0.710) of *Anopheles maculatus* by this method with that (0.761) calculated from regression of mark–recapture data.

In Papua New Guinea Charlwood (1986) concluded that the method of Birley and Rajagopalan (1981) could not be used to estimate mortality rates because in his villages the nullipars of *Anopheles farauti* were either under-sampled or over-sampled. He consequently plotted the numbers of adults caught at bait in the different dilatation classes to estimate survival rates. He found that the data fitted the non-linear Gompertz model (Clements and Paterson 1981) better than the exponential linear model, thus indicating that mortality increased with age.

Reisen et al. (1986) estimated the survivorship of *Culex tritaeniorhynchus, Anopheles culicifacies* and *Anopheles stephensi* collected resting indoors and out of doors, as well as in bovid-bait catches performed for 30 min after sunset, by three different vertical methods. Namely the method of Davidson (1954) based on parity and duration of the gonotrophic cycle, the regression of the log numbers of females showing different numbers of dilatations against time in days from emergence, and by the application of the age-specific method of Service (1973a), that is plotting log numbers in each age class (1, 2, 3 ... × dilatations) as a histogram against the midpoint of the duration of each gonotrophic cycle, a procedure which estimates survival during each gonotrophic cycle. Then the midpoints of the gonotrophic cycles (histogram blocks) are joined up by straight lines (Fig. 13.6). The numbers entering each age class can be estimated graphically and survivorship calculated for each gonotrophic cycle (Table 13.3) (much as larval instar mortalities are calculated). Reisen et al. (1986) concluded that it was technically easy to derive survivorships from the parous rate, and because populations were classified into just nullipars and pars, then statistically reliable estimates can be obtained from moderate sized samples. There are, however, difficulties in sampling the nulliparous population (see above) and with their data Reisen et al. (1986) feared that they were probably overestimating survivorship during the parous period.

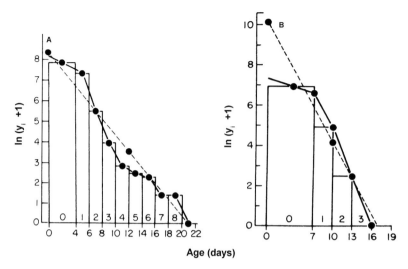

Fig. 13.6. Survivorship curves. A—*Anopheles culicifacies* collected resting during the warm season (April to November), B—*Culex tritaeniorhynchus* collected biting bovid baits after sunset (May–October). y_i = number of females scored into each dilatation class (i) plotted as a function of chronological age at the midpoint of each gonotrophic cycle (t_i). Observed survivorship curve fitted by a straight-edge • — •, and curvilinear regression function (• --- •) fitted through a and the mean of log (y_i + 1) and t_i (Reisen et al. 1986)

The other two methods of estimating survivorship are technically more difficult and much more time consuming. Moreover, the older the females the less are available for dissection, and the slope of the fitted regression line is sensitive to numbers of older females, so a large number of mosquitoes need to be dissected for statistically meaningful results. The age-specific approach is biologically more realistic, because it delineates mortality during those periods of life when death is greatest.

Mark–recapture methods

Gillies (1961) seems to have been one of the first to estimate survival of a mosquito (*Anopheles gambiae*) from the decrease in daily recaptures of marked mosquitoes.

Table 13.3. Stage-specific survivorship tables for *Anopheles culicifacies* and *Culex tritaeniorhynchus*[a] (after Reisen et al. 1986)

Species	i	Y_i	N_i	S_i
Anopheles culicifacies	0	2 591[c]	3 830.5	0.472
(April–November)[b]	1	1 600	1 808.0	0.368
	2	239	665.1	0.174
	3	51	115.6	0.286
	4	16	33.1	0.435
	5	10	14.4	0.764
	6	9	11.0	0.545
	7	3	6.0	0.683
	8	3	4.1	0.488
	9	0	2.0	
Culex tritaeniorhynchus				
Biting	0	1 047	1 047[d]	0.724
(May–October)[b]	1	758	758	0.183
	2	139	139	0.079
	3	11	11	
	4	0	0	

[a] i = dilatation age class; y_i = number of females scored into each i; N_i = antilog of number of females entering each age class i calculated from the curves presented in Fig. 10.15; S_i = stage specific-survivorship calculated as N_{i+1}/N_i.
[b] Months when generations were overlapping and the gonotrophic rhythm constant.
[c] Adjusted by twice the number of pregravids.
[d] First caught when 4 days old then every 3 days thereafter at beginning of gonotrophic cycle, so y_i has to be the same as N_i.

For example, if it can be assumed that the daily survival rate (p) is constant for marked and recaptured females then the number of marked females (M_t) recaptured on the day t after release is given by

$$M_t = M_0 r\, p^t, \text{ or } \log M_t = \log M_0 r + t \log p \qquad (13.85)$$

where M_0 = number marked mosquitoes released, r = sample rate, or percentage of the proportion of the population being sampled, t = days since release and p = daily survivorship rate. There is therefore a linear relationship between $\log M_t$ and t (Fig. 13.7). The shape of the fitted regression line provides an estimate of $\log p$, and its antilog gives the mean daily survival rate (p). In Tanzania Gillies (1961) found, except for the first 2 days, a reasonably good regression line fitted the log numbers of recaptures on days after release, indicating a more or less constant loss rate, which was calculated as about 16%.

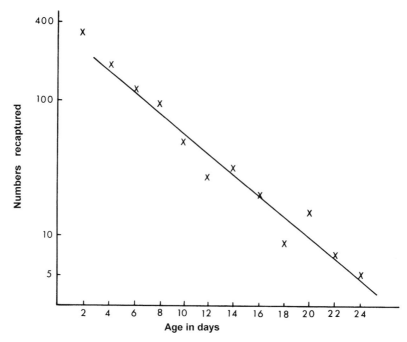

Fig. 13.7. Decrease after release in the numbers of marked adult *Anopheles gambiae* recaptured according to age in days (after Gillies 1961)

Costantini et al. (1996) used the above method to estimate daily survival of *Anopheles gambiae* s.l. in Burkina Faso. In 1991, data for all *Anopheles* species collected were combined and yielded a regression equation using the notation above of $y = 1.290 - 0.133t$, with a coefficient of determination, $r^2 = 0.791$. This equation gave a daily survival rate of 0.736 (95% CI 0.664 – 0.818). In 1992, where only *Anopheles gambiae* s.l. were included in the analysis the regression equation was $y = 1.327 - 0.129t$ and $r^2 = 0.774$. Daily survival rate in 1992 was calculated as 0.743 (95% CI 0.672 – 0.822). The presence of outliers in the 1992 data indicated that survival may not have been constant, with the probability of survival increasing after 9 days post-release.

In Thailand, Day et al. (1994) reared groups of large-bodied (wing length = 3.02 ± 0.1 mm S.E.) and small-bodied (wing length = 2.12 ± 0.1 mm S.E.) *Aedes aegypti* by manipulating larval rearing density and food availability. Newly emerged adults were maintained in cages with access to either sugar or water for approximately 36 hours prior to marking with fluorescent powders and releasing. One half of the sugar-fed group were also given a full blood meal prior to release. Regressions estimating the probability of daily survivorship were calculated from the daily recapture data by calculating the antilog$_{10}$ of the slope of the regression line. Survivorship over the 12 day recapture period did not differ significantly between the groups that had access to sugar or those that had access to water only for either the large-bodied or small-bodied mosquitoes. Taking an initial blood meal also did not appear to increase survivorship.

Lindquist et al. (1967) used the numbers of radioactive *Culex quinquefasciatus* recaptured in Yangon, Myanmar to estimate the numbers of marked mosquitoes surviving. When logarithms of the estimated numbers of marked survivors were plotted against days after release a reasonably good regression line was obtained, indicating that daily mortality (17%) was more or less constant.

In Japan Wada et al. (1969) used this method to calculate that the daily survival rate of *Culex tritaeniorhynchus* was 0.4888. However, Wada et al. (1969) point out this is likely to be an underestimate because some released females probably escaped from the recapture zone. In fact, Reisen et al. (1982*b*) concluded that horizontal estimates of survival rates, such as mark–recapture studies of *Culex tritaeniorhynchus*, are lower than those based on parity because of losses of marked adults due to emigration as well as their removal from the population during sampling.

The same approaches were used by Dow (1971) to estimate daily survival rate (0.81) of *Culex nigripalpus* in Florida, USA and by Yasuno and Rajagopalan (1973) to estimate survival rates of *Culex quinquefasciatus* during different seasons. In California, USA Nelson et al. (1978) used this method to estimate the survival rate of *Culex tarsalis* in different months (Fig. 13.8), and also the numbers of marked mosquitoes remaining in the population (M_t) at time *t* after release, which is the number released minus the numbers recaptured times the daily survival rate *p* raised to the power *t*.

$$M_t = \left(M_0 - \sum_{i=0}^{t} m_i \right) p^t \tag{13.86}$$

They found that survival rates based on recapture experiments were consistently lower than estimates based on the parous rate method of Davidson (1954). In southern Mexico, Fernandez-Salas et al. (1994) used daily recapture data from day 2 onwards and Gillies' (1961) method and calculated the daily survival rate of wild *Anopheles pseudopunctipennis* to be 0.690 and laboratory-reared females to be 0.754, rates which were lower than those estimated for the same populations using Davidson's (1954) method.

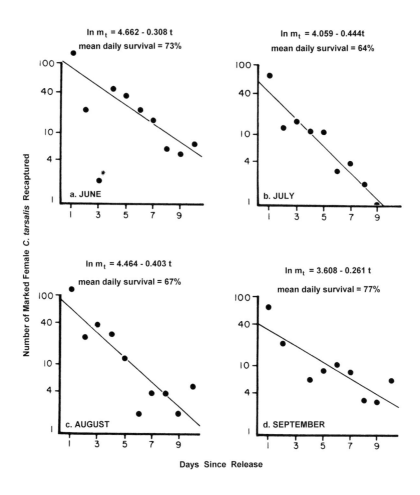

Fig. 13.8. Regressions in logarithms of numbers of female *Culex tarsalis* recaptured on days following release. (The value marked by an asterisk was excluded from calculations) (Nelson et al. 1978)

In Pakistan Reisen and Aslamkhan (1979) estimated survivorship rates of *Anopheles stephensi* by regressing the numbers of recaptured marked females against days from release. They also derived mean longevity ($\bar{L}$) of released adults as

$$\bar{L} = \sum_{t=1}^{n} \log(r_t + 1)t \Big/ \sum_{t=1}^{n} \log(r_t + 1) \qquad (13.87)$$

where r_t = mean recapture rate/day standardised, such as catch/man-hour for age t, t = cohort age in days, and n = last day a marked mosquito was caught.

Also in Pakistan Reisen et al. (1980a) estimated daily survival rates of marked female *Anopheles culicifacies* based on regression of recaptures over 1–5 days was 0.568, whereas over 1–10 days it was 0.738, while plotting numbers of adults with different numbers of dilatations gave a survival estimate of 0.583. There was therefore reasonably good agreement of survivorship rates based on horizontal and vertical methods, except that the regression of recaptures up to 10 days after release gave an unrealistically high daily survivorship rate (0.738). Hii and Vun (1985) estimated the survival rate of *Anopheles balabacensis* from the regression of numbers recaptured against days since their release as 0.7874 for yellow marked mosquitoes, and 0.719 for magenta coloured ones. The survival rate calculated from the proportion parous and the duration of the gonotrophic cycle (3 days) was found to be 0.99, considerably greater than indicated by the regression method. Hii and Vun (1985) thought this might be due to emigration of marked mosquitoes to beyond the sampling areas for recaptures, and possibly because mosquitoes tended to die of old age, that is following the Gompertz function (Clements and Paterson 1981) instead of dying exponentially. In contrast Chiang et al. (1988) estimated a daily survival rate of 0.793 for unfed and 0.867 for blood-fed marked and released *Mansonia uniformis*, which agreed well with survival rates (0.75–0.795) estimated from parous rates.

In Sri Lanka Rawlings et al. (1981) estimated the proportion of 1307 *Anopheles culicifacies* marked with fluorescent dusts remaining in the catch area from the regression obtained by plotting the numbers of marked mosquitoes caught in houses on 9 successive days. The regression coefficient, p, was calculated as 0.829. Successive multiplications of the 1307 marked released mosquitoes by 0.829 gave the numbers of them expected to survive in the area on successive days.

Reisen et al. (1982c) calculated the daily survival rates of *Anopheles culicifacies, Anopheles stephensi* and *Anopheles subpictus* by recapturing marked mosquitoes on different days after release, by regression of the numbers of adults in various dilatation classes against age, and from the parous rate (Davidson 1954). Horizontal estimates based on recapture of marked adults gave the lowest survival estimates because losses due to emigration and 'catching out' were included. Of the two vertical methods, the method of Davidson gave slightly higher estimates than the regression of dilatation classes against age in days, because survivorship during the nulliparous period was measured. Reisen et al. (1982c) concluded that from the epidemiological aspect the regression method involving dilatation classes gave the most meaningful survival estimates.

From seven mark–recapture experiments in Florida, USA Nayar (1982) plotted on a semilog scale the numbers of radioactive adult male and female *Wyeomyia smithii* caught on days after release. Following the suggestion of Gillies (1961) those recaptured on the first 2–3 days were excluded from the plots. From the regression coefficients the daily survival rate was calculated as 0.76–0.87 for males, and 0.74–0.91 for females.

Survival rates based on regression of numbers of marked adults recaptured over time will depend not only on mortality but losses due to migration, so the true survival rate may be greater than estimated by this method. This was indicated in the studies by Nelson et al. (1978) who calculated higher survival rates of female *Culex tarsalis* from parity dissections than from mark–recapture methods. These authors also pointed out that underestimating survival rates will reduce estimates of population size if this parameter enters the calculation. Charlwood et al. (1986a) have also questioned the validity of estimating survival rates from recapture of marked individuals.

Buonaccorsi et al. (2003) have questioned the appropriateness of the commonly used linear regression technique, as recaptured individuals are removed from the population, leading to biased estimates of survival and the inappropriate use of statistical tests to compare survival rates. These authors proposed using a non-linear least squares approach, available in the majority of statistical software packages, as an alternative. In simulations, the nonlinear estimate of survival performed well, exhibiting negligible bias throughout all of the settings used, even in those cases where there was bias in the estimate of the recapture rate. Bootstrap re-sampling methods were used to provide both estimated standard errors and confidence intervals. Buonaccorsi et al. (2003) pointed out that problems associated with the use of *t*-tests to compare survival rates obtained from the simple linear regression method include the assumption that the two slopes in the linear approximation are the same when the two survival rates are the same, which is true only when the recapture rates are also the same. In addition, the *t*-test is only valid when the observations are uncorrelated and have constant variance, which is not the case. In simulations, use of the *t*-test led to many false rejections of the null hypothesis that the two survival rates were the same. The bootstrap method, however, could be used to provide confidence intervals for the difference between survival rates, and to provide an associated test for equal survival rates.

Harrington et al. (2001) present the use of a non-linear regression approach with bootstrapping to estimate survival rates and avoid the objections raised against the use of linear regression approaches. The method assumes constant capture rates (θ) of marked individuals as well as constant survival rates (p_i), and gives the expected number of captures at time tj as:

$$N(1-\theta)^{(j-1)} p_i^{(tj)} \tag{13.88}$$

where N is the number initially marked and released. Standard errors and confidence intervals are generated by boot strapping. Survival rates calculated using this method were significantly higher for younger (released 3 days post-emergence) compared with older cohorts (released 13 days post-emergence and offered human blood meals daily from day 2 to day 13) in Florida, Puerto Rico, but no significant differences were observed between cohorts in a similar study conducted in Thailand. The data presented were considered by the authors as lending further support to the proposition that mosquito survival is not independent of age.

Other examples of the estimation of survivorship rates based on regression of recaptured marked adults against days after release are: *Ochlerotatus communis* (Eldridge and Reeves 1990); *Ochlerotatus triseriatus* (Beier et al. 1982; Haramis and Foster 1983; Pumpuni and Walker 1989; Walker et al. 1987); *Ochlerotatus hendersoni* (Walker et al. 1987); *Ochlerotatus melanimon* (Jensen and Washino 1991); *Ochlerotatus thibaulti* (Copeland 1986); *Culex tarsalis* (Asman et al. 1979; Nelson and Milby 1980); *Culex nigripalpus* (Nayar et al. 1980); *Anopheles punctulatus* (Charlwood and Bryan 1987); *Anopheles albitarsis* (dos Santos et al. 2004); *Anopheles culicifacies* (Rawlings and Davidson 1982; Reisen et al. 1981, 1982*a*) and also *Anopheles stephensi* and *Anopheles subpictus* (Reisen et al. 1981); and *Culex tritaeniorhynchus* (Baker et al. 1979).

Using the method of Fisher and Ford (1947) Macdonald *et al.* (1968) analysed the recaptures of marked *Culex quinquefasciatus* released in a small Myanmar village and calculated that daily survival rate of both sexes was 0.90. Using the same approach Sheppard et al. (1969) estimated the survival rates of male and female *Aedes aegypti* in Bangkok as 0.70 and 0.81, but taking into consideration loss of marked adults through dispersal outside the recapture area a modified Fisher and Ford analysis gave survival rates of 0.72 for males and 0.84 for females. In both instances the survival rates for the sexes were significantly different. Costantini et al. (1996) calculated daily survival rates of *Anopheles gambiae* s.l. in Burkina Faso using the Fisher-Ford and Jolly methods applied to mark-release-recapture data as well as the Macdonald method described below and a computer simulation model. Survival rates calculated using each of the above methods are given in table 13.4.

Table 13.4. Daily survival probabilities obtained using different techniques. Fisher and Ford (a) and (b) methods refer to the mean absolute and square fit respectively. Numbers in parentheses are 95% confidence intervals for the 'decay' model and estimated limits for the Macdonald model (Costantini et al. 1996)

Estimation method	Survival rate	
	1991	1992
Fisher and Ford (a)	0.689	0.724
Fisher and Ford (b)	0.672	0.675
Jolly	0.705	0.383
Recaptures 'decay'	0.736 (0.664–0.818)	0.743 (0.672–0.822)
Simulation Model I	0.817	0.897
Simulation Model II	0.828	0.948
Macdonald model	0.824 (0.731–0.885)	

Saul's method

Estimates of survival rates, and also population size, made from recapture of marked mosquitoes appearing in bait catches, suffer from the problem that feeding is cyclical, therefore the numbers recaptured, as well as the ratio of marked to unmarked adults, will vary greatly from day to day (Charlwood et al. 1986a; Conway et al. 1974). Charlwood et al. (1986a) pointed out that the frequency distribution of marked mosquitoes at bait and thus available for recapture, can be described by a series of renewal equations as a function of the survival throughout the feeding cycle and the frequency distribution of the durations of the feeding cycle. Saul (1987) developed these renewal equations and used them to provide estimates of survival, duration of the feeding cycle and capture efficiency, and from the latter estimates of population size of haematophagous insects. The approach is as follows. The probability that released marked blood-fed mosquitoes will re-feed the next night is P_1, while the probability they will feed for the first time the next night is P_2, so let P_i be the probability they will re-feed for the first time on the ith night following their release as blood-engorged adults. From this the probability of their surviving a feeding cycle (P) and being ready to re-feed is

$$P_f = \sum_{i=1}^{\infty} P_i \qquad (13.89)$$

The model proposed by Saul (1987) has P_f constant throughout the mosquito's life. Following the release of a number of marked mosquitoes there will be more or less synchronous waves of mosquitoes appearing at bait, progressively decreasing in size with time. The numbers feeding after the jth day after release as blood-feds (Nf_j) is given as

$$Nf_j = P_1 Nf_{j-1} + P_2 NF_{j-2} + \dots + P_i Nf_{j-1} + \dots + P_j Nf_0 \qquad (13.90)$$

$$= \sum_{i=1}^{j} P_i Nf_{j-i} \qquad (13.91)$$

Nf_0 = numbers of marked mosquitoes released, and Nf_{-1} and Nf_{-2} etc are zero because there are no marked mosquitoes on days preceding their release. In practice some of these components can be omitted. For example with tropical mosquitoes there are likely to be no mosquitoes feeding again for the first time the day following their release, or as late as say the 5th day. In such a case the above formula simplifies to

$$Nf_j = \sum_{i=2}^{5} P_i Nf_{j-i} \qquad (13.92)$$

The number of marked adults blood-feeding (Nf_j) is estimated from the number of marked ones recaptured on day j (R) and the recapture rate A

$$R_j = ANf_j \qquad (13.93)$$

and substituting eqn (13.92) in eqn (13.91) gives

$$R_j = \sum_{i=1}^{j} P_i R_{j-i} \qquad (13.94)$$

During the days immediately following release eqn (13.90) can be simplified, for example in the situation cited above $P_1 = 0$ and also $P_{>5} = 0$, and let M = the number of marked mosquitoes released so that

$$R_1 = 0 \qquad (13.95)$$

$$R_2 = P_2 MA \tag{13.96}$$

$$R_3 = P_3 MA \tag{13.97}$$

$$R_4 = P_2 R_2 + P_4 MA \tag{13.98}$$

$$R_5 = P_2 R_3 + P_3 R_2 + P_5 MA \tag{13.99}$$

Now by rearranging and substituting eqns (13.96) and (13.97) into (13.98) and (13.99), expressions for P_2 to P_5 can be derived as follows

$$P_2 = R_2 / MA \tag{13.100}$$

$$P_3 = R_3 / MA \tag{13.101}$$

$$P_4 = R_4 / MA - R_2^2 / (MA)^2 \tag{13.102}$$

$$P_5 = R_5 / MA - 2 R_2 R_3 / (MA)^2 \tag{13.103}$$

Again following the specific example the recapture rate (A) is calculated by substituting all above equations for P (i.e. 13.100–13.103) into equations for any of the subsequent days' catches, for example for the catches from days 6, 7, 8 ... In practice, however, the number of marked mosquitoes recaptured decreases rapidly, and so to increase the accuracy, these catches can be summed as follows, again using the example cited above

$$\sum_{j=6}^{n} R_j = P_2 \sum_{j=6}^{n} R_{j-2} + P_3 \sum_{j=6}^{n} R_{j-3} - P_4 \sum_{j=6}^{n} R_{j-4} + P_5 \sum_{j=6}^{n} R_{j-5} \tag{13.104}$$

where there are collections over n days.

Now substitution of eqn (13.104) by eqns (13.100–13.103) gives a quadratic equation in A. For practical values of R_j only one of the two solutions to this equation for A will give a realistic value, namely $0 \leq A \leq 1.0$.

The above method can be used to determine A for any number of values of P_i but only so long as the marked mosquitoes are recaught on more days than the number of values of P_i that are to be estimated. Actual values of P_i can be determined by substituting the estimated value of A into eqns (13.100–13.103)—that is in the example taken to illustrate the method.

Saul (1987) used the following formula proposed by Charlwood et al. (1986a) to estimate the duration of the feeding cycle (FC).

$$FC = \sum_{i=1}^{\infty} i\, P_i \left| \sum_{i=1}^{\infty} P_i \right. \tag{13.105}$$

Using this average value of the feeding cycle (FC) and an estimate of the numbers of mosquitoes feeding each day (M/A), the population of the mosquitoes in the area can be estimated as

$$P = (FCM)/A \tag{13.106}$$

Watson et al. (2000) used this method to estimate survival of *Ochlerotatus notoscriptus* in a mark-release-recapture study in Queensland, Australia. The probability of daily survival was estimated at 0.77 for females marked blue and 0.79 for females marked pink.

Saul (1987) considered his method as having several advantages over regression methods for estimating survival rate or population size from mark–recapture studies, including the fact that median values are obtained of the parameters, and moreover confidence limits can be attached to them. This method also allows the estimation of the recapture rate, which is not obtained by the other methods. The advantages and limitations of the approach are described by Saul (1987).

Hii et al. method

As reviewed by Charlwood et al. (1986a) the frequency distribution of re-captured marked mosquitoes caught at bait can be described by a series of renewal equations as a function of survival through the feeding cycle, and the frequency distribution of lengths of the feeding cycle. In experiments in Sabah, Malaysia with *Anopheles balabacensis* Hii et al. (1990) described a new method for the estimation of adult survival rates and the oviposition interval. The basic underlying relationship in previous models and for the new one they proposed is that if the oviposition cycle is u days then

$$M_t = P_u \times T_{t-u} \tag{13.107}$$

where M_t = parous component, and P_u = survival rate per oviposition cycle. Cross-correlation function of the time series M_t and T_t will have peaks at a lag of u days and the presence of this peak estimates u. Survival rate (P) can be estimated from a time series of M_t and the lagged time series T_{t-u}.

In the simplest type of mark–recapture experiment with just a single release of marked mosquitoes a sample of mosquitoes (C_0) is marked and released on day 0, and on subsequent days marked mosquitoes are recaptured R_t after completion of their oviposition cycle. In this case the equivalent of eqn (13.107), as given by Charlwood et al. (1986a), is

$$R_t = P_u \times R_{t-u} \tag{13.108}$$

where $R_0 = A \times C_0$, with A = the sampling efficiency. However, in the study of Hii et al. (1990) females caught at bait were marked on consecutive nights. For example, those caught at bait were allowed to blood-feed, marked and then released, and on subsequent nights while marked females were counted and removed from the sample those that were unmarked were now marked and released. This procedure was undertaken on 28 consecutive nights, and because of this eqn (13.108) has to be put in the more generalised form of

$$R_t = P_u \times R_{t-u} + A \times P_u \times C_{t-u} \tag{13.109}$$

This eqn (13.109) implies that there will be cross-correlation between the time series of releases of marked mosquitoes and recaptures with a peak at the time lag of u days. The parameters P_u and $A \times P_u$ can be estimated from a multiple linear regression of R_t against R_{t-u} and C_{t-u} with zero intercept. If A is known then a weighted regression estimate of P_u is

$$P_u = \sum R_t / \sum (R_{t-u} + A \times C_{t-u}) \tag{13.110}$$

Now under a steady state condition the number of recaptures should tend to a constant, R, and the number released each day to a constant, C, and then

$$R = A \times P_u \times C / (1 - P_u) \tag{13.111}$$

Hii et al. (1990) pointed out that when $P_u = 0.5$ then $R = A \times C$.

Rawlings et al. (1981) working in Sri Lanka also used a model where marked mosquitoes (*Anopheles culicifacies*) were released on consecutive days. They pointed out that under certain conditions the proportion of marked mosquitoes caught in a daily sample (R/T) could tend to be a constant (Lines et al. 1986). In such a case eqn (13.111) can be divided by T, thus

$$(R/T) = \frac{A \times P_u}{1 - P_u} \times (C/T) \tag{13.112}$$

But Hii et al. (1990) point out that this equation is not consistent with the equation proposed by Rawlings et al. (1981).

In their actual field experiment Hii et al. (1990) marked and released 13 166 *Anopheles balabacensis*, and had a good recapture rate of 11.7% reaffirming previous results of high recapture rates (about 12%) with this species (Hii 1985; Hii and Vun 1985). The oviposition cycle was estimated to be 3 days, while the survival rate/oviposition was 0.48, and the sampling efficiency, A, was 13%. The authors point out that their survival rate is remarkably similar to other estimates for *Anopheles balabacensis* (Hii and Vun 1985), for the *Anopheles punctulatus* group in Papua New Guinea (Charlwood et al. 1985, 1986a) and for Afrotropical vectors (Mutero and Birley 1987). They believe that the survival rate of mosquitoes appears to be relatively independent of the duration of the oviposition cycle. They argue that methods of analysis which concentrate on estimating daily survival rates, instead of survival over a gonotrophic cycle could mask the possibility that survival rates may vary within a species during different phases of its life.

Parasitic infections

Malarial infections

Draper and Davidson (1953) introduced a method of estimating the survival of female *Anopheles* based on sporozoite infections. The procedure relies on the assumption that the percentage of mosquitoes infected with sporozoites on the day of capture, i.e. the immediate sporozoite rate, depends on the proportion of the population that lives sufficiently long for the infection to pass to the salivary glands. If mosquitoes are not dissected on the day of capture, but are kept alive for a period equal to that of the extrinsic cycle of the malaria parasite, then the delayed sporozoite rate is obtained. This will depend on the proportion infected at the time of collection, irrespective of the immaturity of the infection. The ratio between the immediate and delayed sporozoite rates is directly proportional to the chances of survival during the interval between an infective blood-meal and the appearance of sporozoite in the glands. If, as suspected by Macdonald (1952), adult mortality is geometric, i.e. all the adults irrespective of age are subjected to the same mortality risk, then a very simple mathematical relationship exists between the chance of survival and the ratio:

$$\frac{p^{n-1}}{1} = \frac{\text{immediate sporozoite rate}}{\text{delayed sporozoite rate}}, \text{or } \log p = \frac{\log \text{ratio}}{n-1} \quad (13.113)$$

where p = probability of a mosquito surviving through 1 day and n = time in days for completion of the extrinsic cycle of the malaria parasite. Draper and Davidson (1953) used $(n - 1)$ as they considered that mosquitoes were on average at least 1-day old when they took an infective blood-meal.

Although the method estimates survival independently of parity and has been used by a few workers in Africa (Costantini et al. 1996; Davidson 1955b; Davidson and Draper, 1953; Krafsur 1971; Service 1965), it has not been widely used, probably because it is a time consuming method and cannot be applied to all mosquito species.

A somewhat similar approach was proposed by Garrett-Jones (1970). He suggested that as there is a mean difference in age of 8 days between mosquitoes that are gut and gland positive for *Plasmodium falciparum*, then:

$$p^8 = \frac{\text{sporozoite rate}}{\text{total malarial infection rate}} \quad (13.114)$$

From data presented by Shute et al. (1965) on *Anopheles funestus* collected from the Muheza area of Tanzania, Garrett-Jones (1970) calculated p^8 as 0.234, which gives an estimated daily survival rate of 0.835. This agrees well with the value of 0.815 calculated by Garrett-Jones (1970) from data supplied by Gillies and Wilkes (1965) on *Anopheles funestus* collected from the same area.

Saul et al. (1990) presented a modified cyclical feeding model by which oocyst and sporozoite infection rates of *Anopheles*, and human blood indices, can be used to estimate adult: (i) female survival rates through the feeding cycle or through the extrinsic incubation period of malaria parasites; (ii) vectorial capacity; and (iii) the proportion of blood-meals on humans that acquire infection – in other words the probability of infection. Their method does not require knowledge of either the duration of the extrinsic incubation period of malaria or the number of feeds that take place during this period. Their model is best understood by referring to the accompanying paper of Graves et al. (1990), which applies the model to data collected in Papua New Guinea.

Their sampling procedure was as follows. Collections were made in each of three villages on four consecutive nights, followed by an interval of 10 days, and subsequently a further four nights collection. The programme was repeated over about 1 year at 4–6-week intervals. Collections of biting and resting adults were made both inside and outside houses. ELISA tests were undertaken to detect sporozoite rates in all mosquitoes caught biting in one village. In the other two villages mosquitoes caught on two nights were kept alive for 3 days and oocyst rates determined, while a sample of those caught on the other two nights was dissected for parity determination. Bodies of these, and remaining undissected mosquitoes, were used for ELISA detection of sporozoites. Mosquitoes from 1 day's catch/week were used for ELISA determination of blood meal source. Engorged females from the other 3 days of collecting were kept alive for 5 days and oocyst rates determined. From this sampling programme the following parameters were determined.

D_b = Proportion of mosquitoes infected at time of capture at bait, estimated from oocyst rate of those kept alive for 3 days.

D_f = Proportion of mosquitoes infected at time of capture as blood-fed out of door resting mosquitoes, estimated from oocyst rate of those resting out of doors and kept alive for 5 days.

$D_{f'}$ = Proportion of mosquitoes which had become infected by time of capture in collections of blood-fed mosquitoes resting inside houses, estimated from oocyst rate of those caught resting indoors kept alive for 5 days.

S = Proportion of mosquitoes infective at time of capture, estimated by sporozoite rate of those caught biting.

Q = Proportion of blood-feds caught with human blood. In this instance because most of the malaria vectors leave houses after feeding, the best estimate of human feeds (human blood index) is based on ELISA tests on collections of out of door blood-feds.

Q' = Proportion of human feeds on mosquitoes caught inside houses, estimated by human blood index obtained from those caught resting inside houses.

Oocyst rates derived from mosquitoes that were kept alive for 3 or 5 days are regarded as delayed infection rates. The individual vectorial capacity (*IC*) is defined as the number of potentially infective bites arising from each bite on a person infectious to the mosquito. The overall vectorial capacity is obtained by multiplying *IC* by the host biting rate, that is the number of bites/person per night.

The equations generated by Saul et al. (1990) and used by Graves et al. (1990) are given below, together with their approximate variances. P_f = probability of a mosquito surviving through the feeding cycle; P_e = probability of survival through the extrinsic incubation period; IC = individual vectorial capacity; and K = probability of a mosquito becoming infected per bite. The variance of S is estimated approximately as $V(S) = S(1 - S)/n$, where n = the numbers of mosquitoes on which S was estimated; and similarly for the other five parameters ($D_b\, D_f\, D_f'\, Q$ and Q').

$$P_f = D_b / D_f \tag{13.115}$$

$$Var(P_f) \approx P_f^2 [V(D_f)/D_f^2 + V(D_b)/D_b^2]$$

$$P_f = Q'/[QD_f'/D_b) - Q + Q'] \tag{13.116}$$

$$Var(P_f) \approx [P_f^2 Q/Q')^2 \{(D_f'/D_b - 1)^2 [V(Q)/Q^2 + V(Q')/(Q')^2]$$
$$+ (D_f'/D_b)^2 [V(D_f')/(D_f')^2 + V(D_b)/D_b^2]\}$$

$$P_f = [Q(D_f'/D_f) - Q']/(Q - Q') \tag{13.117}$$

$$Var(P_f) \approx [Q'(1 - P_f)/(Q'-Q)]^2 [V(Q)/Q^2 + V(Q')/(Q')^2$$
$$+ [P_f - Q'/(Q'-Q)]^2 [V(D_f)/D_f^2 + V(D_f')/(D_f')^2]$$

$$P_e = S/D_f \tag{13.118}$$

$$Var(P_e) \approx P_e^2 [V(S)/S^2 + V(D_f)/D_f^2]$$

$$P_e = Q'S/[QD_j' - (Q - Q')D_b] \tag{13.119}$$

$$Var(P_e) \approx (P_e/S)^2 \{V(S) + [P_e(D_f' - D_b)Q/Q']^2 [V(Q)/Q^2$$
$$+ V(Q')/(Q')^2] + (P_e Q/Q')^2 [V(D_f')$$
$$+ [(Q'-Q)/Q]^2 V(D_b)]\}$$

Equations for IC. In its simplest form the equation for IC is:

$$IC = QP_e /(1 - P_f) \tag{13.120}$$

Substituting eqns (13.116) and (13.119) into (13.120) gives the following:

$$IC = Q'S/(D'_f - D_b) \tag{13.121}$$

$$Var(IC) \approx (IC)^2 \{V(Q')/Q')^2 + V(S)/S^2 + [1/(D'_f - D_b)]^2 [V(D'_f) + V(D_b)]\}$$

Substituting eqns (13.117) and (13.118) into (13.120) gives the following:

$$IC = S(Q'-Q)/(D_f - D_{f)} \tag{13.122}$$

$$Var(IC) \approx (IC)^2 \{V(S)/S^2 + [1/(Q'-Q)]^2 [V(Q)+V(Q')] \\ + [1/(D'_f - Df)]^2 [V(D_f)+V(D'_f)]\}$$

$$K = (D'_f - D_b)/[Q'(1-D_b)] \tag{13.123}$$

$$Var(K) \approx K^2 \{V(Q')/(Q')^2 + [1/D'_f - D_b)]^2 [V(D'_f) \\ + [(1-D'_f)/(1-D_b)]\}$$

$$K = (D'_f - D_b)/[Q'(1-D_b)] \tag{13.124}$$

$$Var(K) \approx [K/D'_f - D_f)]^4 \{(Q'-Q)^2 [(1-D_f)^2 V(D'_f) + (1-D'_f)^2 V(D_f)] \\ + (D'_f - D_f)^2 [(1-D'_f)^2 V(Q) + (1-D_f)^2 V(Q')]\}$$

The equations chosen to estimate P_e and P_f depend on the information available. The simplest to use are clearly eqns (13.115) and (13.118) if delayed infection rates are known from collections of blood-fed out of door resting mosquitoes, otherwise the human blood index must be obtained for both indoor and outdoor populations, and these may be difficult to obtain without bias. For the estimation of individual vectorial capacity (IC) eqn 13.121 is preferred because the human blood index need only be obtained for indoor resting mosquitoes.

An advantage of the present approach is that the probability of infection (K) can be calculated from just two simple collections, biting and indoor resting. Those interested in this approach should refer to Graves et al. (1990) for a description of methods used to determine appropriate parameters for the *Anopheles punctulatus* complex in Papua New Guinea. Haji et al. (1996) estimated survival rate per gonotrophic cycle of *Anopheles gambiae* s.l. and *Anopheles funestus* in Tanzania from the immediate oocyst rate in mosquitoes coming to feed (R) and the overall probability that a mosquito is infected (D), equivalent to the delayed oocyst rate. R was calculated from the distribution of oocyst sizes from the most recent feed ($\theta(x)$) and from previous infections ($\phi(x)$). The distribution of oocyst sizes in laboratory mosquitoes maintained for three days gives an estimate of $\theta(x)$, and by comparing this with the observed distribution of oocyst size in wild populations, $\phi(x)$ can be estimated and from this K, the probability that a mosquito acquired an infection at the latest feed, can be calculated using maximum likelihood methods and $R = (D - K)/(1 - K)$. The estimated survival rates/gonotrophic cycle for *Anopheles gambiae* s.l. and *Anopheles funestus* calculated using this method were 65.5% and 52.9%, respectively.

Filarial infections

Infections of mosquitoes with filarial worms provide information about the minimum number of days they have survived after ingesting infective blood, and if there is knowledge about the duration of the various parasitic stages in the vector then survivorship estimates can be made. For example, in India Laurence (1963) dissected 651 *Culex quinquefasciatus* and found that 37 adults had microfilariae of *Wuchereria bancrofti*, 22 had sausage-stage worms, 32 had late sausage-stage worms while 10 contained infective 3rd stage larvae. Based on known parasite development times found experimentally, 64 of the 101 infected mosquitoes must have survived at least 2 days, 42 survived at least 4 days and 10 with infective filariae survived at least 10 days. The probability of survival over 1 day (p) is given by:

$$p^2 = 64/101 = 0.634, \quad p^4 = 42/101 = 0.416, \quad p^{10} = 10/101 = 0.099 \quad (13.125)$$

which gives p = 0.769, 0.803 and 0.794. Estimates of daily mortality $100 (1 - p)$ are therefore 20.4%, 19.7% and 20.6%, respectively. Daily mortality rates were calculated also for *Anopheles peditaeniatus*.

   In West Iran van Dijk (1966) found the interval between the appearance of 1st and 2nd stages of *Wuchereria bancrofti* in *Ochlerotatus kochi* was 4 days, and between the appearance of 2nd and 3rd stages was 5 days. He dissected 429 mosquitoes and found that 17 contained 3rd stage larvae, 23 had 2nd stage and 35 had 1st stage sausage forms. So, the survival rates (p) during the first interval of 4 days is

$$p_1^4 = 40/75 = 0.533, \text{ thus } p_1 = 0.852 \quad (13.126)$$

and the survival rate during the second 5 days' interval is

$$p_2^5 = 17/40 = 0.425, \text{ thus } p_2 = 0.853. \tag{13.127}$$

Similar approaches were used on *Culex quinquefasciatus* in Sri Lanka (Samarawickrema 1967), in Trinidad (Nathan 1981) and in Myanmar (De Meillon et al. 1967a).

In Tanzania Krafsur and Garrett-Jones (1977) caught 3289 unfed hungry *Anopheles funestus* in CDC light-traps placed in houses in which the occu_ pants were protected by mosquito nets. Of the 1723 females that were parous 1710 were dissected and 257 had developing larvae of *Wuchereria bancrofti*, and 15 of them harboured 3rd stage infective larvae. The mean intervals between blood-feeding was 3 days, therefore they reasoned that unfed adults caught in light-traps would be, 3, 6, 9, 12 and 15 days old etc. Now, by studying the developmental forms of filarial worms after infection they were able to estimate the numbers of *Anopheles funestus* in each 3-day-old age-group, up to 15 days, since experiments showed that the earliest infective larvae appeared 13 days post-infection. From these data they could calculate the age-distribution of adults and percentage mortality between the age-groups. They then plotted the regression of numbers having filarial infections (y) on age as shown by 3-day age-groups (x).

$$\log_e y = \log_e a - bx \tag{13.128}$$

where b = the slope of the regression and represents mortality rate; a = the origin at zero time of the linear equation, and e = 2.718. They found the regression equation was

$$\log_e y = 5.25 - 0.1646x \tag{13.129}$$

Thus the average daily mortality rate is estimated as 16.46%. This agrees well with the mean value of 15.6% calculated by Krafsur and Garrett-Jones (1977) by plotting log numbers of *Anopheles funestus* having oviposited one or more times (1-, 2-, 3-, 4- or 5-parous) against age in days; these data having been extrapolated from Gillies and Wilkes (1965) who collected adults from the same area.

In Sudan Gad et al. (1989) compared survival rates of *Culex pipiens* by the Davidson (1954) method based on the proportion parous and the duration of the gonotrophic cycle, and by the numbers infected with stages 2 and 3 of *Wuchereria bancrofti*. They considered the latter gave a more reliable indication of survival. However, for this approach to work the development must be synchronised, but in laboratory filarial infections where the parasite numbers may be considerably higher than encountered in the field, there is often asynchronous parasite development. This suggests that the times for various developmental stages to be reached can be variable.

Lardeux and Cheffort (2002) proposed a method of determining the length of the gonotrophic cycle of *Aedes polynesiensis* that depends on measurements of the length and width of larvae of *Wuchereria bancrofti* developing within the body of the mosquito host. In this method, the time since infection is calculated from the age of the developing larvae. Larval dimensions were measured by tracing the length and width of images of the larvae, which were projected onto paper using a *camera lucida* attached to a microscope. A digitizing tablet was used to determine the length and width of the projected images. Larval development is temperature-dependent, and so where development is incomplete, as is the case with field-collected infected mosquitoes,

$$r(T)D_T = A \tag{13.130}$$

where $r(T)$ is the developmental rate at temperature T and D_T is the time elapsed since the beginning of development at temperature $T(D_T < \Gamma_T)$ and Γ_T is total development time at temperature T. A measures the 'fraction of development' of *Wuchereria bancrofti* larvae. If A can be estimated for larvae dissected from wild-caught mosquitoes, then development time D, equivalent to the period between two blood-meals in the case of mosquitoes caught at human bait, can be calculated as

$$D = A / r(T) \tag{13.131}$$

Unfortunately, temperature in the field varies both within and among days, and calculations of development time require use of a more complex computation reflecting the daily sine function oscillation of temperature between T_{min} and T_{max}, such that

$$T = a \sin(t) + b \tag{13.132}$$

where T is temperature, a the amplitude of the sine curve ($a = (T_{max} - T_{min})/2$), t the time of day in radians (1 day = 2π) and b is the mean of the sine curve ($b = (T_{max} + T_{min})/2$). For incomplete larval development, the previous equation becomes

$$\sum_{t=0}^{D} r(f(t))\Delta t \cong A \tag{13.133}$$

where f is the sine function of t and Δt are periods of constant temperature. The time between a mosquito becoming infected and subsequently re-feeding can be calculated using estimates of A obtained from the regression equation below and temperature maxima and minima of preceeding days through a back-calculation process, beginning from the day of capture ($t = 0$).

The relationship between larval length and width, and 'fraction of development' was determined by dissecting mosquitoes held for different periods of time at different constant temperatures using multiple linear regression of the form:

$$A = \alpha + \beta L + \gamma Y \qquad (13.134)$$

where α, β, and γ are regression coefficients. The experiments produced the following relationship: $A = -0.0913(0.005) + 0.238(0.008)\text{Length} + 23.359(0.284)\text{Width}$, with a correlation coefficient $R = 0.96$. Figures in parentheses are standard errors. This regression equation could then be used to reliably estimate the 'fraction of development' of *Wuchereria bancrofti* larvae using measurements of length and width.

Lardeux and Cheffort (2002) used this technique to estimate the blood meal intervals of wild-caught *Aedes polynesiensis*. Larval age was determined using the above equations and a mean age was determined for each of the 59 mosquitoes dissected. A modal analysis was used to decompose the mosquito age histograms into their Gaussian components (Lardeux and Tetuanui 1995), giving estimates of mean and standard deviation. The mean time between Gaussian modes was estimated by observation to be ≈ 3–4 days, agreeing closely with data obtained from previous mark-release-recapture studies.

Miscellaneous methods

In Kenya McDonald (1977) estimated adult survivorship of *Aedes aegypti* by marking newly emerged adults with paint spots, using a different colour and position to date-specific mark over 23 days. Adults were then caught at bait and all those bearing a mark were again colour-coded with the date of capture and released. From the recaptures a survivorship curve was drawn and the daily survival of males was estimated as 0.77 and for females as 0.89.

Also in Kenya Linthicum et al. (1985) calculated the survival rate of female *Aedes lineatopennis* caught in light-traps, at human bait and from out of door collections for 44 days following emergence from a breeding place. They assumed the survival rate was independent of age and estimated it from the regression

$$\ln(N_t) = \ln(N_0) + \ln(p)t \qquad (13.135)$$

where N_t = numbers caught at time t, N_0 = number caught at time 0, p = daily probability of survival and t = number of days. In fact ln (N_0) is the intercept of the regression line and ln (p) is its slope (β). In practice the natural logarithms of total numbers of female *Aedes lineatopennis* caught by all sampling methods were plotted against age in days, starting at the time of maximum collection (in this example day 4) after emergence. Daily probability of survival was estimated as 0.85 (95% confidence limit 0.84–0.86).

In Florida, USA in an *Aedes aegypti*-free area large numbers of adults were released at two experimental sites. The first eggs found in ovitraps were collected 5 days later, thereafter in both localities peak ovipositions occurred at 4-day intervals, which corresponded with the known oviposition cycle. Laboratory experiments determined that the average egg batch was 93 or 96 eggs. Seawright et al. (1977) argued that the numbers of eggs laid in successive ovipositions were directly related to the numbers of surviving released females. So, if mortality was constant and about the same numbers of eggs were produced during each oviposition cycle, then the adult survival rate (S) was

$$\log N_d = \log N_0 + \log S_d \qquad (13.136)$$

where N_d = number of females alive on day d; N_0 = numbers released and d = days after release. Thus the regression of log numbers of eggs collected in ovitraps plotted against days gives the daily survival rate (0.82 and 0.785 for the two experiments). When the regression line is extrapolated back to intercept the egg axis and this egg estimate is divided by the mean number of eggs laid per oviposition then this should give the numbers of adults released. In one experiment there was good agreement, but in the other trial it underestimated the 16 000 released by 83.7%. Possible explanations are presented by the authors.

If a population of mosquitoes is in a steady state with a daily emergence of M mosquitoes with a daily survival rate of p, then the number aged x days is Mp^x. Using this as a basis and letting n = days from adult emergence to the first blood-meal, d = duration of gonotrophic cycle, f = days from oviposition to refeeding, g = proportion of females which are blood-fed or developing eggs, u = proportion of unfed females, and $c = d + f$, Colless (1958) derived the following equations

$$g = \frac{p^n - p^{n+d}}{1 - p^c}, \text{ and } u = 1 - \frac{p^n(1 - p^d)}{1 - p^c}. \qquad (13.137)$$

Consequently, if the proportion of u and g are obtained from sampling resting mosquitoes and if n, d and c are known, this allows the calculation of the survival rate (p).

This approach has rarely been used but Reisen et al. (1983) adopted it to calculate the survival rate (S) of *Culex tarsalis* collected resting as follows

$$S = \frac{(G)^{1/d}}{(U)} \tag{13.138}$$

where G = the number of gravid mosquitoes and U = number of unfeds, and d = duration of the gonotrophic cycle.

References

Allan SA, Surgeoner GA, Helson BV, Pengelly DH (1981) Seasonal activity of *Mansonia perturbans* adults (Diptera: Culicidae) in southwestern Ontario. Can Entomol 113: 133–139

Amalraj DD, Das PK (1996) Life-table characteristics of *Toxorhynchites splendens* (Diptera: Culicidae) cohorts reared under controlled food regimens. J Vector Ecol 21: 136–145

Anufrieva VN, Artem'ev MM (1981) Specification of the method for determining the physiological age of female mosquitoes exemplified by populations of *A. pulcherrimus* and *A. hyrcanus* in north–eastern Afghanistan and fecundity of these species. Medskaya Parazitol 50: 55–62 (In Russian, English summary)

Appawu MA, Baffoe-Wilmot A, Afari EA, Dunyo S, Koram KA, Nkrumah FK (2001) Malaria vector studies in two ecological zones in southern Ghana. Afr Entomol 9: 59–65

Asman SM, Nelson RL, McDonald PT, Milby MM, Reeves WC, White KD, Fine PEM (1979) Pilot release of a sex-linked multiple translocation into a *Culex tarsalis* field population in Kern county, California. Mosquito News 39: 248–258

Bailly-Choumara H (1973) Étude préliminaire d'une récolte d'*Anopheles labranchiae* par piège CDC réaliseé dans la région de Larache, Maroc. Bull World Health Organ 49: 49–55

Baker RH, Reisen WK, Sakai RK, Hayes CG, Aslamkhan M, Saifuddin UT, Mahmood F, Perveen A, Javed S (1979) A field assessment of mating competitiveness of male *Culex tritaeniorhynchus* carrying a complex chromosomal aberration. Ann Entomol Soc Am 72: 751–758

Barr AR, Morrison AC, Guptavanij P, Bangs MJ, Cope SE (1986) Parity rates of mosquitoes collected in the San Joaquin marsh. Proc California Mosq Vector Control Assoc 54: 117–118

Beier JC, Berry WJ, Craig GB (1982) Horizontal distribution of adult *Aedes triseriatus* (Diptera: Culicidae) in relation to habitat structure, oviposition, and other mosquito species. J Med Entomol 19: 239–247

Bellamy RE, Corbet PS (1973) Combined autogenous and anautogenous ovarian development in individual *Culex tarsalis* Coq. (Dipt., Culicidae). Bull Entomol Res 63: 335–346

Bellamy RE, Corbet PS (1974) Occurrence of ovariolar dilatations in nulliparous mosquitoes. Mosquito News 34: 334

Bertram DS, Samariwickrema WA (1958) Age determination for individual *Mansonioides* mosquitoes. Nature 182: 444–446

Birley MH (1977) The estimation of insect density and instar survivorship functions from census data. J Anim Ecol 46: 497–510

Birley MH (1979) The estimation and simulation of variable developmental period, with application to the mosquito *Aedes aegypti* (L.). Researches in Population Ecology 21: 68–80

Birley MH (1984) Estimation, tactics and disease transmission. In: Conway CR (ed) Pest and Pathogen Control: Strategic, Tactical and Policy Models. No. 13, International Series on Applied Systems Analysis, Wiley Interscience, Wiley & Sons, Chichester, pp 272–289

Birley MH (1990) Highly efficient dry season transmission of malaria in Thailand. Trans R Soc Trop Med Hyg 84: 610

Birley MH, Boorman JPT (1982) Estimating the survival and biting rates of haematophagous insects with particular reference to the *Culicoides obsoletus* group (Diptera, Ceratopogonidae) in southern England. J Anim Ecol 51: 135–148

Birley MH, Charlwood JD (1989) The effect of moonlight and other factors on the oviposition cycle of malaria vectors in Madang, Papua New Guinea. Ann Trop Med Parasitol 83: 415–422

Birley MH, Rajagopalan PK (1981) Estimation of the survival and biting rates of *Culex quinquefasciatus* (Diptera: Culicidae). J Med Entomol 18: 181–186

Birley MH, Walsh JF, Davies JB (1983) Development of a model for *Simulium damnosum* s.l. recolonization dynamics at a breeding site in the onchocerciasis control programme area when control is interrupted. J Appl Ecol 20: 507–519

Biswas S, Wattal BL, Tyagi D, Kumar K (1980) Limitation of larval parasitic water mite infestation in age-gradation of adult *Anopheles*. J Commun Dis 12: 214–215

Blackmore JS, Dow RP (1962) Nulliparity in summer and fall populations of *Culex tarsalis* Coq. Mosquito News 22: 291–294

Bockarie MJ, Service MW, Barnish G, Touré YT (1995) Vectorial capacity and entomological inoculation rates of *Anopheles gambiae* in a high rainfall forested area of southern Sierra Leone. Trop Med Parasitol 46: 164–171

Brady J (1963) Results of age-grouping dissections on four species of *Anopheles* from southern Ghana. Bull World Health Organ 29: 147–153

Briegel H, Graf R (1994) Dilatations in mosquito ovarioles. Parasitol Today 10: 356–356

Briët OJT (2002) A simple method for calculating mosquito mortality rates, correcting for seasonal variations in recruitment. Med Vet Entomol 16: 22–27

Bryan JH (1986) Vectors of *Wuchereria bancrofti* in the Sepik provinces of Papua New Guinea. Trans R Soc Trop Med Hyg 80: 123–131

Buéi K, Ito S, Nakamura H, Yoshida M (1980) Field studies on the gonotrophic cycle of *Culex tritaeniorhynchus*. Jap J Sanit Zool 31: 57–62

Buonaccorsi JP, Harrington LC, Edman JD (2003) Estimation and comparison of mosquito survival rates with release-recapture-removal data. J Med Entomol 40: 6–17

Burdick DJ, Kardos EH (1963) The age structure of fall, winter and spring populations of *Culex tarsalis* in Kern county, California. Ann Entomol Soc Am 56: 527–535

Burkot TR, Graves PM, Paru R, Battistutta D, Barnes A, Saul A (1990) Variations in malaria transmission rates are not related to anopheline survivorship per feeding cycle. Am J Trop Med Hyg 43: 321–327

Carnevale P, Le Pont F (1973) Epidemiologie du paludisme humain en République Populaire du Congo. II. Utilisation des pièges lumineux 'C.D.C.' comme moyen d'énchantillonnage des populations anophelines. Cah ORSTOM sér Entomol Méd Parasitol 11: 263–270

Carpenter MJ, Nielsen LT (1965) Ovarian cycles and longevity in some univoltine *Aedes* species in the rocky mountains of western United States. Mosquito News 35: 127–134

Chandra G, Seal B, Hati AK (1996) Age composition of the filarial vector *Culex quinquefasciatus* (Diptera: Culicidae) in Calcutta, India. Bull Entomol Res 86: 223–226

Charlwood JD (1986) Survival rate variation of *Anopheles farauti* (Diptera: Culicidae) between neighbouring villages in coastal Papua New Guinea. J Med Entomol 23: 361–365

Charlwood JD (1997) Vectorial capacity, species diversity and population cycles of anopheline mosquitoes (Diptera: Culicidae) from indoor light-trap collections in a house in southeastern Tanzania. Afr Entomol 5: 93–101

Charlwood JD, Bryan JH (1987) A mark-recapture experiment with the filariasis vector *Anopheles punctulatus* in Papua New Guinea. Ann Trop Med Parasitol 81: 429–436

Charlwood JD, Wilkes TJ (1979) Studies on the age-composition of samples of *Anopheles darlingi* Root (Diptera: Culicidae) in Brazil. Bull Entomol Res 69: 337–342

Charlwood JD, Rafael JA, Wilkes TJ (1980) Métodes de determinar a idade fisiologica em Diptera de importância médica. Uma revisão com especial referência aos vetores de doenças na América do sul. Acta Amaz 10: 311–333 (In Portuguese, English summary)

Charlwood JD, Birley MH, Dagoro H, Paru R, Holmes PR (1985) Assessing survival rates of *Anopheles farauti* (Diptera: Culicidae) from Papua New Guinea. J Anim Ecol 54: 1003–1016

Charlwood JD, Graves PM, Birley MH (1986a) Capture-recapture studies with mosquitoes of the group of *Anopheles punctulatus* Dönitz (Diptera: Culicidae) from Papua New Guinea. Bull Entomol Res 76: 211–227

Charlwood JD, Paru R, Dagoro H, Lagog M (1986b) The influence of moonlight and gonotrophic age on the biting activity of *Anopheles farauti* (Diptera: Culicidae) from Papua New Guinea. J Med Entomol 23: 132–135

Charlwood JD, Kihonda J, Sama S, Billingsley PF, Hadji H, Verhave JP, Lyimo E, Luttikhuizen PC, Smith T (1995) The rise and fall of *Anopheles arabiensis* (Diptera: Culicidae) in a Tanzanian village. Bull Entomol Res 85: 37–44

Charlwood JD, Smith T, Billingsley PF, Takken W, Lyimo EOK, Meuwissen JHET (1997) Survival and infection probabilities of anthropophagic anophelines from an area of high prevalence of *Plasmodium falciparum* in humans. Bull Entomol Res 87: 445–453

Charlwood JD, Vij R, Billingsley PF (2000) Dry season refugia of malaria-transmitting mosquitoes in a dry savannah zone of East Africa. Am J Trop Med Hyg 62: 726–732

Cheke RA, Dutton M, Avissey HSK, Lehane MJ (1990) Increase with age and fly size of pteridine concentrations in different members of the *Simulium damnosum* species complex. Acta Leidensia 59: 307–314

Cheke RA, Garms R, Howe MA, Lehane MJ (1987) Possible use of pteridine concentrations for determining the age of adult *Simulium damnosum* s.l. Trop Med Parasitol 38: 346

Chiang GL, Samarawickrema WA, Cheong WA, Sulaiman I, Yap HH (1984*a*) Biting activity, age composition and survivorship of Mansonia in two ecotypes in Peninsular Malaysia. Trop Biomed 1: 151–158

Chiang GL, Cheong WH, Samarawickrema WA (1984*b*) Filariasis in Bengkoka peninsula, Sabah, Malaysia: bionomics of *Mansonia* sp. Southeast Asian J Trop Med Public Health 15: 294–302

Chiang GL, Loong KP, Mahadevan S, Eng KL (1988) A study of dispersal, survival and gonotrophic cycle estimates of *Mansonia uniformis* in an open swamp ecotype. Southeast Asian J Trop Med Public Health 19: 271–282

Clements AN, Paterson GD (1981) The analysis of mortality and survival rates in wild populations of mosquitoes. J Appl Ecol 18: 373–399

Coene J (1993) Malaria in urban and rural Kinshasa: the entomological input. Med Vet Entomol 7: 127–137

Colless DH (1958) Recognition of individual nulliparous and parous mosquitoes. Trans R Soc Trop Med Hyg 52: 187

Conway GR, Trpis M, McClelland GAH (1974) Population parameters of the mosquito *Aedes aegypti* (L.) estimated by mark–release–recapture in a suburban habitat in Tanzania. J Anim Ecol 43: 289–304

Cook WJ, Smith BP, Brooks RJ (1989) Allocation of reproductive effort in female *Arrenurus* spp. water mites (Acari: Hydrachnidia; Arrenuridae). Oecologia (Berl.) 79: 184–188

Copeland RS (1986) The biology of *Aedes thibaulti* in northern Indiana. J Am Mosq Control Assoc 2: 1–6

Corbet PS (1960) Recognition of nulliparous mosquitoes without dissection. Nature 187: 525–526

Corbet PS (1961) Entomological studies from a high tower in Mpanga forest, Uganda. VIII: The age-composition of mosquito populations according to time and level. Trans R Entomol Soc Lond 113: 336–345

Corbet PS (1962*a*) The Use of External Characters to Age-grade Adult Mosquitoes (Diptera: Culicidae). Proc Int Congr Entomol XIth, vol. II, pp. 387–390

Corbet PS (1962*b*) The age-composition of biting mosquito populations according to time and level: a further study. Bull Entomol Res 53: 409–415

Corbet PS (1963) The reliability of parasitic water-mites (Hydracarina) as indicators of physiological age in mosquitoes (Diptera: Culicidae). Entomol Exp Appl 6: 215–233

Corbet PS (1964) The time elapsing between oviposition and biting in the mosquito *Mansonia (Coquillettidia) fuscopennata* (Theobald). Proc R Entomol Soc Lond (A) 39: 108–110

Corbet PS (1970) The use of parasitic water-mites for age grading female mosquitoes. Mosquito News 30: 436–438

Costantini C, Li SongGang, della Torre A, Sagnon N, Coluzzi M, Taylor CE (1996) Density, survival and dispersal of *Anopheles gambiae* complex mosquitoes in a West African Sudan savanna village. Med Vet Entomol 10: 203–219

Cristesco A (1966) Contributions à l'étude de la composition par âges des populations du complexe *Anopheles maculipennis*, par rapport à l'application des insecticides remanents en Roumaine. Arch Roum Pathol Exp Microbiol 25: 491–502

Davidson G (1954) Estimation of the survival-rate of anopheline mosquitoes in nature. Nature 174: 792–793

Davidson G (1955*a*) Measurement of the ampulla of the oviduct as a means of determining the natural daily mortality of *Anopheles gambiae*. Ann Trop Med Parasitol 49: 24–36

Davidson G (1955*b*) Further studies of the basic factors concerned in the transmission of malaria. Trans R Soc Trop Med Hyg 49: 339–350

Davidson G, Draper C (1953) Field studies of some of the basic factors concerned in the transmission of malaria. Trans R Soc Trop Med Hyg 47: 522–535

Day JF, Curtis GA, Edman JD (1990) Rainfall-directed oviposition behavior of *Culex nigripalpus* (Diptera: Culicidae) and its influence on St. Louis encephalitis virus transmission in Indian River county, Florida. J Med Entomol 27: 43–50

Day JF, Edman JD, Scott TW (1994) Reproductive fitness and survivorship of *Aedes aegypti* (Diptera: Culicidae) maintained on blood, with field observations from Thailand. J Med Entomol 31: 611–617

Deevey ES (1947) Life tables for natural populations of animals. Quart Rev Biol 22: 283–314. Also reprinted in Hazen (1970)

Dégallier N (1979) Le cycle gonotrophique de *Culex portesi* Sénevet et Abonnenc en Guyane française. Cah ORSTOM sér Entomol Méd Parasitol 17: 13–17

De Meillon B, Grab B, Sebastian A (1967*a*) Evaluation of *Wuchereria bancrofti* infection in *Culex pipiens fatigans* in Rangoon, Burma. Bull World Health Organ 36: 91–100

De Meillon B, Paing M, Sebastian A, Khan ZH (1967*b*) Outdoor resting of *Culex pipiens fatigans* in Rangoon, Burma. Bull World Health Organ 36: 67–73

De Meillon B, Sebastian A, Khan ZH (1967*c*) Exodus from a breeding place and time of emergence from pupa of *Culex pipiens fatigans*. Bull World Health Organ 36: 163–167

De Meillon B, Sebastian A, Khan ZH (1967*d*) Cane-sugar feeding in *Culex pipiens fatigans*. Bull World Health Organ 36: 53–65

De Meillon B, Sebastian A, Khan ZH (1967*e*) Time of arrival of gravid *Culex pipiens fatigans* at an oviposition site, the oviposition cycle and the relationship between time of feeding and time of oviposition. Bull World Health Organ 36: 39–46

Desena ML, Clark JM, Edman JD, Symington SB, Scott TW (1999*a*) *Aedes aegypti* (Diptera: Culicidae) age determination by cuticular hydrocarbon analysis of female legs. J Med Entomol 36: 824–830

Desena ML, Clark JM, Edman JD, Symington SB, Scott TW, Clark GG, Peters TM (1999*b*) Potential for aging female *Aedes aegypti* (Diptera: Culicidae) by gas chromatographic analysis of cuticular hydrocarbons, including a field evaluation. J Med Entomol 36: 811–823

Detinova TS (1945) The determination of the physiological age of females of *Anopheles* by changes in the tracheal system of the ovaries. Medskaya Parazitol 14: 45–49 (In Russian)

Detinova TS (1949) Physiological changes in the ovaries of female *Anopheles maculipennis*. Medskaya Parazitol 18: 410–420 (In Russian)

Detinova TS (1953) The duration of the gonotrophic cycle in the mosquito *Anopheles maculipennis*; the interval of time between oviposition and the next blood meal. Medskaya Parazitol 22: 446–449 (In Russian)

Detinova TS (1962) Age-grading methods in Diptera of medical importance. World Health Organ Monogr Ser 47

Detinova TS (1968) Age structure of insect populations of medical importance. Annu Rev Entomol 13: 427–450

Dingle H (1965) The relation between age and flight activity in the milkweed bug, *Oncopeltus*. J Exp Biol 42: 269–283

dos Santos R la C, Forattini OP, Burattini MN (2004) *Anopheles albitarsis* s.l. (Diptera: Culicidae) survivorship and density in a rice irrigation area of the state of São Paulo, Brazil. J Med Entomol 41: 997–1000

Dossou-yovo J, Doannio JMC, Riviere F, Chauvancy G (1995) Malaria in Côte d'Ivoire wet savannah region: the entomological input. Trop Med Parasitol 46: 263–269

Dow RP (1971) The dispersal of *Culex nigripalpus* marked with high concentrations of radiophosphorous. J Med Entomol 8: 353–363

Draper CC, Davidson G (1953) A new method of estimating the survival rate of anopheline mosquitoes in nature. Nature 172: 503

Dye C (1984*a*) Models for the population dynamics of the yellow fever mosquito, *Aedes aegypti*. J Anim Ecol 53: 247–268

Dye C (1992) The analysis of parasite transmission by bloodsucking insects. Annu Rev Entomol 37: 1–19

Eldridge BF, Reeves WC (1990) Daily survivorship of adult *Aedes communis* in a high mountain environment in California. J Am Mosq Control Assoc 6: 662–666

Elissa N, Karch S, Bureau P, Ollomo B, Lawoko M, Yangari P, Ebang B, Georges AJ (1999) Malaria transmission in a region of savanna-forest mosaic, Haut-Ogooué, Gabon. J Am Mosq Control Assoc 15: 15–23

Emord DE, Morris CD (1982) A host-baited CDC trap. Mosquito News 42: 220–224

Feldlaufer MF, Crans WJ (1979) The relative attractiveness of carbon dioxide to parous and nulliparous mosquitoes. J Med Entomol 15: 140–142

Fernandez-Salas I, Rodriguez MH, Roberts DR (1994) Gonotrophic cycle and survivorship of *Anopheles pseudopunctipennis* (Diptera: Culicidae) in the Tapachula foothills of southern Mexico. J Med Entomol 31: 340–347

Fisher RA, Ford EB (1947) The spread of a gene in natural conditions in a colony of the moth *Panaxia dominula* L. Heredity 1: 143–174

Fox AS, Brust BA (1994) Rogue ovarioles and criteria for parity diagnosis in *Culex tarsalis* (Diptera: Culicidae) from Manitoba. J Med Entomol 31: 738–746

Fox AS, Brust RA (1996) Parity diagnosis and ovulation in *Culiseta inornata* (Diptera: Culicidae). J Med Entomol 33: 402–412

Gad AM, Feinsod FM, Soliman BA, El Said S (1989) Survival estimates for adult *Culex pipiens* in the Nile Delta. Acta Trop 46: 173–179

Garrett-Jones C (1968) Epidemiological Entomology and its Application to Malaria. WHO/Mal/68.672; 17 pp. (mimeographed)

Garrett-Jones C (1970) Problems of epidemiological entomology as applied to malariology. Misc Pubs Entomol Soc Am 7: 168–180

Garrett-Jones C (1973) Prevalence of *Plasmodium falciparum* and *Wuchereria bancrofti* in *Anopheles*, in relation to short term female population dynamics. Int Congr Trop Med Malar IXth, vol. 1: 298–300

Garrett-Jones C, Grab B (1964) The assessment of insecticidal impact on the malaria mosquito's vectorial capacity, from the data on the proportion of parous females. Bull World Health Organ 31: 71–86

Garrett-Jones C, Shidrawi GR (1969) Malaria vectorial capacity of a population of *Anopheles gambiae*. Bull World Health Organ 40: 531–545

Germain M, Hervé J-P, Geoffroy B (1974) Evaluation de la durée du cycle trophogonique d'*Aedes africanus* (Theobald), vecteur potentiel de fièvre jaune, dans une galerie forestière du sud de la République Centrafricaine. Cah ORSTOM sér Entomol Méd Parasitol 12: 127–134

Gibb PA, Anderson TJC, Dye C (1988) Are nulliparous sandflies light-shy? Trans R Soc Trop Med Hyg 82: 342–343

Giglioli MEC (1965) The problem of age determination in *Anopheles melas* Theo. 1903, by Polovodova's method. Cah ORSTOM sér Entomol Méd Parasitol 3 & 4: 157–177

Gillett JD (1957) Age analysis in the biting-cycle of the mosquito *Taeniorhynchus (Mansonioides) africanus* Theobald, based on the presence of parasitic mites. Ann Trop Med Parasitol 51: 151–158

Gillies MT (1956) A new character for the recognition of nulliparous females of *Anopheles gambiae*. Bull World Health Organ 15: 451–459

Gillies MT (1961) Studies on the dispersion and survival of *Anopheles gambiae* Giles in East Africa, by means of marking and release experiments. Bull Entomol Res 52: 99–127

Gillies MT (1974) Methods for assessing the density and survival of blood-sucking Diptera. Annu Rev Entomol 19: 345–362

Gillies MT, Wilkes TJ (1963) Observations on nulliparous and parous rates in a population of *Anopheles funestus* in East Africa. Ann Trop Med Parasitol 57: 204–213

Gillies MT, Wilkes TJ (1965) A study of the age-composition of populations of *Anopheles gambiae* Giles and *A. funestus* Giles in north-eastern Tanzania. Bull Entomol Res 56: 237–262

Githeko AK (1992) The behaviour and ecology of malaria vectors and malaria transmission in Kisumu district of Western Kenya. Ph.D. thesis, University of Liverpool

Gottfried KL, Gerhardt RR, Nasci RS, Crabtree MB, Karabatsos N, Burkhalter KL, Davis BS, Panella NA, Paulson DJ (2002) Temporal abundance, parity, survival rates, and arbovirus isolation of field-collected container-inhabiting mosquitoes in eastern Tennessee. J Am Mosq Control Assoc 18: 164–172

Graham JE, Bradley IE (1972) Changes in the age structure of *Culex pipiens fatigans* Wiedemann populations in Rangoon, Burma after intensive larviciding. J Med Entomol 9: 325–329

Graham P (1969) Age grading of mosquitoes from parasitic mites. Mosquito News 29: 259–260

Graves PM, Burkot TR, Saul AJ, Hayes RJ, Carter R (1990) Estimation of *Anopheles* survival rate, vectorial capacity and mosquito infection probability from malaria vector infection rates in villages near Madang, Papua New Guinea. J Appl Ecol 27: 134–147

Gruchet H (1962) Étude de l'âge physiologique des femelles d'*Anopheles funestus* Giles dans la région de Miandrivazo, Madagascar. Bull Soc Pathol Exot 55: 165–174

Gryaznov AI (1995) Age-grading in blackflies (Diptera: Simuliidae) by ovariolar morphology. Bull Entomol Res 85: 339–344

Haeger JS (1960) Behaviour preceding migration in the salt-marsh mosquito *Aedes taeniorhynchus* (Wiedemann). Mosquito News 20: 136–147

Haji H, Smith T, Meuwissen JT, Sauerwein R, Charlwood JD (1996) Estimation of the infectious reservoir of *Plasmodium falciparum* in natural vector populations based on oocyst size. Trans R Soc Trop Med Hyg 90: 494–497

Hájková Z (1966) A study on gonotrophic cycles of the mosquito *Aedes vexans* Meig. in south Moravia. Folia Parasitol (Praha) 13: 361–370

Hamon J, Chauvet G, Thélin L (1961) Observations sur les méthodes d'évaluation de l'âge physiologique des femelles d'anophèles. Bull World Health Organ 24: 437–443

Han-Il Ree, Ui-Wook Hwang, In-Yong Lee, Tae-Eun Kim (2001) Daily survival and human blood index of *Anopheles sinensis*, the vector species of malaria in Korea. J Am Mosq Control Assoc 17: 67–72

Haramis LD, Foster WA (1983) Survival and population density of *Aedes triseriatus* (Diptera: Culicidae) in a woodlot in central Ohio, USA. J Med Entomol 20: 391–398

Haramis LD, Foster WA (1990) Gonotrophic cycle duration, population age structure, and onset of sugar feeding and insemination of *Aedes triseriatus* (Diptera: Culicidae). J Med Entomol 27: 421–428

Harrington LC, Buonaccorsi JP, Edman JD, Costero A, Kittayapong P, Clark GG, Scott TW (2001) Analysis of survival of young and old *Aedes aegypti* (Diptera: Culicidae) from Puerto Rico and Thailand. J Med Entomol 38: 537–547

Hawley WA (1985*b*) Population dynamics of *Aedes sierrensis*. In: Lounibos LP, Rey JR, Frank JH (eds) Ecology of Mosquitoes: Proceedings of a Workshop. Florida Medical Entomology Laboratory, Vero Beach, Florida, pp. 167–184

Hii JLK (1985) Evidence for the existence of genetic variability in the tendency of *Anopheles balabacensis* to rest in houses and to bite man. Southeast Asian J Trop Med Public Health 16: 173–182

Hii JLK, Vun YS (1985) A study of dispersal, survival and adult population estimates of the malaria vector, *Anopheles balabacensis* Baisas (Diptera: Culicidae) in Sabah, Malaysia. Trop Biomed 2: 121–131

Hii JLK, Birley MH, Sang VY (1990) Estimation of survival rate and oviposition interval of *Aedes balabacensis* mosquitoes from mark–recapture experiments in Sabah, Malaysia. Med Vet Entomol 4: 135–140

Hii JLK, Birley MH, Kanai L, Foligeli A, Wagner J (1995) Comparative effects of permethrin-impregnated bednets and DDT house spraying on survival rates and oviposition interval of *Anopheles farauti* No. 1 (Diptera: Culicidae) in Solomon Islands. Ann Trop Med Parasitol 89: 521–529

Hitchcock JC (1968) Age composition of a natural population of *Anopheles quadrimaculatus* Say (Diptera: Culicidae) in Maryland, USA. J Med Entomol 5: 125–134

Hoc TQ (1996*a*) Application of the ovarian oil-injection and ovariolar separation techniques for age-grading haematophagous diptera. J Med Entomol 33: 290–296

Hoc TQ (1996*b*) A method for the rapid recognition of nulliparous and parous females of haematophagous Diptera. Bull Entomol Res 86: 137–141

Hoc TQ, Charlwood JD (1990) Age determination of *Aedes cantans* using ovarian oil injection technique. Med Vet Entomol 4: 227–233

Hoc TQ, Schaub GA (1995) Ovariolar 'basal body' development and physiological age of the mosquito *Aedes aegypti*. Med Vet Entomol 9: 9–15

Hoc TQ, Schaub GA (1996) Improvement of techniques for age grading hematophagous insects: ovarian oil-injection and ovariolar separation techniques. J Med Entomol 33: 286–289

Hoc TQ, Wilkes TJ (1995) Age determination in the blackfly *Simulium woodi*, a vector of onchocerciasis in Tanzania. Med Vet Entomol 9: 16–24

Holmes PR (1986) A study of population changes in adult *Culex quinquefasciatus* Say (Diptera: Culicidae) during a mosquito control programme in Dubai, United Arab Emirates. Ann Trop Med Parasitol 80: 107–116

Holmes PR, Birley MH (1987) An improved method for survival rate analysis from time series of haematophagous dipteran populations. J Anim Ecol 56: 427–440

Ikeshoji T (1985) Age structure and mating status of male mosquitoes responding to sound. Jap J Sanit Zool 36: 95–101

Jalil M, Mitchell R (1972) Parasitism of mosquitoes by water mites. J Med Entomol 9: 305–311

Jensen J, Washino RK (1991) An assessment of the biological capacity of a Sacramento Valley population of *Aedes melanimon* to vector arboviruses. Am J Trop Med Hyg 33: 355–363

Jensen T, Kaiser PE, Barnard DR (1993*a*) Short-term changes in the abundance and parity rate of *Anopheles quadrimaculatus* species C (Diptera: Culicidae) in a central Florida swamp. J Med Entomol 30: 1038–1042

Jensen T, Kramer V, Washino RK (1993*b*) Short-term population dynamics of adult *Aedes dorsalis* (Diptera: Culicidae) in a northern California tidal marsh. J Med Entomol 30: 374–377

Jensen T, Washino RK (1994) Comparison of recapture patterns of marked and released *Aedes vexans* and *Ae. melanimon* (Diptera: Culicidae) in the Sacramento valley of California. J Med Entomol 31: 607–610

Jensen T, Dritz DA, Fritz GN, Washino RK, Reeves WC (1998) Lake Vera revisited: parity and survival rates of *Anopheles punctipennis* at the site of a malaria outbreak in the Sierra Nevada foothills of California. Am J Trop Med Hyg 59: 591–594

Kakitani I, Forattini OP (2000) Paridade e desenvolvimento ovariano de *Anopheles albitarsis* l.s. em área de agroecossistema irrigado. Rev Saúde Pública 34: 33–38

Kardos EH, Bellamy RE (1961) Distinguishing nulliparous from parous female *Culex tarsalis* by examination of the ovarian tracheation. Ann Entomol Soc Am 54: 448–451

Kay BH (1979) Age structure of populations of *Culex annulirostris* (Diptera: Culicidae) at Kowanyama, and Charleville Queensland. J Med Entomol 16: 309–316

Khok CK (=Hoc TQ) (1974) Oogenesis and physiological age of bloodsucking mosquitoes (Culicidae). Ph.D. thesis, University of Moscow (In Russian)

Krafsur ES (1971) Malaria transmission in Gambela Illubabor Province. Ethiop Med J 9: 75–94

Krafsur ES, Garrett-Jones C (1977) The survival in nature of *Wuchereria*-infected *Anopheles funestus* Giles in north–eastern Tanzania. Trans R Soc Trop Med Hyg 71: 155–160

Lanciani CA (1979*a*) Detachment of parasitic water mites from the mosquito *Anopheles crucians* (Diptera: Culicidae). J Med Entomol 15: 99–102

Lanciani CA (1979*b*) Water mite-induced mortality in a natural population of the mosquito *Anopheles crucians* (Diptera: Culicidae). J Med Entomol 15: 529–532

Lanciani CA (1979*c*) The influence of parasitic water mites on the instantaneous death rate of their hosts. Oecologia (Berl.) 44: 60–62

Lanciani CA, (1986) Effect of the water mite *Arrenurus pseudotenuicollis* (Acriformes: Arrenuridae) on the longevity of captive *Anopheles quadrimaculatus* (Diptera: Culicidae). Fla Entomol 69: 436–437

Lanciani CA (1987) Mortality in mite-infested, male *Anopheles crucians*. J Am Mosq Control Assoc 3: 107–108

Lanciani CA, Boyett JM (1980) Demonstrating parasitic water mite-induced mortality in natural host populations. Parasitology 81: 465–475

Lanciani CA, Boyt AD (1977) The effect of a parasitic water mite, *Arrenurus pseudotenuicollis* (Acari: Hydrachnellae), on the survival of and reproduction of the mosquito *Anopheles crucians* (Diptera: Culicidae). J Med Entomol 14: 10–15

Lanciani CA, McLaughlin RE (1989) Parasitsm of *Coquillettidia perturbans* by two water mite species (Acari: Arrenuridae) in Florida. J Am Mosq Control Assoc 5: 428–431

Landry SV, DeFoliart GR, Hogg DB (1988) Adult body size and survivorship in a field population of *Aedes triseriatus*. J Am Mosq Control Assoc 4: 121–128

Lange AB, Khok CK (=Hoc TQ) (1981) Abortive oogenesis and physiological age in blood-sucking mosquitoes (Diptera, Culicidae). Medskaya Parazitol 50: 48–56 (In Russian, English summary)

Lange AB, Khok CK (=Hoc TQ), Sokolova MI (1981) The method of intraovarial oil injection and its use in the determination of the physiological age of females of blood-sucking mosquitoes (Diptera, Culicidae). Medskaya Parazitol 50: 51–53 (In Russian, English summary)

Lardeux F, Cheffort J (2002) Age-grading and growth of *Wuchereria bancrofti* (Filariidea: Onchocercidae) larvae by growth measurements and its use for estimating blood-meal intervals of its Polynesian vector *Aedes polynesiensis* (Diptera: Culicidae). Int J Parasitol 32: 705–716

Lardeux F, Tetuanui A (1995) Larval growth of *Aedes polynesiensis* and *Aedes aegypti* (Diptera: Culicidae). Mosq Syst 27: 118–124

Lardeux F, Ung A, Chebret M (2000) Spectrofluorometers are not adequate for aging *Aedes* and *Culex* (Diptera: Culicidae) using pteridine fluorescence. J Med Entomol 37: 769–773

Laurence BR (1963) Natural mortality in two filarial vectors. Bull World Health Organ 28: 229–234

Lehane MJ (1985) Determining the age of an insect. Parasitol Today 1: 81–85

Lehane MJ, Hargrove J (1988) Field experiments on a new method for determining age in tsetse flies (Diptera: Glossinidae). Ecol Entomol 13: 319–322

Lehane MJ, Laurence BR (1978) Development of the calyx and lateral oviduct during oogenesis in *Aedes aegypti*. Cell Tissue Res 193: 125–137

Lehane MJ, Mail TS (1985) Determining the age of adult male and female *Glossina morsitans morsitans* using a new technique. Ecol Entomol 10: 219–224

Lindquist AW, Ikeshoji T, Grab B, De Meillon B, Khan ZH (1967) Dispersion studies of *Culex pipiens fatigans* tagged with ^{32}P in the Kemmendine area of Rangoon, Burma. Bull World Health Organ 36: 21–37

Lines JD, Lyimo EO, Curtis CF (1986) Mixing of indoor- and outdoor-resting adults of *Anopheles gambiae* Giles s.1. and *A. funestus* Giles (Diptera: Culicidae) in coastal Tanzania. Bull Entomol Res 76: 171–178

Linthicum KJ, Bailey CL, Davies FG, Kairo A (1985). Observations on the dispersal and survival of a population of *Aedes lineatopennis* (Ludlow) (Diptera: Culicidae) in Kenya. Bull Entomol Res 75: 661–670

Loong KP, Chiang GL, Eng KL, Chan ST, Yap HH (1990) Survival and feeding behaviour of Malaysian strain of *Anopheles maculatus* Theobald (Diptera: Culicidae) and their role in malaria transmission. Trop Biomed 7: 71–76

Lord CC, Baylis M (1999) Estimation of survival rates in haematophagous arthropods. Med Vet Entomol 13: 225–233

Lowe RE, Ford HR, Smittle BJ, Weidhaas DE (1973) Reproductive behaviour of *Culex pipiens quinquefasciatus* released into a natural population. Mosquito News 33: 221–227

Macdonald G (1952) The analysis of the sporozoite rate. Trop Dis Bull 49: 569–585

Macdonald G (1957) The Epidemiology and Control of Malaria. Oxford University Press, London

Macdonald WW, Sebastian A, Tun MM (1968) A mark–release–recapture experiment with *Culex pipiens fatigans* in the village of Okpo, Burma. Ann Trop Med Parasitol 62: 200–209

Magnarelli LA (1980). Bionomics of *Psorophora ferox* (Diptera: Culicidae): seasonal occurrence and acquisition of sugars. J Med Entomol 17: 328–332

Mahmood F, Crans WJ (1997) A thermal heat summation model to predict the duration of the gonotrophic cycle of *Culiseta melanura* in nature. J Am Mosq Control Assoc 13: 92–94

Mahmood F, Crans WJ (1998) Ovarian development and parity determination in *Culiseta melanura* (Diptera: Culicidae). J Med Entomol 35: 980–988

Mahmood F, Reisen WK (1981) Duration of gonotrophic cycles of *Anopheles culicifacies* Giles and Anopheles stephensi Liston, with observations on reproductive activity and survivorship during winter in Punjab Province, Pakistsan. Mosquito News 41: 41–50

Mail TS, Lehane MJ (1988) Characterisation of pigments in the head capsule of the adult stablefly *Stomoxys calcitrans*. Entomol Exp Appl 46: 125–131

Mail TS, Chadwick J, Lehane MJ (1983) Determining the age of adults of *Stomoxys calcitrans* (L.) (Diptera: Muscidae). Bull Entomol Res 73: 501–525

Malhotra PR, Sarkar PK, Bhuyan M, Chakraborty BC, Baruah I (1983) Water mites (*Arrenurus* sp.) parasitising mosquitoes in Assam. Indian J Med Res 78: 647–650

McClelland GAH, Conway GR (1971) Frequency of blood feeding in the mosquito *Aedes aegypti*. Nature 232: 485–486

McCrae AWR (1976) The association between larval parasitic water mites (Hydracarina) and *Anopheles implexus* (Theobald) (Diptera, Culicidae). Bull Entomol Res 66: 633–650

McDonald PT (1977) Population characteristics of domestic *Aedes aegypti* (Diptera: Culicidae) in villages on the Kenya coast. I. Adult survivorship and population size. J Med Entomol 14: 42–48

McHugh CP (1989) Ecology of a semi-isolated population of adult *Anopheles freeborni*: abundance, trophic status, parity, survivorship, gonotrophic cycle length, and host selection. Am J Trop Med Hyg 41: 169–176

McHugh CP (1990) Survivorship and gonotrophic cycle length of *Culex tarsalis* (Diptera: Culicidae) near Sheridan, Placer county, California. J Med Entomol 27: 1027–1030

McHugh CP, Washino RK (1986) Survivorship and gonocycle length of *Anopheles stephensi* and *Culex tarsalis* in the Sacramento valley of California. Proc California Mosq Vector Control Assoc 54: 133–135

Mer GG (1932) The determination of the age of *Anopheles* by differences in the size of the common oviduct. Bull Entomol Res 23: 563–566

Mer GG (1936) Experimental study of the development of the ovary in *Anopheles elutus* Edw. (Dipt., Culic.). Bull Entomol Res 27: 351–359

Meyer RP, Washino RK, McKenzie TL, Fukushima CK (1984) Comparison of three methods for collecting adult mosquitoes associated with rice field irrigated pasture habitats in northern California. Mosquito News 44: 315–320

Meyer RP, Hardy JL, Reisen WK (1990) Diel changes in adult mosquito microhabitat temperatures and their relationship to the extrinsic incubation of arboviruses in mosquitoes in Kern County, California, USA. J Med Entomol 27: 607–614

Milby MM, Reisen WK, Reeves WC (1983) Intercanyon movement of marked *Culex tarsalis* (Diptera: Culicidae). J Med Entomol 20: 193–198

Millest AL, Cheke RA, Howe MA, Lehane MJ, Garms R (1992) Determining the ages of adult females of different members of the *Simulium damnosum* complex (Diptera: Simuliidae) by the pteridine accumulation method. Bull Entomol Res 82: 219–226

Mitchell R (1957) Major evolutionary lines in water mites. Syst Zool 6: 137–148

Molineaux L, Gramiccia G (1980) The Garki Project. Research on the Epidemiology and Control of Malaria in the Sudan Savanna of West Africa. World Health Organization, Geneva

Molineaux L, Shidrawi GR, Clarke JL, Boulzaguet R, Ashkar T, Dietz K (1976) The impact of propoxur on *Anopheles gambiae* s.l. and some other anopheline populations, and its relationship with some pre-spraying variables. Bull World Health Organ 54: 379–389

Molineaux L, Dietz K, Thomas A (1978) Further epidemiological evaluation of a malaria model. Bull World Health Organ 56: 565–571

Molineaux L, Shidrawi GR, Clarke JL, Boulzaguet JR, Ashkar TS (1979) Assessment of insecticidal impact on the malaria mosquito's vectorial capacity, from data on the man-biting rate and age-composition. Bull World Health Organ 57: 265–274

Moon RD, Krafsur ES (1995) Pterin quantity and gonotrophic stage as indicators of age in *Musca autumnalis* (Diptera: Muscidae). J Med Entomol 32: 673–684

Moore CG, Reiter P, Xu Jin-Jiang (1986) Determination of chronological age in *Culex pipiens* s.l. J Am Mosq Control Assoc 2: 204–208

Mori A, Wada Y (1977) The gonotrophic cycle of *Aedes albopticus* in the field. Trop Med 19: 141–146

Morris CD (1984) Phenology of trophic and gonobiological states in *Culiseta morsitans* and *Culiseta melanura* (Diptera: Culicidae). J Med Entomol 21: 38–51

Morris CD, DeFoliart GR (1969) A comparison of mosquito catches with miniature light traps & CO_2-baited traps. Mosquito News 29: 424–426

Morris CD, DeFoliart GR (1970) The physiological age of Wisconsin *Aedes* mosquitoes parasitized by water-mites. J Med Entomol 7: 628

Mullen GR (1974) Acarine parasites of mosquitoes. II. Illustrated larval key to the families and genera of mites reportedly parasitic on mosquitoes. Mosquito News 34: 183–195

Mullen GR (1975*a*) Acarine parasites of mosquitoes. I. A critical review of all known records of mosquitoes parasitized by mites. J Med Entomol 12: 27–36

Mullen GR (1975*b*) Acarine parasites of mosquitoes. III. Collection, preservation and rearing techniques used to study water mites (Acarina: Hydrachnellae) parasitic on mosquitoes. Proc New Jers Mosq Exterm Assoc 61: 117–122

Mullen GR (1976) Water mites of the subgenus *Truncaturus* (Arrenuridae, *Arrenurus*) in North America. Search Agric NY State Agric Exp Stn (Ithaca) 6

Mullens BA, Lehane MJ (1995) Fluorescence as a tool for age determination in *Culicoides variipennis sonorensis* (Diptera: Ceratopogonidae). J Med Entomol 32: 569–571

Münchberg P (1954) Zur Kenntnis der an Culiciden (Diptera) schmarotzenden Arrenus-Larven (Hydracarina), sowie über die Bedeutung dieser Parasiten für Wirt und Mensch. Z ParasitKde 16: 298–312

Mutero CM, Birley MH (1987) Estimation of the survival rate and oviposition cycle of field populations of malaria vectors in Kenya. J Appl Ecol 24: 853–863

Nájera JA (1974) A critical review of the field application of a mathematical model of malaria transmission. Bull World Health Organ 50: 449–457

Nasci RS (1986*a*) Relationship between adult mosquito (Diptera: Culicidae) body size and parity in field populations. Environ Entomol 15: 874–876

Nasci RS (1986*b*) The size of emerging and host-seeking *Aedes aegypti* and the relation of size to blood-feeding success in the field. J Am Mosq Control Assoc 2: 61–62

Nasci RS (1988) Biology of *Aedes triseriatus* (Diptera: Culicidae) developing in tires in Louisiana. J Med Entomol 25: 402–405

Nathan MB (1981) Bancroftian filariasis in coastal north Trinidad, West Indies: Intensity of transmission by *Culex quinquefasciatus*. Trans R Soc Trop Med Hyg 75: 721–730

Nayar JK (1982) *Wyeomyia mitchellii*: Observations on dispersal, survival, nutrition, insemination and ovarian development in a Florida population. Mosquito News 42: 416–427

Nayar JK, Knight JW (1981) Occurrence of ovariolar dilatations in nulliparous mosquitoes: *Culex nigripalpus*. Mosquito News 41: 281–287

Nayar JK, Provost MW, Hansen CW (1980) Quantitative bionomics of *Culex nigripalpus* (Diptera: Culicidae) populations in Florida. 2. Distribution, dispersal and survival patterns. J Med Entomol 17: 40–50

Nelson RL (1964) Parity in winter populations of *Culex tarsalis* Coquillett in Kern county, California. Am J Hyg 80: 242–253

Nelson RL, Milby MM (1980) Dispersal and survival of field and laboratory strains of *Culex tarsalis* (Diptera: Culicidae). J Med Entomol 17: 146–150

Nelson RL, Milby MM, Reeves WC, Fine PE (1978) Estimates of survival, population size, and emergence of *Culex tarsalis* at an isolated site. Ann Entomol Soc Am 71: 801–808

Neville AC (1963) Daily growth layers for determining the age of grasshopper populations. Oikos 14: 1–8

Neville AC (1983) Daily growth layers and the teneral stage in adult insects: a review. J Insect Physiol 29: 211–219

Nielsen ET (1958) The initial stage of migration in salt-marsh mosquitoes. Bull Entomol Res 49: 305–313

Packer MJ, Corbet PS (1989) Seasonal emergence, host-seeking activity, age composition and reproductive biology of the mosquito *Aedes punctor*. Ecol Entomol 14: 433–442

Pant CP, Yasuno M (1973) Field studies on the gonotrophic cycle of *Aedes aegypti* in Bangkok, Thailand, J Med Entomol 10: 219–223

Penilla RP, Rodríguez MH, López AD, Viader-Salvadó JM, Sánchez CN (2002) Pteridine concentrations differ between insectary-reared and field-collected *Anopheles albimanus* mosquitoes of the same physiological age. Med Vet Entomol 16: 225–234

Perry EL (1912) Malaria in the Jeypore Hill tract and adjoining coast land. Paludism 5: 32–40

Pfuntner AR (1979) A modified CO_2-baited miniature surveillance trap. Bull Soc Vector Ecol 4: 31–35

Pfuntner AR, Reisen WK, Dhillon MS (1988) Vertical distribution and response of *Culex* mosquitoes to differing concentrations of carbon dioxide. Proc California Mosq Vector Control Assoc 56: 69–74

Polovodova VP (1941) Changes in the oviducts of *Anopheles* with age, and a method of determining the physiological age of mosquitoes. Medskaya Parazitol 10: 387–396 (In Russian)

Polovodova VP (1949) The determination of the physiological age of female *Anopheles* by the number of gonotrophic cycles completed. Medskaya Parazitol 18: 352–355 (In Russian)

Prasad V, Cook DR (1972) The taxonomy of water mite larvae. Mem Am Entomol Inst No. 18

Provost MW (1960) The dispersal of *Aedes taeniorhynchus*. III. Study methods for migratory exodus. Mosquito News 20: 148–161

Pumpuni CB, Walker ED (1989) Population size and survivorship of adult *Aedes triseriatus* in a scrap tireyard in northern Indiana. J Am Mosq Control Assoc 5: 166–172

Rawlings P, Curtis CF (1982) Tests for the existence of genetic variability in the tendency of *Anopheles culicifacies* species B to rest in houses and to bite man. Bull World Health Organ 60: 427–432

Rawlings P, Davidson G (1982) The dispersal and survival of *Anopheles culicifacies* Giles (Diptera: Culicidae) in a Sri Lankan village under malathion spraying Bull Entomol Res 72: 139–144

Rawlings P, Curtis CF, Wickramasinghe MB, Lines J (1981) The influence of age and season on dispersal and recapture of *Anopheles culicifacies* in Sri Lanka. Ecol Entomol 6: 307–319

Reisen WK, Aslamkhan M (1979) A release–recapture experiment with the malaria vector, *Anopheles stephensi* Liston, with observations on dispersal, survivorship, population size, gonotrophic rhythm and mating behaviour. Ann Trop Med Parasitol 73: 251–269

Reisen WK, Mullen GR (1978) Ecological observations on acarine associates (Acari) of Pakistan mosquitoes (Diptera: Culicidae). Environ Entomol 7: 769–776

Reisen WK, Siddiqui TF (1979) Horizontal and vertical estimates of immature survivorship for *Culex tritaeniorhynchus* (Diptera: Culicidae) in Pakistan. J Med Entomol 16: 207–218

Reisen WK, Baker RH, Sakai RK, Aziz-Javed A, Aslam Y, Siddiqui TF (1977) Observations on the mating behavior and survivorship of *Culex tritaeniorhynchus* Giles (Diptera: Culicidae) during late autumn. Southeast Asian J Trop Med Public Health 8: 537–545

Reisen WK, Aslam Y, Siddiqui TF, Khan AQ (1978) A mark release–recapture experiment with *Culex tritaeniorhynchus* Giles. Trans R Soc Trop Med Hyg 72: 167–177

Reisen WK, Mahmood F, Parveen T (1979) *Anopheles subpictus* Grassi: observations on survivorship and population size using mark–release–recapture and dissection methods. Researches in Population Ecology 21: 12–29

Reisen WK, Mahmood F, Parveen T (1980*a*) *Anopheles culicifacies* Giles: release–recapture experiment with cohorts of known age with implications for malaria epidemiology and genetical control in Pakistan. Trans R Soc Trop Med Hyg 74: 307–317

Reisen WK, Sakai RK, Baker RH, Rathor HR, Raana K, Azra K, Niaz S (1980*b*) Field competitiveness of *Culex tritaeniorhynchus* Giles males carrying a complex chromosomal aberration: a second experiment. Ann Entomol Soc Am 73: 479–484

Reisen WK, Mahmood F, Azra K (1981) *Anopheles culicifacies* Giles: adult ecological parameters measured in rural Punjab province, Pakistan using capture–release–recapture and dissection method, with comparative observations on *An. stephensi* Liston and *An. subpictus* Grassi. Researches in Population Ecology 23: 39–60

Reisen WK, Sakai RK, Baker RH, Azra K, Niaz S (1982*a*) *Anopheles culicifacies*: observations on population ecology and reproductive behavior. Mosquito News 42: 93–101

Reisen WK, Hayes CG, Azra K, Niaz S, Mahmood F, Parveen T, Boreham PFL (1982*b*) West Nile virus in Pakistan. II. Entomological studies at Changa Manga national forest, Punjab Province. Trans R Soc Trop Med Hyg 76: 437–448

Reisen WK, Mahmood F, Parveen T (1982*c*) Seasonal trends in population size and survivorship of *Anopheles culicifacies, An. stephensi* and *An. subpictus* (Diptera: Culicidae) in rural Punjab province, Pakistan. J Med Entomol 19: 86–97

Reisen WK, Milby MM, Reeves WC, Meyer RP, Bock ME (1983) Population ecology of *Culex tarsalis* (Diptera: Culicidae) in a foothill environment in Kern county, California: Temporal changes in female relative abundance, reproductive status, and survivorship. Ann Entomol Soc Am 76: 800–808

Reisen WK, Mahmood F, Niaz S, Azra K, Parveen T, Mukhtar R, Aslam Y, Siddiqui TF (1986) Population dynamics of some Pakistan mosquitoes: temporal changes in reproductive status, age structure and survivorship of *Anopheles culicifacies, An. stephensi* and *Culex tritaeniorhynchus*. Ann Trop Med Parasitol 80: 77–95

Reisen WK, Meyer RP, Shields J, Arbolante C (1989) Population ecology of preimaginal *Culex tarsalis* (Diptera: Culicidae) in Kern county, California. J Med Entomol 26: 10–22

Reisen WK, Milby MM, Meyer RP, Pfuntner AR, Spoehel J, Hazelrigg JE, Webb JP (1991) Mark–release–recapture studies with *Culex* mosquitoes (Diptera: Culicidae) in southern California. J Med Entomol 28: 357–371

Reisen WK, Milby MM, Presser SB, Hardy JL (1992) Ecology of mosquitoes and St. Louis encephalitis virus in the Los Angeles basin of California, USA, 1987–1990. J Med Entomol 29: 582–598

Reisen WK, Lothrop HD, Hardy JL (1995) Bionomics of *Culex tarsalis* (Diptera: Culicidae) in relation to arbovirus transmission in southeastern California. J Med Entomol 32: 316–327

Renshaw M (1991) Population dynamics and ecology of *Aedes cantans* (Diptera: Culicidae) in England. Ph.D. thesis, University of Liverpool

Renshaw M, Service MW, Birley MH (1994) Size variation and reproductive success in the mosquito *Aedes cantans*. Med Vet Entomol 8: 179–186

Rosay B (1961) Anatomical indicators for assessing the age of mosquitoes: The teneral adult (Diptera: Culicidae). Ann Entomol Soc Am 54: 526–529

Rosay B (1969) Anatomical indicators for assessing age of mosquitoes: Changes in ovarian follicles. Ann Entomol Soc Am 62: 605–611

Rosenberg R, Maheswary NP (1982) Forest malaria II. Transmission by *Anopheles dirus*. Am J Trop Med Hyg 31: 183–191

Russell RC (1985) The efficiency of various collection techniques for sampling *Culex annulirostris* in southern Australia. J Am Mosq Control Assoc 1: 502–505

Russell RC (1986) Seasonal abundance and age composition of two populations of *Culex annulirostris* (Diptera: Culicidae) at Darwin, northern territory, Australia. J Med Entomol 23: 279–285

Samarawickrema WA (1962) Changes in the ovariole of *Mansonia (Mansonioides)* mosquitoes in relation to age determination. Ann Trop Med Parasitol 56: 110–126

Samarawickrema WA (1967) A study of the age-composition of natural populations of *Culex pipiens fatigans* Wiedemann in relation to the transmission of filariasis due to *Wuchereria bancrofti* (Cobbold) in Ceylon. Bull World Health Organ 37: 117–137

Samarawickrema WA (1968) Biting cycles and parity of the mosquito *Mansonia (Mansonioides) uniformis* (Theo.) in Ceylon. Bull Entomol Res 58: 299–314

Saul A (1987) Estimation of survival rates and population size from mark–recapture experiments of bait-caught haematophagous insects. Bull Entomol Res 77: 589–602

Saul A, Graves PM, Kay BH (1990) A cyclical model of disease transmission and its application to determining vectorial capacity from vector infection rates. J Appl Ecol 27: 123–133

Schlein Y (1975) Further studies on the cuticular daily growth layers of anopheline mosquitos. WHO/MAL/75.857: 7 pp. (mimeographed)

Schlein Y (1979) Age grouping of anopheline malaria vectors (Diptera: Culicidae) by the cuticular growth lines. J Med Entomol 16: 502–506

Schlein Y, Gratz NG (1972) Age determination of some flies and mosquitoes by daily growth layers of skeletal apodemes. Bull World Health Organ 47: 71–76

Schlein Y, Gratz NG (1973) Determination of the age of some anopheline mosquitoes by daily growth layers of skeletal apodemes. Bull World Health Organ 49: 371–375

Schreiber ET, Mulla MS, Chaney JD (1988) Population trends and behavioral attributes of adult mosquitoes associated with dairies in southern California. Bull Soc Vector Ecol 13: 235–242

Schreiber ET, Webb JP, Hazelrigg JE, Mulla MS (1989) Bionomics of adult mosquitoes associated with urban residential areas in the Los Angeles basin, California. Bull Soc Vector Ecol 14: 301–318

Seawright JA, Dame DA, Weidhaas DE (1977) Field survival and ovipositional characteristics of *Aedes aegypti* and their relation to population dynamics and control. Mosquito News 37: 62–70

Self LS, Sebastian A (1971) A high incidence of green coloration in newly-emerged adult populations of *Culex pipiens fatigans* in Rangoon, Burma. J Med Entomol 8: 391–393

Sempala SDK (1981) The ecology of *Aedes (Stegomyia) africanus* Theobald in a tropical forest in Uganda: Mark–release–recapture studies on a female adult population. Insect Science and its Application 1: 211–224

Service MW (1965) Some basic entomological factors concerned with the transmission and control of malaria in northern Nigeria. Trans R Soc Trop Med Hyg 59: 291–296

Service MW (1969) Observations on the ecology of some British mosquitoes. Bull Entomol Res 59: 161–194

Service MW (1973a) Mortalities of the larvae of the *Anopheles gambiae* complex and detection of predators by the precipitin test. Bull Entomol Res 62: 359–369

Service MW (1973b) Flight activities of mosquitoes with emphasis on host seeking behaviour. In: Hudson A (ed) Biting Fly Control and Environmental

Quality. Proc Symp Univ Alberta, Edmonton, 1972. Ottawa No. DR 217: pp. 125–132

Service MW (1993) Mosquito Ecology. Field Sampling Methods. 2nd edn. Chapman & Hall, London

Sheppard PM, Macdonald WW, Tonn RJ, Grab B (1969) The dynamics of an adult population of *Aedes aegypti* in relation to dengue haemorrhagic fever in Bangkok. J Anim Ecol 38: 661–702

Shlenova MF, Bey-Bienko IG (1962) Age composition of mass species populations of mosquitoes genus *Aedes* (according to observations made in Byelrussia). In: Sergiyev PG (ed) Problem Gen Zool Med Parazitol, pp. 589–605

Shute PG, Maryon ME, Pringle G (1965) A method for estimating the number of sporozoites in the salivary glands of a mosquito. Trans R Soc Trop Med Hyg 59: 285–288

Sidavong B, Vythilingam I, Phetsouvanh R, Chan ST, Phonemixay T, Lokman Hakim S, Phompida S (2004) Malaria transmission by *Anopheles dirus* in Attapeu province, Lao PDR. Southeast Asian J Trop Med Public Health 35: 309–315

Singh N, Yasuno M (1972) The Gonotrophic Cycle of *Culex pipiens fatigans* in Nature. WHO/VBC/72.380, 6 pp. (mimeographed)

Slaff M, Crans WJ (1981) The host seeking activity of *Culex salinarius*. Mosquito News 41: 443–447

Slobodkin LB (1962) Growth and Regulation of Animal Populations. Holt, Rinehart & Winston, New York

Slooff R, Herath PRJ (1980) Ovarian development and biting frequency in *Anopheles culicifacies* Giles in Sri Lanka. Trop Geogr Med 32: 306–311

Smiraglia BC, Fahmy MTI, Lavagnino A (1971) Observations on the 'mating plug' in *Anopheles atroparvus* and *Anopheles labranchiae*. Riv Parassitol 32: 105–111

Smith BP (1988) Host-parasite interaction and impact of larval water mites on insects. Annu Rev Entomol 33: 487–507

Smith BP, McIver SB (1984) The impact of *Arrenurus danbyensis* Mullen (Acari: Prostigmata; Arrenuridae) on a population of *Coquillettidia perturbans* (Walker) (Diptera: Culicidae). Can J Zool 62: 1121–1134

Smith IM (1976) A study of the systematics of the water mite family Pionidae (Prostigmata: Parasitengona). Mem Entomol Soc Can No. 98

Smith IM, Oliver DR (1986) Review of parasitic associations of water mites (Acari: Parasitengona; Hydrachnida) with insect hosts. Can Entomol 118: 407–472

Snow WF, Wilkes TJ (1977) Age composition and vertical distribution of mosquito populations in The Gambia, West Africa. J Med Entomol 13: 507–513

Sokolova MI (1981) Age changes and morphological types of ovarioles of females of a northern population of the blood-sucking mosquito *Aedes caspius dorsalis* Mg. Medskaya Parazitol 50: 63–70 (In Russian, English summary)

Sokolova MI (1995) Contributions of female mosquitoes (Diptera: Culicidae) of different reproductive ages to the reproduction of populations. J Vector Ecol 20: 121–128

Spencer M (1979) Age grouping of female *Anopheles farauti* populations (Diptera: Culicidae) in Papua New Guinea. J Med Entomol 15: 555–569

Spencer M, Christian SH (1969) Cyclic ovariole changes in *Anopheles farauti* Laveran (Diptera: Culicidae) in Papua-New Guinea. J Aust Entomol Soc 8: 16–20

Suzuki A, Tsuda Y, Takagi M, Wada Y (1993) Seasonal observations on some population attributes of *Aedes albopictus* females in Nagasaki, Japan, with emphasis on the relation between the body size and the survival. Trop Med (Nagasaki) 35: 91–99

Suzuki T (1977) Preliminary studies on blood meal interval of *Aedes polynesiensis* in the field. WHO/FIL/77.149, 10 pp. (mimeographed)

Suzuki T (1978) Preliminary studies on blood meal interval of *Aedes polynesiensis* in the field. Jap J Sanit Zool 29: 169–174

Thomas DB, Chen AC (1989) Age determination in the adult screwworm (Diptera: Calliphoridae) by pteridine levels. J Econ Entomol 82: 1140–1144

Touré YT, Traore SF, Sankare O, Sow MY, Coulibaly A, Esposito F, Petrarca V (1996) Perennial transmission of malaria by the *Anopheles gambiae* complex in a North Sudan Savanna area of Mali. Med Vet Entomol 10: 197–199

Tsuda Y, Wada Y, Takagi M (1991) Parous rate as a function of basic population parameters of mosquitoes. Trop Med (Nagasaki) 33: 47–54

Tyndale-Biscoe M (1984) Age-grading methods in adult insects: a review. Bull Entomol Res 74: 341–377

Tyndale-Biscoe M, Kitching RL (1974) Cuticular bands as age criteria in the sheep blowfly *Lucilia cuprina* (Weid.) (Diptera: Calliphoridae). Bull Entomol Res 64: 161–174

Ulloa A, Rodriguez MH, Arredondo-Jiménez JI, Fernández-Salas I (2005) Biological variation in two *Anopheles vestitipennis* populations with different feeding preferences in southern Mexico. J Am Mosq Control Assoc 21: 350–354

Ungureanu EM (1972) Methods for dissecting dry insects and insects preserved in fixative solutions or by refrigeration. Bull World Health Organ 47: 239–244

van Dijk WJOM (1966) Age determination of mosquitoes through dissection for filarial larvae. Trop Geogr Med 18: 53–59

Vercruysse J (1985) Estimation of the survival rate of *Anopheles arabiensis* in an urban area (Pikine-Senegal). J Anim Ecol 54: 343–350

Volozina NV (1958) On the fauna and ecology of mosquitoes of the genus *Aedes* in the Ivanovo district. Medskaya Parazitol 27: 670–673 (In Russian)

Vythilingam I, Phetsouvanh R, Keokenchanh K, Yengmala V, Vanisaveth V, Phompida S, Hakim SL (2003) The prevalence of Anopheles (Diptera: Culicidae) mosquitoes in Sekong Province, Lao PDR in relation to malaria transmission. Trop Med Int Health 8: 525–535

Wada Y, Kawai S, Oda T, Miyagi I, Suenaga O, Nishigaki J, Omori N (1969) Dispersal experiment of *Culex tritaeniorhynchus* in Nagasaki area (Preliminary report). Trop Med 11: 37–44

Walker ED, Copeland RS, Paulson SL, Munstermann LE (1987) Adult survivorship, population density, and body size in sympatric populations of *Aedes triseriatus* and *Aedes hendersoni* (Diptera: Culicidae). J Med Entomol 24: 485–493

Walker ED, Torres EP, Villanueva RT (1998) Components of the vectorial capacity of *Aedes poicilius* for *Wuchereria bancrofti* in Sorsogon province, Philippines. Ann Trop Med Parasitol 92: 603–614

Wall RW, Langley PA, Stevens J, Clarke GM (1990) Age-determination in the old-world screw-worm fly *Chrysomya bezziana* by pteridine fluorescence. J Insect Physiol 36: 213–218

Washburn JO, Anderson JR, Mercer DR (1989) Emergence characteristics of *Aedes sierrensis* (Diptera: Culicidae) from California treeholes with particular reference to parasite loads. J Med Entomol 26: 173–182

Watson TM, Saul A, Kay BH (2000) *Aedes notoscriptus* (Diptera: Culicidae) survival and dispersal estimated by mark-release-recapture in Brisbane, Queensland, Australia. J Med Entomol 37: 380–384

Weaver SC, Fashing NJ (1981) Dispersal behavior and vector potential of *Aedes cantator* (Diptera: Culicidae) in southern Maryland. J Med Entomol 18: 317–323

Wharton RH (1959) Age determination in *Mansonioides* mosquitoes. Nature 184: 830–831

World Health Organization (1975*a*) Manual on practical entomology in malaria. Part I. Vector bionomics and organization of anti-malaria activities. WHO Offset Publication No. 13. World Health Organization, Geneva

World Health Organization (1975*b*) Manual on practical entomology in malaria. Part II. Methods and techniques. WHO Offset Publication No. 13. World Health Organization, Geneva

Wu D, Lehane MJ (1999) Pteridine fluorescence for age determination of *Anopheles* mosquitoes. Med Vet Entomol 13: 48–52

Yajima T (1970) A note on the formation of 'false' dilatations in *Culex tritaeniorhynchus summorosus* Dyar. Jap J Sanit Zool 21: 224–225

Yajima T, Yoshida S, Watanabe T (1971) Ecological studies on the population of the adult mosquito, *Culex tritaeniorhynchus summorosus* Dyar: The diurnal activity in relation to physiological age. Jap J Ecol 21: 204–214

Yasuno M, Rajagopalan PK (1973) Population Estimation of *Culex fatigans* in Delhi villages. WHO/VBC/73.431, 18 pp. (mimeographed)

Zaim M, Zahirnia AH, Manouchehri AV (1993) Survival rates of *Anopheles culicifacies* s.l. and *Anopheles pulcherrimus* in sprayed and unsprayed villages in Ghassreghand District, Baluchistan, Iran, 1991. J Am Mosq Control Assoc 9: 421–425

Zalutskaya LI (1959) Comparative data on the biology of *Anopheles minimus* and *Anopheles vagus* in the vicinity of Tay-Nguen (Democratic Republic of Vietnam). Medskaya Parazitol 28: 548–553 (In Russian, English summary)

Chapter 14 Estimating the size of the Adult Population

Several mark–recapture methods for the estimation of population size have been developed as an alternative to counting the number of individuals within a fixed unit of habitat in order to determine absolute population size. Methods have been developed that are applicable to closed populations, in which population size is assumed not to be affected by immigration (including birth) or emigration (including death), and open populations, in which immigration and emigration occur. The application of mark-recapture methods to closed systems requires only one recapture occasion, while the study of open systems requires two or more recaptures. Available methods may be either deterministic, assuming an exact value for survival rate of individuals over the sampling interval, or stochastic, reflecting the more realistic scenario where an individual's survival over the sampling period is described as a probability. The main disadvantage of mark-recapture methods is the requirement to capture a relatively large proportion of the population in order to achieve acceptable levels of accuracy.

Marking methods

A wide variety of methods and techniques for marking insects have been developed, many of which are applicable to the study of mosquitoes, including dyes, inks and paints, dusts, and radioactive isotopes, among others. Hagler and Jackson (2001) have completed a recent review of insect marking methods and they include a discussion of recent developments, some of which could potentially be of value in mosquito studies. The primary requirement of any method of marking insects is that it does not affect the normal behaviours of the insect under study, or at least not the specific behaviour under investigation. Hagler and Jackson (2001) provide a useful discussion of the characteristics of effective markers, which include durability, low cost, ease of application, and ease of identification. In addition, the ideal marker should not be toxic to the insect (or the environment), should not hinder or irritate it, and should not affect its normal behaviour, growth, reproduction or lifespan. In many cases, selection of the most suitable marking method will depend on what specific aspect of ecology or behaviour is being studied and will often involve a degree of compromise regarding which of the ideal characteristics is most important.

Dyes

Southwood and Henderson (2000) provide a useful summary of insect marking methods. Gangwere et al. (1964) review dyes and stains suitable for marking insects and list some of their characteristics, such as solubility. Both Staniland (1959) and Bailey et al. (1962) list suitable fluorescent marking compounds.

Staining insect larvae with dyes

Although several different stains and dyes have been fed to the immature stages of insects to discover whether they persist and are detectable in emerging adults (Heron 1968; Kay and Mottram 1986; Lindig et al. 1980; Vail et al. 1966; Zacharuk 1963), results have usually been disappointing. Stains either do not persist through to the adult stage, or if they do they are either rapidly excreted or difficult to detect. Heron (1968), however, achieved considerable success by feeding rhodamine B and Nile red sulphate to larvae of the larch sawfly (*Pristiphora erichsonii*). Emerging adults had bright green compound eyes and ocelli that strongly fluoresced yellow under ultraviolet light. The dyes were also found in various parts of the adult internal organs.

Adults of pink bollworms, boll weevils and tobacco budworms have been marked successfully by feeding larvae oil-soluble dyes, such as calco red N-1700 (Graham and Magnum 1971; Hendricks and Graham 1970; Lindig et al. 1980; McCarty et al. 1972), oil soluble deep black BB (Hendricks et al. 1971) and oil-soluble blue II (Hendricks 1971).

Almost 70 years ago Weathersbee and Hasell (1938) reported that when mosquito larvae were immersed in a number of stains, the best being Giemsa (1:250 parts water), resultant adults were stained. Vargas and Friere (1940) also apparently marked adults by placing larvae in dilute stains, but until recently most attempts to repeat these results have been unsuccessful (Bailey et al. 1962; Chang 1946; Eyles 1944; Haddow 1942; Reeves et al. 1948). Nielsen and Nielsen (1953) found that with most stains no coloration was imparted to the adult mosquitoes, although when larvae were placed in Giemsa solution (1:250) adults showed some signs of darker coloration on the neck. In India when larvae of *Culex quinquefasciatus* larvae were stained for 8 h in 0.1% methylene blue there was no significant mortality and a bluish-black colour was carried through to the adults, whereas no such colour was apparent when larvae were stained with Giemsa (Rajagopalan et al. 1975).

Joslyn et al. (1985) reported that when larvae of *Ochlerotatus sollicitans* were immersed in a 0.0001% aqueous solution of Giemsa the resultant adults were marked blue. Higher concentrations resulted in larval mortalities, while lower dosages failed to mark the adults; and although another 11 stains including methylene blue and rhodamine B were tested none was suitable as a self-marker. However, the blue coloration from Giemsa staining was not seen unless adults were squashed between a slide and coverslip and examined under a 10–20 × magnification against a white background. Despite adults feeding on sugar and females taking several blood-meals and laying eggs, the stain persisted throughout their laboratory life (maximum 10 days for males and 35 days for females). Preliminary trials involving dosing larval habitats containing 4th instars with 10^{-4}% and 2×10^{-4}% Giemsa showed that most resulting adults were effectively marked. Later Joslyn and Fish (1986) added Giemsa stain to larval habitats of *Ochlerotatus communis* to give a concentration of 2 ppm. Giemsa marks were detected in 7 of the 941 adults, caught in dry ice-baited CDC light-traps, when they were squashed between microscope slides. One marked specimen was trapped 7 weeks after larvae had been marked.

Conrad et al. (1983) attempted to mark adults of *Anopheles quadrimaculatus, Culex pipiens, Ochlerotatus triseriatus* and *Ochlerotatus sollicitans* by staining larvae with Giemsa. Adults of *Ochlerotatus sollicitans* were marked following immersion of early to late 4th instar larvae in aqueous (saline) solutions of stain (10^{-4}%, 4×10^{-4}%, 10^{-3}%, 2×10^{-3}%) buffered with NaH_2PO_4. With *Anopheles quadrimaculatus* early 4th instar larvae had to be immersed in aqueous (non-saline) stain to get the stain to carry over to the adults, and the weakest stain concentration marked only 5% of the adults, although at 10^{-3}% concentration, 92% were marked. There was also some indication that the Giemsa was slightly toxic to the larvae. *Culex pipiens* adults were marked successfully with the lowest concentration, whereas Giemsa proved unsuitable for marking *Ochlerotatus triseriatus* as it caused 86–99% larval mortality.

Paing and Naing (1988) marked adults of both *Anopheles dirus* and *Culex quinquefasciatus* by immersing early 4th instar larvae in aqueous solutions of Giemsa stain. Staining larvae of *Anopheles dirus* for 12 h in 0.03%, or for 3 h in 0.3% effectively marked the adults, but longer exposures caused larval mortality. When *Culex quinquefasciatus* larvae were stained for 12 h in 0.03% Giemsa there was some mortality (12–13.6%). Dissection of adults showed that their fat bodies and Malpighian tubules were stained blue, and in *Culex quinquefasciatus* ovaries and accessory glands were also stained. Paing and Naing (1988) reported that when larvae were immersed in 0.02% rhodamine B, 0.1% fast green or 0.05% Sudan blue only a low percentage of adults became stained, and mortality increased if higher concentrations were used.

In Australia Kay and Mottram (1986) tested 11 stains–Giemsa, methylene blue, Nile blue, rhodamine B, neutral red, fluorescein isiothiocynate, azure A, B, and C, vegetable clothing dyes and cochineal for marking larvae and resultant adults. Larvae were exposed to concentrations of 0.1–500 mg/litre for 1–24 h. Marked adult mosquitoes were tested by crushing them between microscope slides (Joslyn et al. 1985) or by macerating them in the wells of 24-well Lucite plates as used in serology, one drop of alcohol was added to aid detection of the stains. Although methylene blue and azure B (a component of Giemsa) marked adults these stains were not retained beyond 7 days. The best marker was Giemsa at dosages of 6 mg/litre for 24 h or 21 mg/litre for 3 h, and marks could be detected visually without squashing for up to 3 days after emergence, or longer if the adults were squashed. Stain was seen in the intersegmental neck membrane, between abdominal segments I and II, and with microscope examination in the posterior sternites. These retention times are considerably shorter than obtained by Joslyn et al. (1985) with *Ochlerotatus sollicitans* and in *Anopheles* spp. by Weathersbee and Hasell (1938). The formulation of Giemsa stains is apparently variable as is its propensity for staining protozoa. The stain used by Kay and Mottram (1986) was obtained from BDH (Gurr) and contained 26.5% eosin, 10.5% methylene blue and 63.0% azure I (which is A and B plus methylene violet, but with C removed). Larval mortality of *Aedes aegypti, Ochlerotatus vigilax* and *Culex annulirostris* varied between 2–18% with Giemsa at 6 mg/litre for 24 h, to 0–10% with 21 mg/litre for 3 h, mortality was heaviest in the younger instars. Adding 10–20 mg/litre of Giemsa to small-moderate sized pools (10–270 litres) resulted in all larvae of *Ochlerotatus vigilax, Ochlerotatus alternans* and *Culex sitiens* collected the next day being marked, but with a concentration of 6 mg/litre only 0–56% were marked.

Despite the above reported successes, Service (1993) reported having failed to mark adult mosquitoes (*Ochlerotatus cantans, Anopheles gambiae, Anopheles arabiensis, Culex quinquefasciatus*) with Giemsa or rhodamine B by this method, as have a number of other workers. An advantage of marking adults by treating the larvae is that larval habitats can be dosed with suitable chemicals and dispersal and, if possible, longevity of naturally emerging populations studied. The possibility of adverse effects on mosquitoes fed fluorescent dyes should not be overlooked, as both Hayes and Schechter (1970) and Yoho et al. (1971) reported mortalities in colding moths and houseflies due to a photodynamic effect of light on the fluorescent markers. Jones et al. (1972) reported that some fluorescent dyes reduced longevity or flight capacity, as measured on a flight mill, in the corn earworm.

Marking adult insects by feeding them with dyes and trace elements

When adult mosquitoes are fed soluble dyes, usually in sugar solutions, they may become coloured or fluoresce under ultraviolet light. Unfortunately these dyes are usually excreted within a few days (Reeves et al. 1948) and consequently the method is of very limited practical importance. But Bailey et al. (1962) considered that since mosquitoes fed with dyes produced coloured faeces the presence of marked mosquitoes in the field could be detected by placing white paper in artificial resting stations. It seems most unlikely that this will prove very useful, especially as faeces are also only marked for a few days following emergence.

Campbell and Kettle (1976) marked adult *Culicoides brevitarsis* in the laboratory by feeding them 10% sucrose solutions containing 23 dyes. Rhodamine B was detected in the crop of only 20% of adults 8 days after feeding, whereas more than 85% of flies were still stained 8 days after engorging on acid fuchsin, eosin, light green, lissamine green and red and yellow commercial food colourings. Methylene blue disappeared from the midges rapidly (2 days), possibly due to reduction to leucomethylene blue (Fruton and Simmonds 1958). In a semi-arid region of Israel Schlein (1987) successfully marked populations of *Phlebotomus papatasi* by spraying bushes with sucrose solution containing in-digotine or brilliant blue dyes. Sandflies fed on the dried sugary residues and a blue coloration could be detected in them up to 6 days after spraying, either directly or after they had been squashed on filter paper. This method might work with mosquitoes that are known to feed on natural sugary secretions, and it might be possible to use different coloured dyes to study their dispersal from different sites, and other behavioural habits.

Coppedge et al. (1979) marked adult screwworms by feeding them a diet containing fluorescein sodium, a non-toxic freely water-soluble chemical, detectable at 0.02 ppm under short wave (2537 Å) UV light. Marked adults were detected by squashing them on filter paper and examining them under UV light, 97% of those fed on 5% fluorescein remained marked for 14 days.

Another procedure to mark haematophagous flies, is to feed them on animals that have been treated with vital stains. Knight and Southon (1963) for example, marked tsetse flies by feeding them on an ox that had received 200 ml of an intravenous solution of 4 g trypan blue, given over a 20-min period to alleviate distress to the animal. Marked flies were detected by treating them, or their guts, with 0.1 N sodium hydroxide solution and using paper chromatography. Adults fed on the animal shortly after inoculation were positive for trypan blue 8 days later, and could be detected, but less easily, for up to 38 days. With *Aedes aegypti* the dye was only detectable after about 2 days. Clearly the method of retention and excretion in tsetse flies and mosquitoes differs. Cunningham et al. (1963) marked adult tsetses by feeding them on animals labelled with specific agglutinins, which could be detected in the blood-meal of the flies. When chickens were fed on grain stained with oxypren or rhodamine B extra S, or were intravenously injected with stains, they fluoresced, but mosquitoes that fed on birds showed no fluorescence (Bailey et al. 1962).

Topical application of dyes to adult insects

Some of the earliest studies on flight dispersal of mosquitoes involved spraying adults with aniline dyes (Geiger et al. 1919; Le Prince and Griffitts 1917*a,b*; Zetek 1913). Suitable stains consist of aqueous 1–2% solutions of eosin, methylene blue, malachite green, brilliant blue, Congo red, orange G, crystal violet, etc. Dyes are frequently applied to adults confined in small cages as a fine mist produced by a small atomiser, such as a nasal or scent spray. However, as early as 1934 Russell and Santiago found that a commercial paint sprayer produced a finer and more even distribution of stain for marking *Anopheles*. Various other aniline dyes and stains dissolved in acetone, alcohol or mixtures of shellac and alcohol have been used to mark insects. Waterproof inks have also been used (Fales et al. 1964). Porter and Jorgensen (1980) marked ants by spraying them with different coloured fluorescent printing inks or fluorescent inks removed from the canisters of felt-tip marker pens (e.g. 'Magic markers'). Marked individuals were detected by longwave ultraviolet light. Mosquitoes could be similarly marked. Peffly and LaBrecque (1956) marked houseflies by spraying them with 6% phenolphthalein. To minimise mortality, sprayed individuals should be 'dried' soon after marking, by either placing them outside in a relatively windy exposed situation, or indoors in a draft of air.

Recaptured stained mosquitoes can sometimes be recognised without killing them, but a better procedure is to place them on white filter paper and drop on a small quantity of suitable solvent. Marked individuals produce a coloured spot or ring on the paper. A number of solvents can be used, such as ethanol (Edman and Lea 1972), or equal parts of 70% ethanol and chloroform and one-third part of glycerol (Quarterman et al. 1954; Schoof and Siverly 1954). Acetone can be incorporated in this mixture

(Smith et al. 1941), or used alone (MacLeod and Donnelly 1957). Suzuki (1978) marked chloroformed adults of *Aedes polynesiensis* by spraying them lightly with 2% solutions of methylene blue, metanil yellow or crecyl violet in a 1:1 solution of ethanol and water. Recaptured mosquitoes were killed, spread on absorbent paper and a drop of an ethanol, glycerine and chloroform (3:3:1) solution placed on each mosquito, which was then crushed. Houseflies marked with phenolphthalein indicator are detected by the purple coloration produced when they are wetted with 1% sodium hydroxide solutions; the addition of a few drops of detergent improves detection. Edman and Lea (1972) modified the procedures used by Nielsen (1961) and Dalmat (1950) and sprayed *Culex* adults with saturated solutions of fast green, luxol fast blue or safranino in 95% ethanol containing 0.1% acetic acid to enhance the staining properties of these aniline dyes. After filtering, the solutions were placed in a commercial paint sprayer at a pressure of 10–15 lb/in^2 and the insects sprayed for 15–30 s. Some 500 000 adults could be marked within 2 h. There was no apparent mortality. Luxol fast blue was the least satisfactory dye because adults had to be examined under a microscope before staining was apparent, whereas the other two dyes were detected by placing mosquitoes on filter paper and wetting them with a few drops of 95% ethanol.

In Japan, Wada et al. (1969, 1975) used aqueous solutions of fluorescent stains such as 1.0% yellow 8G, 1.0% kaycoll BZ, 0.1% rhodamine 6G and 0.5% crystal violet to spray adults of *Culex tritaeniorhynchus*. Marked individuals were detected by their fluorescence under ultraviolet light.

Schoof and Siverly (1954) labelled all houseflies for release with ^{32}P and then used different coloured dyes for marking those released at different localities. Recaptures were first screened for radioactivity after which positive individuals were inspected more closely for dye marks.

In Thailand, Tsuda et al. (2000) marked 1848 individuals of nine species of wild-caught anophelines with a 0.5% aqueous solution of rhodamine B in order to study the spatial distribution of host-seeking mosquitoes. Marked mosquitoes were identified by placing them on filter paper, adding 80% alcohol, and examining under UV light to check for the presence of dye. The effect of the dyeing procedure on survival was apparently not determined in this study. In China, Tsuda et al. (1999, 2001) marked laboratory-reared *Anopheles minimus* and *Aedes aegypti* respectively, with either a 5% aqueous solution of rhodamine B or a 0.8% solution of Blue AFX + Whitex (Sumitomo Chemicals, Tokyo, Japan).

Fabian et al. (2005) used a 0.5% aqueous solution of rhodamine B to mark male and female *Anopheles saperoi* in a study of dispersal, survival and population size in northern Okinawa, Japan.

Paints and Inks

Different types of paints and inks have been used to apply one or a series of small coloured spots to insects (see Southwood 1978; Southwood and Henderson 2000). Although this procedure can confer date-specific markings, adults usually have to be anaesthetised and individually marked, a procedure that is considerably more time consuming than the simultaneous dusting or spraying of large numbers of mosquitoes. Consequently the method has not been so widely used for mosquitoes as dusting with powders, although tsetse workers have used the method.

A variety of different types of paints have been applied to insects. For instance artist's oil paints have been used on tsetse flies (Jackson 1933) and *Chrysops* (Beesley and Crewe 1963), while poster paints have been used on mosquitoes (Edman et al. 1998; Frank and Curtis 1977; Gillies 1958, 1961; Slooff and Herath 1980), and quick drying nitrocellulose lacquers and paints on a variety of insects. 'Humbrol' enamel paints have proved very useful for marking mosquitoes (Macdonald et al. 1968; McClelland and Conway 1971; McDonald 1977a,b; Renshaw 1991; Renshaw et al. 1994; Sheppard et al. 1969; Trpis 1973; Trpis and Haüsermann 1986). These enamel paints are available in ½-oz quantities and in a great number of colours, including a few fluorescent ones, and are available in shops selling model aircraft and car kits etc.

Paints can be applied with a variety of objects such as artists' fine paint brushes, dissecting needles, bristles, pieces of grass or nylon fishing line. Gillies (1958, 1961) applied poster paints to *Anopheles gambiae* with a micro-loop of 0.0024-gauge plated copper wire fixed to the end of a matchstick with candle wax. If the loop was dipped into dilute detergent (e.g. 'Teepol') prior to dipping in the paint a smaller spot could be placed on the mosquito than if the paint alone was used. Slooff and Herath (1980) working with *Anopheles culicifacies* used this method with poster paints, but modified it so that after dipping the wire loop into the paint it was dipped into 96% ethanol and then applied to only the scutum. Frank and Curtis (1977) applied poster paint to *Wyeomyia vanduzeei* adults using a piece of nylon monofilament (Fisherman's 6-lb strength line) that was passed through a flame to produce a minute bead at the end. A 2.5-cm length of this nylon line with beaded end was taped to a wooden applicator stick so that a 1-cm length protruded. The correct viscosity of water-based poster paints used for marking, was obtained by mixing in a small amount of liquid detergent (about 1% volume). Edman et al. (1998) marked 1000 *Aedes aegypti* with a small spot of water-based fluorescent poster paint applied to the thorax. Paint was applied from a single bristle of a camel's-hair paintbrush that was first coated with liquid detergent. Renshaw et al.

(1994) successfully applied enamel paint to the thorax of *Ochlerotatus cantans* using the tip of a finely-sharpened pencil.

Conway et al. (1974) described a method for marking mosquitoes with paint in which a lightly etherised mosquito is placed dorsal side uppermost on a piece of stockinette glued over a circular hole cut from a sheet of cork. A plastic ring covered on one side with 1-mm nylon mosquito netting is gently lowered on the mosquito (Fig. 14.1) and a spot of paint applied. Marked mosquitoes were detected by a battery operated UV light. A very satisfactory marking method by which up to 100 *Ochlerotatus cantans* can be marked per hour in the field by a single person is as follows. Adults are lightly anaesthetised using diethyl ether in preference to chloroform or ethyl acetate as mosquitoes recover better from the former, while carbon dioxide often results in insufficient anaesthesia. A Pasteur pipette drawn out into a fine point with a very slight dilatation at the end is used to mark the scutum of adults with 'Humbrol' paint. The marker is periodically wiped on a small piece of cotton wool moistened with the appropriate enamel paint thinner. This prevents a blob of paint drying on its tip, which would result in too large a spot on the mosquito.

Anaesthetised and marked mosquitoes are placed in cardboard cups, however Renshaw (1991), also working with *Ochlerotatus cantans*, found that if the paint was still wet when the mosquitoes were placed in the cup, they often became stuck to the surface, hence she preferred to tip anaesthetised adults onto a sheet of white paper and mark them *in situ* with paint, allowing them to recover and fly off. Individuals that did not readily fly off were considered damaged and were killed.

Frank and Curtis (1977) described difficulties in successfully anaesthetising *Wyeomyia vanduzeei*. Chilling adults by placing them in a refrigerator (5°C) then transferring them to a metal block containing ice was unsuccessful as the mosquitoes became wet through condensation. Initial attempts to anaesthetise with chloroform also proved unsuccessful until cotton wool was placed in a plastic wash bottle and moistened with chloroform. The resultant vapour was then blown over the mosquitoes by gently squeezing the wash bottle. The correct concentration to anaesthetise the adults was obtained through trial and error. Recovery of anaesthetised mosquitoes was variable unless they were placed in a high humidity, which was achieved by placing them in plastic containers containing freshly picked green shrub leaves and placing them in a room having 80% R.H.

The idea of using a spot code for marking insects originated with von Frisch (1923), and a short account of the development of the method together with some useful references is given by Rooum (1988).

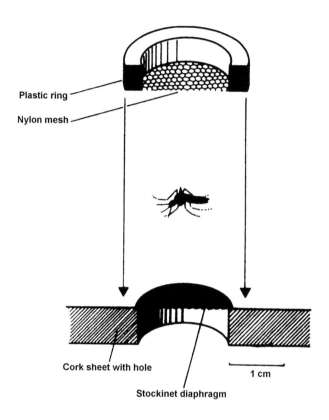

Plastic ring

Nylon mesh

Cork sheet with hole

1 cm

Stockinet diaphragm

Fig. 14.1. Device to hold a mosquito while painting a spot on the mesonotum. In use the mosquito is held gently with slight displacement of the upper mesh and the lower diaphragm (Conway et al. 1974)

Sheppard et al. (1969) obtained unique marking by placing spots of 'Humbrol' paint on eight selected sites on the wings and thorax of *Aedes aegypti*, and by combining this with a binary notation they were able to place individual coded numbers on 255 adults with a single colour (Fig. 14.2). Frank and Curtis (1977) applied paint spots only to the thorax of *Wyeomyia vanduzeei* because they found it hard to believe that spots applied to the wings of mosquitoes, as practised by Sheppard et al. (1969), do not affect their flight performance. Takagi et al. (1995) used three different colours of felt-tip pen to apply individual marks to the wings of *Aedes albopictus* in a study of female dispersal in Nagasaki, Japan.

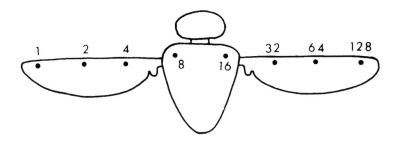

Fig. 14.2. Position and numerical values for a system of marking mosquitoes with paints. By using all combinations of points 255 consecutive numbers can be used with only one colour paint (after Sheppard et al. 1969)

By using three marking positions on each wing and three colours of ink, 4095 ($4^6 - 1$) mosquitoes could be marked individually. In Tanzania (Trpis and Haüsermann, 1986) used three colours of 'Humbrol' paints in a binary notation to date-specific mark *Aedes aegypti*. The first colour represented 1, 2, 4, and 8 when the marks were placed towards the four corners of the scutum, the second colour values were 16, 32, 64 and 128 when marks were placed in the same corners, while a dot in the centre of the scutum represented 256. The maximum number of mosquitoes that can be marked with just three colours applied to a single location is 511. Other three-colour combinations allow a further 511 mosquitoes to be uniquely marked. The number of unique marks obtainable (C) is given by the formula:

$$C = (x+1)^n - 1 \tag{14.1}$$

where x is the number of colours or types of mark and n is the number of sites to which the marks are applied.

Dusts and powders

Gill, in 1925, appears to have been the first to have marked mosquitoes with coloured powders, but the first published account was given by Darling (1925). Aqueous stains, however, remained the most common method of marking mosquitoes until Majid (1937) showed the value of printers 'gold' dusts for labelling mosquitoes. According to Majid one of the principal advantages of these dusts is that they do not stick to the mosquitoes' wings, and are therefore less likely to affect natural mosquito flight activities and behaviour. However, Shapiro et al. (1944) considered these powders did in fact adversely affect mosquitoes. Russell et al. (1944) marked *Anopheles culicifacies* with gold, silver, red and blue printer's dusts in India but found the latter two colours were not so effective as gold or silver. Eyles (1943) used metallic dusts on *Anopheles quadrimaculatus* and in Malaysia Wharton et al. (1963) marked *Anopheles hackeri* with silver and gold bronzing powders. Different coloured bronzing dusts have also been used in Brazil (Causey and Kumm 1948; Causey et al. 1950), in Palestine (Shapiro et al. 1944), in Guyana (Burton 1964), in India (Viswanathan et al. 1945), and later in the USA in conjunction with other coloured powders (Dow 1971). Marked individuals are easily detected by the metallic glint shown when they are examined with good lighting under a stereoscopic microscope. The discovery in Palestine of *Anopheles sergentii* harbouring pollen grains of a plant that grew at the nearest 1–1.5 km away, prompted Shapiro et al. (1944) to speculate on the possibility of marking mosquitoes with exotic pollens. Nielsen and Nielsen (1953) sprayed vegetation on which adult *Ochlerotatus taeniorhynchus* rested soon after emergence with dusts of eosin, methyl violet and Victoria green mixed with three of four parts of flour. Newhouse (1953) also marked mosquitoes by dusting their resting places with coloured powders. In theory this method enables large numbers of newly emerged adults to be marked prior to their dispersal but has been little used.

While several different kinds of coloured dusts and powders have been employed to mark various insects (see MacLeod and Donnelly 1957; Southwood and Henderson 2000), mosquitoes have most frequently been marked using fluorescent powders (Aarons et al. 1951; Bailey et al. 1965; Birley and Charlwood 1989; Dow 1971; Germain et al. 1972; Hervy 1977*b*; Hii and Vun 1985; Howard et al. 1989; Ivanova and Ipatov 1987; Muir and Kay 1998; Pal 1947; Reeves et al. 1948; Reisen et al. 1977, 1981; Renshaw et al. 1994; Self et al. 1971; Sinsko and Craig 1979; Subra 1972; Weathersbee and Meisch 1990; Zukel 1945).

Individuals marked with fluorescent compounds are usually recognised by their fluorescence under ultraviolet light. Consequently they need not be killed, thus making further observations or remarkings possible. However, solvents may be necessary to avoid missing very lightly dusted individuals. Reeves et al. (1948) found that unmarked mosquitoes could fluoresce blue, purple, green, white, yellow, and even orange due to the presence of natural pigments. Service (1993) also observed that mosquitoes may fluoresce blue–green. It may therefore be advisable to use a red fluorescent pigment if powders are not of a strong colour.

A limitation of marking mosquitoes with fluorescent powders is that only a relatively few different colours and hues are available which can be subsequently distinguished. In attempts to overcome this limitation pigments have sometimes been mixed with other marking materials (Dow 1971) including trace amounts of rare earth elements (McClelland et al. 1973a), but these approaches have not been widely adopted. Bennett et al. (1981) proposed a method that allows 42 or more unique marks, but requires the use of a fluorescent microscope at $50\times$ or more magnification and with incident light from above. The system actually used by Bennett et al. (1981) employed six fluorescent powders and could provide up to 42 unique colour codes (including unmarked individuals) when applied in combinations of three colours or less. Using four or more colours may make distinguishing them more difficult. The authors warned that colours may be lost after a long time, or misidentified if too complicated mixtures are used, they therefore suggested that, at least in some cases, it might be best to limit codes to six different colours and using just three at a time. In their trials they used 'Day-Glo' (series A) and Radiant colour fluorescent pigments, which were preferred to 'Helecon' pigments, except that Helecon 2200—a light blue pigment—was used because it could be distinguished in combination from all but the dark blue powders of the other two brands. However, they pointed out that if 'Helecon' phosphorescent pigments, which exhibit an afterglow after removal of blacklight illumination, are used with fluorescent pigments of the same colour, they should be distinguishable thus increasing the variability of the system. They found that powders were more readily lost from wings and legs than from the sides of the thorax.

The same individual insects can be successfully marked on two and sometimes more occasions with contrasting coloured dusts (MacLeod and Donnelly 1957; Quarterman et al. 1954; Reisen et al. 1981; Reuben et al. 1973; Southwood 1978), but the method is not usually very satisfactory for multiple marking, although McClelland et al. (1973b) considered it might be possible to use a combination of several different fluorescent powders. Marked individuals would be recognised in the field from gross fluorescence, while examination of these under an incident-light fluorescent microscope should be able to differentiate between particles of different coloured powders. In general, however, marking with dusts, even fluorescent ones, is unsatisfactory for repeatedly marking the same individuals on different days, and usually not very satisfactory for date-specific marking.

Dusts often stick to mosquitoes without the addition of any gum, but sometimes better adhesion is obtained if one part of marking dust is ground up with 4–6 parts of gum arabic in a little water to form a thick paste. After drying this is pulverised. Trpis (1971) marked *Aedes aegypti* with a mixture of 1 part orcein and 3 parts gum arabic mixed with 96% ethanol and ground to a fine paste. Eddy et al. (1962) marked mosquitoes with 1% different fluorescent powders, 0.05% spreader sticker ('Pylac') and 0.001% emulsifier ('Tween 20'). Sinsko and Craig (1979) and Haramis and Foster (1983) added gum arabic to coloured dusts to ensure they stuck to *Ochlerotatus triseriatus*. However, no adhesive is needed with 'Dayglo' powders, and adults retain their marking for at least 6–8 weeks. With almost any dusts or powders, however, better adhesion is obtained if, prior to marking, the mosquitoes are kept in a humid atmosphere (Reeves et al. 1948; Trpis 1971).

Fluorescent dusts and pigments

Several fluorescent pigments and dusts have been used to mark mosquitoes including: 'Helecon' dusts that have a zinc sulphide base and are made by the U.S. Radium Corporation; and 'Dayglo' powders manufactured from organic dyes that are incorporated into a melamine formaldehyde resin and pounded to form very fine powders, with an average particle size of 10–12 μm. Other powders used for marking mosquitoes are Radiant fluorescent dusts that have a triazine aldehyde amide base and are made by Hercules Inc., U.S.A.; Lumogen™ fluorescent dust made by BASF, Holland; Fiesta™ fluorescent powders manufactured by Swada Ltd., London, UK; and 'Radglo', originally manufactured by Ciba-Geigy (now Novartis).

Mosquitoes are often marked by puffing small quantities of dusts and powders on them from a syringe, a large-bulb pipette or other form of insufflator, or by creating a small dust storm in a cage (e.g. Dunn and Mechalas 1963; O'Donnell et al. 1992; Morrison et al. 1999; Renshaw et al 1994; Sempala 1981; Singh et al. 1975*a*; Trpis 1971).

Pardo et al. (1996) describe a simple apparatus for marking *Lutzomyia* sandflies with fluorescent powder, which would appear to be eminently suitable for marking mosquitoes. The apparatus comprises a plastic insecticide susceptibility test container, to which a smaller container containing the fluorescent powder is attached (Fig. 14.3). The plastic connector section of the apparatus includes a sliding mechanism with a hole in one end and this can be used to open and close the connection between the two sections of the apparatus, keeping the insects separate from the powder, until marking. The smaller compartment (diameter 42 mm), which is used to hold a 5 mm deep layer of fluorescent powder, is fitted with a plastic tube (6 mm diameter), fixed at a height of 7 mm above the bottom of the container. This narrow tube has 3 rows of 2.5 mm diameter holes and a 30 × 2.5 mm slot cut into it and is connected to a length of 6 mm diameter rubber tubing and a rubber bulb. Squeezing and releasing the rubber bulb forces air out of the perforated tube and creates a dust storm within the upper compartment, marking the insects. Advantages of the apparatus described by the authors include: the relatively short period of exposure of insects to the marking powder; the ease with which sand flies could be added to or removed from the device; the ability to rapidly connect successive insect-containing compartments to a single powder-holding compartment; and ease of construction and low cost. Mortality of flies within 1 h of marking was significantly reduced using the new apparatus ($P < 0.0001$), in comparison with a conventional marking method in which sandflies were introduced into large plastic vials containing a 2–3 mm thick layer of fluorescent powder, which was agitated by blowing into the container with an aspirator. All *Lutzomyia* flies introduced into the new apparatus were successfully marked, and none was excessively marked. Excessive marking, associated with increased mortality was observed for 20–98% of sandflies marked using the more conventional method.

No significant difference was observed between the two methods in terms of recapture rates in field trials, but mean survival of those marked using the new apparatus was significantly higher than for those marked with the conventional method, but did not differ from survival of unmarked control flies.

Large numbers of mosquitoes can be marked in the field by placing them in a plastic bag having a small amount of coloured powder at the bottom. The bag is then gently shaken and rotated. For example in Japan Ikeshoji and Yap (1990) marked adult *Aedes albopictus* by shaking them up in plastic bags containing small quantities of rhodamine B or fast green. Recaptured mosquitoes were identified by dropping water on them. Reuben et al. (1973) marked adults of *Aedes aegypti* with either metallic dusts or fluorescent powders by carefully dropping anaesthetised individuals down a funnel into a plastic cylinder containing a thin layer of dust at the bottom. They became marked as they struggled on recovering from the anaesthesia. Moreover, their subsequent flight in the plastic container created a fine spray of dust, which enhanced marking. About 300 adults were marked in × containers measuring 11 mm × 14 cm and about half this number in 8 mm 9.5 cm containers. Laboratory experiments showed that triple-coloured individuals could be recognised up to 15 days after marking. Ulloa et al. (2002) used a similar method to mark adults of *Anopheles vestitipennis* on the coastal plain of Chiapas, Mexico. Mosquitoes were placed in plastic containers lined with paper towels dusted profusely with a fluorescent powder (Lumogen®, BASF Holland).

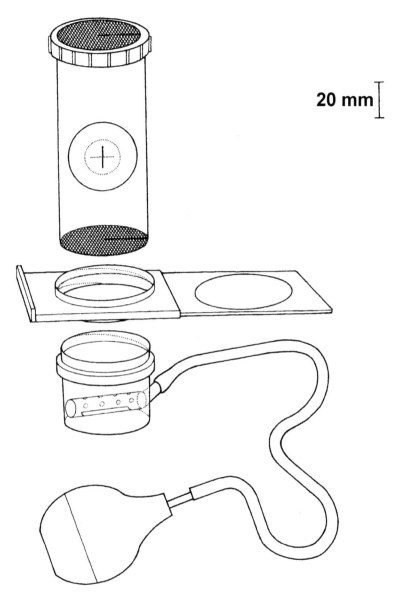

20 mm

Fig. 14.3. Apparatus for marking sand flies or mosquitoes with fluorescent powder (Pardo et al. 1996)

Takken et al. (1998) used 'Dayglo' flourescent powders to mark *Anopheles gambiae* s.l. and *Anopheles funestus* in order to study dispersal and survival rates of these two species in a village in rural southeast Tanzania.

Fluorescent dusts were applied by placing 15 cm cubic cages containing mosquitoes into a plastic bag and blowing small quantities of dust into the bag from a syringe. Yellow, red, orange and pink dusts were used and overall recapture rates were 7.4% for *Anopheles gambiae* s.l. and 4.3% for *Anopheles funestus*. It was reported that marking had no apparent effect on marked mosquitoes, but no details were provided.

In a study of mating behaviour of *Anopheles darlingi* in Brazil, Lounibos et al. (1998) marked virgin females with five distinguishable colours of fluorescent powder. Approximately 1 hour before release, females were confined in 4-litre cardboard ice cream cartons and marked with dust by insufflating it through the screen lid of the cage.

In Papua New Guinea Charlwood et al. (1986) marked blood-fed *Anopheles* mosquitoes by aspirating them into Coleman kerosene-lamp glasses having netting fitted over the two open ends; about 100 mm³ of fluorescent powders were blown into the container with an insufflator. Marked mosquitoes were released by removing the netting. In northern England, Renshaw et al. (1994) marked *Ochlerotatus cantans* with Fiesta™ fluorescent powders (Swada Ltd., London) by squirting powder from a 5 ml plastic syringe into paper cups containing a maximum of 100 individual mosquitoes, or alternatively small, wire cages covered in mosquito netting. This method of filling the barrel of a 5- or 10-ml syringe with fluorescent powders which are puffed onto the mosquitoes contained in plastic drinking beakers, cartons or cages by depressing the plunger is simple and very convenient in many situations.

Weathersbee and Meisch (1990) marked adult *Anopheles quadrimaculatus* with fluorescent pigments by placing them in a mesh holding cage, which was fitted into a slightly larger cardboard box lined with plastic sheeting to form a marking chamber. A 10-cm diameter plastic dryer hose was used to force a stream of air from the exhaust of a backpack powered insect aspirator into the top of the marking chamber. About 3 ml of fluorescent dust was introduced into a 1-cm hole cut at the junction of the dryer hose and the chamber. This procedure proved effective, except that in hot weather (>35°C), mosquitoes appeared rather stressed. Consequently batches of mosquitoes were marked without using the plastic lined chamber, that is by just squirting about 8 ml of powders from a garden duster directly into the holding cage.

Achee et al. (2005) marked *Anopheles darlingi* in Brazil with luminous powder by placing them in 1-gallon cardboard containers with a netting lid and running a ¼-in paintbrush that had been dipped in powder gently around the mesh lid of the holding container. This procedure was repeated four times per container.

Lillie et al. (1985) marked *Culicoides mississippiensis* by placing a plastic bag having several holes in it around a CDC light-trap collection bag containing adult midges. A 5-ml syringe inserted into the bag injected 0.4 ml micronized U.S. Radium fluorescent dust, while the trap fan created a dust storm ensuring that all midges were marked.

Singh et al. (1975*a*) give detailed descriptions of how large numbers of *Culex quinquefasciatus* (approx. million/day) were marked with coloured metallic dusts in a large-scale genetic control programme in India. Basically compressed air was delivered into six large (32-cm high, 26-cm diameter) removable cardboard cartons capable of holding 10 000 mosquitoes each and which were enclosed in aluminium cylindrical cages in which a dust storm was created. About 60 000 mosquitoes can be marked within about 10 min. Details of their subsequent treatment and transportation to the field are described in this paper. A modification of the method allowing marking of mosquitoes in much smaller cartons 8.5 cm high, 4.3 cm in diameter or 14 cm high and 12 cm in diameter was developed for *Aedes aegypti* (Singh et al. 1975*b*).

Moffitt and Albano (1972) used a novel method for marking large numbers of codling moths. These were placed in a small wire cage under a bell jar with the marking powder positioned above them in a watch glass. A partial vacuum was created by a vacuum pump, and then air was suddenly allowed to enter the bell jar to create a small dust storm which marked the moths.

Williams et al. (1979) described and illustrated a dusting chamber for marking large numbers of stable flies (*Stomoxys calcitrans*) that might be suitable for mosquitoes. A 61 × 61 × 122-cm high plywood chamber incorporates a 102-cm hinged drop-down door in the front-facing side. A 0.25 h.p., 120-V squirrel cage motor, with unattached blower, is mounted below the door so that the short blower arm protrudes into the chamber through a hole in the centre of the short (20-cm) panel below the hinged door. The motor, which is outside the chamber, is attached to the chamber on its side thus allowing fluorescent dusts (3 g) to be easily poured into the squirrel cage fan. With the door fastened a 10-s burst from the squirrel cage dusts the insects, which are housed in cages suspended in the chamber. As many as 250 000–300 000 stable flies could be marked in 15 min.

Zinc sulphide based dusts were first used for marking mosquitoes by Reeves et al. (1948). Bailey et al. (1962) found them especially satisfactory because they have excellent adhesive properties without the addition of gum arabic, and even heavily marked mosquitoes appear unharmed and behave normally. In Colorado, USA Mitchell (1979) marked *Culex tarsalis* hibernating in mine shafts with 'Helecon' powders. He then located mosquitoes in the dark with a torch, after which they were examined *in situ* with a portable UV light to determine whether they fluoresced.

In studying the dispersal of *Anopheles culicifacies* in Sri Lanka Curtis and Rawlings (1980) caught mosquitoes in a small hut, marked them with magenta (days 1–3 and 5–7) and yellow (day 4) zinc sulphide fluorescent powders and then released them back in the hut (hut 1). This and another 3 'collecting huts' (2.5 × 1.8 × 2.2 m high) were built of locally woven palm leaves fixed to a wooden frame. For eight nights a cow or calf was placed in hut 1 and removed at dawn. Following this all mosquitoes were collected from the huts with aspirators, placed in plastic beakers which had a netting top and a circle of black paper glued to the bottom. The beakers were placed in a moulded block of ice. The catch was examined with a portable ultraviolet light. Mosquitoes marked with powders were readily identified by their fluorescence against the dark paper disc.

Dow et al. (1965) marked *Culex tarsalis* with green, blue, yellow and red 'Helecon' pigments which had average particle sizes of 2, 2.5, 3.5 and 7 μm, respectively. In later studies, Dow (1971) marked adults of *Culex nigripalpus* with ^{32}P and then dusted them with different combinations of aluminium and bronze powders and six different coloured 'Helecon' dusts and found that of the 180 radioactive *Culex nigripalpus* known to have been marked with dusts, two had no evidence of marking while a few others had only a few particles adhering. He considered that loss of dusts probably occurred due to abrasion during collecting, and that light-trap collections likely suffered most from this. In Florida, USA Kline and Wood (1989) marked adults of *Ochlerotatus taeniorhynchus* and other mosquitoes with yellow, green, blue, red, pink and white 'Helecon' fluorescent powders, which were all readily distinguishable on recaptured adults.

Nowadays mosquitoes are more commonly marked with 'Dayglo' or other daylight fluorescent pigments than with any other dusts, stains, radionuclides or other substances. These fine fluorescent pigments are used by the paint industry and are available in about 13 different colours, but some are very similar and difficult or impossible to distinguish when used to mark mosquitoes. The most useful colours are yellow, blue, magenta, red and arc chrome. Green is also available but this colour can easily be made by mixing blue and yellow powders, and a variety of other colours can also be obtained by judicious mixing of the basic colours. Individual mosquitoes can be successfully marked with multiple colours as long as contrasting ones are used. 'Dayglo' powders fluoresce when activated by visible light at the blue end of the spectrum in addition to fluorescing under UV light. An ultraviolet light is therefore not usually necessary to recognise marked mosquitoes in collected samples, but a battery-operated UV light-tube is useful in detecting marked individuals at close range (Trpis and Haüsermann 1975), such as those resting about 1-m away on house walls. With mains operated UV lamps resting and flying fluorescent mosquitoes can be detected from at least 3 m and Service (1994) successfully used this method to study swarming behaviour of *Culex torrentium* in northern England.

In Tanzania Trpis and Haüsermann (1975) marked feral and domestic forms of *Aedes aegypti*, and in addition to locating marked adults in houses, could, with a portable UV light, detect those resting amongst vegetation and on outside house walls. In the Central African Republic Germain et al. (1977) marked adults with fluorescent powders to study the duration of the gonotrophic cycle of *Aedes africanus*. A similar approach using 'Dayglo' powders was used in Burkina Faso to estimate the gonotrophic cycle in *Aedes aegypti* in different months (Hervy 1977*a*). Also in Burkina Faso, Costantini et al. (1996) marked and released 7260 and 13 854 female *Anopheles gambiae* complex mosquitoes using the following colours of 'Dayglo' A and AX series powders: Aurora Pink, Blaze, Orange, Saturn Yellow, Signal Green and Horizon Blue.

In Uganda Sempala (1981) marked adult *Aedes africanus* on days 1–7 by squirting six different coloured 'Dayglo' dusts and a mixture of two such dusts (corona magneta and horizon blue) from the barrel of a 20-ml plastic syringe into a cage to create a dust storm. All recaptured females were marked with a small spot of quick drying paint ('Duco') which was applied to one of the four corners of the scutum; a total of five different coloured paints were used, thus allowing just a single spot to date-specific mark mosquitoes over 20 days using the method of Conway et al. (1974). Sempala (1981) sometimes found difficulty in distinguishing between mosquitoes marked with signal green and saturn yellow, and between those marked with corona magenta and rocket red, but the problem was resolved by examination under a microscope and using UV light.

Birley and Charlwood (1989) used series T 'Dayglo' powders which differ slightly in physical properties from series A powders that they had previously used (Charlwood et al. 1988). They found that less dust of series T seemed to be retained on the mosquitoes. Loss of dust may have been the reason for their low recapture rate (3.1%) of marked adults, when compared with a recapture rate of about 10% obtained when they used the A series (Charlwood et al. 1988).

Because of the small numbers of mosquitoes usually marked in mark–recapture studies in relation to the size of field populations, Meek et al. (1987) tried to increase the numbers marked by aerial spraying adults with 18.9 litres of 50% water suspension of blaze orange 'Dayglo' pigment. A pigment mixture, which contained a stabiliser to prevent pigment particles settling in the water prior to spraying, and a surfactant, was then poured into the aircraft spray tank (hopper), which had been filled with 397.5 litres of tap water to give an approximate mixture of water/pigment of 21:1. A Gruman AgCat aircraft flying 15.2 m over a marshland pasture sprayed a swath width of 15.2 m. More than 90% of *Culex quinquefasciatus* in cages at 1-m height were marked, which was significantly more than the 69.7% of adults that were marked when held in cages placed in dense vegetation near the ground. In more exposed areas 100 and 98.1% of *Ochlerotatus sollicitans* in cages at 1-m height and on the ground were marked. More than 96% of a natural emerging population of *Ochlerotatus sollicitans* was also marked. In later trials spraying at the rate of 9.8 litre/ha marked more than 98% of *Psorophora columbiae* emerging from flooded cattle hoof-prints in pastures (Meek et al. 1988). It was concluded that aerial spraying was suitable for marking large (but unknown) numbers of mosquitoes provided they were not sheltering in dense vegetation. In later trials Fryer and Meek (1989) found that after aerial application mosquitoes could pick up pigment from vegetation for up to 4 days after spraying, which means that mosquitoes moving into the area can become marked during this time.

Other 'Dayglo' powders that might prove suitable for marking mosquitoes include a series called EBT soluble toners, which are available in about three colours for use with suitable solvents (e.g. acetone and cellosolve) as coloured printing inks.

Nelson et al. (1978) marked adult *Culex tarsalis* with Radiant and with 'Helecon' dusts using an insufflator. Dusts adhered well in laboratory trials, up to 74 days on one female that lived that long. Marked adults caught in light-traps baited with CO_2 were detected under a microscope with UV light (3650 Å). Some recaptured females had dust only on their external genitalia indicating that they had become marked when mating. However, Meek et al. (1988) found there was no transfer of 'Dayglo' pigment when marked *Culex quinquefasciatus* and *Psorophora columbiae* mated with unmarked adults, nor when marked and unmarked adults were confined in cages. In contrast when heavily dusted *Psorophora columbiae* were confined in a cage <1% of unmarked adults became marked within 24 h. In later experiments 3.0% of unmarked mosquitoes kept with 'Dayglo' marked ones became marked after 24 h (Fryer and Meek 1989). Nelson et al. (1978) also found some evidence to suggest that recapture rates of female *Culex tarsalis* were greatest with wild adults that had been trapped and marked (10.9%), than laboratory reared adults (5.3%). In studying dispersal of *Anopheles culicifacies* Curtis and Rawlings (1980) believed that the capture of wild mosquitoes which are marked in the field and returned quickly to the environment likely produces less disturbance to their normal behaviour than either marking reared mosquitoes or transporting wild mosquitoes to the laboratory for marking and then returning them to the field for release. They pointed out that with *Drosophila melanogaster* overcrowding of adults prior to release may stimulate greater than normal dispersal (Wallace 1970). Russell et al. (1944) found some evidence for differential dispersal between reared and wild caught *Anopheles culicifacies*.

Self-marking methods. Singh and Yasuno (1972) developed a useful self-marking technique for emerging mosquitoes. About 2000 pupae were placed in a 14-cm high, 12-cm diameter plastic container. About 18 g of packing material such as thin (1-cm wide) strands of paper were screwed up and liberally dusted with fluorescent powders (e.g. 'Dayglo' pigments) and then placed in a small wire container supported on a small metal platform above the water surface (Fig. 14.4).

Emerging mosquitoes crawled through the loose packing and became marked before they eventually escaped. A wide lid over the apparatus afforded some protection against the weather. In laboratory studies 10 243 *Culex quinquefasciatus* successfully escaped from the automatic marker; only 6.7% were found trapped within the packing material. All of the 1200 adults examined were well marked with fluorescent dusts. This method was used in Pakistan to mark *Culex tritaeniorhynchus* (Reisen et al., 1977, 1978) and *Anopheles culicifacies* (Reisen et al., 1980, 1982*b*). Yasuno (1979), however, modified this self-marking method by increasing the size of the container for holding the pupae (200 cm tall and 175 cm in diameter) so that some 6000 pupae could be introduced. Also, instead of employing strands of paper, two or three layers of coarse hessian mesh dusted with powders were placed in the container on a sponge supported on a stand.

Niebylski and Meek (1989) describe a self-marking device (Fig. 14.5) for mosquitoes emerging from ditches. It consists of four basic components, namely: (1) a $2 \times 2 \times 2$-ft frame made of 0.5-in diameter PVC piping and extending laterally 4 ft at each end (Fig. 14.5 A–C) supporting (2) a heavy duty black fibreglass 16×18-mesh screen tent-like covering; (3) an exit grid comprising a 30×30-in frame supporting ten 24-in horizontal metal tubes (0.25-in diameter) (H, I, K, L) each being inserted through a 4-in casing formed by sewing a double layer of cheese cloth (30×12 in) along the centrefold (J); and finally (4) a horizontal transparent plastic rain shield (N) supported 24 in above the exit grid by four vertical metal rods (M). The ten folds of cheese cloth are impregnated with fluorescent powders so that escaping mosquitoes on flying past them become marked. In each of ten replicates using 100 *Culex quinquefasciatus* pupae 86% of emerging adults flew through the grid and all were marked (5–15 spots/mosquito).

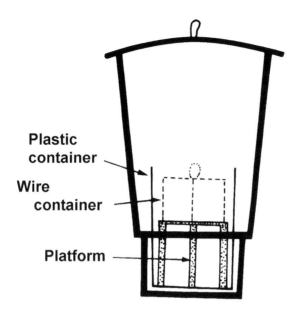

Fig. 14.4. Device for self marking mosquitoes with dusts just after emergence (after Singh and Yasuno 1972)

Most marks were on the tibiae, tarsi, scutellum and abdominal sternites, only rarely were the wings marked. There was no significant mortality in marked and unmarked mosquitoes. The device was originally constructed to mark adults emerging from sewage ditches, but the shape and size of the apparatus could be modified for other breeding places.

A device for self-marking *Simulium venustum* s.l. and *Stegopterna mutata* s.l. adult black flies emerging from streams was designed by Dosdall et al. (1992) and would appear to be equally applicable to marking emerging mosquitoes (Fig. 14.6). The device consists of a styrofoam base that acts as a float and encloses a surface area of 1 m^2, a metal grid with 1 × 2 cm openings onto which were placed a layer of spherical styrofoam chips 3–5cm in diameter, a rain cover, and an anchoring device. The styrofoam chips were coated in powdered fluorescent pigment (Aurora Pink, AR-Moneith Ltd., Vancouver, British Columbia, Canada) by vigorously shaking them with the pigment in a large plastic bag. Emerging black flies crawled through the metal grid and through the layer of pigment coated styrofoam chips before exiting the device.

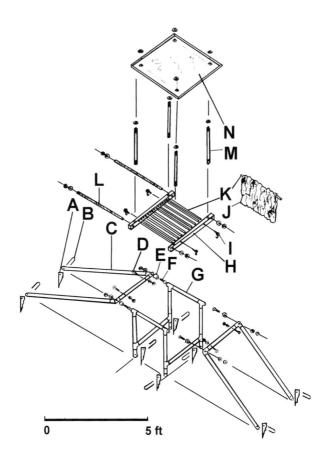

Fig. 14.5. Self marking trap; exploded diagram of trap showing A—Anchoring stake, B — stainless steel wire, C— lateral extension section, D — eyebolt, E — PVC pipe cap, F — machine bolt, G — primary support section, H — wooden piece (30 in long), I—eyebolt with C-hook, J—double-layered cheesecloth partition (30 × 12 in), K—stainless steel tube, L—threaded stainless steel rod (30 in long), M—threaded stainless steel rod (24 in long), N—cast acrylic sheet (30 × 30 in). (Niebylski and Meek 1989)

Over a 3-day period 10 813 adult black flies emerged from three traps and all were successfully marked. The device and the marking pigment used did not appear to adversely affect black fly emergence, survival, or subsequent attraction to human hosts. The authors acknowledged that there

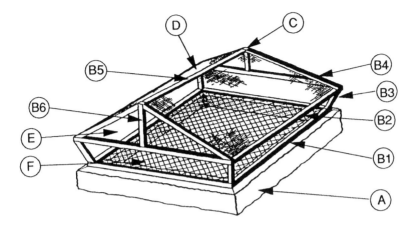

Fig. 14.6. diagram of the self-marking trap of Dosdall et al. (1992). (A) Styrofoam blocks (1.20 m × 0.15 m × 0.15 m; inner area = 1 m²) for flotation, reinforced along their upper surfaces with plywood (thickness = 1.6 cm) glued to the styro-foam, and joined together with bolts (length = 0.20 m); (B) wooden strips com-prising the framework: B1 = 1.00 m × 0.05 m, B2 = 1.10 m × 0.05 m, B3 = 0.20 m × 0.05 m, B4 = 0.60 m × 0.05 m, B5 =1.10 m × 0.05 m, B6 = 0.30 m × 0.05 m; (C) location of an eyebolt for attaching an anchor rope; (D) clear plastic sheets (6mm) for the rain shield; (E) openings for escape of marked adults (the lower opening can be sealed with plastic to prevent wind damage to the styrofoam chips impregnated with fluorescent dust); (F) metal grid with diamond-shaped openings (2 cm × 1 cm) above the water surface for support of styrofoam chips

was large variability in emergence among different traps on different days and assessment over a period of more than three days would be desirable to confirm the preliminary results. Material costs of trap construction were estimated at less than US$ 50.

Trace elements

Marking methods generally suffer from either one or two disadvantages. Although individual labelling with paints imparts most information and allows multiple markings, only comparatively few mosquitoes can be marked. The technique is not suitable for the mass marking and daily releasing of large numbers of individuals. In contrast methods that allow the marking of large numbers of mosquitoes, e.g. dusting with powders, are severely limited by the number of clearly distinguishable marks that can be used. Most optimistically this is below 10 (McClelland et al. 1973a), and usually much lower except where a fluorescent microscope is used to distinguish different fluorescent marks (Bennett et al. 1981). To overcome these difficulties McClelland et al. (1973a) incorporated trace elements into visible markers such as powders. In this method mosquitoes are dusted with a suitable coloured powder such as zinc sulphide 'Helecon' dust, to which an appropriate trace element is added, in an atomic ratio of about 1:100 as this is the minimum ratio for reliable detection. Marked recaptures are recognised by the 'Helecon' pigments and taped to sheets of 0.25-mm 'Mylar' film. They are then mounted in a vacuum chamber, placed in a cyclotron and bombarded with a focused beam of alpha particles. All atoms hit by these particles emit X-rays, and each element has its own characteristic spectrum with peak emissions at certain energies. Possible trace compounds include zirconium oxide, bismuth, lead, cerium oxide, tin, selenium, silver oxide etc..

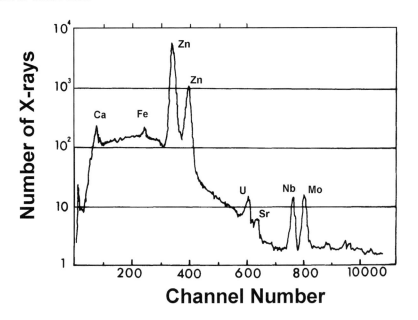

Fig. 14.7. X-ray fluorescent spectrum of mosquitoes dusted with zinc sulphide powders and trace quantities of uranium, niobium and molybdenum (after McClelland et al. 1973a)

Lighter, naturally occurring elements such as hydrogen, carbon, oxygen and calcium, and also the zinc in the 'Helecon' powders, generate low power energy X-rays, but the higher part of the spectrum where peaking of added trace elements occurs is more or less free of peaks, except those of the added trace elements (Fig. 14.7)

In a field trial in California, USA adults of *Ochlerotatus nigromaculis* were dusted with a mixture of 50 parts of a 'Helecon' fluorescent powder containing 1 part of a combination of three of the following non-radioactive trace elements; zirconium oxide, uranium oxide, molybdenum (elemental), niobium oxide, lanthanum carbonate and cerium oxide (McClelland et al. 1973*b*). Recaptured marked mosquitoes were fixed on to 'Mylar' film with an aerosol spray of 'Krylon' clear plastic. Mosquitoes prepared in this way can be kept indefinitely before they are analysed. Although reasonable numbers of marked mosquitoes were recaptured (735) the cyclotron identified only 38. Considerably more dust was lost under field conditions than was expected from preliminary trials, and the amounts of trace elements on the mosquitoes were therefore often reduced below their level of detection, which is about 10^{-7} g. It should be possible to compensate for this loss by using greater quantities of trace elements. An unexpected difficulty was contamination of mosquitoes with lead, which produced such strong X-rays that some of the weaker emissions from trace elements were masked. It was thought that this was caused from lead incorporated in the solder of the aerosol can.

Natural trace elements

Trace elements from the environment accumulate into the tissues of animals and the particular combination of elements present in a specific locality, known as its chemical finger-print or chemoprint, can be used to identify the location of origin of the animal. Mosquito larvae for example take up elements characteristic of their habitat and this natural tagging persists in adults. It has been suggested that the chemical composition of insects (chemoprint) might be used to identify their breeding sites, in other words their place of origin (Bowden et al. 1979; Turncock et al. 1979), and so might serve as a population marker in dispersal studies. Dempster et al. (1986) describe how using this approach, and multivariate analyses of the data using principal component analysis and canonical variate analysis, differences in chemical composition in the Brimstone butterfly (*Gonepteryx rhamni*) from different areas of Britain appear to reflect differences in chemical composition of local soil and food plants. These differences, however, soon became masked with ageing of the adults, and their feeding.

Bowden et al. (1984) present an informative paper on the benefits and limitations of using wavelength dispersive X-ray fluorescence spectrometry to chemoprint insects. They quantitatively studied the chemical composition of a noctuid moth reared on plants grown in different types of soil. They found chemoprinting could sometimes be useful in identifying the breeding source of their insect, but stressed there may be problems with environmental variability in elemental composition.

This method does not appear to have been applied to adult mosquitoes, although it has been applied to simuliid blackflies but with only limited success (Walsh 1977). A potential difficulty is that the chemical composition of larval habitats may change over time thus making it difficult to characterise specific breeding places. It might therefore be better to introduce non-toxic exotic rare elements into the water so as to more specifically mark adults. However, a limitation of this method is that expensive laboratory equipment is needed, such as an X-ray spectrometer.

Rubidium. Agricultural entomologists were among the first to use rubidium to mark adults of several pest species, including moths, beetles, hemipteran bugs, and anthomyid flies. Marking was performed by feeding larvae with diets containing rubidium, or more commonly by spraying food plants with rubidium chloride (Berry et al. 1972; Fleischer et al. 1986; McLean et al. 1979; Pearson et al. 1989; Shepard and Waddill 1976; van Steenwyk et al. 1978). Both Legg and Chiang (1984) and Hayes and Reed (1989) give a number of useful references to the use of rubidium and other trace elements for marking agricultural insects. Marking usually persists for several days, a month in some instances or even longer (Padgham et al. 1984). Rubidium marking of adults has frequently been used to study mosquito dispersal (Honório et al., 2003; Reiter 1996; Reiter et al. 1995).

There are usually no toxic effects associated with rubidium marking, in fact rubidium is biochemically interchangeable with sodium and potassium. Holbrook et al. (1991) reared *Culicoides variipennis* in media having eight concentrations of rubidium chloride (7.8–1000 ppm) but the highest concentrations (500, 1000 ppm) reduced pupal production, adult emergence and adult longevity, whereas concentrations of 250 ppm and below had no noticeable adverse effects. Adults reared from media containing just 15.6 ppm could be detected for at least 13 days after emergence when rubidium levels were measured using an atomic absorption spectrophotometer.

The effects on adult emergence, weight and survival of marking immature *Anopheles stephensi* by rearing them in water containing rubidium at concentrations from 1000 to 7.8 ppm were investigated in the laboratory by Solberg et al. (1999). The amount of Rb detected in mosquitoes increased with increasing concentrations in the rearing water. Concentration more than or equal to 31.2 ppm Rb in the rearing water provided high and consistent detection levels of more than or equal to 3500 ppm Rb/mg of adult mosquito. Concentrations of ≥31.2 ppm were reliably detectable in adults, and were associated successful emergence of 50% or more and a reduction in adult female survival of 29% or less compared with controls. There were no detectable effects of Rb on the weights of adult mosquitoes. It was concluded that 32 ppm was the recommended concentration to maximise detection while minimising mortality.

Rubidium marking of nymphs of *Dermacentor variabilis* ticks (Acari: Ixodidae) by feeding them on mice injected with RbCl had no effect on duration of larval feeding, or on the number of days between detachment and moulting of larvae. Survival of rubidium-marked nymphs was not significantly lower than for un-marked nymphs. Interestingly, the rubidium mark was not transferred between nymph and adult stages (Burg 1994), and in later studies Burg (1997) was unable to infect adult *Dermacentor variabilis* by feeding them on mice injected with RbCl.

In preliminary studies Kimsey and Kimsey (1984) showed that rubidium could be injected into vertebrate hosts to mark blood-fed mosquitoes (see Chapter 8). Rubidium marking by feeding on marked hosts was shown to be feasible in the ixodid tick *Dermacentor variabilis* by Burg (1994).

Holbrook (1995) examined the uptake and persistence of Rubidium by biting midges *Cuilcoides variipennis sonorensis* feeding on a rabbit that had been injected with 4.1 ml of a sterile phosphate-buffered saline solution containing 2.05 g Rb^+ equivalent to a dosage of 500 mg Rb^+ per kg body weight. Adult *Culicoides* females aged 48–72 h from a laboratory colony were placed in 0.24 litre cardboard containers having a fine mesh nylon screen and were exposed to a shaved area of the rabbit's skin for a period of 30 min. *Culicoides* females were fed on days 0, 1, 4, 7 and 14 post-injection. Rb^+ content of batches of 10 females, 100 eggs, and 1.5 μl of rabbit blood were determined by atomic absorption spectrophotometry. Rubidium concentration in rabbit blood samples increased rapidly to a peak at 14 days, after which the levels declined but remained higher than negative controls after 30 days. No notable adverse effects were noted in the test rabbits. There were significant differences in Rb levels among female *Culicoides* that were recently engorged and those maintained for 3 (gravid), 7 and 14 days. 100% of females were marked at engorgement, declining to only 13 of 60 at 14 d post-feeding. All egg batches were marked.

Feeding mosquitoes on labelled hosts is generally more time consuming than other marking methods, and detection of marked flies usually necessitates the collection of blood-fed individuals, which may not be easy to find. The method might, however, be useful if labelled animals are placed in natural habitats to act as normal hosts.

Adult female *Aedes aegypti*, and ultimately their eggs can be marked by feeding them on a blood-meal containing rubidium (Reiter 1996; Reiter et al. 1995). The method used was to feed nulliparous females reared from wild-collected eggs on fresh citrated pig's blood containing a 0.025 M solution of rubidium chloride (RbCl). Honório et al. (2003) used the methods of Reiter et al. (1995) to mark adult *Aedes aegypti* and *Aedes albopictus* to study dispersal in a dengue endemic area of Brazil (see Chapter 15 for details)

Moss and van Steenwyk (1982) marked pink boll worms with another rare earth, caesium, while Anderson et al. (1990) were the first to mark haematophagous insects with caesium, and also the first to simultaneously use two alkali metals as host blood-markers to study mosquito feeding behaviour. Young chickens were labelled by injecting them with 500 mg/kg caesium chloride or rubidium chloride. Blood-engorged mosquitoes that had fed on these hosts were placed in 200 μl of a 50:50 solution of 72% perchloric acid and 70% nitric acid diluted with potassium chloride, to inhibit ionisation of rubidium and caesium, to a final volume of 3.0 ml and a final concentration of 200 ppm KCl. The mosquitoes were kept in this solution at room temperature for 72 h to allow unhydrolysed lipids to coalesce. The clear solutions were then assayed for rubidium and caesium by atomic emission flame spectrophotometry at 780 and 852 nm respectively, and distinguished by their different wavelengths. Mosquitoes were marked with rubidium up to 3 days post-feeding and up to 2 days with caesium; higher injection rates into the chickens extended the detection periods. A similar approach was adopted by Anderson and Brust (1995) in Manitoba, Canada, to investigate the frequency of feeding on multiple hosts by the mosquitoes *Culex tarsalis*, *Culex restuans* and *Culex nigripalpus*, vectors of encephalitis viruses in North America. Blood-fed mosquitoes were collected from box traps, each baited with a pair of quail (*Coturnix japonica*). One quail of each pair was injected with rubidium and the other with caesium in order to allow for differentiation of blood meal sources. The frequency of multiple feeding ranged from 0 to 18.5% for *Culex tarsalis*, 0 to 33.3% for *Culex restuans*, and 0 to 17.6% for *Culex nigripalpus*.

Xue et al. (1995) used rubidium and caesium marking of restrained chickens to investigate incidence of feeding on multiple hosts by *Aedes aegypti* in the laboratory. These authors also studied the effect of mosquito age and body size on frequency of feeding from multiple hosts and determined that larger and older females tended to feed on more than one host more frequently than younger or smaller females.

Hodgson et al. (2001) studied multiple blood-feeding in *Culiseta melanura* by marking robins (*Turdus migratorius*) and European starlings (*Sturnus vulgaris*) with either rubidium or caesium respectively and keeping them in cages in the field overnight. Analysis of blood meals revealed that *Culiseta melanura* appeared to feed preferentially on robins rather than starlings and this was ascribed to differences in defensive behaviour between the two species of bird, however, this was not actually evaluated during the investigation.

Radioactive materials

Radioactive chemicals are now very rarely used to mark mosquito larvae or adults and will therefore not be considered further here. Readers interested in these methods are advised to consult the previous editions of this book (Service 1976; 1993) or the book by Knoche (1991).

Proteins

Zhou et al. (2003) used the method of Hagler & Jackson (1998) to mark aster leafhoppers *Macrosteles quadrilineatus* with rabbit protein in a mark-recapture experiment to determine dispersal. The marking procedure consisted of placing a black cloth over the field collection cage, leaving one end uncovered and oriented toward the sun. Leafhoppers moved towards the light and congregated on the screen, where they were then misted using a compressed air atomiser to produce a fine spray of protein solution, containing 1.0 mg/ml rabbit IgG in distilled water.

Laser etching

An interesting technique to rapidly mark individual carabid beetles with a unique identity number using a laser was recently described by Griffiths et al. (2001). A Synrad Fenix Laser Marker (Synrad Inc., 6500 Harbour Heights Parkway, Mukilteo, WA 98275, USA), commonly used within the engineering sector, was used to mark the elytra of *Pterostichus melanarius*, a carabid beetle with a body length of approximately 12 mm. A 25-W CO_2 laser and a galvo-based marking head set at 20% power with a 370-mm lens and a speed of 380 mm s^{-1} controlled by WinMark 2.0™ software was used. Laser etching with these parameters did not puncture the elytron of the beetles. Beetles were chilled to 4°C and held in a 150 × 150 mm holding plate with sixty 5 mm deep wells. Beetles were held in position against a wire grid using suction. The procedure allowed for the etching of a 3-digit code with a text height of 1.5 mm on the elytra, and this was readable by naked eye or hand lens. Once mounted on the plate 20 beetles could be marked per second. No effect on mortality was observed over a period of 4 weeks and the marks remained readable for at least 3 months on live beetles. The cost of the equipment used was approximately GBP 15,000. Whether this method could be adapted for mosquitoes and other flies with less heavily sclerotised exoskeletons remains to be seen.

Phenotypic mutants

The natural or artificial occurrence of mutant individuals can occasionally be used as a means of identification in studies on dispersal and population size (Peer 1957; Richards and Waloff 1954; Service 1971). Fay and Craig (1969) used two genetic marker strains, spot abdomen and silver mesonotum of *Aedes aegypti* (Craig and Hickey 1967) to study dispersal. They were introduced as homozygotes into a natural population and dispersal studies were based on captures of mutant adults in the traps of Fay, which are described in Chapter 11, and adults reared from eggs collected in ovitraps. In later studies other marker strains of *Aedes aegypti*, such as bronze, black tarsi and black palp, were released into field populations (Bond et al. 1970; Hausermann et al. 1971). It was discovered that not all strains dispersed the same distances. This emphasises the great care that must be paid to the behaviour and longevity of mosquitoes that are marked either genetically, physically or chemically.

Population estimates based on mark-recapture methods

Two main groups of methods for population estimation from mark-recapture techniques have been developed: those for closed populations (where there is assumed to be no immigration or emigration) and those for open populations, where immigration, emigration, or both may occur. In addition, deterministic models have been developed that assume that the survival rate of an individual over an interval of time remains constant (e.g. Lincoln Index, Fisher and Ford's method; Leslie's method). While this is clearly an oversimplification, the attraction of the deterministic approach is that the calculations involved are relatively simple. Stochastic models, e.g. Jolly's (1965) method, make the more realistic assumption that an individual's survival over an interval is best expressed as a probability. When there are a relatively large number of sampling occasions and the recapture rate is relatively high then the stochastic model frequently gives the better population estimate, but if there are few sampling occasions and few recaptures then the deterministic method *may* give the best results. The stochastic method usually gives a more reliable estimate of the variance (Parr 1965). However, it should be remembered that the *estimate* of the variance is influenced by the *estimate* of the population size, consequently if the population estimate is small then the estimated variance tends to be smaller than it should be. The choice between using the various methods for estimating population size depends on the level of accuracy required, the accuracy of estimating survival rates, the feasibility of marking and recapturing individuals on a large number of occasions and not least of all the mathematical inclination of the investigator.

For reliable population estimates to be made from mark-recapture experiments, a number of criteria must be satisfied (Begon 1979; Blower et al. 1981; Jolly 1965; Southwood and Henderson 2000).

(1) the mark should not affect the survival rate of individuals in between sampling dates; (2) it should not cause abnormal behaviour, e.g. altering the chances of it being caught; (3) the mark should be retained; (4) marked individuals must completely mix in the populations before being recaptured; (5) there must be equal chances of recapture of marked and unmarked individuals, and (6) sampling must take place at discrete time intervals and sampling time must be relatively short in relation to the total length of the study. Statistical tests are available to ascertain whether or not these criteria are met.

Effect of marking on survival

Survival of individuals immediately after marking and release can be compared with survival of animals that have been marked for longer periods and can be tested using a method devised by Manly (1971*a*). This method tests only for the effect of the marking process on survival, not the longer term effects of carrying a mark. The difference in survival y between recently marked animals and those marked some time ago is given by:

$$y = \ln\left(\frac{r_i'}{r_i} \times \frac{u_i}{R_i}\right) \qquad (14.2)$$

where r_i' = the number of marked animals in the ith sample that are recaptured again later, r_i = total marked individuals in ith sample, u_i = total unmarked animals in the ith sample, and R_i the total number of animals marked on the ith occasion and subsequently recaptured. If the process of marking has no effect on survival, then y should equal zero. Values of y can be tested for significance by dividing the value of y by its variance to give the statistic g, which behaves as a random normal variate with mean and variance equal to zero. A one-tail test is used as marking is unlikely to increase survival.

$$g = \frac{y}{(\frac{1}{r'_i} + \frac{1}{R_i} - \frac{1}{r_i} - \frac{1}{u_i})^{\frac{1}{2}}} \qquad (14.3)$$

The effects of marking adults with powders or stains, or any other substance, should always be carefully evaluated by comparing mortalities of marked and unmarked mosquitoes of the same species, sex, and if possible same age, over several days or weeks whichever time interval is the more appropriate. It appears from the literature that in many experiments the effects of marking were not evaluated, or inadequately so. In laboratory experiments Reisen et al. (1979) found no differences in survival rates between unmarked males and females of *Anopheles subpictus* and those marked with coloured dusts. Nelson and Milby (1980) found that marking male and female *Culex tarsalis* with Radiant blue dust which were then released into laboratory cages, large out of door cages, and into the field had negligible effect on their survival rates, or insemination rates of females.

Prior to conducting a mark-release-recapture study to estimate survival and dispersal of *Aedes aegypti* in Australia, Muir and Kay (1998) investigated the effect of marking two- or three-day-old colony-reared males and females with powdered Dayglo fluorescent pigment. Approximately 300 males and 300 females were aspirated into two 215-mm diameter waxed containers with gauze tops and bottoms and fluorescent powder was applied using an insufflator constructed from a 1-ml syringe with a 22-gauge needle attached. Controls consisted of groups of 20 males and females maintained in similar waxed containers. Results of the Wilcoxon test of the Kaplan-Meier survival functions for males at 7 days and females at 7 and 42 days revealed significantly higher survival amongst marked mosquitoes compared with unmarked. The mortality rates of female and male controls and marked groups increased with age.

In Sri Lanka to test whether marking had any effect on the survival of *Anopheles culicifacies* Curtis and Rawlings (1980) marked batches of mosquitoes with two coloured dusts (magenta and yellow) while others were left unmarked. Mosquitoes were placed in 30-cm cube cages and provided with blood-meals and water for oviposition, dead individuals were removed daily. The mean ages at death were 18.6, 17.4 and 17.0 days for the magenta, yellow and unmarked adults, and it was concluded that marking had no obvious adverse effect on survival rates. In laboratory experiments Sempala (1981) checked whether etherisation and marking with 'Dayglo' powders or 'Duco' paints affected survival rates or flight activity of *Aedes africanus*. He found that etherisation by itself, and marking unetherised adults with coloured dusts caused little mortality, but females marked with dusts and then etherised and marked with paints suffered a quite drastic mortality—about 30% by day 8 and 93.3% by day 20, whereas control mortality was only 8.7% by day 20.

In laboratory experiments Meek et al. (1988) showed that there was <3% mortality of *Culex quinquefasciatus* and *Psorophora columbiae* bearing a minimum of eight pigment spots; and in aerial spraying of 'Dayglo' water soluble pigments no wild caught mosquitoes had more than five spots of pigment.

In assessing the effects of fluorescent powders on mosquito longevity Chang (1946) marked *Anopheles quadrimaculatus* with either 1 part anthracene mixed with 6 parts of gum arabic, or 1 part rhodamine B or fluorescein and 4 parts of gum arabic. All compounds caused some mortality. For example 80–92% mortality in the third week after marking compared with 38% in unmarked females, and 92–98% mortality in the fourth week compared with 52% in the controls. In contrast no mortality occurred up to six weeks after hibernating *Culex pipiens* were dusted with rhodamine B or with neon red, fire orange, and saturn yellow 'Dayglo' pigments. Moreover all dusts were readily seen on the mosquitoes when examined under a binocular microscope without the addition of any solvents (Service, unpublished). Similarly marking *Ochlerotatus cantans* with many different colours of fluorescent powders has not caused mortality, and marked adults have been recaptured up to 72 days after marking (Renshaw 1991; Renshaw et al. 1994). Nutsathapana et al. (1986) found that much more uniform marking and less mortality was achieved by immobilising *Anopheles minimus* adults with ether and dropping them onto a sheet of paper covered with fluorescent powders (from which excess dust had been shaken off) and gently shaking the paper, than the more usual method of marking mosquitoes in a dust storm.

Marking with paints is usually believed to cause little mortality, but it is likely that applying paint spots to mosquito wings does affect behaviour and possibly survival (Frank and Curtis 1977). Even if paint spots are confined to the scutum there should nevertheless be some check as to whether this causes any mortality.

Even if there is no significant difference between mortalities this does not exclude the possibility that marking might cause atypical behaviour, and thus affect chances of recapture. For example, although Ginsberg (1986) found that there was no significant mortality between unmarked

Ochlerotatus sollicitans and those marked with fluorescent powders, he cautioned that rearing adults for marking and subsequent handling may affect their behaviour and dispersal. Also that capturing adults in traps, for example in emergence traps in which the newly emerged adults are denied nectar feeds before their marking and release, may affect their subsequent behaviour. Some workers believe that creating mini dust storms of pigments in cages covers the wings, legs and bodies of mosquitoes with excessive amounts of dusts (Dunn and Mechalas 1963; Trpis 1971), and that this may affect their flight performance and other behaviours (Chang 1946; Shapiro et al. 1944; Sheppard et al. 1969). Renshaw et al. (1994) found no significant difference in survival rates between unmarked *Ochlerotatus cantans* females and those marked with fluorescent powders when maintained in cages in the field over a period of three weeks. The numbers of females that took a blood-meal from a mouse in the laboratory did not differ when females received 'light' or 'medium' marking with fluorescent powder, however, heavily marked females spent a considerable time grooming and the number taking a blood meal was significantly reduced, particularly on the day of marking ($\chi^2 = 15.45$, $P < 0.001$).

Recent investigations on the dispersal of aster leafhoppers *Macrosteles quadrilineatus* by Zhou et al. (2003) found that marking with fluorescent dusts (DayGlo Rocket red) or rabbit protein (Sigma Chemical Co., St. Louis, MT, USA, No. I5006) did not adversely affect leafhopper flight activity in terms of proportion flying, distance flown, average flight speed, or wingbeat frequencies, as determined in the laboratory by use of flight mills and optical sensors.

Retention of marks

If each individual is marked with two different marks, then the total number of recaptures, r, is given by:

$$r = r_a + r_b + r_{ab} \qquad (14.4)$$

where r_a = number of recaptures with type a mark alone, r_b = number of recaptures having type b mark alone, and r_{ab} = number of captures bearing both marks. If marks are maintained without loss, then the total expected number of marked recaptures is given by:

$$\hat{r} = c(r_a + r_b + r_{ab}) \qquad (14.5)$$

where

$$c = \frac{1}{1-k} \tag{14.6}$$

and

$$k = \frac{r_a r_b}{(r_a + r_{ab})(r_b + r_{ab})} \tag{14.7}$$

The estimate $\hat{r}$ is compared with r above to determine the extent of mark loss.

Equal catchability

Probably the most important conditions concerning the use of these methods on adult mosquitoes is that the markings will not alter their behaviour, and when a series of daily population estimates are made such as in the Fisher and Ford method then marking must not affect survival rates. These conditions should always be checked in preliminary trials. Apart from directly affecting the survival rate coloured paints and powders might conceivably make the mosquito more susceptible to predation.

Closed Populations (no immigration or emigration)

Deterministic models

Petersen-Lincoln Index. Lincoln (1930) is usually credited with developing the method for estimating total population size in which a known number of individuals are marked and returned to the parent population, prior to resampling, and hence the method is most frequently referred to as the Lincoln index. However, Petersen (1896) was the first to use the method to estimate the population size of a fish, *Platichthys platessa*, and as a result some prefer to call it the Petersen index (Begon 1979) or the Petersen-Lincoln index (Southwood and Henderson 2000). The principle behind the method is that a known number of individuals are caught, marked and then returned to the population. After a suitable interval to allow mixing a second population sample is taken and the numbers of marked and unmarked individuals recorded. The ratio of the number of marked individuals in the second sample to the total number in the second sample (both marked and unmarked) will be the same as the ratio of marked individuals released from the original capture to the total population.

The simplest population estimates (P) are obtained from a single marking occasion and the use of the Lincoln Index to analyse the recaptures:

$$P = \frac{an}{r} \qquad (14.8)$$

where P = the estimate of the number of individuals in the population, a = total number marked in the first sample, n = total number of individuals in the second sample, and r = total recaptures. When n is predetermined and approximately equal to a, variance can be calculated as:

$$\text{var } P = \frac{a^2 n(n-r)}{r^3} \qquad (14.9)$$

In many field studies on mosquitoes, marked mosquitoes (a) are added to the natural population whose size is being estimated, and they therefore need to be subtracted from the calculations thus:

$$P = \frac{an}{r} - a \qquad (14.10)$$

This adjustment has sometimes been made by mosquito workers (Nelson et al. 1978; Reisen et al. 1979, 1980).

A less biased estimate has been proposed by Bailey (1952) for small samples ($r < 20$):

$$P = \frac{n(a+1)}{r+1} \qquad (14.11)$$

with an approximate estimate of variance given by:

$$\text{var } P = \frac{a^2 (n+1)(n-r)}{(r+1)^2 (r+2)} \qquad (14.12)$$

Inverse sampling, that is sampling where the number of marked recaptures (r), rather than the total number of individuals in the second sample (n) is predetermined, gives unbiased estimates of population size and allows for the calculation of an exact value of its variance (Bailey 1952):

$$\text{var } P = \frac{(a-r+1)(a+1)n(n-r)}{r^2 (r+1)} \qquad (14.13)$$

When sampling has not been undertaken immediately after release of marked mosquitoes the population remains open and this may result in the loss of some marked individuals from the area (death and/or emigration). So that the numbers of marked mosquitoes available for recapture at say the midpoint in the sampling programme is not a but ax, that is the numbers released have to be multiplied by the proportion still available for capture, and then the population estimate is given as

$$p = \frac{axn}{r} - ax \qquad (14.14)$$

If the overall rate of loss of marked mosquitoes is constant and there is no decrease or increase in the wild population (unmarked individuals), and if sampling removes some marked mosquitoes from the population then x can be calculated as follows (Fletcher et al. 1981).

$$x = \left[\frac{r_2 + (r_2 \times r_1 / a - r_1)}{r_1} \times \frac{n_1 - r_1}{n_2 - r_2} \right]^{(t_1 / t + t_2)} \qquad (14.15)$$

where a = numbers marked mosquitoes released, n_1 = total catch of mosquitoes during first sampling period, n_2 = total catch of mosquitoes during second sampling period, r_1 = numbers of marked mosquitoes recaptured during first sampling period, r_2 = numbers marked mosquitoes recaptured during second sampling period, t_1 = length in days of first sampling period and t_2 length in days of second sampling occasion.

So long as the sampling effort remains constant the assumption that the size of the natural population remains unchanged during the two sampling periods can be checked as follows

$$\frac{n_1 - r_1}{t_1} \cong \frac{n_2 - r_2}{t_2} \qquad (14.16)$$

If there is a significant decrease in population size during the two sampling intervals then t_1/t_2 should be substituted for $(n_i - r_1)/(n_2 - r_2)$, similarly if there is an increase in the wild population the above substitution can still be used, but the population estimate (P) will have a positive bias because n in formula (14.14) will be too large. For further details see Fletcher et al. (1981).

The other assumption of the model, that is a constant rate of decline of marked mosquitoes, can be easily checked by plotting numbers recaptured on three or more successive occasions.

Weighted mean Lincoln Index. In this method data collected over several days are utilised. For example, on any day individuals are caught of which r_i have already been marked on a previous occasion, the unmarked ones are now marked and a_i marked individuals are released back into the population that day. Therefore clearly $(a_i - r_i)$ additional marked individuals are released each day, and the numbers of marked individuals increases each day—so long as we assume there are no losses from the population. Now, we have to introduce the concept of 'marks at risk', (M_i): this is the number of marks in this population that are available for sampling immediately before the sample is taken on day i. This will depend on the numbers of marked individuals that are lost (death, emigration) and the numbers of marked individuals that may be added each day. Hence the numbers of marked individuals 'at risk' on day i – M_i is the number of marked ones released on day (a_i) plus the numbers of all other individuals that have been marked between days 1 and i, in other words that is $(a_2 - r_2) + (a_3 - r_3)... + (a_{i-1} - r_{i-1})$. The population size on any day (P_i) according to the Lincoln Index is

$$P_i = \frac{M_i n_i}{r_i} \tag{14.17}$$

We will have a series of daily estimates of population size, but clearly those based on days when the numbers of recaptured marked individuals is large will be more realistic in estimating what the actual population size is than daily estimates based on low recapture numbers. So it is appropriate to calculate a weighted mean of P_i values, which if we denote the weighted attribute to P_i as w_i we have

$$P = \frac{\sum N_i w_i}{\sum w_i} \tag{14.18}$$

An estimate of P based on a sample containing very few marked recaptures is likely to be greatly influenced by chance effects on the actual size of r_i. On the other hand samples having a large number of r_i values will generally produce more accurate estimates of P. So, it becomes appropriate to use r_i values as the weights (w_i)

$$P = \frac{\sum N_i r_i}{\sum r_i} \tag{14.19}$$

The formula above to estimate P can be reduced, and a virtually unbiased estimate obtained

$$P = \frac{\sum M_i n_i}{(\sum r_i) + 1} \tag{14.20}$$

The standard error is given by

$$SE = P\sqrt{\frac{1}{\sum r_i + 1} + \frac{2}{(\sum r_i + 1)^2} + \frac{6}{(\sum r_i + 1)^3}} \tag{14.21}$$

A more explicit description of using a weighted mean with the Lincoln Index, together with a worked example, is given by Begon (1979).

Skalski's relative abundance methods. Mark-recapture techniques are not only used to estimate absolute population size, but sometimes to measure proportional abundance, such as comparison of population levels in time or place, or to record changes in levels. For instance the ratio (K) of mosquito abundance in one population (N_2) to that of another (N_1) can be obtained as $K = N_2/N_1$. Skalski et al. (1983) presented two alternative methods for obtaining confidence intervals of estimated K values for paired populations obtained by single mark–recapture methods. They also gave methods of analyses that test whether there is equal catchability among the animals in the two populations. They pointed out that no single method of estimating population size was best suited for all situations and that estimates should be obtained by at least two independent methods. Later Skalski et al. (1984) extended the method for use with removal sampling (see below).

A modified Lincoln Index. Gaskell and George (1972) considered that if the Lincoln Index was applied to data when the number of marked individuals recaptured (r) was small, then there were wide intervals between the possible estimates of population size (P) for fixed numbers of individuals originally marked (a) and the size of the sample (n). They concluded that the Lincoln Index was a poor estimator of population size when $r < 10$, and of course could not be applied when $r = 0$. They further pointed out that the modification proposed by Bailey (1952) of adding 1 to both n and r to reduce bias was of limited use because this method really required that a large number of population estimates were made and the average values used, whereas in practice only one estimate is often obtained. They devised a Bayesian modification of the Lincoln Index and derived the following equation:

$$P = \frac{(an + AB)}{r + B} \tag{14.22}$$

where a, n and r are as defined in the Lincoln Index, B = a mathematical variable, and A = a prior guess by the investigator of the probable population size. It was argued that although at first sight this might appear an unreasonable demand, that if pressed most workers would admit to some idea of the size of the population to be measured. In fact previous population estimates might provide information on this. Gaskell and George (1972) concluded that when r is small, B has a value of 2–4 and if there is a good prior estimate of the population size this information will offset the relative error of the Lincoln Index. They suggested that in most cases when r is relatively small B can be taken as 2, but for larger values of r then larger values of B might be used. Graphs were presented to test the efficiency of the method. This method can be criticised on the feasibility of being able to guess population size (A) prior to obtaining any experimental data.

Schnabel censuses. These maximum-likelihood methods for closed populations were developed by Schnabel (1938) and Darroch (1958) and may be applied where there are two or more periods of capture (s)

$$(1 - \frac{M}{N}) = \prod_{i=1}^{s} (1 - \frac{n_i}{N}) \tag{14.23}$$

where M is the total number of different animals captured and n_i is the number of animals captured in sample i. For $s = 2$, the above equation simplifies to the Petersen-Lincoln index. The method can be simplified by carrying out multiple samples and estimating P for each sampling occasion i using:

$$\hat{P}_i = \frac{(M_i + 1)(n_i + 1)}{r_i + 1} - 1 \tag{14.24}$$

Where $M_i = \sum n_i - r_i$ The variance v_i^* of this estimate is

$$v_i^* = \frac{(M_i + 1)(n_i + 1)(M_i - r_i)(n_i - r_i)}{(r_i + 1)^2 (r_i + 2)} \tag{14.25}$$

The average population size is then calculated using:

$$\overline{P} = \frac{\sum \hat{P}_i}{(s-1)} \tag{14.26}$$

Open populations

Deterministic methods

Jackson's methods. In Jackson's pioneering work on tsetse populations he introduced two new concepts, the 'positive' and 'negative' methods (Jackson 1939). In the positive method the loss ratio is calculated for a specific release day across all release groups. In Jackson's positive method a large number of individuals are caught and marked on a single occasion and are subsequently recaptured on a series of occasions. Marked individuals are recorded at each recapture and all, or most of them are then released to avoid reducing the numbers of marked individuals available for future recapture. This gives information on the changes in the proportions of marked to unmarked individuals on each sampling occasion. Extrapolation of mark–recapture data gives an estimate of the number of recaptures that would have been obtained if sampling had occurred directly after marking, assuming of course that marked individuals had mixed with the unmarked population. Begon (1979) provides a readable account of Jackson's (1939) positive method.

The *proportion* of the population that is marked (q_i) on day i is estimated as

$$q_i = \frac{m_i}{n_i} \tag{14.27}$$

where m_i = marked individuals on day i; n_i = numbers of individuals caught on day i.

We wish to estimate the population size (N_0) on day 0, and this can be obtained by firstly estimating the marked population (q_0) in a hypothetical random sample taken on day 0.

$$q_0 = \frac{r_0}{N_0} \tag{14.28}$$

where r_0 = numbers of individuals marked and released on day 0 So,

$$\hat{N}_0 = \frac{r_0}{q_0} \tag{14.29}$$

$$q_i = q_0(1-b)^i \text{ or } \ln q_i = i(\ln(1-b) + \ln q_0) \tag{14.30}$$

where b = birth rate per day.

This is the regression of $\ln q_i$ on i and the two regression parameters $\ln(1-b)$ and $\ln q_n$ can be calculated as follows

$$\ln(1-b) = \frac{\sum m_i(\ln q_i - \overline{\ln q})(i - \bar{i})}{\sum m_i(i - \bar{i})^2} \tag{14.31}$$

$$\ln q_0 = \overline{\ln q} - \ln(1-b)\bar{i} \tag{14.32}$$

and therefore estimates of b and q_0 and thus $N_0 = (r_0/q_0)$ can be calculated. Eyles and Cox (1943) applied this method to estimate populations of *Anopheles quadrimaculatus* in the USA, and Yasuno and Rajagopalan (1977) used it to estimate populations of *Culex quinquefasciatus*, while Reisen and his colleagues used a modified form of Jackson's method with mosquitoes in Pakistan.

Marked and unmarked individuals should theoretically have a common loss rate, due to mortality or emigration. Any difference between the proportional recovery of marked individuals from two successive releases on different dates on a subsequent recapture occasion, should indicate the degree of change that occurred in the first marked group of individuals during the period between the two markings and releases. Working on this principle Jackson (1939) devised his 'negative' method, which estimates the loss ratio from the subsequent daily percentage losses for two release groups. In this method a series of catches are made in which individuals are marked and released, but only on the last catch are the recaptures of marked individuals and the number of marks on them recorded. An advantage of this method is that careful classification of individuals according to marks is only required on the final sampling occasion.

Richards and Waloff (1954) and MacLeod (1958) give worked examples of Jackson's methods. Reuben et al. (1973) used both the 'positive' and 'negative' methods to estimate the population size of *Aedes aegypti* in New Delhi, and to compare the estimates with those obtained by other methods.

Itô (1973) and Hamada (1976) made some modifications to the compution of population estimates by Jackson's method and applied them to estimate population size from just one recapture of marked individuals, this approach was used by Reisen and Aslamkhan (1979), Reisen et al. (1979, 1980, 1982a) to estimate population size of several *Anopheles* species in Pakistan.

The standard errors are calculated from weighted regressions, thus

$$SE_{\ln(1-b)} = \sqrt{\frac{\sum m_i \{\ln q_i - \overline{\ln q} - \ln(1-b)(i-\bar{i})^2\}}{(n-2)\sum m_i (i-\bar{i})^2}} \qquad (14.33)$$

$$SE_{\ln q_0} = \sqrt{\frac{\sum m_i (\ln q_i - \overline{\ln q} + \bar{i}\ln(1-b))^2}{(n-1)\sum m_i}} \qquad (14.34)$$

Antilogs have to be found.

A problem with mark–recapture methods for estimating population size and rates of immigration and emigration is that it is sometimes crucial to separate loss of marked individuals into death and emigration. For example, in determining the probability of survival in the Jolly–Seber method (Jolly 1965; Seber 1965) from one sampling occasion to the next, it is more accurate to define it as the probability not only of remaining alive but of staying within the area for recapture. As Manly (1985) has pointed out if the probability of survival is say 0.8, but the probability of emigration is 0.3, then the true Jolly–Seber survival estimate is 0.8 (1 − 0.3) = 0.56. But the problem is, we rarely know the emigration rate, so are unable to adjust the estimated survival rate.

There have been several attempts to analyse mark–recapture data to allow for emigration, but many are rather complex. The simplest is that proposed by Jackson (1939), which is based on the principle that emigration has less effect in a large area than it does in a small area. Consequently differences between a 'large area' survival estimate and a 'small area' survival estimate can be used to determine an estimate corresponding to no emigration. Manly (1985) using simulation models concluded that Jackson's method, in general, works well and should be more widely used.

Let us suppose that mark–recapture sampling for mosquitoes is undertaken over a relatively large area which can be delineated as a square and that the probability of survival of an adult mosquito = ψs, and that in one of the four smaller square areas into which the large square can be divided, the probability of survival = $\psi 1$. Furthermore, let ϕ = the probability of a mosquito remaining alive per unit time, and ε = probability of it emigrating from the small square per unit time. Then the relationship between survival rates and emigration is

$$\psi s = \phi(1 - \in) \tag{14.35}$$

and

$$\psi 1 = \phi(1 - \in/2) \tag{14.36}$$

The emigration for the large square should be half for the smaller square because half of the immigrants from a small square will move into another small square and thus remain within the large square.

Solving these two equations for ϕ and ε provides

$$\phi = 2\psi 1 - \psi s \tag{14.37}$$

and

$$\in = 2(\psi 1 - \psi s)/(2\psi 1 - \psi s) \tag{14.38}$$

This simple method of adjusting for emigration rates is applicable so long as the amount of movement is not too high. For further details of this approach, and its application where an area cannot be conveniently divided into squares, see Manly (1985).

Fisher and Ford's method. This technique relies on a series of markings and recaptures and was developed by Dowdeswell et al. (1940) and slightly modified by Fisher and Ford (1947) and is frequently referred to as the 'trellis' method, because the mark–recapture data are tabulated as a trellis diagram. Essentially the process involves making a series of Petersen-Lincoln index estimates working backwards from recapture to release date. A constant survival rate that best fits the data must be determined by trial and error. Population size is given by:

$$\hat{N}_t = \frac{n_t a_i \Phi_{i-t}}{r_{ti}} \tag{14.39}$$

where n_t = total sample at time t, a_i = total marked animals released at time i, Φ_{i-t} = survival rate over the period i–t and r_{ti} = recaptures at time t of animals released at time i.

In a situation where any cohort of individuals is only available for recapture, such as mosquitoes attracted to bait only every two or three days, then a model incorporating this cyclical availability β and varying sampling intensity (p) can be developed:

$$\hat{N} = \frac{n_i(1-\Phi^x)}{p_i\Phi^y(1-\Phi)} \qquad (14.40)$$

where x = the period of the cycle of availability, and y = the period between birth and availability for capture. Conway et al. (1974) applied the above method to *Aedes aegypti*.

A simplified account of the Fisher and Ford method is provided by Dowdeswell (1967). Explicit explanations of its workings are given in a paper by Ford (1953), which unfortunately is not readily obtainable. For this reason, a comparatively full account of the method as given by Ford (1953) is presented here. The results concern the recapture data of the moth *Panaxia dominula* for 1941.

Dates on which insects are caught, marked and released are given horizontally along the top of a triangular trellis (Table 14.1) while the total daily captures are entered to the left at the ends of the rows running obliquely downwards, and the numbers released are entered along the right-hand side of the triangle. Recaptures are shown within the body of the triangle and the number of marks they have.

Table 14.1. Trellis diagram of the captures, releases and recaptures of the moth, *Panaxia dominula* in 1941 (Fisher and Ford 1947)

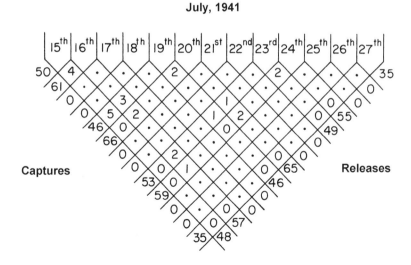

If no recaptures are caught 0 is inserted in the table. It is not necessary to sample the population every day. Consequently, if there is no sampling on a particular day, or no specimens could be caught because none was marked on the previous day, then a dot is entered in the appropriate segment of the table. The scoring is best understood by a few specific examples. From the table it is seen that 50 insects were captured on the 15th July, and that of these 48 were marked and released. On the next day 61 insects were caught, among which were four insects which had a single mark having been caught and released the previous day. The total insects marked and released on the 16th was 57. As a further explanation, consider the recapture on the 20th and 23rd of an individual marked and released on 16th, this will be entered once under the 20th, representing its capture on 16th, and twice on 23rd, representing capture on 16th and 20th. If we now consider the dates 20th and 24th, we see that on 20th 66 insects were caught and 65 were marked and released, and that on 24th 59 insects were caught of which 2 marked ones had been marked and released on the 20th. On the 24th 55 insects were released after marking.

The analysis of the results from this trellis table is best shown in Table 14.2. Column 1 contains in reversed chronological order the sampling days, while column 14 is dated day $n + 1$. The reasons for this are that individuals captured on any given day have reference only to releases on previous days, but those released at the same time have reference only to subsequent days. Entries prior to column 8 are dated relative to column 1, whereas those in columns 8–13 are dated in respect to column 14. In column 2 r is the interval from each day of release to July 27th, while column 3 contains the numbers of releases each day. The survival rate, expressed as a decimal, is entered in column 4 (in this example it is given as 0.84) and raised to successive powers of r. Daily recaptures are entered in column 8 dated relative to column 14; and listed in column 9 are the total number of days which these have survived.

For example, from the trellis diagram it is seen that on the 19th five individuals have survived 4 days, i.e. 20 days of survival, and three have survived 3 days, i.e. 9 days of survival. Consequently a total survival of 29 days is entered in column 9, corresponding to the 19th in the last column. The derivation of a mean survival rate estimated here as 0.84, or in other words a daily mortality of 16%, requires explanation. It is arrived at by reducing approximately to nothing the discrepancies between total days of observed survival (column 9) and the calculated expected values (column 10). This comparison is shown in column 11. Trials are made to determine what survival rate best fits the data.

Table 14.2. An analysis of the data in the trellis diagram (after Ford 1953)

C1	C2	C3	C4	C5	C6	C7
July	r.	Releases	% survival as decimal raised to power of r	C4 × C3	Entries in C5 summed from bottom	Entries in C6 summed from bottom
26	1	—	0.84	—	101.9755	584.0646
25	2	—	—	—	101.8755	482.1891
24	3	55	0.5296	32.5978	101.8755	380.3135
23	4	49	0.4978	24.3957	69.2768	278.4380
22	5	—	—	—	44.8811	209.1612
21	6	—	—	—	44.8811	164.2801
20	7	65	0.2951	19.1809	44.8811	119.3989
19	8	46	0.2479	11.4023	25.7003	74.5178
18	9	—	—	—	14.2980	48.8176
17	10	—	—	—	14.2980	34.5196
16	11	57	0.1469	8.3746	14.2980	20.2217
15	12	48	0.1234	5.9237	5.9237	5.9237

C8	C9	C10	C11	C12	C13	C14
Recaptures	Observed survival in days	Expected survival in days $= \dfrac{C8 \times C7}{C6}$	Excess of observed survival over expected	Captures	Estimated number $= \dfrac{C6 \times C12}{C8 \times C4}$	July
0	—	—	0	35	∞	27
—	—	—	—	—	—	26
—	—	—	—	—	—	25
5	18	20.09	−2.09	59	1379.5	24
4	21	18.64	2.36	53	1194.4	23
—	—	—	—	—	—	22
—	—	—	—	—	—	21
4	10	11.60	−1.60	66	1459.3	20
8	29	27.31	1.69	46	331.7	19
—	—	—	—	—	—	18
—	—	—	—	—	—	17
4	4	4	0	61	614.1	16
			0.36			

The estimated population on 5 days is given in column 13. Sheppard (Ford, 1953) showed how these estimates can be corrected to take into consideration both recruitment of new individuals and loss. Briefly the number of marked individuals surviving on the second day equals the number of marked releases on the first day multiplied by the percentage survival rate, expressed as a decimal. Now, the number of *marks*, not necessarily separate individuals, present on the third day equals that on the second day added to the number marked and released on the third day and then the total multiplied by the survival rate (Table 14.3). The average daily population size over the period in question is obtained by multiplying the number of captures by the numbers of marks present on that day. These products are added and divided by the number of marks recaptured over the sampling period, thus for the 5 days shown in Table 9.5 the average population size based on the 3 days for which data exist is 10 861.0 16 = 678.8.

Table 14.3. Sheppard's procedure for the estimation of the population size from data given in tables 14.1 and 14.2 (Ford 1953)

Date July	No. releases on previous day	Previous marks × survival day	Marks present per day	No. captures	Marks × captures	Recaptures
16	48	48×0.84	40.32	61	2459.5	4
17	57	(40.32+57)×0.84	81.7488	—	—	—
18	—	81.7488×0.84	68.6692	—	—	—
19	—	68.6692×0.84	57.8620	46	2653.4	8
20	46	(57.6820×46)×0.84	87.0929	66	5748.1	4
—	—	—	—	—	10861.0	16

In studying the population dynamics of *Aedes aegypti* in Bangkok Sheppard et al. (1969) used Fisher and Ford's method, and also a modified version which took into account an extra parameter, namely the proportion of live marked mosquitoes which are actually within the study area n days after their release. In other words a correction was made for the number of marked individuals that may have wandered out of the sampling area and not returned. The 'loss' of these marked mosquitoes affects the calculations of both survival rates and population size. Macdonald et al. (1968) also used the Fisher & Ford model to estimate the population size of *Culex quinquefasciatus* in a village in Myanmar. From a series of daily population estimates the average population of *Culex quinquefasciatus* in the village was estimated by the geometric mean of these separate estimates as 2650. Macdonald et al. (1968), however, considered that the harmonic mean might be more appropriate, and this gave an estimated population of 2142 adults.

Southwood and Henderson (2000) recommend the use of more modern methods for population estimation than those of Jackson (1939) and Fisher and Ford (1947), preferring the Jolly-Seber method for open populations. Where mortality is thought to be related to age, then these authors recommend the use of Manly and Parr's (1968) method, but only if 10% or more of the population can be sampled.

Bailey's triple catch method. Bailey (1951, 1952) introduced maximum-likelihood techniques into the analysis of mark-recapture data, facilitating the calculation of variances of estimates. The method and the theory on which it is based are fully described by Bailey (1951, 1952). Essentially, the idea is that the population is sampled on three distinct occasions. On the first occasion adults are caught, marked, and released, on the second occasion adults are caught and both unmarked and marked ones given another distinctive mark and released, and finally on the third occasion adults are caught and the number of types of mark recorded. The intervals between marking must be sufficient for marked individuals to completely mix with the population, but not too long that a high proportion of marked individuals may die. From an analysis of the numbers of marked and unmarked insects caught on the second and third occasions an estimate of the size of the population on the second sampling day (P_2) can be obtained.

$$P_2 = \frac{a_2 n_2 r_{31}}{r_{21} r_{32}} \tag{14.41}$$

The large sample variance of the estimate is:

$$\text{Var } P_2 = (P_2)^2 \left(\frac{1}{r_{21}} + \frac{1}{r_{32}} + \frac{1}{r_{31}} + \frac{1}{n_2} \right) \tag{14.42}$$

where a_2 = number of adults marked and released on the second day, n_2 = the total catch of adults on the second day of which r_{21} were marked on the first day, r_{31} = number of marked adults caught on the third day (i.e. last sampling day) that were marked on day 1, and similarly r_{32} = the number of marked adults recaught on the third day that were marked on day 2. It will be noted that the number of adults marked and released on the first day (a_1) does not enter into any of the calculations for the estimation of the population on the second day (P_2). It should be understood that some of the marked adults caught on the third day (r_{32}) may in fact bear two marks, having also been included in the count of marked ones caught on day 3 and marked on day 1 (r_{31}).

If the recapture rate is small, then the approximately unbiased estimates can be used (Bailey 1952).

$$\hat{P}_2 = \frac{a_2 (n_2 + 1) r_{31}}{(r_{21} + 1)(r_{32} + 1)} \tag{14.43}$$

with variance given by:

$$\text{var } \hat{P}_2 = \hat{P}_2^2 - \frac{a_2^2 (n_2 + 1)(n_2 + 2)(r_{31} - 1) r_{31}}{(r_{21} + 1)(r_{21} + 2)(r_{32} + 1)(r_{32} + 2)} \tag{14.44}$$

It is also possible to calculate any loss (γ) of marked adults, due to mortality or emigration, during the interval between the first and second sampling occasions.

$$\gamma = \ln\left(\frac{a_2 r_{31}}{a_1 r_{32}}\right)^{\frac{1}{t_1}}$$ (14.45)

where t_1 = interval between first and second sampling occasions and a_1 = number of marked adults released on day 1. Likewise any increase (β) in adults due to births or immigration between the second and third sampling occasions is given by:

$$\beta = \ln\left(\frac{r_2 n_3}{n_2 r_{31}}\right)^{\frac{1}{t_2}}$$ (14.46)

where n_3 = total catch on third day of both marked and unmarked adults, and t_2 = the time interval between second and third sampling occasions.

Bailey's triple catch method was used to assess population size of *Aedes aegypti* in Delhi (Reuben et al. 1973), but was found to severely underestimate the population compared with the results obtained by the Lincoln Index and other methods.

Leslie's methods. Three methods (A, B, & C) were proposed and maximum-likelihood equations used to estimate the death-rate of a population which was assumed to remain constant over the sampling period (Leslie 1952; Leslie and Chitty 1951). In Method A individuals are classified on each sampling occasion according to the number of marks they have received, so that a single recaptured individual with multiple marks is entered several times in the recapture table. This results in loss of information, but Leslie (1952) considered the method might be appropriate when the recapture rate is low and a large amount of data has to be analysed. In Method B the recaptures are grouped according to the date on which they were last marked and all other marks are ignored. This approach results in no loss of information and the method should give better estimates of the survival rates. Full accounts of the three methods, which are mathematically laborious and require iterative solutions when there are more than three sampling occasions, are given by Leslie (1952). Reuben et al. (1973) used Method A to estimate the population size of *Aedes aegypti* in Delhi, but found that compared with other methods it gave inconsistent results.

Yamamura et al. (1992) proposed a method that is applicable to the situation where a single release of marked individuals is made, usually of individuals collected from outside the experimental population, or alternatively individuals collected from outside the experimental population, or alternatively reared in the laboratory. This single release is followed by several recaptures, in which the recaptures are not returned to the population. In this situation, the field population is presumed to remain constant.

Later, Yamamura (2003) developed two further models applicable to the situation in which the individuals marked and released originate from the field population. Again, only one release and at least two recaptures are performed. Model A is applicable to closed populations and makes the following explicit assumptions: the total population size is constant between the two consecutive samples; the proportion of marked individuals that survive and remain in the population between the two samples remains constant; marked and unmarked individuals do not differ in their probability of capture; and marked individuals do not lose their marks.

The relationships between the original release and the two subsequent recaptures are as follows:

$$M_1 = \phi M_0 \tag{14.47}$$

$$M_2 = \phi(M_1 - m_1) \tag{14.48}$$

where M_0 = the number of marked individuals released at time 0, M_i = the number of marked individuals remaining in the population at time i, (i = 1,2), m_i = number of marked individuals recaptured at time i (i = 1,2), and ϕ = the proportion of marked individuals that remain in the population between samples.

The expectation of m_1 is

$$E(m_1) = n_1 M_1 / N = n_1 \phi M_0 N \tag{14.49}$$

where N = total number of individuals and n_i = total number recaptured at time i. The conditional expectation of m_2 for a given m_1 is:

$$E(m_2 \mid m_1) = n_2 M_2 / N = n_2 \phi(M_1 - m_1) / N \tag{14.50}$$

Probability distributions of m_1 and m_2 can be described by hypergeometric distributions, but for simplicity binomial distributions can be used to derive the maximum likelihood estimate, L, given by:

$$L = \binom{n_1}{m_1}\left(\frac{M_1}{N}\right)^{m_1}\left(1 - \frac{M_1}{N}\right)^{(n_1 - m_1)} \cdot \binom{n_2}{m_2}\left(\frac{M_2}{N}\right)^{m_2}\left(1 - \frac{M_2}{N}\right)^{(n_2 - m_2)} \tag{14.51}$$

The maximum likelihood estimates of ϕ and N which satisfy $\partial \log L / \partial \phi = 0$ and $\partial \log L / \partial N = 0$ are given by:

$$\hat{\phi} = \frac{n_1 m_2}{m_1 n_2} + \frac{m_1}{M_0} \tag{14.52}$$

$$\hat{N} = \frac{M_0 n_1^2 m_2}{m_1^2 n_2} + n_1 \tag{14.53}$$

The equation for $\hat{N}$ above has a positive bias and an unbiased estimate is obtained through a slight modification to give:

$$\hat{N} = \frac{M_0 (n_1 + 1)(n_1 + 2)m_2}{(m_1 + 1)(m_1 + 2)n_2} + n_1 \tag{14.54}$$

Variances of estimates of ϕ and N must be estimated by the delta method using the variance of hypergeometric distributions.

Model B deals with the situation where there is a constant probability of capture and the total population does not need to remain constant and requires the following assumptions: the probability of an individual being captured in a sample is the same in the two consecutive samples; the proportion of marked individuals that remain in the population is constant between samples; and marked individuals do not lose their marks.

In model B the quantities M_1 and M_2 are as described above for Model A. The expectation of m_1 in this case is:

$$E(m_1) = pM_1 = \phi p M_0 \tag{14.55}$$

and the expectation of m_2 for a given m_1 is:

$$E(m_2 \mid m_1) = pM_2 = p\phi(M_1 - m_1) \tag{14.56}$$

where p = the probability that an individual is recaptured. The expectation for the number of unmarked individuals captured at time time i (u_i) is:

$$Eu_i = pU_i \tag{14.57}$$

where U_i is the total number of unmarked individuals at time i.

The probability distributions of m_1, m_2, u_1 and u_2 are described by multiplicative binomial distributions, but a simplified method based on Poisson distribution is used to obtain the maximum likelihood estimator L:

$$L = \frac{e^{-pM_1}(pM_1)^{m_1}}{m_1!} \cdot \frac{e^{-pU_1}(pU_1)^{u_1}}{u_1!} \cdot \frac{e^{-pM_2}(pM_2)^{m_2}}{m_2!} \cdot \frac{e^{-pU_2}(pU_2)^{u_2}}{u_2!} \quad (14.58)$$

the maximum likelihood estimators for ϕ, p, U_1 and U_2 which satisfy $\partial \log L/\partial \phi = 0$, $\partial \log L/\partial p = 0$, $\partial \log L/\partial U_1 = 0$, and $\partial \log L/\partial U_2 = 0$ are given by:

$$\hat{\phi} = \frac{m_2}{m_1} + \frac{m_1}{M_0} \quad (14.59)$$

$$\hat{p} = \left(\frac{M_0 m_2}{m_1^2} + 1 \right)^{-1} \quad (14.60)$$

$$\hat{U}_i = u_i / \hat{p} = \frac{M_0 m_2 u_i}{m_1^2} + u_i, \, i = (1,2) \quad (14.61)$$

The maximum likelihood estimators of M_1 and N_1 are given by:

$$\hat{M}_1 = m_1 / \hat{p} = \frac{M_0 m_2}{m_1} + m_1 \quad (14.62)$$

$$\hat{N}_1 = n_1 / \hat{p} = \frac{M_0 m_2 n_1}{m_1^2} + n_1 \quad (14.63)$$

and can be obtained for M_2 and N_2 using the same method. Variances of these estimators are calculated by the delta method using the variance of the Poisson distribution.

As with Model A, unbiased estimators can be obtained with slight modifications as follows:

$$\hat{M}_1 = \frac{M_0 m_2}{m_1 + 1} + m_1 \quad (14.64)$$

$$\hat{U}_1 = \frac{M_0 m_2 u_1}{(m_1 + 1)(m_1 + 2)} + u_1 \qquad (14.65)$$

$$\hat{N}_1 = \frac{M_0 (n_1 + 2) m_2}{(m_1 + 1)(m_1 + 2)} + n_1 \qquad (14.66)$$

Numerical simulations indicated that the unbiased estimators for the two models gave closer estimates to the true population than either the Jackson (1939), or Bailey (1952) methods discussed above.

MacLeod's method. MacLeod (1958) gives useful descriptions and comparisons of the methods proposed by Fisher and Ford (1947), Jackson (1939) and Bailey (1952). He concluded that with insects such as *Calliphora* where the recapture rate is low (about 4–7%) and loss rate of marked flies is not necessarily constant, none of these approaches is very useful. Efforts to overcome these difficulties were made by proposing two new formulae, both derived from the Lincoln Index, in which recapture data are pooled from either a single release or serial releases. Both methods suffer the disadvantage that mortality has to be assessed independently by laboratory experiments and moreover the emigration rate of marked flies from the recapture area is calculated from separate experiments.

Stochastic models

Stochastic models make the more realistic assumption of an individual animal having a probability of survival over a given period of time rather than a fixed survival rate.

Stochastic model of Jolly and Seber

Fully stochastic models for open populations in which there are simultaneous births and deaths and possibly immigration and emigration have been proposed by Darroch (1959), and independently by Seber (1965) and Jolly (1965); in all procedures population parameters are estimated by maximum-likelihood. Jolly's method is almost identical with that of Seber but has the advantage that allowances are made for individuals killed after recapture, and not released. The basic assumptions are that: (i) on any sampling occasions, say day i, all animals in the population have the same probability (p_i) of capture and all those caught have the same probability

(η_i) of being released again, irrespective of the previous capture history of the animal; (ii) the survival rate between day i and day $i + 1$ is the same for all individuals in the population, independent of age; and (iii) emigration is permanent and thus is equivalent to death. Pollock (1981) modified the Jolly-Seber model to account for age-dependent survival and capture probabilities. Advantages are: (1) emigration and immigration are allowed; (2) moreover, both of these can vary from day to day; (3) population size and daily survival rates can be estimated separately for almost every day the population is sampled; and (4) standard errors of all estimates can be calculated. The paper by Jolly (1965) should be consulted if the explicit but lengthy algebraic derivations of the essential equations for the population parameters are required. A much simpler account of the basic approach is presented by Cormack (1973), but formulae for obtaining the variances of the parameters are not given. Begon (1979), Blower et al. (1981), Southwood (1978) and Southwood and Henderson (2000) give the basic formulae and procedures for obtaining estimates of population parameters and their variances or standard errors.

The equation for estimated population size on day i is:

$$\hat{N}_i = \frac{\hat{M}_i n_i}{r_i} \tag{14.67}$$

where $\hat{M}_i$ = the estimated total number of marked individuals in the population on day i, r_i = total number of marked individuals recaptured on day i, and n_i = total captures on day i.

The mechanics of the process are illustrated by reproducing the three tables presented by Jolly (1965) and retaining his notation (Tables 14.4–14.6). Marking and recapture extends over several days and the numbers caught (n_t), marked and then released (s_t) are given in Table 14.4 which also shows the days on which the animals were last captured (n_{ij}). The columns are summed to give total numbers of marked and released animals (s_i) subsequently recaptured (R_i). For example, on day 7 a total of 108 individuals were recaptured, i.e. $R_i = 108$. The next step is to sum the values of n_{ij} from the left in each row to give values of a_{ij} in Table 14.5 which represents the total numbers of animals recaptured on day i having already been marked on day j or earlier. The totals in the columns are added, taking care to omit the first entry, to give a series of values $z_i + 1$. These represent the numbers of animals marked before day i which were not caught on day i, but subsequently caught. For example, by adding up the figures in the 6th column $z_7 = 110$. The first figure of 112 in this column, which was omitted from the summation, represents the number of recaptures on day 7, that is $m_7 = 112$.

The estimate of the total numbers of marked animals in the population on any sampling day is given by:

$$\hat{M}_i = \frac{s_i Z_i}{R_i} + m_i \qquad (14.68)$$

So the population estimate ($\hat{M}_7$) on day 7 of *marked* animals is obtained as follows:

$$\hat{M}_7 = \frac{243 \times 110}{108} + 112 = 359.50 \qquad (14.69)$$

and likewise for M_i ($i = 2, 3 \ldots 12$). These estimates are given in Table 14.6. Jolly (Macdonald et al. 1968) has pointed out that if R_i is small this will result in overestimating $\hat{M}_i$ by an estimated magnitude of $(\hat{M}_i - m_i) \times (1/R_i - 1/s_i)$. To reduce this bias $Z_i + 1$ and $R_i + 1$ should be substituted for Z_i and R_i in the formulae for estimating M_i. In Table 14.6 $\hat{\alpha}_i$ is the estimate of the proportion of marked animals in the population on day i when the population is sampled. Values are calculated from:

$$\hat{\alpha}_i = \frac{r_i}{n_i} \qquad (14.70)$$

and

$$\hat{\alpha}_7 = \frac{112}{250} = 0.448 \qquad (14.71)$$

The total daily population estimate is

$$\hat{N} = \frac{\hat{M}_i}{\hat{\alpha}_1} \qquad (14.72)$$

Table 14.4. Tabulation of recapture data set out according to the method of Jolly (1965)

No. captured n_t	No. released s_t	Values of n_{ij} that is number in ith sample last caught in the jth sample											
		1	2	3	4	5	6	7	8	9	10	11	12
54	54												
146	143	10											
169	164	3	34										
209	202	5	18	33									
220	214	2	8	13	30								
209	207	2	4	8	20	43							
250	243	1	6	5	10	34	56						
176	175	0	4	0	3	14	19	46					
172	169	0	2	4	2	11	12	28	51				
127	126	0	0	1	2	3	5	17	22	34			
123	120	1	2	3	1	0	4	8	12	16	30		
120	120	0	1	3	1	1	2	7	4	11	16	26	
142		0	1	0	2	3	3	2	10	9	12	18	35
R_i		24	80	70	71	109	101	108	99	70	58	44	35

Table 14.5. Table of total numbers of marked individuals (a_{ij} recaptured on a given day (i) bearing the mark of day 1 or before (after Jolly 1965)

Values of a_{ij}

	2	3	4	5	6	7	8	9	10	11	12	13
1	10	3	5	2	2	1	0	0	0	1	0	0
2		37	23	10	6	7	4	2	0	3	1	1
3			56	23	14	12	4	6	1	6	4	1
4				53	34	22	7	8	3	7	5	3
5					77	56	21	19	6	7	6	6
6						112	40	31	11	11	8	9
7							86	59	28	19	15	11
8								110	50	31	19	21
9									84	47	30	30
10										77	46	42
11											72	60
12												95
$Z_{(i+1)}$ =	14	57	71	89	121	110	132	121	107	88	60	
	Z_2	Z_3	Z_4	Z_5	Z_6	Z_7	Z_8	Z_9	Z_{10}	Z_{11}	Z_{12}	

Table 14.6. Tabulation of population estimates on different days, together with estimates of Φi and their standard errors (after Jolly 1965)

i	$\hat{a}_i$	$\hat{M}_i$	$\hat{N}_i$	$\hat{\phi}_i$	$\hat{\beta}_i$	$\sqrt{\{V_i(\hat{N}_i)\}}$	$\sqrt{\{V(\hat{\phi}_i)\}}$	$\sqrt{\{V(\hat{\beta}_i)\}}$	$\sqrt{\{V(\hat{N}_i/N_i)\}}$	$\left\{ V(\hat{\phi}_i) - \dfrac{\hat{\phi}_i^2(1-\hat{\phi}_i)}{\hat{M}_{i+1}} \right\}$
1	—	—	—	0·649	—	—	0·114	—	—	0·093
2	0·068 5	35·02	511·2	1·015	263·2	151·2	0·110	179·2	150·8	0·110
3	0·218 9	170·54	779·1	0·867	291·8	129·3	0·107	137·7	128·0	0·105
4	0·267 9	258·00	963·0	0·564	406·4	140·9	0·064	120·2	140·3	0·056
5	0·240 9	227·73	945·3	0·836	96·6	125·5	0·075	111·4	124·3	0·073
6	0·368 4	324·99	882·2	0·790	107·0	96·1	0·070	74·8	94·4	0·068
7	0·448 0	359·50	802·5	0·651	135·7	74·8	0·056	55·6	72·4	0·052
8	0·488 6	319·33	653·6	0·985	−13·8	61·7	0·093	52·5	58·9	0·093
9	0·639 5	402·13	628·8	0·686	49·9	61·9	0·080	34·2	59·1	0·077
10	0·661 4	316·45	478·5	0·884	84·1	51·8	0·120	40·2	48·9	0·118
11	0·626 0	317·00	506·4	0·771	74·5	65·8	0·128	41·1	63·7	0·126
12	0·600 0	277·71	462·8	—	—	70·2	—	—	68·4	—
13	0·669 0	—	—	—	—	—	—	—	—	—

and on day 7,

$$\hat{N}_7 = \frac{359.5}{0.448} = 802.5 \tag{14.73}$$

Two more parameters are measured, namely $\hat{\phi}_i$ which is the probability that an animal at the time of release on day i will survive till the time of recapture on day $i + 1$. This covers emigration and deaths. Estimates are obtained as follows:

$$\hat{\phi}_i = \frac{\hat{M}_{i+1}}{\hat{M}_t - m_i + s_i}. \tag{14.74}$$

The other parameter is $\hat{\beta}_i$ which is the number of new animals entering the population between day i and day $i + 1$ and also still alive on day $i + 1$, and is derived as follows:

$$\hat{\beta}_i = N_{i+1} - \hat{\phi}_i(\hat{N}_t - n_i + s_i). \tag{14.75}$$

Jolly (1965) showed how both the variance and covariances of the three population parameters ($\hat{N}_i$, $\hat{\phi}_i$ and $\hat{\beta}_i$) can be derived. Below are the formulae for the variances.

$$\text{var.} \, \hat{N}_i = N(N_i - n_i)\left\{ \frac{M_i - m_i + s_i}{M_i}\left(\frac{1}{R_i} - \frac{1}{s_i} \right) + \frac{1 - \alpha_i}{m_i} \right\} + N_i - \sum_{j=0}^{i-1} \frac{N_i^2(j)}{B_j}, \tag{14.76}$$

$$\text{var.} \, \hat{\phi}_i = \hat{\phi}_i^2 \left\{ \frac{(M_{i+1} - m_{i+1})(M_{i+1} - m_{i+1} + s_{i+1})}{M_{i+1}^2}\left(\frac{1}{R_{i+1}} - \frac{1}{s_{i+1}} \right) \right. \tag{14.77}$$

$$\left. + \frac{M_i - m_i}{M_i - m_i + s_i}\left(\frac{1}{R_i} - \frac{1}{s_i} \right) + \frac{1 - \phi_i}{M_{i+1}} \right\},$$

$$\text{var.} \, \hat{\beta}_i = \frac{(\beta_i^2(M_{i+1} - m_{i+1})(M_{i+1} - m_{i+1} + s_{i+1})}{M_{i+1}^2}\left(\frac{1}{R_{i+1}} - \frac{1}{s_{i+1}} \right)$$

$$+\frac{M_i - m_i}{M_i - m_i + s_i}\left\{\frac{\phi_i s_i (1-\alpha_i)}{\alpha_i}\right\}^2\left(\frac{1}{R_i}-\frac{1}{s_i}\right)$$

$$+\frac{(N_i - n_i)(N_{i+1} - \beta_i)(1-\alpha_i)(1-\phi_i)}{M_i - m_i + s_i} \qquad (14.78)$$

$$+ N_{i+1}(N_{i+1} - n_{i+1})\frac{1-\alpha_{i+1}}{m_{i+1}} + \phi_i^2 N_i (N_i - n_i)\frac{1-\alpha_i}{m_i}.$$

The mathematics involved in calculating these variances are considerable, and they are best solved with the aid of a computer. However, Jolly (1965) showed that the greater part of the variance of both population size ($\hat{N}_i$) and survival rate ($\hat{\phi}_i$) is due to errors of estimation, and little due to variations in population numbers. Hence the formulae for $V(\hat{N}_i / N_i)$ and $V(\hat{\phi}_i)$ can be simplified to give very good approximate variances (Table 14.6).

$$V(\hat{N}_i / N_i) = N_i(N_i - n_i)\left\{\frac{M_i - m_i + s_i}{M_i}\left(\frac{1}{R_i}-\frac{1}{s_i}\right)+\frac{1+\alpha_i}{m_i}\right\}. \qquad (14.79)$$

In the formula for $V(\hat{\phi}_i)$ only the two end terms do not vanish due to errors of estimation, thus:

$$\text{var. } (\hat{\phi}_i) = \frac{\phi_i^2(1-\phi_i)}{M_{i+1}}. \qquad (14.80)$$

Because no probability distribution was given to β_i its estimated variance ($V(\beta_i)$) is entirely derived from the errors of estimation.

Dempster (1971) used the estimates of new individuals ($\hat{B}_i$) and the survival rate ($\hat{\phi}_i$) to estimate the total numbers of individuals entering the population between different sampling day (B_i^*), and by summation the total numbers entering over the entire sampling period. He also gives a formula for the standard error of this estimate.

Begon (1983) surveyed the use of Jolly's (1965) method in 100 papers published between 1966 and 1980 and concluded that in many cases the assumptions demanded by the method were neither checked nor justified, and estimates were presented with little or no reference to their confidence intervals. Although the Jolly method is mathematically sophisticated it is clearly often abused.

Manly and Parr's method

Manly and Parr (1968) introduced a method for estimating sampling intensity (p_i) in mark–recapture studies, from which population size can be estimated. A number of catches are made on different occasions and individuals are marked with a different coloured paint, or the same colour in a different position, on each day of marking, i.e. date-specific marking. The time interval between the sampling dates need not be constant and the only assumption is that sampling is random. An advantage of the method is that unlike Jolly's (1965) method it does not assume that mortality is independent of age. If the sample size on one of the sampling days, say t_i is n_i then this will represent a certain proportion (p_i) of the total but unknown population (N_i). This proportion is called the sampling intensity and is expressed as follows:

$$p_i = n_i / N_i. \tag{14.81}$$

An estimate of total population (N_i) can be made whenever an estimate of p_i is available, thus:

$$\hat{N}_i = n_i / p_i. \tag{14.82}$$

The individuals present on day t_i, on which a population estimate is to be made, are divided into various categories according to their capture–recapture histories. For the calculation of population size only two categories need be considered here: (1) those individuals caught at time t_i and which were also caught at least once before and after t_i i.e. (A_{1i}) and (2) those alive at t_i although not caught at this time, but which were caught at least once before and after t_i, i.e. (B_{1i}). Such data are obtained from records of marked individuals caught and recorded on the days before and after t_i. It follows that of the individuals known to have been alive before and after the day t_i, i.e. $A_{1i} + B_{1i}$, a proportion, $A_{1i}/(A_{1i} + B_{1i})$, were in fact caught on that day, consequently:

$$\hat{p}_i = A_{1i} /(A_{1i} + B_{1i}), \tag{14.83}$$

since

$$\hat{N}_i = \frac{n_i}{\hat{p}_i}, \tag{14.84}$$

then

$$N_i = n_i (A_{1i} + B_{1i}) / A_{1i}. \tag{14.85}$$

If individuals are caught say on five separate days and marked on the first four occasions, then separate population estimates can be made for each of the three middle days. A restriction on the use of the above technique is that if the estimate of sampling intensity ($\hat{P}_i$) is not to be subjected to sampling fluctuations then A_{1i} should be quite large, in practice more than 10.

Comparison of stochastic and deterministic models

The Lincoln Index and models derived from it (e.g. Dowdeswell et al. 1940; Fisher and Ford 1947) assume a constant survival rate, but more realistically the stochastic models of Seber (1965) and Jolly (1965) assume a probability of survival. Jolly's (1965) method is among the least restrictive, but still assumes that mortality is independent of age. The Manly and Parr (1968) method does not make this assumption, but does not use the data so efficiently as the Jolly method, which is preferred if mortality is in fact independent of age. Morton (1982) pointed out that few ecologists using mark–recapture techniques have bothered to try and test whether marking insects affects their survival or catchability. It can be argued that if the probability of recapture is constant for any sampling occasion the number of repeated recaptures, regardless of their previous history of being recaptured, should fit a Poisson distribution. Hence significant departures from this may indicate that a marked insect is captured more, or less, often than expected. Morton (1982) applied this to mark–recapture studies on a moth, but concluded that even this approach had difficulties, because for example life expectancy during the experiment should remain constant. He found that whereas marking seemed to have no effect on recapture frequencies, repeated handling and disturbance significantly reduced recapture rates.

The stochastic approach has only rarely been used to study mosquito populations, but in Bangkok, Thailand Sheppard et al. (1969) estimated the population size of *Aedes aegypti* by both the Fisher and Ford (1947) model and by Jolly's (1965) method. The latter gave monthly population estimates considerably smaller than those derived from the deterministic model. They concluded that theoretically the stochastic model is far superior when the recapture rate is high, the investigation extends over a long period, and the number of sampling days is large, but when, as in their case, recapture rate is low the deterministic model of Fisher and Ford (1947) appears preferable. Furthermore, the standard errors calculated from Jolly's model appeared unrealistically high; in fact they were greater than the daily population estimate on 31 occasions. Parr (1965) found both models gave comparable results, but in his experiments the recapture rate was relatively large. Sheppard et al. (1969) considered that probably the most important single factor in upsetting population estimates in the deterministic model, and possibly to a lesser extent in the stochastic model, was temporary emigration from the population by marked mosquitoes. The deterministic model is more sensitive to this type of error than that of Jolly. To try to overcome this Sheppard et al. (1969) took into account the proportion of marked *Aedes aegypti* estimated to have remained within the sampling area and available for capture on the different sampling days. With this correction the estimated population size was always considerably lower. This valuable paper of Sheppard et al. (1969) should be consulted for their careful approach to the estimation of dispersal and survival rates as well as population size. Macdonald et al. (1968) estimated the population of *Culex quinquefasciatus* in a village in Myanmar as 2142 by the Fisher and Ford model and 1000 by Jolly's model, again Jolly's method gave a smaller estimate than the deterministic model.

Bishop and Sheppard (1973) compared the relative performances of Jolly's (1965) stochastic model and the deterministic model of Fisher and Ford (1947) by simulating populations of 200, 1000 and 3000 individuals, with probabilities of survival from ith to ith + 1 sampling occasion as 0.5 and 0.9. The proportions of the population collected in the samples (10–20) were set at levels of 0.05, 0.09 and 0.12. Conclusions reached were that Jolly's model gave reliable population estimates when 9% or more of the population was sampled and the survival rate was 0.5. However, when these values were smaller poor estimates were obtained. The stochastic model nearly always overestimated population size, but the Fisher and Ford method also frequently did so. The application of Bailey's correction (1951, 1952) reduced the size of the population estimates, but sometimes resulted in overcorrection. A major fault with Jolly's method is that it nearly always considerably overestimates the survival rates when the arithmetic mean value is calculated over a period and consequently the method should not be used in studies when this is an important factor. It is probably better to estimate survival rates by the geometric mean.

Manly (1970) also compared population estimates obtained from simulated populations by the methods of Fisher and Ford, Jolly, and Manly and

Parr (1968), but as pointed out by Bishop and Sheppard (1973) his results are not very realistic as he inserted exceptionally high sampling intensities (0.1, 0.75) into his models. Carothers (1973b; 1979) has shown that survival estimates obtained by the Jolly (1965) method are robust in the face of an unequal catchability, whereas estimates of population size are much less so: and Manly (1970) has shown that robustness with variable survival is dependent on the pattern of that variability. Another criticism of the method is that the standard errors are of dubious value because they are generated internally (Manly 1971b; Roff 1973), however, as Begon (1983) has pointed out this is unavoidable.

In Delhi, Reuben et al. (1973) marked adults of *Aedes aegypti* with coloured dusts. In the first series of experiments mosquitoes collected at human bait were used to estimate population size by Bailey's (1952) modified Lincoln Index, by the triple catch method (Bailey 1952), by Jackson's positive and negative methods (Jackson 1939) and by Leslie's Method A (Leslie 1952). Table 14.7 shows that Leslie's method seems inapplicable, but there is some degree of similarity between the other methods. In the second experiment adult collections were made of resting adults in houses to sample stages of the gonotrophic cycle not represented at bait. In addition to the above methods for estimating population size the more complicated methods of Fisher and Ford (1947) and the stochastic procedure of Jolly (1965) were used. Analysis by Bailey's triple catch gave the lower estimates, while Fisher and Ford's method gave the highest estimates (Table 14.8). In all instances Jackson's positive method gave lower estimates than the Lincoln Index. In two cases there was almost complete agreement between population size derived from Bailey's and Jolly's methods.

Table 14.7. Population estimates and standard deviations of *Aedes aegypti* adults in Model Basti tyre dump, New Delhi (after Reuben et al. 1973)

June 1974	Bailey's corrected Lincoln Index	Bailey's triple catch	Jackson's neg. (−) and pos. (+) methods	Leslie's type A method
14	4 146 ± 756	—	3 620(+)	14 177 ± 2 536
15	6 129 ± 1 284	—	5 115(+)	12 815 ± 4 050
16	6 156 ± 1 112	3 426 ± 1 342	4 696(+)	7 493 ± 1 184
17	4 108 ± 483	2 948 ± 973	2 548(+)	4 359 ± 571
			1 955(−)	

Reuben et al. (1973) considered that estimates derived from the Lincoln Index and from Jolly's method and Jackson's method agreed but Service (1993) saw no consistent similarity between population size calculated by any of the methods. The loss rate of marked adults (0.32–0.56) and the number of new individuals joining the population (2846–7692) between successive sampling days were calculated from Jolly's method. In addition loss rate was estimated by plotting a straight line regression to the recapture data, but omitting the first day's recaptures as these were consistently higher than later recaptures. It was found that the numbers of recaptured adults decreased geometrically on successive days after release. The regression equation was calculated as $\log y = 2.1373 - 0.2041x$, and from this the loss rate was calculated as 0.375/day.

Table 14.8. Population estimates and standard deviations of *Aedes aegypti* adults in Shastri Nagar, New Delhi (after Reuben et al. 1973)

Date 1971	Bailey's corrected Lincoln Index	Jolly's method	Fisher and Ford's method	Bailey's triple catch	Jackson's neg. (−) and pos. (+) methods
29 June	9 012 ± 1 609	—	11 508	—	7 093(+)
30 June	9 194 ± 1 679	9 935 ± 2 119	12 746	2 919 ± 1 010	4 744(+)
1 July	11 560 ± 2 209	7 195 ± 852	16 995	3 944 ± 3 938	6 593(+)
2 July	13 351 ± 2 755	9 896 ± 5 773	16 710	6 198 ± 5 693	11 569(+)
3 July	13 264 ± 2 966	10 757 ± 5 249	22 960	3 848 ± 1 470	11 536(+)
4 July	14 635 ± 2 839	14 695 ± 2 869	16 410	7 516 ± 7 410	10 094(+)
5 July					11 508(−)
6 July					17 607(−)

Most methods of analysing data from mark–recapture studies to estimate population size only perform well if large samples are obtained. When the numbers of recaptures are small, the lower bounds of the confidence intervals can be less than the numbers of recaptures! There will be a negative bias if the numbers of marked animals and numbers of recaptures are too small. To help overcome the analysis of such data the Bayesian approach has occasionally been used (Carle and Strub 1978; Freeman 1973a,b; Gaskell and George 1972). Gazey and Staley (1986) pointed out that although the approach is intensive in computation, the availability of cheap computing power has made the Bayesian method tractable. Their paper should be consulted by the mathematically brave for the procedures needed to apply this approach.

With the quality of ecological data normally available it seems arguable whether more complex mark–recapture methods give more reliable population estimates than simpler methods such as the Lincoln Index or Bailey's triple catch. One of the intrinsic difficulties in applying any method is the lack of worthwhile biological information, such as survival and dispersal rates and possible effects marking may have on the individuals. There is clearly a need to obtain more ecological data relevant to mark–recapture techniques, and also for more careful comparisons of the efficiencies of the different mathematical procedures for analysing the results. Because the actual size of a population is very rarely known, it is difficult to know which population estimates are near the truth. For this reason computer simulations, where population size is known, are valuable for comparing the relative efficiency of different methods.

Further useful information on mark–recapture methods and procedures is given in the following publications; Alho (1990), Arnason (1972a,b), Barndorff-Nielsen (1972), Carothers (1973a), Chapman and Junge (1956), Cook et al. (1967), Cormack (1968, 1972), Darroch (1959, 1961), Eberhardt (1990), Gaskell and George (1972), Iwao (1963), Krebs and Boonstra (1984), Manly (1969a,b, 1970), Marten (1969), Miller et al. (1987), Parr et al. (1968), Seber (1982, 1986), Southwood (1978), and Southwood and Henderson (2000). Seber (1973) gives an excellent account of estimating population size by various mark–recapture methods and also other methods of estimating animal abundance, but although useful worked examples are given, only those with a good knowledge of mathematics will be able to follow fully his accounts. Manly (1971a) gives a method for estimating the effect, if any, that marking has on survival rates, while White (1970) gives a coding system for marking individuals. Computer software packages that utilise many of the above methods sare freely available on several internet sites, however, they require considerable technical expertise to use them correctly.

In more sophisticated applications of the method population estimates can in fact be obtained even when immigration, emigration and natality occur, but the methods require that individuals are marked and then re-marked on a number of separate occasions. It has also to be assumed that being recaptured and remarked several times does not affect the individuals' chance of survival or subsequent capture. Roff (1973), however, stressed that the assumption that all individuals are equally sampled is rarely, if ever, true. For example, in using Liverpool taxis to assess the efficiency of mark–recapture models Bishop and Bradley (1972) obtained very large differences between two population estimates calculated from sampling localities only 200 m apart, emphasising the need for samples to be taken from homogeneous populations. This may require separate sampling and analysis for sex and age classes, possibly also taking into consideration the genetic variability of the population (Bishop 1973). Conway et al. (1974) showed how important it was that both marked and unmarked adults of *Aedes aegypti* had the same chances of recapture if reliable population estimates were to be made of the biting population of females.

Removal methods

The removal method of Zippin (1956, 1958) assumes catchability remains constant and equal for all individuals in a population. The method of Skalski et al. (1984), however, is equally applicable to the so-called generalised removal method (Otis et al. 1978; Skalski and Robson 1979, 1982; White et al. 1982), which allows some degree of heterogeneity amongst the individuals. The calculation of K, that is the proportional abundance of population N_2 to population N_1 is

$$K_1 = \frac{r_2}{(1-q_2^k)} \bigg/ \frac{r_1}{(1-q_1^k)} = N_2 / N_1 \qquad (14.86)$$

where $r = n_1 + n_2 + ... + n_k$ that is the numbers of individuals caught in k consecutive random removal samples of size n_1, in other words the total catch during the study, and where probabilities of capture ($p = 1 - q$) from the two populations are estimated by iteration from the following formula

$$\frac{t_{2i}}{r_i} = \frac{q_i}{(1-q_i)} - \frac{kq_i^k}{(1-q_i^k)}, \quad \text{for } i = 1, 2 \qquad (14.87)$$

where t_2 = the total number of times the r individuals escaped capture prior to being caught in the sampling programme. That is

$$t_2 = \sum_{i=2}^{k} (i-1)n_i \qquad (14.88)$$

A much simplified estimator of proportional abundance can be derived if probabilities of capture for the two populations are equal ($p_1 = p_2 = p$), namely

$$K_2 = \frac{r_2}{r_1}. \qquad (14.89)$$

Skalski et al. (1984) present formulae for obtaining variances of K_1 and K_2 and it is strongly suggested that field workers with limited mathematical skills use this approach.

Change in ratio methods (Kelker's selective removal)

Change in ratio methods (Kelker 1940; Chapman 1955) exploit natural differences in appearance between different classes of the population, most commonly sexual dimorphism. The proportions of the two sexes in the population is determined, followed by removal of a known number of one of the sexes. The new ratio between the sexes in the population is determined afresh and the total population can then be estimated from the change in ratio between the two sexes, using:

$$\hat{P} = \frac{K_\alpha}{[D_{\alpha 1} - \dfrac{D_{\beta 1} D_{\alpha 2}}{D_{\beta 2}}]}$$ (14.90)

where α and β are the different sexes or other components of the population, with α being the component that was selectively removed from the total population. K_α = the number of individuals of component α that was removed, $D_{\alpha 1}$ = the proportion, expressed as a decimal, of component α in the population prior to removal, $D_{\alpha 2}$ is the proportion of component α in the population following removal, similarly for the proportions of components β before and after removal of individuals of component α. Selective removal of the proportion of the population with a specific dimorphic character could invalidate further ecological studies on that population (Southwood and Henderson 2000). One advantage of this method is that it does not depend on the assumption that initial capture does not alter the likelihood of subsequent recapture.

Chapman (1955) concluded that mark-recapture studies yield more information per unit effort than selective removal approaches and are therefore normally preferred. As selective removal methods do not require the animals to be marked, they may be useful as a means to verify results obtained using Petersen-Lincoln estimators, especially where use of artificial marks is suspected to alter subsequent catchability. Selective removal methods are perhaps most applicable to the study of populations that are subject to selective culling, however they have been used to estimate mosquito population size (e.g. Yasuno and Rajagopalan 1977).

Gouteux et al. (2001) have developed a model for estimating the size of tsetse fly populations from data obtained by removal trapping. This model takes into account reinvasion, which is density-dependent, such that the degree of reinvasion increases as the population size decreases (Glasgow 1953). Population size on day t is denoted N_t and is assumed to be at equilibrium at the commencement of trapping ($t = 0$). The daily capture rate p ($0 < p < 1$) is assumed to be constant and related to trap design, layout, efficiency and the mobility of flies. The number of flies captured on day t is given by pN_t. Birth and death rates are ignored as the experiment is performed over a short period of time. Density-dependent reinvasion is modelled by postulating that the influx of flies each day is proportional to the deficit relative to the initial population N_0. N_{t+1} is given by:

$$N_{t+1} = N_t - pN_t + k(N_0 - N_t) \tag{14.91}$$

The parameter k can be interpreted as a reinvasion rate, equivalent to the number of flies immigrating per unit time for a one fly difference between initial population size N_0 and current population size N_t. The continuous time version of equation 14.91 above is:

$$\frac{dN(t)}{dt} = kN(0) - (p+k)N(t) \tag{14.92}$$

which has the solution

$$N(t) = k\frac{N(0)}{p+k} + p\left(\frac{N(0)}{p+k}\right)e^{-(p+k)t} \tag{14.93}$$

Equation 14.93 describes the situation where when t is large, the population decreases and approaches an equilibrium value $K = kN(0)/(p+k)$, where $k/(p+k)$ is the 'Trapping Survival Proportion' (TSP) or the percentage of the original population at which the population re-stabilises subject to removal of individuals through trapping and their replacement by reinvasion. If the reinvasion rate, k is large compared to the capture rate p, then the TSP approaches 1, trapped flies are rapidly replaced and the population does not decrease much. If k is relatively small the population decreases rapidly because reinvasion cannot balance the effects of removal trapping. If equation 3 above is multiplied by the trapping rate p, then the density of captured flies ($C(t)$) is given by:

$$C(t) = p\left[k\frac{N(0)}{(p+k)} + p\left(\frac{N(t)}{p+k}\right)e^{-(p+k)t}\right] \tag{14.94}$$

Equation 14.94 above can then be fitted to the observed decrease in trapped populations using non-linear least square techniques and the parameters $N(0)$, p and k can be estimated. The model was tested with data from a field collection of *Glossina fuscipes fuscipes* near Bangui, Central African Republic.

Some examples of mosquito population estimates

The Petersen-Lincoln Index has been more frequently used to estimate mosquito population size than any other method. In some studies estimates

derived from it have been compared with those obtained using more complex models. Reisen and colleagues have used this and other recapture methods probably more than any other authors. For example in Pakistan Reisen et al. (1978) marked *Culex tritaeniorhynchus* with coloured dusts and calculated population size by using Bailey's modified Lincoln Index, and Jackson's positive and Jackson's negative methods, using the formula

$$r- = \frac{Y_2 + Y_3}{Y_1 + Y_2} \tag{14.95}$$

for Jackson's negative method, since marked adults were released on just 3 days.

Reisen et al. (1978) concluded that this negative method, based on just three recaptures, seriously underestimated population size ($\male$ 0.165×10^6), whereas the modified Lincoln Index ($\male$ $3.515 \times 10^6 \pm 1.56 \times 10^6$) and Jackson's positive method ($1.726 \times 10^6 \pm 3.578 \times 10^6$) based on the 8-day sampling sequence provided realistic and relatively similar estimates which fell within the fiducial limits of the modified Lincoln Index—despite the fact that the Lincoln Index gave a population estimate 2.04 times greater than the Jackson method. However, one can really only compare methods if one 'knows' the answer—something of a 'catch–22' situation. Just because two methods give similar population estimates this does not mean that they are actually giving the right answer, and the other method the wrong answer, because all three methods are related and therefore share errors.

In Pakistan Reisen and Aslamkhan (1979) estimated the population size of *Anopheles stephensi* from mark–recapture methods using Bailey's modified Lincoln Index and Itô's (1973) modification of Jackson's (1939) positive method and Jackson's negative method. Throughout the ratio of unmarked/marked adults rather than unmarked + marked/marked adults was used in all calculations because the marked released *Anopheles stephensi* were not a part of the natural population. It was concluded that the Lincoln Index gave a more or less realistic population estimate of both males and females, but the Jackson's positive method overestimated the population of males. In similar experiments Reisen et al. (1979) estimated the population size of *Anopheles subpictus*. The daily number of *Anopheles subpictus* added each day (A_t) to the population was also calculated by the Manly and Parr (1968) method, that is

$$A_t = P_t - P_{t-1}Sr. \tag{14.96}$$

where P_t = population size on day t; and S_r = daily survivorship, which is assumed to be constant. The overall addition rate (ar) can be calculated as

$$ar = \overline{A_t} / \overline{P_t} \qquad (14.97)$$

where $\overline{A_t}$ = mean number of additions per day and $\overline{P_t}$ = mean population size. The overall loss rate (L) is given as

$$L = 1 - S_r. \qquad (14.98)$$

The rate of population change (RC) in numbers of individuals per day is given by

$$RC = ar - L. \qquad (14.99)$$

Population estimates of female *Anopheles subpictus* (based over 2–6 days) were, by the modified Lincoln Index 4478.4 ± 1696.8, by the modified Jackson's positive method 5631.6 ± 1248.5, and by Jackson's negative method 4197.1 ± 1486.1. All estimates are in fairly good agreement.

In Pakistan Reisen et al. (1980) used Bailey's (1952) modification of the Lincoln Index in mark–recapture studies with coloured dusts to estimate the population size of *Anopheles culicifacies*. They also calculated the numbers of marked adults remaining in the study area (R_t) on day t as follows

$$R_t = S_r (R_{t-1} + M_{t-1}) \qquad (14.100)$$

where M_t = numbers of marked individuals released on day t and S_r = daily survivorship. As advocated by Nelson et al. (1978) they subtracted the numbers recaptured on previous days (i.e. $-r_{t-1}$) from the calculation.

The numbers added to the population on day $t(A_t)$ and the overall addition rate (immigration and emergence) were calculated as previously by Manly and Parr's (1968) method.

In addition Reisen et al. (1980) calculated the population size of *Anopheles culicifacies* on the day of release by Jackson's (1939) positive and negative methods. Itô's (1973) modification to Jackson's positive method (1939), as presented by Hamada (1976), was also calculated, where the numbers of marked adults recaptured on day t are corrected for numbers marked and released on day $t(z_t)$, so

$$z_t = (10^4 r_t) / (UM_t Mo_{(t)}) \qquad (14.101)$$

where UM = number of unmarked mosquitoes collected on day t, and

$$Mo_{(t)} = Mo - \sum_{j=1}^{t-1} r_j \qquad (14.102)$$

which separates losses due to emigration and mortality from the removal of marked mosquitoes by sampling. Itô's (1973) method fits a regression equation

$$\ln z_t = \ln a = t \ln S_r \qquad (14.103)$$

and population size on day of release $(P_0) = 10^4/a$.

The estimates of population size of females were remarkably similar, with means and 95% confidence intervals as follows, Bailey's Lincoln Index (1048.5 ± 250.1), Jackson's positive method (1045.0 ± 735.2) Itô's modification of this latter method (1049.6 ± 749.2), and Jackson's negative method (1061.1).

Reisen et al. (1980) considered that as the number of marked mosquitoes on a day (R_t) was corrected for constant daily survivorship (S_r), which included the numbers of removed recaptures as well as emigration and mortality, then it was inappropriate to correct R_t, as determined by the modified Lincoln Index, by subtracting r_{t-1} as proposed by Nelson et al. (1978), because this would result in slightly underestimating the population size since it would correct twice by removing r_t.

In later studies Reisen et al. (1982) undertook mark–recapture studies on *Anopheles culicifacies, Anopheles stephensi* and *Anopheles subpictus* to estimate population size of both sexes in different months by the Lincoln Index modified by Bailey (1952), and by Itô's (1973) modification of the Jackson (1939) positive and negative methods. There was no significant difference between the estimates derived by these three methods, but generally the Lincoln Index method seemed more tolerant of low recapture rates because it did not rely on a consistent linear decline in the adjusted recapture rate. Their conclusions agree with previous comparisons of the Lincoln Index and Jackson's methods of population size estimation (Reisen and Aslamkhan 1979; Reisen et al. 1978, 1979, 1980; Reuben et al. 1973). In other experiments Reisen et al. (1981, 1982c) estimated population size of *Anopheles culicifacies*, *Anopheles stephensi* and *Anopheles subpictus* by Bailey's (1952) modification of the Lincoln Index. They also used the Manly and Parr (1968) method to estimate the numbers of daily additions, (emergence + immigrants) to the population.

In Pakistan Reisen and Mahmood (1981) compared Zippin's (1958) removal method and a mark–recapture for estimating the population size of *Anopheles culicifacies* and *Anopheles stephensi*. For the removal method two collectors caught anophelines resting in cattle sheds for 5-min periods until the catch decreased markedly relative to the initial numbers, this usually involved collections for 30–50 min. In mark–recapture experiments the Lincoln Index was used when recaptures were made a few hours after release of dust-marked adults, but Bailey's (1952) modified index was used when recaptures were made over a number of days.

Although in all experiments the numbers removed were greater than 70% of the estimated population size, the removal method seriously underestimated populations, and in fact rarely were more than 40% of the marked mosquitoes released recaptured. In contrast the mark–recapture method provided realistic population estimates. The numbers of marked mosquitoes released (M_1) were regressed as a linear function of population size estimated by the removal method (N_2), and allowed for a corrected population estimate based on the removal method to be obtained. ($M = 203.15 +$ 1.49 N_2 and $r^2 = 0.874$). In addition the numbers of marked mosquitoes caught within 15 min by the two men using aspirators, and referred to as the 0.5 man-hour catch, were plotted as a regression against the corrected population size (N_2) determined by the removal method ($N2 = 254.60 + 3.10$ the 0.5 man-hour catch, and $r^2 = 0.84$). Reisen and Mahmood (1981) concluded that a relative abundance method, such as the 0.5 man-hour catch employing two men provided good population estimates, and compared favourably to the more laborious mark–recapture estimates.

Tietze et al. (2003) used the Petersen-Lincoln index to estimate the population size of *Culex erythrothorax* in Santa Cruz County, California, USA. Mosquitoes were collected using 12 CO_2-baited CDC light traps. Mosquitoes collected were dusted with luminous powder (Shannon Luminous Materials, Inc., Santa Ana, California, USA) applied with a small bulb duster. Recaptures were made with a series of 21 CDC light traps. Marked individuals were identified using ultraviolet light. Approximately 43 000 marked females were released over three separate dates. Only 319 marked females were recaptured. The estimated total population size of *Culex erythrothorax* in the study area, ranged from 3.8 to 9.4 million mosquitoes.

To estimate the population size of *Anopheles albitarsis* s.l. in an irrigated rice-growing area of Vale do Ribeira, São Paulo, Brazil, dos Santos and Forattini (1998) used mark-recapture and Bailey's modification of the Petersen-Lincoln index. The three mark-recapture experiments gave population estimates of 64 560, 50 503 and 22 684.

Lindquist et al. (1967) used a modified Lincoln Index to estimate the population size of *Culex quinquefasciatus* in Yangon. In one experiment adults were released at a central point and 30 recapture stations were sited along five concentric circles separated from each other by 200 yd. The mosquito population in an area enclosed by two concentric circles of radius $(d - 100)$ yd and $(d + 100)$ yd can be estimated from:

$$P_d = \frac{a_d n_d}{r_d} \qquad (14.104)$$

where suffix d = area of ring having a radius d. It was assumed that negligible numbers of *Culex quinquefasciatus* flew < 100 or > 1100 yd, i.e. all released mosquitoes were within the area demarcated by two circles with radii of 100 and 1100 yd. Another assumption was that the numbers of *Culex quinquefasciatus* in the area having a radius d was proportional to $(n_d/k_d)d$, where k_d = number of capture stations in the ring area. It follows that the number of marked adults (a_d) in this area is proportional to $(r_d/k_d)d$, and the *total* number of marked and released adults (a) is proportional to the grand total of this quantity over all the ring areas. An estimate of a_d is therefore derived as follows:

$$a_d = \frac{a(r_d / k_d)d}{\sum_d (r_d / k_d)d} . \qquad (14.105)$$

When the interval between release of marked individuals and subsequent recapture is more than a few days, mortality of the marked adults must be taken into consideration. Let p = daily survival rate and x = the number of days since marking and release, then:

$$a_d(x) = \frac{ap^x (r_d / k_d)d}{\sum_d (r_d / k_d)d} . \qquad (14.106)$$

Substituting this expression of $a_d(x)$ for a_d in the first equation then the estimated total population of *Culex quinquefasciatus* on day x in area d enclosed by two concentric rings is:

$$P_d(x) = \frac{n_d}{r_d} \times \frac{ap^x (r_d / k_d)d}{\sum_d (r_d / k_d)d} \qquad (14.107)$$

and the entire population on day x is given by:

$$P(x) = \sum_d P_d(x). \qquad (14.108)$$

The variance of the population in each ring can be calculated from:

$$\text{var}.P = \frac{(a^2 n - r)}{r^3} \qquad (14.109)$$

while the variance of the entire population in the whole area can be calculated on any one day from:

$$\text{var}.\ P(x) = \sum_d \frac{(n_d - r_d)}{n_d r_d} P_d^2(x). \qquad (14.110)$$

Reuben et al. (1973) compared the Lincoln Index with several other methods for estimating the population size of *Aedes aegypti* in New Delhi, and in Tanzania Trpis (1971) used the index to estimate the population of *Aedes aegypti* in Buguruni. From laboratory data based on 13 different populations of *Aedes aegypti* from various parts of the world supplied by Drs T. J. Crovello & C. S. Hacker, Trpis (1971) postulated a mean daily survival rate of 0.809 which is very similar to the value of 0.806 that Sheppard et al. (1969) had estimated as the mean daily survival of *Aedes aegypti* in Bangkok. Because recaptures of *Aedes aegypti* in Buguruni were made 36 h after marking the mortality rate finally selected was 0.286, and this was used to correct the number of marked adults available for recapture. Trpis (1971) made population estimates when standardised routine human bait catches indicated that the population was at three different population levels. By comparing the man-biting catches with the corresponding three population estimates, obtained by catching unfed females at bait, marking and releasing them and recapturing them at bait as unfed females, it was possible to convert bait catches in terms of absolute densities. A good correlation between the two was found. This is an interesting study but few individuals were marked (391, 275, 488), and very few marked mosquitoes recaptured (6, 2, 2). Moreover no variances are given to the population estimates. It should be realised that it is essential that the females caught at bait, marked and released will have the same chances of recapture in subsequent bait catches as adults that have not previously been collected. Further field experiments are needed to verify this approach of extrapolating bait catches in terms of absolute population size. In the same area Trpis (1973) estimated the population size of *Toxorhynchites brevipalpis* by releasing 195 marked males and females. From the recapture of 19 marked mosquitoes 24 h later the population size in the 1-ha area was estimated by the Lincoln Index as 3459. This high recapture rate indicates that adults dispersed little from the release point.

In a small Kenyan village *Aedes aegypti* adults caught and released in houses were marked with fluorescent powders and estimates made of their population size (Trpis and Haüsermann, 1986). From a total release of 563 female *Aedes aegypti* population size estimated on 14 consecutive days varied from 75.4 ± 28.0 to 728 ± 505.5 with a mean of 331 ± 146.5 using Bailey's modification of the Lincoln Index; from 26.0 ± 19.6 to 1209.6 ± 743.3, mean 270.3 ± 179.5 with Bailey's Triple Catch; and from 88.5 to 594.8, mean 337.3 with the Fisher and Ford model. While using the stochastic approach of Jolly–Seber the population of females was estimated as from 93.0 ± 6.2 to 800.6 ± 234.3 with a mean of 380.1 ± 92.9. There was considerable variation between estimates of minima and maxima populations by the different methods, and there was little agreement in ranking population size over the 14 days by the different methods; the Triple Catch of Bailey gave the highest and lowest values. However, estimated mean population size over the 14-day period by the various methods did not vary too much, although again the Triple Catch method differed most.

Trpis et al. (1995) in a later study used mark-recapture to estimate population size, dispersal and longevity of domestic *Aedes aegypti* in Shauri Moyo village in eastern Kenya. One thousand males and 1000 females reared from locally collected eggs were marked with a fluorescent pigment (DayGlo Rocket Red A-16) and released from a single release point and recaptured via indoor human landing catches. The size of the *Aedes aegypti* population in the village was estimated using three methods: single release and single recapture, single release and repeated recaptures, and removal by daily sampling. Marked recaptured individuals were not returned to the population after sampling. Estimates were calculated using the Petersen-Lincoln index, Jackson's positive method, and the removal methods of Kono (1953) and Zippin (1956, 1958). The total population of females estimated using the Petersen-Lincoln index was 1365, including the 1000 released. Jackson's positive method gave a figure for the total female population, excluding the marked females released, equal to 337. Kono's method gave a population size of 451 and the Moran-Zippin method gave a population size of 467 (SE $\pm$ 18.9). In their discussion, the authors acknowledged that while marked females had reached all houses within 24 h of release, their distribution among the houses was not similar to that of the unmarked population, potentially violating the requirements for complete mixing and equal catchability. The possibility of recruitment to the unmarked population as a result of emergence from domestic water containers was not excluded.

Costantini et al. (1996) working in Burkina Faso analysed mark-release-recapture data using the Peterson-Lincoln Index, the Fisher and Ford method and the Jolly-Seber method to estimate the population density of *Anopheles gambiae* s.l. The overall recapture rate was relatively low (~1%). Population estimates obtained using the different methods are provided in tables 14.8 to 14.10. The authors acknowledged that the three assumptions required for estimating population density, namely that populations are spatially limited; that the entire population is available to be sampled; and sampling is equally effective at all locations, were not fully met, all with potential effects on the reliability of estimates. In addition, no specific tests of the effect of dusting on mosquito survival and behaviour were carried out in this case.

Table14.8. Estimates of population size of endophilic female anophelines in Goundri, Burkina Faso from mark-release-recapture experiments analysed using the Lincoln Index. Median values are shown in italics (Costantini et al. 1996)

Release date	No. released (a)	Total captured (n)	Marked recaptured (r)	Population size (P)	Standard error	95% Confidence limits	
September 1991							
10	1205	1863	21	106,901	23,196	75,003	186,008
11	1628	2130	13	247,805	73,755	172,990	582,335
13	1685	1824	7	384,391	165,631	252,424	1,684,734
16	1455	1671	10	221,160	76,654	150,271	636,382
18	1287	1283	18	86,974	21,470	62,887	169,478
September 1992							
5	1607	2866	20	230,283	51,313	160,282	408,838
8	3580	2635	26	362,819	70,803	262,440	587,549
11	4150	3171	27	487,394	93,399	354,316	780,572
16	3315	2310	17	425,609	108,847	305,676	855,746
21	1202	1283	0	∞	—	—	

Table 14.9. Estimates of population size of endophilic female anophelines in Goundri, Burkina Faso from mark-release-recapture data analysed using Jolly's stochastic model (Costantini et al. 1996)

i	n(i)	s(i)	R(i)	z(i)	m(i)	M(i)	α(i)	N(i)	Φ(i)	β(i)	SE[N(i)]	SE[Φ(i)]
September 1991												
1	1205	1205	33	—	—	—	—	—	—	—	—	—
2	1863	1628	22	12	21	941	0.011	83,946	0.868	141,918	33,888	0.0067
3	2130	1685	9	12	22	2213	0.010	214,210	0.238	89,228	104,107	0.0068
4	1824	1455	15	9	12	921	0.007	140,049	0.497	53,267	69,858	0.0103
5	1671	1287	9	8	16	1174	0.010	122,641	—	—	65,780	—
6	1283	0	—	—	9	24	—	0.019	—	—	—	—
7	1201	0	—	—	11	—	—	—	—	—	—	—
September 1992												
1	1631	1607	29	—	—	—	—	—	—	—	—	—
2	2866	3580	31	9	20	1139	0.007	163,183	0.146	30,949	68,927	0.0052
3	2635	4150	37	5	33	688	0.013	54,957	—	—	24,976	—
4	3171	0	—	10	32	—	0.010	—	—	—	—	—
5	1307	3315	19	—	7	—	0.005	—	—	—	—	—
6	2310	1202	0	2	20	3626	0.009	418,803	—	—	—	—
7	490	0	—	0	3	—	0.006	—	—	—	—	—
8	439	0	—	—	1	—	0.002	—	—	—	—	—

Table 14.10. Estimates of population size of endophilic female anophelines in Goundri, Burkina Faso from mark-release-recapture data analysed using the method of Fisher and Ford. Median values are shown in italics (Costantini et al. 1996)

Recapture date	Population size	
	Mean absolute fit	Mean square fit
September 1991		
11	106,901	106,901
—	135,363	132,675
13	163,825	158,448
16	205,351	190,808
18	171,505	161,154
20	88,984	83,395
23	59,314	51,572
Estimated daily survival rate	0.689	0.672
September 1992		
7	166,679	155,441
9	334,514	325,321
12	568,655	535,405
14	561,329	459,644
18	372,431	317,471
23	246,725	195,430
28	131,734	73,604
Estimated daily survival rate	0.724	0.675

In Uganda Sempala (1981) marked *Aedes africanus* with powders and paints and used the Lincoln Index and Jolly's method to estimate population size. The smallest population was recorded 3 days after release and was estimated at 1104 (Lincoln Index) and 1386 (Jolly) while the largest population, recorded on the 6th day, was 16 320 (Lincoln Index) and 8406 (Jolly), on the other 3 days for which population estimates were made by both methods the Lincoln Index gave 1.5–1.9 times smaller estimates.

In Indiana, USA Sinsko and Craig (1979) located 108 tree-holes in a 10.1-ha isolated wood and counted weekly the numbers of pupae of *Ochlerotatus triseriatus* in them, which when summed gave 4228 pupae of which 1790 were female. This represented a discrete natural cohort. Using a mean daily survival rate of 0.87, derived from the Jolly (1965) method applied to mark–recapture studies on adults (see below), it was estimated that the total population of emerging females should be 825. They then compared this independent direct estimate of population size of females with that derived from mark–recapture studies. Mosquitoes caught biting were marked with seven different coloured 'Dayglo' fluorescent powders, and thereafter with paint spots. Population estimates were derived from the

Schnabel-Thompson method (Schnabel 1938), which is based on the Lincoln Index, the Triple Catch method of Bailey (1951) with and without the correction factor, and by the stochastic Jolly method as modified for computer operation by White (1971a,b). The Jolly method gave a population estimate at the end of the summer of 1225 ± 455 females, which matched well the independent estimate based on pupal counts. The standard error for the Bailey Triple Catch was sometimes very large, and although the standard error of the Schnabel–Thompson method was consistently low, the population estimate was always much too large.

Somewhat similarly Hii and Vun (1985) in Malaysia successively multiplied the 1119 released *Anopheles balabacensis*, which were marked yellow, by their daily survival rate (0.7874) to give expected numbers of survivors in the area on successive days. The actual numbers of recaptured marked mosquitoes caught each day divided by the expected numbers of survivors provided estimates of the proportion of the population in the area that was caught. From this it was estimated that the total population of *Anopheles balabacensis* declined from about 11 000 to about 5500 by the end of the experiment (day 13).

Fabian et al. (2005) used what they termed the Seber (1982) method, but which is actually the simplified version of the Schnabel census (see above) to calculate the population size of *Anopheles saperoi* in a forested area of Okinawa, Japan. Population size (N^*) was estimated as

$$N^* = (n_1 + 1)(n_2 + 1)/(m_2 + 1) - 1 \qquad (14.111)$$

where n_1 is the first sample, n_2 the second sample and m_2 the number marked within the second sample. Variance (v^*) was estimated using:

$$v^* = (n_1 + 1)(n_2 + 1)(n_1 - m_2)(n_2 - m_2)/(m_2 + 1)^2(m_2 + 2) \quad (14.112)$$

Using this method the population size of female *Anopheles saperoi* was estimated in five experiments as: 23 841, 1182, 3514, 5679, 9238. In a sixth experiment population size could not be calculated as the total number of recaptures was less than 10. Survival rate was also estimated by regressing the numbers of *Anopheles saperoi* recaptured (transformed to $\ln(y+1)$) against days post-release. The regression equation calculated using pooled data from all six experiments was $\ln(y+1) = 3.7508 - 0.302n$ ($r^2=0.7907$), equivalent to a daily survival rate of 0.73.

Haramis and Foster (1983) estimated the population size of laboratory-reared *Ochlerotatus triseriatus* females by Bailey's (1952) modified Lincon Index, using the form employed by Yasuno and Rajagopalan (1977) to account for mortality and numbers released into the population. Population estimates from days 4–24 varied from 24–205/ha, this wide range was probably partly due to poor dispersal of marked adults in the wood.

Pumpuni and Walker (1989) used the Lincoln Index, modified for low recapture rates and compensated for daily mortality, to estimate the population size (*P*) of *Ochlerotatus triseriatus* in Indiana, USA.

$$P = [as'(n - r + 1)]/(r + 1) \qquad (14.113)$$

where P = estimated population size on day t; a = number of marked mosquitoes released; s = estimated probability of daily survival; t = sampling day after release; n = total numbers of *Ochlerotatus triseriatus* caught (marked and unmarked) on day t; r = number of marked mosquitoes recaught on day t. Population size from days 3 to 15 after release, based on six recapture days, varied from 4289–11 255 for males, and from recaptures on seven days from 4954–23 680 for female *Ochlerotatus triseriatus*. Recapture rates of males were low after 3 days post-release, and after 4 days post-release for females. These low recapture rates result in the overestimation of population size.

In Bangkok in addition to measuring the population trends of *Aedes aegypti* by routine human landing catches Yasuno and Tonn (1970) estimated adult population size by the collection of pupae from 150 water butts and also during 4 months by mark–recapture studies on adults. Although the number of marked adults recaptured was small, absolute population estimates obtained in the 4 months, and the population size as predicted by pupal counts, more or less agreed with the fluctuations in population size shown by landing catches.

In villages around Delhi, Yasuno and Rajagopalan (1977) estimated the population of *Culex quinquefasciatus* by releasing into the population large numbers (usually about 30 000) of laboratory reared males marked with daylight fluorescent powders. Recaptures were made in houses and from other man-made shelters. Population size was estimated by using Bailey's corrected Lincoln Index, by Jackson's positive and negative methods and also by the CIR (change in ratio) method, in this instance the change in the sex ratio. The population size (*P*) prior to release of males was calculated as follows:

$$P = \frac{d_x s' (1 - p)}{p_1 - p_0} \qquad (14.114)$$

where d_x = number of males introduced, p_0 = fraction of males in samples prior to release of males, p_1 = fraction of males in samples afterwards, s is probability of daily survival and t is time in days between release and recapture. If values of s' are not available then no correction can be made and population size is estimated without this term in the formula. The variance estimate as given by Chapman (1955) is:

$$\text{var}\,P = \frac{(x_0 y_0 / n_0) + (x_1 y_1 / n_1)}{(p_0 - p_1)^2} \qquad (14.115)$$

where p_0 and p_1 are already defined, x_0 and y_0 = numbers of males and females before sex distortion, x_1 and y_1 = numbers of both sexes after distortion, n_0 = total number of both sexes before distortion and n_1 = total number afterwards.

In using the Lincoln Index, the daily loss rate was taken into consideration. Also, because the number of marked and released males (a) did not constitute a part of the real population the numbers of marked recaptures (r) were subtracted from the numbers caught (n). Thus:

$$P = \frac{as'(n - r + 1)}{r + 1} \qquad (14.116)$$

$$\text{var}\,P = \frac{(as')^2 (n - r + 1)(n - 2r)}{(r + 1)^2 (r + 2)} \qquad (14.117)$$

Population estimates on different days in one village varied from 11 707–132 476 (sex distortion method), 11719–132 122 (Lincoln Index) and 11 876–142 653 (Jackson's positive method). There was remarkably good agreement between the estimates by the three different methods. Yasuno and Rajagopalan (1977) concluded that the accuracy of the population estimate derived from the sex distortion method is directly related to the degree of distortion, and that large numbers of males have to be released (or alternatively removed) to reduce the variance of the estimates, as is usually the case with selective removal methods.

In endeavouring to better understand the dry season population dynamics of *Anopheles arabiensis* in West Africa, Taylor et al. (1993) used an indirect/genetic method to calculate effective population size from a comparison of data on gene frequencies over time. This method relies on the assumption that effective population size is defined as the size of a theoretically ideal population that experiences genetic drift at the same average rate as the real population and depends on both population density and movement patterns. When actual population size changes, then the effective population size will be approximately equal to the actual population size of breeding adults at the time when the population density is lowest.

$$\frac{1}{N_e} = \frac{1}{t}\left[\frac{1}{N_1} + \frac{1}{N_2} + \frac{1}{N_3} + ... \frac{1}{N_t}\right]$$ (14.118)

Effective population size, N_e, is calculated from the equation above, where N_1 is population size at time 1, and N_t is population size at time t. The indirect method used by the authors infers the average size of a population from allele or in this case gene arrangement frequencies over several points in time. Changes in allele frequencies, F, can be estimated using the following three methods: Krimbas and Tsakas (1971)

$$\hat{F}_a = \frac{1}{K}\sum_{i=1}^{k}\frac{(x_i - y_i)^2}{x_i(1-x_i)}$$ (14.119)

Nei and Tajima (1981)

$$\hat{F}_c = \frac{1}{K}\sum_{i=1}^{k}\frac{(x_i - y_i)^2}{(x_i + y_i)/2 - x_i y_i}$$ (14.120)

Pollak (1983)

$$\hat{F}_k = \frac{1}{K-1}\sum_{i=1}^{k}\frac{(x_i - y_i)^2}{(x_i + y_i)/2}$$ (14.121)

where allele frequencies in the sample at the start of a time period are represented by x_i where $i=1,....,K$ and in the second period by $y_i = 1, ..., K$. Effective population sizes were calculated as follows: for the Krimbas and Tsakas (1971) method

$$\hat{N}_{e,a} = \frac{t}{2[\hat{F}_a - 1/(2S_0 - 1/(2S_t)]}$$ (14.122)

For the Nei and Tajima (1981) method

$$\hat{N}_{e,c} = \frac{t-2}{2[\hat{F}_c - 1/(2S_0 - 1/(2S_t)]}$$ (14.123)

for the Pollak (1983) method

$$\hat{N}_{e,k} = \frac{t-2}{2[\hat{F}_k - 1/(2S_0 - 1/(2S_t))]}$$ (14.124)

where S_0 and S_t are the sample sizes at the two time points. Confidence intervals for the estimates of N_e should be calculated using the following formula, as they are not normally distributed

$$\hat{F} = \left[\frac{n\hat{F}}{x^2_{0.025[n]}}, \frac{n\hat{F}}{x^2_{0.0975[n]}} \right]$$ (14.125)

where $n = K - 1$. The median effective population sizes calculated according to the above methods ranged from 1902 to 3321, depending on the method used. The median estimates of effective population size generally agreed with expectations based on the numbers of available dry-season breeding sites at the different locations studied. The authors concluded that these dry season estimates were reasonable as long as the population dispersed over several kilometres. If natural selection is acting on the gene arrangements, rather than random genetic drift, then fluctuations in gene frequency may not be directly related to population size, making the method invalid. The authors suggested that while there is some evidence of selection acting on the gene arrangements observed, mathematical analysis suggests that the effects are selectively neutral over a time period.

Near Delhi, India Yasuno et al. (1973) used the removal method of Zippin (1956) to estimate the population of *Culex quinquefasciatus* from houses. To do this four men collected mosquitoes simultaneously from a house until no more could be found, this generally required 90 min of collecting. They also caught mosquitoes from houses by two men using 8-in sweep nets and this usually took 30–35 min. In both methods the catch was divided into 5-min collecting periods so as to apply the removal method. Both methods proved very time consuming and the precision of the population estimates were low.

References

Aarons T, Walker JR, Gray HF, Mezger EG (1951) Studies of the flight range of *Aedes squamiger* (Coquillett). Proc California Mosq Control Assoc 19: 65–69

Achee NL, Grieco JP, Andre RG, Rejmankova E, Roberts DR (2005) A mark-release-recapture study using a novel portable hut design to define the flight behavior of *Anopheles darlingi* in Belize, Central America. J Am Mosq Control Assoc 21: 366–379

Alho JM (1990) Logistic regression in capture–recapture models. Biometrics 46: 623–635

Anderson RA, Brust RA (1995) Field evidence for multiple host contacts during blood feeding by *Culex tarsalis, Cx. restuans*, and *Cx. nigripalpus* (Diptera: Culicidae). J Med Entomol 32: 705–710

Anderson RA, Edman JD, Scott TW (1990) Rubidium and cesium as host blood-markers to study multiple blood feeding by mosquitoes (Diptera: Culicidae). J Med Entomol 27: 999–1001

Arnason AN (1972a) Parameter estimates from mark–recapture experiments on two populations subject to migration and death. Researches in Population Ecology 13: 97–113

Arnason AN (1972b) Prediction Methods and Variance Estimates for the Parameters of the Triple Catch—Two Population Model with Migration and Death. Scientific Report No. 54, Department of Computer Science, University Manitoba, Canada, 31 pp (mimeographed)

Bailey NTJ (1951) On estimating the size of mobile populations from recapture data. Biometrika 38: 293–306

Bailey NTJ (1952) Improvements in the interpretation of recapture data. J Anim Ecol 21: 120–127

Bailey SF, Eliason DA, Iltis WG (1962) Some marking and recovery techniques in *Culex tarsalis* Coq. flight studies. Mosquito News 22: 1–10

Bailey SF, Eliason DA, Hoffman BL (1965) Flight and dispersal of the mosquito *Culex tarsalis* Coquillett in the Sacramento valley of California. Hilgardia 37: 73–113

Barndorff-Nielsen O (1972) Estimation problems in capture–recapture analysis. Danish Rev Game Biol 6, No. 5

Beesley WN, Crewe W (1963) The bionomics of *Chrysops silacea* Austen, 1907 II. The biting-rhythm and dispersal in rain-forest. Ann Trop Med Parasitol 57: 191–203

Begon M (1979) Investigating Animal Abundance: Capture–recapture for Biologists. Edward Arnold, London

Begon M (1983) Abuses of mathematical techniques in ecology: applications of Jolly's capture–recapture method. Oikos 40: 155–158

Bennett SR, McClelland GAH, Smilanick JM (1981) A versatile system of fluorescent marks for studies of large populations of mosquitoes (Diptera: Culicidae). J Med Entomol 18: 173–174

Berry WL, Stimmann MW, Wolf WW (1972) Marking of native phytophagous insects with rubidium: a proposed technique. Ann Entomol Soc Am 65: 236–238

Birley MH, Charlwood JD (1989) The effect of moonlight and other factors on the oviposition cycle of malaria vectors in Madang, Papua New Guinea. Ann Trop Med Parasitol 83: 415–422

Bishop JA (1973) The proper study of populations. In: Geier PW, Clark LR, Anderson DJ (eds) Insects: Studies in Population Management. Mem Ecol Soc Aust 1

Bishop JA, Bradley JS (1972) Taxi-cabs as subjects for a population study. J Biol Educ 6: 227–231

Bishop JA, Sheppard PM (1973) An evaluation of two capture–recapture models using the technique of computer simulation. In: Bartlett MS, Hiorns RW (eds) The Mathematical Theory of the Dynamics of Biological Populations. Academic Press, London, pp 233–244

Blower JG, Cook LM, Bishop JA (1981) Estimating the Size of Animal Populations. Allen & Unwin, London

Bond HA, Craig GB, Fay RW (1970) Field mating and movement of *Aedes aegypti*. Mosquito News 30: 394–402

Bowden J, Brown G, Stride T (1979) The application of X-ray spectrometry to analysis of elemental composition (chemoprinting) in the study of migration of *Noctua pronuba* L. Ecol Entomol 4: 199–204

Bowden J, Digby PGN, Sherlock PL (1984) Studies of elemental composition as a biological marker in insects. I. The influence of soil type and host-plant on elemental composition of *Noctua pronuba* (L.) (Lepidoptera: Noctuidae). Bull Entomol Res 74: 207–225

Burg JG (1994) Marking *Dermacentor variabilis* (Acari: Ixodidae) with Rubidium. J Med Entomol 31: 658–662

Burg JG (1997) Unsuccessful rubidium marking of American dog tick (Acari: Ixodidae) adults. J Med Entomol 34: 29–32

Burton GJ (1964) Attack on the vector of filariasis in British Guiana. US Public Health Rep 79: 137–143

Campbell MM, Kettle DS (1976) Marking of adult *Culicoides brevitarsis* Kieffer (Diptera: Ceratopogonidae). J Aust Entomol Soc 14: 383–386

Carle FL, Strub MR (1978) A new method for estimating population size from removal data. Biometrics 34: 621–630

Carothers AD (1973a) Capture–recapture methods applied to a population with known parameters. J Anim Ecol 42: 125–146

Carothers AD (1973b) The effects of unequal catchability on Jolly–Seber estimates. Biometrics 29: 79–100

Carothers AD (1979) Quantifying unequal catchability and its effect on survival estimates in an actual population. J Anim Ecol 48: 863–869

Causey OR, Kumm HW (1948) Dispersion of forest mosquitoes in Brazil; preliminary studies. Am J Trop Med 28: 469–480

Causey OR, Kumm HW, Laemmert HW (1950) Dispersion of forest mosquitoes in Brazil: further studies. Am J Trop Med 30: 301–312

Chang HT (1946) Studies on the use of fluorescent dyes for marking *Anopheles quadrimaculatus*. Mosquito News 6: 122–125

Chapman DG (1955) Population estimation based on changes of composition caused by a selective removal. Biometrika 42: 279–290

Chapman DG, Junge CO (1956) The estimation of the size of a stratified animal population. Ann Math Stat 27: 375–389

Charlwood JD, Graves PM, Birley MH (1986) Capture–recapture studies with mosquitoes of the group of *Anopheles punctulatus* Donitz (Diptera: Culicidae) from Papua New Guinea. Bull Entomol Res 76: 211–227

Charlwood JD, Graves PM, Marshall TF de C (1988) Evidence for a 'memorized' home range in *Anopheles farauti* females from Papau New Guinea. Med Vet Entomol 2: 101–108

Conrad LB, Joslyn DJ, Stewart BR (1983) Laboratory testing of the Giemsa self-marker in four species of mosquitoes. Proc Ann Mtg New Jers Mosq Control Assoc 70: 84–89

Conway GR, Trpis M, McClelland GAH (1974) Population parameters of the mosquito *Aedes aegypti* (L.) estimated by mark–release–recapture in a suburban habitat in Tanzania. J Anim Ecol 53: 289–304

Cook LM, Brower LP, Croze HJ (1967) The accuracy of a population estimate from multiple recapture data. J Anim Ecol 36: 57–60

Coppedge JR, Spencer JP, Brown HE, Whitten CJ, Snow JW, Wright JE (1979) A new dye marking technique for the screwworm. J Econ Entomol 72: 40–42

Cormack RM (1968) The statistics of capture–recapture estimates. Oceanogr Mar Biol Annu Rev 6: 455–506

Cormack RM (1972) The logic of capture–recapture estimates. Biometrics 28: 337–343

Cormack RM (1973) Commonsense estimates from capture–recapture studies. In: Bartlett MS, Hiorns RW (eds) The Mathematical Theory of the Dynamics of Biological Populations. Academic Press, London, pp 225–234

Costantini C, Li SongGang, della Torre A, Sagnon N, Coluzzi M, Taylor CE (1996) Density, survival and dispersal of *Anopheles gambiae* complex mosquitoes in a West African Sudan savanna village. Med Vet Entomol 10: 203–219

Craig GB, Hickey WA (1967) Genetics of *Aedes aegypti*. In: Wright JW, Pal R (eds) Insect Vectors of Disease. Elsevier, Amsterdam, pp. 67–131

Cunningham MP, Harley JMB, Grainge EB (1963) The Labelling of Animals with Specific Agglutinins and the Detection of these Agglutinins in the Blood Meals of Glossina. East African Trypanosomiasis Research Organisation Report, 1961, pp 23–4

Curtis CF, Rawlings P (1980) A preliminary study of dispersal and survival of *Anopheles culicifacies* in relation to the possibility of inhibiting the spread of insecticide resistance. Ecol Entomol 5: 11–17

Dalmat HT (1950) Studies on the flight range of certain Simuliidae, with the use of aniline dye marker. Ann Entomol Soc Am 43: 537–545

Darling ST (1925). Entomological research in malaria. Sth Med J Nashville 18: 446–449

Darroch JN (1958) The multiple recapture census: I. Estimation of a closed population. Biometrika 45: 343–351

Darroch JN (1959) The multiple-capture census. II. Estimation when there is immigration or death. Biometrika 46: 336–351

Darroch JN (1961) The two-sample capture–recapture census when tagging and sampling are stratified. Biometrika 48: 241–260

Dempster JP (1971) The population ecology of the Cinnabar moth, *Tyria jacobaeae* L. (Lepidoptera, Arctiidae). Oceologia 7: 26–67

Dempster JP, Lakhani KH, Coward PA (1986) The use of chemical composition as a population marker in insects: a study of the Brimstone butterfly. Ecol Entomol 11: 51–65

Dosdall LM, Galloway MM, Gadawski RM (1992) New self-marking device for dispersal studies of black flies (Diptera: Simuliidae). J Am Mosq Control Assoc 8: 187–190

dos Santos R la C, Forattini OP (1999) Marcação-soltura-recaptura para determinar o tamanho da população natural de *Anopheles albitarsis* l.s. (Diptera: Culicidae). Rev Saúde Pública 33: 309–313

Dow RP (1971) The dispersal of *Culex nigripalpus* marked with high concentrations of radiophosphorus. J Med Entomol 8: 353–363

Dow RP, Reeves WC, Bellamy RE (1965) Dispersal of female *Culex tarsalis* into a larvicided area. Am J Trop Med Hyg 14: 656–670

Dowdeswell WH (1967) Practical Animal Ecology. Methuen, London

Dowdeswell WH, Fisher RA, Ford EB (1940) The quantitative study of populations in the Lepidoptera. 1. *Polyommatus icarus* Rott. Ann Eugen Lond 10: 123–136

Dunn PH, Mechalas BJ (1963) An easily constructed vacuum duster. J Econ Entomol 56: 899

Eberhardt LL (1990) Using radio-telemetry for mark–recapture studies with edge effects. J Appl Ecol 27: 259–271

Eddy GW, Roth AR, Plapp FW (1962) Studies on the flight habits of some marked insects. J Econ Entomol 55: 603–607

Edman JD, Lea AO (1972) Sexual behaviour of mosquitoes. 2. Large-scale rearing and marking of *Culex* for field experiments. Ann Entomol Soc Am 65: 267–269

Edman JD, Scott TW, Costero A, Morrison AC, Harrington LC, Clark GG (1998) *Aedes aegypti* (Diptera: Culicidae) movement influenced by availability of oviposition sites. J Med Entomol 35: 578–583

Eyles DE (1943) A method for catching, marking, and re-examining large numbers of *Anopheles quadrimaculatus*. J Natn Malar Soc 2: 85–91

Eyles DE (1944) A Critical Review of the Literature Relating to the Flight and Dispersion Habits of Anopheline Mosquitoes. US Public Health Bull No. 287

Eyles DE, Cox WW (1943) The measurements of a population of *Anopheles quadrimaculatus* Say. J Natn Malar Soc 2: 71–83

Fabian MM, Toma T, Tsuzuki A, Saita S, Miyagi I (2005) Mark-release-recapture experiments with *Anophleles saperoi* (Diptera: Culicidae) in the Yona forest, northern Okinawa, Japan. Southeast Asian J Trop Med Public Health 36: 54–63

Fales JH, Bodenstein OF, Mills GD, Wessel LH (1964) Preliminary studies on face fly dispersion. Ann Entomol Soc Am 57: 135–137

Fay RW, Craig GB (1969) Genetically marked *Aedes aegypti* in studies of field populations. Mosquito News 29: 121–127

Fisher RA, Ford EB (1947) The spread of a gene in natural conditions in a colony of the moth *Panaxia dominula* L. Heredity 1: 143–174

Fleischer SJ, Gaylor MJ, Hue NV, Graham LC (1986). Uptake and elimination of rubidium, a physiological marker, in adult *Lygus lineolaris* (Hemiptera: Miridae). Ann Entomol Soc Am 79: 19–25

Fletcher BS, Kapatos E, Southwood TRE (1981) A modification of the Lincoln index for estimating the population densities of mobile insects. Ecol Entomol 6: 397–400

Ford EB (1953) The Experimental Study of Evolution. Report of the Brisbane Meeting of the Australian and New Zealand Association for the Advancement of Science (1951) 28: 143–154

Frank JH, Curtis GA (1977) On the bionomics of bromeliad-inhabiting mosquitoes. V. A mark–release–recapture technique for estimation of population size of *Wyeomyia vanduzeei*. Mosquito News 37: 444–452

Freeman PR (1973a) Sequential recapture. Biometrika 60: 141–153

Freeman PR (1973b) A numerical comparison between sequential tagging and sequential recapture. Biometrika 60: 499–508

Fruton JS, Simmonds S (1958) General Biochemistry, 2nd edn. John Wiley & Sons, New York

Fryer JC, Meek CL (1989) Further studies on marking an adult mosquito, *Psorophora columbiae*, in situ using fluorescent pigments. SWest Entomol 14: 409–418

Gangwere SK, Chavin W, Evans FC (1964) Methods of marking insects with special reference to Orthoptera (Sens. lat.). Ann Entomol Soc Am 57: 662–669

Gaskell TJ, George BJ (1972) A Bayesian modification of the Lincoln index. J Appl Ecol 9: 377–384

Gazey WJ, Staley MJ (1986) Population estimation from mark–recapture experiments using a sequential Bayes algorithm. Ecology 67: 941–951

Geiger JC, Purdy WC, Tarbett RE (1919) Effective malaria control in a rice-field district with observations on experimental mosquito flights. J Am Med Assoc 72: 844–847

Germain F, Hervé JP, Geoffroy B (1977) Variations du taux de survie des femelles d'*Aedes africanus* (Theobald) dans une galerie forestière de sud de l'Empire Centrafricain. Cah ORSTOM sér Entomol Méd Parasitol 15: 291–299

Germain M, Eouzan JP, Ferrara L (1972) Données sur les facultés de dispersion de deux diptères d'intérrêt médical: *Aedes africanus* (Theobald) et *Simulium damnosum* Theobald dans la domaine montagnard du nord du Cameroun occidental. Cah ORSTOM sér Entomol Méd Parasitol 10: 291–300

Gillies MT (1958) A modified technique for the age-grading of populations of *Anopheles gambiae*. Ann Trop Med Parasitol 52: 261–273

Gillies MT (1961) Studies on the dispersion and survival of *Anopheles gambiae* Giles in East Africa, by means of marking and release experiments. Bull Entomol Res 52: 99–127

Ginsberg HS (1986) Dispersal patterns of *Aedes sollicitans* (Diptera: Culicidae) at the east end of the fire island national seashore, New York, USA. J Med Entomol 23: 146–155

Glasgow JP (1953) The extermination of an animal population by artificial predation and the estimation of population size. J Anim Ecol 22: 32–46

Gouteux J-P, Artzrouni M, Jarry M (2001) A density-dependent model with reinvasion for estimating tsetse fly populations (Diptera: Glossinidae) through trapping. Bull Entomol Res 91: 177–183

Graham HM, Magnum CL (1971) Larval diets containing dyes for tagging pink bollworm moths internally J Econ Entomol 64: 376–379

Griffiths G, Winder L, Bean D, Preston R, Moate R, Neal R, Williams E, Holland J, Thomas G (2001) Laser marking the carabid *Pterostichus melanarius* for mark-release-recapture. Ecol Entomol 26: 662–663

Haddow AJ (1942) The mosquito fauna and climate of native huts at Kisumu, Kenya. Bull Entomol Res 33: 91–142

Hagler JR, Jackson CG (2001) Methods for marking insects: current techniques and future prospects. Annu Rev Entomol 46: 511–543

Hamada R (1976) Density estimation by the modified Jackson's method. Appl Entomol Zool 11: 194–201

Haramis LD, Foster WA (1983) Survival and population density of *Aedes triseriatus* (Diptera: Culicidae) in a woodlot in central Ohio, USA. J Med Entomol 20: 391–398

Haüsermann W, Fay RW, Hacker CS (1971) Dispersal of genetically marked female *Aedes aegypti* in Mississippi. Mosquito News 31: 37–51

Hayes DK, Schechter MS (1970) Survival of codling moth larvae treated with methylene blue under short- and long-day photoperiod. J Econ Entomol 63: 997

Hayes JL, Reed KG (1989) Using rubidium-treated artificial nectar to label adults and eggs of *Heliothis virescens* (Lepidoptera: Noctuidae). Environ Entomol 18: 807–810

Hendricks DE (1971) Oil-soluble blue dye in larval diet marks adults, eggs, and first-stage F1 larvae of the pink bollworm. J Econ Entomol 64: 1404–1406

Hendricks DE, Graham HM (1970) Oil-soluble dye in larval diet for tagging moths, eggs, and spermatophores of tobacco budworms. J Econ Entomol 63: 1019–1020

Hendricks DE, Leal MP, Robinson SH, Hernandez NS (1971) Oil-soluble black dye in larval diet marks, adults and eggs of the tobacco budworm and pink bollworm. J Econ Entomol 64: 1399–1401

Heron RJ (1968) Vital dyes as markers for behavioral and population studies of the Larch sawfly, *Pristiphora erichsonii* (Hymenoptera: Tenthredinidae). Can Entomol 100: 470–475

Hervy JP (1977a) Expérience de marquage–lâcher–recapture portant sur *Aedes aegypti* Linné, en zone de savane soudanienne ouest-africaine. I. Le cycle trophogonique. Cah ORSTOM sér Entomol Méd Parasitol 15: 353–364

Hervy JP (1977b) Expérience de marquage–lâcher–recapture, portant sur *Aedes aegypti* Linné, en zone de savane soudanienne ouest africaine. II. Relations entre habitat, morphologie et comportement. Cah ORSTOM sér Entomol Méd Parasitol 15: 365–372

Hii JLK, Vun YS (1985) A study of dispersal, survival and adult population estimates of the malaria vector, *Anopheles balabacensis* Baisas (Diptera: Culicidae) in Sabah, Malaysia. Trop Biomed 2: 121–131

Hodgson JC, Spielman A, Komar N, Krahforst CF, Wallace GT, Pollack RJ (2001) Interrupted blood-feeding by *Culiseta melanura* (Diptera: Culicidae) on European starlings. J Med Entomol 38: 59–66

Holbrook FR (1995) Rubidium in female *Culicoides variipennis sonorensis* (Diptera: Ceratopogonidae) after engorgement on a rubidium-treated host. J Med Entomol 32: 387–389

Holbrook FR, Belden RP, Bobian RJ (1991) Rubidium for marking adults of *Culicoides variipennis* (Diptera: Ceratopogonidae). J Med Entomol 28: 246–249

Honório NA, SilvaW da C, Leite PJ, Gonçalves JM, Lounibos LP, Lourenço-de-Oliveira R (2003) Dispersal of *Aedes aegypti* and *Aedes albopictus* (Diptera: Culicidae) in an urban endemic dengue area in the State of Rio de Janeiro, Brazil. Mem Inst Oswaldo Cruz 98: 191–198

Howard JJ, White DJ, Muller SL (1989) Mark–recapture studies on the *Culiseta* (Diptera: Culicidae) vectors of eastern equine encephalitis virus. J Med Entomol 26: 190–199

Ikeshoji T, Yap HH (1990) Impact of the insecticide-treated sound traps on an *Aedes albopictus* population. Jap J Sanit Zool 41: 213–217

Itô Y (1973) A method to estimate a minimum population density with a single recapture census. Researches in Population Ecology 14: 159–169

Ivanova VL, Ipatov VP (1987) Migration routes of *Aedes communis* De Geer, *Ae. punctor* Kirby and *Ae. pionips* Dyar mosquito females. Communication I. The method of fluorescent dust labelling applied to follow up the mosquito flying into the settlement in the Middle taiga zone. Med Parasitol Parasitic Dis 56: 55–59 (In Russian, English summary)

Iwao S (1963) On a method for estimating the rate of population interchange between two areas. Researches in Population Ecology 5: 44–50

Jackson CHN (1933) On a method of marking tsetse flies. J Anim Ecol 2: 289–290

Jackson CHN (1939) The analysis of an animal population. J Anim Ecol 8: 238–246

Jolly GM (1965) Explicit estimates from capture–recapture data with both birth and immigration-stochastic model. Biometrika 52: 225–247

Jones RL, Harrell EA, Snow JW (1972) Three dyes as markers for Corn Earworm moths. J Econ Entomol 65: 123–126

Joslyn DJ, Fish D (1986) Adult dispersal of *Aedes communis* using Giemsa self-marking. J Am Mosq Control Assoc 2: 89–90

Joslyn DJ, Conrad LB, Slavin PT (1985) Development and preliminary field testing of the Giemsa self-marker for the salt marsh mosquito *Aedes sollicitans* (Walker) (Diptera: Culicidae). Ann Entomol Soc Am 78: 20–23

Kay BH, Mottram P (1986) In-vivo staining of *Aedes vigilax, Aedes aegypti* and *Culex annulirostris* larvae with Giemsa and other vital dyes. J Am Mosq Control Assoc 2: 141–145

Kelker GH (1940) Estimating deer populations by a differential hunting loss in the sexes. Proc Utah Acad Sci 17: 65–69

Kimsey RB, Kimsey PB (1984) Identification of arthropod blood meals using rubidium as a marker; a preliminary report. J Med Entomol 21: 714–719

Kline DL, Wood JR (1989) Practical assessment and enhancement of some currently used adult surveillance techniques for Florida mosquitoes. Phase II. Final Report to Florida Department of Health and Rehabilitation

Knight RH, Southon HAW (1963) A simple method for marking haematophagous insects during the act of feeding. Bull Entomol Res 54: 379–382

Knoche HW (1991) Radioisotopic Methods for Biological and Medical Research. Oxford University Press, Oxford

Kono T (1953) On the estimation of insect population by time unit collecting. Researches in Population Ecology 2: 85–94

Krebs CJ, Boonstra R (1984) Trappability estimates for mark–recapture data. Can J Zool 62: 2440–2444

Krimbas KB, Tsakas S (1971) The genetics of *Dacus oleae*. V. Changes of esterase polymorphism in a natural population following insecticide control – selection or drift? Evolution 25: 454–460

Legg DE, Chiang HC (1984) Rubidium marking technique for the European Corn Borer (Lepidoptera: Pyralidae) in corn. Environ Entomol 13: 578–583

Le Prince JAA, Griffitts THD (1917*a*) Flight of Mosquitoes. Studies on the distance of flight of *Anopheles quadrimaculatus*. US Public Health Rep 32: 656–659

Le Prince JA, Griffitts THD (1917*b*). Notes from a malaria survey: impounded waters, biting of *A. punctipennis* on porches, distance of flight of *A. quadrimaculatus*. Sth Med J Nashville 10: 642–644

Leslie PH (1952) The estimation of population parameters from data obtained by means of the capture–recapture method. II. The estimation of total numbers. Biometrika 39: 363–388

Leslie PH, Chitty D (1951) The estimation of population parameters from data obtained by means of the capture–recapture method. I. The maximum likelihood equations for estimating the death-rate. Biometrika 38: 269–292

Lillie TH, Kline DL, Hall DW (1985) The dispersal of *Culicoides mississippiensis* (Diptera: Ceratopogonidae) in a salt marsh near Yankeetown, Florida. J Am Mosq Control Assoc 1: 463–467

Lincoln FC (1930) Calculating waterfowl abundance on the basis of banding returns. USDA Circular 118

Lindig OH, Wiygul G, Wright JE, Dawson JR, Roberson J (1980) Rapid method for mass-marking boll weevils. J Econ Entomol 73: 385–386

Lindquist AW, Ikeshoji T, Grab B, De Meillon B, Khan ZH (1967) Dispersion studies of *Culex pipiens fatigans* tagged with ^{32}P in the Kemmendine area of Rangoon, Burma. Bull World Health Organ 36: 21–37

Lounibos LP, Lima DC, Lourenço-de-Oliveira R (1998) Prompt mating of released *Anopheles darlingi* in western Amazonian Brazil. J Am Mosq Control Assoc 14: 210–213

Macdonald WW, Sebastian A, Tun MM (1968) A mark–release–recapture experiment with *Culex pipiens fatigans* in the village of Okpo, Burma. Ann Trop Med Parasitol 62: 200–209

MacLeod J (1958) The estimation of numbers of mobile insects from low-incidence recapture data. Trans R Entomol Soc Lond 110: 363–392

MacLeod J, Donnelly J (1957) Individual and group marking methods for fly-population studies. Bull Entomol Res 48: 585–592

Majid SA (1937) An improved technique for marking and catching mosquitoes. Rec Malar Surv India 7: 105–107

Manly BFJ (1969*a*) On a method of population estimation using capture–recapture data. Entomologist 102: 117–120

Manly BFJ (1969*b*) Some properties of a method of estimating the size of mobile animal populations. Biometrika 56: 407–410

Manly BFJ (1970) A simulation study of animal population estimation using the capture–recapture method. J Appl Ecol 7: 13–39

Manly BFJ (1971*a*) Estimates of a marking effect with capture-recapture sampling. J Appl Ecol 8: 181–189

Manly BFJ (1971*b*) A simulation study of Jolly's method for analysing capture–recapture data. Biometrics 27: 415–424

Manly BFJ (1985) A test of Jackson's method for separating death and emigration with mark–recapture data. Researches in Population Ecology 27: 99–109

Manly BFJ, Parr MJ (1968) A new method of estimating population size, survivorship, and birth rate from capture–recapture data. Trans Soc Br Entomol 18: 81–89

Marten GG (1969) A regression method for mark–recapture estimation of population size with unequal catchability. Ecology 51: 291–295

McCarty JC, Jenkins JN, Parrott WL, Davich TB (1972) Effect of dyes on body fat and eye color of Ebony Pearl boll weevils. J Econ Entomol 65: 370–372

McClelland GAH, Conway GR (1971) Frequency of blood feeding in the mosquito *Aedes aegypti*. Nature 232: 485–486

McClelland GAH, McKenna RJ, Cahill TA (1973*a*) New Approaches for Mark–Release–Recapture Studies on Biting Insects. Proc. Symposium on Biting Fly Control and Environmental Quality. Alberta, Canada (1972), DR 217, pp 49–52

McClelland GAH, McKenna RJ, Jolly DJ, Cahill TA (1973*b*) Results of preliminary mark–release–recapture trials with *Aedes nigromaculis*. Proc California Mosq Control Assoc 41: 107–108

McDonald PT (1977*a*) Population characteristics of domestic *Aedes aegypti* (Diptera: Culicidae) in villages on the Kenya coast. I. Adult survivorship and population size. J Med Entomol 14: 42–48

McDonald PT (1977*b*) Population characteristics of domestic *Aedes aegypti* (Diptera: Culicidae) in villages on the Kenya coast. II. Dispersal within and between villages. J Med Entomol 14: 49–53

McLean JA, Stump IG, d'Auria JM, Holman J (1979) Monitoring trace elements in diets and life stages of the onion maggot, *Hylemya antiqua* (Diptera: Anthomyiidae), with X-ray energy spectrometry. Can Entomol 111: 1293–1298

Meek CL, Broussard BB, Andis MD (1987) Marking adult mosquitoes using aerially applied fluorescent pigment. J Am Mosq Control Assoc 3: 400–402

Meek CL, Fryer JC, Niebylski ML (1988) Marking adult mosquitoes using fluorescent pigments in dispersal studies. Bull Soc Vector Ecol 13: 319–322

Miller SD, Becker EF, Ballard WH (1987) Black and brown bear density estimates using modified capture–recapture techniques. Int Conf Bear Res Mgmt 7: 23–25

Mitchell CJ (1979) Winter survival of *Culex tarsalis* (Diptera: Culicidae) hibernating in mine tunnels in boulder county, Colorado, USA. J Med Entomol 16: 482–487

Moffitt HR, Albano DJ (1972) Vacuum application of fluorescent powders as markers for adult Codling moths. J Econ Entomol 65: 882–884

Morrison AC, Costero A, Edman JD, Clark GG, Scott TW (1999) Increased fecundity of *Aedes aegypti* fed human blood before release in a mark-recapture study in Puerto Rico. J Am Mosq Control Assoc 15: 98–104

Morton AC (1982) The effects of marking and capture on recapture frequencies of butterflies. Oecologia (Berl.) 53: 105–110

Moss JI, van Steenwyk RA (1982) Marking pink bollworm (Lepidoptera: Gelechiidae) with cesium. Environ Entomol 11: 1264–1268

Muir LE, Kay BH (1998) *Aedes aegypti* survival and dispersal estimated by mark-release-recapture in Northern Australia. Am J Trop Med Hyg 58: 277–282

Nei M, Tajima F (1981) Genetic drift and estimation of effective population size. Genetics 98: 624–640

Nelson RL, Milby MM (1980) Effects of fluorescent marker dusts on *Culex tarsalis*, a factor in mark–release–recapture studies. Proc California Mosq Vector Control Assoc 48: 66–68

Nelson RL, Milby MM, Reeves WC, Fine PEM (1978) Estimates of survival, population size, and emergence of *Culex tarsalis* at an isolated site. Ann Entomol Soc Am 71: 801–808

Newhouse BA (1953) A progress report of the Aedes flight range studies in Kern county. Proc California Mosq Control Assoc 22: 80–81

Niebylski ML, Meek CL (1989) A self-marker device for emergent adult mosquitoes. J Am Mosq Control Assoc 5: 86–90

Nielsen ET (1961) On the habits of the migratory butterfly *Ascia monuste*. L. Biol Medd Kgl Dan Vidensk Selsk 23: 1–81

Nielsen ET, Nielsen AT (1953) Field observations on the habits of *Aedes taeniorhynchus*. Ecology 34: 141–156

Nutsathapana S, Sawasdiwongphorn P, Chitprarop U, Cullen JR, Gass RF, Green CA (1986) A mark–release–recapture demonstration of host preference heterogeneity in *Anopheles minimus* Theobald (Diptera: Culicidae) in a Thai village. Bull Entomol Res 76: 313–320

O'Donnell MS, Berry G, Carvan T, Bryan JH (1992) Dispersal of adult females of *Culex annulirostris* in Griffith, New South Wales, Australia. J Am Mosq Control Assoc 8: 159–165

Otis DL, Burnham KP, White GC, Anderson DR (1978) Statistical inferences from capture data on closed animal populations. Wildl Monogr 62: 1–135

Padgham DE, Cook AG, Hutchinson D (1984) Rubidium marking of the rice pest *Nilaparva lugens* (Stål) and *Sogatella furcifera* (Horváth) (Hemiptera: Dephacidae) for field dispersal studies. Bull Entomol Res 74: 379–385

Paing M, Naing TT (1988) Marking mosquito larvae for mark–release–recapture studies on adults. J Commun Dis 20: 276–279

Pal R (1947) Marking mosquitoes with fluorescent compounds and watching them by ultra-violet light. Nature 160: 298–299

Pardo RH, Torres M, Morrison AC, Ferro C (1996) Effect of fluorescent powder on *Lutzomyia longipalpis* (Diptera: Psychodidae) and a simple device for marking sand flies. J Am Mosq Control Assoc 12: 235–242

Parr MJ (1965) A population study of a colony of imaginal *Ischnura elegans* (Van der Linden) (Odonata: Coenagriidae) at Dale, Pembrokeshire. Field Studies 2: 237–282

Parr MJ, Gaskell TJ, George BJ (1968) Capture–recapture methods of estimating animal numbers. J Biol Educ 2: 95–117

Pearson AC, Ballmer GR, Sevacherian V, Vail PV (1989) Interpretation of rubidium marking levels in beet armyworm eggs (Lepidoptera: Noctuidae). Environ Entomol 18: 844–848

Peer DF (1957) Further studies on the mating range of the honey bee, *Apis mellifera* L. Can Entomol 89: 108–110

Peffly RL, LaBrecque GC (1956) Marking and trapping studies on dispersal and abundance of Egyptian house flies. J Econ Entomol 49: 214–217

Petersen CGJ (1896) The early immigration of young plaice into Limfjord from the German sea. Report of the Danish Biological Station 6: 1–48

Pollak E (1983) A new method for estimating the effective population size from allele frequency change. Genetics 104: 531–548

Pollock KH (1981) Capture-recapture models allowing for age-dependent survival and capture rates. Biometrics 37: 521–529

Porter SD, Jorgensen CD (1980) Recapture studies of the harvester ant *Pogonomytmex owyheei* Cole, using a fluorescent marking technique. Ecol Entomol 5: 263–269

Pumpuni CB, Walker ED (1989) Population size and survivorship of adult *Aedes triseriatus* in a scrap tireyard in northern Indiana. J Am Mosq Control Assoc 5: 166–172

Quarterman KD, Mathis W, Kilpatrick TW (1954) Urban fly dispersal in the area of Savannah, Georgia. J Econ Entomol 47: 405–412

Rajagopalan PK, Menon PKB, Mani TR, Brooks GD (1975) Measurement of density of immatures of *Culex pipiens fatigans* in effluent drains. J Commun Dis 7: 327–337

Reeves WC, Brookman B, Hammon WM (1948) Studies on the flight range of certain *Culex* mosquitoes, using a fluorescent-dye marker, with notes on *Culiseta* and *Anopheles*. Mosquito News 8: 61–69

Reisen WK, Aslamkhan M (1979) A release–recapture experiment with the malaria vector, *Anopheles stephensi* Liston, with observations on dispersal, survivorship, population size, gonotrophic rhythm and mating behaviour. Ann Trop Med Parasitol 73: 251–69

Reisen WK, Mahmood F (1981) Relative abundance, removal sampling, and mark–recapture estimates of population size of *Anopheles culicifacies* and *An. stephensi* at diurnal resting sites in rural Punjab province, Pakistan. Mosquito News 41: 22–30

Reisen WK, Baker RH, Sakai RK, Aziz-Javed A, Aslam Y, Siddiqui TF (1977) Observations on the mating behavior and survivorship of *Culex tritaeniorhynchus* Giles (Diptera: Culicidae) during late autumn. Southeast Asian J Trop Med Public Health 8: 537–545

Reisen WK, Aslam Y, Siddiqui TF, Khan AQ (1978) A mark–release–recapture experiment with *Culex tritaeniorhynchus* Giles. Trans R Soc Trop Med Hyg 72: 167–177

Reisen WK, Mahmood F, Parveen T (1979) *Anopheles subpictus* Grassi: observations on survivorship and population size using mark–release–recapture and dissection methods. Researches in Population Ecology 21: 12–29

Reisen WK, Mahmood F, Parveen T (1980) *Anopheles culicifacies* Giles: a mark–release–recapture experiment with cohorts of known age with implications for malaria epidemiology and genetical control in Pakistan. Trans R Soc Trop Med Hyg 74: 307–317

Reisen WK, Mahmood F, Azra K (1981) *Anopheles culicifacies* Giles: adult ecological parameters measured in rural Punjab province, Pakistan using capture–release–recapture and dissection methods, with comparative observations on *An. stephensi* Liston and *An. subpictus* Grassi. Researches in Population Ecology 23: 39–60

Reisen WK, Mahmood F, Parveen T (1982*a*) Seasonal trends in population size and survivorshp of *Anopheles culicifacies, An. stephensi* and *An. subpictus* (Diptera: Culicidae) in rural Punjab province, Pakistan. J Med Entomol 19: 86–97

Reisen WK, Sakai RK, Baker RH, Azra K, Niaz S (1982*b*). *Anopheles culicifacies*: observations on population ecology and reproductive behavior. Mosquito News 42: 93–101

Reiter P (1996) Oviposition et dispersion d'*Aedes aegypti* dans l'environnement urbain. Bull Soc Pathol Exot 89: 120–122

Reiter P, Amador MA, Anderson RA, Clark GG (1995) Dispersal of *Aedes aegypti* in an urban area after blood feeding as demonstrated by rubidium-marked eggs. Am J Trop Med Hyg 52: 177–179

Renshaw M (1991) Population dynamics and ecology of *Aedes cantans* (Diptera: Culicidae) in England. Ph.D. thesis, University of Liverpool

Renshaw M, Service MW, Birley MH (1994) Host finding, feeding patterns and evidence for a memorized home range of the mosquito *Aedes cantans*. Med Vet Entomol 8: 187–193

Reuben R, Yasuno M, Panicker KN, LaBrecque GC (1973) The estimation of adult populations of *Aedes aegypti* at two localities in Delhi, India. J Commun Dis 5: 154–162

Richards OW, Waloff N (1954) Studies on the biology and population dynamics of British grasshoppers. Anti-Locust Bulletin 17

Roff DA (1973) On the accuracy of some mark–recapture estimators, Oecologia (Berl.) 12: 15–34

Rooum D (1988) Identifying individual insects by means of spot codes. Antenna 12: 155–157

Russell PF, Santiago D (1934) Flight range of *Anopheles* in the Philippines. Second experiment with stained mosquitoes. Am J Trop Med 14: 407–424

Russell PF, Knipe FW, Rao TR, Putnam P (1944) Some experiments on flight range of *Anopheles culicifacies*. J Exp Zool 97: 135–163

Schlein Y (1987) Marking *Phlebotomus papatasi* (Diptera: Psychodidae) by feeding on sprayed, coloured sugar bait: a possible means for behavioural and control studies. Trans R Soc Trop Med Hyg 81: 599

Schnabel ZE (1938) The estimation of the total fish population of a lake. Am Math Mon 45: 348–350

Schoof HF, Siverly RE (1954) Multiple release studies on the dispersion of *Musca domestica* at Phoenix, Arizona. J Econ Entomol 47: 830–838

Seber GAF (1965) A note on the multiple-recapture census. Biometrika 49: 249

Seber GAF (1973) The Estimation of Animal Abundance and Related Parameters. Griffin, London

Seber GAF (1982) The Estimation of Animal Abundance and Related Parameters. 2nd edn. Macmillan, New York

Seber GAF (1986) A review of estimating animal abundance. Biometrics 42: 267–292

Self LS, Tun MM, Mathis HL, Abdulcader MHM, Sebastian A (1971) Studies on infiltration of marked *Culex pipiens fatigans* into sprayed areas of Rangoon, Burma. Bull World Health Organ 45: 379–383

Sempala SDK (1981) The ecology of *Aedes (Stegomyia) africanus* Theobald in a tropical forest in Uganda: Mark–release–recapture studies on a female adult population. Insect Science and its Application 1: 211–224

Service MW (1971) Studies on sampling larval populations of the *Anopheles gambiae* complex. Bull World Health Organ 45: 169–180

Service MW (1976) Mosquito Ecology. Field Sampling Methods. Applied Science Publishers, London

Service MW (1993) Mosquito Ecology. Field Sampling Methods. 2nd edn. Chapman & Hall, London

Service MW (1994. Male swarming of the mosquito *Culex (Culex) torrentium* in England. Med Vet Entomol 8: 95–98

Shapiro JM, Saliternik Z, Belferman S (1944) Malaria survey of the Dead Sea area during 1942, including the description of a mosquito flight test and its results. Trans R Soc Trop Med Hyg 38: 95–116

Shepard M, Waddill VH (1976) Rubidium as a marker for Mexican bean beetles, *Epilachna varivestis* (Coleoptera: Coccinellidae). Can Entomol 108: 337–339

Sheppard PM, Macdonald WW, Tonn RJ, Grab B (1969) The dynamics of an adult population of *Aedes aegpyti* in relation to dengue haemorrhagic fever in Bangkok. J Anim Ecol 38: 661–702

Singh KRP, Razdan RK, Vaidyanathan V, Malhotra PR (1975*a*) Caging, marking and transportation of *Culex pipiens fatigans* Wied. for large scale genetic control operations. J Commun Dis 7: 269–279

Singh KRP, Brooks GD, Ansari MA, Razdan RK, Khan QU (1975*b*) Development of equipment for packaging, marking and transportation of *Aedes aegypti* for large scale genetic control operations. J Commun Dis 7: 294–300

Singh N, Yasuno M (1972) A device for self-marking of mosquitoes. Bull World Health Organ 47: 677–679

Sinsko MJ, Craig GB (1979) Dynamics of an isolated population of *Aedes triseriatus* (Diptera: Culicidae). I. Population size. J Med Entomol 15: 89–98

Skalski JR, Robson DS (1979) Tests of homogeneity and goodness-of-fit to a truncated geometric model for removal sampling. In: Cormack RM, Patil GP, Robson DS (eds) Sampling Biological Populations. Co-operative Publishing House, Fairland, Maryland, pp 283–313

Skalski JR, Robson DS (1982) A mark and removal method field procedure for estimating population abundance. J Wildl Mgmt 46: 741–751

Skalski JR, Robson DS, Simmons MA (1983) Comparative census procedures using single mark–recapture methods. Ecology 64: 752–760

Skalski JR, Simmons MA, Robson DS (1984) The use of removal sampling in comparative censuses. Ecology 65: 1006–1015

Slooff R, Herath PRJ (1980) Ovarian development and biting frequency in *Anopheles culicifacies* Giles in Sri Lanka. Trop Geogr Med 32: 306–311

Smith GE, Watson RB, Crowell RL (1941) Observations on the flight range of *Anopheles quadrimaculatus*, Say. Am J Hyg 34: 102–113

Solberg VB, Bernier L, Schneider I, Burge R, Wirtz RA (1999) Rubidium marking of *Anopheles stephensi* (Diptera: Culicidae). J Med Entomol 36: 141–143

Southwood TRE (1978) Ecological Methods with Particular Reference to the Study of Insect Populations. Chapman & Hall, London

Southwood TRE, Henderson PA (2000) Ecological Methods. 3rd edn. Blackwell Science, Oxford

Staniland LN (1959) Fluorescent tracer techniques for the study of spray and dust deposits. J Agric Eng Res 4: 110–125

Subra R (1972) Études ecologiques sur *Culex pipiens fatigans* Wiedemann, 1828 (Diptera: Culicidae) dans une zone urbaine de savane soudanienne ouest-africaine, Longévité et déplacements d'adultes marqués avec des poudres flourescentes. Cah ORSTOM sér Entomol Méd Parasitol 10: 3–36

Suzuki T (1978) Preliminary studies on blood meal interval of *Aedes polynesiensis* in the field. Jap J Sanit Zool 29: 169–174

Takagi M, Tsuda Y, Suzuki A, Wada Y (1995) Movement of individually marked
    Aedes albopictus females in Nagasaki, Japan. Trop Med (Nagasaki) 37: 79–85
Takken W, Charlwood JD, Billingsley PF, Gort G (1998) Dispersal and survival
    of *Anopheles funestus* and *A. gambiae* s.l. (Diptera: Culicidae) during the
    rainy season in southeast Tanzania. Bull Entomol Res 88: 561–566
Taylor CE, Touré YT, Coluzzi M, Petrarca V (1993) Effective population size and
    persistence of *Anopheles arabiensis* during the dry season in West Africa.
    Med Vet Entomol 7: 351–357
Tietze NS, Stephenson MF, Sidhom NT, Binding PL (2003) Mark-recapture of
    Culex erythrothorax in Santa Cruz County, California. J Am Mosq Control
    Assoc 19: 134–138
Trpis M (1971) Seasonal Variation in the Adult Populations of *Aedes aegypti* in
    the Dar es Salaam Area, Tanzania. WHO/VBC 71.291, 29 pp (mimeographed)
Trpis M (1973) Adult population estimate of *Toxorhynchites brevipalpis*. Bull
    World Health Organ 48: 756–757
Trpis M, Haüsermann W (1975) Demonstration of differential domesticity of
    Aedes aegypti (L.) in Africa by mark–release–recapture. Bull Entomol Res 65:
    199–208
Trpis M, Haüsermann W (1986) Dispersal and other population parameters of
    Aedes aegypti in an African village and their possible significance in epidemi-
    ology of vector-borne diseases. Am J Trop Med Hyg 35: 1263–1279
Trpis M, Haüsermann W, Craig GB (1995) Estimates of population size, dispersal,
    and longevity of domestic *Aedes aegypti aegypti* (Diptera: Culicidae) by
    mark-release-recapture in the village of Shauri Moyo in eastern Kenya. J Med
    Entomol 32: 27–33
Tsuda Y, Takagi M, Toma T, Sugiyama A, Miyagi I (1999) Mark-release-
    recapture experiment with adult *Anopheles minimus* (Diptera: Culicidae) on
    Ishigaki Island, Ryukyu Archipelago, Japan. J Med Entomol 36: 601–604
Tsuda Y, Takagi M, Suwonkerd W (2000) A Mark-Release-Recapture study on
    the spatial distribution of host-seeking anophelines in northern Thailand. J Vector
    Ecol 25: 16–22
Tsuda Y, Takagi M, Wang S, Wang Z, Tang L (2001) Movement of *Aedes aegypti*
    (Diptera: Culicidae) released in a small isolated village on Hainan Island,
    China. J Med Entomol 38: 93–98
Turncock WJ, Gerber GH, Bicks M (1979) The applicability of X-ray energy-
    dispersive spectroscopy to the identification of populations of red turnip bee-
    tle, *Entomoscelis americana* (Coleoptera: Chrysomelidae). Can Entomol 111:
    113–125
Ulloa A, Arredondo-Jiménez JI, Rodriguez MH, Fernández-Salas I (2002) Mark-
    recapture studies of host selection by *Anopheles (Anopheles) vestitipennis*.
    J Am Mosq Control Assoc 18: 32–35
Vail PV, Howland AF, Henneberry TJ (1966) Fluorescent dyes for mating and re-
    covery studies with cabbage looper moths. J Econ Entomol 59: 1093–1097
van Steenwyk RA, Ballmer TJ, Page AL, Reynolds HT (1978) Marking pink
    bollworm with rubidium. Ann Entomol Soc Am 71: 81–84

Vargas A, Friere F (1940) Novos processos para avaliar a área de vôos dos anofeles. Anais Paul Med Cirurg 39: 3–7

Viswanathan DK, Ramachandra Rao T, Rama Rao TS (1945) The behaviour of *Anopheles fluviatilis*. Part IV. Experiments on the behaviour of gravid females. J Malar Inst India 6: 243–245

Von Frisch K (1923) Uber die "Sprache" der Bienen. Eine tierpsychologische Untersuchang. Zool Jb Abt Allg Zool Physiol 40: 1–186

Wada Y, Kawai S, Oda T, Miyagi I, Suenaga O, Nishigaki J, Omori N, Takahashi K, Matsuo R, Itoh T, Takatsuki Y (1969) Dispersal experiments of *Culex tritaeniorhynchus* in Nagasaki area (Preliminary Report). Trop Med 11: 37–44

Wada Y, Suenaga O, Miyagi I (1975) Dispersal experiment of *Aedes togoi*. Trop Med 16: 137–146

Wallace B (1970) Observations on the microdispersion of *Drosophila melanogaster*. In: Hecht MK, Steere WC (eds) Essays in Evolution and Genetics Presented in Honour of Theodorius Dobzhansky. North-Holland Publishing Co., Amsterdam, pp 381–399

Walsh JF (1977) Problem of migration of vectors of onchocerciasis in control programmes. Proc 3rd Sci Mtg ICIPE, Lagos, 23–26 November 1976, 70–84

Weathersbee AA, Hasell PG (1938) On the recovery of stain in adults developing from anopheline larvae stained in vitro. Am J Trop Med 18: 531–543

Weathersbee AA, Meisch MV (1990) Dispersal of *Anopheles quadrimaculatus* (Diptera: Culicidae) in Arkansas ricefields. Environ Entomol 19: 961–965

Wharton RH, Eyles DE, Warren McW (1963) The development of methods for trapping the vectors of monkey malaria. Ann Trop Med Parasitol 57: 32–46

White EG (1970) A self-checking coding technique for mark–recapture studies. Bull Entomol Res 60: 303–307

White EG (1971a) A versatile FORTRAN computer program for the capture–recapture stochastic model of G. M. Jolly. J Fish Res Bd Can 28: 443–446

White EG (1971b) A Computer Program for Capture–Recapture Studies of Animal Populations. A FORTRAN Listing for the Stochastic Model of G. M. Jolly. Tussock Grasslands and Mountain Lands Institute, Lincoln College, Christchurch, N.Z. Special Publication, No. 8

White GC, Anderson DR, Burnham KP, Otis DL (1982) Capture–recapture and removal methods for sampling closed populations. LA-8781—NERP, Los Alamos National Laboratory. Los Alamos, New Mexico

Williams DF, Patterson RS, LaBrecque GC (1979) Marking large numbers of stable flies *Stomoxys calcitrans* (L.) for a sterile release program. Mosquito News 39: 146–148

Xue RuiDe, Edman JD, Scott TW (1995) Age and body size effects on blood meal size and multiple blood feeding by *Aedes aegypti* (Diptera: Culicidae). J Med Entomol 32: 471–474

Yamamura K (2003) Population estimation by a one-release, two-capture experiment. Appl Entomol Zool 38: 475–486

Yamamura K, Wakamura S, Kozai S (1992) A method for population estimation from a single release experiment. Appl Entomol Zool 27: 9–17

Yasuno M (1979) An improved device for the self-marking of mosquitoes. Jap J Sanit Zool 30: 292

Yasuno M, Rajagopalan PK (1977) Population estimation of *Culex fatigans* in Delhi villages. J Commun Dis 9: 172–183

Yasuno M, Tonn RJ (1970) A study of biting habits of *Aedes aegypti* in Bangkok, Thailand. Bull World Health Organ 43: 319–325

Yasuno M, Rajagopalan PK, LaBrecque GC (1972a) Migration of *Culex fatigans* from Surrounding Fields to Villages in the Delhi Area. WHO/VBC 72.364 (mimeographed)

Yasuno M, Rajagopalan PK, Russel S, LaBrecque GC (1972b) Dispersal of *Culex fatigans* in Delhi Villages. WHO/VBC 72.352 (mimeographed)

Yasuno M, Russel S, Rajagopalan PK (1973) An application of the removal method to the population estimation of *Culex fatigans* resting indoors. WHO/VBC/73–458 (mimeographed)

Yoho TP, Butler L, Weaver JE (1971) Photodynamic effect of light on dye-fed house flies: preliminary observations of mortality. J Econ Entomol 64: 972–973

Zacharuk RY (1963) Vital dyes for marking living elaterid larvae. Can J Zool 41: 991–996

Zetek J (1913) Determining the flight of mosquitoes. Ann Entomol Soc Am 6: 5–21

Zhou L, Hoy CW, Miller SA, Nault LR (2003) Marking methods and field experiments to estimate Aster leafhopper (*Macrosteles quadrilineaus*) sic dispersal rates. Environ Entomol 32: 1177–1186

Zippin C (1956) An evaluation of the removal method of estimating animal populations. Biometrics 12: 163–189

Zippin C (1958) The removal method of population estimation. J Wildl Mgmt 22: 82–90

Zukel JW (1945). Marking *Anopheles* mosquitoes with fluorescent compounds. Science 102: 157

Chapter 15 Measuring Adult Dispersal

Mark–recapture techniques were principally devised for estimating population size, but have frequently been applied to adult mosquitoes in order to study dispersal, feeding behaviour, duration of the gonotrophic cycle, survival rates and other behaviours. Genetic markers may also be used to study mosquito dispersal. Available methods for marking insects which are applicable to studying the dispersal of marked adult mosquitoes were discussed in Chapter 14.

Dispersal

Flight of insects from one place to another is usually termed dispersal or migration. Williams (1961) used the term migration to describe the movements of insects *en masse* in what seemed to be a purposeful flight in a specific direction. He regarded the more passive and accidental flight of insects as dispersal. Dispersal is commonly held to describe more or less random flights with wind often playing an important factor, while migratory flight is often considered to be linked to the biology and survival of the species and is more controlled and persistent. For example, migration is usually used to describe movements of a whole population of animals, and is often seasonal, with the individuals or their offspring frequently returning to their parent's place of origin. Pielou (1977) recognises migration of animals and much shorter movements, often occurring in a haphazard fashion, which she calls diffusion. This is to avoid confusion with the terms dispersal and dispersion, the latter referring to the spatial pattern of animals such as a negative binomial dispersion. Service (1997) provided a comprehensive review of the different types of dispersal seen in mosquitoes, including unintentional dispersal on human transportation, wind-assisted long-distance dispersal and shorter, daily flights in search of hosts, mates, nectar, oviposition, and resting sites. Southwood and Henderson (2000) refer to two types of dispersal: neighbourhood dispersal, which describes the process of movements of individuals into adjacent areas; and jump dispersal, for longer-distance movements in which individuals move purposefully to a new area or are transported there by an external agent. In this edition, I use Service's (1993) definition of mosquito dispersal, which encompasses the full range of adult mosquito movement, whether self powered or passive, and covering distances of just a few metres or many kilometres. This definition conforms more closely with the definition of migration, dispersal and other associated terms given by Lincoln et al. (1982).

Many flights of mosquitoes are termed *appetitive* or goal-orientated flights. With these flights a suitable stimulus results in cessation of flight and participation in other activities such as nectar-feeding, blood-feeding, mating or oviposition. Typically such flights are short and are under the control of the adults. Provost (1952) termed the 'migratory' flights of *Ochlerotatus taeniorhynchus* from larval habitats as *non-appetential*, because they seemed to serve no special physiological purpose and were mostly windborne. These non-appetential flights are involuntary and are characterised by the lack of control, and adults are not in charge of their destiny—although such flights may be a part of their population biology. Taylor et al. (1973) summed up insect migration as follows. 'Insect migration is a system for searching new territories; linear flight, area, and perhaps distance covered are the relevant parameters, and the particular linear flight mechanism employed is species specific; it uses the same spatial coordinates within which the species lives at other times. Flight becomes linear when released from the inhibition that maintains social cohesion; food, sex, territory and synchronised activity become secondary distractions'.

Nayar and Sauerman (1969) showed that temporary crowding of *Ochlerotatus taeniorhynchus* larvae resulted in a different, and increased, flight activity pattern of adults, than when adults were reared from larvae that were not crowded. They tentatively designated these two flight patterns as non-migrant and migrant phases. Later Nayar and Sauerman (1970) suggested, that eight out of 10 other species investigated also had a potential for a migratory flight phase immediately after emergence. The two exceptions were *Aedes aegypti* and *Culex bahamensis*. However, in later studies none of the 11 species investigated exhibited any well marked migratory phase after larvae were subjected to temporary excessive crowding, although four species showed a tendency for increased flight activity during the first and third days after emergence (Nayar and Sauerman 1973). A summary of appetitive (dispersal) and non-appetential (migration) flights in *Ochlerotatus taeniorhynchus* and *Ochlerotatus sollicitans* is given by Nayar (1985).

Examples of medium and long distance flights of mosquitoes are the mass exodus and dispersal of *Ochlerotatus taeniorhynchus* from salt marshes (Harden and Chubb 1960; Haeger 1960; Nielsen 1958; Nielsen and Haeger 1960; Provost 1952, 1953, 1957, 1960) and the long distance flights of *Anopheles pharoensis* (Garrett-Jones 1962; Kirkpatrick 1925; Low 1925). Other *Anopheles* have been caught 4 miles (Gibson 1923) and 15½ miles (Wright 1918) offshore and 29 km from release sites (Bailey and Baerg 1967). A marked *Culex tarsalis* has been caught 26 km from its release point (Bailey et al. 1965), but flight range of this mosquito is usually much shorter (Reisen and Reeves 1990). Much longer flights have been recorded for salt marsh breeding mosquitoes, for example the discovery of *Ochlerotatus vigilax* more than 60 miles inland and also on a boat 20 miles from the shore (Hamlyn-Harris 1933). *Ochlerotatus sollicitans* has been caught 28 miles out at sea (MacCreary and Stearns 1937) and also as much as 110 miles from the shore (Curry 1939), while *Ochlerotatus melanimon* and *Ochlerotatus nigromaculis* have been caught 47 km from the shore (Smith et al. 1956). Sparks et al. (1986) catalogued the insects caught in light-traps left on unmanned oil rigs off Louisiana, USA. Their list included 7 *Aedes vexans* caught on a rig 32 km from the shore, 1 *Ochlerotatus* (*Ochlerotatus*) sp. caught 74 km offshore, 2 *Ochlerotatus taeniorhynchus* caught 74 km and 2 others caught as far as 106 km from the shore, 1 *Psorophora confinnis*, 1 *Uranotaenia lowii* and 1 *Uranotaenia sapphirina*, all caught in traps sited 32 km from the shore.

As well as wind dispersal mosquitoes sometimes hitch a ride on ships, aircraft, trains, vehicles, camels etc. A classical example is the apparent transportation of *Anopheles gambiae* by fast shipping from West Africa to Brazil in 1930. More recent examples include so-called airport malaria due to infective *Anopheles* being carried as non-paying passengers aboard aircraft to non-malarious countries such as France, Switzerland, Belgium, Britain, Russia and the Netherlands. Russell (1987) found that about 84% of *Culex quinquefasciatus* placed in plastic cups in the wheel bays of a Boeing 747B aircraft survived international journeys, such as from Australia to Singapore, and Singapore to Thailand. Smith and Carter (1984) review cases of international transportation of mosquitoes.

In addition to the excellent book by Johnson (1969) on insect migration and dispersal, Pedgley (1982) has written a useful account of the meteorological factors affecting windborne insects and other pests. A good ecological account of migration and dispersal is given by Begon et al. (1986). The relatively little information available on mosquitoes being transported by frontal systems is briefly reviewed by Service (1980).

General considerations concerning dispersal

Knowledge of the flight range of mosquitoes is of paramount importance in control programmes, as it is essential to know the width of 'barrier zones' needed to prevent the infiltration of adults into an area where control measures are being assessed. Furthermore, the dispersal of sterile mosquitoes, mutants, incompatible forms etc., of a species that has been introduced into a population for control purposes should be compared with the dispersal of normal mosquitoes (Rajagopalan et al. 1973).

In dispersal studies it is not always considered essential to know the numbers of mosquitoes marked and released, and unfortunately there is sometimes little concern over their survival. The survival rate of marked individuals should be assessed, or at least compared with that of unmarked individuals, so that corrections can be applied to dispersal estimates.

Marking and handling may affect mosquitoes so that they do not disperse so far; on the other hand handling may excite them and result in excessive dispersal, especially shortly after marking. This has been observed in both gryllids (Clark 1962) and beetles (Greenslade 1964), and was suspected with *Aedes aegypti* (Sheppard et al. 1969). Reisen et al. (2003) observed no effect on recapture rates of maintaining females at relatively high densities in laboratory cages for several hours prior to release in late afternoon, in contrast to results reported by Wallace (1970) with *Drosophila melanogaster* in which crowded individuals dispersed further. If marking decreases longevity, this will reduce the numbers available for recapture on successive days and hence may result in underestimating dispersal. In some experiments mosquitoes are of known age, e.g. newly emerged laboratory specimens, but often wild caught females of unknown and mixed age composition are marked and released. Laboratory rearing itself may produce adults that have atypical survival rates and dispersal. It is known that overcrowding of mosquito larvae can sometimes lead to a migratory phase in the adults, but little is known of the effects that laboratory diets, rearing conditions, and the subjection to radionuclides have on adult behaviour, although marking with radionuclides is now rarely used in dispersal studies. It is usually assumed that marked mosquitoes are largely confined to the sampling area, which often extends only a relatively short distance from the release point, and yet invariably only few are recaptured. The fate of uncaught marked individuals is not known; they may still be within the sampling area and have escaped capture, or emigrated from it.

Sheppard et al. (1969) estimated that only about 40% of the marked *Aedes aegypti* released in Bangkok remained in the recapture area after 1 day. It is most important to take this type of loss into account in calculating survival rates and dispersal distances. A relatively simple method of estimating the loss of insects from the recapture area was proposed by Gilmour et al. (1946). They compared the theoretical decline in marked blow-flies with distance with actual recaptures. Conceivably some marked mosquitoes are swept upwards by air currents and are wind dispersed outside the experimental area. Such flights even within the recapture zone will be overlooked because in most studies only the aerial population near the ground is sampled. Dispersal rates should theoretically take into account vertical as well as horizontal density gradients, although this is admittedly difficult. Another obvious weakness is that the recovery of very small numbers of marked insects is used to interpret the behaviour and dispersal of exceedingly large field populations. Hocking (1953) appreciated some of the difficulties. He wrote that the method 'yields good results so long as the range of flight is rather limited, communications are good, and the insects are readily trapped in large numbers...it breaks down completely where there is a combination of long flight range and difficult terrain. Even under favourable conditions it is rare for as many as 5% of marked specimens to be recovered—a small proportion on which to base a firm conclusion'. Eddy et al. (1962) for example, failed to recapture any of the 15 000 *Culex tarsalis*, 7500 *Culex stigmatosoma* and 3000 *Ochlerotatus dorsalis* which they estimated they had marked and released. Table 15.1 gives some examples of recapture rates and maximum distances from which marked adults have been recaptured. Clearly for many species recapture rates are extremely small (<1%), but recaptures of 20% to more than 40% have been recorded for some species, such as *Aedes aegypti, Ochlerotatus triseriatus* and *Aedes albopictus*, which indicates that they do not disperse far from their release sites. In some instances this is substantiated by the short maximum distances at which adults have been recaptured, but a problem in interpreting flight range is the diminution of sampling intensity at increasing distances from release sites. Moreover, recapture rates and distances flown will depend not only on intrinsic differences between the behaviour and biology of different species, but according to the type of environment in which the studies were undertaken. For example, mosquitoes will tend to disperse less in woods than those inhabiting more open terrain, and so recapture rates will be greater.

Table 15.1. Examples of recapture rates and maximum distance flown as determined in various marking and recapture experiments[a]

Species	No. marked and released[b]	% recaptures	Max. flight distance	Reference
Anopheles annulipes	11 500 c	0.11	4.2 km	Bryan et al. (1991)
Anopheles annulipes	11 500 c	0.69	5.0 km	Bryan et al. (1991)
Anopheles balabacensis	3 184	8.73	—	Hii and Vun (1985)
Anopheles culicifacies	54 950	1.094	1.5–1.75 miles	Russell et al. (1944)
Anopheles culicifacies	6 789	37.6	—	Rawlings et al. (1981)
Anopheles culicifacies	4 802	16.8	—	Rawlings et al. (1981)
Anopheles culicifacies	3 022	0.73	—	Rawlings et al. (1981)
Anopheles culicifacies	15 933	3.5	—	Rawlings et al. (1981)
Anopheles culicifacies	9 378	33.1	—	Rawlings et al. (1981)
Anopheles culicifacies	4 772	18.96	1.2 km	Rawlings et al. (1981)
Anopheles culicifacies	3 369 ♂	10.74	0.6 km	Reisen et al. (1981)
Anopheles culicifacies	5 123	16.95	9.0 km	Reisen et al. (1981)
Anopheles culicifacies	2 416 ♂	20.44	6.9 km	Reisen et al. (1981)
Anopheles culicifacies	9 323	12.0	—	Reisen et al. (1982c)
Anopheles culicifacies	1 791	4.58	—	Reisen et al. (1982a)
Anopheles culicifacies	967 ♂	1.96	—	Reisen et al. (1982a)
Anopheles culicifacies	489 (wild)	2.04	—	Slooff and Herath (1980)
Anopheles culicifacies	105 (lab)	1.90	—	Slooff and Herath (1980)
Anopheles darlingi	319	14.7	7.2 km	Charlwood and Alecrim (1989)
Anopheles farauti	6 289	3.07	—	Birley and Charlwood (1989)
Anopheles freeborni	54 800 c ♀ ♂	0.032	17.5 miles	Bailey and Baerg (1967)
Anopheles funestus	1 935	7.6	—	Lines et al. (1986)
Anopheles funestus	581 ♂	1.4	—	Lines et al. (1986)

(Continued)

Species	Number released		Distance	Reference
Anopheles gambiae	132 000 c ♀ ♂	0.772	2.25 miles	Gillies (1961)
Anopheles gambiae	12 965	5.5	—	Lines et al. (1986)
Anopheles gambiae	2 840 ♂	2.1	—	Lines et al. (1986)
Anopheles maculatus	419	12.6	—	Loong et al. (1990)
Anopheles minimus	3 526	1.45	—	Nutsathapana et al. (1986)
Anopheles punctulatus	894	9.14	1.8 km	Charlwood and Bryan (1987)
Anopheles quadrimaculatus	16 500 c ♀ ♂	0.19	2.5 miles ♂	Eyles and Bishop (1943)
Anopheles quadrimaculatus	107 200 ♀ ♂ (5 releases)	0.13–0.39 ($\bar{X}$ = 0.24)	3.6–4.3 km	Weathersbee and Meisch (1990)
Anopheles quadrimaculatus	3 800 ♀ ♂	0.116	2 700 ft	Smith et al. (1941)
Anopheles sergentii	larvae 4 260 ♀ ♂	0.87	3.75 km	Abdel-Malek (1966)
Anopheles stephensi	9 227	7.20	—	Reisen et al. (1982a)
Anopheles stephensi	1 866 ♂	4.50	—	Reisen et al. (1982a)
Anopheles stephensi	10 118	7.13	1.78 km	Reisen and Aslamkhan (1979)
Anopheles stephensi	10 863 ♂	4.65	1.78 km	Reisen and Aslamkhan (1979)
Anopheles stephensi	938	12.04	3.1 km	Reisen et al. (1981)
Anopheles stephensi	592 ♂	18.92	1.2 km	Reisen et al. (1981)
Anopheles stephensi	117 000 c ♀ ♂	0.203	4.5 km	Quraishi et al. (1966)
Anopheles subpictus	2 119	6.32	—	Reisen et al. (1979)
Anopheles subpictus	2 527 ♂	1.19	—	Reisen et al. (1979)
Anopheles subpictus	604	12.08	—	Reisen et al. (1981)
Anopheles subpictus	294 ♂	1.70	—	Reisen et al. (1981)
Anopheles subpictus	6 310	6.74	—	Reisen et al. (1982a)
Anopheles subpictus	4 696 ♂	1.96	—	Reisen et al. (1982a)
Aedes aegypti	276 000 c	0.11	0.75 miles	Bugher and Taylor (1949)
Aedes aegypti	9 615	4.70	476–575 ft	Morlan and Hayes (1958)
Aedes aegypti	1 154	0.867	—	Trpis (1971)

(Continued)

Table 15.1. (Continued)

Species	n	%	Distance	Reference
Aedes aegypti	68 548 ♂	3.01	220 m	Reuben et al. (1975b)
Aedes aegypti	68 232 ♂♂	2.76	220 m	Reuben et al. (1975b)
Aedes aegypti	69 232 ♂	6.29	220 m	Reuben et al. (1975b)
Aedes aegypti	1 307	11.8	—	Hervy (1977b)
Aedes aegypti	360	38.33	>0.40 m	Hervy (1977b)
Aedes aegypti	360 ♂	36.94	>0.40 m	McDonald (1977b)
Aedes aegypti	1 251	31.65	800 m	McDonald (1977b)
Aedes aegypti	1 192 ♂	45.64	800 m	McDonald (1977b)
Aedes aegypti	4 050 ♂	7.31	800 m	Trpis and Haüserman (1978)
Aedes aegypti	4 050	10.27	—	Trpis and Haüserman (1978)
Aedes aegypti	828	40.1	—	Trpis and Haüserman (1986)
Aedes aegypti	265 ♂	22.6	154 m	Trpis and Haüserman (1986)
Aedes aegypti	6 000 ♀♀	6.78	113 m	Trpis and Haüserman (1975)
Aedes aegypti	391 ♀ (fed on blood only)	30.0	—	Morrison et al. (1999)
Aedes aegypti	235 ♀ (fed on blood + honey)	23.0		Morrison et al. (1999)
Aedes albopictus	7 050 ♀♂ ♂	2.595	475 yd	Bonnet and Worcester (1946)
Aedes albopictus	3 000	29.4	—	Mori and Wada (1977)
Aedes albopictus	2 400	23.8	—	Mori and Wada (1977)
Aedes albopictus	1 170	2.8	—	Mori and Wada (1977)
Aedes vexans	85 000 ♀	0.18 ± 0.03	620 m	Ba et al. (2005)
Aedes polynesiensis	449	8.9	—	Suzuki (1978)
Aedes polynesiensis	513	13.64	—	Suzuki (1978)
Ochlerotatus canadensis	1 800	2.7	1 mile	Sinsko and Craig (1979)
Ochlerotatus communis	3 million	0.005	♀ 6.6 miles	Jenkins and Hassett (1951)
Ochlerotatus flavescens	larvae 415 000 c	0.020	♂ 1400 yd	Shemanchuk et al. (1955)

Species	Number	Distance	Reference
Ochlerotatus melanimon	5 613	1.04	—	Jensen and Washino (1991)
Ochlerotatus melanimon	5 988 ♂	0.05	—	Jensen and Washino (1991)
Ochlerotatus melanimon	15 487	0.94	—	Jensen and Washino (1991)
Ochlerotatus melanimon	11 679 ♂	0.04	—	Jensen and Washino (1991)
Ochlerotatus nigromaculis	400 000	0.12	1.9 miles	Thurman and Husbands (1951)
Ochlerotatus notoscriptus	1671 ♀ ♂ marked blue	10.6 ♀ only	195 m	Watson et al. (2000)
	1340 ♀ ♂ marked pink	2.3 ♀ only	238 m	
Ochlerotatus sollicitans	83 705 c	0.20	—	Ginsberg (1986)
Ochlerotatus sollicitans	89 657 c	0.74	—	Ginsberg (1986)
Ochlerotatus sticticus and Aedes	1.25 million	0.061	8 km	Brust (1980)
vexans				
Ochlerotatus taeniorhynchus	larvae 2 million c	0.0214	21 miles	Bidlingmayer and Schoof (1957)
Ochlerotatus taeniorhynchus	1 million c	0.0324	20 miles	Provost (1952)
Ochlerotatus triseriatus	726 (wild)	13.9	—	Pumpuni and Walker (1989)
Ochlerotatus triseriatus	2 095 (lab)	11.1	—	Pumpuni and Walker (1989)
Ochlerotatus triseriatus	1 406 (lab)	1.0	—	Pumpuni and Walker (1989)
Ochlerotatus triseriatus	716 (lab)	5.3	362 m	Pumpuni and Walker (1989)
Ochlerotatus triseriatus	1 389	41.2	—	Haramis and Foster (1983)
Ochlerotatus triseriatus	5 600	12.1	—	Haramis and Foster (1983)
Ochlerotatus triseriatus	1 536	29.0	—	Sinsko and Craig (1979)
Ochlerotatus triseriatus	128	30.5	—	Sinsko and Craig (1979)
Ochlerotatus triseriatus	3 734	5.9	—	Beier et al. (1982)
Ochlerotatus triseriatus	4 004 ♂	4.7	—	Beier et al. (1982)
Wyeomyia mitchellii	79 000 c	5.23	—	Nayar (1982)
Wyeomyia vanduzeei	330	9.7	—	Frank and Curtis (1977)
Wyeomyia vanduzeei	5 893	9.54	—	Frank and Curtis (1977)
Wyeomyia vanduzeei	4 947 ♂	12.06	—	Frank and Curtis (1977)

(Continued)

Table 15.1. (Continued)

Haemagogus spegazzinii	7 624	1.3	11.5 km	Causey et al. (1950)
Haemagogus janthinomys	74 and 575	10.8 and 4.1		Dégallier et al. (1998)
Haemagogus leucocelaenus	12 and 12	8.3 and 0		Dégallier et al. (1998)
Sabethes chloropterus	59 and 51	20.3 and 15.6		Dégallier et al. (1998)
Sabethes cyaneus	62 and 18	11.3 and 11.1		Dégallier et al. (1998)
Sabethes belisarioi	34	5.8		Dégallier et al. (1998)
Sabethes amazonicus	32	21.8		Dégallier et al. (1998)
Sabethes quasicyaneus	20	5.1		Dégallier et al. (1998)
Sabethes glaucodaemon	21	0		Dégallier et al. (1998)
Culex erythrothorax	30 000	0.46	mean = 0.57 km/day	Tietze et al. (2003)
Culex erythrothorax	5 000	0.50	Mean = 0.57 km/day	Tietze et al. (2003)
Culex erythrothorax	8 000	0.55	mean = 0.57 km/day	Tietze et al. (2003)
Culex nigripalpus, test 1	141 707 c	0.130	4.8 km	Dow (1971)
Culex nigripalpus, test 2	16 638 c	0.072	—	Dow (1971)
Culex nigripalpus	250 000	1.87	1.2 km	Nayar et al. (1980)
Culex poicilipes	5 800 ♀♂	3.46 ± 0.47	550 m	Ba et al. (2005)
Culex quinquefasciatus	303 ♀♂	5.6	200 m	Macdonald et al. (1968)
Culex quinquefasciatus	28 639	6.648	greatest $\bar{X}$ = 188 m	Yasuno et al. (1972b)
Culex quinquefasciatus	6 868 ♀	7.047	$\bar{X}$ = 105 m	Rajagopalan et al. (1973)
Culex quinquefasciatus	larvae 275 000 c	0.23	3.5 miles	Fussell (1964)
Culex quinquefasciatus, test 1	larvae 583 000 c	0.011	at least 800 m	Lindquist et al. (1967)
Culex quinquefasciatus, test 2	280 000 c ♀♂	0.042	at least 915 m	Lindquist et al. (1967)
Culex quinquefasciatus	37 000 c	0.608	—	Sharma (1977)
Culex quinquefasciatus	100 000 c ♀	0.16	—	Sharma (1977)
Culex quinquefasciatus	39 170 ♀	2.5	—	Yasuno and Rajagopalan (1977)
Culex quinquefasciatus	6 868 ♂	7.0	—	Yasuno and Rajagopalan (1977)
Culex quinquefasciatus	47 750 c	0.341	1.27 km	Schreiber et al. (1988)

Species	Number	Distance	Reference
Culex quinquefasciatus	13 589	3.06	—	Reisen et al. (1991)
Culex quinquefasciatus	8 961 ♂	11.33	—	Reisen et al. (1991)
Culex salinarius	3 800 c	0.158	—	LaSalle and Dakin (1982)
Culex salinarius	18 260 c	0.033	—	LaSalle and Dakin (1982)
Culex stigmatosoma	27 922 c	0.39	4.39 km	Reisen et al. (1991)
Culex stigmatosoma	8 495 ♂	0.30	1.3 km	Reisen et al. (1991)
Culex stigmatosoma	7 500	0.133	1.7 km	Reeves et al. (1948)
Culex tarsalis, test 1	110 301 c	0.17	9.6 miles	Dow et al. (1965)
Culex tarsalis, test 2	40 994 c	0.15		Dow et al. (1965)
Culex tarsalis	253 000 c	0.231	16.75–20 miles	Bailey et al. (1965)
Culex tarsalis	8 612 (wild)	5.4	—	Nelson et al. (1978)
Culex tarsalis	3 481 (lab)	10.9	—	Nelson et al. (1978)
Culex tarsalis	8 400	5.3	362 m	Nelson and Milby (1978)
Culex tarsalis	11 583 ♂	2.12	—	Asman et al. (1979)
Culex tarsalis	31 853	4.08	—	Asman et al. (1979)
Culex tarsalis	5 278 (wild)	3.68	—	Nelson and Milby (1980a)
Culex tarsalis	4 398 ♂ (wild)	0.05	—	Nelson and Milby (1980a)
Culex tarsalis	6 106 (lab)	12.92	—	Nelson and Milby (1980a)
Culex tarsalis	6 590 ♂ (lab)	0.79	—	Nelson and Milby (1980a)
Culex tarsalis	1 446	18.4	—	Reisen et al. (1982b)
Culex tarsalis	3 150	9.9	—	Milby et al. (1983)
Culex tarsalis	1562	9.0	—	Reisen et al. (1983)
Culex tarsalis	1346	3.8	—	Reisen et al. (1983)
Culex tarsalis	7466	1.8	—	Reisen et al. (1983)
Culex tarsalis	9294	3.1	—	Milby et al. (1983)
Culex tarsalis	9224	10.0	—	Milby et al. (1983)
Culex tarsalis	1619	1.11	—	Reisen et al. (1991)

(Continued)

Table 15.1. (Continued)

Culex tritaeniorhynchus, test 1	51300	0.039		Wada et al. (1969)
Culex tritaeniorhynchus, test 2	1900	0.232	8.4 miles	Wada et al. (1969)
Culex tritaeniorhynchus, test 3	86200	0.194	—	Wada et al. (1969)
Culex tritaeniorhynchus	6230	1.12	—	Wada et al. (1975b)
Culex tritaeniorhynchus	7540♀	0.12	—	Wada et al. (1975b)
Culex tritaeniorhynchus	38736	0.54	1.22 km	Reisen et al. (1978)
Culex tritaeniorhynchus	45504♀	0.30	—	Reisen et al. (1978)
Culex tritaeniorhynchus	2626	19.6	—	Reisen et al. (1977)
Culex tritaeniorhynchus	7100♀	8.7	—	Reisen et al. (1977)
Culex tritaeniorhynchus	3412	4.0	71.89 m ($\bar{x}$)	Reisen et al. (1977)
Culex tritaeniorhynchus	3344♀	3.4	64.93 m ($\bar{x}$)	Reisen et al. (1977)
Culiseta melanura	12038	0.889	7.8 km	Howard et al. (1989)
Culiseta morsitans	4764	1.36	9.8 km	Howard et al. (1989)
Mansonia uniformis	23110c	0.21	3.5 km	Chiang et al. (1988)
Mansonia uniformis	22840c	0.07	3.5 km	Chiang et al. (1988)
Psorophora ferox	46720	0.544	10.8 km	Causey et al. (1950)
Psorophora confinnis & Psorophora discolor	larvae & adults 50000c	0.088	Ps. confinnis 6 miles Ps. discolor 1.5 miles	Quarterman et al. (1955)
Toxorhynchites brevipalpis	195 ♀♂	9.743	—	Trpis (1973)

[a]Data presented are not exhaustive and several of the original publications contain much more data on recapture rates and dispersal distances than given here.

[b]Females unless otherwise stated.

[c]= approximately.

Marked mosquitoes are sometimes said to show random dispersal from the release point, or conversely that dispersal is biased in favour of a particular direction, but in most instances such conclusions are statistically unsound. A difficulty pointed out by Johnson (1969) is that the numbers of marked mosquitoes recaptured are usually so small that it would be impossible to test their distribution for randomness. With less mobile insects, however, the number of recaptures may be sufficient to allow either parametric or non-parametric tests for randomness (Clark 1962; Frank 1964; Paris 1965). Another difficulty in interpreting flight range is that initially mosquitoes can only disperse away from the release point, thereafter they can fly either further away or towards it, and then past it in the opposite direction.

Relationship between density and distance

In general population size of dispersing insects, including mosquitoes, decreases with increasing distance from the source of production, such as breeding place or release point. Various simple, and mostly empirical, equations have been proposed and fitted to describe the decline in densities in invertebrates with distance (see Taylor 1978). Most equations for measuring dispersal belong to two distinct groups, namely

$$y = a + bf(x) \text{ and } y = \exp(a + bx^c) \qquad (15.1)$$

where y = the number of insects caught at distance x from the origin of dispersal, and a and b = constants. Equations of the first group relate the rate of change in density with distance independent of the size of y, whereas in the second logarithmic group of equations the rate of change of y is proportional to y. Wadley and Wolfenbarger (1944) proposed the following two formulae, both members of the $y = a + bf(x)$ family:

$$y = a + b \log x \quad \text{and} \quad y = a + b \log x + \frac{c}{x} \qquad (15.2)$$

where y = insect density at distance x, a = 'position on graph', and b and c = 'factors determining the slope of the curve as influenced by the logarithm or reciprocal of the distance'. That is to say a is the value of y at the intercept of the regression line and ordinate, and b is the regression coefficient. Paris (1965) fitted the dispersal of *Isopoda* to the second formula of Wadley and Wolfenbarger (1944), and also to the equation

$$y = a + \frac{c}{x}. \tag{15.3}$$

These equations describe dispersal in which density decreases linearly as a reciprocal of distance, and seem appropriate models for random dispersal in all, or most, directions from the release point.

Taylor (1978) found that the second group of equations ($y = exp(a + bx^c)$) gives the best fit to data for insect dispersal from a point of origin, and also implies that the lengths of individual movements are not random. In the second group of equations, the exponent c was considered by Taylor (1978) as representing some measure of non-randomness, and this often seems to vary from about -1 to $+4$, but when it is $+2$ there is random distribution. When $c > 2$ it results in a flattened curve and may represent repulsion between individuals, leading to a regular distribution, but when $c < 2$ it indicates attraction. Robinson and Luff (1979) applied this formula to data on the dispersal of *Hydrotaea irritans* and O'Donnell et al. (1992) fitted it to their recapture data for *Culex annulirostris* in New South Wales, Australia using maximum likelihood methods. The best fit was obtained for a value of $c = 0.55$, which minimized the deviance of c. Those with a mathematical inclination should consult the paper by Taylor (1978) in which he tests various formulae for the relationship between insect density at various points from their release (or dispersal) site.

The two most common methods for analysing mark–recapture data in studying insect dispersal from a single release point are: (i) to plot population density against log of the distance from the release point, which assumes the population decreases by a constant amount for equal multiples of the distance from the release point (e.g. $y = a + b \log x$); and (ii) to plot log population against distance; this model implies a decrease in population density by a constant proportion for each unit of distance from the release site (e.g. $\log y = a + bx$). Kettle (1951) proposed the following formula:

$$\log y = \log a + bx. \tag{15.4}$$

and Hawkes (1972) proposed a variant of this model incorporating the square root of distance, which was to better fit the dispersal of cabbage root flies, such that

$$\log Y = a - b\sqrt{x} \tag{15.5}$$

Freeman (1977) added a time variable to the above equations, resulting in the following:

$$Y = A(x)e^{-ht}t^{cx} \qquad (15.6)$$

where h and c are constants, and $A(x)$ represents the change in numbers with distance, allowing for an early peak at the centre, where $x = 0$, and later peaks with increasing x. In logarithmic form, the equations become:

$$\log Y = a - bx - ht + cx\log t \qquad (15.7)$$

$$\log Y = a - b\sqrt{x} - ht + cx\log t \qquad (15.8)$$

and represent the generalised equations for catches over a period of time in equally spaced traps. Freeman (1977) used the above equations to describe the data of Gillies (1961) and found rather poor fits on days 1 and 2, while data from days 3–9 and 10+ fitted the equation incorporating the bx term better than that including the $b\sqrt{x}$ term.

Typically recapture stations are arranged along concentric annuli centred on the release site (e.g. Fig. 15.1). Now when marked mosquitoes are allowed to disperse from the centre of a series of concentric annuli the probability (q_i) that a marked mosquito will reach the ith annulus is given by

$$q_i = \frac{A_i r_i / n_i}{\displaystyle\sum_{j=0}^{T} \frac{A_j r_j}{n_j}} \qquad (15.9)$$

where A_i = area of ith annulus *relative* to that of the central annulus; $r = _i$ number released marked mosquitoes; n_i = number of marked individuals recaught in the ith annulus on a recapture day; T = total number of annuli. Now, the mean distance dispersed ($\bar{d}$) from the central release point is

$$\bar{d} = \sum_{i=0}^{T} q_i d_i \qquad (15.10)$$

and an estimate of density (D_1) at the lower limit can be calculated using the total number of flies caught in the central annulus (0) and the next annulus (1) (where efficiency is highest) on both release and recapture days is

$$D_1 = \frac{R + n_0 + n_1}{2(a_0 + a_1)} \qquad (15.11)$$

where R = number of flies caught in annuli 0 and 1 and released. See Begon (1976) for further details and the calculation of a maximum-likelihood estimate of density at the upper limit (D_u).

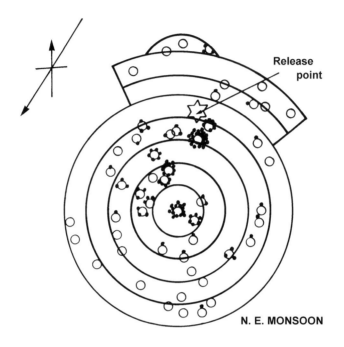

Fig. 15.1. Dispersion of marked *Anopheles gambiae* from a peripheral release point in relation to prevailing wind. Each dot represents a single recapture and each circle a village. Concentric rings delineate 0.25-mile intervals (after Gillies 1961)

When mosquitoes are released from a catch point the decreasing numbers caught at catching stations of increasing distance from the release point reflect not just dispersal but the increasing size of the areas in which mosquitoes are found. For example, if released mosquitoes disperse equally in all directions, each trap at distance r from the release point catches mosquitoes from an annulus of central radius r, with limits of $r + d/2$, and $r - d/2$, where the area is $2\pi r d$, and where d is the distance between traps. To estimate the relative numbers of marked mosquitoes present in each annulus the density at each distance from the release point is multiplied by the area of the annulus at that distance.

The rate of dispersal can be estimated by comparing the changes in variance of the speed of dispersal on successive days (Dobzhansky and Wright

1943). Because the curve of the numbers recaptured on distance is not normal, the variance is estimated as

$$S^2 = \pi \Sigma dp^3 \times \overline{n}p / (dp \times \overline{n}p + C) \qquad (15.12)$$

where dp = distance from release point site, $\overline{n}p$ = mean number of mosquitoes per trap, and C = number in the central trap. If the speed of dispersal is constant then the variance should change by a constant amount. To test this the variances obtained on successive days are plotted as a regression against these days. I am unaware that this procedure has been applied to mosquitoes, but Robinson and Luff (1979) have used this technique in studying *Hydrotaea irritans* in Britain.

Rudd and Gandour (1985) have argued that in all these types of dispersal models there is an *ad hoc* choice of the functional form of the y versus x relationship, and there is no reason to select one model in preference to another—except that it works. As a consequence in a heavily mathematical paper they applied a physical model for the self-diffusion of gases to the analysis of insect dispersal data and concluded that the following formula best described the decline in numbers of insects measured at various distances

$$y_i = A e^{Bx_i^2} \qquad (15.13)$$

where y_i = density at distance x_i from the release point, e = base of natural logarithms and where coefficients A and B are determined by applying a non-linear least squares technique. The authors applied this model to analyse some published dispersal data, and their paper should be consulted if this approach is to be attempted. They also provided equations for calculating mortality of dispersing adults.

Measurement of dispersal

Although mark–recapture methods were primarily designed for the estimation of population size, most studies involving the marking of adult mosquitoes have been to measure dispersal. Whilst dispersal is usually studied using marking methods described in chapter 13, it has occasionally been investigated without physically marking mosquitoes. For example, in an urban area of Savannah, Georgia, USA, Morlan and Hayes (1958) were able to study the dispersal of 9615 unfertilised released *Aedes aegypti* because the town had a negligible natural population of this species. Any damage that might be caused by marking procedures was entirely eliminated in this study. Similarly, in an arid area of Israel where *Aedes aegypti* did not naturally occur Wolfinsohn and Galun (1953) released gravid females and studied their dispersal by collecting eggs in ovitraps placed at varying distances from the release points. In the Delhi area of India dispersal of *Aedes aegypti* was studied both by the release of adults marked with dusts, and by the release of males heterozygous for thoracic genetic silver marker and males having a heterozygous chromosomal translocation. Eggs collected in ovitraps were reared to adults to detect dispersal of the two genetic marker strains. There was little difference between dispersal distances of the wild caught marked individuals and the laboratory reared genetic marker strains (Reuben et al. 1972). Reiter et al. (1995) measured *Aedes aegypti* dispersal in search of oviposition sites by feeding adult females on blood containing rubidium (see Chap. 14 for the specific method used), which subsequently marks the eggs. Ninety marked females were released at dawn at the centre of a 840m diameter circular collection area, that contained 550 ovitrap sites. Marked eggs were recovered from 68 sites at the first collection interval (40–67 h post feeding).

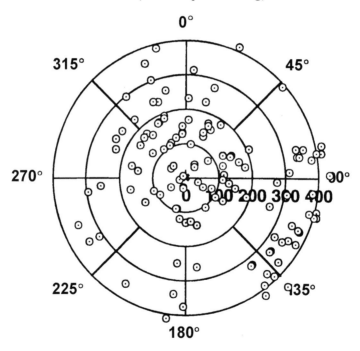

Fig. 15.2. Distribution of sites where rubidium-marked eggs of *Aedes aegypti* were deposited, in relation to the point of release of mosquitoes fed with RbCl-marked blood. The scale is in metres. The study area was roughly circular, with an average diameter of 840 metres. (Reiter et al. 1995)

Mean distance travelled from the release point was 181 metres (range 9–432 m). Oviposition beyond the perimeter of the collection site could not be determined, but was felt likely as eggs were recovered from ovitraps close to the outer perimeter. Despite a consistent wind direction (NE to E) there was no observable pattern in flight direction. Three further collections were made at 24 h intervals. Mean distances travelled and ranges were: 237 m (33–417 m), 221 m (99–417 m), and 279 m (39–441 m) (Fig. 15.2). The authors postulated that the flight range of *Aedes aegypti* is dependent on the availability of oviposition sites, and total flight ranges over a gonotrophic cycle may be considerable, regularly exceeding the often-quoted flight range of 50–100 m (Pan American Health Organization 1994).

Spradberry et al. (1995) studied the dispersal of *Chrysomya bezziana*, the Old World screw-worm fly by releasing laboratory-reared individuals labelled with ^{32}P and collecting the isotope-labelled egg masses from sentinel cattle in Papua New Guinea. The proportions of released females depositing egg masses on sentinel cattle were analysed using a generalised linear model with the specifications of a log link and variance proportional to the mean. The dependence of the proportion (y) on distance (D) was represented by a 'broken stick' model with a node at distance (c):

$$\ln(y) = a + b_n \ln(D/c) \text{ if } D \le c \tag{15.14}$$

$$\ln(y) = a + b_f \ln(D/c) \text{ if } D \le c \tag{15.15}$$

where a is the log-proportion at $D = c$ and b_n, b_f are the slopes for 'near' and 'far' distances. Recapture rates were 0.6–7.4% in the experiment where sentinel cattle were uniformly arranged within a 12.5 km radius, 0.04% for cattle along a 25 km arc, 0.012% along a 50 km arc and 0.002% for one of three 100 km trials. The 'broken stick' model gave an adjusted R^2 of 84.9% indicating a good fit to the data when an estimated node of $c = 11$ km was used, minimising the residual deviance. The authors concluded that while there was no evidence that the break point of the broken stick model had any biological meaning, such that a proportion of the population disperses at a different rate, it did describe the data better than other models and the shape of the curve is not dominated by recaptures at distances close to the release point.

Chapman et al. (1999) used six polymorphic genetic markers to study the population genetic structure and dispersal of *Ochlerotatus vigilax* in coastal south-east Queensland, Australia. The authors detected no evidence of significant genetic differentiation among any of the adult populations collected using carbon dioxide and octenol-baited light traps, suggesting that the population of *Ochlerotatus vigilax* inhabiting this area of Queensland was fully panmictic across all sampling locations, even though these were separated geographically by up to 60km and included both island and coastal mainland habitats. Chapman et al. (1999) therefore concluded that adult *Ochlerotatus vigilax* were capable of travelling between offshore islands and between the islands and mainland and vice versa.

Costantini et al. (1996) used a mark-release-recapture method to determine the dispersal of *Anopheles gambiae* s.l. mosquitoes in the Sudan savannah of Burkina Faso. Relative frequencies of *Anopheles gambiae* s.s. and *Anopheles arabiensis* were approximately 1:1 although there was significant variation in this ratio over the course of the experiment. Mosquitoes for release were hand-collected resting in huts, and marked with a different colour of 'Day-Glo' fluorescent dusts for each release day by gently aspirating them into paper cups that had been lightly dusted with fluorescent powder. All mosquitoes (7260 in 1991 and 13 854 in 1992) were released from the same site and collected from neighbouring villages using pyrethrum spray catches and resting collections. Almost all recaptures were identified to species using PCR techniques and interestingly, a slight but significant excess of *Anopheles arabiensis* was observed among the marked recaptures, compared with the proportion in pyrethrum spray catches. A total of 222 mosquitoes were recaptured, 106 in 1991 and 116 in 1992. Of the 110 *Anopheles gambiae* s.l. mosquitoes identified to species in 1992 30 *Anopheles gambiae* s.s. and 36 *Anopheles arabiensis* were collected within 200 m of the release site and 12 and 32 respectively were captured beyond 200 m. One marked *Anopheles gambiae* s.l. female was recaptured 16 days post release in a village 6 km away. Dispersal was estimated by determining goodness of fit of observed data to two simple computer models of mosquito movement. Model I assumed random movement from release, while Model II allowed for limited directional movement according to physiological status of the mosquitoes in relation to the gonotrophic cycle. Readers interested in knowing more about the specifics of the model designs and computer analysis techniques used are advised to consult the original paper. Model I gave the best fit to the observed data and calculated a mean distance moved of 636 m per day in 1991 and 348 m per day in 1992, suggesting that distances in the order of kilometres could be traversed over the lifetime of an individual mosquito.

Mosquito dispersal has occasionally been studied by the capture of adults at various distances from their natural breeding places, but this is not often a very satisfactory method as it is usually difficult to determine the specific habitats from which the adults originated. However, with coastal species whose immature stages are restricted to saline habitats it may be relatively easy to pinpoint the larval habitats. Nevertheless distance flown by captured adults can only be related to the nearest possible habitats, although they may have originated from more distant ones.

Charlwood et al. (1998) studied the age-dependant dispersal of *Anopheles funestus* and *Anopheles arabiensis* from larval habitats in southern Mozambique by examining the proportions of different age-classes present in indoor-resting collections. Collections were made from 16 randomly selected houses each day for eight days. Mosquitoes were collected by two collectors working for 20 minutes per house using manual aspirators and torches. The decline in numbers of *Anopheles funestus* with increasing distance from the breeding site was fitted to linear and exponential curves, with correlation coefficients of 0.276 and 0.23 respectively. The exponential model predicted that less than five female *Anopheles funestus* would be collected during a 20 min collection period at distances of 300 m or more from the larval habitat. Resting densities of young pre-gravid females were highest close to the larval habitat. Unexpectedly, males tended to disperse further than females, but it was not determined if males return to the proximity of the larval habitat to swarm and mate.

Radar and associated methods have been increasingly used to study the dispersal of relatively large insects such as locusts and moths. Hendricks (1980) described the construction of a low frequency sound-detecting and ranging (sodal) system which is operated like radar and was used for counting moths attracted to a pheromone dispensed from a sonic transmitter. Power supply consisted of three 6-V d.c. batteries connected in series. Schaefer and Bent (1984) developed a relatively inexpensive technique known as Infra-Red Active Determination of Insect Flight Trajectories (IRADIT) that illuminates insects during the day and night with an intense beam of pulsed infrared radiation. A shutter image intensifier connected to a video camera detects and records the flight of insects as small as 1.5 mm^2 from as far away as 15 m at midday. At night, insects can be detected further away, as can larger insects during the day and night. The authors detected a single emerging *Simulium arcticum* from up to 40 m. Gaydecki (1984) used a similar device to observe the behaviour of moths near a 125-W mercury vapour light-trap fixed 5–7 m above the ground.

Boiteau and Colpitts (2004) describe a recently developed prototype portable harmonic radar system for tracking the Colorado beetle and other Coccinellidae, in which an electronic tag with a mass of 23 mg was successfully mounted on insects as small as 6–10 mm in length. For a tag in isolation it was found that a dipole length of 12 mm yielded the largest harmonic radar cross section, but for an antenna attached to a Colorado beetle the optimum length decreased to 8 mm, as the insect forms part of the antenna. The portable harmonic tracking system utilises a pulsed magnetron source to produce high power microwave electromagnetic pulses that illuminate the target. The source operates at a frequency in the marine radar band near 9.4 GHz (fundamental) and is directed through an antenna. The electromagnetic pulse travels to the tagged insect where a portion of the signal energy is captured by the tag, is converted to the second harmonic of the incident signal, and then reradiated away from the tag. A portion of this return harmonic signal is captured by a second antenna, which directs it to a detector. For use in lighter, more delicate insects, smaller and lighter tags will be required.

Calculation of mean distance flown ($\overline{d}$)

In Australia the dispersal of *Dacus tryoni* (Fletcher 1974) and in the USA the dispersal of simuliids (White and Morris 1985) was quantified by estimating the statistic $\overline{d}$, which is the mean distance travelled, such that the density of dispersing marked flies is proportional to the ratio of the total numbers recaptured and the total numbers of traps, thus

$$\overline{d} = \sum_{i=0}^{y} F_i \frac{1}{2}(x_{i+1} + x_i) \qquad (15.16)$$

where F_i = proportional frequency (that is proportion of flies in a particular annulus); x_i = distance from central release point of marked flies to the inner radius of the *i*th annulus; x_{i+1} = distance from central release point to outer radius of *i*th annulus; y = outer annulus.

The statistic ($\overline{d}$) is derived from the proportion of flies that disperse into each annulus, but it cannot be used if significant numbers disperse beyond the outermost annulus (Southwood and Henderson 2000). For this situation a better statistic of the mean distance travelled (*MDT*) is that of Lillie et al. (1981). This is determined by calculating the density of marked flies recaptured in each annulus and then incorporating a correction factor for the recapture rate that would be expected if the recapture site density was equal in each annulus, that is

$MDT = \Sigma$(expected No. recaptures × distance) / Σ total No. recaptures.

(15.17)

Making this correction to obtain the *MDT* is less biased for trap location or size of the study area. Brenner et al. (1984), Bryan et al. (1991), Weathersbee and Meisch (1990), Muir and Kay (1998), and Watson et al. (2000) have all used the methods of Lillie et al. (1981, 1985) and Morris et al. (1991) to calculate *MDT*, where

$$MDT = (\overline{x}_c \times r) / \Sigma \overline{x}_c \qquad (15.18)$$

where $\overline{x}_c$ = corrected mean number per trap ring (subunit) and r = radius of trap ring (km).

In Pakistan Reisen et al. (1977, 1978) estimated the mean dispersal distance ($\overline{d}$), which is the same as MDT, of *Culex tritaeniorhynchus* adults marked with coloured dusts and released, when recapture sites were at 50-m increments from the release point, as follows

$$\overline{d} = \sum_{x=1}^{n} r_x d_x \Big/ \sum_{x=1}^{n} \overline{r}_x \qquad (15.19)$$

where d_x = distance to outer radius of the distance interval x, n = last distance interval flown by a recaptured marked adult, r_x = mean recapture rate of marked adults within distance x, which is calculated as follows

$$r_x = \left(\sum_{i=1}^{f} r_i \right) \Big/ f \qquad (15.20)$$

where r_i = number of recaptures in the field i, and f = number of fields within distance x. In Malaysia Chiang et al. (1988) used the same formula to estimate the mean dispersal distance of *Mansonia uniformis* in mark–recapture experiments.

Jensen and Washino (1994) used the methods of Reisen et al. (1978) to calculate daily cumulative mean dispersal distances for *Aedes vexans* and *Ochlerotatus melanimon* in California, USA. Overall dispersal distances of 113.6 ± 15.9 and 121.0 ± 14.7 m for *Aedes vexans* and *Ochlerotatus melanimon* females, respectively, were not significantly different, however, on day 1 post-release, females of *Aedes vexans* travelled significantly further (138.3 ± 28.2 m) than females of *Ochlerotatus melanimon* (85.8 ± 13.8 m), indicating rapid dispersal from the release site by this species. In addition, cumulative mean dispersal distance of *Ochlerotatus melanimon* increased with time, whereas that of *Aedes vexans* decreased with time.

Morris et al. 1991 developed simple and inexpensive mark-release-recapture methods to study mosquito dispersal in order to determine if mosquitoes emerging from a wastewater treatment plant were the same mosquitoes responsible for nuisance biting in nearby residential areas and to inform control activities. The authors collected, marked and released as many mosquitoes as possible, irrespective of age or physiological condition. They used three release stations situated at varying distances from two residential areas and concentrated collection efforts within these residential areas. Mosquitoes were collected using modified CO_2 baited CDC light-traps. The catching chamber was modified in order to accommodate the large numbers of mosquito captures expected, and consisted of two 1-gallon (3.8 litre) ice cream cartons joined together with duct tape and with holes cut in the sides and covered with screening in order to allow for air circulation. An opening cut in the top was fitted with a 12 inch (30.5 cm) cloth sleeve which was placed over the bottom of the trap body and held in place with rubber bands. Mosquitoes collected in 37 of the 40 traps set were dusted with orange, blue or green fluorescent dust and released from three sites – one colour per site. Dust was applied to the mosquitoes using a 50 ml syringe fitted with a 16 gauge needle broken off at the base. Dust was injected through the sleeve opening of the trap. Mosquitoes from the remaining 3 traps were used to estimate species composition and numbers released. Mosquitoes were collected at 44 sites near Lakeland, Florida. Annuli separated by 0.4km were drawn around the release sites to determine dispersal distance. The correction factor of Lillie et al. (1981, 1985) and White and Morris (1985) was used to correct for unequal trap densities, as follows:

Annulus correction factor = area of annulus/total trapping area × total number of traps

Estimated recaptures ER for each annulus = No. of observed recaptures in annulus/No. of traps in annulus × CF

Annulus distance = inner radius + outer radius/2

Mean distance travelled was calculated as MDT = Sum (ER × distance) for all annuli/Total no. of ER

Corrected data were also used to calculate regressions of the cumulative number of recaptures against log-transformed distance. The regression line estimates the flight range of various proportions of marked populations. Five species (*Anopheles crucians, Culex salinarius, Culex nigripalpus, Mansonia tittilans*, and *Coquillettidia perturbans* accounted for 98.9% of total catch). The linear regression equation revealed 50% and 90% flight ranges of 0.20 km and 2.27 km respectively for all species combined.

O'Donnell et al. (1992) used the method of Hawkes (1972) to calculate mean distance dispersed by *Culex annulirostris*. This method uses the value of *c* from Taylor's (1978) equation above and the 95% confidence limits in the following equation:

$$\text{Mean distance dispersed} = \frac{\Gamma(3d)}{(-b)^d\,\Gamma(2d)} \tag{15.21}$$

where $d = 1/c$ and Γ represents the gamma function. The value of c and its 95% confidence limits obtained by these authors were 0.55, 0.26 and 0.89 respectively. Approximately 75 000 marked females were released over a period 13 days and 215 marked individuals were recaptured in 9 CDC light-traps and four EVS traps. The maximum recorded flight distance was 8.7 km. Using the equation of Hawkes (1972) above, the mean distance dispersed was estimated at 40.9 km for $c = 0.26$, 6.8 km for $c = 0.55$ and 4.1 km for $c = 0.89$. Using the alternative approach of Lillie et al. (1981), O'Donnell et al. (1992) estimated mean distance flown to be 3.5 km ($c = 0.26$), 3.8 km ($c = 0.55$) and 4.2 km ($c = 0.89$)

   To test whether all marked–released and recaptured male *Aedes aegypti* in an experiment in India (Reuben et al. 1975*b*) moved a similar distance, or whether there was heterogeneity in their dispersal the following formula of Dobzhansky and Wright (1943) was applied to detect kurtosis (K_u)

$$K_u = \frac{R \sum_{i=1}^{y} d_i^4 n_i}{\left(\sum_{i=1}^{y} d_i^2 n_i \right)^2} \tag{15.22}$$

where R = number of mosquitoes caught in all traps, d_i = distance moved from release point, n_i = numbers recaptured at a particular distance, y = last equidistant set of traps. It was found that the population was indeed heterogeneous, in that while most *Aedes aegypti* dispersed just short distances, a few adults dispersed much further. Calculations were also made to estimate the 'flight distance 50' (FD$_{50}$) and 'flight distance 90' (FD$_{90}$). That is the radius of a circle centred on the release point within which an estimated 50 and 90% of the released mosquitoes will be at a specified time after release. To calculate these values the mean numbers of mosquitoes captured at each distance was multiplied by a correction factor proportional to the area of each successive annulus expressed as a multiple of the area of the central circle (Reuben et al. 1975*b*).

   To compare mean distances dispersed by simuliids during different time intervals (days, weeks, months) White and Morris (1985) calculated the Vagility Index (V_1)

$$V_1 = \overline{d}_i \bigg/ \sum_{i=1}^{y} -\overline{d}_i. \tag{15.23}$$

   This index represents the relative success of marked flies dispersing from the release site. Howard et al. (1989) used this formula to calculate the vagility of *Culiseta melanura* and *Culiseta morsitans* in mark–recapture studies.

Some factors influencing dispersal and probability of recapture

Dispersal of mosquitoes is likely to be influenced by several extrinsic and intrinsic factors, including among others: meteorological conditions, local topography, physiological status of released mosquitoes, body size, population density, and availability of oviposition and resting sites. In addition, the marking method used may influence dispersal behaviour.

Reisen and Lothrop (1995) reported that topography and in particular location of suitable oviposition sites was more important in determining the direction of dispersal of *Culex tarsalis* in California, USA than wind direction, agreeing with the conclusions of Dow et al. (1965) that *Culex tarsalis* dispersal over desert habitat was independent of wind direction, as well as those of Reisen et al. (1992). Bailey et al. (1965) concluded that short-distance dispersal of *Culex tarsalis* within rice agroecosystems was independent of wind direction, whereas longer distance dispersal (>5 km) was always downwind.

Dobzhansky and Wright (1943, 1947) in their study on *Drosophila pseudoobscura* calculated the change in variance of the numbers recaptured on successive days, and postulated that this should change by a constant amount each day if there was a constant dispersal rate. Convincing evidence was produced that dispersal was affected by temperature, being further during hot weather.

In New Delhi during December the mean dispersal of males and females of this species after 7 days was 54.0 and 90.8 m respectively, but 116.5 and 188.2 m in the hotter month of May (Yasuno et al. 1972*b*). The rate of dispersal also varied in the four release experiments in different months. For example, in December adults did not disperse readily from the release points until the third day, whereas in May they dispersed rapidly soon after emergence. Reisen and Lothrop (1995) reported that for *Culex tarsalis* in California, USA, dispersal seemed to be independent of temperature and humidity, but did vary seasonally, such that dispersal was greater during periods when populations were increasing and new oviposition sites were being created.

Mean dispersal distance of mosquitoes in urban areas is usually considerably less than in open environments. In Yangon, Myanmar Lindquist et al. (1967) found that over 80% of marked *Culex quinquefasciatus* were recaptured within a radius of 660 yd from the release point.

Pumpuni and Walker (1989) found that the recapture rates for small female (1.0%) and male (0.8%) laboratory reared *Ochlerotatus triseriatus* were lower than for laboratory reared large females (5.3%) and males (2.8%). Similarly Mori (1979) reported that recapture rates of small *Aedes albopictus* were less than for larger adults, and Walker et al. (1987) found the same with *Ochlerotatus hendersoni* and *Ochlerotatus triseriatus*. There must either be decreased survival of small mosquitoes or else they disperse further outside the recapture area, or behave in some way that makes their recapture more difficult.

Reisen et al. (2003) undertook a series of mark-release recapture experiments in order to determine the effects of the source of mosquitoes (caught in CO_2 traps within the release area, caught in CO_2 traps located outside the release area, and reared from larvae), the time of release, and the effect of prevailing wind conditions on recapture rates of *Culex tarsalis* in California, USA. Locally captured females were caught more frequently than those that had originally been collected from CO_2 traps located outside of the release area or those that had been reared from larvae. This result differs from that obtained by Reisen and Lothrop (1995) who observed no significant difference in recapture rates between *Culex tarsalis* females collected from outside the study site and those collected within the study site ($\chi2 = 3.2$, d.f. $= 1$, $P > 0.07$). Time of release (just after sunrise, or just before sunset) did not affect recapture rates or distances over which they dispersed (Reisen et al. 2003).

Dow (1971) made the interesting discovery that adult *Culex nigripalpus* which emerged from larvae subjected to a radiation dose of 0.5 *m*Ci ^{32}P/ml, dispersed on average further (1.0 miles) than those (0.73 miles) obtained from larvae exposed to only 0.3 *m*Ci/ml. A tentative explanation given by Dow was that radiation damage, which tended to suppress gonotrophic development, favoured un-distracted migratory flight. Suppression of ovarian development in *Aedes aegypti* due to high radiation levels had already been noted over 20 years earlier (Bugher and Taylor 1949). These observations clearly emphasise the need to evaluate the effect that marking may have on mosquito behaviour.

In Nagasaki, Japan, Takagi et al. (1995) studied the dispersal of 706 laboratory-reared and individually-marked unfed female *Aedes albopictus*, plus 150 field-collected individuals. Ten fixed collecting stations were established in a 0.8 ha plot of land. Females were collected daily between 23 May and 06 June (except 01 and 04 June, due to heavy rain) by human landing catch performed for 7 minutes by a single collector carrying 500g of dry ice. Of the total 856 females released, 116 (13.0%) were recaptured at least once, and 21 (2.5%) were collected twice. Takagi et al. (1995) used the methods of Trpis and Hausermann (1986) to analyse the sequential movements of the females that were recaptured twice. By tabulating the number of individuals against their previous and next recapture points, the authors were able to calculate the frequency of Convergence (C) that is those females arriving from a different collection point, the frequency of Divergence (D) or the numbers dispersing to a different collection point after capture, and the frequency of "stay or return", that is those that either stayed where they were or returned to their initial collection point.

If the dispersal rate of marked mosquitoes is both constant and homogeneous, which is probably an oversimplification in many instances, then the decrease in recaptures with increasing distance can be attributed to (1) a 'dilution' factor, due to their numbers being dispersed over an ever increasing area, thus lowering their density and (2) a 'loss' factor, due to mortality and settling down of the mosquitoes. Now it seems likely that some mosquitoes after initially flying away from the release point will fly back towards it. In the absence of mortality such flight behaviour would not alter the regression equation, but if, as seems probable, there is a steady mortality, there will be decreasing numbers returning to the release point and this would reduce the size of the regression coefficient.

Some examples of mosquito dispersal studies

In their studies in Delhi on the dispersal and survival of chemosterilized, irradiated and cytoplasmically incompatible males of *Culex quinquefasciatus* Rajagopalan et al. (1973) found that the mean distance dispersed by males (y) against time (x) could be fitted to the regression equation

$$\log y = a + b \log x, \qquad (15.24)$$

And for field-collected males

$$\log y = 1.955 + 0.177 \log x, \qquad (15.25)$$

and for laboratory males

$$\log y = 1.831 + 0.199 \log x. \qquad (15.26)$$

No statistical difference was found between these two regression coefficients. Field reared males dispersed a mean of 66.5 m by day 1 and 178.8 m by day 7, and laboratory males a mean of 50.2 and 117.0 m, respectively by these times. The daily decrease in total numbers recaptured of marked males from all recapture points (y) fitted a log–normal regression when the numbers were plotted against time in days (x).

Field males

$$\log y = 1.721 - 0.093x \qquad (15.27)$$

Laboratory males

$$\log y = 2.167 - 0.092x \qquad (15.28)$$

There is no significant difference between these two regressions. Daily loss rates were 19.3 and 19.6% for field and laboratory reared males, respectively. However, there was a significant difference ($P < 0.01$) between the decrease in numbers recaptured daily, and consequently daily loss rate, between normal laboratory males (11.9%) and irradiated males (15.5%).

Dow et al. (1965) were able to fit a regression equation ($y = 0.51 - 0.54$ $\log x$) to the numbers (y) of marked *Culex tarsalis* recaught at various distances (x). However, this relationship only fitted the recapture data for up to about 4 miles, thereafter the slope of the regression line flattened out. This marked change in the regression coefficient may have reflected a true change in dispersal pattern, or resulted from the unreliability of analysing only 12 recaptured adults from five catching stations. Sheppard et al. (1969) found that both male and female *Aedes aegypti* tended to move further during the first 24 h following release, than in subsequent periods. Males in fact dispersed further than females during the initial 24 h, but females had a greater longevity and consequently dispersed at least as far.

Kettle (1951) in studying the dispersal of *Culicoides impunctatus* from larval habitats found that data best fitted the regression equation: $\log y = \log a + bx$, where y = density at distance x, a = density at release point (in this case breeding place), b = regression coefficient, and x = distance from release point in yards. The regression coefficient is always negative because density (y) decreases with increasing distance (x). This relationship implies that log density decreases linearly with increase in distance, and more or less describes the theoretical dispersal by wind of small inanimate objects, and mainly in one direction. Kettle (1951) found a useful value was the reciprocal of the regression coefficient ($1/b$), which he termed the density coefficient and defined as the distance at which density had decreased to 0.1 of its value at the breeding place (release point).

Gillies (1961) initially released *Anopheles gambiae* tagged with ^{32}P or paint at central release points and caught marked adults in huts in all directions up to 1¼ miles. However, it became apparent that dispersal had to be measured over a greater distance, but this was not feasible with existing procedures without drastically reducing the proportion of houses sampled. To avoid this difficulty marked mosquitoes were released at a peripheral release point, situated 1–1¼ miles from the centre of the sampling area. In this way although some mosquitoes escaped from the recapture area almost immediately, those flying towards the centre could be caught up to 2¼ miles from the peripheral release point. A straight line was fitted to the data when the log percentages of the total females recaptured were plotted against distance on an arithmetic scale (Fig. 15.3). The plot of the males still showed evidence of curvature, which might have been overcome if a log–log plot had been used. The mean flight range recorded for males and females released at the centre was 0.52 and 0.64 miles respectively, while for females released at the periphery this was 0.98 miles. A few adults were caught at the maximum range of 2¼ miles from the release point. Gillies (1961) concluded that the dispersal of *Anopheles gambiae* was non-random, but related primarily to the distribution and numbers of houses.

To assess the dispersal of *Culex quinquefasciatus* from wells in irrigated fields into an urban centre, Yasuno et al. (1972a) released marked adults at wells situated 100, 250, 5000 and 1000 m from a village near New Delhi. The numbers of marked–recaptured adults were compared with the recapture rate of mosquitoes released within the village which was standardised as 100. The recapture rate of mosquitoes from wells at 100 m was 98.8%, at 250 m 96.4%, at 500 m 18.1% and as small as 7.7% at wells some 1000 m from the village.

In Kenya Trpis and Haüsermann (1986) marked and released 563 female and 265 male *Aedes aegypti* in a small village and recapture rates were 48.3 and 22.6% respectively, indicating a relatively closed population. Most females were recaptured only once (47.8%), but one was caught on 10 occasions, while 66.7% of males were recaptured once, with the maximum number of recaptures being four. Most mosquitoes remained in the houses in which they were recaptured and released. The maximum distance a female *Aedes aegypti* moved in 24 h was 154 m, 60.0% moved to houses just 11–50 m away, while the maximum dispersal distance of a male *Aedes aegypti* was 113 m, and 66.7% were recaptured in houses 11–50 m away.

The relatively short dispersal distance and its dependence on availability of oviposition sites in *Aedes aegypti* was reported by Edman et al. (1998) in Florida, Puerto Rico. 172 of the 185 individuals recaptured were collected in the house in which they had originally been released. The recapture rate was also significantly higher in houses to which suitable oviposition containers had been added, compared with houses from which containers had been removed.

Thomson et al. (1995) released 125 389 marked female *Anopheles gambiae* s.l. obtained from hand searches of untreated mosquito bed nets or alternatively from exit window traps in a village in The Gambia. Females collected from nets were marked magenta and those from exit traps were marked yellow, using Fiesta™ fluorescent powders.

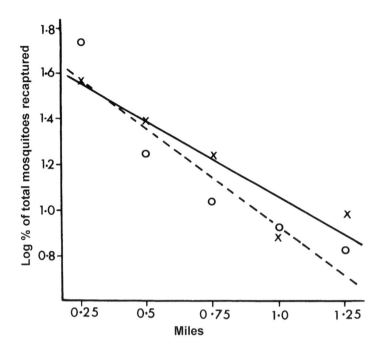

Fig. 15.3. Regression line of recaptured isotope-labelled females ($-\times-\times-$) and males ($-o-o-$) of *Anopheles gambiae* in relation to distance from the central release point (after Gillies 1961)

Recaptures were performed in the release village and two neighbouring villages using a pyrethrum spray collection method. 278 mosquitoes were recaptured (110 magenta and 164 yellow) in the release village and 98 (45 magenta, 53 yellow) in the other two study villages. A 'movement index' for movement between villages was calculated from the recaptures in the pyrethrum spray catches. For example, 13 marked mosquitoes were collected from a total of 12 187 mosquitoes collected by pyrethrum spray catches in the village of Baro Kunda. In Ja Koto village there were 53 recaptures in a total catch of 8272. The index of movement was calculated as $(13/12187)/(53/8272) = 16.6$. Ninety-five percent confidence intervals were assigned based on a binomial distribution. Mosquitoes were observed to readily disperse between the three villages, which were separated by distances of 1.1–1.5 km.

In southeast Tanzania, Takken et al. (1998) studied the dispersal and survival of *Anopheles funestus* and *Anopheles gambiae* s.l. using mark-release-recapture methods. Resting mosquitoes were collected from houses between 0800 and 1000 h using oral aspirators. Mosquitoes were marked by blowing a small quantity of Dayglo™ fluorescent powder into a plastic bag containing a 15 cm cubic cage in which the mosquitoes were held. Following release, resting mosquitoes were collected over a period of two weeks from 11 houses. A CDC light-trap was operated daily in an additional house. A total of 4262 *Anopheles funestus* and 645 *Anopheles gambiae* s.l. were released over two days and 184 and 48 respectively were recaptured over ten days. Dispersal of the marked mosquitoes among the three catching and release sites are presented (Fig. 15.4). For *Anopheles funestus* there was a significant net movement from Area 1 to Area 3, but this was not the case for *Anopheles gambiae* s.l., which moved in each direction equally. The authors assumed that the proximity of Area 3 to rice fields and hence oviposition sites may have been a factor in determining the apparent preference for this area, especially by *Anopheles funestus*.

Ba et al. (2005) marked and released 85 500 *Aedes vexans* females and 5800 *Culex poicilipes* females in Senegal in order to study dispersal and survival. The initial release of *Aedes vexans* comprised newly emerged females from field-collected immatures. Subsequent releases of *Aedes vexans* and all releases of *Culex poicilipes* were of adults captured in CO_2-baited CDC light-traps or in sheep-baited traps. Adults were marked using various colours of Dayglo fluorescent powders and released between 1630 and 1800 h from a single release point. Recaptures were initiated 24 h after release using 44 CDC light-traps and two sheep-baited traps each night at the larval habitat near the release site, additionally four CDC light-traps were operated at each of eight villages. Overall recapture rates were 0.18 ± 0.03 for *Aedes vexans* and $3.46 \pm 0.47\%$ for *Culex poicilipes*. The longest distance traversed in a single day was 400 m for *Aedes vexans* and 550 m for *Culex poicilipes*.

Fabian et al. (2005) studied the dispersal of laboratory-reared *Anopheles saperoi* marked with a 0.5% aqueous solution or rhodamine b in a forested area of Okinawa, Japan. Marked males and females were released at 1300 h, the time at which activity of the wild population was at a maximum. Recaptures were carried out for seven, eight or ten days post-release at nine fixed collection sites using a hand net collection method for 20 minutes/day. Marked individuals were identified by placing the mosquitoes on filter paper, adding a few drops of 80% alcohol and examining under ultra-violet light. The maximum flight distance recorded was 930 m.

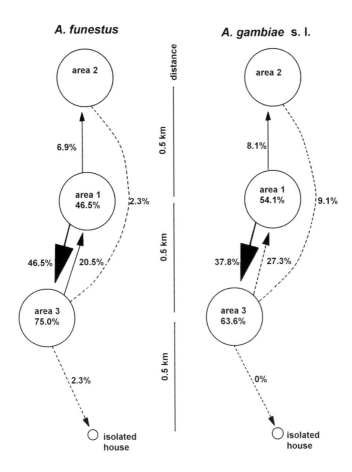

Fig. 15.4. Recaptures and dispersal of marked mosquitoes in Tanzania. Solid arrows: mosquitoes released in area 1. Broken arrows: mosquitoes released in area 3. (Takken et al. 1998)

In India, Yasuno et al. (1973) made recaptures of *Culex quinquefascia-tus* in nearly all houses in villages where marked adults were released, thus avoiding any decrease in sampling intensity with increasing distance from the release points. The daily percentage and cumulative percentage of houses with marked mosquitoes were plotted against days after release. In July the cumulative number of houses having marked mosquitoes had reached nearly 70% on the 7th day, whereas in the cooler month of De-cember only about 20% of houses contained marked mosquitoes by the 7th day. Also in India, Russell et al. (1944) placed series of traps, consisting of thatch-roofed huts, along the circumference of circles ¼, ½, ¾ and 1 mile from the central release point of *Anopheles culicifacies*. One female was caught in the furthest hut (1 mile) just 1 day after release. The distances along the perimeters of these concentric circles were in the proportion of 1, 2, 3 and 4 and the numbers of traps were in the same proportions, namely 8, 16, 24 and 32. The rate of decrease in the proportion of marked *Anopheles culicifacies* recaptured at increasing distances from the release point was not constant but a diminishing one. A good regression line was obtained from a log–log plot of catch against distance from point of release.

The relationship is hyperbolic, and is best described by the following equation:

$$y = ax^{-b}, \tag{15.29}$$

alternatively

$$y = \frac{a}{x^b}, \tag{15.30}$$

where y = number of recaptured mosquitoes, x = distance from release point, and a = a constant, and is the theoretical number of marked individuals to be recaptured 1 mile from the release point, and is consequently dependent on the numbers released. Constant b = the exponent of x, the distance. If random flight had occurred then b would be zero, since the increase in distance and therefore recapture area was compensated for by the larger number of catching stations. However, since b was always calculated as >0 (1.062 for males, 0.681 for females), this indicated that for some reason flight was inhibited and there was theoretically a greater concentration of mosquitoes near the release point than would be expected from random dispersal.

In Sri Lanka Curtis and Rawlings (1980) caught a marked *Anopheles culicifacies* 498 m from its release point within a day after marking. Curtis and Rawlings (1980) were able to derive some information on dispersal by dividing the overall proportions of marked mosquitoes at the furthest recapture points (outlying huts 2, 3 and 4) each day, by the overall proportion found to be marked in hut 1 (release point) on the same day. After doing this the results for each hut were averaged, but because the numbers caught in the outlying huts varied considerably, weighted averages for the huts were calculated. This procedure enabled an estimate to be made of the proportion of both marked and unmarked mosquitoes that moved from hut 1, the release site, to the three outlying huts at distances of 179, 392 and 498 m. A 17% overlap was estimated between populations sampled by huts (1 and 4) 498 m apart, that is an estimated 17% of the population in the vicinity of hut 1 dispersed to hut 4.

In Kenya Linthicum et al. (1985) studied the dispersal of adult *Aedes lineatopennis* from a breeding site using light-traps, human bait collections and collections of out of door resting adults. One collection site for each method (3) was set up 50 m from their emergence site, and four of each sampling method established along concentric annuli having radii of 0.5, 1.0 and 1.5 km. The mean dispersal distance ($\bar{d}$) at any time interval (0–45 days) was determined after applying a correction factor (*CF*) for each annulus as follows

$$CF = \frac{\text{Area of collecting annulus}}{\text{Total collection area}} \times \text{No. of collection sites (in this instance 39)}. \quad (15.30)$$

Mean distance dispersed ($\bar{d}$) is

$$\bar{d} = \Sigma \frac{\text{Corrected number collected} \times \text{distance (km)}}{\text{Total corrected number collected}}. \quad (15.31)$$

For females $\bar{d}$ was 0.11–0.25 km and for males 0.05–0.16 km.

In dispersal studies with dust-marked *Culicoides* and their subsequent recapture in traps in concentric annuli Brenner et al. (1984) pointed out if four traps were used on the first ring 0.5 km distance from the central release point, then for a concentric ring at 3 km from the release point 43 traps would be needed to ensure equal trap density (1 trap/0.2 km^2 in this instance). Although more traps can be located on more distant rings, logistics prevent the correct number being employed, in this instance a total of 141. Therefore a correction factor (*CF*) was computed

$$CF = A_r / A_t \times N_P \quad (15.32)$$

where A_r = area of trap ring, A_t = total trapping area covered by all rings, in other words the area in which recaptures are made, and N_p = number of traps used per sampling period. The correction factor was used to determine the corrected mean number caught per trap ring.

Reisen and Lothrop (1995) studied the population ecology and dispersal of *Culex tarsalis* in the Coachella valley, California, USA using mark-release-recapture methods. Cohorts of known age (obtained by rearing adults from larvae and pupae) or of known physiological status (host-seeking females collected by CO_2-baited CDC light-traps) were marked with fluorescent dust and released on three successive nights in each of five experiments. Mosquitoes were recaptured using 49 CO_2-baited CDC traps within a radius of 6.2 km from the release site. Two walk-in red boxes (Meyer 1985) (see Chapter 5) were also used. Daily dispersal rates were calculated by regression analysis and using a correction factor

(Brenner et al. 1984) to account for decreasing collecting effort with increasing distance. Totals of 22 563 recently emerged females and 55 548 host-seeking females were released. 3758 (6.7%) host-seeking mosquitoes and 37 (0.2%) newly-emerged females were recaptured. The recapture rate of marked females decreased significantly ($R^2 > 0.65$, $P < 0.001$) and varied as a curvilinear function of cohort age in days post-release. Overall, 3172 (84%) of the 3758 marked females were recaptured within 0.5 km of the release site and from 43 of the 49 traps. Dispersion direction appeared to be independent of the prevailing wind (N-NW to S-SE) direction and appeared to be related primarily to topography. The daily dispersal rate was 0.22 km/day and varied between 0.08 in May to 0.30 in September, but the mean distance covered varied significantly between experiments: 2.16 km in February, 1.85 km in May, 1.22 km in July, 2.18 km in September, and 1.30 in November.

Also in California, Kramer et al. (1995) used mark-release-recapture methods to study the dispersal of newly-emerged *Ochlerotatus dorsalis* males and females marked with fluorescent dust (Radiant Colour, Richmond, California, USA). Dispersing mosquitoes were collected using 40 CO_2-baited CDC light-traps positioned within an 8 km radius of the release site and operated continuously over 12 nights (study 1) or 30 traps located within a radius of 2.4 km of the release site and operated for 16 nights (study 2). In study 1, a total of 12 260 females were released, of which only 15 were recaptured. In study 2, a total of 24 070 females were released and 290 were recaptured, with 134 recaptures being made in two traps located within 200 m of the release site. 40 335 males were also released, but none was recaptured. 67% and 81% of recaptures in studies 1 and 2 respectively were made by traps located in a downwind direction.

Tietze et al. (2003) released three batches (30 000, 5000 and 8000 on three consecutive days) of marked *Culex erythrothorax* previously captured using CDC light-traps in Santa Cruz County, California, USA. Released mosquitoes were marked with luminous powder (Shannon Luminous Materials, Inc., Santa Ana, California) applied with a small bulb duster and recaptures were made with a series of 21 CO -baited CDC light-traps. A total of 319 marked mosquitoes were recaptured. The overall average dispersal rate in the study was 0.57 km/day.

In Rio de Janeiro State, Brazil, Honório et al. (2003) marked 3055 *Aedes aegypti* and 2225 *Aedes albopictus* adult females by feeding them on rubidium-marked blood. These marked adults were then released in Nova-Iguaçu, an urban area where dengue is endemic, and dispersal estimated through the identification of Rb-marked eggs collected from 1472 ovitraps situated within an 800m radius from the release point. The ovitraps were distributed in five concentric rings within the 800 m radius circular area as follows: 0–100 m radius (23 ovitraps), 100–200 m (69 ovitraps), 200–400 m (276), 400–600 m (460 ovitraps) and 600–800 m (644 ovitraps). Rb-marked eggs of *Aedes aegypti* were collected up to 800 m from the release site, and interestingly, none was collected from the 0–100 m concentric ring. Rb-marked *Aedes albopictus* eggs were also collected up to 800 m from the release point. 18.2% of ovitrap paddles in the 0–100 m ring were positive for Rb-marked eggs. In the 100–200m area, 16.5% of the 127 paddles were positive. In the 200–400 m area, 50 paddles (9.5%) had Rb-marked eggs. 9.3% of the 875 paddles collected in the 400–600 m area were positive and in the 600–800 m area 11.4% of paddles were positive.

In Arkansas, USA Weathersbee and Meisch (1990) found it too expensive to provide increasing numbers of traps at increasing distance from release points in their study on the dispersal of *Anopheles quadrimaculatus*, so they used the correction factor of Brenner et al. (1984) which allowed the estimation of the expected numbers of recaptures for each subunit (trap ring) under conditions of equal trap density. The expected number of recaptures (Er_s) for each subunit was calculated as follows

$$Er_s = R_s / T_s \times CF_s \qquad (15.33)$$

where R_s = number of recaptures in the subunit, T_s = number of traps in the subunit, and CF_s = the correction factor.

References

Abdel-Malek AA (1966) Study of the dispersion and flight range of *Anopheles sergentii* Theo. in Siwa Oasis using radioactive isotopes as markers. Bull World Health Organ 35: 968–973

Asman SM, Nelson RL, McDonald PT, Milby MM, Reeves WC, White KD, Fine PEM (1979) Pilot release of a sex-linked multiple translocation into a *Culex tarsalis* field population in Kern county, California. Mosquito News 39: 248–258

Ba Y, Diallo D, Kebe CMF, Dia I, Diallo M (2005) Aspects of bioecology of two Rift Valley fever virus vectors in Senegal (West Africa): *Aedes vexans* and *Culex poicilipes* (Diptera: Culicidae). J Med Entomol 42: 739–750

Bailey SF, Baerg DC (1967) The flight habits of *Anopheles freeborni* Aitken. Proc California Mosq Control Assoc 25: 55–59

Bailey SF, Eliason DA, Hoffman BL (1965) Flight and dispersal of the mosquito *Culex tarsalis* Coquillett in the Sacramento valley of California. Hilgardia 37: 73–113

Begon M (1976) Dispersal, density and microdistribution in *Drosophila subobscura* Collin. J Anim Ecol 45: 441–456

Begon M, Harper JL, Townsend CR (1986) Ecology. Individuals, Populations and Communities. Blackwell Scientific Publications, Oxford

Beier JC, Berry WJ, Craig GB (1982) Horizontal distribution of adult *Aedes triseriatus* (Diptera: Culicidae) in relation to habitat structure, oviposition, and other mosquito species. J Med Entomol 19: 239–247

Bidlingmayer WL, Schoof HF (1957) The dispersal characteristics of the saltmarsh mosquito, *Aedes taeniorhynchus* (Wiedemann), near Savannah, Georgia. Mosquito News 17: 202–212

Birley MH, Charlwood JD (1989) The effect of moonlight and other factors on the oviposition cycle of malaria vectors in Madang, Papua New Guinea. Ann Trop Med Parasitol 83: 415–422

Boiteau G, Colpitts B (2004) The potential of portable harmonic radar technology for the tracking of beneficial insects. Int J Pest Manag 50: 233–242

Bonnet DD, Worcester DJ (1946) The dispersal of *Aedes albopictus* in the territory of Hawaii. Am J Trop Med 26: 465–476

Brenner RJ, Wargo MJ, Stains GS, Mulla MS (1984) The dispersal of *Culicoides mohave* (Diptera: Ceratopogonidae) in the desert of southern California. Mosquito News 44: 343–350

Brust RA (1980) Dispersal behaviour of adult *Aedes sticticus* and *Aedes vexans* (Diptera: Culicidae) in Manitoba. Can Entomol 112: 31–42

Bryan JH, Foley DH, Geary M, Carven CTJ (1991) *Anopheles annulipes* Walker (Diptera: Culicidae) at Griffith, New South Wales. 3. Dispersal of sibling species. J Aust Entomol Soc 30: 119–121

Bugher JC, Taylor M (1949) Radiophosphorous and radiostrontium in mosquitoes. Preliminary report. Science 110: 146–147

Causey OR, Kumm HW, Laemmert HW (1950) Dispersion of forest mosquitoes in Brazil: further studies. Am J Trop Med 30: 301–312

Chapman HF, Hughes JM, Jennings C, Kay BH, Ritchie SA (1999) Population structure and dispersal of the saltmarsh mosquito *Aedes vigilax* in Queensland, Australia. Med Vet Entomol 13: 423–430

Charlwood JD, Alecrim WA (1989) Capture–recapture studies with the South American malaria vector *Anopheles darlingi*, Root. Ann Trop Med Parasitol 83: 569–576

Charlwood JD, Bryan JH (1987) A mark–recapture experiment with the filariasis vector *Anopheles punctulatus* in Papua New Guinea. Ann Trop Med Parasitol 81: 429–436

Charlwood JD, Mendis C, Thompson R, Begtrup K, Cuamba N, Dgedge M, Gamage-Mendis A, Hunt RH, Sinden RE, Hogh B (1998) Cordon sanitaire or laissez faire: differential dispersal of young and old females of the malaria vector *Anopheles funestus* Giles (Diptera: Culicidae) in southern Mozambique. Afr Entomol 6: 1–6

Chiang GL, Loong KP, Mahadevan S, Eng KL (1988) A study of dispersal, survival and gonotrophic cycle estimates of *Mansonia uniformis* in an open swamp ecotype. Southeast Asian J Trop Med Public Health 19: 271–282

Clark DP (1962) An analysis of dispersal and movement in *Phaulacridium vittatum* (Siost.) (Acrididae). Aust J Zool 10: 382–399

Costantini C, Li SongGang, della Torre A, Sagnon N, Coluzzi M, Taylor CE (1996) Density, survival and dispersal of *Anopheles gambiae* complex mosquitoes in a West African Sudan savanna village. Med Vet Entomol 10: 203–219

Curry DP (1939) A documented record of a long flight of *Aedes sollicitans*. Proc New Jers Mosq Exterm Assoc 26: 36–39

Curtis CF, Rawlings P (1980) A preliminary study of dispersal and survival of *Anopheles culicifacies* in relation to the possibility of inhibiting the spread of insecticide resistance. Ecol Entomol 5: 11–17

Dégallier N, Sá Filho GC, Monteiro HAO, Castro FC, da Silva OV, Brandão RCF, Moyses M, da Rosa APAT (1998) Release-recapture experiments with canopy mosquitoes in the genera *Haemagogus* and *Sabethes* (Diptera: Culicidae) in Brazilian Amazonia. J Med Entomol 35: 931–936

Dobzhansky T, Wright S (1943) Genetics of natural populations. X. Dispersion rates in *Drosophila pseudoobscura*. Genetics 28: 304–340

Dobzhansky T, Wright S (1947) Genetics of natural populations. XV. Rate of diffusion of a mutant gene through a population of *Drosophila pseudoobscura*. Genetics 32: 303–340

Dow RP (1971) The dispersal of *Culex nigripalpus* marked with high concentrations of radiosphosphorus. J Med Entomol 8: 353–363

Dow RP, Reeves WC, Bellamy RE (1965) Dispersal of female *Culex tarsalis* into a larvicided area. Am J Trop Med Hyg 14: 656–670

Eddy GW, Roth AR, Plapp FW (1962) Studies on the flight habits of some marked insects. J Econ Entomol 55: 603–607

Edman JD, Scott TW, Costero A, Morrison AC, Harrington LC, Clark GG (1998) *Aedes aegypti* (Diptera: Culicidae) movement influenced by availability of oviposition sites. J Med Entomol 35: 578–583

Eyles DE, Bishop LK (1943) An experiment on the range of dispersion of *Anopheles quadrimaculatus*. Am J Hyg 37: 239–245

Fabian MM, Toma T, Tsuzuki A, Saita S, Miyagi I (2005) Mark-release-recapture experiments with *Anopheles saperoi* (Diptera: Culicidae) in the Yona forest, northern Okinawa, Japan. Southeast Asian J Trop Med Public Health 36: 54–63

Fletcher BS (1974) The ecology of a natural population of the Queensland fruit fly, *Dacus tryoni*. V. The dispersal of adults. Aust J Zool 22: 189–202

Frank JH, Curtis GA (1977) On the bionomics of bromeliad-inhabiting mosquitoes. V. A mark–release–recapture technique for estimation of population size of *Wyeomyia vanduzeei*. Mosquito News 37: 444–452

Frank PW (1964) On home range of limpets. Am Nat 98: 99–104

Freeman GH (1977) A model relating numbers of dispersing insects to distance and time. J Appl Ecol 14: 477–487

Fussell EM (1964) Dispersal studies on radioactive-tagged *Culex quinquefasciatus* Say. Mosquito News 24: 422–426

Garrett-Jones C (1962) The possibility of active long-distance migrations by *Anopheles pharoensis* Theobald. Bull World Health Organ 27: 299–302

Gaydecki PA (1984) A quantification of the behavioural dynamics of certain Lepidoptera in response to light. Ph.D. thesis, Cranfield Institute of Technology

Gibson CCG (1923) Malaria and mosquitoes in Belize, British Honduras. J R Army Med. Corps 40: 38–43

Gillies MT (1961) Studies on the dispersion and survival of *Anopheles gambiae* Giles in East Africa, by means of marking and release experiments. Bull Entomol Res 52: 99–127

Gilmour D, Waterhouse DF, McIntyre GA (1946) An account of experiments undertaken to determine the natural population density of the sheep blowfly, *Lucilia cuprina*. Wied. Bull Coun Scient Ind Res Aust 195: 1–39

Ginsberg HS (1986) Dispersal patterns of *Aedes sollicitans* (Diptera: Culicidae) at the east end of the fire island national seashore, New York, USA. J Med Entomol 23: 146–155

Greenslade PJM (1964) The distribution, dispersal and size of a population of *Nebria brevicollis* (F.), with comparative studies on three other Carabidae. J Anim Ecol 33: 311–333

Haeger JS (1960) Behavior preceding migration in the salt-marsh mosquito, *Aedes taeniorhynchus*. Mosquito News 20: 136–147

Hamlyn-Harris R (1933) Some ecological factors involved in the dispersal of mosquitoes in Queensland. Bull Entomol Res 24: 229–232

Haramis LD, Foster WA (1983) Survival and population density of *Aedes triseriatus* (Diptera: Culicidae) in a woodlot in central Ohio, USA. J Med Entomol 20: 391–398

Harden FW, Chubb HS (1960) Observations of *Aedes taeniorhynchus* dispersal in extreme south Florida and the Everglades National Park. Mosquito News 20: 249–255

Hawkes C (1972) The estimation of the dispersal rate of the adult cabbage root fly (*Erioischia brassicae* (Bouche)) in the presence of a brassica crop. J Appl Ecol 9: 617–632

Hendricks DE (1980) Low-frequency sonar device that counts flying insects attracted to sex pheromone dispensers. Environ Entomol 9: 452–457

Hervy JP (1977*b*) Expérience de marquage–lâcher–recapture, portant sur *Aedes aegypti* Linné, en zone de savane soudanienne ouest africaine. II. Relations entre habitat, morphologie et comportement. Cah ORSTOM sér Entomol Méd Parasitol 15: 365–372

Hii JLK, Vun YS (1985) A study of dispersal, survival and adult population estimates of the malaria vector, *Anopheles balabacensis* Baisas (Diptera: Culicidae) in Sabah, Malaysia. Trop Biomed 2: 121–131

Hocking B (1953) The intrinsic range and speed of flight of insects. Trans R Entomol Soc Lond 104: 223–345

Honório NA, Silva W da C, Leite PJ, Gonçalves JM, Lounibos LP, Lourenço-de-Oliveira R (2003). Dispersal of *Aedes aegypti* and *Aedes albopictus* (Diptera: Culicidae) in an urban endemic dengue area in the State of Rio de Janeiro, Brazil. Mem Inst Oswaldo Cruz 98: 191–198

Howard JJ, White DJ, Muller SL (1989) Mark–recapture studies on the *Culiseta* (Diptera: Culicidae) vectors of eastern equine encephalitis virus. J Med Entomol 26: 190–199

Jenkins DW, Hassett CC (1951) Dispersal and flight range of subarctic mosquitoes marked with radiophosphorus. Can J Zool 29: 178–187

Jensen T, Washino RK (1991) An assessment of the biological capacity of a Sacramento valley population of *Aedes melanimon* to vector arboviruses. Am J Trop Med Hyg 44: 355–363

Jensen T, Washino RK (1994) Comparison of recapture patterns of marked and released *Aedes vexans* and *Ae. melanimon* (Diptera: Culicidae) in the Sacramento valley of California. J Med Entomol 31: 607–610

Johnson CG (1969) Migration and Dispersal of Insects by Flight. Methuen & Co. Ltd., London

Kettle DS (1951) The spatial distribution of *Culicoides impunctatus* Goet. under woodland and moorland conditions and its flight range through woodland. Bull Entomol Res 42: 239–291

Kirkpatrick TW (1925) The Mosquitoes of Egypt. Cairo Govt. Press, Cairo

Kramer VL, Carper ER, Beesley C, Reisen WK (1995) Mark-release-recapture studies with *Aedes dorsalis* (Diptera: Culicidae) in coastal northern California. J Med Entomol 32: 375–380

LaSalle MW, Dakin ME (1982) Dispersal of *Culex salinarius* in southwestern Louisiana. Mosquito News 42: 543–550

Lillie TH, Marquardt WC, Jones RH (1981) The flight range of *Culicoides variipennis* (Diptera: Ceratopogonidae). Can Entomol 113: 419–426

Lillie TH, Kline DL, Hall DW (1985) The dispersal of *Culicoides mississippiensis* (Diptera: Ceratopogonidae) in a salt marsh near Yankeetown, Florida. J Am Mosq Control Assoc 1: 463–467

Lincoln RJ, Boxshall GA, Clark PF (1982) A Dictionary of Ecology, Evolution and Systematics. Cambridge University Press, Cambridge

Lindquist AW, Ikeshoji T, Grab B, De Meillon B, Khan ZH (1967) Dispersion studies of *Culex pipiens fatigans* tagged with ^{32}P in the Kemmendine area of Rangoon, Burma. Bull World Health Organ 36: 21–37

Lines JD, Lyimo EO, Curtis CF (1986) Mixing of indoor- and outdoor-resting adults of *Anopheles gambiae* Giles s.l. and *A. funestus* Giles (Diptera: Culicidae) in coastal Tanzania. Bull Entomol Res 76: 171–178

Linthicum KJ, Bailey CL, Davies FG, Kairo A (1985) Observations on the dispersal and survival of a population of *Aedes lineatopennis* (Ludlow) (Diptera: Culicidae) in Kenya. Bull Entomol Res 75: 661–670

Loong KP, Chiang GL, Eng KL, Chan ST, Yap HH (1990) Survival and feeding behaviour of Malaysian strains of *Anopheles maculatus* Theobald (Diptera: Culicidae) and their role in malaria transmission. Trop Biomed 7: 71–76

Low N (1925) Anti-malarial work in Ismailia. J R Army Med Corps 45: 52–54

MacCreary D, Stearns LA (1937) Mosquito migration across Delaware Bay. Proc New Jers Mosq Exterm Assoc 24: 188–197

Macdonald WW, Sebastian A, Tun MM (1968) A mark–release–recapture experiment with *Culex pipiens fatigans* in the village of Okpo, Burma. Ann Trop Med Parasitol 62: 200–209

McDonald PT (1977*b*) Population characteristics of domestic *Aedes aegypti* (Diptera: Culicidae) in villages on the Kenya coast. II. Dispersal within and between villages. J Med Entomol 14: 49–53

Meyer RP (1985) The "walk-in" type red box for sampling adult mosquitoes. Proc New Jers Mosq Control Assoc 72: 104–105

Milby MM, Reisen WK, Reeves WC (1983) Intercanyon movement of marked *Culex tarsalis* (Diptera: Culicidae). J Med Entomol 20: 193–198

Mori A (1979) Effects of larval density and nutrition on some attributes of immature and adult *Aedes albopictus*. Trop Med 21: 85–103

Mori A, Wada Y (1977) The gonotrophic cycle of *Aedes albopictus* in the field. Trop Med 19: 141–146

Morlan HB, Hayes RO (1958) Urban dispersal and activity of *Aedes aegypti*. Mosquito News 18: 137–144

Morris CD, Larson VL, Lounibos LP (1991) Measuring mosquito dispersal for control programs. J Am Mosq Control Assoc 7: 608–615

Morrison AC, Costero A, Edman JD, Clark GG, Scott TW (1999) Increased fecundity of *Aedes aegypti* fed human blood before release in a mark-recapture study in Puerto Rico. J Am Mosq Control Assoc 15: 98–104

Muir LE, Kay BH (1998) *Aedes aegypti* survival and dispersal estimated by mark-release-recapture in Northern Australia. Am J Trop Med Hyg 58: 277–282

Nayar JK (1982) *Wyeomyia mitchellii*: Observations on dispersal, survival, nutrition, insemination and ovarian development in a Florida population. Mosquito News 42: 416–427

Nayar JK (1985) Bionomics and physiology of *Aedes taeniorhynchus* and *Aedes sollicitans*, the salt marsh mosquitoes of Florida. Agr Exp Stn Inst Food Agr Sci University Florida, Bulletin no. 852

Nayar JK, Sauerman DM (1969) Flight behavior and phase polymorphism in the mosquito *Aedes taeniorhynchus*. Entomol Exp Appl 12: 365–375

Nayar JK, Sauerman DM (1970) A comparative study of growth and development in Florida mosquitoes. Part 3. Effects of temporary crowding on larval aggregation formation, pupal ecdysis and adult characteristics at emergence. J Med Entomol 7: 521–528

Nayar JK, Sauerman DM (1973) A comparative study of growth and development in Florida mosquitoes. Part 4. Effects of temporary crowding during larval stages on female flight activity patterns. J Med Entomol 10: 37–42

Nayar JK, Provost MW, Hansen CW (1980) Quantitative bionomics of *Culex nigripalpus* (Diptera: Culicidae) populations in Florida. 2. Distribution, dispersal and survival patterns. J Med Entomol 17: 40–50

Nelson RL, Milby MM (1978) Distribution of *Culex tarsalis* in an isolated field area in California. Proc California Mosq Vector Control Assoc 46: 50–53

Nelson RL, Milby MM (1980*a*) Dispersal and survival of field and laboratory strains of *Culex tarsalis* (Diptera: Culicidae). J Med Entomol 17: 146–150

Nelson RL, Milby MM, Reeves WC, Fine PEM (1978) Estimates of survival, population size, and emergence of *Culex tarsalis* at an isolated site. Ann Entomol Soc Am 71: 801–808

Nielsen ET (1958) The initial stage of migration in salt-marsh mosquitoes. Bull Entomol Res 49: 305–315

Nielsen ET, Haeger JS (1960) Swarming and mating in mosquitoes. Misc Pubs Entomol Soc Am 1: 71–95

Nutsathapana S, Sawasdiwongphorn P, Chitprarop U, Cullen JR, Gass RF, Green CA (1986) A mark–release–recapture demonstration of host preference heterogeneity in *Anopheles minimus* Theobald (Diptera: Culicidae) in a Thai village. Bull Entomol Res 76: 313–320

O'Donnell MS, Berry G, Carvan T, Bryan JH (1992) Dispersal of adult females of *Culex annulirostris* in Griffith, New South Wales, Australia. J Am Mosq Control Assoc 8: 159–165

Pan American Health Organization (1994) Dengue and Dengue Hemorrhagic Fever in the Americas. Guidelines for Prevention and Control. Scientific Publication 584

Paris OH (1965) The vagility of P^{32}-labelled isopods in grassland. Ecology 46: 635–648

Pedgley D (1982) Windborne Pests and Diseases. Meteorology of Airborne Organisms. Ellis Horwood Ltd., Chichester

Pielou EC (1977) Mathematical Ecology. John Wiley, New York

Provost MW (1952) The dispersal of *Aedes taeniorhynchus*. I. Preliminary studies. Mosquito News 12: 174–190

Provost MW (1953) Motives behind mosquito flights. Mosquito News 13: 106–109

Provost MW (1957) The dispersal of *Aedes taeniorhynchus*. II. The second experiment. Mosquito News 17: 233–247

Provost MW (1960) The dispersal of *Aedes taeniorhynchus*. III. Study methods for migratory exodus. Mosquito News 20: 148–161

Pumpuni CB, Walker ED (1989) Population size and survivorship of adult *Aedes triseriatus* in a scrap tireyard in northern Indiana. J Am Mosq Control Assoc 5: 166–172

Quarterman KD, Jensen JA, Mathis W, Smith WW (1955) Flight dispersal of rice field mosquitoes in Arkansas. J Econ Entomol 48: 30–32

Quraishi MS, Faghih MA, Esghi W (1966) Flight range, lengths of gonotrophic cycles and longevity of P^{32}-labeled *Anopheles stephensi mysorensis*. J Econ Entomol 59: 50–55

Rajagopalan PK, Yasuno M, LaBrecque GC (1973) Dispersal and survival in the field of chemosterilized, irradiated, and cytoplasmically incompatible male *Culex pipiens fatigans*. Bull World Health Organ 48: 631–635

Rawlings P, Curtis CF, Wickramasinghe MB, Lines J (1981) The influence of age and season on dispersal and recapture of *Anopheles culicifacies* in Sri Lanka. Ecol Entomol 6: 307–319

Reeves WC, Brookman B, Hammon WM (1948) Studies on the flight range of certain *Culex* mosquitoes, using a fluorescent-dye marker, with notes on *Culiseta* and *Anopheles*. Mosquito News 8: 61–69

Reisen WK, Aslamkhan M (1979) A release–recapture experiment with the malaria vector, *Anopheles stephensi* Liston, with observations on dispersal, survivorship, population size, gonotrophic rhythm and mating behaviour. Ann Trop Med Parasitol 73: 251–269

Reisen WK, Lothrop HD (1995) Population ecology and dispersal of *Culex tarsalis* (Diptera: Culicidae) in the Coachella Valley of California. J Med Entomol 32: 490–502

Reisen WK, Reeves WC (1990) Bionomics and ecology of *Culex tarsalis* and other potential mosquito vector species. In: Reeves WC (ed) Epidemiology and Control of Mosquito-borne Arboviruses in California, 1943–1987. California Mosquito and Vector Control Association, Sacramento, pp. 254–329

Reisen WK, Baker RH, Sakai RK, Aziz-Javed A, Aslam Y, Siddiqui TF (1977) Observations on the mating behavior and survivorship of *Culex tritaeniorhynchus* Giles (Diptera: Culicidae) during late autumn. Southeast Asian J Trop Med Public Health 8: 537–545

Reisen WK, Aslam Y, Siddiqui TF, Khan AQ (1978) A mark–release–recapture experiment with *Culex tritaeniorhynchus* Giles. Trans R Soc Trop Med Hyg 72: 167–177

Reisen WK, Mahmood F, Parveen T (1979) *Anopheles subpictus* Grassi: observations on survivorship and population size using mark–release–recapture and dissection methods. Researches in Population Ecology 21: 12–29

Reisen WK, Mahmood F, Azra K (1981) *Anopheles culicifacies* Giles: adult ecological parameters measured in rural Punjab province, Pakistan using capture–release–recapture and dissection methods, with comparative observations on

An.stephensi Liston and *An. subpictus* Grassi. Researches in Population Ecology 23: 39–60

Reisen WK, Mahmood F, Parveen T (1982*a*) Seasonal trends in population size and survivorship of *Anopheles culicifacies, An. stephensi* and *An. subpictus* (Diptera: Culicidae) in rural Punjab province, Pakistan. J Med Entomol 19: 86–97

Reisen WK, Milby MM, Asman SM, Bock ME, Meyer RP, McDonald PT, Reeves WC (1982*b*) Attempted suppression of a semi-isolated *Culex tarsalis* population by the release of irradiated males: a second experiment using males from a recently colonized strain. Mosquito News 41: 736–744

Reisen WK, Sakai RK, Baker RH, Azra K, Niaz S (1982*c*) *Anopheles culicifacies*: observations on population ecology and reproductive behavior. Mosquito News 42: 93–101

Reisen WK, Milby MM, Meyer RP, Reeves WC (1983) Population ecology of *Culex tarsalis* (Diptera: Culicidae) in a foothill environment in Kern county, California: Temporal changes in male relative abundance and swarming behavior. Ann Entomol Soc Am 76: 809–815

Reisen WK, Milby MM, Meyer RP, Pfuntner AR, Spoehel J, Hazelrigg JE, Webb JP (1991) Mark–release–recapture studies with *Culex* mosquitoes (Diptera: Culicidae) in southern California. J Med Entomol 28: 357–371

Reisen WK, Milby MM, Meyer RP (1992) Population dynamics of adult *Culex* mosquitoes (Diptera: Culicidae) along the Kern river, Kern County, California, 1990. J Med Entomol 29: 531–543

Reisen WK, Lothrop HD, Lothrop B (2003) Factors influencing the outcome of mark-release-recapture studies with *Culex tarsalis* (Diptera: Culicidae). J Med Entomol 40: 820–829

Reiter P, Amador MA, Anderson RA, Clark GG (1995) Dispersal of *Aedes aegypti* in an urban area after blood feeding as demonstrated by rubidium-marked eggs. Am J Trop Med Hyg 52: 177–179

Reuben R, Yasuno M, Panicker KN (1972) Studies on the Dispersal of *Aedes aegypti* at Two Localities in Delhi. WHO/VBC 72.388. 9 pp (mimeographed)

Reuben R, Panicker KN, Brooks GD, Ansari MA (1975*b*) Studies on dispersal and loss rate of male *Aedes aegypti* (L.) in different seasons at Sonepat, India. J Commun Dis 7: 301–312

Robinson J, Luff ML (1979) Population estimates and dispersal of *Hydrotaea irritans* Fallén. Ecol Entomol 4: 289–296

Rudd WG, Gandour RW (1985) Diffusion model for insect dispersal. J Econ Entomol 78: 295–301

Russell PF, Knipe FW, Rao TR, Putnam P (1944) Some experiments on flight range of *Anopheles culicifacies*. J Exp Zool 97: 135–163

Russell RC (1987) Survival of insects in the wheel bays of Boeing 747B aircraft on flights between tropical and temperate airports. Bull World Health Organ 65: 659–662

Schaefer GW, Bent GA (1984) An infra-red remote sensing system for the active detection and automatic determination of insect flight trajectories (IRADIT). Bull Entomol Res 74: 261–278

Schreiber ET, Mulla MS, Chaney JD, Dhillon MS (1988) Dispersal of *Culex quinquefasciatus* from a dairy in southern California. J Am Mosq Control Assoc 4: 300–304

Service MW (1980) Effects of wind on the behaviour and distribution of mosquitoes and blackflies. Int J Biometeorol 24: 347–353

Service MW (1993) Mosquito Ecology. Field Sampling Methods. 2nd edn. Chapman & Hall, London

Service MW (1997) Mosquito (Diptera: Culicidae) dispersal - the long and short of it. J Med Entomol 34: 579–588

Sharma VP (1977) Insemination rate in *Culex pipiens fatigans* Wied. moving from wells to the village. J Commun Dis 9: 128–131

Shemanchuk JA, Fredeen FJH, Kristjanson AM (1955) Studies on flight range and dispersal habits of *Aedes flavescens* (Müller) (Diptera: Culicidae) tagged with radio-phosphorus. Can Entomol 87: 376–379

Sheppard PM, Macdonald WW, Tonn RJ, Grab B (1969) The dynamics of an adult population of *Aedes aegpyti* in relation to dengue haemorrhagic fever in Bangkok. J Anim Ecol 38: 661–702

Sinsko MJ, Craig GB (1979) Dynamics of an isolated population of *Aedes triseriatus* (Diptera: Culicidae). I. Population size. J Med Entomol 15: 89–98

Slooff R, Herath PRJ (1980) Ovarian development and biting frequency in *Anopheles culicifacies* Giles in Sri Lanka. Trop Geogr Med 32: 306–311

Smith A, Carter ID (1984) International transportation of mosquitoes of public health importance. In: Laird M (ed) Commerce and the Spread of Pests and Disease Vectors. Praegar Scientific, New York, pp. 1–17

Smith GE, Watson RB, Crowell RL (1941) Observations on the flight range of *Anopheles quadrimaculatus*, Say. Am J Hyg 34: 102–113

Smith GF, Geib AF, Isaak LW (1956) Investigations of a recurrent flight pattern of flood water *Aedes* mosquitoes in Kern county, California. Mosquito News 16: 251–256

Southwood TRE, Henderson PA (2000) Ecological Methods. 3rd edn. Blackwell Science, Oxford

Sparks AN, Jackson RD, Carpenter JE, Muller RA (1986) Insects captured in light traps in the Gulf of Mexico. Ann Entomol Soc Am 79: 132–139

Spradbery JP, Mahon RJ, Morton R, Tozer RS (1995) Dispersal of the Old World screw-worm fly *Chrysomya bezziana*. Med Vet Entomol 9: 161–168

Suzuki T (1978) Preliminary studies on blood meal interval of *Aedes polynesiensis* in the field. Jap J Sanit Zool 29: 169–174

Takagi M, Tsuda Y, Suzuki A, Wada Y (1995) Movement of individually marked *Aedes albopictus* females in Nagasaki, Japan. Trop Med (Nagasaki) 37: 79–85

Takken W, Charlwood JD, Billingsley PF, Gort G (1998) Dispersal and survival of *Anopheles funestus* and *A. gambiae* s.l. (Diptera: Culicidae) during the rainy season in southeast Tanzania. Bull Entomol Res 88: 561–566

Taylor LR, French RA, Macaulay EDM (1973) Low-altitude migration and diurnal flight periodicity; the importance of *Plusia gamma* L. (Lepidoptera: Plusiidae). J Anim Ecol 42: 751–760

Taylor RAJ (1978) The relationship between density and distance of dispersing insects. Ecol Entomol 3: 63–70

Thomson MC, Connor SJ, Quiñones ML, Jawara M, Todd J, Greenwood BM (1995) Movement of *Anopheles gambiae* s.l. malaria vectors between villages in The Gambia. Med Vet Entomol 9: 413–419

Thurman DC, Husbands RC (1951) Preliminary Report on Mosquito Flight Dispersal Studies with Radioisotopes in California, 1950. Commun Dis Center Bull (US Public Health Service) No. 10

Tietze NS, Stephenson MF, Sidhom NT, Binding PL (2003) Mark-recapture of *Culex erythrothorax* in Santa Cruz County, California. J Am Mosq Control Assoc 19 134–138

Trpis M (1971) Seasonal Variation in the Adult Populations of *Aedes aegypti* in the Dar es Salaam Area, Tanzania. WHO/VBC 71.291, 29 pp (mimeographed)

Trpis M (1973) Adult population estimate of *Toxorhynchites brevipalpis*. Bull World Health Organ 48: 756–757

Trpis M, Haüsermann W (1975) Demonstration of differential domesticity of *Aedes aegypti* (L.) in Africa by mark–release–recapture. Bull Entomol Res 65: 199–208

Trpis M, Haüsermann W (1978) Genetics of house-entering behaviour in East African population of *Aedes aegpyti* (L.) (Diptera: Culicidae) and its relevance to speciation. Bull Entomol Res 68: 521–32

Trpis M, Haüsermann W (1986) Dispersal and other population parameters of *Aedes aegypti* in an African village and their possible significance in epidemiology of vector-borne diseases. Am J Trop Med Hyg 35: 1263–1279

Wada Y, Kawai S, Oda T, Miyagi I, Suenaga O, Nishigaki J, Omori N, Takahashi K, Matsuo R, Itoh T, Takatsuki Y (1969) Dispersal experiments of *Culex tritaeniorhynchus* in Nagasaki area (Preliminary Report). Trop Med 11: 37–44

Wada Y, Oda T, Mogi M, Mori A, Omori N, Fukumi H, Hayashi K, Mifune K, Shichijo A, Matsuo S (1975*b*). Ecology of Japanese encephalitis virus in Japan. II. The population of vector mosquitoes and the epidemic of Japanese encephalitis. Trop Med 17: 111–127

Wadley FM, Wolfenbarger DO (1944) Regression of insect density on distance from center of dispersion as shown by a study of the smaller European elm bark beetle. J Agric Res 69: 299–308

Walker ED, Copeland RS, Paulson SL, Munstermann LE (1987) Adult survivorship, population density, and body size in sympatric populations of *Aedes triseriatus* and *Aedes hendersoni* (Diptera: Culicidae). J Med Entomol 24: 485–493

Wallace B (1970) Observations on the microdispersion of *Drosophila melanogaster*. In: Hecht MK, Steere WC (eds) Essays in Evolution and Genetics Presented in Honour of Theodorius Dobzhansky. North-Holland Publishing Co., Amsterdam, pp. 381–399

Watson TM, Saul A, Kay BH (2000) *Aedes notoscriptus* (Diptera: Culicidae) survival and dispersal estimated by mark-release-recapture in Brisbane, Queensland Australia. J Med Entomol 37: 380–384

Weathersbee AA, Meisch MV (1990) Dispersal of *Anopheles quadrimaculatus* (Diptera: Culicidae) in Arkansas ricefields. Environ Entomol 19: 961–965

White DJ, Morris CD (1985) Bionomics of anthropophilic Simuliidae (Diptera) from the Adirondack mountains of New York State, USA. 1. Adult dispersal and longevity. J Med Entomol 22: 190–199

Williams CB (1961) Studies on the effect of weather conditions on the activity and abundance of insect populations. Phil Trans R Soc Lond B 244: 331–378

Wolfinsohn M, Galun R (1953) A method for determining the flight range of *Aedes aegypti* (Linn). Bull Res Coun Israel 2: 433–436

Wright RE (1918) The distance mosquitoes can fly. J Bombay Nat Hist Soc 25: 511–512

Yasuno M, Rajagopalan PK (1977) Population estimation of *Culex fatigans* in Delhi villages. J Commun Dis 9: 172–183

Yasuno M, Rajagopalan PK, LaBrecque GC (1972*a*) Migration of *Culex fatigans* from Surrounding Fields to Villages in the Delhi Area. WHO/VBC 72.364,6 pp (mimeographed)

Yasuno M, Rajagopalan PK, Russel S, LaBrecque GC (1972*b*) Dispersal of *Culex fatigans* in Delhi Villages. WHO/VBC 72.352, 13 pp (mimeographed)

Yasuno M, Rajagopalan PK, Russel S, LaBrecque GC (1973) Influence of seasonal changes in climate on dispersal of released *Culex pipiens fatigans*. A study in villages in Delhi Union Territory, India. Bull World Health Organ 48: 317–321

Chapter 16 Experimental Hut Techniques

The assessment of the numbers of mosquitoes resting inside huts by aspirator or pyrethrum space spray collections and the use of exit traps fitted to windows, doors or eaves to collect mosquitoes leaving huts have been described in Chap. 5. The present short account concerns the application of these methods for evaluating the efficiency of spraying houses with residual insecticides and using insecticide-treated nets and their effects on mosquito behaviour.

Experimental huts

Normal village huts vary considerably in size and construction and in the furniture and other contents within, and this may make it difficult to collect mosquitoes from them. To try to overcome these difficulties Haddow (1942) and Senior White and Rao (1943) built experimental huts in East Africa and India respectively that were standardised in shape and contained the minimum of furniture, thus facilitating the collection of mosquitoes. The need to evaluate the efficiency of residual insecticides in malaria campaigns gave great impetus to the development of experimental huts, and they have since been used to evaluate insecticide-treated nets.

During the 1940s several workers independently developed window traps to collect mosquitoes either entering or leaving sprayed or unsprayed huts (Hocking 1947; Muirhead-Thomson 1947; Simmons et al. 1945). In Nigeria Muirhead-Thomson (1947) built mud-walled experimental huts with thatched roofs incorporating a 'lobster' type of exit window trap (Fig. 6.8a), further details of the construction of which were presented in later publications (Muirhead-Thomson 1948, 1950). Although these huts were not initially used to study the behaviour of mosquitoes in sprayed huts, they were soon put to this use (Davidson 1953; Muirhead-Thomson 1950). Modifications and improvements were made to both the design of experimental huts and the methods of collecting mosquitoes from them (Burnett 1957; Hocking et al. 1960). The development of the window-type experimental hut is described by Smith (1964).

1425

In many areas it is impossible to assess accurately the numbers of mosquitoes dying overnight in sprayed huts because ants, and other scavengers, remove dead and dying mosquitoes from the floor. In Zimbabwe Mpofu et al. (1988) and Taylor et al. (1981) reported a considerable 'loss' of mosquitoes in both unsprayed and DDT-sprayed houses. Some of this was probably due to predators and scavengers, especially as such losses seemed to be greatest in unsprayed houses, and increased with time in sprayed houses (Mpofu et al. 1988; Taylor et al. 1981), presumably because of increases in scavengers and predators as the insecticidal effects declined. To try to overcome such problems Rapley (1961) designed huts built of burnt brick which had a raised floor supported on concrete pillars which were surrounded by water-filled moats to prevent the entry of scavengers such as ants (Fig.16.1a). Although these huts bear little resemblance to typical village huts, they have proved invaluable for evaluating new insecticides and formulations, despite the criticism of Hamon et al. (1963) that the natural migration of water and insecticidal deposits on the walls are affected by the raised floor. This objection may be overcome, in areas where ant infestations are not large, by building huts not on pillars but by surrounding them by a water-filled concrete moat (Fig 16.1b) or even using tree banding grease to keep out ants. Smith (1964) pointed out that in the uncluttered experimental huts in East Africa, most malaria vectors rested primarily on the roof not on the walls, and that although insecticide migration was important with DDT it was less so with organophosphates and carbamates. He also considered that preventing the removal of mosquitoes by ants was much more important than insecticide migration.

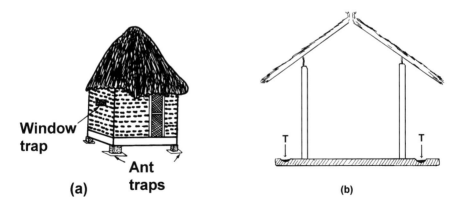

Fig. 16.1. (*a*) Experimental hut of Rapley (1961); (*b*) Experimental hut built on concrete base with T—a trough of oil and water (World Health Organization 1975)

A disadvantage of employing window traps in huts is that when the huts have been treated with a repellent insecticide, species that are normally endophilic, such as *Anopheles gambiae*, are irritated and tend to leave the huts during the night by any available exit. Only a few of these are caught in the window traps, a large number escaping through the eaves. Furthermore, even in unsprayed huts window traps are inefficient in catching mosquitoes such as *Mansonia uniformis* that do not remain long in huts, but escape largely through the eaves during the night. In order to catch mosquitoes flying out through the eaves Smith (1965*a*) developed a veranda trap experimental hut. This type of hut is built of burnt bricks, is supported on concrete pillars surrounded by ant traps, and in fact resembles the hut designed by Rapley (1961), except that verandas incorporating an eave-type exit trap are built on all four walls. During use two veranda exit traps on opposite walls catch about half the mosquitoes leaving the hut via the eaves, while the veranda traps on the other two walls are left open to allow mosquitoes to enter the hut. Alternate opposite pairs of veranda traps are employed as exit traps on different nights. Window-type exit traps can also be used in these huts. This type of experimental hut was used by Lindsay et al. (1992) in The Gambia to study the effect of permethrin-impregnated mosquito nets on house entry by mosquitoes. Mosquito collections were made by searching the room, the two enclosed verandas and the two window traps. Another way of overcoming egress through the eaves is to make it more difficult for mosquitoes to leave via them but not to hinder their entry into huts. Smith and Hudson (1972) fitted a 40-cm wide strip of mosquito wire mesh to the top of the inside of the hut walls and extended this upwards and parallel with the thatch roof (Fig. 16.2*a,b*) to leave an entry gap of 4–5 cm.

Preliminary experiments showed that these baffles greatly reduced the numbers of *Anopheles gambiae* escaping through the eaves, even when pyrethrum smoke was released in the huts, and considerably increased the numbers caught in exit traps (WHO 1975). Hudson (Smith et al. 1972) designed a louvre-type experimental hut that is cheaper and easier to make than the veranda trap and requires a simpler entomological collecting routine. It consists of a typical window trap hut of Rapley (1961) but has a series of entry louvres, based on the type described by Wharton (1951), fitted into one wall (Fig. 6.8*c*). A wooden framework having five 1-ft square horizontal openings is positioned 1 ft from the top of the west wall of the hut. Each opening has 11 aluminium louvres, 2 in wide, set at 53° to the vertical and with 0.8-in entry gaps between them. The louvres and insides of the frames are painted matt black. These louvres allow the entry of mosquitoes but the minimum of light, so that mosquitoes leaving the hut are mainly caught in the window exit trap.

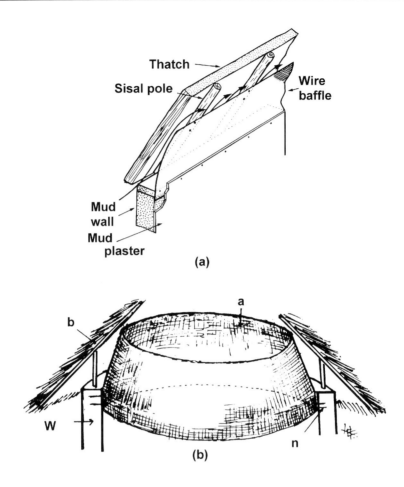

Fig. 16.2. (*a*) Attachment of wire mesh baffle to top of hut wall and sisal poles of roof of an experimental hut to allow the ingress but deter the egress of mosquitoes (after Smith and Hudson 1972); (*b*) modification of eaves showing a — mosquito netting, b — roof pole, n — nail or peg, w — top of hut wall (World Health Organization 1975)

The numbers escaping through the louvres can be estimated by placing an exit trap over one set of louvres and multiplying the numbers caught by five. The eaves' gap surrounding the hut is covered with ½-in mesh to prevent mosquito movement via the eaves. The louvres proved 66–79% efficient in preventing the escape of *Anopheles gambiae* through them, and 74–78% efficient for *Culex quinquefasciatus* and 51–73% efficient for *Mansonia uniformis*. Compared with the egress through the eaves of verandah trap huts, the louvres were much more efficient for *Mansonia uniformis* but slightly less efficient for *Anopheles gambiae*. During the day the louvres are fitted with covers to prevent mosquitoes leaving or entering through them.

The collecting routine includes removing the covers from the louvres in the early evening and replacing them the following morning, before mosquitoes are collected from the exit trap and from the hut.

Smith and Webley (1963) pointed out that when a fumigant insecticide such as dichlorvos (DDVP) is used in huts this will affect not only mosquitoes in the hut but those that have tried to escape but are retained in exit traps. Unrealistically high exit trap mortalities will probably be experienced. To overcome this they suggested that the cone in a 'lobster' type exit trap be made not of netting but of plastic sheeting.

Lines et al. (1987) in evaluating the effectiveness of intact and torn mosquito nets treated with permethrin insecticide in reducing biting by the *Anopheles gambiae* complex in The Gambia used the typical experimental huts described by Smith (1964) and Smith et al. (1972). Similarly in Burkina Faso Darriet et al. (1984) used the same type of experimental huts in their mosquito net trials. In the same country Hervy and Sales (1979) used experimental specially constructed village huts fitted with veranda traps to collect samples of exiting *Anopheles gambiae* and *Anopheles funestus* in sprayed huts. They also collected dead mosquitoes from the hut floors, and live mosquitoes resting on the walls and ceilings. They had two people sleeping in each hut.

In evaluating the effect of insecticide-treated nets on *Anopheles farauti* in the Solomon Islands Ree (1988) used experimental huts (2.5 × 3 m, 2 m high) having a plywood floor, and the walls and a pitched roof made of woven palm fronds. The house was raised on legs 0.6 m from the ground and a door (1.8 × 0.75 m) was fixed in the front. A cube-shaped exit trap, flanked on either side by wooden slats pitched at 30° and set 3.5 cm apart forming entry louvres, was fitted on the walls. Routine sampling consisted of removing dead mosquitoes from the floor sheets, collecting adults from the exit traps followed by aspirating mosquitoes resting in the huts. Darriet et al. (1998) constructed six experimental huts to investigate the impact of pyrethroid resistance in *Anopheles gambiae* s.s. on the efficacy of insecticide-treated nets in Côte d'Ivoire. Dimensions of the huts were not provided by the authors, but the general description was as follows: a wooden framework with walls constructed from planks, a cement floor, and a corrugated iron roof. Wooden baffles were fitted to allow mosquito entry but prevent their escape and a veranda trap was constructed from 'textiglass'.

In Belize, Grieco et al. (2000) constructed three identical huts based on the design of a Mayan house. The walls of the huts measured 7.8 m long × 4.2 m wide × 2.1 m high and were constructed from wooden planks. The floor was of dirt, and a thatched roof was constructed from Cahune palm. A walkway running lengthwise down the centre of each hut was suspended 3 m above the floor to enable examination of the thatch during collections. Each hut had three windows and a north-facing door. Each hut was equipped with a small auxiliary door (0.5 m wide × 1 m high) to allow entry of collectors without needing to remove the traps.

Achee et al. (2005), also in Belize, describe a portable experimental hut that they developed for mark-release-recapture experiments with *Anopheles darlingi*. The huts were constructed in order to match as closely as possible the dimensions, number of rooms and construction materials of local village houses, but using a framework of 1 5/8-in diameter aluminium and 1 ¾-in galvanised pipe, which could be collapsed to facilitate portability (Fig. 16.3). Huts were 13 feet wide × 13 feet long with 6 foot-high side walls. Hut roofs were made from 4 hinged units of corrugated zinc and the height of the roof at its apex was 9 feet. Walls were made from untreated wooden planks 8-in wide and ½-in thick. The aluminium pipes were bolted to the welded joints of the galvanised pipe. Walls were also modular in structure, comprising four panels on each side, three with a hinged window (2 feet × 2 feet), and one with a door (3 feet × 6 feet). Assembly and disassembly could be completed by four individuals in approximately one hour.

Village huts

Specially constructed experimental huts are undoubtedly very useful for testing new insecticides and formulations. They may not, however, be very suitable for evaluating the effect of insecticides in control campaigns, especially if they are relatively costly to make and several are required, but only for relatively short periods. Furthermore, they may not resemble very closely the structure or environment of ordinary village huts and so consequently the behaviour of mosquitoes in them may be atypical. Normal village huts are therefore often used in preference to experimental huts (Bown et al. 1986, 1987; Coz et al. 1966; Kuhlow 1959, 1962; Mpofu et al. 1988; Pant et al. 1969; Service 1964, 1965). Frequently the only modification required is the incorporation of an exit trap. In a Zimbabwean village Mpofu et al. (1988) evaluated the effect of DDT on *Anopheles arabiensis* by fitting lobster-type exit traps to collect blood-engorged mosquitoes that had been released into huts. The numbers trapped each hour after release

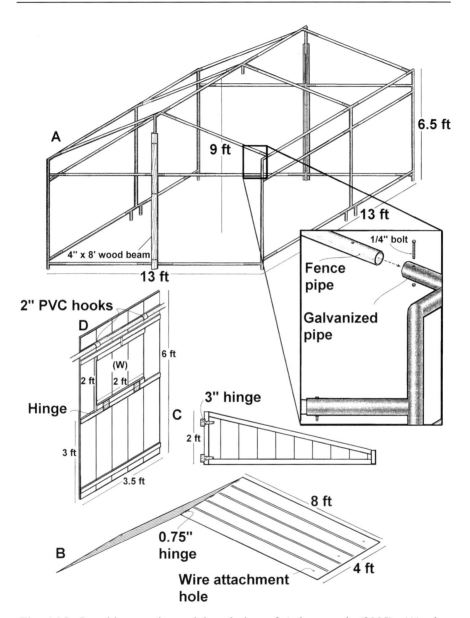

Fig. 16.3. Portable experimental hut design of Achee et al. (2005). (A) pipe framework with inset detail of attachment between aluminium and galvanised tubing. (B) individual hinged roof panel. (C) Inside view of individual gable panel. (D) Individual wall panel with hinged window

gave information on how long *Anopheles arabiensis* rested inside sprayed and unsprayed houses: in the sprayed house most adults left within 2 h of release.

Because of natural openings, cracks and crevices, relatively small proportions of the mosquitoes resting in, or leaving, ordinary village huts will usually be caught, but this may not be a great problem if representative samples are obtained.

There is less control over the use of normal huts and this can present difficulties. For example, a varying number of people may be sleeping in the huts, insecticidal deposits may become covered with soot and tar or even mud due to re-plastering, and there may be difficulties in keeping windows and doors shut during trials. In Nigerian huts, however, heavy soot deposits prolonged the effectiveness of the residual insecticide 'Bayer 39007' because it slowed down the insecticidal adsorption rate (Bar-Zeev et al. 1966). For these reasons it may be preferable to purchase a number of existing village huts and get staff or paid personnel to live or sleep in them, thus ensuring greater control over their operation. Clearly the pros and cons of using normal village huts or specially built experimental ones must be assessed separately for any particular project.

In studying the behaviour of *Anopheles albimanus* in houses in Mexico before and after spraying with bendiocarb or deltamethrin Bown et al. (1987) allowed mosquitoes arriving at a human bait sitting from 1900–2300 h in the doorway of a house to engorge. They were then marked with fluorescent dusts and their movements within the house monitored with a portable ultraviolet lamp. The numbers of landings, types of resting surfaces, and total resting times were recorded. The movements of each marked adult were monitored for an hour, but if a mosquito attempted to leave the house within an hour it was caught.

Elliott (1972) working in Colombia was the first to conceive the idea of surrounding a house with a curtain to study the exodus and ingress of mosquitoes into the house (see Chapter 6). The method was later modified by Bown et al. (1985, 1986) for use in Mexico. Using the modified method Bown et al. (1987) collected *Anopheles albimanus* from the inside and outside of curtains surrounding unsprayed and sprayed houses (bendiocarb and deltamethrin) on alternate hours from 1800–0600 h. For an hour all mosquitoes caught on the outside of the curtain were collected, but only the unfed females were released into the house. During the subsequent hour mosquitoes resting inside were collected and classified according to their physiological condition. The proportion of blood-fed mosquitoes leaving sprayed houses can also be recorded, and the time they rest in sprayed and unsprayed houses determined. Mosquitoes caught exiting can be held for 24 h to assess mortalities; in addition dead and moribund adults can be collected from the floor of the houses and from the space between the curtain and house walls (Bown et al. 1986).

Catching routine in huts

Three types of early morning collection are frequently made.

(1) Dead and dying mosquitoes are removed either directly from the floor or from white calico floor sheets that have been carefully placed over the floor and furniture late in the afternoon of the previous day. This is usually a difficult collection. It will not give a reliable measure of any overnight kill unless hut occupants are prevented from trampling on mosquitoes that have fallen onto the floor or floor sheets, and even when barrier traps are used some mosquitoes may still be removed by predators (Hudson and Esozed 1971). The possibility of this type of loss can be checked by placing known numbers of dead mosquitoes on the floor and recounting them the following morning. If normal village huts are used the collection of mosquitoes from the floor sheets may be so unreliable that it is best omitted.

(2) Exit traps are either removed or emptied and the numbers of dead and live mosquitoes in the different stages of the gonotrophic cycle recorded. Those alive are placed in small cages, supplied with sugar solution, and mortalities assessed after a 24-h holding period.

(3) Finally, the mosquitoes remaining in the hut can be either counted but not removed (Smith 1964), collected by a hand-catch or by pyrethrum space spraying. The collection of these hut resting mosquitoes is not essential for the calculation of overall hut mortalities.

Interpretation of catches

Overall mortality is calculated as the total overnight or 24-h kill. It comprises the dead mosquitoes on the floor (F) together with those dead in the exit traps and those that die within the holding period (D), expressed as a percentage of the total numbers dead on the floor and the total exit trap collection (T) (Smith 1963a) thus:

$$\text{total hut mortality} = \frac{100(D+F)}{T+F} \tag{16.1}$$

A slightly different way of counting and estimating mortality caused by an insecticide was proposed by (WHO 1975) as follows: a = number of dead mosquitoes in exit traps; b = numbers alive in these traps; c = number caught in exit traps still alive after a 24-h holding period in paper cups, or small cages; d = numbers of dead mosquitoes found on floor sheets in the morning and again in the afternoon; e = numbers caught alive resting in the houses in the afternoon; and f = numbers caught resting in the house still alive after a 24-h holding period. Then total dead $(x) = a + (b - c) + d + (e - f)$, and total numbers of mosquitoes caught $(y) = a + b + d + e$. So, percentage mortality is $x/y \times 100$.

WHO (1975) suggest that a better approach is to collect mosquitoes from houses on several consecutive days, but to remove live mosquitoes resting in the house only on the last sampling day, because live mosquetoes in the house will eventually be found dead on the floor or in the exit traps. When veranda-type huts are used the numbers dead in the two veranda traps (V) and the total numbers of mosquitoes caught in these traps (M) are doubled to give the estimated total numbers leaving by the four eaves (Smith 1965b). Thus:

$$\text{total hut mortality} = \frac{100(D + F + 2V)}{T + F + 2M} \tag{16.2}$$

When louvre trap huts are used the formula is modified to take into account the total numbers leaving through the five sets of louvres, as indicated by the numbers in the exit trap (P) fitted to one set of louvres; and, similarly the number in the exit trap that die within the holding period (L) are multiplied by five (Smith et al. 1973). Thus:

$$\text{total hut mortality} = \frac{100(D + 5L + F)}{T + 5P + F} \tag{16.3}$$

Mosquitoes remaining alive in the hut, and which may have been collected in hand-catches or in pyrethrum space spray collections, are omitted from mortality counts, because they constitute individuals that will in future either die within the hut or leave it. However, in Iran in an evaluation of residual house-spraying with bendiocarb against *Anopheles stephensi* in specially constructed experimental huts (Eshghy et al. 1979), the numbers of mosquitoes resting in the hut and their mortality 24 h later, was determined, in addition to the numbers alive and dead in window-type exit traps, those dying after 24 h, and those dead on floor sheets. From these collections the total overall mortality was calculated. Bioassays were also undertaken. Mortalities of males, unfed females, and blood-fed and gravid females combined, can be calculated separately.

The proportions of blood-fed mosquitoes remaining in huts and the proportion caught in exit traps can be compared for treated and untreated huts to measure any repellent effect that causes an increase in the percentage of blood-fed females leaving treated huts. This does not take into account the proportions of mosquitoes that die within the huts overnight, which is likely to be greater in treated huts. A better comparison therefore includes the blood-fed individuals collected in the floor counts. Thus, the numbers of alive and dead blood-fed individuals in the exit traps are expressed as a percentage of the total hut catch of blood-feds, that is the floor, pyrethrum and exit trap catches.

Insecticides may also reduce the frequency of feeding on hut occupants and this can be expressed in the same way, but calculated on unfed individuals. Smith (1963a) proposed a repellency index (RI) to measure this, which is derived by comparing the relative numbers of unfed mosquitoes leaving sprayed and unsprayed huts, and is independent of total catch.

$$RI = \frac{T_1}{T_1 + T_2} \times \frac{C_1 + C_2}{C_1} \tag{16.4}$$

where T_1 = the total numbers alive and dead in exit traps in a treated hut, T_2 = total numbers alive and dead inside a treated hut, i.e. floor sheet and resting collection, C_1 = total numbers alive and dead in exit traps in an untreated hut, C_2 = total numbers alive and dead inside an untreated hut. Smith (1964) considered that only when the repellency index was >1 in the presence of low overall mortalities was the degree of repellency of epidemiological importance.

In addition to having a repellent effect insecticides may exhibit varying degrees of deterrency, causing a decrease in the numbers of mosquitoes entering treated huts (de Zulueta and Cullen 1963; Muirhead-Thomson 1947; Smith 1963b). Smith and Webley (1969) calculated the total number of mosquitoes entering a veranda-type hut (E) over an interval of n days as follows:

$$E = (R_n - R_1) + \Sigma n(F + T + 2M) \tag{16.5}$$

where R_1 = the numbers of mosquitoes resting in the hut on the first day, R_n = the numbers resting in the hut on the last day and F, T and M are as already defined. If the difference between R_n and R_1 is small in comparison with the other terms in the equation, then an approximate but satisfactory estimate of the total numbers of mosquitoes entering a hut is given by:

$$E = \Sigma n(F + T + 2M) \tag{16.6}$$

The expected number of mosquitoes (N) entering a treated hut in the absence of any deterrent effect can be estimated as follows:

$$N = \frac{C \times E_1}{C_1} \tag{16.7}$$

where C = number of mosquitoes entering an untreated control hut, when the treated hut has been treated, C_1 = number of mosquitoes entering this control hut prior to the other hut being treated and E_1 = number of mosquitoes entering the treated hut before spraying. The deterrent effect (DE) which is causing the observed reduction in mosquitoes entering a treated hut is calculated as the difference between the numbers estimated to be entering it (E) and the estimated expected number (N) when assuming there is no deterrency, expressed as a percentage:

$$DE = \frac{100(N - E)}{N} \tag{16.8}$$

Garrett-Jones (Hudson and Esozed 1971) suggested that the reduced numbers of mosquitoes entering a treated hut and the reduction in feeding in these could be considered to give the number of 'feeder–survivors', that is the numbers that enter a treated hut, feed and survive unharmed. This can be assessed by counting the numbers of blood-fed and gravid females caught in exit traps that survive a 24-h holding period. The feeder–survivor density (FSD) is the number of feeder–survivors in a treated hut during a particular period expressed as a percentage of the numbers in the same hut before treatment. Alternatively the feeder–survivor index (FSI) can be used to compare feeder–survivor densities in treated and untreated huts.

$$FSI_n = 100 \left(1 - \frac{T_n C_0}{T_0 C_n} \right) \tag{16.9}$$

where FSI_n = the index for a period of time n after huts have been treated, T_0 = number of feeder–survivors in the treated hut before treatment, T_n = number of feeder–survivors in the treated hut during the period n, C_0 = number of feeder–survivors in a control hut before the treated hut is treated and C_n = number of feeder–survivors in a control hut during period n.

In Sri Lanka Rawlings et al. (1983) assessed the impact of spraying houses with malathion by building a small palm-leaf hut and baiting it with a calf—a host that is very attractive to the principal malaria vector, *Anopheles culicifacies*. The entire hut was enclosed in a large mosquito net. Mosquitoes, either newly emerged unfed females, or parous unfed wild caught adults, or wild blood-fed females, were released into the hut at dusk.

Each batch of released mosquitoes was marked with different coloured powders. Twelve hours later two collectors armed with torches, a small ultra-violet light and aspirators spent 1.5 h searching the hut for released mosquitoes. The leaves comprising the walls and roof of the hut were shaken to flush out mosquitoes. Despite the fact that the hut was surrounded by a net the recovery rates (corrected for the higher mortality of unfed than blood-fed *Anopheles culicifacies* kept in cages) were just 38% for the unfeds and 56% for blood-fed females. These low recapture rates could be due to higher mortalities experienced in the hut than in experimental cages, or more likely due simply to a failure to catch all mosquitoes resting in the hut, emphasising the inefficiency of aspirator collections in collecting mosquitoes in even small huts (1.8 × 1.2 × 1.6 m). For 2 months after malathion spraying there was total mortality of both unfed and blood-fed released adults.

The following formula was used to estimate the likely effect of malathion on mean survival rates of the natural population.

Let p_n = natural daily survival of female *Anopheles culicifacies*.
Let p_i = survival of females resting in a sprayed house.
Let h = proportion of females that rest in houses each day.
Let c = proportion of houses that are sprayed.
Let p_t = mean overall mean daily survival, then we have

$$p_t = p_n (1 - h) + p_n h (1 - c) + p_i h c \qquad (16.10)$$

In Suriname Rozendaal et al. (1989) constructed experimental huts to monitor the effect of DDT spraying on *Anopheles darlingi*. These huts (Rozendaal 1989) were designed to resemble normal village huts, and because these contained so many artefacts (furniture, clothing, plates and other objects on walls) that are not normally sprayed, these were also placed in the experimental huts, to simulate, as far as possible, natural conditions. The huts contained two lobster-type exit traps fitted on the smaller end walls. Mosquitoes entered through the eave gaps and through baffle slits fitted along the two longer walls about 20 cm from the hut floor (Fig. 16.4). No entry traps were fitted as there were very low biting densities in houses, and the collection of adults entering would have further reduced the numbers biting and leaving. The indoor biting population of *Anopheles darlingi* was low, and instead of catching the mosquitoes biting a seated man between 1830–0630 h they were allowed to bite, their numbers recorded each hour and then allowed to fly off either to rest inside the house or to exit.

The probability of *Anopheles darlingi* entering a sprayed hut to feed on a person and then leaving (R) was estimated by applying the formula presented by Hudson (1984).

$$R = 100\left(1 - \frac{Tn \ Co}{To \ Cn}\right) \tag{16.11}$$

where To = mean number mosquitoes (here *Anopheles darlingi*) biting a human bait in the sprayed hut on nights before spraying; Co = mean number biting in control huts before spraying; Tn = mean number biting in sprayed huts after spraying; Cn = mean numbers biting in control hut after the sprayed huts are sprayed.

Voorham (1997) used the same design of experimental hut as Rozendaal et al. (1989) to study the efficacy of wide-mesh gauze treated with lambda-cyhalothrin (11 mg/m^2) as a control method against *Anopheles darlingi* in Suriname. The floor of the huts was covered with light-coloured plastic to facilitate the collection of any dead mosquitoes falling onto it. Human landing-catches were conducted inside the huts from 2100 to 0200 h to determine biting rates.

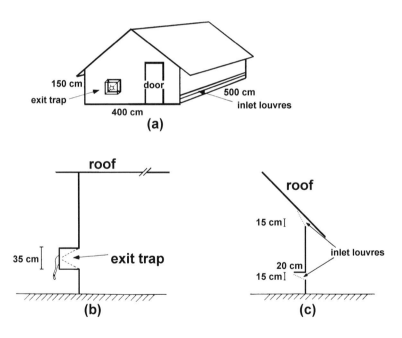

Fig. 16.4. Experimental hut: (*a*) general front view; (*b*) front or back wall in cross section; (*c*) side wall in cross section (Rozendaal 1989)

The mean times mosquitoes rest in sprayed and unsprayed huts can sometimes be measured by marking mosquitoes with paints as they enter houses and recording the times when they are caught in exit traps. In Colombia Elliott (1972) estimated the time *Anopheles albimanus* spent in houses by recording the interval between the time when 50% of the entering population had arrived, and when 50% of the exiting population had been caught in exit traps.

This method depends on a normal distribution of entry and exit times, and in Suriname this simple approach could not be used because the times of entry and exit of *Anopheles darlingi* into huts did not follow a normal distribution (Rozendaal 1989). He assumed there was no difference between times of entry and biting, and no difference between times of departure and collection of adults in exit traps fitted to houses, and finally that the percentiles of *Anopheles darlingi* biting and the percentiles caught in the exit traps represent the same mosquitoes. With these assumptions we can calculate the so-called Surface III (Fig. 16.5), which represents the total number of percent hours for the interval between biting and exiting. The average resting time is then approximately $1/100 \times$ Surface III. This is equal to $1300 - ($Surface I $+$ Surface II$)$, where $1300 = $ total surface area (Fig. 16.5) in percent hours (i.e. $100\% \times 13$ hr). Surface I $=$ total number of percent hours before biting, and Surface II $=$ total number percent hours after departure. Thus the mean indoor resting period $(R) \approx 1/100\ [1300 - ($Surface I $+$ Surface II$)]$ or

$$R \approx 1/100 \left[\sum_{i=1}^{i} O_i - \sum_{i=1}^{i} T_i \right] \tag{16.12}$$

where $\sum_{i=1}^{i} O_i$ = cumulative percentage of all mosquitoes observed biting until observation period i, and $\sum_{i=1}^{i} T_i$ = cumulative percentage of all mosquitoes collected in the exit traps until period i. To calculate the resting period say to 2330 h, the same procedure would be used, but with $i = 5$.

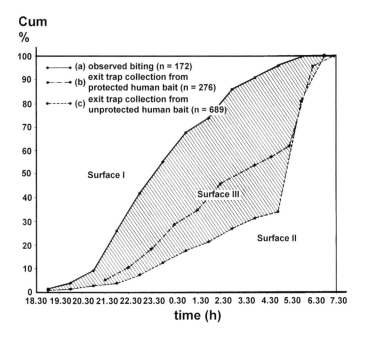

Fig. 16.5. Hourly cumulative percentage of total *Anopheles darlingi*: (*a*) observed biting human bait hourly from 1830 – 0730 h; (*b*) collected hourly in exit traps from 2100–0800 h, human bait within the hut was protected with a mosquito net; (*c*) collected hourly in exit traps from 1830–0730 h, human bait within the hut was not protected with a mosquito net. Also indicated are: Surface I — the total number of percent hours before biting; Surface II — the total number of percent hours after departure; Surface III — the total number of percent hours after biting and before departure (striped) (Rozendaal 1989)

References

Achee NL, Grieco JP, Andre RG, Rejmankova E, Roberts DR (2005) A mark-release-recapture study using a novel portable hut design to define the flight behavior of *Anopheles darlingi* in Belize, Central America. J Am Mosq Control Assoc 21: 366–379

Bar-Zeev M, Bracha P, Self LS (1966) The effect of 'soot' on the residual toxicity of Bayer 39007 on mud surfaces of southern Nigerian huts. Riv Parassitol 27: 33–38

Bown DN, Rios JR, del Angel Cabañas G, Guerrero JC, Méndez JF (1985) Evaluation of chlorphoxim used against *Anopheles albimanus* on the south coast of Mexico: 2. Use of two curtain-trap techniques in a village-scale evaluation trial. Bull Pan-Am Health Organ 19: 61–68

Bown DN, Rios JR, Frederickson C, del Angel Cabañas G, Méndez JF (1986) Use of an exterior curtain-net to evaluate insecticide/mosquito behavior in houses. J Am Mosq Control Assoc 2: 99–101

Bown DN, Frederickson EC, del Angel Cabañas G, Méndez JF (1987) An evaluation of bendiocarb and deltamethrin applications in the same Mexican village and their impact on populations of *Anopheles albimanus*. Bull Pan-Am Health Organ 21: 121–135

Burnett GF (1957) Trials of residual insecticides against anophelines in African-type huts. Bull Entomol Res 48: 631–668

Coz J, Venard P, Attiou B, Somda D (1966) Étude de la remanence de deux nouveaux produits insecticides: O.M.S. 43 et O.M.S. 658. Cah ORSTOM sér Entomol Méd 4: 3–12

Darriet F, Robert V, Tho Vien N, Carnevale P (1984) Evaluation of the efficacy of permethrin-impregnated intact and perforated mosquito nets against vectors of malaria. Document No. 10/RAP/CM/ENT/84 Centre Muraz, Bobo-Dioulasso, Burkina Faso and WHO/VBC/84.899 and WHO/MAL 84.1008, 19 pp. (all mimeographed)

Darriet F, Guillet P, N' Guessan R, Doannio JMC, Koffi A, Konan LY, Carnevale P (1998) Impact de la résistance d' *Anopheles gambiae* s.s. à la perméthrine et à la deltaméthrine sur l' efficacité des moustiquaires imprégnées. Méd Trop (Mars) 58: 349–354

Davidson G (1953) Experiments on the effects of residual insecticides in houses against *Anopheles gambiae* and *A. funestus*. Bull Entomol Res 44: 231–254

de Zulueta J, Cullen JR (1963) Deterrent effect of insecticides on malaria vectors. Nature 200: 860–861

Elliott R (1972) The influence of vector behavior on malaria transmission. Am J Trop Med Hyg 21: 755–763

Eshghy N, Janbakhsh B, Motobar M (1979) Experimental hut trials for the evaluation of bendiocarb (Ficam W) against *Anopheles stephensi* Khesht district, Kazeroun, southern Iran, 1977. Mosquito News 39: 126–129

Grieco JP, Achee NL, Andre RG, Roberts DR (2000) A comparison study of house entering and biting behavior of *Anopheles vestitipennis* (Diptera: Culicidae) using experimental huts sprayed with DDT or deltamethrin in the southern district of Toledo, Belize, C. A. J Vector Ecol 25: 62–73

Haddow AJ (1942) The mosquito fauna and climate of native huts at Kisumu, Kenya. Bull Entomol Res 33: 91–142

Hamon J, Sales S, Eyraud M (1963) Étude biologique de la remanence du DDT dans les habitations de la région de Bobo-Dioulasso, République de Haute Volta. Riv Malar 42: 1–54

Hervy JP, Sales S (1979) Stage IV evaluation of imagocides OMS-43, OMS-1331 and OMS-1394 at the Soumousso experimental station, Upper Volta, during 1978, WHO/VBC/79.727, 19 pp. (mimeographed)

Hocking KS (1947) Assessment of malaria control for mosquito prevalence. Bull Entomol Res 38: 131–136

Hocking KS, Armstrong JA, Downing FS (1960) Gamma-BHC/Cereclor—A new long-lasting Lindane formulation for malaria control. Bull World Health Organ 22: 757–765

Hudson JE (1984) *Anopheles darlingi* Root (Diptera: Culicidae) in the Suriname rain forest. Bull Entomol Res 74: 129–142

Hudson JE, Esozed S (1971) The effects of smoke from mosquito coils on *Anopheles gambiae* Giles and *Mansonia uniformis* (Theo.) in verandah-trap huts at Magugu, Tanzania. Bull Entomol Res 61: 247–265

Kuhlow F (1959) On the behaviour of *Anopheles funestus* in unsprayed and DDT sprayed houses. Z Tropenmed Parasitol 10: 328–333

Kuhlow F (1962) Field experiments on the behaviour of malaria vectors in an unsprayed hut and in a hut sprayed with DDT in Northern Nigeria. Bull World Health Organ 26: 93–102

Lindsay SW, Adiamah JH, Armstrong JRM (1992) The effect of permethrin-impregnated bednets on house entry by mosquitoes (Diptera: Culicidae) in The Gambia. Bull Entomol Res 82: 49–55

Lines JD, Myamba J, Curtis CF (1987) Experimental hut trials of permethrin-impregnated mosquito nets and eave curtains against malaria vectors in Tanzania. Med Vet Entomol 1: 37–51

Mpofu SM, Taylor P, Govere J (1988) An evaluation of the residual lifespan of DDT in malaria control. J Am Mosq Control Assoc 4: 529–535

Muirhead-Thomson RC (1947) The effects of house spraying with pyrethrum and with DDT on *Anopheles gambiae* and *A. melas* in West Africa. Bull Entomol Res 38: 449–464

Muirhead-Thomson RC (1948) Studies on *Anopheles gambiae* and *A. melas* in and around Lagos. Bull Entomol Res 38: 527–558

Muirhead-Thomson RC (1950) DDT and gammexane as residual insecticides against *Anopheles gambiae* in African houses. Trans R Soc Trop Med Hyg 43: 401–412

Pant CP, Joshi GP, Rosen P, Pearson JA, Renaud P, Ramasamy M, Vandekar M (1969) A village-scale trial of OMS-214 (Dicapthon) for the control of *Anopheles gambiae* and *Anopheles funestus* in northern Nigeria. Bull World Health Organ 41: 311–315

Rapley RE (1961) Notes on the construction of experimental huts. Bull World Health Organ 24: 659–663

Rawlings P, Goonatilaka DC, Wickramage C (1983) Assessment of the consequences of the house-spraying of malathion on the interruption of malaria transmission. J Trop Med Hyg 86: 147–151

Ree H-I (1988) Studies on control effect of the pyrethrin-impregnated mosquito net against *Anopheles farauti* in the experimental hut. Jap J Sanit Zool 39: 113–118

Rozendaal JA (1989) Biting behavior of *Anopheles darlingi* in the Suriname rainforest. J Am Mosq Control Assoc 5: 351–358

Rozendaal JA, van Hoof JPM, Voorham J, Oostburg BFJ (1989) Behavioral responses of *Anopheles darlingi* in Suriname to DDT residues on house walls. J Am Mosq Control Assoc 5: 339–350

Senior White R, Rao JVV (1943) On malaria transmission around Vizagapatam. J Malar Inst India 5: 187–205

Service MW (1964) The behaviour of malaria vectors in huts sprayed with DDT and with a mixture of DDT and malathion in northern Nigeria. Trans R Soc Trop Med Hyg 58: 72–79

Service MW (1965) Trials with dichlorvos (DDVP) against malaria vectors in huts in northern Nigeria. Trans R Soc Trop Med Hyg 59: 153–162

Simmons SW et al (1945) Techniques and apparatus used in experimental studies on DDT as an insecticide for mosquitoes. Public Health Rep Wash Suppl 186: 3–20

Smith A (1963*a*) Principles in assessment of insecticides by experimental huts. Nature 198: 171–173

Smith A (1963*b*) Deterrent effect of insecticides on malaria vectors. Nature 200: 861–862

Smith A (1964) A review of the origin and development of experimental hut techniques used in the study of insecticides in East Africa. East Afr Med J 41: 361–374

Smith A (1965*a*) A verandah-trap hut for studying the house-frequenting habits of mosquitos and for assessing insecticides. I.—A description of the verandah-trap hut and of the studies on the egress of *Anopheles gambiae* Giles and *Mansonia uniformis* (Theo.) from an untreated hut. Bull Entomol Res 56: 161–167

Smith A (1965*b*) A verandah-trap hut for studying the house-frequenting habits of mosquitos and for assessing insecticides. II.—The effect of dichlorvos (DDVP) on egress and mortality of *Anopheles gambiae* Giles and *Mansonia uniformis* (Theo.) entering naturally. Bull Entomol Res 56: 275–282

Smith A, Hudson JE (1972) A modification to an Experimental Hut to Reduce Mosquito Eaves-Egress. WHO/MAL/72.775, 6 pp. (mimeographed)

Smith A, Webley DJ (1963) A modified window-trap for assessment of fumigant insecticides in experimental trap-huts. Nature 197: 1227–1228

Smith A, Webley DJ (1969) A verandah-trap for studying the house-frequenting habits of mosquitoes and for assessing insecticides. III.—The effect of DDT on behaviour and mortality. Bull Entomol Res 59: 33–46

Smith A, Hudson JE, Obudho WO (1972) Preliminary louvre-trap hut studies on the egress of *Anopheles gambiae* Giles, *Mansonia uniformis* (Theo.) and *Culex pipiens fatigans* Wied. from untreated huts. Bull Entomol Res 61: 415–419

Smith A, Obudho WO, Esozed S, Myamba J (1973) Louvre-trap hut assessments of mosquito coils, with a high pyrethrin I/pyrethrin II ratio, against *Anopheles gambiae* Giles, *Culex fatigans* Wied. and *Mansonia uniformis* (Theo.). East Afr Med J 50: 352–361

Taylor P, Crees MJ, Hargreaves K (1981) Duration of *Anopheles arabiensis* control in experimental huts sprayed with DDT and decamethrin. Trans Zimbabwe Sci Assoc 61: 1–13

Voorham J (1997) The use of wide-mesh gauze impregnated with lambda-cyhalothrin covering wall openings in huts as a vector control method in Suriname. Rev Saúde Pública 31: 9–14

Wharton RH (1951) The behaviour and mortality of *Anopheles maculatus* and *Culex fatigans* in experimental huts treated with DDT and BHC. Bull Entomol Res 42: 1–20

World Health Organization (1975) Manual on Practical Entomology in Malaria. Part II Methods and Techniques. WHO Offset Publication No. 13. World Health Organization, Geneva

Chapter 17 Indices of Association and Species Diversity Indices

Much has been written on species diversity but it remains hard to define. There has been a proliferation of indices and models for measuring diversity followed by many critiques of these indices. Southwood (1978) and Southwood and Henderson (2000) give good accounts of this topic and provide numerous references to ecological papers, while Pielou's (1975) book on mathematical approaches to species diversity remains a key reference. Nilsson et al. (1988) give a good account of species richness, species area relationships and habitat diversity. Whittaker (1972) reviews the subject of diversity and proposes a useful classification of various types of diversity (α, β, and γ). Magurran (1988) has written an easily understood book on diversity and discusses many of the diversity indices, most of which are accompanied by worked examples.

It is not appropriate to summarise here all the information on species coexistence (association) and diversity, consequently only a few of the more useful, or more commonly used, indices together with those that have been used in mosquito studies, are presented.

Indices of association

Coefficient of interspecific association

The first step in the calculation of this coefficient is to construct a 2×2 contingency table when a, b, c and d are the numbers of occurrence, not the number of larvae collected (Table 17.1). The table should be so arranged that species A is more abundant than species B, in order that $(a + b) < (a + c)$. Cole (1949) proposed the following formulae to measure the degree of association (C_{AB}) between two species.

Table 17.1. 2×2 contingency table for the calculation of the degree of association (C_{AB}) of Cole (1949)

Species B	present	absent (Species A)	
	present	absent	
Present	a	b	$a + b$
Absent	c	d	$c + d$
	$a + c$	$b + d$	$n = a + b + c + d$

When $ad \geq bc$ then:

$$C_{AB} = \frac{ad - bc}{(a+b)(b+d)} \pm \sqrt{\left[\frac{(a+c)(c+d)}{n(a+b)(b+d)}\right]}. \tag{17.1}$$

When $bc > ad$ and $d \geq a$ then:

$$C_{AB} = \frac{ad - bc}{(a+b)(a+c)} \pm \sqrt{\left[\frac{(b+d)(c+d)}{n(a+b)(a+c)}\right]}. \tag{17.2}$$

When $bc > ad$ and $a > d$ then:

$$C_{AB} = \frac{ad - bc}{(b+d)(c+d)} \pm \sqrt{\left[\frac{(a+b)(a+c)}{n(b+d)(c+d)}\right]}. \tag{17.3}$$

Values of C_{AB}, like the correlation coefficient, vary from +1 for complete association to −1 for complete disassociation. The first of these formulae was used to measure the degree of association of *Culex pipiens* and *Culex torrentium* (Service 1968) in both container habitats (+0.52 ± 0.13) and ground pools (+0.35 ± 0.12). Yadav et al. (1997), using Cole's formula, calculated one positive association (0.4 ± 0.16) between larval *Culex brevipalpis* and *Orthopodomyia anopheloides* and three negative associations: *Aedes albopictus* and *Orthopodomyia anopheloides* (−0.52 ± 0.18), *Aedes albopictus* and *Toxorhynchites splendens* (−0.57 ± 0.17), and *Culex brevipalpis* and *Ochlerotatus lophoventralis* (−0.53 ± 0.25) in tree holes in Orissa, India.

Hurlbert (1969), however, considered that the coefficients of Cole (1949) were biased by the species frequencies, but that this could be diminished if the formula was redefined as follows:

$$C_{AB} = \frac{ad - bc}{|ad - bc|} \left|\left(\frac{\text{Obs}\chi^2 - \text{Min}\chi^2}{\text{Max}\chi^2 - \text{Min}\chi^2}\right)^{1/2}\right| \tag{17.4}$$

where the letters are as in the contingency table and where $| ad - bc |$ means that the term is placed in its positive form and where χ^2 is calculated as follows:

$$\chi^2 = \frac{n[| ad - bc | - (n/2)]^2}{(a+c)(b+d)(a+b)(c+d)} \qquad (17.5)$$

If $ad > bc$ then the association is positive, if $bc > ad$ then the association is negative. Calculating χ^2 depends on having a minimum of 5 as any of the expected numbers. Because there is only one degree of freedom the 5% point is 3.84. Therefore if χ^2 is calculated as less than this any apparent association could be due to chance, and further tests are not made.

Returning to formula 17.5, Min χ^2 = value of χ^2 when the observed value of a differs from its expected value ($\hat{a}$) by less than 1.0 (except when $a - \hat{a}$ = 0 or = 0.5; the value of Min χ^2 depends on whether ($ad - bc$) is positive or negative). Thus we have

$$\mathrm{Min}\chi^2 = \frac{n^3(\hat{a} - g[\hat{a}])^2}{(a+b)(a+c)(c+d)(b+d)} \qquad (17.6)$$

where $g(\hat{a}) = \hat{a}$ rounded to the next lowest integer when $ad < bc$; or rounded to next highest integer when $ad \geq bc$; if $\hat{a}$ is an integer then $g(\hat{a}) = \hat{a}$.

Max χ^2 = value of χ^2 when a is as large or small as the marginal totals of the 2 × 2 contingency table will allow to be formulated under the following criteria.

Conditions $ad \geq bc$ $ad < bc$ $ad < bc$
                                  $a \leq d$ $a > d$

$$\mathrm{Max}\chi^2 = \frac{(a+b)(b+d)n}{(a+c)(c+d)} \; \frac{(a+b)(a+c)n}{(a+b)(c+d)} \; \frac{(b+d)(c+d)}{(a+b)(a+c)}. \qquad (17.7)$$

In Pakistan Reisen et al. (1981) calculated the interspecific association index of Cole (1949) to measure larval associations, while in Kenya Lounibos (1981) used the modified index (C_8) of Hurlbert (1969), which ranges from +1 to −1, to measure interspecific associations of mosquitoes breeding in tree-holes and bamboo traps. Statistical significance was tested by the corrected χ^2 formula indicated by Pielou (1977). That is subtracting ½ from the two observed frequencies that exceed expectation, and adding ½ to the two frequencies that are less than expectation. Thus

$$\chi^2 = \frac{[|\,ad - bc\,| - N/2]^2 N}{mnrs} \tag{17.8}$$

where $m = (a + b)$, $n = (c + d)$, $r = (a + c)$ and $s = (b + d)$, and $N = m + n = r + s$. This correction is needed because χ^2 distribution is continuous and in this instance is being used to approximate a discrete distribution. The modified index (C_8) of Hurlbert (1969) has also been used by Suwonkerd et al. (1996) in a study of the seasonal occurrence of *Aedes aegypti* and *Aedes albopictus* in used tyres in Chiangmai, Thailand. These authors found C_8 to equal 0.154, which is significant at the 5% level ($\chi^2 = 12.258$, $P < 0.05$). Bhatt et al. (1993) also calculated values for C_8 for larval and pupal anophelines collected from nine larval habitat types in Gujarat, India. Kant et al. (1998) calculated values for C_8 and H' for anophelines inhabiting several distinct habitats within a rice agroecosystem in Gujarat, India and Kant and Pandey (1999) used the C_8 index to describe species associations in the same rice agroecosystem. Kant and Pandey (1999) observed a significant positive association of *Anopheles culicifacies* with *Anopheles annularis, Anopheles paludis, Anopheles subpictus* and *Culex quinquefasciatus. Anopheles culicifacies* showed a negative association with *Anopheles nigerrimus, Culex tritaeniorhynchus* and the *Culex vishnui* subgroup. Bradshaw and Holzapfel (1983) used the same formula, and factor analysis of Nei et al. (1975), in studying coexistence of tree-hole mosquitoes in North America.

In Kenya Lounibos (1981) found that the highest positive indices of association were between *Aedes aegypti* and *Aedes heischi* (0.241) and between *Aedes heischi* and *Aedes dendrophilus* (0.214), while *Ochlerotatus fulgens* was negatively associated with all four species.

Species dominance using May's (1975) index was calculated for each bamboo trap (Lounibos 1981) as follows

$$D = Y_{max}/Y_t \tag{17.9}$$

where Y_{max} = the numbers of larvae of the most common species in a bamboo or tree-hole, and Y_t = the total numbers of larvae of all species in that habitat. Mean dominance values were calculated for the different ecological zones. Lee (1998) calculated the dominance index of McNaughton (1967) for mosquitoes in organic and non-organic rice fields in southwestern Korea. The index is calculated as follows:

$$DI = \frac{(n_1 + n_2)}{N} \tag{17.10}$$

where n_1 and n_2 are the numbers of the first and second most dominant species, and N is the total individuals of all species. Dominance indices were similar in organic and conventional fields and ranged between 0.48 and 0.96 in the former and between 0.52 and 0.93 in the latter.

Index of association

Many simple formulae, including the above, record only the joint occurrences of species, no weight being given to the numbers of species collected in the samples. Southwood (1978) modified the formula of Whittaker and Fairbanks (1958) to obtain an index of association (I) that takes the numbers of individuals collected into consideration, and has a normal range of +1 to −1. Thus:

$$I = 2\left[\frac{J}{A+B} - 0 \cdot 5\right] \tag{17.11}$$

where J = the number of individuals of both species in samples where they occur together, and A and B = the total numbers of individuals of the two species in all samples. This formula was used to measure the association between larvae of *Culex pipiens* and *Culex torrentium* in containers (+0.59 ± 0.03) and ground pools (+0.38 ± 0.03). In this example a similar high degree of association was found by using this and the formula of Cole (1949). In a study of the insect communities associated with *Ochlerotatus albifasciatus* immatures in floodwater pools in Buenos Aires Province, Argentina, Campos et al. (2004) used the I index of Southwood (1978) and determined that the association between *Liodessus* sp. (Coleoptera: Dytiscidae) and *Ochlerotatus albifasciatus* gave the highest value for I across all seasons ($I = 0.78$). Different species of potential predators showed associations with the different larval and pupal stages of *Ochlerotatus albifasciatus*.

Other indices of association between pairs of species are described by Pielou (1977).

Indices of diversity

Ever since Fisher et al. (1943) introduced an index of diversity (α), ecologists have proposed a variety of mathematical and statistical indices and techniques in efforts to describe the diversity or 'richness' of animal and plant communities. These can also be used to compare the numbers of different species and their frequencies in different habitats and also in the same habitat prior to and after environmental changes (Cancela da Fonseca 1966; Cuba 1981; Elliott and Drake 1981; Gower 1971; Heltshe and Forrester 1983, 1985; MacArthur 1955, 1965; Margalef 1957, 1968; McIntosh 1967; Osborne et al. 1980; Palmer 1990; Pielou 1966, 1967, 1975, 1977; Routledge 1980). Hurlbert (1971) presents a useful critique of the various indices and proposes two alternative parameters, while Southwood (1978), Southwood and Henderson (2000), Wolda (1981) and Magurran (1988) summarise some of the available methods and techniques for expressing species diversity. It is beyond the scope of this book to discuss this aspect of community ecology, but as some of the indices can be useful for comparing the number of species and individuals collected by different sampling methods, a few of the more useful ones are given here.

Fisher et al. (1943) considered that the relationship between numbers of species and numbers of individuals collected was best described by a logarithmic series, and that the constant α in their equation, which is independent of both sample and population size, was in fact a measure of diversity:

$$S = \alpha \log_e \left(1 + \frac{N}{\alpha} \right) \tag{17.12}$$

where S = number of species and N = number of individuals. Thus diversity is obtained simply from these two statistics. The value of α in the equation can be solved by the process of iteration, or more conveniently using the nomograph presented by Williams (1947).

When the number of species is small α is also small, but high when there are large numbers of species. The variance of this estimate, at least for large samples, is

$$\text{Var. } \alpha = \frac{\alpha}{-\log(1 - X)} \tag{17.13}$$

where X = sampling size parameter and is $0 < X < 1$.

The logarithmic series of Fisher et al. (1943) is characterised by the comparatively large numbers of rare species in samples and communities and the comparatively few common species. When the log-series fits, an arithmetic plot with the numbers of different species on the abscissa and numbers of individuals on the ordinate gives a hollow shaped curve with the species of intermediate abundance constituting the majority of individuals in the sample. If the arithmetic plot gives a considerably more hollow curve, that is with more species with very few individuals and a few species with very many individuals, and with less individuals in the intermediate species, then the data will probably best fit the log-normal distribution, or truncated log-normal distribution, suggested by Preston (1948).

The log normal equation has the following three parameters; the numbers of species in the model class, constant a which measures the spread of the distribution curve, and the position of the curve along the x axis, which depends on the ratio of the numbers of individuals to the numbers of species in the sample. Having three instead of two parameters as in the log-series gives the log-normal series more flexibility. So long as the number of species (n_0) occurring in the most common abundance category (model number of species) and the dispersion σ (standard deviation) is known, then the total number of species (N) that occur in a habitat can be predicted:

$$N = n_0 \sqrt{(2\pi\sigma^2)} = 2.5\ \sigma n_0 \qquad (17.14)$$

The α index is strictly speaking only valid when the distribution of individuals among the species mimics the logarithmic series or the log-normal model. However, Taylor et al. (1976), stressed that most spatial distributions in biology are rarely normal, and believed that in most cases data fit a log-normal distribution, and hence α was usually an appropriate statistic of diversity. Taylor (1978) and Kempton and Taylor (1974, 1976) favour this index even when the log series is not the best descriptor of species abundance patterns. Taylor et al. (1976), however, pointed out that the large-sample variance of Fisher used in William's nomograph of species diversity fails to take into account population fluctuations, and hence underestimates the true variability of the diversity index.

Shepard (1984) in a study of benthic animals in streams discusses the application of the log-series distribution and the calculation of its associated diversity index α, and Mountford's (1962) similarity index (I), which in fact can be defined as the reciprocal of α. Forattini et al. (1987) calculated Fisher et al.'s (1943) diversity index for mosquito species caught inside and outside houses in Brazil and later Forattini et al. (1993) calculated it for mosquitoes caught in Shannon traps and CDC light-traps in residual woodland, primitive forest and open land adjacent to rice fields in southeastern Brazil. In the later study, a total of 105 identified species were obtained by the Shannon traps. A comparison of the recorded biodiversities between the three habitats furnished the following indices (with 95% CI): residual woodland (10.0 ± 0.5), open land (7.0 ± 0.35), and primitive forest (8.0 ± 04). Of the 61 species collected by the automatic CDC traps and identified, the indices recorded were: rice paddy (4.0 ± 0.2), drainage trenches (5.0 ± 0.25), and residual ponds (8.0 ± 0.4).

A rather similar index to that of Fisher et al. (1943), but more easily solved, is the index of diversity (D) of Margalef (1957):

$$D = \frac{(S-1)}{\log_e N}.$$
(17.15)

This index was applied by Forattini et al. (1987) to mosquitoes caught both inside houses and out of doors in Brazil.

Sørensen (1948) proposed a very simple index, the quotient of similarity (QS), also known as Czekanowski's index

$$QS = \frac{2j}{a+b}$$
(17.16)

where j = number of species in both collections, and a and b = the numbers of species found in both collections separately.

Mountford (1962) introduced another index of similarity (I) the approximate formula for this being

$$I = \frac{2j}{2ab - (a+b)j}$$
(17.17)

where the symbols are as already defined. Mountford also presents a nomograph for quick estimates of I. Bullock (1971), however, questioned the assumptions made by this index.

The coefficient of similarity of Jaccard (1912) is simple to use and is calculated as

$$C_j = j/(a+b-j)$$
(17.18)

where j = the number of species found in both samples and a and b are the number of species found in the two separate samples. Baroni-Urbani (1980) calculated a slightly modified Jaccard index,

$$Q_j = C/(A+B+C)$$
(17.19)

where A = number of samples with species a, B = number of samples with species b and C = number of samples with both $a + b$ present. Using random samples with different values of N he calculated a series of Q_j values, and presented corresponding probability values in a statistical table.

It will be noted that the coefficients of similarity of Jaccard (1912) and Sørensen (1948) are similar, and that a coefficient of 1 means complete similarity whereas 0 denotes samples are dissimilar with no species in common. Janson and Vegelius (1981) compared a large number of binary indices (presence-absence data) and concluded that only three of them, Ochiai, Jaccard and Sørensen (= Czekanowski) were of real use, although according to Wolda (1983) since the Jaccard index is non-linear it should not be considered a similarity index.

In Russia Masalkina (1979) used the coefficients of similarity of Jaccard (1912) and Shorygin (1939) to study mosquito species caught in light-traps and by different collecting methods on human baits. Khan et al. (1998) applied a hierarchical cluster analysis method (Sneath and Sokal 1973) to Jaccard's (1912) coefficient to analyse the relationships of the larval mosquito faunas of five zones in the upper Brahmaputra valley in northeast India.

Another index of similarity based on relative numbers of each species caught in two different samples uses Euclidean distance. The numbers of each species in each sample is divided by the total numbers of individuals in that sample to give a proportion. If for example the proportion of the ith species in samples A and B are p_{iA} and p_{iB} and there are n species in the samples, then the Euclidean distance E between A and B is

$$E = \sqrt{\sum_{i=n}^{i=1}(p_{iA} - P_{iB})^2} \tag{17.20}$$

Values of zero indicate the samples are identical. Ball (1983) used this similarity coefficient and also Jaccard's coefficient to compare the numbers of flies attracted to Manitoba traps with other sampling methods.

Gower (1971) gave the following General Similarity Index (GSI)

$$\text{GSI}_{if} = \frac{1}{N}\sum_{k=1}^{N}\left(1 - \frac{(x_i - x_j)}{R_k}\right) \tag{17.21}$$

where N = total number of species in the community comparisons; $x_i - x_j$ = absolute difference between abundance of species k in communities i and j; R_k = species range, i.e. differences between the maximum and minimum abundance of species k present in all samples under comparison. Lim and Khoo (1985) found that this index had a high bias when the index was close to unity (high similarity) or zero (low similarity). Most reliable similarity indices were obtained when similarities were moderate. Also, the potential bias increased with the number of species involved in the comparison.

Hill (1973) presented the following diversity index (N_2)

$$N_2 = 1 \bigg/ \sum_{i=1}^{s} p_i^2 \qquad (17.22)$$

where s = number of species and p_i = proportion of ith species in the community. But Gadagkar (1989) has pointed out that caution is needed in using this index when communities have low diversity, and the proportion of one species is close to 1.

The Simpson (1949) index of diversity often denoted by γ, C or D can be written as

$$D = \sum_{j=1}^{s} N_j (N_j - 1) / N(N-1) \qquad (17.23)$$

where N_j = number of individuals of the jth species and N = total number of individuals in the sample. As D increases, diversity decreases and the index is usually therefore expressed as $1 - D$ or $1/D$.

Routledge (1980) concluded that there was, at least sometimes, less bias in measuring species diversity based on sampling by the Simpson (1949) index than by the Shannon–Weaver (1963) index. However, it has been argued that Simpson's index is biased in giving more weight to the common species and less to the rare ones. Pielou (1966, 1967) has suggested that the formula of Brillouin (1962) advocated by Margalef (1957), is better when diversity of samples taken from large populations is being compared.

$$D = \frac{1}{N} \log \frac{N!}{N_1! N_2! N_3! \ldots N_s!} \qquad (17.24)$$

where N = total number of individuals and N_1, N_2, N_3 ... N_s = the total number of individuals of species, 1, 2, 3 ... s.

Originally the logarithms used in these equations were to the base 2 ($\log_2 = \log_{10} \times 3.321928$) as the formulae were devised for analysing information content of a system composed of two different units, but as pointed out by Pielou (1966) there is no special merit in using a binary system when more than two species are involved. See also Pielou (1977) and Magurran (1988) for further discussion of this function.

Another useful but more complicated index is the Shannon–Weaver or Shannon-Wiener function (Shannon and Weaver 1963):

$$H' = -\Sigma p_i \log p_i \qquad (17.25)$$

where p_i = the proportion of the number of individuals of species i in the total sample number (n_i/N). The maximum possible Shannon's Diversity is limited by the number of species collected, and this can be calculated with the formula:

$$H' = \log_{10} k \qquad (17.26)$$

where k is the number of species collected in the sample. The relative proportion of observed diversity to maximum diversity can provide a measure of evenness, such that evenness is low when the majority of the total catch comprises a few species and high when all species in the catch occur in equal proportions. Evenness can be calculated as:

$$J' = \frac{H'}{H'_{max}} \qquad (17.27)$$

de Szalay and Resh (2000) studied the colonization by macroinvertebrates of experimentally manipulated flooded wetlands in California, USA. Emergent vegetation was cut three, five and nine weeks prior to flooding to produce areas with low, medium and high levels of plant cover. Shannon's Diversity index was observed to increase with amount of plant cover. Diversity was lowest in the low plant cover areas and highest in the control areas. Evenness was lowest where plant cover was low and highest in the control areas. Densities of nine of the 12 commonest taxa (including *Culex tarsalis* and *Culiseta inornata*) were correlated with plant cover. Density of some chironomid species was negatively correlated with the amount of plant cover. Reisen (1978) used this index to measure mean diversity of mosquitoes collected by four different sampling techniques in Pakistan, and in later larval studies Reisen et al. (1981) calculated the Shannon–Weaver diversity index (H') as,

$$\bar{D} = -\sum_{i=1}^{n} \left[\left(a_i \bigg/ \sum_{i=1}^{n} a_i \right) \log_2 \left(a_i \sum_{i=1}^{n} a_i \right) \right] \qquad (17.28)$$

where $\bar{D}$ = mean diversity per individual, 1 = the ith species, n = the last species, and a_i = larval abundance in numbers per dip for the ith species. $\bar{D}$ varies from 0 when all larvae are of the same species, to any positive number depending on the number of larvae counted; maximum value is obtained when no two larvae belong to the same species. This index provides an indication of high species equitability and is usually maximised in heterogeneous habitats having large numbers of species, and minimised in homogeneous habitats with only small numbers of species.

Osborne et al. (1980) proposed a hierarchial diversity index (*HDI*) based on the Shannon–Weaver Index (H') to take into account not just species but higher taxonomic levels such as genera and even families. They argued that if a given number of species in a community belongs to different genera and families this increases their diversity, but will not be shown in the normal Shannon–Weaver index. They analysed data from sampling stream fauna to give the normal Shannon–Weaver Index (H') and also a Hierarchial Diversity Index (HDI) calculated as follows:

$$HDI = H'(F) + H'_F(G) + H'_{FG}(S) \qquad (17.29)$$

where $H'(F)$ is the contribution of the families to the total diversity index, $H_F'(G)$ is the generic component of total diversity, and $H_{FG}'(S)$ is the specific component to total diversity. The sum of these three components equals the normal Shannon–Weaver Index (H'). The limitations of this approach and interpretation of the results are discussed in detail by the authors.

Pichon and Gayral (1979) analysed catches of mosquitoes in CDC traps (1968–1969) and in two series of human catches (1968–1969 and 1963–1968) in relict forest in Burkina Faso. They estimated diversity by calculating the Shannon–Weaver and Simpson indices, and concluded that light-traps caught many species (57) but few individuals, whereas bait catches attracted fewer species (42) but greater numbers of mosquitoes.

In Sri Lanka Amerasinghe and Munasingha (1988a) caught 15 243 mosquitoes (both sexes) belonging to 71 species in 9 genera and compared the Shannon–Weaver diversity index for those collected in diurnal human bait collections (0.82), nocturnal human bait collections (0.85), CDC light-traps (1.27) and for catches of resting adults in huts (0.95). The index based on all collecting methods was 1.14. In the same area they collected 38414 immature stages of 64 species of mosquitoes belonging to 14 genera (Amerasinghe and Munasingha (1988b). Shannon–Weaver indices were, ground collections (1.04), rot holes in trees (0.91), tree pans (0.88), tree stumps (0.54), and artificial containers (0.56). The index based on all types of container habitats was 0.99 and for all larval habitats 1.04.

In Gujarat, India, Bhatt et al. (1993) calculated values for the Shannon-Wiener diversity index for larval and pupal samples obtained from nine different habitats using a 'standard' enamel dipper. 25 858 anophelines were collected, comprising 15 species

Zahl (1977) described the process of jackknifing data and pointed out that its advantages are that it makes no assumptions about the distribution of samples, and so it is not necessary to have randomness in the samples, in fact individuals of species can be aggregated. He showed that by applying the jackknife method to sample data that the indices of diversity of both Simpson (1949) and Shannon and Weaver (1963) were considerably improved.

Jackknife estimates are calculated as follows (Heltshe and Forrester 1985). Let us take a collection of $M = 20$ quadrats, from which a diversity index D_0 is calculated. Then remove one quadrat from the number of quadrats and recalculate the index D^{-i}; repeat this process for each of the M quadrats removed, replacing the other quadrats. This will give $M = 20$ so-called pseudo-values defined as

$$D_i = MD_0 - (M-1)D^{-1} \qquad (17.30)$$

where $i = 1, 2, 3, \ldots M$. The jackknife estimate is the average of the pseudovalues, thus

$$\bar{D} = \sum_{i=1}^{M} D_i / M \qquad (17.31)$$

and the variance is

$$\text{Var. } \bar{D} = \sum_{i=1}^{M} (D_i - \bar{D})^2 / M(M-1). \qquad (17.32)$$

Slaff et al. (1983) used the similarity index (Ro_c) of Horn (1966) to compare the numbers of mosquito species caught at human bait, in carbon dioxide-baited CDC traps and in New Jersey light-traps

$$Ro_c = \frac{(\Sigma(x_i + y_i)\log(x_i + y_i) - (\Sigma x_i \log x_i - \Sigma y_i \log y_i)}{(X+Y)\log(X+Y) - X \log X - Y \log Y} \qquad (17.33)$$

where Ro_c = community overlap where species i is represented x times in a community X, and y times in community Y and – where, X and Y = total numbers of individuals in a sampling unit (e.g. in this instance at bait and in two traps). Note using this model only two communities (e.g. traps) can be compared. The index ranges from 0 (completely dissimilar) to 1 (identical species composition). Tallamy et al. (1976) applied this index to compare Malaise trapping and aerial net sampling for Tabanidae.

Wolda (1981) examined the effects of sample size and diversity on 22 different similarity indices. He concluded that only the Morista (1959) index was virtually independent of sample size and independent of diversity, and was usually the best index to use, it can be written as

$$C_A = \frac{2\Sigma n_{1i} n_{2i}}{(\lambda_1 + \lambda_2)N_1 N_2}, \quad \text{where } \lambda_j = \frac{\Sigma n_{ji}(n_{ji}-1)}{N_j(N_j-1)} \qquad (17.34)$$

where nji = number of individuals of species i in sample j; Nj = number of individuals in sample j.

However, as this index is very sensitive to changes in abundance of the more common species a logarithmic transformation of the data [log (n_{ji} + 1)] should be used and then either the Horn (1966), simplified Morista index or the Renkonen percentage similarity (PS) index used.

With the simplified Horn Index, the only modification is that λ is simplified as follows

$$\lambda_j = \frac{\Sigma n^2{}_{ji}}{N^2{}_j} \tag{17.35}$$

The percentage similarity (PS) index of Renkonen (1938) is

$$PS = \Sigma \min(p_{1i}, p_{2i}) \tag{17.36}$$

where p_1 and p_2 are proportions. Sample sizes should not be less than 100, because sampling errors can make the results meaningless.

A disadvantage of the Morista index is that it depends heavily on the most common species. This led Grassle and Smith (1976) to present a modification of the Morista similarity index called the Normalised Expected Species Shared (NESS) that is sensitive to the contribution made by rare species. The measure is based on the expected numbers of species shared between random samples taken from a population. Wolda (1983) pointed out that this modification by Grassle and Smith (1976) which he called the Cm index is unbiased at any sample size and independent of diversity, and is therefore to be preferred.

A diversity index should indicate how heterogeneous an assemblage is within itself. This led Cuba (1981) to propose a new index of diversity (D) based on the assumption that a change in species composition creates greater intra-assemblage variation than does a change in the distribution of the individuals of the species. An assemblage with more species is therefore more diverse than an assemblage with fewer numbers of species.

$$D = S + X \tag{17.37}$$

where S = the number of species in the sample, while X provides a measure of the distribution of individuals among the species. The purpose of calculating X is to determine how close the observed sample is, in its distribution, to a hypothetical sample of equivalent S and N in which individuals are distributed equally among the species.

Now, in the following hypothetical sample all species have exactly the same number of individuals and the hypothetical data (n_{ih}) are calculated as

$$n_{ih} = (N/S) \qquad (17.38)$$

where $n_{1h} = n_{2h} = . \ldots n_{sh}$. Variations from evenness in the observed sample (d_i) are calculated for each species by

$$d_i = | n_{i0} - n_{ih} | \qquad (17.39)$$

where n_{i0} are the observed data. The d_i values are then summed and X calculated as follows

$$X = 1 - \left(\sum_{i=1}^{i=s} d_i \bigg/ 2N \right) \sum \qquad (17.40)$$

and remembering that diversity (D) is calculated as $S + X$.

In Louisiana, USA Schreiber et al. (1988) studied the coexistence of *Ochlerotatus triseriatus, Ochlerotatus hendersoni, Aedes albopictus* and *Aedes vexans* in glass ovijars placed at ground level and on trees at heights of 1, 3, 5, 7 and 9 m. The following formula of Levins (1968) was used to determine niche overlap (αij).

$$\alpha ij = \sum pih \quad pjh(\beta i) \qquad (17.41)$$

where αij = the niche overlap of species i over species j; and pih and pjh = the proportions of each species in the hth of a resource set (in this instance height); and αi = the niche breadth of a species i, which is calculated from

$$\beta = 1/\sum_{i=1}^{n} pi^2 \qquad (17.42)$$

where pi = proportion of a species found in the ith unit of the resource set (here height); n = number of the set (week).

This niche overlap analysis was conducted for each species at each vertical height over the entire study period. Values of niche overlap can range from 1.0 to 0. Low values indicate low ecological similarity, high resource partitioning with respect to habitat and consequently little interspecific competition. High values indicate the opposite, that is ecological similarity and species competition.

In Kenya Lounibos (1980) used the niche overlap index (L) recommended by Hurlbert (1978) to determine whether interspecific encounters among larvae of three sympatric *Eretmapodites* species, *Eretmapodites subsimplicipes, Eretmapodites quinquevittatus* and *Eretmapodites silvestris conchobius*, breeding in available microhabitats (snail shells, tree-holes, leaf pools, fruit husks and fungus caps) were uniform or otherwise.

$$L = \frac{A}{XY}\sum_i \left(\frac{x_i y_i}{a_i}\right)$$
(17.43)

where x_i and y_i are the abundance of species x and y in the ith microhabitat and a_i is the resource abundance which Lounibos considered was the fluid volume (ml) of the ith microhabitat. The total numbers of each species is represented by X and Y, while A is the total volume in all microhabitats. A value of $L = 1$ usually means that each species uses a microhabitat in proportion to its relative abundance, but see Petraitis (1979) for exceptions, and $L < 1$ indicates non-overlap, while $L > 1$ indicates overlap which is proportionally less or more than is to be expected by a uniform distribution.

With the exception of *Eretmapodites quinquevittatus, Eretmapodites silvestris conchobius*, and *Eretmapodites subsimplicipes* which were highly attracted to snail shells (*Achatina fulica*) niche overlap was less than anticipated by a uniform distribution, indicating the species were ecologically separate.

Bradshaw and Holzapfel (1986) studied mosquito larval competition by comparing densities within and between species using Lloyd's (1967) indices of mean interspecific and intraspecific crowding per unit of resource (Hurlbert, 1978). For example mean interspecific crowding between any two (say *Ochlerotatus geniculatus* (*AG*) and *Anopheles plumbeus* (*AP*)) of the three tree-hole breeding mosquitoes *Ochlerotatus geniculatus, Anopheles plumbeus* and *Culex torrentium* is calculated as follows

$$\frac{AG}{AP} = \frac{\sum \frac{AG_i AP_i}{v_i}}{\sum AG_i}$$
(17.44)

where AG_i and AP_i = the numbers of individuals of *Ochlerotatus geniculatus* and *Anopheles plumbeus* respectively in the ith tree-hole, and v_i = volume of water in that tree-hole. To obtain mean intraspecific crowding, $(AG_1 - 1)$ is substituted for AP_i in this formula (as done by Bradshaw and Holzapfel (1986)). Barrera (1996) used the above method to examine interactions within and between several species of aquatic insect, including mosquitoes, inhabiting tree-holes in Pennsylvania, USA. For example, *Ochlerotatus* species encountered scirtids in 95% of tree-holes and ceratopogonids in 91%. Most species in the community encountered other species in the community in proportion to their relative abundance; that is independently. Psychodids demonstrated a negative association with *Ochlerotatus* species.

Disney (1972) proposed an index of uniformity (*IU*) for comparing collections of pre-adult *Simulium* in situations where the numbers of species were too few to allow more conventional indices of diversity, such as α, to be reliably applied:

$$IU = \frac{\Sigma W}{S} \qquad (17.45)$$

where *S* = the number of species; *W* = a value that is given to a percentage range which represents each species' contribution to the total (100%) collection of species (Table 17.2). It progressively weights the highest percentage values. A highly uniform sample would have an index of uniformity of 78, while lower values indicate greater species diversity.

Table 17.2. Arithmetic series of *W* for use in the formulae for the index of uniformity (Disney 1972)

% composition	0.001–0.01	0.011–0.11	0.11–1.0	1.1–10	10.1–20	20.1–30	30.1–40
W	0	1	3	6	10	15	21
% composition	40.1–50	50.1–60	60.1–70	70.1–80	80.1–90	90.1–100	
W	28	36	45	55	66	78	

In Nigeria tree-hole breeding mosquitoes were sampled by using water-filled cylindrical gourds, and Spearman's rank correlation coefficient (r_s), a non-parametric test, used to determined whether the gourds attracted representative samples of the tree-hole breeding mosquitoes (Service 1965):

$$r_s = \frac{1 - 6\Sigma d^2}{N(N^2 - 1)} \qquad (17.46)$$

where *d* = difference in ranking order for each paired observation, and *N* = number of paired observations. The value of r_s ranges from −1 for complete discordance of ranked values to +1 for complete concordance.

Many methods have been proposed to estimate species richness, that is the numbers of species in a community. In a comparison of eight methods of estimating species richness Palmer (1990) concluded that the non-parametric, first-order jackknife estimator was the most precise and least biased.

$$JACK1 = SO + r1(n-1)/n \qquad (17.47)$$

where SO = number of species in n-quadrat; $r1$ = number of species in only one quadrat. Although most of the methods he evaluated either over-estimated or underestimated species richness, they were nevertheless highly correlated with the true species richness, and so can be used for comparing species richness.

Elliott and Drake (1981) in studying the efficiencies of dredges for sampling benthic macroinvertebrates in rivers concluded that the relationship between numbers of taxa (S) and numbers of individuals (N) is best described by a power law $S = a\, Nb$ or $\log_e S = \log_e a + b \log_e N$, where a and b are constants. They thought it likely that this relationship might apply to samples taken with pond nets. They argued that the implications of this relationship casts doubt on the validity of several indices based on species richness, especially when the size of the catches are highly variable.

Whittaker (1972) defined two parameters to describe species association in discrete habitats namely S, which denotes species richness and $\exp(H') = N1$, where H' is the Shannon–Weaver index. Routledge (1977) in a mathematical paper modified these to apply to species in non-discrete habitats.

Bradshaw and Holzapfel (1983) studied species richness in tree-hole mosquitoes in Florida, USA and by plotting the mean numbers of species in tree-holes against their volumes found size becomes a more reliable predicator of species richness as the frequency of drying out increased.

Similarly, Spencer et al. (1999) studied the relationship between habitat size and degree of permanence and the richness of the aquatic community in rock pools in Israel. They observed that larger pools had significantly more species, as did the more permanent pools. In addition, there was a small but significant increase in species richness over time. The proportion of macroscopic predatory species was also significantly higher in larger pools, but there was no significant relationship between proportion of predators and pool permanence.

Buckley et al. (2003) studied species richness in communities inhabiting *Sarracenia purpurea* pitcher plant communities in USA, and observed that species richness tended to increase with increasing latitude, contrary to observations of species richness in individual guilds or taxa within trophic levels, such as birds. Increasing species richness occurred primarily among the bacteria and protozoan elements of the community, whereas invertebrate richness exhibited no significant relationship with latitude. Abundance of the top predator in the system, *Wyeomyia smithii*, decreased significantly with latitude ($P < 0.006$, $R^2 = 0.18$).

References

Amerasinghe FP, Munasingha NB (1988*a*) A predevelopment mosquito survey in the Mahaweli development project area, Sri Lanka: Adults. J Med Entomol 25: 276–285

Amerasinghe FP, Munasingha NB (1988*b*) A predevelopmental mosquito survey in the Mahaweli development project area, Sri Lanka: Immatures. J Med Entomol 25: 286–294

Ball SG (1983) A comparison of the Diptera caught in Manitoba traps with those caught from cattle and other parts of the field ecosystem in northern England. Bull Entomol Res 73: 527–537

Baroni-Urbani C (1980) A statistical table for the degree of coexistence between two species. Oecologia (Berl.) 44: 287–289

Barrera R (1996) Species concurrence and the structure of a community of aquatic insects in tree holes. J Vector Ecol 21: 66–80

Bhatt RN, Sharma RC, Srivastava HC, Gautam AS, Gupta DK (1993) Interspecific associations among anophelines in different breeding habitats of Kheba district Gujarat: part II - non-canal area. Indian J Malariol 30: 91–100

Bradshaw WE, Holzapfel CM (1983) Predator-mediated, non-equilibrium co-existence of tree-hole mosquitoes on southeastern North America Oecologia (Berl.) 57: 239–256

Bradshaw WE, Holzapfel CM (1986) Habitat segregation among European tree-hole mosquitoes. Natl Geogr Res 2: 167–178

Brillouin L (1962) Science and Information Theory, 2nd edn. Academic Press, New York

Buckley HL, Miller TE, Ellison AM, Gotelli NJ (2003) Reverse latitudinal trends in species richness of pitcher-plant food webs. Ecol Letters 6: 825–829

Bullock JA (1971) The investigation of samples containing many species. II. Sample comparison. Biol J Linn Soc 3: 23–56

Campos RE, Fernández LA, Sy VE (2004) Study of the insects associated with the floodwater mosquito *Ochlerotatus albifasciatus* (Diptera: Culicidae) and their possible predators in Buenos Aires Province, Argentina. Hydrobiologia 524: 91–102

Cancela da Fonseca JP (1966) L'outil statistique en biologie du sol. III. Indices d'intérêt écologique. Revue Ecol Biol Sol 3: 381–407

Cole LC (1949) Measurement of interspecific association. Ecology 30: 411–424

Cole LC (1957) The measurement of partial interspecific association. Ecology 38: 226–233

Cuba TR (1981) Diversity: a two-level approach. Ecology 62: 278–279

de Szalay FA, Resh VH (2000) Factors influencing macroinvertebrate colonization of seasonal wetlands: responses to emergent plant cover. Freshwater Biol 45: 295–308

Disney RHL (1972) Observations on sampling pre-imaginal populations of blackflies (Dipt., Simuliidae) in West Cameroon. Bull Entomol Res 61: 485–503

Elliott JM, Drake CM (1981) A comparative study of four dredges used for sampling benthic macroinvertebrates in rivers. Freshwater Biol 11: 245–261

Fisher RA, Corbet AS, Williams CB (1943) The relation between the number of species and the number of individuals in a random sample for an animal population. J Anim Ecol 12: 42–58

Forattini OP, Gomes A de C, Natal D, Kakitani I, Marucci D (1987) Freqüência domicilar e endophilia de mosquitos Culicidae no Vale do Ribeira, São Paulo, Brazil. Rev Saúde Pública 21: 188–192

Forattini OP, Kakitani I, Massad E, Marucci D (1993) Studies on mosquitoes (Diptera: Culicidae) and anthropic environment. 3 - Survey of adult stages at the rice irrigation system and the emergence of *Anopheles albitarsis* in South-Eastern, Brazil. Rev Saúde Pública 27: 313–325

Gadagkar R (1989) An undesirable property of Hills' Diversity index N2. Oecologia (Berl.) 80: 140–141

Gower JC (1971) A general coefficient of similarity and some of its properties. Biometrics 27: 857–871

Grassle JF, Smith W (1976) A similarity measure sensitive to the contribution of rare species and its use in investigation of variation in marine benthic communities. Oecologia (Berl.) 25: 13–22

Heltshe JF, Forrester NE (1983) The jackknife estimate of species richness. Biometrics 39: 1–11

Heltshe JF, Forrester NE (1985) Statistical evaluation of the jackknife estimate of diversity when using quadrat samples. Ecology 66: 107–111

Hill MO (1973) Diversity and evenness: A unifying notation and its consequences. Ecology 54: 427–432

Horn H (1966) Measurement of "overlap" in comparative ecological studies. Am Nat 100: 419–424

Hurlbert SH (1969) A coefficient of interspecific association. Ecology 50: 1–9

Hurlbert SH (1971) The nonconcept of species diversity: a critique and alternative parameters. Ecology 52: 577–586

Hurlbert SH (1978) The measurement of niche overlap and some derivatives. Ecology 59: 67–77

Jaccard P (1912) The distribution of the flora in the alpine zone. New Phytol 11: 37–50

Janson S, Vegelius J (1981) Measures of ecological association. Oecologia (Berl.) 49: 371–376

Kant R, Pandey SD (1999) Breeding preferences of *Anopheles culicifacies* in the rice agro-ecosystem in Kheda District, Gujarat. Indian J Malariol 36: 53–60

Kant R, Pandey SD, Sharma SK, Sharma VP (1998) Species diversity and interspecific associations among mosquitoes in rice agro-ecosystem of Kheda district, Gujarat. Indian J Malariol 35: 22–30

Kempton RA, Taylor LR (1974) Log-series and log-normal parameters as diversity discriminants for the Lepidoptera. J Anim Ecol 43: 381–399

Kempton RA, Taylor LR (1976) Models and statistics for species diversity. Nature 262: 818–820

Khan SA, Handique R, Tewari SC, Dutta P, Narain K, Mahanta J (1998) Larval ecology and mosquito fauna of Upper Brahmaputra Valley, northeast India. Indian J Malariol 35: 131–145

Lee D-K (1998) Effect of two rice culture methods on the seasonal occurrence of mosquito larvae and other aquatic animals in rice fields of southwestern Korea J Vector Ecol 23: 161–170

Levins R (1968) Evolution in Changing Environment. Princeton University Press, Princeton

Lim TM, Khoo HW (1985) Sampling properties of Gower's general coefficient of similarity. Ecology 66: 1682–1685

Lloyd M (1967) Mean crowding. J Anim Ecol 36: 1–30

Lounibos LP (1980) The bionomics of three sympatric *Eretmapodites* (Diptera: Culicidae) at the Kenya coast. Bull Entomol Res 70: 309–320

Lounibos LP (1981) Habitat segregation among African treehole mosquitoes. Ecol Entomol 6: 129–154

MacArthur RH (1955) Fluctuations of animal populations, and a measure of community stability. Ecology 36: 533–536

MacArthur RH (1965) Patterns of species diversity. Biol Rev 40: 510–533

Magurran AE (1988) Ecological Diversity and its Measurement. Croom Helm, London

Margalef DR (1968) Perspectives in Ecological Theory. University of Chicago Press, Chicago

Margalef R (1957) La teoria de la informacion en ecologia. Mems R Acad Cienc Artes Barcelona 3rd ser 32: 373–449. (In Spanish); translated into English by W. Hall (1958) in Gen Syst 3: 36–71

Masalkina TM (1979) A comparative evaluation of methods of capturing blood-sucking mosquitoes. Communication I. Species composition and ratio of mosquito species caught by different methods. Medskaya Parazitol 48: 47–52 (In Russian, English summary)

May RM (1975) Patterns of species abundance and diversity. In: Cody ML, Diamond JM (eds) Ecology and Evolution of Communities. Belknap Press of Harvard University Press, Cambridge, Massachusetts, pp. 81–120

McIntosh RP (1967) An index of diversity and the relation of certain concepts to diversity. Ecology 48: 392–404

McNaughton SJ (1967) Relationship among functional properties of California grassland. Nature 216: 168–169

Morista M (1959) Measuring of interspecific association and similarity between communities. Mem Fac Sci Kyushu Univ Ser E Biol 3: 65–80

Mountford MD (1962) An index of similarity and is application to classificatory problems. In: Murphy PW (ed) Progress in Soil Zoology. Butterworths, London, pp. 45–50

Nei NH, Hull CH, Jenkins JG, Steinbrenner K, Bent DH (1975) SPSS: Statistical package for social sciences. McGraw Hill, New York

Nilsson SG, Bengtsson J, Ås S (1988) Habitat diversity or area per se? Species richness of woody plants, carabid beetles and land snails on islands. J Anim Ecol 57: 685–704

Osborne LL, Davies RW, Linton KJ (1980) Use of hierarchial diversity indices in lotic community analysis. J Appl Ecol 17: 567–580

Palmer MW (1990) The estimation of species richness by extrapolation. Ecology 71: 1195–1198

Petraitis PS (1979) Likelihood measures of niche breadth and overlap. Ecology 60: 703–710

Pichon G, Gayral P (1979) Comparison nomocénologique de deux méthodes de piègeage des moustiques. Cah ORSTOM sér Entomol Méd Parasitol 17: 243–247

Pielou EC (1966) Species-diversity and pattern diversity in the study of ecological succession. J Theor Biol 10: 370–383

Pielou EC (1967) The use of information theory in the study of the diversity of biological populations. Proc Berkeley Symp Math Statist Probab 5: 163–177

Pielou EC (1975) Ecological Diversity. John Wiley & Sons, New York

Pielou EC (1977) Mathematical Ecology. John Wiley & Sons, New York

Preston FW (1948) The commonness and rarity of species. Ecology 29: 254–283

Reisen WK (1978) A quantitative mosquito survey of 7 villages in Punjab province, Pakistan with notes on bionomics, sampling methodology and effects of insecticides. Southeast Asian J Trop Med Public Health 9: 587–601

Reisen WK, Siddiqui TF, Aslamkhan M, Malik GM (1981) Larval interspecific associations and physico-chemical relationships of ground-water breeding mosquitoes of Lahore. Pak J Sci Res 3: 1–23

Renkonen O (1938) Statistisch-ökologische Untersuchungen über der terrestische Käferwelt der finnischen Brunchmoore. Ann Zool Soc Zool-Bot Fenn Vanamo 6: 1–231

Routledge RD (1977) On Whittaker's components of diversity. Ecology 58: 1120–1127

Routledge RD (1980) Bias in estimating the diversity of large, uncensused communities. Ecology 61: 276–281

Schreiber ET, Meek CL, Yates MM (1988) Vertical distribution and species coexistence of tree hole mosquitoes in Louisiana. J Am Mosq Control Assoc 4: 9–14

Service MW (1965) The ecology of the tree-hole breeding mosquitoes in the northern Guinea savanna of Nigeria. J Appl Ecol 2: 1–16

Service MW (1968) The taxonomy and biology of two sympatric sibling species of *Culex, C. pipiens* and *C. torrentium* (Diptera: Culicidae). J Zool Lond 156: 313–323

Shannon CE, Weaver W (1963) The Mathematical Theory of Communication. University of Illinois Press, Urbana

Shepard RB (1984) The logseries distribution and Mountford's similarity index as a basis for the study of stream benthic community structure. Freshwater Biol 14: 53–71

Shorygin AA (1939) Food and food preference of some Gobidae of the Caspian sea. Zool Zh 18: 27–53 (In Russian, English summary)

Simpson EH (1949) Measurement of diversity. Nature 163: 688

Slaff M, Crans WJ, McCuiston LJ (1983) A comparison of three mosquito sampling techniques in northwestern New Jersey. Mosquito News 43: 287–290

Sneath PHA, Sokal RR (1973) Numerical Taxonomy. W.H. Freeman: San Francisco

Sørensen T (1948) A method of establishing groups of equal amplitude in plant sociology based on similarity of species content and its application to analyses of the vegetation on Danish commons. Biol Skr (K Danske Vidensk Selsk N.S.) 5: 1–34

Southwood TRE (1978) Ecological Methods with Particular Reference to the Study of Insect Populations. Chapman & Hall, London

Southwood TRE, Henderson PA (2000) Ecological Methods. 3rd ed. Blackwell Science, Oxford

Spencer M, Blaustein L, Schwartz SS, Cohen JE (1999) Species richness and the proportion of predatory animal species in temporary freshwater pools: relationships with habitat size and permanence. Ecol Letters 2: 157–166

Suwonkerd W, Tsuda Y, Takagi M, Wada Y (1996) Seasonal occurrence of *Aedes aegypti* and *Ae. albopictus* in used tires in 1992–1994, Chiangmai, Thailand. Trop Med (Nagasaki) 38: 101–105

Tallamy DW, Hansens EJ, Denno RF (1976) A comparison of malaise trapping and aerial netting sampling for a horsefly and deerfly community. Environ Entomol 5: 788–792

Taylor LR, Kempton RA, Woiwod IP (1976) Diversity statistics and the log-series model. J Anim Ecol 45: 255–271

Taylor RAJ (1978) The relationship between density and distance of dispersing insects. Ecol Entomol 3: 63–70

Whittaker RH (1972) Evolution and measurement of species diversity. Taxon 21: 213–251

Whittaker RH, Fairbanks CW (1958) A study of plankton copepod communities in the Columbian basin, south eastern Washington. Ecology 39: 46–65

Williams CB (1947) The logarithmic series and the comparison of island floras. Proc Linn Soc Lond 158: 104–108

Wolda H (1981) Similarity indices, sample size and diversity. Oecologia (Berl.) 50: 296–302

Wolda H (1983) Diversity, diversity indices and tropical cockroaches. Oecologia (Berl.) 58: 290–298

Yadav RS, Sharma VP, Chand SK (1997) Mosquito breeding and resting in tree-holes in a forest ecosystem in Orissa. Indian J Malariol 34: 8–16

Zahl S (1977) Jackknifing an index of diversity. Ecology 58: 907–913

Index

1

1-octen-3-ol, 693, 900, 996, 1003, 1006

A

Aedes aegypti indices, 74, 218, 256-287
  Breteau Index, 74, 256, 261, 265, 267, 276, 278, 280, 282, 285
  Container (=Receptacle) Index, 256, 272, 273, 275, 276, 278, 282
  House (=Premise) Index, 74, 256, 271-273, 275, 276, 278, 282
  Premise Condition Index, 140, 264, 265, 274
age-grading, 403, 906, 1161, 1167, 1169, 1171, 1172, 1176
AMSS trap, 393-396
Anderson and Brust box trap, 626
aniline dye, 1278, 1279
aquatic light trap, 221-225
  betalight, 223-225
  chemical light-trap, 224
  sticky light-trap, 226, 227
aquatic mite, 1162, 1164
aquatic net, 163, 179, 195-198, 203, 211, 229, 293
arbovirus surveillance, 151, 458, 630, 959, 966
ascending net, 787

B

backpack aspirator, 257, 261, 390, 444, 445, 569, 680, 686, 727, 1023, 1034

Bailey Triple Catch, 1322, 1324, 1339, 1340, 1350, 1351, 1355
Baited suction trap (Davies), 607-609
basic reproduction rate, 729, 732, 746
Bates' stable trap, 587, 588, 604, 982
bed-net catch (BNC), 522
Bicycle-mounted trap, 788, 797
Bidlingmayer's larval trap, 249, 251
binomial sampling, 113, 114
biting rate, 268, 398, 399, 494, 498, 500, 503, 504, 511, 541, 554, 578, 697, 702, 711, 712, 717, 730-734, 736-741, 743, 745-747, 873, 880, 884, 1211, 1241, 1438
'blacklight' lamp, 852, 905, 908, 909, 922, 1285
Blower trap of de Freitas et al, 605, 606
Briët method, 1218
Buescher et al. bird-baited trap, 604

C

caesium, 682, 683, 715, 1302, 1303
calf-baited trap, 583, 584, 592, 597, 887
cannibalism, 1054, 1058, 1077, 1078
Carbon Dioxide Trap, 105, 597, 771, 921, 947-1011, 1197
Carbon dioxide-baited CDC trap, 916, 953, 956, 957, 960, 1006, 1198, 1457
carboxylic acid, 693
CDC backpack aspirator, 390, 444, 445, 569, 680

CDC light-trap, 17, 108, 111, 418,
    434, 440, 451, 452, 458, 510, 524,
    527-529, 565, 583, 590, 600, 608,
    616, 621-623, 625, 635, 680, 681,
    772, 775, 776, 794, 846, 847, 856,
    857, 859, 864-888, 892, 893, 901,
    906, 908, 915-917, 920, 923, 947,
    949, 968, 973, 974, 999, 1000,
    1003, 1004, 1008, 1009, 1011,
    1032, 1033, 1036-1038, 1177,
    1184, 1191, 1197, 1198, 1214,
    1215, 1221, 1222, 1245, 1275,
    1290, 1348, 1401, 1408, 1411,
    1412, 1451
CDC mechanical aspirator, 436,
    971-976, 978
CFG trap, *see* Counter-flow geome-
    try (CFG) trap
CO_2-baited ABC suction trap, 965
CO_2-baited CDC light-trap, 108, 111,
    451, 680, 956-960, 963, 964,
    1003, 1008, 1011, 1123, 1348,
    1400, 1408, 1411, 1412
Coefficient of Variability, 6-8, 214
Conical trap of DeFoliart and Morris,
    992
Counter-flow geometry (CFG) trap,
    971, 972, 974, 979, 1009, 1010,
    1043
cow-baited net trap, 565-567
crep, 548, 551-553
curtain trap, 538, 540
cuticular growth rings, 1169
cuticular hydrocarbons, 500, 1170,
    1171
cytochrome b, 688-690

D

dengue, 61, 64, 65, 68, 73, 74, 80,
    94, 114, 151, 219, 220, 256, 260,
    262, 264, 268-270, 272, 276, 277,
    282-287, 494, 508, 585
density dependence, 1050-1052,
    1057, 1066, 1075
density dependence, 18

dilatation, 41, 92, 558, 730, 876,
    1173-1181, 1197, 1203, 1223-1227,
    1231, 1232, 1281
dipper, 19, 27, 28, 44, 154-162,
    167-169, 171-174, 176-178,
    180-183, 188, 192, 196, 199, 202,
    203, 211, 212, 214-216, 227, 228,
    240, 246, 251, 252, 289, 292, 294,
    1072, 1077, 1101, 1102, 1114,
    1115, 1121, 1456
dispersal, 18, 94, 190, 456, 467, 510,
    715, 781, 784, 795, 800, 802, 825,
    847, 908, 984, 1032, 1065, 1106,
    1120, 1161, 1197, 1205, 1233,
    1276-1279, 1282, 1284, 1288,
    1291, 1293, 1294, 1299, 1300,
    1302-1304, 1306, 1308, 1338,
    1340, 1351, 1356, 1377-1423
    jump dispersal, 1377
    mean distance dispersed, 1391,
        1400, 1401, 1404, 1411
    neighbourhood dispersal, 1377
Diversity Indices, 1445-1462
DNA 'fingerprinting' method, 684
DNA profiling, 686, 687
DNA tests, 1090
dog-baited trap, 584, 591, 600-603
dominance index, 926, 1448, 1449
drop-net cage, 420, 517
duck-baited trap, 625, 626
duplex cone trap, 969, 1034-1036

E

earth-lined box shelter, 459-461, 467
eastern equine encephalitis, 455, 456,
    589, 960
eastern equine encephalomyelitis,
    458, 589, 623, 628, 689, 727, 875,
    894, 916, 917, 960, 962
eggshell density, 44, 46
EIR, *see* entomological inoculation rate
electric grid, 833, 857
electrophoresis test, 1088
ELISA, *see* enzyme-linked-
    immunosorbent assay

ELISA method, *see* enzyme-linked-immunosorbent assay

emergence trap, 339, 340, 342-349, 351-355, 357-364, 621, 1110, 1308
  Bradley's trap, 342
  Cesspit trap, 355
  Corbet's emergence trap, 351
  Emergence light-trap, 363
  Emergence wheel of Corbet, 363
  Floating trap of Morgan et al, 347
  Mundie's trap, 342, 351
  pit latrine emergence trap, 357-360
  Prism trap, 349
  Pritchard and Scholefield trap, 353
  Pyramidal trap, 343, 346, 348, 364, 1027, 1028
  Sticky emergence trap, 364
  Tree-hole emergence trap, 361

Emord and Morris trap, 622-625

entomological inoculation rate (EIR), 493, 393, 500, 731, 733, 736-740, 743, 744, 747, 880

enzyme-linked-immunosorbent assay (ELISA), 403, 628, 677-680, 688, 689, 703, 706, 710, 722, 723, 726, 727, 734, 1087, 1088, 1141

EVS light-trap, 896-901, 968

EVS trap, 892, 897-903, 916, 963, 966, 968, 979, 1006-1008, 1036, 1401

exit trap, 2, 356, 357, 359, 361, 408, 420, 458, 501, 522, 531-536, 538, 541, 591, 629, 635, 712, 713, 747, 801, 881, 991, 1406, 1425, 1427-1430, 1433-1437, 1439
  Muirhead-Thomson exit trap, 531, 635, 712
  Rachou exit trap, 536

experimental hut, 527, 528, 531, 532, 534, 535, 591, 632, 711, 712, 726, 864

F

Fay-Prince carbon-dioxide trap, 968

Fay-Prince trap, 108, 587, 680, 968-970, 980, 1009, 1010, 1028, 1030-1033, 1037

feeding index, 588, 689, 690, 721, 723, 725-729

Fibre pot resting box of Komar et al, 458

Fisher and Ford method, 1234, 1304, 1309, 1318

floating trap, 216, 217, 340, 342-345, 347

follicular relic, 1173, 1175, 1181

forage ratio, 721-727

funnel trap, 217-221, 229, 339

G

Geographic Information Systems, 17, 18, 146, 147, 150, 151, 153, 260

Giemsa, 300-302, 1177, 1274-1276

Gompertz model, 1182-1185, 1197, 1225

gonotrophic cycle, 556, 559, 585, 687, 691, 692, 718, 719, 721, 730, 731, 744, 747, 1053, 1092, 1124, 1129-1132, 1164, 1173-1175, 1177, 1185, 1186, 1188-1191, 1194-1196, 1199-1249, 1292, 1339, 1377, 1395, 1396, 1433

gonotrophic discordance, 718, 719

gonotrophic dissociation, 719, 719

granular basal body, 1179-1181

gravid trap, 72, 95, 98, 102-112, 680, 794, 795, 921, 959, 962-965, 1011, 1036
  Reiter's gravid trap, 104, 105, 107-111
  Surgeoner and Helson's gravid trap, 103

growth retardant, 1064-1066

H

hand-net collection, 410, 515, 537, 1408

haptoglobin, 683, 711, 719

heteroduplex analysis of cytochrome
    b DNA, 688
Hii et al. method, 1237
horse-baited trap, 468, 569, 1237
host preference, 373, 563, 585, 588,
    590, 591, 595, 615, 617, 683, 701-
    703, 705, 706, 708-710, 713, 716,
    721-723, 727, 776, 887, 1215
host-seeking, 373, 422, 493, 507,
    516, 522, 530, 543, 554, 557, 563,
    577-579, 594, 629, 698, 699, 703,
    775, 823, 832, 859, 860, 875, 916,
    947, 952, 956, 970, 982, 996,
    1010, 1279, 1412
human biting rate (HBR), 494, 702,
    731, 733, 734, 736, 739, 743, 747,
    880
Human Blood Index, 2, 3, 399, 704,
    722, 724, 726, 746, 1240, 1241,
    1243
human landing catch, 60, 108, 145,
    398, 406, 407, 415, 440, 447, 468,
    493, 494, 497, 498, 500-511, 514,
    515, 522, 524, 526, 528-530, 544,
    545, 552, 554, 558, 559, 564, 565,
    567, 568, 574, 575, 595, 603, 633-
    635, 680, 698, 708, 736, 737, 739-
    735, 747, 804, 863, 865, 873, 876,
    877, 879, 881, 882, 887, 888, 910,
    916, 922, 958, 963-966, 969, 970,
    978, 995, 1005, 1006, 1009, 1033,
    1034, 1036, 1043, 1203, 1205,
    1215, 1351, 1356, 1403, 1438
human-baited net, 585, 632-634

I

IMR bait-trap, 609, 610
index of species abundance (ISA),
    513, 514, 1072
index of uniformity, 1461
Indices of Association, 1445-1462
instar duration, 180, 1092-1095,
    1097, 1099-1101, 1103-1105,
    1107-1109, 1112, 1113, 1116,
    1119, 1120, 1131

interspecific competition, 185, 1053-
    1055, 1061, 1062, 1069, 1077, 1459
intraspecific competition, 1053,
    1055, 1056, 1064, 1066, 1123
isokinetic net, 784, 785
Iwao's (1968) model, 186, 188, 190,
    192

J

Jackson's positive method, 1315,
    1339, 1340, 1345-1347, 1351,
    1356, 1357
Jackson's methods, 1315-1317, 1345
Jackson's negative method, 1339,
    1340, 1345-1347, 1356
Japanese encephalitis, 173, 415, 426,
    438, 517, 561, 595, 604, 605, 628,
    783, 892, 908, 909, 979, 1003
Johnson-Taylor exposed cone trap,
    804, 806, 808, 810, 811, 818, 830
Johnson–Taylor suction trap, 808
Johnson–Taylor trap, 804, 808, 810,
    811, 818, 830
Jolly-Seber method, 1317, 1322,
    1351
Jolly-Seber model, 1329

K

kairomone, 692, 694
Katô's dry ice trap, 991
Keg shelter, 448
Kelker's selective removal, 298, 299,
    1342
Kimsey and Chaniotis trap, 901, 902
kite, 782, 783, 785
kriging, 16, 17, 408, 1003
kytoon, 783, 785

L

lactic acid, 693-695, 710, 894,
    995-1011, 1170
lard-can traps, 564, 614-619, 621-
    623, 719, 979, 983, 987, 1198

Larval concentrator, 157-161, 163
latex agglutination test, 677, 1086
Leslie's method, 1304, 1324, 1339
life-table, 173, 180, 187, 299, 1050,
    1051, 1057, 1074, 1090-1095,
    1097-1103, 1105-1107, 1109-
    1132
    age-specific life-table, 1091, 1106,
        1113, 1121, 1122
    generation mean life-table, 1115
    horizontal life-table, 1091, 1092
    time-specific life-table, 1097, 1105,
        1106, 1115
    vertical life-table, 180, 1091
light-emitting diodes (LEDs), 856,
    857, 860
Light-Trap Index (LTI), 925
Lincoln Index, 301, 302, 730, 1304,
    1309, 1310, 1312-1314, 1318,
    1324, 1328, 1337, 1339, 1340,
    1344-1348, 1350-1352, 1354-
    1357
Lloyd's mean crowding, 9, 90, 92,
    186, 188, 190-192, 272, 1126
Lloyd's patchiness index, 186

M

MacLeod's method, 1328
Magoon trap, 568, 580, 583-588,
    591, 593, 594, 600, 627, 707, 722,
    916, 983
Malaise trap, 580, 620, 772-776, 916,
    952, 994, 995, 1036
malaria, 2, 15, 16, 18, 28, 137, 142-
    144, 148-151, 154, 374, 392-394,
    401, 403, 404, 406, 412, 413, 415,
    466, 467, 494, 497, 501, 504, 508,
    509, 514, 524, 541, 559, 563, 567,
    569, 593, 634, 678, 695, 704, 714,
    729-731, 733, 735, 736, 741, 743-
    747, 873, 874, 877, 880, 881, 891,
    892, 906, 978, 1043, 1062, 1188,
    1218, 1239-1241, 1379, 1425,
    1426, 1436
man hour density, 380, 382, 383, 395

man-biting rate, 398, 399, 500,
    731-733, 737, 740, 741
Manly and Parr's (1968) method,
    1332, 1336-1338, 1345-1347
marking
    dye, 300
    fluorescent powder, 392, 400, 499,
        540, 708, 904, 1185, 1203,
        1205, 1228, 1284-1289, 1291,
        192, 1294, 1299, 1306-1308,
        1350, 1354, 1356, 1396, 1406,
        1408
    paint, 57, 68, 362, 582, 616, 896,
        911, 1247, 1278-1283, 1291,
        1292, 1307, 1336, 1354, 1405
    trace elements, 1277, 1297-1300
mark–release-recapture, 300, 392,
    451, 585, 707, 1184, 1200, 1234,
    1237, 1306, 1352-1354, 1396,
    1399, 1403, 1407, 1411, 1412,
    1430
Mbita trap, 527-529
McCreadie et al. trap, 595
mean crowding, 8-10, 90, 92,
    186-188, 190-194, 272, 408, 1126
mean hour density, 395, 440
mechanical auction sweeper
    Davis and Gould's aspirator,
        437-439, 442
mechanical suction sweeper 431,
    436-438
    AFS sweeper, 443, 444
    Arbovirus Field Station (AFS)
        suction sweeper, 442, 969
    Aspirator of Nasci, 439
    CDC Backpack Aspirator, 390,
        444, 445, 569, 680
    CDC sweeper, 435, 436
    Dietrick's machine ('D-vac'), 418,
        431-433, 439-441, 443, 444,
        791, 820, 821
    Perdew and Meek aspirator, 446
    Suction sweeper of de Freitas, 434
    Suction sweeper of Garcia, 437
    Suction sweeper of Kay, 438
Meconium, 366, 1161, 1165, 1166

methylene blue, 300, 301, 1275-1279
Meyer and Bennett trap, 625, 626
migration, 375, 782, 785, 1116,
    1232, 1377-1379, 1426
Monks Wood light-trap, 849, 857,
    873, 906-910, 913, 922, 923
moonlight, 544-546, 548, 832, 845,
    850, 861-863, 865, 874, 1218
Morista index, 1458
Mosquito Breeding Index, 169
Mpofu and Masendu net trap, 576
Mutero and Birley method, 524,
    1216

N

NDVI, 142, 143, 149
nearest-neighbour estimate, 205
negative binomial, 8, 9, 11, 13, 14,
    96, 167, 184, 185, 188, 303, 393,
    407, 408, 882, 1377
negative binomial distribution, 5, 6,
    12, 30, 90, 184, 187-189, 393,
    407, 443, 737
net reproductive rate, 1091, 1092,
    1116, 1125, 1128-1130
New Jersey light-trap, 100, 148, 512,
    776, 820, 821, 846, 847, 854, 858,
    888-893, 897, 913, 915, 916, 919-
    922, 924, 925, 948, 963, 965, 995,
    997, 1030, 1197, 1457
No. 10 Trinidad trap, 606, 610, 612,
    987
No. 17 Trinidad trap, 611, 613
Nozawa-type light-trap, 909, 910

O

o'nyong-nyong, 406, 872
OBET, *see* odour-baited entry trap
octenol, 16, 681, 695, 875, 900, 901,
    917-919, 947, 953, 970-972, 974,
    975, 979, 981, 995-1010, 1396
odour-baited entry trap (OBET), 630,
    632-634
odour-baited trap, 630, 694, 1008

omnidirectional Fay trap, 108, 964
oral aspirator, 376, 378, 380, 382,
    383, 391, 404, 408, 415, 425, 426,
    494, 509, 514, 522, 526, 585, 703,
    707, 727, 1408
ovariolar dilatation, 876, 1173, 1181,
    1223
overcrowding factor, 1065
ovitrap, 60-62, 65-69, 71-78, 88, 90,
    91, 93, 95, 100-102, 112, 113,
    265, 276, 1394, 1413
    autocidal Aedes aegypti ovitrap,
        75, 76
    automatic recording ovitrap, 74, 75
    bamboo pot ovitrap, 80
    car tyre ovitrap, 94, 95
    Culex ovitrap, 99
    enhanced CDC ovitrap, 71, 921
    paddles, 62, 64, 67-70, 75, 77-79,
        86, 87, 90, 91, 96, 1110, 1413
    sticky ovitrap, 73, 74

P

parous rate method, 1186, 1190,
    1191, 1195, 1229
PCR, *see* polymerase chain reaction
peritrophic membrane, 691, 692
Petersen-Lincoln index, 1351
Pfuntner light-trap, 902, 903
Pfuntner trap, 870, 966, 967
Phytotelmata, 231, 233, 239, 1052,
    1056
pipe trap, 365, 462-464, 801
pit shelter, 381, 400, 430, 448,
    464-468, 722
plexiglas trap, 775, 867, 989, 990
Poisson distribution, 5, 9, 11-14, 176,
    184, 188, 189, 303, 882, 1327,
    1337
polarised light, 553, 854
polymerase chain reaction (PCR), 14,
    74, 508, 684-690, 727-729, 736,
    737, 1000, 1090, 1396
precipitin test, 535, 677-679, 591,
    705, 722, 726, 858, 1083-1088

predation, 71, 77, 81, 231, 239, 365,
    1054, 1058, 1060, 1070-1082,
    1085, 1087-1090, 1100, 1103,
    1105, 1114, 1129, 1309
Privy-type shelter, 457, 468-470
pteridine, 1161, 1167-1169
pupal survey, 280, 282, 285, 286
pyrethrum spray collection, 374, 398,
    401, 404, 408, 500, 704, 1407
pyridine haemochromogen test, 681

Q

quadrat, 163, 172, 177, 182,
    201-206, 211, 214-217, 241, 287,
    1102, 1121, 1236, 1457, 1462

R

radar, 144, 150, 785, 811, 1397, 1398
radioimmune assay test, 1088
readionuclides, 223, 300, 1083,
    1291, 1380
ramp trap, 699, 776-669, 824, 947,
    948, 988, 990, 1197
red cloth resting shelter, 453
Relative Variation, 6
remote sensing 18, 138-150, 811
    Advanced Very High Resolution
        Radiometer, 141, 143
    Geostationary satellite, 14
    Landsat Multi-Spectral-Scanner,
        141
    Landsat Thematic Mapper, 18,
        141, 144, 147
removal methods, 174, 206, 209,
    241, 288-290, 293-295, 297-299,
    401, 409, 429, 698, 1342, 1343,
    1347, 1348, 1351, 1357, 1359
    Kelker's selective removal
        method, 298, 299, 1342
    two-catch removal method, 297,
        298
    Zippin's method, 293-295, 297,
        410, 1342, 1351, 1359
repellency index, 1435

resting box of Edman, 394, 454
resting box of Kay, 457
resting box of Morris, 455
resting box of Weathersbee, 456, 457
resting count, 382, 389
rhodamine B, 300, 301, 1274-1279,
    1287, 1307, 1408
rogue ovariole, 1174-1176
Ross River virus, 150, 899-901,
    1000, 1003, 1006
Rothamsted insect survey trap, 810,
    811
rubidium, 682, 683, 715, 1300-1303,
    1394, 1412
rubidium chloride, 682, 1300, 1302

S

Salt–Hollick soil washing machine,
    38, 42
sample size, 3-20, 44, 45, 114, 176,
    262, 390, 632, 721, 740, 1199,
    1336, 1359, 1457, 1458
sampling power, 5
sampling precision, 5, 8
sampling stratification, 2
Saul's method, 1234
Schnabel census, 1314, 1355
sentinel chicken shed, 628-630
sequential sampling, 15, 16, 113,
    114, 168, 185, 302, 303
Shannon's net-trap, 526, 572, 573,
    575, 576, 578
Shannon–Weaver index, 1455, 1456,
    1462
Shannon-Wiener function, 1454
Skalski's relative abundance method,
    1313
skip oviposition, 53
small battery-powered aspirator, 384
small red box shelter, 449
sound trap, 1027, 1037-1044
    Ikeshoji and Ogawa cup trap,
        1040
    Ikeshoji cylinder sound trap, 1039,
        1041-1043

Kanda et al. cylinder and lantern trap, 1041
Nicosia trap, 975, 1043
stable trap, 580-583, 585, 587-595, 604, 725, 982
  Bates' (Egyptian) stable trap, 582, 587
  dawn trap, 591, 592
  Magoon trap, 568, 580, 583-588, 591, 593, 594, 600, 627, 707, 722, 916, 983
  Roberts' stable trap, 589
  Russell's calf-baited hut, 592
  Shannon's stable trap, 591, 592
stationary net, 781-783, 785, 816
sticky CO$_2$-baited trap, 981
sticky trap, 227, 339, 364, 786, 793, 799-804
suction trap, 4, 253, 525, 597, 607, 609, 771, 778, 780, 784, 786, 804-824, 828-833, 846, 951, 965, 980, 997, 1030, 1036, 1038, 1197
  Barnard and Mulla suction trap, 820
  Bidlingmayer suction trap, 805, 813
  directional suction trap, 816-818
  Wainhouse suction trap, 818
sugar-feeding, 556, 557, 699, 710, 1176, 1201, 1204
survival rate, 340, 366, 507, 565, 585, 731, 733, 740, 741, 744, 956, 1044, 1056, 1066, 1081, 1082, 1091, 1093, 1100, 1110, 1116-1119, 1122-1124, 1129-1131, 1161, 1165, 1181-1200, 1202, 1203, 1205-1215, 1217-1219, 1221-1223, 1225, 1227-1229, 1231-1234, 1237-1240, 1243-1245, 1247-1249, 1273, 1288, 1304-1309, 1317, 1318, 1320-1322, 1324, 1328, 1329, 1335, 1337, 1338, 1341, 1349, 1350, 1354, 1355, 1377, 1380, 1381, 1437
survivorship curve, 1057, 1092, 1097-1103, 1105, 1106, 1108, 1111-1114, 1182-1184, 1223, 1226, 1247

sweep net, 199, 274, 279, 401, 410, 416-418, 427, 516, 517, 519, 520, 1359
sweep-netting, 200, 220, 373, 401, 416-419, 421, 422, 516, 517, 519, 520

T

Taylor's (1961) power law, 6, 7, 10-12, 14, 15, 92, 186-188, 190-192
Tracheation, 1161, 1172, 1173, 1176, 1211
trap efficiency, 102, 109, 801, 821, 854
trap hut, 531-560, 592, 593, 684, 712, 1427, 1429, 1434
truck trap, 727, 786, 788-797, 804
Trueman and McIver ramp trap, 988

U

U.S. Army Miniature Solid State light trap (AMSS), 893-896
updraft chemical light-trap, 914
updraft light-trap, 904, 905, 915, 922

V

vectorial capacity, 493, 500, 568, 711, 729-747, 1218, 1240-143
vehicle-mounted aspirator, 448, 797
vent-axia suction trap, 778, 780, 804, 806, 809, 811-813, 817-819, 824, 829, 830
vent-axia trap, 811, 812, 830
veranda trap, 537, 538, 684, 1427, 1429, 1434
Vercruysse method, 1192-1194
visual attraction trap, 968, 994, 1027-1044
  duplex cone trap, 969, 1034-1036
  Fay-Prince trap, 108, 587, 680, 968-970, 980, 1009, 1010, 1028, 1030-1033, 1037
  Haufe and Burgess trap, 1028

Moving-target trap, 980, 981,
 1036, 1037
Wilton and Kloter cylinder trap,
 1033

W

walk-in red box, 105, 450-453, 455,
 963, 965, 1214, 1411
well net, 196, 198, 230

whole-person bag sampler (WPBS),
 520, 521
Wilton and Kloter trap, 970, 1033
window trap, 531, 532, 534-538, 540,
 541, 565, 776, 1406, 1425, 1427
 collapsible window trap, 536

Y

yellow fever, 80, 84, 256, 268, 269,
 273, 418, 494, 508, 873, 955

Species Index

Mosquitoes

Aedeomyia (Aedeomyia) catasticta 610

Aedes (Stegomyia) aegypti 10, 26, 48, 53, 54, 60-71, 73-81, 84, 87, 93, 94, 96, 97, 107, 114, 140, 179, 184, 198-201, 217, 218, 220, 221, 224, 226, 229, 236, 239, 241, 243-246, 256, 257, 260-287, 300, 366, 394, 401, 416, 442, 445, 494, 497, 505, 507, 511, 524, 530, 545, 554, 555, 559, 560, 679, 680, 682, 684, 687, 691, 693-695, 708, 711, 715, 719, 721, 723, 727, 801, 802, 821, 860, 872, 907, 915, 969, 970, 973, 975, 976, 980, 981, 1003, 1007, 1008-1010, 1030-1034, 1036-1039, 1050, 1054, 1058-1065, 1069, 1074, 1077, 1080, 1091, 1093, 1094, 1108, 1109, 1122, 1128, 1131, 1170, 1200-1202, 1233, 1247, 1248, 1276, 1278-1280, 1282, 1283, 1285, 1287, 1290, 1292, 1302, 1304, 1306, 1319, 1322, 1324, 1337-1341, 1350, 1351, 1378, 1380, 1381, 1383, 1384, 1393-1395, 1401, 1403, 1406, 1412, 1413, 1448

Aedes (Stegomyia) africanus 48, 54, 84, 502, 524, 580, 697, 1074, 1100, 1203, 1204, 1292, 1307, 1354

Aedes (Stegomyia) albopictus 26, 62-70, 72, 74, 79, 80, 85, 88, 94-96, 99, 102, 114, 147, 151, 179, 193, 194, 235, 239, 245, 246, 256, 262, 263, 265, 266, 269, 279, 395, 425, 440-442, 459, 494, 503, 510, 511, 516, 530, 555, 575, 678, 680, 693, 694, 698, 701, 714, 723, 727, 731, 869, 950, 955, 964, 970, 980, 981, 1035-1039, 1041, 1042, 1053-1056, 1058-1062, 1065, 1066, 1068, 1069, 1083, 1091, 1109, 1110, 1126, 1162, 1190, 1204, 1282, 1287, 1302, 1381, 1384, 1403, 1412, 1413, 1446, 1448, 1459

Aedes (Lorrainea) amesii 855

Aedes (Stegomyia) apicoargenteus 48, 54, 84, 549, 821

Aedes (Stegomyia) bromeliae 48, 53, 79, 84, 97

Aedes (Stegomyia) calceatus 97

Aedes (Aedes) cinereus 51, 208, 623, 700, 912, 918, 960, 962, 978, 1004, 1093

Aedes (Neomelaniconion) circumluteolus 435, 520, 923

Aedes (Stegomyia) cretinus 921

Aedes (Aedimorphus) culicinus 438

Aedes (Stegomyia) dendrophilus 48, 54, 84, 1448

Aedes (Aedimorphus) dentatus 564

Aedes (Aedimorphus) durbanensis 345

Aedes (Stegomyia) flavopictus 184, 188

Aedes (Aedimorphus) fowleri 54

Aedes (Diceromyia) furcifer 85, 559, 872, 955, 958

Aedes (Stegomyia) guamensis 1063

Aedes (Stegomyia) heischi 97, 1448

Aedes (Stegomyia) hensilli 442

Aedes (Aedimorphus) irritans 872

Aedes (Aedimorphus) jamesi 510

Aedes (Stegomyia) ledgeri 85

Aedes (Neomelaniconion) lineatopennis 173, 700, 864, 894, 1247, 1248, 1411

Aedes (Neomelaniconion) luridus 564

Aedes (Stegomyia) luteocephalus 48, 54, 84, 955

Aedes (Stegomyia) metallicus 85

Aedes (Neomelaniconion) mcintoshi 564

Aedes (Aedimorphus) mixtus 564

Aedes (Diceromyia) micropterus 70

Aedes (Aedimorphus) microstictus 564

Aedes (Levua) mombasaensis = Ochlerotatus (Levua) fryeri 546

Aedes (Stegomyia) novalbopictus 85, 510

Aedes (Aedimorphus) nocturnus 79

Aedes (Aedimorphus) ochraceus 564, 923

Aedes (Stegomyia) pandani 79

Aedes (Skusea) pembaensis 546

Aedes (Stegomyia) polynesiensis 70, 74, 235, 242, 243, 361, 497, 507, 1007, 1008, 1060, 1168, 1205, 1246, 1247, 1279, 1384

Aedes (Stegomyia) pseudoscutellaris 64, 68

Aedes (Stegomyia) riversi 79

Aedes (Aedes) rossicus = ssp. of Aedes (Aedes) esoensis 978, 1004

Aedes (Stegomyia) scutellaris 64

Aedes (Stegomyia) simpsoni 236, 419, 701, 915, 1064, 1067, 1100, 1101

Aedes (Stegomyia) soleatus 97

Aedes (Albuginosus) stokesi 48, 54

Aedes (Diceromyia) taylori 545, 559, 955

Aedes (Stegomyia) unilineatus 54, 70

Aedes (Aedimorphus) vexans 31, 34, 37, 43, 51, 52, 104, 147, 149, 202, 208, 215, 245, 440, 442, 453, 519, 553, 564, 565, 569, 576, 584, 589, 590, 591, 597, 600, 601, 602, 603, 623, 689, 700, 706, 709, 728, 776, 793, 826, 829, 830, 848, 855, 865, 875, 892, 909, 911, 916, 917, 918, 948, 956, 960, 962, 963, 964, 965, 973, 982, 992, 995, 1002, 1004, 1008, 1030, 1069, 1093, 1096, 1384, 1399, 1408, 1459

Aedes (Fredwardsius) vittatus 33, 47, 50, 51, 70, 225, 266, 559, 955

Aedes (Stegomyia) w-albus 510

Anopheles (Cellia) aconitus 351, 468, 501, 535, 543, 566, 567, 568, 699

Anopheles (Nyssorhynchus) albimanus 144, 145, 176, 177, 383, 393, 397, 409, 414, 425, 426, 448, 499, 505, 506, 514, 526, 539, 540, 561, 562, 584, 585, 693, 702, 709, 711, 718, 723, 724, 746, 847, 849, 855, 895, 904, 905, 906, 922, 1005, 1006, 1037, 1092, 1129, 1168, 1169, 1432, 1439

Anopheles (Nyssorhynchus) albitarsis 447, 505, 552, 573-575, 726, 888, 1194, 1233, 1348

Anopheles (Anopheles) algeriensis 1062

Anopheles (Cellia) annularis 380-382, 467, 501, 568, 874, 1163, 1448

Anopheles (Cellia) annulipes 588, 783, 791, 864, 889, 899, 900, 968, 1382

Anopheles (Anopheles) apicimacula 1005

Anopheles (Nyssorhynchus) aquasalis 420, 505, 558, 561, 592, 865, 874, 875, 1005, 1006

Anopheles (Cellia) arabiensis 29, 149, 152, 153, 184, 214, 362, 392, 408, 414, 430, 464-467, 500, 504, 508, 520, 525, 534, 632, 633, 678, 690, 701-703, 705, 707, 712, 713, 734, 736-739, 872, 879, 880, 908, 1054, 1065, 1078, 1092, 1102, 1164, 1183-1185, 1190, 1192-1196, 1222, 1276, 1357, 1396, 1397, 1430, 1432

Anopheles (Anopheles) atratipes 900

Anopheles (Anopheles) atroparvus 179, 224, 255, 1175

Anopheles (Anopheles) atropos 829, 999

Anopheles (Cellia) balabacensis 230, 546, 706, 874, 923, 957, 1231, 1237, 1239, 1355, 1382

Anopheles (Anopheles) bancroftii 588, 883, 885, 886, 968, 1007

Anopheles (Anopheles) barberi 94, 193, 245, 817, 1008, 1072, 1073

Anopheles (Anopheles) barbirostris 230, 383, 426, 566, 910

Anopheles (Kerteszia) bellator 67, 509, 573, 574, 575

Anopheles (Anopheles) bradleyi 963, 1005

Anopheles (Nyssorhynchus) braziliensis 420, 874, 875

Anopheles (Cellia) brohieri 579

Anopheles (Anopheles) campestris 578

Anopheles (Cellia) cinereus 179, 255, 465, 1062

Anopheles (Anopheles) claviger 179, 413, 829, 907

Anopheles (Anopheles) coustani 466, 508, 634, 873, 1009, 1191

Anopheles (Anopheles) crucians 182, 190, 212-215, 348, 431, 449, 453, 469, 470, 580, 583, 589, 722, 790, 793, 828-830, 857, 865, 924-926, 959, 964, 973, 976, 977, 997, 998, 1000-1002, 1036, 1163, 1164, 1400

Anopheles (Kerteszia) cruzii 440, 509, 510, 545, 573, 575

Anopheles (Cellia) culicifacies 55, 163, 166, 173, 229, 230, 380-382, 392, 395, 396, 414, 425, 427, 436, 458, 459, 467, 468, 535, 543, 568, 569, 579, 592, 593, 692, 702, 731, 864, 874, 1092, 1103, 1105, 1163, 1184, 1204, 1217, 1223-1227, 1231, 1233, 1238, 1280, 1284, 1291, 1293, 1294, 1306, 1346, 1347, 1382, 1409, 1410, 1436, 1437, 1448

Anopheles (Nyssorhynchus) darlingi 422, 423, 496, 499, 500, 502, 505, 506, 509, 514, 538, 546, 558, 585, 594, 710, 874, 1198, 1289, 1382, 1430, 1437-1440

Anopheles (Cellia) demeilloni 465

Anopheles (Anopheles) diluvialis 153, 685, 959, 1214

Anopheles (Cellia) dirus 28, 198, 230, 231, 301, 302, 424, 498, 546, 547, 567, 1188, 1189, 1275

Anopheles (Anopheles) donaldi 524, 578, 706, 874

Anopheles (Cellia) d'thali 465, 468, 874

Anopheles (Anopheles) earlei 426, 429, 960

Anopheles (Nyssorhynchus) evansae 505

Anopheles (Cellia) farauti 392, 415,
    458, 503, 507, 524, 546, 565,
    746, 882, 883, 885, 906, 968,
    1000, 1003, 1007, 1180, 1182,
    1184, 1196, 1198, 1200, 1204,
    1215, 1221, 1225, 1382, 1429
Anopheles (Cellia) flavicosta 570,
    873
Anopheles (Cellia) flavirostris 459,
    874, 891
Anopheles (Anopheles) fluminensis
    545
Anopheles (Cellia) fluviatilis 381,
    382, 391, 396, 468, 874, 887
Anopheles (Anopheles) fowleri =
    Anopheles (Anopheles) pallidus
    633
Anopheles (Anopheles) franciscanus
    463, 985
Anopheles (Anopheles) freeborni
    11-13, 57, 150, 154, 192, 211,
    223, 413, 426, 428, 429, 450,
    452, 691, 692, 727, 855, 866,
    915, 916, 919, 920, 921, 984,
    1196, 1213, 1382
Anopheles (Cellia) funestus 14, 15,
    28, 399, 400, 406, 407, 413, 414,
    460, 461, 464-467, 504, 525, 526,
    529, 534, 535, 541, 579, 631-636,
    678, 690, 710, 711, 712, 715,
    716, 718, 719, 731, 734, 736,
    737, 740, 824, 826, 857, 871-873,
    876-881, 894, 908, 915, 1009,
    1062, 1182, 1185, 1191, 1192,
    1195, 1196, 1223, 1240,
    1243-1245, 1288, 1289, 1382,
    1397, 1407-1409, 1429
Anopheles (Nyssorhynchus) galvaoi
    505, 575
Anopheles (Cellia) gambiae 14, 15,
    28, 29, 55, 57, 149, 151, 152,
    153, 157, 180, 184, 224, 227,
    228, 298, 299, 301, 302, 339,
    383, 389, 391, 392, 393, 395,

398, 400, 406, 407, 408, 412,
    413, 414, 415, 422, 430, 460,
    461, 464-467, 504, 507, 508,
    522, 525-527, 529, 534, 535,
    537, 541, 554, 556-559, 562,
    577, 579, 631-636, 678, 684,
    687, 690, 693, 701, 704, 705,
    710-716, 718, 719, 726,
    731-734, 736, 737, 740, 824,
    857, 871-873, 876-881, 884,
    894, 908, 915, 973, 1009, 1054,
    1065, 1069, 1078-1080, 1083,
    1090, 1097-1101, 1104, 1106,
    1119, 1166, 1168, 1169, 1180,
    1181, 1185, 1190, 1191, 1195,
    1198, 1200, 1217, 1222, 1223,
    1226, 1228, 1233, 1243, 1244,
    1276, 1280, 1288, 1289, 1292,
    1351, 1383, 1392, 1396,
    1405-1409, 1427-1429
Anopheles (Cellia) garnhami 465
Anopheles (Anopheles) grabhamii
    855
Anopheles (Cellia) hackeri 563,
    1284
Anopheles (Cellia) hancocki 634,
    873
Anopheles (Cellia) hilli 968
Anopheles (Cellia) hispaniola = ssp.
    of *Anopheles (Cellia) cinereus*
    179, 255, 1062
Anopheles (Anopheles) hyrcanus
    383, 423, 501, 563, 569, 783,
    874, 911, 954, 992
Anopheles (Anopheles) implexus
    424, 502, 548, 697, 1163, 1164
Anopheles (Anopheles) inundatus
    389
Anopheles (Cellia) indefinitus 891
Anopheles (Cellia) jamesii 567
Anopheles (Cellia) jeyporiensis 467,
    1188
Anopheles (Cellia) karwari 501,
    567, 883, 885, 886

Anopheles (Cellia) kochi 978

Anopheles (Cellia) koliensis 504,
701, 883, 885, 886, 1007

Anopheles (Anopheles) labranchiae
397, 922, 1062, 1196

Anopheles (Cellia) leesoni 690

Anopheles (Anopheles) lesteri 566,
576, 731

Anopheles (Anopheles) letifer 578

Anopheles (Cellia) leucosphyrus 524

Anopheles (Cellia) limosus 891

Anopheles (Cellia) longipalpis 466

Anopheles (Cellia) longirostris 883,
885, 886, 1007

Anopheles (Cellia) maculatus 381,
501, 534, 566, 578, 702, 923,
1043, 1188, 1383

Anopheles (Anopheles) maculipennis
27, 28, 201, 347, 405, 569, 912,
956, 1167, 1195

*Anopheles (Nyssorhynchus)
marajoara* 505

Anopheles (Cellia) marshallii 467

Anopheles (Cellia) mascarensis 508,
634

*Anopheles (Anopheles)
mattogrossensis* 502

Anopheles (Anopheles) maverlius
389

*Anopheles (Anopheles)
mediopunctatus* 502, 503, 545

Anopheles (Cellia) melas 28, 55,
148, 522, 543, 777, 778, 824,
832, 871, 947, 1176

Anopheles (Cellia) merus 228, 422,
504, 508, 525, 543, 678, 736

Anopheles (Cellia) minimus 28, 156,
157, 405, 415, 497, 498, 567,
700, 706, 923, 978, 1043, 1188,
1195, 1279, 1383

Anopheles (Cellia) moucheti 740,
1191

Anopheles (Cellia) multicolor 874

Anopheles (Anopheles) messeae
911

Anopheles (Kerteszia) neivai 544

Anopheles (Anopheles) nigerrimus
568, 710, 1448

Anopheles (Cellia) nili 526, 541,
559, 579, 634, 740, 873, 915,
1191, 1198

Anopheles (Cellia) nivipes 567, 700

*Anopheles (Nyssorhynchus)
nuneztovari* 447, 502, 505, 546,
726, 864, 888

Anopheles (Anopheles) obscurus
873

Anopheles (Nyssorhynchus) oswaldoi
447, 502, 545, 726

Anopheles (Cellia) pallidus 382, 633

Anopheles (Anopheles) paludis 873,
1448

Anopheles (Cellia) parensis 546

Anopheles (Nyssorhynchus) parvus
420

Anopheles (Anopheles) peditaeniatus
168, 436, 561, 566, 700, 891,
955, 1074, 1079, 1105, 1106,
1182, 1244

Anopheles (Anopheles) perplexens
959

Anopheles (Anopheles) peryassui
420

Anopheles (Anopheles) petragnani
1062

Anopheles (Cellia) pharoensis 383,
414, 631-634, 824, 826, 871,
877, 908, 923, 1379

Anopheles (Cellia) philippinensis
415, 501, 567, 874, 923

*Anopheles (Nyssorhynchus)
pictipennis* 470

Anopheles (Anopheles) plumbeus
87, 193, 413, 829, 1064, 1460

Anopheles (Cellia) pretoriensis 414,
465, 466

Anopheles (Anopheles)
  pseudopunctipennis 145, 413,
  425, 426, 448, 468, 506, 584,
  585, 722, 1191, 1203, 1229
Anopheles (Cellia) pulcherrimus
  426, 579, 874, 1217
Anopheles (Anopheles) pullus 423
Anopheles (Anopheles) punctimacula
  145, 546, 1005
Anopheles (Anopheles) punctipennis
  147, 245, 427, 449, 453, 455,
  456, 469, 917, 918, 962, 963,
  973, 1002, 1005, 1008, 1177,
  1215
Anopheles (Cellia) punctulatus 410,
  504, 524, 558, 719, 725, 875,
  883, 885, 886, 1182, 1206, 1208,
  1233, 1239, 1243, 1383
Anopheles (Cellia) quadriannulatus
  152, 465, 508, 701, 713, 1090
Anopheles (Anopheles)
  quadrimaculatus 28, 147, 175,
  176, 182, 184, 189, 191, 195,
  203, 205, 216, 217, 245, 342,
  389, 413, 427, 448, 449, 453,
  455, 456, 457, 469, 470, 517,
  580, 583, 589, 685, 692, 790,
  801, 828, 846, 917, 918, 924,
  925, 959, 962, 963, 973, 990,
  995, 997, 998, 1002, 1008, 1069,
  1071, 1164, 1275, 1284, 1289,
  1316, 1383, 1413
Anopheles (Nyssorhynchus) rangeli
  589
Anopheles (Cellia) rivulorum 1062,
  1191
Anopheles (Cellia) rufipes 414, 466,
  631, 634, 877
Anopheles (Cellia) rupicola = ssp. of
  Anopheles (Cellia) rhodesiensis
  465
Anopheles (Anopheles) sacharovi
  414

Anopheles (Anopheles) saperoi
  1279, 1355, 1408
Anopheles (Cellia) sergentii 424,
  874, 1284, 1383
Anopheles (Anopheles) sinensis 9,
  175, 178-181, 192, 204, 423,
  501, 545, 561, 563, 566, 576,
  679, 731, 891, 892, 896, 909,
  982, 992, 1074, 1089, 1106,
  1188
Anopheles (Anopheles) smaragdinus
  389, 685
Anopheles (Cellia) splendidus 381,
  567
Anopheles (Cellia) squamosus 465,
  564, 634, 778, 871, 873, 1191
Anopheles (Cellia) stephensi 230,
  301, 380, 391, 392, 396, 397,
  427, 468, 569, 570, 710, 718,
  731, 856, 874, 904, 1038, 1039,
  1092, 1105, 1131, 1163, 1168,
  1184, 1225, 1230, 1231, 1233,
  1300, 1345, 1347, 1383, 1434
Anopheles (Cellia) subpictus 55, 230,
  380, 382, 383, 392, 396, 397,
  467, 566, 570, 692, 710, 864,
  874, 1225, 1231, 1233, 1306,
  1345-1347, 1383, 1448
Anopheles (Cellia) sundaicus 415,
  723, 874
Anopheles (Cellia) superpictus 874
Anopheles (Anopheles) tenebrosus
  880
Anopheles (Cellia) tessellatus 381,
  467, 543, 561, 874, 948
Anopheles (Cellia) theileri 622
Anopheles (Nyssorhynchus)
  triannulatus 447, 505, 575, 726
Anopheles (Cellia) turkhudi 874
Anopheles (Cellia) vagus 55, 501,
  566, 783, 864, 874, 1195
Anopheles (Cellia) varuna 467, 543,
  593

Anopheles (Anopheles) vestitipennis
145, 568, 569, 584, 707, 708,
727, 855, 1215, 1287

Anopheles (Anopheles) walkeri 149,
413, 414, 601, 625, 917, 918,
948, 958

Anopheles (Cellia) wellcomei 500,
631

Anopheles (*Anopheles*)
yatsushiroensis 576

Anopheles (Anopheles) ziemanni
362, 520, 631, 634, 776, 824,
826, 871, 877, 908, 923

Armigeres (Armigeres) baisasi 188

Armigeres (Armigeres) milnensis
356, 357, 544

Armigeres (Armigeres) subalbatus
67, 85, 110, 418, 422, 559, 605,
701, 950

Armigeres (Armigeres) theobaldi
1056

Chagasia bonneae 509

Chagasia fajardi 575

Coquillettidia (Coquillettidia) aurites
502, 1196

Coquillettidia (Coquillettidia)
crassipes 609, 1003

Coquillettidia (Coquillettidia) fraseri
1162, 1165

Coquillettidia (Coquillettidia)
fuscopennata 550, 578, 908,
1196

Coquillettidia (Coquillettidia)
linealis 17, 889, 968, 1004

Coquillettidia (Coquillettidia)
metallica 1162

Coquillettidia (Coquillettidia)
perturbans 30, 147, 184, 187,
247, 248, 250-252, 254, 341,
348, 350, 351, 364, 418, 449,
453, 584, 589, 600, 601, 623,
625, 688, 689, 695, 729, 790,
800, 828, 846, 875, 913, 918,
955, 956, 958, 960, 963, 973,

976, 977, 989, 990, 997, 998,
1001, 1002, 1027, 1028, 1036,
1164, 1172, 1400

Coquillettidia (Coquillettidia)
pseudoconopas 502, 1162

Coquillettidia (Coquillettidia)
richiardii 249, 413, 549, 551,
552, 615, 829, 892, 907, 911,
977, 987, 1172

Coquillettidia (Rhynchotaenia)
shannoni 575

Coquillettidia (Rhynchotaenia)
venezuelensis 573

Coquillettidia (Coquillettidia)
xanthogaster 900, 1007

Culex (Melanoconion) adamesi 612

Culex (Eumelanomyia) albiventris
80

Culex (Aedinus) amazonensis 612

Culex (Culex) annulirostris 17, 140,
439, 588, 681, 725, 783, 791,
889, 899, 900, 901, 968, 1003,
1004, 1074, 1101, 1102, 1106,
1114, 1165, 1166, 1175, 1177,
1196, 1276, 1390, 1400

Culex (Culex) annulus 697, 891, 985

Culex (Culex) antennatus 362, 435,
611, 634, 824, 872, 873, 923

Culex (Microaedes)
antillummagnorum 93

Culex (Culex) aurantapex 362

Culex (Culex) australicus 783, 889,
899, 968, 1067

Culex (Carrollia) bihaicola 236,
1059

Culex (Microaedes) biscaynensis
239

Culex (Carrollia) bonnei 239

Culex (Eumelanomyia) brevipalpis
425, 1446

Culex (Culex) bahamensis 1378

Culex (Culex) bidens 574

Culex (Culex) bitaeniorhynchus 565,
605, 954

Culex (Melanoconion) chrysonotum = *Culex (Melanoconion) theobaldi* 420

Culex (Lophoceraomyia) cinctellus 585

Culex (Culiciomyia) cinereus 358, 1009, 1063

Culex (Phenacomyia) corniger 96

Culex (Culex) coronator 239, 574

Culex (Culex) decens 777, 778, 1063

Culex (Culex) declarator 603

Culex (Culex) dolosus 65

Culex (Culex) epidesmus 414

Culex (Melanoconion) erraticus 107, 195, 348, 349, 427, 603, 620, 689, 728, 729, 790, 924, 925, 973, 976, 977, 997, 1001, 1175

Culex (Culex) erythrothorax 180, 346, 506, 897, 899, 964, 985, 987, 1348, 1386, 1412

Culex (Culex) fuscocephala 426, 438, 510, 560, 561, 783, 891, 908, 912, 960

Culex (Culex) gelidus 383, 426, 510, 565, 700, 706, 783, 855, 864, 865, 891, 908, 910, 912, 960

Culex (Culex) globocoxitus 899

Culex (Melanoconion) gnomatos 612

Culex (Lutzia) halifaxii 345, 950

Culex (Eumelanomyia) horridus 85

Culex (Neoculex) impudicus 423

Culex (Culex) infula 426

Culex (Eumelanomyia) insignis 435, 911

Culex (Culex) laticinctus 59

Culex (Barraudius) modestus 422, 423, 911, 954

Culex (Culex) neavei 435, 826

Culex (Culiciomyia) nebulosus 85, 97

Culex (Culex) nigripalpus 57, 101, 102, 109, 111, 113, 183, 348, 441, 557, 573, 574, 603, 611, 616, 617, 621, 627, 691, 699, 709, 715, 796, 814, 828, 829, 831, 832, 856, 899, 986, 998, 999, 1001, 1002, 1165, 1176, 1229, 1233, 1291, 1302, 1386, 1400

Culex (Culex) orientalis 423, 978

Culex (Melanoconion) peccator 427

Culex (Melanoconion) pedroi 612

Culex (Culex) perfuscus 873

Culex (Culex) pervigilans 1075

Culex (Culex) pipiens 59, 72, 78, 97, 100, 102-104, 107, 108, 110, 112, 149, 177, 178, 180, 193, 194, 245, 288-291, 347, 355, 389, 390, 423, 449, 453, 563, 564, 568, 569, 584, 595, 601, 603, 604, 615, 616, 618, 623, 626, 709, 772, 789, 794, 823, 829, 846, 848, 855, 875, 892, 907, 909-911, 921, 924, 948, 950, 954, 957, 960, 961, 964, 965, 978, 982, 987, 992, 1004, 1005, 1008, 1037-1039, 1053, 1063, 1065, 1069, 1081, 1122, 1170, 1175, 1245, 1275, 1446, 1449

Culex (Culex) poicilipes 362, 564, 611, 631, 824, 825, 826, 958, 962, 1386, 1408

Culex (Melanoconion) portesi 546, 608, 609, 1204

Culex (Culex) pseudovishnui 415, 438, 510, 561, 700, 710, 783, 982

Culex (Culex) quinquefasciatus Say 58, 71, 73, 85, 99, 100-103, 105, 107-113, 174-176, 180, 198, 204, 218, 219, 221, 224, 229, 239, 266, 301, 346, 355, 356, 358, 360, 380, 382, 390, 395, 398, 401, 409-411, 414, 418, 427, 430, 438, 439, 441-444,

450, 452, 457, 463, 508, 510,
541-543, 558, 559, 565, 579,
585, 588, 591, 601, 603, 610,
620, 621, 630, 631, 633, 686,
690, 709, 711, 727, 747, 783,
792, 801, 802, 830, 846, 855,
859, 861, 876, 878, 884, 897,
908, 910, 915, 919, 921, 922,
955, 959, 965, 966, 974, 983,
987, 995, 1007-1011, 1032,
1033, 1036, 1040, 1053,
1055-1057, 1062, 1063,
1065-1067, 1069, 1080, 1091,
1093, 1094, 1100, 1123, 1129,
1130, 1131, 1165, 1166, 1168,
1175, 1176, 1182, 1185, 1188,
1190, 1195, 1196, 1201, 1203,
1208, 1211-1213, 1229, 1233,
1244, 1245, 1275, 1276, 1279,
1290, 1293, 1294, 1307, 1316,
1348, 1349, 1356, 1359, 1379,
1386, 1387, 1402, 1404, 1406,
1409, 1428, 1448
Culex (Culex) restuans 58-60, 72,
78, 100-104, 107, 108, 110-112,
245, 449, 453, 584, 603, 615,
618, 623, 626, 709, 715, 772,
918, 948, 964, 1002, 1008,
1302
Culex (Melanoconion) ribeirensis
552, 553, 573, 574
Culex (Eumelanomyia) rubinotus
622, 865
Culex (Melanoconion) sacchettae
574
Culex (Culex) salinarius 95, 98,
101-103, 108, 147, 449, 516,
583, 584, 588, 589, 601, 615,
616, 620, 709, 719, 772, 790,
793, 829, 855, 864, 865, 899,
917, 918, 924, 925, 953, 960,
962-964, 973, 976, 977, 990,
993, 997, 998, 1001, 1002, 1005,
1036, 1196, 1387, 1400

Cules (Culex) secutor 93
Culex (Culex) sitiens 17, 900, 979,
1003, 1276
Culex (Melanoconion) spissipes 500
Culex (Culex) squamosus 424
Cules (Culex) stigmatosoma 57, 58,
101, 180, 443, 452, 854, 897,
957, 965, 966, 1190, 1196, 1381,
1387
Culex (Eumelanomyia) subrima
911
Culex (Melanoconion) taeniopus
546, 608, 609
Culex (Culex) tarsalis 6, 11-13, 17,
57, 59, 98, 101, 103, 110, 138,
155, 171, 174, 175, 180, 184,
185, 188, 189, 192, 211, 223,
346, 366, 413, 428, 429, 443,
444, 450-452, 462, 464, 512,
588, 591, 594, 600, 617, 627,
629, 682, 688, 692, 709, 715,
719, 727, 772, 776, 793, 794,
825, 854, 855, 859-861, 866,
897, 915, 916, 919-921, 923,
925, 956, 957, 959, 960, 963,
964, 966, 983-985, 987, 1010,
1039, 1053, 1069, 1072, 1074,
1079, 1092, 1103, 1106, 1114,
1122, 1170, 1174, 1175, 1177,
1190, 1196, 1200, 1204, 1213,
1230, 1232, 1233, 1290, 1293,
1302, 1306, 1379, 1381, 1387,
1402, 1403, 1405, 1411, 1455
Culex (Neoculex) territans 245, 426,
429, 453, 456, 590, 918, 1008
Culex (Culex) thalassius 225, 543,
777, 778, 824, 825, 832, 947,
1196
Culex (Culex) theileri 203, 345, 564,
987
Culex (Lutzia) tigripes 1070, 1119
Culex (Culex) torrentium 87, 193,
417, 829, 846, 961, 1063, 1446,
1449, 1460

Culex (Culex) tritaeniorhynchus
  167-169, 173, 177-179, 181,
  184, 194, 224, 303, 380, 390,
  400, 415, 423, 425, 426, 438,
  541, 558, 561, 563, 565, 568,
  579, 595, 604, 605, 634, 700,
  706, 710, 716, 777, 778, 783,
  785, 798, 847, 848, 855, 864,
  865, 871, 874, 889, 891, 892,
  908-910, 912, 921, 922, 948,
  954, 960, 965, 982, 985, 992,
  1039, 1040, 1041, 1043, 1056,
  1058, 1074, 1080-1082, 1092,
  1100, 1102, 1104, 1106, 1113,
  1115, 1117, 1118, 1122, 1163,
  1170, 1175, 1184, 1198, 1203,
  1225-1227, 1229, 1294, 1345,
  1388, 1399, 1448
Culex (Lophoceraomyia) tuberis
  294
Culex (Culex) univittatus 564, 611,
  612, 709, 872, 923, 987
Culex (Carrollia) urichii 239
Culex (Culex) usquatus 574
Culex (Culex) vishnui 415, 426, 561,
  630, 700, 706, 783, 954, 955,
  960, 983, 995, 1074, 1079, 1105
Culex (Melanoconion) vomerifer
  612
Culex (Culex) weschei 825
Culex (Culex) zombaensis 435
Culiseta (Culiseta) alaskaensis 426,
  519, 588, 912
Culiseta (Culiseta) annulata 87,
  347, 389, 423, 569, 892, 907,
  921, 922, 987
Culiseta (Culiseta) bergrothi 519
Culiseta (Culiseta) impatiens 596,
  597
Culiseta (Culiseta) incidens 101,
  443, 897, 957, 965, 1058, 1064
Culiseta (Culiseta) inornata 59, 98,
  103, 104, 185, 353, 427, 443,
  444, 452, 463, 561, 588, 591,
  597, 772, 793, 821, 859, 897,
  948, 960, 965, 1175, 1176, 1455
Culiseta (Allotheobaldia)
  longiareolata 59, 1073-1076
Culiseta (Climacura) melanura 147,
  243, 362, 448, 449, 454-456,
  458, 620, 622, 623, 682, 683,
  692, 709, 828-830, 846, 857,
  894, 915, 918, 956, 964, 973,
  1000-1002, 1011, 1069, 1166,
  1176, 1200, 1388, 1401
Culiseta (Culicella) minnesotae 426
Culiseta (Culicella) morsitans 30,
  42, 224, 227, 362, 456, 615, 623,
  625, 699, 829, 892, 907, 912,
  948, 956, 961, 987, 1011, 1166,
  1388, 1401
Culiseta (Culiseta) subochrea 423
Deinocerites cancer 829, 891
Eretmapodites chrysogaster 559,
  700, 873
Eretmapodites quinquevittatus 97,
  1460
Eretmapodites silvestris 97, 1460
Eretmapodites subsimplicipes 97,
  1460
Haemagogus (Haemagogus) baresi
  965
Haemagogus (Haemagogus) celeste
  79
Haemagogus (Haemagogus) equinus
  67, 79, 504, 543
Haemagogus (Haemagogus)
  janthinomys 60, 79, 80, 504,
  965, 1386
Haemagogus (Conopostegus)
  leucocelaenus 504, 575, 1386
Haemagogus (Haemagogus) lucifer
  543
Haemagogus (Haemagogus)
  spegazzinii 1386
Hodgesia cyptopous 908
Johnbelkinia longipes 965
Limatus durhami 65, 80, 94

Limatus flavisetosus 239
Limatus pseudomethysticus 239
Malaya genurostris 184, 188
Malaya taeniarostris 908
Mansonia (Mansonioides) africana 345, 520, 611, 777, 824, 872, 908, 915, 1009
Mansonia (Mansonia) amazonensis 965
Mansonia (Mansonioides) annulata 499, 510, 712, 1225
Mansonia (Mansonioides) annulifera 30, 510, 783, 864, 889, 1225
Mansonia (Mansonioides) bonneae 1198, 1225
Mansonia (Mansonioides) dives 1225
Mansonia (Mansonia) dyari 247, 248, 348, 364, 544, 621, 1088
Mansonia (Mansonia) humeralis 575, 965
Mansonia (Mansonioides) indiana 30, 783, 864, 1225
Mansonia (Mansonia) indubitans 573
Mansonia (Mansonia) septempunctata 900, 1003
Mansonia (Mansonioides) titillans 247, 348, 349, 364, 413, 575, 621, 905, 1400
Mansonia (Mansonioides) uniformis 30, 247, 248, 344, 345, 362, 499, 510, 520, 537, 558, 565, 611, 632-634, 690, 700, 777, 824, 864, 872, 889, 900, 908, 915, 923, 955, 962, 1003, 1007, 1041, 1163, 1198, 1388, 1399, 1427, 1428, 1429
Mimomyia (Mimomyia) hybrida 30, 246, 510
Mimomyia (Mimomyia) pallida 246
Ochlerotatus (Ochlerotatus) absserratus 344, 596, 597, 917, 918, 963, 989

Ochlerotatus (Ochlerotatus) albifasciatus 174, 196, 514, 568, 600, 622, 719, 857, 958, 1069, 1070, 1072, 1093, 1102, 1449
Ochlerotatus (Finlaya) alboannulatus 968
Ochlerotatus (Howardina) albonotatus 96
Ochlerotatus (Mucidus) alternans 900, 1276
Ochlerotatus (Ochlerotatus) annulipes 347
Ochlerotatus (Protomacleaya) argyrothorax 965
Ochlerotatus (Ochlerotatus) atlanticus 212, 697, 953, 955, 973, 1000, 1001, 1002, 1036, 1043
Ochlerotatus (Ochlerotatus) atropalpus 78, 245
Ochlerotatus (Ochlerotatus) aurifer 917, 918, 958
Ochlerotatus (Halaedes) australis 1066
Ochlerotatus (Howardina) bahamensis 69, 239, 556, 557, 710, 1063
Ochlerotatus (Geoskusea) baisasi 294
Ochlerotatus (Pseudoskusea) bancroftianus 968
Ochlerotatus (Protomacleaya) berlini 79
Ochlerotatus (Ochlerotatus) caballus 564, 965
Ochlerotatus (Ochlerotatus) canadensis 147, 181, 184, 207, 212, 245, 417, 449, 453, 620, 621, 623, 875, 917, 918, 954, 955, 956, 962, 963, 973, 976, 977, 989, 1000, 1001, 1002, 1384
Ochlerotatus (Ochlerotatus) cantans 29, 31, 48, 184, 188, 224-227, 301, 340-342, 354, 365, 414,

517, 530, 531, 554, 555, 623,
    696, 698, 699, 725, 823, 824,
    829, 1052, 1057, 1074, 1083,
    1085, 1110-1112, 1162, 1177,
    1181, 1276, 1281
Ochlerotatus (Ochlerotatus) cantator
    530, 625, 800, 955, 1028, 1172
Ochlerotatus (Ochlerotatus) caspius
    38, 459, 569, 922, 954, 965
*Ochlerotatus (Ochlerotatus)
    cataphylla* 172, 302, 1096
*Ochlerotatus (Ochlerotatus)
    camptorhynchus* 899
*Ochlerotatus (Ochlerotatus)
    campestris* 957
*Ochlerotatus (Ochlerotatus)
    communis* 29, 169, 183, 293,
    294, 301, 302, 344, 350, 516,
    519, 556, 588, 623, 625, 696,
    847, 912, 913, 956, 989, 1009,
    1173, 1233, 1275, 1384
Ochlerotatus (Ochlerotatus) detritus
    33, 38, 169, 172, 301, 302, 459,
    549, 551, 558, 615, 892, 907
*Ochlerotatus (Ochlerotatus)
    diantaeus* 623, 625, 989
Ochlerotatus (Ochlerotatus) dorsalis
    34, 422, 590, 591, 600, 776, 793,
    916, 921, 922, 957, 960, 1008,
    1087, 1214, 1381, 1412
Ochlerotatus (Ochlerotatus) dupreei
    973, 1001
Ochlerotatus (Ochlerotatus) euedes
    1096
*Ochlerotatus (Ochlerotatus)
    excrucians* 344, 597, 623, 912,
    913, 918, 963
Ochlerotatus (Ochlerotatus) fitchii
    597, 918
*Ochlerotatus (Ochlerotatus)
    flavescens* 597, 957, 1384
*Ochlerotatus (Ochlerotatus)
    fluviatilis* 679
Ochlerotatus (Finlaya) flavipennis
    184, 188, 237
Ochlerotatus (Levua) fryeri 546
Ochlerotatus (Zavortinkius) fulgens
    85, 1448
Ochlerotatus (Ochlerotatus) fulvus
    499, 697
Ochlerotatus (Finlaya) geniculatus
    81, 83, 87, 193, 362, 700, 829,
    1052, 1053, 1064, 1119, 1126,
    1460
*Ochlerotatus (Protomacleaya)
    hendersoni* 69, 78, 87, 90, 99,
    107, 453, 817, 960, 1066, 1233,
    1403, 1459
*Ochlerotatus (Ochlerotatus)
    hexodontus* 350, 519
Ochlerotatus (Ochlerotatus) impiger
    29, 340, 341, 352, 363, 364, 775,
    957, 1030
*Ochlerotatus (Ochlerotatus)
    infirmatus* 43, 211, 973, 1001,
    1002
Ochlerotatus (Finlaya) ingrami 48,
    54, 502
*Ochlerotatus (Ochlerotatus)
    intrudens* 623, 625, 912, 989
Ochlerotatus (Finlaya) japonicus 98,
    99, 109, 110, 112, 151, 245, 698,
    950
Ochlerotatus (Ochlerotatus) juppi
    564
Ochlerotatus (Finlaya) kochi 900,
    1003, 1182, 1244
*Ochlerotatus (Finlaya)
    lophoventralis* 1446
*Ochlerotatus (Gymnometopa)
    mediovittatus* 78, 93
*Ochlerotatus (Ochlerotatus)
    melanimon* 422, 590, 600, 855,
    866, 915, 916, 919, 957, 960,
    1039, 1191, 1233, 1379, 1385,
    1399

Ochlerotatus (Ochlerotatus)
  mercurator 1096
Ochlerotatus (Ochlerotatus)
  mitchellae 964, 973, 995, 1002
Ochlerotatus (Ochlerotatus) nigripes
  29, 340, 341, 352, 363, 364, 775,
  1030
Ochlerotatus (Ochlerotatus)
  nigromaculis 42, 43, 53, 793,
  959, 960, 983, 1165, 1166, 1175,
  1299, 1379, 1385
Ochlerotatus (Ochlerotatus) nubilus
  574
Ochlerotatus (Finlaya) niveus 585
Ochlerotatus (Ochlerotatus)
  normanensis 588, 899, 1003
Ochlerotatus (Finlaya) notoscriptus
  17, 73, 77, 198, 266, 285, 286,
  801, 900, 1003, 1063, 1385
Ochlerotatus (Bruceharrisonius)
  okinawanus 79
Ochlerotatus (Ochlerotatus) pionips
  1009
Ochlerotatus (Finlaya) poicilius 184,
  188, 237, 568, 679, 1188
Ochlerotatus (Rhinoskusea)
  portonovensis 423
Ochlerotatus (Ochlerotatus) procax
  17, 345, 900
Ochlerotatus (Ochlerotatus) pullatus
  519
Ochlerotatus (Ochlerotatus) punctor
  31, 42, 178, 225, 301, 350, 354,
  507, 516, 519, 555, 596, 597,
  800, 913, 956, 957, 989, 1009,
  1027, 1028, 1173
Ochlerotatus (Molpemyia) purpureus
  1003
Ochlerotatus (Rusticoidus) rusticus
  29, 178, 225, 341
Ochlerotatus (Ochlerotatus) sagax
  783, 968
Ochlerotatus (Finlaya) samoanus
  497
Ochlerotatus (Ochlerotatus)
  scapularis 183, 441, 545, 552,
  573, 574, 575, 603, 727, 905,
  1121
Ochlerotatus (Ochlerotatus)
  serratus 183, 440, 574,
  575
Ochlerotatus (Ochlerotatus)
  sierrensis 86, 91, 93, 94, 362,
  389, 511, 559, 587, 602, 696,
  699, 855, 861, 968, 969, 970,
  1032, 1053, 1056, 1059, 1064,
  1067, 1121, 1164
Ochlerotatus (Ochlerotatus)
  sollicitans 32, 36, 38, 97, 139,
  150, 206, 302, 348, 449, 601,
  705, 790, 793, 801, 829, 846,
  865, 956, 963, 990, 1005, 1275,
  1276, 1293, 1308, 1378, 1379,
  1385
Ochlerotatus (Ochlerotatus)
  spencerii 1096
Ochlerotatus (Ochlerotatus) sticticus
  34, 875, 960, 978, 1385
Ochlerotatus (Ochlerotatus)
  stimulans 37, 875, 917, 918,
  1038
Ochlerotatus (Ochlerotatus)
  taeniorhynchus 32, 36, 38, 43,
  44, 79, 97, 180, 183, 204, 348,
  556, 583, 603, 695, 781, 789,
  790, 793, 796, 801, 802, 814,
  845, 846, 863, 905, 963, 972,
  973, 976, 990, 997, 998, 999,
  1005, 1284, 1291, 1378, 1379,
  1385
Ochlerotatus (Protomacleaya)
  terrens 80
Ochlerotatus (Ochlerotatus)
  theobaldi 783, 968
Ochlerotatus (Ochlerotatus) thibaulti
  699, 1233
Ochlerotatus (Finlaya) togoi 79, 224,
  240, 241, 290, 291, 292, 293

Ochlerotatus (Finlaya) rubrithorax
    783
Ochlerotatus (Ochlerotatus)
    tormentor 212, 697, 953
Ochlerotatus (Macleaya) tremulus
    801
Ochlerotatus (Protomacleaya)
    triseriatus 33, 60, 69, 75, 78, 81,
    86-95, 99, 104, 107, 108, 110,
    137, 147, 193, 194, 232, 245,
    440, 444, 555, 559, 584, 587,
    625, 817, 860, 917, 966, 973,
    1008, 1036, 1052, 1056,
    1058-1060, 1066, 1069, 1072,
    1073, 1122, 1173, 1199, 1204,
    1233, 1275, 1286, 1354-1356,
    1381, 1385, 1402, 1403, 1459
Ochlerotatus (Ochlerotatus)
    trivittatus 553, 584, 603, 620,
    772, 793, 875, 917, 918, 960,
    963, 1008
Ochlerotatus (Ochlerotatus)
    varipalpus 559
Ochlerotatus (Ochlerotatus) vigilax
    17, 43-47, 150, 783, 801, 900,
    901, 968, 1004, 1006, 1007,
    1276, 1379, 1395, 1396
Ochlerotatus (Protomacleaya)
    zoosophus 33, 78
Orthopodomyia anopheloides 67,
    1446
Orthopodomyia signifera 78, 86, 93,
    94, 193, 245, 969, 973, 1043,
    1059, 1072, 1073
Psorophora (Psorophora) ciliata 36,
    568, 830, 973
Psorophora (Grabhamia) cingulata
    499, 861
Psorophora (Grabhamia) columbiae
    36, 139, 150, 190, 195, 213, 215,
    417, 418, 431, 583, 588, 589,
    790, 791, 830, 857, 865, 921,
    924, 925, 964, 973, 997, 998,

    1001, 1039, 1043, 1071, 1078,
    1293, 1307
Psorophora (Grabhamia) confinnis
    50, 202, 360, 420, 548, 590, 621,
    776, 801, 828, 856, 990, 995,
    1379, 1388
Psorophora (Grabhamia) discolor
    36, 1388
Psorophora (Janthinosoma) ferox
    212, 418, 453, 584, 697, 953,
    955, 1001, 1173, 1388
Psorophora (Psorophora) howardii
    212
Psorophora (Janthinosoma)
    longipalpus 212
Psorophora (Psorophora) pallescens
    568
Runchomyia paranensis 958
Runchomyia reversa 510
Sabethes (Sabethes) amazonicus
    1386
Sabethes (Peytonulus) aurescens
    236
Sabethes (Sabethes) belisarioi 701,
    1386
Sabethes (Sabethoides) chloropterus
    965, 1386
Sabethes (Sabethes) cyaneus 1386
Sabethes (Sabethoides)
    glaucodaemon 1386
Sabethes (Sabethinus)
    melanonymphe 236
Sabethes (Sabethes) quasicyaneus
    1386
Topomyia (Topomyia) dubitans 188
Topomyia (Suaymyia) yanbarensis
    188
Toxorhynchites (Toxorhynchites)
    amboinensis 66, 78, 1064, 1067,
    1073
Toxorhynchites (Toxorhynchites)
    brevipalpis 69, 80, 184, 188,
    1064, 1080, 1350, 1388

Toxorhynchites (Lynchiella)
  haemorrhoidalis 239, 1068
Toxorhynchites (Lynchiella)
  moctezuma = Toxorhynchites
  (Lynchiella) theobaldi 51, 55,
  66, 79
Toxorhynchites (Lynchiella) rutilus
  69, 94, 193, 245, 1059, 1072,
  1073, 1102
Toxorhynchites (Toxorhynchites)
  splendens 188, 1072, 1073,
  1184, 1446
Toxorhynchites (Toxorhynchites)
  towadensis 1077
Trichoprosopon digitatum 51, 96,
  236, 1056, 1059, 1068
Tripteroides (Tripteroides) bambusa
  67, 235, 698, 950, 1054, 1055,
  1077
Uranotaenia (Uranotaenia) lateralis
  423
Uranotaenia (Uranotaenia) lowii
  793, 829, 830, 1038, 1379
Uranotaenia (Pseudoficalbia)
  ohamai 294
Uranotaenia (Uranotaenia) pygmaea
  345
Uranotaenia (Uranotaenia)
  sapphirina 348, 427, 449, 453,
  590, 829, 830, 856, 857, 1008,
  1379
Uranotaenia (Pseudoficalbia)
  unguiculata 423
Verrallina (Verrallina) butleri 855
Verrallina (Verrallina) carmenti
  1003
Verrallina (Verrallina) funerea 17,
  345, 783, 900, 901, 1003, 1004
Verrallina (Neomacleaya) indica 438
Verrallina (Harbachius) yusafi 438
Wyeomyia (subgenus uncertain)
  bourrouli 603
Wyeomyia (Decamyia) felicia 236,
  1059

Wyeomyia (Wyeomyia) limai 236
Wyeomyia (Wyeomyia) mitchellii
  239, 440, 953, 982, 998, 1385
Wyeomyia (Wyeomyia) smithii 7, 50,
  53, 192, 236, 1052, 1069, 1126,
  1232, 1462
Wyeomyia (Wyeomyia) vanduzeei
  239, 425, 998, 1067, 1280, 1281,
  1385

Other Diptera
Chrysomya bezziana 1167, 1395
Chrysomya putoria 357
Chrysops bicolor 1179
Chrysops caecutiens 804, 825
Cnephia mutata 955
Cochliomyia hominivorax 1167
Culicoides barbosai 974, 975
Culicoides brevitarsis 790, 1277
Culicoides circumscriptus 909
Culicoides furens 788, 974, 966
Culicoides histrio 1006
Culicoides imicola 909
Culicoides impunctatus 1405
Culicoides melleus 974
Culicoides mississippiensis 974,
  975, 995, 1290
Culicoides molestus 1006
Culicoides schultzei 909,
Culicoides subimmaculatus 1006
Culicoides variipennis 34, 1167,
  1300, 1301
Drosophila melanogaster 1294,
  1380
Glossina fuscipes 1344
Glossina morsitans 464
Glossina pallidipes 699
Haematopota pluvialis 804
Hybomitra lundbecki 1009
Hydrotaea irritans 1390, 1393
Lutzomyia intermedia 865
Lutzomyia whitmani 865
Mochlonyx cinctipes 1072
Musca autumnalis 1167

Phlebotomus alexandri 793
Phlebotomus papatasi 1277
Prosimulium hirtipes 955
Sergentomyia clydei 793
Sergentomyia fallax 793
Sergentomyia tiberiadis 793
Simulium arcticum 597, 1397
Simulium damnosum 1083, 1167
Simulium squamosum 908
Simulium venustum 1295
Simulium vittatum 776
Simulium woodi 1179
Stegopterna mutata 1295
Stomoxys calcitrans 1167, 1290

Protozoa
Ascogregarina taiwanensis 1061
Plasmodium elongatum 625

Plasmodium falciparum 558, 730,
  731, 735, 744, 883, 1240
Plasmodium knowlesi 563
Plasmodium vivax 730, 731, 896

Filarial Worms
Cardiofilaria nilesi 610
Dirofilaria immitis 600, 601, 603,
  963
Strelkovimermis spiculatus 1070,
  1102
Wuchereria bancrofti 505, 507, 542,
  568, 1060, 1188, 1205, 1244,
  1245, 1246, 1247
Wuchereria kalimantani 957